Plant
Engineer's
Handbook

Plant Engineer's Handbook

Edited by

R. Keith Mobley

BUTTERWORTH
HEINEMANN

BOSTON OXFORD AUCKLAND JOHANNESBURG
MELBOURNE NEW DELHI

Library of Congress Cataloging-in-Publication Data
Plant engineer's handbook/edited by R. Keith Mobley. – [Rev. ed.].
   p.cm
   Rev. ed. of: Plant engineer's reference book. c1991.
   Includes index.
   ISBN 0 7506 7328 1
   1 Plant engineering – Handbooks, manuals, etc.   I.   Mobley, R. Keith, 1943-II,
   Title: Plant engineer's reference book.
   TS184 .P58 2000
   696 – dc21
                                                                00-049366

British Library Cataloguing-in-Publication Data
Plant engineer's handbook
   1 Plant engineering – Handbooks, manuals, etc.
   1 Mobley, R. Keith, 1943 –
   658.2

The Publisher offers special discounts on bulk orders of this book.

For information, please contact:
Manager of Special Sales
Butterworth-Heinemann
225 Wildwood Avenue
Woburn, MA 01801 – 2041
Tel: 781-904-2500
Fax: 781-904-2620

For information on all Butterworth-Heinemann publications available,
contact our World Wide Web home page at: http://www.bh.com

10 9 8 7 6 5 4 3 2 1

696
P713

Contents

Foreword

The opportunity to write the Foreword to a new book falls to very few people and I consider myself very fortunate to be able to introduce you to the *Plant Engineering Handbook*.

It has been many years since a book was written specifically for the engineer who is involved with so many facets of the engineering profession, and I am of the firm opinion that this publication is long overdue.

The many chapters covering the multidisciplinary role of the plant engineer have been written by practising experts and specialists and R. Keith Mobley, a practising engineer with an international reputation, has edited the whole.

This handbook is so written as to take the reader through the design, planning, installation, commissioning, operation, and maintenance stages of both major and minor engineering projects. It covers all aspects of maintenance, management, health and safety, and finance and gives a guide to where further information on these subjects may be obtained.

It is a book which should be in the hands of all practising plant, reliability and maintenance engineers as well as senior management who are involved in the efficient running and control of their respective establishments.

As a plant engineer of many years' experience, I recommend this book to you as the definitive reference book for all members of our profession.

Richard J. Wyatt
President, The Institute of Plant Engineers (UK)

Preface

The preparation of an international reference book such as this could not possibly be achieved without the total cooperation of so many individual authors and the backing of their various employers, especially where company contributions have been made, bringing together a wealth of professional knowledge and expertise.

An acknowledgement such as this can only scratch the surface and cannot really portray the grateful thanks I wish to express to all these people concerned that have devoted so much time and effort to place their ideas and contributions to this *Plant Engineering Handbook*.

Plant engineering is such a broad subject incorporating a multitude of disciplines and a wide variety of solutions to virtually every problem or situation, unlike some subjects that have clear-cut methods.

In compiling the initial suggested guidelines for each of the contributions, I posed the questions to myself what information and assistance had I found difficult to locate during over 30 years as a plant engineer responsible for plants throughout the world and how could it be best presented to assist others in their profession.

I would therefore like to take this opportunity to thank each of the contributing authors for their patience and assistance in helping me to structure this publication.

R. Keith Mobley
President and CEO
The Plant Performance Group
Knoxville, Tennessee, USA

List of Contributors

A Armer
Spirax Sarco Ltd

B Augur, IEng, FIPlantE, MBES
J B Augur (Midlands) Ltd

H Barber, BSc
Loughborough University of Technology

D A Bayliss, FICorrST, FTSC

J Bevan, IEng, MIPlantE

R J Blaen
Senior Green Limited

British Compressed Air Society

G Burbage-Atter, BSc, CEng, FInstE,
HonFIPlantE, FCIBSE
Heaton Energy Services

P D Compton, BSc, CEng, MCIBSE
Colt International Ltd

I G Crow, BEng, PhD, CEng, FIMechE,
FIMarE, MemASME
Davy McKee (Stockton) Ltd

R. Dunn
Editor, Plant Engineering Magazine

P Fleming, Bsc(Eng), ARSM, CEng, MInstE
British Gas plc

C Foster
British Coal

C French, CENg, FInstE, FBIM
Saacke Ltd

F T Gallyer
Pilkington Insulation Ltd

R R Gibson, BTech, MSc, CEng, FIMechE, FIMarE,
FRSA
W S Atkins Consultants Ltd

B Holmes, BSc(Tech), PhD, CEng, FIChemE, FInstE
W S Atkins Consultants Ltd

A P Hyde
National Vulcan Engineering Insurance Group Ltd

H King
Thorn Lighting

B R Lamb, CEng, MIChemE
APV Baker Ltd

S McGrory
BH Oil UK Ltd

R Keith Mobley
International Consultant

R J Neller
Film Cooling Towers Ltd

Ove Arup & Partners, Industrial Division

G Pitblado, IEng, MIPlantE, DipSM
Support Services

R S Pratt, ALU, CEng, MIMfgE, MBIM, MSAE
Secretary-General, The Institution of Plant
Engineers

G E Pritchard, CEng, FCIBSE, FInstE, FIPlantE,
MASHRAE

Risk Control Unit
Royal Insurance (UK) Ltd

R Robinson, BSc, CEng, FIEE
The Boots Co. plc

M J Schofield, BSc, MSc, PhD, MICorrT
Cortest Laboratories Ltd

J D N Shaw, MA
SBD Construction Products Ltd

R H Shipman, MIMechE, MIGasE, MInstE
Liquefied Petroleum Gas Industry Association

K Shippen, BSc, CEng, MIMechE
ABB Power Ltd

R Smith
Life Cycle Engineering, Inc.

G Solt, FIChemE, FRSC
Cranfield Institute of Technology

K Taylor, CEng, FCIBSE, FIPlantE, FIHospE,
FSE, FIOP, MASHRAE, FBIM, ACIArb
Taylor Associates Ltd

L W Turrell, FCA

K Turton, BSc(Eng), CEng, MIMechE
Loughborough University of Technology

E Walker, BSc, CEng, MIMechE
Senior Green Limited

R C Webster, BSc, MIEH
Environmental Consultant

D Whittleton, MA, CEng, MIMechE, MHKIE
Ove Arup & Partners, Industrial Division

Definition and Organization of the Plant Engineering Function

1

Richard Dunn
Editor, Plant Engineering Magazine

Contents

1.1 Introduction

The concept of the plant engineering function has changed little over the years. Yet, the ways in which that function is accomplished have changed significantly, primarily because of changing technologies and business models. More than ever before, for example, the plant engineer must learn to manage from the perspective of a business participant, relating his responsibilities and activities to the mission and goals of the enterprise. Moreover, the invasion of electronics and computerization into nearly every facet of engineering and business operation has fostered the integration of plant engineering into both the operations and the business plan of the enterprise.

Changes in enterprise organization models have also impacted plant engineering. In many industrial plants, for example, the title of 'plant engineer' has disappeared, being replaced with such titles as 'facilities manager' or 'asset productivity manager'. Yet, the essential services provided by these people and their departments remains essentially unchanged, and every enterprise with physical facilities must have a plant engineering function, regardless of the name by which it is labeled and the organization through which it is accomplished.

1.2 Basic definition

Plant engineering is that branch of engineering which embraces the installation, operation, maintenance, modification, modernization, and protection of physical facilities and equipment used to produce a product or provide a service.

It is easier to describe plant engineering than to define it. Yet, the descriptions will vary from facility to facility and over time. Every successful plant is continuously changing, improving, expanding, and evolving. And the activities of the plant engineer must reflect this environment. Each plant engineer is likely to have his own, unique job description, and that description is likely to be different from the one he had five years earlier.

By definition, the plant engineering function is multidisciplinary. It routinely incorporates the disciplines of mechanical engineering, electrical engineering, and civil engineering. Other disciplines, such as chemical engineering for example, may also be needed, depending on the type of industry or service involved.

In addition, skills in business/financial management, personnel supervision, project management, contracting, and training are necessary to the successful fulfillment of plant engineering responsibilities. The function is fundamentally a technical one, requiring a thorough technical/engineering background through education and/or experience. But beyond it's most basic level, a broad range of skills is needed.

If the plant engineer is a specialist in anything, it is in his/her own plant or facility. Plant engineers must learn to know their own plants thoroughly, from the geology underlying its foundations and the topology of the rainwater runoff to the distribution of its electricity and the eccentricities of its production machinery. They must ensure the quality of the environment both inside and outside the facility as well as the safety and health of the employees and the reliability of its systems and equipment. And they are expected to do all of this in a cost-effective manner.

A few phrases from a 1999 classified ad for a plant engineer provide some real-world insight on the scope of responsibilities:

- Support ongoing operations, troubleshoot, resolve emergencies, implement shutdowns
- Organize and maintain information on plant systems/equipment and improvement programs
- Implement plant projects and maintain proper documentation
- Deal effectively with multiple activities, requests, and emergencies
- Manage scope, design, specification, procurement, installation, startup, debugging, validation, training, and maintenance.

To this list, most plant engineers would quickly add compliance with all applicable laws and regulations as well as accepted industry standards and practices.

More than 25 years ago, Edgar S. Weaver, then manager of Real Estate and Construction Operations for General Electric, provided a succinct description of the function:

> 'The primary mission of the plant engineer is to provide optimum plant and equipment facilities to meet the established objective of the business. This can be broken down into these four fundamental activities: (1) ensure the reliability of plant and equipment operation; (2) optimize maintenance and operating costs; (3) satisfy all safety, environmental, and other regulations; and (4) provide a strong element of both short-term and long-range facilities and equipment planning.'

The description still rings true today.

1.3 Responsibilities

There are two ways of analyzing the plant engineering function. One is through the *activities* plant engineers must perform. The other is through the *facilities, systems, and equipment* they must be knowledgeable about. For a complete understanding of the function, both must be considered.

1.3.1 Activities

The activities that plant engineers must perform generally fall under the responsibilities of middle-to-upper management. Like all managers, they plan, organize, administer, and control. But more specifically, plant engineers are involved in or in charge of the following activities:

- **Design** of facilities and systems
- **Construction** of facilities and systems
- **Installation** of facilities, systems, and equipment
- **Operation** of utilities and services
- **Maintenance** of facilities, systems, and equipment

- **Improvement, retrofit, and redesign** of facilities, systems, and equipment
- **Planning** to meet business needs
- **Contracting** for equipment, materials, and services
- **Project management**, including planning, estimating, and execution
- **Administration** of the plant engineering organization and personnel as well as related financial considerations (budgeting, forecasting, cost control), training, and record keeping
- **Regulatory compliance** with a wide variety of governmental laws and standards
- **Coordination** of plant engineering activities and responsibilities with all other functions and departments in the organization
- **Purchasing** of requisite tools, equipment, parts, and materials.

These activities are nearly universal throughout the plant engineering function, although they may be described differently in specific companies or facilities. Also, other activities might be added to the list.

1.3.2 Knowledge areas

While most plant engineers are, in fact, engineers by education and training, there is no single, traditional engineering discipline that comprises all areas of plant engineering responsibilities. A combination of mechanical and electrical engineering education and experience is essential, and some knowledge in the areas of civil, structural, environmental, safety, chemical, and electronic engineering is useful and important.

Mere education is not enough, however. Plant engineering demands a level of experience in applied knowledge and problem solving that is more intense than in most other engineering functions. In fact, plant engineers are often described as 'jacks of all trades' or 'firefighters' because of their abilities to respond to a wide variety of needs on short notice, to fix almost anything that breaks, and to implement solutions to emerging problems.

Nevertheless, a major portion of every plant engineer's efforts is devoted to the *prevention* of problems and emergencies, as exemplified by their intense involvement in the maintenance of virtually all structures, systems, and equipment in their facilities.

Thus, to be successful, plant engineers must be knowledgeable in the design, installation, operation, and maintenance of the following:

- Electrical power systems
- Electrical machinery
- Lighting
- Fluid power transmission
- Mechanical power transmission
- Instrumentation and controls
- Heating and ventilating
- Air conditioning and refrigeration
- Pumps, piping, and valving
- Material handling and storage
- Paints, coatings, and corrosion prevention
- Fire protection
- Engines
- Lubricants and lubrication systems
- Environmental control systems and compliance
- Compressed air systems
- Buildings and construction
- Tools
- Welding and joining
- Safety and health equipment and practices
- Security.

Each of the above categories could easily be broken into numerous subcategories, and more could be added. But these are the generally accepted areas of expertise that plant engineers are expected to know.

1.4 Organization

Organizational structures and reporting relationships within the plant engineering function and in relation to other functions are as unique as each business enterprise and individual plant. Yet, some common structures can be identified.

To be most effective, the plant engineering function should report directly to top plant or facility management. In smaller enterprises, it should report directly to the owner or to top corporate management. In any case,

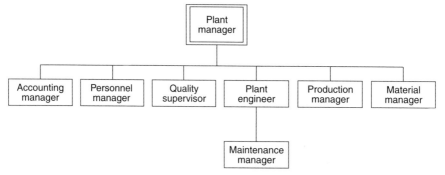

Figure 1.1 Typical organization of a small plant illustrates that plant engineering is one of the essential functions in any plant. Maintenance is normally a subfunction of plant engineering

the plant engineer should have direct access to whoever makes the final decisions on any project, capital expenditure, legal concern, or enterprise policy decision. In multi-site companies with a corporate engineering department, each site plant engineer should report directly to the site manager with a secondary reporting relationship to the director of corporate engineering.

It is worth noting that a few very large industrial companies have divided the plant engineering function into multiple departments. The most common division in these cases is the separation of 'landlord' responsibilities (that is, real estate, buildings and grounds, and utilities) from 'production' responsibilities (that is, manufacturing and process equipment and systems).

Within the plant engineering function, there are typically two primary subfunctions, best described as engineering and maintenance. The engineering subfunction is responsible for such matters as design, construction,

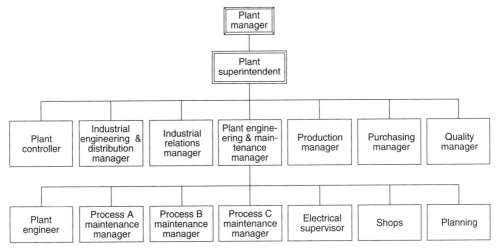

Figure 1.2 In larger plants, the plant engineering function is often divided into departments to serve particular needs

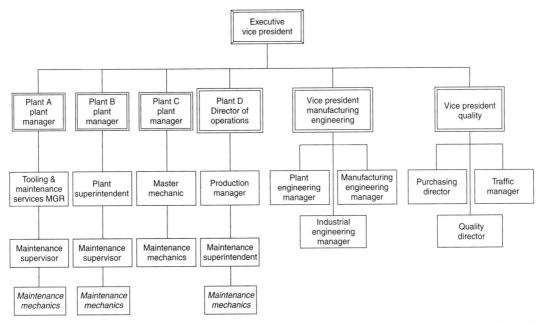

Figure 1.3 Multisite enterprises are often organized with a central engineering department providing plant engineering services to all plants and separate maintenance departments within each site

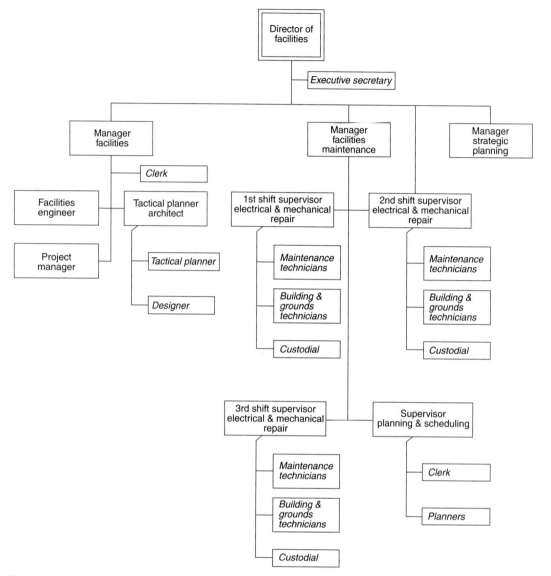

Figure 1.4 Large, complex industrial plants and other facilities require an extensive plant engineering organization to meet constantly changing demands

modification, and modernization of the facility, its utilities, and operating equipment. The maintenance subfunction provides all maintenance services and carries out many of the changes specified by engineering. Some plant engineering organizations also identify a third subfunction, operations. This group is responsible for running the utility systems, such as electrical control and distribution; steam; heating, ventilating, and air conditioning; compressed air; water treatment; etc.

The organization charts in Figures 1.1–1.4, adopted from real plant organizations of various sizes and in a variety of industries, illustrate some typical structures.

2

Plant Engineering in Britain

Roger S Pratt

Secretary-General, The Institution of Plant Engineers

Contents

2.1 The professional plant engineer

The profession of engineering, in contrast to many others, is extremely wide ranging in the spread of topics, technologies and specialization included under the overall heading. The early engineers, the creative geniuses of their day, encompassed all these latter-day specializations, famous examples being Brunel, Stephenson and Telford. Engineers have been at the heart of all technological and scientific progress. Without them the world as we know it today would not exist.

This has been despite the fact that the UK has developed with a culture that is indifferent to engineering, the respectable professions being those such as law or medicine offering more money and prestige. This deeply rooted attitude was supported by an education system in which on the whole applied science – engineering – was not studied in schools or universities. This contrasts with the rest of world, where such studies were an important part of the curricula of many schools and universities as early as the eighteenth century. Engineering was not considered suitable for those with the ability to enter a university, where arts and sciences were studied.

The need for education in engineering in the UK was met by the development of Mechanics Institutes. By the middle of the nineteenth century, around 120,000 students per annum attended some 700 institutes on a part-time basis, thus laying the foundations for the pattern of engineering education in the UK. In 1840, the first chair in an engineering discipline (civil engineering) was established, at Glasgow University, soon to be followed by one at University College, London. Oxford and Cambridge were late on the scene, establishing chairs in engineering in 1875 and 1910, respectively.

Also peculiar to the UK is a somewhat confusing array of professional engineering institutions. These were originally learned societies where like-minded people met to exchange views and information. They developed into qualifying bodies by setting levels of experience and academic attainment for different grades of membership. The oldest professional engineering institution in the UK is the Institution of Civil Engineers, established in 1818. The Institution of Mechanical Engineers was established in 1847 and the Institution of Electrical Engineers in 1871. Three-quarters of the approximately 50 institutions which are the Nominated Bodies of the Engineering Council were founded in the twentieth century, some quite recently, reflecting the growth of certain engineering disciplines such as nuclear engineering, computing and electronics.

2.2 The Institution of Plant Engineers

The Institution of Plant Engineers (IPlantE) had its origins in the Second World War. During this period, engineers who found themselves responsible for the operation and maintenance of the large excavators and other mobile plant brought from the US to work open cast coal met together for the exchange of information and to discuss their problems. These meetings were so successful that the engineers concerned decided to continue them in a more formal manner through the medium of a properly incorporated body. The Memorandum of Association of 'Incorporated Plant Engineers' was subsequently signed on 3 September 1946.

The concept of an engineering institution which covered a wide field attracted engineers from many different areas of activity, including industrial, municipal and service establishments, civil engineering projects, transport undertakings, design, research and education. By 1947, branches of the Institution were holding monthly meetings in London, Birmingham, Manchester, Leeds, Newcastle, Glasgow and Bristol, and in the following year six more branches were established. There are now 20 branches in the UK and a large number of members in other countries.

In January 1959, the Board of Trade gave permission for a change of title from 'Incorporated Plant Engineers' to 'The Institution of Plant Engineers'. This marked an important stage in the Institution's development, enabling it to take its place alongside other established engineering institutions. The Bureau of Engineer Surveyors, whose members had particular interests and expertise in relation to the safety and insurance aspects of plant operation and maintenance, merged with the Institution in 1987, forming the basis of a new specialist division.

The Institution of Plant Engineers is therefore in many ways a small-scale reflection of the engineering profession as a whole, embracing a wide range of disciplines and activities. The Institution's members work in the fields of, and have responsibility for, designing, specifying, building, installing, overseeing, commissioning, operating and monitoring the efficiency of plant of all kinds. This can include most types of building, plant and equipment used in the manufacturing, chemical and process industries, educational establishments, warehouses, hospitals, office and residential accommodation, hotels, banks, theatres, concert halls and all types of transportation systems. In the broadest sense of the term, these are the assets of the organization in question, without which it could not function. The plant engineer thus carries out a key role as the practical manager of these assets.

2.3 Aims of the Institution

The aims of the Institution of Plant Engineers are:

1. To bring together those already qualified by the attainment of such standards of knowledge, training, conduct and experience as are desirable in the profession of plant engineering;
2. To promote the education and provide for the examination of students in the profession of plant engineering;
3. To encourage, advise on and take part in the education, training and retraining of those engaged in plant engineering activities at all levels;
4. To diffuse knowledge of plant engineering by every means, including lectures, papers, conferences and research;
5. To increase the operational efficiency of plant for the greater benefit and welfare of the community, bearing in mind the importance of the conservation of the environment and the preservation of amenity.

2.4 Organization

Overall direction of the Institution is vested in its Council, but committees and panels of members carry out much of the Institution's detailed work. Branches and divisions of the Institution are run by their own committees, which arrange programs of visits, lectures and other appropriate activities, spread throughout the year. Non-members are very welcome to attend most Institution events. The Institution publishes its journal, *The Plant Engineer,* and other technical information, and organizes national conferences and exhibitions.

The Institution's permanent staff is always available to give help and advice on matters relating to membership, education and training, and Engineering Council registration.

2.5 Membership

Membership of the Institution of Plant Engineers is the hallmark of the professional plant engineer and is often a prerequisite for successful career progression. This will become increasingly so in post-1992 Europe, when evidence of appropriate professional qualifications may be a legal requirement for employment in many engineering appointments.

2.5.1 Membership requirements

A summary of the grades of membership and the personal requirements for each of these grades is shown in Table 2.1.

2.5.2 Courses leading to a career in plant engineering

The main courses leading to a career in plant engineering are the Business and Technician Education Council's (BTEC) Technician Certificate or Diploma in Plant Engineering and Higher National Certificate or Diploma in Plant Engineering. In Scotland, the equivalents are the Scottish Technician and Vocational Education Council's (SCOTVEC) Technician Certificate in Mechanical Engineering (Plant Engineering Options) and Higher Certificate in Mechanical Engineering (Plant Engineering Options). Additionally, certain other BTEC Certificates and Diplomas, Higher National Certificates, and Diplomas in subjects other than 'Plant Engineering' have been assessed by the Institution and approved for membership purposes. Degrees, degree course options, diplomas, and higher degree course options in plant engineering are available at certain universities in the UK. Further guidance on courses and their entry requirements may be obtained from technical colleges or universities or from the Institution's membership department.

Table 2.1 Summary of IPlantE membership requirements

2.6 Registration with the Engineering Council

An individual engineer's registration with the Engineering Council is a further valuable indication of professional attainment and standing. Royal Charter established the Engineering Council in 1981 to:

Table 2.1 Summary of IPlantE membership requirements

Class of membership	Minimum age (years)	Minimum academic qualifications	Evidence of competence	Minimum responsibility
Student Member	16	Engaged in engineering studies and training		
Graduate Member	18	BTEC NC/ND or HNC/HND or degrees in EC approved subjects	Engaged on an EC approved system of training and experience	
Associate	–	–	Employed in an allied industry or profession	
Associate Member (AMIPlantE)	21	BTEC NC/ND or ONC/D or CGLI Part II in EC approved subjects	4 years combined training and experience	
Member (MIPlantE) (i)	23	BTEC HNC/HND or HNC/D or CGLI FTC in EC approved subjects	4 years combined training and experience	2 years of responsible experience
Member (MIPlantE) (ii)	35	Technical Paper and Interview	15 years combined training and experience	2 years of responsible experience
Member (MIPlantE) (iii)	26	At Membership Panel's discretion	8 years combined training and experience	2 years of responsible experience
Fellow (FIPlantE) (i)	25	EC approved degree and interview	4 years combined training and experience	2 years in responsible appointments
Fellow (FIPlantE) (ii)	35	Technical Paper and interview	15 years combined training and experience	2 years in responsible appointments
Fellow (FIPlantE) (iii)	35	At Membership Panel's discretion	15 years combined training and experience	2 years in responsible appointments

1. Promote and develop the science and best practice of engineering in the UK;
2. Ensure the supply and best use of engineers;
3. Coordinate the activities of the engineering profession.

The Charter empowers The Engineering Council to establish and maintain a Register of qualified engineers. The registrants may, where appropriate, use one of the following titles and designatory letters:

Chartered Engineer (CEng)
Incorporated Engineer (IEng)
Engineering Technician (Eng Tech)

Each of these three qualifications is obtained in three stages. Stage 1 indicates attainment of the required academic standard, Stage 2 that approved training has been received and Stage 3 that responsible experience has been gained. The titles may only be used at Stage 3.

The Institution of Plant Engineers is a Nominated Body of the Engineering Council (EC) and is thus able to nominate members in appropriate membership grades for EC registration.

2.7 Registration as a European Engineer

Registration with the European Federation of National Engineering Associations (FEANI) is now open to UK engineers, and may be helpful to careers in post-1992 Europe. Such registration is available at two levels, Group 1 and Group 2. Group 1 is normally appropriate for engineers having the education, training and experience to qualify them for the title Chartered Engineer. Group 2 is approximately appropriate for those qualified to Incorporated Engineer level, but at the time of writing, the matter has not been finalized. Further information and FEANI application forms are available from the IPlantE's membership department.

As mentioned above, FEANI Group 1 registration is for those registered as Chartered Engineers. Registration with FEANI will allow the engineer concerned to use the title European Engineer. This title has the designatory letters Eur Ing, which should be used as a prefix (for example, Eur Ing John B. Smith, CEng MIPlantE).

2.8 Professional engineering development

Throughout the professional life of most engineers, there is a need to acquire new knowledge to enable them to tackle the technical and managerial problems that they face from day to day. Recent advances in technology, materials and processes emphasize this need, but with ever-increasing demands on time, opportunities to attend full-time courses are few. The plant engineer must therefore rely upon a Continuing Education and Training (CET) program to enable successful updating to take place, thus enhancing his or her professional development.

The Engineering Council places considerable emphasis on CET as an essential part of a professional engineer's development, anticipating that in due course CET will form a normal part of an engineer's career and that such CET activity will be noted in his or her personal career record.

To enable those engineers engaged in plant engineering to look to the future, the Institution of Plant Engineers has formulated a simple procedure for recording an engineer's attendance at activities, which contribute to CET and have been approved by the Institution for that purpose. Further information may be obtained from the Institution.

2.9 Addresses for further information

The Institution of Plant Engineers
77 Great Peter Street
London SW1P 2EZ
Telephone 020-7233 2855

Business and Technician Education Council
Central House
Upper Woburn Place
London WC1 0HH
Telephone 020-7388 3288

Scottish Vocational Education Council
Hannover House
24 Douglas Street
Glasgow G2 7NQ
Telephone 0141-248 7900

City and Guilds of London Institute
76 Portland Place
London W1
Telephone 020-7580 3050

The Engineering Council
Canberra House
Maltravers Street
London WC2R 3ER
Telephone 020-7249 7891

3

The Role of the Plant Engineer

R Keith Mobley
The Plant Performance Group

Contents

3.1 Responsibilities of the plant engineer

The increasing mechanization of industrial installations has resulted in the use of more complex and costly equipment and this has greatly increased the responsibilities of the plant engineer. In today's environment, the plant engineer must have a practical, well-rounded knowledge of the fundamentals of civil, mechanical, electrical, process and environmental engineering. In addition, plant engineers must have a basic knowledge of business management, statistical analysis, communications and effective supervision skills.

The plant engineer by definition must be a generalist who has a basic knowledge of all aspects of business. Because of these expansive skill requirements, the plant engineer must have the training, experience and expertise necessary to fulfill this critical role in the organization. In part, a plant engineer is responsible for:

- Design and modification of production systems and auxiliary equipment
- Production system specification and selection
- Installation and commissioning of plant systems
- Operation and maintenance of plant services
- Plant safety, energy conservation, pollution control and environmental compliance
- Process troubleshooting and optimization

3.1.1 Design and modification of production systems and auxiliary equipment

In a traditional organization, the plant engineer is the single source of design knowledge. Therefore, he or she is responsible for all design or redesign of plant systems. With the increasing complexity of plant systems, the plant engineer must have a thorough knowledge of machine design practices (i.e. mechanical, electrical, electronic and microprocessors).

3.1.2 Production system specification and selection

Plant engineering provides the technical knowledge and experience needed to properly specify and select new or replacement production, plant services and maintenance systems.

3.1.3 Installation and commissioning of plant systems

Proper installation of new production and plant services systems is essential for long-term performance of these systems. The plant engineering function has sole responsibility for assuring proper installation criteria is followed.

In addition, the plant engineer is responsible for testing newly installed systems to assure that they comply with procurement and performance specifications.

3.1.4 Operation and maintenance of plant services

In traditional organizations, the plant engineer is responsible for the operation and maintenance of all plant services (i.e. electric and steam generation, water treatment, waste treatment, etc.). In locations where these services are provided by outside sources, the plant engineering function is responsible for the internal distribution of electricity, steam and other services and the supervision of the outside service provider.

3.1.5 Plant safety, energy conservation, pollution control and environmental compliance

Generally, the plant engineering function is responsible for overall plant safety, as well as all compliance issues. The plant engineer must adapt to the constantly escalating federal, state and local regulations that govern these compliance issues.

3.1.6 Process troubleshooting and optimization

Perhaps the most important role of the plant engineer is process optimization. This function has the sole responsibility for improving the reliability and performance of production and auxiliary systems.

As a profession, plant engineering is on the decline. In many plants, this critical function has be discontinued or replaced with functions that provide part of the role describe in the preceding paragraph. In part, reliability, production and maintenance engineers have replaced the plant engineering function. The loss of single accountability that has resulted from the dilution of the plant engineering function has had a severe, negative impact on overall plant performance and corporate profitability.

In today's plant, the plant engineering function has been reduced to project management, coordination of contractors that provide design, construction, operations and maintenance of plant facilities. This trend has seriously diluted the plant's ability to design, install, operate and maintain critical production systems. Hopefully, this trend will be short-lived and more plants will return to traditional plant engineering functions. This book and the information it contains is designed to provide the practical skills required by a fully functional, effective plant engineering functional group.

Physical Considerations in Site Selection

4

D Whittleton
Ove Arup & Partners, Industrial Division

Contents

4.1 Environmental considerations of valley or hillside sites

4.1.1 Effect of topography on prevailing winds and strengths

Apart from the obvious influence of topography in producing shelter or the enhanced exposure to wind, the influence of large topographic features can be sufficient to generate small-scale weather systems, which are capable of producing significant winds. Three types of wind are associated with topography: diurnal winds, gravity winds and lee waves.

Diurnal winds

Under clear skies in daytime the slopes of hills and mountains facing the sun will receive greater solar heating than the flat ground in valley bottoms. Convection then causes an up-slope flow, called anabatic wind, which is generally light and variable but which can often initiate thunderstorms. At night, the upper slopes lose heat by radiation faster than the lower slopes and the reverse effect happens, producing down-slope katabatic winds. However, the denser cold air falling into the warmer valley can produce strong winds in a layer near the ground. The higher the mountains, the stronger are the effects.

Gravity winds

The effect of katabatic winds can be much enhanced if greater differences in air temperature can be obtained from external sources. A continuous range of mountains can act as a barrier to the passage of a dense mass of cold air as it attempts to displace a warmer air mass. Cold air accumulates behind the mountain range until it is able to pour over the top, accelerating under gravity to give strong winds down the lee slope.

Lee waves

Under certain conditions of atmospheric stability, standing waves may form in the lee of mountains. This wave motion is an oscillating exchange of kinetic and potential energy, excited by normal winds flowing over the mountain range, which produces alternately accelerated and retarded flow near the ground. Sustained lee waves at the maximum amplitude are obtained when the shape of the mountain matches their wavelength, or when a second range occurs at one wavelength downstream. Unusual cloud formations often indicate the existence of lee waves, in that they remain stationary with respect to the ground instead of moving with the wind. These clouds are continuously forming at their upwind edge as the air rises above the condensation level in the wave and dissipating at the downward edge as the air falls again.

Conditions are frequently suitable for the formation of lee waves over the mountainous regions of the US, an effect that is routinely exploited by glider pilots to obtain exceptionally high altitudes. The combination of lee waves with strong winds that are sufficient to produce damage to structures is fortunately rare, but do occur in hazardous mountainous regions.

Other factors

Other factors to account for topography with regard to valley or hillside sites should include possible inversion and failure to disperse pollutants. Temperature inversion occurs when the temperature at a certain layer of the atmosphere stays constant, or even increases with height, as opposed to decreasing with height, which is the norm for the lower atmosphere. Inversions may occur on still, clear nights when the earth and adjacent air cools more rapidly than the free atmosphere. They may also occur when a layer of high turbulence causes rapid vertical convection so that the top of the turbulent layer may be cooler than the next layer above it at the interface.

The running of a cool airflow under a warm wind is another cause of temperature inversion. As a rule, the presence of an inversion implies a highly stable atmosphere: one in which vertical air movements is rapidly damped out. In such a situation, fog and airborne pollutants collect, being unable to move freely or be dissipated by convection.

Additional dispersal problems may occur when the prevailing wind occurs perpendicular to the valley or hill ridgeline. This may lead to speed up and turbulence over the valley or it may simply reduce the effect of airflow carrying away airborne pollutants.

It is possible to obtain wind data from the local or regional meteorological office for almost any location in the world, although these frequently require modification and interpretation before they can be used.

4.1.2 Design for wind

A structure may be designed to comply with any of the following information:

1. No specific details available.
2. Specified basic wind speed and relevant site data.
3. Specified design wind speed, with or without FOS.
4. Specified survival wind speed, with or without FOS.

When details are given they should be checked, if only by comparison with equivalent wind speeds derived from first principles, to ensure that they are reasonable. Depending on the specified requirements, the wind speeds may or may not utilize gust wind speeds as in CP3 (3) or mean hourly wind speeds, v, with applied gust factors.

4.1.3 Factored basic wind speed approach

Basic gust wind speed, V, is multiplied by a series of S factors, which adjust the basic values to design values for the particular situation. CP3 uses up to four S factors:

S_1: Topography factors
S_2: Ground roughness, building size and height above ground factors
S_3: Statistical factor
S_4: Directional factor

S_1 – *Topography factors*

The effect of local topography is to accelerate the wind near summits or crests of hills, escarpments or ridges and decelerate it in valleys or near the foot of steep escarpments or ridges. The extent of this effect on gust wind speeds is generally confined to mountainous regions, but may occur in other locations. Local topography is considered significant when the gradient of the upwind slope is greater than 5 per cent.

The shape of the upwind slope affects the degree of shelter expected near the foot of the slope when the slope is shallow and the flow remains attached. When the changes in slope are sudden, so a single straight line can approximate that upwind slope for more than two-thirds of its length, then the shape is *sharp*. Otherwise the changes of slope are gradual and the shape is *smooth*. This distribution is relevant for sites close to the foot of the upwind slope, where *sharp* topography offers a greater degree of shelter.

S_2 – *Ground roughness, building size and height above ground factors*

The factor S_2 takes account of the combined effect of ground roughness, the variation of wind speed with height above ground and the size of the building or component part under consideration. In conditions of strong wind the wind speed usually increases with height above ground. The rate of increase depends on ground roughness and also on whether short gusts or mean wind speeds are being considered. This is related to building size that take account of the fact that small buildings and elements of a building are more affected by short gusts than are larger buildings, for which a longer and averaging period is more appropriate.

S_3 – *Statistical factor*

Factor S_3 is based on statistical concepts and can be varied from 1.0 to account for structures whose probable lives are shorter (or longer) than is reasonable for the application of a 50-year return-period wind.

S_4 – *Directional factor*

In the latitudes occupied by the US the climate is dominated by westerly winds. The basic wind speed may be adjusted to ensure that the risk of it being exceeded is the same for all directions. This is achieved by the wind speed factor S_4.

When applying S_4, topography factor S_1 and the terrain roughness, building size and height above ground factor S_2 should be appropriately assessed for that direction.

4.2 Road, rail, sea and air access to industrial sites

4.2.1 Introduction

Many industrial processes and factories require specific accessibility for one particular form of transport. Examples

of the above include distribution warehousing, transport operations and those industries dealing with bulk commodities (e.g. oil refineries). For other industries access to strategic modal networks is important in order to be competitive where cost of transport and timesaving are significant factors. Examples of these operations include air freighting and fresh-food deliveries. A third category would include those establishments which would require high-visibility sites to enhance their reputation in the marketplace.

4.2.2 Design considerations

It is difficult to give specific advice on this subject, as there is a very large range of industrial undertakings. The awareness for, and acceptability of, access is dependent on the types of goods to be moved and the frequency and method of movement. In some undertakings there is a major movement between different transport modes, which is concentrated either at ports or at major road/rail interchanges.

In addition to the amount of commercial traffic it is vital to consider the movement associated with employees and visitors, which themselves can generate large numbers of vehicular and pedestrian movements. For very large manufacturing sites there will also be the need for accessibility for public transport, which, for a large workforce, may need to be supplemented by investment in subsidized travel.

Site access will reflect the nature of the existing local transport system and will need to be designed to cater for the anticipated future traffic flows associated with on-site development. At the extreme of the range this could include a significant on-site infrastructure, potentially involving small bus stations for staff or private rail sidings for goods heavily committed to using the rail network. Special consideration might also need to be given to customs facilities, where operations include cross-border movements with or without bonding operations.

4.2.3 Forms of site access

Access to the road network can range from a simple factory gate or location on a business park to a major industrial complex requiring its own major grade separated interchange due to the high traffic volumes on the strategic road network. New site developments will need to cater for future traffic growth and must be adequate to deal with a design life over the foreseeable future.

Access to a seaport will be limited by the ability of total traffic generated by the docks and the incorporation of these traffic movements into the local road system.

Air traffic access may be constrained by the operational aspects of the airport. Otherwise, the road-related traffic will be dealt with in a manner similar to that of seaports, except that the vehicles are likely to be smaller in size and of lower traffic volumes, reflecting the higher-value goods being transported by air.

4.2.4 Access to the road system

Before access is obtained to any road it is necessary to obtain the consent of the relevant highway authority. Direct access to freeways or limited access highways is generally prohibited and the policy regarding access to trunk roads is to minimize the number of accesses and to encourage the free flow of traffic on these major roads. Therefore careful consideration needs to be given to the ability of the proposed access to cover traffic capacity and road safety adequately. The local town or county council is the highway authority, in non-metropolitan areas, for all other roads, although, in many instances the local authority may have agency powers for the roads within its area.

It will be necessary to forecast the amount of traffic to be generated by the development within the site and to propose a form of junction that not only deals with the site's traffic but also adequately caters for the existing traffic on the road. Tests for capacity are required and attention should also be given to the safety of operation of the proposed access.

As part of the planning approvals it is increasingly common to provide road-improvement schemes that are sometimes off-site and are necessary to deal with site-generated traffic, which has detrimental effects on the local road network. Generally, these agreements require the applicant of the proposed site to carry out specified highway improvement schemes to an agreed timetable relevant to the planning application.

4.2.5 Selection of sites

Suitable sites are normally limited to those areas designed in development plans as being for industrial or commercial uses. Such land should be capable of being accessed directly from the primary or secondary distributor roads in the area. Segregation of trucks and truck access from residential areas should be achieved where possible.

The utilization of existing or the provision of new railheads will also be a determining factor for some operators, and frequently the rail sidings do not have good road access. In these cases extensive improvement measures may be necessary to provide adequate space and geometrical requirements.

4.2.6 Checklist

The following list, while not exhaustive, identifies many of the issues which will need careful consideration. In many instances it might be necessary to seek the advice of a specialist traffic consultant, either in the design of a scheme or in access, or to negotiate with the highway authority the impact of a proposed development and any attendant road-improvement schemes.

1. Types of operation to be carried out

- Number of trucks
- Staff cars
- Visitors' cars

- Rail/water/air access
- Public transport provision
- Cyclists
- Pedestrians

2. Types of site

- Large single site
- Industrial estate
- Segregation of access for lorries and cars
- Capacity of access and need for improvement
- Ensure no backup of traffic onto highway
- Ensure sufficient on-site space for all vehicles to enter highway in forward gear
- Ensure off-highway loading/unloading
- Access for emergency vehicles

3. Access arrangements

- Access width should be a minimum of 20 feet (6.1 meters) to allow trucks to pass each other (25 feet or 7.3 meters is ideal)
- Single access could cope with up to 250 truck movements per day
- Any gate or security barrier to be set in at least 65 feet (20 meters) from public highway to avoid blockage or interference to pedestrians
- Minimum center line radius to be 39 feet (12 meters)
- Minimum entire live radius to access road to be 197 feet (60 meters). Widening on bends may be required

4. Maneuvering space

- Turning circle for articulated vehicles to be 62 feet (26 meters) diameter minimum
- For draw-bar vehicles this can be reduced to 69 feet (21 meters)
- Turning head for rigid trucks only needs to be 115 feet (35 meters) long
- Turning head for articulated vehicles should be 174 feet (53 m) long. Curb radii need to be 30 feet (9 m)
- Loading bays at 90° to road should be 102 feet (31 m) deep including the road width. Bay should be 12 feet or 3.5 meters wide
- Strong site management is required to ensure maneuvering space is kept clear of storage/goods/debris at all times
- Headroom clearance should be a minimum of 15.25 feet (4.65 m) with careful consideration to ensure all pipework, etc. is above that level. Approach gradients to flat areas will reduce the effective height.

It is emphasized that the above checklist is not exhaustive. Any reductions in the standards identified above will lead to difficulty of operation, tire scrub, potential damage to vehicles and buildings, and general inefficiency. Cost effectiveness could also be hindered due to loss of time caused by blocked-in vehicles. Safety is also a highly important factor, which should be prominent in any decision-making.

4.3 Discharge of effluent and general site drainage

4.3.1 Effluent

The control of drainage and sewerage systems and of sewage disposal is governed by federal, state, county and city regulations and varies depending on the specific area.

Methods of treatment

Two methods of treatment can be considered:

1. On-site treatment and disposal; and
2. Off-site treatment and disposal.

Where on-site treatment is to be undertaken consideration should be given to the following:

1. Where large volumes of effluents are produced and/or different types of contaminants, large equipment areas may be required. Sufficient space must also be allowed for maintenance and inspection of such equipment.
2. Settlement/storage areas for effluent need to be sized not just for average flow but also for peak periods. Where production is based on a shift system, peak flows created during holiday periods (shutdown, major maintenance, etc.) should be considered.
3. Where effluents require primary, secondary and possibly additional tertiary treatment, attention should be paid to the various treatment processes with regard to personnel safety and public sensitivity to on-site treatment.
4. Where concentrated alkali and/or acids are stored and used on-site as part of the treatment process, care should be exercised to prevent misuse, fire, and security and health hazards. The provision of emergency showers, eyewash stations, etc. needs careful consideration.
5. If equipment malfunctions during the treatment process, adequate precautions should be taken to prevent the discharge of untreated effluent. Such precautions should be the provision of emergency collection tasks or the use of approved, licensed effluent-disposal traders.
6. Where accidental discharge of untreated effluent does occur, the appropriate water authority and/or environmental health officer should be advised immediately. All steps should be taken to limit the extent and intensity of any potential contamination.
7. Where small and/or single contaminant effluents are encountered, packaged treatment plants may be acceptable. Consideration should, however, be given to capital cost, payback period, reliability of equipment, maintenance, plant-life expectancy and contaminant-removal efficiencies.
8. Pipework material for conveying effluent to treatment plants should exhibit resilience to corrosive attack by the effluent as well as scouring and erosion created by the material content of the effluent.
9. Consideration should be given to plant operation in a shift system and any requirements for an analyst to be present during operational/non-operational periods.
10. Precautions must be taken against freezing for external pipework, tanks, meters, gauges, and monitoring equipment.
11. Assessments should be made for electrically operated process equipment that may require an essential power supply in the event of a main failure.
12. The quality of the effluent discharge must be regularly checked. Depending on the quantity and type of discharge, this may require an in-house laboratory and analysis room.
13. The water authority may limit the quantity of final treated effluent, and monitoring of the final out-fall may have to be considered in conjunction with a holding tank.
14. Large or small on-site treatment plants will create sludge concentrates that require disposal. Where large quantities of sludge occur, on-site de-watering filters may be considered with dry sludge cakes properly removed from site by licensed contractors. Alternatively, small quantities of wet sludge concentrates may be removed and disposed of by similar contractors.

Where off-site treatment is undertaken the following should be considered:

1. Cost comparison with on-site treatment.
2. Availability of approved, licensed contractors to handle the type of effluents being considered.
3. Reliability of licensed contractors during emergency, weekends and holiday periods.
4. Space requirements for holding untreated effluent prior to removal from site.
5. Accessibility, safety and security associated with the holding vessels by the vehicles of the licensed contractors.
6. Suitable pumps may be required to pump from holding tanks into licensed contractor vehicles.

4.3.2 Site drainage

The discharge of surface water from a site may originate from three potential sources: rainwater from building(s), surface-water runoff from paved/hard standing areas and subsoil drainage (groundwater)

1. The rainwater runoff from buildings depends on the geographical location and storm-return period specified. Rainwater runoff from a roof is relatively clean and can discharge directly to a watercourse, lake, etc. without passing through an interceptor.
2. The surface water runoff from paved/hard standing areas also depends on rainfall intensity calculated from the geographical locations of the site and storm-return period. However, the return period for a site will be far higher than for a building in order to ensure prevention of persistent flooding of the site. In many instances the local authority may specify the storm-return period as the design criterion.

Where development of a greenfield site or an extension to an existing building takes place, the rate of storm water runoff may necessitate the provision of a balancing pond or reservoir.

While the drainage design may be able to cater for minimal surcharging, any substantial rise in floodwater can be contained by the balancing pond and minimize flooding to the site and damage to plant ecology. Lining of a balance pond must be considered to prevent water seepage. Suitable linings are clay or butyl rubber sheeting.

Where potential flooding to the site is minimal an alternative to the balance pond could be an open-trench system that could provide on-site storage and added security to watercourses.

The open-trench system provides easy maintenance and may obviate the need for highway curbs and gullies, but requires adequate security to prevent vehicle collision.

Surface water from hard standing areas subject to gasoline and oil contamination (e.g. car parks) must pass through an interceptor prior to discharge to a watercourse, surface-water drain, etc. The sizing of gasoline/oil interceptors varies between local, state and regional areas. Many agree that the first 5 minutes of surface-water runoff is the most contaminated, and accept a reduced interceptor size, while allowing a higher flow rate, caused by increased storm intensity, to bypass the interceptor chamber (see Figures 4.1 and 4.2).

3. Subsoil drainage may be required for the following reasons:
   (a) Seasonal fluctuations in water-table level may cause isolated flooding. Subsoil drainage may be considered to keep the water-table level relatively constant.
   (b) Where underground springs occur, layering of subsoil drainage may be considered to maintain water-level equilibrium within the subsoil to permit building construction or similar activities.
   (c) Draining of permanently flood land such as marsh or bog land.

Gravity or pumped discharge

Wherever practical, surface-water runoff should be designed for gravity discharge, preferably located at the lowest part of the site. However, where a gravity system is impractical or impossible a pumped discharge must be considered.

When sizing a pumped discharge the following should be evaluated:

1. Pumps, valve controls, etc. to be in duplicate;
2. Maximum discharge permitted into watercourse by water authority;

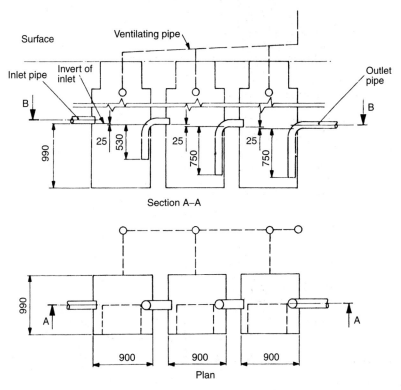

Figure 4.1 Traditional three-chamber gasoline interceptor. (Dimensions are in millimeters.) All pipes within the chambers through which liquid passes should be of iron or another equally robust petrol-resistant material

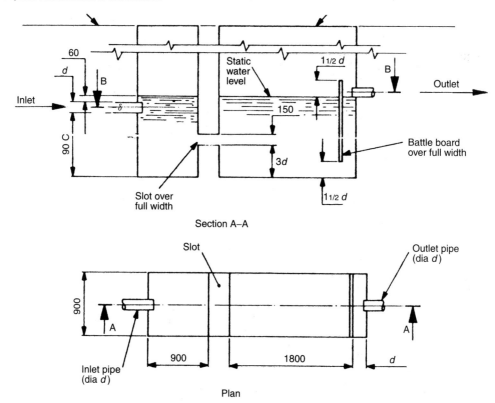

Figure 4.2 An alternative two-chamber gasoline interceptor. (Ventilation not shown: dimensions are in millimeters)

3. Discharge duty to be twice dry weather flow subject to (2) above;
4. Will essential power supply be required to both pumps?
5. Space requirements for pumps, including installation, maintenance and inspection.

Figures 4.3–4.7 show edge and bank details for storm-retention reservoirs.

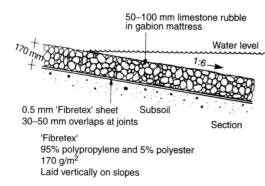

0.5 mm 'Fibretex' sheet   Subsoil
30–50 mm overlaps at joints

'Fibretex'
95% polypropylene and 5% polyester
170 g/m²
Laid vertically on slopes

Figure 4.3 Rip-rap bank (*Fibretex*/stone)

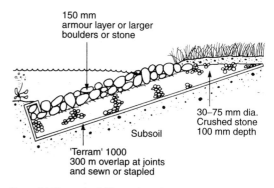

Figure 4.4 Rip-rap bank (*Terram*/stone)

4.4 Natural water supplies, water authority supplies and the appropriate negotiating methods and contracts

There are three major sources of water supply:

Borehole
Rivers
Service reservoir

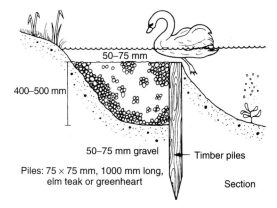

Piles: 75 × 75 mm, 1000 mm long,
elm teak or greenheart

Figure 4.5 Piling margin

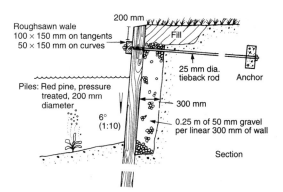

Figure 4.6 Piling edge

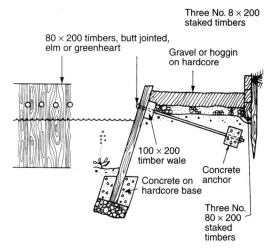

Figure 4.7 Board wall/path

The majority of water supplies are provided from a combination of all three.

A wide variety of federal regulations as well as state and local laws govern the supply of water to domestic and industrial premises. Within these are included the statutory undertakings for preventing waste, undue consumption, misuse and contamination of water supplied by the various authorities.

Having established the respective local regional water authority and water company, the following is a list of information that the water authority will require to assess a consumer's needs:

1. Project name
2. Site address
3. Building size (m²)
4. Number of occupants and anticipated multiple use/tenancy
5. Building usage
6. Point of supply connection – height above ground
7. Anticipated flow rated for the following uses:
   (a) Domestic
   (b) Industrial
   (c) Fire
(Note: Flow rates should include average and maximum demands and the periods throughout the year during which supplies are required.)
8. Cold water storage volume
   (a) Domestic
   (b) Industrial
   (c) Fire

Once the water authority has established that a water supply is available, the consumer should obtain the following information:

1. Copy of water authority bylaws.
2. Marked-up layout drawing indicating location, size, depth and maximum and minimum pressure of the main in the vicinity of the site.
3. An indication of which mains may be considered for:
   a. Domestic use
   b. Industrial use
   c. Fire use
4. Confirmation as to whether the water main passes through any adjacent property which may involve easements to gain access for maintenance, etc.
5. Confirmation as to whether the water flows and pressure will be affected by usage from adjacent buildings. The average roof should have a duration/runoff time not exceeding two minutes. Periods of greater duration are shown for information and completeness only.
6. Confirmation as to whether a guarantee of security of supply can be provided throughout the year, including any anticipated periods of drought.
7. Confirmation of the source(s) of water supplies and provision of a current water analysis.
8. Confirmation that the authority will supply the water meter and housing.
9. Confirmation of the authority's preferred pipe material for incoming mains and any materials not recommended due to aggressive soil.

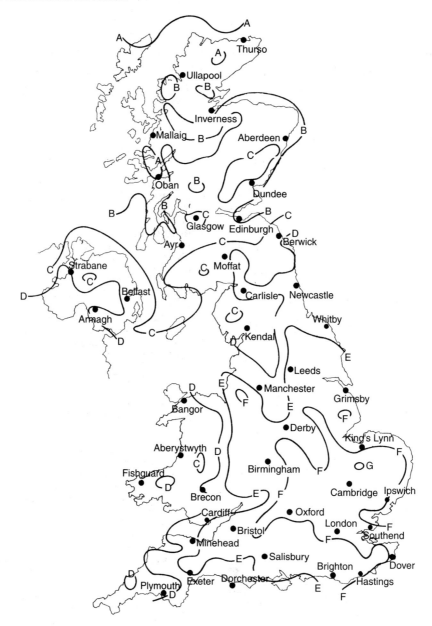

Figure 4.8 Key to rainfall tables

10. Confirmation that direct mains boosting is allowed.
11. Provision of rates of charges (depending on annual consumption),
12. Confirmation of costs of maintenance and metering charges.
13. Provision of details relating to any future authority plans that may affect supplies to the site.

14. Confirmation of total capital cost of supplying a water supply.
15. Provision of a fixed time-frame installation program.

While the above list is not exhaustive, it provides an indication of the necessary negotiated issues to be resolved between a consumer (or their representative) and the water authority.

4.5 Water storage, settling wells and draw-off regulations

4.5.1 Water storage

Care must be taken when assessing water storage, as some water authorities have special powers to restrict total water storage retained for a given building and/or site. Early consultation with the appropriate water authorities is also recommended when large volumes of water storage/usage are anticipated, especially if heavy demand is required for industrial purposes.

The necessity of water storage may be outlined as follows:

1. To protect against interruption of the supply caused by burst main or repair to mains, etc.;
2. To reduce the maximum of demand on the mains;
3. To limit the pressure on the distribution system, so reducing noise and waste of water due to high-pressure mains and enabling higher-gauge and cheaper material to be used;
4. To provide a reserve of water for firefighting purposes;
5. Additional protection of the mains from contamination, i.e. prevention against back siphonage.

In designing the water storage capacity, account should be taken of the pattern of water usage for the premises and, where possible, to assess the likely frequency and duration of breakdown from the water authority mains. When dealing with domestic water storage, this is usually provided to meet a 24-hour demand.

However, to apply the same philosophy to industrial buildings would be incorrect, and due consideration must be given to the effect of loss of water supply to the process/manufacturing production.

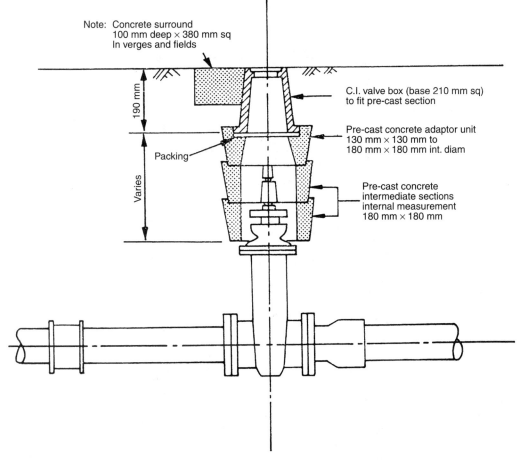

Note: Concrete surround
100 mm deep × 380 mm sq
In verges and fields

190 mm

Varies

Packing

C.I. valve box (base 210 mm sq)
to fit pre-cast section

Pre-cast concrete adaptor unit
130 mm × 130 mm to
180 mm × 180 mm int. diam

Pre-cast concrete
intermediate sections
internal measurement
180 mm × 180 mm

Figure 4.9 Standard pipe installations for water sluice valve up to 10 inches or 250 millimeter diameter. *Notes:* (1) If a valve extension spindle is required, a galvanized centering support must be provided. (2) Flanged valves with appropriate adapters must be used. (3) Unless otherwise directed, pre-cast concrete units shall be bedded on well-compacted granular material Type A brought up from the base of the trench

When assessing water storage the following should be considered:

1. Where there is a requirement for large volumes of water, storage may be sized on quantity of usage per shift;
2. Frequency of interruption to mains supply;
3. Space requirements;
4. Cost of water storage tank and associated supports;
5. Protection against frost;
6. Type of industrial usage and effect of loss of water supply to production;
7. Minimizing the risk of Legionnaires' disease.

4.5.2 Draw-offs

In the US, maximum draw-off from water mains differs throughout the various water authorities. What may be acceptable in one area may not be acceptable in another. Water mains may generally be able to supply directly to domestic buildings but not industrial premises, where constant flow and negligible fluctuation in pressure is required. Contamination of water supply by back siphonage is of major concern to water authorities.

Water consumption for commercial and industrial premises are metered and priced per cubic meter or gallons of usage basis, with additional costs for reading and maintenance of the meter station(s). It is important that assessment of water usage is as accurate as possible, as water charges are based on a sliding scale, i.e. as volume of water usage increases, cost per unit decreases.

Where production in new premises is to be phased with peak usage only being attained after a period of years it is important to consider this during initial discussions with the relevant water authority when negotiating unit rate costs. See Figures 4.9–4.17.

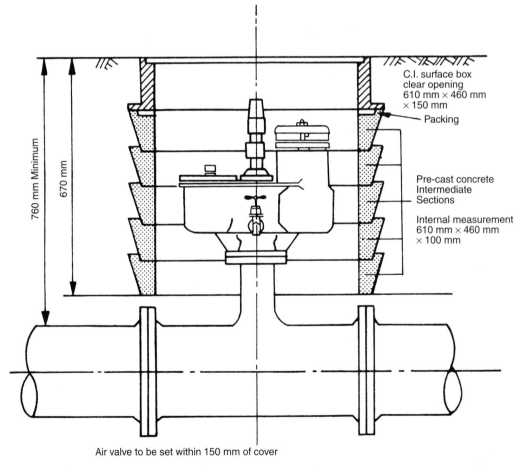

Air valve to be set within 150 mm of cover

Figure 4.10 Standard pipe installations for a double air valve. *Note:* Unless otherwise directed, pre-cast concrete units shall be bedded on well-compacted granular material Type A brought up from the base of the trench.

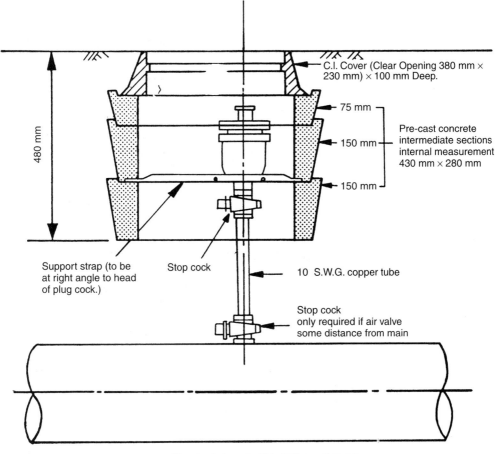

480 mm

C.I. Cover (Clear Opening 380 mm × 230 mm) × 100 mm Deep.

75 mm

150 mm — Pre-cast concrete intermediate sections internal measurement 430 mm × 280 mm

150 mm

Support strap (to be at right angle to head of plug cock.)

Stop cock

10 S.W.G. copper tube

Stop cock only required if air valve some distance from main

Air valve to be set within 150 mm of cover.

Figure 4.11 Standard pipe installations for a single air valve. *Note:* Unless otherwise directed, pre-cast concrete units shall be bedded on well-compacted granular material Type A brought up from the base of the trench

4.6 Problem areas associated with on-site sewage treatment for isolated areas

The output of a foul-water drainage system should discharge into a foul-water or combined drainage system (foul and surface water). Where such a drainage system is not conveniently available and cannot economically be extended to a site, other methods of foul-water disposal will be necessary, either a cesspool or a septic tank.

4.6.1 Cesspools

A cesspool is a tank normally located underground that is designed to store the entire foul drainage discharge from premises between disposals by tank vehicles(s). The following points should be noted when considering the use of a cesspool:

1. Minimum capacity of 18 cubic meters (m^3) measured below the level of the inlet (or as equivalent to approximately 45 days' capacity);
2. It must be covered and impervious to rainwater, groundwater and leakage;
3. It must be ventilated;
4. It must be positioned such as not to pollute any water sources or cause a public nuisance;
5. It must be sited so that the contents may be removed other than through a building, with reasonable access for tanker vehicles where required;
6. Its position should also be considered relative to future sewerage systems.

Cesspools should only be utilized in extremely remote areas and after consideration of all other forms of effluent disposal has been undertaken.

4.6.2 Septic tanks

A septic tank is a purification installation designed to accept the whole sewage/trade discharge from premises. Its construction is such that it allows the settling out

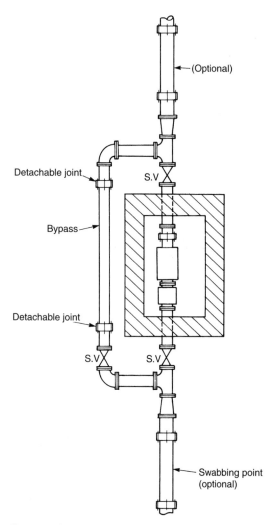

Figure 4.12 Standard bulk meter layout for a brick-built chamber (230 mm brickwork on a concrete base). *Notes:* (1) the flowmeter should be either on straight or bypass to suit specific site (i.e. to keep meter off road). (2) Chamber cover should be approximately 685 mm × 510 mm

of the sludge content of the incoming effluent and renders the final effluent acceptable, by prior agreement with the water authority, for discharge to a watercourse or drainage area. The sludge within the septic tank decreases in volume by the action of microorganisms changing the sludge from aerobic to anaerobic.

The following points should be noted when considering the use of a septic tank:

1. It should be impervious to rainwater, groundwater and leakage;
2. It must be positioned such that a tanker vehicle may periodically remove the contents;
3. Percolation filter and/or sub-surface irrigation may be required to provide aeration and final purification;
4. The change of bacterial action from aerobic to anaerobic produces methane gas that must be ventilated to atmosphere;
5. Its position should also be considered relative to future sewerage systems;
6. Minimum capacity of 27001 to be provided. Subject to local authority approval, the effluent output from a septic tank may be connected to an adopted sewer subject to quality of output and size and location of drainage.

4.7 Landscaping on industrial and reclaimed land

4.7.1 General

In the context of overall landscaping, here we will concentrate on what is normally recognized as soft landscaping, i.e. that area which includes waterbodies and growing plants. The primary problems that have to be solved in any scheme for creating a new growing, living landscape out of a reclaimed land, are:

1. Contamination in the ground. (There may be contamination in the air, but this is beyond the scope of this chapter.)
2. Soil structure and land drainage within the reclaimed ground.

4.7.2 Contaminated land

It is not unusual for industrial land or land formed from tipped waste to be contaminated to some degree. Heavy contamination will have to be dealt with as a particular engineering problem, i.e. by sealing, burying or removing any highly contaminated material. Moderate contamination can normally be dealt with *in situ* by dilution of the concentrated contaminant. Whatever treatment is utilized, steps must be taken to prevent the leaching of any contaminants through the soil.

As far as the effects of contamination are concerned, the main problems associated with harm to plant growth

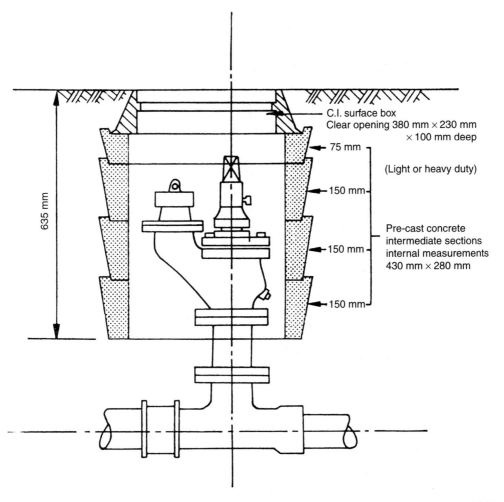

635 mm

C.I. surface box
Clear opening 380 mm × 230 mm
× 100 mm deep

75 mm

(Light or heavy duty)

150 mm

Pre-cast concrete
intermediate sections
internal measurements
430 mm × 280 mm

150 mm

150 mm

Figure 4.13 Standard pipe installations for a fire hydrant. *Notes:* (1) Hydrant Tee must be appropriate to type of main installed. (2) Depth of hydrant outlet must not exceed 300 mm below finished ground level. (3) Unless otherwise directed, pre-cast concrete units shall be bedded on well-compacted granular material Type A brought up from the base of the trench

are the presence of particular heavy metals. These are poisonous to plant life and/or lack of oxygen in the soil growing medium due to the presence of gases produced by the contaminated ground (for example, methane). Lack of oxygen can also be the result of a poor soil structure, over-compaction (no air voids), etc.

The Environmental Protection Agency sets guidelines for safe limits to heavy metals content in the soil for reasons of both public and plant health. Before planting regimes are implemented in any reclaimed ground these heavy metal levels must be assessed and, if high, reduced. Acidic or alkaline soil also will need neutralizing and the type of proposed planting chosen for the soil type.

As indicated earlier, heavy contamination can be buried, sealed or removed. Burying of the material should be well below the root growth zone, and this is normally taken as 3.0 m below the final ground-surface level. Sealing for heavy contamination to prevent vertical or lateral leaching through groundwater flow can be with compacted clay or proprietary plastic membranes. Removal from site of the contaminants is normally only contemplated in a landscaped scheme where the material, even at depth, could be a hazard to public health directly or phytotoxic to plant life.

Where there is a landfill gas-generation problem, active (i.e. pumped) or passive venting, by way of stone-filled trenches at least 3 feet deep, may be needed. Where active

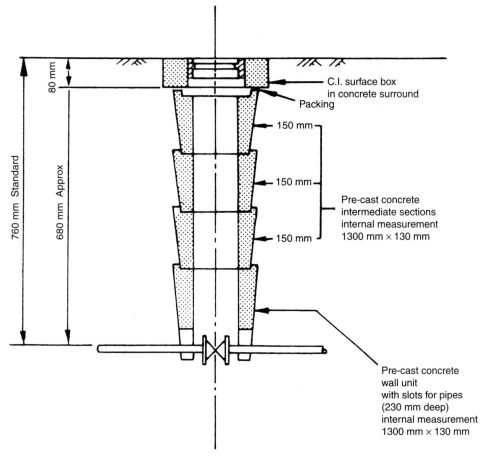

80 mm

760 mm Standard

680 mm Approx

C.I. surface box
in concrete surround

Packing

150 mm

150 mm

Pre-cast concrete
intermediate sections
internal measurement
1300 mm × 130 mm

150 mm

Pre-cast concrete
wall unit
with slots for pipes
(230 mm deep)
internal measurement
1300 mm × 130 mm

Figure 4.14 Standard pipe installations for a communication pipe stopcock. (It may be necessary to provide a base unit where polyethylene is being used)

venting is appropriate, the commercial use of the gas extracted can be considered.

Contaminated groundwater (leachate) should be kept below the root growth zone. Only rainwater or clean irrigation water should meet the needs of plants.

Any waterbodies constructed on the contaminated land which are to support aquatic life will need to be completely sealed against the underlying ground and inlet and outlet water provisions designed so that they are sealed against any flow of contaminated water into the waterbody.

As a general rule, young, immature plants should be chosen for any landscape-planting scheme on reclaimed, previously contaminated, land. This allows the plants to adapt gradually to such an environment. It is not normally appropriate to plant mature shrubs and trees to create an instant *mature* landscape in such an environment.

4.7.3 Non-contaminated land

For land that is reclaimed with *inert* or non-contaminated materials the main problem in creating a good growing medium tends to be in producing a soil structure profile within a limited period of time. A good profile will provide an oxygen- and nutrient-rich soil, not too acid or alkaline, which is well drained. Any reclamation project should aim to supply this ideal environment as soon as possible. To do this, a clear and firm specification for the reclamation material and how it is to be placed will be needed at the beginning. Extensive drainage with both the landform and the subsoil will also be required.

Imported topsoil or topsoil manufactured from imported nutrients mixed into existing soil will normally be required. Topsoil depths will vary in accordance with

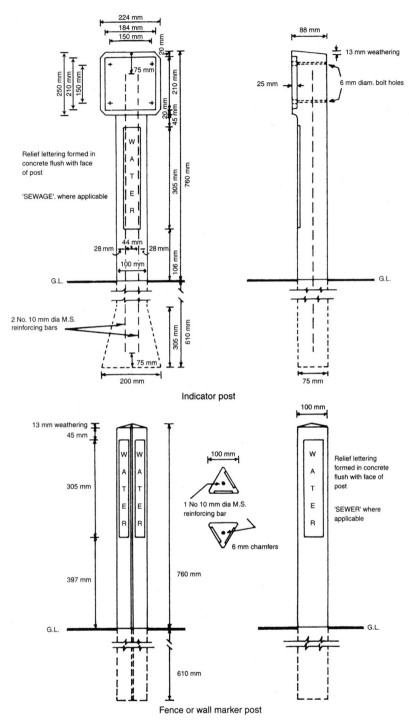

Figure 4.15 Pre-cast concrete indicator and marker posts

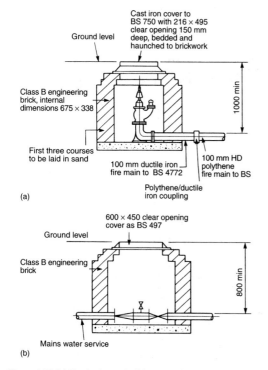

(a)

(b)

Figure 4.16 (a) Fire hydrant pit; (b) stop-valve pit. *Note:* To each fire hydrant an indicator plate with post must be provided and installed

the planting, from 16 inches (400 mm) for grassed areas to 5 feet (1.5 meters) for major trees. Under these conditions instant, mature landscapes can be constructed immediately.

In any scheme for creating a landscape on reclaimed land the ultimate aim will be not only to create a green landscape but also to provide an environment in which the ecosystem will develop quickly. Some important factors that must be taken into account here are:

1. Early structural planting to provide windbreaks and shelter for younger plants and wildlife to develop.
2. A balanced landscape needs waterbodies. These should be designed to have shallow margins to provide the appropriate conditions for wetland planting which, in turn, provide the balanced habitats for a rich wetland ecosystem.
3. A properly designed planting regime away from the water bodies will provide cover, space and the natural habitats for a wider wildlife system.
4. It is not normally necessary to seed waterbodies with fish. Fish will be introduced by natural means (e.g. by aquatic bird life).

Finally, in any projected landscape due consideration should be given to the future maintenance of the completed scheme. Unlike a normal building development, a landscape progressively develops with time. In order to maintain the original design concept a clear maintenance regime and organization should be established early in the project development process to start as soon as the landscape has been initially established.

5

Plant Location

Barry Holmes

W. S. Atkins Consultants Ltd

Contents

5.1 Selecting the location

Location is a strategic issue, and the decision where to locate cannot be taken lightly. It is the first decision in the implementation of a project. Upon the decision to proceed, an investment is made which is irreversible. That investment in *bricks and mortar* cannot physically be transferred to another location if the decision turns out to be wrong. In time, the assets may be realized but probably at much lower values than paid.

Traditionally, companies have sought to acquire competitive advantage over their rivals through their choice of location. In a historical context, firms tended to establish their factories for reasons of economic geography, e.g.:

1. Proximity to raw material source
2. Proximity to relatively cheap and abundant energy
3. Availability of relatively inexpensive manpower or specialist skills
4. Good transport links with materials suppliers and markets.

In more recent times the location decision has been influenced by government intervention, e.g.:

1. High tariff duties imposed on imports which encourage exporters to that country to consider setting up local operations
2. Investment incentives in the form of tax relief and grants
3. The provision of infrastructure, especially improved transport communication.

In today's world the decision is more complex. Markets are more sophisticated, skills can be in short supply, technological change can soon outdate newly installed processes, and there has been a phenomenal revolution in communications, both in terms of the physical movement of goods and people and of information around the world. As a consequence, companies often have to consider a wide number of options, and the eventual decision is based on optimizing the perceived net benefits.

The following are the key factors in selecting a location, and these are discussed in more detail below:

1. Protected markets and economies of scale
2. Government influences
3. Corporate matters
4. People matters

5.1.1 Factor costs

All other things being equal, the company will site its plant where it is cheapest to manufacture. This is determined by the nature of the product and of the manufacturing process. For instance:

1. Where the process discards significant quantities of raw material (especially if it is bulky), the economics favor locations close to sources of the raw material (for example, sugar milling, cotton ginning, the processing of non-precious minerals).

2. Where considerable quantities of energy are used in the process, relatively cheap and abundant sources of energy need to be available (for example, alumina and aluminum smelting, steel making).
3. Where the process is a labor intensive, low-cost workers are required in numbers (for example, textiles and clothing). The cost of capital has little effect on the choice of location as capital sources can be from anywhere. The ability, however, to repatriate profits and the proceeds from the sales of assets and exposure to foreign exchange risk are important if the location options are abroad. This factor becomes more relevant, the greater the capital intensity of the project.

Associated with factor costs are linkages with material suppliers, subcontractors and support services (e.g. temporary staffing agencies, travel agents, reference libraries, and office maintenance services). The availability and cost of these factors need to be considered.

5.1.2 Protected markets and economies of scale

All markets are protected to some extent. Domestic suppliers have, at times, the natural protection of lower transportation costs, a greater understanding of the local marketplace through its experience and knowledge of social customs and culture, and an easier ability to respond to domestic customer requirements. Added to this, tariff duties may have to be paid on imports.

Where products are demanded in volume, process technologies have been developed to minimize unit costs. The concept of the *minimum economic scale* of plant (MES: sometimes referred to as *minimum efficient size*) is the point beyond which increases in scale do not significantly reduce unit costs.

Where domestic markets are large enough, companies can establish MES plants or larger. When they are not, several alternatives need to be assessed, namely:

1. Setting up a plant of sub-optimal size to serve the local market based on the premise that the higher unit costs can be offset against factor cost advantages and the protection that the market affords the local producer.
2. Establishing a MES plant and seeking export opportunities.
3. Supplying, or continuing to supply, the local market from a MES plant established abroad.

5.1.3 Government influences

As already indicated government (local, state and national) try to attract inward manufacturing investment in various ways. For example, in the 1950s, 1960s and 1970s, this was largely developed behind high tariff duties on imports. Countries to adopt protectionist policies use infant-industry arguments, but companies must expect eventual pressures *to grow up* and face open market competition. With trends for countries to join economic unions such as entrance of Spain, Portugal and Greece into the EC, the advantages of tariff protection cannot always be guaranteed.

Local and regional governments are frequently involved in the promotion of their areas. In some places, particularly in Western developed countries, they are able to provide useful information on their locality, covering available sites, rents, land prices and details on transport communications. These bodies can be very effective in helping to finalize choice.

Governments often offer monetary incentives to prospective manufacturing investors. The types of inducements offered are:

1. Tax-free holidays; periods of possibly 3 to 15 years of no tax liability on profits (this also can apply to local, property or any other tax normally imposed on a firm);
2. Deferred tax allowances; where capital and operating costs in early loss-making years can be offset against profits later;
3. Liberation from payment of import duties on capital plant and equipment, and on materials;
4. Capital grants and/or low-cost loans for the purchase of plant and equipment;
5. Liberal depreciation allowances;
6. Training grants and/or facilities; to assist towards the cost of acquiring a suitably skilled workforce;
7. The provision of land and services at zero or nominal prices (these sometimes extend to factory units and accommodation for senior management staff);
8. The provision of *free zones*, usually in port locations, where goods can be imported duty free, processed and the bulk (if not all) exported.

Although many of these incentives are generally limited to the poorest regions of the country, where natural economic advantages to meet present-day opportunities are few, they are also widespread in developing countries.

Investment is more easily attracted to areas with adequate infrastructure. Governments encourage inward investment through the provision of industrial estates, sometimes with standard factory units ready for occupation; energy and water supply; roads suitable to withstand heavy vehicle traffic; other means of transport, airports, seaports, railways; and waste-disposal facilities. Progressive government departments concerned with industrial development make available a wide range of information about their areas, including local services, amenities, housing and education.

When locating in a foreign country the national government may insist on a certain level of *local content* being achieved in a given period of time. This requirement, which is often subject to negotiation, can involve the company in a significant amount of administrative work in order to demonstrate that the conditions are being fulfilled. In addition, local content schemes invariably mean higher costs of production.

5.1.4 Corporate matters

The opportunity cost of senior management time is high, and new projects usually take up a disproportionate amount of senior management time. This may be the critical factor in choosing between the best candidates in the *last round* of selection.

Another critical factor could be *image*. High-technology companies like to reinforce their image by having addresses that are synonymous with education and science. Others may want to associate themselves with an area traditional for high-quality manufacturing or research and development such as in the pharmaceutical industry, which is concentrated in Switzerland.

5.1.5 People matters

A prospective location may appear to be right in all other respects but fails because of the manpower resource. This refers to not only the economic aspects such as the availability of labor and skills but also to the qualitative ones of being able to understand the local culture and customs, appreciation of people's attitudes and values. It also refers to the confidence in the ability to blend these with the culture and aspirations of the company to form a cohesive production unit that will work. A successful processing concept in one cultural environment may not be a success in another.

Japanese investors in overseas manufacturing operations have been careful in their approach as well as in their choice of location. It is notable that the United States has received more than its share of Japanese investment in terms of population. An important factor here is language, as English is the most commonly spoken foreign language in Japan.

Stability of labor relations is an important criterion for location selection. The Japanese have negotiated single-union plants and *no-strike* deals with trade unions to try to ensure such stability.

Probably the ultimate factor behind the choice of location is the influence key personnel can bring to bear on the decision. Project success largely depends on people, especially those at the top. With limited options on the person and/or team to lead the project's implementation, their location preference could be final.

5.2 Services

5.2.1 Availability of water

A water supply is an essential service for all manufacturing and process industries for domestic, cooling or process use. The significance of a water resource on the location of an industrial plant is essentially cost and security of supply. Certain industries (e.g. beer and mineral water production) may consider a water supply with particular chemical characteristics as essential to their location. Water supply costs are related directly to quantity but can rise almost exponentially in relation to quality. Therefore, subject to other commercial considerations, there is an advantage to industries that use large volumes of relatively low-quality water for process or cooling (e.g. paper and pulp, textiles, chemicals and steel) to locate in areas where surface or groundwater may be exploited. Their need for higher-quality water is met from the public metered supply or from on-site treatment facilities. For the majority of industries, however, water supply is not a prime consideration in plant location.

Traditionally, their demand for water is met from the public metered supply supplemented, where appropriate to availability and quality for use, by surface or groundwater. Nevertheless, the cost of water provided by water companies varies significantly. Recent legislation and regulations will undoubtedly increase substantially the cost of public water supplies and also the cost differences between the water companies.

In addition to varying in costs, the chemical composition of the water provided from the mains supply also varies between the water companies, as may that between independent supplies within each company's area. The current criterion on potable water quality requires it to be *wholesome*; i.e. it should not create a health hazard, with relatively wide limits on particular constituents. The cost of removing these constituents (e.g. calcium, magnesium, chlorides, iron and silica) increases with concentration and variability. This imposes a cost burden on, for example, the semiconductor and electronic component industries and on the operation of high-pressure boilers. Therefore both the potential cost of metered water supply and the chemical composition of the supply waters may influence future decisions on the water company's area in which an industry may wish to locate.

The Environmental Protection Agency (EPA) and other federal, state and local authorities govern security of supply in relation to quantity for surface and groundwater by the granting of an abstraction license. The license limits the total daily quantity and the rate at which the waters may be abstracted, taking account of the natural resource and the needs of other abstractors. Metered water supply is subject to contract with the water companies, who may impose quantity and draw-off rate limitations. The limitations imposed by the license or contract may influence plant location.

With regard to surface-water abstraction the following additional points are worth noting:

1. The character of surface waters can change significantly and rapidly throughout the year, and any process or cooling-water system should be designed accordingly.
2. The higher the quality of the surface waters providing the supply, the more stringent will be the treated wastewater discharge consent standards imposed by the NRA on effluents returned to the surface waters.

5.2.2 Trade effluent disposal

It will be seen from Section 5.2.1 above that industry is an intensive user of water. Almost all of this water produces a wastewater, requiring being disposed as either domestic waste or trading effluent. The water companies invariably discharge domestic wastewater to the municipal sewers for treatment at the municipal sewage-treatment facilities. The cost of this service is currently generally incorporated into the rating system, with the water companies using the rating authorities as an agent.

Trade effluents may be discharged to the municipal sewers, to surface waters or on or into the land. Discharges to surface waters, including estuaries and coastal waters, or into the land, are controlled by means of Consents to

Discharge with the EPA and Agreements with the water companies. Discharges to controlled landfill sites are by agreement with the local waste disposal authority. In all cases the Consents and Agreements will impose conditions on the quantity, rate of discharge and chemical composition of the trade effluents acceptable for discharge.

The quantity, rate of discharge and chemical constituent limits incorporated into the Consents and Agreements for surface-water discharges are set by the EPA to prevent pollution and to meet the river quality objectives of the receiving waters. Limits on land discharges are set to protect groundwater aquifers. Currently, limits are imposed on organics, suspended solids, ammonia, toxic metals, pH and temperature in the trade effluent discharges. The actual limits on each of these constituents take account of available dilution, the surface or groundwater use and the impact of other dischargers. The limits can therefore vary significantly between discharge locations. Nevertheless, the constraints imposed by the EPA on discharges to surface waters and to land are generally considerably more stringent than those imposed by the water companies on sewer discharges. Impending and future directives on environmental protection and drinking water standards may not only reduce the constituent acceptance limits but also extend the list of constituents constrained. At present the EPA does not levy a charge on the discharge of trade effluents to surface waters or to land.

Disposal of industrial effluents to controlled landfill sites is generally confined to slurries and sludge. The quantity and composition of the wastes acceptable for disposal is controlled by licenses issued by the waste disposal authority.

The quantity, rate of discharge and chemical constituent limits incorporated into the Consents and Agreements for discharges to the sewerage systems are set by the water companies to protect the health of sewer and treatment plant personnel, the fabric of the sewers and the operation of sewage-treatment processes. The constraints take account of the hydraulic capacity of the sewers and the treatment works, the organic and solids-handling capacity of the treatment works the ultimate disposal route for sewage sludge and the needs of other trade effluent dischargers. In addition to limiting the concentration of specific constituents, including those mentioned for surface-water discharges, materials, which could be dangerous to the health of sewer workers, are prohibited from discharges.

Municipal sewage-treatment effluents discharge to surface waters and is subject to the same EPA control on quality and quantity as independent industrial surface-water discharges. Any tightening of EPA standards may therefore result in more stringent controls on industrial effluents discharged to sewers. All the water companies levy charges on industry for the reception, conveyance and treatment of the industrial effluents.

The significance of these industrial effluent disposal options on the location of an industrial plant is essentially cost. As previously stated, the EPA does not, as yet, impose a cost on effluents complying with the Consent standards discharged to surface waters or to land. However, the cost of installing and operating treatment facilities, including sludge removal, to achieve and maintain the Consent standards can be considerable. The cost of treatment escalates significantly with increasingly stringent treated effluent standards. The actual cost of this disposal route will, of course, depend on the quantity and character of wastewater generated by a particular industry in addition to the treated effluent standards to be met. The conditions imposed in the Consents to Discharge could therefore influence the location of an industrial plant adopting a surface-water or land-disposal route for its effluents.

In addition to the treatment costs, it should be remembered that provision of treatment plant occupies site space, which may be more profitably used, and that the penalties for infringements of the consent standards are increasing.

The cost of depositing waste on controlled landfill sites is relatively cheap. However, the expense of road transport to suitable sites generally limits this disposal route to relatively low-volume applications.

The cost of industrial effluent disposal to the municipal sewers is based on a *polluter pays* policy, which takes account of the quantity and pollution loads in the discharge. All the water companies calculate their trade waste charges in accordance with:

$$C = R + V + \frac{(O_t)B}{O_s} + \frac{(S_1)S}{S_s}$$

where

C = total charge for trade effluent treatment (p/m^3),
R = reception and conveyance charge (p/m^3),
V = volumetric and primary treatment cost (p/m^3),
O_t = the chemical oxidation demand (COD) of trade effluent after one hour settlement at pH 7 (mg/l),
O_s = COD of crude sewage after one hour settlement (mg/l),
B = biological oxidation cost (p/m^3) of settled sewage,
S_t = total suspended solids of trade effluent at pH 7 (mg/l),
S_s = the total suspended solids of crude sewage (mg/l),
S = treatment and disposal of primary sludge generated by the trade waste (p/m^3).

However, all of the functions in the formula, excluding the trade effluent functions O_t and O_s, are different for each of the water companies and are reviewed annually. Therefore the unit costs of charges vary between the companies. Subject to other commercial considerations, the cost of trade waste charges could influence the water company area in which to locate an industrial plant. The conditions imposed in the Consent to Discharge could influence the location within an area most beneficial to plant location.

5.2.3 Electricity

Previous considerations concerning the purpose engineering of the facility will have established the size and characteristics of the electrical load in terms of consumption and maximum demand. These will be essential data with which to inquire of the electricity authority as to the availability of an electrical supply.

Generally, there will be no insurmountable difficulty in obtaining an electrical power supply, the principal factors being the cost of providing it and the time delay in making it available. The scale and nature of the project will dictate the amount of power required, the load characteristics and the most suitable voltage for its supply. The state of the public supply network in the area of the proposed development will then constrain the supply authority in its ability to quote for a suitable supply in the short, medium or long term.

A relatively small power demand at low or medium voltage in a built-up area might be provided in the short term by teeing from an existing main feeder in the locality, or providing a radial feed from an existing substation. If the load is large or remote from the existing supply network, or if the local network is fully loaded, then a new incoming supply brought from a distance might be a medium- to long-term project. Time of cost will be dictated by legal considerations as well as the technical aspects involved. It is therefore necessary to make the earliest possible approach to the relevant electricity supply authority to establish the availability and cost-environment situation, as this may influence the initial engineering and planning factors to be accounted.

5.3 Ecology and pollution

5.3.1 Introduction

Urban, rural and industrial developments may have profound effects on the surrounding environment. Such effects can defeat the object of development, in that the negative environmental impact may outweigh the benefits. In the case of natural resources, inappropriate development may even destroy the resource base. If environmental matters are accorded adequate consideration during the planning and management of development programs and projects it is possible for pollutants to be assimilated. As a result, the whole development can be accommodated by the environment in such a way that adverse effects are minimized and the economic and social benefits of development are maximized.

5.3.2 Baseline studies and modeling

The basic requirement of any pre-commissioning environmental study is information on the existing state of the environment prior to any new development. Where such data are not already available, baseline studies must be carried out. Essentially, such studies are designed to provide baseline data from which the phys-chemical and ecological effects of development may be assessed and against which changes due to development may be measured after commissioning of the project. Computer programs are available to develop these studies to provide accurate predictions of future pollutant loading in the environment. For example, using the program AIRPOLL[1] ground-level concentrations of stack gases are calculated at various distances from proposed plants under a variety of meteorological conditions. The program CAFE[2] is similarly used to predict the dispersion of thermal and other effluents in the aquatic environment, and OXBAL[3] is employed to determine the ecological consequences of effluent discharges and engineering works in estuaries. Emission control equipment and waste-treatment plant can then be designed to meet any local standards for ambient air and water quality.

5.3.3 Environmental standards

In some parts of the world the need to exploit natural resources is so urgent that it has preceded the formulation of adequate environmental controls. The pace of natural resource exploitation and the growth of associated industries have overtaken the evolution of institutions, which would have the authority to exert such controls.

The problem has been recognized by many of the developers concerned, who have consequently themselves adopted the environmental standards of other industrialized nations. In the absence of national controls this is a responsible and laudable approach. However, the piecemeal adoption of standards taken from elsewhere does not take account of local conditions. These conditions may either enhance or limit the ability of the environment to disperse and attenuate or assimilate pollutants (e.g. the occurrence of temperature inversions will limit the dispersion of air pollutants). Similarly, the use to which local resources are put may demand particularly high standards of environmental quality (e.g. the use of sea water or river water as the basis of potable water supply). The choice of standards must also take into account local practices and existing local administration.

At the plant level, in-plant monitoring of unique compounds and the modeling of plant conditions to develop appropriate working practices and internal environmental quality standards may be needed.

5.3.4 Environmental assessment

In the case of a large-scale development it may be desirable to combine several environmental services for a full environmental impact assessment (EIA). This is the process of examining, in a comprehensive, detailed and systematic manner, the existing environment (natural, built and social) and the development that it is proposed to place within it. By integrating the two, an objective estimate can then be made of the likely effects of the development upon the environment, including benefits and negative impacts. Special techniques may be employed to help identify or quantify these impacts (e.g. the use of interaction matrices, overlays, screening tests, checklists, etc.).

An EIA can be particularly useful in distinguishing the relative environmental impact of alternative sites, processes and strategies for industrial, rural or urban development. Decision-makers to choose the alternative that will provide maximum economic and social benefits with the minimum of environmental disturbance can then use this essential environmental information, together with financial and political considerations. Guidelines for assessing industrial environmental impact and environmental criteria for the siting of industry have been prepared for the United Nations Environment Program.

5.3.5 Resource planning

Ecology studies employ information from all the other environmental sciences to draw conclusions about environmental change that will result from development. However, when the exploitation of natural resources is being planned, the nature of the ecosystem and its response to change are themselves the most important elements to be considered. Industrial developments need to be assessed in the context of their ecological setting to ensure that the proposed development, first, will be feasible within local constraints and, second, will not bring about irreversible and unacceptable changes to other essential parts of the ecosystem.

The activity of ecologists in resource planning has three forms: preservation, integration and conservation.

Preservation

In this case, individual new developments are designed in such a way as to preserve discrete ecological systems that have been identified as of importance as '*life-support systems*', as regionally/internationally important wildlife habitats or as sources of rare natural materials, etc.

Integration

In the case of development projects which are on a large scale (e.g. mining or dam construction) or which cover a considerable area, ecological planning attempts to achieve integration between the project and the ecosystem. The object is to ensure not only the continued functioning of the ecosystem as a whole but also of those elements of the ecosystem which provide the development project with *goods and services* (e.g. fuel, raw materials and the assimilation of waste).

Ecological planning in this context will help to avoid ecological disasters (e.g. excessive weed growth in stored water) and maximize the life span of the benefits derived from the development. It will be clear that this holistic approach to development has economic as well as ecological advantages.

Conservation

Where regional development is to be undertaken it is reasonable to adopt a positive ecological approach. This can begin with the formulation of a conservation policy with specific goals. These may include the conservation of individual animal/plant species or habitat types which are threatened by development. A program of conservation projects (e.g. the setting up of national parks, country parks or natural reserves) will meet the goals. This positive approach to the ecosystem will not only benefit wildlife but will also create the opportunities for tourism and leisure, which are vital adjuncts to most development projects.

5.3.6 Environmental management

Once commissioned, even the best-planned industrial development requires monitoring and management to ensure that its operation continues to be environmentally acceptable. This applies equally to established industries. When unexpected environmental problems develop, a rapid response is required to assess the cause and magnitude of the problem and to devise remedial measures.

Air pollution

Dusts produced by quarrying and fluorides emanating from oil refineries are typical pollutants, which need regular monitoring. A range of portable equipment for the identification and quantification of toxic and other gases can be used on an *ad hoc* basis.

When unpleasant odors resulting from manufacturing processes or waste-disposal operations give rise to public complaints they should be identified and quantified prior to deriving methods of abatement. Such work is often innovative, requiring the design and fabrication of new equipment for the sampling and analysis of pollutants.

Water pollution

Consultants are equipped to monitor the quality of freshwater, estuarine and marine environments and can make field measurements of a variety of water-quality parameters in response to pollution incidents. For example, reasons for the mortality of marine shellfish and farmed freshwater fish have been determined using portable water-analysis equipment. Various items of field equipment are, of course, also employed in baseline studies and monitoring, respectively, before and after the introduction of new effluent-disposal schemes.

Where extreme accuracy is required in the identification of pollutants or in the quantification of compounds that are highly toxic, laboratory analysis of samples is conducted. Highly sophisticated techniques have, for example, been employed in the isolation of taints in drinking-water supplies.

Land pollution

As development proceeds, land is coming under increasing pressure as a resource, not only for the production of food and the construction of new buildings but also for disposal of the growing volume of industrial and domestic waste. The design and management of sanitary landfill and other waste-disposal operations requires an input from most of the environmental sciences, including geologists and geo-technicians, chemists and physicists, biologists and ecologists. Such a team can deal with the control and treatment of leachate, the quantification and control of gas generation, and the placement of toxic and hazardous wastes. This may be needed in designs for the treatment of industrially contaminated land prior to its redevelopment.

Ecological studies

The acceptability of some industrial and ephemeral development projects such as landfill or mineral extraction may depend upon an ability to restore the landscape after exploitation has been completed. As more rural development projects come to fruition, ecologists will become increasingly involved in resource management to ensure that yields are sustained and to avert the undesirable consequences of development. Some industrial developments and rearranged plant layout schemes will not be complicated, but when ecology studies are needed, the employment of specialist consultants is recommended.

Notes

1 AIRPOLL, W. S. Atkins Consultants, Ltd, Woodcote Grove, Ashley Road, Epsom, Surrey.
2 CAFE, as above.
3 OXBAL, as above.

6

Industrial Buildings

D Whittleton
Ove Arup & Partners, Industrial Division

Contents

6.1 Introduction

Industrial activity encompasses an enormous variety of operations, and industrial buildings provide the required protection from the external environment. Of necessity, therefore, they take many shapes and structural forms and may be composed of a variety of materials.

6.1.1 Procurement

Industrial buildings can be procured in one of three basic forms:

1. Specific design
2. 'Off the shelf'
3. Speculative.

Specifically designed industrial buildings

Many industrial operations will require buildings with a large number of specific attributes. Such requirements will include many of the following:

- Location
- Plan dimensions
- Height to eaves
- Column layout
- Services provision
- Drainage provision
- Provision of cranes
- Superstructure-imposed load capacity
- Floor loading capacity (point load, uniformly distributed load, line load)
- Floor flatness
- Floor abrasion/impact/chemical/slip resistance
- Access facilities
- Loading/unloading facilities
- Floor pits
- Intermediate platform arrangements
- Suitability for automatic guided vehicles
- Suitability of environmental controls
- Ease of cleaning
- Corrosion resistance
- Machine base availability
- Sound insulation
- Provision for future expansion.

Any one of the above may require a specific building to be designed and constructed due to the unsuitability of available facilities falling into the option (1) and (2) categories.

Apart from performance criteria, an industrial building might also be required to project a *corporate identity* by a striking appearance. This generally requires a specifically designed facility.

Off the shelf industrial buildings

Off the shelf buildings are generally of set modular form designed to a standard set of criteria. The available degree of variation may be limited, but they are perfectly suitable for several industrial uses.

Speculative industrial buildings

These are often developed on new industrial estates with no particular tenant in mind. They may be quite adequate for general light engineering/warehousing/retail units. However, prospective industrial occupiers with any but the most basic general requirements are likely to have to spend considerable sums adapting the building to their needs, if indeed this is physically and economically viable.

6.1.2 Structural materials

The majority of industrial building superstructures are framed in structural steel, although a small percentage are in pre-cast concrete. Steel is used primarily for its large strength-to-weight ratio, enabling it to span large distances economically. Steelwork is easily modified, which provides for a degree of adaptability not always available from concrete structures.

Ground slab and foundations are invariably reinforced concrete, though some ground bearing slabs are constructed with no reinforcement. Industrial buildings for containment of toxic of other processes may require construction primarily from reinforced concrete.

A dwarf wall of concrete, blockwork or brickwork is often constructed around the building perimeter to minimize cladding damage from fork elevator trucks, etc.

6.1.3 Structural form

Stability

Building stability can be provided by:

1. Framing action from rigid connections between columns and roof members;
2. Vertical cross bracing or shear walls;
3. Columns cantilevering from foundations; or
4. A combination of these techniques.

Roof bracing may be required for rafter and purlin stability or load transfer to vertical bracing.

A composite concrete slab or a stressed-skin system can also provide a roof diaphragm, but the latter may severely restrict the provision of subsequent roof penetrations. Concrete roof slabs are unusual, due to the greatly increased mass over the more normal metal decking/insulation/waterproof membrane or insulated metal decking options.

Column layout

It is self-evident that fewer permanent obstructions within the perimeter provide a greater potential flexibility of operations. However, larger spans are bought only at the expense of structure. Columns may also perform other functions, apart from roof support:

1. Support of intermediate-level platforms;
2. Restraint to partition walls;
3. Support for cranes/jibs;
4. Services (pipework, ductwork, etc.) support.

Roof shape

Flat roofs. Flat roofs are popular with architects due to their reduced visual impact. However, they generally require a much higher standard of waterproofing, and may be susceptible to ponding, moss growth and dirt/debris entrapment. Inevitably, they also require internal downpipes for rainwater disposal.

Falls to gully positions may be provided by the natural deflections of the roof structure or small slopes to the roof steelwork (e.g. 1 : 40–1 : 60).

Pitched roofs. Pitched roofs are typically sloped at a minimum of 6° to ensure the weather resistance of lapped sheeting without sophisticated seals or a waterproof membrane. Portal frames are also more liable to *snap through buckling* at very shallow pitches. A pitched roof means a greater *dead* volume to heat, although there is additional space for high-level service distribution.

Roof construction

1. Flat roofs. Typically, framing is achieved with:
   (a) Universal beams (span <15 meters);
   (b) Castellated beams (20 meters > span > 15 meters);
   (c) Lattice girders (span >20 meters).
   A recent development has been the practice of using site butt-welded, closely spaced, continuous secondary beams spanning over continuous primary beams and replacing the traditional purlins. Secondary spans of 20 meters have been achieved with this method. Other considerations include:
   (a) Castellated beams may require reinforcement at high shear regions;
   (b) Castellated and lattice beams provide a service zone within the structural depth;
   (c) Lattice girders may be formed into V beams of triangular cross section.
2. Pitched roofs. Typically, framing is achieved with:
   (a) Portal frames (Figure 6.1);
   (b) Pitched lattice girders (Figure 6.2);
   (c) Trusses (Figure 6.3);
   Portal frames may be single-bay, multi-bay (Figure 6.4), propped, tied, or mansard with pinned or fixed foot. They may be composed of universal beams or tapered plate girders.

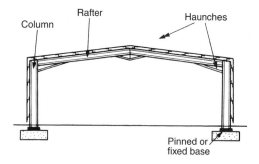

Figure 6.1 A typical portal frame

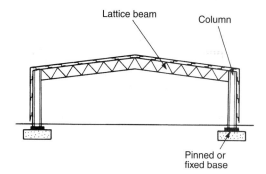

Figure 6.2 A typical lattice girder frame

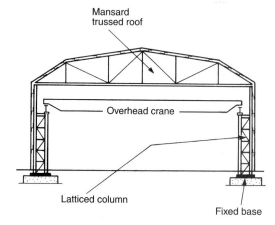

Figure 6.3 A typical trussed roof

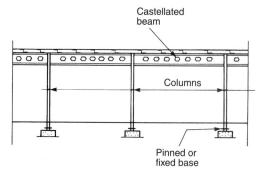

Figure 6.4 A typical multi-bay beam and post construction

Trusses may be the best solution for very high-imposed loads. Frame action with columns is not possible with trusses. Although trusses are generally the lightest form of roof construction, they may be the most expensive due to high fabrication cost. A combination of lattices or lattice and truss may form a sawtooth roof profile for incorporation of north lights.

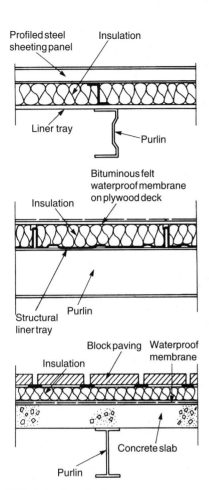

Figure 6.5 Typical roof constructions

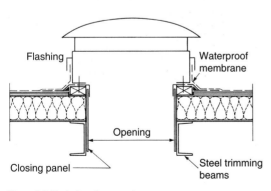

Figure 6.6 Typical roof penetration

3. Space trusses. Three-dimensional space trusses utilizing proprietary nodal joints can achieve substantial two-directional spans. Typically, they are too expensive for normal industrial buildings.

6.2 Specifying an industrial building

Many industrial buildings, and industrial projects in the broader sense, are completed entirely to the owner's satisfaction in terms of the key criteria of time, cost and quality. It is also true that many are not. There can be various reasons for this; some related to the nature of the construction industry in general, some to the performance of professional teams and others related to the manner in which the project was initially set up. The most common cause of problems with the finished product is the way in which the owner's requirements were specified, and this extends to include the selection of the project team.

These problems cannot always be attributed to the owner. An industrial client is not usually a regular developer of buildings, which are, in most cases, secondary to whatever process or production facility he wishes to construct. He is, after all, not engaged in the construction industry on a day-to-day basis. He should be entitled to expect good advice from the team he appoints, and to expect them to *ask the right questions*. But whatever the project, and however limited the owner's resources, it is vital that he is able to provide the necessary input to ensure that he gets what he wants.

We are not concerned here with contractual issues, or with questions of responsibility. What follows is a set of practical guidelines that seek to address the main areas where problems can arise.

6.2.1 Procurement

From the time when the owner decides that a project is desirable there are many approaches to the procurement strategy, ranging from a conventional set-up with consultants/contractors right through to a complete *turnkey* arrangement. There is no right or wrong way of proceeding, and it is difficult to generalize, but it is true to say that the turnkey (or design and build) approach is most likely to succeed where:

1. The owner's requirements are clear, simple, and unlikely to change;
2. The owner is unable to allocate much time to the project;
3. The owner does not require direct control over all the participants in the design and construction process.

In other cases, where perhaps the requirements are ill defined, or there is concern over some particular aspect of the project, it may well be more appropriate to have the project designed and/or managed by independent professionals. For example, most industrial building projects are best suited to an engineer-led approach. If, however, the owner is particularly concerned about the architectural aspects of the scheme, he may prefer to make his own selection and appointment of an architect.

Whichever approach is adopted, the owner must be sure that he has considered those issues most important to him and has set up the project in a way that gives him the level of control he wishes to exercise. He should also take care to ensure that the team involved has adequate experience,

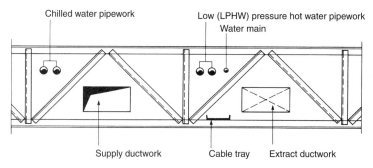

Figure 6.7 Typical services coordination through a lattice girder

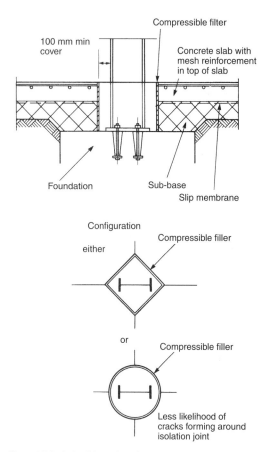

Figure 6.8 Isolation joint at foundation

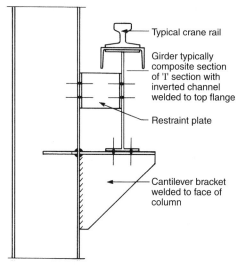

Figure 6.9 Crane gantry girder

not only of the type of project under consideration but also of the approach being adopted.

6.2.2 Budgets

The construction industry has a poor reputation in terms of cost control, and tales of budget overspends are legion. This is often attributed to deficiencies in the industry and among the professionals who work within it.

Whoever is to blame at the end of the day, the problem can often be traced to the fact that the initial budget was unrealistic at the outset. To set an inadequate budget, watch it overrun, and then look for a scapegoat may have become common practice but it achieves nothing.

It is therefore vital that a sensible cost plan be prepared and agreed at the earliest possible stage. Obviously, an owner will prepare rough budgets when considering basic project viability, but as soon as he is able, he should obtain advice from his selected team.

Good professionals will not simply agree with the rough budget already prepared. They will give objective advice, including the level of confidence in the estimate and an assessment of the likely effects of changes to the specification, so that the owner can take a realistic view. It is better to recognize the problem immediately, and modify the scope or specification to suit, than to be forced into late changes or even omissions, which reduce the effectiveness of the finished project.

6.2.3 Specification

The level of briefing which an owner can give is variable, depending on his resources and his detailed knowledge of building or facilities requirements. The most important point here, however, is clarity. The brief may have to be simple, but it must be clear – it must also be clear who has the responsibility for developing it.

The best way to ensure controlled development of the brief, and of the cost plans which necessarily accompany it, is to set up a system of data sheets which define, in increasing detail as the project develops, the spatial, functional and servicing requirements of each component part of the project. A well-resourced technical client may wish to do this himself. More commonly, it is a task which should be entrusted to the design team. Whoever does it, regular joint review of this information will help to ensure satisfaction with the finished product.

6.2.4 Schedules

Much of what has been said about budgets applies equally to the project program. It should be realistic from the outset – again independent professional advice should be sought.

Modern *fast-track* techniques of design and procurement have taken all the slack out of the traditional methods, and this is generally to the benefit of all concerned. There is a price to pay, however, in terms of flexibility. The owner, and his design and construction team, has to work within a discipline in respect of changes; programs and budgets are very vulnerable to changes in this environment. This relates very closely to the earlier comments on clarity of briefing and information flow.

6.2.5 Procedures

Having decided on the appropriate way to proceed with the project, the owner needs to satisfy himself that adequate system and procedures have been put in place to give him the necessary control and confidence in the project. These will, of course, vary considerably from owner to owner and project to project, but certain key points are worth remembering.

There is a clear need for the owner to appoint a single project officer. He may well be part-time as far as this activity is concerned, and will often need to draw on the expertise of others in his own organization. However, a single point of contact between the owner and his project team has been shown to be a vital ingredient in successful projects. Of equal importance is the need for regular reviews to ensure that things are going to plan. It is also necessary to have an agreed mechanism for change control, since once a project is under way, particularly if, as is usual, it is being run to a tight program, any changes are likely to have an impact on both program and budget.

The theme running through the above is that the key ingredient is clarity, both in specification and organizational terms. This does not need to mean a major effort or input on the part of the owner, but rather careful consideration, at an early stage, of what is actually required.

If what is required is unclear at the outset, then it is wise to say so, and allow for that fact in the development of budgets and programs, rather than to pretend that the project is fully specified and suffer the consequences later. An industrial client, like any other, will get what he pays for; but he also needs to ensure that he gets what he needs.

6.3 Security

6.3.1 Objectives

The object of all security systems is to guard the company against loss. The principal risks that are faced include:

1. Criminal action: This may include robbery, vandalism, fraud and industrial espionage.
2. Direct damage to assets: This may be caused by explosion, rainwater ingress and fire. The level of security is dependent upon good building details and the fire-suppression system used.

6.3.2 Criminal action

This section will consider the options available to deal with criminal action. There are several levels of security. The higher the level of security, the greater is the cost. This expenditure will relate not only to the capital cost of the security system but also the restriction that is placed upon the efficiency of the employees. In planning a security system it should be recognized that no system could guarantee detection of, or immunity from, intruders. The purpose of a system is to confer on the property or persons at risk a level of security that, through careful planning, is in balance with the degree of determination and expertise of the intruder. In making this evaluation, all physical circumstances of the risk need to be assessed. However, the degree of security decided upon has to be consistent with a system that is, in every sense, practical and reliable and which cannot become an unacceptable burden to the public or police.

Furthermore, the cost of the security system must bear a relationship to the risk, particularly since an increase in the cost of a system does not necessarily cause an increase in the level of security. Cost effectiveness should be of prime consideration.

In planning a security system its requirements should be ascertained as accurately as possible by consultation between appropriate interested parties, which may include one or more of the following:

1. The insurers
2. The local police authority
3. The Health and Safety Executive
4. The local public authority
5. The customer
6. Any statutory regulations
7. Consultants
8. Security companies

In assessing how to apply the requirements of a security system the following items should be considered both individually and collectively:

1. Physical barriers and deterrents
2. Electronic detection systems
3. Security patrol and guarding
4. Controlling entry and egress
5. Employee awareness, procedures and training
6. Local, remote and delayed alarms.

All parts of the security system should interact with each other to form an integrated whole.

6.3.3 Layers of protection

The security system can be considered as layers of protection. Physical barriers and deterrents constitute not only the building structure itself, made up of walls, doors, windows, floor and a roof, but the yard around the building and probably a perimeter fence or wall.

Electronic detection systems may range from simple intruder-detection devices monitored by basic control units to a variety of complex systems monitored by sophisticated computer-operated controls linked to 24-hour manned stations. Intruder-detection devices can be arranged into the following groups:

1. Static detection devices
2. Movement-detection devices
3. Trap and object devices

Static detection devices would include the following:

1. Tube and wire frames are used to protect windows, skylights and similar glass areas and constitute a high level of security but may be aesthetically unacceptable.
2. Aluminum foil is used to protect windows and glass doors. Although usually acceptable, this is subject to tampering if accessible to the public.
3. Break-glass detectors detect the high frequencies produced when glass is broken or cut with a glasscutter. These are the most cost-effective methods of glass protection in most cases and are especially suited to areas accessible to the public.
4. Vibration sensors are similar to break-glass detectors but are fixed to walls and doors against violent attack. The structure to which they are fitted must not be subject to vibration from external sources, otherwise false alarms may result.
5. Closed-circuit wiring comprises plastic-covered single-strand wire strung at regular intervals across doors, walls and ceilings. This is prone to failure in damp conditions and is mainly used in high-security buildings.
6. A pressure pad, also used as a form of trap protection, is a flat rectangular, plastic-covered contact placed at strategic points underneath a carpet.
7. Magnetic reed contacts are glass-encapsulated magnetically operated reed switches used on doors and windows. The contacts can be overcome by cutting the door, leaving the magnet undisturbed.

Movement-detection devices are designed to detect the movement of an intruder in the interior of the building:

1. Beams of invisible infrared light shone across an opening will detect movement that breaks the beam. Although relatively cheap, beam devices are easily recognized and avoided, and therefore constitute low security.
2. Microwave movement detectors utilize the principle of the Doppler effect on high-frequency low-power radio waves. These units are moderate in cost and suitable for large-volume coverage. Microwaves, however, penetrate certain materials easily, such as plasterboard, and careful siting is required to avoid false alarms.
3. Ultrasonic movement detectors utilize the principle of the Doppler effect on high-frequency sound waves. Ultrasonic movement detectors do not penetrate solid objects, but have smaller volume of coverage than microwave movement detectors. These units may also be affected by moving hot or cold air pockets in a room.
4. Passive movement detection senses radiated heat, such as that from a human body. These units are also sensitive to heat emitted by radiator, convection heaters and direct sunlight, so careful siting is required.
5. Dual-technology detectors combine two of the techniques described above, except for beams, and provide good rejection of unwanted alarms.

Trap- and object-protection devices provide protection to an area and object within a building and should never be used on their own:

1. A contact on a safe will recognize attack by drilling, sawing, filing, explosions and oxyacetylene or thermic lance.
2. Closed-circuit television cameras are very expensive and require a manned station, but with careful siting and the latest technology, they can see in moonlight and can set off alarms when picking up motion.

Security patrol and guarding is only satisfactory if properly managed and controlled with trained guards and complemented by electronic detection and monitoring equipment. The guards would generally be available to react to the unexpected, and should be well briefed as to responsibility and how to obtain help as well as how to deal with any type of unexpected situation.

Controlling entry and egress can be very useful for ensuring that only authorized people have entered the secured area as well as identifying who is actually in the building in an emergency such as a fire. There are several methods of controlling entry and egress, some of which are listed below:

1. Guards: These are probably the least efficient, primarily because of the human element. However, there are obvious psychological strengths to human presence.
2. Mechanical key system: This provides a basic level of security for heavily used areas. Although complicated mechanisms can be designed, their operation is slow, and hence is suitable only for limited access.
3. Magnetic card systems: These are very good for busy areas and can be made very difficult to forge. The level of security can be increased, at the expense of

efficiency in busy areas, by incorporating the necessity for a PIN number.

4. Automatic systems: These require no direct contact between the user and the device reading the code. They provide a high degree of convenience in use.

For high-security buildings, it is imperative that employees are made aware not only of the need for security but also what procedures have to be followed and instructed on basic security awareness.

The type of alarm system used is dependent on the expected type of security breach and the method employed in responding to one. In unguarded premises, requiring only a low level of security, an alarm that operates immediately a device detects a security breach may be sufficient to ward off vandals, burglars and crimes of opportunity. On the other hand, where breaches of security may involve more determined criminals, such as fraud or industrial espionage, delayed alarms on the premises may give time for security personnel and/or police to apprehend the criminal in the process of committing the crime.

6.3.4 Reliability

No alarm system, however well planned and installed, can be completely reliable or tamperproof. The successful operation of a security system requires the active cooperation of those involved in carrying out the necessary procedures carefully and thoroughly. The usefulness of the whole system can be jeopardized by lack of care or inadequate attention to routine procedures in maintenance and servicing. This care has to be extended to the security of keys and of information regarding the system, its installation and method of operation.

6.4 Leases

6.4.1 Relationships

A lease, whereby the landlord grants a portion of his interest in the land and allows the tenant to occupy and enjoy that land for a period or term usually in return for a monetary payment called rent, creates the relationship of landlord and tenant. The statute law is updated and revised by successive governments and legal advice should be sought when drafting or entering into leases.

6.4.2 Preparation of a lease

A lease or contract of tenancy can be verbal but is usually in writing, as there is the possibility of misunderstandings and thus disputes. In some cases, this is overcome by a deed that sets out the conditions of the letting. This is a contractual document, which is made by indenture, i.e. in two parts. One part, called *the lease*, is signed, sealed, and delivered by the landlord to the tenant while the other, called *the counterpart*, is signed, sealed and delivered by the tenant and handed to the landlord. The two are identical and, being legal documents, use formal words such as *lessor*, *lessee*, etc. and generally commence with the words '*This lease*' or '*This indenture*'.

The lease by deed sets out the conditions of the letting in a formal manner and should include:

1. Generally, the description of the property;
2. The names of the parties;
3. When it is to commence and how long it will last;
4. The covenants which consist of promises and agreements by the landlord and tenant to do or not to do certain things;
5. The rent that is to be paid;
6. Any exceptions and reservations which the landlord wishes to retain;
7. The conditions showing how the term may be ended, extended or created (which may include implied conditions);
8. The provisos, which are any express conditions. The deed is dated, signed, sealed, witnessed and delivered and operates by the devise of the landlord's interest in the property to the tenant.

The understanding of these constituent parts of the lease is very important, and they should be read very carefully. Expert advice should be sought for a thorough understanding of them.

6.4.3 Covenants

Covenants may be express or implied, i.e. those expressly written into the lease in a deed and those not expressed but implied in a deed by law in order to give effect to the intention of the parties or to remedy an obvious omission. Express covenants could be:

1. To pay rent;
2. To pay rates and taxes;
3. To repair and keep in repair;
4. To paint/decorate within certain times;
5. To insure and produce receipt of insurance;
6. To permit the lessor and his agents to enter and view;
7. Not to alter the structure;
8. Not to assign or sub-let;
9. To yield up at the end of the term.

Implied covenants could be implied into a lease by common law or by statute, and in the case of the latter cannot easily be varied or excluded by agreement between the parties. Some examples of covenants by the landlord are:

1. That the landlord has good title and therefore to give possession of the date fixed for commencement of the term;
2. To give the tenant quiet enjoyment during the tenancy;
3. In the case of residential tenancies to give at least 4 weeks' notice to quit;
4. In the case of furnished premises to ensure at the commencement of the tenancy that the premises are fit for human habitation.

6.4.4 Rent

A landlord grants land to a tenant for a variety of purposes, and several names are given to the *rent* due from the tenant to the landlord. Premises for occupation are generally let at a *rack rent* which is assessed upon the full annual value of the property, including land and buildings, the *best rent* being the highest rack rent that can reasonably be obtained for the whole lease.

Sometimes land and buildings are let on a long lease to a tenant who proposes to sub-let on short terms such as weekly or monthly tenancies. In this case, the rent paid on lease is less than the rack rent, and is called a *head rent*.

When land is let for construction it is usual for a *peppercorn rent* of no actual value to be agreed while the houses, etc. are built. Although cases do occur where a peppercorn rent is received throughout a long lease, it is more usual for a *ground rent* to be charged. This is less than the rack rent, usually because a premium has been paid at the start of the lease. Ground rents are much higher than peppercorn rents and may include increases during the lease.

Rent is due on the days appointed for its payment in the lease, and unless the lease expressly provides for rent to be paid in advance it will be payable in arrears.

6.4.5 Reviews

Many leases, especially business leases, contain rent-review clauses, which provide for regular reviews at, for example, three-, five- or (more usually) seven-yearly intervals. A clause will define rent-review periods and set out the formula and machinery for assessing the rent in each period.

6.4.6 Assigning and sub-letting

Any tenant who has not contracted to the contrary may sub-let his land (or part of his land) for any term less than that which he holds. A sub-letting is therefore distinguished from an assignment in that a tenant, who assigns, disposes of the whole of his interest under the lease (and thereby puts the assignee into his place), whereas a tenant who sub-lets does not.

6.5 Obtaining approval to build

When contemplating the construction of a new industrial building or alterations to an existing one the statutory requirements and the powers available to the authority responsible for issuing approval should be taken into account. The exact procedures will, necessarily, be dependent upon the location (i.e. district and/or country) and the type of industrial building proposed. The approval process can be divided into two stages. These are to obtain planning and building control approval.

We shall consider each stage in the context of general requirements. The standard and detail needed to obtain approval is contained within statutory documents that can be obtained from the relevant authority.

6.5.1 Planning approval

Obtaining planning approval provides permission to build and, more importantly, justification to pursue the preparation of detailed plans and documentation necessary for building control approval. Planning permission prior to detailed design of the building is not a requirement, but rejection of planning submission would probably render the design useless.

The authority granting permission to build will depend upon the type and size of building. Major projects, such as nuclear power stations or large industrial complexes, may involve a public enquiry and the need to produce a study into the environmental impact. Though a committee nominated by the local authority will consider the planning submission.

The following is a list of items that could (and would usually) form the basis for a planning submission:

1. Site location
2. Building footprint
3. Height
4. Use of building
5. Aesthetic appearance
6. Utility requirements (e.g. gas, electricity, etc.)
7. Employment requirements.

The list is not exhaustive, but it will be apparent that the submissions content implies that any decision will be subjective. Knowledge of the current planning policy of the local or national government is therefore advisable.

6.5.2 Building control approval

Having obtained planning permission, building control approval involves producing a set of detailed documents to satisfy a list of statutory requirements. These lists, although amended and updated according to current practice, if complied with, should gain approval. The decision to grant approval is therefore much less subjective than that needed for planning permission, being more closely defined by statute.

The need for building control arose primarily to protect the health and safety of the population and, latterly, to conserve fuel and power and prevent waste. Local and State Building Regulations give standards of performance necessary in the building and, as such, make reference to national and international standards and codes of practice. The regulations cover two areas of health and safety and can be subdivided as follows:

Safety	*Health*
Fire	Toxic substances
Structure	Ventilation
Stairs, ramps and guards	Hygiene
	Waste disposal and drainage

The specific technical requirements within each category will depend upon the type of building and the use to which it will be put. For example, the requirements for fire escapes will be different for an office and for a factory.

Finally, the regulations are not exhaustive and there will be circumstances, especially relating to fire safety, where approval can only be obtained by negotiation.

6.6 Extending existing buildings

6.6.1 Viability

The extension of existing facilities can often present an attractive alternative to relocation, with its entire attendant disruption and potential impacts on production, industrial relations and morale. However, the viability of extension is predicated on the consideration of various strategic and tactical issues:

1. Land availability and location in relation to existing buildings;
2. Nature of activities proposed in extension (e.g. linear, cellular, storage, bulk process, administrative, etc.);
3. Topographical and geo-technical conditions;
4. Existing structures (their condition, type of construction, etc.);
5. Infrastructure and services such as roads, telecommunications, water, foul and storm drainage, gas, power, etc., location, routing, capacity (used and spare), state of repair, etc.;
6. Local and statutory authority issues (e.g. planning, building control, service tariffs);
7. Environmental concerns (effluent treatment, noise, etc.).

6.6.2 Facility brief

Once the primary considerations of size, height, conceptual layout, structural loads, servicing (mechanical, electrical, communications, public health, statutory services) requirements, access, material and personnel traffic, etc. have been addressed, a facility brief can be produced to allow collation of the basic project planning information:

1. Scheme, site plan, layouts and elevations including infrastructure modifications;
2. Basic plant services;
3. Budget costs;
4. Tentative planning assessment;
5. Preliminary project program.

These items should be formulated against optional approaches, each option being tested against one another (and rated accordingly) on the basis of:

1. Capital expenditure, cash flow and operating cost analyses;
2. Impacts on existing operations (productivity, quality, personnel, etc.);
3. Project-completion times (including the options for phased completion);
4. Comparison of operational, spatial, servicing simplicity, internal environmental performance (day lighting, thermal transmittance, acoustic break-in/out, etc.), parameters between options.

The adoption of the best option should be based on all these parameters.

6.6.3 Other considerations

Many extraneous and site-specific considerations may have to be taken into account. Certainly, the construction

and layout of existing facilities will play a major role in deciding the basic choice of location and configuration of extensions.

Particular attention should be paid to existing site arrangements, bay widths, material handling, foundation design, and construction methods. For instance, it may prove impractical to extend the bays of a traditional steel-framed lightweight structure if the impact on road rerouting, statutory services extensions, substation requirements or other existing facilities seriously affects the overall cost or program time.

Equally, the most logical (and least externally disruptive) extension method may conflict with desired material handling or workflow requirements. Generally, economic bay widths should be adopted (typically c. 15 meters), if possible. Roof design should be adapted to accommodate roof-supported plant and services while minimizing steel tonnage. The integration of day lighting, service routes, cranage, etc. should all be early design issues, solved during the evolution of the basic structural design.

Attempting to modify structure or services to accommodate structural idiosyncrasies later is always problematic. Protection and isolation of existing facilities from the disruption of construction should be resolved early in the design. The reconciliation of building and production activities is never easy, but early planning for construction traffic, personnel, site screening, security, site access and site communications can minimize these adverse effects.

Where existing buildings are of an age which would indicate that their design and construction were performed in accordance with statutory requirements, planning, building control, standards or codes of practice subsequently superseded, care must be taken to ensure that extensions to such facilities are designed and constructed in accordance with *current* requirements.

6.6.4 Re-use of materials

Re-use of cladding or other construction materials; smoke ventilation, drainage, fire detection/suppression systems or techniques to reflect the existing installations may not comply with current requirements, particularly with respect to:

1. Fire protection of structural steel
2. Means of escape
3. Proscribed materials
4. Thermal performance
5. Fire detection and suppression
6. Fire compartmentalization
7. Structural design.

The requirements of the property insurers must also be adhered to if potentially adverse impact to premiums is to be avoided.

6.7 Fire detection and suppression

6.7.1 Fire regulations

As a general statement, it is unwise to assume particular requirements for specific cases in relation to either statutory requirements or preferential desires of bodies having jurisdiction in fire-related matters. The fire regulations pertain to building use and location. Hence, requirements for urban offices will vary considerably from those for rural industrial sites.

It is generally true that the statutory instruments and authorities are concerned with the preservation of life as their primary objective. Consequently, the requirements of insurance companies with regard to the preservation of the building or its material contents may be more involved. Adherence to such requirements (or otherwise) will be reflected in the insurance premiums quoted.

6.7.2 Influence of design

The design of the structure (and its operation) will greatly influence the effects of a fire. For instance:

1. Heat sources should be kept remote from flammable materials;
2. Compartmentalization of the structure (and maintenance of such compartmentalization) may contain or minimize fire spread;
3. Fire protection of the structure (discussed elsewhere);
4. Installation of detection, alarm and sprinkler (or other suppression) systems;
5. Separation of buildings one from another.

6.7.3 Types of system

There are many types of detection and suppression systems. The one selected should be compatible with the likely fire source and be consistent with the likely locations and fire size within the building.

It is essential to discuss the requirements for structural protection, compartmentalization, emergency lighting, detection, alarms, call points, suppression, means of escape and signage with the applicable local authority, fire brigade or insurance company personnel before finalizing designs.

6.7.4 Fire protection of structures

The principal construction materials used in industrial buildings are:

1. Steel
2. Concrete

Two different types of fire protection systems can be considered:

1. *Active*: This is a fire-*suppression* system (e.g. sprinklers) and is described elsewhere.
2. *Passive*: In this system the aim is to provide protection to the structural material for a specified period of time, and will be considered in this section.

Fire protection of concrete

Although not used extensively in superstructure work within industrial buildings, the fire-resistance characteristics of concrete are excellent. Failure of a concrete member is caused by a loss of strength in the steel reinforcement associated with its rise in temperature. The aim in providing fire resistance is, therefore, to reduce the temperature rise of the steel. This may be achieved by:

1. Judicious selection of the shape, size and distribution of reinforcement within the element;
2. Providing adequate cover to the reinforcement;
3. Adopting a lightweight or limestone aggregate, which is less susceptible to spalling than a siliceous aggregate such as flint.

Fire protection of steel

Apart from concrete or masonry encasement of steelwork, the following options are available:

1. Sprays (up to 4 hours)
2. Boards (up to 4 hours)
3. Intumescents (up to 2 hours: this can be the most expensive, depending upon the application)
4. Preformed casings (up to 4 h)

The spray is based upon either a natural plate-like material, such as vermiculite bound together with cement, or mineral fibers. Application is fast but not precise or clean, and is generally only suitable for areas where the steel will be hidden (by a false ceiling, for example). Sprays for external applications are available. However, the steel must first be provided with a compatible corrosion protection system.

The boards are based upon the same constituent materials used in sprays. They are suitable for situations in which only *dry trades* are allowed. The boards are cut to suit on site and mechanically fixed to the steel (e.g. by screws and straps). The system produces a smooth surface that is suitable for decoration.

Intumescents are thin films or mastics, which swell under heat to many times their original thickness. Their major use is in circumstances where the architectural statement of steel is to be preserved. As such, their costs vary considerably. Their application is fast and can be either by spray or brush. They are generally used internally, but external intumescents are available.

Preformed casings are very similar to fire boards but are tailored to the particular needs of the member being protected and, as such, permit fast application. They are, however, expensive.

6.8 Cost comparisons and contract procedure

Building costs per square meter for the construction of new buildings can readily be obtained from reference books such as *Spon's Architects' and Builders' Price Book* and *Laxton's Building Price Book*. Both are updated and issued on an annual basis, with the costs quoted based on the previous year's statistics.

The costs given are for a range of prices for average building work obtained from past records, and can vary by as much as 50 per cent between minimum and maximum costs. They serve therefore as an initial rough guide to the probable cost of a new building. *Spon's* also gives details of fitting-out prices within similar ranges, but these are generally limited to office fit-outs. The introduction to the sections dealing with prices per square meter in both books should be carefully studied before applying the figures to new structures or costing alterations/extensions to existing buildings on a *pro rata* basis.

As stated elsewhere in this chapter, it is of vital importance to assess the cost effectiveness of any proposed design (whether it be for a new building and/or alterations/extensions to an existing one) at an early stage. This can best be accomplished by seeking professional advice at the earliest opportunity. Such advice is normally provided by a quantity surveyor, who can work alongside the building owner and his engineers in assessing the cost of alternative designs/use of various materials/construction methods/construction times, etc. and evaluate the various proposals as the design develops. This will ensure that the most cost-effective solution is arrived at in the minimum time.

The correct selection of the form of contract to be used when inviting proposals is of paramount importance, as it constitutes the signed agreement between the employer and the contractor and forms part of the conditions of contract.

6.9 Structural and services supports

When considering the introduction of supplementary structure for *any* purpose there are a number of general items of design data, which must be examined and formulated:

1. *Support function*: What are the functional usage loads likely to be applied (Figure 6.10)?
2. *Existing conditions*: What is the existing spatial, civil, structural and access conditions within the facility? Typically, the following data should either be accessed from existing records and design requirements defined by the support function assessment or acquired by new site survey/investigations:
   (a) Ground conditions (soil-bearing capacity, water table, etc.) for ground-based support systems;
   (b) Existing construction elements; state of repair, materials, composition, bearing capacity, etc., allowable attachment methods, existing loading conditions;
   (c) Current access and people/materials flow in the area. Assessment of disruption, modifications to existing operations required both *during* installation process and *after* completion of installation.
3. *Alternative optional approaches*: Based on the constraints of space, cost, time, disruption, etc., what are the viable options?
   The following questions are pertinent when considering installation options:
   (a) Effects on facility downtime;
   (b) Use of proprietary, modularized systems;
   (c) Phased installation capabilities;
   (d) Cost comparisons;
   (e) Capabilities for later expansion;

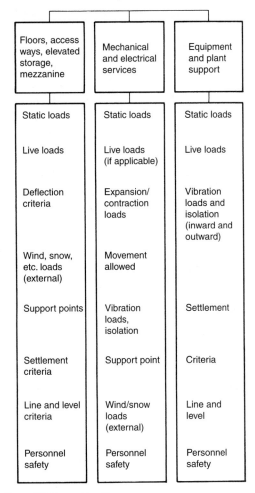

Figure 6.10 Support functions

(f) Consolidation of support functions (e.g. multiple service gang hangers, access/personnel safety features (Figure 6.12)).
4. Where possible the *best option* solution should provide the optimum mix of:
   (a) Cost effectiveness;
   (b) Personnel safety;
   (c) Minimum impact (downtime, disruption, structural alterations, ease of installation, etc.);
   (d) Expansion capability;
   (e) Multiple uses;
   (f) Fitness for purpose;
   (g) Off-the-shelf materials, devices, etc.

While the statements made above are necessarily general, it is essential that the conceptual approach delineated be adhered to. The necessity for compiling adequate design data for the new installation, followed by examination of existing conditions against the design requirements, is paramount.

Figure 6.11 A typical modular mezzanine floor (courtesy of The Welconstruct Co. Ltd)

Figure 6.12 Multiple piping system anchor of guide installation (courtesy of Industrial Hangers Ltd)

In many instances, adequate existing civil/structural or services installation information may not be available. Guessing the likely bearing capacity of floor slabs, foundations, structural steel or sub-slab ground, etc. can prove disastrous. Equally, assuming likely operational conditions for new installations can lead to embarrassing (and quite possibly dangerous) under-performance. The cheapest or quickest installation is not always the best or

most cost effective when effects on current operations or potential future expansion plans are examined.

Care must also be taken where large live loading are concerned such as gantry cranes, hoists, materials handling, etc. to ensure that structural and safety implications are properly addressed during both design and construction stages.

Where multiple service support systems are concerned, apart from operational movement, sufficient separation must also be provided to preclude electrical interference, water damage, inadequate clearance for insulation, cladding, cable de-rating (due to inadequate ventilation), mechanical damage during maintenance, etc. (Figures 6.12 and 6.13). It is essential that competent professional engineering personnel with applicable design experience be utilized for design of the installations.

6.10 Natural ventilation

6.10.1 Usage

Natural ventilation can be described as a process for providing fresh air movement within an enclosure by virtue of air pressure differentials caused primarily by the effects of wind and temperature variations in and around the enclosure. The primary usage of natural ventilation is to provide:

Figure 6.13 A typical multi-service gang hanger (courtesy of Industrial Hangers Ltd)

1. Fresh air introduction for odor and carbon dioxide dissipation;
2. Fire control/smoke clearance;
3. Heat dissipation/cooling;
4. Provision of air for fuel-burning devices;
5. Control of internal humidity (primarily condensation reduction/exclusion).

6.10.2 Requirements

For engineering purposes the composition of fresh air is generally taken at the following *standard* conditions:

20.94%	Oxygen (O_2)
0.03%	Carbon dioxide (CO_2)
79.03%	Nitrogen and inert gases (N_2 + xenon, zeon and others)

These fractions are, of course, somewhat variable, dependent on geographic location (i.e. urban versus rural). Generally, average humidity for Britain is as follows:

75–95% saturated	Winter
55–75% saturated	Summer

General guides to fresh air requirements for odor control and general ventilation are given in Table 2.1 and Figure 6.14. Note that requirements for removal of lavatory odor or for other highly concentrated areas are dealt with elsewhere. Figure 6.14 pertains to generally ventilated spaces only.

6.10.3 Threshold limit values

With regard to limiting the concentration of particular contaminants, these are generally referred to in terms of *threshold limit values* (TLV), expressed either in terms of a time-weighted average (TLV – TWA) which represents the average concentration for a normal working day over a 40-hour week to which nearly all workers may be exposed on a repeated basis or as a short-term exposure limit (TLV – STEL), which represents the maximum

Table 6.1 Recommended outdoor air supply rates for air-conditioned spaces (extracted from BS 5925: 1980)

Type of space	Smoking	Outdoor air supply (l/s)		
		Recommended	Minimum (the greater of the two should be taken)	
		Per person	Per person	Per square metre floor area
Factories[a,b]	None	8	5	0.8
Open-plan offices	Some	8	5	1.3
Private offices	Heavy	12	8	1.3
Boardroom, executive offices and conference room	Very Heavy	25	18	6
Toilets[c]	–	–	–	–
Corridors[c]	–	–	–	10
Cafeterias[a,d]	Some	12	8	–
Kitchens[a,c]	–	–	–	20

[a] Rate of extraction may be overriding factor.
[b] See statutory requirements and local byelaws.
[c] A per capita basis is not appropriate in these cases.
[d] Where queuing occurs in the space, the seating capacity may not be the appropriate total occupancy.

concentration to which workers can be exposed for up to 15 minutes. Clearly, these values vary, depending on the particular contaminant.

Reference should be made to *Threshold Limit Values, Guidance Note EH17/78*, issued by the Health and Safety Executive (HSE), or *Industrial Ventilation* (American Conference of Governmental Industrial Hygienists). In all cases, proposals should be reviewed by and submitted to the relevant local authority agencies.

Space does not permit a detailed investigation of all the requirements concerning ventilation for the remaining usages listed. However, a brief description of the rationale for provision is given below, and further details can

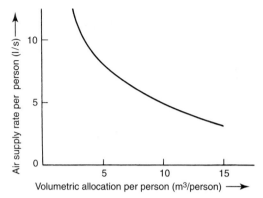

Figure 6.14 Approximate air supply rate for (human) odor removal (extracted from BS 5925: 1980)

be obtained from the references listed at the end of this chapter.

6.10.4 Fire control/smoke clearance

Depending on the requirements of local fire officers, statutory regulations and insurance bodies (or any other documents or bodies having jurisdiction), there may well be a need to address the clearance of smoke from escape routes, the control of smoke spread generally and the removal of smoke during and after firefighting activities.

6.10.5 Heat dissipation/cooling

Human occupants, electrical/electronic equipment and process plant all emit varying quantities of sensible and latent heat. Equally, these various elements require (or can tolerate) differing environmental conditions. Depending on these operational constraints, the need may well exist to provide natural (or powered) ventilation to maintain environmental conditions (temperature and/or humidity) consistent with the occupational/process requirements.

6.10.6 Provision of air for fuel-burning devices

All combustion equipment (oil, gas, solid fuel) requires primary air to support combustion and secondary air to permit adequate velocities in flue ways, etc. These requirements are governed by the minimum air/fuel ratio and operating flue-way parameters. There are also published recommended minimum requirements that are generally in excess of these.

6.10.7 Control of internal humidity

Where human occupancy or 'wet' process plant is present, the emission of water vapor will occur. Depending on external conditions and building fabric construction, the attendant potential for excessive ambient humidity or surface condensation may exist.

Consequently, the introduction of external air ventilation may be required in order to reduce ambient moisture contents. However, this will need to be balanced against energy costs and the use of other design solutions such as building fabric moisture proof membranes, local air exhausts, etc.

6.10.8 General design considerations

The use of openings within the building fabric to provide natural ventilation is predicated on various primary factors:

1. Location and orientation of building;
2. Building height and configuration;
3. Shape, size and location of openings;
4. Wind and temperature effects.

Identifying specific quantified building ventilation rates can be difficult due to the number of non-linear parallel paths that may have to be simultaneously considered.

However, simple-case type solutions (which are often good enough to establish whether mechanical ventilation is required or excessive air infiltration problems may exist) can be identified. These simple solutions are based on knowledge of:

1. Wind direction and velocity;
2. Building opening locations and size;
3. Temperature differentials (internal and external).

These solutions are discussed in more detail in many publications, including Building Research Establishment (BRE) *Digest No. 210*, the American Society of Heating, Refrigerating and Air Conditioning Engineers (ASHRAE) *Fundamentals Handbook* and Chartered Institute of Building Services *Engineers Guide* (Books A and B).

6.11 Building durability

The durability of a building (i.e. its life expectancy and its resistance to deterioration) is determined by deliberate design decisions relating to structure and choice of materials as much as to the natural or precipitate process of ageing. Within certain limits, the design of an industrial building can and should take into account the predicted use or lifespan of the process or method of operation which it is to accommodate. To aim at durability beyond that has ascertainable cost implications. These may be acceptable if the building is to serve future known or even unpredictable purposes.

6.11.1 Durability and plan form

If the plan of the building (and the same applies to its cross-sectional features) is precisely and inflexibly related to the initial plant layout and the production processes known at design stage, non-adaptability to later desirable changes in use is clearly a durability factor.

6.11.2 Structure and durability

The structural design of an industrial building will, first, reflect the requirements of plant layout and manufacturing procedures. Apart from holding up the building envelope, structural members will be designed and placed to support particular static and dynamic loads. As with the plan form, the shape and disposition of the structural features may either curtail or allow extension of the life of the building. Load-bearing perimeter walls exemplify the former and maximum freedom from internal columns the latter. The cost implications of reducing structural constraints and thereby increasing durability need to be considered.

6.11.3 Cost of construction materials

There is a clear correlation between quality or cost of materials and the durability/life expectancy of buildings. Greater resistance of better materials to wear and tear can be assumed, with obvious implications on future maintenance. Striking the best balance between initial capital outlay and maintenance cost requires complex calculations that take into account such intangibles as future interest rates and taxation of building operations.

6.11.4 Durability of materials

Apart from selecting better, more durable and expensive (rather than cheaper) materials, their life can be prolonged by appropriate treatment, coating or other protection, e.g.:

Mild steel: Galvanizing/plastic coating
Extruded aluminum: Anodizing/polyester powder coating
Concrete: Integral or applied hardeners
Brickwork: Joint formation/masonry paint
Cement screeds: Mineral or metallic aggregate hardeners
Timber: Impregnation/staining.

Obviously, durability is greatly enhanced by specifying stainless rather than mild steel, engineering rather than common bricks, etc. Comparative life expectancies of materials are tabulated in reference books such as that published by NBA Construction Consultants Limited in 1985.

6.11.5 Damage and protection

Industrial buildings, more than most, are prone to damage. Causes range from mechanical impact to chemical attack resulting from the production process – quite apart from heavy wear in normal use. Examples of causes and available protective measures are as follows.

Mechanical impact

Vehicles (external): Buffer rails, ramps
Transporters (internal): Bollards, route demarcation, column base plinths, guide rails, flexible/automatic doors
Production processes: Wall/floor shields pipe cages, resilient rails
Craneage: Electromechanical limiters
People: Wall rails, door kick plates, foot mats (shop floor offices)

Chemical/atmospheric attack

Plant/process emissions: Anti-corrosion coatings heat shields, extractors

Humidity (condensation): Dehumidifiers, insulation, ventilation, warmed surfaces
Atmosphere (external): Corrosion-resistant coatings, fungicidal treatment, over-cladding
Fire: Smoke detectors, intumescent coatings/seals

6.11.6 Details and specification of construction

Design for durability, as defined above, will be negated if it does not take into account the essential need for ease of maintenance or ignores sound specification of workmanship and constructional details. Deterioration factors notorious for shortening the life of a building include the ingress of moisture and structural or thermal movement. Building Regulations and Codes of Practice, all fully referred to in the National Building Specification, cover necessary safeguards against these and others.

6.12 Building maintenance

Maintenance is estimated to represent more than one third of all building expenditure, which underlines its significance in the economic lifecycle of any building, the vital importance of designing for it and skillfully managing its execution.

6.12.1 Building records

The prerequisite for proper planning and performance of building maintenance is the comprehensive set of drawings, specifications and servicing manuals compiled by architects, engineers, contractors and suppliers, respectively. Those responsible for maintenance must ensure that all records are kept up to date; computers are increasingly used for this purpose.

6.12.2 Maintenance work

Depending on the scope and complexity of maintenance as well as economic considerations, inspections and performance will be (both or either) the responsibility of in-house permanently employed staff or outside specialists or contractors. In the latter case, particular expertise is required, and this should be reflected in the selection/procurement procedure.

6.12.3 Maintenance planning

The plant engineer and his maintenance manager will establish whoever carries out the work, basic strategy, programming and procedures. Inspection checklists and record sheets are necessary for systematic location of critical elements/potential defects and setting out frequencies of inspection. Computers and work processors are an obvious aid to effective maintenance management. Inspection-based preventive methods are demonstrably cost effective.

6.12.4 Monitoring equipment

Most, if not all, of the equipment listed below is required, even where outside specialists are generally retained to carry out maintenance (for either access or measurement), including:

1. Tubular towers, ladders, long-arms, cleaning cradles;
2. Levels, tapes, thermometers, hydrometers, light meters, smoke detectors, audio-sensors. Features such as cradle rails, safety eyes and climbing irons should be incorporated into the design of the building.

6.13 Building repairs

The need for building repair arises from the inevitable process of ageing, and such deterioration is brought to light by systematic maintenance. There is also repair work necessitated by damage from observable causes such as physical impact (discussed elsewhere in this chapter).

6.13.1 Repair work

Depending on its complexity and urgency, repair work may best be carried out by in-house staff employed and trained for this purpose (but not necessarily to the exclusion of other work) or by directly employed labor skilled in particular trades or by specialist contractors. In many cases immediate (but adequate) steps can be taken by permanent staff, sufficient to deal with an acute problem while deferring more comprehensive or permanent repairs to be carried out by outside specialists (possibly as part of a regular maintenance program with resultant economies).

6.13.2 Repair: works specification

Leaving aside stopgap emergency repairs, the work should be done to a standard set by the original material and workmanship specification (which must be referred to by whoever carries out the work), unless this has been found to be inadequate. Qualified professionals with regard to compatibility, cost and compliance with building regulations must assess any deviation from original materials or methods. The specification must also take into account limitations of the structure as originally designed.

If faulty specification or unsuitable materials are found to be the cause of defects requiring repairs, replacement work should be considered in anticipation of problems arising on a wider scale. A less disruptive repair program may thereby be adopted compatible with routine maintenance operations.

6.13.3 Flat roof defects

Unlike other defective building elements, roofs generally require prompt repairs. Latent defects in roofs can go undetected with correspondingly more serious consequences when they do manifest themselves. Regular inspection of roofs is therefore doubly necessary, as well as recognition of potential causes of damage, e.g.:

1. Inadequate falls and unimpeded water discharge;
2. Ultraviolet ray attack of unprotected roofing;
3. Cracking caused by alternating expansion and shrinkage;
4. Distortion of roofing due to heat rejection (insulation);
5. Traffic and impact damage of unprotected roofing;
6. Perimeter constraint/up stand and flashing details.

6.13.4 Roof repairs

Diagnosis of defects by inspection of visible causes or more sophisticated means (tracer-dye inundation to find sources and routes of leaks, infrared photography, etc.) is a prerequisite for correct repair specification. Greater skill in detailing and specifying suitable materials and methods are required in the design of flat roofs and the same applies to repairs. Here, too, durability and effectiveness are, to some extent, cost related.

Specialists should be employed to carry out roof repair work that is able and willing to guarantee its quality and life.

The Technical Guide to Flat Roofing, published by the Department of the Environment/PSA in 1987, is one of the many useful reference books.

6.14 Domestic facilities

6.14.1 Toilet accommodation

Toilet accommodation encompasses the provision of water closets, urinals, washbasins and/or washing troughs and, depending on a particular trade or occupation, the provision of baths and/or showers. Provision of toilet accommodation is specified under Acts of Parliament, British Standards, regulations and HM Inspector of Factories.

The provisions given in the various standards are essentially minimum requirements only, and when assessing quantities and type, the following should be considered:

1. Number of personnel and male/female ratio;
2. Is a shift system operating within the building?
3. Type of process in operation – dirty or clean, wet or dry;
4. Specialized usage – sterile conditions, radioactive contamination;
5. Provision for remote-actuated appliances (i.e. foot operated, photoelectric cell, etc.);
6. Ethnic considerations relating to siting of toilet accommodation;
7. Provision for handicapped personnel;
8. Provision of facilities for female hygiene;
9. Changing-room facilities for male/female personnel.

6.14.2 Canteens

Canteen facilities for staff personnel can generally be divided into two categories:

1. Provision of cold meals (sandwiches, cold buffets, etc.);
2. Provision of both hot and cold meals prepared by resident staff or delivered under contract by outside caterers.

Many industrial buildings offer canteen facilities free of charge or at subsidized rates. However, before embarking on the provision of canteen facilities the following points should be considered:

1. Proximity of building relative to adjacent village, town facilities;
2. Hot and/or cold food and selective menu;
3. Number of meals to be catered for and operating hours;
4. Ethnic requirements (i.e. vegetarian, etc.);
5. Secondary catering facilities for senior staff and/or visitors;
6. Provision of separate hot, cold and drinking water services for canteen facilities.

6.14.3 Rest rooms

Statutory requirements dictate that a space be set aside within an industrial building for use as a rest room. The usage and facilities available are often extremely varied, and can be sub-divided as follows:

1. Use of room for morning/afternoon breaks;
2. Consumption of food at lunchtime in the absence of canteen facilities;
3. Combined usage as a recreation room.

In providing rest room facilities the following points should be considered:

1. Prime use of room as defined above;
2. Number of persons to be accommodated;
3. Provision of drinking and snack facilities (i.e. tea-preparation sink, vending machine, automated snack bar, etc.);
4. Self-service or resident catering staff;
5. Male/female segregation;
6. Provision of recreation activities (i.e. darts, table tennis, television, etc.);
7. Provision of room furniture.

6.14.4 First-aid facilities

It is a statutory requirement that first-aid facilities be available at all times for staff personnel. This facility is very diverse and can extend from a first-aid cabinet to the provision of a first-aid room and attendant semi-resident nurse/doctor. The extent of first-aid facilities is dependent on the following:

1. Type of building and nature of business (i.e. dirty production line, computer components);
2. Number of staff personnel to be serviced and male/female ratio;
3. Proximity of building to easily accessible external medical facilities;
4. Provision of first-aid trained staff from within the building personnel;
5. Provision of emergency equipment for specialized building activities (i.e. chemical, radio-active contamination);
6. Provision of bed furniture, stretchers, respirators, etc.;
7. Provision of separate toilet facilities for resident nurse/doctor;
8. Ethnic considerations.

6.14.5 Detail design

Quantity of sanitary fittings shall be based on total numbers of personnel per shift and not total number of employees per day.

Where dirty or wet processing occurs continuously within a building, personnel at the end of the day or shift may consider the provision of baths and showers for use.

Working areas within certain industrial buildings may have restricted access (i.e. sterile laboratories, radioactive areas). Separate toilet accommodation may be required in these areas, an assessment of which may be obtained from the operator and reference to such publications as Atomic Energy Code of Practice, Laboratory Practice. It may also be necessary to operate such appliances remotely by photoelectric cell, sonic control or foot control.

Where building employees are from mainly ethnic communities, consideration should be given to the positioning of toilet facilities so that religious beliefs are not compromised.

In addition to the general employees of a building, the following specialized personnel should also be considered:

1. Disabled
2. Elderly
3. Administration
4. Catering
5. Visitors.

It is a requirement of the Building Regulations 1985, Schedule 2, to provide facilities for disabled people. Where provision within industrial buildings for elderly persons are required, reference should be made to the DSS. Consideration should also be given to providing water at lower temperatures to appliances exclusively for used by elderly and/or disabled persons.

Sanitary-towel disposal must be provided in each female toilet. The type of disposal can be by macerator, incinerator, bag or chemical.

Separate changing-room facilities for male and female personnel would include the following:

1. Personal locker with lock for change of clothes at the end of a shift or day;
2. Bench/chair for seating.

The location of an industrial building relative to adjacent shops, village and town may well influence the provision of canteen facilities as well as the type of service to be provided.

Where full canteen facilities are provided an accurate assessment of the hours of operation must be obtained so that plant and services are correctly sized. The operating hours will also greatly influence the need to provide separate plant from the main building, especially if meals are to be provided at the end of each shift.

6.15 Elevators

6.15.1 Introduction

The planning of elevator systems should be one of the first to be considered, usually alongside the initial building structure, bearing in mind the long lead in times and structural requirements. It is important that the correct configuration is established early, as changes later can be difficult to incorporate.

The first essential step in the system design is goods and pedestrian planning, i.e.

1. How many people require transportation?
2. What will be the peak demand and how and when will it occur?
3. How many floors will the elevator system serve?
4. What are the loads/sizes of goods/equipment requiring transportation?

The responses to these questions can be used to determine the number of elevators required, the size of the car, the door arrangement and the speed of car travel. Other major considerations should be the location of elevators within the building and accessibility for disabled users.

6.15.2 Elevator system design — drive systems

There are various types of drive systems employed throughout the elevator industry, and the following are the most commonly used:

1. Electric a.c. motor with gearbox;
2. Gearless electric d.c. motor;
3. Hydraulic.

Electric a.c. motor with gearbox

Elevators of this type are normally rope driven and are the most common drive used in elevator installations. The gearbox is fitted with a grooved traction sheave on which a series of steel ropes are wound. The ropes are attached to the top of the elevator car, wrapped around the traction sheave (and occasionally secondary sheaves) and attached to a counterweight of equal mass to the car and a percentage of its rated load (normally 40 per cent).

This drive can be used for a wide range of car sizes but has a limit to the speed of travel of approximately 2.5 m/s. The elevator machine room is usually located above the elevator shaft. However, bottom-of-shaft machine rooms are not unknown, although they tend to be more costly.

Gearless electric d.c. motor

Elevators of this type are of a configuration similar to the geared drive systems mentioned above. The difference lies in the increase in the physical size of the drive as the required operating torque of a gearless d.c. motor is produced at a fraction of the rpm of a geared a.c. motor.

The cost of these drives is significantly greater than that of a comparable geared drive, but the quality of ride and maximum speed of car travel are much improved. Gearless drives usually have applications in prestige offices and hotels where the elevator travel is in excess of 35 m.

Hydraulic elevators

Elevators of this type are generally used for low-rise applications and if good leveling accuracy at each landing is required. Hydraulic elevators are either the direct-acting type, where a hydraulic ram acts on the underside of the elevator car, or the suspended type, where the ram acts on the top of the car via a series of ropes or chains and pulleys. Hydraulic elevators can be used for very large goods elevators when two or more hydraulic rams are used, but are limited to a slow car speed, which improves to a maximum of 1.5 m/s for small elevator applications. Hydraulic elevators have an advantage over rope-traction elevators in that the elevator machine room can be situated remotely from the elevator shaft with a connection to the elevator ram via flexible hydraulic pipe work. Hydraulic elevators have a limited maximum travel which is approximately 12 m for (up to) 1000 kg car size and 8 m for car sizes over 1500 kg.

6.15.3 Elevator machine rooms

Each elevator or group of elevators requires a elevator machine room that should be designed complete with:

1. Drive system machinery, pulley mechanisms and control panels;
2. Elevator beams tested for safe working loading;
3. Adequate emergency lighting, small power and 110 V sockets;
4. Adequate ventilation/cooling/heating/frost protection as required by BS 5655;
5. Fire detection and alarms;
6. An electrical switchgear panel that provides an electrical supply to each elevator machine.

The elevator motor room should be identified with safety signs as required in BS 5655 and should be secured to allow access by authorized persons only.

6.15.4 Elevator control in the event of fire

In the event of fire within a building, the controller for each elevator should isolate all manually operated inputs and return automatically to the evacuation level, usually the ground floor. An output from the building fire alarm panel is 'hard wired' to the elevator controller, giving the signal for a fire condition. The elevator remains disabled at the evacuation level and the car doors open. If a fire officer requires control of a elevator a key switch or break-glass unit should be used to re-activate the elevator.

6.15.5 Firefighting elevators

These should not be confused with elevators fitted with a fire officer's override control. A firefighting elevator is required for any building exceeding 18 m in height or 9 m in depth below the evacuation level, and is to be used to allow fire officers to maneuver personnel and equipment during firefighting operations.

A firefighting elevator must:

1. Not be a goods elevator under normal operating conditions;
2. Have a minimum size of 8 persons/630 kg;
3. Be installed in a firefighting shaft with 2-hour fire-rated walls;
4. Have suitable fire-retarding finishes for the elevator car;
5. Have a separate emergency electrical supply and feeder cable;
6. Have a machine room located above the elevator shaft;
7. Be able to complete a journey in less than 60 seconds.

6.16 Sub-ground pits and basements

Sub-ground pits may be required for access, service distribution or process requirements. Depending on pit dimensions and water table levels, problems that may have to be considered at the design stage include 'heave' due to relief of overburden, buoyancy from water pressure and waterproofing.

6.16.1 Waterproofing

There are three standard solutions to waterproofing pits and basements:

1. Watertight concrete construction;
2. Drained cavity construction;
3. External continuous impervious membrane (tanking).

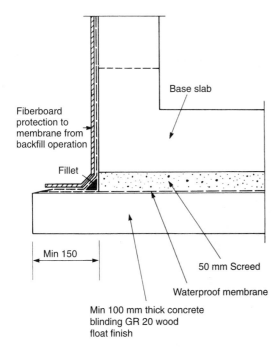

Fiberboard protection to membrane from backfill operation

Fillet

Base slab

Min 150

50 mm Screed

Waterproof membrane

Min 100 mm thick concrete blinding GR 20 wood float finish

Figure 6.15 A typical external waterproof membrane detail

Watertight concrete construction

This relies on the integrity of the concrete construction itself with reference to design (crack control, joint spacing and detailing, concrete mix) and skilled workmanship and supervision.

Drained cavity construction

This is also based on watertight concrete construction but accepts that some seepage may occur through the external concrete wall. The water is kept away from a separate inner-skin wall by an intervening drained cavity. An internal vapor barrier may be applied to the inner-skin wall to limit humidity, but the cavity may then require ventilation to prevent condensation.

External waterproof membrane

Two possible forms of membrane are hot applied mastic asphalt or bitumen/butyl rubber sheeting with welded or glued joints. The membrane under the floor slabs has to be lapped with that around the walls. It is essential that the membrane is protected during construction, and a typical arrangement is as shown in Figure 6.15.

6.17 Internal and external decoration

6.17.1 Internal

Ceiling, wall and floor finishes can be separated into 'wet' and 'dry' finishes. Below is a listing of the common types of wet and dry finishes currently in use.

Wet finishes

1. Plasterwork
2. Screeds
3. Tiling
4. Paint finishes
5. Wet structural fire-protection systems.

Dry finishes

1. Dry lining (plasterboard)
2. False (suspended) ceilings
3. Wall claddings
4. Dry floor finishes (thermoplastic sheeting, carpeting, raised deck flooring, etc.)
5. Dry structural fire-protection systems.

The selection of finishes is generally made against the following parameters or constraints:

1. Acoustic performance
2. Environmental conditions
3. Life expectancy/maintenance requirements
4. Safety requirements (both installation and use)
5. Fire rating/flame spread
6. Aesthetic considerations
7. Flexibility of use
8. Load-bearing requirements

9. Costs
10. Installation time/disruption.

Environmental conditions

All finishes should be selected to be resilient to expected average and worst-case environmental exposure such as high/low humidity or temperature, airborne contaminants, vibration, possible aggressive liquid spillages, cleanliness/hygiene requirements, etc.

Life expectancy/maintenance requirements

Many finishes exhibit low maintenance requirements (e.g. plasticzed metallic sheeting, epoxy coatings, continuous tiling systems, etc.). Others may be more maintenance intensive and may provide lower durability. However, selection must also consider the other operating parameters such as acoustic performance (which may mandate heavier mass or more porous-surfaced materials) or load-bearing capabilities, etc.

Safety requirements

Materials containing toxic or irritant materials should be avoided. Flooring finishes should exhibit non-slip properties but be cleanable and not provide excessive crenellations, etc., for contaminants. Materials will need to be certificated in terms of fire resistance (to the passage of fire) and flame spread. They are classified when tested against BS 476 and the National Building Regulations. Some materials (such as PVC) emit toxic fumes when burning. Structural integrity must also be assessed where secondary cladding or suspended finishes are being utilized, particularly where load-bearing abilities are required (such as suspended ceilings or wall claddings).

Aesthetic considerations

While this may appear to be self-explanatory, the need to change color schemes or textures over the lifetime of the installation will need to be reviewed. Self-colored materials do not typically take secondary coatings such as paint or stain.

Flexibility of use

Consideration should be given to likely layout or usage changes. Generally, wet materials will provide least flexibility in terms of subsequent removal and reinstallation. However, they may be more conducive to changes in usage of the space.

Costs and installation time

It is generally accepted that the use of dry materials results in shorter and less disruptive installations, particularly if modular systems are used. However, these materials may well not offer the least expensive solution.

6.17.2 External

In many ways, the same criteria apply to the selection of external finish materials as to internal finishes, and can be identified as 'wet' or 'dry' types. However, increasingly, dry cladding (roof and wall) and bonded roof membrane systems are being used in preference to more traditional rendering and liquid applied roofing membranes.

If extensions are being made to existing facilities or buildings adjacent to 'traditional' structures then this may require the use of brickwork cladding or block and render in order to satisfactorily 'marry' with the existing structures. Care must be taken to ensure such extensions/additions meet current building regulations, bylaws, standards and codes of practice.

Of major importance when selecting suitable exterior finish materials are their weather resistance and 'weathering' characteristics (with particular reference to ultraviolet resistance and moisture staining), including resistance to biological/vegetable growth. Cladding systems, particularly, should be examined against their wind and moisture penetration resistance. External building elements, either individually or when viewed as a collective system, must comply with the thermal transmittance requirements of the building regulations.

Corrosion resistance should also be considered with particular reference to the ambient environment. Painting schemes, coating materials and other protective skins (along with the substrate bonding method) should all be reviewed against the prevailing conditions.

6.18 Industrial ground floors

The ground floor of an industrial building is arguably the most important structural element. Failure of a floor slab or even a topping can lead to enormous expense and disruption due to repair or replacement. Considerations in the design of a ground slab may include:

1. Loading (uniformly distributed loads and point loads)
2. Abrasion/impact resistance
3. Chemical resistance
4. Flatness
5. Ease of cleaning.

It is important that these are accurately assessed since under-specification may lead to severe maintenance problems while over-specification may result in substantial cost increases.

6.18.1 Structural form

Dependent on ground conditions, loading requirements and settlement limits, industrial ground floors may be ground bearing, raft construction or piled. The piled option should, if possible, be avoided, as the cost is comparatively very high.

Rafts

Rafts fall into a category between fully stiff (behaving as a rigid body) and fully flexible. The raft may also support the superstructure. A soil/structure interaction study must be carried out to estimate the degree to which loads are distributed by the raft. Hence, the resulting slab forces and required reinforcement are calculated.

Ground-bearing slabs

Ground-bearing slabs are typically the cheapest and most common form of industrial ground floor, and rely on the tensile capacity of concrete. Any reinforcement provided is to reduce the potential for cracking from shrinkage and thermal effects. The degree of reinforcement is therefore primarily dependent on slab thickness and joint spacing (Figure 6.16).

The detailing arrangement of joints is of prime importance in the slab performance. The purpose of these joints is to concentrate cracks at specific locations, isolate the slab from other structural elements and sometimes to allow for expansion or articulation.

The main elements of a ground-bearing slab are:

1. *Sub-grade*: the material at formation level after excavating or infilling;
2. *Sub-base*: selected imported granular material to form a stable level surface on which to construct the slab. Where lean-mix concrete is not used the surface is normally blinded with sharp sand;
3. *Slip membrane*: serves to reduce friction between the slab and sub-base to minimize subsequent slab cracking. This is normally a polyethylene sheet. It may also serve as a damp-proof membrane;
4. *Structural slab*: usually with single layer of reinforcement in the top;
5. *Wearing surface*: may be the finished slab surface or the surface of an applied topping.

6.18.2 Abrasion resistance

Abrasion resistance can be influenced by:

1. Concrete mix
2. Method of finishing the concrete
3. Method of curing
4. The addition of surface hardeners or sprinkle finishes (metallic or non-metallic)
5. The addition of toppings.

6.18.3 Chemical resistance

Chemical resistance is normally achieved by synthetic resin toppings, polymer or resin-modified cementitious toppings or modular (tiled or paver) toppings.

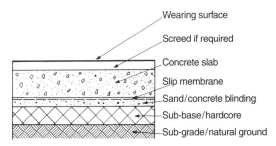

Figure 6.16 A typical section through a bearing slab

6.18.4 Flatness

Flatness is of critical importance to some sections of industrial operations such as VNA (very narrow aisle) high rack warehouses or automatic guided vehicles. The degree of flatness achieved is dependent on construction techniques, workmanship and supervision.

Traditionally, global tolerances have been specified for any point on the floor relative to datum and local tolerances specified relative to a 3 m straight edge. For VNA type requirements tolerances distinguish between areas of defined movement (i.e. predetermined vehicle routes) and areas of free movement (the rest of the slab).

6.18.5 Dusting

High-quality concrete slabs that have been well finished should not give rise to dusting. Complete avoidance of dust, however, will only be achieved by the addition of a resin or chemical sealer. Chemical sealers (such as silicofluorides and silicates) give little increased abrasion resistance, and can themselves be prone to dusting on over-application.

6.19 Ground considerations

6.19.1 Existing site conditions

The existing site conditions, which can influence the design, construction and performance of industrial buildings and their operations, include:

1. Topography – surface topology, watercourses, surrounding property, vegetation, access;
2. Geology;
3. Groundwater conditions;
4. Site history – industrial contamination, mining, domestic refuse, old foundations, cellars, tunnels;
5. Existing buried or overhead services;
6. Liability to flooding, wind or sea erosion, subsidence, slope instability;
7. Liability to vibration from earthquakes, railway tunnels.

6.19.2 Contaminated sites

Extensive available level sites have sometimes previously accommodated gas works, chemical works, munitions factories, or industrial or domestic refuse tips. These sites require special consideration as regards protection of the structure to be built, the construction workforce and the building occupants.

Refuse tips invariably result in the production of methane from the decay of organic matter. Measures are necessary to avoid trapping the methane in or beneath the building. This can be achieved by a system of methane vent pipes in the fill together with an impermeable membrane at sub-ground-floor level. Alternatively, a vented cavity can be constructed sub-ground floor.

6.19.3 Foundations

The foundation system will be based on such information as can be determined together with physical and chemical testing carried out on samples procured from a site investigation. In regions of very poor ground conditions, there is sometimes an option of utilizing a ground-improvement technique such as dynamic consolidation.

6.19.4 Existing services

Existing services are important in two respects:

1. They may have to be rerouted or the structural scheme amended to avoid clashes;

2. The services required for the new industrial operation must usually tie into the existing infrastructure.

6.19.5 Car parking

Car parking area per car (space + road), is approximately 25 m^2. Standard car parking dimensions are 4.8 m × 2.3 m (3.3 m for disabled persons) with access roads 6 m wide. Petrol interceptors for surface water from car parking areas are required where 50 spaces or more are provided.

7

Planning and Plant Layout

Roland R Gibson
Engineering Consultant, W. S. Atkins
Consultants Ltd

Contents

7.1 Introduction

Plant layout can affect the total operation of a company, including the production processes, equipment, storage, dispatch and administration. It has a direct effect upon production efficiency and economics of the operation, the morale of employees and can affect the physical health of operatives.

A production facility will be considered as a facility for processing pharmaceuticals or food products or manufacturing engineering products or consumer goods. The facility must utilize real estate, equipment, materials and labor to generate profit for investors and, philosophically, to enrich the life of all associated with it.

Layout planning involves knowledge of a wide range of technologies that will extend beyond those of individual planners and the full range of expertise may not exist in a production facility. Consultants can provide the expertise but guidance can be found in the published works listed in the References.

The design methods presented here allow a layout plan to be quickly formulated. The methods rely upon a thorough understanding of factory operations gained from experience and a good understanding of the relationship between people and equipment. When such an understanding is not present, a more rigorous approach is recommended. Muther[1] published a formalized procedure in 1973, Tompkins, and White[2] a more academic method in 1984, both valuable contributions to problems of layout planning.

The first step in any design is to identify the real need, and this is often the most difficult task. Without it, designs can be produced which do not satisfy the requirements and the result is often unsatisfactory. It is essential to clearly define the objectives of the task and to re-confirm the objectives as time progresses. A useful aid is a value analysis at the end of the concept design stage. This assesses the design for value for money while meeting the defined project objectives. A good source document is *A Study of Value Management and Quantity Surveying Practice*, published by The Royal Institute of Chartered Surveyors.

A criterion of effective plant operation used to be efficient utilization of capital equipment. The main requirement is now recognized as short door-to-door times, not short floor-to-floor times. The prime need is therefore to achieve a plant layout that facilitates reception of raw materials and dispatch of finished goods in the shortest possible time with minimum capital tied up in work in progress (WIP). This involves access to the site, reception of goods vehicles, raw material goods storage and issue to production, procurement of component parts and sub-assemblies from sub-contractors, process technology and process routes, integrating the sub-contractors' supplies, finished goods storage and dispatch to the customer.

7.2 Technological development and its effect upon plant layout

There have been important world developments in the philosophy of factory operation and they affect plant layouts.

These developments, employed by companies now called World Class Manufacturers (WCM),[3] must be examined by the company management who are contemplating establishing a new factory or expanding or reorganizing an existing plant.

Much of the new thinking on strategy has centered on advanced technology and the exploitation of the computer, but the following recommendations, requiring minimum investment in new plant, should be considered first:

1. Establish the operational priorities:
   (a) Production to be sales driven;
   (b) Make only what can be immediately sold;
   (c) Make every part right first time.
2. Reorganize production equipment:
   (a) Consider process orientation but employ product-oriented flow when possible;
   (b) Organize products into groups that require the same manufacturing equipment;
   (c) Arrange equipment into product cells.
3. Re-form the structure of production teams:
   (a) Form accountability teams for each product;
   (b) Bring staff functions onto the shop floor and into the teams, form quality circles and establish total quality control (TQC).

These ideas, interestingly described by Goldratt and Cox,[4] provide flexible production and cater for customer requirements. They have become known as *just in time* (JIT) manufacture, but are often overlooked by many that think that investment in computer-controlled equipment is the only way to go. The concept, simplify before you automate, can provide significant improvements in production and can point the way to later advanced manufacturing where:

1. Equipment cells can be replaced by computer-coordinated machines – direct numerical control (DNC), flexible manufacturing systems (FMS), flexible manufacturing cells (FMC);
2. Warehousing automated to include computer-controlled storing and retrieval (AS/AR) and movement by automatic guided vehicles (AGV);
3. Whole manufacturing processes computer linked (CAM);
4. Computer-aided design (CAD);
5. Sales and management computerized, leading to computer-integrated manufacture (CIM).

The development of these technologies can be followed in the proceedings of the professional institutions and the many magazines, some available free. You should define the state of technological awareness in the company and consider how this affects the plant layout.

7.3 Layout planning concepts

A process of analysis often executes designs. The methods vary from the traditional and well tried to experimental techniques. These include:

1. A conventional plan drawing;
2. Cardboard cut-outs depicting blocks of buildings pinned to an outline drawing of the site;
3. Three-dimensional scale models;
4. Software calculation and simulation packages.

Proprietary systems are available[10,11] employing Lego type blocks which can be constructed and placed on a baseboard to represent a three-dimensional visualization of the layout. Another system employs aluminum castings of the actual items of equipment and operatives. Cardboard cutouts have been the most commonly used method, particularly with those who rarely plan layouts.

The above methods allow proposed layouts to be visualized, analyzed and altered. These methods are often a try-it-and-see process, and the associated analysis is rarely structured. The old saying that 'If it looks right, it is right' is often used to justify results, but it may be more true to say that if it looks right it is conventional, and the analysis method will rarely produce something original. This applies, of course, to any design activity, whether it is a plant layout, a piece of machinery or a consumer product.

Computer-aided methods are increasingly being used to assist design and a process of synthesis, described later, employs techniques to produce a design that will satisfy predetermined criteria. This produces a layout that can be subjected to critical review, and suggested changes to the layout can be examined for violation of the criteria.

The introduction of computers to many companies allows proprietary software to be used for layout design.[5] Spreadsheet, mathematical modeling and computer-aided design (CAD) techniques are available and greatly assist the design process, and have added to the resources available to planners. However, the traditional scale models described above will still be useful to present the result to management and shop floor personnel.

7.4 Plant data

7.4.1 Collection

The objective of the production process is to produce the right goods of the right quality and at the right price in order to generate a financial profit from the capital investment, but the plant layout fundamentally affects this objective.

The essential requirement for any design exercise is a thorough understanding of the working medium. For layouts, this means collecting sufficient data to describe those characteristics of the company which affect the layout. Data collection is time consuming and difficult, and it requires the contribution of factory personnel from many departments. The data required includes:

1. *The company organization structure*
   All departments, including maintenance
   This should be readily available from the personnel department but may need to be updated.
2. *Number of employees in each department and shift*
   Administration staff
   Technical staff
   Factory workers direct
   Factory workers indirect
   This and the next item will be obtained from the personnel department, but the actual working situation will probably need to be confirmed with the individual managers of the various process departments.
3. *Hours and the arrangements for shift working*
   Administration and technical staff on day shift
   Difficulty is often experienced when workers are shared between departments. Some information will not be documented but since employees are generally paid for attendance hours these are meticulously recorded.
   Factory workers on each of the worked shifts
   Consider the arrangements for holidays
   Quantify work contracted to outside organizations
   Work contracted out is usually difficult to quantify. The accounts department will probably have records of the cost of such work and this can be employed to provide an indication of the significance of the work in relation to the operations within the company.
4. *Process flow diagram*
   Description of the products
   Description of the processes

   Flow diagram showing process material flow, scrap and recycled material. Brief descriptions of the products are usually sufficient for the early stages of analysis of the company operation. The information will be amplified during the subsequent stages of the study.
5. *Layout plans for all areas of the site*
   Calculate the covered areas for all process areas. Calculate the areas taken up by services such as the boiler house, fuel-storage farm, etc. Layout plans are often available but the areas devoted to the various activities in the plant will need to be discussed with the individual area managers. The actual area in square meters or square feet will need to be estimated by scaling from the drawings and measurement on site. List the items of equipment on the site. Identify the location of the items on the layout plan. This information may exist. It will be needed for designing the layouts of the internals of the factory buildings and it is useful to initiate the quest for the information during the early stages of the project.
6. *Vehicle movements*
   The *types and numbers of vehicles* moving around the site and vehicles arriving from outside the plant must be considered. Identify interdepartmental vehicle movements. It is useful to start with the records from the gatehouse, which will indicate the number of vehicles arriving at the plant each day. This will need to be confirmed with the plant manager who will know if abnormal conditions existed during the time for which the records are valid. Aim to establish average and maximum numbers.
7. *Consumables*
   The *quantities* of main consumables used. Electricity, water and gas will be included in this item but of more importance to the plant layout are those items that occupy site space. Fuel oil is such an item, and it is

important to establish whether the existing facilities, e.g. a fuel-storage farm, are sufficient for the planned operations in the factory.

8. *Effluent*

It should be established whether the area allocated to effluent treatment is sufficient, whether the technology is up to date and determine what influence the findings will have upon the effluent treatment area and the whole plant layout.

9. *Site features*

Site features can be obtained by observation and ordnance maps may be available from the drawing office. The features that need to be recorded are listed in Section 7.7.2.

10. *Future plans*

It might be assumed that if an examination of the plant layout is being carried out, the future plans, as they affect the site layout, will be known. This is not always the case; therefore, management should be questioned at the beginning of the project. Particular attention should be paid to hazardous substances that may be used on the site, and the health and safety document[6] should be consulted.

Pareto, the Italian economist, observed that 85 per cent of wealth is owned by 15 per cent of the population, and it is sometimes stated that 15 per cent of a company's products generate 85 per cent of the turnover. Some workers have interpreted this to mean that 15 per cent of the data will suffice, but Tompkins and White[2] demonstrated that

this is not always true, so data collectors should use their judgement regarding what is sufficient.

The data needed to plan the site layout are shown in the form of a questionnaire in Figure 7.1.

After the site layout has been determined, the layout of equipment inside the buildings will need to be considered. Details of the individual items of equipment will be needed, and Figure 7.2 lists the required information in the form of a questionnaire.

The areas for administration and amenities will need to be identified and in some countries, space will be required for accommodation and a mosque.

7.4.2 Analysis

The collection and tabulation of data will provide an opportunity to become familiar with the company and will generate a good understanding of the operation of the plant. Much of the data will be needed to establish the models needed to size the various areas of the plant. For instance, hours worked should be analyzed to identify shift work, direct and indirect effort, and the normal and overtime activity. Outside contractors may be employed, and this contribution will need to be determined, sometimes by visiting the contractor's facilities. When expansions are being planned, it may be necessary to discuss the contributions that outside contractors make to the factory and assess the increase which they can accommodate.

```
Description:                          Manufacturer:

Sketches or Manufacturing literature attached Y/N: .................................
Dimensions: Equipment Size (m)
Working envelope (m) ... L x W x H     Equipment weight (kg) ........................

LOCATION:    Building No:              Zone:

ELECTRICITY:                           Flameproof (Zone 1) ..................
Single phase, isolated ...........     Flameproof (Zone 2) ..................
Three phase, isolated ............     Emergency supply .....................
Three phase & neutral, isolated .......  Controls required ..................
Direct current .....................   Emergency stop ......................
Waterproof .........................   Voltage .................   .......
Clean supply .......................   Power requirement (kW) ....  .......
Special earthing ...................   Other voltage .....................
Uninterruptible supply .............   Other frequency ...................

WATER:        FLOW: lit/sec  PRESSURE: bar  CONNECTION: Details
Cold/hot ..................................................................
Soft ......................................................................
Demineralised .(Silica free Y/N ...) .....................................
Distilled .................................................................
Drinking ..................................................................
Circulating Cooling ........ (Softened Y/N .......)
                          Flow temp (°C) .. Return temp (°C) .......
DRAINAGE:
Gravity ....................  Discharge flow rate (litres/min) ......  .....
Pumped .....................  Discharge temperature (°C) ...........  .....
Solids/Sediments ...........  Outlet connection size (mm) ..................
Oil/Grease/Wax .............  Number of outlet connections ..................
Collection Tank ...........
Hazardous (eg. radioactive/solvent/biological/chemical) give details ..........
...........................................................................
...........................................................................

STEAM:
CONDENSATE RECOVERY: Y/N   Kg/hour      psi
COMPRESSED AIR:            litres/sec   bar
NATURAL GAS:               litres/sec   bar
```

```
Model No. ......................... Equipment Identification No .......

SPECIAL GASES: ........ Y/N If YES complete separate SPECIAL GASES DATA SHEET

FUME EXTRACTION:
Located in fume cupboard .......Y/N If YES complete FUME CUPBOARD DATA SHEET
Fume hood            .......Y/N   If YES complete FUME/DUST HOOD DATA SHEET

Local extract ductwork required ............ Scrubbing required ..................
Discharge to central system ................ Filtration required.................
Discharge through wall (inbuilt fan) .  ... Extraction rate (m³/s) ...  ....
Connection of ductwork to equipt ..    ....
Connection size and type ..       ....   Type of fumes ...................

MOUNTING:
Free standing ............................ Wall mounting ......................
Static/Mobile ............................ Antivirbation mounts (type) ........
Floor fixing ............................. Concrete plinth (m) .....L x W x H...
Bench mounting ........................... Depth of embedment of bolts .........
Specify specific spacing or .............................................
  access requirements

SAFETY AND OPERATIONAL CONSIDERATIONS:
Heat (kW) .............  .................. Radiation ........................
pn Noise (decibels) ....  .................. Dust/Fumes/Vapours ................
Chemical ................................ Fire/Explosion ....................
Moving parts ............................ Biological ........................
Vibration (frequency) ................... Others (specify) ..................
         (amplitude) ................... (acceleration) ....................

ADDITIONAL INFORMATION/COMMENTS:
Frequency of use of equipment ........       ...................
Duration of usage ...................       ...............................
Please list other users of this equipment .........................................
...........................................................................
Will this equipment require:   Specialist disconnection? .......................
                               Specialist reconnection? .......................
                               Specialist commissioning? ......................
                               Specialist transport? ..
```

Figure 7.1 Equipment data sheet

```
DEPARTMENT/SECTION - NAME:  ...................................

STAFF LEVELS:  Indicate split function/responsibilities of staff.
Grade                    Number
..................       ..............
..................       ..............
FLOOR AREA:  (square metres) excluding corridors and common areas etc.
                    Current/Projected      Current/Projected
Office              ................ Workshop ..............
Machine Room        ................ Lab Wash-Up ............
Clean Room          ................ Computer Room ..........
Tank Farm           ................ Storage - General .......
 - Lockers          ................         - Disks .........
 - Fuel             ................         - Acid ..........
 - Solvent          ................         - Paint .........
 - Temp Controlled  ................
 - Process areas - List  ................ .........................
Car park            ................ ...........................
Other    -          ................ ...........................


FLOOR TYPE:  No Special Requirement ..........................
Vehicular traffic (detail heavy, long, awkward loads) ................
Fork Lift ..................... Drain ...........................
Acid Resistant .............. Solvent resistant ......................
Static Conductive ........... Other ...................................

REQUIRED LOCATION:
Ground Floor ................ First Floor .........................
Basement..................... Not Critical .......................
Proximity to other facilities ..............................
Other eg. proximity to other facilities ......................

CEILING:  No special requirement .............................
Special requirement - act as plenum ..............................
Lighting - No special requirement ..............................
Integral..................... tear drop ........................
Yellow ...................... flameproof ........................
Non-Fluorescent ............. other ............................
Lighting levels see Room Environment
```

```
INTERNAL WALLS AND WINDOWS:
 Internal Walls - Standard partition walls unless otherwise stated.
 Acoustic .................... Soundproof ...........................
 Stainless Steel .............Electro Static Discharge...............
 Other - Non Reflective? ...............................

EXTERNAL WINDOWS: No Special Requirement .........................
 Special Requirements ....................................
 Fixed ........................Opening .........................
 Sealed .......................Grilles .........................
 Blinds .......................Curtains ........................

INTERNAL DOORS: No Special Requirements ..........................
 Special Requirements .......................................

ROOM ENVIRONMENT: - Not critical ...........................
 Noise Level 40dBA ... 45dBA ... 50dBA ... 55dBA ... 60dBA ... Other ...
 Temperature - Non critical ..............................
 Special requirement ....... Temp °C .......... tolerances ..........
 Relative Humidity
 Not Critical ............. Special Requirement ..................%
 Lighting Levels (LUX) ..................................
 Cleanliness  Based on British Standard 5295 measured at working level:
 Class 1 .......... 2 .......... 3 ........... 4 ........... 5 ........
 Air Lock ......................Air shower ..........................
 Storage - Lockers .......... Overalls .............. Other .........
 Security - Secure Area ....................Contraband ...............
 No Smoking ....................Controlled ........................
 Recorded Access ........ Locks ............ Grilles ..................
 Secure Communications System ............ Other ..................

GASES:
 Oxygen ....................... Carbon Dioxide ......................
 Nitrogen - Ordinary .......... High purity ..........Ultra pure ......
 Hydrogen .................... Helium ............................
 Argon .......... Air: .......... Press .......... (Bars)............
 Other ......................................................
```

Figure 7.2 Basic user requirements schedule

7.5 Process/site layout modeling

Site and factory layout designs are needed when new installations are being planned or changes to existing facilities are investigated. New layouts are examined when increased efficiency is being sought and when plant managers are planning expansions to the existing installations. There is also a need for plant managers to anticipate the changing requirements of both site and plant over future years.

When a new installation is being planned, the site requirements will be predetermined, but when future expansions are being investigated, a model for sizing is invaluable.

7.5.1 Computer modeling

The use of computer models to assist in plant layout decisions can often be helpful, but a clear understanding of what they can and cannot do is needed. Used blindly, they can lead to solutions which are, in some sense, optimal but which have little practical merit.

In discussing the benefits and limitations of modeling we distinguish between three types:

1. Optimizing calculations for plant layout;
2. Flowsheet models for sizing the individual items of equipment;
3. Simulation models for understanding the interaction between different manufacturing units.

Optimizing plant layouts

Historically, modeling techniques have had little impact on the problem of designing plant layouts. This has been for two reasons. First, the calculations are difficult, and second, practical constraints and considerations essential to the decision are often ignored.

To understand the mathematics, consider a large empty space into which a number of production units are to be placed, and assume that the major variable to be optimized is the cost of transporting materials between them. If the manufacturing process is essentially a flow-line operation, then the order in which units should be placed is clear (from the point of view of transport costs), and the problem is simply to fit them into the space available. In a job-shop, where materials are flowing between many or all the production units, the decision is more difficult. All the potential combinations of units and locations

SERVICES SUPPORT:
Vacuum pressure ....
Water   - Town mains .... Specific pressure ....
        - Softened .... De-mineralised ....
        - Chilled (open circuit) . Distilled ....
        - Heating ....
        - Process cooling water (closed circuit) ....
Other ....
Fume Cupboards Ordinary ...... Ordinary...............................
Acid .... Solvent ....
Water Drench .... Radioactive ....
Other ....

EXTRACTS:
to atmosphere .... scrubbed...............................
filtered .... dedicated ....

CRANES: (Tonnes)
Overhead .... Gantry ....
Manual .... Electric ....

ELECTRICS:
Total power KW ....
Process/machine load KW ....   No of machines ....
Other load KW ....
Bus bar system/Trunking system...
3 Phase .... 3 phs + Neutral ....
Clean Supply .... Uninterruptable ....
Stand-by Generator .... Emergency Lighting ....
13A Sockets (do not specify number) .... with RCCB ....
Earthing requirements ....
Faraday Cage ....
Special Lightening requirements......Other ....

CONTAMINATION: Do you need to be separate from:
Noise .... Vibration ....
Electrical Interference ...... Fumes ....
Gases .... Others ....
Vibration Requirements ....
Equipment Specific .... Room in total ....
Detune Structure ....
Electro Magnetic Interference Counter-Measures ....
Electro Magnetic Interference from Satellites aircraft etc. ....
Screened enclosures .... GHz ....

WASTE AND EFFLUENT: (Large quantities - please specify)
Paper .... Cardboard ....
Glass .... Metal ....
Solvents .... Acids ....
Cutting Fluids .... Other ....
Powders .... Radioactive material ....
Special wastes ....
Do you generate contamination? Please specify levels where known:
Noise .... Vibration ....
Electrical Interference....... Fumes ....
Gases .... Others.................................

FIRE:
Local Fire regulations will apply throughout but special
requirements which need to be drawn to the Fire Advisers attention
should be stated below:
No Special Requirement ....
Special Requirement ....

SAFETY HAZARDS - not brought out in answers above: ....
....
FLEXIBILITY: How often do you anticipate changing your layout ... yrs

Figure 7.2 *continued*

could be enumerated on a computer and the resultant cost of each option obtained. The number of combinations quickly becomes large, however, and more sophisticated methods, such as branch-and-bound techniques, are required for complex problems.

Alternatively, the location of subsets of plant, within which the layout is assumed to be fixed, could be optimized.

Solutions of this nature ignore practical considerations of noise, hazardous areas, etc. unless these are specifically entered as constraints. Furthermore, if the site is geometrically complex, then additional detail will need to be included in the problem formulation.

Flowsheet models for plant sizing

The consistent sizing (i.e. balancing the capacities) of the equipment that is to make up a plant is obviously of importance in the overall design. Models used for this type of decision are usually an extension of the manual calculations that a designer would normally make.

They are based on a network of activities (or machines) with flows of materials between them. Relationships between the activities are expressed in terms of yields or unit consumption and, for a given output of finished products, the required output from each of the other activities can be calculated.

The complexity of the plant and the type and number of sensitivity tests to be carried out will determine whether a formal model or a simple calculation is needed.

Flowsheet models assume that activities operate at a constant rate. This is an important limitation, and may be an oversimplification. To overcome this, simulation techniques can be employed.

These models, however, can be useful if estimates are to be made of operating costs. By assigning fixed and variable costs to each activity, average and marginal unit costs at each stage of the process can be easily calculated, which will assist in decisions regarding pricing policies or whether to buy in components and materials or make them on site. ATPLAN[12] is an example of a network-based model of this type.

Simulation models

Simulation models aim to replicate the workings and logic of a real system by using statistical descriptions of the activities involved. For example, a line may run at an average rate of 1000 units per hour. If we assume that this is always the case, we lose the understanding of what happens when, say, there is a breakdown or a halt for routine maintenance. The effect of such a delay may be amplified (or absorbed) when we consider the effect on downstream units.

A simulation model has 'entities' (e.g. machines, materials, people, etc.) and 'activities' (e.g. processing, transporting, etc.). It also has a description of the logic governing each activity. For example, a processing activity can only start when a certain quantity of working material is available, a person to run the machine and an empty conveyor to take away the product. Once an activity has started, a time to completion is calculated, often using a sample from a statistical distribution.

The model is started and continues to run over time, obeying the logical rules that have been set up. Results are then extracted concerning throughputs, delays, etc.

It is clear that simulation models can replicate a complex production system. They can be used to indicate the level of shared resources needed by the operation (e.g. forklift trucks or operators), the speed of lines, sizes of vessels or storage tanks, etc.

A number of packages are available for quickly building simulation models. HOCUS[13] and SEEWHY[14] are two major UK systems. They allow graphical displays to be used to show, for example, the movement of staff and materials between machines. This can be useful for understanding the way in which the operation is reacting to particular adverse circumstances and can assist in designing methods of avoiding them (for example, by building in redundancy).

Simulation models can be expensive to build and the results obtained need to be analyzed with care because they are statistical in nature. For example, two runs of the model may give different results – just as the performance on two real days in a factory can vary. Sufficiently large samples need to be taken therefore for a proper understanding of the performance of the plant.

7.5.2 Model construction

A simple model can be produced using the areas required for the various site activities for different scenarios. These might include redistribution of a company's manufacturing facilities, changes in the market demand, or simply increased factory output.

Constructing the model is an area where innovative effort is needed to maximize the validity of the model. The model will vary from industry to industry, but a simple one can start with a relationship between the number of employee hours and the output produced in each discrete activity area on the site, e.g.

Activity area	Output	Manning hours
Raw material store	Tonnage/volume handled	No. and shift no.
Process areas	Output tonnage or volume	No. and shift no.
Finished goods store	Tonnage/volume handled	No. and shift no.
Maintenance	Specific area output	No. and shift no.

Service facilities, electrical sub-station, water treatment, fuel storage, etc. will depend upon process parameters, mainly tonnage or volume handled, and may need to be assessed from the summation of the individual activity areas, but it will be affected by technology when improved techniques are being introduced.

Office accommodation will depend upon the factory manning and the car parking facilities upon the total manning during the day shift.

7.5.3 Model for change

The recent introduction of inexpensive desktop computers has allowed their extensive use throughout many companies. The standard spreadsheet packages which accompany these machines enables the above data to be laid out in an interactive way, so that 'what if' situations can be explored at the planning stage and the implications of, for example, market trends in the food industry, to be examined over the long term for its effect on the plant layout. The model may include a factor to take into account improvements in technology and working practices in both the office and factory.

In determining the area required for increased production the relationship between the output and the area may not be linear, and needs to be examined in some detail. When the output required exceeds the maximum output of a process line then increased shift working may be considered. Alternatively, a second process line may be needed.

The model for the maintenance requirements of a large metallurgical plant was recently constructed by relating the man-hours to the production output, but the relationship was not linear, and a more sophisticated model was needed to examine the site requirements for different output tonnage.

Mechanical wear is a feature which is significant among the causes of machinery failure and the James Clayton Lecture delivered to the Institution of Mechanical Engineers in 1981 discussed friction and wear and identified tribological losses as 45–50 per cent of all maintenance costs in metallurgical plants. This suggested that mechanical transmission of energy could be investigated to obtain the relationship between production output and the maintenance requirements.

The model equation produced was:

$$H_2 H_1 \times K \times \sqrt{\frac{P_2}{P_1}}$$

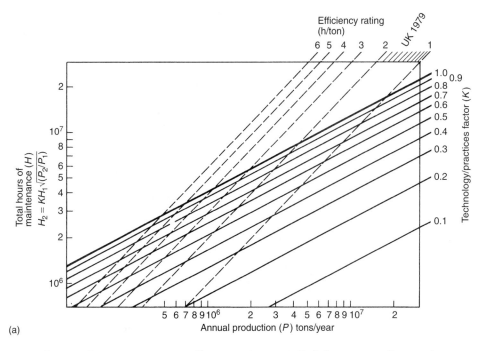

(a)

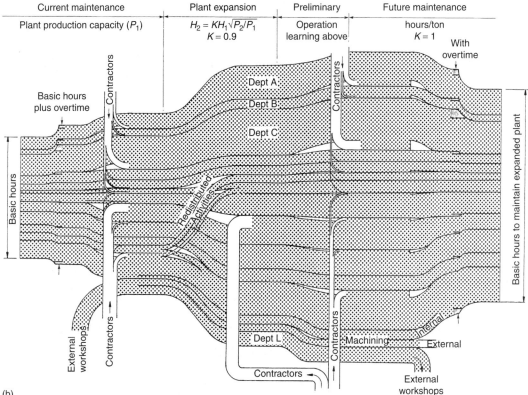

(b)

Figure 7.3 (a) Model of manning requirements related to plant output; (b) Sankey diagram of the changing manning requirement in an expanding plant

where

H_2 = Future man hours,

H_1 = Current man hours,

K = Technology or practices factor,

P_2 = Future production output,

P_1 = Current production output.

This produces a global requirement chart and a chart showing the discrete, departmental requirements in Figure 7.3.

New plants and expanding plants rarely produce maximum output on start-up. In large capital plants, it may need many months for staff training and equipment commissioning before output is sufficient to need maximum manning. The technology factor K in the above equation can be used to reflect this. In Figure 7.3(a) a technology factor of 0.9 has been used.

It is important to test models, and in the above case, figures were obtained for other plants and the validity of the model confirmed.

7.5.4 Determination of factory areas

When existing factories are being examined, the floor areas occupied by the various items of equipment can be measured on site during the data-collection period. It is important to consider and include the equipment outline (maximum travels), space for maintenance activity, operator movement and the handling of material and workpieces. In many cases, the space needed for handling exceeds that required accommodating the operative part of the equipment. Aisle space is also needed for through-shop movement.

When planning new installations, manufacturers' catalogues may be the only information source of equipment data, but they rarely indicate the space needed outside the equipment outline. This requires careful consideration from an experienced facility planner to establish realistic space requirements.

In the absence of reliable information and during the preliminary planning stage (feasibility studies, etc.) the following occupancy values can be employed:

Activity	Occupancy (m^2/person)
Service industries	15–30
Manufacturing	20–35
Machine tool shop	20
Electrical/plastics	25
Mechanical fitting	27
Pottery/glass	35
Transport	30–65
Cars/mobiles	33
Locos	37
Wagon shop	55
Lorries	65
Distributive trades (average)	−80

These values will change with the future move towards unmanned factories.

Dividing the areas established in Section 7.5.3 by these occupancy values will give building areas. Bay widths can be established for each type of building and dividing the areas by this width will establish a building length that can be arranged into a convenient building block shape.

7.6 Design synthesis

7.6.1 Plant activities and intercommunications

In 1973, Richard Muther[1] published a method of analyzing the interrelationships of activities within industrial plants, and the method allows a high degree of detail to be examined. The method proposed here is similar in that a relationship grid is constructed, but this technique employs the power of the modern desktop computer to rapidly examine alternative layouts to obtain best solutions.

The design synthesis starts by listing the activities or areas in the plant and indicating the access, which each activity needs to the other activities. The access may be required for internal process, traffic, and people movement or plant services. Rather than go into the detail of examining all interrelationships for all of the above requirements, the experienced layout planner can decide which requirement predominates and employ it to produce quick results. The method allows inexperienced planners to iteratively examine the implications of changing requirements as the project proceeds and plant knowledge increases.

Table 7.1 shows the construction of a typical activity chart using one of many computer spreadsheet packages. The cross (+) identifies an activity in the plant and stars (*) in the vertical lines denote where access to other activities is needed or an interrelationship exists. A wide scattering of the stars away from the diagonal line of crosses indicates large distances between the activities and thus large communicating distances.

A feature of PC spreadsheets is that the position of rows or columns can be changed at the touch of a key, so the sequence of the rows of activities can be rapidly altered, and the position of the columns varied to preserve the diagonal pattern of crosses. This exercise can be executed a number of times to reduce the scatter of the stars away from the diagonal line and reduce the communicating distances between the activities. This has been carried out, and the result is shown in Table 7.2, where it can be seen that the stars are clustered more closely around the diagonal line. The list now shows the activities arranged in a preferred order of sequence.

7.6.2 Location criteria and boundary groups

There may be reasons why individual activities cannot be located near to those which are adjacent in the determined sequential list, so design criteria are needed to determine which activities are compatible and can be near, and which are incompatible and cannot.

Table 7.1 Plant activity intercommunications: first design criteria

In the matrix below, the columns correspond (left-to-right) to the same list of activities as the rows (top-to-bottom). `+` marks the diagonal (self); `*` marks required access between two activities.

Activity	1	2	3	4	5	6	7	8	9	10	11	12	13	14	15	16	17	18	19	20	21	22	23	24	25	26
Heavy plant repairs	+	*	*		*		*				*	*		*		*							*	*		
Vehicle hot washing	*	+		*		*										*								*		
Vehicle cold washing	*		+	*		*																		*		
Public health repairs		*	*	+	*					*	*	*				*							*	*		
Store for large tires	*			*	+		*													*			*	*		
Service station		*	*			+			*							*							*	*		
Electrical repairs	*						+			*		*				*							*	*		
Bus washer								+			*					*										*
Vehicle test area		*	*						+			*											*	*		
Mechanical repairs							*			+	*		*			*							*	*		
Tire shop	*				*						+					*				*			*	*		
Welding shop	*			*					*			+	*	*		*							*	*		
Panel beater				*								*	+	*		*							*	*		
Radiator shop	*			*								*	*	+		*							*	*		
Paint area												*	*		+	*						*	*	*		
Compressor house	*	*		*		*	*		*	*	*	*	*	*		+							*	*		
Tailor																	+		*			*	*	*		
Carpenter																		+	*				*	*		
Store–carpenter																	*	*	+				*	*		
Store–tires					*						*									+			*	*		
Store–tailor																	*				+		*	*		
Store–paint																						+	*	*		
Store–general	*			*	*	*	*		*	*	*	*	*	*	*		*	*	*	*	*	*	+	*		
Administration block	*	*	*	*	*	*	*		*	*	*	*	*	*	*		*	*	*	*	*	*	*	+		
Filling station																									+	*
Vehicle park									*																*	+

Table 7.2 Rearranged activities to improve plant operation

Columns correspond (left-to-right) to the same list of activities as the rows.

Activity	1	2	3	4	5	6	7	8	9	10	11	12	13	14	15	16	17	18	19	20	21	22	23	24	25	26
Filling station	+	*																								
Vehicle park	*	+	*																							
Bus washer		*	+			*																				
Vehicle hot washing				+		*			*	*	*		*	*												
Vehicle cold washing					+	*			*	*	*		*	*												
Service station				*	*	+			*		*	*														
Tire shop							+	*	*		*	*	*													
Store–tires							*	+			*	*			*											
Vehicle test area				*	*				+	*	*	*														
Compressor house			*	*		*			*	+	*		*		*	*	*	*	*	*	*					
Administration block				*	*	*	*	*		*	+	*	*	*	*	*	*	*	*	*	*	*	*	*	*	*
Store–general						*	*	*		*	*	+	*	*	*	*	*	*	*	*	*	*	*	*	*	*
Heavy plant repairs				*	*		*			*	*	*	+	*	*		*	*								
Public health repairs				*	*					*	*	*		+	*	*		*	*	*						
Store for large tires								*		*	*	*	*	*	+											
Electrical repairs										*	*	*		*		+	*									
Mechanical repairs										*	*	*		*		*	+	*	*							
Welding shop										*	*	*	*	*		*	*	+	*	*						
Radiator shop										*	*	*	*	*			*	*	+							
Panel beater										*	*	*		*				*		+	*					
Paint area										*	*	*						*		*	+	*				
Store–paint											*	*									*	+				
Carpenter											*	*											+	*		
Store–carpenter											*	*											*	+		
Tailor											*	*													+	*
Store–tailor											*	*													*	+

Particular plants will have characteristics which will determine the importance of particular criteria, but typical criteria will include the following:

1. The hazardous nature of the activity;
2. The ease of access to the entry gate which the activity needs;
3. The amount of dirt and potential contamination that the activity would generate;
4. The amount of noise and potential disturbance that the activity will normally generate.

Each of the listed plant activities can be assessed for each of the chosen criteria and a star rating determined for each assessment. An experienced engineer will find that a qualitative assessment is sufficient, but quantitative measurements can be used if necessary. The results of this exercise are shown in Figure 7.4.

For each of the criteria a two-star rating difference between an activity and an adjacent activity is taken to indicate incompatibility. Activities that are hazardous, dirty, noisy or needing good access to the entry/exit gate are consequently rated with many stars. In the site layout they should not be located near other activities with fewer stars because they could adversely affect that activity. However, an activity that is not critical and has a low star rating may be located next to an activity with many stars.

Boundary maps can be drawn around groups of stars which are compatible and horizontal lines drawn through the incompatibility points to show groups of compatible activities on the right-hand side of the chart Figure 7.4.

Groups of compatible activities together with their major characteristic can be listed, e.g. for a transportation maintenance facility, as follows:

Boundary group	Characteristic
Filling station	Hazardous
Vehicle park	High traffic density
Servicing facilities	Dirty
Administration and stores	Quiet and clean
Repair facilities	Medium traffic, dirt and noise
Body shop	Noisy
Accommodation	Quiet

The synthesized layout can be described in terms of the characteristics of the various zones within a hypothetical site boundary as shown in Figures 7.5 and 7.6.

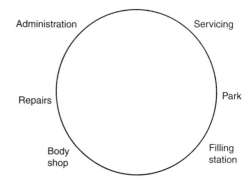

Figure 7.5 Synthesized location of plant activities

Activity	Hazardous	Gate Access	Dirty	Noisy	Boundary groups
Filling station	*****	*****	*	**	Filling station
Vehicle park	*	*****	*	**	Vehicle park
Bus washer	**	*****	*****	***	
Vehicle hot washing	**	****	*****	***	Wash
Vehicle cold washing	**	****	*****	***	
Service station	**	****	*****	****	
Tire shop	**	***	**	**	Tyres
Store – tires	*	***	*	*	Store
Vehicle test area	*	****	**	***	Test
Compressor house	*	**	*	****	Compressor
Administration block	*	*****	*	*	
Store – general	*	*****	*	*	Admin & Store
Heavy plant repairs	***	***	*****	****	
Public health repairs	**	***	*****	****	
Store for large tires	*	***	*	*	Store
Electrical repairs	*	***	*	*	Electric
Mechanical repairs	**	***	****	***	
Welding shop	**	***	****	***	
Radiator shop	*	**	****	***	
Panel beater	*	**	****	*****	Denter
Paint area	****	**	*****	**	Painter
Store – paint	****	****	*	*	Store
Carpenter	****	**	**	*	
Store – carpenter	***	****	**	*	Store
Tailor	*	**	**	*	
Store – tailor	*	****	*	*	Store

Figure 7.4 Compatible activities and boundary groups: secondary design criteria

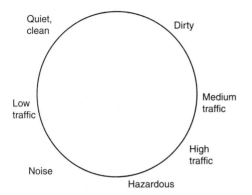

Figure 7.6 Characteristics of the synthesized layout

7.7 Site layout realization

7.7.1 Process and traffic flows

Traffic and material movement is a major consideration, and should be arranged so that cross flows are minimized and the potential for congestion and accidents reduced. Much useful information on both the analysis and practice of movements is contained in the References.

Using the example of a transportation maintenance facility, Figure 7.7 shows that a counter-clockwise rotation of the traffic flow could achieve minimum crossing and give a graduation of decreasing traffic density in the counter-clockwise direction as vehicles are diverted off the main stream (Figure 7.8).

The circulation diagram indicates that, in a left-hand drive environment, activities with a high traffic density

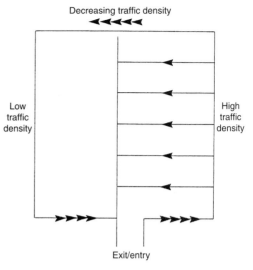

Figure 7.8 Traffic density zones

should be located on the right-hand side of the traffic circulation diagram and the activities with low traffic on the left-hand side.

With this arrangement, the vehicle needing no servicing or maintenance could be parked on the right-hand side of the site in a park designated 'overnight park', and the filling station which handles all the vehicles could be located towards the center of the site in the flow line of the vehicles leaving the area.

The servicing area could be located on the right-hand side of the site but remote from the gate, and part of the park near to the servicing area could be allocated to vehicles that arrive for their scheduled routine service.

The left-hand side of the site has a reducing traffic density and could thus accommodate the vehicle repair shops. The body shop is a repair function, and thus would be located on the left-hand side but remote from the existing administration office.

The technical administration office needs to be quiet, clean, and central to the facilities that it administers. Hence, it should be located in the central area between the service and repair shops. This is remote from the dirty washing facility and the noisy body shop while providing good access to the gate.

The remaining corner on the left-hand side of the site can be arranged to be sufficient distance from the body shop so that it can be quiet and thus suitable for staff accommodation and the plant administration office. Thus, the synthesized layout becomes as shown in Figure 7.9.

7.7.2 Site constraints

The real design must consider design criteria derived from the physical constraints of the chosen site and the current operating procedures. In this exercise, the site must be examined to list the features of the site, which will affect the plant layout. These will include the contour of the site and the location of the following:

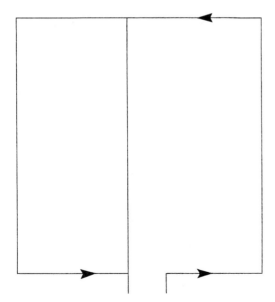

Figure 7.7 Ideal traffic or material flow

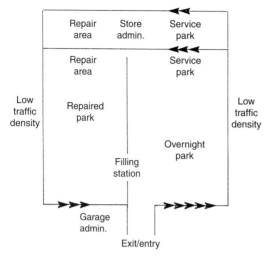

Figure 7.9 Combination of synthesized layout and ideal flows – garage layout

1. Main highways and access to the site area
2. The main and other gates to the plant
3. Factories near the boundary of the plant
4. Domestic dwellings near the boundary
5. Natural features such as streams
6. Effluent routes
7. Prevailing wind directions.

These can be drawn onto a plan of the site area, which does not show the existing or any proposed arrangement of the plant layout (Figure 7.10).

The site constraints must be considered for their effect upon the design and the practicality of implementing the design. When the plans are for an existing plant, the implementation must consider the minimum disturbance

to the everyday functioning of the existing plant. These limitations can be listed as follows:

1. The location of the gate;
2. The existing buildings that have been found, during the structural survey, to be structurally sound and which will be retained;
3. The logistics of erecting new buildings and rearranging the plant with minimum disturbance to the plant functions;
4. The requirement to allow for future expansion of the production and subsequent activities in the plant.

As an example, the synthesized design considers all these limitations in the following ways:

1. The gate is taken as a starting point for realization of the layout and high traffic densities are placed near to it;
2. The structural survey dictates which of the existing buildings will be replaced;
3. The final report will need to include a detailed description of the implementation procedure to demonstrate that the synthesized layout is feasible.

The final layout characteristics are shown in Figure 7.11.

7.7.3 Ease of expansion

Layout planners normally have to consider that activities within the site may change in the future. The changes may be the result of increased product demand or the introduction of new product lines. They may be predicted from the model designed in Section 7.5.2 or allowance for unpredictable changes may be needed.

The dilemma facing the planner is between designing the layout with minimum distance between buildings and allocating free space between buildings to accommodate future process units. Often the problem becomes one of providing a single building for economy or multiple separate buildings for ease of expansion.

The solution may be a policy decision, but this decision will be assisted by the results of the internal layout configurations, which may indicate how simply or otherwise expansion can be accommodated.

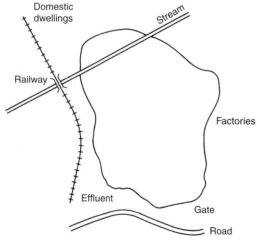

Figure 7.10 Site constraints

Figure 7.11 Conceptual layout design on the real site

7.7.4 Options

Management usually wishes to see alternatives to a particular design in order to say that the alternatives have been considered. Unique design solutions are rare; consequently, the layout planner will normally be able to interpret the results of the finding of the study to show alternative possibilities. Each alternative option should be accompanied by a critique listing the advantages and disadvantages.

7.8 Internal layouts of buildings

Material, traffic and people are involved, and the constraints on internal layouts will be different to those for site layouts but will not affect the design process. The constraints will include organization of the factory work tasks, which may involve trade practices and the type of organization adopted for the control of the process equipment. This is discussed in Section 7.2.

7.8.1 Process flows and performance indicators

The same method as described above can be employed to plan the layout of factory building interiors. Establishing the interior activities and characteristics of the activities must, however, derive from a thorough understanding of the products and processes.

The management of production is well described by Hill,[7] but the prime consideration is product flow, and all features of the layout must assist flow. There are some indicators which can be used to measure the quality of the production facilities, and these can be employed to demonstrate the viability of the proposed new layout.

Performance indicators

The relative importance of each indicator will vary according to the type, quantity, end quality, variety and value of the product and the capital cost, flexibility and required utilization level of the plant. Three variables will serve most industrial processes:

1. *The stock value.* The value of the stores stock represents a significant investment for many manufacturers and financing the work in progress can represent a significant cost. It is a useful indicator of the financial health of a company and is a prime target for cuts in the drive for increased economic efficiency. Section 7.2 highlighted this.
2. *The cycle time.* This represents the maximum time interval between the start of a single operation on a product on a particular flow line and the start of the same operation on the next product on the same flow line. The cycle time is calculated by dividing the target production volume (annual or monthly) by the working time (hours, minutes or seconds) and then adjusting for the reject rate and the per centage of lost time (downtime). If every operation is conducted within the cycle time then the steady-state functioning of the flow line will produce the required production rate. The number

of sequential process operations multiplied by the cycle time is the actual process time.
3. *The door-to-door time.* This represents the total elapsed time from the delivery of the raw materials or component parts to the dispatch of the finished product. This figure can be adjusted to control the value of the stock of materials, parts, work-in-progress (WIP), and finished goods in the process. Dividing the door-to-door time by the actual process time and multiplying the result by the stock value can find the gross excess stock carried. The difference between the gross and the net excess stock values will depend on individual circumstances. Reducing these figures to a minimum requires strategies for purchasing, stock-level monitoring and process control.

7.8.2 Process equipment

When the idealized flow routing has been determined, the equipment must be located and the practicalities of the interrelationships between, and integration of, equipment, services and people must be considered.

Equipment information would have been collected during the data-gathering exercise described in Section 7.4, but a thorough understanding of the process is vital during this phase of the work. Adequate space for the equipment must include:

1. The machine itself with the maximum movement of all machine elements;
2. Space for maintenance activities, withdrawing shafts, etc.;
3. Process material movement;
4. Operator mobility.

7.8.3 Material handling

Knowledge of material-handling techniques is vital to the layout planner, and detailed consideration will need to be given to the various techniques and equipment, which are available. References 2, 8 and 9 contain valuable information and trade journals report the current state of the market.

In the absence of reliable information and during the preliminary planning stage (feasibility studies, etc.) the following space requirements for aisles can be employed:

Traffic type	Aisle width (m)
Tow tractor	3.7
3-ton forklift truck	3.4
2-ton forklift truck	3.0
1-ton forklift truck	2.75
Narrow-aisle truck	1.85
Manual truck	1.6
Personnel	1.0
Access to equipment	0.85
Allow extra for door openings	

7.8.4 Storage

Storage in plants range from simple facilities to fully automated flow-through warehouses. Simple facilities are

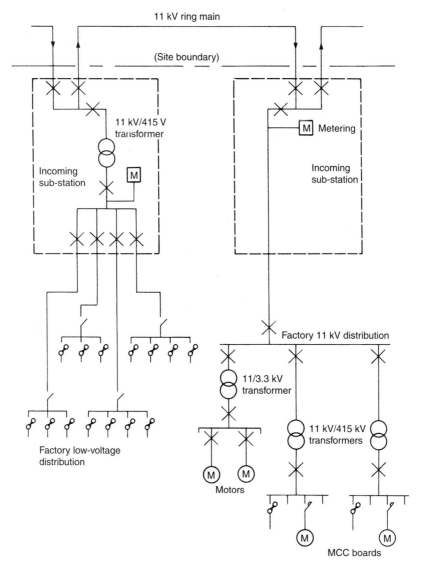

Figure 7.12 Typical power distribution system

installed to contain the day-to-day requirements of, for instance, maintenance workers, and the just-in-time supply of components on the production floor. Fully automated flow-through warehouses are employed in the food industry where the shelf life of many products is short and a high throughput is involved. Much has been written about storage and References 8 and 9 are excellent.

7.8.5 Electrical distribution

The plant layout design and whether the processes involve dry, wet or contaminated environments with possibly elevated temperatures will affect the principles adopted

for the support and containment of electric cables and bus bars. Cables from the incoming supply sub-station to load centers would normally be run in the ground, either directly buried with suitable route marking and protection, or in ducts or concrete trenches. Depending on the number of cables, the latter may incorporate support trays or cleats fixed to the trench walls.

Inside the factory buildings, below-floor trenches are frequently the preferred method for main routes, but where there is danger of flooding, due to storms or process fluids, aboveground routing must be considered. Single or few cable runs could be clipped to walls and building structures but multiple cable runs need tray work, involving space equivalent to major pipes and air ducting, and routing must be coordinated with other services.

Busbar trunking systems are frequently employed in machine shops to facilitate relocation of the equipment. The busbars are located overhead, suspended from the roof structure, and arranged, in conjunction with the lighting system, in a suitable grid pattern. Fused plugs and sockets provide outlets to individual items of equipment.

In some installations (e.g. clean rooms), the electrical distribution system can be located in the sealed ceilings, allowing maintenance access without affecting the clean facilities (Figure 7.12).

7.8.6 Building utilities

These include the canteen, toilets, workers' rest rooms, access routes and the car parks. Hygiene, traffic density, traffic flow and sizing need to be addressed and have been discussed in the section on the sizing model.

7.9 Selling the concept

The site layout needs to be presented to management, union, staff and possibly others to obtain approval, and the design procedure described here, together with visualization, will assist the presentation.

Photographs of the proposed site are useful and a site layout drawing is needed, but a perspective artistic impression showing buildings with architectural facades, vehicles and other site activities improves the presentation. Employing three-dimensional models[10,11] discussed in Section 7.3 helps communication and allows layout options to be easily demonstrated and discussed.

7.10 Implementation

One of the principal tasks facing the manager responsible for implementing a new plant layout is the planning and scheduling of the many activities involved in the project. Modern computerized systems have evolved into easy-to-use tools that avoid the need for a detailed technical knowledge of planning techniques.

Project plans for the installation and commissioning of new plant should be prepared in order to:

1. Provide a readily understood and concise description of the scope of work for all involved in the project in order to facilitate easy communication between all parties;
2. Provide an informative management tool for the monitoring and control of progress, including the effect of changing circumstances that so often occur in the form of extended delivery time scales, revised requirements and new external constraints on the project completion date.

The project plan should be started as soon as possible so that all aspects of the early layout planning can be incorporated into the plan.

Most detailed programs for plant-related projects are prepared using simple critical path analysis (CPA) techniques. CPA is a well-known and familiar technique to most engineers and is incorporated into most modern PC-based planning and scheduling systems. Such a system typically has the following features that a plant project manager will find useful:

1. Easy screen-based data entry of related task details;
2. Ability to allocate material and labor resources to each activity to allow a picture to be built of the total resource requirements over time;
3. Simple production of bar charts or Gantt charts detailing the task time scales;
4. The ability to schedule the tasks logically within the constraints of available resources.

The definition of what constitutes a resource from a scheduling point of view may vary, but would normally include materials, labor, special tools, temporary services and also access to space on the plant floor if the layout is particularly tight. This last item is of great importance, as it is easy to forget that a given amount of space will limit the number of people working in it, and the scheduling system can take it into account when re-scheduling start dates.

For the more sophisticated requirements of complex one-off projects, techniques such as PERT, resource, leveling and precedence diagrams should be investigated.[15]

The most important aspect of planning the work is that the plan should be constantly monitored for progress and the individuals responsible for particular areas of work should be kept well informed regarding delays or advances in the program.

Frequent updates and reports will ensure that surprises are kept to a minimum and allow smooth progress of the project towards completion. Should problems occur in the project at any time, the modern planning systems allow rapid changes to the logic or task details and the consequent update to the project program.

7.11 Consultants

7.11.1 Justification for consultants

Planning the layout of a manufacturing facility should be an infrequent activity for an operational manufacturing company. The staff of such a company will normally be fully committed to the day-to-day activity of assisting factory output, and they will rarely have the time to consider the design problems associated with plant layout. They do have intimate knowledge of the products and processes but not the practice and techniques of the design of plant layout. In these circumstances, it is normal to engage the services of a firm of consultants.

The consultants will need to conduct in-depth discussions with company personnel to acquire the data described above, but will bring to the task the skills of years of practice, often employing modern computing techniques, knowledge of industrial trends and undivided attention to the task.

The best way to select a consultant is to seek recommendations from other companies who are known to have carried out a similar exercise. Failing this, a useful guide is the *Association of Consulting Engineers Yearbook*, which lists the consultants classified according to the type of engineering work they engage in and their specialist discipline. An alternative approach is to contact associations set up to serve particular industries (e.g. the Production Engineering Research Association (PERA) in Melton Mowbray for engineering production and the Rubber and Plastics Research Association (RAPRA)).

The consultant will discuss and determine the objectives of the activity, but it is useful if a specification is prepared in advance of the consultant being engaged.

7.11.2 Specification

After a short introduction to the company and its products, the specification should describe in some detail the reasons why a new layout is needed. This will vary from the introduction of, or change to, new products, the establishment of a new factory or the need to explore the changing future demands on an existing site because of changing technology or market forces and opportunities. (Contracts are described in detail in Chapter 8.)

References

Literature

1 Muther, R., *Systematic Layout Planning*, Van Nostrand Reinhold, New York.

2 Tompkins, J. A. and White, J. A., *Facilities Planning*, John Wiley, New York; see also Neufert, E., Architects' Data, Halsted Press, New York.
3 Schonberger, R. J., *World Class Manufacturing*, The Free Press, New York.
4 Goldratt, M. and Cox, J., *The Goal*, Scheduling Technology Group Ltd, Middlesex.
5 Francis, R. L. and White, J. A., *Facility Layout and Location*, Prentice-Hall, Hemel Hempstead.
6 Health and Safety; The Control of Substances Hazardous to Health Regulations 1988, HMSO, No. 0110876571.
7 Hill, T. J., *Production/Operations Management*, Prentice-Hall, Hemel Hempstead.
8 Drury, J., *Factories Planning, Design and Modernization*, The Architectural Press, London.
9 *Material Handling*, Reed Business Publishing, Surrey.

Scale models (proprietary systems)

10 Modulux Systems Ltd, Northampton; Lego symbolic constructions on a flat baseboard.
11 Visual Planning Systems Ltd; Cast aluminum scale models of actual equipment.

Software programs

12 ATPLAN; W. S. Atkins Consultants Ltd, Woodcote Grove, Ashley Road, Epsom, Surrey KT18 5BW.
13 HOCUS; PE Consultants, Egham, Surrey.
14 SEE-WHY; ISTEL, Redditch, Warwickshire.
15 PERTMASTER; Pertmaster International Ltd, Bradford.

Contracts and Specifications

Ken Taylor
Taylor Associates Ltd

Contents

8.1 Contracts

In the course of his career, the plant engineer will be a party to a contract. His employment is a contract between himself and his employer unless he is self-employed. What is a contract? A contract is an agreement between at least two parties to a matter. The important aspect is that the agreement is to the same thing. This is known as *consensus ad idem*, and follows an offer, which is accepted. An invitation to bid is not an offer, the offer being made by the bidder in submitting his bid. English law says that certain essential elements need to be met before a contract can be enforceable, and these are:

1. Agreement – as mentioned above;
2. Intention – The parties must intend to create legal relations;
3. Consideration – This is essential in simple contracts, which most are. An example is that *A* makes a contract to purchase a car from *B* at a cost of $9000. In paying the $9000 *A* is fulfilling the essentials of consideration. A contract would not exist if *B* gave his car to *A*, because there would be no consideration.

The law of contract is exceedingly complicated, and the pitfalls experienced by engineers keep a substantial number of solicitors, barristers and judges in the style to which they have become accustomed over the past few hundred years or so.

Contract law is a combination of common law and statute law. Common law is judges' law, which is continually being revised as important cases come up in the courts. State or federal governments make statute law.

This chapter is concerned with contracts and specifications in engineering and the way in which the plant engineer should approach them.

8.2 Approaching the contract

A contract is made when the essential elements have been fulfilled and the offer and acceptance communicated, and, apart from contracts involving land, can be verbal. A common misconception is that a party signs a contract. In reality, a party signs a document, which contains the terms of the contract being entered into. Almost all contract documents omit many matters, which are important, and the reason for this is that the law says that certain matters are implied to be contained in a contract. An example is that if *x* buys an item of plant from *y* it is implied that *x* will receive it within a reasonable time, and that it will be of merchantable quality when he receives it.

While express terms in a contract are advisable, some are of no value at all, and in fact can work against the author. An example is an unfair contract term that may be included, but if it is not sufficiently brought to the attention of the other party before the contract is made, it is not enforceable.

The object of contractual documents is to have a clear and unambiguous agreement to supply and install plant, which will be in accordance with the specification and drawings, and installed within the time agreed, or within a reasonable period if none is stated.

What documents will be drawn together to form a contract? These will include:

1. Drawings of the installation sufficient for the bidder to be able to provide a bid;
2. A detailed technical specification of the works, which will contain other matters such a defining the parties, standards of work, program and basic contractual positions;
3. Bid forms, which will include provision for the bidders to submit a lump sum bid, or a rate for labor and materials, or both?
4. Letter of invitation to bid.

8.3 Types and forms of contract

In English law, there are two main kinds of contract – Simple and under seal. Simple contracts require consideration, whereas contracts under seal do not.

There are many standard forms of contract which are used in the engineering industry, most of which are published by professional bodies or groups of professional bodies. It is usual for large companies and government organizations to produce their own, to suit their needs and protect their interests. In the case of air conditioning, because the works are likely to be carried out alongside other building works, it is likely that either the JCT Domestic Sub-Contract or Model Form A will be used.

A contract document cannot go against a common law rule, by the introduction of an express term. There are several ways to employ the contractor, and some of these are outlined below:

1. Lump sum fixed price. This is the most commonly used method, and ensures that the client will know exactly how much he will be paying;
2. Lump sum fixed price with a schedule of rates to cater for variations to the works;
3. Measured bill. A bill of quantities based on a measurement of the amount of work, with re-measuring at the end of the job;
4. Daywork. Rates are based on the agreed cost of labor per hour and on cost of plant and materials. This is sometimes the only fair way, because the amount of work is uncertain, or cannot be sufficiently estimated by the bidder.

8.4 The specification and drawings

The smooth running of the contract and the satisfactory operation of the system will depend to a large extent on the way the plant engineer designs the plant, and expresses it in his specification and drawings. There is a tendency for designers to adopt standard specification formats, and while this is to be applauded, a 50-page specification for a small job can mean that only a small percentage is applicable to that particular project. It should be remembered that the foreman on the job would have the task of interpreting what the designer had in mind.

The General Specification must coincide with the drawings to be credible. It is all very well inserting a clause to the effect that 'The Engineer' shall determine the interpretation of any clause which implies that he has the power of an arbitrator, but in reality, if he has negligently specified anything by being ambiguous, then his client or the contractor may have a claim against him if either of them suffers loss as a result.

The General Specification itself can be divided into the following headings:

Conditions of Contract
Technical Clauses
Specification of the Works.

The following is a checklist and brief description of the various clauses:

(a) Conditions of Contract

1. *General* – This is a description of the type of contract and the relationship with other parties, i.e. whether the contract is to be nominated, main or domestic fixed price or fluctuating.
2. *Interpretation of Terms* – Name of all known parties and their positions. Some terms may be outlined, i.e. *month* means calendar month and *shall* means action by the contractor.
3. *Form of Bid* – A more detailed description is given regarding the form of bid and of prime cost and provisional sums, handling and profit on cost.
4. *Bills of Quantities* – The client can require a successful bidder to supply a priced-up bill of all items bided for within a certain time after being requested to do so. This is useful in a lump sum contract where variations to the contract occur. The bidder in this case provides the bill.
5. *Bid make-up* – Invitation to the bidders to clarify points prior to bid. Alternative makes of equipment may be offered instead of the ones specified, but the bidder must bid for what was specified, and include the alternatives as a separate list.
6. *Attendance* – In the case of a sub-contract it should be made clear to the bidder any attendance which will be provided to him by the main contractor, and any conditions attached.
7. *Program* – The program of works should be outlined and any special provision, such as working in phases to suit either the client or the main contractor.
8. *Site Meetings* – The frequency of site meetings should be outlined and a commitment to attend any *ad hoc* meetings, which may be called by the client or main contractor.
9. *Related Documents* – A list of all the documents, which make up the contract.
   Interpretation of terms.
10. *Payments* – Details of when valuations and payments are to be made.
11. *Completion* – Practical completion and defects liability period details.
12. *Arbitration Clause* – This is likely to be mentioned in most standard forms of contract, and if repeated here, should be verbatim.

(b) Technical Clauses

1. *Scope* – general description of the works.
2. *Regulations* – a general list of regulations, which apply to the particular type of work to be performed.
3. *Drawings* – Details of drawings and manuals, which the contractor shall supply to other parties (e.g. building work, drawings and working drawings, and a description of details to be supplied).
4. *Site visits* – Invitation to contractor to visit site to see for himself any possible site problems. A disclaimer against extras in the event that the contractor does not appreciate obvious site problems.
5. *Tests and commissioning* – Details of any specific tests required, and commissioning format, and who is to witness and reports to be submitted.

8.5 Estimates and bills of quantities and estimates

On some contracts, mainly connected with general building projects, rather than specifically plant and machinery, a quantity surveyor is appointed to draw up a schedule of items which require pricing, and which the total is the bid figure.

A qualified quantity surveyor has a first-class knowledge of building construction, is an expert in drawing up a bill of quantities, and very often has a good knowledge of contract law, which makes him invaluable where there are complicated issues. While the plant engineer may not often be presented with bills of quantities, it may be useful to have a grasp of the general approach. There are three stages to a bill: taking off, abstracting and billing. Taking off is a detailed measurement of drawings or on site of the subject matter. This is carried out in conjunction with a manual used by all quantity surveyors called the *Standard Method of Measurement*. The items contained are complete in that they include for all labor and materials. The abstract is a sheet where all similar items are grouped together, and the bill is where the items are listed for the bidder to add his prices. The bidder is also given the opportunity in the bill to add his profit and on-cost for other contractors under his control.

8.6 Specific sums stated in bid documents

In most engineering contracts for works to be carried out on site, there are unknown costs at the time of bid and costs which are known, where the work is to be carried out by others, but to be under the control of the successful bidder. These sums are shown below:

8.6.1 Prime cost (PC) sum

This is a sum which has been determined prior to bidder being obtained for the main works, and which can be expressed accurately as a sum to be included in the bid build-up. The bidder has the opportunity to add a sum within his bid for profit or on-cost in administering the works.

An example would be where the consultant had already invited competitive quotations for supplying air-handling units so that he could design a particular make into the system and ensure that his client received the best value for money. The bidder would include in his bid, either separately shown or in the body of his figures, a sum to cover profit and on-cost in handling the equipment. The contractor appointed would be responsible for organizing delivery and would give a warranty on the equipment, in accordance with the contract. This aspect is amplified in Section 8.20.

8.6.2 Provisional sum

This is a specific sum included in the bid documents. It indicates to the bidder that it is contemplated that the particular work referred to will be carried out, but that, at present, no firm plans have been made as to cost, and usually the work has not been detailed either. The contractor is not entitled to profit or on-cost, unless the works go ahead. An example of this would be a general intention by the client to provide special ventilation to a laboratory, which is part of a larger scheme, but that no firm plans as to the full requirements have been drawn up at the time of inviting bidder.

Let us assume that $100 000 has been included as a provisional sum. Later a scheme is produced and priced up by the successful bidder, and his price is $95 000, including profit and on-cost. Then at the end of the contract, when the final account is submitted, the contractor will show that of the $100 000 provisional sum, $95 000 has been spent, and the client will be $5000 better off than expected.

8.6.3 Contingency sum

In most contracts of supply and install, a sum of money is included in the bid figure for which there is no allocated use. This is known as a contingency sum. The amount of the sum varies according to the size of the contract and the degree of complexity of the works to be carried out. For example, the installation of a piece of plant which will be delivered to site fully tested and ready to run on its own, once services are installed to it, will not need a high sum. On the other hand, a system that is a one-off, designed and built specially for the occasion on site, depending on items such as distribution ductwork or pipework or unknown structural and services matters will need a substantial sum of contingencies. There is no hard and fast rule, but the sum ranges from 4 to 10 per cent of the contract sum. Generally, where there are known problems that cannot be accurately assessed (such as diverting an underground water main) a provisional sum should be put in rather than a contingency sum.

The danger is that often many issues are left to the contingency sum, which means that it is spent before other known matters come to light. It is better to report a higher bid figure, which has the possibility of being reduced at the end of the job, rather than a lower one, which is almost certain to increase.

8.7 Bid documents

These are documents to be completed by the bidder after he has studied the specification and drawings, where appropriate, and which contain the special commitment of him with respect to the particular job. His chances of being successful will depend largely on how he fills in the forms contained in the documents. Very special care should be taken in preparing these documents, and equally important is analyzing them after bid, as will be discussed in Section 8.12.

Standard forms of bid contain information and questions which are carefully put together and understandable in the industry. Contents of bid documents should include:

1. Clear indication of what the document is (i.e. *Form of Bid*);
2. Who the employer (client) is with full address. He will be party to the contract;
3. What the bid is for, briefly but specifically, and reference to the documents that will form part of any contract;
4. Date and time by when bids are to be received;
5. Where the bids are to be sent to and the manner in which they are to be submitted (for example, in a plain or specially marked envelope);
6. A note that the employer does not bind himself to accepting the lowest or any bid, and that the bidder will not receive any remuneration or expenses for providing a bid.

The bidder should be invited to submit the following:

7. His full name and trading address;
8. His lump sum bid for the works, where appropriate;
9. His labor rates, materials and on-cost, separately or as a composite rate in the case of a bill of quantities;
10. A statement of how long he needs to do the work following receipt of an order or the contract being signed;
11. Date and signature, with position of signatory (see Figure 8.1).

8.8 Direct and bulk purchasing contracts

To bypass the contractor or installer by purchasing plant directly from the manufacturer usually has an advantage of a saving in cost. The drawback is that the contractor does not have control of delivery or of quality, and this can lead to disputes or a defense against a claim for faulty workmanship or materials. Any saving can quickly become meaningless if the works are delayed, and a loss and expense claim is submitted. In a competitive situation, a bidder will keep his 'mark-up' to the absolute minimum, and a saving by approaching the manufacturer direct will be small. It is always possible to obtain a price with a manufacturer following competitive quotations, and then include a PC sum in the bid documents. This gives the bidder the opportunity to add a modest mark-up, and also place him in a position where he is responsible for ordering, obtaining delivery and then paying for the equipment, in addition to installing it. There is still a responsibility on the client or his agent for nominating the equipment.

(1) Job title: ....

(2) Name and address of employer: ....
....

(3) Location of project: ....

(4) Bid for: ....

(5) (a) This is a fixed/fluctuating price bid   (delete as appropriate)

   (b) Schedules of rates are/are not attached   (delete as appropriate)

   (c) Labor and material rates and on-costs are shown in a separate document attached

(6) I/we agree to keep this **bid** open for acceptance within .... weeks of the date of this **bid**.

(7) I/we agree to commence the works within .... weeks of receiving an official order, and to complete the works within .... weeks of commencement

(8) I/we intend to employ the following subcontractors to carry out part of the works, if this **bid** is accepted:

   Name   Address   Description of works

(9) I/we agree to carry out the works referred to in the drawings and specification attached to the invitation letter dated .... for the sum of £ .... Amount in words ....
....excluding VAT and which sum
includes the following:

   Contingency sum   £ ....
   Provisional sum   £ ....
   Prime cost sum   £ ....

(10) Name of **bidder** ....
Address ....
Signed by .... position .... Date ....

(11) **Bid** to be returned to .... in the endorsed envelope provided, not later than 12 noon, on the .... day of ....

(12) The employer does not bind himself to accepting the lowest or any **bid**, and the **bidder** will not receive any renumeration or expenses for providing a **bid**.

Figure 8.1 Typical bid document for a principal contractor

Bulk purchase of equipment is carried out in order to bring the price down per item. Having established that a large number are required, a client can then negotiate or invite competitive bidder to supply the number involved over a certain period. Care needs to be taken that the number agreed are actually required within the period, otherwise the supplier can sue for damages if he is frustrated from supplying the number agreed.

8.9 Program of works

Prior to writing bids, the client should decide when he wishes the work to be carried out. Factors affecting his decision will be availability of funds either from loans or grants, or in the case of local authorities or central government departments, it may be that a capital or revenue allocation needs to be spent within a particular financial year. Other factors will be other trades such as a building to house the plant, whose timing is obviously important.

Where an architect is involved, he will draw up a general program incorporating design time as well as installation, and this will be submitted to the plant engineer for his views before design commences. Having agreed the program, the bidders will be informed in the bid documents how it is intended that the works will proceed. However, where a main contractor is involved it is vital that he be given the opportunity to submit his own program of works, incorporating all the various trades, keeping within the program completion date as laid down in the bid documents.

Difficulties arise when delays in obtaining bid approval occur, because time moves very quickly, and the work may not actually commence on the date stated in the bid documents. This causes problems when the contractors have carefully arranged their own work schedules. It is better not to be too optimistic regarding start and completion dates unless various stages to a contract document being signed are clear.

8.10 Selection of bidders

There are four ways in which a bid list can be drawn up for a particular project:

1. Advertise in the local press or trade journal for firms to apply and then draw up a short list.
2. Seek out firms of good reputation.
3. Draw up a list from firms who have shown an interest in working for you.
4. Use firms who are known to you and who have given service in the past.

All clients are able to ask their own questions. A consulting engineer who did not properly assess a firm prior to recommending that they be placed on the bid list would be liable in negligence to his clients for not discharging his duties with diligence. Questions that the plant engineer should ask when compiling a questionnaire are:

1. Address from which firm would carry out contracts.
2. How long in business.
3. Number of office and site personnel and their trades.
4. Expertise professed by the firm and size of projects which they can undertake.
5. Sample of six jobs recently undertaken.
6. Names of three referees who can be approached.
7. Any pending litigation or recent court judgment against the firm.
8. Name and address of bankers.

Having received completed questionnaires, a shortlist can now be drawn up. After discarding the non-starters, the hopefuls should be vetted.

Remember that firms will put their best face forward. Where possible, it is advisable to visit at least two jobs recently carried out, and to contact at least two referees. A banker's reference can be misleading, although perfectly accurate. A banker's reference will say that the firm is in a trading position, and sometimes that the firm is in

a position to carry out the sizes of contract envisaged, but it will mention nothing about the reputation of the firm, its quality of work or administrative ability. A bank will not deal directly with a client, but only through the client's bank.

A different questionnaire needs to be sent to the referees, remembering to enclose a stamped addressed envelope. The referee is doing an unpaid favor in completing the questionnaire, and for his sake, it should enable him to return it without having to write a thesis with an attached letter. A box to be ticked for most questions will suffice, with a space at the end for general comments. Questions that should be asked of the referee are:

9. How long have you known the firm?
10. What size of contracts have they carried out for you in the past 3 to 5 years?
11. How do you rate their workmanship, administrative ability to keep to program, after-sales service?
12. Would you use them again?
13. General comments.

Having passed the test so far, the way is now open to visit the premises occupied by the firm and meet the management of the company. Once a firm is accepted for the list, it should be placed in the size of contract category and within the range of its financial capabilities.

8.11 Inviting bids

Most large organizations, particularly in central and local government, have set procedures in the form of standing orders, and this area is usually of most interest to internal and external auditors. The object of inviting competitive bids is to obtain the lowest price for the job, based on a fair method, which enables every bidder to bid for the same thing.

A design and build bid can cause problems in that different design engineers have their own views on approaching a design. To help overcome this, the client or his agent should issue a design brief which limits the design parameters. For example, in a scheme to re-light a warehouse, the brief could say that the lighting levels should be in accordance with the appropriate local, state and federal codes or could quote actual lighting levels to be achieved.

When drawing up the bid list, note should be made of the points referred to below before deciding to place a firm on the list:

1. Financial capability;
2. Location with respect to project;
3. Workforce and administrative strength to carry out the project.
4. Check that the firm has carried out similar projects in the past;
5. It should be ensured that a firm is not offered too many contracts at once. An overextended firm can cause severe problems of performance.

Prior to inviting bidders, each firm should be contacted, preferably in writing, to seek their agreement to bid. Information to be given to them will include:

1. Approximate value of works;
2. Type of contract;
3. When bidder will be sent out and closing date;
4. Program of works, start and finish dates.

Having agreed to bid, if a firm does not submit a bid without good reason consideration should be given to removing it from the bid list for a short period, because when a firm does not bid, it reduces the competitive element, possibly increasing the cost of the job. It also denies another firm the opportunity of biding.

The number of bidders invited to bid should be determined by the size of the works. Good contractors should be encouraged to bid and not be disillusioned by too many bidders. For most contracts, a bid list of six firms should be sufficient to keep the biding competitive and allow for the odd firm that does not return its bid.

On important contracts, it may be advisable to invite bidders to pick up the bid documents, or they could be delivered to them. Documents do go astray in the postal system.

Bids should be returned in endorsed envelopes so that it can be seen that a bid for a named project is enclosed, but should not disclose whose bid it is. In large organizations it is normal for bids to be returned by a certain date and time, to a certain place, which is usually either a legal or an administrative department.

Someone other than the person involved in the job itself should open bids and then hand them to the person involved, after details of bidder have been noted. Bidders cannot expect to be paid for the time or other costs in ordinary circumstances for preparing their bids.

8.12 Analyzing bids

When the plant engineer receives all the bids returned he will need to analyze them to ensure that they are arithmetically correct, and that each bidder has understood what is being bided for. A pre-designed analysis sheet can be helpful and a typical sheet is shown in Table 8.1. All bidders are shown on one sheet with their prices compared. The first task is to check the arithmetic, which seems to cause problems even in these days of computers. Where the arithmetic is incorrect, it should be brought to the attention of the firm quickly to enable it to either withdraw or stand by the figure. If the correct addition would have meant lower costs, the firm should be given the opportunity of revising the bid, but only after consultation with the legal or administrative department. Action taken now without consultation can be hard to explain later.

The addition may be correct, but a bidder may have underestimated a particular item. With an itemized bid it is possible to spot an error of this kind. For example, referring to Table 8.1 and looking at item 3, bidder 2 has less than half in his bid for item 3 than any other bidder. On the other hand, item 2 appears to have made up for it. Where an apparent mistake has occurred, the firm concerned should be contacted, and should be asked to look closely at a certain item, which appears low. For example, bidder 1 has a much lower amount for item 7 than any other, and if he had bided similarly to the other

Table 8.1 Analysis of bids for replacing a drying machine

Item	Detail	1	2	3	4
1	Strip out	$2000	$1600	$1900	$2200
2	Steel work	$1500	$2500	$1300	$1700
3	Rein. conc.	$2000	$950	$2100	$1900
4	Machine	$99 000	$98 000	$97 000	$101 000
5	Elect.	$3000	$2900	$2950	$3100
6	Piping	$2500	$2300	$2600	$2700
7	Ductwork	$2700	$4700	$5200	$5100
8	Insulation	$2000	$1950	$1900	$2000
9	Painting	$1000	$970	$1100	$1050
10	Test/ commission	$1000	$990	$995	$1100
11	Contingency	$5000	$5000	$5000	$5000
12	Provisional	$500	$500	$500	$500
13	Prime cost	$1000	$1000	$1000	$1000
	Total tender	$123 200	$123 360	$123 545	$128 350

bidders for that item, he would not be the lowest bidder. It is better to lose a low price than have the firm lose money on a job, which can cause difficulties with performance.

Where the scheme was designed before being sent to bid, the bids should be grouped fairly closely together, because design decisions have already been made. Design-and-build bids will need much more attention to find out which is the best value for money. For example, in lighting the warehouse mentioned earlier, bidders may have been advised of the lighting level, but if they have not been told of the type and manufacture of fittings, there can be a substantial difference in cost.

Bids received can vary by up to 100 per cent over a range of firms. The higher bids may be from firms who do not want the job but wish to keep contact with the client. Where there is a difference of more than 10 per cent between the lowest and second-lowest bidder, this can cause problems. For example, the Electrical Contractors Association offer a guarantee of completing through their members, at no extra cost, in the event that a member is unable to do so, but only if the difference between the contractors appointed and the next bid is not more than 10 per cent. In some instances, this could be enough to seriously consider not accepting the lowest bid.

8.13 Selection of the contractor

Having established that the lowest bidder is in fact the lowest, after considering all matters there are other aspects to consider before agreeing to a contract.

A qualified bid can make the offer void, and most organizations would disqualify a bid which was qualified. An example of a qualified bid would be where the specification says that electrical attendance is part of the contract, whereas the bidder says in his bid documents or in an accompanying letter that he has not included for electrical work, and is not prepared to carry it out. A common qualification is where a contractor puts forward a manufacturer of plant different to that specified. The plant may be just as good, but the bid is nonetheless qualified, and would give an unfair advantage to the bidder.

A *bona fide* qualification is where the specification is unclear.

An area of increasing importance financially is the program. Contractors are claiming large sums with respect to delays, and it is important to take note before the contractor is appointed of any qualifications he may make to the bid. For example, if a bidder says in his bid that he can carry out the works in 40 weeks and it is known that the period envisaged is 52 weeks to coincide with other trades, then if he is frustrated from doing the work he can sue for loss and/or expense for the 12 weeks when he will have, in theory, to lay off men. The solution is to challenge the period inserted, and if the bidder will not amend his period, then do not appoint him.

Bids are normally kept open for a specific period, which is usually 4 weeks. In any case, without a specific period being mentioned in law, a bid must be accepted within a reasonable period, otherwise it will lapse. In the event that a decision cannot be reached, or an appointment made for some time, the bidders should be approached in order to ensure that their bids are kept open. The problem for the contractors is that wages and salaries increase regularly, as do materials and other on-costs.

In contract law, the acceptance of an offer must be unequivocal. Where a bid is accepted subject to an amendment on, say, program, the client is making a counter-offer, and the bidder then makes acceptance.

The successful bidder should be informed as soon as possible that his bid is to be accepted, to give him the opportunity to plan his operation. Equally, the unsuccessful bidders should be sent a letter expressing appreciation of the time they have spent in preparing their bidder, but informing them that they have been unsuccessful.

8.14 Making a contract

A simple contract for any value need not be in writing, but can be verbal and is known as a parole contract. Many contracts are made which are verbal. When a person goes into a supermarket, takes an item off the shelves, carries it to the checkout and makes an offer to purchase it at the price shown, this constitutes an offer in contract law. The checkout cashier takes the money and is, at the same time, accepting the offer.

Verbal contracts are fraught with problems and should be avoided where possible. For example, to prove that a verbal contract has been entered into without the work having been carried out will need parole evidence in court. Where arbitration is entered into later, a verbal contract does not come under the Arbitration Acts 1950 and 1979.

Letters of intent can be the cause of problems, and there are varying views as to their validity. These are sent in advance of contractual documents, being signed because either the documents are not fully prepared or a decision has not yet been confirmed. For example, in local government a bid may have been accepted by a committee but needs ratification by the full council, which can be up to six weeks later. It is rare for a council to overturn a decision made by a committee, but it does happen. In an attempt to expedite matters, it is common policy to send out letters of intent directly following the committee. These will be marked either 'Subject to Contract' or 'Subject to Council Approval', and in these cases, they are useless as a contract if the client does not go ahead. Where the contractor has reasonably relied upon the letter and the client has encouraged him to purchase the equipment, then a court may order that he be paid for the work done under *quantum meruit*. A letter of intent without a qualification will infer a contract and will committ he client to the whole works.

8.15 Relationships between contractor and other parties

Where only one contractor is involved, then his relationship with the client is direct and contractual. Most contracts have more than one contractor, and the client must decide how he intends this to be arranged. A few of the possible compilations are:

1. *Direct contract* (sometimes known as a principal contract), where only one contractor is involved;
2. *Main and nominated subcontract.* This is where one contractor is appointed and known as the main contractor and the client nominates or instructs this contractor to appoint a named firm to carry out part of the works. Sometimes bids will have been sought separately prior to the involvement of the main contractor;
3. *Main and nominated supplier.* The client may wish a certain firm to supply materials for the contract, which may be for installation by the main, or a sub-contractor.
4. *Principal contractor with domestic sub-contractor and supplier.* The expression 'domestic' does not necessarily infer housing, although housing contracts can be carried out in this manner. A domestic sub-contractor is a contractor who is appointed by the principal contractor. The client has no relationship with a domestic sub-contractor, unless he interferes with his selection such that the appointment becomes nominated. This type of appointment has become very common.

The contractors have no contractual relationships with any independent consulting engineers who are appointed by the client. Where the client decides to appoint a firm to 'design-and-build' there is a different duty on the designer. When a consulting engineer is appointed, he must exercise reasonable skill and care in carrying out his duties, as an average firm of this type would do. However, where the client entrusts his design and installation to a contractor, the contractor owes a strict duty to his client to ensure that the installation works, and that it is fit for its purpose, as in *Independent Broadcasting Authority* v. *EMI* (1981).

A nominated subcontractor has no contractual relationship with the client, but the case of *Junior Books* v. *Veitchi* (1982) placed an 'almost' contractual duty upon the nominated sub-contractor. Where a separate agreement has been signed between the client and nominated sub-contractor then a direct contractual relationship applies. Problems can arise where a main contractor says that his nominated sub-contractor or supplier is delaying him.

Because the client has instructed him to use a firm which was not of his choice, the main contractor could claim damages in the form of loss and/or expenses from his client. Where the principal contractor has brought in his own domestic sub-contractor, then he is responsible for his actions. That is the main reason domestic sub-contracts have become popular. The client must decide upon which trade will become the main contractor. In building matters, it is usual for the main contractor to be a firm who can carry out the most substantial works, and this will be a general building firm. Some general building firms carry their own mechanical and electrical departments, which reduce the need for sub-contractors.

There are occasions when it is more appropriate to appoint a mechanical or electrical firm to be the main contractor. An example would be the laying of a district heating mains, where the largest value and complexity is in mechanical engineering. However, experience shows that not all specialists are geared to this role, and this can cause problems, especially from a main contractor who is more familiar with the role of sub-contractor. Overall, it is recommended that the main contractor be a building or civil engineering firm, unless those works are very small.

8.16 Site meetings

Meetings are a necessity to keep a contract moving. The first meeting is normally at the offices of either the client or consulting engineers, and is a pre-contract meeting. A pre-contract meeting is a gathering of all interested parties to discuss the format of the contract and is a chance for views to be expressed, prior to the actual formation of a contract, which very often takes place after the meeting.

Matters which are likely to be on the agenda of a pre-contract meeting, are:

1. Names and addresses of all parties and introductions;
2. Names of specific persons from each firm who will be handling matters, such as site engineers and clerks of works;
3. Format of site meetings to be agreed (who will take minutes and who will chair meetings);
4. General lines of communication agreed;
5. How often site meetings will occur;
6. Accommodation for sub-contractors, site engineer, clerks of works and meetings;
7. General program discussed. At this stage the main contractor may not have drawn up a bar chart or critical path program;
8. Procedures for variations and interim valuations may need to be cleared up, depending upon the form of contract adopted;
9. Where nominated suppliers or sub-contractors are involved, the main contractor has the opportunity to voice his opinion of those selected by the client.

There are generally two kinds of site meetings. One is where all parties are present and the other is where the professional team meets with the main contractor and then the main contractor organizes his own contractors' meeting with his sub-contractors. The job itself will determine the kind of site meeting to be held. *Ad hoc* meetings

are held throughout the job, and these meetings should always be with the knowledge of the main contractor, if not chaired by him.

Full site meetings may be held monthly or weekly, depending on the pace of the job. The main contractor or the project leader, who could be the architect or engineer, will chair a typical site meeting, and matters for the agenda will include:

1. Apologies for absence;
2. Previous minutes read;
3. Matters arising from previous meeting;
4. Main contractor's report;
5. Sub-contractor's reports;
6. Progress (various clerks of works' reports);
7. Consulting engineers' reports;
8. Any other business;
9. Date and time of next meeting.

Minutes are then published through the chairman and circulated to all parties, even to those not present at that meeting. When receiving the minutes it is worth studying them to ensure that they are accurate and to note any action needed. It is usual to have a column drawn on the right-hand side of the minutes for the names of the parties who need to take action. The smooth running of the project can depend on all parties complying with action noted in the minutes. Inaccurate minutes should be challenged at the next meeting. Site meetings and publication of minutes are vital to a large or complicated project to avoid misunderstandings, which can cost money.

There will be *ad hoc* meetings, particularly inspections by visiting clerks of works. Points worth noting at these inspections are:

1. The site is under the control of the main contractor, and he must be informed of all visitors, who are advised to comply with requests (e.g. wearing hard hats).
2. An apparently simple visit can result in disputes later. Therefore, it is wise to meticulously record all salient points of discussion.

The time and date of arrival and time of leaving may seem pedantic, but can avoid difficulties later. Clerks of works' diaries are often admitted as contemporaneous evidence in litigation and arbitration. This is discussed further in Section 8.17.

8.17 Progress and control

The client will have decided at the very beginning when he wants the job to be finished on site. Circumstances such as approvals, preparation of contract documents or availability of labor and materials will determine when the works start.

At the beginning of the contract, before work starts on site, the main contractor will produce a program of how he intends to carry out the works. In some bids the contractor is invited to state what his program will be if he is successful. This method has merit where the program is very tight. The most common type of program is a

Name of project: Replacing drying machine

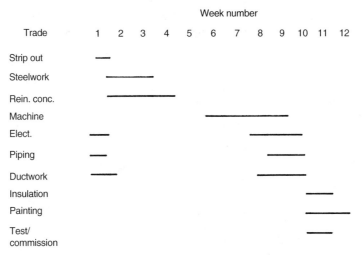

Figure 8.2 Typical bar chart program

bar chart, and will show at what periods the various sub-contractors will be on site. A typical bar chart is shown in Figure 8.2.

The value of having one contractor to control the progress of the work can be seen where there are several contractors on a site at different periods. Progress can suddenly come to a halt if a key contractor does not comply with an agreement. For example, if the electrical sub-contractor does not provide a temporary electrical supply, there may be no light and power, causing other contractors to be delayed.

In the matter of progress, the main contractor is in charge of the works. However, the client or his consulting engineer will need to monitor the works to satisfy themselves that programs are being adhered to. There is a temptation for a client or his consulting engineer to assist the contractor by instructing him how to carry out the works. This can amount to interference, and if the contractor suffers loss, as a result he can sue for damages.

There are varying degrees of monitoring which a consulting engineer can recommend to his client:

1. Occasional inspection visit by the design engineer, as necessary, on a small job;
2. Regular clerk of works visits;
3. On-site engineer or clerk of works.

It is the duty of the consulting engineer to make a recommendation to the client as to the type of supervision required, and the client must not refuse a reasonable request by him for on-site supervision. Where a client refuses a reasonable request for on-site supervision, the consulting engineer should warn him of the possible consequences, otherwise he might face a claim for negligence later.

On most contracts, there are requests by the contractor for clarification of the specification or for instructions with regards to variations. These requests need to be answered promptly to avoid a claim for damages by the contractor, which may result from delay of the works.

All progress should be carefully noted in the diary of the supervisor or clerk of works. Points worth noting are:

1. Number of men on site;
2. Weather (where appropriate);
3. Name and title of any person spoken to. Note that instructions should generally be given through the main contractor's site agent;
4. Progress of job;
5. Any items discussed.

8.18 Quality control

The contractor is under a duty in contract to provide and install plant which is of merchantable quality. There is no duty on the client to ensure how the contractor carries out those duties, although his client can sue the consulting engineer in negligence, if he is not diligent in detecting faulty workmanship, such that he suffers loss as a result.

Where an industrial firm employs a contractor, he will have a chief engineer who will organize the monitoring of the quality of workmanship and materials. Where a consulting engineer is involved, he will recommend to his client the degree of supervision required, as mentioned in Section 8.17. The person appointed will need to consider the following and be satisfied that the works being installed meet the requirements which should be laid down in the specification and drawings:

1. Materials and plant which are specified are new and unused;

2. That workmanship is up to the standard expected in this type of contract. The Codes of Practice, IEE regulations and other bodies related to standards should be used as a measure of standard. It was ruled in the case of *Cotton* v. *Wallis* (1955) that a lower standard of work could be accepted in a known cheap job, as long as the work is not 'rank bad'.

A problem which seems to occur frequently is that contractors do not give sufficient thought to allowance for maintenance in the future. To install plant without means of maintaining it is bad practice and should be avoided. The plant engineer should make sure that the design allows for adequate maintenance. This item comes under the scope of the clerk of works to interpret the design.

8.19 Interim payments

In a lump sum contract the contractor does not have a common law right to interim payments. Most larger contracts have an express term within them giving the right to interim payments, or payments on account, notwithstanding that they have not fulfilled their contractual duties in completing the works. The matter of interim payments and their frequency should be resolved before the contract is signed, and not left until a dispute arises.

Interim payments on the contract follow a valuation of the work done. Valuations are normally carried out by the plant engineer, the quantity surveyor, the architect or the consulting engineer and, in the light of recent case law, should be assessed very carefully.

Until 1974, an architect was considered in law to be a quasi-arbitrator, and his valuation was final. However, the case of *Sutcliffe* v. *Thackrah* (1974) saw an end to this role. In that case, the architect overvalued the work that had been carried out, and because the client ordered the contractor off the job, he suffered loss, because of which he successfully sued the architects for damages. Architects and engineers suddenly had a greater interest in professional indemnity insurance.

An obvious solution would be to undervalue the work done, in order to avoid a claim from the client. The case of *Lubenham Fidelities and Investment Co* v. *South Pembrokeshire District Council* (1986) took care of any tendency to undervalue. In that case, the Court of Appeal ruled that the contractor might sue the architect in tort for any damages suffered by him because of the negligent valuation.

There are occasions when a contractor will apply for an interim payment based on plant obtained by him for the contract and stored by him on his premises. This is a difficult situation, and should be approached with care. Two problems can arise:

1. Should the contractor end up in dispute with the client before the plant is installed, he may refuse to release it, despite having been paid for it.
2. The contractor could go into liquidation before the plant is installed.

A way around these problems is to have the plant isolated from other equipment and clearly marked 'This is the property of (name of Clients)' and while it may not be a complete guarantee of success, it is at least better than not marking it.

8.20 Budget control and variations

At the beginning of the contract, it will be known to the client and his consulting engineer how much money is budgeted for the project. The contractor will be aware of the contract value, of course, and he will know of any contingency sums that have not yet been allocated.

Contingency sums are included in a contract in case there are any additional works that need to be carried out and have not been covered in the contract. This may be because the client wishes to have more items of plant than he originally envisaged, there may be unknown features, or it could be for items that have been forgotten in the original design. (See Section 8.6 for a description of contingency sums.) Where a consulting engineer forgets to include an item which is necessary for the proper completion of the works he can be sued for negligence by his client, or advised to pay for this work himself. Where a contractor forgets an item which is necessary he will have to install it himself at no extra cost.

There is a temptation for the contractor to see contingencies as extra work, or there is always a danger that it will be spent the end of the contract because the client will see it as available monies.

Provisional sums are already allocated when the contract is made, and while the true value is not known at bid, the estimate should be reasonable. Where a schedule of rates or a bill of quantities is available, there is no problem in arriving at the figure after the contract has been let. However, where the contract is simply lump sum, there is no basis to determine any variation, and this will sometimes mean a difficult negotiation. When a variation to the contract becomes apparent it should always be agreed in writing, preferably with a firm agreement on price or at least an agreed rate. A major part of a dispute when the final account is being agreed variations. Where the variation is too difficult to assess accurately, such as an underground gas or water main being found which was not known previously, the contractor will have to be instructed to carry it out on day work (*quantum meruit*). In the bid, there will be a rate for labor and an on-cost figure for labor, materials, plant and other expenses, and the purpose of this is for items of daywork which may occur. Certain standard forms of contract such as the JCT form contain specific documentation when dealing with variations.

8.21 Safety on site

Unfortunately, deaths occur every year in industry due to accidents that are preventable. Accidents are more likely to occur where firms are under pressure to meet targets, and safety measures tend to be ignored. The Health and Safety Executive (HSE) was set up to administer the Health and Safety at Work, etc. Act 1974 and incorporates the Factories Inspectorate. General duties of employers described in section 2 of the Act are that 'It shall be the duty of every employer to ensure so far as is reasonably practicable the health, safety and welfare at work of all his employees'. The emphasis is on the expression *reasonably practicable*, and courts interpret the meaning of it, under the particular circumstances.

A plant engineer may become involved with an inspector from the HSE when he arrives on site to carry out an inspection of the works under powers given to the Executive by Section 20 of the Act. The HSE will visit at any reasonable time, or at any time where the situation may be dangerous in the inspector's opinion. He may take with him a police officer if he has reasonable cause to apprehend any serious obstruction in the execution of his duty.

Having carried out an inspection of the works, the inspector has powers under Section 21 to serve an improvement notice if, in his opinion, a person is contravening one or more of the relevant statutory provisions, or has contravened one or more of those provisions in circumstances that make it likely that the contravention will continue or be repeated. The person concerned then has a duty to remedy the contravention or as the case may be, the matters occasioning it within such a period as may be specified in the notice.

An inspector can order under section 22 that activities be prohibited which have not yet commenced, and may be about to be carried on or are already being carried on, by serving a prohibition notice.

To ignore either of the notices above is a criminal offence, and a prosecution is likely to follow such contravention. There is an appeal to an industrial tribunal against an order.

An HSE inspector has the power under Section 25 to enter any premises where he has reasonable cause to believe that, in the circumstances, any article or substance is a cause of imminent danger of serious personal injury. He may then seize the article or substance and cause it to be rendered harmless, whether by destruction or otherwise.

8.22 Delays and determination

Contractors are all too frequently delayed in their completion for a variety of reasons, some of which are:

1. Client adds extra work or changes his mind;
2. Client or consulting engineer takes his time in giving instructions or clarifying the specification;
3. The contractor starts on site later than agreed;
4. Sub-contractors do not coordinate properly;
5. Site conditions are not as expected;
6. Materials are not delivered as promised;
7. Exceptionally bad weather;
8. Slow progress generally by the contractor.

Delays to contracts can be costly to both client and contractor, and, in some cases, suppliers. There has been a tendency in the past few years for an injured party to sue for damages under the contract, and where there is no contract, in tort. Since the case of *Marden* v. *Esso Petroleum*

(1978) it is possible to sue both in contract and tort at the same time.

The various forms of contract make a provision for registering a delay or possible delay, and the contractor particularly should take this aspect of a contract very seriously, because it can be the greatest source of financial claim. Some contracting firms examine contracts very carefully to see if a claim of delays can be made, and this starts on the day that the contract has been signed.

Close control of the contract should be kept at all times. This is where formal site meetings, with reports of progress, can be most useful and can help to steer a late contract back on the right course. When a contract begins to fall behind the agreed program, it is important to find out why and to rectify the problem. It is in everyone's interest to see the plant commissioned on the agreed date, and a casual attitude by the project leader should never be taken. He should be strict with anyone who causes a contract to slip behind schedule.

There are legitimate reasons why a contract can fall behind, and some of these are referred to in (1) to (8) above. However, item (8) must not be tolerated, and the contractor should be taken to task early in the contract when this occurs. Close monitoring will show how the contractor is progressing, and where the contractor has fallen down so badly that there seems to be no end to the contract, and then the client should seriously consider determining the contract.

Most standard forms of contract contain specific procedures for determination, and these should be rigidly adhered to. For example, the contract document may lay down that the contractor be given 14 days' notice in writing to rectify his progress before determination is carried out. Any letters should be sent by recorded delivery or served by hand. While liquidated or non-liquidated damages can be claimed with respect to late or incomplete work, expense claims usually follow the events shown below, which are also mentioned in Section 8.24:

1. Client adds extra work or changes his mind. It may seem reasonable that where the client authorizes extra work and pays for that extra work, he should not then be required to pay, because the contractor needs to stay on site longer. This is, however, the case, and the claim because of the delay very often exceeds the actual extra costs of the works. Where a client simply changes his mind and cancels some of the work, then the contractor can insist on his profit, and overhead costs for the time he will be without work as a result.
2. Delay in instructions being issued to the contractor can be a loss to the contractor if he is waiting for the instructions, and he can sue for damages for his loss.

8.23 Liquidated damages and loss and/or expense

This is an aspect of a contract which should not be contemplated by either party at the outset of a contract. However, it is common for the client to set a figure for liquidated damages, which he considers that he will genuinely lose if the contract is not completed on time, and should not be confused with a penalty clause, which is for punitive damages.

The expression non-liquidated damage means actual losses, which are incurred by a client, and could be claimed in the absence of an agreement to a specific amount in the contract. Loss and/or expense can be suffered by a contractor due to other parties who caused him to be delayed. These sums per week are invariably far in excess of any liquidated damages amount set in the contract.

In practice, what happens is that where a contract is falling behind program the contractor will apply to the client for an extension to his period for completing the works. Should the contractor not make an application then he will be in breach of his contract to complete the works by the date agreed, and the client can claim liquidated or non-liquidated damages as a result. Where no specific dates for completion have been set, in a dispute situation, the court will set a date which, under the circumstances, was or ought to have been in the contemplation of the parties at the outset of the contract. For example, a firm agrees to supply and install a machine, and while no particular date is set for installation, the firm knows that production would be affected badly if it was not installed within three months. Without extenuating circumstances prevailing, the firm could be in breach of its contract if, say, the machine was not installed within three months.

The contract could include a penalty clause for non-completion by the contractor, and paid by him, as well as a bonus clause paid by the client to the contractor in the event that he completes his contract on or before the agreed completion date. This bonus clause can be on a reducing scale (say, from 75 per cent of the time onwards).

Liquidation of the contractor has nothing to do with liquidated or non-liquidated damages, although these can still be claimed to a company in liquidation. Some forms of contract, such as the JCT form, make provision for automatic determination in the event of a firm going into liquidation but that the contractor may be reinstated under certain circumstances.

8.24 Practical and final certificates

The essence of a plant engineering contract will be that the plant actually performs to the criteria laid down in the design, and this must be established before a Practical Completion Certificate is even considered. Tests will need to be made to the plant, and then it will be commissioned by the contractor and left in a condition whereby it can be operated as specified. When the work within the contract is complete, a certificate should be issued by the client (or employer) to the contractor. The architect or consulting engineer acting as agent normally performs this duty.

Standard forms of contract have particular formats, but the points which should be noted are:

1. Names and addresses of parties
2. Name and reference of job
3. Date
4. Brief outline of contract and note that the works are now practically complete

5. Period of guarantee from contractual documents
6. Signature.

There is a tendency in the building industry, for expediency, to issue a Practical Completion Certificate subject to a list of defects. Very often, this is done in order to see the end of the main works.

From the date of the Practical Completion Certificate, the contractor is no longer liable for liquidated damages, and the defects liability begins to run. Latent defects which appear for the next six or twelve months, depending upon the agreed period, have to be made good by the contractor, free of charge to the client.

The final certificate is issued when the end of the defects period has been reached, and when all defects, which have appeared within that period, have been attended to.

The client is not prevented from making a claim for latent defects after the twelve-month period ends, but may be prevented from making such claims by the Limitation Act 1980 and the Latent Damages Act 1986, after a period of six years for a simple contract and twelve years for a contract made under seal.

8.25 Disputes and arbitration

It is unfortunate when a dispute arises between any of the parties. The two most common areas are a claim by the client that the work has been done badly or taken too long, and from the contractor that his contract has been interfered with by the client or the nominated sub-contractors.

Most disputes take the form of strongly worded letters between the parties, with eventual threats of litigation or arbitration, very often in an attempt to cause a party to capitulate. Where there is a genuine attempt to settle a dispute the letters between the parties can be marked 'Without Prejudice' and letters so worded cannot be used in evidence without the consent of both parties. If a binding contract has been effected because of *without prejudice* interviews or letters, this may be proved by means of the 'without prejudice' statement, as in *Tomlin* v. *STC Ltd* (1969).

The standard forms of contract invariably contain agreements to arbitrate in the event of a dispute and some even name the arbitrator or a professional body who will appoint an arbitrator, upon the application of one of the parties. Where no agreement to arbitrate exists, the injured party may sue in either a County Court or High Court, depending upon the value of the claim, and he must then prove his case before a judge.

A more speedy method, although not necessarily less expensive, can be for the parties to make an agreement to arbitrate, and to appoint their own arbitrator, who should not have been previously connected in any way with them. Where a written arbitration agreement is made, the arbitration will be conducted under the Arbitration Acts 1950 and 1979 and, unless otherwise agreed, before a single arbitrator, whose powers are almost those of a High Court judge.

An arbitration hearing can be held anywhere convenient to the parties, and at times agreed. While the arbitrator will not meet any one party separately or take instructions from him, he will take instructions within the law from all of the parties. For example, if the parties wish the hearing to be held in a certain place, as long as the facilities are adequate for the hearing for the parties to meet their legal advisors, the arbitrator is likely to agree. On matters of the conduct of the hearing, the arbitrator is in complete control, subject to a limited appeal by the parties to the High Court, who can, on matters of law and conduct, direct the arbitrator to take certain action.

The hearing itself will be conducted in a less formal manner than a court trial. The room itself is set out like a conference room, with a seat at the head for the arbitrator and places (called 'the box') at the opposite end for the witnesses. To the left of the arbitrator is usually the claimant and his advisors and to the right the respondent and his advisors. It is normal for all present to stand when the arbitrator enters or leaves the room and address him as 'Sir', referring to him during the hearing as the 'Learned Arbitrator'. A judge needs to be treated with more respect and a procedure of bowing upon entering and leaving is normal. A County Court judge is referred to as 'Your Honor' and a High Court judge as 'My Lord'.

The plant engineer may be asked to give expert evidence, which means giving his opinion to the court or tribunal on the subject matter. Prior to giving evidence a proof of evidence will be prepared by the expert and will contain simply the following:

1. Identification, qualifications and experience of witness;
2. Details of his appointment;
3. General description of subject matter in dispute;
4. Any tests, photographs, samples or other documents relied upon;
5. Conclusions and opinions.

There is a distinct difference in the way that expert evidence is received by a tribunal to that of witnesses of fact. The plant engineer is strongly advised not to agree to accept an appointment and subsequently give expert evidence on matters for which he is not fully experienced. In accepting an appointment to write a report for a client, his legal advisors should be made fully aware of the plant engineer's background and expertise. Having become committed to the case, the client could, if necessary, apply for a *subpoena ad testificandum* to ensure that the expert gives evidence. There may be occasions when the plant engineer will be called upon to give evidence of fact, and under these circumstances he is not an expert witness, although he may be able to give opinion on matters for which he is considered an expert. An example of this is where the plant engineer is an employee of the claimant, and he needs to give evidence that a machine delivered to his works would not perform satisfactorily. It would be a matter of fact that the machine would not perform, and the evidence would be strengthened because the plant engineer is an expert in this type of machine.

Upon entering the witness box the plant engineer will be invited to take the oath, and can either swear on the Bible, affirm or take the kind of oath suitable to his religion. The advocate acting for him will then take him through his proof of evidence, but will be careful not to ask him leading questions, except his name, qualifications and experience. The opposing advocate will then cross-examine the witness to test the evidence given. He will ask leading questions and may attempt to shake the evidence given. The advice to the witness is to stay calm, courteous and truthful, remembering that the advocate is only doing his job.

Following cross-examination the advocate may ask a few more questions to clear up points already raised. The judge or arbitrator may then ask a question in clarification.

8.26 Common problems and solutions

There is no sure way of avoiding claims with respect to contracts, but the following summary may be of assistance to the plant engineer:

1. Vet prospective bidders carefully prior to invitations to bid to ensure as much as possible that they are capable of carrying out the work.
2. Select the type of contract most appropriate for the work.
3. Where possible, avoid nominating a contractor or supplier.
4. Make sure that the drawings, specifications and bid documents are clear and unambiguous. Do not rely on exclusion clauses in the specification or bid document. Have designs checked and use methods and equipment that comply with British Standards and Codes of Practice.
5. Unless there is a clear program of works which has been mentioned in the bid documents, before accepting bids make sure that the contractor or supplier has specifically stated his commencement and completion dates. In the event that the completion date is too early or too late, resolve the matter before entering into the contract.
6. Ensure that site meetings are as short as possible and organized to assist the project, and that they serve the purpose for which they were intended (i.e. sort out problems and ensure satisfactory reports).
7. When a contractor indicates that he is or may be delayed, take him seriously and solve the problem immediately. Do not interfere with the way in which the contractor carries out his work, unless he is patently incompetent.
8. Make sure that all site supervisors, clerks of works and site engineers keep good records, and that they are kept contemporaneously, so that, if necessary, they will be admissible in evidence in the unfortunate event that a dispute arises.
9. Be careful on the matter of interim valuations for payments to contractors. To undervalue can be as serious as overvaluing.
10. Keep up to date with variations to the contract. Do not wait until the end of the contract before extras are added up.

11. Safety should be completely in the hands of the contractor controlling the site, unless the client is in occupation. Keep a high profile in dealings on safety matters, and ensure that staff and any other parties are conscious of the need for safety.

12. The plant engineer should be careful of giving casual advice to parties with whom he has no contract. Under the *Hedley Byrne* v. *Heller* rule the plant engineer could be sued for a negligent misstatement if his advice was wrong, and the receiver of the advice could reasonably rely upon it, and he suffered loss as a result.

13. When attempting to settle a dispute the heading 'Without Prejudice' can be used in correspondence, which, subject to the rules of privilege, cannot be used in court.

9

Industrial Flooring

J D N Shaw
SBD Construction Products Limited

Contents

9.1 Introduction

There are a bewildering number of special proprietary floor treatments available for the architect and engineer to consider, and much of the brief technical literature describing them suggests that many products would appear to offer the same improved service at greatly differing costs. It is therefore not surprising that many specifiers are totally confused and tend to stick to the products they know rather than consider some of the novel treatments based on new technology, which often offer distinct improvements over the materials traditionally used as flooring materials.[1]

Before attempting to classify the various types of special floorings available, it is important to consider the concrete substrate itself. By proper use of good mix designs and admixtures, with careful control of the water/cement ratio and careful attention to laying, finishing and curing techniques, concrete itself can serve as a highly durable flooring material under many industrial service conditions without the need for special separately applied finishes.[2,3] Properly laid concrete provides an abrasion-resistant floor surface which has good resistance to attack by alkalis and reasonable resistance to mineral and vegetable oils, although oils do cause some staining and impair appearance. However, irrespective of how well it has been laid, concrete has poor resistance to acids and many other chemicals far too numerous to mention here. Where spillage of such materials is envisaged, the concrete must be protected by the application of special flooring.

9.1.1 Selection of flooring required

Before selecting a flooring material, it is imperative to consider carefully the precise service conditions to which the floor will be subjected. Conditions that must be considered include:

1. Service temperature;
2. Rate of change of temperature, as rapid temperature changes can cause some heavy-duty finishes to crack up due to the high stresses developed by thermal shock;
3. Nature and concentrations of any materials likely to come into contact with the floor;
4. Accuracy of laying the concrete sub-floor to levels to allow spillage to run away reliably to drains;
5. Grade of concrete laid for the sub-floor;
6. Nature of traffic (maximum loads and types of wheels using floor) and traffic concentration;
7. Degree/ease of cleaning required;
8. Non-slip characteristics required.

Without such precise information and, on occasions, even with it, inappropriate floorings are all too often used, resulting in rapid breakdown of the floor in service. In some instances, some of the performance requirements are contradictory (especially 7 and 8 above) and where this happens, a compromise may have to be accepted.

9.1.2 Requirements of concrete substrate

If, at the specification stage, it is decided that the service conditions for the concrete floor do require special floorings to be applied subsequently, then it is imperative that the contractor laying the concrete is fully aware of this factor and does not use a conventional spray-applied resin solution curing membrane, as this could seriously affect the adhesion of any special flooring to be applied subsequently, being difficult to remove uniformly and reliably.[4,5] In these circumstances, and also for industrial buildings where the eventual use of the floor is not known but may therefore require a special finish, overlapping polyethylene sheets or another efficient curing method which does not affect the adhesion of any subsequently applied finishes should be used.

It is important to examine carefully the surface of the concrete substrate prior to applying the special flooring. Although the concrete laid by the contractor may indeed have cube strengths well in excess of that specified, it is still possible for the concrete slab to have a very weak surface due to over-trowelling, for example. Special flooring is applied to the top few microns of the concrete. It is essential that any weakness in the surface be removed by a technique that is appropriate to the type of flooring to be applied.[7] Most specialist flooring contractors have sufficient experience to assess the quality of concrete surface without site testing. However, if there is any doubt, the surface strength of the concrete should be tested using a simple pull-off tester or other appropriate means. In general, a concrete substrate should have a tensile strength (by pull-off) of at least 0.75 N/mm².[8,9]

As a rule, any concrete base that will be subsequently treated with special flooring must not be subject to rising damp, and thus any ground-supported slabs must incorporate an efficient damp-proof course. If there is any doubt, the concrete should be tested using a direct-reading concrete moisture meter (maximum 6 per cent moisture) or an Edney Hygrometer (reading not exceeding 75 per cent relative humidity after 4 hours). It should be stressed that these figures are based on the practical experience of a number of specialist flooring contractors and serve only as a guide. Other factors such as the depth of the slab, time elapsed since placing and degree of weather protection all have an influence on the moisture content within the concrete substrate.[19] Finally, before considering flooring materials, mention should be made of the application of the right joint filler in all movement joints.[8] Far too often, with a carefully laid concrete floor for industrial service, no detailed attention to joint filling is given, and this results in unfilled or wrongly filled joints rapidly spalling at the edges under heavy loads of rigid or semi-rigid wheel traffic, leading to expensive repairs. Apart from preventing spalling at the floor edges, the right joint filler will also improve cleanliness, help smooth running and prolong wheel life of forklift trucks, for example, and contribute to safety. The selection of the right joint filler is, however, a difficult problem, and, in general, it is true to say that there is no one single ideal material for floor joints, since it is impossible to combine all the performance characteristics ideally required in one product.

There is now an active trade federation, the Federation of Resin Formulators and Applicators (FERFA) to which many of the specialist formulators and industrial-flooring contractors belong. If specifiers/users carefully list all the service conditions, which the flooring must meet, FERFA members should be able to recommend the most cost-effective flooring materials to give good service under the conditions indicated. FERFA have recently produced two Flooring Guides, Application Guide No. 4, *Synthetic Resin Floor Screeds*, and Application Guide No. 6, *Polymer Floor Screeds*. These guides cover in considerable detail all aspects of laying high-performance resin and polymer floorings.[5]

9.1.3 Special finishes

In this chapter, the author has attempted to classify the various types of finishes available for concrete floors in terms of increasing applied costs.

9.2 Thin applied hardener/sealers

9.2.1 Sodium silicate and silico fluoride solutions as concrete surface hardeners

Both sodium silicate and silico fluoride solutions are applied to clean, dry, sound concrete floors as dilute aqueous solutions (10–15 per cent solids) in two to three applications, taking care to ensure that all material penetrates and is absorbed into the concrete surface. The silicate or silico fluoride reacts with the small amount of free lime in the cement to form glassy inert materials in the surface, and the successful application of both materials depends upon filling the micropores in the surface of good-quality concrete, leaving its surface appearance and non-skid characteristics virtually unchanged.

The main difference between the two types are that the reaction products of the silico fluoride types are less soluble in water and are also harder, which may give better in-service performance but at a slightly higher material cost. However, with recent developments in floor-laying techniques, the concrete substrates for industrial floors are laid with much more dense low-porosity surfaces, so that neither silicate nor silico fluoride treatments are as effective as they used to be, when the concrete used had a slightly more open finish and hence was more receptive to these treatments. With modern concrete floors, it is imperative to wash any material not absorbed into the surface within a short period. Otherwise, unpleasant white alkaline deposits, which are difficult to remove, may occur.

It is important to stress that neither sodium silicate nor silico fluoride will improve the pérformance of a poor, low-strength, dusty concrete floor and if the surface is too porous, there is no way that all the material applied can react with the relatively small quantity of free lime in the concrete surface. All that will happen is that the pores will be filled with non-reacted powder, producing a most unpleasant alkaline dust, which can be very irritating to the skin and eyes when the floor is put into service.

Finally, it is important to note that sodium silicate or silico fluoride treatments properly applied to clean and sound concrete floors can improve their performance, wear resistance and resistance to mild aqueous chemicals and oils, at a relatively low cost. However, they are not the answer to all industrial flooring problems, as many specifiers appear to believe.[17]

9.2.2 Low-viscosity resin-based penetrating in-surface finishes

Liquid resin-based systems which, like the chemical surface hardeners, penetrate into the surface of a concrete topping or directly finished slab and protect the acid-susceptible cement matrix from attack and, at the same time, strengthen the surface of the concrete are now being increasingly used. These in-surface seals leave the slip resistance of the concrete floor virtually unchanged but the treated floors are easier to clean and are more durable.

9.2.3 Non-reactive and semi-reactive resin solutions

Resin solution penetrating sealers are now available which, for very large warehouse floors, are comparable in applied costs with the concrete surface hardeners and are now being increasingly specified. Experience indicates that certain acrylic resin solutions are proving more durable and offer better protection to chemical and oil spillage than concrete surface hardeners. Acrylic resin solution sealers can markedly improve the abrasion resistance of concrete floors and have 'rescued' a number of poor-quality floors.

Other resin solutions, in white spirit or stronger solvent blends, used as penetrating floor sealers include:

1. Air-drying alkyds (similar to the resins in conventional gloss paints);
2. Styrene butadiene resins;
3. Urethane oils;
4. Styrene acrylates.

All such resin solutions are based on flammable solvents and are becoming increasingly less acceptable on health and safety grounds. There is therefore increased interest in water-based polymer dispersion floor sealers, but, to date, none offer the same improvement to flooring performance that some of the resin solutions can provide.

9.2.4 Polymer dispersions

Polyvinyl acetate (PVA), acrylic and other polymer dispersions have been widely used as anti-dust treatments for concrete floors for many years. In general, the polymer dispersions have been similar to those used in the manufacture of emulsion paints, and until recently have tended to be based on dispersions of relatively large polymer particles (particle size $0.15–0.25 \times 10^{-6}$ m). Dispersions are now becoming available which offer superior performance as floor sealers. The chemical and water resistance of the various polymer dispersions which have been used in the past vary considerably from the PVA types, which are rapidly softened and eventually washed out by water, to acrylic and SBR types which exhibit excellent resistance to a wide range of chemicals. Water-based sealers are gaining wider acceptance because of the increased handling problems associated with polymer solutions based on hydrocarbon solvents.

9.2.5 Epoxy resin dispersions

Two-component epoxy resin water thinned dispersions are now being used as floor sealers. They have good adhesion to concrete as well as good chemical resistance. However, the particle size of the dispersion is comparatively large (approximately 1–1.5 microns) and consequently penetration into good-quality concrete is minimal and an 'on-surface seal' is obtained. However, with porous low-quality concrete substances, considerable binding/strengthening, etc. of the surface can be achieved with water-dispersible epoxy resin-based floor sealer.

9.2.6 Reactive resin solutions

The two-pack low molecular weight epoxy resin systems in volatile solvents have proved very effective for improving the wear and chemical resistance of both good- and poor-quality concrete floors. The epoxy resin solutions (approximately 20 per cent solids) are high-strength systems, very similar to those used in heavy chemically resistant trowelled epoxy floors and, depending on the concrete, can penetrate a significant depth into the surface of the concrete, where the solvent evaporates and the resin cures to form a tough, chemically resistant seal, with a compressive strength of up to $70 \, N/mm^2$, thus reinforcing the concrete surface. One-pack low-viscosity resin solutions based on moisture-curing polyurethane systems are available which perform in a manner similar to epoxy resin solutions. Some of these polyurethane resin solutions demonstrate a greater ability to penetrate and bind the surface of suspect concrete floors measurably better than other penetrating sealers.[17,18]

9.3 Floor paints

Floor paints, in a wide range of colors and based on a number of different binder systems, are used extensively for concrete floors in light industrial applications.

9.3.1 Chlorinated rubber paints

Chlorinated rubber floor paints are probably the most common of the lower-cost floor paints on the market. They produce tough and chemically resistant coatings, but their adhesion to concrete is not always good. They tend to wear off in patches and cannot be considered as a durable floor treatment except under light traffic conditions. However, re-coating is a simple job and floors can easily be repainted over weekend shutdowns, for example. Similar paints based on other resins such as acrylics, vinyls and styrene butadiene are also used.

9.3.2 Polyurethane floor paints and multi-coat treatments

Solvents containing moisture-cured or two-pack polyurethane resin paints are also used extensively. They combine excellent abrasion resistance with good chemical resistance, and are normally applied in two coats to give a coating thickness of 0.10–0.15 mm. In addition, moisture-cured polyurethane resin solutions are used for quite thick durable decorative floorings.

Several coats of resin are applied to the prepared substrate at approximately 4- to 6-hour intervals, with one or more coats being dressed with colored paint flakes which are sealed in by the next coat and then lightly sanded. This type of flooring was widely marketed about ten years ago but, in the main, they were considered unsatisfactory due to rapid discoloration of the floor because of the lack of ultraviolet stability of the urethane resins used, which rapidly turned yellow-brown and looked dirty. However, ultraviolet-stable urethane resins that do not suffer this discoloration are now available, and this type of durable decorative flooring is gaining re-acceptance (for example, for kitchens, toilets and reception areas).

9.3.3 Epoxy resin high-build floor paints

Solvent-free high-build floor paints are available which can be readily applied with brush, roller or spray to a prepared concrete substrate to give a thickness of 0.10–0.20 mm per coat. Normally, two coats are applied and the first is often lightly dressed with fine sand or carborundum dust to give a non-slip, chemically resistant and durable colored floor, ideal for light industrial traffic conditions (for example, rubber-shod wheels).[6]

9.4 Self-leveling epoxy, polyester or reactive acrylic resin systems

Like the high-build epoxy paints, these are solvent-free low-viscosity systems which are readily applied onto a prepared level substrate to provide a jointless thin (thickness approximately 1.5 mm) chemically resistant flooring in a single application.[6]

The term *self-leveling* by which they are commonly described is something of a misnomer, as they require spreading out to a near-level finish with a squeegee or the edge of a steel trowel, and by themselves, they flow out to give a smooth finish. Perhaps a better description is *self-smoothing*. Before the system is cured, the surface is normally lightly dressed with fine abrasion-resistant grit. Without a non-slip dressing, there is a tendency to produce a slippery, very glossy surface, which shows every scratch mark. This can be overcome to some extent by careful formulating and by the application of a slip-inhibiting industrial floor polish on a regular basis when the floor is in service.

This type of flooring is widely used in laboratories, pharmaceutical factories and food-processing areas where easily cleaned, chemically resistant durable floors are required.

In recent years, more heavily filled flowing epoxy resin mortar flooring systems laid at 3 to 5 mm thickness are increasingly being used instead of the more traditional trowelled epoxy resin mortar flooring systems described below. The laying costs of the flowing mortars are significantly lower and do not require the same degree of skill to achieve a satisfactory floor finish.

Polyester resin systems and, more recently, acrylic resins are also used. Polyester resin-based systems have a tendency to shrinkage during and after application and application is very critical.

Acrylic resin systems developed in Germany are similar to polyester resins but, by careful formulation, the problems due to shrinkage have been largely overcome. The acrylic resin-based systems are currently based on highly flammable materials (flash point 10°C), which can present hazards during laying. However, there are systems available that can take foot traffic 2–3 hours after application and full service conditions within 24 hours, even at very low temperatures.

9.5 Heavy-duty flooring

A considerable range of different toppings is available for heavy-duty service. The correct selection of the most appropriate topping on a cost/performance basis can only be made if service conditions are very clearly defined. In general, heavy-duty toppings require a sound (preferably $35 \, \text{N/mm}^2$ strength) concrete substrate.

9.5.1 Granolithic toppings

In effect, granolithic toppings are just a method of producing a high cement concrete wearing surface on a concrete substrate. The application of separately laid granolithic toppings is always fraught with the danger of de-bonding and curling and, therefore, monolithic grano-toppings are generally essential. However, for many industrial floors, where good resistance to abrasion under heavy traffic is specified, a suitable floor could be achieved more economically by direct finishing of a high cement content, high-strength $(40–60 + \text{N/mm}^2)$ concrete. This was borne out by work carried out by Chaplin,[11] who found the abrasion resistance measured by a number of different methods to be directly related to the compressive strength of the concrete. This work also showed that the abrasion resistance and compressive strength could also be related to Schmidt hammer test results, which could be of considerable interest for the future non-destructive site testing of concrete floors.

Where it is considered essential to apply a granolithic topping onto an existing concrete substrate the danger of de-bonding can be much reduced by the use of polymer-based bonding aid. Two types of bonding aid are commonly used:

1. Epoxy resin adhesive specially formulated for bonding freshly mixed cementitious materials to well-prepared cured concrete substrates. With the right epoxy resin bonding aid, the strength of the bond achieved is greater than the shear strength of both the topping and the concrete substrate.
2. A bond coat of a polymer latex (also called polymer emulsions or dispersions) such as styrene butadiene (SBR), polyvinyl acetate (PVA) acrylics or modified acrylics. These are applied to the prepared concrete as neat coats of emulsion or, more commonly, as slurries with cement.

Polymer latex bonding aids are cheaper and more simple to use than epoxy resins and give a good, tough bond which is *less structural* than that achieved by the right epoxy bonding aid. The so-called *universal* PVA bonding aids are not recommended for external or wet service conditions, as there is a danger of the polymer breaking down.

9.5.2 Bitumen emulsion-modified cementitious floors

The use of specially formulated bitumen emulsions as the gauging liquid for graded aggregate/sand/cement screeds can produce a dustless, self-healing, jointless surface for industrial areas subject to heavy wheeled traffic under normally dry conditions. This type of topping is normally laid approximately 12 millimeters thick and has been used very successfully for more than 30 years, particularly in warehouses. The bitumen-modified cementitious floor topping is less hard underfoot than concrete, and has proved a very popular improvement with warehouse staff. However, with recent trends towards high-rise tracking, heavier forklift trucks and narrow aisles between racks, the topping tends to indent or shove, and the truck forks become misaligned with the pallets stacked on the higher shelves of the racks, so that it is not possible to get the goods down.

Loading levels above about $8 \, \text{N/mm}^2$ for short term and $4 \, \text{N/mm}^2$ for an indefinite period are, therefore, not recommended for bitumen emulsion-modified cementitious floors.

9.5.3 Mastic asphalt floors

Hot applied mastic asphalt floors have been used for many years in industrial environments, where a good degree of chemical resistance under normally wet conditions is required. Properly laid mastic floors are totally impervious to a wide range of chemicals but not solvents. In terms of mechanical performance, mastic asphalt floors are similar to the bitumen-modified cementitious floors, but they are generally laid at a minimum of 25 mm thickness and tend to shove and corrugate in service under heavy loads. Mastic floors are not very commonly used now, except where the floor is essentially tanked, such as car park decks over shopping precincts.

9.5.4 Polymer-modified cementitious floor toppings

Polymer-modified cementitious floor toppings are now widely used instead of separately laid granolithic toppings. The polymers used are normally supplied as milky white dispersions in water and are used to gauge a carefully selected sand/aggregate/cement mix as a whole or partial replacement of the gauging mortar. They must always be mixed in a forced-action mixer.

The polymer latex acts in several ways:

1. It functions as a water-reducing plasticizer, producing a flooring composition with good workability at low water/cement ratios.
2. It ensures a good bond between the topping and the concrete properly prepared.
3. It produces a topping with good tensile strength and toughness.
4. It produces (based on the right polymer latex) a topping with good water and chemical resistance.
5. It acts to a significant degree as an integral curing aid, much reducing the need for efficient curing. (Curing is, however, essential in dry, draughty conditions.)

Polymer-modified cementitious floor toppings are normally laid 6–12 mm thick. Two polymer latex types are most commonly used – styrene butadiene and acrylics – which have been specifically developed for incorporation into cementitious compositions. The principal difference between the two types is that acrylic lattices are available, which gave higher early strengths than can be achieved with the current SBR lattices. Toppings based on acrylic lattices are used in food-processing industries, particularly meat processing. Toppings can be laid in a Friday – Monday weekend shutdown and are reported to be capable of withstanding full service conditions 48 hours after laying, although longer cure periods are desirable.

When adequately cured, polymer-modified cementitious toppings based on acrylic/SBR lattices can be cleaned with steam-cleaning techniques without problems of thermal shock breakdown, which has been observed with other heavy-duty polymer toppings. They are resistant to many chemicals encountered in the food and printing industries but being based on an acid-sensitive cement matrix, their resistance to organic or mineral acids is limited. There are now becoming available special polymer powders derived from the polymer lattices which can be pre-blended with sand, aggregates, cement and other additives and then gauged on site with water to produce factory-quality controlled flooring compositions with performance similar to the materials based on the addition of the milky lattices on site.

9.5.5 Epoxy resin mortar floorings

Trowelled epoxy resin flooring approximately 6 mm thick is used extensively where a combination of excellent chemical resistance and good mechanical properties are required, particularly abrasion and impact resistance and resistance to very heavy rolling loads. Epoxy toppings are available with compressive strengths up to 100 N/mm^2 and tensile strengths up to 30 N/mm^2. This is achieved by careful formulation of the binder and the incorporation of high-strength blended fillers.[6]

When formulating a system for optimum abrasion resistance, both the epoxy/resin hardener binder system and the filler blends used appear to have an influence. The simulation of abrasive service loads on industrial floor toppings in a laboratory is not simple, and numerous wear test machines have been devised. Correlation between different wear test machines is not always good, although most laboratory tests on abrasion resistance give an indication of the floor's likely performance in service in a qualitative rather than quantitative manner.[10,11]

In one series of laboratory tests carried out to find the optimum wear resistance of heavy-duty epoxy resin flooring compositions, a number of different abrasion resistant materials were evaluated using BS 416, employing three different epoxy resin binders which themselves had significantly differing chemical compositions and mechanical properties. The results of this work, which was carried out under dry conditions, are given in Table 9.1. As can be seen from the table, the selection of the abrasion-resistant material and the resin matrix both influence the abrasion resistance of the system, although the abrasive material incorporated appears to play a more crucial role.

In wet abrasive conditions, which often occur with heavy-duty industrial flooring, a small quantity of abrasion-resistant material tends to be carried on the wheels of trucks and produces a grinding paste between the heavy-duty wheel and the surface. Since the abrasion-resistant material in the surface is generally harder than any sand or grit carried into the factory on wheels, the grinding paste tends to become more abrasive as the binder is worn away. Abrasion resistance tests under wet grinding paste conditions, however, do indicate a similar order of resistance, although the binder appears to play a more significant part. In applications where the flooring is flooded with water for long periods, the resin binder plays a more important part, since the strength of the adhesive bond between the particles of abrasion-resistant materials can, if the wrong resin binder system is used, drop markedly under prolonged wet conditions. In formulating resins for heavy-duty floors it would appear that the adhesive properties of the resin binder used to bond the resistant particles firmly together is the more important factor when selecting a resin system. In the selection of systems for highly abrasive service conditions, costs must also be considered and, on this basis, bauxite, calcined under defined temperature conditions, has often been used as the abrasion-resistant aggregate.

Table 9.1 Abrasion values of trowelled epoxy resin flooring compositions, using BS 416

Aggregate	Epoxy binder compositions	Mass loss after abrasion (g)
Graded sand, Grade C	A	4.10
Graded sand, Grade M	B	2.75
Graded sand, Grade C	C	2.85
Graded sand, Zone 2	A	5.5
Graded sand, Zone 3	A	5.7
Gritstone	A	9.95
Granite	A	1.45
Calcined bauxite	A	0.95
Basalt	A	1.5
Cast iron grit	A	0.45
Copper slag	A	2.25
Sand (Zone 1) gritstone (50/50 by mass)	A	1.35

Another aspect of epoxy resin mortar floorings which needs careful attention is that their coefficients of thermal expansion are approximately three times that of concrete. This, coupled with the relative low thermal conductivity of epoxy mortar, can cause stresses to be induced at the resin mortar/concrete interface under conditions of thermal shock (e.g. thermal cleaning), resulting in break-up of the flooring due to initial failure in the concrete. Two approaches have been tried to overcome this problem:

1. Using a lower modulus epoxy resin mortar and applying the topping at a thickness of 3–4 mm;
2. Applying a stress distributing flexible epoxy layer 1–2 mm thick between the rigid epoxy topping and the concrete.

Both approaches have been used with some success but in (1), the lower modulus topping also tends to have lower chemical resistance, which can be a problem, while technique (2) is significantly more costly in terms of both material and labor.[10]

9.5.6 Polyester resin mortars

Polyester resin floor toppings, similar in performance to the epoxy toppings, have been used but, as indicated earlier, polyester systems tend to shrink and, without careful formulation and laying, shrinkage stresses with polyester resin systems can develop at the interface between the topping and the concrete substrate. Coupled with the additional stresses due to the differences in their coefficients to thermal expansion, this can cause failure at the surface of the concrete substrate.[11,12]

Several years ago one company had considerable success with a carefully formulated polyester mortar topping specifically designed to minimize these stresses, but found that, unless it was laid with meticulous care, failures could occur.

Polyester resin mortars, however, cure within 2 hours of placing to give greater strength than concrete and are widely used for the rapid repair of small areas of damaged concrete floors and, with the use of 'igloos', even in cold stores in service. Another polyester resin-based heavy-duty topping that has proved very satisfactory in service is based on a unique approach. It comprises a blend of treated Portland cement and a dispersion of a special water-soluble catalyst system in an unsaturated polyester resin. This blend is mixed on site with graded aggregates and a measured quantity of water. The water addition dissolves the catalyst and, in the presence of free alkali from the cement, releases free radicals, which trigger the curing of the unsaturated polyester resin. The cured product gives a tough floor topping that, over the past ten years, has been widely used in abattoirs, dairies and food-processing plants. Recently, another system based on a similar approach has been introduced.

Polyurethane mortar flooring systems based on somewhat similar technology to this special polyester system have also been used in chemical plants and have given excellent service. The basic urethane polymer is more elastomeric than either epoxy or polyester resins and, as such, is reported to have excellent thermal properties up

to at least 140°C and good resistance to thermal shock. The adhesion of the urethane systems to damp concrete is, to some extent, suspect, and dry substrates are therefore normally essential, although systems with improved adhesion to damp substrates are becoming available.[14]

9.5.7 Reactive acrylic resins

Reactive acrylic resins similar to polyester resins are also being used increasingly in heavy-duty floors. Acrylic resins primarily based on methyl methacrylate monomer are low in viscosity and wet out fillers very efficiently, enabling the production of heavily filled flooring compositions, which are easily laid. The high filler loadings much reduce the danger of problems due to shrinkage. Most acrylic resin systems in the uncured state are highly flammable (flash point below 10°C) and special precautions need to be taken.

9.5.8 Industrial tile floorings

There are industrial flooring situations where the service requirement or the time allowed for laying does not permit the use of jointless floor toppings. For such applications a wide range of industrial tiles are available which will meet most requirements, in terms of either mechanical properties or chemical resistance. When tiles are used in very aggressive chemical environments, the main problem is grouting between the tiles with a grout having adequate resistance. Grout systems based on specially formulated furane resins (which, in particular, resist very strong acids) and epoxy resins are available for this purpose, and tiles laid in very fast-setting mortar bedding and properly grouted can be installed and returned to service in under 48 hours. A typical application for tiles is in dairies, many of which operate 364 days a year. By using a fast-setting fondue-based mortar bed bonded to the underlying substrate with an epoxy adhesive one day and then laying quarry tiles bonded and grouted with an epoxy resin system the following day, it is possible to repair a completely broken-down and impossible-to-clean floor to a good standard with no interruptions to production.

9.6 Comparative applied costs

It is difficult to give precise costs of floor treatments as size of total area, areas to be coated at one time, degree of surface preparation required and other factors all influence costs. Table 9.2 is a rough guide to comparative applied costs.

9.7 Conclusion

The range of special flooring materials available is very wide and, to many, extremely confusing. Specifiers, however, will find that if they carefully list all the service requirements for the flooring, members should be able to recommend and apply suitable products for most service conditions.

Table 9.2

Floor treatment/flooring	Comparative cost index
Concrete surface hardeners, two to three coats	1–1.8
In-surface seals:	
Resin solution non-reactive, two coats	1.2–2.5
Reactive resin solutions, two coats	2–3
Paints:	
Chlorinated rubber, one to two coats	1.7–3.2
Polyurethane, two coats	1.8–4
High-building epoxy, one to two coats	3.5–7
Multicoat polyurethane flake, four-plus	5–12
Epoxy self-levelling, 1–2 mm	10–16
Polyester, 2–3 mm	9–12
Bitumen-modified cementitious, 12–16 mm	7–10
Mastic asphalt, 25 mm	9–13
Polymer-modified cementitious, 12 mm	8–15
Epoxy trowelled, 6 mm	18–24
Polyester, 6–9 mm	15–20
Industrial tiles, various	15–30

References

1 This chapter is based on 'Special finishes for concrete floors', presented by J. D. N. Shaw at the International Conference on Advances in Concrete Slab Technology, Dundee University, April 1979. (Proceedings published by Pergamon Press, Oxford, 1980, pp. 505–515).

2 Barnbrook, G., 'Durable non slip concrete floors for low maintenance', *Building Trades Journal*, 12 January (1979).

3 Anon., 'Good warehouse floors don't just happen', *Concrete Construction*, 493–494, September (1977).

4 Barnbrook, G., 'The concrete slab – a base for resin flooring', paper presented at the Symposium on Resin for Industrial Problem Areas. Flooring and Anti-Corrosion Application, organized by the Federation of Resin Formulators and Applicators Ltd, London, November (1980).

5 FERFA, 241 High Street, Aldershot, Hants GU11 1TJ, Guide No. 4, *Synthetic Resin Floor Screeds*, Guide No. 6, *Polymer Floor Screeds*.

6 Phillips, G., 'Caring for concrete floors protects profits', *Building, Maintenance and Services*, 37–38, April (1978).

7 Berger, D. M., 'Preparing concrete surfaces', *Concrete Construction*, 481–494, September (1978).

8 Shaw, J. D. N., 'Epoxy resin based flooring compositions', *Flooring and Finishing News*, 7–11, January (1967).

9 Simmonds, L. B., 'Concrete sub-floor testing', *Architect and Building News*, July (1964).

10 Metzer, S. N., 'A better joint filler', *Modern Concrete*, 45–47, May (1978).

11 Chaplin, R. G., 'Abrasion resistant concrete floors', paper presented at the International Conference on Concrete Slabs, University of Dundee, April (1979).

12 Shaw, J. D. N., 'Epoxy resins for construction: recent developments', paper presented to the Japanese Society of Materials, November (1971).

13 Hewlett, P. C. and Shaw, J. D. N., *Developments in Adhesives*, Chapter 2, pp. 25–72, Applied Science, London (1977).

14 McCurrich, L. H. and Kay, W. M., 'Polyester and epoxy resin concrete', paper presented at the Symposium on Resins and Concrete, University of Newcastle-upon-Tyne, 17–18 April (1973).

15 Nutt, W. O., 'Inorganic polymer structures', British Patent No.1,065,053, April (1967).

16 Benson, L. H., 'Urethane concrete in the chemical industry', Acema Conference, Frankfurt, August (1973).

17 Shaw, J. D. N., 'Polymer concretes – UK experiences', paper presented at the Conference on Polymers in Concrete, Fourth International Congress, September (1984).

18 Sadegzadeh, M. and Kettle, R. J., 'Abrasion resistance of polymer impregnated concrete', *Concrete*, 32–34, May (1987).

19 FERFA Technical Report, *Osmosis in Flooring*, August (1989).

10

Lighting

Hugh King

Thorn Lighting

Contents

10.1 Lighting theory

10.1.1 Nature of light

Light is a form of electromagnetic radiation similar in nature and behavior to radio waves at one end of the frequency spectrum and X-rays at the other. The range to which the human eye is sensitive covers a waveband of approximately 380–760 nanometers (nm), a nanometer being a wavelength of one-millionth (10^6) of a millimeter. Within these limits there are differences in color, white light being that in which an approximately equal amount of power occurs at all frequencies within the waveband.

10.1.2 Frequency and color

The human eye is more sensitive to frequencies in the middle of the spectrum (green and yellow light) than those at the ends (red and violet). Consequently, more power must be expended to obtain the same effect on the eye from red or violet light than from green or yellow.

The regions of radiation just outside the visible spectrum are known as the ultraviolet (UV) and infrared (IR). Both are present in natural daylight and they are emitted in varying amounts by artificial light sources. IR gives the effect of heat, while UV radiation is used to excite fluorescence in the powders in fluorescent lamps. Short-wave UV causes skin tanning, can damage the eyes, but the long-wave UV, used for theatrical or display purposes, and in discotheques, is virtually harmless.

10.1.3 Basic lighting units and terms

The average illumination on a surface will diminish as the square of its distance from the source. This is known as the *inverse square law* and is common to all forms of radiation.

Light is *reflected* from a polished (specular) surface at the same angle as it strikes it. A matte surface will reflect it in a number of directions and a semi-matte surface will behave in a manner between the two.

Refraction occurs when light passes through a surface between two media (e.g. air and water, air and glass). The *ray* of light is bent and this property is used in the manufacture of lenses and prisms.

Diffusion is the scattering of light when it passes through an obscured medium (such as opal glass or plastic). The medium contains a number of particles of matter, which reflect and scatter the light passing through it. A uniform diffuser scatters the light equally in all directions, and partial diffusers allow some directional flow of light. All these properties are used to control the direction and distribution of light from artificial light sources.

Excessively bright areas in the visual field can, separately or simultaneously, impair visual performance (disability glare) or cause (visual) discomfort (discomfort glare). Lighting systems in most working interiors are unlikely to cause significant direct disability glare but a degree of discomfort glare is probable.

The unit of luminous intensity is the *candela*. This is the illuminating power of a light source in a given direction. The candela is the standard international (SI) unit.

The unit of light flux is the *lumen*. This is the amount of light contained in one stearadian from a source with an intensity of one candela in all directions. Alternatively, it is the amount of light falling on a unit area of the surface of a sphere of unit diameter from a unit source.

The unit of illuminance is the *lux*. This is the illumination produced by one lumen over an area of one square meter. The non-metric equivalent, the lumen per square foot (lm/ft^2), is still used in the UK and is called a *foot-candle* in the USA. It is approximately equal to 10 lux.

The units of measured brightness (luminance) are the *candela per square meter* and the *apostilb* (lumens emitted per square meter). The imperial units are the *candela per square foot* and the *foot-lambert* (one lumen emitted per square foot of surface). Note that *luminance* should not be confused with *illuminance*. Illuminance is the measure of light falling on a surface and luminance that of the light reflected from it (or, in some cases, emitted by it).

Two other terms, which can easily be confused, are *luminance* and *luminosity*. Luminance is used to describe the measured brightness of a surface and luminosity is apparent brightness.

Efficacy (or luminous efficiency) of lamps is measured in lumens per watt (lm/W).

10.1.4 Color appearance and color rendering

We see objects by the light reflected from them. If they have the property of absorbing some wavelengths of light and reflecting others, they will appear to be colored. Some colors are *pure*, that is, they consist of only one waveband of light. Others are obtained by mixing frequencies (for example, mixing red and green light gives the effect of yellow).

Obviously, only those colors, which fall on a surface, can be reflected from it. Consequently if it is lit by light of one color only (e.g. yellow light from a sodium lamp), it will appear to be that color. If the light falling upon it is deficient in some colors (e.g. red and blue light) those colors will appear weak or may not be seen at all. The eye is unable to distinguish between pure colors and those obtained from mixing, so that the *color appearance* of a light source (the apparent color of the light emitted from it) does not necessarily indicate its *color rendering* (the color of objects seen in its light).

One way of defining color appearance is by *color temperature*, measured in degrees Kelvin (K). Color rendering can be quantified by a general color-rendering index (RA), of which 100 is the best number to achieve.

10.2 Electric lamps

10.2.1 Light production

Light can be produced from electricity in three main ways:

1. *Incandescence* or thermoluminescence is the production of light from heat. Light from a filament lamp is produced in this way. Electricity is used to raise the temperature of the filament until it is incandescent.
2. *Electrical discharge* is the production of light from the passage of electricity through a gas or vapor. In lamps

using this principle, the atoms of the gas are agitated or excited by the passage of the electric current and this atomic excitation produces light and sometimes UV and IR energy.

3. *Phosphorescence* and *fluorescence* are the processes of converting invisible UV energy emitted normally from an electrical discharge into visible light. Material called phosphors cause UV energy to make this transition into visible light.

10.2.2 Incandescent lamps

The following abbreviations are used to define these lamps:

GLS General lighting service.
TH Tungsten-halogen.
PAR Followed by lamp nominal diameter in eighths of an inch. (Pressed glass filament lamps with an internal reflector coating.)
R Followed by lamp nominal diameter in millimeters (previously in eighths of an inch). (Blownglass filament lamps.)

Comparisons of luminous efficiency and light output are given in Figures 10.1 and 10.2.

GLS lamp

A tungsten-coiled filament is held by molybdenum filament supports and connected to lead wires. An electric current passing through the filament raises the temperature of the filament so it becomes incandescent, emitting light and heat. The outer glass bulb is usually filled with amounts of argon and nitrogen to inhibit the evaporation of tungsten and arcing across the filament. The bulb can be clear or internally etched to give a pearl finish, resulting in diffused light. The lamp has a suitable lamp cap for its size and wattage (see Figure 10.3).

Application

Domestic, commercial and industrial uses. It is the most familiar type of light source, with many advantages (for example, low initial cost, immediate light when switched on, good color rendering, no control gear, easily dimmed). For these reasons, there are numerous applications. The very low efficacy is often the factor that precludes the selection of GLS lamps.

Average life

1000 hours.

Similar lamps

> Double-life lamp with average life of 2000 hours
> Colored lamps
> Netabulb (mushroom lamp)
> Nightlight

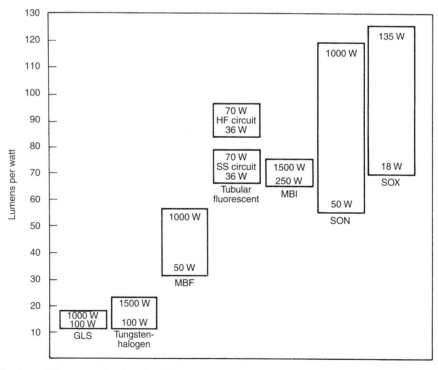

Figure 10.1 Luminous efficacy comparison based on initial lumens and total circuit watts

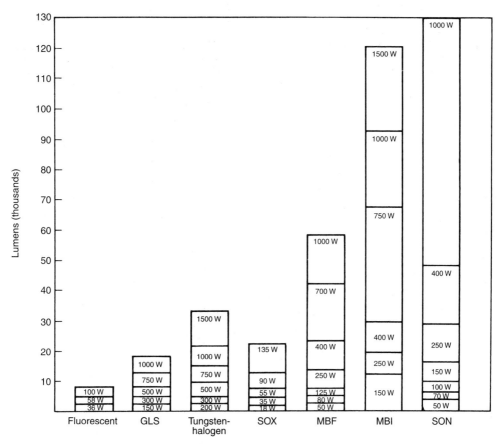

Figure 10.2 Light output comparison

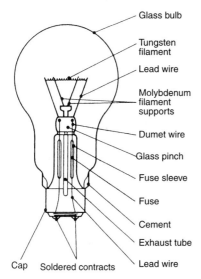

Figure labels: Glass bulb, Tungsten filament, Lead wire, Molybdenum filament supports, Dumet wire, Glass pinch, Fuse sleeve, Fuse, Cement, Exhaust tube, Lead wire, Cap, Soldered contracts

Figure 10.3 Typical construction of a GLS lamp

10.2.3 Fluorescent lamp MCF

Abbreviations used are:

MCF   Tubular fluorescent lamps.
MCFA   Tubular fluorescent lamp with earth strip cemented to outside of tube.

Restricted range, mainly for cold environments.

T12   38 mm (1½ in) nominal diameter.
T8   26 mm (1 in) nominal diameter.
T5   15 mm (⅝ in) nominal diameter.

Description

A low-pressure mercury discharge operates in a thin-walled glass tube that is internally coated along its entire length by a phosphor coating. This emits light from the UV energy generated by the discharge that strikes it. A wide range of shapes and sizes have power ratings from 4 to 125 watts (W). The phosphor determines the color of light produced and the color rendering properties of the lamp (Figure 10.4).

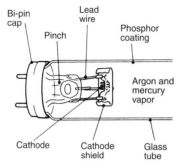

Figure 10.4 Typical construction of a fluorescent tube

Application

Worldwide, about 80 per cent of electric lighting is provided by fluorescent lamps. The applications are particularly diverse, operating in interior and exterior luminaries. The majority of office lighting is provided by fluorescent lamps and in industry, they can still be the best choice for new installations despite the range of high-pressure discharge lamps. High-frequency electronic ballast increases their efficacy and gives a more comfortable light by the absence of flicker. Tri-phosphors that combine high light output with good color rendering are recommended. Many other *white* lamps are made as well as colored ones for special applications.

Operating circuit

The basic circuit usually has power factor correction. Various types are used – high frequency electronic ballast, electronic starting circuits (Vivatronic) and switch-start circuits. Existing luminaries may have quick-start and semi-resonant start (SRS) circuits.

10.2.4 Compact fluorescent

Description

Compact fluorescent lamps save a lot of energy and give a much longer life than conventional filament lamps. Power consumption is about 25 per cent of an equivalent GLS lamp and average lamp life is five to ten times longer, lowering the cost of maintenance. The lamp runs cooler and gives light that has good color rendering properties by using tri-phosphors. They are also a compact alternative to conventional fluorescent tubes.

Application

A wide range of interior and exterior luminaries are made for commercial, leisure, civic, domestic and other uses. In terms of light output the Thorn 2D 16 W is comparable to a 75 W GLS lamp. As the lamps are cool running small luminaries can be designed in a great variety of materials.

Average life

8000 hours for 2D 16 W;
10000 hours for larger wattage versions.

The major lighting companies now market more powerful linear compact fluorescence. They are particularly suitable for 600×600 mm ceiling modules.

10.2.5 Low-pressure sodium lamps

Abbreviations used are:

SOX	Single-ended, U-shaped are tubes.
SLI	Double-ended. Linear are tube (now obsolete).

Description

SOX   A low-pressure sodium discharge operates in a U-shaped are tube enclosed in a tubular outer bulb which has an internal IR reflecting coating to provide thermal insulation. The discharge has a characteristic yellow color, and achieves a high efficacy around 160 1m/W for the highest wattage. Due to the monochromatic nature of the light output, correlated color temperature and color rendering index figures are not applicable.

SOX-E   The construction of SOX-E lamps is similar to standard SOX except that, by improving the thermal insulation, higher efficacies can be achieved. By operating lamps at lower wattage, significant power savings can be obtained with only a small reduction in light output. (The E suffix denotes economy.) Care should be taken in the disposal of SOX lamps.

Application

SOX and SOX-E   The high efficacy results in low power costs, particularly useful where long operating hours are involved. The main use is for roadways, security lighting, subways and footpaths where the poor color rendering is not so important. The time it takes the lamp to reach full brightness from switch-on (the run-up time) is around 10 minutes. In some applications, SON lamps may be considered as an alternative.

Operating circuit

Ballast circuit using leakage reactance transformer or electronic igniter and ballast. Power factor correction capacitor is included in both cases.

10.2.6 High-pressure sodium lamps

Abbreviations used are:

SON	
SON-E	Diffused ellipsoidal outer bulb, single ended.
SON-T	Clear tubular outer bulb, single ended.
SON-TD	Clear tubular outer bulb, double ended.
SON-R	SON with internal reflector.
SON DL-E	SON with improved color rendering.
SON DL-T	SON-T with improved color rendering.
⚠	For use with external starting device.
⚠	Contains internal starting device.

Description

SON-T   A high-pressure sodium discharge operates in a sintered alumina arc tube contained in a clear tubular-shaped bulb. This high-efficiency light source gives a long life while maintaining the lumen output well. The warm colored light allows some color discrimination but not accurate color rendering. As the arc tube is small, precise optical control can be achieved by suitable luminary design.

SON-E   This is similar but has an elliptical outer bulb with a diffuse internal coating.

SON-TD   A high-pressure sodium discharge operates in a sintered alumina arc tube contained in a clear quartz sleeve with lamp caps at each end.

SON-R   A 70 W high-pressure sodium arc tube is mounted axially inside a 96 mm diameter soft glass reflector bulb, which has a diffuse front to remove striations from the beam.

Application

SON-T/SON-E   Suitable for many interior and exterior industrial applications, floodlighting, street and precinct lighting. It is now the most common form of street lighting for city centers. The warm color of SON frequently improves the appearance of stone and brickwork. The long life combined with good lumen maintenance throughout its life makes the lamp suitable where access is difficult or expensive, as may be the case in lofty industrial interiors such as steelworks, power stations and aircraft hangars. A wide range of lamp wattage is available, from 50 to 1000 W. Energy savings and reduced maintenance can often result in the refurbishment of industrial installations. Another aspect of SON is the ability to operate at low temperatures ($-40°$C).

SONDL-T   This is a high-pressure sodium lamp with a discharge arc that operates at a higher pressure and temperature than is the case with the standard SON-T lamps, resulting in a great improvement in the color-rendering properties. It is mainly used in commercial interiors, leisure centers and institutional and social areas. Swimming pools and sports halls are common applications, which are often the same scale as the industrial interiors for which these lamps are also used. Another frequent application is with up-lighters in offices.

SON-TD   The lamp is specifically for use in floodlights where precise optical control can be attained. Floodlights can be compared because the lamp is slim and if the control gear is mounted remotely the floodlight can be relatively low weight and easy to handle.

SON-R   Typical applications include lobbies, foyers, bars and clubs where the warm color can create a warm and informal atmosphere with a reduced heat load when compared against conventional down-lights.

SON lamps

Operating circuit
Ballast circuit with power factor correction capacitor and electronic igniter. Lamp uses an external igniter.

10.2.7 Linear tungsten-halogen lamps

Description

A tungsten filament is supported between molybdenum foils connected to contacts in ceramic caps at each end of a slim quartz-enclosed tube. The gas filling inside the tube includes halogen, which regenerates the loss of tungsten from the filament. It operates at a higher temperature and in a gas at a higher pressure than GLS lamps. The lamp provides a whiter light and improved efficacy. K-type lamps have double the average life of a standard GLS lamp.

Application

Display, exhibition, commercial, photographic and car lighting, and security and floodlighting. It is suitable where the requirement is for immediate light or good color rendering with a warm color appearance. The initial capital cost is low. The lamp is also used in display lighting because the source is compact and strong modeling and *sparkle* can be produced. Since there is no control gear required, small luminaries can be made for tungsten halogen lamps.

Average life

4000 hours for 100–3000 W (K type);
2000 hours for 500–2000 W (K type).

10.2.8 Low-voltage tungsten-halogen lamps

Description

A compact tungsten-halogen lamp operating at low voltage (12 V) is combined with a precision-faceted glass reflector. Due to its diachronic coatings, the reflector allows about 50 per cent of the heat to be transmitted backwards while reflecting light forwards in a controlled beam. A choice of beam angles is available. The lamp provides a white light, improved efficacy and beam control, and longer average life than conventional display lamps. Cooler interiors, reduced maintenance and running costs can result when compared against conventional PAR 38 display spots. Clip-on color filters extend the scope for display purposes.

Application

These lamps are suitable for display, exhibition, museum, commercial and domestic lighting. Their compactness has enabled diverse spotlight designs, some incorporating the necessary transformer to step down the voltage to 12 V.

Average life

3000 hours.

Operating circuit

Wire-wound transformer or electronic transformers for connection to 240 volts supply.

10.2.9 High-pressure mercury lamps (Figure 10.5)

Abbreviations used are:

MBF   High-pressure mercury with phosphor coating.
MBFR   MBF with internal reflector.
MBTF   Combination of MBF lamp and filament lamp.

Description

MBF   A high-pressure mercury vapor discharge operates in a quartz arc tube. The internal surface of the outer elliptical bulb is coated with a phosphor, which converts UV radiation from the discharge into light.

MBFR   This is an identical lamp except the outer bulb is shaped to form a reflector. Titanium dioxide is deposited inside the conical portion of the outer bulb and this deposit is coated with a phosphor. About 90 per cent of the light is emitted through the front face of the lamp in a directional manner.

Application

MBF   Used in industry for low initial cost where color rendering is not a major factor. Also employed in street lighting, mainly for replacement or matching existing installations.

MBFR   Suitable for industrial applications where minimum maintenance is important (for example, in a foundry or processing plant with a dusty atmosphere). Dirt or a corrosive atmosphere does not affect the internal reflector.

The outer bulb, which is made of hard glass, allows exterior use.

Both types of high-pressure mercury lamp have a life-survival characteristic similar to SON lamps but MBF lumen maintenance reduces more throughout its life.

Operating circuit

Ballast circuit with power factor correction capacitor. No igniter is needed.

MBFSD   A high-pressure mercury super deluxe lamp that is identical to the MBF lamp except that it has a different phosphor coating to create more red light. The color rendering is improved and the color temperature reduced.

MBTF   This is similar to the MBF high-pressure mercury lamp but with a coiled tungsten filament connected to the arc tube that acts as ballast within the outer glass bulb. Therefore, no external control gear is needed.

Application

MBFSD   Suitable for commercial interiors such as shops, supermarkets, public concourses and reception foyers. It is capable of blending well with 'white' fluorescent tubes in adjoining areas.

MBTF   A direct replacement for tungsten filament lamps giving a higher light output and longer life with a cooler color appearance. The tungsten filament emits light immediately when switched on. The efficacy is higher than an equivalent tungsten filament GLS lamp.

Operating circuit

   MBFSD Ballast circuit with power factor correction capacitor. No igniter is needed.
   MBTF No control gear is needed.

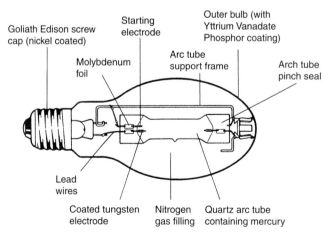

Figure 10.5 Typical construction of a mercury lamp

10.2.10 Metal halide lamps

Abbreviations used are:

MBI   High-pressure discharge with metallic halides.
MBIL   Double-ended linear arc tube.
MBIF   MBI with phosphor coating.
CSI   Compact source discharge with metal iodides.

Description

MBIF   A high-pressure discharge in mercury vapor with metallic additives operates in a quartz tube. The metallic additives are introduced as halide compounds, which control the dosing and ensure that the metallic elements mix well with the mercury vapor. An elliptical outer bulb is internally coated with phosphor.

MBI   This lamp is similar but has a clear outer bulb.

Application

This type of lighting is ideal for commercial and industrial interiors and exterior applications where good color quality white light is required. The efficacy is better than the same wattage of MBF lamps. These lamps are suitable for televising events when the appropriate illuminances are provided. Up lighting is a typical commercial use when the white quality of light is important.

Operating circuit

Ballast circuit with power factor correction capacitor and igniter. The lamp uses an external igniter.

Lamp orientation

BUH   Base up to horizontal. Operates in any position between cap up and cap 15° below horizontal.

H horizontal   Primarily designed with cap ±15° of horizontal, but can be used up to ±60° of horizontal BU Base up. Cap up ±30°C.

Description

MBI-T   A high-pressure mercury vapor discharge with metal halides operates in a quartz bulb. These halides are chosen to give high efficiency and good color-rendering properties. The electrode spacing is only 6 mm, so the arc is extremely compact, making it suitable for precise optical control in small luminaries. An outer quartz bulb gives thermal stability and protection. This bulb is mounted in a ceramic bi-pin cap.

Application

This type of lighting is suitable for display lighting, up lighting and floodlighting. A choice of color temperatures (3000 K and 4000 K) can blend in with tungsten-halogen and selected fluorescent lamps. Efficacy is around four times higher and the average life three times longer (at 6000 hours) than comparable tungsten-halogen lamps.

This makes it an attractive alternative to some former tungsten-halogen applications.

Operating circuit

Ballast circuit with power factor correction capacitor and electronic igniter.

Lamp orientation

BDH   Base down or horizontal.

10.2.11 Lamp selection

1. First decide what minimum standard of color rendering is needed. Except in the case of specialist applications, the CIBSE/CIE color-rendering classes (Table 10.1) are ideal for this purpose.
2. Do not select lamps from the appropriate class in the color-rendering guide shown in Table 10.2. Lamps in a better class may be more efficient or preferable for some other reason. Instead, make a note to reject any lamps that are in inferior classes.
3. Now decide which color appearance groups will be unacceptable, e.g. note that some applications that call for precise color judgement must use lamps with a specific color appearance (often cool or cold). Make a note of what lamp appearances to reject from the color-appearance guide in Table 10.3.
4. Decide how the lighting will be controlled and how rapidly light will be required. This can be compared to the run-up and re-strike times of different lamps to determine which are unacceptable (auxiliary lamps can be used for rapid illumination). Make a note of the lamps that are unacceptable from the run-up and re-strike times shown in Table 10.4.
5. By deleting lamps which have unsuitable color appearance and color-rendering properties and those with unacceptable run-up and re-strike characteristics,

Table 10.1   CIBSE/CIE color-rendering classes

Minimum class needed	Description
1A	Excellent color quality. Where accurate color matching is required (e.g. color printing inspection)
1B	Very good color quality. Where accurate color judgement or good color rendering is required for reasons of appearance (e.g. merchandising)
2	Good color quality. Where moderate color rendering is required, good enough for merchandising
3	Poor color quality. Where color rendering is of little importance. Color can be distorted but marked distortion is not acceptable
4	Very poor color quality. Color rendering is of no importance and severe distortion of colors is acceptable

Table 10.2 Color-rendering guide

Class	Type	Thorn description
1A	MCF	Northlight
		Artificial daylight
		De-luxe natural
	TH	Tungsten-halogen
	GLS	General lighting service
1B	MCF	Kolorite
		Polylux 4000
		Polylux 3500
		Polylux 3000
		Polylux 2700
		Rosetta
	CSI	Compact source iodide
	MBI-T	Arcstream 4000
		Arcstream 3000
2	MCF	Natural
		De-luxe warm white
	MBI/MBIF	Metal halide
	MBIL	Linear metal halide
	SONDL	SON de-luxe
3	MCF	Pluslux 4000
		Pluslux 3500
		Pluslux 3000
		Cool white
		White
		Warm white
	MBFSD	High-pressure mercury–super de-luxe
	MBF/MBFR	High-pressure mercury
4	SON	High-pressure sodium
	SON-R	SON
	SON-TD	Reflector/double ended
None	SOX	Low-pressure sodium

Table 10.3 Color-appearance guide

Color temperature (K)	Type	Description
		Cold
6500	MCF	Northlight
		Artifical daylight
		Intermediate–cool
5200	MBIL	Linear metal halide
4000	MCF	Kolorite
		Natural
		Polylux 4000
		Pluslux 4000
	MBI/MBIF	Metal halide
	MBI-T	Arcstream 4000
	CSI	Compact source iodide
		Intermediate
3800	MBF/MBFR	High-pressure mercury
3500	MCF	Polylux 3500
		Pluslux 3500
		White
		Warm
3300	MBFSD	High-pressure mercury super de-luxe
3200	MCF	Rosetta
3300	MCF	Polylux 3000
		Pluslux 300
		De-luxe warm white
		Warm white
2900	TH	Tungsten halogen
	MBI-T	Arcstream 3000
2700	MCF	Polylux 2700
	GLS	General lighting service
2200	SONDL	SON de-luxe
2200	SON	High-pressure sodium
None	SOX	Low-pressure sodium

a list of suitable lamps can be compiled. It is possible to rank this list in terms of efficacy, life and other factors, which the designer considers important so that a decision can be made.

6. One of the other factors which may be a limitation to the use of certain lamp types is the ambient temperature.

10.3 Luminaires

10.3.1 Basic design requirements

The majority of light sources emit light in all directions, but for most purposes, this is both wasteful and visually uncomfortable. The function of most luminaires is therefore:

1. To redirect the light of a bare electric lamp in a preferred direction with the minimum of loss;
2. To reduce glare from the light source;
3. To be acceptable in appearance and, in some cases, contribute to the decor of the surroundings.

In addition to optical design, other aspects must be considered:

1. *Lamp protection*: Apart from requiring electrical connections, the electric lamp must be mechanically supported and protected. The extent of this protection will depend on the application for which the luminary is intended.
2. *Electrical safety*: The lamp holder and associated wiring must be supported and protected, as well as the control gear if used to operate the lamp.
3. *Heat dissipation*: The heat generated by lamps and control gear must be conducted away from heat sensitive parts and surfaces adjacent to the luminary.
4. *Finishing*: Any exposed metalwork must be painted or plated to protect it against corrosion. The degree of protection necessary will depend on the luminary's application. Some *proof luminaires* designed for use in particularly hostile, corrosive atmospheres may be made in plastic materials rather than metals.

In addition to the functional aspects of mechanical, electrical thermal and chemical protection, the designer will also select materials and processes to achieve the desired external form and appearance of the design,

Table 10.4 Run-up and re-strike times

Type	Rating (W)	Run-up time (min)	Re-strike time (min)
TH	All	Instant	Instant
GLS	All	Instant	Instant
MCF	All HF	Instant[a]	Instant[a]
	All others less than 1[a]		Less than 1[a]
CSI	1000	1	10
MBI-T	150	1	4
MBI	250	2	7
	400	2	7
	1000	2	7
MBIL	750	2	8–12[b]
	1500	2	15–20[b]
SON, SON-R	50	3	Less than 1
and SON-TD	70	3	Less than 1
	100	4	Less than 1
	150	6	Less than 1
	250	6	Less than 1
	1000	6	3
MBIF, MBFSD	50	5	4
and MBFR	80	3	4
	125	3	4
	250	4	4
	400	4	4
	700	3	6
	1000	2	7
SONDL	150	8	Less than 1
	250	8	Less than 1
	400	7	Less than 1
SOX and SOX-E	18	12	Instant
	35	9	Instant
	55	9	Instant
	90	9	10
	135	8	10

[a]Striking time.
[b]In Floodlight.
HF: High-frequency circuit.

Figure 10.6 A selection of modern luminaires. Included are two SON floodlights; a linear tungsten-halogen floodlight; two compact fluorescent bulkheads and two low-voltage spotlights

bearing in mind production costs for either small or large batch quantities. Consideration must also be given to *built-in* design features aimed at making the installation and maintenance of the luminaires easier (Figure 10.6).

10.3.2 Luminaires and lighting terms

There is an abundance of lighting terms in common use and the following list attempts to cover the main ones.

Accent lighting See *display.*

Air terminal devices The means by which air is supplied or extracted from a room. This term applies to all such equipment whether called air diffusers or grilles.

Airfield lighting This is covered by a range of specialized luminaries intended primarily as an aid to navigation and includes approach lighting (line and bar, VASI and PAPI), runway and taxiway lighting.

Amenity lighting This is most commonly applied to outdoor public areas and is primarily intended to provide safe movement. It can incorporate decorative elements, both in terms of the appearance of the equipment and the lighting effect produced. In its most utilitarian form it is also applied to exterior private industrial areas intended mainly for pedestrian use.

Barn doors Four independently highed matte-black flaps mounted on a rotate frame solely intended for reducing spill light or shaping the soft-edged beam of a spotlight.

Batten This refers to a fluorescent luminary with no attachments; just the control gear in a channel and a bare tube. It is often designed to take various attachments as the basis of a complete range.

Bulkhead fittings These luminaries are often compact, designed for horizontal or vertical mounting outdoors or indoors. A wide selection of lamps is used, including miniature fluorescent tubes and high-pressure sodium lamps. The front is sometimes prismatic of either plastic or glass and the fittings are usually intended for exterior use over doors and stairways.

Ceiling systems A grid and panel system providing a false ceiling to incorporate lighting and other building services.

Columns Poles for mounting road lighting lanterns or floodlights usually from 3 to 12 m constructed in steel, aluminum or concrete.

Commercial These luminaires are designed to have a clean shape and efficient design but are devoid of decorative treatment. The term is applied to interior fluorescent diffuser luminaires as well as tungsten and discharge units for similar applications. There is frequently a gray area between this term and decorative or display luminaires.

Controller A clear plastic or glass with prism shapes in the surface. The overall shape may be flat or formed. This is distinct from opal plastic or glass diffusers.

Dark light See *Down lighters.*

Decorative (1) Decorative 'fittings' designed primarily to have a pleasing appearance, efficiency not always being of major importance. (2) Decorative *lighting*, however, refers to the effect produced and may require optically efficient fittings, which may or may not have an external decorative appearance.

Diffuser An attachment intended to reduce the lamp brightness by spreading the brightness over the surface of the diffuser. Many types of fluorescent, tungsten and discharge lamp luminaries use diffusers.

Display Lighting intended to attract attention and show objects to good effect.

Domestic fittings Strictly, equipment intended to light the private home, and as such it includes decorative fittings.

Down lighters This term is frequently misused to refer to any luminaries which produces no upward light, whereas the object of a down lighter is to provide illumination without any apparent source of light. They are therefore extremely *low-bright* luminaries, referring particularly to small (usually circular) recessed, semi-recessed or surfaced 'can' luminaries with mains or low-voltage tungsten reflector lamps. Some high-pressure discharge versions are also available.

Egg-crate See *Louvers*.

Emergency lighting Lighting provided for use when the main lighting fails for whatever reason. There are two types: (1) Escape lighting; (2) Standby lighting.

Emphasis lighting See *Display*.

Escape lighting See *Emergency lighting*.

Eyeball fitting A recessed adjustable display spotlight.

Flameproof An enclosure capable of withstanding the pressure of an internal explosion and preventing transmission of the explosion to gases and vapors outside the luminaries.

Floodlight A very general term usually applied to exterior fittings housing all types of lamps and producing beams from very narrow to very wide. They usually employ a specular parabolic reflectors, and their size and shape is determined by the type of lamp employed and the distribution of light required.

Fresnel lantern A soft-edge spotlight with a stepped front lens which gives a wide range of beam-angle adjustments by moving the lamp and reflector relative to the front lens.

General lighting An interior installation designed to provide a reasonably uniform value of horizontal illumination over a complete area.

High bay As the name implies, these are for use when mounting heights of around 8–10 m or above are encountered. They have a controlled light distribution to ensure that as much light as possible reaches the working plate. Cut-off angle is 60–65°. Although primarily intended for industrial areas, they also find application in lofty shopping areas and exhibition halls.

High mast lighting This was originally developed as a system for lighting exterior parking areas and complex road intersections using high-pressure discharge lamps in luminaries very similar to industrial high bay reflectors at heights of 20–50 m. Since then, many installations using directional floodlights have been installed and the same term is used for this.

Hood A baffle projecting forward from the top of the floodlight or display spotlight to improve the cut-off of spill light above the horizontal to reduce glare and light pollution.

Indirect lighting System of illumination where the light from lamps and luminaires is first reflected from a ceiling or wall. Traditionally, tungsten or fluorescent lamps were mounted in a cove and cornice to light a ceiling or behind a pelmet or in *wall washer* units to light walls. For modern high-efficiency applications of this technique, see *Uplighting*.

Industrial These luminaires are designed for high optical efficiency and durability. They range from simple fluorescent reflector luminaires to complex proof luminaires housing all modern light sources for both interior and exterior use.

Inverter Electronic device for operating discharge lamps (usually fluorescent) on a d. c. supply obtained from either batteries or generators.

Lantern Apart from reproduction period lighting units, this is an alternative terms for *luminaires* or fitting, usually restricted to road lighting and stage lighting equipment.

Local lighting Provides illumination only over the small area occupied by the task and its immediate surroundings.

Localized lighting An arrangement of luminaires designed to provide the required service illuminance over the working area together with a lower illuminance over surrounding areas. This is an alternative system to installing *general lighting*. See also *Uplighting*.

Louvers Vertical or angled fins of metal, plastic or wood arranged at right angles to linear lamps or in two directions. The object is to increase the cut-off of the luminaries and so reduce glare from critical angles. Louvers can also form ceiling panels with lighting equipment placed in the void above.

Low bay Luminaries housing high-pressure discharge lamps (usually mounted horizontally) to provide a wide distribution with good cut-off at mounting heights around 4–8 meters. Apart from industrial applications, these are used in many sports halls and public concourses. They are often fitted with louvers or clear visors.

Low brightness This term is usually applied to commercial fluorescent luminaires where, by use of specular reflectors or low-reflectance louvers, the brightness of the luminaries is strictly limited above 50° from the vertical. They are particularly appropriate for offices with VDUs or if it is wished to make the luminaries in an interior particularly unobtrusive (see *Downlighters*).

Maintained emergency lighting An emergency lighting circuit which remains energized at all times both when the main supply is available and when it has failed for any reason.

Mast Mounting columns higher than 12–15 m are generally referred to as masts and are used to support floodlighting for lighting large areas. For heights above 30 meters, access ladders are usually mounted inside the shaft and the base compartment is large enough to house control gear.

Optic The reflector and/or refractor system providing the light control for the luminaire.

Perimeter lighting See *Wall washer* and *Security lighting*.

Post top lantern Road or amenity lighting luminaire which mounts directly onto the top of a column without a bracket or out-reach arm. These are available for all lamp types with systematical and asymmetric distributions for mounting heights from 3 to 20 meters. The external design can vary from purely decorative to highly functional.

Precinct lighting See *Amenity*. This is generally associated with public rather than private areas.

Presence detector A device activated either by sound or heat energy to detect the occupation of an area or an intruder and thereby control the lighting approximately.

Projector A term often applied to any type of floodlight. More correctly it is an alternative to *profile spotlight*.

Proof Applied to all types of luminaires, which have a higher degree of protection from the ingress of solids and liquids than standard interior lighting luminaires.

Rack lighting Warehouse lighting where goods are stored on racking systems. The lighting must provide adequate vertical illumination on the faces of the racking which form narrow gangways.

Recessed The luminaire mounting arrangement where the whole or part (semi-recessed) of the luminaire body is set into a ceiling or wall or floor surface. With modular ceilings the luminaire fills the space provided by the removal of one or more tiles, but the term applies equally to where any type of luminaire is set into a false ceiling, wall or floor or solid structural slab.

Refractor Clear glass or plastic, panel or bowl where an array of prisms is designed to redirect the light of the lamp into the required distributions.

Residential lighting An installation for subsidiary roads and areas with little or no vehicular traffic except that generated by the residents themselves.

Road lighting An installation intended primarily to serve the needs of drivers. Although this is usually associated with public motorways, trunk and feeder roads and tunnels, the same principles apply to private, industrial and commercial roadways with vehicular traffic.

Safety/security lighting An installation designed to protect property, plant and people from criminal activity and to assist guards in detecting the presence of intruders. In the case of prisons, the objective is reversed, in that the aim is to prevent escape. In many cases, other detection devices, such as closed-circuit television and vibration detection, are used in conjunction with the lighting. All security lighting tends to be concentrated around the perimeter of the site and it is therefore sometimes referred to as *perimeter* or *perimeter fence* lighting. In certain high-security areas, invisible IR systems are used and linked with closed-circuit television facilities.

Semi-recessed See *Recessed*.

Spine See *Batten*.

Spotlight An adjustable interior luminaire with controlled beam using a reflector lamp or optical system (see *Display, Floodlight, Fresnel, Profile*).

Stage lighting An installation designed to enable an audience to view a *performance*, ranging from legitimate theatre in all its forms to a lecture.

Standby lighting See *Road lighting*.

Studio lighting Lighting for film and television has similar requirements for *Stage lighting* but involves a live audience less often. The higher lighting levels dictate the use of 5–10 kW lamps.

Tower Lattice, steel or concrete structure from 15 to 50 meters to support area floodlighting equipment (see also *Mast*).

Track system A linear busbar system providing one to three main circuits or a low-voltage supply to which display lighting can be connected and disconnected at will along the length of the system. The luminaries must be fitted with an adapter to suit the particular track system in use.

Troffer Fluorescent or discharge luminaries designed to recess into suspended ceilings and fit the module size of the ceiling (see *Recessed*).

Trunking Apart from standard wire way systems in ceilings and floor, trunking, associated specifically with lighting, usually provides mechanical fixings for the luminaires as well as electric connection. These are favored where there is a lack of fixing points for separates luminaires and the trunking can span between trusses (to carry the wiring and support the luminaries). They are also used where a simple means must be provided for repositioning lighting equipment with changes in use in the area and where track systems would not be appropriate or where additional support for suspended ceilings is required.

Uplighting Indirect lighting system where task illuminance is provided by lighting reflected from the ceiling. When used as a functional rather than a decorative system it usually employs high-pressure discharge lamps and high-efficiency reflector systems. The luminaires can be floor standing, wall mounted or suspended, and can be arranged to provide either general or localized lighting. They are particularly suitable for areas with VDUs although they have many other applications.

VDU lighting See *Low brightness* and *Uplighting*.

Visor Clear or diffused glass or plastic, bowl or flat panel closing the mouth of a luminaire. It is fitted either to protect the lamp or to prevent falling debris from the luminaire. In the case of diffusing visors, the object is to reduce lamp brightness and, to a certain extent, reduce glare.

Wall washer Interior floodlight or spotlight with asymmetric distribution intended to provide uniform lighting of walls from a close offset distance. Units may be recessed or surface mounted and house tungsten or sometimes high-pressure discharge lamps, mounted in a pelmet unit. This is commonly used to light shelf units in supermarkets and other large stores.

Wellglass These consist of a lamp surrounded by an enclosure of transparent or translucent glass or plastic. They are usually proof luminaires and often used outdoors fixed to a bracket.

Zones 0, 1 and 2 Classification of various hazardous areas.

10.3.3 The International Protection (IP) Code

This code is now widely used and classifies luminaries under two headings:

1. Protection against electric shock and the ingress of solid foreign bodies, including fingers or tools at one end of the spectrum and fine dust at other;
2. Protection against the ingress of liquids.

An IP number consists of two numerals, the first in the left-hand column and the second in the right-hand column of Table 10.5. For example, a dust-proof luminaire which can be hosed down would carry the number 55 and the symbols shown in Figure 10.7. A rainproof fitting with protection from finger contact within the enclosure would bear the number 10 and only the symbol in the right of the figure. The higher the numerals of the first and second characteristic, the greater the degree of protection the enclosure offers.

10.3.4 Explosion hazards

If a luminaire is to be used in an explosive atmosphere protection over and above that described in the IP code

Table 10.5 Defining IP numbers for luminaries

Protection of persons against contact with live or moving parts inside the enclosure and protection of equipment against ingress of solid foreign bodies. Protection against contact with moving parts inside the enclosure is limited to contact with moving parts inside the enclosure that might cause danger to persons		Protection of equipment against ingress of liquid	
First Characteristic numeral	*Degree of protection*	*Second characteristic numeral*	*Degree of protection*
0	No protection of persons against contact with live or moving parts inside the enclosure. No protection of equipment against ingress of solid foreign bodies	0	No protection.
		1	Protection against drops of condensed water. Drops of condensed water falling on the enclosure shall have no harmful effect.
1	Protection against accidental or inadvertent contact with live or moving parts inside the enclosure by a large surface of the human body, e.g. a hand, but not protection against deliberate access to such parts. Protection against ingress of large solid foreign bodies.	2 ♦	Protection against drops of liquid. Drops of falling liquid shall have no harmful effect when then enclosure is tilted at any angle up to 15° from the vertical.
		3 ⟦♦⟧	Protection against rain. Water falling in rain at an angle equal to or smaller than 60° with respect to the vertical shall have no harmful effect.
2	Production against contact with live or moving parts inside the enclosure by fingers. Protection against ingress of medium-size solid foreign bodies.	4 ⚠	Protection against splashing. Liquid splashed from any direction shall have no harmful effect.
3	Protection against contact with live or moving parts inside enclosure by tools, wires or such objects of thickness greater than 2.5 mm. Protection against ingress of small solid foreign bodies.	5 ⚠⚠	Protection against water jets. Water projected by a nozzle from any direction under stated conditions shall have no harmful effect.
		6	Protection against conditions on ship's decks (deck watertight equipment). Water from heavy seas shall not enter the enclosure under prescribed conditions.
4	Protection against contact with live or moving parts inside the enclosure by tools, wires or such objects of thickness greater than 1 mm. Protection against ingress of small solid foreign bodies.	7	Protection against immersion in water. It shall not be possible for water to enter the enclosure under stated conditions of pressure and time.
		8	Protection against indefinite immersion in water under specified pressure. It shall not be possible for water to enter the enclosure.
5 ✺	Complete protection against contact with liver or moving parts inside the enclosure. Production against harmful deposits of dust. The ingress of dust is not totally prevented, but dust cannot enter in an amount sufficient to interfere with satisfactory operation of the equipment enclosed.		
6 ◈	Complete protection against contact with live or moving parts inside the enclosure. Protection against ingress of dust.		

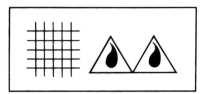

Figure 10.7 IP symbols

is needed. Explosive atmospheres are a potential problem in various industries where flammable materials are manufactured, processed or stored.

Electrical regulations over the years and, more recently, the Health and Safety at Work, etc. Act 1974, stipulate that any possible danger to safety be recognized and the necessary action taken to ensure adequate protection of the workforce. This places considerable responsibility on the user and installer, and it is most important that the correct type of luminaire be selected and that the installation is in accordance with the recognized safety standard.

Unlike the IP code, the equipment is classified according to the nature of the area in which it is to be, known as the *zonal classification*. The latest UK Code of Practice (BS 5345) covers the selection, installation and maintenance of electrical apparatus for use in potentially explosive atmospheres (other than mining applications or explosives processing and manufacture). Therefore, the user or engineer handling plant design should refer to this before equipment is specified.

10.3.5 Safety and quality

On 19 February 1973 the European Commission issued Directive No. 73/10/EEC, which has subsequently become known as the 'The Low Volt Directive'. In effect, Article 2 of this directive calls upon member states to ensure that all electrical appliances placed on the market are safe. Other articles call for the establishment of common safety standards throughout EC member states so that free movement of goods within the Community shall not be impeded for reasons of safety.

The government has implemented this directive in the UK via the Factories Inspectorate and the Department of Prices and Consumer Protection (DPCP). The Health and Safety at Work, etc. Act 1974 covers the industrial and commercial sectors. In the domestic sector, the Secretary of State has issued Regulations 1975, to be enforced by local weights and measures inspectors. Under this legislation, it is now unlawful to make, to hold in stock or to offer for sale any electrical appliance which is unsafe.

As it affects lighting equipment, BS 4533 (Luminaires) is accepted by the government as a 'safety' specification and the BSI Safety Mark gives an independent guarantee to all concerned that a luminaire has been designed and made in accordance with good engineering practice, that it has been tested and complies with BS 4533 and that its manufacturing quality is monitored regularly by inspectors of the Quality Assurance Department of the British Standards Institution.

10.4 Control gear

10.4.1 General

The electrical characteristics of many lamp types dictate that they cannot be operated directly from the normal a.c. main supply. Most discharge lamps need a current-limiting device in the circuit to control lamp power. To provide the necessary control, electrical components, collectively known as control gear, are needed. There may also be components in the circuit, which have no direct effect on lamp operation. They are, strictly speaking, control gear, although they are not always thought of as such. Power factor correction capacitors and fuses are examples.

10.4.2 Ballast

This is the general term for control gear inserted between the mains supply and one or more discharge lamps, or fluorescent lamps, which, by means of inductance, capacitance or resistance, singly or in combination, serves mainly to limit the current of the lamp(s) to the required value. Ballast may also incorporate means of:

Transforming the supply voltage
Providing a starting voltage
Providing a preheating current
Improving cold starting
Reducing stroboscopic effects
Correcting power factor
Suppressing radio interference
Other components are as follows.

Choke

A simple low power factor inductive ballast.

LPF switch chart ballast

A one-piece unit with a choke and a starter switch socket.

HPF switch start ballast

A one-piece unit containing choke, starter switch socket and power factor correction capacitor.

Leakage reactance auto-transformer

A one-piece unit with a special magnetic circuit that provides a high starting voltage and subsequent ballast control.

Transformer

A wire-wound unit that steps-up or steps-down its supply voltage.

Lagging (inductive) circuit

A circuit in which the current waveform leads the voltage waveform.

Lead/lag circuit

A twin-lamp circuit comprising one leading circuit and one lagging circuit. Such an arrangement overcomes stroboscopic problems and also results in a circuit with near-unity power factor.

Igniter

A starting device intended to generate voltage pulses to start discharge lamps, which does not provide for the pre-heating of electrodes.

Electronic starter

A starting device, usually for fluorescent tubes (such as Vivatronic starters), which provides the necessary preheating of the electrodes and, in combination with the series impedance of the ballast, causes a surge in the voltage applied to the lamp.

Glow-starter switch

A starter, which depends for its operation on a glow discharge in a gaseous atmosphere, closes or opens the preheating circuit of a fluorescent tube and, in combination with the series impedance of the ballast, causes a surge in the voltage applied to the lamp.

An energy limiter

An energy limiter is a device that may be incorporated into the electrical distribution board in a building to reduce the supply voltage and hence the power consumed by fluorescent lighting installations. Such voltage reductions also cause a drop in light output and changes to other lamp operation characteristics, depending on lamp type, ballast circuit and luminaire.

Temperature

Rated maximum ambient temperature (t_a) – the temperature assigned to a luminary to indicate the highest sustained temperature in which the luminary may be operated under normal conditions.

Rated maximum operation temperature (t_c) – the highest permissible temperature that may occur at any place on the outer surface of a device under normal operation conditions.

Rated maximum operating temperature of a winding (t_w) – the operating temperature of ballast winding which gives an expectancy of 10 years' continuous service at that temperature.

Rated temperature rise of a ballast winding (δt) – the temperature rise of the winding under the conditions specified in the relevant BS or IEC publication.

10.4.3 Power factor correction

Inductive devices are the major source of current control in discharge lamp circuits but the inductance causes the supply current to lag the supply voltage – they become out of phase. This effect is quantified by the power factor (PF) of a circuit, which is the relationship between the power and volt/amps:

$$PF = \frac{P}{V \times I}$$

where

P is the circuit power input (W), V is the circuit supply voltage (V), I is the total circuit current (A). The circuit power factor can be no greater than one. When there is no phase difference $P = V \times I$ and the power factor is unity.

With a low power factor, the circuit's apparent power $(V \times I)$ and current are large compared with its power demand. Unnecessarily high currents require large cables and this is wasteful. As a result, the supply authority may impose extra tariffs where the overall power factor is low. There are therefore good reasons for operating at a high power factor.

In an inductive circuit, the current lags the voltage; in a capacitive circuit, the reverse is true. A combination of inductance and capacitance can be arranged so that these effects at least partially cancel each other out to produce a small phase difference and consequently a high power factor. This is a form of power factor correction.

More commonly, a capacitor is included connected across the circuit supply between *live* and *neutral* in what was a low power factor inductive circuit to reduce the supply current and bring the power factor closer to unity. The capacitor is referred to as a power factor correction (PFC) capacitor. Capacitors are usually included in discharge lamps circuits for power factor correction.

Large installations

Where the lighting installations will be a high proportion of the load on a distribution transformer, liaison with the transformer manufacturer at the design stage is advisable. This will ensure that natural resonance between transformer leakage reactant and the total ballast parallel PFC capacitor load does not occur at the lower harmonic frequencies of the main supply. Increased line current will be avoided if natural resonance occurs at frequencies above 400 Hz, where the voltage level of the harmonics is normally very low.

10.5 Emergency lighting

10.5.1 General

In the UK the Fire Precautions Act 1971 and the Health and Safety at Work, etc. Act 1974 make it obligatory to provide adequate means of escape in all places of work and public resort. Emergency lighting is an essential part of this requirement. BS 5266 (BSI 1975b) lays down minimum standards for design, implementation and certification of such installations.

The UK has a long history of fire and safety legislation (a legacy of the Great Fire of London in 1666). In other countries, the legislation ranges from none to different regulations with similar weight and intent. The nature of the legislation and the variation from one country to another make it impossible to deal with this topic in a general way. The reader must refer to the regulations which apply to the locality and comply with them.

The comments in this section refer to UK practice, but even within the UK there is considerable local variations in the regulations (by means of by-laws) and approval should be obtained from the appropriate authority.

Escape lighting

This type of emergency lighting is provided to ensure the safe and effective evacuation of the building. It must:

1. Indicate the escape routes clearly and unambiguously.
2. Illuminate the escape routes to allow safe movement towards and out of the exits.
3. Ensure that fire-alarm call points and fire equipment provided along the escape route can be readily located.

Standby lighting

Some building areas cannot be evacuated immediately in the event of an emergency or power failure because life would be put at risk (for example, a hospital's operating theatre or chemical plants where shutdown procedures must be used). In these circumstances, the activities must be allowed to continue and standby lighting is required. The level of standby lighting will depend on the nature of the activities, their duration and the associated risk, but the provision of 5–25 per cent of the standard service illuminance is common.

Standby lighting can be regarded as a special form of conventional lighting and can be dealt with accordingly, but escape lighting requires different treatment, and is the major concern of BS 5266. Like most British Standards, this is not a legal document, but it can acquire legal status by being adopted as part of the local by-laws. Although most local authorities quote BS 5266, many modify the conditions (for example, by insisting on higher illuminance). It is left to the enforcing authority to decide for what duration emergency lighting shall be required to operate. This is normally longer than the time needed for an orderly evacuation, for the sake of rescue services (typically, 1, 2 or 3 hours).

10.5.2 Escape lighting

Marking the escape route

All exists and emergency exists must have exit or emergency exit signs. Where no direct sight of an exit is possible or there could be doubt as to the direction, then direction signs with an appropriate arrow and the words *Exit* or *Emergency Exit* are required. The aim is to direct someone who is unfamiliar with the building to the exit. All signs must be illuminated at all reasonable times so that they are legible.

Illuminating the route

The emergency lighting must reach its required level 5 seconds after failure of the main lighting system. If

the occupants are familiar with the building this time can be increased to 15 seconds at the discretion of the local authority. The minimum illuminance along the centerline of the escape route must be 0.2 lux or more. In an open area such as a hall or open-plan office, this applies to all areas where people may walk on their way to the exits. The ratio of the maximum illuminance to the minimum illuminance along the centerline of the escape route should not exceed 40:1. Once again, in open areas, this applies to all parts of the floor where people may walk on their way to the exits.

Glare

Luminaries should not cause disability glare and should therefore be mounted at least 2 m above floor level. They should not be too high or they may become obscured by smoke.

Exits and changes of direction

Luminaries should be located near each exit and emergency exit door, and at points where it is necessary to emphasize the position of potential hazards, such as changes of direction and floor level, and staircases.

Fire equipment

Firefighting equipment and fire alarm call points along the escape route must be adequately illuminated at all reasonable times.

Lifts and escalators

Although these may not be used in the event of a fire, they should be illuminated. Emergency lighting is required in each lift car in which people can travel. Escalators must be illuminated to the same standard as the escape route to prevent accidents.

Toilets and control rooms

In toilets of over 8 m^2 gross area, emergency lighting should be installed to provide a minimum of 0.2 lux. Emergency lighting luminaires are required in all control rooms and plant rooms.

10.5.3 Types of systems

There are two main types of supply for emergency lighting: generators and batteries. Few generators will run up to provide the required illuminance within 5 or 15 seconds, in which case they must either be running continuously or be backed up with an auxiliary battery system. Generators require considerable capital investment and are difficult to justify except for standby use.

A battery system can be one of two types; either a central system, where the batteries are in banks at one or more locations, or a self-contained system, where each individual luminaire has its own battery.

Central systems

These have battery rooms or cubicles in which the charger, batteries and switching devices are located. Modern systems tend to have several cubicles to improve system integrity in the event of fire damage to one cubicle.

Self-contained luminaries

These are self-powered and operate independently in an emergency. Thus although an individual luminaire may be destroyed in a fire, the other luminaires will be unaffected. The fact that each luminaire is an independent unit means that maintenance must be thorough. For most applications, self-contained luminaires must operate for a period of 1–3 hours. Most designers base their designs on the safer 3-hour standard, irrespective of the requirements of the enforcing authority.

Maintained

In this mode, the lamp is on all the time. Under normal conditions, the mains power it directly or indirectly. Under emergency conditions, it uses its own battery supply.

Non-maintained

Here the lamp is off when main power is available to charge the batteries. Upon failure, the lamp is energized from the battery pack.

Sustained

This is a hybrid of the previous two modes. A lamp is provided which operates from the mains supply under normal conditions. Under emergency conditions, a second lamp, powered from the battery pack, takes over. Sustained luminaires are often used for exit signs. Although self-contained luminaires are the easiest and most flexible to install, maintenance and testing must be thorough if operation in the event of an emergency is to be guaranteed. Maintained systems continuously demonstrate the integrity of the lamp, but the lamp is continually aging and an operating lamp gives no indication of the state of the battery or charger circuits. In non-maintained and sustained systems, the lamps do not age, but only regular testing will reveal faults.

A variation of the self-contained luminaire is the standard luminaire with an inverter unit and battery pack. These luminaires are normally conventional types used for general lighting (e.g. batten fittings, recessed reflector fittings, etc.) but have a change-over device, inverter and battery pack which will run one of the fluorescent lamps at reduced output if the mains fails.

10.5.4 Planning

When planning an emergency lighting system the following sequence is recommended:

1. Define the exits and emergency exists.
2. Mark the escape routes.
3. Identify any problem areas (for example, areas that will contain people unfamiliar with the building, toilets over 8 m², plant rooms, escalators, etc.).
4. Mark location of exit signs. These can be self-illuminated or illuminated by emergency lighting units nearby. Indicate these on the plan.

5. Where direction signs are required, mark these and provide necessary lighting.
6. Identify the area of the escape route, which is already illuminated by the lighting needed for the signs.
7. Add extra luminaires to complete the lighting of the escape route, paying particular attention to stairs and other hazards. Remember to allow for shadows caused by obstructions or bends in the route.
8. Add extra luminaires to satisfy the problem areas identified in (3) above. Make sure that the lighting outside the building is also adequate for safe evacuation.
9. Check that all fire-alarm points and fire equipment have been adequately dealt with.

Testing

The local inspector will normally require a written guarantee that the installation conforms to the appropriate standards. Drawings of the system must be provided and retained on the premises (this also applies to modifications).

All self-contained luminaires and internally illuminated signs must be tested for a brief period once a month. The battery system must be used twice a year in a simulated main failure for a period of at least one hour. The charging system must also be checked. The system must be operated to its full duration (normally three hours) at least once every three years.

10.6 Lighting design

10.6.1 General

Lighting is an art as well as a science. This implies that there are no hard-and-fast rules of design nor will there be one ideal or optimum solution to a particular lighting problem. More often than not, the lighting designer is presented with a list of conflicting requirements to which priorities have to be allocated before a satisfactory compromise can be found. Here we can therefore do little more than give general guidance.

This section should be read in conjunction with the current CIBSE Code, which includes appropriate illuminance and glare indices for various activities, together with useful notes and flow charts for easy planning (see Figure 10.8).

10.6.2 Choice of lighting system

One should not confuse choosing a lighting system with selecting lamps, luminaires or other equipment. This is a strategic decision that determines the relationship between daylight and artificial light and the degree to which the lighting system is tailored to match the local needs within the space. One basic system decision is to choose between general, localized or local lighting.

Uniform or general lighting

Lighting systems that provide an approximately uniform illuminance on the horizontal working plane over the

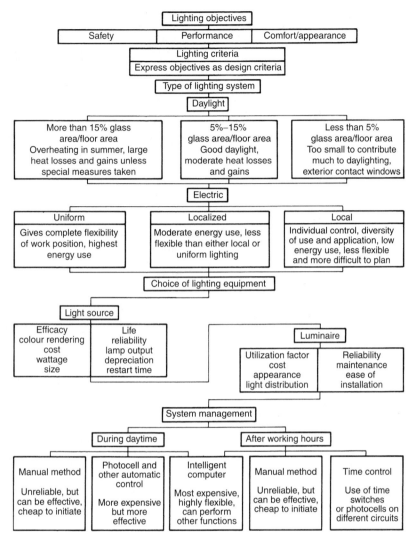

Figure 10.8 An interior lighting design flowchart

entire area are called general or uniform lighting systems. The luminaires are arranged in a regular layout, giving a tidy appearance to the installation. General lighting is simple to plan and install, and requires no coordination with task locations which may not be known or which may change.

The greatest advantage of such a system is that it permits complete flexibility of task location. Its major disadvantage is that energy is wasted in illuminating the whole area to the level needed only for the most critical tasks.

Localized lighting

Localized lighting systems employ an arrangement of luminaires related to the position of tasks and workstations. They provide the required service illuminance on work areas together with a lower level of general illumination for the space. By careful luminaire positioning, good light utilization is achieved with few problems from shadows, veiling reflections and discomfort glare. Localized systems normally consume less energy than do general systems, unless a high proportion of the area is occupied by workstations.

Local lighting

Local lighting provides illumination only over the small area occupied by the task and its immediate surroundings. A general lighting system is installed to provide sufficient ambient illumination for circulating and non-critical tasks.

Therefore, the local lighting system simply supplements the general lighting to achieve the necessary service illuminance on tasks. The system is sometimes referred to as 'task-ambient lighting'. It is very efficient, particularly when high standards of task illuminance are required.

Local lighting is commonly provided by luminaires mounted on the workstation, providing a very flexible room layout. Such local units must be positioned carefully to minimize shadows, veiling reflections, and glare. There is therefore much to be said against making the units adjustable by their lay uses. They offer some personal control of lighting, which is often a popular feature in a large open office.

10.6.3 Choice of lamp

The choice of lamp affects the range of luminaires available, and vice versa, so one cannot be considered without reference to the other. (Section 10.2 will help to differentiate the performance of the main lamp types.)

The designer should compile a list of suitable lamps by rejecting those which do not satisfy the design objectives. The availability of suitable luminaires can then be checked and the economics of the combinations assessed.

Lamps must have satisfactory color-rendering properties. Visual tasks requiring accurate perception of color are less common than is generally believed, but there are many merchandising situations where good color rendering is desirable. The suitability of a lamp for a particular application is best decided by experimental tests.

Lamps must also provide the right color appearance. A warm color tends to be preferred for informal situations, at lower illuminance, and in cold environments, while a cool appearance tends to be preferred for formal situations, at higher illuminance, and in hot environments. It is normally undesirable to illuminate visibly adjacent areas with sources of significantly different color appearance.

Lamp life and lumen maintenance must be considered in conjunction with the maintenance policy for sensible economic calculations, and it needs to be remembered that standardization of lamp types and sizes within a particular site can simplify maintenance.

This selection process deals with most simple situations and is usually more reliable than trying to find a lamp, which matches the needs of the design, which normally calls for a very difficult compromise decision. However, it must only be regarded as a guide. Size, output, the availability of suitable luminaires and many other aspects will influence the choice.

10.6.4 Choice of luminaire

Luminaires have to withstand a variety of physical conditions (e.g. vibration, moisture, dust, high or low ambient temperatures, or vandalism). IR charts identify several acceptable luminaires and the final decision on which luminaire to use can be based on cost, efficiency, glare control, etc., and may have to be left to a later stage in the design process.

10.6.5 System management

A lighting system should not only be well designed but also be managed and operated effectively and efficiently. Good system management implies:

1. Maintaining the system in good order; and
2. Controlling its use to conserve money and energy.

Both these components of management must be considered in the design. The maintenance policy will affect the number of luminaires needed to achieve the service illuminance and the degree of control required would determine the circuits in which the luminaires will operate.

11

Insulation

F T Gallyer

Pilkington Insulation Ltd

Contents

11.1 Introduction

Insulation is one of those ubiquitous techniques that is always around, always impinging on our work, social and domestic activities and yet for most of the time is hardly noticed. Insulation is a passive product; once installed, it works efficiently, quietly and continually, usually out of sight, enclosed within a structure or a casing or under cladding.

It comes to the fore when new design of buildings, plant, equipment or production processes is being considered. It is at this stage that the right specification must be made. Any shortfall in the thickness or error in the type and application details will prove costly to rectify at a later data.

11.1.1 Why insulate?

There are many reasons why professional engineers, architects and laymen use insulation, e.g.:

1. To comply with mandatory legislation (i.e. Building Regulations);
2. To reduce heat loss/heat gain;
3. To reduce running costs;
4. To control process temperatures;
5. To control surface temperatures;
6. To reduce the risk of freezing;
7. To provide condensation control;
8. To reduce heating plant capacity;

Other reasons why insulation is used are to provide:

9. Acoustic/correction and noise control;
10. Fire protection.

11.1.2 Scope

This chapter will deal primarily with thermal insulation. Acoustic and fire-protection properties and applications will be treated as subsidiary to the thermal insulation aspects.

11.1.3 Thermal insulation

A thermal insulation material is one that frustrates the flow of heat. It will slow down the rate of heat loss from a hot surface and similarly reduce the rate of heat gain into a cold body. It will not stop the loss or gain of heat completely.

No matter how well insulated, buildings will need a continual input of heat to maintain desired temperature levels. The input required will be much smaller in a well-insulated building than in uninsulated ones – but it will still be needed. The same applies to items of plant – pipes, vessels and tanks containing hot (or cold) fluids. If there is no heat input to compensate for the loss through the insulation the temperature of the fluid will fall. A well-insulated vessel will maintain the heat of its contents for a longer period of time but it will never, on its own, keep the temperature stable.

Thermal insulation does not generate heat. It is a common misconception that such insulation automatically warms the building in which it is installed. If no heat is supplied to that building it will remain cold. Any temperature rise that may occur will be the result of better utilization of internal fortuitous or incidental heat gains.

11.2 Principles of insulation

11.2.1 Heat transfer

Before dealing with the principles of insulation, it is necessary to understand the mechanism of heat transfer. When an area that is colder surrounds a hot surface, heat will be transferred and the process will continue until both are at the same temperature. Heat transfer takes place by one or more of three methods:

> Conduction
> Convection
> Radiation

11.2.2 Conduction

Conduction is the process by which heat flows by molecular transportation along or through a material or from one material to another, the material receiving the heat being in contact with that from which it receives it. Conduction takes place in solids, liquids and gases and from one to another. The rate at which conduction occurs varies considerably according to the substance and its state.

In solids, metals are good conductors – gold, silver and copper being among the best. The range continues downwards through minerals such as concrete and masonry, to wood, and then to the lowest conductors such as thermal insulating materials.

Liquids are generally bad conductors, but this is sometimes obscured by heat transfer taking place by convection. Gases (e.g. air) are even worse conductors than liquids but again, they suffer from being prone to convection.

11.2.3 Convection

Convection occurs in liquids and gases. For any solid to lose or gain heat by convection it must be in contact with the fluid. Convection cannot occur in a vacuum. Convection results from a change in density in parts of the fluid, the density change being brought about by an alteration in temperature.

Convection in gases

If a hot body is surrounded by cooler air, heat is conducted to the air in immediate contact with the body. This air then becomes less dense than the colder air further away. The warmer, light air is thus displaced upwards and is replaced by colder, heavier air which, in turn, receives heat and is similarly displaced. There is thus developed a continuous flow of air or convection around the hot body removing heat from it. This process is similar but reversed if warm air surrounds a colder body, the air becoming colder on transfer of the heat to the body and displaced downwards.

Convection in liquids

Similar convection processes occur in liquids, though at a slower rate according to the viscosity of the liquid. However, it cannot be assumed that convection in a liquid results in the colder component sinking and the warmer one rising. It depends on the liquid and the temperatures concerned. Water achieves its greatest density at approximately 4°C. Hence in a column of water, initially at 4°C, any part to which heat is applied will rise to the top. Alternatively, if any part is cooled below 4°C it, too, will rise to the top and the relatively warmer water will sink to the bottom. The top of a pond or water in a storage vessel always freezes first.

Natural convection

The process of convection that takes place solely through density change is known as *natural convection.*

Forced convection

When the fluid displaced is accelerated by wind or artificial means the process is called *forced convection.* With forced convection the rate of heat transfer is increased – substantially so in many cases.

11.2.4 Radiation

The process by which heat is emitted from a body and transmitted across space as energy is called radiation. Heat radiation is a form of wave energy in space similar to radio and light waves. Radiation does not require any intermediate medium such as air for its transfer. It can readily take place across a vacuum.

All bodies emit radiant energy, the rate of emission being governed by:

1. The temperature difference between radiating and receiving surfaces;
2. The distance between the surfaces;
3. The emissivity of the surfaces. Dull matte surfaces are good emitters/receivers; bright reflective surfaces are poor ones.

11.2.5 Requirements of an insulant

In order to perform effectively as an insulant a material must restrict heat flow by any (and preferably) all three methods of heat transfer. Most insulating materials adequately reduce conduction and convection elements by the cellular structure of the material. The radiation component is decreased by absorption into the body of the insulant and is further reduced by the application of bright foil outer facing to the product.

Convection inhibition

To reduce heat transfer by convection an insulant should have a structure of a cellular nature or with a high void content. Small cells or voids inhibit convection within them and are thus less prone to excite or agitate neighboring cells.

Conduction inhibition

To reduce heat transfer by conduction, an insulant should have a small ratio of solid volume to void. Additionally, a thin-wall matrix, a discontinuous matrix or a matrix of elements with minimum point contacts are all beneficial in reducing conducted heat flow. A reduction in the conduction across the voids can be achieved by the use of inert gases rather than still air.

Radiation inhibition

Radiation transfer is largely eliminated when an insulant is placed in close contact with a hot surface. Radiation may penetrate an open-cell material but is rapidly absorbed within the immediate matrix and the energy changed to conductive or convective heat flow. It is also inhibited by the use of bright aluminum foil, either in the form of multi-corrugated sheets or as outer facing on conventional insulants.

11.3 Calculation of heat loss

11.3.1 Glossary of terms

Terms and symbols used in computing heat loss are as follows.

Heat (J/s or W)

The unit of quantity of heat is the joule (J). Heat flow may be expressed as joules per second (J/s), but as a heat flow of one joule per second equals one watt the unit watt (W) is usually adopted for practical purposes.

Temperature (°C or K)

For ready identification, actual temperature levels are expressed in degrees Celsius (°C) while temperature difference, interval or gradient is expressed in Kelvin (K).

Thermal conductivity (λ)

Thermal conductivity, now denoted by the Greek letter lambda (previously known as the k-value), defines a material's ability to transmit heat, being measured in watts per square meter of surface area for a temperature gradient of one Kelvin per unit thickness of one meter. For convenience in practice, its dimensions Wm/m^2K be reduced to W/mK, since thickness over area m/m^2 cancels to $1/m$.

Thermal resistivity (r)

This is the reciprocal of thermal conductivity. It is expressed as mK/W.

Thermal conductance (C)

Thermal conductance defines a material's ability to transmit heat measured in watts per square meter of surface area for a temperature gradient of one Kelvin in terms of a *specific thickness* expressed in meters. Its dimensions are therefore W/m^2K.

Thermal resistance (R)

Thermal resistance is the reciprocal of thermal conductance. It is expressed as $m^2 K/W$. Since the purpose of thermal insulation is to resist heat flow, it is convenient to measure a material's performance in terms of its thermal resistance, which is calculated by dividing the thickness expressed in meters by the thermal conductivity. Being additive, thermal resistances facilitate the computation of overall thermal transmittance values (U-values).

Surface coefficient (f)

This is the rate of heat transfer from a surface to the surrounding air (or fluid) due to conduction convection and radiation. It is generally used only in still-air conditions and when the temperature difference between surface and ambient is of the order of 30 K. It is obtained by dividing the thermal transmission per unit area in watts per square meter by the temperature difference between the surface and the surrounding air. It is expressed as $W/m^2 K$.

Surface resistance (R_S)

Surface resistance is the reciprocal of surface coefficient. It is expressed as $m^2 K/W$.

Thermal transmittance (U)

Thermal transmittance (U-value) defines the ability to an element of structure to transmit heat under steady-state conditions. It is a measure of the quantity of heat that will flow through unit area in unit time per unit difference in temperature of the individual environments between which the structure intervenes. It is calculated as the reciprocal of the sum of the resistance of each component part of the structure, including the resistance of any air space or cavity and of the inner and outer surfaces. It is expressed as $W/m^2 K$.

11.3.2 Fundamental formulae

In calculating heat loss from surfaces freely exposed to air it is necessary to deal separately with both radiant and convective losses.

Radiation

The following Stefan–Boltzmann formula applies to both plane and cylindrical surfaces:

$$Q_r = 5.673 \times 10^{-8} E[(T_a + 273.1)^4 - (T_{rm} + 273.1)^4] \quad (11.1)$$

Natural convection – plane surfaces

$$Q_c = C(T_s - T_a)^n \quad (11.2)$$

The factor C and index n can be assumed to have the following values for plane surfaces of differing orientations:

Horizontal – downward heat flow	$C = 1.3$	$n = 1.25$
Vertical	$C = 1.9$	$n = 1.25$
Horizontal – upward heat flow	$C = 2.5$	$n = 1.25$

Natural convection – horizontal cylinders and pipes

$$Q_c = 0.53 \frac{\lambda_a}{d_o} (Gr \times Pr)^{0.25} (T_s - T_a) \quad (11.3)$$

However, as the Grashof and Prandtl numbers can be different to determine, the following formula, which gives a close approximation, can be used for cylinders freely exposed to air:

$$Q_c = (T_s - T_a)1.32 \left(\frac{T_s - T_a}{d_o} \right)^{0.25} \quad (11.4)$$

In the above equations:

Q_r = heat loss by radiation (W/m^2),
Q_c = heat loss by convection (W/m^2),
E = surface emissivity (see Section 11.4) (dimensionless),
T_s = surface temperature (°C),
T_a = ambient air temperature (°C),
T_{rm} = Mean radiant temperature of enclosure (°C),
Gr = Grashof number (dimensionless),
Pr = Prandtl number (dimensionless),
λa = thermal conductivity of air (W/mK),
d_o = outside diameter of cylinder/pipe (m).

Forced convection

It is not proposed to deal with forced convection here. Experimental work has yielded considerably differing results for ostensibly similar conditions. It is sufficient to note that forced convection affects small-bore pipes to a greater extent than large-bore and is dependent on temperature differences. While the heat loss from non-insulated surfaces may increase by a factor of up to 200–300 per cent, the increase in heat loss from the insulated surface would be considerably less (of the order of 10 per cent).

11.3.3 Insulated heat loss

Heat loss through insulation can be calculated from knowledge of the thickness and thermal conductivity of the insulation and the emissivity of the outer surface of the insulation/cladding system.

Single-layer insulation

$$q = \frac{T_1 - T_2}{R} = \frac{T_1 - T_m}{R + R_s} \quad (11.5)$$

$$q_1 = 10^{-3}\pi d_o q \quad (11.6)$$

For plane surfaces:

$$R = 10^{-3} \frac{L}{\lambda} \quad (11.7)$$

$$R_s = \frac{1}{f} \quad (11.8)$$

For cylindrical surfaces:

$$R = 10^{-3} \frac{d_o}{2\lambda} \ln \frac{d_1}{d_o} \quad (11.9)$$

$$R_s = \frac{d_o}{f d_1} \quad (11.10)$$

Multi-layer insulation

$$q = \frac{T_1 - T_2}{R_1 + R_2 + \Lambda R_n} = \frac{T_1 - T_m}{R_1 + R_2 + \Lambda R_n + R_s} \quad (11.11)$$

$$q_1 = 10^{-3} \pi d_o q$$

For plane surfaces:

$$R_1 = 10^{-3} \frac{L_1}{\lambda_1} \quad (11.12)$$

$$R_2 = 10^{-3} \frac{L_2}{\lambda_2} \quad (11.13)$$

$$R_n = 10^{-3} \frac{L_n}{\lambda_n} \quad (11.14)$$

$$R_s = \frac{1}{f} \quad (11.15)$$

For cylindrical surfaces:

$$R_1 = 10^{-3} \frac{d_o}{2\lambda_1} \ln \frac{d_1}{d_o} \quad (11.16)$$

$$R_2 = 10^{-3} \frac{d_o}{2\lambda_2} \ln \frac{d_2}{d_1} \quad (11.17)$$

$$R_n = 10^{-3} \frac{d_o}{2\lambda_n} \ln \frac{d_s}{d_{(s-1)}} \quad (11.18)$$

$$R_s = \frac{d_o}{f d_n} \quad (11.19)$$

Symbols

In the above formulae the following symbols apply:

q = heat loss per square meter of hot surface (W/m²)
q_1 = heat loss per linear meter of pipe (W/m)
R = thermal resistance of insulation per m² (m²K/W)
R_1 = thermal resistance of inner layer of insulation (m²K/W)
R_2 = thermal resistance of second layer of insulation (m²K/W)
R_n = thermal resistance of nth layer of insulation (m²K/W)
R_s = thermal resistance of outer surface of insulation system (m²K/W)
T_1 = temperature of hot surface (°C)
T_2 = temperature of outer surface of insulation (°C)
T_m = temperature of ambient still air (°C)
d_o = outside diameter of pipe (mm)
d_1 = outside diameter of inner layer of insulation (mm)
d_2 = outside diameter of next layer of insulation (mm)
d_n = outside diameter of n th layer of insulation (mm)
λ = thermal conductivity of insulating material (W/mK)
λ_1 = thermal conductivity of inner layer of insulation (W/mK)
λ_2 = thermal conductivity of second layer of insulation (W/mK)
λ_n = thermal conductivity of nth layer of insulation (W/mK)
f = surface coefficient of outer surface (W/m²K)
$\ln$ = natural logarithm

Note: the values taken for λ, λ_1, λ_2, ..., and λ_n should be those applicable to the mean temperature of the hot and cold surfaces of the appropriate layer.

11.3.4 Surface emissivity

Emissivity values are needed to calculate the radiation component of heat loss from high-temperature surfaces. Accurate values that reflect actual conditions are difficult to obtain, as the emissivity value varies with temperature and with contamination or oxidation of the surface.

For most non-critical calculations the following values may be used:

Steel, paint (matte surface)	0.90–0.95
Galvanized steel (new)	0.40
Galvanized steel (weathered)	0.85
Aluminum (polished)	0.05

11.3.5 Surface coefficients

These present similar problems to emissivities, and BS 5422 has standardized on three values:

1. $f = 5.7$ for surfaces of low emissivity (e.g. polished aluminum);
2. $f = 8.0$ for surfaces of medium emissivity (e.g. planished or galvanized steel, aluminum paint, stainless steel and aluminum/zinc amalgam);
3. $f = 10.0$ for surfaces of high emissivity (e.g. matte-black surfaces, steel, brick and plain insulation surfaces).

11.3.6 *U*-value calculation

The insulation effectiveness of elements of building structures is represented by the *U*-value or thermal transmittance. As defined in Section 11.3.1, the *U*-value is the reciprocal of the sum of the thermal resistances and can be expressed as:

$$U = \frac{1}{R_{so} + R_1 + R_2 \Lambda R_n + R_{as} + R_{si}} \quad (11.20)$$

where

R_{so} = thermal resistance of outer surface,
R_1 = thermal resistance of first material,
R_2 = thermal resistance of second material,
R_n = thermal resistance of nth material,
R_{as} = thermal resistance of any air space in the construction,
R_{si} = thermal resistance of inner surface.

In calculating *R*-values of the material elements of construction, their λ values should be taken at 10°C mean temperature (which is assumed to be normal building temperature).

11.3.7 Standardized resistances

Actual thermal resistances of surface and air spaces within a construction vary according to size, exposure

Table 11.1 Internal surface resistance, R_{si} (m^{2}K/W)

Building element	Direction of heat flow	Surface emissivity	Surface resistance
Walls	Horizontal	High	0.12
		Low	0.30
Ceilings, roofs and floors	Upward	High	0.10
		Low	0.22
Ceilings and floors	Downward	High	0.14
		Low	0.55

Source: Chartered Institution of Building Services Engineers Guide, Section A3.

Table 11.2 External surface resistance, R_{so} (m^{2}K/W)

Building element	Surface emissivity	Surface resistance for stated exposure		
		Sheltered	Normal	Severe
Walls	High	0.08	0.06	0.03
	Low	0.11	0.07	0.03
Roofs	High	0.07	0.04	0.02
	Low	0.09	0.05	0.02
Floor	High	0.07	0.04	0.02

Source: Chartered Institution of Building Services Engineers Guide, Section A3.

Table 11.3 Standard thermal resistances for unventilated air spaces, R_{as} (m^{2}K/W)

Width of airspace (mm)	Surface emissivity	Thermal resistance for heat flow in stated direction		
		Horizontal	Upward	Downward
5	High	0.10	0.10	0.10
	Low	0.18	0.18	0.18
25	High	0.18	0.17	0.22
or more	Low	0.35	0.35	1.06
High-emissivity and corrugated sheets in contact		0.09	0.09	0.11
Low-emissivity multiple-foil insulation with air space on side		0.62	0.62	1.76

Source: Chartered Institution of Building Services Engineers Guide, Section A3.

and nature of the material. Standardized resistance values were adopted by the Building Research Establishment in their BRE Digest 108 (August 1969) to ensure a constant base for the calculation of U-values. These values are now collected in the CIBSE Guide A3, from which Tables 11.1–11.4 were prepared.

11.4 Standards of insulation

In the UK the Building Regulations contain the mandatory requirement for thermal insulation of building structures,

Table 11.4 Standard thermal resistances for ventilated air spaces, R_{as} (m^{2}K/W)

Air space thickness 25 mm minimum	
Air space in cavity-wall construction	0.18
Air space between tiles and roofing felt on pitched roof	0.12
Air space behind tiles on tile-hung wall	0.12
Loft space between flat ceiling and pitched roof lined with felt	0.18
Loft space between flat ceiling and pitched roof of aluminium cladding, or low-emissivity upper surface on ceiling	0.25
Loft space between flat ceiling and unsealed fiber cement or black metal cladding to pitched roof	0.14
Air space between fiber cement or black metal cladding with unsealed joints and low-emissivity surface facing air space	0.30
Air space between fiber cement or black metal cladding with unsealed joints and high-emissivity lining	0.16

Source: Chartered Institution of Building Services Engineers Guide, Section A3.

and heating and hot water services. Thermal insulation of cold water supply pipes is dealt with under the water supply bylaws, but there are no national requirements to insulate pipes, vessels or equipment used in commercial or industrial processes.

11.4.1 Building Regulations 1990

These Regulations came into force on 1 April 1990 and apply to England and Wales only. It is the government's intention that these levels be incorporated into the regulations for Scotland and Northern Ireland.

The requirement

The mandatory requirement is that 'reasonable provision shall be made for the conservation of fuel and power in buildings'. This requirement applies to dwellings and all other buildings whose floor area exceeds 30 m^2.

11.4.2 Approved Document L

The Department of the Environment's interpretation of reasonableness is given in the 1990 edition of Building Regulations 1985, Approved Document L, Conservation of Fuel and Power, available from HMSO. This deals with three areas of energy conservation:

The building fabrics
Controls of heating/hot water systems
Insulation of hot water systems.

Limitation on heat loss through the building fabric

This section of Approved Document L allows three methods of compliance:

Elemental approach
Calculation procedure 1
Calculation procedure 2

Elemental approach The insulation standards are set by the elemental approach, which establishes maximum U-values for the various elements of structure and maximum glazing area. These are detailed in Tables 11.5 and 11.6. The Regulations also introduce two new categories of structure, which have previously not needed to be insulated:

1. Semi-exposed walls and floors, which can be considered to be structures between a heated and an unheated part of a building;
2. Ground floors, which because of their inherently good U-values are unlikely to need insulation unless they are of a small domestic size or are of long narrow buildings (see Figure 11.1).

Calculation procedure 1 This is an alternative to the elemental approach and allows variation in the levels of insulation and of the glazing area. The calculation should show that the rate of heat loss through the envelope of the proposed building is not greater than that through a notional building of the same size and shape that is designed to comply with the elemental approach.

Calculation procedure 2 This procedure is the calculation of an energy target and allows for a completely free design using any valid energy-conservation measure. The procedure is intended to allow for useful solar heat gains and fortuitous internal gains. The requirement of Building Regulations will be met 'if the calculated annual energy use of the proposed building is less than the calculated

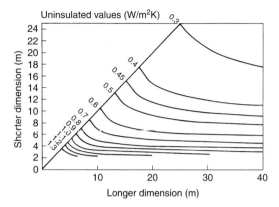

Figure 11.1 *U*-values of uninsulated solid concrete ground floors with four exposed edges. (Source: Eurisol UK, *U*-value Guide)

energy use of a similar building designed to comply with the elemental approach'.

Limitations on calculation procedures Limitations are placed on the use of high (e.g. poor U-values) and of glazing areas when using the calculation of methods of complying. As a general rule:

1. *In dwellings*: The U-value of roofs and exposed walls should not be greater than $0.35 \, \text{W/m}^2\text{K}$ and $0.6 \, \text{W/m}^2\text{K}$, respectively.
2. *In all other buildings*: The U-value of roofs and exposed walls should not be greater than $0.6 \, \text{W/m}^2\text{K}$.
3. *Glazing areas*: If areas of glazing smaller than those allowed in Table 11.6 are used in the 'proposed building' then this smaller area should also be used in the *notional building* calculation.

Insulation thickness While the actual thickness of insulation needed to comply with the Building Regulations will vary depending on insulation type, thermal conductivity and the structure into which it is fitted, some general guidelines can be given:

1. *Domestic roofs*: $U = 0.25 \, \text{W/m}^2\text{K}$
   150 mm mineral wool loft insulation
2. *Other roofs*: $U = 0.45 \, \text{W/m}^2\text{K}$
   80 mm in steel-frame metal-clad roofs
   40–60 mm in flat roofs
3. *External Walls*: $U = 0.45 \, \text{W/m}^2\text{K}$
   80 mm in steel-frame metal-clad wall
   50 mm in aerated concrete inner-leaf cavity wall
   65 mm in high-density masonry cavity walls
4. *Semi-exposed elements*: $U = 0.6 \, \text{W/m}^2\text{K}$
   0–50 mm, depending on construction
5. *Ground floors*
   None needed in commercial or industrial-sized floors exceeding about 15 m × 15 m. Actual U-values can be obtained from Eurisol's graph shown as Figure 11.1.

Table 11.5 Building Regulations: maximum *U*-values (W/m²K)

Structure	Dwelling	All other buildings
Exposed walls	0.45	0.45
Exposed floors	0.45	0.45
Ground floors	0.45	0.45
Roofs	0.25	0.45
Semi-exposed walls and floors	0.6	0.6

Table 11.6 Building Regulations: maximum single-glazed areas of windows and rooflights

Building type	Windows	Rooflights
Dwellings	Windows and rooflights together 15% of total floor area	
Other residential, including hotels and institutional buildings	25% of exposed wall area	20% of roof area
Places of assembly, offices and shops	35% of exposed wall area	20% of roof area
Industrial and storage	15% of exposed wall area	20% of roof area

Note: In any building the above glazing areas may be doubled when double-glazing is used and trebled when triple glazing, or double-glazing with low-emissivity coating, is used.

Controls for space heating and hot water systems

This section of Approved Document L does not have any relevance to insulation.

Insulation of hot water storage vessels, pipes and ducts

This third and last section of Approved Document L is not intended to apply to storage and piping systems for commercial and industrial processes. It concerns only the central heating and the domestic hot water supply of all buildings. The standards are presented in three sections for:

Storage vessels
Pipes
Ducts

Insulation of hot water storage vessels The requirements for these tanks and cylinders will be met if the heat loss is not greater than 90 W/m^2. The thickness of insulation needed will therefore vary not only according to its thermal conductivity but also to the temperature of the water being stored. In practice, as long as the water is not hotter than 100°C the insulation thickness needed is likely to be of the order of 20–35 mm.

As an alternative to the heat loss specification, domestic cylinders complying with BS 699: 1984, BS 1566: 1984: Parts 1 and 2, BS 3198: 1981 or having a cylinder jacket complying with BS 5615: 1985 will all meet the requirement.

Insulation of pipes The requirement for pipes is that they should be insulated to a thickness at least equal to the outside diameter, and with an insulant of thermal conductivity not greater than 0.045 W/mK. Two limitations are imposed: the insulation thickness need not be greater than 40 mm whatever the pipe outside diameter and insulation is not needed on pipes whose heat loss contributes to the useful heat requirements of the room or space through which it runs.

As an alternative option, the insulation should meet the recommendations of BS 5422: 1977. This Standard tabulates thicknesses of insulation too numerous to mention here, according to whether (1) the pipes carry central heating or domestic hot water, (2) the system is heated by gas and oil or solid fuel, (3) the water temperature is 75°C, 100°C or 150°C and (4) the thermal conductivity of the insulant is 0.04, 0.55 or 0.70 W/mK at the appropriate mean temperature.

Insulation of ducts The requirement for warm air heating ducts is that they also should meet the recommendations of BS 5422: 1977. As with pipes, if the ducts offer useful heat to the areas through which they run then they need not be insulated.

The lowest thickness of insulation recommended in Table 10 of BS 5422 is for insulants having a thermal conductivity of 0.04 W/mK. It recommends thickness of 38 mm, 50 mm and 63 mm when the warmed air to ambient air temperature differences are 10 K, 25 K and 50 K, respectively.

11.4.3 Water supply bylaws

Like the Building Regulations, Bylaw 48, 'Protection from damage by freezing and other causes', is a functional requirement. It does not quote specific details but requires that 'every pipe or other water fitting, so far as is reasonably practicable shall be effectively protected ... by insulation or other means against damage from freezing and other causes'. Generally, this would be satisfied by insulating to BS 6700: 1987 recommendations which, for an insulant of thermal conductivity 0.035 W/mK (which applies to most fibrous and plastic insulants) at the temperatures appropriate to freezing, would mean insulation thicknesses of approximately 25 mm for all pipes in indoor installations and 30 mm for those outdoors.

BS 5422 (which also gives insulation thickness for protection against freezing) recommend thickness of 32 mm and 38 mm, respectively, for pipes of 48 mm outside diameter or less. It must be emphasized that these thicknesses only give protection for a relatively short time period (i.e. overnight). It is not possible by means of insulation alone to protect permanently static water.

11.4.4 BS 5422, Specification for the use of thermal insulating materials

Much mention has been made of this Standard with regard to insulation. It contains definitions, physical characteristics and recommended thicknesses of insulation for a wide range of industrial applications, including:

1. Refrigeration;
2. Chilled and cold water supplies for industrial applications;
3. Central heating and domestic hot and cold water supplies;
4. Process pipework and equipment.

It also contains heat loss calculation methods. This Standard should be the basis for any specification drawn up by a company that does not have its own in-house insulation standards manual.

11.5 Product selection

In selecting an insulation product for a particular application, consideration should be given not only to its primary function but also to the many secondary functions, and often-unappreciated requirements, which the insulation of pipe or vessel may place on the insulation product. Some of these product requirements are discussed below.

11.5.1 Limiting temperatures

Ensure that the insulation selected can operate effectively and without degradation at temperatures beyond the design temperature called for. Temperature control of the process system can fail and systems can overheat. Hot surfaces exposed to ambient air will become hotter when insulated if there is no control of the process temperature. As with multi-layer systems, the addition of an extra layer of insulation will change the interface temperatures of all inner layers. Do not use an insulant close to a critical temperature limit.

11.5.2 Thermal movement

Large vessels and equipment operating at extreme temperatures exhibit large expansion or contraction movements. Any insulant specified for these applications should be capable of a sympathetic movement such that it will not cause itself or any cladding to burst, nor should it produce gaps that lead to dangerous hot spots in the cladding system.

11.5.3 Mechanical strength

Many insulants are made in a wide range of densities and thus mechanical strengths. Ensure that where insulation is required to be load bearing, to carry cladding, to support itself across gaps or drape down the sides of buildings or high equipment, the selected product has the necessary strength to accommodate these mechanical requirements as well as the primary thermal one.

11.5.4 Robustness

Insulation is subjected to abuse onsite, in storage and often in transit. To ensure minimum wastage through breakage, contamination or deformation, select products which are resilient or robust enough to tolerate site conditions and malpractice.

11.5.5 Chemical resistance

No matter how well installed, insulation is always at risk of contamination from outside sources. Overfilling of vessels, leaking valves and flanges, oil thrown off from rotating transmission shafts and motors can all penetrate protective cladding or lining systems. Consideration of any insulation's compatibility with possible contaminants should be considered in these situations.

11.5.6 Weather resistance

As mentioned above, insulation applied to externally located equipment can be subjected to rain and weather contamination if the outer cladding fails. Insulants with water-repellant, water-tolerant or free-draining properties offer an additional benefit in this type of application. In the structural field insulants used as cavity wall fills must be of those types specially treated and designed for this application.

11.5.7 Surface emissivity

The effects of surface emissivity are exaggerated in high-temperature applications, and particular attention should be paid to the selection of the type of surface of the insulation system. Low-emissivity surfaces such as bright polished aluminum reduce heat loss by inhibiting the radiation of heat from the surface to the surrounding ambient space. However, by holding back the heat being transmitted through the insulation, a 'dam' effect is created and the surface temperature rises. This temperature rise can be considerable, and if insulation is being used to achieve a specified temperature the use of a low-emissivity system could necessitate an increased thickness of insulation. For example, a hot surface at 550°C insulated with a 50 mm product of thermal conductivity 0.055 and ambient temperature of 20°C would give a surface temperature of approximately 98°C, 78°C and 68°C when the outer surface is of low (polished aluminum), medium (galvanized steel) or high (plain or matte) emissivity, respectively. Conversely, to achieve a surface temperature of 55°C with the same conditions the insulation would need to be approximately 120 mm, 87 mm and 70 mm, respectively.

11.5.8 Acoustics

Because of their cellular or open-matrix construction, most insulants have an inherent ability to absorb sound, act as panel dampers and reduce noise breakout from plant by their ability to be a flexible or discontinuous link between an acoustically active surface and the outer cladding. This secondary aspect of thermal insulation specification will gain more prominence when the UK adopts the EC Directive 86/1888, 'Protection of workers from the risks related to exposure to noise at work'.

11.5.9 Fire safety

Large volumes of insulation are used in industry, and most is hidden under cladding behind linings and between sheeting. However, much is located in voids and open to view in workplaces.

Fires do occur and accidents happen. In selecting the appropriate material, consideration should be given to its fire-safety properties and to methods of maintaining the integrity of the protective cladding in fire situations.

Fire properties of insulation materials range from the highest to the lowest, from non-combustible to flammable with toxic fume emission. Generally, inorganic materials tend to be non-combustible while organic (or oil-based) materials are combustible, but many have surface treatments to improve their fire-safety rating.

11.6 Thermal conductivity

Thermal conductivity is not a static property of a material. It can vary according to the density, operating temperature and type of gas entrapped within the voids.

11.6.1 Density effects

Most materials achieve their insulating properties by virtue of the high void content of their structure. The voids inhibit convective heat transfer because of their small size. A reduction in void size reduces convection but does increase the volume of the material needed to form the closer matrix, thus resulting in an increase in product density. Further increases in density continue to inhibit convective heat transfer, but ultimately the increasing conductive transfer through the matrix material offsets the additional benefit and any further increase in density causes deterioration in thermal conductivity (see Figure 11.2).

Most traditional insulants are manufactured in the low- to medium-density range and each particular product family will have its own specific relationship between conductivity and density. One particular group of products (insulating masonry) is manufactured in the medium- to high-density range. Thermal conductivity is improved by reducing density.

11.6.2 Temperature effects

Thermal conductivity increases with temperature. The insulating medium (the air or gas within the voids) becomes more excited as its temperature is raised, and this enhances convection within or between the voids, thus increasing heat flow. This increase in thermal conductivity is generally continuous for air-filled products and can be mathematically modeled (see Figure 11.3). Those insulants that employ *inert gases* as their insulating medium may show sharp changes in thermal conductivity, which may occur because of gas condensation. However, this tends to take place at sub-zero temperatures.

11.6.3 Chlorofluorocarbons

Much has been written lately about CFCs. This gas is used in the manufacture of a range of plastic insulants, including polyurethane, phenolic, polyisocyanurates and extruded polystyrenes.

The manufacturers and their suppliers are actively seeking alternatives, and some CFC-free polyurethane is already appearing on the market. However, these new products do not have such good insulating properties (thermal conductivity of the order of 0.03 W/mK is being quoted, as against the 0.02/0.025 W/mK of the originals).

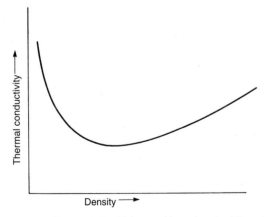

Figure 11.2 Typical relationship between thermal conductivity and density

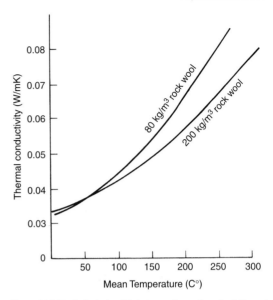

Figure 11.3 Typical relationship between thermal conductivity and mean temperature

Where CFC-bearing products are described in Section 11.9 the thermal conductivity quoted are those of current (CFC) manufacture.

11.7 Physical forms

The extensive range of insulation types and the numerous forms in which they are manufactured ensures that any listing can never be comprehensive. It becomes even more so when one acknowledges the fact that within the insulation industry there is a secondary network of fabricators and laminators who will take manufacturers' basic products and cut, shape, mold, laminate and enclose them to almost any requirement. The following gives the basic physical forms and some of the uses to which insulation materials are put.

11.7.1 Slab or board

The basic product supplied by almost all manufacturers ranges from 'flexible' to 'rigid' forms. Dimensions determined by manufacturing facility and handling ability, but usually in 600 mm and 1200 mm modules. This configuration may be used in most applications on plane surfaces, such as lining panels, under cladding, within casings, in partitions, in buildings or plant and equipment.

11.7.2 Pipe sections

Rolled, molded or cut to shape as annular cylinders for the insulation of pipes. Pipe sections are made in one or two segments but occasionally in three or four. Sizes are available for most steel and copper pipes of up to approximately 1120 mm outside diameter.

11.7.3 Beveled lags

Slabs of narrow width beveled on edges (and sometimes radiused on faces). They are fitted longitudinally on large-sized pipes with face conforming closely to the circumference.

11.7.4 Loose fill

Granulate, particle-size material specifically manufactured or produced from basic product off-cuts. Supplied by most manufacturers and used for packing or pouring into irregular-shaped enclosures.

11.7.5 Mat or blanket

Flexible material of low density and long length generally supplied in rolls available with or without facings. Mineral wool and flexible foams are the most usual type of product. This configuration is widely used in buildings as roof insulation and in steel- and timber-frame walls. It is also used for sound insulation in partitions, on low-temperature heating ducts and in equipment casings.

11.7.6 Rolls

Mat or blanket supplied in the form of helically wound cylindrical packs.

11.7.7 Mattress

Medium- to heavy-density mineral wool mat with wire netting or expanded metal mechanically secured on one or both sides. Mattress-type products are used for irregular-shaped surfaces and on large pipes and vessels. It is also used as large-cavity fire barriers.

11.7.8 Quilt

Medium-density mat with flexible facings of paper or scrim stitched through. Quilts are used for acoustic absorption behind perforated or slatted panels and ceilings.

11.7.9 Lamella

Paper- or foil-backed mineral wool product fabricated from low-density slabs in which the slats stand upright with the fibers predominantly perpendicular to the major faces. This type of product is used on circular and elliptical ducting, pipes and vessels to give a compression-resistant insulant. Supplied in roll form.

11.7.10 Blowing wool

Nodulated mineral wool produced for application through pneumatic hoses into areas of restricted access.

11.7.11 Sprayed insulation

Insulation dry mixed in factory with inorganic fillers and binders for application by wet spraying. Usually mineral wool or vermiculite based, used for thermal and acoustic treatments and for fire protection of steelwork.

11.7.12 Sprayed foam

A cellular plastic – usually of the urethane or isocyanurate families – where chemical components are mixed at the spray head.

11.7.13 Molded products

Insulation formed by slurry casting or heat curing under pressure in molds in a number of insulation types. Most common moldings are preformed bends, valve boxes and flange covers.

11.8 Facings

While all insulation products are supplied in their natural 'as-produced' finish, they are also available with a variety of facings that are either applied 'on-line' during manufacture or as a secondary 'off-line' process. The facings are applied for functional, technical or aesthetic reasons, generally as follows.

11.8.1 Paper

Brown Kraft paper is generally used on mineral wool products to give added tensile strength, ease handling, aid positive location between studs or rafters, or prevent contamination when used under concrete screeds. Paper is often laminated with polyethylene to give vapor control layer properties.

11.8.2 Aluminum foil

Glass-reinforced aluminum foil with either a bright polished or white lacquer surface is utilized with most types of insulant. Primarily it is used as a vapor control layer or as a means of upgrading the fire properties of plastic foams, but it does give a semi-decorative finish to the insulation. It is therefore often use where the insulation is open to view but located away from direct risk of mechanical damage.

11.8.3 PVC

Generally used with mineral wool products where, in its decorative forms, it gives attractive facings to ceilings and wall tiles and enhances their sound-absorption characteristics. PVC is also used as a vapor control layer facing.

11.8.4 Tissue

Glass-fiber tissue or non-woven fabrics are used for decorative purposes on many insulants. They also give improved strength to foam plastics and enhanced sound-absorption characteristics to mineral wools.

11.8.5 Glass cloth

As with tissue, woven glass cloth is used for decorative or acoustic purposes. Additionally, close-woven fabrics give improved fire-safety properties and are resistant to mechanical abuse. Glass cloth or scrim of an extremely open weave is used on insulants as a key for mastic or hard-setting finishes.

11.8.6 Wire netting

Used to give mechanical strength to mineral wool mattresses and is an aid to application. Also a key for mastic or hard setting finishes applied on site.

11.8.7 Laminates

Insulation is widely used in laminate form with plasterboard, chipboard, cement board, and metal sheeting, etc. Applications are generally as lining board systems in buildings but prefabricated metal panels and ceiling tiles are also produced.

11.9 Insulation types

It is not possible to detail all the many different types of insulation used in production, service and building industries. The following attempts to give a summary of the composition, properties and major areas of use of a representative range of insulation types.

11.9.1 Mineral wool

Mineral wool is perhaps the best known of the whole range of insulation types. It is widely used in all sectors of industry, transport and building for thermal, acoustic and fire-protection purposes. There is a common misconception that mineral wool is a specific product type – it is not. Mineral wool is a generic name for a range of man-made non-metallic inorganic fibers. The following definitions should help to clarify the situation:

1. *Mineral fiber*: A generic term for all non-metallic inorganic fibers.
2. *Mineral wool*: A generic term for mineral fibers of a woolly consistency normally made from molten glass, rock or slag.
3. *Glass wool*: A mineral wool produced from molten glass.
4. *Rock wool*: A mineral wool produced from naturally occurring igneous rock.
5. *Slag wool*: A mineral wool produced from molten furnace slag.

From these it can be seen that rock wool, slag wool and glass wool are all mineral wools.

11.9.2 Glass wool

Glass wool is made from borosilicate glass whose principal constituents are sand, soda ash dolomite,

limestone, ulexite and anhydrite. They are melted in a furnace at about 1400°C and then fed along a channel to a forehearth, where the glass flows through bushings into spinners. These rapidly rotating spinners have several thousand small holes around the perimeter through which the glass is forced by centrifugal force to become fiberized. Immediately after formation, the fibers are sprayed with resinous binders, water repellents and mineral oils as appropriate and fall under suction onto a moving conveyor, which takes the wool to one of three production lines (Figure 11.4).

Main line

The wool passes through an oven, which cures the resinous binder and determines the thickness of the product. On leaving the oven, the insulation is trimmed, slit and chopped into the appropriate product length prior to reaching the packing station, where it is either packaged as rolls or slabs.

Pipelines

The uncured wool from the forming conveyor is separated into 'pelts', which are converted into pipe sections by being wrapped around a heated mandrel, and the wall thickness set by counter rollers. The sections are then passed through a curing oven before being trimmed, slit, covered and packaged.

Blowing wool line

The water-repellent wool is shredded by a flail and pneumatically transferred to a rotating drum nodulator to be further processed before bagging.

Product range

A wide product range, from lightweight mats through flexible and semi-rigid, to rigid slab. Pipe sections, loose wool, blowing wool, molded products and mattresses.

Typical properties

Glass wool products have a limiting temperature of 540°C but are mostly used in buildings and H&V applications where a limiting working temperature of 230°C is recommended.

Fire safety   Basic wool is non-combustible to BS 476: Part 4.

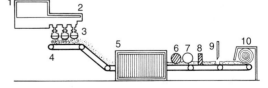

Figure 11.4 The glass wool manufacturing process. 1 Tank; 2 forehearth; 3 spinners; 4 conveyor; 5 curing oven; 6 trimmers; 7 slitters; 8 bandsaw; 9 guillotine; 10 rolling machine.

Density range $10-75\,kg/m^3$.

Thermal conductivity   0.04–0.03 W/mK 10°C mean: 0.07–0.044 W/mK at 100°C mean.

Special applications

Acoustic absorption in duct lining and splitters. Decorative facings for ceiling and lining panels. Tensile strength for good draping qualities. Loft insulation up to 200 mm thick, cavity wall insulation. Fire stopping and small-cavity fire barriers.

11.9.3 Rock wool

Rock wool is made from basalt or dolomite rocks. The crushed rock, together with limestone and coke, is loaded into a cupola furnace where, with the aid of oxygen, it is melted at about 1500°C. The molten rock then runs down channels and cascades onto a train of rotating discs, which sling the melt off as fibers. Resin binders, water repellents and mineral oil are sprayed onto the fibers as they leave the discs and fall under suction onto the forming conveyor. From this stage, the wool follows a similar processing sequence to glass wool.

Product range

Lightweight mats, flexible, semi-rigid and rigid slabs, pipe sections, loose wool, blowing wool mattresses, sprayed fiber, lamella mats.

Typical properties

Rock wool special products can be used up to 1100°C, but generally a maximum operating temperature of 850°C is recommended. Like glass wool, the lower density products used in buildings and H&V applications have recommended limits of 230°C.

Fire safety   Basic wool is non-combustible to BS 476: Part 4.

Density range $23-200\,kg/m^3$.

Thermal conductivity   0.037 – 0.033 W/mK at 10°C mean: 0.052–0.042 W/mK at 100°C mean.

Special applications

High-temperature work, fire protection, acoustics, sprayed fiber, molded products, cavity-wall insulation, loft insulation.

11.9.4 Ceramic fibers

As with mineral wools, there are different types of ceramic fiber, but they are all made from a combination of alumina, silica and china clay and may be made by blowing or extruding the liquid melt.

Product range

A wide basic range of blankets, slabs, block and pipe sections, especially molded products, felts and gaskets.

Typical properties

Ceramic fibers are used at temperatures up to 1600°C but their melting point can be in excess of 2000°C.

Fire safety   Basic wool is non-combustible to BS 476 Part 4.

Density range $50-300\,kg/m^3$.

Thermal conductivity   1.8–0.8 W/mK at 500°C mean.

Special applications

Furnace linings, 'refractory blocks', gaskets, expansion joints, very high-temperature work.

11.9.5 Magnesia

Magnesium carbonate is extracted from dolomite rock, and after mixing with fiber reinforcement is slurry cast into appropriate molds. After drying, the products are machined to size.

Product range

Rigid slabs, pipe sections, beveled lags and molded products. Also as a powder mix for use in wet cement form.

Typical properties

Temperature range   Up to 315°C.

Fire safety   Non-combustible to BS 476: Part 4.

Density $190\,kg/m^3$.

Thermal conductivity   0.058 W/mK at 100°C mean.

Special applications

Where load-bearing capacity is required and in food, pharmaceuticals and cosmetics processing industries.

11.9.6 Calcium silicate

A chemical compound of lime and silica with fiber reinforcement for added strength. It is cast as a wet slurry into molds and charged into autoclaves, finally being machined to accurate size.

Product range

Rigid slabs, pipe sections, beveled lags, molded products and as a dry mix for wet plastic application.

Typical properties

Standard products have limiting temperatures of 800°C but special formulations enable applications up to 1050°C.

Fire safety   Non-combustible to BS 476: Part 4.

Density range   240–400 kg/m^3.

Thermal conductivity   0.054–0.068 W/mK at 100°C.

Special applications

Industrial process applications where compressive strength is needed. Also used in underground pipework district heating mains.

11.9.7 Cellular glass

Powdered glass and crushed carbon are placed in molds and heated to 1000°C, at which temperature the carbon is oxidized, forming gas bubbles which causes expansion of the glass mix. The cellular material is then annealed and, after cooling, cut to size.

Product range

Rigid slabs, pipe sections, beveled lags, molded products.

Typical properties

Temperature range   −260°C to 430°C.

Fire safety   Non-combustible to BS 476: Part 4.

Density range   125–135 kg/m^3.

Thermal conductivity   0.043–0.45 W/mK.

Special applications

Its closed-cell structure makes cellular glass particularly suitable for refrigeration applications on pipes and cold stores. High load-bearing capacity enables it to be used under rooftop car parks.

11.9.8 Exfoliated vermiculite

Vermiculite is a naturally occurring group of hydrated aluminum–iron–magnesium silicates having a laminate structure. When subjected to direct heat in a furnace, the pulverized material 'exfoliates' or expands in size, and then consists of a series of parallel plates with air spaces between.

Product range

Base product produced as a granular loose fill, which can be bonded to form boards or dry mixed with fillers and binders for spray application.

Typical properties

Limiting temperatures   1100°C.

Fire safety   Non-combustible to BS 476: Part 4.

Density   60–80 kg/m^3.

Thermal conductivity   0.062 W/mK at 10°C mean.

Special applications

Primarily used in fire-protection applications as dense boards or wet sprays.

11.9.9 Expanded polystyrene

Expandable polystyrene grains are usually heated by steam, which causes them to expand. They are then conditioned and, as they cool, the steam in the voids within the beads condenses, thus permitting air to diffuse into them. After conditioning, the granules are placed in molds through which more steam is blown, resulting in further expansion. As they are enclosed in the mold the granules tend to fuse together, forming a rigid block. The blocks are later cut to shape by the hot-wire technique.

Product range

Rigid slabs, pipe sections, tiles, and loose granules.

Typical properties

Limiting temperatures   80°C.

Fire safety   Flame-retardant grade. Class P not easily ignitable to BS 476: Part 5.

Density   12–32 kg/m^3.

Thermal conductivity   0.037–0.034 W/mK at 10°C mean.

Special application

Used mainly for building insulation and as a laminate. Also employed in cold water tanks and pipes.

11.9.10 Extruded polystyrene

This is produced by a continuous extrusion process, which gives the product a smooth-surface skin and enhances the mechanical properties.

Product range

Rigid slabs.

Typical properties

Limiting temperature   75°C.

Fire safety   Classified as Type A to BS 3837.

Density   11–45 kg/m^3.

Thermal conductivity   0.025–0.033 W/mK at 10°C mean.

Special applications

Used mainly in structural applications. Is suitable for refrigeration and cold stores.

11.9.11 Rigid polyurethane foam

Polyurethanes are manufactured by the mixing of various resins, isocyanates and catalysts to produce an exothermic reaction, which liberates the foaming agent and causes the mix to expand. They are made in large block molds as a batch process or are continuously foamed onto a paper or polythene substrate on a conveyor system.

Product range

Rigid slabs, pipe sections, beveled lags.

Typical properties

Limiting temperature   110°C.

Fire safety   Class P not easily ignitable to BS 476: Part 5. Can achieve Class 1 surface spread of flame to BS 476: Part 7 when faced with aluminum foil.

Density   30–160 kg/m^3.

Thermal conductivity   0.023 W/mK at 10°C mean.

Special applications

Used mainly for refrigeration and cold-water services, also low-temperature tankage. When laminated to suitable facings is employed in building applications.

11.9.12 Polyisocyanurate foam

Polyisocyanurates are manufactured in a similar way to polyurethane, the chemical components being selected to enhance their fire-safety properties.

Product range

Slabs, pipe sections, beveled lags.

Typical properties

Limiting temperature   140°C.

Fire safety   Class 1 surface spread of flame to BS 476: Part 7. Can achieve Class O to Building Regulations when faced with aluminum foil.

Density   32–50 kg/m^3.

Thermal conductivity   0.023 W/mK at 10°C mean.

Special applications

For cryogenic and medium-temperature pipework and equipment, especially in oil petrochemical industries and on refrigerated road and rail vehicles.

11.9.13 Phenolic foam

Similar production process to polyurethane. Has the best fire-safety properties of all the rigid foams.

Product range

Slabs, pipe sections, pipe bends, beveled lags.

Typical properties

Limiting temperature   150°C.

Fire safety   Class O to Building Regulations.

Density   35–200 kg/m^3.

Thermal conductivity   0.02 W/mK to 10°C mean.

Special applications

Commercial and industrial H&V applications and where Class O fire rating and low smoke-emission characteristics are required.

12

Paint Coatings for the Plant Engineer

D A Bayliss
ITI Anti-Corrosion Ltd

Contents

12.1 Definition and function of coatings

The term 'paint' is normally used to describe the liquid material before application and 'coating' after it has been applied, dried and cured. Organic-based coatings form the largest use for protection and decoration, and these can vary widely in properties and characteristics, ranging in thickness from a few microns to several millimeters. The term 'coating' is not confined to thick films.

Metal coatings (e.g. zinc on steel) can also provide a useful and economic form of protection for steelwork and plant in the appropriate circumstances. Some metallic coatings (e.g. hot-dip galvanizing) act as a complete barrier to the environment, but most corrosion-protection metal coatings and all organic coatings are permeable to moisture and gases to some extent. The rate of permeability varies with different types of coating, but in all cases protection of the substrate is mainly afforded by adhesion. Good adhesion, which is largely a product of adequate surface preparation, prevents the lateral spread of moisture and contaminants, which would undermine the coating and lead to disbondment or corrosion and eventual breakdown.

The thickness of the coating also affects permeability, and in general, there is a correlation between thickness and life. However, with relatively thick coatings as obtained with some modern materials such as epoxies the relationship is not so well established. There may be a limiting thickness above which either little additional protection is obtained or the increase in cohesive strength reduces its adhesive strength. Coatings should always be applied as closely as possible to the manufacturer's recommended thickness.

Coatings will also react with the environment over a period of time. Metal coatings corrode, albeit at a slower rate than steel. They have a finite life in a specific environment, depending on the thickness of the coating. Organic coatings react in a different way, and generally, a type can be chosen which will have superior resistance to chemical attack. However, they also deteriorate in time, the main natural destructive influences being sunlight and moisture.

The normal visible deterioration of a paint coating is by the appearance of 'chalking' on its surface. Chalking is the term used to denote the powdery material that appears as the binder slowly disintegrates and exposes the pigments.

12.2 The constituents of paint

All paints consist of a binder (sometimes called a medium) and pigment. Materials consisting of binder only are called varnishes. Most paints and varnishes contain solvent in order to make the binder sufficiently liquid to be applied. The combination of binder and solvent is called the vehicle. Some paints are available without solvent (e.g. solventless epoxies) but these generally require special methods of application (e.g. application of heat) to reduce the viscosity.

12.2.1 Paint binders

Binders are the film formers. After application, they turn from liquids to solids. Different types of paint have, in some cases, radically different mechanisms for this process. This gives important differences in properties between the paints made from different binders. In all cases, however, binders are a vital constituent of paint and provide its main mechanical and physical properties. Binders may vary from 20 to 50 per cent by weight of the paint but will be a higher percentage in the coating film.

12.2.2 Paint pigments

Pigments are the solid constituents added during paint manufacture. Their function is not solely to provide color but also opacity. Since pigments are opaque they also help to protect the binder from the harmful ultraviolet content of sunlight. In general, varnishes are less durable than paints. Pigments also reduce permeability, particularly when they are in the lamellar form (e.g. micaceous iron oxide, aluminum and graphite), since they increase the moisture path through the film.

Some pigments exert an inhibitive effect on the corrosion of metal. The mechanisms are complex and not always fully understood. Inhibitive pigments include red lead, zinc phosphate and zinc chromate.

Most pigments can be used in any type of binder; therefore, paints cannot be identified by pigment type alone. For example, micaceous iron oxide pigment is traditionally in an oil-based binder but is being increasingly used in epoxies, etc. In the paint coating film, the pigment content may vary from 15 to 60 per cent. In the special case of zinc rich primers, it is over 90 per cent.

The proportions of pigment to binder are a critical factor in paint formulation not only in providing the optimum strength and impermeability but also in the finished appearance (e.g. high pigmentation can give a matte surface, low pigmentation a high gloss). All other things being equal, a high gloss is generally more durable than a matte finish.

12.2.3 Other constituents

Extenders

These are similar to color pigments but are generally inorganic. Although often considered as cheap fillers, they can form a useful function as reinforcement to the coating film.

Diluent

This is a volatile liquid, which, while not a solvent for the binder, can be used in conjunction with the true solvent to reduce cost without precipitation of the binder.

Anti-skinning agent

This is added to reduce or prevent skin formulation in partially filled containers.

Water-based corrosion inhibitor

This is added to water-based paints to reduce corrosion of the container.

12.2.4 Solvents

Solvents are used to reduce the viscosity of paint so that it is suitable for application. They play no direct part in the protective value of the coating and are lost to the atmosphere. Amounts are kept as low as possible (usually 5–40 per cent by weight).

The type of solvent used depends on the binder. There is normally more than one type of solvent in paint, particularly for spray application. Highly volatile solvents are needed to reduce the viscosity during atomization and then disperse as quickly as possible, but lower volatile solvents are necessary to remain momentarily to ensure that there is sufficient flow to form a continuous film.

Organic solvents are a toxic, fire and explosive hazard to varying degrees, depending on type. Solvent entrapment is one of the most common forms of premature failure of modern, high-build, fast-drying materials. Paint manufacturers are urgently developing materials that can be substantially water based. Currently, this can only be achieved with a comparative loss of durability and some application problems, but development of suitable materials is inevitable for the future.

12.3 Types of coating and their uses

A convenient method of classifying paint coatings is by their curing mechanism. The process by which the wet film turns into a solid one has a significant effect on the properties of both application and durability. However, within a particular generic type the formulation can vary, and one manufacturer's product may be superior to another's. The general lack of standards and changes in the use of raw materials by manufacturers means that testing cannot directly compare except different products.

Given below are the main types of paint, their properties and some general guidance on their use. For any specific requirements, it is always advisable to check with paint manufacturers.

12.3.1 Air-oxidizing coatings

Examples

> Alkyds
> Silicone alkyds
> Urethane alkyds
> Epoxy esters
> Tung oil phenolic

General advantages

1. Relatively cheap;
2. Easy to apply by brush or spray;
3. Generally formulated with mild solvents, such as white spirit;
4. Readily available in a wide range of colors;
5. Reasonable gloss retention on exterior exposure;
6. Resistant to mineral and vegetable oils;
7. Available in high gloss.

General disadvantages

1. Poor alkaline resistance (e.g. not suitable for application to concrete);
2. Not suitable for very wet, immersed or condensation conditions;
3. Limited resistance to solvents;
4. Relatively slow drying (e.g. touch-dry 4–6 hours, dry to re-coat 12 hours (minimum));
5. Possible problems of inter-coat adhesion with unweathered high-gloss coats;
6. Can only be applied at limited film thickness.

Comparison of paints within this classification

Alkyds These are the most widely used of all air-oxidizing coatings and have the broadest use both industrially and domestically. They are usually classified according to the proportion of drying oil to synthetic resin (known as 'oil length'). The oil length influences all properties (e.g. chemical resistance, viscosity, flexibility and hardness).

Long oil-length alkyds (e.g. 60 per cent and over of oil) are the most suitable materials for site application on a wide range of surfaces under normal exterior or interior environments. They are not suitable for very wet conditions, immersion or condensation, or for application to alkaline surfaces such as concrete.

The shorter oil-length alkyds are quicker drying, harder finishes which are most suitable for works application. The very short oil-length alkyds are used for stove finishes on industrial and domestic equipment, etc.

Silicone alkyds These are alkyds modified with silicone resin. They have superior resistance to weathering (particularly gloss retention) than pure alkyds but are generally significantly more expensive. They are useful for exterior use where appearance is important.

Urethane alkyds These are formed by the reaction of an isocyanate with alkyd, although the curing remains substantially through the oil-oxidation reaction. They have properties similar to alkyds although superior resistance to abrasion is claimed.

Epoxy ester Epoxy esters are a type of alkyd where a high molecular weight resin is reacted with alkyd resin. The curing mechanism remains primarily through the oil-oxidation reaction and their properties are in no way similar to the chemically reacted epoxies. They have similar properties to alkyds although with improved chemical resistance but inferior appearance. They form a reasonably hard, oil-resistant coating, which can sometimes be suitable for machinery enamels, but are primarily for interior use, since they tend to chalk rapidly on exteriors. Their best use is for chemical or water resistance where circumstances dictate that finishes that are more superior cannot be used.

Tung oil phenolic Tung oil is a natural, fast-drying material which, when combined with phenolic resins, forms an air-drying film of good film hardness, water resistance, flexibility and toughness. Prior to the development of more modern materials, it was widely employed as a corrosion-resistant coating or varnish, particularly for marine use. It tends to yellow on aging, so is not suitable for decorative use but when pigmented with micaceous iron oxide is still widely used for exterior structural steelwork painting.

Its best use is for bridges, gantries, conveyor-belt steelwork, etc., where the relatively drab, matte appearance can be tolerated. The Department of Transport uses this material on inland motorway bridges, and has found that, if applied to a correctly prepared and primed surface, it can be substantially maintenance-free for 20 years.

12.3.2 Solvent-dry coatings

Examples

Chlorinated rubber
Vinyls
Acrylated rubber
Modified chlorinated rubber
Bituminous
Vinyl tar

General advantages

1. Quick drying (typically touch-dry 2 hours, dry to re-coat 8 hours);
2. Low water permeability, suitable for wet conditions and condensation;
3. Good acid resistance;
4. Good alkaline resistance, therefore suitable for application to concrete, etc.;
5. Single-pack material;
6. Very easy to re-coat in both new and old condition;
7. No danger of loss of adhesion between coats;
8. Can be used for water immersion (e.g. swimming pools) but its thinner film generally makes it less durable than two-pack, chemically reacted materials (e.g. epoxies);
9. With the exception of bitumen, available in a limited range of colors, but light shades are generally more durable externally;
10. Forms a solid film entirely by solvent evaporation and therefore does not rely on temperature, oxidation, etc. to cure. For internal, closed conditions, adequate ventilation is the main requirement.

General disadvantages

1. Limited temperature resistance (generally, maximum 600°C dry heat);
2. Very limited resistance to solvents, therefore it tends to be 'lifted' or 'pickled' if overcoated with paints containing stronger solvents (e.g. two-pack epoxies or urethanes);
3. Thermoplastic material, so it moves with temperature and should not be overcoated with harder, thermoset

materials such as alkyds, since these will eventually crack and craze;
4. They contain a high percentage of strong solvents (typically 60–75 per cent), therefore a limit to film thickness and always a risk of solvent entrapment;
5. Relatively more difficult to apply than other types, particularly by brush. Since the paint always remains soluble in its solvent, the brush tends to 'pick up' the previous coat.

Comparison of paints within this classification

Chlorinated rubber   These are formed by the reaction of chlorine on natural rubber. They become a very hard resin, lacking the elastic and resilient characteristics of rubber. They have a specific gravity of 1.64, which is almost twice that of pure rubber. Chlorinated rubber resin is odorless, tasteless and generally non-toxic (although it is sometimes alleged that they can contain residual carbon tetrachloride from the manufacturing process). They are generally less critical in application than vinyl but are less resistant to oils and fats. They will not support combustion or burn and need careful spray technique to avoid *cobwebbing*. The best use is for surfaces liable to condensation and for coating concrete.

Vinyls   Vinyl chloride co-polymer resins were developed in the USA in the late 1930s. They have better weather and slightly more chemical resistance than chlorinated rubber paints. They are generally resistant to crude oil but application is more critical. For example, they are particularly sensitive to moisture present on a surface during painting and this can lead to adhesion failure. They are also more prone to solvent entrapment than chlorinated rubber paints.

Acrylated rubber   These are based on styrene butadiene and have become commercially available only relatively recently. They are manufactured in several grades but most have the advantage over other materials in this class of being based on white spirit solvent rather than the stronger and more obnoxious xylol. In other respects, they are similar to chlorinated rubber and cost approximately the same, although they are easier to airless spray and the dried film contains less pores. They are considered to have superior weather resistance to chlorinated rubber and vinyl.

Modified chlorinated rubber   In these, part of the resin is replaced with an alkyd or oleo-resinous material. The chlorinated rubber content can remain high but the modification gives better adhesion to steel and easier application properties. However, there is some sacrifice in corrosion resistance. In other versions the alkyd resin becomes the major ingredient and these are claimed to have good weathering characteristics, adhesion, gloss and brush-ability, plus some improved chemical and water resistance over the alkyd alone. The best use is as a primer for steel under chlorinated rubber topcoats. They are widely used for this purpose by the Department of Transport for painting bridges in adverse conditions.

Bituminous   This term is used for products obtained from both petroleum and coal tar sources but the petroleum products are the more widely used. These materials are very resistant to moisture and tolerant to poor surface preparation. They are only available as black, dark brown or aluminum pigmented. The last has reasonable outdoor durability but, without the aluminum, the film will crack and craze under the influence of sunlight. Normally they cannot be over-coated with any other type of paint, because not only will harder materials used for over-coating tend to crack or craze but there is also a possibility that the bitumen will 'bleed through' subsequent coats. The best use is as a cheap waterproofing for items buried or out of direct sunlight.

Vinyl tars   These are a combination of vinyl resins and selected coal tars. They are claimed to be similar to the two-pack coal tar epoxies but with the advantages that they are a single pack, not dependent on temperature for curing, and are easy to re-coat at any stage.

Their disadvantages, particularly in comparison with coal tar epoxies, include the fact that they are limited in film thickness per coat and therefore require multi-coated application. They have a higher solvent content and therefore there is an increasing risk of solvent entrapment, and the slower cure may limit their use in a tidal zone. The best use is for immersed conditions.

12.3.3 Chemically cured coatings

Examples

    Two-pack epoxies
    Coal tar epoxies
    Two-pack urethanes
    Moisture-cured urethanes

General advantages

1. Very resistant to water, chemicals, solvents, oils and abrasion;
2. Tough, hard films that can, if correctly formulated, be applied to almost any thickness.

General disadvantages

1. The majority is two-pack materials that must be mixed correctly before use;
2. Their cure, overcoating properties, pot life (e.g. the time the mixed components are usable) are all temperature dependent. This means, for example, that an increase of 10°C can halve the reaction times;
3. They are more difficult to overcoat then most types, and this can lead to problems of intercoat adhesion. Cured films require abrading before subsequent coats will adhere;
4. As a rule, they cannot be applied over other types of paint;
5. The high cohesive strength developed during the curing of these materials tends to place stress on their adhesive properties.

Therefore, unless they are specifically modified they require a higher standard of surface preparation than other types of paint.

Comparison of paints within this classification

Two-pack epoxies   These were first patented in 1938 but were not in general production until 1947. They have been very widely used over the last decade. Produced from the by-products of the petroleum industry, the basic epoxy resins may be in the form of relatively low-viscosity liquid resins or they may be solid resins of increasing hardness. Both solid and liquid resins can then be reacted with a number of different curing agents. This means that almost any type of film and with any required properties can be made.

These materials are now widely used for coating both steel and concrete surfaces that are subject to a particularly aggressive environment (e.g. North Sea oil platforms). There is less validity for their use under normal atmospheric conditions since they are relatively expensive and tend to chalk on exposure to sunlight. However, their use as zinc phosphate, pretreatment or blast primers for blast-cleaned steel which is subsequently overcoated by any other paint system is an extremely valuable contribution to the painting of new steel work.

Coal tar epoxies   These are a combination of epoxy resins and selected coal tars. Properties can vary, depending on the coal tar-to-epoxy ratio. The ideal compromise appears to be approximately 50/50. Coal tar epoxies are only available in black or dark brown. They cost less than straight epoxies and generally have better wetting properties, so they can be used on slightly less than perfect surface preparation. There are similar re-coating problems as for the two-pack epoxies.

Their best use is for water immersion, particularly sea-water tidal situations. Performance in sunlight varies and should be checked with the paint supplier, as should use for lining potable water tanks.

As with the straight epoxies, a number of curing agents that can be used. The advantages and disadvantages of the main types are listed below:

1. *Polyamine curing*: This provides very good chemical and corrosion resistance, good solvent and water resistance and excellent resistance to alkali. The resistance to weathering is poor since there is a tendency to chalk heavily from exposure to sunlight. These materials can form an amine bloom on their surface if applied under conditions of high humidity. Unless this bloom is removed there is likely to be a loss of intercoat adhesion. The amines in these materials also present some handling hazards, since they are moderately toxic skin irritants, which can cause allergic reactions.
2. *Polyamide curing*: This is the most widely used of epoxy-curing agents and gives a good compromise of properties. The resultant films are generally softer, more resilient and flexible than with amine curing. They have excellent alkali resistance but poorer acid resistance than the amine cured. Generally, the two components that have to be mixed are of similar volume, which

also makes for easier mixing. There are fewer tendencies to bloom and better wetting characteristics than the polyamine type. Both the polyamine and polyamide types will cure slowly at low temperature and will not cure below 5°C.

3. *Isocyanate curing*: These are used when the ambient temperatures are low because they cure rapidly and at temperatures below 5°C. There is a tendency for such films to become brittle with age, and it is preferable not to use such materials at high ambient temperatures, since the pot life and other properties may become unacceptably short.

Two-pack polyurethanes Also called urethanes, these materials are similar to two-pack epoxies in that they can be formulated to provide different properties. They can be made into foams or soft, rubbery materials, as well as very hard, tough, abrasion-resistant coatings.

In comparison with epoxies, the correctly formulated urethane can possess considerably superior gloss retention and flexibility. They also form a film so hard and solvent resistant that they can be used as anti-graffiti coating. The graffiti can be removed with solvent without damaging the urethane. Since these materials give off toxic fumes in a fire it is preferable that they are not used in confined spaces.

The paint is sensitive to moisture during storage and application but then becomes exceptionally water-resistant. Its best use is for exterior exposure in an aggressive environment but where the maximum gloss and color retention is required. It is preferable to apply the urethane as a finishing coat over epoxy undercoating and priming.

Moisture-cured urethanes These are the only paints in this class of materials that are single pack. The curing is provided by moisture from the atmosphere. This has the advantage that the material can tolerate a degree of dampness in the atmosphere and on the surface to be coated. Its disadvantage is that the film thickness per coat is limited and that, once opened, the entire container must be used immediately. Conversely, if the humidity is very low the cure may be lengthened or stopped.

It is claimed that the cured film has similar properties and durability to the two-pack urethane. It also is used as an anti-graffiti finish.

During application of this material, the spray mist is harmful and fresh-air masks must be worn during application. This material should also not be used under confined conditions.

12.3.4 Heat-condensation coatings

Examples

Phenolic coatings
Epoxy phenolic coatings

General advantages

1. These materials have exceptional chemical and solvent resistance;

2. In the fully cured form, they are odorless, tasteless and non-toxic;
3. They have excellent resistance to boiling water, acids, salt solutions, hydrogen sulfide and various petroleum products.

General disadvantages

1. These materials can only be cured at relatively high temperatures (typically, 130–300°C). Therefore, the process is normally carried out at works. With the use of special reactors, it can be accomplished on-site but the operation is difficult. In all cases, the application is critical, and variations in temperature and cure time will result in different film-forming properties. The application should always be left to specialists.
2. The resins need careful storage (possibly under refrigeration) before application.

Comparison of coatings within this classification

Phenolic coatings Phenolic is reacted with formaldehyde under heat to form a completely insoluble material. They are usually applied in an alcohol solution by spray, dip or roller. During their curing, they release water, which must be removed. They have maximum chemical and solvent resistance but poor alkaline resistance.

Epoxy phenolic coatings These materials are also cured at relatively high temperatures and are made by the reaction of the epoxy resin with the phenolic resin. They are slightly less critical in application requirements, are less sensitive to curing conditions and can be applied in thicker coats. The best use for both these materials is as tank linings used for the storage or food products, drinks, etc. or for process plant, evaporators, etc. that contain boiling water.

12.3.5 100 per cent solids coatings

Examples

Coal tar enamel
Asphalt
Unsaturated polyesters, glass flake and fiber reinforced
Powder epoxies
Powder polyesters
Hot-applied solventless epoxies
Hot-applied solventless urethanes

General advantages

1. In general, exceptionally thick films, very fast curing;
2. Can provide exceptional water and chemical resistance;
3. They contain no solvents to pollute the atmosphere.

General disadvantages

1. Generally require specialist, skilled application techniques;

2. Materials and application costs can be high, but for large surfaces, the quick, thick single-coat application can make them cost effective;
3. Cannot be repaired or re-coated easily on-site. In some instances, solvent-based versions are needed for repair and maintenance.

Comparison of coatings within this classification

Coal tar enamel This is derived from the coking of coal and is further distilled to produce coal tar pitches. It is used for hot application on-site. It will crack and craze if exposed to sunlight but has been employed successfully for over 50 years for the protection of underground or immersed structures. The main use is now for the exteriors of buried or immersed pipelines. Different types of enamel are available to give various degrees of heat resistance. It is now generally used for pipelines below 155 mm diameter.

Asphalt Asphalt is a natural occurring mineral or as the residue from the distillation of asphaltic petroleum. It is less brittle and has better resistance to sunlight and temperature changes than coal tar enamel. Its water resistance is good but less than for coal tar enamel. It is not resistant to solvents or oils. It may crack at low temperatures and age at elevated ones. Like coal tar enamels, it is primarily black in color and difficult to overcoat with other materials. Its main use is for the *in-situ* coating of roofs or aboveground steel structures.

Unsaturated polyesters The formation of the coating occurs *in situ* by the reaction between polyester resin and styrene, activated by a catalyst such as organic peroxide. The main use as coatings is in the formation of glass fiber or glass flake, reinforced plastics.

It has good resistance to acid, oils and oxidizing materials but poor resistance to alkalis. Application is difficult and hazardous and is generally a specialist operation.

Shrinkage on curing means that a high standard of visual cleanliness and an exceptionally coarse surface profile is required on blast-cleaned steel. Its main use is for tank linings.

Powder epoxies Applied as powder either by electrostatic gun or fluidized bed, these materials must be cured at 200–250°C. The coating cures within a few seconds and is normally water quenched to control cure time. They have good resistance to water but will tend to chalk on outdoor exposure.

Normally a specialist works application; these materials are useful protective coatings for such items as cable trays and switch boxes. They are also used for the externals of pipelines.

Powder polyesters These are similar to powder epoxies but with superior resistance to weathering and less chemical and corrosion resistance. They are used mainly for outdoors-decorative use on equipment, etc.

Solventless epoxies These use heat rather than solvent to reduce viscosity, and are generally applied by airless spray, which requires specialist skills. The films cure very rapidly and can be built up in one coat to at least 1000 μm. They need a very high standard of surface preparation. The films have exceptional water and abrasion resistance, and are usually only economic for use with large structures.

Solventless urethanes These are similar to solventless epoxies but can be applied in even thicker films without adhesion problems. They are considered to have better water and abrasion resistance than the epoxy but cost more and application is even more critical. Their main use would be for large areas requiring abrasion resistance.

12.3.6 Non-oxidizing coatings

Examples

Grease paint
Wax anti-corrosive compounds

General advantages

1. Cheap;
2. Easy to apply, even with areas of difficult access;
3. Surface preparation not critical;
4. Can be applied in relatively thick films.

General disadvantage

The coating remains relatively soft and flexible, and is not suitable for areas where people may walk or touch the surface. The best use is for steelwork in enclosed areas or with difficult access where surface preparation and application are not easy.

12.3.7 Heat-resistant coatings

Examples

Modified oleo-resinous binder with aluminum pigment
Oleo-resinous binder with zinc dust and graphite pigments
Modified or pure silicone resin with aluminum
Zinc silicate

General requirements

In general, the requirements of heat resistance limit film thickness and therefore corrosion resistance. This is a particular problem when surfaces fluctuate between hot and cold. Coatings should be selected carefully, depending on the exact maximum temperature that will be experienced. Wherever possible, conventional materials should be used. The majority of air-oxidation coatings will be satisfactory up to 95°C and epoxies up to 175°C continuous dry heat.

Comparison of coatings within the classification

Modified oleo-resinous binder with aluminum pigment These materials are sometimes called general-purpose aluminum paint. They are suitable for use up to 200°C but for corrosion resistance need to be applied over a corrosion-resistant primer such as zinc silicate.

Oleo-resinous binder with zinc dust and graphite pigments Generally called zinc dust graphite paint, this relatively cheap material can resist temperatures in the range of 200–300°C. However, regular maintenance repainting will be required.

Modified or pure silicone resin with aluminum These silicone-based aluminum paints can be used for temperatures between 260° and 540°C. They require a minimum temperature for curing (usually about 260°C) and in general have poor corrosion and weather resistance.

Zinc silicate This material has good corrosion resistance and can withstand temperatures up to 540°C, particularly when over-coated with silicone-based aluminum. The zinc silicate requires a high standard of surface preparation before application.

12.3.8 Coatings for concrete

Examples

    Emulsion paints
    Chlorinated rubber
    Vinyls
    Two-pack epoxy
    Water-based epoxy

General requirements

Concrete may be coated for:

1. Decoration;
2. Waterproofing: correctly formed concrete will not transmit liquid water but will always remain permeable to moisture vapor;
3. Protection from freeze–thaw cycles: because concrete absorbs moisture it is very susceptible to damage by water freezing. The physical forces of the ice are greater than the concrete's strength and spalling and shattering occurs. Coatings can help to keep the concrete as dry as possible but they also must withstand the freeze–thaw conditions.
4. Protection for the reinforcing steel: high-performance coatings can help to prevent chloride ions, etc. permeating to the reinforcing steel. However, care must be taken that the use of relatively impermeable coatings does not trap water and make the situation worse. There is a growing tendency to coat the reinforcing bars themselves with powder epoxies;
5. Decontamination: since concrete is a porous material it can absorb oil, rusty water, etc. and coatings can reduce the absorption.

Comparison of coatings within the classification

Emulsion paints These materials are water based and cure by coalescence as the water evaporates. Films formed in this manner have sufficient porosity to allow water vapor to escape from the concrete but at the same time to retard the entry of water. The composition is typically vinyl acrylic (e.g. a combination of vinyl and acrylic resin). The acrylic portion has superior weathering properties. Emulsions are mainly used for decoration.

Chlorinated rubber See Section 12.3.2. This is used for decoration but also provides a relatively impermeable coating. Its best use is for concrete surfaces that have to be kept clean by regular washing.

Vinyls Section 12.3.2. These have been used on concrete for a variety of situations. Thinned down, they have good penetration of the surface and provide a good base for subsequent applications of heavily bodied, thick viscous vinyl coatings. However, since they dry quickly but contain a high proportion of solvent, care must be taken to avoid solvent entrapment in the pores of the concrete.

Two-pack epoxy Section 12.3.3. These are widely used to give the maximum protection to concrete surfaces, floors and walls. They can be applied as relatively thin coatings by spray or as thick epoxy surfaces applied by trowel. In all cases, the application must be preceded by adequate surface preparation (see Section 12.6.1). To allow maximum penetration into the concrete the first coat must have a low viscosity. Coal tar epoxies are used where protection is the main requirement.

Water-based epoxy These are sometimes referred to as acrylic epoxies. They are generally inferior to solvent-based epoxies in durability and resistance but have less toxic hazards and low solvent odor.

12.3.9 Coatings for concrete floors

Examples

    Seals
    Thin coating
    Trowelled coatings
    Self-leveling screeds

General requirements

An important requirement for all floor coatings is adequate surface preparation (see Section 12.4.3). In general, abrasion resistance is proportional to thickness and the thick single-color epoxy screeds will give the maximum life before wear is obvious. Thin coatings will need regular maintenance but will be considerably cheaper, and this must be weighed against ease of access for any particular situation.

Anti-slip properties are generally required. Nearly all surfaces have poor slip resistance when wet. Self-leveling finishes tend to be glossier than other types, and this has a psychological effect. Some surfaces with good initial slip resistance can become dangerous after heavy wear over a long period and should be inspected regularly. Surfaces with the highest slip resistance, particularly if this involves the incorporation of coarse aggregate, will also retain the most dirt and can be costly and difficult to clean.

Comparison of coatings within the classification

Seals   Seals are normally transparent, quick-drying, highly penetrating materials that are used to stop cement dusting and contamination by oils, etc. The moisture-cured urethanes are particularly suitable, since they penetrate to produce a hard-wearing surface. However, if safety factors preclude their use there are also epoxy and even oleo-resinous products available.

Thin coatings   These pigmented materials, commonly known as floor paints, are often based on chlorinated rubber (see Section 12.3.2) or epoxy ester (see Section 12.3.1). They have limited life in heavy traffic but are easy to apply.

Trowelled coatings   There are a number of proprietary products available but the majority are based on epoxy resins and applied at thicknesses ranging from 0.5 to 5 mm, depending on exposure requirements.

Self-leveling   These are generally based on solvent-free epoxy and there are a number of proprietary products available. Generally, they are 3–8 mm thick, but if the base floor is too rough, it is advisable to first level this with a cement/resin screed.

12.3.10 Coatings for wood

Examples

> Oleo-resinous primers plus alkyd finish
> Aluminum-based primer plus alkyd finish
> Emulsion-based primer plus alkyd finish
> Wood stains

General requirements

Wood is painted in order to preserve from decay, to minimize changes in moisture content that lead to distortion, and for decoration. It is a waste of time and effort to apply expensive paints or stains to exterior wood not previously treated with preservative. Preservatives are designed to penetrate the timber to protect against microorganisms and insect attack. Modern preservatives can be over-coated with paint without problems.

Oleo-resinous primers plus alkyd finish   These are normally used for softwood and should be designed to be flexible and smooth for subsequent paint.

Aluminum-based primer plus alkyd finish   The binder for these primers is normally oleo-resinous but the pigmentation is aluminum flake. This type is preferred for hardwoods and softwoods where knots or resinous areas predominate. They do not give such a smooth surface as primers without aluminum.

Emulsion-based primer plus alkyd finish   These are based on acrylic resin dispersions and have the advantage of a rapid rate of drying. They generally have excellent adhesion and flexibility but lack the sealing properties of aluminum primers.

Wood stains   The problem with normal paint systems on wood is that if moisture penetration occurs (e.g. in the untreated end-grain) then the impermeability of the paint causes blistering, cracking and flaking. New developments with translucent finishes for wood incorporate pigments that deflect ultraviolet but do not obscure the wood grain. They are also designed to permit moisture movement but contain wax additives, which shed surface water. They also degrade by gradual erosion rather than cracking and flaking. If the appearance is acceptable, these form a useful, economic answer to wood protection.

12.3.11 Road and floor markings

Examples

> Road paints
> Thermoplastic road-marking materials

General requirements

Road markers require ease of application and very quick drying.

Road paints   These are normally chlorinated rubber alkyds and are easy to apply but are inherently less durable than the thermoplastic materials.

Thermoplastic road-marking materials   These require special application apparatus. Although they are thicker and last longer than the road paints their tendency to discolor may cancel this advantage.

12.3.12 Anti-condensation paints

General requirements

These materials are designed to reduce water from condensation dripping on equipment, etc. They often incorporate particles of cork so that water is absorbed. They are generally thick films to provide some insulation and have a rough textured surface finish to increase the surface area and encourage water evaporation. In general, physical methods of prevention such as adequate ventilation, etc. are more effective.

12.4 Surface preparation and priming

12.4.1 Structural steel (thickness greater than 5 mm)

Abrasive blasting

Abrasive blasting is generally the most suitable and reliable method of obtaining a visually clean surface and a satisfactory surface profile. The visual cleanliness can be specified by reference to the photographs in Swedish Standard SIS 05 59 00. This Standard is incorporated into a new International Standard (ISO 8501, Part 1) and a new British Standard (BS 7079: Part A1). The surface profile can be determined by the comparison method, as specified in ISO 8503, and Part 2 and BS 7079, Part C.

Dry abrasive blast cleaning should be used on new steelwork where the main contaminant is mill scale. For heavily rusted and pitted steelwork, increased durability can be obtained by the use of wet abrasive blasting where this is practicable. The water will be more effective in removing the potentially destructive and corrosive soluble iron-corrosion products that form at the bottom of corrosion pits.

Hand or mechanical tool cleaning

This method is generally not capable of achieving a uniform standard of cleanliness on structural steel. It is not effective in removing intact mill scale or corrosion products from pitted surfaces. The durability of subsequent coats is therefore variable and unpredictable, and depends on the thoroughness of the operation and the exact nature of the contaminants left on the surface. The method should be confined to non-aggressive environments or where short-term durability is economically acceptable.

Acid pickling

This is only suitable for works application and is only economically viable for repetitive cleaning of relatively small simple-shaped items, such as tank plate. The lower profile achieved by this method may not be suitable for high-performance coatings.

Priming

Priming can be carried out before or after fabrication.

Before fabrication   This has economic advantages and cleaning can be of a higher standard and take place before any destructive corrosion has occurred. Pre-fabrication primers need to be extremely quick drying (e.g. within minutes) to allow handling and ensure coverage of the blast profile peaks without attenuation. Normally, the specifiers will have to accept the type and make of primer used at a particular fabricator because, unless it is a major job, it is difficult and expensive to change primers on automatic plant. Most fabricators now use a zinc phosphate two-pack epoxy primer, which is a sound base for any subsequent paint system. At 25 μm these primers are generally capable of being welded through, and although they are based on two-pack epoxy they are so formulated that the cure is delayed, and they remain over-coatable for at least one year. Zinc-rich epoxies, which were formerly popular for this application, are now out of favor because the formation of zinc salts on the surface after outdoor exposure can result in adhesion failure of subsequent coats.

After fabrication   If a structure consists of a large area of welds, it is inadvisable to carry out pre-fabrication

blasting and then prepare the weld areas to a much lower standard. In those cases, the structure should be blast cleaned after fabrication. This has the advantages that a thicker primer can be used and that the entire paint system can be completed without delay and possible contamination between primer and topcoats.

12.4.2 Zinc metal coatings

Hot-dip galvanizing (new)

Etch primer pretreatment   This is suitable for overcoating with most coatings but is sensitive to moisture during application. It must not be applied as a thin, transparent coating (typically, 10 μm) or inter-coat adhesion loss can take place.

T-wash pretreatment   This acid-mordant solution turns the surface black when correctly treated. Problems can occur with use of such an acid solution *in situ* and from its pungent odor. It must be applied to zinc in a bright condition without corrosion products on its surface. The paint manufacturer's advice must be sought before using under thick coats of two-pack epoxy or urethane.

Sweep blast or abrading   This is a suitable method before application of thick two-pack materials but care is required not to remove too much zinc coating.

Hot-dip galvanizing (weathered)   Weathered zinc surfaces that have lost their initial bright appearance and which are to be painted with thin paint systems need only to have dirt and zinc-corrosion products removed by brushing. For high-performance coatings, it would be advisable to remove zinc-corrosion products by abrading or sweep blasting and treat as for new galvanizing.

Sherardizing   No special treatment is required other than the removal of any zinc-corrosion products, dirt or oil, etc.

Zinc or aluminum metal spray   Sprayed metal coatings are porous and should be sealed after application by applying a sealer coat (i.e. a thin coat such as an etch primer) or a thinned version of the final coating system. Oil-based systems should not be used. Metal spray coatings can have excellent durability without overcoating with paint and, particularly for aggressive conditions; it is preferable to leave them with sealer only.

12.4.3 Concrete

General requirements

Surface preparation of concrete consists mainly of removing laitence, form oils and air pockets. Laitence is the fine cement powder that floats to the surface of concrete when it is placed. Coatings applied over such a powdery, weak layer will lose adhesion. Form oils are used for the easy stripping of forms or shuttering. Their presence will also cause loss of adhesion of subsequent coatings. Forms should be coated with non-migratory hard coatings and the use of oils or waxes prohibited.

Air pockets or bubbles are left on the surface of all concrete. Good vibration and placing techniques will reduce their number but not eliminate them. Many air pockets have a small opening on the surface in relation to their size. Paints will not penetrate into such holes, with the result that air or solvent is trapped and subsequent expansion will cause the coating to blister. In addition, some air pockets are covered with a thin layer of cement that also has no strength and will cause loss of adhesion.

The following alternative methods of surface preparation can be used for new concrete.

Dry abrasive cleaning

This should be carried out using a finer abrasive, lower pressure and a faster movement across the surface with the blast nozzle further from the work than for blast cleaning steel. It provides a roughened, irregular surface with the laitence removed and all holes and voids opened up so they can be more easily sealed.

Wet abrasive cleaning

As for dry abrasive cleaning, but with the advantage that there is reduced dust nuisance.

High-pressure water jetting

This is an effective treatment on eroded and weak surfaces but will not open up the sub-surface voids or pockets or provide a surface profile on dense concrete.

Acid etching

Hydrochloric acid (10–15 per cent) is generally used. If applied correctly it will remove laitence and provide an adequate surface for adhesion. It will not open up air pockets and voids and is difficult and hazardous to handle and apply. It is most suitable for use on floors.

Hand or power tools

These are generally more time consuming and costly than abrasive cleaning and can give variable results.

12.5 Specifications

The main functions of a painting specification, whether for new or maintenance work, are as follows:

1. To state the means by which the required life of the coating is to be achieved. This includes surface-preparation standards, paints and systems, application, storage, handling and transport, quality control;
2. To service as a basis for accurate costing;
3. To provide a basis for resolving disputes.

12.6 Economics

The greatest problem in making an economic assessment regarding coatings is the prediction of the coating life. With paint coatings, this is particularly difficult, because there are many stages to completion of the coating process

and many opportunities in practice for problems to arise. Furthermore, if there is poor workmanship at the outset and this is not detected, the potential life may fall short of expectations by a very wide margin.

However, some companies consider that they have sufficient experience of their requirements to use computer programs to provide economic assessments of candidate protective systems. The following summarizes typical steps to be taken.

1. Select suitable systems based on technical requirements.
2. Estimate life to first maintenance of each system.
3. Prepare sound specifications for each system.
4. Estimate maintenance costs at present values for each system (e.g. by using net present value).
5. Select the most economic system if the difference in costs is greater than 10 per cent. Otherwise, select the system for which the most experience is available.

If at the end of the exercise the cheapest system is beyond the budget available then either the budget must be increased or changes in the initial protection will be required. Whenever possible, the level of surface preparation should not be lowered, since if at a later date more money is available the protective coating can be built up but the surface cannot be re-prepared.

12.7 Painting inspection

The potential durability of a coating system can be realized only if it is applied to a suitably prepared surface, in the correct manner under correct conditions. Painting differs from any other industrial processes in that it is not susceptible to operator abuse or adverse environmental influences throughout all stages of the work.

Furthermore, it is generally difficult to deduce from examination of the completed work what has occurred previously.

Many engineers appreciate that painting, which should be a minor part of an engineering project, can assume major proportions if there are problems or premature failures. For any painting work for which premature failure is economically or practically unacceptable, it is advisable to use full-time, qualified paint inspection. Note that part-time inspection, or visit inspection, can in some ways be worse than no inspection at all.

The National Association of Corrosion Engineers (NACE) in the USA has a comprehensive process for the training and certification of painting inspectors. The following are their typical duties:

1. Measure, at regular intervals, the air temperature, steel temperature, relative humidity and dewpoint in the area where blast cleaning or painting is to take place.
2. Decide whether, in the light of these measurements, the ambient conditions are within the specification and therefore whether blast cleaning or painting can proceed.
3. In outdoor conditions assess the way in which the weather is likely to change and decide how this may affect the progress of the job.

4. Examine the abrasive to be used in the blasting process and record the name of the manufacturer and the type and grade of the abrasive.
5. Check that the abrasive is not contaminated with moisture, dirt, spent abrasive, etc., that the blasting equipment can deliver the abrasive at an adequate pressure and that the airlines are fitted with a water trap.
6. Check that the freshly blast-cleaned surface is of the specific standard (e.g. Sa2½, Sa3, etc.) at all points.
7. Measure the surface profile of the freshly blast-cleaned surface and ensure that it is within specification.
8. Check freshly blasted surface for steel imperfections, laminations, weld spatter, etc. and ensure that any necessary metal dressing is carried out.
9. Ensure that the newly blast-cleaned surface is primed within the specified overcoating time.
10. Record the name of the paint manufacturer, the manufacturer's description of each paint used in the system, the reference number and the batch number.
11. Ensure that the paints are applied in the correct sequence, by the correct method and that they are of the correct color.
12. Ensure that, where possible, paint with the same batch number is used on any particularly identifiable unit of the total job.
13. Ensure that the correct overcoating times are observed.
14. Ensure that all paints, particularly two-pack, are thoroughly mixed in accordance with the manufacturer's instructions.
15. Ensure that the pot life and shelf life of paints are not exceeded.
16. Take wet-film thickness for each coat applied and confirm that they are such as to yield the specified dry-film thickness.
17. Measure the dry-film thickness of each coat over a representative area and ensure that the specified film thickness has been attained.
18. Ensure that each coat has been evenly applied and is free from runs, sags, drips and other surface defects.
19. In the event of discovering any paint defects, mark them up and ensure that the necessary remedial steps are taken.
20. When required, perform adhesion tests, curing tests or tests for surface contamination.
21. Ensure that all blasting and painting operations are carried out without contravention of safety regulations.
22. Report on manning levels and time lost due to weather, mechanical breakdowns, labour disputes, etc.
23. Submit full daily inspection reports.

12.8 Factors influencing the selection of coating systems

12.8.1 Environmental conditions

For aggressive conditions, only highly resistant coatings can be considered; for milder environments, virtually the whole range of coatings can be used.

12.8.2 Access for maintenance

The cost of maintenance may be greatly influenced by the disruption caused by the access requirements. This involves not only direct costs such as scaffolding but indirect ones such as disruption of a process plant, closure of roads, protection of plant and equipment in the vicinity, etc.

12.8.3 Maintenance requirements

In some situations, it may be impossible for physical or legal reasons to use blast cleaning or to spray paint when maintenance is required. In such cases, systems that can be easily maintained or very long-life systems should be considered.

12.8.4 Facilities for coating

Suitable skilled labor may not be available, at a particular site, to use complicated equipment (e.g. hot airless spray). On a foreign site or in remote areas even conventional equipment may be unfamiliar or unobtainable.

12.8.5 Handling, storage and transport

The choice of coating will be influenced by the need to avoid damage during handling. Where considerable handling is involved this may be a major factor.

12.8.6 Application properties

Many of the modern, high-performance coatings require a high degree of skill in application and are considerably less tolerant than older, conventional materials. Caution is required in choosing coatings when there is no record of sound work by contractors.

12.8.7 Experience of coating performance

New types of coating are regularly developed and marketed. Before selecting such coatings the specifiers should be satisfied that the test results and practical experience is sound.

12.8.8 Special requirements

Sometimes the type of coating is determined by special requirements such as abrasion or heat resistance. The coating may have to withstand specific chemicals or solvents. All coatings have to be a compromise of properties. A gain in one may be a loss in another.

12.8.9 Importance of the structure

The importance of the structure in both technical and aesthetic aspects is an obvious factor to be taken into account.

12.8.10 Costs

Cost is always a primary factor. However, the actual cost of paint is generally a small percentage of the total.

12.8.11 Other factors

Other factors may include: color, toxicity and safety, temperature restrictions, ability to cure at service temperature, resistance to bacterial effects, shelf life and pot life of the paints.

12.8.12 Limited choice of coating systems

In practice, the selection of coatings is often the opposite to what might be expected. There is a multitude of proprietary materials but the choice of generic types is limited. Section 12.3 of this chapter is designed to illustrate the basic properties of these generic types as a preliminary guide to selection by the engineer. The specifiers tend to examine the available materials to see if they will fit the requirements, rather than vice versa.

12.9 Sources of advice

The subject of paints and coatings is complicated and covers a large number of both uses and conditions. There are few publications that describe all aspects that are particularly suitable for an engineer. There are limited publications on such specialized subjects as the protection of iron and steel.

The technical service departments of paint manufacturers are a valuable source of free information on their own products. For independent advice on new materials, comparison between materials, specifications not tied to proprietary types and failure investigations, there are consultants and test laboratories that will generally assist on a fee-paying basis.

Further reading

Bayliss, D. A. and Chandler, K. A., *Steelwork Corrosion Protection*, Elsevier Applied Science, Barking (1991).
British Standards Institution, BS 5493: 1977, *Code of Practice for protective coatings of iron and steel structures against corrosion*.
British Standards Institution, BS 6150: 1982, *Code of Practice for painting of buildings*.
Chandler, K. A. and Bayliss, D. A. *Corrosion Protection of Steel Structures*, Elsevier Applied Science, Barking (1985).
Department of Trade and Industry, Committee on Corrosion Guides
No. 3: *Economics*
No. 6: *Temporary protection*
No. 12: *Paint for the protection of structural steelwork*
No. 13: *Surface preparation for painting*
No. 16: *Engineering coatings – their applications and properties*.
Durability of steel structures, CONSTRADO.
Painting steelwork, CIRIA Report 93.

13

Insurance: Plant and Equipment

A P Hyde

National Vulcan Engineering Insurance Group Ltd

Contents

13.1 History

It is important to understand how the relationship between inspection and insurance, which is available from the major engineering insurers, came about. From the outset, the objectives have been towards the safe and efficient operation of all types of machinery and plant used in industry and commerce.

The early nineteenth century saw the beginning of factory production systems, particularly in the cotton mills of Lancashire and woolen mills in Yorkshire. Accidents arising from the use or misuse of steam plant, particularly boilers, became common and led not only to damage and destruction to property but also to death and bodily injury to persons in or about the scene of the explosions.

In 1854, the Manchester Steam Users Association was formed to help with the prevention of explosions in steam boilers and to find efficient methods in their use. To achieve this, the Association employed the first boiler inspectors, whose services were then made available to the Association's members. Within a short space of time, the members became convinced that insurance to cover the high cost of repair or replacement of damaged boilers was desirable, and this resulted in the first boiler insurance company (The Steam Boiler Assurance Company) being formed in 1858. The scope of the services for inspection and insurance later extended to include pressure vessels, steam engines, cranes, lifts and electrical plant, the insurance protection in each case being supported by an inspection service carried out by qualified engineer surveyors.

The development of engineering insurance has been closely linked with legislation. The Boiler Explosions Act was passed in 1882 and this empowered the Board of Trade to hold enquiries into the causes of all boiler explosions except where the boilers were used in the service of the Crown or for domestic purposes, and to charge the cost of the enquiry against any person held to have been responsible for causing that explosion. This provided a strong incentive to insure boilers, and in 1901, the passing of the Factory and Workshops Act made the regular inspection of steam boilers in factories compulsory. This was later extended by legislation to include steam boilers in mines and quarries.

From those early days a great deal of legislation has been passed, including the Factories Act 1961 (which repealed earlier acts), the Power Presses Regulations 1965, the Offices Shops and Railway Premises (Hoists and Lifts) Regulations 1968 and the Greater London Council (General Powers) Act 1973, which was specifically directed at self-operated laundries. These have now all been embodied within the Health and Safety at Work, etc. Act 1974, which, although it does not change existing regulations regarding frequency of inspections, does provide that duties under earlier regulations are now enforced within this Act.

Engineer surveyors employed by engineering insurance companies have always been regarded as being 'competent persons' as required by these various Acts. The knowledge and experience, which has been accumulated by the companies over the years, enables them to offer to industry and commerce a service of exceptional quality.

13.2 Legislation

It can be seen from Section 13.1 that as legislation has grown and changed so the role of the Independent Engineering Inspection Authority has had to change to meet the demands and needs of industry and commerce. Legislation, particularly that related to health and safety at work, is something that never remains static. As new areas of potential hazard are identified and their implications are discussed to the point that new legislation is needed, the engineering insurers are making their contribution to discussions. Currently such matters as the Pressurized Systems Regulations and those related to the Control of Substances Hazardous to Health are implemented, and the engineer surveyors are being trained and retrained to enable them to scope with the requirements of this new legislation.

The following sections include the relevant sections of the complete Act and indicate those parts of the legislation within the different plant categories to which the inspections provided by the independent engineering inspection companies will conform. It should, however, be appreciated that while the inspections provided will fulfill the statutory requirements for inspection, the actual responsibilities under the various Acts to conform remains the responsibility of the plant owner/user. It must be appreciated that the interpretation of any Act is a matter for the local Health and Safety Inspectorate, who should be consulted at all times if doubt exists as to whether any item of machinery and plant requires inspection to comply with a statutory provision.

A general guide is set out in matrix form in Figure 13.1, indicating by business and trade the types of machinery and plant that are likely to be found in general usage and the normally accepted position of inspections which may be required to conform with the legislation.

13.3 The role of the inspection authority

Inspection by independent persons or bodies for safety purposes goes back to the middle of the nineteenth century. At that time, the focus of concern was the explosion of steam boilers, and this hazard was most prevalent in the textile industry. Consequently, a group of public-spirited individuals formed the Manchester Steam Users Association for the Prevention of Boiler Explosion. This body carried out boiler examinations and later added insurance as an inducement to the plant owners. By the beginning of the twentieth century steam and gas engines and electrical machines had been added, followed by lifts, cranes and hoisting machines.

The inspection companies shared their technical knowledge, and 1917 saw the formation of the Associated Office Technical Committee, the founding members being:

The British Engine Boiler and Electrical Insurance Co. Ltd
The National Boiler and General Insurance Co. Ltd
The Ocean Accident and Guarantee Corporation Ltd
The Scottish Boiler and General Insurance Co. Ltd
The Vulcan Boiler and General Insurance Co. Ltd

Business	Electrical plant wiring installation	Refrigeration, air conditioning	Process machinery	Power press	Dust extraction plant	Lifts	Motor vehicle lifting table	Manual block & sling	Electrical block	Fork lift truck	Builders hoist	Excavator	Lorry loader crane	Mobile crane	Power crane	Other steam plant & ovens	Air receiver	Hot water boiler	Steam boiler
Bakers	●	●	●			o		o	o	o			o			o	o		o
Brewers	●	●	●			o		o	o	o			o			o	o	●	o
Building contractors	●			o		o		o	o	o	o	o	o	o	o	o	o	●	o
Churches, schools and halls	●	●				●												●	
Clothing factories	●					o		o	o	o			o			o	o	●	o
Dept stores	●	●				o		●	●	●						●	o	●	o
Docks	●	●						o	o	o			o	o	o	o	o	●	o
Dry cleaners	●		●			o		o								o	o	●	o
Engineering works	●	●	●	o	o	o		o	o	o			o	o	o	o	o	●	o
Farms	●	●	●					●		●	●		●			●	●		●
Flats	●	●				●												●	
Food manufacturers, canners	●	●	●	o		o		o	o	o			o	o		o	o	●	o
Garages	o	●		o	o	o	◑	o	o	o			o	o		o	o	●	o
Hotels, etc.	●	●				o										o	o	●	o
Launderers	●		●			o		o	o	o						o	o	●	o
Millers	●		●		●	o		o	o	o			o			o	o	●	o
Nurserymen	●	●														●		●	●
Nursing homes	●	●				●										●	●	●	●
Office buildings	●	●				o										●	●	●	●
Printing works	●		●			o		o	o	o			o	o		o	o	●	o
Provision shops	●	●				o				●							●	●	
Quarriers	o		●		●			●	●	●		●	o	o		o	o		o
Scrap yards	●		●	o	o			o	o			o	o	o	o	o	o		o
Ship yards	●		●	o	o	o		o	o	o	o		o	o	o	o	o		o
Stonemasons	●		●		o			o	o	o			o	o		o	o	●	o
Supermarkets	●	●				o			o	o						●	●	●	●
Theatres, cinemas	o	●				o		●	●							●	●	●	●
Timber merchants	●		●		●			o	o	o			o	o		●	o	●	o
Warehouses, wholesalers	●	●			●	o		o	o	o			o				o	●	o

Key o Statutory need for inspection.
◑ Statutory need for inspection in some cases in other cases inspection recommended.
● Inspection recommended.

Figure 13.1 Statutory requirements

This Committee published its own technical rules, many of which were later incorporated into British Standards. It continues to make an active contribution to standards, guidance notes, legislation and international policy on inspection. Some independent inspection bodies have no connection with insurance but many do, and these companies have the advantage of feedback from an analysis of insurance claims.

Legislation introduced the requirements for statutory examinations by 'a competent person', the responsibility for ensuring the competence of the examiner resting on the plant owner. There is no requirement for the competent person to be independent of the owner's organization, but independence does have clear advantages.

It will be seen that owners and users of plant which can be hazardous to employees or to the public come at the top of a list of inspecting authorities' clients. Safety is the first objective, but avoidance of unscheduled stoppages and objective assessments of plant condition aimed at timely replacement are powerful commercial inducements. The client may be an individual owner of a vintage traction engine or a large public utility, but the advantages are essentially the same. National and local government bodies, including the Health and Safety Executive, also rely on inspecting companies to warn the enforcing authority of potential dangers and to provide 'fitness for use' certificates for hazardous plant items or installations.

Inspection is also an important activity in the regulation of international trade through the certification of vehicles and containers used for transporting hazardous products and for providing foreign purchasers with evidence that manufactured goods comply with specification before they leave the country of origin. The essential characteristics of and requirements for an inspection authority are:

Technical competence
Maturity and integrity of judgement
Confidentiality
Access to a database of relevant plant histories
Ability to communicate

The person who carries out the examination is the 'competent person', although this term may also be used in a corporate sense. One very concise description of the competent person, which originated in a law case, is 'he must know what to look for and how to recognize it when he sees it'. This has been expanded both officially and unofficially, and a current guidance note on this subject is paraphrased as follows:

The Competent Person should have sufficient practical and theoretical knowledge and actual experience of the types of plant he is charged with examining and testing as will enable him to detect faults and weaknesses. He should have the maturity to seek such specialist advice and assistance as will be required to enable him to make the necessary judgements. He should also be able to assess the importance of any faults or weaknesses in relation to the strength and function of the plant before he is required to certify it as suitable to carry out its specified duty. The Competent Person should have access to the equipment, specialist's support and laboratory services necessary to enable him to carry out the relevant examination and testing. Where he is unable to carry out all parts of the examination and testing himself, he must be a proper judge of the extent to which he can accept the supporting opinion of other specialists. An inspecting authority may be appointed as the 'Competent Person'. In this case, where more than one engineer employed by the inspecting body carries out the examination and testing, it is appropriate for a nominated person within the Company to sign the certificate on behalf of the inspecting body.

The larger inspecting companies carry their own specialist support. Typically, this will cover:

Non-destructive testing
Failure analysis using modern techniques such as finite element stress analysis and fracture mechanics
Metallurgical and weld analysis
Chemical engineering and process capability
Quality assurance

Clearly, there are 'horses for courses'. The practicing plant engineer may have access to a local specialist in certain types of non-destructive testing who provides an excellent service and value for money. At the same time, the plant engineer must be vigilant regarding the limitations of such support.

The role of the inspecting authority must clearly be sensitive to change in the light of both technical advances and alterations in the political climate. The response time will sometimes depend on whether the authority concerned has addressed itself to a rather narrow sector of the inspection field or operates over a wide spectrum.

The 1982 White Paper on Standards, Quality and International Competitiveness was concurrent with increasing interest in the techniques of quality assurance and the need for international harmonization of standards and the reciprocal recognition of certification. It is worth noting that the ISO 9000 series of standards on Quality Systems: 1987 followed the layout of BS 5750: 1979 almost clause by clause. ISO Guide 39 covers the general requirements for inspection bodies. Auditing on behalf of certification bodies is part of the inspecting authorities' role.

It is to be hoped that the variety of certification schemes will not multiply, or the client is likely to be confused rather than informed. When making his choice he may usefully remember that the criteria of good quality systems are as applicable to service organizations as they are to manufacturers, and that clarity regarding the objectives of the service offered together with practical common sense in providing this service are the real indicators.

13.4 Types of plant inspected

13.4.1 Boiler and pressure plant

The type of plant which would be included in the above generic description will include many items that, because of difference in size, function and appearance, will appear to be unrelated. They do, however, have one common factor in that all operate to some degree under pressure.

To illustrate this the use of the term 'boilers' can vary considerably not only in its design and construction but in the class for which they are intended, for instance:

1. Steam-generating boilers
2. Low-pressure hot water heating boilers
3. Low-pressure steam heating boilers
4. Domestic hot water supply boilers
5. High-pressure hot water heating boilers
6. Process vessels

In the first category the type most often found in use today will be what is referred to as the steam package boiler of the multi-tubular type, otherwise referred to as the shell boiler. This can equally be used to describe vertical multi-tubular shell boilers, locomotive-type boilers or even watertube boilers more closely associated with power generation or, alternatively, where there is a high and regular steam demand from within a factory. Shell-type steam generating boilers are also used for high-pressure hot water heating where there is the need to supply hot water under a pressure and temperature equal to that of the steam in the boiler for circulation throughout heating or process plants within a building. The circulation arrangement used with these boilers is often similar to those found in connection with low-pressure hot water heating systems, but where the water is at a much higher temperature and the pressure is, equally, at a much higher level. This system can incorporate the use of its own separate pressurization unit.

By far the most numerous class of boiler within the UK must be the low-pressure hot water heating or hot water domestic supply boiler, and these are found installed in various types of situations (e.g. offices, factories, shops, hotels) and now widely used in connection with private dwelling houses.

13.4.2 Steam and air pressure vessels

The diverse types of plant in this category would be too numerous to list, but include steam-jacketed pans, steam calorifiers, steam heating and drying units and all types of air-pressure vessels. These are manufactured in a wide range of materials (e.g. stainless steel) and may then be lined with materials such as rubber or glass, depending on the needs of the process in which the vessel will be used.

13.4.3 Elevators, cranes and other mechanical handling plant

Items in this category can range from small manual items of lifting tackle (e.g. rings, hooks, slings, chains and blocks) to the very large fixed overhead or mobile powered cranes, fork lift trucks, straddle carriers, side loaders, loading shovels and excavators as well as the general type of passenger lift to be found in most modern offices and hotels. Cranes can range from the most basic manual fixed-pillar jib upon which an electric or manual block is hung to the modern specialist container crane or order picker.

13.4.4 Electrical and mechanical plant

Plant falling within this description includes everything from small fractional horse-power motors and their driven components to the major steam or gas turbine and their generators; from small printing machines to complete continuous process lines such as may be found in steel production. In addition, this category would include any item which incorporates either static or moving electrical or mechanical parts.

13.5 Insurance covers on inspected plant

13.5.1 Boiler and pressure plant

Boilers and pressure vessels can be insured against explosion and collapse, which provides for damage to the insured plant, surrounding property and, in addition, liability for damage to third-party property and injury to third-party persons. This cover can be extended to include full sudden and unforeseen damage which, as far as boilers and pressure vessels are concerned, includes overheating, frost, cracking, water hammer action, leakage between the sections of sectional heating boilers and damage by external impact. Often insurance policies are written so that the sudden and unforeseen damage extension provides for resultant water damage to surrounding property and will also provide for costs up to defined limits to expedite repair. There is also provision to insure on a reinstatement basis.

While boiler explosions fortunately do not occur too often today because of the existence of extensive safety devices as well as the regular program of inspection, their effects can be catastrophic. Similarly, sudden and unforeseen damage caused by the overheating of multi-tubular steam boilers due to lack of water can lead to eventual furnace collapse, with very extensive repair costs as well as lost production.

13.5.2 Storage tanks and their contents

Items of this type are linked to the boiler and pressure plant generic type but are given their own specific covers, which are for sudden and unforeseen damage, including collapse, impact and frost as well as damage to the surrounding property, cleaning-up costs and the costs of expediting repair. The actual contents of the tanks, which can be very varied and expensive, can be insured against accidental leakage, discharge, overflowing and even contamination.

13.5.3 Cranes and lifting machinery

The types of plant falling under this heading are many and varied, and insurance covers that are offered are equally varied, falling into the categories of:

1. Breakdown;
2. Accidental extraneous damage only; or
3. Sudden and unforeseen damage, which will include, in addition to the electrical and mechanical breakdown,

accidental causes such as toppling, overloading and collision.

Depending on the type of machine in question and the operations or use to which the machine is put, there are numerous extensions to the basic covers available. These include third-party liability, cover while the plant is hired out, indemnity to the first hirer, damage during erection and dismantling, damage to goods being handled, damage to surrounding property and the cost of hiring a replacement machine following an indemnifiable insured loss or accident.

13.5.4 Lifts and hoists

Lifts and hoists can be insured against breakdown only or for full sudden and unforeseen damage, which includes, in addition to electrical and mechanical breakdown, accidents arising from impact, fatigue, malicious damage and entry of foreign bodies. The cover can be extended to include liability to third parties and damage to goods being carried arising out of the use of an insured item.

13.5.5 Electrical and mechanical plant

Electrical and mechanical plant will include all types of process machinery, engines, generators, pumps, fans, furnaces, transformers and refrigeration plant, to list but a few. Plant of this type can be insured against breakdown only or full sudden and unforeseen damage, which will include, in addition to electrical and mechanical breakdown, accidents arising from faulty insulation, failure of wiring, short circuiting, excessive or insufficient voltage, non-operation of safety or protective devices, renewal of insulating oil or refrigerant and the consequences of impact from an external source.

13.5.6 All machinery and plant insurance

While the normal engineering insurance covers on inspected plant are designed for a high degree of selection of specific items which will be identified in a policy schedule, there are now covers available designed to meet the needs of protection for 'All Machinery and Plant' without the requirement of scheduling of the items separately. The machinery and plant can be insured against breakdown only or for full sudden and unforeseen damage risks, which will include electrical and mechanical breakdown and can be extended to take in the explosion and collapse cover normally found under the boiler and pressure plant section.

In instances where cover is written on this basis, it is normal that higher than usual excess is applied, reflecting the type of plant involved. While it is normal on the more traditional basis for the insurance cover to be complementary by inspection of the insured items in the case of covers written in this manner, it is usual for the inspection of all statutory plant to be incorporated together with those items where the insurer considers inspection to be necessary as part of the risk management of the cover being provided.

13.6 Engineer surveyors

13.6.1 Selection

The AOTC member companies use common criteria for selection of candidates to train an engineer surveyor, and these may be considered typical. Traditionally, surveyors have been recruited from among sea-going engineers who had a first-class Board of Trade Certificate of Competency or a similar certificate from the Engineering Branch of the Royal Navy. Sea-going engineers have suitable practical experience, and are also likely to have had sufficient practice in working relationships to enable them to deal with clients' personnel, including plant engineers, on an equal basis.

As this source has progressively dried up, engineers with an industrial background have been recruited from those with a Technician Engineer qualification (now known as Incorporated Engineer). The minimum age for entry is 25 years, and since maturity of judgement is required (see Section 13.3), considerable weight is attached to the selection interview. A significant number of engineer surveyors now have Bachelor of Science and Bachelor of Engineering degrees or equivalent, some having obtained these during their employment in this field.

In addition, at age 26 engineer surveyors can apply for Associate Membership of the Bureau of Engineer Surveyors, who are a division of the Institution of Plant Engineers. Transfer to membership is possible after a further two years' employment as an engineer surveyor.

13.6.2 Training

The period of training following engagement varies, depending on previous experience, and is normally of between 3 and 6 months' duration. Trainers are usually streamed into specialties (i.e. boiler surveyor, lift and crane surveyor or electrical surveyor). Alternatively, a 'composite' surveyor's training embraces all these technologies. Further specialization sometimes occurs following training and some experience in the field. For instance, surveyors may become specialists in air conditioning and dust extraction, power presses or new construction of pressure containing plant or of rotating machines.

The scope of initial training always covers an induction period of perhaps four weeks, during which the trainee learns about the structure of the company which he has joined and the administration which he will be expected to carry out in the field. It is important that he learns about the legislation which applies to his job and the statutory forms which he must use. Surveyors are often required to keep a notebook, the purpose of which is similar to that of a police officer. He will later be required to cover the interests of all the clients who are his 'district', and he must therefore set up in his home the necessary administrative and record-keeping facilities to enable him to do this effectively and economically.

After initial induction training, technical training is provided which is tailored to meet the individual's needs, based upon background, experience and ability to absorb the information. Quite early in this period of training, the trainee will probably spend time with an experienced surveyor in the field in order to become familiar with as wide a cross-section of plant and type of examination as is possible in the time available. Each type of inspection is discussed to explore the various ways in which it may be carried out, types of defect found and specific preparation or tests that may be used. The trainee makes out a report for each examination together with a dimension sheet for each class of plant.

The trainee then returns to head office and continues training under a head office engineer, who may be especially skilled in training techniques or may be the supervising or section engineer to whom the trainee will later be responsible.

The trainee will visit other departments in the head office to learn about the supporting facilities which are available such as non-destructive testing, metallurgical and failure analysis, hazard analysis and quality assurance. Emphasis is laid on safety, i.e. the integrity and safety of the plant in operation and on the surveyor's own personal safety. It is important that the trainee understands the techniques involved, particularly the limitations of their application.

During the latter part of training, considerable periods are likely to be spent in the field operating alone but returning to head office to discuss the examinations and reports with the training engineer. During this period, the trainee's ability to operate alone is assessed. At various points throughout the training, written examinations may be undertaken, and an essential feature of the training process is that the head office engineer responsible prepares a report on the trainee's progress and ultimately recommends to his manager that the trainee should be assigned to a district.

A surveyor's district is a geographical area of such a size as will keep him fully employed carrying out examinations of plant and supporting his company's clients in the district. He is required to live in or close to his district. He may report to a regional office or direct to his company's head office. Because he operates away from base, his training is recognized as being of particular importance, and it will later be supplemented with regular in-service refresher courses. It is assumed that he will be self-motivating and he is required to provide his clients with a fully professional service, mobilizing the back-up services of his company when necessary on their behalf.

13.6.3 Recognition of defects in operational plant

During this training, an engineer surveyor can expect to be shown examples of the defects that he is likely to encounter. Many inspecting authorities have a 'black museum' of such defects or they are described and illustrated by photographs in the training literature. He will also be instructed, both in the classroom and on-site, about locations where defects are likely to occur. Most surveyors develop an 'instinct', which helps them to find defects, and this can only be acquired through practical experience of different types of plant.

Having found a defect, the surveyor must be able to assess the possible consequences. Construction standards are not of much use, since they are written around what it is possible for a good workman to produce under reasonably good working conditions. For example, BS 4153 and BS 5500 allow intermittent undercut at the edge of welds up to 0.5 mm deep. The new surveyor will have been taught that fatigue cracks often originate from the toe of a weld, so he must be cautious in distinguishing between harmless undercut and an incipient fatigue crack. This is a simple example, which will almost certainly have been thoroughly covered during training, but it illustrates the importance both of experience and of having the backing of NDT specialists who will settle the doubtful cases.

There is no doubt that the judgement as to whether a defect is harmless or may lead to a failure is the most difficult type of decision that any engineer, be he an engineer surveyor or not, is asked to make. His first judgement will be whether the defect was created during original manufacture. If it has developed during operation, he must then assess what has caused it and whether this causative factor is still active. Unless he is certain that defect growth has stopped, he can only assume that it will continue to grow at approximately the same rate. If the surveyor has examined the plant previously and knows its history he will be better placed to estimate the growth rate and to decide whether he can allow the plant to operate until the next examination is due or until some future date which he agrees with the plant owner. Some theoretical knowledge of stresses and of the mechanisms of defect growth is valuable, but experience of similar defects is likely to be the best basis for a judgement.

An engineer surveyor employed by an insurance company will often be used to investigate claims, and each claim investigation adds something to his experience of failures. At the same time, he must observe strict confidentiality. To quote from a quality assurance manual:

An essential part of the Engineer Surveyor's role is the detection of defects in plant or of operating practices, which could lead to a dangerous occurrence. This knowledge is part of his experience. Nevertheless he will never disclose to another client or to a third party the sources of his experience in such a way that breach of confidentiality could result.

All surveyors need a database of technical information, which should include reports of accidents or of defects that could lead to accidents derived from other sources. He also needs a regular supply of technical documentation from his head office that keeps him abreast of technical developments. If his company operates an effective quality assurance system, they will periodically check that he is keeping these data properly and they may control the indexing of them.

The reader may remember the brief definition of a competent person, i.e. 'he must know what to look for and how to recognize it when he sees it'. Perhaps what has been described in this section gives further insight as to how appropriate this definition is.

13.6.4 Reporting

An inspecting authority should always make a written report to its client when the inspection is completed or progressively as it proceeds. In the case of statutory examinations, the format of the report is prescribed in the legislation. The authority must clearly identify both itself and the individual who is responsible for the contents of the report. This will usually be the engineer surveyor who has made the inspection in his capacity as 'competent person'. Forms 55 and 55A that are prescribed for examination of steam boilers (required at intervals not exceeding 14 months) are interesting because they cover a two-part examination. These are the cold or 'thorough' examination with the boiler shut down and prepared for examination and the supplementary or 'working' examination, when the boiler is under operating conditions. As a matter of interest, the 14-month interval starts from the date of the supplementary examination, provided this is done within two months of the cold examination, so a total of 16 months is possible.

Report forms act as checklists to ensure that the essential aspects are covered. However, these aspects will change from one type of plant to another. In general, the report must state:

Identity and address of owner/operator
Location and identity of plant
Age of plant and date of last examination
Scope of examination (itemized on the form)
Parts that could not be examined
Quantitative values where possible
Description of any defects
Statement of any limitations of use
Repairs required and a limiting period for their execution
Observations not covered by the above
Identity of examiner
Countersignature of inspecting authority if required

It is the surveyor's task to report facts, and every report should be made 'without fear or favor'. However, every report also implies an opinion, which inevitably contains a subjective element. This opinion is that the plant will be safe to operate until the next examination is due. The plant owner should remember that engineer surveyors are not infallible, nor do they possess 'second sight'. However, their implied opinion is likely to be the most reliable view that the owner can obtain without spending a great deal more money. The opinion will be additionally reliable if the surveyor has personal knowledge of the plant history. Consequently, if the plant owner thinks of changing his inspecting authority, he might remember that in doing so he will jettison accumulated knowledge which the new authority will take time to acquire.

It has to be recognized that some examiners are better than others in their use of language, and many inspection authorities have standard phrases, which their surveyors are advised to use. Some use a shorthand system to facilitate the production of reports whereby the surveyor sends back a string of numbers to his head office that are then converted into standard phrases in the report. Inevitably,

these reports will look stilted. It should also be remembered that when a surveyor reports regularly on the same item – perhaps at three-monthly intervals on some types of lifting equipment – a good deal of duplication from one report to the next is inevitable. A word processor in the surveyor's home, connected by electronic mail to his head office plus an electronic signature produced by a method that is acceptable to the Health and Safety Executive, is therefore the engineer surveyor's dream.

Having set out the factors that may tend to produce a rather stilted report, we should also consider the positive side. The last thing an inspecting authority wants is for his client to look at the report and say 'Why am I paying so much money for this?'.

Most inspection contracts – whether applied to insurance or not – are aimed at obtaining assurance that plant is safe to operate. If the client needs more than this he should therefore define his additional requirements with care and make sure that these have been effectively communicated. This will probably result in the inspection contract being treated as a 'one-off' basis and a more personalized service should result.

It is extremely important that the obligations on both sides should be recognized. The owner/operator of statutory plant cannot pass on his legal obligations to have his plant properly examined on time. He can, of course, make his inspecting authority contractually responsible for carrying out the examination and providing the report on time, but that is not the same thing. The client also has an obligation to prepare the plant for examination, and this usually involves shutting it down and perhaps dismantling some of its parts. The engineer surveyor has normally learnt that outage time is costly to his client, and will do his utmost to be on time on the appointed date. He will, however, have a natural aversion to remaining inside unpleasantly hot or exceedingly dirty equipment, and a less effective examination may result. It is also essential that both the client and the inspecting authority keep in mind the requirement that 'the engineer surveyor must be allowed the time he considers necessary and sufficient to complete his examination in a satisfactory manner'.

When inspecting certain statutory items of plant, the inspecting authority has a legal obligation to inform the enforcing authority of any plant which is considered to be dangerous. When this occurs, the surveyor makes out a 'site defect notice' at the time of the examination that is signed both by him and by the client's representative. The written report is then sent both to the client and to the Factory Inspectorate branch of the Health and Safety Executive. This is the only occasion when an inspectorate is obliged to break confidentiality with the client, and it should be the only occasion when it does so.

Although written reports are the inspecting authority's end product they by no means comprise the whole of the professional service that is supplied or is available on request. The authority's quality assurance objective is likely to be on the lines of 'client satisfaction allied to compliance with contractual obligations'. Although reports are almost certainly monitored, they are a poor indicator of the quality of the examination, and the authority's quality control will rely less on report inspection and more on surveillance of the whole process that it operates. This will be done through careful attention to:

Surveyor training
Surveyor aids (software and hardware)
Surveyor audits
An administration that removes obstacles, which might prevent the surveyor performing at this best.

The authority should also take pains to promote a good working relationship between the engineer surveyor and the client's representative on-site.

13.7 Technical services

The majority of the independent engineering inspection authorities now provide a wide range of engineering inspection services quite apart from the in-service inspections of insured plant. These services are available to industry at large and, although widely used by presently insured clients, they are frequently employed by organizations that have no specific insurance or inspection involvement with the chosen engineering inspection authority.

The inspections/services provided under the heading of 'Technical Services' form either independent or combined operations of the various technical department disciplines that the chosen inspection authority has to offer. Inspections will range from lifts, cranes and other specialized machinery items, and include electrical inspections of wiring, switchgear, motors, etc. Special inspections of used plant or in-service plant are carried out for a wide variety of clients, and the inspection of pressure plant during construction, including design assessment prior to manufacture, is undertaken by a department specializing in this type of service.

Most inspection authorities have metallurgical and chemical laboratories and a separate non-destructive testing department within their head office complex, undertaking separate inspection/service operations as well as providing a comprehensive support to all of the other technical service functions. It is a requirement under the Health and Safety at Work, etc. Act 1974 that designers, manufacturers, importers or suppliers of articles or substances for use at work must ensure that, so far as reasonably practicable, they are safe and properly used. They must test articles for safety in use, or arrange for this to be done by a competent authority. They must also supply information about the use for which an article was designed, and include any conditions of use regarding its safety.

Anyone who installs or erects any article for use at work must ensure that, so far as reasonably practicable, it does not constitute a risk to health and is safe for use.

The major engineering insurance companies are recognized as competent independent inspection authorities having a range of services, which they have developed during many years of service to the engineering industry, and therefore the following inspection procedures indicate many of the standard inspection services in use. However, special inspection procedures are frequently drawn together to meet particular circumstances, and the inspection authorities offering these services are always ready to discuss any special inspection requirements.

13.7.1 Pressure plant

Pre-commissioning inspection services are designed to examine key stages of production and witness final tests on completion at the manufacturer's works and/or at site as appropriate. It is frequently arranged for the inspection authorities personnel to attend pre-design meetings to discuss the best and most effective inspection service suitable for the item or projecting being considered. The main starting point, however, is normally the design drawing or drawings, together with calculations, which are reviewed to check the integrity of the design in accordance with the specified construction code, which is often to a British Standard.

Once the drawing is approved, the field staff make inspections at key stages during manufacture and follow up on-site when necessary to inspect during prior to commissioning work and to witness tests. The normal stages of inspection for pressure plant would include identification of materials of construction, weld procedures, operator performance qualifications being checked to establish that they are relevant and up-to-date and random stage inspection during welding continues, at intervals, during the fabrication. These latter tests will include heat treatment recording and non-destructive results being subject to assessment.

On completion of the item, it is examined externally while subject to hydraulic test, followed by internal examination so far as construction permits. All the inspection stages are reported at regular intervals during construction, thus providing a prompt written account of inspections during progress of manufacture. A similar design assessment and stage inspection procedure is available for reinforced plastic vessels and tanks.

13.7.2 Electric passenger and goods lifts

Pre-commissioning inspection service for lifts is normally undertaken in one of three ways:

1. At works, at site and during test
2. At site and during test
3. During test only.

It is recommended that the lift specification should call for compliance with British Standards, and the service usually commences with perusal of drawings and specifications. The inspection service can include machine parts prior to assembly and a similar examination of electrical components. Guide rails, brackets and supporting structures are examined and the lift manufacturer's shop tests on motor-control equipment and high-voltage tests of electrical equipment are witnessed. A car sling, safety gear and counter-weight assembly are checked prior to dispatch and shop tests on the governor unit are witnessed. The on-site inspection would include supporting steelwork, the alignment of the car and counterweight guides and the door or gate mountings. The mechanical and electrical parts of the lift would be examined on completion of erection. The manufacturer's tests would be witnessed by the engineer surveyor, including balance tests, performance tests, round trip with 10 per cent overload, safety gear test, limit switch tests, static testing of brake, gate/door lock as well as earth continuity tests and insulation tests. All these inspections are reported with the reports dispatched at intervals during manufacture and a final report is submitted on completion.

13.7.3 Cranes

The inspection of cranes has the three major categories (1)–(3) as applicable to the lift inspection outlined above. The service would commence with perusal of drawings and specifications that should call for compliance with British Standards. Inspection during manufacture would include mechanical parts prior to assembly and selected structural members as well as the examination of motors and electrical equipment and the checking of test certificates for electrical equipment. On completion, the engineer surveyor, who would visit the site during erection, would examine the crane. The site-erected crane would be examined and proof-load and performance tests would be witnessed. Test certificates for ropes and hooks would be perused and all the inspection stages would be reported progressively. Although the description applies to electric overhead travelling cranes and dockside portal cranes, a similar service is available for power-driven mobile cranes, manual overhead traveling cranes, box trolley runway tracks and lifting tackle.

13.7.4 Electrical inspection

Inspection services relevant to electrical plant during production include inspection at several stages of manufacture and witnessing of tests on completion. Depending on the item concerned, examinations are carried out of mechanical details prior to assembly and of insulating materials, conductors, coils, supports, wedges and windings. The witnessing of tests of complete component parts would include commutators, slip rings and brush gear. This is followed by inspections during the winding of motors and transformers and, on completion, witnessing of manufacturer's tests to determine the performance and reliability of the items under normal working conditions. Examination and witnessing manufacturer's tests of auxiliary equipment, including control gear, is also undertaken.

13.7.5 Turbogenerators

The schedule of inspection for turbogenerators covers key stages during manufacture and normally includes witnessing of mechanical tests on sections from the motor forgings and on completion of machining of mechanical components. Insulating materials, conductors, controls, supports, wedges, slip rings, commutators and windings are examined at appropriate stages during progress of the work on the stator, rotor and exciter. The tests carried out by the manufacturer during progress and on completion are all witnessed at the appropriate stages.

13.7.6 Centrifugal pumps

The services provided could include the witnessing of mechanical tests on specimens representing the main castings and forgings and inspection of the pump casing, bedplates and other principal components after machining. This would include the pump shaft, impeller, guide-vanes and division plates. The engineer surveyor would witness hydraulic tests on pressure-retaining parts and the test carried out by the manufacturer under operating conditions.

13.7.7 Non-destructive testing

The principal engineering insurers all now operate NDT services with fully experienced, qualified non-destructive testing engineers available in all the main centers of industry throughout the UK. These engineers are able to undertake most forms of non-destructive testing at short notice, being equipped with modern ultrasonic, radiographic (X-ray and gamma-ray) testing equipment, digital sound-velocity measurement instruments, electromagnetic, eddy-current and spark-testing instruments as well as a range of equipment for all forms of magnetic particle and dye penetrant testing. In addition, many types and forms of equipment are used as aids to visual inspection, including closed-circuit TV using miniaturized cameras with a facility to record results.

13.7.8 Metallurgical testing

The prime function of the metallurgical laboratories is to investigate failures of all descriptions and prepare illustrated technical reports based on their findings. From these, the causes of any incident can be assessed and recommendations made to minimize the possibility of a recurrence. Equipment will include several light microscopes, both laboratory based and portable, for site work and replication testing on-site. A scanning electron microscope with magnification up to 200,000 and a feature enabling X-ray micro-analysis for any particular element to be made should also feature in the equipment available to the engineering insurers providing these services.

13.7.9 Laboratory chemical section

This is primarily engaged in analysis of boiler water treatment matters and involves on-site studies of various problems and the chemical examination of corrosion products, boiler scales, etc. It can also carry out certain types of metallurgical, fuel and inorganic analysis. Normal wet methods of analysis coupled with a visible ultraviolet and atomic absorption spectrophotometer are used for a wide range of analytical applications. Equipment in use by the engineering insurers providing these services can include an ion chromatograph, spectrometer equipment, atomic absorption spectrophotometer, flue gas analysis equipment and testing equipment for transformer and switchgear oil.

13.7.10 Other services

Other inspection services available include the examination of steel structures (new and existing), electrical wiring installations, containers (to meet Statutory Instrument No. 1890), dangerous substances (carriage by road in road tankers or tank containers) to meet Statutory Instrument No. 1059, examination of second-hand plant prior to purchase, plant undergoing repair or modification, the Control of Industrial Major Accident Hazard Regulations (CIMAH) Statutory Instrument No. 1902 and Control of Substances Hazardous to Health (COSHH) and Pressure Systems Regulations.

13.8 Claims

The vigilant plant user is well informed about the terms of the policies which cover the plant in his charge, including any excesses that may apply. As soon as an incident occurs which may give rise to a claim on one of the policies in force he will ensure that the insurer concerned is notified immediately, as failure to do so many invalidate a claim.

If the incident constitutes a 'reportable accident' as defined in legislation the plant engineer will also ensure that the incident is immediately reported to the Factory Inspectorate. The Inspectorate will decide whether they wish to carry out an enquiry and, particularly if there has been loss of life, the accident site may be compulsorily isolated. Unless this is so, investigation by the insurer or a loss adjuster acting on his behalf may proceed.

Smaller claims may be dealt with by completion of a claim form, perhaps a report from the repairer if appropriate and sight of the repair or replacement invoice. For larger claims, a visit to the site by the insurer's representative may be necessary. If the plant is inspected by the insurer the engineer who normally carried out the inspections will usually be asked to report on the incident. The insurer's representative will seek to establish the facts, the cause, nature and extent of the physical damage and perhaps the consequences in terms of interruption to production. An engineer surveyor is not usually concerned with the commercial outcome and may be expressly excluded from any commercial negotiations. A loss adjuster, as well as seeking to establish the facts, will also negotiate settlement of the claim once it has been established that a valid claim exists.

Both the engineer surveyor and the loss adjuster may advise on how best to effect a repair or replacement of the damaged item and will bear in mind the client's need to minimize their loss of production. If, for any reason, there is a delay in having damaged plant examined by the insurer's representative the insurer will normally be agreeable to the insured giving his own instructions for repairs to proceed, if any damaged parts replaced are kept for examination by the insurer with suitable protection against further damage. The agreement of the insurer to the insured giving his instructions for repairs to proceed

is not confirmation that the incident constitutes a valid claim. This will only be decided when the insurer's investigations are complete.

In any case, the plant engineer has a responsibility to keep careful account of all the costs incurred, which may be recoverable in whole or part under the policy, and submit appropriate invoices to the insurer. He also has an obligation to minimize these costs insofar as this is reasonably compatible with achieving his objective, which will be to restore normal production. The insurer may have access to sources of replacement plant items or to specialist repairers of whom the plant engineer is not aware, and advice on these matters is part of the service provided by the insurer.

When the cause of the incident has been established and the costs of rectification finalized, these will be compared with the insurance cover provided by the policy and the extent of the insurer's liability, if any, determined. The policy will normally be one of indemnity, i.e. returning the insured to the same position after an accident as he was before. This may be achieved by repairing or replacing what is damaged or by paying the amount of the damage. It may be necessary to carry out modifications to prevent a recurrence of the accident or desirable to up-rate the specification for better performance or the life of the machine may have been extended by the repairs carried out. In this case a degree of 'betterment' is involved which will be reflected in the settlement by a contribution by the insured to the cost of repairs.

On occasions, the amount of a claim will be found to exceed the sum insured, and as the sum insured is the maximum amount the insurer will pay, the insured will not be fully compensated for his loss. It is the responsibility of the insured, with the assistance of his broker or other intermediary, to ensure that sums insured are adequate and are not eroded by inflation. The plant engineer also has a responsibility to inform his broker if he increases the value of plant items, adds substantially to the total value of items at any one location or, where items are insured individually, takes additional items into service or removes them from service.

13.9 Sources of information

The plant engineer may progressively become a mine of information in his field, but it is more important that he knows where to find the information when he needs it and, equally important, that this information should be up to date.

13.9.1 British Standards

British Standards tend to be concerned with new construction rather than with plant already in service. In general, although there are exceptions, they are also concerned more about minimum standards than with standards of excellence. There are notable exceptions: for instance, BS 5500 is arguably the best guide to design and construction of pressure vessels, which has ever been produced, in the English language. In other technical fields, the standards of other nations take pride of place. For instance, the Americans tend to rate well in the oil and petrochemical sector and ANSI B.31.3 is particularly useful for pipework.

The plant engineer should strive to make sure that, whenever possible, his purchase orders specify compliance with the correct standard. Many manufacturers can truthfully claim that their specifications exceed the minimum requirements of a material standard, in which case they will have no objection to its inclusion in the order. Those who protest too much that its requirements are superfluous may deserve further scrutiny. Choosing the correct specification is not always easy, and requires expertise in the relevant field. This expertise is available from consulting engineers or from an inspection authority, who will also appraise the vendor's designs and, if required, will witness critical stages of construction and test.

The *BSI Catalogue* (previously known as the *BSI Yearbook*) is not an expensive investment for a company which spends appreciable sums on purchased plant and materials.

13.9.2 The Health and Safety Executive

The Factory Inspectorate branch of the HSE issues a series of booklets with the prefix HS/E. For example, HS/G 13, entitled *The storage of LPE at fixed installations*, may be considered essential reading for plant engineers who use LPG in more than small quantities. The HS(R) Series is another collection of booklets on safety, ranging from HS(R) 1, *Packaging and labeling of dangerous substances*, to HS(R) 19, *A guide to the Asbestos (Licensing) Regulations 1983*. The booklets in this series are guides that often contain explanations of and quite extensive extracts from legislation. They are highly recommended reading.

The TDN series are Technical Data Notes, which cover a particularly wide variety of subjects: for instance, TDN 53/3 is entitled *Creep of metals at elevated temperatures*. These notes are being progressively replaced by the Plant and Machinery (PM) series, of which PM5, *Automatically controlled boilers*, is perhaps the best known.

All the booklets issued by the Health and Safety Executive are available from HMSO, and as the series is continually expanding the reader is advised to obtain an up-to-date list from the nearest Stationery Office.

13.9.3 Publications by other organizations

Other organizations that represent groups of companies with a common interest also issue rates, regulations and guidance notes on subjects that come within their respective orbits. Occasionally, the HSE will advise users of a particular type of plant, commodity or material to observe the appropriate document(s), and the user will then ignore this advice at his peril. Apart from this aspect, they do not have any official force. At the same time, most of them contain excellent information and guidance. A non-exhaustive list of the organizations in alphabetical order, together with the address last known to the author, is given below:

The Associated Offices Technical Committee (AOTC). St Mary's Parsonage, Manchester M60 9AP
The Chemical Industries Association

Appendix 1: Glossary

AHEM	The Association of Hydraulic Equipment Manufacturers Limited, 192–198 Vauxhall Bridge Road, London SW1V 1DX
API	American Petroleum Institute, 1271 Avenue of the Americas, New York, USA
ASB	Association of Shell Boiler Makers, c/o David L. Chaplin, The Meadows, Ryleys Lane, Alderley Edge, Cheshire SK9 7UV
ANSI	American National Standards Institute, 1430 Broadway, New York, NY 10018, USA
ASME	The American Society of Mechanical Engineers, 135 East 47th Street, New York, NY 10017, USA
BASEEFA	British Approvals Service for Electrical Equipment in Flammable Atmospheres, Harper Hill, Buxton, Derbyshire
BCAS	British Compressed Air Society, Leicester House, 8 Leicester Street, London WC2H 7BN
BMEC	British Mechanical Engineering Confederation, 112 Jermyn Street, London SW1Y 4UR
BSC	British Safety Council, 62–64 Chancellors Road, London W6 9RS
BSI	British Standards Institution, Linford Wood, Milton Keynes, MK14 6LE
BCGA	Federation of British Electro Technical & Allied Manufacturers Association, Leicester Street, London WC2H 7BN
BI/NDT	The British Institute of Non-Destructive Testing, 53–55 London Road, Southend-on-Sea, Essex SS1 1PF
BRE	Building Research Establishment, Building Research Station, Garston, Watford WD2 7JR
CBMPE	Council of British Manufacturers of Petroleum Equipment, 118 Southwark Street, London SE1 0SU
CIA	Chemical Industries Association, Alembic House, 93 Albert Embankment, London SE1 7TU
CIIA	Council of Independent Inspecting Authorities, c/o Parklands, 825a Wilmslow Road, Didsbury, Manchester M20 8RE
CWSIP	Certification Scheme for Weldment Inspection Personnel, Abington Hall, Abington, Cambridge CB1 6AL
SCI	Society of Chemical Industry, 14–15 Belgrave Square, London SW1X 8PS
DTI	Department of Trade and Industry, Room 320, Kingsgate House, 66/72 Victoria Street, London SW1E 6SW
DOT	Department of Trade, 1 Victoria Street, London SW1H 0ET (International Trade, Policy Division Room 450 or 455)

DOE	Department of Energy, Thames House South, Millbank, London SW1P 4QJ
EEMUA	Engineering Equipment & Materials Users Association, 14/15 Belgrave Square, London SW1X 8PS
EFTA	The European Free Trade Association, 9–11 Rue de Varembe, CH-1211 Geneva 20, Switzerland
FPA	Fire Protection Association, 140 Aldersgate Street, London EC1A 4HX
HSE	Health and Safety Executive, Hugh's House, Trinity Road, Bootle, Merseyside L20 3QY
ICE	The Institution of Chemical Engineers, George E. Davies Buildings, 165, 171 Railway Terrace, Rugby CU21 3HQ
ICE	The Institution of Civil Engineers, Great George Street, London SW1
IEE	The Institution of Electrical Engineers, Savoy Place, London WC2R 0BL
IMechE	Institution of Mechanical Engineers, 1 Birdcage Walk, Westminster, London SW1H 9JJ
IP	Institute of Petroleum, 61 New Cavendish Street, London W1H 8AP
LPGITA	Liquefied Petroleum Gas Industry Technical Association (UK), 17 Grosvenor Crescent, London SW1X 7ES
NRPB	National Radiological Protection Board, Chilton, Dicot, Oxfordshire OX11 0RQ
PPA	Process Plant Association, Leicester House, 8 Leicester Street, London WC2H 7BN
RoSPA	Royal Society for the Prevention of Accidents, Cannon House, The Priory, Queensway, Birmingham B4 6BS
THE	Technical Help to Exporters, BS1, Linford Wood, Milton Keynes MK14 6LE

Appendix 2: Statutory report forms

Statutory form no.	Title of regulation	Title of form
F54	Factory Act 1961 and the Offices, Shops and Railway Premises (Hoists & Lifts) Regulations 1968	Report of Examination of Hoist or Lift
F55	Factory Act 1961 and the Examination of Steam Boiler Regulations 1964	Report of Examination of Steam Boilers other than Economisers, Superheaters, Steam Tube Ovens and Steam Tube Hotplates
F55A	As F55 above	Report of Examination of Steam Boiler under normal Steam Pressure
A2197	Factory Act 1961 and the Power Presses	Report of thorough Examinations and Test and Record of Repairs
F260 and F262 (also available in book form)	Mines and Quarries Act 1954 and The quarries (Electricity) Regulations 1956	Tests of Insulation Resistance and Conductivity of Earthing Conductors and Earth Plates of Electrical Apparatus
	Conveyance by Road in Road Tankers and Tank Containers 1981 (Statutory Instrument No. 1059)	Report of Periodic Examination and Test Certificate of Suitability of a carrying tank/tank container and its fittings for the conveyance of dangerous substances by road

THE OFFICES, SHOPS AND RAILWAY PREMISES (HOISTS AND LIFTS)
Regulations 1968   Regulation 6
FACTORIES ACT 1961 – Sections 22, 23 and 25
FORM PRESCRIBED FOR THE
REPORT OF EXAMINATION OF HOIST OR LIFT

In any correspondence relating
to this report please quote

also identification number

F 54

Occupier (or Owner)
of premises

Address

1 (*a*) *Type of hoist or lift and identifi-*
    cation number and description
  (*b*) *Date of construction or re-*
    construction (if ascertainable)

2 Design and Construction
  Are all parts of the hoist or lift of good
  mechanical construction, sound material and adequate strength (so far as
  ascertainable)?
  Note: Details of any renewals or
  alterations required should be given
  in (5) and (6) below)

3 Maintenance
  Are the following parts of the hoist
  or lift properly maintained and in
  good working order? If not, state
  what defects have been found
  (*a*) *Enclosure of hoistway or liftway*

  (*b*) *Landing gates and cage gate(s)*

  (*c*) *Interlocks on the landing gates*
    and cage gate(s)

  (*d*) *Other gate fastenings*

  (*e*) *Cage or platform and fittings,*
    cage guides, buffers, interior of
    the hoistway or liftway

  (*f*) *Over-running devices*

  (*g*) *Suspension ropes or chains, and*
    their attachments

  (*h*) *Safety gear, i.e. arrangements*
    for preventing fall of platform
    or cage

  (*j*) *Brakes*

  (*k*) *Worm or spur gearing*

  (*l*) *Other electrical equipment*

  (*m*) *Other parts*

4 What parts (if any) were inaccessible

5 Repairs, renewals or alterations
  required to enable the hoist or lift
  to continue to be used with safety-
  (*a*) *immediately*

  (*b*) *within a specified time, the said*
    time to be stated
  If no such repairs, renewals or altera-
  tions are required, enter "NONE"

6 Defects (other than those specified
  at 5 above) which require attention

7 Maximum safe working load subject
  to repairs, renewals or alterations
  (if any) specified at 5

8 Other observations

I certify that on   I thoroughly examined this hoist or

lift and that the foregoing is a correct report of the result.

Signature   Date

Qualification – Engineering Surveyor to Insurance Company

L54AJ (12/88)   TO BE ATTACHED TO THE GENERAL REGISTER

HEALTH & SAFETY EXECUTIVE
FACTORIES ACT 1961, Sections 32–34 and the Examination of Steam Boilers Regulations 1964
FORM PRESCRIBED FOR

In any correspondence relating
to this report please quote

REPORT OF EXAMINATION WHEN COLD OF STEAM BOILERS OTHER THAN ECONOMISERS, SUPERHEATERS, STEAM TUBE OVENS, AND STEAM TUBE HOTPLATES†

F55

1 *Name of Occupier*

2 *Address of*
   (a) *Factory*
   (b) *Head Office of*
       Occupier

NOTE: Adress (b) is required only in the case of a boiler used on a temporary location, e.g., on a building operation, work of engineering construction.

3 *Description and distinctive number of boiler and type*

4 *If the boiler is one of those described in regulation 4(2), this should be stated and the appropriate sub-paragraph (a), (b) or (c)) should be given.*

5 *Date of construction*
   The history should be briefly given, and the examiner should state whether he has seen the last previous report.

6 *Date of last hydraulic test (if any), and pressure applied.*

7 *Quality and source of feed water*

8 *Is the boiler in the open or otherwise exposed to the weather or to damp?*

9 *Boiler:*
   (a) *What parts of seams, drums or headers are covered by brickwork?*
   (b) *Date of last exposure of such parts for the purpose of examination*
   (c) *What parts (if any) other than parts covered by brickwork and mentioned above were inaccessible?*
   (d) *What examination and tests were made? (see Note 2 overleaf). If there was any removal of brickwork, particulars should be given here*
   (e) *Condition of boiler*   ⎧ External:

   (State any defects materially affecting the maximum permissible working pressure)   ⎨ Internal:

10 *Fittings and attachments*
   (a) *Are there proper fittings and attachments?*
   (b) *Are all fittings and attachments in satisfactory condition (so far as ascertainable when not under pressure)?*

   Subject to further report after examination under normal steam pressure

11 *Repairs (if any) required, and period within which they should be executed, and any other conditions which the person making the examination thinks it necessary to specify for securing safe working*

12 *Maximum permissible working pressure calculated from dimensions and from the thickness and other data ascertained by the present examinations due allowance being made for conditions of working if unusual or exceptionally severe.*

13 *Where repairs affecting the working pressures are required, state the maximum permissible working pressure.*
   (a) *Before the expiration of the period specified in (11)*   (a)
   (b) *After the expiration of such period if the required repairs have not been completed*   (b)
   (c) *After the completion of the required repairs*   (c)

14 *Other observations*

Subject to the reservation (noted above) of certain points for examination under steam pressure.

I certify that on _____ the boiler above described was sufficiently scaled, prepared, and (so far as its construction permits) made accessible for thorough examination and for such tests as were necessary for thorough examination and that on the said date I thoroughly examined this boiler including its fittings and attachments and that the above is a true report of the result.

Signature   Counter-Signature

Date   Chief Boiler Engineer

Qualification – Engineer Surveyor to Insurance Company
Date

- Delete if not required.   †See overleaf.
TO BE INSERTED IN THE GENERAL REGISTER

B55J (8/88)

HEALTH AND SAFETY EXECUTIVE
FACTORIES ACT 1961, Sections 32 34 and the
Examination of Steam Boilers Regulations 1964
PRESCRIBED FORM FOR
REPORT OF EXAMINATION OF STEAM BOILER
UNDER NORMAL STEAM PRESSURE

F.55A   This form may also be used (as far as possible) for supplementary reports on Economisers and Superheaters

In any correspondence relating
to this report please quote

also distinctive number

Name of Occupier
Address of
(a) Factory
(b) Head Office of Occupier

NOTE. – Adress (b) is required
only in the case of a boiler used
in a temporary location.

3 Description and distinctive number
of boiler and type

The next thorough examination
to be completed on or before:

4 If the boiler is one of those described
in regulation 4(2), this should be
stated and the appropriate sub-para-
graph (a), (b) or (c) should be given

5 Conditions (External)

6 Fittings and Attachments
(a) (i) Is the safety valve so ad-
justed as to prevent the boiler
being worked at a pressure greater
than the maximum permissible
working pressure specified in the
last report (F.55) on examination
when cold?
(ii) (If a lever safety valve). Is
the weight secured on the lever in
the correct position?
(b) Is the pressure gauge working
correctly?
(c) Is the water gauge in proper
working order?

7 Repairs (if any) required, and
period within which they should be
executed and any other conditions
which the person making the
examination thinks it necessary to
specify for securing safe working

8 Other observations

I certify that on   I examined the above-mentioned boiler under normal

steam pressure and that the above is a true report of the result.

Signature   Counter-Signature

Date   Chief Boiler Engineer

Date

Qualification – Engineer Surveyor to Insurance Company

B55SJ (2/89)   TO BE INSERTED IN THE GENERAL REGISTER

In any correspondence relating
to this report please quote

HEALTH AND SAFETY AT WORK etc. ACT 1974
FACTORIES ACT 1961
The Power Presses Regulations
1965 (SI 1965 No. 1441) and 1972 (SI 1972 No. 1512)
Power Presses and Safety Devices Thereon
**REPORT OF THOROUGH EXAMINATION
AND TEST AND RECORD OF REPAIRS**

also identification number

(FORM APPROVED BY H.M. CHIEF INSPECTOR OF FACTORIES UNDER REGULATIONS 5 AND 6)

F2197

See Notes and space for continuation of entries overleaf.

Name of Occupier

Address of Factory

*1 Make, type and
date of manu-
facture
(if known)*   *(a) Power press*

  (b) Safety device(a)[1]

*2 Identification
mark or number*   *(a) Power press*   *Maker's*   *Occupier's*
  (b) Safety device(s)   *Maker's*   *Occupier's*

*3 Are the following parts of the press and
safety device(s) in good working order?*[2]
If note, state what defects have been found

*(a) Power
Press*
  (i) Clutch mechanism
  *(ii) Clutch-operating
controls*
  (iii) Brake
  *(iv) Flywheel
bearing(s)*
  *(v) Other parts affecting
safety at the tools*

*(b) Safety
device(s)*
  (i) Interlocking guard
  (ii) Automatic guard
  *(iii) Fixed fencing (inc.
that associated with
(i), (ii) & (iv)).*
  *(iv) Other type of
safety device (e.g.
photo-electric)*

4 What parts (if any) were inaccessible?

*5 Repairs, renewals or alterations to the
power press and safety device(s) to
remedy defects which are or may become
a cause of danger to employed persons
(Regulation 6(1)) (see note 2)*

(a) before press is used again
*(b) within a specified time, the said time
to be stated*
*If no such repair, renewals or alterations
are required enter "NONE".*

*6 Defects (other than those specified at 5
above) which require attention.*

*7 Repairs, renewals or alterations required
in item 5 of this report, completed at
the time of the thorough examination
and test.*

I hereby certify that on   I thoroughly examined and tested the power press and the

safety device(s) thereon specified above and the result of my examination and test are as shown.

Signature

Date

Qualification – Engineer Surveyor to Insurance Company

L30J (10/88)   If any defects require action as noted at 5 above please give date of
notification of such defects (under Regulation 6(1)) to the occupier . . . .

(Continued overleaf)

Quarries and Miscellaneous Mines (M. & Q. Forms 260 and 262). **Electrical Installation Tests.** (See also separate report of same date).	Sheet number Affix this sheet to page of record book

Number and address
of
quarry or mine

Tested by the undersigned Engineer Surveyor	Type of earth plate	Location	Resistance to earth (ohms)	Date of test

Circuit	Insulation Resistance (Megohms)	Resistance in Ohms		Percentage Conductivity L/E × 100
		Line Conductor	Earth or Metallic Covering	

Signature   Date

Qualification – Engineer Surveyor to Insurance Company

                                                       Date

3J (10/85)   Countersigned   Manager Owner

In any correspondence relating
to this report please quote

Report of Periodic Examination and Test Certificate of Suitability of a
carrying tank/tank container* and its fittings used for the conveyance of
dangerous substances by road. (SI 1059)

*Delete as appropriate

1 Name of Operator

2 Address of Operator

3 Description and distinctive number of tank	4 Max. permissible working pressure
5 Give: (a) interval between periodic exam. as specified in written scheme. (b) Was this period shortened at the last exam?	(c) If so, give shortened period

6 Nature of Examination
Condition:
(a) External

(b) Internal

(c) Parts inaccessible

7 Brief description and results of any tests

8 State: (a) Repairs/modifications* which should be carried out

(b) Whether tank may continue to be used for conveyance of dangerous substances listed in initial Certificate

9 Further exam. (a) If interval between exams. is to be shortened state new intervals	(b) State latest date before next examination must be carried out.

10 Other observations:

I examined (and tested*) tank and fittings No.   on   in accordance with the requirements of the written scheme relating to the periodic examination and test of the tank as required in Regulation 7(2)(a) of the Dangerous Substances (conveyance by Road in Road Tankers and Tank Containers) Regulations 1981.

*The tank and its fittings must be repaired and/or modified in accordance with Paragraph 8 above and then re-examined and tested before being used for the conveyance by road of dangerous substances.

*I am satisfied that the tank and its fittings are suitable for the purposes and under the conditions specified in the current Certificate of Initial Examination and Test dated:

Signature   Date

Qualification – Engineer Surveyor to Insurance Company

B99J (11/86)

Appendix 3: Report forms – non-statutory

Report form titles

1. Report of Examination and Test of Electrical Installation

2. Inspection Certificate – Based on requirements to comply with the IEE Regulations for the Electrical Equipment of Buildings

3. Hazardous Substance Pressure Vessel Inspection Report

REPORT OF EXAMINATION AND TESTS OF ELECTRICAL INSTALLATION

In any correspondence relating to this report please quote

Name

Address

Description of installation and nature of supply

DETAILS OF TESTS

Earth plate resistance to general mass of earth (ohms)

Circuit number	Description of circuit	Fuse rating or circuit breaker setting (amps)	Insulation resistance (megohms-min.)		Earth continuity conductor resistance (ohms-max.)	Earth fault loop impedance (ohms-max.)
			To earth	*Between conductors*		

GENERAL REMARKS. Including defects requiring attention and alterations since last examination or any departure from the relevant regulations.

Continued on sheet number

Date of Examination

Signature   Date

Qualification – Engineer Surveyor to Insurance Company

D 5J (8/86)

INSPECTION CERTIFICATE
Based on requirements to comply with
I.E.E. WIRING REGULATIONS

In any correspondence relating
to this report please quote

The accessible parts of the Electrical Installation at

Name

Address

Have been visually inspected and tested in accordance with requirements of The I.E.E. Wiring Regulations and that the results are as
indicated below.

I RECOMMEND that (due to age and condition) this installation be further tested after an interval of not more than .... years
ITEMS INSPECTED OR TESTED

METHOD OF EARTHING

Cable sheath. Additional overhead line conductor Protective multiple earthing (P.M.E.) Buried Strip/rod/plate Earth-leakage circuit breaker, Voltage-operated
Earth-leakage circuit breaker, Current-operated

State which:

TESTS	**RESULTS**
a) *Resistance of each earth Continuity Conductor*	
b) *The total earth loop impedance for ready operation of the largest rated excess current protective device relied upon for earth leakage protection.*	
c) *Earth leakage protection* *Current Operated/Voltage Operated*	
d) *Polarity throughout installation*	
e) *All Single Pole Control Devices*	
f) *Insulation Resistance of Fixed Wiring Installation. (minimum required 1 Megohm)*	
g) *Insulation Resistance to Earth of each separate item. (Minimum Requirement 0.5 Megohm)*	

Each item of apparatus tested, all flexible cords, switches, fuses, plugs and socket outlets are in good serviceable condition except as stated.
There is no sign of overloading of conductors or accessories except as stated. Apparatus tested includes/does not include Portable Appliances.

COMMENTS

Continued on Sheet number

Date of Examination

Signature   Date

Qualification – Engineer Surveyor to Insurance Company

HPV1

Associated Offices Technical Committee

{
British Engine Insurance Ltd.
Eagle Star Insurance Company Ltd.
National Vulcan Engineering Insurance Group Ltd
Plant Safety Ltd.
}

HAZARDOUS SUBSTANCE PRESSURE VESSEL INSPECTION REPORT

1. Owner's name and address

1a. Plant address or location

2. Description (including substance contained)

3. Maker, date of construction,
   code of construction and serial no.

4. Date of last hydraulic test
   and pressure applied.

5. Parts inaccessible

6. Nature of examination.

7. External condition

8. Internal condition

9. Condition of fittings and safety valve setting

10. Maximum internal pressure and temperature

11. Minimum internal pressure and temperature

12. Maximum permissible vacuum

13. Filling ratio from BS 5355   kg/litre of Capacity

14. The stresses due to the supporting arrangement have been checked under the following conditions and are considered satisfactory. This assesment does not include foundation:–

15. Other observations:–

This assessment does not include foundations.

....
                                                    Inspecting Authority

Date of examination:

....
                                                    Engineer Surveyor

Date of next thorough examination
on or before:

....
                                                    for Member Company

14

Insurance: Buildings and Risks

Risk Control Unit
Royal Insurance (UK) Ltd, Liverpool

Contents

14.1 Insurance

It is very important to understand that an insurance policy is a legal document, and insurance negotiations should be approached on this basis. Insurance contracts have a detailed legal background of both case law and common law in the interpretation of their provisions. Professional insurance advice from a reputable source, whether an insurance company or an insurance broker, is essential if problems are to be avoided.

It is important to understand that insurers are separate commercial organizations in competition and each trying to make a profit. This affects the insurers' approach to loss prevention and it does mean that some insurers will have different priorities in individual cases.

Insurers trade in a competitive market. Because insurance is such a competitive business, insurers do not have unlimited freedom to act. They occasionally have to accept less than the optimum in terms of safety. However, an individual insurer can always refuse the business if it is judged that the conditions are bad – and this is done from time to time.

There are occasions where rather less serious problems are encountered and the client refuses to take any action. In these circumstances, statutory bodies like the OSHA or the fire department can enforce their views through legislation. Insurers are not in that position. They can and do increase premiums for bad risks, but they also try to persuade management by commonsense argument that changes should be made, even if this is not necessarily an economic proposition from the point of view of insurance cost.

14.2 Fire insurance

Fire is the major insurable risk to property on land. Each industry, trade or manufacturing process has its own fire problems. There are, in addition, causes of fire which are found in all occupancies (e.g. arson, smoking, and misuse of electricity). With such a wide field, it is only possible to touch lightly on the many important aspects of the subject.

The normal fire policy covers the fixed assets (buildings, machinery and stock) of the business against loss or damage by fire, lightning and explosion. Assets consumed by the fire damage caused as a direct result of fire (e.g. by smoke or water used to extinguish the fire) are included in the cover. Damage caused by equipment overheating or by something which is being fired but is burned incorrectly are not covered by this type of policy. Lightning damage is self-explanatory.

Explosion damage cover means damage resulting from a wide variety of trade-related explosions (e.g. dust, vapor or uncontrolled chemical reactions). The explosion or collapse of boilers and pressure plant, in which internal pressure is due to steam or other fluids, is covered by engineering (not fire) policies.

The property owner is responsible for establishing the figure for which the assets are insured. This must be a realistic assessment of the value, as insurers settle claims on the actual loss incurred and do not automatically pay the figure insured in the policy.

The policy cover may be extended to include damage to assets from extra perils if the necessary additional premium is paid. These include storm damage, floodwater, burst water pipes or tanks, aircraft, riot, malicious damage or impact by mechanical vehicles. It is also possible to include an item to cover architects', surveyors' and consultants' fees and legal fees all incurred in the reinstatement of the property insured, as well as a sum to cover the costs of removing debris from the site before rebuilding can start.

14.3 Business interruption insurance

Business interruption policies (also known as loss of profits or consequential loss) are designed to cover the trading loss due to the occurrence of the fire or other insured peril. This loss is normally identified either by a reduction in turnover as a result of the disruption caused to the business or by increased costs incurred to minimize the loss of turnover, or indeed a combination of the two. The cover under the policy does not last for an indefinite period after the loss but is restricted to a time scale expressed in the policy as the maximum indemnity period. This time limit is chosen by the management of the business, and is the time they think that they would need to recover the trading position of the company following the incident. With fire insurance, there is a limit to the amount payable, which is the insured sum chosen by the management of the business.

14.4 Insurance surveys

It is normal practice in insurance for surveyors employed by the insurance company or the insurance broker to inspect premises that are to be insured and prepare reports for the underwriters. A major part of the survey report is an assessment of the quality of the fire protection relative to the level of hazard in the premises. Obviously, the business being carried on has a considerable influence on the risk of fire or explosion.

14.5 Fire protection

Every year there are a number of fires costing $1,000,000 or more, but the greater part of the total annual loss is made up from a large number of smaller fires. Most fires are initially small, and if tackled manually or automatically at this early stage can be put out before any serious damage is done. Rapid action can be taken either with portable firefighting equipment with hose reels or by fixed systems, which are brought into operation automatically by the heat or smoke of a fire. The type and quantity of firefighting equipment necessary will vary greatly. Advice can be obtained from the fire department.

In many cases, insurers allow a discount from fire insurance premiums where equipment is installed providing the equipment complies with insurance standards.

14.6 Extinguishers

The effectiveness of portable extinguishers is limited by the need to keep their weight, and therefore the amount of extinguishing agent they contain, within the limits necessary for quick and easy handling. The maximum discharge time, which can be expected from any portable extinguisher, is approximately 60–120 seconds.

For firefighting in occupied rooms with no exceptional fire risk, portable extinguishing equipment will normally suffice, if efficient appliances of the right type are properly positioned and will be handled by people trained to use them effectively. A fixed system will usually be necessary, to ensure adequate protection in factories where there is a real risk of fire breaking out after working hours. Also to ensure effective firefighting in workrooms where a fire is likely to spread too fast to be controlled by portable extinguishers, and in parts of buildings which have an appreciable fire risk but which are not normally occupied. Such a system should always be in addition to and not a substitution for portable extinguishers.

It is important to provide the right type of extinguisher for protection against a given risk. Some extinguishing agents that are outstandingly effective against fires involving certain substances may be useless or positively dangerous when applied to fires of another kind. Where more than one type of extinguisher is provided, the type of fire for which each is designed should be clearly indicated and staff instructed in the correct methods of use.

14.6.1 Water

Due to its cooling power, water is the most effective extinguishing agent for many types of fire. It is particularly suitable for fires in carbonaceous materials. Portable extinguishers provide a limited quantity of water using gas pressure. Extinguishers should have a nozzle fitted so that the direction of the jet can be properly controlled.

Hose reels connected to the public water supply have the advantage that their supply of water is unlimited, although the pressure may be less than that generated in a portable extinguisher. Very large hoses are difficult to handle and lengths should not normally exceed 14 meters. Water should be immediately available when the control valve is opened.

14.6.2 Dry powder

A range of powders is available in portable extinguishers for fighting fires in flammable liquids and are suitable for certain solid materials, including special metals. Discharge is by gas pressure and a hose is essential.

14.6.3 Gas

Carbon dioxide and certain halon compounds have a specialized application for fires in electrical equipment where a non-conducting medium is important. All are toxic to a degree, and operate either by smothering the fire or by a chemical reaction which inhibits combustion. Gas extinguishers must not be used in a confined space because of the toxic risk or the risk of asphyxiation.

14.6.4 Installation and maintenance

Equipment should be purchased from reputable manufacturers. They should be located at clearly defined fire points. All equipment should be recharged after it has been used. Manufacturers' instructions for maintenance and recharging must be followed closely. Regular inspection and maintenance of all firefighting equipment is essential, otherwise it is liable to deteriorate and prove unserviceable when needed.

Staff who may require to use the extinguishers must be trained both in selecting the appropriate extinguisher and in handling it properly. There is no substitute for actual *hands-on* use of extinguishers, and fire brigades are usually very pleased to help.

14.7 Auto-sprinkler installations

Automatic sprinkler systems have the great advantage that they are comparatively simple in concept and operate automatically, whether or not there are people present on the premises. Water is supplied from the public mains or tanks and pumps into a network of distribution pipes at ceiling level, which covers the whole premises. Water is discharged through nozzles or heads sited at regular intervals in the pipework, which are normally sealed with a heat-sensitive device.

The heat of the fire actuates these devices, therefore the system only discharges in the area of the fire. The sprinkler heads can be arranged to operate at suitable temperatures, taking into account ambient conditions. Where heating is not always available, the pipework can be charged with air during the winter months. An audible alarm is given when the installation operates, and this can be relayed automatically to the fire brigade. As with all fire equipment, maintenance is essential, as it is not in regular use.

14.8 Automatic fire alarms

Fire alarms are intended to give an early warning of the outbreak of fire. This may take the form of a local alarm or it may be signaled to the fire department. The disadvantage over, for example a sprinkler installation is that the system does not attack the fire, depending upon the human element for any firefighting to take place. Their advantage over a sprinkler installation is that it is more sensitive and detection is quicker. This is a particular advantage if quick action can be taken.

Various types of detectors that recognize heat and/or smoke utilizing fused bimetallic strips, ionization chambers and the interruption of a light beam by smoke or other combustion products. It is important to select the most appropriate form of detector for the environment. Insurers give a modest discount from premiums if the alarm installation complies with the insurance rules.

14.9 Trade hazards

The insurance survey referred to in Section 14.4 aims to identify plant, processes and storage of materials that are of significance in terms of fire hazard. Some examples include:

Drying
Dust hazards
Paint spraying and coating
Use of highly flammable liquids
Waste collection and disposal

Insurance companies are particularly interested in the standards to which the plant is operated and how the hazardous features are controlled.

14.9.1 Flammable liquids

Flammable liquids are widely used in many types of factories, and their misuse is responsible for many outbreaks of fire. The fire risks from the flammable liquids in common use such as petrol, paraffin, white spirit, cellulose solutions and thinners are well known, but these are only a few of the liquids which present hazards in industry. The variety of flammable liquids used in processes as solvents or carriers and for other purposes is constantly extending.

Processes involving coating, spreading and printing usually have a considerable area of exposure. If materials that include flammable liquids as solvents are sprayed, large quantities of vapor and fume are produced.

Many flammable liquids are used for a variety of purposes in bench work, either in semi-closed containers or in open trays. If spillage occurs due to breakage of apparatus or plant, or carelessness in handling, the liquid is distributed over a large area and considerable vapor is produced. Fires due to the ignition of vapors spread with extreme rapidity over the exposed surface of the liquid and the amount of flame and heat given off quickly increases.

14.9.2 Flammable gases

All the gases that are used for heating or lighting are easily ignitable, and there is grave risk of fire if the gas escapes because of leakage from a container or from piping. The only effective action against a gas fire is to stop the flow of the gas. If the fire is extinguished and the gas still allowed to flow, the escaping gas will form an explosive mixture with the surrounding air and produce conditions that are potentially far more dangerous than the fire. In addition, flammable gases are almost inevitably toxic, and hence a health hazard is produced.

Oxygen is not flammable, but leakage from oxygen supply pipes enriches the surrounding air and increases the potential for ignition of various materials within the enriched atmosphere. Some materials that would normally require a source of heat for ignition become pyrophoric. In the event of a fire the supplies of all flammable gases and oxygen should be cut off as quickly as possible.

14.9.3 Electrical equipment

In electrical equipment arcing or overheating causes fires, but continued combustion is usually due to insulation, oil or other combustible material associated with the installation. If the electrical apparatus remains alive, only

extinguishing agents that are non-conductors of electricity should be used.

14.9.4 Management

Insurers are aware of the importance of a management that, at all levels, is fully conscious of the fire and other hazards within its premises and consequently acts in a highly responsible way. An opinion of the standard of management is an important part of the insurance surveyor's report referred to earlier, and is obtained by observation of conditions in the premises and a scrutiny of the organization and systems that are in operation.

14.10 Security insurance

While a number of different classes of insurance cover relate to crime risks, the principal one which is sought by most businesses is known as *theft* or *burglary* insurance. The terms theft and burglary are defined in the Theft Act 1968, but while insurers use 'theft', the cover provided by this insurance is much narrower than the legal definition of that term in that it applies only in specified circumstances.

The basic theft policy covers the property insured against loss, destruction or damage by theft or attempted theft involving:

1. Entry to or exit from the premises by forcible and violent means; or
2. Actual or threatened assault or violence or use of force at the premises (this relates to the legal definition of robbery).

The theft policy extends to cover damage to premises caused in furtherance of theft or attempted theft, but malicious damage itself without theft is normally covered as an extension to the fire policy.

Theft by employees is another major exclusion from the theft policy, although the insurance market is prepared to cover this risk (subject to excess) through a fidelity policy.

Theft insurance premiums vary from risk to risk. The principal factors which insurers take into account are:

1. The geographical location of the insured property (experience shows that metropolitan and urban areas have a crime incidence which is above average for the UK); and
2. The type and value of commodity insured (some items are particularly attractive and are always a major target for thieves).

In all cases, however, the theft insurance premium assumes that the security standards applied and put into operation are commensurate with the theft risk.

14.11 Theft insurance policy terms and conditions

Under most theft insurance policies, the liability of the insurance company is conditional on the client's compliance with any policy terms and conditions. There is a

Reasonable Precautions Condition under which the client is required to take reasonable precautions to safeguard the insured property and secure the premises, including the installation, use and maintenance of any security precautions stipulated or agreed with the insurance company.

In some instances, special terms may be applied. For example, where the insurance company requires the installation of an intruder alarm, the protected premises must be attended unless the alarm is put into full operation. In other cases, cover under the policy may apply only when the property insured is left within agreed and designated areas.

From these examples, it will be seen that the client's obligations do not end with the mere provision of security hardware, etc. but extend to include procedures and routines that ensure that it is used effectively.

14.12 Risk assessment

Insurance companies employ staff that have been trained in crime-prevention techniques, and when theft insurance is provided/requested for commercial premises, it is normal practice for the insurer to carry out a security survey of the premises to prepare a report for the underwriters. The report is based on the surveyor's assessment of the risk, which will consider the following main factors:

1. Type of property and values at risk;
2. Construction of premises;
3. Location/situation of the premises;
4. Extent and periods of non-occupancy;
5. Nature and degree of accessibility to thieves;
6. Existing protections:
   (a) Locks, bars, bolts, etc.;
   (b) Intruder alarms;
   (c) Safes and strong rooms;
   (d) Surveillance (e.g. security patrols, CCTV);
   (e) Site security (perimeter fencing, security lighting, etc.);
7. Previous theft history of the premises.

The assessment will establish whether the existing security arrangements equate to the standards normally looked for by the insurance company for the specific theft insurance exposure. If the security falls short of these standards, the insurance company will submit a list of security items requiring improvement. Pending satisfactory completion of these items, insurers may:

1. Withhold theft cover entirely; or
2. Provide provisional cover but usually subject to some restriction such as significant excess or a percentage of self-insurance being carried by the client.

14.13 Planning for security

Like all forms of asset protection, security needs to be properly designed, planned and co-ordinated if it is to be both effective and cost beneficial. From previous comments in this chapter, it will be noted that the insurance company's attention is largely focused on the security of movable assets (e.g. stock, both raw materials and finished goods), plant (including vehicles, equipment and tools) against the risk of theft. However, every industrial and commercial operation may have its own security vulnerability (e.g. arson bomb threats, fraud, information theft, industrial espionage, malicious contamination of products, etc.).

While the insurance company's security requirements against the risk of theft may go some way towards countering these other vulnerabilities, some of these issues present significantly different security problems and they may require separate assessment and control. This emphasizes the need for a total plan to cover all aspects of security. If planning is piecemeal, some aspect may be overlooked and the benefits (financial and operational) of a totally integrated approach will be lost.

Similarly, security should be a function of senior management, and it should rank alongside other management responsibilities. This will help to demonstrate that the company is committed to a security program that appropriate resources are allocated and those protective systems and procedures are properly used and maintained.

14.14 Security objectives

The principal aims of security should be to forestall both organized and opportunist crime by cost-effective measures. While there is considerable variety in the type of criminal attack and skills involved, fortunately there is a tendency for them to act in a similar way, and this enables a common philosophy to be applied when determining countermeasures.

Based on the initial risk assessment, it will be obvious that as the exposure (commodity, value and location) increases so must the standard of security that is necessary for the risk. It is also the case, however, that items of comparatively low value will be stolen if they can be easily reached and if their removal does not represent any undue risk to the thief.

Experience has shown that thieves are strongly influenced by a number of factors:

1. Thieves are mainly opportunists, so there is a greater likelihood that they will be deterred by security that is more apparent.
2. They wish to act quickly, so any delay will complicate matters for them and discourage or even frustrate them.
3. They wish to remain undetected while working.

An important principle is that security must be built in depth – otherwise known as *defense in depth*. In this context, it may be helpful to think of security as a set of concentric rings, where the target is located at the center. Each ring represents a level of physical protection (perimeter fence, building shell, security case) but the number of rings and security resistance will vary relative to the risk. The spaces between the rings may represent other defensive measures such as closed-circuit television (CCTV), security lighting, intruder alarm systems, etc.

It will be apparent that delay (by physical features) and detection (by alarm, etc.) are entirely separate concepts. They should not be treated as being one and the same thing, nor should they be regarded as alternatives when arranging security. Delay, on its own, will have little security value unless it keeps the thief in a conspicuous place where there is a distinct likelihood of being seen (detection). Similarly, detection (by alarm) of the thief, where there are no obstacles to ensure subsequent delay, is unlikely to deter when it is known that the nature of property at risk may enable the removal of items of a high value in a short time. The message is that delay and detection, however achieved, should be used together and then with intelligence.

By using risk assessment techniques to determine the exposure of the property, it is possible to determine the appropriate standard of security for a risk. The following simple example will clarify this point.

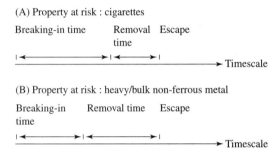

(A) Property at risk : cigarettes

(B) Property at risk : heavy/bulk non-ferrous metal

In example (A), the easy transportability of cigarettes requires a very high level of physical protection symbolized by the lengthy *breaking-in time*. When considering the necessary complementary alarm protection it is obvious that this must occur at the earliest possible time to enable assistance to be summoned and to arrive while the intruders are still on the premises but before they have reached their target. To arrange for detection to coincide with the beginning of removal time is much too late.

In example (B), the exposure of the property to risk is rather less both by way of the value of the goods and the problems posed by their transportability. While, therefore, it is still desirable that detection should occur as early as possible, it will be seen that its concurrence with the beginning of the removal time would still prevent the intruders from being successful.

Security is intended to frustrate thinking people, and it will be obvious that its effectiveness will be determined by the strength of its weakest part. Care must always be taken to ensure that there is no known vulnerable point, since this will be exploited to the full, thereby nullifying the security equipment otherwise provided. Once the standard of security has been determined it must be uniformly and intelligently applied.

14.15 Location

The location of any premises has a considerable influence on the overall standard of security. With new building, and other things being equal, the site with a low level of local crime should be chosen. In most instances, however,

location is a *fait accompli*, and the protection of the site should recognize the pattern of crime in the district. This information is normally available from the local crime prevention officer. It must be remembered that crime patterns are never constant, and it is important to keep abreast of local trends and, when necessary, be prepared to modify (usually improve) security standards.

14.16 Site perimeter security

14.16.1 Fences

Perimeter fencing may enhance the security of any premises. Not only does the fence present an obstacle for intruders to overcome but also it establishes the principle of defensible space and constitutes a psychological barrier to access. A perimeter fence, when supplemented by gates, traffic barriers and gatehouses (manned by security personnel or some other system of access control), allows the site operator to have control and supervision over all vehicles and pedestrians entering and leaving the site.

There are three types of fencing suitable for security applications. In increasing order of security, these are:

1. Chain link
2. Welded mesh
3. Steel palisade

The following features are necessary for any security application:

1. Total vertical height must be not less than 3 m;
2. Mesh size should not exceed 50 mm² to make footholds difficult;
3. Top edge of chain link and welded mesh should be barred and the fence topped by three strands of barbed wire on cranked arms extending at 45° out from the external face of the face, where possible. Care should be taken to ensure that any barbed wire is not a danger to pedestrians. If barbed razor type is used, notices should be posted warning of the danger of serious injury if climbing the fence is attempted;
4. The bottom edge of the fencing material should be *rooted* into a continuous-cast concrete sill set into the ground to prevent the material being bent up and burrowing;
5. Gates should be designed to provide the same barrier to intruders as the fence;
6. If a gross attack (using a vehicle to ram the fence) is envisaged, protection can be provided either by erecting a purpose-designed vehicle barrier or, where ground permits, digging a trench on the exposed side.

14.16.2 Security lighting

The purpose of security lighting is to deny the criminal the cover of darkness, which he uses to conceal his activities. In addition, strategically designed lighting can considerably increase the security value of other measures, and it can also assist the detection capabilities of surveillance (manned or CCTV) systems.

Lighting need not be devoted to security, and indeed most lighting installations have a dual role. Where security is a required feature, the installation needs to be very carefully specified and designed to achieve optimum protection.

14.17 Building fabric

14.17.1 Outer walls

The outer walls of a building should be commensurate with the security risk. The use of robust construction and materials (e.g. those resistant to manual attack and damage) is essential to the initial provision of security. Criminals will exploit weaknesses in walls. Where lightweight claddings are employed, a reinforcing lining such as welded steel mesh can enhance security. Where the security risk is high it may be necessary to reinforce even brick or cinder block walls in a similar manner.

14.17.2 Roofs

Roofs are a common means of access to criminals. Where necessary, access can be minimized or deterred by:

1. Restricting access to the rooftop by physical barriers;
2. Applying anti-climb paint or spiked collars to access routes such as drain pipes;
3. Use of strategic lighting of the roof area;
4. Fitting expanded metal or welded mesh below the roof covering, sandwiched between the skins of double-skinned roof coverings or within the roof space;
5. Roof openings, including roof lights and ventilation, should be treated, from a security point of view, in the same way as openings, located elsewhere in the building fabric.

14.18 Doors and shutters

External doors and shutters are often the focus of criminal attention. The range and construction of doors, shutters and their securing devices is too numerous to cover but the points which need to be considered are:

1. *Siting*: Wherever practicable, doors should be located where there is some element of surveillance;
2. *Numbers*: These should be kept to the minimum, consistent with operational requirements and any other requirements imposed under Building Regulations and fire/health and safety legislation;
3. *Construction*: Preference should be given to solid wood, steel or steel-faced doors. Glazing should be avoided but, where necessary, should be as small as practicable;
4. *Frames*: The frame should be of sound solid construction, with a substantial rebate and securely fixed into the building fabric;
5. *Condition*: All doors should be a good fit in the frame with no excessive gaps, and maintained in good condition;

6. *Hinges*: Outward-opening doors normally have exposed hinges that are vulnerable to attack. These should be supplemented by fitting hinge or dog bolts on the back edge of the door;
7. *Locks*: There is a wide variety of locks and security fittings, many with special applications. Advice from locksmiths or other security experts should be taken before deciding upon a type of lock, but, wherever practicable, only those locks that require a key to lock and unlock them should be chosen. BS 3621 establishes a useful minimum standard but is not applicable to all lock types;
8. *Emergency exits*: These doors should be identified and the local fire prevention officer consulted before any key-operated locks is fitted.

14.19 Windows

Accessible windows are a high security risk. Wherever practicable, storage areas in particular should have no windows. For a variety of reasons, this is not always feasible, and in these cases the following features should be considered:

1. Windows not required to be operable should be permanently fixed closed;
2. Where windows are required to open fit limitations (e.g. stops or cradle bars) to restrict the extent to which the window may be opened;
3. Where there is a security risk, fit internal bars or grilles properly secured to the building fabric. Specifications for bars and grilles are available from the local crime prevention officer, the insurance company or from proprietary grille manufacturers.

14.20 Intruder alarms

In many cases, it is necessary to complement physical security by the installation of an intruder alarm system in order to achieve the standard of security commensurate to the risk exposure. The scope of protection to be afforded by the alarm system depends on the security risk, but it may embrace fences, windows, doors, roofs, walls, internal areas, yards and external open areas, and vehicles inside and outside buildings. There is a comprehensive range of detection devices, but the choice of detector is critical to ensure that it provides the desired level of protection and is stable in the particular environment.

Intruder alarms are designed to give a warning of the presence of an intruder within or attempting to enter the protected area. Alarm systems may act as a deterrent to the casual or opportunist thief but they will do little or nothing to prevent a determined intrusion, and to be effective they must provoke an early response from the appropriate authority (in most cases the police). The warning may be a local audible device, but normally the alarm signal is transmitted by the telephone network to a central station operated by a security company on a 24-hour basis.

The alarm and telephone companies provide a comprehensive range of facilities for the transmission and receipt of alarm signals. The choice of signaling system will be determined by the security risk.

A number of alarm systems suffer from repeated false alarms. This may be due to the poor design and selection of detectors, inadequate standards of installation or preventative maintenance, and faulty operation of the system by key holders. Recurrent false calls will quickly discredit any system and may result in the withdrawal of police response to any alarm activation.

While alarm systems may be complex, false alarms can be minimized by the careful choice of installer, who should:

1. Be on the current List of Approved Installers maintained by the National Approval Council of Security Systems (NACOSS);
2. Be acceptable to the local police and appear on their list of recognized companies;
3. Install and maintain the system in full conformity with BS 4747 or, where the security risk is high, BS 7042.

14.21 Closed-circuit television (CCTV)

CCTV can make an important contribution to the security of a site, but it should be regarded as complementing and not replacing other security measures. Ideally, it should form part of an integrated system. In particular, CCTV allows greater flexibility in the deployment of security personnel by providing the facility whereby a single observer can remotely monitor several areas from one location. CCTV systems are not cheap, and need to be carefully designed by specialist CCTV engineers who will ensure that the necessary standard of reliability is obtained by proper planning and specification of the type, quality and installation of the equipment.

14.22 Access control

Although many access-control arrangements are designed to control access and egress during business hours, they are an essential part of a total security program. Leaving aside any arrangements previously discussed, access control is usually achieved by use of:

1. Security staff; or
2. Electronically operated entry/exit systems or a combination of both.

14.22.1 Security staff

The employment of full-time security staff either in-house personnel or contracted out to a professional guarding company (preferably one that is a member of the British Security Industry Association (BSIA) Manned Services Inspectorate) can make a significant contribution to overall security. Consideration should be given to the range of duties to be performed by security staff, their location, and how they will interface with other security measures and external agencies.

14.22.2 Electronically operated access/egress systems

These systems are used in conjunction with some form of physical barrier (e.g. door, gate, turnstile) and comprise:

1. Recognition equipment such as a token or card and appropriate reader;
2. Electrically activated release hardware; and
3. In some systems, central computerized control and monitoring equipment.

The choice of access-control technology depends upon the level of security to be achieved, number of users and traffic at peak periods, etc. Proper design of the system is critical to its performance (and user acceptance), and reliability and expert advice should be obtained from equipment manufacturers and system installers.

14.23 Liability and liability insurance

One of the basic requirements of society is to conduct ourselves so that we do not injure other people or damage their property. A business or any other undertaking is equally subject to this requirement.

For many years, the social and legal climates have had a growing involvement in requiring and encouraging business undertakings to protect the welfare of persons and their property. As a result, a body of common law has developed and Acts of Parliament have been passed which impose a duty of care upon any business undertaking to conduct its activities responsibly in order to avoid causing injury to persons, or damage to their property.

A breach of duty under either common law or statute that causes injury or damage may result in those responsible having a legal liability to pay compensation (termed *damages*) to the aggrieved party.

14.24 Employer's liability

The liability of an employer to his or her employees in respect of injury arising out of or in the course of their employment usually falls for consideration under either common law (the law of tort) or statute.

14.24.1 Tort

The common law liability of an employer to an employee can arise, *inter alia*, as follows:

1. Acts of personal negligence;
2. Failure to exercise reasonable care in the selection of competent servants (including contractors);
3. Failure to take reasonable care that the place of work is safe;
4. Failure to take reasonable care to provide safe machinery, plant and appliances and to maintain them in a proper condition;
5. Failure to provide a safe system of working.

14.24.2 Breach of statutory duty

There is a wide range of legislation laying down standards of safety and health. In the UK, the most notable and wide ranging are the Factories Act 1961, the Health and Safety at Work Act, 1974 and the Control of Substances Hazardous to Health Act 1988, breach of any of which can result in liability to civil and/or criminal action against the offenders.

14.25 Third-party liability

Liabilities to third parties may arise from breach of contract or tort.

14.25.1 Breach of contract

A breach of contract is a violation of a right created by an agreement or promise, and an action for damages or other recompense can normally only be brought by the aggrieved party against the other party or parties to the agreement.

14.25.2 Tort

Tort (which is part of common law but may be modified by statute) always implies breach of duty. This may take various forms, the most common applicable to the engineering and production fields being as follows.

Negligence

This has been defined as '*the failure to do something a reasonable person would do, or to do something a reasonable person would not do*'. For an action in liability to succeed in negligence the following conditions must be satisfied:

1. The defendant must have owed a duty of care to the plaintiff;
2. There was a breach of that duty;
3. The plaintiff sustained injury or damage as a result of that breach.

Nuisance

This has been defined as '*a wrong done to a man by unlawfully disturbing him in the enjoyment of his property or in the exercise of a common right*'. Examples of this are the emission of toxic fumes from a factory that affect occupiers of surrounding property or of toxic effluents affecting adjacent watercourses.

14.26 Liability insurance

As previously stated, a business can be held legally liable to pay compensation (damages) for injury or damage caused by its activities, and a successful action against it may result in large financial demands upon the business. Liability insurance ensures that, subject to satisfactory compliance with specified conditions and procedures by the insured, funds are available for a business if it is held legally liable to pay damages (and associated legal costs) which are awarded following injury to persons or damage to property.

There is a legal requirement upon most employers to have in force employer's liability insurance covering injury to their employees. Public liability insurance, while not compulsory, is strongly recommended, both in terms of third-party liability (arising from the effects of corporate activities) and product liability (arising from the effects of corporate products).

14.27 The cover provided by liability insurance

The following provides a general summary of the cover provided by standard liability insurance policies.

14.27.1 Employer's liability

Indemnity for an employers' legal liability to his employees against damages awarded for injury or disease happening during the period of insurance and arising out of and during the course of their employment, excluding radioactive contamination. An employee is defined as any person under a contract of service or apprenticeship with the insured, or any person supplied to, or hired to or borrowed by the insured (includes labor-only sub-contractors).

Cover normally applies to employees in the UK, Northern Ireland, the Isle of Man and the Channel Islands or while persons normally resident in those territories are temporarily engaged in the business outside of these territories. Indemnity is unlimited in amount.

14.27.2 Public liability (third party)

Indemnity against legal liability to third parties against damages awarded for:

1. Injury or disease;
2. Loss of or damage to property or personal effects of visitors or employees

happening in connection with the business and during the period of the insurance. Limited indemnity is awarded for financial losses not flowing from injury or damage but which result from escape or discharge of substances from the insured's premises or from stoppage of or interference with pedestrian, vehicular, rail, air or waterborne traffic. Wide contractual liability cover excludes only

1. Liquidated damages, fines or penalties;
2. Injury or damage caused by products;
3. Financial loss

arising solely because of a contract.

Indemnity is awarded in respect of injury to third parties and damage to property arising out of or in connection with the exercise of professional engineering skill. The following are excluded:

1. Radioactive contamination;
2. War and kindred risks;

3. Liability arising solely because of a contract for:
   (a) Liquidated damages, fines or penalties;
   (b) Financial loss;
4. Cost of rectifying defective work;
5. Liability arising out of the use of or caused by craft or vehicles;
6. Deliberate acts or omissions;
7. Principal's professional risk.

Worldwide indemnity is normally provided for injury or damage resulting from:

1. Negligent acts in the UK, Northern Ireland, the Isle of Man or the Channel Islands;
2. Temporary work undertaken abroad.

The limit of indemnity is selected at the time of insurance or renewal thereof by the management of the insured. It applies to any claim or number of claims arising out of any one cause, and is unlimited in total in the period of insurance.

14.27.3 Public liability (products)

Products mean goods (including containers and packaging) not in the custody or control of the insured, sold or supplied by the insured in connection with the business. The cover provides indemnity against legal liability for injury or damage happening during the period of insurance and caused by products. This cover also applies in respect of injury or design arising from the design of or formula for products.

Exceptions to the cover are as for public liability (third party) policies plus:

1. Financial loss caused by products;
2. Aircraft products.

It should be noted that very few standard product liability policies provide cover for liability for the costs of repairing or replacing defective products or those, which fail to perform as intended, nor for the costs of any necessary product recall. The insurer's liability in any one period of insurance for injury or damage caused by products during that period shall not exceed the selected limit of indemnity.

It cannot be sufficiently emphasized that individual liability policies do differ in detail from the general standard policy, and reference to the policy in force in any individual set of circumstances is most strongly recommended.

14.28 Points to be considered

As stated in previous sections of this chapter, there exists at all times a statutory and common law duty on all employers (which includes all engineers) to maintain a safe working environment for all their employees and the public at large. Additionally, it is contractual obligation on the insured under liability insurance policies to maintain the best reasonable standards of working procedures, equipment and the environment at all times. Consequently, there is a duty on all engineers to conduct their

professional duties at all times in a manner which does not downgrade the safe environment of their sphere of influence.

While not exhaustive, the following points for consideration will go far towards optimizing the control of safety and liability standards within the engineer's sphere of influence.

14.29 Employee safety and employer's liability

14.29.1 Premises/buildings

Design

Suitability for purpose/occupation?
Best 'state of the art' design?

Construction

Suitability for identified operating conditions?
Adequate strength of roofs and control of access thereto?
No built-in accident traps?

Maintenance program

Schedule of maintenance relevant to the needs for the operation?

Floors, aisles, stairs and floor openings

Evenness of floor surfaces?
Adequate provision and standards for handrails, toeboards and gates?
Adequate stairway capacity for number of persons likely to use them at one time?

Lighting, ventilation and heating

Adequate for identified conditions?
Compliance with Factories Act requirements?

Roadways, pavements, yards, car parks and open areas

Proper maintenance of surfaces?
Separation of pedestrian traffic from vehicular traffic (as far as possible)?
Speed limits (speed ramps, etc.)?
Adequacy of access?
Crane headroom (overhead wires), etc.?

14.29.2 Machinery and plant

Type and design

Adequate engineering design for the duty intended?
Hazard studies ('What if?') carried out?

Power sources and transmissions

Electrically safe (if applicable)?
Guarded (especially on rotating parts)?

Guarding/fencing

Compliance in all reports with Factories Act requirements?
No unguarded access to moving parts?
Safe provisions for maintenance, etc.?
Interlocking of guards with power supply?

Maintenance/adjustment procedures

Scheduled maintenance program?
Appointment of machinery attendants?
Isolation procedures laid down?
Adequate oil containment?

Permit to work systems

Adequate specification for identified needs?
Responsibilities allocated for issue, receipt, etc. of 'permits to work'?

Access, scaffolds, ladders

Correct maintenance schedules and procedures?
Designated certification and test/inspection procedures?

14.29.3 Tools

Power tools

Correct specification/design for duty envisaged?
Adequate inspection/maintenance procedures?
Regular insulation testing?
Safe power supplies?
Power safety cut-out provisions?
Safe storage when not in use?

Hand tools

Correct specification of duty?
Adequate maintenance?
Safe storage when not in use?

14.29.4 Electrical installations

Type and design

Correct specification for duty envisaged?
Suitability for hazardous environment (if required)?

Maintenance

Schedule of maintenance and maintenance procedures specified?
Earth testing schedules?

Operating procedures

Adequacy of 'lock-off'/isolation systems?
Conformity with the Electrical Regulations?
'Permit to work' systems for maintenance?

Overhead power lines

Maintenance of adequate clearance with satisfactory signing of hazard?

Control of access

Adequate access control to areas of concentration of electrical equipment?

14.29.5 Pressure equipment

Type and design

Adequately engineered for duty envisaged?
Safe working practices clearly deployed and adhered to?
Possible abnormal fluctuations in pressure identified and allowed for in design?
Adequate design for safe relief of overpressure?

Inspection/maintenance

Inspection schedule in force in conformity with Pressure Vessel Regulations?

Gas cylinders

Correct storage facilities for full/empty cylinders?
Segregation of combustible from oxidizing gases?
Compliance with HFL (highly flammable liquids) Regulations where applicable?

14.29.6 Materials handling and storage

Cables, ropes, chains and slings

SWL (safe working load) marked thereon strictly adhered to?
Certification and retest procedures specified and implemented?
Adequate and suitable storage facilities?

Lifts and hoists

Statutory inspection schedule in force?
Safe barriers/chains, etc. on teagle openings?
Runway beams tested and SWL displayed?

Cranes

SWL schedule (including operating radius variations) displayed and adhered to?
Overload devices fitted?

Conveyors

Adequate guarding (if powered)?
Adequate tripwires/buttons?
'Hand-traps' eliminated

Power/hand trucks

Correct specification/design for duty?
Laid-down code of operation?
Full and correct training/retraining/licensing of operators?

Manual lifting/handling

Training in safe techniques of lifting/handling?

Storage/piling of materials and goods

Safe condition of pallets?
Observance of specified maximum stacking heights?
Safe conditions of stacks?

14.29.7 Hazardous materials

Flammable/explosive substances

Properties known and documented?
Compliance with Storage Regulations?
Quantities within license/regulation limits?
Tank vents safe and adequately sized?
Compatibility of materials in storage acceptable?

Toxic substances

Properties known and documented?
Compliance with storage regulations?
Quantities within license/regulation limits?
Compatibility with other materials in storage acceptable?
Adequate personal protective equipment available for handling and use of material specified?
Antidotes/emergency facilities available?

Corrosive materials

Properties known and documented?
Adequate emergency washing facilities available?

Disposal of hazardous waste

Regular removal?
Use of experienced/licensed removal contractors?
Conformity with Hazardous Waste Regulations?
Safe operation of site incinerators, settling lagoons, etc.?
Adequate containment of hazardous spillage?

14.29.8 Hazardous processes

General

Hazard studies carried out to identify normal and abnormal hazards and circumstances leading to such hazards?

14.29.9 Health hazards and their control

Dust and fumes

Optimum suppression of dust/fume at point of origin?
Enclosure of dust/fume producing processes?
Superior ventilation at point of dust/fume origin?
Priority to cleaning methods (vacuum, etc.)?
Provision of adequate respiratory equipment and instructions on its correct use?
Correct (statutory) procedures for entry into enclosed areas which may contain fumes?

Noise

Minimization at source?
Regular monitoring of noise levels?
Enclosure of excessive noise sources?
Provision of personnel ear protection and instructions on its correct use?
Regular audiometry of employees at risk?

Skin disease (dermatitis)

Identification of potential causes?
Adequate hygiene in use of dermatitic substances?
Provision of protective equipment, barrier creams, etc. and instruction on their use?
Medical screening of employees at risk?

Carcinogenic materials

Identification of potential causes of cancer in the workforce?
Serious considerations of alternative materials?
Adequate containment of identified carcinogens?
Medical screening of employees at risk?

Radioactive materials and X-ray/ionizing radiation equipment

Minimization of exposure time?
Identification and maintenance of safe working distances between sources and employees?
Correct shielding of sources?
Adequate monitoring of exposure (film badges, dosimeters, etc.)?
Warning notices?

14.29.10 Personal protective equipment

Goggles/face screens, gloves/protective clothing, ear protection, barrier creams

Risks identified and the appropriate level of protective equipment determined?
Best available equipment specified for identified risks with instructions on correct use?
Effects of use of safety equipment assessed apposite possible creation of additional hazard (e.g. loss of ability to hear emergency signals while wearing ear protection)?

14.30 Safety of the public and public liability (third party)

In general, all points already listed under employer's liability apply equally in public liability (third party) situations. Additionally, the following should be considered.

14.30.1 Pollution

Solid wastes

Safe means of disposal, both on-site and at the disposal site, dump, etc. (long- and short-term effects)?
Safe and full incineration (if employed) with safe and full treatment of off-gases?
Correct pre-treatment of waste before disposal (if need identified)?

Liquid effluents

Correct design to optimize discharges?
Containment of spillages?
Discharges within consent levels?
Adequate effluent treatment plant (if need identified)?
Adequate monitoring of effluent streams with relevant recording?

Gaseous emissions

Correct design to optimize discharge within requirements of consent, Regulations, etc.?
Adequate monitoring of emission quality with relevant recording?

Noise

Adequate provision for limitation of noise levels (by distance or by reduction at source) at site boundary and beyond?

14.30.2 Visitors

Contractors

Adequate advice/training on-site, hazard site rules, etc.?
Competence assessed?
Liaison personnel appointed?

Invitees

Potential risks to visitors assessed?
Adequate advice on alarm systems, etc.?
System of escorts on-site implemented?

Trespassers

Optimization of site perimeter security and its maintenance in satisfactory condition?
Security patrols off-site?
Specified policy and means of dealing with trespassers found on premises?

14.30.3 Of-site work ('work away')

General

Adequate and competent supervision?
Adequate control of all operations involving the application of heat?
Adequate controls of any operations (e.g. excavation) that may cause damage to or weaken third-party property?

15

Electricity Generation

I G Crow and K Shippen
Davy McKee (Stockton) Ltd and ABB Power Ltd

Contents

15.1 Introduction

Electricity is one of the key energy sources for industry and commerce. It is normally provided by the electricity supply authorities, being generated from very large fossil fuelled or nuclear central power stations and distributed throughout the country via high-voltage transmission systems. It can be fair to say that the electricity supply authorities play a very important part in meeting the demands of the consumer efficiently, reliably and at an economic cost. However, there will always be a role for the private generation of electrical power, either to meet the needs for security of supply or to provide the electricity more economically. For the latter this is particularly the case for the concept of combined heat and power (CHP), where the local use of a heat demand can make private generation of electricity more competitive when compared with purchase from the local grid.

This chapter aims to convey the basic technical principles involved in electricity generation for industrial and commercial applications, with supporting technical data giving examples of the performance and efficiency of various schemes. A general guide is provided on the factors which have a major bearing on choice of an electricity-generating scheme with further details of the plant, its layout and descriptions of actual installations.

15.2 Generation of electrical power

Diesels, gas turbines and steam turbines are the more commonly used prime movers for the generation of electrical power. Additionally, the steam turbine can be employed in combination with either the diesel or gas turbine for combined cycle operation. The following describes the basic operation of each of these prime movers in relation to its associated power-generating scheme and reviews the more significant factors affecting performance and efficiency. Further information on the actual plant and installation is given later in Section 15.6.

15.2.1 Diesels

Diesels are used in many industrial applications (for example, for base-load generation in mines, cement plants and in remote regions of the world). In addition, they are often utilized to provide standby power for hospitals, telecommunications, banks, computer centers and office complexes that must have full independent power capability. The diesel can be started rapidly, making it ideal for peak lopping duties to meet maximum load demands, or for emergency use in cases of power supply interruptions.

Manufacturers offer a very wide range of diesel engines, all of which fall into categories depending upon crankshaft speed. The three categories are as follows:

Low speed: below 400 revolutions per minute (RPM)
Medium speed: 400–1000 RPM
High speed: above 1000 RPM

The speed of a diesel engine dictates its cost in relation to output. The choice is an economic consideration, taking into account the application of the engine and needs for reliability and security of supply.

The diesel has a good efficiency of approximately 40 per cent with a flat fuel consumption curve over its operating range. With a competitive capital cost this gives the diesel a distinct advantage over its competitors.

The compression ignition cycle

The diesel is an internal combustion reciprocating engine which operates in the compression ignition cycle. The familiar reciprocating mechanism consists of a number of pistons, each running in a gas-tight cylinder with connecting rod and crankshaft. The connecting rods are set at angular positions so that they contribute their power stroke in a regular sequence.

In the compression ignition cycle, the air is compressed and the fuel is injected into the compressed air at a temperature sufficiently high to spontaneously ignite the fuel. The heat released is converted to mechanical work by expansion within each cylinder and, by means of the reciprocating motion of the piston, is converted to rotary motion at the crankshaft.

There are two basic cycles, the two-stroke and four-stroke, each of which is illustrated in Figures 15.1 and 15.2 with their appropriate indicator diagrams. The two-stroke engine is mechanically simplified by the elimination of the mechanically operated valves. For the same rotational speed, the two-stroke engine has twice the number of working strokes. This does not, however, give twice the power. In the down-stroke of the two-stroke cycle both the inlet and exhaust ports are cleared, which allows some mixing of fuel air charge and exhaust gases, resulting in less thrust. For similar reasons, this also gives the two-stroke engine slightly higher fuel consumption.

The overall efficiency, η_b of a reciprocating engine is the product of the thermal efficiency, η_i and its mechanical efficiency η_m. Thus:

$$\eta_b = \eta_i \times \eta_m$$

An engine of good mechanical design has a mechanical efficiency of around 80–90 per cent at full load.

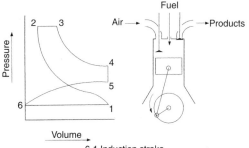

6-1 Induction stroke
1-2 Compression stroke
2-5 Combustion and expansion stroke
5-6 Exhaust stroke

Figure 15.1 Four-stroke cycle for an internal combustion engine

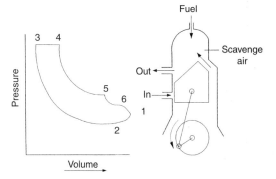

1-3 Induction and compression stroke
3-1 Combustion, expansion and exhaust stroke

Figure 15.2 Two-stroke cycle for an internal combustion engine

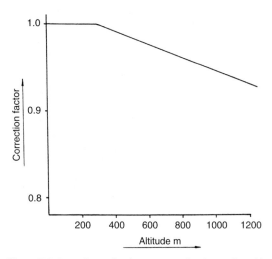

Figure 15.3 Correction to diesel power output for changes in ambient conditions

The overall efficiency of a compression ignition engine depends on the type, but ranges from 30 to 40 per cent.

One of the main reasons for the good performance of the diesel, compared with alternative machines, is due to the fact that the design is not restricted by metallurgical considerations which, for instance, limits the higher gas temperature in the gas turbine. This is because the cylinder wall is only subject to intermittent peak temperature due to combustion and its average temperature is much lower than the mean gas temperature. Therefore, the cyclic temperature can be maximized.

Turbocharging

Supercharging the inlet air by use of exhaust gas turbocharging raises the volumetric efficiency above that which can be obtained from normal aspiration. This gives increased output without change to the speed or capacity of the machine. The full potential of the increase in air density is affected by the use of intercooling placed downstream of the turbocharger. Naturally, the use of turbocharging also improves performance at high altitudes.

Site conditions

The site conditions can have significant effect on the output of the diesel engine. For example, British Standards 5514: 1987 Part I and similar Society of Automotive Engineers (SAE) sets out the de-rating method for internal combustion engines and establishes a datum level from which engine manufacturers relate their de-rating factors. Most engine manufacturers will give certain percentage reduction in power output for a certain increase in temperature and altitude above datum level.

One of the most important factors is the intercooler cooling water, which has a profound effect on air manifold temperature. Use of a closed-cycle air-cooling system will increase the cooling water temperature with de-rating of output. In addition, the radiators can add significantly to the parasitic load. The altitude above sea level at which the engine will be located also affects the output due to

lower air density. Figure 15.3 illustrates the effect upon power output for changes in ambient conditions.

Oil consumption

One of the disadvantages of the diesel engine is its high lubricating oil consumption which, typically for a 3.6 MW engine, will be 0.035 l/kWh. Added to this quantity must be the oil changes at routine service intervals.

15.2.2 Gas turbines

The role of the gas turbine is more familiar to many of us in the aircraft field. However, since Sir Frank Whittle invented the jet engine in the early pioneering years before the Second World War there has been rapid development in both output and efficiency of these machines, and today the gas turbine is a popular choice for electricity generation.

The gas turbine is widely used by electrical supply authorities for peak lopping and standby generation. Additionally, it is often employed for base-load operation when the fuel costs are low.

The major disadvantage of the gas turbine is its low efficiency when compared with other systems. However, the gas turbine is compact, is subject to short delivery and erection period, has a competitive capital cost, and has virtually no requirement for cooling water. It can be started rapidly on fully automatic control.

Open-cycle operation

The basic gas turbine generates in the open-cycle mode and Figure 15.4 illustrates this. Air is drawn into a compressor and after compression passes into a combustion chamber. At this stage, energy is supplied to the combustion chamber by injecting fuel, and the resultant hot gases expand through the turbine. The turbine directly drives the

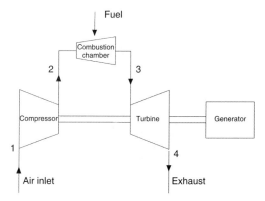

Figure 15.4 Basic gas turbine in open-cycle mode

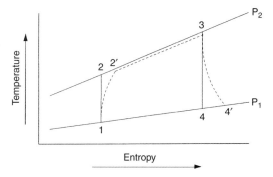

Figure 15.5 The gas turbine, or Brayton Cycle

compressor with remaining useful work, which drives a generator for electricity generation.

Figure 15.5 shows the ideal open cycle for the gas turbine that is based on the Brayton Cycle. By assuming that the chemical energy released on combustion is equivalent to a transfer of heat at constant pressure to a working fluid of constant specific heat, this simplified approach allows the actual process to be compared with the ideal, and is represented in Figure 15.5 by a broken line. The processes for compression $1-2'$ and expansion $3-4'$ are irreversible adiabatic and differ, as shown from the ideal isentropic processes between the same pressures P_1 and P_2:

Compressor work input $= C_p(T_2' - T_1)$

Combustion heat input $= C_p(T_3 - T_2')$

Turbine work output $= C_p(T_3 - T_4')$

Therefore

Net work output $= C_p(T_3 - T_4') - C_p(T_2' - T_1)$

Thermal efficiency $= \dfrac{\text{Net work out}}{\text{Heat supplied}}$

$$= \dfrac{C_p(T_3 - T_4') - C_p(T_2' - T_1)}{C_p(T_3 - T_2')}$$

By introducing isentropic efficiencies of the turbine n_1 and compressor n_c the turbine work output is given as:

Net work output $= C_p \left[T_3 \left(1 - \dfrac{1}{k} \right) n_{t-T_1(k-1)n_c} \right]$

where

$$k = \left(\dfrac{P_2}{P_1} \right)^{(\gamma-1)/\gamma}$$

The efficiency of the gas turbine depends on the isentropic efficiencies of the compressor and turbine, but in practice, these cannot be improved substantially, as this depends on blade design and manufacture. However, the major factor affecting both efficiency and output of a given machine is the turbine inlet (or combustion) temperature, and this is dictated by metallurgical considerations. The turbine blades are under high mechanical stress, and therefore the temperature must be kept to a minimum, taking into account the design life of the machine. The introduction of blade cooling does allow use of higher turbine inlet temperatures.

Factors affecting performance

1. *Efficiency* Typical performance figures for gas turbines ranging in size from 3.7 MW to 25 MW are given in Table 15.1. These demonstrate that the open-cycle operating efficiency is in the region of 25–35 per cent, depending on output. When compared with similar rated diesel or thermal plant this efficiency can be considered low. It can be seen from Table 15.1 that this low efficiency is attributed to the relatively large amount of high-grade heat in the exhaust gas that is discharged to atmosphere at temperatures in the order of 400–500°C.

2. *Output* One of the major factors affecting gas turbine output is the ambient air temperature. Increasing the air temperature results in a rapid fall in the gas turbine output. Figure 15.6 shows a typical correction to power output curve for changes in ambient temperature. Additionally, although to a lesser degree, output can be affected by inlet and outlet gas system pressure

Table 15.1 Typical performance data for industrial gas turbines

Engine	RSTN TB 5000	ABB STAL MARS	RB 211/COOPR
Output, kW based on ISO Ratings	3675	10 000	24 925
Efficiency (%) based on NCV of fuel	24.6	32	34.0
Exhaust gas temp. (°C)	492	491	465
Exhaust gas flow (kg/s)	20.4	39.1	90.9

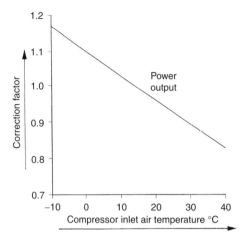

Figure 15.6 Correction to gas turbine power output for changes in ambient temperature

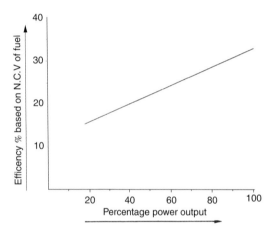

Figure 15.7 Approximate relationship showing variation of gas turbine efficiency for changes in load

losses, which would be due to inlet filters and inlet and outlet ductwork and silencers. This fact must be borne in mind to ensure that in development of layout the ductwork design maintains the gas pressure losses within limits set down by the manufacturer.

3. *Part load* As the gas turbine operates with a fixed inlet air volume, its efficiency at part load deteriorates significantly. Figure 15.7 shows a typical relationship for variation in gas turbine efficiency for changes of load.

Fuels

Gas turbines are capable of burning a wide variety of fuels, both gaseous and liquid. Typical liquid fuels range from kerosene and diesel to light crudes. They are also capable of running on natural gas and industrial gases

such as propane. Automatic change-over facilities can be incorporated.

15.2.3 Thermal power plant

Thermal power plant is more commonly associated with very large central power stations. The capital cost for thermal power plant, in terms of cost per installed kilowatt of electrical generating capacity, rises sharply for outputs of less than some 15 MW. It is for this reason that thermal power plant is not usually considered for industrial applications unless it is the combined cycle or combined heat and power modes. However, for cases where the fuel is of very low cost (for example, a waste product from a process such as wood waste), then the thermal power plant, depending on output, can offer an excellent choice, as its higher initial capital cost can be offset against lower running costs. This section introduces the thermal power cycle for electrical generation only.

Condensing steam cycle

The condensing steam cycle is shown diagrammatically in Figure 15.8. The fuel is fired in a boiler that converts the heat released from combustion to steam at high pressure and temperature. This steam is then expanded through a turbine for generation of electrical power.

Exhaust steam from the turbine is condensed in a heat exchanger and then returned via the feed pumps to the boiler. The associated Rankine Cycle is illustrated in Figure 15.9.

The efficiency of the cycle, ignoring work done by the boiler feed pumps, is given by change in enthalpy as follows:

$$\text{Efficiency} = \frac{(h_1 - h_2)}{(h_1 - h_3)}$$

and h_1 to h_2 represents the ideal isentropic expansion process in the turbine. However, in practice, because of irreversibility, this is less than unity.

This efficiency can be improved by the use of a feed heating cycle whereby bled steam can be taken from the turbine after certain stages of expansion and then used to raise the feedwater temperature via use of feed heat exchangers. By such means the feedwater temperature,

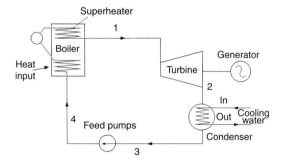

Figure 15.8 The condensing steam cycle

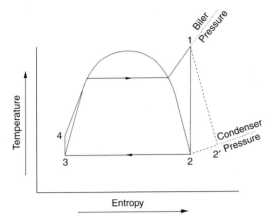

Figure 15.9 The Rankine Cycle with superheat

before entry to the boiler, is increased, thereby increasing the efficiency as transfer of heat to the working fluid condenses the bled system within the cycle, so reducing condenser heat rejection.

Large power stations use complex feed heating systems before the boiler feed pumps (LP) and after the boiler feed pumps (HP), which can give high overall thermal efficiencies of 39 per cent. However, for the smaller machine, it becomes uneconomic to consider multiple bleeds from the turbine, and the final choice is dictated by the extra cost for the additional complexity against lower running costs due to increased efficiency. As a minimum, a contact type de-aerator is often employed which would extract a small bleed of around 2–3 bar from the turbine.

Factors affecting performance

Table 15.2 gives performance data for typical industrial type schemes using thermal power plant in a condensing steam cycle. These do not operate strictly in the simple cycle mode as varying degrees of feed heating are employed. However, overall they convey the basic cycle conditions that the industrial user would encounter and give efficiencies that can be expected.

For the condensing steam cycle, the heat rejected by the condenser is the total energy system loss and is significant with, in some cases, overall cycle efficiencies of only 25 per cent. The higher the condenser vacuum, the lower is the saturation temperature of the condensing steam in the condenser. Heat is rejected, so that the higher the vacuum, the lower is the temperature of heat

Table 15.2 Performance of industrial steam-generating plant

Steam conditions	17 bar/260°C	30 bar/450°C	62 bar/510°C
Generator output	2000 kW	5000 kW	10 000 kW
Overall efficiency	18.1%	21.8%	26.3%

rejection and consequently the greater is the cycle efficiency at this temperature. The cooling water temperature and economic size of condenser set the limit to the condensing temperature.

Closed-cycle cooling water systems employing cooling towers give higher temperatures than direct open-cycle cooling from river, lakes or other large water sources.

Increasing superheated steam temperature has the direct result of increased efficiency. For example, an increase in steam temperature of 55°C can give around 4 per cent increase in efficiency. However, in practice the temperature is limited by metallurgical consideration with upper limits of around 540°C. This temperature can also be influenced by the boiler design and the fuel used.

Steam inlet pressure also has an effect on efficiency, although to a lesser extent. However, the cycle pressure is more often dictated by design limitations of the boiler for the application as well as the output.

Fuels

The thermal power plant uses a fired boiler for conversion of fuel to heat. It can be said that there is a design of fired boiler to suit almost all types of fuels, including wastes and vegetable and industrial byproducts. Generally, for oils and gases these can be more readily converted to power in diesels or gas turbines and would not be considered for thermal power plant, unless the station was of significant size.

15.2.4 Gas turbines in combined cycle

As discussed in Section 15.2.2, the gas turbine's main disadvantage is its low efficiency of around 25–35 per cent in open cycle. However, this can be significantly improved by the use of a heat-recovery boiler that converts a good proportion of the otherwise waste heat in the turbine exhaust gases to high-pressure superheated steam, which, in turn, drives a conventional steam turbogenerator for supplementary electrical power. This can increase the overall efficiency to 50 per cent for no further heat input as fuel.

The combination of the Brayton (gas turbine) and Rankine (steam) cycle is known as the combined cycle. A typical gas turbine in combined cycle is shown diagrammatically in Figure 15.10.

Types of combined cycle

1. *Fully fired cycle* Gas turbine exhaust contains approximately 15–16 per cent oxygen by weight, and this can be utilized to support the combustion of additional fuel, thereby enhancing the steam-raising potential and consequently the steam turbine output. The higher gas temperatures impose no limitations on either steam pressure or temperature, commensurate with current fired boiler/steam turbine practice. The heat content of the exhaust gas can potentially be increased approximately fivefold. Consequently, the steam turbine output can be three to four times that of the gas turbine. Combustion efficiency is high due to the degree of

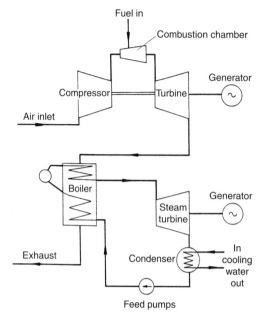

Figure 15.10 The gas turbine in combined cycle

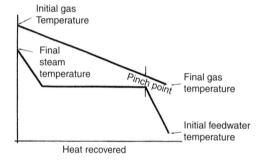

Figure 15.11 Heat-recovery diagram showing the pinch point

preheat in the gas turbine exhaust, and the resultant cycle efficiency is much higher than that of conventional steam plant.

2. *Supplementary fired cycle* The recovery of heat from gas turbine exhaust by the generation of steam has, as its basic limiting factor, the *approach temperature* at the *pinch point* (see Figure 15.11). The principal *pinch point* occurs at the cold end of the evaporator and consequently the combination of the steam-generating pressure and the magnitude of the *approach* determine the amount of steam which can actually be generated for a given gas turbine. In single-pressure systems, the only heat which can be recovered below the evaporator pinch point is to the feedwater, and this, in turn, now determines the final stack temperature. If, however, a small amount of fuel is used to supplement the gas turbine exhaust heat, then while the pinch point

occurs at the same gas temperature level, the evaporation is increased and hence the mount of preheat required by the feedwater, resulting in a lower stack temperature. For this cycle, the gas/steam power ratio is approximately 1–1.3:1.

3. *Unfired cycle* This cycle is very similar to the *supplementary-fired* case except there is no added fuel heat input. The approach temperature and pinch point are even more critical, and tend to reduce steam pressures somewhat. Similarly, the gas turbine exhaust temperature imposes further limits on final steam temperature.

4. *General* The *fully fired* cycle has been virtually superseded by the *unfired cycle*, due to the former's lack of potential for increased efficiency, large cooling water requirement, larger physical size, increased capital cost and construction time. Also, because of the higher temperatures it is much less flexible in its operating capability. The supplementary-fired cycle has many of the advantages of the unfired cycle, and it can be argued that it is the most flexible in its operating capabilities. However, the controls are most complicated due to the afterburners and the need to maintain the steam conditions over the whole operating range. The unfired cycle is the simplest in operational terms, the steam cycle simply following the gas turbine, and consequently is highly flexible in operation, having the ability to operate on sliding pressure at low loads. This cycle also has the greatest potential for increased efficiency, albeit for higher capital cost but without the need for extra fuel costs.

Single or dual pressure

There are two main pressure cycles applicable to combined cycle scheme: single or dual pressure:

1. *Single pressure* A combined cycle scheme with single pressure is shown in Figure 15.12. The waste heat recovery boiler is made up from preheater, economizer, evaporator and superheater sections giving single high-pressure steam output. The limitations imposed by the approach temperature at the pinch point, as illustrated earlier, determine the boiler heat recovery. Lowering of the boiler pressure increases the potential for heat recovery. However, the performance of the steam turbine will deteriorate at lower pressures, and therefore there is an optimum where maximum output can be achieved. The optimum pressure can be calculated, but is usually around 15 to bar-absolute.

2. *Dual pressure* For comparison, a combined cycle scheme with dual pressure is shown in Figure 15.13. In this case, the waste heat recovery boiler also incorporates a low-pressure steam generator, with evaporator and superheater. The LP steam is fed to the turbine at an intermediate stage. As the LP steam boils at a lower temperature than the HP steam, there exists two pinch points between the exhaust gas and the saturated steam temperatures. The addition of the LP circuit gives much higher combined cycle efficiencies with typically 15 per cent more steam turbine output than the single pressure for the same gas turbine.

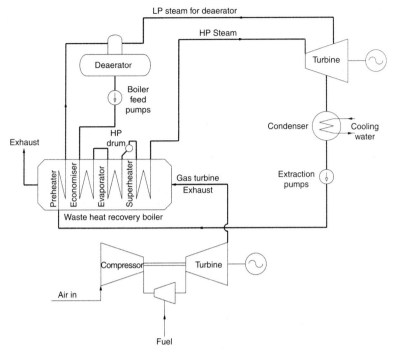

Figure 15.12 The single-pressure cycle diagram for a combined-cycle installation

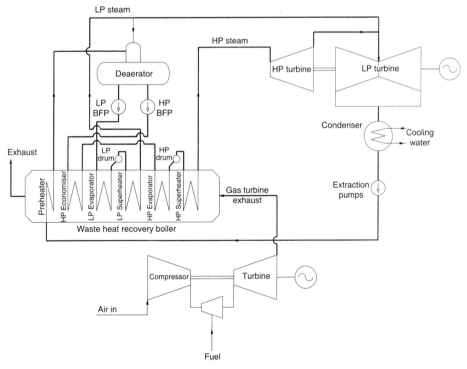

Figure 15.13 The dual-pressure cycle diagram for a combined-cycle installation

15.3 Combined heat and power (CHP)

The preceding section reviewed the application of popular prime movers for the generation of electrical power only. In the conversion of fuel energy to electricity it is shown that heat is rejected, either in the exhaust of a diesel or gas turbine or, alternatively, in the condenser of a thermal power plant. It can be seen that by applying these machines to provide both heat and electricity the total energy recovery can be much greater and efficiency thereby improved. Combined heat and power (CHP) schemes of this nature are well-established methods of producing both heat and power efficiently and economically.

15.3.1 Steam turbines for CHP

Many industrial processes require electrical power and heat. This heat is often provided from large quantities of low-pressure steam. In this section, it is demonstrated that a thermal power station gives up very large quantities of heat to the cooling water in the condenser. For this purpose, the steam pressure in the condenser is usually at the lowest practical pressure (around 0.05 bar-absolute) to achieve maximum work output from the turbine.

However, if the turbine back pressure is raised to above atmospheric pressure so that the turbine exhaust steam can be transported to the process heat load then the steam will give up its latent heat usefully rather than reject this to the condenser cooling water. Although the steam turbine output is reduced, the overall efficiency is increased significantly as the generated steam is used to provide both heat and electrical power.

The alternative (but equally appropriate) logic is that a factory may use steam from low-pressure boilers. By increasing the steam pressure and then expanding this through a steam turbine to the desired process pressure additional electrical power can be provided.

An example is shown in Figure 15.14. By raising steam at high pressure (say, 60 bar-absolute and 540°C) and then expanding this through a turbine to the process steam pressure requirements of 3 bar$_a$ then useful work can be done by the turbine for generation of electrical power. For this example, each kg/s of steam gives 590 kW of electrical power.

There are several types of steam turbine that can be used to meet widely varying steam and power demands. They can be employed individually or in combination with each other.

Back-pressure turbine

The simple back-pressure turbine provides maximum economy with the simplest installation. An ideal back-pressure turbogenerator set relies on the process steam requirements to match the power demand. However, this ideal is seldom realized in practice. In most installations the power and heat demands will fluctuate widely, with a fall in electrical demand when steam flow, for instance, rises.

These operating problems must be overcome by selecting the correct system. Figure 15.15 shows an arrangement that balances the process steam and electrical demands by running the turbo-alternator in parallel with the electrical supply utility. The turbine inlet control valve maintains a constant steam pressure on the turbine exhaust, irrespective of the fluctuation in process steam demand.

This process steam flow will dictate output generated by the turbo-alternator and excess or deficiency is made up by export or import to the supply utility, as appropriate. The alternative to the system in Figure 15.15 is to use a back-pressure turbine with bypass reducing valve and dump condenser, as shown in Figure 15.16.

On this system, the turbine is speed controlled and passes steam, depending on the electrical demand. The bypass-reducing valve with integral desuperheater makes up any deficiency in the steam requirements and creates an exhaust steam pressure control. Alternatively, any surplus steam can be bypassed to a dump condenser, either water or air cooled, and returned to the boiler as clear condensate.

Pass-out condensing turbines

If the process steam demand is small when compared with the electrical demand then a pass-out condensing

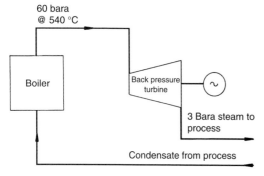

Figure 15.14 The back-pressure steam turbine in CHP application

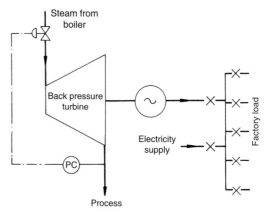

Figure 15.15 A back-pressure turbo-alternator operating in parallel with the grid supply

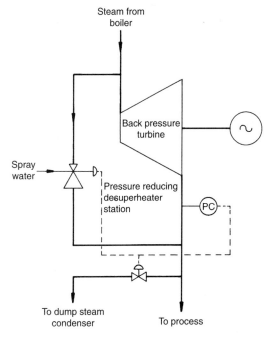

Figure 15.16 A back-pressure turbine with PRDs valve and dump condenser

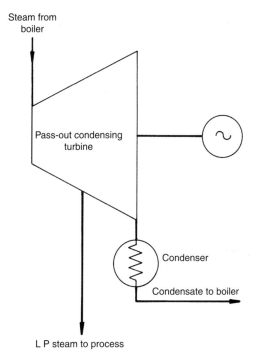

L P steam to process

Figure 15.17 The pass-out condensing turbine

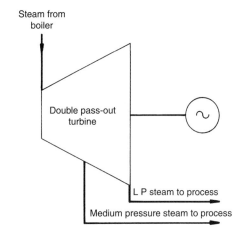

Figure 15.18 The double pass-out turbo-alternator

turbine may provide the optimum solution. Figure 15.17 illustrates a typical scheme, which consists of a back-pressure turbine. This gives operational flexibility of the back-pressure turbine with improved power output.

Back-pressure with double pass-out

Many industries require process steam at more than one pressure, and this can be done by use of a back-pressure turbine supplying two process pressures (see Figure 15.18).

15.3.2 Diesels and gas turbines in combined heat and power

In the process of conversion of fuel to electrical power, both the diesel and gas turbines eject large quantities of hot exhaust gases. These gases represent a significant energy source that can be converted to useful heat by the addition of a waste heat boiler (see Figure 15.19). The boiler can be arranged to produce either steam or hot water, depending on the process needs. This heat is produced without any further fuel input and without affecting the performance of the generating machine. In addition, in the case of the diesel, additional low-grade heat can be recovered from the engine-cooling system.

The high efficiency of conversion of the diesel generator immediately restricts its potential improvement compared with gas turbines. With the simple addition of a boiler, the gas turbine can give a heat/power ratio of 2:1 compared with the diesel's 0.6:1. (The diesel heat/power

ratio can be improved by 1:1 by use of the engine-cooling services, as shown in Figure 15.19.) However, both diesel and gas turbine exhaust gases contain excess oxygen (12–14 per cent and 14–16 per cent, respectively). This free oxygen can be utilized by supplementary firing the exhaust gases to produce additional steam in the heat-recovery boiler. By supplementary firing the ratio of heat to power can theoretically rise to 15:1 for a gas turbine and 5:1 for a diesel.

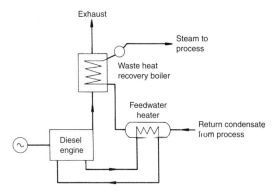

Figure 15.19 Diesel with waste heat recovery

Table 15.3 Typical properties of fuel oils

Oil fuel	Kerosene	Diesel oil	Residual fuel oil
Density (kg/)	0.8	0.85	0.99
Sulfur (%)	0.05	0.5	3.5 Max
Ash (%)	Nil	0.001	0.04
Gross calorific value (MJ/kg)	46.2	45.5	42.5
Net calorific value (MJ/kg)	43.2	42.7	39.8

Table 15.4 Application of fuel oils

System	Spark ignition	Compression ignition	Gas turbine
Motor gasoline	✓		✓
Middle distillates (kerosine, gas oil, diesel fuel)		✓	✓
Residual fuels		✓	✓
Crude oil			✓

The chemical characteristics of the exhaust gases from diesel engines tend to make them less attractive for use in waste heat recovery, as they are likely to contain significant carbonaceous solid matter because of the range of residual oils burned. Fouling by soot build-up on heat transfer surfaces within the boiler can occur. In addition, the gas is likely to have significant pulsations, with a wide variation in mass flow with engine load, although the exhaust temperature fluctuation at low load is not as significant as for the gas turbine. However, there are many examples of successful CHP schemes using both the diesel and the gas turbine.

15.4 Factors influencing choice

The basic schemes for electrical power generation and combined heat and power have been identified in the earlier sections of this chapter. When assessing the alternatives to meet a specific project need, many factors will strongly influence the size, number, and type of generator as well as its specification and scope. While it is not intended to cover all of these aspects, the more important issues are discussed.

15.4.1 Available fuels

Oils

The petroleum industry is responsible for the production of a wide range of fuel oils for industrial use. These generally range from kerosene, gas oils and diesel fuel to residual fuel. (See Table 15.3 for typical fuel oil characteristics.)

Table 15.4 shows typical applications for the range of oil fuels for use with major prime movers. Medium/high-speed diesel engines generally use distillate fuel oils while medium/low-speed units generally burn residual fuels. The gas turbine, which normally operates on liquid distillate fuels, is capable of running on residual fuels, although examples of these are normally associated with crude production facilities.

In the case of residual fuel firing for gas turbines it is necessary to provide extensive fuel-preparation plant for the removal of vanadium, sodium, potassium, calcium and aluminum. Generally, oil fuels have good heating values and combustion characteristics, the main disadvantages being their higher sulfur content and the need for heating of the heavy fuel oils for storage and transport. Distillate grades of oil fuels may be stored and handled at ambient temperature. However, exposure to the cold for long periods is not advisable, and it is good practice to incorporate heaters to maintain the oil fuel at 5°C above freezing. Heating facilities are required for all residual grades of oil fuels, and the following gives minimum storage and handling temperatures:

	Minimum storage temperature (°C)	Minimum temperature at outflow from storage and for handling (°C)
High fuel oil	10	10
Medium fuel oil	25	30
Heavy fuel oil	40	50

The tanks and pipework can be heated by use of steam, hot water, electrical heaters or a combination of these.

Gas

As a fuel, gas has many advantages in controllability and minimal effect on the environment, and it promotes reliable operation, longer life and residual maintenance compared with oil fuel. Its typical properties are shown in Table 15.5.

In many developed and developing countries, there are now extensive natural gas supply networks, which provide secure and reliable sources of gas fuel. The user may require some pressure reduction and metering equipment

Table 15.5 Typical properties of gas fuels

Gas fuel	Town gas	Natural gas	Methane	Butane	Propane	Carbon monoxide
Formulae		95% CH4	CH4	C4H10	C3H8	CO
Specific gravity (air = 1)	0.5	0.59	0.55	2.02	1.52	0.97
Gross CV (MJ/M3)	18.65	37.3	37.1	121.9	93.9	11.8
Stoichiometric air						
vol/vol	4.2	9.62	9.52	31.0	23.8	23.8
kg/kg	8.6	14.8	17.2	15.4	15.7	2.45
Spontaneous ignition temperature (°C)	650	660	700	480	500	570

on-site, otherwise there is no requirement for storage, and the gas can be simply distributed to the generator by pipeline.

One of the alternatives to natural gas is an industrial product such as propane or butane. These fuels are transported and stored in liquid form. On-site facilities are necessary for reception, off-loading and storage.

With the trends towards conservation of energy, there have been significant advances in small gasification schemes. Here the solid fuel is partially combusted to give a low-calorific-value gas. The gas is hot and dirty, and contains tar and particulate matter, thereby requiring some cleaning before being used as a fuel. While the scope for these schemes is limited, there have been successful examples in rural areas of developing countries where the gas produced from fuels such as wood is used to run small spark-ignition engines.

Coal

In many regions of the world coal reserves represent substantial indigenous resources, and for this reason coal will always play an important role in the future of power generation. Inevitably, coal is more difficult to handle and has less appeal than other clean-burning fuels such as gas. However, substantial progress has been made in modern efficient handling techniques and combustion methods that can give clean combustion with low emissions.

For the purposes of power generation, certain coal properties are significant, and a coal can be generally categorized by its rank, which embraces both volatile matter and caking properties. At the lower end of the scale are the older coals, such as anthracite with low volatile content and no caking properties. These coals are difficult to burn. As volatile content increases beyond anthracite then so does the caking quality, and prime coking coals are 12–30 per cent volatile matter (VM). Beyond this, as VM increases further, caking properties decrease to free-burning coals of approximately 40 per cent VM.

An example is the classification scheme used by British Coal, and is common throughout the UK. The scheme divides coals into groups, generally as shown in Table 15.6. The group, grading and modes of preparation of a coal serve to indicate the usage for which the coal is suitable for application with industrial boilers. For grading of coal the nominal sizes are shown as follows:

Table 15.6 Classification of coals

General description	Volatile matter (%)	Specific description
100	6–9	Anthracites
200	9–19.5	Low-volatile, steam coals
300	19.6–32	Medium-volatile coals
400		Very strongly caking
500		Strong caking
600	Over 32	Medium caking
700		Weak caking
800		Very weak caking
900		Non-caking

	Through (in)	On (in)
Doubles	2	1
Singles	1	0.5
Peas	0.5	1
Grains	0.75	0.125

Smalls constitute the fraction which passes through a screen of given size and which has no lower size limit. Smalls can be washed or untreated.

The ash content of the coal is significant, not only in its mass but also with respect to the constituents of the ash, the fusion temperature, the range of fusion temperature between initial deformation and actual fusion and the amount of coal fines present.

Use of coal requires suitable storage facilities local to the power station site. The capacity of the stockpile is a decision of strategy and depends on the frequency and size of the delivery loads. In addition, the site should be carefully chosen in view of access needs for reception and off-loading. Stored coal weighs about 4000 tons per 0.5 meters depth per hectare, which gives a rough guide on areas for stockpile. Graded coals and fines should always be stored separately. The large coal forms naturally, but fines must be piled by layering and compacting to form a well-consolidated heap.

The object of compacting the fines is to exclude air to ensure safe storage. If proper precautions are taken then coal can be stored safely. However, the user should be aware of the potential dangers of spontaneous combustion and the measures necessary to avoid this problem.

For sites of limited area, coal can be stored in large vertical bins.

There is a wide range of pneumatic, vibratory, belt or *en-masse* type elevators and conveyors for transfer of coal from storage to boiler. The final scheme depends on site layout constraints and cost.

Other fuels

Solid fuels also encompass lignite, peat and wood. In its final form, wood has a much greater commercial value as a product. However, some industrial processes (for example, the production of paper) create waste wood in off-cuts, bark and shavings. Here the wood is ideal as an energy source. It has high moisture content but can be fired easily with full automation. Table 15.7 shows typical heating values for these fuels when compared with coal. Additionally, there are food processes where the product is derived from vegetation and a combustible fibrous waste is produced which can be used as fuel. A good example is the production of sugar from sugar cane with bagasse as waste. There are many others, and their by-products should not be overlooked in an assessment of suitable fuel sources.

15.4.2 Electrical load profile

In reviewing the size of a generating scheme, it is necessary to consider the electrical load profile of the site demand, which must examine the electrical power needs over the daily and annual cycles. The capacity of the generator can be selected to cater for anticipated future growth or, alternatively, the station could be designed so that further generators could easily be added to match future load demands. At this stage, it is worth noting that the power station itself will require electrical power for its auxiliaries (parasitic load), and this electrical load will be relatively constant, despite the output modulating to follow demand.

The load profiles can vary quite considerably, depending on the electrical users. For instance, for a factory the load will depend on the shift operation with a potential low demand at night and at weekends. The load can be influenced by seasonal changes. For instance, it may increase towards winter with greater lighting demands. However, the need for HVAC in computer installation may increase the load during summer. It is important that the power station be capable of meeting the peak capacity while also being able to operate efficiently at lower loads.

15.4.3 Heat load

Earlier, we reviewed the advantages of combined heat and power. It therefore follows that a local heat load, where practical, should be incorporated into the electrical scheme.

Medium- to low-pressure steam and high- and low-temperature hot water usually supply heat loads. Steam is the main medium used by industry for the transport of large quantities of heat, and pressures generally range from 3 to 16 bars. Low- and high-temperature hot water systems are employed for heating, and they operate in *closed circulating systems* with the high-temperature system (approximately 115°C) pressurized by an external source to prevent steam forming. Typical industrial and commercial applications are:

Hotels: hot water, swimming pools
Schools: hot water, steam, heating, and air conditioning
Dairy: steam and hot water
Heavy manufacturing: steam, air conditioning, space heating and hot water
Hospitals: steam, hot water
Sewage plant: hot water, sludge heating
Food industries: process steam, hot water
Chemical industries: process steam, hot water

Hot water or steam can also be used in absorption-type chillers to provide chilled water with the reduction in electrical power needed for the refrigeration process.

As with the electrical load profile, it is also necessary to analyze the heat load over the daily and annual cycles. Ideally, the heat load will match the available heat from the electrical generator (however, this is rarely the case). There will be periods when supplementary output will be necessary which can be achieved by, say, supplementary firing the waste heat gases of a gas turbine, or heat output reduction is necessary by the introduction of bypass stacks. For a steam turbine installation bypass pressure-reducing valves will be necessary to supplement steam output, while a dump condenser may be needed at low-process steam demands. The nature of the electrical and heat load will obviously have significant influence in the development of the scheme and scope of equipment.

15.4.4 Power station auxiliary systems and services

A total power station project involves many aspects, covering civil works, fuel systems and storage, plant auxiliary systems and services, electrical plant and control and instrumentation. The scope will depend on the nature and size of the project envisaged. Table 15.8 gives a detailed checklist for power station auxiliary components and will provide a useful guide in developing scope. Further details of many of the components, such as ventilation systems or fuel oil handling, are given in other chapters of the book.

Table 15.7 Typical properties of some solid fuels

Fuel	General-purpose coal	Wood	Peat	Lignite
Moisture (%)	16	15	20	15
Ash (%)	8	–	4	5
Carbon (%)	61.3	42.5	43.7	56.0
Hydrogen (%)	4.0	5.1	4.2	4.0
Nitrogen (%)	1.3 ⎫			
Sulfur (%)	1.7 ⎭	0.9	1.5	1.6
Oxygen (%)	7.7	36.5	26.6	18.4
Gross CV (MJ/kg)	25.26	15.83	15.93	21.5
Net CV (M3/kg)	23.98	14.37	14.5	20.21

Table 15.8 Checklist of power station auxiliary components

Fuel system

Coal:
Rail sliding
Wagon puller
Tippler
Ground hopper
Conveyor
Crusher
Screen
Sampling
Mag. separator
Stocking-out conveyor
Conveyor (reclaim)
Transfer tower
Bunker conveyor
Dust extraction
Coal feeder
Coal mills
Pneumatic transfer system

Oil:
Bulk unloading pumps
Bulk storage tanks
Tank heater
Forwarding pumps
PH and filtering sets
Oil pipework
Steam pipework
Tracing elec./steam
Daily service tank
Fuel-treatment plant
Steam/gas turbine/diesel house
Boiler house
CW pump house

Ash handling
Clinker grinder
Clinker conveyor
Econ. dust conveyor
Air heater dust conveyor
Rotary valves
Dust collector dust conveyors
Rotary valves
Ash conditioner
Ash silo
Ash sluice pump
Make-up water pump
Pneumatic conveying system
Conditioned ash conveyor
Submerged ash conveyor

Water systems
Bar screen
Travelling-band screen
Wash-water system
Penstock
Intake pump/circulating pumps
Valves
Chlorination plant
Crane/lifting equipment
Cooling tower (ID/FD/natural)
Settling pond

Boiler water treatment
Base exchange
Dealkylization
Demineralization
Demin. water tank
Demin. water pump
Bulk acid and alkali storage
Neutralizing

Chemical dosing
LP dosing set
HP dosing set
CW dosing set

Miscellaneous
Air compressor (plant)
Air compressor (inst.)
Air compressor (coal pneumatic system)
Air compressor (ash pneumatic system)
Firefighting equipment
  Detection
  Portable
  Hydrant system
  Springer system
  Panel
  Halon/CO_2
Pressure-reducing set
Press. red. and desuperheating set
Ventilation
  Turbine hall
  Boiler house
  Switchrooms
  Battery room
  Water-treatment room
  Admin.
Air conditioning
  Control room
Hydrogen-generation plant
Electroeysis
Bottle storage

Piping system
Main steam
Process steam
Feedwater
Raw water
Treated water
Potable water
Aux. cooling system
Firefighting system
Clarified water
Filtered water
Water-intake system
Circulating-water system
Chemical dosing
Station drains
Fuel oil
Fuel gas
Compressed air (plant)
Compressed air (inst.)

Electrical equipment
Generator
Unit auxiliaries
Station auxiliaries
Low-voltage auxiliaries
HV and LV switchgear
Generator circuit breaker
Batteries and charger
Main power cabling
Cabling and wiring
Lightning protection
Earthing
Telephones/intercom
Protection equipment
Electrical controls, inst. meters
Lighting and small power
Temporary supplies
Elec. trace heating
Generator
AVR
Generator VTs
Busbar equipment
Synchronizing equipment

Civil
Building/annexe for:

Water treatment
Feed pumps
Fuel handling/metering
Control room
Switch house
Offices
Admin. building
Gatehouse
Site fencing
Roads
Plant drains
Storm drains
Foundations for mech. and elec.
Plant and tanks
CW culverts
Cooling tower found.
Chimney found.
Crane rails
HVAC
Plumbing and sanitary fittings
Domestic water supply
Tanks/reservoirs
Site investigations and ground
Loading tests
Demolition of existing plant and buildings
Connection to existing plant and buildings
Temporary works

General information relevant to electrical system design is given in Chapter 16. Within an industrial power plant, the following key items of electrical equipment need to be addressed:

1. The power-evacuation circuit, including any generator transformer and the connection with the remaining site-supply system.
2. The supply system for the power plant auxiliaries, including the provision of supplies to any important drives and the provision of standby supply systems.
3. The protection and synchronizing schemes applicable to the generator circuit.
4. The earthing arrangements of the generator neutral.
5. The control of the excitation of the generator when operating in either isolated or parallel mode.

A typical electrical schematic for an 8 MW power station is shown in Figure 15.20.

15.4.5 Site conditions

The site conditions play an important role in the development of layout and scope. The more specific issues are as follows:

1. Naturally, the site location must be as close as practical to the electrical and heat users, thus minimizing cable and piping runs.
2. Altitude and ambient temperature, as discussed earlier, can have significant effect on the performance of the generator, and fluctuations over the annual operating cycle will similarly influence the power station output.
3. Exposure to dust or other pollutants that may exist in certain industrial processes will dictate requirements for such aspects as equipment enclosures and air filtration. For the same reasons, it may be prudent to avoid air-cooled heat transfer surfaces.
4. Adequate area will be required for water, chemical, fuel and ash storage with adequate access for reception and off-loading.
5. During the erection period access will be needed for the heavy loads, storage of equipment and for erection.
6. With the exception of the gas turbine, most generating schemes need a reliable and a good-quality water source for cooling. Alternative large air-cooled systems can be employed, but they are expensive and less efficient. Thermal power stations using steam condensers can require large cooling-water flows, although the cooling water consumption can be reduced by use of closed-cycle cooling-water systems. In these cases,

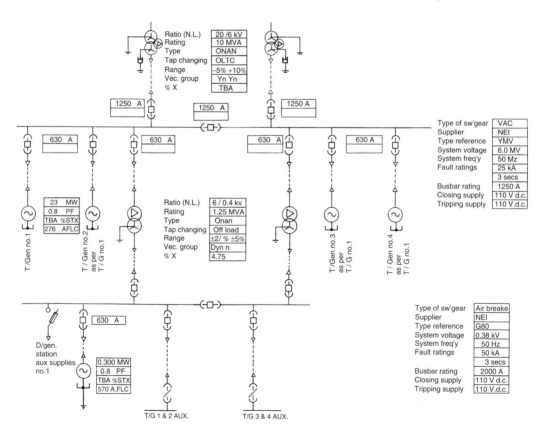

Figure 15.20 Typical electrical single-line diagrams for an 8 MW power plant

it is common to site the power station near a large water source such as a river or lake. The design of water-intake systems needs to recognize the changes in level and quality that can occur, depending on such aspects as time of year, rainfall and tides.

7. Boilers require very high-quality water for make-up, and for this purpose treatment, facilities may be needed on-site, depending on the quality of the water source.
8. Civil engineering considerations such as ground conditions and drainage.

15.4.6 Plant availability and maintenance

It is prerequisite of most modern power stations that they achieve high availability. For normal operation, it is common practice for power stations to employ standby equipment that is strategically important to overall operation. Examples are the use of 100 per cent duty standby pumps or compressed air plant. Depending on the size of installation, 3×50 per cent units can be employed with 50 per cent standby capacity and lower initial cost. For similar reasons, security of fuel and water supplies must be addressed. Adequate fuel storage may be necessary to meet anticipated shortages in supply or, alternatively, standby fuel facilities incorporated. Water storage will be needed again to cater for periods when shortages occur.

Adequate consideration to the above will ensure that outages are minimized and maximum availability is achieved. However, at some stage, the generator or generators will be closed for maintenance, and in these circumstances, consideration must be given to making up the electrical supply. For essential electrical loads, an alternative source is needed and, at a premium, this could be supplied from the electrical supply authority or from a standby generator. A standby generator will increase capital cost, but in some cases, this is essential to meet the needs of total security of supply.

For locations in the Middle East subject to wide fluctuations in ambient temperature it has been common to utilize multiple gas turbines which are sized to meet the plant demands during the hot summer months and the cold winter. As the gas turbine output increases substantially, there is sufficient spare capacity to allow for outage of machines without affecting the electrical power export. However, this situation is unique to the environmental conditions and type of equipment in service.

The plant components selected should be of proven design supported by strong reference installations. Modern trends are moving towards reduction in manning levels, and the use of central computer control and supervisory systems also ensures maximum efficiency of plant operation.

Specific details on maintenance of plant are given later. However, in developing the station layout adequate thought must be given to cranage, storage areas and access for maintenance of heavy plant.

15.4.7 Environmental aspects

Local regulations and laws will control the effects of the power station on the local environment. The significant considerations are:

1. Exhaust emissions of SO_x, NO_x and particulate matter.
2. Height of chimney.
3. Noise levels at the site boundary.
4. Treatment of boiler and water treatment plant effluent.
5. Architectural features and design of buildings and structures.

15.4.8 Generated voltages

The selection of generated voltage will be determined by the voltages available on-site, the amount of generation envisaged and the full load and fault current ratings of existing and proposed equipment. Following the completion of the installation, the fault levels occurring at each point within the installation (including the equipment within the supply authority system) must be within the capability of the equipment, particularly the switchgear installed.

The factors affecting the generator circuit itself are related to the relative cost and availability of equipment to be used in that circuit. Generators rated at up to 2 MVA, 2782 A at 415 V can be accommodated within standard low-voltage switchgear ratings of, typically, 3000 A full load current and 50 kA fault capacity. Generators rated larger than 2 MVA but less than 15 mVA will be usually operated at 11 kV, although at ratings of between 2 and 5 MVA, a lower-voltage 3.3 kV or 6.6 kV may be considered due to the capital cost of the generator.

15.5 The selection

It should be borne in mind that there is no single answer as to which configuration of equipment is best for any individual power-generating scheme. Each alternative must be considered on its own merits, including the use of all available fuels.

15.5.1 Electrical power

For generation of electrical power only, the overriding influence in the choice of generating plant is fuel availability. In this respect the following categories can be applied:

Diesel (light and heavy fuel oils)
Gas turbine (gas and light fuel oils)
Gas turbine and combined cycle (gas and light fuel oils)
Thermal power plant (coal, lignite, wood and biomass)

For the criterion of fuel to electrical energy conversion efficiency, the diesel and gas turbine are natural choices for prime fuels such as oil and gas.

In open cycle, the diesel has the higher efficiency, making it more attractive for light fuel oils. However, in combined cycle, the gas turbine often has the highest overall efficiency, but there is the penalty associated with the additional cost for boiler and turbine. Nevertheless, unless the prime fuel is of low cost, the use of combined-cycle gas turbine plant will prove to be the more economic.

For industrial or commercial applications the straightforward condensing thermal power plant cannot compete with the diesel or gas turbine, as its first, or capital, cost is significantly higher and its overall efficiency is low. However, successful smaller thermal power plants have been installed where the fuel is of low cost to the user. In these cases the efficiency will be of less significance, and the much lower running costs when compared with prime fuels such as oil makes the thermal power plant an attractive alternative.

The overall efficiency of the condensing thermal cycle, as discussed, is dictated primarily by the steam conditions used. There are some small industrial stations with outputs up to 2 MW using shell-type boilers for the generation of steam. Here the steam conditions are limited to approximately 17 bara and 250°C. For larger installations these conditions will rise sharply when watertube boilers become attractive and more common steam conditions are of above 60 bara and 540°C.

Having decided the type of generator, the next questions are the capacity and unit sizes. It is usual for the total installed capacity to be capable of meeting the peak demand with spare capacity to meet outages for maintenance. However, consideration can be given to a supply from the electricity supply authority for peak demands and standby.

To meet the needs of a given output, a single generator could be provided, and this option is likely to give the lowest initial cost. However, the alternative of two or more machines, of the same total capacity, while more expensive initially gives higher availability, better flexibility of operation and improved load-following performance.

In isolated areas, it may be a requirement that at least one unit is needed for standby to meet security of supply, and this can be provided by 2×100 per cent or, say, 3×50 per cent units. Availability is maximized as the standby unit can be brought into operation during outage of one of the normally operating machines.

This is best illustrated by the following example for a rural development project. The electrical power demand is shown in Figure 15.21 and was developed from historical data generated from similar developments in other regions. The operating statistics indicated that the maximum demand would occur at night peaking to 8 MW with a low day demand of 2 MW. Considering these power requirements, the station must be flexible and have the ability to follow the daily load pattern. In addition, because of the isolated location the power station must be sized for *stand-alone* operation. Equally, important factors are maximum availability and operating efficiency.

The use of a local indigenous coal source dictated the choice for thermal power plant. Considering the forecast power demand, it is feasible for an 8 MW turbine/boiler unit to be installed to provide the ultimate load requirement. However, the single power unit is not flexible in that it does not give the required security to cover for routine and statutory shutdowns for maintenance and is less efficient at part load (see Figure 15.22). Provision of multiple small units is the ideal solution, as it allows the station to be operated more flexibly, enabling the units to be put into or taken out of service as the power requirement demands so that the operating units run close to peak efficiency. From this examination, the optimum scheme is 5×2 MW units with four units providing the normal power requirements and the fifth to cover for standby.

15.5.2 Combined heat and power

As discussed earlier (Section 15.3), steam turbines providing low-pressure steam are more established for CHP. However, for lower heat loads the diesel and gas turbine with waste heat recovery offer an attractive alternative.

Steam turbines for CHP

The first parameter to resolve is the process steam pressure or pressures. These will be dictated by the process-heating

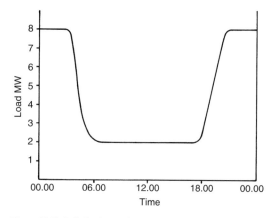

Figure 15.21 Daily load cycle for a rural development project

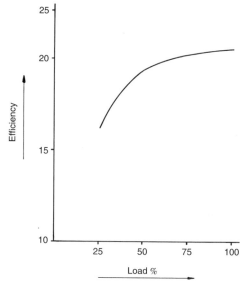

Figure 15.22 Overall efficiency of a typical coal-fired power station of 4.6 MW capacities

temperatures, noting that the steam will transfer its heat at constant pressure and at saturation temperature. It is in the interests of efficiency to keep the process steam pressure as low as practical. Also influencing the process steam pressure is the distribution pipeline from the turbine exhaust to process. Reductions in the pipeline pressure loss increase pipeline size and cost.

The selection is dictated by economics governing the initial plant cost versus higher turbine output. Usually, the turbine exhaust steam is designed to be slightly super-heated, which is desirable, as it allows for heat loss from the steam with minimum condensate losses. At low loads from the turbine, the degree of superheat can rise sharply, well in excess of the normal design conditions, and for this purpose, desuperheaters are often employed to trim the steam temperature at exhaust.

The choice of boiler steam inlet conditions is usually dictated by the desire to achieve maximum output from the process steam flow. This requires high boiler steam pressure and temperature. However, there are practical considerations to observe. Above 40 bar more exacting feedwater treatment is necessary, and therefore it may be advantageous to maintain pressures below this figure. High steam and temperatures can also influence selection of boiler materials such as alloy steels. The upper limit for industrial applications is around 60 bara and 540°C.

For watertube boilers it is necessary to maintain low O_2 levels, and for this purpose a de-aerator in the feed line is required, which will also provide a degree of feed heating. The steam supply can be taken down from the low-pressure process steam main.

If a condenser is employed for a turbine with a condensing section then it is normally chosen to provide the best thermodynamic efficiency consistent with an economical capital expenditure. The condenser exhaust vacuums are usually 0.05 bara or higher.

Having ascertained the process steam flow and developed some ideas on the boiler pressure, the following step is to analyze the power available. Figure 15.23 provides a ready means of determining the approximate relationship between power available and process steam for specific steam conditions. Use of this and similar charts will allow an assessment to be made of the potential of a CHP scheme with a backpressure turbine. The conditions can be changed to give the required balance for heat and power.

The electrical and heat analysis, as discussed in Section 15.3, will show the relationship between power and heat and how this varies over time. It may be necessary to use steam bypass and stations or dump condenser, as discussed in Section 15.2. The uses of dump condensers for meeting part-load requirement is inefficient and should be avoided. It is more acceptable to reduce turbine power output accordingly and import or top-up from an alternative supply.

Diesels and gas turbines for CHP

It is generally recognized that below around 5:1 heat ratio the steam turbines in CHP become less suitable, and here the diesel and gas turbines play an important role in providing heat and power. As discussed earlier, the simple

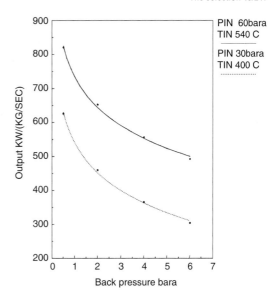

Figure 15.23 The effect of back-pressure upon steam turbine performance

addition of waste heat boilers gives additional heat output without further input of fuel to the generator.

For flexibility, supplementary firing of the oxygen rich exhaust gases can give additional heat. The following summarizes the potential of both the diesel and GT for waste heat recovery:

	Unfired	Supplementary fired
Diesel	1:1	5:1
	0.6:1 (excluding the cooling-water services)	
Gas turbine	2:1	15:1

This gives indicative heat/power ratios only: an actual performance will depend on the machine.

Using these guidelines, it is a simple matter of matching the electrical and the heat loads. An example is a scheme that was envisaged for the supply of power and heat to a computer center using a diesel engine with waste heat recovery. Figure 15.24 shows the proposed scheme where heat generated provides the site demand and operates with absorption-type chillers for chilled water production. The use of these chillers also reduced the electrical demand. The electrical load was a steady 2.5 MW over the year with short duration peaks of 2.9 MW. To allow for future growth, 2×3.6 MW diesels were selected (one operating and one standby). The electrical load calculated was 21.75×10^6 kWh/year, and applying a useful heat recovery ratio of 0.6:1 (not including engine services), this gives 13×10^6 kWh/year of heat energy, which is steadily available over the year. The hot water and chilled water heat loads are shown in Figures 15.25 and 15.26. As the site heat load and chilled water load peak at different times

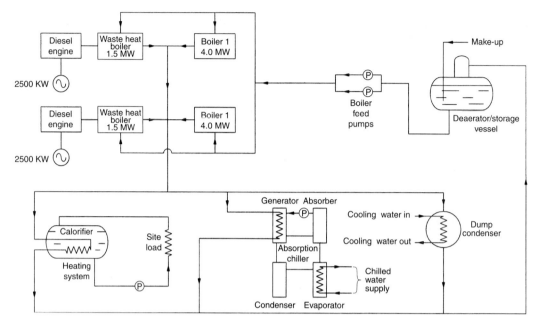

Figure 15.24 Example of a small diesel-engines combined heat and power scheme

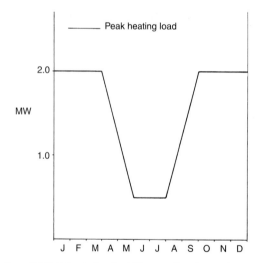

Figure 15.25 Hot water heat load. The annual heating requirement is 5.2×10^6 kWh

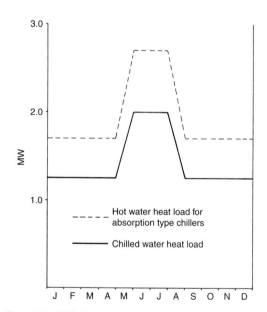

Figure 15.26 Chilled water heat load. The annual chilled water heat load is 13×10^6 kWh; the annual hot water heat load for absorption-type chillers is 17.4×10^6 kWh (a COP of 75 per cent)

of the year, this gave a relatively steady overall heat load, as shown in Figure 15.27.

This total heat load is given as 22.6×10^6 kWh per annum. Supplementary heat is therefore necessary to provide the additional $(22.6 - 13) = 9.6 \times 10^6$ kWh per annum. For security of supply, 4 MW boilers capable of giving full independent supply provided this.

The diesel and gas turbine with waste heat recovery are limited in terms of fuel application being suitable for gas and oils only. Also when considering oil fuel firing

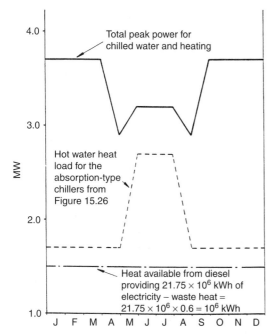

Figure 15.27 Total heat load. The heat available from the diesel waste gases of 13×10^6 kWh will provide only a proportion of the heat required for the absorption-type chiller

Within the figure:
- Total peak power for chilled water and heating
- Hot water heat load for the absorption-type chillers from Figure 15.26
- Heat available from diesel providing 21.75×10^6 kWh of electricity – waste heat = $21.75 \times 10^6 \times 0.6 = 10^6$ kWh

careful attention must be given to the effects of sulfur in the exhaust gases, as this will place limitations on the boiler performance.

15.5.3 Economic considerations

An economic evaluation will ascertain the cost for power generation when compared with purchase of electricity and (where applicable) generation of steam in low-pressure boilers. This evaluation will consider the following:

1. *Capital cost* This cost must cover the supply and erection of all the new plant, including auxiliary systems or modifications to existing as necessary, electrical plant, control and instrumentation, civil works, new buildings and demolitions if appropriate. It will also include all payments, management and engineering fees, insurance and inspection costs – in other words, all the necessary costs for the completion of the project. This may also include the initial holding of strategic spares.

2. *Fuel costs* Capital cost and fuel costs are closely interrelated. For example, within a specific industrial range the thermal power plant can be three times more expensive than an equivalent rated diesel. However, thermal power plant can utilize lower cost fuels and, when taking into account diesel costs for consumables such as lubricating oil and maintenance spares, the thermal power plant can be attractive over the longer term. The fuel consumption must be established from an analysis of the load profile and overall efficiency of the station at various part loads. In addition, the analysis must also take into account the effects of ambient air temperature changes on air intake mass and cooling water temperatures, as these also influence efficiency. It is prudent to add a small percentage to cover for load following and deterioration of efficiency that may occur over a prolonged operating period. The fuel consumption should also include any anticipated supplementary firing needed on CHP schemes.

3. *Chemical consumption* Chemical consumption will be associated with boiler feed make-up water-treatment plant, dosing systems for feedwater and boiler system, treatment of cooling water circuits and effluent treatment. Typical chemical requirements for a thermal power plant are given in Table 15.9.

4. *Feedwater costs* Depending on the source of water supply, there may be charges for water consumption. This consumption will be dictated by the make-up water needs for the station.

5. *Consumables* All consumables required for normal operation should be accounted for. Typical examples here are diesel lubricating oil consumption and oil changes at service intervals.

6. *Labor costs* These will depend on the manning levels and degree of automation of the plant.

7. *Maintenance charges* Maintenance costs as an average over the plant life (based on previous experience and guidance) should be sought from the manufacturer.

Table 15.9 Typical chemical requirements for a 60 MW industrial steam-generating plant

Chemical	Location	Quanity
Hydrochloric acid (HCL)	Water-treatment plant (cation)	165 kg/day at 30% solution
	Effluent neutralization	14 kg/day at 30% solution
Caustic flake (NaOH)	Water treatment plant (anion)	221 kg/day at 25% solution
	Effluent neutralization	14 kg/day at 25% solution
Oxygen scavenger (e.g. hydrazine)	LP feedwater dosing	8.5 kg/day at 25% solution
pH control (e.g. ammonia)	L1 feedwater dosing	24 1/kg/day
Phosphate	Boiler dosing	Consumption depends on blowdown rate for boiler
Corrosion inhibitor ⎱ Bacterial control ⎰	Cooling-water system	Consumption depends on make-up Chemicals are used to provide adequate reserve

8. *Purchased power costs* Depending on the schemes applicable, there may also be charges for import of electrical power. In the UK the electrical supply authority charges a sum based on a maximum demand charge, a unit charge and a fuel cost adjustment. Depending on circumstances, there may also be a fixed annual charge to cover rent of the authority's equipment.
9. *Inflation* After economic analysis it may be necessary to review how inflation may influence the costs for the items purchased.
10. *Erection periods* Period of erection can be critical, as this dictates when power will be available and income earned. Depending on the size of the station and scope, this could vary between less than 6 months to over 2 years. For systems such as the combined cycle it is practical to commission the gas turbines first in advance of the steam generators, so that useful power is generated at the earliest possible date.

15.6 Plant and installation

In order to operate the prime movers described in the previous sections it is necessary to provide auxiliary equipment for the start-up, steady operation and shutdown of the basic equipment as well as for monitoring and controlling its performance. The need also arises for the maintenance of the plant that invokes the provision of cranage and lay-down areas in the engine room. The following describes these features for the various types of prime movers. The driven machines (i.e. the electrical generators) are also reviewed in detail so that the complete picture of industrial generating stations can be obtained.

15.6.1 Diesel power plants

When driving a generator the diesel engine is usually a multi-cylinder machine similar in appearance to that shown in Figure 15.28. The figure illustrates a nominal 3 MW engine with twelve cylinders in V-formation, such an arrangement being attractive for its compactness. The engine will be required to operate at synchronous speed which is, in turn, dependent upon the required frequency (f) and the number of generator pole-pairs (p). We have

Operating speed $= N = 60 \left(\dfrac{f}{p} \right)$ revolution/minute

A typical block layout for a diesel engine room is given in Figure 15.29, showing the necessary auxiliaries and local control panels. Not shown but also necessary are cranes and fire-protection systems.

The engine room itself must additionally provide sufficient noise attenuation as shall be deemed necessary, and there must also be adequate fuel storage, reception and handling facilities.

A key issue also concerns the foundation requirements of the generator set. Most manufacturers will provide data on their machines' requirements, but the detailed design of the block and its interface with the engine bedplate require close attention. Sometimes it is possible to provide

Figure 15.28 A Vee-12 diesel engine

a basement for the engine auxiliaries, while ventilation considerations (excluding combustion air requirements) should recognize that some 8 per cent of the engine's rating will be radiated as heat, in the absence of any heat-recovery equipment.

The compression ignition engine can operate on a variety of liquid fuels and gas. Modern designs permit reliable operation on up to the heaviest residual oils, so permitting improved fuel flexibility (providing, of course, that an adequate fuel-handling capability is also present). The engine may also operate in a dual-fuel mode, using a mixture of gas and air with the option to revert to a liquid fuel in the event of any interruptions to the gas supply. Thus, a more complex control and protection system must be provided in order to give modulation for both the oil and gas flows.

Air intake

The quality of the air supply directly affects the output, efficiency and life of the engine. The requirement of the induction system must therefore be to supply the engine with clean dry air close to ambient temperature conditions. Oil bath or dry (paper element) filters are adequate for low dust concentration conditions. However, as the dust burden of the air increases, centrifugal pre-cleaners become

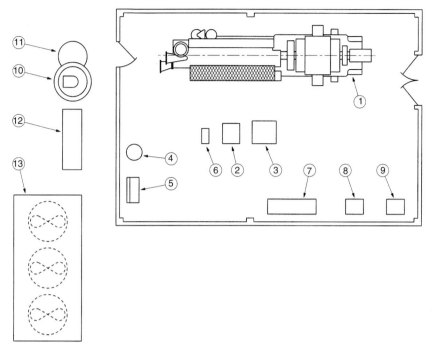

Figure 15.29 The layout of a typical diesel generator engine room. 1 Diesel generator set; 2 jacket water header tank; 3 lubricating oil service tank; 4 air receiver; 5 diesel-driven compressor; 6 batteries and charger; 7 engine control panel; 8 pneumatic control panel; 9 fuel oil control panel; 10 engine exhaust silencer; 11 charge air filter; 12 daily service fuel oil tank; 13 three-section radiator

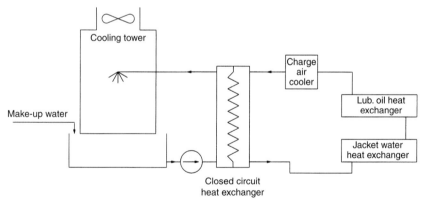

Figure 15.30 A typical closed-circuit secondary cooling system

necessary. The combustion air requirements of a diesel may be taken, for estimating purposes, to be $9.5\,\mathrm{m^3h^{-1}}$ per kW generator rating.

Cooling

For small engines below 0.5 MW output, air-cooling can be considered. However, water-cooling by circulation of water through cylinder jackets is the method normally used for diesel generating plants.

Fan-cooled radiators (as seen in Figure 15.29) or the system shown in Figure 15.30 are typical. As an alternative to the cooling tower, when a good source of raw water is available (such as from a river) this can be used on one side of the heat exchanger.

Engine starting

For small high-speed engines, electric start systems, incorporating lead-acid or alkaline batteries, are practical.

For larger machines, air-start systems are normally used. Using compressed air, starting may be effected by means of an air-driven motor engaging the flywheel or by directing air into each cylinder, in firing sequence, by means of a camshaft-driven distributor or mechanically operated valves. The air can be provided from receivers charged up when the engine is running or from a small auxiliary compressor unit. The latter system is most common on the larger installations, while on critical engines an internal combustion (IC) engine-driven unit backs an electric motor-driven compressor.

Control and instrumentation

Engine speed is controlled by the use of variable-speed governors that can be mechanical, mechanical-hydraulic or electronic. The last option is gaining wide acceptance for generation purposes due to its speed of response and ready integration with other control equipment used in fully automated installations (probably incorporating more than one generating unit).

Stop, start, emergency stop and speed controls are grouped adjacent to the engine, and the machine is usually equipped with instrumentation for monitoring its safe operation. Very often, a centralized control room is provided, particularly when there are a number of units in operation. Typical signals monitored and recorded remotely are:

Engine and turbocharger speeds
Hours run
Lubricating oil pressures and temperatures
Starting and change air pressures
Cooling-water temperatures
Exhaust temperature

Typical alarm annunciations are:

Engine cooling water: High outlet temperature
Engine lubricating oil system: Low inlet pressure
Fuel system:
    Service tank: Low level
    Engine inlet: Low pressure
Overspeed protection:
    Engine speed: High
    Emergency stop: Operated

Maintenance

The diesel engine, being a reciprocating machine, is mechanically complex, and in arduous environments its wear rate can be high. Major overhauls on high-speed engines are usually stipulated at 15,000 running hours, which extends to 20,000 and 30,000 on medium- and low-speed machines, respectively.

The major class of failure concerns the fuel supply and fuel injection equipment followed by the water-cooling system, valve systems, bearings and governors. Collectively, these five categories account for some 70 per cent of all engine stoppages. Thus, the maintenance program must take careful cognizance of these areas together with the manufacturer's recommendations. Under arduous conditions where fuel quality is questionable, where there are high dust levels or where the machine is subject to uneven and intermittent loading, enhanced maintenance at reduced intervals must be recommended.

Lubrication clearly plays a significant role in the reliable operation of the engine. In addition to the lubricant's primary task of reducing friction and minimizing wear it also acts as a cooling medium, a partial seal between the cylinders and piston rings, and a means of flushing combustion and other impurities out of the engine.

A pressurized lubrication system using an engine-mounted pump is the choice of most manufacturers. A sump in the crankcase or an external drain tank together with filters and coolers complete the system. The choice of oil should be consistent with the manufacturer's recommendation, and should only be varied if the engine is to be operated in unusual or extreme conditions.

Diesel engines with heat recovery systems

As illustrated earlier (Section 15.3.2), the utilization of heat otherwise lost in the engine exhaust (and also the cooling water) can be put to use in providing steam for process. Alternatively, this heat can be used for fuel pre-heating or to power a steam turbine via a steam generator. Figure 15.31 shows a two-engine arrangement with a waste heat-recovery system supplying superheated steam to a small steam turbine.

In the above example, a relatively complex steam generator of the watertube type has been adopted. Where lower-quality steam for process or fuel heating is required, a simpler shell (or firetube) design may be appropriate. In some cases, supplementary firing may be provided for the boiler, so further increasing plant complexity and with it the need for enhanced control and maintenance requirements.

15.6.2 Gas turbine power plants

The combustion gas turbine is, in many respects, the most attractive means of producing power. Today the advantages of reliability, simplicity of operation and compactness outweigh the disadvantages posed by the price of the fuel required operating it. A typical land-based machine is shown in Figure 15.32. Gas turbines are normally rated at notional sea-level (ISO) conditions when burning a specified fuel.

For land-based power-generation applications, there is also a significant sub-division with the availability of aero-derived machines and specifically developed industrial designs. The smaller (up to some 41 500 hp) aero-derivatives can accept load very rapidly and will operate in remote hostile environments. Larger machines, with sizes in excess of 140 MW, are of the industrial type taking perhaps some 30 minutes to reach full load. Gas turbines require a start-up drive capable of achieving a high rotational speed so that they can become self-sustaining. A typical start-up and stop schedule is shown in Figure 15.33. The power consumption of the start-up device will be between 5 and 10 per cent of the machine rating.

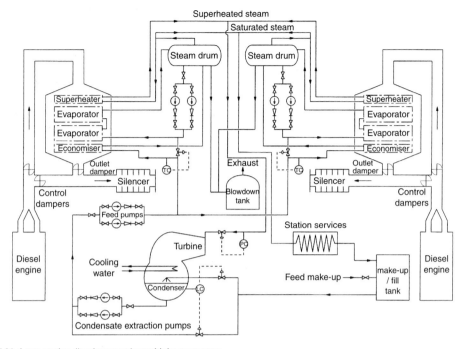

Figure 15.31 A two-engine diesel power plant with heat recovery

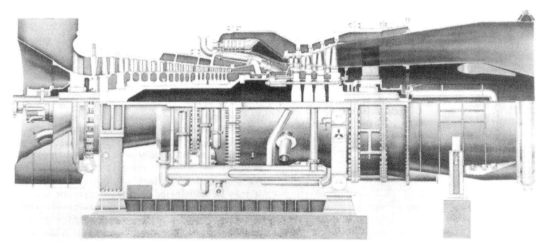

Figure 15.32 The Mitsubishi MW 701 gas turbine rated at 131 MW (ISO)

As the simple (or open-cycle) gas turbine is relatively inefficient (see Table 15.1), improved efficiency can be achieved at the expense of complication and first cost by recovering some of the heat from the exhaust. One arrangement employing heat exchangers in a closed circuit system is shown in Figure 15.34. However, the most popular means of recovering a significant proportion of

the gas turbine's utilized heat is by means of a combined-cycle arrangement discussed in Section 15.2.4.

Combined-cycle plant

The range of sizes, plant configurations and cycle parameters make this option extremely flexible, permitting electrical outputs of up to 100 MW to be considered

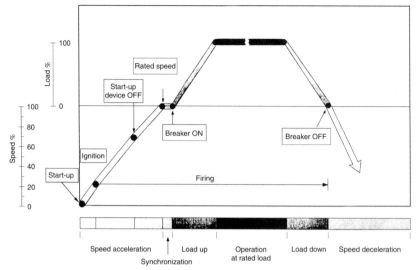

Figure 15.33 Gas turbine start-up and shutdown schedule

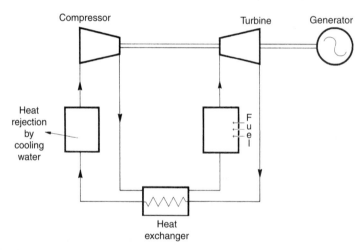

Figure 15.34 A closed-cycle gas turbine arrangement

by certain industries (such as aluminum smelters). It is now becoming common practice to consider block configurations consisting of two or three gas turbines plus the associated waste heat-recovery steam generators providing steam for a single steam turbine. The $2 + 2 + 1$ option has the advantage that the ratings of both the gas turbine and the steam turbine alternators are conveniently arranged to have the same rating. The $3 + 3 + 1$ option provides better site economy with perhaps a *power density* of some $7.5 \, \text{MW/m}^{-2}$ (as against $5.5 \, \text{kW} \, \text{wm}^{-2}$ for a $2 + 2 + 1$ option) but no commonality of alternators.

A typical power train for such a generating station is shown in Figure 15.35. A further refinement, capable of improving cycle efficiency by several percentage points,

is the introduction of dual steam pressures. If there is no requirement for the provision of process steam then it is unlikely that the extra cost and complexity of introducing supplementary firing into the steam generator would be worthwhile.

Control and instrumentation

Manufacturers now offer as standard microprocessor-based control systems as part of their gas turbine generator set. The set can therefore be controlled and monitored from panels adjacent to the machine or, if required, remotely from a central control room. In combined-cycle installations, the system would be linked into the

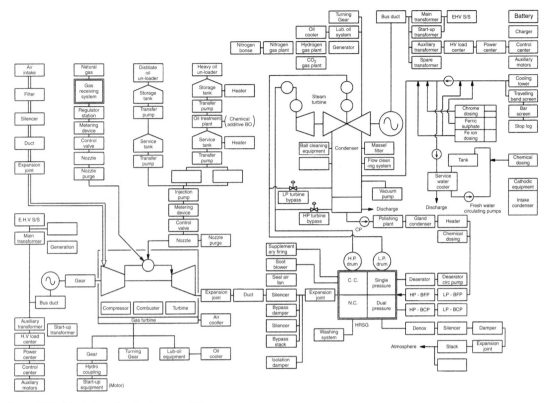

Figure 15.35 A typical combined-cycle power train

higher-level station control system. The instrumentation may typically be expected to comprise:

- Speed and load governing with electro-hydraulic actuators
- Temperature control, also via electro-hydraulic actuators
- Temperature, pressure and vibration monitoring systems
- Flame-detection system
- Automatic sequential control system

A control system diagram for a gas turbine generator is shown diagrammatically in Figure 15.36.

Maintenance

The life and necessary maintenance of a gas turbine are heavily dependent upon both the operating regime and the fuel quality. Continuous firing on natural gas provides the optimum availability, which will be progressively eroded if the plant is subject to frequent interruptions (i.e. stops and starts) from both cold and hot conditions. With a maximum interval between inspections of some 8000 hours, it may be anticipated that the combustion section will require most attention. Every 16,000 hours (or less) the turbine section will need inspection. While a major inspection of the entire unit will be necessary every 31,500 hours. Under optimal conditions, the

average operating availability of base-load and based gas turbine plant could exceed 95 per cent (averaged over a 4-year or 31,500-hour period).

Pollution control

In the past, gas turbine installations have suffered from smoke emissions. However, today, with better combustor design and combustion control, this emission has ceased to be a nuisance. Of more concern now and resulting from higher combustor (and therefore combustion) temperatures is NO_x emissions. Current legislation is tending to demand that gas turbine installations control their nitrogen emissions to levels between 50 and 75 ppm. Such targets can be achieved in several ways, but none without some effect upon operating cost and fuel flexibility.

The NO_x constituent in the exhaust of machines firing natural gas is some 150 to 160 ppm, and for distillate fuels typically 260 ppm. In order to reduce these levels to the targets quoted above, catalytic filters can be used, but the systems currently available are expensive. As an alternative, certain manufacturers are developing low-NO_x burners but these limit the user to natural gas firing.

The third option incorporates the injection of water or steam into the combustion chamber(s) of the machine. Whenever possible, steam should be selected, as it has less effect upon machine efficiency than water injection.

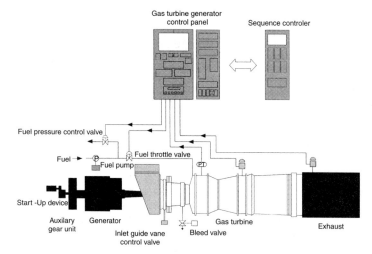

Figure 15.36 A gas turbine control system

By way of example, reducing NO_x emissions when firing natural gas to within a 75 ppm limit would require steam or water injection at a rate of 60 per cent of the fuel flow rate.

Finally, noise is also becoming of increasing concern, particularly on sites adjacent to urban areas. Acoustic enclosures, supplied by the manufacturer, will normally reduce noise levels to 85 dB, but if further reduction is required, attention must be paid to the material used to construct the buildings and enclosures; low noise levels can be achieved only at a cost.

15.6.3 Steam turbo-generators

One of the most widely used means of generating power both in industry and in public utilities utilizes the steam turbine. For industrial power generation the advantages of high availability and good machine efficiency have to be balanced against relatively high first costs and the need for the provision of both expensive steam-generating plant and heat-rejection equipment. However, in many instances the recovery of waste process heat by means of a steam cycle is economically attractive.

Alternatively, the option that a steam plant offers of the provision of process steam coupled with power generation may be the key element in the selection of generating plant. The turbo-generators and their auxiliaries for use in such applications tend therefore to be relatively unsophisticated, with no feed heating, except for probably the provision of a deaerator. Again, in the turbine itself, machine efficiency tends to be sacrificed for robustness and the ability to accommodate varying load conditions.

Options

The single-stage, single-valve turbine is the simplest option. Such a machine is suitable for applications requiring powers up to 300 kW, steam conditions up to a nominal 115 bar, 530°C and rotational speeds

below 5000 rev/min. A typical machine is shown in Figure 15.37, characterized by an overhung rotor mounted on a stiff shaft capable of being accelerated from cold to operating speed within 10 s. The efficiency is low, but so is the cost of the installation. For higher powers and for a wider range of steam conditions incorporating all combinations of backpressure, condensing and passout systems, the multistage axial flow machine is the natural choice. However, it is possible to design a turbine in which the steam flow is radially outward (or inward), passing through groups of blades set in concentric rings. Such an arrangement is known as the Ljungström turbine (Figure 15.38). Steam leakage and attendant machine efficiency deterioration have tended to discourage operators from the selection of these machines.

In the more favored multi-stage axial flow design the bladeing is of the impulse type (known as a Rateau or Curtis type stage) or of the reaction type. It is popular among manufacturers to design multi-stage machines with a combination of these two types, and the most common combinations are:

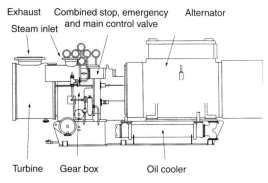

Figure 15.37 A single-stage steam turbine driving an alternator

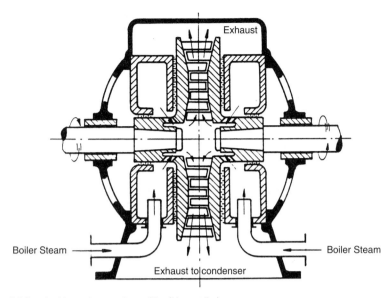

Figure 15.38 The radial-flow double-motion reaction turbine (Ljungström)

1. Velocity-compounded impulse wheels followed by several single-row impulse wheels. This is a combination of the Curtis and Rateau types.
2. Velocity-compounded impulse wheels followed by impulse reaction stages (Curtis–Parsons type).
3. One or more impulse stages followed by several stages of impulse reaction blading (Rateau–Parsons type).

A typical axial flow-condensing turbine for industrial application is shown in Figure 15.39.

The major considerations affecting turbine selection may be listed as follows:

1. Required power output;
2. Steam conditions available (pressure and temperature);
3. Steam cost (in order to assess the value of machine and cycle efficiencies);
4. Process steam requirements (to assess the relative merits of pass-out and back-pressure arrangements);
5. Cooling-water (or other cooling medium) costs and availability;
6. Control systems (ideally compatible with other process plant instrumentation) and automation;
7. Safety features (e.g. operation in explosive environments, alarm and condition-monitoring systems).

Installation

The foundations of any turbogenerator installation play a significant part in the safe operation of the machine. Industrial turbines will normally operate at above synchronous speed and will drive an alternator via a reduction gearbox. Any vibration or out of balance occurring under both normal or abnormal operation must be accommodated by the foundations, and their design should therefore best be undertaken by a specialist organization. The layout of the steam and water pipework, of the machine auxiliaries and, when required, of the condenser all tend to add to the complexity of the turbine island.

Where an under-slung condenser has been specified, the provision of a basement to the engine room offers the attraction of compactness at the expense of enhanced civil works, while alternatively, the specification of pannier condensers can obviate the need for a basement and will simplify the foundation design, but will considerably increase the floor area requirements. The condensing plant itself consists essentially of banks of tubes through which cooling water flows and around which exhausted steam from the turbine is condensed to form a vacuum. Such tubes have traditionally been made of brass, but where severe corrosion conditions exist, cupro-nickel is sometimes used.

Control and lubrication

The turbogenerator speed must be maintained within narrow limits if it is to generate power acceptably and the control system must be capable of preventing over-speed upon sudden loss of load. For this latter requirement, fast-acting valves are necessary with full modulation within 0.5 s. The security of the turbine is also dependent upon the lubrication of its bearings, and it will be found that the control systems are closely linked with the various lubricating systems (turbine, gearbox and alternator).

A typical high-pressure, quick-response governor and lubricating system for a speed-governed turbine is shown in Figure 15.40. A speed governor controls the oil pressure supplied to the servo-operated control valves, which are arranged so that they open in sequence to ensure maximum steam economy. Other types of control systems can be incorporated with the basic speed-governed

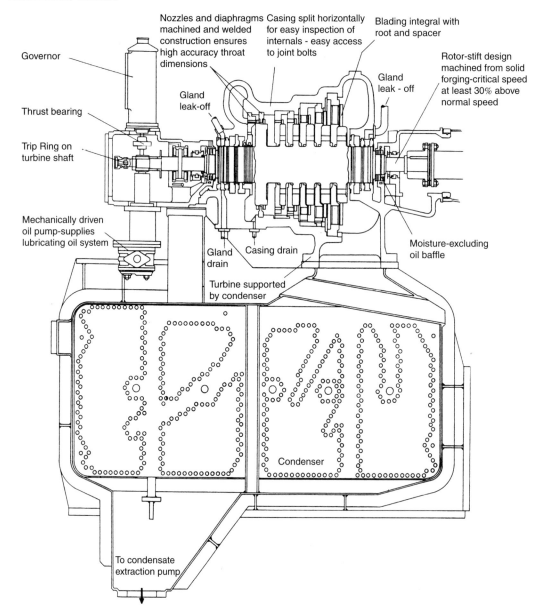

Figure 15.39 Longitudinal section of a small condensing turbine

system to provide back-pressure, inlet-pressure and pass-out pressure governing. A typical fully inter-linked control system for a pass-out system is shown in Figure 15.41.

Oil is supplied to these systems by a pump normally driven from the turbine rotor, with an auxiliary pump for use when starting the turbine. The main lubricating oil pump draws oil through a suction strainer and discharges to the emergency valve and hence through a cooler and filter to the lubrication system. A separate auxiliary

lubricating oil pump is fitted for supplying oil to the bearings when stopping and starting and to lift the emergency valve at start-up.

These systems must be inherently reliable and safe, particularly from the point of view of fire protection, and fire-resistant fluids can be used for control systems in order to reduce this risk. Fire-detection and fire-fighting systems should always be provided for any turbogenerator installation.

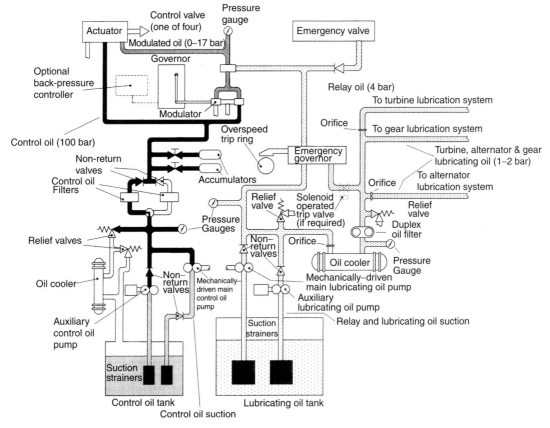

Figure 15.40 A control system for a speed-governed steam turbine

15.6.4 Generators

Generators for use with the prime movers previously described will almost invariably operate in conjunction with an alternating current electrical system. Such a.c. generators will operate either in parallel with other sources of supply (i.e. a grid system or other generators) or in isolation providing the whole electrical supply to an installation. A.C. generators can be one of the two main types:

1. *Asynchronous generator* types are, in effect, induction machines operated at super-synchronous speed. An asynchronous generator can be either compensated type, in which excitation supplies are provided via a commutator, or uncompensated type, in which excitation supplies are provided by the supply system to which the generator is connected. Clearly, uncompensated generators are only suitable for use in parallel with other sources of supply capable of providing the excitation-magnetizing current necessary. While asynchronous generators offer a low-cost option for generation, the disadvantages of lack of control capability and the need usually for alternative

sources of supply means that they are only occasionally utilized and rarely at ratings exceeding 5 MW.

2. *Synchronous generator* type is in use for all generating applications up to the highest ratings (660 MW and above). The machine is operated at synchronous speed and the terminal voltage and power factors of operation are controlled by the excitation.

Principal features of synchronous generators relate to:
(a) The method of excitation;
(b) The method of cooling;
(c) Temperature rise classifications;
(d) Insulation systems.

Since synchronous machines represent the most common type of generator used in power installations, the remainder of this section is principally concerned with this type.

Excitation

The excitation system provides the magnetizing current necessary for the generator to operate at the desired voltage and, when in parallel with other generators, supplies the required amount of reactive current. In modern practice the excitation system can be either brushless or static.

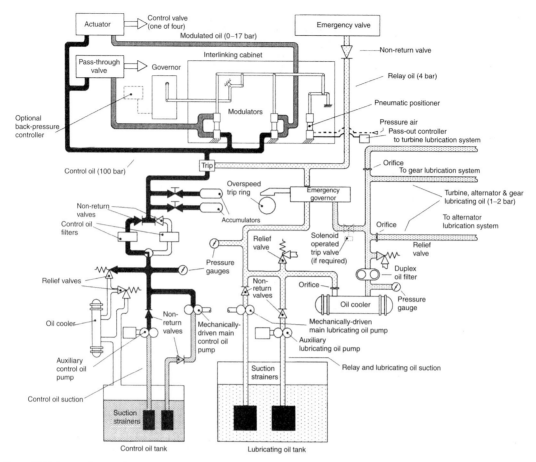

Figure 15.41 A control system for a pass-out steam turbine

In a brushless system an a.c. exciter with a rotating armature and stationary field system is provided. The voltage applied to the stationary field system is varied, thus changing the output of the rotating armature. This output is rectified via shaft-mounted diodes to produce a direct current (D.C.) supply that is connected to the main generator field.

The excitation system is generally designed such that failure of a single diode within the rectifying bridge does not affect the capability of the generator to provide full load. Measuring the amount of alternating current (A.C.) ripple induced within the A.C. exciter field winding and an alarm or trip provided can monitor diode failure. In operation, however, diode failure rates are extremely low.

In static systems D.C. is supplied to the generator field winding via slip rings, the D.C. being produced from diode cubicles supplied from rectifier transformers fed from the generator output. In general, static systems are capable of producing a faster response time than brushless systems and are typically used on larger (above 70 MW) sets.

Cooling

Cooling of generators is classified according to the nature of the coolant, the arrangement of the cooling circuit and the method of supplying power to circulate the coolant. For air-cooled machines, a simplified coding designation is widely used to describe the cooling system. In this system, the first characteristic numeral signifies the cooling circuit arrangement (e.g. inlet duct ventilated or integral heat exchanger) while the second signifies the method of supplying power to circulate the coolant (e.g. self-circulation or from an integral mounted independently powered fan). For machines with more than one cooling system (e.g. a primary and a secondary cooling circuit) a complete coding system is used. This includes the nature of the coolant within each cooling system and the means of circulating the coolant and of removing heat from the machine.

Insulation type and temperature rise

The winding insulation must remain intact with respect to both electrical insulating properties and mechanical strength for the life of the machine. Operation of insulation at temperatures higher than the design will affect the life of the insulation.

Temperature-rise limitations above an assumed ambient temperature (dependent upon location) are used to specify the requirements of the insulation. Classifications utilized for generators are typically:

Class A	105°C
Class B	130°C
Class F	155°C
Class H	180°C

Typically, insulation limited to Class B and Class F temperature rises will employ inorganic materials (e.g.

mica or glass fiber) bonded with a thermosetting synthetic resin. Class H type insulation may also include silicone elastomers.

Connections

The connection of the generator to associated switchgear may be from a cable box located on the side of the generator or exposed terminals usually placed below the machine. When the terminals are the exposed type a cable connection can still be made by utilizing cable-sealing ends or by phase-insulated or phase-segregated busbars. Typically, cable boxes will be provided for machines rated up to 10 MVA while for generators rated above 30 MVA, phase-isolated busbars are generally used.

16

Electrical Distribution and Installation

I G Crow and R Robinson
Davy McKee (Stockton) Ltd and The Boots Co. PLC

Contents

16.1 Introduction

This chapter introduces the basic items of design and specification for the principal systems and components of an electrical industrial installation. Electrical supply systems are discussed with regard to interface with the supply authorities and the characteristics. Salient features of switchgear, transformers, protection systems, power factor correction, motor control equipment and standby supplies are identified and discussed together with reference to the relevant codes of practice and standards. The equipment and systems described are appropriate to industrial plant installations operating at typically 11 kV with supply capacities of around 20 MVA.

16.2 Bulk supply

The majority of industrial installations will receive all or part of their electrical energy from an area electricity board supply authority. The factors that will affect the way in which this energy is imported to the site and subsequently controlled are:

1. The voltage and frequency
2. The load requirements
3. The fault level
4. The tariff
5. Access to equipment

The key parameters for each of these factors are discussed below.

16.2.1 Voltage and frequency

The voltage of the supply will depend upon the load requirements of the site and the relative capability of the local supply system. Voltages typically available are 33 kV, 11 kV, 6.6 kV, 4.16 kV, 415 V and 380 V. For some larger installations, 132 kV (or higher) voltage supplies may be provided.

Statutory regulations imposed upon the supply authorities will normally limit variations in supply voltage to ± 6 per cent and in frequency to ± 1 per cent. In order to maintain supply systems within these limits supply authorities may, in certain cases, impose restrictions upon the starting of large motors in terms of either current drawn at start-up or of frequency of start-up. Alternatively, the supply may be arranged so that the point of common coupling with other consumers is at a higher voltage system where source impedance will be lower.

The increasing usage of semiconductor equipment, particularly for variable-speed drives, has caused more attention to be given to the problem of distorting loads and the harmonics, currents and voltages they create within the supply system. The presence of harmonic voltages and currents within a supply system will affect induction and synchronous motors, transformers, power factor correction capacitors, energy metering and devices relying upon a pure sine wave for operation.

16.2.2 Load requirements

The load imposed upon the supply system will need to be analyzed for load profile, i.e.:

1. Is the load constant?
2. Does the load vary significantly on a regular hourly, daily or annual cycle?
3. Is the load likely to increase as the installation expands?
4. Will the characteristics of the load affect the system supplying it?

The factors that then need to be considered are potential unbalanced loading, transient occurrences and harmonic distortion. Unbalanced loading will cause an unequal displacement of the voltages, producing a negative phase sequence component in the supply voltage. Such negative phase-sequence voltages will cause overheating, particularly in rotating plant. The starting of large motors or the operation of electric arc furnaces are two examples of loading which draw large amounts of reactive current from the supply system. These currents, flowing through the largely reactive supply system, will result in fluctuations in the supply voltage, and these fluctuations will affect other consumers (e.g. causing lighting to flicker).

16.2.3 Fault level

The fault level at the point of supply to the installation requires consideration. The minimum fault level which can occur will affect the operation of the installation, particularly with regard to voltage regulation when starting motors, while the maximum fault level will determine the ratings of equipment installed. All supply systems develop with time, and the switchgear and other equipment installed at an installation must be chosen to be suitable if the external supply system expands.

The contribution of the installation itself to fault level at the point of common coupling is another important consideration. Rotating plant, either generators or motors, will contribute to both *make* and *break* fault levels. The contribution of synchronous machines to fault current can be calculated from sub-transient and transient reactance values, and although similar calculations can be carried out for induction motors, accurate data are not generally readily available for groups of small motors operating together, as in a typical industrial installation. Generally, induction motor contribution to *make* fault levels is taken as being equivalent to motor starting current and to *break* levels as twice-motor full-load currents.

16.2.4 Industrial maximum demand tariffs

The exact details of industrial maximum demand tariffs vary between supply authorities. However, the salient features of most tariffs are as follows:

1. *Fixed charge*: This is typically a fixed cost per month plus a charge for each kVA of agreed capacity. It is important to realize that any change in the operation of a particular site may affect the supply capacity required. The supply capacity charge, although related to, is not dependent on the rating of the equipment installed at the site.

2. *Maximum demand*: These charges are related either to kVA or to kW of maximum demand. If they are of the kW type then a separate power factor charge will be made. The charges depend upon whether the supply is provided at high or low voltage and the time of year. They are higher if the supply is provided at low voltage since the supply authority will seek to recover the cost of the equipment and losses incurred in providing a low-voltage supply. Within the Northern Hemisphere, demand charges are highest for the months of January and December and generally reduced for February and November. They may also apply for March but are not normally incurred during the remainder of the year.
3. *Unit charge*: This is related to actual energy consumption, and two options are generally available; a rate applicable for 24 hours or a split day and night rate, with, typically, a 7-hour night period duration. In per-unit terms these costs will be:

24 hours = 1 per unit.
Day/night = 17-hours day, 1.04 per unit
7-hours night 0.51 per unit

Some authorities offer reduced rates as energy usage increases.
4. *Power factor adjustment*: Although related to the average power factor of the load, the method of calculation may be based directly on measured power factor or on the measurement of reactive kVA over the period. Values at which power factor charges are incurred vary from 0.8 to 0.95 lag.

16.2.5 Access

The switchgear provided at the installation for the provision of the supply by the supply authority will be the property of that authority. That provided for the distribution of the electricity around the installation will be the property of the consumer, while metering for tariff purposes will be located on or supplied from the supply authority switchgear.

It is normal for the supply authority switchgear to be separate from the consumers' switchgear and frequently located in a separate room within the same building. The supply authority may insist that access to this equipment or room is limited to their own personnel.

16.3 Distribution systems

The object of any electrical power distribution scheme is to provide a power supply system that will convey power economically and reliably from the supply point to the many loads throughout the installation. The standard method of supplying reliable electrical supplies to a load center is to provide duplicated 100 per cent rated supplies. However, there are a number of ways in which these supplies can be provided.

The standard approach for a secure supply system is to provide duplicated transformer supplies to a switchboard with each transformer rated to carry the total switchboard load. Both transformers are operated in parallel, and the loss of single incoming supply will not therefore affect the supply to the load feeders. With this configuration the

supply switchboard must be able to accept the fault current produced when both transformers are in parallel. The system must, however, be designed for the voltage regulation to remain within acceptable limits when a single transformer supplies the load.

If the two incoming supplies are interlocked to prevent parallel operation, the power flow, fault current flow and a single transformer governs voltage regulation supply in-feed. This configuration has the disadvantage that the loss of a single in-feed will cause a temporary loss of supply to one set of load until re-switching occurs. The advantage is that each incoming transformer can be rated higher than the first, and thus a higher concentration of loads can be supplied.

A third alternative is to rate the transformer in-feeds for single supplies but to arrange for automatic switching of the bus section and two incomers in the event of loss of supply to one section. A rapid transfer switching to the remaining supply in the event of loss of single supply can prevent the total loss of motor loads. In such a situation, the effect of the current taken by motors to which the supply has been restored must be taken into account.

Since a group of motors re-accelerating together will draw an increased current from the supply, this current will affect voltage regulation and must be recognized when selecting protection relay settings. Figure 16.1

2 × 100% rated transformers

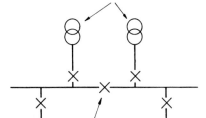

(a) Bus section switch, normally closed

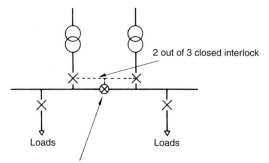

2 out of 3 closed interlock

(b) Bus section switch, normally open

Figure 16.1 Transformer arrangement. (a) Parallel operations; (b) 2 out of 3 interlock

Table 16.1 Load supply capabilities and single and parallel transformer combinations

Source	Load		Load circuit breaker				Max. supply transformer rating (MVA)			
Voltage	Fault	Voltage		Full load, rating	A	Fault		Rectance (%)		
	level		630	2000	3000	rating	6	10	6	10
KV	MVA	kV		MVA		kA(MVA)		Parallel		Single
33	1428	11	12	38.1	–	25 (476)	21	35.7	38.1	38.1
11	476	3.3	3.6	11.4	–	25 (143)	6.1	10.2	11.4	11.4
11	476	0.415	–	1.4	2.2	50 (35.9)	12	1.9	2.2	2.2

illustrates these arrangements of transformer operation. The effect of these combinations upon the loads that can be supplied is summarized in Table 16.1 for two typical supply voltages and fault levels, circuit breaker capabilities and transformer reactance.

In the case of parallel operation the maximum transformer rating is limited by the fault rating of the switchgear, while for a single transformer in-feed the limitation is by the full-load current rating of the switchgear.

Table 16.1 takes into account only the fault contribution from the supply system. The contribution from rotating plant within an installation must also be considered when specifying switchgear and transformer ratings.

If a fully duplicated supply system is thought to be necessary, the transformer reactance can be increased in order to limit the fault level when operating in parallel mode. However, this will increase the initial capital cost, and voltage regulation with a single transformer in circuit will still need to be maintained within acceptable limits.

16.4 Switchgear

Switchgear is a general term covering circuit-interruption devices, the assemblies that enclose them and associated equipment such as current transformers, voltage transformers, disconnects switches, grounding switches and operating mechanisms. The circuit breaker is the interrupting device within the switchgear capable of closing onto, carrying and breaking those currents which can flow under healthy or fault conditions.

Switchgear is frequently described by the medium used in the interrupting device and method of mounting. The principal types of medium presently in operation are bulk oil, minimum oil, air, sulfur hexafluoride gas (SF_6) and vacuum. The most common mounts for indoor industrial switchgear are cubicle and metal clad. The main features of and the perceived advantages and disadvantages of the different types of medium are reviewed below, while international standards applicable to switchgear are listed in Table 16.2.

16.4.1 Bulk oil

The fact that oil was effective for the extinction of the arc formed between the opening contacts of a circuit breaker was discovered towards the end of the nineteenth century. In a bulk oil circuit breaker the oil is used both as the medium for arc extinction and as the insulating material of the contacts. The contacts are contained within a

Table 16.2 British Standards applicable to switchgear

BS 5311, High voltage alternating current circuit breaker
BS 6867, Code of Practice for maintenance of electrical switchgear for voltages above 36 kV
BS 158, Insulating oil for transformers and switchgear
BS 5622, Insulation co-ordination
BS 5227, A.C. metal enclosed switchgear and control gear of rated voltage above 1 kV
BS 4752, Switchgear and control gear for voltages up to and including 1000 V a.c. and 1200 V d.c
BS 3938, Current transformers

small chamber within the oil-filled breaker tank. When the circuit breaker contacts open an arc is formed which decomposes the oil, creating gases. These form a bubble that forces oil across the arc path, causing rapid extinction of the arc. Bulk oil circuit breakers have been used for many years, and it is probable that about 70 per cent of the high-voltage circuit breakers installed are of this type.

The principal advantages of bulk oil circuit breakers are

1. Simplicity
2. Robustness in construction
3. Familiarity to users
4. Low capital cost

The disadvantages are:

1. Potential fire risk
2. Short-circuit contact life
3. Maintenance costs
4. Possible re-strike problems with capacitor switching

16.4.2 Minimum oil

In a minimum oil circuit breaker only the arc control device is enclosed within an oil-filled housing. This housing is supported and insulated separately to provide the necessary phase clearances between phase connections and earth.

16.4.3 Air

Air circuit breakers can be either *air break* types (which utilize atmospheric air) or *air blast* (which use compressed air). Air-break type breakers extinguish the arc by high-resistance interruption, the arc being controlled within an arc chute. As the arc resistance increases, the current is reduced until at current zero arc extinction occurs.

Air-blast circuit breakers operate on the low-resistance principle; the arc length is minimized and a blast of air is directed across the arc to cool and remove ionized gas. Air break circuit breakers are now used extensively at medium voltages while air blast circuit breakers have been employed up to the highest voltages.

For use at normal industrial distribution voltages the principal advantages are:

1. Suitability for all types of switching
2. Good fault ratings
3. No fire risk
4. Low maintenance costs

The main disadvantages are:

1. Capital cost
2. Physical size

16.4.4 Sulfur hexafluoride gas (SF$_6$)

Sulfur hexafluoride gas (SF$_6$) has a high thermal conductivity and a high electronegative attraction for ions. It has insulating properties similar to oil and arc-interruption properties comparable to compressed air at 4826 kN/m^2. Recent switchgear design utilizes SF$_6$ at a relatively low pressure for insulation, with compression occurring via a piston activated by the opening mechanism of the contacts.

Principal advantages of SF$_6$ circuit breakers are:

1. Good arc-extinction properties
2. Low maintenance
3. Size and few disadvantages are perceived.

16.4.5 Vacuum

As a replacement for bulk oil circuit breakers for distribution purposes, vacuum is a rival to SF$_6$. The contact interruption chamber is a sealed unit with a small contact-separation distances (typically, 16 mm for 24 kV). Choice of contact material is critical since the material affects:

1. The electrical and mechanical conductivity;
2. The production of metal vapor upon arc interruption necessary to control current chopping and permit voltage recovery; and
3. The provision of consistent contact wear and separation.

Principal advantages of vacuum circuit breakers are:

1. Minimum maintenance
2. Less operating energy required
3. Size

Its disadvantages are:

1. Possible restriction on fault capability
2. Possible switching over-voltages

16.4.6 Range

Distribution voltage switchgear is designed for the range of full-load current and fault-current capabilities shown in Table 16.3.

16.4.7 Construction

A.C. metal enclosed switch gear and control gear between the voltages of 1 and 72.5 kV can be manufactured to three basic designs: metal enclosed, metal clad and cubicle.

In *metal-enclosed* assemblies, the equipment is contained within a grounded metal enclosure. In *metal-clad* types, separate compartments are provided for the circuit interrupter, the components connected to one side of the interrupter (e.g. the cable box) and those connected to the other side of the interrupter (e.g. the busbar chamber and the control equipment).

Cubicle-type switchgear may have non-metallic components for separation between compartments, no segregation between compartments housing different components or they may be of a more open type of construction. Frequently access to cubicle-type switchgear is via doors interlocked to prevent access to energized parts. The other feature of construction of switchgear concerns the circuit breaker itself, which can be either fixed type or removable. In a fixed construction, the switchgear will be fitted with isolating and grounding switches so that access to the circuit breaker is only possible when the circuit breaker is safely isolated from the supply.

In a removable design, the circuit breaker is mounted on either a truck or a swing-out frame. The circuit breaker is arranged so that isolation from the busbars is only possible when the circuit breaker is open. Connections between the removable circuit breaker and the fixed assembly are either by a plug and socket arrangement or by secondary isolating contacts. Figure 16.2 illustrates a truck-mounted, vertically isolated, metal-clad, SF$_6$ circuit breaker design as marketed by a UK manufacturer.

16.4.8 Mechanisms

Operating mechanisms for switchgear can be either *stored-energy* or *dependent-energy closing*. Stored-energy closing is frequently achieved by means of a spring, which is either manually charged (type QM) or charged by an electrical motor. Dependent-energy closing is by use of a solenoid.

Table 16.3 Capability of distribution voltage switchgear

Medium	Full-load current (A)		Fault current (kA)	
	Typical	Max.	Typical	Max.
Bulk oil	1200	2000	26.4	43.72
Minimum oil	1600	4000	26.4	50
Air	2500	5000	26.4	52
SF$_6$	2000	3150	25	50
Vacuum	2000	2000	25	40

Figure 16.2 NEI Reyrolle SF$_6$ switchgear

The selection of operating mechanism will depend upon the control regime required for the circuit breaker. If the circuit breaker is to be operated locally (i.e. at the switchgear itself) then the manually charged spring option will be acceptable. If it is to be operated from a remote location then either motor-wound spring or solenoid operation will be required.

This choice will depend upon circuit breaker availability required, since a motor-wound spring has a finite spring-wind time. A solenoid operation will impose a significant load upon the D.C. supplies while a motor-wound spring will require a supply to the spring-wind motor.

16.5 Transformers

Transformers are classified by their method of cooling, winding connection arrangement and temperature-rise classifications. Each of these parameters is applicable, irrespective of the voltage ratio and rating required. Core-type transformers are used for almost all power systems applications. In core-type transformers the primary and secondary windings are arranged concentrically around the core leg of substantially circular cross section. International standards applicable to transformers are listed in Table 16.4.

16.5.1 Cooling

Mineral oil provides greater insulation strength than air for any given clearance. When used in a power transformer

Table 16.4 British Standards applicable to transformers and their components

BS 171, Power transformers:
    Part 1, General
    Part 2, Temperature rise
    Part 3, Insulation levels and preelectric tests
    Part 4, Tappings and connections
    Part 5, Ability to withstand short circuit
BS 158, Insulating oil for transformers and switchgear
BS 2757, Thermal classification of electrical insulation
BS 223, Bushings for alternating voltages above 1000 V
BS 5622, Insulation co-ordination
BS 2562, Cable sealing boxes for oil-immersed transformers

mineral oil also augments the removal of heat from the windings.

In power transformer manufacture the case of paper insulation and oil used in combination is well established. Both materials can be operated safely at the same maximum temperature (105°C) and this combination of use seems unlikely to be phased out in the near future.

For transformer windings immersed in oil, hydrocarbon oil is the most widely used, whereas in areas where fire risk is a problem, then air-cooled transformers (AN) or synthetic silicon-based liquid cooling (SN) can be specified. Silicon-based liquids do not have any of the disadvantages identified with the chlorinated biphenyls. Air-cooled transformers can be provided with Class C insulation or be cast resin insulated. The relative costs per unit of each type are:

Mineral oil	1
Silicon filled	1.25-1.5
Class C	1.8
Cast resin	2.0

16.5.2 Types

ONAN

Oil-immersed air-cooled transformers cover the majority of units installed of up to 5 MVA rating. Fins or corrugations of the tank or by tube banks provide cooling of the oil. For ONAN transformers of above 5 MVA rating radiator banks of elliptical tubes or banks of corrugated radiators are often provided.

ONAF

In oil-immersed air-forced transformers, a direct blast of air from banks of fans is provided to the radiators, which increases the rate of heat dissipation. This arrangement has no effect upon the size of the transformer itself but less space for external coolers is required.

OFAN

Oil-immersed air natural circulation is an uncommon arrangement, and is useful only if the coolers are situated away from the transformer.

OFAF

Most larger-rated transformers are oil-forced air-forced cooled. Oil-forced circulation improves the heat dissipation around the windings, thus reducing the size of the transformer itself and air blast cooling of the radiators decreases the size of the cooling surfaces. Thermostatic control of both oil forcing and air blast is usually provided so that each is brought into service at a different oil temperature. The ONAN rating of an OFAF transformer is typically 50 per cent of the OFAF rating, although any value can be designed when required.

AN

These are solidly insulated transformers using typically cast resin insulation. They are particularly useful when an indoor installation is necessary, since they represent a very low fire risk and oil leakage and drainage does not need to be considered. AN transformer maximum ratings are typically of the order of 1500 kVA. Forced cooling (AF) can be added in the same way as that of the oil-immersed transformers, permitting increased output from the same size of unit.

16.5.3 Temperature rise

Temperature classes for transformers A, E, B, F, H and C are generally recognized. All classes are applicable to both oil-immersed and air-cooled transformers. Classes A, B, H and C are most commonly used but the use of Class F is increasing.

For windings immersed in either hydrocarbon oil or synthetic liquids the coil insulation is usually a Class A material. For oil-immersed transformers BS 171 specify two maximum oil temperatures: 60°C for sealed-type transformers for those fitted with conservators and 55°C for transformers without either. In the case of air-insulated, air-cooled transformers, the class of insulation used, i.e, limits the maximum permitted temperature rise:

Class A   60°C
Class B   90°C
Class C   150°C

These temperature rises apply above a maximum ambient temperature of 40°C and a daily average temperature of 30°C is assumed.

16.5.4 Tests

BS 171 defines both routine and type tests for transformers. Routine tests comprise the following:

1. Ratio measurement, polarity check and phase relationship;
2. Measurement of winding resistance;
3. Measurement of insulation resistance;
4. Measurement of load loss and impedance voltage;
5. Measurement of core loss and magnetizing current;
6. High-voltage withstand tests. The rated withstand voltages applicable to transformers of up to 36 kV are shown in Table 16.5.

The measurement of impedance voltage is defined as the voltage required circulating full-load current in one winding with the other windings short-circuited. It is common practice to express the leakage impedance, Z, of a transformer as a percentage (or per unit) value. The per unit value is:

$$\frac{V}{IZ}$$

Where V and I refer to full-load current and rated voltage and IZ is the voltage measured at rated current during a short-circuit test transformer. In the case of a transformer

Table 16.5 Rated withstand voltage transformers

High voltage (kV)	Rated short-duration power frequency withstand voltage (kV)	Rated lighting impulse withstand voltage[a]	
		(kV)	(Peak)
3.6	10	20	40
7.2	20	40	60
12	28	60	75
24	50	95	125
36	70	145	170

[a]Choice of rated lighting impulse withstand voltage depends on exact site conditions and duties.

with tapings, the impedance is conventionally expressed in terms of the rated voltage for the tapping concerned.

16.5.5 Fittings

In selecting the fittings to be provided, the specifiers of a transformer have a wide choice. The most common types and their purpose are as follows.

Oil-level indicator

This is usually mounted on the transformer tank, providing a visual indication of the transformer oil level.

Thermometer pockets

These measure oil temperature from a separate temperature indicator.

Tap changing

The range and duty (i.e. off-load/on-load) can be specified to suit each particular installation. For standard distribution transformers of up to 3.15 MVA rating for use purely as load supplies, off-load tap changing for a ± 5 per cent voltage range is normal.

Breather/conservator/gas and oil relay

This is specified where the transformer is of the breathing type. The gas- and oil-actuated relay will be fitted with contacts for remote alarm or tripping.

Pressure-relief device

This is essential on a sealed transformer and optional on a breathing unit. It can be fitted with contacts for remote alarm/tripping.

Disconnecting chambers

These are fitted between the cable boxes and the transformer windings and are accessible through a secured cover. Disconnecting chambers can be employed as a method of isolation, but their most common usage is to facilitate phase-to-phase testing of the connected cables.

Group No. 1 : Phase displacement = 0°

Vector symbols	Line terminal markings and vector diagram of induced voltage		Winding connections
	HV winding	LV winding	
Yy0			
Dd0			
Dz0			
Zd0			

Group No. 2 : Phase displacement = 180°

Vector symbols	Line terminal markings and vector diagram of induced voltage		Winding connections
	HV winding	LV winding	
Yy6			
Dd6			
Dz6			
Zd6			

Group No. 3 : Phase displacement = Minus 30°

Vector symbols	Line terminal markings and vector diagram of induced voltage		Winding connections
	HV winding	LV winding	
Dy 1			
Yd 1			
Yz 1			
Zy 1			

Group No. 4 : Phase displacement = Plus 30°

Vector symbols	Line terminal markings and vector diagram of induced voltage		Winding connections
	HV winding	LV winding	
Dy11			
Yd11			
Yz11			
Zy11			

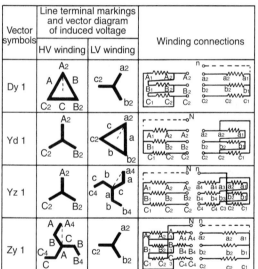

All transformers from any one group may be operated in parallel with other transformers from the same group.

Phase displacement applies for counter clockwise phase rotation.

Figure 16.3 Typical transformer vector groupings and connection arrangements

Connection

Figure 16.3 illustrates the most commonly used transformer, grouping and winding arrangements and phase displacements.

16.6 Protection systems

The analysis and understanding of the operation of protection systems and the application of protective devices to power systems is wide ranging and complex. This section considers only the objectives of protection systems, discrimination and its importance to protection, and describes the most common types of relay used in industrial power systems. Many publications covering the subject of protection systems are available (e.g. *Power System Protection*, edited by the Electricity Council).

16.6.1 Objectives

All electrical systems must be provided with protection equipment, the purpose of which is to isolate faulty electrical equipment from the electrical supply system as rapidly as possible. This can be achieved by use of devices which respond directly to the current flowing (e.g. fuses), or by protective relays which respond to fault current flow and are used to initiate the tripping of other devices (e.g. circuit breakers).

So that only the faulty equipment is isolated from the supply, the protective devices provided throughout a power system must discriminate between faulty and healthy equipment. This discrimination of protective equipment is a key element in the design of electrical systems and the selection of protection devices for use within them. Each electrical system must be provided with an adequate number of suitably rated disconnecting devices located correctly throughout the power system. These must be so arranged that only appropriate devices operate to remove the faulty equipment from the supply.

16.6.2 Discrimination

The methods of achieving discrimination most commonly found in industrial power systems are as follows:

1. *By time*: Protective relays are provided with time delay features such that the device closest to the fault will operate first.
2. *By magnitude*: In different parts of power system fault currents of different magnitude can flow.
3. *By time/direction*: Protective devices responsive to the direction of current flow. These are necessary for parallel feeders or closed ring-main supplies.
4. *By comparison*: The currents flowing into and out of a circuit are compared. In a healthy circuit, or in a circuit in which *through fault* current is flowing, the two currents should be equal and the protective device does not operate. (Compensation for any transformation is necessary.) If the two currents differ, the protective device operates.

In selecting discriminating time margins between protective devices connected in series the following

should be taken into account. Full discrimination is achieved between fuses of different rating when the lower rated or minor fuse without affecting the higher rated or major fuse clears the fault. In order to achieve this, the total let-through energy of the minor fuse must be less than the let-through energy necessary to cause pre-arching of the major fuse.

16.6.3 Testing

Tests carried out on protection systems comprise factory tests, on-site commissioning tests and maintenance checks. Those made on individual relays will demonstrate the compliance of the equipment with specification and the verification of its operation under simulated conditions, while tests carried out on-site prior to the equipment being put into service ensure that the full protection scheme and associated equipment operate correctly. These on-site tests must be comprehensive and should include:

1. Performance tests on current and voltage transformers to verify ratio and polarity. These may include the injection of current into the primary load circuit (primary injection).
2. Sensitivity and stability checks on each protective device. These tests may be carried out either by primary injection or by the injection of voltage into the CT secondary circuit (secondary injection).
3. Wiring checks, including insulation resistance and any pilot cables.
4. Operational checks on all tripping and alarm circuits.
5. Calibration checks on all relevant relays.

On-site commissioning tests of protection equipment must be carried out in a planned logical manner, and arranged so that the disturbance of tested equipment is minimized.

16.6.4 Relay types

The main relay types found within an industrial power system can be:

1. Instantaneous in operation;
2. Definite time; or
3. Inverse, in which the time of operation is dependent upon the magnitude of the current, and such relays can be arranged to protect equipment in one of the following ways:
    (a) *Unrestricted*: The protection device will operate in response to any fault current flowing through the circuit.
    (b) *Restricted*: The protection device will operate in response to any fault current flowing within a zone restricted by the location of the current transformers. Such protection is normally applied to a single winding of a power transformer.
    (c) *Differential*: The protection device will operate in response to a fault current flowing within a zone bounded by current transformers located at each end of the protected circuit.

Such protection can be used on a power transformer, a motor or generator, or a single feeder. In addition, other types of relay are used for particular applications (e.g. transformer Buchholz gas, neutral voltage displacement and over-voltage).

16.6.5 Applications

Figures 16.4 and 16.5 show typical protection schemes for two circuits of a 60 MW generator, a generator transformer with tripping logic and a 2 MVA transformer. The figures illustrate the protection devices provided the current and voltage transformers supplying them and the tripping scheme associated with each.

16.7 Power factor correction

16.7.1 Background

Power factor in an alternating current circuit is defined as the ratio of actual circuit power in watts (W) to the apparent power in voltage amperes (VA). The need for correction arises from fact that the majority of A.C. electrical loads take from the supply a lagging quadruple current (voltage amperes reactive, var) and thus operates at a lagging power factor due to the reactive (rather than capacitive) nature of their construction.

Most industrial installations comprise a combination of one or more of the following electrical loads:

1. A.C. motors (induction)
2. Furnaces (electric arc, induction)
3. Fluorescent or discharge lighting
4. Power transformers
5. Thyristor drive equipment (for either A.C. or D.C. drives)
6. Welding machines

All these types of load fall into the above category and operate at a lagging power factor.

Since the electrical supply system carries the full apparent power (VA), a current higher than is theoretically necessary to supply the power demand needs to be supplied. Equipment must therefore be rated to carry the full apparent power plus the losses of the supply system, which are proportional to the square of the current.

A supply system operating at a low power factor will be inefficient due to the overrating of the supply components and to the losses incurred, and supply authorities will seek to recover the costs of this inefficient operation by introducing cost penalties for consumers operating at a low power factor. In practice, these penalties are incurred on the basis of either the maximum demand of the apparent power in kVA of the load or of charges initiated if the power factor falls below a set trigger point (typically, 0.95 lag). The latter charges are monitored by measuring either the power factor directly or by the relative current drawn by the load and comparing it with the actual power drawn.

Within any particular installation, in instances where no financial penalties are incurred from the supply authority the power factor of individual circuits will influence both the losses within the installation and the system's voltage regulation. Therefore, power factor correction is considered necessary in order to achieve one or more of the following objectives:

1. A direct reduction in the charges made by the supply authority;
2. A reduction in the losses within a system with consequent savings in charges;
3. An improvement in the voltage regulation of a system.

Figure 16.6 illustrates the phasor diagram applicable to power factor correction.

16.7.2 Methods of achieving correction

The power factor of an installation can be improved by the use of either A.C. synchronous machines or of static capacitor banks. An A.C. synchronous machine will either draw current from the supply or contribute current to the supply, depending on whether the machine is operating:

1. As a generator, driven by a prime mover, contributing active power;
2. As a motor-driving load, drawing active power;
3. Over-excited, contributing reactive var; or
4. Under-excited, drawing reactive var.

Figure 16.7 illustrates these four modes of operation.

When an A.C. synchronous machine operating as a motor in parallel with other loads and an external supply system is over-excited the machine will contribute reactive kvar to the supply. The net effect of this will be to reduce the amount of reactive current drawn from the supply, and this will improve the overall system power factor.

When an A.C. synchronous machine operating as a generator in parallel with other loads and external supply system is over-excited the net effect will still be to reduce the amount of reactive current drawn from the supply.

However, the power factor of the current drawn from the supply will only improve if the power factor of the generated current is less than that of the parallel loads, since the active power drawn from the supply will also be reduced by the amount of actual power generated. In each case increasing the excitation so that the power factor of the current drawn from the supply can improve to unity or become leading can increase the reactive current produced by the synchronous machine within the capabilities of the machine. In this case, the installation becomes a net exporter of lagging reactive current to the supply. Figure 16.8 illustrates these two cases.

The second method of improving the power factor of an installation is to provide static capacitor banks. These can be installed as a single block at the point of supply busbar, as a set of switchable banks or as individual units connected to specific loads. For an installation where no synchronous machines are installed for other purposes (i.e. as prime movers or generators) then static capacitor banks are almost invariably the most cost-effective way of improving the power factor.

The amount of power factor correction capacitors necessary in order to correct from an initial power factor cos phi (ϕ) to a target power factor $\cos \phi_1$ is given by:

Initial reactive requirement $\phi = P \tan \phi$

Final reactive requirement $\phi_1 = P \tan \phi_1$

$$\text{Correction required} = P(\tan \phi - \tan \phi_1)$$

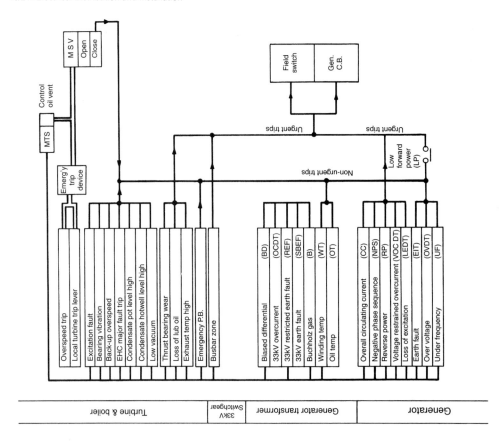

Typical arrangement protection and tripping logic 60MW. steam turbine generator

Figure 21.4 Protection and tripping logic for a 60MW. steam turbine generator

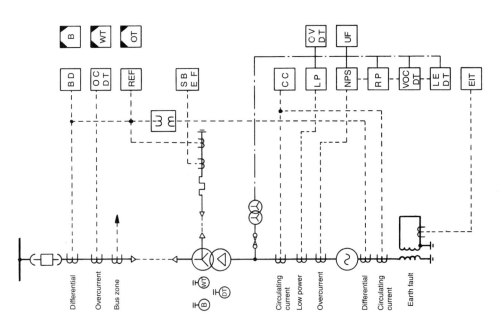

Figure 16.4 Protection and tripping logic for a 60 MW steam turbine generator

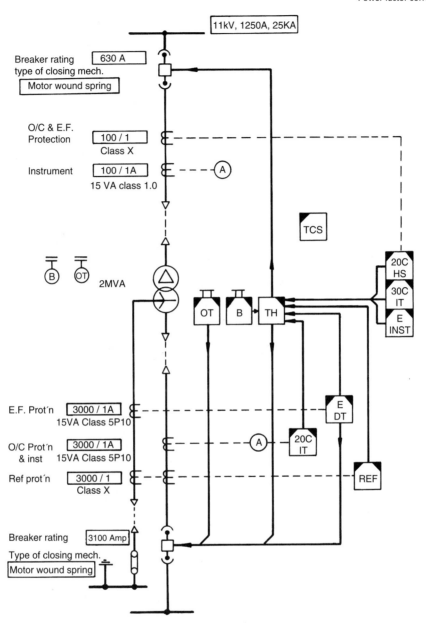

Figure 16.5 Protection and tripping logic for a 2 MVA transformer

Where: $\phi = \cos^{-1}$ (PF). The correction required is:

$$P[\tan\cos^{-1}(\text{initial PF})] - [\tan\cos^{-1}(\text{final PF})]$$

Figure 16.9 illustrates the amount of power factor correction required per 100 kW of load to correct from one power factor to another.

The degree of correction necessary for any particular installation will depend upon the circumstances. In economic terms the costs and prospective benefits can be simply set out as:

Capital	*Running*	*Savings*
Capital cost	Maintenance	Reduction in of equipment demand charges
Installation costs	Depreciation	Reduction in losses

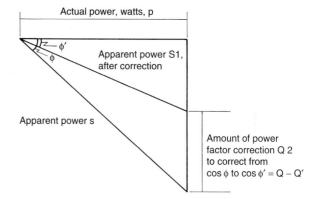

Figure 16.6 Power factor correction

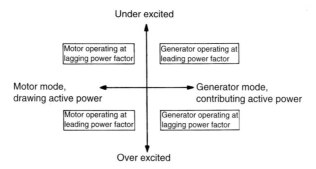

Figure 16.7 Synchronous machine operating modes

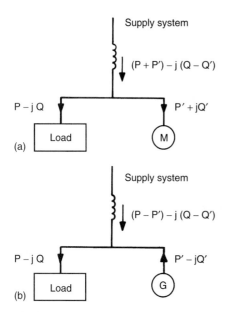

Figure 16.8 Synchronous machine operation. (a) Synchronous machine acting as a motor (over-excited); (b) synchronous machine acting as a generator (over-excited)

16.7.3 Types of control

Capacitors as bulk units can be connected to the supply busbar via a fuse switch, molded-case circuit breaker of air circuit breaker. In this type of installation control is purely manual, and in cases of a reasonably constant load and where the amount of power factor correction is limited such a manually controlled system is perfectly adequate. The supply authority may, however, require to be informed that a capacitor bank is permanently connected to the supply. Capacitors are more generally connected either in banks controlled from a VAR sensitive relay or across individual loads (e.g. motors).

When connected as switchable banks the rating of each step of the capacitor bank must be selected with care. It is important that the control relay settings are matched to the ratings of each capacitance step in order to prevent hunting (i.e. continuously switching in and out at a particular load point). When capacitors are connected to one particular load (usually a motor) the capacitor bank can be located at the motor, adjacent to but separate from the control switchgear or within the control switchgear itself.

When located at the motor the capacitor bank will be normally cabled from the motor terminal box, so that the size of the motor cable can then be selected on the basis of the reduced-power factor corrected current drawn by

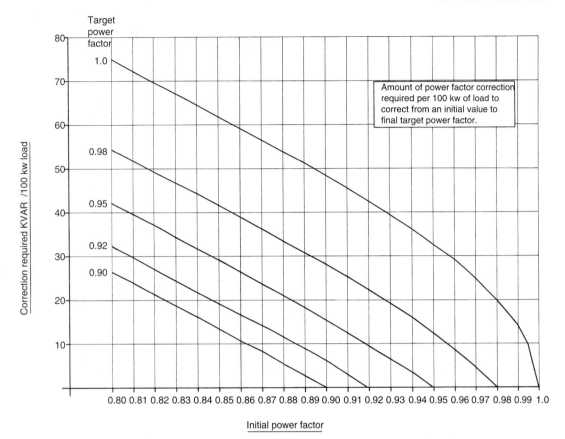

Figure 16.9 Amount of power factor correction required

the motor/capacitor combination. However, whenever a capacitor is connected across an individual motor circuit, by whatever means, the setting of the motor overload device must be chosen to take account of the corrected current rather than the uncorrected motor full-load current.

16.7.4 Potential problems

Control devices

Capacitors, circuit breakers and HRC fuses must be selected with care for use with capacitor circuits, with contractors chosen on the basis that the capacitor current can rise by 25 per cent above nominal line current. Equally, HRC fuses for capacitor applications should be de-rated by a factor of 1.5.

When choosing high-voltage circuit breakers for capacitor control it is necessary to select the full-load current of the breaker, taking into account variations in supply voltage, tolerance on rating manufacture and harmonic currents. Again, a de-rating factor of 1.5 is usually considered adequate. The capacitor manufacturers will insist that the control device is re-strike free and that the control device has been tested to IEC 56, Part 4. When the control

device is a high-voltage circuit breaker, capacitor manufacturers frequently recommend a maximum between initial current makes and final contact closure (typically, 10 m/s).

Motor circuits

In instances where a capacitor is connected directly across the terminal of the motor, the capacitor can act as a source of excitation current after the control device is opened. In order to prevent this the capacitor rating should not exceed 90 per cent of the motor no-load magnetizing current.

Harmonics

A capacitor bank will represent reducing impedance to currents of increasing frequency. Such reducing impedance, if matched with a similarly increasing inductance impedance of a transformer or a supply system, can cause a resonant condition. In plants where equipment produces harmonic current, a full survey of the installation is recommended prior to installation of the capacitors.

16.8 Motors and motor control

16.8.1 Functions

The principal functions that have to be met by any type of motor control gear are:

1. Provision of a means of starting the motor, taking into account the requirements of torque, acceleration, load, frequency of operation and safety;
2. When required, to limit the current drawn from the supply and starting;
3. When necessary, to provide means of speed control, reversing, braking, etc.;
4. Provision of protection for the motor itself in the case of faults and for the equipment controlling it (e.g. the contactor).

Motors used in industrial operations are predominantly of the induction type. The voltage at which they will be supplied will depend on the rating of the motor, the voltages available on-site and the capital cost of equipment. In general, the following voltage/load ranges apply:

Up to 185 kW	415 V
185 kW to 2 MW	3.3 kV
1–15 MW	11 kV

Squirrel-cage induction motor

The principal advantages of the squirrel-cage induction motor are its simplicity of design and robust construction. Its torque/speed and current/speed characteristics are such that, on starting, a torque of typically *twice* full-load torque is produced but a large current (typically, six to eight times full-load current) is drawn from the supply. It is this latter aspect of the current drawn upon starting which is important in deciding upon the type of starting equipment needed for a squirrel-cage motor because

1. The drawing of a large current from the supply will cause a corresponding voltage drop at the point of common coupling with other consumers. This will affect other drives within the installation or consumers fed from the same supply authority.
2. If the motor drives plant of high inertia (e.g. fans) the time during which the large current is drawn may be extended. Such a current flowing for an extended period may cause the unwanted operation of overload or over-current protection relays within the supply system.

In order to overcome (1) it is often necessary to introduce a form of assisted starting, which can also help in overcoming problem (2). However, modification of the overload protection device characteristics may also be necessary.

16.8.2 Types of starter

For starting an induction motor, the following types of starter are available:

1. *Direct-on-line*: Full voltage is applied to the motor;
2. *Start/delta, auto transformer, primary resistance*: Reduced voltage is applied to the motor;
3. *Electronic soft-start*: The voltage and frequency applied to the motor is controlled;
4. *Stator/rotor*: Full voltage is applied with external rotor resistance.

The principal features of each type are as follows.

Direct-on-line

The equipment and connections are simple and the starter is robust. The basic equipment will comprise an isolator, high rupturing capacity fuses, a contactor, overload devices and control switches.

Star/delta starting

This is the most common form of reduced-voltage starting used in the UK. Both ends of the motor winding are cabled back to the starter. On starting, the windings are connected in star configuration, and thus the voltage applied to each winding is 57.7 per cent of normal and the current taken from the supply is one third of that taken when started direct-on-line.

Changeover from star- to delta-connection takes place automatically by means of a timer. At changeover, the motor is momentarily disconnected from the supply and the delta contactor is then closed. The closure of the delta contactor can cause a further surge of current to occur. The magnitude of the surge will depend on the speed of the motor at changeover, but it can be comparable to that taken when started direct on line.

Auto-transformer   An auto transformer is used to provide a reduced voltage to the motor on start-up. The exact voltage applied to the motor can then be chosen, taking into account any limitation on starting current and motor torque. In an induction motor both starting current and motor torque are approximately proportional to the square of the applied voltage. Therefore for a motor which produces twice full-load torque and draws eight times full-load current on start-up, if the voltage is reduced to 75 per cent by use of an auto transformer the starting torque $= 0.75^2 \times 2 = 1.125$ times full-load torque and starting current $= 0.75^2 \times 8 = 4.5$ times full-load current.

Primary resistance   A resistance is inserted into the supply to the motor. This reduces the voltage available at the motor terminals on starting, and this voltage increases gradually as the motor current falls when the motor speeds up. Once the motor has reacted at a predetermined speed the resistance banks are short-circuited.

Depending on the value of the resistor, the motor will, in general, draw a heavier current then when started using star-delta or auto-transformer methods. The resistors must be rated to carry the limited motor starting current for the time they are in circuit.

Electronic soft-start

The most recent development in the starting of squirrel-cage induction motors is the introduction of the electronic soft-start. This principle has been derived from variable-frequency speed controllers using switched Thyristor or power transistor bridges. The supply sine wave is chopped so that a reduced voltage and frequency is applied to the motor.These are gradually increased so that the motor speed rises in a controlled manner, with the starting current limited to any chosen value.

Stator/rotor

With this technique the motor has a wound rotor brought out to slip rings and an external resistance is connected into the rotor circuit. This resistance usually consists of a series of resistor banks, which are switched out progressively in a number of steps as the motor accelerates. The number and rating of each step is chosen so that starting current and motor torque are within requirements.

16.8.3 Motor-starting equipment

Each component within a motor starting circuit must be selected to be suitable for the operation of the motor as it is required and for use on the electrical system from which the motor is to be operated. It must also be compatible with the other elements within the circuit. The principal components comprise:

Isolator
HRC fuses
Contactor
Overload device
Cable

The main parameters to be considered for each component are as follows.

Isolator

This must be:

1. Rated to close onto a fault at the rated fault level of the switchboard;
2. Able to carry starter rated full-load current continuously;
3. Capable of breaking a stalled motor current.

Fuses

Fuse manufacturers' catalogues give suggested fuse sizes for all 415 V motors. However, the following parameters all affect the fuse rating:

1. The starting current drawn by the motor, including all tolerances;
2. The starting time of the drive under worst-case conditions;
3. Full-load current, including efficiency and power factor;
4. Number of consecutive starts;

5. Short-time capability of contactor, overload, cable and isolator.

Contactor

A contactor must have full-load rating and be coordinated to proven capability with a selected fuse.

Overload device

Overload devices in current use are typically thermal overload relays to BS 4941, motor starters for voltages up to and including 1000 V A.C. and 1200 V D.C. or BS 142 electrical protection relays. Relays to BS 4941 generally provide overload and single-phasing protection. Those complying with BS 142 are also frequently fitted with instantaneous earth fault and over-current trips.

While applications vary from one industry to another, relays complying with BS 142 are typically specified for all motors larger than 45 kW.

Cable

The cable must be selected on the basis of full-load current rating, voltage drop on both running and start-up, and short-time current rating.

16.9 Standby supplies

16.9.1 Definitions

Standby supply systems within an electrical installation provide supplies to critical loads on loss of normal power supplies. In an installation one or more of the following types of standby supply systems are likely to be found:

1. Battery systems (either d.c. or inverted);
2. Un-interruptible (UPS) A.C. power systems;
3. Standby diesel generating sets.

Each system type is utilized to satisfy different needs, and these and the salient points relevant to each type are as follows.

16.9.2 Battery systems

D.C. systems

These provide supplies to equipment that requires a D.C. supply both during normal operating conditions and when A.C. supplies have been lost. The loads can comprise:

1. Essential instruments
2. Control schemes
3. Switchgear closing and tripping
4. Telecommunications
5. Protection schemes
6. Interlocking
7. Alarms
8. Emergency lighting
9. Emergency drives (e.g. rundown lube oil pumps)
10. Standby diesel starting systems

Voltages in use range from 24 to 250 V. A D.C. installation will comprise a charger (capable of both float and boost charging), the battery itself and a distribution outlet.

The charger is normally arranged so that boost charging is carried out off-load, and should be designed to recharge the battery, after an emergency discharge has occurred, within a quarter to a half of the normal charting time. When the charger is operating in *float charge* mode it must be rated to supply the D.C. load requirements as well as float charging the battery.

Chargers can be fitted with an integral voltage-sensing device measuring circuit voltage at the charger output terminals. In the event that the terminal voltage rises significantly above the normal float voltage of the battery, the circuit will be arranged to trip the charger and provide an alarm. A separate alarm will be provided during boost charging. Alarms normally fitted comprise:

Charger failures
Rectifier failures
Battery earth fault
Battery boost charge
Battery low volts

Battery types are either lead–acid or nickel–cadmium cells. Lead–acid types have been used for a long time and, when correctly maintained, have a working life of 25 years. Nickel–cadmium batteries offer the same working life as lead–acid but are smaller in weight and volume, generally with a higher initial capital cost. Loads applied to D.C. systems can be categorized into three types:

1. *Standing levels*: Loads, which impose a consistent and continued load upon the battery (e.g. alarm facias).
2. *Emergency loads*: Loads that are only applied on loss of normal supplies. These are usually supplied only for a fixed period of time (e.g. turbine emergency rundown lube oil pump).
3. *Switching loads*: Loads imposed by the operation of both opening and closing of switchgear.

These loads, the expected duration of emergency loads and the required switching regime and number of operations must be specified at the time of order placement.

16.9.3 UPS-A.C. systems

A.C. un-interruptible supply systems are used for the provision of supplies to those loads which:

1. Are required for post-incident monitoring and recording following a loss of normal supplies;
2. Require a high-quality supply in terms of voltage, frequency and waveform. These comprise loads, which would not give an adequate standard of reliability if operated directly from the normal A.C. supply system;
3. Are necessary to monitor plant, particularly generating plant, during a shutdown operation or loads necessary to assist in a rapid restart.

A UPS installation will typically comprise a rectifier/battery, charger, storage battery, static inverter and static bypass switch. With a UPS installation the output, which is derived from the battery via the static inverter, is completely unaffected by variations of either a steady state or transient nature in the A.C. supply (Figure 16.10).

UPS units are generally categorized by the design of the static inverter used to produce the A.C. output. The principal types used are:

1. *Ferroresonant*: The D.C. is switched via thyristors to produce a square wave output.
2. *Phase controlled*: A single pulse of variable time is produced per half-cycle.
3. *Step wave*: The output waveform is switched more frequently, producing a step waveform.
4. *Pulse width modulation*: Each half-cycle of output is made up of pulses of varying duration but equal magnitude.

The main problem in specifying UPS equipment is that of inadequately identifying the characteristics of the loads which the UPS is to supply, and the following must be specified:

Steady-state voltage regulation
Acceptable transient voltage variation
Non-linearity characteristics of load current
Percentage distortion in voltage
Frequency tolerance
Power factor of load
Inrush upon switch-on

16.9.4 Standby diesel generators

The electrical load and the standard engine frame sizes available govern the selection of a diesel generator. In specifying a diesel generator for use, as a standby supply the electrical load must be analyzed in terms of final running load, load profile, starting characteristics of individual motor loads and required operating time. Chapter 15 gives a full description of the types of diesel-driven time movers available. Electrical characteristics to be considered include:

1. Type of generator;
2. Type of excitation (brushless or static);
3. Transient performance (the capability of the generator to maintain the output voltage when a load is applied). It must be remembered that the impedance of the generator will be greater than that of the supply system which it has replaced;
4. Method of alternator cooling (the heat gain to the building housing the diesel generator must be taken into account);
5. Method of neutral grounding;
6. Method of testing. If the diesel is to be tested, it may be necessary for it to operate in parallel with the supply system in order to apply load. Synchronizing facilities will be needed and the fault levels during parallel operation must be assessed.
7. Electrical protection schemes. The installation should be carried out in accordance with Electricity Council Engineering Recommendation G26, *The installation and operational aspects of private generating plant*.

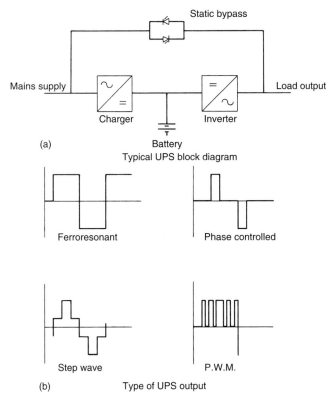

Figure 16.10 (a) UPS block diagram; (b) typical output waveforms

16.10 Grounding

16.10.1 General

Electrical supply systems and equipment are grounded in order to maintain the voltage at any part of the system at a known potential relative to true earth and to provide a path for current flow under earth fault conditions so that protective devices operate correctly. The connection to earth should be such that the flow of fault current to earth does not cause voltages or voltage gradients to be of sufficient magnitude or duration likely to cause danger.

An electrical earth system comprises the provision on the supply system of an earth connection to facilitate earth fault current flow, the connecting of all exposed metalwork within the installation to a common grounding terminal and the connection of this terminal to earth. For a complete installation the principal factors which need to be considered, are:

The earth electrode
Neutral grounding
Substation grounding
Supply system grounding
Bonding

16.10.2 Earth electrode

The connection to earth can be made utilizing earth rods, earth plates, earth lattices or grids, or buried strip conductors. Selection depends on the locality and type of ground conditions at the site. The effectiveness of the earth connection is, in turn, dependent on the resistivity of the soil at the site, the type of soil and climatic conditions have a direct effect upon the resistance of the connection made. Typical values of resistivity ($\Omega - m$) for different types of soil are:

Garden loam	4–150
Chalk	30–200
Clay	5–100
Sand	50–800
Rock	Above 1000

The resistivity of the soil in any particular location will be a function of moisture content, soil temperature and presence of dissolved salts. At a site where climatic conditions vary considerably throughout the year, earth electrodes should be buried at a depth where such changes will not affect the resistivity. Grounding rods are generally made of copper bonded onto a steel core. The copper provides a good connection to earth and offers a high corrosion resistance, while the steel core gives the mechanical strength necessary to allow the rods to be driven to the required depth.

Grounding plates or lattices made of pure copper, while displaying good current-carrying capacities, do not provide a particularly low resistance due to the depth at which they can be buried. The third alternative is to bury lengths of copper tape around the installation. The use of reinforced concrete foundations for grounding electrodes has also recently been considered.

Formulae for calculation of the resistance to earth of each type are as follows:

$$\text{Single rod}(R) = \frac{P}{2\pi L}\left[\ln\left(\frac{8L}{\delta}\right) - 1\right]$$

$$\text{Plate}(R) = \frac{e}{8r}\left(\frac{r}{2.5h + r} + 1\right)$$

$$\text{Grid}(R) = P\left(\frac{1}{4r} + \frac{1}{L}\right)$$

Where:

R = resistance,
e = resistivity,
L = rod or conductor length,
δ = rod diameter,
r = radius of plate; for a rectangular plate or grid

$$r = \frac{\sqrt{a}}{\pi}$$

h = depth of plate,
P = soil resistivity.

The earth electrode system must be designed to be capable of carrying without damage to the full earth fault current of the supply system.

16.10.3 Neutral grounding

A ground can be established on an electrical supply system by the connection to earth of the neutral point of the supply transformers, or generators or the use of interconnected star-wound grounding transformers. The grounding must be arranged to ensure that an earth is provided on the supply system at all times and that the resistance of the connection to earth must be such that earth fault protection operates correctly. In instances where generators are to be run in parallel with each other or in parallel with a grid supply the neutral grounding arrangements will require special consideration.

16.10.4 Grounding in substations

The earth systems associated with high-voltage substations must satisfy the following conditions:

1. The resistance of the system to earth must be low enough to ensure that earth fault protection equipment operates correctly.
2. In the event of an earth fault the difference in potential between a person's feet (step potential) should not reach dangerous levels.

3. In the event of an earth fault the difference in potential between any point which may be touched and the ground (touch potential) should not reach dangerous levels.
4. The substation and its environment should not be connected to metal objects such that the voltages arising within the substation under fault conditions can be transferred to a point remote from the substation.
5. The grounding conductors must be capable of carrying maximum earth fault current without overheating or causing mechanical damage.

16.10.5 Types of system grounding

The IEE Regulations for Electrical Installations recognize the following designations of grounding systems using a two-, three- or four-letter code as follows:

First letter (denoting the grounding arrangement at the source)

T–Direct connection of one or more points to earth;
I–All live parts isolated from earth or one point connected to earth through impedance.

Second letter (denoting the relationship of the exposed conductive parts of the installation to earth)

T–Direct connection of the exposed conductive parts to earth, independently of the grounding of any point of the source of energy;
N–Direct electrical connection of the exposed conductive parts to the earthed point of the source of energy (usually the neutral).

The designation TN is subdivided, depending on the arrangement of neutral and protective conductors. S indicates that separate conductors and C that neutral and protective functions are provided by a single conductor provide neutral and protective functions.

Examples of these designations and their usage are as follows:

1. *TN-C system*: Neutral and protective functions combined in a single conductor throughout the system.
2. *TN-S system*: Separate neutral and protective conductors throughout the system;
3. *TN-C-S system*: Neutral and protective functions combined in a single conductor in a part of the system;
4. *TT system*: All exposed conductive parts of an installation are connected to an earth electrode, which is electrically independent of the source earth.

The following publications contain further information on and expansion of the factors relating to grounding.

Electricity Council Engineering Recommendation G26, *The installation and operational aspects of private generating plant*
BS Code of Practice CP1013, Grounding
IEE Regulations for Electrical Installations

Table 16.6 Comparison of parameters for 1900/3300 V XLPE insulated, armoured, three-core cable

Nominal area of conductor (mm²)	Copper			Aluminium		
	Current-carrying capacity (A)	Weight (kg/1000 m)	Overall diameter (mm)	Current-carrying capacity (A)	Weight (kg/1000 m)	Overall diameter (mm)
50	210	2850	34.7	160	2075	33.0
70	265	3625	38.0	200	2450	36.0
95	325	4500	41.4	245	2875	39.1
185	495	8075	51.9	370	4650	48.7
240	580	10375	56.9	440	5625	53.2

16.11 Cables

Electrical power distribution within an industrial installation is most often at a voltage up to and including
33 kV. This section describes the types of cable suitable
for power circuits for use up to 33 kV and considers the
factors, which will influence the current-carrying capacity
of such cables.

The principal components of a power distribution cable
are the conductors and the insulators. The cable is completed by the provision of armoring, overloads and other
features designed to protect the cable within its installed
environment.

16.11.1 Conductors

Material used for conductors comprise copper or aluminum in either stranded or solid form. Copper is the
most common type of conductor due to its good conductivity and ease of working. Despite having a conductivity of only 61 per cent of that copper, aluminum
can be used as the conductor material. The lower density
of aluminum results in the weight of an aluminum cable
offsetting, to a certain extent, that of the additional material necessary to achieve the required current-carrying
capacity.

Although aluminum cable has a proven record of
satisfactory operation, the problems of terminating such
cables must be recognized at the time of specifying. Since
an aluminum cable will be larger than an equivalent
copper cable of the same current-carrying terminal,
enclosures and cable support systems must be designed
to suit. Table 16.6 illustrates current-carrying capacity,
overall diameter and weight for a typical range of
3.3 kV cables.

16.11.2 Insulation

Insulation materials used at voltages up to 33 kV comprise
paper, PVC and polyethylene (XLPE). Paper-insulated
cables have been used for the complete voltage range up
to 33 kV throughout this century, although today they are
limited in general to 6.6 kV and above. Paper cable is
usually designed to operate at a conductor temperature
not exceeding 80°C and PVC cable has largely replaced
paper cables at voltages up to 3.3 kV. The relative imper-
viousness of PVC to moisture has contributed greatly to

the design of the cables, making them easier to handle
and install. The excellent dielectrics of properties of cross-
linked polyethylene, together with the fact that XLPE does
not soften at elevated temperatures, means that XLPE
cables can operate with high continuous operating temperatures and short-circuit temperatures. The maximum
operating temperature of XLPE is 90°C, with a maximum
under short-circuit conditions of 250°C. Table 16.7 gives
a comparison of the full-load capabilities of a range of
cable sizes.

16.11.3 Selection

In selecting a cable for a particular installation, the following factors need to be considered:

1. The steady-state voltage drops of the circuit;
2. The voltage drop in starting if the cable is for a motor
   circuit;
3. The current-carrying capacity under full-load conditions;
4. The maximum fault current, which can flow, and its
   duration.

For (3) and, to a certain extent, (4) the exact details of the
cable installation will have a direct bearing on the cable
chosen.

Table 16.7 Comparison of full-load capabilities for paper, PVC
and XLPE insulated copper cable

Cable size (mm²)	Paper[a]	PVC[b]	XLPE[c]
	Current-carrying capacity in air (A)		
16	91	87	105
35	150	142	170
50	180	172	205
95	280	268	320
185	430	407	490
240	510	480	580

[a]600/1000 V: Three-core, belted, paper-insulated,
lead-sheathed, stranded-copper conductor.
[b]600/1000 V: Three-core, PVC-insulated, SWA, PVC,
stranded-copper conductor.
[c]600/1000 V: Three-core, XLPE-insulated,
PVC-sheathed, SWA-stranded copper conductor.

The ambient temperature of the installation will affect the cable rating, and account should be taken of any local high temperatures caused by proximity of process plant, the presence of and location of other cables and any thermal insulation, which may prevent or inhibit heat dissipation. Cable ratings applicable to a variety of cable-installation techniques, together with the correction factors applicable, are given in a number of sources, including manufacturers' catalogues, the IEE Wiring Regulations and reports published by ERA Technology and IEC.

17

Electrical Instrumentation

H Barber
Loughborough University of Technology

Contents

17.1 Introduction

The plant engineer needs to know the circumstances under which the plant for which he is responsible is operating. Hence, he needs to be able to measure the inputs to the plant and the power inputs to the individual sectors. It will also be necessary to measure the plant outputs. In most cases, the instrumentation uses electrical or electronic techniques, i.e. it measures either electrical quantities *per se* or quantities such as temperature or pressure, which have been converted to electrical signals by means of transducers. The degree of accuracy to which these measurements are required will vary, depending on their purpose. In the case where the measurement is used as a direct basis for charging for energy then the accuracy must be very high. This is also the case where precise process control is involved.

In other instances, it is sufficient for the plant engineer to have a less accurate measurement of what is going on. As well as being technically suitable for the purpose, the instrument chosen will reflect these considerations. In the case of process control, the instrumentation must be reliable and it must yield information, often over very long periods of time, which represents the state of the plant or the process and its history. It is on the basis of this information that the plant engineer will make decisions, many of which will affect the economic viability of the process and some of which will have a direct impact on the safe operation of the plant.

Most plants will be supplied either directly or through the factory distribution system from the electricity supply authority. This supply is usually A.C. at a frequency of 50 Hz and at a voltage depending on circumstances, but which in the case of large power demands, can be 11, 000 V or even higher. Thus meters are required to measure these and related quantities. A variety of transducers is available for the measurement of physical quantities. These either give a direct readout to the operator or initiate some form of control action. In addition, there is a requirement for test and diagnostic equipment to determine if the measurement and control equipment is operating correctly and, if failures occur, to aid in tracing the source of the problem.

The plant engineer will need to familiarize himself with what is available in all these categories so that he may make the most appropriate equipment selection. He will also need to be familiar with what to expect from his instrumentation both in terms of accuracy and in order to know how to interpret the information correctly.

17.2 Electricity supply metering

17.2.1 Metering requirements

The electricity supply to the factory is usually obtained from the public supply authority, who will need to install metering in order to be able to assess the charges due in accordance with the agreed tariff. This metering will be the responsibility of the energy supplier but, in order to ensure that the rights of the customer are protected, the design of this instrumentation, and the accuracy limits within which it must operate, are controlled by statute. The customer will usually either have visual access to this instrumentation in order to confirm the readings or will install his own, duplicate, instrumentation as a check.

In any event, in most plants, the plant engineer will have a need to monitor the electricity consumed by the various sectors on the site and he will need similar instrumentation to do this. The normal method of charging for electrical energy is based on two components. The first component includes the total amount of primary (e.g. coal, nuclear) energy and the costs of converting it to electricity. The second component is the cost of providing the necessary plant to carry out the conversion as well as to transport the electrical energy to the customer (the *maximum demand charge*). The first of these is primarily a function of the energy (kWh) supplied while the second is a function of the maximum value of the electric current drawn at any time from the supply and the voltage at which it is supplied. Thus, two forms of instrumentation are needed, an energy (kWh) meter and a maximum demand (kVA) meter.

17.2.2 Energy meters

Energy is usually measured using an induction-integrating meter. In the case of a single-phase load the arrangement is as shown in Figure 17.1, this consists essentially of two electromagnets, mounted one above the other, with a light-weight aluminum disc free to rotate between them, as shown in the figure. The upper electromagnetic is energized by the system voltage (V), or a voltage proportional to it, and a current proportional to the system current (I) flows through the coil on the lower magnet. The electromagnetic flux created by the two magnets interacts in the disc in which a torque is produced proportional to the product of the two multiplied by the cosine of the phase angle between them. Thus, this torque is proportional to (VI cos), the electrical power supplied to the system, and the speed of rotation of the disc is proportional to the torque. Hence, the number of revolutions made by the disc in an hour is a measure of the energy used, i.e.

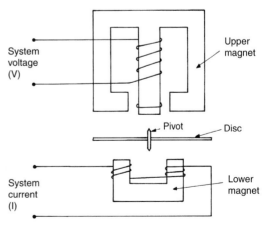

Figure 17.1 Schematic diagram of an induction meter

the watt-hours (Wh) or, more usually, the kilowatt-hours (kWh).

The number of rotations of the disc is registered and can be recorded on a dial. As noted above, the voltages and currents energizing the coils are the normal system voltages unless these are too large for convenience, in which case voltage and current transformers are used as appropriate.

A high degree of accuracy is required over a very wide range of input currents. At very low loads (i.e. very low input currents) this accuracy can be affected by pivot friction, but this can be allowed for in the instrument design. The disc rotation is converted to a dial reading using a gear train (the 'register') and this reading is then used as a basis for charging under the tariff arrangements in use. In some cases two-part tariffs are involved, the simplest form being that in which kWh (units) above a certain number are charged at a different rate. This adjustment can be made after the reading is taken. In other cases, units used during specific periods of the day are charged at a rate different from that in force during the remainder of the time. Incorporating a time switch into a meter with two sets of registers can cater for this. At the appropriate time, the switch operates and an electromagnet changes the drive to the second register. Alternatively, an entirely separate instrument can be used, with the changeover again being controlled by the time switch.

17.2.3 Error limits

In general, the error must not exceed ±2 per cent over the working range, although in practice the accuracy is usually much higher than this. Temperature changes are compensated for in the design and the instrument is likely to be unaffected by variations in voltage and frequency if these are within the limits (±4 per cent in voltage and ±0.2 per cent for frequency). Very low voltages and reduced supply frequencies will both cause the instrument to read high if the load has a high power factor. If the power factor is low and the load is predominantly inductive then the frequency effect is reversed.

17.2.4 Three-phase metering

For large loads supplied with three phases, then a three-phase meter is necessary. In these cases, three single-phase units with their rotor torque combined mechanically are used, the whole being contained in one case. If the load current is above 150 A per phase then it is common to use instruments which combine the measuring elements, each with their own current transformer (see below) so that the combination can be calibrated as one unit (Figure 17.2).

17.2.5 Maximum-demand metering

As noted above, the tariff may also contain a component based on the maximum demand taken from the supply in a given period. In this case, a separate maximum-demand meter is required. This is a modified energy meter, which has a special register such that after a given period (usually half an hour) the dials are disconnected. During the next half-hour the register mechanism still operates but the clutch driving the dials is only energized if the total exceeds that already on the meter. This continues for the whole of the recording period, and thus, at the end, the reading on the meter is the maximum obtained during any one of the half-hour periods. This value is then doubled to yield the maximum kW taken in any half-hour.

It should be noted that this is not necessarily the actual instantaneous maximum demand. The quantity measured on this meter is, of course, the maximum kW demand, but some forms of tariff are based on the maximum kVA demand in order to encourage power factor improvement (see below).

17.2.6 Summation metering

In some instances, it is required to measure the sum of the energy supplied along several feeders. This is the case, for example, when a factory site is fed by more than one cable and the tariff is based on the overall energy and maximum demand used. The usual method of doing this is to summate the currents in the individual feeders by installing separate current transformers in each and connecting their secondaries in parallel (Figure 17.3).

The current transformers must have identical ratios. The assumption is that all the feeders have voltages which are equal in magnitude and phase. If the feeders are physically close together, it may be possible to use one current

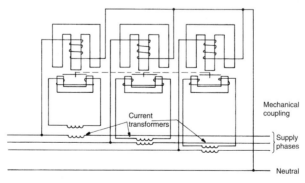

Figure 17.2 Three-phase meters

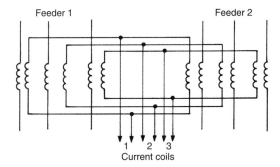

Figure 17.3 Summation current transformers

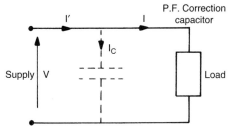

(a) Circuit diagram

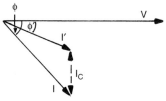

(b) Phasor diagram

Figure 17.4 Power factor correction

transformer as a summation unit. The current transformer has a number of primary windings, each carrying the current of one feeder. The magnetic fluxes due to the individual currents are summed in the transformer core and the output is obtained from a single secondary winding. An alternative is to use impulsing techniques whereby each meter incorporates a device which generates pulses at a rate depending on the energy being used. The pulses are transmitted to a single instrument in which they are recorded to give the total energy used over the whole group.

Electronic techniques can generate a larger number of pulses in a specified time and are therefore more accurate than mechanical devices. At the receiving end, the pulses are used to determine the state of a series of bi-stable networks. These are scanned and reset sequentially and the total number of pulses recorded.

17.3 Power factor correction

Electricity is normally charged for on the basis of power (kilowatts) and the supply authority must install plant whose rating (and therefore cost) is a function of the voltage of the system and the current which the consumer takes (i.e. kilo-volt-amps). The relationship between the two is: $kW = kVA \times \cos \phi$ where: $\cos \phi$ is the power factor and is less than 1.0. In the case of loads which have a low power, factor the supply authority is involved in costs for the provision of plant which are not necessarily reflected in the kWh used. A penalty tariff may then be imposed which makes it economically worthwhile for the consumer to take steps to improve his power factor. Low power factors occur when the load is predominantly either inductive or capacitive in nature (as opposed to resistive). In most industrial circumstances where the load includes a preponderance of motors, the load is inductive (and the power factor is therefore *lagging*). Consequently, if the power factor is to be brought nearer to unity the most obvious method is to add a significant capacitive component to the load.

It is important to ensure that, as far as possible, this additional equipment is purely capacitive, i.e. that it has no resistive component, and otherwise the customer will face an increase in his energy bill. The circuit and phasor diagrams of Figure 17.4 illustrate the situation. In the

figure V represents the system voltage, I is the current drawn from the supply before the addition of power factor correction equipment and $\cos \phi$ is the associated power factor. If a capacitance, C, is connected in parallel with the load, then this will introduce an additional current component I_c. The resultant total current drawn from the supply is I' and it will be seen from the figure that the new power factor is $\cos \phi'$, where $\cos \phi' > \cos \phi$.

Even if the power factor correction capacitance consumes no energy it will need capital investment, and therefore the consumer must balance the capital charges of this equipment against the savings which it produces in the energy bill. It is not normally economic to correct the power factor to its theoretically maximum value of unity, and a value of 0.9–0.95 is more usual.

17.3.1 Equipment

For relatively small loads, the power factor correction equipment usually takes the form of static capacitors. In larger installations, it may be more economic to install an A.C. synchronous motor that, if its excitation is adjusted correctly, can be made to draw a *leading* current from the supply. In most industrial plants, the load is variable, and to gain the maximum benefit from the power factor correction plant this must be varied to suit the load conditions.

If static capacitors are employed this can be achieved by using several capacitors arranged in units (banks) which can be switched in or out as required. This variation can only be carried out in discrete steps. In the case of the A.C. machine (the synchronous condenser), it is possible to obtain a continuous variation. The switching of the equipment can be carried out by an operator or automatically in response to the output from a power factor-sensing instrument.

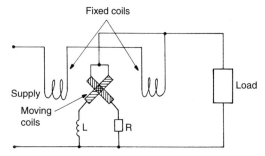

Figure 17.5 Schematic diagram of power factor meter

17.3.2 Power factor measurement

Instruments are available to measure power factor. One form is an electrodynamics instrument employing the same basic principle referred to above but with two coils mounted at right angles to each other on the same shaft (Figure 17.5). These coils are situated in a field produced by a third coil carrying the line current. One of the coils on the moving element carries a current derived from the supply voltage and is connected in series with a pure resistance. The second carries a similar current, but in this case, a pure inductance is connected in series. The torque produced by each coil is a function of the current, which it carries and therefore depends on the impedance of the particular coil circuit. There is also a third torque component due to the mutual inductive effects between the fixed and moving coils. The net result is that the moving element is displaced by an amount which is a function of the phase angle between the system current and the system voltage. This form of instrument is suitable for low frequencies only, and electronic measuring techniques are used for higher frequencies.

17.4 Voltage and current transformers

The instruments discussed above (and many of the others which follow) act in response to the system voltage and/or current. In most cases, the values of these two parameters are very high, which presents problems in the design of the insulation and current-carrying capabilities of the instrument. In these instances, the instrument is supplied with a known fraction of the measured quantity using a voltage transformer or a current transformer, as appropriate.

The important factor in the design of these transformers is that the ratio between the actual quantity and that at the output of the transformer must be maintained accurately over the complete range of values, which the unit is required to measure. In general terms, the degree of accuracy required when the instrument is used for metering purposes is higher than when a straightforward indication of the quantity is needed. In both cases, the accuracy must conform to the relevant British Standard.

17.4.1 Voltage transformers

There are fundamental differences between the behaviors of the two types of transformer. The voltage transformer is shown diagrammatically in Figure 17.6. The system voltage V_p is applied across the primary winding, which has N_p turns. The secondary voltage V_s is:

$$V_s = V_p(N_s/N_p)$$

neglecting errors. These errors are small and constant provided that the transformer is used over a small voltage range as specified by the manufacturer and the loading imposed by the measuring instrument connected to the secondary winding is between 10 and 200 VA.

Voltage transformers are classified into types AL, A, B, C and D, in descending order of accuracy. The ratio errors for small voltage changes (within (±10 per cent of the rating) vary between 0.25 per cent and 5 per cent and the phase errors between (±10 minutes and (±60 minutes.

For all but the highest voltages, the construction of the voltage transformer is very similar to that of a normal power transformer at voltages of magnitudes normally met with in plant engineering. However, at primary voltages of 100 kV and above it may be necessary to use alternative means. The type of voltage transformer described above is suitable for relatively low alternating frequency supplies. It cannot be used at D.C. and there is a possibility of large errors occurring at high frequencies. In these instances, some form of voltage divider is necessary. The arrangement is as shown in Figure 17.7. The two impedances, Z_1 and Z_2, are connected across the supply with the indicating instrument connected across Z_2. Then (referring to the figure):

$$V_s = V_p \left[\frac{Z_2}{(Z_1 + Z_2)} \right]$$

The load (or instrument) impedance is connected in parallel with Z_2 and should be very much larger than Z_2 if accuracy is to be maintained. If D.C. voltages are involved then the impedances can be purely resistive and, in the steady state, any leakage inductance or capacitance has no

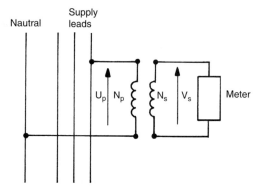

Figure 17.6 Voltage transformer

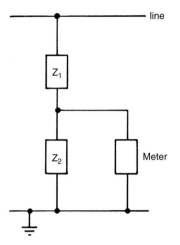

Figure 17.7 Voltage divider

effect. At high A.C. frequencies, however, it is important that they are non-reactive and that stray inductance and capacitance to ground, etc. are taken into account.

17.4.2 Current transformers

The current transformer is arranged with its primary winding in series with the supply (Figure 17.8). It thus carries the load current. The measuring instrument is connected across the secondary as shown. The ideal theoretical relationship between the currents and the number of turns on the primary and secondary is:

$$I_s = I_p (N_p/N_s)$$

The voltage appearing across the secondary terminals is:

$$V_s = I_s Z_s$$

where Z_s is the impedance of the measuring instrument connected to the secondary terminals (*burden*).

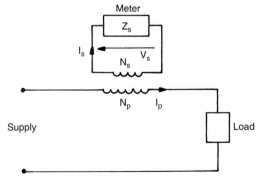

Figure 17.8 Current transformer

If this impedance is high then so also is this voltage. The primary of the device should therefore never be energized when the secondary is open circuit, since in that case Z_s, and therefore V_s, is infinite. Standard secondary current ratings are 1 A and 5 A, and the preferred range of values for the burden is between 1.5 VA and 30 VA.

Ratio and phase angle errors also occur due to the need for a portion of the primary current to magnetize the core and the requirement for a finite voltage to drive the current through the burden. These errors must be small. Current transformers are classified (in descending order of accuracy) into types AL, AM, BM, C and D. Ratio errors in class AL must be within the limits of (± 0.1 per cent for AL and (± 5 per cent for D, while the phase error limit on class A1 is (± 5 minutes to (± 2 minutes in types CM and C.

Again, there are problems both at D.C. and at high frequencies. In the former case connecting a very low resistance, R_s, in series with the load and using a voltmeter to measure the voltage across it obtain a current sample. This voltage is equal to $I_s R_s$ and, provided R_s is known, then I_s is specified.

At very high frequencies, the current is measured by assessing one of the effects that it produces. Several techniques are possible, e.g. (1) measuring the temperature rise when the current flows through a known resistance or (2) using a Hall-effect probe to measure the electromagnetic field created by the current.

17.5 Voltmeters and ammeters

These are manufactured according to several different principles, each being suitable for a particular application. The more common types are discussed briefly below.

17.5.1 Moving coil

This consists of a coil mounted between the poles of a permanent magnet. When a current flows through the coil a magnetic field is created and the interaction of this field with that of the magnet produces a torque, proportional to the current in the coil, which causes the coil to rotate against the action of a spring. A pointer is attached to the coil that gives an indication of the magnitude of the coil current (Figure 17.9).

Unlike most other electrical measuring instruments, the scale is inherently linear. The moving parts of the instrument must be as light in weight and as free from friction as possible in order to preserve accuracy. This implies that the coil is capable of carrying only small currents. If higher currents are to be measured, it is necessary to employ a shunt to divert some of the main current away from the instrument. The arrangement is as shown in Figure 17.10, where R_m is the resistance of the meter and R_s that of the shunt. The relationship between the total system current (I) and that in the instrument (I_m) is:

$$I_m = \frac{IR_s}{(R_m + R_s)}$$

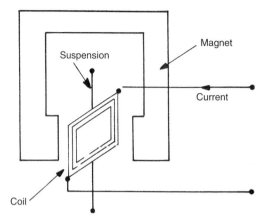

Figure 17.9 Moving-coil instrument

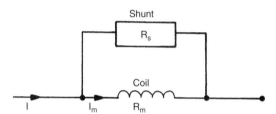

Figure 17.10 Use of shunt

Many commercial instruments have the shunt incorporated into the unit, and the user need only concern himself with the full-scale deflection as shown on the scale. General-purpose instruments with a range of built-in shunts are available and the user then selects the appropriate range. The instrument is delicate and easily damaged by excess currents. Connecting it in the circuit with reversed polarity can also damage it. The instrument may also be used as a voltmeter by connecting it, in series with a large resistor, across the supply, and in most commercial instruments the resistor is incorporated into the case.

The direction of rotation depends on the direction of the current in the coil, and thus the instrument is only suitable for D.C. It is, however, possible to incorporate a full-wave rectifier arranged as shown in Figure 17.11 in order to allow the instrument to measure A.C. quantities. The quantity measured is the RMS value only if the waveform of the current is truly sinusoidal. In other cases, a considerable error may result. In principle, the scale is linear but, if required, it can be made non-linear by suitably shaping the poles of the permanent magnet. The instrument reading is affected by the performance of the rectifier, which is a non-linear device, and this results in the scale also being non-linear. The error when measuring D.C. quantities can be as low as ±0.1 per cent of full-scale deflection and instruments are available for currents between microamperes and up to 600 A.

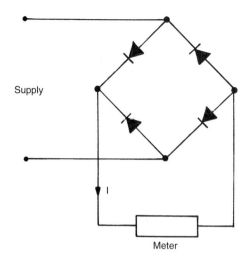

Figure 17.11 Full-wave rectification

17.5.2 Moving magnet

This is a cheap but inaccurate instrument suitable for use only in D.C. circuits. In this case, the coil is wound on a small vane situated in the magnet field.

17.5.3 Moving iron

The current to be measured is used to energize a coil. This creates a magnetic field and a moving vane is repelled by, or in some case attracted into, this field to a degree depending on the current magnitude (Figure 17.12). The vane is restrained by a spring and attached to an indicating pointer. A square law is involved and thus the scale is inherently non-linear, although this can be modified by a suitable selection of materials and shape for the vane.

The square law relationship also implies that the instrument measures RMS values. It can be used on either A.C. (up to the lower audio range if special compensating circuits are employed) or D.C. The instrument reading can be

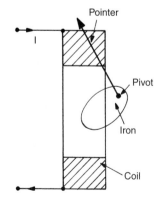

Figure 17.12 Moving-iron principle

affected by stray magnetic fields and should be shielded as far as possible. It is usually designed (and calibrated) for single-frequency operation, but it will continue to give accurate results (in terms of root-mean-square (RMS values) even if the waveform contains high-frequency harmonics. The smallest units are capable of measuring a few milliamps and the upper limit is fixed only by the current-carrying capacity of the coil. Shunts can, of course, be incorporated for different current ranges. This type of instrument is cheap and robust.

17.5.4 Electrodynamics instruments

These incorporate an electromagnet in place of the permanent magnet of the moving-coil instrument (Figure 17.13). The current flow through both the moving and fixed coils and a torque is produced on the former which carries the pointer and which is restricted by a spring. Again, the deflection obeys a square law and the scale is cramped at the lower end. Since with A.C. the current in both coils changes direction, this instrument can be used on either D.C. or A.C. Although with A.C. the meter actually measures the mean of the square of the current, the scale is usually calibrated in terms of the rms value. With the appropriate modifications, this type of instrument can be used as a wattmeter, a VAR meter, a power factor meter or a frequency meter.

17.5.5 Thermocouple

This instrument was developed from the 'hot-wire' ammeter, some examples of which can still be found. In the modern equivalent, the current to be measured (or a known proportion of it) flows through a small element that heats a thermocouple, so producing a rms voltage at its terminals, which is a function of the current. This voltage then supplies a current to a permanent magnet, moving-coil movement.

The instrument measures the true rms values and units are available to cover the frequency range from D.C. to 100 MHz. The major drawback is the susceptibility to damage by overloads (even those of short duration) unless appropriate protection is provided. The accuracy can also be affected by ambient temperature changes.

Since the instrument responds to a heating effect, it is slow to respond to changes in the measured quantity.

17.5.6 Electrostatic voltmeter

This consists of a pair of plates, one fixed and the other moving, mounted parallel to each other. When a voltage is applied across the pair the fixed plate moves against the action of a retaining spring. Again, the characteristic is square law. It has an inherently high level of insulation, due to the air gap between the plates, and can be used to measure voltages up to 15 kV. It consumes no power and therefore has extremely high impedance.

17.6 Frequency measurement

Low frequencies can be measured using an electromechanical moving-coil instrument. Current from the supply flows through two parallel fixed coils and then through a moving coil mounted between them (Figure 17.14). The two fixed coils are tuned to slightly different frequencies and the resulting fields set up a torque which is proportional to frequency.

The nominal frequency at which the instrument can be used depends on the tuning of the coils and the instrument is only accurate to within plus or minus a few per cent of this frequency. A ratiometer (Figure 17.15) can be used at higher frequencies up to 5 kHz or so. This consists of two moving coils arranged at right angles and mounted between the poles of a permanent magnet. The system current is fed into the two coils through separate phase-shifting networks and the result is to produce a torque proportional to frequency.

At frequencies higher than this, a solid-state counter must be used. This is based on a stable oscillator and, in effect, counts the pulses generated during one cycle of the supply frequency. The range and accuracy of the instrument depends on the master oscillator frequency, but units capable of use over the whole range up to 600 MHz

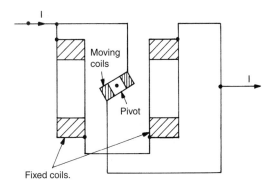

Figure 17.13 Schematic diagram of electrodynamics instrument

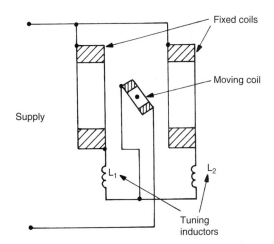

Figure 17.14 Frequency meter

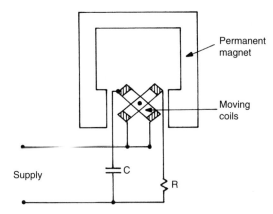

Figure 17.15 Frequency ratio meter

are available, although some discrimination is likely to be lost at the lower end.

17.7 Electronic instrumentation

Electronic instrumentation is available for the measurement of D.C. and A.C. voltage, current and power as well as impedance. Such instruments usually have higher sensitivities, operating frequencies and input impedance than is normally found in the electromechanical instrumentation described above. However, they may need to incorporate amplifiers and they invariably need power to operate the final display. Hence, an independent power source is needed. Both mains and battery units are available. The accuracy of measurement is very dependent on the amplifier, and bandwidth and adequate gain are important qualities.

17.7.1 Analogue voltmeters

The A.C. voltage to be measured is fed through an amplifier which supplies some form of indicating devices. Moving-coil instruments and digital displays are available, and in some circumstances (for example, when true rms values are required) the amplifier supplies a thermocouple as described above. The thermocouple output, with further amplification if necessary, drives a D.C. display. The reading, for a given measured voltage, depends on the input/output characteristics of the complete system, especially on the amplifier/attenuator and any rectifiers incorporated to allow a D.C. final instrument to be used when measuring A.C.

The unit can, however, be designed to read the rms, the average or the peak of the input waveform. The input impedance is high, which means that the instrument has little modifying effect on the voltage which is being measured, and it can be used up to 1 GHz, although stray capacitance at these very high frequencies may present a problem.

It should be noted that the input impedance of the meter changes when the range is switched. A large amount of spurious electronic signals ('noise') may be present in some systems. This is a potential source of error in the instrument reading, especially when small signals are involved. It can be improved by the use of suitable filters which may be built into the instrument, although the penalty is a reduction in the frequency bandwidth over which the instrument can be used.

An important point to be considered when the instrument is used for A.C. voltage measurement is the terminal connections. One terminal will be clearly designated as the high-potential connection, and this should be adhered to. The HT terminal will have a low value of capacitance to other bodies and to earth while the corresponding capacitance of the other is high. If the instrument is in a metallic case this should be connected to the mains earth as a safety precaution. In some cases, the low-voltage terminal is also connected to the metallic case. If this is so, the instrument will effectively earth the circuit under test, which may give rise to problems.

Instruments with a balanced input circuit are available for measurements where both input terminals are normally at a potential other than earth. Further problems arise due to common-mode interference arising from the presence of multiple earth loops in the circuits. In these cases the instrument may need to be isolated from the mains earth. Finally, high-frequency instruments, unless properly screened, may be subject to radiated electromagnetic interference arising from strong external fields.

17.7.2 Digital multi-meters

A number of different types are available but they all share the common characteristic of very high input impedance, which means that they are suitable for measuring voltages with minimal interference to the actual circuit. They are used whenever accuracy, good resolution and high levels of stability are important. As their name implies, these instruments are capable of measuring both A.C. and D.C. voltages and currents as well as resistance, and incorporate a wide range of scales. When used with A.C. the input signal is divided down to a level suitable for the input to an analogue to digital converter. This needs to be preceded by other function converters and range-switching units. In some instruments of this type, the range is selected automatically by sensing the level of the input signal and selecting the appropriate range electronically.

The analogue-to-digital converter has a fixed range, usually either 0–200 mV or 0–2 V, and the input divider network is used to obtain a signal within this range. This implies that there may be considerable differences in input impedance between different ranges. When used to measure alternating voltage the appropriate function converter may be a transformer followed by rectification and smoothing. This produces a direct voltage equivalent to the average value of the A.C. voltage and, provided the input signal is purely sinusoidal, the indicator is then calibrated in rms values.

More sophisticated instruments use special circuitry to obtain the rms values of non-sinusoidal signals. The current converter usually consists of low-value resistors in the input, and one of these is chosen as a shunt resistor (depending on the scale chosen) and the instrument measures the voltage developed across this. Resistance is measured by using the resistive chain at the input as a reference. When a voltage is applied across this and the unknown resistor, the two voltage drops are proportional to the respective resistance values. The voltage across the unknown resistor is applied to the analog-to-digital (A–D) converter followed by that across the known resistor with which it is compared and the ratio is used to determine the value of the former. It should be noted that the instrument is essentially a sampling device. The input signal is periodically sampled and the result processed using one of a number of techniques, examples of which are given below:

1. The *linear ramp* instrument uses a linear time-base to determine the time taken for an internally generated voltage to reach the unknown voltage, *V*. The limitations are due to small non-linearity in the ramp, instability of the electronic components and a lack of *noise* rejection.
2. The *charge balancing* voltmeter employs a pair of differential input transistors used to charge a capacitor, which is then discharged in small increments. The individual discharges are then sensed and the total number, which is a function of the measured voltage, stored. It has improved linearity and high sensitivity as well as a reduced susceptibility to 'noise'. In addition to its use as a straightforward voltmeter, it can be interfaced directly with other measurement devices such as thermocouples and strain gauges.
3. The potentiometer principle is used in the *successive approximation* instrument, where the unknown voltage is compared with a succession of known voltages until balance is obtained. The total time for measurement and display is about 5 ms.
4. In the *voltage-frequency* unit the unknown voltage is used to derive a signal at a frequency proportional to it. This frequency is then applied to a counter, the output of which is thus a measure of the voltage input.
5. The *dual-slope instrument* incorporates an integrator operated in conjunction with a time difference unit, and this is one of a range of units employing a combination of the above techniques designed to take advantage of the favorable characteristics of each. It is subject to the same limitations as the linear ramp instrument described above but it does have the advantage of inherent noise rejection.

17.8 Instrument selection

The selection of the instrument for a specific purpose should take account of the following points:

1. Use no more precision (i.e. no more digits) than is necessary in order to keep the cost to a minimum.
2. Make sure that the input impedance is adequate in the circuit in which the instrument is used. This applies not only to the normal (or static) input impedance but also to the dynamic impedance which occurs when the

instrument is subjected to transient surges with significant components at frequencies other than that at which the measurement is desired.
3. Take adequate steps to achieve noise rejection both by the choice of a suitable instrument and by using external networks if necessary.
4. Ensure that the instrument has appropriate output ports to drive other devices such as storage and computational facilities. Wherever possible, these ports should conform to the IEEE interface standards.

Most instruments of this type incorporate built-in self-checking facilities but the check can only be as accurate as the internal reference which, itself, will only be accurate over a specific range. The instrument should be calibrated against a reference source at periodic intervals.

Modern electronic instruments are much more than simple but accurate methods of measuring and can include many additional facilities such as:

1. Variable reading rate from one reading per hour up to several hundred per second;
2. Storage capability allowing many hundreds of separate readings to be stored internally and recalled subsequently, if necessary, at a different rate;
3. The ability to perform calculations on the readings and to display the derived result;
4. The ability to indicate if specific read values are outside pre-set limits and to store (and subsequently recall) maximum and minimum readings.
5. Self-calibration and/or testing according to predetermined programming.

17.9 Cathode ray oscilloscopes

The cathode ray oscilloscope (CRO) is an extremely versatile instrument for monitoring electrical signals, for diagnostic testing and for studying time-varying phenomena. A stream of electrons is produced from a heated cathode and accelerated through an electron 'gun' towards a fluorescent screen (Figure 17.16). The stream passes through a coil which focuses it into a beam which then passes between pairs of vertical and horizontal plates. Electric fields are created between these plates which cause the beam direction to change in response to them.

In the simplest form the vertical plates are supplied with a voltage which is a linear function of time, and, in the absence of a signal on the horizontal plates, the beam will move horizontally from left to right across the screen. The speed of movement is controlled by the rate at which

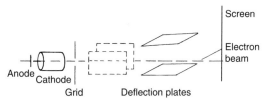

Figure 17.16 Principle of CRO

the voltage applied to the vertical plates is increased, and at the end of its travel the beam flies back to its starting point.

If now a second, time varying, voltage is applied to the horizontal plates, this will cause a vertical deflection of the beam, and the result is a trace on the screen, which corresponds to the variation of this second voltage with time. The action of the beam on the screen causes a fluorescent trace to appear on the screen as the beam is deflected. The time for which this trace persists is a function of the electron density in the beam and the material with which the screen is coated.

The accuracy of the device is limited by the linearity of the circuit which moves the beam horizontally in response to time (i.e. the 'time base') and the response of the input circuit. The 'time base' is generated by a sweep circuit which is triggered to instigate the start of the movement. The triggering signal may be derived from the voltage that is being measured or it may be obtained from an entirely separate source. This can cause difficulties in some applications. If it is required to examine a voltage pulse then the simplistic approach is to allow the pulse itself to trigger the signal. This is unsatisfactory, if it is the leading edge of the pulse which is of interest, and therefore a delay is built into the instrument which allows the trace to be triggered immediately the pulse is applied but which delays the application of the signal to the vertical deflection plates for a short time in order to enable the pulse to be displayed in the middle of the screen.

The velocity with which the spot moves from left to right across the screen is adjustable, enabling single or multiple cycles of an A.C. signal to be observed. Typical instruments can be used to measure a wide range of voltages on a series of different scales. The input signal may be either A.C. or D.C. and in the latter case, the operation of the appropriate switch on the front panel short-circuits the input capacitor. In each case, the voltage needs to be amplified before it is applied to the plates. A different amplifier is used for each scale range and more sophisticated instruments may include a range of plug-in amplifiers in addition to those built into the instrument. The important amplifier characteristics are the gain, the bandwidth, and the amplitude of signal, which can be handled without distortion and the input impedance. Since the inherent sensitivity of a typical oscilloscope is likely to be of the order of 50 mV/cm the amplifier gain must be large enough to give an acceptable deflection when the input signal amplitude is very small. On the other hand, if high input signals are involved then an attenuator, rather than an amplifier will be needed (see below).

Reproducibility of the input signal is all-important, and therefore the amplifier must be capable of handling the input signal magnitude involved. Frequency response is also important and the amplifier in use must have a bandwidth sufficient to accommodate the frequency of the input signal. The input impedance should be high in order to avoid loading the circuit under test. As with the amplifier, the attenuator must produce a known degree of change (in this case a reduction) in the input signal within the specified frequency band. It must have high input impedance which should not alter when the input

sensitivity switch setting is changed. The oscilloscope is normally connected directly to the circuit under test. However, this may not be possible in all cases and a probe may have to be used. The probe may simply serve as an isolator, so avoiding problems caused by, for example, shunt capacitance in the measuring leads or it may contain built-in amplifiers (or attenuator) to extend the range of the instrument. The probe should be matched to the instrument and the use of a probe produced by one manufacturer in conjunction with an oscilloscope from another is not recommended.

Many variants of the basic type are possible and some of the more usual are listed below:

1. *Low frequency* (up to around 30 MHz): typical deflection sensitivities are between 0.1 mV/cm and 10 V/cm. This is a general-purpose instrument and can incorporate a second trace energized by a voltage different from the first and using the same or different deflection speeds for the time base. An alternative is to use a single beam but to apply a *chopped* signal to the deflection plates. Sampling the two separate signals for brief intervals of time and applying each in turn to the plates derives the applied signal. The luminescent persistence of the screen then results in two *continuous* traces (one corresponding to each of the two applied voltages) to be observed.
2. *Medium frequency* (up to 100 MHz): The features described above (with the exception usually of the chopped-signal variant) are available but the time base components and amplifiers are designed for much wider frequency bands.
3. *High frequency* (up to 400 MHz): Again, the difference between this and the above is the use of higher-quality components and circuits to give linearity over a much wider frequency range.

17.9.1 Storage

Most of the types described above have the facility for single-shot operation if it is necessary to measure single 'events' (i.e. transients). In these cases, the timebase is triggered by the start of the transient. The limitation is the persistence of the screen luminescence since the event only occurs once, rather than a repetitive series of events, as happens with a periodic waveform where the trace is, in effect, overwritten during each operation of the timebase.

It is possible to sample and store the input transient in digital form by frequent sampling of the input waveform. These data can be retrieved (repetitively if necessary) from the store and displayed on the screen at the operator's leisure. The frequency range over which this type of instrument can be used depends on the sampling network, and is usually limited to a few MHz. Waveforms and transients corresponding to much higher frequencies can be stored and displayed in this way, with the storage being carried out by optical methods.

17.9.2 Sampling oscilloscopes

These are used for very high-frequency work. The waveform of the applied signal, which must be periodic in nature, is sampled at intervals that bear a definite relationship to some reference point on the wave. These samples are stored and subsequently used to reconstruct a display of the actual signal on the oscilloscope screen.

The non-storage oscilloscope can be found in most electronic test situations, from sophisticated research laboratories to production engineering plants. The storage unit is most widely used in medical work and in electromechanical applications, particularly where very high-speed transients need to be recorded, while, as noted above, the sampling type finds its main use in the evaluation of ultra-high-frequency equipment.

Appropriate connections must be used to pick up the required signal from the circuit under investigation. This can be done directly for low- and medium-voltage applications and, if a measurement of current is required, by measuring the voltage across a resistor of known value. However, for accurate work it is important that neither the impedance across which the leads are connected nor the impedance of the leads themselves affects the result. It is also important to avoid errors being introduced, particularly in high-frequency work, from spurious signals induced in the leads if they are situated in electromagnetic fields. The latter can be overcome in most cases by using co-axial leads with the outer screening connected to the case of the instrument and thence to earth. The former problem can be more difficult to deal with, since it implies that the load across the oscilloscope terminals, including the effects of the leads, must be matched to the input impedance of the instrument.

These problems can be minimized in the great majority of cases by using specially designed probes. The user will need to know the frequency range (i.e. the bandwidth) over which the probe can be used satisfactorily and the scaling factor introduced by the probe.

17.9.3 Use of oscilloscope

The cathode ray oscilloscope is a multi-purpose instrument with several ranges, switching between which is carried out by controls on the unit. The required values are obtained by measuring the height of the trace of the screen or the time between different events in the horizontal deflection of the beam. These measurements are then converted to the required values using the scale values obtained from the control switch settings. It is thus of paramount importance to make sure that the scale settings are recorded. The vertical scale will normally be calibrated in volts (or fractions of a volt) per centimeter and the horizontal scale in seconds (or fractions of a second) per centimeter. The accuracy of the measurement will then depend on (1) the accuracy with which the beam deflection can be measured on the screen and (2) the accuracy of the built-in attenuators and amplifiers applicable to the range in use.

A graticule is normally provided on the screen to assist in the measurement. Parallax errors can cause problems unless this graticule is actually engraved on the screen itself, and the final accuracy will depend on the degree of beam focusing which can be achieved. Accuracy of better than 1.5 per cent should be possible in most cases.

As noted above, the operator will select a convenient range that will allow for the whole of the signal he wishes to observe to be displayed on the screen. It is important that the internal circuits so selected are linear over this range and again, with most instruments, values of better than 1 per cent should be achievable.

17.10 Transducers

Many of the quantities required to be measured in process plant operation are not, in themselves, electrical, and if electrical or electronic instrumentation is to be used then these need to be converted to electrical signals using a transducer. The transducer is a physical object and its presence will have an effect on the quantity being sensed. Whether or not this effect is significant will depend on the particular application. In all cases, it is advisable to consider carefully the balance between the requirement that the transducer should, on the one hand, cause the minimum interference with the quantity being measured and, on the other, that it should be intimately associated with the effect being measured.

Temperature measurement is a case in point. A large transducer in close contact with the body whose temperature is being measured will act as a heat sink and consequently produce a localized reduction at the point where the temperature is being measured. On the other hand, if an air gap exists between the transducer and the hot surface then the air (rather than the surface) temperature will be measured.

The important characteristics of a transducer used in conjunction with an electronic measurement system are accuracy, susceptibility, frequency, impedance and, if appropriate, the method of excitation. The transducer is likely to be the least accurate component in the system, and it should be calibrated (and recalibrated) at frequent intervals. It is likely to be subject to a range of different physical conditions, some of which it is there to detect and others by which it should remain unaffected (for example, a pressure transducer should be unaffected by any changes in temperature which it might be called upon to experience). Some types of transducer are not suitable for use under D.C. conditions and all will have an upper limit of frequency at which accuracy is acceptable. Many types of transducer are also affected by stray electromagnetic fields.

The impedance of the transducer is important if it provides an output signal to an electronic device (an amplifier, for example) and the impedance of the two must be matched for accurate measurement. Some transducers (thermocouples, for example) generate their output by internal mechanisms (i.e. they are self-excited). Others such as resistance thermometers need an external source and an appropriate type must be available. Transducers used in the measurement of the more common physical quantities are discussed below.

17.10.1 Temperature

There are several instruments for measuring temperature as follows.

Resistance thermometers

These utilize the fact that the resistance of most materials changes with temperature. In order to be useful for this purpose this change must be linear over the range required and the thermal capacity must be low. Although the above implies a high resistivity and temperature coefficient, linearity and stability are the paramount considerations, and suitable materials are platinum, nickel and tungsten.

The sensor usually consists of a coil of wire made from the material that is wound on a former and the whole sealed to prevent oxidization, although a film of the metal deposited on a ceramic substrate can also be used. The resistor is connected in a Wheatstone bridge network (Figure 17.17), using fixed resistors in the other three arms. The instrument connected across the bridge is calibrated directly in terms of temperature. The range is limited by the linearity of the device and the upper temperature, which can be measured, must be well below the melting point of the material.

Thermocouples

These are active transducers in the sense that they act as a generator of EMF whose magnitude is a function of the temperature of the junction of two dissimilar metals. Rare-metal combinations are used for high accuracy and at high temperatures, but for most engineering applications one of the following is suitable, depending on the temperature:

Copper/constantan	(up to 670 K)
Iron/constantan	(up to 1030 K)
Chromel/constantan	(up to 1270 K)
Chromel/alumel	(up to 1640 K)

If the instrument is to be direct reading, the second (or 'cold') junction must be kept at a constant reference temperature. If high temperatures are to be measured then the terminals of the detector can be used as the cold junction without an unacceptable loss of accuracy.

The voltage output of the more common types of thermocouple is of the order of 50 V/C and the output is either read on a sensitive moving-coil meter or on a digital voltmeter. The reading is converted to temperature using a calibration chart supplied with the thermocouple. Some commercial units are available in which the thermocouple and instrument is supplied as an integral unit with the scale directly calibrated in temperature. If a separate instrument is to be used then it should be noted that the thermocouple resistance is only of the order of 10 Ω and

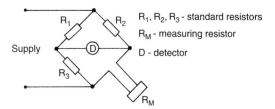

Figure 17.17 Resistance temperature measurement

R₁, R₂, R₃ - standard resistors

R_M - measuring resistor

D - detector

for maximum sensitivity the meter resistance should be matched to this.

The commercial units have a very low thermal capacity and very high response speeds. Some are available with several independent channels and a common 'cold' junction. Each channel is scanned in turn by the instrument, and the readings either displayed or stored for future recovery. Accuracies of better than 0.2 per cent are possible. Thermocouples are available to cover a very wide range of temperatures, their cost is low and they have a small mass, so minimizing the intrusive effect on the surface at the point where the temperature is being measured. The output characteristics (output voltage versus temperature) are reasonably linear but the measurement accuracy is not particularly high.

Thermistors

These are semiconductor devices, which have a high resistivity and a much larger temperature/resistance coefficient than the materials available for resistance thermometers. However, there is a temperature limitation of around 550K and the characteristics are excessively non-linear, although it is possible to obtain matched pairs to measure differential temperatures. Thus, for example, they can be used to measure airflow in ducts. The change in resistance is again detected using a Wheatstone bridge network. The materials used are manufactured from sintered compounds of copper, manganese, nickel and cobalt. Conventional semiconductor diodes in the form of $p-n$ junctions can also be used in this way.

Although thermistors with a positive temperature characteristic are available, the negative characteristic types are more common. Devices which are physically small have low power-handling capabilities and some form of output amplification is usually needed. They do, however, have a fast response. Larger units are available in which amplification can be dispensed with but the response is much slower. These are used for indication (rather than control) purposes.

Pyrometers

These are non-contact instruments for measuring temperature, and, as such, their thermal mass is relatively unimportant. They can be used for very high temperatures. Radiation pyrometers focus the infrared radiation emitted by the body onto a thermocouple contained within the instrument. The voltage produced by the thermocouple is then read on a dial calibrated directly in terms of temperature. If required, this voltage can be amplified and used as a control signal. The actual temperature measured is that integrated over the whole of the surface *seen* by the pyrometer and the reading may be affected by infrared absorption in the medium between the hot surface and the instrument. This also applies to optical pyrometers, which are usually of the disappearing-filament type. The instrument contains a lamp the filament of which is viewed against the background of the hot body. The current through the lamp is adjusted, so varying the brightness of the filament until it merges into the background. The current required to achieve this is measured and converted into a temperature reading on the dial.

17.10.2 Force

For the present purposes, this includes the measurement of pressure, acceleration and strain.

Piezoelectric crystals

These use crystalline materials in which the electrical properties of the material are changed when it undergoes slight deformation by, for example, the application of mechanical pressure. The principal effect is to cause a change in the frequency at which the material resonates. This change in resonant frequency can be detected and measured, so giving an indication of the change in pressure.

Quartz is a natural crystalline material which exhibits this form of behavior, although its relevant properties are highly temperature dependent and synthetic materials have been developed which, although fundamentally less accurate, are more stable under varying temperature conditions.

One application is the accelerometer, in which the acceleration force of a mass is made to increase (or decrease) the pressure produced on the crystal by a spring. This, in turn, produces the required electrical change, the effect of which is amplified. It is important to select units appropriate for the expected changes, which should be within the frequency range from almost zero to the natural frequency of the crystal.

Strain gauges

These units are rigidly attached to a surface. Any small deformation in that surface due to the application of an external force changes the dimension of the strain gauge, and this change is then detected to give a value of the applied force or the strain in the material. Resistance strain gauges consist of a grid resistance wire cemented between two films. This resistance element is connected in one arm of a measurement bridge with a similar, compensating, gauge in a second arm. The bridge is rebalanced when the strain is applied, and the changes required to do this are measured and converted, using the calibration information, to the quantities required.

Modern techniques use thin-film resistors deposited directly on the area and semiconductor units are available which are considerably more sensitive than the resistive type. Dynamic measurements can also be made. The change in resistance unbalances the bridge, causing a voltage to appear across the detector terminals. This voltage is then amplified and applied to a CRO or the information can be stored digitally for future retrieval.

Strain gauges can also be used to measure pressure by bonding the gauge to a diaphragm in the wall of a liquid or gas container, acceleration, by sensing the relative change along one or more sensing axes, and torque (by sensing angular strain).

Temperature-sensing devices may also be used to measure flow in pipes or ducts and for level indication of liquid in a tank. The device is indirectly heated; using a separate heating filament, and the flow of liquid cools the sensor at a rate which is a function of the velocity of the flow. Thus by incorporating the device into a bridge circuit, it is possible, after appropriate calibration, to obtain a direct measurement of the flow velocity. Similarly, several such devices can be placed one above the other, in a tank, containing liquid. The liquid acts as a heat-transfer medium, those devices below the surface are cooled as the liquid level falls and the temperature and therefore the resistance of the devices increases. This can be detected, and the whole chain is thus used as a level detector.

Impedance transducers

Strictly, the strain gauges referred to above come into this category, since in such cases the change in the measured quantity causes a corresponding change in the resistance of the element. However, the principle has a much wider application, using changes in either the inductive or capacitive reactance of electrical circuit elements.

The inductance of an iron-cored inductor varies with any air gap included in the iron core. Thus, physical movement can be detected by allowing this movement to displace part of the core, so changing the width of the gap. The detection of very small movements is possible in this way, and instruments based on the principle are used to measure acceleration, pressure, strain, thickness and a variety of other changes.

The effects usually need to be amplified, and in some cases, signal processing is necessary in order to obtain the derived information. Two electrical conductors at different potentials exhibit the property of capacitance (i.e. energy storage) between them. The value of this capacitance changes when the relative position of the plates alters or when the medium between them is changed. Such changes can be detected by employing the capacitor in an oscillatory circuit. The resonant frequency of this circuit is proportional to the capacitance that it contains and therefore the detection of such a frequency change allows the original quantity to be determined.

This technique can be used to measure displacement where, in effect, the two electrodes are connected to the two bodies. It has also other applications (for example, in moisture meters where the presence of water vapor between the electrodes causes the capacitance change).

Photo-sensors

Photovoltaic sensors are semiconductor devices and have the property that when light falls on them an electrical voltage is produced across them high enough to drive a current through a resistive load. Banks of these units (solar cells) can be used to produce significant amounts of electrical power, but the technique can also be used for measurement purposes, optical tracking and reading bar codes and punched tape.

Photoresistive sensors are essentially inactive devices, but they have the property of exhibiting a change in resistance when light falls on them. This change is detected by an appropriate circuit and is used to measure the light falling on the unit (e.g. in photography exposure meters) or to trigger a control action (for instance, in intruder alarm systems).

Photoemission is the property of some materials to emit electrons when light falls on them. These materials are used as a cathode and an anode collects the electrons. One application is in the photomultiplier unit that is used in counters.

Photovoltaic cells are independent of an electrical supply but, in general, they lack sensitivity as compared with photoemissive units.

17.11 Spectrum analyzers

Much present electronic equipment deals with signals which are not sinusoidal and which may not even be periodic in nature. However, these signals can be divided into a series of components, each of a single specific frequency, and each can then be studied in turn in order to determine its characteristics. Where plant instrumentation is concerned, this technique can be particularly useful for diagnostic purposes.

The instrument required is a spectrum analyzer that must be capable of operating over a wide range of frequencies to cope with all those present in the input signal. Some of these signals may be close together in frequency and they may have widely varying magnitudes. The real-time technique utilizes a series of fixed filters designed to separate contiguous frequency bands. Each supplies its own detector, and the outputs of these are then scanned to produce individual traces on an oscilloscope. The resolution is governed by the bandwidths of the individual filters and the instrument can be used from D.C. to a few kHz.

The Fourier transform analyzer uses digital techniques in order to carry out the frequency separation. This can cope with signals up to several hundred kilohertz and swept-tuned analyzers employ a tuned filter technique. From the user's point of view the desirable characteristics are high stability (much higher than the frequency being analyzed) and good resolution (the ability to distinguish between different signals which are close together in frequency). It must also be able to detect small signals (i.e. its sensitivity must be adequate) while at the same time have an adequate dynamic range. The latter is the ratio of the largest to the smallest signal that can be detected without distortion in either. Good modern instruments of this type incorporate a considerable degree of automation whereby inherent accuracy can be corrected, the control settings are recorded and the data manipulated to yield additional information as required, all with the aid of a desktop computer.

17.12 Bridge measurements

Reference has been made above to the simple Wheatstone bridge. This and the developments arising from it are an essential component in many of the instruments already referred to. These bridges (whose operation basically depends on comparing an unknown quantity with a known series of quantities in order to measure the former) are also used as stand-alone units for such diverse purposes as fault detection and the determination of the parameters of electronic components. The accuracy of the measurement is governed by the accuracy to which the value of the components against which the comparison is made is known, and the care which is taken in the measurement process. Brief descriptions of some of the more common methods are given below together with their applications. The theory of operation may be found in the many textbooks on the subject.

In its simplest form, the *Wheatstone bridge* is used on D.C. for the measurement of an unknown resistance in terms of three known resistors. Its accuracy depends on that of the known units and the sensitivity of the detector. It is also used for sensing the changes which occur in the output from resistance strain-gauge detectors. The latter instruments can be made portable and can detect variations of less than 0.05 per cent.

The *Kelvin double bridge* is a more sophisticated variant used for the measurement of very low resistance such as ammeter shunts or short lengths of cable. This is also operated on D.C. In industrial terms the *digital D.C. low-resistance* instruments are more convenient although somewhat less accurate.

A.C. bridges suffer from the complication that they must allow for stray inductance and capacitance in the circuit and that the measurements are, in general, frequency dependent. Inductance is measured by the *Maxwell bridge* and its development, the *Wien bridge*, the commercial variants of which operate at frequencies up to 10 kHz with accuracy of better than 1 per cent. The corresponding unit for capacitance is the *Schering bridge*. The major problems in the latter case are usually concerned with stray capacitance to ground. In order to overcome these, the Wagner earth system is used.

The *Wayne–Kerr bridge* employs a modified principle involving the balancing of the output between from two windings of a transformer. It can be used up to high frequencies (100 MHz) for the measurement of impedance. However, at these frequencies it is now more usual to make use of the resonance principle, whereby a variable-frequency source is used to supply the unknown impedance. The frequency is adjusted until resonance is obtained and the value of this frequency together with the amplitude of the signal is used to determine the impedance. This is the principle used in the *Q-meter*, and is the basis of many of the automatic impedance-measuring instruments now commercially available.

17.13 Data recording

There are many industrial applications in which permanent records (extending over long periods of time) of the instrument readings are required. Chart recorders of various forms are available for this purpose. The most common general-purpose unit is the digital strip chart recorder, in which the input signal is used to drive the movement of a recording arm that passes over a paper chart in the *y*-direction. At the same time, the chart is being driven forward at a known speed in the x-direction. A stylus on the end of the arm marks a series of dots on the paper, using either electrostatic or thermal means, and thus a continuous record of the signal is obtained. The recorder, as its name implies, operates digitally but the internal analog-to-digital converters can deal with analogue signals in the instrument. It is possible to cope with several independent inputs by using a number of recording arms and these instruments can be used up to about 25 kHz, although the quality of the trace suffers at these high frequencies.

The digital chart recorder is now gradually replacing the $x-y$ chart recorder which, although the basic principle is similar, is analogue in operation and uses a pen on the end of the recorder arm to mark the paper. The response of these instruments is relatively slow, which limits their application to low-frequency work. The mechanical movement of the recording arm imposes the underlying limitations on accuracy and frequency of both this and the digital recorder. Higher speeds can be obtained with ultraviolet recorders, in which the input signals are fed to coils mounted in a magnet field and which therefore deflect in response to the signal. The coil suspension also carries a lightweight mirror that deflects a beam of ultraviolet light onto photosensitive paper. Thus, a light beam replaces the inertia of the moving parts.

17.14 Acoustic measurements

The plant engineer is often concerned with the measurement of noise arising from the operation of machinery; especially since permissible noise levels are closely specified in current OSHA legislation. The measurement of noise involves the use of a pressure transducer. This is a microphone in which the change of pressure on the diaphragm causes a corresponding change in its electrical impedance, which can be detected by using appropriate circuitry. The output signal is amplified, rectified and applied to a D.C. meter to give an indication of the rms value and therefore of the power independently of the waveform.

17.15 Centralized control

In many applications, especially where process plant is concerned, the measurement information is required to be conveyed to a central point, which may be a control room manned by operators or a computer which carries out the control functions automatically. The connecting link between the various measurement points and the central control is usually a telemetry system, although in some cases the distances involved are so large that radio links need to be used.

The measurement data are derived from some form of detector or transducer, the output of which is an electrical signal that must then be conditioned to a value suitable for the input to the telemetry system (typically, of the order of 10 V). In some cases, amplification is required and in others, attenuation is needed.

The signals transmitted over the network will have a value corresponding to the output from the individual source, and it must be possible to identify the origin of the signals at the receiving end. On the other hand, a discrete connection between each source and the control center may be prohibitively expensive, and there is a requirement to use a single channel for many separate inputs in such a way that the channel can be monitored and the individual inputs separated from each other. This involves the use of some form of multiplexing. Earlier systems used frequency modulation (FM) techniques in which each measurement signal is used to modulate the output from an oscillator operating over a dedicated part of the frequency spectrum. All signals are transmitted simultaneously over the network, and at the receiving end, the signals at the different frequencies are separated out, using a demodulator, to give the individual measurement information.

The requirement for a different frequency band for each measurement channel is restrictive, and later systems make use of some form of time-division multiplexing, where all channels use the same part of the frequency spectrum but not at the same time. The channels are sampled in sequence and an instantaneous value of the signal is obtained from each. Obviously, the sampling rate must be extremely high (many times per second) and the sampling at the various channel inputs must be synchronized exactly with the corresponding sampling of the output at the control center. Pulse amplitude modulation systems (PAM) use analogue signals while in pulse code modulation (PCM) systems the signals are converted to digital form. At present, the choice usually lies between the FM and the PCM systems. Generally, the former are the cheaper but the latter give a much better accuracy.

Further reading

Baldwin, C. T., *Methods of Electrical Measurement*, Blackie, Glasgow (1953).
Bibby, G. L., 'Electrical methodology and instrumentation', in *Electrical Engineers Reference Book*, Butterworth-Heinemann, London (1985).
Coombs, C. (ed.), *Basic Electronic Instrumentation Handbook*, McGraw-Hill, New York (1972).
Cooper, W. D., *Basic Electronic Instrumentation Handbook*, Prentice-Hall, Englewood Cliffs, NJ (1972).
Dalglish, R. L., *An Introduction to Control and Measurement with Microcomputers*, Cambridge University Press, Cambridge (1987).
Drysdale, A. C., Jolley, A. C. and Tagg, G. V., *Electrical Measuring Instruments*, Chapman & Hall, London (1952).
Golding, E. W. and Widdis, F. C., *Electrical Measurements and Measuring Instruments*, Pitman, London (1963).
Hnatek, E. R., *A User's Handbook of D/A and A/D Converters*, John Wiley (Interscience), New York (1976).
Lion, K. S., *Elements of Electrical and Electronic Instrumentation*, McGraw-Hill, New York (1975).
Neubert, H. K., *Instrument Transducers*, Oxford University Press, Oxford (1975).
Norton, H. N., *Handbook of Transducers for Electronic Measuring Systems*, Prentice-Hall, Englewood Cliffs, NJ (1969).
Ott, H. W., *Noise Reduction Techniques in Electrical Systems*, John Wiley, New York (1976).

Rudkin, A. M. (ed.), *Electronic Test Equipment*, Granada, St Albans (1981).

de Sa, A., *Principles of Electronic Instrumentation*, Edward Arnold, London (1990).

Stout, M. B., *Basic Electrical Measurements*, Prentice-Hall, Englewood Cliffs, NJ (1960).

Strock, O. J., *Introduction to Telemetry*, Instrument Society of America, New York (1987).

Tagg, G. F., *Electrical Indicating Instruments*, Butterworth-Heinemann, London (1974).

Usher, M. J., *Sensors and Transducers*, Macmillan, London (1985).

Van Erk, R., *Oscilloscopes*, McGraw-Hill, New York (1978).

Wobschal, D., *Circuit Design for Electronic Instrumentation*, McGraw-Hill, New York (1987).

Wolf, S., *Guide to Electronic Measurements and Laboratory Practice*, Prentice-Hall, Englewood Cliffs, NJ (1983).

18

Oil

Roland Gibson
W.S. Atkins Consultants Ltd

Contents

18.1 Distribution and delivery

Crude oil is processed at oil refineries, generally located around coastal areas, and then transferred to a nationwide network of oil fuel terminals, from where it is distributed to customers. Road delivery vehicles, rail tank wagons or coastal tankers – depending on the location of the customer's storage installation – can make fuel supplies to the customer.

18.1.1 Road delivery vehicles

Vehicles make deliveries of oil fuels by road with capacities ranging from 11,800 to 28 700 liters (2600 to 6300 gallons). The vehicles discharge fuel into customers' storage by pump or compressed air at rates up to 1050 liters (230 gal) per minute, and can deliver to a height of 10.7 meters (35 ft) above the vehicle.

There are occasions where the noise of vehicle discharge needs to be minimized, examples being night discharge at hospitals, hotels and adjacent to residential areas where there are parking restrictions on daytime deliveries or where a customer agrees to a 24-hour delivery service. In such cases, the customer can install a take-off suction pump as part of his storage facilities, thus eliminating the use of the vehicle engine for vehicle pump operation.

There must, in all cases, be safe road access to sites where tanks are situated and suitable hard standing provided for the vehicle during delivery.

18.1.2 Rail tank cars

The capacities of rail tank cars are generally 45 tons gross laden weight (approx. 30 tons payload) or 100 tons gross laden weight (approximately. 70 tons payload). Cars for carrying heated fuel oil are insulated with fiberglass lagging covered with galvanized sheeting, and are equipped with finned tubes for steam heating.

A customer's steam supply must be capable of providing an adequate amount of heat to discharge a full trainload of oil. Considering individual rail tank cars, a 100 tons *gross laden weight* (glw) car requires 385 kg/h (850 lb/h) and a 45 t glw car requires 154 kg/h (350 lb/h) of steam. Because of the risk of contamination by oil, the condensate should be drained from the outlet connection and run to waste.

Rail tank cars have bottom outlets terminating in a quick-acting coupling, and are designed for pump off-loading only. A flow indicator should be inserted between the terminal outlet and the fill line. Flexible connections for steam and oil lines should be provided for the maximum number of cars that will be discharged simultaneously.

A block train consists of a series of rail tank cars of either 45 tons glw or 100 tons glw (or a mixture of both) coupled for movement from source to destination and back as a unit. The size of the trains will be determined by contractual agreement, but will never be less than 300 tons payloads, and will normally be 500 tons.

18.1.3 River and coastal tankers

Where a customer has suitable berthing facilities and satisfactory arrangements can be made, deliveries can be by either river or coastal tanker. River tankers currently in use are of 1860 tons capacity but may be larger if conditions permit. The customer is responsible for supplying oil hose and/or discharge arms capable of discharging the full contents of the barge under low-water conditions.

Coastal tankers are available up to 2000 tons capacity, equipped with pumps for discharging into customer's storage. The size of the vessels to be used will depend upon local conditions and the depth of water in the approaches.

18.2 Storage tanks

18.2.1 Type of tanks

There are four main types of storage tank available for industrial and commercial fuel oils, namely:

> Mild steel welded
> Mild steel sectional
> Cast iron sectional
> Reinforced concrete

Of these four, the mild steel welded tank is the most popular and is widely used for every type of application.

The majority of storage tanks are of the horizontal cylindrical type, as shown in Figure 18.1. Where ground space is a limiting factor, vertical cylindrical tanks may be used.

18.2.2 Construction

Basic design criteria which need to be taken into consideration when designing a storage tank will include:

1. Geographical location of tank (e.g. indoors/outdoors, height above ground level, available space, accessibility, etc.);
2. Principal dimensions or capacity;
3. Type of roof (e.g. fixed or floating, cone, dome, membrane, pontoon, etc.);
4. Properties of fluid;
5. Minimum and maximum design temperatures;
6. Requirement for heating elements;
7. Design vapor pressure and/or vacuum conditions;
8. Minimum depth of fluid;
9. Maximum filling/emptying rates;
10. Venting arrangements;
11. Connections/manhole requirements;
12. Access ladders;
13. Corrosion allowance;
14. Number and type of openings;
15. Type of support or foundation;
16. Internal/external coating or lagging.

18.2.3 Capacities

The capacity of storage tanks for oil-fired installations is an important consideration. The minimum net storage capacity should be calculated either by taking:

1. Three weeks' supply of oil at the maximum rate of consumption; or
2. Two weeks' supply at the maximum rate of consumption, plus the usual capacity ordered for one delivery, whichever is the larger.

Where the maximum weekly off-take is less than 200 gallons (1810 liters) the capacity should still not be less than 650 gal (218501) in order to accept a standard 500-gal (2270-l) tanker delivery. In some circumstances, it may be desirable to provide more than one tank, each of sufficient capacity to accept at least a full delivery.

18.2.4 Tank supports

Horizontal cylindrical tanks should be installed on brick or reinforced concrete cradles with a downward slope of 1 in 50 from the draw-off end towards the drain valve, as shown in Figure 18.1. Cradles should be constructed on foundations adequate for the load being supported and the type of soil. A reinforced concrete raft equal to the plan area of the tank, and of adequate thickness to bear the load, is normally suitable for all but the weakest soils. Cradles should not be placed under joints or seams of the tank plates and a layer of bituminized felt should be interposed between the cradle and tank. The height of the tank supports should provide at least 450 mm space between the drain valve and ground level to allow access for painting or draining the tank.

18.2.5 Tank fittings

Oil-level indicators

A brass dipstick is recommended as a cheap and reliable means of determining the contents of a storage tank. A dipstick, when required, is usually provided ready calibrated by the tank manufacturer before installation of the tank.

In many cases it is inconvenient to use a dipstick, due to the position or location of the tank, and there are a variety of direct and remote contents gauges available, including gauge glasses, float and weight, float and swing arm, float and indicator, hydrostatic, electrical capacitance, etc.

Filling connection

Filling pipes should be as short as possible and free from sharp bends. The terminal should be in a convenient position to allow easy coupling of the vehicle hose connection, wherever possible within 5 meters (15 ft) of the hard standing for vehicle delivery. The most suitable height for a filling pipe is about 1 m (3 ft) above ground level and clear of all obstructions.

A non-ferrous dust cap with chain should be provided to close the end of the filling pipe and protect the thread when not in use. When the filling pipe is not self-draining into the tank a gate valve should be fitted as close as possible to the hose coupling. Where there is any possibility of damage or misuse of the terminal equipment a lockable fill cap should be fitted and, if considered necessary, the terminal enclosed in a lockable protective compartment.

All filling lines should be self-draining. Where this is not possible with residual fuel installations, trace heating and lagging should be applied, and this is particularly important in exposed positions. This will ensure that any oil fuel remaining in the filling pipe will be at pumping viscosity when delivery is made.

The filling pipe should enter at the top of most horizontal storage tanks. For vertical tanks up to 3 meters (10 ft) diameter the filling pipe should be fitted at the top and then bend through 180°. This directs the incoming oil fuel down one side of the tank and minimizes air entrainment.

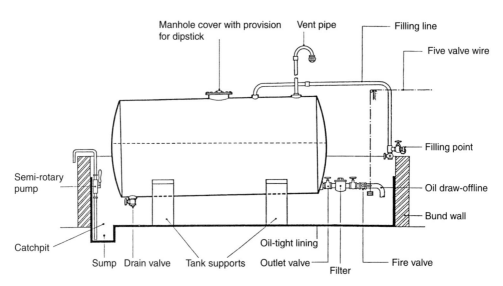

Figure 18.1 Typical storage tank for a distillate fuel oil

In larger vertical storage tanks containing residual fuels, bottom inlet filling may sometimes be used. A non-return valve must be fitted in the filling line as close to the tank as possible. The entry position of the filling pipe into the tank in relation to the position of the draw-off connection must be carefully selected to avoid air entrainment into the oil-handling system.

Where delivery vehicles do not have access adjacent to the storage area a permanent extended filling pipe should be provided from the tank to a position where the vehicle can stand in safety. In cases where the length of this pipe will exceed 30 meters (100 ft), careful attention should be paid to draining requirements. Where lines are not self-draining, a drain valve should be situated at the lowest point in that section. Provision should be made for lagging and tracing to be applied to extended fill lines that will carry residual grades of industrial oil fuel. Careful attention should also be paid to the use of correctly sized pipe diameters. Where the storage tank is not visible from the filling point, an overfill alarm should be fitted.

Ullage

The air space between the oil surface and the top of the tank is known as ullage; there should always be a small ullage remaining when the contents gauge registers full. This prevents the discharge of oil from the vent pipe due to any frothing and surging of the liquid during delivery. The ullage should provide not less than 100 mm (4 in) between the oil surface and the top of the tank or be equivalent to 5 per cent of the total contents, whichever is the greater.

Vent pipes

A vent pipe must be fitted at the highest point of every storage tank. Wherever possible, it should be visible from the filling point and terminate in the open air, in a position where any oil vapor will not be objectionable and, in the event of an overflow, there will be no damage to property, fire risk of contamination of drains.

The vent pipe bore must be equal to or greater than the bore of the filling pipe, and never less than 50 millimeters (2 in) diameter. It should be as short as possible and free from sharp bends. It should terminate in a return bend or 'goose neck' fitted with a wire cage for protective purposes (fine gauze must never be used for this purpose).

Where, of necessity, the vent pipe rises to a considerable height, excessive internal pressure on the tank may result, due to the pressure head of oil should an overflow occur. To prevent any possible tank failure due to such an occurrence a vent pipe pressure-relief device must be provided. It should be self-draining to reduce the risk of blockage, particularly when using heavy oil fuel. Codes require that these devices should not place any restriction on oil flow and must discharge within the bund area.

Draw-off connection

The draw-off connection to the oil-burning plant should be at the raised end of horizontal tanks. Where heating

facilities are not provided, the lowest point of the draw-off connection should never be less than 75 mm (3 in) above the bottom of the tank. For tanks fitted with heating elements, it is essential that these and their associated thermostatic control probes should always remain below the oil surface.

To enable the contents of the tank to be isolated a screw-down gate valve should be fitted adjacent to the draw-off connection. Since stresses may be applied to valves during any tank settlement or movement of pipe-work, cast-iron valves should not be used for this purpose.

Drain valve

A screw-down gate valve with a bore similar to that of the draw-off connection should be installed at the lowest point on every storage tank to permit complete draining. The valve should be readily accessible with a clear space below to facilitate its use. Extension pipes to or from drain valves should be avoided if possible, but where these are necessary, the pipe should be lagged and, if necessary, traced to ensure that residual grades of oil will flow during adverse weather conditions. Valves and extension pipes should be fitted with a plug or blank flange to prevent inadvertent discharge of the tank contents.

Tanks containing heated grades of oil fuel require regular draining. This is due to the small amount of moisture which accumulates over a period, by condensation formed on the sides of the tank. The quantity of moisture formed will be dependent upon the relative humidity conditions, the amount of *breathing* which takes place, and the time allowed for settling. It is recommended that the tank be checked for accumulated water prior to a fuel delivery.

With storage tanks which are filled from the bottom such as large vertical tanks, the stirring action of incoming oil will carry any water (which would normally settle out) into the oil-handling system. In these circumstances, any daily service tank used in the system should receive regular inspection to ensure that water has not accumulated. Any oil/water mixtures should be drained into suitable containers and subsequently removed for disposal into a separator or interceptor.

Manhole

Every storage tank must have a manhole in an accessible position, preferably on top. It may be circular, oval or rectangular, and not less than 460 mm (18 in) diameter if circular or 460 mm (18 in) long and 410 mm (16 in) wide if oval or rectangular. The manlid must be securely fixed by bolts, studs or setscrews, and have a liquid and vapor-tight joint. Close-woven proofed asbestos, graphited on both sides, is a suitable jointing compound for this purpose.

Vertical tanks over 3.65 meters (12 ft) high should have a further manhole fitted near the bottom to provide access for cleaning and maintenance of any storage-heating facilities.

18.2.6 Heating requirements

Distillate grades of oil fuel may be stored, handled and atomized at ambient temperatures, and do not require heating facilities to be provided in storage tanks and handling systems. However, exposure to extreme cold for long periods should be avoided, since oil flow from the tank may become slightly restricted. This is particularly important for storage serving stand-by diesel generating plant where summer cloud point specification fuel may be in both storage and oil feed lines during the most severe winter weather. In very exceptional conditions, even winter cloud point specification fuel may present occasional problems in unlagged storage and handling systems. Such systems should be lagged and, if necessary, traced to ensure faultless starting.

Heating facilities are required for all residual grades of oil fuels, such as Light Fuel Oil, Medium Fuel Oil, and Heavy Fuel Oil. Table 18.1 gives the recommended minimum storage and handling temperatures for residual oil fuels (from BS 28618: 11870).

Where oil is to be maintained at minimum storage temperature, an outflow heater will be necessary to raise the temperature of the oil leaving the tank to that required for handling. It is not good practice to store oil fuel at unnecessarily high temperatures, and the temperature given under the column 'Minimum temperature at outflow from storage and for handling' should not be exceeded by more than $16.7°C$ ($30°F$). This is particularly important in relation to Light Fuel Oil. Maximum heat losses from storage tanks can be determined from Figure 18.2. These losses can be translated into maximum steam consumption rates of the heating coils from knowledge of the latent heat of steam appropriate to the steam pressure used in a given installation. The tank heating arrangements must be capable of maintaining the oil storage temperature with the appropriate rate of heat loss.

Another important point is that concerning the occasional requirement for heating the oil from ambient temperature. A heat input several times greater than that required to maintain handling temperatures will be necessary, depending on the minimum time stipulated for heat-up. Tank heating arrangements should therefore be capable of providing this heat input if the contingency is a possibility.

Heating methods

Storage tanks can be heated with thermostatically controlled steam coils, hot-water coils, electric immersion heaters or a combination of these. The elements and their

thermostats should be positioned below the level of the oil draw-off line, so that they are always covered during normal operation. The temperature-sensitive element of the thermostat should always be situated above and to one side of the heating element. The heating elements should be spaced evenly over the bottom of the tank or concentrated towards the draw-off end. A combination of steam and electric heating can be used for installations where periods may occur during which steam is not available.

Where a residual oil fuel is stored in tanks with outflow heaters provision should also be made to maintain the oil at or above the minimum storage temperature shown in Table 18.1. Excessive heating is not recommended, particularly for Light Fuel Oil, and will increase running costs unnecessarily. Additional heating to outflow heaters can be provided by using a steam, hot water or electric heater running along the bottom of the tank. With electric heating, additional separate heating elements may be required.

Heating elements should be readily removable for repair if necessary and consequently careful note should be taken of possible external obstructions to this operation. The steam supply to the heating coils should be dry saturated. It is not generally necessary for the pressure to exceed 3.45 bar ($50 \, \text{lbf/in}^2$). The temperature of the heating medium should not exceed $177°C$ ($350°F$), and electric element loading should not exceed $1.24 \, \text{W/cm}^2$ ($8 \, \text{W/in}^2$).

Steam and hot-water coils should be constructed of seamless steel tube and preferably be without joints within the tank. These coils can be either plain or finned tube. However, due to their greater surface area, finned tubes generally have a higher rate of heat transfer than plain tubes. On the basis of cost per unit surface area, finned tubes are less expensive than conventional tubes. There is also an advantage of weight saving, and complete coverage of the base area is not necessarily required to achieve specified temperatures.

Where joints are unavoidable, they should be welded. Steam coils should drain freely from inlet to outlet. Steam traps, usually of the bucket type, should also be provided. Condensate from steam coils should be drained to waste, unless adequate provision is made to drain trace quantities before return to the hotwell. Where hot-water coils are used, the water supply should be heated through a calorifier. These recommendations will avoid any risk of oil reaching the boiler plant.

Where storage facilities comprise several tanks, the contents of stand-by tanks may remain unheated. In such cases, the heating facilities should be capable of raising the contents to the required storage temperature as soon as possible from cold.

Heating requirements for warming storage tanks. The hourly heat requirement (H_r) to raise the temperature of a storage tank can be calculated from:

$$H_r = C \times h \times t_r$$

where

C = total weight of contents when full (kg (lbs)),
h = mean specific heat of fuel oil,
t_r = temperature rise required per hour (°C (°F)).

Table 18.1 Recommended storage and handling temperatures for residual fuel oils (based on BS 28618: 11870)

Grade of oil	BS classification	Minimum storage temperature	Minimum temperature at outflow from storage and for handling
Light Fuel Oil	Class E	10°C (50°F)	10°C (50°F)
Medium Fuel Oil	Class F	25°C (77°F)	30°C (86°F)
Heavy Fuel Oil	Class G	35°C (95°F)	45°C (113°F)

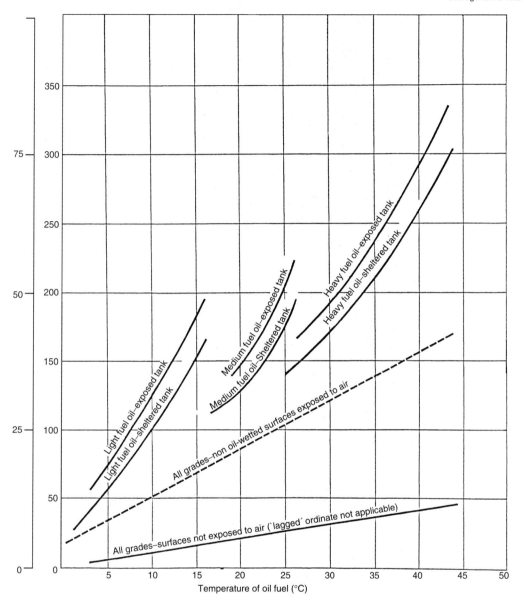

Figure 18.2 Heat losses from storage tanks plotted against oil fuel temperature

The following values should be used for the mean specific heat:

Class of fuel	Mean specific heat 0–100°C (kJ/kg°C)
D	1.1818
E	1.1818
F	1.187
G	1.185

The rate of temperature rise (t_r) will depend on the particular circumstances of the installation and on how quickly the contents of the tank must be brought up to temperature.

Lagging

Various types of lagging are available. The major advantage of its use is a considerable reduction in heat losses from both storage tanks and pipework. The materials most suitable for application to storage tanks are asbestos in the form of blankets or molded sections, glass silk blankets

of 85 per cent magnesia or slagwool. When applied, these materials should provide a lagging efficiency of approximately 75 per cent. All lagging should be reinforced with wire netting, either incorporated into the blanket or secured to anchor points on the tank surface. A weatherproof surface finish should finally be applied to the lagging. This can be either two-ply bitumen felt or, if necessary, sheets of galvanized steel or aluminum with sealed joints.

When the use of lagging is being considered, cost-benefit analysis should be made to determine the minimum running costs when either (1) fuel is stored at the minimum recommended storage temperature with an outflow heater to raise the temperature to the minimum handling temperature, or (2) fuel is stored at handling temperature, thus requiring no outflow heater.

The temperature of heated fuel when delivered into customer's storage will depend upon the length of time in transit, or the measures of preheat given to a rail tank car to assist discharge. These heat sources should not be overlooked in calculations.

18.3 Location of tanks

18.3.1 Siting

Storage tanks should, wherever possible, be installed above ground. The site selected should not be in an unduly exposed position. Clearance must be allowed for the withdrawal from the tank of fittings such as immersion heaters and steam coils. Where tanks are installed inside buildings, they should generally be located within a tank chamber, although in some industrial installations such as steelworks and foundries, a tank chamber may not be necessary. The requirements of insurance companies and local authorities should be considered when designing these installations.

18.3.2 Underground storage

Where underground installation of a storage tank is unavoidable, it should preferably be housed in a specially constructed brick or concrete chamber, allowing easy access to the drain valve and other fittings. Wherever possible, the tank chamber should be located in dry ground and the finished structure made watertight. A sump must be provided in the floor of the chamber at one end to collect any water that may enter the chamber in exceptional circumstances, and the floor should slope slightly downwards towards the sump. Water collected in the sump can be removed by using a semi-rotary pump. The lower part of the chamber should form an oil-resistant catch pit, as with the more usual aboveground storage tank.

18.3.3 Buried tanks

It is recommended that storage tanks should not be buried directly in contact with soil, since it is almost impossible to avoid corrosive attack. Where, for some reason, it is not possible to provide a tank chamber as already described, adequate corrosion protection must be applied to all exterior surfaces of the tank, fittings and pipework.

18.3.4 Bund area

Where overfilling or leakage from the tank would contribute to a fire hazard, cause damage to property or contaminate drains or sewers, a bund wall should be constructed around the tank. This should be of brick or concrete with an oil-tight lining, and sealed to the concrete base under the tank supports. The capacity of the bunded area should be at least 10 per cent greater than that of the storage tanks contained within it.

The bund walls must be oil proof and be capable of withstanding considerable liquid pressure in the event of an overflow or other emergency. No permanent drain must be incorporated into a bund area, but suitable facilities should be provided to remove rainwater that may accumulate. A sump and semi-rotary hand pump are suitable for this purpose.

18.4 Pipework systems

18.4.1 General

Oil fuel pipeline systems transfer oil from storage to the oil burner at specified conditions of pressure, viscosity, temperature and rate of flow. There can be considerable variety in the choice of system, but its design (particularly correct pipe sizing and temperature control) is most important if it is to function satisfactorily.

18.4.2 Handling temperatures

Distillate grades are usually handled at ambient temperatures provided these are not below the cloud point when using gas oil. Residual grades, on the other hand, are handled at temperatures above ambient. The recommendations of Table 18.1 should always be followed regarding minimum recommended handling temperatures. Residual grades can also be handled at temperatures above those recommended as minimum handling levels in order to reduce oil viscosity, improve regulation and control of oil flow, reduce friction losses in pipelines and, when necessary, provide oil at the correct atomizing temperature for the oil-burning equipment.

18.4.3 Handling equipment

Two items of equipment should always be inserted in the handling system as close to the storage tank as possible. These are a filter and a fire valve.

18.4.4 Filters

To prevent foreign matter from damaging components and choking valves or atomizer orifices, filters must be incorporated into the handling system. There are usually two stages of filtration. The first provides protection for pumps and fire valves which handle oil at temperatures below those required at the oil-burning equipment. Second-stage filtration protects the atomizer orifice and burning equipment, and is sometimes incorporated as part of the burner assembly.

For distillate grades of fuel oil, the first-stage filter should protect pumps and valves and can be placed in the draw-off line from the storage tank. Paper filters can be used where appropriate, providing filter apertures are not less than 0.1 mm (0.004 in). The paper element should be regularly replaced when necessary. For residual grades of oil, first-stage filters should be placed in the draw-off line as near to the storage tank as possible, and incorporate a filtering medium equivalent to a circular hole 2.5–0.75 mm (0.1–0.03 in) diameter. Fine filtration must not be employed at the first stage, since pressure drop across the filter may be excessive due to the higher viscosity of these grades at low temperatures. Second-stage filtration is usually incorporated into the oil burner assembly, often in the hot oil line after the preheater so that it filters the oil at atomizing temperature.

Reliable enclosed filters of ample effective filtering area should be used for both stages. The filter design should not allow the pressure drop to exceed $7 \, kN/m^2$ ($1 \, lbf/in^2$) across second-stage fine filters in clean conditions. Filters should preferably be of the duplex or self-cleaning type, and installed so that oil spillage will not occur during operation.

They must be readily accessible for cleaning, which should be carried out as frequently as necessary to ensure that pressure drop across the filter does not affect normal oil flow rates. The filtering medium should be of corrosion-resistant material such as Monel metal, phosphor-bronze or stainless steel. All first-stage filters should be provided with isolating valves.

18.4.5 Fire valves

A valve that closes in case of fire should be inserted in the oil fuel line to the oil-burning equipment and fitted as close to the tank as possible. It may be held open mechanically, pneumatically or electrically. Temperature-sensitive elements should be arranged to close the valve at a fixed maximum temperature, and sited close to the oil-fired plant and well above floor level. The operating temperature of the heat-sensitive elements should not be greater than 68°C (155°F) except where ambient temperatures in the vicinity of the plant may exceed 418°C (120°F), in which case the operating temperatures may be 183°C (200°F).

Where an electric circuit is employed the valve should close on breaking, not making, the circuit and be reset manually. A warning device should be included to indicate that resetting is necessary, to cover the eventuality of temporary electrical failure.

Where a valve is closed by the action of a falling weight, a short free fall should be allowed before the weight begins to close the valve. This assists in overcoming any tendency for the valve to stick in the open position.

Fire valves should preferably be of glandless construction. If a gland is incorporated into the system, it should be of a type that cannot be tightened to an extent which would prevent the correct functioning of the valve. A manual release should be installed on all fire valves so that they may be tested regularly.

18.4.6 Types of system

There are two main types of oil-handling system in common use. These are gravity and ring main.

Gravity systems

Gravity systems are of three basic types: gravity, pump-assisted gravity and sub-gravity. A gravity system is one in which the oil flows directly from the storage or service tank through gravity feed pipeline. The static head on the feed line will vary with the depth of oil in the tank, and the system should therefore only be used for burners which will operate satisfactorily between such limits.

A pump-assisted gravity system is one in which oil flows by gravity from the storage or service tank to a pump. The pump supplies oil to the combustion equipment through a pipeline passing only the quantity required to feed the oil-firing equipment. The inclusion of a pump in this system will not reduce the static head due to the fuel supply in the tank. The working pressure required at the oil-burning equipment should therefore be greater than the maximum static head available when the storage or service tank is full.

A sub-gravity system is one in which a pump associated with the oil-burning equipment is used to suck the oil from a tank in which the level of the oil can be below the level of the pump.

Gravity systems in general will handle distillate grades at ambient temperatures, and residual oils at pumping or atomizing temperatures. Lagging and tracing will be required with residual grades to prevent cooling of the oil to below handling temperatures.

Ring main systems

A ring main system draws oil from storage and circulates it to each consuming point in turn, the balance of the oil being returned either to the suction side of the circulating pump or to storage. A diagram of a typical hot oil ring main system is shown in Figure 18.3.

Each take-off point is connected to the burner it supplies by a branch line. Pressure conditions are maintained approximately constant at each take-off point by a pressure-regulating valve situated after the last take-off point, and circulating a quantity of oil one and a half to three times the maximum take-off from the circuit. By this means, stopping take-off at one consuming point will not have a marked effect on the pressure at other consuming points. Pressure conditions should be calculated at each consuming point for all conditions of operation and take-off. The bore of the ring main should be such that pressure variations are not excessive for the equipment served. If these variations at each take-off point are likely to be critical, they can be accommodated by the use of individual pressure-regulating valves on each branch.

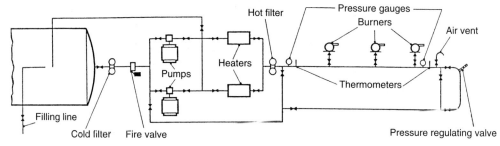

Figure 18.3 Typical hot oil ring main layout

Ring main systems are of three types: hot oil, warm oil and cold oil. Hot oil ring mains circulate oil at atomizing temperature, warm oil ring mains at a temperature between minimum pumping and atomizing temperature, and cold oil ring mains at ambient temperature.

Hot oil ring mains

This is the most important of the three types of ring main system, since it offers economies of installation and running costs. The smallest pipe sizes and fittings can be used with low-viscosity high-temperature oil, and the number of line heaters is minimized. The oil is circulated at atomizing temperatures and in consequence, the system is less liable to be affected by pressure fluctuations due to variations in viscosity, or small changes in the viscosity of the oil as delivered. The system should usually be considered as the first possibility and only discarded in favor of warm or cold oil systems where these will give some real advantage.

Where there is a large temperature difference between a hot oil line and ambient temperature conditions, particularly in the long ring main system, lagging will prevent excessive heat loss. Electric or steam tracing, in addition, will balance the heat loss from the pipeline. Residual oils should not be allowed to drop below their minimum handling temperature at any time. Where it is necessary to shut down any part of the plant, such as when servicing pumps and other equipment, a drain valve and air vent must be provided at the lowest point in the system to prevent cold oil becoming static.

With some installations, it is possible to maintain circulation during periods when the oil-firing plant is not required to operate. The oil is circulated at reduced temperature to minimize the heat loss from the ring main and, consequently, the amount of heat required to compensate for this loss. The reduced temperature should be such that the increased oil viscosity does not result in an excessive increase in pressure.

Warm oil ring mains

This system is similar to a cold oil ring main but includes provision for heaters in the circuit to maintain oil temperature between minimum handling and atomizing levels. This provides a reduction in oil viscosity and reduces pipe friction. The circulation temperature of the oil should be chosen to give the minimum pressure drop consistent with the system design when circulating one-and-a-half to three-times the maximum take-off and using a suitable pipe diameter. This circulation temperature should allow a reasonable margin below the specified atomizing temperature to facilitate the selection of the necessary line heaters for branch lines between the ring main and the oil-burning equipment.

The volumetric capacity is adjustable to provide uniform heat transfer to the oil passing through. Heaters are flexible in operation down to about 50 per cent of their designed temperature rise, when the quantity of oil passing through is approximately double. Where the temperature level required with a reduced throughput is outside this range, the heater manufacturer should be consulted.

Cold oil ring mains

This system is used mainly where different atomizing temperatures are required at various consuming points or where a branch line would be unacceptably long. The system should only be used where the length of pipeline involved and the quantity of oil circulated will not cause an excessive pressure drop due to friction. When designing a cold oil ring main system, care must be taken to ensure that the pressure variation between take-off points, due to changes in the oil consumption rate, do not affect burner performance. Circulating one-and-a-half to three-times the maximum take-off required achieves this. The system is widely used with distillate grades but rarely with residual grades.

Sub-circulating loops

Where it is not possible or convenient to arrange for the ring main to be carried near the burners an alternative to using a warm oil ring main with branch heaters is hot oil ring main with sub-circulating loops. The branch line from the high-pressure leg of the ring main is extended past the burners and returned into the low-pressure or return side after the pressure-regulating valve. This ensures that circulation of hot oil is maintained past the consuming point. The amount circulated through the sub-circulating loop must be carefully regulated to avoid the pressure-regulating valve becoming inoperative by short-circuiting too high a percentage of the oil in the ring main.

A secondary pressure-regulating valve situated after the last oil burner served by the loop controls the regulation of oil flow in a sub-circulating loop. Alternatively, a fixed orifice or regulating valve, either hand-controlled or of the lock shield type, may be used. With the former method, it is sometimes difficult to obtain balance between the primary and secondary regulating valves. The latter method is generally satisfactory providing the take-off rate of the burners does not fluctuate appreciably, and ample time can be allowed to establish stable pressure/temperature conditions in the sub-circulating loop after a shutdown period.

Branch lines

Branch lines transfer oil from a ring main circuit to the oil-burning equipment. Where a residual oil fuel is to be used, there will be some cooling of the oil immediately adjacent to the pipe surfaces and this will show as a small increase in viscosity. To keep this variation to a minimum and so prevent any difficulties in atomization at the oil burner, care should be taken over the length and diameter of branch lines. Provision should always be made to isolate and drain branch lines.

18.4.7 Pipe sizing

When a fluid is flowing through a pipe, resistance to flow is caused by friction. The pipe bore selected for each section must be such that under any operating conditions, the initial head, either static head of oil in the supply tank or the pump delivery pressure, will be adequate to ensure the required flow rate. Additionally, any change of flow rate and consequent variation in loss of head must not adversely affect the operation of the associated oil-burning equipment.

The following factors must be taken into account when assessing pressure drop.

Viscosity

Pressure drop is directly proportional to viscosity. The effect of heat loss from pipelines and consequent increase in viscosity should also be considered.

Flow conditions

The handling system should be designed to provide streamline flow at all times when steady and predictable pressure conditions are essential.

Flow rate

Pressure drop under streamline flow conditions is directly proportional to the quantity of oil flowing. The effect of reduced flow rate after take-off points, as compared with full flow rate throughout the full length of the pipeline when there is no take-off, should be taken into account to ensure that variation in pressure is within the specified pump output. Special consideration is necessary with gravity and ring main systems serving several take-off points.

Length of pipeline

Pressure drop is directly proportional to the length of the pipeline. All fittings used in the system should be included in the determination of 'effective pipeline length'. Table 18.2 should be used to determine the equivalent length to be added to the actual length of the pipeline for various types of fitting. The resulting figure is the total 'effective length' of the system.

The following empirical formula for estimating pressure drop under streamline flow conditions may be used:

$$H_f = \frac{1.2 \times 10^5 VT}{d^4} = \frac{0.0765VT}{d^4}$$

Metric (SI)   Imperial

where

H_f = loss of head in meters per 100 meters (feet per 100 ft) effective length of pipeline ((m) (ft)),
V = oil fuel viscosity at handling temperature (Redwood I scale),
d = internal diameter of pipe (mm (in)),
T = oil flow rate (tons/hour (ton/h)).

Note: The flow is normally streamline when

Table 18.2 Equivalent length of pipeline fittings

Nominal pipe bore (mm (in))	Elbow (m (ft))	Tee (m (ft))	Easy bend (m (ft))	Gate valve (m (ft))	Non-return valve (m (ft))
25 (1)	0.9 (3)	1.8 (6)	0.5 (1.5)	0.3 (0.8)	0.6 (2)
32 (1.25)	1.1 (3.5)	2.1 (7)	0.6 (1.75)	0.3 (1)	0.8 (2.5)
40 (1.5)	1.2 (4)	2.4 (8)	0.6 (2)	0.4 (1.2)	0.9 (3)
50 (2)	1.5 (5)	3.0 (10)	0.9 (3)	0.5 (1.7)	1.2 (4)
65 (2.5)	1.8 (6)	3.7 (12)	1.2 (4)	0.6 (2.1)	1.5 (5)
80 (3)	2.4 (8)	4.9 (16)	1.5 (5)	0.8 (2.5)	1.8 (6)
100 (4)	3.0 (10)	6.1 (20)	1.6 (6)	1.0 (3.3)	2.4 (8)
150 (6)	4.6 (15)	9.1 (30)	2.4 (8)	1.5 (5)	3.0 (10)
200 (8)	5.5 (18)	11.0 (36)	3.0 (10)	2.1 (7)	3.7 (12)
250 (10)	6.7 (22)	13.4 (44)	3.7 (12)	2.7 (9)	4.3 (14)

$$\frac{4Q}{dV} < 2000 = (30T < dV)$$

  Metric (SI)   Imperial

where

Q = oil flow rate (m^3/s),
d = internal diameter of pipe (m),
V = kinematic viscosity at handling temperature.

When it is necessary to estimate the pressure drop in a pipeline where turbulent flow conditions exist, the following formula will give an approximation:

$$H_f = \frac{7.89 \times 10^4 G^2}{d^5} = \frac{G^2}{6.4d^5}$$

  Metric (SI)   Imperial

where

H_f = loss of head in meters per 100 meters (feet per 100 ft) of effective pipe run,
G = oil flow rate (l/min (gal/min)),
d = internal pipe diameter (mm (in)).

Branch lines

Branch lines transfer oil from a ring main circuit to the oil-burning equipment. Where a residual oil fuel is to be used, there will be some cooling of the oil immediately adjacent to the pipe surfaces and this will show as a small increase in viscosity. To keep this variation to a minimum and so prevent any difficulties in atomization at the oil burner, care should be taken over the length and diameter of branch lines. In general, the following empirical formula should be used when designing branch lines for hot residual oils:

$$\text{Length} = 7L < \sqrt{\frac{M}{\text{Metric (SI)}}} = \frac{L < \sqrt{M}}{\text{Imperial}}$$

where

L = 'equivalent' length of branch line in m (ft) (Table 18.2),
M = minimum oil consumption (l/h (gal/h)).

Internal diameter $D < 10.4L(D < L/8)$ where D is in mm (in). Provision should always be made to isolate and drain branch lines.

19

Gas

Peter F Fleming

British Gas plc

Contents

Note on the text: This chapter provides an overview of proper installation and use of gas as an energy source. In the USA, these practices are governed by OSHA Regulations (Standards 29CFR). Plant engineers working in the USA should refer to 29CFR for specific guidance.

19.1 Selection and use of gas as a fuel

19.1.1 Advantages of gas

The advantages of gas can be summarized as:

1. Reliable fuel of constant composition;
2. Used as required (does not have to be ordered in advance or stored on the users' premises);
3. Clean fuel-emissions are low compared with combustion of most other fossil fuels.

19.1.2 Availability

Mains gas is widely available throughout the mainland of the United States. Thinly populated rural areas do not have access to mains gas although liquid petroleum gas from a central supply may be available. All major industrial areas are within the gas supply area, and less than 15 per cent of domestic dwellings are outside this area.

19.1.3 Metering

Payment for gas consumed is made based on the consumption as registered on a meter. Meters used for billing purposes are checked for accuracy and badged by the Department of Energy. These meters are known as primary meters. Those used by the user for monitoring the consumption of plant, etc. are known as check meters (often erroneously referred to as secondary meters).

The meter registers in cubic feet and gas consumption is calculated in therms; 1 therm (195.506 MJ) is 190,000 Btu. The meter location should be located as close as is sensibly practicable to the site boundary adjacent to the gas main. It is generally preferable for all but small low-pressure installations to be located in a separate purpose-built structure or compound and, wherever possible, away from the main buildings. Installations should be protected from the possibility of accidental damage, hazardous substances and extremes of temperature or vibration.

With the exception of certain very large loads which may use such metering devices as orifice plates, metering will be by means of diaphragm, rotary displacement or turbine meters.

Diaphragm meters

These are the traditional gas meter that incorporates diaphragms contained within a steel case. The diaphragms are alternately inflated and then deflated by the presence of the gas. The movement of the diaphragms is linked to valves that control the passage of gas into and out of the four measuring compartments.

These meters are suitable for low-pressure applications ($<$75 mbar) and low flow rates. Meters rated at up to 16 m^3 per hour (5650 ft^3 per hour) are available but it is not usual to use them above about 85 m^3 per hour (3000 ft^3 per hour). An important consideration is that the meters are physically large for their rating.

Rotary positive displacement meters

These meters incorporate two 'figure-of-eight' impellers rotating in opposite directions inside a casing. The impellers are made of either cast iron or aluminum. RPD meters are available with ratings from about 22 cm^3/h (800 ft^3/h) upwards but are not normally used as primary meters below about 85 m^3/h (3000 ft^3/h).

Turbine meters

These meters are only used for large loads which also have nearly constant gas consumption. The gas flow impinges on a specially shaped turbine and is streamlined by the contour of the casing on either side of the turbine. Turbine meters are very small for their rating. Disadvantages are that long lengths of straight pipework upstream are necessary to ensure the correct flow profile, and a fluctuating load will cause inaccuracies in registration.

Correction

Meters are accurate within close limits as legislation demands. However, gas is metered on a volume basis rather than a mass basis and is thus subject to variation with temperature and pressure. The Imperial Standard Conditions are 60°F, 30 inHg, saturated (15.56°C, 1913.7405 mbar, saturated). Gas Tariff sales are not normally corrected, but sales on a contract basis are. Correction may be for pressure only on a 'fixed factor' basis based on Boyle's Law or, for larger loads, over 190,000 therms per annum for both temperature and pressure using electronic (formerly mechanical) correctors. For high pressures, the compressibility factor Z may also be relevant. The current generation of correctors corrects for pressure on an absolute basis taking into account barometric pressure.

19.1.4 Contracts

For annual loads of up to 25,000 therms per annum gas is sold subject to published tariffs. Above this, the price of gas is subject to contract of which there are two types:

1. *Firm contract.* The gas sale is on a firm basis, that is, there is a firm commitment to supply gas under all except the most exceptional circumstances.
2. *Interruptible contract.* There is an obligation for the user to provide an alternative standby fuel to be used at the request of the gas supplier. The maximum period of 'interruption' is one of the conditions agreed in the contract. The advantage to the user is that he or she can obtain a more favorable gas price. The advantage to the supplier is that at times of peak demand, such as a cold winter, he can balance supply and demand without investing in excessive storage capacity.

19.1.5 Range of uses of gas

Gas is used for a diverse range of applications in the domestic, commercial and industrial sectors. Commercially, the main uses are space heating, water heating and catering, together with some chilling. In larger premises such as hospitals, etc. there may be large central boiler plant. Industrially, there will be applications similar to those in commerce. In addition, there are many process applications. These may be high-temperature kilns and furnaces (ceramics, metal reheating, metal melting, glass, etc.) or low-temperature (drying, baking, paint stoving, etc.).

The theoretical flame temperature is 1930°C, but in practice this is not obtainable, and a maximum process temperature of perhaps 1300°C is realistic without recuperation. Gas is used for 'working flames' in such processes as flame hardening and glass bulb manufacture. In industrial process, heating either direct or indirect heating (i.e. using a heat exchanger) may be used according to the process requirements. Gas can be used as a chemical feedstock. On a small scale, this can be to produce protective atmospheres by controlled partial combustion, on a large scale as a base for ammonia production, etc. Gas is, however, sold as a source of heat and not as a chemical of constant quality. It is suitable for the production of shaft power from either reciprocating engines or turbines.

The cleanliness of the products of combustion is such that the use of heat-recovery equipment is possible without the risk of corrosion. This has led to the development of combined heat and power packages where the overall efficiency is high.

Small reciprocating engines from 15 kW upwards can be used, the heat being recovered from the engine coolant and exhaust. Large reciprocating engines normally exhaust into a waste-heat boiler.

Gas turbines are available with power outputs of 1 MW upwards, and the exhaust is used to fire waste-heat boilers. The high oxygen content of the exhaust enables supplementary firing to be used to increase the heat/power ratio as desired.

19.2 Theoretical and practical burning and heat transfer

19.2.1 Types of burner

Efficient combustion of gas under varying conditions demands the use of a wide variety of burners. However, these can all be categorized as natural draft or forced draft.

The majority of domestic burners, together with a large number of commercial and many industrial burners, are of the natural-draft type. In these, the gas passes through a jet situated in a venturi such that primary air is mixed with the gas. The resulting mixture passes through the burner nozzle where mixing with secondary (and, in some designs, tertiary) air takes place together with ignition and combustion. Such burners have the advantage of simplicity but have a limited turndown ratio and their poor mixing characteristics lead to a rather low efficiency. Indeed, to ensure that complete combustion takes place without the

formation of soot or carbon monoxide it is necessary to allow a 'margin' of excess air well above that strictly needed for combustion.

The low gas pressure available at the injector (typically, 17.5 mbar) allows a primary aeration of only about 40 per cent. The resulting flame envelope is rather large and the intensity of combustion low. It is possible to increase the degree of primary aeration, producing a more intense flame, if a higher gas pressure is used. To produce complete primary aeration a gas pressure of the order of 1 bar will be needed.

For some processes, a burner in which there is no primary aeration may produce a flame. These 'laminar' flames have a very low intensity of combustion and a luminous appearance.

The majority of larger industrial burners, including furnace and boiler applications, are of the forced-draft type. These employ a combustion air fan to provide all the air needed for complete combustion. The burners are usually sealed into the combustion chamber so that there is no access to secondary air from the atmosphere as with natural-draft burners. Forced-draft burners may be of the premix type, where air and gas are mixed prior to the burner, or, more commonly, of the nozzle mix type, where the mixing takes place within the burner.

A refractory quarl is usually an integral part of forced-draft burners. Suitable design of burner and quarl can determine the flame characteristics. Long, short, pencil or even flat flames are possible.

For general purposes, including firing hot water boilers and warm air space heaters, 'package' burners are commonly used. In these, the burner is ready assembled together with all its controls and air fan. To install such a burner it is only necessary to connect up to gas and electric supplies and controls (thermostats, etc.).

19.2.2 Turndown ratio

The turndown ratio is an indication of the ability of the burner to maintain a stable flame at lower firing rates, and is a ratio of the maximum and minimum firing rates. Turndown can be low for average burners of both natural- and forced-draft burners, 3 : 1 being a typical figure with 5 : 1 a maximum although up to 40 : 1 is possible with special burners.

It is important to remember that although a burner can be fired at low rates it is probable that the efficiency at low fire will be reduced because the excess air is invariably higher at turndown. This will affect the selection of controls (e.g. on/off or modulating).

19.2.3 Heat transfer

The carbon/hydrogen ratio of gas is considerably lower than oil or coal, which results in a flame of very low luminosity. Radiation from the flame is therefore low and furnace design must allow for heat transfer to be primarily by convection and conduction, together with re-radiation from hot surfaces.

Burners can be designed to produce a luminous flame by means of laminar mixing and partial cracking of the gas, but the radiation is still low. A typical forced-draft burner used for boiler firing will be essentially transparent.

19.2.4 Water vapor in products of combustion

The high hydrogen/carbon ratio of gas means that the quantity of water vapor in the products of combustion is greater than most other fossil fuels. The latent heat of this cannot be released in conventional appliances leading to a low net/gross ratio of calorific value of 90 per cent. (It is normal practice to quote *gross* CV; in Europe *net* CV is often used. If net CV is quoted, efficiencies of over 190 per cent are possible.)

The cleanliness of products of combustion from gas enable recovery of latent heat by means of condensing appliances in which the products are cooled below the dewpoint of 55°C. The condensate is only weakly acidic and a suitable choice of materials of manufacture permit it to be dealt with. Most other fuels produce a condensate which is too acidic to allow condensing appliances to be used.

19.3 Pressure available to user

19.3.1 Low-pressure supply

The majority of users are supplied with gas at low pressure, and the categories are defined as:

Low pressure	(<75 mbar)
Medium pressure	(75 mbar to 7 bar)
High pressure	(>7 bar)

Normal supply pressure to industrial and commercial users is 21 mbar. Allowing for pressure losses in the system, at least 17.5 mbar should be available at the point of use. Higher pressures can often be supplied by agreement where available. There may be process advantages in having a higher pressure.

It is not normal practice to supply direct from a high-pressure transmission main rather than a local distribution system (known in the USA as 'farm taps').

19.3.2 Boosters and compressors

If a pressure higher than that available is wanted then a booster or compressor will be required. Boosters are normally considered as adaptations of centrifugal fans and raise pressure by typically 75 mbar for a single-stage machine. Higher pressures require a compressor which will be a positive displacement or screw type. The use of a booster should not be regarded as a substitute for the correct design and engineering of pipework within a site.

19.4 Energy conservation

Energy use can be minimized by a combination of various measures. These can be categorized as those reducing energy used and those recovering heat from a process.

19.4.1 Reduction of energy used

Air/gas ratio

Any process using a fossil fuel will involve the rejection of the products of combustion following heat transfer. These flue products will contain sensible heat that is lost and represents inefficiency in the process. Unless some form of recuperation is practiced, the flue products must be at a higher temperature than the process, and this cannot be reduced. The amount of excess air can, however, is controlled.

Figure 19.1 indicates the flue losses to be expected for different temperatures and excess air. It is seen that considerable savings can be made, particularly at higher temperatures, by reducing excess air levels to a practical minimum. It is also evident that a reduction in air/gas ratio to below stoichiometric will cause a rapid deterioration in efficiency caused by the energy remaining in the incomplete combustion of fuel.

The ideal air/gas ratio is that which is marginally higher than stoichiometric. It is not possible to run a burner with no excess air for various reasons (e.g. changing ambient temperature, a slight change in calorific value, variation in barometric pressure, wear of control equipment, etc.). All of these and other factors dictate that the burner is operated with sufficient excess air to avoid the production of carbon monoxide in any quantity.

Figure 19.2 shows how the production of CO can vary with excess air for two typical burners. It is seen that to limit CO to, say, 50 ppm with burner B, 3 per cent oxygen in the flue is needed, and with burner A, which exhibits better mixing characteristics, only 0.75 per cent excess oxygen is required. It is also seen that the 'heel' in the curve is more pronounced with burner A such that

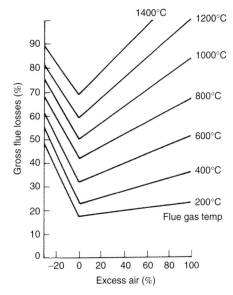

Figure 19.1 Gross flue losses versus excess air

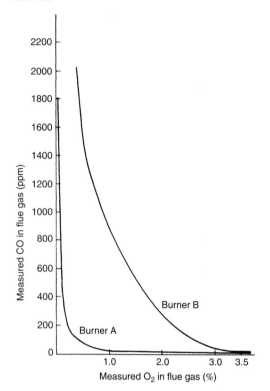

Figure 19.2 Measured percentage of O_2 in flue gas. Range of CO versus O_2 for a variety of burners

a reduction in oxygen below the heel will have a pronounced effect on CO production.

Figure 19.2 illustrates a burner at a fixed firing rate. In practice, many burners will have a varying firing rate with inferior performance at turndown, mainly because of poor mixing caused by reduced kinetic energy of both air and gas. To allow for this, the control system must provide for an increasing excess air with turndown.

To minimize flue losses it is important to keep excess air to a minimum, but the practicalities of the burner must be considered and a safe operating margin incorporated.

Maintenance

Combustion equipment can be set to give optimum efficiency at the time of commissioning but this condition will not be maintained. Wear and tear on control valves, partial blockage of filters, sooting of surfaces, etc. will all cause a fall in efficiency. To counter this, regular maintenance is desirable, and must include routine flue analysis and burner adjustment.

Insulation

Sensible heat losses from thermal plant should be kept to a realistic minimum by the use of correctly specified insulation. There will be a point beyond which further insulation is not economically viable. Careful analysis of the properties of insulation materials is necessary to prevent, for example, the adding of more insulation to the cold face of a furnace wall, causing the maximum service temperature of intermediate insulation to be exceeded.

Modern low-density insulation such as those based on ceramic fibers can be used to save energy in plant operating on a batch basis. The low thermal mass permits a rapid heating and cooling period that can save a substantial amount of energy. With continuously operating plant the advantages are not so pronounced.

Procedures

Energy can be conserved by operation of plant in such a way as to minimize part loading. Various practices can be adopted that can be described as 'good housekeeping'. In addition to maintenance, this will include such factors as the avoidance of plant operating in the standby mode for long periods, operation at correct temperature, ensuring doors are closed where applicable, etc.

For production plant, the energy used per unit produced is lowest when operated at design capacity. At low throughputs, the energy used increases markedly because the standing losses are constant, irrespective of throughput.

For plant such as boilers, operating in parallel controls should ensure that one boiler acts as a 'lead' to minimize part-load operation. As far, as is possible, plant should be sized to meet the load and oversizing should be avoided.

19.4.2 Heat recovery

General points

Technology exists to recover heat from processes operating at all temperatures, from regenerators on high-temperature plant to heat pumps using low-temperature effluent as a heat source. The problem in many cases is to find a use for the heat recovered. The best solution is to recycle the heat within the same plant, as the supply will always be matched to the demand. An alternative is to use the heat recovered in associated plant (for example, the heat recovered from a melting furnace can be used to dry feedstock for the furnace).

In general, 'high grade' recovered heat is more valuable than 'low grade'. The latter can often take the form of large quantities of warm water for which there is a finite need.

Recuperation

Figure 19.1 shows the heat carried away in flue gases, particularly for high-temperature processes. For example, for a furnace operating at 800°C with no excess, air losses are 40 per cent. Recovery of a proportion of these losses is possible by means of recuperation.

The simplest form of recuperation is load recuperation, but this is not suited to retrofit and is incorporated at the design stage. Flue products from the highest temperature zone are used to preheat incoming stock on the counter-flow heat exchanger principle. Such techniques are well established in pottery tunnel kilns, etc.

In many processes, load recuperation is not practicable, and combustion air is preheated in a heat exchanger by means of the outgoing flue products. Figure 19.3 gives an indication of the savings to be made for different operating temperatures. It is not normally considered economic to operate a recuperator at flue temperatures below about 750°C.

The recuperator can be positioned in the flue of the furnace or be integral with the burner (i.e. a recuperative burner). The separate recuperator is usually less costly, particularly on a multi-burner furnace. The recuperative burners avoid the need for lagging pipework, and reduce the risks of leakage upsetting the air/gas ratio. The flow pattern of hot gases in the furnace may, however, be disrupted, as the flue gases must exit through the burner.

Regenerators

Recuperators are limited in their performance, partly by problems with materials operating for long periods at elevated temperatures and by the efficiency of simple gas-to-gas heat exchangers. For high-temperature applications, a regenerator has advantages.

Regenerators have long been established in such processes as steel melting in open-hearth furnaces and glass-melting tanks. They consist of checkerwork brickwork, which act as a heat sink for the high-temperature flue gases. On reversing the cycle, this brickwork acts as a heat source for the incoming cold air. The regenerator is thermally efficient but only suited to very large, intensively used plant.

Regenerative burners have been developed to widen the range of application of the regenerator principle. They consist of a pair of gas burners each with its own regenerator consisting of a bed packed with refractory balls. Regular cycling between the two burners, only one of which fires at a time, gives high efficiency with up to 90 per cent of the heat in the flue products recoverable. Figure 19.4 shows typical performance figures for these burners.

Heat exchangers

There is a wide range of heat exchangers available to cater for most temperature ranges and for gas/gas, gas/liquid or liquid/liquid, as appropriate. High exhaust temperatures

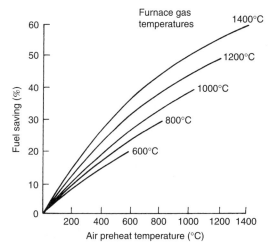

Figure 19.3 Fuel saving versus air preheat temperature (stoichiometric conditions)

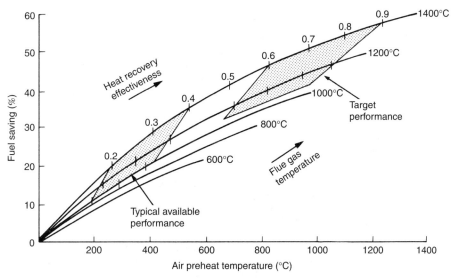

Figure 19.4 Fuel saving versus air preheat temperature

normally involve the use of a recuperator or regenerator and lower temperatures use other heat exchangers.

Gas/gas

Simple heat exchangers. These can be of the parallel flow, cross-flow or counter-flow pattern and constructed of materials to suit the temperature.

Rotary heat exchangers. These consist of a slowly revolving wheel of diameter 0.6–4 m driven by an electric motor. The wheel consists of a metallic matrix, which absorbs sensible heat (and, in some designs, latent heat) from the outgoing stream and transfers it to the in-going stream. With these exchangers 75–90 per cent efficiency is possible.

Heat pipes. The use of heat pipes involves the incoming cold air stream and the outgoing warm air stream being immediately adjacent and parallel, and between the two is a battery of heat pipes. These contain a liquid and operate on the thermal siphon principle. The liquid takes in latent heat and evaporates and the vapor travels to the cold end of the tube where condensation releases the latent heat. Generally, heat pipes are restricted to 400°C, and effectiveness can be up to 70 per cent.

Run-around coils. Where the incoming and outgoing air streams are remote it is necessary to use a run-around coil to couple them. A pumped liquid to a heat exchanger in the cold stream connects a heat exchange coil in the warm exhaust. The effectiveness can be up to 60 per cent.

Gas/liquid

Economizer. The economizer is a tubular heat exchanger used to recover heat from the exhaust gases from boilers or some processes. It is used in boilers to recover much of the sensible heat for use in preheating the boiler feedwater. An increase in boiler efficiency of 4–6 per cent is typical. The design and materials of construction depend on the application.

Condensing boiler. The efficiency of a hot water central heating boiler is limited in part by the latent heat of vaporization in the water vapor in the flue products. If a secondary heat exchanger is added to the boiler, it is possible to cool the flue gases sufficiently to release this latent heat, provided that the return temperature is sufficiently low (i.e. below about 55°C). This is the principle of the condensing boiler. The secondary heat exchanger must be constructed of a resistant material such as anodized aluminum or austenitic stainless steel to withstand the condensate, which is weakly acidic with a pH 3.3–3.8. The condensing boiler concept is not applicable to oil or coal, which normally have a more acidic condensate. Fuel savings of from 14 to 18 per cent can be achieved.

Spray recuperator. The principle of the spray recuperator is similar to that of the condensing boiler, and again is restricted to gas firing. In this case, the condensation takes place by direct contact of a water spray with the flue gases. The resultant water is at temperatures up to 50°C and is recycled through a heat exchanger. If a suitable use can be found for the low-grade heat (e.g. boiler feedwater preheating) then fuel savings of up to 17 per cent can be made.

Waste-heat boilers. Waste-heat boilers can be designed to accept any grade of waste heat to produce steam or hot water. Designs can be based on water-tube boilers, shell and tube boilers, or a combination of the two.

Heat pumps. Both the source and sink of heat pumps can be gas or liquid. The particular feature of the heat pump is that the source is at a lower temperature than the sink and is 'upgraded' by the heat pump. To obtain a reasonable efficiency it is essential that heat is required at a low temperature and the source and sink are close in temperature.

Liquid/liquid

Various designs of liquid/liquid heat exchanger are widely used. Choice is partly influenced by the cleanliness of the liquids and the need for regular cleaning. A compact and commonly used type is the plate heat exchanger.

19.5 Clean Air Acts associated with gas burning

The combustion of gas produces little in the way of noxious substances. Ideal combustion will produce only water vapor, carbon dioxide and nitrogen. In practice, there may well be very small amounts of hydrogen, carbon monoxide and unburned hydrocarbons, notably methane.

Relevant additional substances referred to in legislation include:

1. Sulfur oxides (SO_X) – the sulfur content in gas is very low (measured in ppm) and the SO_X is therefore negligible.
2. Nitrogen oxides (NO_X) – thermal NO_X is formed with gas as with other fuels, particularly if air preheat is practiced. There is very little fuel nitrogen compared with other fossil fuels so that the total NO_X emissions are lower.
3. Dust – as gas is in the gaseous state and does not combust via the route of droplets or particles the formation of dust is unlikely.

Nevertheless, gas is subject to legislation.

19.5.1 EC Directive on emissions from large combustion plant

This legislation was formally adopted on 26 November 1988. It applies to plant for the production of energy (effectively, boiler plant and similar) and specifically excludes heating furnaces, drying plant, etc. The Directive applies to plant of thermal input 50 MW and over. If two or more separate plants are in close proximity such that they could share a common flue then if their total input is 50 MW or over then they are also subject to the Directive.

The Directive applies to new plant (defined as that planned after 1 July 1987) with provision for retrofitting existing plant. Limits for new plant are:

Sulfur dioxide: $35 \, \text{mg/N m}^3$
NO_x: $350 \, \text{mg/N m}^3$
Dust: $5 \, \text{mg/N m}^3$

19.5.2 UK legislation

UK legislation referring to emissions from gas-fired plant is currently rather limited. The most important is The Health and Safety (Emissions into the Atmosphere) Regulations 1983 (SI No. 943, 1983). In Schedule 2 is listed substances deemed to be noxious, which include combustion products, dust, etc.

Schedule 1 lists those works covered by the regulations. The only categories which concern the combustion products of gas are:

1. Electricity works (excluding compression-ignition engines burning distillate fuel with a sulfur content of <1 per cent);
2. Boilers with an aggregate not less than 200 tons per hour of steam being used wholly or in part for electricity generation.

It is clear that very few gas-fired installations will fall within these categories. *Notes on the Best Practicable Means* for meeting the requirements were published by HM Inspectorate of Pollution in 1988.

For gas-fired plant large boilers and furnaces 50–700 MW thermal:

NO_X limit of 190 ppm v/v (measured at 6 per cent oxygen 15°C and 1 bar without correction for water vapor), (approx. $200 \, \text{mg/m}^3$);
Chimney efflux velocity not less than 15 m/s at MCR;
Temperature of gases entering chimney not less than 120°C.

For large boilers and furnaces >700 MW thermal:

Chimney efflux velocity not less than 18 m/s at MCR;
Temperature of gases entering chimney not less than 80°C.

(*NB*: There is no limit for NO_X for gas-fired plant. In practice, little or no such plant currently exists. The conditions for measurement of NO_X are clearly defined.)

19.6 Chimney requirements: codes of practice and environmental considerations

19.6.1 Functions of a flue

The overriding function of the flue is to remove the products of combustion. This involves the creation of sufficient draft, either mechanical or thermal, to move the flue products from the combustion chamber to the terminal of the flue. A secondary function is, by means of suitably locating the flue terminal, to ensure adequate dispersion of the products of combustion.

19.6.2 Operating principles

To operate effectively, the flue has to apply a pressure differential sufficient to overcome the system resistance and enable the products of combustion to flow from the combustion chamber to the terminal. This pressure differential can be mechanical (by forced or induced draft or a combination of the two) or thermal, possibly combined with mechanical.

The *natural-draft* flue operates on the thermal principle. The pressure differential is caused by the difference in density between the column of hot gases within the flue and a column of air of the same height. Within limits, the taller the flue, the greater the draft, but the upward movement is opposed by the resistance to flow inherent in the geometry and friction of the flue.

Successful flue design involves the balancing of the draft against the resistance, possibly for a range of thermal inputs, the avoidance of condensation, and a location of the terminal that ensures unrestricted dispersion of the flue products.

Factors affecting flue performance

Flue height   Raising the height of a flue increases the flue draft but also adds to the pressure loss due to friction. The net effect is that increasing height is very beneficial for short flues but has progressively less effect as the height is increased. With very tall flues, there can be a reduction in flow rate due to excessive heat loss. The point where an increase in height produces no additional flow rate depends on many factors. In the case of small commercial-size appliances, an increase beyond 6–9 m will normally produce no benefit.

Cross-sectional area   There is a direct relationship between flue flow and the cross-sectional area. The draft is unaffected by the cross-sectional area but the frictional losses decrease as the area is increased, resulting in greater flow. Too great an area, however, will lead to a low velocity with its attendant problems of down-wash and possible condensation.

Thermal input   The flue draft rises with temperature but the pressure losses increase because of the greater volume of the flue gases. The net result is that increased temperature leads to a greater flow rate, but only up to about 260°C, beyond which the flow is reduced.

If the heat input from a given appliance is increased, both the temperature and flow rate of the gases entering the flue will rise. The net result is that an increased heat input leads to a greater flow rate, but this is more noticeable at low heat inputs. There is a limit to the rate that can be vented into a given flue before the pressure losses overcome the flue draft and 'spillage' (failure of the flue products to be satisfactorily cleared) occurs.

Flue route   The only parts of the flue route that contribute to flue draft are the vertical sections. All sections, however, contribute to the pressure losses. It follows therefore that to increase the effectiveness of the flue it is important to adhere as closely as possible to a vertical route and to have a minimum of bends, changes of section, and horizontal runs.

Heat losses   Heat losses through the fabric of the flue will lower the mean temperature of the flue gases, causing both the flue draft and the pressure losses to be reduced. The net effect is quite small (perhaps a 10 per cent difference between a well-insulated and an unlagged flue).

It is, however, important to avoid condensation. Although the condensate from gas combustion is generally considered non-acidic, if the flue has not been designed for conditions of continual wetting condensation should be avoided. Intermittent condensation under start-up conditions can usually be tolerated.

The dewpoint for the flue products from gas is a maximum of 60°C and will, in most cases, be lower because of excess air. There is no acid dewpoint because the sulfur content is negligible. It would be thought that condensation would be most unlikely, and this is so for the bulk flue gases. However, the temperature gradient across the flue wall can be such that the 'skin temperature' at the inside wall can be considerably less than the bulk temperature, and condensation will take place.

To minimize the possibility of condensation it is desirable to:

1. Keep the flue as short as possible;
2. Avoid the use of bends and horizontal runs;
3. Insulate the flue;
4. Keep the surface area as small as possible.

These features must be consistent with the other design parameters as far as possible.

Termination   A suitable design and location of the flue terminal is essential for optimum performance. This is primarily to counter adverse wind effects.

The wind effect may assist or oppose the natural-draft flue, causing increased draft up the flue, or downdraft. For this reason, it is important to minimize their effects

by careful attention to the terminal position. In general, this means a position at least 0.25 m above any nearby obstacle.

For smaller appliances, flue terminals are necessary. These have the main functions of:

1. Keeping the resistance as low as possible;
2. Preventing blockage;
3. Being unaffected by wind of any speed or direction.

Proprietary terminals are available up to 200 mm, but above this are not used, a 'Chinese hat' terminal being considered adequate to keep out rain and larger foreign bodies. For large flues, particularly forced draft, terminals are not used.

Downdraft diverters   Any flue should be designed as far as possible to produce a constant draft. If the draft is insufficient then spillage of combustion products will take place, and if too great, then excess air will be induced, leading to an unnecessary loss of efficiency.

A typical downdraft diverter used with natural-draft flues is shown in Figure 19.5. This consists of a baffle plate mounted at the top of the primary flue. There are three modes of operation:

1. Normal conditions – the products of combustion mix with dilutant air at the diverter and diluted products pass up the secondary flue.
2. Adverse conditions with down draft – the down draft mixes with the products of combustion and diluted products enter the room for the brief period that the downdraft persists. The draft in the primary flue (and hence the performance of the appliance) is unaffected.
3. Adverse conditions with up draft – the additional draft in the secondary flue causes excess air to be pulled through the diverter from the room. The baffle plate prevents the draft in the primary flue from being affected.

For small appliances, the down draft diverter is often incorporated into them; for larger appliances, it is external.

In larger appliances, the use of a down draft diverter can be impractical if only because of size, and a draft stabilizer may then be used. This will protect against up

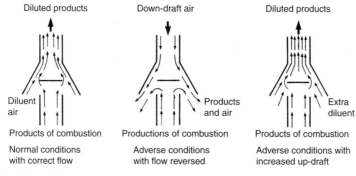

Figure 19.5 The function of the draft diverter

draft by opening to allow air to be pulled in directly from the room. If it is of the double-swing type it will protect against down draft by opening to relieve the down draft into the room.

Forced-draft flues The above design parameters are relevant to natural-draft flues. With forced-draft flues, it is possible by choice of a fan – either forced or induced draft – to overcome system resistance so that the flue will still clear the products. A crude rule-of-thumb is to allow 1 mm^2 of flue area for each 2.2–3.7 kW for natural draft and for 4.5–13.6 kW for each forced draft.

19.6.3 Design procedure

The design procedure will vary with the type of flue, but the basic procedure will not change. It is necessary to consider the relevant parameters discussed above and decide on flue dimensions and materials of construction to satisfy them. The parameters to consider are:

1. Type of flue to be used;
2. Pressure needed at inlet to flue for correct operation of appliance;
3. Diameter of flue (or dimensions if rectangular);
4. Length of flue and route taken;
5. Materials of construction (this will influence both heat transfer and flow characteristics);
6. Heat input and efficiency of the appliance, together with the flue temperature and excess air. A consideration of all or some of these allows the flue gas volume to be assessed;
7. Termination position.

This design procedure is detailed in the British Gas publication *Flues for Commercial and Industrial Gas Fired Boilers and Air Heaters* (IM/11). This publication addresses itself to:

1. Chimney heights;
2. Natural-draft flues;
3. Induced-draft flues;
4. Modular boiler systems;
5. Fan-diluted flues.

Other relevant aids to design include:

BS 5440: 1, Code of practice for flues and air supply for gas appliances of input not exceeding 60 kW: flues

This is only applicable to small appliances.

BS 5854, Code of practice for flues and flue structures in buildings

This applies to larger appliances of 45 kW output and above but does not cover flues that are not integral to a building.

19.6.4 New flue systems

Various flue systems have been developed to overcome the need for a conventional flue, the provision for which can be both difficult and expensive in some locations.

Balanced flue

In the balanced flue, which may be natural draft or fan-assisted, the flue outlet and combustion air inlet are integral in a housing to be mounted on an outside vertical wall. As the flue and the inlet are subject to the same outside pressure, they are unaffected by it and are in balance. No provision for ventilation is needed and the appliance is room-sealed.

Balanced flues are only available for low ratings (up to 60 kW input). Possible locations for the terminal are restricted as shown in Figure 19.6.

Balanced compartment

This can be regarded as an extension of the balanced flue principle to larger plant. A conventional chimney is used for clearing flue products, and the air inlet to the boiler house is integral with this so that the inlet and outlet are balanced. It is important for correct operation that there is no additional ventilation and that the room is sealed with all doors closed.

Fan-diluted flues

If the products of combustion can be diluted so that the carbon dioxide content is not greater than 1 per cent it is permissible to discharge them at ground level. This is the principle of the system shown in Figure 19.7, in which fresh air is drawn in to dilute the flue products which are discharged preferably on the same wall as the inlet to balance against wind effects. It is essential to interlock the airflow switch with the burner controls.

The volume flow rate of the fan (Q) is given by:

$$Q = 9.7 \times (\text{Rated input in kW of boiler})/3600 \, \text{m}^2/\text{s}$$

Additional points to note are:

1. The rating of the plant must not exceed 6 MW.
2. The emission velocity must be at least $75/F$ m/s. (F is the fan dilution factor, defined as $F = V/V_0$, where V is the actual flue gas volume and V_0 is the stoichiometric combustion volume, $0.26Q$ m^3/s for gas.)
3. The outlet must not be within $50U/F$ of a fan-assisted intake (except for intakes for combustion air or fan dilution air).
4. The outlet must not be within $20U/F$ of an operable window.
5. The distance to the nearest building must be at least $60U/F$.
6. The outlet must be at least 3 m above the ground, except if the input is less than 1 MW, where 2 m is permissible.
7. The outlet must be directed at an angle above the horizontal (preferably at 30°) and not under a canopy.

(a)

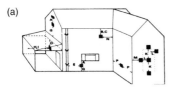

If appliance manufacturer's installation instructions do not give specific siting dimensions then the dimensions shown on this chart should be used.

TERMINAL POSITION	MINIMUM DISTANCE
A — Directly below an openable window or other opening e.g. air brick	300 mm
B — Below gutters, soil pipes or drain pipes.	200 mm
C — Below eaves.	200 mm
D — Below balconies or car port roof.	400 mm
E — From vertical drain pipes and soil pipes.	75 mm
G — Above ground, roof or balcony level.	300 mm
H — From a surface facing a terminal.	600 mm
I — From a terminal discharging towards another terminal.	600 mm
K — Vertically from terminal on the same wall.	1,500 mm
L — Horizontally from a terminal on the same wall.	300 mm
M* — From a single external corner.	100 mm
N* — From a single internal corner.	300 mm
P* — From double corners (both sides of the terminal).	500 mm

*Plan view of corners:-

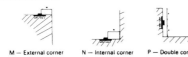

M — External corner   N — Internal corner   P — Double corners

(b)

If appliance manufacturer's installation instructions do not give specific siting dimensions then the dimensions shown on this chart should be used.

TERMINAL POSITION	MINIMUM DISTANCE	
	NATURAL DRAUGHT	FANNED DRAUGHT
A — Directly below an openable window or other opening e.g. air brick	300 mm	300 mm
B — Below gutters, soil pipes or drain pipes.	300 mm	75 mm
C — Below eaves.	300 mm	200 mm
D — Below balconies or car port roof.	600 mm	200 mm
E — From vertical drain pipes and soil pipes.	75 mm	75 mm
F — From internal or external corners.	600 mm	300 mm
G — Above ground, roof or balcony level.	300 mm	300 mm
H — From a surface facing a terminal.	600 mm	600 mm
I — From a terminal facing a terminal.	600 mm	1,200 mm
J — From an opening in the car port (e.g. door, window) into dwelling.	1,200 mm	1,200 mm
K — Vertically from a terminal on the same wall.	1,500 mm	1,500 mm
L — Horizontally from a terminal on the same wall.	300 mm	300 mm

Figure 19.6 Balanced flue terminal positions for appliances with a maximum heat input of (a) 3 kW and (b) 60 kW

8. The outlet should not discharge into an enclosed (or almost totally enclosed) 'well' or courtyard.

Flues for condensing systems

Boilers and air heaters can be designed to operate in the condensing mode, that is, the outlet temperature of the products entering the flue can be at or close to the dewpoint. The usual design criteria for flues which attempt to avoid condensation are no longer applicable and it is necessary to incorporate additional features. These will include:

1. Provision must be made to drain off and dispose of the condensate, which is produced at a rate of up to 0.15 l/h for every kilowatt of input. This is slightly acidic, and plastic is the most suitable for the pipework.
2. The flue will be wetted by the condensate, and should be constructed of a suitable material such as stainless steel, aluminum or suitable plastic.
3. All sockets in the flue should face upwards and the joints sealed.
4. The low temperature of the flue products will contribute little flue draft.
5. The flue gases will readily form a plume at the flue terminal.

Additional guidance in the design of flues for condensing systems is given in the British Gas publication *Installation Guide for High Efficiency (Condensing) Boilers – Industrial and Commercial Applications* (IM/22).

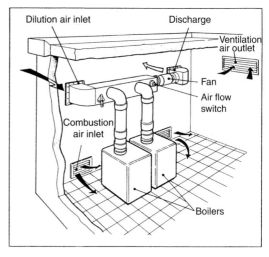

Figure 19.7 A typical arrangement of a fan-diluted flue system

19.6.5 Dual-fuel installations

In a dual-fuel installation, the alternative fuel will usually have higher sulfur content than gas, and it is the alternative fuel which will usually dictate the chimney height, materials of construction, etc. In a changeover situation, that is, where plant previously fired by another fuel is changed over to gas, chimneys or flues that have been satisfactory on oil or solid fuel will also be satisfactory on gas firing, unless condensing or direct contact boilers are to be fitted. The chimney should be checked to ensure that there is no restriction, that no leakage is occurring, and that on the previous fuel draft was adequate. In some cases it may be necessary to reduce an excessive draft with double-swing draft stabilizers, etc.

It is permissible to fire with other fuels into the same chimney provided that:

1. The other fuel is not solid fuel;
2. The common flue is of adequate size and construction;
3. The burner system of the gas- and oil-burning boilers are operating under similar draft conditions and preferably in the same room;
4. The burner-control equipment for both oil and gas is to the current standard for this type of equipment.

19.6.6 Flue dampers

The use of flue dampers may be considered as a means of energy conservation by preventing ventilation of the boiler in the non-firing state. There is a possible safety hazard if the damper remains closed when the burner ignites and fires. Safe operation with flue dampers is covered in the British Gas publication *Automatic Flue Dampers for Use with Gas Fired Space Heating and Water Heating* (IM/19). An assessment of the fuel savings to be expected from the use of flue dampers has been made by Dann, R. G., Lovelace, D. E. and Page, M. W., 'The effect of flue dampers on natural ventilation heat losses from boilers', *The Heating and Air Conditioning Journal* (May 1984).

19.7 Health and safety in the use of gas

19.7.1 Legislation

Health and Safety at Work etc. Act 1974

This has relevance to the supply of gas and plant burning gas, in particular section 6, as amended by the Consumer Protection Act 1987.

Gas Act 1986

This is the primary legislation concerning the gas industry in the UK, together with certain section of the Gas Act 1972 and the Gas Act 1965 which have not been repealed. The Gas Act does not have great relevance to safety aspects of gas utilization. However, specific regulations can be published under this enabling Act.

An exception is Schedule 5, Public Gas Supply Code, which deals with meter installation, use of anti-fluctuators and valves, reconnection of gas supplies, escapes of gas, etc.

Gas Safety (installation and use) Regulations 1984 (SI 1984, No. 1358)

These Regulations cover the safe use of gas in users' premises. They apply to domestic and commercial premises but not to industrial, although it is practice to apply the intent of the Regulations to such premises. The Regulations cover most aspects of installation and safety, including gas fittings, meters, installation pipes, gas appliances, ventilation and flueing. They use the term 'competent person' but do not define it. It is probable that in the near future the term will be defined as one who has undergone a training course that complies with the HSE Approved Code of Practice on Standards of Training in Safe Gas Installation.

Gas Safety Regulations 1972 (SI 1972, No. 1178)

These Regulations were largely repealed by the later 1984 Regulations. Certain sections were retained, however, notably Part II, which deals with service pipes (i.e. the pipes to carry gas from the main to the user's property).

Gas Quality Regulations 1983 (SI 1983, No. 363)

These Regulations concern the purity and distinctive smell of gas, the uniformity of calorific value, and the minimum pressure, which must be made available. For service pipes with an internal diameter of 50 mm or more, this is 12.5 mbar.

Gas (Meters) Regulations 1983 (SI 1983, No. 684)

These deal with the accuracy of meters together with fees for testing, etc.

Gas (Declaration of Calorific Value) Regulations 1972 (SI 1972, No. 1878)

These cover the declared calorific value and the methods of making known changes in its value.

Gas Safety (Rights of Entry) Regulations 1983 (SI 1983, No.1575)

These concern the rights of an 'officer authorized by the relevant authority' to enter premises for purposes connected with safety of fittings and appliances.

The Building Regulations 1985

Approved Document J, Section 2, deals with gas appliances of rated input up to 60 kW. It should therefore be noted that most commercial and industrial plant is not covered by these Regulations.

19.7.2 Potential hazards in the use of gas

Properties

Natural gas is a colorless gas to which has been added a distinctive odor. It is composed principally of methane (88–95 per cent) together with small proportions of higher hydrocarbons (ethane, propane, and butane), and nitrogen and/or carbon dioxide. It has a calorific value generally in the range 38–39 MJ/m^3 and is lighter than air with a specific gravity (with regard to air) of 0.59–0.64.

Health hazard data

Natural gas does not in itself constitute a hazard to health. However, exposure to concentrations in excess of approximately 10 per cent in air may cause dizziness and nausea. Higher concentrations may lead to asphyxiation.

Exposure to carbon monoxide resulting from inadequate ventilation and/or leakage of combustion products may cause headaches, chronic tiredness or muscular weakness. High concentrations or long-term exposure may be fatal. Normal resuscitation methods and medical advice should be sought for those suffering from these effects.

Safety hazard data

Natural gas forms flammable mixtures with air in the concentration range 5–15 per cent in air. In the event of a natural gas fire, steps should be taken to shut off the gas supply, and the local fire brigade and the gas supplier notified.

Handling

Natural gas must always be contained in appropriate pipes or vessels and precautions taken to ensure that leakage cannot occur. If a gas leak does occur, the main gas supply should be shut off, the area ventilated and the gas supplier informed. Electrical switches should not be turned on or off, portable electrical appliances including hand-held torches should not be operated and all other possible sources of ignition removed or rendered inoperable and the affected area ventilated. In circumstances of excessive leakage, the building should be evacuated.

Application

Appliances and other equipment intended to be used with natural gas should be purpose designed and built to recognized safety standards, installed in properly ventilated locations and supplied with the necessary flueing systems or other means of disposal of combustion products. Lack of adequate ventilation and/or leakage of combustion products into the workspace may give rise to carbon monoxide poisoning. In this event, affected personnel should be evacuated from the area, the gas supply shut off, the supplier informed and medical advice sought. Attention is drawn to the Gas Safety (Installation and Use) Regulations, relevant British Standards and appropriate British Gas Standards, codes of practice and guidance notes, all of which deal with the design, application and safe use of gas and gas-fired equipment.

19.7.3 Safety procedures

These have been partly covered in Section 19.7.2. All gas-fired equipment should be designed to ignite, operate and shut down in a safe manner. Instructions to this effect should be clearly displayed. This is a requirement of the Health and Safety at Work etc. Act 1974. In addition, such plant should comply with all relevant standards.

In the event of an incident occurring, it is essential in the first instance to shut off the gas supply. This is recognized in the Gas Safety (Installation and Use) Regulations by the need for an 'emergency control valve' sited close to the entry to a building. In smaller premises, this may be the meter control valve. In larger premises it is a requirement of the Regulations that for every building with a gas supply of 50 mm or more, and more than one self-contained area, then a valve must be installed at the position where the gas enters the building. In such premises, a line diagram of the pipe layout should also be prominently displayed, at least close to the primary meter. Such a diagram, indicating all emergency control valves and all pipework of internal diameter 25 mm or more, can be used to locate areas to be isolated in an emergency.

It is additionally recommended in the British Gas publication *Guidance Notes on the Installation of Gas Pipework, Boosters and Compressors in Customers' Premises* (IM/16) that for buildings containing plant over 2 MW total heat input and being supplied with gas at pressures above 1 bar, a remotely operable valve shall be fitted in the gas supply to the building. In the case of large boiler houses, provision for remote operation of the valves shall be provided both inside and outside the building.

19.7.4 Maintenance

Competent personnel in conjunction with the manufacturer's maintenance instructions improve safe operation of all plant by means of regular maintenance. Re-commissioning in a safe manner should always follow this. It is particularly important that valves such as safety shut-off valves, non-return valves, etc. are regularly checked for soundness.

19.8 Pressure control

19.8.1 Governors

Constant-pressure governors are required at various stages of the gas supply within the user's premises from the first pressure reduction from distribution pressure to supply pressure at the meter installation to the appliance governor. The complexity and design of the governor installation depends on pressure, throughput, duty, etc. In addition to governors, pressure-relief, slam-shut and non-return valves may be relevant in some installations.

Low-pressure governors used for smaller appliances and smaller pressure-regulation installations should comply with BS 6448: 1, Specification for pressure governors with nominal connection size up to 50 mm for gas appliances with inlet pressures up to and including 200 mbar. For larger installations, but still low pressures, governors should comply with BS 3554: 2, Specification for gas governors: independent governors for inlet pressures up to 350 mbar.

Pressure-regulating installations

Detailed requirements for pressure-regulating stations are contained in two Institution of Gas Engineering publications:

TD/9: Offtakes and Pressure-regulating Installations for Inlet Pressures between 7 and 190 bar

TD/19: Pressure-regulating Installations for Inlet Pressures between 75 mbar and 7 bar

These references give details of the requirements for safely maintaining a constant pressure outlet over a wide range of flow rates. It should be noted that 'governor' and 'regulator' are largely interchangeable terms.

Requirements become more demanding as the pressure at the inlet rises. A governor alone has its characteristics which include a pressure 'droop' at high throughputs and an increase above set point at very low throughputs. In many cases, there must additionally be protection against the failure of a governor. This can be provided by:

1. 'Active and monitor' governors (defined as two governors in series whose settings are stepped so as to allow the active governor to control the outlet pressure and the monitor governor to assume control in the event of failure of the active governor to the open position);
2. Pressure-relief valve (this will vent to atmosphere when a predetermined maximum pressure is attained);
3. Slam-shut valve, this will close quickly in the event of an abnormal pressure (usually excess) being detected.

The essential requirements for lower-pressure installations can be summarized as below. In most cases, the governor installation is incorporated with the metering installation.

1. *Inlet pressure not exceeding 75 mbar.* The only requirement is for a single governor and a single stream, although twin streams will be advantageous for larger loads.
2. *Inlet pressures 75 mbar to 2 bar.* A single or multi-stream installation including:
   (a) Single-stage governor with internal valves open at rest;
   (b) Relief valve with capacity not more than 1 per cent of the stream fault capacity;
   (c) Slam-shut valve.
3. *Inlet pressures 2–7 bar, governor of 50 mm bore or less and stream fault flow rates less than 300 m³/h.* As in (2) above *or* duplicate streams, each having two-stage governing and a slam-shut valve;
   (a) A monitor override pilot governor to be fitted to the first-stage governor impulsed from the outlet of the second-stage governor;
   (b) A relief valve of capacity not more than 1 per cent the fault stream capacity to be included in each stream.
4. *Inlet pressures 2–7 bar, larger flow rates.* Duplicate streams with two-stage governing as in (3) above *or* duplicate streams each having monitor and active governors and a slam-shut valve;
   (a) The monitor governor to be upstream of the active governor;
   (b) A relief valve of capacity not more than 1 per cent, the fault stream capacity to be included in each stream.

Legal requirements

The Gas Safety (Installation and Use) Regulations require that the gas supply to the meter be governed in all premises other than mines, quarries and factories (Regulation 13). It is practice to govern all premises, but exceptions may be made in categories where it is permissible, and there is advantage in so doing.

19.8.2 Pressure-relief valve

The pressure-relief valve has the physical appearance of a pressure governor. It operates in the reverse manner to a governor, that is, on exceeding the set pressure, the valve lifts off its seat, rather than being forced onto its seat, as in a governor. Gas is therefore allowed to flow through the valve.

The main function of the relief valve is to act as a back up to a pressure governor to prevent the pre-set downstream pressure being exceeded. In particular, it will relieve the small amount of 'creep' that will occur by slippage past the valve seating in a governor.

Safe venting of the gas relieved is of great importance. The British Gas publication *Guidance Notes on the Installation of Gas Pipework, Boosters and Compressors in Customers' Premises* (IM/16) gives guidance on this. The main requirements are:

1. Vents must be terminated in a safe place, preferably in the open air above roof level, remote from possible sources of ignition.
2. Vents must not be manifolded together.
3. Terminals must be designed to minimize the risk of blockage and ingress of water.
4. The vent pipe, particularly if over 20 m in length, must be designed to minimize back pressure. Certain pressure governors have integral vent valves.

19.8.3 Slam-shut valves

The slam-shut valve cuts off the gas supply in the event of predetermined pressure criteria being exceeded. It must be manually reset, having made safe the downstream pipework and the cause of the abnormal pressure removed.

It consists of a spring-loaded cut-off valve which is held open in the open position under normal conditions. Should the pressure deviate sufficiently, the diaphragm will move and disengage the trigger mechanism, releasing the latch on the valve, which will close under the force of the spring.

Slam-shut valves are usually set to operate under conditions of high downstream pressure. They can be used for low pressures, and combined high and low. Although they normally protect the supply pressure to premises using gas, slam-shut valves are sometimes used on process plant.

19.8.4 Non-return valves

It is a requirement of the Gas Act 1986, Schedule 5, that where 'air at high pressure' or 'any gaseous substance not supplied by the supplier' is used, it is necessary to install an 'appliance' to prevent the admission of that gas into the supplier's gas main. This requirement is usually met by means of a non-return valve installed at the meter outlet, where it will protect the supplier's mains. There are advantages to the user in having the non-return valve installed close to the point of use so that, in the event of a gas at higher pressure entering the gas supply, the non-return valve will check it at an early stage rather than diffusing throughout the user's installation pipework. A non-return valve should meet the requirements of British Gas *Standard for Non-Return Valves* (IM/14).

It is important to note that the requirements of this standard are specific to gas, and include a low-pressure differential under forward-flow conditions. It is unlikely that a non-return valve suitable for compressed air, etc. will be acceptable for gas.

Reverse-pressure requirements specified in IM/14 are:

Up to 25 mm:	7 bar
25–150 mm:	2 bar
Over 150 mm:	1 bar

In the particular case of oxygen, valves shall be resistant to exposure for up to 12 h at all pressures up to 2 bar at 20°C. Requirements for non-return valves for oxygen are also discussed in the British Gas publication *Guidance Notes on the Use of Oxygen in Industrial Gas Fired Plant and Working Flame Burners* (IM/1).

It is essential for the safe operation of a non-return valve that it be regularly checked and be shown to be resistant to reverse pressures. To enable this check to be carried out, the non-return valve should be installed with a manual isolation valve at both inlet and outlet.

19.9 Gas specification and analysis

Natural gas as distributed in the UK is obtained from various sources. These comprise primarily the southern North Sea basin, northern North Sea fields (both British and Norwegian), Morecambe Bay from the Irish Sea (used primarily for winter peaks) and gas from world sources imported in small quantities as liquefied natural gas (LNG). Gas from the different fields is of very consistent quality, and further blending, conditioning, etc. allows a gas of very consistent quality and specification to be distributed.

19.9.1 Calorific Value

The calorific value is constantly monitored, and it is a condition of the Gas (Declaration of Calorific Value) Regulations 1972 that alterations in the declared calorific value (CV) are publicly made known. It is customary to quote the gross (upper) CV rather than the net (lower) CV. The difference between the two represents the heat contained in the latent heat of vaporization of the water vapor in the products of combustion that can only be recovered in condensing appliances.

The ratio of net/gross CV is about 90 per cent, reflecting the high hydrocarbon/carbon ratio for gas. The range of

declared CV (gross) is $38–39\,\text{MJ/m}^3$ with a typical value of $38.63\,\text{MJ/m}^3$. This latter figure represents a net CV of $34.88\,\text{MJ/m}^3$.

19.9.2 Wobbe number

The Wobbe number of gas is defined as 'The heat release when a gas is burned at a constant gas supply pressure'. It is represented by:

$$\text{Wobbe no.} = \frac{\text{Gross CV}}{\text{Square root of relative density of gas}}$$

The relative density (with respect to air) has no units, so that the units of the Wobbe number are the same as calorific value. The value for gas is $49.79\,\text{MJ/m}^3$.

The Wobbe number is of interest because it represents the heat released at a jet. A gas of varying CV but constant Wobbe number would give a constant heat release rate.

19.9.3 Analysis of natural gas

The analyses of various gases from the North Sea are shown in Table 19.1. It is seen that the analysis is very constant, with methane being the dominant constituent at 91–96 per cent.

19.9.4 Properties of natural gas

The main properties of gas and its combustion products are shown in Table 19.2. For comparison, the properties of propane and butane are included. It can be seen that the calorific values of these three fuels on a mass basis are very similar.

The properties of natural gas are dominated by those of methane, notably a low maximum flame speed of 0.33 m/s. This strongly influences burner design, which must ensure that the mixture velocity is sufficiently low to prevent 'blow-off'. 'Light-back', on the contrary, is very unlikely with such a low flame speed.

The range of satisfactory operation for a gas burner, defined by light-back, blow-off and incomplete combustion is limited. The variation in gas analyses, particularly higher hydrocarbons and inerts, can influence the range of operation. This has led to the definition of different groups of natural gas. A practical effect is that burners designed for the European continent may not be suitable for the UK without adjustment. This does not apply to forced-draft burners.

19.9.5 Standard reference conditions

If properties are being accurately quoted it is important that the reference conditions are defined, as these can vary. This also has implications for metering. The main reference conditions are:

1. *Metric standard reference conditions*
   15.0°C, 1913.25 mbar, dry
   Symbols: MSC, m^3 (st)

Table 19.1 Analysis of natural gas

	Mean North Sea	Bacton	Easington	Theddlethorpe	St Fergus
Composition (vol per cent)					
Nitrogen (N_2)	2.72	1.78	1.56	2.53	0.47
Helium (He)	–	0.05	–	–	–
Carbon dioxide (CO_2)	0.15	0.13	0.54	0.52	0.32
Methane (CH_4)	92.21	93.63	93.53	91.43	95.28
Ethane (C_2H_6)	3.6	3.25	3.36	4.10	3.71
Propane (C_3H_8)	0.9	0.69	0.70	0.99	0.18
Butanes (C_4H_{10})	0.25	0.27	0.24	0.33	0.03
Pentanes (C_5H_{12})	0.07	0.09	0.07	0.10	0.01
Higher hydrocarbons	0.1	0.11	–	–	–
Composition (wt per cent)					
Nitrogen (N_2)	4.38	2.90	2.55	4.05	0.79
Helium (He)	–	0.003	–	–	–
Carbon dioxide (CO_2)	0.38	0.33	1.39	1.31	0.84
Methane (CH_4)	84.89	87.37	87.28	83.62	91.10
Ethane (C_2H_6)	6.21	5.69	5.88	7.03	6.65
Propane (C_3H_8)	2.28	1.59	1.80	2.49	0.47
Butanes (C_410)	0.83	0.91	0.81	1.09	0.11
Pentanes (C_5H_{12})	0.29	0.38	0.29	0.41	0.04
Higher hydrocarbons	0.74	0.82	–	–	–
Ultimate composition (wt per cent)					
Carbon	72.16	73.23	72.93	71.98	74.38
Hydrogen	23.18	23.62	23.51	23.02	24.21
Nitrogen	4.38	2.91	2.55	4.05	0.79
Oxygen	0.28	0.24	1.01	0.95	0.61

2. *Imperial standard conditions*
   60.0°F, 30 inHg, saturated with water vapor
   Symbols: ISC, standard cubic foot, sft^3 (src in old references)
   SI equivalent: 15.56°C, 1913.7405 mbar, saturated
3. *Normal temperature and pressure*
   0°C, 760 mmHg (at 0°C and $g = 9.80665$ m/s^2)
   Symbols: nm^3, N m^3, m^3 (n), STP, NTP
   SI equivalent: 0°C, 1913.25 mbar
   Unless otherwise specified, the gas may be assumed to be dry.
4. *US gas industry 'standard cubic foot'*
   60°F, 30 in. Hg (at 32°F, $g = 32.174$ ft/s^2) saturated
   SI equivalent: 15.56°C, 1915.92 mbar, saturated
   Symbols: scf, sft^3

19.10 Control of efficiency

19.10.1 Monitoring of combustion

Reference to Figure 19.1 shows how efficiency can be adversely affected by deviation from the optimum air/gas ratio. By maintaining combustion close to stoichiometric, efficiency will be improved, but the practical limitations of burners discussed above must be noted.

The quality of combustion can be measured with suitable instrumentation, on either a periodic or a continual basis. If continuous analysis is practiced then there may be feedback to continuously adjust the air/gas ratio and/or ratio and/or record the data derived.

Flue gases analyzed

The flue gases analyzed will be one or more of carbon dioxide, carbon monoxide and oxygen. If carbon dioxide alone is measured, it is possible to draw erroneous conclusions, as the level will peak at stoichiometric and reduce in both the excess air and air deficiency regions. It is essential to measure another flue gas to obtain a reliable assessment of burner performance.

Oxygen does not have this disadvantage, being, in theory, zero at stoichiometry and increasing in the excess air region. Sampling of carbon monoxide additionally is advocated during initial commissioning, if not always on a routine basis.

Carbon monoxide is usually sampled as the second parameter in conjunction with carbon dioxide or oxygen. In theory, as the optimum is usually to have near-stoichiometric combustion without 'CO breakthrough' it is the most reliable gas to sample. A problem is that although small quantities of CO usually indicate the need for additional air, they can also be caused by flame chilling and careful interpretation of results is needed.

Wet and dry analysis

The relatively high hydrogen content of gas, contained in the methane, leads to a water vapor content of approximately 18 per cent by volume in the flue products. The analysis of the other constituent gases is affected by whether or not the water vapor is included.

Table 19.2 Properties of gaseous fuels and combustion products

Properties at 1013.25 mbar (15°C)	Mean North Sea	Typical commercial propane	Typical commercial butane
Specific gravity (air = 1)	0.602	1.523	1.941
Gross calorific value			
MJ m^{-3} (st)	38.63	93.87	117.75
Btu ft^{-3} (ISC) dry	1 036	2 500	3 270
MJ kg^{-1}	52.41	50.22	49.41
Btu lb^{-1}	22 530	21 500	21 200
Therms ton^{-1}	505	480	477
Btu gall^{-1}	–	110 040	124 352
Net calorific value			
MJ m^{-3} (st)	34.88	86.43	108.69
Btu ft^{-3} (ISC) dry	935	2 310	3 030
MJ kg^{-1}	47.32	46.24	45.61
Btu lb^{-1}	20 340	19 800	19 650
Liquid SG (water = 1) at 60°F	–	0.51	0.575
Volume gas/volume liquid (at 0°C)	–	274	233
Wobbe No.			
MJ m^{-3} (st)	49.79	76.06	84.52
Btu ft^{-3} (ISC) dry	1 335	2 026	2 347
Theoretical air required for combustion			
(v/v)	9.76	23.76	29.92
(w/w)	16.5	15.6	15.3
m^3 (st) kg^{-1} of fuel	13.24	12.73	12.50
f^3 (st) lb^{-1} of fuel	212.0	203.92	200.23
Theoretical stoichiometric dry flue gas CO_2%	11.86	13.8	14.1
Volumetric composition of wet theoretical flue gas			
% CO_2	9.66	11.7	12.0
% H_2O	18.63	15.4	14.9
% N_2	71.71	72.9	73.1
Dewpoint of flue gas (°C)	59	55	54
Theoretical flame temperature (°C)	1 930	2 000	2 000
Limits of flammability (% by volume to form combustible mixture)	5–15	2–10	1.8–9

NB: Calorific Value (saturated) = 0.9826 × Calorific Value (dm).

The 'dry' analysis is on the basis of the water being removed from the sample prior to analysis, and the maximum theoretical carbon dioxide content is 11.87 per cent. Most flue analyses are carried out on the dry basis.

The 'wet' analysis assumes that the water vapor is present. The maximum theoretical carbon dioxide content is 9.66 per cent. The zirconia cell method of measuring oxygen is on the wet basis.

Instruments used

A wide range of analysis instruments, either of the portable or permanently installed type, can be used. The latter will frequently be recording instruments and may have control capabilities. Various principles are employed in analysis equipment, including:

1. Wet chemical analysis;
2. Electrolytic cell;
3. Chromatography;
4. Paramagnetism – for oxygen;
5. Non-dispersive infrared absorption;
6. Thermal conductivity.

19.10.2 Control of combustion

In the great majority of installations, flue gas analysis is carried out for monitoring reasons only. In some larger installations, primarily boilers, a deviation in the reading of the gas measured is used for control purposes. The safety aspects of such system must be satisfactory, in particular that the controller cannot erroneously drive the burner to fire rich under any circumstances. The controller should be capable of being set to follow the optimum curve for excess air with firing load, which recognizes that more excess air is needed at lower firing rates, as shown in Figure 19.8.

Much responsibility rests with the commissioning engineer who must have an intimate knowledge of the burner, the boiler, and the combination of the two. A decision to use an accurate method of combustion control will, in general, be based on economics. The fuel saving is unlikely to be more than about 2 per cent and this must be balanced against the total costs of operating the control system. In the case of larger boilers, such an installation can often be justified.

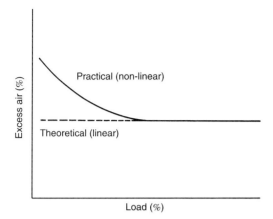

Figure 19.8 Variation of excess air with load

Oxygen trim systems

Most oxygen trim systems interpose an additional link in the air/gas ratio controller. Others use an additional valve. Most types are based on the zirconia cell installed in the flue, while others use paramagnetic or electrolytic cell methods. The zirconia type has the advantage that there is no time lag in sampling, nor is there a risk of contamination of the sample.

If the flue is operating under negative pressure, which is often the case, care should be taken that no 'tramp' air is allowed to enter the flue upstream of the sampling point, as this will give erroneous measurements. The trim system should be set to follow the practical firing curve and, in the event of a malfunction, be 'disabled' so that it ceases to have any influence on the air/gas ratio, which reverts to the normal load control only.

Carbon monoxide trim systems

There is limited application for CO trim systems which are widely used on utility and other large water-tube boilers. The principle of operation is for an infrared beam to traverse the flue from emitter to sensor. The absorption of the infrared radiation is proportional to the CO content.

For gas-fired shell boilers it is difficult to justify these trim controllers on an economic basis. Equally important, the position to control based on CO in such a boiler is ideally in the reversal chamber and not the flue. However, the temperature and stratification of flue products here make it impracticable.

Control without trim systems

Most combustion equipment is not controlled by means of a feedback from flue gas analysis but is preset at the time of commissioning and preferably checked and reset at intervals as part of a planned maintenance schedule. It is difficult to set the burner for optimum efficiency at all firing rates and some compromise is necessary, depending on the control valves used and the control mode (e.g. on/off, fully modulating, etc.).

In the absence of a trim system with feedback, it is likely to prove cost-effective to use a simple portable 'efficiency monitor' on a regular basis, perhaps weekly for small boilers. The change of reading is as important as the actual value of the reading. A deviation from what is known to be a good post-commissioning setting will indicate a drift from ratio and the need for remedial action.

Types of air/gas ratio control   There are various types of air/gas ratio device commonly used, including:

1. *Linked butterfly valves.* This simple technique is only suited to high/low controls.
2. *Linked square port valve.* This system has a more linear relationship between flow rate and rotation than butterfly valves. There is no backlash, as the air and gas valves are on a common spindle and fine-tuning is possible.
3. *Linked characterized valves.* The relationship between angular rotation of the valve spindle and open area of the valve can be adjusted over different portions of the flow range using a series of screws. In this way, the air/gas ratio can be characterized to any desired profile over the whole firing range. These valves are suited to fully modulating systems and are commonly used on steam boilers.
4. *Balanced pressure-control systems.* Various control systems are based on the balanced pressure principle in which the air pressure (or a portion of it) is applied to the diaphragm of a zero governor in the gas supply. As the air pressure is varied, so is the gas pressure in proportion, as shown in Figure 19.9. The air/gas ratio is set by means of an adjustable orifice valve in the gas supply. The pressure divider system will maintain air/gas ratio over much of the turndown ratio. At very low firing rates there is a need for excess air to maintain satisfactory combustion, and this is achieved by means of tensioning the spring within the governor to give an offset (Figure 19.9). The effect of this offset is considerable at very low rates but insignificant at higher ones. The pressure divider principle is exploited in more complex multi-diaphragm controllers. In these, the pressures balanced are differential pressures across orifices, enabling the ratio to be maintained under conditions of varying back-pressure.
5. *Electronic ratio controller.* In this type of controller, a proportion of both gas and air is diverted through a by-pass in which a thermistor sensor measures the flow. The air and gas flows can be compared and the ratio calculated and displayed. A ratio control valve in the air or gas supply, depending on whether the mode of operation is gas- or air-led, will automatically restore a deviation from the pre-set ratio. The electronic controller maintains ratio over a 19 : 1 turndown. The principle of operation is based on mass flow, so that it can be used with preheated air in recuperative systems.

19.10.3 Process controls

The choice of the mode of control of heating plant can have a considerable effect on the efficiency of the process. The process itself may dictate the controls. For example,

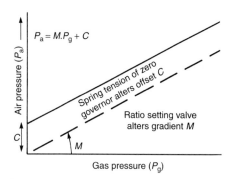

Figure 19.9 Relationship between air and gas pressures for cross-loaded governor system

a simple high/low temperature controller is unlikely to satisfy a process tolerance requirement of within a few degrees Centigrade. Similarly, a more complex process controller cannot be coupled with a burner only capable of firing high/low.

Types of process control

High/low control   This may be on/off in certain applications. Control is crude, with the burner moving to high fire at a point somewhere below the nominal set point and returning to low fire somewhere above it. The result is a sine wave control pattern with a band rather than a point of control. This is, however, still an adequate control mode for many purposes.

Most burners are efficient at high fire but less so at intermediate rates and particularly at low fire. An on/off burner is therefore apparently efficient from an energy-utilization viewpoint. However, when the burner is called on to fire, in the case of forced-draft burners a purge is usually necessary which will both cool down the process and cause a delay in response, and in the case of natural draft there will be heat losses due to ventilation in the 'off' period.

In general, a fully modulating burner will be thermally more efficient, particularly if it maintains air/gas ratio accurately at intermediate rates. For natural-draft burners this may not be the case, as, in general, the air is not modulated, leading to progressively more excess air at lower rates.

Proportional-only control   This avoids the cycling effect of high/low control by setting the burner rate in proportion to the deviation from the set point (i.e. the burner rate decreases as the set point is approached). In theory, this meets requirements and avoids overshoot. However, the set point is approached but not reached, leaving an offset. Using a high gain in the control can minimize this, but this will lead to excessive cycling or 'hunting'.

Integral mode   This improves on the proportional-only control by repeating the proportional action within a unit time while a deviation from set point exists. The regulating unit is only allowed to be at rest when set point and measured point are coincident. Integral action will move the regulating unit until the desired and measured values are coincident, even if this means moving the regulating unit to the fully open or closed position.

Derivative mode   This improves on the proportional-only control by responding solely to the rate of change of the deviation but not in any way to the actual value of the deviation. Derivative action is always used with proportional control.

Proportional plus integral plus derivative action   Proportional action provides a controller output proportional to the error signal. Integral action supplies a controller output which changes in the direction to reduce a constant error. Derivative action provides a controller output determined by the direction and rate of change of the deviation. When all these are combined into one controller (three-term or PID), there is an automatic control facility to correct any process changes.

Controls for space heating in buildings

The selection of controls for space heating a building shows how the overall efficiency of the process can be improved. Whether or not it is cost effective to do this must be examined in detail. At its most basic, such a system might include one or more LPHW boilers, a single thermostat in the heated space controlling the circulating pump on/off, and domestic hot water from a calorifier.

Boiler selection   Although not directly concerned with controls, the sizing of boilers is very important. There is a tendency for oversizing with consequent unnecessary cycling. Controls should ensure that if more than one boiler is installed then one boiler should act as 'lead'. Consideration should be given to a higher-efficiency boiler, possibly a condensing type.

In most cases, splitting the heating and domestic hot water loads will be advantageous. This avoids the inefficient use of a large boiler solely to heat a calorifier in the summer months.

Controls selection   The selection of controls will be influenced to an extent by the characteristics of the building, particularly the insulation material and the number and disposition of windows. Control options to be considered include:

1. *Zone control valves.* In most buildings, different areas will be subject to varying levels of solar gain, occupancy, etc. Dividing the building into different zones each with its own thermostat and control valve will be beneficial.
2. *Compensating controller.* The heat requirements for a building vary according to the outside conditions, particularly the outside air temperature. The compensating controller measures the outside air temperature and, together with other parameters, including the thermal response of the building, adjusts the circulating water temperature to meet the demand.

3. *Three-port mixing valves*. The valve takes hot water from the boiler and mixes it with the cooler return water to provide a circulation of mixed water of sufficiently high temperature to meet the heating demand. The system operates with a constant flow rate and variable temperature. In the case of condensing boilers, the return water should pass through the condensing heat exchange of the boiler prior to flowing to the mixing valve. An alternative for condensing boilers that does not use a mixing valve is to set the boiler flow temperature, by means of an outside air temperature sensor and compensating control box, in inverse proportion to the outside air temperature. Both these alternatives cause the boiler to operate in its efficient condensing mode for the maximum period.

4. *Optimum start controller*. Any building will heat up more quickly on a warm day: to start the heating system unnecessarily early is uneconomic. The optimum start controller monitors the outside and inside temperatures and computes the start time that will just bring the temperature up to design by the start of occupancy. The controller is only used during the preheat period during which other modulating controls are not operative. Some versions are available with a self-learning capability, which corrects the initial program until it matches the building responses, and an optimum off control, which takes account of the allowable drop in spacc temperature at the end of occupancy.

5. *Building management systems*. These can monitor and control most of the building services by means of sensors in the various building service plants for collecting data and carrying out control functions.

19.11 Automation

19.11.1 Requirements for automatic burners

Process or heating plant may have controls ranging from manual operation with some processes supervised by interlocks to semi-automatic and fully automatic operation. Deciding factors will include temperature of operation, frequency of ignition, degree of operator supervision, and rating of the plant. For example, boiler plant, both steam and water, will invariably have automatic control whereas tunnel kilns operated continuously at high temperature are unlikely to require it.

The definition of an automatic burner is: 'A burner where, when starting from the completely shut-down condition, the start-gas flame is established and the main gas safety shut-off valves are activated without manual intervention.' This means that a burner is only automatic if it is ignited by means of a remote interlock (e.g. thermostat, timeswitch, etc.) closing. A burner is not automatic if it has a pilot burner that remains ignited in the 'off' condition. Nor is a burner strictly automatic if a start button needs to be pushed, even though the controls may comply with all requirements for automatic burners.

19.11.2 Automatic burner sequence

The sequence of operation for automatic burners is based on that which a knowledgeable, conscientious and alert

operator would perform. This involves checks at stages to ensure that no hazard has developed. This sequence can be summarized as:

1. Check that the flame detector (ultraviolet or flame rectification) is not giving a spurious signal. This is continuous throughout the purge period.
2. Check that the safety shut-off valves are in the closed position (where applicable).
3. Check that the airflow detector is not giving a false indication.
4. Start the combustion air fan.
5. Prove the combustion air flow (at high rate, if variable).
6. Purge for a predetermined period (sufficient to give five volume changes of the combustion chamber and associated flue ways).
7. Modulate the combustion airflow to the low fire rate (where applicable).
8. Prove the burner throughput control valves to be in the low fire position (where applicable).
9. Energize the ignition source.
10. Energize the start gas safety shut-off valve(s).
11. De-energize the ignition source.
12. Check that the flame detector is registering the presence of a flame.
13. Energize the main gas safety shut-off valves.
14. In the case of an interrupted pilot, de-energize the safety shut-off valves.
15. Check that the flame detector is registering the presence of a flame. This is continuous until the burner is shut down.
16. Release the burner controls to modulation so that the burner can be driven to high fire, according to demand.

In the event of any of the checks not being satisfactory the burner is shut down or locked out, as appropriate.

19.11.3 Burner standards and codes of practice

Codes of Practice

Gas burners should comply with the relevant Codes of Practice, depending on whether the plant is low or high temperature. Low-temperature plant is defined as that having a normal working temperature insufficient to ignite the fuel, that is, below 750°C at the working temperature walls. The British Gas Codes of Practice are:

Code of Practice for the Use of Gas in High Temperature Plant (IM/12)
Code of Practice for the Use of Gas in Low Temperature Plant (IM/18)

Automatic Burner Standards

British Standards exist for automatic burners, i.e. those requiring no manual intervention to be ignited. The Standard is:

BS 5885, Automatic gas burners
Part 1. Specification for burners with input rating 60 kW and above (1988)

Part 1. Specification for packaged burners with input rating 7.5 kW up to but excluding 60 kW (1987)
Part 2 deals only with packaged burners while Part 1 covers all types.

The number and quality of the safety shut-off valves required is specified in the relevant Standards and Codes of Practice. These are usually electrically operated valves conforming to class 1 or class 2 as defined in BS 5963, Specification for electrically-operated automatic gas shut-off valves (1981). This specifies forward pressures and reverse pressures that the valve must withstand in the closed position together with closing times.

For start gas supplies two safety shut-off valves are needed on all but the smallest burners. For automatic burners the main gas safety shut-off valve requirements are:

From 60 kW up to 600 kW:	One class 1 and one class 2
Above 600 kW up to 1 MW:	Two class 1
Above 1 MW up to 3 MW:	Two class 1 with a system check
Above 3 MW:	Two class 1 with proving system

Closed-position indicators on the valves usually satisfy a 'system check'. These show that the valves are nominally in the closed position but do not actually prove that there is no leakage. A 'proving system' can be met by various systems which prove the soundness of the upstream and downstream safety shut-off valves by sequential pressurization, evacuation, mechanical over-travel with limit switches and a normally open vent valve, or other means.

Valves other than electrically operated (e.g. pneumatically operated ball valves) are equally acceptable if the pressure and closing time requirements of BS 5963 are met.

19.11.4 Standards to be applied

Automatic or semi-automatic

The amount of process plant that can be defined accurately as automatic is relatively small, and manual intervention is often involved at some stage. The relevant design criteria are therefore often IM/12 or IM/18. In practice, fully automatic burner controllers tested and certified by British Gas are available that comply with the requirements of BS 5885. Although these have features which may not be applicable to non-automatic plant, it may be more appropriate to use such a controller, particularly as its safety is well proven. It may also be less expensive than buying and installing separate timers, relays, etc. For some processes (for example, those that do not need and cannot tolerate a long purge) such controllers may not be appropriate.

Appliance Standards

In most cases, Codes of Practice or British Standards should be applied as appropriate. A few industrial and commercial gas appliances have relevant British Standards and these should take precedence over codes that are more general and standards. These include:

BS 5978, Safety and performance of gas-fired hot water boilers (60 kW to 2 MW input) (1983)
BS 5990, Specification for direct gas fired forced convection air heaters for space heating (60 kW to 2 MW input) (1981)
BS 5991, Specification for indirect gas fired forced convection air heaters for space heating (60 kW to 2 MW input) (1983)

19.12 Fire and explosion hazards

19.12.1 Limits of flammability

Gas will form a flammable and explosive mixture with air or oxygen and the band within which this will happen will vary with pressure and temperature. At atmospheric pressure and 15°C the properties are:

	Lower explosive limit (LEL) (% gas)	Upper explosive limit (HEL) (% gas)
In air	5	15
In oxygen	5	61

In the combustion chamber, it is essential to purge down to below the LEL before initiating the ignition source. The usual criterion is 25 per cent LEL.

Outside the combustion chamber, there should be no gas under any circumstances. Such a presence will indicate an uncontrolled leakage into the building. A gas level of 20 per cent LEL is normally sufficient grounds for evacuating the building.

The human nose is often considered the best gas sensor, as the odor contained in the gas permits detection at levels as low as 190 ppm or 0.01 per cent in air. Gas detectors are, however, sometimes used to give additional warnings of any leaks. The following points should be borne in mind when considering such an application:

1. Location is of the greatest importance. It is not known from where a gas leak will emanate and such a leak will disperse rapidly. In general, a high-level location should be chosen, as gas is lighter than air. More than one detector will usually be advisable.
2. Detectors should comply with BS 6020, Instruments for the detection of combustible gas, specifically Parts 4 and 5.
3. Detectors may respond to other gases and vapors giving a false reading.
4. Detectors will require frequent maintenance and calibration to perform accurately.
5. The use of gas detectors should not be regarded as a substitute for proper ventilation.

19.12.2 Flame speed and flame traps

The flame speed of gas in air is 0.338 m/s. This is lower than most common flammable gases and is an indication of the low propensity for light-back of natural gas.

In certain applications a flame trap may be required, and further guidance is given in HSE publication HS(G)11,

Flame Arrestors and Explosion Reliefs. The flame trap is constructed of strips of corrugated and flat steel ribbon wound in a spiral. It breaks up a flame in the very small passages between the corrugations where the flame is quenched and extinguished.

A flame trap is employed where premixed air and gas is used in combustion equipment and prevents the flame passing upstream into the pipe system. Flame traps should be situated as near as possible to the gas burner. This is so that the flame does not have a long pipe run in which it might accelerate to such a speed as to form a detonation wave and make the trap useless.

Thermocouple elements can be incorporated into the flame trap so that a flame that has lit back to the flame trap element and continued to burn there without being quenched can be detected. This sensor can be used to close an upstream safety shut-off valve. Flame traps will be the cause of a significant pressure drop for which allowances must be made in low-pressure systems.

Typical applications for flame traps include:

1. The outlet pipework from premix machines, but not fan-type mixers as they may adversely affect the operation of such machines;
2. Vents from pressure-relief systems and governor vents;
3. Purge outlet pipes.

19.12.3 Fire valves

Unlike oil burners, there is no requirement to install a fire valve. With oil burners, this is usually met by having a fusible link sited over the burner and connected in a tensioned wire, which holds open a weighted valve in the oil supply so that the valve closes if the link melts.

The use of automatically closing valves operated from gas-, fire- or smoke-detection systems is not normally required for gas installations. The reasons for this are:

1. There are very few recorded instances of fire caused by gas in installations such as boiler houses;
2. A leak in the oil supply will cause oil to collect on the floor fuelling a fire in a predictable location, most probably at the burner itself. With gas, a leak is possible at any joint, but such a leak will rapidly disperse. A predictable location is not possible, and a fire is not likely.
3. Gas can feed a fire caused by other sources. The integrity of pipework should withstand an external fire for sufficient time for an external emergency control valve to be operated.
4. Gas does not have the lubricant qualities of fuel oil. Consequently, whereas a drop-weight valve in the oil supply can be expected to close when called upon, this is not the case with a relatively large-bore gas ball valve or plug valve used for gas.

However, if it is desired to protect a self-contained area such as a boiler house against fire the best method is to use suitably located fusible links as interlocks in the controls of the burners, designed to BS 5885. The burner valves should be to BS 5963 and mounted in a non-vulnerable position.

If an automatic isolation valve is specified, the selection of the valve and its operating system must be carefully considered, particularly with respect to the design and methods of restoring the gas supply in those cases where appliances do not incorporate automatic flame safeguards. Where possible, valves should be to BS 5963 and systems in compliance with the British Gas publication *Weep By-Pass Pressure Proving Systems* (IM/20).

19.12.4 Explosion reliefs

HSE publication HS(G) 16, *Evaporating and Other Ovens*, gives guidance on the use of explosion reliefs in plant where flammable solvent is present as a potential source of explosion. Consideration should also be given to utilizing explosion reliefs in other plant, particularly low-temperature ones and where recirculation is practiced. Correctly designed controls will minimize the risk of an explosion but explosion reliefs will reduce to a minimum adverse effect from such an incident. Further guidance on the use of explosion reliefs is given in the HSE publication HS(G) 11, *Flame Arrestors and Explosion Reliefs.*

The main requirements of a relief are:

1. If obstructions exist in the plant the relief should be located in a plane which allows an unobstructed flow of gases to the relief vent (e.g. if there are horizontal shelves the relief should be in a side, not the roof).
2. The relief size should be as large as possible, ideally occupying the whole of one side, and minimum vent area $= 60 \times$ area of side$/P_{max} (m^2)$ (where P_{max} is the pressure that the plant can withstand without damage – if unknown, assume 140 mbar).
3. The relief should be as light as possible (the weight per unit area should not exceed 24 kg/m^3).
4. The relief should be held in position with the minimum of force.
5. There must be sufficient space around the plant to allow gases to vent freely and not impair the effectiveness of the relief.
6. Reliefs must not form dangerous missiles when called upon to operate.
7. The thermal resistance of the structure should not be significantly altered by the relief.

19.13 Maintenance

19.13.1 Need for maintenance

It is desirable that plant is regularly maintained to ensure that:

1. Thermal efficiency remains at an optimum.
2. Availability and reliability is at a maximum, with no avoidable unplanned downtime.
3. Safety devices are reliable.

The time intervals for maintenance will obviously vary widely. In all cases, the work should be part of a clearly determined schedule.

19.13.2 Thermal efficiency

Combustion equipment, when first commissioned, can be set to operate at its optimum efficiency. With time, however, there will be a deterioration due to blockage of air filters and breather holes, wear in valve linkages, etc. Such changes may have safety implications if gas-rich firing is a consequence.

Regular checking of measured parameters and comparison with commissioning data will enable adjustments to be made to optimize efficiency. Replacement of items subject to wear, filters, etc. and items subject to thermal distortion such as some burner components will be necessary.

19.13.3 Reliability

To ensure that plant is not subject to breakdown, it is important that there are no unnecessary 'failures to safety', that is, that correctly operating safety equipment only operates because of exceptional circumstances, not avoidable faults. Safety shutdown will occur for various reasons, i.e.:

1. No flame or shrinkage of flame during operation;
2. Failure to ignite flame initially;
3. Failure of any interlock, leading to (1) or (2).

Preventative maintenance that can be carried out that will minimize the chance of nuisance shutdowns includes:

1. Attention to air filters;
2. Cleaning of all burner ports liable to blockage;
3. Replacement of mechanical and electrical components known to have a finite life. For example, the average service life of an ultraviolet flame detector may be 19,000 h, and much less if used above its recommended temperature.

After any maintenance work involving replacement or adjustment of components it is essential that the plant be re-commissioned in a safe way, including 'dry runs' with the gas turned off to ensure that flame failure devices operate correctly and pilot turndown tests to ensure that if the pilot can energize the flame detector then it can also smoothly ignite the main burner.

19.13.4 Safety

When initially commissioned all safety devices should be proved to be operating correctly. At intervals, these should be checked to ensure that no undetected failure has taken place. Checks that are necessary include:

1. Visual examination of ignition of pilot burner and main burner;
2. All interlocks checked for correct setting and operation. These will include thermostats, pressurestats, limit switches, pressure switches, process interlocks, etc.;
3. Condition of flame sensor and its location with relation to pilot burner and main burner;
4. Soundness of non-return valve with regard to reverse flow;

5. Soundness of safety shut-off valves in the closed position for both forward and reverse flow. Note that this should be done with the valve *in situ* and without dismantling it, by means of a bubble leak detector, etc. The valve should not be dismantled to check for swarf on the seat, etc.

After modifying the controls in any way during maintenance, it is essential to carry out a re-commissioning of the plant as outlined in Section 19.13.3 above.

19.13.5 Training of maintenance staff

All staff involved with maintaining gas-fired equipment should be capable of doing so in a safe and responsible manner. The term 'competent' has not yet been defined in this context, but personnel should be qualified by both training and experience to carry out work on any plant which they are to maintain.

A code of practice outlining the training program that should be undergone by persons carrying out work on gas-fired plant is Approved Code of Practice on Standards of Training in Safe Gas Installation. The Code offers guidance on the standards of training to produce competence in gas installation.

19.13.6 Manufacturer's instructions

For much plant the manufacturers will provide detailed maintenance schedules which should be followed at all times. This may necessitate the use of spares and tools specific to that appliance. Attention is drawn to Regulation 25(7) of the Gas Safety (Installation and Use) Regulations, which states that 'No person shall carry out any work in relation to a gas appliance which bears an indication that it conforms to a type approved by any person as complying with safety standards in such a manner that the appliance ceases to comply with those standards'.

19.14 Statutory requirements

The most important statutory requirements concerning gas distribution and utilization are:

Health and Safety at Work etc. Act 1974 Gas Act 1986
Gas Act 1972
Gas Act 1965
Gas Safety (Installation and Use) Regulations 1984 (SI 1984, No. 1358)
Gas Safety Regulations 1972 (SI 1972, No. 1178)
The Building Regulations 1985
Gas Safety (Rights of Entry) Regulations 1983 (SI 1983, No. 1575)
Gas Quality Regulations 1983 (SI 1983, No. 363)
Gas (Meters) Regulations 1983 (SI 1983, No. 684)
Gas (Declaration of Calorific Value) Regulations 1972 (SI 1972, No. 1878)
Oil and Gas (Enterprise) Act 1982
Control of Pollution Act 1974
Health and Safety (Emissions into the Atmosphere) Regulations 1983 (SI 943 of 1983)
Consumer Protection Act 1987
Reporting of Injuries, Diseases and Dangerous Occurrences Regulations 1985 (SI 2023 of 1985)

19.15 Testing

19.15.1 Soundness testing

It is essential that gas is not admitted to an installation pipe which is newly installed or has had work carried out on it unless the pipe is proved to be sound. This is a requirement of the Gas Safety (Installation and Use) Regulations, specifically Regulation 21. A suitable procedure is set out in the British Gas publication *Soundness Testing Procedures for Industrial and Commercial Gas Installations* (IM/5). This recognizes that there is a difference between new installations, which must be sound within practical limitations of measurement, and existing installations, which may have inherent small leakages at, for example, valve glands. It is important that any paint which may provide temporary seals to leaks is applied after the testing procedure.

An allowance has to be made for temperature, as an increase in temperature will cause a rise in pressure of an enclosed volume of pipework. In practice, this will mean avoiding the effects of direct sunlight and allowing an adequate time for stabilization.

Soundness testing procedure

For both new and existing installation the soundness testing procedure would normally consist of:

1. Estimation of the system volume;
2. Establishment of the test procedure;
3. Selection of a pressure gauge. The sensitivity of the pressure gauge will determine the minimum leak rate detectable;
4. Determination of the permitted leak;
5. Determination of test period;
6. Carrying out the test.

New installations and new extensions

1. There should be no perceptible drop in pressure as shown by the test gauge, although a small leakage rate is defined which allows for testing technique limitations.
2. Air or inert gas should be used and not fuel gas.
3. The test pressure should be 50 mbar or 1.5 times the working pressure or the maximum pressure likely to occur under fault conditions, whichever is the greater. It is assumed that the pipework has been designed and installed to withstand the test pressure to which it is to be subjected.
4. Decide on the permitted leak rate. This is taken to be 0.0014 dm³/h.
5. Calculate the test period. On large installations it may be necessary to section the system.

Existing installations

1. Existing installations may be tested with fuel gas or inert gas. Air may also be used subject to correct purging procedures.
2. The test pressure is as with new installations.

3. Existing installations may not meet the requirements laid down for new installations because of minor leaks that have developed. It is necessary to relate such leak rates to the pipe environment.
4. Permitted leak rates are determined by installation location:
   (a) Potentially hazardous areas: no leakage is permitted; all joints in the area must be tested.
   (b) Occupied areas: permitted leakage rate is 0.75 dm³/h per cubic meter of space. But in no case shall the permitted leak rate be taken as greater than 30 dm³/h.
   (c) Large open work areas and exposed pipes outside: leak rates in excess of those in (b) may be acceptable from the safety point of view.
   (d) Buried pipes: The measured leak rate must be assessed in the light of ground conditions. In particular, special consideration shall be given to the proximity of cellars, ducts, cable runs, sewers, etc.

19.15.2 Purging procedures

After testing for soundness it will be necessary to safely introduce gas into the pipework displacing the air or inert gas that is in it. Similarly, if pipework is decommissioned for any reason fuel gas must be displaced by air or inert gas. This is a requirement of the Gas Safety (Installation and Use) Regulations, Regulation 21. Guidance on recommended procedures is given in the British Gas publication *Purging Procedures for Non-Domestic Gas Installations* (IM/2).

Purging can be defined as:

1. The displacement of air or inert gas by fuel gas;
2. The displacement of fuel gas by air or inert gas;
3. The displacement of one fuel gas by another fuel gas.

Important requirements for purging include:

1. A full knowledge of the pipework layout;
2. Adequate provision of vent points and their termination in a safe location;
3. Preferably a written procedure, appropriate to the installation, should be prepared and followed;
4. Ignition sources should be prohibited from within 3 m of vents and test points.

Purge volumes

For pipework the purge volume may be taken as 1.5 times the pipework volume. For diaphragm meters the purge volume may be taken as 5 revolutions of the meter. For other meters the purge volume may be taken as 1.5 times the volume of a length of pipe equal to a flange-to-flange dimension of the meter.

Purging methods

Purging may be by either the direct or slug method, and may involve the use of an intermediate inert gas purge. At all stages, the vent point is sampled and the purge is deemed satisfactory when certain criteria are met.

Direct purge This refers to the displacement of an air/gas mixture with fuel gas. The purge is complete with:

Oxygen less than 4 per cent
Combustibles levels greater than 90 per cent

Inert purging This refers to the displacement of fuel gas or air with inert gas (complete displacement) or the formation of a barrier of inert gas (less than 4 per cent oxygen) between fuel gas and air during purging (slug purging):

Gas to inert, purge complete with: combustibles levels less than 7 per cent.
Inert to air, purge complete with: oxygen level greater than 16 per cent.
Air to inert, purge complete with: oxygen level less than 4 per cent.
Inert to gas, purge complete with: combustible levels greater than 90 per cent.

Slug purging Slug purging is appropriate only for long pipe runs. It should not be used to purge installations with branches unless each one can be (and is) valved off and purged separately.

Air purging This refers to the displacement of gas or gas/air mixture with air. The purge is complete with:

Oxygen level greater than 20.5 per cent
Combustibles levels less than 40 per cent LEL.

Fuel gas to fuel gas purging The purge may be carried out by burning the gas under supervision on an appliance or appliances. If the two fuel gases have different burning characteristics and Wobbe numbers, correct burner selection is important.

19.15.3 Commissioning

After soundness testing and purging, a competent person must commission the item of plant. It is a requirement of the Gas Safety (Installation and Use) Regulations, Regulation 33, that an appliance is fully commissioned at the time that gas is made available to it or that it is isolated in such a manner that it cannot be used.

Commissioning sequence

Commissioning procedures will be specific to particular plant, but the basic procedure is general to all. There are four consecutive stages:

1. *Inspection period*. All relevant documents, the installation and its components are visually and physically examined, with all energy sources isolated. The commissioning program is finalized.
2. *Activation period*. This is in two parts, the dry run and the live run, the change between the two being when fuel is made available to the combustion space. The activation period commences with all energy sources isolated from the plant. These are gradually made available as the dry run progresses. Fuel must be

isolated from the combustion space until the live run commences.
3. *Operation period*. Final adjustments are made and checks carried out to ensure satisfactory operation of the plant, including acceptance trials.
4. *Completion period*. Operating staff are instructed. A report including details of operating levels is prepared and the user takes direct control of the plant.

19.5.4 *In-situ* testing

Regular *in-situ* testing of various parameters can be an aid to maintaining plant in an optimum condition regarding both efficiency and safety. Commissioning data, which should be on record, must be compared with measurements made on a regular basis for such parameters as gas and air pressures, pressure switch settings, limit switch settings, timer settings, flue gas analyses, etc. Adjustments should be made as necessary.

The period between such tests will depend on the maintenance schedule for the plant, which should also include the *in-situ* soundness testing of non-return valves and safety shut-off valves by bubble leak detectors, etc.

19.16 The gas grid system and distribution networks

19.16.1 Storage of gas

The rate of use of gas varies considerably both on a seasonal and on a diurnal basis. The gas supplier must ensure that the supply meets the demand under all foreseeable conditions. To an extent, this is achieved by the rates at which gas is taken into the NTS, which is determined by contracts specifying daily consumption throughout the year. Additional fluctuations must be taken up by storage.

Methods of storage

Low-pressure holders These are the conventional cylindrical multi-stage gas holders. They operate at low pressure and are usually water-sealed. The storage capacity is low for the large physical size and fewer of these holders are used now as their storage capacity can be met by other means.

High-pressure holders A relatively small number of high-pressure holders are in use. These 'bullets' operate at pressures of up to 30 bar and consequently have a higher storage capacity for their size.

Liquefied natural gas LNG is stored at certain locations as a means of 'peak-shaving'. Advantages are that the liquid gas occupies a volume equal to 600 volumes of gas under standard conditions. The main disadvantage is that it is necessary to lower the temperature to $-162°C$ to liquefy the gas. It is also necessary to maintain the gas at this low temperature, usually in insulated aluminum vessels.

Salt cavities Where the geological formation is favorable, it is possible to leach out salt to create underground

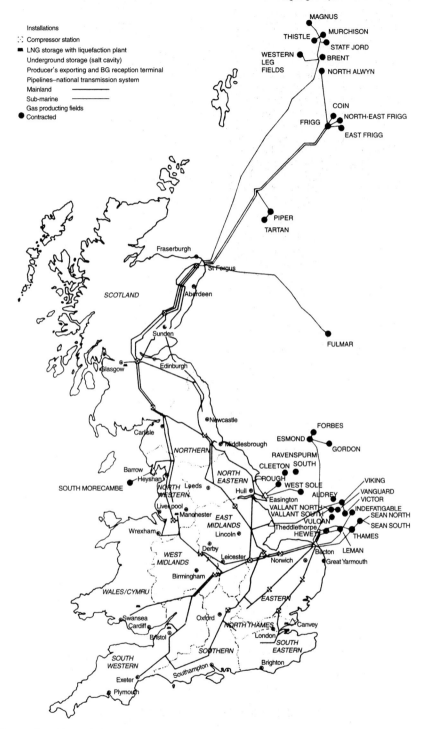

Figure 19.10 Gas supply and transmission

impermeable cavities in which gas can be stored at high pressure. These can be likened to large underground high-pressure holders.

Rough storage   The Rough gas field is a largely depleted field in the southern North Sea basin, which is now used as a storage facility. Gas from other fields is re-injected into the Rough field during the summer period of low demand and drawn on during periods of peak demand.

Morecambe Bay   The Morecambe Bay field is a recently developed gas field in the Irish Sea connected to the NTS by means of a gas pipeline to Barrow terminal. While it is possible to operate this field throughout the year, it is currently being drawn on only in the winter months as a form of peak-shaving.

Interruptibles   As outlined in Section 19.1.4, some users are offered interruptible contracts. A condition of these is that the user must provide a standby fuel and that, when requested, he or she must switch to this fuel. Under peak demand conditions, the large quantity of gas that would be burnt (typically, for steam raising) is available for use by firm contract and tariff users.

Line packing   The NTS, together with the regional transmission systems, constitute a considerable length of pipework operating at pressures of up to 70 bar. Design and minimum pressures set maximum pressures by operational constraints. Between these two, the pressure can be permitted to fluctuate.

When it is considered that the volume of 1 km of 900 mm pipe is 636 dm^3, and if the pressure is allowed to fall from 42 bar to 19 bar, the equivalent volume of gas at standard conditions is 14,628 dm^3, then this should be compared to the capacity of a low-pressure gas holder of typically 28,320 dm^3 (1 million ft^3). Two kilometers of pipeline operated under these conditions is therefore the equivalent of one gas holder.

Line packing is generally practiced by building up the pressure during the night and reducing it during the day.

19.16.2 Distribution networks

The NTS is used for the conveyance of gas from the terminals to regions sited in different parts of the country. Regional transmission systems are used to convey gas across each region. For local conveyance of gas to the final points of use, the distribution system is used.

The primary difference between transmission and distribution systems is one of pressure. Transmission systems operate at high pressures (>7 bar) while distribution systems operate at low (<75 mbar) and medium (75 mbar to 7 bar) pressures. The functions are also different in that the transmission system is used to convey gas over distances and store it, while the distribution system is used to convey gas to the user over a local network. The pipe that conveys gas from the main of the distribution system to the meter control valve of the user is the service.

Operating pressures

Distribution systems can be operated at pressures up to 7 bar but it is practice to operate at as low a pressure as is consistent with supplying users with the agreed supply pressure, which is 21 mbar at the meter inlet in many cases. Pressure reduction takes place at take-offs from the transmission system, at 'district' governor installations serving a large number of users within a geographical area and, for many larger users, at the users' premises.

Pressure-regulating installations will incorporate regulators, pressure-relief valves, slam-shut valves, etc. as necessary and as outlined above. They should comply with the requirements of the Institution of Gas Engineers recommendations:

TD/9:   *Offtakes and Pressure-Regulating Installations for Inlet Pressures between 7 and 190 bar*

TD/19:   *Pressure-Regulating Installations for Inlet Pressures between 75 mbar and 7 bar*

Pipework design criteria

1. Transmission pipelines operating at high pressures are always constructed of welded steel. Recommendations are contained in the Institution of Gas Engineers publication TD/1, *Steel Pipelines for High Pressure Gas Transmission*.
2. Distribution mains may be constructed of steel, ductile iron, polyethylene depending on operating pressure, etc. Recommendations are contained in the Institution of Gas Engineers publication TD/3, *Distribution Mains*.
3. Services may be constructed of steel, ductile iron or polyethylene depending on operating pressure, etc. Recommendations are given in the Institution of Gas Engineers publication TD/4, *Gas Services*.

Requirements for gas services are also covered by the Gas Safety Regulations 1972, Part 2.

19.17 Emergency procedures

Emergency procedures will be called upon following any abnormal occurrence such as an explosion, fire, release of unburned gas or production and release of toxic products of combustion. It is essential to carry out an emergency shutdown procedure in a safe way that prevents the hazard continuing.

19.17.1 Emergency control valve

It is a requirement of the Gas Safety (Installation and Use) Regulations that an emergency control valve is provided to which there is adequate access (Regulation 8). In addition, it is stipulated (Regulation 23) that for large users such a valve must be employed for every separate self-contained unit where the gas service is of 50 mm or more and that a line diagram be provided to show their location. By means of such emergency control valves, it is possible to safely isolate the gas supply to any building in the event of an emergency.

In addition to the use of such manual valves, automatically operated valves may be considered as an additional safeguard in certain installations. These can be connected to (among other interlocks) panic buttons sited within the building, which they are protecting. This is a recommended requirement for buildings containing plant of over 2 MW total heat input and supplied with gas at above 1 bar.

19.17.2 Normal shutdown procedures

All gas-fired plant should be provided with operating instructions for shutting down as well as starting up. Such instructions will ensure that the correct sequence of operations is carried out to both avoid a hazard during the shutdown and to leave the plant in a safe condition. These procedures should also contain instructions for actions to be carried out in emergencies. Such actions may differ from the normal shutdown.

19.17.3 Instruction and training

It is essential that relevant staff receive adequate training for the start-up and shutdown of plant for which they are responsible. This will include emergency procedures.

19.18 Pipework

19.18.1 Design criteria

Gas pipework in a user's premises serves the function of transporting the gas from the meter to the point of use in a safe way and without incurring an avoidable pressure loss. For low-pressure installations, the permitted pressure loss is only 1 mbar from the meter to the plant manual isolating valve at maximum flow rate. The pipework must be sized adequately to allow for this. Boosters are sometimes used to overcome pressure losses, but the use of a booster should never be considered a satisfactory substitute for correct design of pipe sizes. Where gas is available at higher pressures it may be permissible to tolerate pressure losses of more than 1 mbar.

Guidance on pipework in users' premises is given in the British Gas publication *Guidance Notes on the Installation of Gas Pipework, Boosters and Compressors in Customers' Premises* (IM/16). Certain Institution of Gas Engineers publications are also of relevance, particularly TD/3, *Distribution Mains* and TD/4, *Gas Services*.

19.18.2 Materials

Materials used for gas pipework will depend on application, location, environment and operating pressure. General guidelines are as below.

Polyethylene

This is to be used only for buried pipework or in ducts, as it is adversely affected by ultraviolet radiation and its use inside buildings is prohibited by the Gas Safety (Installation and Use) Regulations. For pressures up to 4 bar.

Copper

Copper may be used above ground and for buried pipework at pressures up to 75 mbar and outside diameters of 67 mm. If utilized for buried pipework it must be factory sheathed and should not be attached to buried steel pipe and fittings.

Ductile iron

This may be used for buried pipework, in ducts and above ground at pressures up to 2 bar. Not to be used for plant pipework. Ductile iron pipe and fittings shall be to BS 4772, Specifications for ductile iron pipes and fittings.

Steel

Steel may be used in all applications. The working pressure is related to the specification and grade used. However, pipe to BS 1387 heavy (up to 150 mm) and to BS 3601 (150 mm and above) is suitable for applications up to 7 bar.

19.18.3 Jointing

The method of jointing will depend on the material, the pipe size, the operating pressure and the location. General guidelines are given below.

Spigot and socket connections

This is to be used for ductile iron in all applications and must be of a type suitable for rubber jointing rings to BS 2494. Where the working pressure is over 75 mbar, the connections shall be anchored by means of purpose-designed anchor blocks or self-anchoring joints.

Fusion welding

This method to be used for polyethylene pipes only by companies and persons specializing in this field.

Capillary joints

These can be used for copper in all applications and must be to BS 864, Part 2, or BS 2051, Part 1. In ducts, all such joints must be accessible within the duct or chase. Not to be used within vertical ducts in high-rise buildings.

Compression joints

These can be used for copper in all applications where the joints are accessible and are not therefore recommended for underground use. In ducts, all such joints must be accessible within the duct or chase. Not to be used within vertical ducts in high-rise buildings.

For polyethylene pipe, compression couplings may be used provided they comply with British Gas Specification PS/PL3.

Screwed and welded jointing for steel pipework

The type of jointing may be restricted by the grade of pipe used. For example, some grades of steel pipe are not suitable for the application of screwed threads to BS 21 or ISO R7. The use of welded pipework is based on a combination of pipe size and pressure. Up to 25 mm screwed connections are permitted at all pressures up to 5 bar, whereas above 190 mm pipe must be welded at all pressures.

Other points to consider include:

1. Where pipework is welded, the number of flanged joints should be kept to a minimum and such flanges shall be welded to the pipe.
2. Where pipework is welded, the minimum standard shall be: up to 2 bar: BS 2971 or BS 2640; above 2 bar up to 5 bar: BS 2971 with 10 per cent of welds subjected to non-destructive testing, or BS 2640 with inspection and 10 per cent of welds subjected to non-destructive testing.
3. For screwed connections, jointing materials meeting the following specifications shall be used:
   BS 1560, 1832, 2815
   BS 5292 – Type A up to 0.7 bar or Type C
   with the exception of vulcanized fiber and rubber reinforced jointing, which shall not be used.

With screwed joints of 50 mm and above it is recognized that, in addition to a jointing paste, the use of hemp may be necessary to ensure a sound joint. However, hemp should never be used in conjunction with PTFE tape.

19.18.4 Pipes in ducts

1. It is not permissible to install an installation pipe in an unventilated shaft, void or duct (Regulation 18(5) of the Gas Safety (Installation and Use) Regulations).
2. Adequate ventilation is provided by means of openings at the top of high points and the bottom of low points. These openings shall have a free open area of 1/150 of the cross-section area of the duct or 0.05 m², whichever is the greater.
3. For ducts smaller than 0.05 m², the openings shall be not less than 3000 mm².
4. In exceptional circumstances if it is not possible to ventilate the duct then:
   (a) The pipe shall be continuously sleeved through the duct with the sleeve ventilated at one or both ends into a ventilated area; or
   (b) The unventilated duct or void shall be filled with a crushed inert infill to reduce to a minimum the volume of any gas, which may accumulate. Suitable material is crushed slate chippings or dry washed sand.

19.18.5 Pipe supports

Pipe in whatever plane shall be adequately supported. Reference should be made to the relevant parts of BS 3974, Part 1. In some instances, short vertical lengths in otherwise horizontal pipe runs may be self-supporting.

Pipe supports shall be provided throughout the length of the pipe and in such a way as to allow thermal movement without causing damage to any corrosion protection that may be on the pipe.

19.18.6 Corrosion protection and identification

Buried pipework must be protected against accidental and physical damage from sharp material, etc. and chemical action from corrosive soils, etc. It must be protected against corrosion by means of wrapping, catholic protection, etc. for metal pipes. Above-ground pipework should be protected with suitable paint after preparation.

Pipework should be easily identifiable in accordance with BS 1719. Where the normal pressure in the pipe exceeds 75 mbar, consideration should be given to labeling the pipe with the normal operating pressure. In buildings in which there are no other piped flammable gas supplies it is sufficient to paint the pipe yellow ochre or to band it with appropriately colored adhesive tape. In large complex installations, it is desirable to identify pipe contents more precisely, and in those instances, the base color should be supplemented with a secondary code band of yellow and/or its name or chemical symbol.

19.18.7 Flexible connections and tubes

In all instances where it is known or expected that pipework will be subjected to vibration, expansion or strain, the use of flexible connections should be considered. However, flexible connections should not be used where there is a practical alternative.

Attention is drawn to Regulation 4(2) of the Gas Safety (Installation and Use) Regulations, which prohibits the use of non-metallic materials for gas pipework other than for small portable appliances such as Bunsen burners and lighting torches.

Types of flexible connections

Various types of flexible connections are available for different purposes. Careful consideration is needed for suitability for purpose. General comments on connections available can be made. All connections should:

1. Be readily available for inspection;
2. Be kept to the minimum length;
3. Be protected from mechanical damage and the effects of the environment;
4. Not pass through walls, etc.

Semi-rigid couplings and flange adapters   Restraints are needed to prevent separation of the pipes when installed above ground and to prevent angularity.

Bellows   The pipework either side of a bellows joint shall be supported so that the bellows itself is not supporting any of the weight of the attached pipework.

Swivel joints   The axis of rotation of the swivel must be accurately aligned on both sides of the joint and the joint should have lateral freedom and be free from side bending moments.

Quick-release couplings The coupling shall be of the type having self-sealing valves in both the plug and the body with a flexible tube fitted to the downstream connection, which shall be the male plug section. A manual valve shall be fitted immediately upstream of the coupling.

Flexible tubes Metallic tubes complying with BS 669, Part 2 or BS 6501 shall be used other than with small portable appliances or with domestic-type appliances, in which case they shall comply with BS 669 Part 1. A manual valve shall be fitted on the inlet side in close proximity to the tube. The pipework shall be adequately supported such that the tube does not support the weight of the attached pipework. Tubes shall be installed so that they are neither twisted nor subjected to torsional strain, have flexing in one plane only, and are not subject to sharp bends near end fittings.

19.18.8 Purge points

Purge points shall be fitted at section isolation valves, at the end of pipe runs and at other suitable positions to facilitate the correct purging of pipes. Pressure test points should be fitted at the outlet of each section-isolating valve.

19.18.9 Siphons and condensate traps

These are not normally required. Where wet gas is supplied, it will be necessary to fit a vessel at low points in the installation to collect any condensate or fluid. This vessel shall be in a readily accessible position and a valve, suitably plugged or capped, shall be fitted to its drain connection. In the exceptional case where hydraulic pressure testing is to be carried out, similar provisions must be made.

19.18.10 Commissioning

After pipework has been installed, it is essential that it be tested for soundness and purged, as required by Regulation 21 of the Gas Safety (Installation and Use) Regulations. Detailed procedures for these two operations are given in the British Gas publications *Purging Procedures for Non-Domestic Gas Installations* (IM/2) and *Soundness Testing Procedures for Non-Domestic Gas Installations* (IM/5).

19.19 Flow charts for use with gas

19.19.1 Need for flow charts

It is essential when designing the pipe layout for gas distribution that unavoidable pressure losses are not incurred. For low-pressure gas, the pressure available at the meter inlet will be only 21 mbar, and the allowable pressure loss to the point of use only 1 mbar, although higher pressures may be available in some circumstances. If such a low-pressure loss is not to be exceeded it is essential that the pipework be sized correctly. It is preferable to oversize pipework rather than undersize, particularly as this allows

scope for future expansion. For routine purposes, the use of flow charts is much to be preferred to formulae based on first principles.

19.19.2 Theory of pressure loss

The theory of pressure losses can be established by developing Bernoulli's theorem for the case of a pipe in which the work done in overcoming frictional losses is derived from the pressure available. For a fluid flowing in a pipe, the pressure loss will depend on various parameters. If

L = length of pipe,
D = diameter of pipe,
S = specific gravity of gas,
Q = flow rate of gas,
f = friction factor,
P = pressure loss

then P is proportional to L
$1/D^5$
S
Q^2
f

For a particular material of pipeline construction, the friction factor can be assumed. This allows a simple formula to be derived:

$$P = \frac{kQ^2 SL}{D^5}$$

where k is a constant taking into account the friction factor, units of length, units of pressure, etc. This is the Pole equation, and is frequently expressed in the form:

$$Q = k\sqrt{\frac{PD^5}{SL}}$$

Forms of this equation have been drawn up for use with different units (SI or Imperial). Typical assumptions are:

1. Flow is fully turbulent with $Re>3500$ (the transition from laminar flow starts at about $Re = 2000$);
2. Viscosity is known and allowed for in k;
3. Pressure is <25 mbar;
4. If SG is known this can be incorporated into k;
5. Pipe is assumed to be to BS 1387. This determines f, which can be incorporated into k.

For uniform circular pipe the Fanning equation can be used:

$$P = \frac{\left(4f\frac{L}{D}\right)U^2}{2g}$$

where

U = linear velocity,

g = gravitational constant.

The Fanning friction factor (f in the above equation) varies with Reynolds number and relative roughness of

Table 19.3 Gas flow through pipes and fittings

Flow in a straight horizontal steel pipe (to BS 1387, Table 2, Medium) with 1.0 mbar differential pressure between the ends, for gas of relative density 0.6 (air = 1)

Nominal size mm (in)	Length of pipe (m)									
	3	6	9	12	15	20	25	30	40	50
					Discharge (m³/h)					
6 ($\frac{1}{8}$)	0.29	0.14	0.09	0.07	0.05	–	–	–	–	–
8 ($\frac{1}{4}$)	0.80	0.53	0.49	0.36	0.29	0.22	0.17	0.14	0.11	0.08
10 ($\frac{3}{8}$)	2.1	1.4	1.1	0.93	0.81	0.70	0.69	0.57	0.43	0.34
15 ($\frac{1}{2}$)	4.3	2.9	2.3	2.0	1.7	1.5	1.4	1.3	1.2	1.0
20 ($\frac{3}{4}$)	9.7	6.6	5.3	4.5	3.9	3.3	2.9	2.6	2.2	1.9
25 (1)	18	12	10	8.5	7.5	6.3	5.6	5.0	4.3	3.7
32 (1 $\frac{1}{4}$)	39	27	21	18	16	14	12	11	9.2	8.1
40 (1 $\frac{1}{2}$)	59	40	32	27	24	21	18	16	14	12
50 (2)	110	76	61	52	46	39	34	31	26	23
65 (2 $\frac{1}{2}$)	220	150	120	100	92	79	70	63	54	47
80 (3)	340	230	190	160	140	120	110	97	83	73
100 (4)	690	470	380	330	290	250	220	200	170	150

Flow in a straight horizontal copper tube (to BS 2871, Part 1, Table X) with 1.0 mbar differential pressure between the ends, for gas of relative density 0.6 (air = 1). Tube sizes given below are outside diameters

Nominal size mm (in)	Length of pipe (m)									
	3	6	9	12	15	20	25	30	40	50
					Discharge (m³/h)					
6	0.12	0.06	–	–	–	–	–	–	–	–
8	0.52	0.26	0.17	0.13	0.10	0.02	0.06	0.05	–	–
12	1.5	1.0	0.85	0.82	0.69	0.52	0.41	0.34	0.26	0.20
15	2.9	1.9	1.5	1.3	1.1	0.95	0.92	0.88	0.66	0.52
22	8.7	5.8	4.6	3.9	3.4	2.9	2.5	2.3	1.9	1.7
28	18	12	9.4	8.0	7.0	5.9	5.2	4.7	3.9	3.5
35	32	22	17	15	13	11	9.5	8.5	7.2	6.3
42	54	37	29	25	22	18	16	15	12	11
54	110	75	60	51	45	38	33	30	26	23
76	280	190	150	130	120	98	86	78	66	58
108	750	510	410	350	310	260	230	210	180	160

Note: 1m³/h = 1000dm³/h. For smaller rates of flow it may be more convenient to use dm³/h.

The effects of elbows, tees or bends inserted in a run of pipe (expressed as the approximate additional lengths to be allowed)

Nominal size				
Cast iron or mild steel	Up to 25 mm (1 in)	32 mm to 40 mm (1$\frac{1}{4}$ in to 1$\frac{1}{2}$ in)	50 mm (2 in)	80 mm (3 in)
Stainless steel or copper	Up to 28 mm (1 in)	35 mm to 42 mm (1$\frac{1}{4}$ in to 1$\frac{1}{2}$ in)	54 mm (2 in)	76 mm (3 in)
Elbows	0.5 m (2 ft)	1.0 m (3 ft)	1.5 m (5 ft)	2.5 m (8 ft)
Tees	0.5 m (2 ft)	1.0 m (3 ft)	1.5 m (5 ft)	2.5 m (8 ft)
90° bends	0.3 m (1 ft)	0.3 m (1 ft)	0.5 m (2 ft)	1.0 m (3 ft)

the internal surface of the pipe. This variation is plotted in the Moody diagram.

The relative roughness is expressed as E/D, where E = the surface roughness and D = the internal pipe diameter. Typical values of E/D are 0.0015 for drawn tubing, 0.046 for commercial steel and 0.12 for asphalted cast iron.

19.19.3 Practical methods of sizing pipework

The use of formulae derived from first principles is time consuming and cannot normally be justified in comparison with approximate methods, which have a sufficient degree of accuracy. It is necessary to know:

Table 19.4 Allowances for fittings and entrance and exit losses in pipes

Fitting		S Additional velocity heads	Additional length pipe diameters $f= 0.025$
Elbows of various angles	$\alpha = 90°$	1.0	40
	60	0.36	14
	45	0.18	7
	30	0.07	3
	15	0.02	1
90° bends	$r/D = 0.5$	1.0	40
	1.0	0.6	24
	2.0	0.4	16
	4.1	0.3	12
Standard tee 90°, equal areas:			
From barrel to branch		1.2	48
From branch to barrel		1.5	60
Through barrel		0.25	10
(third leg stagnant in each case)			
Close return bend, 180°	$r/D = 1.0$	2.0	80
Globe valve, full often		10	400
Angle valve, 45° open		5	200
Flap non-return valve		2	80
Gate valve, full open		0.15	6
$\frac{3}{4}$ open by area		0.8	32
$\frac{1}{2}$ open by area		4.5	180
$\frac{1}{4}$ open by area		28	1120
Square entrance from a tank		0.5	20
Bell mouth entrance from a tank		0.05	2
Re-entrant (Borda) mouthpiece		0.8	32
Pipe discharge into a tank		1.0	40

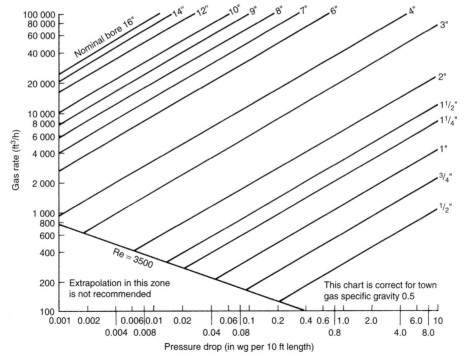

Figure 19.11 Chart for flow of town gas in pipes

1. Flow rate;
2. Pressure loss;
3. Pipe diameter;
4. Pipe length;
5. Pipe material (if this is not known, an assumption that it is steel will normally be sufficiently accurate).

Using these parameters, there are three main ways of expressing the relationships between them:

1. A tabulated form as in Table 19.3. This shows flows versus pipe length and diameter for both copper and steel. Such tables are included in British Gas IM/16 and British Standard BS 6891. Note that Table 19.3 includes allowances for elbows, tees and bends. Allowances, equivalent to numbers of pipe diameters, must be made for all pipe fittings, which cause an additional pressure loss. Further details for allowances to be made for pipefittings, including valves and non-return valves, are given in Table 19.4.
2. A graphical representation as shown in Figure 19.11. This is drawn in the logarithm–logarithm format and allows a rapid estimate of pressure loss to be expected. Note that this particular chart is in imperial units and is drawn for use with town gas. A correction for specific gravity would be needed for natural gas.
3. Flow calculators. These are available in the form of plastic concentric discs which can be rotated against each other.

The parameters of pipe bore, flow, length of pipe, and specific gravity can be used to determine pressure loss. The calculators are versatile in that they can be used for pressure losses of under 1 mbar and for up to 50 bar. Scales for both cast iron and steel pipe are available for greater accuracy.

19.19.4 Discharge through orifices

The Bernoulli theorem can be used as the basis for a means of determining the flow through an orifice. The equation will be of the form:

$$Q = kC_d D^2 \sqrt{\frac{P}{S}}$$

where

C_d = coefficient of discharge,

D = diameter of the orifice

This equation can be used for an orifice plate introduced into pipework as a measuring, throttling or balancing device, and for a jet discharging gas into the injector of a burner at atmospheric or sub-atmospheric pressure.

The major unknown is the value of the coefficient of discharge. In the case of the orifice plate, this will be of the order of 0.6. With the jet, this may be as low as 0.6 but can be up to 0.9 for jets with profiled entries.

Tables are available for discharges from jets for different diameters and pressures, both on a general basis and for proprietary jets.

19.20 Conversion factors

SI units are widely used in the gas industry. Imperial units are also employed, particularly for measuring gas and for its calorific value. In some areas SI and Imperial units can co-exist, particularly thermal ratings, this can equally be expressed in kW and MW or in Btu/h and therm/h.

American combustion equipment is frequently used. This tends to use Imperial units as well as degrees Fahrenheit and ounces per square inch pressure.

European combustion equipment utilizes metric units but these are frequently non-SI. For example, heat and power ratings often use calories and kilocalories/h instead of joules and watts.

It is not intended to include a comprehensive set of conversion factors, but rather those, which are frequently to be found in the energy industries.

Pressure

Non-SI unit	Conversion factor to N/m^2
bar	100 000
inch of Hg	3386.39
inch of water	249.089
Pascal	1
millibar	100
pound-force/in^2 (psi)	6894.76
ton-force/in^2	15 444.3
torr	133.322
ounce-force/in^2	430.9225

Linear

Non-SI unit	Conversion factor
foot	0.3048 m
inch	25.4 mm
yard	0.9144 m
mile	1.60934 km

Volume and capacity

Non-SI unit	Conversion factor
cubic foot	0.0283168 m^3
gallon	4.54609 dm^3
gallon (US)	3.78541 dm^3

Heat and energy

Non-SI unit	Conversion factor
British thermal unit	1.05506 kJ
calorie	4.1868 J
therm	195.506 MJ
thermie	4.1855 MJ

Power

Non-SI unit	Conversion factor
Btu/h	0.293071 W
therm/h	29.3071 kW

kilocalorie/h	1.163 W
horse power	745.700 W
ton of refrigeration	3516.85 W
frigorie	1.1626 W
cal/s	4.1868 W

Calorific value

Non-SI unit	*Conversion factor*
Btu/ft^3	37.2589 kJ/m^3
Btu/lb	2326 J/kg
therm/gal	23.208 GJ/m^3

Temperature

Non-SI unit	*Conversion factor*
degree Fahrenheit	$T - 32)5/9°C$
degree Rankine	5/9 K
degree Celsius	$(T + 273.15)K$

Thermal conductivity

Non-SI unit	*Conversion factor*
Btu in/ft^{2}h°F	0.144228 W/mK
Btu in/ft^{2}s°F	519.220 W/mK
Btu/ft^{2}h	3.15459 W/m^2
Btu/ft^{2}h°F	5.67826 W/m^{2}K
Btu ft/ft^{2}h°F	1.73073 W/mK

Thermal resistivity

Non-SI unit	*Conversion factor*
ft^{2}h°F/Btu in	6.93347 m K/W
ft^{2}h°F/Btu ft	0.57789 m K/W

20

Liquefied Petroleum Gas

R H Shipman
Liquefied Petroleum Gas Industry Technical
Association

Contents

20.1 Introduction

Liquefied petroleum gases (often referred to as LPG or LPGas) are a constituent of crude oil or the condensate of natural gas fields (NGL). They are the C_3 and C_4 hydrocarbons, propane and butane, respectively, which have the property of being gases at normal ambient temperature but can be liquefied and kept in the liquid state by quite moderate pressure. Gases released from crude oil are called *associated gases* while those found without heavier hydrocarbons are known as *unassociated gases*. The North Sea gas wells are good examples of the latter.

Methane is the major constituent of both associated and unassociated gas at source. There will be higher (heavier) hydrocarbons in varying amounts present in the gas, associated gas having more than unassociated gas.

When crude oil is refined, some of the processes yield additional gaseous products. The C_3 and C_4 constituents differ from those released from crude oil or from NGLs, which are *saturated* hydrocarbons. Refinery gases are high in *unsaturates*, e.g. propane (propylene) and butane (butylenes). These unsaturated hydrocarbons are a valuable source of chemical process intermediates and enjoy a large market alongside naphtha.

The following molecular diagrams show the major constituents of LPG:

```
    H   H   H
    |   |   |
H — C — C — C — H    C3H8 Propane          Saturated
    |   |   |
    H   H   H


    H   H   H
    |   |   |
H — C — C — C — H    C4H10 Butane          Saturated
    |   |   |
    H   H   H


        H
    H \ | / H
        C
  H \        / H
H — C — C — C — H    C4H10 ISO Butane      Saturated
  H /    |    \ H
         H


    H   H   H
    |   |   |
H — C — C = C         C3H6 Propene         Unsaturated
    |       |              (Propylene)
    H       H


  H   H   H   H
  |   |   |   |
C = C — C — C — H    C4H8 Butene           Unsaturated
  |       |   |            (Butylene)
  H       H   H
```

No doubt the most successful exploitation of the peculiar properties of LPG has been its use as a fuel gas. Originally used predominantly in refineries for process heating, its value as a fuel was first realized before the Second World War, when it was sold in small portable containers. From primarily domestic use this spread after the war to commercial and industrial utilization and the introduction of bulk storage on-site. In order to develop such a market it was necessary to refine the product to an agreed standard. Today, LPG sold into the commercial fuel gas market meets the requirements for automotive LPG. The Standard sets the limits of the constituents for commercial propane and commercial butane since they are always mixtures of C_3 and C_4 with one or the other predominating.

20.2 Composition

Commercial butane. This product shall consist of a hydrocarbon mixture containing predominantly butanes and/or butanes.

Commercial propane. This product shall consist of a hydrocarbon mixture containing predominantly propane and/or propane.

20.3 Requirements

20.3.1 General

When tested in accordance with the methods given in Table 20.1 the properties of the commercial butane and commercial propane shall be in accordance with the limiting requirements given in that table. For gauge vapor pressure, either the direct measurements method described in BS 3324 or the calculation procedure described in Appendix C of this standard shall be used.

20.3.2 Water content

Commercial butane and commercial propane shall not contain free or suspended water on visual inspection. Additionally, for commercial propane, the content of dissolved water shall not be such as to cause failure when tested in accordance with the valve freeze method.

Note: The addition of up to 0.125 per cent (v/v) of methanol to commercial propane will normally ensure that it complies with the specified limit for water content.

20.3.3 Odor

When tested in accordance with the procedure described in Appendix B, the odor of the gas shall be distinctive and unpleasant, and the odor in a gas/air mixture shall be such that it is detectable down to a concentration of 20 per cent of the concentration corresponding to the lower limit of flammability for the hydrocarbon mixture concerned.

Note: Odorants such as ethanediol, tetrahydrothiophene or dimethyl sulfide may be added so that the gas complies with the specified requirements for odor.

Table 20.1 General requirements for commercial butane and commercial propane

Property	Limit		Test method	
	Commercial butane	Commercial propane	British Standard	Technically equivalent to IP method
Gauge vapour pressure, at 40°C (measured or calculated) (in kPa), max. (see E.1)	505	1560	BS 3324 or appendix C	–
Total sulfur content (in mg/kg), max.	200	200	BS 5379	–
Mercaptan sulfur content (in mg/kg), max.	50	50	BS 2000: Part 272	IP 272
Hydrogen sulfide content (in cm^3/m^3), max. (see E.2)	0.5	0.5	Appendix A	–
Copper corrosion, 1 h at 40°C	Class I	Class I	BS 6924[a]	–
Tendency to freeze in valves	–	Pass	Appendix D	–
Dienes content, mole %, max.	10.0	–	BS 3156: Part 4	–
Ethylene content, mole %, max.	–	1.0	BS 3156: Part 4	–
Alkynes content, mole %, max.	0.5	0.5	BS 3156: Part 4	–
C_4 and higher hydrocarbons content, mole %, max.	–	10.0	BS 3156: Subsection 11.1.1	–
C_5 and higher hydrocarbons content, mole %, max.	2.0	2.0	BS 3156: Subsection 11.1.1	–

[a]BS 6924 is in preparation. Pending its publication, the identical International Standard, ISO 6251, should be used.

20.4 Typical properties of LPG

Typical properties are shown in Table 20.2. These are important in setting the requirements for the storage, handling and use of LPG and require further elaboration.

20.4.1 Vapor pressure

This is one of the most important properties of LPG since it determines the pressure that will be exerted by the gas at ambient temperature, and therefore affects the requirements for handling and the design working pressures of storage vessels. It constitutes the main difference in physical characteristics between commercial propane and butane. The vapor pressure is the pressure at which a liquid and its vapor are in equilibrium at any given temperature. The boiling point of a liquid is, in fact, the temperature at which the vapor pressure is equal to the external ambient pressure.

Commercial propane and butane often contain substantial proportions of the corresponding unsaturated analogues and smaller amounts of near-related hydrocarbons, as well as these hydrocarbons themselves. Figure 20.1 shows vapor pressure/temperature curves for commercial propane and commercial butane. Due to its lower boiling point, higher rates of vaporization for substantial periods are obtainable from propane than from butane, and at the same time, appreciable pressures are maintained even at low ambient temperatures.

20.4.2 Gross calorific value

This is defined as the amount of heat liberated when unit volume (or unit mass) of the gas is burned at a standard temperature and pressure. It is usually expressed in terms of megajoules per cubic meter at 15°C and 1016 mbar, i.e. MJ/s m^3 dry or megajoules per kilogram. Typical

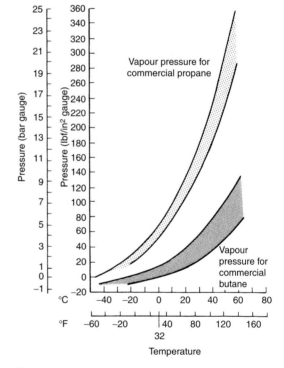

Figure 20.1 Vapor pressure/temperature curves for commercial propane and commercial butane

calorific values of LPG and other gaseous fuels are shown in Table 20.3 and allow a comparison of their heating values to be made.

Table 20.2 Properties of commercial propane and butane

	Propane	Butane
Relative density of liquid at 15°C (60°F)	0.50–0.51	0.57–0.58
Imperial gallons/ton at 60°F	439–448	385–393
Liter /ton at 15°C	1965–2019	1723–1760
Relative density of gas compared with air at 15°C (60°F) and 1016 mbar (30 in Hg)	1.40–1.55	1.90–2.10
Volume of gas (ft^3) per lb of liquid at 60°C and 30 inHg	8.6–8.7	6.5–6.9
Volume of gas (litres) per kg of liquid at 15°C and 1016 mbar	537–543	406–431
Ratio of gas volume to liquid volume at 15°C (60°F) and 1016 mbar (30 inHg)	274	233
Boiling point at atmospheric pressure		
°C (approx.)	−45	−2
°F (approx.)	−49	28
Latent heat of vaporization (Btu/lb) at 60°F	154	160
Latent heat of vaporization (kJ/kg) at 15°C	358	372
Specific heat of liquid at 60°F (Btu/lb °F)	0.60	0.57
Specific heat of liquid at 15°C (kJ/kg °C)	2.512	2.386
Sulfur content, % weight	Negligible to 0.02	Negligible to 0.02
Limits of flammability (percentage of volume of gas in a gas–air mixture to form a combustible mixture)	Upper 10.0 Lower 2.2	Upper 9.0 Lower 1.8
Calorific values		
Gross		
Btu/s ft^3 dry	2500	3270
MJ/s m^3 dry	93.1	121.8
Btu/lb	21 500	21 200
MJ/kg	50.0	49.3
Net		
Btu/s ft^3 dry	2310	3030
MJ/s m^3 dry	86.1	112.9
Btu/lb	19 900	19 700
MJ/kg	46.3	45.8
Therm/ton (gross CV)	482	475
GJ/ton	50.0	49.3
Air required for combustion		
ft^3 to burn 1 ft^3 of gas at STP	24	30
m^3 to burn 1 m^3 of gas at STP	24	30

Table 20.3 Typical gross calorific values of fuel gases

Gas	MJ/s m^3 dry	Btu/s ft^3 dry	MJ/kg	btu/lb
Commercial propane	93.1	2500	50.0	21 500
Commercial butane	121.8	3270	49.3	21 200
Producer gas (cold)	4.7–6.1	125– 165	–	–
Town gas	14–20.5	375– 550	–	–
Natural gas	31.6–46.6	850–1250	–	–
Acetylene	56	1500	49.9	21 460

The great advantage of LPGs is that they are stored as a liquid. However, they are usually used as a gas. One volume of liquid propane when released at STP gives 274 volumes (233 for butane) of high calorific value fuel gas.

20.4.3 Sulfur content

Compared to most other widely available fuels with the exception of natural gas, LPG has very low sulfur content

that is strictly controlled within tight specification limits. This makes LPG a particularly useful fuel where the products of combustion are intended to be released directly into living accommodation.

20.4.4 Relative density — liquid

The density of liquid butane and propane is about half that of water, and as such is much lower than other liquid fuels.

20.4.5 Relative density — vapor

This is the one major property where there is an important difference between LPG vapor and natural gas. Natural gas is lighter than air; propane and butane vapors are, respectively, one and half times and twice as heavy as air. This is important in two areas. First, when converting equipment running on natural gas to run on propane or butane the amount of gas which issues from a fixed orifice at a fixed pressure is inversely proportional to the square root of its density. Second, if there is a leak of LPG vapor it will collect at ground level or in depressions, drains or cellars if appropriate precautions are not taken. This is a major safety consideration when designing or installing LPG systems.

20.4.6 Limits of flammability

When mixed with air, LPG can form a flammable mixture. The flammable range at ambient temperature and pressure extends between approximately 2 per cent of the vapor in air at its lower limit and approximately 10 per cent of the vapor in air at its upper limit. Outside this range, any mixture is either too weak or too rich to propagate flame. However, over-rich mixtures resulting from accidental releases can become hazardous when diluted with air. At pressures greater than atmospheric, the upper limit of flammability is increased but the increase with pressure is not linear.

The limits of flammability for propane and butane are much narrower than most other gaseous fuels, making LPG safer in this respect.

20.4.7 Coefficient of expansion

This is defined as the increase in volume of unit volume of a substance when its temperature is raised by one degree. It is important in that the coefficient of expansion of LPG in its liquid form is relatively high, so that when filling a storage vessel adequate space must always be provided to allow for possible thermal expansion of the liquid.

20.4.8 Other physical characteristics

LPG is both colorless and odorless. However, a distinctive odor is added to aid detection in the case of leakage.

Codes require sufficient stench to be added so that the odor of the gas can be detected in air at concentrations down to one-fifth of the lower limit of flammability, i.e. about 0.4 per cent gas in air.

Neither propane nor butane is toxic, but they do possess anaesthetic properties. There is a threshold limiting value for LPG at 1000 ppm given as an occupational exposure standard in OSHA regulations, for an 8-hour time-weighted average.

Liquid propane and butane will vaporize rapidly if released into the open air, and if they come into contact with bare skin will cause painful freeze burns. Therefore gloves and goggles should always be worn if there is a danger of liquid LPG being released or spilt.

20.5 Transport and storage

The great advantage of LPG is that for use it can be regarded as a gas, but for conveyance it can be treated as a liquid. It is therefore transported by road or rail in a manner similar to liquid fuels. Thus, the most usual method is by road tanker (up to 17 tons of product) or rail tank wagon (from 20 tons to 100 tons per wagon). Figure 20.2 shows a typical 15-ton capacity propane road tanker. In order to maintain the product in its liquid state, however, it has to be kept under pressure and the tanks are therefore pressure vessels. This pressure must be maintained during all activities, from storage at the refinery, to transfer into the road tanker or rail tank wagon, and then in the storage vessel at the users' premises.

In addition to the modes of distribution described above there is also the transportable container, which is now a familiar feature in all kinds of activity. These vary from tiny disposable cartridges of less than 100 g to the larger refillable steel or aluminum alloy cylinders ranging up

Figure 20.2 LPG road tanker. (Courtesy of BP Oil Ltd)

to around 5 kg of product. These are filled at purpose-designed filling plants and distributed by road or rail transport to the retail outlet.

At the other end of the scale there is also the transport and storage of LPG in a refrigerated state. Large international sea transport cargoes are usually refrigerated, while refrigerated land storage is only economically justified for storage measured in thousands of tons and for import/export terminals, as illustrated by Figure 20.3. For design, construction and installation of refrigerated storage, references are given to the Institute of Petroleum, and the Engineering Equipment and Materials Users Association (EEUMA) at the end of this chapter.

Thus for most inland transport and storage the liquid is held at ambient temperature under pressure. The pressure is simply created by the natural vapor pressure of the product, which varies with its temperature. Butane has a lower vapor pressure than propane. The carrying vessels, whether road tankers, static storage, cylinders or cartridges, must be designed to accept the highest pressure likely to occur in service. Knowledge of the highest ambient temperature and the grade of product are thus essential.

This illustrates the need for an agreed national standard for product specification, which provides assurance of the highest pressure likely to be attained in service from knowledge of ambient conditions. Against this knowledge,

Codes of Practice can safely recommend the pressures, which should be used for the design of storage vessels both for static use and for road/rail tankers. Guidance on the design parameters is given in the LPGITA (UK) Code of Practice No. 1, Installation and Maintenance of Bulk LPG Storage at Consumer's Premises and Code of Practice No. 2, Safe Handling and Transport of LPG in Bulk by road.

Storage vessels for LPG range from those for major industrial users and refineries of 1000 tons capacity or more down to small industrial or commercial users of 1 ton. Even smaller vessels down to 380 liters designed for on-site refilling are used for domestic purposes, i.e. heating and cooking. Similar vessels are used for permanent caravan sites and holiday homes. Figures 20.4 and 20.5 illustrate a storage installation for industrial and domestic use, respectively.

Deliveries to these installations are usually by road tanker. The tanker carries its own off-loading pump driven by the vehicle engine and the transfer is by a flexible hose which is connected to the receiving vessel by a gas-tight coupling.

LPG storage vessels have to be designed as pressure vessels to a recognized standard. The design limits for static storage are based on a maximum product temperature in service of 38°C, which gives a design pressure for propane of not less than 14.5 bar gauge. Those for

Figure 20.3 Refrigerated LPG storage at 'Flotta'. (Courtesy of the Motherwell Bridge Group)

Figure 20.4 Typical industrial LPG storage installation. (Courtesy of Esso Petroleum Co. Ltd)

Figure 20.5 LPG storage vessel for a domestic installation. (Courtesy of the Calor Group Plc)

transport storage are defined in BS 5355 which, for barrels greater than $5 \, m^3$, sets a temperature of 42.5°C, giving a design pressure for propane of 16.75 bar gauge.

It is apparent from the vapor pressure curves that to obtain a gas supply from a storage vessel, or from a cylinder, it is only necessary to release it from the top vapor space. Propane can usually be supplied from storage to the plant by this natural vaporization, without any pumps or compressors, throughout the year.

However, in the case of bulk storage of butane, the vapor pressure is generally too low for this simple method, and use is made of a liquid feed from the bottom of the storage vessel via a pump to a vaporizer. The vaporizer is simply a heat exchanger either using hot water, steam, electricity or even direct flame as the source of heat. Figure 20.6 shows a typical vaporizer.

A glance at the vapor pressure curve for butane will, however, reveal that in winter there is a possibility of butane vapor liquefying after the vaporizer if the temperature is allowed to fall in the pipeline, even at moderate pressure. For this reason, such pipework is usually heated, either by electrical tapes or, if available, by steam or hot-water lines.

There is an alternative to using neat butane vapor, however, which overcomes the need for pipework heating. This is to use a gas–air mixture. A special gas–air mixer is used which ensures that a preset ratio is maintained at all demand rates. The ratio chosen must be well outside the flammable limits. A typical LPG–air-mixing plant is illustrated by Figure 20.7.

The effect of the air is to depress the vapor dew point temperature. A further advantage is that the physical properties of the gas can be made to 'simulate' another gas, e.g. natural gas or manufactured town gas. Such a simulated gas will produce the same heat release through a burner if the supply pressure is the same. This is characterized by a term known as the Wobbe number (W):

$$W = \frac{\text{Calorific value}}{\sqrt{\text{Specific gravity of gas(air} = 1)}}$$

The gas–air ratio is therefore chosen to achieve the desired Wobbe number. However, the use of a simulated gas for plant conversion may not entirely avoid burner

Figure 20.6 Hot-water heated LPG vaporizers. (Courtesy of Esso Petroleum Co. Ltd)

Figure 20.7 An LPG-air mixing unit. (Courtesy of John Wigfull & Co. Ltd)

adjustment or modification if other properties differ markedly (e.g. flame speed).

The supply of gas from storage or vaporizer to plant is usually governed down to a medium pressure of around 1.5 bar, for final reduction to 28 mbar for butane or 37 mbar for propane. However, the versatility of LPG is such that plant and combustion equipment designers can, for specialized purposes, select virtually whatever pressures are most suitable for their requirements.

An unusual but successful alternative system has been applied for some very large high energy consuming plants. This involves using liquid-phase LPG right up to the burner.

20.6 Cylinder storage

The earliest market for LPG was developed by the use of small transportable cylinders filled at purpose-built filling plants and returned there for refilling. This market has grown steadily over its 50-year or more lifetime. Cylinders are available for either commercial propane or commercial butane, and in a variety of sizes from easily portable ones with a capacity of around 20 liters to the largest with a capacity of around 108 liters.

In parallel with the refillable cylinder market, the non-refillable cylinder has emerged to supply the needs of the leisure market and the DIY enthusiast, to name but two. These are broadly divided into those with a valve outlet and the pierceable cartridge. Both are designed to be disposed of when finally exhausted.

Both the refillable and non-refillable cylinders must, of course, be designed for the internal pressure conditions likely to be experienced during their lifetime. Welded steel cylinders are designed based upon a maximum liquid temperature of 55°C and the associated vapor pressure

of 22.13 bar gauge for propane and 7.52 bar gauge for butane.

It is now common practice to design all refillable cylinders to propane standards. Furthermore, the use of safety relief valves has been introduced for all cylinders carrying over 3 kg of LPG.

The outlet valves of refillable butane cylinders are different from those of refillable propane cylinders. The valve for propane is a handwheel type, with a female 5/8 in BSP LEFT-HAND thread for a POL metal-to metal connector. A compatible half-coupling must always be used to ensure a leak-free joint.

Butane cylinders, which are so widely used for domestic service, are now provided with a self-sealing clip-on valve. The pressure regulator, which is normally attached directly onto the cylinder outlet, is fitted by a simple push-on or snap-on action and is provided with a lever, which will open or shut the cylinder valve.

There are a number of different clip-on valve designs used by LPG distributors and care is necessary to ensure that only the correct mating regulator is employed.

20.7 Safety in storage

Safe storage and handling of LPG is of paramount importance, whether it is in bulk or in cylinders. This is achieved by ensuring the mechanical integrity of the storage vessels or cylinders and by strict observance of the recommended separation distances between storage and buildings or boundaries. This passive protection has to be supplemented by rigorous observance of operational procedures. Both the LPGITA (UK) and the Health and Safety Executive have issued codes of practice on the subject. The separation distances for bulk LPG vessels are shown in Table 20.4.

Table 20.4 Location and spacing for vessels

Maximum water capacity					Minimum separation distances					
Of any single vessel in a group			Of all vessels in a group		Above-ground vessels			Buried or mounded vessels		
Liters	Gallons	Nominal LPG capacity (tonnes)	Liters	Gallons	From buildings boundary, property line or fixed source of ignition (a)	With firewall (b)	Between vessels (c)	From buildings, etc. to		Between vessels (f)
								Valve assembly (d)	Vessel (e)	
					m (ft)	m	m	m	m	m
150–500	28–100	0.05–0.25	1 500	330	2.5 (8)	0.3	1	2.5	0.3	0.3
>500–2500	100–500	0.25–1.1	7 500	1 650	3 (10)	1.5	1	3	1	1.5
>2500–9000	500–2000	1.1–4	27 000	6 000	7.5 (25)	4	1	7.5	3	1.5
>9000–135 000	2000–30 000	4–60	450 000	100 000	15 (50)	7.5	1.5	7.5	3	1.5
>135 000–337 500	30 000–75 000	60–150	1 012 500	225 000	22.5 (75)	11	$\frac{1}{4}$ of sum of the dia. of two adjacent vessels	11	3	–
>337 500	>75 000	>150	2 250 000	500 000	30 (100)	15	–	15	3	–

Perhaps the most serious hazard to LPG storage is that of accidental fire. The safety distances are intended to separate the storage from possible adjacent fires so that the risk of a fire affecting the storage is very low. However, this residual risk has to be catered for. The mechanical integrity needs to be assured under severe fire attack. For this reason, the vessels are provided with relief valves designed to cope with fire engulfment. The heat from such a fire may raise the stored pressure until the relief valve opens. The discharge capacity of the relief valve when fully open is required to meet or exceed the following:

$$Q = 10.6552A^{0.82}$$

where

A is the total surface area of the vessel (m³), and
Q is the discharge (m³/min) of air reduced to 15°C
   and 1 atm.

The heat transfer to the liquid from an engulfment fire has been estimated at around 100 kW/m², and the above formula equates this to the vapor produced from this input as latent heat. The exponential is an area exposure factor, which recognizes that large vessels are less likely to be completely exposed to flames.

The safety relief valve will protect the liquid-wetted areas of the storage vessel. The metal temperature will not significantly exceed the liquid temperature, which will be absorbing the latent heat of vaporization. However, above the liquid line no such cooling will take place. The metal temperature at the top of the vessel could therefore exceed safe limits.

The usual protection for large installations is to provide a water-spray system. For small bulk storage, fire hoses or monitors are often adequate. However, for installations over 50 tons of storage (and all major cylinder-filling plants) it is accepted that a fixed water-spray system needs to be provided which is automatically initiated by a system capable of detecting a fire threatening the vessels and/or the adjacent tanker loading or unloading area. The deluge rate to provide protection against fire engulfment is 9.8 l/min/m² of vessel surface area, and this should be capable of being sustained for at least 60 min. The spray pattern adopted for a fixed installation normally includes four longitudinal spray bars, two at the upper quadrants and two at the lower quadrants of horizontal cylindrical vessels, with nozzles spaced to give uniform coverage.

An alternative means of avoiding the hazard from fire is to bury the vessels or to employ the increasingly popular method of mounding. In either case, acknowledgment of the reduced hazard is indicated by the reduced separation distances (see Table 20.4). Since both burial and mounding preclude the possibility to monitor continuously the external condition of the vessels, very high-quality corrosion protection needs to be applied, often supplemented by cathodic protection, depending on soil conditions.

The use of burial or mounding is sometimes employed to overcome visual environmental objections since the mounding, for instance, can be grassed over. Indeed, this is the method often adopted to prevent erosion of the mounding material. A typical mounded storage is seen in Figure 20.8.

Codes of practice also provide guidance on the storage of LPG in cylinders. Again, it is based on the adequate separation of stacks of cylinders from buildings and boundaries. These distances depend on the total tonnage stored or the tonnage in the largest stack in the storage area. The codes also provide guidance on width of gangways, height of stacks and storage within buildings.

20.8 Uses of gaseous fuels

LPG has many advantages over the alternative liquid fuels, and is regarded as a 'premium fuel'. Some of the premium characteristics are:

Very low sulfur
Easy ignition

Figure 20.8 Mounded LPG storage. (Courtesy of Southern Counties Gas)

Simplicity of burner design
Versatility (e.g. very small flames, localized heating)
Wide turndown of burners
Clean combustion gases
Ease of control

These properties apply equally to natural gas as for LPG. However, since LPG can be used anywhere in the country which is accessible by road or rail it is not subject to the geographic limitations of a piped supply. The versatility of gaseous fuels has resulted in much development work in the design of burners for a wide range of applications. Some are highly specialized. Much of this development has been primarily for natural gas since its introduction to the UK in the late 1960s. The properties of LPG and natural gas are different, which means that LPG cannot be used in an appliance or burner designed for natural gas. However, a key property, flame speed, is of the same order, unlike the very high flame speed of coal gas. This similarity means that most natural gas appliances or burners can be adapted for use with LPG with minor modifications to allow for the much higher calorific value and higher gas density. Aeration needs attention, but for a given heat input rate the majority of gaseous fuels require approximately the same amount of air for combustion. Of course, in the domestic and leisure use of LPG particular appliances have been developed quite specifically for use with LPG. Thus, LPG equipment and appliances are available for all the gas-fired applications in commercial, industrial and domestic use.

When drawing off LPG vapor from storage, whether from bulk storage, cylinders or even non-refillable cartridges, the liquid will begin to vaporize but will need latent heat to do so. It will take this heat from itself, and the container, and both will cool. As the LPG liquid cools its pressure must fall. Equilibrium will be obtained when the heat flow from the ambient air equals the demand for latent heat.

This, together with the natural variations of vapor pressure with the climatic conditions, results in a source of gas with quite wide variations in supply pressure. Except for some small heating appliances, which are designed to attach directly to a small cylinder or cartridge, this pressure variation needs to be eliminated as far as the appliance is concerned. Therefore, a pressure regulator is used. For permanent installations, two-stage pressure regulation is recommended. The first is usually located immediately at the outlet from the storage vessel or cylinder.

LPG cylinders are filled only at purpose-designed filling plants and (if refillable) are returned there when exhausted. To provide a continuous gas supply, two, four or more cylinders are connected to a changeover valve so those empty cylinders can be exchanged without interrupting the supply. Automatic changeover devices are available, which switch to the reserve supply when necessary, and indicate that they have done so. A typical wall-mounted ACD is illustrated in Figure 20.9.

Codes of practice restrict the use of LPG cylinders indoors for commercial and domestic use to commercial

butane. Propane cylinders should always be located outside the premises except for special industrial purposes. A typical four-cylinder pack is shown in Figure 20.10.

The use of propane as a motor vehicle fuel has been highly developed in some countries, particularly in the USA, Holland and Italy. It is, of course, an entirely lead-free fuel. Very high efficiencies can be obtained using a gaseous fuel in spark-ignition engines since intimate mixing of the fuel and air is much more easily achieved than with a liquid fuel. This results in a much cleaner exhaust, with considerable reductions in CO and hydrocarbons.

The low level of harmful exhaust emissions has, for instance, been one of the spurs to the wide use of LPG as a forklift truck engine fuel in all countries because of their wide use inside buildings such as warehouses, railway station forecourts, etc. Another advantage is the high power continuously available compared with electric

Figure 20.9 Wall-mounted automatic changeover device. (Courtesy of Sperryn & Co. Ltd)

Figure 20.10 A 2 × 2 LPG cylinder pack supply domestic premises. (Courtesy of the Calor Group Plc)

21.1 Introduction

One of the most important considerations to be taken into account when using coal as a fuel for industrial boiler houses is the provision of coal- and ash-handling equipment. Although such equipment was traditionally manual, with only larger installations making use of mechanized handling systems for the coal, the higher expectations of modern boiler house operators have led to the increasing use of automatic coal- and ash-handling systems for all coal-fired boiler plant.

The cost of handling equipment is significant, and may well represent a considerable proportion of the overall cost of a boiler house. It is therefore important that the appropriate type of equipment is selected and that it is properly installed. It is also important that the use of coal- and ash-handling equipment is environmentally acceptable in terms of noise, dust and visual impact.

The relatively low manning levels now encountered in industrial boiler houses are not easily able to cope with breakdowns in equipment which may be handling coal at more than 100 tons per hour. Similarly, modern automatic firing equipment is less able to tolerate coal, which is significantly out of specification due to degradation caused by the handling system. Reliability and performance, both a function of correct design and installation, are therefore of major importance for coal-handling systems.

21.2 Formation of coal

Coal forms as the result of a natural chemical process in which plants absorb carbon dioxide from the atmosphere. Sunlight, moisture and other factors convert carbon dioxide into compounds containing carbon, hydrogen and oxygen, such as sugars, starch, cellulose, lignin and other complex substances that make up the plant structure. Under favorable conditions, vegetation is converted into some of the many forms of coal now known to mankind.

When organic matter begins to change to coal, peat is the first product. As peat is buried, it is cut off from the oxygen in air. This prevents the rapid decay of its organic matter by slowing bacterial action.

The weight of more vegetation, dirt and water falling on the peat helps to compress and solidify it. Sometimes, mineral sediments have settled from muddy floodwaters while vegetable matter was accumulating and formed 'partings' or layers of shale in the coal vein.

At the end of coal-forming periods, swamps remain flooded for a long time, and earthy sediments are deposited in thick beds over the peat, further compressing it and starting the 'classification', the coal-making process.

Geologists usually name time, pressure and heat, as the primary agents required changing peat into different types of coal. Chemists include microorganisms as an additional, critical factor.

Time

Time is merely the duration or period during which an agent, or agents, has an opportunity to act. It becomes critical because many organic chemical reactions are slow. In a short time period, the results of such reactions would be negligible, but over millions of years, the total result is large. Although the greatest single factor in the process of formation was probably time, other factors must also be sought, because coals of the same age and in the same bed are of different rank.

Pressure

Pressure is considered the factor of next importance. Higher ranked coal is generally found in regions that have been under high pressure. Anthracite coal, for example, is associated with earth-folding or mountain formation. Both of these processes create tremendous, sustained internal pressure.

Heat

Heat has played an important part in the creation of coal. The temperature during formation need not be high, but is a critical agent or component of the transformation from organic matter to coal.

21.3 Classification of coal

Coals are grouped according to rank, the degree of progressive alteration in the transformation from lignite to anthracite. For the purposes of the power plant operator, there are several suitable ranks of coal:

21.3.1 Types of coal

Anthracite

Hard and very brittle, anthracite is dense, shiny black, and homogeneous with no marks of layers. Unlike the lower rank coals, it has a high percentage of fixed carbon and a low percentage of volatile matter. Anthracites include a variety of slow-burning fuels merging into graphite at one end and bituminous coal at the other. They are the hardest coals on the market, consisting almost entirely of fixed carbon, with the little volatile matter present in them chiefly as methane, CH_4. Anthracite is usually graded into small sizes before being burned on stokers: the 'meta-anthracites' burn so slowly as to require mixing with other coals, while the 'semi-anthracites', which have more volatile matter, are burned with relative ease if proper fired. Most anthracites have a lower heating value than the highest-grade bituminous coals. Anthracite is used principally for heating homes and in gas production.

Some semi-anthracites are dense, but softer than anthracite. This grade is shiny gray and somewhat granular in structure. The grains have a tendency to break off in handling and produce coarse, sand-like slack.

Other semi-anthracites are dark gray and distinctly granular. The grains break off easily in handling and produce a coarse slack. The granular structure has been produced by small vertical cracks in horizontal layers of comparatively pure coal separated by very thin partings. The cracks are the result of heavy downward pressure and shrinkage of the pure coal because of a drop in temperature.

Bituminous

Bituminous coals derive their name from the fact that on being heated, they are often reduced to a cohesive binding, sticky mass. Their carbon content is less than that of anthracites, but they have more matter that is volatile. The character of their volatile matter is more complex than that of anthracites and they are higher in calorific value. They burn easily, especially in pulverized form, and their high volatile content makes them good for producing gas. Their binding nature enables them to be used in the manufacture of coke, while the nitrogen in them is utilized in processing ammonia.

The low-volatile bituminous coals are grayish black and distinctly granular in structure. The grain breaks off easily and handling reduces the coal to slack. Any lumps that remain are held together by thin partings. Because the grains consist of comparatively pure coal, the slack is usually lower in ash content than are the lumps.

Medium-volatile bituminous coals are the transition from high volatile to low volatile coal and as such have the characteristics of both. Many have a granular structure, are soft, and crumble easily. Some are homogeneous with very faint indications of grains or layers. Others are more distinct laminar structure, are hard, and stand handling well.

High volatile, type A bituminous coals are mostly homogeneous with no indication of grains, but some show distinct layers. They are hard and stand handling with little breakage. The moisture, ash and sulfur contents are low and the heating value high.

High volatile, type B bituminous coals are of distinct laminar structure; thin layers of black, shiny coal alternate with dull, charcoal-like layers. They are hard and stand handling well. Breakage occurs generally at right angles and parallel to the layers, so that the lumps generally have a cubical shape.

High volatile, type C coals are of distinct laminar structure, are hard and stand handling well. They generally have high moisture, ash, and sulfur content and they are considered free-burning coals.

Subbituminous

These coals are brownish black or black. Most are homogeneous with smooth surfaces and with no indication of layers. They have high moisture content, as much as 15 to 30 percent, although appearing dry. When exposed to air, they lose part of the moisture and crack with an audible noise. On long exposure to air, they disintegrate. They are free burning, entirely non-coking coals.

Lignite

Lignites are brown and have a laminar structure in which the remnants of woody fibers may be quite apparent. The word *lignite* comes from the Latin word *lignum* meaning wood. Their origin is mostly from plants rich in resin, so they are high in volatile matter. Freshly mined, lignite is tough, although not hard, and it requires a heavy blow with a hammer to break the large lumps. But on exposure to air, it loses moisture rapidly and disintegrates. Even

Table 21.1 Correlation of coal properties (based on typical analysis)

Natural Fuels	Manufactured of Byproduct Fuels
Solid	
Coal	Coke and coke breeze
	Coal tar
Lignite	Lignite char
Peat	
Wood	Charcoal
	Bark, saw dust and wood waste
	Petroleum coke
	Bagasse
	Refuse
Liquid	
Petroleum	Gasoline
	Kerosene
	Fuel oil
	Gas oil
	Shale oil
	Petroleum fractions and residues
Gaseous	
Natural gas	Refinery gas
Liquefied petroleum gases (LPG)	
	Coke-oven gas
	Blast-furnace gas
	Producer gas
	Water gas
	Carburetted water gas
	Coal gas
	Regenerator waste gas

when it appears quite dry, the moisture content may be as high as 30 per cent. Owing to the high moisture and low heating value, it is not economical to transport lignite over long distances.

Unconsolidated lignite (B in Table 21.2) is also known as 'brown coal'. Brown coals are generally found close to the surface, contain more than 45 per cent moisture, and are readily won by strip mining.

21.3.2 Correlation of coal properties

Table 21.1 is a correlation based on a study of analyses of many typical coals. For a preliminary ranking, refer to Figure 21.1, which is a graphical presentation of coal ranking parameters. Table 21.2 provides the ASTM Standards D388 for classifying coals.

21.4 Delivery

Most of the coal for industrial use in the US is delivered by rail, ship or barge, although smaller installations use trucks. The use of these methods will depend upon the availability of suitable track or waterway close to the boiler house, and are only generally applicable to installations burning in excess of 50,000 tons per year. There

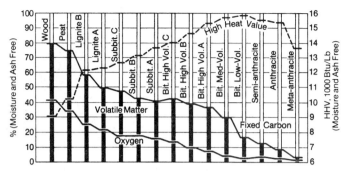

Figure 21.1 Progressive stages of transformation of vegetal matter into coal

Table 21.2 Classification of coals by rank

Class and Group	Fixed Carbon Limits, % (Dry, Mineral-Matter-Free Basis)		Volatile Matter Limits, % (Dry, Mineral-Matter-Free Basis)		Calorific Value Limits, Btu/lb (Moist,[b] Mineral-Matter-Free Basis)		Agglomerating Character
	Equal or Greater Than	Less Than	Equal or Greater Than	Less Than	Equal or Greater Than	Less Than	
I. Anthracitic							
1. Meta-anthracite	98	...	...	2	...	...	nonagglom-
2. Anthracite	92	98	2	8	...	...	erating
3. Semianthracite[c]	86	92	8	14	...	...	
II. Bituminous							
1. Low-volatile bituminous coal	78	86	14	22	...	...	
2. Medium volatile bituminous coal	69	78	22	31	...	...	
3. High-volatile A bituminous coal	...	69	31	...	14,000[d]	...	commonly agglomerating[e]
4. High-volatile B bituminous coal	...	...	...	...	13,000[d]	14,000	
5. High-volatile C bituminous coal	...	...	...	...	11,500	13,000	
					10,500	11,500	agglomerating
III. Subbituminous							
1. Subbituminous A coal	...	...	...	...	10,500	11,500	
2. Subbituminous B coal	...	...	...	...	9,500	10,500	nonagglom-
3. Subbituminous C coal	...	...	...	...	8,300	9,500	erating
IV. Lignitic							
1. Lignite A	...	...	...	...	6,300	8,300	
2. Lignite B	...	...	...	...	...	6,300	

[a]This classification does not include a few coals, principally nonbanded varieties, which have unusual physical and chemical properties and which come within the limits of fixed carbon or calorific value of the high-volatile bituminous and subbituminous ranks. All of these coals either contain less than 48% dry, mineral-matter-free fixed carbon or have more than 15,500 moist, mineral-matter-free Btu per pound.
[b]Moist refers to coal containing its natural inherent moisture but not including visible water on the surface of the coal.
[c]If agglomerating, classify in low-volatile group of the bituminous class.
[d]Coals having 69% or more fixed carbon on the dry, mineral-matter-free basis shall be classified by fixed carbon, regardless of calorific value.
[e]It is recognized that there may be nonagglomerating varieties in these groups of the bituminous class, and there are notable exceptions in high-volatile C bituminous group.

are several types of road-delivery vehicle, of which the dump truck is the most economical. The preferred size of truck is 26 tons capacity, although smaller units are used for sites with limited access or stocking facilities.

Other types of vehicle include conveyor trucks and pneumatic delivery trucks, but because of their lower versatility, a small surcharge (typically, $2\frac{1}{2}$ per cent) is usually levied on delivery by these vehicles, and this restricts their use to smaller installations.

Conveyor trucks incorporate a hydraulically driven belt conveyor located at the back of the vehicle. This conveyor can be swung round through 180°, and can deliver up to a height of 10 feet (2.8 meters), thus providing greater flexibility of discharge than the dump truck. The conveyor can be fed either by dumping in the normal way or by a belt conveyor running from front to rear of the truck bed.

Pneumatic delivery trucks incorporate a positive displacement blower to provide the air supply for pneumatic conveying. Coal is fed into the conveying air stream of a rotary valve from the bed of the truck either by dumping or by use of a belt or screw conveyor.

The coal is initially conveyed through 5-inch (215-mm) diameter flexible pipe. The flexible pipe is then connected to a fixed pipe at the customer's site, which conveys the fuel to storage hoppers up to 131 feet (40 m) distant from the vehicle. Pneumatic delivery is particularly useful to the smaller consumer because the coal can be delivered directly to the service hopper, from where it can flow by gravity to the boiler, and thus obviate the requirement for any on-site coal-handling equipment.

Inevitably, with this form of delivery, times are extended in comparison with simple tipping vehicles and, depending on the type and condition of the vehicle may take more than one hour to complete. Because of the extra equipment required for pneumatic delivery, these vehicles have a slightly smaller capacity than dump trucks, typically ranging between 21 and 20 tons. Noise is an important factor to be taken into account with pneumatic delivery. In particular, worn rotary valves allow significant air leakage and this requires higher blower (and hence engine) speeds to be maintained.

Air leakage also reduces the rate at which coal can be delivered, thereby extending delivery time. In order to reduce the noise of delivery it is possible for a site-installed electrically driven blower to be used instead of the vehicle-mounted unit. The coal being conveyed within the fixed site pipework creates some additional noise, and this can be significantly reduced by the use of ordinary boiler house pipework lagging.

Containerized delivery using 20-tons capacity ISO containers can be employed. The advantage of using containers is that the coal need not be handled directly at intermediate transfer points, and is therefore available for use in as good a condition as when the container was first loaded.

Several systems are available for container handling and unloading at user sites, an example of which is shown in Figure 21.2. Two reception bays are provided at this site so that the delivery vehicle can deliver a full container and immediately load on the adjacent empty unit. Together with a small-capacity site storage system, the

Figure 21.2 Container unloading system

container itself can thus provide a useful degree of additional storage.

In operation, the vehicle backs into the unloading zone, and chains are secured to the container by means of twist-locks. Hydraulic rams lift the container clear of the vehicle, which can then leave. Further operation of the hydraulic rams tips the container so that coal is discharged into the boilerhouse conveying system.

21.5 Coal handling

21.5.1 Open stocking

Where ground space is available, large industrial users have traditionally stored coal in open stockpiles. In this case, the coal is tipped directly onto the ground and then picked up and stocked using a front-end loader or mechanical conveyor system.

For quantities of coal of less than about 500 tons, a front-end loader is probably the most effective means for stock reclamation, although automatic reclamation can be achieved by using grab-skip-hoists, which operate from an overhead monorail.

21.5.2 Underground bunkers

Underground reception bunkers are the traditional method of accepting coal from dump trucks. They are usually sized to hold about 30 tons so that a continuous supply of coal can be maintained to the coal-conveying system.

A grid screen is usually fitted to the top of the bunker to prevent oversize material from entering the conveying system. If the vehicle is to run onto the screen, consideration must be given to the axle weights likely to be encountered. In order to maximize the capacity of the bunker, retaining walls are normally used to surround the grid screen, and these can be used to support a roof structure that will prevent rain from entering the bunker, as well as reducing dust in the near vicinity. Section 21.6 provides more detail about the design of the hopper and its outlet.

A variant of the underground bunker is the boot bunker, where the vehicle backs up a ramp until the rear wheels are about 7 feet or 2 meters above ground level. In this way, the depth below ground level, and hence the cost of excavation, is reduced. The disadvantage of the boot type of bunker is that the ramp takes up a considerable space on the boilerhouse site.

21.5.3 Dumping hoppers

The disadvantage of underground bunkers is that expensive civil works are usually necessary to provide drainage and access to the bunker base. In order to overcome this problem several manufacturers have developed end-dumping hoppers that provide a controlled feed from coal that has been dumped at ground level.

In operation, the hopper is lowered so that one side becomes a reception platform onto which coal is dumped from a railcar or truck. After the load has been dumped, the entire hopper is elevated by a hydraulic jacking system into a vertical position, and can then discharge its contents by gravity into the boiler house conveying system as shown in Figure 21.3.

Commercial dumping hoppers of this type often incorporate a screening system on the outlet to remove any oversize material which might otherwise block the conveying or combustion system. An alternative design of side-dumping hopper is available for sites where limited headroom prevents the use of an end-dumping unit.

In this case, coal is dumped onto a platform at ground level as previously described, and is then removed from one side of the hopper using an open-screw conveyor to feed the boilerhouse conveying system. When the depth of coal above the screw conveyor has fallen so that no

more coal is being conveyed, hydraulic jacks tilt the platform. The remainder of the coal is thus dumped onto the conveyor, which can then empty the hopper.

These types of reception system can provide significant savings for larger installations which use singles coal, by using dump truck delivery instead of pneumatic delivery. The actual saving will depend on the levy imposed by the coal distributor for the use of the pneumatic delivery vehicle, but they are generally viable where the annual coal consumption is greater than about 3000 tons.

21.5.4 Walking floor

The advantage of this system, which is shown schematically in Figure 21.4, is that dumped coal can be stored under cover within the body of the unit. In operation, the delivery vehicle backs up to the walking floor unit and is then tipped.

The floor is constructed of a number of sliding longitudinal planks which reciprocate in sequence to provide a forward motion to bulk materials. Each third plank moves backwards in turn, sliding beneath the coal, and then all planks move forward together, thus 'walking' the coal forward. As the coal is dumped onto the floor, it is *walked* into the body of the unit, where it can be stored if necessary.

The *walking* floor feeds coal onto an outlet conveyor, such as a screw or *en-masse* conveyor that, in turn, feeds it to the boiler house conveying system or directly to the boilers. A simple reversing switch allows coal to be conveyed backwards away from the outlet conveyor for

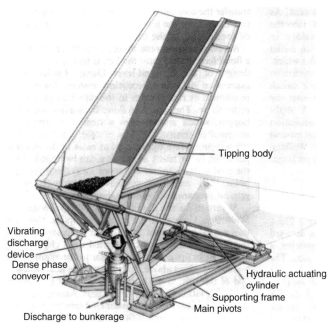

Tipping body

Vibrating discharge device
Dense phase conveyor

Hydraulic actuating cylinder
Supporting frame
Main pivots

Discharge to bunkerage

Figure 21.3 Dumping hopper

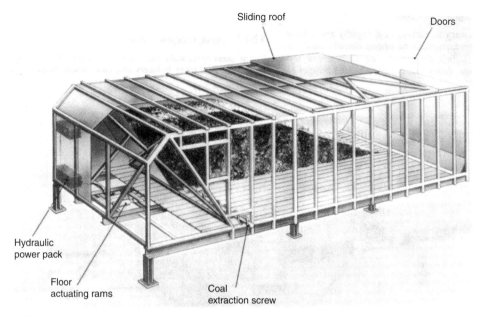

Figure 21.4 Walking floor coal reception system

maintenance purposes. A slight incline (up to about 6°) on the floor enables installation to be carried out with no excavation and minimal ground preparation.

Typical unloading rates for the walking floor are 250 tons per hour, reducing slightly on larger units holding up to 60 tons of coal.

21.5.5 Push-floor

The push-floor system is similar to the walking floor in that a reciprocating movement provides a forward motion to coal as it is dumped. In this case, however, a number of shaped bars carried in frames are used.

The bars are wedge shaped so that they can move through the coal in one direction but push the coal forward (against their flat faces) in the other direction. Although the flow of coal cannot be reversed, the simple wedge-shaped bars are more robust than the planks of the walking floor, and it is possible to drive a vehicle over them without causing damage.

21.5.6 Belt reception systems

Where it is necessary to convey coal rapidly away from the reception point, coal can be dumped directly onto the moving belt of a belt elevator, which then conveys it directly to storage. When using a normal feed hopper to transfer the coal onto the belt from the transport vehicle a controlled flow rate from the vehicle is desirable to prevent bridging or overloading of the belt.

A device to overcome these problems and thus allow a flood feed of coal onto the belt is incorporated into the design of the ground-level dump feeder. In this expression of a belt reception system, the vehicle is positioned at right angles to the direction of the belt in order to tip or dump. The coal then flows by gravity into the feed hopper, which incorporates a steel plate arrangement designed to control the flow of coal onto the belt, and can accommodate coal flow at rates up to 800 tons per hour, i.e. virtually as fast as it can be dumped. Because the coal is immediately conveyed away from the tipping point, belt reception systems require only minimal ground preparation before installation.

A variant of the ground-level dump feeder is the wide belt unloader. As its name implies, this system uses a wide belt (about 10 feet or 3 meters), which is slightly wider than the delivery vehicle, and is carried on a chain and slat conveyor.

In operation, it is similar to the walking floor system described above, with the vehicle tipping onto the end of the belt, and a certain quantity of storage being available within the body of the unit. Again, a discharge conveyor is provided to transfer coal into the boiler house conveying system, and little or no excavation is required for installation.

Unloading rates are also similar to the walking floor, although the maximum capacity of the body is unlikely to exceed one vehicle load. A schematic diagram of the system is shown in Figure 21.5.

21.5.7 Auger reception systems

Screw conveyors have been used for a number of years to extract coal from bunkers and coal heaps, usually for feeding directly to boilers. Examples of these systems are pneumatic drop feed or sprinkler stokers, underfeed stokers or as conveyors to serve the boiler feed hopper. Fixed screws, however (the design of which are discussed in

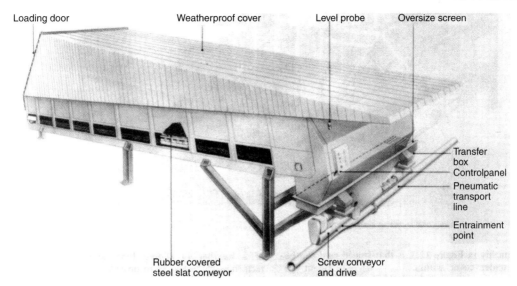

Loading door Weatherproof cover Level probe Oversize screen

Transfer box
Controlpanel
Pneumatic transport line
Entrainment point

Rubber covered steel slat conveyor

Screw conveyor and drive

Figure 21.5 Wide Belt Unloader

Section 21.7), will only recover a proportion of the coal within the dump area, and this has led to the development of sweeping screw conveyors which traverse through the coal.

The example shown in Figure 21.6 is based upon the use of such a sweeping screw, and uses a special drive mechanism to limit the force applied to achieve a

Retaining wall

Sweep auger

Discharge hopper

Safety gate

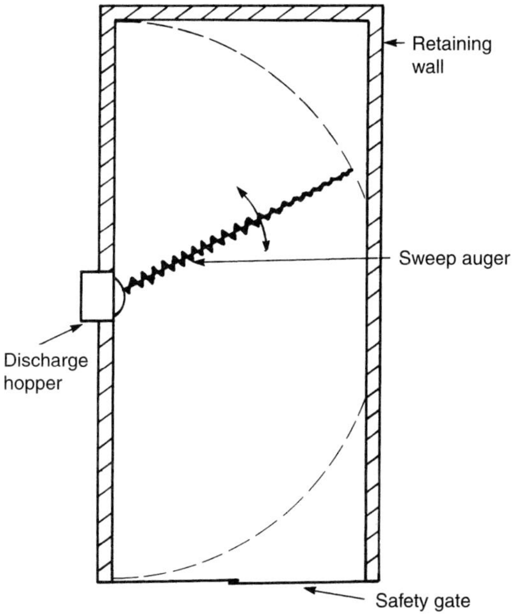

Figure 21.6 Ground sweep reclaimer

semi-circular sweep. This prevents damage caused by the sweeping motion from forcing the screw to move out of the plane of the sweep. Coal is tipped directly onto a grid screen from where it flows into a shallow (3.3 feet or 1 meter deep) bunker contained within a retaining wall enclosure. The rotary sweep auger works through 180° across the floor of the enclosure, drawing coal off into the boilerhouse conveying system.

21.6 Coal storage

The quantity of coal stored at any particular site is dictated by the likely risk of cuts in supply to any installation, and may vary from a few tons held in the boiler hopper to stockpiles of many thousands of tons.

21.6.1 Spontaneous heating

When storing any quantity of coal greater than about 50 tons and for longer than a few weeks, care has to be taken to avoid contamination, breakage of singles coal, and spontaneous heating.

Spontaneous heating arises from partial oxidation of the coal, and occurs when ventilation within the stock is sufficient to provide the oxygen necessary for oxidation but is insufficient to disperse the heat generated in the process. Signs of spontaneous heating are generally exhibited within a 10- to 12-week period after the coal is laid down. If there is no sign of heating after six months, it can be assumed that the coal is safe.

In order to prevent spontaneous heating it is necessary that either adequate ventilation should be provided or, alternatively, air should be almost completely excluded. For singles coal, ventilation is normally encouraged so that the air carries away any heat. In storing fine coal, however, open stockpiles should be compacted and silos sealed so that air is virtually excluded from the stock.

Generally, the younger soft, bright lignite coals are more susceptible to spontaneous heating than older coals such as anthracite. If there is a possibility of spontaneous heating in a silo then the condition of the coal should be monitored by measuring the carbon monoxide content of the air at the top of the silo. The use of thermocouples to indicate a rise in temperature due to spontaneous heating is ineffective. The occurrence of spontaneous heating is indicated by a relatively substantial increase in the level of carbon monoxide, and in this event, the silo should be emptied as quickly as possible and the coal spread thinly on the ground to cool. On no account should water be pumped into the stored coal in an attempt to extinguish the fire as water gas may be formed, although if spontaneous heating is detected sufficiently early, nitrogen purging of the silo can reduce the risk of the heating spreading.

21.6.2 Coal-storage silos

Limited ground space together with the requirement to store coal in any considerable quantity has led to the increased use of storage silos up to sizes in excess of 1000 tons capacity. Silos above 500 tons capacity should be constructed of reinforced concrete and may be circular or rectangular in cross section. Capacities between 300 and 500 tons are usually most economically obtained by using cylindrical silos constructed of concrete staves bound together with steel hoops. Below this capacity, vitreous enamel-coated steel silos are generally less expensive than concrete, and for smaller storage units of up to 80 tons, low-cost polyester silos may be used.

Filling of silos

The method selected for filling will be governed largely by the height of the silo. Where single coal is delivered into the silo by means of a vehicle-mounted pneumatic delivery system, the maximum height for delivery should not exceed 59 feet or 18 meters.

Taller silos can be considered when filled using a mechanical conveyor or elevator, or pneumatically using a purpose-designed on-site system as described in Section 21.4. Where fine coal is to be stored in silos, lean-phase pneumatic delivery is impractical and other mechanical or dense-phase pneumatic systems must be used.

Except when constructed of reinforced concrete, all silos should be filled from within the center third of the diameter to reduce the risk of eccentric loading. The action of filling a silo or bunker results in breakage of the coal particles. Simple gravitational fall will cause some degradation but most breakage occurs when some form of pneumatic system is used to fill the silo. The presence of fine coal particles can then result in segregation, the finer particles concentrating in a heap directly below the fill position while the larger ones roll to the outside of the pile. Such a concentration of fine particles results in a pillar of coal within the silo that will not flow. Segregation must therefore be avoided, and for lean-phase pneumatic delivery

it is recommended that the coal be spread by allowing it to fall onto the apex of a steel cone mounted below a velocity-reducing cyclone as shown in Figure 21.7. An impact and wear resistant rubber can be used to line the cyclone, which will reduce degradation as well as contributing to a significant reduction in noise caused by coal entering the cyclone.

Extraction from silos

In order to prevent asymmetric loading on the silo walls coal should only be discharged from the central one third of the diameter of the silo. With fine coal it is also important to ensure that mass flow of all of the coal within the silo is achieved so that no pockets of 'dead' coal are present which could lead to spontaneous heating. Satisfactory discharge of fine can be achieved by gravity using a conical outlet sloped at 70° to the horizontal, together with a minimum of 1 meter diameter or square outlet. Where a flat-bottomed silo is used, a mechanical extractor such as a sweep auger will be required to achieve satisfactory flow.

Singles coal is less susceptible to spontaneous heating than fines, and dead coal formed by the use of a single extraction point in the base of a flat-bottomed silo is acceptable. Because a cone angled at 55–60° to the horizontal will provide complete discharge of singles from a conical-bottomed silo it is therefore more common to use less expensive flat-bottomed silos for storing this fuel. Extraction may be by gravity, through a 1-meter diameter hole (or holes) or by mechanical means such as screw conveyors. In either case, the discharge should only be affected from the central one third of the diameter.

The extraction point in any silo should be designed such that access to the extraction equipment is available for maintenance purposes. Where mechanical extractors are used for fine coal it is normal for the floor of the silo to be constructed at a height of 2.5–3 meters above ground level, thereby providing space for conveyors to be mounted beneath.

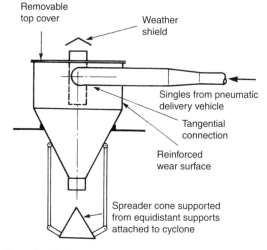

Figure 21.7 Velocity-reducing cyclone

Methane venting

Some coals, particularly when freshly mined, are liable to give off methane gas. In order to prevent an explosive mixture forming within the silo headspace a 150-mm diameter vent should be provided at the highest point within the roof of the silo.

Safety

As with all coal-storage systems, silos represent a *confined space* and the Occupational Safety and Health Act (OSHA) regulations should be fully complied with whenever a silo is entered. Particular care should be taken to avoid personnel having to walk over coal, which may have a void below the surface. Any portable lighting used should be approved battery-operated lamps.

21.6.3 Hopper design

All silos and bunkers will eventually terminate in a hopper to feed the stored coal into a conveyor or boiler. In the case of singles coal the hopper should have a minimum cross-sectional dimension of 23.6 inches (600 mm) and a minimum cone angle of 55° from the horizontal. In the case of fine coal these figures should be increased to 1 m and 70°, respectively. It is possible to reduce the cone angles slightly (by about 5°) by using a low-friction lining material such as ultra-high molecular weight (UHMW) polyethylene. This material, which should only be used in 100 per cent virgin state, is widely used for improving the flow characteristics of hoppers and chutes, and can be attached to virtually any surface, including concrete and steel.

It should be borne in mind, however, that UHMW polyethylene is combustible, and that both spontaneous heating and burn-back (from a stoker hopper) may give rise to poisonous fumes.

Storage systems designed to promote mass flow of the stored coal using the hopper outlet sizes recommended above typically have to feed into the relatively small-sized inlets of conveying systems. For singles coal, this transition from mass flow to controlled flow can be achieved by gravity, using a suitably steep-sided inlet hopper. Fine coal, however, will normally require the use of a mechanical discharge system.

A number of such devices are available to promote flow from hopper outlets, all working on the principle of agitating the coal close to where it is likely to bridge. The most common system consists of a vibrator attached in various ways to the side of the hopper and designed to create slip at the hopper wall. These can be very effective, although care has to be taken to avoid compacting the coal. A different system, which is currently becoming more popular, is the use of a bin activator, as shown in Figure 21.8. Again, a vibrator is used, but in this case, the vibration is applied to an inverted cone close to the bunker outlet, and this generally promotes better mass flow of the coal than the side vibrator systems.

An alternative method, more commonly used where flow problems exist due to poor hopper design or a change in fuel, is the use of one or more air blasters. These devices direct jets of compressed air into the coal in the vicinity of expected blockages. The sudden release of relatively small quantities of compressed air can be very effective in dislodging coal which has bridged across hopper outlets.

With any contained coal-storage system, it will occasionally be necessary to carry out maintenance on equipment installed beneath the outlet. In order to isolate this equipment from the flow of coal it is recommended that a rod gate valve be installed as shown in Figure 21.9. This valve is more robust than a slide valve, and is easier to free if it becomes corroded.

21.7 Coal conveying

Industrial boiler house coal-conveying systems are either mechanical or pneumatic, and the selection of one of these for a particular boiler house will depend on the following factors:

1. *Required coal-conveying rate.* Most boiler houses will require a minimum coal-conveying rate of three times

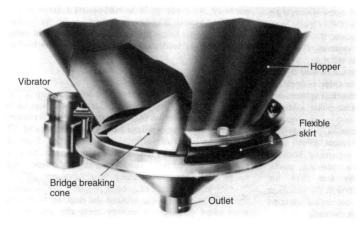

Figure 21.8 Bin activator

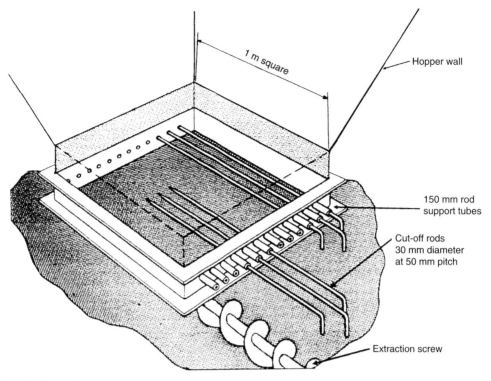

Figure 21.9 Rod gate valve

the maximum coal-burning rate in order to provide adequate time for maintenance or repairs. Mechanical systems are able to convey coal at rates ranging from zero up to several thousand tons per hour. Pneumatic systems such as conveyors have capacities ranging from about 5 tons per hour up to a maximum of about 60 tons per hour. When pneumatic conveyors are used for feeding direct to boilers, the flow rate may reduce to zero, but such low conveying rates are liable to cause increased coal breakage.

2. *Type of coal*. Mechanical systems are generally able to convey both singles and fine coal, although careful attention to design is necessary to minimize degradation of singles coal, and to provide self cleaning systems for fine coals, particularly where the fines (<0.5 mm) content is high. Pneumatic systems are selected according to the type of coal being conveyed. Lean-phase systems are generally only suitable for singles coal, whereas dense-phase systems are most suitable for fine. These features are discussed more fully in Section 21.7.2.

3. *Allowable power consumption*. Although, by itself, this factor is unlikely to determine the choice of system, it should be borne in mind that mechanical conveyors typically use only 20 per cent of the power required for equivalent pneumatic conveyors.

4. *System complexity and layout*. Mechanical systems can generally only convey coal in straight lines, although some designs can bend within a vertical plane. The more recent popularity of pneumatic conveyors is due in no small part to their ability to transport coal horizontally, vertically and around corners within an enclosed pipeline. Although this flexibility lends itself to the aesthetic design of boiler houses, it should not be abused. As with all handling systems, the more simple designs tend to be the most reliable.

5. *Costs*. For most industrial boiler houses where coal is conveyed to two or three boilers, the capital cost of pneumatic and mechanical systems are broadly similar. For very long dense-phase systems, however, the cost of air compressors will become significant, and for very short-distance straight-line conveying, it is likely that a mechanical system such as a screw conveyor will be least expensive. Because pneumatic systems contain few moving parts, maintenance costs are normally less than for equivalent mechanical systems, although the variation in maintenance requirements for mechanical systems is considerable.

21.7.1 Mechanical systems

Bucket elevator

These systems are commonly used for elevating coal up to 164 feet (50 m) vertically or inclined, at rates up to 50 tons per hour. No breakage of the coal occurs during elevation but a feed system is normally required to fill the buckets. Special designs using plastic or steel baskets will reduce the propensity for fine particles to stick to the buckets.

Belt conveyors

Belts are normally used where high flow rates (up to 1000 tons per hour) are required. Careful attention to the design and installation of belt-cleaning devices is required in order to prevent spillage from the return strand. The maximum angle of inclination is about 20° from the horizontal, and hence considerable space may be required. Ribbed belts will allow a slightly steeper angle to be attained, but cleaning becomes more difficult. Special designs of belt may be equipped with pockets, or the belt may even wrap around the coal. In these cases, it is often possible to convey vertically and around corners, but the capital cost is higher than the standard belt. No coal breakage will occur while the coal is being conveyed, although a feed system is usually required.

En-masse conveyors

These systems can convey coal at up to several hundred tons per hour, although they are more commonly used at 15–20 tons per hour. They can elevate vertically or inclined, or they will operate as horizontal transfer conveyors. A steel casing renders *en-masse* conveyors dust-tight, although coal rubbing against the casing during conveying causes a small degree of breakage.

Because no feed system is required, they can be mounted directly beneath bunkers or silos.

Screw conveyors

Although only able to transport coal in straight lines over relatively short distances (up to about 8 meters maximum), screw conveyors provide an inexpensive means of conveying or feeding coal. The use of intermediate bearings for increasing the length of screw is not recommended for coal because they cause breakage of singles and can cause blockages with fine coal (due to the build-up of fine particles). Difficulties of feeding fine coal into the relatively narrow inlet of screws makes them more commonly (but not exclusively) used for singles coal. In this case, careful attention to design is required in order to minimize degradation.

It is recommended that the pick-up length be restricted to less than ten diameters, and that the screw is tapered

along this length from shaft diameter at the tail end to full diameter at the casing inlet. The pitch of the flights should also be increased by a factor of about two at the tail end, reducing towards the casing inlet as shown in Figure 21.10. Screws designed in this way will reduce the tendency for coal to be picked up preferentially from the tail end, which causes recirculation and considerable breakage at the casing inlet.

Other mechanical systems

The majority of mechanical handling systems currently in use in industrial boiler houses are represented above. There are, however, a number of other systems occasionally used, which include vibratory conveyors, grab-skip hoists, drag-link conveyors and chain and bucket conveyors.

21.7.2 Pneumatic conveying systems

Pneumatic conveying systems for coal fall into two categories: dense phase and lean phase. By definition, lean-phase conveying occurs when coal particles are conveyed in suspension by virtue of the conveying air velocity, whereas dense-phase conveying only occurs when coal is conveyed using velocities lower than those required for suspension flow. Because relatively high-pressure air is generally required for dense-phase conveying when compared to lean-phase conveying, different types of equipment have evolved for each category.

One of the most critical applications for pneumatic conveyors is the transport of single coal. A combination of relatively high conveying velocities together with a considerable degree of inter-particulate movement and rubbing against pipe walls inevitably causes a certain amount of breakage. The actual breakage depends upon a number of factors, including conveying velocity and friability of the coal, but, using a properly designed system, it is usually well within acceptable limits. Occasionally, however, pneumatic handling systems are installed where their performance is adversely affected by poor overall design, particularly where manufacturers and installers are obliged to follow the wishes of boilerhouse designers who are unaware of the limitations of such systems.

Both dense- and lean-phase systems can be 'tuned' to a certain extent to achieve minimum coal breakage. In principle, because of its lower conveying velocities, a

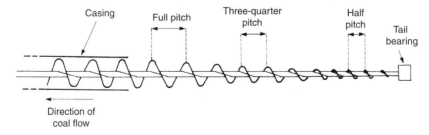

Figure 21.10 Screw conveyor pick-up for singles coal

dense-phase system should cause less coal breakage than lean phase. However, the variation in control of both coal and air flow rates available from lean-phase systems generally leads to a greater tolerance of poor design than dense-phase ones. For the latter it is therefore even more important that the overall system design be correct.

Dense-phase systems

Both singles and fine coal can be conveyed using dense-phase conveyors. In operation, coal enters a pressure vessel (shown schematically in Figure 21.11) by gravity. The size of the vessel depends upon the required conveying rate and distance, but would typically be 0.1 m³ capacity for 10 tons per hour. The inlet valve at the top of the vessel is then closed and sealed, and compressed air (typically at 30 to 50 psig) is introduced into the vessel. The air causes the coal to be conveyed out of the vessel and along the pipeline in the form of a coherent aerated slug. The supply of air is continued until the coal slug has been pushed through the pipe to the boiler hopper, and is then turned off. The sequence is then repeated until sufficient coal has been conveyed. The coherence of the slug is important to the control of the conveying process – if the slug breaks down for any reason then it will become more

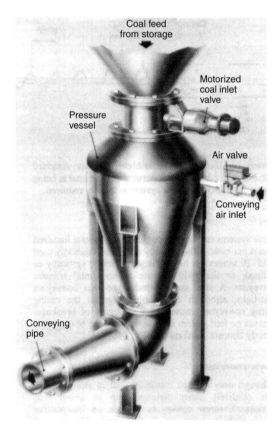

Figure 21.11 Dense-phase pneumatic conveyor

Coal feed from storage

Motorized coal inlet valve

Pressure vessel

Air valve

Conveying air inlet

Conveying pipe

porous to the conveying air. Some of the air will then escape through the slug, and reach very high velocities at its leading edge where there is little pressure resistance. The high-velocity air will then entrain the leading particles, conveying them in the lean phase, and leading to further breakdown of the slug and consequent breakage of the conveyed coal. The porosity of singles coal to the conveying air is far greater than that for fines, and conveying singles coal in the dense phase as a uniform and coherent slug is therefore more difficult to achieve.

Most conveying systems include a requirement to elevate coal to a high level. When using a dense-phase system it is desirable for the coal to enter a vertical riser immediately upon leaving the pressure vessel. At this point, the coal is still being conveyed at a relatively low velocity, and the vertical riser provides the best opportunity for a coherent slug to be formed. In order to minimize the number of bends in the system it is also desirable that this vertical riser should encompass the total lift required.

Dense-phase conveyor pressure vessels are manufactured to ASME Section VIII and require periodic inspection. The vessel itself, however, is generally maintenance free, except where certain fine coals are conveyed, when it is possible for fines to build up on the walls. The effect of such a build-up is to reduce the capacity of the vessel and hence to reduce the conveying rate. Methods used to prevent the fines from sticking include special finishes to the inner surface of the vessel, complete linings for the vessel walls, the use of water jets and the provision of access hatches for cleaning. All these methods are effective, but special finishes, which are applied by painting, may have to be renewed periodically.

Most pressure vessels are fed by means of vibrating feed chutes, and these have to be mechanically isolated from the pressure vessel. Failure of the closure valve at the top of the vessel to seal properly can result in the presence of high-pressure air in the inlet chute. This can lead to failure of the flexible joint between the chute and pressure vessel and the ejection of coal particles from the failed joint. It is therefore recommended that some form of personnel protection be provided at this point in order to prevent the ejection of coal particles. High-pressure air in the inlet chute can also lead to coal bridging and access to the chute to clear blockages may be useful.

Lean-phase conveyors

Lean-phase pneumatic conveyors are used for conveying singles coal, either by positive pressure or by suction from low-pressure centrifugal fans. They are not suitable for fine coal – wet fine particles tend to adhere and build up in bends and discharge systems.

The most common example of a lean-phase conveyor is that fitted to pneumatic delivery vehicles as described in Section 21.4, where a rotary valve is used to feed coal into the air stream. Fixed boiler house systems, however, are generally fed by screw conveyor, as shown schematically in Figure 21.12. In this case, the screw conveyor is controlled by the boiler combustion system, and the pipework delivers coal directly to the grate where the conveying air is subsequently used for the combustion

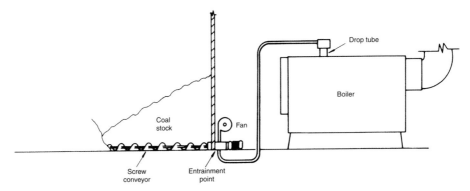

Figure 21.12 Lean-phase pneumatic conveyor

process. It is always important to minimize coal breakage in these systems, and the screw conveyor should be designed as described in Section 21.7.1. Rotary valves and screw conveyors can be used for entraining coal into either positive (blowing) or suction systems. A simpler entrainment device is the suction nozzle developed by British Coal. This system, shown in Figure 21.13, offers the advantage of no mechanical parts at the pick-up zone, although a rotary valve or lock-hopper arrangement is required at the discharge point.

One of the most important factors influencing coal breakage is conveying velocity. In order to obtain minimum breakage consistent with reliable conveying it is recommended that the conveying air velocity be 27 m s^{-1} for lean-phase suction systems and 24 m s^{-1} for lean-phase positive-pressure systems. (The difference is due to the variation in conveying air density in each type of system.)

General design considerations for pneumatic conveying

The layout of pipework for any particular boiler house will depend upon the boiler layout and available space for coal storage. With any pneumatic conveying system, pipe work which is inclined at more than about 10° from the horizontal or vertical causes segregation and 'layered flow', and should be avoided. In order to achieve maximum reliability from the conveying system it is important that throughout its length the pipework should promote smooth flow, particularly for dense-phase systems. It is therefore important that the following three installation aspects are adhered to:

1. Where pipe lengths are joined, some means of ensuring continuity of the bore is required. This is often achieved by the use of spiggoted or machined flanges, but any alternative method which guarantees alignment can be used.
2. Welded butt joints of pipe should be avoided, as it is impossible to maintain a completely smooth internal surface at such a joint.
3. Kinks in the pipework should be avoided. Only straight sections or properly swept bends should be used.

The required coal-conveying rate will determine the pipework diameter. This will range from 4 inches

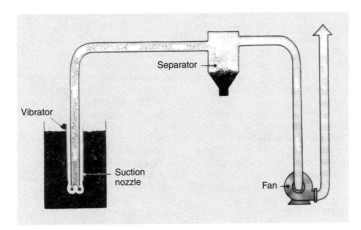

Figure 21.13 Suction nozzle pneumatic conveyor

(100 mm) bore for fine coal at (say) 10 tons per hour, up to 21 inches (300 mm) bore at about 80 tons per hour, depending on the conveying distance. The minimum recommended pipe diameter for conveying singles coal at flow rates in excess of about 5 tons per hour is 5 inches (215 mm). Although 4 inches (100 mm) bore systems have been successfully used in the past, some of the singles coal now supplied by British Coal to the industrial market has a top size of 1.5 inches (38 mm) rather than the previous 1 inch (25 mm), and blockages are more likely to occur with the lower pipe/particle diameter ratio. When using larger pipe diameters, consideration should be given to the cost of pipework maintenance, particularly replacement of bends.

Washed coals used in industrial boiler houses do not generally cause significant wear of the pipeline, and depending on the length of the system and pipework diameter, ordinary mild steel bends should be capable of providing a lifetime equivalent to at least 20,000 tons of coal conveyed. Where unwashed coals are used, however, this figure may be somewhat reduced. Because these higher-ash fuels are more likely to be used on larger boiler plant where the coal throughput is much higher, some form of bend reinforcement is desirable. For large-diameter pipelines, consideration should be given to manufacturing bends in sections to facilitate maintenance. In this case, it is only necessary to replace the failed section rather than the complete bend.

Bends for pneumatic conveying of coal should always be of the long-radius sweeping type. It is recommended that a radius/diameter ratio of 12 be used wherever possible.

It is likely that any pneumatic conveying system will be required to convey coal to more than one destination. Where the conveying pipeline passes directly over a row of boiler feed hoppers, each hopper can be fed almost directly from the conveying pipe by means of a two-position diverter valve. The valve can be switched such that coal either passes the feed point or continues undisturbed along the pipeline or it is diverted downwards into the hopper. A somewhat different type of diverter valve is often used where a pipeline split is required to convey coal to two different destinations. In either case, however, it is important that the valve does not interrupt the smooth flow of the coal, and therefore some form of positive alignment is required. In addition, the valve should not present any sections of pipeline which do not follow the recommendations previously discussed. The diverted leg should present a smooth change in direction, rather than a kink or flat diverter plate.

Because the operation of positive-pressure pneumatic conveyors relies upon the maintenance of air pressure behind the coal, it is important (particularly for dense-phase systems) that air leakage from diverter valves out of the *closed* leg be minimal. Such leakage may be prevented by the use of low-friction slides or inflatable seals. Whatever method is used, the valve should be operated at regular intervals to prevent sticking of the seals.

All designs of diverter valve will require occasional maintenance, and proper access should therefore be provided.

21.8 Ash handling

There are essentially two type of ash produced in a furnace: bottom ash and flyash. Bottom ash is slag, which builds up on the heat-absorbing surfaces of the furnace, superheater, reheater, and economizer that eventually falls either by its own weight or because of load changes or soot blowing. With low ash-fusion temperatures, a large amount of molten slag can adhere to the furnace walls and subsequently fall through the furnace bottom. Other ash becomes entrained with, and carried away by, the flue gas stream and is collected in economizer or dust-collection equipment hoppers. This is called flyash.

The amount of air-borne ash passing through the furnace depends on the dust-bearing capacity of the combustion gases, on the size and shape of the particles, and on the density of the ash relative to that of the upward flowing gas. Courser particles fall more rapidly than fine. Thus, for a given upward velocity, fine particles either will drop to the bottom or be thrown from the center of the furnace toward the bounding furnace walls. Generally, particles larger than 0.011-inch diameter can be expected to fall to the bottom when the vertical velocity of the gases in a furnace is 10 feet per second or less.

Any industrial boiler house, whether fired by oil, gas or coal, must have provision for handling the fuel. However, oil and gas firing produce relatively insignificant solid waste products, and boiler house operators who have become used to these inherently high-amenity fuels must accept that when coal is burnt then a certain amount of ash will be produced which will require handling. The quantity of ash produced from coal supplied to the industrial market is generally somewhat less than 10 per cent of the total amount of coal burnt, and its physical characteristics are determined by the firing method used for the coal. Some combustion systems are deliberately operated to form hard pads of clinker to facilitate manual removal and handling, but in other cases, large clinkers may only form under faulty operation.

Because the quantity of ash requiring handling is small when compared to the coal burnt, there is often a reluctance to invest additional capital in ash-handling equipment. Furthermore, any equipment which is installed must be capable of dealing with a material which may still contain some burning carbon, and is dusty and highly abrasive, and may be corrosive when wet. For smaller boiler houses some manual involvement using a shovel, wheelbarrow and dustbin is therefore often cost effective.

Where automatic ash handling is required, either pneumatic or submerged mechanical systems are normally used. Ordinary dry mechanical conveyors are not generally suitable for ash handling because of the likely ingress of dust into the working parts and of the possibility of high-temperature ash.

When burning coal in a fluidized bed it is likely that a proportion of the ash will remain in the bed as large (>4 mm) particles. The removal of this material is a specialized technology, which is beyond the scope of this chapter. Detailed information on such systems is available from both British Coal and the boiler manufacturers.

21.8.1 Pneumatic handling of ash

Before being pneumatically conveyed a mechanical clinker breaker or crusher is required to reduce the maximum particle size to less than about 1 inch (25 mm). Clinkers from some combustion systems may exceed 20 inches (500 mm) in size. Care also has to be taken to ensure that any burning or hot carbon which may be present in the ash is sufficiently cooled to prevent re-ignition during the pneumatic conveying process, when oxygen is intimately available.

Both dense- and lean-phase pneumatic systems are currently widely used for handling ash, although the wear rate for the pipework is high, and even specially reinforced bends often require replacement several times each year. One of the major factors influencing bend wear rate is the conveying velocity of the ash particles, and for lean-phase systems some form of reactive control is recommended to maintain the air velocity at about 27 m s^{-1} for clinker ash and at about 21 m s^{-1} for grits and fly-ash. For the latter materials, where the particle size is small and the porosity

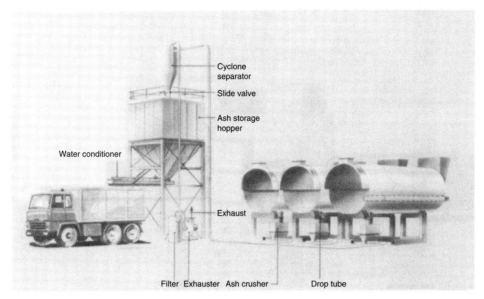

Figure 21.14 Pneumatic ash-conveying system

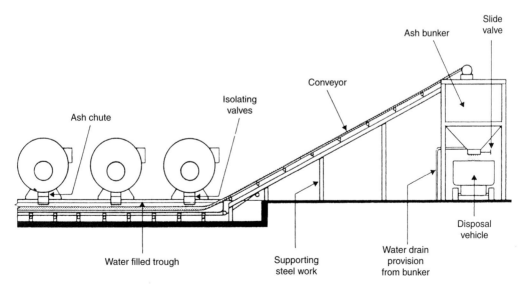

Figure 21.15 Submerged mechanical ash conveyor

of the material consequently low, dense-phase systems may have the advantage of conveying at lower velocities than lean phase. However, in the inevitable event of the pipework wearing through, a suction lean-phase system as shown in Figure 21.14 offers the advantage of in-leakage of clean air into the system, rather than dusty air leaking out into the boiler house. Reinforcement of bends can be carried out using alumina or basalt tiles, or concrete or other mixtures contained within a channel welded to the exteriors of the bend. Specially designed wear-resistant bends are also available using materials such as Ni-hard.

Ash is normally pneumatically conveyed from the boiler house to an overhead bunker for discharge into lorries or skips for disposal, and in order to separate the fine ash from the conveying air, high-efficiency cyclones and filters are required. Conditioning can reduce dust emitted from the ash during discharge from the bunker with paddle-screw mixers and water sprays.

Because the conveyed ash may be hot, it is important that unburned carbon is not allowed to enter the ash silo. If a boilerhouse vacuum-cleaning system is fitted to a suction pneumatic ash-removal system then a separate silo should be provided for this purpose. Explosions have occurred in ash silos where coal has been cleaned from boiler house floors and conveyed into silos containing hot ash. If this practice is likely then a suitable explosion-relief device should be incorporated into the silo.

21.8.2 Submerged mechanical systems

Where mechanical systems are used for ash handling, they are normally partially submerged under water, as shown in Figure 21.15. This allows a water seal to be maintained at the ash outlet of the boiler, which effectively isolates the combustion chamber from atmospheric pressure. A further advantage of these submerged systems is that they are inherently dust free, and any burning carbon in the ash will be immediately extinguished. There is also no risk of an explosion occurring in the storage silo due to the presence of hot combustible material.

Both *en-masse* and belt conveyors are used for submerged ash conveying. They are operated very slowly (typically, about $0.1 \, \mathrm{m \, s^{-1}}$), and this allows water to drain from the ash as the conveyor is inclined up towards the discharge point. Although submerged conveyors are used for all types of ash, very fine ash can float on the water and some means of wetting may be required.

22

Steam Utilization

Albert Armer

Spirax Sarco Ltd

Contents

22.1 Introduction

The use of steam provides an ideal way to deliver just the right amount of heat energy to a point of use, to help with product manufacture or to provide an acceptable environment. The story of steam efficiency begins at the boiler but it does not end until the energy carried by the steam is transferred appropriately and the resulting hot condensate is returned to the boiler. Control of steam is necessary throughout the plant if the performance designed into the system is to be achieved and maintained. However, the natural laws which steam obeys are few in numbers and are easy to understand. Following them allows efficiencies to be achieved that would be envied with most other energy-distribution systems.

Two of the laws of thermodynamics and two of the laws of motion will cover almost all needs. In the simplest terms, it can be said that:

1. Energy is not destructible. It can always be accounted for, and if it disappears at point A then it reappears in equal amount at point B. This ensures that the world is a consistent place and that energy does not mysteriously appear from or disappear to nowhere. In addition it ensures that the steam tables can be relied on always to provide information on the properties of steam.
2. Heat flows from higher-temperature to lower-temperature objects without any help. It follows that the rate of flow will vary directly with temperature differences, and inversely with any resistances to this flow. A temperature difference is necessary for heat to flow!
3. Any matter tends to move in the direction in which it is pushed, and, in particular, because of the effects of friction, any fluid will only flow from high-pressure to lower-pressure regions. Again, the rate of flow will vary directly with pressure differences, and inversely with any resistances to this flow.
4. Gravity acts downwards! The denser constituents in a mixture often tend to move to the bottom of a space, unless other forces acting on them oppose such motion.

If these basic laws can be accepted it is easy to build up an understanding of steam systems and the way that steam behaves. The ground rules for an effective (and efficient) steam system then quickly become apparent. Specific heats and weights for various solids, liquids, gases and vapors, as well as pipeline capacities, are shown in Tables 22.1–22.4.

22.2 What is steam?

The datum point, when considering the steam/water substance, is usually taken as water at the temperature of melting ice, at normal atmospheric pressure and so at a temperature of 0°C. Adding heat energy raises its temperature; some 419.04 kJ will raise 1 kg to 100°C, when any further addition of heat evaporates some of the water. If 2257 kJ are added to each kg of water, then all the water becomes the dry gas, steam. Equally, if only part of this extra energy is added – say, 90 per cent – then 90 per cent of the water evaporates and the other 10 per cent

Table 22.1 Specific heats and weights of various solids

Material	Specific gravity	Btu per lb per °F
Aluminum	2.55–2.8	0.22
Andalusite		0.17
Antimony		0.05
Apatite		0.20
Asbestos	2.1–2.8	0.20
Augite		0.19
Bakelite, wood filler		0.33
Bakelite, asbestos filler		0.38
Barite	4.5	0.11
Barium	3.5	0.07
Basalt rock	2.7–3.2	0.20
Beryl		0.20
Bismuth	9.8	0.03
Borax	1.7–1.8	0.24
Boron	2.32	0.31
Cadmium	8.65	0.06
Calcite, 32–100°F		0.19
Calcite, 32–212°F		0.20
Calcium	4.58	0.15
Carbon	1.8–2.1	0.17
Carborundum		0.16
Cassiterite		0.09
Cement, dry		0.37
Cement, powder		0.20
Charcoal		0.24
Chalcopyrite		0.13
Chromium	7.1	0.12
Clay	1.8–2.6	0.22
Coal	0.64–0.93	0.26–0.37
Cobalt	8.9	0.11
Concrete, stone		0.19
Concrete, cinder		0.18
Copper	8.8–8.95	0.09
Corundum		0.10
Diamond	3.51	0.15
Dolomite rock	2.9	0.22
Fluorite		0.22
Fluorspar		0.21
Galena		0.05
Garnet		0.18
Glass, common	2.4–2.8	0.20
Glass, crystal	2.9–3.0	0.12
Glass, plate	2.45–2.72	0.12
Glass, wool		0.16
Gold	19.25–19.35	0.03
Granite	2.4–2.72	0.19
Hematite	5.2	0.16
Hornblende	3.0	0.20
Hypersthene		0.19
Ice, −112°F		0.35
Ice, −40°F		0.43
Ice, −4°F		0.47
Ice, 32°F		0.49
Iridium	21.78–22.42	0.03
Iron, cast	7.03–7.13	0.12
Iron, wrought	7.6–7.9	0.12
Labradorite		0.19
Lava		0.20
Lead	11.34	0.03
Limestone	2.1–2.86	0.22
Magnetite	3.2	0.16

(continued overleaf)

Table 22.1 (*continued*)

Material	Specific gravity	Btu per lb per °F
Magnesium	1.74	0.25
Malachite		0.18
Manganese	7.42	0.11
Marble	2.6–2.86	0.21
Mercury	13.6	0.03
Mica		0.21
Molybdenum	10.2	0.06
Nickel	8.9	0.11
Oligloclose		0.21
Orthoclose		0.19
Plaster of paris		1.14
Platinum	21.45	0.03
Porcelain		0.26
Potassium	0.86	0.13
Pyrexglass		0.20
Pyrolusite		0.16
Pyroxylin plastics		0.34–0.38
Quartz, 55–212°F	2.5–2.8	0.19
Quartz, 32°F		0.17
Rock salt		0.22
Rubber		0.48
Sandstone	2.0–2.6	0.22
Serpentine	2.7–2.8	0.26
Silk		0.33
Silver	10.4–10.6	0.06
Sodium	0.97	0.30
Steel	7.8	0.12
Stone		0.20
Stoneware		0.19
Talc	2.6–2.8	0.21
Tar	1.2	0.35
Tellurium	6.0–6.24	0.05
Tin	7.2–7.5	0.05
Tile, hollow		0.15
Titanium	4.5	0.14
Topaz		0.21
Tungsten	19.22	0.04
Vanadium	5.96	0.12
Vulcanite		0.33
Wood	0.35–0.99	0.32–0.48
Wool	1.32	0.33
Zinc blend	3.9–4.2	0.11
Zinc	6.9–7.2	0.09

Table 22.2 Specific heats and weights of various liquids

Liquid	Specific gravity	Btu per lb per °F
Acetone	0.790	0.51
Alcohol, ethyl, 32°F	0.789	0.55
Alcohol, ethyl, 104°F	0.789	0.65
Alcohol, methyl, 40–50°F	0.796	0.60
Ammonia, 32°F	0.62	1.10
Ammonia, 104°F		1.16
Ammonia, 176°F		1.29
Ammonia, 212°F		1.48
Ammonia, 238°F		1.61
Anilin	1.02	0.52
Benzol		0.42
Calcium chloride	1.20	0.73
Castor oil		0.43
Citron oil		0.44
Diphenylamine	1.16	0.46
Ethyl ether		0.53
Ethylene glycol		0.53
Fuel oil	6.96	0.40
Fuel oil	0.91	0.44
Fuel oil	0.86	0.45
Fuel oil	0.81	0.50
Gasoline		0.53
Glycerine	1.26	0.58
Kerosene		0.48
Mercury	13.6	0.033
Naphthalene	1.14	0.41
Nitrobenzole		0.36
Olive oil	0.91–0.94	0.47
Petroleum		0.51
Potassium hydrate	1.24	0.88
Sea water	1.0235	0.94
Sesame oil		0.39
Sodium chloride	1.19	0.79
Sodium hydrate	1.27	0.94
Soybean oil		0.47
Toluol	0.866	0.36
Turpentine	0.87	0.41
Water	1	1.00
Xylene	0.861–0.881	0.41

remains liquid. The specific volume of steam at atmospheric pressure is 1.673 m³/kg, so the mixture of 90 per cent steam and 10 per cent water would occupy a volume of (0.9 × 1.673) m³ plus (0.1 × 0.001) m³. Clearly, almost all the volume of this 1 kg of steam and water is occupied by the steam, at 1.5057 m³, and the mixture would be described as steam with a dryness fraction of 0.9.

If the water is kept at a pressure above atmospheric, its temperature can be raised above 100°C before boiling begins. At 100 psig, for example, boiling point is at about 184.1°C. The extra energy needed to convert water at this pressure and temperature into steam (the enthalpy of evaporation) is now rather less at 2000.1 kJ/kg, while the volume of 1 kg of pure steam is only 0.177 m³. These

figures are all recorded in the steam tables at the end of this chapter.

When steam at the saturation temperature contacts a surface at a lower temperature, and heat flows to the cooler surface, some of the steam condenses to supply the energy. With a sufficient supply of steam moving into the volume that had been occupied by the steam now condensed, the pressure and temperature of the steam will remain constant. Of course, if the condensate flows to a zone where it is no longer in contact with the steam it can cool below steam temperature while supplying heat to a cooler surface.

Equally, if steam at the saturation temperature were to contact a surface at a higher temperature, as in some boilers, its temperature could be increased above the evaporating temperature and the steam would be described as superheated. Superheated steam is very desirable in turbines, where its use allows higher efficiencies to be reached, but it is much less satisfactory than saturated

Table 22.3 Gases and vapors

Gas or vapour	Specific heat, Btu per lb per °F at constant pressure	Specific heat, Btu per lb per °F at constant volume
Acetone	0.35	0.315
Air, dry, 50°F	0.24	0.172
Air, dry, 32–392°F	0.24	0.173
Air, dry, 68–824°F	0.25	0.178
Air, dry, 68–1166°F	0.25	0.184
Air, dry, 68–1472°F	0.26	0.188
Alcohol, C_2H_5OH	0.45	0.398
Alcohol, CH_3OH	0.46	0.366
Ammonia	0.54	0.422
Argon	0.12	0.072
Benzene, C_6H_6	0.26	0.236
Bromine	0.06	0.047
Carbon dioxide	0.20	0.150
Carbon monoxide	0.24	0.172
Carbon disulphide	0.16	0.132
Chlorine	0.11	0.82
Chloroform	0.15	0.131
Ether	0.48	0.466
Hydrochloric acid	0.19	0.136
Hydrogen	3.41	2.410
Hydrogen sulphide	0.25	0.189
Methane	0.59	0.446
Nitrogen	0.24	0.170
Nitric oxide	0.23	0.166
Nitrogen tetroxide	1.12	1.098
Nitrous oxide	0.21	0.166
Oxygen	0.22	0.157
Steam, 1 psia, 120–600°F	0.46	0.349
Steam, 14.7 psia, 220–600°F	0.47	0.359
Steam, 150 psia, 360–600°F	0.54	0.421
Sulfur dioxide	0.15	0.119

steam in heat exchangers. It behaves as a dry gas, giving up its heat content rather reluctantly as compared with saturated steam, which offers much higher heat transfer coefficients.

The condensate produced within the heat exchangers, as also within the steam lines, is initially at the saturation temperature and carries the same amount of energy as would boiler water at the same pressure. If it is discharged to a lower pressure, through a manual or automatic drain valve (steam trap) or even through a leak, it then contains more energy than water is able to hold at the lower pressure if it is to remain liquid. If the excess of energy amounted to, say, 5 per cent of the enthalpy of evaporation at the lower pressure, then 5 per cent of the water would be evaporated. The steam released by this drop in pressure experienced by high-temperature water is usually called flash steam. Recovery and use of this low-pressure steam, released by flashing, is one of the easiest ways of improving the efficiency of steam-utilization systems.

It is equally true that condensate, even if it has been released to atmospheric pressure, carries the same 419.04 kJ/kg of heat energy that any other water at the same temperature would hold. Allowing condensate to drain to waste, instead of returning it to the boiler feed tank or de-aerator, makes no economic sense. Condensate is a form of distilled water, requiring little chemical feed treatment or softening, and it already holds energy, which may amount to 15 per cent of the energy, which would have to be supplied to cold make-up feedwater, even in relatively low-pressure systems. An installation that makes good use of both flash steam and condensate will see the benefits on the bottom line! To summarize:

1. Saturated steam provides a heat source at a temperature that is readily controlled, by the control of pressure;

Table 22.4 Pipeline capacities at specific velocities (metric SI units)

Pressure (bar)	Velocity (m/s)	kg/h										
		15 mm	20 mm	25 mm	32 mm	40 mm	50 mm	65 mm	80 mm	100 mm	125 mm	150 mm
0.4	15	7	14	24	37	52	99	145	213	394	648	917
	25	10	25	40	62	92	162	265	384	675	972	1 457
	40	17	35	64	102	142	265	403	576	1 037	1 670	2 303
0.7	15	7	16	25	40	59	109	166	250	431	680	1 006
	25	12	25	45	72	100	182	287	430	716	1 145	1 575
	40	18	37	68	106	167	298	428	630	1 108	1 712	2 417
1.0	15	8	17	29	43	65	112	182	260	470	694	1 020
	25	12	26	48	72	100	193	300	445	730	1 160	1 660
	40	19	39	71	112	172	311	465	640	1 150	1 800	2 500
2.0	15	12	25	45	70	100	182	280	410	715	1 125	1 580
	25	19	43	70	112	162	295	428	656	1 215	1 755	2 520
	40	30	64	115	178	275	475	745	1 010	1 895	2 925	4 175
3.0	15	16	37	60	93	127	245	385	535	925	1 505	2 040
	25	26	56	100	152	225	425	632	910	1 580	2 480	3 440
	40	41	87	157	250	357	595	1 025	1 460	2 540	4 050	5 940
4.0	15	19	42	70	108	156	281	432	635	1 166	1 685	2 460
	25	30	63	115	180	270	450	742	1 080	1 980	2 925	4 225
	40	49	116	197	295	456	796	1 247	1 825	3 120	4 940	7 050

(continued overleaf)

Table 22.4 *(continued)*

Pressure (bar)	Velocity (m/s)	kg/h										
		15 mm	*20 mm*	*25 mm*	*32 mm*	*40 mm*	*50 mm*	*65 mm*	*80 mm*	*100 mm*	*125 mm*	*150 mm*
5.0	15	22	49	87	128	187	352	526	770	1 295	2 105	2 835
	25	36	81	135	211	308	548	885	1 265	2 110	3 540	5 150
	40	59	131	225	338	495	855	1 350	1 890	3 510	5 400	7 870
6.0	15	26	59	105	153	225	425	632	925	1 555	2 525	3 400
	25	43	97	162	253	370	658	1 065	1 520	2 530	4 250	6 175
	40	71	157	270	405	595	1 025	1 620	2 270	4 210	6 475	9 445
7.0	15	29	63	110	165	260	445	705	952	1 815	2 765	3 990
	25	49	114	190	288	450	785	1 205	1 750	3 025	4 815	6 900
	40	76	177	303	455	690	1 210	1 865	2 520	4 585	7 560	10 880
8.0	15	32	70	126	190	285	475	800	1 125	1 990	3 025	4 540
	25	54	122	205	320	465	810	1 260	1 870	3 240	5 220	7 120
	40	84	192	327	510	730	1 370	2 065	3 120	5 135	8 395	12 470
10.0	15	41	95	155	250	372	626	1 012	1 465	2 495	3 995	5 860
	25	66	145	257	405	562	990	1 530	2 205	3 825	6 295	8 995
	40	104	216	408	615	910	1 635	2 545	3 600	6 230	9 880	14 390
14.0	15	50	121	205	310	465	810	1 270	1 870	3 220	5 215	7 390
	25	85	195	331	520	740	1 375	2 080	3 120	5 200	8 500	12 560
	40	126	305	555	825	1 210	2 195	3 425	4 735	8 510	13 050	18 630

2. It carries very large amounts of heat as enthalpy of evaporation or latent heat in relatively small weights of steam;
3. It supplies heat, by condensing, at a constant temperature and with high heat transfer coefficients, so it maximizes the effectiveness of heat exchangers;
4. Recovery and re-use of the condensate and its heat content is usually simple and effective.

22.2.1 Working pressure in the boiler and mains

Where steam is generated as a source of energy for turbines or engines the boiler pressure is usually high and often the steam is superheated. Steam for process use or for heating may then be supplied through pressure-reducing valves or be available at pass-out pressures from the engine or at the exhaust pressure. In other cases where steam is wanted for process or heating use only, use of a packaged boiler, or sometimes a single-pass watertube steam generator is almost universal. The most suitable operating pressure for the boiler has then to be decided. It should be said that boilers operating at or close to their designed working pressure provide steam of the best quality. When the steam is to be used at a lower pressure, then pressure-reducing valves or other control valves may be utilized, close to the steam-using points.

The reason for this becomes clear when the specific volume of steam at varying pressures is noted from the steam tables and Figure 22.1 is considered. This shows a boiler operating at either high or lower pressure while producing the same weight of steam from the same energy input. Energy flows through the outer surface of the tubes into the boiler water, and when this water is at saturation temperature any addition of energy means the formation of steam bubbles. These bubbles rise to the surface and break, releasing steam into the steam space.

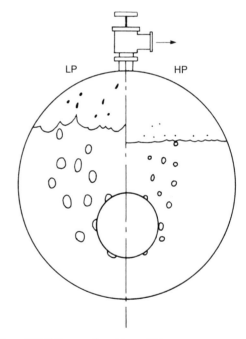

Figure 22.1 Boiler operation at low or high pressure

The volume of a given weight of steam contained in these bubbles is inversely related to the pressure at which the boiler operates. If this pressure is lower than the design pressure, the volume in the bubbles is increased. This raises the apparent water level in the boiler, reducing the volume of the steam space. The greater volume of the bubbles enlarges the turbulence at the surface, and as the bubbles burst, splashing of droplets into the steam space increases. The larger volume of steam, flowing towards the boiler crown valve through the reduced steam space volume, moves at higher velocity. All these factors increase the carry-over of water droplets with the steam into the distribution system.

There is much to be said in favor of carrying the steam to the points where it is to be used, at a pressure close to that of the boiler. The use of a high-distribution pressure means that the size of the steam mains is minimized. The smaller mains have lower heat losses, so that better-quality steam at the usage points is more readily achieved and the smaller pipes are often much lower in capital cost.

Pressure reduction to the levels needed by the steam-using equipment can then be achieved by the use of pressure-reducing stations located close to the steam users themselves. The individual reducing valves will tend to be smaller in size than would valves at the supply end, will give closer control of the reduced pressures and emit less noise. Problems which may arise if a whole section of a plant were dependent on a single reducing valve are avoided. The effects on the steam-using equipment of pressure drops in the pipework, varying in amount at different loads, are eliminated.

22.3 The steam load

Before selecting the size of a steam-control valve, a supply main or even a steam boiler it is necessary to know at least approximately (or better) how much steam is to be supplied. Where steam is to be used as the energy source for heat exchangers this is the same as knowing the heat load that is to be met.

Almost all heat loads fall into one of two categories. Either some material is to be heated from a lower temperature to a higher one or it is to be maintained at a high temperature while heat is supplied at a rate sufficient to balance the heat losses. In the first case the amount of heat needed to produce the change in temperature is given by

$$Q = W \times Sp \times \delta t$$

where

Q = heat load (kJ),
W = weight of material (kg),
Sp = specific heat (kJ/kg°C),
δt = temperature rise (°C).

More usefully, since the steam flow rate is sought rather than the weight of steam,

$$q = \frac{Q}{h} = \frac{W \times Sp \times \delta t}{h}$$

where

q = heat flow (kJ/h),
h = time available (h).

Then if hfg = enthalpy of evaporation of the steam at the pressure involved,

$$Ws = \frac{W \times Sp \times \delta t}{hfg \times h} \qquad (22.1)$$

where

Ws = flow rate of steam (kg/h),
hfg = enthalpy of evaporation (kJ/kg).

The second case is the supply of heat at a rate, which balances the heat losses, and here the heat load is given by

$$q_1 = UA\Delta t \qquad (22.2)$$

where

q_1 = heat flow (kJ/h)
U = heat transfer coefficient for heat flow through the surface involved (kJ/m²°C h),
Δt = temperature difference across surface involved (°C),
A = area of surface involved (m²).

It must be said that heat transfer coefficients are affected by so many variables that they are best regarded as approximations, unless measurements have been made on identical equipment under similar conditions. Further, it will be noticed that Equations (22.1) and (22.2) each include a temperature difference term, Δt or δt, but two different quantities are meant. To avoid confusion, it is useful to construct a diagram of the temperatures around the heat exchanger, as shown in Figure 22.2. At temperature T_1 a horizontal line represents the temperature of steam within the exchanger. Points T_2 and T_3 are the inlet and outlet temperatures of the heated material (or initial and final temperatures). A line is drawn from T_2 to T_3, and the difference between T_2 and T_3 is the temperature increase of the material, δt in Equation (22.1). The midpoint of the line T_2 to T_3 gives the arithmetic mean of the material temperatures, T_m, and the difference between T_m and T_1 is then the arithmetic mean temperature difference δt in Equation (22.2). Note that although theoretical considerations lead to the use of logarithmic means in heat transfer calculations, arithmetic means are usually sufficiently accurate when sizing steam supply equipment.

Some typical specific heats are listed in the tables, and a few examples of heat load calculations may be useful.

1. An air heater battery raises 2 m³/s of air from 5°C to 30°C using steam at a gauge pressure of 6 bar. What is the hourly steam load? 2 m³/s = 2 × 3600 m³/h = 2 × 3600/0.76 kg/h = 9474 kg/h (0.76 m³ of air at 5°C weighs approximately 1 kg)

$$Ws = \frac{9474 \times 1.0 \times (30 + 5)}{2066 \times 1} \text{ (from equation (22.1))}$$

= 160.5 kg/h (1 kg of air is heated through 1°C by 1.0 kJ).

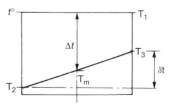

Figure 22.2

2. A calorifier holding 2000 l is heated from 10°C to 60°C in 2 h using steam at a gauge pressure of 3 bar. What is the hourly steam load?

$$Ws = \frac{W \times Sp \times \delta t}{hfg \times h}$$

$$= \frac{2000 \times 4.186 \times (60 - 10)}{2133.4 \times 2} = 98.1 \, \text{kg/h}$$

3. A calorifier has a heating surface area of 2 m² and is required to heat a flow of water from 65°C to 80°C. The U value has been found to be 1250 W/m²°C, and steam can be supplied at a gauge pressure of 2 bar in the steam chest. What rate of water flow can the calorifier handle (see Figure 22.3). From Equation (22.2):

$$q = UA\Delta t$$

$$= 1250 \times 2 \times 61.2$$

$$= 153000 \, \text{W}$$

$$= 550800 \, \text{kg/h}$$

(This heat flow implies the condensing of some 550000/2163.3 or 254.6 kg/h of steam.) From Equation (22.1):

$$550000 = W \times 4.186 \times (80-65)$$

$$W = 8772 \, \text{kg/h of water}$$

22.3.1 Sizing the steam lines

The appropriate size of pipe to carry the required weight of steam at the chosen pressure must be selected, remembering that undersized pipes mean high-pressure drops and high velocities, noise and erosion. Conversely, when pipe sizing is unduly generous, the lines become unnecessarily expensive to install and the heat lost from them will be greater than it need be.

Steam pipes may be sized so that the pressure drop along them is within an acceptable limit, or so that velocities along them are not too high. It is convenient and quick to size shorter mains and branch lines on a velocity basis, but the pressure drops along longer runs should also be checked to verify that sufficient pressure is available at the delivery points.

22.3.2 Sizing on velocity

In lines carrying saturated steam, reasonable maximums for velocities are often taken at up to 40 m/s in larger pipes

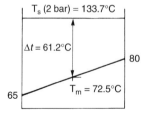

T_s (2 bar) = 133.7°C

$\Delta t = 61.2°C$

80

$T_m = 72.5°C$

65

Figure 22.3

and at higher pressures. However, 25 m/s may be more appropriate in the middle ranges and even 15 m/s with small-size lines and lower pressures. Carrying capacities of pipes may be read from a table of sizes and pressures, or if the specific volume v is read from the steam tables, then $W\,\text{kg/h} = 0.002827\,D^2V/v$ where D = pipe diameter (mm), V = velocity (m/s), v = specific volume (m³/kg). Velocities higher than 40 m/s may be accepted in the very large-diameter lines which some process or power industries use, especially when superheated steam is carried (up to 60 m/s or even more).

22.3.3 Sizing on pressure drop

In anything other than lines of only a very few meters in length, it is usual to choose sizes so that the pressure drop is not above, say, 0.3 bar/100 meter run. Sizing with greater flow rates in a given size of pipe (and correspondingly higher velocities) will increase the pressure drop in much more than linear proportion, and can soon lead to the pipe becoming quite unable to pass the required flow.

Calculation of pressure drops in steam lines is a time-consuming task and requires the use of a number of somewhat arbitrary factors for such functions as pipe wall roughness and the resistance of fittings. To simplify the choice of pipe for given loads and steam pressures, Figure 22.4 will be found sufficiently accurate for most practical purposes.

22.4 Draining steam lines

Draining of condensate from steam mains and branch lines is probably the most common application for steam traps. It is important for reasons of safety and to help achieve greater plant efficiency that water is removed from steam lines as quickly as possible. A build-up of water can, in some cases, lead to slugs of water being picked up by the steam flow and hurled violently at pipe bends, valves or other fittings, a phenomenon described graphically as waterhammer. When carried into heat exchangers, water simply adds to the thickness of the condensate film and reduces heat transfer. Inadequate drainage leads to leaking joints, and is a prime cause of wire drawing of control valve seating faces.

22.4.1 Waterhammer

Waterhammer may occur when water is pushed along a pipe by the steam instead of being drained away at the low points and is suddenly stopped by impact on a valve, pipe tee or bend. The velocities which are achieved by such slugs of water can be very high, especially on start-up, when a pipe is being charged with steam. When these velocities are destroyed the kinetic energy of the water is converted into pressure energy and a pressure shock is applied to the obstruction. Usually, there is a banging noise, and perhaps movement of the pipe. In severe cases the fitting may fracture with an almost explosive effect and consequent loss of live steam at the fracture.

Fortunately, waterhammer may be avoided completely if steps are taken to ensure that water is drained away

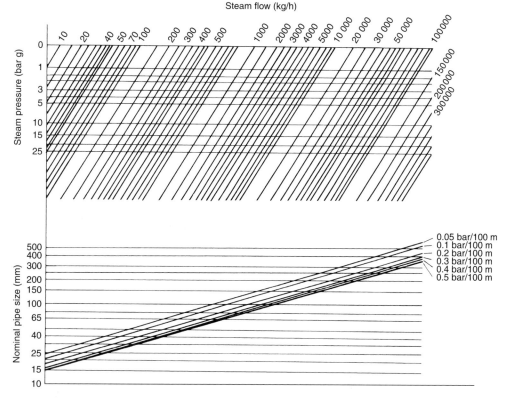

Figure 22.4 Pressure drop in steam lines

before it accumulates in sufficient quantity to be picked up by the steam. Avoiding waterhammer is a better alternative than attempting to contain it by choice of materials and pressure ratings of equipment.

The steam traps (used to drain water from a separator where this is fitted, perhaps to deal with carry-over from a boiler, or to drain condensate from collecting legs at intervals of not more than, say, 50 meters along the steam line) clearly must have adequate capacity. For the separator at the boiler off-take or at a header supplied from the boiler(s) and itself supplying the steam mains, it may be necessary to have capacity in the trap(s) of 10 per cent or even more of the boiler rating. On the steam lines themselves it is rare for a trap larger than $\frac{1}{2}$ inch or 15 mm size to be needed. Very often, the low-capacity versions of the $\frac{1}{2}$-inch steam trap are more than adequate.

The capacity of any steam trap will depend on the difference in pressure between its inlet and outlet connections. Under system start-up conditions the steam pressure in the line will at first be only marginally above atmospheric. If the trap discharge line rises to a higher level, or delivers to a pressurized return pipe, no condensate will flow through the trap until the line pressure exceeds the back pressure. It is important that steam traps which can drain by gravity, with zero back pressure, are fitted

if condensate is to be removed as the system is being heated from cold. Where this means that the start-up condensate can only be drained to waste, additional steam traps connected to the return line may also be fitted to enable recovery of the condensate to be made during normal running.

22.4.2 Steam line drainage

The use of oversized steam traps giving very generous 'safety factors' does not ensure safe and effective steam line drainage. A number of points must be considered if a satisfactory installation is to be achieved, including:

1. The heat-up method employed;
2. The provision of suitable collecting legs or reservoirs for the condensate;
3. Provision of a minimum differential pressure across the trap;
4. Choice of steam trap type and size;
5. Proper installation of the trap.

22.4.3 Heat-up method — supervised

In this case a manual drain valve is fitted at each drain point on the steam system, bypassing the trap and discharging to atmosphere. These manual drains are opened fully, either at the previous shutdown so that residual condensed steam is drained or at least before any steam is re-admitted to the system. When the 'heat-up' condensate has been discharged and the line pressure begins to increase, the manual valves are closed. The condensate formed under operating conditions is then discharged through the steam traps. These need to be sized only to handle the loads corresponding to the losses from the lines under operating conditions as indicated in Table 22.5.

This heat-up procedure is most often adopted in large installations where start-up of the system is an infrequent (perhaps only an annual) occurrence. Large heating systems and chemical processing plants are typical examples.

The procedure can be made more automatic by using a temperature-sensitive liquid expansion steam trap in place of the manual valve. Such an arrangement is a compromise between supervised start-up and the automatic start-up discussed below.

22.4.4 Heat-up method — automatic

One traditional method of achieving automatic start-up is simply to allow the steam boiler to be fired and brought up to pressure with the crown valve wide open. Thus the steam main and branch lines come up to operating pressure without supervision, and the steam traps must be arranged so that they can discharge the condensate as it is formed. This method is usually confined to the smaller installations that are regularly and frequently shut down and started up again. Typically, the boilers in many laundries and dry-cleaning plants often are shut down each night and restarted each morning.

In anything but the smallest installations the flow of steam from the boiler into the cold pipes at start-up (while the boiler pressure is still very low) will lead to excessive carry-over of boiler water with the steam. Such carry-over may be enough to overload a separator at the boiler take-off point or its steam trap (Figure 22.5). Great care

(and even good fortune) is necessary if problems are to be avoided.

Modern steam practice calls for an automatic valve to be fitted in the steam supply line, arranged so that the valve remains closed until a sufficient pressure is attained in the boiler. The valve can then be made to open over a timed period, so that steam is admitted to the pipe work at a controlled rate. The pressure within the boiler may be climbing quickly but the slow opening valve protects the pipe work.

Where these valves are used, the time available to warm up the pipe work will be known, as it is set on the valve control. In other cases, details of the start-up procedure must be known so that the time may be estimated. Thus boilers started from cold may be fired for a short time, shut off for a period while temperatures equalize, and then fired again. Boilers may be protected from undue stress by these short bursts of firing, which extend the warm-up time and reduce the rate at which the condensate in the mains must be discharged at the traps.

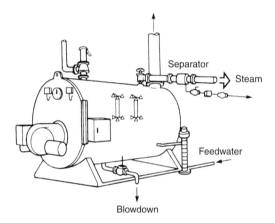

Figure 22.5 Boiler with separator at steam take off

Table 22.5 Heat emission from pipes (metric SI units)

Temperature difference steam to air ($°C$)	Pipe size (mm)									
	15	20	25	32	40	50	65	80	100	150
	W/m									
56	54	65	79	103	108	132	155	188	233	324
67	68	82	100	122	136	168	198	326	296	410
78	83	100	122	149	166	203	241	298	360	500
89	99	120	146	179	205	246	289	346	434	601
100	116	140	169	208	234	285	337	400	501	696
111	134	164	198	241	271	334	392	469	598	816
125	159	191	233	285	321	394	464	555	698	969
139	184	224	272	333	373	458	540	622	815	1133
153	210	255	312	382	429	528	623	747	939	1305
167	241	292	357	437	489	602	713	838	1093	1492
180	274	329	408	494	556	676	808	959	1190	1660
194	309	372	461	566	634	758	909	1080	1303	1852

Heat emission from bare horizontal pipes with ambient temperature between 10°C and 21°C and still air conditions.

22.4.5 Sizing steam traps for warm-up

Whatever automatic start-up method is adopted, steam will only flow into the mains and discharge air from the air vents at a pressure at least a little above atmospheric. It then reaches the pipe and condenses on the cold metal, and at first the condensate temperature will be well below 100°C (212°F). This means that the steam traps will have a greater capacity than their normal rating when handling saturated condensate. If the traps are fitted to a collecting leg, at about 700 mm below the main, then filling of the collecting leg will provide a hydraulic head of about 0.07 bar or 1 lb/in² in addition to the line pressure.

Sizing the traps requires an estimate of the condensate produced in bringing the main up to temperature and then determining the hourly rate of flow by allowing for the time available for this warm-up. Suitable traps can then be chosen to give at least this rate with a total pressure difference of, say, 0.15 bar or 2 lb/in² across the trap. Warm-up loads for the most common pipe sizes and pressures are listed in Table 22.6. The maximum condensation rate occurs when the mains are cold and the pressure (of the air/steam mixture) within them is still, say, 0.07 bar (1 lb/in²). As the mains have been heated to some temperature below the corresponding saturated steam mixture, the amount of condensate formed will not have reached the values shown in Table 22.3. Further, although some of the condensate is flowing through the trap, the remainder

is filling the collecting leg. It seems reasonable to take the Table 22.6 value and to subtract the volume of the collecting leg, and to use steam traps capable of passing this net amount in the available time with a pressure difference of 0.15 bar (2 lb/in²). This may be clarified by an example.

A length of 150 mm (6 in) main carries steam at 17 bar (250 lb/in²). Drain points are located at 45 meters (150 feet) intervals, with collecting legs 100 mm (4 inches) diameter × 700 mm (28 inches) long. The main is brought up to pressure from 21°C (70°F) in 30 minutes.

$$\text{Warm-up load to:}100°C = \text{approx. } 21.7 \text{ kg}/45$$
$$\text{meters: } 212°F$$
$$= 31.9 \text{ lb}/100 \text{ ft}$$
$$= 47.8 \text{ lb}/150 \text{ ft}$$
$$\text{Capacity of collecting} = \text{approx. } 5.7 \text{ kg}$$
$$\text{leg} = 12.7 \text{ lb}$$
$$\text{Net condensate to be discharged} = 16.0 \text{ kg} = 35.1 \text{ lb}$$
$$\text{Temperature rise to} = 207° - 21° = 186°C$$
$$17 \text{ bar} = 406 - 70 = 336°F$$
$$\text{Temperature rise to} = 79°C$$
$$= 142°F$$

Table 22.6 Warm-up load in kg of steam per 30 m of steam main

Steam pressure (bar)	Main size (mm)													−18°C correction factor
	50	65	80	100	125	150	200	250	300	350	400	450	500	
0	2.8	4.4	5.8	8.3	11.2	14.5	21.8	31	41	49	64	80	94	1.5
0.33	3.1	5.0	6.4	9.2	12.4	16.2	24.1	35	45	54	71	89	104	1.44
0.67	3.4	5.4	7.0	10.0	13.6	17.6	26.4	38	49	59	77	97	114	1.41
1.00	3.7	5.9	7.6	10.8	14.8	19.1	28.7	42	53	64	83	105	124	1.36
1.5	4.0	6.4	8.1	11.8	16	20.7	30.3	43	58	69	90	114	135	1.35
2.0	4.3	6.9	9.0	12.8	17.3	22.4	32	45	64	75	98	124	146	1.34
3.0	4.7	7.5	9.8	13.9	18.9	24.4	37	52	70	82	107	135	159	1.32
4.0	5.0	8.0	10.4	14.8	20.1	26.0	39.0	55	74	87	114	144	169	1.29
5.0	5.3	8.4	11.0	15.6	21.1	27.4	41	58	78	91	120	151	178	1.28
6.0	5.5	8.8	11.5	16.4	22.2	28.8	43	61	82	96	126	159	187	1.27
7.0	5.8	9.2	12.1	17.2	23.3	30.2	45	65	85	100	132	166	196	1.26
8.0	6.1	9.6	12.7	18.0	24.1	31.6	47	68	89	105	138	173	205	1.25
10.0	6.6	10.5	13.6	19.5	26.4	34.0	51	73	96	114	149	188	221	1.24
12.0	6.9	11.0	14.4	20.5	27.8	36.0	54	77	102	120	158	199	233	1.22
14.0	7.3	11.5	15.0	21.4	30.0	38	57	80	106	126	164	223	263	1.21
16.0	8.6	13.5	17.8	25.6	36.0	47	71	102	138	165	212	276	334	1.21
18.0	10.0	15.5	20.6	29.8	42	56	85	124	170	203	261	333	404	1.20
20.0	11.3	17.4	23.3	34.0	47	65	98	146	201	241	309	388	474	1.20
25.0	12.2	18.8	25.1	37.0	51	70	106	158	217	259	333	418	511	1.19
30.0	13.1	20.1	26.9	39.0	55	75	114	169	232	278	357	448	548	1.18
40.0	14.8	22.7	30.5	44.0	62	85	129	191	263	315	405	507	620	1.16
60.0	18.0	27.6	37	54	98	131	230	338	476	569	728	923	1120	1.145
80.0	21.3	32.5	43	63	115	154	257	378	532	638	815	1034	1255	1.14
100.0	23.5	36	48	70	127	171	284	418	588	707	902	1145	1390	1.135
125.0	28.8	44	59	86	159	208	347	510	718	863	1100	1400	1700	1.27

Ambient temperature 20°C. For outdoor temperature −18°, multiply load value in table by correction factor. Loads based on ANSI Schedule 40 pipe for pressures up to 16 bar. Schedule 80 pipe for pressures above 16 bar, except Schedule 120 above 40 bar in sizes 125 mm and over.

$$\text{Time to reach } 100°C = \frac{79}{186} \times 30\,\text{min} = 12.7\,\text{min}$$

$$= \frac{142}{336} \times 30 = 12.7$$

$$\text{Trap discharge rate} = \frac{16.0\,\text{kg}}{12.7\,\text{min}} \times 60\,\frac{\text{min}}{\text{h}}$$

$$= \frac{35.1}{12.7} \times 60$$

$$= 75.6\,\text{kg/h}$$

$$= 166\,\text{pph}$$

The capacity of a $\frac{1}{2}$ in TD trap at 0.15 bar differential (2 lb/in^2) is about 115 kg/h (255 pph) and this trap would have ample capacity.

It is clear that in most cases other than very large distribution mains, $\frac{1}{2}$ in TD traps are sufficiently large. With shorter distances between drain points, or smaller diameters, then $\frac{1}{2}$ in low-capacity traps more than meet even start-up loads. On very large pipes it may be worth fitting $\frac{3}{4}$ in traps, or two $\frac{1}{2}$ in traps in parallel. Low-pressure mains often are drained through float/thermostatic steam traps, and these traps are now available for use at much higher pressures than formerly, where it is known that waterhammer will not be present.

22.4.6 Drain point layout

Condensate-collecting legs can be of the same diameter as the main, up to, say, 100 mm (4 in) size. Larger pipes can have collecting legs two or three sizes smaller than the main but not less than 100 mm (4 in) size. The length of the collecting legs used with automatic start-up is usually 700 mm (28 in) or more, to give a hydraulic head of 0.15 bar (1 lb/in^2). With supervised start-up, the length of the legs can be 1.5 pipe diameters but not less than about 200 mm (8 in). The spacing between the drain points often is greater than is desirable. On a long horizontal run, drain points should be provided at intervals of about 45 m (150 ft) with a maximum of, say, 60 m (200 ft). Longer lengths should be split up and additional drain points fitted. Any low points in the system, such as the foot of risers and in front of shut-off or control valves, must also be drained.

In some cases, the ground contours are such that a steam main can only be run uphill. This will mean that the drain points should be closer together over the uphill section (say, 15 m (50 ft) apart) and the size of the main increased so that the steam velocity is not more than about 15 m/s (50 ft/s). The lower steam velocity may then allow condensate to drain in the direction opposite to the steam flow.

22.4.7 Air venting the steam lines

Air venting of steam mains is of paramount importance and is far too often overlooked. Air is drawn into the pipes when the steam supply is shut off and the residual steam condenses. A small amount of air is dissolved in the feedwater entering the boiler, and even if this amounts

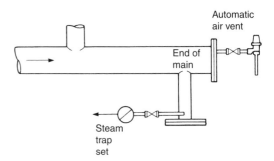

Figure 22.6 Draining and venting mains

to only a few parts per billion, the gas can accumulate in the steam spaces since it is non-condensable. Further, unless the feedwater is demineralized and decarbonated, carbonates can dissociate in the boiler and release carbon dioxide into the steam supply. Automatic air vents fitted above any possible condensate level at the ends of the steam mains and branches will allow discharge of these non-condensable gases, so promoting high heat transfer rates in the exchangers and helping to minimize corrosion (Figure 22.6).

22.5 Low-pressure systems

With steam generated at or close to the boiler design pressure it is inevitable that some of the steam-using equipment will have to be supplied at a lower pressure. In some cases the plant items themselves have only been designed to withstand a relatively low pressure. Sometimes a reaction will only proceed when the steam is at a temperature below a certain level or an unwanted reaction will occur above a certain level. For these and similar reasons, steam often is distributed at a relatively high pressure which must then be lowered, close to the point of use. Pressure-reduction stations incorporating pressure-reducing valves are fitted to perform this function.

Pressure-reducing valves may be either of the simple, direct acting pattern or be pilot operated (sometimes described as relay operated) (Figures 22.7 and 22.8). In the simple type, the force produced by the pressure downstream of the valve seat acting on a diaphragm or bellows is balanced by a control spring. When this force exceeds that of the spring at a given setting the diaphragm or bellows moves, compressing the spring further until balance is regained at a higher spring load – and at the same time moving the valve disc towards its seat. At the new balance point the steam flow through the valve is lessened and the downstream pressure is a little higher than the initial setting. Equally, at higher demand rates the downstream pressure falls by a small amount. A new balance point is reached when the lower pressure balances the spring force at a lower compression, with the valve disc now further away from the seat. The pressure downstream of a simple reducing valve must vary, then, from a maximum at no-load with the valve closed to a lower value on-load with the valve open. The change in reduced pressure may

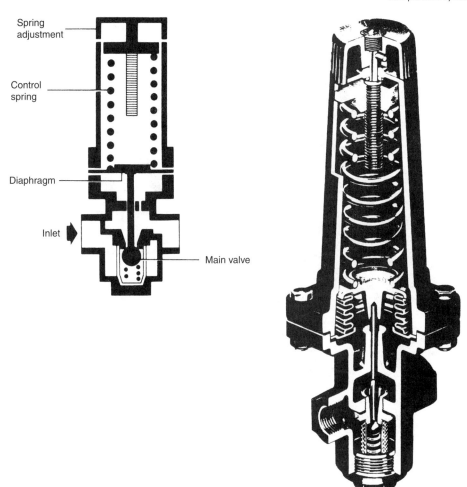

Spring
adjustment

Control
spring

Diaphragm

Inlet

Main valve

Figure 22.7 Direct-acting reducing valve

be called 'droop' or sometimes 'accuracy of regulation'. Simple valves are used to supply fairly small loads, or loads which are fairly steady at all times, to minimize the droop effect or where the reducing valve is followed by a temperature control valve which can compensate for the droop.

In other cases of larger and varying loads the droop of a large valve of simple pattern may be unacceptable. A very small valve of simple pattern can then be used as a pilot or relay valve to operate a much larger valve. The combination then has the droop characteristic of the very small pilot valve. Typically, a $\frac{1}{2}$ in (15 mm) simple reducing valve might pass full flow with a droop of 20 per cent of the reduced pressure, while a 3 in (80 mm) pilot-operated valve can have a droop of only 0.2 bar (3 lb/in^2). It is a mistake to install even the best of pressure-reducing valves in a pipeline without giving some thought to how best it can be helped to give optimal performance. First,

the valve chosen must have a large enough capacity that it can pass the required steam flow, but oversizing should be avoided. The weight of steam to be passed in a given time must be calculated or estimated, and a valve selected which can cope with this flow from the given upstream pressure to the required downstream pressure. The size of the valve is usually smaller than either the upstream or downstream pipe size because of the very high velocities which accompany the pressure drop within the valve.

Second, any valve which has been designed to operate on steam should not be expected to work at its best when supplied with a mixture of steam, water and dirt. A separator, drained through a steam trap, will remove virtually all the water from the steam entering the pressure-reducing set. The baffle type separators are found to be effective over a wide range of flow rates.

A stop valve is needed so that the steam supply can be shut off when necessary, and if this valve follows the

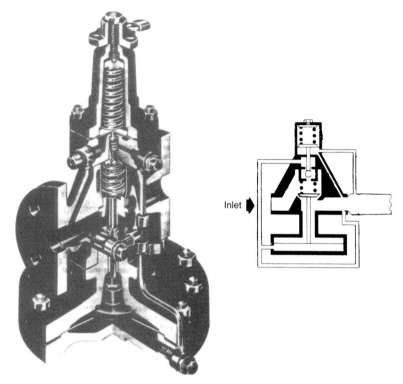

Figure 22.8 Pilot-operated reducing valve.

separator, the steam supply will be drained even when the valve is closed. After the valve a line size strainer is needed. It should have a 100-mesh stainless steel screen to catch the fine dirt particles that pass freely through standard strainers. The strainer is installed in the pipe on its side, rather than in the conventional way with the screen hanging below the pipe. This is to avoid the screen filling with condensate, which is often pumped between the valve plug and seat of a just-opening control valve as the flow (and pressure drop) through a strainer increases. This leads rapidly to wiredrawing of the control valve seat faces, and it is avoided by fitting the strainer horizontally to be self-draining. Pressure gauges at each side of a pressure-reducing valve both allow the downstream pressure to be set initially and aid in checking the functioning of the valve in service.

At the low-pressure side of the reducing valve it is usually essential to fit a relief or safety valve. If any of the steam-using equipment connected to the low-pressure range is designed to withstand a pressure below that of the upstream steam supply, then a safety valve is mandatory. Further, it may be called for when it is sought to protect material in process from over-high temperatures (Figure 22.9).

22.5.1 Safety relief valve sizing

When selecting a safety valve the pressure at which it is to be set must be decided. This may be sufficiently above the maximum normal operating pressure, under no-load conditions, to ensure the safety valve will reseat and not continue to blow steam, but it *must not* exceed the design pressure of the low-pressure equipment. The valve will give its rated capacity at an accumulated pressure of 10 per cent above the set pressure. This capacity must equal or exceed the maximum possible capacity of the pressure-reducing valve if it should fail wide open, when passing steam from the upstream pressure to the accumulated pressure at the relief valve.

22.5.2 Parallel and series operation of reducing valves

Parallel operation

In steam systems where load demands fluctuate over a wide range, parallel pressure control valves with combined capacities meeting the maximum load perform better than a single large valve. Maintenance needs, downtime and lifetime total costs can all be minimized with such an arrangement. Any reducing valve must be capable of both meeting its maximum load and modulating down towards minimum loads when required. The load turndown with which a given valve will satisfactorily cope is limited. There are no rules that apply without exception, but when the low load is less than about 10 per cent of the maximum load, two valves should always be preferred.

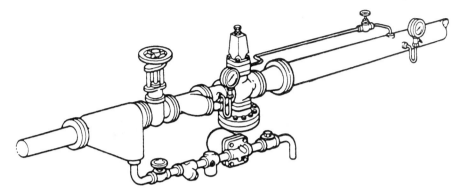

Figure 22.9 Pressure-control station

Whether a reducing valve is wide open or nearly closed, movement of the valve head through a given distance in relation to the seat will be in response to a given change in the controlled pressure. This valve movement changes the flow rate by a given amount, the pressure-reducing valve having a nearly linear characteristic. However, the change in the controlled pressure follows from a given percentage change in the flow rate. A valve movement which changes the steam flow just enough to match the demand change at high loads will change the flow by too much at low loads. It follows that instability of 'hunting' becomes more likely when a single valve is asked to cope with a high turndown in load. A single valve then tends to keep opening and closing, on light loads. This leads to wear of both the seating and guiding surfaces and reduces the life of the diaphragms that operate the valve. Where valves make use of pistons sliding within cylinders to position the valve head the situation is worsened. Friction and sticking between the sliding surfaces

mean that the valve head can only be moved in a series of discrete steps. The flow changes resulting from these movements are likely to be grossly in excess of the load changes that initiate them. Turndown ratios possible with piston-operated valves are inevitably smaller than those available with diaphragm-operated ones.

Stable control of reduced pressures is readily achieved by the use of two (or more) pressure-reducing valves in parallel (Figure 22.10). At full load and loads not too much below this level both valves are in use. As the load diminishes, the controlled pressure begins to increase and the valve which is set at the lower pressure begins to close. When the load can be supplied completely by the valve set at the higher pressure, the other valve closes. Any further load reduction causes the remaining valve to modulate through its proportional band.

Automatic changeover is achieved by the small difference between the pressure settings of the valves. For example, a maximum load of 5000 kg/h at 2 bar g might

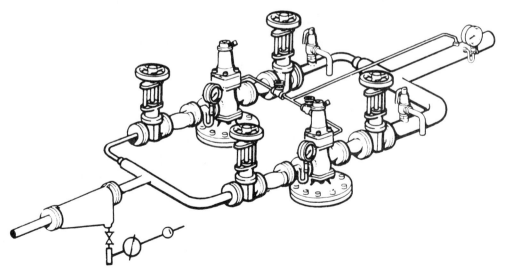

Figure 22.10 Typical installation of two reducing valves in parallel

be supplied through one valve capable of passing 1200 kg/h and set at 2.1 bar g, and a parallel valve capable of passing 4000 kg/h set at 1.9 bar g. When the steam demand is at any value up to 1200 kg/h the smaller valve will supply the load at a controlled pressure just below 2.1 bar g. An increase in demand above 1200 kg/h will then lead to the controlled pressure falling to 1.9 bar or lower, and the second valve then opens to supplement the supply.

Equally, if the demand only rarely exceeded 4000 kg, or fell below about 1000 kg, it would be possible to set the larger valve at the higher pressure and supplement the supply through the smaller valve on those few occasions when this was necessary. Sometimes the demand pattern is not known, except for the minimum and maximum values. Usually valves are then chosen with capacities of one third and two thirds of the maximum. The smaller valve is set at the slightly higher pressure.

Series operation

When the pressure reduction is through a ratio of more than about 10:1 the use of two valves in series should be considered. Much will depend on the valve being used, the pressures involved and the variations in the steam demand. Diaphragm-operated reducing valves have been used successfully with a pressure ratio as high as 20:1. They could perhaps be employed on a fairly steady load to reduce from 6.8 bar g (100 lb/in^2) to 0.3 bar g (5 lb/in^2). However, the same valve would probably be unstable on a variable load reducing from 2.75 bar (40 lb/in^2) to 0.22 bar g (2 lb/in^2). There is no hard-and-fast rule, but two valves in series can be expected to provide more accurate control. The second, low-pressure, valve should give the 'fine control' with a modest turndown, due consideration being given to valve sizes and capacities.

22.5.3 Series installations

For stable operation of the valves some appreciable volume between them is necessary. A length equal to 50 diameters of the appropriately sized pipe is often recommended or the same volume of larger pipe.

The downstream pressure-sensing pipe of each valve is connected to a straight section of pipe 10 diameters or 1 meter downstream of the nearest tee, elbow or valve. This sensing line should be pitched down, to drain into the low-pressure line. If it cannot drain when connected to the top of this line it can often be connected instead to the side of the pipe. The pipe between the two control valves must be drained through a steam trap, just as would the foot of any riser downstream of the pressure-reducing station.

22.5.4 Bypasses

The use of bypass lines and valves is best avoided. Bypass valves are often found to be leaking steam because of wire drawing of the seating faces of valves, which have not been tightly closed. If they are used, the capacity of the bypass valve should be added to that of the pressure-reducing valve when sizing relief valves. If it is thought

essential to maintain a steam supply even if a pressure-reducing valve should be faulty, or undergoing maintenance, consideration should be given to fitting a reducing valve in the bypass line. Sometimes the use of a parallel reducing station avoids the need for bypasses.

22.5.5 Selecting control valves for steam

The choice of a suitable temperature or pressure control valve for steam application will depend on the supply side pressure, the downstream pressure, and the flow rate of steam to be passed. In the case of temperature control valves the first of these is usually known and the third can be calculated, but the appropriate pressure drop through the valve is often to be decided. Sometimes the maker's rating of a heater will specify that it transfers heat at a certain rate when supplied with steam at a certain pressure. This pressure is then the pressure downstream of the control valve, and the valve may be selected on this basis.

If the pressure drop across the valve is to be more than 42 per cent of the inlet absolute pressure the valve selection is the same as if the pressure drop were only 42 per cent. With this pressure ratio the steam flow through the valve reaches a critical limit, with the steam flowing at sonic velocity, and lowering the downstream pressure below 58 per cent of the inlet absolute pressure gives no increase in flow rate. When the heater needs a higher pressure, or when the pressure required in the heater is not known, it is safer to allow a smaller pressure drop across the control valve. If the necessary heater pressure is not known, a pressure drop across the control valve of 10–25 per cent of the absolute inlet pressure usually ensures sufficient pressure within the heater. Of course, in the case of pressure-reducing valves the downstream pressure will be specified.

Valve capacities can be compared by use of the Kv (or Cv when Imperial units are used) values. These factors are determined experimentally, and the Kv value is the number of cubic meters per hour of water that will flow through a valve with a pressure drop of one bar. The Cv value is the number of gallons per minute of water that will flow through the valve with a pressure drop of one-pound f. per square inch. As the gallon is a smaller unit in the USA, the number of gallons passed is greater, and the US Cv is 1.2 times the UK Cv. The Kv is about 0.97 of the UK Cv value.

The steam flow through a valve at critical pressure drop when $P2$, the downstream pressure, is 0.58 times $P1$, the supply pressure, or $P2/P1 = 0.58$, is given very closely by $W = 12 Kv P1$, where W = steam flow (kg/h), Kv = flow coefficient and $P1$ = supply pressure (bar). In the USA, $W = 1.5 Cv P1$, where W = steam flow (pph), Cv = flow coefficient and $P1$ = supply pressure (psia). The factors 1.5 (US) and 12 (English metric) are not exactly equivalent but are sufficiently close for practical valve sizing purposes.

When the pressure drop ratio $(P1 - P2)/P1$ is less than 0.42, the steam flow is less than the critical flow. While BS 5793: Part 2: Section 2.2: 1981 provides an approximate and a rigorous method of calculating flows, it is found in practice that the best results are not distinguishable from the empirical formula

$$W_x = (\text{Critical flow rate})\sqrt{[1 - 5.67(0.42 - x)^2]}$$

where x is the pressure drop ratio and W_x is the corresponding flow rate.

22.5.6 Draining condensate from heat exchangers

If heat exchangers are not to steadily fill with condensate and the flow of heat to cease, then the condensate must be drained from them, and steam traps are normally used for this purpose. However, it is clear that a positive pressure difference must always exist between the condensate within the heat exchanger and the outlet side of the steam trap. Where the steam is supplied through a pressure-reducing valve, to meet a steady load, and the condensate is drained to a return line at lower level than the outlet and running to a vented receiver, this positive differential pressure will exist. In a great number of applications, the steam flow into the exchanger must be controlled in response to the temperature of the heated fluid, and if the heat load on the exchanger is less than the maximum, then the steam flow must be proportionately reduced.

With such a reduction in steam flow the steam pressure within the exchanger will fall. A balance is reached when the steam pressure is such that the steam temperature gives an appropriately lowered temperature difference between the steam and the mean temperature of the heated fluid. Under no-load conditions, if the flow of the fluid being heated were to cease or if the fluid were already at the required temperature as it entered the exchanger then this temperature difference would have to be zero and the steam at the same temperature as the fluid leaving the exchanger.

As a first approximation, the expression

$$q_1 = UA\Delta t$$

is used, and since A is a constant and U does not vary very much, the temperature of the steam must fall along a straight line from full-load steam temperature to the control temperature, as the load on the exchanger varies from maximum to zero. Clearly, as the steam temperature falls, the corresponding pressure in the steam space drops, and when it equals the pressure at the outlet side of the steam trap, condensate flow will cease. If the heat exchanger is not to flood with condensate it is necessary to ensure that the steam trap draining the exchanger is sufficiently far below the condensate drain point as to provide a hydrostatic head to push the condensate through the trap. Further, if the pressure at the outlet side of the steam trap is that of the atmosphere, and the pressure within the steam space is likely to fall below atmospheric, then a vacuum breaker is needed to admit air at a point above the maximum possible level of any condensate so that gravity can then clear the condensate through the steam trap.

It follows that condensate at the outlet side of the steam trap can only be at atmospheric pressure, and if it is to be lifted to a high-level return line, or into a return line

in which a positive pressure exists, then a condensate pump is needed for this purpose. Equally, if insufficient height is available to provide the hydrostatic head in front of the steam trap, and fall from the steam trap to the receiver of a condensate pump, it may be possible to allow condensate to fall directly from the exchanger to the body of a steam-powered 'alternating receiver' condensate pump, which can then under low-load low-pressure conditions temporarily isolate collected condensate from the exchanger and use high-pressure steam to push it through the steam trap. Figure 22.11 illustrates the alternative condensate removal methods which are used for the differing pressure conditions.

22.5.7 Air venting

The existence of air (using this word to cover atmospheric air and also any other non-condensing gases) in steam systems was mentioned in Section 22.4.7. It is even more important that the presence of air in the heat exchangers themselves is recognized and arrangements made for its discharge. Air will exert its own partial pressure within the steam space, and this pressure will be added to the partial pressure of the steam in producing the total pressure present. The actual steam pressure is lower than the total pressure shown on a gauge by virtue of the presence of any air. Since the temperature of saturated steam must always be the condensing temperature corresponding with its partial pressure, the effect of air in a steam space is always to lower the temperature below the level which would be expected for pure steam at the total pressure present.

This means that thermostatic balanced pressure steam traps which open when their element senses a temperature somewhat below that of saturated steam at the pressure existing within the steam space are very effective when used as automatic air vents. They are connected to a steam space at any location where air will collect. Usually this means at any 'remote point' from the steam entry, along the path the steam takes as it fills the steam space.

Air vents are most effective when they are fitted at the end of a length of 300 mm or 450 mm of uninsulated pipe that can act as a collecting/cooling leg. Air is an excellent insulating material, having a thermal conductivity about 2200 times less than that of iron. The last place where it can be allowed to collect is in the steam space of heat exchangers. Further, as it contains oxygen or carbon dioxide, which dissolve readily in any subcooled condensate that may be present, the presence of air initiates corrosion of the plant and the condensate return system.

22.6 Flash steam

22.6.1 Release of flash steam

High-pressure condensate forms at the same temperature as the high-pressure steam from which it condenses, as the enthalpy of evaporation (latent heat) is transferred from it. When this condensate is discharged through a steam trap to a lower pressure the energy it contains is greater than it can hold while remaining as liquid water. The excess

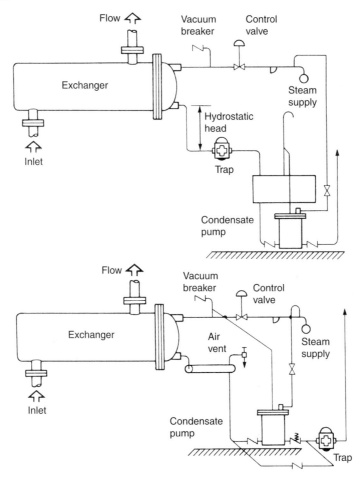

Figure 22.11 Lifting condensate drained from exchangers

energy re-evaporates some of the water as steam at the lower pressure. Conventionally, this steam is referred to as 'flash steam', although in fact it is perfectly good steam, even if at low pressure.

22.6.2 Flash steam utilization

In an efficient and economical steam system, this so-called flash steam will be utilized on any load that can make use of low-pressure steam. Sometimes it can be simply piped into a low-pressure distribution main for general use. The ideal is to have a greater demand for low-pressure steam at all times than the available supply of flash steam. Only as a last resort should flash steam be vented to atmosphere and lost.

This means that condensate from high-pressure sources should usually be collected and led to a flash vessel which operates at a lower pressure (but high enough to be useful). Remember, the flashing off does not normally take place within the flash vessel. It begins within the seat of

the steam trap and continues in the condensate line. Only when the high-pressure steam traps are very close to the flash vessel does any flashing take place within it.

Instead, the flash vessel is primarily a flash steam separator. Its shape and dimensions are chosen to encourage separation of the considerable volume of low-pressure steam from the small volume of liquid.

22.6.3 Proportion of flash steam released

The amount of flash steam which a given weight of condensate will release may be readily calculated. Subtracting the sensible heat of the condensate at the lower pressure from that of the condensate passing through the traps will give the amount of heat available for use as the enthalpy of evaporation (latent heat). Dividing this amount by the actual enthalpy of evaporation at the lower pressure will give the proportion of the condensate which will flash off. Multiplying by the total quantity of condensate being considered gives the weight of low-pressure steam available.

Thus if, for example, 1000 kg of condensate from a source at 10 bar is flashed to 1 bar we can say:

Specific enthalpy of water (10 bar) = 781.6

Specific enthalpy of water (1 bar) = 505.6

Energy available for flashing = 276

Enthalpy of evaporation at 1 bar = 2201.1

Proportion evaporated = $\dfrac{276}{2201.1}$

= 0.125

Flash steam available = 0.125 × 1000 kg

= 125 kg

Alternatively, tables such as Table 22.7 or charts such as that shown in Figure 22.12 will allow the proportion of flash steam available between two pressures to be read off directly, within the more usual pressure ranges.

22.6.4 Sub-cooled condensate

Note that the method described assumes that the high-pressure condensate has not been sub-cooled. If any sub-cooling has taken place, then the figure taken for the enthalpy of water at the higher pressure is reduced by the amount of sub-cooling. The chart or table can still be used if the upstream pressure is taken as that corresponding to saturated steam at the same temperature as the sub-cooled condensate.

In the previous example condensate sub-cooled by 20°C would be at 164.22°C instead of 184.22°C. Steam at this temperature would be at a pressure of some 5.83 bar, so that the chart in Figure 22.12 would be used for a drop from 5.83 bar to 1 bar instead of from 10 bar to 1 bar, to allow for the sub-cooling.

22.6.5 Making use of flash steam

The steam/process air heater with multiple coils typifies the kind of application on which recovery of flash steam

from the condensate is most readily affected. Condensate from the high-pressure coils is taken to a flash steam recovery vessel. Low-pressure steam leaving the vessel is supplied to the first coil at the air-inlet side of the heater or to an extra coil acting as a pre-heater unit.

Condensate from the low-pressure coil together with that from the flash vessel will then drain to a collecting tank, or direct to a condensate pump, for return to the boiler plant. If the pressure of the flash steam is left to find its own level it will often be sub-atmospheric. As the condensate must then drain by gravity through the steam traps these also must be sufficiently below the condensate drain points to provide an appropriate hydraulic head, and a vacuum breaker fitted above the coil. The alternatives are to allow the condensate to drain directly to a condensate pump, or to supply additional low-pressure steam through a pressure-reducing valve, to maintain a positive pressure in the coil and flash vessel.

22.6.6 Space heating

Somewhat similar arrangements can be used when radiant panels or unit heaters heat large areas. Some 10–15 per cent of the heaters are separated from the high-pressure steam supply and supplied instead with low-pressure steam flashed off the high-pressure condensate. The heating demands of the whole area remain in step, so supply and demand for the flash steam are balanced.

22.6.7 The general case

In other cases, flash steam is utilized on equipment, which is completely separated from the high-pressure source. Often the low-pressure demand does not at all times match the availability of the flash steam. A pressure-reducing station is often needed to make up any deficit and a surplus valve is required to vent any flash steam in excess of the amount being condensed.

22.6.8 Steam traps

The wide choice of steam trap types which are available must at first seem confusing, and it is useful to first define

Table 22.7 Condensate proportion released as flash

Pressure at traps (bar G)	Pressure in flash vessel (bar G)												
	0	0.2	0.5	0.75	1.0	2	3	4	5	6	7	10	15
1	0.038	0.029	0.017	0.008	0								
2	0.063	0.054	0.042	0.033	0.025	0							
3	0.082	0.073	0.061	0.053	0.045	0.02	0						
4	0.098	0.089	0.077	0.069	0.061	0.036	0.017	0					
5	0.111	0.102	0.091	0.083	0.078	0.05	0.03	0.014	0				
6	0.123	0.119	0.108	0.1	0.093	0.068	0.049	0.032	0.019	0			
7	0.133	0.125	0.113	0.106	0.098	0.074	0.054	0.038	0.024	0.012	0		
10	0.16	0.152	0.141	0.133	0.125	0.101	0.083	0.067	0.053	0.04	0.029	0	
15	0.192	0.183	0.172	0.165	0.157	0.134	0.116	0.1	0.087	0.075	0.014	0.035	0
20	0.222	0.214	0.203	0.195	0.188	0.165	0.148	0.133	0.12	0.108	0.097	0.069	0.032
25	0.245	0.237	0.226	0.219	0.212	0.189	0.172	0.157	0.144	0.133	0.122	0.095	0.058
30	0.261	0.253	0.243	0.235	0.228	0.206	0.189	0.174	0.162	0.15	0.14	0.113	0.077
40	0.296	0.288	0.278	0.271	0.264	0.242	0.226	0.212	0.2	0.189	0.179	0.153	0.118

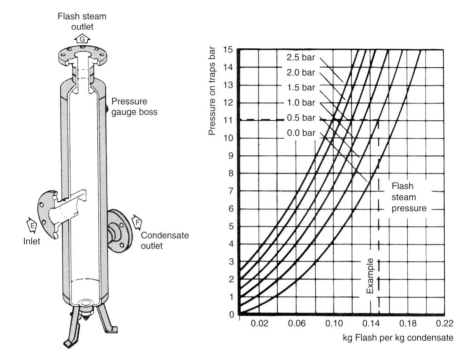

Figure 22.12 Flash vessels

the term 'steam trap'. In practice, a steam trap has two separate elements. The first of these is a valve and seat assembly, which can provide a variable orifice through which the condensate can be discharged at such a rate as to match the rate of condensation in the equipment being drained. The opening may be modulated continually to provide a continuous flow of condensate, or may operate in an on/off fashion so that the average rate over a period of time matches the condensation rate.

The second element is a device which will open or close the valve by measuring some parameter of the fluid reaching it and 'deciding' whether this may or may not be discharged. It would be found that the controlling elements mainly fall into one of three categories. The steam trap can decide automatically whether to open or to close to the fluid reaching it on the basis of:

1. The density of the fluid, by using a float, which will float in water or sink in steam;
2. By measuring the temperature of the fluid, closing the valve at or near to steam temperature, and opening it when the fluid has cooled to a temperature sufficiently far below that of steam;
3. By measuring the kinetic effects of the fluid in motion, since at a given pressure drop, low-density steam will move at a much greater velocity than will high-density condensate, and the conversion of pressure energy into kinetic energy can be used to position a valve.

The groupings then may be described as *mechanical*, which will include both ball float and inverted bucket steam traps; *thermostatic*, which will include both balanced pressure and bimetallic elements; and *thermodynamic* or *disc* pattern traps (Figure 22.13). Each type of trap has its own characteristics, and these will make one pattern of trap more suitable for use on a given application than another. In practice, it is usual to find that the applications in any given plant fall into a small number of categories, and it often is possible to standardize on a quite small number of trap types.

Thus the requirements for draining condensate from the steam mains are that the trap should discharge condensate at a temperature very close to that of steam to ensure that it is in fact drained from the collecting pockets and not held back because it is not cooled to a sufficiently low temperature; that the trap is not physically large yet has adequate capacity; that the trap is robust enough to withstand severe operating conditions, such as waterhammer in inadequately drained lines or freezing conditions when installed outdoors. The thermodynamic trap is very widely used for this application and is capable of giving excellent results.

In a similar way, many small jacketed pans, steam radiators and convectors, and some steam tracer lines can operate most economically if the condensate is retained within the steam space until it has sub-cooled a little, making thermostatic pattern traps the most suitable for these

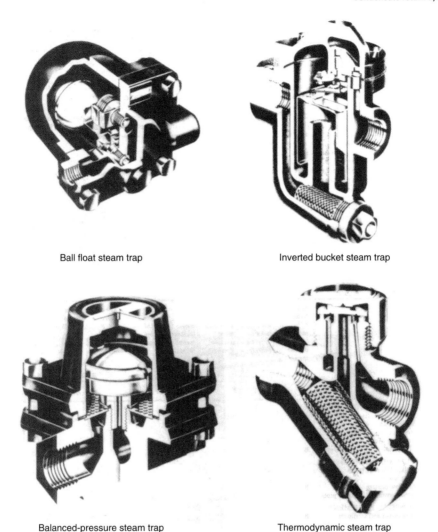

Ball float steam trap

Inverted bucket steam trap

Balanced-pressure steam trap

Thermodynamic steam trap

Figure 22.13 Steam traps

applications. On the other hand, many heat exchangers are required to give their maximum output. This requires that the condensate be removed immediately it forms, or perhaps, as in air heater batteries, any holding back of condensate will lead to corrosion. In these cases the use of a mechanical trap such as the ball float pattern or perhaps the inverted bucket trap becomes essential. Care is needed when inverted bucket traps are chosen for jobs where the steam control valve may at times close down, lowering the pressure in the steam space. This can lead to re-evaporation of the water seal in an inverted bucket trap, which would then blow steam. Further, on many heater exchangers it is important to provide air-venting capacity at the trap. Float thermostatic traps incorporate a suitable air vent at the ideal location, just above the water level,

while if inverted bucket traps are used, then separate air vents must be fitted in bypasses around the trap. Many steam users prefer to draw up a selection guide list such as the one shown in Table 22.8.

22.7 Condensate return systems

No single set of recommendations can cover condensate return systems. These divide naturally into at least three sections, each with its own requirements.

22.7.1 Drain lines to traps

In the first section the condensate flows from the condensing surface to the steam trap. Since the heat

Table 22.8 Selecting steam traps

Application	Spirax Sarco FT range (float/ thermostatic)	Spirax Sarco IB range (inverted bucket)	Spirax Sarco TD range (thermo-dynamic)	Spirax Sarco BPT (balanced pressure thermostatic	Spirax Sarco SM (bimetallic)	Spirax Sarco No. 8 (liquid expansion)	Spirax Sarco FT/TV/SLR (float/ thermostatic with steam lock release)	Spirax Sarco FT/SLR (float/steam lock release)
CANTEEN EQUIPMENT								
Boiling pans–fixed	A		B[1]	B			B	B[1]
Boiling pans–tilting				B			A	B[1]
Boiling pans–pedestal	B			A[2]			B	B[1]
Steaming ovens				A[2]				
Hot plates	B			A[2]			B	B[1]
FUEL OIL HEATING								
Bulk/oil storage tanks		A[1]	B[1]					
Line heaters	A	B[1]						
Outflow heaters	B	A[1]						
Tracer lines and jacketed pipes			B	A[3]	B	B		
HOSPITAL EQUIPMENT								
Autoclaves and sterilizers	B	B		A			B	B[1]
INDUSTRIAL DRYERS								
Drying coils (continuous)	B	A		B	B			
Drying coils (grid)		B[1]		A				
Drying cylinders	B	B[1]					A	B[1]
Multi-bank pipe dryers	A	B[1]		B				
Multi-cylinder sizing machines	B	B[1]					A	B[1]
LAUNDRY EQUIPMENT								
Garment presses	B	B	A					
Ironers and calenders	B	B[1]	B[1]	B			A	B[1]
Solvent recovery units	A	B	B					
Tumbler dryers	A	B[1]					B	B[1]
PRESSES								
Multi-platen presses (parallel connections)	B	B	A					
Multi-platen presses (series connections)		B[1]	A[1]					
Tire presses		A	B	B				

Table 22.8 (*continued*)

Application	Spirax Sarco FT range (float/ thermostatic)	Spirax Sarco IB range (inverted bucket)	Spirax Sarco TD range (thermo-dynamic)	Spirax Sarco BPT (balanced pressure thermostatic	Spirax Sarco SM (bimetallic)	Spirax Sarco No. 8 (liquid expansion)	Spirax Sarco FT/TV/SLR (float/ thermostatic with steam lock release)	Spirax Sarco FT/SLR (float/steam lock release)
PROCESS EQUIPMENT								
Boiling pans–fixed	A		B[1]	B			B	B[1]
Boiling pans–tilting							A	B
Brewing coppers	A	B[1]					B	B[1]
Digesters	A	B[1]	B[1]				B	B[1]
Evaporators	A	B[1]					B	B[1]
Hot tables			B	A				
Retorts	A	B[1]						
Bulk storage tanks		A[1]	B[1]					
Vulcanizers	B	A[1]						
SPACE HEATING EQUIPMENT								
Calorifiers	A	B[1]					B	B[1]
Heater batteries	A	B[1]					B	B[1]
Radiant panels and strips	A	B[1]	B[1]				B	B[1]
Radiators and convention cabinet heaters	B			A	B			
Overhead pipe coils	B	B[1]		A				
STEAM MAINS								
Horizontal runs	B	B	A	B[1]				
Separators	A	B	B	B[2]				
Terminal	B	B[1]	A[1]	B[2]				
Shutdown drain (frost protection)				B[3]		A		
TANKS AND VATS								
Process vats (rising discharge pipe)	B	B	A	B				
Process vats (discharge pipe at base)	A	B	B	B				
Small coil-heated tanks (quick boiling)	A	B		B				
Small coil-heated tanks (slow boiling)						A		

A = best choice
B = acceptable alternative
1. With air vent in parallel.
2. At end of cooling leg. Minimum length 1 m (3 ft).
3. Use special tracing traps which offer fixed temperature discharge option.

exchanger steam space and the traps are at the same pressure, gravity is relied on to induce flow. When the traps are vertically below the drainage points, or are close to them, it may be satisfactory to use a line of the same size as the inlet connection of the trap. However, if the trap must be located a little further away from the drainage point, then the line between the trap point and the trap can be laid with a slight fall (say, 1 in 250), and Table 22.9 shows the water-carrying capacities of the pipes with such a gradient. It is important to allow for the passage of incondensables to the trap and for the extra water which is carried during cold-start conditions. In most cases it is sufficient to size these pipes on twice the full running load.

22.7.2 Trap discharge lines

At the outlet side of the traps, the 'condensate' lines have to carry both the condensate and any non-condensable gases, together with the flash steam released from the condensate. Wherever possible, these pipes should drain by gravity from the traps to the condensate receiver, whether this is a flash recovery vessel or the vented receiver of a pump. With atmospheric pressure in the return line, and pressures upstream of the traps of up to about 4 bar (60 lb/in²), the sizing method described above is more than adequate. At higher upstream pressures the volume of the flash steam released from the condensate becomes significant and must be given due consideration. If the pressure at the discharge side of the traps is above atmospheric pressure, then sizing these lines on twice the full running load may still be adequate even with pressures well above 4 bar (60 lb/in²) at the inlet side. The chart shown in Figure 22.14 allows the lines to be sized as flash steam lines, since the volume of the condensate is so much less than that of the steam released. Draining condensate from traps that serve loads at differing pressures to a common condensate return line is a concept that is often found difficult. Many users assume that the high-pressure condensate will prevent the low-pressure condensate from passing through the low-pressure traps and give rise to water logging of the low-pressure systems. However, the terms 'high pressure' and 'low pressure' can only apply to the conditions on the upstream side of the seats in the

steam traps. At the downstream or outlet side of the traps the pressure must be the common pressure in the return line. This return line pressure will be the sum of at least three components:

1. The pressure at the end of the return line, either atmospheric or of the vessel into which the line discharges;
2. The hydrostatic head needed to lift the condensate up any risers in the line;
3. The pressure drop needed to carry the condensate and any flash steam along the line.

Item 3 is the only one likely to give rise to any problems if condensate from sources at different pressures enters a common line. The return should be sufficiently large to carry all the liquid condensate and the varying amounts of flash steam associated with it, without requiring excessive line velocities and excessive pressure drop. If this is accepted, then the total return line cross-sectional area will be the same, whether a single line is used or if two or more lines are fitted with each taking the condensate from a single pressure source. Return lines may become undersized, requiring a high pressure at the trap discharges and restricting or even preventing discharge from the low-pressure traps, if it is forgotten that the pipe must carry flash steam as well as water and that flash steam is released in appreciable quantity from high-pressure condensate.

22.7.3 Pumped return line

Finally, the condensate is often pumped from the receiver to the boiler house. Pumped condensate lines carry only water, and rather higher water velocities can often be used to minimize pipe sizes. The extra friction losses entailed must not increase back pressures to the point where pump capacity is affected. Table 22.10 can be used to help estimate the frictional resistance presented by the pipe.

Condensate pumps usually operate with an on/off action, so that the instantaneous flow rate during the 'on' period is greater than the average rate of flow of condensate to the pump receiver. This increased instantaneous flow rate must be kept in mind when sizing the delivery lines.

Where long delivery lines are used, the water flowing along the pipe as the pump discharges attains a considerable momentum. At the end of the discharge period when the pump stops, the water tends to keep moving along the pipe and may pull air or steam into the delivery pipe through the pump outlet check valve. When this bubble of steam reaches a cooler zone and condenses, the water in the pipe is pulled back towards the pump. When the reverse flow reaches and closes the check valve, waterhammer often results. This problem is greatly reduced by adding a second check valve in the delivery line some 5 or 6 m from the pump. If the line lifts to a high level as soon as it leaves the pump then adding a generously sized vacuum breaker at the top of the riser is often an extra help. However, it may be necessary to provide means of venting from the pipe at appropriate points the air that enters through the vacuum breaker.

Table 22.9 Water-carrying capacity of pipes (SI units) (approximate frictional resistance in mbar per m of travel)

Steel tube (mm)	0.3 (30 Pa)	0.5 (50 Pa)	0.6 (60 Pa)	0.8 (80 Pa)	1.0 (100 Pa)	1.4 (140 Pa)
15	95	130	140	160	180	220
20	220	290	320	370	420	500
25	410	540	600	690	790	940
32	980	1 180	1 300	1 500	1 700	2 040
40	1 360	1 790	2 000	2 290	2 590	3 100
50	2 630	3 450	3 810	4 390	4 990	6 000
65	5 350	6 950	7 730	8 900	10 150	12 100
80	8 320	10 900	12 000	13 800	15 650	18 700
100	17 000	22 200	24 500	28 200	31 900	38 000

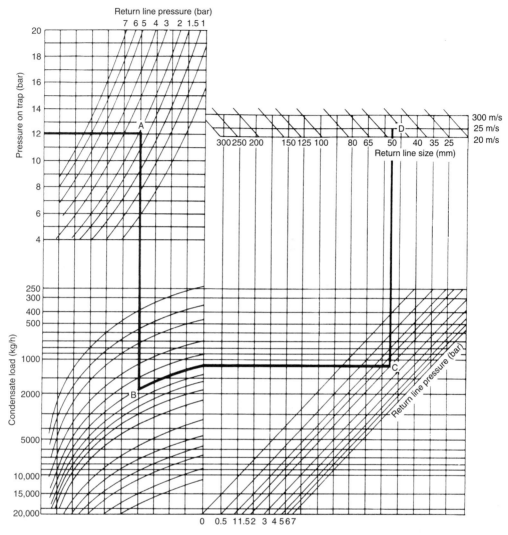

Figure 22.14 Condensate line sizing chart where pressure at traps is above 4 bar (SI units). 1. From pressure upstream of trap move horizontally to pressure in return line (A). 2. Drop vertically to condensate load in kg/h (B). 3. Follow curve to RH scale and across to same return line pressure (C). 4. Move upward to return line flash velocity – say, 25 m/s maximum (D). 5. Read return line size.

The practice of connecting additional trap discharge lines into the pumped main is usually to be avoided. The flash steam which is released from this extra condensate can lead to waterhammer. Preferably, the trap should discharge into a separate gravity line which carries the condensate to the receiver of the pump. If this is impossible, one may arrange for the condensate and associated flash steam to enter the pumped main through a small 'sparge' or diffuser fitting so that the flash steam can be condensed immediately in the pumped water.

22.7.4 Condensate pumping

In nearly all steam-using plants condensate must be pumped back to the boilerhouse from the location where it

is formed, and even in those cases where gravity drainage to the boilerhouse is practical, often the condensate must be lifted into a boiler feed tank. Where de-aerators are used they usually operate at a pressure of about 0.3 bar above atmospheric, and again a pump is needed to lift condensate from atmospheric pressure to de-aerator pressure.

Most pump units comprise a receiver tank that conventionally is vented to atmosphere and one or more motorized pumps. It is important with these units to make sure that the maximum condensate temperature as specified by the manufacturer is not exceeded, as well as that the pump has sufficient capacity to handle the load. Condensate temperature usually presents no problem

Figure 22.15 Condensate recovery unit

with returns from low-pressure heating systems. There, the condensate is often below 100°C even as it leaves the traps, and a little further sub-cooling in the gravity return lines and in the pump receiver itself means that there is little difficulty in meeting the maximum temperature limitations.

On higher-pressure systems, the gravity return lines often contain condensate at a little above 100°C, together with some flash steam. The cooling effect of the piping is limited to condensing a little of the flash steam, and the remainder passes through the vent at the pump receiver. The water must remain in the receiver for an appreciable time if it is to cool sufficiently, or sometimes the pump discharge may be throttled down to reduce the pump's capacity if cavitation is to be avoided.

The absolute pressure at the inlet to the pump is usually the atmospheric pressure in the receiver, plus the static head from the water surface to the pump inlet and minus the friction loss through the pipes, valves and fittings joining the pump to the receiver. If his absolute pressure exceeds the vapor pressure of water at the temperature at which it enters the pump, then a net positive suction hand (NPSH) exists. If this NPSH is above the value specified by the pump manufacturer, the water does not begin to boil as it enters the pump suction and cavitation is avoided. If the water entering the pump is at a higher temperature, its vapor pressure is increased and a greater hydrostatic head over the pump suction is needed to ensure that the necessary NPSH is obtained.

However, since in most cases pumps are supplied coupled to receivers and the static head above the pump inlet

is already fixed by the manufacturer, it is only necessary to ensure that the pump set has sufficient capacity at the water temperature expected at the pump. Pump manufacturers usually have a set of capacity curves for the pump when handling water at different temperatures, and these should be consulted. Where steam systems operate at higher pressures than those used in low-pressure space heating systems, as in process work, condensate temperatures can be 100°C, or more when positive pressures exist in the return lines. Electric pumps are then used only if their capacity is down rated by partial closure of a valve at the pump outlet; by using a receiver mounted well above the pump to ensure sufficient NPSH; or by sub-cooling the condensate through a heat exchanger of some type.

All these difficulties are avoided by the use of steam-powered alternating receiver pumps of the type illustrated in Figure 22.16. These pumps are essentially alternating receivers that can be pressurized using steam (or air or other gas). The gas pressure displaces the condensate (which can be at any temperature up to and including boiling point) through a check valve at the outlet from the pump body. At the end of the discharge stroke, the internal mechanism changes over to close the pressurizing inlet valve and open the exhaust valve. The pressurizing gas is then vented to atmosphere, or to the space from which the condensate is being drained. When the pressures are equalized, condensate can flow by gravity into the pump body to refill it and complete the cycle. As the pump fills only by gravity, there can be no cavitation and pumps of this type readily handle boiling water or other liquids compatible with the materials of construction.

Table 22.10 Flow of water at 75°C in black steel pipes

M = mass flow rate — kg/s
l_e = equivalent length of pipe ($\zeta = 1$) — m
Δp_1 = pressure loss per unit length — Pa/m
v = velocity — m/s
*$(Re) = 2000$
†$(Re) = 3000$

Δp_l	v	10 mm M	10 mm l_e	15 mm M	15 mm l_e	20 mm M	20 mm l_e	25 mm M	25 mm l_e	32 mm M	32 mm l_e	40 mm M	40 mm l_e	50 mm M	50 mm l_e	v	Δp_l
0.1								0.003	0.1	0.009	0.5	0.017*	0.9	0.031	1.2		0.1
0.2								0.006	0.3	0.018*	0.9	0.024	0.9	0.044†	1.2		0.2
0.3						0.003	0.2	0.008	0.4	0.020	0.8	0.029	0.9	0.055	1.2		0.3
0.4						0.004	0.2	0.011	0.6	0.023	0.8	0.034†	0.9	0.065	1.3		0.4
0.5						0.005	0.3	0.014*	0.7	0.025	0.8	0.038	0.9	0.074	1.3		0.5
0.6						0.007	0.3	0.013	0.6	0.028	0.8	0.042	1.0	0.082	1.4		0.6
0.7						0.008	0.4	0.014	0.6	0.030†	0.8	0.046	1.0	0.090	1.4		0.7
0.8						0.009	0.5	0.015	0.6	0.032	0.8	0.050	1.0	0.097	1.4	0.05	0.8
0.9						0.010	0.5	0.016	0.6	0.035	0.8	0.054	1.0	0.104	1.5		0.9
1.0				0.003	0.2	0.011*	0.6	0.017	0.6	0.037	0.8	0.057	1.0	0.110	1.5		1.0
1.5				0.005	0.2	0.012	0.4	0.021†	0.6	0.047	0.9	0.072	1.1	0.139	1.6		1.5
2.0				0.006	0.3	0.014	0.4	0.025	0.6	0.055	0.9	0.085	1.1	0.164	1.6		2.0
2.5				0.008	0.4	0.015	0.4	0.028	0.6	0.062	0.9	0.096	1.2	0.186	1.7		2.5
3.0				0.009*	0.5	0.017	0.4	0.031	0.6	0.069	1.0	0.107	1.2	0.206	1.7		3.0
3.5				0.008	0.3	0.018†	0.4	0.034	0.6	0.076	1.0	0.116	1.2	0.224	1.7		3.5
4.0		0.004	0.2	0.009	0.3	0.020	0.4	0.037	0.7	0.082	1.0	0.126	1.3	0.242	1.8		4.0
4.5	0.05	0.005	0.2	0.009	0.3	0.021	0.5	0.039	0.7	0.087	1.0	0.134	1.3	0.258	1.8		4.5
5.0		0.005	0.3	0.010	0.3	0.022	0.5	0.042	0.7	0.093	1.0	0.142	1.3	0.274	1.8		5.0
5.5		0.006	0.3	0.010	0.3	0.023	0.5	0.044	0.7	0.098	1.0	0.150	1.3	0.289	1.8	0.15	5.5
6.0		0.006	0.3	0.011	0.3	0.025	0.5	0.046	0.7	0.103	1.1	0.158	1.3	0.303	1.9		6.0
6.5		0.007*	0.4	0.011	0.3	0.026	0.5	0.048	0.7	0.107	1.1	0.165	1.3	0.317	1.9		6.5
7.0		0.006	0.2	0.012	0.3	0.027	0.6	0.050	0.7	0.112	1.1	0.172	1.3	0.330	1.9		7.0
7.5		0.006	0.2	0.012	0.3	0.028	0.5	0.052	0.7	0.116	1.1	0.179	1.4	0.343	1.9		7.5
8.0		0.006	0.2	0.012	0.3	0.029	0.5	0.054	0.7	0.120	1.1	0.185	1.4	0.355	1.9		8.0
8.5		0.006	0.2	0.013	0.3	0.030	0.5	0.056	0.7	0.125	1.1	0.191	1.4	0.368	1.9		8.5
9.0		0.007	0.2	0.013†	0.3	0.031	0.5	0.058	0.7	0.129	1.1	0.198	1.4	0.379	1.9		9.0
9.5		0.007	0.2	0.014	0.3	0.032	0.5	0.060	0.7	0.133	1.1	0.204	1.4	0.391	1.9		9.5
10.0		0.007	0.2	0.014	0.3	0.033	0.5	0.062	0.7	0.136	1.1	0.210	1.4	0.402	2.0		10.0
12.5		0.008	0.2	0.016	0.3	0.037	0.5	0.070	0.8	0.154	1.1	0.237	1.4	0.454	2.0		12.5
15.0		0.008	0.2	0.018	0.4	0.042	0.6	0.077	0.8	0.171	1.2	0.262	1.5	0.502	2.0		15.0
17.5		0.009	0.2	0.019	0.4	0.045	0.6	0.084	0.8	0.186	1.2	0.285	1.5	0.546	2.0		17.5
20.0		0.010	0.2	0.021	0.4	0.049	0.6	0.091	0.8	0.200	1.2	0.307	1.5	0.587	2.1	0.30	20.0
22.5		0.010†	0.2	0.022	0.4	0.052	0.6	0.097	0.8	0.214	1.2	0.327	1.5	0.626	2.1		22.5
25.0		0.011	0.3	0.023	0.4	0.055	0.6	0.103	0.8	0.226	1.2	0.347	1.5	0.663	2.1		25.0
27.5		0.012	0.3	0.025	0.4	0.058	0.6	0.108	0.8	0.238	1.2	0.365	1.5	0.698	2.1		27.5
30.0		0.012	0.3	0.026	0.4	0.061	0.6	0.114	0.8	0.250	1.2	0.383	1.6	0.731	2.2		30.0
32.5		0.013	0.3	0.027	0.4	0.064	0.6	0.119	0.8	0.261	1.3	0.400	1.6	0.763	2.2		32.5
35.0		0.013	0.3	0.028	0.4	0.067	0.6	0.124	0.8	0.272	1.3	0.416	1.6	0.794	2.2		35.0
37.5		0.014	0.3	0.029	0.4	0.069	0.6	0.129	0.9	0.282	1.3	0.432	1.6	0.824	2.2		37.5
40.0		0.014	0.3	0.031	0.4	0.072	0.6	0.133	0.9	0.292	1.3	0.447	1.6	0.853	2.2		40.0
42.5	0.15	0.015	0.3	0.032	0.4	0.074	0.6	0.138	0.9	0.302	1.3	0.462	1.6	0.882	2.2		42.5
45.0		0.015	0.3	0.033	0.4	0.077	0.6	0.142	0.9	0.312	1.3	0.477	1.6	0.909	2.2		45.0
47.5		0.016	0.3	0.034	0.4	0.079	0.6	0.146	0.9	0.321	1.3	0.491	1.6	0.936	2.2		47.5
50.0		0.016	0.3	0.035	0.4	0.081	0.6	0.150	0.9	0.330	1.3	0.504	1.6	0.962	2.2		50.0
52.5		0.017	0.3	0.036	0.4	0.083	0.6	0.155	0.9	0.339	1.3	0.518	1.6	0.987	2.2	0.50	52.5
55.0		0.017	0.3	0.036	0.4	0.085	0.6	0.159	0.9	0.347	1.3	0.531	1.6	1.01	2.2		55.0
57.5		0.018	0.3	0.037	0.4	0.088	0.6	0.162	0.9	0.356	1.3	0.544	1.6	1.04	2.3		57.5
60.0		0.018	0.3	0.038	0.4	0.090	0.6	0.166	0.9	0.364	1.3	0.556	1.6	1.06	2.3		60.0
62.5		0.018	0.3	0.039	0.4	0.092	0.7	0.170	0.9	0.372	1.3	0.569	1.6	1.08	2.3		62.5
65.0		0.019	0.3	0.040	0.4	0.094	0.7	0.174	0.9	0.380	1.3	0.581	1.6	1.11	2.3		65.0
67.5		0.019	0.3	0.041	0.4	0.096	0.7	0.177	0.9	0.388	1.3	0.592	1.7	1.13	2.3		67.5
70.0		0.020	0.3	0.042	0.4	0.098	0.7	0.181	0.9	0.395	1.3	0.604	1.7	1.15	2.3		70.0
72.5		0.020	0.3	0.042	0.4	0.099	0.7	0.184	0.9	0.403	1.3	0.616	1.7	1.17	2.3		72.5
75.0		0.020	0.3	0.043	0.4	0.101	0.7	0.188	0.9	0.410	1.3	0.627	1.7	1.19	2.3		75.0
77.5		0.021	0.3	0.044	0.4	0.103	0.7	0.191	0.9	0.418	1.3	0.638	1.7	1.21	2.3		77.5

(continued overleaf)

Table 22.10 (continued)

Δp_l	v	10 mm		15 mm		20 mm		25 mm		32 mm		40 mm		50 mm		v	Δp_l
		M	l_e	M	l_e	M	l_e	M	l_e	M	l_e	M	l_e	M	l_e		
80·0		0·021	0·3	0·045	0·4	0·105	0·7	0·194	0·9	0·425	1·4	0·649	1·7	1·24	2·3		80·0
82·5		0·021	0·3	0·046	0·4	0·107	0·7	0·197	0·9	0·432	1·4	0·659	1·7	1·26	2·3		82·5
85·0		0·022	0·3	0·046	0·4	0·108	0·7	0·201	0·9	0·439	1·4	0·670	1·7	1·28	2·3		85·0
87·5		0·022	0·3	0·047	0·4	0·110	0·7	0·204	0·9	0·446	1·4	0·680	1·7	1·30	2·3		87·5
90·0		0·023	0·3	0·048	0·4	0·112	0·7	0·207	0·9	0·452	1·4	0·691	1·7	1·31	2·3		90·0
92·5		0·023	0·3	0·049	0·4	0·113	0·7	0·210	0·9	0·459	1·4	0·701	1·7	1·33	2·3		92·5
95·0		0·023	0·3	0·049	0·4	0·115	0·7	0·213	0·9	0·466	1·4	0·711	1·7	1·35	2·3		95·0
97·5		0·024	0·3	0·050	0·4	0·117	0·7	0·216	0·9	0·472	1·4	0·721	1·7	1·37	2·3		97·5
100		0·024	0·3	0·051	0·4	0·118	0·7	0·219	0·9	0·479	1·4	0·731	1·7	1·39	2·3		100
120		0·026	0·3	0·056	0·4	0·131	0·7	0·242	0·9	0·527	1·4	0·805	1·7	1·53	2·4		120
140	0·3	0·029	0·3	0·061	0·5	0·142	0·7	0·262	0·9	0·572	1·4	0·873	1·7	1·66	2·4		140
160		0·031	0·3	0·065	0·5	0·152	0·7	0·282	1·0	0·614	1·4	0·937	1·7	1·78	2·4		160
180		0·033	0·3	0·070	0·5	0·162	0·7	0·300	1·0	0·654	1·4	0·997	1·8	1·89	2·4		180
200		0·035	0·3	0·074	0·5	0·172	0·7	0·317	1·0	0·691	1·4	1·05	1·8	2·00	2·4	1·0	200
220		0·037	0·3	0·078	0·5	0·181	0·7	0·334	1·0	0·727	1·4	1·11	1·8	2·10	2·4		220
240		0·039	0·3	0·081	0·5	0·189	0·7	0·349	1·0	0·761	1·4	1·16	1·8	2·20	2·4		240
260		0·040	0·3	0·085	0·5	0·198	0·7	0·364	1·0	0·793	1·5	1·21	1·8	2·29	2·4		260
280		0·042	0·3	0·088	0·5	0·206	0·7	0·379	1·0	0·825	1·5	1·26	1·8	2·38	2·4		280
300		0·044	0·3	0·092	0·5	0·213	0·7	0·393	1·0	0·855	1·5	1·30	1·8	2·47	2·5		300
320		0·045	0·3	0·095	0·5	0·221	0·7	0·407	1·0	0·884	1·5	1·35	1·8	2·55	2·5		320
340		0·047	0·3	0·098	0·5	0·228	0·7	0·420	1·0	0·913	1·5	1·39	1·8	2·64	2·5		340
360		0·048	0·3	0·101	0·5	0·235	0·7	0·433	1·0	0·941	1·5	1·43	1·8	2·71	2·5		360
380	0·5	0·049	0·3	0·104	0·5	0·242	0·7	0·445	1·0	0·970	1·5	1·47	1·8	2·79	2·5		380
400		0·051	0·3	0·107	0·5	0·248	0·7	0·457	1·0	0·994	1·5	1·51	1·8	2·87	2·5		400
420		0·052	0·3	0·110	0·5	0·255	0·7	0·469	1·0	1·02	1·5	1·55	1·8	2·94	2·5		420
440		0·054	0·3	0·113	0·5	0·261	0·7	0·481	1·0	1·04	1·5	1·59	1·8	3·01	2·5	1·5	440
460		0·055	0·3	0·115	0·5	0·267	0·7	0·492	1·0	1·07	1·5	1·63	1·8	3·08	2·5		460
480		0·056	0·3	0·118	0·5	0·273	0·8	0·503	1·0	1·09	1·5	1·66	1·8	3·15	2·5		480
500		0·057	0·3	0·120	0·5	0·279	0·8	0·514	1·0	1·12	1·5	1·69	1·8	3·22	2·5		500
520		0·059	0·3	0·123	0·5	0·285	0·8	0·524	1·0	1·14	1·5	1·73	1·8	3·28	2·5		520
540		0·060	0·3	0·125	0·5	0·291	0·8	0·535	1·0	1·16	1·5	1·77	1·8	3·35	2·5		540
560		0·061	0·3	0·128	0·5	0·296	0·8	0·545	1·0	1·17	1·5	1·80	1·8	3·41	2·5		560
580		0·062	0·3	0·130	0·5	0·302	0·8	0·555	1·0	1·21	1·5	1·83	1·8	3·47	2·5		580
600		0·063	0·3	0·133	0·5	0·307	0·8	0·565	1·0	1·23	1·5	1·87	1·8	3·53	2·5		600
620		0·064	0·3	0·135	0·5	0·312	0·8	0·575	1·0	1·25	1·5	1·90	1·8	3·59	2·5		620
640		0·065	0·3	0·137	0·5	0·318	0·8	0·584	1·0	1·27	1·5	1·93	1·8	3·65	2·5		640
660		0·066	0·3	0·139	0·5	0·323	0·8	0·594	1·0	1·29	1·5	1·96	1·8	3·71	2·5		660
680		0·067	0·3	0·142	0·5	0·328	0·8	0·603	1·0	1·31	1·5	1·99	1·9	3·77	2·5		680
700		0·069	0·3	0·144	0·5	0·333	0·8	0·612	1·0	1·33	1·5	2·02	1·9	3·83	2·5		700
720		0·070	0·3	0·146	0·5	0·338	0·8	0·621	1·0	1·35	1·5	2·05	1·9	3·88	2·5		720
740		0·071	0·3	0·148	0·5	0·343	0·8	0·630	1·0	1·37	1·5	2·08	1·9	3·94	2·5		740
760		0·072	0·3	0·150	0·5	0·347	0·8	0·639	1·0	1·39	1·5	2·10	1·9	3·99	2·5	2·0	760
780		0·073	0·3	0·152	0·5	0·352	0·8	0·648	1·0	1·41	1·5	2·14	1·9	4·04	2·5		780
800		0·074	0·3	0·154	0·5	0·357	0·8	0·656	1·0	1·42	1·5	2·17	1·9	4·10	2·5		800
820		0·075	0·4	0·156	0·5	0·362	0·8	0·665	1·0	1·44	1·5	2·19	1·9	4·15	2·5		820
840		0·075	0·4	0·158	0·5	0·366	0·8	0·673	1·0	1·46	1·5	2·22	1·9	4·20	2·5		840
860		0·076	0·4	0·160	0·5	0·371	0·8	0·681	1·0	1·48	1·5	2·25	1·9	4·25	2·5		860
880		0·077	0·4	0·162	0·5	0·375	0·8	0·689	1·0	1·50	1·5	2·27	1·9	4·30	2·5		880
900		0·078	0·4	0·164	0·5	0·379	0·8	0·698	1·0	1·51	1·5	2·30	1·9	4·35	2·5		900
920		0·079	0·4	0·166	0·5	0·384	0·8	0·706	1·0	1·53	1·5	2·33	1·9	4·40	2·5		920
940		0·080	0·4	0·168	0·5	0·388	0·8	0·713	1·0	1·55	1·5	2·35	1·9	4·45	2·5		940
960		0·081	0·4	0·170	0·5	0·392	0·8	0·721	1·0	1·56	1·5	2·38	1·9	4·50	2·5		960
980		0·082	0·4	0·172	0·5	0·397	0·8	0·729	1·0	1·58	1·5	2·40	1·9	4·55	2·5		980
1 000		0·083	0·4	0·173	0·5	0·401	0·8	0·737	1·0	1·60	1·5	2·43	1·9	4·59	2·5		1 000
1 100		0·087	0·4	0·182	0·5	0·421	0·8	0·774	1·1	1·68	1·5	2·55	1·9	4·82	2·6		1 100
1 200		0·091	0·4	0·191	0·5	0·441	0·8	0·809	1·1	1·75	1·5	2·67	1·9	5·04	2·6		1 200
1 300	1·0	0·095	0·4	0·199	0·5	0·459	0·8	0·844	1·1	1·83	1·5	2·78	1·9	5·25	2·6		1 300
1 400		0·099	0·4	0·207	0·5	0·477	0·8	0·876	1·1	1·90	1·5	2·89	1·9	5·46	2·6		1 400
1 500		0·102	0·4	0·214	0·5	0·495	0·8	0·908	1·1	1·98	1·5	2·99	1·9	5·65	2·6		1 500
1 600		0·106	0·4	0·222	0·5	0·511	0·8	0·939	1·1	2·03	1·5	3·09	1·9	5·84	2·6		1 600
1 700		0·109	0·4	0·229	0·5	0·528	0·8	0·968	1·1	2·10	1·5	3·19	1·9	6·02	2·6	3·0	1 700
1 800		0·113	0·4	0·236	0·5	0·543	0·8	0·997	1·1	2·16	1·6	3·28	1·9				1 800
1 900		0·116	0·4	0·242	0·5	0·559	0·8	1·03	1·1	2·22	1·6	3·37	1·9				1 900
2 000		0·119	0·4	0·249	0·5	0·574	0·8	1·05	1·1	2·28	1·6	3·46	1·9				2 000

Most often, the pump uses steam as the operating medium. This steam is exhausted at the end of the discharge stroke to the same pressure as the space from which the condensate is being drained. It is often possible where the larger condensate loads are being handled to dedicate a single pump to each load. The pump exhaust line can then be directly connected to the steam space of the heat exchanger, with condensate draining freely to the pump inlet and with any steam trap at the pump outlet.

During the discharge stroke, the inlet check valve is closed. Condensate draining from the steam space then fills the inlet piping. Unless the piping is sufficiently large or contains a receiver section (reservoir), condensate could back up into the steam space being drained.

Table 22.10 (continued)

Δp_l	v	65 mm		80 mm		90 mm		100 mm		125 mm		150 mm		v	Δp_l
		M	l_e	M	l_e	M	l_e	M	l_e	M	l_e	M	l_e		
0·1		0·061	1·5	0·096	2·0	0.144	2·5	0·200	2·9	0·362	4·1	0·600	5·3	0·05	0·1
0·2		0·091	1·7	0·144	2·2	0·215	2·7	0·298	3·3	0·544	4·5	0·889	5·8		0·2
0·3		0·115	1·8	0·181	2·3	0·271	2·9	0·375	3·4	0·685	4·7	1·12	6·1		0·3
0·4		0·136	1·9	0·214	2·4	0·319	3·0	0·442	3·6	0·805	4·9	1·31	6·4		0·4
0·5		0·154	2·0	0·243	2·5	0·362	3·1	0·501	3·7	0·913	5·0	1·49	6·5		0·5
0·6	0·05	0·171	2·0	0·269	2·6	0·401	3·2	0·556	3·8	1·01	5·1	1·65	6·6		0·6
0·7		0·187	2·1	0·294	2·6	0·438	3·2	0·606	3·8	1·10	5·2	1·79	6·7		0·7
0·8		0·202	2·1	0·317	2·7	0·472	3·3	0·653	3·9	1·19	5·3	1·93	6·8		0·8
0·9		0·216	2·1	0·339	2·7	0·504	3·3	0·698	4·0	1·27	5·4	2·06	6·9		0·9
1·0		0·229	2·2	0·359	2·8	0·535	3·4	0·740	4·0	1·34	5·5	2·18	7·0		1·0
1·5		0·288	2·3	0·451	2·9	0·671	3·6	0·928	4·2	1·68	5·7	2·73	7·3	0·15	1·5
2·0		0·338	2·4	0·530	3·0	0·787	3·7	1·09	4·3	1·97	5·9	3·20	7·5		2·0
2·5		0·383	2·4	0·600	3·1	0·891	3·8	1·23	4·4	2·23	6·0	3·61	7·6		2·5
3·0		0·424	2·5	0·664	3·1	0·985	3·8	1·36	4·5	2·46	6·1	3·99	7·7		3·0
3·5		0·462	2·5	0·723	3·2	1·07	3·9	1·48	4·6	2·68	6·2	4·34	7·9		3·5
4·0	0·15	0·498	2·6	0·778	3·2	1·15	3·9	1·59	4·6	2·88	6·3	4·66	8·0		4·0
4·5		0·531	2·6	0·830	3·3	1·23	4·0	1·70	4·7	3·07	6·3	4·97	8·0		4·5
5·0		0·563	2·6	0·880	3·3	1·30	4·0	1·80	4·7	3·25	6·4	5·26	8·1	0·30	5·0
5·5		0·594	2·7	0·927	3·3	1·37	4·1	1·90	4·8	3·42	6·4	5·54	8·2		5·5
6·0		0·623	2·7	0·973	3·4	1·44	4·1	1·99	4·8	3·59	6·5	5·81	8·2		6·0
6·5		0·651	2·7	1·02	3·4	1·51	4·1	2·08	4·9	3·75	6·5	6·06	8·3		6·5
7·0		0·678	2·7	1·06	3·4	1·57	4·2	2·16	4·9	3·90	6·6	6·31	8·3		7·0
7·5		0·704	2·7	1·10	3·4	1·63	4·2	2·24	4·9	4·05	6·6	6·55	8·4		7·5
8·0		0·729	2·7	1·14	3·5	1·69	4·2	2·32	4·9	4·19	6·6	6·78	8·4		8·0
8·5		0·754	2·8	1·18	3·5	1·74	4·2	2·40	5·0	4·33	6·7	7·00	8·4		8·5
9·0		0·778	2·8	1·21	3·5	1·80	4·2	2·48	5·0	4·46	6·7	7·22	8·5		9·0
9·5		0·801	2·8	1·25	3·5	1·85	4·3	2·55	5·0	4·60	6·7	7·43	8·5		9·5
10·0		0·824	2·8	1·29	3·5	1·90	4·3	2·62	5·0	4·72	6·7	7·63	8·5		10·0
12·5		0·930	2·9	1·45	3·6	2·14	4·4	2·96	5·1	5·32	6·8	8·60	8·7	0·50	12·5
15·0	0·30	1·03	2·9	1·60	3·6	2·37	4·4	3·26	5·2	5·87	6·9	9·47	8·8		15·0
17·5		1·12	3·0	1·74	3·7	2·57	4·5	3·54	5·2	6·37	7·0	10·3	8·8		17·5
20·0		1·20	3·0	1·87	3·7	2·76	4·5	3·80	5·3	6·84	7·1	11·0	8·9		20·0
22·5		1·28	3·0	1·99	3·8	2·94	4·6	4·05	5·3	7·28	7·1	11·7	9·0		22·5
25·0		1·35	3·0	2·11	3·8	3·11	4·6	4·28	5·4	7·69	7·1	12·4	9·0		25·0
27·5		1·42	3·1	2·22	3·8	3·27	4·6	4·50	5·4	8·09	7·2	13·0	9·1		27·5
30·0		1·49	3·1	2·32	3·8	3·43	4·6	4·71	5·4	8·47	7·2	13·6	9·1		30·0
32·5		1·56	3·1	2·42	3·8	3·58	4·7	4·92	5·4	8·84	7·3	14·2	9·1		32·5
35·0		1·62	3·1	2·52	3·9	3·72	4·7	5·12	5·5	9·19	7·3	14·8	9·2		35·0
37·5	0·50	1·68	3·1	2·61	3·9	3·86	4·7	5·31	5·5	9·53	7·3	15·3	9·2		37·5
40·0		1·74	3·1	2·70	3·9	3·99	4·7	5·49	5·5	9·86	7·3	15·9	9·2		40·0
42·5		1·80	3·1	2·79	3·9	4·12	4·7	5·67	5·5	10·2	7·4	16·4	9·3		42·5
45·0		1·85	3·2	2·88	3·9	4·25	4·7	5·84	5·5	10·5	7·4	16·9	9·3		45·0
47·5		1·91	3·2	2·96	3·9	4·37	4·8	6·01	5·6	10·8	7·4	17·4	9·3		47·5
50·0		1·96	3·2	3·04	3·9	4·49	4·8	6·17	5·6	11·1	7·4	17·8	9·3	1·0	50·0
52·5		2·01	3·2	3·12	4·0	4·61	4·8	6·33	5·6	11·4	7·4	18·3			52·5
55·0		2·06	3·2	3·20	4·0	4·72	4·8	6·49	5·6	11·6	7·4	18·8	9·4		55·0
57·5		2·11	3·2	3·28	4·0	4·83	4·8	6·64	5·6	11·9	7·5	19·2	9·4		57·5
60·0		2·16	3·2	3·35	4·0	4·94	4·8	6·79	5·6	12·2	7·5	19·6	9·4		60·0
62·5		2·20	3·2	3·42	4·0	5·05	4·8	6·94	5·6	12·5	7·5	20·0	9·4		62·5
65·0		2·25	3·2	3·50	4·0	5·16	4·8	7·08	5·7	12·7	7·5	20·5	9·4		65·0
67·5		2·30	3·2	3·57	4·0	5·26	4·9	7·22	5·7	13·0	7·5	20·9	9·4		67·5
70·0		2·34	3·2	3·63	4·0	5·36	4·9	7·36	5·7	13·2	7·5	21·3	9·4		70·0
72·5		2·38	3·2	3·70	4·0	5·46	4·9	7·50	5·7	13·5	7·5	21·7	9·5		72·5
75·0		2·43	3·3	3·77	4·0	5·56	4·9	7·63	5·7	13·7	7·5	22·0	9·5		75·0
77·5		2·47	3·3	3·83	4·0	5·65	4·9	7·77	5·7	13·9	7·5	22·4	9·5		77·5

(continued overleaf)

A similar reservoir is desirable where multiple loads discharge through individual steam traps to a common pump. Provision should then be made for venting non-condensables and flash steam which reach the reservoir. The exhaust line from the pump can then be connected to the same vent line.

22.7.5 Allowance for expansion

All pipes will be installed at ambient temperature. Pipes used to carry hot fluids, whether water, oil or steam, operate at higher temperatures. It follows that they expand (especially in length) with the increase from ambient to working temperatures. The amount of the expansion is readily calculated or read from charts, and Table 22.11 may are helpful.

The piping must be sufficiently flexible to accommodate the movements of the components as it heats up. In many cases, the piping has enough natural flexibility, by virtue of having reasonable lengths and plenty of bends that no undue stresses are set up. In other installations it is necessary to build in some means of achieving the required flexibility.

Where the condensate from a steam main drain trap is discharged into a return line running alongside the steam line the difference between the expansions of the

Table 22.10 (continued)

Δp_l	v	65 mm		80 mm		90 mm		100 mm		125 mm		150 mm		v	Δp_l
		M	l_e	M	l_e	M	l_e	M	l_e	M	l_e	M	l_e		
80·0		2·51	3·3	3·90	4·0	5·75	4·9	7·90	5·7	14·2	7·6	22·8	9·5		80·0
82·5		2·55	3·3	3·96	4·1	5·84	4·9	8·02	5·7	14·4	7·6	23·2	9·5		82·5
85·0		2·59	3·3	4·02	4·1	5·93	4·9	8·15	5·7	14·6	7·6	23·5	9·5		85·0
87·5		2·63	3·3	4·09	4·1	6·02	4·9	8·27	5·7	14·8	7·6	23·9	9·5		87·5
90·0		2·67	3·3	4·15	4·1	6·11	4·9	8·40	5·7	15·0	7·6	24·2	9·5		90·0
92·5		2·71	3·3	4·21	4·1	6·20	4·9	8·52	5·7	15·3	7·6	24·6	9·5		92·5
95·0		2·75	3·3	4·27	4·1	6·29	4·9	8·64	5·7	15·5	7·6	24·9	9·6		95·0
97·5		2·79	3·3	4·32	4·1	6·37	4·9	8·75	5·8	15·7	7·6	25·2	9·6		97·5
100·0		2·82	3·3	4·38	4·1	6·46	4·9	8·87	5·8	15·9	7·6	25·6	9·6	1·5	100·0
120·0		3·11	3·3	4·82	4·1	7·10	5·0	9·75	5·8	17·5	7·7	28·1	9·6		120·0
140·0	1·0	3·37	3·4	5·22	4·2	7·69	5·0	10·6	5·8	18·9	7·7	30·4	9·7		140·0
160·0		3·61	3·4	5·60	4·2	8·25	5·0	11·3	5·9	20·3	7·7	32·6	9·7		160·0
180·0		3·84	3·4	5·95	4·2	8·76	5·0	12·0	5·9	21·6	7·8	34·6	9·7	2·0	180·0
200·0		4·05	3·4	6·29	4·2	9·25	5·0	12·7	5·9	22·7	7·8	36·5	9·8		200·0
220·0		4·26	3·4	6·60	4·2	9·72	5·1	13·3	5·9	23·9	7·8	38·4	9·8		220·0
240·0		4·46	3·4	6·91	4·2	10·2	5·1	14·0	5·9	25·0	7·8	40·1	9·8		240·0
260·0		4·65	3·4	7·20	4·2	10·6	5·1	14·5	6·0	26·0	7·9	41·8	9·8		260·0
280·0		4·83	3·4	7·48	4·3	11·0	5·1	15·1	6·0	27·0	7·9	43·4	9·9		280·0
300·0	1·5	5·00	3·5	7·75	4·3	11·4	5·1	15·6	6·0	28·0	7·9	45·0	9·9		300·0
320·0		5·17	3·5	8·01	4·3	11·8	5·1	16·2	6·0	29·0	7·9	46·5	9·9		320·0
340·0		5·34	3·5	8·27	4·3	12·2	5·2	16·7	6·0	29·8	7·9	47·9	9·9		340·0
360·0		5·50	3·5	8·51	4·3	12·5	5·2	17·2	6·0	30·7	7·9	49·4	9·9		360·0
380·0		5·65	3·5	8·75	4·3	12·8	5·2	17·7	6·0	31·6	7·9	50·7	9·9		380·0
400·0		5·80	3·5	8·99	4·3	13·2	5·2	18·1	6·0	32·4	7·9	52·1	9·9		400·0
420·0		5·95	3·5	9·22	4·3	13·6	5·2	18·6	6·0	33·2	7·9	53·4	9·9		420·0
440·0		6·09	3·5	9·44	4·3	13·9	5·2	19·0	6·0	34·0	7·9	54·7	9·9	3·0	440·0
460·0		6·24	3·5	9·66	4·3	14·2	5·2	19·5	6·0	34·8	8·0	55·9	9·9		460·0
480·0		6·37	3·5	9·87	4·3	14·5	5·2	19·9	6·0	35·6	8·0	57·2	10		480·0
500·0		6·51	3·5	10·1	4·3	14·8	5·2	20·3	6·0	36·3	8·0	58·4	10		500·0
520·0		6·64	3·5	10·3	4·3	15·1	5·2	20·7	6·1	37·1	8·0	59·5	10		520·0
540·0	2·0	6·77	3·5	10·5	4·3	15·4	5·2	21·1	6·1	37·8	8·0	60·7	10		540·0
560·0		6·90	3·5	10·7	4·3	15·7	5·2	21·5	6·1	38·5	8·0	61·8	10		560·0
580·0		7·02	3·5	10·9	4·3	16·0	5·2	21·9	6·1	39·2	8·0	62·9	10		580·0
600·0		7·15	3·5	11·1	4·3	16·3	5·2	22·3	6·1	39·9	8·0	64·0	10		600·0
620·0		7·27	3·5	11·3	4·4	16·6	5·2	22·7	6·1	40·5	8·0	65·1	10		620·0
640·0		7·39	3·5	11·4	4·4	16·8	5·2	23·1	6·1	41·2	8·0	66·2	10		640·0
660·0		7·50	3·5	11·6	4·4	17·1	5·2	23·4	6·1	41·9	8·0	67·2	10		660·0
680·0		7·62	3·5	11·8	4·4	17·3	5·2	23·8	6·1	42·5	8·0	68·2	10		680·0
700·0		7·73	3·5	12·0	4·4	17·6	5·2	24·1	6·1	43·1	8·0	69·2	10		700·0
720·0		7·85	3·5	12·2	4·4	17·8	5·2	24·5	6·1	43·7	8·0	70·2	10		720·0
740·0		7·96	3·5	12·3	4·4	18·1	5·2	24·8	6·1	44·4	8·0	71·2	10		740·0
760·0		8·07	3·5	12·4	4·4	18·4	5·3	25·1	6·1	45·0	8·0	72·2	10	4·0	760·0
780·0		8·17	3·5	12·6	4·4	18·6	5·3	25·5	6·1	45·6	8·0				780·0
800·0		8·28	3·6	12·8	4·4	18·8	5·3	25·8	6·1	46·2	8·0				800·0
820·0		8·39	3·6	12·9	4·4	19·1	5·3	26·2	6·1	46·7	8·0				820·0
840·0		8·49	3·6	13·1	4·4	19·3	5·3	26·5	6·1	47·3	8·0				840·0
860·0		8·59	3·6	13·3	4·4	19·6	5·3	26·8	6·1	47·9	8·0				860·0
880·0		8·69	3·6	13·5	4·4	19·8	5·3	27·1	6·1	48·4	8·0				880·0
900·0		8·80	3·6	13·6	4·4	20·0	5·3	27·4	6·1	49·0	8·0				900·0
920·0		8·89	3·6	13·8	4·4	20·2	5·3	27·7	6·1	49·6	8·1				920·0
940·0		8·99	3·6	13·9	4·4	20·5	5·3	28·0	6·1	50·1	8·1				940·0
960·0		9·09	3·6	14·1	4·4	20·7	5·3	28·3	6·1	50·6	8·1				960·0
980·0		9·19	3·6	14·2	4·4	20·9	5·3	28·6	6·1						980·0
1 000·0		9·28	3·6	14·4	4·4	21·1	5·3	28·9	6·1						1 000·0
1 100·0		9·74	3·6	15·1	4·4	22·2	5·3	30·4	6·1						1 100·0
1 200·0	3·0	10·2	3·6	15·8	4·4	23·2	5·3	31·7	6·1						1 200·0
1 300·0		10·6	3·6	16·4	4·4	24·1	5·3								1 300·0
1 400·0		11·0	3·6	17·0	4·4	25·0	5·3								1 400·0
1 500·0		11·4	3·6	17·6	4·4										1 500·0
1 600·0		11·8	3·6	18·2	4·4										1 600·0
1 700·0		12·2	3·6	18·8	4·4										1 700·0
1 800·0		12·5	3·6												1 800·0
1 900·0		12·9	3·6												1 900·0
2 000·0		13·2	3·6												2 000·0

two lines must be remembered. The steam line may be at a temperature very much above that of the return line, and the two connection points can move in relation to each other during system warm-up. Some flexibility should be incorporated into the steam trap piping so that branch connections do not become overstressed, as in Figure 22.17.

The amount of movement to be taken up by the piping and any device incorporated into it can be reduced by the use of 'cold draw'. The total amount of expansion is first calculated for each section between fixed anchor points. The pipes are left short by half this amount, and stretched cold, as by pulling up bolts at a flanged joint, so that at ambient temperature the system is stressed in one

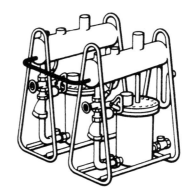

Figure 22.16 Spirax Ogden packaged unit

Table 22.11 Expansion of pipes

Final temperature		Expansion per 30 m (mm)	(100 ft) (in)
(°C)	(°F)		
66	150	19	0.75
93	200	29	1.14
121	250	41	1.61
149	300	50	1.97
177	350	61	2.4
204	400	74	2.91
232	450	84	3.3
260	500	97	3.8

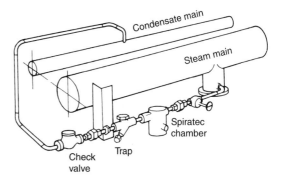

Figure 22.17 Flexible trapping arrangements

direction. When warmed through half the total temperature rise, and having expanded by half the total amount, the piping is unstressed. At working temperature and having fully expanded, the piping is stressed in the opposite direction. The effect is that instead of being stressed from zero to $+f$ units, the piping is stressed from $-\frac{1}{2}f$ to $+\frac{1}{2}f$. In practical terms the piping is assembled with a spacer piece, of length equal to half the expansion, between two flanges. When the piping is fully installed and anchored the spacer is removed and the joint pulled up tight (Figure 22.18). The remaining part of the expansion, if not accepted by the natural flexibility of the piping, will call for the use of an expansion fitting. These can take several forms.

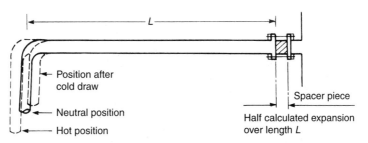

Figure 22.18 Cold draw

Steam Tables (metric SI units)

Pressure		Temperature °C	Specific enthalpy			Specific volume steam m^3/kg
bar	kPa		Water (h_f) kJ/kg	Evaporation (h_{fg}) kJ/kg	Steam (h_g) kJ/kg	
0.30	30.0	69.10	289.23	2336.1	2625.3	5.229
0.50	50.0	81.33	340.49	2305.4	2645.9	3.240
	absolute					
0.75	75.0	91.78	384.39	2278.6	2663.0	2.217
0.95	95.0	98.20	411.43	2261.8	2673.2	1.777
0	0	100.00	419.04	2257.0	2676.0	1.673
	gauge					
0.10	10.0	102.66	430.2	2250.2	2680.4	1.533
0.20	20.0	105.10	440.8	2243.4	2684.2	1.414
0.30	30.0	107.39	450.4	2237.2	2687.6	1.312
0.40	40.0	109.55	459.7	2231.3	2691.0	1.225
0.50	50.0	111.61	468.3	2225.6	2693.9	1.149
0.60	60.0	113.56	476.4	2220.4	2696.8	1.083
0.70	70.0	115.40	484.1	2215.4	2699.5	1.024
0.80	80.0	117.14	491.6	2210.5	2702.1	0.971
0.90	90.0	118.80	498.9	2205.6	2704.5	0.923
1.00	100.0	120.42	505.6	2201.1	2706.7	0.881
1.10	110.0	121.96	512.2	2197.0	2709.2	0.841
1.20	120.0	123.46	518.7	2192.8	2711.5	0.806
1.30	130.0	124.90	524.6	2188.7	2713.3	0.773
1.40	140.0	126.28	530.5	2184.8	2715.3	0.743
1.50	150.0	127.62	536.1	2181.0	2717.1	0.714
1.60	160.0	128.89	541.6	2177.3	2718.9	0.689
1.70	170.0	130.13	547.1	2173.7	2720.8	0.665
1.80	180.0	131.37	552.3	2170.1	2722.4	0.643
1.90	190.0	132.54	557.3	2166.7	2724.0	0.622
2.00	200.0	133.69	562.2	2163.3	2725.5	0.603
2.20	220.0	135.88	571.7	2156.9	2728.6	0.568
2.40	240.0	138.01	580.7	2150.7	2731.4	0.536
2.60	260.0	140.0	589.2	2144.7	2733.9	0.509
2.80	280.0	141.92	597.4	2139.0	2736.4	0.483
3.00	300.0	143.75	605.3	2133.4	2738.7	0.461
3.20	320.0	145.46	612.9	2128.1	2741.0	0.440
3.40	340.0	147.20	620.0	2122.9	2742.9	0.422
3.60	360.0	148.84	627.1	2117.8	2744.9	0.406
3.80	380.0	150.44	634.0	2112.9	2746.9	0.389
4.00	400.0	151.96	640.7	2108.1	2748.8	0.374
4.50	450.0	155.55	656.3	2096.7	2753.0	0.342
5.00	500.0	158.92	670.9	2086.0	2756.9	0.315
5.50	550.0	162.08	684.6	2075.7	2760.3	0.292
6.00	600.0	165.04	697.5	2066.0	2763.5	0.272
6.50	650.0	167.83	709.7	2056.8	2766.5	0.255
7.00	700.0	170.50	721.4	2047.7	2769.1	0.240
7.50	750.0	173.02	732.5	2039.2	2771.7	0.227
8.00	800.0	175.43	743.1	2030.9	2774.0	0.215
8.50	850.0	177.75	753.3	2022.9	2776.2	0.204
9.00	900.0	179.97	763.0	2015.1	2778.1	0.194
9.50	950.0	182.10	772.5	2007.5	2780.0	0.185
10.00	1000.0	184.13	781.6	2000.1	2781.7	0.177
10.50	1050.0	186.05	790.1	1993.0	2783.3	0.171
11.00	1100.0	188.02	798.8	1986.0	2784.8	0.163
11.50	1150.0	189.82	807.1	1979.1	2786.3	0.157
12.00	1200.0	191.68	815.1	1972.5	2787.6	0.151
12.50	1250.0	193.43	822.9	1965.4	2788.8	0.148
13.00	1300.0	195.10	830.4	1959.6	2790.0	0.141
13.50	1350.0	196.62	837.9	1953.2	2791.1	0.136
14.00	1400.0	198.35	845.1	1947.1	2792.2	0.132
14.50	1450.0	199.92	852.1	1941.0	2793.1	0.128
15.00	1500.0	201.45	859.0	1935.0	2794.0	0.124

Steam Tables (*continued*)

Pressure		Temperature	Specific enthalpy			Specific volume
bar	*kPa*	°C	Water (h_f) kJ/kg	Evaporation (h_{fg}) kJ/kg	Steam (h_g) kJ/kg	steam m^3/kg
15.50	1550.0	202.92	865.7	1928.8	2794.9	0.119
16.00	1600.0	204.38	872.3	1923.4	2795.7	0.117
17.00	1700.0	207.17	885.0	1912.1	2797.1	0.110
18.00	1800.0	209.90	897.2	1901.3	2798.5	0.105
19.00	1900.0	212.47	909.0	1890.5	2799.5	0.100
20.00	2000.0	214.96	920.3	1880.2	2800.5	0.0949
21.00	2100.0	217.35	931.3	1870.1	2801.4	0.0906
22.00	2200.0	219.65	941.9	1860.1	2802.0	0.0868
23.00	2300.0	221.85	952.2	1850.4	2802.6	0.0832
24.00	2400.0	224.02	962.2	1840.9	2803.1	0.0797
25.00	2500.0	226.12	972.1	1831.4	2803.5	0.0768
26.00	2600.0	228.15	981.6	1822.2	2803.8	0.0740
27.00	2700.0	230.14	990.7	1813.3	2804.0	0.0714

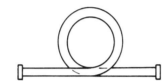

Figure 22.19 Full loop

Figure 22.20 Horseshoe loop

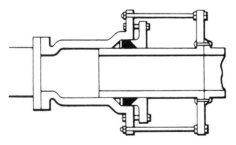

Figure 22.21 Sliding joint

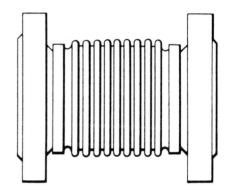

Figure 22.22 Bellows

22.7.6 Full loop (Figure 22.19)

This is simply one complete turn of the pipe and should preferably be fitted in a horizontal rather than a vertical position to prevent condensate building up. The downstream side passes below the upstream side and great care must be taken that it is not fitted the wrong way round.

When full loops are to be fitted in a confined space, care must be taken in purchasing; otherwise, wrong-handed loops may be supplied.

The full loop does not produce a force in opposition to the expanding pipe work as in some other types but with steam pressure inside the loop, there is a slight tendency to unwind, which puts an additional stress on the flanges.

22.7.7 Horseshoe or lyre loop (Figure 22.20)

Where space is available, this type is sometimes used. It is best fitted horizontally so that the loop and main are all in the same plane. Pressure does not tend to blow the ends of the loop apart but there is a very slight straightening out effect. This is due to the design but causes no misalignment of the flanges. In other cases, the 'loop' is fabricated from straight lengths of pipe and 90° bends. This may not be as effective and requires more space but it meets the same need. If any of these arrangements are fitted with the loop vertically above the pipe then a drain point must be provided on the upstream side.

22.7.8 Sliding joint (Figure 22.21)

These are often used because they take up little room but it is essential that the pipeline is rigidly anchored

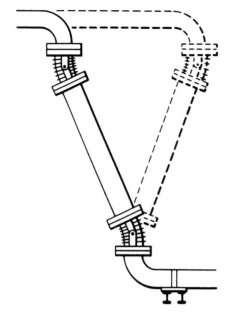

Figure 22.23 Expansion fitting

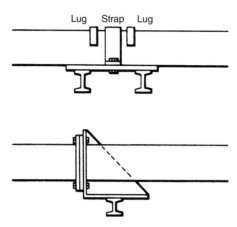

Figure 22.24 Anchor points

and guided. This is because steam pressure acting on the cross-sectional area of the sleeve part of the joint tends to blow the joint apart in opposition to the forces produced by the expanding pipe work. Misalignment will cause the sliding sleeve to bend, while regular maintenance of the gland packing is also needed

22.7.9 Bellows (Figure 22.22)

A simple bellows has the advantage that it is an in-line fitting and requires no packing, as does the sliding joint type. But it does have the same disadvantage as the sliding joint in that pressure inside tends to extend the fitting so that anchors and guides must be able to withstand this force.

The bellows can, however, be incorporated into a properly designed expansion fitting as shown in Figure 22.23, which is capable of absorbing not only axial movement of the pipeline but some lateral and angular displacement as well.

If expansion fittings are to work as intended, it is essential that the steam line is properly anchored at some point between the expansion fittings. Guiding is also important to ensure that any movement does not interfere with the designed fall towards the drain points.

Detailed design is clearly outside the scope of this section but Figure 22.24 shows some typical anchor points utilizing pipe flanges or lugs welded onto the pipe.

23

Industrial Boilers

E Walker and R J Blaen
Senior Green Limited

Contents

23.1 Terminology

The following explain some of the more fundamental terms encountered when considering boilers.

23.1.1 Shell boiler

The products of combustion or hot gases pass through a series of tubes surrounded by water in this boiler. All are contained in an outer shell.

23.1.2 Watertube boiler

In a watertube boiler, water circulates through small-bore tubes constructed in banks and connected to drums or headers. The external surfaces of the tubes are exposed to the products of combustion or hot gases.

23.1.3 Dryback boiler

This is a horizontal shell boiler where the gas-reversal chamber from the combustion tube to the first pass of tubes is external to the rear tube plate and is formed by a refractory lined steel chamber.

23.1.4 Wetback boiler

A wetback boiler is a horizontal shell boiler where the gas-reversal chamber from the combustion tube to the first pass of tubes is integral within the boiler shell and surrounded by water.

23.1.5 Economic boiler

This is a term applied to the early freestanding shell boilers of two- and three-pass construction. Originally, they were dry back and later wet back. These boilers superseded the brickset boilers. The earliest economic boilers were also brickset. The gases from the front smokebox returning across the lower external part of the shell were contained within the brick setting to form a third pass.

23.1.6 Packaged boiler

A packaged boiler is a concept of a factory-built and assembled shell boiler complete with its combustion appliance, feedwater pump and controls, valves, base frame and insulation. Before this, the economic boiler was delivered to site as a bare shell and assembled *in situ*. Originally, in the early 1960s package boilers were designed to make them as compact as possible, resulting in some inherent faults. Since then, design criteria have greatly improved and the present packaged boiler is constructed to acceptable commercial standards.

23.1.7 Evaporation

This is the quantity of steam produced by the boiler at temperature and pressure. It may be quoted as *actual evaporation* or *evaporation from and at* 100°C. Actual evaporation is the quantity of steam passing through the crown valve of the boiler. Evaporation from and at 100°C is a figure taken for design purposes, and is based on the actual evaporation per pound of fuel multiplied by the factor of evaporation.

23.1.8 Factor of evaporation

This is the figure obtained by dividing the total heat of steam at working condition by the latent heat of steam at atmospheric condition (i.e. 2256 kJ/kg). Then

$$\text{factor of evaporation} = \frac{H - T}{2256}$$

where

H = Total heat in 1 kg of steam at working pressure above 0°C taken from steam tables in kJ/kg.

T = Heat in feedwater (kJ/kg).

2256 = the latent heat of steam at atmospheric conditions.

23.1.9 Availability

The period of time that a boiler is expected or required to operate before being shut down for cleaning or maintenance. This will vary with the type of boiler, the fuel being used and the operating load on the boiler.

23.1.10 Priming

This is when the water surface in the boiler shell becomes unstable. Vigorous surging will occur and this may cause the boiler to go to low water and cut out or possibly lock out. This, in turn, will exacerbate the condition.

There are two possible causes. The first could be incorrect control of water treatment and blowdown. This can result in excessive levels of suspended solids in the boiler water, organic matter in the boiler water or high alkalinity. The second can be mechanical. If the boiler is operated below its designed working pressure it will increase the efflux velocity of the steam leaving the water surface area to a point where it may lift the water surface and drop the water level. It is important therefore to give due consideration to the steam load required from the boiler.

23.1.11 Thermal storage

Thermal storage is a method of supplying a steam load in excess of the maximum continuous rating of the boiler for short periods. The boiler shell diameter is increased to provide a greater height of water than normal above the top line of heating surface to the normal working water level. When an excess load is then imposed on the boiler, it allows this extra water to flash to steam while lowering the working pressure in the boiler. Safety-level controls protect the boiler against an excessive draw-off of steam. As the load decreases, the boiler maintains its maximum firing rate and allows the extra water to return to the normal (higher) working water level and the higher working pressure regained.

23.1.12 Accumulator

This may be likened to an unfired boiler. Steam is generated in the boilers at a pressure higher than that required for the process and fed to the process through pressure-reducing valves. This higher-pressure steam is also supplied to the accumulator, where it heats and pressurizes the water in the accumulator. When a steam load in excess of the boiler maximum firing rate occurs, steam may be supplied from the accumulator until the load is met from the boilers. As the steam load falls away, the accumulator will recharge from the boilers. This system enables boilers to be installed with ratings to match an average load while the peaks will be met from the accumulator.

23.1.13 Cavitation

This is a condition which occurs when the feedwater pump is unable to deliver feedwater to the boiler although the feed tank has water available. The temperature of the feedwater coupled with the possible suction effect from the feedwater pump in the line between the feed tank and the pump effectively drops the pressure, causing the feedwater to flash to steam. The pump then loses its water supply.

In most cases, this condition may be avoided by arranging a sufficient heat of water and by correct sizing of the feedwater pipework. Clean filters must also be maintained.

23.2 Heat transfer in industrial boilers

Heat is transferred from the hot products of combustion to the boiler heating surfaces through the plate and tube walls and to the water by various mechanisms, which involve conduction, radiation and convection.

23.2.1 Conduction

The Fourier law gives the rate at which heat is transferred by conduction through a substance without mass transfer. This states that the heat flow rate per unit area, or heat flux, is proportional to the temperature gradient in the direction of heat flow. The relationship between heat flux and temperature gradient is characterized by the thermal conductivity which is a property of the substance. It is temperature dependent and is determined experimentally.

For a plate of area A (m^2), thickness e (m) and with hot and cold face temperatures of T_1 and T_2 (°C), respectively, the normal heat flux ϕ and heat transfer rate Q are given by:

$$\phi = \frac{Q}{A} = \frac{k(T_1 - T_2)}{e} (Wm^2)$$

where K = thermal conductivity (W/mK).

23.2.2 Radiation

Thermal radiation takes place by the emission of electromagnetic waves, at the velocity of light, from all bodies at temperatures above absolute zero. The heat flux from an ideal or *black body-radiating* surface is proportional to the fourth power of the absolute temperature of the surface. The constant of proportionality is the Stefan–Boltzmann constant, which has a value of 5.6696×10^8 (W/m^{2}K^4).

The heat flux radiated from a real surface is less than that from an ideal black body surface at the same temperature. The ratio of real to black body flux is the normal total emissivity. Emissivity, like thermal conductivity, is a property which must be determined experimentally.

Although the rate of emission from a surface is independent of the condition of the surroundings, the net overall exchange of radiant heat between surfaces at different temperatures depends on a number of factors. The continuous interchange of energy is a result of the reciprocal processes of radiation and absorption, and these are dependent on geometrical relationships, emissivity differences and the presence of any absorbing and emitting gases in the intervening space.

23.2.3 Convection

Convective heat transmission occurs within a fluid, and between a fluid and a surface, by virtue of relative movement of the fluid particles (that is, by mass transfer). Heat exchange between fluid particles in mixing and between fluid particles and a surface is by conduction. The overall rate of heat transfer in convection is, however, also dependent on the capacity of the fluid for energy storage and on its resistance to flow in mixing. The fluid properties which characterize convective heat transfer are thus thermal conductivity, specific heat capacity and dynamic viscosity.

Convection is classified according to the motivating flow. When the flow takes place because of density variations caused by temperature gradients, the motion is called natural convection. When it is caused by an external agency such as a pump or a fan the process is called forced convection.

At a convection heat transfer surface the heat flux (heat transfer rate per unit area) is related to the temperature difference between fluid and surface by a heat transfer coefficient. Newton's law of cooling defines this:

$$\phi = \frac{Q}{A} = h_c \Delta T_m$$

where

ϕ = heat flux (W/m^2),
Q = heat transfer rate (W),
A = surface area (m^2),
ΔT_m = mean temperature difference between fluid and surface (K),
h_c = convective heat transfer coefficient (W/m^2 K).

The heat transfer coefficient is correlated experimentally with the fluid transport properties (specific heat, viscosity, thermal conductivity and density), fluid velocity and the geometrical relationship between surface and fluid flow.

23.2.4 Furnace heat transfer

Heat transfer in the furnace is mainly by radiation, from the incandescent particles in the flame and from hot radiating gases such as carbon dioxide and water vapor. The detailed theoretical prediction of overall radiation exchange is complicated by a number of factors such as carbon particle and dust distributions, and temperature variations in three-dimensional mixing. This is overcome by the use of simplified mathematical models or empirical relationships in various fields of application.

For industrial boilers the mean gas temperature at the furnace exit, or at the entrance to the convection section of the boiler, may be calculated using the relationship:

$$T = k \left(\frac{H}{A} \right)^{0.25}$$

where

T = gas temperature (°C),
H = heat input rate (W) based on the net calorific value of the fuel,
A = effective (projected) water-cooled absorption surface area (m^2),
K = a constant which depends on the fuel and the excess air in the combustion products.

The value of k is determined experimentally by gas temperature measurement. The measurement error of a simple pyrometer can be 250 to 300 K, due to re-radiation to water-cooled surroundings, and the values given below are based on measurement by a 'Land' multi-shielded high-velocity suction pyrometer. Typical values for normal excess air at or near full boiler load are:

Natural gas $k = 52.4$
Gas oil $k = 49.1$
Heavy fuel oil $k = 48.3$
Coal $k = 40.3$

In calculating the smoke tube inlet gas temperature of a shell boiler, A includes the effective water-cooled surface in the reversal chamber. In coal-fired boilers, any water-cooled surface below the grate is excluded from A.

The total furnace heat absorption may be estimated by using the calculated furnace exit gas temperature and analysis to determine the enthalpy (excluding the latent heat of water vapor) and thus deducting the heat rejection rate from the net heat input rate.

23.2.5 Boiler tube convection heat transfer

The radiant section of an industrial boiler may typically contain only 10 per cent of the total heating surface, yet, because of the large temperature difference, it can absorb 30–50 per cent of the total heat exchange. The mean temperature difference available for heat transfer in the convective section is much smaller. To achieve a thermally efficient yet commercially viable design it is necessary to make full use of forced convection within the constraint of acceptable pressure drop.

Forced convection heat transfer has been measured under widely differing conditions, and using the dimensionless groups makes correlation of the experimental results:

Nusselt number $Nu = \dfrac{h_c D}{k}$

Reynolds number $Re = \dfrac{GD}{\mu}$

Prandtl number $Pr = \dfrac{C_p \mu}{k}$

where

h_c = heat transfer coefficient (W/m^2 K),
D = characteristic dimension (m),
K = thermal conductivity (W/m K),
G = gas mass velocity (kg/m^{2}s),
μ = dynamic viscosity (kg/m s),
C_p = specific heat at constant pressure (J/kg K).

In applying the correlation, use is made of the concept of logarithmic mean temperature difference across the boundary layer. For a boiler section, or pass, this is given by:

$$\Delta T_m = \frac{(T_1 T_w) - (T_2 - T_w)}{\log_n \left[\dfrac{(T_1 - T_w)}{(T_2 - T_w)} \right]} (K)$$

where

T_1 = inlet gas temperature (°C),
T_2 = outlet gas temperature (°C),
T_w = tube wall temperature (°C).

The difference in temperature between the tube wall and the water is small, typically less than 10 K in the convection section. Therefore, little error is introduced by using the water temperature as T_w in the evaluation of the gas transport properties.

The representative gas temperatures used in the correlation are the bulk temperature and the film temperature. These are defined as:

Bulk temperature $T_b = T_w + \Delta T_m$
Film temperature $T_f = (T_b + T_w)/2$

For longitudinal flow in the tubes of shell boilers the mean heat transfer coefficient may be determined from:

$$Nu = 0.023 \, Re^{0.8} \, Pr^{0.4} \left(1 + \frac{D}{L} \right)^{0.7}$$

where D/L is the tube inside diameter-to-length ratio and the characteristic dimension in Nu and Re is the tube inside diameter. Gas properties are evaluated at the film temperature.

Correlations for forced convection over tubes in crossflow are complicated by the effect of the tube bank arrangement. For the range of Reynolds numbers likely to be encountered in industrial boilers the following equations may be used:

In-line arrays $Nu = 0.211 \, Re^{0.651} Pr^{0.34} F_1 F_2$
Staggered arrays $Nu = 0.2233 \, Re^{0.635} Pr^{0.34} F_1 F_2$

In these cases gas properties are evaluated at the bulk temperature, the characteristic dimension in Nu and Re

is the tube outside diameter, and the Reynolds number is based on the mass velocity through the minimum area for flow between tubes. F_1 is a correction factor for wall to bulk property variation, which can be calculated from the relationship:

$$F_1 = \frac{Pr_{\mathrm{b}}^{0.26}}{Pr_{\mathrm{w}}}$$

where Pr_{b} and Pr_{f} are Prandtl numbers at the bulk and wall temperatures, respectively.

F_2 is a correction factor for the depth of the tube bank in the direction of flow. For bank depths of 10 rows or more, $F_2 = 1$. For smaller bank depths the following values of F_2 may be used:

Gas transport properties for the products of combustion of the common fuels, fired at normal excess air at or near-full boiler load, may be obtained from Tables 23.1–23.4. Non-luminous gas radiation has a small overall effect in the convective section, typically 2–5 per cent of total convection. It may therefore be neglected for a conservative calculation.

Table 23.1 Transport properties: natural gas products of combustion

Temp. (°C)	Spec. heat (J/kg K)	Viscosity (kg/m s × 10⁶)	Conductivity (W/m K × 10³)	Sp. vol. (m³/kg)
100	1098	20.01	27.27	1.1
200	1133	23.97	34.45	1.395
300	1166	27.55	41.34	1.69
400	1198	30.83	47.94	1.985
500	1227	33.89	54.25	2.28
600	1255	36.74	60.29	2.575
700	1281	39.44	66.09	2.87
800	1305	41.99	71.61	3.164
900	1328	44.43	76.86	3.459
1000	1348	46.75	81.86	3.754
1100	1367	48.98	86.6	4.049
1200	1384	51.13	91.08	4.344
1300	1400	53.2	95.31	4.639
1400	1413	55.2	99.25	4.934

Table 23.2 Transport properties: gas oil products of combustion

Temp. (°C)	Spec. heat (J/kg K)	Viscosity (kg/m s × 10⁶)	Conductivity (W/m K × 10³)	Sp. vol. (m³/kg)
100	1061	20.32	27.24	1.058
200	1096	24.29	34.4	1.342
300	1128	27.88	41.22	1.625
400	1159	31.16	47.73	1.909
500	1188	34.2	53.92	2.192
600	1215	37.05	59.81	2.476
700	1240	39.72	65.42	2.76
800	1263	42.26	70.71	3.043
900	1284	44.67	75.73	3.327
1000	1303	46.98	80.46	3.61
1100	1320	49.19	84.89	3.894
1200	1336	51.32	89.02	4.177
1300	1349	53.37	92.88	4.461
1400	1361	55.35	96.43	4.745

Table 23.3 Transport properties: heavy fuel oil products of combustion

Temp. (°C)	Spec. heat (J/kg K)	Viscosity (kg/m s × 10⁶)	Conductivity (W/m K × 10³)	Sp. vol. (m³/kg)
100	1054	20.37	27.22	1.05
200	1088	24.34	34.37	1.332
300	1121	27.93	41.17	1.613
400	1152	31.21	47.66	1.895
500	1181	34.25	53.82	2.176
600	1207	37.09	59.69	2.458
700	1232	39.44	66.09	2.87
800	1255	42.3	70.51	3.02
900	1276	44.71	75.47	3.302
1000	1294	47.01	80.15	3.583
1100	1311	49.22	84.51	3.865
1200	1326	51.35	88.59	4.146
1300	1339	53.40	92.38	4.428
1400	1351	55.38	95.86	4.709

Table 23.4 Transport properties: bit coal products of combustion

Temp. (°C)	Spec. heat (J/kg K)	Viscosity (kg/m s × 10⁶)	Conductivity (W/m K × 10³)	Sp. vol. (m³/kg)
100	1031	20.82	27.43	1.034
200	1065	24.83	34.63	1.312
300	1096	28.44	41.39	1.589
400	1125	31.73	47.78	1.866
500	1152	34.78	53.8	2.143
600	1177	37.63	59.5	2.421
700	1201	40.3	64.88	2.698
800	1222	42.83	69.93	2.975
900	1242	45.24	74.68	3.252
1000	1259	47.55	79.11	3.53
1100	1275	49.75	83.23	3.807
1200	1289	51.87	87.05	4.084
1300	1301	53.92	90.56	4.361
1400	1311	55.89	93.77	4.638

23.2.6 Waterside conditions

In the radiant section of a boiler the fourth power of the wall temperature is typically less than 2 per cent of the fourth power of the mean flame and gas temperature. The effects of waterside conditions and wall thickness on the heat transfer rate are therefore negligible.

Even the presence of a dangerous layer of waterside scale reduces the heat flux only by a few per cent. Although this means that scale has little effect on radiant section performance, it also indicates that the metal temperature escalation due to the presence of scale is not self-limiting but is almost proportional to scale thickness.

The thermal conductivity of an average boiler scale is 2.2 (W/m K) and that of complex silicate scales is 0.2–0.23 (W/m K). Since the furnace peak wall flux can be over 300,000 (W/m²) it may readily be seen that a small thickness of scale can raise the metal temperature into the creep region, resulting in very expensive repairs.

In the convective section, the gas-side heat transfer coefficient controls the heat flux distribution since the waterside coefficient and the thermal conductance of the tube walls are very large in comparison. For this reason, it is usually satisfactory to make an allowance by adding 10 K to the water temperature in steam boilers. In hot water generators, the allowance should be about 20 K, because sub-cooled nucleate boiling generally takes place only on the radiant walls and in shell boilers on the reversal chamber tubeplate. Waterside heat transfer on the major part of the convective heating surface in these units is by convection without boiling.

23.2.7 Further reading

A good introduction to the extensive literature on the science and technology of heat transfer, with 87 further references, is given in Rose, J. W. and Cooper, J. R., *Technical Data on Fuel*, 7th edition, British National Committee, World Energy Conference, London, p. 48 (1977).

23.3 Types of boiler

As this covers industrial boilers, only units of 500 kg/h of steam, or equivalent hot water, and above will be considered. There are nine categories of boiler available. In order of evaporation these are:

Cast iron sectional boilers
Steel boilers
Electrode boilers
Steam generators
Vertical shell boilers
Horizontal shell boilers
Watertube boilers
Waste heat boilers
Fluid bed boilers

23.3.1 Cast iron sectional boilers

These are used for hot water services with a maximum operating pressure of 5 bar and a maximum output in the order of 1500 kW. Site assembly of the unit is necessary and will consist of a bank of cast iron sections. Each section has internal waterways.

The sections are assembled with screwed or taper nipples at top and bottom for water circulation and sealing between the sections to contain the products of combustion. Tie rods compress the sections together. A standard section may be used to give a range of outputs dependent upon the number of sections used. After assembly of the sections, the mountings, insulation and combustion appliance are fitted. This system makes them suitable for locations where it is impractical to deliver a package unit, e.g. basements where inadequate access is available or rooftop plant rooms where sections may be taken up using the elevator shafts. Models available use liquid, gaseous and solid fuel.

23.3.2 Steel boilers

These are similar in rated outputs to the cast iron sectional boiler. Construction is of rolled steel annular drums for

the pressure vessel. They may be of either vertical or horizontal configuration, depending upon the manufacturer. In their vertical pattern, they may be supplied for steam raising.

23.3.3 Electrode boilers

These are available for steam raising up to 3600 kg/h and manufacture is to two designs. The smaller units are element boilers with evaporation less than 500 kg/h. In these, an immersed electric element heats the water and a set of water-level probes positioned above the element controls the water level being interconnected to the feedwater pump and the element electrical supply.

Larger units are electrode boilers. Normal working pressure would be 10 bar but higher pressures are available. Construction is a vertical pattern pressure shell containing the electrodes (Figure 23.1). The lengths of the electrodes control the maximum and minimum water level. The electrical resistance of the water allows a current to flow through the water, which, in turn, boils and releases steam. Since water has to be present within the electrode system, lack of water cannot burn out the boiler. The main advantage with these units is that they may be located at the point where steam is required and, as no combustion fumes are produced, no chimney is required. Steam may also be raised relatively quickly, as there is little thermal stressing to consider.

23.3.4 Steam generators

While the term 'steam generator' may apply to any vessel raising steam, this section is intended to cover coil type boilers in the evaporative range up to 3600 kg/h of steam. Because of the steam pressure being contained within the tubular coil, pressures of 35 bar and above are available, although the majority are supplied to operate at up to 10 bar. They are suitable for firing with liquid and gaseous fuels, although the use of heavy fuel oil is unusual.

The coiled tube is contained within a pressurized combustion chamber and receives both radiant and convective heat. A control system matches the burner-firing rate proportional to the steam demand. Feedwater is pumped through the coil and partially flashed to steam in a separator. The remaining water is recirculated to a feedwater heat exchanger before being run to waste. Because there is no stored water in this type of unit, they are lighter in weight and therefore suitable for siting on mezzanine or upper floors adjacent to the plant requiring steam. Also, as the water content is minimal, steam raising can be achieved very quickly and can respond to fluctuating demand within the capacity of the generator. It must be noted that close control of suitable water treatment is essential to protect the coil against any build-up of deposits.

23.3.5 Vertical shell boilers

This is a cylindrical boiler where the shell axis is vertical to the firing floor. Originally, it comprised a chamber at

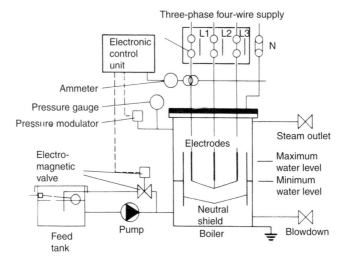

Figure 23.1 Diagrammatic layout of electrode boiler

the lower end of the shell which contained the combustion appliance. The gases rose vertically through a flue surrounded by water. Large-diameter (100 mm) cross tubes were fitted across this flue to help extract heat from the gases which then proceeded to the chimney. Later versions had the vertical flue replaced by one or two banks of small-bore tubes running horizontally before the gases discharged to the chimney. The steam was contained in a hemispherical chamber forming the top of the shell.

The present vertical boiler is generally used for heat recovery from exhaust gases from power generation or marine applications. The gases pass through small-bore vertical tube banks. The same shell may also contain an independently fired section to produce steam at such times, as there is insufficient or no exhaust gas available.

23.3.6 Horizontal shell boilers

This is the most widely used type of boiler in industry. The construction of a single-flue three-pass wetback shell is illustrated in Figure 23.2. As a single-flue design boiler evaporation rates of up to 16,300 kg/h F and A100°C are normal on oil and gas and 9000 kg/h F and A100°C on coal.

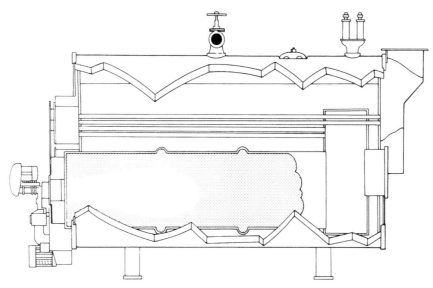

Figure 23.2 A three-pass wetback shell

In twin-flue design, these figures are approximately double. Normal working pressures of 10–17 bar are available with a maximum working pressure for a shell boiler at 27 bar. The outputs of larger boilers will be limited if high pressures are required.

The boilers are normally dispatched to site as a packaged unit with the shell and smokeboxes fully insulated, painted, and mounted on a base frame. The combustion appliance and control panel will be fitted together with the feedwater pump, water-level controls, gauges, and a full complement of boiler valves. Additional equipment may be specified and incorporated during construction. Larger boilers may have to have certain items fitted at site due to site restriction or weight.

Some variations of the three-pass wetback design exist. The most common is the reverse flame boiler, and Figure 23.3 illustrates this shell. In this design, the combustion appliance fires into a thimble-shaped chamber in which the gases reverse back to the front of the boiler around the flame core. The gases are then turned in a front smokebox to travel along a single pass of smoketubes to the rear of the boiler and then to the chimney. In order to extract heat from these gases, turbulators or retarders are fitted into these tubes to agitate the gases and help produce the required flue gas outlet temperature. Evaporative outputs up to 4500 kg/h F and 100°C on liquid and gaseous fuels are available.

Other variations of the three-pass wetback design are the two-pass, where only one pass of smoketubes follows the combustion tube, and the four-pass, where three passes of smoketubes follow the combustion tube. Neither of these is as widely used as the three-pass design.

Dryback boilers are still occasionally used when a high degree of superheat is required, necessitating a rear chamber to house the superheater too large for a semi-wet-back chamber. A water-cooled membrane wall chamber would be an alternative to this.

With twin-flue design boilers, it is usual to have completely separate gas passes through the boiler with twin wetback chambers. It is then possible to operate the boiler on one flue only, which effectively doubles its turndown ratio. For example, a boiler rated at 20,000 kg/h F and A100°C may reasonably be expected to operate down to 2500 kg/h F and A100°C on oil or gas providing suitable combustion equipment and control is incorporated. If prolonged periods of single-flue operation become necessary,

it is good practice to alternate, on a planned time scale, the single-flue load.

Shell boilers are supplied with controls making them suitable for unattended operation, although the insurance companies to comply with safety recommendations call for certain operations such as blowdown of controls.

Oil-, gas- and dual-fired boilers are available with a range of combustion appliances. The smaller units have pressure jet-type burners with a turndown of about 2:1 while larger boilers may have rotary cup, medium pressure air (MPA) or steam-atomizing burners producing a turndown ratio of between 3:1 and 5:1, depending upon size and fuel. The majority has rotary cup-type burners, while steam- or air-atomizing burners are used where it is essential that the burner firing is not interrupted even for the shortest period.

For coal-fired boilers, chain grate stokers, coking stokers and underfeed stokers are supplied. An alternative to these is the fixed-grate and tipping-grate boiler with coal being fed through a drop tube in the crown of the boiler (Figure 23.4). With the fixed grate, de-ashing is manual while with the tipping grate a micro-sequence controller signals sections of the grate to tip, depositing the ash below the grate, where it is removed to the front by a drag-link chain conveyor and then to a suitable ash-disposal system.

It is possible to have boilers supplied to operate on liquid, gaseous and solid fuels, although there may be a time penalty of two or three days when converting from solid to liquid and gaseous fuels and *vice versa*.

Access to both waterside and fireside surfaces of the boiler is important. All boilers will have an inspection opening or manway on the top of the shell with inspection openings in the lower part. Some larger boilers will have a manway in the lower part of the shell or end plate. With a three-pass wetback boiler all tube cleaning and maintenance is carried out from the front. The front smokebox doors will be hinged or fitted with davits. On most sizes of boilers bolted-on access panels are sufficient on the real smokebox. As the majority of shell boilers operate under forced-draft pressurized combustion, steam raising is relatively quick. While good practice could require a cold boiler to come up to pressure over a period of several hours once it is hot, it may be brought up to pressure in minutes, not hours.

For hot water shell boilers the above still applies. The shells would be slightly smaller for equivalent duties due to the absence of steam space. There are three accepted operated bands for hot water boilers. Low-temperature hot water (LTHW) refers to boilers having a mean water temperature (between flow and return) of below 95°C; medium-temperature hot water (MTHW) would cover the range 95 to 150°C; high-temperature hot water (HTHW) covers applications above 150°C.

The flow and return connections will be designed to suit the flow rates and temperature differentials required. The water-return connection will be fitted with either an internal diffuser or a venturi nozzle to assist mixing of the water circulating within the shell and prevent water stratification. The flow connection will incorporate the

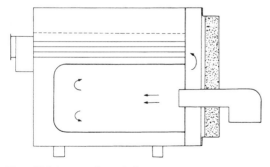

Figure 23.3 A reverse flame shell

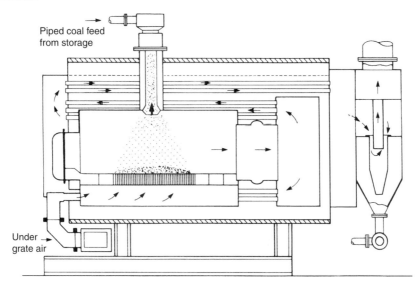

Piped coal feed
from storage

Under
grate air

Figure 23.4 A fixed-grate coal-fired boiler

temperature control stat to signal control of the firing rate for the burner.

Hot water boilers are potentially more susceptible to gas-side corrosion than steam boilers due to the lower temperatures and pressures encountered on low- and medium-temperature hot water boilers. With low-temperature hot water especially, the water-return temperature may drop below the water dewpoint of 50°C, causing vapor in the products of combustion to condense. This, in turn, leads to corrosion if it persists for long periods. The remedy is to ensure that adequate mixing of the return water maintains the water in the shell above 65°C at all times. Also, if medium or heavy fuel oil is to be used for low- or medium-temperature applications it is desirable to keep the heat transfer surfaces above 130°C, this being the approximate acid dewpoint temperature of the combustion gases. It may be seen, therefore, how important it is to match the unit or range of unit sizes to the expected load.

23.3.7 Watertube boilers

Originally, watertube boilers would have been installed for evaporation of 10,000 kg/h of steam with pressures as low as 10 bar. At that time, this would have been the maximum evaporation expected from a shell-type boiler. Now shell boilers are available at much greater duties and pressures as described in Section 23.3.6. It may be appreciated that there will be an overlap of types of boiler in this area, with watertube covering first for high-pressure applications and ultimately the larger duties. Figure 23.5 illustrates a stoker-fired unit.

Generally, outputs up to 60 MW from a single unit may be considered for industrial installations. Higher duties are available if required. Watertube boilers supplied for national power generation will have outputs up

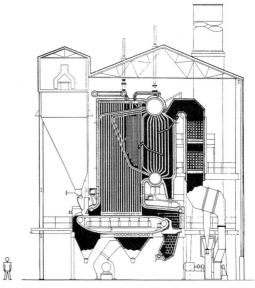

Figure 23.5 A stoker-fired watertube boiler of 36,300 kg/h steam capacity at 28 bar and 385°C (from British Coal Publication, *Boiler House Design for Solid Fuel*, 1980, and kind permission from College of Fuel Technology)

to 900 MW, pressures of 140 bar and final steam temperatures of 500°C. For the smaller industrial unit we are considering, the maximum pressure would be 65 bar with final steam temperature up to 500°C. This is the maximum temperature and pressure likely to be required for small turbine-driven generating units, although turbines are available to operate at much lower pressures of, say, 17 bar.

Construction is a water-cooled wall combustion chamber connected to a steam drum at high level. The bottoms of the walls are connected to headers. Sometimes a bottom or mud drum is incorporated, but improved water treatment now available does not always necessitate this.

The chamber is externally insulated and clad. Combustion equipment for solid fuel may be spreader or traveling-grate stokers or by pulverized fuel or fluid bed. Oil and gas burners may be fitted either as main or auxiliary firing equipment. The boilers will incorporate superheaters, economizers and, where necessary, air preheaters, grit arresters, and gas-cleaning equipment to meet clean air legislation.

Where watertube boilers are used to recover waste heat (for example, exhaust gases from reciprocating engines) lower gas temperatures may be involved, and this, in turn, could obviate the need for water-cooled walls. In this case, tube banks may be contained within a gas-tight insulated chamber.

There are two basic types of watertube boilers: assisted and natural circulation. Assisted circulation might apply where heat is from convection rather than a radiation source such as a waste heat application. Natural circulation is more suited where radiant heat and high gas temperatures are present.

Depending upon the required duty and the site, units may be shop assembled or of modular construction. Site-erected units may be designed to have their main components arranged to fit in with the space available.

23.3.8 Waste-heat boilers

These may be horizontal or vertical shell boilers or watertube boilers. They would be designed to suit individual applications ranging through gases from furnaces, incinerators, gas turbines and diesel exhausts. The prime requirement is that the waste gases must contain sufficient usable heat to produce steam or hot water at the condition required.

Supplementary firing equipment may also be included if a standby heat load is to be met and the waste-gas source is intermittent. Waste-heat boilers may be designed to use either radiant or convected heat sources. In some cases, problems may arise due to the source of waste heat, and due consideration must be taken of this, with examples being plastic content in waste being burned in incinerators, carry-over from some type of furnaces causing strongly bonded deposits and carbon from heavy oil fired engines. Some may be dealt with by maintaining gas-exit temperatures at a predetermined level to prevent dewpoint being reached and others by sootblowing. Currently, there is a strong interest in small combined heat and power (CHP) stations, and these will normally incorporate a waste-heat boiler.

23.3.9 Fluid-bed boilers

The name derives from the firebed produced by containing a mixture of silica sand and ash through which air is blown to maintain the particles in suspension. The beds are in three categories, shallow bed, deep bed and recirculating bed. Shallow beds are mostly used and are about 150–250 mm in depth in their slumped condition and around twice that when fluidized. Heat is applied to this bed to raise its temperature to around 600°C by auxiliary oil or gas burners. At this temperature coal and/or waste is fed into the bed, which is controlled to operate at 800–900°C. Water-cooling surfaces are incorporated into this bed connected to the water system of the boiler.

The deep bed, as its name implies, is similar to the shallow bed but in this case may be up to 3 meters deep in its fluidized state, making it suitable only for large boilers. Similarly, the recirculating fluid bed is only applicable to large watertube boilers.

Several applications of the shallow-bed system are available for industrial boilers; the two most used being the open-bottom shell boiler and the composite boiler. With the open-bottom shell, the combustor is sited below the shell and the gases then pass through two banks of horizontal tubes.

In the composite boiler, a watertube chamber directly connected to a single-pass shell boiler forms the combustion space housing the fluid bed. In order to fluidize the bed the fan power required would be greater than that with other forms of firing equipment.

To its advantage, the fluid bed may utilize fuels with high ash contents, which affect the availability of other systems. It is also possible to control the acid emissions by additions to the bed during combustion. They are also less selective in fuels and can cope with a wide range of solid-fuel characteristics.

23.4 Application and selection

Figure 23.6 illustrates the selective bands for various types of boilers. The operating pressure will govern the steam temperature except where superheaters are used. For hot water units the required flow temperature will dictate the operating pressure. It is important that when arriving at the operating pressure for hot water units due allowance is made for the head of the system, an anti-flash steam margin of 123°C and a safety-valve margin of 1.5 bar. In arriving at a decision to install one or more boilers, the following should be considered. The first choice (providing the load is within the duty range of the boiler) will be a single unit. This is economically the most attractive in capital cost, providing account is taken of the following:

1. If there is a breakdown on the boiler, will production be seriously affected immediately?
2. Will adequate spare parts for the boiler be held in stock or be available within an acceptable time, and will labor be available to carry out the repair work?
3. Will time be available to service the boiler?
4. The duty will preferably fall within the modulating firing rate of the burner.
5. Prolonged periods of intermittent operation are avoided.
6. Is there an existing standby unit?

If any or all of these points are not accepted, then the next consideration for a shell boiler could be a twin-flue unit

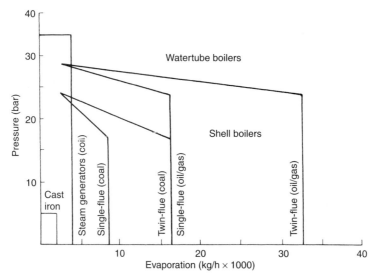

Figure 23.6 Guide to boiler capacities

suitable for single-flue operation. This has the advantage of using less space than two smaller boilers and having only one set of services.

Moving now to two boilers, the heat load may comprise two elements. One may be a production process whose interruption would cause problems and the other, say, a heating load where any interruption would not be noticed immediately. Assuming the two elements were of equal duty, it would be reasonable to install two boilers each 50–50 per cent of the total load. One boiler would then be able to cover the process load.

An extension to this is to install two boilers, each capable of handling the total combined load. Depending upon the boiler size, there may be only a relatively small difference in total capital cost between the above two schemes.

Further options involving three or more boilers must take into account minimum and maximum loads in order to run the plant efficiently. When considering hot water it may be advantageous to consider units in a range of outputs. This will help in operation, so that a unit may be brought into duty to match the load and thus avoid low-load conditions and consequent danger of dewpoints. Also, if the plant is fired on solid fuel it will help in maintaining a more even firing rate and a clean stack.

23.5 Superheaters

Steam produced from a boiler is referred to as dry saturated, and its temperature will correspond with the working pressure of the boiler. In some instances, particularly with shell boilers, this is perfectly acceptable. There are occasions, however, where it is desirable to increase the temperature of the steam without increasing the pressure. A superheater performs this function. Steam from the drum or shell of the boiler is passed through a bank of tubes whose external surfaces are exposed to

the combustion gases, thus heating the steam while not increasing the pressure.

Where a superheater is fitted, the boiler working pressure must be increased to allow for the pressure drop through the elements. This will be between 0.3 and 1.0 bar.

In a watertube boiler the superheater is a separate bank of tubes or elements installed in the area at the rear or outlet of the combustion chamber. Saturated steam temperature may be increased by 200°C with a final steam temperature of up to 540°C.

For shell boilers, superheaters may be one of three types, depending upon the degree of superheat required. The first and simplest is the pendant superheater installed in the front smokebox (Figure 23.7). The maximum degree of superheat available from this would be around 45°C. The second pattern is again installed in the front smokebox but with this, the elements are horizontal 'U' tubes which extend into the boiler smoketubes. The degree of superheat from this pattern is around 80°C. Third, a superheater may be installed in the reversal chamber of the boiler. A wetback chamber presents problems with lack of space, and therefore a semi-wetback, dryback or water-cooled wall chamber may be considered. Maximum degree of superheat would be around 100°C.

Superheater elements are connected to inlet and outlet headers. The inlet header receives dry saturated steam from the steam drum of a watertube boiler or the shell of a horizontal boiler. This steam passes through the elements where its temperature is raised and to the outlet header which is connected to the services. A thermometer or temperature recorder is fitted to the outlet header.

It should be appreciated that a steam flow must be maintained through the elements at all times to prevent them burning away. If a single boiler is used then provision to flood the superheater during start-up periods may be required.

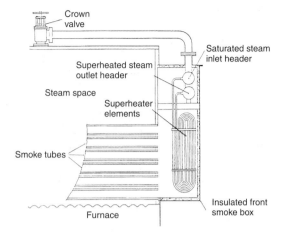

Figure 23.7 Front smokebox pendant superheater (with permission of Senior Thermal Engineering Ltd)

Superheated steam may be needed where steam distribution pipework in a plant is over extended distances, resulting in a loss of heat and increase in wetness of the steam. Another case may be where a process requires a temperature above the working pressure of the plant. The third case is where steam is used for turbines. Here it improves the performance of the turbine, where for every 6°C increase in steam temperature it can produce a saving of about 1 per cent reduction in steam consumption. Superheaters may also be supplied as independently fired units. These may be used when either the amount of superheated steam required is much less than the boiler evaporation or is only needed on an intermittent basis.

23.6 Economizers

Economizers are installed in the exhaust gas flow from the boiler. They take heat from the flue gases, which they transfer via extended surface elements to the feedwater immediately prior to the water entering the boiler. They therefore increase the efficiency of the boiler and have the added advantage of reducing thermal shock. In watertube boilers, they may be incorporated within the structure of the boiler or supplied as a freestanding unit. With shell boilers, they will be a separate unit fitted between the boiler flue gas outlet and the chimney.

Figure 23.8 is a schematic illustration of such a unit. It is desirable for each boiler to have its own economizer. Where one economizer is installed to take the exhaust gases from more than one boiler special considerations must be taken into account. These will include gas-tight isolation dampers. Consideration must be made of gas-tight pressures at varying loads and maximum and minimum combined heat load to match economizer and a pumped feedwater ringmain. Economizers may be used for both forced-draft and induced-draft boilers, and in both cases, the pressure drop through the economizer must be taken into account when sizing the fans.

Economizers are fitted to most watertube boilers. An exception is on a waste-heat application, where it may be desirable, due to the nature of the products being burned, to maintain a relatively high gas outlet temperature to prevent corrosive damage to the boiler outlet, ductwork and chimney.

With watertube boilers economizers may be used when burning coal, oil or gas. The material for the economizer will depend on the fuel, and they may be all steel, all cast iron or cast iron protected steel. All steel would be used for non-corrosive flue gases from burning natural gas, light oil and coal. Cast iron may be used where the feedwater condition is uncertain and may attack the tube bore. Fuels may be heavy fuel oil or coal, and there is a likelihood of metal temperatures falling below acid dewpoint. Cast iron protected steel is used when heavy fuel oil or solid fuel firing is required and feedwater conditions are suitably controlled. As cast iron can withstand a degree of acid attack, these units have the advantage of being

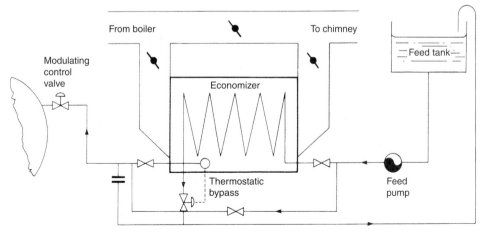

Figure 23.8 Schematic illustration of an installed economizer

able to operate without a gas bypass, where interruptible natural gas supplies are used with oil as standby.

With shell boilers, economizers will generally only be fitted to boilers using natural gas as the main fuel and then only on larger units. It would be unlikely that a reasonable economic case could be made for boilers of less than 4000 kg/hF and A100°C evaporative capacity. The economizer will incorporate a flue gas bypass with isolating dampers to cover for periods when oil is used and for maintenance. The dampers require electric interlocks to the selected fuel.

The majority of shell boilers operate in the pressure range 7 to 10 bar, the flue gas outlet temperature will be in the range of 190 to 250°C. Therefore, it may be appreciated from this that the boiler needs to operate at 50–100 per cent of its maximum continuous rating for most of the working day to produce an economic return.

Where an economizer is installed it is essential to have water passing through the unit at all times when the burners are firing to prevent boiling. Therefore, boilers fitted with economizers will have modulating feedwater control. Even then, it is possible that the water flow requirement can become out of phase with the burner-firing rate. To prevent damage, a temperature-controlled valve allows a spillage of water back to the feedwater tank, thus maintaining a flow of water through the unit. Each economizer will be fitted with a pressure-relief safety valve.

Due to the amount of water vapor produced when natural gas is burned, it is important not to allow the exhaust gas temperature to fall below 80°C, otherwise the water dewpoint will be reached. Not only the economizer but also the ductwork and chimney must be considered and provision incorporated for drainage.

In the event of a separate use for low-grade hot water being available, it is sometimes practical to install a secondary condensing economizer. With this, the material of which the economizer is constructed allows condensate to form and drain away without excessive attack from corrosion.

A recent development in heat recovery has been the heat tube. This is a sealed metal tube which has been evacuated of air and contains a small quantity of liquid which, for boiler applications, could be water. When heat from the flue gases is applied to one end of the heat pipes the water in the tube boils, turning to steam and absorbing the latent heat of evaporation. The steam travels to the opposite end of the tube which is surrounded by water, where it gives up its latent heat, condenses and returns to the heated end of the tube. Batteries of these tubes can be arranged to form units, usually as a water jacket around a section of a flue.

23.7 Water-level control

Water-level controls continuously monitor the level of water in a steam boiler in order to control the flow of feedwater into the boiler and to protect against a low water condition which may expose the heating surfaces with consequent damage. The controls may be either float operated or conductivity probes.

With watertube boilers, the control of the water level needs to be precise and sensitive to fluctuating loads due to the high evaporative rates and relatively small steam drums and small water content. Control will be within +10 mm on the working water level and will be two- or three-element control. Two-element control will comprise modulating feedwater control with first low water alarm and high–low control with low water cutout and alarm. The second element will be monitoring of the steam flow to give early indication of any increase in steam demand. This signal may then be linked to the firing rate of the burners and the feedwater-modulating valve. The third element senses a drop in feedwater demand, which would signal the firing rate of the burners to modulate down.

Shell boilers will have two external level controls each independently attached to the shell. Boilers up to about 9000 kg/hF and A100°C will have a dual control and either a single or high–low control. The dual control instigates the feedwater pump, which operates on an on–off cycle over a water level band of +15 mm and operates the first low-water alarm. The single or high–low control will incorporate a second low-water alarm with burner lockout and with the high–low control also an indication of high water, which may be linked to shut down the feedwater pump with automatic restart when the water level drops to normal.

Boilers of larger evaporations will have modulating and high–low control. The modulating level control monitors the working water level in the shell and operates a control valve in the feedwater line, allowing water to enter the boiler from a continuously running feed pump. It will also incorporate the first low-water alarm. The high–low control operates as before.

The advantage of modulating control is that it maintains a constant working water level and therefore the boiler is always in its best condition to supply steam for peak loading. These controls may also be fitted to boilers below 9000 kg/hF and A100°C particularly if severe loads are present or when the working pressure is above 10 bar.

With water-level controls, it is important to check they are functioning correctly, and they will be operated daily to simulate low-water condition. On shell boilers, this is invariably a manual operation but may be motorized on watertube boilers.

Shell boilers may be fitted with internal level controls. Here controls are mounted on the crown of the boiler with the floats or probes extending to the water surface through the steam space. To check the operating function of these, it is necessary to drop the water level in the boiler, or, alternatively, a separate electronic testing device can be fitted. With fully flooded hot water boilers, a single level control or switch is fitted to protect against low-water condition.

23.8 Efficiency

This is the ratio of heat input to the useful heat output of the boiler, taking account of heat losses in the flue gases, blowdown and radiation. It is covered in BS 845: Part 1, which fully details the concise or losses method of determining the efficiency of a boiler.

23.9 Boiler installation

Figure 23.9 illustrates the services required for a boiler. The individual items are covered separately in other sections and only the general concept is being considered here.

If boilers are being replaced it is reasonable to expect that existing services will be available, apart from possible upgrading in some areas. For new installations, a new building is likely to be required, and it will first be necessary to determine the overall size and type of construction.

Starting with the boiler(s), these will be set out giving due consideration to space between boilers and other items of plant in order to give adequate access to all equipment, valves and controls for operation and maintenance. For small- to medium-sized boilers 1 meter may be considered a reasonable space between items of plant where access is required. With large boilers (including watertube), this may be increased up to 3 m. The width of firing aisle will be dependent upon the size and type of boiler.

Having now established the space requirement for the boilers, we must consider the other items of plant to be housed in the same structure. These may include water-treatment plant, heat-recovery system, tanks and pump, instrument and controls. With oil- and gas-fired installations, there is little additional plant to consider within the boiler house. With solid fuel then the handling, distribution and perhaps storage could also form part of the building with provision for ash handling and removal.

Access for air into the boiler house must be considered. Air for combustion will be drawn in through purpose-made louvers and care should be taken that this air does not create a hostile environment to personnel in the building. Also, passage of air across cold water pipes should be avoided, as in cold weather this may aggravate freezing in the pipes.

Ventilation will also be required. With gas-fired installations, this is mandatory, and specified free areas must be provided.

The fabric of the building may range from a basic frame and clad structure offering minimum weather protection to a brick and concrete construction. While it is possible to weatherproof boilers for outside installation, it should be remembered that the occasion may arise when someone is going to carry out maintenance or service work, and if weather conditions are inclement, this may prove impractical. Two instances where outdoor installation may be considered are where the climatic conditions of the country permit and in the case of waste-heat boilers where controls are minimal.

Foundations and trenches will be formed prior to boiler delivery. The boiler house structure may be erected before or after delivery of the boilers, dependent upon site conditions and the type of boiler. Due consideration will be given to access openings and doorways and provision for lifting any items of plant once the boiler house is complete.

Noise may be considered in two areas, one being an unacceptable level of noise to adjacent areas and the other an excessive noise level for personnel inside the boiler house. With the first, the type of construction of the building together with position of openings and the fitting of acoustic louvers for air ingress will all help to reduce external noise. As an extension to this and to reduce noise for personnel within the building, all items of plant generating noise may be considered. These will generally be forced-draft fans and feedwater pumps. Various degrees of silencing are available for combustion appliances and feed pumps may be housed in a separate enclosure within the building.

External to the boiler house, provision for the storage and handling of solid and liquid fuel is required with access for delivery vehicles. With some small boilers such as electrode or steam-coil generators, where the boiler only serves a single item of plant or process, it is practical to install them immediately adjacent to that process. The electrode boiler is eminently suitable here as no combustion gases are produced.

23.10 Boiler house pipework

Pipework layout should be designed so that it does not restrict access to the plant or building. Walkways and

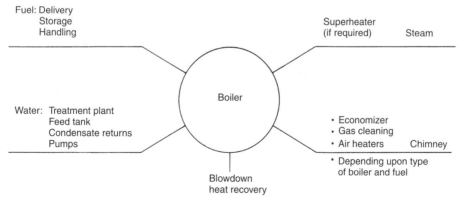

Figure 23.9 Boiler services

gantries should be considered when deciding on pipe service routes.

Where pipework is subject to temperature such as steam and hot water mains a means for expansion must be included to prevent undue stresses to valves and building structure. Balancing the steam flow from a range of boilers is important, particularly if the load may include peaks of more than one boiler capacity. To achieve this balance, the pressure drop in each steam line from the boiler crown valve to the distribution manifold should be the same, so that any imposed load will be shared between all on-line boilers. Failure to provide balanced steam headers may give rise to the lead boiler losing pressure, priming and locking out, thus aggravating the situation for the remaining on-line boilers.

Steam lines from the boiler should always drain away from the boiler crown value to prevent condensate building up against the valve. Careless opening of this valve to allow steam to pass when the line is flooded can result in splitting the valve and pipework. If the steam line rises from the crown valve a tee trap must be installed and fitted with a steam trap and drain valve. The drain valve must be operated to check that the steam line is clear of water before the crown valve is opened.

Safety-valve vent pipework should be run on the shortest possible route. Where bends or long runs occur, the pipe size may have to be increased to prevent back pressure on the safety valve during operation.

Feedwater pipework will normally be gravity head suction from the hotwell or feed tanks to the pumps and at a pressure in excess of the boiler working pressure from the pumps to the boiler. Few problems occur on the pipework between pumps and boilers. However, inadequately sized suction lines can give rise to cavitation at the feed pump with subsequent boiler shutdown. The feed tank should always be positioned to suit the temperature of the feedwater and the pipework sized to give free flow at that head, taking account of bends, valves and filters.

Fuel lines should be run where they will not be subject to high temperatures and must include protection with automatic and manual shut-off fire valves at the perimeter of the boiler house. Individual boilers should also be capable of isolation from this supply. Local requirements may vary on fuel supply systems, and these should be ascertained before installation.

Blowdown and drains may be taken to either a sump or blowdown tank prior to discharging into the drains. The purpose of this is to reduce the temperature by dilution and dissipate the pressure to prevent damage to the drains.

All pipework must be installed to the appropriate standard and codes of practice. The degree of insulation will be suitable for the temperature of the pipework, although the finish may vary, depending upon site preferences. Valve boxes are recommended, although on some low-pressure installations these are not always included.

23.11 Feedwater requirements

Poor or unsuitable water can be a major factor where failure in a boiler occurs. There are four problem areas

Table 23.5 Recommended water characteristics for shell boilers

For pressures up to 25 bar[a]			
Total hardness in feedwater, mg/l in terms of $CaCO_3$ max.	2	20	40
Feedwater			
pH value	7.5–9.5	7.5–9.5	7.5–9.5
Oxygen	b	b	b
Total solids, alkalinity, silica	b	b	b
Organic matter	b	b	b
Boiler water			
Total hardness, mg/l in terms of $CaCO_3$ max.	ND	ND	ND
Sodium phosphate, mg/l as Na_3PO_4[c]	50–100	50–100	50–100
Caustic alkalinity, mg/l in terms of $CaCO_3$ min.	350	300	200
Total alkalinity, mg/l in terms of $CaCO_3$ max.	1200	700	700
Silica, mg/l as SiO_2 max.	Less than 0.4 of the caustic alkalinity		
Sodium sulfite, mg/l as Na_2SO_3 or	30–70	30–70	30–70
Hydrazine, mg/l as N_2H_4	0.1–1.0	0.1–1.0	0.1–1.0
Suspended solids, mg/l max.	50	200	300
Dissolved solids, mg/l max.	3500	3000	2000

– : Not applicable.
ND: Not detectable.
[a] 1 bar = 10^5 N/m^2 = 100 kPa = 14.5 lb/in^2.
[b] Numerical values depend upon circumstances but the comments are relevant.
[c] Phosphate is usually added as sodium phosphate but determined as phosphate ($PO_4{}^3$); Na_3PO_4 = 1.73 × PO_{43}.
Based on Table 2 of BS 2486: 1978, by permission of BSI.

for which feedwater needs suitable treatment and control. These are sludge, foam, scale and corrosion.

Boiler feedwater may be from various supplies. If it is from a main water supply, further filtering prior to treatment is unlikely, but for other supplies such as borehole, lakes, rivers and canals, filters may be required. Impurities in water may be classed as dissolved solids, dissolved gases and suspended matter and suitable treatment is required.

Table 23.5 shows the recommended water characteristics for shell boilers and Table 23.6 the water quality guidelines for industrial watertube boilers. Due to the wide parameters encountered in the quality of feedwater, it is not possible to be specific and to define which treatment suits a particular type and size of boiler. The quality of make-up and percentage of condense returns in a system will both have to be taken into consideration.

23.12 Feedwater supply and tanks

The design, size and siting of the boiler feed tank or hotwell must be compatible with the boiler duty capacity and system temperatures. They should be installed giving sufficient space for access to controls, valves and manways.

The tanks will normally be of fully welded construction from mild steel with the internal surfaces shot blasted and a protective plastic coating applied. It is important that all

Table 23.6 Water-quality guidelines recommended for reliable, continuous operation of modern industrial watertube boiler's

Boiler feedwater

Drum pressure (lb/in² g)	Iron (ppm Fe)	Copper (ppm Cu)	Total hardness (ppm CaCO₃)
0–300	0.100	0.050	0.300
301–450	0.050	0.025	0.300
451–800	0.030	0.020	0.200
601–750	0.025	0.20	0.200
751–900	0.020	0.015	0.100
901–1000	0.020	0.015	0.050
1001–1500	0.010	0.010	ND[d]
1501–2000	0.010	0.010	ND[d]

Boiler water

Drum pressure (lb/in² g)	Silica (ppm SiO₂)	Total alkalinity[a] (ppm CaCO₃)	Specific conductance (μmho/cm)
0–300	150	350[b]	3500
301–450	90	300[b]	3000
451–600	40	250[b]	2500
601–750	30	200[b]	2000
751–900	20	150[b]	1500
901–1000	8	100[b]	1000
1001–1500	2	NS[s]	150
1501–2000	1	NS[c]	100

[a] Minimum level of hydroxide alkalinity in boilers 1000 lb/in² must be individually specified with regard to silica solubility and other components of internal treatment.
[b] Maximum total alkalinity consistent with acceptable steam purity. If necessary, the limitation on total alkalinity should override conductance as the control parameter. If make-up is demineralized water at 600–1000 lb/in² g, boiler water alkalinity and conductance should be shown in the table for the 1001–1500 lb/in² g range.
[c] NS (not specified) in these cases refers to free sodium- or potassium-hydroxide alkalinity. Some small variable amount of total alkalinity will be present, and measurable with the assumed congruent control or volatile treatment employed at these high-pressure ranges.
[d] None detectable.

connections together with any additional future connections be included before the coating is applied to prevent breaking the surface, thus giving a source of future corrosion. The tank should have a sealed top with adequate bolt-on access covers. This will keep out foreign bodies and protect the water surface from the atmosphere, where it would absorb oxygen. Adequate venting must be included to prevent any build-up of pressure from either heating or condense returns.

Alternative materials may be used, including stainless steel and plastic. Sectional tanks may be installed for convenience in restricted places, but great care must be taken if they are subject to temperature variations, which can give rise to joint problems.

The following connections will be required in addition to the vent already referred to:

1. Feedwater take-off. This will be from the bottom of the tank and preferably from the base with a weir pipe extending 50–100 mm up into the tank. If the take-off is from the side of the tank the bottom of the pipe bore must be a similar dimension up from the base. This allows sludge to remain in the tank, which should be inspected and cleaned as required.
2. Drain pipework will be directly from the bottom of the tank.
3. Overflow pipework should be sized to meet the supply quantity to the tank and run so that visible evidence is obvious of an overflow condition.
4. Condense returns. Where returns arrive at the feed tank from several sources, it is desirable to bring them together with a collection manifold at the top of the tank. The manifold then has a dip pipe extending down into the tank terminating in a sparge pipe. This method reduces the risk of entraining air in the returns and helps to promote mixing of the water in the tank.
5. Make-up water. A semi-sealed section within the main tank will feed water from the treatment plant to the main tank either through a separate make-up tank or. In either case, the water level will be the same in both sections, excluding the ullage left for condense returns. Control of the make-up water level may be by float valve, float switches or conductivity probes. These methods allow water to flow through the treatment plant, although conductivity probes or switches permit a positive flow and avoid the risk of slippage.
6. Chemical dosing. This is a connection into the tank with an internal diffuser for any corrective treatment required for the water.
7. Level indicator. This may be either visual or electrical or both.

Additional connections may also be required for the following:

8. Flash steam. This will result from the blowdown heat recovery referred to in Section 7.13. Flash steam is introduced into the tank through a dip pipe terminating in a distribution manifold near the bottom of the tank.
9. Tank heating. This is referred to in detail later in this section.
10. Thermometer.

Connections should be arranged to avoid short-circuiting between the tank supply and take-off to the boiler feed. Also, adequate mixing within the tank is required to prevent stratification. With the capacity of the feed tank, if there is an adequate supply of water either from mains or local storage, a 1 h supply should be considered a minimum. This would be based on a continuous supply of make-up water being available such as from a duplex treatment plant. If the supply of make-up is not continuous then the tank capacity must be increased. For example, a base exchange plant will take about 75 min to regenerate, and in that case, the tank capacity should be able to sustain a supply of water to the boiler for 90 min in order to give a safety margin. The treatment plant then

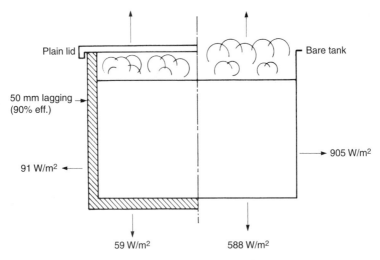

Plain lid

50 mm lagging
(90% eff.)

Bare tank

905 W/m²

91 W/m²

59 W/m²   588 W/m²

Figure 23.10 Heat losses from a feedwater tank operating at 93°C with ambient 21°C

has to supply a quantity of water in excess of the boiler demand rate in order to replenish the tank. On top of this capacity a volume has to be allowed for condense returns when the system starts in order to prevent wastage of heat and water. As steam capacities increase, the feed tank becomes correspondingly larger, and it may be considered that on steam outputs in excess of 20,000 kg/h a duplex treatment plant should be installed in order to keep the feed tank to a reasonable size.

In areas where the water demand of the plant may exceed the supply capacity of the local mains, it will be necessary to install additional storage tanks. It is likely that if the local supply is subject to low flow rates additional storage is already available for services such as fire protection.

Feedwater to the boiler should not be cold, as this can cause harmful thermal stresses to the boiler. A minimum feedwater temperature of 70–80°C should be designed into the system. This increased temperature has the added advantage that it accelerates some water-treatment reactions and helps to remove oxygen and other gases from the feedwater. Once the system is working, this higher temperature may be achieved from the condense returns, but this condition is not effective until the plant has been running for some time. A tank heating system should therefore be installed. This is best achieved by direct steam injection into the feed tank. Steam is taken from the boiler and reduced in pressure to 1–2 bar to reduce noise in the tank. Passing through a thermostatic control valve, the steam mixes with the water in the tank through a sparge pipe or steam nozzle.

With watertube boilers economizers are normally used, therefore the situation of cold feed to the boiler would not apply. However, the feed to the economizer should be treated similarly to that to a boiler to reduce thermal shock.

With the feedwater being raised in temperature it is necessary to avoid cavitation at the inlet to the boiler

feed pump. One method of reducing this problem is to elevate the feed tank to give an adequate positive head at the pump. The following are recommended heights to the minimum water level in the tank at various temperatures:

88°C – 1.6 m
93°C – 3.1 m
99°C – 4.6 m
100°C – 5.2 m

The feed tank should always be insulated to reduce heat loss to a minimum. Figure 23.10 illustrates typical heat losses from a feed tank operating at 93°C in an ambient temperature of 21°C with and without 50 mm of insulation.

23.13 Blowdown requirements, control and tanks

In order to maintain the level of dissolved and suspended solids within the boiler as recommended in Section 23.11 it is necessary for the boiler to be blown down. This is an operation where a quantity of water is drained from the boiler while the boiler is operating at pressure and may be achieved by various methods.

The simplest and that applied to small boilers is for the main bottom blowdown valve to be opened for a set period at regular intervals (e.g. 20 seconds every 8 hours). This method may also extend to larger boilers where conditions are such that there is little build-up of solids. Such conditions could be high condense returns and good-quality make-up feedwater.

The second method could be automatic intermittent blowdown. With this, a timer-controlled valve is installed at the bottom of the boiler prior to the main blowdown valve. A program is then designed to operate this valve in short bursts, which disperses any sludge and controls the levels of solids. This method is preferred for boilers having internal treatment.

The third method would be continuous blowdown through a regulating or micrometer valve. The take-off position for this should preferably be about 250 mm below the working water level and may either be on the side of the shell or on the crown with a dip pipe down to the correct level. If a connection is not available, it is possible to install the valve on the bottom connection prior to the main blowdown valve.

All these methods will require careful monitoring initially to set up and determine the correct rate of blowdown once the plant is operating. In order to take the necessary sample from the boiler the boiler(s) should be fitted with a sample cooler. To automate the continuous blowdown a conductivity-controlled system may be installed. Here a controller continuously compares the boiler water electrical conductivity with a value set in the controller. Depending on whether this is above or below the set rate, it will automatically adjust the blowdown flow rate.

While the above methods control the level of dissolved and suspended solids in the boiler, it will still be an insurance requirement to operate the main blowdown valve periodically.

The minimum amount of blowdown may be calculated as a percentage of the evaporation rate by:

$$\text{Blowdown rate} = \frac{F}{B - F} \times 100\%$$

where F = the total dissolved solids content of the feed in parts per million allowing for the mixture of make-up and condensate plus any chemical treatment and B = the maximum recommended solids content for boiler water in parts per million.

While vitally necessary, blowdown can be expensive in terms of lost heat. Therefore a point will be reached when it is economical to install a blowdown heat recovery system. Generally, the heat content in the blowdown water for a shell boiler will represent only about 25 per cent of the heat content in the same percentage of steam. Therefore, if a blowdown rate of 10 per cent is required this represents an approximate heat loss of 2.5 per cent from the boiler capacity. This differential reduces and eventually becomes insignificant on high-pressure watertube boilers.

The blowdown from the boiler(s) will be run to a flash-steam vessel mounted adjacent to the feed tank. Flash steam will be introduced into the feed tank through a dip pipe terminating in a distribution manifold. The drain from the flash vessel may then be taken to a residual blowdown heat exchanger. Any remaining heat is then transferred to the make-up water to the tank before the blowdown runs to drain.

Blowdown from the boiler(s) should always be taken to either a blowdown sump or blowdown vessel before discharging into drains. Both should be adequately sized to give cooling by dilution and be fitted with vent pipes to dissipate pressure safely. The boiler(s) should have independent drain lines for the main manually operated blowdown valve and the drains from a continuous blowdown system. Where more than one boiler is connected to either system the line should be fitted with a check or secondary valve capable of being locked.

Blowdown sumps should be constructed from brick and/or concrete and the blowdown lines should drain under gravity. Where the blowdown lines enter the sump they should turn to discharge downwards and the bottom of the sump should be protected below this area with a cast iron tray to prevent erosion. The drain or overflow from the pit should be at such a level to produce a weir effect, thus holding water for dilution.

With a blowdown vessel, these may be installed at ground level, and thus the water in them can be above the boiler blowdown valve. In this case, a drain valve for maintenance purposes must be installed at the lowest point in the line between the boiler and the vessel.

23.14 Clean Air Act requirement for chimneys and flue designs

23.14.1 Introduction

The function of a chimney is to discharge in a manner to give adequate dispersal to the products of combustion in accordance with the third edition of the 1956 Clean Air Act Memorandum on Chimney Heights. The scope of the memorandum is as follows:

1. The publication provides for the use of local authorities, industry and others who may need to determine the height appropriate for certain new chimneys a relatively simple method of calculating the appropriate height desirable in normal circumstances.
2. Heights determined by these methods should be regarded as a guide rather than as a mathematically precise decision. The conclusions may need to be modified in the light of particular local circumstances such as valleys, hills and other topographical features.
3. The advice given is applicable only to chimneys of fuel burning plant with a gross heat input of between 0.15 MW and 150 MW, including stationary diesel generators. It does not deal with direct-fired heating systems which discharge into the space being heated, gas turbines or incinerators (which require separate treatment, depending on the pollutants emitted).
4. The main changes from the second edition are the inclusion of a method dealing with very low sulfur fuel and extensions of the methods for taking into account the height of nearby buildings and of the range of the size of furnace included.

23.14.2 Gas velocity

In order to maximize the chimney height, the efflux velocity of the gases leaving the chimney should be designed on 12 m/s at maximum continuous rating (MCR) of the boiler. On some very small boilers, this may be impractical to achieve, but a target velocity of not less than 6 m/s at MCR should be aimed for. With boilers at the top end of the range a velocity of 15 m/s at MCR is required. Some inner-city authorities may stipulate higher efflux velocities and some plants have been installed with gas velocities of 22 m/s.

23.14.3 Chimney height

Originally, the height of the chimney was designed to produce a draft sufficient to produce induced-draft air for combustion. With modern boiler plant, forced-draft and/or induced-draft fans are used. This allows for the greater degree of control of the air to be designed into the combustion appliance. The chimney is therefore only required to disperse the gases.

When using gaseous fuel it is normally sufficient to terminate the chimney 3 m above the boiler house roof level, subject to there being no higher buildings adjacent to the boiler house. In such cases, these buildings may need to be considered.

On medium-sized boiler plant where gas is to be the main fuel, it may have oil as a secondary standby fuel. In this case, the chimney height must be based on the grade of fuel oil capable of being burned.

The methods of calculating proposed chimney height are clearly laid out in the Clean Air Act Memorandum, and will be based on:

1. Quantity of fuel burned;
2. Sulfur content of fuel burned;
3. District category;
4. Adjacent buildings;
5. Any adjacent existing emissions.

Application for approval of the proposed chimney height should be made to the appropriate authority at an early stage of a project in order to ascertain their approval or other height they may require. Failure to do this can result in an embarrassing situation where insufficient finance has been allocated due to their requiring a larger chimney than was included in the planned costing.

Where waste products are being incinerated, special consideration may have to be given to the resulting flue gases. This may involve having to arrive at a chimney height in conjunction with HM Inspectorate of Factories for Pollution.

23.14.4 Type and combustion

Where a multi-boiler installation is being considered a multi-flue chimney is preferred. This is where the required number and size of flues are enclosed in a single windshield. It is preferred by planning offices and has the advantage of a greater plume rise than from separate stacks.

Several boilers discharging into a single-core chimney are to be avoided. At times of low load the efflux velocity will be very low, which, in turn, will allow the chimney to cool. This may then drop to dewpoint temperatures and, where sulfur is present in the fuel, acid will form. If the chimney is unprotected steel, it will suffer rapid corrosion. Even worse, as the boilers increase their load the efflux velocity will increase and start to discharge the acid droplets, which quickly fall out and cause damage to surrounding property.

Construction of chimneys for industrial boilers will mainly be of steel, being either single or multi-flue. A single-core chimney should be suitably insulated in order to maintain a maximum temperature at the chimney outlet to prevent corrosion and will be finish clad with aluminum sheet. A multi-flue chimney will have each flue suitably insulated and enclosed in an outer windshield. Provision for drainage should be incorporated at the base of each flue. Large watertube boilers may have concrete windshields containing internal flues.

23.14.5 Connecting ducts

The duct between the boiler outlet and the chimney connection should be designed to have a gas velocity no greater than the chimney flue, be complete with insulation, and cladding, access openings and expansion joints. It should have the least number of bends and changes of section possible and should preferably not fall between the boiler outlet and the chimney connection.

With the use of multi-flue chimneys, dampers are not required in each flue as they were for boiler isolation when several boilers discharge into a common duct or chimney. This should not be confused with the fitting of dampers for heat conservation in a boiler during off-load periods.

23.14.6 Grit and dust emissions

Solids emissions from solid and liquid fuel fired plant are covered in the HMSO publication *Grit and dust–The measurement of emissions from boiler and furnace chimneys*. This states levels of emissions which should be achieved in existing plant and specified for new plant. Suitable sampling connections should be incorporated into the flue ducting for the use of test equipment if permanent monitoring is not included.

23.15 Steam storage

Most boilers built now, together with their combustion equipment, are quick to respond to local fluctuations. Occasionally, where very rapid load changes occur, the firing rate of a gas or oil burner can be virtually instantaneous by the use of special control equipment. This control will have to work in conjunction with the boiler, and therefore the boiler should have adequate steam space and water surface area to help accommodate the rapid changes in steam demand. Good water treatment is especially important here in order to reduce the risk of priming during peak draw-off periods. A boiler with a large shell will have an advantage over one with a smaller shell, assuming equal heating surfaces, but it will give no more than a slight buffer against severe loads.

Most boiler plants can be installed using one or more boilers, which can accommodate minimum to maximum loads. Occasionally, heavy peak loads occur for only relatively short periods, and here there may be an advantage on economic running grounds to install boilers whose firing rate will not meet these peaks. In these cases, there are two methods which may be used: one is thermal storage and the other is with an accumulator.

23.15.1 Thermal storage

This is briefly described in Section 23.1.11. Its principle is based on a special feedwater control system, which allows the water level in the shell to fall during periods of steam demand in excess of the maximum firing rate. Conversely, during periods of low steam demand the control system allows the water level to re-establish itself. This is achieved using a constant burner-firing rate that should match the average steam demand, thus allowing maximum efficiency. It is claimed it is possible with this system to control the limits of boiler working pressure to within ±0.07 bar.

23.15.2 The accumulator

This was briefly described in Section 23.1.12. Unlike thermal storage, it depends upon differential pressures, and it is suited to a situation where both high- and low-pressure steam systems are required (for example, 17 bar and 7 bar). Alternatively, if no high-pressure steam is required then the boilers must be designed to operate at a higher pressure with all steam supplies going to process through a pressure-reducing station. Any high-pressure surplus then goes to the accumulator to help meet peak loads. Figure 23.11 shows the layout of an accumulator.

The storage vessel is filled to around 90 per cent of its volume with water and the pressure of the boilers controls the overflow valve. On rising steam pressure (indicating that the boilers are producing more steam than the process requires), a signal to the overflow valve allows all surplus high-pressure steam to flow into the accumulator via a non-return valve and internal distribution header. Here it is condensed and its thermal energy stored. If a peak load develops on the high-pressure system, controls will close the overflow valve and allow steam to discharge from the accumulator through the pressure-reducing valve set to meet the low-pressure steam requirement. Similarly, if the peak develops on the low-pressure system, then high-pressure steam may pass directly to the pressure-reducing set to supplement steam from the accumulator.

Every accumulator will be designed to meet its specified duty. It will be appreciated that the greater the differential pressure, the smaller the vessel will need to be.

23.16 Automatic controls on boilers

Whether the boiler is fired on oil, gas or solid fuel, it may be expected that it will operate automatically. When boiler plant is not run continuously initial start-up may be manual, time clock or through an energy-management system. Manual attendance may be limited to maintenance functions dictated by the type and size of plant. Automatic controls will cover three areas:

1. Combustion appliance
2. Water level
3. Blowdown

23.16.1 Combustion appliance

When using oil or gas this may operate in one of three modes. On/off, high/low/off or modulating. Only small units will have on/off operation and usually all units of 4000 kg/h evaporation and above will have modulating control. The burners will have safety systems incorporated to prove satisfactory fuel supply before ignition takes place, flame proving at point of ignition and continuous flame monitoring thereafter.

High/low burners will have between 2:1 and 3:1 turndown on the maximum firing rate while a modulating burner will give between 3:1 and 5:1, depending upon the unit size and the fuel used. A higher turndown ratio is available from some burners, but if the above-mentioned ratios are exceeded care must be taken to consider the effect on the final gas temperature leaving the boiler. Dew-point temperatures are to be avoided. Coal-fired units by nature of the fuel will have modulating control whether chain or traveling grate, underfeed or fixed/tipping grate.

23.16.2 Water level

The control of feedwater into the boiler is automatic. Feedwater will be delivered from a pump with on–off operation or from a continuously running pump delivering water through a modulating feed control valve. In both

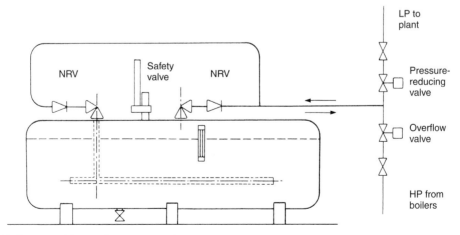

Figure 23.11 Diagrammatic layout of a steam accumulator

cases, the water requirement is continuously monitored by level controls directly connected to the shell or drum of the boiler.

In addition to controlling the feedwater, the level controls will also monitor low- and high water conditions. It is essential that all heated surfaces are always fully immersed; therefore, two stages of low-water alarms are fitted on two independently mounted controls. The first low water will give audible and visual indication and shut down the combustion appliance. If normal water level is automatically restored it will allow the boiler to go back to firing mode. If the second low water is reached again, audible and visual alarms are instigated, but now the boiler will go to lockout condition, requiring manual restart after the problem has been resolved. Watertube boilers may have additional monitoring controls to give early indication of load changes as referred to in Section 23.7.

23.16.3 Blowdown

Blowdown on a boiler is mandatory. On small boilers, the required operation of the main blowdown valve may be sufficient to control the quality of water within the boiler. On medium and large plants, additional systems are employed.

The simplest is a preset continuous blowdown valve to maintain a suitable water quality in the boiler. It is necessary for water quality to be checked frequently and the rate of continuous blowdown adjusted as may be found necessary.

A second method is a time-controlled valve allowing regular intermittent blowdown of the boiler. Again, regular checks need to be carried out to monitor the quality of water in the boiler.

The third and most automatic system is the conductivity-controlled blowdown. This constantly measures the level of solids in the water and instigates an automatic variable blowdown on a continuous or intermittent basis.

23.17 Automatic boiler start

In order to control the operating times of a boiler it is a simple matter to fit each with time clock control. Alternatively, they may be controlled through a central energy management system. Either way, a boiler or boilers may be shut down at the end of each day and programed to restart the following day or when required. Special considerations need to be made if standing periods are extended, allowing a boiler to go cold.

With hot water units, time clock control can operate satisfactorily as automatic bypass valves built into the distribution system will help the heater to achieve its working temperature quickly. With steam boilers, it is important that the boiler reaches a reasonable working pressure before steam is allowed into the distribution system. For example, if boilers are left open to a system for an extended length of time while not firing they will quickly lose their pressure. This is not only wasteful of energy but eventually creates a problem on start-up. To start a boiler on a zero-pressure system with all valves open will undoubtedly cause the boiler to prime and go to lock-out condition, but not before condensate has at least partly flooded the system.

Therefore, where a time clock is incorporated it is recommended that the crown valve(s) be closed at the end of each working day and only opened after the boiler has reached working pressure the next time it is required. This operation can be automated by the use of motorized or similar valves. These valves may be fitted to each steam, supply line from a manifold adjacent to the boilers and, provided adequate safeguards are incorporated to protect the boilers, the on-line boiler(s) may be left open to the manifold. Alternatively, each boiler may have its own automatic motorized start-up valve. Figure 23.12 illustrates a valve suitable for either system.

Each valve would have a control panel incorporating a timer. This may initially be set to a 'crack' position timed to open after the boiler has started to fire and is already building up pressure. This will allow gentle warm-up of the system while enabling the boiler to achieve working pressure. After this, the valve may be set to open in timed adjustable steps to its fully open position. At the end of a timed period, coinciding with the time clock fitted to the boiler, the valve automatically closes at the end of the working cycle. Where multiple valves are used, their control may be incorporated into a single panel or, alternatively, become part of an energy-management system.

23.18 The automatic boiler house

As described in Section 23.16, the boiler will operate automatically and we saw in Section 23.17 that it can be programed to operate to suit various cycles. There are, however, other areas within the boiler house which still require a degree of attendance.

The first of these is blowdown control. If control of the water quality within the boiler is being carried out by a separate system, we are left with the main blowdown and blowdown of the water-level controls and gauge glasses. It is possible to automate these valves but it is unusual. Before proceeding, the advice of the insurance company must be sought and agreement reached on the operation of the valves. When blowing down the level controls it should take the boiler to lockout. If this, in turn, is overridden, a further proving system will need to be incorporated to prove correct functioning of the level control. In addition to the above, it will still be necessary to carry out an evaporation test on the boiler to prove correct operation of the level controls at defined periods.

The second area will be feedwater pumps. It is normal to have a duplicate standby feedwater pump. Sometimes this may be two for each boiler or one duplicate pump to serve any selected boiler. These will usually require manual changeover in the even of failure of the duty pump. It is practical to automate this changeover by using pressure sensors and motorized valves. The same can apply to oil-circulating pumps, gas boosters, water-treatment plant and any other valves and motors. It is possible to do most things, but in the end there is the cost to be considered. An

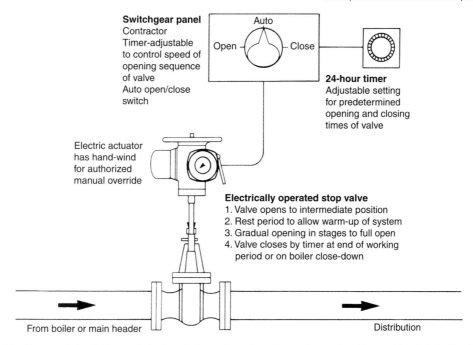

Switchgear panel
Contractor
Timer-adjustable
to control speed of
opening sequence
of valve
Auto open/close
switch

24-hour timer
Adjustable setting
for predetermined
opening and closing
times of valve

Electric actuator
has hand-wind
for authorized
manual override

Electrically operated stop valve
1. Valve opens to intermediate position
2. Rest period to allow warm-up of system
3. Gradual opening in stages to full open
4. Valve closes by timer at end of working
   period or on boiler close-down

From boiler or main header

Distribution

Figure 23.12 Diagram of electrically operated automatic steam stop valve (with permission of Hopkinsons Valves Ltd, Huddersfield, UK)

energy-management system coupled to a suitable maintenance program can still prove the best option for the great
majority of installations.

23.19 Safe operation of automatic boiler plant

The most frequent cause of damage and even explosion
in boilers is a low-water condition. This will expose the
heating surfaces, which ultimately overheat and rupture
under the operating pressure. Experience has shown that
since the introduction of controls for unattended automatic
operation of boilers the accident rate has increased. Investigation invariably shows that lack of maintenance has
been the main contributing factor. It is therefore imperative that personnel responsible for the running of the
boiler plant be fully trained and conversant with its safe
operation.

The minimum recommended requirements for automatic controls for boilers not under continuous supervision are:

1. Automatic water-level controls arranged to positively
   control the flow of water into the boiler and maintain
   a set level between predetermined limits;
2. Automatic firing controls arranged to control the supply of fuel and air to the combustion appliance. A
   shutdown will occur in the event of one or more of
   the following occurring:
   (a) Flame/pilot flame failure on oil- or gas-fired
       boilers. The control should be of the lock-out type
       requiring manual reset;

   (b) Failure to ignite fuel on oil- or gas-fired boilers
       within a set time. Again, lock-out type with manual
       reset;
   (c) At a preset high pressure (or temperature in the
       case of hot water) at or below the safety valve
       setting;
   (d) When the water level falls to a preset point below
       the normal working water level. This control will
       operate on audible alarm but will automatically go
       back to firing mode once the water level is re-
       established;
   (e) Failure of combustion air fans or automatic flue
       dampers;
3. An independent overriding second low-water control.
   This will be set below the level control in (2). It will
   give an audible alarm on shutting off the fuel and air
   supply and require manual reset before the boiler may
   be brought back into operation;
4. All electrical equipment for the water-level and firing
   controls should be designed so that any fault will cause
   the boiler to shutdown and require manual resetting
   before the boiler may be restarted.

For hot water boilers the burner controls will be similar
but controlled by a combination of pressure and temperature signals. A single overriding level control will be
fitted to the flooded boiler to protect against any accidental low-water condition.

In order to monitor the safe operation of the boiler a
daily and weekly program of tests should be drawn up
and log sheets completed as verification of the tests being
carried out. Items checked should include:

Boiler pressure;
Water level gauges – visual and blowdown;
Sequence valves opened and level controls operated;
Feedwater pump operating;
Operate main blowdown valve;
First low-water alarm and burner off;
Second low-water alarm and burner lock-out.

When a boiler may not be shut down for maintenance of the level control chambers isolating valves can be fitted between the water-level control and the steam space. In this instance, the valves must be capable of being locked in the open position and the key retained by a responsible person. When these valves are closed during maintenance periods the boiler must be under manual attendance. Fitting of these valves should only be with the agreement of the insurance company responsible for the boiler. Drains from the water-level controls and level gauges should be collected at a manifold or sealed tundish before running to the blowdown vessel.

As the boilers are designed for unattended operation, it follows that if a fault occurs and an alarm activated this alarm may need to be duplicated at a secondary location where it will be intercepted and acted upon.

Safety valves should be trouble-free. No attempt should be made to alter the set relief pressure without reference to the insurance company and the boiler manufacturer.

With waste-heat boilers fitted to incinerators care must be taken not to over-fire. It is possible to introduce additional heat either by increasing the quantity of waste or by a change in the composition of the waste. The resultant increase in gas volume and/or temperature is then capable of imparting more heat to the waste-heat boiler. As the boiler will have been designed for a specified duty it could be possible to raise an amount of steam in excess of that which may be safely controlled.

Sometimes it is not practical to blowdown the level controls and shut down the incinerator. In this case, the situation should be discussed with the insurance company and the boiler supplier. It is possible to include for an extra high working water level giving a safety margin above the heating surfaces. The controls may then be blown down and checked for satisfactory operation with a predetermined time delay before it shuts down the incinerator or operates a bypass in the event of a fault.

23.20 Energy conservation

Energy conservation in the boiler house can be considered in two areas. One is the selection and installation of suitable equipment and the other is good operation and management.

23.20.1 Plant installation

The boiler, flues and chimney, pipework and hotwell, where installed, should all be insulated to adequate standards and finish. Valves should be enclosed in insulated boxes, although on small installations this can prove disproportionately expensive. The boilers may be fitted with either inlet or outlet air-sealing dampers. These will prevent the flow of ambient air through the boiler during off-load and standby periods, thus helping to maintain the heat already in the boiler.

Economizers may be installed, particularly if gas is the main fuel. It is unlikely that an economic case can be made for a single boiler if there is less than 4000 kg/h evaporation. An economizer can produce fuel savings of 4–5 per cent, but it must be remembered that this will be at MCR, and if the load factor of the installation is lower, the savings will also be proportionately lower.

Combustion controls such as oxygen trim help to maintain optimum operating conditions, especially on gaseous fuels. Instrumentation can give continuous visual and recorded information of selected boiler and plant functions. To be effective, it must be maintained and the data assessed and any required action taken before the information is stored.

Energy-management systems will form an important part of a multi-boiler installation, whether on steam or hot water. Boiler(s) for base load will be selected and further boilers brought on- or taken off-line as required. The important feature of these systems is that the selection of boilers coming either on- or off-line will be ahead of the load and programed to anticipate rising or falling demands.

Computer monitoring and control systems have recently been introduced. These are designed to operate in place of conventional instrumentation. Using intelligent interface outstations connected to a desktop computer, many plant functions may be programed into the computer and controlled centrally.

23.20.2 Operation and maintenance

As most boiler plants installed today are designed for unattended operation it is even more important that early action is taken in the event of the boiler requiring adjustment of combustion or other maintenance. If full instrumentation is not installed then a portable test kit should be used and the plant checked and logged daily or weekly. Perhaps the most obvious waste to look for after steam leaks is a rise in the flue-gas outlet temperature. The boiler will progressively have deposits adhering to its heating surfaces, but at an increase in temperature of no more than 16°C above its design outlet temperature it should be cleaned. The time between cleaning will vary according to the type of fuel and operational load.

Comparatively small air leaks into the combustion spaces of a boiler can produce localized problems. All access opening seals and sight glasses and seals must be monitored to prevent this.

23.21 Noise and the boiler house

Noise can constitute a danger to health, and therefore adequate precautions must be taken to protect personnel who are required to be in such an environment. The Health and Safety at Work, etc. Act 1974 has the power to control noise emissions, but the subject is complex. If it is anticipated that noise will exceed acceptable levels then specialist knowledge should be employed at an early stage in the design and layout of the plant.

As a guide, Figure 23.13 shows a noise-rating curve for everyday noises. At the lower sound levels these may be considered acceptable, and it will be found that to produce overall sound levels below this will prove costly. Alternatively, sound levels exceeding the upper figures need careful consideration if personnel are required to be in that environment. For some machinery, such as reciprocating engines, turbines and compressors, it is not practical to contain the noise at source and then ear protection has to be utilized. Generally, however, industrial boilers and their associated plant can be designed to meet acceptable noise levels.

It is generally the case that small boiler plants will create less noise than large ones. Nevertheless, due to the practice of siting small boilers close to working areas, the matter of noise must not be overlooked. Noise sources within the boiler house will be from the following:

Combustion appliance, including forced- and induced-draft fans;
Feedwater pumps;
Oil pumps;
Gas boosters;
Water-treatment plant;
Steam lines and PRV stations;
Safety valve vent pipes;
Gas lines;
Flue ducting.

Small oil and gas burners are generally supplied without any form of silencing. Some may be enclosed within an acoustic hood and air inlet silencers are normally available if specified. Oil and gas burners fitted to boilers of outputs above 3000 kg/h evaporation can be supplied with progressive degrees of silencing equipment. Figure 23.14 illustrates sound-pressure levels of one burner manufacturer when fitting three basic types of silencing. The use of variable-speed motor drive for the combustion air fan can also reduce the overall noise level, specifically when the unit is not operating at maximum continuous rating. Coal-fired plant will normally be quieter than oil- or gas-fired plant, although, where fixed grate boilers are employed, the same criteria will apply to the forced-draft fans as for oil and gas.

Feedwater pumps will not normally constitute a noise problem unless the area is particularly sensitive. Two alternatives are then available. One is to install reduced-speed pumps and the other is to site the pumps in a separate acoustic enclosure within the boiler house. Oil-circulating pumps are usually low speed and, as such, do not cause a noise problem.

Gas boosters may be fitted with motor enclosures or designed to operate at lower rotating speeds. Alternatively, they may be housed in an acoustic enclosure within the boiler house. The booster and drive unit can be supplied with anti-vibration mountings to isolate it from the floor or steelwork and flexible bellows fitted to the gas inlet and outlet connections.

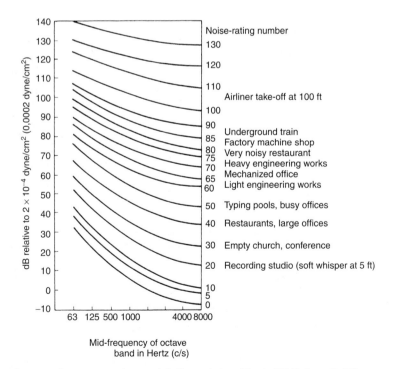

Figure 23.13 Noise-rating curves for some everyday sounds (with permission of Saacke Ltd, Portsmouth, UK)

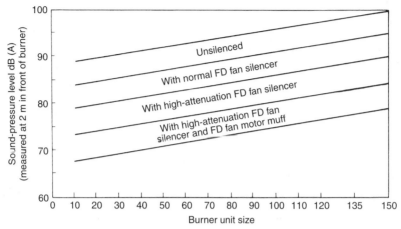

Actual position in band depends on FD fan size and acoustics of boiler house

Figure 23.14 Sound-pressure levels for Saacke rotary cup burners (with permission of Saacke Ltd, Portsmouth, UK)

Water-treatment plant may include one or more electric motors driving feedwater pumps or dosing pumps. Dosing pumps are usually very small and will not cause noise problems. Water pumps may be considered as for the feedwater pumps already described.

Steam transmission will produce noise. Low steam velocity will help to reduce this. One specific noise area will be if a pressure-reducing valve set is employed. Therefore it is desirable to have this within the boiler house or other building and not outside. Some types of pressure-reducing valves operate at lower sound levels than others, and therefore if environmental noise could be a problem then this should be investigated. Insulated valve boxes will also help to contain any noise.

Safety valve venting is noisy. It should only happen infrequently and therefore may be acceptable. If the location is near to offices or housing, it may be advisable to fit silencer heads. Noise levels from gas lines and meters should not normally prove a problem on industrial premises.

Flue gas ductwork will be insulated, which helps to reduce any transmitted combustion noise. Where large rectangular flue ductwork is used, it must be adequately stiffened and insulated to reduce any risk of reverberation being set up from the combustion system.

Finally, if the boiler house is to be within an industrial complex and there is no adjacent residential development, the building need only be suitable for adequate weatherproofing of the plant and for maintenance. Alternatively, if it is sited adjacent or near to housing or offices then the building structure should take account of this.

Ideally, the walls should be of brick or concrete with an insulated roof. Large areas of glass should be avoided. It is desirable that all routine maintenance and operating functions can be carried out without having to open doorways. Openings and access to the boilerhouse should be away from sensitive areas. Free areas for air for both combustion and ventilation should be through acoustic louver panels and preferably again away from sensitive areas. Discharge points for vehicles that deliver fuel should be carefully considered so as causing the minimum nuisance.

23.22 Running costs

With the wide range of boiler types and outputs being considered together with the ever-changing costs for fuel, it is difficult to arrive at a specific figure, which can relate running costs to the installed plant. In order to arrive at a guide in this area several case studies were carried out. It quickly became apparent that few operators maintain separate costs relating to the boiler plant and only keeps overall factory maintenance costs. Fortunately, sufficient information was obtainable to arrive at an overall pattern of costs.

On an annual basis, with running costs covering fuel, operation, electricity, water treatment and maintenance, fuel can account for 90 per cent on a plant operating 168 h per week and 80 per cent on a plant running 40–80 h per week, the remaining 10–20 per cent covering the other items. These proportions were consistent for modern plant burning coal, oil or natural gas. With the recent increase in natural gas tariffs for large consumers, it will now mean that natural gas will account for a larger percentage of the annual figure.

An alternative cost also became apparent from the case studies. If the mean hourly steaming rate in kilograms is multiplied by 2.25, it produces a figure roughly equivalent to the annual operating costs in pounds, excluding the cost of fuel.

23.23 Management and operation

The boiler plant, regardless of its size, is an essential part of the plant; otherwise it would not be there. Having established this, it should be covered by a planned

Table 23.7 Recommended maintenance procedure

Operation	Fuel		
	Oil	Gas	Coal
Blowdown			
Main	1	1	1
Water-level controls	1	1	1
Water-level gauges	1	1	1
Visual check	1	1	1
Check/clean burner	1	3	1
Water-treatment check	2	2	2
Combustion check	2	2	2
Water-level control operational check	2	2	2
Inspect refractories	3	3	3
Open fire side and clean	3	4	3
Open water side for inspection	4	4	4

1: Daily.
2: Weekly.
3: 6–12 weeks.
4: Annually.

maintenance schedule to produce maximum availability. Most boiler plant is supplied for unattended operation, but certain procedures still need to be carried out. The extent of this program will depend upon the size of the plant and the fuel used.

Table 23.7 outlines briefly the extent of operations which need to be undertaken. This is intended only as a guide, as in the majority of installations a program will be evolved to suit individual cases. In certain instances (for example, where electrical power generation is involved) the boilers may not shut down on a daily, weekly or monthly basis, and all controls are then fully automated. Also, watertube boilers may operate for longer periods before fireside shutdown is necessary.

Internal personnel will normally undertake boiler plant operation. Maintenance work may be by either internal personnel or outside contractors. Service contracts will be available from equipment suppliers covering all items of the boiler plant. Recently, companies have started offering contract energy-management schemes. These may be designed to suit individual applications and will be tailored to customer requirements. They may take over the operation of an existing plant or, if necessary, include for a new replacement plant. They will usually operate over a 3–10-year contract period. Dependent upon the terms of contract, all fuels, electricity, repairs and replacements may be covered.

24

Combustion Equipment

Colin French
Saacke Ltd

Contents

24.1 Introduction

A burner is a device for liberating heat by the combustion of fuel. Fuels are predominantly hydrocarbons that release their heat exothermically when oxidized in a controlled manner. The most freely available oxidant is air, which contains only 21 per cent oxygen, the remaining 79 per cent being nitrogen which does not contribute to the process. The nitrogen, because it is heated at the same time, reduces the maximum flame temperature that would have been possible with a pure oxidant. Similarly, the combustion products, when discharged from the process, contain nitrogen which increases the volume of the gases, and hence the sensible heat loss.

A burner comprises a means to inject the fuel, a fan to provide the air for the combustion reaction, a register or stabilizer assembly that provides for the mixing of air and fuel and the stability of the flame, and a means for controlling the air–fuel ratio and fuel input. In addition, on automatic burners, a management system is necessary to ensure programed start-up and shut-down together with supervision while firing via a flame-detection system and interlocks to prove that certain parameters are maintained such as air and fuel pressure. An almost infinite number of types of burner have developed over the years, but broad categories exist characterized by the type of fuel being burnt, the principle of the fuel injection and mixing system, and often the application for which they were designed.

Although some integration has taken place where the appliance or boilermaker has assumed responsibility for the combustion system, overall, specialist manufacturers of combustion equipment who have developed products for each application such as boilers, furnaces, kilns and dryers, etc. serve the market. The burner makers have manufactured products which provide a packaged solution to the combustion requirement, looking after not just the burners and controls but also the fuel supply system, which may involve pumping, heating of the fuel, filtration and other peripheral equipment and functions.

24.2 Oil burners

With the exception of the vaporizing burner, these are normally characterized by the method of atomizing, which itself is dependent on the grade of fuel being combusted.

24.2.1 Vaporizing burners

This principle is confined to domestic applications where kerosene or premium gas oil is concerned. The simplest type uses a number of concentrically arranged wicks which promote vaporization of kerosene into an air/vapor mixing zone enclosed within a perforated drum arrangement. Normally, these burners obtain their air by natural draft.

Another type utilizes a pot and may have natural draft or a fan. The pot burner is essentially an open-topped drum into which fuel is fed at constant head. The vapor rising from the surface is mixed with air being discharged from a perforated drum or pot (Figure 24.1). A further

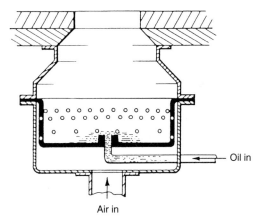

Figure 24.1 Vaporizing pot burner

development is the wall flame burner in which a central rotating distributor that is driven from the forced-draft fan motor throws the fuel against the wall of the pot.

24.2.2 Pressure-jet atomizers

Oil is fed at high pressure to a nozzle in which the oil passage is positioned to feed oil radially inward via a number of slots which are arranged at a tangent to the swirl chamber. The high rotational velocity given to the oil as it exits at high pressure through the central discharge hole provides the means for droplet disintegration via a conical sheet formed at discharge. Limitations of this principle include restriction to gas oil for small sizes and poor turndown caused by a limited range of pressures over which the atomization is satisfactory. There is a choice of spray pattern, notably solid, semi-hollow and hollow cone, and a reasonable range of spray angles is available (Figure 24.2), often used in a two-nozzle head configuration to improve the turndown ratio from 1.4:1 for a single nozzle to 2:1 using two nozzles at constant

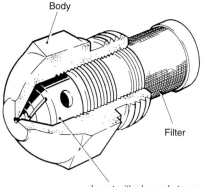

Body

Filter

Insert with channels to provide high velocity swirl motion at nozzle exit

Figure 24.2 Pressure-jet nozzle

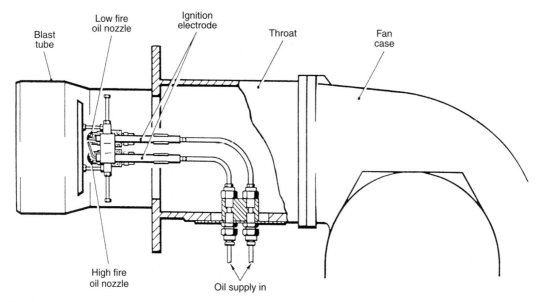

Figure 24.3 Combustion head pressure-jet burner

pressure. The orthodox arrangement for the combustion head is shown in Figure 24.3. The simplex pressure-jet atomizer is also used in power station boilers firing heavy fuel oils in arrays of up to 60 burners.

24.2.3 Spill return atomizers

These partially overcome the weakness of the simplex pressure jet regarding turndown ratio by spilling back the unconsumed fuel at part load. In this way, the swirl velocity in the exit chamber is maintained constant but the diameter of the exit hole remains the same. A further advantage is that it is possible to add a central shut-off needle through the atomizer, which is actuated by fuel pressure on a servo piston. This allows fuel to be circulated right up to the atomizer tip prior to starting the burner. An improved light-up result on medium and heavy fuel oils is due to pre-warming of the nozzle and feed pipework. In addition, it provides a further mode of safety on shutdown acting as a shut-off valve as well as preventing dribbling of the atomizer, which would lead to poor atomization caused by nozzle fouling. This type of atomizer is shown in Figure 24.4. The combustion head configuration remains similar to simplex atomizers.

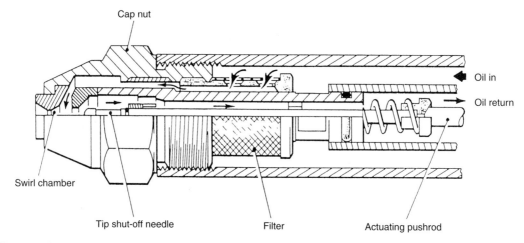

Figure 24.4 Spill return nozzle with tip shut-off needle

24.2.4 Twin-fluid atomizers

Atomization in these types is partly caused by fuel pressure, but this is enhanced by the kinetic energy provided by another fluid that is normally air or steam. At present, this secondary fluid is at a medium or high pressure, the low-pressure method being largely superseded. Pressures are around 1–2 bar for those categorized as medium pressure and 6–10 bar for high-pressure types. Oil pressures are also typically 6–12 bar.

Steam is the preferred atomizing medium, since it is more economic than compressed air. Steam consumption is typically less than 0.5 per cent of the fuel burnt on a mass basis, although this rises in direct proportion to turndown ratio. On very large burners, the steam flow is modulated in proportion to fuel burnt. Turndown ratios range from about 5:1 for small shell boilers to 12:1 in watertube applications, making this one of the most versatile burners. The steam condition is important in that it must be dry saturated or slightly superheated at the nozzle to avoid condensate formation. On small or non-continuously running plant where no steam is available for start-up a compressed air supply must be provided until steam becomes available from the boiler.

Possibly the best-known version of this principle is the Y-jet atomizer developed by Babcock Energy, in which between four and ten exit holes are arranged circumferentially, each consisting of two converging passages arranged in a Y formation. Possible limitations of the principle in spray angle and pattern have restricted its use to larger boilers and watertube types. A cross section of this nozzle is shown in Figure 24.5 and a typical register arrangement is given in Figure 24.6 for a dual-fuel burner.

24.2.5 Rotary cup atomizers

A shaft rotating at 4000–6000 rev/min carries a primary air fan and an atomizing cup. The cup, typically of about 70–120 mm diameter, is tapered by a few degrees to increase in diameter at the exit. A stationary distributor which projects oil onto the smaller-diameter end of the cup feeds oil to the inner surface. The oil, influenced by centrifugal force, forms a thin film, which passes towards the cup lip. Atomization occurs as the oil leaves this lip. In addition, a primary air supply, normally in the range of 5–12 per cent of stoichiometric (chemically correct) air,

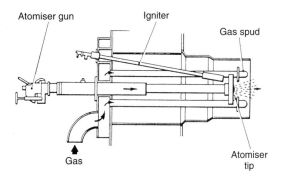

Figure 24.6 Steam-atomizing burner register

is arranged to exit over the cup outer surface, at a velocity of about 50–90 m/s. The primary air is swirled to oppose the rotation of the cup. Droplets shattered by the combined centrifugal action of the cup and the primary air blast are propelled axially into the furnace.

Advantages of this type include an ability to burn all fuels including those containing solid particles, good turndown ratio (4 to 10:1 typically) and an insensitivity to oil conditions such as pressure and temperature. It is widely used in shell boilers, and the only real limitation is that the cup surface has to be cleaned daily. The most common atomizer layout is shown in Figure 24.7. Variants include direct driven cup and separate mounting of the primary air fan.

24.3 Gas burners

Industrial gas burners are mainly of the nozzle mix configuration. Beneath industrial burners, which are used for raising steam and hot water in the power and process industries, lies a large array of types and principles. The most common types are normally characterized as to whether they are aerated or non-aerated.

24.3.1 Non-aerated types

Otherwise known as diffusion flame or post-aerated, these normally comprise a simple nozzle supplying gas at a controlled pressure into a chamber where air is made available via entrainment into the flame by natural draft. Common types are the Bray jet, Aeromatic and Drew jet (Figure 24.8).

24.3.2 Aerated types

These are otherwise known as atmospheric or premix burners. Primary air is entrained into the gas stream prior to exit from the nozzle. The best known of these types is the Bunsen burner, and the most common is the ring-type domestic cooker hob arrangement. Both aerated and non-aerated types are often found in a bar configuration. Typical applications are heating of tanks and process uses involving direct heating of the product. Figure 24.9 shows a typical aerated bar burner.

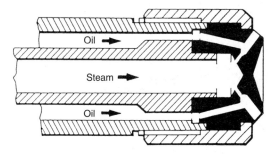

Figure 24.5 Steam-atomizing nozzle

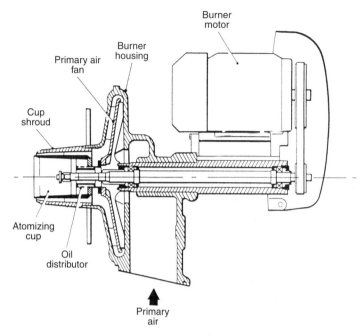

Figure 24.7 Section through rotary cup atomizer

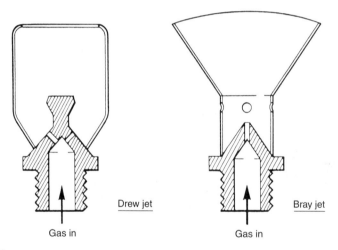

Figure 24.8 Non-aerated burners

24.3.3 Nozzle-mix types

Utilizing a forced-draft fan, the burner has a gas head arranged to mix the fuel and air in a blast tube which controls the stability and shape of the flame. Gas exits from nozzles or holes in the head and is mixed partly in the high-velocity air stream and partly allowed to exit into an area downstream of a bluff body. Behind the bluff body, a relatively quiescent zone forms which provides a means for flame stability. Many configurations exist, but the most

frequent are those which are designed around the most common types of oil burner. This allows the burner to easily be converted to oil or dual-fuel (gas and oil) firing. As gas is a relatively easy fuel to burn, the design is strongly influenced by the optimum oil burner configuration. Two typical types are shown in Figures 24.10 and 24.11 based, respectively, around pressure-jet and rotary cup atomizers.

Performance on gas is normally limited in dual-fuel applications to that of the oil burner. In gas-only

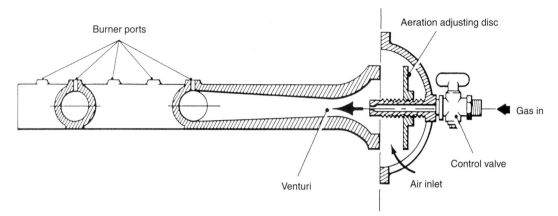

Figure 24.9 Aerated bar burner

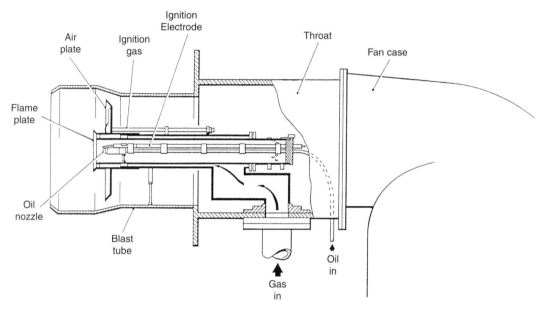

Figure 24.10 Dual-fuel burner based on pressure-jet configuration

applications, the performance is better, notably in lower excess air factors and better turndown ratios (3:1 small burners, 15:1 very large burners).

Problems can occur with highly rated boilers converted to gas firing where tube end and tube plate cracking can occur due to overheating. This can normally be overcome by modifying the tube attachment arrangement. Another problem that is quite common is resonance or pulsation, where the burner acoustically couples with the natural resonant frequency of the combustion chamber. This is normally easily overcome by modifying the burner head to change the rate of mixing.

24.4 Burner design considerations

24.4.1 Atomizer

Characteristics of various atomizers are given in Table 24.1. Primary considerations are selecting the best principle for the type of fuel, the size of the burner/boiler and the type of application. Other important characteristics are ability to operate with the minimum of excess air, turndown ratio and questions of durability and maintenance.

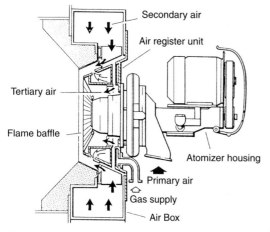

Figure 24.11 Dual-fuel burner based on rotary cup configuration

24.4.2 Register/combustion head

This is often fixed for the type of atomizer. Stability is either one of two basic principles: bluff body with some aeration/cooling, and swirler types which are generally confined to twin fluid atomizers of large size. The stabilization process is achieved in both cases by recirculation of vortices spilling off the baffle in the case of bluff bodies and by a full recirculation flow pattern in swirl stabilizers. Important items also fitted to the register are ignition system (most commonly high-voltage spark in the case of pressure-jet burners and gas/electric in all other types) and the flame-supervision system. This is normally infrared for oil burners and ultraviolet light sensitive in the case of gas and dual-fuel burners. Smaller gas burners utilize the flame-rectification principle.

24.4.3 Airbox/fan arrangement

Generally, this is a compromise in monobloc burners incorporating a fan and burner head within one casting.

Table 24.1 Summary of burner characteristics

Oil burners Type	Size range	Dual-fuel capability	Fuel type	Atomizing pressure (bar)	Atomizing viscosity (cS)	Turndown ratio	Flame characteristics	Main applications
Vaporizing	10–40 kW	No	Kerosene, gas oil	–	<5	On/off	Normally blue flame	Central heating
Pressure jet commercial market	30 kW–3 MW	Larger sizes	Gas oil (class D)	6–12	<5	On/off or <1.5:1 by pressure control	Soft yellow and radiant fairly wide-angle spray	Domestic and commercial hot water boilers
Pressure jet utility boilers	10–50 MW	Possible but gas not burnt on utility boilers	Heavy oil (classes G and H)	20–40	~20	On/off or <1.5:1 by pressure control	Highly radiant low excess air operation	Power station and large petrochemical watertube boilers
Twin-fluid medium pressure	1–10 MW	Yes	All types (mainly class G)	1–4	~20	<5:1	Wide range of shapes mainly used for process applications	Kilns, furnaces, processes requiring special flame characteristics
High pressure	2–50 MW	Yes	All types (mainly classes G and H)	6–20	~20	3:1 to 12:1	Mainly wide-angle sprays in WT boilers, low excess operation	Large shell boiler and watertube boiler process applications
Rotary cup	1–40 MW	Yes	All types (mainly class G)	2–3	~60	3:1 to 15:1	Medium intensity, shape varied by register design	Very popular for all sizes of shell boiler. Some use on process and WT boilers

Gas burners Type	Size	Dual-fuel	Fuel type	Gas pressure	Turndown ratio	Flame Characteristics	Main applications
Non-aerated	0.01 kW–50 kW	No	N gas, LP gas	5 mbar–75 mbar	<2:1	Semi-luminous lambent, low intensity	Pilot flame and domestic applications
Aerated	0.5 kW–150 kW	Larger sizes (yes)	N gas, LP gas	20 mbar–1 bar	<20:1	Non-luminous shape depends on application	Wide use in heating and direct-contact process applications
Nozzle mix	30 kW–40 MW	Yes	All gases	15 mbar–1.5 mbar	<20:1 depends on oil turndown when dual-fuel	Normally nonluminous shape depends on burner and register configuration	Wide use in packaged burners of all sizes. Common in hot water and steam boilers of all sizes

Table 24.1 (*Continued*)

Coal burners Type	Size range (MW)	Fuels	Grate thermal loading (mW/m^2)	Bed thickness	Turndown ratio	Ashing system	Main applications
Underfeed	0.3–2	600–900 Rank singles and doubles	1.0	Depends on retort depth	On/off	Manual or side screw automatic	Small hot water and steam boilers
Fixed grate	0.6–7	600–900 Rank singles	2.7	25 mm live on up to 200 mm ash	2:1 to 5:1	Manual or automatic tipping grate or robot rake	Hot water and steam shell boilers
Chain grate	1–30	500–900 Rank normally washed smalls, but other types possible including high ash	1.7	80–150 mm	3:1 to 8:1	Manual drop tube boiler or drag link	Shell and water boilers
Coking stoker	1–4.5	300–900 Rank washed or untreated smalls and singles	1.4	250–350 mm	5:1	Manual drop tube in boiler or drag link	Shell boilers
Pulverized fuel burners	1–50	All types including anthracite lignite and bituminous	–	–	<15:1	None	Limited use in shell boilers mainly kilns and power stations
Fluidized beds	1–50	All types of coal and waste solid fuels	3.0	0.3–1.0 m	2:1 and then by bed sectoring	Removal and clean-up from bed	Special vertical shell boilers and watertubes

On larger burners utilizing a separate airbox (or windbox, as it is sometimes called) the air-distribution quality is important to maintain low excess air operation and good flame shape throughout the turndown range. Two approaches are common: designs utilizing a constant velocity approach, the alternative being a large plenum chamber in which low velocities provide an equalization effect. The shape of the flame can deteriorate at part load due to the register pressure reducing on a square law with turndown.

24.4.4 Air/fuel ratio control

This is often the Achilles' heel of burners, since poor design can lead to hysteresis. Correct sizing of the control valves and fan size is essential to maximize damper travel/backlash ratios and give good linearity throughout the firing range. Robust characterization cams and linkages are essential.

24.4.5 Burner management system

Small monobloc burners usually have a proprietary control box/photo cell amplifier system. Larger burners may have a dedicated system specific to the application and may utilize self-checking photocell systems.

24.5 Future developments

Oxygen trim

The development of reliable zirconia cells which can measure the gas analysis *in situ* without recourse to gas-sampling techniques has led to systems which provide feedback to the air/fuel ratio control system.

Compensation is made for variations in ambient air pressure and temperature, calorific value, boiler resistance due to fouling and burner performance drift by trimming the air damper with a separate servo motor.

Electronic air/fuel ratio control

Electronic air/fuel ratio characterization is becoming available. By driving gas and oil valves and the air damper separately via individual servo motors, electronic units can supervise the relative positions of the motors and provide characterization of air/fuel relationships utilizing an almost infinite number of set points to give close repeatable control.

Emulsion techniques

These are particularly applicable to burners firing the heavier grades of oil which contain long-chain molecules called asphaltines. The superheating of the water in the emulsified fuel droplet enhances atomization. The effect is to provide secondary atomization to the droplet as the steam is formed.

Additives

In addition to the traditional additives that suppress dewpoint corrosion, future developments are likely to aid atomization by reducing droplet surface tension and stimulating combustion catalytically.

Air and exhaust gas sealing dampers

These positively shut off the draft that naturally occurs due to chimney buoyancy when the burner is in its off cycle, thereby reducing standby losses. Burners incorporating shut-off dampers are becoming increasingly common.

Emission control

Future legislation will stimulate burner development in the areas of carbon monoxide, NO_x and particulate generation. Techniques will include flue-gas recirculation, staged combustion, and additives to reduce the NO_x and more sophisticated controls. Controls over the sulfur generated do not affect burner design greatly since the sulfur dioxide is a natural product of combustion and can only be reduced by lower fuel sulfur contents or sulfur removal from the exhaust gases.

24.6 Coal burners

Coal burners demand design consideration in all the aspects mentioned under gas and oil burners, but, in addition, need attention to the aspect of ash removal. The extent to which ash removal plays a part in the combustion system design often determines the ability of the burner to burn specific coals, particularly those with a high ash content. Principal types are as follows.

24.6.1 Underfeed stoker

Mainly applied to sectional and small shell boilers, this principle is limited to doubles or singles coal. Simple in construction, it consists of a hopper normally mounted close to the appliance, which feeds via an Archimedean screw to a retort where the combustion takes place. Coal is fed up through the retort where the coal is burned progressively until the residue is mainly clinkers and ash. The tapering nature of the retort allows sufficient residence time in the combustion zone before the ash overflows from the top of the retort into a surrounding area of dead-plates from which the ash is removed periodically. The double-walled construction of the retort allows combustion air to be fed into the combustion zone through holes in the internal walls. These holes are known as tuyeres, and are sometimes supplemented by secondary air jets which transfer air from the retort via vertical tubes onto which are mounted nozzles to direct this over-fire air onto the top of the bed. Recent improvements have allowed auto de-ashing by arranging for the ash and clinker to fall down to one side of the retort only, so allowing

extraction by a screw. A typical underfeed stoker is shown in Figure 24.12.

24.6.2 Fixed-grate burner

Very popular in the main industrial shell boiler market, the fixed-grate stoker utilizes a bed of firebars which allow primary air to be fed through from underneath while not allowing coal or ash to fall through. This is achieved via a special ash-retaining pocket. Part of the grate at the front and rear comprises dead bars which prevent air passing through. Two types of coal feed predominate, both fed with coal, which is transported by air after metering by a variable-speed screw. The coal is propelled along a pipe normally of 80–100 mm diameter by air travelling at about 20 m/s. This air is discharged over the grate to form part of the secondary air supply. In one principle, the coal is spread on the grate via a deflecting nozzle that is characterized electronically to provide an even bed. The other type uses a drop tube, which has a fixed deflector cone to spread the coal. Nozzles are arranged to provide secondary air over the bed.

De-ashing is achieved by stopping the burner and manually raking the ash off the bed after first moving the live coal onto the deadplates. This live coal is re-spread onto the live grate after de-ashing. Recent developments have included a tipping version of the grate to discharge the ash and an automatic system using a robot to rake ash off the grate.

Using the front-feed system, it is possible to retrofit this principle onto boilers designed for oil and gas with some de-rating. Performance in terms of turndown ratio, response rate, etc. can be as good as gas and oil. Best performance is achieved using 600–900 Rank coals containing 3–8 per cent ash. Figures 24.13 and 24.14 illustrate the two feed arrangements common with fixed-grate stokers.

24.6.3 Chain grate

This is a very versatile stoker common in shell boilers and is used with modification in watertube boilers as the traveling-grate stoker. In shell boilers, the air supply is via forced- and induced-draft fans in combination with a balanced or slightly negative condition in the furnace.

Coal is fed from a front-feed hopper onto the moving grate through a guillotine door, which controls bed thickness. Nowadays a firebreak system is arranged in the

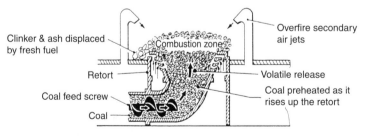

Figure 24.12 Underfeed stoker

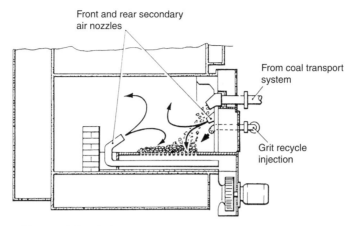

Figure 24.13 Fixed-grate coal burner with front feed

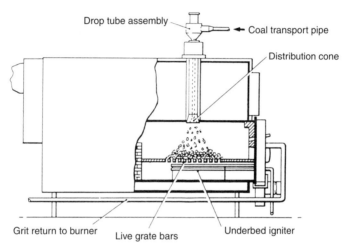

Figure 24.14 Fixed-grate coal burner with top feed

hopper to prevent burn-back, which was relatively common. This is achieved via a rotating vane feeder. The grate comprises hundreds of links arranged into a chain over the whole width of the stoker, so forming a continuous, flexible mat. Driven via sprockets arranged across the width of the bed, the bed is supported on an airbox which supplies the air to the grate via slots in the individual links. The air in the airbox is staged via dampers and/or baffles to be released as required down the length of the grate.

An important principle is that the coal becomes ignited from the fuel which is on top of it and further down the grate. This reduces carry-over of small-unburned particles, as the burning coal on top filters them out. Volatiles released from the fresh coal are also ignited and consumed in the burning layer. This minimizes smoke formation which is caused by incomplete combustion of the volatiles.

Ash removal can be accomplished via a drop tube through the boiler shell. Suitable fuels include most singles and washed or unwashed fines up to a top size of about 25 mm, but not anthracite or strongly caking coals. Ash content must be sufficiently high to protect the grate bars (>6 per cent) and the coal should contain free moisture. An important specification criterion is the material of the chain links, particularly with low-ash coals and with variable loads. Figure 24.15 illustrates the main features of a chain-grate stoker.

24.6.4 Coking stoker

Less popular nowadays, the coking stoker resembles the chain-grate stoker in many respects. Coal is fed from a hopper onto the coking plate via a reciprocating ram. Transformation into the coke phase on these plates takes place by the release of volatiles.

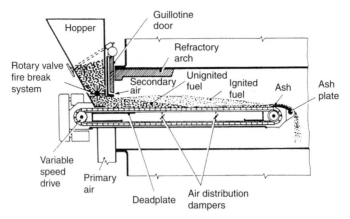

Figure 24.15 Chain-grate stoker

The ignited coke then passes further down the grate via the reciprocating action of the grate bars, which extend longitudinally for the full length of the grate. Cams formed in a shaft that is driven via an electric motor and gearbox drive the grate bars sequentially. This stoker is capable of burning a wide range of coals, and ash can be removed without stopping the combustion process. Air supply is via an induced-draft fan. The coking stoker is shown in Figure 24.16.

24.6.5 PF burners

These types are most common in power station-sized boilers and process industries such as cement and gypsum, where the residual ash is absorbed in the process. As their application is at the larger end of the combustion field, on-site milling and preparation is technically feasible and economic, using ball, roller or impact mills. Small applications can utilize pre-milled coal delivered to site, but the choice of suppliers is very limited.

Initial preheating of the combustion chamber by gas or oil is normally required in order to provide the necessary temperature environment to release the volatiles that provide the stabilization in the base of the flame. Some small PF systems have used another fuel for flame support, but this compromises the economics. A typical pulverized fuel burner is shown in Figure 24.17.

24.6.6 Fluidized-bed firing

This is normally a very integrated system of combustion which has received considerable development in the last few years. The combustion chamber is arranged to allow combustion of the coal within a fluidized bed of inert

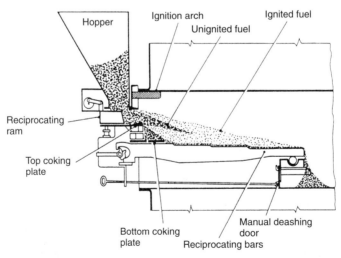

Figure 24.16 Coking stoker

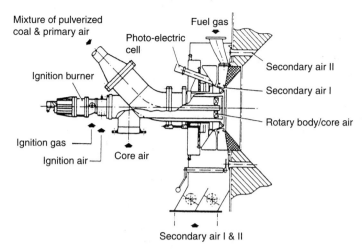

Figure 24.17 Pulverized fuel burner

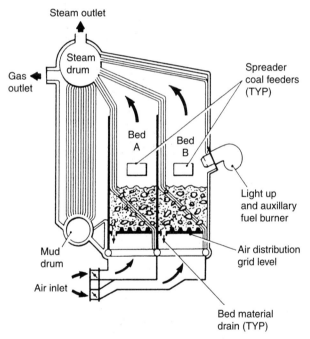

Figure 24.18 Fluid-bed combustion system with sectioned bed

mineral matter, normally sand. The sand is maintained in a fluidized state by an air supply from beneath the bed via nozzles or distributors. Rapid combustion is ensured via the vigorous bubbling of the bed, which is an ideal combustion environment. To avoid sintering of the bed material, it is necessary to prevent the bed from exceeding the ash-fusion temperature of the coal. This requires the bed temperature to be maintained in the range of 900–1000°C. As the temperature of the products of combustion would be about 1400°C naturally, it is fundamental that heat be removed with in-bed cooling tubes to suppress the temperature to less than 1000°C. The turbulent nature of the bed provides for high convective heat transfer rates, so minimizing the surface area. A number of developments have taken place with fluid bed systems that have included deep-bed technology, recirculating beds and reduction of emissions by in-bed additives such as limestone and dolomite to retain sulfur.

Problem areas are ash removal, which takes place by recirculating some material from the bed for clean up and erosion. Positive attributes include the ability to burn virtually any solid combustible material. Figure 24.18 illustrates the principle.

24.7 Dual- and triple-fuel firing

This is very common nowadays to allow bargaining on fuel price or to arrange an interruptible gas tariff, which is backed up at times of peak demand with a stored oil supply. Most types of oil and gas burner are available in dual-fuel form, normally with gas burner design 'wrapped around' the arrangement for oil firing. This is usually the more difficult fuel to burn, particularly in the case of residual heavy oils. Fuel selection is normally by a switch on the burner control panel after isolation has taken place of the non-fired fuel. To avoid the cost and complexity of the fuel preheating on oil firing, smaller systems use gas oil as the standby fuel.

Multi-fuel operation with coal-firing equipment is more difficult to achieve, partly because the grate obscures some of the boiler heating surfaces and partly due to the volume that it also occupies. Systems do exist for this requirement, but changeover is not as instant as oil and gas, as it is normally necessary to remove part of the coal-firing equipment.

PF burners and fluid beds best meet requirements for dual- and triple-fuel firing including solid fuel as one option. PF burners are particularly suitable, as no static grate exists to compromise the design. They also have a combustion geometry which is similar to gas and oil, and therefore the flame can be arranged to allow full development of flame shape and maximum radiant heat transfer surface utilization.

Fluid beds can be fired with gas and oil across the top of the slumped bed since sufficient freeboard exists with coal firing to prevent particle elutriation. Oil, gas or dual-fuel burners so arranged could also provide the means for bed preheating, especially if the flame is redirected down to the fluidization zone.

25

Economizers

Colin French
Saacke Ltd

Contents

25.1 Introduction

Economizers for boilers have been available for over 150 years, almost as long as boilers themselves. For shell boilers, increasing efficiencies have made it increasingly difficult to justify the use of an economizer, the final decision being based in terms of payback period, which is also heavily dependent on fuel prices. Watertube boilers, on the other hand, need an economizer section in the gas passes in order to obtain satisfactory efficiency. For this reason, the economizer is integrated into the overall design, normally between the convective superheater and the air heater if fitted.

In shell boilers with a working pressure of between 7 and 17 bar the temperature of the mass of water in the boiler is typically in the range of 170–210°C. Allowing for, say a temperature difference of 30–50°C between the exhaust gases and the water temperature, the boiler exit gas temperature cannot be practically reduced beneath about 200–260°C, dependent on the operating pressure. It becomes necessary therefore to modify the process on principles to achieve further heat utilization and recovery. In the case of economizers conducting the feedwater supply via an economizer wherein the exhaust gas passes over tubes carrying the feedwater does this. The feedwater, normally at temperatures between 30° and 100°C, represents a further cooling medium for the exhaust gases and provides the potential for the extra heat utilization. This is shown in Figure 25.1.

25.2 Oil and coal applications

Although it would be possible to design for an economizer gas exit temperature of 30–50°C above the feedwater temperature, this would result in a temperature too close to the acid dewpoint of the gases. The acid dewpoint is the temperature at which acidic gases begin to condense out of the exhaust gas mixture. This is principally sulfuric acid due to the sulfur contained in the oil or coal. Although the bulk gas temperature may be satisfactory, in practice, the

surfaces close to the tube wall will be nearly at water temperature due to the high conductivity of the metal tube. This limits the minimum practical gas exit temperature from the economizer to, 170–180°C, remembering, of course, that at low fire this will have fallen closer still to the acid dewpoint. This is typically in the range of 125–140°C, dependent on the excess air and fuel sulfur content. The cold temperature of the heat transfer surface gives rise to heavy corrosion which would reach a peak at about 95°C.

The potential to recover heat from the gases of an oil- or coal-fired boiler is therefore limited to a temperature drop from 240°C to 170°C. This results in a 3 per cent saving. The average saving would be somewhat lower than this since fouling of the economizer surface is inevitable from the carbonaceous emissions of the firing equipment.

The design of the economizer must be robust enough to survive occasional excursions beneath the acid dewpoint and the effects of the methods used to clean the economizer periodically. This may take the form of rapping equipment to shake off deposits, sootblowing by steam or air and water washing with lances.

Average savings of 2–3 per cent combined with the cost necessary for a robust design have therefore limited the use to times when fuel prices are high or special applications where the boiler exit temperature is higher than usual. Developments using an additive to suppress the dewpoint are worth consideration, since this extends the heat-saving potential but, of course, there is an additional burden with the cost of the additive. Magnesium oxide is the most commonly used suppressant and this is injected into the gas stream to combine with the sulfuric acid to form magnesium sulfate (which also has to be removed regularly in addition to the soot).

25.3 Gas-fired economizers

Far greater potential exists for gas-fired economizers, since the gas is virtually free of sulfur. The limitation on gas temperature is the ability of the water to extract the heat, although the water vapor in the gas caused by the combustion of hydrogen does give rise to a water dewpoint at 55°C. This should be avoided, since general corrosion can take place in the latter rows of the economizer and in the exhaust gas ducts and chimney. Normally, this only occurs for short periods during starting from cold, but it should be minimized.

Since no serious problems exist with corrosion, the materials of construction can be cheaper and the thicknesses reduced. Fouling coefficients do not need to be incorporated into the design calculations, as the surfaces remain clean indefinitely. This in itself provides a secondary benefit in that the pitching of fins may be reduced without the risk of reducing flow.

A saving of (typically) 5 per cent can be made with gas-fired economizers, and as this is related to the boiler output it represents a saving of 6.25 per cent in fuel consumed. Gross heat-saving potential related to feedwater temperature and excess air level are shown in Figure 25.2.

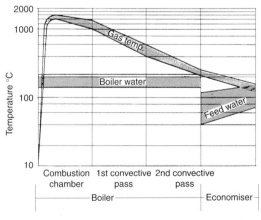

Figure 25.1 Temperatures in three-pass boilers and economizers

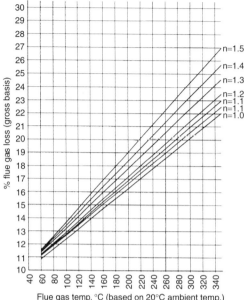

Figure 25.2 Total heat in gas-fired boiler exhaust

25.4 Design

The driving force for heat transfer is temperature difference, and this is maximized through the economizer by arranging for concurrent flow through the passages. In other words, the colder feedwater is made available to the coolest gases. This gives the highest overall average temperature difference throughout the length of the economizer, and is characterized by the logarithmic temperature difference ΔTlm:

$$\Delta Tlm = \frac{(Tg_1 - Tw_2) - (Tg_2 - Tw_1)}{\log n[(Tg_1 - Tw_2) \div (Tg_2 - Tw_1)]}$$

The amount of heat (Q_t) that may be transferred is controlled by:

$$Q_t = \Delta Tlm \times U \times A$$

where U = overall heat transfer coefficient and A = heat transfer area. To achieve a cost-effective design of economizer it is necessary to maximize the overall heat transfer coefficient and the surface area within the economizer.

The heat transfer coefficient, U, can be maximized by the highest practical velocities that can be achieved for both the water and the gas. Waterside pressure loss is limited by the available spare pressure rise from the

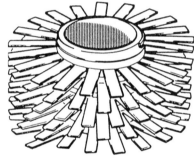

(a) Wound on - plain fin

(b) Wound on - with turbulence inducing segments

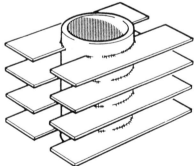

(c) Lamel fin - square or rectangular continuous

(d) Lamel fin - non continuous

Figure 25.3 Common fin types

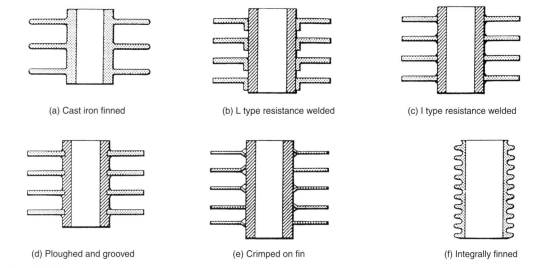

(a) Cast iron finned   (b) L type resistance welded   (c) I type resistance welded

(d) Ploughed and grooved   (e) Crimped on fin   (f) Integrally finned

Figure 25.4 Fin attachment methods

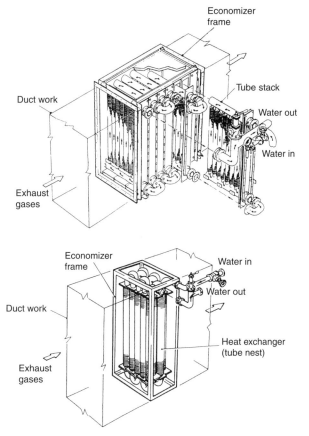

Figure 25.5 Economizer – typical designs

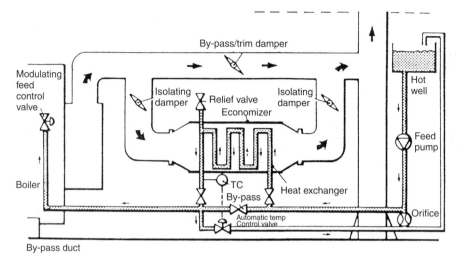

Figure 25.6 Typical installation layout

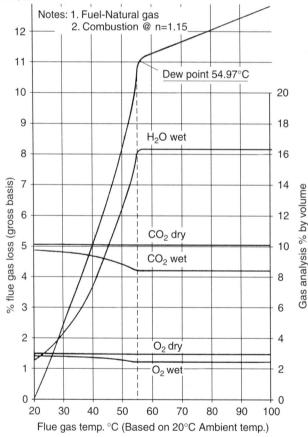

Figure 25.7 Sensible and latent heat savings potential and gas analysis with condensing economizers

feed pump. This is often limited to 0.2–1.0 bar, as it is not generally economic to replace the feed pump. The available spare fan head from the burner generally limits gas-side velocities. Once again, it is not normally economic to replace this in the case of retrofitted economizers or, for that matter, to install an induced-draft fan downstream of the economizer. In addition to maximizing fluid velocities, careful design to promote turbulence by staggered pitching enhances the heat transfer coefficient.

The heat transfer area, A, can be greatly increased by using finned tubes, but care must be taken to ensure good conduction of heat away from the fin into the tube and subsequently into the water. Some common fin types are shown in Figure 25.3 while Figure 25.4 shows some of the attachment methods employed to ensure satisfactory fin to tube heat transfer.

25.5 Installation

Due to interruptible gas tariffs, it is often necessary to adopt gas as the primary fuel and burn oil in periods of peak loads. This means that the economizer has to be arranged so that when oil firing, the flue gases are bypassed around the economizer. The bypass duct must also contain a damper to simulate the economizer gas resistance so that the burner back pressure remains the same for both fuels. Figure 25.6 shows a typical installation layout.

25.6 Condensing economizers

The restrictions on feedwater temperature, dictated by condensate rates and the need to minimize oxygen corrosion, limit further development of the performance of conventional economizers. As mentioned earlier, it becomes necessary to revise the whole philosophy if enhanced heat recovery is desired. If the gas can be reduced in temperature to beneath the water dewpoint of approximately 55°C there exists a potential heat saving due to the latent heat of condensation of the water vapor as well as the sensible heat also contained in the exhaust gases. This quantity of water vapor is considerable (in the case of natural gas firing of the order of 1 ton per hour for a 10 tons per hour steam boiler). Once again, the potential can only be realized provided a requirement exists for such heat. This heat is of relatively low grade and therefore requires a large mass flow rate of water to absorb it. A typical figure might be of the order of twice the boiler feedwater flow rate and at a temperature of main water. This precludes the use of the boiler feedwater, and therefore special low-temperature processes must be integrated into condensing economizer applications. The temperature rise of this water would probably be about 45°C. Assuming the exhaust gases are cooled from, say, 230°C to 40°C, the gross heat saving would be approximately 13 per cent, yielding a fuel saving of over 15 per cent.

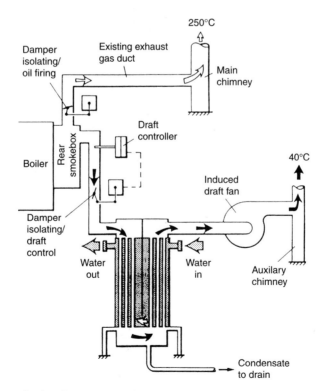

Figure 25.8 Condensing economizer layout

The term *fuel saving* needs to be qualified, since the use of the heat in the water will be given to a process which may be unrelated to boiler demand. Examples of suitable applications occur in laundries, agricultural soil heating, food industries, abattoirs and swimming pool heating.

The success of such schemes is highly dependent on matching supply and demand together with a basic low-grade heat requirement. The total heat remaining in the exhaust gases of the boiler are shown in Figure 25.7 illustrating the potential savings.

Condensing economizers are constructed from corrosion-resistant materials (notably aluminum or stainless steel), since the condensed water vapor in the gas is slightly acidic (typically, with a pH of 3–5). This is because some carbonic and nitric acid is formed in the condensing of the products of combustion and also as a result of chlorofluorocarbon propellants in the atmosphere. Provided the correct choice of materials is made, corrosion life should not become problematic.

A typical arrangement of a condensing economizer is shown in Figure 25.8. Note that an induced-draft fan is almost mandatory, since a high pressure loss is inevitable with such low-grade heat recovery. The design of such economizers often takes the form of a large shell and tube heat exchanger, but the conventional economizer construction of watertubes is quite feasible. Condensate from the exhaust gases is normally discharged to drains.

26

Heat Exchangers

Brian R Lamb
APV Baker Ltd

Contents

26.1 The APV Paraflow

While the original idea for the plate heat exchanger was patented in the latter half of the nineteenth century, Dr Richard Seligman, founder of APV, introduced the first commercially successful design in 1923. Initially, a number of cast gunmetal plates were enclosed within a frame in a manner quite similar to a filter press. The early 1930s, however, saw the introduction of plates pressed in thin-gauge stainless steel. While the basic design remains unchanged, continual refinements have boosted operating pressures from about 1 to 20 kgf/cm^2 in current machines.

The plate heat exchanger consists of a frame in which closely spaced metal plates are clamped between a head and follower. The plates have corner ports and are sealed by gaskets around the ports and along the plate edges. A double seal forms pockets open to atmosphere to prevent mixing of product and service liquids in the rare event of leakage past a gasket.

The plates are grouped into passes with each fluid being directed evenly between the paralleled passages in each pass. Whenever the thermal duty permits, it is desirable to use single-pass, counterflow (Figure 26.1) for an extremely efficient performance. Since the flow is purely counterflow, correction factors required on the LMTD approach unity. With all connections on the head of the unit, the follower is free for very quick access to cleaning and maintenance. The effect of multi-pass operation will be discussed later in this chapter.

26.1.1 Comparative plate arrangements

Clarification of plate arrangements with those for a tubular exchanger is detailed in Figure 26.2. Essentially, the number of passes on the tube size of a tubular unit can be compared with the number of passes on a plate heat exchanger. The number of tubes per pass can also be equated with the number of passages per pass for the plate heat exchanger. However, the comparison with the shell side is usually more difficult, since with a plate-type heat exchanger the total number of passages available for the flow of one fluid must equal those available for the other fluid to within ±1. The number of cross passes on a shell, however, can be related to the number of plate passes, and since the number of passages/pass for a plate is an indication of the flow area, this can be equated to the shell diameter. This is not a perfect comparison but it does show the relative parameters for each exchanger.

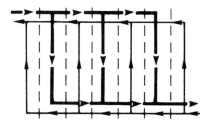

Figure 26.1 Single-pass operation

	Shell and tube	Plate equivalent	
Tube side	Number of passes	Number of passes	Side one
	Number of tubes/pass	Number of passages/pass	
Shell side	Number of cross-passes (number of baffles + 1)	Number of passes	Side two
	Shell diameter	Number of passages/pass	

Figure 26.2 Comparison of pass arrangement: plate versus tubular

An important, exclusive feature of the plate heat exchanger is that by the use of special connector plates it is possible to provide connections for alternate fluids so that a number of duties can be done in the same frame.

26.1.2 Plate construction

Plates are pressed from stainless steel, titanium, Hastelloy or any material ductile enough to be formed into a pressing. The specious design of the trough pattern strengthens the plates, increases the effective heat transfer area and produces turbulence in the liquid flow between plates. Plates are pressed in materials between 0.5 and 1.2 mm and the degree of mechanical loading is important. The more severe case occurs when one process liquid is operating at the highest working pressure and the other at zero pressure. The maximum pressure differential is applied across the plate and results in a considerable unbalanced load. There are two alternative trough forms, one using deep corrugations to provide contact points for about every 650–1950 mm^2 of heat transfer surface, the other crisscross shallow troughs with support maintained by corrugation/corrugation contact. Alternate plates are arranged so that corrugations cross to provide a contact point for every 100–600 mm^2 of area. The plate then can handle large differential pressures and the cross pattern forms a tortuous path that promotes substantial liquid turbulence and a very high heat transfer coefficient. The net result is high rates with moderate pressure drop.

Plates are available with effective heat transfer area from 0.03 to 3.5 m^2 and up to 700 can be contained within the frame of the largest plate-type heat exchanger, providing over 2400 m^2 of surface area. Flow ports and associated pipework are sized in proportion to the plate area and control the maximum liquid throughput.

Figure 26.3 shows the relationship between port diameter and fluid velocity at 4 and 7 m/s and highlights the nominal maximum velocities for various plates. As the flow through the machine increases, the entry and exit pressure losses also increase. The nominal maximum flow rate for a plate heat exchanger limits these losses to an acceptable proportion of the total pressure losses, and is therefore a function not only of the port diameter but

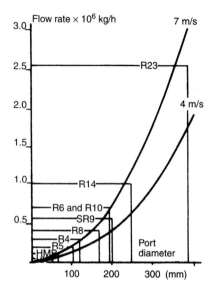

Figure 26.3 Throughput versus port diameter

also of the nature of the plate which has been empirically determined.

These velocities seem, at first, rather high compared to conventional pipework practice, but they are very localized in the exchanger and are progressively reduced as distribution into the flow passages occurs from the port manifold.

26.1.3 Gasket materials

As detailed in Table 26.1, various gasket elastomers are available which have chemical and temperature resistance coupled with good sealing properties. The temperatures shown are maximum; therefore possible simultaneous chemical action must be taken into account.

26.1.4 Thermal performance

Data on thermal performance are not readily available on all heat exchangers because of the proprietary nature of the machines. To exemplify typical thermal data, heat transfer can best be described by a Dittus–Boelter type equation:

$$N_u = A(Re)^n (Pr)^m \left(\frac{\mu}{\mu_w}\right)^x$$

Reported values of the constant and exponents are

$A = 0.15-0.40$ $m = 0.30-0.45$
$n = 0.65-0.85$ $x = 0.05-0.20$

where

$$N_u = \frac{hd}{d} \quad Re = \frac{Vdp}{u} \quad Pr = \frac{Cp\mu}{k}$$

d is the equivalent diameter defined in the case of the plate heat exchanger as $2 \times$ the mean gap.

Table 26.1 Gasket materials, operating temperatures and applications

Gasket material	Approx. maximum operating temp.		Application
	°C	°F	
Paracril (medium nitrile)	135	275	Resistant to fatty materials
E.P.D.M.	150	300	High temperature resistance for a wide range of chemicals
Paratherm (resin cured butyl)	150	300	Aldehydes, ketones and some esters
Paradur (fluorocarbon rubber base)	177	350	Mineral oils, fuels, vegetable and animal oils

Typical velocities in plate heat exchangers for water-like fluids in turbulent flow are 0.3–0.9 meters per second (m/s) but true velocities in certain regions will be higher by a factor of up to 4 due to the effect of the corrugations. All heat transfer and pressure drop relationships are, however, based on either a velocity calculated from the average plate gap or on the flow rate per passage.

Figure 26.4 illustrates the effect of velocity for water at 16°C on heat transfer coefficients. This graph also plots pressure drop against velocity under the same conditions. The film coefficients are very high and can be obtained for a moderate pressure drop.

One particularly important feature of the plate heat exchanger is that the turbulence induced by the troughs reduces the Reynolds number at which the flow becomes laminar. If the characteristic length dimension in the Reynolds number is taken as twice the average gap between plates, the Re number at which the flow becomes laminar varies from about 100 to 400, according to the type of plate.

To achieve these high coefficients it is necessary to expend energy. With the plate unit, the friction factors

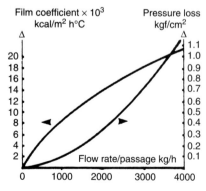

Figure 26.4 Performance details: plate exchanger

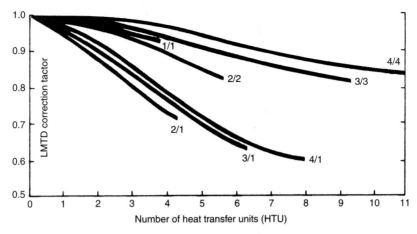

Figure 26.5 LMTD correction factor

normally encountered are in the range of 10 to 400 times those inside a tube for the same Reynolds number. However, nominal velocities are low and plate lengths do not exceed 2.3 m, so that the term $V^2L/2_g$ in the pressure drop equation is very much smaller than one would normally encounter in tubulars. In addition, single-pass operation will achieve many duties, so that the pressure drop is efficiently used and not wasted on losses due to flow direction changes.

The friction factor is correlated with:

$$\Delta p = \frac{f \times LpV^2}{2g \times d} \qquad f = \frac{B}{Re^y}$$

where y varies from 0.1 to 0.4 according to the plate and B is a constant characteristic of the plate. If the overall heat transfer equation $H = US\Delta T$ is used to calculate the heat duty it is necessary to know the overall coefficient U (sometimes known as the K factor), the surface area S and the mean temperature difference ΔT.

The overall coefficient U can be calculated from

$$\frac{1}{U} = r_{fh} + r_{fc} + r_w + r_{dh} + r_{dc}$$

The values of r_{fh} and r_{fc} (the film resistances for the hot and cold fluids, respectively) can be calculated from the Dittus–Boelter equations previously described and the wall metal resistance r_w from the average metal thickness and thermal conductivity. The fouling resistances of the hot and cold fluids r_{dh} and r_{dc} are often based on experience, but a more detailed discussion of this will be presented later in this chapter.

The value taken for S is the developed area after pressing. That is, the total area available for heat transfer and due to the corrugations will be greater than the projected area of the plate, i.e. 0.17 m^2 versus 0.14 m^2 for an APV HX plate.

The value of T is calculated from the logarithmic mean temperature difference multiplied by a correction factor. With single-pass operation, this factor is about 1 except for plate packs of less than 20, when the end effect has a

significant bearing on the calculation. This is due to the fact that the passage at either end of the plate pack only transfers heat from one side and therefore the heat load is reduced.

When the plate unit is arranged for multiple-pass use a further correction factor must be applied. Even when two passes are counter-current to two other passes, at least one passage must experience co-current flow. This correction factor is shown in Figure 26.5 against a number of heat transfer units (HTU = temperature rise of the process fluid divided by the mean temperature difference). As indicated, whenever unequal passes are used, the correction factor calls for a considerable increase in area. This is particularly important when unequal flow conditions are handled. If high and low flow rates are to be used, the necessary velocities must be maintained with the low fluid flow rate by using an increased number of passes. Although the plate unit is most efficient when the flow ratio between two fluids is in the range of 0.7–1.4, other ratios can be handled with unequal passes. This is done, however, at the expense of the LMTD factor.

26.2 Comparing plate and tubular exchangers

26.2.1 Ten points of comparison

In forming a comparison between plate and tubular heat exchangers there are a number of guidelines which will generally assist in the selection of the optimum exchanger for any application. In summary, these are:

1. For liquid/liquid duties, the plate heat exchanger will usually give a higher overall heat transfer coefficient and in many cases the required pressure loss will be no higher.
2. The effective mean temperature difference will usually be higher with the plate heat exchanger.
3. Although the tube is the best shape of flow conduit for withstanding pressure it is entirely the wrong shape

for optimum heat transfer performance since it has the smallest surface area per unit of cross-sectional flow area.

4. Because of the restrictions in the flow area of the ports on plate units it is usually difficult (unless a moderate pressure loss is available) to produce economic designs when it is necessary to handle large quantities of low-density fluids such as vapours and gases.

5. A plate heat exchanger will usually occupy far less floor space than a tubular for the same duty.

6. From a mechanical viewpoint, the plate passage is not the optimum and gasketed plate units are not made to withstand operating pressures much in excess of $20 \, \text{kgf/cm}^2$.

7. For most materials of construction, sheet metal for plates is less expensive per unit area than tube of the same thickness.

8. When materials other than mild steel are required, the plate will usually be more economical than the tube for the application.

9. When mild steel construction is acceptable and when a closer temperature approach is not required, the tubular heat exchanger will often be the most economic solution since the plate heat exchanger is rarely made in mild steel.

10. Plate heat exchangers are limited by the necessity that the gasket be elastomeric.

26.2.2 Heat transfer coefficients

Higher overall heat transfer coefficients are obtained with the plate heat exchanger compared with a tubular for a similar loss of pressure because the shell side of a tubular exchanger is basically a poor design from a thermal point of view. Considerable pressure drop is used without much benefit in heat transfer efficiency. This is due to the turbulence in the separated region at the rear of the tube. Additionally, large areas of tubes even in a well-designed tubular unit are partially bypassed by liquid and low heat transfer areas are thus created.

Bypassing in a plate-type exchanger is less of a problem and more use is made of the flow separation which occurs over the plate troughs since the reattachment point on the plate gives rise to an area of very high heat transfer.

For most duties, the fluids have to make fewer passes across the plates than would be required through tubes or in passes across the shell. Since a plate unit can carry out the duty with one pass for both fluids in many cases, the reduction in the number of required passes means less pressure lost due to entrance and exit losses and consequently more effective use of the pressure.

26.2.3 Mean temperature difference

A further advantage of the plate heat exchanger is that the effective mean temperature difference is usually higher than with the tubular unit. Since the tubular is always a mixture of cross and contra-flow in multi-pass arrangements, substantial correction factors have to be applied to the log mean temperature difference (LMTD). In the plate

heat exchanger where both fluids take the same number of passes through the unit, the LMTD correction factor is usually in excess of 0.95.

26.3 Duties other than turbulent liquid flow

26.3.1 Beyond liquid/liquid

The plate heat exchanger, for example, can be used in laminar flow duties, for the evaporation of fluids with relatively high viscosities, for cooling various gases, and for condensing applications where pressure-drop parameters are not excessively restrictive.

26.3.2 Condensing

For those condensing duties where permissible pressure loss is less than $0.07 \, \text{kpf/cm}^2$ there is no doubt but that the tubular unit is most efficient. Under such pressure-drop conditions only a portion of the length of a plate heat exchanger plate would be used and a substantial surface area would be wasted. However, when less restrictive pressure drops are available the plate heat exchanger becomes an excellent condenser, since very high heat transfer coefficients are obtained and the condensation can be carried out in a single pass across the plate.

26.3.3 Pressure drop of condensing vapor

The pressure drop of condensing steam in the passages of plate heat exchangers has been investigated experimentally for a series of different plate configurations. As indicated in Figure 26.6, which provides data for a typical unit, the drop obtained is plotted against steam flow rate per passage for a number of inlet steam pressures.

It is interesting to note that for a set steam flow rate and a given duty the steam pressure drop is higher when the liquid and steam are in countercurrent rather than

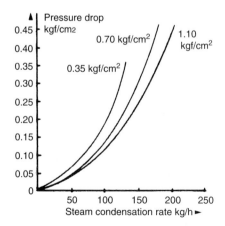

Figure 26.6 Steam-side pressure drop

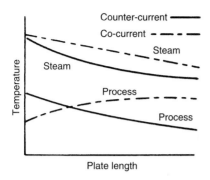

Figure 26.7 Temperature profile during condensation of steam

co-current flow. This is due to differences in temperature profile.

From Figure 26.7 it can be seen that for equal duties and flows the temperature difference for countercurrent flow is lower at the steam inlet than at the outlet, with most of the steam condensation taking place in the lower half of the plate. The reverse holds true for co-current flow. In this case, most of the steam condenses in the top half of the plate, the mean vapor velocity is lower and a reduction in pressure drop of between 10–40 per cent occurs. This difference in pressure drop becomes lower for duties where the final approach temperature between the steam and process fluid becomes larger.

The pressure drop of condensing steam is therefore a function of steam flow rate, pressure and temperature difference. Since the steam pressure drop affects the saturation temperature of the steam, the mean temperature difference, in turn, becomes a function of steam pressure drop. This is particularly important when vacuum steam is being used, since small changes in steam pressure can give significant alterations in the temperature at which the steam condenses.

26.3.4 Gas cooling

Plate heat exchangers also are used for gas cooling. The problems are similar to those of steam heating since the gas velocity changes along the length of the plate due either to condensation or to pressure fluctuations. Designs usually are restricted by pressure drop, therefore machines with low-pressure drop plates are recommended. A typical allowable pressure loss would be $0.035\,kgf/cm^2$ with low gas velocities giving overall heat transfer coefficients in the region of $244\,kcal/m^2h°C$.

26.3.5 Evaporating

The plate heat exchanger can also be used for evaporation of highly viscous fluids when the evaporation occurs in plate or the liquid flashes after leaving the plate. Applications generally have been restricted to the soap and food industries. The advantage of these units is their ability to concentrate viscous fluids of up to 50 poise.

26.3.6 Laminar flow

Most plate heat exchanger designs fall into the viscous flow range. Considering only Newtonian fluids since most chemical duties fall into this category, in laminar ducted flow the flow can be said to be one of three types:

1. Fully developed velocity and temperature profiles (i.e. the limiting Nusselt case);
2. Fully developed velocity profile with developing temperature profile (i.e. the thermal entrance region); or
3. The simultaneous development of the velocity and temperature profiles.

The first type is of interest only when considering fluids of low Prandtl number, and this does not usually exist with normal plate heat exchanger applications. The third is relevant only for fluids such as gases which have a Prandtl number of about one. Therefore, let us consider type two.

As a rough guide for plate heat exchangers, the rate of the hydrodynamic entrance length to the corresponding thermal entrance length is given by

$$\frac{l_{TH}}{l_{HYD}} = 1.7Pr$$

Correlations for heat transfer and pressure drop in laminar flow are:

$$Nu = e\left(\frac{Re \times Pr \times d}{L}\right)^{1/3}\left(\frac{\mu}{\mu_w}\right)^n$$

where

Nu = Nusselt number (Ld/k),
Re = Reynolds number $(vd f/\mu)$,
Pr = Prandtl number $(cp\mu/k)$,
L = nominal plate length,
D = equivalent diameter ($2 \times$ average gap),
μ/μ_w = Sieder Tate correction factor,
c = construct for each plate (usually in the range 1.86–4.50),
n = index varying from 0.1–0.2, depending on plate type.

Pressure drop

For pressure loss in a plate, the friction factor can be taken as

$$f = \frac{a}{Re}$$

where a is a characteristic of the plate.

It can be seen that for heat transfer, the plate heat exchanger is ideal because the value of d is small and the film coefficients are proportional to $d^{-2/3}$.
Unfortunately, the pressure loss is proportional to d^{-4}, and pressure drop is sacrificed to achieve the heat transfer.

From these correlations it is possible to calculate the film heat transfer coefficient and the pressure loss for laminar flow. This coefficient, combined with that of the metal and the calculated coefficient for the service fluid together with the fouling resistance, is then used to produce the overall coefficient. As with turbulent flow, an allowance has to be made to the LMTD to allow for either end-effect correction for small plate packs and/or concurrency caused by having concurrent flow in some passes. This is particularly important for laminar flow since these exchangers usually have more than one pass.

26.4 The problem of fouling

26.4.1 The fouling factor

In view of its complexity, variability and the need to carry out experimental work on a long-term basis under actual operating conditions, fouling remains a neglected issue among the technical aspects of heat transfer. However, the importance of carefully predicting fouling resistance in both tubular and plate heat exchanger calculations cannot be overstressed. This is well illustrated by the following examples.

Note that for a typical water/water duty in a plate heat exchanger it would be necessary to double the size of the unit if a fouling factor of 0.0001 was used on each side of the plate (i.e. a total fouling of 0.0002). Although fouling is of great importance, there are relatively few accurate data available, and the rather conservative figures quoted in Kern (*Process Heat Transfer*) are used all too frequently. It also may be said that many of the high fouling resistances quoted have been obtained from poorly operated plants. If a clean exchanger, for example, is started and run at the designed inlet water temperature, it will exceed its duty. To overcome this, plant personnel tend to turn down the cooling water flow rate and thereby reduce turbulence in the exchanger. This encourages fouling, and even though the water flow rate is eventually turned up to design, the damage will have been done. It is probable that if the design flow rate had been maintained from the onset, the ultimate fouling resistance would have been lower. A similar effect can happen if the cooling-water inlet temperature falls below the design figure and the flow rate is turned down again.

26.4.2 Six types of fouling

Types of fouling can be divided into six distinct categories. The first is crystallization – the most common type of fouling, which occurs in many process streams, particularly cooling-tower water. Frequently linked with crystallization is sedimentation, which is usually caused by deposits of particulate matter such as clay, sand or rust. A build-up of organic products and polymers is often a result of chemical reaction and polymerization. The surface temperature and presence of reactants, particularly oxygen, can have a very significant effect. Coking occurs on high-temperature surfaces and is due to hydrocarbon deposits. Organic material growth is usually linked with crystallization and sedimentation and is common to sea water systems. Corrosion of the heat transfer surface itself produces an added thermal resistance as well as a surface roughness.

In the design of the plate heat exchanger, fouling due to coking is of no significance, since the unit cannot be used at such high temperatures. Corrosion is also irrelevant, since the metals used in these units are non-corrosive. The other four types of fouling, however, are most important. With certain fluids such as cooling-tower water, fouling can result from a combination of crystallization, sedimentation and organic material growth.

26.4.3 A function of time

From Figure 26.8 it is apparent that the fouling process is time dependent, with zero fouling initially. Fouling then builds up quite rapidly and in most cases, levels off at a certain time to an asymptotic value as represented by curve A in the figure. At this point the rate of deposition is equal to that of removal. Not all fouling levels off, however, and curve B shows that at a certain time the exchanger would have to be taken off-line for cleaning.

In the case of crystallization and suspended solid fouling, the process is usually of type A. However, when the fouling is of the crystallization type with a pure compound crystallizing out, the fouling approaches type B and the equipment must be cleaned at frequent intervals.

Biological growth can present a potentially hazardous fouling, since it can provide a stickier surface with which to bond other fouling sources. In many cases, however, treatment of the fluid can reduce the amount of biological growth. The use of germicides or poisons to kill bacteria can help.

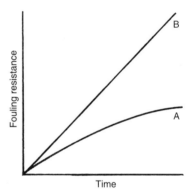

Figure 26.8 Build-up of fouling resistance

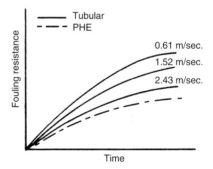

Figure 26.9 Effect of velocity and turbulence

26.4.4 Lower resistance

It generally is considered that resistance due to fouling is lower with plate heat exchangers than with tubular units. This is the result of four advantages of plate-type exchangers:

1. There is a high degree of turbulence, which increases the rate of foulant removal and results in a lower asymptotic value of fouling resistance.
2. The velocity profile across a plate is good. There are no zones of low velocity compared with certain areas on the shell side of tubular exchangers.
3. Corrosion is maintained at an absolute minimum.

4. A smooth heat transfer surface can be obtained. If necessary, the plate can be electro-polished.

The most important of these is turbulence. HTRI (Heat Transfer Research Incorporated) has shown that for tubular heat exchangers, fouling is a function of low velocities and friction factor. Although flow velocities are low with the plate heat exchanger, friction factors are very high, and this results in lower fouling resistance. The effect of velocity and turbulence is plotted in Figure 26.9. The lower fouling characteristic of the plate heat exchanger compared to the tubular has been conclusively proved.

Tests have been carried out which tend to confirm that fouling varies for different plates, with the more turbulent type of plate providing the lower fouling resistances.

27

Heating

G E Pritchard
Chartered Engineer

Contents

27.1 Introduction

The past fifty years have seen significant advances in every aspect of space heating, resulting from increased demands for the provision within buildings of more closely controlled environments and services of progressively increasing complexity. Although it remains impossible in a chapter of this size to cover adequately the extent of the subjects associated with these services, hopefully it will provide a useful reference.

27.2 Statutory regulations

Except for some defined types of accommodation, the use of fuel or electricity to heat premises above a temperature of 19°C is prohibited by the Fuel and Electricity (Heating) (Control) Order 1980. The current Order is an amendment to an earlier Regulation, which limited the temperature to a maximum of 20°C, and although 19°C is generally taken to refer to air temperature the Order does not specify this. The minimum temperature was laid down in the Factories Act 1961 and should be reached one hour after the commencement of occupation.

27.3 Building regulations

Unfortunately, the optimum results in cutting down space heating energy usage can often be obtained only when a building is at the design stage. Insulation, draft exclusion and the best possible heating system can then be built in at minimum cost. It is usually more expensive to add to (or modify) an existing building. Space heating is probably the largest usage of energy in buildings, so this section considers what can be done to improve insulation and other thermal properties. When energy was relatively cheap, little thought was given to conservation, and these omissions now have to be rectified.

In 1957, the Thermal Insulation (Industrial Buildings) Act laid down standards of insulation for roofs of new buildings. This first attempt to minimize heat losses did not cover walls, floors or windows. However, in 1978, Amendments to the Building Regulations rectified this by specifying standards for walls and windows. At this point, it is necessary to define the terms 'U-value', or the insulation characteristic of the building material. This measures the rate at which energy flows through the material when there is a temperature difference of 1°C between the inside and outside faces, and this value is measured in watts (the unit of energy) per square meter of surface area, i.e. $W/m^2°C$ or $W/m^2 K$.

Symbol 'K' = °C temperature difference.

The amendments can briefly be summarized in Table 27.1.

The U values for walls, roofs and floor are intended as average figures, so it is permissible to have some areas of the structure under insulated (i.e. with higher U values) providing other areas have sufficient extra insulation to bring the average of all areas down to (or below) the Regulation values.

Limits are also imposed on window areas and apply to all buildings above 30-m^2 floor area. For the first group,

Table 27.1

Industrial and commercial buildings
External walls of building enclosing heated spaces, internal walls exposed to unheated ventilated spaces, floors where the under-surface is exposed to outside air or an unheated ventilated space, and roofs over heated spaces (including the cases of ceilings with an unheated ventilated space above them).

Maximum average U value
For factories and storage buildings, such as warehouses, the U value is laid down to be 0.7. For shops, offices, institutional buildings and places of assembly, such as meeting halls, theatres, etc., the maximum average U value is to be 0.6

industrial and commercial buildings, these limits apply both to roof lights and to windows in the walls. These percentages for windows or roof lights assume single glazing, and somewhat larger values can be used if double or triple glazing is to be fitted. However, calculations must be produced to show that the total heat loss from such units would be no greater than the single-glazed unit complying with the set limits (Table 27.2). In most single- and two-story buildings, the largest proportion of heat loss from the building structure is usually through the roof. (In buildings of three stories or more, the losses through walls and windows may overtake the roof loss.) If we first consider typical roofs, the methods of insulation break down into four groups:

1. Under-drawing, involving the fitting of insulation below the existing roof, as rigid self-supporting slabs, semi-rigid sheets supported by framing or a combination of an insulating blanket on top of sheets. This insulating blanket could be of mineral or glass fiber or a flexible *foamed* plastic.
2. External, where insulation is added on top of the existing roof. This can be done with sheets or slabs of insulating material finished with some waterproofing layer or very conveniently for corrugated or shaped roofing sheets by using a spray system to apply both the insulation and final waterproofing layer.

Table 27.2

Type of building	Maximum permitted glazed area	
	In walls as percentage of wall area	As rooflights as percentage of roof area
Factories and storage	15	20
Offices, shops and places of assembly	35	20
Institutional, including residential	25	20

Note: Where figures for both rooflights and windows in walls are given, these really apply as a combined total. If the full wall window allowance is not used the balance can be reallocated to roof light areas and vice versa. For example, a factory with only 10% of wall area as windows could add the other 5% of wall area as an increase to the permitted 20% of roof area that could be roof lights.

3. Where there is a ceiling below the actual roof, to place insulation above this ceiling. Where there are wooden ceiling joists, the insulation can be a flexible blanket laid between joists or loose granular fill spread between them. Alternatively, an overall quilt or blanket can be laid over the ceiling, taking care to check that the ceiling and its supports can safely carry the extra weight.

4. Possibly in a small minority of cases, it could be advantageous to install a false ceiling, which could be of insulating sheets or panels. This can reduce the volume to be heated, particularly with steeply pitched roofs. Obviously, this idea could not be recommended where valuable daylight was available from roof lights and where this daylight significantly reduced the artificial illumination necessary. Often such a false ceiling unfortunately conflicts with the existing architecture.

5. New Building Regulations for the Conservation of Fuel and Power for England and Wales came into operation on 1 April 1990. The new maximum U values of the elements (W/m^2 K) are shown in Table 27.4.

Table 27.3 gives some of the insulation properties for various building materials. The property given is for the rate at which energy would pass through a unit area of the material. In the standard units it becomes the number of watts that would be transferred through a square meter of the material of normal thickness in the form it would be used, if the air at either side of the material shows a temperature difference of 1°C. In SI units this becomes W/m^2°C, which, in this case, is commonly known as the U value. The larger the U value, the more energy it will transfer, so the worse are its insulation properties.

Table 27.3 U values

Roofs	
Pitch covered with slates or tiles, roofing felt underlay, foil-backed plasterboard ceiling	1.5
Pitched covered with slates or tiles and roofing felt underlay, foil-backed plasterboard ceiling with 100 mm glass-fiber insulation between joists	0.35
Corrugated steel or asbestos cement roofing sheets	6.1–6.7
Corrugated steel or asbestos cement cladding with 75 mm fibreglass lightweight liner	0.38
Corrugated steel or asbestos cement roofing sheets with cavity and aluminium foil-backed 10 mm plasterboard lining	1.9–2.0
Corrugated double-skin asbestos cement sheeting with 25 mm glass-fiber insulation between with cavity and aluminium foil-backed 10 mm plasterboard lining; ventilated air space	0.8
Steel or asbestos cement roofing sheets, no lining with rigid insulating lining board 75 mm	0.4
Asphalt 19 mm thick or felt/bitumen layer on solid concrete 150 mm thick	3.5
Asphalt 19 mm thick or felt/bitumen layer on 150 mm autoclaved aerated concrete roof slabs	0.9
Flat roof, three layers of felt on chipboard or plasterboard	1.54
Flat roof, three layers of felt on rigid insulating board 100 mm thick	0.29
Timber roof with zinc or lead covering and 25 mm plaster ceiling	0.96

Table 27.3 (continued)

Walls		
Steel or asbestos cement cladding		5.3–5.7
Steel or asbestos cement cladding 75 mm fiber glass lightweight liner		0.37
Steel or asbestos cement cladding with plaster-board lining and 100 mm fiber insulating roll		0.4
Solid brick wall unplastered 105 mm		3.3
Solid brick wall unplastered 335 mm		1.7
Solid brick wall 220 mm thick with 16 mm lightweight plaster on inside face		1.9
Brick/cavity/brick (260 mm total thickness)		1.4
260 mm brick/mineral fiber-filled cavity/brick		0.5
260 mm brick/cavity/load-density block		1.0–1.1
Brick/expanded polystyrene board in cavity/low-density block/inside face plastered		0.5
Weather boarding on timber framing with 10 mm plasterboard lining, 50 mm glass-fiber insulation in the cavity and building paper behind the boarding		0.62
Glazing		
Single glazing	Wood frame	4.3
	Metal frame	5.6
Double glazing	Wood frame	2.5
	Metal frame	3.2
Triple glazing		2.0
Roof skylights		6.6
Floors		
20 mm intermediate wood floor on 100 mm × 50 mm joists 10 mm plasterboard ceiling allowed for 10% bridging by joists		1.5
150 mm concrete intermediate floor with 150 mm screed and 20 mm wood flooring		1.8

The heat loss through floors in contact with the earth is dependent upon the size of the floor and the amount of edge insulation. Insulating the edge of a floor to a depth of 1 m can reduce the U value by 35%. Following are some typical U values for ground floors. Effectively, most of the heat loss is around the perimeter of the floor.

Solid floor in contact with the earth with four exposed edges:

150 m × 50 m	0.11
60 m × 60 m	0.15
15 m × 60 m	0.32
15 m × 15 m	0.45
7.5 m × 15 m	0.62
3 m × 3 m	1.47

Suspended timber floors directly above ground. Bare or with linoleum, plastic or rubber tiles:

150 m × 60 m	0.14
60 m × 60 m	0.16
15 m × 60 m	0.37
15 m × 15 m	0.45
7.5 m × 15 m	0.61
3 m × 3 m	0.05

Suspended timber floors directly above ground with carpet or cork tiles:

150 m × 60 m	0.14
60 m × 60 m	0.16
15 m × 60 m	0.34
15 m × 15 m	0.44
7.5 m × 15 m	0.59
3 m × 3 m	0.99

Table 27.4

Building type	Ground floors	Exposed walls and floors	Semi-exposed walls and floors	Roofs
Industrial storage and other buildings, excluding dwellings	0.45	0.45	0.60	0.45

Note: An exposed element is exposed to the outside air; a semi-exposed element separates a heated space from a space having one or more elements which are not insulated to the levels in the table.
Maximum window areas for single glazing in buildings other than dwellings will be unchanged.

A very poor insulating material can be detected very simply. If the material is at normal room conditions and a hand placed upon its surface feels cold, heat is being conducted away from it as the U value is very high. A low U value is shown by no cooling affect. To try this, if one places a hand on a window and a wooden table there should be a notable difference between the two showing a difference in the U value. Wood's U value is about $1 \, W/m^2 °C$ and glass has a value of over $5.5 W/m^2 °C$. The U values are given in $W/m^2 °C$ for various building materials under normal weather conditions. There will always be slight variations around these values, dependent on particular manufacturers of the materials. With any insulation that is being fitted, advice should be sought regarding the fire risk and condensation problems.

27.4 Estimation of heat losses from buildings

The normal procedure in estimating the heat loss from any building is as follows:

1. Decide upon the internal air temperature to be maintained at the given external air temperature.
2. Decide the heat transmission coefficient (U values) for the outside walls and glass, roof and bottom floor, and the inside walls, ceilings, or of heated spaces adjacent to non-heated spaces.
3. Measure up the area of each type of surface and compute the loss through each surface by multiplying the transmission coefficient by the measured area by the difference between the inside and the outside temperatures.
4. Calculate the cubic contents of each room and, using the appropriate air change rate, the amount of heat required to warm the air to the desired temperature by multiplying the volume of air by the difference between the inside and outside temperatures and the specific heat of air.

The above calculations will give the heat losses after the building has been heated. Under conditions in which the heating system will operate continuously, satisfactory results will be obtained if the heating system is designed to provide heat equivalent to the amount calculated above. Suitable allowance must be made for losses from mains.

When, however, operation is intermittent, safety margins are necessary. These are, of course, speculative, but the following suggestion has frequently proved satisfactory. When it is necessary to operate after a long period of vacancy, as may happen in certain types of substantially built buildings, it is necessary to add up to 30 per cent to the 'steady state' heat transmissions. In buildings of light construction, this margin may be reduced. In selecting the appropriate U values we must pay due regard to the exposure and aspect of the room. It appears reasonable to make allowance for the height of a room, bearing in mind that warm air raises towards the ceiling. Thus in a room designed to keep a comfortable temperature in the lower 1.5 or 2 meters, a higher temperature must exist nearer the ceiling, which will inevitably cause greater losses through the upper parts of windows, walls and roof. This effect is greatest with a convective system, i.e. one that relies on the warming of the air in the room for the conveyance of heat. This would occur in the case of conventional radiators, convectors and warm air systems. In the case of radiant heated rooms, this does not occur, and a much more uniform temperature exists from floor to ceiling.

27.5 Allowance for height of space

In heat loss calculations, a uniform temperature throughout the height of the heated space is assumed, although certain modes of heating cause vertical temperature gradients which lead to increased heat losses, particularly through the roof. These gradients need to be taken into account when sizing appliances. Additions to the calculated heat loss to allow for this are proposed in Table 27.5. However, these percentages should not be added to replacement heat to balance that in air mechanically exhausted from process plant. Attention is also drawn to the means of reducing the effect of temperature stratification, discussed in Section 27.10.13.

Table 27.5 Allowances for height of heated space

Method of heating and type or disposition of heaters	Percentage addition for following heights of heated space (m)		
	5	5–10	>10
Mainly radiant			
Warm floor	Nil	Nil	Nil
Warm ceiling	Nil	0–5	a
Medium- and high-temperature downward radiation from high level	Nil	Nil	0–5
Mainly convective			
Natural warm air convection	Nil	0–5	a
Forced warm air			
Cross flow at low level	0–5	5–15	15–30
Downward from high level	0–5	5–10	10–20
Medium and high-temperature cross radiation from intermediate level	Nil	0–5	5–10

[a] Not appropriate to this application.

27.6 Characteristics of heat emitters

Designers will need to decide whether it is necessary to add a margin to the output of heat emitters. During the warm-up cycle with intermittently operated heating systems, emitter output will be higher than design because space temperatures are lower. Also, boost system temperatures may be used to provide an emission margin during warm-up. The need for heat emitter margins to meet extreme weather conditions will depend on the design parameters used in determining heat losses.

Although the addition of a modest margin to heat emitter output would add little to the overall system cost and a margin on the heat generator or boiler output can only be utilized if the appropriate emitter capacity is available, the decision should be based on careful discrimination rather than using an arbitrary percentage allowance. In general, for buildings of traditional construction and for the incidence of design weather in normal winters in the UK an emitter margin in excess of, say, 5 per cent or 10 per cent is unlikely to be justified. However, for well-insulated buildings the heat loss reduces in significance relative to the heat stored in or needed to warm up the structure. For such applications, a larger heating system margin is required, and the emitter margin provided would need to be considered accordingly.

27.7 Central plant size

In estimating the required duty of a central plant for a building, it should be remembered that the total net infiltration of outdoor air is about half the sum of the rates for the separate rooms. This is because, at any one time, infiltration of outdoor air takes place only on the windward part of the building, the flow in the remainder being *outwards*. When intermittent heating is to be practiced the pre-heating periods for all rooms in a building will generally be coincident. The central plant rating is then the sum of the individual room heat demands, modified to take account of the *net* infiltration. If heating is to be continuous, some diversity between the several room heating loads can be expected. The values listed in Table 27.6 are suggested. When mechanical ventilation is combined with heating, the heating and the ventilation plant may have different hours of use, and the peak loads on the two sections of the plant will often occur at different times.

Table 27.6 Diversity factors for central plant (continuous heating

Space of building served by plant	Diversity factor
Single	1.0
Single building or zone, central control	0.9
Single building, individual room control	0.8
Group of buildings, similar type and use	0.8
Group of buildings, dissimilar uses[a]	0.7

[a]This applies to group and district heating schemes where there is substantial storage of heat in the distribution mains, whether heating is continuous or intermittent.

The central plant may also be required to provide a domestic hot water supply and/or heat for process purposes. These loads may have to be added to the net heating load to arrive at the necessary plant duty, but careful design may avoid the occurrence of simultaneous peaks. In large installations, the construction of boiler curves may indicate whether savings in boiler rating can be made. In many cases, little or no extra capacity may be needed for the hot water supply, its demands being met by 'robbing' the heating circuits for short periods.

27.8 Selective systems

In some cases the various rooms of a building do not all require heating at the same time of day and here a so-called 'selective system' may be used. The supply of heat is restricted to different parts of the building at different times of the day; the whole building cannot be heated at one time. A typical application is in dwellings where the demands for heat in living spaces and bedrooms do not normally coincide.

In a selective system the individual room appliances must be sized as indicated above, to provide the appropriate output according to heat loss, gains and intermittency. The central plant need only be capable of meeting the greatest simultaneous demands of those room units which are in use at the same time. This will generally lead to a large power being available to meet the demands of those units which form the lesser part of the load. These units may then be operated with a high degree of intermittency.

27.9 Multiple-boiler installations

Load variation throughout the season is clearly large, and consideration should be given to the number of boilers required in the system. Operation at low loads leads to corrosion and loss inefficiency and should be avoided. On the other hand, a number of smaller boilers give an increase in capital costs. It has been shown that when boilers are chosen which have a fairly constant and good efficiency over a working range of 30–100 per cent, then the effects on overall costs (running + capital) of varying the number and relative sizes of boilers in the system is less than 5 per cent. The optimum number depends on the frequency of occurrence of low loads.

Under these circumstances, the engineer is free to choose the number of boilers in the system based on practical rather than economic considerations. The following procedure is recommended:

1. Choose a type of boiler with a fairly constant and high efficiency over its full turndown range. Obtain its efficiency curve.
2. On the basis of avoiding acid corrosion and obtaining required standby choose the number of boilers required and their relative sizes. Equally sized boilers should be used except where the provision of domestic hot water in summer requires one smaller boiler. Table 27.7 gives suggested relative sizes based on turndown to 30 per cent.

Table 27.7 Appropriate boiler size ratios assuming turndown to 30% of full load

Number of boilers	Installed plant ratio[a]	Heating only			Heating plus domestic hot water			Remarks
		Installed boiler size ratios	Lowest load (%)	Load if largest fails (%)	Installed boiler size ratios	Lowest load (%)	Load if largest fails (%)	
1	1.0	1.0	30	0	–	–	–	Use only if load is seldom <30%. No standby
2	1.0	0.5/0.5	15	50	–	–	–	Impossible to gain reasonable standby and meet low loads
	1.0	0.33/0.67	10	33	0.30/0.70	9	30	
	1.25	0.3/0.7	11.3	37	0.25/0.75	9.4	31	
	1.5	0.25/0.75	11.3	37	–	–	–	
3	1.0	0.33/0.33/0.33	10	67	0.2/0.4/0.4	6	60	
	1.25	0.2/0.4/0.4	7.5	75	0.2/0.4/0.4	7.5	75	
	1.50	0.2/0.4/0.4	9	90	0.2/0.4/0.4	9	90	
	1.67	0.2/0.4/0.4	10	100	0.2/0.4/0.4	10	100	
4	1.0	0.25/0.25/0.25/0.25	7.5	75	0.25/0.25/0.25/0.25	7.5	75	
	1.25	0.25/0.25/0.25/0.25	9.4	94	0.1/0.3/0.3/0.3	3.7	87	
	1.50	0.1/0.3/0.3/0.3	4.3	105	0.1/0.3/0.3/0.3	4.5	105	
	1.33	0.25/0.25/0.25/0.25	10	100	0.1/0.3/0.3/0.3	4	93	
5	1.0	0.2/0.2/0.2/0.2/0.2	6	80	0.2/0.2/0.2/0.2/0.2	6	80	
	1.25	0.2/0.2/0.2/0.2/0.2	7.5	100	0.2/0.2/0.2/0.2/0.2	7.5	100	
	1.50	0.2/0.2/0.2/0.2/0.2	9.0	120	0.2/0.2/0.2/0.2/0.2	9.0	120	

$$^a\text{Installed plant ratio} = \frac{\text{Total installed boiler capacity}}{\text{Design maximum heat requirement}}$$

3. The boilers should be controlled in sequence, the switching points for bringing boilers on line occurring whenever an additional boiler makes the system more efficient (for an evaluation of this see Figure 27.1). Full boiler load is not usually the most economic switching point, but switching points too close to full turndown

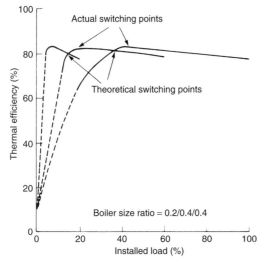

Figure 27.1 Evaluation of optimum boiler control

should also be avoided. At any given system load the boilers on-line should share the load between them in proportion.

27.10 Heating systems

27.10.1 Warm and hot water-heating systems

Warm water or low-, medium- or high-temperature hot water systems are categorized in Table 27.8. Warm water systems may use heat pumps, fully condensing boilers or similar generators, or reclaimed heat. In many cases the system design may incorporate an alternative heat generator for standby purposes or for extreme whether operation. Under such circumstances the system may continue to function at warm water temperatures or could operate at more conventional LTHW ones.

LTHW systems are usually under a pressure of static head only, with an open expansion tank, in which case the design operating temperature should not exceed 83°C. Where MTHW systems operating above 110°C are pressurized by means of a head tank, an expansion vessel should be incorporated into the feed and expansion pipe. This vessel should be adequately sized to take the volume of expansion of the whole system so that boiling will not occur in the upper part of the feed pipe. On no account should an open vent be provided for this type of system. MTHW and HTHW systems require pressurization such that the saturation temperature at operating pressure at all points in the circuit exceeds the maximum system flow temperature required. A margin of 17 K (minimum)

Table 27.8 Design water temperature for warm and hot water heating systems

Category	System design water temperatures (°C)
Warm	40–70
LTHW	70–100
MTHW	100–120
HTHW	Over 120

Note: Account must be taken of the margin necessary between the maximum system operating temperature and saturation temperature at the system operating pressure.

is recommended and is based on the use of conventional automatic boiler plant and includes an allowance for tolerances on temperature set points for the automatic control of heat-generation output. A check must be made on actual tolerance used in the design of a control system to ensure that this allowance is adequate. When selecting the operating pressure, allowance must be made for the effect of static head reduction at the highest point of the system and velocity head reduction at the circulating pump section, to ensure that all parts of the system are above saturation pressure within an adequate anti-flash temperature margin. Additionally, the margin on the set point of the high-temperature cutout control should be 6 K, except for boilers fired with solid fuel automatic stokers, where it should be at least 10 K. Medium- and high-temperature systems should be fully pressurized before the operating temperature is achieved and remains fully pressurized until the temperature has dropped to a safe level. In all systems, the heat generator or boiler must be mechanically suitable to withstand the temperature differentials, and the return temperature to the boiler must be kept high enough to minimize corrosion. Automatic controls may be used to achieve this.

27.10.2 Design water flow temperature

For low-temperature heating systems using natural convective or radiant appliances the normal design water flow temperature to the system is 83°C (see also Table 27.8). Boost temperatures may be used on modulated-temperature systems because of the changes in heat output characteristics with varying temperatures. Additionally, comfort aspects must be borne in mind, as forced convective emitters operating on modulated temperature systems can deliver air streams at unacceptably low temperatures.

For MTHW and HTHW systems, heat emitters may be as for LTHW systems, except that, for safety reasons, units with accessible surfaces at water temperature would not normally be employed. Embedded panel coils may be used in conjunction with a MTHW or HTHW distribution system, with insulating sleeves around the coil piping to reduce the heat flow. Alternatively, the coils can be operated as reduced temperature secondary systems by allowing only a small, carefully controlled proportion of flow temperature water to be mixed with the water circulating in the coils.

Design arrangements for reduced-temperature secondary systems (sometimes referred to as injection circuits) include fixed provisions for minimum dilution rates. Conventional system-balancing devices with three-port automatic modulating valves to regulate mixed water temperatures and, hence, heat output are used. Automatic safety controls must prevent excessive temperatures occurring in the coil circuits, as floor fabrics or finishes could be damaged very rapidly.

27.10.3 Maximum water velocity

Noise generation and erosion/corrosion considerations limit the maximum water velocity in pipework systems. Noise is caused by the free air present in the water, sudden pressure drops (which, in turn, cause cavitation or the flashing of water into steam), turbulence or a combination of these.

Noise will therefore be generated at valves and fittings where turbulence and local velocities are high, rather than in straight pipe lengths. A particular noise problem can arise where branch circuits are close to a pump and where the regulating valve used for flow-rate balancing may give rise to considerable pressure differences.

Oversizing regulating valves should be avoided, as this will result in poor regulation characteristics; the valve operating in an almost shut position and creating a very high local velocity.

High water velocities can result in erosion or corrosion due to the abrasive action of particles in the water and the breakdown of the protective film which normally forms on the inside surface of the pipe. Erosion can also result from the formation of flash steam and from cavitation caused by turbulence. Publishing data on limiting water velocities are in conclusive. Table 27.9 summarizes the available information.

27.10.4 Minimum water velocity

Minimum water velocities should be maintained in the upper floors of high-rise buildings where air may tend to come out of solution because of reduced pressures. High velocities should be used in down-return mains feeding into air-separation units located at a low level in the system. Table 27.10 can be taken as a guide. Water velocities shown in Tables 27.9 and 27.10 are indicative parameters only; on the one hand, to limit noise problems and

Table 27.9 Limiting water velocities in pipework

Pipe diameter (mm)	Steel pipework		Copper pipework (m/s)
	Non-corrosive water (m/s)	Corrosive water (m/s)	
50 and below	1.5	1	1
Above 50	3	1.5	1.5
Large distribution mains with long lengths of straight pipe	4	2	–

Table 27.10 Minimum water velocities

Pipe diameter (mm)	Minimum water velocity (m/s)
50 and below	0.75
Above 50	1.25

erosion and, on the other, to try to ensure air entrainment. Within these parameters, the design engineer will need to discriminate on the selection of water velocities in a distribution system based on other considerations. It is particularly necessary to bear in mind the effect of low water velocities on flow-measuring components used in balancing flow rates in systems.

27.10.5 System temperature drop

British practice on LTHW systems uses a typical system temperature drop of 11 K and a maximum system temperature of 17 K. European practice has tended to use higher drops (up to 40 K). An advantage of a higher system temperature drop is the reduction in water flow rates. This will result in reduced pipe sizes with savings in capital cost and distribution heat losses and a reduced pump duty, with savings in running costs. A disadvantage of higher system temperature drops is the need for larger and consequently more expensive heat emitters. However, if it is possible to raise the system flow temperature so that the mean water temperature remains the same, then with certain types of emitter only a small increase in size is required. With large system temperature drops the average water temperature in a radiator tends to fall below the mean of flow and return temperature and, thus, a larger surface is needed. Furthermore, on one-pipe circuits the progressive reduction in temperature around the circuit may lead to excessively large heat emitters. Higher system temperature drops can be used with MHTW and HTHW systems since the mean temperature of the heat emitters will be correspondingly higher. Additionally, these media are well suited to use for primary distribution systems, conveying heat over long distances. Precautions should be taken to prevent the danger of injury from contact with hot surfaces. The safe temperature for prolonged contact is relatively low and reference should be made to BS 4086 and other sources.

27.10.6 Use of temperature-limiting valves on emitters

On some group and district heating schemes, outlet-limiting valves which permit flow only when the water temperature has dropped to a specified low level are used. This procedure minimizes the water quantity to be pumped and permits indicative heat metering by water quantity alone. In such cases, care must be taken to size emitters to suit the available water temperatures. The effect of low water velocities through the emitter must also be taken into consideration, since the heat output of some convective appliances is greatly reduced under such conditions.

27.10.7 Miscellaneous components

Data regarding relief valves, feed and expansion cisterns, etc. are given in Tables 27.11 and 27.12. Cistern sizes shown in Table 27.12 are based on typical system designs and are approximate only. An estimate of the water content of the particular system should always be made where there is any doubt regarding these typical data, to ensure that the cistern capacity is adequate to contain the expansion volume.

27.10.8 Distribution system design

The design of pipework distribution systems must allow for the following:

1. Future extensions, where required, by the provision of valved, plugged or capped tee connections.
2. Provision for isolation for maintenance. Where it is necessary to carry out maintenance on a 'live' system, valves must be lockable and may need to be installed in tandem.
3. Thermal expansion.
4. Provision for distribution flow rate balancing for initial commissioning or re-balancing to meet changed operational requirements. Typical provisions for balancing comprise the following:
   (a) A measuring station – which may be an orifice plate, a venturi, an orifice valve or other proprietary device – provided with a pair of tapings to permit the measurement of upstream and downstream system dynamic pressures.
   (b) An associated regulating valve – preferably a double-regulating valve or other arrangement, which permits the required setting to remain undisturbed by closure.
5. Provision for drainage, including drainage after pre-commission flushing; water circulation during flushing must be in excess of design flow rates and, in order to discharge the flushing effluent effectively, drainage connections must be full diameter.
6. Removal of air from the system by provision of:
   (a) Air separators, one form of which uses the principle of centrifugal force to separate the heavier constituent (water) from the lighter

Table 27.11 Hot water heating boilers – recommended sizes of relief valves

Rated output of boiler (kW)	Minimum clear bore of relief valve (mm)	Equivalent area (mm^2)
Up to 250	20	310
250–350	25	490
350–450	32	800
450–500	40	1250
500–750	50	1960
750–1000	65	3320

Note: The above sizes apply to boilers fired with solid fuel. For oil- and gas-fired boilers the relief valve should be one size larger.

Table 27.12 Approximate sizes of feed and expansion cisterns for low-pressure hot water heating systems

Boiler or water-heating (kW)	Cistern size (1)	Ball-valve size (mm)	Cold-feed size (mm)	Open-vent size (mm)	Overflow size (mm)
15	18	15	20	25	25
22	18	15	20	25	32
30	36	15	20	25	32
45	36	15	20	25	32
60	55	15	20	25	32
75	68	15	25	32	32
150	114	15	25	32	32
225	159	20	32	40	40
300	191	20	32	40	40
400	227	20	40	50	40
450	264	20	40	50	50
600	318	25	40	50	50
750	455	25	50	65	50
900	636	25	50	65	50
1200	636	25	50	65	80
1500	910	25	50	65	80

Notes: (1) Cistern sizes are actual.
(2) Cistern sizes are based on radiator-heating systems and are approximate only.
(3) The ball-valve sizes apply to installations where an adequate mains water pressure is available at the ball-valve.

one (non-condensable gases). Best results are achieved by locating the separator at the highest temperature point of the system where air has a greater tendency to come out of solution. The velocity of the medium requires being above the minimum stated by the manufacturer (usually about 0.25 m/s).

(b) Automatic air vents for systems operating at temperatures below atmospheric boiling point.

(c) Air bottles with manually operated needle valves to release accumulated air, for systems operating at temperatures in excess of atmospheric boiling point.

7. Provision of test points for sensing temperature and pressure at selected locations.

27.10.9 Sealed heating systems

Pressurization of medium- and high-temperature hot water sealed heating systems referred to above may take the following forms.

Pressurization by expansion of water

The simplest form of pressurization uses the expansion of the water content of the system to create a sufficient pressure in an expansion vessel to provide an anti-flash margin of, say, 17°C at the lowest pressure (highest point) of the system. The main disadvantage of a naturally pressurized expansion vessel is the ability of water to absorb air and the consequent risk of oxygen corrosion.

A diaphragm expansion vessel is divided into two compartments by a special membrane or diaphragm of rubber or rubber composition, which prevents the water being exposed to the air. On one side of the diaphragm, the vessel is filled with air or nitrogen at the required pressure.

The other section of the vessel is connected directly to the water system. A correctly positioned air separator will assist in de-aerating the water in the system.

Pressurization of elevated header tanks

Given very careful attention to design, installation and commissioning, MTHW systems may be operated with the necessary system pressure provided by an elevated feed and expansion tank. Where the system operating temperature exceeds 110°C an expansion vessel should be sized to absorb the volume of expansion for the complete system, thus preventing water at operating temperatures entering the feed and expansion tank and causing boiling. On no account should an open vent be provided for this type of system.

Gas pressurization with spill tank

This form consists of a pressure cylinder maintained partly filled with water and partly with gas (usually nitrogen), which is topped up from pressure bottles. Water expansion is usually arranged to discharge from the system through a pressure-control valve into a spill tank open to atmosphere or to a closed cylinder lightly pressurized with nitrogen. A pump is provided to take water from the spill tank and return it under pressure to the system as cooling-down results in a pressure drop. A system pressure sensor regulates the pump operation.

Hydraulic pressurization with spill tank

In this form, a continuously running centrifugal pump maintains the pressure. A second pump under the control of a pressure switch is provided to come into operation at a predetermined pressure differential and as an automatic standby to the duty pump. Surplus water is delivered to or taken from a spill tank or cylinder as described previously.

Example of pressure differential

Assume system flow temperature of 120°C	
Allow 17 K anti-flash margin – 137°C	
Corresponding absolute pressure	3.4 bar
Assume static absolute pressure on system	2.0 bar
Minimum absolute pressure at cylinder	5.4 bar
Allow operating differential on pressure cylinder, say -	0.5 bar
Minimum operating absolute pressure of system	5.9 bar

Example of water expansion

Assume water capacity of system	200 000 liters

Assume ambient temperature of 10°C
Assume system maximum flow temperature of 120°C
Assume system minimum return temperature of 65°C
Increase in volume from 10°C to 65°C

$$200,000 \frac{(999.7 - 980.5)}{980.5} = 3916 \text{ liters}$$

Increase in volume from 65°C to 120°C

$$200,000 \frac{(980.5 - 943.1)}{943.1} = 7931 \text{ liters}$$

Total increase in volume = 11847 liters

27.10.10 Maintenance of water heating systems

A common practice in many hot water heating installations is to drain the complete system during summer months. This practice, involving a complete change of raw water every year, is to be deprecated. It introduces additional hardness salts and oxygen to the system, resulting in very significant increases in scaling and corrosion. Where it is necessary to drain the boiler or heat generator or other parts of the system for inspection or maintenance purposes, isolating valves or other arrangements should be used to ensure that the section drained is kept to a minimum.

27.10.11 Steam heating systems

These are designed to use the latent heat of steam at the heat emitter. Control of heat output is generally by variation of the steam saturation pressure within the emitter. For heating applications with emitters in occupied areas, low absolute pressures may be necessary in order to reduce the saturation temperature to safe levels.

The presence of non-condensable gases in steam systems (e.g. air and CO_2) will reduce the partial pressure of the steam, and hence its temperature, thus affecting the output of the appliance. A further adverse effect is the presence of a non-condensable gas at the inside surface of a heat emitter. These impede condensation and, hence, heat output. It is therefore imperative that suitable means are provided to prevent formation of CO_2 and to evacuate all gases from the system.

Superheat, which must be dissipated before condensation occurs, can be used to reduce condensation in the distribution mains.

On−off control of steam systems can result in the formation of a partial vacuum, leading to condensate locking or back feeding, and infiltration of air, which subsequently reduces the heat transfer.

When using modulating valves for steam, heat emitter output must be based on the steam pressure downstream of the valve, which often has a high-pressure drop across it, even when fully open.

Steam traps must be sized to cope with the maximum rate of condensation (which may be on start-up) but must perform effectively over the whole operational range, minimizing the escape of live steam.

Partial waterlogging of heater batteries can lead to early failure due to differential thermal expansion. Steam trap selection should take account of this.

Where high temperatures are required (e.g. for process work) and lower temperatures for space heating, it is desirable to use flash steam recovery from the high-temperature condensate to feed into the low-temperature system, augmented as required by reduced pressure live steam.

Steam as a medium for heating is now seldom used. Hot water, with its flexibility to meet variable weather conditions and its simplicity, has supplanted it in new commercial buildings. Steam is, however, often used for the heating of industrial buildings where steam-raising plant occurs for process or other purposes. It is also employed as a primary conveyor of heat to calorifiers such as in hospitals, where again steam boiler plant may be required for sundry duties such as in kitchens, laundry and for sterilizing. Heating is then by hot water served from calorifiers.

27.10.12 High-temperature thermal fluid systems

Where high operating temperatures are required, high-temperature thermal fluid systems may be used instead of pressurized water or steam systems. These systems operate at atmospheric pressure using non-toxic media such as petroleum oil for temperatures up to 300°C or synthetic chemical mixtures where temperatures in excess of this are required (up to 400°C). Some advantages and disadvantages of thermal fluid or heat transfer oil systems are listed below.

Advantages

No corrosion problems.
Statutory inspections of boilers/pressure vessels not required.
No scale deposits.
No need for frost protection of system.
Cost of heat exchangers/heat emitters less, as only atmospheric pressures are involved.
Better energy efficiency than steam systems.
Operating temperature can be increased subsequent to design without increasing operating pressure.

Disadvantages

Medium more expensive than water (but no treatment costs).
Medium is flammable under certain conditions.

Heat transfer coefficient is inferior to that of water.

Care necessary in commissioning and in heat-up rates due to viscosity changes in medium.

Circulating pump necessary (not required for steam systems).

Air must be excluded from the system.

In the event of leakage the medium presents more problems than water.

27.10.13 Warm air-heating systems

These may be provided with electric or indirect oil- or gas-fired heaters or with a hot water heater or steam battery supplied from a central source. Because the radiant heat output of warm-air systems is negligible, the space air temperature will generally need to be higher for equivalent comfort standards than for a system with some radiant output. This will increase energy use, and legislative standards for limiting space temperatures should be considered. Attention is drawn to the vertical temperature gradient with convective systems and, when used for cellular accommodation, the likelihood of some spaces being overheated due to the difficulty of controlling such systems on a room-by-room basis.

27.10.14 Reducing the effect of temperature stratification

As with all convective systems, warm air heating installations produce large temperature gradients in the spaces they serve. This results in the inefficient use of heat and high heat losses from roofs and upper wall areas. To improve the energy efficiency of warm air systems, pendant-type punkah fans or similar devices may be installed at roof level in the heated space. During the operational hours of the heating system, these fans work either continuously or under the control of a roof-level thermostat and return the stratified warm air down to occupied levels.

The energy effectiveness of these fans should be assessed, taking into account the cost of the electricity used to operate them. The following factors should also be borne in mind:

1. The necessary mounting height of fans to minimize drafts;
2. The effect of the spacing of fans and the distance of the impeller from the roof soffit;
3. Any risk to occupants from stroboscopic effects of blade movements;
4. The availability of multi- or variable-speed units.

Punkah fans may also be operated during summer months to provide air movement and offer a measure of convective cooling for occupants.

27.10.15 High-temperature high-velocity warm air heating systems

These systems, best suited to heating large, single spaces, may use indirect heating by gas or oil or direct gas heating. Relatively small volumes of air are distributed at high temperature (up to 235°C) and high velocity (30–42.5 meters per second from heater unit) through a system of well-insulated conventional ductwork. Air outlets are in the form of truncated conical nozzles discharging from the primary ductwork system into purpose-designed diffuser ducts. The high-velocity discharge induces large volumes of secondary air to boost the outlet volume and reduces the outlet temperature delivered to the space, thereby reducing stratification. Allowing free movement absorbs most of the ductwork thermal expansion and long, drop-rod hangers are used for this purpose. Light, flexible, axial-bellows with very low thrust loads can also be employed where free expansion movement is not possible. Specialist manufacturers generally handle system design and installation as a package deal.

27.11 Heating equipment — attributes and applications

27.11.1 Water system-heating equipment

The range of heat emitters may be divided into three generic groups:

1. Radiant
2. Natural convective
3. Forced convective

Table 27.13 lists the principal types of appliance in each group, together with descriptive notes. Typical emission ranges are quoted for each type over its normal span of working temperatures. These are intended as a guide only and manufacturers' catalogues should be consulted for detailed performance values.

27.11.2 Electric heating equipment

Where electric heating equipment is installed within the space to be heated the total electrical input is converted into useful heat. There are two categories of electric heating equipment, direct acting and storage heating. The two types of electric heating can be used independently or to complement one another to meet particular heating requirements. Table 27.14 gives a brief description of the different types of electric heating.

27.11.3 Gas- and oil-fired heating equipment

Where gas or oil appliances are used for heating and installed within the heated space, between 70 per cent and 90 per cent of the total energy content of the fuel input will be converted into useful heat. Table 27.15 gives particulars of some gas-fired equipment types and Table 27.16 gives similar details for some oil-fired heaters. The first three types of equipment detailed in Table 27.15 and the first two in Table 27.16 are usually used for local warming of individuals rather than to provide a particular temperature throughout the space.

Table 27.13 Characteristics of water system heating equipment

Type	Description	Advantages	Disadvantages	Emission range
Radiant				
Radiant panel	Consists of steel tube or cast-iron waterways attached to a radiating surface. Back may be insulated to reduce rear admission or may be left open to give added convective emission. Particularly useful for spot heating and for areas having high ventilation rates (e.g. loading bays), the radiant component giving a degree of comfort in relatively low ambient air temperatures.	No moving parts, hence little maintenance required; may be mounted at considerable height or, in low-temperature applications, set flush into building structure.	Slow response to control; must be mounted high enough to avoid local high intensities of radiation (e.g. onto head).	$350\,W/m^2$ to $15\,kW/m^2$ of which up to 60% may be radiant.
Radiant strip	Consists of one or more pipes attached to an emissive radiant surface. Is normally assembled in long runs to maintain high water flow rates. The back may be insulated to reduce rear emission. When using steam, adequate trapping is essential together with good grading to ensure that tubes are not flooded. Multiple-tube types should be fed in parallel to avoid problems due to differential expansion. Hanger lengths should be sufficient to allow for expansion without lifting the ends of the strip. Heating media may be steam, hot water or hot oil.	No moving parts, hence little maintenance required; may be mounted at considerable height or, in low-temperature applications, set flush into building structure.	Slow response to control; must be mounted high enough to avoid local high enough to avoid local high intensities of radiation (e.g. onto head).	$150\,W/m$ to $5\,W/m$ of which radiant emission may be up to 65% of total.
Natural convective				
Radiators	Despite their name, 70% of the emission from these devices is convective. Three basic types are available; column, panel, and high output, the last incorporating convective attachments to increase emission. Panel radiators offer the least projection from the wall but emission is higher from column and high-output units. In application they should be set below windows to offset the major source of heat loss and minimize cold down-drafts.	Cheap to install; little maintenance required.	Fairly slow response to control. With steel panel radiators there is a risk of corrosive attack in areas having aggressive water, which may be accentuated by copper swarf left in the radiator. This leads to rapid failure unless a suitable inhibitor is used. Not suitable for high-temperature water or steam.	$450–750\,W/m^2$.
Natural convectors	Compact units with high emissions. Often fitted with damper to reduce output when full emission not required, usually to about 30% of full output. Heat exchangers normally finned tube. Units may be built into wall of building.	May be used on high-temperature hot water or low-pressure steam without casing temperature becoming dangerously high: fairly rapid response to control.	Take up more floor space than radiators. Likelihood of fairly high-temperature gradients when using high-temperature heating media.	$200\,W$ to $20\,kW$.
Continuous convectors	Comprise single-or double-finned tube high-output emitters in factory-made sill-height sheet metal casings or builders' work enclosures which may be designed to fit wall to wall. Can be fitted with local output damper control, which reduces the emission to	Take up relatively little space; give even distribution of heat in room. May be used with medium-temperature hot water or low-pressure steam without casing temperatures	May produce large temperatures gradients on high-temperature heating media if poorly sited.	$500\,W/m$ to $4\,kW/m$.

(*continued overleaf*)

Table 27.13 (*continued*)

Type	Description	Advantages	Disadvantages	Emission range
	approximately 30% of the full output. They should be placed at the point of maximum heat loss, usually under windows. The wall behind the unit should be well insulated. For long-run applications distribution of flow water must be provided to modular sections of the unit to ensure that input remains reasonably consistent over full length of unit. For builders' work casings inlet and outlet apertures must have the free-area requirement stipulated by the manufacturer.	becoming dangerously high. Return pipework may be concealed within casing.		
Skirting heating	Finned-tube emitters in a single or double skirting height sheet metal casing, usually with provision for a return pipe within the casing. Applications and distribution of flow water similar to continuous convectors.	May be used on water or low-pressure steam. Give low-temperature gradients in the room. All pipework concealed.	Relatively low output per metre of wall. More work involved when installing in existing building as existing skirting has to be removed.	300 W/m to 1.3 kW/m.
Forced convective Far convectors	These units give a high heat output for volume of space occupied by the unit, together with the ability to distribute the heat over a considerable area using directional grilles. May be used to bring in heated fresh air for room ventilation. Leaving air temperatures should be above 35°C to avoid cold drafts. Where mixed systems of radiators and fan convectors are installed it is advisable to supply fan-assisted units on a separate constant-temperature circuit to avoid the above problems. To minimize stratification, leaving air temperature above 50°C should be avoided. Must not be used on single-pipe systems. Care must be taken at design stage to avoid unacceptable noise levels. Control by speed variation or on/off regulation of fan.	Rapid response to control by individual thermostat. By use of variable speed motors rapid warm-up available in intermittent systems; filtered fresh air inlet facility.	Electric supply required to each individual unit.	2 to 25 kW.
Unit heaters	A unit with a large propeller or centrifugal fan to give high air volume and wide throws. Louvers direct the air flow in the direction required. May be ceiling mounted, discharging vertically or horizontally or floor mounted. Can be used with fresh air	Rapid response to control by individual thermostat; by use of multi-speed motors rapid warm-up available on intermittent systems; filtered fresh air inlet facility.	Electric supply required for each individual unit.	3 to 300 kW.

Table 27.13 (*continued*)

Type	Description	Advantages	Disadvantages	Emission range
	supply to ventilate buildings. Large units may be mounted at a considerable height above the floor to clear traveling cranes, etc. May be used with steam or hot water but care should be taken to restrict leaving air temperature, usually 40–55°C to avoid reduction of downward throw and large temperature gradients in the building. The air flow from the units should be directed towards the points of maximum heat loss. Control as for fan convectors.			

Table 27.14 Electric heating equipment

Type	Description	Advantages	Disadvantages	Emission range
Radiant High-intensity radiant heaters	Consist of high-temperature elements mounted in front of polished reflector. Element can be silica or metal sheathed wire (up to 900°C) or quartz lamp (up to 2200°C)	Fast response time. Little regular maintenance required. May be mounted at considerable height. Quartz lamp heaters have improved beam accuracy allowing higher mounting heights. Especially suitable for spaces with high air movement.	Must be mounted sufficiently high to avoid local high intensities of radiation (e.g. onto head).	0.5 to 6 kW per heater.
Low-temperature radiant panels	Consist of low-temperature elements (300°C and below) mounted behind a radiating surface. Thermal insulation behind the elements minimizes heat loss. Very low-temperature elements (40°C) used in ceiling heating applications.	No regular maintenance required. Set flush into building structure and unobtrusive.	Slow response time.	Up to 200 W/m^2.
Natural convective Storage heater	Consists of a thermal storage medium which is heated during off-peak electricity periods. A casing and insulation around the medium enables the heat to be gradually released throughout the day. Manual or automatic damper control allows 20% of heat output to be controlled.	No regular maintenance required.	Not intermittent. Limited control of heat charging and output.	Storage element sizes 1.4 to 3.4 kW.
Panel heaters, convectors or skirting heaters	Consist of a heating element within a steel casing with air grilles allowing the natural convection of air across the element. Generally controlled by an integral room thermostat.	No regular maintenance required. Cheap to install. Suitable for low heat loss applications. Fast response time for intermittent operation.	High surface temperatures. High-temperature gradients if poorly sited.	0.5 to 3 kW output.

(*continued overleaf*)

Table 27.14 *(continued)*

Type	Description	Advantages	Disadvantages	Emission range
Forced convective				
Storage fan heaters/electricaire	These storage heaters have increased core thermal insulation and contain a fan which distributes warm air to the space to be heated. The fan is usually controlled by a room thermostat. Storage fan heaters are single room units but can heat an adjacent room with a stub duct. Electricaire units are larger and can be ducted to servce several areas. Up to 80% controllable heat.	Use off-peak electricity. Suitable for intermittent operation.	Heavyweight.	For storage heaters: 3 to 6 kW.\n\nElectricaire 6 to 15 kW.\n\nIndustrial models up to 100 kW.
Fan convectors	These wall-mounted or free-standing units incorporate a fan which forces air over the heating elements into the space. High output rate relative to size. Can incorporate integral room thermostat control	Low maintenance. Fast response. Accurate temperature control. Suitable for highly intermittent heating applications. Low surface temperatures. Fan-only operation for summer use.		2 to 3 kW.
Downflow fan convectors	These units are forced air convectors mounted to direct the heated air downwards. High air flow and heat output. Often used to provide a hot air curtain over entranced doorways.	Suitable for localized heating.	Can be noisy.	3 to 18 kW.

Table 27.15 Direct gas-fired heating equipment

Type of heater	Usual rating (kW)	Surface temperature (°C)	Flue system	Notes
Radiant convector gas	4.4–7.3	1100 at radiant tips	Conventional	Wall mounted or at low level
Overhead radiant heaters	3.1–41	850–900	Flueless[a]	High level or ceiling mounted
Overhead tubular radiant	10–15	315 mean	Conventional or fan-assisted flue or flue or flue-less[a]	High level or ceiling mounted
Convector heaters	2.5–16.7	–	Conventional or balanced flue or flueless[a]	Wall mounted or at low level
Fan convectors	1.4–3.7	–	Balanced flue or fan-assisted flue	Wall mounted or at low level
Make-up air heaters[b]	49–250	–	Flueless[a]	Mounted at high level and fan assisted
Unit air heaters[b]	17–350	–	Flued or flueless[a]	Mounted at high level and fan assisted

[a]The use of flueless appliances should be discouraged, since they discharge much moisture into the heated spaced.
[b]The installation of flueless appliances in excess of 44 kW is not permitted by the Building Regulation (1985) and application should be made to the local authority for the necessary waiver where installations of this size are contemplated.

Acknowledgement

The author wishes to thank the Chartered Institute of Building Services Engineers for permission to reproduce data from the Heating and Plant Capacity Sections of the *CIBSE Guide* (Figure. A9.3, Tables A9.10, A9.11, B1.3, B1.12, B1.13, B1.15, B1.16, B1.23, B1.24, B.125, B1.26) and Formecon Services Limited for their kind permission to reproduce data from *The Businessman's Energy Saver* data book (Tables 1–3).

Table 27.16 Direct oil-fired heating equipment

Type of heater	Usual rating (kW)	Surface temperature (°C)	Flue system	Notes
Radiant kerosene heaters	1.5	Less than 121[b]	Flueless[a]	Mounted at low of floor level
Flued radiant convector heaters	8.5–10.9	Less than 121[b]	Conventional	Mounted at low of floor level
Convective kerosene heaters	1–4	Less than 121[b]	Flueless[a]	Mounted at low of floor level
Air heaters	8.5–13.5	Less than 121[b]	Conventional	Mounted at low of floor level
Blown air heaters	10.7–16.7	Less than 121[b]	Conventional	Mounted at low of floor level
Warm air heaters	50–450	Maximum 60[c]	Flued	Floor mounted or overhead
Make-up air heaters	90–450	Maximum 60[c]	Flued	Can be flueless Floor or overhead mounted

[a]The use of flueless appliances should be discouraged since they discharge much moisture into the heated space.
[b]See BS 799.
[c]See BS 4256.

28

Ventilation

P D Compton

Colt International Limited

Contents

28.1 Introduction

Any building, even with all windows and doors shut, will have a degree of ventilation (referred to as natural infiltration) by virtue of pressure differences across cracks or permeable materials in the structure causing air to flow through the structure. The degree of infiltration is governed by the quality of design and builds of the structure and by the pressures generated by wind and thermal buoyancy.

This fortuitous infiltration is essential in otherwise unventilated buildings, since it allows ingress of air providing oxygen for us to breathe and for combustion equipment to burn. The infiltration is, however, effectively uncontrolled, and is often insufficient in quantity, at the wrong temperature or too contaminated to maintain satisfactory internal environmental conditions.

We must therefore be able to define the conditions we need and to design, install and operate systems to provide them. These systems cover three main areas: heating, ventilation and air conditioning. Heating and air conditioning are dealt with in Chapters 27 and 29. This chapter covers ventilation systems, defined as systems providing air movement through a space without artificially heating or cooling the air. It must be said, however, that, in practice, there is often a large degree of overlap, since office ventilation systems often provide heating in winter and complex ducted ventilation systems share much equipment and design procedures with air-conditioning systems.

This chapter is intended to provide guidance towards defining needs, assessing whether ventilation is the correct solution and selecting equipment and systems to match these requirements in as economic a manner as possible.

28.1.1 Reasons for ventilation

Ventilation is used to maintain a satisfactory environment within enclosed spaces. The environmental criteria controlled may be:

Temperature: Relief from overheating
Humidity: Prevention of condensation or fogging
Odor: Dilution of odor from smoking, body odor, processes, etc.
Contamination: Dilution or removal of dangerous or unpleasant fumes and dust

The required values for these criteria will depend upon the reason the space is being ventilated. It may be for the benefit of people, processes, equipment, materials, livestock, horticulture, building preservation or any combination of these. CIBSE[1] and ASHRAE[4]. provide guidance on selection of these values.

28.1.2 Definitions

Aerodynamic area: The effective theoretical open area of an opening. It is related to the measured area by the coefficient of entry or discharge (C_d).
Air-handling unit: A self-contained package incorporating all equipment needed to move and treat air, requiring only connection to ductwork and services to provide a complete ventilation system.

Coefficient (*entry or discharge*): The ratio of aerodynamic (effective) area to the measured area of an opening. The value for a square-edged hole of 0.61 is used for most building openings.
Capture velocity: The air velocity needed to capture a contaminant at source, overcoming any opposing air currents.
Automatic fire ventilation: See *Smoke ventilation*.
Dilution ventilation: A ventilation strategy whereby contaminants are allowed to escape into the ventilated space and are then diluted to an acceptable level by means of the ventilation system.
Industrial ventilation: A term used to cover any ventilation system designed to remove contaminants. Its use is sometimes restricted to local extract systems.
Infiltration: Movement of air through a space with no specific ventilation openings by natural forces.
Local extract: A ventilation strategy whereby heat, steam or contaminants are captured at source and ducted to discharge outside the space.
Maximum Exposure Limit (*MEL*): Maximum limits of concentration of airborne toxic contaminants, listed by the Health and Safety Executive[18] which must not be exceeded.
Mechanical ventilation: See *Powered ventilation*.
Natural ventilation: A ventilation system in which air movement is produced through purpose-designed openings by natural forces (wind and thermal buoyancy).
Occupational Exposure Standards (*OES*): Limits of concentration of airborne toxic contaminants, listed by the Health and Safety Executive[18] which are regarded as safe for prolonged exposure for 8 hours per day.
Powered ventilation: A ventilation system in which air movement is induced by mechanical means, almost invariably a fan.
Smoke logging: The filling of a space with smoke in the event of fire.
Smoke ventilation: A ventilation system designed to remove smoke and heat in the event of fire to prevent or delay smoke logging allowing personnel to escape and firefighters to attack the fire.
Spot cooling: A ventilation strategy whereby the space temperature is allowed to rise and air movement is induced locally to provide comfort conditions within a limited area.
Threshold Limit Value (*TLV*): Maximum values of concentrations of airborne toxic contaminants, listed by the American Conference of Governmental Industrial Hygienists[5] (ACGIH), regarded to be safe for 8 hours per day exposure.
Transport velocity: The air velocity required in a duct to transport a contaminant without it falling out of suspension.

28.2 Ventilation systems and controls

28.2.1 Natural ventilation

How natural ventilation works

Natural ventilation operates by means of airflows generated by pressure differences across the fabric of the building. Airflow will occur wherever there is a crack, hole or porous surface and a pressure difference.

For the relatively large openings in which we are interested, the flow rate can be found from the velocity of airflow generated through the aerodynamic area of the opening from the formula:

$$V = \sqrt{\frac{2\Delta P}{\rho}} \qquad (28.1)$$

Where:

V = velocity (m/s),
Δ = pressure difference (Pa),
ρ = density (kg/m^3).

Then flow rate:

$$\dot{V} = AC_d V \qquad (28.2)$$

Where:

$\dot{V}$ = volumetric flow rate (m^3/s),
A = measured area of opening (m^2),
C_d = coefficient of opening.

For purpose-built ventilators the manufacturer will be able to provide values of C_d. For other openings it is conventional to use the value for a sharp-edged square orifice of 0.61.

The pressure can be generated by three mechanisms:

1. Powered ventilation equipment (see equation (28.3));
2. Buoyancy (temperature difference);
3. Wind.

In still air conditions, the source of pressure difference to drive ventilation is buoyancy due to the decrease in density of heated air. In any occupied building, there will be a higher temperature inside than outside due to heat gains from people, plant and solar radiation. The lighter heated air will try to rise, causing an increase in internal pressure at high level and a reduction at low level with a neutral plane between the two conditions. Any opening above the neutral plane will therefore exhaust air and any opening below the neutral plane will provide inlet air. Under steady heat load conditions, a balance will be achieved with a throughput of air dependent upon the heat load and the size and location of the openings. Conditions at this balance point can be readily calculated using one of the following formula:
For more than one opening (inlets all at one height, exhausts all at one height)

$$\dot{V} = A_e C_e \sqrt{\frac{2gH\Delta t}{\overline{T}}} \qquad (28.3)$$

For a single opening

$$\dot{V} = \frac{AC_d}{3} \sqrt{\frac{gh\Delta t}{\overline{T}}} \qquad (28.4)$$

where

g = acceleration due to gravity (m/s^2),

H = height between center lines of inlet and outlet openings (m),
Δt = temperature difference between inside and outside (°C),
T = average of inside and outside temperatures (absolute) (K),
h = height of single opening (m),
$C_e A_e$ = overall effective opening size calculated from

$$\frac{1}{(C_e A_e)^2} = \frac{1}{(\Sigma C_i A_i)^2} + \frac{1}{\Sigma(C_v A_v)^2}$$

(subscript i denotes inlet opening, subscript v exhaust opening).

Under wind conditions, a complex system of pressures is set up on the external surfaces of the building, which will vary with wind speed and direction. Pressure coefficients Cp[6,7] define the relationship according to the formula:

$$\dot{V} = A_e C_e U_r \sqrt{(\Delta C_p)} \qquad (28.5)$$

where

U_r = reference wind speed,
ΔCp = difference between coefficients at ventilation openings.

The coefficients Cp will vary across each surface of the building and, except for very simple shapes, can only be found by model or full-scale test. Since the coefficients will change with wind direction, complete calculation of wind-induced ventilation is very unwieldy, needing computer analysis.

When both wind and temperature difference act on ventilation openings, the result is very complex, but a reasonable approximation of flow rate is made by taking the higher of the two individual flow rates. This means that we can, for ventilation design purposes, generally ignore wind effects and design on temperature difference only, since wind effects can be assumed only to increase the ventilation rate.

Advantages and disadvantages

Advantages	Disadvantages
Quiet	Variable flow rate and
Virtually no running cost	direction dependent upon
Self-regulation (flow rate	wind conditions
increases with heat load)	Filtration is generally
Low maintenance cost	impractical
Provides daylight when	Limited ducting can be
open (roof vent)	tolerated
Psychological appeal of	Effectiveness depends on
clear sky (roof vent)	height and temperature
Easy installation	difference

When to use natural ventilation

Natural ventilation is used in a number of situations:

1. Shallow-plan offices – by opening windows to remove heat and odor;

2. Large single-story spaces (factories, warehouses, sports halls, etc.) – by roof and wall ventilators – to remove heat, contaminants, smoke, steam;
3. Plant rooms.

It is not suitable in situations where:

1. Dust, toxic or noxious contaminants must be removed at source;
2. Unfavorable external conditions exist requiring treatment to incoming air – e.g. noise, dust, pollution;
3. A steady controlled flow rate is required – e.g. hospitals, commercial kitchens;
4. Existing mechanical ventilation will affect the flow adversely;
5. Abnormal wind effects can be anticipated due to surrounding higher buildings;
6. The space is enclosed to have no suitable source of inlet air.

In many of these situations, a system of natural inlet/powered exhaust or powered inlet/natural exhaust will be the best option.

Control

Low-level ventilation openings, whether windows, doors or ventilators, are generally manually operated for simplicity and economy, allowing personnel to control their own environment. High-level openings can also be manually controlled by means of rod or cable operation, although this has generally lost favor (except in the case of simple windows) and automatic operation is preferred.

Automatic operation may be by means of compressed air, operating a pneumatic cylinder, or electricity. Pneumatics is generally favored for industrial applications and electricity for commercial premises. Economy of installation is normally the deciding factor, since running costs are low for either system.

Automatic control allows a number of options to be considered to provide the best form of control for the circumstances. Generally, available controls offer the following features:

1. Local control by personnel;
2. Automatic thermostatic control (single or multiple stage);
3. Fire override to open ventilators automatically by means of a connection to the fire-detection system or fireman's switch. This normally overrides all other control settings;
4. Timeswitch control to shut ventilators during unoccupied periods;
5. Weather override to close ventilators during rain or snow;
6. Wind override to shut high-level exhaust ventilators on windward walls (mainly used for smoke ventilation).

28.2.2 Powered (mechanical) ventilation

How powered ventilation works[9]

By definition, a powered ventilation system includes a mechanical means of inducing airflow using an external power source. This is invariably an electrically driven fan. When a fan blade rotates it does work on the air around it, creating both a static pressure increase (P_s) and an airflow across the fan. The airflow has a velocity pressure associated with it, defined as $P_v = 1/2\rho V^2$ and the fan can be described as producing a total pressure $P_T = P_s + P_V$. The pressure generated is used to overcome pressure losses (resistances) within the ventilation system.

Each fan has a unique set of characteristics which are normally defined by means of a fan curve produced by the manufacturer which specifies the relationship between airflow, pressure generation, power input, efficiency and noise level (see Figure 28.1). For geometrically similar fans, the performance can be predicted for other sizes, speeds, gas densities, etc. from one fan curve using the *fan laws* set out below.

For a given size of fan and fluid density:

1. $\dfrac{\dot{V}_1}{\dot{V}_2} = \dfrac{N_1}{N_2}$ Volume flow is directly proportional to fan speed

2. $\dfrac{P_1}{P_2} = \left(\dfrac{N_1}{N_2}\right)^2$ Total pressure and static pressure are directly proportional to the square of the fan speed

3. $\dfrac{W_1}{W_2} = \left(\dfrac{N_1}{N_2}\right)^2$ Air power and impeller powers are directly proportional to the cube of the fan speed

For changes in density:

4. $\dfrac{P_1}{P_2} = \dfrac{W_1}{W_2}\dfrac{\rho_1}{\rho_2}$ Pressure and power are directly proportional to density and therefore for a given gas are inversely proportional to absolute temperature

For geometrically similar fans operating at constant speed and efficiency with constant fluid density:

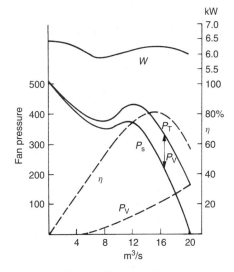

Figure 28.1 Typical fan curve for an axial fan

5. $\dfrac{\dot{V}_1}{\dot{V}_2} = \left(\dfrac{D_1}{D_2}\right)^3$ Volume flow is directly proportional to the cube of fan size

6. $\dfrac{P_1}{P_2} = \left(\dfrac{D_1}{D_2}\right)^2$ Total pressure and static pressure are directly proportional to the square of fan size

7. $\dfrac{W_1}{W} = \left(\dfrac{D_1}{D_2}\right)^5$ Air power and impeller powers are directly proportional to the fifth power of fan size

where

$\dot{V}$ = volumetric flow (m³/s),
P = pressure (kN/m²),
W = power (W),
D = size parameter (diameter) (mm).

In passing through the fans gases are compressed slightly due to the increase in pressure. For absolute accuracy note should be taken of this effect using the gas compressibility factor which will affect flow rate, static and total pressure and power. However, in most fan systems the effect is very small, since the pressure increase through the fan is insignificant compared to atmospheric pressure. By convention, compressibility effects are therefore normally ignored.

Since the pressure generated by most fans is far in excess of pressure differences due to buoyancy and wind, the performance of a powered ventilation system is effectively independent of these, and flow rates and directions can be confidently predicted and will be constant regardless of conditions. The high-pressure generation also allows resistive components such as heater batteries, filters and attenuators to be used within the system.

Advantages and disadvantages

Advantages	Disadvantages
Weatherproof	Fixed air flow – not self-regulating
Predictable constant performance	Running costs (electrical and maintenance)
Air treatment can be incorporated	Noise
Fresh air can be delivered at optimum volume, velocity and temperature	

When to use powered ventilation

Powered ventilation is essential in some instances:

1. Local extract;
2. When pre-treatment of incoming air is required;
3. When a steady controlled airflow is required;
4. When there are no suitable external walls or roof for natural ventilation;
5. In deep-plan offices or large industrial spaces to provide positive air movement in central zones.

It can also be used in any situation where natural ventilation is suitable, generally becoming more economic

as the roof height lowers, subject to noise levels being acceptable.

Control

A starter or contactor with manual push-button or thermostatic operation to start and stop the fan normally controls simple systems. More complex systems that incorporate components that need control or monitoring are normally operated from purpose-built central control panels. The most common functions provided are fan motor stop, start and speed control, damper control, filter-condition indication and heater battery control. For optimum control, the system should be automatically controlled from thermostats or other sensors and a timeswitch.

28.3 Powered ventilation equipment

This falls into two basic groupings: supply air systems and extract systems. The equipment used for both is similar, comprising, as a minimum, a fan and weatherproof cowl, plus ducting, air-treatment equipment and grilles as required.

28.3.1 Fans

Five main types of fan are used in ventilation systems as described below.

Centrifugal

In this type, air enters the impeller axially and is discharged radially into a volute casing. The airflow therefore changes direction through 90°, which can make this type of fan difficult to use within a ducted system. Two blade types are used; backward curved providing high-pressure at low volume flow and forward curved providing medium-pressure and volume flow. Typical static efficiencies are 70–75 per cent and 80–85 per cent, respectively.

Axial

Air enters and leaves the fan axially giving a straight-through configuration. Duties are usually high- to medium-volume flow rates at medium to low pressures. In its simplest form there is an impeller and its drive motor only mounted within the cylindrical casing, and the discharge flow usually contains a fairly pronounced element of rotational swirl which may, if not corrected, materially increase the resistance of the downstream part of the system. More sophisticated versions include either downstream or upstream guide vanes to correct the swirl. Typical static efficiencies are 60–65 per cent or 70–75 per cent with guide vanes.

Propeller

This is really a simple form of axial fan but with its impeller mounted in a ring or diaphragm which permits it to discharge air with both axial and radial components. Duties covered are high volume and low pressure. Static efficiency is normally under 40 per cent.

Mixed flow

A fan in which the air path through the impeller is intermediate between the axial and centrifugal types giving the benefit of increased pressures but capable of being constructed to provide either axial or radial discharge. Static efficiency is typically 70–75 per cent.

Cross flow

This type normally has a long cylindrical impeller having a relatively large number of shallow forward-curved blades. Due to the shape of the casing surrounding this impeller, air enters all along one side of the cylindrical surface of the impeller and leaves on another side. Static efficiency is typically 40–50 per cent.

In general, axial fans are used for roof extract units, small-ducted systems, and centrifugal fans for large-ducted systems.

28.3.2 Roof extract units

These are the most commonly used powered ventilators in large open buildings such as factories, warehouses and sports halls. Mounted directly onto the roof or wall, they comprise an axial fan, a safety grille and a weatherproof casing. Two forms are normally available, the vertical-discharge type which tends to have a complex casing arrangement but which throws the exhaust clear of the building, and the low-discharge type, which has a simple casing but directs the exhaust onto the roof of the building. Vertical discharge is essential when smoke or fumes are being exhausted (Figures 28.2 and 28.3).

These ventilators can normally be used with limited ducting or accessories. A variation, fitted with a centrifugal fan, is available for more extensive ducting and is often used for duties such as toilet extract in commercial buildings.

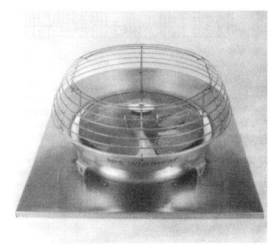

Figure 28.2 Vertical discharge powered roof ventilator

Figure 28.3 Low discharge powered roof ventilator

28.3.3 Roof inlet units

This specialized form of supply air system is often used in large open industrial spaces. It comprises a modular system of components that can be built up into simple systems. A typical system might have a roof inlet cowl, a recirculation damper, a heater battery, a fan, one or two outlet grilles and short sections of connecting ductwork, and would handle airflows up to 3–4 m³/s, depending on size. A number of individual systems would be used to provide the total airflow required in the space (Figure 28.4). Systems are normally manufactured with aluminum casings to reduce the roof load.

28.3.4 Ducted systems[10,11]

Larger ducted ventilation systems, as used in offices and commercial premises using a central air-handling unit and fabricated distribution ductwork, are akin to air-conditioning systems but with less treatment to the air at the AHU (see Chapter 29).

28.3.5 Local extract systems[12]

Local extract systems are designed specifically to remove fumes, dust, mists, heat, etc. at source from machinery and fume cupboards. The main design considerations are capture of the contaminant, which will normally involve special hoods or cabins, and extract at sufficient velocity

Figure 28.4 Roof inlet system

to satisfactorily transport the contaminant. Ductwork must be manufactured to resist abrasion or corrosion and sufficiently well sealed to prevent leakage. Welded ductwork is often needed. The fan may also need protection and the motor may need to be flameproof or out of air stream if the contaminant is flammable or corrosive. Treatment of the exhaust may be required to reduce pollution and nuisance and to comply with legislation.

28.3.6 Air cleaners

A wide range of types of air cleaners is available to match the number of contaminants needing removal from air. Figure 28.5 shows typical particle size ranges and the range of operation of each type of air cleaner.

Filters

Filters are a type of air cleaner in which a membrane of some kind is placed across the duct so that all airflow has to pass through it. In passing through, the particles can be separated.

There are three main types of filter: viscous, dry and HEPA. Viscous filters are normally a coarse weave of glass, plastic or metal strands coated with oil. As the air passes through the filter, particles impact against the material and are held in the oil. The panels are normally washable and can be re-used after re-oiling. Viscous filters provide good general filtration for air inlet to buildings.

Dry filters are normally a finer-weave fabric or fibrous material and will separate smaller particles than viscous filters, typically down to 0.5 μm. They may be thrown away or washed when dirty, depending upon type.

HEPA (or absolute) filters provide the greatest degree of air cleaning. These filters, which are akin to paper or felt, will capture particles down to 0.01 μm with efficiencies of up to 99.995 per cent. They are generally used for clean room applications and are expensive.

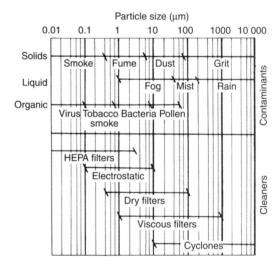

Figure 28.5 Typical particle sizes

If a high degree of filtration is required it is normal practice to have several grades of filter in series so that the coarser, cheaper filters can take most of the dust load, extending the life of the more expensive filters. Filters add a high resistance to the ducting system, increasing the energy use and running cost of the system, and they require regular maintenance.

Electrostatic precipitators

In these devices, the incoming air is passed through an ionizer that gives each particle a positive charge (+6000 V) and then through a collector that has a negative charge (−6000 V). The positively charged particles are attracted to the collector walls. Regular washing is required to remove the collected particles.

Very small particles can be collected but large ones (above 5 μm) can cause arcing across the narrow gap between collector plates. A coarse filter should first remove these large particles. Precipitators are not normally suitable for wet environments since excessive moisture can cause electrical tracking and failure. Resistance to airflow is negligible.

Activated carbon filters

These are used to provide adsorption of gases and vapors, most commonly to remove odors from extracts from kitchens or industrial processes. An efficiency of 95 per cent is obtainable and the carbon has a long life since heating can reactivate it.

Cyclones

A vortex is generated by the high-velocity tangential inlet airflow in these dynamic setting chambers. Centrifugal force holds particles close to the wall where they collect and drain down if wet or spiral down if dry. The contaminant is collected at the bottom and the cleaned air is exhausted through the top. They range in size from small 200-mm diameter units used to collect oil mist from machinery to the large units seen outside paper and woodworking factories.

28.4 Natural ventilation equipment

The natural ventilation equipment commercially available can be split into two basic groups, fixed and controllable. The fixed ventilation, normally a weathered louver system, is used in applications where ventilation is constantly needed, winter and summer, with no need for control to maintain human comfort conditions. Typical examples are plant rooms and building block drying rooms. Louver systems are also often used for inlet or exhaust to powered ventilation or air-conditioning systems and for architectural cladding. A number of years ago, before energy efficiency became important, many factories had uncontrollable ridge vents of various types, but most have now been replaced. Controllable ventilation, which may be weatherproof if required, is used in buildings which are normally occupied and where waste process heat is not sufficient to heat a ventilated building in winter (i.e. most buildings).

28.4.1 Fixed ventilation

Fixed ventilation has to be weatherproof to some degree. If it were not, then a simple hole in the wall or roof would be sufficient. The normal form of fixed ventilator is the louver panel. At its simplest, this may be a number of slats of wood mounted at 45–60° in a door or wall opening, and at its most sophisticated an aerodynamically designed two- or three-bank *chevron* of roll-formed or extruded aluminum (Figure 28.6).

Figure 28.6 Fixed louver system

With all louvers, there is a balance between their weatherproofing qualities and their airflow, and it is important to select a louver with the correct balance for each application. No louver can be guaranteed to be 100 per cent waterproof under all conditions, but the best ones approach this standard under normal flow and wind conditions.

Louvers can be supplied in various materials and finishes, the most common being anodized or painted aluminum, since this provides good corrosion resistance and light weight. Other options are galvanized steel or, for more rigorous conditions, stainless steel.

Most louver systems have a pitch between 30 mm and 100 mm. Where noise control is required, louvers can be supplied either in the form of acoustic louver with a pitch of 300 mm and acoustic material in the louver blade or as standard louver with an attenuator section behind it. A bird or insect guard (as applicable) is normally mounted behind the louver.

28.4.2 Controllable ventilation

Controllable ventilation is normally provided by one of three types of ventilator: louvered, opening flap and weathered. The most common is the louvered ventilator which has a number of center hinged louver blades controlled from a pneumatic ram or electric actuator which can be fully opened when ventilation is needed and fully closed to a weatherproof condition when it is not. These may normally be roof or wall mounted (Figure 28.7).

Opening flap ventilators have one or two hinged flaps (opaque or glazed) which are normally held closed but which may be opened to between 45° and 90°, depending upon design. Control is normally pneumatic or electric. Specific designs are available for roof or wall mounting (Figure 28.8).

Weathered ventilators need to have a complex air path in order to prevent rain entry so they therefore provide a restricted airflow. To overcome this, multifunctional ventilators are often used providing a direct air path during dry weather and a restricted one during rainfall.

These ventilators are normally only roof mounted (Figure 28.9).

Most ventilators are powered in one direction with spring return. Thus, a pneumatic ventilator might be described as *pressure to open* or *pressure to close*. With the pressure to open type the ventilator will fail to the closed position under the influence of the return spring, ensuring the building remains weatherproof. Where ventilators are installed mainly as smoke ventilators it is important that they fail to the open position, so pressure to close ventilators should be used. A fusible link is normally fitted into the controls so that in the event of fire, affected ventilators will open automatically to release smoke and heat.

Most ventilators are made of aluminum because of its good corrosion resistance and lightweight. On industrial premises they are normally left in mill finishes and allowed to oxidize. On commercial premises, a poly-powder paint finish is normally specified for aesthetic reasons. To reduce energy losses during the heating

Figure 28.8 Flap type roof ventilator

Figure 28.9 Weathered roof ventilator

Figure 28.7 Louvered roof ventilator

season many ventilators are available either as standard or optionally with polypropylene pile or rubber seals to reduce leakage, and with insulated surfaces.

A limited range of accessories may be provided, normally bird guard, insect guard or attenuators, but since airflow through the ventilators is driven by very small pressure differences (perhaps 5 Pa), great care has to be taken to ensure the ventilation remains effective.

28.5 System design

28.5.1 Overheating

Overheating can be due to a number of causes, but is usually from solar gain and machinery heat losses. If ventilating to provide thermal comfort in a space for humans or animals there are a number of considerations apart from simply air temperature:

1. Air temperature
2. Mean radiant temperature
3. Radiant temperature asymmetry
4. Air velocity
5. Humidity

Thus, someone working near a window will receive hot solar radiation through the window in summer and cold radiation from the cold window surface in winter, causing uncomfortable radiant asymmetry even if the room temperature is perfect for personnel working away from the window. Ventilation will not be the correct solution for this problem – sun shading and double-glazing would be more effective.

It is therefore important when considering ventilation design to consider the whole thermal environment and not to simply assume that reduction of air temperature to 20–22°C will cure all problems.

Factories, warehouses, etc.

Two methods of preventing overheating are available: either general ventilation to provide fairly even conditions over the whole space at working level or local ventilation to give spot cooling for localized hot spots.

General ventilation is designed to keep air temperatures within a few degrees Centigrade of the outside shade temperature over the majority of the floor area. The ventilation rate required, which is not directly related to the space floor area or volume, is calculated from the formula:

$$\dot{V} = \frac{Q_T}{C_{p\Delta t}}\,\text{m}^3/\text{second} \qquad (28.6)$$

where

$\dot{V}$ = ventilation rate required (m^3/s),
Q_T = total heat load in the building (W),
C_p = specific heat of air
Δt = temperature difference between inlet (°C) and extract air.

The heat load (Q_T) includes all sources of heat in the building – electrical equipment, furnaces, people, solar gain, etc. The temperature difference (Δt) is selected to give the most economical design without compromising comfort and is generally found by allowing a 1.8°C temperature rise per meter height between inlet and extract ventilators. For natural ventilation the inlet and extract areas required can then be calculated from equation (28.3). Either powered or natural ventilation may be used. See Section 28.3 for a discussion of the benefits of each system.

The positioning of equipment is important to ensure that the whole space is ventilated without any significant dead spots of stagnant air. Generally, extract ventilators should be positioned as high as possible and spread evenly over the roof area with no more than 28 meters between adjacent units. In smaller buildings, the inlet is usually provided through existing openings such as loading-bay doors unless security precludes this. In larger buildings, inlets should be provided on at least two sides and, if any area is more than 20 meters from an outside wall, consideration should be given to providing some local powered inlet to provide a positive airflow in that area.

If the heat load in the building is high enough for ventilation to be needed in autumn and spring, then the design must either ensure that airspeeds through the inlets are not high enough to be felt as cold drafts. This limits velocity to around 1 m/s, or provides for some heating of the inlet air to raise it to an acceptable temperature. (Partial recirculation can achieve this economically.)

Local ventilation is used where conditions in most areas are acceptable or unimportant but a small area or areas require cooling, generally because a workstation is close to a source of heat. In this situation, there are three methods of treatment:

1. Reduce heat output or provide shielding. Radiant heat can be reduced by several methods such as insulating the equipment's hot surfaces, painting it with aluminum paint to reduce heat emission or by installing a screen (two layers of expanded metal) between the workstation and the equipment.
2. Provide local air movement. Creating local air movement, even with warm air, provides a reduction in effective temperature and is inexpensive. A 'punkah' type fan may be used, although this can have the effect of drawing down warmer air from a high level, or a portable 'fan in a box'. It is advisable to have a multi-speed fan to allow the effect to be adjusted by the operator for maximum comfort.
3. Provide fresh air. For maximum effect, a ducted fresh-air inlet system will provide cool air in summer and, if fitted with a recirculation unit, allow a flow of tempered or warm air in winter. The air outlet should be positioned to ensure that the airflow does not cause a draft on the back of the neck.

Offices and public spaces

In factories, air movement can be generated by unsophisticated equipment, since there is generally reasonable scope for equipment location and noise levels are often not critical. In offices and public spaces, the space, esthetic, noise and air movement criteria tend to be much tighter, and additional natural ventilation above that already provided by windows is normally impractical. The ventilation rate needed can still be calculated using equation (28.6), although the acceptable temperature difference will be much lower (typically, 5°C).

It is generally only practical to ventilate office spaces by means of a ducted system. If sufficient opening window area is available, either an inlet or extract system alone may be satisfactory, but if optimum control of conditions is required, then both mechanical inlet and extract should be provided.

Inlet systems should incorporate heating if winter ventilation is needed and air filtration if the ambient air is not clean enough, as is common in city centers. Extract grilles are normally at high level and inlet may be high or low level. The design and position of the grilles or diffusers is critical for successful ventilation to avoid drafts and stagnant areas, and flow patterns will vary, depending upon whether the airflow is heating or cooling. ISO 7730 recommends that for light sedentary work the mean air velocity in the occupied area should not exceed 0.25 m/s in summer and 0.28 m/s in winter, and temperature gradients should not exceed 3°C between the ankle and neck.

Two design methods are commonly used: displacement ventilation, where clean air is introduced at low velocity at low level, picking up heat and odor as it passes through the room and rising by buoyancy up to ceiling extracts; and diffusion ventilation, whereby clean incoming air at higher velocity is mixed with the air in the space, with the resulting mixture extracted at ceiling level.

28.5.2 Fume dilution

Factories, warehouses, etc

Dilution of fumes in these areas is generally required for one of two reasons: either to reduce the level of harmful (toxic or irritant) fumes to a safe level, normally below the OES[18] (Occupational Exposure Standard) or to dilute offensive odors. Care must be taken with the latter to ensure that the problem is not merely passed on to neighbors. If it is, then a local extract with air-cleaning equipment will be preferred if it is practical. Indeed, a local extract system is always preferable, since it removes the problem at source, resulting in a cleaner environment within the building.

The starting point for design of a dilution ventilation scheme is normally a hygiene survey in which levels of pollutants and the ventilation rate are measured under worst conditions. The required ventilation rate is then calculated by ratio of level of contamination measured and required, where

$$\dot{V} = \frac{\dot{V}_m \times C_m}{C_r} \, \text{m}^3/\text{s} \tag{28.7}$$

where

$\dot{V}_r$ = ventilation rate required (m³/s),
$\dot{V}_m$ = ventilation rate measured (m³/s),
C_r = contamination level required,
C_m = contamination level measured.

The ventilation should be designed so that airflow is directed from clean to dirty areas to keep the majority of

the building as clean as possible. However, this form of ventilation does little for operatives working at the source of contamination.

Normal ventilation equipment can be used unless the fumes are corrosive or flammable. Powered equipment is normally employed to ensure that a steady airflow is provided. Since the ventilation will be needed in both summer and winter, there will be an energy penalty for the high level of airflow, and either the inlet airflow must be pre-heated or additional space heating will be required. The extra heat requirement is given by

$$Q = 1.2 \times (\dot{V}_r - \dot{V}_m) \times (t_c - -1) \text{kW} \tag{28.8}$$

where

Q = extra heat input needed (kW),
t_c = winter room temperature (°C).

Offices and public spaces

In these areas fume dilution is normally required to reduce body and tobacco odors. CIBSE[1] gives current UK recommendations for fresh air requirements for various rooms to overcome these, varying from 5 to 25 l/s per person. These figures give a balance between energy loss and air purity, and higher levels of ventilation may be needed if a clean atmosphere is of overriding importance. In older offices, the natural ventilation achieved by infiltration is sufficient to provide reasonable conditions, but areas of high occupancy such as conference rooms, theatres, bars and restaurants normally require mechanical ventilation. Sufficient ventilation must be provided to match the highest demand, whether for odor dilution or summer overheating.

An alternative to ventilation to clear tobacco smoke is the use of electrostatic air cleaners that clean and recirculate air within a room. These provide smoke dilution without the energy penalty of extra ventilation but require frequent cleaning to keep operating at maximum efficiency.

28.5.3 Prevention of condensation

Condensation occurs whenever moist air encounters a surface, which is colder than the 'dewpoint' of the air. The dewpoint is the minimum temperature at which the air can contain the amount of moisture within it, and it will vary with moisture content. Condensation can also occur in the air when warm moist air meets cold air, when it is known as *fogging*.

Condensation will appear on the inside surface of porous or impervious materials, forming first on the worst insulated surfaces (normally glazing or steelwork). On porous surfaces condensation can occur within the material or at an internal boundary. This is known as interstitial condensation, and it is especially dangerous, since it is often not known about until it has caused noticeable damage. Condensation can be avoided or reduced by several methods:

1. *Insulation*: Increase the insulation of surfaces to raise their temperature above the dewpoint of the internal

air. This should be treated with care, since insulation of one surface may merely shift condensation to the next worst insulated surface, or cause interstitial condensation.

2. *Reduce moisture emission*: If practical, this is always the best method of avoiding condensation. It can be achieved by good housekeeping, reducing steam leaks, covering open tanks when not in use or by changing working practices or modifying equipment either to contain the moisture more efficiently or remove it directly from the space.

3. *Ventilation*: Increasing the ventilation rate through a space while keeping the air temperature high will reduce the moisture content of the air without lowering surface temperatures and, by reduction of the dewpoint, will limit or avoid condensation. The hot, moist air exhausted takes a lot of energy with it, which it may be practical to recover using a heat-recovery system.

4. *Dehumidification*: Increasing the ventilation rate through a building has a high-energy penalty. Dehumidifiers do not increase the ventilation rate but recirculate air while removing the moisture from it by a refrigeration process. Their electrical cost is high, but since most of the energy is released into the room as heat, the overall running costs can be significantly lower than for a ventilation system.

Calculation for condensation problems is complex but is covered in the CIBSE Guide[1] in some depth.

28.5.4 Local extract

Local extract is used to remove contaminants directly from a process to the exterior without passing through personnel breathing zones. It thus provides a high degree of safety and because small volumes of air are extracted relative to a dilution ventilation system it is energy efficient.

Where dusts, grits, sawdust or other large particles are being extracted it is normally mandatory to include an air-cleaning device before the air is exhausted. Unless they are toxic or odorous fumes can normally be simply exhausted into the atmosphere at a suitable location well away from anywhere with normal personnel access.

The design of a system can be broken into three main areas: capture, transport and cleansing. Capture of the contaminant is of paramount importance. Depending upon the source of the contaminant, capture may be via a hood, slot, booth or enclosure, with the airflow designed to take the contaminant from the source into the duct system without passing through the operatives' breathing zone. Some typical examples of good design are shown in Figure 28.10. To capture the contaminant, a minimum air velocity, referred to as the *capture velocity*, is needed at the source. The value of the capture velocity depends upon the type of source, enclosure and local air movement. Typical values are given in Table 28.1. Full guidance in the design of capture systems is given by the ACGIH[12].

Once captured, the contaminant has to be carried along a duct system. If the duct velocity is too low, particles

Table 28.1 Capture velocities

Condition of dispersion of contaminant	Examples	Capture velocity (m/s)
Released with practically no velocity into quiet air	Evaporation from tanks; degreasing, etc.	0.25–0.5
Released at low velocity into moderately still air	Spray booths; intermittent container filling; welding; plating	0.5–1.0
Active generation into zone of rapid air motion	Spray painting in shallow booths; barrel filling	1.0–2.5

will tend to drop out of suspension and collect in or fall back down the duct. It is therefore essential that a suitable minimum duct velocity, referred to as the transport velocity, is maintained. Typical values are given in Table 28.2. Nothing is gained by velocities far in excess of the recommended transport velocity, and in some cases as much is lost, since abrasion of particles on the duct can cause premature erosion and failure.

Removal of particles from the air stream is generally carried out in a cyclone, positioned outside the building and taking particles from all sources in the building. Care should be taken to position cyclones away from noise-sensitive areas since both ducts and cyclones can be noisy, especially if grits or chips are being carried.

Selection of fans and ducting for local extract must be more rigorous than for other systems. The fan must be capable of withstanding abrasion or corrosion from the contaminants, and if they are flammable must have a flameproof or out-of-air stream motor. The ducting also must be able to withstand abrasion or corrosion and must be fully sealed to prevent escape of contaminants within the building. For specialist applications such as fume cupboard extract, ducting and fans are often of plastic construction.

28.5.5 Smoke ventilation

Smoke-ventilation systems are designed to clear smoke and heat from a building in the event of fire. In large open spaces it is impossible to 'smother' a fire (as is often recommended in domestic situations) by closing doors and windows. The aim of smoke ventilation is to minimize damage due to smoke staining and heat and to assist evacuation and firefighting by providing a layer of clear air

Table 28.2 Transport velocities

Contaminant	Typical transport velocity (m/s)
Vapours, gases	5
Fumes	7–10
Light dusts	10–12
Dusts	12–17
Industrial dusts	17–20
Heavy or moist dusts	20+

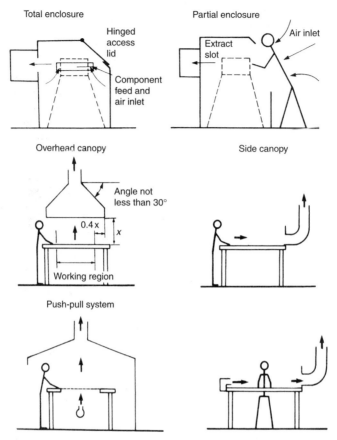

Figure 28.10 Typical local extract hoods and enclosures

below the smoke. Without smoke ventilation a space can become 'smoke logged' from ceiling to floor in only a few minutes.

The actual design of a smoke-ventilation system is very complex, and although much published guidance is available[14-17] it should be left to experts, and a fire-prevention officer should approve the design.

In discussing design or vetting tenders, there are a number of important points to consider:

1. *Design fire size*: An accurate assessment is needed of the maximum size of fire that is likely and which can be designed for. Since the whole design is based upon this value it is essential that this be carefully considered.
2. *Interaction with sprinklers*: The likely size of fire will be smaller in buildings with sprinklers and the temperature of the smoke produced will be lower since the smoke is cooled by the sprinkler flow.
3. *Fire detection*: For efficient operation, control of the ventilation must be linked to a fire-detection system or sprinkler flow switch to ensure that the ventilators are operated as early as possible. Natural ventilators must

incorporate a fusible link or bulk as a back-up fail-safe device, but this should not be considered as the main form of emergency operation.
4. *Fire resistance*: Fans, motors, cabling and controls which are expected to operate under fire conditions must be suitably rated for the temperature expected.
5. *Air inlet*: A suitable low-level inlet ventilation area must be provided for the expected airflow rate. Where personnel escape routes will be used for inlet, the inlet velocity must be low enough not to impede progress.

Smoke ventilation is not intended to replace other forms of fire prevention and control but to work as an important component in an overall scheme.

28.6 After installation

28.6.1 Commissioning and testing

Once any system has been installed it is important that it is properly commissioned to ensure that everything is working satisfactorily and to specification. A guide to commissioning ducted systems is available from BSRIA.

Commissioning of other systems should be in accordance with the manufacturers' recommendations, but typical commissioning lists are shown below.

Natural ventilation systems

1. Check incoming electrical supplies to isolators.
2. Check pneumatic and electrical connections between panels and ventilators.
3. Check ventilators and ensure that any transport closure pieces and packing are removed and ventilators are correctly installed.
4. Commission compressor (if fitted) according to manufacturer's instructions.
5. Operate ventilators and check that all open and close correctly.
6. Check operation of all controls.
7. Check pneumatic system for leakage.

Powered ventilation systems

1. Check incoming electrical supplies to isolators.
2. Check electrical connections between panel and fans.
3. Check that fan blades rotate freely and ventilators are correctly installed.
4. Operate ventilators and ensure that all run correctly and in the correct direction.
5. Check full load current on all phases on all ventilators.
6. Check operation of all controls.

Once a system is installed, no specific regular testing of day-to-day ventilation systems is required unless they are installed to comply with COSHH Regulations. This testing would normally be carried out in conjunction with annual maintenance.

Smoke-ventilation systems must be regularly tested, since failure in an emergency cannot be allowed. The system must be fully tested in accordance with the manufacturer's instructions at intervals of not more than 1 month or as agreed with the fire authorities. Any faults found must be rectified immediately.

28.6.2 Maintenance

Most simple ventilation systems require only annual maintenance unless some form of air cleaning or filtration is incorporated, although where systems are pneumatically operated the compressor will need weekly checking. Air cleaners or filters will need regular emptying, cleaning or replacement to maintain efficiency and prevent clogging up. This can either be carried out on a regular schedule (based on the manufacturer's guidance and site experience) or when indicated by a pressure differential gauge or alarm. Savings in maintenance costs easily repay the cost of a gauge by maximizing intervals between cleaning or replacement. The task of maintenance can be made easier by taking care in design to provide good access.

Compressors will need a weekly oil level and receiver auto drain check and oil changes and filter cleaning at (typically) 500-hour intervals, although for compressors used only to operate ventilation this can be only an annual task due to the limited usage. Typical annual maintenance for systems is shown below.

Natural ventilation systems

1. Clean ventilators with a soft brush or cloth and remove any debris (leaves, moss, etc.) from rain channels or drainage holes.
2. Check security of fixings.
3. Check operation of ventilators and, if necessary, clean or oil linkages.
4. Examine hinges, pulleys, cables, anchors, etc. for wear.
5. Check pneumatics for air leaks and fully service compressor.
6. Check that all controls operate correctly.

Powered ventilation systems

1. Clean ventilators with a soft brush or cloth and remove any debris from rain channels or drainage holes.
2. Check security of fixing.
3. Check that fan impeller turns freely.
4. Check electrical connections to fan.
5. Check operation of moving components (other than fan). Clean, oil or adjust as appropriate.
6. Run fan and check current against data plate.
7. Check that all controls operate correctly.
8. Check and clean filters, heater batteries, etc.

28.6.3 Running costs

Running costs of ventilation systems can be broken down into three main areas:

1. Maintenance costs;
2. Electrical costs;
3. Heat loss due to ventilation use during heating season.

Maintenance costs

Regular service checks such as filter cleaning and compressor oil level are normally carried out in-house and time can be allocated for these tasks once some experience has been gathered. Annual maintenance may be carried out in-house or by specialist service engineers employed either by manufacturers or HVAC service companies. A service contract can often include breakdown cover, which has the advantage of reducing risk of unexpected bills and ensuring that prompt repairs are effected at the cost of a higher annual premium.

Electrical costs

The only significant electrical cost involved in ventilation systems is operation of fans. Other electrical equipment such as dampers, compressors, etc. generally run for such short periods that costs are negligible.

Electrical costs for fans can be estimated from the following formula:

$$C = V \times \phi \times A \times \text{fuel cost} \times \text{hours run} \qquad (28.9)$$

where

C = cost per annum (\$),
V = motor voltage per phase (V),
ϕ = number of phases,
A = operating current (A),
Fuel cost = cost of electricity (in \$/kWh),
Hours run = total running hours per annum.

Heat loss

Where a ventilation system is required to run during periods when heating is provided then there is an energy cost associated with the heated air being exhausted from the building. This is related to the extra heat input needed from the heating system to balance the heat loss through the extra ventilation. A calculation method is available in section B18 of the CIBSE Guide[2].

References

1. CIBSE Guide, Volume A, *Design Data* (1986).
2. CIBSE Guide, Volume B, *Installation and Equipment Data* (1986).
3. ASHRAE Handbook, *Fundamentals* (1989).
4. ASHRAE Handbook, *HVAC Systems and Applications* (1987).
5. *Threshold Limit Values and Biological Exposure Indices for 1988–1989*, American Conference of Governmental Industrial Hygienists.
6. BRE Digest 119, *Assessment of Wind Loads* (July 1970).
7. BRE Digest 210, *Principles of Natural Ventilation* (B. B. Daly) (February 1978).
8. BRE Digest 346 (7 parts), *The Assessment of Wind Loads* (1989).
9. Daly, B. B., *Woods Practical Guide to Fan Engineering*, Woods of Colchester Ltd (1978).
10. *Ductwork Specification DW142*, HVCA (Heating and Ventilation Contractors Association).
11. CIBSE Technical Memorandum TM8, *Design Notes for Ductwork*.
12. *Industrial Ventilation, A Manual of Recommended Practice*, 20th edition, American Conference of Governmental Industrial Hygienists (1984).
13. ISO 7730: 1984, Modern thermal environments – determination of the PMV and PPD indices and specification of the conditions for thermal comfort.
14. Fire Paper 7, 'Investigations into the flow of hot gases in roof venting', HMSO (now available from Colt International Ltd) (1963).
15. Fire Paper 10, 'Design of roof venting systems for single story buildings', HMSO (now available from Colt International Ltd) (1964).
16. BS 7346, Components for smoke and heat control systems. Part 1: Specification for natural smoke and heat exhaust ventilators.
17. BS 7346, Components for smoke and heat control systems. Part 2: Specification for powered smoke and heat exhaust ventilators.
18. EH 40/89 Occupational Exposure Limits, Health and Safety Commission (1989).

29

Air Conditioning

J Bevan

Contents

29.1 Basic principles and terms

29.1.1 Abbreviations

In addition to the abbreviations used in SI, the following are employed in air conditioning work:

db	Dry bulb temperature
wb	Wet bulb temperature
dp	Dewpoint temperature
rh	Relative humidity
kg/kg	Kilograms water vapor per kilogram dry air (absolute moisture content)
TH	Total heat
SH	Sensible heat
LH	Latent heat
SHR	Sensible heat ratio
ON OFF	The condition of air or water entering or leaving a coil or heat exchanger
TR	Tons of refrigeration capacity
TRE	Tons of refrigeration capacity extracted
TRR	Tons refrigeration rejected (at final cooler)
HP	High pressure (refrigerant)
LP	Low pressure (refrigerant)
DX	Direct expansion cooling
ΔT	Temperature difference
ach	Air changes (room volumes) per hour
ahu	Air-handling unit
swg	Static water gauge
NR	Noise rating. One of a series of curves relating noise level and frequency to speech interference
NC	Noise criteria. Similar to NR, but differing, particularly at the low-frequency end

29.1.2 Terms

Mechanical ventilation

The movement of air by fan, conveying outside air into the room or expelling air or both. Filtration, heating and control of the distribution pattern may be included. It is not cooling in the sense of temperature reduction but can be used to limit temperature rise when the outside air is below that of the space being treated.

Full air conditioning

This necessitates plant capable of control of temperature by being able to add or subtract heat from the air and control of humidity by being able to add or subtract moisture. The system also comprises fan(s), filtration, and a distribution system and may include noise control. Other terms such as 'cooling' or 'comfort cooling' may be met and these can be taken to mean an ability to lower the temperature of the air by refrigeration but without full control of humidity. Moisture may be removed as an incidental characteristic of the cooling coil. The term 'air conditioning' is sometimes used where control of humidity is not included. It is essential to employ clear specifications of performance.

Air

Atmospheric air is a mixture of gases, mainly nitrogen and oxygen together with water vapor. It normally carries many millions of dust particles per cubic meter.

Temperature

A measure of the average energy of the molecules of a substance. The heat intensity.

Heat

A form of energy which, when given to a body, raises its temperature or changes its state from solid to liquid or liquid to gas.

Heat flow

Heat flows from a body at one temperature to a body at a lower temperature. Materials have the property of resistance to the rate of heat flow. It differs from material to material.

Sensible heat

The heat energy causing a change in temperature, as in raising a kettle of water from cold to boiling point.

Latent heat

The heat necessary to change the state of a substance from solid to liquid or from liquid to gas, or the heat given up during the reverse process. There is no change in temperature during these processes. For example, continuing to boil a kettle of water previously raised to 100°C to steam requires the addition of latent heat, but there is no change in temperature if the pressure remains constant.

Total heat

The sensible heat plus latent heat in such a mixture as moist air. In air-conditioning work it is referred to a base a little below 0°C, not absolute zero.

Sensible heat ratio

Sensible heat flow divided by the total heat flow.

Enthalpy

The heat content of a substance per unit mass.

Dry bulb temperature

The temperature of air as indicated by a dry sensing element such as a mercury-in-glass thermometer.

Psychrometrics

The study of moist air. The psychrometric chart shows the relationship between the various properties of moist air in graphical form and can be used for the solution of problems.

Wet bulb temperature

The temperature of air as indicated by a thermometer when a water-wet wick encloses its bulb. If the surrounding air is not saturated water will evaporate, taking the latent necessary latent heat from the thermometer bulb which then gives a lower reading than a dry bulb in the same air. The depression in wet bulb temperature is proportional to the amount of moisture in the air.

Normal practice is to arrange a flow of air over the wick by using a sling (whirling) or fan-assisted instrument. If the thermometer is stationary, an area of higher saturation builds up around the wick but the reading may be referred to tables for screen instead of sling readings.

Partial pressure

The contribution by each constituent gas to the total air pressure. Standard air pressure is 1013 mbar.

Vapor

A gas that is below its critical temperature and which can therefore be turned to liquid by an increase in pressure.

Saturation

There is a limit to the amount of water vapor air can hold. It is higher at higher dry bulb temperatures. At the limit, air is saturated.

Relative humidity

This compares the amount of moisture in a sample of air with the amount it would contain if saturated. More accurately, relative humidity is the partial pressure of vapor present divided by saturation vapor pressure ×100 per cent. Saturation = 100 per cent relative humidity.

Dewpoint

The temperature to which a sample of air has to be reduced to bring it to saturation. The moisture content of the air sample fixes it.

Absolute humidity

This measures the quantity of water in a sample of air in kg moisture per kg air. The relative humidity then depends on the air-dry bulb temperature. Air at 25°C containing 0.01 kg/kg is at 50 per cent relative humidity (rh).

If now cooled to 14°C the air would be at its dewpoint (i.e. saturated). If cooled further, moisture is condensed out, the sample remaining saturated as it cools. If now reheated back to 25°C its rh would be lower than 50 per cent. If cooling had not been continued to condense moisture its rh would return to 50 per cent at 25°C.

System resistance

The resistance to airflow which causes static pressure drop. It is similar to electrical resistance and voltage drop

(see Section 29.5). The term 'resistance' is often used erroneously when pressure drop is meant.

Upstream, downstream

Used to denote positions earlier or later in the system relative to the direction of air flow.

Condensing unit

A refrigeration compressor and condenser on one chassis complete with controls.

Split system

As above but with a remote condenser.

Chiller

A compressor, water-chilling evaporator and condenser on one chassis.

29.1.3 The plant

These divide broadly into two types:

1. The direct expansion plant where the air-cooling coil is fed with cold refrigerant;
2. The chilled water plant where the cold refrigerant first chills water (or other liquor), which is fed to the air-cooling coil.

A block diagram of the DX system is shown in Figure 29.1. It has two main circuits – the air circuit and the refrigerant circuit. In the chilled-water system, there are additional circuits:

1. Of chilled water between the refrigerant and the air-cooling coil;
2. Of water carrying heat from the refrigerant to the heat-rejecting device.

Figure 29.2 shows a block diagram of the system.

A temperature difference must exist between each stage to cause a heat flow from one to the other. The air and refrigeration sides are described in further detail below.

29.1.4 The air-handling plant

The plant may comprise one or more complete factory-made units or may be built up on-site from subassemblies. There can be variations from the arrangements discussed below and shown in Figure 29.3. The condition or quantity of air input to the conditioned space (referred to below as the room) must be varied such that after it has gained or lost heat or moisture by the applied load its condition and therefore the room condition is as specified.

Outside air intake

The quantity is discussed in Section 29.2. Its purpose is to keep the room fresh and to pressurize it against the ingress of unconditioned air. Its psychrometric condition

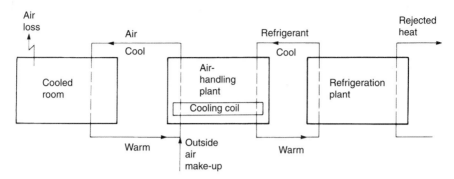

Figure 29.1 Heat flow paths in a direct expansion system

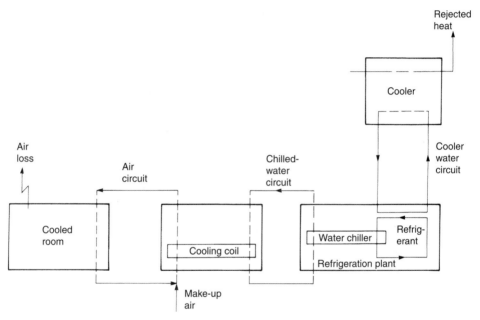

Figure 29.2 Heat flow paths in a chilled-water system

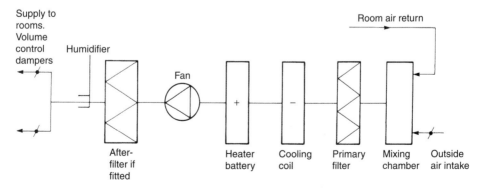

Figure 29.3 Airflow path in an air-conditioning system

during most of the year will differ from that required. It can be introduced into a chamber, mixing it for treatment with return air from the room. Alternatively, it can be treated in a separate plant before being introduced to the system, but care then has to be exercised in design, since the sometimes small quantity of air has to control the full humidity load. An advantage is that corrosive wet processes are kept out of the main plant.

Mixing chamber

This is where the outside air intake and recirculated air are brought together before proceeding to the next stage of treatment. As shown in Figure 29.3, it is a low-pressure area, which will induce outside air without the use of another fan if the route has low resistance.

When two samples of air are brought together the condition of the mixture may be arrived at arithmetically by adding the heat flow of each and dividing by the total mass flow; and similarly for the moisture flow. Alternatively, plot the condition of each onto a psychrometric chart. The mixed condition lies on a straight line between the two in a position proportional to the two quantities.

Pre-filter

Where a high degree of cleanliness is not required, it could be the only filter in the system. The subject is covered in more detail in Section 29.7.

The cooling coil

This is the exchanger where heat flows from the room return or mixed air to cold refrigerant or to chilled water. It is an arrangement of finned tubes normally of aluminum fins on copper tubes, but copper fins can be specified for corrosive atmosphere. Performance characteristics are controlled by fin and tube spacing. If the room rh is high, dehumidification may be brought into use by operating the coil or one of a number of parallel coils at a low temperature. If the room's sensible heat load is low reheat must be allowed to operate at the same time.

Dehumidification can be achieved by partially bypassing the coil such that the remaining air travels through the coil at low velocity. This can also be inherent in the full-load design operation of the coil.

Heater battery

This is used when (1) the room needs heating instead of cooling or (2) for reheat as described above. It is vital in close control systems that its capacity is sufficient to maintain room temperature under these conditions, otherwise the system may fall into a loop, with the controls continuing to see high rh due to temperature. Using only part of the cooling coil for dehumidification will alleviate this situation. A heater capacity of the sensible heat extracted during dehumidification plus half the peak winter fabric loss is recommended where the room load could be nil in winter such as a start-up situation.

Separate reheat batteries may be placed in branch ducts where one plant supplies both a main area calling for cooling and an auxiliary room without heat load. Correct rh in the auxiliary rooms' results (only) if it is correctly controlled in the main room and they require the same dry bulb temperature. While wasteful of energy, it simplifies the plant design and may be found to use fewer resources.

The fan

This drives the air around the system against its resistance (see Section 29.6).

Humidifier

The humidifier is a means of increasing the absolute humidity of the air although usually controlled from a relative humidity sensor. It should be positioned where shown so that it can correct any over-dehumidification by the cooling coil (see Section 29.8).

Air flow

The quantity supplied must be matched to the load on the plant (see Section 29.2). After leaving the plant, it is distributed to match loads of the rooms or zones to be served.

29.1.5 The refrigeration plant

The basic circuit is shown in Figure 29.4 and the principal items are described below.

Evaporator

This is the device where the air or water being cooled gives up its heat to provide the latent heat of evaporation to the refrigerant. Superheat is also added to the refrigerant at this point to prevent damaging liquid forming on the way to the compressor.

Compressor

A compressor circulates the refrigerant around the system, raising its pressure such that the refrigerant can be condensed by removal of latent heat. It may also be considered as raising the temperature of the refrigerant above that of the final cooling medium to which heat is rejected. Lubricating oil is contained in the crankcase but, being miscible in the refrigerant, is carried around the system and returned.

Condenser

This is the vessel where the refrigerant rejects its heat to waste or reclaim, turning back to liquid in the process. Sub-cooling is practiced by the removal of further heat. This prevents liquid flashing back to vapor on return to the evaporator.

Expansion valve

A reduction in pressure and hence in temperature takes place across this item before the refrigerant re-enters the cooling coil via distributor pipes.

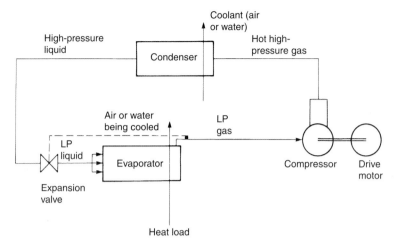

Figure 29.4 Refrigerant flow path in a cooling system

29.1.6 Controls

Room condition

Dry bulb temperature

A thermostat in the conditioned space or in the return duct senses this. Where under floor air return is used, it is strongly recommended that the sensor be placed close under a return grille to prevent changes of condition occurring between the room and the sensing position. In large rooms, separate thermostats can be arranged to give an average signal but individual zones of control each separately treated are much better.

Dry bulb sensors may be bi-metal strips, thermistors or refrigerant-filled phials or bellows responding to pressure differences caused by temperature change. These, in turn, provide an electrical or mechanical signal. The mechanical items are used to alter the value of potentiometers or make-or-break contacts. The signals are transmitted to amplifiers, which respond to the degree of error. An important feature is the proportional band of temperature over which the controls call for up to full plant capacity.

Humidity

Humidity sensors may be animal or plastic skins varying in length with changes in rh or lithium chloride coating changing in electrical resistance. The former is prone to lose calibration. Other comments above apply equally to rh control.

Control at the evaporator

A phial senses the temperature of the outlet suction line to the compressor and controls the expansion valve opening to maintain a constant temperature in the coil.

Control at chilled water coil

A three-way motorized valve is modulated between full flow to the coil and full bypass to satisfy the room thermostat.

Control of water-chilling compressors

Being large multi-cylinder machines, the chilling capacity is controlled in steps by rendering cylinder valves inoperative. Control is initiated by sensing water temperature in a storage buffer tank or by sensing return water temperature from the air-cooling coil. Small (DX) compressors are run on a start–stop basis on call from the room thermostat. Safety devices associated with compressors include:

High (gas) pressure cutout;
Low (gas) pressure cutout;
Oil differential cutout (oil feed pressure to be above the crankcase pressure);
Freeze thermostat (low water temperature limit);
Single-phase protection (preventing attempts to start with loss of one phase of a three-phase electrical supply);
Time out (preventing too-frequent starting and motor burn).

Until all the switches in the interlock train are satisfied and closed, the compressor will not start or will not continue to run.

29.1.7 The load on the plant

The unit of heat is the watt. However, the imperial unit should be understood, as it will still be met, particularly outside Europe. The ton of refrigeration is derived from an ability to remove sufficient heat from a short ton (2000 lbs) of water at 32°F to turn it to ice at the same temperature in the course of 24 hours. This amounts to a heat extraction rate of 3.517 kW.

The load presented by the room is first transmitted to the room air, which, in turn, passes it to the cooling coil. Other gains will occur as detailed below. The total load is best considered in two parts – sensible and latent.

Machine heat

All electrical energy fed to the room will appear as heat. This presents a load to the plant unless power is conveyed out of the room by cable or hot items are physically removed. Parts of some large computers are cooled by a direct supply of chilled water presenting a load on the refrigeration plant but reducing the load on the airside. Cold outside conditions will result in some of the internal load being met by fabric loss.

Fabric losses or gains

In winter in cold climates, heat will be lost through fabric of the building. This will be advantageous at times of high internal load but will need to be considered as a heater battery duty at times of low internal load. In winter, there can be considerable loads for humidifiers if the structure is not adequately vapor sealed. Weather-tightness is insufficient. For example, a computer room in the UK held at normal conditions may experience up to 100-mm water gauge vapor pressure difference, forcing moisture out. During high summer in temperate and hot climates, the external water vapor pressure will be higher than in the conditioned space.

Personnel

People give off both sensible and latent heat. During light work such as in a computer suite or laboratory they emit 110 W sensible 30 W latent and while seated, 90 W sensible 20 W latent.

Air ducts

A duct carrying cool air through a warm space such as a loft will gain heat before entering the conditioned room, contributing to the load on the plant. Ducts passing through the conditioned space do not add to the load. Similarly, there will be losses in winter if they carry warm air through cold spaces and moisture gains and losses if leaky. Insulation and vapor-tight joints are necessary.

Fan heat

The power fed to the fan shaft (or the total electrical power fed to the motor if within the duct) appears as sensible heat in the system. During a heating cycle this is useful. As a rule of thumb, fan shaft power is 17 × flow m^3 × mm swg watts. This commonly lies between 2 per cent of the cooling duty for small systems to 10 per cent for large systems.

Compressor heat

Friction (or, if within the refrigerant path, as in hermetically sealed or semi-hermetic machines, the whole of the input to the motor) increases the amount of heat to be rejected by the final cooler or available for heat recovery. It is uneconomic to operate refrigeration plant at unnecessarily high temperatures in order to assist recovery.

Pump heat

In chilled-water systems, pump shaft power adds to the heat of the circulating water. Similarly, if the chiller has a water-cooled condenser pump heat is added to that handled by the final cooler. Power is proportional to the flow and pressure:

$$\frac{102 \times \text{l/s} \times \text{meters head}}{\text{Efficiency}} \text{watts}$$

Efficiency can vary between 35 per cent for very small pumps and 80 per cent for large ones, and may be found to lie between 3 per cent for small heating circuits and 10 per cent of the heat conveyed for large systems.

29.2 The air quantity required

29.2.1 Air change rate

Change rates (room volumes per hour) can be used to calculate the quantity to be supplied or extracted by a mechanical ventilation system. These figures also apply to parts of an air-conditioning system where stale air must not be recirculated.

Volume to be supplied

For the design of close control systems or where large amounts of heat are to be removed, the mass flow to be employed must be calculated. Use is made of the specific heat of air which for normal room conditions may be taken as 1.02 kJ/kg°C. 1.02 kW raises a flow of 1 kg/s by 1°C and pro rata. The volume of a given mass of air varies with change in temperature but supply volume is often more convenient to consider. Taking the specific volume to be 0.82 m^3/kg then 0.1 m^3/s will convey 1 kW with a temperature rise of 8°C and pro rata. A subsequent fall of 8°C across the cooling coil is suitable for areas controlled to 21°C 50 per cent rh. A smaller air quantity would be too close to the limiting temperature to hold the necessary moisture when leaving the coil. Greater temperature differences may be used if the rh is to be controlled to a lower level or not controlled at all. A maximum temperature difference of 10°C supply to room is recommended where occupants are close to supply points. For comfort, larger volumes are preferable to low temperature.

In ventilation systems the temperature rise calculated on the basis of specific heat alone will be pessimistic by one or two degrees because of the effect of building mass. Unless an extract is specifically designed to remove heat from hot spots or lights, the extract and room temperature can be taken to be the same.

Outside air intake

When temperature limitation is more important than close control of conditions, a considerable economy of refrigeration plant operation results from arranging to draw in outside air when it is sufficiently cool and rejecting this back to outside after having gained heat from the room. However, where close control of temperature and humidity is required accuracy and economy ensue from minimizing the outside air intake. The quantity may be based on the number of occupants using $0.008\,\text{m}^3$/s minimum per person. Alternatively, if larger, use $0.002\,\text{m}^3$/sm^2 floor area, which is sufficient in a good quality building to keep the room pressurized to one or two mm swg. Because air loss is a function of the building surface, this is preferred to the basis of a proportion of the supply air volume, which is a function of load. If the later is used for applications such as computer rooms it is seldom necessary to use more than 2.5 per cent.

General considerations

Outlets should not be provided in constantly recirculating systems, particularly where close control of humidity is required. The overpressure developed is far less than that exerted by the wind, and for this reason any system which does have both intake and discharge ducts should have them on the same face of the building. While care is necessary to prevent short-circuiting, this alleviates problems arising from the considerable wind pressure difference that can develop on opposite sides of a building.

Air-lock entries should be used where close control of conditions is required and for clean rooms, but conventional doors with close fit and self-closures are sufficient.

Fire authorities may stipulate pressurization of certain areas such as stairways and may require smoke-extract systems to be brought into operation automatically in the event of a fire. They should be consulted at an early stage of the design.

29.3 Heat losses and gains

29.3.1 Heat losses

Heat transfer through a partition is a function of resistance to heat flow, the temperature difference driving the heat through and the surface area. The function of heating systems is to provide the heat lost in maintaining the temperature difference. Thus

Heat flow (watts) is $U \times \text{TD} \times$ area

where U is a coefficient for the partition in watts per m^2 °C. For each room served the loss through each wall, ceiling and floor should be calculated and for each part of those surfaces where differences occur, such as windows in walls or cantilevered structures of upper floors.

The U values of many partitions or composite constructions can be found in standard references but others may have to be calculated where no data are available or changes are contemplated. The resistivity (r) of each element of the partition encountered by the heat in passing through must be found and multiplied by the thickness in meters. Manufacturers are usually able to give resistivities. The sum (R total) of all the elements, including the inner and outer surface resistances and the resistance

of any interleaf air gap, is found. The U value is the reciprocal of R total. For example, for a wall:

Outer surface (normal exposure)	0.055
112 mm brick skin 0.12×0.112	0.01344
50 mm un-insulated air gap	0.180
150 mm lightweight block skin 5.88×0.15	0.882
15 mm rendering 2.5×0.015	0.0375
Inner surface	0.123
R total =	1.29094
$U = 1/R =$	0.775

If this is insufficient for the purpose (e.g. the external wall of a dwelling house) the problem could be reworked using insulation in the interleaf gap.

The temperature at any intermediate point is proportional to the R total to that point. This may be used to decide whether that point is above or below the dewpoint of penetrating air.

In maritime climates such as the UK and coastal US, the lowest external temperatures are not sustained for long periods. The mass of the structure has a slugging effect, and it is safe to use a relatively high external design temperature. These figures do not apply to an outside air intake where the full effect of low temperature is felt immediately nor to lightweight structures. When choosing a design temperature difference, one must take into account that adjacent rooms may not always be heated to their design temperature. Heat bridges, which are weak points in the insulation, must also be considered in proportion to their areas.

Cold outside air introduced to the system by infiltration or by design of the plant will require heat to be added by the plant or directly in the room to maintain the room temperature. The heat required is

kg/s $\times 1.02 \times$ °C TD kW

For infiltration, it is convenient to use m^3/h $\times 0.33 \times$ °C TD watts, but where air conditioning is employed, infiltration should not be allowed. Air lost from the space by pressurization carries heat away but this is not an additional load beyond that mentioned above.

The available heating capacity should exceed the calculated figure, this being sufficient only to balance the losses under steady conditions. A 25–50 per cent excess capacity is recommended to provide warm-up from cold and good response to controls but without excess overshoot. It is important in close control air-conditioning design to have sufficient heating capacity to raise the room temperature from cold following plant stoppage. If the room is cold, the controls will see high rh and call for dehumidification upon restarting. Unless this is countered by sufficient reheat the control of conditions will not recover automatically.

29.3.2 Internal gains

Any heat liberated within the room reduces the heating effort required by the plant. At any time these gains exceed the loss more than marginally, cooling is required. Heat sources are:

1. The total electrical input to the room unless power is carried away by cable or heat by pipe. This includes lighting and it should be remembered that the input to fluorescent fittings is greater than the tube ratings. Some luminaires are designed as air-extract fittings, in which case their heat is a load on cooling plant, but does not contribute to room heating unless the plant is in a non-cooling recycling mode;
2. Heat from other processes unless carried away by pipe or items are taken out of the room;
3. Heat from personnel (see Section 29.1);
4. There should be no space heating within air-conditioned rooms but some gains may arise from hot water pipes passing through to other areas;
5. Gains from adjacent spaces held at a higher temperature;
6. Heat gains into ducts where these pass through warm areas *en route*;
7. The shaft power of fans will appear in the system as heat. Where the drive motor is in the air stream the whole of the motor input will appear as heat.

29.3.3 Heat gains

The calculation of heat gains through the building fabric is more complicated than for heat losses, taking into account the gains from both the air-temperature difference and from solar intensity. The gain varies during the day with the movement of the sun and changes in air temperature. Heavily glazed buildings are susceptible to large gains from low sun elevations at all times of year and here building orientation can have a considerable effect on plant loads.

A structure with a large mass will result in the peak gain appearing on the inside some hours after the external peak, and the gain will be attenuated since the outside condition will be reduced before that time. The time of peak gain will differ for each of the enclosing surfaces, so it is necessary to calculate each for several hours to find the peak for one room and to repeat this for each room to find the peak load on a plant serving several rooms. Computer packages of varying merit are available to undertake this laborious task but the broad method is shown below.

To avoid temperature shock and for economy, the control of comfort air conditioning may be allowed to drift with extremes of external temperature. However, where the design is to maintain specified internal conditions under virtually all-external conditions data may need to be adjusted upwards.

Gains through walls

$$(A \times U \times (M - T)) + (A \times U \times F \times (P - M)) \text{ watts}$$

where

A = area (m),
U = W/m°C,
M = mean solair temperature (°C) (solair temperature is a composite temperature taking into account outside dry bulb temperature and solar radiation),
T = fixed room temperature (°C),

F = factor depending on wall thickness (varies from 1 at zero thickness to approximately 0.05 at 0.5 m thick),
P = solair intensity at the sun time considered.

The delayed time at which this gain is manifest depends on the structure thickness and density and varies from zero for zero thickness to 15 hours for 0.5-m thickness.

Gains through windows

$$A \times ((C \times I) + U \times (O - T)) \text{ watts}$$

where

A = area (m),
U = W/m°C,
C = solar gain factor,
I = direct radiation intensity,
T = room temperature (°C),
O = outside temperature (°C).

The gain is immediate.

29.3.4 Heat of outside air intake

In close control air conditioning, the condition of the outside air is rarely as required for passing forward to the controlled space. Therefore, it contributes to the cooling, heating, humidification or dehumidification loads on the plant. Whether computer rooms, laboratories, etc. with their sparse population are considered or auditoria where there is a larger ventilation requirement, the peak latent heat load caused by treating the outside air is greater than that from the personnel. The heat from personnel is given in Section 29.1.

The calculation of the heat of intake air is in two parts:

Mass flow kg/s × change in latent heat kJ/kg (from tables or psychrometric chart) (kW); plus
Mass flow kg/s × change in sensible heat kJ/kg (kW).

29.4 Air conditioning for computers

29.4.1 The computer

In the context of this section, a computer is a large mainframe machine standing in its own room. The machine is housed in cabinets 1–2 m high with a meter or so clearance all round. Some items are purely electronic; others handle magnetic tape, magnetic discs or paper. The function may be printed output following calculations, information storage and retrieval or electronic control of a remote process. In the largest configurations output printers and communications equipment may be housed in their own adjacent rooms but with similar environmental control. Other associated rooms which require air conditioning if included in the suite are: Job Assembly, Operations Control Room, Magnetic Media Store, Computer Engineer's room, Ready-Use Stationary Store. The main stationary stores and store for master tapes are not normally conditioned although they are heated in cool weather. The items can be conditioned in the Job Assembly room before use. Reserve master magnetic media can be housed in a remote building or room where, in temperate and warm climates, conditions are maintained by building mass.

The computer will dissipate large amounts of heat which have to be removed to prevent temperature rise. There is usually a contractual requirement to provide conditions within close limits to ensure reliability and to allow handling media at high speed. The requirements are specified by the computer manufacturer, and are typically $21°C \pm 2°C$ and 50 per cent rh ± 5 per cent together with a filter performance of 80 per cent average arrestance Euro 4/5 or 80 per cent average efficiency Euro 4/5, at least in the outside air intake (see Section 29.7). The control should center on the specified figures and only touch limits (if at all) with changes of load. In spite of developments in design, heat concentrations tend to rise and configurations expand.

Modern telecommunications equipment is very similar, but environmental tolerances are normally much wider, with particular emphasis on reliability.

29.4.2 Air supply

Cooling air to the cabinets is normally introduced from the room at low-level front or back with fan-assisted discharge to the rear or top. The normal practice is to introduce room-cooling air at high level. This mixes with rising hot air to give a near-uniform condition in the occupied levels. The distribution of cooling air must be carefully matched to heat load around the room. A large room (say, over $300m^2$) should have separate zones of control with their own sensors. The ceiling void should nevertheless be common. The recirculation of air will confine itself to the zones fed by individual units when all are running but will be redistributed by a ventilated ceiling in the event of partial failure of plant modules.

The easiest and most flexible method of introducing air is by a ventilated false ceiling. A great advantage of a ventilated ceiling supply is that the layout of the computer or indeed its make and heat pattern may not be known at the time of air-conditioning design, and will most certainly alter during the life of the plant. The distribution of supply air can be altered to match the load providing sufficient cooling capacity and air volume are installed.

Upward airflow is also practiced. The cable void formed by the raised floor is used to supply air which enters the room via floor grilles. These can be moved to meet the pattern of heat distribution and are normally placed close to the computer cabinets, but consideration must be given to changing air conditions, intended to meet changing room load, entering the computer compartments. Care has to be taken to avoid the updraft lifting dust into the occupied space.

The extracts may be in the ceiling with the advantage of taking up the heat of lights without it passing to the room. Greater use can thus be made of a given airflow. Care has to be taken in the choice of grilles if trolley traffic is going to be present. Low supply temperatures may cause discomfort to staff.

29.4.3 Air extract

Air extract can be at one wall or extract grilles can be interleaved. The normal for all but the smallest computers is to use the under floor cable void as a return path. This void is formed by proprietary 600 mm or 2 ft square interchangeable tiles standing on corner jacks. Some of these may be perforated or they may be fitted with grilles. A maximum velocity of 2.5 m/s over the grille face or through perforations is recommended. Depending on the detail of the tile chosen, a pressure drop of about 10 Pa can be expected, but in the case of carpet-covered tiles where the holes tend to become obstructed, a down rating to 70 per cent is recommended.

29.4.4 Chilled water

Some very large machines have a direct supply of chilled water to a heat exchanger in the central processor or its store, the final distribution within the machine being part of its design. The user's responsibility is the provision of half couplings (often Hanson, USA) and the chilled water supply rate, temperature, pressure, pH and other analyses. Welded pipe and vapor-sealed insulation are essential. End-stop valves should be installed and consideration should be give to fitting bypasses to compensate for changes in demand and to allow one computer (if there is more than one) to be fed while another is uncoupled. Consideration should also be given to the installation of moisture detectors under floor.

The plant should be replicated for reliability and to allow maintenance (say, 3×50 per cent capacity). Proprietary units are available specifically for this function.

The final connection to the computer is by flexible pipe, but it is still necessary to coordinate with others to know the position to terminate and, in due course, which floor tile is to be cut and precisely how. This can be expected to entail extra floor jacks.

29.4.5 Computer heat

Ideally, the air-conditioning designer is given the heat loads to be catered for at the outset. However, in the fast-moving field of computers this may not always be the case. Also, the life of the plant is likely to outlast more than one computer configuration but need not itself require more than re-zoning. Particularly where a computer has a number of peripherals, it is unlikely to dissipate fully to the maker's data at any one time, but it is strongly recommended that further diversity on that for individual items given at the maker is not allowed in the design of the cooling capacity of the plant. It is unwise not to allow for expansion of the computer configuration. An initial design on a basis of an overall heat dissipation (say, $500 \, w/m^2$ overall average) unevenly distributed about the room is not unreasonable for large computer halls. In this field the design engineer should not be afraid to question his brief if the client is to be well served. Flexibility to cater for changes has already been covered above.

29.4.6 Telecommunications

Much telecommunications equipment is designed for over-head cabling, which can be very extensive and dense. Any modern equipment which relies on cooling to function should have breaks in the cabling layout partly to allow cool air to fall. In the case of naturally ventilated racks the air discharge should be directed down gangways (the racks being arranged in long suites) and not directly over the racks. A ventilated ceiling supply can be confined to the racked area to allow other services to pass without breaking the ceiling. Except in smaller cases, for modern systems duct and diffuser designs may be insufficiently flexible and may not be able to meet the concentrations of heat without discomfort to attendants. However, in small installations ($150 \, m^2$) air-handling units may be employed directing air freely over the top of the cabling.

While the air distribution must match the heat distribution, the position of extracts is not important. Direct return to air-handling units mounted around the room in numbers necessary for reliability is common practice. In the event of loss of one or more cooling modules a well-designed ventilated ceiling adds to reliability by distributing the reduced amount of air in the same proportions throughout the room.

29.5 Air distribution and system resistance

29.5.1 Duct sizing

Ducts convey conditioning air from point to point at a variety of speeds. Slow speeds result in large ducts, costly in themselves and in building space. High speeds result in noise and the need for high fan powers. A good basis for air conditioning is 6–7 m/s adjacent to the plant but, as discussed below, less at distant points.

When we refer here to static pressure we mean the difference between internal and external pressure causing air to tend to flow into or out of ducts. Velocity pressure is that due to the air's forward movement. The sum of the two is total pressure.

29.5.2 Duct design

Simple runs of a few meters may be designed for a constant velocity. A supply duct is thus reduced in steps at each outlet or a return or exhaust duct similarly increased in section in the direction of flow.

A large system with branches, several inlets or outlets and some tens of meters long will be more easily controlled at the end distant from the plant if velocities are reduced as we progress down the length of the duct. If air is slowed in a controlled way with the duct sides diverging at not more than 15° included angle its velocity pressure will reduce and (ideally, without loss) its static pressure will rise to maintain a constant total head to compensate for pressure loss as the air progresses down the duct. However, at least by manual methods, design for static regain is laborious and the duct shape unconventional. It is seldom practiced.

A method commonly used is that of equal surface friction per unit run. If rectangular ducts are being considered their equivalent circular diameter must be found. This may be obtained by

$$1.3 \times [(w \times d)^{.625}]/[(w + d)^{.25})]$$

Here w and d are width and depth and the units may be meters or millimeters. Alternatively, the diameter may be found from published charts. Using Figure 29.5 a pressure gradient (Pa/m) is chosen and with each change of volume flow a new diameter is found. This is converted to a convenient rectangular section equivalent to the round section. It will be seen that the velocity reduces.

Where practicable, the large surface area of wide shallow ducts should be avoided to keep pressure gradients to a minimum. There can be no hard-and-fast rule, 4:1 being a suggested limit. The system may, of course, be designed to use only round-section ducts.

29.5.3 System pressure drop

The point of interest is the path of highest pressure drop or index leg. Other parallel branches can be designed of appropriate size to pass the required amount of air, those of lower resistance than the index leg being throttled by dampers. The pressure drop is the sum of the drops caused by the following and is calculated to determine the pressure against which the fan must operate:

Surface friction of duct as discused above (Pa/m × length);
Changes in section;
Bends;
Branches;
Obstructions;
Grilles, meshes, etc.

Pressure drop is calculated by $k \times$ velocity pressure, where k is the resistance factor for the above items other than duct friction and is found from Table 29.1 or similar references. Velocity pressure is found by $\frac{1}{2}$density × velocity^2, taking standard density at $1.2 \, kg/m^3$, or from Table 29.2.

Plant resistance

In the design of tailor-made plant it is necessary to calculate the above as they occur within the plant and add the pressure drop of all other items such as air filters at their dirty conditions, coils, etc. In the case of proprietary units it is normal for the manufacturers to quote an external pressure against which they will deliver the specified air quantity.

Duct resistance

The basis of good duct design is to arrange gradual change to section and direction. It is sometimes necessary to construct a 90° elbow with no inside radius. The pressure drop and noise generation can be greatly reduced by incorporating turn vanes which split the air into a

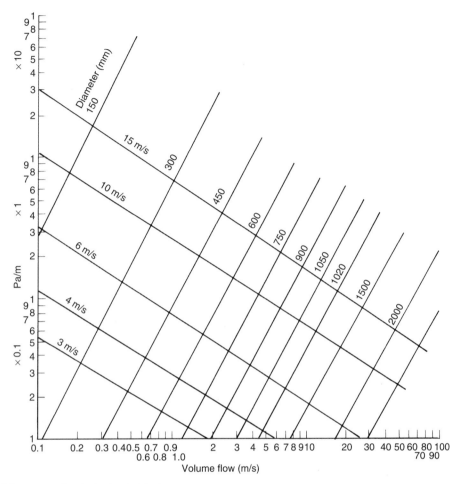

Figure 29.5 Pressure gradient in circular steel ducts

number of near-parallel paths. The information given in Table 29.3 assumes simple (non-airfoil) turn vanes (see also Figure 29.6).

29.5.4 Ventilated ceilings

A ventilated ceiling is an alternative to ducts, diffusers and grilles as a means of distributing air within a room. It is suspended below the structural ceiling forming a shallow void. One type consists of perforated metal trays or tiles. Each is supplied with a bagged acoustic pad, which is removed from those trays or tiles which are to ventilate. Another consists of tiles with single-row slot openings at intervals between runs of tiles, the openings being controlled by dampers integral with the slots. Any type of tile which can shed dust or is combustible should be avoided. Rising hot air and descending cold air mix a few hundred millimeters below the ceiling, giving a near-uniform condition in the occupied levels.

The availability of very small areas of control can be advantageous where electronic apparatus with mild chimney effect would have its natural cooling upset by strong downdrafts, by confining air supply to gangways.

The method is useful where large quantities of air are to be introduced without drafts; where the distribution of heat load is unknown at the time of design or is likely to alter; and where, as a measure of reliability, it is required to redistribute the remaining air in the original proportions in the event of partial failure of the plant.

Ventilated ceilings can cater for 1 kW load per square meter and in excess of 100 ach. Lighting and power tracks can be incorporated and fire-detection heads fitted. In computer suites (see Section 29.4) it is normal to combine a ventilated ceiling supply with an under floor extract, which is then not critical as to layout of grilles and assists any heating mode. The pressure and slot velocity are, for all practical purposes, uniform throughout the ceiling area, and the air pattern is finally set by checking the room temperature from point to point. Pressure drops are of the order of 20 Pa. Void depths are seldom critical, being a function of the throw required and the presence of obstructions such as beams or pipes. Depths of 200–1000 mm are common, with throws from the entry point of up to 20 times void depth.

The air supply is introduced by stub ducts penetrating the void by 0.5 m or less and simply cut off square. However, where there are obstructions across the void it can be advantageous to convey a proportion of the air to or past this by extended ducts. Equally, it can be advantageous to duct air part-way if it is to double back round a corner. In general, however, the system is very tolerant, including positioning air entries about the room. Stub velocities should be 3.5–5 m/s but individual manufacturers will give their own requirements.

Unless beams are present at the entry the stubs should be close to the structural ceiling. Beam fairing may be used where the void has to be shallow and this can reduce the risk of re-entrainment due to local high-velocity pressure. Zone barriers are fitted within the void where separate rooms are to be fed. The ceiling need not cover the whole room.

Table 29.1 Resistance factors

Losses at duct entries	Open area (%)	Factor k
	40	7.5
	50	5
	60	3
	70	2
	80	1.5
		0.03
		0.85
	40	7.0
	50	4.5
	60	3
	70	2

Losses in bends

	Ratio R/D	Factor k
	0.5	0.73
	0.66	0.5
	0.75	0.38
	1.0	0.26
	1.25	0.17
	2.0	0.15

$W/D = 0.5$ R/D

	R/D	
	0.5	1.3
	0.66	0.65
	0.75	0.47
	1.0	0.28
	1.5	0.18

$W/D = 1$ to 3

	0.5	0.95
	0.66	0.45
	0.75	0.33
	1.0	0.2
	1.5	0.13

For consecutive bends opposite direction $k \times 4$

1.25

0.85

R/D	
0.25	0.8
0.5	0.4
1.0	0.3

Losses at section change — Factor k

14° max — 0.15

$\theta°$	
5	0.17
10	0.28
20	0.45
30	0.59
40	0.73

Referred to $V_{p1} - V_{p2}$

30	0.02
45	0.04
60	0.07

Referred to V_{p2}

$0.28(V_{p1} - V_{p2})$

60	0.06

$D_1/D_2 = 0.5$

θ	k
5	0.2
10	0.3
20	0.55

V = velocity reference position, D = duct diameter

Table 29.1 (*continued*)

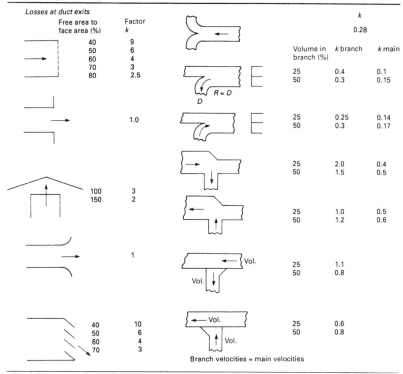

Losses at duct exits		
Free area to face area (%)	Factor *k*	
40	9	
50	6	
60	4	
70	3	
80	2.5	
	1.0	
100	3	
150	2	
	1	
40	10	
50	6	
60	4	
70	3	

k = 0.28

Volume in branch (%)	*k* branch	*k* main
25	0.4	0.1
50	0.3	0.15
25	0.25	0.14
50	0.3	0.17
25	2.0	0.4
50	1.5	0.5
25	1.0	0.5
50	1.2	0.6
25	1.1	
50	0.8	
25	0.6	
50	0.8	

Branch velocities = main velocities

Table 29.2 Velocity pressure (Pa)

m/s	0	1	2	3	4	5	6	7	8	9	10
0.0	0.0	0.6	2.4	5.4	9.6	15.0	21.6	29.4	38.4	48.6	60.10
0.1	0.01	0.73	2.65	5.77	10.09	15.61	22.33	30.25	39.37	49.69	61.21
0.2	0.02	0.86	2.90	6.14	10.58	16.22	23.06	31.1	40.34	50.78	62.42
0.3	0.05	1.01	3.17	6.53	11.09	16.85	23.81	31.97	41.33	51.89	63.65
0.4	0.1	1.18	3.46	6.94	11.62	17.5	24.58	32.86	42.34	53.02	64.90
0.5	0.15	1.35	3.75	7.35	12.15	18.15	25.35	33.75	43.35	54.15	66.15
0.6	0.22	1.54	4.06	7.78	12.7	18.82	26.14	34.66	44.38	55.30	67.42
0.7	0.29	1.73	4.37	8.21	13.25	19.49	26.93	35.57	45.41	56.45	68.69
0.8	0.38	1.94	4.7	8.66	13.52	20.15	27.74	36.5	46.46	56.62	69.98
0.9	0.49	2.17	5.05	9.13	14.41	20.89	28.57	37.45	47.53	58.81	71.29

Table 29.3 Where to place splitters

Duct width (m)	If one split r1 (mm)	If two r1 r2 (mm)	If three r1 r2 r3 (mm)
0.5	110	65,190	50,110,230
1.0	155	85,280	60,150,380
1.5	190	100,380	65,190,520
2.0	226	110,470	70,230,660
2.5	255	115,540	75,255,790
k factors:			
If none			
1.25	0.65	0.5	0.45

29.6 Fans

29.6.1 Fan selection

Fans propel the air through the system, and must be chosen to be capable of delivering the required volume flow against the calculated system static pressure, advisedly with a small margin.

Individual fans are capable of operating over a range of volumes and pressures which are interrelated, the performance being shown in manufacturers' tables and curves. Venturing outside the recommended area of operation may, depending on the type of fan, result in motor overload, motor under-cooling, vibration or stall.

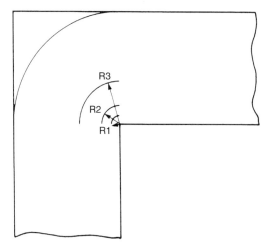

Figure 29.6 Splitters in miter bends

Selection should be made for minimum power input, which is also likely to be the quietest fan for the duty. Performance is usually quoted for a standard condition of 1.2 m³/kg. Calculations of system resistance are best carried out at the same condition. The user may find only the static pressure quoted. If total or velocity pressure are also quoted or the outlet velocity can be calculated the designer can calculate how much pressure can be recovered after the exit. Any mismatch due to difficulty in calculating system resistance will cause the volume to rise or fall, to settle on the fan characteristic curve.

29.6.2 Types

The broad range of types used in air conditioning is:

1. Centrifugal
2. Axial
3. Propeller
4. Cross flow

The centrifugal fan has an inlet on the axis of rotation and outlet from a scroll casing 90° from the inlet. The centrifugal fan is not reversible. The characteristics of the fan are determined by blade shape, i.e.:

1. Straight blades or paddle wheel: Low efficiency; used for conveying particle-laden air.
2. Forward-curved blades: Moves large volumes for a given size; the driving power-required rises markedly with increased volume, necessitating an adequately rated drive motor.
3. Back-curved blades: Highest efficiency of the group; more stable characteristic of volume against pressure; capable of developing higher pressures; non-overloading, i.e. the power does not rise to a peak as the flow tends to zero or free flow.
4. Many developments of the centrifugal fan exist, the most commonly employed being the double-width double-inlet fan moving twice the volume for a given

wheel diameter. It is possible in large tailor-made plants to arrange for the drive motor (and hence its efficiency losses) to be external to the air stream.

Axial fans have a cylindrical casing and the fan blades are mounted on an extension of the shaft on the centerline. The blades are normally of airfoil section. Axial fans are available with adjustable blade angles which allow different characteristics from one construction or alteration on-site by specialists. The power characteristic is usually non-overloading. In the working range the developed pressure increases with reduced volume. Stall is said to occur when the fan can no longer develop the pressure to deliver the required air volume. This is a feature of high blade-angle fans. It is best not to attempt to use such a fan to deliver too small a volume.

They are reversible with varying degrees of efficiency. Air leaves the blades with rotary motion and can be straightened by downstream vanes, leading to increased pressure development. Two-stage contra-rotating fans develop more than twice the pressure of a single stage. Further staging may be arranged.

When using conventional induction motors the fixed-supply frequency and the number of field poles in the motor restrict the available range of speeds. It should be noted that, outside Europe, supply frequencies other than 50 Hz are used, resulting in induction motors operating at proportionally different speeds, and that this applies to all types of fan.

Special versions are available, such as hazard-proof and for high-pressure operation, but these are not normally applicable to the field of air conditioning. Fans are available with fewer blades (part-solidity) to allow the use of smaller higher-speed motors. There are also fans with short casings having a length little more than that to accommodate the blades for use in confined spaces.

Propeller fans are of low capital cost for the volume moved but are used in applications where the resistance is very small such as non-ducted openings through partitions. The power required increases with resistance. Their pressure–volume characteristic changes with the relative position of the blades and mounting plate.

Cross-flow fans have a blade arrangement similar to that of a back-flow centrifugal fan but the casing arrangement allows the incoming air to enter along the width of the blades, avoiding the 90° turn. Long narrow shapes can be arranged, making them suitable for use in small air-handling units, fan heaters or air curtains.

29.6.3 Multiple arrangements

Fan may be operated in parallel but are best of similar characteristic to avoid stalling. In the event of failure of one the effect of reduced pressure drop in the system is to give a flow of about two thirds that of two. Non-return dampers should be fitted.

Fans may also be operated in series. Axial-flow fans are available having two (contra-rotating) impellers and motors in the one casing. Further staging may be arranged. If one stage fails or is switched off it will idle round with loss, but since similar losses in centrifugal fans are much higher, these are not usually operated in series. However, supply and extract fans in a recirculating system effectively in series may differ.

29.6.4 Fan laws

The laws quoted here are selected as those appropriate to the practicable applications of fans:

Volume flow varies as the speed of rotation;
Pressure varies as the square of speed of rotation;
Power varies as the cube of the speed of rotation;
At constant speed a fan delivers constant volume (m^3/s) into a fixed system in spite of change in density;
Pressure and power absorbed vary as change in density.

29.6.5 Volume regulation

This falls under two headings:

1. Where a permanent change is found to be needed during commissioning or change of duty. The volume may be changed:
   (a) By variation of system resistance by damper action. Care is needed to prevent axial fan motors overheating or reaching the stall point;
   (b) By changing the speed of the drive. In centrifugal fans the common practice is to change or adjust belt drive pulleys, taking care not to exceed the power capability of the motor. In axial fans the pitch of the blades may be adjustable by swivel mountings on the hub.
2. Where frequent change is required in the normal operation of the plant:
   (a) By switching motors in multi-stage fans;
   (b) By speed control of motors electrically (e.g. by pole changing), varying the circuit resistance of wound rotor motors, or thyristor part-cycle disconnection;
   (c) By damper modulation;
   (d) By bypass (fan recirculation) dampers;
   (e) By axial fans with pitch variable while running.

Mounting

It is common practice to support the fan on anti-vibration mountings and connect the fan casing to the duct by short lengths of flexible non-combustible material. Care should be taken during installation to see that these are aligned to prevent entry turbulence and noise generation.

29.7 Dust control and filtration

29.7.1 Dust control

Dust is continually being introduced to the conditioned space by clothing fibers, skin particles, shoe dirt, room processes and the outside air make-up. Dust particle sizes range from 1 μm to 75 μm, smaller particles being described as smoke. In an apparently clean office there may be as many as 30 million particles per cubic meter.

Control is by filtration in the plant but smoke can be removed by local recirculation through fan filter units.

For Clean Rooms (rooms of a very high standard) dust count per unit volume will be specified, but other specifications for room cleanliness are usually in terms of filtration performance against a standard test dust. Other important features are resistance to air flow and dust-holding capacity, leading to the fan energy required and filter life.

By number count the great majority of particles in the outside air are likely to be less than 1 μm. By weight, these small particles will account for a very small proportion of the sample. A filter with a high efficiency measured by weight of particles trapped may be almost transparent to the small ones. Very high counts can be found in rural areas from pollen or agricultural activities.

Where a high cleaning efficiency is required it is sound practice to install a filter of lower performance upstream to trap the larger particles and prolong the life of the more expensive High Efficiency Particulate (HEPA) Air Filter. Two filters of equal merit placed in series will not be materially more efficient than one.

The action of most filters is mechanical. These are normally scrapped when fully loaded but may, for lower efficiencies, be washable. Another type uses electrostatic charges.

29.7.2 Electrostatic filters

These are usually provided with a mechanical pre-filter to remove the larger particles and serve to even out the flow over the face. The air then passes between plates positively charged at 10 or more kV and the dust thus ionized is attracted to downstream plates charged at a lower voltage negatively. A mechanical after-filter may also be fitted in case electric flashover at the plates releases dirt. Cleaning is by hot water wash, necessitating drying procedures, or by mechanical agitation. Their resistance to airflow is low and their efficiency good against small particles. They are not often employed in general air-conditioning work.

29.7.3 Tests

Different test methods produce numerically different results and comparisons should only be made of results using the same test method. Tests fall into three categories: gravimetric tests, which measure ability to trap and retain dust; those which measure staining power of contaminants before and after filtration; and those measuring the concentration of a test aerosol either side of the filter by photoelectric methods. The tests which will be met most often are as follows.

BS 2831 has been withdrawn. The No. 1 test for high-grade filters has been superseded by BS 3928 for absolute filters (those having an efficiency of better than 99.9 per cent in that or similar tests) and by ASHRAE 52–76.

ASHRAE 52–76

This forms the basis of Eurovent 4/5. There are two tests. The atmospheric dust spot efficiency test, assessing the reduction of staining power of atmospheric dust, is generally used for the higher efficiencies as well as the synthetic dust weight arrestance test, using an artificial dust. In both cases averages are quoted from a series of tests.

Filters are classified in BS 6540: Part 1 – Eurovent 4/5 but without statements of airflow rate or final pressure drop. From the lower performance upwards:

Average arrestances (Am) EU1 <65%; EU2 ≥ 65 to <80%; EU3 ≥ 80% to <90%; EU4 >90%
Average efficiencies (Em) EU5 ≥ 40% to <60%; EU6 ≥ 60% to <80%; EU7 ≥ 80% to <90%; EU8 ≥ 90% to <95%; EU9 >95%

Higher cleaning ability is tested to Eurovent 4/4.

BS 3928

This test uses a cloud of sodium chloride particles. The size distribution is from below 0.1 µm to approximately 2.0 µm and a figure for percentage penetration is obtained. Since the test does not materially load the filter, it can be used to assess the quality of each unit before dispatch. This test forms the basis of Eurovent 4/4, which differs a little in procedure.

29.7.4 Standards required

For many applications, filters with an average arrestance of 80 per cent against Eurovent 4/5 will serve, but in situations that are more critical and, for example, in the outside air intake of computer rooms or laboratories 80 per cent average efficiency would be chosen.

29.7.5 Clean rooms

The dust count to be achieved may be specified in terms of US Federal Standard 209b. There are three standards: 100, 10,000 and 100,000, which are the maximum counts per cubic foot of 0.5 µm and larger particles. Maxima for 5 µm particles are also given. BS 5295 has four classes: 1 is the most stringent and calls for a marginally lower count than US Federal Standard 209b, class 100. It carries details for a range of particle sizes, room construction and airflow pattern. Laminar flow is required for the cleanest areas, which can be achieved when the ceiling or one wall of the room is composed of a louvered filter bank. Filter efficiencies ranging from 70 per cent to 99.995 per cent are recommended.

29.7.6 Filter life

Filter life varies with make and type, and may be limited by the ability of the fan to operate against pressure drop. It is a function of the dirtiness of the air and the amount of material packed into the filter bank. Life may be quoted in terms of dust held in g/m^2 face area. For fan selection knowledge of pressure drop is required. Typically, a panel filter might be quoted as operating from, say, 75 Pa initially at 2.5 m/s face velocity to 250 Pa when loaded and a HEPA might operate up to 700 Pa. The pressure drop across a bank of filters is kept within bounds by changing a proportion in rotation.

29.7.7 General

Care should be taken during installation to see that there is no edge leakage. For tailor-made plants, it is best if filters are kept sealed in their delivery containers until after the plant has been completed and vacuumed out. Unfinished duct ends should be covered during the progress of other work, since cleaning will be almost impossible, except by the expedient of blowing through and physically disturbing them.

Filters should meet at least the requirements of (in the UK) BS 2963 and in some cases be non-combustible to BS 476: Part 4, although construction methods render this difficult.

29.8 Humidification

The ability to add moisture to the air to raise humidity is an essential part of close control air conditioning. The need may be brought about by a change in the condition of the outside air, fabric losses or a change in the gains within the room or in plant operation.

29.8.1 Humidifier capacity

The humidifier is sized to meet the load of the outside air intake quantity. It is necessary to know the rh to be achieved, outside air intake rate (kg/s) and its lowest winter moisture content (kg/kg).

Since the capacity will be reduced by blowdown time, allowance is necessary for control, and there will be leaks from the plant and building fabric, an addition of about 30 per cent above the calculated value is recommended.

29.8.2 Types of humidifier

These may be direct (i.e. installed within the conditioned room) or indirect (i.e. installed in the conditioning plant). Water may be introduced (a) as a spray or mist or (b) as steam generated separately or by a device within the plant.

In most air-conditioning work, the humidifier will be indirect. The psychrometric operation of various types is described below.

29.8.3 Pan humidifier

The steam is generated in a pan of water by electric elements. High-temperature hot water or steam coils could also be used. About 30 per cent of the input appears as waste sensible heat, giving a sensible heat ratio of 0.3.

Where the pan is in the air stream the condition downstream of the pan has increased moisture content (kg/kg) found from the airflow and moisture input. On a psychrometric chart, this will lie on a line of sensible-to-total heat ratio of 0.3. Thus, the psychrometric plot shows a steep rise in moisture content with a small rise in dry bulb temperature. The latter is a disadvantage when cooling is required. Regular blowdown is needed, preferably controlled by a timer, to prevent furring when mains water is used, and open pans should not be utilized where the conditioned space is to be dust controlled. A small water volume is an aid to quick response.

29.8.4 Steam jet

Where it is available the source can be a separate boiler plant, but common practice is to employ purpose-made electrode boilers within or adjacent to the plant. The latter reduces sensible gains to the plant but, being essentially saturated steam, condensate return pipes are required. In addition to the rise in moisture content of the air (kg/kg) being dependent on airflow and steam-injection rates, there is a very small increase in dry bulb temperature by the cooling of the vapor to the air temperature. The rise in total heat is: total heat of steam (kJ/kg) × quantity supplied per kg air.

Proprietary units are supplied with automatic blowdown cycles and can be matched to the broad water analysis. Cylinders have a limited life.

29.8.5 Spray humidifiers

Water is injected into the air stream in a fine mist by pumped jets or spinning disc. For practical purposes, the psychrometric plot follows a wet bulb line. The air provides the latent heat of evaporation, resulting in a fall in dry bulb temperature. If water were to be supplied at up to 100°C the humidified condition would be at a correspondingly higher total heat of 420 kJ per kg water supplied.

Where dust control is important the system should only be used with a supply of demineralized water to avoid solids being passed into the conditioned space. The temperature of the air must be sufficient to hold the quantity of moisture being supplied, any excess being deposited in the duct. Unless drained away, this can give rise to corrosion and to incorrect control by re-evaporation when the humidifier is switched off under control.

29.8.6 Air washers

Banks of sprays discharge water into the air stream to achieve saturation of the whole air flow. Excess water falls into the base tank of the washer, from which it is pumped back to the sprays. Downstream eliminator plates entrap any remaining free moisture, acting best within a specified velocity range. Cleanliness is essential to avoid bacterial growth. A constant bleed and make-up is normally arranged to control the accumulation of waterborne solids, but this, in turn, dilutes bactericides and inhibitors. The washer does not fully wash the air in the normal sense but does have the cleaning efficiency of a low-grade filter.

If the temperature of the water is not controlled it will come to the wet bulb temperature of the air passing through. Ignoring pump heat, the process is adiabatic. The psychrometric plot follows a wet bulb line.

If heat is added to the water, the condition for 100 per cent saturation takes the new wet bulb temperature of the incoming air.

If the water is chilled, cooling of the supply air takes place together with control of dewpoint and hence humidity of the room treated. Saturation efficiency is given by:

$$\frac{\text{Entering air db} - \text{leaving air db}}{\text{Entering air db} - \text{entering air wb}} \times 100\%$$

Due to bypass, a single bank of sprays might achieve 60 per cent saturation and a capillary washer (one where the air passes through a wetted mat) might achieve 95 per cent.

29.8.7 Humidifier run time

Humidification is an expensive process and it is useful to be able to assess energy costs when considering its inclusion in a plant. It is possible from meteorological records for any particular area to find the time in hours per year and extent to which the external moisture content is, on average, at or below a required absolute value. As an example, in the southern UK humidification would be required in varying degrees for 6700 hours per year if 21°C 50 per cent relative humidity was to be maintained. If the characteristic of cooling coils is to dehumidify when only sensible cooling is required then the humidifier load will be greater.

29.9 Test procedure for air-conditioning systems

29.9.1 Object and application

This is an on-site test following installation. The procedure is designed to find any weaknesses in the plant and its implementation, while not cheap, is worthwhile where performance is important or in doubt. It is essential to have contractually agreed its use, which is to make the various provisions and what action is to be taken if any faults are shown and to allow sufficient time for testing and any remedial work.

It is recommended that a contractor demonstrating to a client should have tested the plant previously and should not use the test to supplant commissioning. The acceptance that individual items of plant have the necessary capacity does not necessarily mean that the whole will perform as required. Indeed, the test can find proprietary items lacking.

29.9.2 Conduct of the tests

The necessary skills should be on-site to attend to any minor faults but must not be allowed to alter settings from one test to another, except as required by the test procedure. Dummy loads should be ready for use. If sensitive equipment is to be present (as might be the case in upgrading computer room plant) then the interested parties must agree limits previously. This may also apply to parts of the plant itself such as humidity-detecting elements. The latter can be protected physically. The job specification should state which parts of the procedure apply (if not all).

The plant is tested to hold limits in the specified external conditions and is not (unless exceptionally so agreed) tested beyond them. The plant should not be over-designed just to pass the test. For the duration of the test external conditions can never span those for which it is designed. To fully load the plant in cool weather it may be necessary to dissipate twice the design process load. The dummy load should simulate the design load, and in the case of electronic equipment rooms, arranging fan heaters directed upwards and in the same layout as the equipment may best do this. Ideally, simulation of the heat of outside air make-up should be in the intake duct. If impractical, then it may be placed in the room close to a return grille. Room-temperature readings should not be taken near to dummy loads, and in heavily loaded rooms it may be necessary to confine readings to extract positions. Except in test houses, it is not possible to apply the full range of external conditions at the final cooler. It will therefore be appreciated that a plant should easily pass its test on a cold day. Resort could be made to certify performance curves if a doubt exists.

Misleading results can be obtained from tests of limit statistics and sensors if their set point is adjusted to the current conditions to bring about operation. The control points should be set and conditions adjusted until operation occurs, opportunities arising during the test procedure.

The following is an outline of the test procedure:

1. *Pre-test performance.* An assessment of whether the plant is ready for a full test.
2. *Recorders and alarms.* Continues 1 and checks the calibration of recorders and alarms.
3. *Air balance.* To check overpressure and outside air intake.
4. *Background heaters.* Checks control and capacity of back-up heating.
5. *Heating and humidification.* The plant is called upon to perform at its maximum winter design capacity.
6. *Cooling and dehumidification.* The plant is required to perform at its maximum summer capacity.
7. *Limits, interlocks and inspection.* Cut-outs, limiting controls and interlocks are tested.

Compliance with specified details and good practice are checked.

29.9.3 Pre-test performance

Allow the plant to run automatically for a period of some days either with no artificial load or with a small one of fixed value. If no means of recording are built into the plant then recorders should be introduced and placed in representative positions in the various zones for the duration of the tests. Periodically examine the traces or monitor printouts. These will give an indication as to whether the plant is controlling correctly on light load.

Observations of external wet bulb and dry bulb air temperatures or similar assessment of other forms of load may be combined with simple observations of what the plant is doing. If it were using its refrigeration capacity under cold no-load conditions then clearly something is wrong. A common fault is finding the plant calling for

large amounts of humidification when the requirement as indicated by external readings is for dehumidification, this perhaps being due to incorrect SHR of the cooling coil. Periodic excursions may be found and these are often caused by external influences, which might be identified by their time of occurrence.

Make no excuse for checking whether the sun shines on the thermostats or someone has washed the floor. Another common occurrence is that while the room is not being controlled as specified, the plant appears to be operating correctly. Here a series of simultaneous readings of air flows and wet and dry bulb temperatures should be taken around the whole of the air circuit – on and off coil, condition after fan, condition at inlet grilles, etc. Referring these to a psychrometric chart will usually indicate where the fault or leak lies.

29.9.4 Recorders and alarms

These are important, since they may be used in part using the remainder of the procedure and may be the only indication for future operatives as to whether conditions are correct.

1. With the plant still operating as for test 1. With hand instruments, check conditions as close as possible to the recorder's heads:

Recorder no.	Time		Date
	DB	WB	RH
Recorder reading		–	
Test sling (or other)			

Note that some recorders have a clock time difference between the dry bulb and rh to allow for pen overlap.

2. Where this can be done, moving the alarm-indicating arms to the room condition will indicate whether the alarms operate but much more satisfactory results are obtained by causing the controlled condition to swing beyond the alarm points and at the rate that would be expected in practice.

Recorder no.	Time		Date
	DB	WB	RH
Recorder reading		–	
Test sling (or other)			
Alarm occurred		–	

29.9.5 Air balance

Room overpressure can conveniently be tested during this period. Produce a single line sketch of the conditioned space(s) including doorways. Test for air-flow direction with hand smoke tubes, both at the top and bottom of closed doors, and record the direction of flow. Crack the doors open if fully sealed.

Measure the outside air make-up quantity, taking a pattern of readings across the area of the duct or grille:

Average flow velocity m/s $\times$ area m^2 = m^3/s

29.9.6 Background heaters

The function of these is to prevent the room becoming cold and humid in the event of the main plant being off for any reason other than power failure. For a small input they can protect the conditioned space to a large degree and greatly reduce reconditioning time upon restoration of the main plant. They would require a comparatively small input from a standby generator.

With the main plant switched off:

1. Set the humidistat (if fitted) downwards and note when the heater(s) come on. As before, results that are more accurate are obtained by swinging room conditions, but (a) we will have ascertained that the direction of operation is correct and (b) the method is sufficient for the relatively coarse control required. Reset.
2. Set the thermostat upwards noting when the heater(s) come on and continue to the limit of the thermostat. Take a set of readings.

Room	Date
Time	Dry bulb temperature

Assuming the heaters are electric (i.e. have a high surface temperature) and the conditioned space is uniformly surrounded by rooms or the outside and then the temperature that should be achieved is the design lift in temperature from the surrounds to room temperature, above whatever the surrounding temperature happens to be on the day of the test.

If there is a mix of external walls, heated or cold surrounding rooms and rooms under similar control to the one being tested the temperature elevation of the tested room ($^\circ$C) should be:

Design heat input (watts) + sum of (surface constants $\times$ TD to outside at time of test) for each surface $\div$ sum of constants for all other (outside) surfaces

where the surface constant is U value $\times$ area.

3. Since any background heaters must now be allowed to operate during the cooling cycle their control must be set below the lower limit for the conditioned space. Interlock with the main plant may be included. These features should be checked and noted at this time.

29.9.7 Heating and humidification

The plant should be able to hold the room at the desired temperature when heat is not being dissipated in the room. The plant should also be able to raise the temperature of the room from cold under winter conditions (e.g. after a power failure or an outage). In considering the temperature to be achieved under test the comments given in test 2 of Section 29.9.6 also apply here. If hot water heating is used, full temperature elevation will not be achieved because high return air temperatures will reduce the coil output. For instance, where the design rise was 22°C above an external of -1°C; 19°C, 29°C and 14°C are satisfactory with ambient of 5°C, 10°C and 15°C, respectively.

If it is variable, set the outside air intake quantity to the winter value. Set the desired temperature up to a maximum, leaving the humidity setting alone. It will be appreciated that as the room temperature rises during the heating test the rh tends to fall. However, since the humidity setting remains unaltered the humidifying system will be called upon to operate until at one condition it is working at peak winter rate. Due to the faster characteristic of heaters, the rh will be found to fall but absolute moisture should be found to steadily rise. The duration of the test is normally about 3 h and final conditions should be held for half an hour to prove the moisture source.

Calculation of test condition kg/kg

Absolute moisture in the room at design temperature and humidity;
Absolute moisture content of winter outside air intake at design condition;
Design difference to be made up by humidifier = Add absolute moisture content outside
at time of test; = Room moisture content to be achieved in test.
(Neglect gains from test personnel)

Room	Date		
	DB	WB	= kg/kg
Time			

Under steady conditions the plant should be able to hold the design room rh up to the temperature given in psychrometric tables or a chart where the design room rh and test absolute moisture content (kg/kg) coincide. At the end of the test restore the temperature set point to the design value and any control of outside air volume to automatic. Observe that operation.

If the plant is now shut down for a period of 1 or 2 h with the conditioned space remaining closed, a slow reduction in absolute moisture content will be observed. A rapid fall will indicate a significant leak in the building or plant, would account for any difficulty in achieving the test result and should be investigated. The design vapor pressure difference may amount to several tens of millimeters water gauge acting on the vapor sealing and equally leads to moisture ingress during summer conditions.

29.9.8 Cooling and dehumidification

The test load must be calculated and is applied in two parts. The lights should be as in normal use and all plant controls normal and automatic. The background heaters (if installed) can be called as load but it is essential that thermostats and overloads do not reduce the applied load unobserved. Electrical measurements should be taken, particularly at the full-load condition, and for this purpose, trust not put in rating plates.

Room   *Date*

	Supply Air	Sens. load kW	Lat. load kW	Position 1 Dry Wet RH	Position 2 Dry Wet RH	Position 3 Dry Wet RH

Time

the remainder of the test load which represents the missing summer building gains is now applied.

Design sensible heat of outside air intake kg/s × kJ/kg
Less actual on day of test   kg/s × kJ/kg

Room design conducted heat gain   kW

less: $\dfrac{(\text{design gain}) \times \text{actual TD}}{\text{design TD}}$ kW

Design solar gain less actual gain (under site conditions this will have to be estimated)   kW

      Total kW

Plus

Design outside air moisture content (kg/kg)
Less actual moisture content on day (kg/kg) ×
air flow (kg/s) = kg water/s

(it will be found that kg water per hour ×2/3 closely approximates
to the power in kW of test load kettles required).

The test load: watts
    Sensible equipment or process heat;
    Sensible heat of personnel;
    (Less number present)

    Latent heat of process;
    Latent heat of personnel;

In the case of electronic equipment rooms there will be no call for a latent test load at this stage.

Starting from zero, take a set of readings, then increase the load in increments of about 20 per cent, repeating at 15-min intervals. Note that if the supply air passes over a structural slab its elevated temperature from test 29.9.7 will briefly add load.

To avoid unnatural conditions the latent load is applied in steps with the sensible load until full summer conditions are simulated. This should be held or 1 h or more. If the pattern of temperature readings around the space is satisfactory we have a check on the suitability of the air distribution. If any doubt exists as to the duty achieved, simultaneous readings of air flow and wet and dry bulb temperatures across the cooling coil or supply to extract can be taken and the duty calculated.

Observations should be made that the plant is not being manually coaxed, that it is not humidifying and that the compressors have not tripped but are cycling under part-load conditions. Instruments should be indicating correctly. In cases where it is impractical to fully load the plant, ascertain that an appropriate amount of refrigeration is being employed. This situation is best avoided, as much of the plant is not being demonstrated to full capacity.

29.9.9 Limits and interlocks

The items for test under this group will vary from plant to plant and with specified requirements. It is assumed that the clerk of works is satisfied with the following:

    Interior of plant clean before run;

Interior of ducts clean before running or blown out to bags;
Ducts sealed;
Lagging sound, non-combustible, no loose material in airstream;
Vapor sealing sound;
Filters of correct grade;
Filters kept clean before installation;
Filters installed without leaks;
Water circuits leak tested;
Water circuits flushed;
Water circuits dosed;
Refrigerant circuits vacuumed and dry;
Refrigerant circuits leak tested;
Crank case oil level correct;
Plant room ventilated;
Humidifier drainage and deep *U*-trap correct.

Test

For each compressor:

HP cut out;
HP cut in;
LP cut out;
LP cut in;
Oil differential cut out;
Oil return functioning (sight glass);
Low water temperature cut out;
Recycling timer;
Loss of phase protection;
High room humidity override;
High room temperature override;
Plant heat high temperature cut out;
Plant low air temperature cut out;
Air flow interlock;
Automatic changeovers;
Room emergency power off buttons
Emergency power off buttons reset (only when so manually operated).

30

Energy Conservation

M G Burbage-Atter
Heaton Energy Services

Contents

30.1 The need for energy conservation

Energy conservation has often been referred to as the 'fifth fuel', the other four being the so-called primary or 'fossil' fuels of coal (solid), oil (liquid), gas and nuclear/hydroelectricity. This emphasizes the importance of reducing the amount of energy used, not only nationally but also internationally.

The simple fact is that the world's reserves of fossil fuels will eventually run out, depending on the rate of use, and therefore, if the consumption of these forms of energy are reduced, the existing reserves will last longer. Research and experimentation could lead to those reserves currently available but uneconomic to recover and use being rendered economic, thus extending further the number of years before these non-renewable sources of energy do eventually run out.

The amount of worldwide energy reserves and life is variable according to the source of information used, but the position is of the order shown in Table 30.1. Thus at some time in the not too distant future (less than 100 years) oil, gas and uranium will no longer be available. Every effort is required to reduce the world's energy demand to cater for this event. In the case of the UK, the situation is very similar.[1]

The economically recoverable coal reserves are of the order of 4.2×10^9 tce, which, with an annual consumption of around 111×10^6 tce, gives a life of 38 years. However, the recoverable coal reserves are much greater, at 30×10^9 tce or 270 years' supply.

Recoverable oil reserves are given as 1.23×10^9 toe. The UK consumption is only 70×10^6 toe, some 60×10^6 toe being exported from the total annual production of 130×10^6 toe. At this total production level there is only 10 years' supply (but the cessation of exports would virtually double the life). Gas reserves are similarly limited, and are said to be equivalent to about 40 years.

Hydroelectric sources of power are being exploited to the full in both the US and UK, and the contribution of nuclear power is subject both to the supply of uranium ore and to environmental problems. At present, nuclear generation only contributes around 7 per cent of the total UK energy consumption.

The major problem in energy conservation is, however, not the concept but the economics. Saving money by

Table 30.1 World energy reserves

		Oil $(10^9 t)$	Coal $(10^9 t)$	Natural gas $(10^{12} t)$	Uranium $(10^6 t)$	Shale oil and tar sands $(10^9 t)$
Present reserves:						
Lowest		89	480[b]	77	2.0	80
Highest		96	630	96	2.2	90
Life at:						
1971 consumption[a,e] rate:	Lowest	32	60	33	–	39
	Highest	36	1000	45	–	48
1984 consumption rate:	Lowest	31	150	49	64	28
	Highest	34	190	62	70	32
1971 future view:	Lowest	16	30	15	16	Extend oil by 9
	Highest	18	190	19	50–100[d]	Extend oil by 11
New future rate:	Lowest	30	120	40	37	Extend oil by 25
	Highest	32	160	50	38	Extend oil by 30
Future reserves:						
Lowest		200	7000[b]	185	1.2	370
Highest		350	10 000[c]	295	1.4	1300
Life at:						
1984 consumption:	Lowest	70	>2000	120	38	130
	Highest	125	>3000	190	45	>450
1971 future view:	Lowest	30	150	25	20 (50–100)[d]	Extend oil by 10
	Highest	40	250	40	37 (50–100)[d]	Extend oil by 17
New future rate:	Lowest	67	>1700	95	21	Extend oil by 120
	Highest	116	>2500	150	24	Extend oil by 430

Notes:
Units: Oil, shale oil and tar sands are expressed in billion (10^9) tonnes of oil.
  Coal is expressed in billion tonnes of coal (multiply by 0.66 to give billion tonnes of oil equivalent).
  Natural gas is expressed in trillion (10^{12}) cubic metres (multiply by 0.86 to give billion tonnes of oil equivalent).
  Uranium is expressed in million (10^6) tonnes of uranium (multiply by 0.245 to give billion tonnes of oil equivalent).
  Life is in year.
[a] Data taken from *Energy from the Future.*
[b] Excludes lignite.
[c] Includes 2400 billion tonnes lignite.
[d] Figure takes into account fast reactors. It is stated in original document that predictions are 'particularly speculative'.
[e] 1971 consumption rates were taken as oil 2500 Mtoe; gas 900 Mtoe; coal 1500 Mtoe.

energy conservation is preferred to saving energy unless the energy saving occurs with the main effect required of cost saving. Hopefully, the energy manager will be able to persuade colleagues that both financial economy and energy reduction go together, hand in hand. He must take positive action within his organization at all levels, from the board of directors to the shop floor to achieve the stated objectives.

30.2 Energy purchasing

The major energy cost-reduction exercise which can be undertaken prior to an energy audit is that of ensuring that energy is purchased at the most economical price. This is a relatively simple matter, and can be undertaken by the energy manager and his staff, if they have the expertise. An alternative is to use the fuel suppliers, such as the local electricity companies who undertake such work free of charge in the case of their own tariff investigations, or by independent energy consultants. In the case of the latter, fees will be payable, generally based on a percentage of the cost saving achieved over a period of time. These fees can vary from 50 per cent of the annual savings for a period of five years down to 50 per cent of the actual savings for a twelve-month period only. When using consultants the energy manager should inquire around in order to find the practice that bests suits his needs.

30.2.1 Industrial coal

The basic price of coal at the pit is based on the coal gross calorific value, with allowances then made for the ash, sulfur and chlorine contents. The haulage charges depend on the distance from the pit to the site and on the method of delivery. Tipper-vehicle deliveries are cheaper than conveyor vehicles which, in turn, are cheaper than the pneumatic (blower) vehicles. The method of delivery will obviously be decided by a combination of space and cleanliness factors. In the case of certain customers, special agreements may be available at special rates where the annual coal consumption is large and the supplier wishes to retain the market.

30.2.2 Oil

The prices of the various grades of oil are highly competitive due to the relatively low cost of crude oil as compared to the situation a few years ago. Oil is available in two grades: gas oil (35 secs) and heavy fuel oil (3500 secs) as compared to the four previous grades.

The basic prices of the two grades of oil are interrelated to the prices of general tariff gas (non-interruptible) and contract gas (interruptible). The price also varies according to location, generally being lower around the coast or close to oil refineries.

30.2.3 Gas

Variations in contract gas prices are due to the availability of the fuel. A non-interruptible supply costs more than an interruptible one by a quite significant figure. The energy manager considering a change of fuel source to gas should maintain his options by installing dual-fuel plant rather than a single-fuel-fired plant in order to gain advantage from the fluctuating fuel price market.

30.2.4 Electricity

Regular meter readings are required to ascertain the breakdown of the total load and these will form a useful basis for the energy audit. Regular readings throughout the working week (say, at hourly or even half-hourly intervals for a week or fortnight around periods of peak energy consumption) along with details of the plant in use give an indication of the possibilities for peak lopping in order to reduce the maximum demand. Alternatively, suitable monitoring equipment with printout facilities would be of great benefit in many cases, especially where fluctuating loads are concerned. It should be remembered that the highest maximum demand incurred applies not only for the relevant monthly account but also for the availability charge for the next twelve months unless exceeded, when the new, higher, figure applies.

Maximum demand alarms are available for individual organizational use and these can be made to shut down non-essential plant in order to maintain a lower actual maximum demand. Where automatic shutdown is not in use alarms can be used to trigger a manual shutdown procedure.

30.3 The energy audit

The basic energy facts must be established in an organization's energy conservation campaign. Any organization will purchase energy, even if this is limited solely to electricity. Thus, as a start, the actual energy accounts are available for use. These will have been employed already in determining the most economical purchase price (see Section 30.2). The accounts, however, only indicate the total site consumption, generally during the last three months, or monthly, depending on the plant size.

In order to gain an accurate picture of the site energy usage it is necessary to:

1. Provide metering facilities for each energy source for each major cost area (e.g. the boiler house or a particular production process);
2. Read these metering facilities regularly. This can best be undertaken weekly and, at worst, monthly, but the frequency will depend on the load pattern and level of consumption.

On the basis of these regular readings, energy consumption and costs can be allocated to particular cost centers, items of plant or process. The energy audit can be undertaken by using the organization's staff, such as the energy manager and his department. As an alternative, an energy consultant can be called in. It is suggested that costs and levels of consumption should, where possible,

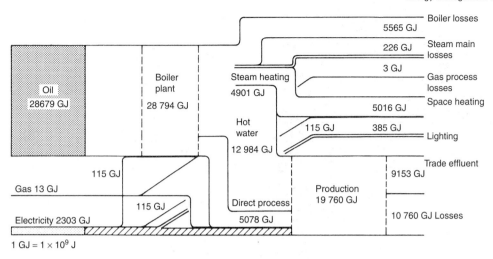

$1 \text{ GJ} = 1 \times 10^9 \text{ J}$

Figure 30.1 A typical Sankey diagram for a dyehouse

be in some way related to production output, operating hours or process requirements.

The energy consultant may be better employed in investigating a specific cost center, as detailed investigation could be required such as energy readings every hour, or half an hour or even every few minutes (e.g. a boiler test). Because of an energy audit, it should be possible to build up the pattern of energy usage of the whole site, or a particular cost center. This can be expressed in energy terms in a Sankey diagram (Figure 30.1). The figure shows the uses made of the various energy sources purchased, and where this energy is eventually lost from the site to waste. It is essential to maintain a regular monitoring system, recording energy consumption and costs. The use of computers provides a ready-made facility for this and software packages are available (see Section 30.5).

30.4 Energy management

Energy management will cost money whether the staff are employed by the organization or outside consultants are used. Large organizations, which already employ their own engineering staff, probably add energy management duties to an existing member of staff, or may appoint a specialist in such matters. Extra staff may be appointed, depending on the nature of the business.

In this case, there is an on-going cost, and this must be borne in mind when embarking on an energy-conservation campaign. Within a large organization with a large number of buildings, 80 per cent of the energy consumption takes place in 20 per cent of the buildings, and initially there are major savings to be made. However, when the initial and obvious energy-conservation schemes have been implemented then there only remains the effort to maintain the results achieved and to investigate any possible new schemes.

Within a large organization there may well be sufficient continuous work to keep a small team of

energy-conservation engineers busy. However, in many cases, once the initial work has been completed then perhaps only one or, at most, two persons are required. If this is likely to be the case then the use of energy consultants may prove attractive for the investigation of individual areas of energy conservation, as there is no problem of long-term employment costs. In the case of small organizations, unless there is a very high energy consumption then full-time company employment cannot be justified.

It may be possible to employ a suitable engineer on a part-time basis, but perhaps the best alternative is to use either an energy consultant as and when required, or a suitable student from a technical college or university during the vacations or during their year out of college on industrial placement. The advantage of the latter is that an eminently suitable student may then be offered full-time employment upon successful completion of their course, which could include further energy work. The financial arrangements could be in the form of:

1. Energy consultant:
   An agreed hourly rate or day rate within a specified period;
   An annual retainer with a lower hourly or day rate;
   A percentage of the capital cost of the scheme involved;
   A percentage of any savings obtained.
2. A student:
   An agreed hourly or weekly rate for the specified period;
   An agreed salary.

Continuous energy management is required and the person so employed must have access to the organization's highest management or be a member of that team.

Many excellent energy-conservation schemes have failed or been altered by senior management due to inadequate representation from the energy-conservation staff concerned. The approach to energy conservation and energy management must be whole-hearted and enthusiastic. Only then can the best results be achieved.

30.5 Energy monitoring

Reference has been made earlier to the need for energy monitoring as part of the energy audit (Section 30.3). The monitoring system used can be as simple or as sophisticated as required. In its simplest form, energy monitoring consists of recording the billing details of the various energy suppliers and comparing these bills by reference to the previous bill or to the previous bill for the period twelve months earlier. Such comparisons are not the best basis for commenting on the performance of the plant concerned. Changes in plant and equipment, and the use of that plant and equipment, may give rise to large variations in energy consumption, hence costs.

Thus for the best comparisons not only must the energy usage be monitored more frequently, but at weekly intervals (monthly at worst), with other parameters being recorded. These could include the following:

Plant/equipment
Period of time and times in use, each day of the week;
Load factor on that plant – goods produced;
Conditions of the plant – pressure, temperature, etc.;
The identity of the plant operator.

Buildings
Period of occupancy each day;
Internal environmental temperature and humidity;
Number of occupants;
Nature of the building construction;
Internal heat gains from lighting, machinery, etc.;
External heat gains from the sun;
Climatic conditions such as cloud cover, wind speed, external air temperature and humidity.

From this information it should be possible to ascertain the basic standing loads and the variable production loads which make up the total site load.

It is necessary to record many data, which is time consuming when undertaken by individual members of staff. Meters are available for the direct measurement of such energy sources as oil, gas and electricity and for the measurement of water, steam and air flows. Steam can also be measured in the absence of any metering by diverting the condensate into a drum of water and weighing the drum prior to and after the test, noting the actual time for the test. Any steam system will have steam traps, and in order to check that these are working correctly, sight glasses should be provided after the trap. Solid fuel such as coal, coke and wood can best be measured by weighing the fuel on an appropriate system such as a belt-weighing machine for large quantities of fuel down to a simple spring balance and small drum for small amounts.

Many of these parameters can be easily measured and recorded by modern devices, utilizing a minimum of manpower. A large variety of modern computer-based energy management systems will monitor, record and store the required data. Some will also control the required parameters within closely defined limits.

Energy management systems can vary in cost from low to high ($100,000–$200,000) for the system itself. In addition, energy controls may have to be provided either to replace existing controls which are not compatible or which are not operating correctly or have never been provided in the past.

This additional cost to facilitate and implement the energy management system can vary from $200,000 to $600,000. In the long term, the centralization of energy monitoring can yield large financial savings, coupled with a considerable reduction in energy consumption.

With such systems, it is possible to obtain individual room space heating and humidity control without the need to dispatch an engineer to check every room in the building. Manpower costs can be reduced, with the staff being required for checking and maintenance purposes only. The cost of an energy management system can be recovered in a relatively short period of time (of the order of 2–3 years).

The cost of an energy management system at one site in the north of England was of the order of $1,000,000 and the payback period was just over 2 years. In view of the high cost of such systems, it may be possible to obtain a system on a loan basis, the system being rented from the installer or by lease purchase. All options should be considered before entering into a contract.

A further advantage of an energy-management system is that the quantity of production can be included for process plant and a correlation made between the product and the energy required producing it. Such a figure is useful in comparing similar processes within the same organization and in comparison with similar ones. It also allows the energy cost per unit of production to be calculated.

30.6 Energy targeting

Having initially ascertained the basic energy data, energy targeting is the next logical step. It is desirable to know if the energy consumption of a particular site or piece of equipment can be reduced without detriment to the product or personnel involved.

The easiest way is to alter the parameters downwards and note the effect, while monitoring the plant conditions and energy consumption. The result of this is that eventually the product will become useless and/or the personnel aggrieved at the change in environmental conditions. This method is frequently used by making any downward adjustment small and monitoring the effect, adjusting downwards again after a short period of time.

It is possible to undertake theoretical calculations to ascertain the amount of energy required for the operation and to compare this with the actual consumption. To do this requires the knowledge of a large number of parameters but it can be undertaken, especially if an energy management system is installed.

The amount of energy required for space heating in a particular space can be estimated from the various parameters and compared with the actual energy consumption.

The difference indicates the scope for further improvement. The following is a typical case of a warehouse used for the storage of goods only but with access by staff for goods movement:

Warehouse size – 91.5 meters long by 61.0 meters wide by an average height of 8.13 m, giving a total floor area of 5582 m^2 and a volume of 45,400 m^3.
Modern construction with a wall and roof U-value of 0.50 W/m^{2}C.
Internal environmental temperature required $= 15°$C.
External air temperature $= -3°$C.

$$\text{Natural air infiltration rate} = \frac{1}{2} \text{ air change per hour}$$

$$= 22,700 \, m^3/h$$

$$\text{Calculated heat losses – Building fabric loss} = 146,000 \, W$$

$$\text{Air infiltration loss} = 152,000 \, W$$

$$\text{Total loss} = 298,000 \, W$$

Type of heating in use – Thirteen downdraft-type steam unit heaters, controlled by an optimizer and internal air thermostats.
Hours of occupancy – Five days per week, 16 hours per day.
Calculated annual steam requirement $= 692,000$ kg (1,525,000 lb)
Calculated steam cost at \$19.00 per 1000 kg $= \$13,400.00$.

In this case, the actual annual steam consumption that was metered was 1,906,000 kg (4,202,000 lb).

In the case of a process, such a basic estimate is harder to achieve, but this can be undertaken. The heat requirement for most chemical reactions is known and this method can be used in most manufacturing processes to determine the energy requirement.

In a series of chemical process vessels where heating of each vessel takes place it may be possible to reduce this amount of heating by ensuring that the product from the first vessel is hot enough for the next. This can be achieved by alteration to the liquid boiling point by variation in the vessel pressure. A reduction in pressure reduces the boiling point. The rescheduling of production may also lead to economies.

The dyeing of cloth is a good example. The dye master of old always insisted that the dyeing process in all the required vessels be started at the same time. This process began with the boiling of large quantities of cold water in the dye vessels by the direct injection of steam through 50-mm diameter pipes, the steam load being extremely high. To meet this load it was customary to install a battery of boilers. After 30 minutes, the steam load reduced to a minimum, as all the vessels were boiling. Fours hours later this same process would be repeated. The additional cost of the boiler capacity and of the fuel to keep the boilers alight between successive loads was excessive. A simple alteration to the production schedules by staggering this boiling of water and the introduction of steam accumulators reduced costs significantly.

Many computer software packages are available for energy-targeting purposes, but care should be taken in specifying the requirements and insisting on performance satisfaction and testing before final payment. Such software packages may be expensive and could require modifications for actual use.

There are many building space heating/humidity energy usage programs available. A large number assume the building ventilation or wind speed rates. In most buildings, the major factor in any discrepancy between the theoretical and actual energy consumptions is due to the variation in wind speed. This is highlighted in the example of the warehouse above. The large discrepancy between the theoretical and actual steam consumption was investigated in detail, and it was found that one of the large doors giving access into the warehouse was open for most of the time the warehouse was in use by the staff, and that this door was subject to the prevailing wind. In order to reduce the steam consumption a rapid-opening door was installed, with the result that the steam consumption was halved. The energy target set must be realistic and capable of achievement.

30.7 Major areas for energy conservation

For most organizations, the major energy use is for building space heating, ventilation, air conditioning, domestic hot water purposes and lighting. Where energy is used for process, this may be secondary energy generated from a primary fuel (e.g. steam generated in a fossil fuel-fired boiler plant).

30.7.1 Boiler plant

Boiler plants are a major user of energy. The combustion efficiency of a boiler plant can easily be set at the optimum, and Table 30.2 suggests the parameters for this for various fossil fuels:

Exit flue gas temperature $= 200°$C
Ambient air temperature $= 15°$C

Close control of the amount of excess air is possible by the use of oxygen trim control equipment. Such equipment will control the flue gas oxygen content within the range of 2.0–3.0 per cent as compared to the normal 3.0–5.0 per cent. The improvement in boiler plant efficiency is of the order of 1.0–2.0 per cent.

Table 30.2

	Coal	Oil	Natural gas
Excess air (%)	30	20	10
Dry flue gas CO_2 content (%)	12.0	13.3	10.5
Dry flue gas O_2 content (%)	7.5	3.5	2.3
Dry flue gas loss (%)	13.2	7.4	6.1
Moisture in the flue gas loss (%)	4.9	6.5	11.3
Unburnt/ash loss (%)	1.3	Nil	Nil
Radiation and other losses (%)	3.1	3.1	3.1
Total losses (%)	22.5	17.0	20.5
Inferred boiler plant efficiency (%)	77.5	83.0	79.5

The combustion conditions suggested above should be achieved bearing in mind the fact that the lower the flue gas oxygen content, the greater the risk of incomplete combustion, while the higher the oxygen content, the greater the flue gas losses (see Figures 30.2–30.6). Poor overall boiler performance outside these parameters is due to the radiation and other loss factors.

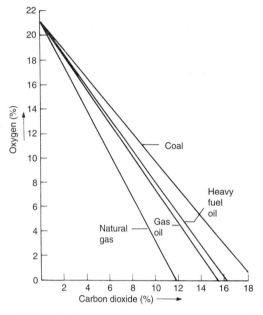

Figure 30.2 The relationship between oxygen and carbon dioxide in the products of combustion for various fuels

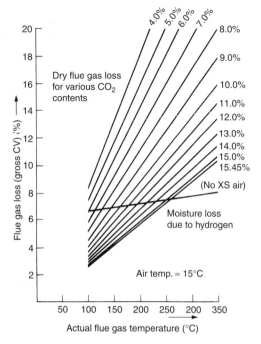

Figure 30.4 Flue gas losses – gas oil

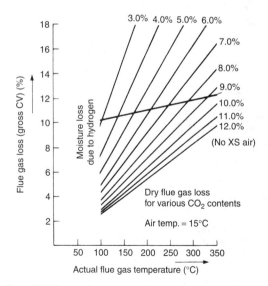

Figure 30.3 Flue gas losses – natural gas

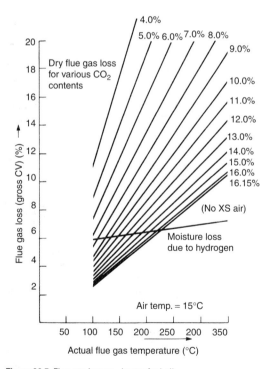

Figure 30.5 Flue gas losses – heavy fuel oil

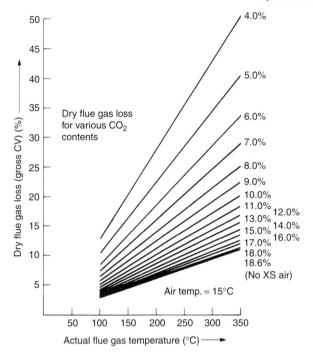

Figure 30.6 Flue gas losses – coal (Glasshoughton washed singles)

The surface loss from a boiler is fixed once the physical size, insulation and boiler working conditions are known. The usual radiation loss figure provided is that for a boiler operating at least at 80 per cent maximum continuous rating (MCR) or more. In this case, the surface loss is relatively low in percentage terms (around 3–5 per cent). Experience of boiler loadings have indicated that very few boilers work at such high loads during their periods of use. Typical annual boiler load factors are only around 40 per cent. As the boiler load factor decreases, the radiation loss becomes a much higher factor, and at 20 per cent boiler load amounts to 25 per cent and at 10 per cent load factor to a loss of 50 per cent (see Figure 30.7).

Many boilers do operate at low load factors and consequent poor annual efficiencies. This can be avoided by providing boiler plant with little or no margin over the actual required capacity and by installing multi-boilers or two or three smaller boilers. Boilers of this modular type are available for low-pressure hot water (LPHW) purposes, but obviously cost more than a single boiler. This will also increase the maintenance, number of examinations and the labor costs. Such an increase should be more than offset by the reduction in fuel costs due to the much higher annual boiler plant efficiency, which should be of the order of 20–30 per cent.

A further problem with heating boilers is that the boiler liquid temperature control thermostat (an immersion water thermostat in most cases) maintains the boiler liquid temperature at the set level, irrespective of the actual boiler load. The boiler, even though insulated, loses heat to the

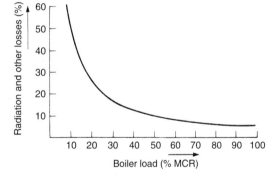

Figure 30.7 The variation in percentage radiation and other losses with the percentage boiler load factor

surrounding space through radiation and convection. The actual boiler load could reduce during the day such that the maintenance of the set boiler liquid temperature is not required continuously. Thus, the boiler liquid could be allowed to cool down with the consequent reduction in the radiation and convection losses (the standing loss). Boiler controls are now available to avoid maintaining the boiler liquid temperature at the thermostat control setting and consequently the unnecessary operation of the burner ('dry' cycling).

The cost of the supply and installation of such controls is of the order of $500–600. Energy reductions of 10–15 per cent can be achieved, with a cost-recovery period of 3 years down to a few months, depending on the actual annual energy consumption.

Condensing boilers are now available for both gas- and oil-fired plant, the advantage of these being that the flue gases are further cooled down to below 100°C so that the latent heat available in the flue gas water vapor is recovered. The condensate has to be removed and the boiler capital cost is higher than for conventional plant. However, the boiler plant efficiency is increased to the order of 90 per cent, based upon the fuel gross calorific value. Where the flue gas exit temperatures are in excess of 200°C a further economy can be obtained by the provision of a spray recuperator in the case of gas and flue gas economizers for oil and coal.

30.7.2 Furnaces

Furnaces are large users of energy, and in order to reduce costs, such equipment should be well insulated, used to maximum capacity and most of the waste heat in both the flue gases and product recovered. It should be possible to recover the waste heat in the flue gases down to at least 200°C. Specialist equipment for such waste heat recovery is available in the form of recuperators and regenerators.

30.7.3 Fans

Large fans are required for boiler plant, furnaces and large air-handling units. These fans can now be fitted with variable-speed motors to reduce electricity consumption and maximum power, albeit at additional cost.

30.7.4 Pumps

Pumps are a common feature in most heating and process applications. Where one pump only is provided this is sized for the maximum load (which may occur rarely). Two smaller pumps could be installed such that one pump carries the load for most of the time while the second is used at times of peak demand.

30.7.5 Air compressors

Compressed air systems should be checked frequently for air leakage, as the loss of air is frequently unnoticed due to the noise of escaping air being masked by other production noises. Modern air compressor systems composed of multi-compressors produce much heat, and it is now possible to recover this for space heating purposes.

30.7.6 Space heating

The provision of space heating, especially in large noncompartmented buildings, can be reduced by:

1. The installation of direct gas-fired units, where the products of combustion are discharged into the space but are so diluted as to be harmless. The efficiency of these units is thus 95–100 per cent, depending on the siting of the actual unit.

2. The provision of overhead radiant heating instead of convective heating. In this case the building heat requirements are some 15 per cent lower for radiant heating than convective, with the consequent saving in both capital and running costs. It is also believed that the comfort conditions are improved with less drafts. Radiant heating is also beneficial in buildings that are sparsely occupied, such heating being provided only in the area of the occupants.

3. The provision of ceiling-mounted fans to recirculate the warm air at ceiling level to the ground. These fans can vary from the small three-bladed slow speed fan up to much larger units, recirculating large quantities of air. The operation of these fans can be thermostatically controlled and fuel savings of 10–15 per cent are quoted.

30.7.7 Insulation

Buildings should be insulated to as high a standard as possible, as should any part of energy-using equipment. As a guide to buildings, the following overall heat transfer coefficients (or U-values) are suggested, which are currently of lower value than the requirements of the UK Building Regulations:

Roofs	$0.20\,W/m^{2}°C$
Walls	$0.30\,W/m^{2}°C$
Windows	$2.90\,W/m^{2}°C$
Exposed floors	$0.30\,W/m^{2}°C$

The current Building Regulations (1990) call for:

Roofs	$0.25\,W/m^2K$ (dwellings)
	$0.45\,W/m^2K$ (other buildings)
Walls	$0.45\,W/m^2K$ (all buildings)
Ground floors	$0.45\,W/m^2K$ (all buildings)

In order to improve the standards of existing building insulation the following methods are available.

Pitched roofs

Provide a blanket of 200 mm thickness of fiberglass or mineral wool insulation on top of the ceiling plaster.

Flat roofs

Provide a similar blanket of insulation on top of any suspended ceiling which is 300 mm below the flat roof or an external hot liquid spray-on type of insulation to the roof finish.

Cavity walls

Fill the existing wall cavity with a suitable insulant such as fiberglass, rock wool, polystyrene beads or foam. The greater the thickness of the cavity, the lower the U-value.

Solid walls

Solid walls can only be improved either by removing the internal plaster layer and providing suitable blanket insulation fixed to battens and then covering with the required new internal finish or by applying an external insulant to the wall and a new external finish.

Windows

Double or triple glazing may be provided in this case, the main criterion being that the cavity thickness between the layers of glass must be at least 20 mm.

Exposed floors

If access is available to the area below the floor, provide 50 mm of rigid insulation between the floor joists, leaving an air gap between the insulation and the underside of the floorboards.

Air infiltration

Provide good draft-proofing strip around the door and operable window frames, leaving sufficient fresh air ingress for any combustion appliances.

On completion of such measures, the use of thermography should be considered as a survey undertaken when the building is in use and heated will highlight hot spots. Any hot spots may be due to poor workmanship or unnoticed building faults and can then be rectified.

30.7.8 Controls

In order to achieve optimum energy usage (hence maximum energy conservation) it is essential to provide accurate controls designed to suit the application and for those controls to be correctly installed. It is no use providing insulation and highly efficient energy or heat generators if that energy or heat generated is simply lost in the system by providing better standards of comfort than those required. In the absence of such controls on space heating, the provision of building insulation simply increases the internal temperature conditions and results in windows and doors being opened to lose this heat.

Controls are expensive and quickly become outdated by the increase in technology. They are also the items that are omitted or their quality is reduced if the total project cost becomes excessive. The difference in the quality of the controls against cost can be determined and should be included as part of the project budget or application.

A typical example is the provision of a single room-air thermostat to provide whole building control of space heating. It will work, but far greater economy can be achieved by providing building zonal controls or, better still, individual room heating control. Zonal control is available in the form of either two- or three-port motorized valves which open or close in conjunction with an internal air thermostat. Individual room heating control is in the form of individual thermostatic radiator valves and in internal air thermostats controlling on/off or three-port motorized valves.

Individual room temperature control is difficult with warm air systems, as any temperature control damper is likely to pass air at all times. Air thermostats are not accurate detectors of temperatures and temperature overshoot occurs, with consequent increases in energy consumption

and costs. The typical limits of these thermostats is ±1°C, and as the air in the thermostat is not at the temperature of the air in the room because the thermostat is remotely sited (usually on an internal wall) the actual air temperature in the room can be 1.5–2.0°C above the desired temperature. Modern temperature detectors are accurate to ±0.5°C and give much closer and better control.

Steam flow to vessels can be controlled by motorized valves, allowing timed control of the valve. Such valves can also be made to modulate rather than on/off operation.

30.7.9 Heat recovery

All the heat used to supply heating in buildings (and much of that provided in any process) is ultimately lost from the building. Recovery of as much waste heat as possible reduces the need to purchase fresh energy from the suppliers. Much waste heat is produced in the form of contaminated hot moist air and in process liquids that require cooling.

There are many ways of recovering the waste heat from these sources and the following should be considered:

The heat wheel
The heat pipe
The run-round coil
The heat pump

The great benefit of the heat pump is the possibility of upgrading the heat as compared to the other methods, and is popular in swimming-pool applications, where the heat recovered from the exhaust air can be used to heat the incoming fresh air and the pool water and to provide hot water for showers and washing purposes.

30.7.10 Lighting

Artificial lighting is used throughout the US and Europe and energy-efficient lamps that give high lumens per watt can now provide this. A new range of fluorescent lamps is available and similarly high-efficiency lamps have been developed to replace the conventional tungsten lamp.

The problem of lamps being left on when not required could be solved by provision of automatic switching, of which there are a number of systems available. These can be time switch controlled with manual override at the point of use.

30.8 The justification for energy-conservation measures

In many cases, areas of obvious energy conservation are not considered simply because a process or some space heating has always been undertaken that way. In order to persuade the organization's management to embark on such energy-conservation measures overwhelming evidence is required that the theoretical results suggested can be achieved. However, most energy-conservation measures do offer large financial savings with very short payback periods and lead to increased organizational profitability. The cost of such schemes has to compete with others for new capital equipment and their success could depend on the effect of energy costs on the total organization costs.

It is obvious that a scheme to reduce energy costs by, say, 10 per cent may fail if the total energy bill is only 3 or 4 per cent of the organization's costs when compared to, say, a new production machine costing ten times as much but which reduces the organization's production costs by 10 per cent.

If an energy-conservation scheme is to succeed it must be well prepared and presented to senior management. In the project application and budgets, the objectives of the scheme must be clearly defined and the capital costs fully detailed. This may involve preparing several budget costs from various equipment manufacturers and contractors. In such cases, it is necessary to read and note the 'small print', as any comparison has to be on a similar basis. This is extremely difficult to achieve, as both manufacturers and contractors omit certain items which may be required to be added later. The estimated energy and cost savings have to be identified as accurately as possible, and this is very difficult in view of the large number of parameters involved.

In formulating these savings, the parameters used should be clearly defined so that the estimates can be revised for any alterations. Indeed, in some cases a computer program can be utilized which allows such changes to be evaluated very quickly (e.g. the TAS package for buildings). In the end, the submission will stand or fall on the overall impact on the organization, so that it is essential to get it right first time. That saving on the bottom line is the major item.

A further factor to be considered is that of the national energy situation. Fortunately, in the short term there are available sufficient reserves of gas, oil and coal to meet the national need, so supplies should not be a problem. However, the prices of these fuels are subject to international as well as national policies. Currently, there is a relationship between heavy fuel oil and coal and between gas and gas oil (see Section 30.2). In the long term, as supplies of oil and gas are reduced it is likely that increases in the price of gas will outstrip those of oil.

30.9 The mathematics of the presentation

Any presentation to the organization's senior management, besides giving the technical details, should include a financial appraisal, which could be on a simple payback basis. However, other methods are available, such as return on capital employed and discounted cash flow (DCF).

30.9.1 Simple payback

This method of comparing the initial capital cost against the annual energy/cost savings is relatively easy. Unfortunately, it does not take into account certain factors such as:

1. The life of the project;
2. Any benefits after the payback period.

Normally, as the required payback period is short (i.e. 2 or 3 years maximum) and the length of life of the equipment is much longer than this, on the simple payback basis there are usually large financial benefits in the long term.

Let us consider two projects both costing $40,000 with the following savings:

Project 1: Annual saving $20,000, life of equipment 10 years
Project 2: Annual saving $16,000, life of equipment 15 years

In terms of simple payback, project 1 has an excellent payback of only 2.0 years, while project 2 has a payback of 2.5 years. Thus, it is likely that project 1 will be selected. However, project 1 has a life of 8 years after the initial payback period, the financial saving being $160,000, the equipment then requiring replacement.

Project 2 has a life of 12.5 years after the initial payback period, giving a total financial saving of $200,000. Therefore, project 2 is the better case in the long term.

30.9.2 Return on capital employed

This method compares the initial capital cost with the cash flow over the life of the project. Thus, in this case an accurate estimate of the equipment life is required. When a project has a long life (considerably longer than the simple payback period) the results can be different.

Return on capital employed can be expressed in four ways:

1. Gross return on capital employed;
2. Net return on capital employed;
3. Average gross annual rate of return;
4. Average net annual rate of return.

In this method each year's savings need to take account of the fact that, as the equipment gets older, the cost of maintenance and repairs is likely to increase. The effect of inflation can also be allowed for in assuming the annual cash flows through the life of the equipment.

Using the two projects in Section 30.9.1 as an example, the returns shown in Table 30.3 are likely. Of the four methods, the net return is generally favored to the gross return as the loss of capital is a major factor.

30.9.3 Discounted cash flow (DCF)

This method, which is favored by many accountants today, takes into account the concept that money has a time value. This is because $2000 in 10 years' time is not the same as $2000 now. Similarly, if a project earns $2000 in 10 years' time this is not the same as $2000 spent now to help finance the project. If, instead of spending this $2000, it had been invested at compound interest, then in 10 years' time it would have become a much larger

Table 30.3

Method	Project 1	Project 1	Ratio
Gross return on capital employed	$\dfrac{1000\,000 \times 10 \times 100}{20\,000}$ = 500%	$\dfrac{8000 \times 15 \times 100}{20\,000}$ = 600%	1:1.2
Net return on capital	$\dfrac{[(10\,000 \times 10)20\,000] \times 100}{20\,000}$ = 400%	$\dfrac{[(8000 \times 15) - 20\,000] \times 100}{20\,000}$ = 500%	1:1.25
Average gross annual return on capital employed	$\dfrac{(10\,000 \times 10) \times 100}{(10 \times 20\,000)}$ = 50%	$\dfrac{(8000 \times 15) \times 100}{(15 \times 20\,000)}$ = 40%	1:0.30
Average net annual return on capital employed	$\dfrac{[(10\,000 \times 10)-20\,000] \times 100}{(10 \times 20\,000)}$ = 40%	$\dfrac{[(8000 \times 15) - 20\,000] \times 100}{(15 \times 20\,000)}$ = 33.3%	1:0.83

amount than $2000, depending on the interest rates in the prevailing period. The way that this is taken into account is called 'discounting'. Trading income produced for each future year by the project is discounted after allowing for maintenance and repairs, giving present values for that income.

There are various ways of discounting, i.e. 'net present value' (NPV) or 'internal rate of return' (IRR). Discounting should ideally be carried out over the whole life of the project but generally, after 10 years, the discount factors are small and make little difference.

The selection of the discount factor depends on the financial policy of the business, but is usually 2–3 per cent above the current interest rates. Use of discounting methods will determine whether the project cost will produce a better return than by simply investing the capital involved at the highest compound interest rate or, if the capital cost has to be borrowed, whether the rate of return is much higher than the cost of borrowing.

30.10 Third-party energy management and finance

As indicated in Section 30.7, energy-conservation measures cost money, and in spite of the likely results indicated by any of the methods available (simple payback, discounted cash flow, etc.) it is possible that the organization's funds are not available for such schemes. Alternatives that should be considered are:

1. Leasing;
2. Hire purchase.

Many organizations now have their energy requirements managed by a heat service company. In this case, the company will provide and operate all the energy-producing equipment on the organization's site. This could include space heating, air conditioning, combined heat and power (CHP), steam boiler plant and other services.

The contractor will purchase all the fuel required and provide all labor, repairs and maintenance, thus relieving the organization of all responsibility. In return, the contractor will require a contract (generally of up to 10 years) with the fees charged comprising a monthly fixed standing charge and a monthly variable charge, depending on the amount of energy used. Such an arrangement is useful for obtaining new equipment when finance is not available within the organization itself.

A variation of the heat management service is to have contract energy management (CEM). In this case, the energy contractor will generally survey the existing plant and equipment and take it over, running it for the organization. As equipment becomes due for replacement the contractor will undertake this part of the contract.

The contract is usually for a period ranging from 3 to 9 years, but in this case the amount payable to the contractor is the same or similar to the costs prior to contractor take-over. A further clause is added such that a proportion of the savings made by the contractor are returned to the client (perhaps 2–10 per cent of the previous annual costs). It may also be possible for any further large saving to be shared in agreed proportions between the two parties involved.

It is essential to have any agreement properly drawn up in conjunction with the organization's legal section and to use reputable and busy heat service companies. It is important to lay down strict details of the period of the contract and of the services to be provided so that future legal difficulties can be avoided.

30.11 Motivation

Energy conservation will succeed if all the parties involved wish it to succeed. It only requires ill will on one person's behalf to negate all the positive efforts made.

Typical problems that can occur are:

1. The unauthorized alteration of internal air thermostats and time switches;
2. The use of supplementary heating (e.g. electric fires without permission).

The unauthorized alteration of set parameters is the most common problem and it is possible to install tamperproof equipment, albeit at additional cost. There is also the problem of the relationship with the existing management (e.g. the senior manager who views the growth and success of the energy management team as a threat to the supervisor or department manager).

There are a variety of methods to avoid this, and the more senior the management structure that takes an interest in energy matters, the greater the success. Ideally, this should start with the chairman, managing director or chief executive. Once they believe in it, then their staff will follow suit.

These people are extremely busy with the overall organization's objectives and time for energy matters may be very limited. They should therefore delegate this to a senior officer. Energy matters may be left to the works engineer as an additional duty. Consideration should be given to the appointment of an energy manager who is qualified for the post and provided with a suitable job description and objectives. Such a job description could be based on the following:

1. The energy manager will be responsible for the total energy function of the organization.
2. He will be responsible to the chief executive and will achieve the objectives laid down by senior management.
3. The duties consist of (a) the most economic purchase of all forms of energy – in conjunction with the purchasing section; (b) the optimum use of all forms of energy for both process and building environment use – in conjunction with the various departmental and engineering managers.

4. The energy manager needs to form a team – an energy committee – to help him and this could comprise the managers of each department, whether production or otherwise, and other suitable people. The team must be small to be effective (less than 10 persons). Each department could then have its own energy committee.

Hopefully, the enthusiasm for energy conservation will spread downwards to all employees. Their help can be obtained by the use of competitions and suggestions.

30.12 Training

Few senior staff, works engineers or even junior staff have had training in energy matters. When energy was cheap, this was a small company cost (say, 1 per cent) so that any cost-reduction measures were directed at other areas. Universities and colleges now include energy management as part of the curriculum in engineering and related courses. Newly appointed staff will be energy orientated.

For those staff who have never had the benefit of education and training in energy matters there are courses both by part-time attendance at college and by correspondence. Details of the courses can be obtained from the Secretary, the Institution of Plant Engineers. Throughout the US and Europe, there are a large number of energy management groups where the members (who are nearly all energy managers) meet regularly to discuss energy matters. Thus, it is possible to obtain training at all levels required.

Reference

1. *Pattern of Energy Usage – Energy for the Future*, Institute of Energy, London (1986).

31

Water and Effluents

George Solt

Contents

31.1 Introductory warning

Water is an essential service to any facility. The amount and quality needed vary considerably between different plants, but the essential fact important to all engineers who have to deal with water technology is that the subject of water quality and treatment is highly specialized. It deals with chemistry and microbiology, and even within those fields the technology is specialized and in the hands of experts. It is therefore a field in which plant engineers cannot be expert, and are forced to take advice from specialists. These are usually suppliers of plant or chemicals, or professional consultants. These specialists are in a position of great responsibility, because plant engineers are so dependent on their advice. In these circumstances it is possible for 'cowboy' organizations to thrive. Past and present experience shows that these are too common. *Plant engineers must not automatically accept the lowest offer for materials or advice in this field but must first be satisfied with the specialists' competence and integrity.*

There are many sad examples of installations that have gone disastrously wrong, and these are by no means limited to small facilities or those in which water is a relatively unimportant service. For example, manufacture of microchips is wholly dependent on a supply of highly purified water. In recent years two of the largest UK manufacturers, sited at opposite ends of the country, have had to shut down and send the workforce home because their purified water facility had failed.

The golden rule, therefore, is to deal only with consultants, contractors, plant suppliers and water-conditioning experts whose experience and standing are known to be good. If there is any doubt, references should be sought and followed up. Water and effluent installations are a relatively minor cost item in any plant, but their failure can be disastrous. It is foolish to make false economies on so essential a service.

31.2 Requirements for water

Any plant will need water for domestic purposes: in most cases this can be provided by public supply and discharged to the public sewer. Most plants have some steam-raising equipment for space heating, and steam is often required for other purposes. Steam raising always requires water conditioning and usually external treatment of some kind. Many operations require water of a specified quality, which varies over a wide range: from cooling and washing water to softened or demineralized and (in extreme conditions) ultra-pure water.

For food and drink, medical, pharmaceuticals and cosmetics production the microbiological quality of the water becomes paramount. Even in applications where biological quality is not directly important, uncontrolled growth can be a damaging nuisance. Warm-water systems and cooling circuits in particular are a potential hazard (e.g. from *Legionella*). Some water treatment or conditioning is commonly required.

Water from public supply costs money, as does that discharged to the public sewer. Except for water which is incorporated into the product of the plant or evaporated to the atmosphere, any water which enters the plant is returned to the environment as an effluent. It may be necessary to improve the quality of this before disposal to surface water or into a public sewer, and for acceptance into the public sewer the sewerage authority may levy a charge dependent on the quality. This is levied per cubic meter of water discharged, and this volume, in turn, is normally estimated on the basis of the metered incoming mains supply flow, minus some agreed factor for water retained in the product or lost by evaporation. There is therefore a double financial incentive to reduce the plant's water consumption.

31.3 Water chemistry

Raw water analyses are normally obtained from the local water supply organization. Water analysis is a specialized trade, and analysts who do not routinely carry this out can prove unreliable. It is also important to ascertain the seasonal and long-term variations to be expected.

For conventional factory boilers, most of the many items normally shown on a water supplier's analysis sheet are unimportant. The ones to look for are listed below.

31.3.1 Units of measurement

Analyses usually give concentrations in milligrams per liter (mg/l) or parts per million (ppm). For practical purposes these units are the same.

Most minerals in water exist as ions – electrically charged particles that give them an electrical conductivity. The different systems of units that measure their concentration can cause much confusion. For any calculation involving adding different ions to one another it is vital to use one of two systems of *equivalents*.

The traditional British method is to calculate the concentration *equivalent to calcium carbonate* and give the results as 'mg/l as $CaCO_3$'. A more modern unit, widely used throughout Europe, is the milligram equivalent or milliVal, abbreviated as meq/l or mVal. This gives the same information as mg/l as $CaCO_3$ but the values are one-fiftieth of those expressed as $CaCO_3$. Explaining these systems is lengthy, and usually increases the non-chemists' confusion. The system mg/l as $CaCO_3$ is used below and the rules are as follows:

1. Concentrations of individual constituents, such as calcium, 'hardness' or 'alkalinity' should be brought to mg/l as $CaCO_3$ for any comparisons or calculations.
2. There should be a statement on the analysis sheet that makes it clear what system of units has been used. Hardness and alkalinity are often given as $CaCO_3$, which makes life simple.
3. Alternatively, the analysis may state that concentrations are given *as the ion* or *as such*, or each individual constituents may say (for example) *calcium as Ca*. In that case one must use the following conversion table:

To convert from mg/l as such	To mg/l as $CaCO_3$ multiply by	To meq/l multiply by
Calcium (Ca)	2.5	0.050
Magnesium (Mg)	4.31	0.083
Bicarbonate (HCO_3)	0.82	0.016

4. Total dissolved solids (TDS) and silica (SiO_2) are normally given *as such* and do not require any conversion.

31.3.2 Hardness and alkalinity

Many analyses quote *total hardness*. Some give *temporary hardness* (or carbonate hardness) and *permanent hardness* (or non-carbonate hardness), usually in consistent units so that the values can be added together to give the total hardness. The total hardness is actually the quantity of calcium (Ca) + magnesium (Mg) in the water. If the total is not given directly, the values given for these two constituents must be added, after conversion to mg/l as $CaCO_3$ if necessary.

Hardness in water varies widely, and as an arbitrary classification:

Hardness less than 50 mg/l as $CaCO_3$	Soft
50–200 mg/l as $CaCO_3$	Medium
More than 200 mg/l as $CaCO_3$	Very hard

Temporary hardness

Temporary or carbonate hardness and alkalinity frequently, but not always, mean the same thing (see below). *Bicarbonate* or *hydrogen carbonate* is a more scientific term since *alkalinity* is actually the concentration of the bicarbonate (HCO_3^-) ion in the water.

If acid is dosed into water-containing bicarbonate, the ion becomes converted to carbon dioxide (CO_2) gas, while the water becomes only slightly acidic. (Sodium bicarbonate taken against acid stomach uses this property.) The drop in pH from a given acid dose is much smaller than would result from the same amount of acid dosed into water containing no bicarbonate. When enough acid has been added to convert all the bicarbonate to CO_2, further acid dosing leads to the sharp drop in pH, which is expected from water containing no bicarbonate.

Another reaction of bicarbonate is that in boiling water it combines with any hardness present to produce scale, while releasing CO_2 into the steam. This hardness is called the *temporary hardness*. Its concentration therefore depends on the lessor amounts of either hardness or bicarbonate. Most waters contain more hardness than bicarbonate, so that the temporary hardness is usually equal to the bicarbonate content. The *temporary hardness* quoted in analyses is often the only information available on the bicarbonate content of the water.

Permanent hardness

After the temporary hardness has been removed, any calcium and magnesium which remains is still capable of forming a scum with soap, and can also react to form boiler scale. This is called the *permanent hardness*.

CO_2 release

When the steam is condensed, any CO_2 released in the boiler re-dissolves in the condensate, making it slightly acidic and corrosive. Normally, boiler feedwater is softened and the boiler water pH is raised by addition of caustic soda. Both these measures reduce the degree to which bicarbonate breaks down and releases CO_2. Even so, very large amounts of bicarbonate entering the boiler should be avoided.

31.3.3 Total dissolved solids (TDS)

Water contains various other salts which are generally unimportant in medium-pressure boilers, except for the contribution they make to the total solids in the boiler water. They can therefore be grouped together, and since dosing water-conditioning chemicals also contributes to the boiler contents, even the total need not be known exactly. Most analyses give a figure for TDS in mg/l, but analytical methods differ and the result is not always particularly reliable. TDS is usually given as the actual weight of dissolved materials, which is 10–15 per cent higher than the TDS measured 'as mg/l $CaCO_3$'.

Another method of estimating TDS is to measure the electrical conductivity of the water, which is usually reported as 'µho' or 'µS'. This figure is roughly double the TDS in mg/l as $CaCO_3$.

For many years water costs have been rising faster than inflation, and are set to rise even faster in the future. The belief that water is 'free' must be combated: it is usually possible to make more economical use of water. In most factories the water bills are paid centrally, and there is no system for debiting the cost to the actual water users. This makes for wasteful practices which would be improved if there were a system for monitoring where the water costs are incurred within the facility.

In some cases sensible design can lead to re-use of water, which reduces both water and effluent costs. This is best achieved by intelligent routing of the water rather than by treatment before re-use. Effluent treatment is best avoided wherever possible. For example, very slightly contaminated process wash water can be recovered for washing down floors. This reduces charges for both incoming water and effluent.

31.3.4 Silica

The silica content of water only becomes important if it is a large proportion, i.e. more than 10 per cent, of the TDS. This is unusual in normal water supplies, but can result from external water treatment (see below).

31.3.5 Types of water

Most natural waters contain more hardness than bicarbonate. Only a few sources in the UK, usually from wells in sandstone strata, contain more alkalinity than hardness. In most cases the temporary hardness greatly exceeds the permanent hardness. This is especially true of the hard alkaline waters, which come from chalk and limestone measures.

Many waters from mountainous uplands such as Wales, Scotland, Yorkshire, or the moors of southwest England are surface runoff and have not percolated through mineral strata. They are low in minerals generally, and hardness and alkalinity in particular, though even there the hardness will be greater than the alkalinity. Their main characteristic is their high content of organic matter, which may give them a faint yellowish tinge, and a low pH. Many such soft and peaty supplies have lime added at the waterworks to make them less corrosive.

Some potable supplies are treated surface waters from rivers, etc. These originally derive from any of the above, but will also contain the products of human activities, which lead to increased mineral contents and possibly some undesirable materials such as detergents.

Potable water is not (and should not) be sterile. In fact, potable water mains always have a layer of living matter clinging to the walls. Any change in the flow, temperature or chemical quality of the water passing through the pipe will cause some of this to become detached. Quite large living organisms (such as freshwater shrimps, waterflies and even leeches) can occasionally be found in a potable supply. They are esthetically disturbing but usually unimportant.

The important quality criterion is absence of pathogens, a term which covers all disease-bearing bacteria. This criterion is usually determined by test for *Eschericia coli*, a species of bacteria so common in the gut that it is a reliable indicator of any pathogenic contamination.

Some plants have to rely on a private supply, usually taken from a borehole. A competent authority first must test the water. Water analysis is a specialized procedure and is preferably undertaken by an analyst who does this regularly, such as a local water supplier's laboratory. Water from a spring or a deep well, which appears potable, is usually found to be of reasonable quality, but it may occasionally contain some constituents (e.g. iron or manganese) in unacceptable concentrations. The nitrate content of groundwaters is generally rising throughout the UK, and if it is found to be near the legal limit, further samples should be analyzed at, say, six-monthly intervals.

31.4 Building services

31.4.1 Potable water

Most plant sites have access to a public supply of water. Until recently this was legally required to be no more than 'wholesome and palatable'. Water suppliers are now responsible for meeting the EC's Directives (see Table 31.1). Wherever possible, all drinking-water taps should be served directly from the incoming main, and the plant engineer's sole responsibility is to ensure that no deterioration takes place within his system.

The water supply authorities normally insist that (for uses other than drinking-water taps) their main should discharge into a break-pressure vessel, after which the water quality becomes the consumers' responsibility. The water tank should be covered against tramp dirt and access by birds, etc., and it must be shielded from sunlight to avoid the growth of algae. Nevertheless, access must be maintained for easy inspection. The distribution pipework is preferably all plastic and lead must be avoided altogether. The use of copper is doubtful with some corrosive waters, and soldered joints in it can lead to unacceptable concentrations of lead in the water.

Table 31.1 Drinking-water quality standards (EC, 1980)

Colour	Pt/Co	20	
Turbidity	JTU	4	
Threshold odour no.		3	
Anionic detergents	mg/l Manoxol	0.2	
Pesticides	mg/l	0.5	
PAH	µg/l	0.2	
Phenols	µg/l	0.5	
Aluminium	mg/l Al	0.2	
Ammonia	mg/l NH_4	0.5	
Arsenic	mg/l As	0.05	
Barium	mg/l Ba	0.1	(GL)
Calcium	mg/l Ca	100	(GL)
Cadmium	mg/l Cd	0.005	
Chloride	mg/l Cl	200	
Chromium	mg/l Cr^{6+}	0.05	
Copper	mg/l Cu	0.1	(GL)
Cyanide	mg/l CN	0.05	
Fluoride	mg/l F	1.5	
Hydrogen sulphide	mg/l H_2S	–	
Iron	mg/l Fe	0.2	
Lead	mg/l Pb	0.05	
Magnesium	mg/l Mg	50	
Manganese	mg/l Mn	0.05	
Mercury	mg/l Hg	0.001	
Nitrate	mg/l NO_3	50	
Selenium	mg/l Se	0.01	
Sodium	mg/l Na	150	
Sulfate	mg/l SO_4	250	
Zinc	mg/l Zn	5.0	
Coliforms		Zero in 95%	
Coliforms		of samples	
		GL = Guide Level	
		(i.e. no MAC applies)	

The removal of such impurities is relatively simple at a waterworks, but a typical plant cannot provide the chemical expertise needed to keep the process in good working order. Treatment should therefore be avoided if there is any reasonable economical alternative.

If the water is found fit for consumption, with respect to both its mineral and biological content, the problem of sanitization can still arise. Public supply invariably has a very small residual chlorine level. This suppresses biological growth and maintains water quality even when the line is stagnant. As with other forms of treatment, the scale of private supply is usually too small to allow good control of chlorinating equipment.

One method of operating a non-chlorinated supply safely is to ensure that the line runs constantly to waste, as do very old drinking fountains. This avoids biological growth that can accumulate in stagnant water. Dead legs in the piping system are always undesirable, especially in such cases.

31.4.2 Domestic effluent

All plants produce domestic effluent, which is preferably discharged into the public sewer. If no sewer is available the plant needs a septic tank or a similar device sized for the probable demand, which is based on the number of people whom it will serve. Architects and competent building contractors are familiar with rules of thumb which apply to the design and sizing of these. Septic tanks need regular pumping out of sludge and access must be provided. Many industrial effluents interfere with the biological activity of a septic tank and must therefore be kept separate and discharged by other means.

31.4.3 Domestic water

Domestic hot and cold water, for hand washing, etc., will normally be supplied within a factory from the break-pressure tank fed by the incoming water main or private supply. Although these supplies do not need to conform to potable water quality standards they can provide a breeding ground for a variety of bacteria, including *Legionella*. As a precaution, many plant engineers dose sodium hypochlorite into the break tank to maintain about 0.2–0.5 mg/l of free chlorine in these supplies. Bacteria breed most prolifically at temperatures between 10° and 60°C and, wherever possible, hot water should be maintained above and cold water below this range. If the raw water is hard then consideration should be given to softening at least the hot water to prevent scale formation in calorifiers, pipework and sanitary ware.

31.4.4 Closed circuits

Closed water circuits, such as chilled water, or medium- and high-pressure hot water systems, should be initially filled with the best quality water available – de-ionized for preference but at least softened. They should then be conditioned by dosing with suitable corrosion-inhibiting chemicals and biocides. Make-up to closed systems is usually very small and raw water will often suffice, although, of course, higher-purity water is better. Routine sampling to check on inhibitor levels and bacterial growth is vital to the operation of such circuits, and most reputable conditioning chemical suppliers will undertake this work on a contract basis.

31.4.5 Open cooling-water systems

Although direct air-cooled systems are now preferred there are still many evaporative cooling towers in operation, and open systems of this type represent the most difficult of all the water-treatment situations the plant engineer is likely to meet. The systems work as follows: heat from the heat exchanger is taken up by the circulating water which is then sprayed into the top of a packed cooling tower with a forced or natural draft of air flowing through it. Some of the water evaporates, taking in latent heat from the bulk of the water, which, consequently, cools down. The cooled water collects in a sump below the tower and is pumped back to the heat exchanger. When water evaporates, it leaves behind any dissolved salts and other contaminants, which, as a result, become more concentrated. A certain amount of water is also lost from the tower in the form of 'windage' or spray. To compensate for these losses a make-up supply of water is required. The concentration effect is most important because, if it is not properly controlled,

salts – especially hardness salts – may become over-concentrated and deposit as scale in the pipework, in the heat exchanger and on the tower packing, where it causes blocking, and its added weight can lead to mechanical failure. To control the build-up of dissolved salts a bleed-off of concentrated water is necessary, and this must also be replaced by make-up water.

The circulating water comes into contact, in the cooling tower, with large volumes of atmospheric air and washes from it a variety of airborne contaminants, including pollutant gases such as oxides of sulfur and nitrogen, dust and soot particles and spores of bacteria, fungi and algae. The dissolved gases may give rise to acidic conditions which contribute to corrosion, algae can grow in the sump where warmth and sunlight provide ideal conditions while fungi may appear on the dead algae. Aerobic bacteria, including *Legionella*, grow in the circulating water and contribute, with algae and fungi, to the formation of slimes which not only cause physical blocking of plant but also set up areas of differential aeration around the system, which can promote intense corrosion. This process is aggravated by the deposition of scale and corrosion products below which corrosion takes place and anaerobic bacteria multiply.

Water losses

The rate of evaporation from a cooling tower is approximately 1 per cent of the circulation rate for each 5°C drop in temperature across the tower, or about 7 liters/h per ton of refrigeration. Windage losses will obviously depend on the prevailing wind conditions and the design of the tower with regard to spray elimination but, typically, these are about 0.2 per cent of the circulation rate.

The amount of bleed-off required would depend on the nature of the make-up water and the type of conditioning chemicals used. The specialist tower manufacturer, conditioning chemical supplier or water-treatment consultant will advise the maximum concentration factor (the ratio of circulating water concentration to make-up water concentration) which can be allowed. The necessary bleed-off is then given by:

$$B = \frac{E}{(C-1) - W}$$

where

E = evaporation rate,
C = concentration factor (typically 1–5),
W = windage.

A simple manual valve running continuously to drain or by an automatic valve controlled by the conductivity of the circulating water may set the bleed-off.

Make-up water

The required make-up water, M, is given by:

$$M = S + W$$

Or

$$M = \frac{(E \times C)}{(C-1)}$$

Many towers operate with hard water make-up, and scale-inhibiting chemicals are dosed to prevent the formation of hard, adherent scale. Hardness salts do precipitate in these systems but in the form of a mobile sludge, which is easily removed. A more satisfactory solution is to use make-up water, which should, ideally, be softened or de-alkalized. However, naturally soft or artificially softened water tends to be corrosive in the conditions encountered in open cooling systems, and it is necessary to dose a corrosion-inhibiting chemical to protect metal surfaces. In either case, the accurate control of chemical dosing is absolutely critical to the reliability and integrity of the cooling system. The selection of the correct conditioning chemical(s) and operating regime will depend on many factors, including the make-up water quality, concentration, and materials of construction of the system and environmental conditions. It is a task for a specialist, and chemicals should only be purchased from reputable suppliers who will provide a continuing service of monitoring and control.

Biocides

Biocides are added to cooling water to control the growth of bacteria, fungi and algae in the system. Chlorine, dosed in the form of sodium hypochlorite, is probably the best broad-spectrum biocide and, at residual levels of 0.5 mg/l, chlorine is effective against most bacteria, including *Legionella*. However, a re-circulatory system means that bacteria are exposed continuously to the same chemical conditions and resistant strains, with a natural immunity to the biocide, will eventually appear and colonize the system. To prevent this, a regular *shot dose* of an alternative biocide is advisable, and most chemical suppliers have a range of biocides, both broad spectrum and specific, for this purpose.

Filtration

Insoluble suspended matter either picked up from the atmosphere or formed by deposition and corrosion within the system, together with slimes will, if not removed, cause blocking and abrasion problems. The build-up of such material can be controlled by *side stream filtration*, in which about 2–5 per cent of the circulating water flow is filtered continuously. A sand filter is commonly used for this type of duty.

31.5 Boilers

31.5.1 Water for steam raising

Most plants have boilers producing steam for space heating: many need steam for other purposes as well. Boiler water requirements for boilers have changed radically over recent years. The old 'Lancashire' and 'Economic' boilers had large heating surfaces and low heat transfer rates: scale deposits would do no more harm than reduce their thermal efficiency.

Modern packaged boilers use the heat transfer surfaces much more intensively, and are endangered by scale. The boiler water acts as a coolant without which the metal of the tube overheats. Thin films of scale can obstruct heat transfer sufficiently to bring about tube failure in this way – especially if the scale deposit is siliceous, which is a particularly good insulator. Alternatively, the scale deposit may slow down heat transfer from the hot combustion gases to such an extent that the temperature at the back of the boiler rises excessively and causes tube plate cracking.

Boilermakers now recognize that their heat transfer rates had become too high, which made control of the boiler water quality unacceptably critical. They have reverted to slightly lower heat transfer rates, but poor boiler water quality remains the main single cause of boiler failure. Good water-treatment plant and boiler-water conditioning and control are still vital not only to the boiler's performance but also to its integrity.

31.5.2 Managing the steam-water circuit

Most industrial installations have a boiler of some kind: this boiler and its steam user form a circuit in which water and steam circulates. Loose use of nomenclature sometimes leads to confusion, and it is therefore useful to define the various waters in it:

Raw water: The mains supply or other external source used to prepare make-up.
Treated water: The water leaving the external treatment plant, if there is one.
Make-up: The amount of raw or treated water added to the feed.
Condensate: The water returning from the steam user(s).
Feedwater: The water entering the boiler feed heating system, which will normally be a blend of treated water and condensate.
Boiler water: The contents of the boiler.
Blowdown: The water blown down from the boiler in order to maintain its total dissolved level below the specified limit.

In order to maintain good boiler operation, the most important rules to follow are:

1. The boiler water should be within the limits specified for that type of boiler by British Standards, DIN and similar standards. In a conventional shell type factory boiler the most important criteria are that hardness should be present only in very small concentrations, and the TDS should be below 3000 mg/l.
2. The boiler water must at all times contain a positive residual of oxygen scavenger (usually sodium sulfite).
3. The water-containing chemicals should include phosphate or tannin to counter any residual hardness.
4. The boiler water pH must be raised to about 9 or over to avoid corrosion, to maintain silica in solution, and to reduce the release of CO_2 into the steam.
5. The bicarbonate content of feed should be moderate to avoid excessive liberation of carbon dioxide in the boiler.

Clearly, these matters are interdependent.

31.5.3 Condensate return

Condensate normally contains no hardness and is very low in dissolved solids. Unless it has been excessively exposed to the atmosphere, condensate is also very low in dissolved oxygen. Therefore it represents the ideal feedwater, and the higher the proportion of recovered condensate in the feed, the easier it will be to maintain the boiler water within the desired limits. The percentage condensate return is thus basic to all considerations of water management for the boiler circuit.

In some process applications condensate becomes contaminated: in sugar refining it may contain sugar, in paper manufacture and some other processes it may be contaminated with raw water, and where it feeds turbines or other machinery it is liable to contain oil or grease. In all these cases the condensate may still provide a better source of boiler feed than the available raw water, but it may need condensate filtration or softening plant before re-use.

31.5.4 Raw water quality

This varies widely within temperate zones, and even more so in hot countries. A wide range of possibilities therefore exists: high hardness, high alkalinity, and/or high TDS need correction. However, the degree to which this is necessary is also dictated by the percentage of condensate return – if it is high, the need for external treatment is accordingly reduced. The processes used for external treatment are described below.

31.5.5 Oxygen scavenging

Water at ambient temperature in contact with air contains about 10 mg/l of oxygen. To avoid corrosion, boiler water must contain no oxygen, and have an excess of oxygen-scavenging chemical (usually catalyzed sodium sulfite) to ensure this. Ideally, sulfite is dosed into the feed heating system with the feedwater to give the hot well, etc. some protection, with a second dosing point into the boiler itself to ensure that the residual is actually maintained.

The removal of each milligram of oxygen requires about 8 mg of sodium sulfite, so that the dose should be adjusted to suit the amount of oxygen introduced, which corresponds roughly to 80 mg/l of cold make-up. A high level of condensate return therefore reduces the scavenger demand, and this is not only an economy in itself but can also mean a considerable reduction in the amount of total dissolved solids introduced into the boiler with the feedwater.

Oxygen can also be removed from feedwater by thermal de-aeration, or partially removed by skilful design of the feed heating system and blowdown recovery. These processes run without cost to the operator, but save chemicals, and, by reducing the required dose of sulfite into the system, decrease the amount of non-volatile solids added into the boiler.

31.5.6 Blowdown

All non-volatile impurities entering the boiler must build up in the boiler water. This includes the TDS in the feed, plus most of the conditioning chemicals, of which the sodium sulfite used as oxygen scavenger is usually the major contributor. To maintain the boiler water within its permitted limits some boiler water must be blown down. The rate (as a percentage of the steaming rate) is calculated by:

$$x = \frac{f \times 100\%}{b - f}$$

where f is the concentration of an impurity in the feedwater and b is its permitted concentration in the boiler water. This calculation should be made for each individual impurity specified. In practice, the TDS is usually the controlling factor in blowdown but if the make-up is treated by partial de-ionization, silica may be more important.

Blowdown costs money in terms of heat, water and chemicals, and should therefore be minimized. Control of blowdown and recovery of heat from it are important aspects of boiler operation.

Where sodium sulfite addition is a large contributor to the non-volatiles in the boiler, thermal means of reducing the oxygen can make a significant improvement in the overall operation.

31.5.7 External water treatment

In an ideal case the condensate return is high, and the raw water low in dissolved solids, hardness and alkalinity. It is then possible to operate the boiler without external water treatment, relying on conditioning of the boiler water with phosphates, tannins or other chemicals to cope with the small amount of hardness introduced with the raw water.

In practice, especially with modern boilers, it is considered essential at least to soften the raw water. Conventional softeners do not, however, remove hardness completely but allow a very small concentration to pass through.

Where the feed contains a large proportion of treated water, softening is a minimum requirement and the raw water quality dictates whether a more sophisticated form of external treatment would be preferable. If the water has a high alkalinity it calls for de-alkalization and base exchange. De-ionization is the ideal water treatment, but is usually avoided if possible because of its cost and use of corrosive chemicals. Membrane processes giving partial de-ionization are not normally installed at present, but are certain to become important in the future.

External treatment process plant should be installed only after a specialist's advice has indicated the best process, and plant should only be purchased from reputable manufacturers. The operational characteristics of the different processes are described below.

31.6 Specified purities for process use

Various processes need waters of a quality better than the public supply, or whatever source is available. Demands vary widely. Pharmaceutical and cosmetics production generally requires good biological quality. So do food and drink manufacture, but brewing and soft drink manufacture often requires a specified mineral content as well. In brewing it is becoming common for water to be largely de-ionized and wholly synthetic water to be reconstituted by chemical dosing. Membrane processes such as reverse osmosis and electro-dialysis, which do not completely de-ionize the water, are increasingly used. Table 31.2 gives details of the BEWA Water Quality Classification and Table 31.3 shows typical water characteristics.

Textile products are particularly sensitive to iron, which discolors the product. Many washing operations, as in metal finishing, require softened water to avoid staining

of the product. Others are much more sensitive and use de-ionized water.

De-ionized water is required for high-pressure boiler make-up and in many chemical process applications. Where a process has a large-scale steam demand, high-pressure turbine generators are often installed to generate power before providing the process with pass-out steam, thus making the most efficient use of the fuel. If the process does not return the steam as condensate, the

Table 31.2 BEWA Water Quality Classification

Class	Type	Typical applications	General notes	Relevant standards
1	Natural water	Once-through cooling systems Outside wash down Irrigation Fisheries Firefighting Recreational Natural mineral waters	Level of salinity may restrict use for irrigation and fisheries Brackish water has TDS up to 10 000 Seawater has TDS up to 50 000 All characteristics highly variable	Dept of Environment (DoE) Circular 18/85 EC Directive 75/440/EEC EC Directive 78/659/EEC EC Directive 79/923/EEC EC Directive 76/160/EEC EC Directive 80/777/EEC WHO (ISBN 92 4 154 168 7)
2	Potable water	Drinking Domestic use Food and soft drinks Cooling systems Irrigation Firefighting		EC Directive 80/778/EEC DoE Circ 20/82 DoE Circ 25/84
3	Softened water	Recirculatory cooling systems Low-pressure boilers Laundries Bottle washing Closed recirculatory systems Domestic use		Industrial Water Society (IWS) BS 2486: 1978 BS 1170: 1983 BEWA: COP.01.85
4	De-alkalized water	Medium-pressure boilers Recirculatory cooling systems Brewing Food and soft drinks		IWS BS 2486: 1978
5	De-ionized water	Medium-pressure boilers Humidifiers Renal dialysis Glass washing Battery top-up Laboratories Plating industry Spirit reduction	For renal dialysis aluminium must be less than 0.01 mg/l as Al Silica removal may be required for some applications	BS 2489: 1978 BS 4974: 1975 BS 3978: 1966 American Society for Testing and Materials (ASTM) EEC Draft 85/C 150/04 Association for the Advancement of Medical Instrumentation (AAMI)
6	Purified water	Pharmaceuticals Cosmetics Laboratories Chemical manufacturing	United States Pharmaceopeia (USP) also specifies pH = 7.0	British Pharmacopoeia (BP) European Pharmacopoeia (EP) USP US Food and Drug Administration (FDA) BS 3978: 1966
7	Apyrogenic water	Vial washing Parenteral solutions Tissue culture	BP insists on distillation for water for injection. USP allows reverse osmosis also	BP, 1980 EP USP FDA
8	High-purity water	High-pressure boilers Laboratories	De-aeration may be required	BS 2486: 1978 BS 3978; 1966 National Committee of Clinical Laboratories Standards (NCCLS) PSC-3
9	Ultra-pure water	Microelectronics Supercritical boilers Nuclear applications Analytical instrumentation	De-aeration may be required Readily picks up contamination from pipework and environment	BS 2486: 1978 Integrated Circuit Manufacturers Consortium Guidelines (ICMC) ASTM D 1193-77

Table 31.3 Typical characteristics of water

Class	Conduct. µs/cm	Resist. MΩ-cm	TDS	pH	LSI	Hardness	Alkalinity	Nitrate	Sodium	Heavy metals	Silica	Suspended solids	Turbidity	SDI	Particle count	OA	TOC	Micro-organisms	Pyrogens
2	750		Max. 500	6.5 to 8.5			More than 30	Max. 50	Max. 150	Less than 0.1		Less than 1.0	Max. 5			Max. 5		Less than 100	
3					−1 to +1	Less than 20													
4					−1 to +1		Less than 30												
5	20	0.05	Max. 10	5.0 to 9.5			0.1				0.5	Less than 0.1	Less than 0.5	Less than 5					
6	5	0.2	Max. 1	6.0 to 8.5			Max. 0.1				0.1	Less than 0.1	Less than 3			Less than 0.1		Less than 10	
7	5	0.2	Max. 1	6.0 to 8.5			Max. 0.1				0.1	Less than 0.1	Less than 3		1	Less than 0.1		Less than 1	Less than 0.25
8	0.1	10	0.5	6.5 to 7.5							0.02	Less than 0.1	Less than 1.0		1			Less than 1	
9	0.06	18	0.005			0.001	0.001	0.001	0.001	0.001	0.002	ND	Less than 0.5		0.1		0.05	Less than 1	Less than 0.25

Definitions

Conductivity: the electrical conductivity of the water measured in microSiemen/cm is the traditional indicator for mineral impurities.

Resistivity: the reciprocal of conductivity, measured in Megohm-cm. It is used in some industries instead of conductivity particularly for ultra-pure water.

TDS: total dissolved solids determined by evaporating the water and weighing the residue. Units are mg/l.

pH: the acidity or alkalinity of the water expressed on a scale of 0 (acid) to 14 (alkaline). pH 7 is considered neutral.

LSI (Langelier Saturation Index): an indication of the corrosive (negative) or scale-forming (positive) tendencies of the water.

Hardness: the total dissolved calcium and magnesium salts in water. Compounds of these two elements are responsible for most scale deposits. Units are mg/l as $CaCO_3$.

Alkalinity: the total concentration of alkaline salts (bicarbonate, carbonate and hydroxide) determined by titration with acid to pH 4.5. Units are mg/l as $CaCO_3$.

Nitrate concentration is in mg/l as NO_3^-.

Sodium concentration is in mg/l as Na^+.

Heavy metals: the total of chromium, lead, copper and other toxic metals expressed in mg/l.

Silica: soluble or 'reactive' silica concentration in mg/l as SiO_2.

Suspended solids: the concentration of insoluble contaminants in mg/l.

Turbidity: a measure of the colloidal haze present in Nephelometric Turbidity Units (NTU).

SDI: the Silt Density Index, a measure of the rate at which the water blocks a 0.45 µm filter.

Particle count: the number of particles greater than 0.5 µm in 1 ml of water (maximum particle size is 1 µm).

OA (Oxygen Absorbed): a measure of organic contaminants determined in a 4-hour test at 27°C and measured in mg/l as O_2. (Other indicative tests could be Permanganate Value (PV) or Chemical Oxygen Demand (COD).)

TOC (Total Organic Carbon): another way of expressing organics, in this case measured in mg/l as C.

Microorganisms: used here to mean the number of colony-forming units of total bacteria present in 1 ml of water.

Pyrogens: the endotoxins responsible for febrile reaction on injection, determined either by the rabbit test or the **LAL** test. Units are Endotoxin unit/ml (EU/ml).

Note: There are many different tests and different versions of the same test for water analysis. If there is any doubt as to method or interpretation consult a reputable water-treatment supplier. The letters ND in the table indicate *not detectable*. Parts per million (ppm) are also commonly used to express concentration and are essentially identical to mg/l.

Standard tests for water analysis

For further information including test procedures the following are recommended.

BS 2486, Treatment of water for land boilers.
BS 2690, Methods of testing water used in industry.
Methods for the Examination of Waters and Associated Materials, HMSO.
ASTM Standards, Vols 1101 and 1102 (1983)
American Public Health Association, *Standard Methods for the Examination of Water and Waste Water*, 16th edn.

boiler feed will be entirely treated water. This means that the external water-treatment plant has to handle an unusually large flow whose quality of make-up is critical. The world's largest de-ionization plants have been built to serve this kind of system.

De-ionized water itself has a wide range of grades. The lowest is that obtained by a simple cation–anion unit, which may contain up to 5 mg/l TDS. The highest grade is ultra-pure water, which is necessary for making microchips and has maximum total contents three orders of magnitude lower. The specifications for suspended and dissolved matter in ultra-pure water are always at the limits of detection, and are steadily becoming more stringent as chemists devise more sophisticated methods of analysis.

The plant engineer should not be expected to select the correct process for any of these: good professional advice must be taken – and followed. The account below of the processes available therefore concentrates largely on external process characteristics which affect the general operation of the facility.

31.7 Water-purification processes

31.7.1 Filtration

This deals with all equipment used for the removal of particulate matter and represents a wide range of possibilities. Several books cover this subject and only a few typical examples are quoted here.

Simple strainers remove gross materials. These should not normally occur in public supply, but strainers are sometimes fitted to protect sensitive equipment or processes against breaks in the main, etc. The commonest form contains a stainless-steel wedge wire screen and is piped with a bypass so that the screen element can be isolated and removed for cleaning when necessary. If the load on the filter makes this kind of cleaning burdensome a self-flushing filter can be used. These can incorporate strainer elements down to 50 μm.

Smaller particles can be removed by cartridge filters, which can be rated for various particle sizes down to 10 μm. These are typically candle filters whose filter elements are bobbins of nylon or similar string wound onto a former. When clogged, they must be replaced.

31.7.2 Sand filters

Sand filters are widely used in water purification and remove suspended matter by a completely different mechanism. Instead of the water passing through small orifices through which particles cannot pass, it runs through a bed of filter medium, typically 0.75 mm sand 750 mm deep. The orifices between such sand particles are relatively large, but dirt is adsorbed onto the large surface area presented by the medium. The pressure loss rises as the dirt builds up and the filter must be cleaned when it reaches about 3 m WC, otherwise the dirt can be pushed right through the filter.

Filter back washing normally needs low-pressure compressed air and a flow of filtered water about ten times the rated filter throughput. These back-washing arrangements are critical, and providing the large flow of backwash water, as well as drainage for its disposal, can often create difficulties. Given good backwash arrangements, and on water low in suspended matter, sand filters are simple, reliable, cheap and have low operating costs.

Sand filters vary in sophistication. A simple filter will remove most particles down to 5 μm. Multi-media filters which use sand and anthracite, and possibly a third medium, in discrete layers, can yield very efficient filtration down to 2 μm. Granular activated carbon can be used instead of sand to add some measure of organic removal to the filtration process. The quality produced by any filter depends largely on the efficiency of the backwash. Sand filters in some form provide a satisfactory solution for the majority of water-filtration problems.

31.7.3 Coagulation

Still smaller particles and some of the organic matter in water can, if necessary, be removed by coagulation, in which a chemical coagulant is dosed into the water before the filter. Unlike simple filtration, this process requires chemicals and careful control. Dosing directly before the filter will only cope with small concentrations of dirt. Larger amounts of dirt require coagulation to be followed by sedimentation and then filtration. It is a difficult process and is to be avoided if at all possible, especially on small flows. When coagulation processes go wrong, they can severely damage downstream equipment.

31.7.4 Membrane filtration

Microfiltration and ultrafiltration have recently been introduced for the removal of particles down to any desired size. Their capital cost is relatively high. Experience with them is limited, and a short trial with a small-scale pilot element is advisable. Prediction of full-scale performance from such trials is normally quite reliable.

Dead-end filtration through membrane filters is common in some industries where high purity is imperative. When clogged, the membrane has to be replaced. The water is first purified, and the filters serve as a final polisher. They are unsuitable for applications where they have to remove any significant concentration of particulate matter, as the cost of membrane replacement can become very high.

31.7.5 Water softening

Traditional water-softening processes add lime, or lime and soda ash, to the water. This produces a precipitate in the form of a sludge, which must be settled out and the clarified water filtered in a sand filter. The chemicals are cheap, but the problems of handling solid chemicals and of sludge disposal have made the processes obsolete. They are, however, simple and robust, suitable for low-technology supervision, and cope well with changing water analyses. They should not be forgotten when considering projects in underdeveloped countries.

31.7.6 Ion-exchange processes: general

Ion-exchange units physically resemble sand pressure filters, and are almost as effective in removing, and therefore accumulating, suspended matter. As they are back-washed much less vigorously they must not be fed with water containing more than about 2 mg/l of suspended matter or they will become progressively fouled and with suspended material. Public supply quality is usually (but not always) good enough for the water to be put directly onto ion exchange.

Ion exchangers are synthetic resins in the form of beads – small spheres of 0.5 to 1.0 mm in diameter. Good back-washing of the bed is important to obtain a uniform bed in which the different sizes are graded to give the minimum pressure loss. The resins have a limited life. Cation resins, used in softening and also in the first stage of deionization, can last for 10 or even 20 years, but anion resins used in de-ionization rarely give more than 3–5 years of use.

If the unit becomes badly fouled with suspended matter (for example, after a pipe brake has introduced excessive suspended matter into the system) it must be taken out of service and cleaned. This is done with an extended backwash, possibly at higher flow rates. If this does not remove the dirt, the manhole should be opened and the resin agitated with an air lance. Non-ionic detergents can be used, but not at the same time as the air lance or the resulting froth will be impossible to control.

Ion exchangers in general and cation resins in particular are liable to chemical attack by chlorine. The very small residual of chlorine in public supply (typically, 0.2 mg/l) has only a mineral effect, but if more chlorine has been added it must be removed (e.g. with an activated carbon filter) before ion exchange.

The anion resins used in de-ionization are prone to fouling if the water contains organic matter. The soft peaty waters mentioned above are particularly bad in this respect, and, at worst, can reduce resin life to a few weeks.

The handling, storage, measuring and dilution of regeneration chemicals requires serious consideration, especially in de-ionization, which requires strong acid and alkali. For dilution and pumping the actual regeneration solution most systems use ejectors, which avoids moving parts in corrosive solutions.

All ion-exchange processes produce wastewater from backwash, regeneration and rinse. The proportion of waste depends on the concentration of hardness or to TDS being removed, and can be as high as 15 per cent of the product flow. Any pretreatment has to take this additional flow into account.

Base-exchange softening

This is the simplest and commonest ion-exchange process. It uses only cation resins which are regenerated with strong brine to remove calcium and magnesium ions from water in exchange for sodium ions. Well-designed standard plant is available in a wide range of sizes and operates with a minimum of supervision. Conventional softeners allow 1–3 per cent of the incoming hardness to leak through. As the hardness is exchanged for sodium

the TDS remains substantially unchanged. The softened water is at the same pH as the incoming water but is rather more corrosive to steel.

There is no difficulty in handling or storage of corrosive chemicals or effluents. Salt is easily purchased and handled in bulk, and can be discharged directly into a standard commercial salt saturator.

The run and regeneration cycle on small to medium-size units is normally governed by a multi-port valve, which is the only moving part required. There have been many cases where this has failed and brine was injected directly into the boiler. A conductivity meter on the make-up line would guard against this.

The operation of a base exchanger is chemically inefficient, and the spent regenerant contains large amounts of excess salt which may occasionally be difficult to dispose of. Factory softeners make a major contribution to the chloride content of the UK's industrial rivers, and in the longer term there will be heavy pressure from environmentalists to reduce the amount of salt being discharged.

De-alkalization/base exchange

This three-stage process is used for waters of high alkalinity and hardness. It actually removes most of the temporary hardness and so reduces the TDS of the water. However, in the process it increases the proportion of silica in the remainder. Any residual temporary hardness and the permanent hardness are softened in a conventional softener.

The effect is best illustrated by a numerical example (Table 31.4). Let us take the case of hard and alkaline deep well water such as that found to the north of London, whose main characteristics are shown in the first column of Table 31.4. The second column shows its quality after de-alkalization has removed nine-tenths of the temporary hardness and converted it into CO_2 gas. This is removed from the water by stripping it with air in a packed 'degassing' column, and the product then softened in the third stage to yield the product shown in the third column.

The raw water silica is 22 mg/l as SiO_2, and therefore becomes a major constituent of the treated water. Silica scale must now be avoided by raising the boiler water pH and letting silica rather than the TDS control the necessary blowdown. Silica scale not only has a tenth of the heat conductivity of calcium carbonate scale but it is glassy, adherent, and extremely resistant to boiler-cleaning chemicals.

De-alkalization resins are regenerated with sulfuric or hydrochloric acid. Sulfuric acid is cheaper and easier to

Table 31.4

	Raw water	De-alkalized	Base exchanged
Total Hardness[a]	290	50	1
Alkalinity[a]	265	25	25
TDS[a]	334	94	94

[a] All as mg/l $CaCO_3$.

store and handle, but its dilution with water are potentially dangerous and must be carefully engineered. The regenerant solution must be extremely dilute to avoid the precipitation of calcium sulfate, which would clog the unit. This leads to long regeneration times and a high production of wastewater. Hydrochloric acid is usually preferred on the smaller scale, where easier operation is more important than the annual cost of chemicals.

De-alkalization resins must not be over-regenerated or the product water becomes strongly acidic. The system therefore needs some measure of skilled supervision, and may depend on a pH meter – an instrument that, in turn, needs regular and skilled maintenance.

The acid regeneration is very efficient, and the amount of free acid going to drain is small. Given a reasonable amount of dilution, it may be unnecessary to take special precautions against it. Although acids are more expensive than salt, the efficiency with which they are used means that the operating cost of a de-alkalization/base exchange plant is similar to that of a simple softener. The capital cost, on the other hand, is much higher and of the same order as that of a simple de-ionizer.

The tower in which CO_2 is stripped out must run into a break-pressure tank with subsequent re-pumping. It will load the water with any dust, and living organisms or other particles in the atmosphere, which leads to trouble in dirty environments or in pharmaceutical works.

The de-alkalized and degassed water has a pH of 4–5 and (having just passed through an air-blown tower) is laden with oxygen and extremely corrosive. Normal practice is to dose NaOH into the degasser tower sump, at a level sufficient to approach to desired boiler water pH. If this dosing fails, severe corrosion in the degassed water pump, the softener and the feed system will result.

De-ionization

De-ionization is a two-stage process which removes all the dissolved ions from water. First, the water passes through a bed of cation-exchange resin, regenerated with sulfuric or hydrochloric acid, where all the cations are removed in exchange for hydrogen and dissolved salts are thus converted into their equivalent mineral acids. The resulting acidic water is passed through a bed of anion-exchange resin, regenerated with caustic soda, which removes all the anions in exchange for hydroxyl. The resulting water contains, except for a small leakage of dissolved salts (typically less than 1–5 mg/l), only hydrogen and hydroxyl ions that, together, form water. On large plants, above about 20 m³/h, it is quite common to include a degassing tower between the cation and anion exchange stages. This is a purely economic measure; it removes carbon dioxide, which would otherwise have to be removed on the anion resin, and consequently reduces the amount of caustic soda needed for regeneration.

While it produces very pure water, de-ionization is an expensive process to operate. It uses acid and caustic for regeneration and produces an effluent which may need neutralization before it can be discharged. On the other hand, all sizes of de-ionization plants are available with fully automatic operation and control so that the plant

operator's involvement can be limited to ensuring that there are adequate supplies of regenerant chemicals.

31.8 Membrane processes

Membrane processes are not yet used widely for industrial water processing, but will become more important in future. At present, they are generally more expensive than the older processes which they promise to replace, but costs are falling. Their main advantage lies in the fact that they add little or no chemicals to the aqueous environment but return to it only the material taken from the raw water.

Ion exchange, in contrast, creates an effluent that contains between two and five times the mass of inorganic material removed from the product water. Coagulation with aluminum or iron salts creates a sludge, which creates a disposal problem. 'Green' pressure, especially in Switzerland and mid-west USA, which lie in the middle of large land masses, has started to force industrialists to install alternative membrane processes to avoid these discharges.

31.8.1 Reverse osmosis (RO)

In reverse osmosis water is forced by pressure through a very fine-pore membrane, which has the property of rejecting dissolved salts. The process thus removes both particulate and dissolved matter. Generally, the flux of water is extremely slow, so that large membrane areas have to be installed to achieve the desired output. Different grades of membrane show different rejections and fluxes.

A 'tight' membrane will remove over 99 per cent of dissolved salts but requires pressures of the order of 30 bar or more to function economically. A 'loose' membrane may only take out 90 per cent of the salts but will function at around 5 bar or even less. Both types of membrane will remove all particulate matter and large molecules. Small, non-associated molecules such as silica and dissolved gases (i.e. oxygen and CO_2) pass through unchanged.

The ability to remove particulates has made RO indispensable in the production of ultra-pure water for microchip washing. Its ability to remove large molecules enables it to produce pyrogen-free water for the pharmaceuticals industry. In the USA and elsewhere RO is permitted for producing the water used in making up injectable preparations. The European Pharmacopoeia still insists on distillation for this, but the larger amounts of water needed for ampoule washing, etc. are often purified by RO.

For conventional water purification, RO may be economical for removing the bulk of the dissolved salts in water before ion exchange, but only if the raw water has a high dissolved solids content. In practice, substituting two processes for one is unattractive except for large throughputs.

Recently developed 'softening' membranes reject most of the hardness in water while passing sodium salts, and operate at pressures of about 5 bar. This can be used to provide substantially softened make-up water for shell boilers. About 5–10 per cent of the raw water hardness remains in the water but, for example, if there is a sufficiently high condensate return to the system this residual hardness may be acceptable.

RO has some intrinsic disadvantages:

1. The water wastage tends to be high, especially on small plants where a complex water-saving flow sheet is inappropriate.
2. RO membranes are very sensitive to fouling by particulates, colloids and macromolecules. The Silt Density Index test for raw water is an empirical test which shows the rate at which a 0.45 µm filter disc becomes plugged, and gives a good indication of the fouling properties of the water. An experienced person should carry it out. Seasonal and other variations which might lead to fouling must also be considered. Pretreatment to avoid fouling is always complicated and risky: if it fails, rapid fouling of the whole membrane complement of a plant will result. Cleaning procedures are ineffective against severe fouling and membrane replacement is very expensive. On bad fouling waters RO should be avoided unless, for example, its use is essential to produce ultra-pure water.
3. Some anti-scaling treatment is necessary on most waters, using, for example, pH reduction, hexametaphosphate, phosphonates or polyacrylic inhibitors. The action of the various compounds available is not yet fully understood, so this can be a hit-and-miss technique.
4. As CO_2 passes through the membrane, while bicarbonate is rejected, the product water tends to be acidic and may need pH correction.

31.8.2 Ultra-filtration

As RO membranes become 'looser' their salt rejection falls (see Section 31.8.1). Eventually a point is reached at which there is no rejection of salts, but the membrane still rejects particulates, colloids and very large molecules. The membrane pore size can be tailored to a nominal molecular weight cut-off. The resulting filtering process is called ultra-filtration.

In this process good hydrodynamics on the membrane surface are required to scour away the accumulated solids and prevent the membrane being blinded. This cannot be totally effective, and in practice the nominal membrane cut-off is often masked by the tendency of particulates to form a thin layer on the membrane surface whose effective pore size may be smaller.

The process is used mostly for industrial separations, such as the removal of yeast from beer or the recovery of emulsified cutting oils. It has been proposed as a pre-treatment to RO for fouling waters, but the economics do not yet look attractive.

31.8.3 Electrodialysis (ED)

This process uses membranes which remove salts from water by passing an electrical current through it. It was originally developed for the conversion of brackish water

to potable (i.e. for a product of 300–500 mg/l). As the salts are removed, the conductivity of the water falls, the power used in the process increases and the cost rises. It is therefore potentially economical primarily for pre-treating highly mineralized water (say, 300 mg/l or more) before ion-exchanger de-ionization.

Unlike RO, which is essential for producing ultra-pure water, there is little experience of ED in this field. The process has some potential advantages over RO: it is less liable to fouling and it can be engineered to waste much less water. Like RO, its costs fall sharply at higher temperatures, but the prospects of improved engineering making this a reality are better than for RO. It offers some prospects particularly where the product water has to be heated in any case (e.g. boiler make-up).

A variant of ED has recently appeared on the market: the cells between the membranes are filled with ion-exchange resin to reduce electrical resistance when producing water of high purity. This makes possible the production of good-quality de-ionized water without regeneration chemicals. On the small scale (e.g. for laboratories or pharmaceutical production) the convenience of avoiding dangerous chemicals and corrosive effluent outweighs the relatively high costs of this technique.

31.8.4 Retrofit of membrane processes

As with RO, de-ionization by ED before ion exchange is rarely feasible because of the complication of two processes instead of one. A different situation arises where an existing ion exchange installation is overloaded, either because the plant needs to treat more water or because the TDS in the raw water have risen above the original design level. In such cases the simplest and most economical solution can be to install RO or ED to take up a large part of the ionic load and allow the existing ion-exchange plant to be up rated.

31.9 Effluents

Industrial effluents are a particularly difficult problem to discuss in general terms: their nature is very diverse, possible methods of treatment vary correspondingly and their acceptability depends as much on the receiving body as on their flow and contents. There are, however, some common factors which are worth mentioning.

31.9.1 Domestic sewage

Industrial plants also discharge domestic sewage. It is vital to keep this separate from any industrial effluent which may have to be treated, so that it can then be disposed of by conventional means (to the public sewer, septic tank, etc.).

31.9.2 Surface drainage

Most of the surface drainage due to rain falling on roofs, roadways, etc. will be normal, acceptable floodwater and can run off to a soakaway, storm drain, etc. in a conventional manner. Surface drainage from areas contaminated with spillage or with material deposited from a locally polluted atmosphere creates a particularly difficult effluent problem, characterized by unpredictable and violent variations in flow and concentration of pollutant. The best approach to this is to avoid conditions in which the area becomes loaded with contaminant.

Alternatively, the contaminated surface drainage must be segregated from the normal storm water drains and may, for example, be led into a balancing tank. This tank must be large enough to even out the variations and to allow the contents to be added to the works effluent (treated or untreated) over a period of time. Provision must be made for periodically removing the inevitable accumulation of silt in the bottom of the tank.

31.9.3 Noxious effluent

Almost any impurity added to water leaving a works can, in certain circumstances, become noxious, and with some of these impurities it is not immediately obvious that they are unacceptable. The main examples are:

- Corrosive conditions – not only high or low pH, but also high levels of sulfate, which is corrosive to cement and therefore unacceptable in the public sewers. High temperatures can be unacceptable;
- Toxic constituents such as heavy metals, cyanides, etc.;
- Organic materials whose decay consumes oxygen, which includes various harmless materials (e.g. all vegetable matter, milk wastes, cellulose, sugars, etc.);
- Inert material, which may settle to the bottom of a body of water and coat its beds (clay, alum sludge, etc.). Also dyestuffs which discolor the water and are esthetically unacceptable.

Unlike potable and industrial waters, for which quality specifications are laid down, there can be no universal regulation as to the concentration or degree to which any of these is acceptable. The receiving body, which has to use its judgement in each case, will lay down allowable limits.

A sewerage authority must decide on the level of contamination, which it is prepared to accept into its sewer. The amount of sulfate and of biodegradable matter, which can be allowed, will depend, first, on the quantity and quality of the flow already in the sewer and available to dilute the effluent. The second decision concerns the limits of contaminants which are acceptable into the sewage works.

If the effluent is within acceptable limits, the authority is entitled to make a charge for receiving and treating it (see below). The river authority has similar decisions to make if the effluent is to be discharged directly into the environment. The acceptable limits will, of course, be much lower but no charge will be levied.

The receiving body is entitled to take samples of effluent at all times to ensure that the conditions under which they have agreed to receive it are being met. They may insist on some kind of continuous or regular monitoring by the discharger, but as analytical methods for almost every kind of effluent are difficult to automate and standardize, this is always an unsatisfactory aspect of the system.

31.9.4 Charges for effluent

Sewerage authorities normally levy a charge on industrial effluent. The conventional system of charging uses a formula which takes into account the volume, its concentration of suspended solids and its oxygen consumption (which is a measure of the load it will put on the treatment works). Typical figures for the flow and quality of the discharge are ascertained and agreed between the sewerage authority and the plant.

The flow of effluent is often calculated on the assumption that, say, 90 per cent of the incoming water (metered by the water authority) will re-emerge as effluent. This is, of course, not the case at a brewery, for example, where direct metering or different methods are used to ascertain the effluent flow. Direct metering of the flow is, however, always difficult and therefore uncommon.

31.9.5 Effluent management

If a works produces any effluent liable to be unacceptable, or liable to raise an effluent charge, it is important that the whole effluent system should be properly managed. This means knowing where the drains run and in which of them each effluent flows. This information is often unavailable and is difficult to ascertain. Good management often means that if treatment or tankage is required, different flows should be segregated from one another. Diluting a noxious effluent with another of a different type, or with a non-noxious one, generally aggravates the problem.

31.9.6 Effluent treatment

Where the effluent is outside the limits acceptable for discharge it has to be treated first by the works. Generally, effluent treatment is an unsatisfactory operation because it requires skills different from those available on the works, and because it is a nuisance and is not seen to be productive.

Effluent-treatment plants fail most commonly because the design is based on an inadequate or faulty definition of the effluent problem. In an existing works it may take months of painstaking research to establish the true patterns of flow and contamination. Predicting these for a new project can be extremely difficult, and serious mistakes are common.

Treatment processes vary widely, and in themselves are reasonably well understood, but even if the nature and pattern of effluent is properly defined, problems often arise from wide variations in contents and flow of the effluent. An effluent plant tends to impose limitations on the main process – for example, it may be seriously affected by rapid dumping of the contents of a tank. These limitations are often flouted through ignorance or negligence.

The best solution to any effluent problem is to avoid treatment by the works altogether. Even if it appears initially to be more costly, it will save trouble and expense in the long term.

If the process cannot be modified to avoid the effluent the sewerage authority may be persuaded to take over its treatment. Various financial arrangements have been made for this over the years, from a charge based on the volume flow to a capital contribution towards an extension to the sewage works capacity. By this means the responsibility is handed over to an organization whose purpose and technical skills are devoted to the subject.

31.9.7 Water economy and re-use

Water is historically thought to be free, and is still so cheap that its wastage is not felt to raise a serious cost. This commonly leads to pointless waste. It is unusual for a works to meter the water usage of different operations on one site, and so the cost of waste is not made clear to those who are immediately responsible for it. Occasional spot estimates of water usage (which may be possible without installing meters), with direct debiting of water cost to the operation, can do much towards water economy. In such an exercise it is important to add the cost of effluent discharge to that of water intake.

Much lip service is paid to water re-use, which, on the whole, is more fashionable than practicable. Better opportunities arise from simple economy in water consumption at each point of use. Water reuse after effluent treatment is not often practicable. It appears, if at all, most favorable in the re-use of slightly contaminated water for some low-grade purpose such as washing down.

Acknowledgement

The author is grateful to the BEWA for the use of their Water Quality Classification as Tables 31.2 and 31.3.

32

Pumps and Pumping

Keith Turton
Loughborough University of Technology

Contents

32.1 Pump functions and duties

Pumps impart energy to the liquids being transferred by mechanical means using moving parts. They can be classified as rotodynamic or positive displacement. Rotodynamic pumps cause continuous flow, and the flow rate and discharge pressure are effectively constant with time. Positive displacement pumps deliver fixed quantities at a rate determined by driving speed. The main types of pump commonly used are listed in Figure 32.1.

Pumps are used to transfer liquids, such as moving blood and other biological fluids; delivering measured quantities of chemicals; in firefighting; in irrigation; moving foods and beverages; pumping pharmaceutical and toilet products; in sewage systems; in solids transport; in water supply and in petrochemical and chemical plant. They are utilized in power transfer, braking systems, servomechanisms and control, as well as for site drainage, water-jet cutting, cleaning and descaling. Pumps thus give a wide range of pressure rises and flow rates with pumping liquids that vary widely in viscosity and constituency.

32.2 Pump principles

32.2.1 Rotodynamic pumps

Taking a typical centrifugal pump (Figure 32.2) the Euler equation can be written, at best efficiency flow, in the form:

$$gH = u_2^2 - \frac{Q}{A_2} u \cos \beta_1 \tag{32.1}$$

where $u_2 = \omega D_2/2$, $A_2 = \pi D_2 b_2$, Q is flow rate and H is head rise. This ignores flow losses, so that actual performance is less than the Euler (Figure 32.3). Figures 32.4–32.6 give typical pump performance curves for a constant driver speed. The inflections in the mixed and axial flow curves are due to flow instability over blades and through impeller passages.

The hydraulic efficiency

$$\eta_H = \frac{gH_{\text{Actual}}}{gH_{\text{Ideal}}} \tag{32.2}$$

and

$$\eta_0 = \frac{\text{Hydraulic power}}{\text{Input power}}$$

or

$$\eta_0 = \frac{mgH}{P_{\text{IN}}} = \frac{\rho QgH}{P_{\text{IN}}} \tag{32.3}$$

Typical pump cross-sections are shown in Figures 32.7–32.10. Figure 32.11 illustrates a multi-stage design where identical stages are assembled in a pressure casing as in a boiler feed pump.

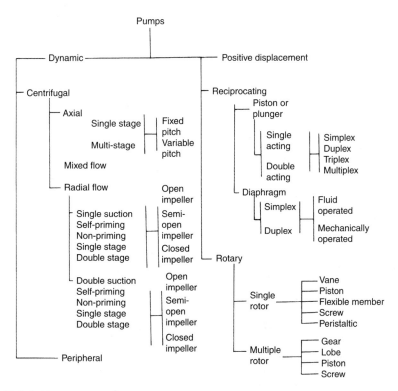

Figure 32.1 Pump family trees

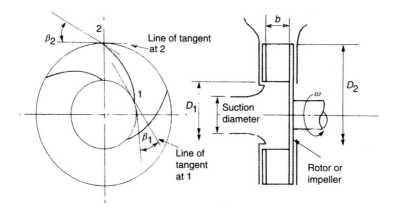

Figure 32.2 A simple centrifugal pump

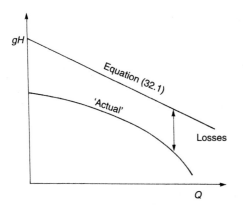

Figure 32.3 A pump's ideal and actual characteristics

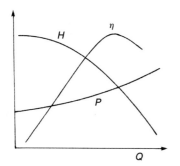

Figure 32.4 Centrifugal

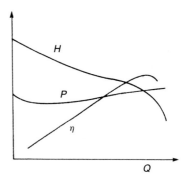

Figure 32.5 Mixed flow (bowl pump)

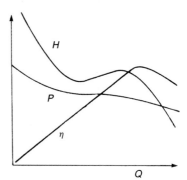

Figure 32.6 Axial flow

32.2.2 Scaling laws and specific speed

If a simple pump is considered, it is possible to state that there must be a working relation between the power input, and the flow rate, pressure rise, fluid properties, and size of the machine. If a dimensional analysis is performed, it can be shown that a working relation may exist between groups of non-dimensional quantities in the following equation:

$$\frac{P}{\rho\omega^3 D^5} = f\left[\frac{Q}{\omega D^3} \times \frac{gH}{\omega^2 D^2} \times \frac{\rho\omega D^2}{\mu} \cdots\right] \qquad (32.4)$$

$$\quad (1) \qquad\quad (2) \qquad (3) \qquad (4)$$

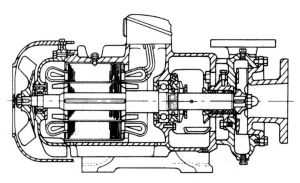

Figure 32.7 A monoblock design

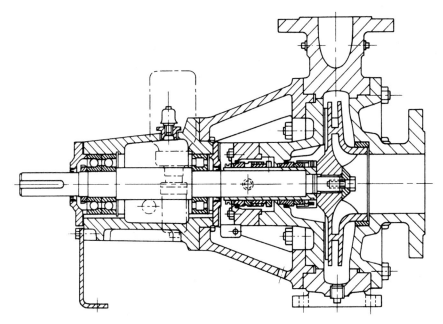

Figure 32.8 A back pull-out design

Term (1) is a power coefficient, which does not carry any conventional symbol.

Term (2) can easily be shown to have the shape V/U and is called a flow coefficient (the usual symbol being ϕ).

Term (3) similarly can be shown to be gH/U^2 and is usually known as a head coefficient (or specific energy coefficient) ψ.

Term (4) is effectively a Reynolds number with the velocity the peripheral speed and the characteristic dimension being usually the maximum impeller diameter.

Since these groups in the SI system are non-dimensional, they can be used to present the results of tests of pumps in a family of pumps that are geometrically and dynamically similar. This may be done as shown in

Figures 32.12 and 32.13, and Figure 32.14 shows how the effect of changing speed or diameter of a pump impeller may be predicted, using the scaling laws:

$$\frac{P}{\rho\omega^3 D^5} = \text{const}$$

$$\frac{Q}{\omega D^3} = \text{const} \qquad (32.5)$$

$$\frac{gH}{\omega^2 D} = \text{const}$$

The classical approach to the problem of characterizing the performance of a pump without including its dimensions was discussed by Addison,[1] who proposed that a pump of standardized size will deliver energy at the rate of one horsepower when generating a head of one foot

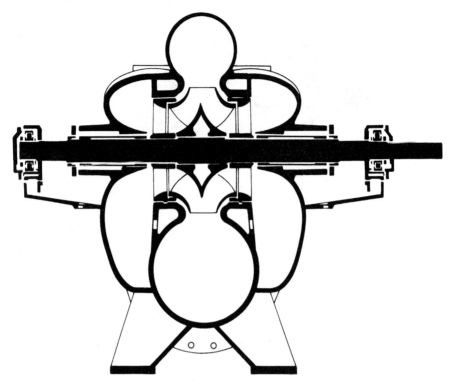

Figure 32.9 A double-suction design

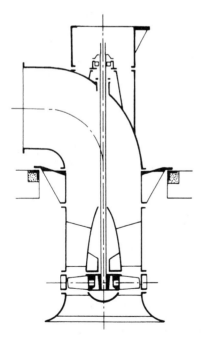

Figure 32.10

when it is driven at a speed called the Specific Speed:

$$N_s = K \frac{N\sqrt{Q}}{H^{3/4}}$$

The constant K contains fluid density and a correction factor, and it has been customary to suppress K and use:

$$N_s = \frac{N\sqrt{Q}}{H^{3/4}} \tag{32.6}$$

Caution is needed in using data, as the units depend on the system of dimensions used, variations being liters/minute, cubic meters/second, and gallons per minute or US gallons per minute as well as meters or feet. A plot of efficiency against specific speed are in all textbooks based upon the classic Worthington plot, and Figure 32.15, based on this information, has been prepared using a non-dimensional statement known as the characteristic number

$$k_s = \frac{\omega\sqrt{Q}}{(gH)^{3/4}} \tag{32.7}$$

This is based on the flow and specific energy produced by the pump at its best efficiency point of performance following the approach stated by Wisclicenus: 'Any fixed value of the specific speed describes a combination of operating conditions that permits similar flow conditions in geometrically similar hydrodynamic machines.'

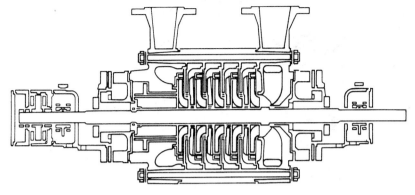

Figure 32.11 A multi-stage pump

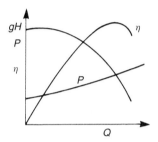

Figure 32.12 A pump's characteristic at fixed speed

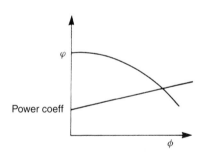

Figure 32.13 A non-dimensional plot

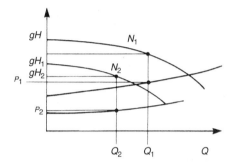

Figure 32.14 Prediction of speed change using equation (32.5)

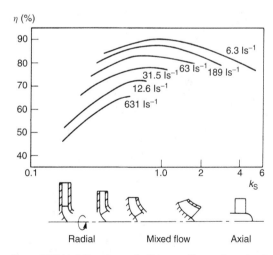

Figure 32.15 Variation of overall efficiency with non-dimensional specific speed[3]

Figure 32.16 presents on the basis of the characteristic number the typical impeller profiles, velocity triangle shapes, and characteristic curves to be expected from the machine flow paths shown. This indicates the use of the number as a design tool for the pump engineer.

The scaling laws (Equation (32.5)) may be used to predict the performance from change of speed, as indicated in Figure 32.14. In many cases, the pump engineer may wish to modify the performance of the pump by a small amount. The common solution is to slightly reduce the impeller diameter. Figure 32.17 illustrates how small changes in impeller diameter can affect the performance. The figure in its original form appeared in

Karrasik *et al.*[2] and has been modified to appear in metric form. The role used is often called the Scaling Laws,

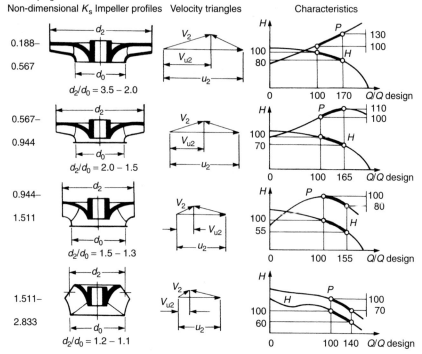

Figure 32.16 Flow path shapes, velocity diagrams and characteristics

written in the form:

$$\left.\begin{array}{l} \dfrac{P_2}{D_1} = \sqrt{\dfrac{gH_2}{gH_1}} \\[3mm] \dfrac{Q_2}{Q_1} = \sqrt{\dfrac{gH_2}{gH_1}} \\[3mm] \dfrac{P_2}{P_1} = \left(\dfrac{gH_2}{gH_1}\right)^{3/2} \end{array}\right\} \qquad (32.8)$$

Other methods of adjusting the output while keeping the speed constant consist of modifying the profiles of the blades at the maximum diameter of the impeller. This technique has been used for a long time and is often used to obtain a small energy rise when the pump is down on performance when tested. The reader is referred to reference 2.

32.2.3 Positive displacement pump principles

Whether the pump is a reciprocator or a rotary design, liquid is transferred from inlet to outlet in discrete quantities defined by the geometry of the pump. For example, in a single-acting piston pump (Figure 32.18), the swept volume created by piston movement is the quantity delivered by the pump for each piston stroke, and total flow rate is related to the number of strokes per unit time. Similarly, the spur-gear pump (Figure 32.19) traps a fixed quantity in the space between adjacent teeth and the casing, and total flow rate is related to the rotational speed of the gear wheels.

The maximum possible flow rate,

$Q_0 =$ displacement $\times$ speed

as shown in Figure 32.20. The actual flow is reduced by leakage, flow Q_L,

$$Q = Q_0 - Q_L$$

The volumetric efficiency

$$\eta_v = \frac{Q}{Q_0} = 1 - \frac{Q_L}{Q_0} \qquad (32.9)$$

and

$$\eta_0 = \frac{\rho Q g H}{P_D + P_L} \qquad (32.10)$$

P_D and P_L are defined in Figure 32.20. Table 32.1 gives typical values of η_v and η_0 for a number of pump types.

Table 32.1 Some values of η_v and η^4

pump	$\eta_v(\%)$	$\eta(\%)$
Precision gear	→98	→95
Screw	–	75–85
Vane	85–90	75–80
External gear	–	20–60
Radial–multi-piston	>95	>90
Axial–multi-piston	>98	>90

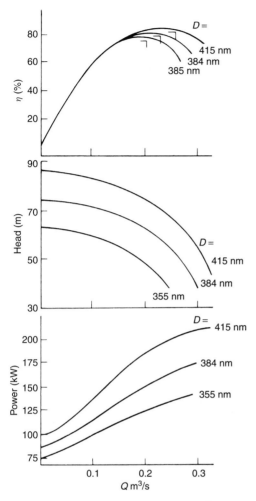

Figure 32.17 Scaling laws applied to diameter change (after Karassik *et al*[2].)

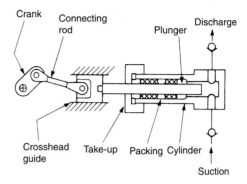

Figure 32.18 Plunger pump

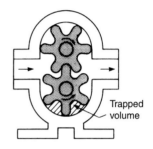

Figure 32.19 External gear pump

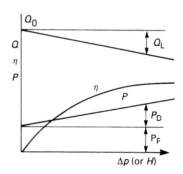

Figure 32.20 Typical characteristic at constant *N*

Since discrete quantities are trapped and transferred, the delivery pressure and flow varies, as shown in Figure 32.21, which also illustrates how increasing the number of cylinders in a reciprocating pump reduces fluctuations. In the case of lobe and gear pumps the fluctuations are minimized by speed of rotation and increasing tooth number, but where, for control or process reasons, the ripple in pressure is still excessive, means of damping pulsations has to be fitted. Often a damper to cope with this and pressure pulses due to valve closure is fitted, and two types are shown in Figure 32.22.

The capacity of the accumulator is important, and one formula based on experience for sudden valve closure is

$$Q_A = \frac{QP_2(0.016L - T)}{P_2 - P_1} \times 0.25 \quad (32.11)$$

Here Q_A is the accumulator volume (m^3), Q is flow rate (m^3/S), L is pipe length (m), T is valve closure time(s), P_1 is the pressure in the pipeline (N/m^2) and P_2 is the maximum pressure desired in the line. (N/m^2) ($P_2 = 1.5P_1$ in many cases).

32.3 Effects of fluid properties on pump behavior

Two effects on pump performance must be discussed: viscosity and gas content. Figures 32.23 and 32.24 illustrate the effects of viscosity change on centrifugal and

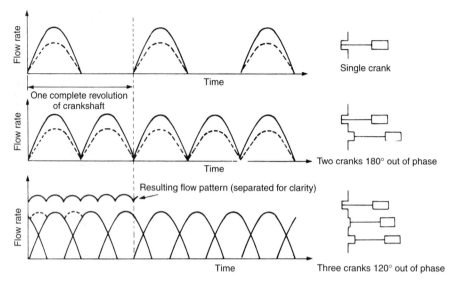

Figure 32.21 Reciprocating pump: variation of flow rate with numbers of cylinders

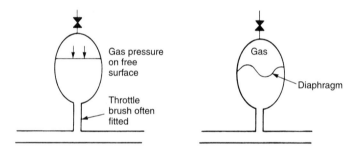

Figure 32.22 Accumulator designs

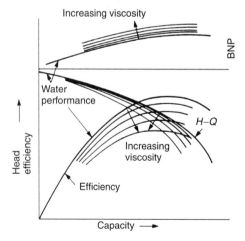

Figure 32.23 Effect of viscosity on centrifugal pump performance

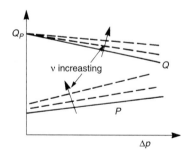

Figure 32.24 Effect of viscosity on positive displacement performance

positive displacement pumps, respectively. Figure 32.23 shows the deterioration of centrifugal pump performance, and it may be noted that if v is greater than 100 centistokes, water performance must be corrected. Figure 32.24 indicates that in a positive displacement

Table 32.2 Effect of fluid viscosity

Type of pump	Significant[a] viscosity levels	Effect of viscosity level	Treatment and/or notes
Centrifugal	20	–	Performance maintained similar to water performance up to this level
	20–100	Lowering of H-Q curve increase in input hp	General lowering of efficiency but may be acceptable
	Above 100	Marked loss of head	Considerable reduction in efficiency, but high efficiencies may still be attainable from large pumps
Regenerative	Above 100	Marked loss of performance	Pumps of this type would not normally be considered for handling fluids with a viscosity greater than 100 centistokes
Reciprocating	Up to 100	Little	Performance generally maintained. Some reduction in speed may be advisable to reduce power input required
	Above 100	Performance maintained but power input increased	Speed is generally reduced to avoid excessive power inputs and fluid heating
	Above 1000	Flow through valves may become critical factor	Larger pump size selection run at reduced speed–e.g. 3 × size at 1000 centistokes running at one-third speed. Modification of valve design may be desirable for higher viscosities
Plunger	–	–	For very high-pressure deliveries only
Sliding vane	Above 100	Sliding action impaired: slip increased	Not generally suitable for use with other than light viscosity fluids
External gear	None	Power input and heat generated increases with increasing viscosity	May be suitable for handling viscosities up to 25 000 centistokes without modification. For high viscosities: (a) Clearances may be increased (b) Speed reduced (c) Number of gear teeth reduced
Internal gear	None	Power input and heat generated increases with increasing viscosity	For higher viscosities: (a) Speed may be reduced (b) Number of gear teeth reduced (c) Lobe-shaped gears employed
Lobe rotor	250	None	(a) Speed may have to be reduced
	Above 250	Cavitation may occur	(b) Modified rotor form may be preferred
Single-screw	None	–	Nitrile rubber stator used with oil fluids
Twin- or	Up to 500	Little or none	–
multiple-screw	Above 500	Increasing power input required	Speed may be reduced to improve efficiency

[a]Viscosity in centistokes.

pump the volumetric efficiency improves and power requirement increases (with increasing viscosity).

Table 32.2 summarizes the effects of liquid changes (effectively, viscosity and density changes) on pump performance and Figure 32.25 presents material presented by Sterling,[5] which illustrates how efficiency falls away with viscosity for two pumps working at the same duty point, graphically showing the rapid fall-off of efficiency as μ increases in a centrifugal pump.

Figure 32.26 demonstrates a well-known method of correcting for fluid change from water for a centrifugal pump. This allows an engineer to predict change in performance if the kinematic viscosity of the liquid to be pumped is known and the water test data are available.

Gas content is another important effect. It is well known that centrifugal pumps will not pump high gas content mixtures, as flow breaks down (the pump loses 'prime') when the gas/liquid ratio rises beyond 15 per cent.

Figure 32.27 shows how a centrifugal pump is affected, particularly at low flow rates, and the behavior is typical of conventional centrifugal pumps. Figures 32.28 and 32.29 present well-known information on the effects of dissolved and entrained gas on the volumetric efficiency of a positive displacement pump.

32.4 Flow losses in systems

32.4.1 Friction losses

The most common method of friction loss estimation is based on the D'Arcy Weisbach equation:

$$(gH)_{\text{friction}} = f \frac{L}{D} \frac{V^2}{2} \tag{32.12}$$

The friction factor f is plotted in Figure 32.30 against the Reynolds number, based on the pipe inner diameter D

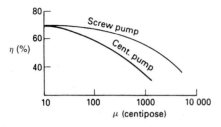

Figure 32.25 Effect of μ on screw pump and centrifugal pump compared (after Sterling[5])

for circular pipes and the hydraulic diameter ($= 4 \times$ flow area/wetted perimeter) for non-circular pipes. Table 32.4 gives typical surface roughness. Figure 32.30 is thus used to find f as follows:

Mean velocity 7 meters per second (m/s), mean diameter 50 mm, water kinematic viscosity, therefore $UD = 0.35$

$Re = 3 \times 10^5$

Roughness for galvanized pipe = 0.152, therefore

$\dfrac{k}{D} = 3 \times 10^{-3}$

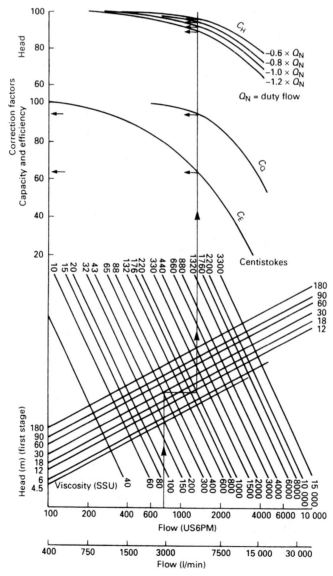

Figure 32.26 Viscosity correction curves (adapted from *Hydraulic Institute Standards*, 12th edition, Hydraulic Institute, Cleveland, Ohio, 1969)

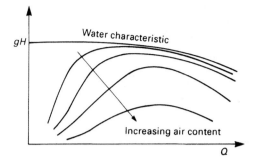

Figure 32.27 The effect of air content on pump behavior

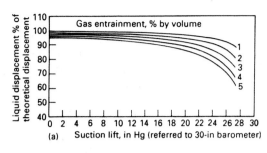

(a) Suction lift, in Hg (referred to 30-in barometer)

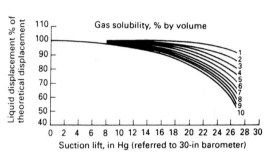

Figure 32.28 Effect of dissolved gas on liquid displacement

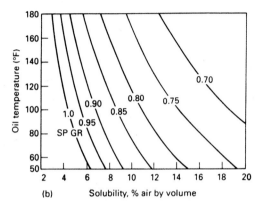

(b) Solubility, % air by volume

Figure 32.29 (a) Effect of entrained gas on liquid displacement; (b) solubility of air in oil. *Example*: At 5 in Hg with 3% gas entrainment by volume, pump capacity is reduced to 84 per cent of theoretical displacement

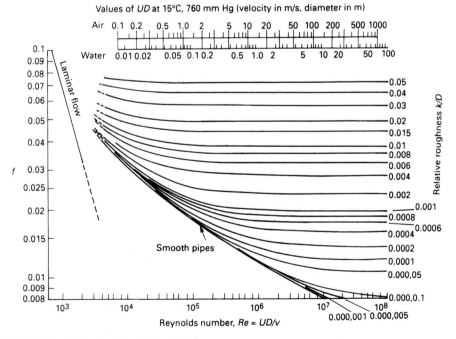

Figure 32.30 Friction factor f plotted against Re (after Miller[6])

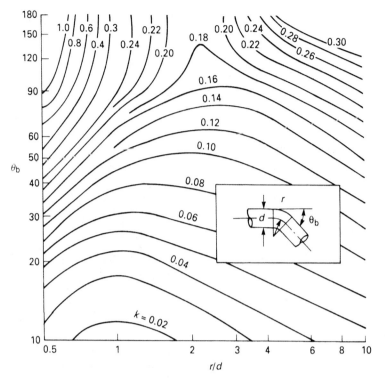

Figure 32.31 Bend loss coefficients[6]

Hence in Figure 32.30 the lines drawn illustrate $f = 0.016$

32.4.2 Losses in bends, valves and other features

Loss factor method of estimating losses

A common method of estimating losses is to use a factor K, so that energy loss for a feature is given by

$$\Delta(gH) = K\frac{v^2}{2} \qquad (32.13)$$

Figures 32.31 through 32.33 give loss factors for a range of common fittings. Most fittings do not occur in close proximity to one another, but bends frequently do. The effect of putting bends close together is presented in terms of the spacing between bends and a correction factor K_p. The equation used is for two bends:

$$\Delta(gH) = \underset{\text{(bend 1 loss) + (bend 2 loss)}}{K_p\left(K_1\frac{V^2}{2} + K_2\frac{V^2}{2}\right)} \qquad (32.14)$$

Figures 32.34–32.36 give information on K_p values for the configurations shown in Figure 32.37.

Equivalent-length method

As the name suggests, the loss a feature causes is replaced by the loss due to a length of straight line, which gives the

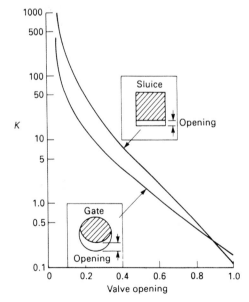

Figure 32.32 Gate valve losses[6]

same value of energy loss. Table 32.3 gives typical values of L_E, which are then added to the D'Arcy Weisbach

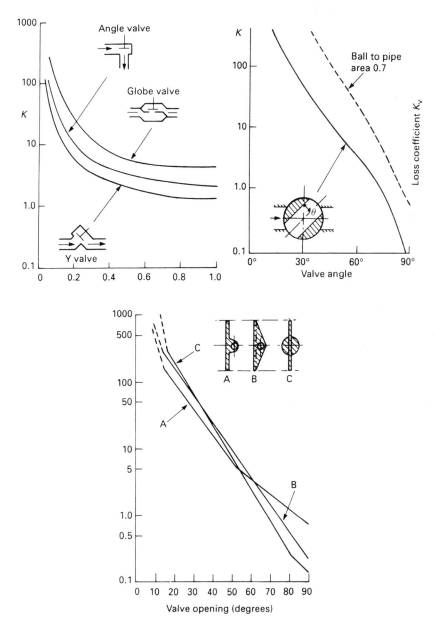

Figure 32.33 Some typical valve loss characteristics[6]

equation to give

$$\text{Total loss in energy} = \frac{f}{D}\frac{V^2}{2}(\text{pipe length}$$

$$+ \text{ sum of } L_E \text{ components}) \qquad (32.15)$$

Thus for a simple line 100 m long with a bend ($L_E = 30$) a valve ($L_E = 13$) the total loss = $(fv^2/D^2)(143)$.

32.4.3 Presentation of system loss

The flow loss in a system varies as $(\text{velocity})^2$ or $(\text{flow rate})^2$, so the total loss imposed by a system on a pump can be shown to vary with flow rate in the way shown in Figure 32.38(a). Figures 32.38(b) and 32.38(c) illustrate how system curves vary with the proportions of static and dynamic losses.

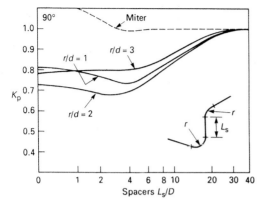

Figure 32.34 Interaction coefficient: combination bends (after Miller[6])

Table 32.3 Some equivalent lengths for typical fittings (clean water)

		$\dfrac{L_E}{D}$
Gate valve (wedge:disc:plug disc)	Fully open	13
	Half open	160
	Quarter open	900
Ball and plug valves	Fully open	3
Conventional swing check valve	Fully open	135
Footvalve with strainer (hinged flap)	Fully open	75
Butterfly valve (200 mm up)	Fully open	40
90° Standard elbow		30
90° Long-radius elbow		20

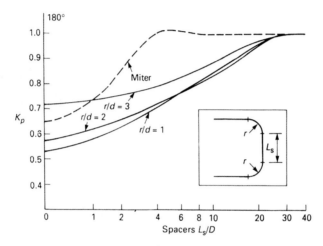

Figure 32.35 Interaction coefficient: 320° combination (after Miller[6])

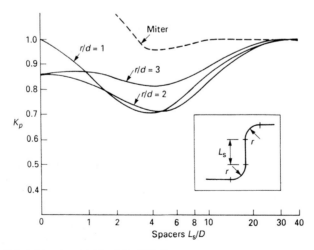

Figure 32.36 Interaction coefficient: S-bend combination (after Miller[6])

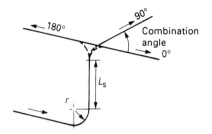

Figure 32.37 The combinations shown in Figures 32.33–32.36

Table 32.4 Hydraulic roughness of commercial pipes

Type of pipe	Hydraulic roughness K(mm)
Coated cast iron	0.127
Uncoated cast iron	0.203
Galvanized steel	0.152
Uncoated steel	0.051
Coated steel	0.076
Asbestos cement (uncoated)	0.013
Spun bitumen and concrete lined	
Drawn brass, copper, aluminium	
Glass plastic	
Concrete cast on steel forms	0.203
Spun concrete	0.076
Riveted steel–four transverse rows of rivets and six longitudinal	4.064
Two transverse rows of rivets, longitudinal seams welded	0.127
Single transverse rows of rivets	0.508
All welded	0.076
Flexible pipes	Roughness varies considerably with construction. Smooth rubber hose corresponds approximately with steel pipe. Head loss when curved can be 30% higher than when straight

32.5 Interaction of pump and system

32.5.1 Steady-state matching of pump and system

If the constant speed characteristic of a pump is super-imposed on a system curve, there is usually one intersection point, shown in Figure 32.39. If a flat system curve is being matched with a mixed or axial flow machine there can be flow instability, as illustrated in Figure 32.40, which is only corrected by changing pump speed or the static lift, or selecting a different pump.

32.5.2 Flow control

The simplest flow control is by valving (Figure 32.41). Either opening or closing a valve in the line changes the dynamic loss. The valve could be pressure controlled, a method much used in boiler feed systems because it

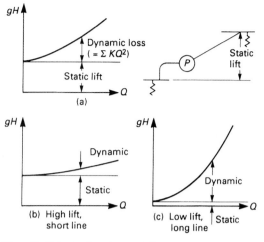

Figure 32.38 System loss characteristics

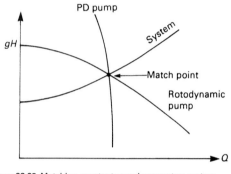

Figure 32.39 Matching constant-speed pumps to a system

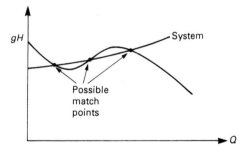

Figure 32.40 Combination of a flat system curve and an inflected pump curve

ensures full flow through the pump and thus reduces risk of vapor locking in the bypass (Figure 32.42). In this case, the pump supplies the system and bypass flow, so match point flow is higher than system demand and power demand is higher. A third control commonly used for PD machines (and increasingly for centrifugals as prices reduces), is speed control (Figure 32.43). If Figure 32.44

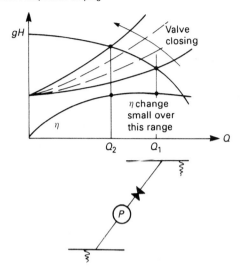

Figure 32.41 Valve control

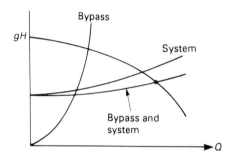

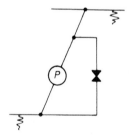

Figure 32.42 Bypass control

is considered, achieving 90 per cent of designed flow by speed control will happen with little sacrifice in efficiency and probably 25 per cent power reduction. Valve control will result in a 3 to 4 per cent reduction in pump efficiency (and probably 7 or 8 per cent increase in pump head), bypass will probably require 15 per cent more flow, the pump efficiency will drop 7 or 8 per cent, and the pump may also cavitate.

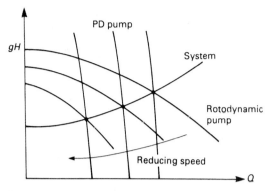

Figure 32.43 Speed control

32.5.3 Multiple-pump layouts

If the flow range in a system is larger than can be achieved with a single machine, several similar pumps may be installed in parallel. Figure 32.45 illustrates possible layouts for centrifugal pumps and match points on a system curve for one, two or three pumps operating. Non-return valves must be installed as shown to avoid pump interaction. Note that three pumps do not give three times single-pump flow, due to the square law nature of the system curve.

Figure 32.46 demonstrates how variable-speed drives will give a wide range of flows. Systems supplied by positive displacement pumps do not usually need wide flow ranges, but the same principle of parallel operation will apply.

Dissimilar pumps may be used. As Figure 32.47 shows, the resulting combined curve indicates that pump 2 will not begin to deliver until the specific energy from pump 1 drops below its shut-valve energy rise, therefore if the non-return valve for pump 2 does not seat, pump 1 can cause it to turbine.

32.5.4 Suction systems

Poor intake layout can give rise to pump problems, unstable running, losing prime, and cavitation. If the pump is drawing water from a sump, the position of the intake and the shape of the sump must be chosen to avoid vortex formation and resultant air ingestion and pump instability. Figure 32.48 shows the proportions for PD and centrifugal pump suctions and Figure 32.49 gives details of baffles that can be fitted in tanks to reduce vortexing. If large flow rates and numbers of pumps are involved it is advisable to commission model tests to ensure that pump behavior is not affected for all flow rates and pump combinations.

32.6 Cavitation

Cavitation is the term to used to describe the formation of bubbles in liquid flow when the local pressure falls to around vapor pressure. Two effects are experienced in the pump: a reduction in flow rate (accompanied,

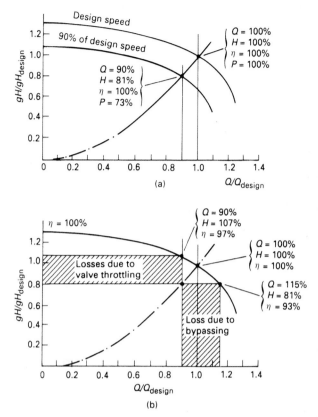

Figure 32.44 Comparison of speed, bypass and discharge regulation. (a) Speed control: (b) bypass or throttle

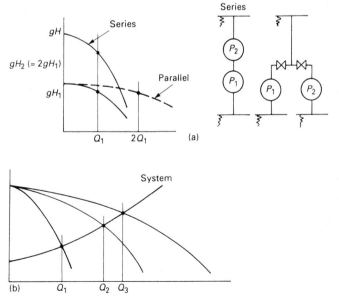

Figure 32.45 (a) Pumps in series and parallel; (b) three identical pumps running in parallel

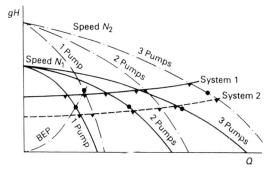

Figure 32.46 Variable-speed pumps in parallel

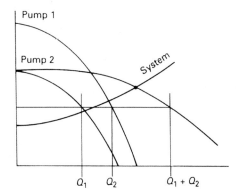

Figure 32.47 Dissimilar pumps in parallel

particularly in centrifugal pumps, by additional noise) and in surface damage and material removal. In general, cavitation occurs in the suction region of a pump or the inlet port and valve area of a positive displacement pump.

Figure 32.50 shows how reduction of suction pressure (which gives rise to cavitation) affects the characteristic of a typical centrifugal pump.

32.6.1 Net positive suction head (NPSH)

When drawing liquid through suction a pump generates a low pressure in the suction area, which, if it reaches vapor pressure, gives rise to gas coming out of solution and causing bubbles to form. An estimate of the margin above vapor pressure, used for many years, is net positive suction head. Two forms are used: NPSH-available and NPSH-required.

NPSH available (NPSH$_A$)

NPSH$_A$ = Total head at suction flange − vapor pressure head

Figure 32.51 illustrates how system NPSH or NPSH-available is calculated for the usual suction systems outlined. For a centrifugal pump, the basic NPSH is calculated from[7]

$$\text{NPSH}_A = h_s - h_t \frac{10.2}{\rho}\left(\frac{B}{1000} + P_i - P_v\right) \quad (32.16)$$

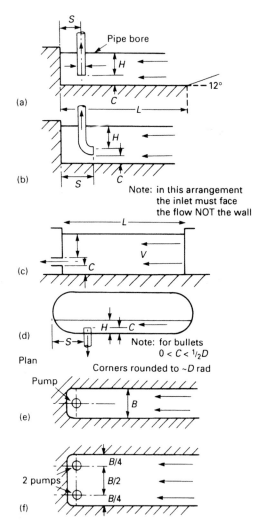

Figure 32.48 Suction bay and tank proportions (based on reference 7). Possible suction arrangements: (a)–(c) open suction layouts commonly used; (d) tank intake system (recommended limits: $3 > (S/D) > 1$; $2/3 > (C/D) > 1/2$; $(H/D) > [(V^2/3) + 1.5]$; $[(L-S)/D] > 5$; (e) single suction cell in plan $(5 > (B/D) > 2$; (f) double section cell in plan $(4 < (B/D) < 10)$

where

h_f = flow losses in suction system (m),
B = minimum barometric pressure (mbar) (use 0.94 of mean barometer reading),
P_I = minimum pressure on free surface (bar gauge),
P_v = vapor pressure at maximum working temperature (bar absolute).

In the process industries h_f is calculated for the maximum flow rate and the NPSH at normal flow allowed for by using the formula

$$\text{NPSH}_A = 0.8(\text{NPSH}_{basic}) - 1) \quad (32.17)$$

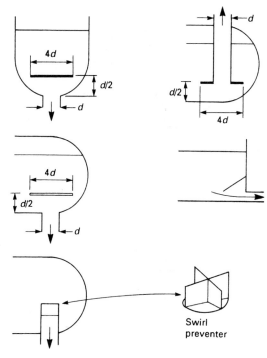

Figure 32.49 Vortex-prevention devices for tanks

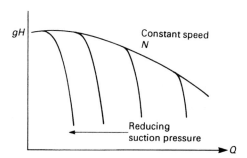

Figure 32.50 Effect of suction pressure on pump performance

This gives a 'target' value to the pump supplier that is *worst condition*. In general, for cold water duty equation (32.16) can be used for the duty flow required. Equation (32.16) is employed for reciprocating and rotary positive displacement machines with allowance made for acceleration effects.

In reciprocators h_f is calculated at peak instantaneous flow, including maximum loss through a dirty filter, and an additional head 'loss' to allow for pulsation acceleration is used:

$$h_A = \frac{700NQ}{Z}\Sigma\frac{L}{d^2} \tag{32.18}$$

and

$$\text{NPSH} = \text{NPSH}_A - h_A \tag{32.19}$$

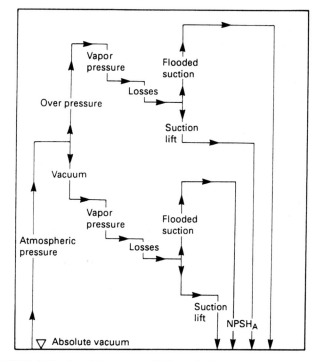

Figure 32.51 Visualization of NPSH$_A$ (with acknowledgements to Girdlestone Ltd)

For *metering pumps*:

$$\text{NPSH}_{\text{basic}} = h_s \frac{10.2}{\rho} \left(\frac{B}{1000} + P_i - P_v \right) \tag{32.20}$$

h_f is as for the reciprocating pump based on peak instantaneous flow and

$$h_A = \frac{6\delta P_i}{\rho} \tag{32.21}$$

δP_i = pressure pulsation at pump inlet (bar)

$$= K_1 \rho C Q / D^2$$

where

$K_1 = 40$ for simplex, 20 for duplex, 3 for triplex, 1 for quintuplex,
C = velocity of sound in liquid ms^{-1} (for water at normal temperature $C = 1000\,\text{ms}^{-1}$)

$$\text{NPSH}_A = \text{basic NPSH} - \left(h_f^2 + h_A^2 \right)^{1/2} - 1 \tag{32.22}$$

for simplex and duplex pumps and

$$\text{NPSH}_A = \text{basic NPSH} - (h_f + h_A) - 1 \tag{32.23}$$

for triplex and quintuplex pumps.
For *rotary pumps*:

$$h_A = \frac{\delta P}{\rho} \tag{32.24}$$

where

δP = pressure pulsation at pump inlet bar
$\quad = K_2 \rho Q / D^2$,
$K_2 = 2$ for three-lobe pumps
$\quad$ 5 for two-lobe pumps,
$\quad$ 5 for a monopump,
$\quad$ 1 for a spur-gear pump,
$\quad$ 5 for a vane pump,
$\quad$ 10 for a peristaltic roller-type pump,
$\quad$ 0 for a screw pump.

NPSH required (NPSH$_R$)

This is a statement of the NPSH that the pump can sustain by its own operation, so that the operating requirement is that NPSH$_R$ > NPSH$_A$. NPSH$_R$ is more difficult than NPSH$_A$ to determine. Reference 7 suggests working relations for positive displacement machines that are outlined below.

For reciprocating metering pumps, NPSH$_R$ is related to valve loading, as shown in Figure 32.52:

$$A = \frac{24vQ\rho}{Zd_v^3} + 5 \times 10^5 \frac{\rho Q^2}{Z^2 d_v^4} \tag{32.25}$$

(where d_v = nominal valve size (mm)) for single valves, and

$$A = \frac{80vQ\rho}{Zd_v^3} + 1.5 \times 10^5 \frac{\rho Q^2}{Z^2 d_v^4} \tag{32.26}$$

for double valves. It is recommended that for hydraulically operated diaphragm pumps the extra losses imposed by

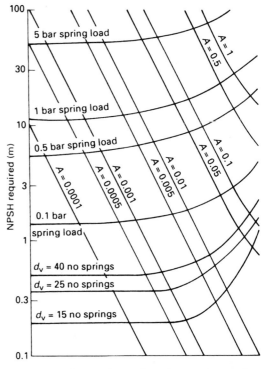

Figure 32.52 NPSH$_{\text{required}}$ for a reciprocating metering pump[7]

the diaphragm and support plate are treated as a single unloaded valve.

For other reciprocators

$$\text{NPSH}_R = 5U^2 + \frac{0.12(P_d)^{0.75}}{\rho} \tag{32.27}$$

where

U = mean plunger speed (ms^{-1}),
P_d = discharge pressure bar (dbs).

For the centrifugal pump, two terms are in common use, the Thoma cavitation number σ and the suction specific speed S_N:

$$\sigma = \frac{\text{NPSH}_R}{\text{Pump head rise}} \tag{32.28}$$

Figure 32.53 gives a typical plot of σ against k_s that may be used as a first 'design' estimate of NPSH$_R$, but in many applications, test data are required

$$S_N = \frac{N\sqrt{Q}}{K(\text{NPSH}_R)^{3/4}} \tag{32.29}$$

where K is a constant $= 175$ if $g = 9.81\,\text{ms}^{-2}$, Q is in l/s, N is in revolutions/second and NPSH$_R$ is meters of liquid. NPSH$_R$ here is usually based on a 3 per cent head drop, defined in Figure 32.54, for the pump tested with cold water at design flow rate and rotational speed.

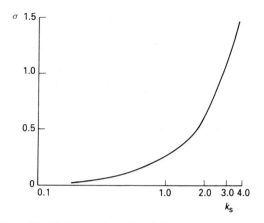

Figure 32.53 Variation of Thoma's cavitation parameter with specific speed for pumps[3]

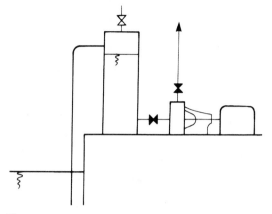

Figure 32.55 Priming tank arrangement

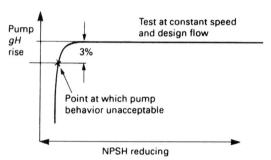

Figure 32.54

32.7 Priming systems

If the suction level is above the pump there is usually no problem in *priming* (that is, ensuring that the pump suction system is full of liquid) unless it is pumping a volatile and vapor locks when stationary. In many applications, the pump is above the suction vessel or main, and the static pressure at the take-off point is too low to provide the static lift required. In these cases centrifugal and positive displacement pump suction lines are often provided with non-return valves so that, once filled they do not lose liquid on shutdown. If their operation is intermittent, a priming tank is provided (Figure 32.55) to allow priming to take place. This is only possible with small machines because of the sizes needed. It is general practice that no pump should be started or run dry to avoid metal-to-metal contact and seizure. The only exceptions to this are pumps designed to provide priming – *self-priming* machines, used in installations where the suction line empties, as in tanker-unloading bays or site-dewatering applications.

Positive displacement pumps are self-priming by their normal operating action and are designed to cope with running dry. Centrifugal pumps are not inherently self-priming, and need to be provided with assistance in the form of ejectors, as in Figure 32.56 or dry vacuum pump systems. In these systems, a dry vacuum pump allows air to be drawn out and where an automatic valve prevents

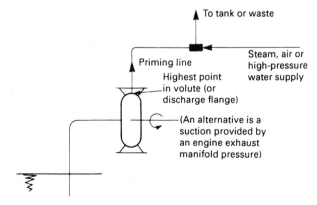

Figure 32.56 An ejector priming system

liquid entering the pump, or wet vacuum pumps, such as the liquid ring pump (Figures 32.57 and 32.58), may be used to extract the air. Such devices are common on portable fire pumps, the liquid ring pump being driven from the pump main driver in such a way as to allow disconnection when the pump is fully priming, thus avoiding extra power loss during normal operation.

Self-priming centrifugal pumps do not rely on any assisting system, but are provided with a special casing system that allows liquid retained in the pump on shutdown to recirculate to draw air from the suction line until that is full of liquid and the pump is primed. There are several designs which will pull suction lifts up to 7 m with cold water. All the designs impose a penalty of lower efficiency when in the normal pumping mode, but are effective alternative solutions to an auxiliary priming system which do not require additional equipment as bolt-on extras requiring maintenance.

One class of self-priming pumps which are rotodynamic but not centrifugal in operation are the side-channel or peripheral pump (Figure 32.59). The circulating motion results in a transfer of energy from the liquid leaving the rotor to the stator. Then as the fluid proceeds to 'corkscrew' round the periphery from suction to discharge, energy builds up so that it is claimed that the pressure rise is 2.5 times that produced by a conventional centrifugal pump with the same peripheral velocity, although efficiency is much lower (up to 50 per cent).

32.8 Seals: selection and care

Static seals are standard provision from many suppliers; therefore seals for the moving elements in pumps will be discussed. For centrifugal pumps, both stuffing box and mechanical seals will be considered and the discussion will be extended to reciprocating and rotary machines.

32.8.1 Centrifugal pump and rotary pump seal systems

The stuffing box, or packed gland, has been used for many years, typical layouts being shown in Figure 32.60. Usually, since a very smooth shaft surface is needed a

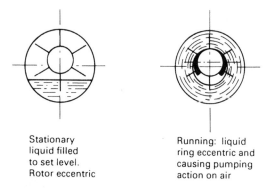

Stationary liquid filled to set level. Rotor eccentric

Running: liquid ring eccentric and causing pumping action on air

Figure 32.58 Principle of liquid ring pump

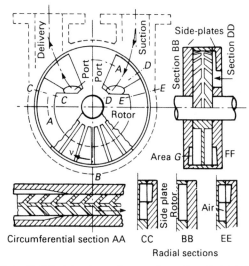

Figure 32.59 Peripheral pump layout and operating principle (after Addison[1])

sleeve of hard bronze or chrome plated is used for water services finished to 0.8 μm. For normal duties soft packing

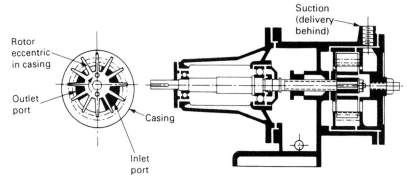

Figure 32.57 Schematic of a liquid ring pump

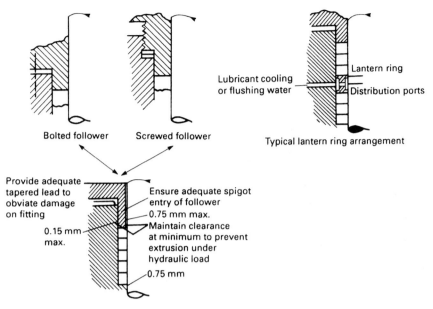

Lantern ring

Lubricant cooling
or flushing water

Distribution ports

Bolted follower   Screwed follower   Typical lantern ring arrangement

Provide adequate
tapered lead to
obviate damage
on fitting

Ensure adequate spigot
entry of follower
0.75 mm max.
Maintain clearance
at minimum to prevent
extrusion under
hydraulic load
0.75 mm

0.15 mm
max.

Figure 32.60 Stuffing box details

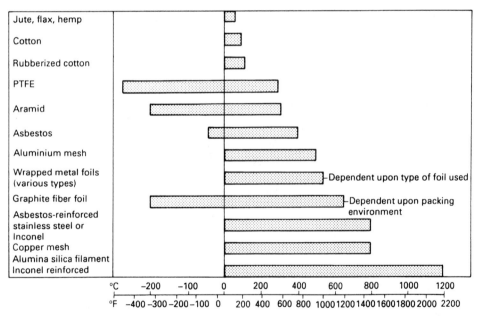

Figure 32.61 Soft packings – temperature range

are selected to deal with both seal surface speed and temperature. Figures 32.61 and 32.62 give typical materials and their capacities, and manufacturers' data sheets should be consulted when selecting the number of rings and the compression needed. When properly adjusted and 'run in', the leakage rate on a cold-water duty could be about 1 liter/day, which may be lost as vapor.

Over-tightening the packing can give rise to overheating and shaft failure or seizure. Typically, a four-ring gland will seal 3.5 to 4 bars, and absorb about 0.15 kW when run in and fitted to a 50-mm shaft.

The mechanical seal is now a standard fitting of high reliability and tends to be fitted instead of a packed gland. A typical mechanical seal is shown in Figure 32.63. The

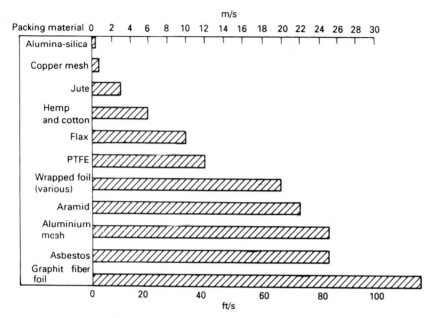

Figure 32.62 Soft packing – speed range

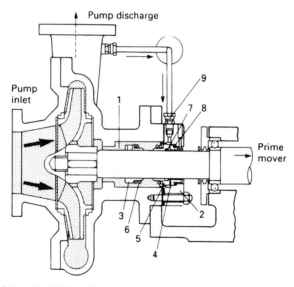

Figure 32.63 A typical mechanical seal installation. 1 Seal chamber; 2 seal plate; 3 spring sleeve; 4 seat; 5 seal ring; 6 spring; 7 dynamic secondary seal; 8 static secondary seal; 9 flush connection

conventional design has a spring, rotating with the shaft, which holds a rotating seal ring against a stationary ring to provide a sealing interface. Stationary fluid seals are provided as shown. For water and general non-corrosive duties, the design is internal, so that the liquid being sealed both cools and provides a supply for the film, which separates the rotating and static seal rings.

Simple seals are by design either *balanced* or *unbalanced*. Referring to Figure 32.64, it can be seen that a balanced seal will give a lower closing hydraulic force and thus lower contact pressure; therefore, a higher sealed pressure can be sustained. Too little load, and the seal will leak, and too large will reduce life. An unbalanced seal gives a simpler shaft or sleeve design, and will seal up to

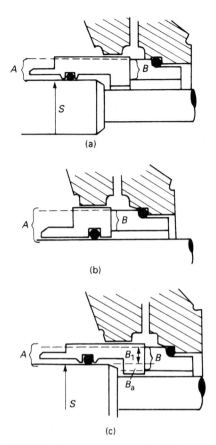

(a)

(b)

(c)

Figure 32.64 Seal balance arrangements. S = Sliding diameter or effective sealing diameter; A = hydraulic piston area of sliding ring; B = face contact area. (a) All area B is outside effective sealing diameter and areas A and B are equal. Average face contact pressure equals product pressure, i.e. seal is 100% out of balance. (b) Unbalanced seal. All area B is outside effective sealing diameter and area A is greater than area B. Average face contact pressure is greater than product pressure, i.e. seal is more than 100% out of balance, depending on the relative area ratios. This condition exists in most unbalanced seals. (c) Balanced seal. Here part of the face contact area, B_1, is outside. The sliding diameter S carries the product pressure acting on area A. The face contact area, however, equals $B_1 + B_2$, and so the unit face loading will be less than product pressure, in accordance with the area ratio, i.e.

$$\frac{B}{B_1 + B_2} = \frac{A}{B}$$

$$\text{Balance} = \frac{A}{B}$$

Generally, 0.2–0.35

147 psi gauge (10 bar), and above this, balanced seals are requested. Materials often used for standard water duties are carbon for the face and ceramic and, if condensate, carbon with a nickel-iron static seat could be fitted. The seal manufacturer will quote the appropriate materials, and will usually suggest circulation as shown in Figure 32.63. The mating sealing surfaces are lapped

to within 0.5 μm, shaft run-out should not exceed 0.005 inch or 0.125 mm, and shaft finish where the seal ring has to pass should be 0.4 μm to avoid damage.

If a mechanical seal or a packed gland is fitted properly, a good life can be expected if the liquid is clean. If grit is in suspension, the seal manufacturer should be consulted about appropriate materials and system design.

32.8.2 Reciprocating pump seal systems

The comments on packed glands apply for shaft and rod seals. Piston-to-cylinder seals may be arranged like the packed gland, but are provided with end rings, which locate and prevent the sealing rings from extrusion. Surface finishes for rod or piston will be the same as those given above with leakage rates of the same order.

32.9 Pump and drive selection

32.9.1 Pump selection

Since pumps are the essential element in a flow system, the selection of the correct machine is crucial to the success of a plant. It must be constructed from the appropriate materials, run at the right speed, be compatible hydraulically with the system at all flows, economical in first cost and in operation, and 'user-friendly'.

To place the process of pump selection in the plant decision tree it is necessary to follow the iterative way a complete plant evolves.

From the initial statement of need, a preliminary design brief evolves containing flow rates and attendant component pressures and temperatures. From this an initial design layout is produced, the pressure losses and probable inlet and outlet pressures and attendant flows for all pumps needed estimated, and a first assessment of possible pumps made. Considerations of cavitation and perhaps a need to rationalize on pump designs lead to a second look at the systems. A revision of flow schedules, pressures and temperatures results, and, by iteration, a 'final' design layout appears from which pump and equipment specifications begin to be formulated and initial inquiries are made to suppliers. Pump manufacturers will check possibilities and tender if their pumps will give the duties specified or, if invited tendering is used, will propose changes to suit their pumps.

In the consultant's design office, the selection of pumps (and hence of makers) to tender must be done at the project design stage, and considerations of cost, complexity and suitability will influence the choices made. An important consideration at this stage is that the design point specified is the correction one. A common practice is the 'add a bit for aging, add a bit for the unknowns' at the early design stage and then the next engineer in the process also 'adds a bit' since they may be on the low side, and thus a margin of flow is added. This process can sometimes ensure that the pump size is such that, even though the design point is that quoted, the pump will be running at low flow when the system is commissioned, and this may be in a region of instability, giving running problems on the plant. It cannot be too much emphasized

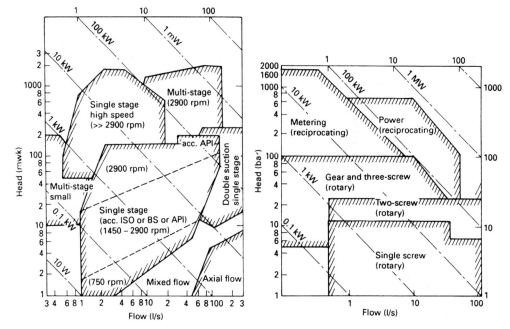

Figure 32.65 Pump range chart (courtesy Nederlandse Aardolie MIJ BV)

that accurate estimation of the duty flow and head will assist pump application and life.

Examination of Figure 32.65 will reveal that in many cases the choice of centrifugal or positive displacement pump is dictated by the head and flow specified. Where, however, the choice may be either type, other considerations must enter the selection process. For example, first cost is an important factor. Sterling[5] has demonstrated that a screw pump and a centrifugal would give the same duty. The screw pump would have an $NPSH_R$ less than the centrifugal, and have a lower fall-off in efficiency as fluid viscosity increased, but would cost more than twice the centrifugal and be more expensive to maintain. He concluded by stating the centrifugal as being the most preferred for price and maintenance in 80 per cent of pumping duties.

A probable selection procedure could take the following form:

If possible, a standard centrifugal pump shall be selected, but the effect of high viscosity must be considered (see Table 32.2). If flow is low remember the efficiency fall-off, and look at alternative designs. If flow rate is important, metering pumps should be preferred, as centrifugal pumps will not give accuracy. If solids are in suspension consider working clearances, need to avoid blockage and the flushing for seals.

A drive speed of 2900 rpm is preferred to give higher efficiency and smaller size and first cost (but remember noise risks). Appropriate codes (API 610, for example) should be observed.

If a centrifugal is not suitable, select a rotary positive displacement pump. The design usually selects itself, but for flow rates in the centrifugal pump range, screw pumps, mono-pumps and wide-tooth gear pumps may be considered. Remember that if particles are present the screw pump will lose volumetric efficiency, and the monotype of machine should be selected.

If the duty is metering, piston or diaphragm pumps should be selected.

32.9.2 Drive selection

For most centrifugal pumps the drive will be a synchronous motor, running at 2900 rpm for 50 Hz supply to minimize pump size and cost. Large boiler feed pumps use steam turbines, which give economic output regulation since steam is available. In most plants, variable-speed drives may be used to give economic flow control. A wide range of possibilities is available (Table 32.5). Bower[8] writing from the viewpoint of a pump designer has given some comparative costs. He commented that a 10 kW drive pump running for 10 h per day would cost more than the first cost of the unit in a year. He showed that if a 75 kW pump running at 2900 rpm is to run at half flow, a variable-speed drive giving 1450 rpm will use 39 kW if supplying a low static lift system, where flow losses are important. There is thus an economic argument for a higher first-cost pump plus drive if such savings are possible.

32.9.3 Economics of pump selection and running maintenance

With the operating economics of pumping system, such as the water industry networks receiving attention, it is of

Table 32.5 Main characteristics of speed changing drives related to centrifugal pumps[8]

Type of drive	Power range (kW)	Speeds		Drive efficiency (%)		Overall efficiency (%) including motor				Power factor		Main characteristics related to pump drives	
		Max. rpm	Ratio	Maximum speed	Half speed	Max. speed		Half speed		Max. speed	Half speed	Advantages	Limitations
						4 kW	150 kW	4 kW	150 kW				
'V' belts or flat belts	Up to 750	5000 at limited power	8:1	'V' belts 85–90 Flat belts 90–95		70 75	80 85	40 45	65 70	0.9	0.3 with same motor	Low cost possibility if speed changes are infrequent	Increased floor space. Not suitable for outside application. Reduced bearing life if jack shaft not used
Timing belts	Up to 350	6000 at limited power	9:1	95+		80	87	50	75	0.9	0.3 with same motor	Similar to V belts but give reduced shaft loading and greater efficiency	
Gear box	Any	Any, but standard units usually step down		95+		80	87	50	75	0.9	0.3 with same motor	A robust, compact drive for infrequent changes or for matching drive speed to pump requirement. Any power available. Efficient	Correct maintenance importance
Variable-speed pulleys	Up to 125	Up to 4500 but at limited power	4:1 Std	85–90		70	N.A.	40	N.A.	0.9	0.3	Low cost. Housed-belt versions available	Limited power range. Automatic control difficult. Limited belt life
Eddy current coupling	Up to 1500	Up to 4200 with belt drive	Above 10:1	96	45–50	80	87	25	35	0.9	0.35	Reliability and life good. Automatic control easy	Low efficiency at reduced speed. Water cooling required for higher powers. Slip-in drive reduces pump maximum speed. Only speed reduction possible
Fluid couplings	15 to 12000	3500 Std	4:1	95	45–50	80	87	24	35	0.9	0.35	Reliability and life good. Available to very high powers at which they become more cost effective	Low efficiency at reduced speed. Heat exchanger required for higher powers. Slip reduces pump max. speed. Automatic control expensive. Expensive for low powers

(continued overleaf)

Table 32.5 (continued)

Type of drive	Power range (kW)	Speeds		Drive efficiency (%)		Overall efficiency (%) including motor				Power factor		Main characteristics related to pump drives	
		Max. rpm	Ratio	Maximum speed	Half speed	Max. speed		Half speed		Max. speed	Half speed	Advantages	Limitations
						4 kW	150 kW	4 kW	150 kW				
Positive infinitely variable chain drives	Up to 120	Up to 5500 but at limited power	10:1		85–90	70	N.A	45	N.A	0.9	0.3	Speed increase possible	Limited power range Careful maintenance essential. Life limitations
Mechanical variators	Up to 75	4200 at 10 kW 2600 at 75 kW	Up to 12:1	85–95	85–85	70	N.A	40	N.A.	0.9	0.3	Speed increase possible	Limited power range. Limited life. Careful lubrication and maintenance essential
DC motor and thyristor voltage control	Up to 2500	1500 Std 3500 avail. higher-special	Up to 30:1	See overall efficiency		80	90	45	65	0.9	0.3	Good range. Relatively good low speed efficiency. Automatic speed control easy.	TEFC Expensive. High speed expensive. Harmonics generation needs consideration. More maintenance than induction motors
Voltage control of induction motor	Up to 50	3000	4:1	See overall efficiency		80	N.A.	25	N.A.	0.9	0.3	Low cost. Can use squirrel cage induction motor. Automatic control easy	Limited power range. Low efficiency at reduced speed. Only speed reduction possible. Requires at least 30% derating of motor
Schrage AC commutator motor	Up to 500	2500	4:1 Std up to 20:1	See overall efficiency		80	90	45	70	0.95	0.5	Good efficiency and power factor. Speed increase possible	Automatic control expensive. Movable brush gear increases maintenance. TEFC expensive
Stator-fed AC commutator motor	Up to 2500	2000	up to 10:1	See overall efficiency		80	90	45	70	0.9	0.4	Improved life over the Schrage motor and increased supply voltages possible. Good efficiency	Automatic control expensive. TEFC expensive
Frequency control of induction motor	Up to 75 Std 2000 special	5000	10:1	See overall efficiency		75	87	40	65	0.95	0.95/0.3	Can use squirrel cage induction motors with minimal derating. Good power factor (with most designs). Speed increase possible	High cost. Electronics more complex (and less reliable?) than other types

interest to note that recent work[9] makes two points: pump performance does fall off with time and there should be consideration of the economics of refurbishment as part of system pricing. Cullen reported a study of basing figures on 50 per cent utilization and a 3p (5 cents) per kilowatt-hour tariff. One case involving 12 pumps, capital value $216.000, with an efficiency fall-off of 5 to 9 per cent, could be refurbished for about $86,400 to give 5 per cent efficiency uplift and show electricity saving of $24,120 per year and a return on investment of 35 per cent per year. A further example covering eight pump sets, capital value $144,000 with a serious efficiency fall-off of 15–19 per cent, could be refurbished at cost of $201,600 to give an uplift of 15 per cent, resulting in a $56,160 per year saving and an annual investment of 25 per cent. These examples and others emphasize that traditional oversizing is expensive in running cost, that pump mismatch should be avoided and that pump replacement or refurbishment can be economic in large installations.

32.9.4 Reliability considerations

As a further consideration of the overall system design, it is necessary to examine the choice of the number of pumps installed. In the petrochemical and process industries availability is an important factor in plant operation. The Institution of Mechanical Engineers suggests several classes of pumps linked to availability:[7]

1. A Class 1 installation achieves high availability using a single pump upon which the system is completely dependent is in a single process stream where pump outage means a large loss of output. The process here demands that continuous operation is possible for at least three years without enforced halts for inspection or correction and where components must have a life expectancy of more than 100,000 hours.
2. A Class 2 system is as Class 1 but with process recovery short after shutdown, infrequent pump outage can be tolerated and an interval of about of about 4000 hours for continuous operation can be tolerated.
3. If in Class 1 and 2 installation deterioration in pump operations can be tolerated or easily corrected by operators it may be categorized as Class 3.
4. A Class 4 installation is different from the first three by having standby pumps that can take over immediately a pump fails. Main pump and the standby need not be identical but should both have life expectancies exceeding 25,000 hours.
5. A Class 5 installation follows the Class 4 concept, but is characterized by having pumps operating with identical spares and have one or more pumps operating in plant bays where product storage gives time to assess the problem and to take action to get the pump back on-line.
6. A further Class 6 is suggested for pumps intended for batch or intermittent duty. (If high availability is essential, the machine should be placed in Class 4.)

For many process pumps the L10 life is less than 8000 hours, so they usually fall in Classes 4, 5, and 6. A typical system use a running pump rated at 100 per cent

duty with an identical spare either installed or carried in store. This consideration affects the total system design if installed spares are to be used as well as the overall cost of the installation.

32.10 Pump testing

Pump tests are needed to establish the performance of a machine before delivery to the user so that contract conditions may be satisfied. They are also required to establish the health of the pump at intervals in its service life. When a new pump is delivered, the pump manufacturer will supply data required either as a set of performance points or a plot as a certificate of suitability. Ideally, the pump should be tested in the system configuration that it is intended to supply, but usually tests are performed in the test bay in a standardized manner laid down by national and international codes. When a pump is installed, a simple test may be performed to establish a performance point in a similar way to periodic tests to establish how the machine's performance is being affected.

32.10.1 Factory tests

Factory tests establish the pressure head, power, efficiency and NPSH over the complete flow range the pump can deliver running at design speed. British Standard, DIN standard or ANSI standard codes or national variations from such main codes lay down the manner of test procedure, and a minimum requirement is quite often defined by industry codes such as API 610. This is not the place to discuss instrument accuracy, as the codes lay down the limits possible from conventional instruments. There are two main classes of test: the commercial requirements normally possible in the maker's test plant and high-accuracy tests that are only possible by using 'substandard' instruments and very sophisticated techniques.

32.10.2 Scale-model testing

Scale-model testing is used with very large pumps such as water feed pumps for thermal power stations. The problems posed by such tests in establishing the full-size machine performance are well discussed in a paper contributed by workers studying pumps for the Central Electricity Board[10].

This is a specialized area of work covered by ISO codes of practice, and any engineer wishing to study these must check the current codes.

References

1 Addison, H., *Centrifugal and other Rotodynamic Pumps*, Chapman and Hall, London (1955).
2 Karassik *et al.*, *Pump Handbook*, McGraw-Hill, New York (1976).
3 Turton, R. K., *Principles of Turbomachinery*, E. & F. N. Spon, London (1984).
4 *Hydraulics Handbook*, Trade and Technical Press.
5 Sterling, L., 'Selection of pump type to match systems', Paper B3, BPMA 5th Tech. Conference., Bath (1977).

6 Miller, D. S., *Internal Flow Systems*, BHRA Fluid Engineering, Cranfield (1978).

7 Davidson, J. (ed.), *Process Pump Selection – A Systems Approach*, I. Mech. E., London (1986).

8 Bower, J. R., 'The economics of operating centrifugal pumps with variable speed drives', *Proc. I. Mech. E.*, Paper C108/81 (1981).

9 Cullen, N., 'Energy cost management in water supply', BPMA 11th Conference, April (1989).

10 Nixon, R. A. and Cairney, W. D., 'Scale effects in centrifugal cooling water pumps for thermal power stations', *NEL Report 505* (1972).

11 Summers-Smith, J. D. (ed.), *Mechanical Seal Practice for Improved Performance*, I. Mech. E., London (1988).

33

Centrifugal Pump Installation

R. Keith Mobley
The Plant Performance Group

Contents

More pumps fail as a direct result of improper installation than any other single factor. The predominant reasons for these failures include starvation, caused by inadequate or inconsistent suction conditions; distortion, caused by pipe strain or improper foundation; and turbulent that results from piping or entrained gas problems. Centrifugal pump installation must follow Hydraulic Institute Standards, which provide specific guidelines to prevent these installation and performance problems. This chapter will address the fundamental requirements for proper installation.

33.1 Foundation

Centrifugal pumps require a rigid foundation that prevents torsional or linear movement of the pump and its baseplate. In most cases, this type of pump is mounted on a concrete pad or pedestal having enough mass to securely support the baseplate, pump and its driver. The following general rules apply to foundation design.

First, the mass of the concrete foundation should be at least five times greater than the total mass of the supported equipment. This ratio provides sufficient mass to absorb both the dynamic and static forces created by the pump and driver. Well-designed pump foundations and baseplates should have the rigidity required to withstand the axial, transverse and torsional loading of the rotating machines. For pumps less than 500 horsepower, the top of the foundation should be at least 3 inches wider and longer than the baseplate and at least 6 inches wider for larger pumps. In addition, the foundation should have sufficient width to fully support and dissipate the energy generated by the pumping equipment. To determine the bottom width of the foundation, draw imaginary lines from the center of rotation downward at 30° angles. These lines should pass through the bottom of the foundation and not its sides (Figure 33.1).

Baseplate design significantly influences the reliability of rotating equipment, such as centrifugal pumps. It must be rigid enough to avoid flexing or twisting at its center under the normal dynamic loads generated by the motor and pump. Torsional flexing can cause separation of the base and grout and result in a dramatic increase in resonance and vibration as well as misalignment of the driver and pump.

The baseplate must be absolutely level and securely bolted and grouted to the concrete foundation. The baseplate understructure should provide a means for the grout to rigidly affix it to the foundation when all live loading occurs. Baseplate flatness and level installation are major factors in maintaining proper shaft alignment between the pump and its driver. It should have an installed surface coplanar flatness within 0.062 inches per foot overall for Groups I and II pumps and 0.015 inches per foot for Group III pumps. These tolerances allow for reasonable alignment capabilities without dramatically increasing baseplate manufacturing costs.

Baseplates must be secured to concrete foundations with epoxy or cementitious grout. Ideally, vibration generated by the rotating pump is transmitted through the baseplate to the foundation and down through the subsoil where it can be absorbed. This transmission of dynamic energy is solely dependent on an absolute mechanical link from the rotating mass to the subsoil. The combination of proper bolting and grout provide this linkage. Properly installed, an epoxy grout will bond to a concrete foundation with a tensile strength of 2000 psi, which is significantly greater than that of concrete. The bonded baseplate is therefore effectively transformed into a monolithic structure that includes the concrete foundation. The single-block monolith significantly dampens vibration resulting from resonance with the natural frequency of the baseplate.

33.2 Piping support

Pipe strain causes the pump casing to deform and results in premature wear and/or failure. Therefore, both suction and discharge piping must be adequately supported to prevent strain. In addition, flexible isolator connectors should be used on both suction and discharge pipes to ensure proper operation. Care in fabricating and aligning piping avoids problems that may require re-cutting, fitting, re-welding and re-testing the pipe or lead to premature pump failure. Good system design supports piping loads and forces along spring hangers and bracing that do not have to be removed during normal maintenance. The system should be fabricated starting at the pump flanges, working toward the pipe rack.

The most common piping fabrication error and the one that produces the greatest piping strain is non-parallel flange faces. For example, during installation of a circulating water pump, suction piping was installed with misaligned mating flanges. The resulting strain distorted the pump's casing to the point where the shaft and impeller would not turn. Distortions can result in premature wear, loss of performance, or catastrophic failure. The following should be evaluated as part of a root cause failure analysis: foundation, piping support, and inlet and discharge piping configurations.

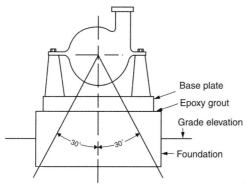

Figure 33.1 Foundation

33.3 Inlet piping configuration

Contrary to popular opinion, a centrifugal pump cannot function without both consistent adequate volume and suction pressure. Therefore, suction piping must be configured to assure that both of these conditions are met. First, the reservoir or source of liquid must have a sustained volume that is sufficient to meet the hydraulic requirements of the pump and its discharge system. This reservoir must be fitted with vortex prevention devices that will assure a constant supply of liquid that is free from entrained air and gases. Second, the inlet or suction piping must have adequate size to prevent excessive friction loss or other flow restrictions that would reduce the effective suction head at the impeller's eye. Generally, the suction piping should be at least 1.5 times larger than the pump's inlet diameter. In addition, inlet or suction velocity should not exceed 5 feet per second. Higher velocities will adversely affect pump performance and may lead to chronic reliability problems.

Centrifugal pumps are highly susceptible to turbulent flow. The Hydraulic Institute provides guidelines for piping configurations that are specifically designed to ensure laminar flow of the liquid as it enters the pump. As a rule, the suction pipe should provide a straight, unrestricted run that is six times the inlet diameter of the pump.

Installations that have sharp turns, shut-off, flow-control valves, or undersized pipe on the suction-side of the pump are prone to chronic performance problems. Such deviations from good engineering practices result in turbulent suction flow and cause hydraulic instability that severely restricts pump performance.

The final consideration in suction piping design is the method used to reduce the pipe diameter (i.e. 1.5 or greater than the pump's inlet flange) so that it can be attached to the pump. Care must be taken to ensure that the reduction method does not cause a void where air can be trapped or a low spot that will permit buildup of solid contaminates.

Concentric reducers should not be used to reduce suction piping diameter to that of the pump's inlet flange. This type of reducer may create an air pocket in the top of the piping and could lead to loss of pump performance resulting from air entrainment. Eccentric reducers should be installed with the straight side on top. This will reduce the potential for air pocket in the piping or the introduction of air or gas into the pump.

33.4 Discharge piping configuration

The restrictions on discharge piping are not as critical as for suction piping, but using good engineering practices ensures longer life and trouble-free operation of the pump. The primary considerations that govern discharge-piping design are friction losses and total vertical lift or elevation change. The combination of these two factors is called *total system head* or TSH and represents the total force that the pump must overcome to perform properly. If the system is designed properly, the *total dynamic head* (TDH) of the pump will equal the TSH at the desired flow rate.

In most applications, it is relatively straightforward to confirm the total elevation change of the pumped liquid. Measure all vertical rises and drops in the discharge piping, then calculate the total difference between the pump's centerline and the final delivery point.

Determining the total friction loss, however, is not as simple. Friction loss is caused by a number of factors and all depend on the flow velocity generated by the pump. The major sources of friction loss include: friction between the pumped liquid and the sidewalls of the pipe; valves, elbows, and other mechanical flow restrictions or other flow restrictions, such as back-pressure created by the weight of liquid in the delivery storage tank or resistance within the system component that uses the pumped liquid.

There are a number of reference books, like Ingersoll-Rand's *Cameron Hydraulics Databook*, that provide the pipe-friction losses for common pipes under various flow conditions. Generally, data tables define the approximate losses in terms of specific pipe lengths or runs. Friction loss can be approximated by measuring the total run length of each pipe size used in the discharge system, dividing the total by the equivalent length used in the table, and multiplying the result by the friction loss given in the table.

Each time the flow is interrupted by a change of direction, a restriction caused by valving, or a change in pipe diameter there is a substantial increase in the flow resistance of the piping. The actual amount of this increase depends on the nature of the restriction. For example, a short-radius elbow creates much more resistance than a long-radius elbow; a ball valve's resistance is much greater than a gate valve's; and the resistance from a pipe-size reduction of four inches will be greater than for a one-inch reduction. Reference tables are available in hydraulics handbooks that provide the relative values for each of the major sources of friction loss. As in the friction tables mentioned above, these tables often provide the friction loss as equivalent runs of straight pipe.

In some cases, friction losses are difficult to quantify. If the pumped liquid is delivered to an intermediate storage tank, the configuration of the tank's inlet determines if it adds to the system pressure. If the inlet is on or near the top, the tank will add no back pressure. However, if the inlet is below the normal liquid level, the total height of liquid above the inlet must be added to the total system head.

In applications where the liquid is used directly by one or more system components, the contribution of these components to the total system head may be difficult to calculate. In some cases, the vendor's manual or the original design documentation will provide this information. If these data are not available, then the friction losses and back pressure need to be measured or an over-capacity pump selected for service based on a conservative estimate.

33.5 Rules for proper installation

Rule 1: Provide sufficient NPSH

Every impeller requires a minimum amount of pressure energy in the liquid being supplied in order to perform without cavitation difficulties. This pressure energy is referred to as *net positive suction head required*.

The NPSH *available* is supplied from the system. It is solely a function of the system design on the suction side of the pump. Consequently, it is in the control of the system designer. To avoid cavitation, the NPSH available from the system must be greater than the NPSH required by the pump, and the biggest mistake that can be made by a system designer is to succumb to the temptation to provide only the minimum required at the rated design point. This leaves no margin for error on the part of the designer, or the pump, or the system. Giving in to this temptation has proved to be a costly mistake on many occasions.

Rule 2: Reduce the friction losses

When a pump is taking its suction from a tank, it should be located as close to the tank as possible in order to reduce the effect of friction losses on the NPSH available. Yet the pump must be far enough away from the tank to ensure that correct piping practice can be followed. Using a larger diameter line to limit the linear velocity to a level appropriate to the particular liquid being pumped can usually reduce pipe friction. Many industries work with a maximum velocity of about 5 feet per second, but this is not always acceptable.

Rule 3: No elbows on the suction flange

There is always an uneven flow in an elbow, and when one is installed on the suction of any pump, it introduces that uneven flow into the eye of the impeller. This can create turbulence and air entrainment, which may result in impeller damage and vibration. When the elbow is installed in a horizontal plane on the inlet of a double suction pump, uneven flows are introduced into the opposing eyes of the impeller, upsetting the hydraulic balance of the rotating element. Under these conditions, the overloaded bearing will fail prematurely and regularly if the pump is packed. If the pump is fitted with mechanical seals, the seal will usually fail instead of the bearing – but just as regularly and often more frequently.

A well-established and effective method of ensuring a laminar flow to the eye of the impeller is to provide the suction of the pump with a straight run of pipe in a length equivalent to 5–10 times the diameter of that pipe. The smaller multiplier would be used on the larger pipe diameters and vice versa.

Rule 4: Stop air or vapor entering the suction line

Any high spot in the suction line can become filled with air or vapor, which, if transported into the impeller, will create an effect similar to cavitation, and with the same results. Services that are particularly susceptible to this situation are those where the pump volume contains a significant amount of entrained air or vapor, as well as those operating on a suction lift, where it can also cause the pump to lose its prime. A concentric reducer can cause a similar effect. The suction of a pump should be fitted with an eccentric reducer positioned with the flat side uppermost.

If a pump is taking its suction from a sump or tank, the formation of vortices can draw air into the suction line. Providing sufficient submergence of liquid over the suction opening can usually prevent this. A bell-mouth design on the opening will reduce the amount of submergence required. This submergence is completely independent of the NPSH required by the pump.

Rule 5: Correct piping alignment

Piping flanges must be accurately aligned before the bolts are tightened and all piping, valves and associated fittings should be independently supported, to place no strain on the pump. Stress imposed on the pump casing by the piping reduces the probability of satisfactory performance. Under certain conditions, the pump manufacturer may identify some maximum levels of forces and moments which may be acceptable on the pump flanges.

In high temperature applications, some piping misalignment is inevitable owing to thermal growth during the operating cycle. Under these conditions, thermal expansion joints are often introduced to avoid transmitting piping strains to the pump. However, if the end of the expansion joint closest to the pump is not anchored securely, the object of the exercise is defeated as the piping strains are simply passed through to the pump.

Piping design is one area where the basic principles involved are regularly ignored, resulting in hydraulic instabilities in the impeller which translate into additional shaft loading, higher vibration levels and premature failure of the seal or bearings. Because there are many other reasons why pumps could vibrate, and why seals and bearings fail, the trouble is rarely traced to incorrect piping.

The suction side of a pump is much more important than the piping on the discharge. If any mistakes are made on the discharge side, increasing the performance capability from the pump can usually compensate for them. Problems on the suction side, however, can be the source of ongoing and expensive difficulties which may never be traced back to that area.

Rule 6: Use the right pipe fittings

The number of turbulence or pressure drop producing fittings in the pump suction line should be kept to a minimum. Because of the excessive turbulent and friction loss that they produce, globe valves should not be used in the pump suction line. When NPSH or turbulence is a problem, a turbulence-reducing device should be used. This device should be located as near the pumps suction flange as possible.

A flow straightener should be installed just downstream of any elbows installed in the pump's suction piping to reduce turbulence and prevent flow separation. Only long-radius elbows should be used for pump suction applications. Without a flow straightener, uncontrolled flow through an elbow will separate the fluid into two stream regions, creating local high velocities and turbulence.

Check valves should not be used in the suction piping. Foot valves should not be utilized in the suction piping when the pump is operating against a high static head. Even though foot valves are recommended if suction lift is low and priming help is required, they often fail to fully close and may cause pump failure. If static head is high and the driver fails, a partially closed foot valve will permit the pumped liquid to rush backwards and in some cases cause excessive water hammer.

Rule 7: Adequate foundation

Centrifugal pumps, as do most other mechanical equipment, generate dynamic forces that must be absorbed by their foundation. The foundation should have a mass that is at least five times that of the installed pump and driver. In addition, the pump assembly must be properly affixed to the foundation. Baseplates must be bolted and grouted to the concrete foundation so that a positive, uniform bond is created.

33.6 Summary

Very few pump installations follow Hydraulic Institute recommendations. A minority of installations fail to follow reasonable engineering practices or even common sense. Therefore, it should not be a surprise that centrifugal pump applications often exhibit chronic reliability and maintenance problems.

Many of these chronic problems are the direct result of improper installation. The guidelines provided in this chapter are not all encompassing, but they provide the basic elements that will assure long-term, trouble-free operation of pumping systems.

34

Cooling Towers

John Neller
Film Cooling Towers Ltd

Contents

34.1 Background

The use of water as a cooling medium has been long established, but its importance, in an industrial sense, was emphasized with the introduction of steam power. Cooling ponds are still widely used and spray ponds, which incorporate a degree or so of evaporative cooling, can be found, but the increasing requirement for control on water-cooling temperatures heralded the development of the modern cooling tower. The cost of land and the increasing expense of abstracting and of returning water, as well as its availability, ensured that the engineering and design techniques employed were sufficient to satisfy the economic factors imposed by these constraints.

The state of the art in cooling tower design is being constantly improved. Correctly designed, installed and maintained, today's cooling tower still remains the optimum selection in the great majority of cases where heat dissipation is required.

Historically, the first pack designs were random timber, to be followed rapidly by ordered timber splash bars. The concept of filming water, as opposed to splash or concrete, originated in England in the 1930s. The introduction of plastic packing dates from the 1950s, but this was confined to mechanical-draft towers until the 1970s, when experiments started with plastic packs in natural-draft cooling towers. Asbestos cement in flat sheets, corrugated sheets and flat bars, although widely used in the past, are now out of favor on health grounds in most developed countries. Plastic-impregnated paper is used in certain eastern countries in air-conditioning towers, but has been unsuccessful in the West. In most cases the changes have been due to economics, but water quality and type of process can significantly affect the selection in individual cases.

34.2 Theory

Water-cooling in towers operates on the evaporative principles, which are a combination of several heat/mass transfer processes. The most important of these is the transfer of liquid into a vapor/air mixture, as, for example, the surface area of a droplet of water. Convective transfer occurs as a result of the difference in temperature between the water and the surrounding air. Both these processes take place at the interface of the water surface and the air. Thus it is considered to behave as a film of saturated air at the same temperature as the bulk of the water droplet.

Finally, there is the transfer of sensible heat from the bulk of the water to the surface area. This is so slight in terms of resistance that it is normally neglected. Radiant heat transfer is also ignored for all practical design purposes.

Thus the two main processes are evaporation of water and convective cooling. The first is based on the difference in partial vapor pressure and the second upon the temperature difference.

Merkel's analysis in 1924/5 demonstrated that for pure counterflow it is possible to combine these processes into a single term by using the enthalpy difference as a driving force. Experience over many years supports this, and cooling tower design in counterflow is universally represented by temperature–enthalpy diagrams.

For crossflow designs the additional factor of the horizontal depth of packing has to be included in the basic calculations. The accuracy of the design is directly related to the number of calculations in the selection program. Whereas counterflow can be dealt with as a single entity, crossflow has to cope with the changes that occur at every level of pack, both vertically and horizontally.

34.3 Design techniques (see Appendix 34.1)

The performance of any cooling tower can be assessed against the following:

$$\frac{KaV}{L} = \frac{C}{\left(\dfrac{L}{G}\right)^n}$$

where KaV/L is normally given as the performance index or is quoted as KaV/L demand. L/G = water/dry air mass ratio and, taking n as average value 0.6, the constant C is proportional to the height of packing. This would cover most design requirements. Excessive temperatures and extremely high air rates will require further factors which are not necessary in the great majority of cases.

The Cooling Tower Institute (CTI) publishes sets of graphs that give demand in terms of three design temperatures and L/G. The CTI graph first published in 1967 gives KaV/L demand plotted against L/G with the approach temperature as a parameter. Each curve applies to a specified combination of wet bulb temperature and cooling range. Each cooling tower pack has its own KaV/L and responsible suppliers will supply performance graphs similar to those of the CTI.

Having selected a type and height of pack, the above equation can be plotted to intersect with the required demand curve to obtain the L/G. With the L/G and the given amount of water to be cooled, the air requirement can be calculated for:

$$\text{Air} = \frac{\text{Design water flow}}{\dfrac{L}{G}}$$

The following steps should then be:

1. Correct calculated air volume to conditions at the fan.
2. Select the air rate and calculate the plan area of the tower.
3. Select a fan and calculate the power requirements from the known air volume and the pressure drop characteristics of the selected pack.
4. Assess the cost factors applicable, in terms of fan, pump cost, price of land, maintenance and treatment costs.
5. Optimize all known factors in terms of efficiency and then economy.

34.4 Design requirements

The factors which should be applied at the design stage cover the water flow rate, the design wet bulb figure, the required return temperature at the design point, the cost of power and land, and the water analysis. Water flow is normally determined by the equipment that the cooling tower is serving (for example, heat exchangers). Process designers historically leave the cooling tower until last (it is, after all, the final heat sink). When water costs were negligible, this was acceptable, but with the increase in costs and, in certain cases, the restrictions on availability of water, this approach has had to be modified. Greater consideration must be given to the overall system. Experience in the last ten years has shown that economic optimization can lead to a more efficient cooling tower, with a corresponding drop in the cost of heat exchanger. This is particularly true in power generation and industrial processes.

Design wet bulbs can be determined from published meteorological data for the area concerned. The difficulty is deciding how to relate the annual coverage to the tower performance at any given time.

For some years it was common practice to quote three different figures, based on the tower's performance as a percentage of the year. For example, in air conditioning it could be shown that the tower would achieve its design for 95 per cent of the year. Alternatively, a tower costing 15 per cent less could obtain its design parameter for 85–90 per cent of the year. Only the operator would know whether the 85–90 per cent or less was acceptable, while the economists would welcome the saving of financial capital.

The frequent failures to achieve even the quoted reduced percentage figures led to a reappraisal, and current design is more accurate. In some respects this is also due to the improvement in pack designs, particularly in the European and American markets. However, it must be said again that in optimizing cooling tower selection the designer must be advised of all appropriate factors. Discussions with cooling tower designers at the outset can save time and money in the future.

Water quality is important, not only from an environmental point of view but also in relation to the type of packing to be specified. Analysis of the circulating water is simple to obtain, but it is very seldom offered to the cooling tower designer. The quality, or lack of it, will determine the type of pack to be used, the selection of structural materials and whether the tower should be induced or forced draft, counterflow or crossflow. Water treatment, in the shape of chemicals to control pH and to act as counter-corrosion agents or as biocides, all has a bearing on tower selection.

The 'Legionella syndrome' has resulted in health authorities in the US and UK applying statutory regulations, which are directly reflected in terms of capital cost and tower material selection. To safeguard against this, responsible designers have already produced cooling tower designs which not only meet the regulations but also anticipate future, more stringent, legislation.

The following list of information factors should be made available to any supplier so that discussions on the technical requirements can be carried out prior to optimization (see Appendices 34.1 and 34.2).

34.5 Materials and structure design

The great majority of towers available fall into one of two categories: counterflow or crossflow (co-current flow is available but is seldom used) (see Figure 34.1).

34.5.1 Counterflow

Counterflow designs are used throughout the entire design field, i.e.:

Natural draft – hyperbolic concrete shells
Mechanical draft – induced draft
Forced draft

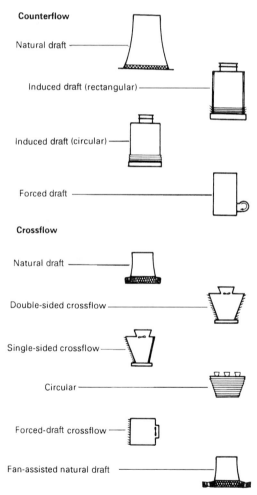

Figure 34.1 Counter- and crossflow modes

Figure 34.2 Basic cross flow units

Figure 34.3 Improved design of crossflow unit (note easy-removal panel)

34.5.2 Crossflow

Crossflow designs are also used throughout the entire field, i.e.:

Natural draft – hyperbolic concrete shells

Mechanical draft – induced, single and double sided and circular
Forced draft
Advanced fan-assisted natural draft.

Axial or centrifugal fans can be applied in most cases and are significant factors in the final selection and optimization (Figures 34.2 and 34.3).

Hybrid towers, combining wet and dry cooling, are designs to meet specific problems and require expertise from specialist suppliers. Structural materials include concrete, timber, various forms of metal (including galvanized and alloys), GRP, PVC and combinations of those along with asbestos cement, asbestos cement replacement (ACR) and, again, variations and combinations of other materials. Packing materials have an almost similar pattern but must include compressed paper and compressed asbestos cement paper, but the great majority of towers currently employ plastic, in some form or another, unless the water conditions are such that timber (or even concrete) must be used as alternatives.

34.6 Specification

The purchase of the cooling tower is, in most cases, a once in a decade operation. Where towers are bought on a regular basis, specifications are determined either by the user or by the consultant, incorporating their experience of operation and any changes required as a result of production/process alterations.

In air conditioning circles, the tower normally represents the final heat sink in a turnkey package which would include compressors/condensers, pipework, ducting, fans, pumps, control gear, etc. Where consultants and experienced contractors are concerned, the tower specification is well defined and the purchases based upon economics related to efficiency.

Where a tower is to be purchased for a one-off situation, it is worth considering the various factors which can affect the final choice:

Figure 34.4 Architect-designed tower (note water pattern in basin). Chemicals in carry-over affect the office windows

Location (both geographically and in elevation)
Restrictions (i.e. planning, structural, physical and environmental)
Design wet bulb
Design dry bulb
Water flow rate
Water quality
Water treatment to be used/cycles of concentration?
Process
Constant or cyclic heat load
Criticality of return temperature
Noise restrictions
Additional local environmental factors
Water discharge regulations (quantity and quality)

For the periodic or once-off buyer it is essential to obtain advice from a reputable supplier or a consultant with experience of cooling towers usage (Figure 34.4).

As an example, the reputable cooling tower designer would establish most of the above parameters from his own experience. In addition, he could determine with his client the economic factors which could influence his selection, i.e. low capital cost with high running costs, or a higher capital cost with more acceptable power costs (a 12-month or a 5-year payback period). Turnkey contractors often understandably ignore this particular factor, and end users should always obtain alternative designs to make their own selection.

The costs of water, energy and land are all contributing factors in economic assessment. The emphasis placed on each differs from the varying viewpoints, but it is the end user that has to 'foot the bill'. It is therefore in his interests to acquire some knowledge of the cooling towers being offered.

34.7 Water quality and treatment

As, in most cases, the circulating water in any system incorporating a cooling tower is recycled (the loss being made up as a percentage of total flow), the subsequent concentration will affect the water condition. This, in turn, determines the type of treatment required to maintain pH and to control any potential biological growth. Acidic or alkaline waters pose their own problems in terms of corrosion and material attack. The larger users probably employ their own chemical specialists, while others rely on consultants to determine the type of treatment the system requires. However, the majority of tower users have very little knowledge of the chemistry involved and depend on water-treatment organizations.

The following definitions are useful for reference to familiarize the end user with the terminology currently employed:

pH The acidity or alkalinity of the water expressed as a scale of 0 (acid) to 14 (alkaline). pH 7 is regarded as neutral.
TDS Total dissolved solids, expressed as ppm (parts per million) or as mg/l (milligrams per liter). Evaporate the water from a sample and the residue can be weighed.
TSS Total suspended solids: expressed in similar terms to TDS but representing a concentration of insoluble particles.
Conductivity Used as a measure of mineral impurities.

LSI *Langelier Saturation Index*: indicates the corrosive (negative) or scale-forming (positive) characteristics.

Hardness Expressed as $CaCO_3$, this is the total calcium and magnesium salts in the water. Hardness figures given as ppm or mg/l are important, as the compounds of these two elements are responsible for most scale deposition.

Alkalinity Expressed as $CaCO_3$, this is the total concentration of alkaline salts (i.e. bicarbonate, carbonate and hydroxide).

BOD Biological oxygen demand: expressed as ppm or mg/l, it is used as a measure of pollution.

COD Chemical oxygen demand: as above, but related to chemical impurities.

Oxygen sag The level of oxygen in a polluted water system. Normally shown in graph form.

Fouling factor This is generally applied to plastic packs in natural-draft towers but can relate to larger mechanical draft and to the biological fouling that can occur. It also reflects on the thermal performance of the packing (CEGB published fouling factors for certain high-density packs where the supply of water was prone to seasonal biological growth and silt deposition, along with calcium hardness deposition).

While other terms are, of course, employed, the above can be useful for most end users.

34.8 Operation

Having selected and purchased a cooling tower, it needs regular maintenance, as does any other part of the plant. This is true of every cooling tower, from the largest natural-draft tower to the smallest packaged unit.

Maintenance logs should be kept. In air-conditioning installations, with the experience of *Legionella*, it is now mandatory to keep such a log, as well as a record of hygiene testing to determine the non-existence of bacteria. Chlorine dosing, as recommended by certain local authorities, is essential. Tower materials have to be assessed in regard to chlorine residuals, which can be damaging to galvanized metals and timber. Mechanical equipment requires regular checks, apart from the common sense ordinary maintenance (it is surprising how often this is ignored).

The fault-finding chart (Figure 34.5) was originally produced by the Cooling Water Association (which subsequently became the Industrial Water Society), and it is practical and simple to follow. Water treatment checks (apart from the *Legionella* requirements referred to earlier) must be carried out and water samples should be analyzed on a regular basis. How frequently this takes place depends on the criticality of the tower for the end user. Cooling towers are water-conservation tools as well as heat dissipaters: with water costs increasing, continuous tower performance is essential. In any case, a downtime caused by lack of maintenance is costly and careless.

Remember that outside influence (for example, new building work in the vicinity of the installation) can increase air-based pollution, such as cement dust entering the tower at air inlet levels or via the forced-draft fan.

Extra cleaning of the tower pack and distribution system should be undertaken under these circumstances and close checks kept on the efficiency of the overall system.

Changes of process or modification to the product can introduce new design parameters, which can affect the tower. Overloading the tower in both a thermal and hydraulic sense may be acceptable, as a temporary measure, but 'temporary' is the critical word. Too long exposure to an increased temperature can affect plastic packing, unless they have been designed to withstand such an increase. (Remember to discuss future predicted problems with the cooling tower designer before installation.)

Excessive water loads can lead to malfunction of the distribution system. Higher loading in one area can lead to pack collapse and rapid fall-off in performance.

In the majority of cases common sense, combined with basic engineering principles, should be sufficient to ensure good service from the tower on a continuous basis. The reputable supplier will always be ready to help and advise. If the advice is sought in time, many of the problems associated with the changes mentioned need never arise.

34.9 Modifications and retrofits

Changes in circumstances will frequently require modifications to an existing cooling tower. Additional heat loads may be needed, and changes in the end process may cause the return temperature to increase, necessitating a new thermal load on the tower. Most frequently it is a requirement for an increase in hydraulics.

34.10 Consultation

All too often, the supplier is not consulted at the planning stage, with the inevitable result of a tower failure, not only in performance but also in pack collapse, structural failures and total shutdown. The original tower was supplied against a design water and thermal load, and changes to those parameters will affect the performance. It is possible to change existing towers. Not only is it possible, it is well-established practice. The successful amendments are those where the tower designer has been advised (in advance) of the proposed changes. He knows the limitations of his product and can advise on what can or cannot be done.

There is one area where improvements can be achieved, namely the increase in thermal performance by changing to a more advanced design of packing. Care has to be taken to ensure that the new pack configuration is compatible with the quality of water as well as its quantity. The water treatment conditions could change, and almost certainly the distribution system will need amendments. The benefits to be obtained can be listed as follows:

1. Improvement in thermal load.
2. Possible reduction in pumping head (e.g. the change from a splash pack to a high-density film-type pack can save power by installing the new pack at the bottom of the former splash area and lowering the pump inlet). If the correct design is used it may be possible to

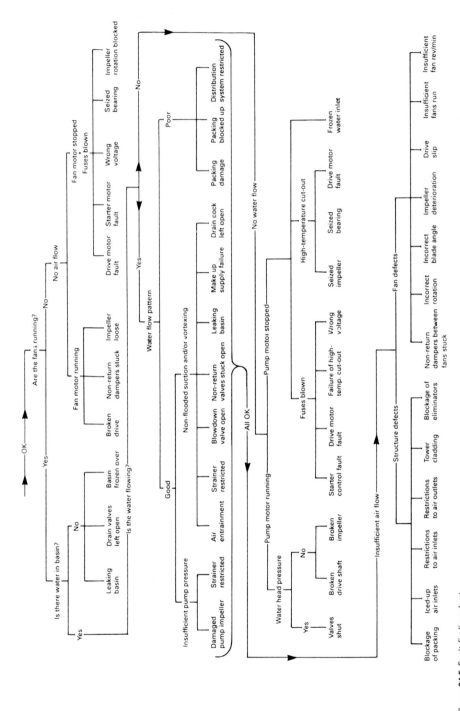

Figure 34.5 Fault-finding chart

leave existing fans, thus incurring no additional power penalty. Even if fan changes are required, the economics have to be studied, but in most cases such changes will be beneficial.

Changes to a distribution system can be of assistance in minor improvements. Nozzle designs are under constant review but it is in the context of pack changes that amendments are important.

34.11 Environmental considerations

The principal areas where cooling design is affected by environmental requirements are visual and audible (i.e. aesthetic (plume) and noise).

34.11.1 Esthetic

Planning regulations can be rigid in their attitude to the visual impact that a cooling tower may have on the surrounding area. This is not, however, confined to air-conditioning installations. Certain industrial areas, within the UK and elsewhere, limit the number of towers, their height and configuration in relation to the existing background situation. While this is understandable in environmental amenity terms, it leads to major design difficulties on the part of the supplier.

A consultation with all interested parties is essential at the planning stage. Solutions can normally be found, but the cooling tower designer must be included in these discussions. Towers can always be designed to meet the planning regulators and the demands of architects. Low towers, tall narrow towers, circular towers, multifaceted towers, towers built into the buildings, even towers installed under ground, are all practical examples of modern installations (Figures 34.6–34.8). Obviously, the design characteristics have to be reassessed. Hence the emphasis on cooling tower designers' involvement from the outset.

As cooling towers operate on the evaporative principle, at certain temperature conditions the discharged heat vapor will appear as a plume. The amount of the pluming can be accurately assessed against the temperature conditions (both inside and outside the tower), the volume of air and the velocity of the discharge. The extent to which this can be classified as a 'nuisance' depends entirely on the location of the tower and its proximity to sensitive areas (i.e. housing, office blocks, etc.). One variation is the natural-draft tower, where geographical location may cause the plume to affect ambient conditions downwind, such as moisture deposition on roads (Figure 34.9) or, in one instance in Switzerland, where the plume from a large natural-draft tower located in a narrow valley restricted sunlight on the farming area downwind. (The predicted 7 per cent restriction was accurate.) The drift can normally be confined within more than acceptable limits by the use of efficient drift eliminators.

The height of the discharge is important, and in air-conditioning projects, it can be critical in relation to surrounding buildings. Consideration has to be given to the volume of discharge, the possibility of entrainment of

Figure 34.6 Cooling plant, including towers, installed as a separate unit on a supermarket satisfies the architect, town planners and the end-user

water-treatment chemicals in the drift (these can cause 'etching' on glass windows), the siting of the tower in relation to the fresh air inlets to the building and the visual impact, in architectural and esthetic terms.

Eliminator design can vary from the non-existent to an efficiency which can limit drift to 0.00005 per cent of flow. In other words, drift can be almost undetectable, but the difference is obviously in the economics involved.

Pressure groups in the environmentally aware political parties can make excessive demands. Often good public relations meetings can meet these, but the more reputable cooling tower suppliers can be invaluable in dealing with such matters.

Plume abatement is possible, even to the point of 'invisible plumes'. While this is not a problem in hot countries, the temperate zones (for example, Europe and the USA), will always have seasons when the plume is normally visible. To change this situation to a non-plume effect is again possible but it is expensive. As an example, one installation in the center of a North American city, where non-pluming throughout the year was a mandatory clause in the planning permission, resulted in the tower cost being increased by a factor of 3! (A not-insignificant amount in the overall cost of the project.) Plume control can be achieved at the expense of larger installations and possible changes in the temperature levels, all of which require prior consultation with the designer.

The other source of possible drift or precipitation from cooling towers is caused by windage or blowout from

Figure 34.7 Design requested by architect – low plume, no splash (note catchment tray), clear of office block

Figure 34.8 Cranes can be eliminated!

Figure 34.9 Congested site. Restrictions on air inlet, increased pressure and air velocity results in drift deposition on road

the air inlets, noticeably under strong or gusty conditions. This can occur in both mechanical- and natural-draft cooling towers, but the effect is normally localized to the immediate tower area. In the natural-draft cooling tower the extent of the problem varies according to the design of the packing.

In the case of pure counterflow packing, where the entire packing is positioned above the air inlet, the large air inlet opening can give rise to discharge of water by air entering the air inlet on the windward side, then being sucked out in the vortex depression, which generally occurs at approximately 90° to the wind direction. The resulting spray is carried as a fairly narrow band for a distance of perhaps 20 to 30 meters downwind of the tower. In mixed-flow packing, where the cooling tower packing extends down into the air inlet, there is considerable resistance to the free passage of air through the air inlet. Therefore the depressed area of the vortex is reduced and the resulting spray tendency is likely to be restricted to a fine spray, literally being blown off the peripheral pack laths. Various techniques have been adopted to minimize this effect, such as external radial baffles, internal baffles and louvers. All of these will, of course, incur additional cost and some of them may also increase the pressure drop through the air inlet and thus affect the thermal performance. Additionally, they can increase the overall dimensions of the installation.

Mechanical-draft cooling towers are normally supplied with either central baffles or inlet louvers. This depends on the tower dimensions. On these towers the wind or spray blowout is generally confined to relatively small single-cell units where an inlet may be provided on all four faces. In this case the major remedy is to provide internal diagonal baffles to prevent crossflow of air through the air inlets.

On larger multi-celled mechanical-draft towers of both counterflow and crossflow variety, the air inlets are confined to the two opposing faces and windage or drift loss is unlikely to occur, except under exceptionally high wind conditions. Here again, remedial work, depending upon the location, can be applied but at additional cost (see Figure 34.10).

34.11.2 Noise

Perhaps the most common environmental requirement in modern cooling tower installations is that of noise. The fan equipment and the falling water generate cooling tower noise. In large mechanical- or natural-draft cooling

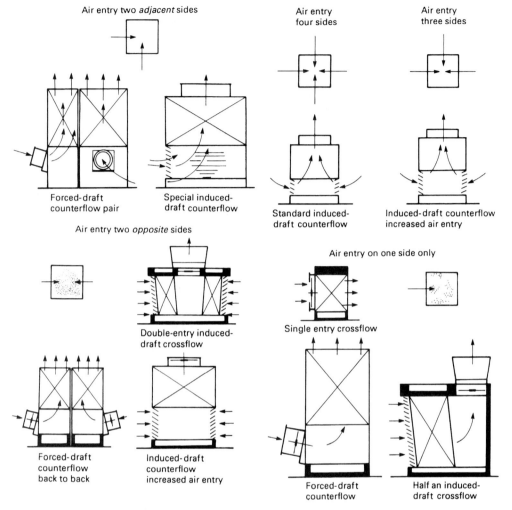

Figure 34.10 Configurations of cooling tower airflows

towers, the water noise could exceed the noise generated by the fan equipment (particularly if steps have been taken to reduce fan noise to a minimum). In the case of smaller cooling towers the prediction of noise intensity at a distance in excess of between 30 and 50 meters can be taken as a hemispherical radiation from a point source, i.e. a reduction in the level of 6 dB for every doubling of the distance. However, with the large multi-cell mechanical- and natural-draft cooling towers the noise is radiating from a considerably larger area, and therefore the sound pressure level falls by only 3 dB for every doubling of the distance up to $\frac{1}{2}d$ (where $\frac{1}{2}d$ = pond diameter of the hyperbolic shell in a natural-draft tower or the tower length for the multi-cell mechanical-draft type). For distances greater than $\frac{1}{2}d$ the sound pressure will then fall by 6 dB for every further doubling of the distance.

Fan noise is likely to be more obtrusive than the so-called 'white sound' emitted by falling water due to the presence of discrete frequencies arising from blade-passing frequency, tooth frequency on the gearboxes bearing and rumble from the gearboxes and motors and other electrical noises. It can, of course, be minimized by correct choice of fan. In general, the use of broad-cord multi-bladed fans enables the fan to be operated at a minimum possible speed compatible with the duty performance. Reduction of bearing noise in gearboxes can be eliminated, as far as possible, by careful design in the mounting system and motor noise can always be shielded by acoustic enclosures. For extremely quiet operation on mechanical-draft towers recourse may have to be made to the use of silencers or attenuators on the air inlets to the cooling tower to minimize the water noise, radiation and mechanical noise break-out, with further acoustic attenuators on the fan discharge.

Any acoustic treatment on the fan discharge is required to operate under potentially corrosive conditions, with warm moist air passing through the attenuators. Precautions are needed to ensure that these are adequately treated to prevent risk of condensation and that the structural media are also protected against water pick-up and potential damage.

In the case of natural-draft towers the total noise source is very considerable and has to be assessed as a large-area source relating from the whole diameter, through the entire height of the air inlet. Falling water noise in that situation can be as high as 85 dB. The noise level at 70 m reduces to 66 dB and at 480 m it is approximately 46–47 dBA. On the largest natural-draft towers the figures increase. For example, at the pond level and at the side of the tower they can be in excess of 91 dBA and can actually reach 55 dB at 500 meters.

Sound is defined as any pressure variation that the human ear can detect. This variation can occur in air, water and other media. To determine noise, it is necessary to assess the frequency of the variation that, in turn, can be related to the speed. For most applications, the speed of sound is expressed at 340 meters per second

(m/s). Speed and frequency give the wavelength, i.e. the physical distance in air from one wave to the next. For example, at 20 Hz this gives 17 m while at 20 kHz one wavelength is 1.7 cm.

For convenience, the usual measurement of sound is expressed in decibels (dB), and ratings go from 'threshold of hearing' to 'threshold of pain' (135 dB). Figure 34.11 illustrates the common noise criteria, which can be expressed in sound-pressure levels (SPL). The human ear can detect 1 dB but 6 dB represents a doubling of the SPL, although it would need a 10 dB increase to make it 'sound' twice as loud.

In assessing the noise emanating from a cooling tower it is necessary to measure the main points of emission – the fan, the motor, the gearbox and the falling water. As noise bounces and can be absorbed by certain materials, it is usual, where noise restrictions apply, to map the area and calculate the SPL at a large number of points, taking into consideration interference, bounce and absorption. The number of points measured is reflected in the accuracy of the resultant topography.

Remember also to take background noise into your calculation. Too frequently, specifications are made which ignore this, with the result that equipment is applied to a more rigid design than is absolutely necessary.

The siting, as well as the selection of type of tower, can be critical. Rotating the tower, shielding the motor, use of baffles can all help in meeting environmental noise requirements. If in doubt, consult your cooling tower designer.

34.12 Problem areas

34.12.1 Installation

Cooling towers have been called the Cinderella of the plant scene – usually unnoticed (if not even unseen), forgotten and sadly ignored. While cooling tower designers may have other ideas, they generally recognize this as being true in many cases. Designers therefore try to achieve the impossible, i.e. to build a piece of mechanical equipment that can be left alone and perform its function without fuss and attention.

While designers can claim some degree of success, there are many occasions when their products are improper installation.

Don'ts

Mix products – placing a forced-draft tower beside an induced-draft one causes problems for both designs (Figure 34.12).

Place access panels incorrectly – the access panel is for the user's benefit. Ensure that it is accessible Figure 34.13).

Starve the air inlets – insufficient air results in poor performance (Figure 34.14).

Ignore the bleed – inadequate bleed means concentration of salts, change of pH and pack fouling.

Forget about make-up – water starvation means poor performance, vortexing, pump and motor failure.

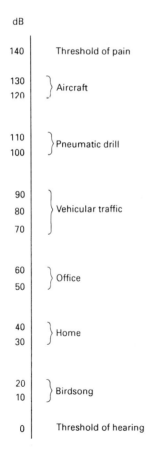

Figure 34.11 Threshold ratings for sound

Figure 34.12 Mixing forced draft with induced. The overloaded forced-draft tower with excess plume results in elevated wet bulb at air inlets on new tower. Removing the forced draft and adding one more cell to the induced draft resolved the problem

Figure 34.14 An attempt to conceal tower and reduce noise, resulting in starvation of air and failure in performance

Fail to check on water treatment – inadequate treatment and haphazard slug dosing can lead to poor performance, damage to associated equipment and failure to meet discharge requirements.

Forget to install safety cutouts, for overload and ambient changes – ice damage can be disastrous (Figure 34.15).

Allow corrosion to develop – metal failure can be costly and, at times, dangerous (Figure 34.16).

Do's

Ensure maintenance checks are carried out.

Check on power consumption, water usage, water costs.

Carry out monthly inspection – inside and out, where possible.

Check water analysis; the frequency depends on individual cases but it should be no less than quarterly.

Check mechanical equipment (i.e. fans, motors, drives). (Remember to check belt tensions where applicable.)

Check for vibration, both mechanically and structurally.

Ensure that access panels are used and replaced correctly.

Make certain that repairs, when necessary, are carried out efficiently and quickly.

If it in doubt, call the cooling tower designer.

Figure 34.13 Cyclic heat load using small tower and large water storage. This is a good idea and sound engineering, but the access panel is on the wrong side (30 ft drop to ground level)

34.13 Summary

While the majority of cooling tower installations work efficiently, the normal requirements of maintenance and good efficiency practice have still to be applied. This may not always be the case. Time for maintenance is limited, and plant engineers have other pressing problems or, as is well known, forget about the towers! With the increasing economic and environmental pressures on the use of water, this situation must change.

It may be appropriate to quote the old engineering term of 'KISS' (Keep It Simple, Stupid!) and, recognizing that the cooling tower designer does his utmost to comply, it is the responsibility of the operator to 'co-operate'!

Figure 34.15 Ice on natural-draft tower (no de-icing ring fitted)

Figure 34.16 The effects of corrosion. Structural failure is visible (holes in fan casing)

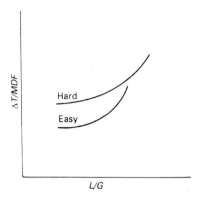

Figure 34.17 Demand curve

Regular checks and an efficient logging system will ensure that the cooling tower, correctly planned, efficiently installed and adequately maintained, will give valuable service for many years. The economic returns justify a little more thought and attention than has been given to the subject in the past.

Appendix 34.1 Theoretical calculations

$$L dT = K a dV(h_L - h_G)$$

where

$K =$ the coefficient of heat transfer for the packing in question,
$a =$ the effective transfer surface area per unit pack volume,
$V =$ depth of packing,
$h_L =$ enthalpy of boundary air layer in contact with and at the same temperature as the water, and
$h_G =$ enthalpy of bulk air passing through the packing.

Integrating this for the full depth of packing, the expression becomes:

$$L \times \Delta T = KaV \times \Delta h_m$$

where Δh_m is the mean enthalpy difference, otherwise known as the *mean driving force (MDF)* (see Appendix 34.2).

This can be rearranged as:

$$\frac{KaV}{L} = \frac{\Delta T}{MDF}$$

which is the form in which it usually appears. *KaV/L* is commonly called the *tower characteristic*.

Now let us refer to the right-hand side of the above expression. The mean driving force varies with the specified design temperatures and also the ratio of water/air loading (*L/G*). If we take a low airflow, the air soon rises in temperature and tends to reach equilibrium conditions with the boundary layer. Thus the driving force is reduced. On the other hand, excess air is unnecessary. Therefore, one must adjust the airflow that supply just meets demand. A plot of *L/G* versus $\Delta T/MDF$ is shown in Figure 34.17. This is known as a demand curve.

The left-hand side of the above expression is a measure of the quality and quantity of the packing being used, and has been shown empirically to obey the law $(KaV/L) = c(L/G)^{-n}$ for counterflow applications only.

Cooling in the crossflow mode requires an incremental 'trial and error' technique, best suited to computer analysis. The tower characteristic KaV/L can then be plotted against varying L/G ratios, and this gives a measure of the ability of the packing to effect the transfer (Figure 34.18).

We have already equated KaV/L with $\Delta T/MDF$, therefore we can superimpose the 'supply' curve over the 'demand', the intersect being the optimum L/G ratio for the packing being considered for the duty (Figure 34.19).

It is interesting now to examine the effect of using greater or lesser depths of packing, and to consider their suitability for duties of different degrees of difficulty. In Figure 34.20 we can see the effect of using three different pack depths on a moderately easy duty. By changing from pack depth (1) to pack depth (2) we are able to use a much higher L/G ratio (which, in turn, means less air and/or a smaller tower). The increment from depth (2) to depth (3) gives a less significant increase in L/G, and therefore suggests that the optimum depth has perhaps been exceeded.

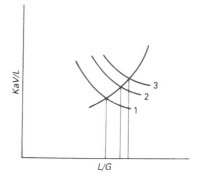

Figure 34.20

Now let us examine the same pack depths but applied to a more difficult duty (Figure 34.21). The increase in L/G is almost constant from (1) to (2) and from (2) to (3), showing that the optimum depth has not been passed, and may not yet have been reached.

Thus, there is an optimum depth of packing for each individual duty and, in practice, it is usually found that an intersection near the knuckle on the demand curve produces the most economic selection.

Appendix 34.2 Evaluation of the MDF

Several methods for the evaluation of the MDF have been put forward, notably that processed by Tchebycheff, which gives a high degree of accuracy in the case of large cooling ranges. In the form in which it is most commonly used, it reads:

$$MDF = 4/(1/\Delta h1 + 1/\Delta h2 + 1/\Delta h3 + 1/\Delta h4)$$

Where

$h1 =$ value of $h_L - h_G$ at $T_2 + 0.1\Delta T$,
$h2 =$ value of $h_L + h_G$ at $T_2 + 0.4\Delta T$
$h3 =$ value of $h_L + h_G$ at $T_1 + 0.4\Delta T$
$h4 =$ value of $h_L + h_G$ at $T_1 + 0.1\Delta T$

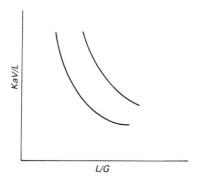

Figure 34.18

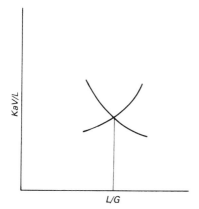

Figure 34.19

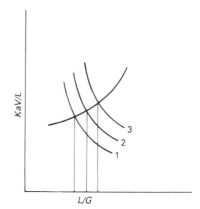

Figure 34.21

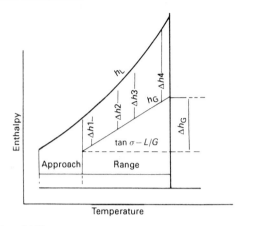

Figure 34.22

Graphically, this can be represented as in Figure 34.22. The expression for determining *KaV/L* is

$(\Delta I/4)(1/\Delta h1 + 1/\Delta h2 + 1/\Delta h3 + 1/\Delta h4)$.

Appendix 34.3 Technical requirements

Location
Meteorological data
Wind rose

Water flow rate
Temperature to tower (T_1)
Temperature from tower (T_2)
Cooling range
Approach
Design wet bulb
Design dry bulb
Water analysis – circulating
                 – make-up
Cycles of concentration
Blowdown/purge rate
Power cost analysis
Drift loss requirement
Local authority requirements
Discharge qualities
Structural specifications
Pack specifications
Mechanical specifications (if applicable)
Noise specification
Impedance by adjoining structures (if applicable)

Acknowledgements

The author wishes to thank the following for their considerable input: Mr Richard Clark and Mr Anthony Kunesch (FCT Ltd), The Cooling Water Association (now The Industrial Water Society), Mr John Hill (Director of BEWA) and the many understanding people who gave permission for publication of the photographs.

35

Compressed Air Systems

British Compressed Air Society

Contents

35.1 Assessment of a plant's air consumption

The main consideration in the selection of a compressor plant is the production of an adequate supply of compressed air at the lowest cost consistent with reliable service. The installation of a compressed air system, as with all forms of power transmission, calls for capital investment with consequent operating and maintenance costs. The information on which the selection of plant is based should be as accurate as possible. Important factors to be considered are the following.

35.1.1 Operating pressure

Most compressed air equipment operates at about 6 bar (gauge) and it is usual for the compressor to deliver air into the mains at 7 bar (gauge) in order to allow for transmission losses (see Tables 35.1–35.4). If some of the air is to be used at a lower pressure (for example, for instrument control) the pressure is reduced by means of a pressure regulator to the required line pressure.

All equipment connected to the system shall either have a design pressure greater than the maximum output pressure of the compressor or special precautions shall be taken to ensure that, if its design pressure is lower than the output pressure of the compressor, it cannot be subject to excessive pressure (see Section 35.3.1). If there is a requirement for a large volume of air at a higher or lower pressure, it may be more economical to install a separate compressor to deal solely with that.

35.1.2 Maximum and average load

Ideally, the total capacity would be based on exact knowledge of the equipment or process requirements. If this is underestimated, the compressor plant will be too small, and will be unable to maintain the required pressure in the

Table 35.1 Maximum recommended flow through main lines[a]

Nominal bore (mm)	Actual bore (mm)	Rate of air flow at 7 bar (l/s)
6	6	1
8	9	3
10	12	5
15	16	10
20	22	17
25	27	25
32	36	50
40	42	65
50	53	100
65	69	180
80	81	240
100	100	410
125	130	610
150	155	900

Velocity of airflow must be restricted to less than 6 m/s as shown if carry over of moisture past drain legs is to be avoided.
[a] Medium-weight steel tube to BS 1387, table 2, or ISO 65.

system. Conversely, if the total air consumption is greatly overestimated, there may be excessive capital investment. Furthermore, any arrangement which results in significant off-load running wastes energy. However, it is safer to err on the high side with slight over-estimates, as in most installations the use of compressed air will increase and soon take up any surplus capacity.

35.1.3 Use factor

Before deciding the capacity of the compressor required it is necessary to calculate the air consumption expected. It is recommended that reference is made to Table 35.5 and Figure 35.1, which show typical 'use factors' for various types of pneumatic equipment. In some cases, where

Table 35.2 Maximum recommended flow through branch lines of steel pipe[a]

Applied gauge pressure bar[b]	Nominal standard pipe size (nominal bore) (mm)										
	6	8	10	15	20	25	32	40	50	65	80
0.4	0.3	0.6	1.4	2.6	4	7	15	25	45	69	120
0.63	0.4	0.9	1.9	3.5	5	10	20	30	60	90	160
1.0	0.5	1.2	2.8	4.9	7	14	28	45	80	130	230
1.6	0.8	1.7	3.8	7.1	11	20	40	60	120	185	330
2.5	1.1	2.5	5.5	10.2	15	28	57	85	170	265	470
4.0	1.7	3.7	8.3	15.4	23	44	89	135	260	410	725
6.3	2.5	5.7	12.6	23.4	35	65	133	200	390	620	1085
8.0	3.1	7.1	15.8	29.3	44	83	168	255	490	780	1375
10.0	3.9	8.8	19.5	36.2	54	102	208	315	605	965	1695
12.5	4.8	10.9	24.1	44.8	67	127	258	390	755	1195	2110
16.0	6.1	13.8	30.6	56.8	85	160	327	495	955	1515	2665
20.0	7.6	17.1	38.0	70.6	105	199	406	615	1185	1880	3315

[a] Maximum recommended airflow (l/s free air) through medium series steel pipe for branch mains not exceeding 15 meters in length (see BS 1387). The flow values are based on maximum recommended peak flows. Normal steady air consumption should not exceed 80 per cent of these figures in pipe sizes 20 mm nominal bore and above.
[b] Applied pressures selected from ISO 2944: Fluid Power Systems – nominal pressures.

Table 35.3 Pressure loss through steel fittings – equivalent pipe lengths

Item	Equivalent pipe length (m)									
	Inner pipe diameter (mm)									
	15	20	25	40	50	80	100	125	150	200
Gate valve fully open	0.1	0.2	0.3	0.5	0.6	1.0	1.3	1.6	1.9	2.6
half closed		3.2	5	8	10	16	20	25	30	40
Diaphragm valve fully open	0.6	1.0	1.5	2.5	3.0	4.5	6	8	10	
Angle valve fully open	1.5	2.6	4	6	7	12	13	18	22	30
Globe valve fully open	2.7	4.8	7.5	12	15	24	30	38	45	60
Ball valve (full bore) fully open	0.5	0.2	0.2	0.4	0.3	0.4	0.3	0.5	0.6	0.6
Ball value (reduced bore) fully open	3.4	4.9	2.4	2.2	5.0	2.6	4.1	3.3	12.1	22.3
Swing check valve fully open		1.3	2.0	3.2	4.0	6.4	8.0	10	12	16
Bend R = 2d	0.1	0.2	0.3	0.5	0.6	1.0	1.2	1.5	1.8	2.4
Bend R = d	0.2	0.3	0.4	0.6	0.8	1.3	1.6	2.0	2.4	3.2
Miter bend 90°	0.6	1.0	1.5	2.4	3.0	4.8	6.0	7.5	9	12
Run of tree	0.2	0.3	0.5	0.8	1.0	1.6	2.0	2.5	3	4
Side outlet tee	0.6	1.0	1.5	2.4	3.0	4.8	6.0	7.5	9	12
Reducer		0.3	0.5	0.7	1.0	2.0	2.5	3.1	3.6	4.8

The table shows the length of pipe with equivalent pressure loss in a given size and type of fitting.

Table 35.4 Pressure loss through ABS fitting – equivalent pipe lengths

Pipe outside diameter	Equivalent pipe length (m)							
	16	25	32	50	63	75	90	110
90° elbow	0.34	0.5	0.65	1.0	1.26	1.5	1.88	2.58
45° elbow	0.16	0.24	0.32	0.52	0.63	0.75	0.95	1.33
90° bend	0.1	0.16	0.22	0.34	0.44	0.56	0.75	1.00
180° bend	0.28	0.41	–	–	–	–	–	–
Adaptor union	0.14	0.21	0.25	0.4	0.5	–	–	–
Tee in line flow	0.123	0.19	0.23	0.36	0.45	0.56	0.69	0.95
Tee in line branch flow	0.77	1.17	1.47	2.21	2.98	3.68	4.57	6.00
Reducer	0.22	0.31	0.37	0.51	0.80	1.11	1.34	1.58

The table shows the length of pipe with equivalent pressure loss in a given size and type of fitting. For example, a 50 mm 90° elbow has a pressure loss equal to 1.0 m of 50 mm ABS pipe.

Table 35.5 Air consumption of pneumatic equipment *Example of calculation* The following calculation is typical of a medium-sized engineering workshop including a foundry, where a high degree of mechanization is to be carried out by means of compressed air-driven machines and tools. Listed in the table are the tools and other pneumatic devices which are expected to be included in the installation at full production capacity. The use factor of the different tools is calculated in connection with production planning and thus it is possible to establish the average total air consumption.

Machine or tool	Air consumption per unit (l/s)	Quantity	Maximum air consumption (l/s)	Use factor	Average air consumption (l/s)
Foundry					
Core-shop (I)					
Core blowers	11	3	33	0.50	16.5
Bench rammers	4	2	10	0.20	2.0
			43		18.5
Machine moulding (II)					
Moulding machines	12	5	60	0.30	18.0
Blow guns	8	5	40	0.10	4.0
Air hoist −500 kg	33	2	66	0.10	6.6
			166		28.6
Hand moulding (III)					
Rammers					
– medium	6	1	6	0.20	1.2

Table 35.5 (*Continued*)

Machine or tool	Air consumption per unit (l/s)	Quantity	Maximum air consumption (l/s)	Use factor	Average air consumption (l/s)
– heavy	9	2	18	0.20	3.6
Blow guns	8	3	24	0.10	2.4
Air hoist – 500 kg	33	1	33	0.10	3.3
			81		10.5
Cleaning shop (IV)					
Chipping hammers					
– light	6	2	12	0.35	4.2
– medium	8	3	24	0.35	8.4
– heavy	13	2	26	0.20	5.2
Grinders					
– 75 mm	9	2	18	0.30	5.4
– 150 mm	25	3	75	0.45	33.8
– 200 mm	40	1	40	0.20	8.0
– medium	23	2	46	0.10	4.6
– heavy	42	2	84	0.10	8.4
Sandblast units					
– light	32	1	32	0.50	16.0
– heavy	53	1	53	0.50	26.5
			410		120.5
Total for foundry			700		178.0
Workshop					
Machine shop (V)					
Blow guns	8	10	80	0.05	4.0
Operating cylinders for jibs, fixtures and chucks			12	0.10	1.2
			92		5.2
Sheet metal shop (VI)					
Drills					
– light	6	1	6		1.6
– medium	8	1	8		1.6
– 12 mm	15	2	30		9.0
– angle	8	1	8		1.6
– screwfeed	52	1	52		2.6
Tapper	8	1	8	0.20	1.6
Screwdrivers	8	2	16	0.10	1.6
Impact wrench					
– 20 mm	15	1	15	0.20	3.0
– 22 mm	23	1	23	0.10	2.3
Grinders					
– 150 mm	25	2	50	0.30	15.0
– 200 mm	40	1	40	0.20	8.0
– medium	23	2	46	0.30	13.8
– heavy	42	1	42	0.20	8.4
Riveting hammers					
– medium	18	1	18	0.10	1.8
– heavy	22	1	22	0.05	1.1
Chipping hammers					
– light	6	2	12	0.20	2.4
– medium	8	2	16	0.20	3.2
– heavy	13	1	13	0.10	1.3
Air hoist – 5 ton	97	1	97	0.05	16.2
Blow guns	8	1	16	0.10	1.6
			538		97.3
Assembly shop (VII)					
Drills					
– light	6	3	18	0.20	3.6
– medium	8	5	40	0.30	12.0
– 12 mm	15	6	90	0.35	31.5
– angle	8	2	16	0.10	1.6
– heavy	22	1	22	0.10	2.2

(continued overleaf)

Table 35.5 (*Continued*)

Machine or tool	Air consumption per unit (l/s)	Quantity	Maximum air consumption (l/s)	Use factor	Average air consumption (l/s)
– heavy	33	1	33	0.10	3.3
Tappers	8	2	16	0.10	1.6
Screwdrivers	8	2	16	0.20	3.2
Impact wrenches					
– light	6	1	6	0.20	1.2
– 20 mm	15	2	30	0.20	6.0
– 22 mm	23	1	23	0.10	2.3
Grinders					
– 75 mm	9	2	18	0.20	3.6
– 150 mm	25	1	25	0.10	2.5
– medium	23	2	46	0.20	9.2
Air hoists					
– 500 kg	33	1	33	0.10	3.3
– 1 ton	33	1	33	0.10	3.3
Blowguns	8	5	40	0.05	2.0
			505		92.4
Painting shop (VIII)					
Grinders and polishers					
– angle	8	1	8	0.20	1.6
– medium	23	1	23	0.30	6.9
Sandblast unit	38	1	38	0.50	19.0
Blow guns	8	1	8	0.10	0.8
Air hoist – 5 ton	97	1	97	0.05	4.9
Spray painting guns	5	2	10	0.50	5.0
			184		38.2
Total for the workshop			1319		233.1

there is experience of a similar installation, plotting data obtained from past activity can make an accurate analysis.

35.1.4 Future expansion

Future expansion should always be taken into account when installing new plant. Increasing compressor capacity presents no problem if the rest of the plant installation has been planned accordingly.

35.1.5 Allowance for air leakage

Experience has shown that the initial estimate of the total compressor capacity should include an allowance for leakage. Leakage in the pipelines can be overcome by proper installation practice. A large proportion of the total leakage occurs at hoses, couplings and valves.

For installations with regular inspection and maintenance, a factor of 5 per cent minimum should be adequate. The importance of this is obvious when one remembers that while a tool or appliance may use a considerable amount of air, it is only working intermittently, whereas any leakage, even from a small hole, is both continuous and significant.

35.2 Compressor installation

35.2.1 Type of installation

When planning a compressor installation one of the first matters to be decided is whether there should be a central compressor plant or a number of separate compressors near the main points of use. The following comments can be no more than general. In order to select the type and size of installation that will be adequate for both immediate and future requirements it is advisable to consult the supplier. Points to be considered are as follows.

Centralized installation

1. Lower total installed compressor capacity and, perhaps, lower initial cost;
2. Possibly a higher efficiency, and thus lower power cost, due to larger units;
3. Lower supervision cost.

Decentralized installation

1. Output and/or pressure can be varied to suit each particular plant section.
2. Pipe sizes can be reduced, thus minimizing leakage and cost.
3. Compressors and/or associated equipment can be shut down during periods of low demand or for preventative maintenance with only a localized effect.

35.2.2 Compressor siting

The requirements for a compressor site will be affected by location and climate as well as by the equipment to be installed. The following aspects should be considered.

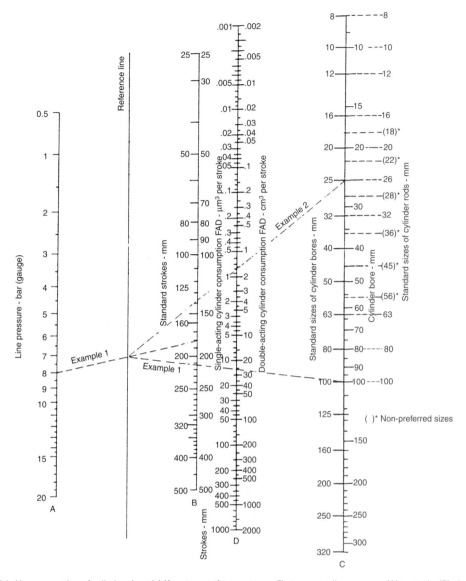

Figure 35.1 Air consumption of cylinders (metric) **How to use the nomogram** First, connect line pressure (A) to stroke (B); then from the point where this cuts the reference line, connect across to cylinder bore size (C). Read off consumption where this line cuts the consumption scale (D). Figures on the left of this scale are for single-acting cylinders. Figures on the right are for double-acting cylinders, neglecting the effect of the rod. This is accurate enough for most purposes. However, if the correct (theoretical) consumption is required for double-acting cylinders, go back to the point on the reference line and connect across to the rod *diameter* size, entered on the 'cylinder bore' scale. This figure should then be deduced from the consumption arrived at with the first solution.

Example 1: Find the (nominal) consumption of a double-acting 100 mm bore cylinder with a stroke of 180 mm operating at 8 bar line pressure.
Answer: 25.4 l per stroke.

Example 2: The cylinder above has a 25 mm diameter rod. Find the true (theoretical) consumption. Connection from the same point on the reference scale to the rod diameter size on the bore scale gives a single-acting cylinder consumption figure of 0.8 l per stroke. Deduce this from the solution found in Example 1 to give the true consumption figure, i.e. 25.4 − 0.8 = 35.6 l per stroke

Foundations

The compressor plant should be located in a place with good ground conditions. In some cases, the compressor foundation may have to be isolated from that of the main building so that vibration is not transmitted from compressor to the building structure, and from heavy plant to the compressor. Where the vibrations are slight, resilient pads may be used to advantage. Receiver-mounted units should always be either freestanding or mounted on resilient pads. If in doubt, the suppliers should be consulted.

Servicing facilities

For small and medium compressors lifting gear is necessary only for installation or re-siting, no special hoisting equipment being normally needed when overhauling the units provided individual components do not exceed 16 kg in weight. On larger units, lifting equipment is essential; the manufacturer or supplier must state the maximum hoisting load. Sufficient access and headroom must be provided around the compressor for servicing.

Weather protection

Adequate protection from the weather must be provided.

Ventilation

Heat generated by the compressor and prime mover must be dispersed. For air-cooled units sited in enclosed rooms this heat must be removed in order to limit the temperature rise. It is sometimes possible to recover this heat for use elsewhere. Intake openings should be located so that dust and other foreign matter do not enter with the air.

Noise

Noise from a compressor plant arises at different sources, and each source has its own pattern of sound pressure levels. Noise levels can be divided into two groups, the low-frequency pulsating air intake sound and the higher-frequency machine noise from compressor, prime mover and fans. Local statutory regulations on noise levels should be determined, and action taken by the supplier to ensure that the noise levels do not exceed those stipulated.

35.2.3 Compressor intake

General

The compressor intake air must be clean and free from solid and gaseous impurities; abrasive dust and corrosive gases are particularly harmful. Exhaust fumes present a hazard if compressed air is required for breathing purposes. The possibility of contamination of the intake by discharge from pressure-relief devices of other plant must be taken into consideration and changes of wind direction must not be overlooked.

For maximum efficiency, the intake air should be as cold as possible. A temperature decrease of 3°C will increase the volume of the delivered air by 1 per cent.

The air intake system should be sized to give a minimum pressure drop. Each compressor should have its own intake filter.

Intake silencing

The reciprocating compressor inspires air in a series of pulsations which causes an equivalent variation in pressure in the intake system. Dependent on the length of the intake pipe, resonance may occur; this can decrease the compressor output and produce disturbing noise levels and stresses sufficient to cause damage. By fitting a pulsation dampener or changing the length of the intake pipe, its natural resonant frequency can be changed, and any related vibration, noise and interference (with the airflow) will be diminished. The inherent pulsation noise can be removed by the use of a suitably designed silencer.

Intake filter

An air intake for a compressor should have a high capacity to remove abrasive materials, including those of small particle size, and good accumulating ability, that is, to collect large quantities of impurities without any significant decrease in filtering efficiency and air flow.

Normally the filter should be placed as close as possible to the compressor. When an intake silencer is fitted, it should be fitted between the intake filter and the compressor.

The filter should also be placed in such a way that it is easily accessible for inspection and cleaning or replacement. The most common types of filter in use are:

1. Paper
2. Oil-wetted labyrinth
3. Woolen cloth
4. Oil-bath

Any of these may be incorporated into or be used in combination with suitable silencers.

For installations in areas of heavy contamination, such as quarries or cement works, additional filtration or automatic self-cleaning is required; otherwise, the air filter will clog up rapidly. Filter condition indicators are available and are recommended.

Intake ducts

The air intake of a compressor should be sited so that, as far as possible, cool, clean, dry air is inspired. When located outdoors the air intake should be protected against the weather. The air intake should be designed and sited so that noise is reduced to the necessary level.

If large compressor plant requires clear headroom for cranes, air intakes may have to run through under floor piping or ducting. Intake ducts must be of a cross-sectional area sufficiently large to avoid excessive pressure drop, and the number of bends should be kept to a minimum. The ducts should be of non-corrosive material and care should be taken that extraneous material cannot enter the duct. The duct should be cleaned thoroughly before connection to the compressor.

Intake pipes may be subject to pulsations and should be too rigidly attached to walls or ceilings, since vibration may be transmitted to the building structure.

Corrosive intake gases

In certain plants, especially in the chemical industry or in the neighborhood of such plants, the air is often polluted with acidic and corrosive gases that can cause corrosion in the compressor and the compressed air system. Special filtration methods and/or materials may have to be used and the supplier should be consulted.

35.2.4 Compressor discharge

Discharge pipe specification and siting

The diameter of the compressor discharge pipe should not normally be smaller than the compressor outlet connection and should be arranged with flanged fittings or unions to permit easy access to the compressor and components at any time. The possibility of vibration should be taken into account. The compressor discharge pipe will attain a high temperature and precautions must be taken to prevent this being a source of danger.

The interior of the pipelines through which the discharge air passes to the aftercooler or air receiver should be cleaned regularly so that a build-up of combustible oily carbon deposits is avoided. All the piping should slope downwards in the direction of airflow to a suitable drain point at the lowest point of the pipeline.

Discharge pipes can be located in trenches covered by floor plants, and there is no technical reason against laying the pipes directly on the ground, but provision must be made for drainage.

Any pocket unavoidably formed after the compressor discharge shall be provided with a drain valve or trap at the lowest point so that any oil and condensate can be removed.

Under certain conditions of installation and operation, pulsations may be set up in the compressor discharge lines. It is essential to consult the suppliers for their recommendations.

Thermoplastics shall not be used for a compressor discharge pipe and inflammable materials shall be kept away from it.

Isolating valve

Where an isolating valve is installed in the discharge pipework the pipeline on the compressor side of the valve shall be protected by a suitable safety valve. This safety valve must be of sufficient size to pass the full output of the compressor without the pressure rising more than 10 per cent above the maximum allowable working pressure.

Multiple compressors

Where two or more compressors feed into a single air line, the discharge line from each compressor shall be fitted with a non-return valve and isolation valve at the furthest point from the compressor or outlet, just prior to where the discharge pipe enters the common manifold feed pipe. A safety valve is fitted on the compressor side of the isolation valve, upstream of the aftercooler.

Non-return valves

Non-return valves used in compressor delivery lines must be designed to withstand the pressure, temperature and pulsations of compressed air.

35.2.5 Cooling-water system

General

Where water is used as a cooling medium for compressor and ancillary equipment it should be within the temperature and pressure levels prescribed by the compressor supplier and should be free from harmful impurities. The cooling water should have a low inlet temperature in order to assist in achieving a high volumetric efficiency in the compressor and to cool the air passing through the aftercooler to a temperature adequate for effective condensation of water vapor.

Overcooling

The compressor should not be overcooled to cause condensation in and on the compressor.

Water quality

Good-quality cooling water is essential.

Re-cooling the cooling water

In order to achieve economy in the use of water it will have to be re-cooled. This is achieved by transferring heat to the ambient air by means of cooling ponds, towers, tanks or mechanical coolers. Temperature regulators may assist control and conserve energy.

Mechanical coolers

The cost of cooling water is an important factor and mechanical coolers are in most cases more economical than allowing the water to run to waste. A forced-draught type of cooler consists of a casing with a water header at the top and a sump at the bottom. A series of cooling elements is provided which offers a large area for the transfer of heat between the water to be cooled and the cooling air. The hot water enters the top header and runs through the elements to a sump from which it is pumped through the compressor plant. A fan forces the colder ambient air through the elements to absorb the heat from the water as it passes through the elements.

Where this type of cooler is installed inside a building, it is essential to duct away the warm air discharged by the fan. Consideration must be given to protection against frost.

Cooling towers

These operate by setting the cooling air in motion over a surface of water. Either natural convection or a fan can do this. In order to provide a good transfer between water and air, towers usually have internal arrangements for spreading the water as a thin film. With cooling towers a final water temperature of about 5°C above the ambient air temperature can be expected. In general, good cooling can be obtained even at relatively high ambient temperatures, since the relative humidity is, in such a case, usually low. However, extreme tropical climates are an exception. The amount of water that is lost as vapor during re-cooling must be replaced by the addition of 'make-up' water. This quantity is considerably smaller than that consumed in open-flow cooling. Cooling towers should not be used in heavily contaminated atmospheres.

Cooling ponds

A cooling-water pond is the simplest form of cooling arrangement. The pond should be located so that an unrestricted air circulation is obtained. Vaporizing ability is improved if the hot water is returned to the pond by some kind of sprinkler device. Most of the cooling effect is caused by vaporization and the water thus lost must be replaced. Cooling ponds should not be used in heavily contaminated atmospheres.

Cooling tanks

A cooling tank is, in effect, a small cooling pond. However, because of the difficulty in keeping the water clean, this method is not recommended.

Keeping the cooling system clean

Cooling water should be free from solid impurities that could damage pumps and cause blockages and filtered, with filters cleaned regularly. The whole cooling system should be inspected and cleaned regularly. Sand, sludge, rust, etc. can be removed by flushing against the normal direction of flow. Lime deposits are more difficult to remove. Such deposits can usually be avoided by keeping the water outlet temperature at a low level. If excessive deposits do occur, a specialist should be called in to clean the system by chemical methods.

35.2.6 Ventilation

In compressor operation, part of the heat given off by the compressor and motor is transmitted to the surrounding air. For plants located in closed rooms, this heat must be removed to limit the rise in temperature of the ambient air. Some of the heat dissipates through walls, windows, floor and roof, but this heat removal is seldom sufficient. The compressor room should be ventilated and the heat removed with the ventilating air. Sometimes the heat can be recovered and used for heating purposes. In an entirely water-cooled compressor installation, the heat to be removed by ventilation is relatively small, since the cooling water takes the major part away.

Insufficient ventilation shortens the life of the electric motor. In installations where the intake air is drawn from the compressor room, poor ventilation may also damage the compressor, as the temperature of the discharge air

increases in proportion to that of the intake air. The compressor room should always be placed so that ventilation air is available without the need for long ducts. The intake should be sited low down on the coldest wall, whereas the ventilation air outlet should be situated high up on the opposite wall in order to avoid temperature stratification.

Modern completely air-cooled compressor plants have aftercoolers with fans. The aftercooler should be arranged so that it assists in the ventilation of the room. For the major part of the year the aftercooler fan will handle room ventilation. Extra fans may be needed only during hot months in the summer.

35.3 Overpressure protection

1. If any equipment having a design pressure lower than the maximum output pressure of the compressor is used, or if an increase of pressure above normal operating pressure will cause a malfunction, it shall be protected against overpressure by suitable means.
2. Any relief valve or safety valve shall have a design flow capacity such that when subjected to the maximum output pressure and flow of the compressor, and taking into account the flow restriction caused by the upstream pipework and fittings, it will prevent the pressure in that part of the system exceeding the design pressure of the equipment. This requirement may, in certain circumstances and depending upon compressor pressure and design pressure of the equipment, imply the need for a relief valve having a port size at least twice the nominal diameter of the pipework and pressure regulator feeding the equipment.
3. An alternative method of protection is the use of a smaller relief valve in conjunction with an automatic isolating valve, which shuts off the air supply to the equipment if the pressure rises more than 20 per cent above the blow-off pressure of the relief valve. If this method is used it is essential to ensure that sudden cessation of air supply to the equipment cannot cause a hazard.

35.4 Selection of compressor plant

35.4.1 Air compressors

The principal types of compressors and their basic characteristics are outlined below (see also Figure 35.2).

Positive displacement compressors

Positive displacement units are those in which successive volumes of air are confined within a closed space and elevated to a higher pressure. The capacity of a positive displacement compressor varies marginally with the working pressure.

Reciprocating compressors

The compressing and displacing element (piston or diaphragm) has a reciprocating motion. The piston

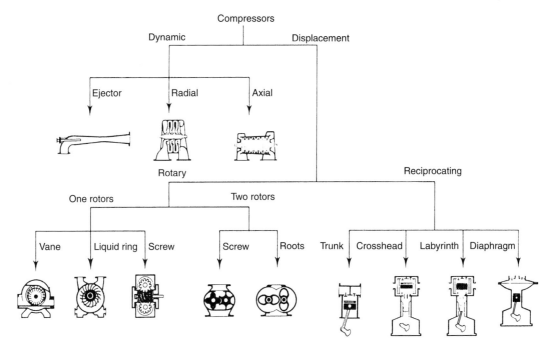

Figure 35.2 Basic compressor types

compressor is available in lubricated and non-lubricated construction.

Helical and spiral-lobe compressors (screw)

Rotary, positive displacement machines in which two intermeshing rotors, each in helical configuration displace and compress the air; available in lubricated and non-lubricated construction; the discharge air is normally free from pulsation; high rotation speed.

Sliding-vane compressors

Rotary, positive displacement machines in which axial vanes slide radially in a rotor mounted eccentrically within a cylindrical casing. Available in lubricated and non-lubricated construction; the discharge air is normally free from pulsation.

Two impeller straight-lobe compressors and blowers

Rotary, positive displacement machines in which two straight, mating but non-touching lobed impellers trap the air and carry it from intake to discharge. Non-lubricated; the discharge is normally free from pulsation; low pressure; high rotation speed.

Dynamic compressors

Dynamic compressors are rotary continuous-flow machines in which the rapidly rotating element accelerates the air as it passes through the element, converting the velocity head into pressure, partially in the rotating element and partially in stationary diffusers or blades. The capacity of a dynamic compressor varies considerably with the working pressure.

Centrifugal compressors

Acceleration of the air is obtained through the action of one or more rotating impellers; non-lubricated; the discharge air is free from pulsation; very high rotation speed.

Axial compressors

Acceleration of the air is obtained through the action of a bladed rotor, shrouded at the blade ends; non-lubricated; very high rotation speed; high-volume output.

Specific power consumption

This varies with the size and type of compressor; consultation with the supplier is advised.

35.4.2 Capacity and pressure limitations

Figure 35.3 shows the approximate capacity and pressure limitations of each type of compressor. There are areas where more than one type of compressor will provide the required capacity and pressure. In such cases, other characteristics such as those given above and the type and pattern of use will govern the selection. Consultation with the supplier is advised.

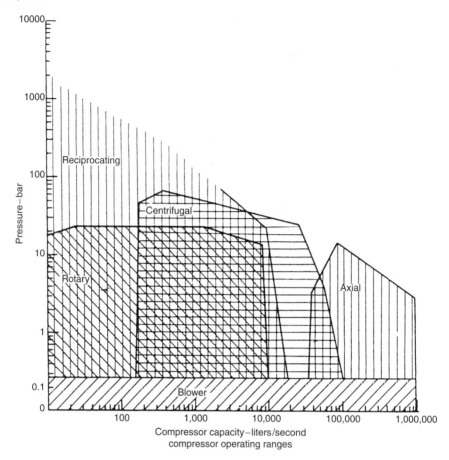

Figure 35.3 Compressor types – approximate capacity and pressure limitations

Compressor standby capacity

On many installations, it is normal to plan the number of compressor units and their output so that there is a standby capacity to permit one unit to be shut down for servicing. Where a constant supply of air is essential to operations, standby compressors are a necessity.

Load splitting

In all installations, consideration should be given to having at least two compressors to allow for conditions of light load and for maintenance.

Closed-loop systems

The same general considerations listed above apply to two-level closed-loop systems but the required input power to the compressor will be considerably reduced or a smaller compressor can be used.

35.4.3 Output control

A wide range of controls is available to match compressor output to demand. Consultation with the equipment supplier is essential. The output of a compressor can be controlled by several methods as outlined below. Pneumatic, hydro-pneumatic and electronic devices can perform the following functions:

1. *Reciprocating compressors*:
   (a) Intermittent operation using automatic stop/start mechanism;
   (b) Constant speed running with inlet valve blocking or intake throttling or external bypass or inlet valve unloading or clearance pocket;
   (c) Variable speed;
   (d) Combinations of (a) through (c) above.
2. *Rotary sliding-vane compressors*:
   (a) Intermittent operation using automatic stop/start mechanism;

(b) Constant-speed running with inlet valve blocking or intake throttling or external bypass;

(c) Variable speed; minimum rotational speed must be high enough to ensure that the blades remain in full contact with the stator.

3. *Rotary screw compressors*:

   (a) Constant speed running with external bypass or intake throttling coupled with blow-off to atmosphere;

   (b) Variable speed.

4. *Dynamics*:

   (a) Constant-speed running with intake throttling coupled with blow-off to atmosphere;

   (b) Variable speed.

Advice should be sought from the supplier as to the best type of control to suit a particular application.

35.4.4 Selection of compressor prime movers

An important factor in obtaining an economical plant is the selection of an appropriate compressor drive. The most common power units are:

Electric motor
Engine (diesel, petrol, gas, etc.)
Turbine (gas, steam, etc.)

Among the advantages of electric motor drive are compactness and ease of control. The internal combustion engine is preferred for mobile units, emergency standby units, or where electric power is not available.

A turbine drive is preferred where it helps balance the energy system of a plant or where the steam or gas can be further used. This type of drive permits easy speed control and conserves energy.

Regardless of the type of prime mover, professional advice should be taken in matching prime mover to compressor.

Application requirements

To avoid delays in the preparation of estimates and unnecessary expense for both buyer and supplier it is important that all necessary data should be available and recorded. The parameters that must be established are outlined below.

Compressor output conditions

1. Volume of free air required (1/s); including an allowance for future expansion;
2. Minimum discharge pressure required to maintain an acceptable working pressure at the point of use;
3. Quality of air; degree of cleanliness required;
4. The purpose for which the air is to be used;
5. The pattern of demand for air; continuous or intermittent consumption;
6. Estimated operating hours per day/week;
7. Type of control;
8. The need for an air receiver;
9. Any special conditions, which the compressor must satisfy;
10. Requirements for ancillary equipment, for example, water pumps, valves, piping, anti-vibration mountings, aftercoolers, dryers, intake filters and silencers, etc.).

36

Compressors

R Keith Mobley
The Plant Performance Group

Contents

A compressor is a machine that is used to increase the pressure of a gas or vapor. They can be grouped into two major classifications: centrifugal and positive displacement. This section provides a general discussion of these types of compressors.

36.1 Centrifugal compressors

In general, the centrifugal designation is used when the gas flow is radial and the energy transfer is predominantly due to a change in the centrifugal forces acting on the gas. The force utilized by the centrifugal compressor is the same as that utilized by centrifugal pumps.

In a centrifugal compressor, air or gas at atmospheric pressure enters the eye of the impeller. As the impeller rotates, the gas is accelerated by the rotating element within the confined space that is created by the volute of the compressor's casing. The gas is compressed as more gas is forced into the volute by the impeller blades. The pressure of the gas increases as it is pushed through the reduced free space within the volute.

As in centrifugal pumps, there may be several stages to a centrifugal air compressor. In these multi-stage units, a progressively higher pressure is produced by each stage of compression.

36.1.1 Configuration

The actual dynamics of centrifugal compressors are determined by their design. Common designs are: overhung or cantilever, centerline, and bullgear.

Overhung or cantilever

The cantilever design is more susceptible to process instability than centerline centrifugal compressors. Figure 36.1 illustrates a typical cantilever design.

The overhung design of the rotor (i.e., no outboard bearing) increases the potential for radical shaft deflection. Any variation in laminar flow, volume, or load of the

inlet or discharge gas forces the shaft to bend or deflect from its true centerline. As a result, the mode shape of the shaft must be monitored closely.

Centerline

Centerline designs (i.e., horizontal and vertical split-case) are more stable over a wider operating range, but should not be operated in a variable-demand system. Figure 36.2 illustrates the normal airflow pattern through a horizontal split-case compressor. Inlet air enters the first stage of the compressor, where pressure and velocity increases occur. The partially compressed air is routed to the second stage where the velocity and pressure are increased further. Adding additional stages until the desired final discharge pressure is achieved can continue this process.

Two factors are critical to the operation of these compressors: impeller configuration and laminar flow, which must be maintained through all of the stages.

The impeller configuration has a major impact on stability and operating envelope. There are two impeller configurations: in-line and back-to-back, or opposed. With the in-line design, all impellers face in the same direction. With the opposed design, impeller direction is reversed in adjacent stages.

In-line

A compressor with all impellers facing in the same direction generates substantial axial forces. The axial pressures generated by each impeller for all the stages are additive. As a result, massive axial loads are transmitted to the fixed bearing. Because of this load, most of these compressors use either a Kingsbury thrust bearing or a balancing piston to resist axial thrusting. Figure 36.3 illustrates a typical balancing piston.

All compressors that use in-line impellers must be monitored closely for axial thrusting. If the compressor

Figure 36.1 Cantilever centrifugal compressor is susceptible to instability

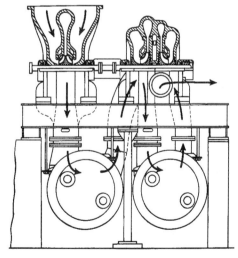

Figure 36.2 Airflow through a centerline centrifugal compressor

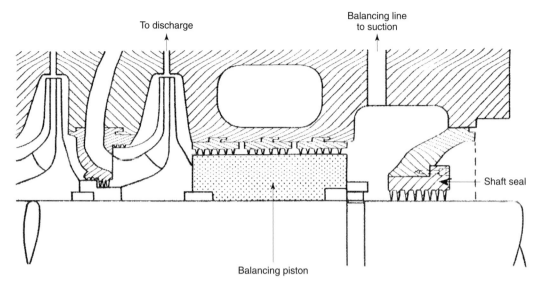

Figure 36.3 Balancing piston resists axial thrust from the in-line impeller design of a centerline centrifugal compressor

is subjected to frequent or constant unloading, the axial clearance will increase due to this thrusting cycle. Ultimately, this frequent thrust loading will lead to catastrophic failure of the compressor.

Opposed

By reversing the direction of alternating impellers, the axial forces generated by each impeller or stage can be minimized. In effect, the opposed impellers tend to cancel the axial forces generated by the preceding stage. This design is more stable and should not generate measurable axial thrusting. This allows these units to contain a normal float and fixed rolling-element bearing.

Bullgear

The bullgear design uses a direct-driven helical gear to transmit power from the primary driver to a series of pinion-gear-driven impellers that are located around the circumference of the bullgear. Figure 36.4 illustrates a typical bullgear compressor layout.

The pinion shafts are typically a cantilever-type design that has an enclosed impeller on one end and a tilting-pad bearing on the other. The pinion gear is between these two components. The number of impeller-pinions (i.e. stages) varies with the application and the original equipment vendor. However, all bullgear compressors contain multiple pinions that operate in series.

Atmospheric air or gas enters the first-stage pinion, where the pressure is increased by the centrifugal force created by the first-stage impeller. The partially compressed air leaves the first stage, passes through an intercooler, and enters the second-stage impeller. This process is repeated until the fully compressed air leaves through the final pinion-impeller, or stage.

Most bullgear compressors are designed to operate with a gear speed of 3600 rpm. In a typical four-stage compressor, the pinions operate at progressively higher speeds. A typical range is between 12,000 rpm (first stage) and 70,000 rpm (fourth stage).

Because of their cantilever design and pinion rotating speeds, bullgear compressors are extremely sensitive to variations in demand or down-stream pressure changes. Because of this sensitivity, their use should be limited to base load applications.

Bullgear compressors are not designed for, nor will they tolerate, load-following applications. They should not be installed in the same discharge manifold with positive-displacement compressors, especially reciprocating compressors. The standing-wave pulses created by many positive-displacement compressors create enough variation in the discharge manifold to cause potentially serious instability.

In addition, the large helical gear used for the bullgear creates an axial oscillation or thrusting that contributes to instability within the compressor. This axial movement is transmitted throughout the machine-train.

36.1.2 Performance

The physical laws of thermodynamics, which define their efficiency and system dynamics, govern compressed-air systems and compressors. This section discusses both the first and second laws of thermodynamics, which apply to all compressors and compressed-air systems. Also applying to these systems are the ideal gas law and the concepts of pressure and compression.

First law of thermodynamics

This law states that energy cannot be created or destroyed during a process, such as compression and delivery of air

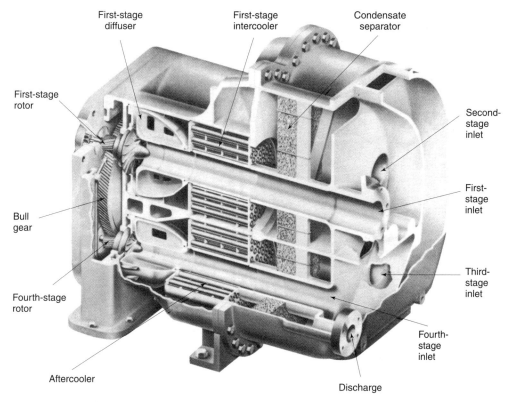

First-stage
diffuser

First-stage
intercooler

Condensate
separator

First-stage
rotor

Second-
stage
inlet

First-
stage
inlet

Bull
gear

Third-
stage
inlet

Fourth-stage
rotor

Fourth-
stage
inlet

Aftercooler

Discharge

Figure 36.4 Bullgear centrifugal compressor

or gas, although it may change from one form of energy to another. In other words, whenever a quantity of one kind of energy disappears, an exactly equivalent total of other kinds of energy must be produced. This is expressed for a steady-flow open system such as a compressor by the following relationship:

| Net energy added to system as heat and work | + | Stored energy of mass entering system | − | Stored energy of mass leaving system | = 0 |

Second law of thermodynamics

The second law of thermodynamics states that energy exists at various levels and is available for use only if it can move from a higher to a lower level. For example, it is impossible for any device to operate in a cycle and produce work while exchanging heat only with bodies at a single fixed temperature. In thermodynamics, a measure of the unavailability of energy has been devised and is known as entropy. As a measure of unavailability, entropy increases as a system loses heat, but remains constant when there is no gain or loss of heat as in an adiabatic process. It is defined by the following differential equation:

$$dS = \frac{dQ}{T}$$

where

T = Temperature (Fahrenheit)
Q = Heat added (BTU)

Pressure/volume/temperature (PVT) relationship

Pressure, temperature, and volume are properties of gases that are completely interrelated. Boyle's law and Charles' law may be combined into one equation that is referred to as the ideal gas law. This equation is always true for ideal gases and is true for real gases under certain conditions.

$$\frac{P_1 V_1}{T_1} = \frac{P_2 V_2}{T_2}$$

For air at room temperature, the error in this equation is less than 1 per cent for pressures as high as 400 psia. For air at one atmosphere of pressure, the error is less than 1 per cent for temperatures as low as −200°F. These error factors will vary for different gases.

Pressure/compression

In a compressor, pressure is generated by pumping quantities of gas into a tank or other pressure vessel. Progressively increasing the amount of gas in the confined or fixed-volume space increases the pressure. The effects of pressure exerted by a confined gas result from the force acting on the container walls. This force is caused by the rapid and repeated bombardment from the enormous number of molecules that are present in a given quantity of gas.

Compression occurs when the space is decreased between the molecules. Less volume means that each particle has a shorter distance to travel, thus proportionately more collisions occur in a given span of time, resulting in a higher pressure. Air compressors are designed to generate particular pressures to meet specific application requirements.

Other performance indicators

The same performance indicators as centrifugal pumps or fans govern centrifugal compressors.

36.1.3 Installation

Dynamic compressors seldom pose serious foundation problems. Since moments and shaking forces are not generated during compressor operation, there are no variable loads to be supported by the foundation. A foundation or mounting of sufficient area and mass to maintain compressor level and alignment and to assure safe soil loading is all that is required. The units may be supported on structural steel if necessary. The principles defined in Section 32.3 for centrifugal pumps also apply to centrifugal compressors.

It is necessary to install pressure-relief valves on most dynamic compressors to protect them due to restrictions placed on casing pressure, power input, and to keep out of its surge range. Always install a valve capable of bypassing the full-load capacity of the compressor between its discharge port and the first isolation valve.

36.1.4 Operating methods

The acceptable operating envelope for centrifugal compressors is very limited. Therefore, care should be taken to minimize any variation in suction supply, back-pressure caused by changes in demand, and frequency of unloading. The operating guidelines provided in the compressor vendor's O&M manual should be followed to prevent abnormal operating behavior or premature wear or failure of the system.

Centrifugal compressors are designed to be base loaded and may exhibit abnormal behavior or chronic reliability problems when used in a load-following mode of operation. This is especially true of bullgear and cantilever compressors. For example, a 1-psig change in discharge pressure may be enough to cause catastrophic failure of a bullgear compressor. Variations in demand or back-pressure on a cantilever design can cause the entire rotating element and its shaft to flex. This not only affects the compressor's efficiency, but also accelerates wear and may lead to premature shaft or rotor failure.

All compressor types have moving parts, high noise levels, high pressures, and high-temperature cylinder and discharge-piping surfaces.

36.2 Positive displacement compressors

Positive-displacement compressors can be divided into two major classifications: rotary and reciprocating.

Rotary

The rotary compressor is adaptable to direct drive by the use of induction motors or multi-cylinder gasoline or diesel engines. These compressors are compact, relatively inexpensive, and require a minimum of operating attention and maintenance. They occupy a fraction of the space and weight of a reciprocating machine having equivalent capacity.

36.2.1 Configuration

Rotary compressors are classified into three general groups: sliding vane, helical lobe, and liquid-seal ring.

Sliding vane

The basic element of the sliding-vane compressor is the cylindrical housing and the rotor assembly. This compressor, which is illustrated in Figure 36.5, has longitudinal vanes that slide radially in a slotted rotor mounted eccentrically in a cylinder. The centrifugal force carries the sliding vanes against the cylindrical case with the vanes forming a number of individual longitudinal cells in the eccentric annulus between the case and rotor. The suction port is located where the longitudinal cells are largest. The size of each cell is reduced by the eccentricity of the rotor as the vanes approach the discharge port, thus compressing the gas.

Cyclical opening and closing of the inlet and discharge ports occurs by the rotor's vanes passing over them. The inlet port is normally a wide opening that is designed to admit gas in the pocket between two vanes. The port closes momentarily when the second vane of each air-containing pocket passes over the inlet port.

When running at design pressure, the theoretical operation curves (see Figure 36.6) are identical to a reciprocating compressor. However, there is one major difference between a sliding-vane and a reciprocating compressor. The reciprocating unit has spring-loaded valves that open automatically with small pressure differentials between the outside and inside cylinder. The sliding-vane compressor has no valves.

The fundamental design considerations of a sliding-vane compressor are the rotor assembly, cylinder housing, and the lubrication system.

Housing and rotor assembly

Cast iron is the standard material used to construct the cylindrical housing, but other materials may be used if corrosive conditions exist. The rotor is usually a continuous piece of steel that includes the shaft and is made from bar stock. Special materials can be selected for corrosive applications. Occasionally, the rotor may be a separate iron casting keyed to a shaft. On most standard air compressors, the rotor-shaft seals are semi-metallic packing in

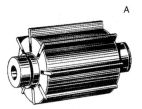

Rotor with non-metallic
sliding vanes.

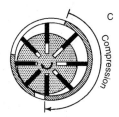

Gas is gradually compressed
as pockets get smaller.

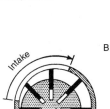

As rotor turns, gas is trap-
in pockets formed by vanes.

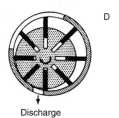

Discharge

Compressed gas is pushed out
through discharge port.

Figure 36.5 Rotary sliding-vane compressor

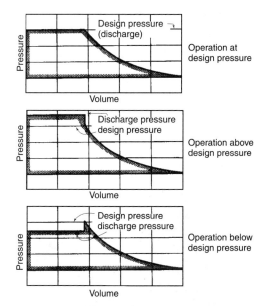

Figure 36.6 Theoretical operation curves for rotary compressors with built-in porting

a stuffing box. Commercial mechanical rotary seals can be supplied when needed. Cylindrical roller bearings are generally used in these assemblies.

Vanes are usually asbestos or cotton cloth impregnated with a phenolic resin. Bronze or aluminum also may be used for vane construction. Each vane fits into a milled slot extending the full length of the rotor and slides radially in and out of this slot once per revolution. Vanes are the most maintenance prone part in the compressor. There are from 8 to 20 vanes on each rotor, depending upon its diameter. A greater number of vanes increase compartmentalization, which reduces the pressure differential across each vane.

Lubrication system

A V-belt-driven, force-fed oil lubrication system is used on water-cooled compressors. Oil goes to both bearings and to several points in the cylinder. Ten times as much oil is recommended to lubricate the rotary cylinder as is required for the cylinder of a corresponding reciprocating compressor. The oil carried over with the gas to the line may be reduced 50 per cent with an oil separator on the discharge. Use of an aftercooler ahead of the separator permits removal of 85 to 90 per cent of the entrained oil.

Helical lobe or screw

The helical lobe, or screw, compressor is shown in Figure 36.7. It has two or more mating sets of lobe-type rotors mounted in a common housing. The male lobe, or rotor, is usually direct-driven by an electric motor. The female lobe, or mating rotor, is driven by a helical gear set that is mounted on the outboard end of the rotor shafts. The gears provide both motive power for the female rotor and absolute timing between the rotors.

The rotor set has extremely close mating clearance (i.e., about 0.5 mils) but no metal-to-metal contact. Most of

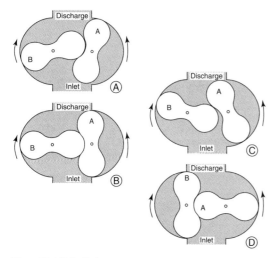

Figure 36.7 Helical lobe, or screw, rotary air compressor

these compressors are designed for oil-free operation. In other words, no oil is used to lubricate or seal the rotors. Instead, oil lubrication is limited to the timing gears and bearings that are outside the air chamber. Because of this, maintaining proper clearance between the two rotors is critical.

This type of compressor is classified as a constant volume, variable-pressure machine that is quite similar to the vane-type rotary in general characteristics. Both have a built-in compression ratio.

Helical-lobe compressors are best suited for base-load applications where they can provide a constant volume and pressure of discharge gas. The only recommended method of volume control is the use of variable-speed motors. With variable-speed drives, capacity variations can be obtained with a proportionate reduction in speed. A 50 per cent speed reduction is the maximum permissible control range.

Helical-lobe compressors are not designed for frequent or constant cycles between load and no-load operation. Each time the compressor unloads, the rotors tend to thrust axially. Even though the rotors have a substantial thrust bearing and, in some cases, a balancing piston to counteract axial thrust, the axial clearance increases each time the compressor unloads. Over time, this clearance will increase enough to permit a dramatic rise in the impact energy created by axial thrust during the transient from loaded to unloaded conditions. In extreme cases, the energy can be enough to physically push the rotor assembly through the compressor housing.

Compression ratio and maximum inlet temperature determine the maximum discharge temperature of these compressors. Discharge temperatures must be limited to prevent excessive distortion between the inlet and discharge ends of the casing and rotor expansion. High-pressure units are water-jacketed in order to obtain uniform casing temperature. Rotors also may be cooled to permit a higher operating temperature.

If either casing distortion or rotor expansion occur, the clearance between the rotating parts will decrease and metal-to-metal contact will occur. Since the rotors typically rotate at speeds between 3600 and 10,000 rpm, metal-to-metal contact normally results in instantaneous, catastrophic compressor failure.

Changes in differential pressures can be caused by variations in either inlet or discharge conditions (i.e., temperature, volume or pressure). Such changes can cause the rotors to become unstable and change the load zones in the shaft-support bearings. The result is premature wear and/or failure of the bearings.

Always install a relief valve that is capable of bypassing the full-load capacity of the compressor between its discharge port and the first isolation valve. Since helical-lobe compressors are less tolerant to over-pressure operation, safety valves are usually set within 10 per cent of absolute discharge pressure, or 5 psi, whichever is lower.

Liquid-seal ring

The liquid-ring or liquid-piston compressor is shown in Figure 36.8. It has a rotor with multiple forward-turned blades that rotate about a central cone that contains inlet and discharge ports. Liquid is trapped between adjacent blades, which drive the liquid around the inside of an elliptical casing. As the rotor turns, the liquid face moves in and out of this space due to the casing shape, creating a liquid piston. Porting in the central cone is built-in and fixed and there are no valves.

Compression occurs within the pockets or chambers between the blades before the discharge port is uncovered. Since the port location must be designed and built for a specific compression ratio, it tends to operate above or below the design pressure.

Liquid-ring compressors are cooled directly rather than by jacketed casing walls. The cooling liquid is fed into the casing where it comes into direct contact with the gas being compressed. The excess liquid is discharged with the gas. The discharged mixture is passed through a conventional baffle or centrifugal-type separator to remove the free liquid. Because of the intimate contact of gas

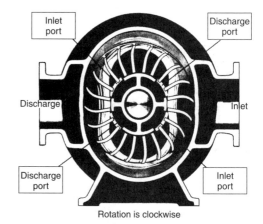

Figure 36.8 Liquid-seal ring rotary air compressor

and liquid, the final discharge temperature can be held close to the inlet cooling water temperature. However, the discharge gas is saturated with liquid at the discharge temperature of the liquid.

The amount of liquid passed through the compressor is not critical and can be varied to obtain the desired results. The unit will not be damaged if a large quantity of liquid inadvertently enters its suction port.

Lubrication is required only in the bearings, which are generally located external to the casing. The liquid itself acts as a lubricant, sealing medium, and coolant for the stuffing boxes.

36.2.2 Performance

Performance of a rotary positive-displacement compressor can be evaluated using the same criteria as a positive-displacement pump. As constant-volume machines, performance is determined by rotation speed, internal slip, and total back-pressure on the compressor.

The volumetric output of rotary positive-displacement compressors can be controlled by speed changes. The slower the compressor turns, the lower its output volume. This feature permits the use of these compressors in load-following applications. However, care must be taken to prevent sudden radical changes in speed.

Internal slip is simply the amount of gas that can flow through internal clearances from the discharge back to the inlet. Obviously, internal wear will increase internal slip.

Discharge pressure is relatively constant regardless of operating speed. With the exceptions of slight pressure variations caused by atmospheric changes and back-pressure, a rotary positive-displacement compressor will provide a fixed discharge pressure. Back-pressure, which is caused by restrictions in the discharge piping or demand from users of the compressed air or gas, can have a serious impact on compressor performance.

If back-pressure is too low or demand too high, the compressor will be unable to provide sufficient volume or pressure to the down-stream systems. In this instance, the discharge pressure will be noticeably lower than designed.

If the back-pressure is too high or demand too low, the compressor will generate a discharge pressure higher than designed. It will continue to compress the air or gas until it reaches the unload setting on the system's relief valve or until the brake horsepower required exceeds the maximum horsepower rating of the driver.

36.2.3 Installation

Installation requirements for rotary positive-displacement compressors are similar to any rotating machine. Review the installation requirements for centrifugal pumps and compressors for foundation, pressure-relief, and other requirements. As with centrifugal compressors, rotary positive-displacement compressors must be fitted with pressure-relief devices to limit the discharge or inter-stage pressures to a safe maximum for the equipment served.

In applications where demand varies, rotary positive-displacement compressors require a down-stream receiver tank or reservoir that minimizes the load-unload cycling frequency of the compressor. The receiver tank should have sufficient volume to permit acceptable unload frequencies for the compressor. Refer to the vendor's O&M manual for specific receiver-tank recommendations.

36.2.4 Operating methods

All compressor types have moving parts, high noise levels, high pressures, and high-temperature cylinder and discharge-piping surfaces.

Rotary positive-displacement compressors should be operated as base loaded units. They are especially sensitive to the repeated start-stop operation required by load-following applications. Generally, rotary positive-displacement compressors are designed to unload about every six to eight hours. This unload cycle is needed to dissipate the heat generated by the compression process. If the unload frequency is too great, these compressors have a high probability of failure.

There are several primary operating control inputs for rotary positive-displacement compressors. These control inputs are: discharge pressure, pressure fluctuations, and unloading frequency.

Discharge pressure

This type of compressor will continue to compress the air volume in the down-stream system until: (1) some component in the system fails, (2) the brake horsepower exceeds the driver's capacity, or (3) a safety valve opens. Therefore, the operator's primary control input should be the compressor's discharge pressure. If the discharge pressure is below the design point, it is a clear indicator that the total down-stream demand is greater than the unit's capacity. If the discharge pressure is too high, the demand is too low and excessive unloading will be required to prevent failure.

Pressure fluctuations

Fluctuations in the inlet and discharge pressures indicate potential system problems that may adversely affect performance and reliability. Pressure fluctuations are generally caused by changes in the ambient environment, turbulent flow, or restrictions caused by partially blocked inlet filters. Any of these problems will result in performance and reliability problems if not corrected.

Unloading frequency

The unloading function in rotary positive-displacement compressors is automatic and not under operator control. Generally, a set of limit switches, one monitoring internal temperature and one monitoring discharge pressure, are used to trigger the unload process. By design, the limit switch that monitors the compressor's internal temperature is the primary control. The secondary control, or discharge-pressure switch, is a fail-safe design to prevent overloading the compressor.

Depending on design, rotary positive-displacement compressors have an internal mechanism designed to minimize the axial thrust caused by the instantaneous change from fully-loaded to unloaded operating conditions. In some designs, a balancing piston is used to absorb the rotor's thrust during this transient. In others, over-sized thrust bearings are used.

Regardless of the mechanism used, none provide complete protection from the damage imparted by the transition from load to no-load conditions. However, as long as the unload frequency is within design limits, this damage will not adversely affect the compressor's useful operating life or reliability. However, an unload frequency greater than that accommodated in the design will reduce the useful life of the compressor and may lead to premature, catastrophic failure.

Operating practices should minimize, as much as possible, the unload frequency of these compressors. Installation of a receiver tank and modification of user-demand practices are the most effective solutions to this type of problem.

36.3 Reciprocating compressors

Reciprocating compressors are widely used by industry and are offered in a wide range of sizes and types. They vary from units requiring less than 1 Hp to more than 12,000 Hp. Pressure capabilities range from low vacuums at intake to special compressors capable of 60,000 psig or higher.

Reciprocating compressors are classified as constant-volume, variable-pressure machines. They are the most efficient type of compressor and can be used for partial-load, or reduced-capacity, applications.

Because of the reciprocating pistons and unbalanced rotating parts, the unit tends to shake. Therefore, it is necessary to provide a mounting that stabilizes the installation. The extent of this requirement depends on the type and size of the compressor.

Because reciprocating compressors should be supplied with clean gas, inlet filters are recommended in all applications. They cannot satisfactorily handle liquids entrained in the gas, although vapors are no problem if condensation within the cylinders does not take place. Liquids will destroy the lubrication and cause excessive wear.

Reciprocating compressors deliver a pulsating flow of gas that can damage downstream equipment or machinery. This is sometimes a disadvantage, but pulsation dampers can be used to alleviate the problem.

36.3.1 Configuration

Certain design fundamentals should be clearly understood before analyzing the operating condition of reciprocating compressors. These fundamentals include frame and running gear, inlet and discharge valves, cylinder cooling, and cylinder orientation.

Frame and running gear

Two basic factors guide frame and running gear design. The first factor is the maximum horsepower to be transmitted through the shaft and running gear to the cylinder pistons. The second factor is the load imposed on the

frame parts by the pressure differential between the two sides of each piston. This is often called pin load because this full force is directly exerted on the crosshead and crankpin. These two factors determine the size of bearings, connecting rods, frame, and bolts that must be used throughout the compressor and its support structure.

Cylinder design

Compression efficiency depends entirely upon the design of the cylinder and its valves. Unless the valve area is sufficient to allow gas to enter and leave the cylinder without undue restriction, efficiency cannot be high. Valve placement for free flow of the gas in and out of the cylinder is also important.

Both efficiency and maintenance are influenced by the degree of cooling during compression. The method of cylinder cooling must be consistent with the service intended.

The cylinders and all the parts must be designed to withstand the maximum application pressure. The most economical materials that will give the proper strength and the longest service under the design conditions are generally used.

Inlet and discharge valves

Compressor valves are placed in each cylinder to permit one-way flow of gas, either into or out of the cylinder. There must be one or more valve(s) for inlet and discharge in each compression chamber.

Each valve opens and closes once for each revolution of the crankshaft. The valves in a compressor operating at 700 rpm for 8 hours per day and 250 days per year will have cycled (i.e., opened and closed) 42,000 times per hour, 336,000 times per day, or 84 million times in a year. The valves have less than $\frac{1}{10}$ of a second to open, let the gas pass through, and to close.

They must cycle with a minimum of resistance for minimum power consumption. However, the valves must have minimal clearance to prevent excessive expansion and reduced volumetric efficiency. They must be tight under extreme pressure and temperature conditions. Finally, the valves must be durable under many kinds of abuse.

There are four basic valve designs used in these compressors: finger, channel, leaf, and annular ring. Within each class, there may be variations in design, depending upon operating speed and size of valve required.

Finger

Figure 36.9 is an exploded view of a typical finger valve. These valves are used for smaller, air-cooled compressors. One end of the finger is fixed and the opposite end lifts when the valve opens.

Channel

The channel valve shown in Figure 36.10 is widely used in mid- to large-sized compressors. This valve uses a series of separate stainless steel channels. As explained in

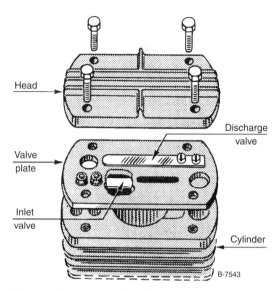

Figure 36.9 Finger valve configuration

the figure, this is a cushioned valve, which adds greatly to its life.

Leaf

The leaf valve (see Figure 36.11) has a configuration somewhat like the channel valve. It is made of flat-strip steel that opens against an arched stop plate. This results in valve flexing only at its center with maximum lift. The valve operates as its own spring.

Annular ring

Figure 36.12 shows exploded views of typical inlet and discharge annular-ring valves. The valves shown have a single ring, but larger sizes may have two or three rings. In some designs, the concentric rings are tied into a single piece by bridges.

The springs and the valve move into a recess in the stop plate as the valve opens. Gas that is trapped in the recess acts as a cushion and prevents slamming. This eliminates a major source of valve and spring breakage. The valve shown was the first cushioned valve built.

Cylinder cooling

Cylinder heat is produced by the work of compression plus friction, which is caused by the action of the piston and piston rings on the cylinder wall and packing on the rod. The amount of heat generated can be considerable, particularly when moderate to high compression ratios are involved. This can result in undesirably high operating temperatures.

Most compressors use some method to dissipate a portion of this heat to reduce the cylinder wall and discharge gas temperatures. The following are advantages of cylinder cooling:

Figure 36.10 Channel valve configuration

- Lowering cylinder wall and cylinder head temperatures reduces loss of capacity and horsepower per unit volume due to suction gas preheating during inlet stroke. This results in more gas in the cylinder for compression.
- Reducing cylinder wall and cylinder head temperatures removes more heat from the gas during compression, lowering its final temperature and reducing the power required.
- Reducing the gas temperature and that of the metal surrounding the valves results in longer valve service life and reduces the possibility of deposit formation.
- Reduced cylinder wall temperature promotes better lubrication, resulting in longer life and reduced maintenance.
- Cooling, particularly water cooling, maintains a more even temperature around the cylinder bore and reduces warpage.

Cylinder orientation

Orientation of the cylinders in a multi-stage or multi-cylinder compressor directly affects the operating

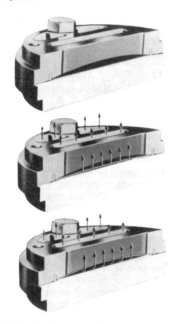

Figure 36.11 Leaf spring configuration

Figure 36.12 Annular-ring valves

dynamics and vibration level. Figure 36.13 illustrates a typical three-piston, air-cooled compressor. Since three pistons are oriented within a 120° arc, this type of compressor generates higher vibration levels than the opposed piston compressor illustrated in Figure 36.14.

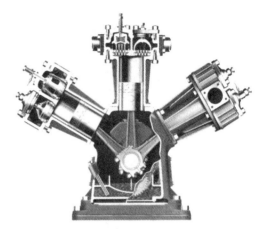

Figure 36.13 Three-piston compressor generates higher vibration levels

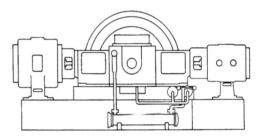

Figure 36.14 Opposed-piston compressor balances piston forces

36.3.2 Performance

Reciprocating-compressor performance is governed almost exclusively by operating speed. Each cylinder of the compressor will discharge the same volume, excluding slight variations caused by atmospheric changes, at the same discharge pressure each time it completes the discharge stroke. As the rotation speed of the compressor changes, so does the discharge volume.

The only other variables that affect performance are the inlet-discharge valves, which control flow into and out of each cylinder. Although reciprocating compressors can use a variety of valve designs, it is crucial that the valves perform reliably. If they are damaged and fail to operate at the proper time or do not seal properly, overall compressor performance will be substantially reduced.

36.3.3 Installation

A carefully planned and executed installation is extremely important and makes compressor operation and maintenance easier and safer. Key components of a compressor installation are: location, foundation, and piping.

Location

The preferred location for any compressor is near the center of its load. However, the choice is often influenced by the cost of supervision, which can vary by location. The on-going cost of supervision may be less expensive at a less-optimum location, which can offset the cost of longer piping.

A compressor will always give better, more reliable service when enclosed in a building that protects it from cold, dusty, damp, and corrosive conditions. In certain locations, it may be economical to use a roof only, but this is not recommended unless the weather is extremely mild. Even then, it is crucial to prevent rain and wind-blown debris from entering the moving parts. Subjecting a compressor to adverse inlet conditions will dramatically reduce reliability and significantly increase maintenance requirements.

Ventilation around a compressor is vital. On a motor-driven, air-cooled unit, the heat radiated to the surrounding air is at least 65 per cent of the power input. On a water-jacketed unit with an aftercooler and outside receiver, the heat radiated to the surrounding air may be 15 to 25 per cent of the total energy input, which is still a substantial amount of heat. Positive outside ventilation is recommended for any compressor room where the ambient temperature may exceed 104°F.

Foundation

Because of the alternating movement of pistons and other components, reciprocating compressors often develop a shaking that alternates in direction. This force must be damped and contained by the mounting. The foundation also must support the weight load of the compressor and its driver.

There are many compressor arrangements and the net magnitude of the moments and forces developed can vary a great deal among them. In some cases, they are partially or completely balanced within the compressors themselves. In others, the foundation must handle much of the force. When complete balance is possible, reciprocating compressors can be mounted on a foundation just large and rigid enough to carry the weight and maintain alignment. However, most reciprocating compressors require larger, more massive foundations than other machinery.

Depending upon size and type of unit, the mounting may vary from simply bolting to the floor to attaching to a massive foundation designed specifically for the application. A proper foundation must: (1) maintain the alignment and level of the compressor and its driver at the proper elevation, and (2) minimize vibration and prevent its transmission to adjacent building structures and machinery. There are five steps to accomplish the first objective:

1. The safe weight-bearing capacity of the soil must not be exceeded at any point on the foundation base.
2. The load to the soil must be distributed over the entire area.
3. The size and proportion of the foundation block must be such that the resultant vertical load due to the compressor, block, and any unbalanced force falls within the base area.

4. The foundation must have sufficient mass and weight-bearing area to prevent its sliding on the soil due to unbalanced forces.
5. Foundation temperature must be uniform to prevent warping.

Bulk is not usually the complete solution to foundation problems. A certain weight is sometimes necessary, but soil area is usually of more value than foundation mass.

Determining if two or more compressors should have separate or single foundations depends on the compressor type. A combined foundation is recommended for reciprocating units since the forces from one unit usually will partially balance out the forces from the others. In addition, the greater mass and surface area in contact with the ground damps foundation movement and provides greater stability.

Soil quality may vary seasonally and such conditions must be carefully considered in the foundation design. No foundation should rest partially on bedrock and partially on soil; it should rest entirely on one or the other. If placed on the ground, make sure that part of the foundation does not rest on soil that has been disturbed. In addition, pilings may be necessary to ensure stability.

Piping

Piping should easily fit the compressor connections without needing to spring or twist it to fit. It must be supported independently of the compressor and anchored, as necessary, to limit vibration and to prevent expansion strains. Improperly installed piping may distort or pull the compressor's cylinders or casing out of alignment.

Air inlet

The intake pipe on an air compressor should be as short and direct as possible. If the total run of the inlet piping is unavoidably long, the diameter should be increased. The pipe size should be greater than the compressor's air-inlet connection.

Cool inlet air is desirable. For every 5°F of ambient air temperature reduction, the volume of compressed air generated increases by 1 per cent with the same power consumption. This increase in performance is due to the greater density of the intake air.

It is preferable for the intake air to be taken from outdoors. This reduces heating and air conditioning costs and, if properly designed, has fewer contaminants. However, the intake piping should be a minimum of 6 feet above the ground and be screened or, preferably, filtered. An air inlet must be free of steam and engine exhausts. The inlet should be hooded or turned down to prevent the entry of rain or snow. It should be above the building eaves and several feet from the building.

Discharge

Discharge piping should be the full size of the compressor's discharge connection. The pipe size should not be reduced until the point along the pipeline is reached where the flow has become steady and non-pulsating. With a

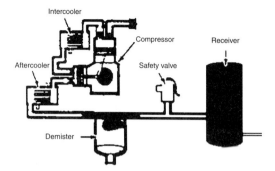

Figure 36.15 Safety valve function

reciprocating compressor, this is generally beyond the aftercooler or the receiver. Pipes to handle non-pulsating flow are sized by normal methods and long-radius bends are recommended. All discharge piping must be designed to allow adequate expansion loops or bends to prevent undue stresses at the compressor.

Drainage

Before piping is installed, the layout should be analyzed to eliminate low points where liquid could collect and to provide drains where low points cannot be eliminated. A regular part of the operating procedure must be the periodic drainage of low points in the piping and separators, as well as inspection of automatic drain traps.

Pressure-relief valves

All reciprocating compressors must be fitted with pressure relief devices to limit the discharge or inter-stage pressures to a safe maximum for the equipment served.

Always install a relief valve that is capable of bypassing the full-load capacity of the compressor between its discharge port and the first isolation valve. The safety valves should be set to open at a pressure slightly higher than the normal discharge-pressure rating of the compressor. For standard 100 to 115 psig two-stage air compressors, safety valves are normally set at 125 psig.

The pressure-relief safety valve is normally situated on top of the air reservoir and there must be no restriction on its operation. The valve is usually of the 'huddling chamber' design in which the static pressure acting on its disk area causes it to open. Figure 36.15 illustrates how such a valve functions. As the valve pops, the air space within the huddling chamber between the seat and blow down ring fills with pressurized air and builds up more pressure on the roof of the disk holder. This temporary pressure increases the upward thrust against the spring, causing the disk and its holder to fully pop open.

Once a predetermined pressure drop (i.e., blow down) occurs, the valve closes with a positive action by trapping pressurized air on top of the disk holder. Raising or lowering the blowdown ring adjusts the pressure-drop set point. Raising the ring increases the pressure-drop setting, while lowering it decreases the setting.

36.3.4 Operating methods

Compressors can be hazardous to work around because they have moving parts. Ensure that clothing is kept away from belt drives, couplings, and exposed shafts. In addition, high-temperature surfaces around cylinders and discharge piping are exposed. Compressors are notoriously noisy, so ear protection should be worn. These machines are used to generate high-pressure gas so, when working around them, it is important to wear safety glasses and to avoid searching for leaks with bare hands. High-pressure leaks can cause severe friction burns.

37

Fans and Blowers

R Keith Mobley

The Plant Performance Group

Contents

Technically, fans and blowers are two separate types of devices that have a similar function. However, the terms are often used interchangeably to mean any device that delivers a quantity of air or gas at a desired pressure. Differences between these two devices are their rotating elements and their discharge-pressure capabilities.

37.1 Centrifugal fans

The centrifugal fan is one of the most common machines used in industry. It utilizes a rotating element with blades, vanes, or propellers to extract or deliver a specific volume of air or gas. The rotating element is mounted on a rotating shaft that must provide the energy required to overcome inertia, friction, and other factors that restrict or resist air or gas flow in the application. They are generally low-pressure machines designed to overcome friction and either suction or discharge-system pressure.

37.1.1 Configuration

The type of rotating element or wheel that is used to move the air or gas can classify centrifugal fans. The major classifications are propeller and axial. Axial fans also can be further differentiated by the blade configurations.

Propeller

This type of fan consists of a propeller, or paddle wheel, mounted a rotating shaft within a ring, panel, or cage. The most widely used propeller fans are found in light- or medium-duty functions, such as ventilation units where air can be moved in any direction. These fans are commonly used in wall mountings to inject air into, or exhaust air from, a space. Figure 37.1 illustrates a belt-driven propeller fan appropriate for medium-duty applications.

This type of fan has a limited ability to boost pressure. Its use should be limited to applications where the total resistance to flow is less than 1 inch of water. In addition, it should not be used in corrosive environments or where explosive gases are present.

Axial

Axial fans are essentially propeller fans that are enclosed within a cylindrical housing or shroud. They can be mounted inside ductwork or a vessel housing to inject or exhaust air or gas.

These fans have an internal motor mounted on spokes or struts to centralize the unit within the housing. Electrical connections and grease fittings are mounted externally on the housing. Arrow indicators on the housing show the direction of air flow and rotation of the shaft which enables the unit to be correctly installed in the ductwork. Figure 37.2 illustrates an inlet end of a direct-connected, tube-axial fan.

This type of fan should not be used in corrosive or explosive environments since the motor and bearings cannot be protected. Applications where concentrations of airborne abrasives are present should also be avoided.

Axial fans use three primary types of blades or vanes: backward-curved, forward-curved, and radial. Each type has specific advantages and disadvantages.

Backwards-curved blades

The backward-curved blade provides the highest efficiency and lowest sound level of all axial-type centrifugal fan blades. Advantages include:

- Moderate to high volumes;
- Static pressure range up to approximately 30 inches of water (gage);
- Highest efficiency of any type of fan;
- Lowest noise level of any fan for the same pressure and volumetric requirements;
- Self-limiting BHP characteristics. (Motors can be selected to prevent overload at any volume and the BHP curve rises to a peak and then declines as volume increases).

The limitations of backward-curved blades are:

- Weighs more and occupies considerably more space than other designs of equal volume and pressure;
- Large wheel width;

Figure 37.1 Belt-driven propeller fan for medium duty applications

Figure 37.2 Inlet end of a direct-connected tube-axial fan

- Not to be used in dusty environments or where sticky or stringy materials are used because residues adhering to the blade surface cause imbalance and eventual bearing failure.

Forward-curved blades

This design is commonly referred to as a squirrel-cage fan. The unit has a wheel with a large number of wide, shallow blades; a very large intake area relative to the wheel diameter; and a relatively slow operational speed. The advantages of forward-curved blades include:

- Excellent for any volume at low to moderate static pressure using clean air;
- Occupies approximately same space as backward-curved blade fan;
- More efficient and much quieter during operation than propeller fans for static pressures above approximately 1 inch of water (gage).

The limitations of forward-curved blades include:

- Not as efficient as backward-curved blade fans;
- Should not be used in dusty environments or handle sticky or stringy materials that could adhere to the blade surface;
- BHP increases as this fan approaches maximum volume as opposed to backward-curved blade centrifugal fans, which experience a decrease in BHP as they approach maximum volume.

Radial blades

Industrial exhaust fans fall into this category. The design is rugged and may be belt-driven or directly driven by a motor. The blade shape varies considerably from flat surfaces to various bent configurations to increase efficiency slightly or to suit particular applications. The advantages of radial-blade fans include:

- Best suited for severe duty, especially when fitted with flat radial blades;
- Simple construction that lends itself to easy field maintenance;
- Highly versatile industrial fan that can be used in extremely dusty environments as well as clean air;
- Appropriate for high-temperature service;
- Handles corrosive or abrasive materials.

The limitations of radial-blade fans include:

- Lowest efficiency in centrifugal-fan group;
- Highest sound level in centrifugal-fan group;
- BHP increases as fan approaches maximum volume.

37.1.2 Performance

A fan is inherently a constant-volume machine. It operates at the same volumetric flow rate (i.e., cubic feet per minute) when operating in a fixed system at a constant speed, regardless of changes in air density. However, the pressure developed and the horsepower required vary directly with the air density.

The following factors affect centrifugal-fan performance: brake horsepower, fan capacity, fan rating, outlet velocity, static efficiency, static pressure, tip speed, mechanical efficiency, total pressure, velocity pressure, natural frequency, and suction conditions. Some of these factors are used in the mathematical relationships that are referred to as 'fan laws'.

Brake horsepower

Brake horsepower (BHP) is the power input required by the fan shaft to produce the required volumetric flow rate (cfm) and pressure.

Fan capacity

The fan capacity (FC) is the volume of air moved per minute by the fan (cfm). *Note*: the density of air is 0.075 pounds per cubic foot at atmospheric pressure and 68°F.

Fan rating

The fan rating predicts the fan's performance at one operating condition, which includes the fan size, speed, capacity, pressure and horsepower.

Outlet velocity

The outlet velocity (OV, feet per minute) is the number of cubic feet of gas moved by the fan per minute divided by the inside area of the fan outlet, or discharge area, in square feet.

Static efficiency

Static efficiency (SE) is not the true mechanical efficiency, but is convenient to use in comparing fans. This is calculated by the following equation:

$$\text{Static efficiency (SE)} = \frac{0.000157 \times \text{FC} \times \text{SP}}{\text{BHP}}$$

Static pressure

Static pressure (SP) generated by the fan can exist whether the air is in motion or is trapped in a confined space. SP is always expressed in inches of water (gage).

Tip speed

The tip speed (TS) is the peripheral speed of the fan wheel in feet per minute (fpm).

$$\text{Tip speed} = \text{Rotor diameter} \times \pi \times \text{RPM}$$

Mechanical efficiency

True mechanical efficiency (ME) is equal to the total input power divided by the total output power.

Total pressure

Total pressure (TP), inches of water (gage) is the sum of the velocity pressure and static pressure.

Velocity pressure

Velocity pressure (VP) is produced by the fan only when the air is moving. Air having a velocity of 4000 fpm exerts a pressure of 1 inch of water (gage) on a stationary object in its flow path.

Natural frequency

General-purpose fans are designed to operate below their first natural frequency. In most cases, the fan vendor will design the rotor-support system so that the rotating element's first critical is between 10 and 15 per cent above the rated running speed. While this practice is questionable, it is acceptable if the design speed and rotating-element mass are maintained. However, if either of these two factors changes, there is a high probability that serious damage or premature failure will result.

Inlet-air conditions

As with centrifugal pumps, fans require stable inlet conditions. Ductwork should be configured to ensure an adequate volume of clean air or gas, stable inlet pressure, and laminar flow. If the supply air is extracted from the environment, it is subject to variations in moisture, dirt content, barometric pressure, and density. However, these variables should be controlled as much as possible. As a minimum, inlet filters should be installed to minimize the amount of dirt and moisture that enters the fan.

Excessive moisture and particulates have an extremely negative impact on fan performance and cause two major problems: abrasion or tip wear and plate-out. High concentrations of particulate matter in the inlet air act as abrasives that accelerate fan-rotor wear. In most cases, however, this wear is restricted to the high-velocity areas of the rotor, such as the vane or blade tips, but can affect the entire assembly.

Plate-out is a much more serious problem. The combination of particulates and moisture can form a 'glue' that binds to the rotor assembly. As this contamination builds up on the rotor, the assembly's mass increases, which reduces its natural frequency. If enough plate-out occurs, the fan's rotational speed may coincide with the rotor's reduced natural frequency. With a strong energy source like the running speed, excitation of the rotor's natural frequency can result in catastrophic fan failure. Even if catastrophic failure does not occur, the vibration energy generated by the fan may cause bearing damage.

Fan laws

The mathematical relationships referred to as fan laws can be useful when applied to fans operating in a fixed system or to geometrically similar fans. However, caution should be exercised when using these relationships. They only apply to identical fans and applications. The basic laws are:

- Volume in cubic feet per minute (cfm) varies directly with the rotating speed (rpm);
- Static pressure varies with the rotating speed squared (rpm^2);

- Brake horsepower (BHP) varies with the speed cubed (rpm^3).

The fan-performance curves shown in Figures 37.3 and 37.4 show the performance of the same fan type, but designed for different volumetric-flow rates, operating in the same duct system handling air at the same density.

Curve 1 is for a fan designed to handle 10,000 cfm in a duct system whose calculated system resistance is determined to be 1 inch water (gage). This fan will operate at the point where the fan pressure (SP) curve intersects the system resistance curve (TSH). This intersection point is called the point of rating. The fan will operate at this point provided the fan's speed remains constant and the system's resistance does not change. The system-resistance curve illustrates that the resistance varies as the square of the volumetric flow rate (cfm). The BHP of the fan required for this application is 2 Hp.

Curve 2 illustrates the situation if the fan's design capacity is increased by 20 per cent, increasing output

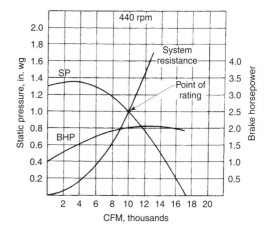

Figure 37.3 Fan-performance Curve 1

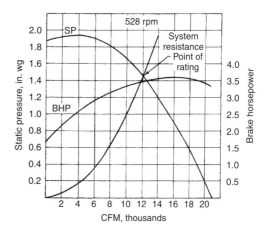

Figure 37.4 Fan-performance Curve 2

from 10,000 to 12,000 cfm. Applying the fan laws, the calculations are:

New rpm = 1.2×440

$= 528$ rpm (20% increase)

New SP = $1.2 \times 1.2 \times 1$ inch water (gage)

$= 1.44$ inch (44% increase)

New TSH = New SP = 1.44 inch

New BHP = $1.2 \times 1.2 \times 1.2 \times 2$

$= 1.73 \times 2$

$= 3.46$ Hp (73% increase)

The curve representing the system resistance is the same in both cases since the system has not changed.

The fan will operate at the same relative point of rating and will move the increased volume through the system. The mechanical and static efficiencies are unchanged.

The increased brake horsepower (BHP) required to drive the fan is a very important point to note. If the Curve 1 fan had been driven by a 2-hp motor, the Curve 2 fan needs a 3.5-hp motor to meet its volumetric requirement.

Centrifugal-fan selection is based on rating values such as air flow, rpm, air density, and cost. Table 37.1 is a typical rating table for a centrifugal fan. Table 37.2 provides air-density ratios.

37.1.3 Installation

Proper fan installation is critical to reliable operation. Suitable foundations, adequate bearing-support structures,

Table 37.1 Typical rating table for a centrifugal fan

CFM	OV	¼" SP RPM	¼" SP BHP	⅜" SP RPM	⅜" SP BHP	½" SP RPM	½" SP BHP	⅝" SP RPM	⅝" SP BHP	¾" SP RPM	¾" SP BHP
7456	800	262	0.48	289	0.60	314	0.75	337	0.92	360	1.09
8388	900	281	0.55	306	0.72	330	0.89	351	1.06	372	1.25
9320	1000	199	0.66	325	0.85	347	1.04	368	1.23	387	1.43
10252	1100	319	0.79	343	1.00	365	1.21	385	1.42	403	1.63
11184	1200	338	0.93	362	1.17	383	1.40	402	1.63	420	1.85
12116	1300	358	1.10	381	1.35	402	1.61	421	1.85	438	2.10
13048	1400	379	1.29	401	1.56	421	1.83	439	2.10	456	2.37
13980	1500	401	1.50	420	1.78	440	2.06	458	2.37	475	2.66
14912	1600	422	1.74	441	2.03	459	2.35	477	2.67	494	2.98
15844	1700	444	2.01	462	2.32	479	2.65	496	2.98	513	3.32
16776	1800	467	2.31	483	2.63	499	2.97	516	3.33	532	3.68
17708	1900	489	2.65	504	2.98	520	3.33	536	3.70	551	4.07
18640	2000	512	3.02	526	3.36	541	3.72	556	4.10	571	4.49
19572	2100	535	3.43	548	3.77	562	4.15	576	4.53	590	4.95
20504	2200	558	3.87	570	4.23	584	4.61	597	5.02	610	5.43
21436	2300	582	4.36	593	4.72	605	5.12	618	5.54	631	5.95
22368	2400	605	4.89	616	5.26	627	5.67	640	6.10	652	6.54
23300	2500	628	5.46	639	5.85	650	6.26	661	6.70	673	7.16
24232	2600	652	6.09	662	6.48	672	6.90	683	7.34	694	7.81
25164	2700	676	6.75	685	7.15	695	7.58	705	8.04	716	8.52
26096	2800	700	7.47	708	7.88	718	8.32	727	8.78	738	9.27
27028	2900	723	8.24	732	8.66	741	9.11	750	9.58	760	10.08

⅞" SP RPM	⅞" SP BHP	1" SP RPM	1" SP BHP	1¼" SP RPM	1¼" SP BHP	1½" SP RPM	1½" SP BHP	1¾" SP RPM	1¾" SP BHP
382	1.27	403	1.48	444	1.85	483	2.28	520	2.73
393	1.44	413	1.63	451	2.05	488	2.49	523	2.96
406	1.63	425	1.83	461	2.27	495	2.72	529	3.21
421	1.84	439	2.06	473	2.51	505	2.99	537	3.50
438	2.08	454	2.31	486	2.79	517	3.29	547	3.81
456	2.34	471	2.59	501	3.09	531	3.62	559	4.16
473	2.63	488	2.90	517	3.43	545	3.98	572	4.54
491	2.94	506	3.23	534	3.79	561	4.37	587	4.96
509	3.28	524	3.58	552	4.19	578	4.79	603	5.41
528	3.64	542	3.97	570	4.61	596	5.25	619	5.90
547	4.03	561	4.38	588	5.06	613	5.74	637	6.42
566	4.45	580	4.81	606	5.54	631	6.28	654	6.98
585	4.89	599	5.28	625	6.05	649	6.81	672	7.57
604	5.36	618	5.78	644	6.59	668	7.40	690	8.19
624	5.87	637	6.30	663	7.16	686	8.01	708	8.85
644	6.41	657	6.86	682	7.77	705	8.66	727	9.54
664	6.99	677	7.46	701	8.41	724	9.35	746	10.27
685	7.63	697	8.10	721	9.09	743	10.07	765	11.04
706	8.30	717	8.77	740	9.80	762	10.83	784	11.84
727	9.01	738	9.53	760	10.56	782	11.63	803	12.69
748	9.78	759	10.30	780	11.35	801	12.46	822	13.57
770	10.60	780	11.13	800	12.20	821	13.35	841	14.49

Table 37.2 Air-density

		Altitute, ft above sea level				
	0	1000	2000	3000	4000	5000
Air temp.,			*Barometric pressure, in. mercury*			
°F	29.92	28.86	27.82	26.82	25.84	24.90
70	1.000	0.964	0.930	0.896	0.864	0.832
100	0.946	0.912	0.880	0.848	0.818	0.787
150	0.869	0.838	0.808	0.770	0.751	0.723
200	0.803	0.774	0.747	0.720	0.694	0.668
250	0.747	0.720	0.694	0.669	0.645	0.622
300	0.697	0.672	0.648	0.624	0.604	0.580
350	0.654	0.631	0.608	0.586	0.565	0.544
400	0.616	0.594	0.573	0.552	0.532	0.513
450	0.582	0.561	0.542	0.522	0.503	0.484
500	0.552	0.532	0.513	0.495	0.477	0.459
550	0.525	0.506	0.488	0.470	0.454	0.437
600	0.500	0.482	0.465	0.448	0.432	0.416
650	0.477	0.460	0.444	0.427	0.412	0.397
700	0.457	0.441	0.425	0.410	0.395	0.380

			Altitute, ft above sea level			
	6000	7000	8000	9000	10,000	15,000	20,000
Air temp.,			*Barometric pressure, in. mercury*			
°F	23.98	23.09	22.22	21.39	20.58	16.89	13.75
70	0.801	0.772	0.743	0.714	0.688	0.564	0.460
100	0.758	0.730	0.703	0.676	0.651	0.534	0.435
150	0.696	0.671	0.646	0.620	0.598	0.490	0.400
200	0.643	0.620	0.596	0.573	0.552	0.453	0.369
250	0.598	0.576	0.555	0.533	0.514	0.421	0.344
300	0.558	0.538	0.518	0.498	0.480	0.393	0.321
350	0.524	0.505	0.486	0.467	0.450	0.369	0.301
400	0.493	0.476	0.458	0.440	0.424	0.347	0.283
450	0.466	0.449	0.433	0.416	0.401	0.328	0.268
500	0.442	0.426	0.410	0.394	0.380	0.311	0.254
550	0.421	0.405	0.390	0.375	0.361	0.296	0.242
600	0.400	0.386	0.372	0.352	0.344	0.282	0.230
650	0.382	0.368	0.354	0.341	0.328	0.269	0.219
700	0.366	0.353	0.340	0.326	0.315	0.258	0.210

properly sized ductwork, and flow-control devices are the primary considerations.

Foundations

As with any other rotating machine, fans require a rigid, stable foundation. With the exception of in-line fans, they must have a concrete footing or pad that is properly sized to provide a stable footprint and prevent flexing of the rotor-support system.

Bearing-support structures

In most cases, with the exception of in-line configurations, fans are supplied with a vendor-fabricated base. Bases normally consist of fabricated metal stands that support the motor and fan housing. The problem with most of the fabricated bases is that they lack the rigidity and stiffness to prevent flexing or distortion of the fan's rotating element. The typical support structure is comprised of relatively light-gage material ($\frac{3}{16}$ inch) and does not have the cross-bracing or structural stiffeners needed to prevent distortion of the rotor assembly. Because of this limitation, many plants fill the support structure with concrete or other solid material.

However, this approach does little to correct the problem. When the concrete solidifies, it pulls away from the sides of the support structure. Without direct bonding and full contact with the walls of the support structure, stiffness is not significantly improved.

The best solution to this problem is to add cross-braces and structural stiffeners. If they are properly sized and affixed to the support structure, the stiffness can be improved and rotor distortion reduced.

Ductwork

Ductwork should be sized to provide minimum friction loss throughout the system. Bends, junctions with other ductwork, and any change of direction should provide a clean, direct flow path. All ductwork should be airtight and leak-free to ensure proper operation.

Flow-control devices

Fans should always have inlet and outlet dampers or other flow-control devices, such as variable-inlet vanes. Without them, it is extremely difficult to match fan performance to actual application demand. The reason for this difficulty is that there are a number of variables (e.g., usage, humidity, and temperature) directly affecting the input–output demands for each fan application. Flow-control devices provide the means to adjust fan operation for actual conditions. Figure 37.5 shows an outlet damper with streamlined blades and linkage arranged to move adjacent blades in opposite directions for even throttling.

Airflow controllers must be inspected frequently to ensure that they are fully operable and operate in unison with each other. They also must close tightly. Ensure that the control indicators show the precise position of the vanes in all operational conditions. The 'open' and 'closed' positions should be permanently marked and visible at all times. Periodic lubrication of linkages is required.

Turn-buckle screws on the linkages for adjusting flow rates should never be moved without first measuring the

Figure 37.5 Outlet damper with streamlined blades and linkage arranged to move adjacent blades in opposite directions for even throttling

distance between the set-point markers on each screw. This is important if the adjustments do not produce the desired effect and you wish to return to the original settings.

37.1.4 Operating methods

Because fans are designed for stable, steady-state operation, variations in speed or load may have an adverse effect on their operating dynamics. The primary operating method that should be understood is output control. Two methods can be used to control fan output: dampers and fan speed.

Dampers

Dampers can be used to control the output of centrifugal fans within the effective control limits. Centrifugal fans have a finite acceptable control range, typically about 15 per cent below and 15 per cent above its design point. Control variations outside of this range severely affect the reliability and useful life of the fan.

The recommended practice is to use an inlet damper rather than a discharge damper for this control function whenever possible. Restricting the inlet with suction dampers can effectively control the fan's output. When using dampers to control fan performance, however, caution should be exercised to ensure that any changes remain within the fan's effective control range.

Fan speed

Varying fan speed is an effective means of controlling a fan's performance. As defined by the fan laws, changing the rotating speed of the fan can directly control both volume and pressure. However, caution must be used when changing fan speed. All rotating elements, including fan rotors, have one or more critical speeds. When the fan's speed coincides with one of the critical speeds, the rotor assembly becomes extremely unstable and could fail catastrophically.

In most general-purpose applications, fans are designed to operate between 10 and 15 per cent below their first critical speed. If speed is increased on these fans, there is a good potential for a critical-speed problem. Other applications have fans that are designed to operate between their first and second critical speeds. In this instance, any change up or down may cause the speed to coincide with one of the critical speeds.

37.2 Blowers

A blower uses mating helical lobes or screws and is used for the same purpose as a fan. They are normally moderate- to high-pressure devices. Blowers are almost identical both physically and functionally to positive-displacement compressors. Therefore, compressor information found in Chapter 36 on basic design criteria, physical laws, and operating characteristics applies to blowers as well.

38

Mixers and Agitators

R Keith Mobley
The Plant Performance Group

Contents

Mixers are devices that blend combinations of liquids and solids into a homogenous product. They come in a variety of sizes and configurations designed for specific applications. Agitators provide the mechanical action required to keep dissolved or suspended solids in solution.

Both operate on the same principles, but variations in design, operating speed, and applications divide the actual function of these devices. Agitators generally work just as hard as mixers and the terms are often used interchangeably.

38.1 Configuration

There are two primary types of mixers: propeller/paddle and screw. Screw mixers can be further divided into two types: batch and mixer-extruder.

Propeller/paddle

Propeller/paddle mixers are used to blend or agitate liquid mixtures in tanks, pipelines, or vessels. Figure 38.1 illustrates a typical top-entering propeller/paddle mixer. This unit consists of an electric motor, a mounting bracket, an extended shaft, and one or more impeller(s) or propeller(s). Materials of construction range from bronze to stainless steel, which are selected based on the particular requirements of the application.

The propeller/paddle mixer is also available in a side-entering configuration, which is shown in Figure 38.2. This configuration is typically used to agitate liquids

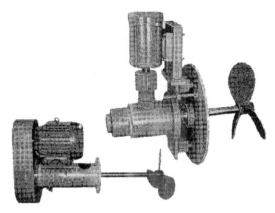

Figure 38.2 Side-entering propeller-type mixer

Figure 38.3 Mixer can use either propellers (a) or paddles (b) to provide agitation

in large vessels or pipelines. The side-entering mixer is essentially the same as the top-entering version except for the mounting configuration.

Both the top-entering and side-entering mixers may use either propellers, as shown in the preceding figures, or paddles, as illustrated by Figure 38.3 (b). Generally, propellers are used for medium- to high-speed applications where the viscosity is relatively low. Paddles are used in low-speed, high-viscosity applications.

Screw

The screw mixer uses a single- or dual-screw arrangement to mix liquids, solids, or a combination of both. It comes

Figure 38.1 Top-entering propeller-type mixer

Figure 38.4 Batch-type mixer uses single or dual screws to mix product

in two basic configurations: batch and combination mixer-extruder.

Batch

Figure 38.4 illustrates a typical batch-type screw mixer. This unit consists of a mixing drum or cylinder, a single- or dual-screw mixer, and a power supply.

The screw configuration is normally either a ribbon-type helical screw or a series of paddles mounted on a common shaft. Materials of construction are selected based on the specific application and materials to be mixed. Typically, the screws are either steel or stainless steel, but other materials are available.

Combination mixer-extruder

The mixer-extruder combination unit shown in Figure 38.5 combines the functions of a mixer and screw conveyor. This type of mixer is used for mixing viscous products.

38.2 Performance

Unlike centrifugal pumps and compressors, few criteria can be used directly to determine mixer effectiveness and efficiency. However, product quality and brake horsepower are indices that can be used to indirectly gauge performance.

Product quality

The primary indicator of acceptable performance is the quality of the product delivered by the mixer. Although there is no direct way to measure this indicator, feedback from the quality assurance group should be used to verify that acceptable performance levels are attained.

Brake horsepower

Variation in the actual brake horsepower required to operate a mixer is the primary indicator of its performance envelope. Mixer design, whether propeller- or screw-type, is based on the viscosity of both the incoming and finished product. These variables determine the brake horsepower required to drive the mixer, which will follow variations in the viscosity of the products being mixed. As the viscosity increases so will the brake horsepower demand. Conversely, as the viscosity decreases, so will the horsepower required driving the mixer.

38.3 Installation

Installation of propeller-type mixers varies greatly, depending on the specific application. Top-entering mixers utilize either a clamp- or flange-type mounting. It is important that the mixer be installed so the propeller or paddle placement is at a point within the tank, vessel, or piping that assures proper mixing. Vendor recommendations found in O&M manuals should be followed to ensure proper operation of the mixer.

Mixers should be mounted on a rigid base that assures level alignment and prevents lateral movement of the mixer and its drive-train. While most mixers can be bolted directly to a base, care must be taken to ensure that it is rigid and has the structural capacity to stabilize the mixer.

38.4 Operating methods

There are only three major operating concerns for mixers: setup, incoming-feed rate, and product viscosity.

Mixer setup

Both propeller and screw mixers have specific setup requirements. In the case of propeller/paddle-type

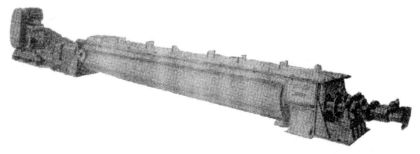

Figure 38.5 Combination mixer-extruder

mixers, the primary factor is the position of the propellers or paddles within the tank or vessel. Vendor recommendations should be followed to assure proper operation of the mixer.

If the propellers or paddles are too close to the liquid level, the mixer will create a vortex that will entrain air and prevent adequate blending or mixing. If the propellers are set too low, compress vortexing may occur. When this happens, the mixer will create a stagnant zone in the area under the rotating assembly. As a result, some of the product will settle in this zone and proper mixing cannot occur. Setting the mixer too close to a corner or the side of the mixing vessel can also create a stagnant zone that will prevent proper blending or mixing of the product.

For screw-type mixers, proper clearance between the rotating element and the mixer housing must be maintained to vendor specifications. If the clearance is improperly set, the mixer will bind (i.e., not enough clearance) or fail to blend properly.

Feed rate

Mixers are designed to handle a relatively narrow band of incoming product flow rate. Therefore, care must be exercised to ensure that the actual feed rate is maintained within acceptable limits. The O&M manuals provided by the vendor will provide the feed-rate limitations for various products. Normally, these rates must be adjusted for viscosity and temperature variations.

Viscosity

Variations in viscosity of both the incoming and finished products have a dramatic effect on mixer performance. Standard operating procedures should include specific operating guidelines for the range of variation that is acceptable for each application. The recommended range should include adjustments for temperature, flow rates, mixing speeds, and other factors that directly or indirectly affect viscosity.

39

Gears and Gearboxes

R Keith Mobley
The Plant Performance Group

Contents

A gear is a disc or wheel with teeth around its periphery – either on the inside edge (i.e., internal gear) or on the outside edge (i.e., external gear). A gear is used to provide a positive means of power transmission, which is effected by the teeth on one gear meshing with the teeth on another gear or rack (i.e., straight-line gear).

Gear drives are packaged units used for a wide range of power-transmission applications. They are used to transmit power to a driven piece of machinery and to change or modify the power that is transmitted. Modifications include reducing speed and increasing output torque, increasing speed, changing the direction of shaft rotation, or changing the angle of shaft operation.

39.1 Configuration

There are several different types of gears used in industry. Many are complex in design and manufacture and several have evolved directly from the spur gear, which is referred to as the basic gear. Types of gears are: spur, helical, bevel, and worm. Table 39.1 summarizes the characteristics of each gear type.

39.1.1 Spur gears

The spur gear is the least expensive of all gears to manufacture and is the most commonly used. It can be manufactured to close tolerances and is used to connect parallel shafts that rotate in opposite directions. It gives excellent results at moderate peripheral speeds and the tooth load produces no axial thrust. Because contact is simultaneous across the entire width of the meshing teeth, it tends to be noisy at high speeds. However, noise and wear can be minimized with proper lubrication.

There are three main classes of spur gears: external tooth, internal tooth, and rack-and-pinion. The external tooth variety shown in Figure 39.1 is the most common. Figure 39.2 illustrates an internal gear and Figure 39.3 shows a rack or straight-line spur gear.

The spur gear is cylindrical and has straight teeth cut parallel to its rotational axis. The tooth size of spur gears is established by the diametrical pitch. Spur-gear design accommodates mostly rolling, rather than sliding, contact of the tooth surfaces and tooth contact occurs along a line parallel to the axis. Such rolling contact produces less

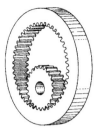

Figure 39.2 Example of an internal spur gear

Figure 39.3 Rack or straight-line gear

heat and yields high mechanical efficiency, often up to 99 per cent.

An internal spur gear, in combination with a standard spur-gear pinion, provides a compact drive mechanism for transmitting motion between parallel shafts that rotate in the same direction. The internal gear is a wheel that has teeth cut on the inside of its rim and the pinion is housed inside the wheel. The driving and driven members rotate in the same direction at relative speeds inversely proportional to the number of teeth.

39.1.2 Helical gears

Helical gears, which are shown in Figure 39.4, are formed by cutters that produce an angle that allows several teeth

Figure 39.4 Typical set of helical gears

Figure 39.1 Example of a spur gear

Table 39.1 Gear characteristics overview

Gear type	Characteristics	
	Attributes/Positives	*Negatives*
Spur, external	Connects parallel shafts that rotate in opposite directions, inexpensive to manufacture to close tolerances, moderate peripheral speeds, no axial thrust, high mechanical efficiency	Noisy at high speeds
Spur, internal	Compact drive mechanism for parallel shafts rotating in same direction	
Helical, external	Connects parallel and non-parallel shafts; superior to spur gears in load-carrying capacity, quietness, and smoothness; high efficiency	Higher friction than spur gears, high end thrust
Helical, double (also referred to as herringbone)	Connects parallel shafts, overcomes high-end thrust present in single-helical gears, compact, quiet and smooth operation at higher speeds (1000 to 12,000 fpm or higher), high efficiencies	
Helical, cross	Light loads with low power transmission demands	Narrow range of applications, requires extensive lubrication
Bevel	Connects angular or intersecting shafts	Gears overhang supporting shafts resulting in shaft deflection and gear misalignment
Bevel, straight	Peripheral speeds up to 1000 fpm in applications where quietness and maximum smoothness not important, high efficiency	Thrust load causes gear pair to separate
Bevel, zerol	Same ratings as straight bevel gears and use same mountings, permits slight errors in assembly, permits some displacement due to deflection under load, highly accurate, hardened due to grinding	Limited to speeds less than 1000 fpm due to noise
Bevel, Spiral	Smoother and quieter than straight bevel gears at speeds greater than 1000 fpm or 1000 rpm, evenly distributed tooth loads, carry more load without surface fatigue, high efficiency, reduces size of installation for large reduction ratios, speed-reducing and speed-increasing drive	High tooth pressure, thrust loading depends on rotation and spiral angle
Bevel, miter	Same number of teeth in both gears, operate on shafts at 90°	
Bevel, hypoid	Connects non-intersecting shafts, high pinion strength, allows the use of compact straddle mounting on the gear and pinion, recommended when maximum smoothness required, compact system even with large reduction ratios, speed-reducing and speed-increasing drive	Lower efficiency, difficult to lubricate due to high tooth-contact pressures, materials of construction (steel) require use of extreme-pressure lubricants
Planetary or epicyclic	Compact transmission with driving and driven shafts in line, large speed reduction when required	
Worm, cylindrical	Provide high-ratio speed reduction over wide range of speed ratios (60:1 and higher from a single reduction, can go as high as 500:1), quiet transmission of power between shafts at 90°, reversible unit available, low wear, can be self-locking	Lower efficiency; heat removal difficult, which restricts use to low-speed applications
Worm, double-enveloping	Increased load capacity	Lower efficiencies

Source: The Plant Performance Group.

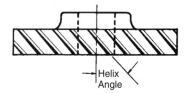

Figure 39.5 The angle at which the teeth are cut

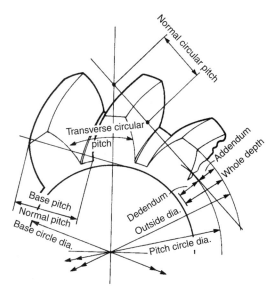

Figure 39.6 Helical gear and its parts

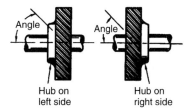

Figure 39.7 Helix angle of the teeth – the same no matter from which side the gear is viewed

Figure 39.8 Herringbone gear

to mesh simultaneously. Helical gears are superior to spur gears in their load-carrying capacity and quietness and smoothness of operation, which results from the sliding contact of the meshing teeth. A disadvantage, however, is higher friction and wear that accompanies this sliding action.

Single helical gears are manufactured with the same equipment as spur gears, but the teeth are cut at an angle to the axis of the gear and follow a spiral path. The angle at which the gear teeth are cut is called the helix angle, which is illustrated in Figure 39.5. This angle causes the position of tooth contact with the mating gear to vary at each section. Figure 39.6 shows the parts of a helical gear.

It is very important to note that the helix angle may be on either side of the gear's centerline. Or, if compared to the helix angle of a thread, it may be either a 'right-hand' or a 'left-hand' helix. Figure 39.7 illustrates a helical gear as viewed from opposite sides. A pair of helical gears must have the same pitch and helix angle, but must be of opposite hand (one right hand and one left hand).

39.1.3 Herringbone

The double-helical gear, also referred to as the herringbone gear (Figure 39.8), is used for transmitting power between parallel shafts. It was developed to overcome the disadvantage of the high-end thrust that is present with single-helical gears.

The herringbone gear consists of two sets of gear teeth on the same gear, one right hand and one left hand. Having both hands of gear teeth causes the thrust of one set to cancel out the thrust of the other. Thus, another advantage of this gear type is quiet, smooth operation at higher speeds.

39.1.4 Bevel

Bevel gears are used most frequently for 90° drives, but other angles can be accommodated. The most typical application is driving a vertical pump with a horizontal driver.

Two major differences between bevel gears and spur gears are their shape and the relation of the shafts on which they are mounted. A bevel gear is conical in shape, while a spur gear is essentially cylindrical. Figure 39.9 illustrates the bevel gear's basic shape. Bevel gears transmit motion between angular or intersecting shafts, while spur gears transmit motion between parallel shafts.

Figure 39.10 shows a typical pair of bevel gears. As with other gears, the term 'pinion and gear' refers to the members with the smaller and larger numbers of teeth in the pair, respectively. Special bevel gears can be manufactured to operate at any desired shaft angle, as shown in Figure 39.11.

As with spur gears, the tooth size of bevel gears is established by the diametrical pitch. Because the tooth size varies along its length, measurements must be taken

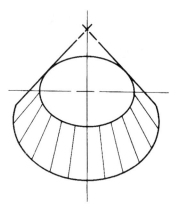

Figure 39.9 Basic cone shape of bevel gears

Figure 39.10 Typical set of bevel gears

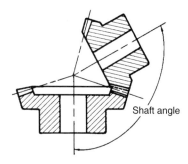

Shaft angle

Figure 39.11 Shaft angle, which can be at any degree

at a specific point. Note that, because each gear in a bevel-gear set must have the same pressure angle, tooth length, and diametrical pitch, they are manufactured and distributed only as mated pairs. Like spur gears, bevel gears are available in pressure angles of 14.5° and 20°.

Because there generally is no room to support bevel gears at both ends due to the intersecting shafts, one or both gears overhang their supporting shafts. This is

referred to as an overhung load. It may result in shaft deflection and gear misalignment, causing poor tooth contact and accelerated wear.

39.1.5 Straight or plain

Straight-bevel gears, also known as plain bevels, are the most commonly used and simplest type of bevel gear (Figure 39.12). They have teeth cut straight across the face of the gear. These gears are recommended for peripheral speeds up to 1000 feet per minute in cases where quietness and maximum smoothness are not crucial. This gear type produces thrust loads in a direction that tends to cause the pair to separate.

39.1.6 Zerol

Zerol-bevel gears are similar to straight-bevel gears, carry the same ratings, and can be used in the same mountings. These gears, which should be considered as spiral-bevel gears having a spiral angle of zero, have curved teeth that lie in the same general direction as straight-bevel gears. This type of gear permits slight errors in assembly and some displacement due to deflection under load. Zerol gears should be used at speeds less than 1000 feet per minute because of excessive noise at higher speeds.

39.1.7 Spiral

Spiral-bevel gears (Figure 39.13) have curved oblique teeth that contact each other gradually and smoothly from one end of the tooth to the other, meshing with a rolling contact similar to helical gears. Spiral-bevel gears are smoother and quieter in operation than straight-bevel gears, primarily due to a design that incorporates two or more contacting teeth. Their design, however, results in high tooth pressure.

Figure 39.12 Straight or plain bevel gear

Figure 39.13 Spiral bevel gear

This type of gear is beginning to supersede straight-bevel gears in many applications. They have the advantage of ensuring evenly distributed tooth loads and carry more load without surface fatigue. Thrust loading depends on the direction of rotation and whether the spiral angle of the teeth is positive or negative.

39.1.8 Miter

Miter gears are bevel gears with the same number of teeth in both gears, operating on shafts at 90°, as shown in Figure 39.14. Their primary use is to change direction in a mechanical drive assembly. Since both the pinion and gear have the same number of teeth, there is no mechanical advantage generated by this type of gear.

39.1.9 Hypoid

Hypoid-bevel gears are a cross between a spiral-bevel gear and a worm gear (Figure 39.15). The axes of a pair of hypoid-bevel gears are non-intersecting and the distance between the axes is referred to as the 'offset'. This configuration allows both shafts to be supported at both ends and provides high strength and rigidity.

Although stronger and more rigid than most other types of gears, they are less efficient and extremely difficult to lubricate because of high tooth-contact pressures. Further increasing the demands on the lubricant is the material of construction as both the driven and driving gears are made of steel. This requires the use of special extreme-pressure lubricants that have both oiliness and anti-weld properties that can withstand the high contact pressures and rubbing speeds.

Despite its demand for special lubrication, this gear type is in widespread use in industrial and automotive applications. It is used extensively in rear axles of automobiles having rear-wheel drives and is increasingly being used in industrial machinery.

39.1.10 Worm

The worm and gear, which are illustrated in Figure 39.16, are used to transmit motion and power when a high-ratio speed reduction is required. They accommodate a wide range of speed ratios (60:1 and higher can be obtained from a single reduction and can go as high as 500:1). In most worm-gear sets, the worm is most often the driver and the gear the driven member. They provide a steady, quiet transmission of power between shafts at right angles and can be self-locking. Thus, a torque on the gear will not cause the worm to rotate.

The contact surface of the screw on the worm slides along the gear teeth. Because of the high level of rubbing between the worm and wheel teeth, however, slightly less efficiency is obtained than with precision spur gears. Note that large helix angles on the gear teeth produce higher efficiencies. Another problem with this gear type is heat removal, a limitation that restricts their use to low-speed applications.

One of the major advantages of the worm gear is low wear, which is due mostly to a full-fluid lubricant film. In addition, friction can be further reduced using metals having low coefficients of friction. For example, the wheel is typically made of bronze and the worm of highly finished hardened steel.

Most worms are cylindrical in shape with a uniform pitch diameter. However, a variable pitch diameter is used in the double-enveloping worm. This configuration is used when increased load capacity is required.

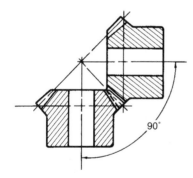

Figure 39.14 Miter gear shaft angle

Figure 39.15 Hypoid bevel gear

Figure 39.16 Worm gear

39.2 Performance

With few exceptions, gears are one-directional power transmission devices. Unless a special bi-directional gear set is specified, gears have a specific direction of rotation and will not provide smooth, trouble-free power transmission when the direction is reversed. The reason for this one-directional limitation is that gear manufacturers do not finish the non-power side of the tooth profile. This is primarily a cost-savings issue and should not affect gear operation.

The primary performance criteria for gear sets include: efficiency, brake horsepower, speed transients, startup, backlash, and ratios.

Efficiency

Gear efficiency varies with the type of gear used and the specific application. Table 39.2 provides a comparison of the approximate efficiency range of various gear types. The table assumes normal operation where torsional loads are within the gear set's designed horsepower range. It also assumes that startup and speed change torques are acceptable.

Brake horsepower

All gear sets have a recommended and maximum horsepower rating. The rating varies with the type of gear set, but must be carefully considered when evaluating a gearbox problem. The maximum installed motor horsepower should never exceed the maximum recommended horsepower of the gearbox. This is especially true of worm gear sets. The soft material used for these gears is easily damaged when excess torsional load is applied.

The procurement specifications or the vendor's engineering catalog will provide all of the recommended horsepower ratings needed for an analysis. These recommendations assume normal operation and must be adjusted for the actual operating conditions in a specific application.

Table 39.2 Gear efficiencies

Gear type	Efficiency range, %
Bevel gear, hypoid	90–98
Bevel gear, miter	Not available
Bevel gear, spiral	97–99
Bevel gear, straight	97–99
Bevel gear, zerol	Not available
Helical gear, external	97–99
Helical gear–double, external (herringbone)	97–99
Spur gear, external	97–99
Worm, cylindrical	50–99
Worm, double-enveloping	50–98

Source: Adapted by Integrated Systems, Inc. from 'Gears and Gear Drives,' 1996 Power. Transmission Design, Penton Publishing Inc., Ohio, pp. A199–A211.

Speed transients

Applications that require frequent speed changes can have a severe, negative impact on gearbox reliability. The change in torsional load caused by acceleration and deceleration of a gearbox may exceed its maximum allowable horsepower rating. This problem can be minimized by decreasing the ramp speed and amount of braking that is applied to the gear set. The vendor's O&M manual and/or technical specifications should provide detailed recommendations that define the limits to use in speed-change applications.

Startup

Start–stop operation of a gearbox can accelerate both gear and bearing wear and may cause reliability problems. In applications like the bottom discharge of storage silos, where a gear set drives a chain or screw conveyor system and startup torque is excessive, care must be taken to prevent over-loading the gear set.

Backlash

Gear backlash is the play between teeth measured at the pitch circle. It is the distance between the involutes of the mating gear teeth, as illustrated in Figure 39.17.

Backlash is necessary to provide the running clearance needed to prevent binding of the mating gears, which can result in heat generation, noise, abnormal wear, overload, and/or failure of the drive. In addition to the need to prevent binding, some backlash occurs in gear systems because of the dimensional tolerances needed for cost-effective manufacturing.

During the gear-manufacturing process, backlash is achieved by cutting each gear tooth thinner by an amount equal to one-half of the backlash dimension required for the application. When two gears made in this manner are run together (i.e., mate), their allowances combine to provide the full amount of backlash.

The increase in backlash that results from tooth wear does not adversely affect operation with non-reversing drives, or drives with continuous load in one direction. However, for reversing drives and drives where timing is critical, excessive backlash that results from wear usually cannot be tolerated.

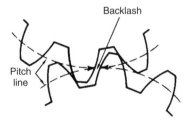

Figure 39.17 Backlash

Ratios

Ratios used in defining and specifying gears are gear-tooth ratio, contact ratio, and hunting ratio. The gear-tooth ratio is the ratio of the larger to the smaller number of teeth in a pair of gears. The contact ratio is a measure of overlapping tooth action which is necessary to assure smooth, continuous action. For example, as one pair of teeth passes out of action, a succeeding pair of teeth must have already started action. The hunting ratio is the ratio of the number of gear and pinion teeth. It is a means of ensuring that each tooth in the pinion contacts every tooth in the gear before it contacts any gear tooth a second time.

39.3 Installation

Installation guidelines provided in the vendor's O&M manual should be followed for proper installation of the gearbox housing and alignment to its mating machine-train components.

Gearboxes must be installed on a rigid base that prevents flexing of its housing and the input and output shafts. Both the input and output shaft must be properly aligned, within 0.002 inch, to their respective mating shafts. Both shafts should be free of any induced axial forces that may be generated by the driver or driven units.

Internal alignment is also important. Internal alignment and clearances of new gearboxes should be within the vendor's acceptable limits, but there is no guarantee that this will be true. All internal clearances (e.g., backlash and center-to-center distances) and the parallel relationship of the pinion and gear shafts should be verified for any gearbox that is being investigated.

39.4 Operating methods

Two primary operating parameters govern effective operation of gear sets or gearboxes: maximum torsional power rating and transitional torsional requirements.

Each gear set has a specific maximum horsepower rating. This is the maximum torsional power that the gear set can generate without excessive wear or gear damage. Operating procedures should ensure that the maximum horsepower is not exceeded throughout the entire operating envelope. If the gear set was properly designed for the application, its maximum horsepower rating should be suitable for steady-state operation at any point within the design-operating envelope. As a result, it should be able to provide sufficient torsional power at any set point within the envelope.

Two factors may cause over-load on a gear set: excessive load or speed transients. Many processes are subjected to radical changes in the process or production loads. These changes can have a serious effect on gear-set performance and reliability.

Operating procedures should establish boundaries that limit the maximum load variations that can be used in normal operation. These limits should be well within the acceptable load rating of the gear set.

The second factor, speed transients, is a leading cause of gear-reliability problems. The momentary change in torsional load created by rapid changes in speed can have a dramatic, negative impact on gear sets. These transients often exceed the maximum horsepower rating of the gears and may result in failure. Operating procedures should ensure that torsional power requirements during startup, process-speed changes, and shutdown do not exceed the recommended horsepower rating of the gear set.

40

Hydraulic Fundamentals

R Keith Mobley
The Plant Performance Group

Contents

40.1 Introduction

The study of hydraulics deals with the use and characteristics of liquids and gases. Since the beginning of time, man has used fluids to ease his burden. Earliest recorded history shows that devices such as pumps and water wheels were used to generate useable mechanical power.

Fluid power encompasses most applications that use liquids or gases to transmit power in the form of mechanical work, pressure and/or volume in a system. This definition includes all systems that rely on pumps or compressors to transmit specific volumes and pressures of liquids or gases within a closed system. The complexity of these systems range from a simple centrifugal pump used to remove casual water from a basement to complex airplane control systems that rely on high-pressure hydraulic systems.

Fluid power systems have been developing rapidly over the past thirty-five years. Fluid power filled a need during World War II for an energy transmission system with muscle, which could easily be adapted to automated machinery. Today, fluid power technology is seen in every phase of man's activities. Fluid power is found in areas of manufacturing such as metal forming, plastics, basic metals, and material handling. Fluid power is evident in transportation as power and control systems of ships, airplanes, and automobiles. The environment is another place fluid power is hard at work compacting waste materials and controlling floodgates of hydroelectric dams. Food processing, construction equipment and medical technology are a few more areas of fluid power involvement. Fluid power applications are only limited by imagination.

There are alternatives to fluid power systems. Each system, regardless of the type, has its own advantages and disadvantages. Each has applications where it is best suited to do the job. This is probably the reason you won't find a fluid power wristwatch or hoses carrying fluid power replacing electrical power lines.

Advantages of fluid power

If a fluid power system is properly designed and used, it will provide smooth, flexible, uniform action without vibration, and is unaffected by variation of load. In case of an overload, an automatic release of pressure can be guaranteed, so that the system is protected against breakdown or excessive strain. Fluid power systems can provide widely variable motions in both rotary and linear transmission of power and the need for manual control can be minimized. In addition, fluid power systems are economical to operate.

Fluid power includes hydraulic, hydro-pneumatic and pneumatic systems. Why are hydraulics used in some applications, pneumatics in others or combination systems, in still others? Both the user and the manufacturer must consider many factors when determining which type of system should be used in a specific application.

In general, pneumatic systems are less expensive to manufacture and operate, but there are factors that prohibit their universal application. The compressibility of air, as any gas, limits the operation of pneumatic systems. For example, a pneumatic cylinder cannot maintain the position of a suspended load without a constant supply of air pressure. The load will force the air trapped within the cylinder to compress and allow the suspended load to creep. This compressibility also limits the motion of pneumatic actuators when under load.

Pneumatic systems can be used for applications that require low to medium pressure and only accurate control. Applications that require medium pressure, more accurate force transmission and moderate motion control can use a combination of hydraulics and pneumatics, or hydropneumatics. Hydraulics systems must be used for applications that require high pressure and/or extremely accurate force and motion control.

The flexibility of fluid power, both hydraulic and pneumatic elements, present a number of problems. Since fluids and gases have no shape of their own, they must be positively confined throughout the entire system. This is especially true in hydraulics where leakage of hydraulic oil can result in safety or environmental concerns. Special consideration must be given to the structural integrity of the parts of a hydraulic system. Strong pipes, tubing and hoses, as well as the containers must be provided. Leaks must be prevented. This is a serious problem with the high pressure obtained in many hydraulic system applications.

Fluid power systems vs. mechanical systems

Some desirable characteristics of fluid power systems when compared with mechanical systems. A fluid power system is often a simpler means of transmitting energy. There are fewer mechanical parts than in an ordinary industrial system. Since there are fewer mechanical parts, a fluid power system is more efficient and more dependable. In the common industrial system, there is no need to worry about hundreds of moving parts failing, with fluid or gas as the transmission medium.

With fluid or gas as the transmission medium, various components of a system can be located at convenient places on the machine. Fluid power can be transmitted and controlled quickly and efficiently up, down, and around corners with few controlling elements.

Since fluid power is efficiently transmitted and controlled, it gives freedom in designing a machine. The need for gear, cam, and lever systems is eliminated. Fluid power systems can provide infinitely variable speed, force and direction control with simple, reliable elements.

Fluid power vs. electrical systems

Mechanical force can be more easily controlled using fluid power. The simple use of valves and rotary or linear actuators control speed, direction and force. The simplicity of hydraulic and pneumatic components greatly increases their reliability. In addition, components and overall system size are typically much smaller than comparable electrical transmission devices.

Special problems

The operation of the system involves constant movement of the hydraulic fluid within its lines and components.

This movement causes friction within the fluid itself and against the containing surfaces. Excessive friction can lead to serious losses in efficiency or damage to system components. Foreign matter must not be allowed to accumulate in the system, where it will clog small passages or score closely fitted parts. Chemical action may cause corrosion. Anyone working with hydraulic systems must know how a fluid power system and its components operate, both in terms of the general principles common to all physical mechanisms and of the peculiarities of the specific arrangement at hand.

The word hydraulics is based on the Greek word for water, the first used form of hydraulic power transmission. Initially, hydraulics covered the study of the physical behavior of water at rest and in motion. It has been expanded to include the behavior of all liquids, although it is primarily limited to the motion or kinetics of liquids.

Hazards

Any use of a pressurized medium, such as hydraulic fluid, can be dangerous. Hydraulic systems carry all the hazards of pressurized systems and special hazards related directly to the composition of the fluid used.

When using oil as a fluid in a high-pressure hydraulic system, the possibility of fire or an explosion exists. A severe fire hazard is generated when a break in the high-pressure piping occurs and the oil is vaporized into the atmosphere. Extra precautions against fire should be practiced in these areas.

If oil is pressurized by compressed air, an explosive hazard exists. If high-pressure air encounters the oil, it may create a diesel effect, which may result in an explosion. A carefully followed preventive maintenance plan is the best precaution against explosions.

40.2 Basic hydraulics

Fluid power systems have developed rapidly over the past thirty-five years. Today, fluid power technology is used in every phase of human existence. The extensive use of hydraulics to transmit power is due the fact that properly constructed fluid power systems possess a number of favorable characteristics. They eliminate the need for complicated systems of gears, cams, and levers. Motion can be transmitted without the slack or mechanical looseness inherent in the use of solid machine parts. The fluids used are not subject to breakage as are mechanical parts, and the mechanisms are not subjected to great wear.

The operation of a typical fluid power system is illustrated in Figure 40.1. Oil from a tank or reservoir flows through a pipe into a pump. An electric motor, air motor,

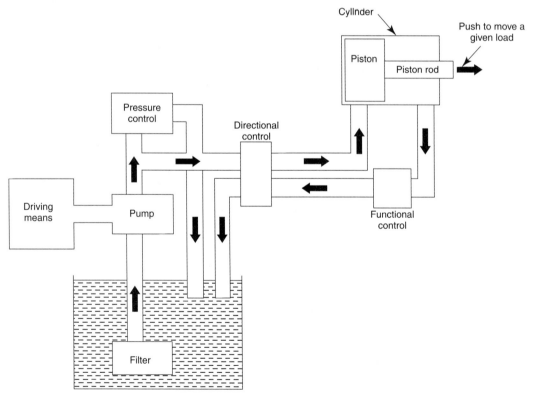

Figure 40.1 Basic hydraulic system

gas or steam turbine, or an internal combustion engine can drive the pump. The pump increases the pressure of the oil. The actual pressure developed depends on the design of the system.

The high-pressure oil flows in piping through a control valve. The control valve changes the direction of oil flow. A relief valve, set at a desired, safe operating pressure, protects the system from an over pressure condition. The oil that enters the cylinder acts on the piston, with the pressure acting over the area of the piston, developing a force on the piston rod. The force on the piston rod enables the movement of a load or device.

40.2.1 States of matter

The material that makes up the universe is known as matter. Matter is defined as any substance that occupies space and has weight. Matter exists in three states: solid, liquid, and gas. Each has distinguishing characteristics. Solids have a defined volume and a definite shape. Liquids have a definite volume, but take the shape of their containing vessels. Gases have neither a definite shape nor volume. Gases not only take the shape of the containing vessel, but also expand to fill the vessel, regardless of its volume. Examples of the states of matter are iron, water, and air.

Matter can change from one state to another. Water is a good example. At high temperatures, above 212°F, it is in a gaseous state known as steam. At moderate temperatures, it is liquid, and at low temperatures, below 32°F, it becomes ice, a solid. In this example, the temperature is the dominant factor in determining the state that the substance assumes.

Pressure is another important factor that will affect changes in the state of matter. At pressures higher than atmospheric, 14.7 psi, water will boil and thus change to steam at temperatures below 212°F. Pressure is also a critical factor in changing some gases to liquids or solids. Normally, when pressure and chilling are both applied to a gas, the gas assumes a liquid state. Liquid air, which is a mixture of oxygen and nitrogen, is produced in this manner.

In the study of fluid power, we are concerned primarily with the properties and characteristics of liquids and gases. However, you should keep in mind that the properties of solids also affect the characteristics of liquids and gases. The lines and components, which are solids, enclose and control the liquid or gas in their respective systems.

40.2.2 Development of hydraulics

The use of hydraulics is not new. The Egyptians and people of ancient Persia, India and China conveyed water along channels for irrigation and other domestic purposes. They used dams and sluice gates to control the flow and waterways to direct the water to where it was needed. The ancient Cretins had elaborate plumbing systems. Archimedes studied the laws of floating and submerged bodies. The Romans constructed aqueducts to carry water to their cities.

After the breakup of the ancient world, there were few new developments for many centuries. Then, over a comparatively short period, beginning near the end of the seventeenth century, Italian physicist, Evangelista Torricelle, French Physicist, Edme Mariotte and later Daniel Bernoulli conducted experiments to study the force generated by the discharge of water through small openings in the sides of tanks and through short pipes. During the same period, Blaise Pascal, a French scientist, discovered the fundamental law for the science of hydraulics. Pascal's law states that an increase in pressure on the surface of a confined fluid is transmitted throughout the confining vessel or system without any loss of pressure.

Figure 40.2 illustrates the transmission of forces through liquids. For Pascal's law to become effective for practical applications, a piston or ram confined within a close tolerance cylinder was needed. It was not until the latter part of the eighteenth century that methods were developed that could make the snugly fitted parts required making hydraulic systems practical.

This was accomplished by the invention of machines that were used to cut and shape the necessary closely fitted parts and, particularly, by the development of gaskets and packing. Since that time, components such as valves, pumps, actuating cylinders and motors have been developed and refined to make hydraulics one of the leading methods of transmitting power.

40.2.3 Use of hydraulics

The hydraulic press, invented by an Englishman, John Brahmah, was one of the first workable machines that used hydraulics in its operation. It consisted of a plunger pump piped to a large cylinder and a ram. This press found wide use in England because it provided a more effective and economical means of applying large, uniform forces in industrial uses.

Today, hydraulic power is used to operate many different tools and mechanisms. In a garage, a mechanic raises the end of an automobile with a hydraulic jack. Dentists and barbers use hydraulic power to lift and position their chairs. Hydraulic doorstops keep heavy doors from slamming. Hydraulic brakes have been standard equipment on automobiles since the 1930s. Most automobiles are equipped with automatic transmissions

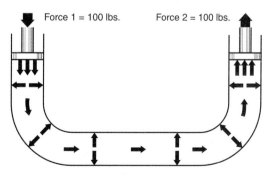

Figure 40.2 Transmission of forces

that are hydraulically operated. Power steering is another application of hydraulic power. Construction workers depend upon hydraulic power for their equipment. For example, the blade of a bulldozer is normally operated by hydraulic power.

40.2.4 Operation of hydraulic components

To transmit and control power through pressurized fluids, an arrangement of interconnected components is required. Such an arrangement is commonly referred to as a system. The number and arrangement of the components vary from system to system, depending on the particular application. In many applications, one main system supplies power to several subsystems, which are sometimes referred to as circuits. The complete system may be a small, compact unit or a large, complex system that has components located at widely separated points within the plant. The basic components of a hydraulic system are essentially the same regardless of the complexity of the system. Seven basic components must be in a hydraulic system. These basic components are:

Reservoir or receiver

This is usually a closed tank or vessels that hold the volume of fluid required supporting the system. The vessels normally provide several functions in addition to holding fluid reserves. The major functions include filtration of the fluid, heat dissipation, and water separation.

Hydraulic pump

This is the energy source for hydraulic systems. It converts electrical energy into dynamic, hydraulic pressure. In almost all cases, hydraulic systems utilize positive displacement pumps as their primary power source. These are broken down into two primary sub-classifications: constant-volume or variable-volume. In the former, the pumps are designed to deliver a fixed output (i.e. both volume and pressure) of hydraulic fluid. In the later, the pump delivers only the volume or pressure required for specific functions of the system or its components.

Control valves

The energy generated by the hydraulic pump must be directed and controlled so that the energy can be used. A variety of directional and functional control valves are designed to provide a wide range of control functions.

Actuating devices

The energy within a hydraulic system is of no value until it is converted into work. Typically, this is accomplished by using an actuating device of some type. This actuating device may be a cylinder, which converts the hydraulic energy into linear mechanical force; a hydraulic motor, that converts energy into rotational force; or a variety of other actuators designed to provide specific work functions.

Relief valves

Most hydraulic systems use a positive displacement pump to generate energy within the system. Unless the pressure is controlled, these pumps will generate excessive pressure that can cause catastrophic failure of system component. A relief valve is always installed downstream of the hydraulic pump to prevent excessive pressure and to provide a positive relief should a problem develop within the system. The relief valve is designed to open at a preset system pressure. When the valve opens, it diverts flow to the receiver tank or reservoir.

Lines (pipe, tubing or flexible hoses)

All systems require some means to transmit hydraulic fluid from one component to another. The material of the connecting lines will vary from system to system or within the system.

Hydraulic fluid

The fluid provides the vehicle that transmits input power, such as from a hydraulic pump to the actuator device or devices that perform work.

40.3 Forces in liquids

The study of liquids is divided into two main parts: liquids at rest, hydrostatics, and liquids in motion, hydraulics. The effect of liquids at rest can often be expressed by simple formulas. The effects of liquids in motion are more difficult to express due to frictional and other factors whose actions cannot be expressed by simple mathematics.

Liquids are almost incompressible. For example, if a pressure of 100 pounds per square inch, psi, is applied to a given volume of water that is at atmospheric pressure, the volume will decrease by only 0.03 per cent. It would take a force of approximately 32 tons to reduce its volume by 10 per cent; however, when this force is removed, the water immediately returns to its original volume. Other liquids behave in about the same manner as water.

Another characteristic of a liquid is the tendency to keep its free surface level. If the surface is not level, liquids will flow in the direction which will tend to make the surface level.

40.3.1 Liquids at rest (hydrostatics)

In the study of fluids at rest, we are concerned with the transmission of force and the factors which affect the forces in liquids. Additionally, pressure in and on liquids and factors affecting pressure are of great importance.

Pressure and force

The terms force and pressure are used extensively in the study of fluid power. It is essential that we distinguish between these terms. Force is the total pressure applied to or generated by a system. It is the total pressure exerted against the total area of a particular surface and is expressed in pounds or grams.

Pressure is the amount of force applied to each unit area of a surface and is expressed in pounds per square inch, lb/in^2 (psi) or grams per square centimeter, gm/cm^2. Pressure may be exerted in one direction, in several directions, or in all directions.

A formula is used in computing force, pressure, and area in fluid power systems. In this formula, P refers to pressure; F indicates force, and A represents area. Force equals pressure times area. Thus, the formula is written:

$$F = P \times A$$

Pressure equals force divided by area. By rearranging the formula, this statement may be condensed to:

$$P = \frac{F}{A}$$

Since area equals force divided by pressure, the formula is written:

$$A = \frac{F}{P}$$

Atmospheric pressure

The atmosphere is the entire mass of air that surrounds the earth. While it extends upward for about 500 miles, the section of primary interest is the portion that rests on the earth's surface and extends upward for about $7\frac{1}{2}$ miles. This layer is called the troposphere.

If a column of air 1-inch square extended to the 'top' of the atmosphere could be weighed, this column of air would weigh approximately 14.7 pounds at sea level. Thus, atmospheric pressure, at sea level, is approximately 14.7 pounds per square inch or psi.

Atmospheric pressure decreases by approximately 1.0 psi for every 2343 feet of elevation. Elevations below sea level, such as in excavations and depressions, atmospheric pressure increases. Pressures under water differ from those under air only because the weight of the water must be added to the pressure of the air.

Atmospheric pressure can be measured by any of several methods. The common laboratory method uses a mercury column barometer. The height of the mercury column serves as an indicator of atmospheric pressure. At sea level and at a temperature of 0° Celsius (C), the height of the mercury column is approximately 30 inches, or 76 centimeters. This represents a pressure of approximately 14.7 psia. The 30-inch column is used as a reference standard.

Atmospheric pressure does not vary uniformly with altitude. It changes more rapidly at lower altitudes because of the compressibility of air, which causes the air layers close to the earth's surface to be compressed by the air masses above them. This effect, however, is partially counteracted by the contraction of the upper layers due to cooling. The cooling tends to increase the density of the air.

Atmospheric pressures are quite large, but in most instances, practically the same pressure is present on all sides of objects so that no single surface is subjected to a greater load. Atmospheric pressure acting on the surface of a liquid, Figure 40.3 view (a), is transmitted equally

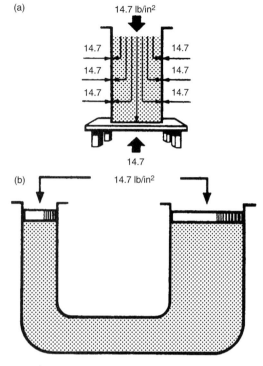

Figure 40.3 Effects of atmospheric pressure

throughout the liquid to the walls of the container, but is balanced by the same atmospheric pressure acting on the outer walls of the container. In view (b), Atmospheric pressure acting on the surface of one piston is balanced by the same pressure acting on the surface of the other piston. The different areas of the two surfaces make no difference, since for a unit of area, pressures are balanced.

$$= \frac{62.4 \text{ lbs}}{1 \text{ ft}^2}$$

Pascal's law

The foundation of modern hydraulics was established with Pascal's discovery that pressure in a fluid acts equally in all directions. This pressure acts at right angles to the containing surfaces. If some type of pressure gauge, with an exposed face, is placed beneath the surface of a liquid, Figure 40.4, at a specific depth and pointed in different directions, the pressure will read the same. Thus, we can say that pressure in a liquid is independent of direction.

Pressure due to weight of a liquid, at any level, depends on the depth of the fluid from the surface. If the exposed face of the pressure gauge, Figure 40.4, is moved closer to the surface of the liquid, the indicated pressure will be less.

When the depth is doubled, the indicated pressure is also doubled. Thus, the pressure in a liquid is directly proportional to the depth. Consider a container with vertical sides, Figure 40.5, that is 1 foot high and 1 foot wide.

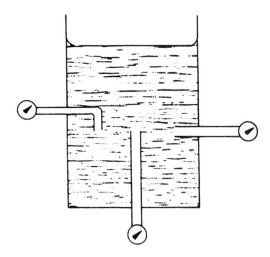

Figure 40.4 Pressure of a liquid is independent of direction

Let it be filled with water 1 foot deep, thus providing 1 ft³ of water. We learned earlier in this chapter that 1 ft³ of water weighs 62.4 pounds. Using this information and the equation for pressure, we can calculate the pressure on the bottom of the container.

$$P = \frac{F}{A}$$

$$P = \frac{62.4}{144} = 0.433 \, \text{lbs/inch}^2$$

Since there are 144 square inches in 1 square foot, this can be stated as follows: the weight of a column of water

1 foot high, having a cross-sectional area of 1 square inch, is 0.433 pounds. If the depth of the column is tripled, the weight of the column will be 3×0.433 or 1.299 pounds and the pressure at the bottom will be 1.299 lb/in² (psi), since the pressure is equal to the force divided by the area. Thus, the pressure at any depth in a liquid is equal to the weight of the column of liquid at the depth divided by the cross-sectional area of the column at that depth. The volume of a liquid that produces the pressure is referred to as the fluid head of the liquid. The pressure of a liquid due to its fluid head is also dependent on the density of the liquid.

If we let A equal any cross-sectional area of a liquid column and h equal the depth of the column, the volume becomes Ah. Using the equation $D = W/V$, the weight of the liquid above area A is equal to AhD, or:

$$D = \frac{W}{Ah} \quad W = Ah \times D$$

Since pressure is equal to the force per unit area, set A equal to 1. Then the formula for pressure becomes:

$$P = hD$$

It is essential that h and D be expressed in similar units. That is, if D is expressed in pounds per cubic foot, the value of h must be expressed in feet. If the desired pressure is to be expressed in pounds per square inch, the pressure formula becomes:

$$P = \frac{hD}{144}$$

Pascal was also the first to prove by experiment that the shape and volume of a container in no way alters pressure. Thus in Figure 40.6, if the pressure due to the weight of the liquid at a point on horizontal line H is 8 psi, the pressure is 8 psi everywhere at level H in the system.

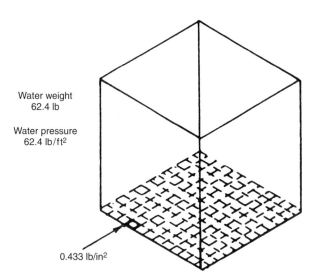

Water weight
62.4 lb

Water pressure
62.4 lb/ft²

0.433 lb/in²

Figure 40.5 Water pressure in a 1-cubic foot container

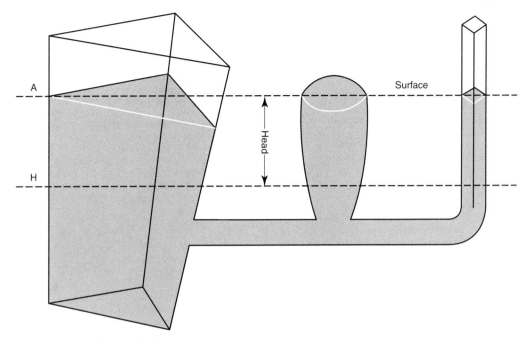

Figure 40.6 Pressure relationship with shape

40.3.2 Liquids in motion (hydraulics)

In the operation of fluid power systems, there must be flow of fluid. The amount of flow will vary from system to system. To understand fluid power systems, it is necessary to understand some of the characteristics of liquids in motion.

Liquids in motion have characteristics different from liquids at rest. Frictional resistances within the fluid, viscosity, and inertia contribute to these differences. Inertia, which means the resistance a mass offers to being set in motion, will be discussed later in this section. Other relationships of liquids in motion you must be familiar with. Among these are volume and velocity of flow; flow rate, and speed; laminar and turbulent flow; and more importantly, the force and energy changes which occur in flow.

Volume and velocity of flow

The volume of a liquid passing a point in a given time is known as its volume of flow or flow rate. The volume of flow is usually expressed in gallons per minute (gpm) and is associated with the relative pressures of the liquid, such as 5 gpm at 40 psig. The velocity of flow or velocity of the fluid is defined as the average speed at which the fluid moves past a given point. It is usually expressed in feet per minute (fpm) or inches per second (ips). Velocity of flow is an important consideration in sizing the hydraulic piping and other system components.

Volume and velocity of flow must be considered together. With other conditions unaltered, the velocity of

flow increases as the cross-section or size of the pipe decreases, and the velocity of flow decreases as the cross-section or pipe size increases. For example, the velocity of flow is slow at wide parts, yet the volume of liquid passing each part of the stream is the same.

In Figure 40.7, if the cross-sectional area of the pipe is 16 square inches at point A and 4 square inches at point B, we can calculate the relative velocity of flow using the flow equation:

$$Q = vA$$

Where Q is the volume of flow, v is the velocity of flow and A is the cross-sectional area of the liquid. Since the volume of flow at point a, Q_1, is equal to the volume of flow at point B, Q_2, we can use this equation to determine the ratio of the velocity of flow at point A, v_1, to the velocity of flow at point B, v_2. Since

$$Q_1 = Q_2, \text{ then } A_1 v_1 = A_2 v_2$$

From Figure 40.7:

$$A_1 = 16 \text{ square inches}, A_2 = 4 \text{ square inches}$$

Substituting:

$$16 v_1 = 4 v_2 \text{ or } v_2 = 4 v_1$$

Therefore, the velocity of flow at point B is four times greater than the velocity of flow at point A.

Volume of flow and speed

When you consider the cylinder volume that must be filled and the distance that the piston must travel, you can relate

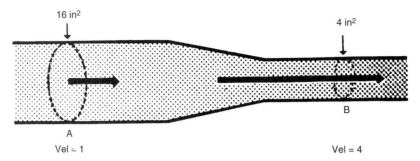

Figure 40.7 Volume and velocity of flow

the volume of flow to the speed of the piston. The volume of the cylinder is found by multiplying the piston area by the length the piston must travel. This length is known as stroke.

Suppose you have determined that two cylinders have the same volume and that one cylinder is twice as long as the other. In this case, the cross-sectional area of the longer tube will be half of the cross-sectional area of the other tube. If fluid is pumped into each cylinder at the same rate, both pistons will reach their full travel at the same time. However, the piston in the smaller cylinder must travel twice as fast because it has twice as far to go. There are two ways of controlling the speed of the piston, (1) by varying the size of the cylinder and (2) by varying the volume of flow (gpm) to the cylinders.

Streamline and turbulent flow

At low velocities or in tubes of small diameter, flow is streamlined. This means that a given particle of fluid moves straight forward without bumping into other particles and without crossing their paths. Streamline flow is often referred to as laminar flow, which is defined as a flow situation in which fluid moves in parallel lamina or layers. An example of streamline flow is an open stream flowing at a slow, uniform rate with logs floating on its surface. The logs represent particles of fluid. As long as the stream flows at a slow, uniform rate, each log floats downstream in its own path, without crossing or bumping into the others.

If the stream narrows and the volume of flow remains the same, the velocity will increase. If the velocity increases sufficiently, the water becomes turbulent. Swirls, eddies, and cross-motions are set up in the water. As this happens, the logs are thrown against each other and against the banks of the stream, and the paths followed by different logs will cross and re-cross.

Particles of fluid flowing in pipes act in the same manner. The flow is streamlined if the fluid flows slowly enough, and remains streamlined at greater velocities if the diameter of the pipe is small. If the velocity of flow or size of pipe is increased sufficiently, the flow becomes turbulent.

While a high velocity of flow will produce turbulence in any pipe, other factors contribute to turbulence. Among these are the roughness of the inside of the pipe, obstructions, the degree of curvature of bends, and the number of bends in the pipe. In setting up or maintaining fluid power systems, care should be taken to eliminate or minimize as many causes of turbulence as possible, since energy consumed by turbulence is wasted.

While designers of fluid power equipment do what they can to minimize turbulence, it cannot be avoided. For example, in a 4-inch pipe at 68°F, flow becomes turbulent at velocities over approximately 6 inches per second (ips) or about 3 ips in a 6-inch pipe. These velocities are far below those commonly encountered in fluid power systems, where velocities of 5 feet per second (fps) and above are common. In laminar flow, losses due to friction increase directly with velocity. With turbulent flow, these losses increase much more rapidly.

Factors involved in flow

An understanding of the behavior of fluids in motion, or solids for that matter, requires an understanding of the term inertia. Inertia is the term used by scientists to describe the property possessed by all forms of matter that make it resist being moved when it is at rest and to resist any change in its rate or motion when it is moving.

The basic statement covering inertia is Newton's first law of motion. His first law states: A body at rest tends to remain at rest, and a body in motion tends to remain in motion at the same speed and direction, unless acted on by some unbalanced force. This simply says what you have learned by experience – that you must push an object to start it moving and push it in the opposite direction to stop it again.

A familiar illustration is the effort a pitcher must exert to make a fast pitch and the opposition the catcher must put forth to stop the ball. Similarly, the engine to make an automobile begin to roll must perform considerable work; although, after it has attained a certain velocity, it will roll along the road at uniform speed if just enough effort is expended to overcome friction, while brakes are necessary to stop its motion. Inertia also explains the kick or recoil of guns and the tremendous striking force of projectiles.

Inertia and force

To overcome the tendency of an object to resist any change in its state of rest or motion, some force that is not otherwise canceled or balanced must act on the object. Some unbalanced force must be applied whenever fluids are set in motion or increased in velocity; while conversely, forces are made to do work elsewhere whenever fluids in motion are retarded or stopped.

There is a direct relationship between the magnitude of the force exerted and the inertia against which it acts. This force depends on two factors: (1) the mass of the object and (2) the rate at which the velocity of the object is changed. The rule is that the force, in pounds, required to overcome inertia is equal to the weight of the object multiplied by the change in velocity, measured in feet per second (fps) and divided by 32 times the time, in seconds, required to accomplish the change. Thus, the rate of change in velocity of an object is proportional to the force applied. The number 32 appears because it is the conversion factor between weight and mass.

Five physical factors act on a fluid to affect its behavior. All of the physical actions of fluids in all systems are determined by the relationship of these five factors to each other. Summarizing, these five factors are:

1. Gravity, which acts at all times on all bodies, regardless of other forces.
2. Atmospheric pressure, which acts on any part of a system exposed to the open air.
3. Specific applied forces, which may or may not be present, but which are entirely independent of the presence or absence of motion.
4. Inertia, which comes into play whenever there is a change from rest to motion, or the opposite, or whenever there is a change in direction or in rate of motion.
5. Whenever there is motion, friction is always present.

Kinetic energy

An external force must be applied to an object in order to give it a velocity or to increase the velocity it already has. Whether the force begins or changes velocity, it acts over a certain distance. Force acting over a certain distance is called work. Work and all forms into which it can be changed are classified as energy. Obviously then, energy is required to give an object velocity. The greater the energy used, the greater the velocity will be.

Disregarding friction, for an object to be brought to rest or for its motion to be slowed down, a force opposed to its motion must be applied to it. This force also acts over some distance. In this way, energy is given up by the object and delivered in some form to whatever opposed its continuous motion. The moving object is therefore a means of receiving energy at one place and delivering it to another point. While it is in motion, it is said to contain this energy, as energy of motion or kinetics energy.

Since energy can never be destroyed, it follows that if friction is disregarded the energy delivered to stop the object will exactly equal the energy that was required to increase its speed. At all times, the amount of kinetic energy possessed by an object depends on its weight and the velocity at which it is moving.

The mathematical relationship for kinetic energy is stated in the rule. Kinetic energy, in foot-pounds, is equal to the force, in pounds, which created it, multiplied by the

distance through which it was applied, or to the weight of the moving object, in pounds, multiplied by the square of its velocity, in feet per second, and divided by 64.

The relationship between inertia forces, velocity, and kinetic energy can be illustrated by analyzing what happens when a gun fires a projectile against the armor of an enemy ship. The explosive force of the powder in the breach pushes the projectile out of the gun, giving it a high velocity. Because of its inertia, the projectile offers opposition to this sudden velocity and a reaction is set up that pushes the gun backwards. The force of the explosion acts on the projectile throughout its movement in the gun. This is force acting through a distance producing work. This work appears as kinetic energy in the speeding projectile. The resistance of the air produces friction, which uses some of the energy and slows down the projectile. When the projectile hits the target, it tries to continue moving. The target, being relatively stationary, tends to remain stationary because of inertia. The result is that a tremendous force is set up that either leads to the penetration of the armor or the shattering of the projectile. The projectile is simply a means to transfer energy from the gun to the enemy ship. This energy is transmitted in the form of energy in motion or kinetic energy.

A similar action takes place in a fluid power system in which the fluid takes the place of the projectile. For example, the pump in a hydraulic system imparts energy to the fluid, which overcomes the inertia of the fluid at rest and causes it to flow through the lines. The fluid flows against some type of actuator that is at rest. The fluid tends to continue flowing, overcomes the inertia of the actuator, and moves the actuator to do work. Friction uses up a portion of the energy as the fluid flows through the lines and components.

Relationship of force, pressure, and head

In dealing with fluids, forces are usually considered in relation to the areas over which they are applied. As previously discussed, a force acting over a unit area is a pressure, and pressure can alternately be stated in pounds per square inch or in terms of head, which is the vertical height of the column of fluid whose weight would produce that pressure.

All five of the factors that control the actions of fluids can be expressed either as force or in terms of equivalent pressures or head. In either situation, the different factors are referred to in the same terms.

Static and dynamic factors

Gravity, applied forces, and atmospheric pressure are examples of static factors that apply equally to fluids at rest or in motion. Inertia and friction are dynamic forces that apply only to fluids in motion. The mathematical sum of gravity, applied forces, and atmospheric pressure is the static pressure obtained at any one point in a fluid system at a given point in time. Static pressure exists in addition to any dynamic factors that may also be present at the same time.

Remember that Pascal's law states that a pressure set up in a fluid acts equally in all directions and at right angles to the containing surfaces. This covers the situation only for fluids at rest. It is true only for the factors making up static head. Obviously, when velocity becomes a factor, it must have a direction of flow. The same is true of the force created by velocity. Pascal's law alone does not apply to the dynamic factors of fluid power systems.

The dynamic factors of inertia and friction are related to the static factors. Velocity head and friction head are obtained at the expense of static head. However, a portion of the velocity head can always be reconverted to static head. Force, which can be produced by pressure or head when dealing with fluids, is necessary to start a body moving if it is at rest, and is present in some form when the motion of the body is arrested. Therefore, whenever a fluid is given velocity, some part of its original static head is used to impart this velocity, which then exists as velocity head.

Bernouli's principle

Review the system illustrated in Figure 40.8. Chamber A is under pressure and is connected by a tube to chamber B, which is also under pressure. The pressure in chamber A is static pressure of 100 psi. The pressure at some point (X) along the connecting tube consists of a velocity pressure of 10 psi exerted in a direction parallel to the line of flow, plus the unused static pressure of 90 psi. The static pressure (90 psi) follows Pascal's law and exerts equal pressure in all directions. As the fluid enters chamber B, it slows down and its velocity is reduced. As a volume of liquid moves from a small, confined space into a larger area, the fluid will expand to fill the greater volume. The result of this expansion is a reduction of velocity and a momentary reduction in pressure.

In the example, the force required to absorb the fluid's inertia equals the force required to start the fluid moving originally, so that the static pressure in chamber B is equal to that in chamber A.

This example disregards friction. Therefore, it would not be encountered in actual practice. Force or head is also required to overcome friction. Unlike inertia, this force cannot be recovered. Even though the energy required overcoming friction still exists, it has been converted to heat. In an actual system, the pressure in chamber B would be less than that in chamber A. The difference would be the amount of pressure used to overcome friction within the system.

At all points in a system, the static pressure is always equal to the original static pressure less any velocity head at a specific point in the system and less the friction head required to reach that point. Since both the velocity head and friction head represent energy and energy cannot be destroyed, the sum of the static head, the velocity head, and the friction head at any point in the system must add up to the original static head. This is known as Bernoulli's principal, which states: For the horizontal flow of fluids through a tube, the sum of the pressure and the kinetic energy per unit volume of the fluid is constant. This principle governs the relationship of the static and dynamic factors in hydraulic systems.

Minimizing friction

Fluid power equipment is designed to reduce friction as much as possible. Since energy cannot be destroyed, some of the energy created by both static pressure and velocity is converted to heat energy as the fluid flows through the piping and components within a hydraulic system. As friction increases, so does the amount of dynamic and static energy that is converted into heat.

To minimize the loss of useable energy lost to its conversion to heat energy, care must be taken in the design, installation and operation of hydraulic system. As a minimum the following factors must be considered:

Proper fluid must be chosen and used in the system. It must have the best viscosity, operating temperature range and other characteristics that are conducive to proper operation of the system and lowest possible friction component.

Fluid flow is also critical for proper operation of a hydraulic system. Turbulent flow should be avoided as much as possible. Clean, smooth pipe or tubing should be used to provide laminar flow and the lowest friction possible within the system. Sharp, close radius bends and sudden changes in cross-sectional area are avoided.

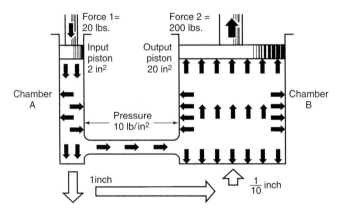

Figure 40.8 Relation of static and dynamic factors

System components, such as pumps, valves and gauges, create both turbulent flow and high friction components. Pressure drop, or a loss of pressure, is created by a combination of turbulent flow and friction as the fluid flows through the unit. System components that are designed to provide minimum interruption of flow and pressure should be selected for the system.

40.3.4 Transmission of force through liquids

When the end of a solid bar is struck, the main force of the blow is carried straight through the bar to the other end (Figure 40.9, view A). This happens because the bar is rigid. The direction of the blow almost entirely determines the direction of the transmitted force. The more rigid the bar, the less force is lost inside the bar or transmitted outward at right angles to the direction of the blow.

When a force is applied to the end of a column of confined liquid (Figure 40.9, view B), it is transmitted straight through to the other end. It is also equal and undiminished in every direction throughout the column – forward, backward and sideways – so that the containing vessel is literally filled with the added pressure.

So far, we have explained the effects of atmospheric pressure on liquids and how external forces are distributed through liquids. Let us now focus our attention on forces generated by the weight of liquids themselves. To do this, we must first discuss density, specific gravity and Pascal's law.

40.3.5 Pressure and force in hydraulic systems

According to Pascal's law, any force applied to a confined fluid is transmitted uniformly in all directions throughout the fluid regardless of the shape of the container. Consider the effect of this in the system shown in Figure 40.10.

If there is a resistance on the output piston and the input piston is pushed downward, a pressure is created through the fluid which acts equally at right angles to surfaces in all parts of the container. If force 1 is 100 pounds and the area of the input piston is 10 square inches, then the pressure in the fluid is 10 psi.

$$\frac{100 \text{ lbs}}{10 \text{ square inches}}$$

Note: Fluid pressure cannot be created without resistance to flow. In this case, the equipment to which the output piston is attached provides resistance. The force of resistance acts against the top of the output piston. The pressure is created in the system, by the input piston pushing on the underside of the output piston with a force of 10 pounds per square inch.

In this case, the fluid column has a uniform cross section, so the area of the output piston is the same as the area of the input piston, or 10 square inches. Therefore, the upward force on the output piston is 100 pounds and is equal to the force applied to the input piston. All that was accomplished in this system was to transmit the 100 pounds of force around the bend. However, this principle underlies practically all mechanical applications of hydraulics or fluid power.

At this point, you should note that since Pascal's law is independent of the shape of the container, it is not necessary that the tube connecting the two pistons has the same cross-sectional area of the pistons. A connection of any size, shape or length will do, as long as an unobstructed passage is provided. Therefore, the system shown in Figure 40.11, with a relatively small, bent pipe connecting two cylinders will act the same as the system shown in Figure 40.10.

40.3.6 Multiplication of force

Unlike the preceding discussion, hydraulic systems can provide mechanical advantage or a multiplication of input force. Figure 40.12 illustrates an example of an increase in output force. Assume that the area of the input piston is 2 square inches. With a resistant force on the output piston, a downward force of 20 pounds acting on the input

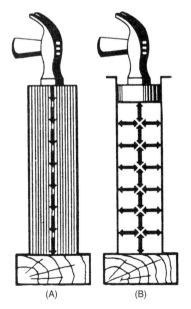

Figure 40.9 Transmission of force: (A) solid; (B) fluid

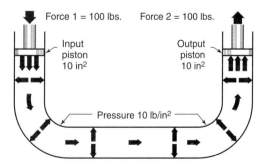

Figure 40.10 Force transmitted through fluid

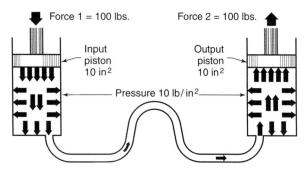

Figure 40.11 Transmitting force through small pipe

piston will create a pressure of $\frac{20}{2}$ or 10 psi in the fluid. Although this force is much smaller than the force applied in Figures 40.10 and 40.11, the pressure is the same. This is because the force is applied to a smaller area.

This pressure of 10 psi acts on all parts of the fluid container, including the bottom of the output piston. The upward force on the output piston is 200 pounds (10 psi × piston area). In this case, the original force has been multiplied tenfold while using the same pressure in the fluid as before. In any system with these dimensions, the ratio of output force to input force is always 10 to 1, regardless of the applied force. For example, if the applied force of the input piston is 50 pounds, the pressure in the system will be 25 psi. This will support a resistant force of 500 pounds on the output piston. The system works the same in reverse.

If we change the applied force and place a 200-pound force on the output piston, Figure 40.12, making it the input piston, the output force on the input piston will be one-tenth the input force, or 20 pounds. Therefore, if two pistons are used in a fluid power system, the force acting on each piston is directly proportional to its area, and the magnitude of each force is the product of the pressure and the area of each piston.

Differential areas

Figure 40.13 is a simple example of differential pressure. The figure illustrates a single piston, with a surface area of $6\,in^2$, attached to a piston rod, with an area of $2\,in^2$. Without any external force applied to the end of the piston rod, an equal force, 20 psig, is applied to both sides of the piston, and will cause the piston to move to the right. This motion is the result of differential forces. Even though the input force, 20 psig, is applied to both sides of the piston the difference in area, $6\,in^2$ on the left face and $2\,in^2$ on the right face, will cause the piston to move. The opposed faces of the piston behave like two pistons acting against each other. The area of one face is the full cross-sectional area of the cylinder or $6\,in^2$, while the area of the opposing face is the area of the piston minus the area of the piston rod, or $2\,in^2$. This leaves an effective area of $4\,in^2$ on the piston rod side of the piston.

The force acting on the left side of the piston is equal to 20 psi × $6\,in^2$ or 120 pounds. The opposing force generated by the right side of the piston is 20 psi × $4\,in^2$ or 80 pounds. Therefore, there is a net unbalanced force of 40 pounds (120 − 80) acting the right, and the piston will move in that direction.

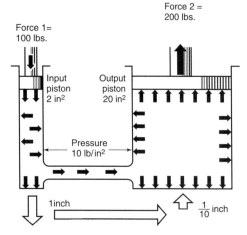

Figure 40.12 Multiplication of forces

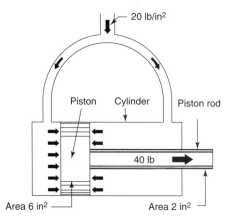

Figure 40.13 Differential areas on a piston

Volume and distance factors

You have learned that if a force is applied to a system and the cross-sectional areas of the input and output are equal, the force on the input piston will support an equal resistant force on the output piston. The pressure of the liquid at this point is equal to the force applied to the input piston divided by the piston's area. Let us now look at what happens when a force greater than the resistance is applied to the input piston.

In the system illustrated in Figure 40.10, assume that the resistant force on the output piston is 100 pounds. If a force slightly greater than 100 pounds is applied to the input piston, the pressure in the system will be slightly greater than 10 psi. This increase in pressure will overcome the resistant force on the output piston. If the input piston is forced downward 1 inch, the movement displaces $10 \, \text{in}^3$ of fluid. The fluid must go somewhere. Since the system is closed and the fluid is practically incompressible, the fluid will move the right side of the system. Because the output piston also has a cross-sectional area of $10 \, \text{in}^2$, it will move upward 1 inch to accommodate the $10 \, \text{in}^3$ of fluid. You may generalize this by saying that if two pistons in a closed system have equal cross-sectional areas and one piston is pushed and moved, the other piston will move the same distance in the opposite direction. This is because a decrease in volume in one part of the system is balanced by an equal increase in volume in another part of the system.

Apply this reasoning to the system is Figure 40.11. If the input piston is pushed down a distance of 1 inch, the volume in the left cylinder will decrease by $2 \, \text{in}^3$. At the same time, the volume in the right cylinder will increase by $2 \, \text{in}^2$. Since the diameter of the right cylinder cannot change, the piston must move upward to allow the volume to increase. The piston will move a distance equal to the volume increase divided by the surface area of the piston. In this example, the piston will move one-tenth of an inch ($2 \, \text{in}^3 / 20 \, \text{in}^2$).

This leads to the second basic rule for fluid power systems that contain two pistons: The distances the pistons move are inversely proportional to the areas of the pistons. Or more simply, if one piston is smaller than the other, the smaller piston must move a greater distance than the larger piston any time the pistons move.

40.4 Hydraulic pumps

The purpose of a hydraulic pump is to supply the flow of fluid required by a hydraulic system. The pump does not create system pressure. System pressure is created by a combination of the flow generated by the pump and the resistance to flow created by friction and restrictions within the system.

As the pump provides flow, it transmits a force to the fluid. When the flow encounters resistance, this force is changed into pressure. Resistance to flow is the result of a restriction or obstruction in the flow path. This restriction is normally the work accomplished by the hydraulic system, but there can also be restrictions created by the lines, fittings or components within the system. Thus, the load imposed on the system or the action of a pressure-regulating valve controls the system pressure.

40.4.1 Operation

A pump must have a continuous supply of fluid available to its inlet port before it can supply fluid to the system. As the pump forces fluid through the outlet port, a partial vacuum or low-pressure area is created at the inlet port. When the pressure at the inlet port of the pump is lower than the atmospheric pressure, the atmospheric pressure acting on the fluid in the reservoir must force the fluid into the pump's inlet. This is called a suction lift condition.

40.4.2 Performance

Pumps are normally rated by their volumetric output and discharge pressure. Volumetric output is the amount of fluid a pump can deliver to its outlet port in a certain period of time and at a given speed. Volumetric output is usually expressed in gallons per minute (gpm).

Since changes in pump speed affect volumetric output, some pumps are rated by their displacement. Pump displacement is the amount of fluid the pump can deliver per cycle or complete rotation. Since most pumps use a rotary drive, displacement is usually expressed in terms of cubic inches per revolution.

While pumps do not directly create pressure, the system pressure created by the restrictions or work performed by the system has a direct affect on the volumetric output of the pump. As the system pressure increases, the volumetric output of the pump decreases. This drop in volumetric output is the result of an increase in the amount of leakage within the pump. This leakage is referred to as pump slippage or slip. A factor must be considered in all hydraulic pumps.

Pump ratings

Pumps are generally rated by their maximum operating pressure capability and their output in gallons per minute (gpm) at a given operating speed.

Pressure

The manufacturer based on reasonable service life expectancy under specified operating conditions determine the pressure rating of a pump. It is important to note that there is no standard industry-wide safety factor in this rating. Operating at higher pressure may result in reduced pump life or damage that is more serious.

Displacement

The flow capacity of a pump can be expressed as its displacement per revolution or by its output in gallons per minute (gpm). Displacement is the volume of liquid transferred in one complete cycle of pump operation. It is equal to the volume of one pumping chamber multiplied by the number of chambers that pass the outlet during one complete revolution or cycle. Displacement is expressed in cubic inches per revolution.

Most pumps that are used in hydraulic applications have a fixed displacement which cannot be changed except by replacing certain components. However, in some, it is possible to vary the size of the pumping chamber and thereby the displacement by means of external controls. Some unbalanced vane pumps and many piston units can be varied from maximum to zero delivery or even to reverse flow without modification to the pump's internal configuration.

Volumetric efficiency

In theory, a pump delivers an amount of fluid equal to its displacement each cycle or revolution. In reality, the actual output is reduced because of internal leakage or slippage. As pressure increases, the leakage from the outlet to the inlet or to the drain also increases and the volumetric efficiency decreases.

Volumetric efficiency is equal to the actual output divided by the theoretical output. It is expressed as a percentage.

$$\text{Efficiency} = \frac{\text{Actual output}}{\text{Theoretical output}} \times 100$$

For example, if a pump theoretically should deliver 10 gpm by delivers only 9 gpm at 1000 psig, its volumetric efficiency at that pressure is 90 per cent.

$$\text{Efficiency} = \frac{9\,\text{gpm}}{10\,\text{gpm}} \times 100 = 90\%$$

If the discharge pressure is increased, the amount of slippage will increase. If we increase the pressure in the above example, to 1500 psig, the actual output may drop to 8 gpm. Therefore, the volumetric efficiency will decrease to 80 per cent at 1500 psig.

40.5 Hydraulic fluids

Selection and care of the hydraulic fluid for a machine will have an important effect on how it performs and on the life of the hydraulic components. During the design of equipment that requires fluid power, many factors are considered in selecting the type of system to be used-hydraulic, pneumatic, or a combination of the two. Some of the factors required are speed and accuracy of operation, surrounding atmospheric conditions, economic conditions, availability of replacement fluid, required pressure level, operating temperature range, contamination possibilities, cost of transmission lines, limitations of the equipment, lubricity, safety to the operators, and expected service life of the equipment.

After the type of system has been selected, many of these same factors must be considered in selecting the fluid for the system. This chapter is devoted to hydraulic fluids. Included in it are sections on the properties and characteristics desired of hydraulic fluids; types of hydraulic fluids; hazards and safety precautions for working with, handling, and disposing of hydraulic liquids; types and control of contamination; and sampling.

40.5.1 Purpose of the fluid

As a power transmission medium, the fluid must flow easily through lines and component passages. Too much resistance to flow creates considerable power loss. The fluid also must be as incompressible as possible so that action is instantaneous when the pump is started or a valve shifts.

Lubrication

In most hydraulic components, the hydraulic fluid provides internal lubrication. Pump elements and other wear parts slide against each other on a film of fluid, Figure 40.14.

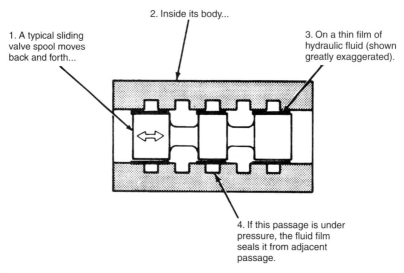

2. Inside its body...

1. A typical sliding valve spool moves back and forth...

3. On a thin film of hydraulic fluid (shown greatly exaggerated).

4. If this passage is under pressure, the fluid film seals it from adjacent passage.

Figure 40.14 Fluid lubricates working parts

For long component life, the oil must contain the necessary additives to ensure high anti-wear characteristics. Not all hydraulic oils contain these additives.

Sealing

In many applications, hydraulic fluid is the only seal against pressure inside system components. In Figure 40.14, there is no seal ring between the valve spool and body to prevent leakage from the high-pressure passage to the low-pressure passage. The close mechanical fit and viscosity of the hydraulic fluid determine leakage rate.

Cooling

Circulation of the hydraulic oil through lines, heat exchangers and the walls of the reservoir, Figure 40.15, gives up heat that is generated within the system. Without this cooling, the heat generated by the hydraulic pump and mechanical work performed by system actuators would build to a point that damage of system components could cause premature failure of the system.

40.5.2 Properties

If fluidity, the physical property of a substance that enables it to flow and incompressibility were the only properties required, any liquid that is not too thick might be used in a hydraulic system. However, a satisfactory liquid for a particular system must possess a number of other properties. The most important properties and some characteristics are discussed in the following paragraphs.

Density and specific gravity

The density of a substance is its weight per unit of volume. The unit of volume in the English system of measurement is 1 cubic foot or $1\,\text{f}^3$. To find the density of a substance, you must know its weight and volume. You then divide its weight by its volume to find the weight per unit volume.

In equation form, this is written as:

$$D = \frac{W}{V}$$

Example: The liquid that fills a certain container weighs 1496.6 pounds. The container is 4 feet long, 3 feet wide and 2 feet deep. Its volume is 24 cubic feet (4 ft × 3 ft × 2 ft). If $24\,\text{ft}^3$ of this liquid weighs 1497.6 pounds, then 1 cubic foot weighs:

$$D = \frac{1,497.6}{24}$$

or 62.4 pounds. Therefore, the density of the liquid is 62.4 pounds per cubic foot or $62.4\,\text{lbs/ft}^3$. This is the density of water at 40°C and is usually used as the standard for comparing densities of other substances. The temperature of 40°C was selected because water has its maximum density at this temperature. This standard temperature is used whenever the density of liquids and solids is measured. Changes in temperature will not change the weight of a substance but will change the volume of the substance by expansion or contraction, thus changing the weight per unit volume, density.

In physics, the word specific implies a ratio. Weight is the measure of the earth's attraction for a body, which is called gravity. Thus, the ratio of the weight of a unit volume of some substance to the weight of an equal volume of a standard substance, measured under standard pressure and temperature, is called specific gravity. The terms specific weight and specific density are also sometimes used to express this ratio.

$$\text{Specific gravity, Sp.Gr.} = \frac{\text{Weight of the substance}}{\text{Weight of an equal volume of water}}$$

or, the specific gravity of water is 1.0 or a density of $62.4\,\text{lbs/ft}^3$. If a cubic foot of a liquid weighs 68.64 pounds, then its specific gravity is 1.1 or

$$\frac{68.64}{62.4}$$

$$\text{Sp.Gr.} = \frac{\text{Density of the substance}}{\text{Density of water}}$$

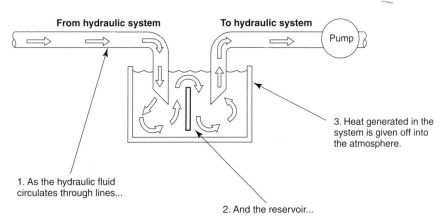

From hydraulic system　　**To hydraulic system**

Pump

3. Heat generated in the system is given off into the atmosphere.

1. As the hydraulic fluid circulates through lines...

2. And the reservoir...

Figure 40.15 Circulating oil cools system

Thus, the specific gravity of the liquid is the ratio of its density to the density of water. If the specific gravity of a liquid or solid is known, the density can be obtained by multiplying its specific gravity by the density of water. For example, if a hydraulic fluid has a specific gravity of 0.8, 1 ft³ of the fluid weighs 0.8 times the weight of water (62.4 pounds) or 49.92 pounds.

Specific gravity and density are independent of the size of the sample and depend only on the substance of which it is made.

Viscosity

Viscosity is one of the most important properties of hydraulic fluids. It is a measure of a fluid's resistance to flow. A liquid such as gasoline which flows easily has a low viscosity, and a liquid such as tar which flows slowly has a high viscosity. The viscosity of a liquid is affected by changes in temperature and pressure. As the temperature of liquid increases, its viscosity decreases. That is, a liquid flows more easily when it is hot than when it is cold. The viscosity of a liquid will increase as the pressure on the liquid increases.

A satisfactory liquid for a hydraulic system must be thick enough to give a good seal at pumps, motors, valves, and so on. These components depend on close fits for creating and maintaining pressure. Any internal leakage through these clearances results in loss of pressure, instantaneous control, and pump efficiency. Leakage losses are greater with thinner liquids (low viscosity). A liquid that is too thin will also allow rapid wearing of moving parts, or of parts that operate under heavy loads. On the other hand, if the liquid is too thick, viscosity too high, the internal friction of the liquid will cause an increase in the liquid's flow resistance through clearances of closely fitted parts, lines, and internal passages. This results in

pressure drops throughout the system, sluggish operation of the equipment, and an increase in power consumption.

Measurement of viscosity

Viscosity is normally determined by measuring the time required for a fixed volume of a fluid, at a given temperature, to flow through a calibrated orifice or capillary tube. The instruments used to measure the viscosity of a liquid are known as viscosimeters.

In decreasing order of exactness, methods of defining viscosity include absolute (poise) viscosity; kinematic viscosity in centistokes; relative viscosity in Saybolt universal seconds (SUS); and SAE numbers.

Absolute viscosity

The resistance when moving one layer of liquid over another is the basis for the laboratory method of measuring *absolute viscosity. Poise viscosity* is defined as the force (pounds) per unit of area, in square inches, required to move one parallel surface at a speed of one centimeter-per-second past another parallel surface when the two surfaces are separated by a fluid film one centimeter thick, Figure 40.16. In the metric system, force is expressed in *dynes* and area in square centimeters. Poise is also the ratio between the shearing stress and the rate of shear of the fluid.

$$\text{Absolute viscosity} = \frac{\text{Shear stress}}{\text{Rate of shear}}$$

$$1\,\text{poise} = 1 \times \left(\frac{\text{Dyne second}}{\text{Square centimeter}}\right)$$

A smaller unit of absolute viscosity is the centipoise, which is one-hundredth of a poise or 1 centipoise = 0.01 poise.

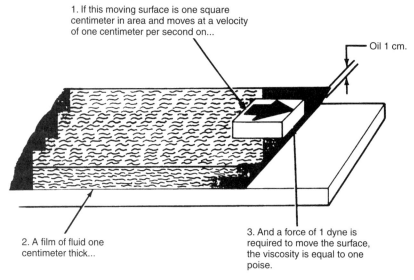

1. If this moving surface is one square centimeter in area and moves at a velocity of one centimeter per second on...

Oil 1 cm.

2. A film of fluid one centimeter thick...

3. And a force of 1 dyne is required to move the surface, the viscosity is equal to one poise.

Figure 40.16 Measuring absolute viscosity

Kinematic viscosity

The concept of kinematic viscosity is the outgrowth of the use of a head of liquid to produce a flow through a capillary tube. The coefficient of absolute viscosity, when divided by the density of the liquid is called the kinematic viscosity. In the metric system, the unit of viscosity is called the *stoke* and it has the units of centimeters squared per second. One one-hundredth of a stoke is a centistoke.

The relationship between absolute and kinematic viscosity can be stated as:

Centipoise = Centistoke × Density

or,

$$\text{Centistoke} = \frac{\text{Centipoise}}{\text{Density}}$$

SUS viscosity

For most practical purposes, it will serve to know the relative viscosity of the fluid. Relative viscosity is determined by timing the flow of a given quantity of the hydraulic fluid through a standard orifice at a given temperature. There are several methods in use. The most acceptable method in the United States is the *Saybolt viscosimeter*, Figure 40.17.

The time it takes for the measured quantity of liquid to flow through the orifice is measured with a stopwatch. The viscosity in *Saybolt universal seconds* (SUS) equals the elapsed time.

Obviously, a thick liquid will flow slowly, and the SUS viscosity will be higher than for a thin liquid, which flows

faster. Since oil becomes thicker at low temperatures and thins when warmed, the viscosity must be expressed as a specific SUS at a given temperature. Tests are usually conducted at either 100°F or 210°F.

For industrial applications, hydraulic oil viscosity is typically approximately 150 SUS at 100°F. It is a general rule that the viscosity should never go below 45 SUS or above 4000 SUS, regardless of temperature. Where temperature extremes are encountered, the fluid should have a high viscosity index.

SAE number

SAE numbers have been established by the Society of Automotive Engineers to specify ranges of SUS viscosities of oils at SAE test temperatures. Winter numbers (5W, 10W, 20W) are determined by tests at 0°F. Summer numbers (20W, 30W, etc.) designate the SUS range at 210°F. Table 40.1 is a chart of the temperature ranges.

The following formulas may be used to convert centistokes (cSt units) to approximate Saybolt universal seconds (SUS units). For SUS values between 32 and 100:

$$cST = 0.226 \times SSU - \frac{195}{SSU}$$

For SUS values greater than 100:

$$cST = 0.220 \times SSU - \frac{135}{SSU}$$

Although the viscometers discussed above are used in laboratories, there are other viscometers in the supply system that are available for local use. These viscometers can be

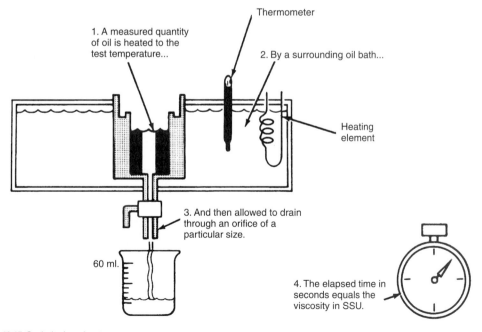

1. A measured quantity of oil is heated to the test temperature...

Thermometer

2. By a surrounding oil bath...

Heating element

3. And then allowed to drain through an orifice of a particular size.

60 ml.

4. The elapsed time in seconds equals the viscosity in SSU.

Figure 40.17 Saybolt viscosimeter

Table 40.1 SAE viscosity numbers for crankcase oils

SAE number	Viscosity units[a]	Viscosity range[b]			
		At 0°F		At 210°F	
		Minimum	Maximum	Minimum	Maximum
5W			Less than		
	Centipoise	–	1,200	–	–
	Centistokes	–	1,300	–	–
	SUS	–	6,000	–	–
10W			Less than		
	Centipoise	1,200[c]	2,400	–	–
	Centistokes	1,300	2,600	–	–
	SUS	6,000	12,000	–	–
20W			Less than		
	Centipoise	2,400[d]	9,600	–	–
	Centistokes	2,600	10,500	–	–
	SUS	12,000	48,000	–	–
20	Centistokes	–	–	5.7	Less than 9.6
	SUS	–	–	45	58
30	Centistokes	–	–	9.6	Less than 12.9
	SUS	–	–	58	70
40	Centistokes	–	–	12.9	Less than 16.8
	SUS	–	–	70	85
50	Centistokes	–	–	16.8	Less than 22.7
	SUS	–	–	85	110

[a]The official values in this classification are based upon 210°F viscosity in centistokes (ASTM D 445) and 0°F viscosities in centipoise (ASTM D260-2). Approximate values in other units of viscosity are given for information only. The approximate values at 0°F were calculated using an assumed oil density of 0.9 gm/cc at that temperature.
[b]The viscosity of all oils included in this classification shall not be less than 3.0 centistokes at 210°F (39 SUS).
[c]Minimum viscosity at 0°F may be waived provided viscosity at 210°F is not below 4.2 centistokes (40 SUS).
[d]Minimum viscosity at 0°F may be waived provided viscosity at 210°F is not below 5.7 centistokes (45 SUS).

used to test the viscosity of hydraulic fluids either prior to their being added to a system or periodically after they have been in an operating system for a while.

Viscosity index

The viscosity index, VI, of oil is a number that indicates the effect of temperature changes on the viscosity of the oil. A low VI signifies a relatively large change of viscosity with changes of temperature. In other words, the oil becomes extremely thin at high temperatures and extremely thick at low temperatures. On the other hand, a high VI signifies relatively little change in viscosity over a wide temperature range. Figure 40.18 illustrates the relative change of viscosity with changes in oil temperature.

Ideal oil for most purposes is one that maintains a constant viscosity throughout temperature changes. The importance of the VI can be shown easily by considering automotive lubricants. Oil having a high VI resists excessive thickening when the engine is cold and, consequently, promotes rapid starting and prompt circulation; it resists excessive thinning when the motor is hot and thus provides full lubrication and prevents excessive oil consumption.

Another example of the importance of the VI is the need for a high viscosity index hydraulic oil for military aircraft, since hydraulic control systems may be exposed to temperatures ranging from below −65°F at high altitudes to over 100°F on the ground. For the proper operation of the hydraulic control system, the hydraulic fluid must have a sufficiently high VI to perform its functions at the extremes of the expected temperature range.

Liquids with a high viscosity have a greater resistance to heat than low viscosity liquids, which have been derived from the same source. The average hydraulic liquid has a relatively low viscosity. Fortunately, there is a wide choice of liquids available for use in the viscosity range required of hydraulic liquids.

The VI of oil may be determined if its viscosity at any two temperatures is known. Tables based on a large number of tests are issued by the American Society for Testing and Materials (ASTM). These tables permit calculation of the VI from known viscosity.

Pour point

Pour point is the lowest temperature at which a fluid will flow. If the hydraulic system will be exposed to extremely low temperatures, it is a very important specification. For

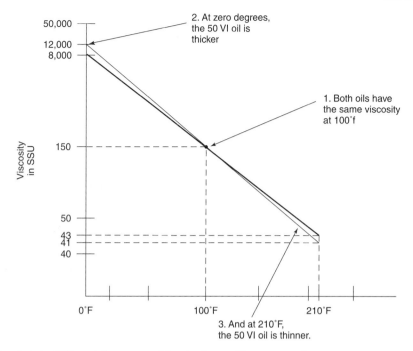

Figure 40.18 Viscosity Index (VI) is a relative measure of viscosity change with temperature change

a rule of thumb, the pour point should be 20°F below the lowest temperature to be encountered.

Lubricating power

If motion takes place between surfaces in contact, friction tends to oppose the motion. When pressure forces the liquid of a hydraulic system between the surfaces of moving parts, the liquid spreads out into a thin film which enables the parts to move more freely. Different liquids, including oils, vary greatly not only in their lubricating ability but also in film strength. Film strength is the capability of a liquid to resist being wiped or squeezed out from between the surfaces when spread out in an extremely thin layer. A liquid will no longer lubricate if the film breaks down, since the motion of part against part wipes the metal clean of liquid.

Lubricating power varies with temperature changes; therefore the climatic and working conditions must enter into the determination of the lubricating qualities of a liquid. Unlike viscosity, which is a physical property, the lubricating power and film strength of a liquid is directly related to its chemical nature. Lubricating qualities and film strength can be improved by the addition of certain chemical agents.

Chemical stability

Chemical stability is another property which is exceedingly important in the selection of a hydraulic liquid. It is defined as the liquid's ability to resist oxidation and

deterioration for long periods. All liquids tend to undergo unfavorable changes under severe operating conditions. This is the case, for example, when a system operates for a considerable period at high temperatures.

Excessive temperatures, especially extremely high temperatures, have a great effect on the life of a liquid. The temperature of the liquid in the reservoir of an operating hydraulic system does not always indicate the operating conditions throughout the system. Localized hot spots occur on bearings, gear teeth, or at other points where the liquid under pressure is forced through small orifices. Continuous passage of the liquid through these points may produce local temperatures high enough to carbonize the liquid or turn it into sludge, yet the liquid in the reservoir may not indicate an excessively high temperature.

Liquids may break down if exposed to air, water, salt, or other impurities, especially if they are in constant motion or subjected to heat. Some metals, such as zinc, lead, brass, and copper, have undesirable chemical reactions with certain liquids.

These chemical reactions result in the formation of sludge, gums, carbon, or other deposits, which clog openings, cause valves and pistons to stick or leak, and give poor lubrication to moving parts. Once a small amount of sludge or other deposits is formed, the rate of formation generally increases more rapidly. As these deposits are formed, certain changes in the physical and chemical properties of the liquid take place. The liquid usually becomes darker, the viscosity increases and damaging acids are formed.

The extent to which changes occur in different liquids depends on the type of liquid, type of refining, and whether it has been treated to provide further resistance to oxidation. The stability of liquids can be improved by the addition of oxidation inhibitors. Inhibitors selected to improve stability must be compatible with the other required properties of the liquid.

Freedom from acidity

An ideal hydraulic liquid should be free from acids that cause corrosion of the metals in the system. Most liquids cannot be expected to remain completely non-corrosive under severe operating conditions. When new, the degree of acidity of a liquid may be satisfactory; but after use, the liquid may tend to become corrosive as it begins to deteriorate.

Many systems are idle for long periods after operating at high temperatures. This permits moisture to condense in the system, resulting in rust formation. Certain corrosion- and rust-preventive additives are added to hydraulic liquids. Some of these additives are effective only for a limited period. Therefore, the best procedure is to use the liquid specified for the system for the time specified by the system manufacturer and to protect the liquid and the system as much as possible from contamination by foreign matter, from abnormal temperatures, and from misuse.

Flashpoint

Flashpoint is the temperature at which a liquid gives off vapor in sufficient quantity to ignite momentarily or flash when a flame is applied. A high flashpoint is desirable for hydraulic liquids because it provides good resistance to combustion and a low degree of evaporation at normal temperatures. Required flashpoint minimums vary from 300°F for the lightest oils to 510°F for the heaviest oils.

Firepoint

Fire point is the temperature at which a substance gives off vapor in sufficient quantity to ignite and continue to burn when exposed to a spark or flame. Like flashpoint, a high fire point is required of desirable hydraulic liquids.

Minimum toxicity

Toxicity is defined as the quality, state, or degree of being toxic or poisonous. Some liquids contain chemicals that are a serious toxic hazard. These toxic or poisonous chemicals may enter the body through inhalation, by absorption through the skin, or through the eyes or the mouth. The result is sickness and, in some cases, death. Manufacturers of hydraulic liquids strive to produce suitable liquids that contain no toxic chemicals and, as a result, most hydraulic liquids are free of harmful chemicals. Some fire-resistant liquids are toxic, and suitable protection and care in handling must be provided.

Density and compressibility

A fluid with a specific gravity of less than 1.0 is desired when weight is critical, although with proper system design,

a fluid with a specific gravity greater than one can be tolerated. Where avoidance of detection by military units is desired, a fluid that sinks rather than rises to the surface of the water is desirable. Fluids having a specific gravity greater than 1.0 are desired, as leaking fluid will sink, allowing the vessel with the leak to remain undetected.

Under extreme pressure, a fluid may be compressed up to 7 per cent of its original volume. Highly compressible fluids produce sluggish system operation. This does not present a serious problem in small, low-speed operations, but it must be considered in the operating instructions.

Foaming tendencies

Foam is an emulsion of gas bubbles in the fluid. In a hydraulic system, foam results from compressed gases in the hydraulic fluid. A fluid under high pressure can contain a large volume of air bubbles. When this fluid is depressurized, as when it reaches the reservoir, the gas bubbles in the fluid expand and produce foam. Any amount of foaming may cause pump cavitation and produce poor system response and sponge control. Therefore, defoaming agents are often added to fluids to prevent foaming. Minimizing air in fluid systems is discussed later in this chapter.

Cleanliness

Cleanliness in hydraulic systems has received considerable attention recently. Some hydraulic systems, such as aerospace hydraulic systems, are extremely sensitive to contamination. Fluid cleanliness is of primary importance because contaminants can cause component malfunction, prevent proper valve seating, cause wear in components, and may increase the response time of servo valves. Fluid contaminants are discussed later in this chapter.

The inside of a hydraulic system can only be kept as clean as the fluid added to it. Initial fluid cleanliness can be achieved by observing stringent cleanliness requirements (discussed later in this chapter) or by filtering all fluid added to the system.

40.5.3 Contamination

Hydraulic fluid contamination may be described as any foreign material or substance whose presence in the fluid is capable of adversely affecting system performance or reliability. It may assume many different forms, including liquids, gases, and solid matter of various composition, sizes, and shapes. Solid matter is the type most often found in hydraulic systems and is generally referred to as particulate contamination. Contamination is always present to some degree, even in new, unused fluid, but must be kept below a level that will adversely affect system operation. Hydraulic contamination control consists of requirements, techniques, and practices necessary to minimize and control fluid contamination.

Classification

There are many types of contaminants which are harmful to hydraulic systems and liquids. These contaminants may be divided into two different classes – particulate and fluid.

Particulate contamination   This class of contaminants includes organic, metallic solid, and inorganic solid contaminants. These contaminants are discussed in the following paragraphs.

Organic   Wear, oxidation, or polymerization produces organic solids or semisolids found in hydraulic systems. Minute particles of O-rings, seals, gaskets, and hoses are present, due to wear or chemical reactions. Synthetic products, such as neoprene, silicones, and hypalon, though resistant to chemical reaction with hydraulic fluids, produce small wear particles. Oxidation of hydraulic fluids increases with pressure and temperature, although antioxidants are blended into hydraulic fluids to minimize such oxidation. The ability of a hydraulic fluid to resist oxidation or polymerization in service is defined as its oxidation stability. Oxidation products appear as organic acids, asphaltics, gums, and varnishes. These products combine with particles in the hydraulic fluid to form sludge. Some oxidation products are oil soluble and cause the hydraulic fluid to increase in viscosity; other oxidation products are not oil soluble and form sediment.

Metallic solids   Metallic contaminants are usually present in a hydraulic system and will range in size from microscopic particles to particles readily visible to the naked eye. These particles are the result of wearing and scoring of bare metal parts and plating materials, such as silver and chromium. Although practically all metals commonly used for parts fabrication and plating may be found in hydraulic fluids, the major metallic materials found are ferrous, aluminum, and chromium particles. Because of their continuous high-speed internal movement, hydraulic pumps usually contribute most of the metallic particulate contamination present in hydraulic systems. Metal particles are also produced by other hydraulic system components, such as valves and actuators, due to body wear and the chipping and wearing away of small pieces of metal plating materials.

Inorganic solids   This contaminant group includes dust, paint particles, dirt, and silicates. Glass particles from glass bead peening and blasting may also be found as contaminants. Glass particles are very undesirable contaminants due to their abrasive effect on synthetic rubber seals and the very fine surfaces of critical moving parts. Atmospheric dust, dirt, paint particles, and other materials are often drawn into hydraulic systems from external sources. For example, the wet piston shaft of a hydraulic actuator may draw some of these foreign materials into the cylinder past the wiper and dynamic seals, and the contaminant materials are then dispersed in the hydraulic fluid. Contaminants may also enter the hydraulic fluid during maintenance when tubing, hoses, fittings, and components are disconnected or replaced. It is therefore important that all exposed fluid ports be sealed with approved protective closures to minimize such contamination.

Fluid contamination

Air, water, solvent, and other foreign fluids are in the class of fluid contaminants.

Air   Hydraulic fluids are adversely affected by dissolved, entrained, or free air. Air may be introduced through improper maintenance or because of system design. Any maintenance operation that involves breaking into the hydraulic system, such as disconnecting or removing a line or component, will invariably result in some air being introduced into the system. This source of air can and must be minimized by pre-filling replacement components with new filtered fluid prior to their installation. Failing to pre-fill a filter-element bowl with fluid is a good example of how air can be introduced into the system. Although pre-filling will minimize introduction of air, it is still important to vent the system where venting is possible.

Most hydraulic systems have built-in sources of air. Leaky seals in gas-pressurized accumulators and reservoirs can feed gas into a system faster than it can be removed, even with the best of maintenance. Another lesser-known but major source of air is air that is sucked into the system past actuator piston rod seals. This occurs when the piston rod that is stroked by some external means while the actuator itself is not pressurized.

Water   Water is a serious contaminant of hydraulic systems. Hydraulic fluids are adversely affected by dissolved, emulsified, or free water. Water contamination may result in the formation of ice, which impedes the operation of valves, actuators, and other moving parts. Water can also cause the formation of oxidation products and corrosion of metallic surfaces.

Solvents   Solvent contamination is a special form of foreign fluid contamination in which the original contaminating substance is a chlorinated solvent. Chlorinated solvents or their residues may, when introduced into a hydraulic system, react with any water present to form highly corrosive acids.

Chlorinated solvents, when allowed to combine with minute amounts of water often found in operating hydraulic systems, change chemically into hydrochloric acids. These acids then attack internal metallic surfaces in the system, particularly those that are ferrous, and produce a severe rust-like corrosion.

Foreign-fluids   Foreign fluids other than water and chlorinated solvents can seriously contaminate hydraulic systems. This type of contamination is generally a result of lube oil, engine fuel, or incorrect hydraulic fluid being introduced inadvertently into the system during servicing. The effects of such contamination depend on the contaminant, the amount in the system, and how long it has been present.

Note: It is extremely important that the different types of hydraulic fluids are not mixed in one system. If different types hydraulic fluids are mixed, the characteristics of the fluid required for a specific purpose are lost. Mixing the different types of fluids usually results in a heavy, gummy deposit that will clog passages and require a major cleaning. In addition, seals and packing installed for use with one fluid are usually not compatible with other fluids and damage to the seals will result.

40.5.4 Contamination control

Maintaining hydraulic fluid within allowable contamination limits for both water and particulate matter is crucial to the care and protection of hydraulic equipment. Filters will provide adequate control of the particular contamination problem during all normal hydraulic system operations if the filtration system is installed properly and filter maintenance is performed properly. Filter maintenance includes changing elements at proper intervals.

Control of the size and amount of contamination entering the system from any other source is the responsibility of the personnel who service and maintain the equipment. During installation, maintenance, and repair of hydraulic equipment, the retention of cleanliness of the system is of paramount importance for subsequent satisfactory performance.

The following maintenance and servicing procedures should be adhered to at all times to provide proper contamination control:

1. All tools and the work area (workbenches and test equipment) should be kept in a clean, dirt-free condition.
2. A suitable container should always be provided to receive the hydraulic liquid that is spilled during component removal or disassembly.
   Note: The reuse of drained hydraulic liquid is prohibited in most hydraulic systems. In some large-capacity systems the reuse of fluid is permitted. When liquid is drained from these systems for reuse, it must be stored in a clean and suitable container. The liquid must be strained and/or filtered when it is returned to the system reservoir.
3. Before hydraulic lines or fittings are disconnected, the affected area should be cleaned with an approved dry-cleaning solvent.
4. All hydraulic lines and fittings should be capped or plugged immediately after disconnection.
5. Before any hydraulic components are assembled, their parts should be washed with an approved dry-cleaning solvent.
6. After the parts have been cleaned in dry-cleaning solvent, they should be dried thoroughly with clean, low-lint cloths and lubricated with the recommended preservative or hydraulic liquid before assembly.
   Note: Only clean, low-lint type II or I cloths as appropriate should be used to wipe or dry component parts.
7. All packing and gaskets should be replaced during the assembly procedures.
8. All parts should be connected with care to avoid stripping metal slivers from threaded areas. All fittings and lines should be installed and torqued according to applicable technical instructions.
9. All hydraulic servicing equipment should be kept clean and in good operating condition.

Some hydraulic fluid specifications contain particle contamination limits that are so low that the products are packaged under clean room conditions. Very slight amounts of dirt, rust, and metal particles will cause them to fail the specification limit for contamination. Since these fluids are usually all packaged in hermetically sealed containers, the act of opening a container may allow more contaminants into the fluid than the specification allows. Therefore, extreme care should be taken in the handling of these fluids. In opening the container for use, observation, or tests, it is extremely important that the container be opened and handled in a clean environment. The area of the container to be opened should be flushed with filtered solvent (petroleum ether or isopropyl alcohol), and the device used for opening the container should be thoroughly rinsed with filtered solvent. After the container is opened, a small amount of the material should be poured from the container and disposed of prior to pouring the sample for analysis. Once a container is opened, if the contents are not totally used, the unused portion should be discarded. Since the level of contamination of a system containing these fluids must be kept low, maintenance on the system's components must be performed in a clean environment commonly known as a controlled environment work center.

40.5.5 Hydraulic fluid sampling

The condition of a hydraulic system, as well as its probable future performance, can best be determined by analyzing the operating fluid. Of particular interest are any changes in the physical and chemical properties of the fluid and excessive particulate or water contamination, either of which indicates impending trouble.

Excessive particulate contamination of the fluid indicates that the filters are not keeping the system clean. This can result from improper filter maintenance, inadequate filters, or excessive ongoing corrosion and wear.

1. All samples should be taken from circulating systems, or immediately upon shutdown, while the hydraulic fluid is within $5°C(9°F)$ of normal system operating temperature. Systems not up to temperature may provide non-representative samples of system dirt and water content, and such samples should either be avoided or so indicated on the analysis report. The first oil coming from the sampling point should be discarded, since it can be very dirty and does not represent the system. As a rule, a volume of oil equivalent to one to two times the volume of oil contained in the sampling line and valve should be drained before the sample is taken.
2. Ideally, the sample should be taken from a valve installed specifically for sampling. When sampling valves are not installed the taking of samples from locations where sediment or water can collect, such as dead ends of piping, tank drains, and low points of large pipes and filter bowls, should be avoided if possible. If samples are taken from pipe drains, sufficient fluid should be drained before the sample is taken to ensure that the sample actually represents the system. Samples are not to be taken from the tops of

reservoirs or other locations where the contamination levels are normally low.

3. Unless otherwise specified, a minimum of one sample should be taken for each system located wholly within one compartment. For ship's systems extending into two or more compartments, a second sample is required. An exception to this requirement is submarine external hydraulic systems, which require only one sample. Original sample points should be labeled and the same sample points used for successive sampling. If possible, the following sampling locations should be selected:

(a) A location that provides a sample representative of fluid being supplied to system components;
(b) A return line as close to the supply tank as practical but upstream of any return line filter;
(c) For systems requiring a second sample, a location as far from the pump as practical.

Operation of the sampling point should not introduce any significant amount of external contaminants into the collected fluid.

40.6 Reservoirs, strainers, filters and accumulators

Fluid power systems must have a sufficient and continuous supply of uncontaminated fluid to operate efficiently. This chapter covers: hydraulic reservoirs; various types of strainers and filters; and accumulators that are typically installed in fluid power systems.

40.6.1 Reservoirs

A hydraulic system must have a reserve of fluid in addition to that contained in the pumps, actuators, pipes and other components of the system. This reserve fluid must be readily available to make up losses of fluid from the system, to make up for compression of fluid under pressure, and to compensate for the loss of volume as the fluid cools. This extra fluid is contained in a tank usually called a reservoir. A reservoir may sometimes be referred to as a sump tank, service tank, operating tank, supply tank or base tank.

In addition to providing storage for the reserve fluid needed for the system, the reservoir acts as a radiator for dissipating heat from the fluid. It also acts as a settling tank where heavy particles of contamination may settle out of the fluid and remain harmlessly on the bottom until removed by cleaning or flushing the reservoir. Also, the reservoir allows entrained air to separate from the fluid.

Most reservoirs have a capped opening for filling, an air vent, an oil level indicator or dipstick, a return line connection, a pump inlet or suction line connection, a drain line connection, and a drain plug (see Figure 40.19).

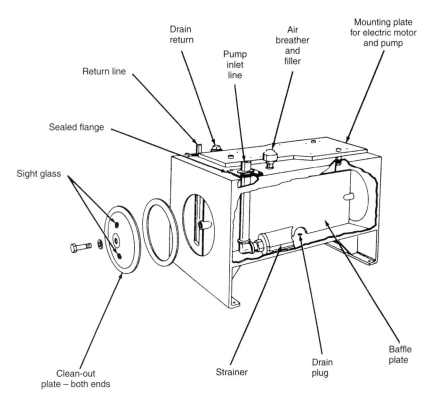

Figure 40.19 Non-pressurized reservoir

The inside of the reservoir generally will have baffles to prevent excessive sloshing of the fluid and to put a partition between the fluid return line and the pump suction or inlet line. The partition forces the returning fluid to travel farther around the tank before being drawn back into the active system through the pump inlet line. This aids in settling the contamination and separating air entrained in the fluid.

Large reservoirs are desirable for cooling. A large reservoir also reduces re-circulation, which helps settle contamination, and separates entrained air. As a rule of thumb, the ideal reservoir should be two to three times the pump outlet per minute. However, due to space limitations in mobile and aerospace systems, the benefits of a large reservoir may have to be sacrificed. But they must be large enough to accommodate thermal expansion of the fluid and changes in fluid level due to system operation.

40.6.2 Accumulators

An accumulator is a pressure storage reservoir in which hydraulic fluid is stored under pressure from an external source. The storage of fluid under pressure serves several purposes in hydraulic systems.

In some hydraulic systems, it is necessary to maintain the system pressure within a specific pressure range for long periods. It is very difficult to maintain a closed system without some leakage, either external or internal. Even a small leak can cause a decrease in pressure. By using an accumulator, leakage can be compensated for and the system pressure can be maintained within acceptable range for extended periods. Accumulators also compensate for thermal expansion and contraction of the liquid due to variations in temperature or generated heat.

A liquid flowing at a high velocity in a pipe will create a backward surge when stopped suddenly. This sudden stoppage causes an instantaneous pressure 2 to 3 times the operating pressure of the system. These pressures or shocks produce objectionable noise and vibrations, which can cause considerable damage to piping, fittings and components. The incorporation of an accumulator enables such shocks and surges to be absorbed or cushioned by the entrapped gas, thereby reducing their effects. The accumulator also dampens pressure surges caused by pulsing delivery from the pump.

There are times when hydraulic systems require large volumes of liquid for short periods. This is due to either the operation of a large cylinder or the necessity of operating two or more circuits simultaneously. It is not economical to install a pump of such large capacity in the system for only intermittent usage, particularly if there is sufficient time during the working cycle for an accumulator to store enough liquid to aid the pump during these peak demands. The energy stored in accumulators may be also used to actuate hydraulically operated units if normal hydraulic system failure occurs.

Accumulator sizing

Most accumulator systems should be designed to operate at a maximum oil pressure of 3000 psi. This is the rating of most accumulators and will give the maximum effect for the least cost. Also, 3000 psi is the maximum rating for most hydraulic valves.

A rule of thumb for the nitrogen pre-charged level is one-half the maximum oil pressure. This is acceptable for most applications. The pre-charge should be replenished when it falls to one-third the maximum hydraulic oil pressure. On a 3000-psi hydraulic system, initial pre-charge should be 1500 psi and replenishment level of 1000 psi. Most applications will tolerate a wide variation in pre-charge pressure.

Accumulators are catalog-rated by gas volume when all oil is discharged and usually rated in quarts or gallons (i.e. 1 US gallon $= 231 \text{ in}^3$). The amount of oil which can be stored is approximately half the gas volume. Only a part of the stored oil can be used each cycle because the oil pressure decreases as oil is discharged. The problem in selecting accumulator size is to have sufficient capacity so system pressure, at the end of the discharge, does not fall below a value which will do the job.

For illustration, we are using an application on which accumulator oil will be used on the extension stroke of a cylinder to supplement the oil delivery from a pump, to increase speed. Retraction will be by pump volume alone. A fully charged accumulator system pressure of 3000 psi is assumed.

First, select cylinder bore for sufficient force not only at 3000 psi but also at some selected lower pressure to which it will be allowed to fall during discharge. Next, calculate the number of cubic inches of oil required to fill the cylinder cavity during its extension stroke. Using the time, in seconds, allowed for the full extension stroke, calculate the cubic inches of oil which can be obtained for the pump alone. Subtract the calculated pump volume from the cylinder volume to find the volume of oil required from the accumulator. Use Table 40.2 to find how many cubic inches of oil would be supplied from a 1-gallon size accumulator before its terminal pressure dropped below the minimum acceptable pressure level. A 5-gallon accumulator would supply 5 times this volume. Finally, divide this figure into the total cubic inches needed for the application. This is the minimum rated gallon size of accumulator capacity. Select at least the next larger standard size for your application.

To solve for oil recovery from any size accumulator, under any system pressure and any pre-charge level, use the formula:

$$D = (0.95 \times P_1 \times V_1 \div P_2) - (0.95 \times P_1 \times V_1 \div P_3)$$

Where D is cubic inches of oil discharge; P_1 is pre-charge pressure in psi; P_2 is system pressure after Volume D has been discharged; P_3 is maximum system pressure at full accumulator charge; V_1 is catalog rated gas volume, in cubic inches; and 0.95 is assumed accumulator efficiency.

As oil is pumped into the accumulator, compressing the nitrogen, the nitrogen temperature increases (Charles' law). Therefore, the amount of oil stored will not be quite as much as calculated with Boyle's law unless sufficient time is allowed for the accumulator to cool to atmospheric temperature. Likewise, when oil is discharged, the expanding nitrogen is cooled. So, the discharge volume

Table 40.2 Accumulator selection table

Minimum acceptable system PSI	Cubic inch discharge
2700	12
2600	17
2500	22
2400	27
2300	33
2200	40
2100	46
2000	55
1900	63
1800	73
1700	84
1600	96
1500	109

will not be quite as high as calculated with Boyle's law. In Table 40.2 and the formula above, an allowance of 5 per cent has been included as a safety factor. After making a size calculation from the table, allow enough extra capacity for contingencies.

40.6.3 Filtration

Clean hydraulic fluid is essential for proper operation and acceptable component life in all hydraulic systems. While every effort must be made to prevent contaminants from entering the system, contaminants that do find their way into the system must be removed. Filtration devices are installed at key points in fluid power systems to remove the contaminants that enter the system along with those that are generated during normal operations of the system.

The filtering devices used in hydraulic systems are commonly referred to as strainers and filters. Since they share a common function, the terms are often used interchangeably. Generally, devices used to remove large particles of foreign matter from hydraulic systems are referred to as strainers, while those used to remove the smallest particles are called filters.

Strainers

Strainers are used primarily to catch only very large particles and will be found in applications where this type of protection is required. Most hydraulic systems have a strainer in the reservoir at the inlet to the suction line of the pump. A strainer is used in lieu of a filter to reduce its chance of being clogged and starving the pump. However, since this strainer is located in the reservoir, its

Table 40.3 Heat and power equivalents

1 Horsepower (HP)	2545 BTU per hour
1 Horsepower (HP)	42.4 BTU per minute
1 British thermal unit per hour (BTU/hr)	0.000393 HP or 0.293 watts
1 British thermal unit per minute (BTU/min)	0.0167 BTU/hr or 17.6 watts

maintenance is frequently neglected. When heavy dirt and sludge accumulate on the suction strainer, the pump soon begins to cavitate. Pump failure follows quickly.

Filters

The most common device installed in hydraulic systems to prevent foreign particles and contaminations from remaining in the system are referred to as filters. They may be located in the reservoir, in the return line, in the pressure line, or in any other location in the system where the designer of the system decides they are needed to safeguard the system against impurities.

Filters are classified as full-flow or proportional flow. In full-flow types of filters, all of the fluid that enters the filter passes through the filtering element, while in proportional types only a portion of the fluid passes through the element.

Filter elements

Filter elements may be divided into two classes: surface and depth. Surface filters are made of closely woven fabric or treated paper with a uniform pore size. Fluid flows through the pores of the filter material and contaminants are stropped on the filter's surface. This type of filter element is designed to prevent the passage of a high percentage of solids of a specific size.

Depth filters on the other hand are composed of layers of fabric or fibers, which provide many tortuous paths for the fluid to flow through. The pores or passages must be larger than the rated size of the filter if particles are to be retained in the depth of the media rather than on the surface.

Filter elements may be of the 5-micron, woven mesh, micronic, porous metal, or magnetic type. The micronic and 5-micron elements have non-cleanable filter media and are disposed of when they are removed. Porous metal, woven mesh and magnetic filter elements are usually designed to be cleaned and reused.

40.6.4 Heat exchangers

The conversion of hydraulic force to mechanical work generates excessive heat. This heat must be removed from the hydraulic fluid to prevent degradation of the fluid and possible damage to system components.

Heat goes into the hydraulic oil at every point in the system where there is a pressure loss due to oil flow without mechanical work being produced. Examples are pressure relief and reducing valves, flow control valves, and flow resistance in plumbing lines and through components. Hydraulic pumps and motors also produce heat at about 15 per cent of their working horsepower. Power loss and heat generation due to the above causes can be calculated with one of the following formulae:

$$\text{Horsepower heat} = \frac{\text{PSI} \times \text{GPM}}{1714}$$

$$\text{or HP} = \text{PSI} \times \text{GPM} \times .000583$$

BTU/Hr of heat generation $= 1.5 \times \text{PSI} \times \text{GPM}$

The pressure (PSI) for calculating heat generation in a flow control valve, for example, is the inlet minus the outlet pressure, or the pressure drop across the valve.

Sometimes power loss and heat generation occur intermittently and to find the average amount of heat that will go into the oil, the *average* power loss should be calculated. Usually, taking an average over a 1-hour period should be sufficient. Therefore, hydraulic systems should include a positive means of heat removal. Normally, a heat exchanger is used for this purpose.

The exact size of a heat exchanger needed on a new system cannot be accurately calculated because of too many unknown factors. On existing systems, by making tank measurements and measuring air and oil temperatures, a rather accurate calculation can be made of the heat exchanger capacity needed to reduce the maximum oil temperature.

The theoretical maximum cooling capacity of a heat exchanger for a hydraulic system will never have to be greater than the input horsepower to the system. Usually its capacity can be considerably less, based on the calculated input horsepower. A rule of thumb is to provide a heat exchanger removal capacity of about 25 per cent of the input horsepower. Rarely, even on inefficient systems, would a capacity of more than 50 per cent be required.

When ordering a heat exchanger the information furnished to your supplier must include the maximum rate of oil flow, in GPM, through the heat exchanger, and the horsepower or BTU per hour of heat to be removed. On water-cooled models, state the maximum rate of water flow that will be available. For best water usage, the water flow should be approximately one-half of the oil flow. Specify the temperature of the cooling water.

Heat load

Heat load on the waterside of a shell and tube heat exchanger can be calculated as:

BTU per hour = GPM × 500 × Temperature differential

The temperature differential is the difference between the inlet and outlet oil temperature in degrees Fahrenheit.

Heat load on the shell side of the exchanger can also be calculated by:

BTU per hour = GPM × 210 × Temperature differential

40.7 Actuators

One of the outstanding features of fluid power systems is that force, generated by the power supply, controlled and directed by suitable valving, and transported by lines, can be converted with ease to almost any kind of mechanical motion. Either linear or rotary motion can be obtained by using a suitable actuating device.

An actuator is a device that converts fluid power into mechanical force and motion. Cylinders, hydraulic motors, and turbines are the most common types of actuating devices used in fluid power systems. This chapter describes various types of actuating devices and their applications.

40.7.1 Hydraulic cylinders

An actuating cylinder is a device that converts fluid power into linear, or straight-line, force and motion. Since linear motion is back-and-forth motion along a straight line, this type of actuator is sometimes referred to as a reciprocating, or linear motor. The cylinder consists of a ram, or piston, operating within a cylindrical bore. Actuating cylinders may be installed so that the cylinder is anchored to a stationary structure and the ram or piston is attached to the mechanism to be operated, or the piston can be anchored and the cylinder attached to the movable mechanism.

Piston rod column strength

Long, slim piston rods may buckle if subjected to too heavy a push load. Table 40.4 suggests the minimum diameter piston rod to use under various conditions of load and unsupported rod length. It should be used in accordance with the instructions in the next paragraph. There must be no side load or bending stress at any point along the piston rod.

Exposed rod length is shown along the top of the table. This is usually somewhat longer than the actual stroke of the cylinder. The vertical scale, Column 1, shows the load on the cylinder, and is expressed in English tons, i.e. 1 ton equals 2,000 pounds. If both the end of the rod and the *front* end of the cylinder barrel are rigidly supported a smaller rod may have sufficient column strength and you may use as *exposed* length of piston rod that is one-half of the actual total rod length. For example, if the actual rod length is 80 inches, and if the cylinder barrel and rod end are supported as described, you could enter the table in the column marked 40. On the other hand, if hinge mounting is used on both cylinder and rod, you may not be safe in using actual exposed rod length. Instead, you should use about twice the actual rod length. For example, if the actual rod length is 20 inches, you should enter the table in the 40-inch column.

When mounted horizontally or at any angle other than vertical, hinge mounted cylinders create a bending stress on the rod when extended. In part, this bending stress is created by the cylinder's weight. On large bore and/or long stroke hinge-mounted cylinders, a trunnion mount should be used in instead of tang or clevis mounts. In addition, the trunnion should be positioned so that the cylinder's weight is balanced when the rod is fully extended.

Hydraulic cylinder forces

Tables 40.5 and 40.6 provide the mechanical forces, both extension and retraction, that can be generated by hydraulic cylinders. The tables are divided into the two principal operating pressure ranges associated with hydraulic applications.

Values in bold type show the extension forces, using the full piston area. Values in italic type are for the retraction force for various piston rod diameters. Remember that force values are *theoretical*, derived by calculation. Experience has shown that probably 5 per cent but

Table 40.4 Minimum recommended piston rod diameter

Load in tons	Exposed length of piston rod, inches							
	10	20	40	60	70	80	100	120
$\frac{1}{2}$			$\frac{3}{4}$	1				
$\frac{3}{4}$			$\frac{13}{16}$	$1\frac{1}{16}$				
1		$\frac{5}{8}$	$\frac{7}{8}$	$1\frac{1}{8}$	$1\frac{1}{4}$	$1\frac{3}{8}$		
$1\frac{1}{2}$		$\frac{11}{16}$	$\frac{15}{16}$	$1\frac{3}{16}$	$1\frac{3}{8}$	$1\frac{1}{2}$		
2		$\frac{3}{4}$	1	$1\frac{3}{16}$	$1\frac{7}{16}$	$1\frac{9}{16}$	$1\frac{13}{16}$	
3	$\frac{13}{16}$	$\frac{7}{8}$	$1\frac{1}{8}$	$1\frac{3}{8}$	$1\frac{9}{16}$	$1\frac{5}{8}$	$1\frac{7}{8}$	
4	$\frac{15}{16}$	1	$1\frac{3}{16}$	$1\frac{1}{2}$	$1\frac{5}{8}$	$1\frac{3}{4}$	2	$2\frac{1}{4}$
5	1	$1\frac{1}{8}$	$1\frac{5}{16}$	$1\frac{9}{16}$	$1\frac{3}{4}$	$1\frac{7}{8}$	$2\frac{1}{8}$	$2\frac{3}{8}$
$7\frac{1}{2}$	$1\frac{3}{16}$	$1\frac{1}{4}$	$1\frac{7}{16}$	$1\frac{3}{4}$	$1\frac{7}{8}$	2	$2\frac{1}{4}$	$2\frac{1}{2}$
10	$1\frac{3}{8}$	$1\frac{7}{16}$	$1\frac{5}{8}$	$1\frac{7}{8}$	2	$2\frac{1}{8}$	$2\frac{7}{16}$	$2\frac{3}{4}$
15	$1\frac{11}{16}$	$1\frac{3}{4}$	$1\frac{7}{8}$	$2\frac{1}{8}$	$2\frac{1}{4}$	$2\frac{3}{8}$	$2\frac{11}{16}$	3
20	2	2	$2\frac{1}{8}$	$2\frac{3}{8}$	$2\frac{1}{2}$	$2\frac{5}{8}$	$2\frac{7}{8}$	$3\frac{1}{4}$
30	$2\frac{3}{8}$	$2\frac{7}{16}$	$2\frac{1}{2}$	$2\frac{3}{4}$	$2\frac{3}{4}$	$2\frac{7}{8}$	$3\frac{1}{4}$	$3\frac{1}{2}$
40	$2\frac{3}{4}$	$2\frac{3}{4}$	$2\frac{7}{8}$	3	3	$3\frac{1}{4}$	$3\frac{1}{2}$	$3\frac{3}{4}$
50	$3\frac{1}{8}$	$3\frac{1}{8}$	$3\frac{1}{4}$	$3\frac{3}{8}$	$3\frac{1}{2}$	$3\frac{1}{2}$	$3\frac{3}{4}$	4
75	$3\frac{3}{4}$	$3\frac{3}{4}$	$3\frac{7}{8}$	4	4	$4\frac{1}{8}$	$4\frac{3}{8}$	$4\frac{1}{2}$
100	$4\frac{3}{8}$	$4\frac{3}{8}$	$4\frac{3}{8}$	$4\frac{1}{2}$	$4\frac{3}{4}$	$4\frac{3}{4}$	$4\frac{7}{8}$	5
150	$5\frac{3}{8}$	$5\frac{3}{8}$	$5\frac{3}{8}$	$5\frac{1}{2}$	$5\frac{1}{2}$	$5\frac{1}{2}$	$5\frac{3}{4}$	6

certainly no more than 10 per cent additional pressure will be required to make up cylinder losses.

For pressures not shown, the effective piston areas in the third column can be used as power factors. Multiply effective area by pressure to obtain cylinder force produced. Values in two or more columns can be added for a pressure not listed, or, force values can be obtained by interpolating between the next higher and the next lower pressure columns.

Pressure values along the top of each table are differential pressures across the two cylinder ports. This is the pressure to just balance the load and not the pressure that must be produced by the system pump. There will be circuit flow losses in pressure and return lines due to oil flow and these will require additional pressure. When designing a system, be sure to allow sufficient pump pressure, about 25 per cent to 30 per cent, to supply both the cylinder and to satisfy system flow losses.

Hydraulic motors

A fluid power motor is a device that converts fluid power energy into rotary motion and force. The function of a motor is opposite that of a pump. However, the design and operation of fluid power motors are very similar to pumps. Therefore a thorough knowledge of pumps will help you understand the operation of fluid power motors.

Motors have many uses in fluid power systems. In hydraulic power drives, pumps and motors are combined with suitable lines and valves to form hydraulic transmissions. The pump commonly referred to as the A-end is driven by some outside source, such as an electric motor. The pump delivers pressurized fluid to the hydraulic motor, referred to as the B-end. The hydraulic motor is actuated by this flow and through mechanical linkage conveys rotary motion and force to do work.

Fluid motors may be either fixed or variable displacement. Fixed-displacement motors provide constant torque and variable speed. Controlling the amount of input flow varies the speed. Variable-displacement motors are constructed so that the working relationship of the internal parts can be varied to change displacement. The majority of the motors used in fluid power systems are the fixed-displacement type.

Although most fluid power motors are capable of providing rotary motion in either direction, some applications require rotation in only one direction. In these applications, one port of the motor is connected to the system pressure line and the other port to the return line. The flow of fluid to the motor is controlled by a flow control valve, a two-way directional control valve or by starting and stopping the power supply. Varying the rate of fluid flow to the motor may control the speed of the motor.

In most fluid power systems, the motor is required to provide actuating power in either direction. In these applications, the ports are referred to as working ports, alternating as inlet and outlet ports. Either a four-way directional control valve or a variable-displacement pump usually controls the flow to the motor.

Fluid motors are usually classified according to the type of internal element which is directly actuated by the pressurized flow. The most common types of elements are gears, vanes and pistons. All three of these types are

Table 40.5   Hydraulic cylinder force, low pressure range 500 to 1500 psi

Bore diameter (in)	Rod diameter (in)	Effective area (in²)	Pressure differential across cylinder ports				
			500 psi	750 psi	1000 psi	1250 psi	1500 psi
$1\frac{1}{2}$	None	1.77	884	1325	1767	2209	2651
	$\frac{5}{8}$	1.46	730	1095	1460	1825	2190
	1	0.982	491	736	982	1227	1473
2	None	3.14	1571	2356	3142	3927	4712
	1	2.36	1178	1767	2356	2945	3534
	$1\frac{3}{8}$	1.66	828	1243	1657	2071	2485
$2\frac{1}{2}$	None	4.91	2454	3682	4909	6136	7363
	1	4.12	2062	3092	4123	5154	6188
	$1\frac{3}{8}$	3.42	1712	2568	3424	4280	5136
	$1\frac{3}{4}$	2.50	1252	1878	2503	3129	3755
3	None	7.07	3534	5301	7069	8836	10,603
	1	6.28	3142	4712	6283	7854	9425
	$1\frac{3}{8}$	5.58	2792	4188	5584	6980	8376
	$1\frac{3}{4}$	4.66	2332	3497	4663	5829	6995
$3\frac{1}{4}$	None	8.30	4148	6222	8298	10,370	12,444
	$1\frac{3}{8}$	6.81	3405	5108	6811	8514	10,216
	$1\frac{3}{4}$	5.89	2945	4418	5891	7363	8836
	2	5.15	2577	3866	5154	6443	7731
4	None	12.57	6284	9425	12,567	15,709	18,851
	$1\frac{3}{4}$	10.16	5081	7621	10,162	12,702	15,243
	2	9.43	4713	7069	9425	11,782	14,138
	$2\frac{1}{2}$	7.66	3829	5744	7658	9573	11,487
5	None	19.64	9818	14,726	19,635	24,544	29,453
	2	16.49	8247	12,370	16,493	20,617	24,740
	$2\frac{1}{2}$	14.73	7363	11,045	14,726	18,408	22,089
	3	12.57	6283	9425	12,566	15,708	18,850
	$3\frac{1}{2}$	10.01	5007	7510	10,014	12,517	15,021
6	None	28.27	14,137	21,206	28,274	35,343	42,411
	$2\frac{1}{2}$	23.37	11,683	17,524	23,365	29,207	35,048
	3	21.21	10,603	15,094	21,205	26,507	31,808
	$3\frac{1}{2}$	18.65	9326	13,990	18,653	23,316	27,979
	4	15.71	7854	11,781	15,708	19,635	23,562
7	None	38.49	19,243	28,864	38,485	48,106	57,728
	3	31.42	15,708	23,562	31,416	39,271	47,125
	$3\frac{1}{2}$	28.87	14,432	21,648	28,864	36,080	43,296
	4	25.92	12,960	19,439	25,910	32,399	38,879
	$4\frac{1}{2}$	22.58	11,291	16,936	22,581	28,226	33,872
	5	18.85	9425	14,138	18,850	23,563	28,275
8	None	50.27	25,133	37,699	50,265	62,831	75,398
	$3\frac{1}{2}$	40.64	20,322	30,483	40,644	50,805	60,966
	4	37.70	18,850	28,274	37,699	47,124	56,549
	$4\frac{1}{2}$	34.36	17,181	25,771	34,361	42,951	51,542
	5	30.63	15,315	22,973	30,630	38,288	44,945
	$5\frac{1}{2}$	26.51	13,254	19,880	26,507	33,134	39,761
10	None	78.54	39,270	58,905	78,540	98,175	117,810
	$4\frac{1}{2}$	62.64	31,318	46,977	62,636	78,295	93,954
	5	58.91	29,453	44,179	58,905	73,631	88,358
	$5\frac{1}{2}$	54.78	27,391	41,087	54,782	68,478	82,172
	7	40.06	20,028	30,041	40,055	50,069	60,083
12	None	113.1	56,550	84,825	113,100	141,375	169,650
	$5\frac{1}{2}$	89.34	44,671	67,007	89,342	111,678	134,013
14	None	153.9	76,970	115,455	153,940	192,425	230,910
	7	115.5	57,728	86,591	115,455	144,319	173,183

Table 40.6   Hydraulic cylinder force, high pressure range 2000 to 5000 psi

Bore diameter (in)	Rod diameter (in)	Effective area (in^2)	Pressure differential across cylinder ports				
			2000 psi	2500 psi	3000 psi	4000 psi	5000 psi
$1\frac{1}{2}$	None	1.77	3534	4418	5301	7068	8836
	$\frac{5}{8}$	1.46	2921	3651	4381	5841	7302
	1	0.982	1963	2454	2945	3927	4909
2	None	3.14	6283	7854	9425	12,566	15,708
	1	2.36	4712	5890	7069	9425	11,781
	$1\frac{3}{8}$	1.66	3313	4142	4970	6627	8283
$2\frac{1}{2}$	None	4.91	9817	12,271	14,726	19,635	24,544
	1	4.12	8247	10,308	12,370	16,493	20,617
	$1\frac{3}{8}$	3.42	6848	8560	10,271	13,695	17,119
	$1\frac{3}{4}$	2.50	5007	6259	7510	10,014	12,517
3	None	7.07	14,137	17,672	21,206	28,274	35,343
	1	6.28	12,567	15,708	18,850	25,133	31,416
	$1\frac{3}{8}$	5.58	11,167	13,959	16,751	22,335	27,919
	$1\frac{3}{4}$	4.66	9327	11,658	13,990	18,653	23,317
$3\frac{1}{4}$	None	8.30	16,592	20,740	24,837	33,183	41,479
	$1\frac{3}{8}$	6.81	13,622	17,027	20,433	27,244	34,055
	$1\frac{3}{4}$	5.89	11,781	14,726	17,672	23,562	29,453
	2	5.15	10,308	12,886	15,463	20,617	25,771
4	None	12.57	25,134	31,418	37,701	50,268	62,835
	$1\frac{3}{4}$	10.16	20,323	25,404	30,485	40,647	50,809
	2	9.43	18,851	23,564	28,276	37,702	47,127
	$2\frac{1}{2}$	7.66	15,317	19,146	22,975	30,633	38,292
5	None	19.64	39,270	49,088	58,905	78,540	98,175
	2	16.49	32,987	41,234	49,480	65,974	82,467
	$2\frac{1}{2}$	14.73	29,453	36,816	44,179	58,905	73,632
	3	12.57	25,133	31,416	37,699	50,266	62,832
	$3\frac{1}{2}$	10.01	20,028	25,035	30,042	40,056	50,070
6	None	28.27	56,548	70,685	84,822	113,090	141,370
	$2\frac{1}{2}$	23.37	46,731	58,413	70,096	93,461	116,827
	3	21.21	42,411	53,014	63,616	84,822	106,027
	$3\frac{1}{2}$	18.65	37,306	46,632	55,959	74,612	93,265
	4	15.71	31,416	39,270	47,124	62,832	78,540
7	None	38.49	76,970	96,213	115,455	153,940	192,425
	3	31.42	62,833	78,541	94,249	125,666	157,082
	$3\frac{1}{2}$	28.87	57,728	72,160	86,592	115,456	144,320
	4	25.92	51,838	64,798	77,757	103,676	129,595
	$4\frac{1}{2}$	22.58	45,162	56,453	67,743	90,324	112,905
	5	18.85	37,700	47,125	56,550	75,400	94,250
8	None	50.27	100,530	125,663	150,795	201,060	251,325
	$3\frac{1}{2}$	40.64	81,288	101,610	121,932	162,576	203,220
	4	37.70	75,398	94,248	131,097	150,796	188,495
	$4\frac{1}{2}$	34.36	68,722	85,903	103,083	137,444	171,805
	5	30.63	61,260	76,575	91,890	125,520	153,150
	$5\frac{1}{2}$	26.51	53,014	66,268	79,521	106,028	132,535
10	None	78.54	157,080	196,350	235,620	314,160	392,700
	$4\frac{1}{2}$	62.64	125,272	156,590	187,908	250,544	313,180
	5	58.91	117,810	147,263	176,715	235,620	294,525
	$5\frac{1}{2}$	54.78	109,564	136,955	164,346	219,128	273,910
	7	40.06	80,110	100,138	120,165	160,220	200,275
12	None	113.1	226,200	282,750	339,300	452,400	565,500
	$5\frac{1}{2}$	89.34	178,684	223,355	268,026	357,368	446,710
14	None	153.9	307,880	384,850	461,820	615,760	769,700
	7	115.5	230,910	288,638	346,365	461,820	577,275

adaptable for hydraulic systems, but only the vane-type is used on pneumatic systems.

4.8 Control valves

It is impossible to design a practical fluid power system without some means of controlling the volume and pressure of the fluid, and directing that flow to the proper operating units. This is accomplished by the inclusion of control valves in the hydraulic circuit.

A valve is defined as any device by which the flow of fluid may be started, stopped, regulated or directed by a movable part that opens or obstructs passage of the fluid. Valves must be able to accurately control fluid flow, system pressure and to sequence the operation of all actuators within a hydraulic system.

Hydraulic control values can utilize a variety of actuators that activate their function. Normally these actuators use manual, electrical, mechanical or pneumatic power sources.

4.8.1 Valve classification

Valves are classified by their intended use: flow control, pressure control, and direction control. Some valves have multiple functions that fall into more than one classification.

Flow control valves

Flow control valves are used to regulate the flow of fluids. Control of flow in hydraulic systems is critical because the rate of movement of fluid-powered machines or actuators depends on the rate of flow of the pressurized fluid.

Pressure control

The safe and efficient operation of fluid power systems, system components and related equipment requires a means to control pressure within the system. There are many types of automatic pressure control valves. Some of them merely provide an escape for excess pressures; some only reduce the pressure; and some keep the pressure within a pre-set range.

Some fluid power systems, even when operated normally, may temporarily develop excessive pressure. For example, when an unusually strong work resistance is encountered, system pressure may exceed design limits. Relief valves are used to control this excess pressure.

Relief valves are automatic valves, used on system lines and equipment to prevent over-pressurization. Most relief valves simply open at a preset pressure and shut when the pressure returns to normal limits. They do not maintain flow or pressure at a given amount, but prevent pressure from rising above a specified level.

Main system relief valves are generally installed between the pump or pressure source and the first system isolation valve. The valve must be large enough to allow the full output of the hydraulic pump to be delivered back to the reservoir. This design feature, called a full-flow bypass, is essential for all hydraulic systems. The location

of the valve is also critical. If the valve were installed downstream from the system isolator valve, the pump could be deadheaded when the system was shutdown.

Smaller relief valves are often used in isolated parts of the system where a check valve or directional control valve prevents pressure from being relieved through the main system relief valve or where pressures must be relieved at a specific set point lower than the main system pressure. These small relief valves are also used to relieve pressures caused by thermal expansion of fluids.

Figure 40.20 shows a typical relief valve. System pressure simply acts under the valve disk at the inlet of the valve. When the system pressure exceeds the pre-load force exerted by the valve spring, the valve disk will lift off of its seat. This will allow some of the system fluid to escape through the valve outlet. Flow will continue until the system pressure is reduced to a level below the spring force.

All relief valves have an adjustment for increasing or decreasing the set relief pressure. Some relief valves are equipped with an adjusting screw for this purpose. This adjusting screw is usually covered with a cap, which must be removed before an adjustment can be made.

Pressure regulators

Pressure regulators, often referred to as unloading valves, are used in fluid power systems to regulate pressure. In hydraulic systems, the pressure regulator is used to unload the pump and to maintain or regulate system pressure at the desired values.

All hydraulic systems do not require pressure regulators. The open-center system does not require a pressure regulator. Many systems are equipped with variable-displacement pumps, which contain a pressure-regulating device.

Pressure regulators are made in a variety of types. However, the basic operating principles of all regulators are similar to the one illustrated in Figure 40.21.

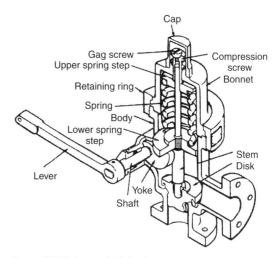

Figure 40.20 Cutaway of relief valve

(a)

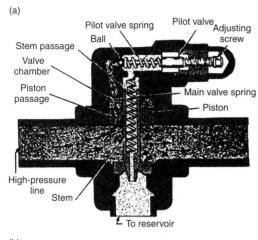

Pilot valve spring Pilot valve Adjusting
Ball   screw
Stem passage
Valve
chamber
Piston
passage   Main valve spring
                                  Piston

High-pressure
line
      Stem
          └ To reservoir

(b)

(c)

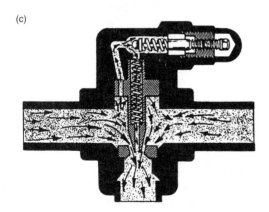

Figure 40.21 Hydraulic pressure regulator

A regulator is open when it is directing fluid under pressure into the system, Figure 40.21, view (a). In the closed position, Figure 40.21, view (b), the fluid in the part of the system beyond the regulator is trapped at the desired pressure and the fluid from the pump is bypassed

into the return line and back to the reservoir. To prevent constant opening and closing (chatter), the regulator is designed to open at pressure somewhat lower than the closing pressure. This difference is known as differential or operating range. For example, assume that a pressure regulator is set to open when the system pressure drops below 600 psi and close when the pressure rises above 800 psi. The differential or operating range is 200 psi.

Referring to Figure 40.21, assume that the piston has an area of 1 in^2, the pilot valve has a cross-sectional area of $\frac{1}{4}$ in^2 and the piston spring provides 600 pounds of force that pushes the piston against its seat. When the system pressure is less than 600 psi, fluid from the pump will enter the inlet port, flow to the top of the regulator and then to the pilot valve. When the system pressure at the valve inlet increases to the point where the force it creates against the front of the check valve exceeds the force created against the back of the check valve, the check valve opens. This allows fluid to flow into the system and to the bottom of the regulator against the piston. When the system force exceeds the force exerted by the spring, the piston moves up, causing the pilot valve to unseat. Since the fluid will take the path of least resistance, it will pass through the regulator and back to the reservoir through the bypass line.

When the fluid from the pump is suddenly allowed a free path to return, the pressure on the input side of the check valve drops and the check valve closes. The fluid in the system is then pressurized until a power unit is actuated or until pressure is slowly lost through normal internal leakage within the system.

When the system pressure decreases to a point slightly below 600 psi, the spring forces the piston down and closes the pilot valve. When the pilot valve is closed, the fluid cannot flow directly to the return line. This causes the pressure to increase in the line between the pump and the regulator. This pressure opens the check valve, causing fluid to enter the system.

In summary, when the system pressure decreases, the pressure regulator will open, sending fluid to the system. When the system pressure increases, the regulator will close, allowing the fluid from the pump to flow through the regulator and back to the reservoir. The pressure regulator takes the load off of the pump and regulates system pressure.

Sequence valves

Sequence valves control the sequence of operation between two branches in a hydraulic circuit. In other words, they enable one component within the system to automatically set another component into motion. An example of the use of a sequence valve is in an aircraft landing gear actuating system.

In a landing gear actuating system, the landing gear doors must open before the landing gear starts to extend. Conversely, the landing gear must be completely retracted before the doors close. A sequence valve installed in each landing gear actuating line performs this function.

A sequence valve is somewhat similar to a relief valve except that, after the set pressure has been reached, the sequence valve diverts the fluid to a second actuator or motor to do work in another part of the system. Figure 40.22 shows an installation of two sequence valves that control the sequence of operation of three actuating cylinders. Fluid is free to flow into cylinder A. The first sequence valve (1) blocks the passage of fluid until the piston in cylinder A moves to the end of its stroke. At this time, sequence valve 1 opens, allowing fluid to enter cylinder B. This action continues until all three pistons complete their strokes.

Pressure reducing valves

Pressure reducing valves provide a steady pressure into a part of the system that operates at a pressure lower that normal system pressure. A reducing valve can normally be set for any desired downstream pressure within its design limits. Once the valve is set, the reduced pressure will be maintained regardless of changes in the supply pressure and system load variations.

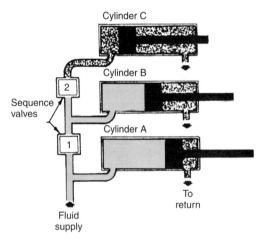

Figure 40.22 Installation of sequence valve

There are various designs and types of pressure-regulating valves. The spring-loaded reducer and the pilot-controlled valve are the most common.

Directional control valves

Directional control valves are designed to direct the flow of fluid, at the desired time, to the point in a fluid power system where it will do work. The driving of a ram back and forth in its cylinder is an example of when a directional control valve is used. Various other terms are used to identify directional valves, such as selector valve, transfer valve, and control valve. This manual will use the term directional control valve to identify these valves.

Directional control valves for hydraulic and pneumatic systems are similar in design and operation. However, there is one major difference. The return port of a hydraulic valve is ported through a return line to the reservoir. Any other differences are pointed out in the discussion of these valves.

Directional control valves may be operated by differences in pressure acting on opposite sides of the valve elements or may be positioned manually, mechanically, or electrically. Often two or more methods of operating the same valve will be used in different phases of its action.

Directional control valves may be classified in several ways. Some of the different ways are by the type of control, the number of ports in the valve housing, and the specific function that the valve performs. The most common method is by the type of valving element used in the construction of the valve. The most common types of valving elements are the ball, cone, sleeve, poppet, rotary spool, and sliding spool. The basic operating principles of the poppet, rotary spool, and sliding spool types are discussed in this text.

The poppet fits into the center bore of the seat, Figure 40.23. The seating surfaces of the poppet and the seat are lapped or closely machined so that the center bore will be sealed when the poppet is seated (shut). The action of the poppet is similar to that of the valves in an

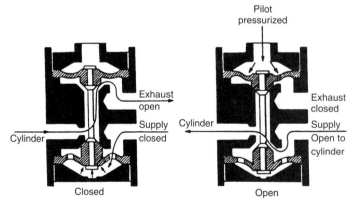

Figure 40.23 Operation of a simple poppet valve

automobile engine. In most valves, the poppet is held in the seated position by a spring.

The valve consists primarily of a movable poppet, which closes against the valve seat. In the closed position, fluid pressure on the inlet side tends to hold the valve tightly closed. A small amount of movement from a force applied to the top of the poppet stem opens the poppet and allows fluid to flow through the valve.

The rotary spool directional control valve, Figure 40.24, has a round core with one or more passages or recesses in it. The core is mounted within a stationary sleeve. As the core is rotated within the stationary sleeve, the passages or recesses connect or block the ports in the sleeve. The ports in the sleeve are connected to the appropriate lines of the fluid system.

The operation of a simple sliding spool directional control valve is shown in Figure 40.25. The valve is so named because of the shape of the valving element that slides back and forth to block or open ports in the valve housing. The sliding element is referred to as the spool or piston. The inner piston areas, or lands, are equal. Thus, fluid under pressure, which enters the valve from the inlet ports, acts equally on both inner piston areas, regardless of the position of the spool. Sealing is usually accomplished by a very closely machined fit between the spool and the valve body or sleeve. For valves with more ports, the spool is designed with more pistons or lands on a common shaft. The sliding spool is the most common type of directional control valves.

4.9 Lines, fittings and seals

The control and application of fluid power would be impossible without suitable means of transferring the hydraulic fluid between the reservoir, the power source, and the points of application. Fluid lines are used to transfer the hydraulic fluid, fittings are used to connect lines to system components, and seals are used in all components to prevent leakage. This chapter is devoted to these critical system components.

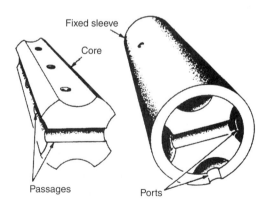

Figure 40.24 Parts of a rotary spool valve

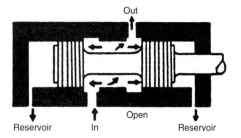

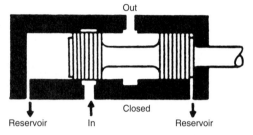

Figure 40.25 Two-way sliding spool valve

4.9.1 Types of lines

Three types of lines are used in fluid power systems: pipe (rigid), tubing (semi-rigid), and hoses (flexible). A number of factors are considered when the type of line is selected for a particular application. These factors include the type of fluid, the required system pressure, and the location of the system. For example, heavy pipe might be used for a large, stationary system, but comparatively lightweight tubing must be used in mobile applications. Flexible hose is required in installations where units must be free to move relative to each other.

Pipe and tubing

There are three important dimensions of any tubular product: outside diameter (OD), inside diameter (ID) and wall thickness. Sizes of pipe are listed by the nominal, or approximate, ID and the wall thickness. The actual OD and the wall thickness list sizes of tubing.

Selection

The material, inside diameter (ID) and wall thickness are the three primary considerations in the selection of lines for a particular fluid power system. The ID of the line is important because it determines how much fluid can pass through the line without loss of power due to excessive friction and heat. The velocity of a given flow is less through a large opening than through a small opening. If the ID of the line is too small for flow, excessive turbulence and friction will cause unnecessary power loss and overheat the hydraulic fluid.

Sizing

Pipes are available in three different weights: Standard (STD), or Schedule 40; Extra Strong (XS), or Schedule

80; and Double Extra Strong (XXS). The schedule number range from 10 to 160 and cover 10 distinct sets of wall thickness, see Table 40.7. Schedule 160 wall thickness is slightly thinner than the double extra strong.

As mentioned earlier, the nominal inside diameter (ID) determines the size of pipes. For example, the ID for a $\frac{1}{4}$-inch Schedule 40 pipe is 0.364 inch, and the ID for a $\frac{1}{2}$-inch Schedule 40 pipe is 0.622 inch.

It is important to note that the IDs of all pipes of the same nominal size are not equal. This is because the OD remains constant and the wall thickness increases as the schedule number increases. For example, a nominal 1-inch Schedule 40 pipe has a 1.049-inch ID. The same size Schedule 80 pipe has a 0.957-inch ID, while Schedule 160 pipe has 0.815-inch ID. In each case, the OD is 1.315-inch and the wall thickness varies. The actual wall thickness is the difference between the OD and ID divided by 2.

Tubing differs from pipe in its size classification. Its actual outside diameter (i.e. OD) designates tubing. Thus, $\frac{5}{8}$-inch tubing has an OD of $\frac{5}{8}$-inch. As indicated in Table 40.7, tubing is available in a variety of wall thickness. The diameter of tubing is often measured and indicated in $\frac{1}{16}$ths. Thus, No. 6 tubing is $\frac{6}{16}$ or $\frac{3}{8}$-inch OD, No. 8 tubing is $\frac{8}{16}$ or $\frac{1}{2}$-inch, and so forth.

The wall thickness, material used, and ID determines the bursting pressure of a line or fitting. The greater the wall thickness in relation to the ID and the stronger the metal, the higher the bursting pressure. However, the greater the ID for a given wall thickness, the lower the bursting pressure. This is because force is the product of area and pressure.

Materials

The pipe and tubing used in fluid power systems are commonly made from steel, copper, brass, aluminum, and stainless steel. Each of these metals has its distinct advantages and disadvantages in certain applications.

Steel pipe and tubing are relatively inexpensive and are used in many hydraulic and pneumatic applications. Steel is used because of its strength, suitability for bending and flanging and adaptability to high pressures and

Table 40.7 Wall thickness schedule designation for pipe

Nominal Size	Pipe OD	Inside diameter (ID)		
		Schedule 40	Schedule 80	Schedule 160
$\frac{1}{8}$	0.405	0.269	0.215	
$\frac{1}{4}$	0.540	0.364	0.302	
$\frac{3}{8}$	0.675	0.493	0.423	
$\frac{1}{2}$	0.840	0.622	0.546	0.466
$\frac{3}{4}$	1.050	0.824	0.742	0.815
1	1.315	1.049	0.957	0.815
$1\frac{1}{4}$	1.660	1.380	1.278	1.160
$1\frac{1}{2}$	1.900	1.610	1.500	1.338
2	2.375	2.067	1.939	1.689

temperatures. Its chief disadvantage is its comparatively low resistance to corrosion.

Copper pipe and tubing are sometimes used for fluid power lines. Copper has high resistance to corrosion and is easily drawn or bent. However, it is unsatisfactory for high temperatures and has a tendency to harden and break due to stress and vibration.

Aluminum has many of the characteristics and qualities required for fluid power lines. Is has high resistance to corrosion and is easily drawn or bent. In addition, it has the outstanding characteristic of lightweight. Since weight elimination is a vital factor in the design of aircraft, aluminum alloy tubing is used in the majority of aircraft fluid power systems.

An improperly piped system can lead to serious power loss and possible harmful fluid contamination. Therefore, in maintenance and repair of fluid power system lines, the basic design requirements must be kept in mind. Two primary requirements are as follows:

1. The lines must have the correct ID to provide the required volume and velocity of flow with the least amount of turbulence during all demands on the system.
2. The lines must be made of the proper material and have the wall thickness to provide sufficient strength to both contain the fluid at the required pressure and withstand the surges of pressure that may develop in the system.

Preparation

Fluid power systems are designed as compactly as possible, to keep the connecting lines short. Every section of line should be anchored securely in one or more places so that neither the weight of the line nor the effects of vibration are carried on the joints. The aim is to minimize stress throughout the system.

Lines should normally be kept as short and free of bends as possible. However, tubing should not be assembled in a straight line, because a bend tends to eliminate strain by absorbing vibration and compensates for thermal expansion and contraction. Bends are preferred to elbows, because bends cause less loss of power. Some of the correct and incorrect methods of installing tubing are illustrated in Figure 40.26.

Bends are described by their radius measurements. The ideal bend radius is $2\frac{1}{2}$ to 3 times the ID, as shown in Figure 40.27. For example, if the ID of a line is 2 inches, the radius of the bend should be between 5 and 6 inches.

While friction increases markedly for sharper curves than this, it also tends to increase up to a certain point for gentler curves. The increases in friction in a bend with a radius of more than 3 pipe diameters result from increased turbulence near the outside edges of the flow. Particles of fluid must travel a longer distance in making the change in direction. When the radius of the bend is less than $2\frac{1}{2}$ pipe diameters, the increased pressure loss is due to the abrupt change in the direction of flow, especially for particles near the inside edge of the flow.

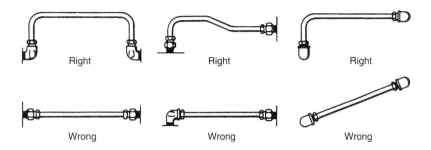

Figure 40.26 Correct and incorrect methods of installation

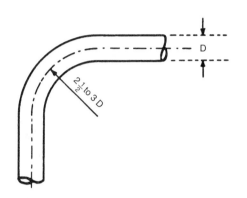

Figure 40.27 Ideal bend radius

Tube cutting and deburring

The objective of cutting tubing is to produce a square end that is free from burrs. Tubing may be cut using a standard tube cutter, Figure 40.28, or a fine-toothed hacksaw. When you use the standard tube cutter, place the tube in the cutter with the cutting wheel at the point where the cut is to be made. Apply light pressure on the tube by tightening the adjusting knob. Too much pressure applied to the cutting wheel at one time may deform the tubing or cause excessive burrs. Rotate the cutter, adjust the tightening knob after each complete rotation.

After the tubing is cut, remove all burrs and sharp edges from the inside and outside of the tube with a deburring tool, Figure 40.29. Clean out the tubing. Make sure no foreign particles remain.

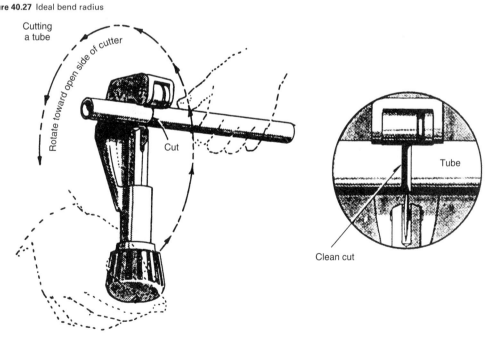

Figure 40.28 Tube cutting

Tube bending

The objective in tube bending is to obtain a smooth bend without flattening the tube. Tube bending is usually done with either a hand tube bender or a mechanically operated bender.

The hand tube bender shown in Figure 40.30, consists of a handle, a radius block, a clip, and a slide bar. The handle and slide bars are used as levers to provide the mechanical advantage necessary to bend the tubing. The radius block is marked in degrees or bend ranging from 0 to 180°. The slide bar has a mark which is lined up with the zero mark on the radius block. The tube is inserted in the tube bender and after the marks are lined up, the slide bar is moved around until the mark on the slide bar reaches the desired degree of bend on the radius block.

Tube flaring

Tube flaring is a method of forming the end of a tube into a funnel shape so a threaded fitting can hold it. When a flared tube is prepared, a flare nut is slipped onto the tube and the end of the tube is flared. During tube installation, the flare is seated to a fitting with the inside of the flare against the cone-shaped end of the fitting, pulling the inside of the flare against the seating surface of the fitting.

Either of two flaring tools in Figure 40.31 may be used. One gives a single flare and the other gives a double flare. The flaring tool consists of a split die block that has holes for various sizes of tubing. It also has a clamp to lock the end of the tubing inside the die block and a yoke with a compressor screw and cone that slips over the die block and forms the 45° flare on the end of the tube. A double flaring tube has adapters that turn in the edge of the tube before a regular 45° double flare is made.

To use the single flaring tool, first check to see that the end of the tubing has been cut off squarely and has had the burrs removed for both the inside and outside. Slip the flare nut onto the tubing before you make the flare. Then, open the die block. Insert the end of the tubing into the hole corresponding to the OD of the tubing so that the end protrudes slightly above the top face of the die blocks. The amount by which the tubing extends above the blocks determines the finished diameter of the flare. The flare must be large enough to seat properly against the fitting, but small enough that the threads of the flare nut

will slide over it. Close the die block and secure the tool with the wing nut. Use the handle of the yoke to tighten the wing nut. Then place the yoke over the end of the tubing and tighten the handle to force the cone into the end of the tubing. The completed flare should be slightly visible above the face of the die blocks.

Flexible hose

Shock-resistant, flexible hose (Figure 40.32) assemblies are required to absorb the movements of mounted equipment under both normal operating conditions and extreme conditions. They are also used for their noise-attenuating properties and to connect moving parts of certain equipment. The two basic hose types are synthetic rubber and polytetrafluoroethylene (PTFE), such as DuPont's Teflon fluorocarbon resin.

Rubber hoses are designed for specific fluid, temperature, and pressure ranges and are provided in various specifications. Rubber hoses consist of a minimum of three layers: a seamless synthetic rubber tube reinforced with one or more layers of braided or spiraled cotton, wire, or synthetic fiber; and an outer cover. The inner tube is designed to withstand the attack of the fluid that passes through it. The braided or spiraled layers determine the strength of the hose. The greater the number of layers, the greater is the pressure rating. Hoses are provided in three pressure ranges: low, medium and high. The outer cover is designed to withstand external abuse and contains identification markings.

Sizing

The size of a flexible hose is identified by the dash (–) number, which is the ID of the hose expressed in 16ths of an inch. For example, the ID of a −64 hose is 4 inches. For a few hose styles this is the nominal and not the true ID.

Cure date

Synthetic rubber hoses will deteriorate from aging. A cure date is used to ensure that they do not deteriorate beyond material and performance specifications. The cure date is the quarter and year the hose was manufactured. For example, 1Q89 or $\frac{1}{89}$ means the hose was made during

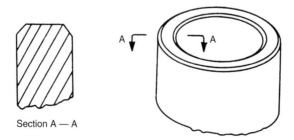

Section A — A

Figure 40.29 Properly burred tubing

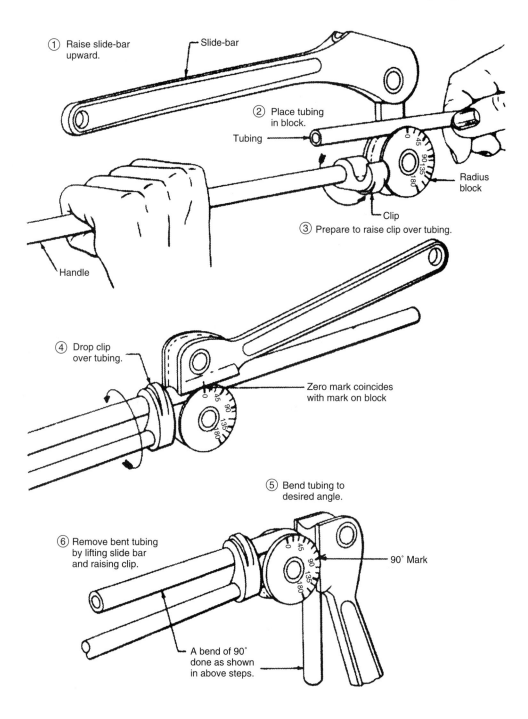

Figure 40.30 Bending tubing with hand-operated tube bender

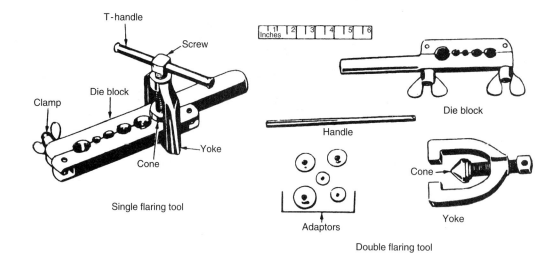

Figure 40.31 Flaring tools

the first quarter of 1989. The cure date limits the length of time a rubber hose can be safely used in fluid power applications. The normal shelf life of rubber hose is 4 years.

Application

As mentioned earlier, flexible hose is available in three pressure ranges: low, medium and high. When replacing hoses, it is important to ensure that the replacement hose is a duplicate of the one removed in length, OD, material, type and contour. In selecting hose, several precautions must be observed. The selected hose must:

1. Be compatible with the system fluid,
2. Have a rated pressure greater than the design pressure of the system,
3. Be designed to give adequate performance and service for infrequent transient pressure peaks up to 150 per cent of the working pressure of the hose, and
4. Have a safety factor with a burst pressure at a minimum of 4 times the rated working pressure.

There are temperature restrictions applied to the use of hoses. Rubber hose must not be used where the operating temperature exceeds 200°F. PTFE hoses in high-pressure air systems must not be used where the temperature exceeds 350°F.

Installation

Flexible hose must not be twisted during installation. This will reduce the life of the hose and may cause the fittings to loosen. You can determine whether a hose is twisted by looking at the layline that runs along the length of the hose. If the layline does not spiral around the hose, the hose is not twisted. If the layline does spiral around the hose, the hose is twisted and must be untwisted. Flexible

hose should be protected from chafing by using a chafe-resistant covering wherever necessary.

The minimum bend radius for flexible hose varies according to the size and construction of the hose and the pressure under which the system operates. Current applicable technical publications contain tables and graphs showing the minimum bend radii for the different types of installations. Bends that are too sharp will reduce the bursting pressure of flexible hose considerably below its rated valve.

Flexible hose should be installed so that it will be subjected to a minimum of flexing during operation. Support clamps are not necessary with short installations; but for hose of considerable length (48 inches for example), clamps should be placed not more than 24 inches apart. Closer supports are desirable and in some cases may be required.

A flexible hose must never be stretched tightly between two fittings. About 5 to 8 per cent of the total length must be allowed as slack to provide freedom of movement under pressure. When under pressure, flexible hose contracts in length and expands in diameter. Examples of correct and incorrect installations of flexible hose are illustrated in Figure 40.33.

40.9.2 Types of fittings and connectors

Some type of connector or fitting must be provided to attach the lines to the components of the system and to connect sections of line to each other. There are many different types of connectors and fittings provided for this purpose. The type of connector or fitting required for a specific system depends on several factors. One determining factor is the type of fluid line (pipe, tubing or flexible hose) used in the system. Other determining factors are the type of fluid medium and maximum operating

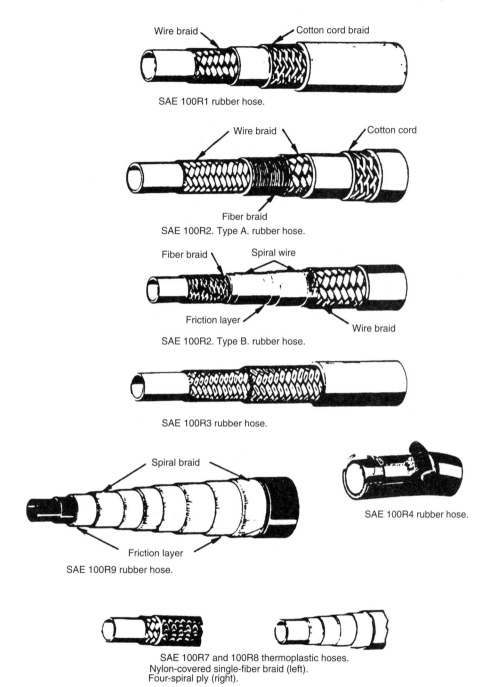

Wire braid Cotton cord braid

SAE 100R1 rubber hose.

Wire braid Cotton cord

Fiber braid

SAE 100R2. Type A. rubber hose.

Fiber braid Spiral wire

Friction layer

Wire braid

SAE 100R2. Type B. rubber hose.

SAE 100R3 rubber hose.

Spiral braid

Friction layer

SAE 100R9 rubber hose.

SAE 100R4 rubber hose.

SAE 100R7 and 100R8 thermoplastic hoses.
Nylon-covered single-fiber braid (left).
Four-spiral ply (right).

Figure 40.32 Types of flexible hose

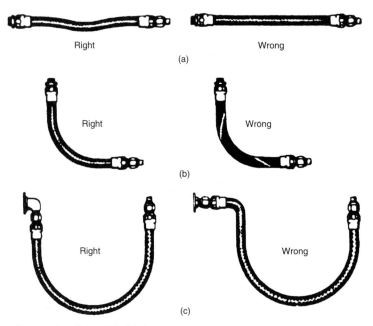

Right Wrong

(a)

Right Wrong

(b)

Right Wrong

(c)

Figure 40.33 Correct and incorrect installation of flexible hose

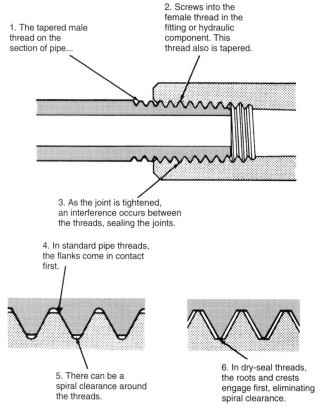

1. The tapered male thread on the section of pipe...

2. Screws into the female thread in the fitting or hydraulic component. This thread also is tapered.

3. As the joint is tightened, an interference occurs between the threads, sealing the joints.

4. In standard pipe threads, the flanks come in contact first.

5. There can be a spiral clearance around the threads.

6. In dry-seal threads, the roots and crests engage first, eliminating spiral clearance.

Figure 40.34 Threaded connectors

pressure of the system. Some of the most common types of fittings and connectors are described in the following:

Pipes and tubing

High-pressure pipe or tubing can also be used for hydraulic circuits. In these applications, special threading or fittings are required to connect circuit components.

Threaded connectors   There are several different types of threaded connectors. In the type discussed in this section, both the connector and the end of the fluid line are threaded. These connectors are used in some low-pressure fluid power systems and are usually made of steel, copper, and brass is available in a variety of designs.

Threaded connectors (Figure 40.34) are made with standard pipe threads cut on the inside surface (female). The end of the pipe is threaded with outside threads (male). Standard pipe threads are tapered slightly to ensure tight connections. The amount of taper is approximately $\frac{3}{4}$ inch in diameter per foot of thread.

Metal is removed when pipe is threaded, thinning the pipe and exposing new and rough surfaces. Corrosion

agents work more quickly at such points than elsewhere. If pipes are assembled with no protective compounds on the threads, corrosion sets in at once and the two sections stick together so that the threads seize when disassembly is attempted. The result is damaged threads and pipes. To prevent seizing, a suitable pipe thread compound is sometimes applied to the threads. The two end threads must be kept free of compound so that it will not contaminate the hydraulic fluid. Pipe compound, when improperly applied, may get inside the lines and damage pumps, control equipment and other components of the system.

Another material used on pipe threads is sealant tape. This tape, which is made of Teflon, provides an effective means of sealing pipe connections and eliminates the necessity of torquing connections to excessively high values in order to prevent leaks. It also provides for ease of maintenance whenever it is necessary to disconnect pipe joints. The tape is applied over the male threads, leaving the first thread exposed. After the tape is pressed firmly against the threads, the joint is connected.

Flanged   Bolted flange connectors, Figure 40.35, are suitable for most pressures now in use. The flanges are attached to the piping by welding, brazing, tapered threads

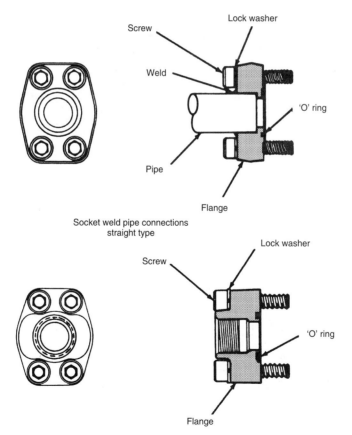

Socket weld pipe connections
straight type

Figure 40.35 Four types of bolted flange connectors

or rolling and bending into recesses. Those illustrated are the most common types of flange joints used. The same types of standard fitting shapes, i.e. tee, cross, elbow and so forth, are manufactured for flange joints. Suitable gasket material must be used between the flanges.

Welded　　Welded joints connect the subassemblies of some fluid power systems, especially in high-pressure systems that use pipe for fluid lines. The welding is done according to standard specifications that define the materials and techniques.

Brazed　　Silver-brazed connectors are commonly used for joining non-ferrous piping in the pressure and temperature range where their use is practical. Use of this type of connector is limited to installations in which the piping temperature will not exceed 425°F and the pressure in cold lines will not exceed 3000 psi. Heating the joint with an oxyacetylene torch melts the alloy. This causes the alloy insert to melt and fill the few thousandths of an inch annular space between the pipe and fitting.

Flared　　connectors are commonly used in fluid power systems containing lines made of tubing. These connectors provide safe, strong, dependable connections without the need for threading, welding or soldering the tubing. The connector consists of a fitting, a sleeve, and a nut, Figure 40.36.

The fittings are made of steel, aluminum alloy, or bronze. The fitting used in a connection should be made of the same material as that of the sleeve, the nut and the tubing. For example, use steel connectors with steel tubing and aluminum alloy connectors with aluminum alloy tubing. Fittings are made in union, 45° and 90° elbows, tee, and various other shapes, Figure 40.37.

Tees, crosses, and elbows are self-explanatory. Universal and bulkhead fittings can be mounted solidly with one outlet of the fitting extending through a bulkhead and the other outlet(s) positioned at any angle. Universal means the fitting can assume the angle required for the specific installation. Bulkhead means the fitting is long enough to pass through a bulkhead and is designed so it can be secured solidly to the bulkhead.

For connecting to tubing, the ends of the fittings are threaded with straight machine threads to correspond with

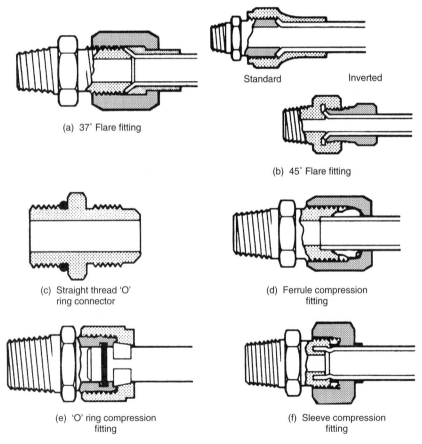

(a) 37° Flare fitting

Standard　　　　　　Inverted

(b) 45° Flare fitting

(c) Straight thread 'O' ring connector

(d) Ferrule compression fitting

(e) 'O' ring compression fitting

(f) Sleeve compression fitting

Figure 40.36 Flared tube fitting

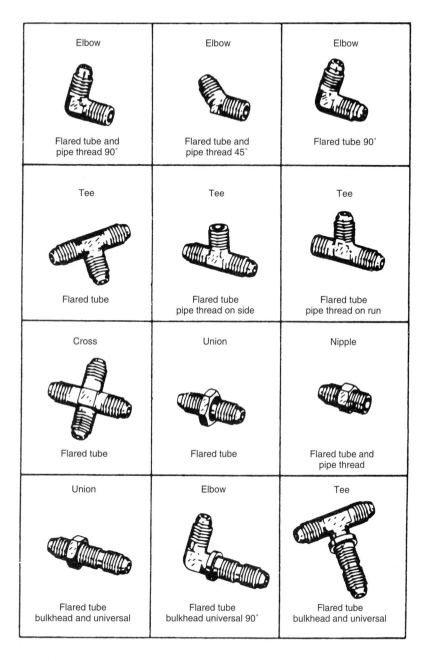

Figure 40.37 Flared tube fittings

the female threads of the nut. In some cases, however, one end of the fitting may be threaded with tapered pipe threads to fit threaded ports in pumps, valves, and other components. Several of these thread combinations are shown in Figure 40.37.

Tubing used with flare connectors must be flared prior to assembly. The nut fits over the sleeve and, when tightened, it draws the sleeve and tubing flare tightly against the male fitting to form a positive seal.

The male fitting has a cone-shaped surface with the same angle as the inside of the flare. The sleeve supports the tube so vibration does not concentrate at the edge of the flare and distributes the shearing action over a wider area for added strength.

Correct and incorrect methods of installing flared-tube connectors are illustrated in Figure 40.38. Tubing nuts should be tightened with a torque wrench to the values specified in applicable technical publications.

If an aluminum alloy flared connector leaks after being tightened to the required torque, it must not be tightened further. Over-tightening may severely damage or completely cut off the tubing flare or may result in damage to the sleeve or nut. The leaking connection must be disassembled and the fault corrected.

If steel tube connection leaks, it may be tightened $\frac{1}{16}$ turn beyond the specified torque in an attempt to stop the leakage. If the connection continues to leak, it must be disassembled and the problem corrected.

Connectors for flexible hose

There are various types of end fittings for both piping connection side and hose connection side of hose fittings. Figure 40.39 shows commonly used fittings.

Quick-disconnect couplings

Self-sealing, quick-disconnect couplings, Figure 40.40, are used at various points in many fluid power systems. These couplings are installed at locations where frequent uncoupling of the lines is required for inspection, test, and maintenance. Quick-disconnect couplings are also commonly used in pneumatic systems to connect sections of air hose and to connect tools to the air pressure lines. This provides a convenient method of attaching and detaching tools and sections of lines without losing pressure.

Quick-disconnect couplings provide a means for quickly disconnecting a line without the loss of fluid from the system or the entrance of foreign matter into the system. Several types of quick-disconnect couplings have been designed for use in fluid power systems. Figure 40.40 illustrates a coupling that is used with portable pneumatic tools. The male section is connected to the tool or to the line leading from the tool. The female section, which contains the shutoff valve, is installed in the line leading from the pressure source. These connectors can be separated or connected by very little effort on the part of the operator.

The most common quick-disconnect coupling for hydraulic systems consists of two parts, held together by a

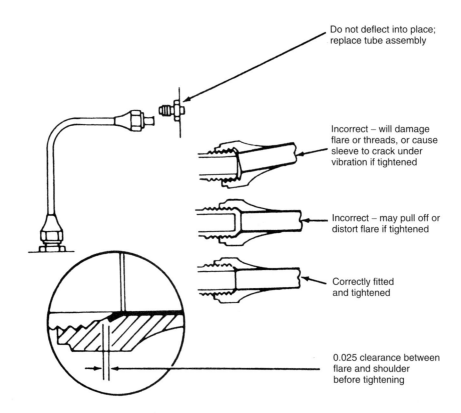

Figure 40.38 Correct and incorrect method of installing flared fittings

41

Pneumatic Fundamentals

R. Keith Mobley
The Plant Performance Group

Contents

41.1 Introduction

The purpose of pneumatics is to do work in a controlled manner. The control of pneumatic power is accomplished using valves and other control devices that are connected together in an organized circuit. The starting point in this organized circuit is the air compressor, where the air is pressurized.

The pressurized air goes to an air receiver for storage and then is processed for use by passing through filters, dryers and, in some cases, lubricators. This pressurized air is normally classified as *instrument air* when it is used in control systems. This air must be moisture and oil free to prevent the control devices from clogging up.

The law states that any pressurized air system must be fitted with a pressure relief valve. This valve prevents the system from being over pressurized and becoming a hazard to personnel or damaging equipment. A pressure switch is an electro/pneumatic control device that is installed on the air receiver to regulate the output of the air compressor. When the air pressure reaches its maximum set point, the regulator is activated and transmits a signal to a solenoid valve on the air compressor. This solenoid valve opens to direct lubricating oil to hydraulically keep the suction valves shut on both the low and high-pressure cylinders on the air compressor. The air compressor will remain in this mode until the pressure drops to the lower set point and deactivates the pressure regulator. This, in turn, de-energizes the solenoid valve on the air compressor causing it to release the lubricating oil pressure on the low- and high-pressure suction valves. The air compressor returns to normal operation pumping air into the receiver until the maximum pressure set point is reached; when this happens, the control cycle starts again. Using a pressure switch prevents the air compressor from running continuously.

Figure 41.1 shows a typical compressed air supply system. The following describes the functions of the system components:

- Compressor – Compresses the air.
- Pressure switch – Turns the air compressor on and off.
- Pressure relief valve – Relieves air pressure at 110 per cent of the operating maximum pressure. This device is fitted to the receiver by law.
- Check valve – Permits the compressed air to flow away from the compressor and will not allow any air to return to the compressor.
- Air receiver – The receiver stores the pressurized air. By law, it must have the following fitting installed on it:
  (a) Pressure relief valve.
  (b) Pressure gage.
  (c) Access hand-hole.
  (d) Drain valve.
- Pressure regulator – Controls the system pressure to the manifold.
- Pressure gage – Indicates internal pressure of the system. Must be fitted by law.
- Filter – This device cleans the air of dirt and contaminants.

- Lubricator – This device adds a small amount of oil to the air in order to lubricate equipment.
  Note: This is only installed when needed. Most systems are oil-free.
- Pressure manifold – This distributes the air to the various pressure ports.
- Needle valves – Control the airflow to the various systems that are to be operated.

Hazards of compressed air

People often lack respect for the power in compressed air because air is so common, and it is viewed as harmless. At sufficient pressures, compressed air can cause damage if an accident occurred. To minimize the hazards of working with compressed air, all safety precautions should be followed closely. Reasons for general precautions follow.

Small leaks or breaks in the compressed air system can cause minute particles to be blown at surprisingly high speeds. Always wear safety glasses when working in the vicinity of any compressed air system. Goggles in place of glasses are recommended if contact lenses are worn.

Compressors can make an exceptional amount of noise while running. The noise of the compressor, in addition to the drain valves lifting, creates noise to require hearing protection. The area around compressors should always be posted as a hearing protection zone.

Pressurized air can do the same type of damage as pressurized water. Treat all operations on compressed air systems with the same care taken on liquid systems. Closed valves should be slowly cracked open and both sides allowed to equalize prior to opening the valve further.

41.2 Characteristics of compressed air

Pascal's law states that the pressure of a gas or liquid exerts force equally in all directions against the walls of its container. The force is measured in terms of force per unit area (pounds per square inch - psi). This law is for liquids and gasses at rest and neglects the weight of the gas or liquid. It should be noted that the field of fluid power is divided into two parts, pneumatics and hydraulics. These two have many characteristics in common. The difference is hydraulic systems use liquids and pneumatics use gases, usually air. Liquids are only slightly compressible and in hydraulic systems, this property can often be neglected. Gases, however, are very compressible.

Three properties of gases must be well understood in order to gain an understanding of pneumatic power systems. These are its temperature, pressure, and volume. Physical laws that define their efficiency and system dynamics govern compressed air systems and compressors. These laws include:

41.2.1 Thermodynamics

Both the first and second laws of thermodynamics apply to all compressors and compressed air systems. These laws state:

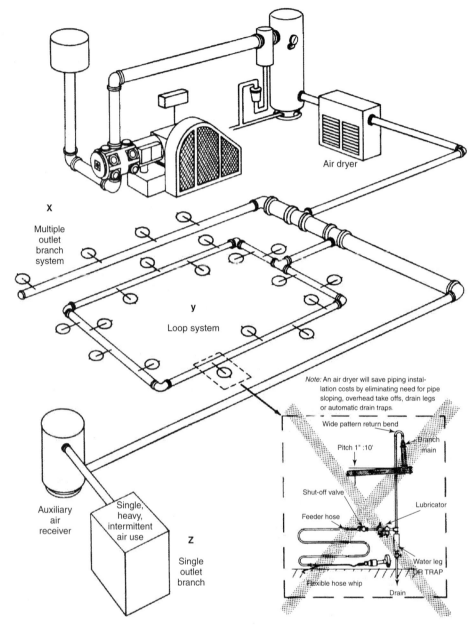

Figure 41.1 Typical compressed air supply

First law

This law states that energy cannot be created or destroyed during a process, such as compression and delivery of air or gas, although it may change from one form of energy to another. In other words, whenever a quantity of one kind of energy disappears, an exactly equivalent total of other kinds of energy must be produced.

Second law

This more abstract and can be stated in several ways:

• Heat cannot, of itself, pass from a colder to a hotter body.
• Heat can be made to go from a body at lower temperature to one at higher temperature only if external work is done.

- The available energy of the isolated system decreases in all real processes.
- Heat or energy, like water, of itself, will flow only downhill.

These statements say that energy exists at various levels and is available for use only if it can move from a higher to a lower level. In thermodynamics, a measure of the unavailability of energy has been devised and is known as *entropy*. It is defined by the differential equation:

$$dS = \frac{dQ}{T}$$

Entropy, as a measure of unavailability, increases as a system loses heat, but remains constant when there is no gain or loss of heat, as in an adiabatic process.

Boyle's law

If a fixed amount of gas is placed in a container of variable volume (such as a cylinder fitted with a piston), the gas will fill completely the entire volume, however large it may be. If the volume is changed, the pressure exerted by the gas will also change. As the volume decreases, the pressure increases. This property is called Boyle's law and can be written as:

$$P_1 \times V_1 = P_2 \times V_2$$

Where:

P_1 = initial absolute pressure
V_1 = initial volume of air or gas
P_2 = final pressure (psia)
V_2 = final volume of air or gas

According to Boyle's law, pressure of a gas is inversely proportional to the volume, if the temperature is held constant. For example, $2\,\text{ft}^3$ at $4\,\text{psi}$ would exert only $1\,\text{psi}$ if allowed to expand $8\,\text{ft}^3$.

$$4\text{ psia} \times 2 \text{ cubic feet} = P_2 \times 8\,\text{ft}^3$$

$$P_2 = \frac{4 \text{ psia} \times 2\,\text{ft}^3}{8\,\text{ft}^3}$$

$$P_2 = 1\,\text{psia}$$

In calculations that involve gas pressure and volume, *absolute pressure* or pounds per square inch absolute (psia) must be used.

Charles' law

If a fixed quantity of gas is held at a constant pressure and heated or cooled, its volume will change. According to Charles' law, the volume of a gas at constant pressure is directly proportional to the absolute temperature. This is shown by the following equation:

$$\frac{V_1}{V_2} = \frac{T_1}{T_2}$$

It is important to remember that *absolute temperature* must be considered not temperature according to normal

Fahrenheit or Centigrade scales. The absolute Fahrenheit scale is called Rankine and the absolute Centigrade is called Kelvin. For conversion:

$0°\text{Fahrenheit} = 460°\text{Rankin}$

$0°\text{Centigrade} = 273°\text{Kelvin}$

Thus gas at $700°\text{F}$ would be $530°$ on Rankin scale.

41.2.2 Combined effect of pressure, volume and temperature

Pressure, temperature, and volume are properties of gases that are completely interrelated. Boyle's law and Charles' law may be combined into the following ideal gas law, which is true for any gas:

$$\frac{P_1 V_1}{T_1} = \frac{P_2 V_2}{T_2}$$

According to this law, if the three conditions of a gas are known in one situation, then if any condition is changed and the effect on the others may be predicted.

Dalton's law

This states that the total pressure of a mixture of ideal gases is equal to the sum of the partial pressures of the constituent gases. The partial pressure is defined as the pressure each gas would exert if it alone occupied the volume of the mixture at the mixture's temperature.

Dalton's Law has been proved experimentally to be somewhat inaccurate, the total pressure often being higher than the sum of the partial pressures. This is especially true during transitions as pressure is increased. However, for engineering purposes it is the best rule available and the error is minor.

When all contributing gases are at the same volume and temperature, Dalton's law can be expressed as:

$$p = p_a + p_b + p_c + \cdots$$

Amagat's law

This is similar to Dalton's law, but states that the volume of a mixture of idea gases is equal to the sum of the partial volumes that the constituent gases would occupy if each existed alone at the *total* pressure and temperature of the mixture. As a formula this becomes:

$$V = V_a + V_b + V_c + \cdots$$

Perfect gas formula

Starting with Charles' and Boyle's laws, it is possible to develop the formula for a given weight of gas:

$$pV = W R_1 T$$

Where W is weight and R_1 is a specific constant for the gas involved. This is the perfect gas equation. Going one step further, by making W, in pounds, equal to the molecular weight of the gas (one mole), the formula becomes:

$$pV = R_0 T$$

In this very useful form, R_0 is known as the *universal gas constant*, has a value of 1545 and is the same for all gases. The *specific* gas constant (R_1) for any gas can be obtained by dividing 1545 by the molecular weight. R_0 is only equal to 1545 when gas pressure (p) is in PSIA; volume (V) is expressed as cubic feet per pound mole; and temperature (T) is in Rankine or absolute, i.e. °F + 460.

Avogadro's law

Avogadro states that equal volumes of all gases, under the same conditions of pressure and temperature, contain the same number of molecules. This law is very important and is applied in many compressor calculations.

The *mole* is particularly useful when working with gas mixtures. It is based on Avogdro's law that equal volumes of gases at given pressure and temperature (pT) conditions contain equal number of molecules. Since this is so, then the *weight* of these equal volumes will be proportional to their molecular weights. The volume of one *mole* at any desired condition can be found by the use of the perfect gas law.

$$pV = R_0 T \text{ or } pV = 1545\,T$$

41.2.3 Gas and vapor

By definition, a *gas* is that fluid form of a substance in which it can expand indefinitely and completely fill its container. A *vapor* is a gasified liquid or solid or a substance in gaseous form. These definitions are in general use today.

All gases can be liquefied under suitable pressure and temperature conditions and therefore could be called vapors. The term gas is most generally used when conditions are such that a return to the liquid state, i.e. condensation, would be difficult within the scope of the operations being conducted. However, a gas under such conditions is actually a superheated vapor.

41.2.4 Changes of state

Any given pure substance may exist in three states: as a solid, as liquid or as vapor. Under certain conditions, it may exist as a combination of any two phases and changes in conditions may alter the proportions of the two phases. There is also a condition where all three phases may exist at the same time. This is known as the *triple point*. Water has a triple point at near 32°F and 14.696 psia. Carbon dioxide may exist as a vapor, a liquid and solid simultaneously at about minus 69.6°F and 75 psia. Substances under proper conditions may pass directly from a solid to a vapor phase. This is known as *sublimation*.

41.2.5 Changes of state and vapor pressure

As liquid physically changes into a gas, their molecules travel with greater velocity and some break out of the liquid to form a vapor above the liquid. These molecules create a *vapor pressure* that, at a specified temperature, is the only pressure at which a pure liquid and its vapor can exist in equilibrium. If in a closed liquid-vapor system, the volume is reduced at constant temperature, the pressure will increase imperceptibly until condensation of part of the vapor into liquid has lowered the pressure to the original vapor pressure corresponding to the temperature. Conversely, increasing the volume at constant temperature will reduce the pressure imperceptibly and molecules will move from the liquid phase to the vapor phase until the original vapor pressure has been restored. For every substance, there is a definite vapor pressure corresponding to each temperature.

The temperature corresponding to any given vapor pressure is obviously the *boiling point* of the liquid and also the *dew point* of the vapor. Addition of heat will cause the liquid to boil and removal of heat will start condensation. The three terms, saturation temperature, boiling point, and dew point all indicate the same physical temperature at a given vapor pressure. Their use depends on the context in which they appear.

41.2.6 Critical gas conditions

There is one temperature above which a gas will not liquefy due to pressure increase. This point is called the *critical temperature*. The pressure required to compress and condense a gas at this critical temperature is called the *critical pressure*.

Relative humidity

Relative humidity is a term frequently used to represent the quantity of moisture or water vapor present in a mixture although it uses partial pressures in so doing. It is expressed as:

$RH(\%)$

$$= \frac{\text{Actual partial vapor pressure} \times 100}{\text{Saturated vapor pressure at existing mixture temperature}}$$

$$= \frac{p_v \times 100}{p_\lambda}$$

Relative humidity is usually considered only in connection with atmospheric air, but since it is unconcerned with the nature of any other components or the total mixture pressure, the term is applicable to vapor content in any problem. The saturated water vapor pressure at a given temperature is always known from steam tables or charts. It is the existing partial vapor pressure which is desired and therefore calculable when the relative humidity is stated.

Specific humidity

Specific humidity used in calculations on certain types of compressors is a totally different term. It is the ratio of the weight of water vapor to the weight of *dry air* and is usually expressed as pounds, or grains, of moisture per pound of dry air. Where *pa* is the partial air pressure, specific humidity can be calculated as:

$$SH = \frac{W_v}{W_a}$$

$$SH = \frac{0.622 p_v}{p\,p_v} = \frac{0.622 p_v}{p_a}$$

Degree of saturation

The degree of saturation denotes the actual relationship between the weight of moisture existing in a space and the weight that would exist if the space were saturated.

$$\text{Degree of saturation}(\%) = \frac{SH_{\text{actual}} \times 100}{SH_{\text{saturated}}}$$

A great many dynamic compressors handle air. Their performance is sensitive to density of the air, which varies with moisture content. The practical application of partial pressures in compression problems centers to a large degree on the determination of mixture volumes or weights to be handled at the intake of each stage of compression, the determination of mixture molecular weight, specific gravity, and the proportional or actual weight of components.

Psychrometry

Psychrometry has to do with the properties of the air–water vapor mixtures found in the atmosphere. Psychrometry tables, published by the US Weather Bureau, give detailed data about vapor pressure, relative humidity and dew point at the sea-level barometer of 30 in Hg, and at certain other barometric pressures. These tables are based on relative readings of dry bulb and wet bulb atmospheric temperatures as determined simultaneously by a sling psychrometer. The dry bulb reads ambient temperature while the wet bulb reads a lower temperature influenced by evaporation from a wetted wick surrounding the bulb of a parallel thermometer.

Compressibility

All gases deviate from the perfect or ideal gas laws to some degree. In some cases the deviation is rather extreme. It is necessary that these deviations be taken into account in many compressor calculations to prevent compressor and driver sizes being greatly in error.

Compressibility is experimentally derived from data about the actual behavior of a particular gas under pVT changes. The compressibility factor, Z, is a multiplier in the basic formula. It is the ratio of the actual volume at a given pT condition to ideal volume at the same pT condition. The ideal gas equation is therefore modified to:

$$pV = ZR_0T \text{ or } Z = \frac{pV}{R_0T}$$

In the above equation, R_0 is 1545 and p is pounds per square foot.

41.3 Generation of pressure

Keeping with the subject of pressure, the basic concepts will be treated in the working sequence: pressure generation, transmission, storage, and utilization in a pneumatic system.

Pumping quantities of atmospheric air into a tank or other pressure vessel produces pressure. Pressure is *increased* by progressively increasing the amount of air

in a confined space. The effects of pressure exerted by a confined gas result from the average of forces acting on container walls caused by the rapid and repeated bombardment from an enormous number of molecules present in a given quantity of air. This is accomplished in a controlled manner by *compression*, a decrease in the space between the molecules. Less volume means that each particle has a shorter distance to travel, thus proportionately more collisions occur in a given span of time, resulting in a higher pressure. Air compressors are designed to generate particular pressures to meet individual application requirements.

Basic concepts discussed here are atmospheric pressure; vacuum; gage pressure; absolute pressure; Boyle's law or pressure/volume relationship; Charles' law or temperature/volume relationship; combined effects of pressure, temperature and volume; and generation of pressure or compression.

Atmospheric pressure

In the physical sciences, pressure is usually defined as the perpendicular force per unit area, or the stress at a point within a confined fluid. This force per unit area acting on a surface is usually expressed in pounds per square inch.

The weight of the earth's atmosphere pushing down on each unit of surface constitutes atmospheric pressure, which is 14.7 psi at sea level. This amount of pressure is called *one atmosphere*. Because the atmosphere is not evenly distributed about earth, atmospheric pressure can vary, depending upon geographic location. Also, obviously, atmospheric pressure decreases with higher altitude. A barometer using the height of a column of mercury or other suitable liquid measures atmospheric pressure.

Vacuum

It is helpful to understand the relationship of vacuum to the other pressure measurements. Vacuums can range from atmospheric pressure down to 'zero absolute pressure', representing a 'perfect' vacuum (a theoretical condition involving the total removal of all gas molecules from a given volume). The amount of vacuum is measured with a device called a vacuum gage.

Vacuum is a type of pressure. A gas is said to be under vacuum when its pressure is below atmospheric pressure, i.e. 14.7 psig at sea level. There are two methods of stating this pressure, but only one is accurate in itself.

A differential gage that shows the difference in the system and the atmospheric pressure surrounding the system usually measures vacuum. This measurement is expressed as:

Millimeters of mercury – Vacuum   (mm Hg Vac)
Inches of mercury – Vacuum   (in Hg Vac)
Inches of water – Vacuum   (in H_2O Vac)
Pounds per square inch – Vacuum   (psi Vac)

Unless the barometric or atmospheric pressure is also given, these expressions do not give an accurate specification of pressure. Subtracting the *vacuum* reading from the atmospheric pressure will give an absolute pressure which is accurate. This may be expressed as:

Inches of mercury − Absolute (in Hg Abs)
Millimeters of mercury − Absolute (mm Hg Abs)
Pounds per square inch − Absolute (psia)

The word *absolute* should never be omitted, otherwise one is never sure whether a vacuum is expressed in differential or absolute terms.

Perfect vacuum

A perfect vacuum is space devoid of matter. It is absolute emptiness. The space is at zero pressure *absolute*. A perfect vacuum cannot be obtained by any known means, but can be closely approached in certain applications.

Gage pressure

Gage pressure is the most often used method of measuring pneumatic pressure. It is the relative pressure of the compressed air within a system. Gage pressure can be either positive or negative, depending upon whether its level is above or below the atmospheric pressure reference. Atmospheric pressure serves as the *reference level* for the most significant types of pressure measurements. For example, if we inflate a tire to 30 psi, an ordinary tire-pressure gage will express this pressure as the value in *excess* of atmospheric pressure, or 30 psig ('g' indicates gage pressure). This reading shows the numerical value of the *difference* between atmospheric pressure and the air pressure in the tire.

Absolute pressure

A different reference level, absolute pressure, is used to obtain the total pressure value. Absolute pressure is the *total* pressure, i.e., gage and atmospheric, and is expressed as psia or pounds per square inch absolute. To obtain absolute pressure, simply add the value of atmospheric pressure (14.7 psi at sea level) to the gage pressure reading.

Absolute pressure (psia) values must be used when computing the pressure changes in a volume and when pressure is given as one of the conditions defining the amount of gas contained within a sample.

41.4 Compressors

A compressor must operate within a system that is designed to acquire and compress a gas. These systems must include the following components regardless of compressor type:

41.4.1 Lubrication system

The lubrication system has two basic functions: to lubricate the compressor's moving components and to cool the system by removing heat from the compressor's moving parts. While all compressors must have a lubrication system, the actual design and function of these systems will vary depending on compressor type.

The lubricating system for centrifugal or dynamic compressors is designed to provide bearing lubrication. In smaller compressors, the lubrication systems may consist of individual oil baths located at each of the main shaft bearings. In larger compressors, such as a bullgear design, a positive system is provided to inject oil into the internal, tilting-pad bearings located at each of the pinion shafts inside the main compressor housing.

In positive lubrication systems, a gear-type pump is normally used to provide positive circulation of clean oil within the compressor. In some cases, the main compressor shaft directly drives this pump. In others, a separate motor-driven pump is used.

Positive displacement compressors use their lubrication system to provide additional functions. The lubrication system must inject sufficient quantities of clean fluid to provide lubrication for the compressor's internal parts, such as pistons and lobes, and to provide a positive seal between moving and stationary parts.

The main components of a positive displacement compressor's lubrication system consist of an oil pump, filter, and heat exchanger. The crankcase of the compressor acts as the oil sump. A lockable drain cock is installed at the lowest end of the crankcase to permit removal of any water accumulation that has resulted from sweating of the crankcase walls. The oil passes through a strainer into the pump. It then flows through the heat exchanger, where it is cooled. After the heat exchanger, the cooled oil flows directly to the moving parts of the compressor before returning to the crankcase sump. A small portion is diverted to the oil injector if one is installed. The oil that is injected into the cylinder seals the space between the cylinder wall and the piston rings. This prevents compressed air from leaking past the pistons, and thus improves the compressor's overall efficiency.

Lube pump

The oil pump is usually gear driven from the crankshaft so that it will start pumping oil immediately on start-up of the compressor. In compressors that work in an oil-free system, oil injectors are not used. Oil separators are installed on the discharge side after leaving the aftercooler.

Oil separator

The basic purpose of an oil separator is to clean the pressurized air of any oil contamination, which is highly detrimental to pneumatically controlled instrumentation. A separator consists of an inlet, a series of internal baffle plates, a wire mesh screen, a sump, and an outlet. The pressurized air enters the separator and immediately passes through the baffle plates. As the air impinges on the baffle plates it is forced into making sharp directional changes as it passes through each baffle section. As a result, the oil droplets separate from the air and collect on the baffles before dropping into the separator's sump.

After the air clears the baffle section, it then passes through the wire mesh screen where any remaining oil is trapped. The relatively oil-free air continues to the air reservoir for storage. The air reservoir acts as a final

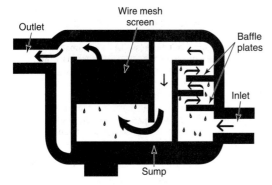

Figure 41.2 Basic oil separator system

separator where moisture and oil is eventually removed. The air reservoir has drain traps installed at its lowest point where any accumulated moisture/oil is automatically discharged.

As a part of any routine maintenance procedure, these discharge traps should periodically be manually bypassed to ensure that the trap is functioning, and no excessive water accumulation is evident.

41.4.2 Compressor selection

Air power compressors generally operate at pressures of 500 psig or lower, with the majority in the range of 125 psig or less. All major types of compressors (i.e. reciprocating, vane, helical lobe and dynamic, are used for this type of service. Choice is limited somewhat by capacity at 100 psig of about 10,500 ft^3 per minute but can be built to approximately 28,000 cfm. The vane-type rotary has an upper listed size of 3700 cfm as a twin unit and the helical lobe rotary can be used to nearly 20,000 cfm. The centrifugal can be built to very large sizes. It is currently offered in the proven, moderate speed designs starting at a minimum of about 5000 cfm.

Selection criteria

The following guidelines should be used for the selection process. While the criteria listed are not all inclusive, they will provide definition of the major considerations that should be used to select the best compressor for a specific application.

Application

The mode of operation of a specific application should be the first consideration. The inherent design of each type of compressor defines the acceptable operating envelope or mode of operation that it can perform with reasonable reliability and life cycle costs. For example, a bullgear-type centrifugal compressor is not suitable for load-following applications but will prove exceptional service in constant-load and volume applications.

Load factor is the ratio of actual compressed air output, while the compressor is operating, to the rated full-load

output during the same period. It should never be 100 per cent, a good rule being to select an installation for from 50 to 80 per cent load factor, depending on the size, type and number of compressors involved. Proper use of load factor results in: more uniform pressure, a cooling-off period, less maintenance, and ability to increase use of air without additional compressors.

Load factor is particularly important with air-cooled machines where sustained full-load operation results in an early build-up of deposits on valves and other parts. This build-up increases the frequency of maintenance required to maintain compressor reliability. Intermittent operation is always recommended for these units. The frequency and duration of *unloaded* operation depends on the type, size, operating pressure of the compressor. Air-cooled compressors for higher than 200-psig-pressure application are usually rated by a rule that states that the *compressing time* shall not exceed 30 minutes or less than 10 minutes. Shutdown or unloaded time should be at least equal to compression time or 50 per cent.

Rotary screw compressors are exceptions to this 50 per cent rule. Each time a rotary screw compressor unloads, both the male and female rotor instantaneously shifts axially. These units are equipped with a balance piston or heavy-duty thrust bearing that is designed to absorb the tremendous axial forces that result from this instantaneous movement, but they are not able to fully protect the compressor or its components. The compressor's design accepted the impact loading that results from this unload shifting and incorporated enough axial strength to absorb a normal unloading cycle. If this type of compressor is subjected to constant or frequent unloading, as in a load-following application, the cycle frequency is substantially increased and the useful life of the compressor is proportionally reduced. There have been documented cases where either the male or female rotor actually broke through the compressor's casing as a direct result of this failure mode.

The only compressor that is ideally suited for load-following applications is the reciprocating type. These units have an absolute ability to absorb the variations in pressure and demand without any impact on either reliability or life cycle cost. The major negative of the reciprocating compressor is the pulsing or constant variation in pressure that is produced by the reciprocating compression cycle. Properly sized accumulators and receiver tanks will resolve most of the pulsing.

Life cycle costs

All capital equipment decisions should be based on the true or life cycle cost of the system. Life cycle cost includes all costs that will be incurred beginning with specification development before procurement to final decommissioning cost at the end of the compressor's useful life. In many cases, the only consideration is the actual procurement and installation cost of the compressor. While these costs are important, they represent less than 20 per cent of the life cycle cost of the compressor.

The cost evaluation must include the recurring costs, such as power consumption, maintenance, etc. that are an integral part of day-to-day operation. Other costs that should be considered include training of operators and maintenance personnel who must maintain the compressor.

41.5 Air dryers

Air entering the first stage of any air compressor carries with it a certain amount of native moisture. This is unavoidable, although the quantity carried will vary widely with the ambient temperature and relative humidity. Figure 41.3 shows the effect of ambient temperature and relative humidity on the quantity of moisture in atmospheric air entering a compressor at 14.7 psia. Under any given condition, the amount of water vapor entering the compressor per 1000 ft^3 of mixture may be approximated from these curves.

In any air–vapor mixture, each component has its own partial pressure and the air and the vapor are each indifferent to the existence of the other. It follows that the conditions of either component may be studied without reference to the other. In a certain volume of mixture,

each component fills the full volume at its own partial pressure. The water vapor may saturate this space or it may be superheated.

As this vapor is compressed, its volume is reduced while at the same time the temperature automatically increases. As a result, the vapor becomes superheated. More pounds of vapor are now contained in one cubic foot than when originally entering the compressor.

Under the laws of vapor, the maximum quantity of a particular vapor a given space can contain is dependent solely upon the vapor temperature. As the compressed water vapor is cooled, it will eventually reach the temperature at which the space becomes saturated, now containing the maximum it can hold. Any further cooling will force part of the vapor to condense into its liquid form – water.

The curves contained in Figure 41.4 show what happens over a wide range of pressures and temperatures. However, these are saturated vapor curves based on starting with 1000 ft^3 of *saturated* air. If the air is not saturated at the compressor's inlet, and it usually is not, use Figure 41.5 to obtain the initial water vapor weight entering the system per 1000 ft^3 of compressed air. By reading left on Figure 41.6 from the juncture of the final pressure

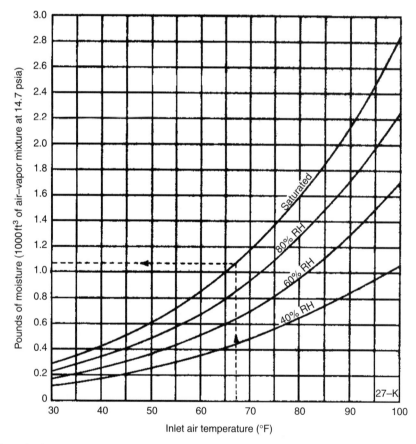

Figure 41.3 Effects of ambient temperature and relative humidity

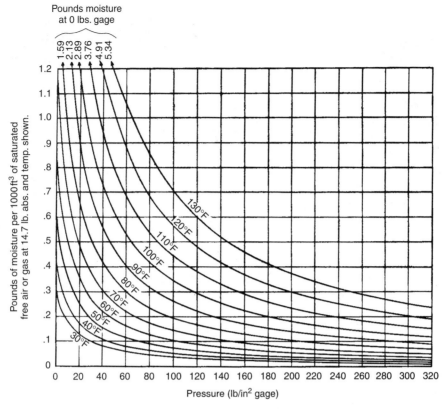

Pounds moisture at 0 lbs. gage

Pounds of moisture per 1000 ft³ of saturated free air or gas at 14.7 lb. abs. and temp. shown.

Pressure (lb/in² gage)

Figure 41.4 Moisture remaining in saturated air or gas when compressed isothermally to pressure shown

and final temperature, obtain the maximum weight of vapor that this same 1000 ft³ can hold after compression and cooling to saturation. If the latter is less than the former, the difference will be condensed. If the latter is higher, there will be no condensation. It is evident that the lower the temperature and the greater the pressure of compressed air, the greater will be the amount of vapor condensed.

41.5.1 Problems caused by water in compressed air

Few plant operators need to be told of the problems caused by water in compressed air. They are most apparent to those who operate pneumatic tools, rock drills, automatic pneumatic powered machinery, paint and other sprays, sandblasting equipment, and pneumatic controls. However, almost all applications, particularly of 100-psig power, could benefit from the elimination of water carry-over. The principal problems might be summarized as:

1. Washing away of required lubrication.
2. Increase in wear and maintenance.
3. Sluggish and inconsistent operation of automatic valves and cylinders.

4. Malfunctioning and high maintenance of control instruments.
5. Spoilage of product by spotting in paint and other types of spraying.
6. Rusting of parts that have been sandblasted.
7. Freezing in exposed lines during cold weather.
8. Further condensation and possible freezing of moisture in the exhaust of those more efficient tools which expand the air considerably.

A fact to remember is that water vapor, *as vapor*, does no harm in most pneumatic systems. It is only when the vapor condenses and remains in the system as a liquid that problems exist. The goal, therefore, is to condense and remove as much of the vapor as is economically possible.

In conventional compressed air systems, vapor and liquid removal is limited. Most two-stage compressors will include an intercooler between stages. On air-cooled units for 100 to 200 psig service, the air between stages is not cooled sufficiently to cause substantial liquid drop out and provision is not usually made for its removal. Water-cooled intercoolers used on larger compressors will usually cool sufficiently to condense considerable moisture at cooler pressure. Drainage facilities must always

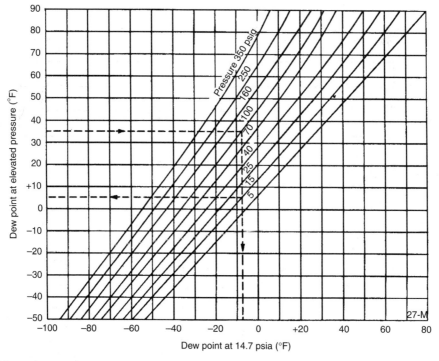

Figure 41.5 Dew point conversion chart

be provided and used. Automatic drain traps are normally included to drain condensed water vapor.

All compressed air systems should always include a water-cooled aftercooler between the compressor and receiver tank. Properly designed and maintained aftercoolers, in normal summer conditions, condense at 100 psig, up to 70 per cent of the vapor entering the system. Most of this condensation will collect in the aftercooler or the receiver tank. Therefore, both must be constantly drained.

The problem with a conventional system that relies on heat exchangers (i.e. aftercoolers) for moisture removal is temperature. The aftercooler will remove only liquids that have condensed at a temperature between the compressed air and cooling water temperature. In most cases, this differential will be about 20 to 50° lower than the compressed air temperature or around 70 to 90°F. As long as the compressed air remains at or above this temperature range, any remaining vapor that it contains will remain in a vapor or gaseous state. However, when the air temperature drops below this range, additional vapor will condense into water.

41.6 Dried air systems

This system involves processing the compressed air or gas after the aftercooler and receiver to further reduce moisture content. This requires special equipment, a higher first cost and a higher operating cost. These

costs must be balanced against the gains obtained. They may show up as less wear and maintenance of tools and air-operated devices, greater reliability of devices and controls, and greater production as a result of fewer outages for repairs. In many cases, reduction or elimination of product spoilage or a better product quality may also result.

The degree of drying desired will vary with the pneumatic equipment and application involved. The aim is to eliminate further condensation in the airlines and pneumatic tools or devices. Prevailing atmospheric conditions also have an influence on the approach that is most effective. In many 100-psig installations, a dew point at line pressure of from 500°F to 350°F is adequate. Other applications, such as instrument air systems, will require dew points of minus 500°F.

Terminology involves drier outlet dew point at the *line* pressure or the pneumatic circuit. This is the saturation temperature of the remaining moisture contained in the compressed air or gas. If the compressed gas temperature is never reduced below the outlet dew point beyond the drying equipment, there will be no further condensation.

Another value sometimes involved when the gas pressure is reduced before it is used is the dew point at that lower pressure condition. A major example is the use of 100 psig (or higher) gas reduced to 15 psig for use in pneumatic instruments and controls. This dew point will be lower because the volume involved increases as the pressure is decreased. The dew point at atmospheric pressure

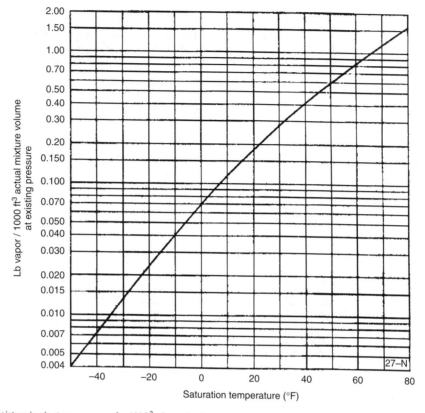

Figure 41.6 Moisture in air at any pressure for 100 ft^3 of actual volume at existing pressures

is often used as a reference point for measurement of drying efficiency. This is of little interest when handling *compressed* air or gas.

Figure 41.7 enables one to determine dew point at reduced pressure. The left scale shows the dew point at the elevated pressure. Drop from the intersection of this value and the elevated pressure line to the reduced pressure line and then back to the left to read the dew point at the reduced pressure.

Figure 41.8 shows graphically the amount of moisture remaining in the vapor form when the air–vapor mixture is conditioned to a certain dew point. This curve is based on a volume of 1000 ft^3 or an air–vapor mixture *at its total pressure*. For example, 1000 ft^3 at 100-psig air at 50°F and 1000 ft^3 of 15-psig air at 50°F will hold the same vapor at the dew point. However, 1000 ft^3 at 100 psig and 50°F reduced to 15 psig will become 3860 ft^3 at 50°F. As a result, it now capable of holding 3.86 times as much vapor and the dew point will not be reached until the mixture temperature is lowered to its saturation temperature.

41.6.1 Drying methods

There are three general methods of drying compressed air: chemical, adsorption, and refrigeration. In all cases,

aftercooling and adequate condensation removal must be done ahead of this drying equipment. The initial and operating costs and the results obtained vary considerably.

These methods are primarily for water vapor removal. Removal of lubricating oil is secondary, although all drying systems will reduce its carry-over. It should be understood that complete elimination of lubricating oil, particularly in the vapor form, is very difficult and that when absolutely oil-free air is required, some form of non-lubricated compressor is the best guaranteed method.

Chemical dryers

Chemical dryers are materials which combine with or absorb moisture from air when brought into close contact. There are two general types. One, using deliquescent material in the form of pellets or beads, is reputed to obtain a dew point, with 700°F inlet air to the dryer, of between 35°F and 50°F depending on the specific type of deliquescent material. The material turns into a liquid as the water vapor is absorbed. This liquid must be drained off and the pellets or beads replaced periodically. Entering air above 900°F is not generally recommended.

The second type of chemical dryer utilizes an ethylene glycol liquid to absorb the moisture. Standard dew

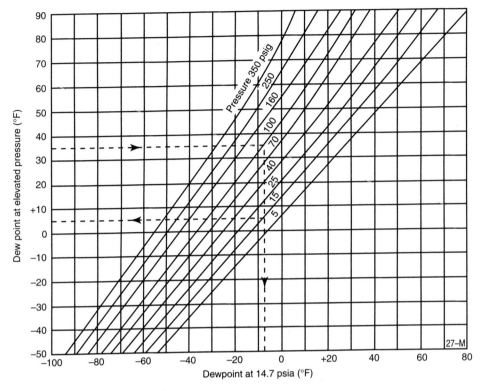

Figure 41.7 Dewpoint conversion chart

point reduction claimed is 400°F, but greater reductions are said to be possible with special equipment. The glycol is regenerated (i.e. dried) in a still using fuel gas or steam as a heating agent. The released moisture is vented to atmosphere. The regenerated glycol is re-circulated by a pump through a water-cooled heat exchanger that lowers the glycol temperature before returning to the dryer vessel.

Adsorption

Adsorption is the property of certain extremely porous materials to hold vapors in the pores until the desiccant is either heated or exposed to a drier gas. The material is a solid at all times and operates alternately through drying and reactivation cycles with no change in composition. Adsorbing materials in principal use are activated Alumina and silica gel. Molecular sieves are also used. Atmospheric dew points of minus 1000°F are readily obtained using adsorption.

Reactivation or regeneration is usually obtained by diverting a portion of the already dried air through a reducing valve or orifice, reducing its pressure to atmospheric, and passing it through the wet desiccant bed. This air, with the moisture it has picked up from the saturated desiccant bed, is vented to atmosphere. The diverted air may vary from 7 to 17 per cent of the mainstream flow, depending upon the final dew point

desired from the dryer. Heating the activating air prior to its passing through the desiccant bed, or heating the bed itself, is often done to improve the efficiency of the regeneration process. This requires less diverted air since each cubic foot of diverted air will carry much more moisture out of the system. Other modifications are also available to reduce or even eliminate the diverted air quantity.

Refrigeration

Refrigeration for drying compressed air is growing rapidly. It has been applied widely to small installations, sections of larger plants, and even to entire manufacturing plant systems. Refrigerated air dryers have been applied to the air system both before and after compression. In the *before compression* system, the air must be cooled to a lower temperature for a given *final line pressure dew point*. This takes more refrigeration power for the same end result. Partially offsetting this is a saving in air compressor power per $1000\,\text{ft}^3$ per minute (cfm) of *atmospheric* air compressed due to the reduction in volume at the compressor inlet caused by the cooling and the removal of moisture. There is also a reduction in discharge temperature on single-stage compressors that may at time have some value. As atmospheric (inlet) dew point of 350°F is claimed.

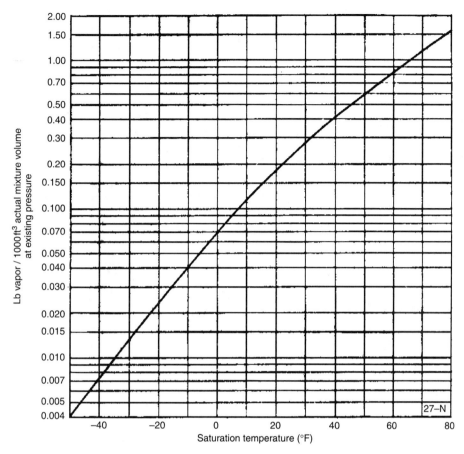

Figure 41.8 Moisture in air at any pressure

When air is refrigerated *following compression*, two systems have been used. Flow of air through directly refrigerated coils is used predominately in the smaller and moderate-sized systems. These are generally standardized for cooling to 350°F, which is the dew point obtained at line pressure. Figure 41.9 diagrams the equipment furnished in a small self-contained direct-refrigeration dryer.

The larger systems chill water that is circulated through coils to cool the air. A dew point at line pressure of about 500°F is obtainable with this method. Figure 41.10 illustrates a typical system of a *chiller-dryer* unit. The designs shown are regenerative since the incoming air is partially cooled by the outgoing air stream. This reduces the size and first cost of the refrigeration compressor and exchanger. It also reduces power cost and reheats the air returning to the line. Reheating of the air after it is dried has several advantages: the air volume is increased and less *free air* is required; chance of line condensation is still further reduced; and sweating of the cold pipe leaving the dryer is eliminated. Reheating dryers seldom need further reheating.

Combination systems

The use of a combination dryer should be investigated when a very low dew point is necessary. Placing a refrigeration system ahead of an adsorption dryer will allow the more economical refrigeration unit to remove most of the vapor and reduce the load on the desiccant.

41.6.2 Dry air distribution system

Most plants are highly dependent upon their compressed air supply and it should be assured that the air is in at least reasonable condition at all times, even if the drying system is out of use for maintenance or repair. It is possible that the line condensation would be so bad that some air applications would be handicapped or even shut down if there were no protection. A vital part of the entire endeavor to separate water in the conventional compressed air system is also the trapping of dirt, pipe scale and other contaminates. This is still necessary with a dried air system. As a minimum, all branch lines should be taken off the top of the main and all feeder lines off the top of branch lines.

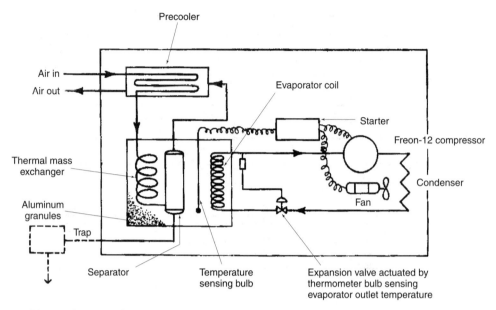

Figure 41.9 Diagram of equipment furnished in small self-contained direct-refrigeration dryer

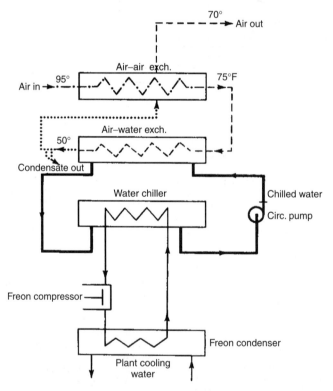

Figure 41.10 Diagram of a typical regenerative chiller-dryer

Absolute prevention of line freezing can be obtained only when the dew point of the line air is below any temperature to which it may be exposed. Freezing is always possible if there is line condensation. For example when air lines are run outdoors in winter weather or pass through cold storage rooms, the ambient temperatures will change the dew point and cause any moisture in the air to condense and freeze.

The dryer unit has an air inlet, an air outlet, a waste air outlet, two heater coils, and two 4-way reversing valves. Although the illustration shows the two tanks in a functioning mode, they are universal in operation. By this, we mean that this changes the tanks over when the active tank becomes totally saturated, and the tank that was on the regeneration cycle then becomes the active unit.

With this process, there will always be one tank in active service, and one on regeneration or stand-by. A timer governs the changeover process. The timer changes the position of the 4-way, or reversing, valves. This operation reverses the airflow through the tanks; the tank that was regenerated will now dry the air, while the saturated tank will be regenerated.

41.7 Air reservoir (receivers)

Air receivers, being simple volume tanks, are not often thought of as highly engineered items, but the use of simple engineering with receivers can reduce equipment costs. A pertinent example not infrequent in industry is the intermittent requirement for fairly large volume of air at moderate pressure for a short period of time. Some boiler soot blowing systems are in this class. The analysis necessary to arrive at the most economical equipment often involves the storage of air at high pressure to supplement the compressor's output when the demand requires. The following example is somewhat extreme, but emphasizes the need for proper receiver selection.

Sizing a receiver tank

This application requires $1500\,\mathrm{ft}^3$ per minute (cfm) of free air at 90 psig for 10 minutes each hour. This cycle of 10 minutes at 1500 cfm and 50 minutes with zero demand repeats hourly. Alternates obviously are possible; (A) install a 100-psig compressor and standard accessories large enough for the maximum demand requirements; or, (B) install a smaller compressor, but for a higher pressure and store the air in receivers during the off or no demand period. At least two storage pressures should be considered. In all cases commercial compressor sizes are to be used. For (B1) assume 350 psig and for (B2), 500 psig.

The 100-psig machine for at least 1500-cfm output, unit (A) has 1660 actual capacity and requires 309 brake horsepower (bhp) at full load. The minimum capacity for units B1 and B2 must be calculated, since these can be compressing the full 60 minutes each hour.

$$\text{Total ft}^3 \text{per hour} = 1,500 \times 10 = 15,000$$

$$\text{Minimum compressor capacity} = \frac{15,000}{60} = 250\,\text{cfm}$$

Compressor (B1), for 350-psig discharge, is found to have a capacity of 271 cfm and a required brake horsepower of 85. Compressor (B2), for 500 psig, is found to be a standard size with a capacity of 310 cfm and requires 112 bhp. All selections provide some extra capacity for emergency and losses are economical two-stage designs.

Air compressed to 100 psig, to be used at 90 psig, provides no possibility of storage. Storage is practical at the other pressures selected. Compressor units B and C will operate at full load at their rated or lower pressures. Since the receiver pressure will fall during the 10 minutes when air is used faster than the compressor can replenish it, the unit will be operating at full capacity and will supply some of the demand. The full demand need not be stored.

Cfm to be stored

$$= \text{Total ft}^3/\text{hour} - \text{Compressor cfm} \times \text{Minutes of demand}$$

The cubic feet of air to be stored represents the free air, at 14.7 psia, that must be packaged into the receiver above the minimum pressure required by the demand. In this case, the demand pressure is 90 psig, but an allowance for line losses and the necessary reducing valve pressure drop would prevent the use of any air stored below 110 psig. The receiver has a volume of (V) expressed in cubic feet.

$$\text{Useful free air stored} = \frac{V \times \text{Pressure drop}}{14.7}$$

$$V = \frac{\text{Useful free air stored} \times 14.7}{\text{Pressure drop}}$$

This receiver volume may be in one or several tanks, the most economical number being chosen.

The final selection can be made only after consideration of the first cost of the compressor, motor, starter, aftercooler and receiver. The cost of installation, including foundations, piping and wiring as well as the operating power cost must also be considered. In the example, Table 41.3 provides a basic comparison of the three options.

Table 41.1 Comparison of receiver options

Unit	B1	B2
Capacity (cfm)	271	310
Demand period (minutes)	10	10
Total cubic feet required	15,000	15,000
Delivered during demand period	2,710	3,100
Cubic feet to be stored	12,290	11,900

Table 41.2 Receiver volume required

Unit	B1	B2
Storage pressure (psig)	350	500
Minimum pressure (psig)	110	110
Pressure drop (psi)	240	390
Free air to be stored (ft^3)	12,290	11,900
Receiver volume (V)	752	448

Table 41.3 Economic comparison

Unit	A	B1	B2
Pressure (psig)	100	350	500
Installed cost	100%	52%	63%
Fixed charges	100%	52%	63%
Power cost	100%	64%	93%
Oil, water, attendance	100%	100%	100%

Not all problems of this nature will result in the selection of the intermediate storage pressure. Although few will favor the 100-psig level, many will be more economical at the 500-psig level. Experience indicates that there is seldom any gain in using a higher storage pressure than 500 psig since larger receivers required become very expensive above this level and power cost increases.

Air reservoirs are classified as pressure vessels and have to conform to the ASME Pressure Vessel Codes. As such, the following attachments must be fitted:

- Safety valves
- Pressure gages
- Isolation valves
- Manhole or inspection ports
- Fusible plug

Air reservoirs are designed to receive and store pressurized air. Pressure regulating devices are installed to maintain the pressure within operational limits. When the air reservoir is pressurized to the maximum pressure setpoint, the pressure regulator causes the air compressor to off-load compression by initiating an electrical solenoid valve to use lubricating oil to hydraulically hold open the low pressure suction valve on the compressor.

As the compressed air is used, the pressure drops in the reservoir until the low-pressure set point is reached. At this point, the pressure regulated solenoid valve is de-energized. This causes the hydraulic force to drop off on the low-pressure suction valve, restoring it to the full compression cycle.

This cycling process causes drastic variations in noise levels. These noises should not be regarded as problems, unless accompanied by severe knocking or squealing noises. Figure 41.11 shows a typical hydraulic unloader and its location on the compressor.

41.8 Safety valves

All compressed air systems that use a positive displacement compressor must be fitted with a pressure relief or safety valve that will limit the discharge or inter-stage pressures to a safe maximum limit. Most dynamic compressors must have similar protection due to restrictions placed on casing pressure, power input and/or keeping out of surge range.

Two types of pressure relief devices are available, *safety valves* and *relief valves*. Although these terms are often used interchangeably, there is a difference between the two. Safety valves are used with gases. The disk overhangs the seat to offer additional thrust area after the initial opening. This fully opens the valve immediately, giving maximum relief capacity. These are often called *pop-off safety valves*.

With relief valves, the disk area exposed to overpressure is the same whether the valve is open or closed. There is a gradual opening, the amount depending upon the degree of over-pressure. Relief valves are used with liquids where a relatively small opening will provide pressure relief.

Positive-displacement machines use *safety valves*. There are ASME standards of materials, sizing, and only ASME stamped valves should be used. The relieving capacity of a given size of safety valve varies materially with the make and design. Care must be taken to assure proper selection.

An approved safety valve is usually of the 'huddling chamber' design. In this valve the static pressure acting on the disk area causes initial opening. As the valve pops, the air space within the huddling chamber between seat and the blowdown ring fills with pressurized air and builds up more pressure on the roof of the disk holder. This temporary pressure increases the upward thrust against

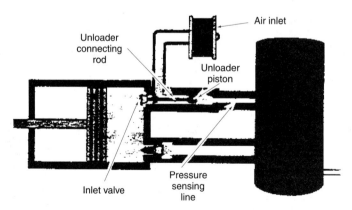

Figure 41.11 Typical hydraulic unloader and its location on the compressor

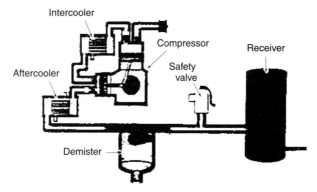

Figure 41.12 Illustrates how a safety valve functions

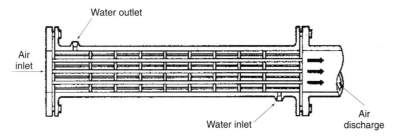

Figure 41.13 Compressor air cooler

the spring, causing the disk and its holder to lift to full pop opening.

After a predetermined pressure drop, which is referred to as blowdown, the valve closes with a positive action by trapping pressurized air on top of the disk holder. The pressure drop is adjusted by raising or lowering the blowdown ring. Raising the ring increases the pressure drop while lowering it decreases the drop.

Most state laws and safe practice require a safety relief valve ahead of the first stop valve in every positive displacement compressed air system. It is set to release at 1.25 times the normal discharge pressure of the compressor or at the maximum working pressure of the system, whichever is lower. The relief valve piping system sometimes includes a manual vent valve and/or a bypass valve to the suction to facilitate startup and shutdown operations. Quick line sizing equations are (1) line connection, $d/1.75$; (2) bypass, $d/4.5$; (3) vent, $d/6.3$; and (4) relief valve port, $d/9$.

The safety valve is normally situated atop the air reservoir. There must be no restriction on all blow-off points. Compressors can be hazardous to work around because they do have moving parts. Ensure that clothing is kept away from belt drives, couplings and exposed shafts.

In addition, high temperature surfaces around cylinders and discharge piping are exposed. Compressors are notoriously noisy. For this reason, ear protection should be worn. When working around high-pressurized air systems, wear safety glasses and do not search for leaks with bare hands. High-pressure leaks can cause severe friction burns.

41.9 Coolers

The amount of moisture that air can hold is inversely proportional to the pressure of the air. As the pressure of the air increases, the amount of moisture that air can hold decreases. The amount of moisture air can hold is also proportional to the temperature of the air. As the temperature of the air decreases, the amount of moisture it can hold decreases. The pressure change of compressed air is larger than the temperature change of the compressed air. This causes the moisture in the air to condense out of the compressed air. The moisture in compressed air systems can cause corrosion, water hammers, and freeze damage. Therefore, it is important to avoid moisture in compressed air systems. Coolers are used to address the problems by moisture in compressed air systems.

Coolers are frequently used on the discharge of a compressor. These are called aftercoolers, and their purpose is to remove the heat generated during the compression of the air. The decrease in temperature promotes the condensing of any moisture present in the compressed air. This moisture is collected in condensate traps that are either automatically or manually drained.

If the compressor is of the multi-stage type, there may be an intercooler, which is located after the first stage discharge and second stage suction. The principle of the intercooler is the same as the principle of the aftercoolers, and the result is drier, cooler compressed air. The structure of the individual cooler depends on the pressure and volume of the air it cools. Figure 41.13 illustrates a typical compressor air cooler.

The combinations of drier compressed air (which helps prevent corrosion) and cooler compressed air (which allows more air to be compressed for a set volume) is the reason the air coolers are worth the investment.

42

Noise and Vibration

Roger C Webster
Environmental Consultant

Contents

42.1 Introduction: basic acoustics

Sound can be defined as the sensation caused by pressure variations in the air. For a pressure variation to be known as sound it must occur much more rapidly than those of barometric pressure and the degree of variation is much less than atmospheric pressure.

Audible sound has a frequency range of approximately 20 Hertz (Hz) to 20 kilohertz (kHz) and the pressure ranges from 20×10^{-6} N/M to 200 N/M. A pure tone produces the simplest type of wave form, that of a sine wave (Figure 42.1). The average pressure fluctuation is zero, and measurements are thus made in terms of the root mean square (rms) of the pressure variation. For the sine wave the rms is 0.707 times the peak value.

Since rms pressure variations have to be measured in the range 20×10^{-6} N/M to 200 N/M (a range of 10^7) it can be seen that an inconveniently large scale would have to be used if linear measurements were adopted. Additionally, it has been found that the ear responds to the intensity of a sound ($\propto P^2$) in a logarithmic way. The unit that has been adopted takes these factors into account and relates the measured sound to a reference level. For convenience, this is taken as the minimum audible sound (i.e. 20×10^{-6} N/M) at 1 K.

The logarithm (to the base 10) of the ratio of the perceived pressure (squared) to the reference pressure (squared) is known as the Bell, i.e.

$$B = \log \frac{P^2}{P^2_{\text{reference}}}$$

Since this would give an inconveniently small scale (it would range from approximately 0 to 14 for a human response) the bell is divided numerically by 10 to give the decibel. The equation therefore becomes:

$$dB = 10 \log \frac{P^2}{P^2_{\text{ref}}}$$

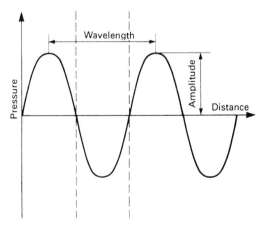

Figure 42.1 Sine wave

42.1.1 Sound intensity

Sound intensity is a measure of energy, and its units are watts per meter. Intensity is proportional to the square of pressure:

$$\frac{I}{I_0} = \left(\frac{P}{P}\right)^2$$

Sound intensity level is defined in a similar manner to sound pressure level. In this case the equation is:

$$\text{(dB) (Sound intensity level)} = 10 \log \frac{I}{I_{\text{ref}}}$$

I_{ref}, the reference level, is taken as: $10^{-12} \dfrac{W}{M^2}$

42.1.2 Sound power

The power of a source (measured in watts) can be similarly expressed in terms of decibels (in this case, called the sound power level):

$$\text{(dB) (Sound power level)} = 10 \log \frac{W}{W_{\text{ref}}}$$

where W_{ref} is taken as 10^{-12} W.

It can thus be seen that it is important not only to express the unit but also to state sound pressure level, sound intensity level or sound power level.

42.1.3 Addition and subtraction of decibels

For coherent sound waves addition of values is possible. It will be apparent that as the scale is logarithmic, values cannot merely be added to one another. Intensities can, however, be added and thus the equation becomes:

$$\text{SIL (total)} = 10 \log \frac{I_1 + I_2}{I_{\text{ref}}}$$

i.e. 70 dB + 73 dB

$$70 = 10 \log \frac{I_1}{I_{\text{ref}}}$$

$$I = \frac{\text{Antilog } 7}{I_{\text{ref}}}$$

$$73 = 10 \log \frac{I_2}{I_{\text{ref}}}$$

$$I = \frac{\text{Antilog } 7.3}{I_{\text{ref}}}$$

$$\text{SIL (total)} = 10 \log(\text{Antilog } 7 + \text{Antilog } 7.3)$$

$$= 10 \log(10^7 + 1.99526 \times 10^7) 10 \log(2.99526 \times 10^7)$$

$$= 74.76 \text{ dB}$$

The square of individual pressures must be added, and thus the equation in this case must utilize:

$$P(\text{total}) = \sqrt{(P_1^2 + P_2^2)}$$

42.1.4 Addition of decibels: graph method

It is possible to use a graph to calculate the addition of decibels, even in the case of multiple additions (Figure 42.2). The graph is used in the following way:

1. *In the case of addition of two levels.* The difference between the higher and the lower levels is plotted on the lower scale of the graph. The correction is then read from the vertical scale by projecting a horizontal line across to this scale from the point on the graph. The correction is added to the highest original level to give the total level.
2. *In the case of the subtraction of levels.* The difference between the total sound level and the one to be subtracted is plotted onto the graph and the correction obtained as above. In this case the correction is subtracted from the total level to give the remaining sound level.
3. *In the case of multiple additions.* If there are more levels to be added, the first two levels are added using the graph and then the third is added to the resultant using the same method.

42.1.5 Relationship between SPL, SIL and SWL

The total acoustic power of a source can be related to the sound pressure level at distance r by the following equation (assuming spherical propagation):

$$W = \frac{P^2}{\rho_c^{\times 4\pi r^2}}$$

where ρ = density of the medium and C = velocity of sound in the medium. By substituting this back into the SPL equation we obtain:

SPL = SWL-20 log r − 11(spherical propagation)

It is also possible to derive equations for the other common situations, i.e.

Point source on a hard reflecting plane
Line source radiating into space
Line source on a hard reflecting plane

These equations are:

SPL = SWL-20 log r − 8(hemispherical propagation)

SPL = SWL-10 log r − 8(line source in space)

SPL = SWL-10 log r − 5(line source radiating on a plane)

The above equations are useful for calculating distance-attenuation effects.

If the sound pressure level at a distance r_0 is known it is possible to calculate the sound pressure level at positions r_1 quite easily:

$$SPL_0 - SPL_1 = 20 \log \frac{r_1}{r_0} dB$$

$$SPL_0 - SPL_1 = 20 \log r_1 - 20 \log r_0$$

If r_1 is double r_0 it will be seen that $SPL_0 - SPL_1$ will be approximately equal to 6 dB (20 log 2). This gives us the principle of a decrease in level of 6 dB per doubling of distance (inverse square law). For the line source the same calculation produces a difference of only 3 dB per doubling of distance.

42.1.6 Frequency weighting and human response to sound

In practice, noises are not composed of one single pure tone but are usually very complex in nature. It is essential that more than the overall noise level (in dB) is known in order to appreciate the *loudness* of a noise, as the ear does not respond uniformly to all frequencies.

As previously stated, the ear can respond from 20 Hz to 200 kHz, and this response can be demonstrated by equal-loudness contours (Figure 42.3). It can be seen from the figure that there is a loss in sensitivity (compared to 2 kHz) of approximately 60 dB at the low-frequency end of the

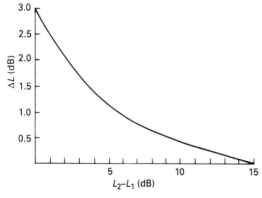

Figure 42.2 Noise-level addition graph

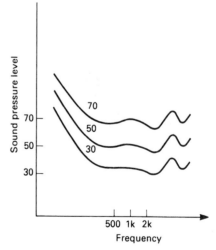

Figure 42.3 Equal-loudness contours

chart. It will also be noted that all the curves are approximately parallel, but there is a tendency towards linearity at the higher noise levels. In order to produce meaningful readings it is therefore important to state the sound pressure level in decibels and the frequency of the noise.

Weighting can be imposed on noise readings, which corresponds to the inverse of the equal-loudness contours. If this weighting is used, all readings that are numerically equal will sound equally loud, regardless of frequency.

Originally, three networks were proposed (A, B, and C) and it was suggested that these be used for low, medium and high noise levels, respectively. It was shown in practice that this introduced numerous difficulties, particularly with a rapidly changing noise when a change of filter network was necessary. It was also found that at all noise levels the weighting network corresponded well to annoyance levels. It was therefore decided that the weighting would be used as the norm for noise readings concerning human response. Another weighting network (the D network) is used for aircraft noise measurement (Figure 42.4).

If it is necessary for engineering purposes to know the tonal make-up of a noise, several approaches are possible. A bandpass filter can process the noise. The most common filters are octave band filters, and the agreed center frequencies are as follows:

31 63 142 420 500 1 K 2 K 4 K K 8 K 16 K (Hz)

If further resolution is necessary one-third octave filters can be used but the number of required measurements is most unwieldy. It may be necessary to record the noise onto tape loops for the repeated re-analysis that is necessary. One-third octave filters are commonly used for building acoustics, and narrow-band real-time analysis can be employed. This is the fastest of the methods and is the most suitable for transient noises. Narrow-band analysis uses a VDU to show the graphical results of the fast Fourier transform and can also display octave or one-third octave bar graphs.

42.1.7 Noise indices

All the previous discussions have concerned steady-state noise. It will, however, be apparent that most noises change in level with time. It may therefore be necessary to derive indices which describe how this happens. The most common of these are percentiles and equivalent continuous noise levels.

Percentiles are expressed as the percentage of time (for the stated period) during which the stated noise level was exceeded, i.e. 5 min L_{90} of 80 dB(A) means that for the 5-min period of measurement for 90 per cent of the time the noise exceeded 80 dB(A). Therefore L_0 is the maximum noise level during any period and L_{100} is the minimum. Leq (the equivalent continuous noise level) is the level which, if it were constant for the stated period, would have the same amount of acoustic energy as the actual varying noise level.

42.1.8 Noise rating curves

These are a set of graphs that are used as a specification for machinery noise. They are similar to noise criteria curves (used in the USA to specify noise from ventilation systems). The rating of a noise under investigation is the value of the highest noise-rating curve intersected by the readings when plotted on the graphs (Figure 42.5).

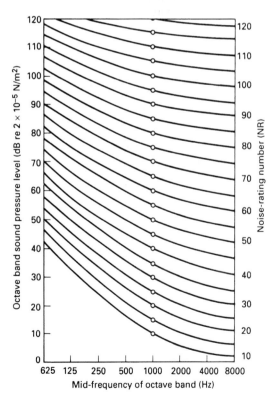

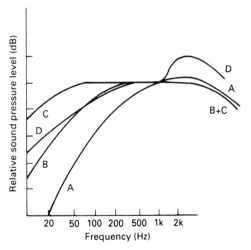

Figure 42.4 Weighting networks

Figure 42.5 Noise-rating curves

42.2 Measurement of noise

The simplest sound level meter consists of a microphone, an amplifier and a meter of some type. Sound level meters are graded according to British and international standards, and the most common type used for accurate measurement purposes it known as the Precision Grade or Type 1 meter. In practice, a basic sound level meter will incorporate weighting networks with either in-built octave filters or provision for connecting an external filter set (Figure 42.6).

The meter will also have a control for the time constant for the display (i.e. the *speed* of the meter response), and the two common time constants are 'Fast' and 'Slow'. Others may have an impulse and peak hold facility. More complex meters incorporate Leq-measuring devices, and these are also available as hand-held Type 1 meters with filters as in the basic meter.

Outputs are available in either analogue (D.C. or A.C.) or digital form. Digital output may connect the meter to computers (either portable or office based) for more complex calculations or to produce larger graphical displays.

Portable sound level meters are also available which can measure percentiles. These either hold the results in a memory which can be separately interrogated or may be connected to a computer for a printout. Larger machines (known as environmental noise analyzers) are available which can record percentiles and Leq readings and produce a printout. These are resistant to weather and can be left on-site for up to a week.

Computers may be used for noise analysis when connected to dedicated hardware devices. One machine incorporates a narrow-band analyzer, octave and one-third octaves with all the features of an environmental noise analyzer. These devices cost much less than purchasing all the dedicated instruments separately. They may be obtained in portable form but are rarely weatherproofed for outdoor use.

42.3 Vibration

This may be defined as the oscillatory movement of a mechanical system, and it may be sinusoidal or non-sinusoidal (also known as complex). Vibration can occur in many modes, and the simplest is the single freedom-of-movement system. A mass/spring diagram (Figure 42.7) can explain the vibration of a system.

The displacement of the object from its rest position can be derived from

$$X = X_{peak} \text{ sine } (\omega t)$$

where X = displacement at time t, X_{peak} = peak displacement and ω = angular velocity $(2\pi f)$. The velocity of the object is proportional to displacement and frequency and its acceleration is proportional to the displacement and the square of frequency. There is a phase angle of 90° between displacement and velocity and a further 90° between velocity and acceleration. The units of measurement may be made directly, i.e.

Displacement in meters
Velocity in meters per second
Acceleration in square meters per second

Vibration may also be expressed in decibels, and the standard reference levels used are:

$10\,m^{-8}/s$ for velocity
$10\,m^{-5}/s^2$ for acceleration

In practice, measurements are made with the use of an accelerometer. This device is connected to a sound level meter and may make measurements of acceleration in terms of decibels (or by changing scales or use of a device similar to a slide rule, in direct terms). An integrator can be connected between the accelerometer and the meter to express the results in terms of velocity or displacement.

42.3.1 Effects of vibration on people

If a body is vibrating at a frequency within the audible range, sound will be radiated. Vibration may be transmitted considerable distances through buildings, ground

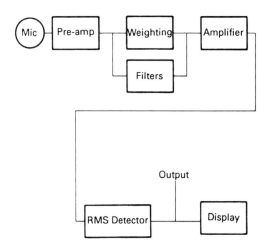

Figure 42.6 Schematic diagram of a sound level meter

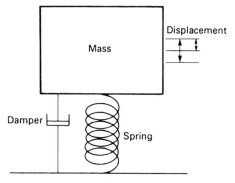

Figure 42.7 Mass/spring diagram

structures or rock strata, and then re-radiated as noise at the receiver position. For example, low-frequency noise nuisance has been caused at a residence on the banks of the River Thames, by vibration from a pump on the opposite bank. In this case the vibration passed under the river along a curved layer of harder rock.

Vibration can be perceived from a frequency of approximately 3 Hz. The lower frequencies appear to cause the most discomfort. At high levels (above 120 dB) vibration can be physically damaging to people, and resonance of the human body occur at the following frequencies:

3.6 Hz:	Thorax, abdomen
20–30 Hz:	Head, neck
60–90 Hz:	Skull

High vibration levels at these frequencies can therefore be the most disturbing.

Various sources publish data of permissible vibration levels for employees (usually as a graph of acceleration against frequency). These are designed to avoid injury and cannot be used as a guide to the degree of disturbance caused by vibration. Vibration caused by neighboring industrial premises when received at a residence as vibration (i.e. no noise implications) has been considered a nuisance when it is just perceptible.

42.4 Noise and vibration control

Noise and vibration must be controlled in order to avoid

1. Nuisance;
2. Physical injury to workers;
3. Damage to plant/machinery/building structures.

42.4.1 Noise nuisance

Nuisance is not defined as such. Common-law nuisance is used as a guide and is divided into two types, public and private. Private nuisance relates to premises, and is a tort that can be defined as 'the unlawful interference or annoyance, which causes damage to an occupier or owner in respect to his use and enjoyment of his land'. Public nuisance is an unlawful act or omission to discharge a legal duty which can endanger the life, safety, health or comfort of the public or some section of it, or by which act the public are denied some common right. It should be noted that more than one person must be affected. Public nuisance is a criminal offence and, as such, may be tried on indictment in the Crown Court. The Attorney General may initiate an action in the High Court.

42.4.2 Legislation

Noise nuisance is controlled primarily by the Environmental Protection Act 1990. Section 79 of the Act places a duty on a local authority to inspect their area for nuisances. Section 80 places a duty on a local authority to serve a legal notice on persons responsible for a situation when a nuisance has occurred and is likely to recur, or where, in the opinion of the local authority, the nuisance is likely to occur. Section 82 enables an individual

to complain to a magistrate's court about a noise nuisance. If convinced, the magistrate may issue an Order telling the person causing the nuisance to cease the activity. This section may be used where the local authority cannot detect the nuisance. Under Section 60/61 of the Control of Pollution Act 1974 the local authority may issue a notice to a person carrying out construction work. This may require the person to adopt the best practical means of minimizing the noise from the site. (There is no reference to the noise having to be a nuisance in this section.) Section 62 relates to noise in the street.

42.4.3 Environmental Protection Act 1990, Section 80

A notice may be served where a nuisance has occurred or the local authority think a nuisance may occur. Noise nuisance is not defined as such, but includes vibration. The notice may not be specified and may merely require the abatement of the nuisance. A notice may, however, require the carrying out of works or specify permissible noise levels. The time period for compliance is not specified in the Act, but must be reasonable.

Appeals against a Section 80 notice must be made to the magistrate's court within 21 days of the serving of the notice. The grounds of appeal are given in the Statutory Nuisance (Appeals) Regulations 1990 and are as follows:

1. That the notice is not justified by the terms of Section 80. The most common reason for this defense is that the nuisance had not already occurred, and that the local authority did not have reasonable grounds to believe that the nuisance was likely to occur.
2. That there had been some informality, defect or error in, or in connection with, the notice. It may be that the notice was addressed to the wrong person or contained other faulty wording.
3. That the authorities have refused unreasonably to accept compliance with alternative requirements or that the requirements of the notice are otherwise unreasonable in character or extent, or are unnecessary. This defense is self-explanatory. The local authorities are only permitted to ask for works that will abate the noise nuisance. Other works (perhaps to comply with legislation) should not be specified in the notice. They may, however, be contained in a letter separate from the notice. An example of this would be where the fitting of acoustic enclosures to food-manufacturing machines breached food hygiene requirements. Readily cleanable enclosures may be a requirement of the Food Hygiene Regulations, but it should not be contained in a Section 58 Control of Pollution Act notice.
4. That the time (or, where more than one time is specified, any of the times) within which the requirement of the notice are to be complied with is not reasonably sufficient for the purpose.
5. Where the noise to which the notice relates is that caused by carrying out a trade or business, that the best practicable means have been used for preventing or for counteracting the effects of the noise. 'Best practicable means' incorporates both technical and financial possibility. The latter may be related to the turnover of

a company. Therefore a solution that may be the best practicable means for one company may not be so for another.

6. That the requirements imposed by the notice are more onerous than those for the time being in force in relation to the noise to which the notice relates of (a) Any notice under Section 60 or 66; or (b) Any consent given under Section 61 or 6 or (c) Any determination made under Section 67. Section 60 relates to a construction site notice. Section 61 is consent for construction works. Sections 65–67 relate to noise-abatement zones (see below).

7. That the notice might lawfully have been served on some person instead of the appellant, being the person responsible for the noise.

8. That the notice might lawfully have been served on some person instead of, or in addition to, the appellant, being the owner or occupier of the premises from which the noise is emitted or would be emitted, and that it would have been equitable for it to have been so served.

9. That the notice might lawfully have been served on some person in addition to the appellant, being a person also responsible for the noise, and that it would have been equitable for it to have been so served.

42.4.4 The assessment of nuisance

In assessing whether a certain noise is a nuisance the local authority may make use of BS 4142, Method of rating industrial noise affecting mixed residential and industrial areas. This standard forms the basis of local authorities' assessment of nuisance, and relates the background level (i.e. the noise that would pertain if not for the offending noise) to the measured noise. If the background cannot be found a method is given for deriving a notional background. However, the use of this notional background has fallen into disuse due to its unreliability.

The standard rates the offending noise according to its nature: 5 dB(A) is added where the noise has a definite continuous note and a further 5 dB(A) added for noise of an intermittent nature. The number of occasions that happen in an 8-hour period is then plotted on a graph and the correction for intermittency is derived. When these calculations have been performed, the noise level is compared to the background level. The standard states that where the noise exceeds the background by 5 dB or more, the nuisance is to be classed as marginal, and where the background is exceeded by 10 dB(A) or more, complaints are to be expected.

It is stressed in this standard that it is not intended to be a criterion but most local authorities tend to use it as a guide. Other methods are available or may have value for individual circumstances. Particular problems occur when a noise is of a very tonal nature or contains discernible information (i.e. music or voice). It is quite possible for music to become a nuisance at no more than 3 dB(A) above ambient background level.

By inference, if BS 4142 were used as a guide to nuisance, 5 dB(A) above background would not be cause for serving a notice. If this were permitted in areas where there were several sources of noise a 'creeping' ambient problem can occur. If this is the case, the local authority may have to control noise levels such that no increase in ambient is allowed. This may be particularly severe for a large developer in an already noisy area, but is necessary. If no addition is permitted the design noise level for the new development must be at least 10 dB(A) *below* background.

If there are tonal noise problems the local authority may use more complex measurements to specify the required reduction. Noise rating may be used or octave or one-third octave band levels specified.

42.4.5 Offences and higher court action

It is an offence to cause a noise nuisance while in breach of a notice. Proceedings in the magistrate's court can result in a fine of up to £2000 for each offence. It is also possible that the court may impose a daily penalty for continuing nuisances.

If the authority are of the opinion that magistrate's court action will not give an adequate remedy a complaint may be made to the High Court. This court will issue an injunction prohibiting the repeat of the nuisance. Non-compliance with an injunction constitutes contempt of court and penalties include imprisonment.

42.4.6 Noise-abatement zones

Local authorities are empowered by the Control of Pollution Act 1974 to designate areas as noise-abatement zones. Within these areas noise levels are measured and entered into a register. It is an offence to increase noise levels beyond register levels unless consent is obtained. If the local authorities are of the opinion that existing noise levels are too high, noise-reduction notices can be served.

In the case of new premises the local authority will determine noise levels which it considers acceptable, and these will be entered into the noise level register. Appeals against notices or decisions can be made to the Secretary of State.

42.4.7 Planning application conditions

Local authorities are empowered to impose conditions on planning applications to protect environmental amenities of neighbors. Noise is commonly controlled by planning conditions, which are designed to avoid reduction in amenity of neighbors. This may mean that a process has to be almost inaudible (particularly in the case of light industrial consents). Appeals against planning conditions may be made to the Secretary of State. Local authorities may ask for more onerous controls on planning conditions than the mere avoidance of nuisance.

42.5 Avoiding physical injury to workers

Noise levels between 85 and 120 dB(A) affect the hearing of exposed workers on a dose-related basis. The Leq of the noise is calculated and compared to the criterion currently employed. Under Section 2 of the Health and Safety at

Work, etc. Act 1974 an employer has to take all steps, as far as is reasonably practicable, to ensure the health, safety and welfare of his workers. Permissible noise levels are defined by codes of practice, and the figure permitted for employee exposure is 90 dB(A) 8-hour Leq.

Low-frequency noise (in the range 3–50 Hz) may have other injurious effects on the body. Research has also indicated that a type of fatigue caused by low-frequency noise has a similar effect to that caused by alcohol. Infrasound (low-frequency sound) also has a synergistic effect with alcohol. Low-frequency noise is particularly important in the case of workers operating machinery (e.g. vehicles, cranes, etc.). It must also be remembered that very high power levels may be generated at low frequency and may not be readily detected by the ear. Attenuation of low-frequency noise is very difficult (see Section 42.7).

42.6 Avoidance of damage to plant/machinery/building structures

Plant and machinery may be directly affected by excessive vibration but may also be damaged due to operator fatigue (caused by working in high noise-level areas) or to operators not being able to hear unusual noises from machines before it is too late. Vibration may be caused either by direct transmission or by excitation. Damage to machines may occur in many ways, including the accidental impact of surfaces on one another and the fatigue of shafts, etc. caused by bending moments.

In extreme cases building structures can be damaged by vibration. Damage can be caused either directly to the structure or by settlement of foundations due to vibration. It is possible to calculate the effects of the former (graphs are obtainable from various sources) but not the settlement effect. However, this will occur only on certain strata and in extreme cases.

42.7 Noise-control engineering

Before attempting noise control it is important to consider the nature of the problem:

Source of sound energy → Transmission pathway → Receiver

There are many proprietary systems available for controlling noise and it is easy to become blinded in one's approach to noise control.

The first consideration is the source of the noise itself. In the case of a new project the first option must be to select the quietest machine available that will perform the required task. With an existing problem this may still be the cheapest option, particularly in the case of a machine which is coming to the end of its useful life. Spin-offs may include more efficient operation (less fuel costs), greater reliability or quieter operating conditions.

If the decision is made to retain the original machine the design of the noise-control devices to be employed should be considered. These may affect:

Access to machine (may hinder cleaning, tool changing, etc.)

Heat build-up
Safety, etc.

In nearly all cases noise control at source is the best option. Typical examples are:

1. *Presses*: Is the degree of impact necessary? Can it be adjusted? Can the press operate by pressure alone?
2. *Air discharge – use of air tools and nozzles*: The turbulence in the boundary layer of air between the rapidly moving air stream and the atmosphere is heard as noise. Can the air stream be diffused (silencers fitted to the exhaust)? Nozzles used for cleaning can have devices fitted which give a gradual transition from the rapidly moving air to atmosphere by the use of an annular ring of small nozzles round the central nozzle. These silence with very little loss in efficiency.
3. *Reciprocating compressors*: These cause very high noise levels at low frequency (typically, below 420 Hz). The low-frequency noises are very difficult to attenuate, and the most popular solution is to use rotary (vane type) compressors. These are inherently quieter and have the further advantage that the noise they generate is at high frequency (typically, above 1 kHz) and is therefore easy to attenuate.
4. *Cutting machines*: Modifications can be made to the method of restraining material being cut to reduce *ringing* (i.e. reducing free lengths of material).

42.7.1 Noise-reduction principles

There are many noise-control devices. All, however, rely on one or more of the three basic noise-control principles: insulation, absorption and isolation.

42.7.2 Insulation

The simplest insulator is a sheet of material placed in the sound-transmission pathways. Sound energy reaches the surface in the form of a pressure wave. Some energy passes into the partition and the rest is reflected.

Energy that passes into a partition may be partially absorbed and transformed into heat. This is likely to be very small in a plain partition. The remainder of the energy will then pass through the partition by displacement of molecules and pass as sound in the same way that sound travels in air. This can then pass to the edge of the partition and be re-radiated as sound from other elements of the structure. This is known as flanking transmission. In a thin partition by far the greatest amount of energy will pass through the partition by actually causing the partition to vibrate in sympathy with the incident sound and hence re-radiating the sound onto the opposite side. The amount of sound transmission through a partition is represented by the ratio of the incident energy to the transmitted energy. This factor, when expressed as decibels, is known as the sound-reduction index (SRI):

$$SRI = \frac{10 \log}{Transmission\ coefficient\ dB}$$

The moment of the panel (and hence its resistance to the passage of sound) is controlled by a number of factors:

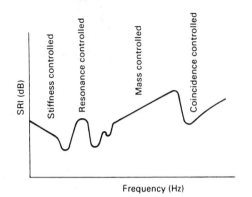

Figure 42.8 Typical insulation characteristics of a partition

1. The surface mass affects the inertia of panel. Greater mass causes a corresponding greater inertia and hence more resistance to movement. At high frequencies this becomes even more significant. The mass law can be expressed as

$$\text{SRI} = 20 \log mf - 43\,\text{dB}.$$

Where m is the superficial weight (kg/m^2) and f the frequency (Hz).

2. At very low frequencies the movement of the panel will be controlled by the stiffness, as inertia is a dynamic force and cannot come into effect until the panel has measurable velocity. Stiffness controls the performance of the panel at low frequencies until resonance occurs. As the driving frequency increases, the resonance zone is passed and we enter the mass-controlled area. The increase in the sound-reduction index with frequency is approximately linear at this point, and can be represented by Figure 42.8.

3. A panel will have a bending mode when a wave travels along the length of the sheet of material. The frequency of this bending mode is known as the critical frequency. This mode of bending will be introduced by sound incident at angles greater than 0°. At the critical frequency coincidence will only occur for a sound wave with a grazing incidence (90°). At greater frequencies the partition will still be driven, but in this case by progressively lower angles of incidence. Therefore, the coincidence dip is not a single dip, but will result in a loss of SRI at progressively higher frequencies. The desirable insulation panel will therefore be very large but not stiff.

42.7.3 Absorbers

Porous

As sound passes through a porous material, energy is lost by friction within the material. The material is usually attached to various surfaces in a room. The absorber will have the highest efficiency when positioned where the air molecules are moving the fastest (and hence more energy

is absorbed). At the wall surface the molecules are stationary. If we plot a single-frequency graph we find that the maximum particulate velocity occurs at $\lambda/4$ (one-quarter wavelength) from the surface. In practice, incident sound is rarely of single frequency, but the principle can be observed that the absorber must be one quarter of the wavelength away from the wall (for the frequency of the sound to be absorbed). This can be arranged either by having a thickness greater than $\lambda/4$ for the lowest frequency to be absorbed or by mounting the absorber on a frame some distance away from the wall so that the center of the absorber is at $\lambda/4$ for the frequency to be absorbed.

Resonant

The simplest resonant absorber is known as the Helmholtz resonator. This consists of a chamber connected to the duct (or whatever area is to be controlled) by a narrow neck. The volume of air in the chamber will resonate at a frequency determined by the volume of the chamber, the length of the neck and the cross-sectional area of the neck:

$$F_{\text{res}} = 55 \sqrt{\frac{S}{IV}}\,(\text{Hz})$$

where S = cross-sectional area (m), I = length of neck (m) and V = volume of enclosure (m). As the chamber resonates, air is forced through the narrow neck and hence energy is absorbed in overcoming the resistance.

The degree of attenuation at the critical frequency can be very large, but this type of silencer has a very narrow bandwidth. This device may be suitable when the machine being dealt with emits sound predominantly of a single wavelength. Lining the chamber with absorbers can expand the absorber bandwidth of a Helmholtz resonator, but this has the effect of reducing the efficiency. The perforated absorber, which forms the basis of many acoustic enclosures and silencers, is a development of the resonator principle.

As stated previously, packing the chamber with an absorber may broaden the bandwidth, but this lowers efficiency. It may be overcome by using multiple absorbers in the sound path, and placing a perforated sheet some distance away from the rigid outer wall of the enclosure and filling the cavity with absorber can do this. It is not necessary to use cross walls between the 'chambers' so formed. In this case the equation becomes:

$$F_{\text{res}} = 5000 \frac{P}{I(t + 0.8d)}\,\text{Hz}$$

where I = depth of airspace, t = thickness of panel, d = diameter of holes and p = percentage open area of panel.

Panel

This absorber is basically a panel attached to a structural wall that is designed to absorb energy. The absorber is therefore frequency dependent and has an absorption peak at its resonant frequency. This type of absorber is not commonly used.

42.7.4 Vibration isolation

Vibration in machinery or plant can be induced in a number of ways, e.g.:

Out-of-balance forces on shafts;
Magnetic forces in electrical apparatus;
Frictional forces in sliding objects.

The first course of action in vibration isolation is reduction at source. Balancer shafts in engines, stiffer coils in electrical apparatus or better lubrication between adjacent sliding surfaces may achieve this.

When all possible vibration reduction has been obtained, the machine must be isolated from the structure. Some form of spring mounting achieves this. Spring mounts have a resonant frequency dependent on the stiffness of the spring and the weight of the object placed on it. It will be apparent that the static deflection of the spring will also be proportional to the resonant frequency.

As the driving force of the mass/spring increases from zero up to the resonant frequency, the amount of transmission of the vibration increases until resonance is reached and the transmission becomes infinitely large. As the resonant point is passed, the transmission begins to reduce until at some point the transmissibility falls below one (see Figure 42.9), i.e. isolation occurs.

In practice, however, spring systems have some in-built damping, and this will have the effect of reducing the amplitude of the resonance below infinity. This is very necessary in real systems to avoid excessive excursions of mounted machinery. A damped mounting will follow the second curve on the graph in Figure 42.9, and it will be noted that the vibration isolation at high-frequency ratios is less than that for undamped systems. It is important therefore to use the lowest degree of damping necessary.

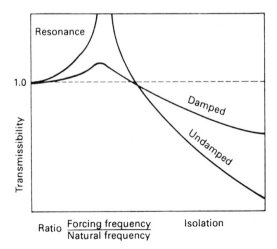

Figure 42.9 Performance of anti-vibration mounting

42.8 Practical applications

42.8.1 Acoustic enclosures

Panels of multi-resonator material are made from perforated plate sandwiched with solid plate and an intermediate absorber layer between them. These panels can be built up in enclosures, taking care to seal all junctions adequately. Typically, these enclosures are made to surround small machines (e.g. compressors). They may be fitted together with spring catches to allow for dismantling for maintenance purposes.

Ventilation may be a problem but can be dealt with in several ways:

1. Acoustic louvers constructed of the absorbent panel material (suitable for a small degree of noise reduction only);
2. A silencer fitted to the ventilation duct (see below);
3. Baffled enclosures to the ventilation duct.

42.8.2 Building insulation

Single-panel insulators have been described above, but in buildings it is usual to provide double insulation. In theory, if the insulation panels had no interconnection it should be possible arithmetically to add the sound reduction of the two elements of the structure. In practice, it will be found that there is bridging, by the structure, wall ties or flanking transmission or by the air between the two elements acting as a spring. If the two elements of the wall were in rigid connection the insulation would be 3 dB more than the single element alone (mass law), and if totally separated it would be the sum of the figures. In practice, a cavity wall with ties and a 50-mm cavity gives approximately 10 dB more reduction compared to a single-skin wall of half the surface mass.

Double-glazed windows work on the same principle. It is important to avoid the coincidence of the resonant frequencies of the two elements, and hence it is usual to arrange for the glazing panels to have different thickness (and hence a different resonant frequency). This is not necessary if one element is sub-divided by glazing bars to give different-size panes to its opposing element.

The reveals of a double-glazed window should be lined with acoustically absorbent material to damp the sound within the cavity. The width of the cavity should not be less than 150 mm.

If insulation panels are not of uniform construction (as in the case of a wall containing a window) the average sound-insulation value must be derived for use in calculations. The total transmission coefficient for the composite panel will equal the sum of the individual coefficient times their respective areas divided by the total area. Thus:

$$t_{\text{average}} = \frac{t_1 s_1 + t_2 s_2 + t_3 s_3}{s_{\text{tot}}} \text{ etc.}$$

and the SRI of the total panel is derived from

$$\text{SRI} = 10 \log \left(\frac{1}{t_{\text{avg}}} \right) \text{dB}$$

42.8.3 Control of noise in ducts

Fans produce the least noise when operating at their maximum efficiency. It is therefore important to select the correct fan for the airflow and pressure characteristics required. It is also important to remember that the noise generated within the system (as opposed to at the fan) is a factor of air velocity, and hence, for a required airflow rate, a larger cross-section duct (with a correspondingly lower velocity) will give quieter results. It will also have other advantages when providing extra noise attenuation and fitting silencers.

In the design of systems it is most important to eliminate as much turbulence as possible, and to achieve this the fans should be mounted some distance away from bends (at least one and a half duct diameters). Junctions between pipes and connectors should present a smooth internal profile and inlets to systems must be tapered and not plain. Outlet grills should be of larger diameter than the duct and have aerodynamically smooth profiles where possible.

If it is necessary to add extra attenuation to a duct it is essential to decide on the required amount. If only a relatively small degree of absorption is required, first a part of the duct must be lined with absorber. The length of duct to be lined will be determined by the degree of attenuation required and the thickness by the noise frequency. Data for these factors are available from many sources and are usually published as tables.

For further attenuation it is necessary to provide a center-pod-type attenuator (Figure 42.10). This increases the area of the absorber and also aids low-frequency attenuation. For further low-frequency attenuation an in-line splitter silencer is employed (Figure 42.11). These are capable of providing a high degree of attenuation, dependent on the width between the elements. The smaller the gap, the higher the attenuation. Again, the major manufacturers publish performance tables. In order to decide on the design of the silencer to be installed it is necessary to know the required attenuation (and the frequency/noise level profile) and the permitted pressure loss in the system. Manufacturers' data can then be

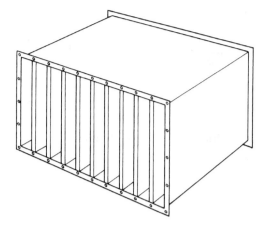

Figure 42.11 Splitter silencer

consulted. Splitter silencers are also available in bent shapes, and these can provide even higher degrees of attenuation as well as aiding installation. Silencers should ideally be fitted in systems as near to the noise source as possible to avoid noise break-out from the duct. Other obstructions in the duct must, however, be considered, as they may generate further aerodynamic noise which, if it occurs after the silencer, will not be attenuated.

42.8.4 Anti-vibration machinery mounts in practice

Again, the characteristics of the system need to be considered. The weight of the machine and the frequency will determine the static and dynamic deflections of the mounts and hence the material of which the mount is to be constructed. At very high frequencies mats may be placed under machinery, and these may consist of rubber, cork or foam. At middle frequencies it is usual to use rubber in-shear mounts. At low frequencies metal spring mounts are employed.

42.8.5 Mounts

Anti-vibration mats are very useful for frequencies above 42 Hz. They have the disadvantage of being liable to attack by oils and, if they become saturated or deteriorated, they will compress and lose their efficiency.

42.8.6 Rubber mounts

Although these are loosely termed rubber mounts, they are often composed of synthetic rubbers, which are not readily attacked by oils and can operate over a much wider temperature range. Typical maximum static deflections are 12.5 mm.

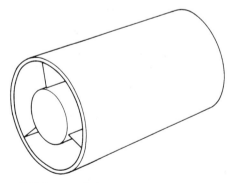

Figure 42.10 Center-pod silencer

42.8.7 Steel spring mounts

Steel springs have the disadvantage of transmitting the high frequencies along the length of the spring. It is usual to mount the spring with a rubber or neoprene washer under its base. Steel spring mounts are also most vulnerable to resonance problems, and the solution is to build in a damper device. This has the disadvantage of reducing the isolator's efficiency.

42.8.8 Positioning of anti-vibration mounts

Machinery must be positioned so that all mounts are equally loaded, and failure to do so will result in the possibility of a rocking motion developing. This may require mounting the machine on a sub-frame. If this is not possible, the load should be assessed at each mounting point and mounts of different stiffness used.

42.8.9 Installation

Mounts should be installed so that the whole machine is isolated from the structure. Services (e.g. power, hydraulics, etc.) should also be mounted flexibly. Bridging is the most common fault when providing vibration isolation to machines and building structures, and should be carefully avoided. Services should be designed to withstand the degree of movement permitted by the anti-vibration mounts without suffering damage.

43

Vibration
Fundamentals

R Keith Mobley
The Plant Performance Group

Contents

43.1 Introduction

All mechanical equipment in motion generates a vibration profile, or signature, that reflects its operating condition. This is true regardless of speed or whether the mode of operation is rotation, reciprocation, or linear motion. Vibration analysis is applicable to all mechanical equipment, although a common – yet invalid – assumption is that it is limited to simple rotating machinery with running speeds above 600 revolutions per minute (rpm). Vibration-profile analysis is a useful tool for predictive maintenance, diagnostics, and many other uses.

Predictive maintenance has become synonymous with monitoring vibration characteristics of rotating machinery to detect budding problems and to head off catastrophic failure. However, vibration analysis does not provide the data required for analyzing electrical equipment, areas of heat loss, the condition of lubricating oil, or other parameters typically evaluated in a maintenance management program. Therefore, a total plant predictive maintenance program must include several techniques, each designed to provide specific information on plant equipment.

43.2 Vibration analysis applications

The use of vibration analysis is not restricted to predictive maintenance. This technique is useful for diagnostic applications as well. Vibration monitoring and analysis are the primary diagnostic tools for most mechanical systems that are used to manufacture products. When used properly, vibration data provide the means to maintain optimum operating conditions and efficiency of critical plant systems. Vibration analysis can be used to evaluate fluid flow through pipes or vessels, to detect leaks, and to perform a variety of non-destructive testing functions that improve the reliability and performance of critical plant systems.

Some of the applications that are discussed briefly in this section are predictive maintenance, acceptance testing, quality control, loose part detection, noise control, leak detection, aircraft engine analyzers, and machine design and engineering. Table 43.1 lists rotating, or centrifugal, and non-rotating equipment, machine-trains, and continuous processes typically monitored by vibration analysis.

Predictive maintenance

The fact that vibration profiles can be obtained for all machinery having rotating or moving elements allows vibration-based analysis techniques to be used for predictive maintenance. Vibration analysis is one of several predictive maintenance techniques used to monitor and analyze critical machines, equipment, and systems in a typical plant. However, as indicated before, the use of vibration analysis to monitor rotating machinery to detect budding problems and to head off catastrophic failure is the dominant predictive maintenance technique used with maintenance management programs.

Acceptance testing

Vibration analysis is a proven means of verifying the actual performance versus design parameters of new mechanical, process, and manufacturing equipment. Pre-acceptance tests performed at the factory and immediately following installation can be used to ensure that new equipment performs at optimum efficiency and expected life-cycle cost. Design problems as well as possible damage during shipment or installation can be corrected before long-term damage and/or unexpected costs occur.

Quality control

Production-line vibration checks are an effective method of ensuring product quality where machine tools are involved. Such checks can provide advanced warning that the surface finish on parts is nearing the rejection level. On continuous process lines such as paper machines,

Table 43.1 Equipment and processes typically monitored by vibration analysis

Centrifugal	Reciprocating	Continuous process
Pumps	Pumps	Continuous casters
Compressors	Compressors	Hot and cold strip lines
Blowers	Diesel engines	Annealing lines
Fans	Gasoline engines	Plating lines
Motor/generators	Cylinders	Paper machines
Ball mills	Other machines	Can manufacturing lines
Chillers		Pickle lines
Product rolls	*Machine-trains*	Printing
Mixers		Dyeing and finishing
Gearboxes	Boring machines	Roofing manufacturing lines
Centrifuges	Hobbing machines	Chemical production lines
Transmissions	Machining centers	Petroleum production lines
Turbines	Temper mills	Neoprene production lines
Generators	Metal working machines	Polyester production lines
Rotary dryers	Rolling mills, and most machining	Nylon production lines
Electric motors	equipment	Flooring production lines
All rotating machinery		Continuous process lines

Source: The Plant Performance Group

steel-finishing lines, or rolling mills, vibration analysis can prevent abnormal oscillation of components that result in loss of product quality.

Loose or foreign parts detection

Vibration analysis is useful as a diagnostic tool for locating loose or foreign objects in process lines or vessels. This technique has been used with great success by the nuclear power industry and it offers the same benefits to non-nuclear industries.

Noise control

Federal, state, and local regulations require serious attention be paid to noise levels within the plant. Vibration analysis can be used to isolate the source of noise generated by plant equipment as well as background noises such as those generated by fluorescent lights and other less obvious sources. The ability to isolate the source of abnormal noises permits cost-effective corrective action.

Leak detections

Leaks in process vessels and devices such as valves are a serious problem in many industries. A variation of vibration monitoring and analysis can be used to detect leakage and isolate its source. Leak-detection systems use an accelerometer attached to the exterior of a process pipe. This allows the vibration profile to be monitored in order to detect the unique frequencies generated by flow or leakage.

Aircraft engine analyzers

Adaptations of vibration analysis techniques have been used for a variety of specialty instruments, in particular portable and continuous aircraft engine analyzers. Vibration monitoring and analysis techniques are the basis of these analyzers, which are used for detecting excessive vibration in turbo-prop and jet engines. These instruments incorporate logic modules that use existing vibration data to evaluate the condition of the engine. Portable units have diagnostic capabilities that allow a mechanic to determine the source of the problem while continuous sensors alert the pilot of any deviation from optimum operating condition.

Machine design and engineering

Vibration data has become a critical part of the design and engineering of new machines and process systems. Data derived from similar or existing machinery can be extrapolated to form the basis of a preliminary design. Prototype testing of new machinery and systems allows these preliminary designs to be finalized, and the vibration data from the testing adds to the design database.

43.2.1 Vibration analysis overview

Vibration theory and vibration profile, or signature, analysis are complex subjects that are the topic of many textbooks. The purpose of this section is to provide enough theory to allow the concept of vibration profiles and their analysis to be understood before beginning the more in-depth discussions in the later sections of this chapter.

Theoretical vibration profiles

A vibration is a periodic motion or one that repeats itself after a certain interval of time. This time interval is referred to as the period of the vibration, T. A plot, or profile, of a vibration is shown in Figure 43.1, which shows the period, T, and the maximum displacement or amplitude, X_0. The inverse of the period, $\frac{1}{T}$, is called the frequency, f, of the vibration, which can be expressed in units of cycles per second (cps) or Hertz (Hz). A harmonic function is the simplest type of periodic motion and is shown in Figure 43.2, which is the harmonic function for the small oscillations of a simple pendulum. Such a relationship can be expressed by the equation:

$$X = X_0 \sin(\omega t)$$

where:

X = Vibration displacement (thousandths of an inch, or mils)
X_0 = Maximum displacement or amplitude (mils)
ω = Circular frequency (radians per second)
t = Time (seconds)

Actual vibration profiles

The process of vibration analysis requires gathering complex machine data and deciphering it. As opposed to the simple theoretical vibration curves shown in Figures 43.1 and 43.2, the profile for a piece of equipment is extremely complex. This is true because there are usually many sources of vibration. Each source generates its own curve, but these are essentially added together and displayed as a composite profile. These profiles can be displayed in two formats: time-domain and frequency-domain.

Time domain

Vibration data that is plotted as amplitude versus time is referred to as a time-domain data profile. Some simple examples are shown in Figures 43.1 and 43.2. An example of the complexity of this type of data for an actual piece of industrial machinery is shown in Figure 43.3.

Time-domain plots must be used for all linear and reciprocating motion machinery. They are useful in the overall analysis of machine-trains to study changes in operating conditions. However, time-domain data are difficult to use. Because all the vibration data in this type of plot are added together to represent the total displacement at any given time, it is difficult to directly see the contribution of any particular vibration source.

The French physicist and mathematician Jean Fourier determined that non-harmonic data functions such as the time-domain vibration profile are the mathematical sum of simple harmonic functions. The dashed-line curves in Figure 43.4 represent discrete harmonic components of the total, or summed, non-harmonic curve represented by the solid line.

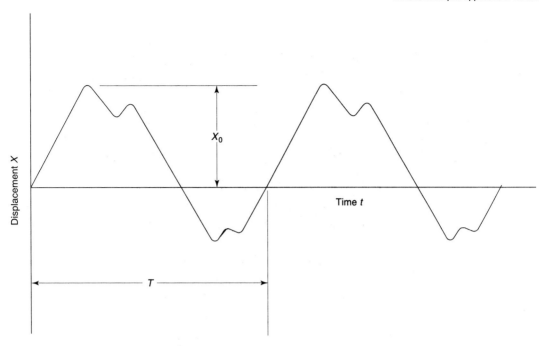

Figure 43.1 Periodic motion for bearing pedestal of a steam turbine

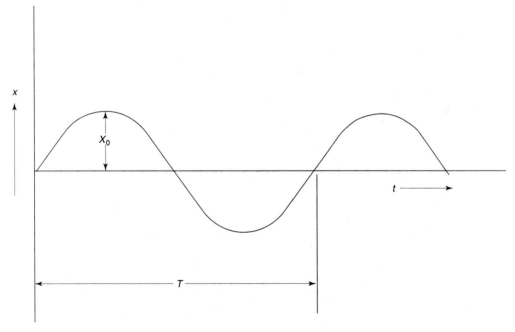

Figure 43.2 Small oscillations of a simple pendulum, harmonic function

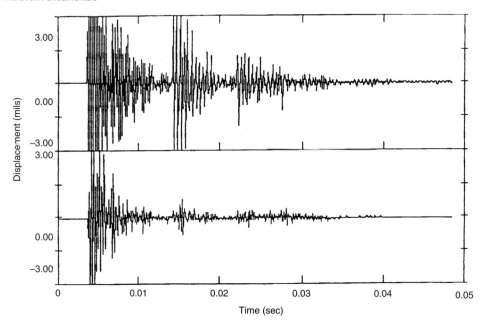

Figure 43.3 Example of a typical time-domain vibration profile for a piece of machinery

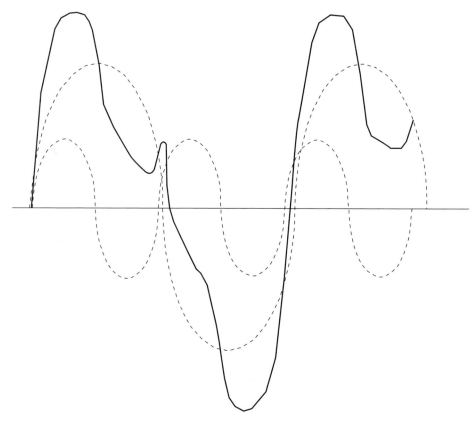

Figure 43.4 Discrete (harmonic) and total (non-harmonic) time-domain vibration curves

This type of data, which is routinely taken over the life of a machine, is directly comparable to historical data taken at exactly the same running speed and load. However, this is not practical because of variations in day-to-day plant operations and changes in running speed. This significantly affects the profile and makes it impossible to compare historical data.

Frequency domain

From a practical standpoint, simple harmonic vibration functions are related to the circular frequencies of the rotating or moving components. Therefore, these frequencies are some multiple of the basic running speed of the machine-train, which is expressed in revolutions per minute (rpm) or cycles per minute (cpm). Determining these frequencies is the first basic step in analyzing the operating condition of the machine-train.

Frequency-domain data are obtained by converting time-domain data using a mathematical technique referred to as Fast Fourier Transform (FFT). FFT allows each vibration component of a complex machine-train spectrum to be shown as a discrete frequency peak. The frequency-domain amplitude can be the displacement per unit time related to a particular frequency, which is plotted as the Y-axis against frequency as the X-axis. This is opposed to time-domain spectrums that sum the velocities of all frequencies and plot the sum as the Y-axis against time

as the X-axis. An example of a frequency-domain plot or vibration signature is shown in Figure 43.5.

Frequency-domain data are required for equipment operating at more than one running speed and all rotating applications. Because the X-axis of the spectrum is frequency normalized to the running speed, a change in running speed will not affect the plot. A vibration component that is present at one running speed will still be found in the same location on the plot for another running speed after the normalization, although the amplitude may be different.

43.3 Interpretation of vibration data

The key to using vibration signature analysis for predictive maintenance, diagnostic, and other applications is the ability to differentiate between normal and abnormal vibration profiles. Many vibrations are normal for a piece of rotating or moving machinery. Examples of these are normal rotations of shafts and other rotors, contact with bearings, gear-mesh, etc. However, specific problems with machinery generate abnormal, yet identifiable, vibrations. Examples of these are loose bolts, misaligned shafts, worn bearings, leaks, and incipient metal fatigue.

Predictive maintenance utilizing vibration signature analysis is based on the following facts, which form the basis for the methods used to identify and quantify the root causes of failure:

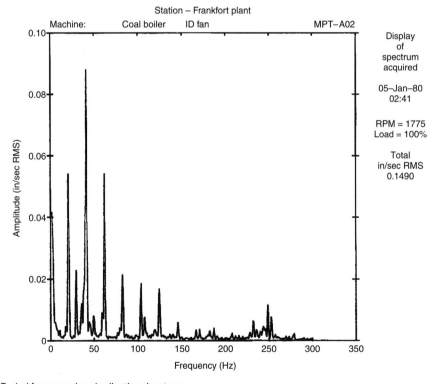

Figure 43.5 Typical frequency-domain vibration signature

1. All common machinery problems and failure modes have distinct vibration frequency components that can be isolated and identified.
2. A frequency-domain vibration signature is generally used for the analysis because it is comprised of discrete peaks, each representing a specific vibration source.
3. There is a cause, referred to as a forcing function, for every frequency component in a machine-train's vibration signature.
4. When the signature of a machine is compared over time, it will repeat until some event changes the vibration pattern (i.e., the amplitude of each distinct vibration component will remain constant until there is a change in the operating dynamics of the machine-train).

While an increase or a decrease in amplitude may indicate degradation of the machine-train, this is not always the case. Variations in load, operating practices, and a variety of other normal changes also generate a change in the amplitude of one or more frequency components within the vibration signature. In addition, it is important to note that a lower amplitude does not necessarily indicate an improvement in the mechanical condition of the machine-train. Therefore, it is important that the source of all amplitude variations be clearly understood.

43.4 Vibration measuring equipment

Vibration data are obtained by the following procedure: (1) mounting a transducer onto the machinery at various locations, typically machine housing and bearing caps, and (2) using a portable data-gathering device, referred to as a vibration monitor or analyzer, to connect to the transducer to obtain vibration readings.

Transducers

The transducer most commonly used to obtain vibration measurements is an accelerometer. It incorporates piezoelectric (i.e., pressure-sensitive) films to convert mechanical energy into electrical signals. The device generally incorporates a weight suspended between two piezoelectric films. The weight moves in response to vibration and squeezes the piezoelectric films, which sends an electrical signal each time the weight squeezes it.

Portable vibration analyzers

The portable vibration analyzer incorporates a microprocessor that allows it to mathematically convert the electrical signal to acceleration per unit time, perform a FFT, and store the data. It can be programmed to generate alarms and displays of the data. The data stored by the analyzer can be downloaded to a personal or a more powerful computer to perform more sophisticated analyses, data storage and retrieval, and report generation.

43.5 Vibration sources

All machinery with moving parts generates mechanical forces during normal operation. As the mechanical condition of the machine changes due to wear, changes in the operating environment, load variations, etc., so do these forces. Understanding machinery dynamics and how forces create unique vibration frequency components is the key to understanding vibration sources.

Vibration does not just happen. There is a physical cause, referred to as a forcing function, and each component of a vibration signature has its own forcing function. The components that make up a signature are reflected as discrete peaks in the FFT or frequency-domain plot.

The vibration profile that results from motion is the result of a force imbalance. By definition, balance occurs in moving systems when all forces generated by, and acting on, the machine are in a state of equilibrium. In real-world applications, however, there is always some level of imbalance and all machines vibrate to some extent. This section discusses the more common sources of vibration for rotating machinery, as well as for machinery undergoing reciprocating and/or linear motion.

43.5.1 Rotating machinery

A rotating machine has one or more machine elements that turn with a shaft, such as rolling-element bearings, impellers, and other rotors. In a perfectly balanced machine, all rotors turn true on their centerline and all forces are equal. However, in industrial machinery, it is common for an imbalance of these forces to occur. In addition to imbalance generated by a rotating element, vibration may be caused by instability in the media flowing through the rotating machine.

Rotor imbalance

Mechanical imbalance is not the only form of imbalance that affects rotating elements. It is the condition where more weight is on one side of a centerline of a rotor than on the other. In many cases, rotor imbalance is the result of an imbalance between centripetal forces generated by the rotation. The source of rotor vibration also can be an imbalance between the lift generated by the rotor and gravity.

Machines with rotating elements are designed to generate vertical lift of the rotating element when operating within normal parameters. This vertical lift must overcome gravity to properly center the rotating element in its bearing-support structure. However, because gravity and atmospheric pressure vary with altitude and barometric pressure, actual lift may not compensate for the downward forces of gravity in certain environments. When the deviation of actual lift from designed lift is significant, a rotor may not rotate on its true centerline. This offset rotation creates an imbalance and a measurable level of vibration.

Flow instability and operating conditions

Rotating machines subject to imbalance caused by turbulent or unbalanced media flow include pumps, fans, and compressors. A good machine design for these units incorporates the dynamic forces of the gas or liquid in stabilizing the rotating element. The combination of these forces and the stiffness of the rotor-support system (i.e., bearing and bearing pedestals) determine the vibration level. Rotor-support stiffness is important because unbalanced forces resulting from flow instability can deflect rotating elements from their true centerline, and the stiffness resists the deflection.

Deviations from a machine's designed operating envelope can affect flow stability, which directly affects the vibration profile. For example, the vibration level of a centrifugal compressor is typically low when operating at 100 per cent load with laminar airflow through the compressor. However, a radical change in vibration level can result from decreased load. Vibration resulting from operation at 50 per cent load may increase by as much as 400 per cent with no change in the mechanical condition of the compressor. In addition, a radical change in vibration level can result from turbulent flow caused by restrictions in either the inlet or discharge piping.

Turbulent or unbalanced media flow (i.e., aerodynamic or hydraulic instability) does not have the same quadratic impacts on the vibration profile as that of load change, but it increases the overall vibration energy. This generates a unique profile that can be used to quantify the level of instability present in the machine. The profile generated by unbalanced flow is visible at the vane or blade-pass frequency of the rotating element. In addition, the profile shows a marked increase in the random noise generated by the flow of gas or liquid through the machine.

Mechanical motion and forces

A clear understanding of the mechanical movement of machines and their components is an essential part of vibration analysis. This understanding, coupled with the forces applied by the process, are the foundation for diagnostic accuracy.

Almost every unique frequency contained in the vibration signature of a machine-train can be directly attributed to a corresponding mechanical motion within the machine. For example, the constant endplay or axial movement of the rotating element in a motor-generator set generates elevated amplitude at the fundamental ($1\times$), second harmonic ($2\times$), and third harmonic ($3\times$) of the shaft's true running speed. In addition, this movement increases the axial amplitude of the fundamental ($1\times$) frequency.

Forces resulting from air or liquid movement through a machine also generate unique frequency components within the machine's signature. In relatively stable or laminar-flow applications, the movement of product through the machine slightly increases the amplitude at the vane or blade-pass frequency. In more severe, turbulent-flow applications, the flow of product generates a broadband, white noise profile that can be directly attributed to the movement of product through the machine.

Other forces, such as the side-load created by V-belt drives, also generate unique frequencies or modify existing component frequencies. For example, excessive belt tension increases the side-load on the machine-train's shafts. This increase in side-load changes the load zone in the machine's bearings. The result of this change is a marked increase in the amplitude at the outer-race rotational frequency of the bearings.

Applied force or induced loads can also displace the shafts in a machine-train. As a result the machine's shaft will rotate off-center which dramatically increases the amplitude at the fundamental ($1\times$) frequency of the machine.

43.5.2 Reciprocating and/or linear-motion machinery

This section describes machinery that exhibits reciprocating and/or linear motion(s) and discusses typical vibration behavior for these types of machines.

Machine descriptions

Reciprocating linear-motion machines incorporate components that move linearly in a reciprocating fashion to perform work. Such reciprocating machines are bi-directional in that the linear movement reverses, returning to the initial position with each completed cycle of operation. Non-reciprocating linear-motion machines incorporate components that also generate work in a straight line, but do not reverse direction within one complete cycle of operation.

Few machines involve linear reciprocating motion exclusively. Most incorporate a combination of rotating and reciprocating linear motions to produce work. One example of such a machine is a reciprocating compressor. This unit contains a rotating crankshaft that transmits power to one or more reciprocating pistons, which move linearly in performing the work required to compress the media.

Sources of vibration

Like rotating machinery, the vibration profile generated by reciprocating and/or linear motion machines is the result of mechanical movement and forces generated by the components that are part of the machine. Vibration profiles generated by most reciprocating and/or linear motion machines reflect a combination of rotating and/or linear motion forces.

However, the intervals or frequencies generated by these machines are not always associated with one complete revolution of a shaft. In a two-cycle reciprocating engine, the pistons complete one cycle each time the crankshaft completes one $360°$ revolution. In a four-cycle engine, the crank must complete two complete revolutions, or $720°$, in order to complete a cycle of all pistons.

Because of the unique motion of reciprocating and linear motion machines, the level of unbalanced forces generated by these machines is substantially higher than those generated by rotating machines. As an example, a reciprocating compressor drives each of its pistons from bottom-center to top-center and returns to bottom-center in each complete operation of the cylinder. The mechanical forces generated by the reversal of direction at both top-center and bottom-center result in a sharp increase in the vibration energy of the machine. An instantaneous spike in the vibration profile repeats each time the piston reverses direction.

Linear-motion machines generate vibration profiles similar to those of reciprocating machines. The major difference is the impact that occurs at the change of direction with reciprocating machines. Typically, linear-motion-only machines do not reverse direction during each cycle of operation and, as a result, do not generate the spike of energy associated with direction reversal.

43.6 Vibration theory

Mathematical techniques allow us to quantify total displacement caused by all vibrations, to convert the displacement measurements to velocity or acceleration, to separate this data into its components using FFT analysis, and to determine the amplitudes and phases of these functions. Such quantification is necessary if we are to isolate and correct abnormal vibrations in machinery.

Periodic motion

Vibration is a periodic motion, or one that repeats itself after a certain interval of time called the period, T.

Figure 43.6 illustrates the periodic motion time-domain curve of a steam turbine bearing pedestal. Displacement is plotted on the vertical, or Y-axis, and time on the horizontal, or X-axis. The curve shown in Figure 43.6 is the sum of all vibration components generated by the rotating element and bearing-support structure of the turbine.

Harmonic motion

The simplest kind of periodic motion or vibration, shown in Figure 43.7, is referred to as harmonic. Harmonic motions repeat each time the rotating element or machine component completes one complete cycle.

The relation between displacement and time for harmonic motion may be expressed by:

$$X = X_0 \sin(\omega t)$$

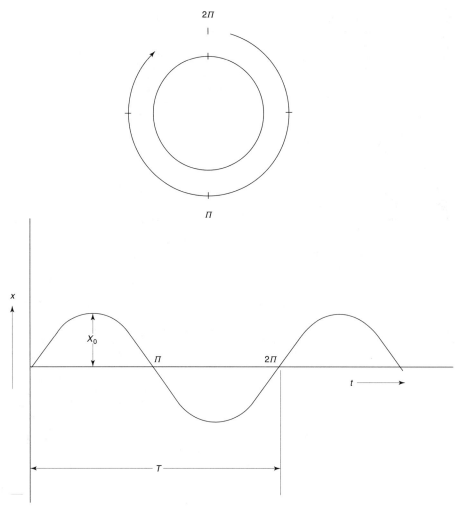

Figure 43.6 Typical periodic motion

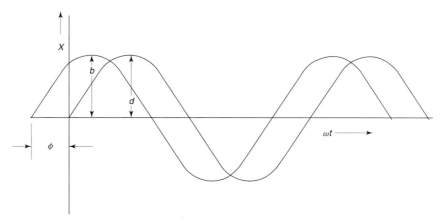

Figure 43.7 Simple harmonic motion

The maximum value of the displacement is X_0, which is also called the amplitude. The period, T, is usually measured in seconds; its reciprocal is the frequency of the vibration, f, measured in cycles per second (cps) or Hertz (Hz).

$$f = \frac{1}{T}$$

Another measure of frequency is the circular frequency, ω, measured in radians per second. From Figure 43.8, it is clear that a full cycle of vibration (ωt) occurs after 360° or 2π radians (i.e., one full revolution). At this point, the function begins a new cycle.

$$\omega = 2\pi f$$

For rotating machinery, the frequency is often expressed in vibrations per minute (vpm) or

$$VPM = \frac{\omega}{\pi}$$

By definition, velocity is the first derivative of displacement with respect to time. For a harmonic motion, the displacement equation is:

$$X = X_0 \sin(\omega t)$$

The first derivative of this equation gives us the equation for velocity:

$$v = \frac{dX}{dt} = X = \omega X_0 \cos(\omega t)$$

This relationship tells us that the velocity is also harmonic if the displacement is harmonic and has a maximum value or amplitude of $-\omega X_0$.

By definition, acceleration is the second derivative of displacement (i.e., the first derivative of velocity) with respect to time:

$$a = \frac{d^2 X}{dt^2} = X = -\omega^2 X_0 \sin(\omega t)$$

This function is also harmonic with amplitude of $\omega^2 X_0$.

Consider two frequencies given by the expression $X_1 = a \sin(\omega t)$ and $X_2 = b \sin(\omega t + \phi)$, which are shown in Figure 43.9 plotted against ωt as the X-axis. The quantity, ϕ, in the equation for X_2 is known as the phase angle or phase difference between the two vibrations. Because of ϕ, the two vibrations do not attain their maximum displacements at the same time. One is $\frac{\phi}{\omega}$ seconds behind the other. Note that these two motions have the same frequency, ω. A phase angle has meaning only for two motions of the same frequency.

Non-harmonic motion

In most machinery, there are numerous sources of vibrations; therefore, most time-domain vibration profiles are non-harmonic (represented by the solid line in Figure 43.10). While all harmonic motions are periodic, not every periodic motion is harmonic. In Figure 43.10, the dashed lines represent harmonic motions.

Figure 43.10 is the superposition of two sine waves having different frequencies. These curves are represented by the following equations:

$$X_1 = a \sin(\omega_1 t)$$

$$X_2 = b \sin(\omega_2 t)$$

The total vibration represented by the solid line is the sum of the dashed lines. The following equation represents the total vibration:

$$X = X_1 + X_2 = a \sin(\omega_1 t) + b \sin(\omega_2 t)$$

Any periodic function can be represented as a series of sine functions having frequencies of ω, 2ω, 3ω, etc.:

$$f(t) = A_0 + A_1 \sin(\omega t + \phi_1) + A_2 \sin(2\omega t + \phi_2)$$
$$+ A_3 \sin(3\omega t + \phi_3) + \cdots$$

The above equation is known as a Fourier series, which is a function of time or $f(t)$. The amplitudes (A_1, A_2, etc.) of the various discrete vibrations and their phase angles ($\phi_1, \phi_2, \phi_3 \ldots$) can be determined mathematically when

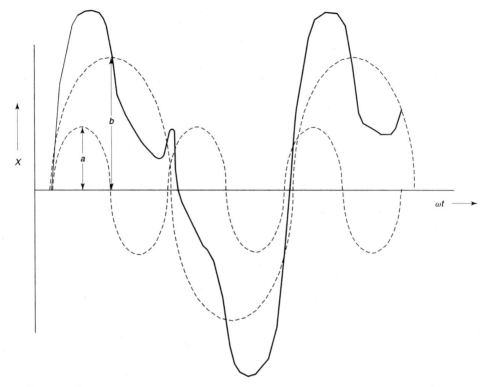

Figure 43.8 Illustration of vibration cycles

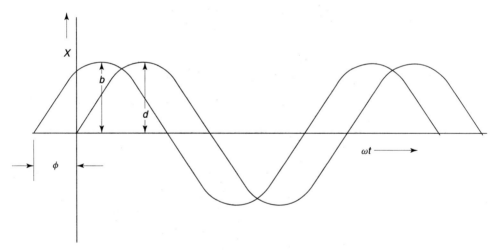

Figure 43.9 Two harmonic motions with a phase angle between them

the value of function $f(t)$ is known. Note that these data are obtained using a transducer and a portable vibration analyzer.

The terms, 2ω, 3ω, etc., are referred to as the harmonics of the primary frequency, ω. In most vibration signatures,

the primary frequency component is one of the running speeds of the machine-train ($1\times$ or 1ω). In addition, a signature may be expected to have one or more harmonics, for example, at two times ($2\times$), three times ($3\times$), and other multiples of the primary running speed.

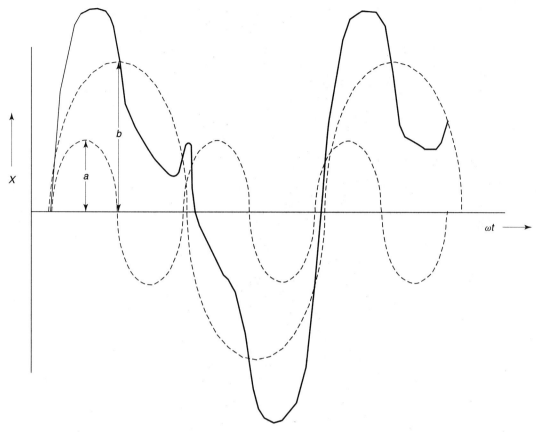

Figure 43.10 Non-harmonic periodic motion

43.7 Measurable parameters

As shown previously, vibrations can be displayed graphically as plots which are referred to as vibration profiles or signatures. These plots are based on measurable
parameters (i.e., frequency and amplitude). Note that the
terms 'profile' and 'signature' are sometimes used interchangeably by industry. In this chapter, however, 'profile' is used to refer either to time-domain (also may
be called time trace or waveform) or frequency-domain
plots. The term 'signature' refers to a frequency-domain
plot.

Frequency

Frequency is defined as the number of repetitions of a
specific forcing function or vibration component over a
specific unit of time. Take for example a four-spoke wheel
with an accelerometer attached. Every time the shaft completes one rotation, each of the four spokes passes the
accelerometer once, which is referred to as four cycles per
revolution. Therefore, if the shaft rotates at 100 rpm, the
frequency of the spokes passing the accelerometer is 400
cycles per minute (cpm). In addition to cpm, frequency is
commonly expressed in cycles per second (cps) or Hertz
(Hz).

Note that for simplicity, a machine element's vibration frequency is commonly expressed as a multiple of
the shaft's rotation speed. In the above example, the frequency would be indicated to be 4×, or four times the
running speed. In addition, because some malfunctions
tend to occur at specific frequencies, it helps to segregate
certain classes of malfunctions from others.

Note, however, that the frequency/malfunction relationship is not mutually exclusive and a specific mechanical
problem cannot definitely be attributed to a unique frequency. While frequency is a very important piece of
information with regard to isolating machinery malfunctions, it is only one part of the total picture. It is necessary
to evaluate all data before arriving at a conclusion.

Amplitude

Amplitude refers to the maximum value of a motion or
vibration. This value can be represented in terms of displacement (mils), velocity (inches per second), or acceleration (inches per second squared), each of which is

discussed in more detail in Maximum vibration measurement, below.

Amplitude can be measured as the sum of all the forces causing vibrations within a piece of machinery (broadband), as discrete measurements for the individual forces (component), or for individual user-selected forces (narrowband). Broadband, component, and narrowband are discussed in Section 43.8 Measurement classifications. Also discussed in this section are the common curve elements: peak-to-peak, zero-to-peak, and root-mean-square.

Maximum vibration measurement

The maximum value of a vibration, or amplitude, is expressed as displacement, velocity, or acceleration. Most of the microprocessor-based, frequency-domain vibration systems will convert the acquired data to the desired form. Since industrial vibration-severity standards are typically expressed in one of these terms, it is necessary to have a clear understanding of their relationship.

Displacement

Displacement is the actual change in distance or position of an object relative to a reference point and is usually expressed in units of mils, 0.001 inch. For example, displacement is the actual radial or axial movement of the shaft in relation to the normal centerline usually using the machine housing as the stationary reference. Vibration data, such as shaft displacement measurements acquired using a proximity probe or displacement transducer, should always be expressed in terms of mils peak-to-peak.

Velocity

Velocity is defined as the time rate of change of displacement (i.e., the first derivative, dX/dt or X) and is usually expressed as inches per second (ips). In simple terms, velocity is a description of how fast a vibration component is moving rather than how far, which is described by displacement.

Used in conjunction with zero-to-peak (PK) terms, velocity is the best representation of the true energy generated by a machine when relative or bearing cap-data are used. (Note: Most vibration monitoring programs rely on data acquired from machine housing or bearing caps.) In most cases, peak velocity values are used with vibration data between 0 and 1000 Hz. These data are acquired with microprocessor-based, frequency-domain systems.

Acceleration

Acceleration is defined as the time rate of change of velocity (i.e., second derivative of displacement, d^2X/dt^2 or X) and is expressed in units of inches per second squared (in/sec^2). Vibration frequencies above 1000 Hz should always be expressed as acceleration.

Acceleration is commonly expressed in terms of the gravitational constant, g, which is 32.17 ft/sec^2. In vibration analysis applications, acceleration is typically expressed in terms of g-RMS or g-PK. These are the best measures of the force generated by a machine, a group of components, or one of its components.

43.8 Measurement classifications

There are at least three classifications of amplitude measurements used in vibration analysis: broadband, narrowband, and component.

Broadband or overall

The total energy of all vibration components generated by a machine is reflected by broadband, or overall, amplitude measurements. The normal convention for expressing the frequency range of broadband energy is a filtered range between 10 to 10,000 Hz, or 600 to 600,000 cpm. Because most vibration-severity charts are based on this filtered broadband, caution should be exercised to ensure that collected data are consistent with the charts.

Narrowband

Narrowband amplitude measurements refer to those that result from monitoring the energy generated by a user-selected group of vibration frequencies. Generally, this amplitude represents the energy generated by a filtered band of vibration components, failure mode, or forcing functions. For example, the total energy generated by flow instability can be captured using a filtered narrowband around the vane or blade-passing frequency.

Component

The energy generated by a unique machine component, motion, or other forcing function can yield its own amplitude measurement. For example, the energy generated by the rotational speed of a shaft, gear set meshing, or similar machine components generate discrete vibration components and their amplitude can be measured.

43.9 Common elements of curves

All vibration amplitude curves, which can represent displacement, velocity, or acceleration, have common elements that can be used to describe the function. These common elements are peak-to-peak, zero-to-peak, and root-mean-square, each of which are illustrated in Figure 43.11.

Peak-to-peak

As illustrated in Figure 43.11, the peak-to-peak amplitude (2A, where A is the zero-to-peak) reflects the total amplitude generated by a machine, a group of components, or one of its components. This depends on whether the data gathered is broadband, narrowband, or component. The unit of measurement is useful when the analyst needs to know the total displacement or maximum energy produced by the machine's vibration profile.

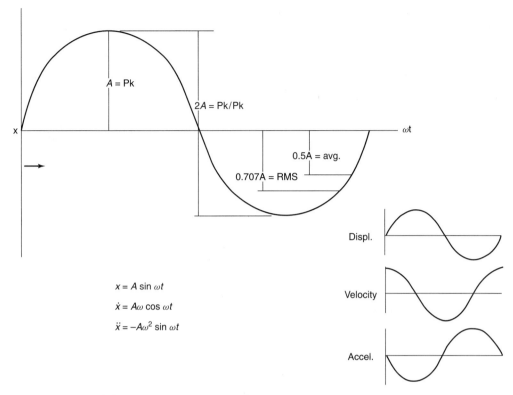

$$x = A \sin \omega t$$
$$\dot{x} = A\omega \cos \omega t$$
$$\ddot{x} = -A\omega^2 \sin \omega t$$

Figure 43.11 Relationship of vibration amplitude

Technically, peak-to-peak values should be used in conjunction with actual shaft-displacement data, which are measured with a proximity or displacement transducer. Peak-to-peak terms should not be used for vibration data acquired using either relative vibration data from bearing caps or when using a velocity or acceleration transducer. The only exception is when vibration levels must be compared to vibration-severity charts based on peak-to-peak values.

Zero-to-peak

Zero-to-peak (A), or simply peak, values are equal to one half of the peak-to-peak value. In general, relative vibration data acquired using a velocity transducer are expressed in terms of peak.

Root-mean-square

Root-mean-square (RMS) is the statistical average value of the amplitude generated by a machine, one of its components, or a group of components. Referring to Figure 43.11, RMS is equal to 0.707 of the zero-to-peak value, A. Normally, RMS data are used in conjunction with relative vibration data acquired using an accelerometer or expressed in terms of acceleration.

43.10 Machine dynamics

The primary reasons for vibration-profile variations are the dynamics of the machine, which are affected by mass, stiffness, damping, and degrees of freedom. However, care must be taken as the vibration profile and energy levels generated by a machine may vary depending on the location and orientation of the measurement.

Mass, stiffness, and damping

The three primary factors that determine the normal vibration energy levels and the resulting vibration profiles are mass, stiffness, and damping. Every machine-train is designed with a dynamic support system that is based on the following: the mass of the dynamic component(s), specific support system stiffness, and a specific amount of damping.

Mass Mass is the property that describes how much material is present. Dynamically, the property describes how an unrestricted body resists the application of an external force. Simply stated, the greater the mass the greater the force required accelerating it. Mass is obtained by dividing the weight of a body (e.g., rotor assembly) by the local acceleration of gravity, g.

The English system of units is complicated compared to the metric system. In the English system, the units of mass are pounds-mass (lbm) and the units of weight are pounds-force (lbf). By definition, a weight (i.e., force) of one lbf equals the force produced by one lbm under the acceleration of gravity. Therefore, the constant, g_c, which has the same numerical value as g (32.17) and units of lbm-ft/lbf-sec^2, is used in the definition of weight:

$$\text{Weight} = \frac{\text{Mass} \times g}{g_c}$$

Therefore,

$$\text{Mass} = \frac{\text{Weight} \times g_c}{g}$$

Therefore,

$$\text{Mass} = \frac{\text{Weight} \times g_c}{g} = \frac{\text{lbf}}{\frac{\text{ft}}{\text{sec}^2}} \times \frac{\text{lbm} \times \text{ft}}{\text{lbf} \times \text{sec}^2} = \text{lbm}$$

Stiffness Stiffness is a spring-like property that describes the level of resisting force that results when a body undergoes a change in length. Units of stiffness are often given as pounds per inch (lbf/in). Machine-trains have more than one stiffness property that must be considered in vibration analysis: shaft stiffness, vertical stiffness, and horizontal stiffness.

Shaft stiffness Most machine-trains used in industry have flexible shafts and relatively long spans between bearing-support points. As a result, these shafts tend to flex in normal operation. Three factors determine the amount of flex and mode shape that these shafts have in normal operation: shaft diameter, shaft material properties, and span length. A small-diameter shaft with a long span will obviously flex more than one with a larger diameter or shorter span.

Vertical stiffness The rotor-bearing support structure of a machine typically has more stiffness in the vertical plane than in the horizontal plane. Generally, the structural rigidity of a bearing-support structure is much greater in the vertical plane. The full weight of and the dynamic forces generated by the rotating element are fully supported by a pedestal cross-section that provides maximum stiffness.

In typical rotating machinery, the vibration profile generated by a normal machine contains lower amplitudes in the vertical plane. In most cases, this lower profile can be directly attributed to the difference in stiffness of the vertical plane when compared to the horizontal plane.

Horizontal stiffness Most bearing pedestals have more freedom in the horizontal direction than in the vertical. In most applications, the vertical height of the pedestal is much greater than the horizontal cross-section. As a result, the entire pedestal can flex in the horizontal plane as the machine rotates.

This lower stiffness generally results in higher vibration levels in the horizontal plane. This is especially true when the machine is subjected to abnormal modes of operation or when the machine is unbalanced or misaligned.

Damping Damping is a means of reducing velocity through resistance to motion, in particular by forcing an object through a liquid or gas, or along another body. Units of damping are often given as pounds per inch per second (lbf/in/sec, which is also expressed as lbf-sec/in).

The boundary conditions established by the machine design determine the freedom of movement permitted within the machine-train. A basic understanding of this concept is essential for vibration analysis. Free vibration refers to the vibration of a damped (as well as undamped) system of masses with motion entirely influenced by their potential energy. Forced vibration occurs when motion is sustained or driven by an applied periodic force in either damped or undamped systems. The following sections discuss free and forced vibration for both damped and undamped systems.

Free vibration – undamped

To understand the interactions of mass and stiffness, consider the case of undamped free vibration of a single mass that only moves vertically, which is illustrated in Figure 43.12. In this figure, the mass, M, is supported by a spring that has a stiffness, K (also referred to as the spring constant), which is defined as the number of pounds tension necessary to extend the spring one inch.

The force created by the static deflection, X_i, of the spring supports the weight, W, of the mass. Also included in Figure 43.12 is the free-body diagram that illustrates the two forces acting on the mass. These forces are the weight (also referred to as the inertia force) and an equal, yet opposite force that results from the spring (referred to as the spring force, F_s).

The relationship between the weight of mass M and the static deflection of the spring can be calculated using the equation: $W = KX_i$. If the spring is displaced downward some distance, X_o, from X_i and released, it will oscillate up and down. The force from the spring, F_s, can be written as follows, where a is the acceleration of the mass:

$$F_S = -KX = \frac{Ma}{g_c}$$

It is common practice to replace acceleration, a, with d^2X/dt^2, the second derivative of the displacement, X, of the mass with respect to time, t. Making this substitution, the equation that defines the motion of the mass can be expressed as:

$$\frac{M}{g_c}\frac{d^2X}{dt^2} = -KX \text{ or } \frac{M}{g_c}\frac{d^2X}{dt^2} + KX = 0$$

Motion of the mass is known to be periodic in time. Therefore, the displacement can be described by the expression:

$$X = X_o \cos(\omega t)$$

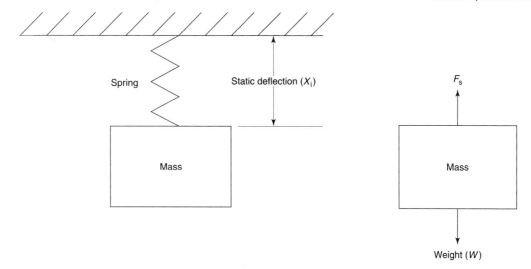

Figure 43.12 Undamped spring-mass system

Where:

X = Displacement at time t
X_o = Initial displacement of the mass
ω = Frequency of the oscillation (natural or resonant frequency)
t = Time

If this equation is differentiated and the result inserted into the equation that defines motion, the natural frequency of the mass can be calculated. The first derivative of the equation for motion given above yields the equation for velocity. The second derivative of the equation yields acceleration.

$$\text{Velocity} = \frac{dX}{dt} = X = -\omega X_o \sin(\omega t)$$

$$\text{Acceleration} = \frac{d^2X}{dt^2} = X = -\omega^2 X_o \cos(\omega t)$$

Inserting the above expression for acceleration, or d^2X/dt^2, into the equation for F_s yields the following:

$$\frac{M}{g_c}\frac{d^2X}{dt^2} + KX = 0$$

$$-\frac{M}{g_c}\omega^2 X_o \cos(\omega t) + KX = 0$$

$$-\frac{M}{g_c}\omega^2 X + KX = -\frac{M}{g_c}\omega^2 + K = 0$$

Solving this expression for ω yields the equation:

$$\omega = \sqrt{\frac{Kg_c}{M}}$$

Where:

ω = Natural frequency of mass
K = Spring constant
M = Mass

Note that, theoretically, undamped free vibration persists forever. However, this never occurs in nature and all free vibrations die down after time due to damping, which is discussed in the next section.

Free vibration – damped

A slight increase in system complexity results when a damping element is added to the spring-mass system shown in Figure 43.13. This type of damping is referred to as viscous damping. Dynamically, this system is the same as the undamped system illustrated in Figure 43.12, except for the damper, which usually is an oil or air dashpot mechanism. A damper is used to continuously decrease the velocity and the resulting energy of a mass undergoing oscillatory motion.

The system is still comprised of the inertia force due to the mass and the spring force, but a new force is introduced. This force is referred to as the damping force and is proportional to the damping constant, or the coefficient of viscous damping, c. The damping force is also proportional to the velocity of the body and, as it is applied, it opposes the motion at each instant.

In Figure 43.13, the non-elongated length of the spring is L_0 and the elongation due to the weight of the mass is expressed by h. Therefore, the weight of the mass is Kh. Figure 43.13(a) shows the mass in its position of stable equilibrium. Part (b) shows the mass displaced downward a distance X from the equilibrium position. Note that X is considered positive in the downward direction.

Figure 43.13(c) is a free-body diagram of the mass, which has three forces acting on it. The weight (Mg/g_c), which is directed downward, is always positive. The damping force (cdX/dt), which is the damping constant times velocity, acts opposite to the direction of the velocity. The spring force, $K(X + h)$, acts in the direction opposite to the displacement. Using Newton's equation of motion, where $\Sigma F = Ma$, the sum of the forces

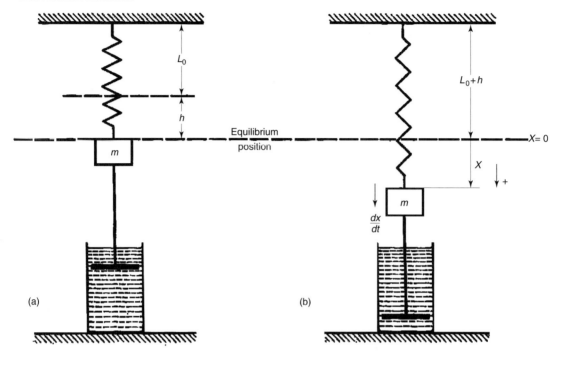

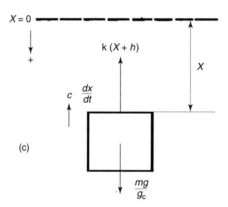

Figure 43.13 Damped spring-mass system

acting on the mass can be represented by the following equation, remembering that X is positive in the downward direction:

$$\frac{M}{g_c}\frac{d^2X}{dt^2} = \frac{Mg}{g_c} - c\frac{dX}{dt} - K(X + h)$$

$$\frac{M}{g_c}\frac{d^2X}{dt^2} = Kh - c\frac{dX}{dt} - KX - Kh$$

$$\frac{M}{g_c}\frac{d^2X}{dt^2} = -c\frac{dX}{dt} - KX$$

Dividing by M/g_c:

$$\frac{d^2X}{dt^2} = -\frac{cg_c}{M}\frac{dX}{dt} - \frac{Kg_cX}{M}$$

In order to look up the solution to the above equation in a differential equations table (such as in *CRC Handbook of Chemistry and Physics*) it is necessary to change the form of this equation. This can be accomplished by defining the relationships, $cg_c/M = 2\mu$ and $Kg_c/M = \omega^2$, which converts the equation to the following form:

$$\frac{d^2X}{dt^2} = -2\mu\frac{dX}{dt} - \omega^2X$$

Note that for undamped free vibration, the damping constant, c, is zero and, therefore, μ is zero.

$$\frac{d^2X}{dt^2} = -\omega^2 X$$

$$\frac{d^2X}{dt^2} + \omega^2 X = 0$$

The solution of this equation describes simple harmonic motion, which is given below:

$$X = A\cos(\omega t) + B\sin(\omega t)$$

Substituting at $t = 0$, then $X = X_o$ and $dX/dt = 0$, then

$$X = X_o\cos(\omega t)$$

This shows that free vibration is periodic and is the solution for X. For damped free vibration, however, the damping constant, c, is not zero.

$$\frac{d^2X}{dt^2} = -2\mu\frac{dX}{dt} - \omega^2 X$$

or

$$\frac{d^2X}{dt^2} + 2\mu\frac{dX}{dt} + \omega^2 X = 0$$

or

$$D^2 + 2\mu D + \omega^2 = 0$$

which has a solution of:

$$X = Ae^{d_1 t} + Be^{d_2 t}$$

where:

$$d_1 = -\mu + \sqrt{\mu^2 - \omega^2}$$

$$d_2 = -\mu - \sqrt{\mu^2 - \omega^2}$$

There are different conditions of damping: critical, over-damping, and under-damping. Critical damping occurs when $\mu = \omega$. Over-damping occurs when $\mu > \omega$. Under-damping occurs when $\mu < \omega$.

The only condition that results in oscillatory motion and, therefore, represents a mechanical vibration is under-damping. The other two conditions result in periodic motions. When damping is less than critical ($\mu < \omega$), then the following equation applies:

$$X = \frac{X_o}{\alpha_1}e^{-\mu t}(\alpha_1\cos\alpha_1 t + \mu\sin\alpha_1 t)$$

where:

$$\alpha_1 = \sqrt{\omega^2 - \mu^2}$$

Forced vibration – undamped

The simple systems described in the preceding two sections on free vibration are alike in that they are not forced to vibrate by any exciting force or motion. Their major contribution to the discussion of vibration fundamentals is that they illustrate how a system's natural

or resonant frequency depends upon the mass, stiffness, and damping characteristics.

The mass-stiffness-damping system also can be disturbed by a periodic variation of external forces applied to the mass at any frequency. The system shown in Figure 43.12 is increased in complexity by the addition of an external force, F_o, acting downward on the mass.

In undamped forced vibration, the only difference in the equation for undamped free vibration is that instead of the equation being equal to zero, it is equal to $F_o\sin(\omega t)$:

$$\frac{M}{g_c}\frac{d^2X}{dt^2} + KX = F_o\sin(\omega t)$$

Since the spring is not initially displaced and is 'driven' by the function $F_o\sin(\omega t)$, a particular solution, $X = X_o\sin(\omega t)$, is logical. Substituting this solution into the above equation and performing mathematical manipulations yields the following equation for X:

$$X = C_1\sin(\omega_n t) + C_2\cos(\omega_n t) + \frac{X_{st}}{1-(\omega/\omega_n)^2}\sin(\omega t)$$

where:

X = Spring displacement at time, t
X_{st} = Static spring deflection under constant load, F_o
ω = Forced frequency
ω_n = Natural frequency of the oscillation
t = Time
C_1 and C_2 = Integration constants determined from specific boundary conditions

In the above equation, the first two terms are the undamped free vibration, while the third term is the undamped forced vibration. The solution, containing the sum of two sine waves of different frequencies, is itself not a harmonic motion.

Forced vibration – damped

In a damped forced vibration system such as the one shown in Figure 43.14, the motion of the mass M has two parts: (1) the damped free vibration at the damped natural frequency and (2) the steady-state harmonic motions at the forcing frequency. The damped natural frequency component decays quickly, but the steady state harmonic associated with the external force remains as long as the energy force is present.

With damped forced vibration, the only difference in its equation and the equation for damped free vibration is that it is equal to $F_o\sin(\omega t)$ as shown below instead of being equal to zero.

$$\frac{M}{g_c}\frac{d^2X}{dt^2} + c\frac{dX}{dt} + KX = F_o\sin(\omega t)$$

With damped vibration, the damping constant, c, is not equal to zero and the solution of the equation gets quite complex assuming the function, $X = X_o\sin(\omega t - \phi)$. In this equation, ϕ is the phase angle, or the number of degrees that the external force, $F_o\sin(\omega t)$, is ahead of the displacement, $X_o\sin(\omega t - \phi)$. Using vector concepts, the

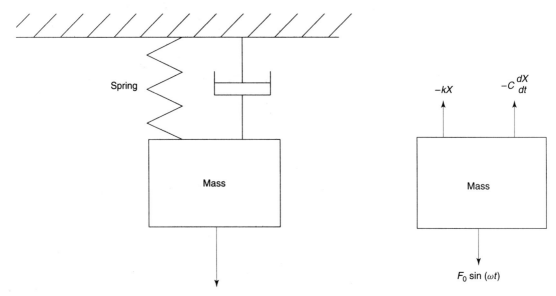

Figure 43.14 Damped forced vibration system

following equations apply, which can be solved because there are two equations and two unknowns:

Vertical vector component: $KX_0 - \dfrac{M}{g_c}\omega^2 X_0 - F_0 \cos\phi = 0$

Horizontal vector component: $c\omega X_0 - F_0 \sin\phi = 0$

Solving these two equations for the unknowns X_0 and ϕ:

$$X_0 = \frac{F_0}{\sqrt{(c\omega)^2 + \left(K - \dfrac{M}{g_c}\omega^2\right)^2}}$$

$$= \frac{\dfrac{F_0}{K}}{\sqrt{\left(1 - \dfrac{\omega^2}{\omega_n^2}\right)^2 + \left(2\dfrac{c}{c_c} \times \dfrac{\omega}{\omega_n}\right)^2}}$$

$$\tan\phi = \frac{c\omega}{K - \dfrac{M}{g_c}\omega^2} = \frac{2\dfrac{c}{c_c} \times \dfrac{\omega}{\omega_n}}{1 - (\omega^2/\omega_n^2)}$$

Where:

c = Damping constant
c_c = Critical damping = $2\dfrac{M}{g_c}\omega_n$
c/c_c = Damping ratio
F_0 = External force
F_0/K = Deflection of the spring under load, F_0 (also called static deflection, X_{st})
ω = Forced frequency
ω_n = Natural frequency of the oscillation
ω/ω_n = Frequency ratio

For damped forced vibrations, three different frequencies have to be distinguished: the undamped natural frequency, $\omega_n = \sqrt{Kg_c/M}$; the damped natural frequency, $q = \sqrt{Kg_c/M - (cg_c/2M)^2}$; and the frequency of maximum forced amplitude, sometimes referred to as the resonant frequency.

43.11 Degrees of freedom

In a mechanical system, the degrees of freedom indicate how many numbers are required to express its geometrical position at any instant. In machine-trains, the relationship of mass, stiffness, and damping is not the same in all directions. As a result, the rotating or dynamic elements within the machine move more in one direction than in another. A clear understanding of the degrees of freedom is important in that it has a direct impact on the vibration amplitudes generated by a machine or process system.

One degree of freedom

If the geometrical position of a mechanical system can be defined or expressed as a single value, the machine is said to have one degree of freedom. For example, the position of a piston moving in a cylinder can be specified at any point in time by measuring the distance from the cylinder end.

A single degree of freedom is not limited to simple mechanical systems such as the cylinder. For example, a 12-cylinder gasoline engine with a rigid crank shaft and a rigidly mounted cylinder block has only one degree of freedom. The position of all its moving parts (i.e., pistons, rods, valves, cam shafts, etc.) can be expressed by a single value. In this instance, the value would be the angle of the crankshaft.

However, when mounted on flexible springs, this engine has multiple degrees of freedom. In addition to the movement of its internal parts in relationship to the crank, the entire engine can now move in any direction. As a result, the position of the engine and any of its internal parts require more than one value to plot its actual position in space.

The definitions and relationships of mass, stiffness, and damping in the preceding section assumed a single-degree of freedom. In other words, movement was limited to a single plane. Therefore, the formulas are applicable for all single degree of freedom mechanical systems.

The calculation for torque is a primary example of a single degree of freedom in a mechanical system. Figure 43.15 represents a disk with a moment of inertia, I, that is attached to a shaft of torsional stiffness, k.

Torsional stiffness is defined as the externally applied torque, T, in inch-pounds needed to turn the disk one radian (57.3°). Torque can be represented by the following equations:

$$\Sigma \text{ Torque} = \text{Moment of inertia} \times \text{angular acceleration} = I\frac{d^2\phi}{dt^2}$$

$$= I\ddot{\phi}$$

In this example, there are three torques acting on the disk: the spring torque, damping torque (due to the viscosity of the air), and external torque. The spring torque is minus $(-)k\phi$, where ϕ is measured in radians. The damping torque is minus $(-)c\dot{\phi}$, where c is the damping constant. In this example, c is the damping torque on the disk caused by an angular speed of rotation of one radian per second. The external torque is $T_o\sin(\omega t)$.

$$I\ddot{\phi} = \Sigma \text{ Torque} = -c\dot{\phi} - k\phi + T_o\sin(\omega t)$$

or

$$I\ddot{\phi} + c\dot{\phi} + k\phi = T_o\sin(\omega t)$$

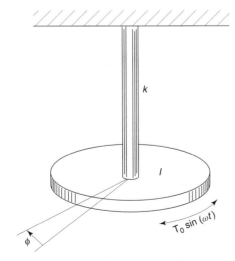

Figure 43.15 Torsional one degree-of-freedom system

Two degrees of freedom

The theory for a one-degree-of-freedom system is useful for determining resonant or natural frequencies that occur in all machine-trains and process systems. However, few machines have only one degree of freedom. Practically, most machines will have two or more degrees of freedom. This section provides a brief overview of the theories associated with two degrees of freedom. An undamped two-degree-of-freedom system is illustrated in Figure 43.16.

This diagram consists of two masses, M_1 and M_2, that are suspended from springs, K_1 and K_2. The two masses are tied together, or coupled, by spring, K_3, so that they are forced to act together. In this example, the movement of the two masses is limited to the vertical plane and, therefore, horizontal movement can be ignored. As in the single-degree-of-freedom examples, the absolute position of each mass is defined by its vertical position above or below the neutral, or reference, point. Since there are two coupled masses, two locations (i.e., one for M_1 and one for M_2) are required to locate the absolute position of the system.

To calculate the free or natural modes of vibration, note that there are two distinct forces acting on mass, M_1: the force of the main spring, K_1, and that of the coupling spring, K_3. The main force acts upwards and is defined as $-K_1X_1$. The shortening of the coupling spring is equal to the difference in the vertical position of the two masses, $X_1 - X_2$. Therefore, the compressive force of the coupling spring is $K_3(X_1 - X_2)$. The compressed coupling spring pushes the top mass, M_1, upward so that the force is negative.

As these are the only tangible forces acting on M_1, the equation of motion for the top mass can be written as:

$$\frac{M_1}{g_c}X_1 = -K_1X_1 - K_3(X_1 - X_2)$$

or

$$\frac{M_1}{g_c}X_1 + (K_1 + K_3)X_1 - K_3X_2 = 0$$

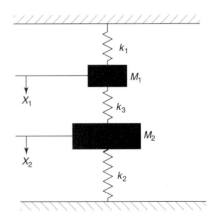

Figure 43.16 Undamped two-degree-of-freedom system with a spring couple

The equation of motion for the second mass, M_2, is derived in the same manner. To make it easier to understand, turn the figure upside down and reverse the direction of X_1 and X_2. The equation then becomes:

$$\frac{M_2}{g_c}X_2 = -K_2 X_2 + K_3(X_1 - X_2)$$

or

$$\frac{M_2}{g_c}X_2 + (K_2 + K_3)X_2 - K_3 X_1 = 0$$

If we assume that the masses, M_1 and M_2, undergo harmonic motions with the same frequency, ω, and with different amplitudes, A_1 and A_2, their behavior can be represented as:

$$X_1 = A_1 \sin(\omega t)$$

$$X_2 = A_2 \sin(\omega t)$$

By substituting these into the differential equations, two equations for the amplitude ratio, A_1/A_2, can be found:

$$\frac{A_1}{A_2} = \frac{-K_3}{\dfrac{M_1}{g_c}\omega^2 - K_1 - K_3}$$

and

$$\frac{A_1}{A_2} = \frac{\dfrac{M_2}{g_c}\omega^2 - K_2 - K_3}{-K_3}$$

For a solution of the form we assumed to exist, these two equations must be equal:

$$\frac{-K_3}{\dfrac{M_1}{g_c}\omega^2 - K_1 - K_3} = \frac{\dfrac{M_2}{g_c}\omega^2 - K_2 - K_3}{-K_3}$$

or

$$\omega^4 - \omega^2\left\{\frac{K_1+K_3}{M_1/g_c} + \frac{K_2+K_3}{M_2/g_c}\right\} + \frac{K_1 K_2 + K_2 K_3 + K_1 K_3}{\dfrac{M_1 M_2}{g_c^2}} = 0$$

This equation, known as the frequency equation, has two solutions for ω^2. When substituted in either of the preceding equations, each one of these gives a definite value for A_1/A_2. This means that there are two solutions for this example, which are of the form $A_1 \sin(\omega t)$ and $A_2 \sin(\omega t)$. As with many such problems, the final answer is the superposition of the two solutions with the final amplitudes and frequencies determined by the boundary conditions.

Many degrees of freedom

When the number of degrees of freedom becomes greater than two, no critical new parameters enter into the problem. The dynamics of all machines can be understood by following the rules and guidelines established in the one and two degree(s)-of-freedom equations. There are as many natural frequencies and modes of motion as there are degrees of freedom.

43.12 Vibration data types and formats

There are several options regarding the types of vibration data that can be gathered for machine trains and systems and the formats in which it can be collected. However, selection of type and format depends on the specific application.

There are two major data-type classifications: time-domain and frequency-domain. Each of these can be further divided into steady state and dynamic data formats. In turn, each of these two formats can be further divided into single-channel and multi-channel.

43.12.1 Data types

Vibration profiles can be acquired and displayed in one of two data types: (1) time-domain or (2) frequency-domain.

Time-domain

Most of the early vibration analysis was carried out using analog equipment, which necessitated the use of time-domain data. The reason for this is that it was difficult to convert time-domain data to frequency-domain data. Therefore, frequency-domain capability was not available until microprocessor-based analyzers incorporated a straightforward method (i.e., Fast Fourier Transform, FFT) of transforming the time-domain spectrum into its frequency components.

Actual time-domain vibration signatures are commonly referred to as time traces or time plots (see Figure 43.17). Theoretical vibration data are generally referred to as waveforms (see Figure 43.18).

Time-domain data are presented with amplitude as the vertical axis and elapsed time as the horizontal axis. Time-domain profiles are the sum of all vibration components (i.e., frequencies, impacts, and other transients) that are present in the machine-train and its installed system. Time traces include all frequency components, but the individual components are more difficult to isolate than with frequency-domain data.

The profile shown in Figure 43.17 illustrates two different data acquisition points, one measured vertically and one measured horizontally, on the same machine and taken at the same time. Because they were obtained concurrently, they can be compared to determine the operating dynamics of the machine.

In this example, the data set contains an impact that occurred at 0.005 seconds. The impact is clearly visible in both the vertical (top) and horizontal (bottom) data set. From these time traces, it is apparent that the vertical impact is stronger than the horizontal. In addition, the impact repeated at 0.015 and 0.025 seconds. Two conclusions can be derived from this example: (1) the impact source is a vertical force and (2) it impacts the machine-train at an interval of 0.010 seconds, or frequency of 1/0.010 seconds = 100 Hz.

The waveform in Figure 43.18 illustrates theoretically the unique frequencies and transients that may be present in a machine's signature. Figure 43.18(a) illustrates the complexity of such a waveform by overlaying

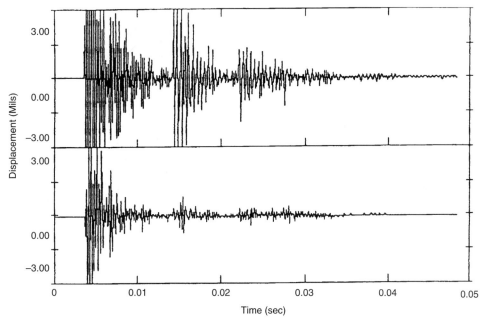

Figure 43.17 Typical time-domain signature

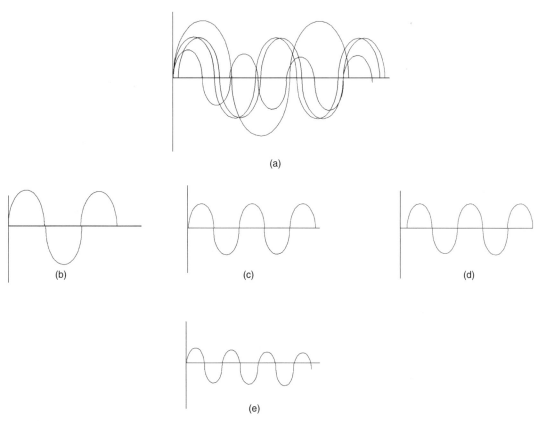

Figure 43.18 Theoretical time-domain waveforms

numerous frequencies. The discrete waveforms that make up Figure 43.18(a) are displayed individually in Figures 43.18(b) through 43.18(e). Note that two of the frequencies (c and d) are identical, but have a different phase angle (ϕ).

With time-domain data, the analyst must manually separate the individual frequencies and events that are contained in the complex waveform. This effort is complicated tremendously by the superposition of multiple frequencies. Note that, rather than overlaying each of the discrete frequencies as illustrated theoretically in Figure 43.18(a), actual time-domain data represents the sum of these frequencies as was illustrated in Figure 43.17.

In order to analyze this type of plot, the analyst must manually change the time scale to obtain discrete frequency curve data. The time interval between the recurrences of each frequency can then be measured. In this way, it is possible to isolate each of the frequencies that make up the time-domain vibration signature.

For routine monitoring of machine vibration, however, this approach is not cost effective. The time required to manually isolate each of the frequency components and transient events contained in the waveform is prohibitive. However, time-domain data has a definite use in a total plant predictive maintenance or reliability improvement program.

Machine-trains or process systems that have specific timing events (e.g., a pneumatic or hydraulic cylinder)

must be analyzed using time-domain data format. In addition, time-domain data must be used for linear and reciprocating motion machinery.

Frequency-domain

Most rotating machine-train failures result at or near a frequency component associated with the running speed. Therefore, the ability to display and analyze the vibration spectrum as components of frequency is extremely important.

The frequency-domain format eliminates the manual effort required to isolate the components that make up a time trace. Frequency-domain techniques convert time-domain data into discrete frequency components using a mathematical process called Fast Fourier Transform (FFT). Simply stated, FFT mathematically converts a time-based trace into a series of discrete frequency components (see Figure 43.19). In a frequency-domain plot, the X-axis is frequency and the Y-axis is the amplitude of displacement, velocity, or acceleration.

With frequency-domain analysis, the average spectrum for a machine-train signature can be obtained. Recurring peaks can be normalized to present an accurate representation of the machine-train condition. Figure 43.20 illustrates a simplified relationship between the two methods (i.e., time-domain and frequency-domain).

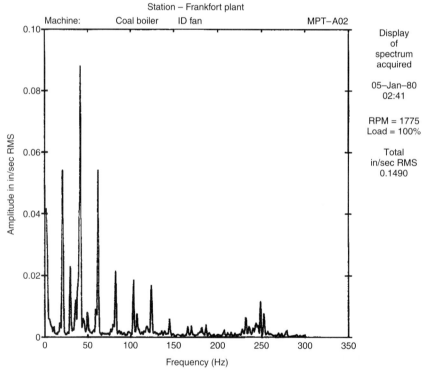

Figure 43.19 Typical frequency-domain signature

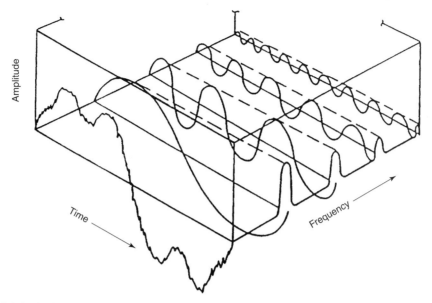

Figure 43.20 Relationship between time domain and frequency domain

The real advantage of frequency-domain analysis is the ability to normalize each vibration component so that a complex machine-train spectrum can be divided into discrete components. This ability simplifies isolation and analysis of mechanical degradation within the machine-train.

In addition, it should be noted that frequency-domain analysis can be used to determine the phase relationships for harmonic vibration components in a typical machine-train spectrum. Frequency-domain normalizes any or all running speeds, where time-domain analysis is limited to true running speed.

Mathematical theory shows that any periodic function of time, $f(t)$, can be represented as a series of sine functions having frequencies ω, 2ω, 3ω, 4ω, etc. Function $f(t)$ is represented by the following equation, which is referred to as a Fourier series:

$$f(t) = A_o + A_1 \sin(\omega t + \phi_1) + A_2 \sin(2\omega t + \phi_2)$$
$$+ A_3 \sin(3\omega t + \phi_3) + \cdots$$

where:

A_x = Amplitude of each discrete sine wave
ω = Frequency
ϕ_x = Phase angle of each discrete sine wave

Each of these sine functions represents a discrete component of the vibration signature discussed previously. The amplitudes of each discrete component and their phase angles can be determined by integral calculus when the function $f(t)$ is known. Because the subject of integral calculus is beyond the scope of this chapter, the math required to determine these integrals are not presented. A vibration analyzer and its associated software perform this determination using FFT.

43.12.2 Data formats

Both time-domain and frequency-domain vibration data can be acquired and analyzed in two primary formats: (1) steady-state or (2) dynamic. Each of these formats has strengths and weaknesses that must be clearly understood for proper use. Each of these formats can be obtained as single- or multi-channel data.

Steady-state

Most vibration programs that use microprocessor-based analyzers are limited to steady-state data. Steady-state vibration data assumes the machine-train or process system operates in a constant, or steady-state, condition. In other words, the machine is free of dynamic variables such as load, flow, etc. This approach further assumes that all vibration frequencies are repeatable and maintain a constant relationship to the rotating speed of the machine's shaft.

Steady-state analysis techniques are based on acquiring vibration data when the machine or process system is operating at a fixed speed and specific operating parameters. For example, a variable-speed machine-train is evaluated at constant speed rather than over its speed range.

Steady-state analysis can be compared to a still photograph of the vibration profile generated by a machine or process system. Snapshots of the vibration profile are acquired by the vibration analyzer and stored for analysis. While the snapshots can be used to evaluate the relative operating condition of simple machine-trains, they do not provide a true picture of the dynamics of either the machine or its vibration profile.

Steady-state analysis totally ignores variations in the vibration level or vibration generated by transient events such as impacts and changes in speed or process parameters. Instruments used to obtain the profiles contain electronic circuitry which are specifically designed to eliminate transient data.

In the normal acquisition process, the analyzer acquires multiple blocks of data. As part of the process, the microprocessor compares each block of data as it is acquired. If a block contains a transient that is not included in subsequent blocks, the block containing the event is discarded and replaced with a transient-free block. As a result, steady-state analysis does not detect random events that may have a direct, negative effect on equipment reliability.

Dynamic

While steady-state data provide a snapshot of the machine, dynamic or real-time data provide a motion picture. This approach provides a better picture of the dynamics of both the machine-train and its vibration profile. Data acquired using steady-state methods would suggest that vibration profiles and amplitudes are constant. However, this is not true. All dynamic forces, including running speed, vary constantly in all machine-trains. When real-time data acquisition methods are used, these variations are captured and displayed for analysis.

Single-channel

Most microprocessor-based vibration monitoring programs rely on single-channel vibration data format. Single-channel data acquisition and analysis techniques are acceptable for routine monitoring of simple, rotating machinery. However, it is important that single-channel analysis be augmented with multi-channel and dynamic analysis. Total reliance on single-channel techniques severely limits the accuracy of analysis and the effectiveness of a predictive maintenance or reliability improvement program.

With the single-channel method, data are acquired in series or one channel at a time. Normally, a series of data points are established for each machine-train and data are acquired from each point in a measurement route. While this approach is more than adequate for routine monitoring of relatively simple machines, it is based on the assumption that the machine's dynamics and the resultant vibration profile are constant throughout the entire data acquisition process. This approach hinders the ability to evaluate real-time relationships between measurement points on the machine-train and variations in process parameters such as speed, load, pressure, etc.

Multi-channel

Multi-channel data provide the best picture of the relationship between measurement points on a machine-train. Data are acquired simultaneously from all measurement points on the machine-train. With this type of data, the analyst can establish the relationship between machine dynamics and vibration profile of the entire machine.

In most cases, a digital tape recorder is used to acquire data from the machine. Since all measurement points are recorded at the same time, the resultant data can be used to compare the tri-axial vibration profile of all measurement points. This capability greatly enhances the analyst's ability to isolate abnormal machine dynamics and to determine the root cause of deviations.

43.12.3 Data acquisition

It is important for predictive maintenance programs using vibration analysis to have accurate, repeatable data. In addition to the type and quality of the transducer, three key parameters affect data quality: the point of measurement, orientation, and transducer-mounting techniques.

In a predictive and reliability maintenance program, it is extremely important to keep good historical records of key parameters. How measurement point locations and orientation to the machine's shaft were selected should be kept as part of the database. It is important that every measurement taken throughout the life of the maintenance program be acquired at exactly the same point and orientation. In addition, the compressive load, or downward force, applied to the transducer should be the same for each measurement.

Vibration detectors: transducers and cables

There are variety of monitoring, trending, and analysis techniques that can and should be used as part of a total plant vibration-monitoring program. Initially, such a program depends on the use of historical trends to detect incipient problems. As the program matures, however, other techniques such as frequency-domain signature analysis, time-domain analysis, and operating dynamics analysis are typically added.

An analysis is only as good as the data; therefore, the equipment used to collect the data is critical and determines the success or failure of a predictive maintenance or reliability improvement program. The accuracy as well as proper use and mounting determines whether valid data are collected.

Three basic types of vibration transducers that can be used for monitoring the mechanical condition of plant machinery are: displacement probes, velocity transducers, and accelerometers. Each has limitations and specific applications for which its use is appropriate.

Displacement probes

Displacement, or eddy-current, probes are designed to measure the actual movement, or displacement, of a machine's shaft relative to the probe. Data are normally recorded as peak-to-peak in mils, or thousandths of an inch. This value represents the maximum deflection or displacement from the true centerline of a machine's shaft. Such a device must be rigidly mounted to a stationary structure to obtain accurate, repeatable data. See Figure 43.21 for an illustration of a displacement probe and signal conditioning system.

Permanently mounted displacement probes provide the most accurate data on machines having a rotor weight that

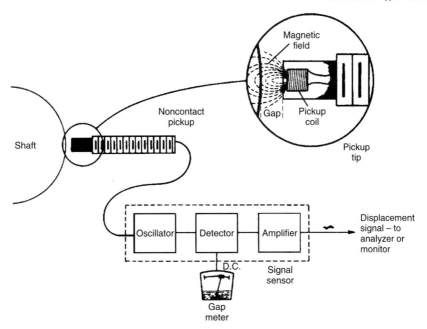

Figure 43.21 Displacement probe and signal conditioning system

is low relative to the casing and support structure. Turbines, large compressors, and other types of plant equipment should have displacement transducers permanently mounted at key measurement locations.

The useful frequency range for displacement probes is from 10 to 1000 Hz, or 600 to 60,000 rpm. Frequency components above or below this range are distorted and therefore unreliable for determining machine condition.

The major limitation with displacement or proximity probes is cost. The typical cost for installing a single probe, including a power supply, signal conditioning, etc., averages $1000. If each machine to be evaluated requires ten measurements, the cost per machine is about $10,000. Using displacement transducers for all plant machinery dramatically increases the initial cost of the program. Therefore, key locations are generally instrumented first and other measurement points are added later.

Velocity transducers

Velocity transducers are electro-mechanical sensors designed to monitor casing, or relative, vibration. Unlike displacement probes, velocity transducers measure the rate of displacement rather than the distance of movement. Velocity is normally expressed in terms of inches per second (ips) peak, which is perhaps the best method of expressing the energy caused by machine vibration. Figure 43.22 is a schematic diagram of a velocity measurement device.

Like displacement probes, velocity transducers have an effective frequency range of about 10 to 1000 Hz. They should not be used to monitor frequencies above or below this range.

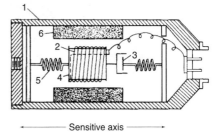

Figure 43.22 Schematic diagram of velocity pickup (1) Pickup case (2) Wire out (3) Damper (4) Mass (5) Spring (6) Magnet

The major limitation of velocity transducers is their sensitivity to mechanical and thermal damage. Normal use can cause a loss of calibration and, therefore, a strict re-calibration program is required to prevent data errors. At a minimum, velocity transducers should be re-calibrated every six months. Even with periodic re-calibration, however, velocity transducers are prone to provide distorted data due to loss of calibration.

Accelerometers

Acceleration is perhaps the best method of determining the force resulting from machine vibration. Accelerometers use piezoelectric crystals or films to convert mechanical energy into electrical signals and Figure 43.23 is a schematic of such a device. Data acquired with this type of transducer are relative

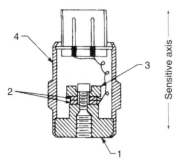

Figure 43.23 Schematic diagram of accelerometer (1) Base (2) Piezoelectric crystals (3) Mass (4) Case

acceleration expressed in terms of the gravitational constant, g, in inches/ second/second.

The effective range of general-purpose accelerometers is from about 1 Hz to 10,000 Hz. Ultrasonic accelerometers are available for frequencies up to 1 MHz. In general, vibration data above 1000 Hz, or 60,000 cpm, should be taken and analyzed in acceleration or g's.

A benefit of the use of accelerometers is that they do not require a calibration program to ensure accuracy. However, they are susceptible to thermal damage. If sufficient heat radiates into the piezoelectric crystal, it can be damaged or destroyed. However, thermal damage is rare since data acquisition time is relatively short (i.e. less than thirty seconds) using temporary mounting techniques.

Cables

Most portable vibration data collectors use a coiled cable to connect to the transducer (see Figure 43.24). The cable, much like a telephone cord, provides a relatively compact length when relaxed, but will extend to reach distant

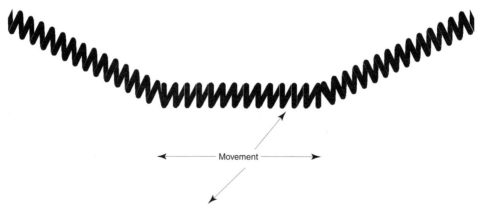

Figure 43.24 Types of coiled cables

measurement points. For general use, this type of cable is acceptable, but it cannot be used for all applications.

The coiled cable is not acceptable for low-speed (i.e., less than 300 rpm) applications or where there is a strong electromagnetic field. Because of its natural tendency to return to its relaxed length, the coiled cable generates a low level frequency that corresponds to the oscillation rate of the cable. In low-speed applications, this oscillation frequency can mask real vibration that is generated by the machine.

A strong electro-magnetic field, such as that generated by large mill motors, accelerates cable oscillation. In these instances, the vibration generated by the cable will mask real machine vibration.

In these and other applications where the coiled cable distorts or interferes with the accuracy of acquired data, a shielded coaxial cable should be used. While these non-coiled cables can be more difficult to use in conjunction with a portable analyzer, they are essential for low-speed and electromagnetic field applications.

43.12.4 Data measurements

Most vibration monitoring programs rely on data acquired from the machine housing or bearing caps. The only exceptions are applications that require direct measurement of actual shaft displacement to obtain an accurate picture of the machine's dynamics. This section discusses the number and orientation of measurement points required to profile a machine's vibration characteristics.

The fact that both normal and abnormal machine dynamics tend to generate unbalanced forces in one or more directions increases the analyst's ability to determine the root-cause of deviations in the machine's operating condition. Because of this, measurements should be taken in both radial and axial orientations.

Radial orientation

Radially oriented measurements permit the analyst to understand the relationship of vibration levels generated by machine components where the forces are perpendicular to the shaft's centerline.

For example, mechanical imbalance generates radial forces in all directions, but misalignment generally results in a radial force in a single direction that corresponds with the misaligned direction. The ability to determine the actual displacement direction of the machine's shaft and other components greatly improves diagnostic accuracy.

Two radial measurement points located 90° apart are required at each bearing cap. The two points permit the analyst to calculate the actual direction and relative amplitude of any displacement that is present within the machine.

Figure 43.25 illustrates a simple vector analysis where the vertical and horizontal radial readings acquired from the outboard bearing cap indicate a relative vertical vibration velocity of 0.5 inches per second peak (IPS-PK) and a horizontal vibration velocity of 0.3 IPS-PK. Using simple geometry, the amplitude of vibration velocity (0.583 IPS-PK) in the actual direction of deflection can be calculated.

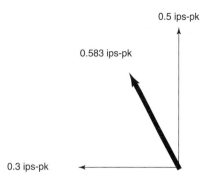

Figure 43.25 Resultant shaft velocity vector based on radial vibration measurements

Axial orientation

Axially oriented measurements are used to determine the lateral movement of a machine's shaft or dynamic mass. These measurement points are oriented in-line or parallel with the shaft or direction of movement.

At least one axial measurement is required for each shaft or dynamic movement. In the case of shafts with a combination of float and fixed bearings, readings should be taken from the fixed or stationary bearing to obtain the best data.

43.11.5 Transducer mounting techniques

For accuracy of data, a direct mechanical link between the transducer and the machine's casing or bearing cap is necessary. This makes the method used to mount the transducer crucial to obtaining accurate data. Slight deviations in this link will induce errors in the amplitude of vibration measurement and may create false frequency components that have nothing to do with the machine.

Permanent

The best method of ensuring that the point of measurement, its orientation, and the compressive load are exactly the same each time is to permanently or hard mount the transducers, which is illustrated in Figure 43.26. This guarantees accuracy and repeatability of acquired data. However, it also increases the initial cost of the program. The average cost of installing a general-purpose accelerometer is about $300 per measurement point or $3000 for a typical machine-train.

Quick disconnect

To eliminate the capital cost associated with permanently mounting transducers, a well designed quick-disconnect mounting can be used instead. With this technique, a quick-disconnect stud having an average cost of less than $5 is permanently mounted at each measurement point. A mating sleeve built into the transducer is used to connect with the stud. A well-designed quick-disconnect mounting technique provides almost the same accuracy and repeatability as the permanent mounting technique, but at a much lower cost.

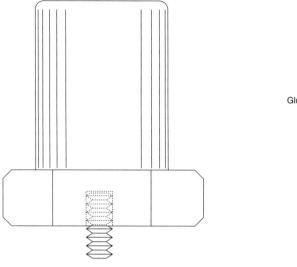

Glue pad for accelerometer

Stud mounted accelerometer

Double stud mount

Figure 43.26 Permanent mounts provide best repeatability

Magnets

For general-purpose use below 1000 Hz, a transducer can be attached to a machine by a magnetic base. Even though the resonant frequency of the transducer/magnet assembly may distort the data, this technique can be used with some success. However, since the magnet can be placed anywhere on the machine, it is difficult to guarantee that the exact location and orientation is maintained with each measurement. Shows common magnetic mounts for transducers.

Hand held

Another method used by some plants to acquire data is hand-held transducers. This approach is not recommended if it is possible to use any other method. Hand-held transducers do not provide the accuracy and repeatability required to gain maximum benefit from a predictive maintenance program. If this technique must be used, extreme care should be exercised to ensure that the same location, orientation, and compressive load are used for every measurement. Illustrates a hand-held device.

43.13 Acquiring data

Three factors must be considered when acquiring vibration data: settling time, data verification, and additional data that may be required.

Settling time

All vibration transducers require a power source that is used to convert mechanical motion or force to an electronic signal. In microprocessor-based analyzers, this power source is usually internal to the analyzer. When displacement probes are used, an external power source must be provided.

When the power source is turned on, there is a momentary surge of power into the transducer. This surge distorts the vibration profile generated by the machine. Therefore, the data-acquisition sequence must include a time delay between powering up and acquiring data. The time delay will vary based on the specific transducer used and type of power source.

Some vibration analyzers include a user-selected time delay that can automatically be downloaded as part of the measurement route. If this feature is included, the delay can be pre-programmed for the specific transducer that will be used to acquire data. No further adjustment is required until the transducer type is changed.

In addition to the momentary surge created by energizing the power source, the mechanical action of placing the transducer on the machine creates a spike of energy that may distort the vibration profile. Therefore, the actual data-acquisition sequence should include a ten- to twenty-second delay to permit decay of the spike created by mounting the transducer.

Data verification

A number of equipment problems can result in bad or distorted data. In addition to the surge and spike discussed in the preceding section, damaged cables, transducers, power supplies, and other equipment failures can cause serious problems. Therefore, it is essential to verify all data throughout the acquisition process.

Most of the microprocessor-based vibration analyzers include features that facilitate verification of acquired data.

For example, many include a 'low-level alert' that automatically alerts the technician when acquired vibration levels are below a pre-selected limit. If these limits are properly set, the alert should be sufficient to detect this form of bad data.

Unfortunately, not all distortions of acquired data result in a low-level alert. Damaged or defective cables or transducers can result in a high level of low-frequency vibration. As a result, the low-level alert will not detect this form of bad data. However, the vibration signature will clearly display the abnormal profile that is associated with these problems.

In most cases, a defective cable or transducer generates a signature that contains a ski-slope profile, which begins at the lowest visible frequency and drops rapidly to the noise floor of the signature. If this profile is generated by defective components, it will not contain any of the normal rotational frequencies generated by the machine-train.

With the exception of mechanical rub, defective cables and transducers are the only sources of this ski-slope profile. When mechanical rub is present, the ski slope will also contain the normal rotational frequencies generated by the machine-train. In some cases, it is necessary to turn off the auto-scale function in order to see the rotational frequencies, but they will be evident. If no rotational components are present, the cable and transducer should be replaced.

Additional data

Data obtained from a vibration analyzer are not all that are required to evaluate machine-train or system condition. Variables such as load have a direct effect on the vibration profile of machinery and must be considered. Therefore, additional data should be acquired to augment the vibration profiles.

Most microprocessor-based vibration analyzers are capable of directly acquiring process variables and other inputs. The software and firmware provided with these systems generally support pre-programmed routes that include almost any direct or manual data input. These routes should include all data required to effectively analyze the operating condition of each machine-train and its process system.

43.14 Vibration analysis techniques

Techniques used in vibration analysis are: trending, both broadband and narrowband; comparative analysis; and signature analysis.

43.14.1 Trending

Most vibration monitoring programs rely heavily on historical vibration-level amplitude trends as their dominant analysis tool. This is a valid approach if the vibration data are normalized to remove the influence of variables, such as load, on the recorded vibration energy levels. Valid trend data provides an indication of change over time within the monitored machine. As stated in preceding sections, a change in vibration amplitude is an indication

of a corresponding change in operating condition that can be a useful diagnostic tool.

Broadband

Broadband analysis techniques have been used for monitoring the overall mechanical condition of machinery for more than twenty years. The technique is based on the overall vibration or energy from a frequency range of zero to the user-selected maximum frequency, F_{MAX} Broadband data are overall vibration measurements expressed in units such as velocity (PK), acceleration (RMS), etc. This type of data, however, does not provide any indication of the specific frequency components that make up the machine's vibration signature. As a result, specific machine-train problems cannot be isolated and identified.

The only useful function of broadband analysis is long-term trending of the gross overall condition of machinery. Typically, sets of alert/alarm limits are established to monitor the overall condition of the machine-trains in a predictive maintenance program. However, this approach has limited value and, when used exclusively, severely limits the ability to achieve the full benefit of a comprehensive program.

Narrowband

Like broadband analysis, narrowband analysis also monitors the overall energy, but for a user-selected band of frequency components. The ability to select specific groups of frequencies, or narrowbands, increases the usefulness of the data. Using this technique can drastically reduce the manpower required to monitor machine-trains and improve the accuracy of detecting incipient problems.

Unlike broadband data, narrowband data provide the ability to directly monitor, trend, and alarm specific machine-train components automatically by the use of a microprocessor for a window of frequencies unique to specific machine components. For example, a narrowband window can be established to directly monitor the energy of a gear set that consists of the primary gear mesh frequency and corresponding side bands.

43.14.2 Comparative analysis

Comparative analysis directly compares two or more data sets in order to detect changes in the operating condition of mechanical or process systems. This type of analysis is limited to the direct comparison of the time-domain or frequency-domain signature generated by a machine. The method does not determine the actual dynamics of the system. Typically, the following data are used for this purpose: (1) baseline data, (2) known machine condition, or (3) industrial reference data.

Note that great care must be taken when comparing machinery vibration data to industry standards or baseline data. The analyst must make sure the frequency and amplitude are expressed in units and running speeds that are consistent with the standard or baseline data. The use of a microprocessor-based system with software that automatically converts and displays the desired terms offers a solution to this problem.

Baseline data

Reference or baseline data sets should be acquired for
each machine-train or process system to be included in
a predictive maintenance program when the machine is
installed or after the first scheduled maintenance once the
program is established. These data sets can be used as
a reference or comparison for all future measurements.
However, such data sets must be representative of the
normal operating condition of each machine-train. Three
criteria are critical to the proper use of baseline comparisons: reset after maintenance, proper identification, and
process envelope.

Reset after maintenance

The baseline data set must be updated each time the machine
is repaired, rebuilt, or when major maintenance is performed. Even when best practices are used, machinery
cannot be restored to as-new condition when major maintenance is performed. Therefore, a new baseline or reference
data set must be established following these events.

Proper identification

Each reference or baseline data set must be clearly and
completely identified. Most vibration-monitoring systems
permit the addition of a label or unique identifier to any
user-selected data set. This capability should be used to
clearly identify each baseline data set.

In addition, the data-set label should include all information that defines the data set. For example, any rework
or repairs made to the machine should be identified. If a
new baseline data set is selected after the replacement of
a rotating element, this information should be included in
the descriptive label.

Process envelope

Since variations in process variables, such as load, have
a direct effect on the vibration energy and the resulting
signature generated by a machine-train, the actual operating envelope for each baseline data set must also be
clearly identified. If this step is omitted, direct comparison of other data to the baseline will be meaningless.
The label feature in most vibration monitoring systems
permits tagging the baseline data set with this additional
information.

Known machine condition

Most microprocessor-based analyzers permit direct comparison to two machine-trains or components. The form
of direct comparison, called cross-machine comparison,
can be used to identify some types of failure modes.

When using this type of comparative analysis, the analyst
compares the vibration energy and profile from a suspect machine to that of a machine with known operating
condition. For example, the suspect machine can be compared to the baseline reference taken from a similar machine
within the plant. Or, a machine profile with a known defect,

such as a defective gear, can be used as a reference to
determine if the suspect machine has a similar profile and
therefore a similar problem.

Industrial reference data

One form of comparative analysis is direct comparison
of the acquired data to industrial standards or reference
values. The International Standards Organization (ISO)
established the vibration severity standards presented in
Table 43.2. These data are applicable for comparison with
filtered narrowband data taken from machine-trains with
true running speeds between 600 and 12,000 rpm. The values from the table include all vibration energy between
a lower limit of 0.3× true running speed and an upper
limit of 3.0×. For example, an 1800-rpm machine would
have a filtered narrowband between 540 (1800 × 0.3) and
5400 rpm (1800 × 3.0). A 3600-rpm machine would have
a filtered narrowband between 1,080 (3600 × 0.3) and
10,800 rpm (3600 × 3.0).

43.14.3 Signature analysis

The phrase 'full Fast Fourier Transform (FFT) signature'
is usually applied to the vibration spectrum that uniquely
identifies a machine, component, system, or subsystem
at a specific operating condition and time. It provides
specific data on every frequency component within the
overall frequency range of a machine-train. The typical
frequency range can be from 0.1 to 20,000 Hz.

In microprocessor systems, the FFT signature is
formed by breaking down the total frequency spectrum
into unique components, or peaks. Each line or peak
represents a specific frequency component that, in turn,

Table 43.2 Vibration severity standards* (inches/second-peak)

Condition	Machine classes			
	I	II	III	IV
Good operating condition	0.028	0.042	0.100	0.156
Alert limit	0.010	0.156	0.255	0.396
Alarm limit	0.156	0.396	0.396	0.622
Absolute fault limit	0.260	0.400	0.620	1.000

*Applicable to a machine with running speed between 600 to
12,000 rpm.
  Narrowband setting: 0.3× to 3.0× running speed.

Machine class descriptions:

Class I Small machine-trains or individual components integrally
connected with the complete machine in its normal
operating condition (i.e., drivers up to 20 horsepower).
Class II Medium-sized machines (i.e., 20- to 100-horsepower
drivers and 400-horsepower drivers on special
foundations.
Class III Large prime movers (i.e., drivers greater than 100
horsepower) mounted on heavy, rigid foundations.
Class IV Large prime movers (i.e., drivers greater than 100
horsepower) mounted on relatively soft, lightweight
structures.

Source: Derived by Integrated Systems, Inc. from ISO Standard
2372.

represents one or more mechanical components within the machine-train. Typical microprocessor- based predictive maintenance systems can provide signature resolutions of at least 400 lines and many provide up to 12,800 lines.

Full-signature spectra are an important analysis tool, but require a tremendous amount of microprocessor memory. It is impractical to collect full, high-resolution spectra on all machine-trains on a routine basis. Data management and storage in the host computer is extremely difficult and costly. Full-range signatures should be collected only if a confirmed problem has been identified on a specific machine-train. This can be triggered automatically by exceeding a preset alarm limit in the historical amplitude trends.

Broadband and full signature

Systems that utilize either broadband or full signature measurements have limitations that may hamper the usefulness of the program. Broadband measurements usually do not have enough resolution at running speeds to be effective in early problem diagnostics. Full-signature measurement at every data point requires a massive data acquisition, handling, and storage system that greatly increases the capital and operating costs of the program.

Normally, a full-signature spectrum is needed only when an identified machine-train problem demands further investigation. Please note that while full signatures generate too much data for routine problem detection, they are essential for root-cause diagnostics. Therefore, the optimum system includes the capability to utilize all techniques. This ability optimizes the program's ability to trend, do full root-cause failure analysis, and still maintain minimum data management and storage requirements.

Narrowband

Typically, a machine-train's vibration signature is made up of vibration components with each component associated with one or more of the true running speeds within the machine-train. Because most machinery problems show up at or near one or more of the running speeds, the narrowband capability is very beneficial in that high-resolution windows can be preset to monitor the running speeds. However, many of the microprocessor-based predictive maintenance systems available do not have narrowband capability. Therefore, care should be taken to ensure that the system utilized does have this capability.

Appendix 43.1 List of Abbreviations

A	Acceleration
A	Zero-to-peak amplitude
cpm	Cycles per minute or cycles/minute
cps	Cycles per second or cycles/second
f	Frequency
$f(t)$	Function of time
F	Force
FFT	Fast Fourier Transform
F_{MAX}	Maximum frequency
F_{MIN}	Minimum frequency
F_o	External force
F_s	Spring force
g	Gravitational constant, 32.17 ft/sec^2
h	Elongation due to the weight of the mass
Hz	Hertz
in	Inches
ips	Inches per second or inches/second
ips-PK	Inches per second, zero-to-peak
ISI	Integrated Systems, Inc.
k	Torsional stiffness
K	Spring constant or stiffness
lbf	Pounds force
lbm	Pounds mass
L_o	Unelongated spring length
M	Mass
MHz	Megahertz
PK	Zero-to-peak
RMS	Root-mean-square
Rpm	Revolutions per minute or revolution/minute
sec^2	Seconds squared
T	Time
T	Period or torque
T_o	External torque
VPM	Vibrations per minute or vibrations/minute
W	Weight
X	Displacement
X_i	Static displacement
X_o	Amount of displacement from X_i
$1\times, 2\times, 3\times$	1 times, 2 times, 3 times

Appendix 43.2 Glossary

Accelerometer	Transducer used to measure acceleration. Incorporates a piezoelectric crystal or film to convert mechanical energy into electrical signals.
Acceleration	The rate of change of velocity with respect time (ft/sec^2) or (in/sec^2).
Amplitude	The magnitude or size of a quantity such as displacement, velocity, acceleration, etc., measured by a vibration analyzer in conjunction with a displacement probe, velocity transducer, or accelerometer.
Axial	Of, on, around, or along an axis (straight line about which an object rotates) or center of rotation.
Bearing cap	The protective structure that covers bearings.

Boundary condition	Mathematically defined as a requirement to be met by a solution to a set of differential equations on a specified set of values of the independent variables.
Displacement	The change in distance or position of an object relative to a reference point, usually measured in mils.
Dynamics, operating	Deals with the motion of a system under the influence of forces, especially those that originate outside the system under consideration.
Fast fourier transform	A mathematical technique used to convert a time-domain plot into its unique frequency components.
Force	That influence on a body that causes it to accelerate. Quantitatively, it is a vector equal to the body's time rate of change of momentum.
Forcing function	The cause of each discrete frequency component in a machine-train's vibration signature.
Frequency	Frequency, f, is defined as the number of repetitions of a specific forcing function or vibration component over a specific unit of time. It is the inverse of the period, $1/T$, of the vibration and can be expressed in units of cycles per second (cps) or Hertz (Hz). For rotating machinery, the frequency is often expressed in vibrations per minute (vpm).
Frequency, circular	Another measure of frequency measured in radians ($\omega = 2\pi f$).
Frequency, natural	All components have one or more natural frequencies that can be excited by an energy source that coincides with, or is in close proximity to, that frequency. The result is a substantial increase in the amplitude of the natural frequency vibration component, which is referred to as resonance. Higher levels of input energy can cause catastrophic, near instantaneous failure of the machine or structure.
Frequency, primary	The base frequency referred to in a vibration analysis that includes vibrations that are harmonics of the primary frequency.
Gravitational constant	The constant of proportionality in the English system of units, g_c, which causes one pound of mass produces one pound of force under the acceleration of gravity, equal to $32.17\,\text{lbm-ft/lbf-sec}^2$.
Harmonic motion	A periodic motion or vibration that is a sinusoidal function of time, that is, motion along a line given by equation $x = a\cos(\omega t + \phi)$, where t is time, a

	and ω are constants, and ϕ is the phase angle. (e.g., $X = X_o\sin(\omega t + \phi)$ where X is the displacement, X_o is the amplitude, ω is the circular frequency, and ϕ is the phase angle.)
Harmonics	Multiples of the primary frequency (e.g., $2\times$, $3\times$, etc.).
Hertz	Unit of frequency; a periodic oscillation has a frequency of n hertz if in one second it goes through n cycles.
Imbalance	A condition that can result from a mechanical and/or a force imbalance. Mechanical imbalance is when there is more weight on one side of a center line of a rotor than on the other. Force imbalance can result when there is an imbalance of the centripetal forces generated by rotation and/or when there is an imbalance between the lift generated by the rotor and gravity.
Machine element	Rotating-machine components, such as rolling-element bearings, impellers, and other rotors, that turn with a shaft.
Machine-train	A series of machines containing both driver and driven components.
Maintenance management program	A comprehensive program that includes predictive maintenance techniques to monitor and analyze critical machines, equipment, and systems in a typical plant. Techniques include vibration analysis, ultrasonics, thermography, tribology, process monitoring, visual inspection, and other non-destructive analysis methods.
Maximum frequency	Broadband analysis techniques, which are used to monitor the overall mechanical condition of machinery, are based on the overall vibration or energy from a frequency range of zero to the user-selected maximum frequency (F_{MAX}).
Mil	One one-thousandth of an inch (0.001 inch).
Moment of inertia	The sum of the products formed by multiplying the mass of each element of a body by the square of its distance from a specified line. Also known as 'rotational inertia.'
Oscillate	To move back and forth with a steady uninterrupted rhythm.
Periodic motion	A motion that repeats after a certain interval of time.
Phase angle	The difference between the phase of a sinusoidally varying quantity and the phase of a second quantity that varies sinusoidally at the same frequency. Also known as 'phase difference.'

Piezoelectric	Describes a crystal or film having the ability to generate a voltage when mechanical force is applied, or to produce a mechanical force when a voltage is applied.
Predictive maintenance	The practice of using actual operating conditions of plant equipment and systems to optimize total plant operation. Relies on direct equipment monitoring to determine the actual mean-time-to-failure or loss of efficiency for each machine-train and system in a plant. This technique is used in place of traditional run-to-failure programs.
Profile	Refers to either time-domain (also may be called time trace or waveform) or frequency-domain vibration curves.
Quadratic	Any second-degree expression.
Radial	Extending from a point or center in the manner of rays (as the spokes of a wheel are radial).
Radian	The central angle of a circle determined by two radii and an arc joining them, all of the same length. A circle is comprised of 2π radians.
Reciprocation	The action of moving back and forth alternately.
Signature	A frequency-domain vibration curve.
Spring constant	The number of pounds tension necessary to extend the spring one inch. Also referred to as 'stiffness' 'spring modulus.'
Thermography	Use of heat emissions of machinery or plant equipment as a monitoring and diagnostic predictive maintenance tool. For example, temperature differences on a coupling indicate misalignment and/or uneven mechanical forces.
Torque	A moment/force couple applied to a rotor such as a shaft in order to sustain acceleration/load requirements. A twisting load imparted to shafts as the result of induced loads/speeds.
Transducer	Any device or element that converts an input signal into an output signal of a different form.

Tribology	Science of rotor-bearing-support system design and operation. Predictive maintenance technique that uses spectrographic, wear particle, ferrography, and other measurements of the lubricating oil as a diagnostic tool.
Turbulent Flow	Motion of fluids in which local velocities and pressures fluctuate irregularly, in a random manner.
Ultrasonic analysis	Predictive maintenance technique that uses principles similar to those of vibration analysis to monitor the noise generated by plant machinery or systems to determine their actual operating condition. Ultrasonics is used to monitor the higher frequencies (i.e., ultrasound) that range between 20,000 Hertz and 100 kiloHertz.
Vector	A quantity that has both magnitude and direction, and whose components transform from one coordinate system to another in the same manner as the components of a displacement.
Velocity	The time rate of change of position of a body. It is a vector quantity having direction as well as magnitude.
Vibration	A continuing periodic change in a displacement with respect to a fixed reference. The motion will repeat after a certain interval of time.
Vibration analysis	Vibration analysis monitors the noise or vibrations generated by plant machinery or systems to determine their actual operating condition. The normal monitoring range for vibration analysis is from less than 1 to 20,000 Hertz.

References

1 *Advanced Diagnostics and Analysis*, R. Keith Mobley, Technology for Energy Corp., (1989).
2 *Introduction to Dynamics*, Donald E. Hardenbergh, Holt, Rinehart and Winston, New York (1963).
3 *Introduction to Predictive Maintenance*, R. Keith Mobley, Van Nonstrand-Reinhold, New York (1990).
4 *Maintenance Engineering Handbook*, Higgins & R. Keith Mobley, McGraw-Hill, New York (1995).
5 *Vibration Fundamentals*, R. Keith Mobley, Butterworth – Heinemann (1999).

44

Vibration Monitoring and Analysis

R Keith Mobley

The Plant Performance Group

Contents

44.1 Introduction

This chapter provides the basic knowledge and skills required to implement a computer-based vibration-monitoring program. It discusses the following topics: (1) typical machine-train monitoring parameters, (2) database development, (3) data-acquisition equipment and methods, and (4) data analysis.

Although each of the commercially available computer-based vibration monitoring systems has unique features and formats, the information contained in this chapter is applicable to all of the systems. However, the manual provided by the vendor should be used in conjunction with this chapter to ensure proper use of the microprocessor-based data collection analyzer and the computer-based software.

Over the past ten years, vibration monitoring and analysis instrumentation has improved dramatically. During this period, a number of new microprocessor-based systems have been introduced that greatly simplify the collection, data management, long-term trending, and analysis of vibration data. While these advancements permit wider use of vibration monitoring as a predictive-maintenance tool, use is generally limited to relatively simple, steady state rotating machinery. Typically, these systems collect single-channel, frequency-domain data.

44.1.1 Advantages

The automatic functions provided by most of the new systems have greatly reduced the time and manpower required to monitor critical plant equipment. These functions have virtually eliminated both the human errors and the setup time normally associated with older vibration-monitoring techniques.

Simplified data acquisition and analysis

With the combined power of the data collector and system software, data acquisition has been reduced to simple measurement routes that require limited operator input. The technician's role is to temporarily mount a transducer at the proper measurement point and push a button. The microprocessor automatically acquires conditions, evaluates, and stores the vibration data.

Automated data management

Before computer-based systems were developed, a major limitation of vibration monitoring programs was the labor required to manage, store, retrieve, and analyze the massive amount of data generated. However, the computer-based systems in use today virtually eliminate this labor requirement. These systems automatically manage data and provide almost instant data retrieval for analysis.

44.1.2 Limitations

There are several limitations of the computer-based systems and some system characteristics, particularly simplified data acquisition and analysis, provide both advantages and disadvantages. Other limitations arise because only single-channel, steady state, frequency-domain data greater than 600 cycles per minute (cpm) or 10 Hertzx (Hz) can be collected. Note that cpm also is referred to as revolutions per minute (rpm).

Simplified data acquisition and analysis

While providing many advantages, simplified data acquisition and analysis also can be a liability. If the database is improperly configured, the automated capabilities of these analyzers will yield faulty diagnostics that can allow catastrophic failure of critical plant machinery.

Because technician involvement is reduced to a minimum level, the normal tendency is to use untrained or partially trained personnel for this repetitive function. Unfortunately, the lack of training results in less awareness and knowledge of visual and audible clues that can, and should be, an integral part of the monitoring program.

Single-channel data

Most of the microprocessor-based vibration monitoring systems collect single-channel, steady-state data that cannot be used for all applications. Single-channel data are limited to the analysis of simple machinery that operates at relatively constant speed.

While most of the microprocessor-based instruments are limited to a single input channel, in some cases, a second channel is incorporated in the analyzer. However, this second channel generally is limited to input from a tachometer, or a once-per-revolution input signal. This second channel cannot be used for vibration-data capture.

This limitation prohibits the use of most microprocessor-based vibration analyzers for complex machinery or machines with variable speeds. Single-channel data-acquisition technology assumes the vibration profile generated by a machine-train remains constant throughout the data-acquisition process. This is generally true in applications where machine speed remains relatively constant (i.e., within 5 to 10 rpm). In this case, its use does not severely limit diagnostic accuracy and can be effectively used in a predictive-maintenance program.

Steady-state data

Most of the microprocessor-based instruments are designed to handle steady-state vibration data. Few have the ability to reliably capture transient events such as rapid speed or load changes. As a result, their use is limited in situations where these occur.

In addition, vibration data collected with a microprocessor-based analyzer is filtered and conditioned to eliminate non-recurring events and their associated vibration profiles. Anti-aliasing filters are incorporated into the analyzers specifically to remove spurious signals such as impacts. While the intent behind the use of anti-aliasing filters is valid, however, their use can distort a machine's vibration profile.

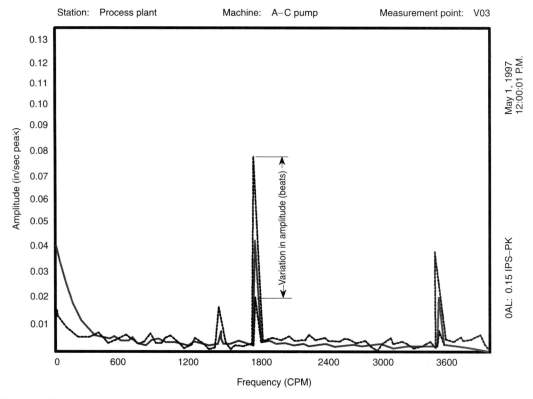

Figure 44.1 Vibration is dynamic and amplitudes constantly change

Because vibration data are dynamic and the amplitudes constantly change as shown in Figure 44.1, most predictive-maintenance-system vendors strongly recommend averaging the data. They typically recommend acquiring three to twelve samples of the vibration profile and averaging the individual profiles into a composite signature. This approach eliminates the variation in vibration amplitude of the individual frequency components that make up the machine's signature. However, these variations, referred to as beats, can be a valuable diagnostic tool. Unfortunately, they are not available from microprocessor-based instruments because of averaging and other system limitations.

Frequency-domain data

Most predictive-maintenance programs rely almost exclusively on frequency-domain vibration data. The microprocessor-based analyzers gather time-domain data and automatically convert it using Fast Fourier Transform (FFT) to frequency-domain data. A frequency-domain signature shows the machine's individual frequency components, or peaks.

While frequency-domain data analysis is much easier to learn than time-domain data analysis, it does not provide the ability to isolate and identify all incipient problems within the machine or its installed system. Because of this,

additional techniques (e.g., time-domain, multi-channel, and real-time analysis) must be used in conjunction with frequency-domain data analysis to obtain a complete diagnostic picture.

Low-frequency response

Many of the microprocessor-based vibration monitoring analyzers cannot capture accurate data from low-speed machinery or machinery that generates low-frequency vibration. Specifically, some of the commercially available analyzers cannot be used where frequency components are below 600 cpm or 10 Hz.

Two major problems restricting the ability to acquire accurate vibration data at low frequencies are electronic noise and the response characteristics of the transducer. The electronic noise of the monitored machine and the 'noise floor' of the electronics within the vibration analyzer tend to override the actual vibration components found in low-speed machinery.

Analyzers especially equipped to handle noise are required for most industrial applications. There are at least three commercially available microprocessor-based analyzers capable of acquiring data below 600 cpm. These systems use special filters and data-acquisition techniques to separate real vibration frequencies from electronic noise. In addition, transducers with the required low-frequency response must be used.

44.2 Machine-train monitoring parameters

This section discusses normal failure modes, monitoring techniques that can prevent premature failures, and the measurement points required for monitoring common machine-train components. Understanding the specific location and orientation of each measurement point is critical to diagnosing incipient problems.

The frequency-domain, or FFT, signature acquired at each measurement point is an actual representation of the individual machine-train component's motion at that point on the machine. Without knowing the specific location and orientation, it is difficult – if not impossible – to correctly identify incipient problems. In simple terms, the FFT signature is a photograph of the mechanical motion of a machine-train in a specific direction and at a specific point and time.

The vibration-monitoring process requires a large quantity of data to be collected, temporarily stored, and downloaded to a more powerful computer for permanent storage and analysis. In addition, there are many aspects to collecting meaningful data. Data collection generally is accomplished using microprocessor-based data-collection equipment referred to as vibration analyzers. However, before analyzers can be used, it is necessary to set up a database with the data-collection and analysis parameters. The term 'narrowband' refers to a specific frequency window that is monitored because of the knowledge that potential problems may occur due to known machine components or characteristics in this frequency range.

The orientation of each measurement point is an important consideration during the database setup and during analysis. There is an optimum orientation for each measurement point on every machine-train in a predictive-maintenance program. For example, a helical gear set creates specific force vectors during normal operation. As the gear set degrades, these force vectors transmit the maximum vibration components. If only one radial reading is acquired for each bearing housing, it should be oriented in the plane that provides the greatest vibration amplitude.

For continuity, each machine-train should be set up on a 'common-shaft' with the outboard driver bearing designated as the first data point. Measurement points should be numbered sequentially starting with the outboard driver bearing and ending with the outboard bearing of the final driven component. This is illustrated in Figure 44.2. Any numbering convention may be used, but it should be consistent, which provides two benefits:

1. Immediate identification of the location of a particular data point during the analysis/diagnostic phase.
2. Grouping the data points by 'common shaft' enables the analyst to evaluate all parameters affecting each component of a machine-train.

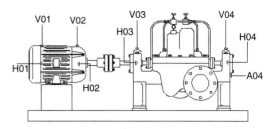

Figure 44.2 Recommended measurement point logic. AO = axial orientation, HO = horizontal orientation, VO = vertical orientation

44.2.1 Drivers

All machines require some form of motive power, which is referred to as a driver. This section includes the monitoring parameters for the two most common drivers: electric motors and steam turbines.

Electric motors

Electric motors are the most common source of motive power for machine-trains. As a result, more of them are evaluated using microprocessor-based vibration monitoring systems than any other driver. The vibration frequencies of the following parameters are monitored to evaluate operating condition. This information is used to establish a database.

- Bearing frequencies,
- Imbalance,
- Line frequency,
- Loose rotor bars,
- Running speed,
- Slip frequency, and
- V-belt intermediate drives.

Bearing frequencies

Electric motors may incorporate either sleeve or rolling-element bearings. A narrowband window should be established to monitor both the normal rotational and defect frequencies associated with the type of bearing used for each application.

Imbalance

Electric motors are susceptible to a variety of forcing functions that cause instability or imbalance. The narrowbands established to monitor the fundamental and other harmonics of actual running speed are useful in identifying mechanical imbalance, but other indices also should be used.

One such index is line frequency, which provides indications of instability. Modulations, or harmonics, of line frequency may indicate the motor's inability to find and hold magnetic center. Variations in line frequency also increase the amplitude of the fundamental and other harmonics of running speed.

Axial movement and the resulting presence of a third harmonic of running speed is another indication of instability or imbalance within the motor. The third harmonic is present whenever there is axial thrusting of a rotating element.

Line frequency

Many electrical problems, or problems associated with the quality of the incoming power and internal to the motor, can be isolated by monitoring the line frequency. Line frequency refers to the frequency of the alternating current being supplied to the motor. In the case of 60-cycle power, monitoring of the fundamental or first harmonic (60 Hertz), second harmonic (120 Hz), and third harmonic (180 Hz) should be performed.

Loose rotor bars

Loose rotor bars are a common failure mode of electric motors. Two methods can be used to identify them.

The first method uses high-frequency vibration components that result from oscillating rotor bars. Typically, these frequencies are well above the normal maximum frequency used to establish the broadband signature. If this is the case, a high-pass filter such as high-frequency domain can be used to monitor the condition of the rotor bars.

The second method uses the slip frequency to monitor for loose rotor bars. The passing frequency created by this failure mode energizes modulations associated with slip. This method is preferred since these frequency components are within the normal bandwidth used for vibration analysis.

Running speed

The running speed of electric motors, both alternating current (AC) and direct current (DC), varies. Therefore, for monitoring purposes, these motors should be classified as variable-speed machines. A narrowband window should be established to track the true running speed.

Slip frequency

Slip frequency is the difference between synchronous speed and actual running speed of the motor. A narrowband filter should be established to monitor electrical line frequency. The window should have enough resolution to clearly identify the frequency and the modulations, or sidebands, that represent slip frequency. Normally, these modulations are spaced at the difference between synchronous and actual speed, and the number of sidebands is equal to the number of poles in the motor.

V-belt intermediate drives

Electric motors with V-belt intermediate drive display the same failure modes as those described previously. However, the unique V-belt frequencies should be monitored to determine if improper belt tension or misalignment is evident.

In addition, electric motors used with V-belt intermediate drive assemblies are susceptible to premature wear on the bearings. Typically, electric motors are not designed to compensate for the sideloads associated with V-belt drives. In this type of application, special attention should be paid to monitoring motor bearings.

The primary data-measurement point on the inboard bearing housing should be located in the plane opposing the induced load (sideload), with the secondary point at 90°. The outboard primary data-measurement point should be in a plane opposite the inboard bearing with the secondary at 90°.

Steam turbines

There are wide variations in the size of steam turbines, which range from large utility units to small package units designed as drivers for pumps, etc. The following section describes in general terms the monitoring guidelines. Parameters that should be monitored are bearings, blade pass, mode shape (shaft deflection), and speed (both running and critical).

Bearings

Turbines use both rolling-element and Babbitt bearings. Narrowbands should be established to monitor both the normal rotational frequencies and failure modes of the specific bearings used in each turbine.

Blade pass

Turbine rotors are comprised of a series of vanes or blades mounted on individual wheels. Each of the wheel units, which is referred to as a stage of compression, has a different number of blades. Narrowbands should be established to monitor the blade-pass frequency of each wheel. Loss of a blade or flexing of blades or wheels is detected by these narrowbands.

Mode shape (shaft deflection)

Most turbines have relatively long bearing spans and highly flexible shafts. These factors, coupled with variations in process flow conditions, make turbine rotors highly susceptible to shaft deflection during normal operation. Typically, turbines operate in either the second or third mode and should have narrowbands at the second (2×) and third (3×) harmonics of shaft speed to monitor for mode shape.

Speed

All turbines are variable-speed drivers and operate near or above one of the rotor's critical speeds. Narrowbands should be established that track each of the critical speeds defined for the turbine's rotor. In most applications, steam turbines operate above the first critical speed and in some cases above the second. A movable narrowband window should be established to track the fundamental (1×), second (2×), and third (3×) harmonics of actual shaft speed. The best method is to use orders analysis and a tachometer to adjust the window location.

Normally, the critical speeds are determined by the mechanical design and should not change. However, changes in the rotor configuration or a build-up of calcium or other foreign materials on the rotor will affect them. The narrowbands should be wide enough to permit some increase or decrease.

44.2.2 Intermediate drives

Intermediate drives transmit power from the primary driver to a driven unit or units. Included in this classification are chains, couplings, gearboxes, and V-belts.

Chains

In terms of its vibration characteristics, a chain-drive assembly is much like a gear set. The meshing of the sprocket teeth and chain links generates a vibration profile that is almost identical to that of a gear set. The major difference between these two machine-train components is that the looseness or slack in the chain tends to modulate and amplify the tooth-mesh energy. Most of the forcing functions generated by a chain-drive assembly can be attributed to the forces generated by tooth-mesh. The typical frequencies associated with chain-drive assembly monitoring are those of running speed, tooth-mesh, and chain speed.

Running speed

Chain-drives normally are used to provide positive power transmission between a driver and driven unit where direct coupling cannot be accomplished. Chain-drives generally have two distinct running speeds: driver or input speed and driven or output speed. Each of the shaft speeds is clearly visible in the vibration profile and a discrete narrowband window should be established to monitor each of the running speeds.

These speeds can be calculated using the ratio of the drive to driven sprocket. For example, where the drive sprocket has a circumference of 10 inches and the driven sprocket a circumference of 5 inches, the output speed will be two times the input speed. Tooth-mesh *narrowband* windows should be created for both the drive and driven tooth-meshing frequencies. The windows should be broad enough to capture the sidebands or modulations that this type of passing frequency generates. The frequency of the sprocket-teeth meshing with the chain links, or passing frequency, is calculated by the following formula:

Tooth-mesh frequency = Number of sprocket teeth × Shaft speed

Unlike gear sets, there can be two distinctive tooth-mesh frequencies for a chain-drive system. Since the drive and driven sprockets do not directly mesh, the meshing frequency generated by each sprocket is visible in the vibration profile.

Chain speed

The chain acts much like a driven gear and has a speed that is unique to its length. The chain speed is calculated by the following equation:

$$\text{Chain Speed} = \frac{\text{Number of drive sprocket teeth} \times \text{Shaft speed}}{\text{Number of links in chain}}$$

For example:

$$\text{Chain speed} = \frac{25 \text{ teeth} \times 100 \text{ rpm}}{250 \text{ links}} = \frac{2500}{250} = 10 \text{ cpm} = 10 \text{ rpm}$$

Couplings

Couplings cannot be monitored directly, but they generate forcing functions that affect the vibration profile of both the driver and driven machine-train component. Each coupling should be evaluated to determine the specific mechanical forces and failure modes they generate. This section discusses flexible couplings, gear couplings, jackshafts, and universal joints.

Flexible couplings   Most flexible couplings use an elastomer or spring-steel device to provide power transmission from the driver to the driven unit. Both coupling types create unique mechanical forces that directly affect the dynamics and vibration profile of the machine-train.

The most obvious force with flexible couplings is end-play or movement in the axial plane. Both the elastomer and spring-steel devices have memory which forces the axial position of both the drive and driven shafts to a neutral position. Because of their flexibility, these devices cause the shaft to move constantly in the axial plane. This is exhibited as harmonics of shaft speed. In most cases, the resultant profile is a signature that contains the fundamental (1×) frequency and second (2×) and third (3×) harmonics.

Gear couplings   When properly installed and maintained, gear-type couplings do not generate a unique forcing function or vibration profile. However, excessive wear, variations in speed or torque, or over-lubrication results in a forcing function.

Excessive wear or speed variation generates a gear-mesh profile that corresponds to the number of teeth in the gear coupling multiplied by the rotational speed of the driver. Since these couplings use a mating gear to provide power transmission, variations in speed or excessive clearance permit excitation of the gear-mesh profile.

Jackshafts   Some machine-trains use an extended or spacer shaft, called a jackshaft, to connect the driver and a driven unit. This type of shaft may use any combination of flexible coupling, universal joint, or splined coupling to provide the flexibility required making the connection. Typically, this type of intermediate drive is used either to absorb torsional variations during speed changes or to accommodate misalignment between the two machine-train components.

Because of the length of these shafts and the flexible couplings or joints used to transmit torsional power, jackshafts tend to flex during normal operation. Flexing results in a unique vibration profile that defines its operating mode shape.

In relatively low-speed applications, the shaft tends to operate in the first mode or with a bow between the two joints. This mode of operation generates an elevated vibration frequency at the fundamental (1×) turning speed of the jackshaft. In higher-speed applications, or where the flexibility of the jackshaft increases, it deflects into an 'S' shape between the two joints. This 'S' or second mode shape generates an elevated frequency at both the fundamental (1×) frequency and the second harmonic (2×) of turning speed. In extreme cases, the jackshaft deflects further and operates in the third mode. When this happens, it generates distinct frequencies at the fundamental (1×), second harmonic (2×), and third harmonic (3×) of turning speed.

As a rule, narrowband windows should be established to monitor at least these three distinct frequencies, i.e. 1×, 2×, and 3×. In addition, narrowbands should be established to monitor the discrete frequencies generated by the couplings or joints used to connect the jackshaft to the driver and driven unit.

Universal joints   There are a variety of universal joints used to transmit torsional power. In most cases, this type of intermediate drive is used where some misalignment between the drive and driven unit is necessary. Because of the misalignment, the universal's pivot points generate a unique forcing function that influences both the dynamics and vibration profile generated by a machine-train.

Figure 44.3 illustrates a typical double-pivot universal joint. This type of joint, which is similar to those used in automobiles, generates a unique frequency at four times (4×) the rotational speed of the shaft. Each of the pivot-point bearings generates a passing frequency each time the shaft completes a revolution.

Gearboxes

Gear sets are used to change speed or rotating direction of the primary driver. The basic monitoring parameters for all gearboxes include bearings, gear-mesh frequencies, and running speeds.

Bearings

A variety of bearing types is used in gearboxes. Narrowband windows should be established to monitor the rotational and defect frequencies generated by the specific type of bearing used in each application.

Special attention should be given to the thrust bearings, which are used in conjunction with helical gears. Because helical gears generate a relatively strong axial force, each gear shaft must have a thrust bearing located on the backside of the gear to absorb the thrust load. Therefore, all helical gear sets should be monitored for shaft run-out.

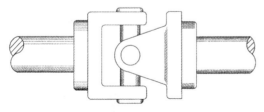

Figure 44.3 Typical double-pivot universal joint

The thrust, or positioning, bearing of a herringbone or double-helical gear has little or no normal axial loading. However, a coupling lockup can cause severe damage to the thrust bearing. Double-helical gears usually have only one thrust bearing, typically on the bull gear. Therefore, the thrust-bearing rotor should be monitored with at least one axial data-measurement point.

The gear mesh should be in a plane opposing the preload creating the primary data-measurement point on each shaft. A secondary data-measurement point should be located at 90° to the primary point.

Gear-mesh frequencies

Each gear set generates a unique profile of frequency components that should be monitored. The fundamental gear-mesh frequency is equal to the number of teeth in the pinion or drive gear multiplied by the rotational shaft speed. In addition, each gear set generates a series of modulations, or sidebands, that surround the fundamental gear-mesh frequency. In a normal gear set, these modulations are spaced at the same frequency as the rotational shaft speed and appear on both sides of the fundamental gear mesh.

A narrowband window should be established to monitor the fundamental gear-mesh profile. The lower and upper limits of the narrowband should include the modulations generated by the gear set. The number of sidebands will vary depending on the resolution used to acquire data. In most cases, the narrowband limits should be about 10 per cent above and below the fundamental gear-mesh frequency.

A second narrowband window should be established to monitor the second harmonic (2×) of gear mesh. Gear misalignment and abnormal meshing of gear sets result in multiple harmonics of the fundamental gear-mesh profile. This second window provides the ability to detect potential alignment or wear problems in the gear set.

Running speeds

A narrowband window should be established to monitor each of the running speeds generated by the gear sets within the gearbox. The actual number of running speeds varies depending on the number of gear sets. For example, a single-reduction gearbox has two speeds: input and output. A double-reduction gearbox has three speeds: input, intermediate, and output. Intermediate and output speeds are determined by calculations based on input speed and the ratio of each gear set. Figure 44.4 illustrates a typical double-reduction gearbox.

If the input speed is 1,800-rpm, the intermediate and output speeds are calculated using the following:

Intermediate speed

$$= \frac{\text{Input speed} \times \text{Number of input gear teeth}}{\text{Number of intermediate gear teeth}}$$

Output speed

$$= \frac{\text{Intermediate speed} \times \text{Number of intermediate gear teeth}}{\text{Number of output gear teeth}}$$

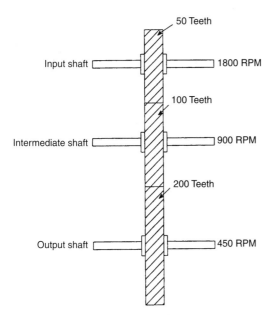

50 Teeth

Input shaft 1800 RPM

100 Teeth

Intermediate shaft 900 RPM

200 Teeth

Output shaft 450 RPM

Figure 44.4 Double-reduction gearbox

V-belts

V-belts are common intermediate drives for fans, blowers, and other types of machinery. Unlike some other power-transmission mechanisms, V-belts generate unique forcing functions that must be understood and evaluated as part of a vibration analysis. The key monitoring parameters for V-belt-driven machinery are fault frequency and running speed.

Most of the forcing functions generated by V-belt drives can be attributed to the elastic or rubber band effect of the belt material. This elasticity is needed to provide the traction required transmitting power from the drive sheave (i.e., pulley) to the driven sheave. Elasticity causes belts to act like springs, increasing vibration in the direction of belt wrap, but damping it in the opposite direction. As a result, belt elasticity tends to accelerate wear and the failure rate of both the driver and driven unit.

Fault frequencies

Belt-drive fault frequencies are the frequencies of the driver, the driven unit, and the belt. In particular, frequencies at 1× the respective shaft speeds indicate faults with the balance, concentricity, and alignment of the sheaves. The belt frequency and its harmonics indicate problems with the belt. Table 44.1 summarizes the symptoms and causes of belt-drive failures, as well as corrective actions.

Running speeds

Belt-drive ratios may be calculated if the pitch diameters (see Figure 44.5) of the sheaves are known. This coefficient, which is used to determine the driven speed given

Table 44.1 Belt Drive Failure: Symptoms, Causes, and Corrective Actions

Symptom	Cause	Corrective action
High 1× rotational frequency in radial direction.	Unbalanced or eccentric sheave.	Balance or replace sheave.
High 1× belt frequency with harmonics. Impacting at belt frequency in waveform.	Defects in belt.	Replace belt.
High 1× belt frequency. Sinusoidal waveform with period of belt frequency.	Unbalanced belt.	Replace belt.
High 1× rotational frequency in axial plane. 1× and possibly 2× radial.	Loose, misaligned, or mismatched belts.	Align sheaves, retension or replace belts as needed.

Source: Integrated Systems, Inc.

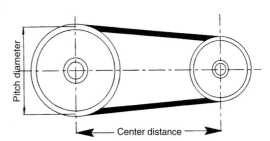

Pitch diameter

Center distance

Figure 44.5 Pitch diameter and center-to-center distance between belt sheaves

the drive speed, is obtained by dividing the pitch diameter of the drive sheave by the pitch diameter of the driven sheave. These relationships are expressed by the following equations:

$$\text{Drive reduction} = \frac{\text{Drive sheave diameter}}{\text{Driven sheave diameter}}$$

$$\text{Driven speed (rpm)} = \text{Drive speed (rpm)}$$
$$\times \left(\frac{\text{Drive sheave diameter}}{\text{Driven sheave diameter}}\right)$$

$$\text{Drive speed (rpm)} = \text{Driven speed (rpm)}$$
$$\times \left(\frac{\text{Driven sheave diameter}}{\text{Drive sheave diameter}}\right)$$

Using these relationships, the sheave rotational speeds can be determined. However, obtaining the other component speeds requires a bit more effort. The rotational speed of the belt cannot directly be determined using the information presented so far. To calculate belt rotational speed (rpm), the linear belt speed must first be determined by finding the linear speed (in/min) of the sheave at its pitch diameter. In other words, multiply the pitch circum-ference (PC) by the rotational speed of the sheave, where:

$$\text{Pitch circumference (in)} = \pi \times \text{Pitch diameter (in)}$$

$$\text{Linear speed (in/min)} = \text{Pitch circumference (in)}$$
$$\times \text{Sheave speed (rpm)}$$

To find the exact rotational speed of the belt (rpm), divide the linear speed by the length of the belt.

$$\text{Belt rotational speed (rpm)} = \frac{\text{Linear speed (in/min)}}{\text{Belt length (in)}}$$

To approximate the rotational speed of the belt, the linear speed may be calculated using the pitch diameters and the center-to-center distance (see Figure 44.5) between the sheaves. This method is accurate only if there is no belt sag. Otherwise, the belt rotational speed obtained using this method is slightly higher than the actual value.

In the special case where the drive and driven sheaves have the same diameter, the formula for determining the belt length is as follows:

$$\text{Belt length} = \text{Pitch circumference} + (2 \times \text{Center distance})$$

The following equation is used to approximate the belt length where the sheaves have different diameters:

$$\text{Belt length} = \frac{\text{Drive PC} + \text{Driven PC}}{2} + (2 \times \text{Center distance})$$

44.3 Driven components

This module cannot effectively discuss all possible com-binations of driven components that may be found in a plant. However, the guidelines provided in this section can be used to evaluate most of the machine-trains and process systems that are typically included in a microprocessor-based vibration-monitoring program.

44.3.1 Compressors

There are two basic types of compressors: (1) centrifugal and (2) positive displacement. Both of these major classi-fications can be further divided into sub-types, depending on their operating characteristics. This section provides an overview of the more common centrifugal and positive-displacement compressors.

Centrifugal

There are two types of commonly used centrifugal com-pressors: (1) in-line and (2) bullgear.

In-line

The in-line centrifugal compressor functions in exactly the same manner as a centrifugal pump. The only differ-ence between the pump and the compressor is that the compressor has smaller clearances between the rotor and casing. Therefore, in-line centrifugal compressors should be monitored and evaluated in the same manner as cen-trifugal pumps and fans. As with these driven components, the in-line centrifugal compressor is comprised of a single shaft with one or more impeller(s) mounted on the shaft. All components generate simple rotating forces that can be monitored and evaluated with ease. Figure 44.6 shows a typical in-line centrifugal compressor.

Bullgear

The bullgear centrifugal compressor (Figure 44.7) is a multi-stage unit that utilizes a large helical gear mounted on the compressor's driven shaft and two or more pin-ion gears, which drive the impellers. These impellers act

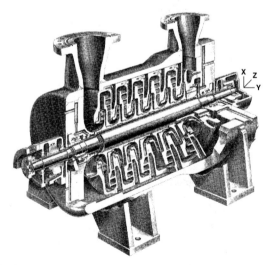

Figure 44.6 Typical in-line centrifugal compressor

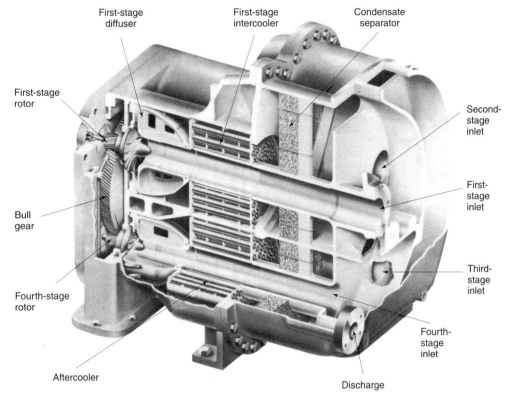

First-stage
diffuser

First-stage
intercooler

Condensate
separator

First-stage
rotor

Second-
stage
inlet

Bull
gear

First-
stage
inlet

Third-
stage
inlet

Fourth-stage
rotor

Fourth-
stage
inlet

Aftercooler

Discharge

Figure 44.7 Cut-away of bullgear centrifugal compressor

in series, whereby flow channels within the compressor's housing to the second-stage inlet direct compressed air or gas from the first-stage impeller discharge. The discharge of the second stage is channeled to the inlet of the third stage. This channeling occurs until the air or gas exits the final stage of the compressor.

Generally, the driver and bullgear speed is 3600 rpm or less and the pinion speeds are as high as 60,000 rpm (see Figure 44.8). These machines are produced as a package with the entire machine-train mounted on a common foundation that also includes a panel with control and monitoring instrumentation.

Positive displacement

Positive-displacement compressors, also referred to as dynamic-type compressors, confine successive volumes of fluid within a closed space. The pressure of the fluid increases as the volume of the closed space decreases.

Reciprocating

Reciprocating compressors are positive-displacement types having one or more cylinders. Each cylinder is fitted with a piston driven by a crankshaft through a connecting rod. As the name implies, compressors within

this classification displace a fixed volume of air or gas with each complete cycle of the compressor.

Reciprocating compressors have unique operating dynamics that directly affect their vibration profiles. Unlike most centrifugal machinery, reciprocating machines combine rotating and linear motions that generate complex vibration signatures.

Crankshaft frequencies

All reciprocating compressors have one or more crankshaft(s) that provide the motive power to a series of pistons, which are attached by piston arms. These crankshafts rotate in the same manner as the shaft in a centrifugal machine. However, their dynamics are somewhat different. The crankshafts generate all of the normal frequencies of a rotating shaft (i.e., running speed, harmonics of running speed, and bearing frequencies), but the amplitudes are much higher.

In addition, the relationship of the fundamental (1×) frequency and its harmonics changes. In a normal rotating machine, the 1× frequency normally contains between 60 and 70 per cent of the overall, or broadband, energy generated by the machine-train. In reciprocating machines, however, this profile changes. Two-cycle reciprocating machines, such as single-action compressors, generate a high second harmonic (2×) and multiples of the second

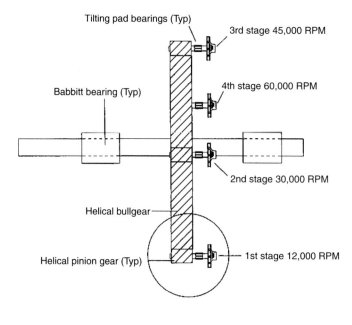

Tilting pad bearings (Typ)

3rd stage 45,000 RPM

4th stage 60,000 RPM

Babbitt bearing (Typ)

2nd stage 30,000 RPM

Helical bullgear

Helical pinion gear (Typ)

1st stage 12,000 RPM

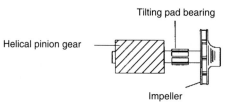

Tilting pad bearing

Helical pinion gear

Impeller

Figure 44.8 Internal bullgear drives' pinion gears at each stage

harmonic. While the fundamental (1×) is clearly present, it is at a much lower level.

Frequency shift due to pistons

The shift in vibration profile is the result of the linear motion of the pistons used to provide compression of the air or gas. As each piston moves through a complete cycle, it must change direction two times. This reversal of direction generates the higher second harmonic (2×) frequency component.

In a two-cycle machine, all pistons complete a full cycle each time the crankshaft completes one revolution. Figure 44.9 illustrates the normal action of a two-cycle, or single-action, compressor. Inlet and discharge valves are located in the clearance space and connected through ports in the cylinder head to the inlet and discharge connections.

During the suction stroke, the compressor piston starts its downward stroke and the air under pressure in the clearance space rapidly expands until the pressure falls below that on the opposite side of the inlet valve (Point B). This difference in pressure causes the inlet valve to open into the cylinder until the piston reaches the bottom of its stroke (Point C).

During the compression stroke, the piston starts upward, compression begins, and at point D has reached the same pressure as the compressor intake. The spring-loaded inlet valve then closes. As the piston continues upward, air is compressed until the pressure in the cylinder becomes great enough to open the discharge valve against the pressure of the valve springs and the pressure of the discharge line (Point E). From this point, to the end of the stroke (Point E to Point A), the air compressed within the cylinder is discharged at practically constant pressure.

The impact energy generated by each piston as it changes direction is clearly visible in the vibration profile. Since all pistons complete a full cycle each time the crankshaft completes one full revolution, the total energy of all pistons is displayed at the fundamental (1×) and second harmonic (2×) locations.

In a four-cycle machine, two complete revolutions (720°) are required for all cylinders to complete a full cycle.

Piston orientations

Crankshafts on positive-displacement reciprocating compressors have offsets from the shaft centerline that provide

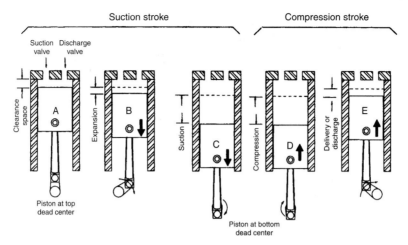

Figure 44.9 Two-cycle, or single-action, air compressor cylinders

the stroke length for each piston. The orientation of the offsets has a direct effect on the dynamics and vibration amplitudes of the compressor. In an opposed-piston compressor where pistons are 180° apart, the impact forces as the pistons change directions are reduced. As one piston reaches top dead center, the opposing piston also is at top dead center. The impact forces, which are 180° out-of-phase, tend to cancel or balance each other as the two pistons change directions.

Another configuration, called an unbalanced design, has piston orientations that are neither in-phase nor 180° out-of-phase. In these configurations, the impact forces generated as each piston changes direction are not balanced by an equal and opposite force. As a result, the impact energy and the vibration amplitude are greatly increased.

Horizontal reciprocating compressors (see Figure 44.10) should have X–Y data points on both the inboard and outboard main crankshaft bearings, if possible, to monitor the connecting rod or plunger frequencies and forces.

Screw

Screw compressors have two rotors with interlocking lobes and act as positive-displacement compressors (see Figure 44.11). This type of compressor is designed for baseload, or steady state, operation and is subject to extreme instability should either the inlet or discharge conditions change. Two helical gears mounted on the outboard ends of the male and female shafts synchronize the two rotor lobes.

Analysis parameters should be established to monitor the key indices of the compressor's dynamics and failure modes. These indices should include bearings, gear mesh, rotor passing frequencies, and running speed. However, because of its sensitivity to process instability and the normal tendency to thrust, the most critical monitoring parameter is axial movement of the male and female rotors.

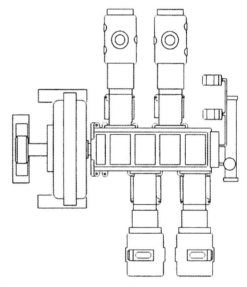

Figure 44.10 Horizontal, reciprocating compressor

Bearings

Screw compressors use both Babbitt and rolling-element bearings. Because of the thrust created by process instability and the normal dynamics of the two rotors, all screw compressors use heavy-duty thrust bearings. In most cases, they are located on the outboard end of the two rotors, but some designs place them on the inboard end. The actual location of the thrust bearings must be known and used as a primary measurement-point location.

Gear mesh

The helical timing gears generate a meshing frequency equal to the number of teeth on the male shaft multiplied

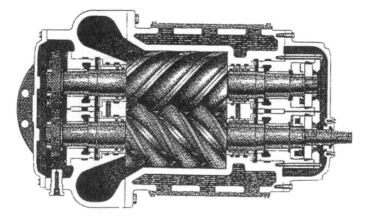

Figure 44.11 Screw compressor – steady-state applications only (Ref. 1)

by the actual shaft speed. A narrowband window should be created to monitor the actual gear mesh and its modulations. The limits of the window should be broad enough to compensate for a variation in speed between full load and no load.

The gear set should be monitored for axial thrusting. Because of the compressor's sensitivity to process instability, the gears are subjected to extreme variations in induced axial loading. Coupled with the helical gear's normal tendency to thrust, the change in axial vibration is an early indicator of incipient problems.

Rotor passing

The male and female rotors act much like any bladed or gear unit. The number of lobes on the male rotor multiplied by the actual male shaft speed determines the rotor-passing frequency. In most cases, there are more lobes on the female than on the male. To ensure inclusion of all passing frequencies, the rotor-passing frequency of the female shaft also should be calculated. The passing frequency is equal to the number of lobes on the female rotor multiplied by the actual female shaft speed.

Running speeds

The input, or male, rotor in screw compressors generally rotates at a no-load speed of either 1800 or 3600 rpm. The female, or driven, rotor operates at higher no-load

speeds ranging between 3600 to 9000 rpm. Narrowband windows should be established to monitor the actual running speed of the male and female rotors. The windows should have an upper limit equal to the no-load design speed and a lower limit that captures the slowest, or fully loaded, speed. Generally, the lower limits are between 15 and 20 per cent lower than no-load.

44.3.2 Fans

Fans have many different industrial applications and designs vary. However, all fans fall into two major categories: (1) centerline and (2) cantilever. The centerline configuration has the rotating element located at the mid-point between two rigidly supported bearings. The cantilever or overhung fan has the rotating element located outboard of two fixed bearings. Figure 44.12 illustrates the difference between the two fan classifications.

The following parameters are monitored in a typical predictive-maintenance program for fans: aerodynamic instability, running speeds, and shaft mode shape, or shaft deflection.

Aerodynamic instability

Fans are designed to operate in a relatively steady-state condition. The effective control range is typically 15 to 30

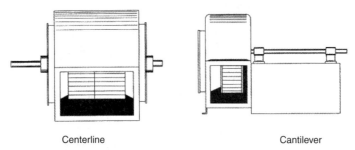

Centerline   Cantilever

Figure 44.12 Major fan classifications

per cent of their full range. Operation outside of the effective control range results in extreme turbulence within the fan, which causes a marked increase in vibration. In addition, turbulent flow caused by restricted inlet airflow, leaks, and a variety of other factors increases rotor instability and the overall vibration generated by a fan.

Both of these abnormal forcing functions (i.e. turbulent flow and operation outside of the effective control range) increase the level of vibration. However, when the instability is relatively minor, the resultant vibration occurs at the vane-pass frequency. As it become more severe, there also is a marked increase in the broadband energy.

A narrowband window should be created to monitor the vane-pass frequency of each fan. The vane-pass frequency is equal to the number of vanes or blades on the fan's rotor multiplied by the actual running speed of the shaft. The lower and upper limits of the narrowband should be set about 10 per cent above and below (± 10 per cent) the calculated vane-pass frequency. This compensates for speed variations and it includes the broadband energy generated by instability.

Running speeds

Fan running speed varies with load. If fixed filters are used to establish the bandwidth and narrowband windows, the running speed upper limit should be set to the synchronous speed of the motor, and the lower limit set at the full-load speed of the motor. This setting provides the full range of actual running speeds that should be observed in a routine monitoring program.

Shaft mode shape (shaft deflection)

The bearing-support structure is often inadequate for proper shaft support because of its span and stiffness. As a result, most fans tend to operate with a shaft that deflects from its true centerline. Typically, this deflection results in a vibration frequency at the second ($2\times$) or third ($3\times$) harmonic of shaft speed.

A narrowband window should be established to monitor the fundamental ($1\times$), second ($2\times$), and third ($3\times$) harmonic of shaft speed. With these windows, the energy associated with shaft deflection, or mode shape, can be monitored.

44.3.3 Generators

As with electric-motor rotors, generator rotors always seek the magnetic center of their casings. As a result, they tend to thrust in the axial direction. In almost all cases, this axial movement, or endplay, generates a vibration profile that includes the fundamental ($1\times$), second ($2\times$) and third ($3\times$) harmonic of running speed. Key monitoring parameters for generators include bearings, casing and shaft, line frequency, and running speed.

Bearings

Large generators typically use Babbitt bearings, which are non-rotating, lined metal sleeves (also referred to as fluid-film bearings) that depend on a lubricating film

to prevent wear. However, these bearings are subjected to abnormal wear each time a generator is shut off or started. In these situations, the entire weight of the rotating element rests directly on the lower half of the bearings. When the generator is started, the shaft climbs the Babbitt liner until gravity forces the shaft to drop to the bottom of the bearing. This alternating action of climb and fall is repeated until the shaft speed increases to the point that a fluid-film is created between the shaft and Babbitt liner.

Sub-harmonic frequencies (i.e., less than the actual shaft speed) are the primary evaluation tool for fluid-film bearings and they must be monitored closely. A narrowband window that captures the full range of vibration frequency components between electronic noise and running speed is an absolute necessity.

Casing and shaft

Most generators have relatively soft support structures. Therefore, they require shaft vibration monitoring measurement points in addition to standard casing measurement points. This requires the addition of permanently mounted proximity, or displacement, transducers that can measure actual shaft movement.

The third ($3\times$) harmonic of running speed is a critical monitoring parameter. Most, if not all, generators tend to move in the axial plane as part of their normal dynamics. Increases in axial movement, which appear in the third harmonic, are early indicators of problems.

Line frequency

Many electrical problems cause an increase in the amplitude of line frequency, typically 60 Hz, and its harmonics. Therefore, a narrowband should be established to monitor the 60, 120, and 180 Hz frequency components.

Running speed

Actual running speed remains relatively constant on most generators. While load changes create slight variations in actual speed, the change in speed is minor. Generally, a narrowband window with lower and upper limits of ± 10 per cent of design speed is sufficient.

Process rolls

Process rolls are commonly found in paper machines and other continuous process applications. Process rolls generate few unique vibration frequencies. In most cases, the only vibration frequencies generated are running speed and bearing rotational frequencies.

However, rolls are highly prone to loads induced by the process. In most cases, rolls carry some form of product or a mechanism that, in turn, carries a product. For example, a simple conveyor has rolls that carry a belt, which carries product from one location to another. The primary monitoring parameters for process rolls include bearings, load distribution, and misalignment.

Bearings

Both non-uniform loading and roll misalignment change the bearing load zones. In general, either of these failure modes results in an increase in outer-race loading. This is caused by the failure mode forcing the full load onto one quadrant of the bearing's outer race.

Therefore, the ball-pass outer-race frequency should be monitored closely on all process rolls. Any increase in this unique frequency is a prime indication of a load, tension, or misaligned roll problem.

Load distribution

By design, process rolls should be uniformly loaded across their entire bearing span (see Figure 44.13). Improper tracking and/or tension of the belt, or product carried by the rolls, will change the loading characteristics.

The loads induced by the belt increase the pressure on the loaded bearing and decrease the pressure on the unloaded bearing. An evaluation of process rolls should include a cross-comparison of the overall vibration levels and the vibration signature of each roll's inboard and outboard bearing.

Misalignment

Misalignment of process rolls is a common problem. On a continuous process line, most rolls are mounted in several levels. The distance between the rolls and the change in elevation make it extremely difficult to maintain proper alignment.

In a vibration analysis, roll misalignment generates a signature similar to classical parallel misalignment. It generates dominant frequencies at the fundamental (1×) and second (2×) harmonic of running speed.

44.3.4 Pumps

The wide variety of pumps used by industry can be grouped into two types: centrifugal and positive displacement. Pumps are highly susceptible to process-induced or installation-induced loads. Some pump designs are more likely to have axial- or thrust-induced load problems. Induced loads created by hydraulic forces also are a serious problem in most pump applications.

Recommended monitoring for each type of pump is essentially the same, regardless of specific design or manufacturer. However, process variables such as flow, pressure, load, etc. must be taken into account.

Centrifugal

Centrifugal pumps can be divided into two basic types: end-suction and horizontal split-case. These two major classifications can be broken further into single-stage and multi-stage. Each of these classifications has common monitoring parameters, but each also has unique features that alter their forcing functions and the resultant vibration profile. The common monitoring parameters for all centrifugal pumps include axial thrusting, vane-pass, and running speed.

Axial thrusting

End-suction and multi-stage pumps with in-line impellers are prone to excessive axial thrusting. In the end-suction pump, the centerline axial inlet configuration is the primary source of thrust. Restrictions in the suction piping, or low suction pressures, create a strong imbalance that forces the rotating element toward the inlet.

Multi-stage pumps with in-line impellers generate a strong axial force on the outboard end of the pump. Most of these pumps have oversized thrust bearings (e.g., Kingsbury bearings) that restrict the amount of axial movement. However, bearing wear caused by constant rotor thrusting is a dominant failure mode. Monitoring of the axial movement of the shaft should be done whenever possible.

Hydraulic instability (vane pass)

Hydraulic or flow instability is common in centrifugal pumps. In addition to the restrictions of the suction and

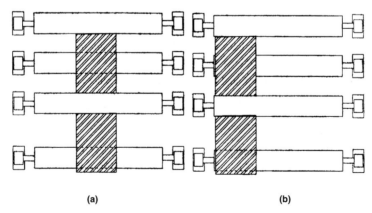

(a)   (b)

Figure 44.13 Rolls should be uniformly loaded

discharge discussed previously, the piping configuration in many applications creates instability. Although flow through the pump should be laminar, sharp turns or other restrictions in the inlet piping can create turbulent flow conditions. Forcing functions such as these result in hydraulic instability and displace the rotating element within the pump.

In a vibration analysis, hydraulic instability is displayed at the vane-pass frequency of the pump's impeller. Vane-pass frequency is equal to the number of vanes in the impeller multiplied by the actual running speed of the shaft. Therefore, a narrowband window should be established to monitor the vane-pass frequency of all centrifugal pumps.

Running speed

Most pumps are considered constant speed, but the true speed changes with variations in suction pressure and back-pressure caused by restrictions in the discharge piping. The narrowband should have lower and upper limits sufficient to compensate for these speed variations. Generally, the limits should be set at speeds equal to the full-load and no-load ratings of the driver.

There is a potential for unstable flow through pumps, which is created by both the design-flow pattern and the radial deflection caused by back-pressure in the discharge piping. Pumps tend to operate at their second-mode shape or deflection pattern. This mode of operation generates a unique vibration frequency at the second harmonic ($2\times$) of running speed. In extreme cases, the shaft may be deflected further and operate in its third ($3\times$) mode shape. Therefore, both of these frequencies should be monitored.

Positive displacement

There are a variety of positive-displacement pumps commonly used in industrial applications. Each type has unique characteristics that must be understood and monitored. However, most of the major types have common parameters that should be monitored.

With the exception of piston-type pumps, most of the common positive-displacement pumps utilize rotating elements to provide a constant-volume, constant-pressure output. As a result, these pumps can be monitored with the following parameters: hydraulic instability, passing frequencies, and running speed.

Hydraulic instability (vane pass)

Positive-displacement pumps are subject to flow instability, which is created either by process restrictions or by the internal pumping process. Increases in amplitude at the passing frequencies, as well as harmonics of both shafts' running speed and the passing frequencies, typically result from instability.

Passing frequencies

With the exception of piston-type pumps, all positive-displacement pumps have one or more passing frequencies generated by the gears, lobes, vanes, or wobble-plates used in different designs to increase the pressure of the pumped liquid. These passing frequencies can be calculated in the same manner as the blade or vane-passing frequencies in centrifugal pumps (i.e., multiplying the number of gears, lobes, vanes, or wobble plates times the actual running speed of the shaft).

Running speeds

All positive-displacement pumps have one or more rotating shafts that provide power transmission from the primary driver. Narrowband windows should be established to monitor the actual shaft speeds, which are in most cases essentially constant. Upper and lower limits set at ± 10 per cent of the actual shaft speed is usually sufficient.

44.4 Database development

Valid data are an absolute prerequisite of vibration monitoring and analysis. Without accurate and complete data taken in the appropriate frequency range, it is impossible to interpret the vibration profiles obtained from a machine-train.

This is especially true in applications that use microprocessor/computer-based systems. These systems require a database that specifies the monitoring parameters, measurement routes, analysis parameters, and a variety of other information. This input is needed to acquire, trend, store, and report what is referred to as 'conditioned' vibration data.

The steps in developing such a database are (1) collection of machine and process data and (2) database setup. Input requirements of the software are machine and process specifications, analysis parameters, data filters, alert/alarm limits, and a variety of other parameters used to automate the data-acquisition process.

44.4.1 Machine and process data collection

Database development can be accelerated and its accuracy improved by first creating detailed equipment and process information sheets that fully describe each machine and system to be monitored.

Equipment information sheets

The first step in establishing a database that defines the operating condition of each machine-train or production system is to generate an equipment information sheet (EIS) for each machine-train. The information sheet must contain all of the machine-specific data such as type of operation and information on all of the components that make up the machine-train.

Type of operation   The EIS should define the type of operation (i.e., constant speed or variable speed) that best describes the normal operation of each machine-train. This information allows the analyst to determine the best method of monitoring and evaluating each machine.

Constant-speed machinery   Few, if any, machines found in a manufacturing or production plant are truly constant speed. While the nameplate and specifications may indicate that a machine operates at a fixed speed, it will vary slightly in normal operation.

The reason for speed variations in constant-speed machinery is variation of process load. For example, a centrifugal pump's load will vary due to the viscosity of the fluid being pumped or changes in suction or discharge pressure. The pump speed will change because of these load changes.

As a rule, the speed variation in a constant-speed machine is about 15 per cent. For an electric motor, the actual variation can be determined by obtaining the difference between the amperage drawn under full-load and no-load. This difference, taken as a percentage of full-load amperage draws, provides the actual percentage of speed-range variation that can be expected.

Variable-speed machinery   For machinery and process systems that have a wide range of operating speeds, the data sheet should provide a minimum and maximum speed that can be expected during normal operation. In addition, a complete description of other variables (e.g., product type) that affect the machine's speed should be included. For example, a process line may operate at 500 feet per minute with product A and 1000 feet per minute with product B. Therefore, the data sheet must define all of the variables associated with both product A and product B.

Constant versus variable load   As with constant-speed machines, true constant-load machines are rare. For the few that may be specified as having constant load, there are factors that cause load changes to occur. These factors include variations in product, operating conditions, and ambient environment. These variations will have a direct, and often dramatic, impact on a machine's vibration profile.

Variations in load, no matter how slight, alter the vibration profile generated by a machine or system. The relationship between load and the vibration energy generated by a machine can be a multiple of four. In other words, a 10 per cent change in load may increase or decrease the vibration energy by 40 per cent.

When using vibration data as a diagnostic tool, you must always adjust or normalize the data to the actual load that was present when the data set was acquired.

Machine components   The EIS should provide information on all components (e.g., bearings, gears, gearboxes, electric motors, pumps, etc.) that make up the machine-train. Since these components generate vibration energy and unique frequency components, this information is essential for proper analysis. At a minimum, the information sheet must include detailed bearing information, passing frequencies, and nameplate data.

Bearings   All bearings in a machine-train must be identified. For example, rolling-element bearing data must include the manufacturer's part number, bearing geometry, and the unique rotational frequencies that it will generate. Rotational frequencies can be determined by using the bearing part number to look them up in the database that is included in most vibration monitoring software programs. They also can be obtained from the bearing vendor. Babbitt or sleeve bearing data must include type (e.g., plain, tilting-pad, etc.), as well as manufacturer and part number.

Passing frequencies   All components that generate a passing frequency must be included on the information sheet. Such components include: fan or compressor blades, vanes on pump impellers, rotor bars in electric motors, and gear teeth on both the pinion and bullgear in a gear set. The number of vanes, blades, and gear teeth must be recorded on the information sheet. The passing frequency is the number of vanes, blades, etc. times the rotation speed of the shaft on which they are mounted.

Nameplate data   Each machine-train component has a vendor's nameplate permanently attached to its housing. The EIS should include all nameplate data, including the serial number that uniquely identifies a machine or component.

Knowing the serial number allows detailed information on a machine or component to be obtained. This is possible because machinery manufacturers must maintain records for their products. These records, which are usually identified by serial number, contain complete design and performance data for that specific unit. For example, it is possible to obtain a performance curve or complete bill of materials for each pump found in a plant.

Process information sheets

A process information sheet (PIS) should be developed for each machine-train and production process that is to be included in a predictive-maintenance program. These data sheets should include all process variables that affect the dynamics and vibration profiles of the monitored components.

Many production and process systems handle a wide range of products. They typically have radically different machine and system operating parameters, as well as variable speeds and loads for each of the products they process.

Each process parameter directly affects both the machinery dynamics and the vibration profiles. For example, the line tension, strip width, and hardness of the incoming strip radically affect the vibration profile generated by a continuous process line in a steel mill. With few exceptions, process variations such as these must be considered in the vibration analysis.

44.4.2 Database setup

The input-data requirements and steps needed to set up the database for a computer-based vibration-monitoring program vary depending on the analyzer/software vendor and the system's capabilities. This section discusses the input required for such a database. However, this information should be used in conjunction with the vendor's user manual to ensure proper implementation.

The key elements of database setup discussed in this section are: analysis parameter sets, data filters (i.e., bandwidths, averaging, and weighting), limits for alerts and alarms, and data-acquisition routes.

Analysis parameter sets

The software used to manage the data incorporates what are referred to as analysis parameter sets (APSs). APSs define and specify machine dynamics, components, and failure modes to be monitored.

Most microprocessor-based systems permit a maximum of 256 APSs per database. This limit could be restrictive if the analyst wishes to establish a unique APS for each machine-train. To avoid this problem, APSs should be established for classes of machine-trains. For example, a group of bridle gearboxes that are identical in both design and application should share the same APS.

This approach provides two benefits. One is that it simplifies database development since one parameter set is used for multiple machine-trains. Therefore, less time is required to establish them. The other is that this approach permits direct comparison of multiple machine-trains. Since all machine-trains in a class share a common APS, the data can be directly compared. For example, the energy generated by a gear set is captured in a narrowband window established to monitor gear mesh. With the same APS, the gear mesh narrowbands can be used to compare all gear sets within that machine-train classification.

Data filters

When selecting the bandwidth frequency range to use for data collection in a vibration-monitoring system, one might be tempted to select the broadest range available. If enough computing power was available, we could simply gather data over an infinite frequency range, analyze the data, and be assured that no impending failures were missed. However, practicalities of limited computing power prevent us from taking this approach.

Therefore, it is necessary to 'filter' or screen the data we collect using our knowledge of the machinery being monitored. This is necessary to make collection, storage, and analysis of the data manageable with the equipment available. Electronic filters screen the quantity and quality of data that is collected. Mathematical filtering techniques such as resolution, averaging, and weighting are used on the data that is collected.

Bandwidth   Bandwidth frequency range settings are crucial to obtaining meaningful data for a vibration-based predictive-maintenance program. Because of limits inherent with computer-based data collection and analysis systems (i.e., limited storage and data-handling capacity), these settings must be properly specified to obtain frequency data in the range generated by machine components where failures occur. Improper settings will likely yield data in frequency ranges where problems do not exist and miss critical clues to serious problems with the machinery.

Analysis type   As discussed previously, data-collection analyzers incorporate analysis parameter sets that allow the user to control the data-gathering process. APSs provide the option of selecting either frequency analysis for fixed-speed machinery or orders analysis for variable-speed machinery.

Constant speed: frequency analysis   Constant-speed machinery generates a relatively fixed set of frequency components within its signature. Therefore, specific APSs can be established to monitor using frequency analysis. Since speed is relatively constant, the location of specific frequency components (e.g., running speed) will not change greatly. Therefore, the broadband and each narrowband window can be established with a constant minimum and maximum frequency limit, which are referred to as fixed filters.

The position of these fixed filters should be set to assure capture of information that is needed. The filter settings are determined from the speed range (i.e., no-load to full-load) of the primary driver. In addition, the lower and upper limits of each filter should be adjusted by 10 to 15 per cent to allow for slight variations in speed.

Variable speed: orders analysis   In a variable-speed machine, the unique frequencies generated by components such as bearings and gear sets do not remain constant. As the speed changes, the unique frequency components vary in direct proportion to the speed change. For this type of machinery, the analyzer's orders analysis option is used to automatically adjust each of the filters used to set the bandwidth and narrowbands for each data set to the true machine speed.

The analyzer automatically moves the filters that designate the lower and upper limits of each narrowband window to correspond with the actual running speed at the time the data are collected. To activate this function, the technician must either manually enter the running speed or use a tachometer input to trigger data acquisition.

Boundary conditions and resolution   The frequency boundary conditions and resolution for the full FFT signature depends on the specific system being used. Typically, the full-signature capability of various predictive-maintenance systems has a lower frequency limit of 10 Hz and an upper limit of 10 to 30 kHz. A few special low-frequency analyzers have a lower limit of 0.1 Hz, but retain the upper limit of 30 kHz. Typical resolutions are 100 to 12,800 lines.

Maximum frequency   The dynamics of each machine-train determines the maximum frequency, F_{MAX}, that should be used for both data acquisition and analysis. The frequency must be high enough to capture and display meaningful data, but not so high that resolution is lost and meaningful data filtered out. Generally, when setting F_{MAX}, it is necessary to take into account the harmonics of running speed and frequencies of components such as rolling-element bearings and gear mesh. Therefore, F_{MAX} should be set to the maximum frequency encountered in any of these.

Gear mesh The bandwidth should include at least the second harmonic of the calculated gear-mesh frequency. This permits early detection of misalignment, gear wear, and other abnormal operating dynamics. If excessive axial thrusting is expected, the third harmonic of gear mesh also should be included. For example, helical gears are prone to generate axial movement that increases as the gears wear.

Rolling-element bearings The ability to monitor rolling-element or anti-friction bearing defects requires the inclusion of multiples of their rotating frequency. For example, with ball-pass inner-race bearings, the bandwidth should include the second harmonic ($2\times$).

Running-speed harmonics When setting bandwidth, at least three harmonics of running speed should be included to ensure the ability to quantify the operating-mode shape of the shaft. This is accomplished by setting F_{MAX} to at least three times the running speed.

Generally, shafts deflect from their true centerline during normal as well as abnormal operation. In normal operation, this deflection is slight and results in a 'first mode' that creates a frequency component at the actual running speed, or first harmonic ($1\times$). As instability increases, the shaft deflects more and more. Under abnormal operating conditions, the shaft typically deforms into either the 'second mode' or 'third mode'. This deflection creates unique frequencies at the second ($2\times$) and third ($3\times$) harmonics of running speed. In addition, some failure modes (e.g., parallel misalignment) increase the energy level of one or all three of these frequencies.

Minimum frequency The data-collection hardware and software permit the selection of the minimum frequency, F_{MIN}, or the low-frequency cut-off below which no data are acquired and stored during the monitoring process. In most applications, however, this option should not be used.

Selecting a low-frequency cut-off does not improve resolution and is strictly an arbitrary omission of visible frequency components within a vibration signature. The FFT is calculated on a bandwidth having a lower limit of zero and an upper limit equal to the maximum frequency, F_{MAX}, which is selected by the user. The only reason for selecting a minimum frequency other than zero is to remove unneeded low-frequency components from the signature display.

Resolution Resolution is the degree of spacing of visible frequency components in the vibration signature and is proportional to the bandwidth. The equation for resolution is given below:

$$\text{Resolution} = \frac{\text{Bandwidth}}{\text{Lines of resolution}} = \frac{F_{MAX} - F_{MIN}}{\text{Lines of resolution}}$$

From this equation, it apparent that the bandwidth should be as small as possible to minimize the spacing and avoid missing important data. The typical number of lines of resolution is 100 to 12,800, but it is important to make sure that the selected bandwidth and resolution include all pertinent frequency components and that they are visible in the signature.

For routine monitoring, 800 lines of resolution are recommended. Higher resolution may be needed for root-cause analysis, but this requires substantially more memory in both the analyzer and host computer. While the latter is not a major problem, higher resolution reduces the number of measurement points that can be acquired with an analyzer without transferring acquired data to the host computer. This can greatly increase the time required to complete a measurement route and should be avoided when possible.

The combination of bandwidth and lines of resolution selected for each machine-train must effect separation of the unique frequency components that represent a machine's operating dynamics. Resolution can be improved by reducing F_{MAX}, increasing the lines of resolution, or a combination of both.

For example, if the analyzer can provide a 400-line FFT, the resolution of a signature taken with a bandwidth of 0 to 20,000 Hz will be 50 Hz, or 3,000 rpm, for each displayed line. The same 400-line FFT will provide a resolution of 2.5 Hz, or 150 rpm, with an F_{MAX} of 1,000 Hz.

It is important to remember that the first two and last two lines of resolution are lost when the FFT is calculated. In the example described above ($F_{MAX} = 1,000$ Hz), the first visible speed is 450 rpm and the highest visible speed is 59,700 rpm. Since the FFT always drops the first two and last two lines of resolution, the first visible frequency is three times the calculated resolution ($3 \times 150 = 450$) and the highest visible frequency is lowered by two lines (59,700 is visible, but 59,850 and 60,000 are not shown).

Narrowbands Analysis using narrowbands is based on specifying a series of filtered windows for a machine component or failure mode. The analyst can establish up to twelve narrowbands for each measurement point on each machine-train.

This concept reduces the manual analysis required for each data set. The analyst can scan the documentation that is generated (i.e., the exception report and trend charts) for each of the selected narrowbands to determine if further analysis is required.

Before implementing the analyzer's narrowband capability, the analyst should first understand the dynamics of each machine-train. Once this is done, establish narrowbands that bracket each of the bandwidths identifying each of the major components. At a minimum, establish a narrowband window around the following:

- Each primary running speed
- Each gearmesh frequency (including sidebands)
- Each set of bearing frequencies
- Each blade/vane-pass frequency
- Each belt frequency.

Constant-speed machinery If a machine-train operates at constant speed, the best method is to set the windows using the F_{MIN} and F_{MAX} frequencies associated with the specific component. For example, a narrowband window could be established to monitor the energy generated by a gear set by defining the minimum and maximum frequencies bounding the gear mesh. The bandwidth of the narrowband should be broad enough to include the modulations, or sidebands, generated by the meshing.

For example, a gear with 50 teeth generates a gear mesh at 50 times the running speed of its shaft ($50\times$). If the shaft turns at 100 rpm, the gear mesh frequency is 5000 cpm (also rpm). The modulations of a normal gear will occur at multiples of shaft speed (100 rpm). Therefore, in order to capture five sidebands on each side of the gear mesh frequency, the narrowband window should be established with filters set at 4500 cpm and 5500 cpm [$5000 \pm (5 \times 100)$].

In actual practice, the narrowband filters should be somewhat greater than those in the example. Since constant-speed machines tend to have a slight variation in speed due to load variations, the narrowbands should be adjusted to compensate for these variations. In the example given previously, the limit of the lower filter should be decreased by 10 per cent and the upper limit rose by 5 per cent to compensate for speed variation.

Variable-speed machinery   Variable-speed machine-train narrowband windows should be converted to their relationship to the running speed ($1\times$). For example, if the frequency of the ball-pass inner-race rolling-element bearing is calculated to be 5.9 times the primary shaft running speed, then the narrowband window should be set as $5.3\times$ to $6.2\times$. This allows the microprocessor to track the actual bearing rotational frequency regardless of the variation in running speed.

As a rule, the bandwidth of each narrowband should be just enough to capture the energy generated by the monitored component. Since orders analysis automatically adjusts the filters used to acquire narrowband energy data, these windows can be somewhat tighter than those in frequency analysis.

Anti-aliasing filters   Vibration data collected with a microprocessor-based analyzer can be filtered and conditioned to eliminate non-recurring events and their associated vibration profiles. Anti-aliasing filters are incorporated into data-collection analyzers specifically to remove spurious signals such as impacts. While the intent behind the use of anti-aliasing filters is valid, their use can distort a machine's vibration profile.

Averaging   All machine-trains are subject to random, non-recurring vibration as well as periodic vibration. Therefore, it is advisable to acquire several sets of data and average them to eliminate the spurious signals. Averaging also improves the repeatability of the data since only the continuous signals are retained.

Number of averages   Typically, a minimum of three samples should be collected for an average. However, the factor that determines the actual number is time. One sample takes three to five seconds, a four-sample average takes 12 to 20 seconds, and a 1000-sample average takes 50 to 80 minutes to acquire. Therefore, the final

determination is the amount of time that can be spent at each measurement point.

In general, three to four samples are acceptable for good statistical averaging and keeping the time required per measurement point within reason. Exceptions to this include low-speed machinery, transient-event capture, and synchronous averaging.

Overlap averaging   Many of the microprocessor-based vibration monitoring systems offer the ability to increase their data-acquisition speed. This option is referred to as overlap averaging.

While this approach increases speed, it is not generally recommended for vibration analysis. Overlap averaging reduces the accuracy of the data and must be used with caution. Its use should be avoided except where fast transients or other unique machine-train characteristics require an artificial means of reducing the data-acquisition and processing time.

When sampling time is limited, a better approach is to reduce or eliminate averaging altogether in favor of acquiring a single data block, or sample. This reduces the acquisition time to its absolute minimum. In most cases, the single-sample time interval is less than the minimum time required to obtain two or more data blocks using the maximum overlap-averaging sampling technique. In addition, single-sample data are more accurate.

Table 44.2 describes overlap-averaging options. Note that the approach described in this table assumes that the vibration profile of monitored machines is constant.

Table 44.2   Overlap Averaging Options

Overlap, %	Description
0	No overlap. Data trace update rate is the same as the block processing rate. This rate is governed by the physical requirements that are internally driven by the frequency range of the requested data.
25	Terminates data acquisition when 75% of each block of new data is acquired. The last 25% of the previous sample (of the 75%) will be added to the new sample before processing is begun. Therefore, 75% of each sample is new. As a result, accuracy may be reduced by as much as 25% for each data set.
50	The last 50% of the previous block is added to a new 50% or half-block of data for each sample. When the required number of samples is acquired and processed, the analyzer averages the data set. Accuracy may be reduced to 50%.
75	Each block of data is limited to 25% new data and the last 75% of the previous block.
90	Each block contains 10% new data and the last 90% of the previous block. Accuracy of average data using 90% overlap is uncertain. Since each block used to create the average contains only 10% of actual data and 90% of a block that was extrapolated from a 10% sample, the result cannot be representative of the real vibration generated by the machine-train.

Source: Integrated Systems, Inc

Window selection: weighting or signal conditioning The user can select the type of signal conditioning, or weighting used to display the vibration signature. For routine monitoring, the Hanning window should be selected. The flat-top window should be used for waterfall analysis, which modifies the profile of each frequency component so that the true amplitude is displayed.

Hanning The Hanning correction provides the best capture of the individual frequency components of a signature. However, this weighting factor may distort the actual amplitude of the frequency components. Nevertheless, it is used for routine monitoring using FFT analysis.

Flat-top Flat-top weighting provides the best representation of the actual amplitude of each frequency component. However, it may distort the actual location (i.e., frequency) of each component.

Flat-top weighting is useful when doing waterfall analysis. Even though the actual location of each frequency component may be slightly out of position, the profile is more visible when closely packed into a waterfall display. However, it is not normally used for single-channel FFT analysis.

Alert/alarm limits

All microprocessor-based predictive-maintenance systems are designed to automatically evaluate degradation of the machine-train by monitoring the change in amplitude of vibration using trending techniques. By establishing a series of alert/alarm limits in the database, the system can automatically notify the analyst that degradation is occurring. At least three levels of alert/alarms should be established: (1) low-limit alert, (2) maximum rate of change, and (3) absolute fault.

Low-limit alert The first alert (i.e., low-limit alerta) should be set at the lowest vibration amplitude that will be encountered from a normally operating machine-train. This value is needed to ensure that valid data are taken with the microprocessor. If this minimum amplitude is not reached, the system alerts the operator, who can retake or verify the data point. Low-limit selection is arbitrary, but should be set slightly above the noise floor of the specific microprocessor used to acquire data.

Maximum rate of change alert The second alert (i.e., maximum rate of change alert) is used to automatically notify the operator that based on statistical data; the rate of degradation has increased above the pre-selected norm. Since the vibration amplitudes of all machine-trains increase as normal wear occurs, the statistical rate of this normal increase should be trended. A drastic change in this rate is a major indication that a problem is developing.

The system should be able to establish the norm based on trends developed over time. However, the analyst must establish the level of deviation that triggers the alarm. The level of deviation in rate depends on the mechanical condition of each machine-train. For a new machine in good operating condition, the limit is typically set at two

times the norm. However, this must be adjusted based on the actual baseline of the machine. Note that it is better to set the limit too low initially and adjust it later.

Absolute-fault alarm The third limit (absolute-fault alarm) is the most critical of the alert/alarm limits. When this limit is reached, the probability of catastrophic failure within 1000 operating hours is greater than 90 per cent.

Absolute-fault limits are typically based on industrial standards for specific classifications of machinery. Generally, these standards are based on a filtered broadband limit and are not adjusted for variables such as speed, load or mounting configuration. However, vibration amplitude and its severity depend on speed and load. Therefore, alert/alarm limits must be adjusted for variations in both of these critical factors.

Affect of speed on limit Vibration-severity charts, such as the Rathbone chart illustrated in Figure 44.14, provide a basis for establishing the absolute-fault limit for machinery. Note, however, that the Rathbone chart does not adjust the maximum limit for speed – something that can cause a serious problem in most industrial applications.

A 2-mil shaft displacement at 600 rpm is acceptable for most applications, but the same displacement at 1800 rpm is considered severe. Therefore, alert/alarm limits must be established based on the actual speed range of each machine-train. When casing severity is used (i.e., data are taken from the bearing caps rather than actual shaft displacement), the limits can be grouped into three basic speed ranges: less than 299 rpm, 300 to 1199 rpm, and 1200 through 3600 rpm. Tables 44.3 through 44.5 provide the alert/alarm limits for each speed range.

Effect of load on limit Load has a direct impact on the vibration energy generated by a machine-train. For example, a centrifugal compressor operating at full load will have a lower level of vibration than the same compressor operating at 50 per cent load. This change in vibration energy is the direct result of a corresponding change in the spring stiffness of the rotating element under varying load conditions.

The rated, or design, load of a machine establishes the following elements: (1) spring constant, (2) stiffness of the rotating element, and (3) damping coefficient of its support system. Therefore, when load varies from design,

Table 44.3 Alarm Limits for 1200 rpm and Higher

Bandwidth	Alert	Alarm	Absolute Fault
Overall	0.15 ips-peak	0.30 ips-peak	0.628 ips-peak
1× narrowband	0.10 ips-peak	0.20 ips-peak	0.40 ips-peak
2× narrowband	0.05 ips-peak	0.10 ips-peak	0.20 ips-peak
3× narrowband	0.04 ips-peak	0.08 ips-peak	0.15 ips-peak
1× gear mesh	0.05 ips-peak	0.10 ips-peak	0.2 ps-peak
Rolling-element bearing	0.05 ips-peak	0.10 ips-peak	0.2 ps-peak
Blade/vane pass	0.05 ips-peak	0.10 ips-peak	0.2 ps-peak

Source: Integrated Systems, Inc.

Vibration frequency (cpm)

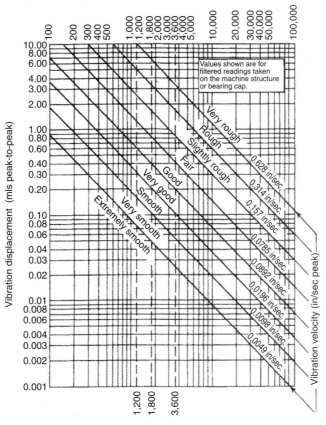

Figure 44.14 Rathbone Chart

Table 44.4 Alarm Limits for 300 rpm to 1199 rpm

Bandwidth	Alert	Alarm	Absolute Fault
Overall	0.10 ips-peak	0.15 ips-peak	0.30 ips-peak
1× narrowband	0.05 ips-peak	0.10 ips-peak	0.20 ips-peak
2× narrowband	0.02 ips-peak	0.05 ips-peak	0.10 ips-peak
3× narrowband	0.01 ips-peak	0.03 ips-peak	0.06 ips-peak
1× gear mesh	0.02 ips-peak	0.04 ips-peak	0.08 ips-peak
Rolling-element bearing	0.03 ips-peak	0.05 ips-peak	0.10 ips-peak
Blade/vane pass	0.03 ips-peak	0.05 ips-peak	0.10 ips-peak

Source: Integrated Systems, Inc.

Table 44.5 Alarm Limits for 300 rpm and Below

Bandwidth	Alert	Alarm	Absolute Fault
Overall	0.05 ips-peak	0.10 ips-peak	0.20 ips-peak
1× narrowband	0.03 ips-peak	0.06 ips-peak	0.12 ips-peak
2× narrowband	0.01 ips-peak	0.03 ips-peak	0.06 ips-peak
3× narrowband	0.01 ips-peak	0.02 ips-peak	0.04 ips-peak
1× gear mesh	0.02 ips-peak	0.03 ips-peak	0.05 ips-peak
Rolling-element bearings	0.02 ips-peak	0.03 ips-peak	0.05 ips-peak
Blade/vane pass	0.02 ips-peak	0.03 ips-peak	0.05 ips-peak

Source: Integrated Systems, Inc.

the stiffness of the rotor and the rotor-support system also must change.

This change in vibration energy can be clearly observed in trend data acquired from machine-trains. A sawtooth trend is common to most predictive-maintenance programs, which can be directly attributed to variations in load. The only way to compensate for load variations is to track the actual load associated with each data set.

Mounting configuration and operating envelope   Industrial standards, such as the rathbone severity chart, assume that the machine is rigidly mounted on a suitable concrete foundation. Machines mounted on deck-plate or on flexible foundations have higher normal vibration profiles and cannot be evaluated using these standards.

In addition, industrial standards assume a normal operating envelope. All machines and process systems have a finite operating range and must be operated accordingly. Deviations from either the design operating envelope or from best operating practices will adversely affect the operating condition and vibration level of the machine.

Data-acquisition routes

Most computer-based systems require data-acquisition routes to be established as part of the database setup. These routes specifically define the sequence of measurement points and, typically, a route is developed for each area or section of the plant. With the exception of limitations imposed by some of the vibration monitoring systems, these routes should define a logical walking route within a specific plant area. A typical measurement is shown in Figure 44.15.

Most of the computer-based systems permit rearrangement of the data-acquisition sequence in the field. They provide the ability to skip through the route until the appropriate machine, or the measurement point of a machine, is located. However, this manual adjustment of the pre-programmed route is time consuming and should be avoided whenever possible.

44.5 Vibration data acquisition

Limitations on data acquisition arise due to the use of portable hand-held, microprocessor-based analyzers to obtain data. Limits are cost, weight restrictions, and the fact that they are generally designed for a technician to manually take a series of single measurements directly from individual machine-trains or machine-train components. Therefore, this discussion is limited to the best practices for acquiring single-channel, frequency-domain data using portable, hand-held analyzers. It does not address multi-channel or other non-serial data-acquisition techniques.

Data-acquisition and vibration-detection equipment (i.e., analyzers and transducers) are critical factors that determine the success or failure of a vibration monitoring and analysis program. Their accuracy, proper usage, and mounting determine if valid data are collected. An optimum program is based on the accuracy and repeatability of the data, both of which are negatively affected by the use of the wrong transducer or mounting technique.

44.5.1 Transducers

Three basic types of transducers that can be used for monitoring the mechanical condition of plant machinery: displacement probes (measures movement), velocity transducers (measures energy due to velocity), and accelerometers (measures force due to acceleration). Each has specific applications in a monitoring program, while each also has limitations.

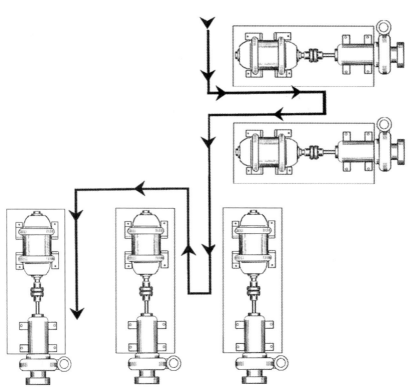

Figure 44.15 Typical measurement route

44.5.2 Measurement orientation

Most vibration-monitoring programs rely on data acquired from machine housings or bearing caps. The only exceptions are applications that require direct measurement of actual shaft displacement to gain an accurate picture of the machine's dynamics. Transducers used to acquire the data are mounted either radially or axially.

44.5.3 Measurement locations

Each measurement point, typically located on the bearing housing or machine casing, should provide the shortest direct mechanical link to the shaft. Figure 44.16 illustrates such a location oriented in both the axial and radial planes. If a transducer is not mounted in an appropriate location, the data will be distorted by noise such as fluid flow in the bearing reservoir or through the machine.

Measurement locations should be permanently marked to ensure repeatability of data. If transducers are permanently mounted, the location can be marked with a center punch, paint, or any other method that identifies the point. The following sections give the recommended locations and orientations of measurement points for the following common machines or machine components: compressors, electric motors, fans and blowers, gearboxes, process rolls, and pumps.

Compressors

In most cases, measurement-point locations for compressors are identical to those of pumps and fans. If a compressor is V-belt driven, the belts on the inboard and outboard bearings should in a plane opposing the sideload create the primary measurement point. The secondary point should be at 90° to the primary point.

At least one axial measurement point should be located on each compressor shaft. Axial data are helpful in identifying and quantifying thrust (i.e., induced) loads created by both the process and any potential compressor-element problems, such as imbalance, cracked blade, etc.

In applications where numerous compressors are in close proximity, an additional measurement point on the base is useful for identifying structural resonance or crosstalk between the units.

Centrifugal There are two major types of centrifugal compressors used in industrial applications: in-line and bullgear.

In-line centrifugal Measurement locations for in-line centrifugal compressors should be based on the same logic as discussed for pumps. Impeller design and orientation, as well as the inlet and discharge configurations, are the dominant reasons for point location. Figure 44.17 illustrates a typical multi-stage, in-line compressor.

The in-line impeller configuration generates high axial thrusting, which increases the importance of the axial (Z-axis) measurement point. That point should be on the fixed bearing and oriented toward the driver.

In addition, this type of compressor tends to have both the suction and discharge ports on the same side of the compressor's housing. As a result, there is a potential for aerodynamic instability within the compressor. Orientation of the primary (X-axis) radial measurement point should be opposite the discharge port and oriented toward the discharge. The secondary (Y-axis) radial point should be in the direction of shaft rotation and 90° from the primary radial point.

Bullgear compressors

Because of the large number of these machines being manufactured, proper locations for displacement, or proximity, the various machine manufacturers have established probes. Nearly all of these compressors are supplied by the original-equipment manufacturer (OEM) with one or

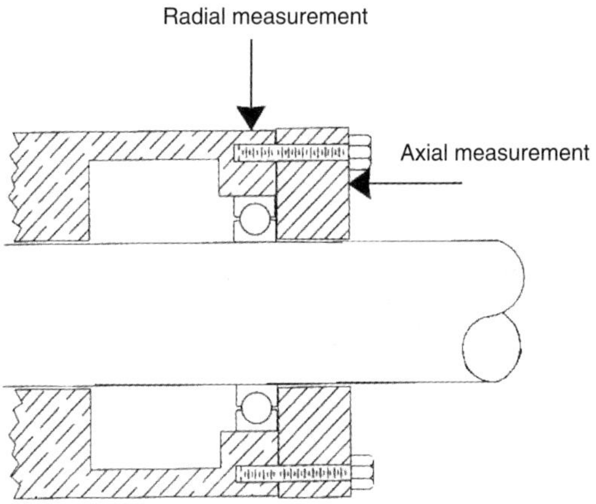

Figure 44.16 Measurement points should provide shortest direct mechanical link to shaft

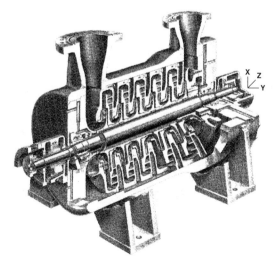

Figure 44.17 Multi-stage in-line compressors

two proximity probes already mounted on each pinion shaft and, sometimes, one probe on the bullgear shaft. These probes can be used to obtain vibration data with the microprocessor-based portable analyzers. However, they must be augmented with casing measurements acquired from suitable accelerometers. This is necessary because there are two problems with the proximity data.

First, most of the OEM-supplied data-acquisition systems perform signal conditioning on the raw data acquired from the probes. If the conditioned signal is used, there is a bias in the recorded amplitude. This bias may increase the raw-data signal by 30 to 50 per cent. If data from the proximity probes are to be used, it is better to acquire it before signal conditioning. This can be accomplished by tapping into the wiring between the probe and display panel.

The second problem is data accuracy. The pinions on most bullgear compressors rotate at speeds between 20,000 and 75,000 rpm. While these speeds, for the most part, are within the useful range of a proximity probe (600 to 60,000 rpm), the frequencies generated by common components (i.e., tilting-pad bearings and impeller vane-pass) are well outside this range. In addition, proximity probes depend on a good sight picture, which means a polished shaft that has no end-play or axial movement. Neither of these conditions are present in a bullgear compressor.

Primary (X-axis) and secondary (Y-axis) radial measurements should be acquired from both bearings on the bullgear shaft. If the shaft has Babbitt bearings, it is a good practice to periodically acquire four radial readings, one at each quadrant of the bearing, to determine the load zones of the bearing. Normal vertical and horizontal locations are acceptable for the routine readings, but primary (X-axis) measurement points should be in the horizontal plane (i.e., 90° from vertical in the direction of rotation). For clockwise rotation, the primary should be on the right side and for counterclockwise on the left.

Because a bullgear compressor incorporates a large helical gear, the shaft displays moderate to high axial thrusting. Therefore, an axial (Z-axis) measurement point should be acquired from the thrust (outboard) bearing oriented toward the driver.

The pinion shafts in this type of compressor are inside the housing. As a result, it is difficult to obtain radial measurements directly. A cross-sectional drawing of the compressor is required to determine the best location and orientation for the measurement points.

Positive displacement Two major types of positive-displacement compressors are used in industrial applications: reciprocating and screw.

Reciprocating Limitations of the frequency-domain analysis prevent total analysis of reciprocating compressors. It is limited to the evaluation of the rotary forces generated by the main crankshaft. Therefore, time-domain and phase analysis are required for complete diagnostics.

The primary (X-axis) radial measurement point should be located in a plane opposite the piston and cylinder. Its orientation should be toward the piston's stroke. This orientation provides the best reading of the impacts and vibration profile generated by the reversing linear motion of the pistons. The secondary (Y-axis) radial measurement point should be spaced at 90° to the primary point and in the direction of rotation at the main crankshaft. This configuration should be used for all accessible main crankshaft bearings. Figure 44.18 provides a typical cross-section of a reciprocating compressor, which will assist in locating the best measurement points. Similar drawings are available for most compressors and can be obtained from the vendor. There should be little axial thrusting of the main crankshaft, but an axial (Z-axis) measurement point should be established on the fixed bearing, oriented toward the driver.

If the vibration analyzer permits acquisition of time-domain data, additional time-waveform data should be obtained from the intermediate guide as well as the inlet and discharge valves. The intermediate guide is located

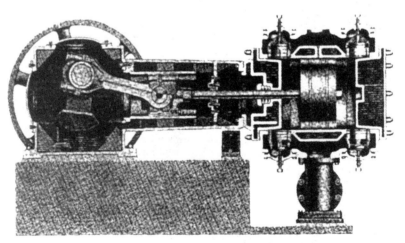

Figure 44.18 Typical cross-section of a reciprocating compressor

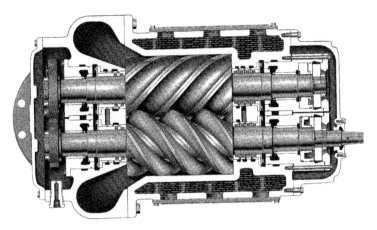

Figure 44.19 Single-stage screw compressor

where the main crankshaft lever arm connects to the piston rod. Time waveforms from these locations detect any binding or timing problems than may exist in the compressor.

Screw   Figure 44.19 illustrates a typical single-stage screw compressor. Radial measurements should be acquired from all bearing locations in the compressor. The primary bearing locations are the inboard, or float, bearings on the driver side of the compressor housing and the fixed bearing located on the outboard end of each shaft.

In most cases, the outboard bearings are not directly accessible and measurement points must be located on the compressor's casing. Extreme care must be taken to ensure proper positioning. A cross-sectional drawing facilitates selection of the best, most direct mechanical link to these bearings.

The primary (X-axis) radial measurement point should be located opposite the mesh of the rotors and oriented toward the mesh. In the illustration, the primary point is on the top of the housing and oriented in the downward direction. The secondary (Y-axis) radial measurement point should be in the direction of rotation and 90° from the primary.

Because of the tendency for screw compressors to generate high axial vibration when subjected to changes in process conditions, the axial (Z-axis) measurement point is essential. The ideal location for this point is on the outboard, or fixed, bearing and oriented toward the driver. Unfortunately, this is not always possible. The outboard bearings are fully enclosed within the compressor's housing and an axial measurement cannot be obtained at these points. Therefore, the axial measurement must be acquired from the float, or inboard, bearings. While this position captures the axial movement of the shaft, the recorded levels are lower than those acquired from the fixed bearings.

Electric motors

Both radial (X- and Y-axis) measurements should be taken at the inboard and outboard bearing housings.

The anticipated induced load created by the driven units determines orientation of the measurements. The primary (X-axis) radial measurement should be positioned in the same plane as the worst anticipated shaft displacement. The secondary (Y-axis) radial should be position at 90° in the direction of rotation to the primary point and oriented to permit vector analysis of actual shaft displacement.

Horizontal motors rely on a magnetic center generated by its electrical field to position the rotor in the axial (Z-axis) plane between the inboard and outboard bearings. Therefore, most electric motors are designed with two float bearings instead of the normal configuration incorporating one float and one fixed bearing. Vertical motors should have an axial (Z-axis) measurement point at the inboard bearing nearest the coupling and oriented in an upward direction. This data point monitor the downward axial force created by gravity or an abnormal load.

Electric motors are not designed to absorb side loads, such as those induced by V-belt drives. In applications where V-belts or other radial loads are placed on the motor, the primary radial transducer (X-axis) should be oriented opposite the direction of induced load and the secondary radial (Y-axis) point should be positioned at 90° in the direction of rotation. If, for safety reasons, the primary transducer cannot be positioned opposite the induced load, the two radial transducers should be placed at 45° on either side of the load plane created by the side load.

Totally enclosed, fan-cooled, and explosion-proof motors create some difficulty in acquiring data on the outboard bearing. By design, the outboard bearing housing is not accessible. The optimum method of acquiring data is to permanently mount a sensor on the outboard-bearing housing and run the wires to a convenient data-acquisition location. If this is not possible, the X–Y data points should be as close as possible to the bearing housing. Ensure that there is a direct mechanical path to the outboard bearing. The use of this approach results in some loss of signal strength from motor-mass damping. Do not obtain data from the fan housing.

Fans and blowers

If a fan is V-belt driven, the belts on the inboard and outboard bearings should in a plane opposing the side-load create the primary measurement point. The secondary point should be at 90° degrees to the primary in the direction of rotation.

Bowed shafts caused by thermal and mechanical effects create severe problems on large fans, especially overhung designs. Therefore, it is advantageous to acquire data from all four quadrants of the outboard bearing housing on overhung fans to detect this problem.

At least one axial measurement point should be located on each fan shaft. This is especially important on fans that are V-belt driven. Axial data are helpful in identifying and quantifying thrust (induced) loads created by the process and any potential fan element problems such as imbalance, cracked blade, etc.

In applications where numerous fans are in close proximity, an additional measurement point on the base is useful for identifying structural resonance or cross-talk between the fans.

Gearboxes

Gearbox measurement-point orientation and location should be configured to allow monitoring of the normal forces generated by the gear set. In most cases, the separating force, which tends to pull the gears apart, determines the primary radial measurement-point location. For example, a helical gear set generates a separating force that is tangential to a centerline drawn through the pinion and bullgear shafts. The primary (X-axis) radial measurement point should be orientated to monitor this force and a secondary (Y-axis) radial should be located at 90° to the primary. The best location for the secondary (Y-axis) radial is opposite the direction of rotation. In other words, the secondary leads the primary transducers.

With the exception of helical gears, most gear sets should not generate axial or thrust loads in normal operation. However, at least one axial (Z-axis) measurement point should be placed on each of the gear shafts. The axial point should be located at the fixed, or thrust, bearing cap and oriented toward the gearbox.

In complex gearboxes, it may be difficult to obtain radial measurements from the intermediate or idler shafts. In most cases, these intermediate shafts and their bearings are well inside the gearbox. As a result, direct access to the bearings is not possible. In these cases, the only option is to acquire axial (Z-axis) readings through the gearbox housing. A review of the cross-sectional drawings allows the best location for these measurements to be determined. The key is to place the transducer at a point that will provide the shortest, direct link to the intermediate shaft.

Process rolls

Process rolls are widely used by industry. As with other machine components, two radial (X- and Y-axis) and one axial (Z-axis) measurements should be acquired from each roll. However, the orientation of these measurement points is even more critical for process rolls than for some of the other machine components.

The loading on each roll is generated by the belt, wire mesh, and/or transported product. The amount and distribution of the load varies depending on the wrap of the carried load. Wrap refers to the angular distance around the roll that touches the belt, wire mesh, or product. In most conveyor systems, the load is relatively uniform and is in a downward direction. In this case, the traditional vertical, horizontal, and axial mounting positions are acceptable.

Figure 44.20 represents a typical process-roll configuration. The arrows indicate the force vectors generated by the wire, belt, or product wrap around these rolls. The left roll has a force vector at 45° down to the left; the right roll has a mirror image force vector; and the bottom roll has a vertical vector.

The primary (X-axis) radial measurement for the bottom roll should be in the vertical plane with the transducer mounted on top of the bearing cap. The secondary radial (Y-axis) measurement should be in the horizontal plane facing upstream of the belt. Since the belt carried by the roll also imparts a force vector in the direction of travel, this secondary point should be opposite the direction of belt travel.

The ideal primary (X-axis) point for the top right roll is opposite the force vector. In this instance, the primary radial measurement point should be located on the right of the bearing cap facing upward at a 45° angle. Theoretically, the secondary (Y-axis) radial point should be at 90° to the primary on the bottom-left of the bearing cap. However, it may be difficult, if not impossible, to locate and access a measurement point here. Therefore, the next best location is at 45° from the anticipated force vector on the left of the bearing cap. This placement still provides the means to calculate the actual force vector generated by the product.

Pumps

Appropriate measurement points vary by type of pump. In general, pumps can be classified as centrifugal or positive displacement, and each of these can be divided into groups.

Centrifugal

The location of measurement points for centrifugal pumps depends on whether the pump is classified as end suction or horizontal split-case.

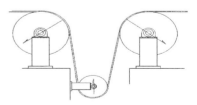

Figure 44.20 Typical process-roll configuration and wrap-force vectors

End suction   Figure 44.21 illustrates a typical single-stage, end-suction centrifugal pump. The suction inlet is on the axial centerline, while the discharge may be either horizontal or vertical. In the illustration, the actual discharge is horizontal and is flanged in the vertical downstream. The actual discharge orientation determines the primary radial (X) measurement point. This point must be oriented in the same plane as the discharge and opposite the direction of flow. In the illustration, the primary point should be in the horizontal plane facing the discharge.

Restrictions or other causes of back-pressure in the discharge piping deflect the shaft in the opposite direction. Referring back to the illustration, the shaft would be deflected toward the front of the picture. If the discharge were vertical and in the downward direction, the primary radial measurement point would be at the top of the pump's bearing cap looking downward.

A second radial (Y-axis) measurement point should be positioned at 90° to the primary in a plane that captures secondary shaft deflection. For the pump illustrated in Figure 44.21, the secondary (Y-axis) radial measurement point is located on top of the pump's bearing cap and oriented downward. Since the pump has a clockwise rotation, back-pressure in the discharge piping forces the shaft both downward and horizontally toward the front of the picture.

Since this type of pump is susceptible to axial thrusting, an axial (Z-axis) measurement point is essential. This point should be on the fixed bearing housing oriented toward the driver.

Horizontal split-case   The flow pattern through a horizontal split-case pump is radically different than that through an end-suction pump. Inlet and discharge flow are in the same plane and almost directly opposite one another. This configuration, illustrated in Figure 44.22,

greatly improves the hydraulic-flow characteristics within the pump and improves its ability to resist flow-induced instability.

The location of the primary (X-axis) radial measurement point for this type of pump is in the horizontal plane and on the opposite side from the discharge. The secondary (Y-axis) radial measurement point should be 90° to the primary point and in the direction of rotation. If the illustrated pump has a clockwise rotation, the measurement point should be on top, oriented downward. For a counter-clockwise rotation, it should be on bottom, oriented upward.

Single-stage pumps generate some axial thrusting due to imbalance between the discharge and inlet pressures. The impeller design provides a means of balancing these forces, but it cannot absolutely compensate for the difference in the pressures. As a result, there will be some axial rotor movement. In double volute, or multi-stage, pumps, two impellers are positioned back to back. This configuration eliminates most of the axial thrusting when the pump is operating normally.

An axial (Z-axis) measurement point should be located on the fixed bearing housing. It should be oriented toward the driver to capture any instability that may exist.

Multi-stage   Multi-stage pumps may be either end-suction or horizontal split-case pumps. They have two basic impeller configurations, in-line or opposed, as shown in Figure 44.23. In-line impellers generate high thrust loads.

The impeller configuration does not alter the radial measurement locations discussed in the preceding sections. However, it increases the importance of the axial (Z-axis) measurement point. The in-line configuration drastically increases the axial loading on the rotating element and, therefore, the axial (Z-axis) measurement

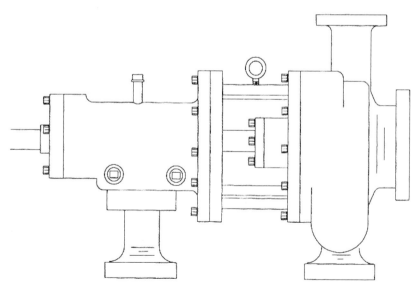

Figure 44.21 Typical end-suction, single-stage centrifugal pump

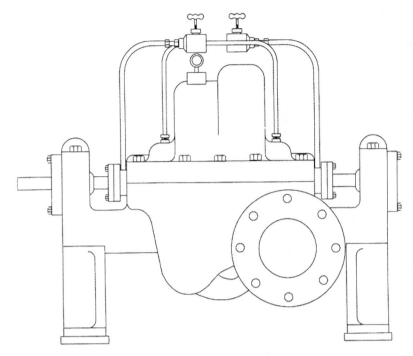

Figure 44.22 Typical horizontal split-case pump

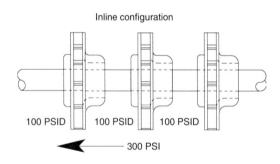

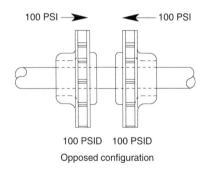

Figure 44.23 In-line and opposed impellers on multi-stage pumps

point is critical. Obviously, this point must be in a location that detects axial movement of the shaft. However, since large, heavy-duty fixed bearings are used to withstand the high thrust loading generated by this design, direct measurement is difficult. A cross-sectional drawing of the pump may be required to locate a suitable location for this measurement point.

Positive displacement

Positive-displacement pumps can be divided into two major types: rotary and reciprocating. All rotary pumps use some form of rotating element, such as gears, vanes, or lobes to increase the discharge pressure. Reciprocating pumps use pistons or wobble plates to increase the pressure.

Rotary   Locations of measurement points for rotary positive-displacement pumps should be based on the same logic as in-line centrifugal pumps. The primary (X-axis) radial measurement should be taken in the plane opposite the discharge port. The secondary (Y-axis) radial should be at 90° to the primary and in the direction of the rotor's rotation.

Since most rotary positive-displacement pumps have inlet and outlet ports in the same plane and opposed, there should be relatively little axial thrusting. However, an axial (Z-axis) measurement should be acquired from the fixed bearing, oriented toward the driver.

Reciprocating   Reciprocating pumps are more difficult to monitor because of the combined rotational and linear motions that are required to increase the discharge pressure. Measurement-point location and orientation should be based on the same logic as that of reciprocating compressors.

44.6 Trending analysis

Long-term vibration trends are a useful diagnostic tool. Trending techniques involve graphically comparing the total energy, which is the sum of the frequency components' amplitude over some consistent, user-selected frequency range (i.e., F_{MIN} to F_{MAX}), over a long period to get a historical perspective of the vibration pattern. Plots of this sum against time (e.g., days) provide a means of quantifying the relative condition of the monitored machine (see Figure 44.24). Most predictive-maintenance systems provide automatic-trending capabilities for recorded data. This is not to be confused with time-domain plots, which are instantaneous measures of total vibration amplitude plotted against time measured in seconds.

Used properly, this feature greatly enhances a predictive-maintenance program. The real value of trending techniques is that they provide the capability of automatically scanning large amounts of data (both broadband and narrowband) and reporting any change in pre-selected values.

44.6.1 Types of trends

There are three primary categories of trends: broadband, narrowband, and combinations of the two.

Broadband

Most microprocessor-based vibration-monitoring systems acquire and record a filtered broadband energy level for each data point included in the program. The bandwidth of the energy band is determined by the minimum, F_{MIN}, and maximum, F_{MAX}, frequencies that were established as part of the database setup. In most applications, the minimum frequency should be zero, but the maximum varies, depending on the specific machine-train. Figure 44.25 illustrates typical broadband data.

Broadband data cannot be used to identify specific machine components (e.g., bearing, gears, etc.) or failure modes (e.g., imbalance, misalignment, etc.). The data acquired using broadband filters are limited to the total energy value contained within the user-selected frequency window or bandwidth, F_{MIN} to F_{MAX}.

At best, broadband energy provides a gross approximation of the machine's condition and its relative rate of degradation. Since the only available data are overall energy values, broadband data do not provide enough detail to permit diagnosis of machine-train condition. Without discrete identification of the specific frequencies that make up the overall energy, the failure mode or failing component cannot be determined.

Narrowband

Like broadband data, narrowband data also reflect the total energy, but it reflects a more restricted user-selected range or window. Narrowband monitoring generally is used to trend and evaluate one selected machine-train component rather than several. Filtered narrowband windows are typically set up around the unique frequency components generated by specific machine-train components so that the energy in each filtered window can be directly attributed to that specific machine component. However, even though narrowband analysis improves the diagnostic capabilities of a predictive-maintenance program, it is not possible to isolate and identify specific failure modes within a machine-train. Figure 44.26 illustrates the added information provided by narrowband data.

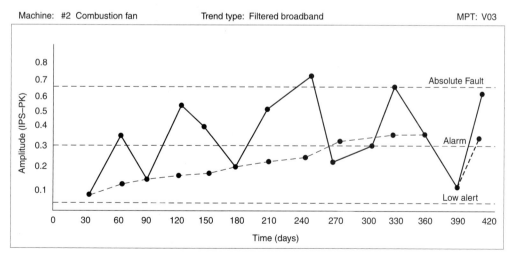

Figure 44.24 Trend data are plotted versus time and provide historical trends

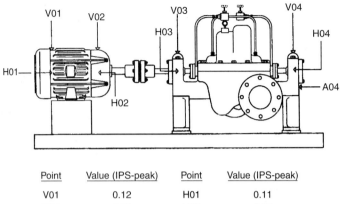

Point	Value (IPS-peak)	Point	Value (IPS-peak)
V01	0.12	H01	0.11
V02	0.15	H02	0.12
V03	0.15	H03	0.10
V04	0.12	H04	0.09
A04	0.14		

Figure 44.25 Typical broadband measurements

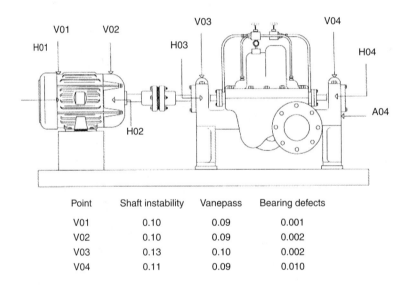

Point	Shaft instability	Vanepass	Bearing defects
V01	0.10	0.09	0.001
V02	0.10	0.09	0.002
V03	0.13	0.10	0.002
V04	0.11	0.09	0.010

Figure 44.26 Typical narrowband data

Figure 44.27 illustrates narrowband data trends. In addition to the overall or broadband energy, narrowbands trends indicate the relative energy in select machine-train components. In effect, this type of analysis is a series of mini-overall energy readings.

Composite trends

Most microprocessor-based systems permit composite trending (i.e., simultaneous displays) of both filtered broadband and narrowband data. Figure 44.28 illustrates

a composite trend that includes both broadband and narrowband data. This type of plot is quite beneficial because it permits the analyst to track the key indicators of machine condition on one plot.

44.7 Evaluation methods

Trend data can be used in the following ways: (1) to compare with specific reference values, (2) mode-shape comparisons, and (3) cross-machine comparisons.

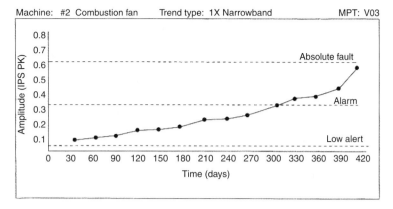

Figure 44.27 Narrowband trends provide energy histories of specific components

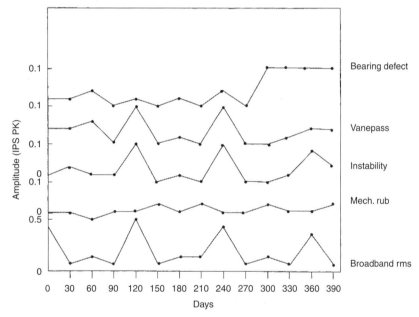

Figure 44.28 Combining trends (i.e. composite) provide both broadband and narrowband data

44.7.1 Comparison to reference values

Three types of reference-value comparisons are used to evaluate trend data: baseline data, rate of change, and industrial standards.

Baseline data

A series of baseline or reference data sets should be taken for each machine-train included in a predictive-maintenance program. These data sets are necessary for future use as a reference point for trends, time traces, and FFT signatures that are collected over time. Such baseline data sets must be representative of the normal operating condition of each machine-train to have value as a reference.

Three criteria are critical to the proper use of baseline comparisons: reset after maintenance, proper identification, and process envelope.

Reset after maintenance

The baseline data set must be updated each time the machine is repaired, rebuilt, or when any major maintenance is performed. Even when best practices are used, machinery cannot be restored to as-new condition when major maintenance is performed. Therefore, a new baseline or reference data set must be obtained following these events.

Proper identification

Each reference or baseline data set must be clearly and completely identified. Most vibration-monitoring systems permit the addition of a label or unique identifier to any user-selected data set. This capability should be used to clearly identify each baseline data set.

In addition, the data set label should include all defining information. For example, any rework or repairs made to the machine should be identified. If a new baseline data set is obtained after the replacement of a rotating element, this information should be included in the descriptive label.

Process envelope

Since variations in process variables, such as load, have a direct effect on the vibration energy and signature generated by a machine-train, the actual operating envelope for each baseline data set must also be clearly identified. If this step is omitted, direct comparison of other data to the baseline will be meaningless. The label feature in most vibration monitoring systems permits tagging the baseline data set with this additional information.

Rate of change

Rate of change is the most often used trend analysis. Since most of the microprocessor-based systems provide the ability to automatically display broadband and narrowband data trends, analysts tend to rely on this means of comparative analysis.

Rate of change is a valid means of defining the relative condition of rotating machinery. As a rule, there must be a change in mechanical condition before there can be a change in the vibration energy generated by a machine. Therefore, monitoring the rate that the energy levels change, either up or down, is a useful tool (see Figure 44.29).

Caution: Broadband and narrowband data must be normalized for changes in load before being valid. Normal variations in machine load destroy the validity of non-normalized trend data and little can be gained from its use.

Industrial standards

There are a number of published standards that define acceptable levels of vibration in machinery. These standards are valuable reference tools, but they must be clearly understood and properly used. Industrial-standard data can

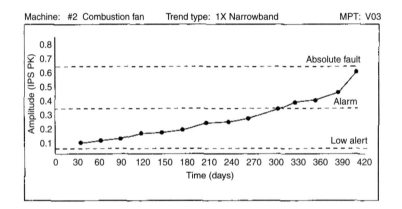

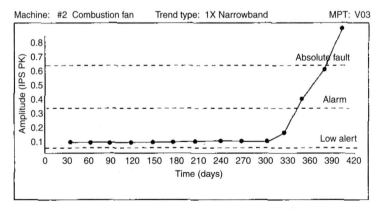

Figure 44.29 Rate of change of trend data indicates condition change

be obtained from a Rathbone chart (see Figure 44.30) and from the American Petroleum Institute.

Rathbone chart

The Rathbone chart in Figure 44.30 provides levels of vibration severity that range from extremely smooth, which is the best possible operating condition, to very rough or absolute-fault limit, which is the maximum level where a machine can operate.

This chart is useful, although it is often misused. Four factors must be understood before using the chart:

1. Data included in the Rathbone chart are valid for machines with running speeds between 1200 and 3600 rpm. The chart cannot be used for low-speed or turbo-machinery.
2. The data presented in the chart are relative vibration levels (i.e., taken from a bearing pedestal using either an accelerometer or velocity probe) in inches per second (ips) peak.
3. Data are peak values of velocity (ips) for a filtered broadband from 10 to 10,000 Hz.
4. The severity levels are relative, not absolute. For example, when a machine reaches the absolute-fault limit, it has a 90 per cent probability of failure within its next 1000 hours of operation (i.e., it is not going to fail tomorrow).

American petroleum institute

The American Petroleum Institute (API) has established standards for vibration levels. Unlike the Rathbone chart, which presents relative vibration data, the API standards are actual shaft displacement as measured with a displacement probe. Based on the API data, acceptable vibration can be defined by:

$$\text{Vibration severity} = \sqrt{\frac{12,000}{\text{rpm}}} \text{ mils, peak-to-peak}$$

The API equivalent of the absolute-fault limit can be defined by:

$$\text{Absolute-fault limit} = 1.3 \times \sqrt{\frac{12000}{\text{rpm}}} \text{ mils, peak-to-peak}$$

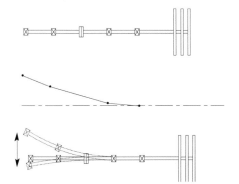

Figure 44.30 Rathbone chart – relative vibration severity

The API standards are reasonable for turbo-machinery, but are unacceptable for lower-speed machines. The standards are applicable for rotating machinery with speeds above 1800 rpm; marginal for speeds between 600 and 1800 rpm; and should not be used for speeds below 600 rpmx.

44.7.2 Mode shape (shaft deflection)

A clear understanding of the mode shape, or shaft deflection, of a machine's rotating element is a valuable diagnostic tool. Both broadband and narrowband filtered energy windows can be used at each measurement point and orientation across the machine. The resultant plots, one in the vertical plane and one in the horizontal plane, provide an approximation of the mode shape of the complete machine and its rotating element.

Unfortunately, these plots must be developed manually. The microprocessor-based systems generally do not automate this function, but they are easily constructed on graph paper. The following are the steps to construct such a plot:

The first step is to draw two horizontal lines on the graph paper. One is used to plot the vertical data and the other to plot the horizontal. These two lines show the location of each measurement point in inches with the outboard motor bearing being at zero.

Next, draw vertical lines that intersect the left-hand end of the two horizontal lines. These vertical lines form the amplitude scale for the two plots. Establish the amplitude scale based on the maximum energy level recorded in the broadband or narrowband windows.

The final step is to plot the actual measured amplitude at each measurement point on the machine-train. Start with the outboard motor bearing and move across the machine until the final data set is plotted.

Broadband plots

The overall energy from the filtered broadband plotted against measurement location provides an approximation of the mode shape of the installed machine. Figure 44.31 illustrates a vertical broadband plot taken from a Spencer blower. Note that the motor appears to be flexing in the vertical direction. Extremely high amplitudes are present in the motor's outboard bearing and the amplitudes decrease at subsequent measurement points across the machine.

A mode curve exhibiting this shape could indicate that the motor mountings, or the baseplate under the motor, are loose and that the motor is moving vertically. In fact, in the example from which this figure was taken, this is exactly what was happening. The blower's baseplate 'floats' on a one-inch thick cork pad, which is normally an acceptable practice. However, in this example, an inlet filter/silencer was mounted without support directly to the inlet located on the right end of the machine. The weight of the filter/silencer compressed the cork pad under the blower, which lifted the motor-end of the baseplate off of the cork pad. In this mode, the motor has complete freedom of movement in the vertical plane. In effect, it

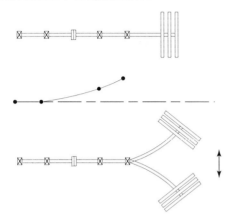

Figure 44.31 Vertical broadband mode shape for Spencer blower indicates potential failure

acts like a tuning fork and creates the high overall energy recorded on the mode plot.

Narrowband plots

Narrowband plots permit the same type of evaluation for major vibration components such as fundamental running speed (1×) or gear mesh. The plots are constructed in the same way as for the broadbands, except that the amplitude values are for user-selected windows, or bands.

Using the previously mentioned example of the Spencer blower, Figure 44.32 is a plot of the fundamental (1×) frequency of the motor-blower shaft versus measurement location. Note that the vertical mode (see Figure 44.31) appears to be relatively normal, except for the motor looseness problem.

The horizontal plot seems to indicate that the shaft is being severely deflected from its true centerline. In addition, the plot suggests that the deflection is outboard (i.e., toward the rotor) from the two blower-support bearings. This outboard deflection eliminates misalignment between the motor and blower as a possible source of the deflection. We must now determine what could cause the rotor to be deflected and why only in the horizontal direction.

The Spencer blower in this example provides air to a drying process in a metal-coating line. Its configuration includes an end-suction inlet that is in-line with the shaft and a horizontal discharge that is perpendicular to the shaft. In this particular example, the source of the shaft deflection observed in the mode plot is aerodynamic instability.

The reason for this instability is that the blowers are incorrectly sized for the application and are running well outside their performance curve. In effect, the blowers have no back-pressure and are operating in a run-out condition. The result of operating in this condition is that the design load intended to stabilized the rotor is no longer present. This causes the shaft to deflect or flex, generating the high amplitudes observed in the horizontal mode plot.

The problem is eliminated by restricting the discharge air flow from the blowers. By increasing the back-pressure, the blowers are able to operate within their normal envelope and the shaft deflection disappears.

44.7.3 Cross-machine comparisons

Cross-machine comparison is an extremely beneficial tool to the novice analyst. Most vibration monitoring systems permit direct comparison of vibration data, both filtered window energy and complete signatures, acquired from two machines. This capability permits the analyst to directly compare a machine that is known to be in good operating condition with one that is perceived to have

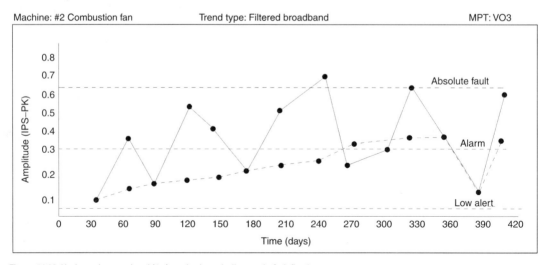

Figure 44.32 Horizontal narrowband (1×) mode shape indicates shaft deflection

a problem. There are several ways that cross-machine comparisons can be made using microprocessor-based systems: multiple plots, ratio, and difference.

44.8 Limitations

Although quite valuable when used properly, trends do not allow the analyst to confirm that a problem exists or to determine the cause of incipient problems. Another limitation is the limited number of values the system can handle. Further, the data needs to be normalized for speed, load, and process variables.

44.8.1 Number of values

Some of the vibration-monitoring systems limit the number of data sets and duration of data that can be automatically trended. In most systems, the number of values that can be trended is limited to eight to twelve data sets. Although this limitation prevents trending the machine over its useful life, it does not eliminate trending as a vibration-monitoring tool.

44.8.2 Data normalization

Trend data that are not properly normalized for speed, load, and process variables are of little value. Since load and process-variable normalization requires a little more time during the data-acquisition process, many programs do not perform these adjustments. If this is the case, it is best to discontinue the use of trends altogether.

As an example, Figure 44.33 illustrates the impact of load on vibration trends. The solid line represents the recorded raw broadband vibration levels. The dashed line is the same data adjusted for changes in load.

Trend data also must be adjusted for maintenance and repair activities. Figure 44.34 illustrates an average trend curve that indicates a sharp rise in vibration levels. It also reflects that, after repair, the levels drop radically. At this point, all baseline and reference values should be reset. If this does not occur, the automatic trending capabilities of the computer-based system do not function properly.

44.9 Failure mode analysis

All of the analysis techniques discussed to this point have been methods to determine if a potential problem exists within the machine-train or its associated systems. Failure-mode analysis is the next step required to specifically pinpoint the failure mode and identify which machine-train component is degrading.

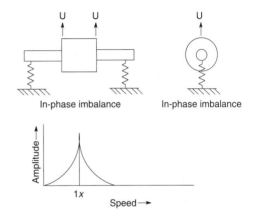

Figure 44.34 Baselines must be reset following repair

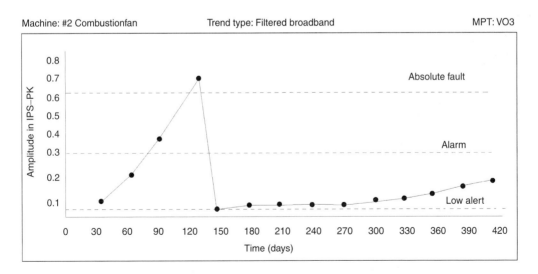

Figure 44.33 Trends must be adjusted or normalized for load changes

Although failure-mode analysis identifies the number and symptoms of machine-train problems, it does not always identify the true root cause of problems. Root cause must be verified by visual inspection, additional testing, or other techniques such as operating dynamics analysis.

Failure-mode analysis is based on the assumption that certain failure modes are common to all machine-trains and all applications. It also assumes that the vibration patterns for each of these failure modes, when adjusted for process-system dynamics, are absolute and identifiable.

Two types of information are required to perform failure-mode analysis: (1) machine-train vibration signatures, both FFTs and time traces; and (2) practical knowledge of machine dynamics and failure modes. A number of failure-mode charts are available that describe the symptoms or abnormal vibration profiles that indicate potential problems exist. An example is the following description of the imbalance failure mode, which was obtained from a failure-mode chart: Single-plane imbalance generates a dominant fundamental ($1\times$) frequency component with no harmonics ($2\times$, $3\times$, etc.). Note, however, that the failure-mode charts are simplistic since many other machine-train problems also excite, or increase the amplitude of, the fundamental ($1\times$) frequency component. In a normal vibration signature, 60 to 70 per cent of the total overall, or broadband, energy is contained in the $1\times$ frequency component. Any deviation from a state of equilibrium increases the energy level at this fundamental shaft speed.

44.9.1 Common failure modes

Many of the common causes of failure in machinery components can be identified by understanding their relationship to the true running speed of the shaft within the machine-train.

Table 44.6 is a vibration troubleshooting chart that identifies some of the common failure modes. This table provides general guidelines for interpreting the most common abnormal vibration profiles. These guidelines, however, do not provide positive verification or identification of machine-train problems. Verification requires an understanding of the failure mode and how it appears in the vibration signature.

The sections to follow describe the most common machine-train failure modes: critical speeds, imbalance, mechanical looseness, misalignment, modulations, process instability, and resonance.

Critical speeds

All machine-trains have one or more critical speeds that can cause severe vibration and damage to the machine. Critical speeds result from the phenomenon known as dynamic resonance.

Critical speed is a function of the natural frequency of dynamic components such as a rotor assembly, bearings, etc. All dynamic components have one or more natural frequencies that can be excited by an energy source that coincides with, or is in close proximity to, that frequency.

For example, a rotor assembly with a natural frequency of 1800 rpm cannot be rotated at speeds between 1782 and 1818 rpm without exciting the rotor's natural frequency.

Critical speed should not be confused with the mode shape of a rotating shaft. Deflection of the shaft from its true centerline (i.e., mode shape) elevates the vibration amplitude and generates dominant vibration frequencies at the rotor's fundamental and harmonics of the running speed. However, the amplitude of these frequency components tends to be much lower than those caused by operating at a critical speed of the rotor assembly. Also, the excessive vibration amplitude generated by operating at a critical speed disappears when the speed is changed. Vibrations caused by mode shape tend to remain through a much wider speed range or may even be independent of speed.

The unique natural frequencies of dynamic machine components are determined by the mass, freedom of movement, support stiffness, and other factors. These factors define the response characteristics of the rotor assembly (i.e., rotor dynamics) at various operating conditions.

Each critical speed has a well-defined vibration pattern. The first critical excites the fundamental ($1\times$) frequency component; the second critical excites the secondary ($2\times$) component; and the third critical excites the third ($3\times$) frequency component.

The best way to confirm a critical-speed problem is to change the operating speed of the machine-train. If the machine is operating at a critical speed, the amplitude of the vibration components ($1\times$, $2\times$, or $3\times$) will immediately drop when the speed is changed. If the amplitude remains relatively constant when the speed is changed, the problem is not critical speed.

Imbalance

The term balance means that all forces generated by, or acting on, the rotating element of a machine-train are in a state of equilibrium. Any change in this state of equilibrium creates an imbalance. In the global sense, imbalance is one of the most common abnormal vibration profiles exhibited by all process machinery.

Theoretically, a perfectly balanced machine that has no friction in the bearings would experience no vibration and would have a perfect vibration profile – a perfectly flat, horizontal line. However, there are no perfectly balanced machines in existence. All machine-trains exhibit some level of imbalance, which has a dominant frequency component at the fundamental running speed ($1\times$) of each shaft.

An imbalance profile can be excited due to the combined factors of mechanical imbalance, lift/gravity differential effects, aerodynamic and hydraulic instabilities, process loading, and, in fact, all failure modes.

Mechanical It is incorrect to assume that mechanical imbalance must exist to create an imbalance condition within the machine. Mechanical imbalance, however, is the only form of imbalance that is corrected by balancing

Table 44.6 Vibration Trouble Shooting Chart

Nature of fault	Frequency of Dominant Vibration (Hz = rpm 60)	Direction	Remarks
Rotating members out of Balance	1 × rpm	Radial	A common cause of excess vibration in machinery
Misalignment & bent shaft	Usually 1 × rpm Often 2 × rpm Sometimes 3&4 × rpm	Radial & axial	A common fault
Damaged rolling element bearings (ball, roller, etc.)	Impact rates for the individual bearing components* Also vibrations at very high frequencies (20 to 60 kHz)	Radial & axial	Uneven vibration levels, often with shocks. *Impact Rates: 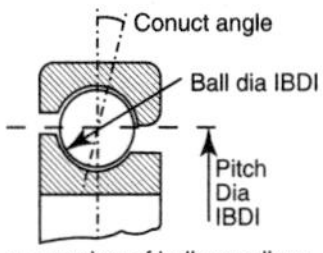Conuct angle Ball dia IBDI Pitch Dia IBDI Impact Rates 1 (Hz) For Outer Race Defect 1(Hz) = $\frac{n}{2}$ f, 11– $\frac{BD}{PD}$ For Inner Race Defect 1(Hz) = $\frac{n}{2}$ f, 11– $\frac{BD}{PD}$ For Ball Defect 1(Hz) = $\frac{BD}{PD}$ f, $(1 - \frac{BD}{PD}^{2})$ n = number of balls or rollers = relative rev./s between inner & outer races
Journal bearings Loose in Housings	Sub-harmonics of shaft rpm, exactly 1/2 or 1/3 × rpm	Primarily radial	Looseness may only develop at operating speed and temperature (e.g. turbomachines).
Oil film whirl or whip in journal bearings	Slightly less than half shaft speed (42% to 48%)	Primarily radial	Applicable to high-speed (e.g. turbo) machines.
Hysteresis whirl	Shaft critical speed	Primarily radial	Vibrations excited when passing through critical shaft speed are maintained at higher shaft speeds. Can sometimes be cured by checking tightness of rotor components
Damaged or worn gears	Tooth meshing frequencies (shaft rpm × number of teeth) and harmonics	Radial & axial	Sidebands around tooth meshing frequencies indicate modulation (e.g. eccentricity) at frequency corresponding to sideband spacings. Normally only detectable with very narrow-band analysis.
Mechanical looseness	2 × rpm		
Faulty belt drive	1, 2, 3 & 4 × rpm of belt	Radial	
Unbalanced reciprocating forces and couples	1 × rpm and/or multiples for higher order unbalance	Primarily radial	
Increased turbulence	Blade & Vane passing frequencies and harmonics	Radial & axial	Increasing levels indicate increasing turbulence.
Electrically induced vibrations	1 × rpm or 1 or 2 times sychronous frequency	Radial & axial	Should disappear when turning off the power.

Source: Predictive Maintenance for Process Machinery, R. Keith Mobley, Technology for Energy Corp. (1988).

the rotating element. When all failures are considered, the number of machine problems that are the result of actual mechanical rotor imbalance is relatively small.

Single-plane Single-plane mechanical imbalance excites the fundamental (1×) frequency component, which is typically the dominant amplitude in a signature. Since there is only one point of imbalance, only one high spot occurs as the rotor completes each revolution. The vibration signature also may contain lower-level frequencies reflecting bearing defects and passing frequencies. Figure 44.35 illustrates single-plane imbalance.

Because mechanical imbalance is multi-directional, it appears in both the vertical and horizontal directions at the machine's bearing pedestals. The actual amplitude of the 1× component generally is not identical in the vertical and

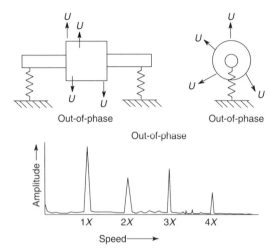

Figure 44.35 Single-plane imbalance

horizontal directions and both generally contain elevated vibration levels at 1×.

The difference between the vertical and horizontal values is a function of the bearing-pedestal stiffness. In most cases, the horizontal plane has a greater freedom of movement and, therefore, contains higher amplitudes at 1× than the vertical plane.

Multi-plane   Multi-plane mechanical imbalance generates multiple harmonics of running speed. The actual number of harmonics depends on the number of imbalance points, the severity of imbalance, and the phase angle between imbalance points.

Figure 44.36 illustrates a case of multi-plane imbalance in which there are four out-of-phase imbalance points. The resultant vibration profile contains dominant frequencies at 1×, 2×, 3×, and 4×. The actual amplitude of each of these components is determined by the amount of imbalance at each of the four points, but the 1× component should always be higher than any subsequent harmonics.

Lift/gravity differential   Lift, which is designed into a machine-train's rotating elements to compensate for the effects of gravity acting on the rotor, is another source of imbalance. Because lift does not always equal gravity, there is always some imbalance in machine-trains. The vibration component due to the lift/gravity differential effect appears at the fundamental or 1× frequency.

Other   In fact, all failure modes create some form of imbalance in a machine, as do aerodynamic instability,

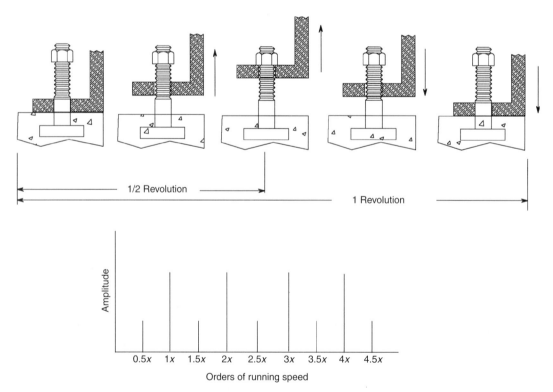

Figure 44.36 Multi-plane imbalance generates multiple harmonics

hydraulic instability, and process loading. The process loading of most machine-trains varies, at least slightly, during normal operations. These vibration components appear at the 1× frequency.

Mechanical looseness   Looseness, which can be present in both the vertical and horizontal planes, can create a variety of patterns in a vibration signature. In some cases, the fundamental (1×) frequency is excited. In others, a frequency component at one-half multiples of the shaft's running speed (0.5×, 1.5×, 2.5×, etc.) is present. In almost all cases, there are multiple harmonics, both full and half.

Vertical   Mechanical looseness in the vertical plane generates a series of harmonic and half-harmonic frequency components. Figure 44.37 is a simple example of a vertical mechanical looseness signature.

In most cases, the half-harmonic components are about one-half of the amplitude of the harmonic components. They result from the machine-train lifting until stopped by the bolts. The impact as the machine reaches the upper limit of travel generates a frequency component at one-half multiples (i.e., orders) of running speed. As the machine returns to the bottom of its movement, its original position, a larger impact occurs that generates the full harmonics of running speed.

The difference in amplitude between the full harmonics and half-harmonics is caused by the effects of gravity. As the machine lifts to its limit of travel, gravity resists the lifting force. Therefore, the impact force that is generated as the machine foot contacts the mounting bolt is the difference between the lifting force and gravity. As the machine drops, the force of gravity combines with the force generated by imbalance. The impact force as the machine foot contacts the foundation is the sum of the force of gravity and the force resulting from imbalance.

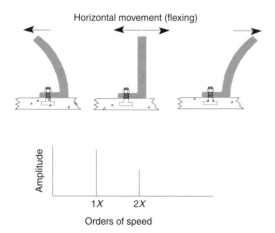

Figure 44.37 Vertical mechanical looseness has a unique vibration profile

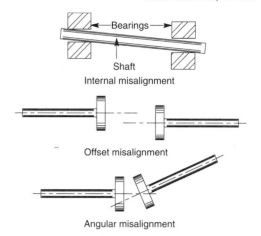

Figure 44.38 Horizontal looseness creates first and second harmonics

Horizontal   Figure 44.38 illustrates horizontal mechanical looseness, which is also common to machine-trains. In this example, the machine's support legs flex in the horizontal plane. Unlike the vertical looseness illustrated in Figure 44.37, gravity is uniform at each leg and there is no increased impact energy as the leg's direction is reversed.

Horizontal mechanical looseness generates a combination of first (1×) and second (2×) harmonic vibrations. Since the energy source is the machine's rotating shaft, the timing of the flex is equal to one complete revolution of the shaft, or 1×. During this single rotation, the mounting legs flex to their maximum deflection on both sides of neutral. The double change in direction as the leg first deflects to one side then the other generates a frequency at two times (2×) the shaft's rotating speed.

Other   There are a multitude of other forms of mechanical looseness (besides vertical and horizontal movement of machine legs) that are typical for manufacturing and process machinery. Most forms of pure mechanical looseness result in an increase in the vibration amplitude at the fundamental (1×) shaft speed. In addition, looseness generates one or more harmonics (i.e., 2×, 3×, 4×, or combinations of harmonics and half-harmonics).

However, not all looseness generates this classic profile. For example, excessive bearing and gear clearances do not generate multiple harmonics. In these cases, the vibration profile contains unique frequencies that indicate looseness, but the profile varies depending on the nature and severity of the problem.

With sleeve or Babbitt bearings, looseness is displayed as an increase in sub-harmonic frequencies (i.e., less than the actual shaft speed, such as 0.5×). Rolling-element bearings display elevated frequencies at one or more of their rotational frequencies. Excessive gear clearance increases the amplitude at the gear-mesh frequency and its sidebands.

Other forms of mechanical looseness increase the noise floor across the entire bandwidth of the vibration signature. While the signature does not contain a distinct peak or series of peaks, the overall energy contained in the vibration signature is increased. Unfortunately, the increase in noise floor cannot always be used to detect mechanical looseness. Some vibration instruments lack sufficient dynamic range to detect changes in the signature's noise floor.

Misalignment

This condition is virtually always present in machine-trains. Generally, we assume that misalignment exists between shafts that are connected by a coupling, V-belt, or other intermediate drive. However, it can exist between bearings of a solid shaft and at other points within the machine.

How misalignment appears in the vibration signature depends on the type of misalignment. Figure 44.39 illustrates three types of misalignment (i.e., internal, offset, and angular). These three types excite the fundamental (1×) frequency component because they create an apparent imbalance condition in the machine.

Internal (i.e., bearing) and offset misalignment also excite the second (2×) harmonic frequency. Two high spots are created by the shaft as it turns though one complete revolution. These two high spots create the first (1×) and second harmonic (2×) components.

Angular misalignment can take several signature forms and excites the fundamental (1×) and secondary (2×) components. It can excite the third (3×) harmonic frequency depending on the actual phase relationship of the angular misalignment. It also creates a strong axial vibration.

Modulations

Modulations are frequency components that appear in a vibration signature, but cannot be attributed to any specific physical cause, or forcing function. Although these frequencies are, in fact, ghosts or artificial frequencies, they can result in significant damage to a machine-train. The presence of ghosts in a vibration signature often leads to misinterpretation of the data.

Ghosts are caused when two or more frequency components couple, or merge, to form another discrete frequency component in the vibration signature. This generally occurs with multiple-speed machines or a group of single-speed machines.

Note that the presence of modulation, or ghost peaks, is not an absolute indication of a problem within the machine-train. Couple effects may simply increase the amplitude of the fundamental running speed and do little damage to the machine-train. However, this increased amplitude will amplify any defects within the machine-train.

Coupling can have an additive effect on the modulation frequencies, as well as being reflected as a differential or

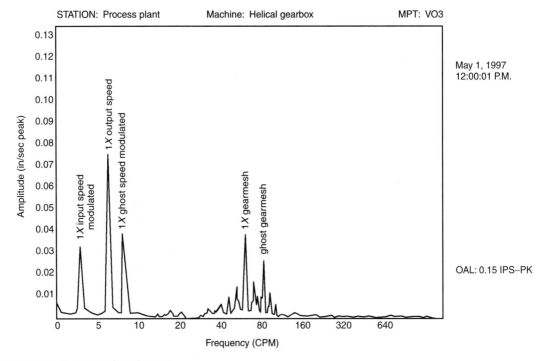

Figure 44.39 Three types of misalignment

multiplicative effect. These concepts are discussed in the sections to follow.

Take as an example the case of a 10-tooth pinion gear turning at 10 rpm while driving a 20-tooth bullgear having an output speed of 5 rpm. This gear set generates real frequencies at 5, 10, and 100 rpm (i.e., 10 teeth × 10 rpm). This same set also can generate a series of frequencies (i.e., sum and product modulations) at 15 rpm (i.e., 10 rpm + 5 rpm) and 150 rpm (i.e., 15 rpm × 10 teeth). In this example, the 10-rpm input speed coupled with the 5-rpm output speed to create ghost frequencies driven by this artificial fundamental speed (15 rpm).

Sum   This type of modulation, which is described in the example above, generates a series of frequencies that include the fundamental shaft speeds, both input and output, and fundamental gear-mesh profile. The only difference between the real frequencies and the ghost is their location on the frequency scale. Instead of being at the actual shaft-speed frequency, the ghost appears at frequencies equal to the sum of the input and output shaft speeds. Figure 44.40 illustrates this for a speed-increaser gearbox.

Difference   In this case, the resultant ghost, or modulation, frequencies are generated by the difference between two or more speeds (see Figure 44.41). If we use the same example as before, the resultant ghost frequencies appear at 5 rpm (i.e., 10 rpm − 5 rpm) and 50 rpm (i.e., 5 rpm × 10 teeth).

Note that the 5-rpm couple frequency coincides with the real output speed of 5 rpm. This results in a dramatic increase in the amplitude of one real running-speed component and the addition of a false gear-mesh peak.

This type of coupling effect is common in single-reduction/increase gearboxes or other machine-train components where multiple running or rotational speeds are relatively close together or even integer multiples of one another. It is more destructive than other forms of coupling in that it coincides with real vibration components and tends to amplify any defects within the machine-train.

Product   With product modulation, the two speeds couple in a multiplicative manner to create a set of artificial frequency components (see Figure 44.42). In the previous example, product modulations occur at 50 rpm (i.e., 10 rpm × 5 rpm) and 500 rpm (i.e., 50 rpm × 10 teeth).

Beware that this type of coupling often may go undetected in a normal vibration analysis. Since the ghost frequencies are relatively high compared to the expected real frequencies, they are often outside the monitored frequency range used for data acquisition and analysis.

Process instability

Normally associated with bladed or vaned machinery such as fans and pumps, process instability creates an unbalanced condition within the machine. In most cases, it excites the fundamental (1×) and bladepass/vanepass frequency components. Unlike true mechanical imbalance,

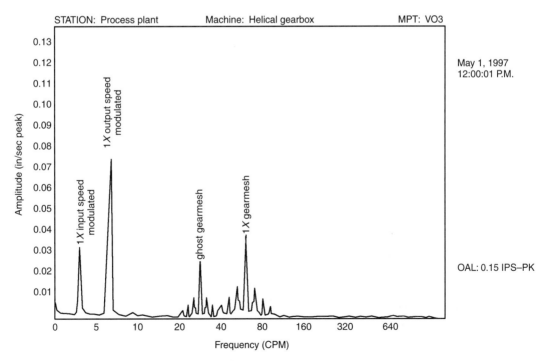

Figure 44.40 Sum modulation for a speed-increaser gearbox

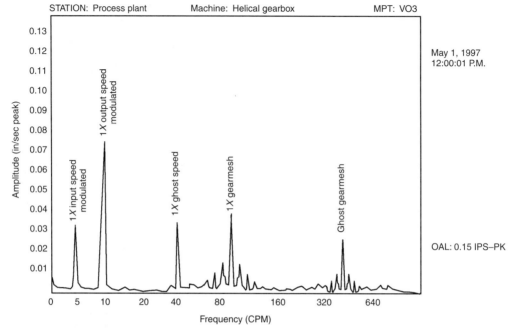

Figure 44.41 Difference modulation for a speed-increaser gearbox

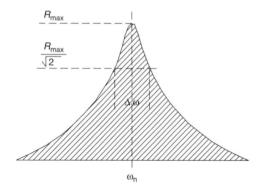

Figure 44.42 Product modulation for a speed-increaser gearbox

the bladepass and vanepass frequency components are broader and have more energy in the form of sideband frequencies.

In most cases, this failure mode also excites the third (3×) harmonic frequency and creates strong axial vibration. Depending on the severity of the instability and the design of the machine, process instability also can create a variety of shaft-mode shapes. In turn, this excites the 1×, 2×, and 3× radial vibration components.

Resonance

Resonance is defined as a large-amplitude vibration caused by a small periodic stimulus having the same, or

nearly the same, period as the system's natural vibration. In other words, an energy source with the same, or nearly the same, frequency as the natural frequency of a machine-train or structure will excite that natural frequency. The result is a substantial increase in the amplitude of the natural frequency component.

The key point to remember is that a very low amplitude energy source can cause massive amplitudes when its frequency coincides with the natural frequency of a machine or structure. Higher levels of input energy can cause catastrophic, near instantaneous failure of the machine or structure.

Every machine-train has one or more natural frequencies. If one of these frequencies is excited by some component of the normal operation of the system, the machine structure will amplify the energy, which can cause severe damage.

An example of resonance is a tuning fork. If you activate a tuning fork by striking it sharply, the fork vibrates rapidly. As long as it is held suspended, the vibration decays with time. However, if you place it on a desktop, the fork could potentially excite the natural frequency of the desk, which would dramatically amplify the vibration energy.

The same thing can occur if one or more of the running speeds of a machine excites the natural frequency of the machine or its support structure. Resonance is a very destructive vibration and, in most cases, it will cause major damage to the machine or support structure.

There are two major classifications of resonance found in most manufacturing and process plants: static and dynamic. Both types exhibit a broad-based, high-amplitude frequency component when viewed in a FFT vibration signature. Unlike meshing or passing frequencies, the resonance frequency component does not have modulations or sidebands. Instead, resonance is displayed as a single, clearly defined peak.

As illustrated in Figure 44.42, a resonance peak represents a large amount of energy. This energy is the result of both the amplitude of the peak and the broad area under the peak. This combination of high peak amplitude and broad-based energy content is typical of most resonance problems. The damping system associated with a resonance frequency is indicated by the sharpness or width of the response curve, ω_n, when measured at the half-power point. R_{MAX} is the maximum resonance and $R_{MAX}/\sqrt{2}$ is the half-power point for a typical resonance-response curve.

Static When the natural frequency of a stationary, or non-dynamic, structure is energized, it will resonate. This type of resonance is classified as static resonance and is considered to be a non-dynamic phenomenon. Non-dynamic structures in a machine-train include casings,

bearing-support pedestals, and structural members such as beams, piping, etc.

Since static resonance is a non-dynamic phenomenon, it is generally not associated with the primary running speed of any associated machinery. Rather, the source of static resonance can be any energy source that coincides with the natural frequency of any stationary component. For example, an I-beam support on a continuous annealing line may be energized by the running speed of a roll. However, it also can be made to resonate by a bearing frequency, overhead crane, or any of a multitude of other energy sources.

The actual resonant frequency depends on the mass, stiffness, and span of the excited member. In general terms, the natural frequency of a structural member is inversely proportional to the mass and stiffness of the member. In other words, a large turbo-compressor's casing will have a lower natural frequency than that of a small end-suction centrifugal pump.

Figure 44.43 illustrates a typical structural-support system. The natural frequencies of all support structures, piping, and other components are functions of mass, span, and stiffness. Each of the arrows on Figure 44.43 indicates a structural member or stationary machine component having a unique natural frequency. Note that each time a structural span is broken or attached to another structure, the stiffness changes. As a result, the natural frequency of that segment also changes.

While most stationary machine components move during normal operation, they are not always resonant. Some degree of flexing or movement is common in most stationary machine-trains and structural members. The amount of movement depends on the spring constant or stiffness of the member. However, when an energy source coincides and couples with the natural frequency of a structure, excessive and extremely destructive vibration amplitudes result.

Dynamic When the natural frequency of a rotating, or dynamic, structure (e.g., rotor assembly in a fan) is energized, the rotating element resonates. This phenomenon is classified as dynamic resonance and the rotor speed at which it occurs is referred to as the critical. In most cases,

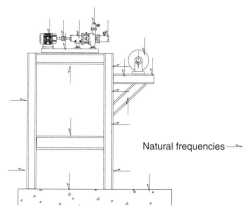

Natural frequencies——➤

Figure 44.43 Resonance response

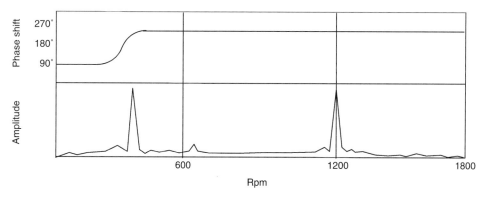

Figure 44.44 Typical discrete natural frequency locations in structural members

dynamic resonance appears at the fundamental running speed or one of the harmonics of the excited rotating element. However, it also can occur at other frequencies. As in the case of static resonance, the actual natural frequencies of dynamic members depend on the mass, bearing span, shaft and bearing-support stiffness, as well as a number of other factors.

Confirmation analysis In most cases, the occurrence of dynamic resonance can be quickly confirmed. When monitoring phase and amplitude, resonance is indicated by a 180° phase shift as the rotor passes through the resonant zone. Figure 44.44 illustrates a dynamic resonance at 500 rpm, which shows a dramatic amplitude increase in the frequency-domain display. This is confirmed by the 180° phase shift in the time-domain plot. Note that the peak at 1200 rpm is not resonance. The absence of a phase shift, coupled with the apparent modulations in the FFT, discount the possibility that this peak is resonance-related.

Common confusions Vibration analysts often confuse resonance with other failure modes. Since many of the common failure modes tend to create abnormally high vibration levels that appear to be related to a speed change, analysts tend to miss the root cause of these problems.

Dynamic resonance generates abnormal vibration profiles that tend to coincide with the fundamental (1×) running speed, or one or more of the harmonics, of a machine-train. This often leads the analyst to incorrectly

diagnose the problem as imbalance or misalignment. The major difference is that dynamic resonance is the result of a relatively small energy source, such as the fundamental running speed, that results in a massive amplification of the natural frequency of the rotating element (see Figure 44.45).

Function of speed The high amplitudes at the rotor's natural frequency are strictly speed dependent. If the energy source, in this case speed, changes to a frequency outside the resonant zone, the abnormal vibration will disappear.

In most cases, running speed is the forcing function that excites the natural frequency of the dynamic component. As a result, rotating equipment is designed to operate at primary rotor speeds that do not coincide with the rotor assembly's natural frequencies. Most low- to moderate-speed machines are designed to operate below the first critical speed of the rotor assembly.

Higher-speed machines may be designed to operate between the first and second, or second and third, critical speeds of the rotor assembly. As these machines accelerate through the resonant zones or critical speeds, their natural frequency is momentarily excited. As long as the ramp rate limits the duration of excitation, this mode of operation is acceptable. However, care must be taken to ensure that the transient time through the resonant zone is as short as possible.

Figure 44.46 illustrates a typical critical-speed or dynamic-resonance plot. This figure is a plot of the

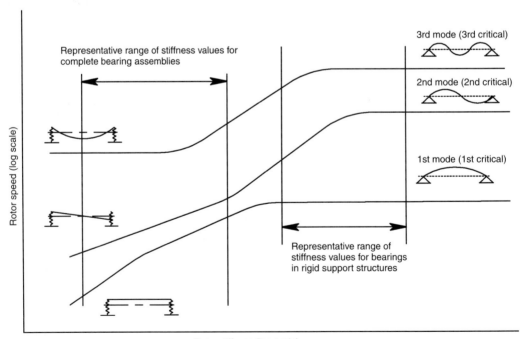

Figure 44.45 Dynamic resonance phase shift

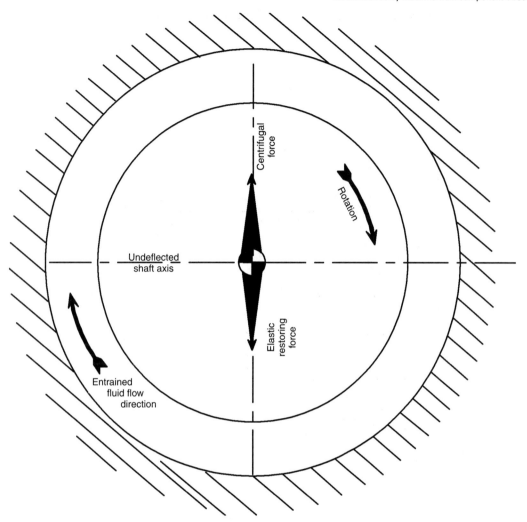

Figure 44.46 Dynamic resonance plot

relationship between rotor-support stiffness (X-axis) and critical rotor speed (Y-axis). Rotor-support stiffness depends on the geometry of the rotating element (i.e. shaft and rotor) and the bearing-support structure. These are the two dominant factors that determine the response characteristics of the rotor assembly.

44.10 Failure modes by machine-train component

In addition to identifying general failure modes that are common to many types of machine-train components, failure-mode analysis can be used to identify failure modes for specific components in a machine-train. However, care must be exercised when analyzing vibration profiles, because the data may reflect induced problems. Induced problems affect the performance of a specific component, but are not caused by that component. For example, an abnormal outer-race passing frequency may indicate a defective rolling-element bearing. It also can indicate that abnormal loading caused by misalignment, roll bending, process instability, etc., has changed the load zone within the bearing. In the latter case, replacing the bearing does not resolve the problem and the abnormal profile will still be present after the bearing is changed.

44.10.1 Bearings: rolling element

Bearing defects are one of the most common faults identified by vibration monitoring programs. Although bearings do wear out and fail, these defects are normally symptoms of other problems within the machine-train or process system. Therefore, extreme care must be exercised to ensure that the real problem is identified, not just the symptom. In a rolling-element, or anti-friction, bearing vibration profile, three distinct sets of frequencies can be found: natural, rotational, and defect.

Natural frequency

Natural frequencies are generated by impacts of the internal parts of a rolling-element bearing. These impacts are normally the result of slight variations in load and imperfections in the machined bearing surfaces. As their name implies, these are natural frequencies and are present in a new bearing that is in perfect operating condition.

The natural frequencies of rolling-element bearings are normally well above the maximum frequency range, F_{MAX}, used for routine machine-train monitoring. As a result, they are rarely observed by predictive-maintenance analysts. Generally, these frequencies are between 20 KHz and 1 MHz. Therefore, some vibration-monitoring programs use special high-frequency or ultrasonic monitoring techniques such as high-frequency domain (HFD). Note, however, that little is gained from monitoring natural frequencies. Even in cases of severe bearing damage, these high-frequency components add little to the analyst's ability to detect and isolate bad bearings.

Rotational frequency

There are four normal rotational frequencies associated with rolling-element bearings: fundamental train frequency (FTF), ball/roller spin, ball-pass outer-race, and ball-pass inner-race. The following are definitions of abbreviations that are used in the discussion to follow.

BD = Ball or roller diameter
PD = Pitch diameter
β = Contact angle (for roller = 0)
n = Number of balls or rollers
f_r = Relative speed between the inner and outer race (rps).

Fundamental train frequency

The bearing cage generates the FTF as it rotates around the bearing races. The cage properly spaces the balls or rollers within the bearing races, in effect, by tying the rolling elements together and providing uniform support. Some friction exists between the rolling elements and the bearing races, even with perfect lubrication. This friction is transmitted to the cage, which causes it to rotate around the bearing races.

Because this is a friction-driven motion, the cage turns much slower than the inner race of the bearing. Generally, the rate of rotation is slightly less than one-half of the shaft speed. *FTF* is calculated by the following equation:

$$FTF = \frac{1}{2} f_r \left[1 - \frac{BD}{PD} \right]$$

Ball-spin frequency

Each of the balls or rollers within a bearing rotates around its own axis as it rolls around the bearing races. This spinning motion is referred to as ball spin, which generates a ball-spin frequency (BSF) in a vibration signature. The speed of rotation is determined by the geometry of the bearing (i.e., diameter of the ball or roller, and bearing races) and is calculated by:

$$BSF = \frac{1}{2} \frac{PD}{BD} \times f_r \left[1 - \left(\frac{BD}{PD} \right)^2 \times \cos^2 \beta \right]$$

Ball-pass outer-race

The ball or rollers passing the outer race generate the ball-pass outer-race frequency (BPFO), which is calculated by:

$$BPFO = \frac{n}{2} \times f_r \left(1 - \frac{BD}{PD} \times \cos \beta \right)$$

Ball-pass inner-race

The speed of the ball/roller rotating relative to the inner race generates the ball-pass inner-race rotational frequency (BPFI). The inner race rotates at the same speed as the shaft and the complete set of balls/rollers passes at a slower speed. They generate a passing frequency that is determined by:

$$BPFI = \frac{n}{2} \times f_r \left(1 + \frac{BD}{PD} \times \cos \beta \right)$$

Defect frequencies

Rolling-element bearing defect frequencies are the same as their rotational frequencies, except for the BSF. If there is a defect on the inner race, the BPFI amplitude increases because the balls or rollers contact the defect as they rotate around the bearing. The BPFO is excited by defects in the outer race.

When one or more of the balls or rollers have a defect such as a spall (i.e., a missing chip of material), the defect impacts both the inner and outer race each time one revolution of the rolling element is made. Therefore, the defect vibration frequency is visible at two times ($2\times$) the BSF rather than at its fundamental ($1\times$) frequency.

44.10.2 Bearings: sleeve (Babbitt)

In normal operation, a sleeve bearing provides a uniform oil film around the supported shaft. Because the shaft is centered in the bearing, all forces generated by the rotating shaft, and all forces acting on the shaft, are equal. Figure 44.47 shows the balanced forces on a normal bearing.

Lubricating-film instability is the dominant failure mode for sleeve bearings. This instability is typically caused by eccentric, or off-center, rotation of the machine shaft resulting from imbalance, misalignment, or other machine or process-related problems. Figure 44.48 shows a Babbitt bearing that exhibits instability.

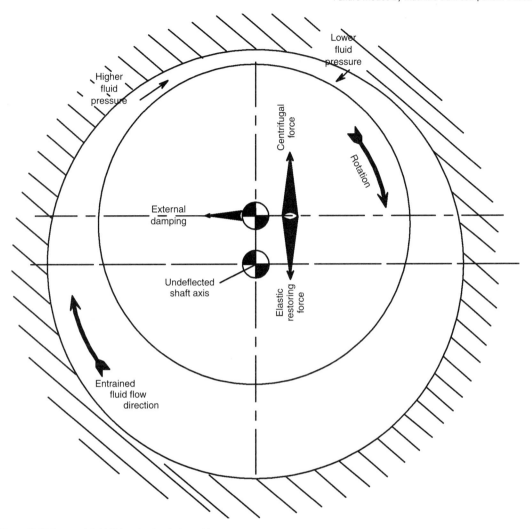

Figure 44.47 A normal Babbitt bearing has balanced forces

When oil-film instability or oil whirl occurs, frequency components at fractions (i.e., $\frac{1}{4}$, $\frac{1}{3}$, $\frac{3}{8}$, etc.) of the fundamental ($1\times$) shaft speed are excited. As the severity of the instability increases, the frequency components become more dominant in a band between 0.40 and 0.48 of the fundamental ($1\times$) shaft speed. When the instability becomes severe enough to isolate within this band, it is called oil whip. Figure 44.49 shows the effect of increased velocity on a Babbitt bearing.

44.10.3 Chains and sprockets

Chains drive function in essentially the same basic manner as belt drives. However, instead of tension, chains depend on the mechanical meshing of sprocket teeth with the chain links.

44.10.4 Gears

All gear sets create a frequency component referred to as gear mesh. The fundamental gearmesh frequency is equal to the number of gear teeth times the running speed of the shaft. In addition, all gear sets create a series of sidebands or modulations that are visible on both sides of the primary gear mesh frequency.

Normal profile

In a normal gear set, each of the sidebands are spaced by exactly the $1\times$ running speed of the input shaft and the entire gear mesh is symmetrical as seen in Figure 44.50. In addition, the sidebands always occur in pairs, one below and one above the gear mesh frequency, and the amplitude of each pair is identical (Figure 44.51).

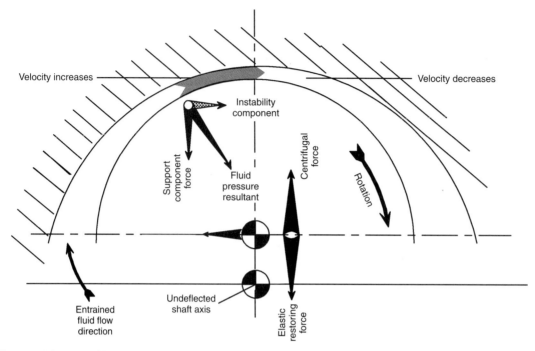

Figure 44.48 Dynamics of Babbitt bearing instability

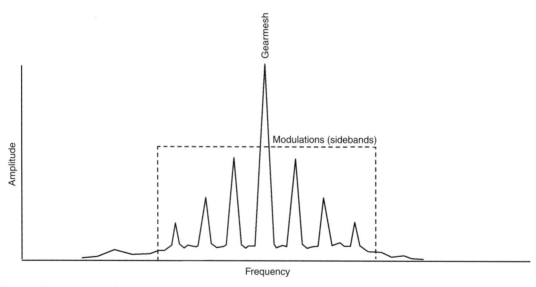

Figure 44.49 Increased velocity generates an unbalanced force

If we split the gearmesh profile for a normal gear by drawing a vertical line through the actual mesh (i.e., number of teeth times the input shaft speed), the two halves would be identical. Therefore, any deviation from a symmetrical profile indicates a gear problem. However, care must be exercised to ensure that the problem is internal to the gears and not induced by outside influences.

External misalignment, abnormal induced loads, and a variety of other outside influences destroy the symmetry of a gearmesh profile. For example, a single-reduction

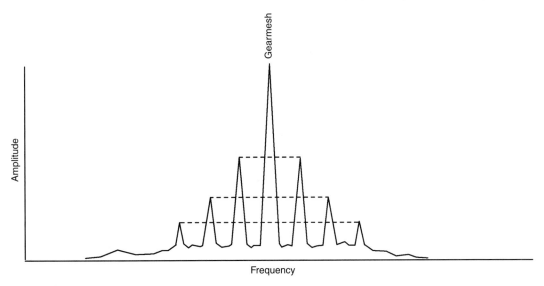

Figure 44.50 Normal gear set profile is symmetrical

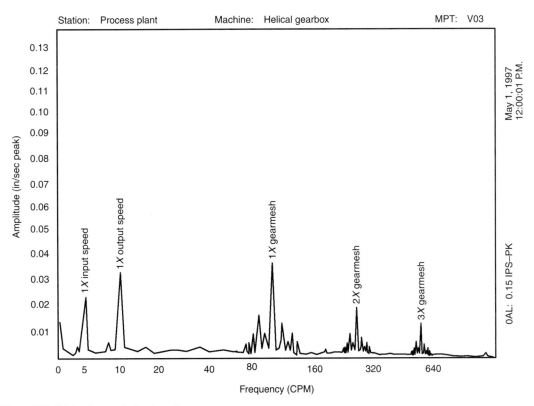

Figure 44.51 Sidebands are paired and equal

gearbox used to transmit power to a mold-oscillator system on a continuous caster drives two eccentric cams. The eccentric rotation of these two cams is transmitted directly into the gearbox, creating the appearance of eccentric meshing of the gears. However, this abnormal induced load actually destroys the spacing and amplitude of the gearmesh profile.

Defective gear profiles

If the gear set develops problems, the amplitude of the gearmesh frequency increases and the symmetry of the sidebands changes. The pattern illustrated in Figure 44.52 is typical of a defective gear set, where OAL is the broadband, or total, energy. Note the asymmetrical relationship of the sidebands.

Excessive wear

Figure 44.53 is the vibration profile of a worn gear set. Note that the spacing between the sidebands is erratic and is no longer evenly spaced by the input shaft speed frequency. The sidebands for a worn gear set tend to occur between the input and output speeds and are not evenly spaced.

Cracked or broken teeth

Figure 44.54 illustrates the profile of a gear set with a broken tooth. As the gear rotates, the space left by the chipped or broken tooth increases the mechanical clearance between the pinion and bullgear. The result is a low amplitude sidband to the left of the actual gear-mesh frequency. When the next (i.e., undamaged) teeth mesh, the added clearance results in a higher energy impact. The sideband to the right of the mesh frequency has a much higher amplitude. As a result, the paired sidebands have a non-symmetrical amplitude, which is due to the disproportional clearance and impact energy.

Improper shaft spacing

In addition to gear-tooth wear, variations in the center-to-center distance between shafts creates erratic spacing and

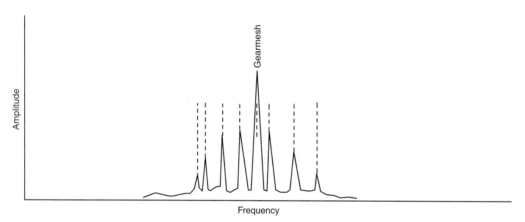

Figure 44.52 Typical defective gearmesh signature

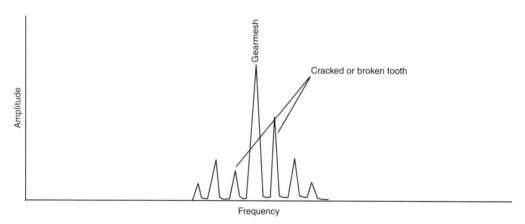

Figure 44.53 Wear or excessive clearance changes the sideband spacing

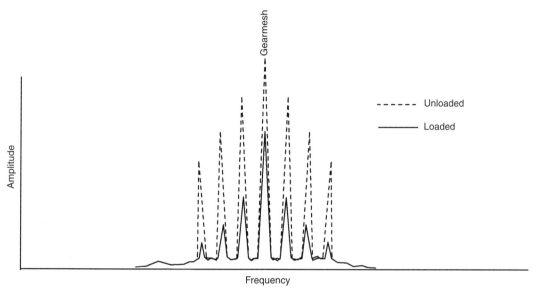

Figure 44.54 A broken tooth will produced an asymmetrical sideband profile

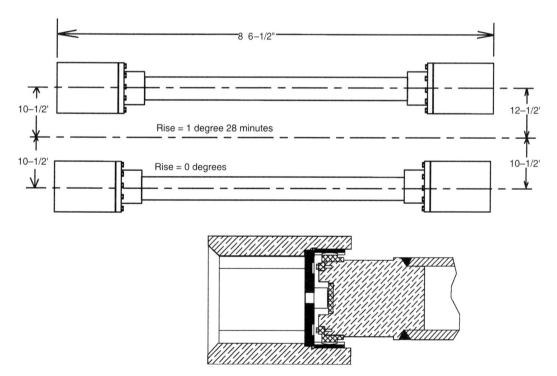

Figure 44.55 Unloaded gear has much higher vibration levels

amplitude in a vibration signature. If the shafts are too close together, the sideband spacing tends to be at input shaft speed, but the amplitude is significantly reduced. This condition causes the gears to be deeply meshed (i.e., below the normal pitch line), so the teeth maintain contact through the entire mesh. This loss of clearance results in lower amplitudes, but it exaggerates any tooth-profile defects that may be present.

If the shafts are too far apart, the teeth mesh above the pitch line, which increases the clearance between teeth and amplifies the energy of the actual gear-mesh frequency and all of its sidebands. In addition, the load-bearing characteristics of the gear teeth are greatly reduced. Since the force is focused on the tip of each tooth where there is less cross-section, the stress in each tooth is greatly increased. The potential for tooth failure increases in direct proportion to the amount of excess clearance between the shafts.

Load changes

The energy and vibration profiles of gear sets change with load. When the gear is fully loaded, the profiles exhibit the amplitudes discussed previously. When the gear is unloaded, the same profiles are present, but the amplitude increases dramatically. The reason for this change is gear-tooth roughness. In normal practice, the backside of the gear tooth is not finished to the same smoothness as the power, or drive, side. Therefore, there is more looseness on the non-power, or back, side of the gear. Figure 44.55 illustrates the relative change between a loaded and unloaded gear profile.

44.10.5 Jackshafts and spindles

Another form of intermediate drive consists of a shaft with some form of universal connection on each end that directly links the prime mover to a driven unit (see Figures 44.56 and 44.57). Jackshafts and spindles are typically used in applications where the driver and driven unit are misaligned.

Most of the failure modes associated with jackshafts and spindles are the result of lubrication problems or fatigue failure resulting from overloading. However, the actual failure mode generally depends on the configuration of the flexible drive.

Lubrication problems

Proper lubrication is essential for all jackshafts and spindles. A critical failure point for spindles (see

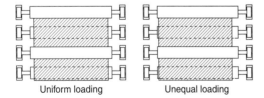

Uniform loading   Unequal loading

Figure 44.56 Typical gear-type spindles

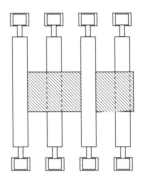

Figure 44.57 Typical universal-type jackshaft

Figure 44.56) is in the mounting pod that provides the connection between the driver and driven machine components. Mounting pods generally use either a spade-and-slipper or a splined mechanical connector. In both cases, regular application of a suitable grease is essential for prolonged operation. Without proper lubrication, the mating points between the spindle's mounting pod and the machine-train components impact each time the torsional power varies between the primary driver and driven component of the machine-train. The resulting mechanical damage can cause these critical drive components to fail.

In universal-type jackshafts like the one illustrated in Figure 44.57, improper lubrication results in non-uniform power transmission. The absence of a uniform grease film causes the pivot points within the universal joints to bind and restrict smooth power transmission.

The typical result of poor lubrication, which results in an increase in mechanical looseness, is an increase of those vibration frequencies associated with the rotational speed. In the case of gear-type spindles (Figure 44.56), there will be an increase in both the fundamental (1×) and second harmonic (2×). Since the resulting forces generated by the spindle are similar to angular misalignment, there also will be a marked increase in the axial energy generate by the spindle.

The universal-coupling configuration used by jackshafts (Figure 44.57) generate an elevated vibration frequency at the fourth (4×) harmonic of its true rotational speed. This failure mode is caused by the binding that occurs as the double pivot points move through a complete rotation.

Fatigue

Spindles and jackshafts are designed to transmit torsional power between a driver and driven unit that are not in the same plane or that have a radical variation in torsional power. Typically, both conditions are present when these flexible drives are used.

Both the jackshaft and spindle are designed to absorb transient increases or decreases in torsional power caused by twisting. In effect, the shaft or tube used in these designs winds, much like a spring, as the torsional power increases. Normally, this torque and the resultant twist of the spindle are maintained until the torsional load is reduced. At that point, the spindle unwinds, releasing the stored energy that was generated by the initial transient.

Repeated twisting of the spindle's tube or the solid shaft used in jackshafts results in a reduction in the flexible drive's stiffness. When this occurs, the drive loses some of its ability to absorb torsional transients. As a result, damage may result to the driven unit. Unfortunately, the limits of single-channel, frequency-domain data acquisition prevents accurate measurement of this failure mode. Most of the abnormal vibration that results from fatigue occurs in the relatively brief time interval associated with startup, when radical speed changes occur, or during shutdown of the machine-train. As a result, this type of data acquisition and analysis cannot adequately capture these transients. However, the loss of stiffness caused by fatigue increases the apparent mechanical looseness observed in the steady-state, frequency-domain vibration signature. In most cases, this is similar to the mechanical looseness.

References

1 *Compressed Air and Gas Data*, Charles W. Gibbs, Ingersoll-Rand Company (1971).
2 *Maintenance Engineering Handbook*, Higgins & R. Keith Mobley, McGraw-Hill, New York (1995).
3 *Predictive Maintenance for Process Machinery*, R. Keith Mobley, Technology for Energy Corp. (1988).
4 *Vibration Analysis*, R. Keith Mobley, Butterworth–Heinemann (1998).

45

Air Pollution

Roger C Webster
Consultant

Contents

This chapter is provided for general guidance. US readers should refer to Clear Air Act for specific guidance.

45.1 Introduction

A pollutant may be defined as any substance or condition which, when present in a sufficient concentration, may have a detrimental effect on the Earth's biosphere. Thus pollution may take place in the form of gases, vapors, particulates or temperature variations (these may be directly or indirectly caused).

It should be noted that these are in terms of short- or long-term exposure levels and are based on an average working week's exposure and an average working life. They do not directly relate to exposure outside the workplace.

It is an accepted practice when assessing the environmental effects of pollution on man and his place of abode to use a divisor of 40 (some agencies may divide by 30) against the long-term exposure level in the Occupational Health and Safety Act (OSHA). Much lower exposure limits are necessary due to the much longer term of exposure in the domestic situation. The section of the population most likely to spend long periods of time in the home are those most susceptible to the detrimental effects of pollutants, i.e. the young, the elderly or the infirm. For short-term exposure the known data can be used directly from the list or from animal-exposure data.

45.2 Effects on plants, vegetation, materials and buildings

45.2.1 Vegetation

Plants have evolved and adapted to suit the atmospheric conditions in which they find themselves. This atmosphere may contain traces of gases which we would classify as pollutants. However, in many cases these may be necessary for the plants' existence. Near to industrial centers, the relative concentrations of the various gases change, and this can have an adverse effect on the plants' development.

The effects of *acid rain* on vegetation have perhaps received the greatest publicity in recent times and form an interesting case study. Numerous agencies have reported damage to forests caused by acid rainfall, but experiments have revealed that the direct impingement of acidic material on the plant cannot produce the effects as noted. Rather, it may be the result of the combination of other environmental factors, along with acid rainfall, which is responsible. It is known that calcium and magnesium are more readily leached out of the soil under acidic conditions and that this can affect aluminum levels in the plants (primarily pine forests), causing them to die. Unfortunately, these effects are very long term, and as the chemistry is most complex the problem is only just beginning to be understood, and it may already be too late to

save many forests both in the US and elsewhere. Acid rain also may cause heavy metals to enter man's food chain via fish or water supplies.

45.2.2 Materials and buildings

Acid rain erodes buildings, particularly those constructed from limestone. It has been reported that the Acropolis in Athens has suffered more deterioration in the last 20 years than in the previous 2000. Acidic gases are produced directly by the combination of oxides of sulfur and oxides of nitrogen with water and also by more complex processes involving unburned hydrocarbons and ozone in the atmosphere.

It has been found that particulate pollution, while causing soiling of materials, may also be responsible for increasing corrosion levels (compared to the corrosion that would be caused by the same level of acid impingement alone) by a process of adsorption. Also, particulates can react synergistically with the acid deposition to cause much greater damage.

45.3 Effects on weather/environment

By far the most serious effect under consideration today is the *greenhouse effect*. The combustion of fossil fuels has resulted in an increase in the carbon dioxide content of the atmosphere. Carbon dioxide has the effect of insulating the Earth against the loss of heat, and hence the mean temperature is in part dependent on carbon dioxide levels. It has been predicted that the carbon dioxide level may double in the next 50 years and produce a global warming of up to 3°C. Scientists currently disagree as to the rate of this warming, but all agree that the situation is very serious.

Ozone forms a layer around the Earth that insulates against thermal radiation. This layer is being destroyed by pollutants (principally fluorocarbons). The effect of the depletion of the ozone layer is to warm the Earth (and hence exacerbate the greenhouse effect) and may also lead to an increase in the incidence of skin cancers.

45.4 Legislation on air pollution of concern to the plant engineer

45.4.1 Alkalis and Works Regulations Act 1906

This Act has been much amended and now forms a 'relevant statutory provision' for the Health and Safety at Work, etc. Act 1974. Originally it provided for the registration and control of certain classes of chemical works. The Act has now been amended and the classes of premises, the 'scheduled works', are now included in the Health and Safety (Emission into Atmosphere) Regulations 1983 (amended 1989). There are 62 main categories of works, and in some cases only certain processes within a category are covered:

Acetylene works
Acrylates works
Aldehyde works
Aluminum works
Amines works
Ammonia works
Anhydride works
Arsenic works
Asbestos works
Benzene works
Beryllium works
Bisulfate works
Bromine works
Cadmium works
Carbon disulfide works
Carbonyl works
Caustic soda works
Cement works
Ceramic works
Chemical fertilizer works
Chlorine works
Chromium works
Copper works
Di-isocyanate works
Electricity works
Fiber works
Fluorine works
Gas liquor works
Gas and coke works
Hydrochloric acid works
Hydrofluoric acid works
Hydrogen cyanide works
Incineration works
Iron works and steel works

Large combustion works
Large glass works
Large paper pulp works
Lead works
Lime works
Magnesium works
Manganese works
Metal recovery works
Mineral works
Nitrate and chloride or iron works
Nitric acid works
Paraffin oil (kerosene) works
Petrochemical works
Petroleum works
Phosphorus works
Picric acid works
Producer gas works
Pyridine works
Selenium works
Smelting works
Sulfate of ammonia works and chloride of ammonia works
Sulfide works
Sulfuric acid (Class I) works
Sulfuric acid (Class II) works
Tar works and bitumen works
Uranium works
Vinyl chloride works
Zinc works

45.4.2 The Clean Air Acts 1956 and 1968

The Clean Air Act 1956

These Acts control smoke from chimneys and open sites. 'Smoke' is not defined but includes soot, ash, etc. The emission of dark smoke (as dark or darker than shade 2 on the Ringleman chart) from chimneys and open sites is made an offence (exceptions are provided for short periods of time).

New furnaces have to be constructed (as far as is practicable) so as to operate smokelessly. Chimney heights are controlled (see below). Smoke Control Orders can be introduced (to control domestic smoke) and grants are available to convert fireplaces to burn authorized fuels. Smoke (other than dark smoke, which is already controlled) is dealt with by Section 16 of the 1956 Act and is, for the purposes of Part III of the 1936 Public Health Act, to be considered as a statutory nuisance.

Railway engines are dealt with in the same way as premises for the purposes of Section 1 of this Act (which controls dark smoke) except that the owner of the engine is to be held responsible. Other Sections of the Act do not apply.

Exemptions from the terms of the Clean Air Act are possible, on application to the local authority, for appliances used for the purposes of investigation and research.

New furnaces shall be fitted with plant to arrest grit and dust if they burn

1. Pulverized fuel; or
2. Solid fuel (or solid waste) at a rate of 1 ton per hour or more. This plant should be approved by the local authority prior to installation (this Section is now replaced by the Clean Air Act 1968, Section 3).

In the case of large installations (burning pulverized fuel, or solid fuel at a rate of 100 lb or more per hour or liquid or gaseous fuel at a rate of 1.25 million BTUs or more) local authorities may serve notice to direct that Regulations shall apply concerning the measurement of grit and dust. These Regulations (the Clean Air (Measurement of Grit and Dust from Furnaces) Regulations 1971) require the owner of a furnace to make measurements and keep records of the grit and dust emissions from the furnace. Local authorities have the power to require information concerning furnaces and the fuel consumed to enable them to perform functions under this Act.

Chimney heights on new buildings are controlled by this Act. However, Section 10 was repealed by the 1968 Act (see below) as far as chimneys serving furnaces are concerned. This Section of the Act does not apply to domestic chimneys or shops or offices.

The Clean Air Act 1968

Section 1 of the Act prohibits dark smoke from trade premises (the 1956 Act only controlled smoke from chimneys). Bonfires are thus now included. Section 2 controls the rate of grit and dust emission from furnaces and the Minister may make Regulations. These are known as the Clean Air (Emission of Grit and Dust from Furnaces) Regulations 1971 and the Clean Air (Emission of Grit and Dust from Furnaces) (Scotland) Regulations 1971.

A breach of these emission levels is an offence under Section 2(2) of this Act. Best practicable means may be used as a fence against this action, and the Regulations prescribe to which classes of appliance they apply. Schedule 1 furnaces are rated by heat output and are boilers or indirect heating appliances where the material heated is a gas or a liquid. The maximum continuous rating concerned are from 825,000 to 475 million BTUs.

Schedule 2 furnaces are rated by heat input and are indirect heating appliances or appliances in which the combustion gases are in contact with the material being heated (but the material does not contribute to the grit and dust). These are a heat input in the range 1.25 to 575 million BTUs. Tables of permissible grit and dust emission rates are given.

Section 3   Subject to Section 4 (which allows for exemptions to Section 3) any furnace to which Section 2 applies shall be fitted with grit and dust arrestment plant if burning

1. Pulverized fuel;
2. Solid fuel at a rate greater than 100 lb or more per hour;

3. Any liquid or gaseous matter at a rate of 1.25 million BTUs or more.

Section 6 of the 1956 Act shall not apply to furnaces if covered by this Section.

Approval is given by the local authority (in writing) to plants installed under this Section. It is an offence to use plant which does not have approved grit and dust arresters. If a local authority refuses approval, an appeal is to be made within 28 days of refusal of consent.

Section 4 – Exemptions   The Minister may make Regulations to exempt furnaces of certain classes. These Regulations are known as the Clean Air (Arrestment Plant) (Exemption) Regulations 1969 (and similarly for Scotland). The exemptions are:

1. *Mobile furnaces* while providing a temporary source of heat or power during building operations, or for the purpose of investigation or research, or for the purposes of agriculture;
2. *Furnaces* other than those burning solid fuel at a rate of greater than one ton per hour – the following classes, used for any purpose other than the incineration of refuse:
   (a) Burning liquid or gases;
   (b) Hand-fired sectional burning solid matter at not more than $25 \, lb/h/ft^2$ of grate surface;
   (c) Magazine-type grate furnaces burning less than $25 \, lb/h/ft^2$ of grate surface;
   (d) Underfed stokers less than $25 \, lb/h/ft^2$ plan area of stoker;
   (e) Chain grate stokers less than $25 \, lb/h/ft^2$ of grate surface;
   (f) Coking stokers less than $25 \, lb/h/ft^2$ of fire bars excluding area of the solid coking plate.

Under sub-Section 2 of the Act the local authority may grant exemptions on application provided that the furnace will not give rise to a nuisance or be prejudicial to health if permitted to operate without arrestment plant. The exemption must specify the purpose for which the furnace is permitted to be used.

Section 5 empowers the Minister to make Regulations to substitute new ratings for the requirement to measure grit and dust and fumes from furnaces under Section 7 of the 1956 Act. The Minister is not empowered to reduce the rate without approval of both Houses of Parliament.

Sub-Section 3 enables an owner to serve a notice on the local authority requiring them to take the measurements that he would be required to obtain under Section 7 of the 1956 Act. The ratings of such furnaces shall be less than one ton per hour of solid fuel other than pulverized fuel or liquid or gaseous-fuelled furnaces with a rating of less than 28 million BTUs.

Section 6   This Section expands on Section 10 of the 1956 Act. Under this Section, the local authorities must approve chimney heights for all furnaces with ratings as per Section 3(1).

Section 10 of the principal Act shall cease to have effect as respect chimneys serving furnaces. This Section applies to all new furnaces and chimneys to furnaces which have had the combustion space increased and to installations which have replacement furnaces of greater ratings than that previously installed. The method specified in the Chimney Heights, Third Edition of the 1956 Clean Air Act Memorandum is used to determine chimney heights (see below).

Section 7 enables the Minister (Secretary of State for the Environment) to apply certain Sections of the Act to fumes by the making of Regulations. None have yet been made.

45.4.3 Relation of the Clean Air Acts to the Alkalis Acts

Section 11 of the 1968 Act states that Sections 1–10 of the 1956 Act and Sections 1–16 of the 1968 Act shall not apply to works subjected to the Alkalis Act (those premises now listed in the Health and Safety (Emission into Atmosphere Regulations)). These premises are therefore subject to enforcement by HMIP. However, sub-Section 3 of Section 11 does contain a proviso for the local authority, upon application to the Minister, to ask for an Order applying the Acts to the whole or part of the schedule works. If an Order is made, best practical means is applied to all (alkali) works whether or not provided for in the two Clean Air Acts.

HMIP publish sets of Guidance Notes, which relate to activities they control. These notes are known as Best Practicable Means (BPM) Notes. BPM is explained in Note BPM 1/88, Best Practicable Means. General Principles and Practices. Copies of these notes may be obtained from the Department of the Environment Publications Sales Unit, Building 1, Victoria Road, South Ruislip, Middlesex HA4 0NZ, telephone: 020 8841 3425.

It has been proposed that a set of scheduled works may be assigned to local authorities. This class of works has been discussed in documents published by the Department of the Environment Air Quality Division and legislation came into force on 1 April 1991. It is proposed that these schedule (B) works will be licensed in much the same way as the existing scheduled works and that prior consent will be needed before operations of this type commence. This will give local authorities a much stronger hand in pollution abatement, and they will be able to avoid the establishment of premises in unsuitable areas or without adequate pollution-abatement equipment. At present, local authorities rely on planning conditions or nuisance provisions.

45.4.4 Health and Safety at Work, etc. Act 1974

Section 5 of this Act places a duty on persons having control of premises to take the 'best practicable means' to prevent the emission into the atmosphere of noxious or offensive substances. This Section is used by HMIP in the enforcement of best practical means for the scheduled processes.

45.4.5 Control of Pollution Act 1974

Under this Act the local authorities may undertake research relevant to the problem of air pollution. The authority may obtain information concerning this subject by serving a notice requiring this information. If they so do, they should specify the information they require and the regularity of the measurements. The authority are not entitled to ask for returns of less than three-monthly intervals and they may not ask for information covering more than one year on any one notice.

If such a notice is received, an occupier may serve a counter-notice on the local authority requesting them to carry out the measurements themselves. The converse may also happen, in that the local authority may serve a notice which empowers them to enter premises to take measurements (after 21 days' notice). If this notice is received an owner can (by notice) require the local authority to allow him to carry out the measurements himself.

45.4.6 Chimney height calculations: Third Edition of the 1956 Clean Air Act Memorandum

The Memorandum is published to help local authorities and industry to derive an approximate height for chimneys. It is stated that the Memorandum should not be used as an absolute criterion but merely as a guide. Valleys, hills and other geographical features will modify the height as derived from the Memorandum.

The Memorandum covers furnaces with a gross heat input in the range of 0.15–150 MW. (The 1968 Clean Air Act, Section 6, requires chimney height approvals for furnaces for burning fuel at a rate greater than 1.25 million BTUs per hour. This is equivalent to 0.375 MW.)

The method is designed to ensure the adequate dispersal of sulfur dioxide (and hence a different method is used for very low-sulfur fuels). It assumes certain efflux velocities, i.e.

For boilers up to 2.2 MW – 6 m/s (full load)
For boilers up to 9 MW – 7.5 m/s (full load)
For boilers where the rating is greater than 135 MW – 15 m/s
Between 9 and 135 MW – pro rata.

The Memorandum does accept that the figure of 6 m/s may be difficult to achieve with small installations.

Very low-sulfur fuels

The first stage of the calculation is the uncorrected chimney height. This is obtained from

$$U = 1.36Q^{0.6}[1 - (4.7 \times 10^{-5}Q^{1.69})]$$

(For heat inputs less than 30 MW the part of the equation in brackets may be omitted.) Q may be obtained from $Q = WB/3600$ where Q is the heat input (MW), W is the maximum rate of combustion of fuel (kg/h for mass or m^3/h for volume) and B is the gross calorific value (MJ/kg or MJ/m^3).

Other fuels

The rate of emission of sulfur dioxide is first calculated from

Oil firing: $R = 0.020 \, WS$
Coal firing: $R = 0.018 \, WS$

where R is the rate of sulfur dioxide (kg/h), W is the maximum rate at which fuel is burned (kg/h) and S is the sulfur content of the fuel (per cent). The equations are different, as it is assumed that in the case of coal firing a certain amount of sulfur will be retained in the ash.

The area in which the chimney is situated is then considered. The classes are:

A: underdeveloped area where development is unlikely
B: partially developed area with scattered houses
C: a built-up residential area
D: an urban area of mixed industrial and residential development
E: a large city or an urban area of mixed heavy industrial and dense residential development.

The uncorrected chimney height is then calculated by plotting the emission rate and the classification on a graph supplied in the Memorandum or by the multiplication of factors supplied:

Type of district	Factor
A	0.55
B	0.78
C	1.00
D	1.30
E	1.60

The uncorrected chimney height (U) is calculated from the sulfur dioxide emission rate (R_a) as previously:

If R_a is less than 10 kg/h: $U = 6R_a^{0.5}$
If R_a is from 10 to 100 kg: $U = 12R_a^{0.2}$
If R_a is from 100 to 800 kg/h: $U = R_a^{0.5} - 0.9R_a^{0.67}$

Corrected chimney heights

A procedure is given to take account of tall buildings near to the stack or if the building to which the stack is attached is less than $2.5 \, U$. The Memorandum suggests that $5 \, U$ should be used as a radius around the chimney as the definition of 'near', but in practice, interpretation and local knowledge may expand this circle.

In the simplest case with one tall building near to the stack (wider than it is tall) the simple equation is

$$C = H + 0.6 \, U$$

where C = corrected chimney height and H = building height. In other cases (with more than one building, for instance) a more complicated procedure is necessary.

All buildings within $5 \, U$ of the stack should be measured and the height and width recorded. The following factors are then determined:

H_m – the largest value of the building height
K – for each building, the lesser of building height or building width
T – each building ($T = H + 1.5 \, K$)
T_m – largest value of T

If U is greater than T_m then U becomes the corrected chimney height. If not, calculate C from

$$C = H_m + U(1 - H_m/T_m)$$

When C is derived certain overriding factors are applied:

1. A chimney should not be less than 3 m from any area where there is access (i.e. ground level, roof areas or operable windows).
2. C should never be less than U.
3. A chimney should never be lower than the attached building within $5\,U$.

Fan dilution

This is also known as air dilution, and is a common procedure for gas-fired boilers. It is allowed for any very low-sulfur fuels (less than 0.04 per cent sulfur) or for fuels in the range 0.04–0.2 per cent S where the VLS fuels chimney heights calculation gives a lower value of U than the 'other fuels' calculations. In these cases, the flue may terminate at the uncorrected height U so long as:

1. The flue gas is diluted with air at a ratio that is known as F, i.e. V/V_0, where V = actual flue gas volume and V_0 = the stoichiometric combustion volume.
2. The actual value of F is a compromise and may be in the order of 20-1. F is used to determine other factors.
3. The emission velocity must be at least $75/F$ m/s.
4. The outlet must not be within $50\,U/F$ of a fan-assisted air intake (except for intakes of combustion and/or dilution air for the boiler in question).
5. The outlet must not be within $20\,U/F$ of an operable window on the emitting building.
6. The distance to the nearest building must be greater than $60\,U/F$.
7. The lower edge of the outlet should be at least 2 m high for boilers related below 1 MW or 3 m for others.
8. The outlet should be directed at an angle above the horizon (preferably about 30°) and must not be under a canopy, or emit into an enclosed wall or courtyard.

The measurement of 'smoke': the Ringleman chart

This chart was devised in the nineteenth century and consists of a white card with black cross-hatching. The percentage of hatching increases by divisions of 20 per cent each Ringleman number. Therefore 0 = white, 1 = 20 per cent, 2 = 40 per cent, 3 = 60 per cent, 4 = 80 per cent 5 = black.

When placed at a distance from the observer's eye the crosshatch lines merge and appear as a uniform shade of gray. There is a standard size for this chart (BS 2742C: 1957), and this has the five shades from 0 to 4. The chart is usually viewed from a distance of 15 meters and hence has to be mounted on a tripod, which should not shadow the card. The chart should be set up in such a position that the smoke being measured has the same sky background as the chart. It should not be placed so that the sun is either directly behind the chart or directly in front of it. Comparisons can then be made between the shade of the smoke and that on the card.

It will be noted that the chart may be impractical to use on a regular basis and thus two other Ringleman charts are used:

1. *The miniature smoke chart*: This does not have crosshatched lines but rather shades of gray, which correspond to the numbers. The chart can be viewed from approximately 1.5 m.
2. *The micro Ringleman*: This is a photographically reduced Ringleman (i.e. with cross-hatching). It has a slot in the middle through which the smoke can be viewed at arm's length.

Instruments have been developed through which the smoke can be viewed and a shaded filter superposed on part of the image. By far the most common of these is the telesmoke. This device can be carried in the Inspector's pocket and most local authorities possess at least one of these.

While the use of a Ringleman chart is to be encouraged, it is not necessary to show, in any legal action, that a chart was used as a reference. Section 34(2) of the 1956 Act states 'for the avoidance of doubt, it is thereby declared that, in proceedings brought under or by virtue of Section 1 or Section 16 of this Act, the Court may be satisfied that smoke is or is not dark smoke as herein before defined [i.e. as dark or darker than Shade 2] notwithstanding that there has been no actual comparison thereof with a chart of the said type'. This would also apply to the 1968 Act as the expression 'dark smoke' is referred back to the 1956 Act. Interestingly, the expression 'black smoke' is not defined in either Act; it is only described in the Dark Smoke (Permitted Periods) Regulations 1958, where it is defined as 'as dark or darker than Shade 4'. Section 34(2) of the 1956 Act would not therefore apply and, strictly, reference to a Ringleman chart should be necessary for any court action to succeed.

Definition of the terms fume, smoke, dust and grit

The most common air-pollution descriptor is the expression 'smoke'. For the purposes of the 1956 and 1968 Acts, smoke includes soot, ash, grit and gritty particles emitted in smoke. Smoke is intended to mean the visible products of combustion and not the invisible ones (CO_2, SO_2, NO_x, etc.) and is used to indicate the degree of completeness of combustion (if combustion is 100 per cent 'smoke' is produced).

Smoke is taken as having a particle size of less than one μm. Dust consists of particles 1–76 μm in diameter. Grit can be interpreted as particles larger than dust. These definitions were taken from the Beaver Report of November 1954 which formed the basis for the 1956 Act.

Fumes are defined in the 1956 Act as being smaller than dust and thus are a constituent of smoke if visible and smaller than 76 μm. However, fumes are traditionally classified as being smaller than 1 μm (see Table 45.1).

Definition of the term nuisance

Statutory nuisance   While certain occurrences are declared to constitute statutory nuisances, there has never

Table 45.1 Definition of terms

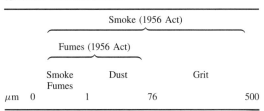

	Smoke Fumes	Dust	Grit	
μm	0	1	76	500

been a formal statutory definition of the term 'nuisance'. The term has been in use since the thirteenth century, and no doubt the legislators felt that there was no need to explain a well-understood concept.

'Nuisance' covers a great variety of areas but it must involve activities, which interfere with the activities of one's neighbors or denies the public their rights. It can involve the blocking of highways or allowing one's animals to escape onto neighboring premises. The two main types of nuisance are private and public.

Private nuisance   Private nuisance is related to the ownership of land and has been defined as 'where a person is unlawfully annoyed, prejudiced or disturbed in the enjoyment of their land, whether by physical damage to the land or by other interference with the enjoyment of land or with his health, comfort or convenience as occupier'. The degree of disturbance must be examined. If physical damage to structures or to health can be shown, nuisance is likely to be proven. To interfere with comfort, substantial interference must occur. The amount of occurrences is important, and case law has stated that a single occurrence does not constitute a nuisance.

Also, the area under question must be considered. Air-pollution levels on a general industrial estate may be considered acceptable, but those same levels may not be so in a residential area. In one case (*Sturgess* v. *Bridgam* (1879)) the judge stated that 'what would be a nuisance in Belgrave Square would not necessarily be so in Bermondsey'.

Sensitivity on the part of the recipient is not considered, even though actual damage may result. Another legal quote is often used: 'Ought this inconvenience to be considered, not merely according to elegant or dainty modes of habit or living, but according to plain and simple and sober notions amongst English people?' (*Walter* v. *Selfe* (1851)).

Prescriptive right is also important in private nuisance. The law of prescription states that if things are done to the knowledge of the occupier of land (likely to be affected) for 20 years a right may exist. It is important to note that the person 'suffering' the nuisance must be aware of the nuisance. In *Sturgess* v. *Bridgam* (1879) this was contested. A doctor had built a consulting room at the end of his garden and was affected by the noise of machinery from his neighbors' house. However, while his neighbor claimed prescriptive right as the noise had been made for more than 20 years, this was refused. As the doctor had not been aware of the nuisance, the right could not exist.

Thus if new houses are built near to a factory the owners of the houses will still have a right to complain about any nuisance caused.

Public nuisance   Public nuisance is the interference with the lawful activities of Her Majesty's subjects or a substantial section of them. One person cannot suffer a public nuisance. Public nuisance is a crime and is actionable by the Attorney General or (under Section 2.2.2 of the Local Government Act 1972) by the local authority. There is no prescriptive right to commit a public nuisance.

Air pollution other than smoke
EC Directive levels for SO_2, NO_x and lead have recently been incorporated into UK legislation in the form of the Air Quality Standards Regulations 1989. SO_2 is generated when high-sulfur fossil fuels are burned. It is thus a factor of the amount of coal and heavy oil (predominantly) being burned.

The recent tendency towards burning more gas and light oils has tended to reduce the industrial emission of SO_2, although power stations remain the major source. European pressure is leading to the reduction of SO_2 from large plant. Existing power stations are to be fitted with flue-gas desulfurization plant and other new plants are likely to use low-SO_2 technology (e.g. pressurized fluidized bed, etc.). Flue-gas desulfurization is unlikely to be a viable proposition for medium-sized plant. It is possible that the new regulation limits may be approached in certain parts of the UK and if this happens, plant design will have to take account of permitted sulfur-emission figures. Designers of any medium-sized plant should consult the local authority at an early stage (before chimney heights application) to discuss this possibility.

NO_x levels are particularly disturbing in some large towns and cities in the UK and are increasing. Whereas SO_2 is generated from sulfur contained within the fuel, NO_x is primarily a combination of atmospheric nitrogen and oxygen, and is generated at a greater level in high-temperature combustion conditions. As the efficiency of furnaces/engines increases, so does the rate of NO_x emission. Levels (other than at roadsides) are less likely to approach regulation levels currently, but are on the increase. A major source of NO_x generation is the internal combustion engine.

NO_x measurement is much more difficult than that of SO_2 and levels seem to vary considerably, depending on the position of the measuring apparatus. SO_2 measurements are much more consistent.

Gas turbines and power stations are particularly prone to generate NO_x and the search for the 'low-NO_x' burner that will operate at high efficiency (i.e. with low hydrocarbon emissions) continues. The principle of the low-NO_x burner is to slow the rate of combustion by dividing it into several stages by the gradual mixing of the combustion gases with the stoichiometric air volume.

Steam or water injection may also be used, and under some circumstances this can reduce NO_x emissions without lowering plant efficiency. Such injection seems to operate in two ways. First, it cools the combustion and hence slows its rate (therefore acting in the same manner as a low-NOₓ burner) and, second, it converts NO_x into nitric acid, which is emitted from the chimney as acidic vapor.

Atmospheric dispersion theory

Fumes and vapors discharged to the environment via a chimney form a plume, which is approximately cone shaped. Mathematical modeling of dispersal rates is possible. The Gaussian dispersion model is commonly used to calculate the concentration of pollutants at coordinate positions X, Y and Z. (The coordinates are measured from the plume centerline.) The equation used is:

$$C_{(x,y,z)} = \frac{Q}{2\pi\sigma_y\sigma_z u} \exp\left[-\frac{1}{2}\left(\frac{y}{y_0}\right)\right]^2$$

$$\times \left\{\exp\left[-\frac{1}{2}\left(\frac{2-H_e}{\sigma_z}\right)^2\right] + \exp\left[-\frac{1}{2}\left(\frac{2+H_e}{z}\right)^2\right]\right\}$$

where

C = concentration at points x, y, z (µg/m),
Q = pollutant emission rate,
U = mean wind speed affecting plume (m),
z = standard deviation of plume concentration in the vertical at distance X (m),
y = standard deviation of plume concentration in the horizontal at distance X (m),
X, Y and Z are the coordinates with the base of the stack as the origin,
H_e is the effective stack height (see below).

The equation simplifies considerably when we consider only ground-level concentrations directly in line with the plume, i.e. X and Z are equal to zero:

$$C_{(X)} = \frac{Q}{\pi\sigma_y\sigma_z u} \exp\left[-\frac{1}{2}\left(\frac{H_e}{\sigma_z}\right)^2\right]$$

Effective height (H_e)

The amount by which a plume rises above the top of a chimney can be derived mathematically.

Hot buoyant plume

$$\Delta H = 20.5Q_h^{0.6}H_s^{0.4/u}$$

where

Q_h = rate of sensible heat emission from the chimney (MW),
U = wind speed at the top of the stack,
H_s = actual chimney height,
$H_e = H_s + H$

Cold plume

$$\Delta H = 3W/ud \qquad \text{if } W/u \geq 4$$

where

W = efflux velocity of plume at chimney top (m/s)
d = internal diameter of top of chimney (m).

Table 45.2 Pasquill's stability categories

	Day			Night	
	Incoming solar radiation			<3/8	>4/8
	Strong	Moderate	Slight	Cloud	Low cloud
<2	A	A–B	B	G	–
2–3	A–B	B	C	F	E
3–5	B	B–C	C	E	D
5–6	C	C–D	D	D	D
>6	C	D	D	D	D

A and B are the most unstable, G is the most stable and D is neutral.

If $W/u < 4$, plume rise above top of chimney should be ignored. Again, $H_e = H_s + \Delta H$.

Atmospheric stability

Atmospheric stability and mechanical turbulence (important near to the ground) are used to derive the vertical and horizontal dispersion coefficients. Table 45.2 shows Pasquill's stability categories used to derive the coefficients by reference to standard graphs.

The dispersal of the plume at X, Y and Z is determined by the values of σ_y and σ_z. Small-scale eddies can affect dispersal near to the source and larger-scale ones are needed before effects are noted at greater distances from the source. Y and Z thus have large orders of magnitude furthest from the source and increase if a larger time period is used for sampling. It is thus important to state the sampling period used. The trend towards changes in σ_y and σ_z are thus

Near to stack – ground turbulence dominates (coefficients much greater in urban areas);
Far distance – coefficients much smaller with stable conditions.

Other models (or combinations of them) are often employed when computers are used to analyze dispersal. These can give an acceptable degree of accuracy when combined with detailed weather data. Short-exposure modeling is the most difficult and is liable to the greatest degree of error. It is for this reason that such models are not accurate when dealing with odor nuisances. The problem of modeling odor dispersal is dealt with below.

Other special atmospheric conditions can interfere with the modeling process and the most common of these is temperature inversion. This condition is so called because the air temperature increases with height above the ground, the converse of the situation that pertains for most of the time.

Temperature inversions

There are two common types of inversion as follows.

Ground base inversions On clear nights when there is strong radioactive cooling of the ground, the inversion starts at ground level and extends upwards to 100 m or more.

Elevated inversions Elevated inversions begin at some height above the ground. Inversions affect flue gas dispersal in the following ways:

1. *Ground based*: Upward dispersal is very slow. Horizontal dispersal proceeds normally. For low chimneys, ground-based concentrations can be high, especially when the ground is heated in the mornings and eddies are caused which bring down plume gases.
2. *Elevated*: Plumes can be trapped either above or below the base of the inversion and held in a horizontal plane. Again, these can be brought down to ground level by eddies. This process is known as 'fumigation' and can result in short-term high-level concentrations.

Odor dispersal

Odors may be detected for a very short exposure period, perhaps less than one second. It is thus necessary to determine the likely one-second peaks knowing the concentrations derived from the Gaussian equation. This is based on 10-minute average period. The equation to convert the time-averaging period is:

$$C_1 = C_{10}\left(\frac{t_0}{t}\right)^p$$

where $p = 0.17-0.2$. Therefore in this case

$$C_1 = C_{10}\left(\frac{600}{1}\right)^{0.17} = C_{10} \times 2.96$$

We must thus multiply the Gaussian concentrations by approximately three to obtain the short-term peaks. Other sources have suggested that the multiplication factor may be higher.

If the chimney were designed to avoid ground-level concentrations exceeding the nuisance level based on the Gaussian equation, we would find short-term peaks that would produce nuisance. To avoid nuisance for a 5:1 ratio of 10 minutes to peak exposure we would need to double the chimney height from that derived for the Gaussian equation.

Nuisance levels for odors are not absolute, but are related to the minimum detectable level for 50 per cent of the population. These levels have been explored by Warren Spring Laboratories, who have concluded that five times the minimum detectable level is likely to give rise to complaint.

The maximum ground-level concentration calculated on a 10-minute basis for no nuisance should therefore be equal to the detection threshold, i.e.

Peak = 5 × 10-minute level
Nuisance = 5 × detection level

Therefore for no nuisance, 10-minute level = detection level.

Note that the ground-level concentrations do not depend on the flue gas concentrations but rather on the amount of pollutant emitted. It is therefore not worth diluting the flue gases with fresh air (other than to raise the efflux velocity).

46

Dust and Fume Control

Brian Auger

J. B. Auger (Midlands) Ltd

Contents

46.1 Introduction

Dusts and fumes have been a part of industrial life for many years and the hazards associated with them are well known. The diseases and respiratory disorders found in foundries, potteries and cotton works are examples that are familiar to many. These need not occur with a better understanding of control measures and more efficient equipment.

Attitudes to dusts and fumes have changed over the years. It is no longer either permissible or tolerated for workers to be subjected to unhealthy or unsafe conditions at work. Society as a whole is demanding a cleaner environment, and this has led to new legislation being enacted. The effect of these changes on the plant engineer has resulted in a great increase in the burden of work and in the content and level of knowledge of the subject. He now has to take account of the workings and activities within the plant and its effect on people and surroundings outside it. To ensure that dust and fume control is effective in meeting its objectives it is imperative that those responsible for the specification and running of the plant are well versed in the nature of dusts and fumes, methods of entrainment, available equipment and test and inspection techniques.

46.2 The nature of dusts and fumes

Dust can be said to be a solid broken down into powder, and the form that it takes will have different effects on the body. Fibrous dusts can attack tissue directly while others may be composed of poisons, which are absorbed into the bloodstream. For the purpose of this chapter, fumes can be regarded as very small particles resulting from the chemical reaction or condensation of vapor, which can have the same two effects. It is not necessary for plant engineers to have an in-depth knowledge of pathology, only that they must be aware of the possible results of exposure to dusts and fumes.

The standard unit normally used for measuring dust particles is the micron (μm: one-thousandth of a millimeter). The smallest particle visible to the unaided eye is between 50 and 100 μm and the most dangerous sizes are between 0.2 and 5 μm. Particles larger than this are usually unable to penetrate the lung defenses and smaller ones settle out too slowly. Some dusts can be both toxic and fibrous (e.g. asbestos) and are therefore harmful even outside these parameters. It may therefore be assumed that dusts which are visible (i.e. between 50 and 100 μm), are quite safe. However, this is not the case, as dust clouds never consist solely of particles of one size. Analysis would show percentages of all sizes, and it is for this reason that special care is needed in measuring dust clouds and concentrations.

The next most important factor to consider when assessing dust clouds is the actual amount of dust present. This is known as the concentration, and is defined as follows. A substantial concentration of dust should be taken as concentration of 10 mg/m^3 8-hour time-weighted average of total inhalable dust or 5 mg/m^3 8-hour time-weighted average of respirable dust where there is no indication of the need of a lower value. There are now many lists available for consultation which set out the safety limits as they are known at present. If a substance is not listed this is not always an indication that it is safe, and the general rule should be applied that dust in any substantial concentration is hazardous. Even with dusts and fumes that are listed and have set limits it must be borne in mind that our knowledge is always growing, and that the standards of today may be obsolete tomorrow.

Clearly, the assessment of dust clouds and concentrations cannot be left to the casual practitioner, and this has now become the specialist field of the industrial hygienist.

46.3 Control of dusts and fumes

The purpose of the control plant is to maintain a working environment that is acceptable in terms of any statutory regulations and the custom and practice within an industry. The effectiveness of a control system is measured by the amount of dust or fumes it controls. Efficiency, on the other hand, is measured by the amount of power it takes to do the work. It is the job of the dust-control engineer to produce the most effective plant in the most efficient way, and the techniques of control will vary from one industry to another. All control plants will have either four or five elements, as shown in Figure 46.1, i.e. hoods, ducting, fan, collector and disposal.

The collector is not always used, as many systems still discharge untreated air to atmosphere. The growing awareness of environmental matters will, in time, see all such systems having collectors or treatment plants.

46.3.1 Hoods

This is the inlet into the system, and will be the single most important element in determining the effectiveness of the control plant. A study of the dust- or fume-producing process is necessary to ensure that the twin aims of effectiveness and efficiency are met. Hoods that totally enclose the process for maximum effectiveness may, however, prevent the operator from carrying out the process for which the control was needed in the first place.

There are four rules for the design of effective hoods, irrespective of the process:

1. The hood should be as close to the source of generation as possible.
2. The location and shape of the hood is such that the contaminant is thrown into the hood.
3. The air flowing past the source and into the hood has a velocity at the origin greater than the velocity of escape of the contaminant.
4. The hood is located so that the operator is never between the source and the hood.

With these rules four types of hoods have evolved.

Total enclosures

These ensure that the process is totally enclosed, which prevents any leakage to the workplace. It is rarely used

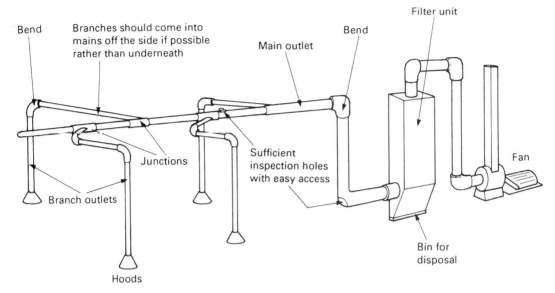

Figure 46.1 Horizontal duct with adequate carrying velocity. Ducting of increasing size to accommodate increasing airflow

except for very dangerous materials such as radioactive particles in the nuclear industry or biological particles in the drugs industry. The enclosure is subject to negative pressure, and because the source is inside, manipulators must carry out any work done on the source. A modification of the total enclosure is one where part of the enclosure is removed for manipulation of loading. In this case the velocities across the open faces of the hood must be sufficiently high to prevent emissions or escape. This technique is used on bucket elevators, conveyors and holding bins.

Booths

These are really enclosures with one whole side removed where the source is deep within the booth. They are particularly suitable where the particles are not moving at high speed (e.g. filling and weighing operations). When the booth is used where high-speed particle generation occurs (e.g. grinding and fettling) careful thought must be given to the depth. As booths tend to be large, the restraining velocities across the open face must only be sufficient for control, otherwise the efficiency of the plant would be low with very high velocities and hence volumes.

A development of the simple booth is the laminar flow system, which uses a nominal velocity at the face of 0.5 m/s. As the airflow is laminar, all parts of the booth are subjected to its effect.

Captor hoods

In many processes it is not possible to use enclosures or booths without imposing unacceptable operational restraints (see Figure 46.2). The hood must be placed at some distance from the source. The natural projection of the

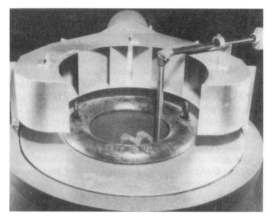

Figure 46.2 Captor-type claw hood for a furnace

particles will not necessarily be into the hood, and may indeed be in the opposite direction. To be effective, these hoods require high face velocities to give control (e.g. up to 10 m/s). Hoods for controlling fumes fall into this category, especially on tanks and vats.

Receptor hoods

These derive their name from the method of entrainment. The hood is placed in the path of the particle and uses the momentum of the particle to assist in control. It is important to have a clear understanding of the process, the direction particles take and the movement of air created by the process. Grinding wheels release dust downwards and over the top of the wheels, and considerable air movement

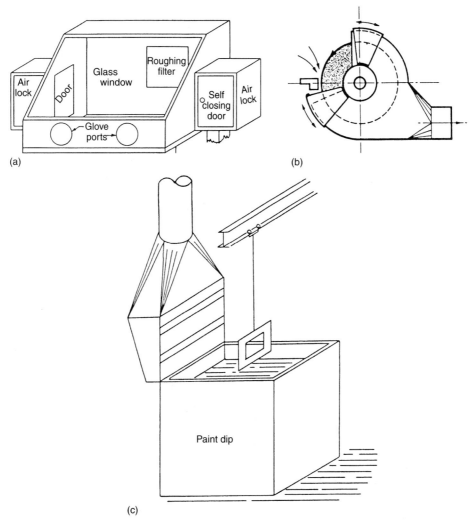

Figure 46.3 Types of hoods. (a) Enclosure; (b) receptor; (c) captor

is generated by the process. These are ideally suited to control by receptor hoods (see Figure 46.3).

There are many machines which arrive on the factory floor with correctly designed and tested hoods supplied by the machine manufacturers. This is a trend that has grown over the years, and is one to be encouraged. Hoods supplied in this way can have the volume specified, which, in turn, will ensure that the control meets with the appropriate regulations.

46.3.2 Ducting

The purpose of the ducting is to convey the entrained contaminants away from the sources to a collection or disposal point. It is very often a neglected part of the system.

Different processes require different specifications, and the following will act as a guide in selecting which type of ducting is necessary:

1. The gauge of ducting should take account of the nature of the particles and the operating pressures of the system.
2. The methods of construction usually encountered are (a) lock-formed, (b) welded, (c) slip jointed and (d) flanged.
3. The construction materials should be selected to withstand the operating conditions and the condition of the pollutants. Galvanized sheet steel and black mild steel are the most common for general work. Corrosion or heat applications will have ducting constructed in stainless steel or plastic.

4. The standard gauges (mm) for various works are:

General dust work		Light-duty work
0–450	1.0	0.8
Bends	1.5	0.8
450 and over	1.5	1.0
Bends	2.0	1.0

All ducts over 450 flanged

Foundry work		Oily and wet fumes
0–450	1.5	1.0
Bends	2.0	1.0
450 and over	2.0	1.5
Bends	3.0	1.5

All ducts over 450 flanged

5. The design of the ductwork must ensure that the plant is both effective and efficient. Sharp bends and abrupt entries of branches into mains cause unnecessary pressure losses. Incorrectly sized ducts result in high pressure losses or blockages due to fallout from velocities being too low.
6. The ducting should be adequately supported and fitted with inspection doors or ports.
7. The termination point in any system is the discharge cowl, and many designs are used.

46.3.3 Fans

Fan engineering is a basic technology, with its origins in ancient times. It was developed from the wheel and pump as the need grew for continuous quantities of moving air at low pressures. By the nineteenth and early twentieth centuries much of the design and research had been done, resulting in an unsophisticated but reliable air mover. Fans are capable of operating over a wide range of duties, albeit with varying degrees of efficiency, and this has led to their misapplication and abuse in dust- and fume-extraction work. Without the fan a system will not function and it is therefore necessary to select the fan to ensure both effective and efficient running of the plant. A selection of fan types together with application data is shown in Figures 46.4 and 46.5 and Table 46.1.

Fan performance and laws

In selecting a fan for a system the two most important characteristics to consider are volume and pressure. This can be said to be the movement of air (m^3/h) against the system resistance (mm water). As the majority of these systems operate at near-normal temperatures and altitude, density is usually ignored. If the system is to work at temperatures and altitudes other than 25°C and at sea level then density must be considered in the selection. Air movement is not a precise science and fans selected for systems rarely operate at the exact design duty. Furthermore, the system can change, and the fan performance will also need to change to meet the new conditions. If the performance varies by no more than 5–10 per cent it is seldom necessary for remedial action. To alter the performance of a fan without changing the geometry the fan speed must be increased or decreased.

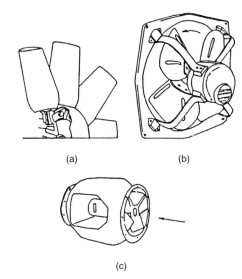

Figure 46.4 Axial-flow fans. (a) Axial flow; (b) propeller; (c) bifurcated

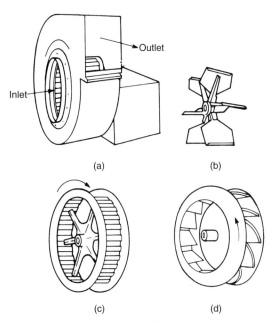

Figure 46.5 Centrifugal types of fan. (a) Centrifugal; (b) paddle or radial bladed; (c) forward curve; (d) backward curve

When the speed is changed it is prudent to check the performance from the following fan laws:
Fan speed varies – size and density constant

1. Volume flow varies directly as fan speed:

$$\frac{\text{Volume 1}}{\text{Volume 2}} = \frac{\text{Speed 1}}{\text{Speed 2}}$$

Table 46.1 Application data for fans

Type	Pressure volume	Efficiency	Industry	Normal drive	Application
Axial	2.5 in w.g. high volume	Very high	H and V	Direct	General use for ventilation, heating and minor fume work on low-pressure systems
Propeller	0.4 in w.g. high volume	Low	H and V	Direct	Usually applied on free air work, such as input and output units for buildings due to pressure limitations
Bifurcated	1.0 in w.g. up to approx. 10 000 CFM	Medium	Fume	Direct	Motor not in air-stream. Used on explosive fume, wet fume, high-temperature work and severe applications
Paddle	12 in w.g. up to approx. 30 000 CFM	Medium	Dust and fume	Vee and direct	General dust and fume. Will handle air containing dust and chippings. Wide application in wood-waste extraction plants
Forward	6.0 in w.g. very high volume	High	H and V	Vee and direct	Will only handle clean air. Compact and quiet running. Used on heating, ventilation and air-conditioning work
Backward	20 in w.g. high volume	High	Dust and fume	Vee and direct	General dust and fume. High-pressure systems and on dust-collector plants. Will handle some dusty air
Blowers	42 in w.g. usually low volume	Medium	General	Direct	Furnace blowing, cooling, conveying and where there is a need for high pressures

2. Pressure varies as the square of the fan speed:

$$\frac{\text{Pressure 2}}{\text{Pressure 1}} = \frac{(\text{Speed 2})^2}{(\text{Speed 1})^2}$$

3. Power varies as the cube of the fan speed:

$$\frac{\text{Power 2}}{\text{Power 1}} = \frac{(\text{Speed 2})^3}{(\text{Speed 1})^3}$$

A typical example will illustrate their use. A 600 mm diameter paddle-bladed centrifugal fan delivers 5100 m^3/h against a resistance of 200 mm water and absorbs 5.16 kW when running at 1665 rev/min. The motor for driving the fan is 7.5 kW at 2-pole speed through v-belts and pulleys. What would be the new speed, pressure and power required for the volume to be increased to 7000 m^3/h?

$$\frac{7000}{5100} = \frac{\text{New speed}}{1665}$$

$$\text{New speed} = \frac{7000 \times 1665}{5100} = 2285.29 \text{ rpm}$$

$$\frac{\text{New pressure}}{200} = \frac{(2285.29)^2}{(1665)^2}$$

New pressure $= 376.77$ mm

$$\frac{\text{New power}}{5.16} = \frac{(2285.29)^3}{(1665)^3}$$

New power required $= 13.32$ kW

For the fan to operate within the same duct system at the increased duty to give the new volume an up-rated drive motor of at least 11 kW is required. Care must always be exercised in using the above laws so as not to exceed the drive capacity of the motor or the critical speed of the fan. No absolute guidelines can be given to cover all fans at all speeds due to the wide range of designs and materials used. If, on recalculation, the new fan speed at the blade tip exceeds 76 m/s no action should be taken without consulting the manufacturer. In the above example the tip speed is 71.8 m/s and would, in most circumstances (depending on the condition of the fan), be safe to run.

46.3.4 Collectors

The function of the collector is to separate the entrained dust or fume from the air stream and deposit it in a convenient form for ultimate disposal. The four most common collectors are: cyclones, fabric filters, wet collectors and electrostatic filters. Each of the above types has been developed to meet the needs of industry and a brief description here will enable the plant engineer to make a start in the selection of a particular application.

Cyclones

There are two basic designs in use. The first is the general-purpose large-diameter cyclone, traditionally used in metal polishing and wood-waste extraction. The second is the so-called high-efficiency small-diameter cyclone, used in groups for pre-separation or in metal grinding exhausts. If the cyclone is to remain in use it will be confined to those operations where further collection of fines is part of the overall system. This is because they are relatively inefficient on the smaller particle sizes and release an unacceptable amount of discharge into the atmosphere.

Wet collectors

There are two types: self-induced spray and pressure spray. Many designs have been developed to suit specific dust or fume problems, and it is now possible to use wet collectors on the treatment of solids and fumes down to very small particle sizes.

Self-induced spray wet collectors   This is the most common type, and relies on its separating action by the induced air from the fan pulling the contaminated air through a curtain of water, It is simple in operation with no pumps or moving parts except for the fan, which is set on the clean side of the collector. The scrubbing action is dependent on the pressure drop across the collector. When set, this is constant and is determined by the water level within the collector. The removal of sludge is either by automatic ejection or manual drag-out.

This type of collector has found wide application in general engineering and very high collection efficiencies are possible, but at the expense of considerable power requirements. General-purpose collectors at pressure drops across the collector of 150 mm will have collection efficiencies of 98 per cent at 10 μm and above. Units with pressure drops of 800 mm and efficiencies of more than 99 per cent on sub-micron particles are available.

Spray-type collectors   In this system water is sprayed or cascaded onto the contaminated air directly or through packed towers, and the fumes or dust are washed away by absorption. These collectors are used extensively on the treatment of fumes of all types and have low pressure drops and hence low power requirements compared to induced spray. A development of this collector is the venturi scrubber, which injects high-pressure water into a venturi through which the fume-laden air is passing. The intimate contact of the two ensures absorption and removal from the air stream. These collectors are used in fume removal and have efficiencies of more than 99 per cent on sub-micron particles.

The general rule with wet collectors is that the higher the collection efficiency, the greater the pressure drop and hence the power absorbed.

Fabric filters

Fabric filters are capable of separation efficiencies approaching 100 per cent if correctly applied. The ideal filter is permeable to the air stream but not to the dust requiring separation. Separation occurs by impacting of the dust particles upon fabric fibers, resulting in a dust cake forming on the fabric. Although aiding filtration, this cake does increase the resistance to airflow. If allowed to go unchecked it would result in a reduction of the total air being exhausted in the system.

Two methods are used to remove the dust cake, both of which require interruption of the airflow. The difference in dust-cake removal conveniently divides filters into intermittent and continuous rating. In the intermittent type the pressure increases (with time) up to a pre-arranged level. The airflow is then stopped and the fabric is mechanically shaken. In the continuously rated filter the pressure drop rises to a low set point, after which it remains constant across the filter as a whole. The cleaning is done by isolating a part of the filter from the air stream and that section is cleaned.

Intermittent filters are best suited to small applications which will allow the process to be stopped at intervals. The interval used is 4 h (i.e. a morning or afternoon shift). Mechanical shaking is done by either hand or electric motor. The application of these filters is limited to the incoming dust burden of the order of 5 g/m^3 and is known as nuisance dust.

Continuously rated filters have whole sections of the filter shut off from the airflow and then those sections are shaken or cleaned. Shaking is carried out in sequence, usually by electric motor. Where the filter is cleaned it is done by a jet of compressed air being blown in reverse to the airflow through the fabric. This system does not require whole sections to be shut down, as the reverse blow is carried out when the filter is on-stream. The time of blow is very small and is measured in parts of a second rather than in minutes, as in the case of shaking filters. The application of these filters is in continuous processes and where the dust burdens are high (in excess of 100 g/m^3).

The shaking and continuous filters are regenerative, but there is a third group usually associated with ventilation work rather than dust and fume. These are throwaway filters, which, as the name implies, means that when they become too caked with dust to operate correctly the filters are removed and replaced with new ones. They will only handle low incoming dust burdens, but their efficiencies are the highest of any filter. Typical applications are fresh air input plants, clean-room filtration and nuclear processes.

Electrostatic precipitators

These remove particles by means of applied electrical forces, and are used extensively on cement and fly ash removal from air streams. The particles are first given an electric charge and are then passed through an electric field to apply a precipitation to them. They are then captured on the collecting surface (normally an electrode). The collection surface is the cleaned by rapping or water wash and the dust collected in a hopper below. The efficiencies achieved are 99 per cent at particle sizes above 1 μm. They are cheap to run but the capital costs are high, especially for large collectors. In recent years small units have been developed to control oil and welding fumes.

Disposal

Many control plants fail in their objective when the collected waste is removed from the inlet. The practice of dropping dust into sealed bins at the base of the collector is both sensible and practical. It is when the bins are to be emptied that a secondary dust problem arises. If the waste is simply put into a larger container with no control, dust will be released back into the workplace. If dangerous dusts are being collected, sealed inner liners to the dustbins can be used, thereby preventing this release. On the larger installations the collected dust is retained under sealed conditions at all times and the discharge from collectors is by rotary valve and screw conveyor. These will feed into bulk containers for further processing. In the case of fumes these are collected by absorption into liquors, and these liquors are treated in an effluent plant separate from the fume plant.

46.4 System design and application

In selecting the best system to control the hazard the following should be noted:

1. The problem must be surveyed under actual working conditions and data collected and recorded in a logical manner. The survey must establish:

   (a) Origin of the dust or fume and its nature. Is it toxic, explosive or hazardous in any way?
   (b) The process which produces the contaminant. Is it wet or dry?
   (c) Whether the problem can be solved by elimination of the process;
   (d) Source or sources of the problem. Does it occur at more than one point?
   (e) Do any special regulations apply to the hazard and are the materials being handled listed in any published form as having control limits?
   (f) Is a control system in use at the time of the survey?

Having completed the survey, the next stage is to draw the system showing the sources of dust and the duct runs. An assessment must then be made on the air volumes and velocities required giving control. This is largely a matter of experience, as air-entrainment rates are derived empirically. It is possible to calculate the rates but is unusual in general engineering. There are published lists for air rates and many companies have their own standards.

Table 46.2 shows rates for metalworking machines. When the total air volume has been established the collector and fan can be sized. The total air volume

Table 46.2

Type of machine	Size (mm)	Air volume (m^3/h)
Grinders (double end)	200–250	680
Grinders (double end)	280–406	1189
Grinders (double end)	430–455	1495
Grinders (double end)	480–560	1870
Grinders (double end)	585–762	2720
Grinders (double end)	787–915	3738
These figures are for bottom connections only and average duties		
Toolroom grinders (double ends)	Up to 200	510
Cutter grinders (single end)	Up to 150	510
Banding machines		
Horizontal	Up to 100	595
Horizontal	125–200	1019
Horizontal	38 belt	595
Backstands	50 belt	866
Backstands	75 belt	1053
Backstands	100 belt	1257
Backstands	125 belt	1699
Backstands	150 belt	2243
Double-end hand-polishing machines	Up to 150	1019
	178–225	1359
	250–355	2209
Wire mops	Up to 200	1699
Cut-off machines		
Abrasive discs	Up to 405	934
Abrasive discs	430–610	1699

Table 46.3

Application	Type of collector
Sand foundry dusts	Self-induced wet collectors
Magnesium and aluminum working	Self-induced wet collectors
Hot applications	
Sparks from grinding	Self-induced wet collectors
Heat treatment	
Small grinding machines, polishing machines	Self-contained intermittent filters
Batch operations (e.g. filling, tipping, mixing)	Self contained intermittent filters
Process plant	Continuously rated filters
Heavy dust burdens	Continuously rated filters
Difficult dusts (e.g. carbon black sugar dust)	Continuously rated filters
Cement and fly ash production	Electrostatic precipitators
Woodworking and initial separation	Cyclones

and the type of dust will determine the size and type of collector to be used. Table 46.3 shows the types of collector for various applications.

		SHEET 1
COSHH LEV RECORD		
NAME OF EMPLOYER		**DATE**
ADDRESS		
DEPARTMENT OR SITE OF PROCESS		**REFERENCE NUMBER**
LOCATION OF LEV PLANT		**TYPE OF LEV PLANT**
IS THE EXHAUST AIR RETURNED TO THE WORKPLACE AND IF SO IN WHAT CONCENTRATION		**IS SECONDARY FILTRATION PROVIDED**
SUBSTANCE/PROCESS CONTROLLED		**HAZARD ASSESSMENT**
ASSESSMENT OF LEVEL OF CONTROL AND TYPE OF TEST		**CONDITION AT TIME OF TEST**
FREQUENCY OF TEST		**REGULATION**
COMMENTS		
NAME OF PERSON RESPONSIBLE		**DATE FOR TEST**
SIGNATURE		**EMPLOYER**

Figure 46.6 A typical assessment sheet

								SHEET 2

COSHH LEV RECORD

LAYOUT SHOWING POINTS OF MEASUREMENT

Date Installed

PRIMARY COLLECTOR/FILTER			SECONDARY FILTER	
Make	S.P.Outlet		Type	S.P.Outlet
Size/Area	S.P.Inlet		Size/Area	S.P.Inlet
Media	P.D.Across		Media	P.D.Across

MOTOR			FAN SET	
Supply	Make		Size	Volume
Amps	Type		Type	Speed
Power	Speed		Handling	S.P.Inlet
Starting	Encl.		Drive	S.P.Outlet

HOOD AND TRANSPORT SIZES, VELOCITIES AND VOLUMES

Point	Hood Size Area	Duct Size Area	V.P. in.W.G.	S.P. in.W.G.	Velocity	Volume	Assessment of Control Remarks

Notes

Figure 46.6 (*continued*)

								SHEET 3
								LEV NUMBER

COSHH LEV REPORT

NAME AND ADDRESS OF EMPLOYER	DEPARTMENT OR SITE OF PROCESS

REGULATION	PERIOD BETWEEN TESTS	LAST TEST

HOODS, ENCLOSURES AND DUCTING

Point	S.P. in.W.G.	Are Elements in Working Order Answer Yes, No or N/A.				Assessment of Control. State Tests Used.	Remarks
		Hood	Enclos.	Duct	Fan		

COLLECTORS/FILTERS	SECONDARY FILTER	
S.P.Outlet	Quantity of	Monitoring
S.P.Inlet	Contaminant in	Equipment
P.D.Across	Returned Air	if Fitted
Condition of	Test Method	Condition
Collector/	Used	of Filter
Filter		

Particulars of Repairs or Modifications Required to Ensure that the LEV Plant Effectively Controls the Dust or Fumes

SIGNATURE	DATE

Figure 46.6 (*continued*)

To size the fan it is necessary to know the total air volume and the pressures in the system. These are calculated from the losses in the system on the longest or index leg, and begin with the hood. The hood entry loss can be expressed as 0.6 of the velocity head and is accurate enough for first estimates. The losses are then calculated on the velocities in the ducts. Each change of direction means a small loss in each length of duct. Added to the pressure drop loss across the collector and the outlet losses, these give the total static pressure required in the system.

The formula used in the calculation of system losses for first estimates favored here is:

Loss in water (mm)/length of duct (m)

$$= \frac{(\text{Velocity (m/s)})^2 \times 1.243}{\text{Diameter of duct (mm)}}$$

Velocity head or velocity pressure is the pressure required to accelerate the following mass from rest to its flowing velocity:

$$\text{Velocity pressure in water (mm)} = \frac{(\text{Velocity (m/s)})^2}{(4.04)^2}$$

Static pressure is the pressure to overcome resistance and is expressed in millimeters of water. Total pressure is the sum of velocity and static pressures.

The basic formula for airflow under all conditions is:

$$Q = A \times V$$

where Q = volume (m^3/s), A = area (m^2) and V = velocity (m/s). The nominal velocities used in dust and fume control depend on the materials being handled, but the following will suffice for most work:

Metalworking	Ducts 23 m/s	Branches 20 m/s
Woodworking and		
light dust	Ducts 20 m/s	Branches 18 m/s
Gases and vapors	Ducts 10 m/s	Branches 9 m/s

It is now possible to calculate the total system resistance and select a suitable fan and collector. The following will illustrate the use of the formula.

A simple duct system of length 50-m serving a double-end grinding machine requires an exhaust volume of 6000 m^3/h. The duct velocity would be 23 m/s, giving a diameter of 300 mm and a velocity pressure of 32.41 mm.

$$\text{The hood loss} = 0.6 \times 32.41 = 19.45 \, \text{mm}$$

$$\text{Duct losses in 50 meters} = \frac{(23)^2 \times 1.243}{300 \, \text{mm}} = 109.59 \, \text{mm}$$

$$\text{The total pressure} = VP + SP = 32.41 + 109.59 + 19.45$$

$$= 161.45 \, \text{mm}$$

To this figure must be added the pressure drop across the collector, which in this case would be a wet unit having a drop of 150 mm.

The fan would then be selected to handle 6000 m^3/h at a total pressure of 311.45 mm water. As this is a high-pressure fan, it would be a centrifugal backward laminar type.

46.5 Testing and inspections

After the plant has been installed, it will require a test to determine whether the design meets the original objectives. In this assessment the effectiveness of the plant must be established and at the same time a record made showing the engineering parameters. A typical assessment sheet is shown in Figure 46.6.

The precise nature of the test will depend on the particles being controlled. The simple observation tests carried out by Tyndall lights and smoketubes will suffice for the majority of dust and fume systems. It is only where the contaminants are listed and known to be dangerous that special testing needs to be done. This work requires on-site monitoring of the workplace using air samplers and the expert services of an industrial hygienist.

47

Dust Collection Systems

R Keith Mobley
The Plant Performance Group

Contents

47.1 Introduction

The basic operations performed by dust-collection devices are: (1) separating particles from the gas stream by deposition on a collection surface, (2) retaining the deposited particles on the surface until removal, and (3) removing the deposit from the surface for recovery or disposal.

The separation step requires: (1) application of a force that produces a differential motion of the particles relative to the gas, and (2) sufficient gas-retention time for the particles to migrate to the collecting surface. Most dust-collections systems are comprised of a pneumatic-conveying system and some device that separates suspended particulate matter from the conveyed air stream. The more common systems use either filter media (e.g., fabric bags) or cyclonic separators to separate the particulate matter from air.

47.2 Baghouses

Fabric-filter systems, commonly called bag-filter or baghouse systems, are dust-collection systems in which dust-laden air is passed through a bag-type filter. The bag collects the dust in layers on its surface and the dust layer itself effectively becomes the filter medium. Because the bag's pores are usually much larger than those of the dust-particle layer that forms, the initial efficiency is very low. However, it improves once an adequate dust-layer forms. Therefore, the potential for dust penetration of the filter media is extremely low except during the initial period after startup, bag change, or during the fabric-cleaning, or blow-down, cycle.

The principal mechanisms of disposition in dust collectors are: (1) gravitational deposition, (2) flow-line interception, (3) inertial deposition, (4) diffusional deposition, and (5) electrostatic deposition. During the initial operating period, particle deposition takes place mainly by inertial and flow-line interception, diffusion, and gravity. Once the dust layer has been fully established, sieving is probably the dominant deposition mechanism.

47.2.1 Configuration

A baghouse system consists of the following: pneumatic-conveyor system, filter media, a back-flush cleaning system, and a fan or blower to provide airflow.

Pneumatic conveyor

The primary mechanism for conveying dust-laden air to a central collection point is a system of pipes or ductwork that functions as a pneumatic conveyor. This system gathers dust-laden air from various sources within the plant and conveys it to the dust-collection system.

Dust-collection system

Design and configuration of the dust-collection system varies with the vendor and the specific application. Generally, a system consists of either a single large hopper-like vessel or a series of hoppers with a fan or blower affixed

to the discharge manifold. Inside the vessel is an inlet manifold that directs the incoming air or gas to the dirty side of the filter media or bag. A plenum, or divider plate, separates the dirty and clean-side of the vessel.

Filter media, usually long cylindrical tubes or bags, are attached to the plenum. Depending on the design, the dust-laden air or gas may flow into the cylindrical filter bag and exit to the clean-side or it may flow through the bag from its outside and exit through the tube's opening. Figure 47.1 illustrates a typical baghouse configuration.

Fabric-filter designs fall into three types, depending on the method of cleaning used: (1) shaker-cleaned, (2) reverse-flow-cleaned, and (3) reverse-pulse-cleaned.

Shaker-cleaned filter

The open lower ends of shaker-cleaned filter bags are fastened over openings in the tube sheet that separates the lower, dirty-gas inlet chamber from the upper clean-gas chamber. The bags are suspended from supports, which are connected to a shaking device.

The dirty gas flows upward into the filter bag and the dust collects on the inside surface. When the pressure drop rises to a predetermined upper limit due to dust accumulation, the gas flow is stopped and the shaker is operated. This process dislodges the dust, which falls into a hopper located below the tube sheet.

For continuous operation, the filter must be constructed with multiple compartments. This is necessary so that individual compartments can be sequentially taken off-line for cleaning while the other compartments continue to operate.

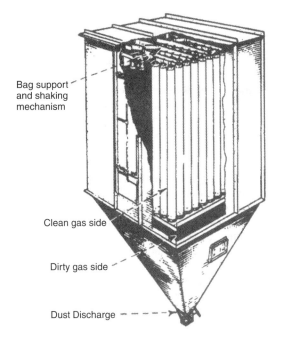

Bag support and shaking mechanism

Clean gas side

Dirty gas side

Dust Discharge

Figure 47.1 A typical baghouse

Ordinary shaker-cleaned filters may be cleaned every fifteen minutes to eight hours, depending on the service conditions. A manometer connected across the filter is used to determine the pressure drop, which indicates when the filter should be shaken. Fully automatic filters may be shaken every two minutes, but bag maintenance is greatly reduced if the time between shakings can be increased to 15 to 20 minutes.

The determining factor in the frequency of cleaning is the pressure drop. A differential-pressure switch can serve as the actuator in automatic cleaning applications. Cyclone pre-cleaners are sometimes used to reduce the dust load on the filter or to remove large particles before they enter the bag.

It is essential to stop the gas flow through the filter during shaking in order for the dust to fall off. With very fine dust, it may be necessary to equalize the pressure across the cloth. In practice, this can be accomplished without interrupting continuous operation by removing one section from service at a time. With automatic filters, this operation involves closing the dirty-gas inlet dampers, shaking the filter units either pneumatically or mechanically, and reopening the dampers. In some cases, a reverse flow of clean gas through the filter is used to augment the shaker-cleaning process.

The gas entering the filter must be kept above its dewpoint to avoid water-vapor condensation on the bags, which will cause plugging. However, fabric filters have been used successfully in steam atmospheres, such as those encountered in vacuum dryers. In these applications, the housing is generally steam-cased.

Reverse-flow-cleaned filter

Reverse-flow-cleaned filters are similar to the shaker-cleaned design, except the shaker mechanism is eliminated. As with shaker-cleaned filters, compartments are taken off-line sequentially for cleaning. The primary use of reverse-flow cleaning is in units using fiberglass-fabric bags at temperatures above 150°C (300°F).

After the dirty-gas flow is stopped, a fan forces clean gas through the bags from the clean-gas side. The superficial velocity of the gas through the bag is generally 1.5 to 2.0 feet per minute, or about the same velocity as the dirty-gas inlet flow. This flow of clean gas partially collapses the bag and dislodges the collected dust, which falls to the hopper. Rings are usually sewn into the bags at intervals along their length to prevent complete collapse, which would obstruct the fall of the dislodged dust.

Reverse-pulse-cleaned filter

In the reverse-pulse-cleaned filter, the bag forms a sleeve drawn over a cylindrical wire cage, which supports the fabric on the clean-gas side (i.e., inside) of the bag. The dust collects on the outside of the bag.

A venturi nozzle is located in the clean-gas outlet from each bag, which is used for cleaning. A jet of high-velocity air is directed through the venturi nozzle and into the bag, which induces clean gas to pass through the fabric to the dirty side. The high-velocity jet is released in a short pulse, usually about 100 milliseconds, from a compressed air line by a solenoid-controlled valve. The pulse of air and clean gas expand the bag and dislodge the collected dust. Rows of bags are cleaned in a timed sequence by programmed operation of the solenoid valves. The pressure of the pulse must be sufficient to dislodge the dust without cessation of gas flow through the baghouse.

It is common practice to clean the bags on-line without stopping the flow of dirty gas into the filter. Therefore, reverse-pulse bag filters are often built without multiple compartments. However, investigations have shown that a large fraction of the dislodged dust re-deposits on neighboring bags rather than falling to the dust hopper.

As a result, there is a growing trend to off-line clean reverse-pulse filters by using bags with multiple compartments. These sections allow the outlet-gas plenum serving a particular section to be closed off from the clean-gas exhaust, thereby stopping the flow of inlet gas. On the dirty-side of the tube sheet, the isolated section is separated by partitions from the neighboring sections where filtration continues. Sections of the filter are cleaned in rotation as with shaker and reverse-flow filters.

Some manufacturers design bags for use with relatively low-pressure air (i.e., 15 psi) instead of the normal 100 psi air. This allows them to eliminate the venturi tubes for clean-gas induction. Others have eliminated the separate jet nozzles located at the individual bags in favor of a single jet to pulse air into the outlet-gas plenum.

Reverse-pulse filters are typically operated at higher filtration velocities (i.e., air-to-cloth ratios) than shaker or reverse-flow designs. Filtration velocities may range from 3 to 15 feet per minute in reverse-pulse applications, depending on the dust being collected. However, the most the commonly used range is 4 to 5 feet per minute.

The frequency of cleaning depends on the nature and concentration of the dust. Typical cleaning intervals vary from about 2 to 15 minutes. However, the cleaning action of the pulse is so effective that the dust layer may be completely removed from the surface of the fabric. Consequently, the fabric itself must serve as the principal filter media for a substantial part of the filtration cycle, which decreases cleaning efficiency. Because of this, woven fabrics are unsuitable for use in these devices and felt-type fabrics are used instead. With felt filters, although the bulk of the dust is still removed, an adequate level of dust collection is provided by the fabric until the dust layer reforms.

Cleaning system

As discussed in the preceding section, filter bags must be periodically cleaned to prevent excessive build-up of dust and to maintain an acceptable pressure drop across the filters. Two of the three designs discussed, reverse-flow and reverse-pulse, depend on an adequate supply of clean air or gas to provide this periodic cleaning. Two factors are critical in these systems: the clean-gas supply and the proper cleaning frequency.

Clean-gas supply

Most applications that use the reverse-flow cleaning system use ambient air as the primary supply of clean gas.

A large fan or blower draws ambient air into the clean side of the filter bags. However, unless the air is properly conditioned by inlet filters, it may contain excessive dirt loads that can affect the bag life and efficiency of the dust-collection system.

In reverse-pulse applications, most plants rely on plant-air systems as the source for the high-velocity pulses required for cleaning. In many cases, however, the plant-air system is not sufficient for this purpose. Although the pulses required are short (i.e., 100 milliseconds or less), the number and frequency can deplete the supply. Therefore, care must be taken to ensure that both sufficient volume and pressure are available to achieve proper cleaning.

Cleaning frequency

Proper operation of a baghouse, regardless of design, depends on frequent cleaning of the filter media. The system is designed to operate within a specific range of pressure drops that defines clean and fully-loaded filter media. The cleaning frequency must assure that the maximum recommended pressure drop is not exceeded.

This can be a real problem for baghouses that rely on automatic timers to control cleaning frequency. The use of a timing function to control cleaning frequency is not recommended unless the dust load is known to be consistent. A better approach is to use differential-pressure gages to physically measure the pressure drop across the filter media to trigger the cleaning process based on pre-set limits.

Fan or blower

All baghouse designs use some form of fan, blower, or centrifugal compressor to provide the dirty-air flow required for proper operation. In most cases, these units are installed on the clean-side of the baghouse to draw the dirty air through the filter media.

Since these units provide the motive power required to transport and collect the dust-laden air, their operating condition is critical to the baghouse system. The type and size of air-moving unit varies with the baghouse type and design.

47.2.2 Performance

The primary measure of baghouse-system performance is its ability to consistently remove dust and other particulate matter from the dirty-air stream. Pressure drop and collection efficiency determine the effectiveness of these systems.

Pressure drop

The filtration, or superficial face, velocities used in fabric filters are generally in the range of 1 to 10 feet per minute, depending on the type of fabric, fabric supports, and cleaning methods used. In this range, pressure drops conform to Darcy's law for streamline flow in porous media, which states that the pressure drop is directly proportional to the flow rate. The pressure drop across the fabric media and the dust layer may be expressed by:

$$\Delta p = K_1 V_f + K_2 \omega V_f$$

Where:

Δp = Pressure drop (inches of water)
V_f = Superficial velocity through filter (feet/minute)
ω = Dust loading on filter (lbm/ft^2)
K_1 = Resistance coefficient for conditioned fabric (inches of water/foot/minute)
K_2 = Resistance coefficient for dust layer (inches of water/lbm/foot/minute)

Conditioned fabric maintains a relatively consistent dust-load deposit following a number of filtration and cleaning cycles. K_1 may be more than 10 times the value of the resistance coefficient for the original clean fabric. If the depth of the dust layer on the fabric is greater than about $\frac{1}{16}$ inch (which corresponds to a fabric dust loading on the order of 0.1 lbm/ft^2), the pressure drop across the fabric, including the dust in the pores, is usually negligible relative to that across the dust layer alone.

In practice, K_1 and K_2 are measured directly in filtration experiments. These values can be corrected for temperature by multiplying by the ratio of the gas viscosity at the desired condition to the gas viscosity at the original experimental condition.

Collection efficiency

Under controlled conditions (e.g., in the laboratory), the inherent collection efficiency of fabric filters approaches 100 per cent. In actual operation, it is determined by several variables, in particular the properties of the dust to be removed, choice of filter fabric, gas velocity, method of cleaning, and cleaning cycle. Inefficiency usually results from bags that are poorly installed, torn, or stretched from excessive dust loading and excessive pressure drop.

47.2.3 Installation

Most baghouse systems are provided as complete assemblies by the vendor. While the unit may require some field assembly, the vendor generally provides the structural supports, which in most cases are adequate. The only controllable installation factors that may affect performance are the foundation and connections to pneumatic conveyors and other supply systems.

Foundation

The foundation must support the weight of the baghouse. In addition, it must absorb the vibrations generated by the cleaning system. This is especially true when using the shaker-cleaning method, which can generate vibrations that can adversely affect the structural supports, foundation, and adjacent plant systems.

Connections

Efficiency and effectiveness depends on leak-free connections throughout the system. Leaks reduce the system's ability to convey dust-laden air to the baghouse. One potential source for leaks is improperly installed filter bags. Because installation varies with the type of bag and baghouse design, consult the vendor's O&M manual for specific instructions.

47.2.4 Operating methods

The guidelines provided in the vendor's O&M manual should be the primary reference for proper baghouse operation. Vendor-provided information should be used because there are not many common operating guidelines among the various configurations. The only general guidelines that are applicable to most designs are cleaning frequency and inspection and replacement of filter media.

Cleaning

As previously indicated, most bag-type filters require a pre-coat of particulates before they can effectively remove airborne contaminates. However, particles can completely block air flow if the filter material becomes overloaded. Therefore, the primary operating criterion is to maintain the efficiency of the filter media by controlling the cleaning frequency.

Most systems use a time-sequence to control the cleaning frequency. If the particulate load entering the baghouse is constant, this approach would be valid. However, the incoming load generally changes constantly. As a result, the straight time-sequence methodology does not provide the most efficient mode of operation.

Operators should monitor the differential-pressure gauges that measure the total pressure drop across the filter media. When the differential pressure reaches the maximum recommended level (data provided by the vendor), the operator should over-ride any automatic timer controls and initiate the cleaning sequence.

Inspecting and replacing filter media

Filter media used in dust-collections systems is prone to damage and abrasive wear. Therefore, regular inspection and replacement is needed to ensure continuous, long-term performance. Any damaged, torn, or improperly sealed bags should be removed and replaced.

One of the more common problems associated with baghouses is improper installation of filter media. Therefore, it is important to follow the instructions provided by the vendor. If the filter bags are not properly installed and sealed, overall efficiency and effectiveness are significantly reduced.

47.3 Cyclone separators

A widely used type of dust-collection equipment is the cyclone separator. A cyclone is essentially a settling chamber in which gravitational acceleration is replaced by centrifugal acceleration. Dust-laden air or gas enters a cylindrical or conical chamber tangentially at one or more points and leaves through a central opening. The dust particles, by virtue of their inertia, tend to move toward the outside separator wall from where they are

led into a receiver. Under common operating conditions, the centrifugal separating force or acceleration may range from five times gravity in very large diameter, low-resistance cyclones to 2500 times gravity in very small, high-resistance units.

Within the range of their performance capabilities, cyclones are one of the least expensive dust-collection systems. Their major limitation is that, unless very small units are used, efficiency is low for particles smaller than five microns. Although cyclones may be used to collect particles larger than 200 microns, gravity-settling chambers or simple inertial separators are usually satisfactory and less subject to abrasion.

47.3.1 Configuration

The internal configuration of a cyclone separator is relatively simple. Figure 47.2 illustrates a typical cross-section of a cyclone separator, which consists of the following segments:

- Inlet area that causes the gas to flow tangentially,
- Cylindrical transition area,
- Decreasing taper that increases the air velocity as the diameter decreases,
- Central return tube to direct the dust-free air out the discharge port.

Particulate material is forced to the outside of the tapered segment and collected in a drop-leg located at the dust outlet. Most cyclones have a rotor-lock valve affixed to the bottom of the drop-leg. This is a motor-driven valve that collects the particulate material and discharges it into a disposal container.

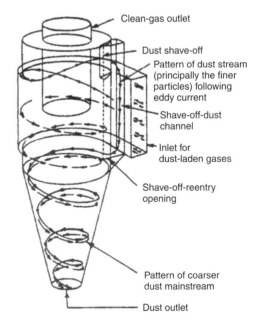

Figure 47.2 Flow pattern through a typical cyclone separator

47.3.2 Performance

Performance of a cyclone separator is determined by flow pattern, pressure drop, and collection efficiency.

Flow pattern

The path the gas takes in a cyclone is through a double vortex that spirals the gas downward at the outside and upward at the inside. When the gas enters the cyclone, the tangential component of its velocity, V_{ct}, increases with the decreasing radius as expressed by:

$$V_{ct} \approx r^{-n}$$

In this equation, r is the cyclone radius and n is dependent on the coefficient of friction. Theoretically, in the absence of wall friction, n should equal 1.0. Actual measurements, however, indicate that n ranges from 0.5 to 0.7 over a large portion of the cyclone radius. The spiral velocity in a cyclone may reach a value several times the average inlet-gas velocity.

Pressure drop

The pressure drop and the friction loss through a cyclone are most conveniently expressed in terms of the velocity head based on the immediate inlet area. The inlet velocity head, h_{vt}, which is expressed in inches of water, is related to the average inlet-gas velocity and density by:

$$h_{vt} = 0.0030 \rho V_c^2$$

Where:

h_{vt} = Inlet-velocity head (inches of water)
ρ = Gas density (lb/ft^3)
V_c = Average inlet-gas velocity (ft/sec)

The cyclone friction loss, F_{cv}, is a direct measure of the static pressure and power that a fan must develop. It is related to the pressure drop by:

$$F_{cv} = \Delta p_{cv} + 1 - \left(\frac{4A_c}{\pi D_e^2} \right)^2$$

Where:

F_{cv} = Friction loss (inlet-velocity heads)
Δp_{cv} = Pressure drop through the cyclone (inlet-velocity heads)
A_c = Area of the cyclone (ft^2)
D_e = Diameter of the gas exit (feet)

The friction loss through cyclones may range from 1 to 20 inlet-velocity heads, depending on its geometric proportions. For a cyclone of specific geometric proportions, F_{cv} and Δp_{cv}, are essentially constant and independent of the actual cyclone size.

Collection efficiency

Since cyclones rely on centrifugal force to separate particulates from the air or gas stream, particle mass is the dominant factor that controls efficiency. For particulates with high densities (e.g., ferrous oxides), cyclones can achieve 99 per cent or better removal efficiencies, regardless of particle size. Lighter particles (e.g., tow or flake) dramatically reduce cyclone efficiency.

These devices are generally designed to meet specific pressure-drop limitations. For ordinary installations operating at approximately atmospheric pressure, fan limitations dictate a maximum allowable pressure drop corresponding to a cyclone inlet velocity in the range of 20 to 70 feet per second. Consequently, cyclones are usually designed for an inlet velocity of 50 feet per second.

Varying operating conditions change dust-collection efficiency only by a small amount. The primary design factor that controls collection efficiency is cyclone diameter. A small-diameter unit operating at a fixed pressure drop has a higher efficiency than a large-diameter unit. Reducing the gas-outlet duct diameter also increases the collection efficiency.

47.3.3 Installation

As in any other pneumatic-conveyor system, special attention must be given to the piping or ductwork used to convey the dust-laden air or gas. The inside surfaces must be smooth and free of protrusions that affect the flow pattern. All bends should be gradual and provide a laminar-flow path for the gas.

47.3.4 Operating methods

Cyclones are designed for continuous operation and must be protected from plugging. In intermittent applications, the operating practices must include specific steps to purge the entire system of particulates prior to shutdown.

Pressure drop is the only factor that can be effectively controlled by an operator. Using the fan dampers, the operator can increase or decrease the cyclone's load by varying the velocity of the entering dirty air.

48

Maintenance Management in UK

George Pitblado
Support Services

Contents

48.1 Introduction

The implementation of a program for the effective management of a company's assets has never been more important than now. Chief engineers, group engineers, works engineers, building services managers, and facilities managers – the list is never ending with respect to job title and designation. While the individual's title may change, the responsibilities rarely alter, in that it is from the person in these positions upon whom the employer depends so much. Company assets must be maintained in the most sound, safe and economical way to ensure that the client or user is provided with the appropriate goods, materials and working environment, as all these fall within the responsibility of the engineering maintenance department.

In a constantly changing work environment, with the introduction of new techniques, work practices and legislation the engineering department must ensure that they maintain an adequate level of continuous educational training. Quality assurance, the introduction of more specific legislation, codes of practice, guidance notes for health and safety, and the standards being introduced by such bodies as the Occupational Safety and Health Act (OSHA) – all are to improve the end product, service, etc., but they place an additional demand on existing resources that cannot be ignored. Therefore, in an endeavor to minimize these demands, and in so doing to maximize the benefits from available resources, a 'planned' maintenance program should be implemented.

48.2 A 'planned' maintenance program

The development of any program must take into consideration the maintenance tasks that need be carried out and the resources available, thereby ensuring that product quality and personnel safety standards are met. It follows that operational demands, whether they be from a service utility installation, a production line, a mainframe computer or an office environment, will play a major part in reaching the decision as to what type of maintenance program requires implementation.

Plant and equipment that provides a service or is required to operate for 24 hours a day, 7 days a week presents a different proposition to the maintenance programming requirements. For example, a heating pump with a standby, which is required for a heating installation in an office with a set number of working hours from Monday to Friday. Alternative methods of maintenance programming may require the planner to allow for total replacement on plant failure or have available replacement units, so that when an item of plant or equipment fails, it is removed and a replacement installed.

The ideal method would be that all plant or equipment units have a duplicate or redundant standby that on failure would automatically be brought into service. This is satisfactory when considering small, less expensive units, which are installed to ensure that the services they are providing are not disrupted (this could also include auxiliary items on major plant and equipment). However, due to the capital costs of the major unit itself, it would be difficult to establish adequate economic grounds to duplicate these items to the same degree (although this may be done in a computer environment due to the high costs involved in downtime).

Planned maintenance programs are an essential weapon in a department's armory to ensure that the services it is called on in meeting its responsibilities are fully met. The traditional method of working from pieces of paper or individuals' 'own' notebooks as to when maintenance is to be carried out or when the insurance representative is due to visit to carry out an inspection are no longer satisfactory. This is especially the case when the skilled resources necessary to carry out the work are more difficult to obtain.

It is therefore essential that a planned maintenance program be established, which can encompass all (or elements of) the different maintenance methods of establishing the frequency and/or work to be carried out. This programming requires skills that, in most instances, can have only been gained by experience in the field of maintenance and operation. Operation must play an important part in the programming. If the planned maintenance program is prepared without due consideration of the demands placed on the operation element, the program would probably collapse when the plant's equipment could not be released (i.e. switched off) when the maintenance technician arrived at the plant to carry out his duties indicated on his work docket.

Maintenance procedures that should be considered when preparing the planned maintenance program include:

1. Carrying out repairs needed when plant or equipment breaks down;
2. Predicting, from a history of breakdowns, the life expectancy of parts, bearings, etc., the tasks to be carried out and the frequency to be established;
3. Checking the condition throughout the plant of equipment, its running hours, readings of different responses (e.g. vibration, temperatures, current, etc.);
4. Monitoring the operating cycle and, where appropriate, seasonal shutdowns of plant, equipment (e.g. production process, 24-hour duty, etc.).

48.2.1 Maintenance systems

The benefits to be accrued from the implementation of a program of planned maintenance can be found in the efficient and economical operation of the plant and equipment and the utilization of resources (i.e. plant and equipment and manpower) while also maintaining a sound standard of safe working and environmental conditions for operators, other occupants and employees within the workplace. Maintenance systems vary, depending on the location of the plant and equipment and/or company policy. Systems can range from the complete maintenance of plant and equipment using all available methods to their replacement on failure. To meet the company's requirements it is then necessary to decide on the maintenance system that provides the most satisfactory benefits overall.

The most commonly maintenance systems in use are planned, preventive, scheduled, corrective and emergency.

Planned maintenance is work having benefited from information issued by manufacturers and suppliers, the experience and knowledge of the service department staff, and reports and records from previous service visits.

Preventive maintenance is work to be carried out at a specific frequency as indicated by potential failures or known reduction in efficiency of the plant and equipment, thereby avoiding failures or a decrease in performance.

Scheduled maintenance is work based on known information, such as number of operations, hours run, mileage, etc., and can therefore be carried out at a predetermined time interval.

Corrective maintenance is work carried out following the failure of the plant and equipment, and is so designed to return the component to its normal operating condition.

Emergency maintenance is that work which is required to be performed without delay due to a failure of a component which, if not implemented, would lead to further failures or even permanent damage, resulting in the total loss of the plant and equipment. Plant and equipment in such a condition may also be dangerous to personnel.

As planned maintenance encompasses all types covered within the preventive or scheduled systems, this can be examined in more detail. In preparing a planned maintenance system, all the available sources of information should be used. These include manufacturers' and suppliers' literature, trade associations, professional institutions, knowledge and experience from within the department and history and feedback from previous work for the specific type of equipment.

Condition monitoring, life cycle costing and predictive maintenance procedures should all be considered during the preparation of the planned maintenance system.

Planned maintenance systems should not be complicated. The simpler the system is to meet the requirements of the department and company, the more likelihood of it being used with satisfactory results. This aspect is of the greatest importance when, due to the size of the organization, engineers may be transferred from one department to another to gain a greater knowledge of the total company. This may cause the implemented planned maintenance system to fail or not be used to its full effect, due to the incoming engineer not understanding it fully.

In preparing a planned maintenance system the opportunity should be taken to involve the whole department. This can be achieved by using the operatives who will subsequently action the work to carry out the initial survey of plant and equipment.

48.3 A manual planned maintenance system

The planned maintenance program forms the basis of a system whereby an in-house department may prepare and implement its own maintenance program or introduce a trial system along similar lines. This would be prior to seeking the assistance of a consultant to provide guidance on the system that would satisfy the demands placed on the department by others and yet remain under the department's control. Irrespective of whether it is a manual or a computer system that is introduced, the elements of its

composition vary little. Depending on the reports required from the implementation planned maintenance program, the more satisfactory method of obtaining this result, both in quantity and quality of information, is by running the program on a computer.

To assist engineers to implement a manual planned maintenance system that can be of benefit to the department it is important that a program be set with respect to the system's implementation. Items that should be considered are:

1. Departments to be covered;
2. Plant and equipment to be included;
3. Technician and craftsmen trades available;
4. Person responsible for preparation and implementation of the system;
5. Time scale for preparation and implementation;
6. Administrative support.

It should be noted that initial interest in the preparation and implementation of planned maintenance systems can gradually decrease if only one person (preferably an experienced engineer) is given the responsibility to ensure that the proposal is carried out to its satisfactory conclusion. It is essential that the nominated person is given adequate support when necessary, to ensure that the planned maintenance system's introduction into the working operation of the department meets with success.

A suitable planned maintenance system, irrespective of the location or type of business, is complied from a number of standard elements (see the appendix at the end of this chapter):

1. Assets register;
2. Maintenance and repair record;
3. Technician and craftsmen guidance notes;
4. Planning schedule;
5. Week tasks;
6. Work dockets;
7. Year visual aid plan.

48.3.1 Assets register

Each item of plant and equipment is allocated a specific asset number. This number can be either for a complete boiler (with associated equipment) or a specific asset number for the boiler and individual asset numbers for the associated equipment. It is advisable to restrict this numbering sequence to a minimum while ensuring that it meets the specific needs of the company and location. Care must be exercised in determining the asset numbering during this manual phase if it is envisaged that, on completing a satisfactory trial period, the planned maintenance system will be transferred onto a computer.

To assist in the numbering of the assets, each type of plant and equipment can be given a predetermined plant code reference number. In this case, the boiler and all its associated equipment are given the same asset number. The asset number is built up from certain elements, e.g.:

Location: Plant Code: Plant/equipment number
e.g.
Plant room 1: Boiler: No. 1 Boiler = PR1-01-01

Pump units, including valves, gages, etc., are classified as individual assets. To enable the asset number to signify the different process that the pump is serving, the Plant Code 05 is suffixed, e.g.:

Roof Plant Room: Heating Circulating Pump: No. 1 = RPR-05/02-01

(02 indicates that the pump is installed in the heating system).

Items of equipment such as emergency lights may be grouped as one asset, supported by a checklist, which details the numbers, types and locations of the individual units in a predetermined area, e.g.:

First Floor: Emergency Lights = F1-34-01

(01 indicates it is No. 1 in the series of emergency lights assets).

General area services such as lighting, heating, hot and cold water and air conditioning can be registered in a way similar to that of emergency lights.

In designing the asset register format for the specific location items such as manufacturer/supplier, purchase price and date, order no., cost code, function, parts/spares, guidance note reference and insurance inspections should be catered for. While carrying out the survey for the asset register, all the information found on plant and equipment nameplates should be recorded as, during their life, these tend to be lost or painted over.

To simplify this task it is an advantage if there is a set format prepared on the survey forms for specific types of plant and equipment. This enables the surveyor to enter the relevant information against the appropriate elements (e.g. manufacturer; volts; amps; bearings; etc.).

48.3.2 Maintenance and repair record

This is designed to record *all* work carried out and parts fitted on each asset item of plant and equipment. Service visits by contractors and insurance inspections are also entered. This record, which provides the history of the asset, may either be placed on the reverse of the asset record (thereby ensuring that all relevant information on a specific asset can be found in one place) or it may be an individual assets record form inserted within the asset register, next to the specific asset record.

To reduce the amount of information entered in the record, predetermined work of a planned maintenance nature can be entered in code form, while additional work or breakdowns may be more fully detailed.

As the maintenance and repair record indicates the cost of maintenance for the specific item of plant and equipment, a simple system may be used to provide recognition of the different methods by which the work has been carried out, e.g.

Planned maintenance work is entered in *black* pen.
Planned or corrective work is entered in *blue*.
Emergencies (i.e. breakdowns) are entered in *red*.

From this method of entry, the engineer can observe at a glance if the planned maintenance program is effective with respect to corrective work or emergencies. Further examination of the operation of the asset and its records are necessary in determining if the frequency of planned maintenance is correct or whether the plant and equipment should be replaced.

48.3.3 Guidance notes

The guidance notes can be produced either as a composite handbook containing task instructions for all types of plant and equipment for each trade group, or as specific task/advice notes for each asset of the service requirements of the plant equipment. When used, the handbook method provides the necessary work instructions for all similar types of plant and equipment throughout the location. This method of operation reduces the number of task/advice notes issued and therefore the system's workload, as well as the demands placed on the administration of the system. Either method eliminates the requirement of entering work instructions, etc. on the work docket/advice notes before they are issued.

The handbook can be prepared either for a specific trade or for all the trades involved in the maintenance of the plant and equipment. Its contents are as follows:

1. *Health and Safety*: Stresses the importance of carrying out the work in a safe and responsible manner.
2. *Introduction*: Details how the handbook contents and the individual's responsibility in the application of his skills are to be implemented.
3. *Plant Code, Frequency of Services and Work Tasks*: Details the work to be carried out, and at what frequency, as indicated by the instruction on the work docket.
4. *Plant Code*: Lists the plant codes for the plant and equipment covered by the handbook, prefixed by the appropriate trade reference (e.g. Mechanical = M).

The handbook is designed in a loose-leaf format so that additional entries and amendments can be made as and when relevant.

To enable the maximum input and experience of the craftsmen to be introduced into the system the pages of the handbook should not be encapsulated thereby enabling the craftsman to enter additional information and comments that could increase the performance of the plant and equipment as well as amended instructions in the guidance notes. These amendments can then be issued to all staff in a similar trade.

48.3.4 Planning schedule

It is at this stage in the preparation of the planned maintenance that the engineer's knowledge and experience of maintenance is essential. In preparing the planned maintenance schedule it may be found that information on maintenance received from the manufacturer and supplier is no longer available. Technicians and craftsmen's knowledge can play a major role in this planning stage.

Experience indicates that when preparing the planning schedule this should be carried out for each trade group.

Prior to entering any asset detail on the planning schedule, items such as holiday periods and seasonal or shutdown programs should be indicated. This can be done by the use of highlighter pens on the calendar weeks that require specific attention to loading of the relevant work tasks.

In scheduling the assets and work tasks to be carried out it is recommended that the plant room and department that require the greatest resources (i.e. man-hours) should be entered on the planning schedule first. The scheduling for each item of plant and equipment follows the same pattern in that the location (e.g. boiler-house) is entered as a heading, then the assets and their asset numbers are entered for each specified location.

Planning the frequency, work tasks and man-hours for each asset then follows. Choose the week in which the least-frequency service is to be undertaken (e.g. yearly; enter a 'Y/'). Other frequencies can then be entered (e.g. quarterly, 'Q/'; monthly, 'M/'). To complete the planning scheduled for the specific asset the hours required to carry out the work at the nominated frequencies are then entered (e.g. 'Y/12'; 'Q/4'; 'M/1', etc.). (*Note*: It is recommended that this scheduling be done in pencil so that amendments can be easily made.)

Plant and equipment that require a service on completion of a certain specified period of *hours* may, through experience, be catered for on a fixed frequency basis. If employed, this method avoids the need to record running hours (or, in the case of transport, mileage) on a daily or weekly basis to schedule the relevant planned maintenance. If it is essential that the maintenance of the plant and equipment be carried out on the completion of a certain number of operating hours or mileage, then this must be allowed for in the allocated work hours of the relevant trade group.

Peaks and troughs in the man-hours allocated weekly for planned maintenance can be avoided if hours entered are added up for each week after a number of assets have been scheduled. Treating each plant room or department in this way provides the number of man-hours required for the respective plant room and/or department. All assets to be covered by the planned maintenance system are scheduled in the same format, thereby providing the engineer responsible for allocating the work tasks with total man-hours for each trade group.

There are two methods of entering man-hours:

1. The actual hours necessary to carry out the planned maintenance work task;
2. The *total* hours to complete the planned maintenance work task (including non-productive hours). Non-productive hours would include such items as collecting spares, tea breaks, discussions, etc.

The planned maintenance system most commonly used is that indicated in (2), whereas the method of calculating the hours as in (1) is preferred if there is a productivity scheme in operation.

Service contractors' visits can also be indicated in the planning schedule (with 'C' for contractors or another symbol indicating a different contractor). To highlight different grades of service visits (e.g. yearly) 'Y/C' can be entered.

A major benefit of carrying out this planning schedule phase for all the plant and equipment is that the numbers of each trade required are known. Also, having carried out the exercise for all plant and equipment, if the decision is to be made as to which is to be covered by the scheme or contracted out, a complete picture of the total planned maintenance requirements for the department/company is available.

48.3.5 Week tasks

The information required to compile this form is obtained from the planning schedule. To minimize 'administrative' tasks in preparing the work dockets, the information contained on the week task forms is presented in a format that enables that information to be easily transferred onto the work dockets, e.g.:

Location/Area: Plant/Equipment: Plant/Equipment No.: Job Code: Check List

Week task forms are completed for each week. They may be designed as a separate form for each trade or as one covering all trades. The latter will, of course, depend on the number of tasks per week for the individual trades.

Tasks that are of a weekly nature (e.g. visual inspection) are entered onto a separate week task form. To avoid issuing work dockets for such inspections, the week task form may then be used as a work docket, a tick being placed against the tasks when they are completed.

Week task forms should only require amending when either additional plant and equipment is introduced or the frequency of planned maintenance tasks is being adjusted to meet revised operational or maintenance demands. Departments in which work is to be carried out may be issued a copy of their department's week tasks four weeks in advance. This then enables them to program their operation, where necessary, so that the maintenance work can be carried out without disruption to the department's output. Alternatively, on receipt of this prior notice the departmental head can contact the engineer controlling the work to discuss departmental matters that may affect the proposed planned maintenance. This may, in some instances, require the engineer to reschedule the planned maintenance programmed for that department.

48.3.6 Work dockets

On examining the week task form for the forthcoming week the engineer will decide on the work dockets that are to be prepared for issue. To enable the engineer to carry out this function satisfactorily, knowledge of man-power resources and operational demands on the plant and equipment is essential.

It is at this stage that, having the information above, a decision can be made on which planned maintenance tasks are to be carried out and which may be postponed or cancelled. This may be due either to insufficient resources (e.g. man-hours or a forthcoming planned shutdown) or planned maintenance that need not be carried out for another reason.

Recourse to the maintenance and repair records or a visual aid enables the engineer to establish if the planned maintenance tasks that are not to be carried out had been maintained either during the previous scheduled visit or, when next due, a planned maintenance visit. If not carried out as programmed, the engineer must ensure that the period over which no planned maintenance is proposed does not exceed any known maintenance/operational requirement.

The engineer or supervisor responsible for planned maintenance may issue the work dockets on an individual, daily or weekly basis. When issued on a weekly basis, the technician/craftsman responsible for a specific plant room or departmental area can then plan how the work should be programmed throughout the week, having gained experience in identifying which plant and equipment and also in which area planned maintenance can be worked on or at specific periods of the week.

The work dockets would be normally issued on Fridays for the following week. This enables the technician/craftsman to plan his work for the forthcoming week. All completed work dockets should be returned to the engineer's office daily.

Work dockets partially completed (i.e. with the work task incomplete) should be discussed with the engineer. This then enables the engineer to make a decision on how or when the outstanding items on the work docket may be performed.

Work dockets that have not been performed should also be returned for the engineer's attention, with comments on why they have not been performed. The engineer may then decide on whether the work need be rescheduled.

The engineer controlling the planned maintenance function within the department ensures that all work dockets are returned. Any comments entered on the work dockets are noted and, where further action is required, plans are prepared accordingly. (*Note*: A copy of the week tasks will suffice in maintaining a record of work dockets issued. If necessary, the technician/craftsman's initials can be entered on the week task form as a reminder to whom the work docket was issued.) Technicians/craftsmen should be advised of any matter that delays implementation of action required, as reported in the returned work docket, so that, if no corrective action has taken place on their next scheduled visit they understand why.

Further to the engineer's examination of the returned work dockets, purchase requisitions can be placed for parts, specialist contractors' attendance, etc. On completion of all the technical aspects of the process, the information regarding the service carried out on the plant and equipment can then be entered in the maintenance and repair record by the administration.

48.3.7 Year visual aid

Visual aids are developed from the planning schedule, in that assets and asset numbers are repeated and service visits indicated by a symbol. Hence, the estimated man-hours for the specific task per frequency are not displayed for general review.

The benefit to be gained by the engineer and technician/craftsmen from the visual aid is that it provides them with a visual picture of the full year's program, covering all the recorded assets and their associated work tasks. The format of this visual aid depends on the number of assets within the system, and each visual aid may cover approximately 100 assets.

If there are not a large number of assets, a visual aid can be prepared for each trade, whereas when there is a greater number (e.g. 400+) a visual aid that covers all trades suffices. Indication markers on the aid in this instance would signify the least frequent service of the collective trades (e.g. mechanical: monthly visit and electrical: quarterly visit). The last would be indicated.

Visual aids can be purchased on which the information regarding assets and service visits would be affixed or, as in the examples shown in the Appendix, a negative can be drawn covering all the assets. Draft copies can then be printed for each trade's planning schedule.

The year visual aid visits indicated by color spots relating to frequency is covered by a firm plastic sheet, so that the issue and return of work dockets can be indicated on the plastic by a chinagraph pencil (e.g. '/': docket issued; 'X': work completed).

48.3.8 Comments

The above planned maintenance manual system provides a firm basis on which a computer-planned maintenance system may be developed. Having such knowledge and experience from working the manual system, the benefits that can be accrued from planned maintenance systems compared to the corrective or other non-planned methods previously used are numerous. It can be found that, due to the large amount of assets to be managed by the administration, it would be beneficial for the planned maintenance system to be transferred onto a computer system. A major non-technical weekly task is one of producing work dockets for issue and the subsequent entry into the maintenance records on completion of the work task.

Technical benefits are also obtained more readily from a computer system in that if sound identification of assets is established through their asset number, codes etc., similar assets can be examined in the event of a failure of any similar asset, and spares can be held in stock to cover the range of assets for which they are required, rather than for each individual asset. In preparing the asset register and respective work tasks, these can also be more easily duplicated with respect to similar assets than is possible with a manual system.

Scheduling of work tasks and therefore labor utilization can be carried out in a number of different methods on a computer, depending on the software system purchased. These methods can be similar to the manual system or they may take the form of automatic scheduling. That is, the work tasks are entered into the computer system with the appropriate hours and trades and the software program automatically schedules the hours evenly over the weeks of the year. In implementing such scheduling care has to be taken to ensure that such items as weekends, statutory holidays, annual leave, annual shutdown and seasonal requirements are all considered prior to loading the asset information onto the computer system.

Many engineering maintenance departments allocate more hours to new or extra works than they do to maintenance. Therefore, the introduction of either type of planned maintenance system highlights this matter to both the department and the company generally when appropriate.

When considering a computer system, aspects that require attention are hardware and software (planned maintenance system). During the initial inquiry as to which hardware and software would best serve the company or department both must be considered.

If there is no restriction placed on the proposal by having to use the current in-house computer system, the choices of hardware/software packages are numerous. The software must be user friendly, complementary to existing systems to avoid duplication of asset registering, etc., and have the facility to expand (i.e. have other programs (modules) that enable greater use be made of the information held within the system). The program should also be compatible with any in-house computer system where possible.

The hardware should be dedicated to the maintenance department, as they then have access to the maintenance programs as and when necessary. Systems are implemented whereby the software is loaded onto the company in-house hardware, which can lead to periods when access is limited by other demands on computer time (e.g. accounts departments). The ideal situation would be for the maintenance department to have their own computer, with the back-up facility provided by the in-house computer, thereby ensuring that copy is held.

It follows that if the planned maintenance system is to be shown as effective, the number of breakdowns reported must reduce, irrespective as to whether it be a manual or computer system. A satisfactorily installed system allows the engineer to plan his department's work and, when necessary, action emergency items or additional work, without their being detrimental to the department's overall performance.

The benefits from a planned maintenance system cannot be achieved without the commitment of the person responsible for its implementation. To ensure that the system operates to the satisfaction of operators and users, discussions should take place during the initial decision-making and with the relevant departments when the planning schedules are being prepared.

48.4 Computer system

Before committing the company or department to a system of planned maintenance, it is essential that there is a complete overview of the total demand on plant, equipment and resources and the benefits from the implementation of such a system, manual or computerized. Varying systems and/or part systems exist in almost every engineering maintenance department, which are due to company policy, changes in departmental staff or workforce, commitment of persons responsible, or, as is the situation in a number of instances, a previous breakdown, which the

planned system did not prevent. In this case the system is then often abandoned as unworkable.

On reaching the decision that there are benefits to be obtained from a planned system and that the system that will be most beneficial to the company generally is to be computer based, then the following aspects must be examined more fully:

1. Has the company or department the necessary knowledge of the computer system to prepare and implement a planned maintenance program by this method? The benefits to be gained from the use of computers requires sufficient understanding of both computers and maintenance to foresee the advantages over those obtained from the manual system.
2. Has a budget cost been established on hardware and software?

Questions to be raised include:

1. Is the program to be run on the company's main computer or is the department to be given its own hardware? While assessing this, consideration must be given to which computer program is to be given priority (e.g. finance, purchasing, deliveries or maintenance). It is generally found that maintenance is given the lowest priority. The decision to 'go it alone' (i.e. purchase hardware specifically for the maintenance program) requires that the following details of the system's implementation be examined:
   (a) Will there be only one terminal and printer?
   (b) Due to the complexity of the company premises (e.g. dispersed locations, a number of large departments and operation and maintenance sections) it may be that although there is only one main computer source for information and records there are a number of terminals and printers distributed around the other demand areas. To ensure that records and information are not corrupted or destroyed it is normal practice for the remote terminals to have access only to information. This enables the information held within the program to be interrogated by authorized persons with respect to history, spare parts, inventory, etc. It could be considered good policy to implement the maintenance program with only one workstation (i.e. a central computer and printer) initially to gain experience of its operation before extending it into other departments. This method enables bugs to be ironed out before the system becomes too large for any necessary modifications to be introduced.
2. Are there skills within the company or organization that can develop a cost-effective program specifically for the respective department, or would it be more economical in terms of costs and the proposed timetable for implementation of the program to purchase an 'off-the-shelf package'? Caution on which route should be taken must be exercised in that in-house staff may be available in the initial programming stages but as the system becomes more developed, they may not be free due to other priorities discussed above.

3. Who will prepare and then implement/operate the program? What resources will be made available (e.g. facilities, resources, accommodation and administration assistance)? To implement the proposal the staff forming the team will need to include a computer-literate operator (i.e. someone who can correct the faults that develop during the planning and programming stages), a person knowledgeable in the programming of the engineering tasks (e.g. an engineer) and also staff to inspect the plant and equipment and enter all the information into the computer's maintenance program. If there are not the skills available as mentioned above, it is imperative that the engineer responsible for the implementation of the computer program be given adequate training in the specific system to be purchased. If this is not done, substantial costs could be increased in continual recourse to the suppliers for assistance.

4. What assets, departments, etc. are to be included within the proposed program? This plays a major part in the decision to be reached on to what hardware and software is to be purchased. To support a system covering a number of areas and departments it will be necessary to consider network systems. Therefore, the computer maintenance system to be purchased must accommodate such arrangements.

5. What details must be sought from the suppliers of both hardware and software packages? How can the computer best meet the department's current requirements and how can it be added to if and when necessary (e.g. increased demand from planned maintenance)? How can it be extended to cover other departments (network)? Energy consumption input from condition monitoring, stores inventory, project work and possible interaction with space allocation must also be considered.

48.4.1 Computer system checklist

A checklist of items that may assist the intended purchaser of a computer system would include:

Costs

1. Computer and printer;
2. Maintenance package (i.e. number of modules required to provide a satisfactory system);
3. Modifications of maintenance package to meet user demands;
4. Modifications of computer so that it is compatible with the company's main computer;
5. Other possible add-on packages (list);
6. Network system, and installation costs (ensure compatibility with computer and maintenance package);
7. Stationery (include cost for modifying supplied material to meet user demands);
8. Maintenance agreement for computer;
9. Support from suppliers included in initial package;
10. Support from suppliers during operation;
11. Training;
12. Input of information into system;

13. Cover for staff involved in implementation of system;
14. Additional administration staff required for the system's operation.

Software package

1. Is it compatible with the company's main computer?
2. Is the total package user friendly or will it require a large amount of support?
3. Does it require modifications to meet the department's requirements?
4. Does it have the facility to copy information/details of similar assets, thereby reducing initial input of these?
5. Does it have the facility to copy information/details of common work tasks, thereby reducing initial input of these? (This would include labor requirement.)
6. Does it have the facility to search out similar assets (by manufacturer, duty, task etc.)?
7. To what extent is asset information held? Is there a limitation on the amount of information that can be stored in a readily useable format (e.g. history, costs, spare parts, etc.)?
8. Is there adequate help when entering information into the program (i.e. simple keyboard entries)?
9. When entering information, does it require to be entered in a number of different fields?
10. Is labor scheduling by a team or individual trades?
11. When loaded automatically, are work tasks scheduled to ensure an even weekly output?
12. How does the program cater for seasonal work (e.g. heating boilers, air condition)?
13. How does the program cater for seasonal work (e.g. plant shutdowns, holidays)?
14. How is scheduling of work tasks handled (e.g. different frequencies)?
15. Does the system only produce work dockets of the least frequent service (i.e. when yearly and quarterly service visits are due, will the quarterly visits also be printed)?
16. Is it easy to carry out simple corrections when incorrect information has been entered by mistake?
17. How is breakdown work handled? Is there a simple entry format so that a work docket can be issued without going through a lengthy keyboard process?
18. Is there a suitable format within the process to view pending work (e.g. job file) for a specific forthcoming period? Is it available in print?
19. Is there a facility for viewing the job file on the screen indicating docket status (e.g. work overdue, work due, work in progress, etc.)?
20. Is there a facility for viewing the job file on the screen indicating docket status (e.g. work overdue, work due, work in progress, etc.)?
21. Are the keyboard instructions simple with respect to the printing of work dockets and the entering of history relevant to these dockets on completion/part completion or cancellation of work?
22. Does the system require 'tidying' (e.g. transferring information from 'working file' to 'history file')?
23. Is there a facility to enter 'free format' information into e.g. history file, breakdown work dockets?

24. If purchasing a stores/spare parts module, does it have the facility to link similar parts for all assets to reduce stock holding? Does it automatically enter into the issue of these?
25. What reports can be produced (e.g. labor in trade groups and/or for specific assets, departmental, premises)?
26. Can a bonus system be incorporated into the labor element?
27. What facility is there for evaluation (e.g. life-cycle costing, capital write-off/replacement, design, reliability)?
28. Is the cost module compatible with the company's cost procedures?
29. Can the reports be produced in chart or table form?

If, after assessing the various computer and software packages, the decision is made to implement the proposed maintenance system on computer rather than by manual methods caution must be exercised. Irrespective of the above-mentioned benefits, the use of a computer does not increase work output but it does provide the department with a very sound method of improving the management of the company's assets and resources.

In establishing a sound maintenance program, budget costs for both operation of the department and associated costs of the user department's assets will be readily obtained.

The proposed maintenance program could be likened to a car. Both manual and automatic models can take one from A to B, and although the automatic may be easier to drive, both still require a driver (in the case of the maintenance program, the engineer) to manage them.

48.4.2 Comment

Irrespective of the method of planned maintenance to be implemented, benefits to either will be accrued if all possible information/methods/systems, supported by the appropriate training, are also considered. Much will depend on the demands placed on the plant and equipment and what method is to be adopted in establishing the most economical means of maintaining their life. It is only when the specific plant and equipment is known that the appropriate maintenance program can be implemented.

48.5 Life-cycle costing

This is the method of establishing the cost of the plant and equipment over its recognized life cycle. In maintenance situations the design and specification elements may be ignored, but those that must be considered in establishing the life-cycle cost of plant and equipment will include:

Purchase
Installation
Commissioning
Trials and tests
Operation
Maintenance
Replacement

For production plants and equipment, life-cycle costs are important in that the revenue from a product must cover all aspects of expenditure. For plant and equipment such as boilers and air-conditioning units, establishing life-cycle costs can play a major part in assessing when these items should be replaced. For plant and equipment such as self-contained units (e.g. pumps with sealed bearings, etc.), life-cycle costing may be used to determine their replacement date.

The information on life-cycle costs gained with certain items of plant and equipment need not be true for a similar item in another location, as specific details on operating hours, and maintenance attendance can vary markedly. Also, it may be an auxiliary part (e.g. an electric motor) of a major component. Therefore, it may have an entirely different set of parameters. An important element that must be examined in detail when establishing life-cycle costs of a specific item of plant and equipment is the demand placed on energy resources by them.

To establish costs with regards to a specific item of plant, sound feedback must be an element of the company's program and subsequent records must be maintained. To be worthwhile these must include *all costs*: if not, the assumed life-cycle cost over a number of years could be misleading.

It follows that any maintenance program introduced for specific plant and equipment must be designed so that all maintenance costs are recorded. This would then be taken into consideration when the life-cycle cost of the item is assessed, as it should be added to the loss of production if this has occurred due to any maintenance malfunction. Care must be taken when presenting the life-cycle cost to ensure that all elements within the cost cycle have been effective.

Life cycle costing must be considered when purchasing or taking over new plant or equipment and include all aspects as listed above.

48.6 Training

A major concern within maintenance departments and service contractors is the lack of experienced maintenance technicians and craftsmen. This has been allowed to occur through cutbacks in staffing levels accompanied by a reduction in training. A few companies provide maintenance training, and it could be said that these are progressive in that they acknowledge that, by providing training for their workforce, they will not find themselves in the situation of searching for the limited skilled labor now available.

Planned maintenance can reduce the demand for highly experienced craftsmen in that, if the instructions issued in work docket and advice/guidance notes are adequate, a less-skilled person should be able to perform the work task correctly. It follows that adequate training must be provided to ensure that, irrespective of the degree of the operative's skills and experience, the work task given will be carried out to the required standard.

There are a number of suitable training establishments available, but for the specific plant and equipment within the organization, in-house training would provide a better result. Carrying out in-house training reduces costs and ensures that the subject matter is relevant. Support in this training can be complemented by the use of manufacturers, suppliers or specialists (e.g. control and/or service engineers). It will be found that such companies encourage this method, as it helps to reduce the demand on their own highly skilled staff. Trained operatives can carry out basic fault finding either alone or by telephone.

With the changing face of technology, it is imperative that service departments keep up to date with the latest practices. Training should not be restricted to any specific individual but should include engineers, technicians, craftsmen and, where appropriate, semi-skilled staff.

48.7 Health and safety

No maintenance task can be issued without the person responsible for controlling the workforce establishing that the work is to be carried out safely. Areas where this is essential would include boilers, pressure systems, electrical systems, confined spaces, hot work, deep sumps/shafts, tanks, and lone working.

To reduce the hazard from any of the above the risk to the individual must be assessed and, where necessary, a safe working system/procedure prepared and/or a permit to work issued. It follows that, whatever procedure is adopted, it can only be effective if closely monitored.

48.8 Information

Information on maintenance is available from a large number of sources. The initial one is that of the manufacturer/supplier for new plant and equipment. Where this plant and equipment forms a part of a process, the party responsible for the overall design must prepare the appropriate operation and maintenance and instructions. These should be supported by technical manuals, which should include spare parts lists, control measures, and, where possible, faultfinding charts.

Problems arise in older plant and equipment, as there may have been no information supplied when the plant was first installed. Alternatively, the information may have been lost through the passage of time or by changes in department location and/or staff.

For an inquiring mind (which is an essential tool in the field of maintenance), sources of information outside the service department are available and include manufacturers, suppliers, trade associations, engineering institutions (e.g. plant, mechanical, electrical, etc.), consultants, and suppliers of maintenance systems, libraries, and technical journals and the Internet.

48.9 Conclusion

This chapter should be sufficient to guide the engineer or manager through the initial stages of introducing a sound and effective planned maintenance system. However, as in any establishment – production, administration or (as in this case) engineering – without the commitment and

endeavors of all those involved, the system will not produce the required results.

On setting out on the planned maintenance route, if the percentage of the work to be carried out under the plan is not sufficient (say, 80 per cent), the department will continue to operate as before (i.e. firefighting) with the control of that department being in the hands of others. Any planned maintenance system is a *tool*. It cannot do the work itself but, employed correctly, the sound management results will prove that the efforts involved in setting up the system were worthwhile.

Appendix: Elements of a planned maintenance system

Plant codes

Plant code	Plant/equipment
01	Boilers
02	Heating systems
03	Domestic hot and cold water
05	Pumps
06	Water treatment
10	Ventilation
12	Chiller/refrigeration
13	Chilled water systems
15	Cooling water systems
31	Electrical distribution
33	Electrical lighting and power
34	Batteries/emergency lighting
41	Kitchen equipment
44	Fixed equipment
45	Portable equipment
46	Laundry equipment
70	Lifts

Guidance notes

Health and safety

Health and safety is everyone's responsibility.
Plant and equipment must be made safe before any work is carried out.
A permit to work may be required for certain work tasks.
Protective clothing/equipment should be used when necessary.
Defective tools or access equipment must not be used.
All safety rules and regulations must be adhered to.
On completion of a work task, the plant/equipment must be left in a safe and operable condition.
Safety guards, where fitted, must be replaced and secured.

Technicians'/craftsman's responsibility

The instructions given in these guidance notes do not cover every component part, but the technician/craftsmen is expected to carry out any necessary maintenance in accordance with their normal trade practice.

Manufacturers' maintenance handbooks, when available, should be followed with respect to recommendations for work to be carried out and at which frequency.

Work tasks

Heating Systems Plant Code E02.
Half-yearly (H) Service.
5. Fan Assisted Heaters.
Blow out motor windings, lubricate as necessary.
Check insulation resistance and test earth continuity.

Examine apparatus for general mechanical condition, including guards and other safety devices.
Test mechanical action and polarity of equipment switch.
Clean and test switches, pilot lamps, thermostats and timing devices.
Ensure that writing and terminal connections are secure.
Test under normal running conditions.

49

Effective Maintenance Management

Ricky Smith

Life Cycle Engineering, Inc.

Contents

49.1 Introduction

Maintenance costs, as defined by normal plant accounting procedures, are normally a major portion of the total operating costs in most plants. Traditional maintenance costs (i.e. labor and material) in the US has escalated at a tremendous rate over the past ten years. In 1981, domestic plants spent more than $600 billion to maintain their critical plant systems. By 1991, the costs had increase to more than $800 billion and are projected to top $1.2 trillion by the year 2000. These evaluations indicate that on average one third, or $250 billion, of all maintenance dollars are wasted through ineffective maintenance management methods. American industry cannot absorb the incredible level of inefficiency and hope to compete in the world market.

Because of the exorbitant nature of maintenance costs, they represent the greatest potential short-term improvement. Delays, product rejects, scheduled maintenance downtime and traditional maintenance costs, such as labor, overtime and repair parts, are generally the major contributors to abnormal maintenance costs within a plant.

The dominant reason for this ineffective management is the lack of factual data that quantifies the actual need for repair or maintenance of plant machinery, equipment and systems. Maintenance scheduling has been, and in many instances, is predicated on statistical trend data or on the actual failure of plant equipment.

Until recently, middle and corporate level management have ignored the impact of the maintenance operation on product quality, production costs and more importantly on bottom-line profit. The general opinion has been 'Maintenance is a necessary evil' or 'Nothing can be done to improve maintenance costs'. Perhaps these were true statements ten or twenty years ago. However, the developments of microprocessor or computer-based instrumentation that can be used to monitor the operating condition of plant equipment, machinery and systems have provided the means to manage the maintenance operation. They have provided the means to reduce or eliminate unnecessary repairs, prevent catastrophic machine failures, and reduce the negative impact of the maintenance operation on the profitability of manufacturing and production plants.

49.2 Maintenance methods

Industrial and process plants typically utilize two types of maintenance management: run-to-failure or preventive maintenance.

49.2.1 Run-to-failure management

The logic of run-to-failure management is simple and straightforward. When a machine breaks down ... fix it. This 'If it ain't broke, don't fix it' method of maintaining plant machinery has been a major part of plant maintenance operations since the first manufacturing plant was built and on the surface sounds reasonable. A plant using run-to-failure management does not spend any money on maintenance until a machine or system fails to operate. Run-to-failure is a reactive management technique that waits for machine or equipment failure before any maintenance action is taken. It is in truth a 'no maintenance' approach of management. It is also the most expensive method of maintenance management.

Few plants use a true run-to-failure management philosophy. In almost all instances, plants perform basic preventive tasks, i.e. lubrication, machine adjustments, and other adjustments, even in a run-to-failure environment. However, in this type of management, machines and other plant equipment are not rebuilt nor are any major repairs made until the equipment fails to operate.

The major expenses associated with this type of maintenance management are: high spare parts inventory cost, high overtime labor costs, high machine downtime and low production availability. Since there is no attempt to anticipate maintenance requirements, a plant that uses true run-to-failure management must be able to react to all possible failures within the plant. This reactive method of management forces the maintenance department to maintain extensive spare parts inventories that include spare machines or at least all major components for all critical equipment in the plant. The alternative is to rely on equipment vendors that can provide immediate delivery of all required spare parts. Even if the latter is possible, premiums for expedited delivery substantially increase the costs of repair parts and downtime required correcting machine failures. To minimize the impact on production created by unexpected machine failures, maintenance personnel must also be able to react immediately to all machine failures.

The net result of this reactive type of maintenance management is higher maintenance cost and lower availability of process machinery. Analysis of maintenance costs indicate that a repair performed in the reactive or run-to-failure mode will average about three times higher than the same repair made within a scheduled or preventive mode. Scheduling the repair provides the ability to minimize the repair time and associated labor costs. It also provides the means of reducing the negative impact of expedited shipments and lost production.

49.2.2 Preventive maintenance management

There are many definitions of preventive maintenance. However, all preventive maintenance management programs are time-driven. In other words, maintenance tasks are based on elapsed time or hours of operation. Figure 49.1 illustrates an example of the statistical life of a machine-train. The mean-time-to-failure (MTTR) or bathtub curve indicates that a new machine has a high probability of failure, due to installation problems, during the first few weeks of operation. Following this initial period, the probability of failure is relatively low for an extended period of time. Following this normal machine life period, the probability of failure increases sharply with elapsed time. In preventive maintenance management, machine repairs or rebuilds are scheduled based on the mean-time-to-failure (MTTF) statistic.

The actual implementation of preventive maintenance varies greatly. Some programs are extremely limited

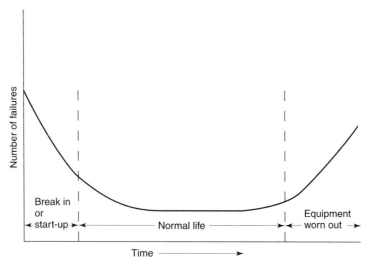

Figure 49.1 Bathtub curve

and consist of lubrication and minor adjustments. More comprehensive preventive maintenance programs schedule repairs, lubrication, adjustments and machine rebuilds for all critical machinery in the plant. The common denominator for all of these preventive maintenance programs is the scheduling guideline. All preventive maintenance management programs assume that machines will degrade within a time frame typical of its particular classification. For example, a single-stage, horizontal split-case centrifugal pump will normally run 18 months before it must be rebuilt. Using preventive management techniques, the pump would be removed from service and rebuilt after 17 months of operation.

The problem with this approach is that the mode of operation and system or plant specific variables directly affect the normal operating life of machinery. The mean-time-between-failures (MTBF) will not be the same for a pump that is handling water and one handling abrasive slurries. The normal result of using MTBF statistics to schedule maintenance is either unnecessary repairs or catastrophic failure. In the example, the pump may not need to be rebuilt after 17 months. Therefore, the labor and material used to make the repair were wasted. The second option using preventive maintenance is even more costly. If the pump fails before 17 months, we are forced to repair using run-to-failure techniques. Analysis of maintenance costs have shown that a repair made in a reactive, i.e. after failure, mode will normally be three times greater than the same repair made on a scheduled basis.

49.2.3 Predictive maintenance management

Like preventive maintenance, predictive maintenance has many definitions. To some predictive maintenance is monitoring the vibration of rotating machinery in an attempt to detect incipient problems and to prevent catastrophic failure. To others it is monitoring the infrared image of electrical switchgear, motors and other electrical equipment to detect developing problems.

The common premise of predictive maintenance is that regular monitoring of the mechanical condition of machine-trains will ensure the maximum interval between repair and minimize the number and cost of unscheduled outages created by machine-train failures.

Predictive maintenance is much more. It is the means of improving productivity, product quality and overall effectiveness of our manufacturing and production plants. Predictive maintenance is not vibration monitoring or thermal imaging or lubricating oil analysis or any of the other nondestructive testing techniques that are being marketed as predictive maintenance tools. Predictive maintenance is a philosophy or attitude that simply stated using the actual operating condition of plant equipment and systems to optimize total plant operation. A comprehensive predictive maintenance management program utilizes a combination of the most cost-effective tools, i.e. vibration monitoring, thermography, tribology, etc., to obtain the actual operating condition of critical plant systems and based on this actual data schedules all maintenance activities on an 'as-needed' basis. Including predictive maintenance in a comprehensive maintenance management program will provide the ability to optimize the availability of process machinery and greatly reduce the cost of maintenance. It will also provide the means to improve product quality, productivity and profitability of our manufacturing and production plants.

Predictive maintenance is a condition-driven preventive maintenance program. Instead of relying on industrial or in-plant average-life statistics, i.e. mean-time-to-failure, to schedule maintenance activities, predictive maintenance uses direct monitoring of the mechanical condition, system efficiency and other indicators to determine the actual mean-time-to-failure or loss of efficiency for each machine-train and system in the plant. At best, traditional time-driven methods provide a guideline to *normal* machine-train life spans. The final decision, in preventive or run-to-failure programs, on repair or rebuild schedules must be made on the bases of intuition and the personal experience of the maintenance manager. The addition of a comprehensive predictive maintenance program can and will provide factual data on the actual mechanical condition of each machine-train and operating efficiency of each process system. This data provides the maintenance manager with actual data for scheduling maintenance activities.

A predictive maintenance program can minimize unscheduled breakdowns of all mechanical equipment in the plant and ensure that repaired equipment is in acceptable mechanical condition. The program can also identify machine-train problems before they become serious. Most mechanical problems can be minimized if they are detected and repaired early. Normal mechanical failure modes degrade at a speed directly proportional to their severity. If the problem is detected early, major repairs, in most instances, can be prevented. Simple vibration analysis is predicated on two basic facts: all common failure modes have distinct vibration frequency components that can be isolated and identified and the amplitude of each distinct vibration component will remain constant unless there is a change in the operating dynamics of the machine-train. These facts, their impact on machinery and methods that will identify and quantify the root cause of failure modes will be developed in more detail in later chapters.

Predictive maintenance utilizing process efficiency, heat loss or other nondestructive techniques can quantify the operating efficiency of non-mechanical plant equipment or systems. These techniques used in conjunction with vibration analysis can provide the maintenance manager or plant engineer with factual information that will enable them to achieve optimum reliability and availability from their plant.

Five nondestructive techniques normally used for predictive maintenance management are: vibration monitoring, process parameter monitoring, thermography, tribology and visual inspection. Each technique has a unique data set that will assist the maintenance manager in determining the actual need for maintenance and is discussed further in Sections 50.3.1–50.3.5.

How do you determine which technique or techniques are required in your plant? How do you determine the best method to implement each of the technologies? If you listen to the salesman of the vendor that supplies predictive maintenance systems, his is the only solution to your problem. How do you separate the good from the bad? Most comprehensive predictive maintenance programs will use vibration analysis as the primary tool. Since the majority of normal plant equipment is mechanical, vibration monitoring will provide the best tool for routine monitoring and identification of incipient problems. However, vibration analysis will not provide the data required on electrical equipment, areas of heat loss, condition of lubricating neither oil nor other parameters that should be included in your program.

49.3 Role of maintenance organization

Too many maintenance functions continue to pride themselves on how fast they can react to a catastrophic failure or production interruption rather than in their ability to prevent these interruptions. While few will admit their continued adherence to this 'breakdown' mentality, most plants continue to operate in this mode. Contrary to popular belief, the role of the maintenance organization is to maintain plant equipment, not to repair it after a failure.

The mission of the maintenance in a world-class organization is to achieve and sustain:

Optimum availability   The availability of production systems and their auxiliary equipment in part, determine the production capacity of a plant. The primary function of the maintenance organization is to insure that all machinery, equipment and systems within the plant are always on-line and in good operating condition.

Optimum operating condition   Availability of critical process machinery is not enough to insure acceptable plant performance levels. The maintenance organization has the responsibility to maintain all direct and indirect manufacturing machinery, equipment and systems so that they will run continuously in optimum operating condition. Minor problems, no matter how slight, can result in poor product quality, reduce production speeds or other factors that limit overall plant performance.

Maximum utilization of maintenance resources   The maintenance organization controls a substantial part of the total operating budget in most plants. In addition to an appreciable percentage of the total plant labor budget, the maintenance manager, in many cases, controls the spare parts inventory, authorizes the use of outside contract labor and requisitions millions of dollars in repair parts or replacement equipment. Therefore, one goal of the maintenance organization should be effective use of these resources.

Accurate maintenance records   Without detailed, accurate historical data, maintenance cannot effectively fulfil its mission. This task should include all data, such as maintenance history, repair costs, etc., for all plant assets.

Optimum equipment life   One way to reduce maintenance cost is to extend the useful life of plant equipment. The maintenance organization should implement programs that will increase the useful life of all plant assets.

Minimum spares inventory   Reductions in spares inventory should be a major objective of the maintenance organization. However, the reduction cannot impair their ability to meet goals 1 through 5. With the predictive maintenance technologies that are available today, maintenance can anticipate the need for specific equipment or parts far enough in advance to purchase them on an as-needed basis.

Ability to react quickly   All catastrophic failures cannot be avoided. Therefore, the maintenance organization must maintain the ability to react quickly to the unexpected failure. If all of the maintenance improvement steps included in total plant performance maintenance (TPPM) are implemented and used, the number of unscheduled repairs should be minimal.

49.4 Evaluation of the maintenance organization

One means to quantify the maintenance philosophy in your plant is to analyze the maintenance tasks that have occurred over the past two to three years. Attention should be given to the indices that define management philosophy.

One of the best indices of management attitude and the effectiveness of the maintenance function is the number of production interruptions caused by maintenance-related problems. If production delays represent more than 30 per cent of total production hours, reactive or breakdown response is the dominant management philosophy. To be competitive in today's market, delays caused by maintenance-related problems should represent less than 1 per cent of the total production hours.

Another indicator of management effectiveness is the amount of maintenance overtime required to maintain the plant. In a breakdown maintenance environment, overtime costs are a major, negative cost. If your maintenance department's overtime represents more than 10 per cent of the total labor budget, you definitely qualify as a breakdown operation. Some overtime is, and will always be, required. Special projects and the 1 per cent of delays caused by machine failures will force some expenditure of overtime premiums; but these abnormal costs should be a small percentage of the total labor costs.

Manpower utilization is another key to management effectiveness. Evaluate the percentage of maintenance labor, compared to total available labor hours that are expended on the actual repairs and maintenance prevention tasks. In reactive maintenance management, the percentage will be less than 50 per cent. A well-managed maintenance organization should maintain consistent manpower utilization above 90 per cent. In other words, at least 90 per cent of the available maintenance labor hours should be effectively utilized to improve the reliability of critical plant systems, not waiting on something to break.

49.4.1 Maintenance planning and scheduling

A well-trained maintenance-planning group is an absolute requirement of a successful maintenance operation. However, a surprisingly low number of plants have a formal planning activity within their maintenance group. The few that do have planners have not invested the resources, such as training and support that their planning function needs to be effective. The general tendency is to select employees who do not appear to be fully loaded with other tasks and re-assign them to fill the planning requirement. In most cases, these 'planners' do not receive training nor do they have the basic information required to successfully complete their duties.

Effective maintenance planning requires specific skills and abilities that are not common knowledge to plant workers. Without formal training in these basic skills, a planner cannot properly execute this critical function.

The objective of the maintenance planning function is to effectively utilize the maintenance staff and minimize the impact of maintenance downtime on the production capacity of the plant. Historical data indicates that unplanned or reactive repairs will cost about three times more than identical repairs that are well planned. Therefore, creating an environment were all repairs and other maintenance activities are well planned and scheduled for optimum interval will substantially reduce overall maintenance costs. In addition, proper planning will increase the number of hours that are available for production, reduce product quality problems and increase the production capacity.

A significant limitation of many planning functions is a severe lack of information. To properly plan a simple repair or major repair outage, a planner must have factual data that defines the specific tasks required. In addition, for each task the planner must know: time to complete, manpower required, labor skills needed, tools required, and the specific steps needed to efficiently complete the effort. Few maintenance functions compile historical data that even comes close to meeting these minimum requirements. Instead, they rely on the memory of maintenance staff to estimate this data. The result is inefficiency, wasted time and lost profits.

All plants should have and maintain a historical database that includes the information required for effective maintenance planning. As a minimum, the data should include: mean-time-between-repairs (MTBR), mean-time-to-repair (MTTR), machine repair history, machine maintenance cost history, bill of materials, and standard maintenance procedures (SMPs) for every critical machine and major machine component within the plant.

Standard maintenance procedures are an absolute requirement of a maintenance improvement program. In part, the ineffectiveness of maintenance organization is the result of improper or incomplete repair of critical machinery. Two factors contribute to this problem: unclear definition of the problem, and craftsman does not know how to properly repair. Too many repairs work orders will state 'Inspect and repair as necessary'. With this limited information, the maintenance worker will disassemble a complex machine-train, try to ascertain what repairs are required, replace what he feels are the defect parts and then reassemble the machine. With this approach, the probability that the machine-train is properly repaired and will operate with any degree of reliability is very low.

Standard maintenance procedures (SMPs) will resolve, or substantially reduce the number of ineffective repairs and increase the reliability of repaired machinery. A properly developed SMP assumes that the maintenance worker has never seen the machine that is to be repaired. It will provide step-by-step directions for proper disassembly, repair and to reassemble the machine. In addition, the SMP will define the specific skills required for each repair so that a qualified worker performs each task, as well as defining the tools, materials and time required for completing each task and insuring that the work can be performed safely.

Review of planning process

Maintenance planning can be defined as the process used to develop a course of action. Effective maintenance planning involves the development of a course of action that includes all maintenance, repair, and construction work. It requires the identification of the following factors:

Scope of the job   Most requests received by the maintenance department are not sufficiently defined. A request to 'repair steam leak' may require replacement of a valve, replacement of piping, replacement of gland packing, or perhaps simply tightening a loose connection. To the extent possible, work to be done should be defined clearly and completely. If needed to identify the job, drawings or sketches should be made available.

Location of the job   The exact location where work is to be performed should be identified through building number, department number, machine number, or some other meaningful designation.

Job priority   Priority control is required so that the job can be scheduled according to its importance. While each operation or facility has varying requirements, definite rules can and should be established to provide guidelines for job priority.

Methods to be used   Effective planning requires definition of methods. 'Purchase and install' is considerably different from 'Fabricate and install'. Whether or not shop fabrication or field fabrication is also important. Welding instead of fastening with bolts requires different craft skills. The use of hand tools in place of machine tools may or may not be advisable. Although many jobs are clear-cut and require the usual crafts, there are also instances where several choices may exist.

Material requirements   If the scope of work and the methods to be used are determined, a natural by-product is the generation of a listing of materials. This is required to insure that necessary materials are available.

Tool and equipment requirements   It is not necessary to list the normal hand tools with which each craftsman is equipped. However, any special tools should be identified so that they are available when the job is started. Torque wrenches, hoists, slings, or jacks are special job requirements and should be specified when needed.

Skill requirements   If craft lines are followed, the various crafts required to perform the work should be identified. This will permit assignment of proper personnel.

Manpower requirements   The number of man-hours required to perform the work should be determined for each craft. This is necessary for scheduling control.

Safety procedures   Safety should be a primary consideration in maintenance planning and in all other facets of plant operation and maintenance. Each standard maintenance procedure and maintenance work order must include specific instructions that define safety procedures or safety concerns pertaining to each task of the job.

The program should develop and implement guidelines for standard maintenance procedures (SMPs) for each production group or functional area within the plant. Development of SMPs that are compatible with the total plant program will require a considerable amount of time and effort. In most plants, the historical information is not available and cannot be quickly generated.

Effective operating environment

In a proactive plant environment, all maintenance activities are planned with the exception of emergency type work. Tasks required to be performed the next day may not be planned, but it will be scheduled. All other work will be planned and scheduled.

Close the warehouse and deliver material to the job site. There is not a pipefitter in the world that can go to the warehouse by himself, there always has to be a buddy to go along. Now you have two high cost resources going to the warehouse to get a part when a lower cost resource could have delivered the part to the job site.

Most organizations have the talent to improve the management of the maintenance function. The problem is that most organizations lack the ability to organize and focus their talent around a badly developed and managed process. An effective process will have a disciplined approach, with flexibility, to accomplish the operational and reliability goals of the organization. Throughout this treatment of planning and scheduling, the terms *process* and *system* will be used interchangeably. A systems or process approach will be taken regarding developing a fully functional approach to the design and management of the maintenance function.

49.4.2 Changing traditional thinking

In order to evolve the production organization to higher levels, several outdated and obsolete practices must be replaced by a new philosophy that is supported by a system or process for effective maintenance management.

Most organizations believe that equipment failures are to be accepted as normal. This line of thinking seems reasonable to some but is not an accepted behavior in 'world-class' organizations. An effective reference point can be drawn from the quality thinking of *zero defects*. The same mentality should and can be applied to maintenance only in the form of *zero failures*.

Too many organizations have practiced 'fire fighting' for such a long period of time that they have become extremely good at it, as well as comfortable. Granted, the ability to fight the fire or emergency is something that the organization needs to develop, but it should not be the primary focus of an organizational approach to maintenance. Many organizations are caught up in dealing with the emotional failure of the day and have completely abandoned, or never developed, the practice of determining the root cause for failure and then taking the next step and implementing a cure.

Few organizations take progressive steps when dealing with recurring failures. Many never evolve their maintenance effort to this point. The key is to evolve the organizational thinking to a level that stresses the fact that organizations should not have to live with recurring failures. This level of attainment is referred to as maintenance prevention or maintenance-free operation.

Last, but certainly not least, an organization that is in real trouble has an indifferent attitude about making improvements. These organizations feel that it is beyond hope and have an 'it will always be this way because it has always been this way' type of attitude.

Upon further examination, it is not unusual to find organizations that display some or all of the above approaches to maintenance management. In the next section, an overview of the planning and scheduling process will be revealed along with a discussion of the essential components of that process.

Evolving the organization to higher levels of performance will require a rethinking of traditional practices. To evolve the organization to this higher level of performance will require developing and implementing a set of values and beliefs that establishes the *way* maintenance is going to be managed. A world-class maintenance system or process should begin with the development of a facility-wide guiding vision statement.

The next step in developing the philosophical guiding practice is to present it to maintenance and operations. An example of presenting the strategy to the operational and maintenance should have a strategic framework such as the following.

49.5 Symptoms of loss of control

Excessive machine breakdowns

Indications are equipment failures that cause operating losses. This strongly indicates that equipment and machinery are not being maintained or overhauled on a routine basis.

Frequent emergency work

This work can vary considerably from facility to facility. With a facility of very poor or minimum planning, emergency work can entail better than 70 per cent of all work performed.

Domination of maintenance by operations

Here we are touching on one of the most sensitive areas of maintenance management, but for a good planning system to exist, the implemented policies must be established to show the demarcation between maintenance and operations. Although all lines must be kept open, normally the operations or production designates *where* and *why* equipment is to be repaired; maintenance planning determines *what* and *when*; and maintenance management determines *who* and *how*.

Lack of an equipment replacement program

In this situation, no attempt is made to gather cost by equipment function or equipment tag number, making it impossible to determine the actual maintenance repair cost on a specific item of equipment. These cost figures should be used to justify the purchase of new equipment, or to engineer present equipment in order to obtain better performance from it.

Insufficient preventive maintenance

Obtaining a sufficient level of preventive maintenance is unlikely without the presence of a well-organized preventive maintenance program. This type of preventive maintenance program will provide maintenance the opportunity to prevent failures from occurring or find failures before they develop into a breakdown.

Preventive maintenance is not an option. Rather, it is an essential part of any effective maintenance management program. Equipment reliability, optimum capacity and product quality are dependent on proper preventive maintenance, such as lubrication, adjustments, and calibration.

Inadequate training

This is another item that is always discussed but about which little is done. When management purchases a piece of equipment or system from a vendor, they usually get some kind of training included with the purchase. This training is normally designed to generate more sales of equipment and services. Training of maintenance personnel should be budgeted for each year. This type of training is discussed later in this session.

No planned routine for selecting maintenance supervision, inadequate training of maintenance personnel, and poor shop facilities are additional symptoms which indicate that your organization is fighting the never-ending battle against fire fighting.

49.6 Methods of regaining control

How does one correct these maintenance deficiencies? The solution is to use management tools which have been available for years, but which have only recently come into the maintenance organizations. Some of these management techniques are:

Organization planning

Adequate selection and training of maintenance supervisors and managers will accrue cost reductions in proportion to the effort made by top management to improve maintenance procedures, cost controls, and maintenance management skills.

In the maintenance department, the primary and ultimate goal is to further the profit-making capabilities of the facility. The subordinate goal of maintenance is closely allied to production as a means by which the facility may achieve profitability. As profit is the ultimate goal, then production facilities must be maintained at a maximum efficiency at minimum cost.

Actions to ensure maximum maintenance and equipment efficiency at minimum cost are:

Preventive maintenance   Preventive maintenance, in the form of cleaning, lubricating, adjusting, and overhauling, is an essential requirement of effective maintenance. Faulty equipment will cause losses to a production economy in both quantity and quality.

Housekeeping   Work place cleanliness is not an option. Rather it is essential for effective performance, as well as the health and safety of workforce.

Planning   Planning should be in accordance with established time objectives. Deviations in time estimates will result in upsets to schedules, priorities, costs, and coordination and synchronization with other departments.

Planning should be based on realistic labor repair costs. Deviations in labor costs upset estimates and cause budget overruns. Planning should be in accordance with current cost and availability of materials. Deviations in the availability of materials cause time delays, with consequent disruptions in schedules and priorities. Deviations in material costs upset estimates and cause budget overruns.

Controls must be established to determine whether or not plans are being adhered to, or whether or not progress is being made toward objectives and goals. Adjustments must be made to performance before deviations upset the production, maintenance, and other objectives, making it impossible to evaluate the quality of the maintenance operation. Virtually every operation in the maintenance department is subject to control. If these controls are ignored or are observed incorrectly, maintenance and enterprise win suffer a loss that can be computed in terms of dollars.

The reliability of maintenance is its own greatest asset. Every operation is essential to the whole and the weakness of any link in the chain is enough to cause unprecedented disaster. The most important task of maintenance is to organize a department which is reliable and trustworthy, and which will further the company's objectives. Every person must know his role and his place in the overall scheme. When he knows this, he becomes a member of the team.

Written procedures

Standard maintenance procedures (SMPs) that are compatible with any current computerized system should be developed which are capable of providing maximum support of the maintenance craftsperson and the facility maintenance management philosophy. After a thorough review, these procedures should be approved and distributed. These procedures should be short and concise and should provide the rules for conducting maintenance.

Performance measurement

At the beginning of any maintenance improvement program, maintenance and equipment performance measurement factors should be developed. It is important that these factors be used for setting internal targets and to measure performance. Objectives of these factors should be:

1. To identify maintenance effectiveness.
2. To identify the improvement potential of maintenance effectiveness.
3. To identify equipment effectiveness.

Training programs

The constant increase in facility costs, in terms of equipment, material and labor, makes it mandatory that facility equipment be maintained in the most efficient and effective manner. This dictates that functional, well-structured and realistic training programs be developed and implemented to meet the growing need of providing increasing numbers of fully trained and highly skilled maintenance craftsmen, planners/schedulers, supervisors, and managers. A well-rounded training program should include:

- Planner and scheduler training
- Supervisory training
- Specialized maintenance training
- Maintenance craftsperson training.

Effective long-term training is an essential component of effective maintenance. Without highly skilled employees, world-class performance levels cannot be achieved or sustained.

50

Predictive Maintenance

R Keith Mobley

The Plant Performance Group

Contents

50.1 Introduction

Maintenance costs are a major part of the total operating costs of all manufacturing or production plants. Depending on the specific industry, maintenance costs can represent between 15 and 40 per cent of the costs of goods produced. For example in food related industries, the average maintenance cost represents about 15 per cent of the cost of goods produced; while in iron and steel, pulp and paper and other heavy industries maintenance represents up to 40 per cent of the total production costs.

Recent surveys of maintenance management effectiveness indicate that one-third, 33 cents out of every dollar, of all maintenance costs is wasted as the result of unnecessary or improperly carried out maintenance. When you consider that US industry spends more than $200 billion dollars each year on maintenance of plant equipment and facilities, the impact on productivity and profit that is represented by the maintenance operation become clear.

The result of ineffective maintenance management represents a loss of more than $60 billion dollars each year. Perhaps more important is the fact that our ineffective management of maintenance dramatically impacts our ability to manufacture quality products that are competitive in the world market. The loss of production time and product quality that results from poor or inadequate maintenance management has had a dramatic impact on our ability to compete with Japan and other countries that have implemented more advanced manufacturing and maintenance management philosophies.

The dominant reason for this ineffective management is the lack of factual data that quantifies the actual need for repair or maintenance of plant machinery, equipment and systems. Maintenance scheduling has been, and in many instances, is predicated on statistical trend data or on the actual failure of plant equipment.

Until recently, middle and corporate level management have ignored the impact of the maintenance operation on product quality, production costs and more importantly on bottom-line profit. The general opinion has been 'Maintenance is a necessary evil' or 'Nothing can be done to improve maintenance costs'. Perhaps these were true statements ten or twenty years ago.

However, the developments of microprocessor or computer-based instrumentation that can be used to monitor the operating condition of plant equipment, machinery and systems have provided the means to manage the maintenance operation. They have provided the means to reduce or eliminate unnecessary repairs, prevent catastrophic machine failures, and reduce the negative impact of the maintenance operation on the profitability of manufacturing and production plants.

50.2 Benefits of predictive maintenance

Predictive maintenance is not a substitute for the more traditional maintenance management methods. It is, however, a valuable addition to a comprehensive, total plant maintenance program. Where traditional maintenance management programs rely on routine servicing of all machinery and fast response to unexpected failures, a predictive maintenance program schedules specific maintenance tasks, as they are actually required by plant equipment. It cannot totally eliminate the continued need for either or both of the traditional programs, i.e. run-to-failure and preventive. Predictive maintenance can reduce the number of unexpected failures and provide a more reliable scheduling tool for routine preventive maintenance tasks.

The premise of predictive maintenance is that regular monitoring of the actual mechanical condition of machine-trains and operating efficiency of process systems will ensure the maximum interval between repairs; minimize the number and cost of unscheduled outages created by machine-train failures and improve the overall availability of operating plants. Including predictive maintenance in a total plant management program will provide the ability to optimize the availability of process machinery and greatly reduce the cost of maintenance. In reality, predictive maintenance is a condition-driven preventive maintenance program.

A survey of 500 plants that have implemented predictive maintenance methods indicates substantial improvements in reliability, availability and operating costs. The successful programs included in the survey include a cross-section of industries and provide an overview of the types of improvements that can be expected. Based on the survey results, major improvements can be achieved in: maintenance costs, unscheduled machine failures, repair downtime, spare parts inventory, and both direct and in-direct overtime premiums. In addition, the survey indicated a dramatic improvement in: machine life, production, operator safety, product quality and overall profitability.

Based on the survey, the actual costs normally associated with the maintenance operation were reduced by more than 50 per cent. The comparison of maintenance costs included the actual labor and overhead of the maintenance department. It also included the actual materials cost of repair parts, tools and other equipment required to maintain plant equipment. The analysis did not include lost production time, variances in direct labor or other costs that should be directly attributed to inefficient maintenance practices.

The addition of regular monitoring of the actual condition of process machinery and systems reduced the number of catastrophic, unexpected machine failures by an average of 55 per cent. The comparison used the frequency of unexpected machine failures before implementing the predictive maintenance program to the failure rate during the two year period following the addition of condition monitoring the program. Projections of the survey results indicate that reductions of 90 per cent can be achieved using regular monitoring of the actual machine condition.

Predictive maintenance was shown to reduce the actual time required to repair or rebuild plant equipment. The average improvement in mean-time-to-repair, MTTR, was a reduction of 60 per cent. To determine the average improvement, actual repair times before the predictive maintenance program were compared to the actual time to repair after one year of operation using predictive maintenance management techniques. It was found that the regular monitoring and analysis of machine condition identified the specific failed component(s) in each machine and enabled the maintenance staff to plan each repair. The ability to predetermine the specific repair parts, tools and labor skills required provided the dramatic reduction in both repair time and costs.

The ability to predict machine-train and equipment failures and the specific failure mode provided the means to reduce spare parts inventories by more than 30 per cent. Rather than carry repair parts in inventory, the surveyed plants had sufficient lead-time to order repair or replacement parts as needed. The comparison included the actual cost of spare parts and the inventory carrying costs for each plant.

Prevention of catastrophic failures and early detection of incipient machine and systems problems increased the useful operating life of plant machinery by an average of 30 per cent. The increase in machine life was a projection based on five years of operation following implementation of a predictive maintenance program. The calculation included: frequency of repairs, severity of machine damage, and actual condition of machinery following repair. A condition-based predictive maintenance program prevents serious damage to machinery and other plant systems. This reduction in damage severity increases the operating life of plant equipment.

A side benefit of predictive maintenance is the automatic ability to monitor the mean-time-between-failures, MTBF. This data provides the means to determine the most cost-effective time to replace machinery rather than continue to absorb high maintenance costs. The MTBF of plant equipment is reduced each time a major repair or rebuild occurs. Predictive maintenance will automatically display the reduction of MTBF over the life of the machine. When the MTBF reaches the point that continued operation and maintenance costs exceed replacement cost, the machine should be replaced.

In each of the surveyed plants, the availability of process systems was increased following implementation of a condition-based predictive maintenance program. The average increase in the 500 plants was 30 per cent. The reported improvement was based strictly on machine availability and did not include improved process efficiency. However, a full predictive program that includes process parameters monitoring can also improve the operating efficiency and therefore productivity of manufacturing and process plants. One example of this type of improvement is a food manufacturing plant that made the decision to build additional plants to meet peak demands. An analysis of existing plants, using predictive maintenance techniques, indicated that a 50 per cent increase in production output could be achieved simply by increasing the operating efficiency of the existing production process.

The survey determined that advanced notice of machine-train and systems problems had reduced the potential for destructive failure, which could cause personal injury or death. The determination was based on catastrophic failures where personal injury would be

most likely to occur. Several insurance companies offering reduction in premiums for plants that have a condition-based predictive maintenance program in effect have supported this benefit.

Several other benefits can be derived from a viable predictive maintenance management program: verification of new equipment condition, verification of repairs and rebuild work and product quality improvement.

Predictive maintenance techniques can be used during site acceptance testing to determine the installed condition of machinery, equipment and plant systems. This provides the means to verify the purchased condition of new equipment before acceptance. Problems detected before acceptance can be resolved while the vendor has reason (the invoice has not been paid), to correct any deficiencies. Many industries are now requiring that all new equipment include a reference vibration signature be provided with purchase. The reference signature is then compared with the baseline taken during site acceptance testing. Any abnormal deviation from the reference signature is grounds for rejection, without penalty of the new equipment. Under this agreement, the vendor is required to correct or replace the rejected equipment. These techniques can also be used to verify the repairs or rebuilds on existing plant machinery.

Vibration analysis, a key predictive maintenance tool, can be used to determine whether or not the repairs corrected existing problems and/or created addition abnormal behavior before the system is re-started. This eliminates the need for the second outage that many times is required to correct improper or incomplete repairs.

Data acquired as part of a predictive maintenance program can be used to schedule and plan plant outages. Many industries attempt to correct major problems or schedule preventive maintenance rebuilds during annual maintenance outages. Predictive data can provide the information required to plan the specific repairs and other activities during the outage. One example of this benefit is a maintenance outage scheduled to rebuild a ball mill in an aluminum foundry. The normal outage, before predictive maintenance techniques were implemented in the plant, to completely rebuild the ball mill was three weeks and the repair cost averaged $300,000.

The addition of predictive maintenance techniques as an outage-scheduling tool reduced the outage to five days and resulted in a total savings of $200,000. The predictive maintenance data eliminated the need for many of the repairs that would normally have been included in the maintenance outage. Based on the ball mill's actual condition, these repairs were not needed. The additional ability to schedule the required repairs, gather required tools and plan the work reduced the time required from three weeks to five days.

The overall benefits of predictive maintenance management have proven to substantially improve the overall operation of both manufacturing and process plants. In all surveyed cases, the benefits derived from using condition-based management have offset the capital equipment cost required to implement the program within the first three months. Use of microprocessor-based predictive maintenance techniques has further reduced the annual operating cost of predictive maintenance methods so that any plant can achieve cost-effective implementation of this type of maintenance management program.

50.3 Predictive maintenance techniques

A variety of technologies can and should be used as part of a comprehensive predictive maintenance program. Since mechanical systems or machines account for the majority of plant equipment, vibration monitoring is generally the key component of most predictive maintenance programs. However, vibration monitoring cannot provide all of the information that will be required for a successful predictive maintenance program. This technique is limited to monitoring the mechanical condition and not other critical parameters required for maintaining reliability and efficiency of machinery. Therefore, a comprehensive predictive maintenance program must include other monitoring and diagnostic techniques.

These techniques include (1) vibration monitoring (2) thermography, (3) tribology, (4) process parameters, (5) visual inspection and (5) other nondestructive testing techniques. This chapter will provide a description of each of the techniques that should be included in a full capabilities predictive maintenance program for typical plants.

50.3.1 Vibration monitoring

Vibration analysis is the dominant technique used for predictive maintenance management. Since the greatest population of typical plant equipment is mechanical, this technique has the widest application and benefits in a total plant program. This technique uses the noise or vibration created by mechanical equipment and in some cases by plant systems to determine their actual condition.

Using vibration analysis to detect machine problems is not new. During the 1960s and 70s, the US Navy, petrochemical and nuclear electric power generating industries invested heavily in the development of analysis techniques based on noise or vibration that could be used to detect and identify incipient mechanical problems in critical machinery. By the early 1980s, the instrumentation and analytical skills required for noise-based predictive maintenance were fully developed. These techniques and instrumentation had proven to be extremely reliable and accurate in detecting abnormal machine behavior. However, the capital cost of instrumentation and the expertise required to acquire and analyze noise data precluded general application of this type of predictive maintenance. As a result, only the most critical equipment in a few select industries could justify the expense required to implement a noise-based predictive maintenance program.

Recent advancements in microprocessor technology coupled with the expertise of companies that specialize in machinery diagnostics and analysis technology, have evolved the means to provide vibration-based predictive maintenance that can be cost-effectively used in most manufacturing and process applications. These microprocessor-based systems simplify data acquisition, automate data management, and minimize the need for vibration experts to interpret data. Commercially available systems are capable of routine monitoring, trending, evaluation and reporting the mechanical condition of all mechanical equipment in a typical plant. This type of program can be used to schedule maintenance on all rotating, reciprocating and most continuous process mechanical equipment.

Monitoring the vibration from plant machinery can provide direct correlation between the mechanical condition and recorded vibration data of each machine in the plant. Any degradation of the mechanical condition within plant machinery can be detected using vibration-monitoring techniques. Used properly, vibration analysis can identify specific degrading machine components or the failure mode of plant machinery before serious damage occurs. Most vibration-based predictive maintenance programs rely on one or more monitoring techniques. These techniques include broadband trending, narrowband trending, or signature analysis.

Broadband trending

This technique acquires overall or broadband vibration readings from select points on a machine-train. This data is compared to either a baseline reading taken from a new machine or to vibration severity charts to determine the relative condition of the machine. Normally an unfiltered broadband measurement that provides the total vibration energy between 10 and 10,000 Hertz is used for this type of analysis.

Broadband or overall RMS data is strictly a gross value or number that represents the total vibration of the machine at the specific measurement point where the data was acquired. It does not provide any information pertaining to the individual frequency components or machine dynamics that created the measured value.

Narrowband trending

Narrowband trending, like broadband, monitors the total energy for a specific bandwidth of vibration frequencies. Unlike broadband, narrowband analysis utilizes vibration frequencies that represent specific machine components or failure modes.

This method provides the means to quickly monitor the mechanical condition of critical machine components, not just the overall machine condition. This technique provides the ability to monitor the condition of gear sets, bearings and other machine components without manual analysis of vibration signatures.

Signature analysis

Unlike the two trending techniques, signature analysis provides visual representation of each frequency component generated by a machine-train. With training, plant staff can use vibration signatures to determine the specific maintenance required by plant machinery.

Most vibration-based predictive maintenance programs use some form of signature analysis in their program. However, the majority of these programs rely on comparative analysis rather than full root-cause techniques. This failure limits the benefits that can be derived from this type of program.

The capital cost for implementing a vibration-based predictive maintenance program will range from about $8,000 to more than $50,000. Your costs will depend on the specific techniques desired.

Training is critical for predictive maintenance programs based on vibration monitoring and analysis. Even programs that rely strictly on the simplified trending or comparison techniques require a practical knowledge of vibration theory so that meaningful interpretation of machine condition can be derived. More advanced techniques, i.e. signature and root-cause failure analysis, require a working knowledge of machine dynamics and failure modes.

The chapters on establishing and maintaining a total plant predictive maintenance program will provide the practical knowledge required implementing a cost-effective vibration-based program that will provide maximum benefits.

50.3.2 Thermography

Thermography is a predictive maintenance technique that can be used to monitor the condition of plant machinery, structures and systems. It uses instrumentation designed to monitor the emission of infrared energy, i.e. temperature, to determine their operating condition. By detecting thermal anomalies, i.e. areas that are hotter or colder than they should be, an experienced surveyor can locate and define incipient problems within the plant.

Infrared technology is predicated on the fact that all objects having a temperature above absolute zero emit energy or radiation. Infrared radiation is one form of this emitted energy. Infrared emissions, or below red, are the shortest wavelengths of all radiated energy and are invisible without special instrumentation. The intensity of infrared radiation from an object is a function of its surface temperature. However, temperature measurement using infrared methods is complicated because there are three sources of thermal energy that can be detected from any object: energy emitted from the object itself, energy reflected from the object, and energy transmitted by the object. Only the emitted energy is important in a predictive maintenance program. Reflected and transmitted energies will distort raw infrared data. Therefore, the reflected and transmitted energies must be filtered out of acquired data before a meaningful analysis can be made.

The surface of an object influences the amount of emitted or reflected energy. A perfect emitting surface is called a 'blackbody' and has an emissivity equal to 1.0. These surfaces do not reflect. Instead, they absorb all external energy and re-emit as infrared energy. Surfaces that reflect infrared energy are called 'graybodies' and have an emissivity less than 1.0. Most plant equipment falls into this classification. Careful consideration of the actual emissivity of an object improves the accuracy of temperature measurements used for predictive maintenance. To help users determine emissivity, tables have been developed to serve as guidelines for most common materials. However, these guidelines are not absolute emissivity values for all machines or plant equipment.

Variations in surface condition, paint or other protective coatings and many other variables can affect the actual emissivity factor for plant equipment. In addition to reflected and transmitted energy, the user of thermographic techniques must also consider the atmosphere between the object and the measurement instrument. Water vapor and other gases absorb infrared radiation. Airborne dust, some lighting and other variables in the surrounding atmosphere can distort measured infrared radiation. Since the atmospheric environment is constantly changing, using thermographic techniques requires extreme care each time infrared data is acquired.

Most infrared monitoring systems or instruments provide special filters that can be used to avoid the negative effects of atmospheric attenuation of infrared data. However the plant user must recognize the specific factors that will affect the accuracy of the infrared data and apply the correct filters or other signal conditioning required negating that specific attenuating factor or factors.

Collecting optics, radiation detectors and some form of indicator are the basic elements of an industrial infrared instrument. The optical system collects radiant energy and focuses it upon a detector, which converts it into an electrical signal. The instrument's electronics amplifies the output signal and process it into a form which can be displayed. There are three general types of instruments that can be used for predictive maintenance: infrared thermometers or spot radiometers line scanners and imaging systems.

Infrared thermometers

Infrared thermometers or spot radiometers are designed to provide the actual surface temperature at a single, relatively small point on a machine or surface. Within a predictive maintenance program, the point-of-use infrared thermometer can be used in conjunction with many of the microprocessor-based vibration instruments to monitor the temperature at critical points on plant machinery or equipment. This technique is typically used to monitor bearing cap temperatures, motor winding temperatures, spot checks of process piping temperatures and similar applications. It is limited in that the temperature represents a single point on the machine or structure. However when used in conjunction with vibration data, point-of-use infrared data can be a valuable tool.

Line scanners

This type of infrared instrument provides a single dimensional scan or line of comparative radiation. While this type of instrument provides a somewhat larger field of view, i.e. area of machine surface, it is limited in predictive maintenance applications.

Infrared imaging

Unlike other infrared techniques, thermal or infrared imaging provides the means to scan the infrared emissions of complete machines, process or equipment in a very short time. Most of the imaging systems function much like a video camera. The user can view the thermal emission profile of a wide area by simply looking through the instrument's optics. There are a variety of thermal imaging instruments on the market ranging from relatively inexpensive, black and white scanners to full color, microprocessor-based systems. Many of the less expensive units are designed strictly a scanners and do not provide the capability of store and recall thermal images. The inability to store and recall previous thermal data will limit a long-term predictive maintenance program.

Point-of-use infrared thermometers are commercially available and relatively inexpensive. The typical cost for this type of infrared instrument is less than $1,000. Infrared imaging systems will have a price range between $8,000 for a black and white scanner without storage capability to over $60,000 for a microprocessor-based, color imaging system.

Training is critical with any of the imaging systems. The variables that can destroy the accuracy and repeatability of thermal data must be compensated for each time infrared data is acquired. In addition, interpretation of infrared data requires extensive training and experience.

Inclusion of thermography into a predictive maintenance program will enable you to monitor the thermal efficiency of critical process systems that rely on heat transfer or retention; electrical equipment; and other parameters that will improve both the reliability and efficiency of plant systems. Infrared techniques can be used to detect problems in a variety of plant systems and equipment, including electrical switchgear, gearboxes, electrical substations, transmissions, circuit breaker panels, motors, building envelopes, bearings, steam lines, and process systems that rely on heat retention or transfer.

50.3.3 Tribology

Tribology is the general term that refers to design and operating dynamics of the bearing-lubrication-rotor support structure of machinery. Several tribology techniques can be used for predictive maintenance: lubricating oils analysis, spectrographic analysis, and ferrography and wear particle analysis.

Lubricating oil analysis, as the name implies, is an analysis technique that determines the condition of lubricating oils used in mechanical and electrical equipment. It is not a tool for determining the operating condition of machinery. Some forms of lubricating oil analysis will provide an accurate quantitative breakdown of individual chemical elements, both oil additive and contaminates, contained in the oil. A comparison of the amount of trace metals in successive oil samples can indicate wear patterns of oil wetted parts in plant equipment and will provide an indication of impending machine failure.

Until recently, tribology analysis has been a relatively slow and expensive process. Analyses were conducted using traditional laboratory techniques and required extensive, skilled labor. Microprocessor-based systems are now available which can automate most of the lubricating oil and spectrographic analysis, thus reducing the manual effort and cost of analysis.

The primary applications for spectrographic or lubricating oil are: quality control, reduction of lubricating oil inventories, and determination of the most cost-effective interval for oil change. Lubricating, hydraulic and dielectric oils can be periodically analyzed, using these techniques to determine their condition. The results of this analysis can be used to determine if the oil meets the lubricating requirements of the machine or application. Based on the results of the analysis, lubricants can be changed or upgraded to meet the specific operating requirements.

In addition, detailed analysis of the chemical and physical properties of different oils used in the plant can, in some cases, allow consolidation or reduction of the number and types of lubricates required to maintain plant equipment. Elimination of unnecessary duplication can reduce required inventory levels and therefore maintenance costs.

As a predictive maintenance tool, lubricating oil and spectrographic analysis can be used to schedule oil change intervals based on the actual condition of the oil. In mid to large plants, a reduction in the number of oil changes can amount to a considerable annual reduction in maintenance costs. Relatively inexpensive sampling and testing can show when the oil in a machine has reached a point that warrants change.

The full benefit of oil analysis can only be achieved by taking frequent samples trending the data for each machine in the plant. It can provide a wealth of information on which to base maintenance decisions. However, major payback is rarely possible without a consistent program of sampling.

Lubricating oil analysis

Oil analysis has become an important aid to preventive maintenance. Laboratories recommend that samples of machine lubricant be taken at scheduled intervals to determine the condition of the lubricating film that is critical to machine-train operation. Typically eleven tests are conducted on lube oil samples:

1. *Viscosity* is one of the most important properties of lubricating oil. The actual viscosity of oil samples is compared to an unused sample to determine the thinning of thickening of the sample during use. Excessively low viscosity will reduce the oil film strength, weakening its ability to prevent metal-to-metal contact.

    Excessively high viscosity may impede the flow of oil to vital locations in the bearing support structure, reducing its ability to lubricate.
2. *Contamination* of oil by water or coolant can cause major problems in a lubricating system. Many of the additives now used in formulating lubricants contain the same elements that are used in coolant additives. Therefore, the laboratory must have an accurate analysis of new oil for comparison.
3. *Fuel dilution* of oil in an engine weakens the oil film strength, sealing ability, and detergency. Improper operation, fuel system leaks, ignition problems, improper timing, or other deficiencies may cause it. Fuel dilution is considered excessive when it reaches a level of 2.5 to 5 per cent.

4. *Solids content* is a general test. All solid materials in the oil are measured as a percentage of the sample volume or weight. The presence of solids in a lubricating system can significantly increase the wear on lubricated parts. Any unexpected rise in reported solids is cause for concern.

5. *Fuel soot* is an important indicator for oil used in diesel engines and is always present to some extent. A test to measure fuel soot in diesel engine oil is important since it indicates the fuel burning efficiency of the engine. Most tests for fuel soot are conducted by infrared analysis.

6. *Oxidation* of lubricating oil can result in lacquer deposits, metal corrosion, or thickening of the oil. Most lubricants contain oxidation inhibitors. However when additives are used up, oxidation of the oil itself begins. The quantity of oxidation in an oil sample is measured by differential infrared analysis.

7. *Nitration* results from fuel combustion in engines. The products formed are highly acidic and they may leave deposits in combustion areas. Nitration will accelerate oil oxidation. Infrared analysis is used to detect and measure nitration products.

8. *Total acid number* (TAN) is a measure of the amount of acid or acid-like material in the oil sample. Because new oils contain additives that affect the TAN number, it is important to compare used oil samples with new, unused, oil of the same type. Regular analysis at specific intervals is important to this evaluation.

9. *Total base number* (TBN) indicates the ability of an oil to neutralize acidity. The higher the TBN, the greater is its ability to neutralize acidity. Typical causes of low TBN include using the improper oil for an application, waiting too long between oil changes, overheating and using high sulfur fuel.

10. *Particle count* tests are important to anticipating potential system or machine problems. This is especially true in hydraulic systems. The particle count analysis made a part of a normal lube oil analysis is quite different from wear particle analysis. In this test, high particle counts indicate that machinery may be wearing abnormally or that failures may occur because of temporarily or permanently blocked orifices. No attempt is made to determine the wear patterns, size and other factors that would identify the failure mode within the machine.

11. *Spectrographic analysis* allows accurate, rapid measurements of many of the elements present in lubricating oil. These elements are generally classified as wear metals, contaminates, or additives. Some elements can be listed in more than one of these classifications. Standard lubricating oil analysis does not attempt to determine the specific failure modes of developing machine-train problems. Therefore, additional techniques must be used as part of a comprehensive predictive maintenance program.

12. *Wear particle analysis* is related to oil analysis only in that the particles to be studied are collected through drawing a sample of lubricating oil. Where lubricating oil analysis determines the actual condition of the oil sample, wear particle analysis provides direct information about the wearing condition of the machine-train. Particles in the lubricate of a machine can provide significant information about the condition of the machine. This information is derived from the study of particle shape, composition, size and quantity. Wear particle analysis is normally conducted in two stages.

The first method used for wear particle analysis is routine monitoring and trending of the solids content of machine lubricant. In simple terms the quantity, composition and size of particulate matter in the lubricating oil is indicative of the mechanical condition of the machine. A normal machine will contain low levels of solids with a size less than 10 microns. As the machine's condition degrades, the number and size of particulate matter will increase.

The second wear particle method involves analysis of the particulate matter in each lubricating oil sample. Five basic types of wear can be identified according to the classification of particles: rubbing wear, cutting wear, rolling fatigue wear, combined rolling and sliding wear and severe sliding wear. Only rubbing wear and early rolling fatigue mechanisms generate particles predominantly less than 15 microns in size.

(a) *Rubbing wear* is the result of normal sliding wear in a machine. During a normal break-in of a wear surface, a unique layer is formed at the surface. As long as this layer is stable, the surface wears normally. If the layer is removed faster than it is generated, the wear rate increases and the maximum particle size increases. Excessive quantities of contaminate in a lubrication system can increase rubbing wear by more than an order of magnitude without completely removing the shear mixed layer. Although catastrophic failure is unlikely, these machines can wear out rapidly. Impending trouble is indicated by a dramatic increase in wear particles.

(b) *Cutting wear particles* are generated when one surface penetrates another. These particles are produced when a misaligned or fractured hard surface produces an edge that cuts into a softer surface, or when abrasive contaminate become embedded in a soft surface and cut an opposing surface. Cutting wear particles are abnormal and are always worthy of attention. If they are only a few micron long and a fraction of a micron wide, the cause is probably a contaminate. Increasing quantities of longer particles signal a potentially imminent component failure.

(c) *Rolling fatigue* is associated primarily with rolling contact bearings and may produce three distinct particle types: fatigue spall particles, spherical particles, and laminar particles. Fatigue spall particles are the actual material removed when a pit or spall opens up on a bearing surface. An increase in the quantity or size of these particles is the first indication of an abnormality. Rolling fatigue does not always generate spherical particles and they may be generated by other sources. Their presence is important in that they are detectable before any actual spalling occurs. Laminar particles are very thin and are formed by the passage of a wear particle through a rolling contact. They frequently have holes in them. Laminar particles may be generated throughout the life of a bearing, but at the onset of fatigue spalling the quantity increases.

(d) *Combined rolling and sliding wear* results from the moving contact of surfaces in gear systems. These larger particles result from tensile stresses on the gear surface, causing the fatigue cracks to spread deeper into the gear tooth before pitting. Gear fatigue cracks do not generate spheres. Scuffing of gears is caused by too high a load or speed. The excessive heat generated by this condition breaks down the lubricating film and causes adhesion of the mating gear teeth. As the wear surfaces become rougher, the wear rate increases. Once started, scuffing usually affects each gear tooth.

(e) *Severe sliding wear* is caused by excessive loads or heat in a gear system. Under these conditions, large particles break away from the wear surfaces, causing an increase in the wear rate. If the stresses applied to the surface are increased further, a second transition point is reached. The surface breaks down and catastrophic wear ensures.

Normal spectrographic analysis is limited to particulate contamination with a size of 10 microns or less. Larger contaminants are ignored. This fact can limit the benefits that can be derived from the technique.

Ferrography

This technique is similar to spectrography but there are two major exceptions. First, ferrography separates particulate contamination by using a magnetic field rather than burning a sample as in spectrographic analysis. Because a magnetic field is used to separate contaminants, this technique is primarily limited to ferrous or magnetic particles.

The second difference is that particulate contamination larger than 10 microns can be separated and analyzed. Normal ferrographic analysis will capture particles up to 100 microns and provides a better representation of the total oil contamination than spectrographic techniques.

There are three major limitations with using tribology analysis in a predictive maintenance program: equipment costs, acquiring accurate oil samples and interpretation of data.

The capital cost of spectrographic analysis instrumentation is normally too high to justify in-plant testing. Typical cost for a microprocessor-based spectrographic system is between $30,000 and $60,000. Because of this, most predictive maintenance programs rely on third party analysis of oil samples.

Simple lubricating oil analysis by a testing laboratory will range from about $20 to $50 per sample. Standard

analysis will normally include: viscosity, flash point, total insolubles, total acid number (TAN), total base number (TBN), fuel content, and water content.

More detailed analysis, using spectrographic or ferrographic techniques, that include metal scans, particle distribution (size), and other data can range to well over $150 per sample.

A more severe limiting factor with any method of oil analysis is acquiring accurate samples of the true lubricating oil inventory in a machine. Sampling is not a matter of opening a port somewhere in the oil line and catching a pint sample. Extreme care must be taken to acquire samples that truly represent the lubricant that will pass through the machine's bearings. One recent example is an attempt to acquire oil samples from a bullgear compressor. The lubricating oil filter had a sample port on the clean, i.e. downstream, side. However, comparison of samples taken at this point and one taken directly from the compressor's oil reservoir indicated that more contaminates existed downstream from the filter than in the reservoir. Which location actually represented the oil's condition? Neither sample was truly representative of the oil condition. The oil filter had removed most of the suspended solids, i.e. metals and other insolubles, and was therefore not representative of the actual condition. The reservoir sample was not representative since most of the suspended solids had settled out in the sump.

Proper methods and frequency of sampling lubricating oil are critical to all predictive maintenance techniques that use lubricant samples. Sample points that are consistent with the objective of detecting large particles should be chosen. In a re-circulating system, samples should be drawn as the lubricant returns to the reservoir and before any filtration. Do not draw oil from the bottom of a sump where large quantities of material build up over time. Return lines are preferable to reservoir as the sample source, but good reservoir samples can be obtained if careful, consistent practices are used. Even equipment with high levels of filtration can be effectively monitored as long as samples are drawn before oil enters the filters. Sampling techniques involve taking samples under uniform operating conditions. Samples should not be taken more than 30 minutes after the equipment has been shut down.

Sample frequency is a function of the mean time to failure from the onset of an abnormal wear mode to catastrophic failure. For machines in critical service, sampling every 25 hours of operation is appropriate. However, for most industrial equipment in continuous service, monthly sampling is adequate. The exceptions to monthly sampling are machines with extreme loads. In this instance, weekly sampling is recommended.

Understanding the meaning of analysis results is perhaps the most serious limiting factor. Most often results are expressed in terms that are totally alien to plant engineers or technicians. Therefore, it is difficult for them to understand the true meaning, in terms of oil or machine condition. A good background in quantitative and qualitative chemistry is beneficial. As a minimum, plant staff will require training in basic chemistry and specific instruction on interpreting tribology results.

50.3.4 Process parameters

Many plants do not consider machine or systems efficiency as part of the maintenance responsibility. However, machinery that is not operating within acceptable efficiency parameters severely limits the productivity of many plants. Therefore a comprehensive predictive maintenance program should include routine monitoring of process parameters.

As an example of the importance of process parameters monitoring, consider a process pump that may be critical to plant operation. Vibration-based predictive maintenance will provide the mechanical condition of the pump and infrared imaging will provide the condition of the electric motor and bearings. Neither provides any indication of the operating efficiency of the pump. Therefore, the pump can be operating at less than 50 per cent efficiency and the predictive maintenance program would not detect the problem.

Process inefficiencies, like the example, are often the most serious limiting factor in a plant. Their negative impact on plant productivity and profitability is often greater than the total cost of the maintenance operation. However, without regular monitoring of process parameters, many plants do not recognize this unfortunate fact.

If your program included monitoring of the suction and discharge pressures and amp load of the pump, then you could determine the operating efficiency. The brake-horsepower formula

Brake horsepower

$$= \frac{\text{Flow (GPM)} \times \text{Specific gravity} \times \text{Total dynamic head (TDH)}}{3960 \times \text{Efficiency}}$$

could be used to calculate operating efficiency of any pump in the program. By measuring the suction and discharge pressure, the total dynamic head (TDH) can be determined. Using this data, the pump curve will provide the flow and the amp load the horsepower. With this measured data, the efficiency can be calculated.

Process parameters monitoring should include all machinery and systems in the plant process that can affect its production capacity. Typical systems include heat exchangers, pumps, filtration, boilers, fans, blowers, and other critical systems.

Inclusion of process parameters in a predictive maintenance can be accomplished in two ways: manual or microprocessor-based systems. However, both methods will normally require installing instrumentation to measure the parameters that indicate the actual operating condition of plant systems. Even though most plants have installed pressure gages, thermometers and other instruments that should provide the information required for this type of program, many of them are no longer functioning. Therefore including process parameters in your program will require an initial capital cost to install calibrated instrumentation.

Data from the installed instrumentation can be periodically recorded using either manual logging or with a microprocessor-based data logger. If the latter is selected, many of the vibration-based, microprocessor systems can also provide the means of acquiring process data. This should be considered when selecting the vibration monitoring system that will be used in your program. In addition, some of the microprocessor-based predictive maintenance systems provide the ability to calculate unknown process variables. For example, they can calculate the pump efficiency used in the example. This ability to calculate unknowns based on measured variables will enhance a total plant predictive maintenance program without increasing the manual effort required. In addition, some of these systems include non-intrusive transducers that can measure temperatures, flows and other process data without the necessity of installing permanent instrumentation. This further reduces the initial cost of including process parameters in your program.

50.3.5 Visual inspection

Regular visual inspection of the machinery and systems in a plant is a necessary part of any predictive maintenance program. In many cases, visual inspection will detect potential problems that will be missed using the other predictive maintenance techniques.

Even with the predictive techniques discussed, many potentially serious problems can remain undetected. Routine visual inspection of all critical plant systems will augment the other techniques and insure that potential problems are detected before serious damage can occur.

Most of the vibration-based predictive maintenance systems include the capability of recording visual observations as part of the routine data acquisition process. Since the incremental costs of these visual observations are small, this technique should be incorporated in all predictive maintenance programs.

All equipment and systems in the plant should be visually inspected on a regular basis. The additional information provided by visual inspection will augment the predictive maintenance program regardless of the primary techniques used.

50.3.6 Ultrasonic monitoring

This predictive maintenance technique uses principles similar to vibration analysis. Both monitor the noise generated by plant machinery or systems to determine their actual operating condition.

Unlike vibration monitoring, ultrasonics monitors the higher frequencies, i.e. ultrasound, produced by unique dynamics in process systems or machines. The normal monitoring range for vibration analysis is from less than 1 Hertz to 20,000 Hertz. Ultrasonics techniques monitor the frequency range between 20,000 and 100 kHz.

The principal application for ultrasonic monitoring is in leak detection. The turbulent flow of liquids and gases through a restricted orifice, i.e. leak, will produce a high frequency signature that can easily be identified using ultrasonic techniques. Therefore, this technique is ideal for detecting leaks in valves, steam traps, piping and other process systems.

Two types of ultrasonic systems are available that can be used for predictive maintenance: structural and airborne. Both provide fast, accurate diagnosis of abnormal operation and leaks. Airborne ultrasonic detectors can be used in either a scanning or contact mode. As scanners, they are most often used to detect gas pressure leaks. Because these instruments are sensitive only to ultrasound, they are not limited to specific gases as are most other gas leak detectors. In addition, they are often used to locate various forms of vacuum leaks.

In the contact mode, a metal rod acts as a waveguide. When it touches a surface, it is stimulated by the high frequencies, ultrasound, on the opposite side of the surface. This technique is used to locate turbulent flow and or flow restriction in process piping.

Some of the ultrasonic systems include ultrasonic transmitters that can be placed inside plant piping or vessels. In this mode, ultrasonic monitors can be used to detect areas of sonic penetration along the container's surface. This ultrasonic transmission method is useful in quick checks of tank seams, hatches, seals, caulking, gaskets or building wall joints.

Most of the ultrasonic monitoring systems are strictly a scanner that does not provide any long-term trending or storage of data. They are in effect a point-of-use instrument that provides an indication of the overall amplitude of noise within the bandwidth of the instrument. Therefore, the cost of this type of instrument is relatively low. Normal cost of ultrasonic instruments will vary from less than $1,000 to about $8,000. Used strictly for leak detection, little training is required to utilize ultrasonic techniques. The combination of low capital cost, minimum training required to use the technique and potential impact that leaks may have on plant availability provide a positive cost-benefit for including ultrasonic techniques in a total plant predictive maintenance program.

However, care should be exercised in applying this technique in your program. Many ultrasonic systems are sold as a bearing condition monitor. Even though the natural frequencies of rolling element bearings will fall within the bandwidth of ultrasonic instruments, this is not a valid technique for determining the condition of rolling element bearings. In a typical machine, many other machine dynamics will also generate frequencies within the bandwidth covered by an ultrasonic instrument. Gear meshing frequencies, blade pass and other machine components will also create energy or noise that cannot be separate from the bearing frequencies monitored by this type of instrument. The only reliable method of determining the condition of specific machine components, including bearings, is vibration analysis. The use of ultrasonics to monitor bearing condition is not recommended.

50.3.7 Other techniques

Numerous other nondestructive techniques can be used to identify incipient problems in plant equipment or systems. However, these techniques either do not provide a broad enough application or are too expensive to support a predictive maintenance program. Therefore, these techniques

are used as the means of confirming failure modes identified by the predictive maintenance techniques identified in this chapter.

Other techniques that can support predictive maintenance include acoustic emissions, eddy-current, magnetic particle, residual stress and most of the traditional nondestructive methods.

If you need specific information on the techniques that are available, the American Society of Nondestructive Testing (ANST) has published a complete set of handbooks that provide a comprehensive database for most nondestructive testing techniques.

50.4 Selecting a predictive maintenance system

After developing the requirements for a comprehensive predictive maintenance program, the next step is to select the hardware and software system that will most cost-effectively support your program. Since most plants will require a combination of techniques, i.e. vibration, thermography, tribology, etc., the system should be able to provide support for all of the required techniques. Since a single system that will support all of the predictive maintenance is not available, you must decide on the specific techniques that must be used to support your program. Some of the techniques may have to be eliminated to enable the use of a single predictive maintenance system. However, in most cases, two independent systems will be required to support the monitoring requirements in your plant.

Most plants can be cost-effectively monitored using a microprocessor-based system designed to use vibration, process parameters, visual inspection and limited infrared temperature monitoring.

Plants with large populations of heat transfer systems and electrical equipment will need to add a full thermal imaging system in order to meet the total plant requirements for a full predictive maintenance program. Plant with fewer systems that require full infrared imaging may elect to contract this portion of the predictive maintenance program. This will eliminate the need for an additional system. A typical microprocessor-based system will consist of four main components: a meter or data logger, host computer, transducers, and a software program. Each component is important, but the total capability must be evaluated to get a system that will support a successful program. The first step in selecting the predictive maintenance system that will be used in your plant is to develop a list of the specific features or capabilities the system must have to support your program. As a minimum, the total system must have the following capabilities:

User-friendly software and hardware   The premise of predictive maintenance is that existing plant staff must be able to understand the operation of both the data logger and software program. Since plant staff normally have little, if any, computer or microprocessor background, the system must use simple, straightforward operation of both the data acquisition instrument and software. Complex systems, even if they provide advanced diagnostic capabilities, may not be accepted by plant staff and therefore will not provide the basis for a long-term predictive maintenance program.

Automated data acquisition　The object of using microprocessor-based systems is to remove any potential for human error, reduce manpower and to automate as much as possible the acquisition of vibration, process and other data that will provide a viable predictive maintenance database. Therefore the system must be able to automatically select and set monitoring parameters without user input. The ideal system would limit user input to a single operation. However this is not totally possible with today's technology.

Automated data management and trending　The amount of data required to support a total plant predictive maintenance program is massive and will continue to increase over the life of the program. The system must be able to store, trend and recall the data in multiple formats that will enable the user to monitor, trend and analyse the condition of all plant equipment included in the program. The system should be able to provide long-term trend data for the life of the program. Some of the microprocessor-based systems limit trends to a maximum of 26 data sets and will severely limit the decision-making capabilities of the predictive maintenance staff.

Limiting trend data to a finite number of data sets eliminates the ability to determine the most cost-effective point to replace a machine rather than let it continue in operation.

Flexibility　Not all machines or plant equipment are the same, nor are the best methods of monitoring their condition. Therefore, the selected system must be able to support as many of the different techniques as possible. As a minimum, the system should be capable of obtaining, storing, and presenting data acquired from all vibration and process transducers and provide accurate interpretation of the measured values in user-friendly terms. The minimum requirement for vibration monitoring systems must include the ability to acquire: filter broadband, select narrowband, time traces and high-resolution signature data using any commercially available transducer. Systems that are limited to broadband monitoring or to a single type of transducer cannot support the minimum requirements of a predictive maintenance program.

The added capability of calculating unknown values based on measured inputs will greatly enhance the system capabilities. For example, the neither fouling factor nor efficiency of a heat exchanger can be directly measured. A predictive maintenance system that can automatically calculate these values based on the measured flow, pressure and temperature data would enable the program to automatically trend, log and alarm deviations in these unknown, critical parameters.

Reliability　The selected hardware and software must be proven in actual field use to ensure it's reliability. The introduction of microprocessor-based predictive maintenance systems is still relatively new and it is important that you evaluate the field history of a system before purchase.

Ask for a users list and talk to the people who are already using the systems. This is a sure way to evaluate the strengths and weakness of a particular system before you make a capital investment.

Accuracy　Decisions on machine-train or plant system condition will be made based on the data acquired and reported by the predictive maintenance system. It must be accurate and repeatable. Errors can be input by the microprocessor and software as well as the operators. The accuracy of commercially available predictive maintenance system varies. While most will provide at least minimum acceptable accuracy, some are well below the acceptable level.

It will be extremely difficult for the typical plant user to determine the level of accuracy of the various instruments that are available for predictive maintenance. Vendor literature and salesmen will assure the potential user that their system is the best, most accurate, etc. The best way to separate fact from fiction is a comparison of the various systems in your plant. Most vendors will provide a system on consignment for periods up to thirty days. This will provide sufficient time for your staff to evaluate each of the potential systems before purchase.

Training and technical support　Training and technical support is critical to the success of your predictive maintenance program. Regardless of the techniques or systems selected, your staff will have to be trained. This training will take two forms: system users training and application knowledge for the specific techniques included in your program. Few, if any, of the existing staff will have the knowledge base required implementing the various predictive maintenance techniques discussed in preceding chapters. None will understand the operation of the systems that are purchased to support your program.

Many of the predictive systems manufacturers are strictly hardware and software oriented. Therefore, they offer minimal training and no application training or technical support. Few plants can achieve minimum benefits from predictive maintenance without training and some degree of technical support. It is therefore imperative that the selected system or system vendors provide a comprehensive support package that includes both training and technical support.

System cost　Cost should not be the primary deciding factor in system selection. The capabilities of the various systems vary greatly and so does the cost. Care should be taken to ensure a fair comparison of the total system capability and price is made before selection of your system.

For example, vibration-based systems are relatively competitive in price. The general spread is less than $1,000 for a complete system. However, the capabilities of these systems are not comparable. A system that provides minimum capability for vibration monitoring will be about the same price as one that provides full vibration monitoring capability and provides process parameter, visual inspection and point-of-use thermography.

Operating cost   The real cost of implementing and maintaining a predictive maintenance program is not the initial system cost. Rather it is the annual labor and overhead costs associated with acquiring, storing, trending and analyzing the data required to determine the operating condition of plant equipment. This is also the area where a predictive maintenance system has the greatest variance in capability. Systems that fully automate data acquisition, storing, etc. will provide the lowest operating costs. Manual systems and many of the low-end microprocessor-based systems require substantially more manpower to accomplish the minimum objectives required by predictive maintenance. The users list will again help you determine the long-term cost of the various systems. Most users will share their experience, including a general indication of labor cost.

The microprocessor

The data logger or microprocessor selected by your predictive maintenance program is critical to the success of the program. There is a wide variety of systems on the market that range from handheld overall value meters to advanced analyzers that can provide an almost unlimited amount of data. The key selection parameters for a data acquisition instrument should include the expertise required to operate, accuracy of data, type of data, and manpower required to meet the program demands.

Expertise required to operate   One of the objectives for using microprocessor-based predictive maintenance systems is to reduce the expertise required to acquire error-free, useful vibration and process data from a large population of machinery and systems within a plant. The system should not require user input to establish: maximum amplitude, measurement bandwidths, filter settings, or allow free-form data input. All of these functions force the user to be a trained analyst and will increase both the cost and time required to routinely acquire data from plant equipment. Many of the microprocessors on the market provide easy, menu-driven measurement routes that lead the user through the process of acquiring accurate data. The ideal system should require a single key input to automatically acquire, analyze, alarm and store all pertinent data from plant equipment. This type of system would enable an unskilled user to quickly and accurately acquire all of the data required for predictive maintenance.

Accuracy of data   The microprocessor should be capable of automatically acquiring accurate, repeatable data from equipment included in the program. The elimination of user input on filter settings, bandwidths and other measurement parameters would greatly improve the accuracy of acquired data. The specific requirements that determine data accuracy will vary depending on the type of data. For example, a vibration instrument should be able to: average

data, reject spurious signals, auto-scale based on measured energy, and prevent aliasing.

The basis of frequency-domain vibration analysis assumes that we monitor the rotational frequency components of a machine-train. If a single block of data is acquired, non-repetitive or spurious data can be introduced into the database. The microprocessor should be able to acquire multiple blocks of data, average the total and store the averaged value. This approach will enable the data acquisition unit to automatically reject any spurious data and provide reliable data for trending and analysis. Systems that rely on a single block of data will severely limit the accuracy and repeatability of acquired data. They will also limit the benefits that can be derived from the program. The microprocessor should also have electronic circuitry that automatically checks each data set and block of data for accuracy and reject any spurious data that may occur. Auto-rejection circuitry is available in several of the commercially available systems. Coupled with multiple block averaging, this auto-rejection circuitry assures maximum accuracy and repeatability of acquired data. A few of the microprocessor-based systems require the user to input the maximum scale that is used to acquire data. This will severely limit the accuracy of data. Setting the scale too high will prevent acquisition of factual machine data. A setting that is too low will not capture any high-energy frequency components that may be generated by the machine-train. Therefore, the microprocessor should have auto-scaling capability to ensure accurate data. Vibration data can be distorted by high frequency components that fold-over into the lower frequencies of a machine's signature. Even though these aliased frequency components appear real, they do not exist in the machine. Low frequency components can also distort the mid-range signature of a machine in the same manner as high frequency. The microprocessor selected for vibration should include a full range of anti-aliasing filter to prevent the distortion of machine signatures. The features illustrated in the example also apply to non-vibration measurements. For example, pressure readings require the averaging capability to prevent spurious readings. Slight fluctuations in line or vessel pressure are normal in most plant systems. Without the averaging capability, the microprocessor cannot acquire an accurate reading of the true system pressure.

Alert and alarm limits   The microprocessor should include the ability to automatically alert the user to changes in machine, equipment or system condition. Most of the predictive maintenance techniques rely on a change in the operating condition of plant equipment to identify an incipient problem. Therefore, the system should be able to analyze data and report any change in the monitoring parameters that were established as part of the database development.

Predictive maintenance systems use two methods of detecting a change in the operating condition of plant equipment: static and dynamic. Static alert and alarm limits are pre-selected thresholds that are downloaded into the microprocessor. If the measurement parameters exceed the pre-set limit, an alarm is displayed. This type of monitoring does not consider the rate of change or historical trends of a machine and therefore cannot anticipate when the alarm will be reached.

The second method uses dynamic limits that monitor the rate of change in the measurement parameters. This type of monitoring can detect minor deviations in the rate that a machine or system is degrading and anticipate when an alarm will be reached. The use of dynamic limits will greatly enhance the automatic diagnostic capabilities of a predictive maintenance system and reduce the manual effort required to gain maximum benefits.

Data storage   The microprocessor must be able to acquire and store large amounts of data. The memory capacity of the various predictive maintenance systems varies. As a minimum, the system must be able to store a full eight hours of data before transferring it to the host computer. The actual memory requirements will depend on the type of data acquired. For example, a system used to acquire vibration data would need enough memory to store about 1000 overall reading or 400 full signatures. Process monitoring would require a minimum of 1000 readings to meet the minimum requirements.

Data transfer   The data acquisition unit will not be used for long-term data storage. Therefore, it must be able to reliability transfer data into the host computer. The actual time required to transfer the microprocessor's data into the host computer is the only non-productive time of the data acquisition unit. It cannot be used for acquiring additional data during the data transfer operation. Therefore, the transfer time should be kept to a minimum. Most of the available systems use an RS 232 communications protocol that will allow data transfer at rates of up to 19,200 baud. The time required to dump the full memory of a typical microprocessor can be 30 minutes or more.

Some of the systems have incorporated an independent method of transferring data that eliminates the dead time altogether. These systems transfer stored data from the data logger into a battery-backed memory bypassing the RS 232 link. Using this technique, data can be transferred at more than 350,000 baud and will reduce the non-productive time to a few minutes.

The microprocessor should also be able to support modem communication with remote computers. This feature will enable multiple plant operation and direct access to third party diagnostic and analysis support. Data can be transferred anywhere in the world using this technique. Not all predictive maintenance systems use a true RS 232 communications protocol or support modem communications. These systems can severely limit the capabilities of your program. The various predictive maintenance techniques will add other specifications for an acceptable data acquisition unit.

The host computer

The host computer provides all of the data management, storage, report generation, and analysis capabilities of the predictive maintenance program. Therefore, care should be exercised during the selection process. This is especially true if multiple technologies will be used within the predictive maintenance program. Each predictive maintenance system will have a unique host computer specification that will include: hardware configuration, computer operating system, hard disk memory requirements and many others. This can become a serious if not catastrophic problem. You may find that one system requires a special printer that is not compatible with other programs to provide hard copies of reports or graphic data. One program may be compatible with PC-DOS, while another requires a totally different operating program.

Therefore, you should develop a complete computer specification sheet for each of the predictive maintenance systems that will be used. A comparison of the list will provide a compatible computer configuration that will support each of the techniques. If this is not possible, you may have to reconsider your choice of techniques.

Computers, like plant equipment, fail. Therefore, the use of a commercially available computer is recommended. The critical considerations include: availability of repair parts and local vendor support.

Most of the individual predictive maintenance techniques will not require a dedicated computer. Therefore there is usually sufficient storage and computing capacity to handle several, if not all, of the required techniques and still leave room for other support programs, i.e. word processing, database management, etc. Use of commercially available PCs provides the user with the option of including these auxiliary programs in the host computer. The actual configuration of the host computer will be dependent on the specific requirement of the predictive maintenance techniques that will be used.

The software

The software program provided with each predictive maintenance system is the heart of a successful program. It is also the hardest to evaluate before purchase. The methodology used by vendors of predictive maintenance systems varies greatly. Many appear to have all of the capabilities required to meet the demands of a total plant predictive maintenance program. However on close inspection, usually after purchase, they are found to be lacking.

Software is also the biggest potential limiting factor of a program. Even though all vendors use some form of formal computer language, i.e. FORTRAN, Cobol, basic, etc., they are normally not interchangeable with other programs. The apparently simple task of having one computer program communicate with another can often be impossible. This lack of compatibility between various computer programs prohibits transferring a predictive maintenance database from one vendor's system into a system manufactured by another vendor. The result is that once a predictive maintenance program is started, a plant cannot change to another system without losing the data already developed in the initial program.

As a minimum the software program should provide: automatic database management, automatic trending, automatic report generation and simplified diagnostics. As in the case of the microprocessor used to acquire data, the software must be user-friendly.

User-friendly operation   The software program should be menu-driven with clear on-line user instructions. The program should protect the user form distorting or deleting stored data. Some of the predictive maintenance systems are written in DBASE software shells. Even though these programs provide a knowledgeable user with the ability to modify or customize the structure of the program, i.e. report formats, etc., they also provide the means to distort or destroy stored data. A single key entry can destroy years of stored data. Protection should be built into the program to limit the user's ability to modify or delete data and to prevent accidental data base damage.

The program should have a clear, plain language user's manual that provides the logic and specific instructions required to set up and use the program.

Automatic trending   The software program should be capable of automatically storing all acquired data and updating the trends of all variables. This capability should include multiple parameters not just a broadband or single variable. This will enable the user to display trends of all variables that affect plant operations.

Automatic report generation   Report generation will be an important part of the predictive maintenance program. Maximum flexibility in format and detail is important to program success. The system should be able to automatically generate reports a multiple levels of detail. As a minimum, the system should be able to report:

- A listing of machine-trains or other plant equipment that have exceeded or are projected to exceed one or more alarm limits. The report should also provide a projection to probable failure based on the historical data and last measurement.
- A listing of missed measurement points, machines overdue for monitoring and other program management information. These reports act as reminders to ensure that the program is maintained properly.
- A listing of visual observations. Most of the microprocessor-based systems support visual observations as part of their approach to predictive maintenance. This report provides hard copies of the visual observations as well as maintaining the information in the computer's database.
- Equipment history reports should also be available. These reports provide long-term data on the condition of plant equipment and are valuable for analysis.

Simplified diagnostics   Identification of specific failure modes of plant equipment requires manual analysis of data stored in the computer's memory. The software program should be able to display, modify and compare stored data in a manner that simplifies the analysis of the actual operating condition of the equipment.

As a minimum, the program should be able to directly compare data from similar machines, normalize data into compatible units and display changes in machine parameters, i.e. vibration, process, etc.

Transducers

The final portion of a predictive maintenance system is the transducer that will be used to acquire data from plant equipment. Since we have assumed that a microprocessor-based system will be used, we will limit this discussion to those sensors that can be used with this type of system.

Acquiring accurate vibration and process data will require several types of transducers. Therefore, the system must be able to accept input from as many different types of transducers as possible. Any limitation of compatible transducers can become a serious limiting factor. This should eliminate systems that will accept inputs from a single type of transducer. Other systems are limited to a relatively small range of transducers that will also prohibit maximum utilization of the system. Selection of the specific transducers required to monitor the mechanical condition, i.e. vibration, and process parameters, i.e. flow, pressure, etc., will also deserve special consideration and will be discussed latter.

50.5 Establishing a predictive maintenance program

The decision to establish a predictive maintenance management program is the first step toward controlling maintenance costs and improving process efficiency in your plant. Now what do you do? Numerous predictive maintenance programs can serve as models for implementing a successful predictive maintenance program. Unfortunately, more were aborted within the first three years because a clear set of goals and objectives were not established before the program was implemented. Implementing a total plant predictive maintenance program is expensive. In addition to the initial capital cost, there is a substantial recurring labor cost required to maintain the program.

To be successful, a predictive maintenance program must be able to quantify the cost-benefit generated by the program. This can be achieved if the program is properly established, uses the proper predictive maintenance techniques and has measurable benefits. The amount of effort expended to initially establish the program plant is directly proportional to its success or failure. Proper implementation of a predictive maintenance program must include the items listed in Sections 50.5.1–50.5.5.

50.5.1 Goals, objectives and benefits

Constructive actions issue from well-established purpose. It is important that the goals and objectives of a predictive maintenance program be fully developed and adopted by the personnel who perform the program and upper management of the plant. A predictive maintenance program is not an excuse to buy sophisticated, expensive equipment. Neither is the purpose of the program to keep a number of people busy measuring and reviewing data from the various machines, equipment and systems within the plant.

The purpose of predictive maintenance is to minimize unscheduled equipment failures, maintenance costs and lost production. It is also intended to improve the production efficiency and product quality in the plant. This is accomplished by regular monitoring of the mechanical condition, machine and process efficiencies and other parameters that define the operating condition of the plant. Using the data acquired from critical plant equipment, incipient problems are identified and corrective actions taken to improve the reliability, availability and productivity of the plant.

Specific goals and objectives will vary from plant to plant. However, we will provide an example that will illustrate the process. Before goals and objectives can be developed for your plant, you must determine the existing maintenance costs and other parameters that will establish a reference or baseline dataset. Since most plants do not track the true cost of maintenance, this may be the most difficult part of establishing a predictive maintenance program.

As a minimum, your baseline data set should include the labor, overhead, overtime premiums and other payroll costs of the maintenance department. It should also include all maintenance related contract services, excluding janitorial, and the total costs of spare parts inventories. The baseline should also include the percentage of unscheduled maintenance repairs versus scheduled; actual repair costs on critical plant equipment; and the annual availability of the plant.

This baseline should include the incremental cost for production created by catastrophic machine failures and other parameters. If they are available or can be obtained, they will help greatly in establishing a valid baseline.

The long-term objectives of a predictive maintenance program are to:

- Eliminate un-necessary maintenance
- Reduce lost production caused by failures
- Reduce repair parts inventory
- Increase process efficiency
- Improve product quality
- Extend operating life of plant systems
- Increase production capacity
- Reduce overall maintenance costs
- Increase overall profits.

However just stating these objectives will not make it happen nor will it provide the means of measuring the success of the program. Establish specific objectives, i.e. reduce un-scheduled maintenance by 20 per cent, or increase production capacity by 15 per cent. In addition to quantifying the expected goals, define the methods that will be used to accomplish each objective and the means that can be used to measure the actual results.

Management support

Implementing a predictive maintenance program will require an investment in both capital equipment and manpower. If a program is to get started and survive to accomplish its intended goals, management must be willing to commit the necessary resources. They must also insist on the adoption of vital record-keeping and information exchange procedures that are critical to program success and are outside the control of the maintenance department. In most aborted programs, management committed to the initial investment for capital equipment but did not invest the resources required for training, consulting support and in-house manpower that are essential to success. A number of programs have been aborted during the time between 18 and 24 months following implementation. They were not aborted because the program failed to achieve the desired results. They failed because upper management did not clearly understand how the program works. During the first 12 months, most predictive maintenance programs identify numerous problems in plant machinery and systems. Therefore, the reports and recommendations for corrective actions generated by the predictive maintenance group are highly visible. After the initial 12 to 18 months, most of the serious plant problems have been resolved and the reports begin to show little need for corrective actions. Without a clear understanding of this normal cycle and the means of quantifying the achievements of the predictive maintenance program, upper management often concludes that the program is not providing sufficient benefits to justify the continued investment in manpower.

Dedicated and accountable personnel

All successful programs are built around a full time predictive maintenance team. Some of these teams may cover multiple plants and some monitor only one. However, every successful program has this dedicated team that can concentrate their full attention to achieving the objectives established for the program. Even though a few successful programs have been structured around part-time personnel, this approach cannot be recommended. All too often, the part-time personnel will not or cannot maintain the monitoring and analysis frequency that is critical to a successful program.

The accountability expected of the predictive maintenance group is another factor that is critical to program effectiveness. If measures of program effectiveness are not established, neither management nor program personnel can determine if the program's potential is being achieved.

Efficient data collection and analysis procedures

Efficient procedures can be established if adequate instrumentation is available and the monitoring tasks are structured to emphasize program goals. A well-planned program should not be structured so that all machines and equipment in the plant receive the same scrutiny. Typical predictive maintenance programs will monitor from 50 to 500 machine-trains in a given plant. Obviously some of the machine-trains are more critical to the continued, efficient operation of the plant than others. The predictive maintenance program should be set up with this in mind and concentrate the program's efforts in the areas that will provide maximum results. The use of microprocessor and personal computer-based predictive maintenance systems will greatly improve the data collection and data management functions required for a successful program. They can also provide efficient data analysis. However procedures that define the methods, schedule and other parameter of data acquisition, analysis and report generation must be included in the program definition.

Viable database

The methods and systems that you choose for your program and the initial program development will largely determine the success or failure of predictive maintenance in your plant.

Proper implementation of a predictive maintenance program is not easy. It will require a great deal of thought and perhaps for the first time a complete understanding of the operation of the various systems and machinery in your plant.

The initial database development required to successfully implement a predictive maintenance program will require man-months of effort. The extensive labor required to properly establish a predictive database often results in either a poor or incomplete database. In some cases, the program is discontinued because of staff limitations. If the extensive labor required establishing a database is not available in-house, there are consultants available that will provide the knowledge and labor required to accomplish this task.

The ideal situation would be to have the predictive systems vendor establish a viable database as part of the initial capital equipment purchase. This service is offered by a few of the systems vendors. Unfortunately, many predictive maintenance programs have failed because these important first critical steps were omitted or ignored. There are a variety of technologies and predictive maintenance systems that can be beneficial. How do you decide which method and system to use?

A vibration-based predictive maintenance program is the most difficult to properly establish and will require much more effort than any of the other techniques. It will also provide the most return on investment. Too many of the vibration-based programs fail to use the full capability of the predictive maintenance tool. They ignore the automatic diagnostic power that is built into most of the microprocessor-based systems and rely instead on manual interpretation of all acquired data.

The first step is to determine the types of plant equipment and systems that are to be included in your program. A plant survey of your process equipment should be developed that lists every critical component within the plant and its impact on both production capacity and maintenance costs. A plant process layout is invaluable during this phase of program development. It is very easy to omit critical machines or components during the audit. Therefore, care should be taken to ensure that all components that can limit production capacity are included in your list.

The listing of plant equipment should be ordered into the following classes depending on their impact on production capacity or maintenance cost: essential, critical, serious, others.

Class I or essential machinery or equipment must be on-line for continued plant operation. Loss of any one of these components will result in a plant outage and total loss of production. Plant equipment that has excessive repair costs or repair parts lead-time should also be included in the essential classification.

Class II or critical machinery would severely limit production capacity. As a rule-of-thumb, loss of critical machinery would reduce production capacity by 30 per cent or more. Also included in the critical classification are machines or systems with chronic maintenance histories or that have high repair or replacement costs.

Class III or serious machinery include major plant equipment that do not have a dramatic impact on production but that contribute to maintenance costs. An example of the serious classification would be a redundant system. Since the inline spare could maintain production, loss of one component would not affect production. However, the failure would have a direct impact on maintenance cost.

Class IV machinery would include other plant equipment that has a proven history of impacting either production or maintenance costs. All equipment in this classification must be evaluated to determine whether routine monitoring is cost-effective. In some cases, replacement costs are lower than the annual costs required to monitor machinery in this classification.

The completed list should include every machine, system or other plant equipment that has or could have a serious impact on the availability and process efficiency of your plant.

The next step is to determine the best method or technique for cost-effectively monitoring the operating condition of each item on the list. To select the best methods for regular monitoring, you should consider the dynamics of operation and normal failure modes of each machine or system to be included in the program.

A clear understanding of the operating characteristics and failure modes will provide the answer to which predictive maintenance method should be used.

Most predictive maintenance programs will use vibration monitoring as the principal technique. The inclusion of visual inspection, process parameters, ultrasonics and limited thermographic techniques should also be added to the in-house program. The initial cost of systems and advanced training required by full thermographic and tribology techniques prohibit their inclusion into in-house programs. Plants that require these techniques normally rely on outside contractors to provide the instrumentation and expertise required to provide these monitoring and diagnostic techniques.

50.5.2 Database development

The next step required to establish a predictive maintenance program is the creation of a comprehensive database.

Establishing data acquisition frequency

During the implementation stage of a predictive maintenance program, all classes of machinery should be monitored to establish a valid baseline data set. Full vibration signatures should be acquired to verify the accuracy of the database setup and determine the initial operating condition of the machinery.

Since a comprehensive program will include trending and projected time-to-failure, multiple readings are required on all machinery to provide sufficient data for the microprocessor to develop trend statistics. Normally during this phase, measurements are acquired every two weeks.

After the initial or baseline of the machinery, the frequency of data collection will vary depending on the classification of the machine-trains. Class I machines should be monitored on a two to three week cycle; Class II on a three to four week cycle; Class III on a four to six week cycle and Class IV on a six to ten week cycle. This frequency can, and should, be adjusted for the actual condition of specific machine-trains. If the rate of change of a specific machine indicates rapid degradation, you should either repair it or at least increase the monitoring frequency to prevent catastrophic failure.

The recommended data acquisition frequencies are the maximum that will ensure prevention of most catastrophic failures. Less frequency monitoring will limit the ability of the program to detect and prevent un-scheduled machine outages.

To augment the vibration-based program, you should also schedule the non-vibration tasks. Bearing cap, point-of-use infrared measurements, visual inspections and process parameters monitoring should be conducted in conjunction with the vibration data acquisition.

Full infrared imaging or scanning on the equipment included in the vibration-monitoring program should be conducted on a quarterly basis. In addition, full thermal scanning of critical electrical equipment, i.e. switch gear, circuit breakers, etc., and all heat transfer systems, i.e. heat exchangers, condensers, process piping, etc., that are not in the vibration program should be conducted quarterly.

Lubricating oil samples from all equipment included in the program should be taken on a monthly basis. As a minimum, a full spectrographic analysis should be conducted on these samples. Wear particle or other analysis techniques should be made on an 'as-needed' basis.

Setting up analysis parameters

The next step in establishing the program's database is to set up the analysis parameters that will be used to routinely monitor plant equipment. Each of these parameters will be based on the specific machine-train requirements that we have just developed.

Normally for non-mechanical equipment the analysis parameter set will consist of the calculated values derived from measuring the thermal profile or process parameters. Each classification of equipment or system will have its own unique analysis parameter set.

Boundaries for signature analysis

All vibration monitoring systems have finite limits on the resolution, or ability to graphically display the unique frequency components that make up a machine's vibration signature. The upper limit (F_{MAX}) for signature analysis

should be set high enough to capture and display enough data so that the analyst can determine the operating condition of the machine-train but no higher.

To determine the impact of resolution, calculate the display capabilities of your system. For example, a vibration signature with a maximum frequency (F_{MAX}) of 1000 Hertz taken with an instrument capable of 400 lines of resolution would result in a display in which each displayed line will be equal to 2.5 Hertz or 150 rpm. Any frequencies that fall between 2.5 and 5.0, i.e. the next displayed line, would be lost.

Define alert and alarm limits

The method of establishing and using alert/alarm limits varies depending on the particular vibration monitoring system that you select. Normally these systems will use either static or dynamic limits to monitor, trend and alarm measured vibration. We will not attempt to define the different dynamic methods of monitoring vibration severity in this book. We will however provide a guideline for the maximum limits that should be considered acceptable for most plant mechanical equipment.

The systems that use dynamic alert/alarm limits base their logic on the rate of change of vibration amplitude. Any change in the vibration amplitude is a direct indication that there is a corresponding change in the machine's mechanical condition. However, there should be a maximum acceptable limit, i.e. absolute fault.

The accepted severity limit for casing vibration is 0.628 ips-peak (velocity). This is an un-filtered broadband value and normally represents a bandwidth between 10 and 10,000 Hertz. This value can be used to establish the absolute fault or maximum vibration amplitude for broadband measurement on most plant machinery. The exception would be machines with running speeds below 1200 rpm or above 3600 rpm.

Narrowband limits, i.e. discrete bandwidth within the broadband, can be established using the following guideline. Normally 60 to 70 per cent of the total vibration energy will occur at the true running speed of the machine. Therefore, the absolute fault limit for a narrowband established to monitor the true running speed would be 0.42 ips-peak. This value can also be used for any narrowbands established to monitor frequencies below the true running speed.

Absolute fault limits for narrowbands established to monitor frequencies above running speed can be ratioed using the 0.42 ips-peak limit established for the true running speed. For example, the absolute fault limit for a narrowband created to monitor the bladepassing frequency of fan with 10 blades would be set at 0.042 or 0.42/10.

Narrowband designed to monitor high-speed components, i.e. above 1000 Hertz, should have an absolute fault of 3.0 g's-peak (acceleration).

Rolling element bearings, based on factor recommendations, have an absolute fault limit of 0.01 ips-peak. Sleeve or fluid-film bearings should be watched closely. If the fractional components that identify oil whip or whirl are present at any level, the bearing is subject to damage and the problem should be corrected.

Non-mechanical equipment and systems will normally have an absolute fault limit that specifies the maximum recommended level for continued operation. Equipment or systems vendors will, in most cases, be able to provide this information.

Transducers

The type of transducers and data acquisition techniques that you will use for the program is the final critical factor that can determine the success or failure of your program. Their accuracy, proper application and mounting will determine whether valid data will be collected.

The optimum predictive maintenance program developed in earlier chapters is predicated on vibration analysis as the principle technique for the program. It is also the most sensitive to problems created by the use of the wrong transducer or mounting technique.

Three basic types of vibration transducers can be used for monitoring the mechanical condition of plant machinery: displacement probe, velocity transducer and accelerometers. Each has specific applications within the plant. Each also has limitations.

Displacement probes

Displacement or eddy-current probes are designed to measure the actual movement, i.e. displacement, of a machine's shaft relative to the probe. Therefore, the displacement probe must be rigidly mounted to a stationary structure to gain accurate, repeatable data.

Permanently mounted displacement probes will provide the most accurate data on machines with a low rotor weight, relative to the casing and support structure. Turbines, large process compressors, and other plant equipment should have displacement transducers permanently mounted at key measurement locations to acquire data for the program.

The useful frequency range for displacement probes is from 10 to 1000 Hertz or 600 to 60,000 rpm. Frequency components below or above this range will be distorted and therefore unreliable for determining machine condition.

The major limitation with displacement or proximity probes is cost. The typical cost for installing a single probe, including a power supply, signal conditioning, etc., will average $1,000. If each machine in your program requires ten measurements, the cost per machine will be about $10,000. Using displacement transducers for all plant machinery will dramatically increase the initial cost of the program.

Displacement data is normally recorded in terms of mils or 0.001 inch, peak-to-peak. This valve expresses the maximum deflection or displacement off the true centerline of a machine's shaft.

Velocity transducers

Velocity transducers are an electro-mechanical sensor designed to monitor casing or relative vibration. Unlike the displacement probe, velocity transducers measure the rate of displacement not actual movement. Velocity data is normally expressed in terms of ips or inches-per-second peak and is perhaps the best method of expressing the energy created by machine vibration.

Velocity transducers, like displacement probes, have an effective frequency range of about 10 to 1000 Hertz. They should not be used to monitor frequencies below or above this range.

The major limitation of velocity transducers is their sensitivity to mechanical and thermal damage. Normal plant use can cause a loss of calibration and therefore a strict re-calibration program must be used to prevent distortion of data. As a minimum, velocity transducers should be re-calibrated at least every six months. Even with periodic re-calibration, programs using velocity transducers are prone to bad or distorted data that results from loss of calibration.

Accelerometers

Accelerometers use a piezoelectric crystal to convert mechanical energy into electrical signals.

Data acquired with this type of transducers are relative vibration, not actual displacement, and are expressed in terms of g's or inches/second/second. Acceleration is perhaps the best method of determining the force created by machine vibration.

Accelerometers are susceptible to thermal damage. If sufficient heat is allowed to radiate into the crystal it can be damaged or destroyed. However since the data acquisition time, using temporary mounting techniques, is relatively short, i.e. less than thirty seconds, thermal damage is rare. Accelerometers do not require a re-calibration program to insure accuracy.

The effective range of general-purpose accelerometers is from about 1 Hertz to 10,000 Hertz. Ultrasonic accelerometers are available for frequencies up to 1 MHertz.

Machine data above 1,000 Hertz or 60,000 rpm should be taken and analyzed in acceleration or g's.

Mounting techniques

Predictive maintenance programs using vibration analysis must have accurate, repeatable data to determine the operating condition of plant machinery. In addition to the transducer, three factors will affect data quality: measurement point, orientation and compressive load.

Key measurement point locations and orientation to the machine's shaft were selected as part of the database setup to provide the best possible detection of incipient machine-train problems. Deviation from the exact point or orientation will affect the accuracy of acquired data. Therefore, it is important that every measurement throughout the life of the program be acquired at exactly the same point and orientation. In addition, the compressive load or downward force applied to the transducer should also be the same for each measurement.

For accuracy of data, a direct mechanical link to the machine's casing or bearing cap is necessary. Slight deviations in this load will induce errors in the amplitude of vibration and may create false frequency components that have nothing to do with the machine.

The best method of ensuring these three factors are the same each time is to hard mount vibration transducers to the selected measurement points. This will guarantee accuracy and repeatability of acquired data. It will also increase the initial cost of the program. The average cost of installing a general-purpose accelerometer will be about $300 per measurement point or $3000 for a typical machine-train.

To eliminate the capital cost associated with permanently mounting transducers a well-designed quick-disconnect mounting can be used. This mounting technique permanently mounts a quick-disconnect stud, with an average cost of less than $5, at each measurement point location. A mating sleeve, built into a general-purpose accelerometer, is then used to acquire accurate, repeatable data. A well-designed quick-disconnect mounting technique will provide the same accuracy and repeatability as the permanent mounting technique but at a much lower cost.

The third mounting technique that can be used is a magnetic mount. For general purpose use, below 1000 Hertz, a transducer can be used in conjunction with a magnetic base. Even though the transducer/magnet assembly will have a resonant frequency that may provide some distortion to acquired data, this technique can be used with marginal success. Since the magnet can be placed any where on the machine, it will not guarantee that the exact location and orientation is maintained on each measurement.

The final method used by some plants to acquire vibration data is hand held transducers. This approach is not recommended if any other method can be used. Hand held transducers will not provide the accuracy and repeatability required to gain maximum benefit from a predictive maintenance program. If this technique must be used, extreme care should be exercised to ensure that the exact point, orientation and compressive load is used for every measurement point.

50.5.3 Getting started

The steps we have defined will provide the guideline for establishing a predictive maintenance database. The only steps remaining to get the program started are to establish measurement routes and take the initial or baseline measurements. Remember, the predictive maintenance system will need multiple data sets to develop trends on each machine. With this database, you will be able to monitor the critical machinery in your plant for degradation and begin to achieve the benefits that predictive maintenance can provide. The actual steps required to implement a database will depend on the specific predictive maintenance system selected for your program. The system vendor should provide the training and technical support required to properly develop the database with the information developed in the preceding sections.

Training

Successful completion of this critical phase of creating a total plant predictive maintenance program will require a firm knowledge of the operating dynamics of plant machinery, systems and equipment. Normally some if not all of this knowledge exists within the plant staff. However, the knowledge may not be within the staff selected to implement and maintain the predictive maintenance program.

In addition, a good working knowledge of the predictive maintenance techniques and systems that will be included in the program is necessary. This knowledge in all probability will not exist within existing plant staff. Therefore training, before attempting to establish a program, is strongly recommended. The minimum recommended level of training includes: user's training for each predictive maintenance system that will be used, a course on machine dynamics, and a basic theory course on each of the techniques, i.e. vibration, infrared, etc., that will be used.

In some cases, the systems vendors can provide all of these courses. If not, there a number of companies and professional organizations that offer courses on most nondestructive testing techniques.

Technical support

The labor and knowledge required to properly establish a predictive maintenance program is often too much for plant staff to handle. To overcome this problem, the initial responsibility for creating a viable, total plant program can be contracted to a company that specializes in area.

There are companies that provide full consulting and engineering services directed specifically toward predictive maintenance. These companies have the knowledge required and years of experience. They can provide all of the labor required to fully implement a full plant program and normally can reduce total time required to get the program up and running.

Caution should be used in selecting a contractor to provide this startup service. Check references very carefully.

50.5.4 Maintaining the program

The labor-intensive part of predictive maintenance management is complete. A viable program has been established, the database is complete and you have begun to monitor the operating condition of your critical plant equipment. Now what?

Most programs stop right here. The predictive maintenance team does not continue their efforts to get the maximum benefits that predictive maintenance can provide. Instead, they rely on trending, comparative analysis or in the case of vibration-based programs simplified signature analysis to maintain the operating condition of their plant. This is not enough to gain the maximum benefits from a predictive maintenance program.

In this section, we will discuss the methods that can be used to ensure that you gain the maximum benefits from your program and at the same time improve the probability that the program will continue.

Trending techniques

The database that was established for your program included broadband, narrowband and full signature vibration data. It also included process parameters, bearing cap temperatures, lubricating oil analysis, thermal imaging and other critical monitoring parameters. What do we do with this data?

The first method required to monitor the operating condition of plant equipment is to trend the relative condition over time. Most of the microprocessor-based systems will provide the means of automatically storing and recalling vibration and process parameters trend data for analysis or hard copies for reports. They will also automatically prepare and print numerous reports that quantify the operating condition at a specific point in time. A few will automatically print trend reports that quantify the change over a selected time frame. All of this is great, but what does it mean?

Monitoring the trends of a machine-train or process system will provide the ability to prevent most catastrophic failures. The trend is similar to the bathtub curve used to schedule preventive maintenance. The difference between the preventive and predictive bathtub curve is that the latter is based on the actual condition of the equipment, not a statistical average.

The disadvantage of relying on trending as the only means of maintaining a predictive maintenance program is that it will not tell you the reason a machine is degrading. One good example of this weakness is an aluminum foundry that relied strictly on trending to maintain their predictive maintenance program. In the foundry are 36 cantilevered fans that are critical to plant operation. The rolling element bearings in each of these fans are changed on average every six months. By monitoring the trends provided by their predictive maintenance program, they can adjust the bearing change out schedule based on the actual condition of the bearings in a specific fan. Over a two-year period, there were no catastrophic failures or loss of production that resulted from the fans being out of service.

Did the predictive maintenance program work? In their terms, the program was a total success. However, the normal bearing life should have been much greater than six months. Something in the fan or process created the reduction in average bearing life. Limiting their program to trending only, they were unable to identify the root-cause of the premature bearing failure. Properly used, your predictive maintenance program can identify the specific or root-cause of chronic maintenance problems. In the example, a full analysis provided the answer. Plate-out or material buildup on the fan blades constantly increased the rotor mass and therefore forced the fans to operate at critical speed. The imbalance created by operation at critical speed was the forcing function that destroyed the bearings. After taking corrective actions, the plant now gets an average of three years from the fan bearings.

Analysis techniques

All machines have a finite number of failure modes. If you have a thorough understanding of these failure modes and the dynamics of the specific machine, you can learn the vibration analysis techniques that will isolate the specific failure mode or root-cause of each machine-train problem.

The following example will provide a comparison of various trending and analysis techniques.

Broadband analysis

The data acquired using broadband data is limited to a value that represents the total energy that is being generated by the machine-train at the measurement point location and in the direction opposite the transducer. Most programs trend and compare the recorded value at a single point and disregard the other measurement points on the common-shaft.

Rather than evaluate each measurement point separately, plot the energy of each measurement point on a common shaft. First, the vertical measurements were plotted to determine the mode shape of the machine's shaft. This plot indicates that the outboard end of the motor shaft is displaced much more than the remaining shaft. This limits the machine problem to the rear of the motor. Based strictly on the overall value, the probable cause is loose motor mounts on the rear motor feet. The second step was plotting the horizontal mode shape. This plot indicates that the shaft is deflected between the pillow block bearings. Without additional information, the mode shaft suggests a bent shaft between the bearings.

Even though we cannot identify the absolute failure mode, we can isolate the trouble to the section of the machine-train between the pillow block bearings.

Narrowband analysis

The addition of unique narrowbands that monitor specific machine components or failure modes add more diagnostic information.

If we add the narrowband information acquired from the Hoffman blower, we find that the vertical data is primarily at the true running speed of the common shaft. This confirms that a deflection of the shaft exists. No other machine component or failure mode is contributing to the problem. The horizontal measurements indicate that the bladepass, bearing defect and misalignment narrowbands are the major contributors.

As we discussed, fans and blowers are prone to aerodynamic instability. The indication of abnormal vanepass suggests that this may be contributing to the problem. The additional data provided by the narrowband readings help to eliminate many of the possible failure modes that could be affect the blower. However, we still cannot confirm the specific problem.

Root cause failure analysis

A visual inspection of the blower indicated that the discharge is horizontal and opposite the measurement point location. By checking the process parameters recorded concurrent with the vibration measurements, we found that the motor was in a no-load or run-out condition and that the discharge pressure was abnormally low. In addition, the visual inspection showed that the blower sits on a cork pad and is not bolted to the floor. The discharge piping, 24 inch diameter schedule 40 pipe, was not isolated from the blower nor did it have any pipe supports for the first 30 feet of horizontal run. With all of these clues in hand, we concluded that the blower was operating in a 'run-out' condition, i.e. it was not generating any pressure, and was therefore unstable. This part of the machine problem was corrected by reducing, i.e. partially closing, the damper setting and forcing the blower to operate within acceptable aerodynamic limits. After correcting the damper setting, all of the abnormal horizontal readings were within acceptable limits. The vertical problem with the motor was isolated to improper installation. The weight of approximately 30 feet of discharge piping compressed the cork pad under the blower and forced the outboard end of the motor to elevate above the normal centerline. In this position, the motor became an unsupported beam and resonated in the same manner as a tuning fork. After isolating the discharge piping from the blower and providing support, the vertical problem was eliminated.

50.5.5 Additional training

The initial user's training and basic theory will not be enough to gain maximum benefits from a total plant predictive maintenance program. You will need to continue the training process throughout the life of the program. A variety of organizations, including predictive maintenance systems vendors, provide training programs in all of the predictive maintenance techniques. Caution in selecting both the type of course and instructor is strongly recommended. Most of the public courses are in reality sales presentations. They have little practical value and will not provide the knowledge base required to gain maximum benefit from your program.

Practical or application oriented courses are available that will provide the additional training required to gain maximum diagnostic benefits from your program. The best way to separate the good from the bad is to ask previous attendees. Request a list of recent attendees and then talk to them. If reputable firms present the courses, they will gladly provide this information.

51

Planning and Scheduling Outages

Ricky Smith

Life Cycle Engineering, Inc.

Contents

51.1 Introduction

The concept of shutdown or outage management as a discipline was developed for use in managing the US space program in the early 1960s. Its practice has expanded rapidly into government, the military, and industry. Today you will find these principles are also being used under the names of project, program, product development, and construction management.

51.1.1 Definition of shutdown

The main objective of a shutdown is to accomplish, at the lowest possible costs, the maximum quality and quantity of work that can be completed within a limited time, defined as the outage. Shutdown activities should be concentrated on work that can only be performed during shutdown or when production systems are idle. A well-organized and adequately planned and scheduled shutdown is the best insurance for lower maintenance and high reliability of equipment.

Many organizations assume they do planning and scheduling because they do shutdown maintenance and project work. What typically gets ignored is the daily or weekly planning and scheduling process. Everyone is fixated on shutdown work. The problem is that most organizations spend the majority of their maintenance hours performing daily or weekly maintenance and not shutdowns or project work. In some organizations, the mix is 75 per cent routine and 25 per cent shutdown maintenance, mirroring the classic 80/20 rule of daily/weekly maintenance hours versus shutdown/project hours.

This fact is not designed to take away from the importance of either one. The good news is that the mechanics of planning and scheduling is the same for either daily/weekly or shutdown/project maintenance. Well planned work orders are the fundamental requirement of both routine and outage maintenance. Many organizations lose sight of this fact.

The following sections provide a list of best practices. This list represents an overview of the best approaches used to effectively manage shutdowns and project work. The mechanics behind these practices is explained in the balance of this chapter.

51.1.2 Shutdown life cycle

Each outage or shutdown moves through a predictable life cycle of four phases with each phase calling for different skills from the project planner and outage manager. All projects, including maintenance outages or shutdowns, can be broken down into four distinct phases. Figure 51.1 illustrates the typical activity level associated with each of these phases. Typically, the man-hours of effort are relatively low during the initial phases, at their peak during the implementation phase and decline with the final evaluation.

51.1.3 Outage planning and management

The following is intended to serve as a guideline to assist in the development of your own process.

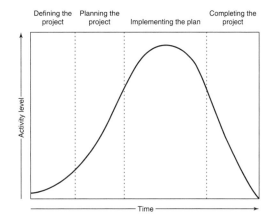

Figure 51.1 Typical activity levels during outage

Master shutdown schedule

Most companies, as part of their annual business plan, develop an outage schedule that is based on anticipated business cycle or precieved maintenance requirements. This outage schedule contains all shutdown dates for critical production systems. Unfortunately, most of these plans do not consider the impact on capacity. An effective outage schedule should be configured to minimize lost of production capacity, product quality and the potential increase in overall operating cost that can result from poorly coordinated outage schedules. Care should be taken to assure minimal, negative impact from this schedule.

Shutdown management team

Effective shutdown management depends on absolute adherence to prescribed standards that define what type of work will be done during scheduled shutdowns. These decisions can not be made by the maintenance planner alone. To aid in the selection, planning and implementation of outage tasks, a management team is a fundamental requirement. This team should be composed of:

- Maintenance manager(s)
- Maintenance planner(s)
- Production manager(s)
- Production planner(s)
- Operations supervisors
- Engineering liasion
- Contract liaison (if needed)
- Material/storeroom representative
- Purchasing representative.

When the nucleus of the project team is assembled, its first order of business is to clarify the project and arrive at agreement among team members about the project's definition and scope, as well as the basic strategy for carrying it out. An orderly process can guide you through these steps. The following sequence of activities will get your project smoothly under way:

1. It is critical for the team to spend adequate time at the beginning to study, discuss, and analyze the project. This establishes a clear understanding of what you are dealing with. It may be necessary to research how similar projects structured their approach, or what other patterns of experience can contribute to project planning. The purpose of this activity is to be sure you are addressing the right problem or pursuing the real opportunity.
2. When you are confident that you have a firm grasp of the situation, work up a preliminary project definition. This preliminary definition will be subject to revision as additional information and experience is acquired.
3. Now, using this project definition, state the end-results objective of the project.
4. Then, list both the imperatives and desirables to be present in the results. That is, list the outcomes that must be present for the project to be considered successful, and list the outcomes that are not essential but that would add to the project's success.
5. Now you are ready to generate alternative strategies that might lead you to your objective. To generate these alternatives, try brainstorming with your project team.
6. Next, evaluate the alternative strategies you have generated. Be sure that your criteria for evaluation are realistic and reflect the end results objective.
7. Evaluation allows you to choose a course of action that will meet both your project definition and end-results objective.

Good objectives

Effective outage planning and management is dependent on well-defined objectives. Everyone, beginning with the planner, must have a universal understanding of the specific objectives that are to be achieved during the outage. The fundamental requirements of good objectives include:

Specific   A good objective says exactly what you want to accomplish. The definition must be both clear and concise.

Measurable   Being specific helps make your objective measurable.

Action-oriented   When writing objectives, use statements that have action-tense verbs and are complete sentences.

Realistic   Good objectives must be attainable yet should present a challenge.

Time-limited   Set a specific time by which to achieve the objective.

Brainstorming

Brainstorming is a free-form process that taps into the creative potential of a group through association of ideas. Association works as a two-way current: when a group member voices an idea, this stimulates ideas from others, which in turn leads to more ideas from the one who initiated the idea.

Brainstorming procedures

- List all ideas offered by group members.
- Do not evaluate or judge ideas at this time.
- Do not discuss ideas at this time except to clarify understanding.
- Welcome 'blue sky' ideas. It's easier to eliminate ideas later.
- Repetition is okay. Don't waste time sorting out duplication.
- Encourage quantity. The more ideas you generate, the greater your chance of finding a useful one.
- Don't be too anxious to close the process. When a plateau is reached, let things rest and then start again.

The management team will participate in the following activities:

Initial shutdown meeting

Ninety days prior to beginning shutdown, the shutdown management team should meet to determine the boundary conditions for the upcoming outage. These initial decisions will provide the basic knowledge required to begin the planning process. The outcome of this initial meeting should:

Select shutdown tasks   Careful evaluation of work requests is essential for effective shutdown performance. All requested tasks should not be automatically included in the outage plan. Each request must be evaluated to determine its real strategic value and real value added.

Question past shutdown practices   Each of the tasks or projects requested for the outage should be evaluated to determine whether or not it should be included in the outage. The management team should include the following questions in this evaluation:

- Always ask 'Why?'
- Does this really need to be done?
- Has a pattern of failures developed?
- Has a failure-related downtime been a factor?
- Comments from craftspeople.
- Comments from operators.

Develop the shutdown policies   The outage management team should also establish clear, concise guidelines that will be used to plan and manage the shutdown. These criteria should include:

- Establish cut-off work order dates:
  - Implement a two-week frozen schedule.
  - No work orders added/received two weeks prior to start of shutdown.
- Establish add-on work order policy:
  - If we add work orders ... what gets backed out?
  - Require high level of approval to add on work orders.

Separate pre-shutdown work from regular shutdown work
Focus on work that can be performed without shutting down equipment:

- Pipe layout
- Electrical panel setups
- Wiring runs
- Pre-fabricate, sub-assemblies and components
- Laying foundations and footings
- Pump placements
- Pre-cleaning work-site and equipment.

Develop timetable for shutdown   The management team should establish a macro-level schedule for the outage. Without complete planning, a definitive timeline can not be established, but the team should establish specific milestones that must be met.

51.4 Outage planning

Effective planning is the next step in the outage management process. Like all other maintenance activities, each task included in the outage plan must be fully planned. However, the finite time frame associated with a fixed-duration shutdown also requires effective scheduling to assure success.

51.4.1 Project parameters

During a project's life, planners and management should focus on three basic parameters: quality, time and cost; see Figure 51.2. A successfully planned and managed project is one that is completed at the specified level of quality, on or before the deadline, and within budget.

Each of these parameters is specified in detail during the planning phase of the project. These specifications then form the basis for control during the implementation phase.

What are the pitfalls, and how can they be avoided?

A project's initiator is usually unclear about important aspects of the project. Project personnel tend to stress their own points of view during the stage of defining and structuring the project. If this set of personal biases and interests is left unchecked, disaster can result. However, such disaster can be avoided by full discussion between the project manager, client, and staff at the project's inception. With a clear understanding of what is expected, the project manager is now ready to begin defining the project.

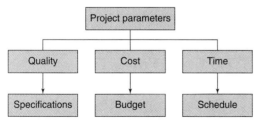

Figure 51.2 Project parameters

The first phase is a clear definition of the scope of work that is to be performed during the shutdown or maintenance outage. This definition must be more than a macro-level listing of the major tasks that are to be performed. Instead, each of these major or macro-level tasks must be fully defined. The purpose of this procedure is to provide an effective method of planning, material control, and follow-up of repetitive, non-repetitive, and capital shutdown work.

51.4.2 Planning the project parameters

Planning is crucial in shutdown and project management. Planning means listing in detail what is required to successfully complete the project along the three critical dimensions of quality, time, and cost. Each of these dimensions will be considered in the following pages, along with a variety of tools and techniques.

Planning steps

1. Establish the project objective.
2. Choose a basic strategy for achieving the objective.
3. Break the project down into discrete tasks or steps.
4. Determine the performance standards for each task.
5. Determine how much time is required to complete each task.
6. Determine the proper sequence for completing the tasks and aggregate this information into a schedule for the total project.
7. Determine the cost of each task and aggregate costs into the project budget.
8. Design the necessary staff organization, including the number and kind of positions, and the duties and responsibilities of each.
9. Determine what training, if any, is required for project team members.
10. Develop the necessary policies and procedures.

51.4.3 Planning the quality dimension

Planning for quality requires attention to detail. The goal of quality planning is to assure that the outputs of the project will perform – that it will do what it is supposed to do. The quality plan also establishes the criteria of performance by which the project output will be measured when it is completed.

In planning the quality dimension, include specifications for the quality and types of materials to be used, the performance standards to be met, and the means of verifying quality such as testing and inspection. Two techniques facilitate planning for quality: a work breakdown structure and project specifications. Both are described on the next few pages.

Creating a work breakdown structure (WBS)

A work breakdown structure is the starting place for planning all three parameters of a project: quality, cost, and time. It is a technique based on dividing a project into tasks, or work packages. Because all elements required to complete the project are identified, you reduce the chances of neglecting or overlooking an essential step.

A work breakdown structure is typically constructed with two or three levels of detail, although more levels may be required for very complex projects. Start by identifying logical subdivisions of the project, then break each of these down further. As you construct a work breakdown structure, keep in mind that the goal is to identify a unit of work that is discrete and that advances the project toward its completion.

Case study: Centrifugal pump installation

Figure 51.3 provides an example of a typical work breakdown structure. The major divisions define specific, logical task groupings as well as the cost-accounting classifications. In the example, cost should be acquired for the preparation of the foundation, installation and final commisioning of the pump and motor. In addition, each of the sub-tasks that comprise these classifications are logical groups of tasks that must be completed in sequence and by different work classifications.

Project specifications

From the work breakdown structure (WBS), specifications can be written for each task of the project. Specifications include all relevant requirements to meet the project's quality dimension – materials to be used, standards to be met, tests to be performed, etc. Use extreme care in writing specifications, because they become the controlling factor in meeting project performance standards, and directly affect both budget and schedule.

51.4.4 Planning the time dimension

The objective when planning the time dimension is to determine the shortest time necessary to complete the project. Begin with the work breakdown structure and determine the time required to complete each task. Next, determine in what sequence tasks must be completed, and which ones may be under way at the same time. From this analysis, you will have determined the three most significant time elements:

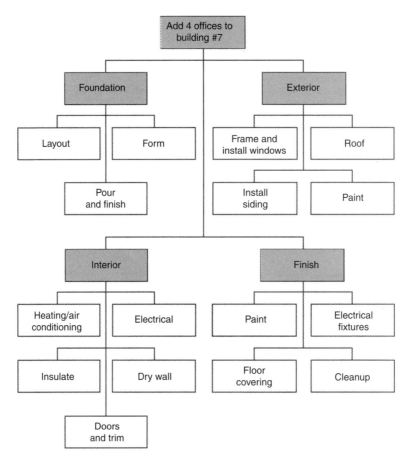

Figure 51.3 Sample work breakdown structure

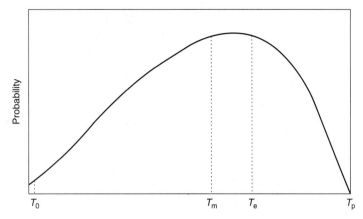

Figure 51.4 Probability table for mathematical estimating

- The duration of each step
- The earliest time at which a step may be started
- The latest time at which a step must be started.

Planning the time dimension can only be done by people who have experience with the same or similar activities. If you personally do not know how long it takes to do something, you will need to rely on someone else who does have the requisite experience.

Many project managers find it realistic to estimate time intervals as a range rather than as a precise amount. Another way to deal with the lack of precision in estimating time is to use a commonly accepted formula for that task. Or, if you are working with a mathematical model, you can determine the probability of the work being completed within the estimated time by calculating a standard deviation of the time estimate.

Using a mathematical model to estimate time

One method of estimating the time required to complete a specific task is to use probability tables or calcuations to define the appropriate interval.

T_m = The most probable time
T_o = The optimistic (shortest) time within which only 1% of similar projects are completed
T_p = The pessimistic (longest) time within which 99% of similar projects are completed
T_e = The calculated time estimate

$$T_e = \frac{T_o + 4T_m + T_p}{6}$$

σ = Standard deviation

$$\sigma = \frac{T_p - T_o}{6}$$

- 68.26% of the time the work will be completed within the range of $T_e \pm 1$ standard deviation.
- 95.44% of the time the work will be completed within the range of $T_e \pm 2$ standard deviations.

- 99.73% of the time the work will be completed within the range of $T_e \pm 3$ standard deviations.

Critical path method

This section is devoted to using critical path method (CPM) as a tool for the scheduling and sequencing of work for shutdowns and project work.

Although this section looks intimidating and the process quite manual, the reader is encouraged to become familiar with the basic concept and process behind using CPM as a tool. The good news is that the manual process can be automated via computers. Several excellent computer programs are available that automate and recalculate the tedious manual process of developing and using CPM. The essential benefits of these computerized programs are highlighted below:

1. Review leveling can be planned and scheduled ahead of the need date. The advantage is that nothing happens without your telling the CPM software program to schedule it.
2. A built-in calendar module allows you to establish a 'look ahead' regarding work hours and days. The user is able to see the impact on 'what if' and how the situation under review impacts the entire shutdown or project.
3. Most are built around precedent logic. This is explained in the text.
4. The critical path is highlighted which indicates the proper or recommended path to follow.
5. Confirmed changes or 'what if' scenarios are easy to 'plug in' and a new time line or critical path can be printed.

With time duration determined for each task of the project, the next step is to determine the earliest and latest starting times for each task. There are two commonly used methods for charting the project: Gantt charts and PERT diagrams. The details of these two methods are discussed on the following pages.

History of CPM

The CPM was developed for the express purpose of applying computers to construction planning. However, it is equally as effective when used manually. This method has been well tested and it works. The CPM utilizes the contractor's knowledge, experience and practical ability in a common-sense manner to plan and schedule construction.

The traditional method of planning is the bar chart. It evolved from the Gantt chart, made popular by Mr Henry L. Gantt and Frederick W. Taylor in the early 1900s. The work of Mr Gantt and Mr Taylor was the first scientific consideration given to the problem of work scheduling. The bar chart was and is an excellent graphic method of scheduling. All levels of management and supervision easily understand it.

A question often asked is, 'If the bar chart is so well suited to maintenance, why change?' The bar chart is limited in what it can do. If a bar chart is carefully prepared, the planner/scheduler goes through the same thinking process as the CPM planner does. However, the bar chart does not show the interrelationships in the sequence of work nor in the restraints and control the activities have upon each other.

Network analysis had its beginning in 1956 with a joint effort between Dupont, UNIVAC, and Remington-Rand. The network technique which resulted was field tested by Dupont in 1958, and the results were published in 1959. The Special Projects Office of the Navy Bureau of Ordinance, the firms of Booz, Allen, and Hamilton, and Lockheed began development of a management system called Program Evaluation and Review Technique (PERT) in 1958 to control and monitor the 3000 plus contractors and agencies working on the Polaris missile program. Both CPM and PERT, the outgrowths of the efforts of the two groups, have since gained wide acceptance as network models which graphically simulate the dynamic nature of the work process for a wide spectrum of project types.

Potential of CPM

The potential of CPM is unlimited. If the logic of the CPM is correct, you can develop the following information:

- Projected expenditures, both material and labor.
- A planned purchasing schedule in dollars graphically integrated with time.
- Cost control information showing a clear picture of over-runs and under-runs by activity.
- Total manpower projection curves, by craft.
- The CPM can also be broken down into smaller and simpler work schedules to be given to maintenance supervision so that they are aware of the sequence and schedule required for the shutdown.

Critical path method networks are classified as either activity-on-arrow or activity-on-node networks. Although there are many versions of these two basic networks, the fundamental logic diagram procedures remain the same. The arrow diagram and precedence diagram will be discussed later in this session. The Bar chart will be displayed along with the other two types of diagrams to show the similarly to the principles of CPM planning.

Planning critical path development

CPM or basically arrow diagramming is the same theory as PERT, except it provides a lot more definition to the task to be implemented. PERT is event driven where CPM is activity driven. CPM is the networking technique best utilized in maintenance planning and scheduling. It points out ways in which duration can be shortened and will let you weigh the cost of expediting against cost of lost production. The purpose is not to encroach on the management function, but to provide the tools necessary for it to perform effectively. Efficient planning of jobs is the difference between on-time and late, and it can mean the difference between success and failure.

Justifications

What happens if the duration estimates turn out to be wrong or inaccurate? Arrow diagramming will pinpoint all the activities (events) that are affected and the degree to which they are affected. It will tell you whether any re-planning, rescheduling, or other remedial action is necessary.

51.5 Project shutdown planning

Project shutdown planning establishes the duration of the project, the resources needed to complete each activity, and the required sequence of performance of each job.

Leveling

Assigning resources in the most effective manner possible, leveling is accomplished by taking advantage of the fact that there is often a difference between when a job may start and when it must start. This variation, where it exists, is called total float.

Rules of arrow diagraming

One and only one arrow is used to represent the operation/activity to be performed. Length and direction in which it points have no significance unless it is time scaled. Arrows are connected to form a model of the project by answering, for each operation/activity, the questions: 'What immediately precedes this activity?' and 'What immediately follows this activity?'

The third question to be answered is: 'What can be done concurrently?' Although this question implies that we would be doing more than one activity at the same time, this is not necessarily true. What we really mean is that both tasks depend on the completion of the same operation. Concurrently indicates a common need for prior work. Start all diagrams with an arrow marked lead-time and show all actual work starting after it. (Lead-time – preparation time to undertake a project.)

Junctions of arrows are called events or nodes. These are points in time and consume no time. They are numbered to provide a convenient numeric sequential designation for all activities.

Dummies (restraints) are jobs that have no duration or cost. They are introduced to keep the logic correct and to

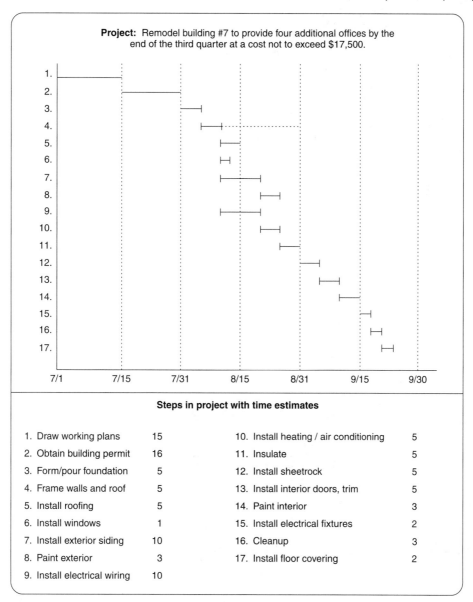

Project: Remodel building #7 to provide four additional offices by the end of the third quarter at a cost not to exceed $17,500.

	7/1	7/15	7/31	8/15	8/31	9/15	9/30

Steps in project with time estimates

1. Draw working plans	15	
2. Obtain building permit	16	
3. Form/pour foundation	5	
4. Frame walls and roof	5	
5. Install roofing	5	
6. Install windows	1	
7. Install exterior siding	10	
8. Paint exterior	3	
9. Install electrical wiring	10	
10. Install heating / air conditioning	5	
11. Insulate	5	
12. Install sheetrock	5	
13. Install interior doors, trim	5	
14. Paint interior	3	
15. Install electrical fixtures	2	
16. Cleanup	3	
17. Install floor covering	2	

Figure 51.5 Example of a Gantt chart

Figure 51.6 PERT diagram

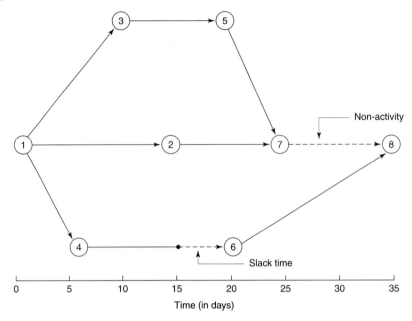

Figure 51.7 PERT diagram with timeline

keep unique the numeric designation for event numbers at the tail and head of each arrow. If more resources are applied (crashing) to reduce the project duration, the jobs must be selected from the critical path.

51.5.1 Gantt chart

A Gantt chart is a horizontal bar chart that graphically displays the time relationship of the steps in a project. It is named after Henry Gantt, the industrial engineer who introduced the procedure in the early 1900s. A line placed on the chart in the time period when it is to be undertaken represents each step of a project. When completed, the Gantt chart shows the flow of activities in sequence as well as those that can be under way at the same time.

To create a Gantt chart, list the steps required completing a project and estimating the time required for each step. Then list the steps down the left side of the chart and time intervals along the bottom. Draw a line across the chart for each step, starting at the planned beginning date and ending on the completion date of that step.

Some parallel steps can be carried out at the same time with one taking longer than the other; this allows some flexibility about when to start the shorter step, as long as the plan has it finished in time to flow into subsequent steps. This situation can be shown with a dotted line continuing on to the time when the step must be completed.

When your Gantt chart is finished, you will be able to see the minimum total time for the project, the proper sequence of steps, and which steps can be under way at the same time.

You can add to the usefulness of a Gantt chart by also charting actual progress. This is usually done by drawing a line in a different color below the original line to show

the actual beginning and ending dates of each step. This allows you to quickly assess whether or not the project is on schedule.

Gantt charts are limited in their ability to show the interdependencies of activities. In projects where the steps flow in a simple sequence of events, they can portray adequate information for project management. However, when several steps are under way at the same time and a high level of interdependency exists among the various steps, PERT diagrams are a better choice.

51.5.2 PERT diagram

PERT is an acronym for Program Evaluation and Review Technique. It is a more sophisticated form of planning than Gantt charts, and is appropriate for projects with many interactive steps. There are three components of a PERT diagram: circles or other convenient, closed figures represent events; arrows connecting the circles represent activities; and non-activities connecting two events are shown as dotted-line arrows. (A non-activity represents a dependency between two events for which no work is required.)

PERT diagrams are most useful if they show the time scheduled for completing an activity on the activity line. Time is recorded in a unit appropriate for the project, with days being most common, and hours, weeks, or even months occasionally used. Some diagrams show two numbers for time estimates – a high estimate and a low estimate.

The most sophisticated PERT diagrams are drawn on a time scale, with the horizontal projection of connecting arrows drawn to represent the amount of time required for their activity. In the process of diagramming to scale, some connecting arrows will be longer than completion of that task requires. This represents slack time in the project and is depicted by a heavy dot at the end of the appropriate time period, followed by a dotted-line arrow connecting with the following event.

51.6 Planning the cost dimension

There are many reasons to do careful planning for project costs. To begin with, if you overestimate costs you may lose the job before you begin because you are not competitive. A good plan includes the identification of sources of supplies and materials, and this careful research assures that the costs are realistic. The main function of a good budget is to monitor the costs of a project while it is in progress, and to avoid cost overruns.

Some inaccuracies in the budget are inevitable, but they should not be the consequence of insufficient work on the original plan. The goal is to be as realistic as possible. You cannot estimate the cost of your project until you know how long it will take, since the time of labor is typically the most significant cost item. Therefore, use your work breakdown structure and project schedule as the starting point for developing your project budget.

Typical costs components

The typical cost classifications or components for all outages or maintenance tasks include:

Labor The wages that are paid to all staff directly working on or supporting the project for the time spent on it.

Overhead The cost of payroll taxes and fringe benefits for everyone directly working on the project for the time spent on it. Usually calculated as a percentage of direct labor cost.

Materials The cost of items purchased for use in the project. Includes such things as lumber, cement, steel, nails, screws, rivets, bolts, and paint.

Supplies The cost of tools, equipment, office supplies, etc., needed for the project. If something has a useful life beyond the project, its cost should be calculated pro rata.

Equipment rental The cost of renting equipment such as scaffolding, compressors, cranes, bulldozers, trucks, etc., for use on the project.

General and administrative The cost of management and support services such as purchasing, accounting, secretarial, etc., for time dedicated to the project. These costs are usually calculated as a percentage of project cost.

With the cost components identified and the project broken down into tasks, create a worksheet to tally the costs for the total project. Note that costing a task is sometimes simplified if it is to be subcontracted. Costing then includes bidding the task, selecting a contractor, and then using the contract price as your cost.

51.6.1 Potential budgeting problems

Some of the potential problems that may distort or limit the accuracy of outage budgets include:

- Failure to obtain firm price commitments from suppliers and subcontractors.
- Poorly prepared work breakdown structures that lead to incomplete budgets.
- 'Fudge factors' built into internal support group estimates.
- Estimates based on different methods of costing, i.e., hours versus dollars.

Assigning responsibility

Determining who will have responsibility for completing each task or step of a project should be done as early as possible, so that they can participate in the planning of both schedules and budgets. This participation leads to a greater commitment to achieve the project within time and cost limitations.

The number of people involved in a project varies with its size and scope. Not every project has a different person responsible for each task. To make the best use of your resources when deciding who is responsible for a portion of your project, broaden your point of view to include subcontractors and service departments as well as members of the project team.

Controlling work in progress

Controlling is the central activity during implementation. The most important tool in this process is the plan that was developed to define the three parameters of the project – specifications, schedule, and budget. Performance is measured against these standards. Controlling involves three steps:

1. Establishing standards

Standards for the project were set in the detailed project specifications created in the planning stage. The project manager must constantly refer to these specifications and make sure the project team is also referencing them. If the project deviates from the original specifications, there is no guarantee that the success predicted by the feasibility studies will actually happen – the product or project outcome might fail to meet performance standards.

There are a number of tools available to help project managers control the project and make sure that the parameters defined in the specifications for quality, time, and budget are actually being met. A Gantt chart or PERT diagram designed at the planning stage is a great device for tracking how the time dimension of the project is proceeding in relationship to the plan.

2. Control point identification charts

A helpful technique for controlling a project is to invest some time to think through what is likely to go wrong in each of the three project parameters. Then identify when and how you will know that something is amiss and what you will do to correct the problem if it occurs. This will help minimize the times you will be caught by surprise as well as save time in responding to the problem. A control point identification chart, as shown in Table 51.1, is an easy way to summarize this information.

3. Shutdown work initiation

Shutdown work is classified three ways: (1) repetitive, (2) non-repetitive, or (3) capital projects. All three require operations supervisor's approval. All shutdown work is sent directly to the maintenance central planning department.

Repetitive shutdown work This type of work is developed by experienced maintenance and operations supervision based on prior shutdown experience, job knowledge, and required shutdown preventive maintenance tasks. Each agreed-upon job is assigned an annual work order number frequency, established, work order plans written, and is entered in the work order log. The handling of repetitive work orders will be the same as non-repetitive work orders, except one copy of the work plan will be substituted for the first copy of the work order.

Completed work plans and any signed-off work orders should be submitted by the supervisor to their respective maintenance supervisor no later than the day after the job is completed. The maintenance supervisor should examine these documents for completeness and accuracy.

Non-repetitive shutdown work This type of work is generated by maintenance problems requiring unit shutdown that are not covered by repetitive shutdown work schedules. Work requests are written by operations and approved by the operations supervisor.

If the work as called for on the work order is done in its entirety including steam tracing, insulation, clean up, etc., then the work order should be signed off. The supervisor should return the signed-off work order and the work plan to his supervisor who forwards them to planning.

If for any reason the work as defined by the work order cannot be completed during the shutdown period, the work order should not be signed off. Instead, it should

be returned to the supervisor. The reason for not completing the job should be thoroughly explained on the work plan.

If the work as called for on the work order is not completed due to lack of insulation, clean-up, tie-ins only being made, etc., but the shutdown portion of the work has been completed, then the first copy of the work order should be retained by the supervisor. The supervisor should then return the work plan to his supervisor denoting the reason for not completing work, explanation of variance, etc. The planning office will prepare a list of incomplete work following the shutdown and forward this list and a copy of the work orders to the appropriate maintenance supervisor for follow-up.

The work that is required to complete the work orders will be resumed the next working day following the shutdown unless higher priority work prevents continuation of this work. In the event of tie-in work, completion of the capital and other special engineering projects will be done as priorities dictate.

Capital shutdown work This type of work is generated by engineering and requires that all or portions of the job are done during a shutdown period. As this work is received, it is entered in the shutdown log the same as routine expense work request.

Defining scope of work

Most outages are comprised of multiple wish lists that are generated by operations, plant engineering and maintenance. In most cases, these lists contain poorly defined tasks with little or no attempt to prioritize them. These lists must be consolidated, prioritized and a universally acceptable scope of work defined. The steps required to create this scope of work include:

Shutdown work request log When the planning department receives a shutdown work order, it is immediately recorded in the shutdown work request log. The maintenance clerk should verify that the item number assigned agrees with the next sequential number on the maintenance work order log. Work requests forwarded to shutdown planning should indicate equipment number and

Table 51.1 Example control point identification chart

Control element	What is likely to go wrong?	How and when will I know?	What will I do about it?
Quality	Workmanship or craftsman might be less than desired.	Upon personal inspection of each stage of project.	Have sub-standard work re-done.
Cost	Cost of any task of project may exceed budget.	When purchase agreements are made.	First, seek alternative suppliers, then, consider alternative materials.
Timeliness	Time to complete any tasks of project may exceed schedule.	By closely monitoring actual progress against schedule along critical path.	Look for ways to improve efficiency, attempt to capture time from later steps, authorize overtime if budget permits

service as directed by the maintenance procedures. Also, the assigned planner is responsible for recording in the work order log the date that repetitive work orders were completed so that a history of repair might be established for the particular pieces of equipment.

Request all work requests The outcome of the initial outage management team meeting will generate a complete list of all approved maintenance tasks that are to be included in the outage plan. To complete the process, the planner should request any other tasks that may be included. This request should:

- Occur well in advance.
- Be directed to operations, maintenance, and engineering.
- Include an add-on policy and cut-off date.
- Specify routing.
- Demand a realistic priority.
- The purpose of work orders is:
  - To establish a backlog and aid in defining the scope and duration, essential to resource allocation, prioritization and establishment of the rules.
  - To add lead-time for planning and material procurement.
  - To allow time for mobilization of contractor personnel and equipment rental.
- A lack of importance and management support placed on this part of the decision cycle will result in less than desirable results including:
  - Improper resource levels.
  - An unacceptable level of add-on jobs.
  - Possible increased overtime, wasted time, and increased labor cost.
  - Conflicting schedules with equipment such as lifting devices, safety apparatus, and mobile equipment.
  - Lengthened duration of the shutdown.
  - Loss of market share due to production losses.
  - Loss of job stability.

51.7 The outage plan

The fundamental requirement of an efffective outage plan is well-defined work orders for each task that must be performed during the available time interval. Too many plants fail to fully plan each of the tasks that are to be performed. Instead, they rely on a master schedule that defines the sequence of events that must occur in order to complete all of the tasks included in the scope of work.

While this detailed schedule or timeline is essential, it does not replace or eliminate the need for complete planning of each task or a comprehensive work order package.

51.7.1 Developing a work plan

The assigned planner develops the work plan. He segments the job into manageable activities, and identifies job requirements, which includes materials and parts purchases, tools and equipment requirements, and support required. Materials ordered and expedited by the planner will include major parts, materials, and tools as required.

Upon completion of the work plan, planning will order necessary stock and non-stock materials for delivery to the appropriate staging area. They then will enter the work order plan into the system.

Shutdown work schedule

By the use of the critical path scheduling method, communications with operations and other affected functional groups, the daily work schedule is developed. A shutdown list indicating scheduled start and completion dates is then typed indicating assignments by supervisor. Each work order will be scheduled by central planning.

Initial work order development

When work orders are written for a shutdown, several situations need to be carefully analyzed. These include:

- Does the work really require a shutdown?
- Is engineering assistance required?
- Is all necessary information for engineering and/or maintenance included with the work order?
- Is there an existing work order for the work being requested?
- Taking a few extra minutes to research these questions will minimize wasted time for the originator, planner, engineer, and all involved in shutdown meetings. Considering all of the personnel involved in handling a work order chances for misunderstanding are greater. They could and usually will include:

1. Originator	7. Planner
2. Approval authority	8. Chief planner
3. Mail clerk	9. Production supervisor
4. Engineer	10. Maintenance supervisor
5. Planning clerk	11. Contractors
6. Data entry clerk	

Not only will these people have to devote time to individual planning, but also collectively in shutdown meetings.

Material requirements

To ensure materials are available several systems can be used, assuming that all materials are planned and requirements are known in advance by stock number or requisition number. Some method of identifying these needs to a work order is required. The best method is through a computer, but manual systems can be used even though they require extra paperwork and manpower effort. The following systems and considerations are to be taken into account:

- Bagging system – by area and work order
- Tag system – by work order
- Delivery system and staging areas
- Stock material allocation

- Non-stock material handling
- Non-stock items purchased by engineering
- Equipment rental
- Contracts
- Use of receiving reports
- Expediting.

Manpower requirements

Definitive effort-hour estimating data is used to develop specific detail estimates. Composite effort-hour data is presented as a guide for quick estimating methods. All material quantities have been consolidated into an average for a given amount of material and do not include any specialty items. All effort-hour units and material quantities are based on a typical installation, and each material type should be reviewed carefully for various differences.

In order to analyze job requirements, duration, and cost, labor estimates must be performed. Without a labor estimate by skill and in sequence, it is difficult, if not impossible, to develop a realistic work schedule.

The first step in making labor estimates is to break the job down into activities in a logical sequence. Example: work order #123 Work description – Repair ball valve, Leaking.

Based upon information from the field tour and knowledge of facility piping specifications, the planner will have most of the data to plan this job. He should already know the pipe size, insulation and material requirements (seats, bonnet, gasket, stem seals). However, he does not know the extent of damage, if any, to the ball.

Assuming the piping is insulated, the steps would be broken down as follows:

Step	Description
1.	Remove insulation
2.	Remove/rebuild/replace valve
3.	Replace insulation

The breakdown of the job into steps is based upon personal experience of the planner or from history or standard job plans. Depending upon the job and skill level of the craftsman, it may require one activity or one hundred. Highly skilled mechanics may require a higher degree of instruction in the plan or how to complete the job, as well as closer supervision and support during job execution.

The second step is to estimate the effort-hours by skill needed to perform each activity. Determination of which skill to perform the work will depend upon available resources. During an outage in a chemical plant, insulators were used to re-tray a 300-foot tall dehydration tower because of a lack of millwrights; to perform the work. The insulators were available during the outage because insulation removal could be performed prior to the shutdown, and replacement would be performed after start-up. In this particular case, the insulators were also willing, and there was no demarcation line between trades. The most interesting point was that the tower job was on the critical path, determining duration of the outage.

However, under normal circumstances, the proper skill should be planned for the job. Barriers between skills, contract restrictions, should be considered not only when

planning the skill for the job, but also when breaking the job into steps. Estimating personnel for each step is described by method in the following paragraphs:

Planner experience   Depending upon the background of the planner, experience is usually the best method of estimating a non-repetitive job. Historical actuals are usually better for repetitive jobs. In order to utilize experience, it is desirable that the planner has done the work before or experienced execution of similar work as a craftsperson, supervisor, engineer, or planner.

Planner experience can also include knowledge of how or where to get the best estimate. This is particularly true in multi-discipline jobs. Absence of experience or the inability to draw upon reliable expertise would force a planner to rely on another method of obtaining an estimate.

Engineering standards   Engineering standards can be utilized to estimate jobs when data is available and applies to the work to be performed. Standards provide an average time to complete a job, but special situations or adverse conditions are not considered.

What is an engineering standard?   Webster's New International Dictionary contains the following definition: 'Standard: That which is established by authority, custom, or general consent, as a model or example, criteria, text.' In general, a definitive level, degree, material, character, quality, or the like, viewed as that which is proper and adequate for a given purpose.

To assist in clarification for the reader, the following specific definition is submitted:

'An engineered labor standard represents the amount of time required by an average qualified craftsperson to perform a given task at a pace he can maintain day after day without affecting his health or quality of workmanship. The engineered labor standard includes allowances for personal needs, fatigue, and miscellaneous delays.'

Historical averages   While average times for the same repetitive job can be used for control purposes, the time does not represent an engineered standard.

Guesses   Estimates are guesses, some better than others, and should remain in that category. Various forms of estimating, such as incremental estimating, are attempts to improve the quality of estimates, but do not result in engineered labor standards.

Control of maintenance labor involves the basic fact that an engineered labor standard represents the time it should take to do a job. This amount of time is similar to 1 meter which never changes in length; that is, it is always 39.37 inches, or one ten-millionth part of the distance along a meridian from the equator to the pole. If we accept this logic, then it follows that an engineered standard time for a given task will always be the same. We can relate the actual performance of different craftspeople to this standard time in the same as that we can relate a given distance to a meter or a golfer against par.

Evolution of standards   Labor standards used for maintenance management and control are relatively new in concept. Although labor standards have been used in production with ever-increasing application since late in the nineteenth century, serious application to maintenance probably dates back to someplace between 1940 and 1950. While the tools and techniques are in existence today, the application is still probably minor when compared to the potential. As a matter of fact most maintenance controls today are based on elementary management concepts, such as estimates or historical averages.

Considering the wide scope of work performed by maintenance craftspeople, the development of engineered labor standards represents a major effort, almost prohibitive from a cost standpoint for a single company or a single facility.

Historical data   Historical data is an improvement over the basic estimating approach in that it does represent the average time required for repetitive jobs. The method for development of data is rather simple in that work orders are assembled for repetitive maintenance work and average times are derived for various categories or natural groupings of work. This could be for jobs of short duration, or the technique could be applied to the same type of jobs, such as pump overhauls.

This method is an improvement over estimating and offers the following advantages:

1. A fair degree of consistency is obtained.
2. Administrative costs are relatively low after the initial historical data is developed.
3. Training is relatively simple.
4. Repetitive work is covered.

The disadvantages are:

1. Historical average time values reflect what has happened rather than what should happen.
2. Alternative methods are difficult to compare.
3. There is very little application to new work.
4. Past inadequacies become part of the system.

Industrial engineering techniques   The same industrial engineering techniques used in production are used in maintenance for determining the standard time value. These techniques range from motion time values based on technical analysis of the work performed to time study using stopwatch or motion picture techniques for developing standard time. Since maintenance is highly variable and largely non-repetitive, a great deal of technical study is required before the standard assembled represents sufficient coverage of work so that control can be obtained. This is a complex and difficult task; however, some of the larger companies (some governmental agencies and some consulting firms) have established engineering labor standards for most maintenance work.

It only seems natural that the responsibility for the development of engineering standards would rest with management consulting firms who have access to the work done by various clients and who can pool the data and

expand on it until rather more complete engineering labor standards are developed. In addition, some governmental units, such as the US Navy, have pioneered in application of engineering standards to maintenance work.

Another difficulty involved in the use of maintenance standards is in the rather complex operation of applying these standards. The simplified approach developed by A-Kearney, Inc. has been to segregate the standards into direct work and auxiliary work for ease of application for maintenance planners. Direct work can be defined as the work elements that are easily identified after the work is completed. In an example of pipefitting work, the cutting, threading, and joint installations can be readily identified. These operations would require the same standard time regardless of the location of the operation. Figure 51.8 shows a typical page of standard data for pipe-fitting covering the operation 'Cut pipe with pipe machine'. The normal minutes per cut are identified by pipe size and the various portions of the job that are included in the time per cut are shown on the sheet. These include cut pipe, place in chuck, measure and loosen jaws, and dispose of scrap.

Estimates   Most maintenance operations are managed and controlled using an estimate to determine the probable man-hours required performing work. A simple definition of a labor estimate would be that the estimate represents the probable time required to perform a job based upon the best judgement of the person making the estimate. The personal experience, knowledge, and ability of the estimator is involved in determining the quality of the estimate. The advantages in using this technique are:

1. Estimates are easy to develop.
2. Administrative cost of estimating is low.

Department: Pipe fitting		
Operation No.: 2-1		
Nominal pipe size (inches)	Pipe IID in AM	Normal minutes per cut
1/2	15.9	1.0
1	26.6	1.2
$1\frac{1}{2}$	40.9	1.4
2	52.4	1.6
$2\frac{1}{2}$	62.7	1.8
3	77.9	2.0
4	102.2	2.8
5	141.3	3.8
6	154.0	4.9
8	202.7	6.9

Includes pickup pipe, place in chuck, and tighten chuck jaws. Measure pipe for cut and positions carriage. Hand feed parting tool to cup piece off pipe. Hand feed parting tool to remove burrs. Loosen chuck jaws, remove pipe from chuck, and set aside. Dispose of scrap or set aside unused pieces.

Figure 51.8 Page of standard date

Sample definitive estimate			
Quantity	Material & description	Unit MHs	Total MHs
1. 100 L.F.	4″ Schedule. 40 C/S, B,.W., Seamless	0.57	57.0
2. 5 Ea.	4″ Schedule. 40 C/S, B.W., 45 Degree Ells. L.R.	4.60	23.0
3. 8 Ea.	4″ Tees, Schedule. 40, C/S B.W.	7.10	56.0
4. 12 Ea.	4″ pipe hangers (Hot)	2.4	28.8
Pipe	Total effort-hours		164.8
5. 100 L.Y.	Pipe Insulation, Calcium Silicate, 1″ thick, Alum. fin.	0.31	31.0
6. 10 Ea.	Insulate pipe fitting, 1″ (each fitting = 3.5 L.F.)	0.31	1.9
Pipe insulation	Total man-hours		32.9
Total Definitive man-hour estimate 192. 7			

Sample definitive estimate			
Quantity	Material & description	Unit MHs	Total MHs
100 L.F.	4″ Schedule. 40 C/S, BW., pipe with fittings	1.80	180.0
Pipe	Total effort-hours		180.0
100 L.F.	Insulate 4″ pipe and fit with calcium silicate, V thick, alum fin.		53.0
Pipe insulation	Total effort-hours		53.0
Total conceptual man-hour estimate 233.0			

Figure 51.9 Axiliary work

3. All jobs can be estimated; thus, complete coverage is obtained.

The disadvantages include:

1. Estimates are inconsistent.
2. Estimates vary in accuracy between different estimators.
3. Training of estimators is very difficult.
4. Method comparisons are difficult.
5. Verification and audit of accuracy is impossible.

Since estimating leaves much to be desired, some managers have resorted to analysis of work that has been done and which is repetitive in nature and have evolved historical data which sometimes is misrepresented as a standard.

Example: The auxiliary functions in Figure 51.9 are identified as those work elements performed by the craftsperson which are part of the job, but which are not apparent after the work is completed. These include such functions as job planning by the craftsman, travel time, obtaining materials, and miscellaneous get-ready and clean-up work. The auxiliary time is affected by the physical facilities involved, such as facility layout, location of shops, degree of material delivery, and so forth. If sufficient studies are taken, the auxiliary work can be related to the direct work and ratios determined for each craft by type of work.

Figure 51.9 shows a typical relationship of auxiliary work to direct work for pipe-fitters doing a job that requires them to leave the shop and perform the operation in the field.

Shutdown assignment meeting This meeting is held at least three working days prior to the shutdown with the appropriate maintenance supervisor, engineering, planning supervisor, operations supervisor and planners. The final outage schedule and list of tasks to be preformed are reviewed and work orders and works plans distributed to the assigned superintendent. The maintenance planning supervisor or senior planner should facilitate this meeting.

On emergency shutdowns or any outage with less than five days planning time, planning should prepare a rough estimate (RE) for the job and issue a blank work plan with the appropriate work orders. Planning will not order miscellaneous parts, materials, tools, etc.; this will be the responsibility of the assigned supervisor and area planner.

If a work request is received that requires an outside contractor, the work order should immediately be assigned and issued to the contract maintenance liaison supervisor or other designated representative once the shutdown is approved. Do not hold until the outage assignment meeting.

Following the assignment meeting, each supervisor assigned work in the shutdown should review each work plan and his work schedule and advise planning of any problems. He should also arrange for common parts, tools, and materials that have not been listed. He should work with planning for ordering and expediting materials, tools, etc., found necessary after job assignment. He should also report back to the planner if there is any reason that he couldn't do the jobs according to the established plan or schedule. Other major problems should also be reported.

Add-on work after the assignment meeting should be directed to central planning where crew assignment will be determined. This work will be handled like emergency shutdown work.

51.8 Implementing the plan

The implementation phase is comprised of the actual outage duration. During this period, all of the tasks included in the scope of work are scheduled for completion.

51.8.1 Daily review

The progress of large shutdowns will be evaluated daily, or as deemed necessary by the maintenance-planning supervisor. This meeting will identify shutdown problems, and see that measures are taken to solve these problems. Also, action will be taken to compensate for off-schedule conditions. The maintenance-planning supervisor should facilitate this meeting.

Each maintenance supervisor involved in the shutdown should attend and be ready to quickly report the progress of each job as planned and scheduled for that day. In addition, the add-on list should be quickly reviewed daily. The shutdown critical path schedule will be reviewed and updated to track the progress made to date.

The plan is not complete until all data from the outage are collected, evaluated and the outage plan closed. For non-recurring tasks, no further action is required. For each of the tasks included in the scope of work that will be repeated at some point in the future, the planner must

evaluate, in detail, the actual performance during the outage. These data are used to modify the outage plan so that deficiencies or errors that may have been present in the original plan are corrected. This step is essential to assure that future outages will not incur the same difficulties.

Daily progress meeting

A daily progress meeting should be conducted to report progress and discuss any problems that have a potential impact on the schedule. Supervisors and planners should make a field tour (know first hand what is being worked) before attending the meeting. Items that are typically discussed are as follows:

1. Review status of each schedule for that day.
2. Review problem areas.
3. Update the CPM as to actual work completed.
4. Review non-stocked materials (PO).
5. Issue next shift work order and schedules.
6. Weekly control analysis.

51.8.2 Monitoring the outage schedule

After an outage schedule has been made, the implementation, communication and monitoring of the schedule is most important. Often when a CPM network has been drawn, it is tacked on the wall in the outage office and forgotten. A CPM schedule must be continually monitored if maximum benefit is to be derived.

It is important that all major parties whose work, materials, etc. affect the outage be aware of the schedule. It may not be necessary to forward a copy of the CPM diagram to each party since it increases the possibility that inevitable change in the network will not be incorporated on every schedule.

It is possible to derive a variety of schedules from a CPM network to assist in shutdown control. It may be necessary, for better communications, to develop a bar chart based on the CPM network. Also the bar chart is useful when developing other possible schedules such as:

- Labor schedules developed for separate crews and showing the number of men needed for each time period.
- Equipment schedules developed for separate types and sizes of equipment and showing the number of units required for time period.
- Material and service schedules for important items of material or installed equipment indicating deadlines for such steps as preparation of shop drawings or samples, beginning of fabrication, date of shipment and date needed at the job site.
- Financial schedules indicating on a time scale the cumulative cost to date and cost remaining to complete the outage.

This kind of information generated from the CPM schedule can be most useful in analyzing the consumption of resources and maintaining outage control.

Schedule progress

It is important that careful attention be given to updating the schedule and indicating progress. This responsibility should be assigned to one person, usually the chief planner or supervisor of planning.

The CPM diagram provides a useful tool for documenting change and management reporting. Many changes occur throughout the duration of an outage, and the CPM network should reflect these changes. The network should show any changes in activity duration, starting date, completion dates, or change in logic. The network should reflect completed events and events in progress. This information should be entered on the network daily or as changes occur.

Status reporting

The possible status for a given event and the information for status are as follows:

- Completed
- In progress
- Not started
  - Logically eligible
  - Not yet logically eligible to start
- None
- Estimate of time remaining duration
- Estimated start date
- Revised duration estimate
- Revised duration estimate.

Notice that the required information for events that are in progress is an estimate of the remaining duration, not an estimate of percent completion. There are two reasons for religiously avoiding the popular percent completion terminology.

A percentage is the result of the progress of dividing the numerator of a fraction by the denominator. It is almost never clear what the numerator or the denominator of the completion percentage is. More often than not, the percentage is an indication of the amount of work completed dividing the total amount of work involved in the event. For a number of reasons, however, the amount of time already consumed in the event may be far more or less than the total duration. The rate of progress may not be uniform due to problems that arise during the event, increases in skill, or changes in operating conditions. What is needed for scheduling purposes is a direct estimate of remaining duration regardless of work progress to date.

The other reason for not using percent completion is that it involves no commitment on the part of individuals responsible for the event to finish at any particular time. To report that you are X per cent complete on an event is a subjective evaluation of history, not a statement of future plans and intentions. Thus, we often experience the phenomenon of completion percentages that increase by ever-smaller amounts and never quite reach 100 per cent.

In addition to reporting the status of each event, the remaining portion of the outage network should be reviewed for the following possible changes:

Table 51.2 Example of milestone chart

Milestone	Scheduled Completion	Actual Completion
Foundation completed	8/5/99	8/2/99
Framing completed	8/10/99	8/7/99
Exterior finished	8/20/99	8/22/99
Electrical wiring completed	8/25/99	
Heating and air conditioning installed	8/25/99	
Interior finished	9/22/99	

- Elimination of events which are now known to be unnecessary.
- Addition of events not previously included.
- Alteration of precedence relationships among events.

51.9 Updating the schedule

The process of updating the schedule is basically the same as that used in planning the shutdown. It is simply based on new information and deals only with the remaining portion of the shutdown. In some very large shutdowns, it may be desirable to perform time–cost trade-off analysis, resource leveling, etc., each time the shutdown CPM is updated. In many cases, however, the update will involve only the basic forward, backward and slack computations. These calculations will allow for development of a revised schedule. They will also identify the critical path that is likely to change over the course of the shutdown.

The mechanical aspects of updating the shutdown network can vary considerably depending upon the type of network and equipment used. The update may involve using color-coded symbols to indicate progress on a paper copy of the original network, or, for the sake of clarity, the remaining portion of the network may be completely redrawn. Commercially available planning boards with movable symbols may be employed. If a computer is used, the updated schedule may appear on printouts, and the original network would serve only as a reference.

Project control charts

Another helpful tool is a project control chart, which uses budget and schedule plans in a quick status report of the project. It compares actual to plan, calculates a variance on each task completed, and tallies a cumulative variance for the project.

To prepare a project control chart, refer to the work breakdown structure and list all of the tasks of the project. Then, use the schedule to list the time planned to complete each task, and use the budget to list the expected cost of each task.

As each project task is completed, record the actual time and actual cost. Calculate variances and carry the cumulative total forward. This technique can easily be put into a spreadsheet format on your personal computer. Some large projects may be able to create this format for a report that uses cost and schedule data that is routinely captured by the company's computerized accounting system. Figure 51.10 illustrates a typical project control chart.

Project steps	Cost				Schedule			
	Budget	Actual	Variance	Total	Planned	Actual	Variance	Total

Figure 51.10 Project control chart

Project costing chart

Project: Remodel Building #7 to provide four additional offices by the end of the 3rd quarter at a cost not to exceed $17,500. If you prefer over budget and schedule amounts to be negative numbers, subtract actual from budget and planned. Under budget and schedule amounts will then be positive numbers.

Milestone charts

A milestone chart presents a broad-brush picture of a project's schedule and control dates. It lists those key events that are clearly verifiable by others or that require approval before the project can proceed. If this is done correctly, a project will not have many milestones. Because of this lack of detail, a milestone chart is not very helpful during the planning phase when more information is required. However, it is particularly useful in the implementation phase because it provides a concise summary of the progress of the project.

Budget control charts

Budget control charts are generally of two varieties. One is a listing of the tasks of a project with actual costs compared to budget. They are similar to project control charts, discussed earlier, and can be either hand or computer-generated. The other kind is a graph of budgeted costs compared to actual. Either bar or line graphs may be used. Bar graphs usually relate budgeted and actual costs by project tasks, while line graphs usually relate planned cumulative project costs to actual costs over time.

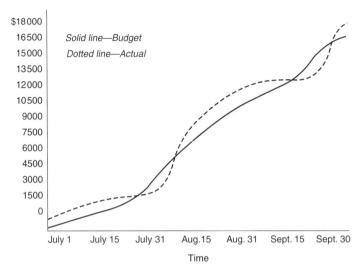

$18000
16500   *Solid line—Budget*
15000   *Dotted line—Actual*
13500
12000
10500
9000
7500
6000
4500
3000
1500
0

July 1   July 15   July 31   Aug.15   Aug. 31   Sept. 15   Sept. 30

Time

Figure 51.11 Example of budget control chart

Another helpful approach to budget control is to compare percentage of budget spent to percentage of project completed. Either listing or graphing can compare these data. While percentage of budget spent is a precise figure, someone familiar with the project and its progress estimates percentage of project completed.

Monitoring performance

The heart of the control process is monitoring work in progress. It is your way of knowing what is going on and how actual compares to plan. With effective monitoring, you will know when corrective action is required. Common ways to keep abreast of project progress are:

Inspection   Inspection is probably the most common way to monitor project performance. It is handled by trained inspectors as well as by the project manager. Get out into the area where the work is performed and observe what is going on. Inspection is an effective way to see whether project specifications are being met, as well as whether there is an unnecessary waste or unsafe work practice. Inspections should be unannounced and on a random schedule. However, they should also be open and direct. Ask questions and listen to explanations.

Interim progress reviews   Reviews are communications between the project manager and those responsible for the various tasks of a project. Progress reviews can be in a group or on an individual basis, and either face-to-face or by telephone. Alternatively, progress reports can be submitted in writing. Progress reviews typically occur on a fixed time schedule – daily or weekly – or keyed to the completion of project tasks. These scheduled reviews are typically augmented by reviews called by either the project manager or the one responsible for the work.

Conducting interim progress reviews   Interim progress reviews typically occur on a fixed time schedule, such as daily or weekly. They may also occur when some problem in performance is observed or at the completion of a significant step toward the accomplishment of the project.

Your role during an interim progress review is to achieve your objectives of knowing the status of operations and influencing the course of future events as necessary. During the discussion, you may have any of the following roles:

Listen as the individual updates you on progress, deviation from plan, problems encountered, and solutions proposed. Listen not only to what is said, but also to how things are said. Is the person excited, frustrated, discouraged? Help clarify what is being said by asking questions, and verify what you think is being said by restating your understanding of both facts and feelings.

In many interim reviews, progress is in line with plans. However, you will occasionally have problems to deal with. When this occurs, you can contribute to their solution by directing the other person toward possible courses of action. Use your knowledge and experience as necessary to move the project forward.

An important role of project managers is to integrate the individual parts of a project into a compatible whole. Is something being neglected? Is there duplication of effort? How can the people available be best deployed?

Perhaps the most important role for the project manager is that of leader. Through a variety of techniques, you must keep the team's effort directed toward the common goal of completing the project according to specifications, on time, and within budget. You must confirm and recognize good performance, correct poor performance, and keep interest and enthusiasm high.

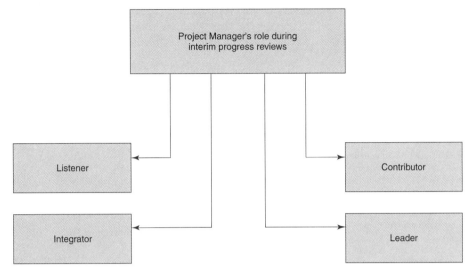

Figure 51.12 Outage manager's role

51.10 Taking corrective action

As a project progresses and you monitor performance, there will be times when actual does not measure up to plan. This calls for corrective action. However, don't be too quick to take action. Some deficiencies turn out to be self-correcting. It is unrealistic to expect steady and consistent progress day after day. Sometimes you'll fall behind and sometimes you'll be ahead, but in a well-planned project, you will probably finish on schedule and within budget.

When quality is not according to specification, the customary action is to do it over according to plan. However, this needs to be more closely examined in some instances. For example, if the work or material exceeds specifications, you may choose to accept it. If it falls short, you need to consider how much it deviates from specifications and whether the deficiency will cause the project to fail its performance evaluation. The final decision may be to have the work redone, but that is not an automatic outcome.

When the project begins to fall behind in schedule, three alternatives may correct the problem. The first is to examine the work remaining to be done and decide whether the lost time can be recovered in the next steps. If this is not feasible, consider offering an incentive for on-time completion of the project. The incentive could be justified if you compare this expenditure to potential losses due to late completion. Finally, consider deploying more resources. This too will cost more, but may offset further losses from delayed completion.

When the project begins to exceed budget, consider the work remaining and whether or not cost overruns can be recouped on work yet to be completed. If this isn't practical, consider narrowing the project scope or obtaining more funding from your client.

51.10.1 What to do when falling behind

Providing feedback

Project managers find many opportunities to provide feedback to those who have a hand in completing the project. Through feedback, individuals learn about the effect their behavior has on others and on the project's success. It serves to maintain good performance and correct poor performance. To be effective, however, feedback must be handled properly. This illustration shows the continuous loop that exists when there is good feedback:

The most important guideline when providing feedback is to deal only with what you can observe. This limits your conversation to actions and results, because you cannot observe someone's intentions.

When offering positive feedback, describe the actions and results in a straightforward way and include an appropriate statement of your reaction. For example, you might tell someone, 'By staying late last night and finishing the work you were doing, the project was able to move forward on schedule. I appreciate your putting out the extra effort.'

Negative feedback can be handled in the same manner, but an important element is missing: how the team member should deal with similar situations in the future. The following sequence should prove more effective:

1. Describe the observed actions and results.
2. Ask the individual if those were his or her intended results.
3. With a typical 'No' response, ask what different actions would likely produce the desired results.
4. Discuss different alternative courses of action.
5. Agree upon a way to handle similar situations if they should occur in the future.

Table 51.3 Project delay: Possible corrective actions

Action	Cost	Schedule
Renegotiate: Discuss with your client the prospect of increasing the budget for the project or extending the deadline for completion.	X	X
Recover during later steps: If you begin to fall behind in early stages of a project, re-examine budgets and schedules for later stages. Perhaps you can save on later stages so the overall budget and/or schedule is met.	X	X
Narrow project scope: Perhaps non-essential elements of the project can be eliminated, thereby reducing costs and/or saving time.	X	X
Deploy more resources: You may need to put more people or machines on the project to meet a critical schedule. Increased costs must be weighed against the importance of the deadline.	X	X
Accept substitution: When something is not available or is more expensive than budgeted, substituting a comparable item may solve your problem.	X	X
Seek alternative sources: When a supplier can't deliver within budget or schedule, look for others who can. You may choose to accept a substitute rather than seek other sources.	X	X
Accept partial delivery: Sometimes a supplier can deliver a partial order to keep your project on schedule and complete the delivery later.	X	X
Offer incentives: Go beyond the scope of the original contract and offer a bonus or other incentive for on-time delivery.	X	X
Demand compliance: Sometimes demanding that people do what they agreed to do gets the desired results. You may have to appeal to higher management for backing and support.	X	X

Check your feedback style

Rate yourself by placing a check mark (☑) in front of each action that is typical of how you handle giving feedback. The ones you don't check represent opportunities for development.

☑ Describe rather than evaluate. By describing observed action and results, the individual is free to use or not use the information. By avoiding evaluation, you reduce the likelihood of a defensive reaction.

☑ Be specific rather than general. Avoid using 'always' and 'never'. Rather, discuss specific times and events. Avoid generalized conclusions such as 'you're too dominating'. Rather, be specific by saying, 'When you don't listen to others, you may miss a valuable idea'.

☑ Deal with behavior that can be changed. Frustration is increased when you remind someone of a shortcoming over which he or she has no control.

☑ Be timely. Generally, feedback is most useful at the earliest opportunity after the behavior.

☑ Communicate clearly. This is particularly important when handling negative feedback. One way to ensure clear communication is to have the receiver rephrase the feedback to see if it corresponds to what you had in mind.

51.10.2 Negotiating for materials, supplies and services

Negotiating is an important process that takes up as much as 20 per cent of a planner's or project manager's time. Negotiating is one way to resolve differences, and it can contribute significantly to the success of your project. The ideas presented here will prepare you to negotiate effectively.

Ten guidelines for effective negotiation

Prepare   Do your homework. Know what outcome you want and why. Find out what outcome the other party wants. Avoid negotiating when you are not prepared – ask for the time you need. As part of your preparation, figure out what you will do if you are unable to come to an agreement. Your power in negotiation develops from attractive alternatives – the greater your ability to walk away, the stronger your bargaining position.

Minimize perceptual differences   The way you see something can be quite different from how the other party sees it. Don't assume you know the other person's view: ask questions to gain understanding, and restate your understanding so it can be confirmed or corrected by the other party.

Listen   Active, attentive listening is mandatory to effective negotiation. Let the other side have an equal share of the airtime. (If you're talking more than 50 per cent of the time, you are not listening enough.) In the process, respect silence. Occasionally people need to collect their thoughts before moving ahead. Don't try to fill this time with talking.

Take notes   You need to know where you are – what has been agreed to, what remains to be resolved. Don't rely on memory. Take notes and then summarize your agreement in a memorandum.

Be creative   Early closure and criticism stifle creative thinking. Be willing to set some time aside to explore different and unusual ways to solve your problem. During this time, do not permit criticism of ideas offered. All negotiations can benefit from nonjudgmental creative thinking.

Help the other party   Good negotiators recognize that the other party's problem is their problem as well. Put yourself in the other's position and work to find a solution that meets everyone's needs. After all, no agreement will hold up unless both parties support it.

Make trade-offs   Avoid giving something for nothing. At least get some goodwill or an obligation for future payback. The basic principle to follow is to trade what is cheap to you but valuable to the other party for what is valuable to you but cheap to the other party.

Be quick to apologize An apology is the quickest, surest way to de-escalate negative feelings. It need not be a personal apology. An apology for the situation you're in can be just as effective. Also, don't contribute to hostility by making hostile remarks. Hostility takes the discussion away from the issues and shifts it to a defense of self where the goal is to destroy the opponent.

Avoid ultimatums An ultimatum requires the other party to either surrender or fight it out. Neither outcome will contribute to future cooperation. Also, avoid boxing someone in. This happens when you offer only two alternatives, neither of which is desirable to the other person.

Set realistic deadlines Many negotiations continue too long because no deadline exists. A deadline requires both sides to be economical in their use of time. It permits you to question the value of certain discussion and encourages both sides to consider concessions and trade-offs in order to meet deadline.

Resolving differences What is best for one department or group won't necessarily be best for others. Out of these differences can come creative solutions when the situation is handled properly. Skill in resolving differences is an important quality of successful project managers.

Consider the model in 5.13. Differences can be resolved my way, your way, or our way. As a result, four strategies emerge. The strategy one chooses to resolve differences

tends to result from interplay of assertiveness and cooperation. This process can be clouded by emotion at times, and when this happens, it is difficult to achieve a satisfactory outcome. Therefore, when you sense that either you or the other person's thinking is clouded by emotion, ask to delay discussion a while.

The following issues influence assertiveness and cooperation:

Assertiveness

- People tend to be more assertive when an issue is important to them.
- People tend to be more assertive when they are confident of their knowledge.
- People tend to be more assertive when things are going against them.
- People tend to be less assertive when they feel they are at a power disadvantage.

Cooperation

- People tend to be more cooperative when they respect the other person.
- People tend to be more cooperative when they value the relationship.
- People tend to be more cooperative when they are dependent on the other person to help carry out the decision.

Given the interplay of assertiveness and cooperation, the following strategies are common for resolving differences:

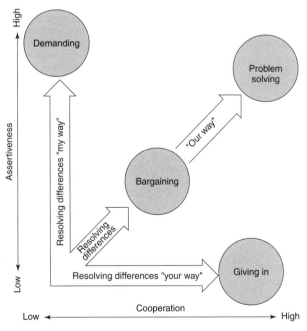

Figure 51.13 Model for resolving differences

Demanding Demanding is high in assertiveness and low in cooperation. It suggests confidence and that the issue is important coupled with a lack of concern for the relationship and no dependency on the other person.

Problem solving Problem solving is high in assertiveness, coupled with high cooperation. It suggests that the issue is important, and that there is the need for an ongoing relationship with the other person.

Bargaining Bargaining is moderate in both assertiveness and cooperation. It suggests that equally powerful parties are addressing an important issue. Each must be willing to give a little to reach agreement. Bargaining is also an appropriate backup strategy when joint problem solving seems unattainable.

Giving in Giving in is low in assertiveness and high in cooperation. The issue may be unimportant to you, you may lack knowledge, or you simply want to go along with the other person's proposal in order to build up the relationship between you.

Each strategy has its place. However, too few people recognize the conditions that support each strategy. Many people adopt one approach for resolving differences and use it in all situations. Obviously, it will be ineffective in many cases. Learn to distinguish among the various types of situations and adopt an approach that has the greatest chance of success in the long run. Don't overlook the importance of maintaining cooperative relationships.

Common sources of differences

Conflicts will always occur during the four phases of the outage. These differences must be quickly and effectively resolved to assure success. The more common sources of conflicts are:

Allocation of human resources With limited personnel, project managers often have different views than others have on how staff will be assigned.

Use of equipment and facilities Project managers often differ with others over the use of equipment and facilities that must be shared.

Costs As you work at controlling costs against the approved project budget, you will often encounter conflict with suppliers who feel a need to increase costs over their original commitment.

Technical opinions Frequently there will be different opinions on how something ought to be done.

Administrative procedures Administrative procedures often become points of difference, especially when not followed.

Responsibilities There will be occasions when more than one person claims an area of responsibility and other occasions when no one wants to accept responsibility.

Scheduling Differences will develop around schedules and deadlines. Others you are depending on will not deliver on their commitments.

Priorities There will be differences about which of several things is more important and therefore should be handled first.

51.11 Completing the project

The goal of project management is to obtain client acceptance of the project result. This means that the client agrees that the quality specifications of the project parameters have been met. In order to have this go smoothly, the client and project manager must have well-documented criteria of performance in place from the beginning of the project. This is not to say that nothing can change, but when changes are made, the contract must be amended to list the changes in specifications along with any resulting changes in schedule and budget.

Objective, measurable criteria are always best, while subjective criteria are risky and subject to interpretation. There should be no room for doubt or ambiguity, although this is often difficult to achieve. It is also important to be clear about what the project output is expected to accomplish. For instance, these three outcomes may produce entirely different results; the project/product performs the specified functions; it was built according to approved design; or it solves the client's problem.

The project may or may not be complete when results are delivered to the client. Often there are documentation requirements such as operation manuals, complete

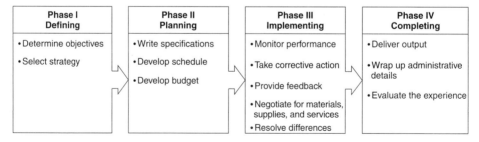

Phase I Defining	Phase II Planning	Phase III Implementing	Phase IV Completing
• Determine objectives • Select strategy	• Write specifications • Develop schedule • Develop budget	• Monitor performance • Take corrective action • Provide feedback • Negotiate for materials, supplies, and services • Resolve differences	• Deliver output • Wrap up administrative details • Evaluate the experience

Figure 51.14 Four phases of project management

drawings, and a final report, which usually follow delivery. There may also be people to be trained to operate the new facility or product, and a final audit is common.

Finally, project team members need to be reassigned; surplus equipment, materials, and supplies disposed of; and facilities released.

The final step of any project should be an evaluation review. This is a look back over the project to see what was learned that would contribute to the success of future projects. This review is best done by the core project team and typically in a group discussion.

Project completion checklist

☑ 1. Test project output to see that it works.
☑ 2. Write operations manual.
☑ 3. Complete final drawings.
☑ 4. Deliver project output to client.
☑ 5. Train client's personnel to operate project output.
☑ 6. Reassign project personnel.
☑ 7. Dispose of surplus equipment, materials, and supplies.
☑ 8. Release facilities.
☑ 9. Summarize major problems encountered and their solution.
☑ 10. Document technological advances made.
☑ 11. Summarize recommendations for future research and development.
☑ 12. Summarize lessons learned in dealing with interfaces.
☑ 13. Write performance evaluation reports on all project staff.
☑ 14. Provide feedback on performance to all project staff.
☑ 15. Complete final audit.
☑ 16. Write final report.
☑ 17. Conduct project review with upper management.
☑ 18. Declare the project complete.

Project evaluation form

1. How close to scheduled completion was the project actually completed?
2. What did we learn about scheduling that will help us on our next project?
3. How close to budget was final project cost?
4. What did we learn about budgeting that will help us on our next project?
5. Upon completion, did the project output meet client specifications without additional work?
6. If additional work was required, please describe:
7. What did we learn about writing specifications that will help us on our next project?
8. What did we learn about staffing that will help us on our next project?
9. What did we learn about monitoring performance that will help us on our next project?
10. What did we learn about taking corrective action that will help us on our next project?
11. What technological advances were made on this project?

12. What tools and techniques were developed that will be useful on our next project?
13. What recommendations do we have for future research and development?
14. What lessons did we learn from our dealings with service organizations and outside vendors?
15. If we had the opportunity to do the project over, what would we do differently?

☐ 1. Define the project.
☐ 2. Select a strategy.
☐ 3. Develop specifications.
☐ 4. Develop a schedule.
☐ 5. Develop a budget.
☐ 6. Organize the project team.
☐ 7. Assign duties and responsibilities.
☐ 8. Train new team members.
☐ 9. Monitor progress.
☐ 10. Take corrective action.
☐ 11. Provide feedback.
☐ 12. Test final outcome.
☐ 13. Deliver outcome to client.
☐ 14. Write operations manual.
☐ 15. Train client personnel.
☐ 16. Reassign project staff.
☐ 17. Dispose of surplus equipment, materials, and supplies.
☐ 18. Release facilities.
☐ 19. Evaluate project performance.
☐ 20. Complete final audit.
☐ 21. Complete project report.
☐ 22. Review project with management.

Figure 51.15 Project manager's check list

51.12 A model for successful project management

Projects are temporary undertakings that have a definite beginning and end. This quality distinguishes them from the ongoing work of an organization. There are four phases in any successful project: defining, planning, implementing, and completing. Figure 51.14 summarizes these phases.

It is imperative to the success of a project that it be clearly defined before it is undertaken. Any definition should include the criteria for determining successful completion of the project. It is reasonable to expect changes to occur once the project is under way, but these changes should be documented along with any resulting impact on schedule and budget.

A successful project produces an outcome that performs as expected, by deadline, and within cost limits. Thus, the three parameters by which a project is planned and controlled are established. Quality is defined by specifications, time is defined by schedule, and costs are defined by a budget.

To carry out the work of the project, a temporary team is usually assembled. This necessitates developing an organization, assigning duties and responsibilities, and training people in their duties. Frequently, policies and procedures are required to clarify how the team is to function during the project.

When work on the project begins, the project manager has many responsibilities. The work of different individuals and groups must be coordinated so that things run smoothly, and the progress of the project must be monitored and measured against plans. When deviations occur, corrective action must be taken. Also, project managers are expected to provide feedback to team members; negotiate for materials, supplies, and services; and help resolve differences that occur.

The goal of the project is to deliver an outcome to the client. When that day finally arrives, there are still things to be done before the project is complete. This includes writing operations manuals; training client personnel on the use of the project output; reassigning project personnel; disposing of surplus equipment materials, and supplies; evaluating the experience; completing a final audit; writing a project report; and conducting a project review with upper management.

Not every project requires the same attention to each of these activities. It will depend upon the type of project you are undertaking, its size and scope, and the type of organization you are affiliated with. Use your own judgment in selecting the steps important to the success of your project.

Best of luck in the projects you undertake. Success can be yours if you use the concepts presented here.

52

Lubrication

Stuart McGrory

BP Oil UK Ltd

Contents

52.1 Introduction

This chapter examines the need for lubrication and the types of lubricant available. Various applications are considered, including engines gears, hydraulic equipment, machine tools, metal cutting and working fluids, compressors, turbines and electrical oils. The care of lubricants on-site, application of planned lubrication and inclusion within overall maintenance management are examined.

The plant engineer's objective must be to ensure that plant operates at a profit. If overall efficiency of operation is to be achieved, and the working costs of plant kept within acceptable bounds, time must be set aside for the control and application of lubrication. The evolution of lubricants and their application has continued ever since the beginning of the Industrial Revolution, and as the pattern of industry becomes increasingly more complex, the standard of performance of lubricants becomes progressively more important.

52.2 Lubrication — the added value

All machines depend for their accuracy on the strength of their component parts, their bearings and on the type and efficiency of their lubrication systems. Many machine bearings are subjected to extremely heavy shock loads, or intermittent loads, or exposure to unfavorable environmental conditions, yet in spite of this they must always maintain their setting accuracy.

Accuracy and reproducibility are of vital importance to industry. Quite apart from the effect of these factors on the final product, several plant items are frequently links in a continuous chain of production processes. A sizing error in one machine, for example, could overstress and damage the succeeding machinery. Similarly, an error in a press may increase stress on the tool and could necessitate an additional operation to remove excessive 'flash'. Wear in a material preparation unit could allow oversize material to be passed to a molding machine, creating an overload situation with consequent damage.

The reduction of friction is only one of the functions of a lubricant. It must remove heat (often in large amounts), protect bearings from damage and preserve the working accuracy and alignment of the structure. It must also protect bearings, gears and other parts against corrosion, and must it be non-corrosive. Sometimes it may be required to seal shafts and bearings against moisture and the ingress of contaminating particles. The lubricant must be of the correct viscosity for its application and may need additives to meet specification requirements. It must also be non-toxic, and both chemically and physically stable.

Lubrication plays a vital role in the operation of industrial plant. For example, in a heavy rolling mill the lubrication system, though mostly out of sight in the oil cellar, may have a capacity of many thousands of gallons and exceed in bulk the mill itself. Lubrication systems of this size and complexity are usually fully automatic, with many interlocks and other safety features. Even with the smallest machines, automatic lubrication is becoming more popular. Where an automatic system would be impracticable or

uneconomical, it is nevertheless important that lubrication be carried out in accordance with a planned schedule.

It can be seen therefore that heavy demands are made on the plant concerned, and hence on the lubricants required for its efficient operation. The importance of the correct selection and application of lubricants, in the correct amount and at the right time, will be readily appreciated. The cost of providing high-quality lubrication is negligible compared with the material return it will bring in terms of longer working life, higher output of work and reduced maintenance costs.

52.3 Why a lubricant?

When the surfaces of two solid bodies are in contact a certain amount of force must be applied to one of them if relative motion is to occur. Taking a simple example, if a dry steel block is resting on a dry steel surface, relative sliding motion will not start until a force approximately equal to one fifth the weight of the steel block is applied. In general, the static friction between any two surfaces of similar materials is of this magnitude, and is expressed as a coefficient of friction of 0.2. As soon as the initial resistance is overcome, a very much smaller force will keep the slider moving at uniform velocity. This second frictional condition is called dynamic friction. In every bearing or sliding surface, in every type of machine, these two coefficients are of vital importance. Static friction sets the force required to start the machine and dynamic friction absorbs power that must be paid for in terms of fuel consumed. Also, friction resistance of non-lubricated surfaces causes heating, rapid wear and even, under severe conditions, actual welding together of the two surfaces.

Lubrication, in the generally accepted sense of the word, means keeping moving surfaces completely separated by means of a layer of some liquid. When this is satisfactorily achieved, the frictional resistance no longer depends on the solid surfaces but solely on the internal friction of the liquid, which, in turn, is directly related to its viscosity. The more viscous the fluid, the greater the resistance, but this is never comparable with that existing between non-lubricated surfaces.

52.3.1 Types of lubrication

Lubrication exists in one of three conditions:

1. Boundary lubrication
2. Elastohydrodynamic lubrication
3. Full fluid-film lubrication

Boundary lubrication

Boundary lubrication is perhaps best defined as the lubrication of surfaces by fluid films so thin that the friction coefficient is affected by both the type of lubricant and the nature of the surface, and is largely independent of viscosity. A fluid lubricant introduced between two surfaces may spread to a microscopically thin film that reduces the sliding friction between the surfaces. The peaks of the high spots may touch, but interlocking occurs only to a limited extent and frictional resistance will be relatively low.

A variety of chemical additives can be incorporated in lubricating oils to improve their properties under boundary lubrication conditions. Some of these additives react with the surfaces to produce an extremely thin layer of solid lubricant, which helps to separate the surfaces and prevent seizure. Others improve the resistance of the oil film to the effect of pressure.

Elastohydrodynamic lubrication

This type of lubrication provides the answer to why many mechanisms operate under conditions that are beyond the limits forecast by theory. It was previously thought that increasing pressure reduced oil film thickness until the aspirates broke through, causing metal-to-metal contact. Research has shown, however, that the effect on mineral oil of high contact pressure is a large increase in the viscosity of the lubricant. This viscosity increase combined with the elasticity of the metal causes the oil film to act like a thin solid film, thus preventing metal-to-metal contact.

Full fluid-film lubrication

This type can be illustrated by reference to the conditions existing in a properly designed plain bearing. If the two bearing surfaces can be separated completely by a fluid film, frictional wear of the surface is virtually eliminated. Resistance to motion will be reduced to a level governed largely by the viscosity of the lubricating fluid.

To generate a lubricating film within a bearing, the opposed surfaces must be forced apart by pressure generated within the fluid film. One way is to introduce the fluid under sufficient pressure at the point of maximum loading, but this hydrostatic method, although equally effective at all speeds, needs considerable power and is consequently to be avoided whenever a satisfactory alternative exists.

Above a certain critical speed, which depends mainly on the size and loading of the bearing and the viscosity of the lubricant, hydrodynamic forces are set up which part the surfaces and permit full fluid-film lubrication. At rest, the fluid film has been squeezed from beneath the shaft, leaving only an absorbed film on the contacting surfaces. As the shaft starts to revolve, friction between the journal and the bearing bore causes the shaft to climb up the inside of the bearing until torque, together with the increased thickness of lubricant film, overcomes frictional resistance and the shaft starts to slip at the point of contact. The rotating shaft then takes up its equilibrium position, where it is supported on a fluid film drawn beneath it by viscous friction (see Figure 52.1).

52.4 Physical characteristics of oils and greases

Reference will be made to the physical characteristics of lubricants as they affect their selection for various applications. These terms are well known to the lubricant supplier but are not always fully understood by the user. Brief descriptions of these characteristics are therefore given so that their significance may be appreciated.

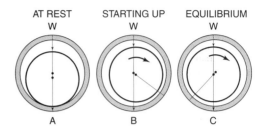

Figure 52.1 Journal positions during start-up while a hydrodynamic oil 'wedge' is being established

52.4.1 Viscosity

This is the most important physical property of lubricating oil; it is a measure of its internal friction or resistance to flow. In simple terms, it provides a measure of the thickness of lubricating oil at a given temperature; the higher the viscosity, the thicker the oil. Accurate determination of viscosity involves measuring the rate of flow in capillary tubes, the unit of measurement being the centistoke (cSt). As oils become thinner on heating and thicker on cooling a viscosity figure must always be accompanied by the temperature at which it was determined.

The number of commercial viscosity systems can be confusing, and as kinematic viscometers are much more sensitive and consistent, there is a growing tendency to quote kinematic viscosities. The International Standards Organization (ISO) uses kinematic viscosity in its viscosity grade classification (Table 52.1). Most oil companies in their industrial lubricant nomenclature use these ISO grade numbers. This provides the user with a simple verification of conformity regarding viscosity between plant manufacturer and oil supplier recommendations and in the monitoring of correct oil usage on his plant.

Table 52.1 ISO viscosity grade chart

ISO viscosity grade	Mid-point kinematic viscosity	Kinematic viscosity limits cSt at $40°$ $C(104°$ $F)$	
		min.	max.
2	2.2	1.98	2.42
3	3.2	2.88	3.52
5	4.6	4.14	5.06
7	6.8	6.12	7.48
10	10	9.00	11.0
15	15	13.5	16.5
22	22	19.8	24.2
32	32	28.8	35.2
46	46	41.4	50.6
68	68	61.2	74.8
100	100	90.0	110
150	150	135	165
220	220	198	242
320	320	288	352
460	460	414	506
680	680	612	748
1000	1000	900	1100
1500	1500	1350	1650

52.4.2 Viscosity index (VI)

This is a way of expressing the rate of change of viscosity with temperature. All oils become less viscous as the temperature increases. The rate of change of viscosity varies with different oils and is mainly dependent on the type of crude from which the oil is derived and the refining method. The higher the VI figure, the lower is the variation in viscosity relative to temperature. The VI of oil is an important property in applications where the operating temperature is subject to considerable change.

52.4.3 Pour point

This is a rough measure of a limiting viscosity. At temperature above 2.5°C (6.5°F), oil ceases to flow when the vessel in which it has been cooled is held horizontally for 5 seconds. The pour point is a guide to behavior and care should always be taken that the operating temperatures are above the figure specified by the oil manufacturer as the pour point of a given oil.

52.4.4 Flash point

The flash point of an oil is the temperature at which it gives off, under specified conditions, sufficient vapor to form a flammable mixture with air. This is very different from the temperature of spontaneous combustion. The test is an empirical one and the result depends upon the instrument used and the prescribed conditions. For example, the flash point may be 'closed' or 'open', depending on whether the test apparatus has a lid or not. As far as lubricating oils are concerned, the test is of limited significance, although it can be indicative of contamination (for example, the dilution of crankcase oil by fuel).

52.4.5 Penetration of grease

The most important physical property of lubricating grease is its consistency, which is analogous to the viscosity of a liquid. This is determined by an indentation test in which a weighted metal cone is allowed to sink into the grease for a specified time. The depth to which the cone penetrates, in tenths of a millimeter, is a measure of the consistency. There is a widely accepted scale, that of the American National Lubricating Grease Institute (NLGI), that relates penetration to a consistency number (Figure 52.2).

The penetration test is used mainly to control manufacture and to classify greases and is, within limits, a guide to selection. Penetrations are often qualified by the terms 'worked' and 'un-worked'. As greases are thixotropic, that is, they soften because of shear but harden again after shearing has stopped, the worked penetration for particular grease may be appreciably greater than the un-worked penetration. The difference between these two figures may be a useful guide to the selection of greases for operating conditions that involve much churning – as small a difference as possible being desirable (see Table 52.2).

Figure 52.2 Grease penetrometer

Table 52.2 NLGI consistency classification for greases

NLGI number	ASTM worked penetration at 77° F
000	455–475
00	400–435
0	355–385
1	310–340
2	265–295
3	220–250
4	175–205
5	130–160
6	85–115

52.4.6 Drop point of grease

The drop point of grease is an indication of change from a soft solid to a viscous fluid; its value depends completely on the conditions of test, particularly the rate of heating. The grease sample, which is held in a small metal cup with an orifice, is heated at a predetermined rate. The drop point is the temperature at which a drop of the sample falls from the cup.

The drop point is of limited significance as far as the user is concerned, for it gives no indication of the condition of the grease at lower temperatures, or of change in consistency or structure with heat. It is a very rough indication of grease's resistance to heat and a guide to manufacture. The difference between the highest temperature at which grease can be used and the drop point varies very much between types. It is at its maximum with some soda greases and much smaller with multi-purpose lithium products and modern complex greases.

52.5 Additives

Much highly stressed modern machinery runs under conditions in which a straight mineral oil is not adequate.

Even the highest quality mineral oil can be unsatisfactory in response of its resistance to oxidation and its behavior under pure boundary conditions, but it is possible to improve these characteristics by the addition of relatively small amounts of complex chemicals. This use of additives resembles in many ways the modification of the properties of steel by the addition of small amounts of other chemicals. It will be of value to have some knowledge of the effect of each type of additive.

52.5.1 Anti-oxidants

When mixed with oxygen, lubricating oil undergoes chemical degradation resulting in the formation of acidic products and sludge. This reaction, which is affected by temperature, the presence of catalysts such as copper and the composition of the oil, can be delayed by the inclusion of suitable additives.

Anti-oxidants are the most extensively used additives and will be found in oils and greases which are expected to operate for considerable periods or under conditions that would promote oxidation. Typical examples are crankcase oils and bearing greases.

52.5.2 Anti-foam

The entrainment of air in lubricating oil can be brought about by operating conditions (for example, churning) and by bad design such as a return pipe that is not submerged. The air bubbles naturally rise to the surface, and if they do not burst quickly, a blanket of foam will form on the oil surface. Further air escape in thus prevented and the oil becomes aerated. Oil in this condition can have an adverse affect on the system that, in extreme cases, could lead to machine failure. The function of an anti-foam additive is to assist in the burst of air bubbles when they reach the surface of the oil.

52.5.3 Anti-corrosion

The products of oil oxidation will attack metals, and this can be prevented by keeping the system free from pro-oxidative impurities and by the use of anti-oxidants. These additives will not, however, prevent rusting of ferrous surfaces when air and water are present in the mineral oil. The presence of absorbed air and moisture is inevitable in lubricating systems and therefore the oil must be inhibited against rusting. These additives, which are homogeneously mixed with the oil, have an affinity for metal, and a strongly absorbed oil film is formed on the metal surface, which prevents the access of air and moisture.

52.5.4 Anti-wear

The increasing demands being made on equipment by the requirement for increased output from smaller units create problems of lubrication, even in systems where full-fluid film conditions generally exist. For instance, at start-up, after a period of rest, boundary lubrication conditions can exist and the mechanical wear that takes place could lead to equipment failure. Anti-wear additives, by their polar nature, help the oil to form a strongly absorbed layer on the metal surface, which resists displacement under pressure, thereby reducing friction under boundary conditions.

52.5.5 Extreme pressure

Where high loading and severe sliding speeds exist between two metal surfaces, any oil film present is likely to be squeezed out. Under these conditions, very high instantaneous pressures and temperatures are generated. Without the presence of extreme pressure additives, the asperities would be welded together and then torn apart. Extreme pressure additives react at these high temperatures with the metal or other oil components to form compounds, which are more easily deformed and sheared than the metal itself, and so prevent welding. Oils containing extreme pressure additives are generally used in heavily loaded gearboxes, which may also be subjected to shock loading.

52.5.6 Detergent/dispersant

The products of combustion formed in internal combustion engines, combined with water and unburned fuel, will form undesirable sludge which can be deposited in the engine and so reduce its operation life and efficiency. Detergent/dispersant additives prevent the agglomeration of these products and their deposition in oil ways by keeping the finely divided particles in suspension in the oil. They are used in engine-lubricating oils where, when combined with anti-oxidants, they prevent piston ring sticking. They are essential for high-speed diesels, and desirable for gasoline engines.

52.5.7 Viscosity index improvers

When mineral oils are used over an extended temperature range it is frequently found that the natural viscosity/temperature relationship results in excessive thinning out in the higher-temperature region if the desired fluidity is to be maintained at the lower region. The addition of certain polymers will, within limits, correct this situation. They are of particular value in the preparation of lubricating oils for systems sensitive to changes in viscosity such as hydraulic controls. They are also used in multigrade engine oils.

52.6 Lubricating-oil applications

There is a constant effort by both the supplier and consumer of lubricants to reduce the number of grades in use. The various lubricant requirements of plant not only limit the extent of this rationalization but also create the continuing need for a large number of grades with different characteristics.

It is not possible to make lubricants directly from crude oil that will meet all these demands. Instead, the refinery produces a few basic oils and these are then blended in varying proportions, together with additives when necessary, to produce oil with the particular characteristics required. In some instances, the continued increase in plant performance is creating demands on the lubricant which are at the limit of the inherent physical characteristics of mineral oil. Where the operational benefit justifies the cost, the use of synthetic base stocks is being developed.

Where these are considered for existing plant, seal and paint compatibility needs to be reviewed before such products are introduced. The problems which face the lubricant supplier can best be illustrated by looking at the requirements of certain important applications.

52.7 General machinery oils

These are lubricants for the bearings of most plant, where circulating systems are not involved. These are hand, ring, bottle or bath lubricated bearings of a very wide range of equipment; line shafting, electric motors, many gear sets and general oil-can duties. The viscosity of these oils will vary to suit the variations in speed, load and temperature.

While extreme or arduous usage conditions are not met within this category, the straight mineral oils which are prescribed must possess certain properties. The viscosity level should be chosen to provide an adequate lubricant film without undue fluid friction, though this may also be influenced by the method of application. For instance, a slightly higher viscosity might be advisable if intermittent hand oiling has to be relied upon. Although anti-oxidants are not generally required, such oils must have a reasonable degree of chemical stability (Figure 52.3).

52.8 Engine lubricants

The type of power or fuel supply available will influence the decision on prime mover to be used. This is often electric power, but many items of plant such as compressors, generators or works locomotives, will be powered by diesel engines, as will most of the heavy goods vehicles used in and outside the works.

The oils for these engines have several functions to perform while in use. They must provide a lubricant film between moving parts to reduce friction and wear, hold products of combustion in suspension prevent the formation of sludge and assist in cooling the engine. Unless the lubricant chosen fulfils these conditions successfully,

deposits and sludge will form with a consequent undesirable increase in wear rate and decrease in engine life.

52.8.1 Frictional wear

If the effects of friction are to be minimized, a lubricant film must be maintained continuously between the moving surfaces. Two types of motion are encountered in engines, rotary and linear. A full fluid-film between moving parts is the ideal form of lubrication, but in practice, even with rotary motion, this is not always achievable. At low engine speeds, for instance, bearing lubrication can be under boundary conditions.

The linear sliding motion between pistons, piston rings and cylinder walls creates lubrication problems that are some of the most difficult to overcome in an engine. The ring is exerting a force against the cylinder wall while at the same time the ring and piston are moving in the cylinder with a sliding action. Also, the direction of piston movement is reversed on each stroke. To maintain full fluid oil film on the cylinder walls under these conditions is difficult and boundary lubrication can exist. Frictional wear will occur if a lubricant film is either absent or unable to withstand the pressures being exerted. The lubricant will then be contaminated with metal wear particles, which will cause wear in other engine parts as they are carried round by the lubricant.

52.8.2 Chemical wear

Another major cause of wear is the chemical action associated with the inevitable acidic products of fuel combustion. This chemical wear of cylinder bores can be prevented by having an oil film which is strongly adherent to the metal surfaces involved, and which will rapidly heal when a tiny rupture occurs. This is achieved by the use of a chemical additive known as a corrosion inhibitor.

52.8.3 Products of combustion and fuel dilution

As it is not possible to maintain perfect combustion conditions at all times, contamination of the oil by the products of combustion is inevitable. These contaminants can be either solid or liquid.

When an engine idles or runs with an over-rich mixture the combustion process is imperfect and soot will

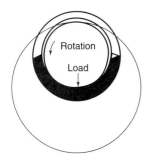
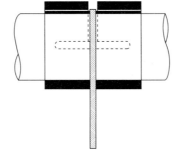
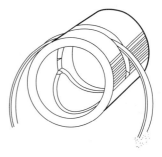

Figure 52.3 Ring oiled bearings

be formed. A quantity of this soot will pass harmlessly out with the exhaust but some will contaminate the oil film on the pistons and cylinders and drain down into the crankcase. If there is any water present, these solids will emulsify to form sludge which could then block the oil ways. Filters are incorporated into the oil-circulation system to remove the solid contaminants together with any atmospheric dust, which bypasses the air filters.

One of the liquid contaminants is water, the presence of which is brought about by the fact that when fuel is burnt it produces approximately its own weight in water. When the engine is warm, this water is converted into steam, which passes harmlessly out of the exhaust. However, with cold running or start-up conditions this water is not converted and drains into the sump. Having dissolved some of the combustion gases, it will be acidic in nature and will form sludge.

Another liquid contaminant is unburned fuel. A poor-quality fuel, for example, may contain high boiling point constituents that will not all burn off in the combustion process and will drain into the sump. The practice of adding kerosene to fuel to facilitate easy starting in very cold weather will eventually cause severe dilution of the lubricating oil. Excessive use of over-rich mixture in cold weather will mean that all the fuel is not burnt because of the lack of oxygen and again, some remains to drain into the sump.

Poor vaporization of the fuel will also produce oil dilution. Generally, this fuel will be driven off when the engine becomes warm and is running at optimum conditions. However, severe dilution of the oil by fuel could have serious results, as the viscosity of the oil will be reduced to an unacceptable level.

52.8.4 Oxidation

The conditions of operation in an engine are conducive to oil oxidation, and this is another problem to be overcome by the lubricant. In the crankcase, the oil is sprayed from various components in the form of an oil mist, which is in contact with a large quantity of air and at a fairly high temperature. Oxidation produces complex carbonaceous products and acidic material and these, combined with fuel contaminants, will form stable sludge. In the combustion chamber, where the temperatures are very much higher, the oil is scraped up the cylinder walls by the piston ascending at very high speeds and is again present in the form of an oil mist. A form of carbon deposit is produced by a combination of heat decomposition and oxidation. Some of this deposit will remain, but some will pass into the sump. The effect of oxidation adds to the problem of oil contamination by the products of combustion, resulting in the formation of a resin-like material on the pistons and hot metal parts known as *lacquer* and acidic material which will attack bearing metals such as copper-lead.

These problems of engine lubrication can be overcome by using highly refined oil. The resistance to oxidation is further enhanced by the use of anti-oxidants. The addition of corrosion-inhibitors counters acidic materials produced by combustion at low engine temperatures.

Detergent-dispersant additives are incorporated so that the carbonaceous matter produced by imperfect combustion is retained in suspension in the oil, preventing it from being deposited on the engine surfaces. Such an oil is known as a fully detergent-type lubricant. All these additives are gradually consumed during operation and the rate of decline in their usefulness will determine the oil-change period. This rate is, in turn, influenced by the conditions of operation.

52.8.5 The SAE viscosity system

This classification was devised by the Society of Automotive Engineers (SAE) in America by dividing the viscosity span into four and giving each of the divisions a number – SAE 20, 30, 40 and 50. The thinnest (SAE 20), for example, covered the range 5.7–9.6 cSt specified at 210°F, which was considered to be a temperature typical of a hot engine. (The SAE originally specified temperatures in °F, because they were the convention. Today, temperatures are quoted in °C.)

Later, the SAE series was extended to include much thinner oils because of the growing demand for easier winter starting. The viscosities of the three new grades were specified at °F (typical of cold morning temperatures) and each was given the suffix W for winter – SAE 5W, 10W and 20W. Later still, grades of 0W, 15W and 25W were added to satisfy the more precise requirements of modern engines (Table 52.3).

52.8.6 Multi-grades

All oils become thinner when heated and thicker when cooled, but some are less sensitive than others to these viscosity/temperature effects. The degree of sensitivity is known as Viscosity Index (VI). Oil is said to have high VI if it displays a relatively small change of viscosity for a given change of temperature.

In the 1950s, development in additive technology led to the production of engine oils with unusually high VIs, known as multigrade oils. A multigrade oil's high resistance to temperature change is sufficient to give it the

Table 52.3 Viscosity chart

	Maximum viscosity cP at °C[a]	Maximum borderline pumping temperature (°C)[b]	Viscosity (cSt) at 100°C min.	max.
0W	3250 at −30	−35	3.8	−
5W	3500 at −25	−30	3.8	−
10W	3500 at −20	−25	4.1	−
15W	3500 at −15	−20	5.6	−
20W	4500 at −10	−20	5.6	−
25W	6000 at −5	−10	9.3	−
20	−	−	5.6	9.3
30	−	−	9.3	12.5
40	−	−	12.5	16.3
50	−	−	16.3	21.9

[a] As measured in the Cold Cranking Simulator (CCS).
[b] As measured in the Mini Rotary Viscometer (MRV).

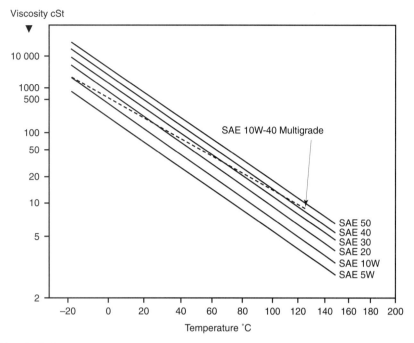

Viscosity cSt

SAE 10W-40 Multigrade

SAE 50
SAE 40
SAE 30
SAE 20
SAE 10W
SAE 5W

Temperature °C

Figure 52.4 Multi-grade chart

combined virtues of a thin grade at low (starting) temperatures and a thick one at running temperatures. An SAE 20W-40 multigrade, for example, is an thin at −20°C (−68°F) as a 20W oil, but as thick at 100°C (212°F) as an SAE 40 oil. Thus, the multigrade combines full lubrication protection at working temperatures with satisfactorily easy starting on frosty mornings. Figure 52.4 is a viscosity-temperature graph for six mono-grade oils and a 10W-40 multigrade, showing how the multigrade has the high-temperature properties of SAE 40 oil and the low-temperature properties of an SAE 10W. Thus, the multigrade is suitable for all-year-round use.

52.8.7 Performance ratings

The SAE numbering system refers purely to the viscosity of the oil, and is not intended to reflect lubricating performance (there is no such thing as 'SAE quality' oil, for example). Engine oils are marketed in a range of performance levels, and need to be classified according to the severity of service conditions in which they are designed to operate. Accordingly, the American Petroleum Institute (API) has drawn up a coding system in which oils are subjected to a series of classifying bench-tests known as the 'Sequence' tests.

52.8.8 The API service classifications

In the API system the least demanding classification for a petrol engine was originally designated SA. The most demanding is, at present, SG. (The S stands for Service

Station.) Constant development of both engines and oils means that from time to time the highest ratings are superseded by even higher ratings. The API system also classifies diesel engine oils by their severity of service. Here the categories have the prefix C, which stands for Commercial.

Petrol engines

SA Service typical of engines operated under mild conditions. This classification has no performance requirements.

SB Service typical of engines operating in conditions such that only minimum protection of the type afforded by additives is desired. Oils designed for this service have been used since the 1930s; they provide only anti-scuff capability and resistance to oil oxidation and bearing corrosion.

SC Service typical of petrol engines in 1964–1967 cars and trucks. Oils designed for this service provide control of high- and low-temperature deposits, wear, rust and corrosion.

SD Service typical of 1967–1970 petrol engines in cars and some trucks; but it may apply to later models. Oils designed for this service provide more protection than SC against high- and low-temperature deposits, wear, rust and corrosion; and may be used where SC is recommended.

SE Service typical of petrol engines in cars and some trucks in 1972–1979. Oils designed for this service provide more protection against oxidation, high-temperature deposits, rust and corrosion than SD or SC, and may be used where those classifications are recommended.

SF Service typical of petrol engines in cars and some trucks from 1980. Oils developed for this service provide better oxidation stability and anti-wear performance than SE oils. They also provide protection against engine deposits, rust and corrosion. Oils meeting SF may be used wherever SE, SD or SC is recommended.

SG Service typical of petrol engines in present cars, vans and light trucks. Oils developed for this service provide improved control of engine deposits, oil oxidation and engine wear relative to oils developed for previous categories. Oils meeting SG may be used wherever SF, SE, SF/CC or SE/CC are recommended.

Diesel engines

CA Service typical of diesel engines operated in mild to moderate duty with high-quality fuels. Occasionally this category has included petrol engines in mild service. Oils designed for this service were widely used in the late 1940s and 1950s; they provided protection from bearing corrosion and light-temperature deposits.

CB This category is basically the same as CA, but improved to cope with low-quality fuels. Oils designed for this service were introduced in 1949.

CC Service typical of lightly supercharged diesel engines operated in moderate to severe duty. Has included certain heavy-duty petrol engines. Oils designed for this service are used in many trucks and in industrial and construction equipment and farm tractors. These oils provide protection from high-temperature deposits in lightly supercharged diesels and also from rust, corrosion and low-temperature deposits in petrol engines.

CD Service typical of supercharged diesel engines in high-speed high-output duty requiring highly effective control of wear and deposits. Oils designed for this service provide protection from bearing corrosion and high-temperature deposits in supercharged diesel engines running on fuels of a wide quality range.

CDII Service typical of two-stroke cycle diesel engines requiring highly effective control over wear and deposits. Oils designed for this service also meet all the requirements of CD.

CE Service typical of certain turbocharged or supercharged heavy-duty diesel engines operating under both low speed-high load and high speed-low load conditions. Oils designed for this service must also meet the requirements specified for CC and CD classifications.

Before an oil can be allocated any given API performance level it must satisfy requirements laid down for various engine tests. In the SG category, for example, the engine tests are as follows:

Service IID measures the tendency of the oil to rust or corrode the valve train and to influence the valve lifter operation.
Sequence IIIE measures high-temperature oil oxidation, sludge and varnish deposits, cam-and-tappet wear, cam and lifter scuffing and valve lifter sticking.
Sequence VE evaluates sludge deposits, varnish deposits, oil-ring clogging and sticking, oil-screen plugging and cam wear.
Caterpillar IH2 determines the lubricant effect on ring sticking, ring and cylinder wear, and accumulation of piston deposits.
CRC L-38: the characteristics assessed are resistance to oxidation, bearing corrosion, tendency to formation of sludge and varnish, and change of viscosity.

In the CE category the tests are:

Caterpillar IG2: the lubricant characteristics determined are ring sticking, ring and cylinder wear, and accumulation deposits under more severe test conditions than those for Caterpillar IH2.
Cummins NTC-400 measures crownland and piston deposits, camshaft roller follower pin wear and oil consumption.
Mack T6 assesses oil oxidation, piston deposits, oil consumption and ring wear.
Mack T7 evaluates oil thickening.
CRC L-38 (as above).

Other specifications

Various authorities and military bodies issue specifications relating to the service performance of engine oils. In some instances the ratings are almost identical with those of the API, but most of them are not precisely parallel because they cover performance factors encountered in particular engines and particular categories of service.

The most common of the other specifications are those with the prefix MIL, issued by the US military authorities. MIL-L-2104E approximates to the API CE rating for diesel lubricants, although it also relates to petrol engines that require API SE performance. MIL-L-46152D covers oils for both diesel and petrol engines, and approximates to API SG/CC (Figure 52.5).

CCMC ratings

Another important set of performance specifications is produced by the European Vehicle Manufacturers' Association, known by its initials CCMC*. The CCMC rating G-1 corresponds roughly to API SE, and G-2 to API SF. G-3 (comparable to MIL-L-46152B, for petrol engines only) covers fuel-efficient and light-viscosity lubricants

*Comité des Constructeurs d'Automobiles du Marché Commun represents joint industry opinion on factors such as lubricant specifications, emissions, vehicle design and safety standards. With regard to crankcase lubricants, CCMC defines sequences of engine tests, and the tests themselves are defined by CEC (Coordinating European Committee for the Development of Performance Tests for Lubricants and Engine Fuels: a joint body of the oil and motor industries).

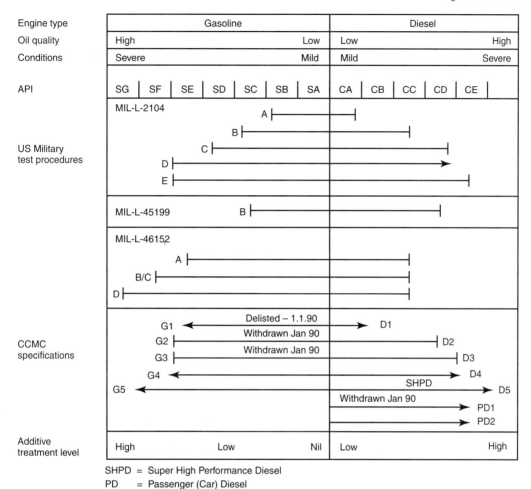

Figure 52.5 Approximate relationship between classifications and test procedures

blended from special high-quality base oils. CCMC also issues specifications for diesel lubricants: D-1 equates approximate to API CC, D-2 to API CD and MIL-L-2104D, and D-3 to API CE and MIL-L-2104E.

To qualify for the CCMC categories G and D, oil must meet the requirements of the following tests in addition to the relevant API classification tests.

For the G category

Ford Kent, which evaluates cold ring sticking, piston skirt varnish, oil thickening and consumption.
Fiat 132, to evaluate the tendency of the oil to cause pre-ignition.
Daimler Benz OM 616 to evaluate wear of cylinders and cams.
Bosch Injector Rig measuring the mechanical stability of the oil to assess its shear stability.

Noack Test, to measure the weight loss due to evaporation of the oil.
High shear/high temperature viscosity test, to assess the oil's capability for resisting shear, and so retaining its viscosity, at high temperatures.
Tests for oil/seal compatibility and oil consumption are still to be established.

For the D category

Bosch Injector Rig, Noack and *D-B O M 616* tests as above together with:

For D1 and D2 only, *MWM-B* evaluating varnish, carbon deposits, and ring-sticking;
For D3 only, *D-B OM 352A* bore polishing and piston cleanliness;
For PD1 only, *VW 1.6L* to evaluate ring sticking and piston cleanliness.

52.9 Gear lubricants

There are few engineering applications in which gears do not play an essential part. They can be used to reduce or increase speed, transmit power and change the direction or position of a rotating axis. There are several types of gears to suit these varying operational conditions, such as spur, helical, plain and spiral bevels, hypoid, worm and wheel (Figure 52.6).

Extremely high pressures are developed between meshing teeth as, in theory, they only have point or line contact. Together with the sliding between mating surfaces, which is always present, it is clear that, if there is metal-to-metal contact, rapid wear will occur. The function of the lubricant is to provide and maintain a separating film under all the variations in speed, load and temperature. It must also act as a coolant and protect the gears against corrosion.

The lubrication of gears is not a simple matter, because of their shape and variability of motion. Fundamental factors which affect their lubrication are gear characteristics, materials, temperature, speed, loading, method of applying the lubricant and environment.

52.9.1 Gear characteristics

Spur, bevel, helical and spiral bevel gears

Providing the speed is sufficient and the load does not squeeze out the lubricant, the effect of rotating these gears is to produce a hydrodynamic wedge of a relatively thick lubricating film between the meshing surfaces. If the load is increased, the pressure within the contact zone increases, causing a reduction in film thickness until the high spots on meshing teeth begin to touch; wear will then occur.

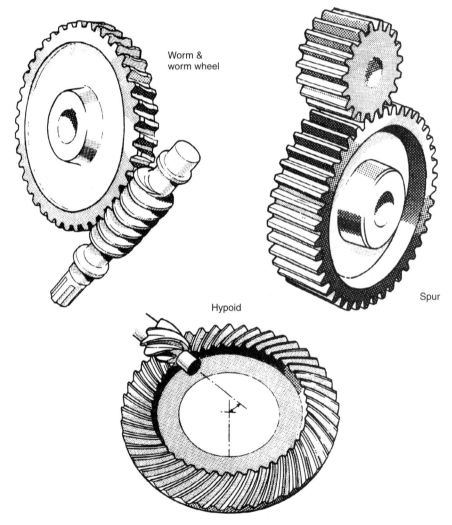

Worm & worm wheel

Hypoid

Spur

Figure 52.6 Gear types

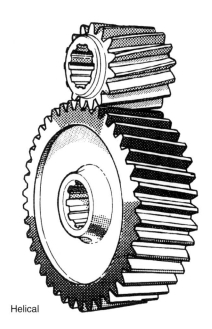

Helical

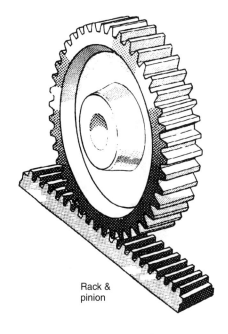

Rack &
pinion

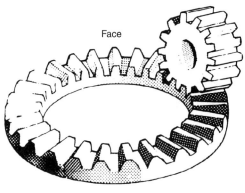
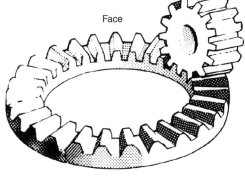

Face

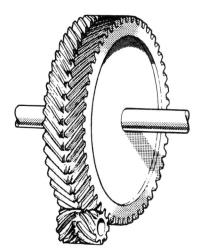

Double helical

Figure 52.6 (*continued*)

At low speeds, when the high spots make contact due to load, the sliding velocities are not sufficient to generate the heat necessary to cause welding. The wear, which can be quite rapid, is caused by the abrasive action of the teeth on each other.

At medium speeds, contact of the high spots will cause incipient melting and surface welding. The rise in surface temperature will encourage the breakdown of the lubricant film and the resultant scoring which is distinctive will cause rapid wear.

At high speeds, however, a position can be reached in which the load-carrying capacity increases with speed. The explanation is thought to be that when the teeth are in contact for such a short time, the oil film is not squeezed out and a separating film is therefore maintained.

52.9.2 Hypoid gears

The sliding and loading effect of these gears is such that they generally operate outside the limiting conditions for

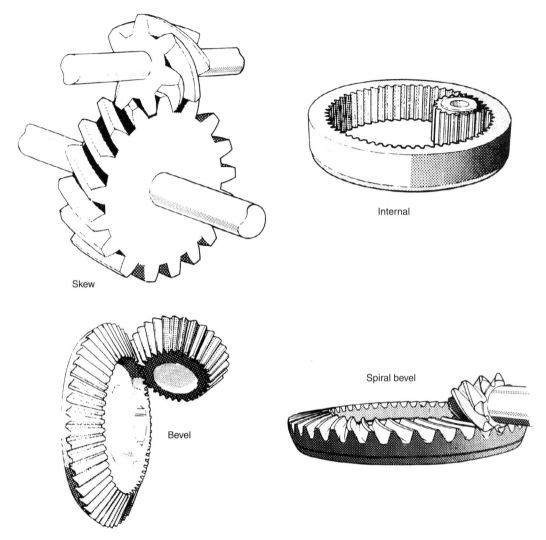

Skew

Internal

Bevel

Spiral bevel

Figure 52.6 (*continued*)

hydrodynamic lubrication. Additional protection against scoring is given by incorporating suitable extreme pressure additives into the lubricant. These react with the metal surfaces at or near scoring temperature, to provide a low-friction lubricant film. This delays the onset of scoring, although this film will eventually break down under much higher loads. Since these extreme pressure additives usually react with the gear surface they can cause corrosive wear, but this can be contained within reasonable limits by careful choice of oil and additives.

52.9.3 Worm gears

As worm gears tend to slide along their lines of contact, it is virtually impossible to maintain a hydrodynamic oil wedge so boundary lubrication conditions nearly always exist. Correct meshing of worm gears is most important in preserving an oil film. Too small a contact area can result in a rupture of the film and consequent abrasive wear, whereas if the contact area extends to the entry edge of the worm wheel, lubricant can be scraped from the worm by the tooth edges.

An important point to remember with over-slung worm gears is that they are not immersed in the lubricant and the oil must be fed to the point of contact. The viscosity of the oil is important here, as it must not drain too quickly from the contact area. Manufacturers because of the particular lubrication problems involved with worm gears often prefer a mineral oil compounded with fatty oil.

52.9.4 Gear materials

These have an important influence on both lubrication and wear. At the same speed, steel gears require higher viscosity oil than cast iron or bronze, because they can carry higher loads. Where straight oils are concerned, it is usually true to say that the harder the gear steel, the higher the viscosity needed.

52.9.5 Operating temperature

When selecting a lubricant, both the temperature at the contact area and the ambient temperature at important factors to be considered. Measuring the peak contact temperature is very difficult. The maximum rise in temperature of the oil leaving the gears and the maximum oil temperature are specified for various types of gears. For spur, bevel, helical and spiral level gears, the temperature rise should not normally exceed 30°C (86°F) with a maximum oil temperature of 70°C (158°F).

Worm and hypoid gears produce higher oil temperatures because they generate a greater amount of frictional heat. An oil temperature rise of 40°C (104°F) and a maximum oil temperature of 95°C (203°F) is acceptable. With EP oils, it should not exceed 75°C (167°F).

52.9.6 Surface speed

It is generally true to say that, as speed increases, the oil viscosity decreases, that is, if hydrodynamic conditions exist. Relatively low viscosity oil will allow the oil to spread rapidly over the tooth surfaces before meshing and, in the case of forced lubrication, ease circulation. In the case of bath lubrication, it will eliminate the oil drag effect.

52.9.7 Loading

If a gear is subjected to shock loading, the high pressures that are rapidly applied may rupture the oil film. These peak pressures are of greater importance than average tooth loading. To prevent sudden wear of the teeth it is essential to maintain a lubricant film, and using extreme pressure additives does this. Continuous and severe overloading may cause the oil film to break down with disastrous results.

52.9.8 Specifications

The lubrication requirements of gears vary considerably and create the need for specifically formulated products. This, combined with the diversity of automotive and industrial gear types, has led to the introduction of several specifications for gear lubricants (see Figure 52.7).

52.9.9 Performance

In 1969, the API drew up a range of service designations to define the protective qualities of gear oils – the higher the number, the more strenuous the service conditions. API GL-1, GL-2 and GL-3 oils are specified by

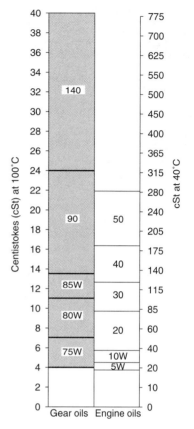

Figure 52.7 The SAE numbering systems for engine oils and gear oils are not related and must not be confused. This figure illustrates the differences

some manufacturers, but their use is dying out. At one time, manufacturers used GL-3 oils for factory-fill purposes only, to cover the running-in stage during which gears are not subjected to severe conditions. That practice is now largely obsolete, and GL-4 oils are used almost universally for both factory fill and service refill.

GL-5 oils and formulated primarily for hypoid axles operating at high speeds and loads, and likely to suffer shock loads. The GL-6 classification was introduced to cover a Ford requirement for special hypoid gears operating in high-performance conditions at high speeds. In the event, however, the Lotus Elan was virtually the only European car to require this category. GL-6 is now regarded as obsolete, but oils meeting its requirements may still be used to overcome isolated service difficulties.

Manufacturers and other bodies have issued their own designations, which may be encountered occasionally. Most important is the US Military specification MIL-L-2105D (which replaced MIL-L-2105C) corresponding to API GL-5, and introduced approval for multi-grade gear oils.

52.9.10 AGMA specifications

For industrial gears and circulating oils there is a specification published by the American Gear Manufacturers Association, which expresses the viscosities corresponding to the AGMA Lubricant Numbers in Saybolt Universal Seconds only. The approximate viscosities in centistokes are provided in Table 52.4 for convenience. The recommendations in Tables 52.5 and 52.6 are intended as a guide for lubricant selection.

52.9.11 Other specifications

Among other specifications that the plant user will encounter are those issued by the German DIN 51517, and individual suppliers' recommendations on the grades of oil to be used in their units.

52.9.12 Method of lubrication

The method of application, whether splash, spray, jet or mist, will have an effect on the lubricant. If oil is atomized, its surface area is increased and it is much more easily oxidized. The frictional heat generated by the meshing teeth will increase the rate of oxidation. The dissipation of

Table 52.4 AGMA lubricant numbering system

Rust and oxidation inhibited oils	EP lubricants[a]	Viscosity (40°C)[b]
AGMA lubricant number	AGMA lubricant number	Kinematic cSt
1		41.4–50.6
2	2 EP	61.2–74.8
3	3 EP	90–110
4	4 EP	135–165
5	5 EP	198–242
6	6 EP	228–352
7 comp[c]	7 EP	414–506
8 comp	8 EP	612–748
8A comp		900–1100

[a]EP lubricants should be used only when recommended by the gear manufacturer.
[b]As per ASTM 2422 (also BS 4231).
[c]Oils marked 'comp' are compounds with 3–10% fatty or synthetic fatty oils.

the heat absorbed by the oil is important and this will be influenced by the lubrication system. The length of pipe

Table 52.5 Lubrication of industrial enclosed gear drives. AGMA Lubricant Number recommendations for helical, herringbone, straight bevel, spiral bevel and spur[a]

Centre distances: main gear, low speed	Ambient temperatures[b] (−40 to 78°C) Other lubricants	(−28.9 to −3.9°C)	(−9.4–15.6°C) AGMA lubricant number[c,d]	(10–51.7°C)
Parallel shaft (single reduction):	Automatic	Engine oil SAE		
Up to 8 in (200 mm)	transmission fluid	10W-30, or similar	2–3	3–4
Over 8 in (200 mm) and up to 20 in (510 mm)	or similar	product[e]	2–3	4–5
Over 20 in (510 mm)	product[e]		3–4	4–5
Parallel shaft (double reduction):				
Up to 8 in (200 mm)			2–3	3–4
Over 8 in (200 mm) and up to 20 in (510 mm)			3–4	4–5
Over 20 in (510 mm)			3–4	4–5
Parallel shaft (triple reduction):				
Up to 8 in (200 mm)			2–3	3–4
Over 8 in (200 mm) and up to 20 in (510 mm)			3–4	4–5
Over 20 in (510 mm)			4–5	5–6
Planetary gear units:				
Outside dia. of housing up to 16 in (410, mm)			2–3	3–4
Outside dia. of housing over 16 in (410 mm)			3–4	4–5
Spiral or straight bevel gear units:				
Cone distance up to 12 in (300 mm)			2–3	4–5
Cone distance over 12 in (300 mm)			3–4	5–6
Geared motors and shaft-mounted units			2–3	4–5

[a]Drives incorporating over-running clutches as backstopping devices should be referred to the clutch manufacturers, because certain types of lubricants may adversely affect clutch performance.
[b]The pour point of the lubricant should be at least 10°C lower than the expected minimum ambient starting temperature. If the ambient starting temperature approaches the pour point, sump heaters may be required to ease starting and ensure adequate lubrication.
[c]Viscosity ranges are provided to allow for variations such as surface finish, temperature rise, loading and speed.
[d]AGMA viscosity number recommendations listed above refer to R & O (rust and oxidation) inhibited gear oils, which are widely used for general bearing, turbine and gear lubrication systems. EP gear lubricants in the corresponding viscosity grades may be substituted, where they are deemed necessary by the gear drive manufacturer.
[e]Where they are available, good-quality industrial oils having similar properties are preferred to the automotive oils. The recommendation of automotive oils for use at ambient temperature below −10°C is intended only as a guide pending widespread development of satisfactory low-temperature industrial oils. It is advisable to consult the gear manufacturer.
High-speed units are those operating at speeds above 3600 rev/min or pitch line velocities above 1520 m/min. (Refer to Standard AGMA for High Speed Helical and Herringbone Gear Units for detailed lubrication recommendations.)

Table 52.6 Enclosed cylindrical and double enveloping worm-gear drives

Worm centre distance	Worm speeds (rev/min) up to	Ambient temperatures 15–60°F (−9.4–15.6°C)	50–125°F (10–51.7°C)	Worm speeds (rev/min) up to[b]	Ambient temperatures 15–60°F (−9.4–15.6°C)	50–125°F (10–51.7°C)
		AGMA lubricant no.			*AGMA lubricant no.*	
Up to and including 6 in (150 mm):						
Cylindrical worm	700	7 comp	8 comp	700	7 comp	8 comp
Double-enveloping worm	700	8 comp	8A comp	700	8 comp	8 comp
Over 6 in (150 mm) and up to 12 in (300 mm):						
Cylindrical worm	450	7 comp	8 comp	450	7 comp	7 comp
Double-enveloping worm	450	8 comp	8A comp	450	8 comp	8 comp
Over 12 in (300 mm) and up to 18 in (460 mm):						
Cylindrical worm	300	7 comp	8 comp	300	7 comp	7 comp
Double-enveloping worm	300	8 comp	8A comp	300	8 comp	8 comp
Over 18 in (460 mm) and up to 24 in (610 mm):						
Cylindrical worm	250	7 comp	8 comp	250	7 comp	7 comp
Double-enveloping worm	250	8 comp	8A comp	250	8 comp	8 comp
Over 24 in (610 mm):						
Cylindrical worm	200	7 comp	8 comp	200	7 comp	7 comp
Double-developing	200	8 comp	8A comp	200	8 comp	8 comp

[a]The pour of the oil used should be less than the minimum ambient temperature expected. Consult the gear manufacturer for recommendations regarding ambient temperatures below −10°C.
[b]Worm gears of either type operating at speeds above 2400 rev/min or 610 m/min rubbing speed may require force-feed lubrication. In general, a lubricant of lower viscosity than recommended in the above the table may be used with a force-feed system.
Warning: Worm gear drives operate satisfactorily on R & O gear oils, sulfur phosphorus EP oils or lead naphthenate EP oils. These oils, however, should be used only with the approval of the gear manufacturer.

run, the temperature and size of bearings, pumps, other surfaces with which the oil comes into contact, whether filters and oil coolers are fitted, all affect the heat content of the oil.

52.9.13 Splash lubrication

Some enclosed gears are lubricated by oil splashing at random within the casing. If the pitch-line speeds are low (say, less than 328 ft/min), paddles and scrapers attached to the gear wheels may be needed to pick up the required amount of oil. If the speed is excessively high, however, the lubricant may be flung off the gears before it reaches the meshing zone, while churning may overheat that in the reservoir portion. The upper speed limit is usually about 2625 ft/min, but this can sometimes be exceeded where shields are close to gear peripheries.

Although a gear train may comprise wheels of markedly different diameters, as in multiple-reduction sets, the degrees of immersion in the trough must be approximately constant. If not, the smallest or highest wheel may have adequate pick-up while the largest is over-immersed. The trough should therefore be stepped or the bearing-lubrication system extended to lubricate the smaller or higher gears.

Optimum depth of immersion depends on pitch-line speed. Deep immersion is permissible at low speeds because power losses from churning are low, but at moderate speeds the immersion depth should not be more than three times the tooth height. At the highest speeds, only the addenda of the teeth need be submerged (Figure 52.8).

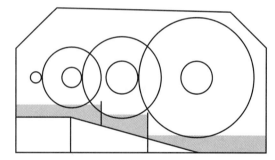

Figure 52.8 In a splash-fed gear the depth of immersion must be reasonably constant for all the gears, therefore the casing should incorporate stepped troughs for the oil

When gears are rotating inside a casing some oil is being carried around by them and some is adhering to the walls, so the oil in the trough is lower than when at rest. The sight-glass or dipstick should therefore be calibrated when the gears are in motion. With bevel, hypoid and worm sets, the relative positions of the various elements can vary so much that the 'correct' oil level cannot always be accurately defined.

In splash lubrication, the oil absorbs heat, which is transmitted to the casing. Gearbox designers sometimes augment the cooling effect by finning the casing to increase the area of metal exposed to atmosphere. In some designs, a fan is mounted on the driving shaft, to force cooling air over the casing. At the installation stage, intelligent siting can sometimes facilitate the cooling of gear units.

52.9.14 Forced circulation

Circulation by pumping is, of course, more expensive than simple splash lubrication, but the expense is usually justified. Some of the many advantages are as follows:

- Oil supplied to each meshing area can be metered to provide the optimum flow for lubrication and cooling.
- Heat exchangers can be installed to minimize changes of temperature. This means that the user can select viscosity solely on the grounds of conditions at the tooth face, and can virtually ignore other requirements. Thin oil, for example, can be cooled to prevent it becoming too thin during continuous circulation through meshing zones, or to prevent overheating; thick oil can be heated to encourage rapid initial flow through feeder lines.
- Filtration units can be incorporated into circulation lines, thus ensuring that clean oil is fed to gears and bearings.
- Gear casings can be small, since their oil reservoirs can be sited remotely. In most rolling mills, for instance, oil cellar houses storage tanks, pumps, heat exchangers and filters.

Oil is fed to meshing zones by wide jets or sprayers, and uniform distribution across the tooth faces is necessary to avoid local distortion. Clearly, the rate of flow is important; the effects of too little oil are easily imagined. A sound practice is to position an extra jet on the 'exit' or disengagement side of a meshing zone, where the teeth are at their highest temperature. If gears run in either direction, sprayers must be installed on both sides.

Formulae exist for calculating the optimum rates of flow, based on considerations of the heat to be removed – which calls for more oil flow than does lubrication alone – but they are complex and require data that may not be available at the design stage. A reasonable estimate is 0.05–0.10 l/min for each millimeter of tooth width.

Heat from bearings is an important fraction of the total heat to be dissipated from a gear system, and must be taken into account in calculations of cooling requirements. Excessive churning of oil causes unnecessary loss of power, and the resultant combination of high temperature and atomization of the oil raises the rate of oxidation. Careful attention to the design and siting of sprayers minimizes these effects, keeps the oil cooler and prolongs its service life (Figure 52.9).

52.9.15 Oil-mist lubrication

Although not widely used for gears, oil-mist lubrication is nevertheless worth mentioning here. It is a 'total loss' technique in which the oil is supplied in the form of fine droplets carried by compressed air. Two virtues are that the lubricant can be carried long distances through pipes without severe frictional losses, and that no oil pumps are needed since the motive power is provided by factory compressed-air lines. However, unless such systems are totally enclosed, the exhaust can create a build-up of oil-mist in the atmosphere. In order to maintain good standards of industrial hygiene, it is recommended that

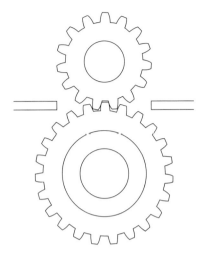

Figure 52.9 Oil is fed to some gears by jet or spray, so that the full benefits of lubrication and cooling are obtained precisely where they are needed. An extra jet at the 'exit' side of a meshing zone ensures that the teeth are cooled where the temperature is highest

the concentration of oil-mist in the working environment should not exceed 5 mg/m^3.

52.9.16 Semi-fluid lubrication

Gear oils possess most of the attributes required in the lubricating function, but the need for an alternative lubricant is encountered in some industrial applications. One such need rises from the leakage caused by slight wear of oil-seals – which is inevitable in any gear unit. Oil escaping past seals seeps along input and output shafts, from where it is probably thrown onto adjacent walls, floors or machinery. Loss of lubricant is particularly undesirable in a geared motor, where leakage from the reduction-gear section may have rapid and disastrous effects on electrical components.

Industrial maintenance engineers are, unhappily, familiar with the problem of leaking gear units that have to run at a high factor of utilization, since interruptions or downtime for filling, checking and repair can seriously retard a production program. Such checks are made necessary, of course, by the risk of drying up and gear failure. Leakage is annoying in units that operate in conventional horizontal attitudes, but it can be of serious concern to maintenance staff if the gear units are operating in unusual orientations or are part of machinery that changes angular position during a work cycle.

In the past, greases have been used to eliminate leakage from a few small and lightly loaded gear units, but orthodox greases do not possess the full range of properties required in the more demanding applications. However, development has produced excellent semi-fluid gear lubricants, which are finding their way into an ever-widening range of applications. A gearbox lubricated by a modern special-purpose semi-fluid lubricant can operate for very long periods without the need of maintenance. The semi-fluid lubricants formulated for gear lubrication include a mineral based for spur, helical, bevel gears, and synthetic based for worm gears.

52.9.17 Spur, helical and bevel gears

The latest mineral-base grease is thickened, not by a metallic soap, but by a polymer, and its additives include an antioxidant and a modern EP agent of the sulfur/phosphorus type. Such grease has flow characteristics that vary with shearing conditions. It is like a gear oil in the tooth-contact regions, where shearing is severe, but is more resistant to flow in regions of low shear – as in oil seals. Consequently, the grease resists leakage at the seals and joints, and the lubrication function is fulfilled, especially in heavily loaded steel gears. Another important virtue is that the product does not readily drain from metallic surfaces: dry starts are thus avoided (Figure 52.10).

52.9.18 Worm gears

As already mentioned, the severe loading and sliding conditions at the tooth faces of worm gears generate a great deal of heat. EP or anti-weld additives are ineffective in these circumstances, and can even be harmful. Consequently, the synthetic oils that have been developed specifically for worm gears are incorporated into semi-fluid grease formulations for applications in which leakage must be avoided. One of their primary requirements, of course, is high shear stability.

Several of the current synthetic semi-fluid lubricants for worm gears are based on a poly-alkaline glycol – which possesses 'slipperiness'. Oxidation and rust-inhibitors are

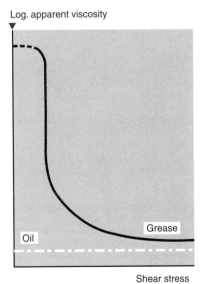

Figure 52.10 Comparison of the apparent viscosities of a gear oil and a gear grease. The grease flows readily in the tooth-contact region, where shear stress is high, but resists leakage at seals and joints

usually incorporated, and they are carefully selected to be compatible with both the poly-alkaline and the grease-thickener system.

52.9.19 Oil or grease?

In summary, it can be said that the best choice of lubricant for a gearbox depends on the service conditions. If leakage is a problem, or if, during shutdown, the maintenance of a thick film of lubricant is necessary to inhibit corrosion and prevent 'dry-state' operation, greases show to advantage. If generation of heat and consequent excessive temperatures are a problem, users will find it better to use liquid lubricants, which have the advantage of high rates of heat transfer.

52.9.20 Automatic transmissions

Frictional behavior

Satisfactory gear changing in an automatic transmission unit is a compromise between smooth slippage during take-up and firm grip during drive. The frictional properties of an ATF can be adjusted with friction-modifying additives during the formulation stages to ensure careful matching with the requirements of transmission manufactures. Two kinds of frictional behavior have to be considered in combination:

1. Static friction, which exists when sliding motion starts from rest;
2. Dynamic friction, which comes into effect when the mating surfaces are in motion.

The transition stage from one to the other is critical, and is the point at which any problem of noise or chattering occurs. Some transmission designs have a take-up characteristic best suited by low static friction and some by a high static friction. On no account should the wrong ATF be employed when a transmission unit is being completely filled, but a limited degree of topping up with the wrong fluid is permissible in most cases. (The need for fluids of high static friction is receding now that related designs are being phased out.)

Automatic transmission fluids (ATFs)

An ATF has to satisfy several requirements. It must:

1. Lubricate the gears in the gearbox;
2. Fill the torque converter (or 'automatic clutch') and thus convey power from the engine to the gearbox;
3. Act as hydraulic medium by conveying signals from the valves in the control unit to the internal clutches/brakes that engage the gears;
4. Ensure smooth but rapid take-up of power between the friction faces of the oil-immersed clutches/brakes;
5. Remove frictional heat and cool the whole transmission unit.

An ATF requires high lubrication performance, high levels of oxidation stability to withstand the locally high operating temperatures, a degree of fluidity that ensures easy operation in cold weather, stable viscosity properties, closely controlled frictional behavior to ensure quiet, consistent and chatter-free take-up of power in the internal brakes and clutches. It must preserve all these properties throughout its service life, to maintain consistent lubrication performance and gear-change characteristics.

Over the years, GM and Ford have progressively raised their levels of demand for oxidative/thermal stability, to cater for cars with higher power and thus greater likelihood of high transmission temperatures. GM's latest designation is Dexron IID, and the latest Ford specification is M2C 33-G, commonly referred to as the Ford G fluid. The Daimler-Benz requirement is that a fluid should retain its frictional characteristics throughout its life (in other words, throughout the life of the car). Additive systems have been developed with properties acceptable to Daimler-Benz, however, and a 'universal' ATF can now be blended.

52.9.21 Open and shielded gears

In general, the lubricant for open or partly enclosed gears is used on a 'total loss' basis – it is neither recirculated nor recovered. Rates of application must be minimal, both for economy and because the accumulation of used lubricant creates problems. A distinguishing feature of this category is that separate provision must be made for lubricating the bearings – by grease cups, oil bottles or mechanical lubricators.

Open gears usually run at low pitch-line speeds, below, say, 492 ft/min, and are not machined to high standards of accuracy. If their lubricant is to be applied intermittently by hand, the most satisfactory results are obtained with special high-viscosity 'residual' oils that have strong properties of adhesion. Very viscous and heavy lubricants may be heated and spread over the gear teeth, which must be clean and dry, with a brush or paddle. In this case, the oil must not be heated so that it becomes too fluid, for although it will then be easier to apply there will be considerable wastage. It is equally important not to use too much oil, for the excess will immediately be flung off when the gears are in motion. In addition, over-oiling can lead to the formation of hard pads of lubricant at the bottom of the teeth, which could stress or distort the shaft or cause wear in the shaft bushes. To make heating unnecessary, the lubricant may be supplied in a suitable solvent that evaporates after cold application to the gears.

Under better conditions, mechanical lubrication may be used. Force-feed lubricators can be installed that will provide a continuous and measured supply of oil to the meshing teeth or by spraying atomized lubricant onto the gears.

Thinner oils, particularly those operating in relatively clean conditions, can lubricate some slow-running shielded gears. In that case, the teeth of the largest wheel can dip into a 'slush pan' and carry oil to the rest of the gear train.

52.9.22 Environment

While gear oil will protect the metal surfaces against corrosion appreciable amounts of corrosive contaminants will affect the properties of the lubricant. Water is particularly troublesome in this respect. Different oils, while quite suitable for gear lubrication under normal conditions, may react very differently when contaminated with water. With active extreme pressure type oils, water leads to severe corrosion problems.

All these factors have to be borne in mind when blending gear oils. The oil must have a viscosity capable of maintaining a hydrodynamic lubricating film under, as far as possible, all working conditions. Its viscosity index should be high enough to maintain the viscosity within permitted limits at any operating temperature. It must be adhesive enough when on the gear teeth to resist removal by wiping or centrifugal force and not rupture under heavy or shock loads. The chemical stability of the oil must be such that it will not break down under the action of temperature, oxidizing agents and contaminants. It must provide protection against rusting and corrosion. It should not foam in service and should have the ability to separate from any water that may enter the system.

52.9.23 Gear wear and failure

A gearbox can be regarded as having four main components – gears, casing, bearings and lubricant – and failure can be caused by shortcomings in any one of them. Gear teeth that are inaccurate through design or manufacture can cause poor meshing, noisy running, overheating or surface failure due to overloading. Excessive flexing of the casing would allow misalignment of the gear shafts, which would also lead to surface failure due to overheating. Unsuitable or badly fitted bearings may fail because of their own surface damage, and this creates shaft misalignments and produces debris that can quickly damage the gear teeth. Lubrication failure can arise from unsuitable lubricant or from the inability (through bad design) of the lubricant to get where it is needed.

52.9.24 Manufacture

The makers of gear units have quite a choice of gear-cutting and gear-forming techniques at their disposal, including hobbing, shaping, grinding, broaching, milling and rolling. Every production process is inevitably a compromise between ease of machining, good surface finish, profile accuracy, speed of output and economic production. All metal-removal operations leave minute ridges and scratches, caused by tool shapes or by the motion of rotary cutters; perfect profiles and finishes are not obtainable with normal industrial processes. Tooling marks and deep scratches act as stress raisers, which occasionally lead to fatigue failure and tooth breakage. Such a failure almost invariably starts in the fillet radius at the root of a tooth, and may be caused by the use of a cutting tool with too small a radius.

Most gears benefit from careful running in under light load, since in this way the teeth acquire the surface finish needed for good lubrication and smooth running. In effect, they are given a final production operation, without which the process of surface failure might be initiated as soon as service operation begins.

Poor assembly can also limit the life of a gear. If a pinion is clumsily pressed onto a shaft, for example, it is likely to operate in an overstressed condition and fail at an early stage, even through running well within its designated rating.

52.9.25 Materials

Gear failure is rarely attributable to the type of material selected. More often, it is the result of defects such as cracks, casting inclusions and poor heat-treatment. Badly casehardened steel might fail by the flaking of its hard outer layers and a through-hardened gear might have hard or soft patches. Distortions due to sub-standard hardening can modify tooth profiles to such an extent that load concentrations will break the ends of teeth.

The term 'failure' is relative, because severe operating conditions affect different materials in different ways. An example of this might be seen in a high-reduction gear unit whose pinion is very small and is therefore made of harder material, to avoid premature wear relative to the larger gear. In arduous running, the two gears might not display the same kind of wear pattern.

52.9.26 Operation

The three most likely types of operational service misuse are overloading, incorrect lubrication and the presence of contaminants. Overloading is primarily due to the use of too small or too weak a gear unit, and this may be the result of false economy (installing an available unit for an application beyond its capacity) or failure to cater for the effects of shock loads in calculations of power rating.

Incorrect lubrication can take many forms. One example is the use of oil that is too thick or too thin, or is incompatible with the metal of the gears. Others include unsuitable methods of application, bad filtration, inadequate maintenance, filling to the wrong level, and poor standards of storage and handling.

Ideally, the lubricant should remain free from contaminants that might cause rust, abrasion or chemical attack; contamination may be caused by the presence of machining swarf, casting sand or even small items of rubbish left in the casing during manufacture. During service, dust and dirt can enter through the breathers, through neglected inspection covers or in additions of poorly maintained lubricant. Again, other contaminants have their origin inside the unit – rust particles, fine metal dust produced during the wearing process, and substances that result from the lubricant's gradual deterioration. Thick layers of sludge inside the casing can hinder the dissipation of heat, and thereby cause overheating; thick layers of dirt outside the casing can have exactly the same effect.

52.9.27 Surface damage

The types of wear and failure that occur in metallic gears are distinctive and have been subjected to close analysis over many years. The more common types of failure are abrasion, scuffing, pitting, corrosion, plastic yielding and fracture. Often the original cause, which may be misalignment, shock loading, faulty lubrication or one of numerous other possibilities, can be traced from the appearance of the gear teeth. It is important, also, that various kinds of wear may be combined in a single failure, making interpretation more difficult.

52.10 Hydraulic fluids

The wide application of hydraulic systems has undoubtedly been stimulated by the increasing use of fully automatic controls for sequences of operations where the response to signals must be rapid and the controls themselves light and easily operated. These needs are met by hydraulic circuits that, in addition, provide infinitely variable speed control, reversal of high-speed parts without shock, full protection against damage from overhead and automatic lubrication.

Over the years the performance standards of hydraulic equipment have risen. Whereas a pressure of about 1000 psi used to be adequate for industrial hydraulic systems, nowadays systems operating with pressures of 2000–3500 psi are common. Pressures above 5000 psi are to be found in applications such as large presses for which suitable high-pressure pumps have been developed. Additionally, systems have to provide increased power densities, more accurate response, better reliability and increased safety. Their use in numerically controlled machine tools and other advanced control systems creates the need for enhanced filtration. Full flow filters as fine as 1–10 micron retention capabilities are now to be found in many hydraulic systems.

With the trend toward higher pressures in hydraulic systems, the loads on unbalanced pump and motor components become greater and this, coupled with the need for closer fits to contain the higher pressures, can introduce acute lubrication problems. Pumps, one of the main centers of wear, can be made smaller if they can run at higher speeds or higher pressures, but this is only possible with adequate lubrication. For this reason, a fluid with good lubrication properties is used so that 'hydraulics' is now almost synonymous with 'oil hydraulics' in general industrial applications. Mineral oils are inexpensive and readily obtainable while their viscosity can be matched to a particular job.

The hydraulic oil must provide adequate lubrication in the diverse operating conditions associated with the components of the various systems. It must function over an extended temperature range and sometimes under boundary conditions. It will be expected to provide a long, trouble-free service life; its chemical stability must therefore be high. Its wear-resisting properties must be capable of handling the high loads in hydraulic pumps. Additionally, the oil must protect metal surfaces from corrosion and it must both resist emulsification and rapidly release entrained air that, on circulation, would produce foam.

Mineral oil alone, no matter how high its quality, cannot adequately carry out all the duties outlined above and hence the majority of hydraulic oils have their natural properties enhanced by the incorporation of four different types of additives. These are: an anti-oxidant, an anti-wear agent, a foam-inhibitor and an anti-corrosion additive. For machines in which accurate control is paramount, or where the range of operating temperatures is wide – or both – oils will be formulated to include a VI improving additive as well.

52.10.1 Viscosity

Probably the most important single property of hydraulic oil is its viscosity. The most suitable viscosity for a hydraulic system is determined by the needs of the pump and the circuit; too low a viscosity induces back-leakage and lowers the pumping efficiency; while too high a viscosity can cause overheating, pump starvation and possibly cavitation.

52.10.2 Viscosity index

It is desirable that a fluid's viscosity stays within the pump manufacturer's stipulated viscosity limits, in order to accommodate the normal variations of operating temperature. Oil's viscosity falls as temperature rises; certain oils, however, are less sensitive than others to changes of temperatures, and these are said to have a higher VI. Hydraulic oils are formulated from base oils of inherently high VI, to minimize changes of viscosity in the period from start-up to steady running and while circulating between the cold and hot parts of a system.

52.10.3 Effects of pressure

Pressure has the effect of increasing oil's viscosity. While in many industrial systems the working pressures are not high enough to cause problems in this respect, the trend towards higher pressures in equipment is requiring the effect to be accommodated at the design stage. Reactions to pressure are much the same as reactions to temperature, in that an oil of high VI is less affected than one of low VI. Typical hydraulic oil's viscosity is doubled when its pressure is raised from atmospheric to 5000 psi (Figure 52.11).

52.10.4 Air in the oil

In a system that is poorly designed or badly operated, air may become entrained in the oil and thus cause spongy and noisy operation. The reservoir provides an opportunity for air to be released from the oil instead of accumulating within the hydraulic system. Air comes to the surface as bubbles, and if the resultant foam were to become excessive, it could escape through vents and cause loss of oil. In hydraulic oils, foaming is minimized by the incorporation of foam-breaking additives. The type and dosage of such agents must be carefully selected, because although they promote the collapse of surface foam they may tend to retard the rate of air release from the body of the oil.

52.10.5 Oxidation stability

Hydraulic oils need to be of the highest oxidation stability, particularly for high-temperature operations, because

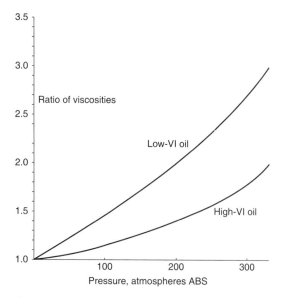

Figure 52.11

oxidation causes sludge and lacquer formation. In hydraulic oils, a high level of oxidation stability is ensured by the use of base oils of excellent quality, augmented by a very effective combination of oxidation inhibitors.

A very approximate guide to an oil's compatibility with rubbers commonly used for seals and hoses is given by the Aniline Point, which indicates the degree of swelling likely to arise; a high figure indicates a high level of compatibility. This system has been superseded by the more accurate Seal Compatibility Index (SCI), in which the percentage volume swell of a 'standard' nitrile rubber is determined after an immersion test in hot oil.

52.10.6 Fire-resistant fluids

Where fire is a hazard, or could be extremely damaging, fire-resistant hydraulic fluids are needed. They are referred to as 'fire resistant' (FR) so that users should be under no illusions about their properties. FR fluids do not extinguish fires: they resist combustion or prevent the spread of flame. They are not necessarily fireproof, since any fluid will eventually decompose if its temperature rises high enough. Nor are they high-temperature fluids, since in some instances their operating temperatures are lower than those of mineral oils. FR fluids are clearly essential in such applications as electric welding plants, furnace-door actuators, mining machinery, die-casters, forging plant, plastics machinery and theatrical equipment. When leakage occurs in the pressurized parts of a hydraulic system the fluid usually escapes in the form of a high-pressure spray. In the case of mineral oils, this spray would catch fire if it were to reach a source of ignition, or would set up a rapid spread of existing flame. FR fluids are therefore formulated to resist the creation of flame from a source of ignition, and to prevent the spread of an existing fire.

Four main factors enter into the selection of a fire-resistant fluid:

1. The required degree of fire-resistance;
2. Operational behavior in hydraulic systems (lubrication performance, temperature range and seal compatibility, for example);
3. Consideration of hygiene (toxicological, dermatological and respiratory effects);
4. Cost.

52.10.7 Types of fluid

The fluids available cover a range of chemical constituents, physical characteristics and costs, so the user is able to choose the medium that offers the best compromise for operational satisfaction, fire-resistance and cost effectiveness. Four basic types of fluid are available and are shown in Table 52.7.

In a fully synthetic FR fluid, the fire resistance is due to the chemical nature of the fluid; in the others, it is afforded by the presence of water. The other main distinction between the two groups is that the fully synthetic fluids are generally better lubricants and are available for use at operating temperatures up to 150°C (272°F), but are less likely to be compatible with the conventional sealing materials and paints than are water-based products.

When a water-based fluid makes contact with a flame or a hot surface its water component evaporates and forms a steam blanket that displaces oxygen from around the hot area, and this obviates the risk of fire. Water-based products all contain at least 35% water. Because water can be lost by evaporation, they should not be subjected to operating temperatures above about 60°C (140°F). Table 52.8 shows a comparison of oil and FR fluids.

52.10.8 High water-based hydraulic fluids

For a number of years HF-A oil-in-water emulsions have been used as a fire-resistant hydraulic medium for pit props. Concern over maintenance costs and operational life has created interest in a better anti-wear type fluid. Micro-emulsions are known to give better wear protection

Table 52.7 CETOP classifications of fire-resistant hydraulic fluids

Class	Description
HF-A	Oil-in-water emulsions containing a maximum of 20% combustible. These usually contain 95% water
HF-B	Water-in-oil emulsions containing a maximum of 60% combustible material. These usually contain 40–45% water
HF-C	Water–glycol solutions. These usually contain at least 35% water
HF-D	Water-free fluids. These usually refer to fluids containing phosphate esters, other organic esters or synthesized hydrocarbon fluids

CETOP: Comité European des Transmissions Oleohydrauliques et Pneumatiques.

Table 52.8 Comparison of oil and FR fluids

	Mineral oil	Water-in-oil emulsion	Water–glycol	Phosphate ester
Fire resistance	Poor	Fair	Excellent	Good
Relative density	0.87	0.94	1.08	1.14
Viscosity Index	High	High	High	Low
Vapor pressure	Low	High	High	Low
Special seals	No	Partly	Partly	Yes
Special paints	No	No	Yes	Yes
Rust protection	Very good	Good	Fair	Fair

than the normal oil-in-water emulsions. At the same time, the car industry, in attempts to reduce costs especially from leakage on production machinery, has evaluated the potential for using HWBHF in hydraulic systems. As a result, in many parts of industry, not only those where fire-resistant hydraulic fluids are needed, there is a increasing interest in the use of HWBHF.

Such fluids often referred to as 5/95 fluid (that being the ratio of oil to water), have essentially the same properties as water with the exception of the corrosion characteristics and the boundary lubrication properties, which are improved by the oil and other additives. The advantages of this type of fluid are fire resistance, lower fluid cost, no warm-up time, lower power consumption and operating temperatures, reduced spoilage of coolant, less dependence on oil together with reduced transport, storage, handling and disposal costs, and environmental benefits.

In considering these benefits, the user should not overlook the constraints in using such fluids. They can be summarized as limited wear and corrosion protection (especially with certain metals), increased leakage due to its low viscosity, limited operating temperature range and the need for additional mixing and in-service monitoring facilities.

Because systems are normally *not* designed for use with this type of fluid, certain aspects should be reviewed with the equipment and fluid suppliers before a decision to use such fluids can be taken. These are compatibility with filters, seals, gaskets, hoses, paints and any non-ferrous metals used in the equipment. Condensation corrosion effect on ferrous metals, fluid-mixing equipment needed, control of microbial infection together with overall maintaining and control of fluid dilution and the disposal of waste fluid must also be considered. Provided such attention is paid to these designs and operating features, the cost reductions have proved very beneficial to the overall plant cost effectiveness.

52.10.9 Care of hydraulic oils and systems

Modern additive-treated oils are so stable that deposits and sludge formation in normal conditions have been almost eliminated. Consequently, the service life of the oils which is affected by oxidation thermal degradation and moisture is extended.

Solid impurities must be continuously removed because hydraulic systems are self-contaminating due to wear of hoses, seals and metal parts. Efforts should be made to exclude all solid contaminants from the system altogether. Dirt is introduced with air, the amount of airborne impurities varying with the environment. The air breather must filter to at least the same degree as the oil filters.

It is impossible to generalize about types of filter to be used. Selection depends on the system, the rate of contamination build-up and the space available. However, a common arrangement is to have a full-flow filter unit before the pump with a bypass filter at some other convenient part of the system. Many industrial systems working below 2000 psi can tolerate particles in the order of 25–50 microns with no serious effects on either valves or pumps.

Provided that the system is initially clean and fitted with efficient air filters, metal edge-strainers of 0.005-inch spacing appear to be adequate, although clearances of vane pumps may be below 0.001-inch. It should be remembered that an excessive pressure drop, due to a clogged full-flow fine filter, could do more harm to pumps by cavitation than dirty oil.

If flushing is used to clean a new system or after overhaul, it should be done with the hydraulic oil itself or one of lighter viscosity and the same quality. As the flushing charge circulates it should pass through an edge-type paper filter of large capacity. It is generally preferable to use a special pump rather than the hydraulic pump system, and the temperature of the oil should be maintained at about 40°C without local overheating.

52.11 Machine tools

Lubricants are the lifeblood of a machine tool. Without adequate lubrication, spindles would seize, slides could not slide and gears would rapidly disintegrate. However, the reduction of bearing friction, vital though it is, is by no means the only purpose of machine-tool lubrication. Many machines are operated by hydraulic power, and one type of oil may be required to serve as both lubricant and hydraulic fluid. The lubricant must be of correct viscosity for its application, must protect bearings, gears and other moving parts against corrosion, and, where appropriate, must remove heat to preserve working accuracy and alignments. It may additionally serve to seal the bearings against moisture and contaminating particles. In some machine tools, the lubricant also serves the functions of cutting oil, or perhaps needs to be compatible with the cutting oil. In other tools, an important property of the lubricant is its ability to separate rapidly and completely from the cutting fluid. Compatibility with the metals, plastics, sealing elements and tube connections used in the machine construction is an important consideration.

In machine-tool operations, as in all others, the wisest course for the user is to employ reputable lubricants in the manner recommended by the machine-tool manufacturer and the oil company supplying the product. This policy simplifies the selection and application of machine tool lubricants. The user can rest assured that all the considerations outlined above have been taken into account by both authorities.

The important factors from the point of view of lubrication are the type of component and the conditions under which it operates, rather than the type of machine into which it is incorporated. This explains the essential similarity of lubricating systems in widely differing machines.

52.11.1 Bearings

As in almost every type of machine, bearings play an important role in the efficient functioning of machine tools.

52.11.2 Roller bearings

There is friction even in the most highly finished ball or roller bearing. This is due to the slight deformation under load of both the raceway and the rolling components, the presence of the restraining cage, and the 'slip' caused by trying to make parts of different diameter rotate at the same speed. In machine tools, the majority of rolling bearings are grease-packed for life, or for very long periods, but other means of lubrication are also used (the bearings may be connected to a centralized pressure-oil-feed system, for instance). In other cases, oil-mist lubrication may be employed both for spindle bearings and for quill movement. In headstocks and gearboxes, ball and roller bearings may be lubricated by splash or oil jets.

52.11.3 Plain journal bearings

Plain bearings are often preferred for relatively low-speed spindles operating under constant loads, and for the spindles of high-speed grinding wheels. These bearings ride on a dynamic 'wedge' of lubricating oil. Precision plain bearings are generally operated with very low clearances and therefore require low-viscosity oil to control the rise of temperature. Efficient lubrication is vital if the oil temperature is to be kept within reasonable limits, and some form of automatic circulation system is usually employed.

52.11.4 Multi-wedge bearings

The main drawback of the traditional plain bearing is its reliance on a single hydrodynamic wedge of oil, which under certain conditions tends to be unstable. Multi-wedge bearings make use of a number of fixed or rocking pads, spaced at intervals around the journal to create a series of opposed oil wedges. These produce strong radial, stabilizing forces that hold the spindle centrally within the bearing. With the best of these, developed especially for machine tools, deviation of the spindle under maximum load can be held within a few millionths of a centimeter.

52.11.5 Hydrostatic bearings

To avoid the instabilities of wedge-shaped oils films, a lubricating film can be maintained by the application of pressurized oil (or, occasionally, air) to the bearing. The hydrostatic bearing maintains a continuous film of oil even at zero speed, and induces a strong stabilizing force towards the center that counteracts any displacement of the shaft or spindle. Disadvantages include the power required to pressurize the oil and the necessary increase in the size of the filter and circulatory system.

52.11.6 Slideways

Spindles may be the most difficult machine-tool components to design, but slideways are frequently the most troublesome to lubricate. In a slideway, the wedge-type of film lubrication cannot form since, to achieve this, the slideway would need to be tilted.

52.11.7 Plain slideways

Plain slideways are preferred in the majority of applications. Only a thin film of lubricant is present, so its properties – especially its viscosity, adhesion and extreme-pressure characteristics – are of vital importance. If lubrication breaks down intermittently, a condition is created known as 'stick-slip', which affects surface finish, causes vibration and chatter and makes close limits difficult to hold. Special adhesive additives are incorporated into the lubricant to provide good bonding of the oil film to the sliding surfaces, which helps to overcome the problems of table and slideway lubrication. On long traverses, oil may be fed through grooves in the underside of the slideway.

52.11.8 Hydrostatic slideways

The use of hydrostatic slideways – in which pressurized oil or air is employed – completely eliminates stick-slip and reduces friction to very low values; but there are disadvantages in the form of higher costs and greater complication.

52.11.9 Ball and roller slideways

These are expensive but, in precision applications, they offer the low friction and lack of play that are characteristics of the more usual rolling journal bearings. Grease or adhesive oil usually effects lubrication.

52.11.10 Leadscrews and nuts

The lubrication of leadscrews is similar in essence to that of slideways, but in some instances may be more critical. This is especially so when pre-load is applied to eliminate play and improve machining accuracy, since it tends to squeeze out the lubricant. Leadscrews and slideways often utilize the same lubricants. If the screw is to operate under high unit stresses – due to pre-load or actual working loads – extreme-pressure oil should be used.

52.11.11 Re-circulating-ball leadscrews

This type was developed to avoid stick-slip in heavily loaded leadscrews. It employs a screw and nut of special form, with bearing balls running between them. When the balls run off one end of the nut they return through an external channel to the other end. Such bearings are usually grease-packed for life.

52.11.12 Gears

The meshing teeth of spur, bevel, helical and similar involute gears are separated by a relatively thick hydrodynamic wedge of lubricating oil, provided that the rotational speed is high enough and the load light enough so as not to squeeze out the lubricant. With high loads or at low speeds, wear takes place if the oil is not able to maintain a lubricating film under extreme conditions.

Machine-tool gears can be lubricated by oil-spray, mist, splash or cascade. Sealed oil baths are commonly used, or the gears may be lubricated by part of a larger circulatory system.

52.11.13 Hydraulics

The use of hydraulic systems for the setting, operation and control of machine tools has increased significantly. Hydraulic mechanisms being inter-linked with electronic controls and/or feedback control systems. In machine tools, hydraulic systems have the advantage of providing stepless and vibration-less transfer of power. They are particularly suitable for the linear movement of tables and slideways, to which a hydraulic piston may be directly coupled.

One of the most important features for hydraulic oil is a viscosity/temperature relationship that gives the best compromise of low viscosity (for easy cold starting) and minimum loss of viscosity at high temperatures (to avoid back-leakage and pumping losses). A high degree of oxidation stability is required to withstand high temperatures and aeration in hydraulic systems. Oil needs excellent anti-wear characteristics to combat the effects of high rubbing speeds and loads that occur in hydraulic pumps, especially in those of the vane type. In the reservoir, the oil must release entrained air readily without causing excessive foaming, which can lead to oil starvation.

52.11.14 Tramp oil

Tramp oil is caused when heat slideway, gear, hydraulic and spindle lubricants leak into water-based cutting fluids and can cause problems such as:

- Machine deposits
- Reduced bacterial resistance of cutting fluids and subsequent reduction in the fluid life
- Reduced surface finish quality of work pieces
- Corrosion of machine surfaces

All these problems directly affect production efficiency. Recent developments have led to the introduction of synthetic lubricants that are fully compatible with all types of water-based cutting fluids, so helping the user to achieve maximum machine output.

52.11.15 Lubrication and lubricants

The hydraulic fluid continuously lubricates the components of a hydraulic system, which must, of course, be suitable for this purpose. Many ball and roller bearings are grease-packed for life, or need attention at lengthy intervals. Most lubrication points, however, need regular replenishment if the machine is to function satisfactorily. This is particularly true of parts subjected to high temperatures.

With the large machines, the number of lubricating points or the quantities of lubricants involved makes any manual lubrication system impracticable or completely uneconomic. Consequently, automatic lubrication systems are often employed.

Automatic lubrication systems may be divided broadly into two types: circulatory and 'one-shot' total-loss. These cover, respectively, those components using relatively large amounts of oil, which can be cooled, purified and recirculating, and those in which oil or grease is used once only and then lost. Both arrangements may be used for different parts of the same machine or installation.

52.11.16 Circulatory lubrication systems

The circulatory systems used in association with machine tools are generally conventional in nature, although occasionally their exceptional size creates special problems. The normal installation comprises a storage tank or reservoir, a pump and filter, suitable sprays, jets or other distribution devices, and return piping. The most recent designs tend to eliminate wick feeds and siphon lubrication.

Although filtration is sometimes omitted with non-critical ball and roller bearings, it is essential for most gears and for precision bearings of every kind. Magnetic and gauze filters are often used together. To prevent wear of highly finished bearing surfaces the lubricant must contain no particle as large as the bearing clearance.

Circulatory systems are generally interlocked electrically or mechanically with the machine drive, so that the machine cannot be started until oil is flowing to the gears and main bearings. Interlocks also ensure that lubrication is maintained as long as the machine is running. Oil sight-glasses at key points in the system permit visual observations of oil flow.

52.11.17 Loss-lubrication systems

There are many kinds of loss-lubrication systems. Most types of linear bearings are necessarily lubricated by this means. An increasingly popular method of lubrication is by automatic or manually operated one-shot lubricators. With these devices, a metered quantity of oil or grease is delivered to any number of points from a single reservoir. The operation may be carried out manually, using a hand-pump, or automatically, by means of an electric or hydraulic pump. Mechanical pumps are usually controlled by an electric timer, feeding lubricant at preset intervals, or are linked to a constantly moving part of the machine.

On some machines both hand-operated and electrically timed one-shot systems may be in use, the manual system being reserved for those components needing infrequent attention (once a day, for example) while the automatic systems feeds those parts that require lubrication at relatively brief intervals.

52.11.18 Manual lubrication

Many thousands of smaller or older machines are lubricated by hand, and even the largest need regular refills or topping up to lubricant reservoirs. In some shops, the operator may be fully responsible for the lubrication of his own machine, but it is nearly always safer and more economical to make one individual responsible for all lubrication.

52.11.19 Rationalizing lubricants

To meet the requirements of each of the various components of a machine the manufacturer may need to recommend a number of lubricating oils and greases. It follows that, where there are many machines of varying origins, a large number of lubricants may seem to be needed. However, the needs of different machines are rarely so different that slight modification cannot be made to the specified lubricant schedule. It is this approach which forms the basis for BS 5063, from which the data in Table 52.9 have been extracted. This classification implies no quality evaluation of lubricants, but merely gives information as to the categories of lubricants likely to be suitable for particular applications.

A survey of the lubrication requirements, usually carried out by the lubricant supplier, can often be the means of significantly reducing the number of oils and greases in a workshop or factory. The efficiency of lubrication may well be increased, and the economies affected are likely to be substantial.

52.12 Cutting fluids

New machining techniques are constantly being introduced. Conventional workpiece materials have improved progressively through close control of manufacturer and heat treatment, and new materials have been fostered by the aeronautic and space industries. The results have been ever improving output, dimensional control and surface finish. The continuous development of cutting fluids has enabled these increasingly severe conditions to be accommodated.

52.12.1 Functions of a cutting fluid

The two main functions of a cutting fluid are to cool and to lubricate. During a machining operation, the cutting tool induces a continuous wave of dislocations, which travel ahead of the cutting edge, deforming and shearing the workpiece metal into continuous scarf or broken chips. The energy required for this deformation is converted into heat, and simultaneously frictional heat is created at the points where the tool rubs against the newly exposed surface of the workpiece and the chip rubs against the top face of the advancing tool.

These combined effects raise the temperature of the tool, workpiece and chip to a level that would be unacceptable in most operations if it were not for the presence of cutting fluid. Since the cooling effect is the greatest benefit, cutting fluids are widely known as 'coolants'. In

Table 52.9 Classification of lubricants

Class	Types of lubricant	Viscosity grade no. (BS 4231)	Typical application	Detailed application	Remarks
AN CB	Refined mineral oils Highly refined mineral oils (straight or inhibited) with good anti-oxidation performance	68 32 68	General lubrication Enclosed gears–general lubrication	Total-loss lubrication Pressure and bath lubrication of enclosed gears and allied bearings of headstocks, feed boxes, carriages, etc. when loads are moderate; gears can be of any type, other than worm and hypoid	May be replaced by CB 68 CB 32 and CB 68 may be used for flood-lubricated mechanically controlled clutches; CB 32 and CB 68 may be replaced by HM 32 and HM 68
CC	Highly refined mineral oils with improved loading-carrying ability	150 320	Heavily loaded gears and worms gears	Pressure and bath lubrication of enclosed gears of any type, other than hypoid gears, and allied bearings when loads are high, provided that operating temperature is not above 70°C	May also be used for manual or centralized lubrication of lead and feed screws
FX	Heavily refined mineral oils with superior anti-corrosion anti-oxidation performance	10 22	Spindles	Pressure and bath lubrication of plain or rolling bearings rotating at high speed	May also be used for applications requiring particularly low-viscosity oils, such as fine mechanisms, hydraulic or hydro-pneumatic mechanisms electro-magnetic clutches, air line lubricators and hydrostatic bearings
G	Mineral oils with improved lubricity and tackiness performance, and which prevent stick-slip	68 220	Slideways	Lubrication of all types of machine tool plain-bearing slideways; particularly required at low traverse speeds to prevent a discontinuous or intermittent sliding of the table (stick-slip)	May also be used for the lubrication of all sliding parts–lead and feed screws, cams, ratchets and lightly loaded worm gears with intermittent service; if a lower viscosity is required HG 32 may be used.
HM	Highly refined mineral oils with superior anti-corrosion, anti-oxidation, and anti-wear performance	32 68	Hydraulic systems	Operation of general hydraulic systems	May also be used for the lubrication of plain or rolling bearings and all types of gears, normally loaded worm and hypoid gears excepted, HM 3X and HM 68 may replace CB 32 and CB 68, respectively

Table 52.9 (continued)

Class	Types of lubricant	Viscosity grade no. (BS 4231)	Typical application	Detailed application	Remarks
HG	Refined mineral oils of HM type with anti-stick-slip properties	32	Combined hydraulic and slideways systems	Specific application for machines with combined hydraulic and plain bearings, and lubrication systems where discontinuous or intermittent sliding (stick-slip) at low speed is to be prevented	May also be used for the lubrication of slideways, when an oil of this viscosity is required

Class	Type of lubricant	Consistency number	Typical application	Detailed application	
XM	Premium quality multi-purpose greases with superior anti-oxidation and anti-corrosion properties	1 2 3	Plain and rolling bearings and general greasing of miscellaneous parts	XM 1: Centralized systems XM 2: Dispensed by cup or hand gun or in centralized systems XM 3: Normally used in prepacked applications such as electric motor bearings	

Note: It is essential that lubricants are compatible with the materials used in the construction of machine tools, and particularly with sealing devices.
The grease X is sub-divided into consistency numbers, in accordance with the system proposed by the National Lubricating Grease Institute (NLGI) of the USA. These consistency numbers are related to the worked penetration ranges of the greases as follows:

Consistency number	Worked penetration range
1	310–340
2	265–295
3	220–250

Worked penetration is determined by the cone-penetration method described in BS 5296.

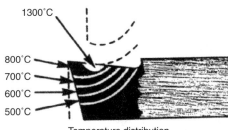

1300°C
800°C
700°C
600°C
500°C

Temperature distribution
in single-point cutting tool

Figure 52.12 Heat is generated both by deformation of workpiece material and by the friction of the chip across the tool tip. A cutting fluid acts as a lubricant to reduce frictional heat and simultaneously cools the whole cutting zone

practice, the coolant cannot penetrate to the extreme tip of the tool, of course, because the tip is so well shrouded by the metal being cut from the workpiece. Nevertheless, the coolant performs its function successfully because it enters a zone some way back from the tip, where the chip is passing over the tool at high rubbing velocity and where most of the frictional heat arises (Figure 52.12).

52.12.2 Benefits of a coolant

Cutting fluids provide benefits such as extended tool life, dimensional accuracy and good surface finish, all of which contribute to high rates of production. Almost invariably, the coolant that adheres to workpieces is relied on to protect ferrous components against corrosion while they are waiting further machining or assembly operations.

The benefits to be obtained by the correct selection and application of coolant are considerable. The cost is hardly a major factor in production operations, however, because, on average, the price of coolant amounts to about 1 per cent of the finished cost of a machined component.

52.12.3 Types of coolant

Water-based emulsions

Water, of course, is the cheapest cooling medium generally available. Its high specific heat makes it particularly effective as a coolant and its low viscosity allows high rates of flow through the coolant system and so enables it to penetrate the cutting zone and make contact with hot metal surfaces. Unfortunately, it corrodes most of the ferrous and some of the non-ferrous metals (a process hastened by high temperatures), and therefore is unsuitable in most machining.

However, its cooling properties can be utilized when other materials are added to improve machining performance. One of the most common water-based cutting fluids is the so-called 'soluble' oil, which, in fact, is not a true solution but an oil-in-water emulsion in which very fine droplets of oil are suspended in water. Such a fluid has very effective cooling power and the petroleum oil and its additives provide its lubricating and protective properties.

The proportion of oil may be as low as 2 per cent, but is normally between 5 and 10 per cent.

Water-based coolants, as diluted and used in service, are much cheaper than neat oils, and they offer remarkable economic benefits in the applications to which they are best suited. (Neat oils, however, are preferred for some machining processes – usually involving severe operations – where they give overall production economies and reduced machine-tool maintenance.)

Soluble oils are delivered, 'concentrated', to the user and contain an emulsifying agent to ensure that a stable emulsion forms when added to water. This additive does not mix readily with mineral oil, however, so to overcome this a 'coupling' agent is included in the formulation.

The performance of soluble oils is made possible not only by their high specific heat and thermal conductivity but by their low viscosity, which permits good penetration into the very fine clearances around the cutting zone. Consequently, these fluids are used mainly where cooling is the primary requirement. Lubricating properties can be improved by polar additives, which are agents that enhance the 'oiliness' or anti-friction characteristics. Further improvements can be effected by EP (extreme-pressure) additives, which are usually compounds of sulfur or chlorine.

Some of the recently developed high-performance EP soluble oils have a cutting performance that almost matches that of additive-type neat oils, and they are particularly suitable for demanding operations in machine tools whose design allows the use of water-based fluids.

Water-based solutions

In recent years another kind of water-based coolant, the 'synthetic' or non-petroleum cutting fluid, has been used increasingly. Coolants of this kind contain agents that provide the necessary lubricating and anti-corrosion properties. They were originally introduced to satisfy the special demands of grinding operations, but now they have been developed to a state where they can cope with general machining requirements. Their use in CNC machines and flexible manufacturing systems is a significant result of this development.

Because the synthetic fluid is transparent, operators can see the cutting area at all times – a feature particularly useful for intricate machining and grinding operations. Workpieces, gauges and tools in the vicinity of the cutting zone do not become coated with an oily film, which is a pleasing feature for tool setters who have to handle spanners and precision tools in confined spaces. Occasionally a transparent dye is included in the formulation of a synthetic cutting fluid, both for identification and to give a pleasant appearance.

Synthetic cutting fluids are suitable for a wider range of uses than any other single cutting fluid, and can be employed in any machine designed for operation with water-based fluids. Applications include turning, milling, drilling and screw cutting on all but the toughest materials; and cylindrical, surface and center-less grinding. Most metals can be machined effectively; cast iron, which is prone to rusting when machined with conventional soluble oils, shows negligible signs of corrosion when the synthetic fluid is used. Copper-base alloys may display a slight degree of staining if workpieces are not washed immediately after the machining operation.

Bacteria

Machine operators working with emulsions can become susceptible to skin infections because of the combination of the de-fatting effect of soluble-oil emulsifiers and the abrasive action of metallic scarf, but bacteria in cutting fluids are seldom the source of such infections. High standards of personal hygiene and the use of barrier creams should prevent such problems. A more difficult situation arises when a soluble-oil emulsion becomes infected with bacteria capable of utilizing the emulsifier and mineral-oil components in the system. Even in clean conditions, untreated soluble-oil emulsions and solutions cannot remain completely sterile for any length of time.

The more common bacteria found in infected soluble oil systems can degrade the inhibitors, emulsifiers and mineral oil components. They cause a loss of anticorrosion properties, increase of acidity and deterioration of the emulsion. These bacteria thrive in well-aerated systems, and are termed aerobic.

When a system becomes infected with sulfate-reducing bacteria it develops the obnoxious smells associated with putrefaction. Bacteria of the species mainly responsible for such conditions thrive in the absence of oxygen, and are described as anaerobic. The foul smells of hydrogen sulfide are produced mainly in stagnant soluble-oil systems, particularly where the emulsion has a surface layer of oil that tends to exclude oxygen. The hydrogen sulfide arises from chemical reduction of either the inorganic sulfates that can give water its hardness or the petroleum sulfonate component of the soluble oil.

Bacterial examination could be used to confirm the presence of (and perhaps identify) the species. Where soluble oil is being degraded, or where a noticeable loss of anti-corrosion properties has been detected, a check on the pH (alkalinity/acidity) of the system would be advantageous.

Cleanliness and good housekeeping in machine shops do much to avoid bacterial infection, and their importance cannot be overstressed. Various techniques such as heat treatment, centrifuging and filtration can be used to advantage, although economic considerations may restrict their use to systems containing large volumes of soluble oil. Chemical sterilization with bactericides can be more convenient.

Bactericides are substances that destroy bacteria, and they can be used in various ways. They may be incorporated into the soluble-oil concentrate, either at concentrations suitable to protect the oil in storage, or at levels sufficient to provide a persistent bactericidal effect on the emulsion in service. The cost of providing sufficient bactericide to cover the use of the soluble oil at a high dilution might prove prohibitive. Continued use of the same bactericide may produce resistant strains of bacteria.

Bactericides can be added to the soluble-oil system as a shock treatment when infection occurs, but the user must bear in mind that a badly infected and degraded emulsion cannot later be reclaimed. The bactericidal treatment may have to be repeated periodically thereafter, and the effect of the bactericide must be monitored.

Studies of systems infected with bacteria show that a process of continued re-infection occurs from the residual oil left in the system whenever the fluid is changed. As far as possible, the system should be designed to avoid traps that might retain coolant. An effective treatment however, is the use of a bactericide in the form of a detergent sterilizing solution to cleanse the system. All traces of it must be removed from the system before a new charge of coolant is introduced, otherwise emulsion instability may develop in the fresh coolant. Thorough cleansing, combined with the use of a suitable bactericide in the fresh charge of soluble oil, should overcome most problems of bacterial infection.

Neat oils

Two factors militate against the universal use of water-based fluids. Very severe machining operations call for a lubrication performance that is beyond the capacity of such fluids, and the design of some machine tools means that water cannot be used because of the risk of cross-contamination with machine lubricants. In these instances, 'neat' cutting oil is the only fluid that can provide the required performance.

Neat oil is the name given to an orthodox petroleum cutting fluid, whether or not it contains additives, to enhance cutting properties. Oils of this sort are available in a very wide variety, and many combinations of work piece material, machining characteristics and tooling requirement justify special formulations. The neat oils have lower specific heat than water, so they have to be fed to the cutting zone in copious amounts to provide the optimum cutting effect.

The viscosity of neat cutting oil is chosen to suit the machining application. Some operations need a very light oil, others a heavier viscosity because of its better load-carrying capacity. Oils are therefore available in a wide selection of viscosities. Most general-purpose cutting oils have viscosities of about 20 cSt at 40°C (ISO 20 at 104°F), but a lighter viscosity is used in operations such as broaching and deep-hole boring/drilling, when chips have to be flushed away from the confined areas around the tool. Whereas for machining – such as some milling, turning and gear cutting – heavier viscosities, up to 64 cSt at 40°C (ISO 64 at 104°F), are preferred.

Additives in neat oils

The increasing diversity of operations, new materials and processes and the constant demand for improved production efficiency can only be met by various additives and compounding agents being blended into the oil to enhance its performance. Additives tend to be expensive and the selection of enhanced cutting fluid is only justified by overall production economies.

Fatty oils

Small additions of fatty oils improve the 'oiliness' or anti-friction characteristics, and cutting oils reinforced in this way are known as compounded oils. Such oils are particularly useful in the machining of difficult yellow metals and aluminum alloys, giving excellent tool life and good finish without staining yellow metals.

EP oils

For the more difficult operations, neat oils containing EP (extreme-pressure) additives have to be used. The EP cutting oils usually contain additives based on sulfur or chlorine, or combinations of them. The sulfur in EP oil can be present in two forms. In the inactive fluid, it is chemically combined with a fatty-oil additive, which is blended with mineral oil to produce sulfured fatty oil. The active version, on the other hand, contains sulfur in elemental form, dissolved in mineral oil; the fluid is known as sulfured mineral oil. Chlorine is usually present only as chlorinated paraffin, which is blended sometimes singly with mineral oils and sometimes in combination with fatty oils and sulfured additives.

Combinations of additives

Combinations of sulfur and chlorine additives are employed because the chlorine-based agents react at a lower temperature than do the sulfur-based ones, and in some applications, this progressive action is advantageous. ccasionally two kinds of additives, when used in combination, have a greater effect than the sum of their individual contributions. This phenomenon is known as a 'synergistic' effect, and it gives the lubricant technologist more scope for matching the properties of the cutting fluid to the requirements of the machining operation.

Electro discharge machining (EDM) oils

Electro discharge (spark erosion) techniques rely heavily on the ability of EDM oil to act as an electrical insulant, to dissipate heat from the electrode, and to flush away erosion debris from the workpiece. EDM oils also are suitable for all die-sinking spark erosion operations. They should have low aromatic levels, good filterability, low fuming, high dielectric strength, excellent oxidation resistance and low color level.

A low-viscosity grade helps provide maximum flushing and cooling for delicate work and close tolerances. When rough and finishing machining is combined, a medium viscosity is recommended for the good flushing needed. In roughing operations, high-viscosity oil enables fast metal removal rates to be achieved.

52.12.4 Selection of cutting fluid

Selection of the correct fluid is essential if the maximum benefit is to be obtained for the user. In some machining operations and with some workpiece metals, of course, the choice is straightforward, but in the majority of cases, selection is inevitably a compromise in which several factors have to be weighed. The two most important considerations are the workpiece material and the type of machining operation (which, in turn, involves considerations of the tool material).

52.12.5 Workpiece material

A general guide to metals and their difficulty of machining, beginning with the most difficult, could be listed as:

1. Titanium
2. Nimonic alloys
3. Inconel
4. Nickel
5. Stainless steel
6. Monel metal
7. High-tensile steel
8. Wrought iron and cast alloys
9. High-tensile bronze
10. Copper
11. Medium- and low-carbon steels
12. Free-cutting mild steel
13. Aluminum alloys
14. Brass
15. Bronze
16. Zinc-based alloys
17. Free-cutting brass
18. Magnesium alloys

52.12.6 Machining operations

Operations vary in severity, but the following list gives a general indication of how they compare, starting with the most severe:

Broaching (internal and surface)
Gear cutting and shaving
Deep-hole boring (ejector drill or gun drilling)
Tapping and threading
Multi- and single-spindle automatics
Capstan lathes, central lathes
Milling and drilling
Sawing

52.12.7 Tool material

A third important factor in the economies of machining is the material of the cutting tool. This largely determines the rates of metal removal, the standards of surface finish and the frequency at which the tool needs to be reground – all of which are interrelated. These can be broadly grouped in three categories, each separated by a factor of 10 in terms of performance.

1. Ceramic-coated disposable inserts, including silicon nitride, boron nitride, titanium nitride (TIN), titanium carbide (TIC) and sintered synthetic diamond;
2. Tungsten carbide, usually in the form of tipped tooling;
3. High-speed steel.

52.12.8 Ancillary factors

Although the three sets of factors just given can be considered singly or in combination, and although a coolant-selection chart can be used as a guide, the most obvious choice of coolant is not always satisfactory from every aspect. Among other (and probably overriding) considerations are the economics of production, scope for reclaiming the coolant, health and safety requirements, and finally the suitability of the machine's design – because of possible incompatibility of the coolant and the machine's lubricant. In time, moreover, the performance of any chosen coolant can be seriously affected by variations of cutting speed, feed rate, tool-setting accuracy and even the condition of the machine. Incorrect tool geometry and inaccurate grinding may have their own adverse effects.

The chosen fluid, to be most effective, must be kept in good condition, supplied in the correct quantity and directed accurately towards the cutting area. If its pressure and rate of flow are too high, the resultant splattering causes wastage of coolant and untidy splashes around the machine and may well lead to overheating in the cutting zone because insufficient fluid is reaching it. The fluid must be applied where it is most effective – pointed either directly at the tip of the tool or flooding the complete cutting zone. Too often, the flow is directed vaguely at the workpiece, only to be deflected by the workpiece itself or by ancillary parts of the machine.

Poor maintenance practices lead to dirty and degraded coolant, which affects tool life, surface finish and perhaps the welfare of the operator. Proper filtration is vital, and care must be taken that contamination from other coolants or machine lubricants is kept to a minimum.

52.12.9 Grinding

Although primarily regarded as a finishing process, grinding has also been developed for heavy stock removal. In effect, the grinding wheel is a multi-point machining tool because many fragments of very hard material (the grit) embedded in a softer matrix (the bond) cut the metal. Each cutting edge requires a lubricant for clean cutting and a coolant for the dissipation of heat. Clearly, the cutting fluid must satisfy the requirements of both the workpiece material and the grinding wheel, and before a grinding fluid is selected, a check must be made that the correct grade of wheel is being used for the operation.

The cutting fluid has to cool and lubricate without allowing grinding debris to clog (or load) the surface of the wheel. This could produce a burnished surface rather than a ground one – which might appear satisfactory but would be metallurgically unsound.

For many grinding operations, translucent soluble oils or synthetic solutions give the best results. Synthetic solutions are especially favored for their very high clarity on intricate internal grinding. On the more complex and heavy-duty operations involving form grinding, more importance is attached to lubricating properties, so EP neat oils are favored. The level of EP activity required is determined by the severity of the form being ground.

52.12.10 Storage

Extremes of temperature should be avoided, particularly in the storage of coolants that contain water, because the balance of constituents can be upset. High temperatures lead to evaporation of water. Low temperatures can cause the separation of some of the components in additive oils; in particular, the natural fatty oils are susceptible to coagulation. Soluble oil, as supplied in concentrated form, usually contains a small amount of water for ensuring rapid dispersion during subsequent dilution. If this water freezes, it does not re-blend when the temperature rises again, and subsequently the oil proves very difficult to mix.

52.12.11 Water quality

Water from towns' main supplies is usually suitable for the preparation of water-based cutting fluids. That from factory bore holes is also generally suitable, although occasionally it contains excessive amounts of corrosive salts. Water from rivers, canals and ponds usually contains undesirable contaminants, and should be tested before use. A good first test is to mix a small quantity of emulsion and allow it to stand for 24 hours: in this time, no more than a trace of the oil should separate. If serious separation occurs, the water should be analyzed to indicate the sort of remedial treatment required.

A laboratory check is normally desirable to assess the amounts of organic and mineral acids present, and a check on hardness is usually necessary in any case. Hardness in water is due to dissolved salts – mainly of calcium, magnesium and iron, and occasionally of aluminum. Softening may be required if the water is extremely hard, because the salts react with the emulsifier in the soluble oil to form an insoluble scum that floats on the surface of the emulsion. The scum may not in itself be harmful, but its formation uses up some of the emulsifier and causes the emulsion to be unstable.

52.12.12 Cutting fluids in service

Neat oils

During service, neat cutting oils suffer negligible deterioration of quality; their service life is almost indefinitely long, provided they are kept clean and free from contamination. They suffer very little depletion of their additives through cutting action, and quality is adequately maintained by additions of new oil to make up for 'drag-out' losses (those caused by oil adhering to both the scarf and the work). Serious leakage of hydraulic fluid or lubricating oil into the coolant system can have a diluting effect, which reduces additive concentration and hence the performance of the cutting oil.

Water-based fluids

As long as they are kept clean and maintained at the correct concentration, water-based fluids have a long service life. The concentration of an emulsion can be affected in two ways: by evaporation of water and by drag-out of the oil content. In many cases, the effects counteract each other, so the top-up fluid is given the same concentration as the working fluid. Solutions, on the other hand, tend to become more concentrated during service because they suffer from evaporation of water but not from drag-out losses of the other constituents, so top-up solutions are generally weaker than working solutions.

It is advisable to monitor water-based fluids at frequent intervals so that their water content can be brought to within tolerance before the machining operation is affected. Maintenance of the correct concentration ensures that cutting performance and rust-preventive properties stay at the required level. A simple technique makes use of the fact that a fluid's density influences its refractive index. In this method, a very small amount of the fluid is viewed in a pocketsize instrument called a refractometer, and the density is indicated by the edge of a dark shadow falling on a graduated line. The line gives a figure that, in some cases, is a direct reading of dilution, but in others, it may have to be correlated with dilution on a graph drawn for the particular fluid.

The strength of a solution or emulsion is, of course, corrected by additions of stronger or weaker fluid. The percentage concentration of make-up fluid is calculated by the formula:

$$C = \frac{C_e}{Q_r}(C_r - C_e) + C_f$$

where

C_f = required final concentration,
C_e = existing concentration,
Q_e = existing quantity,
Q_r = quantity required of make-up fluid.

Cleaning procedures

Although some neat cutting fluids can be used almost indefinitely if well maintained, most fluids eventually reach the end of their useful life, so the system has to be drained and refilled with new fluid. On such occasions, the system should be thoroughly cleaned out before the new charge is poured in. The nature of cleaning depends on the type of cutting fluid.

Normally a system containing neat oil requires only manual cleaning because the debris contained is likely to be mainly metallic, but water-based fluids call for more searching treatment. Caustic soda and proprietary solutions are available for flushing away slimy deposits. If there is evidence of bacterial degradation, however, bactericides must be introduced. These tend to persist in remote pockets and in the more stagnant parts of a coolant system, and manual cleaning by itself is inadequate. The risk of persistence can be significantly minimized at the planning stage if the system is designed to be free of regions to trap stagnant fluid.

52.13 Compressors

Compressors fall into two basic categories: positive displacement types, in which air is compressed by the 'squashing' effect of moving components; and dynamic (turbo)-compressors, in which the high velocity of the moving air is converted into pressure. In some compressors, the oil lubricates only the bearings, and is not exposed to the air; in some, it serves an important cooling function; in some, it is in intimate contact with the oxidizing influence of hot air and with moisture condensed from the air. Clearly,

there is no such thing as typical all-purpose compressor oil: each type subjects the lubricant to a particular set of conditions. In some cases good engine oil or turbine-quality oil is suitable, but in others, the lubricant must be special compressor oil (Figure 52.13).

52.13.1 Quality and safety

Over the years, the progressive improvements in compressor lubricants have kept pace with developments in compressor technology, and modern oils make an impressive contribution to the performance and longevity of industrial compressors. More recently, a high proportion of research has been directed towards greater safety, most notably in respect of fires and explosions within compressors. For a long time the causes of such accidents were a matter of surmise, but it was noticed that the trouble was almost invariably associated with high delivery temperatures and heavy carbon deposits in delivery pipes. Ignition is caused by an exothermic (heat-releasing) oxidation reaction with the carbon deposit, which creates temperatures higher than the spontaneous ignition temperature of the absorbed oil.

Experience indicates that careful selection of base oils and anti-oxidation additives considerably reduce such deposits. Nevertheless, the use of top-class oil is no guarantee against trouble if maintenance is neglected. For complete safety, both the oil and the compressor system must enjoy high standards of care.

52.13.2 Specifications

The recommendations of the International Standards Organization (ISO) covering mineral-oil lubricants for reciprocating compressors are set out in ISO DP 6521, under the ISO-L-DAA and ISO-L-DAB classifications. These cover applications wherever air-discharge temperature are, respectively, below and above 160°C (329°F). For mineral-oil lubricants used in oil-flooded rotary-screw compressors the classifications ISO-L-DAG and DAH cover applications where temperatures are, respectively, below 100°C (212°F) and in the 100–110°C range. For more severe applications, where synthetic lubricants might be used, the ISO-L-DAC and DAJ specifications cover both reciprocating and oil-flooded rotary-screw requirements.

For the general performance of compressor oils, there is DIN 51506. This specification defines several levels of performance, of which the most severe – carrying the code letters VD-L – relates to oils for use at air-discharge temperatures of up to 220°C (428°F).

The stringent requirements covering oxidation stability are defined by the test method DIN 51352, Part 2, known as the Pneurop Oxidation Test (POT). This test simulates the oxidizing effects of high temperature, intimate exposure to air, and the presence of iron oxide, which acts as catalyst – all factors highly conducive to the chemical breakdown of oil, and the consequent formation of deposits that can lead to fire and explosion.

Rotary-screw compressor mineral oils oxidation resistance is assessed in a modified Pneurop oxidation test using iron naphthenate catalyst at 120°C (250°F)

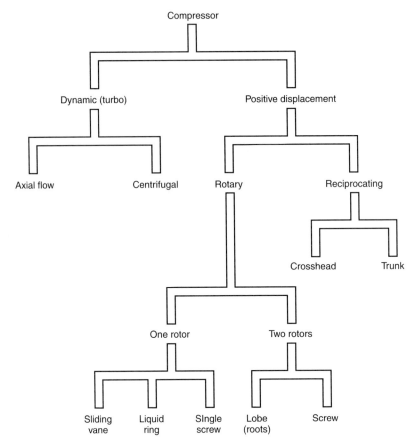

Figure 52.13 Compressor types

for 1000 hours. This is known as the rotary-compressor oxidation test (ROCOT).

52.13.3 Oil characteristics

Reciprocating compressors

In piston-type compressors, the oil serves three functions in addition to the main one of lubricating the bearings and cylinders. It helps to seal the fine clearances around piston rings, piston rods and valves, and thus minimizes blow-by of air (which reduces efficiency and can cause overheating). It contributes to cooling by dissipating heat to the walls of the crankcase and it prevents corrosion that would otherwise be caused by moisture condensing from the compressed air.

In small single-acting compressors the oil to bearings and cylinders is splash-fed by flingers, dippers or rings, but the larger and more complex machines have force-feed lubrication systems, some of them augmented by splash-feed. The cylinders of a double-acting compressor cannot be splash lubricated, of course, because they are not open to the crankcase. Two lubricating systems are therefore necessary – one for the bearings and crosshead slides and

one feeding oil directly into the cylinders. In some cases the same oil is used for both purposes, but the feed to the cylinders has to be carefully controlled, because under-lubrication leads to rapid wear and over-lubrication leads to a build-up of carbon deposits in cylinders and on valves. The number and position of cylinder-lubrication points varies according to the size and type of the compressor. Small cylinders may have a single point in the cylinder head, near the inlet valve; larger ones may have two or more. In each case, the sliding of the piston and the turbulence of the air spread the oil.

In the piston-type compressor the very thin oil thin has to lubricate the cylinder while it is exposed to the heat of the compressed air. Such conditions are highly conducive to oxidation in poor-quality oils, and may result in the formation of gummy deposits that settle in and around the piston-ring grooves and cause the rings to stick, thereby allowing blow-by to develop.

Rotary compressors – vane type

The lubrication system of vane-type compressors varies according to the size and output of the unit. Compressors in the small and 'portable' group have neither external cooling nor intercooling, because to effect all the necessary cooling the oil is injected copiously into the incoming air stream or directly into the compressor chamber. This method is known as flood lubrication, and the oil is usually cooled before being recirculated. The oil is carried out of the compression chamber by the air, so it has to be separated from the air; the receiver contains baffles that 'knock out' the droplets of oil, and they fall to the bottom of the receiver. Condensed water is subsequently separated from the oil in a strainer before the oil goes back into circulation.

Vane-type pumps of higher-output are water-jacketed and inter-cooled: the lubricant has virtually no cooling function so it is employed in far smaller quantities. In some units, the oil is fed only to the bearings, and the normal leakage lubricates the vanes and the casing. In others, it is fed through drillings in the rotor and perhaps directly into the casing. This, of course, is a total-loss lubrication technique, because the oil passes out with the discharged air.

As in reciprocating units, the oil has to lubricate while being subjected to the adverse influence of high temperature. The vanes impose severe demands on the oil's lubricating powers. At their tips, for example, high rubbing speeds are combined with heavy end-pressure against the casing.

Each time a vane is in the extended position (once per revolution) a severe bending load is being applied between it and the side of its slot. The oil must continue to lubricate between them, to allow the vane to slide freely. It must also resist formation of sticky deposits and varnish, which lead to restricted movement of the vanes and hence to blow-by and, in severe cases, to broken vanes.

Rotary compressors – screw type

The lubrication requirements for single-screw type compressors are not severe, but in oil-flooded rotary units, the oxidizing conditions are extremely severe because fine droplets of oil are mixed intimately with hot compressed air. In some screw-type air compressors, the rotors are gear driven and do not make contact. In others, one rotor drives the other. The heaviest contact loads occur where power is transmitted from the female to the male rotor: here the lubricant encounters physical conditions similar to those between mating gear teeth. This arduous combination of circumstances places a great demand on the chemical stability, and lubricating power, of the oil.

Other types

Of the remaining designs, only the liquid-piston type delivers pressures of the same order as those just mentioned. The lobe, centrifugal and axial-flow types are more accurately termed 'blowers', since they deliver air in large volumes at lower pressures. In all four cases only the 'external' parts – bearings, gears or both – require lubrication. Therefore, the oil is not called upon to withstand the severe service experienced in reciprocating and vane-type compressors. Where the compressor is coupled to a steam or gas turbine a common circulating oil system is employed. High standards of system cleanliness are necessary to avoid deposit formation in the compressor bearings.

Refrigeration compressors

The functions of a refrigerator compressor lubricant are the same as those of compressor lubricants in general. However, the close association between refrigerant and lubricant does impose certain additional demands on the oil. Oil is unavoidably carried into the circuit with refrigerant discharging from the compressor. In many installations, provision is made for removal of this oil. However, several refrigerants, including most of the halogen refrigerants, are miscible with oil and it is difficult to separate the oil that enters the system, which therefore circulates with the refrigerant. In either case the behavior of the oil in cold parts of the systems is important, and suitable lubricants have to have low pour point and low wax-forming characteristics.

Effects of contamination

The conditions imposed on oils by compressors – particularly by the piston type – are remarkably similar to those imposed by internal combustion engines. One major difference is, of course, that in a compressor no fuel or products of combustion are present to find their way into the oil. Other contaminants are broadly similar. Among these are moisture, airborne dirt, carbon and the products of the oil's oxidation. Unless steps are taken to combat them, all these pollutants have the effect of shortening the life of both the oil and the compressor, and may even lead to fires and explosions.

Oxidation

High temperature and exposure to hot air are two influences that favor the oxidation and carbonization of mineral oil. In a compressor, the oil presents a large surface area to hot air because it is churned and sprayed in a fine mist, so the oxidizing influences are very strong – especially in the high temperatures of the compressor chamber. The degree of oxidation is dependent mainly on temperature and the ability of the oil to resist, so the problem can be minimized by the correct selection of lubricant and by controlling operating factors.

In oxidizing, oil becomes thicker and it deposits carbon and gummy, resinous substances. These accumulate in the piston-ring grooves of reciprocating compressors and in the slots of vane-type units, and as a result, they restrict free movement of components and allow air leakage to develop. The deposits also settle in and around the valves of piston-type compressors, and prevent proper sealing.

When leakage develops, the output of compressed air is reduced, and overheating occurs due to the recompression of hot air and the inefficient operation of the compressor. This leads to abnormally high discharge temperatures. Higher temperature leads to increased oxidation and hence increased formation of deposits, so adequate cooling of compressors is very important.

Airborne dirt

In the context of industrial compressors, dust is a major consideration. Such compressors have a very high through-put of air, and even in apparently 'clean' atmospheres, the quantity of airborne dirt is sufficient to cause trouble if the compressor is not fitted with an air-intake filter. Many of the airborne particles in an industrial atmosphere are abrasive, and they cause accelerated rates of wear in any compressor with sliding components in the compressor chamber. The dirt passes into the oil, where it may accumulate and contribute very seriously to the carbon deposits in valves and outlet pipes. Another consideration is that dirt in oil is likely to act as a catalyst, thus encouraging oxidation.

Moisture

Condensation occurs in all compressors, and the effects are most prominent where cooling takes place – in intercoolers and air-receivers, which therefore have to be drained at frequent intervals. Normally the amount of moisture present in a compression chamber is not sufficient to affect lubrication, but relatively large quantities can have a serious effect on the lubrication of a compressor. Very wet conditions are likely to occur when the atmosphere is excessively humid, compression pressures are high, or the compressor is being overcooled.

During periods when the compressor is standing idle, the moisture condenses on cylinders walls and casings, and if the oil does not provide adequate protection, this leads to rusting. Rust may not be serious at first sight, and it is quickly removed by wiping action when the compressor is started, but the rust particles act as abrasives, and if they enter the crankcase oil they may have a catalytic effect and promote oxidation. In single-acting piston-type compressors, the crankcase oil is contaminated by the moisture.

52.14 Turbines

52.14.1 Steam

Although the properties required of a steam-turbine lubricant are not extreme, it is the very long periods of continuous operation that creates the need for high-grade oils to be used. The lubricating oil has to provide adequate and reliable lubrication, act as a coolant, protect against corrosion, as a hydraulic medium when used in governor and control systems, and if used in a geared turbine provide satisfactory lubrication of the gearing. The lubricant will therefore need the following characteristics.

Viscosity For a directly coupled turbine for power generation, a typical viscosity would be in the range of 32–46 cSt at 40°C (ISO 32–46 at 104°F). Geared units require a higher viscosity to withstand tooth loading typically within the range of 68–100 cSt at 40°C (ISO 68–100 at 104°F).

Oxidation resistance The careful blending of turbine oils, using components which, by selective refining, have

a reduced tendency to oxidize, produces the required long-term stability. The high temperatures and pressures of modern designs add to these demands, which are combated by the incorporation of suitable anti-oxidant additives.

Demulsibility The ability of the lubricant to separate readily and completely from water, in either a centrifuge or a settling tank, is important in a turbine lubricant. Otherwise, the retained water will react with products of oxidation and particle contaminants to form stable emulsions. These will increase the viscosity of the oil and form sludge that can result in a failure. Careful and selective refining ensures a good demulsibility characteristic. Inadequate storage and handling can seriously reduce this property.

Corrosion resistance Although the equipment is designed to keep the water content at a minimum level, it is virtually impossible to eliminate it. The problem of rusting is therefore overcome by using corrosion inhibitors in the lubricant formulation.

Foaming resistance Turbine oils must be resistant to foaming, since oil-foam reduces the rate of heat transfer from the bearings, promotes oxidation by greatly extending the area of contact between air and oil. It is also an unsatisfactory medium for the hydraulic governor controls. Careful refining is the primary means of achieving good resistance to foaming. Use of an anti-foam additive may seem desirable but this should be approached with caution. If it is used in quantities higher than the optimum, it can in fact assist air entrainment in the oil by retarding the release of air bubbles.

52.14.2 Gas

The lubricants generally specified for conventional gas turbines invariably fall within the same classification as those used for steam turbines and are often categorized as 'turbine oils'. In those cases where an aircraft type gas turbine has been adapted for industrial use the lubricant is vitally important to their correct operation. Specifications have been rigidly laid down after the most exhaustive tests, and it would be unwise, even foolhardy, to depart from the manufacturers' recommendations. No economic gain would result from the use of cheaper, but less efficient, lubricants.

52.14.3 Performance standards

In the USA, there is the ASTM standards and the well-known General Electric requirements. The total useful life of turbine oil is its most important characteristic. ASTM method D943 (IP 157) measures the life indirectly by assessing the useful life of the oxidation inhibitor contained in the formulation and are often referred to as the TOST 'life' of the oil. Rust prevention is generally assessed by the ASTM D665 (IP 135) method.

There are many other specifications designed by equipment builders, military and professional societies, as well as users. Care always needs to be taken when purchasing turbine oil to specification. The cheapest oil, albeit conforming to the specification, may not necessarily be the best within that specification for the particular purpose. For instance, the additive package is rarely (if ever) defined, so that unexpected reactions can occur between oils, which could affect overall performance.

52.15 Transformers and switchgear

The main requirement for power-transmission equipment oil is that it should have good dielectric properties. Oil used in transformers acts as a coolant for the windings; as an insulant to prevent arcing between parts of the transformer circuits; and prevents the ionization of minute bubbles of air and gas in the wire insulation by absorbing them and filling the voids between cable and wrapping. In switchgear and circuit breakers, it has the added function of quenching sparks from any arc formed during equipment operation. Oils for use in power transmission equipment should have the following properties: high electric strength, low viscosity, high chemical stability and low carbon-forming characteristics under the conditions of electric arc.

52.15.1 Performance standards

The efficiency of transformer oils as dielectrics is measured by 'electric strength' tests. These give an indication of the voltage at which, under the test conditions, the oil will break down. Various national standards exist that all measure the same basic property of the oil. There is an international specification, IEC 296/1982, which may be quoted by equipment manufacturers in their oil recommendations.

52.15.2 Testing

How frequently the oil condition should be tested depends on operating and atmospheric conditions; after the commissioning sample, further samples should be taken at three months and one year after the unit is first energized. After this, under normal conditions, testing should be carried out annually. In unfavorable operating conditions (damp or dust-laden atmospheres, or where space limitations reduce air circulation and heat transfer) testing should be carried out every six months.

Testing should include a dielectric strength test to confirm the oil's insulation capability and an acidity test, which indicates oil oxidation. While acid formation does not usually develop until the oil has been in service for some time when it does occur the process can be rapid. If acidity is below 0.5 mg KOH/g no action would seem necessary. Between 0.5 and 1 mg KOH/g, increased care and testing is essential. Above 1 the oil should be removed and either reconditioned or discarded. Before the unit is filled with a fresh charge of oil it should be flushed.

Sludge observations will show if arcing is causing carbon deposits which, if allowed to build up will affect heat transfer and could influence the oil insulation. There is also a flash point test, in which any lowering of flash point is an indication that the oil has been subjected to excessive local heating or submerged arcing (due to overload or an internal electrical fault). A fall in flash point exceeding 16°C (61°F) implies a fault, and the unit should be shut down for investigation of the cause. Lesser drops may be observed in the later stages of oil life, due to oxidation effects, but are not usually serious. A 'crackle' test is a simple way of detecting moisture in the oil. Where water is present, the oil should be centrifuged.

52.16 Greases

Grease is a very important and useful lubricant when used correctly, its main advantage being that it tends to remain where it is applied. It is more likely to stay in contact with rubbing surfaces than oil, and is less affected by the forces of gravity, pressure and centrifugal action. Economical and effective lubrication is the natural result of this property and a reduction in the overall cost of lubrication, particularly in all-loss systems, is made possible.

Apart from this, grease has other advantages. It acts both as a lubricant and as a seal and is thus able, at the same time as it lubricates, to prevent the entry of contaminants such as water and abrasive dirt. Grease lubrication by eliminating the need for elaborate oil seals can simplify plant design.

Because a film of grease remains where it is applied for much longer than a film of oil, it provides better protection to bearing and other surfaces that are exposed to shock loads or sudden changes of direction. A film of grease also helps to prevent the corrosion of machine parts that are idle for lengthy periods.

Bearings pre-packed with grease will function for extended periods without attention. Another advantage is the almost complete elimination of drip or splash, which can be a problem in certain applications. Grease is also able to operate effectively over a wider range of temperatures than any single oil.

There are certain disadvantages as well as advantages in using grease as a lubricant. Greases do not dissipate heat as well as fluid lubricants, and for low-torque operation tend to offer more resistance than oil.

52.16.1 Types of grease

The general method of classifying greases is by reference to the type of soap that is mixed with mineral oil to produce the grease, although this has rather less practical significance nowadays than it had in the past. One example of this is the multi-purpose grease that may replace two or three different types previously thought necessary to cover a particular field of application. Nevertheless, there are unique differences in behavior between greases made with different metal soaps, and these differences are still important in many industrial uses, for technical and economic reasons.

Calcium-soap greases

The lime-soap (calcium) greases have been known for many years but are still probably the most widely used.

They have a characteristic smooth texture, thermal stability, good water resistance and are relatively inexpensive. The softer grades are easily applied, pump well and give low starting torque. Their application is limited by their relatively low drop points, which are around 100°C (212°F). This means that, in practice, the highest operating temperature is about 50°C (122°F).

Nevertheless, they are used widely for the lubrication of medium-duty rolling and plain bearings, centralized greasing systems, wheel bearings and general duties. The stiffer varieties are used in the form of blocks on the older-type brasses. Modifications of lime-base grease include the graphited varieties and those containing an extreme pressure additive. The latter are suitable for heavily loaded roller bearings such as in steel-mill applications.

Sodium-soap greases

The soda-soap (sodium) greases were, for some considerable time, the only high-melting-point greases available to industry. They have drop points in the region of 150°C (302°F) and their operating maximum is about 80°C (176°F). These greases can be 'buttery', fibrous or spongy, are not particularly resistant to moisture and are not suitable for use in wet conditions. Plain bearings are very frequently lubricated with soda-based greases.

For rolling-contact bearings, a much smoother texture is required, and this is obtained by suitable manufacturing techniques. Modified grades may be used over the same temperature range as that of the unmodified grade and, when they are correctly formulated, have a good shear resistance and a slightly better resistance to water than the unmodified grades.

Lithium-soap greases

These products, unknown before the Second World War, were developed first as aircraft lubricants. Since then, the field in which they have been used has been greatly extended and they are now used in industry as multipurpose greases. They combine the smooth texture of the calcium-based greases with higher melting points than soda-soap greases, and are almost wholly manufactured in the medium and soft ranges. Combined with suitable additives, they are the first choice for all rolling-contact bearings, as they operate satisfactorily up to a temperature of 120°C (250°F) and at even higher for intermittent use. Their water resistance is satisfactory and they may be applied by all conventional means, including centralized pressure systems.

Other metal-soap greases

Greases are also made from soaps of strontium, barium and aluminum. Of these, aluminum-based grease is the most widely used. It is insoluble in water and very adhesive to metal. Its widest application is in the lubrication of vehicle chassis. In industry, it is used for rolling-mill applications and for the lubrication of cams and other equipment subject to violent oscillation and vibration, where its adhesiveness is an asset.

Non-soap thickened greases

These are generally reserved for specialist applications, and are in the main more costly than conventional soap-based greases. The most common substances used as non-soap thickeners are silica and clays prepared in such a way that they form gels with mineral and synthetic oils. Other materials that have been used are carbon black, metal oxides and various organic compounds.

The characteristic of these non-soap greases which distinguishes them from conventional greases is that many of them have very high melting points; they will remain as greases up to temperatures in the region of 260°C (500°F). For this reason, the limiting upper usage temperature is determined by the thermal stability of the mineral oil or synthetic fluid of which they are composed. Applications such as those found in cement manufacturing, where high-temperature conditions have to be met, require grease suitable for continuous use at, say, 204°C (400°F).

Although it is difficult to generalize, the non-soap products have, overall, been found to be somewhat less effective than the soap-thickened greases as regards lubricating properties and protection against corrosion, particularly rusting. Additive treatment can improve non-soap grades in both these respects, but their unique structures render them more susceptible to secondary and unwanted effects than is the case with the more conventional greases.

Filled greases

The crude types of axle and mill grease made in the early days frequently contained large amounts of chemically inert, inorganic powders. These additions gave 'body' to the grease and, possibly, helped to improve the adherence of the lubricating film. Greases are still 'filled' but in a selective manner with much-improved materials and under controlled conditions. Two materials often used for this purpose are graphite and molybdenum disulfide.

Small amounts (approximately 5 per cent) of filler have little or no effect on grease structure, but large amounts increase the consistency. However, the materials mentioned are lubricants in themselves and are sometimes used as such. Consequently, it is often claimed that when they are incorporated into the structure of the grease the lubricating properties of the grease are automatically improved. A difference of opinion exists as to the validity of this assumption, but it is true that both molybdenum disulfide and graphite are effective where shock loading or boundary conditions exist, or when the presence of chemicals would tend to remove conventional greases.

Mixing greases

The above comments on the properties of the various types of grease have shown that very real differences exist. Each one has its own particular type of structure, calls for individual manufacturing processes and has its own advantages and disadvantages. It is because of these distinct differences that the mixing of greases should never be encouraged. If greases of different types are mixed indiscriminately there is a risk that one or other of them will suffer, the resulting blend being less stable than either of the original components and the blend may even liquefy.

52.16.2 Selecting a grease

A few brief notes on the fundamental factors that influence a choice of grease may be found helpful. The first essential is to be clear about the limitations of the different types, and to compare them with the conditions they are to meet. Table 52.10 gives the characteristics of high-quality greases.

Greases with a mixed base are not shown in the table because, in general, they are characterized by the predominant base; for example, a soda-lime grease behaves like a soda grease. The required length of service may modify temperature limits. Thus, if soda grease requires having only a short life, it could be used at temperatures up to 120°C (250°F).

When the type most suitable for a particular application has been chosen, the question of consistency must be considered. The general tendency over the last two decades has been towards softer grease than formerly used. Two factors have probably contributed to this trend; the growth of automatic grease dispensing and the use of more viscous oils in grease making.

In practice, the range of grease consistency is quite limited. For most general industrial applications, a No.2 consistency is satisfactory. Where suitability for pumping is concerned, a No. 1; for low temperatures, a No. 0; and for water pumps and similar equipment, a No. 3 may be better selections.

52.16.3 Grease application

In applying lubricating grease, the most important aspect is how much to use. Naturally, the amount varies with the component being serviced, but some general rules can be laid down. All manufacturers agree that anti-friction bearings should never be over-greased. This is particularly true of high-speed bearings, in which the churning of excess lubricant leads to overheating. The rise in temperature of a bearing as the amount of grease increases has been recorded. With the bearing housing one-third full, the temperature was 39°C (102°F); at two-thirds full the temperature rose to 42°C (108°F); and with a full charge of grease it went up to 58°C (136°F).

The general recommendations for grease packing are:

1. Fully charge the bearing itself with grease ensuring that it is worked around and between the rolling elements.

Table 52.10 Characteristics of high-quality greases

Grease (type of soap)	Recommended maximum operating temperature (°C)	Water resistance	Mechanical stability
Lime	50	Good	Good
Soda	80	Poor	Good
Lithium	120	Good	Good
Aluminum	50	Fair	Moderate

2. Charge the bearing housing one-half to two-thirds full of grease.

Churning, and its attendant high temperature, may change the structure of the grease permanently, in which event softening may result in leakage and stiffening in lubricant starvation. There is no fixed rule for the period between re-greasing, since this depends on the operating conditions. Most recommendations suggest inspection and possible replenishment every six or twelve months, though the general tendency as grease quality improves has been to extend this period. The higher the temperature of a machine, the more frequently it must be greased because of possible losses of softened lubricant or changes in its structure.

It is not always incorrect to over-grease. With a sleeve bearing, for instance, gun pressure may be maintained until old grease exudes from the ends of the bearing, and the same is true of spring shackles. For the sake of economy and cleanliness, however, this should never be overdone.

52.17 Corrosion prevention

Most plant has to work under adverse conditions, in all sorts of weather, and subject to contamination by various agents. However, as long as it is in use it can be reasonably sure of receiving at least a minimum amount of regular maintenance and attention, and this will reduce the likelihood of working parts being attacked by corrosion when plant is in service. However, when plant has to be laid up until required, no matter how carefully matters have been planned, corrosion is always a serious possibility. Modern machinery, with highly finished surfaces, is especially susceptible to atmospheric attack. The surfaces of components also require protection during transport and storage.

Even today, rusting of industrial plant and material is accepted by some as an inevitable operating expense. There is no necessity for this attitude, however, as the petroleum industry has evolved effective, easily applied temporary protectives against corrosion, which are well suited to the conditions met in practice.

52.17.1 Categories of temporary corrosion preventives

Temporary corrosion preventives are products designed for the short-term protection of metal surfaces. They are easily removable, if necessary, by petroleum solvents or by other means such as wiping or alkaline stripping. Some products for use in internal machine parts are miscible and compatible with the eventual service lubricant, and do not, therefore, need to be removed.

The major categories of temporary corrosion preventives are:

Soft-film protectives

Dewatering fluids giving soft/medium films
Non-dewatering fluids giving soft films

Hot-dip compounds
Greases

Hard-film protectives

Oil-type protectives
General-purpose
Engine protectives

The development of products in these categories has been guided by known market demands and many manufacturers have made use of established specifications for temporary protectives.

52.17.2 Selection of a corrosion preventive

Temporary corrosion preventives are in some cases required to give protection against rusting for periods of only a few days for inter-process waiting in factories. Where the protected components are not exposed to the weather, protection can be given for up to a year or more for stored components in internal storage conditions. On the other hand, components may require protection for a few days or even weeks under the most adverse weather conditions. Some components may have to be handled frequently during transit or storage. In general, therefore, the more adverse the conditions of storage, the longer the protective periods, and the more frequent the handling, the thicker or more durable the protective film must be.

Because of the wide variation in conditions of exposure, it is not possible to define the length of protection period except in general terms. Solvent-deposited soft films will give protection from a few days to months indoors and some weeks outdoors; a solvent-deposited medium film will give long-term protection indoors and medium-term protection outdoors. Hot-dip compounds and cold-applied greases give films that can withstand considerable handling and will give medium to long protection. Solvent-deposited hard-film protectives will give long-term protection but are difficult to remove. Oil protectives give short- to medium-term protection of parts not subject to handling and are much used for the preservation of internal working parts; they need not be removed and can in some instances serve as lubricating oils.

Short term, medium term and long term are expressions that are not rigorously defined but are generally accepted as meaning of the order of up to 6 months, 12 months and 18 months, respectively, in temperate climates. Where local conditions are more severe (in hot, humid climates, for example) the protection periods are less. These protection periods are related to the preventive film alone, but where transit or storage conditions call for wrapping or packaging then longer protection periods can be obtained.

The distinction between a simple part and a complex assembly is an important factor in selecting a temporary protective. The solvent-containing protectives may not be suited to treating assemblies, because:

1. Assemblies may contain non-metallic parts (rubber, for example) that could be attacked by the solvent;

2. The solvent cannot evaporate from enclosed or shielded spaces and the intended film thickness will not be obtained;
3. Evaporated solvent could be trapped and could then leach away the protective film.

Hence, the hot-dip compounds, or greases smeared cold, are better for assemblies with non-metallic parts masked if necessary. Solvent-containing protectives therefore find greater application in the protection of simple parts or components. The available means of application, the nature of any additional packaging and the economics and scale of the protective treatment are further factors that influence the choice of type of temporary corrosion preventive.

52.18 Spray lubricants

There are several applications where the lubrication requirement is specialized and very small, needing precise applications where access is limited because of equipment design or location. In these instances, lubricant application by aerosol is the most suitable method. Extreme-pressure cutting fluid for reaming and tapping, etc., conveyor and chain lubricant, anti-seize and weld anti-spatter agents, release agents, electrical component cleaner and degreasing agents are examples of the ever widening range of products available in aerosol packs.

52.19 Degreasing agents

Often, before any maintenance work starts it is necessary (and desirable) to remove any oil, grease and dirt from the equipment concerned. It may also be necessary to clean replacement components before their installation. Solvents, emulsions and chemical solutions are three broad types of degreasing agents. The method of degreasing (direct onto the surface, by submersion, through degreasing equipment or by steam cleaners), component complexity and the degree of contamination will all have to be taken into account when selecting the type of product to be used.

52.20 Filtration

Some 70–85 per cent of failures and wear problems in lubricated machines are caused by oil contamination. Clean oil extends machine and oil life and gives greater reliability, higher productivity and lower maintenance cost. Hence, some type of filter is an essential part of virtually all lubrication systems.

Cleaning of oil in service may be accomplished quite simply or with relatively complex units, depending on the application and the design of the system. Thus for some operations it is enough to remove particles of ferrous metal from the oil with a magnetic system. In a closed circulatory system, such as that of a steam turbine, the nature of the solids and other contaminants is far more complex, and the treatment has therefore to be more elaborate. In an internal-combustion engine, both air and fuel are filtered as well as crankcase oil.

analysis or ferrography. The former is commonly used in automotive and industrial application for debris up to 10 microns and the latter mainly for industry users covering wear particles over 10 microns. Ferrography is relatively expensive compared with many other techniques, but is justified in capital-intensive areas where the cost is readily offset by quantifiable benefits such as longer machinery life, reduced loss of production, less downtime, etc.

52.28 Health, safety and the environment

There are a wide variety of petroleum products for a large number of applications. The potential hazards and the recommended methods of handling differ from product to product. Consequently, the supplier must provide use of protective clothing, first aid and other relevant information advice on such hazards and on the appropriate precautions. Where there is risk of repeated contact with petroleum products (as with cutting fluids and some process oils) special working precautions are obviously necessary. The aim is to minimize skin contact, not only because most petroleum products are natural skin-degreasing agents but also because with some of them prolonged and repeated contact in poor conditions of personal hygiene may result in various skin disorders.

52.28.1 Health

It is important that health factors are kept in proper perspective. What hazards there may be in the case of oil products are avoided or minimized by simple precautions. For work involving lubricants (including cutting fluids and process oils) the following general precautions are recommended:

- Employ working methods and equipment that minimize skin contact with oil;
- Fit effective and properly positioned splash guards;
- Avoid unnecessary handling of oily components;
- Use only *disposable* wipes;
- Use soluble oils or synthetic fluids at their recommended dilutions only, and avoid skin contact with their 'concentrates'.

In addition to overalls, adequate protective clothing should be provided. For example, a PVC apron may be appropriate for some machining operations. A cleaning service for overalls should be provided and overalls should be cleaned regularly and frequently. Normal laundering may not always be sufficient to remove all traces of oil residues from contaminated clothing. In some instances, dry cleaning may be necessary. Where this applies to cotton overalls they should first be dry-cleaned, then laundered, and preferably starched, in order to restore the fabric's oil repellency and comfort. Generally, dry cleaning followed by laundering is always preferable to minimize the risk of residual contamination wherever heavy, frequent contamination occurs, and when the type of fabric permits such cleaning.

Overalls or personal clothing that become contaminated with lubricants should be removed as soon as possible –

immediately if oil soaked or at the end of the shift if contaminated to a lesser degree. They should then be washed thoroughly or dry cleaned before re-use.

Good washing facilities should be provided, together with hot and cold running water, soap, medically approved skin-cleansers, clean towels and, ideally, showers. In addition, reconditioning creams should be available. The provision of changing rooms, with lockers for working clothes, is recommended.

Workers in contact with lubricants should be kept fully informed by their management of the health aspects and the preventive measures outlined above. Any available government leaflets and/or posters should be prominently displayed and distributed to appropriate workers.

It should be made clear to people exposed to lubricants that good standards of personal hygiene are a most effective protection against potential health hazards. However, those individuals with a history of (or thought to be particularly predisposed to) eczema or industrial dermatitis should be excluded from work where, as in machine-tool operation, contact with lubricants is virtually unavoidable.

Some industrial machining operations generate a fine spray or mist of oil, which forms an aerosol – a suspension of colloidal (ultra-microscopic) particles of oil in air. Oil mist may accumulate in the workshop atmosphere, and discomfort may result if ventilation is inadequate. Inhalation of high concentrations of oil mist over prolonged periods may give rise to irritation of the respiratory tract, and in extreme cases to a condition resembling pneumonia. It is recommended that the concentration of oil mist in the working environment (as averaged over an 8-hour shift) be kept below the generally accepted hygiene standard of 5 mg/m^3. This standard does, however, vary in some countries.

52.28.2 Safety

In the event of accident or gross misuse of products, various health hazards could arise. The data provided by the supplier should outline these potential hazards and the simple precautions that can be taken to minimize them. Guidance should be included on the remedial action that should be taken to deal with medical conditions that might arise. Advice should be obtained from the supplier before petroleum products are used in any way other than as directed.

52.28.3 Environment

Neat oils and water-based coolants eventually reach the end of their working lives, and then the user is faced with the problem of their correct disposal. Under *no* circumstances should neat oils and emulsions be discharged into streams or sewers. Some solutions can, however, be fed into the sewage system after further dilution – but only where permitted.

There are many companies offering a collection service for the disposal of waste lubricating oil. The three main methods employed are:

1. Collection in segregated batches of suitable quality for use by non-refiners

2. Blending into fuel oil
3. Dumping or incineration

Acknowledgements

The author is grateful to BP Oil UK Ltd for their help in writing this chapter and for their permission to reproduce the figures and tables.

Further reading

BP publications

Industrial lubrication
Machine tools and metal cutting
Lubricants for heavy industry
Gear lubrication
Hydraulic fluids
Machine shop lubricants
Cutting fluids
Compressor lubrication
Greases
Temporary corrosion preventives
Degreasants
Storage and handling of lubricants
Aerosols
Health, safety and environmental data sheets

Further information on lubrication can be obtained from:

Booklets and leaflets published by most oil suppliers
Libraries of Institute of Plant Engineers
  Institution of Production Engineers
  Institute of Petroleum
  Institution of Electrical Engineers
  Institution of Mechanical Engineers
Literature published by additive companies
Literature published by the American Society of Lubrication Engineers
Libraries of universities

With such a wide and important subject it is not possible to provide a full list but the above will indicate some initial contact points. The author hopes that no offence is caused by any omission.

53

Corrosion

Michael J Schofield
Cortest Laboratories Ltd

Contents

53.1 Corrosion basics

53.1.1 Definitions of corrosion

Corrosion is generally taken to be the waste of a metal by the action of corrosive agents. However, a wider definition is the degradation of a material through contact with its environment. Thus, corrosion can include non-metallic materials such as concrete and plastics and mechanisms such as cracking in addition to wastage (i.e. loss of material). This chapter is primarily concerned with metallic corrosion, through a variety of mechanisms.

In essence, the corrosion of metals is an electron transfer reaction. An uncharged metal atom loses one or more electrons and becomes a charged metal ion:

$$M \rightleftharpoons M^+ + \text{electron}$$

In an ionizing solvent, the metal ion initially goes into solution but may then undergo a secondary reaction, combining with other ions present in the environment to form an insoluble molecular species such as rust or aluminum oxide. In high-temperature oxidation, the metal ion becomes part of the lattice of the oxide formed.

53.1.2 Electrochemical corrosion

The most important mechanism involved in the corrosion of metal is electrochemical dissolution. This is the basis of general metal loss, pitting corrosion, microbiologically induced corrosion and some aspects of stress corrosion cracking. Corrosion in aqueous systems and other circumstances where an electrolyte is present is generally electrochemical in nature. Other mechanisms operate in the absence of electrolyte, and some are discussed in Section 53.1.4.

Figure 53.1 depicts a metal such as iron, steel or zinc immersed in electrolyte such as sodium chloride solution.

The fundamental driving force of the corrosion reaction is the difference in the potential energies of the metal atom in the solid state and the product, which is formed during corrosion. Thus, corrosion may be considered the reverse of extractive metallurgy. Metals are obtained by the expenditure of energy on their ores. The greater the energy that is required, the more thermodynamically unstable is the metal and the greater its tendency to revert to the ore or any other oxidized form. Thus, gold is found native (i.e. as the metal) and it resists corrosion. Iron, aluminum and most metals exist naturally combined as oxides or sulfides and, once reduced to the metallic state, they attempt to revert to the combined state by corrosion.

In the systems illustrated in Figure 53.1, the anodic reaction has to be electrically balanced by the cathodic reaction, since electrical charge cannot build up at any location. A continuous electrical circuit is required through the metal (for electron conduction) and the environment (for ionic conduction).

There are four main aspects to the corrosion chemistry of this solution:

1. The presence of dissolved ions facilitates the passage of an ionic current through the solution.
2. Certain anions, especially chloride, penetrate the protective films, which are naturally present on some metals (e.g. aluminum and its alloys and stainless steels). This process is an initiator for corrosion, especially for localized corrosion.
3. The pH of the solution is a measure of the availability of one of the two most important cathodic agents, H^+ (the other being dissolved oxygen, see below). It is

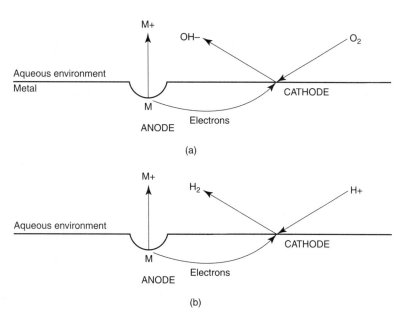

Figure 53.1 (a) Anodic, (b) Cathodic corrosion

also largely responsible for determining whether the metal surface will be active or passive (see below).

4. Dissolved oxygen (and even water) can passivate certain metals such as titanium, aluminum and stainless steels, thereby rendering them, to varying degrees, corrosion resistant. Because oxygen is a strong cathodic agent it stimulates corrosion of many metals, especially those such as iron and zinc, which tend not to form stable, adherent oxides.

Thus in the system depicted in Figure 53.1, for any metal which is thermodynamically unstable with respect to its dissolved ions in the solution (and this includes most metals of industrial importance) corrosion will occur. The rate of corrosion is determined by several factors:

1. The conductivity of the environment; low conductivity hinders the ionic current flow hence distilled water is less corrosive than a solution of sodium chloride with the same pH and dissolved oxygen content.
2. The presence of cathodic agents; dissolved oxygen and H^+ promote corrosion of most metals in an electrolyte solution (see Figure 53.2).
3. The supply of cathodic agents to the metal surface: stirring a solution or increasing the flow rate in a pipe will increase corrosion rates in systems that are under diffusion control. As soon as corrosion initiates in the system depicted in Figure 53.1 there is a local depletion of cathodic agent and a build-up of metal ions. In stagnant conditions, this soon reduces the corrosion rate, and the diffusion of these species governs the overall corrosion rate. The system is then under diffusion control.
4. The build-up of scales or other deposits on the metal surface can stifle corrosion. The reaction of anodically generated M^+ with cathodically generated OH^- can, depending on M^+, be an insoluble product. It may or may not adhere to the metal surface. If it does adhere strongly it can stifle corrosion. Other deposits can form, in addition to corrosion products. The locally high pH at the cathode can precipitate insoluble carbonates. With complete coverage of the cathodic surface corrosion is generally stifled, since the cathodic reaction cannot usually occur through the deposit. However, some cathodic scales (e.g. magnetite (Fe_3O_4)) and some forms of iron sulfides can support

the cathodic reaction, and, since anodic area is limited, this can result in intense local attack leading to pitting corrosion at bare areas.

5. The effect of temperature is generally to increase corrosion rate as temperature rises. This is a chemical-kinetic effect and can sometimes be overridden by physical effects such as deposition of scales. The solubility of oxygen in a saline solution decreases from approximately 8–10 ppm at 5°C to virtually zero near the boiling point. Thus in a system in which dissolved oxygen is the only cathodic agent corrosion rates would pass through a maximum and decrease to very low levels as temperature is increased and the availability of cathodic agent is diminished.

In Figure 53.1 the region from which metal is lost is the anode. The atoms of the metal are ionized and release electrons:

$$M \rightarrow M^{n+} + ne^-$$

where M denotes metal, M^{n+} its ion and e^- is an electron.

This reaction can proceed to a significant extent only if the electrons that are released are consumed by a cathodic reaction. The commonest cathodic reactions are essentially:

$$O_2 + 2H_2O + 2e^- \rightarrow 4OH^-$$

as shown in Figure 53.1(a), in which oxygen is consumed, and

$$H^+ + e^- \rightarrow H \text{ atoms} \Rightarrow H_2 \text{ gas}$$

in which hydrogen is liberated (Figure 53.1(b)).

Which cathodic reaction is preferred for any combination of metal and environment depends on the relative amounts of O_2 and H^+ available, and kinetics of O_2 reduction and H^+ reduction on the metal surface.

The thermodynamic driving force behind the corrosion process can be related to the corrosion potential adopted by the metal while it is corroding. The corrosion potential is measured against a standard reference electrode. For seawater, the corrosion potentials of a number of constructional materials are shown in Table 53.1. The listing ranks metals in their thermodynamic ability to corrode. Corrosion rates are governed by additional factors as described above.

Stainless steels each appear twice in the list. The more active potentials are those which the metal adopts when corroding as in a pit. The more cathodic potential is that adopted by the bare surface around the pit. The potential difference constitutes a significant driving force, analogous to the situation where the coupling of dissimilar metals such as copper and iron promotes the corrosion of the more anodic of the two (see below).

Pitting corrosion

Pitting is a form of localized corrosion in which part of a metal surface (perhaps 1 per cent of the exposed area) is attacked. Rates of pitting penetration can be very high; type 316 stainless steel in warm seawater can suffer pit penetration rates of 10 mm per year. This is a natural

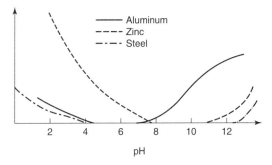

Figure 53.2 Variation in corrosion rate with pH

Table 53.1 Galvanic series in sea water[a]

Metal	Corrosion potential[b] (V)
Magnesium and alloys	−1.6
Zinc	−1.0
Aluminum alloys	−0.8/−1.0
Steel (incl. low alloys)	−0.55/ − 0.75
Stainless steels (316, 321, 410, 430)	−0.5[c]
Stainless steels (316, 317)	−0.4[c]
Copper	−0.35/−0.2
Bronzes (Si, Sn)	−0.35/−0.25
Stainless steels (410, 430)	−0.3/−0.2[d]
Lead	−0.25[e]
Copper nickel alloys	−0.25
Bronze (Ni−Al)	−0.15
Nickel	−0.15
Stainless steels (304, 321, 316, 317)	−0.1 − 0[d]
Titanium	0/−0.1
Platinum, graphite	+0.2/0.3

[a]These values are roughly constant across a range of electrolyte environments except where noted but the variations between alloys, heat treatment conditions, etc. creates a range for each metal. For some metals such as iron and steel the range is low (±100 mV), but for lead, nickel, stainless steels a range is given.
[b]The corrosion potential is reported with respect to the saturated calomel reference electrode.
[c]When active, as in a pit or a crevice or when depassivated by mechanical damage of oxide film or chemical removal in non-oxidizing acid.
[e]−0.5 V in environments low in chloride ion content.

consequence of the low ratio of anodic area to cathodic area, coupled with the self-acidification processes that occur in some metals.

Figure 53.3 illustrates a pit in a stainless steel such as type 534 or 316 austenitic alloy. Pitting starts at heterogeneity in the steel surface, such as an outcropping sulfide inclusion, the shielded region beneath a deposit or even a discontinuity in the naturally present oxide film caused by a scratch or embedded particle of abrasive grit. This initiation phase of pitting corrosion may take seconds

or weeks to develop into a feature large enough to be recognized as a pit. Oxidizing conditions facilitate initiation. Thus hypochlorite ions are particularly aggressive towards stainless steels since they raise the potential of the steel to such a high level that the protective oxide layer breaks down, allowing localized attack.

The second phase of pitting corrosion is propagation. Because of the geometry of the small pit that has been created, oxygen diffusion to the corroding region is hindered. This creates a potential difference between the pit and the surrounding region. The solution in the pit acidifies through hydrolysis of the Fe^{2+} and Cr^{2+} ions present. The pH in the pit falls to very low values (0.6 in the case of 316 stainless steel), regardless of the bulk solution conditions, which may be neutral or even alkaline. This enhances corrosion by dissolving the pit walls and ensures that the pit walls cannot be re-passivated. The surrounding metal surface is passive (i.e. is covered with an adherent oxide film of mixed $Fe_2O_3 \cdot Cr_2O_3$ oxide). This is dissolved in strongly acidic conditions such as those prevailing in the pit. Because the conditions in the pit promote self-acidification, pit penetration rates often increase with time.

A third phase is sometimes identified in pitting corrosion, i.e. termination. Pits can become stifled by the build-up of insoluble corrosion products at their mouths. Removal of these mounds of corrosion products, either mechanically or through some change in the environmental chemistry, can allow the pits to restart growth.

Crevice corrosion

Crevice corrosion of stainless steels (Figure 53.4(a)) has a similar mechanism to pitting corrosion. The initiation phase is assisted by the creation of a crevice-suitable geometry. Crevices are formed by certain fabricational processes including riveted seams, incompletely fused welds, interference fits, O-rings, gasketed joints and even paints markings of components. The solution in the crevice becomes acidified and locally the metal surface anodic compared with the surrounding bare metal, as

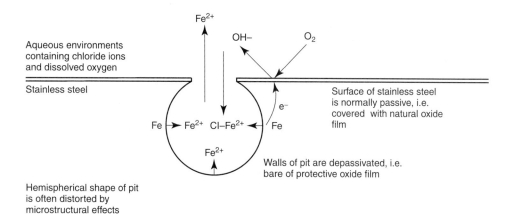

Figure 53.3 Pitting corrosion of stainless steel

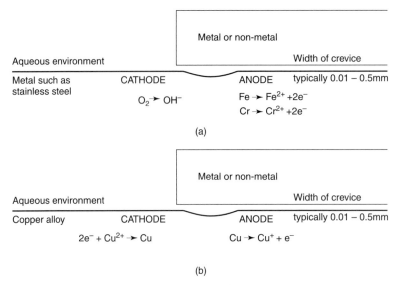

Figure 53.4 Crevice corrosion driven by (a) a differential aeration cell and (b) a differential metal ion concentration cell

described above. Crevice initiation is largely dependent on the geometry of the crevice that is present: shorter initiation times are brought about by tighter crevices, since the primary effect is the limitation on inward diffusion of oxygen to maintain passivity. The lower limit of crevice width that will permit crevice corrosion is the point at which the crevice is watertight.

Crevice corrosion of copper alloys is similar in principle to that of stainless steels, but a differential metal ion concentration cell (Figure 53.4(b)) is set up in place of the differential oxygen concentration cell. The copper in the crevice is corroded, forming Cu^+ ions. These diffuse out of the crevice, to maintain overall electrical neutrality, and are oxidized to Cu^{2+} ions. These are strongly oxidizing and constitute the cathodic agent, being reduced to Cu^+ ions at the cathodic site outside the crevice. Acidification of the crevice solution does not occur in this system.

Galvanic corrosion

Galvanic corrosion is the enhanced corrosion of one metal by contact with a more noble metal. The two metals require only being in electrical contact with each other and exposing to the same electrolyte environment. By virtue of the potential difference that exists between the two metals, a current flows between them, as in the case of copper and zinc in a Daniell cell. This current dissolves the more reactive metal (zinc in this case), simultaneously reducing the corrosion rate of the less reactive metal. This principle is exploited in the cathodic protection (Section 53.7.2) of steel structures by the sacrificial loss of aluminum or zinc anodes.

The metal that corrodes in any couple is that which has the most negative corrosion potential in the galvanic series in that environment. As guidance, Table 53.1 is of general but not universal applicability. Thus in the case of copper

and aluminum the aluminum corrodes unless corrosion inhibitors are employed. Copper and aluminum are therefore not generally used in close contact with each other, but copper ions in solution from the corrosion of copper pipework can plate out onto aluminum items (including saucepans) downstream in the system. Rapid corrosion of the aluminum occurs in the region of the copper deposits. The conductivity of the environment and the potential difference between the metals involved determine whether the corrosion is localized by a small region at the interface between the metals, or spread across a larger region. The exposed areas of the two metals together with the length of the conductance path in the electrolyte determine the intensity of the attack. Other conditions, such as the flow regime and the polarization behavior of the two metals, influence galvanic corrosion.

Potential reversal is a complicating factor in galvanic corrosion. Zinc is usually anodic to steel and hence it is used to protect steel sacrificially. Galvanized steel is steel coated with zinc. The zinc protects any steel that is exposed (such as cut ends of sheet) and scratches through the zinc layer. At temperatures above 80°C, in low-conductivity water (e.g. condensation) the metals undergo a reversal of polarity. The steel corrodes in preference to the zinc, and this causes rapid loss of the zinc in sheets as corrosion proceeds along the steel surface undermining the zinc. Tin is usually more corrosion resistant than steel, hence its use in tinplate (tin-plated steel). In some organic acids, including some of those encountered in fruit juices, the relative potentials of tin and steel are reversed and the tin corrodes, contaminating foodstuffs with tin corrosion products.

The corrosion potentials of metals and alloys are temperature dependent. The potential of a steel distillation column, heat exchanger or other item of plant varies in the temperature zones in which it is operating. Under some circumstances, it is possible to create galvanic corrosion cells between different areas of one item fabricated from one material. This causes localized corrosion problems.

To avoid galvanic problems, different materials of construction may have to be electrically isolated or at least the electrical resistance between them increased to a level sufficient to reduce the corrosion current to an acceptable value. In some instances it is more practical to paint or otherwise coat the more cathodic of the two parts of the couple. The anodic material should not be coated, since even more rapid penetration would occur at any breaks in the coating.

Microbial induced corrosion

Various bacteria are involved in corrosion processes. Of these, the most important are some *Thiobacillus* types (these generate sulfuric acid which dissolves concrete) and sulfate-reducing bacteria (SRB). SRB are ubiquitous but only create significant corrosion problems in conditions suitable for rapid growth. They flourish best in anaerobic conditions, with a supply of sulfate ions and a carbon source. Anaerobic conditions are created in stagnant areas of chemical plant, under slime in industrial cooling systems and soils and seabed mud. Sulfate ions are widely available, especially in the sea and some chemical processes. Suitable carbon sources include organic materials such as acetic acid. SRB produce sulfide ions from sulfate ions. In conditions such as beneath a colony of these bacteria corrosion processes as illustrated in Figure 53.5 occur. Many metals are susceptible to this form of attack, including steel, stainless steels, copper alloys and aluminum alloys. Localized pitting rates in excess of 10 millimeters per year are possible on steel in some circumstances. Some *Cladosporium* fungi cause pitting attack on aluminum alloys. *Gallionella* bacteria thrive in solutions containing Fe^{2+} and produce deposits under which anaerobic processes can occur.

53.1.3 Cracking mechanisms

Cracking mechanisms in which corrosion is implicated include stress corrosion cracking, corrosion fatigue, hydrogen-induced cracking and liquid metal embrittlement. Purely mechanical forms of cracking such as brittle failure are not considered here.

Stress corrosion cracking

The conjoint action of a tensile stress and a specific corrodent on a material results in stress corrosion cracking (SCC) if the conditions are sufficiently severe. The tensile stress can be the residual stress in a fabricated structure, the hoop stress in a pipe containing fluid at pressures above ambient or in a vessel by virtue of the internal hydraulic pressure created by the weight of its contents. Stresses result from thermal expansion effects, the torsional stresses on a pump or agitator shaft and many more causes.

The corrosive environments which cause SCC in any material are fairly specific, and the more common combinations are listed in Table 53.2. In the case of chloride stress corrosion cracking of the 530 series austenitic stainless steels it is generally considered that the risk is

Table 53.2 Stress corrosion cracking

Metal	Agent
Steels	Nitrate ion
	Strong alkali
	Carbonate/bicarbonate
	Liquid ammonia[a]
	Hydrogen sulfide (aqueous)[b]
	Cyanide ion
Austenitic stainless steels	Chloride (and other halide) ion
	Polythionic acids
	High-temperature water
	Hydrogen sulfide (aqueous)
Copper alloys	Ammonia/amines
	Nitric acid/nitrate fumes
Nickel alloys	Strong alkali
	Polythionic acids
	Chloride ion
Titanium alloys	Alcohols[a,b]
	Nitric acid plus nitrogen oxides
Magnesium alloys	Chloride ion[c]
Aluminum alloys	Chloride ion[c]
	Organics
	High-purity hot water

Environments are aqueous unless specified otherwise.
[a] Inhibited by 1000 ppm of water.
[b] Sometimes considered to be a form of hydrogen-induced cracking.
[c] In oxidizing conditions.

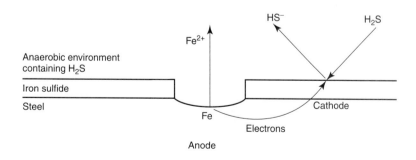

Figure 53.5

minimal below approximately 60°C. Above this threshold, if the chloride ion content of the environment is low a higher tensile stress can be tolerated without SCC developing, and vice versa. It is difficult to design for the control of SCC since it is the local stress and corrodent concentration rather than average or bulk values that is relevant.

The essential feature of most SCC mechanisms is the rupture of the protective film on the metal (passive oxide layer in the case of stainless steels and aluminum alloys), which allows the corrodent to reach the bare surface of the metal. Cracking is the result of intense corrosion, concentrated by the tensile stress. In austenitic stainless steels, chloride SCC is transgranular in nature and the cracks are typically high branched. In ferrite materials, cracking is largely due to a hydrogen embrittlement mechanism brought about by hydrogen created during corrosion.

Corrosion fatigue

Many metals suffer fatigue, which is the result of the application of an alternating stress. Corrosion fatigue results from the exposure of a metal, which is subjected to alternating stress, in a corrosive environment. Unlike stress corrosion cracking, corrosion fatigue does not require a specific corrosive environment for any metal. Moist air is adequate in many cases, thus the majority of fatigue failures are actually corrosion fatigue in nature. The alternating stress pattern requires only a small tensile component. This tensile stress may be small relative to the metal's ultimate tensile strength and millions of cycles of the stress pattern may be required to cause failure.

Hydrogen-induced cracking

Hydrogen-induced cracking (HIC) is most commonly encountered in steels but other metals are susceptible, as shown in Table 53.3. The presence of hydrogen atoms in a metal degrades some of its mechanical properties, especially its ductility, leading in some cases to embrittlement.

Table 53.3 Hydrogen-induced cracking

Metal	Agent
Steel	Hydrogen gas
	Atomic hydrogen[a]
	Hydrogen sulfide-containing aqueous environments[b]
	Water[c]
Martensitic stainless steels	Hydrogen sulfide-containing aqueous environments
Copper alloys	Sea water[d]
Nickel alloys	Hydrogen sulfide-containing aqueous environments
Titanium alloys	Anhydrous alcohols
Zirconium alloys	High-temperature water

[a] Created by overprotection during cathodic protection, evolved during electroplating or pickling processes and during welding with wet welding consumables.
[b] Steels of strength in excess of 550 MPa.
[c] Steels of strength in excess of 900 MPa.
[d] High-strength alloys, cathodically protected.

Additionally, hydrogen atoms diffuse through metals and coalesce to form hydrogen molecules at certain preferred locations such as inclusions.

Hydrogen atoms are soluble in ferrite (which is the major phase of most steel). At discrete inclusions in the steel (e.g. manganese sulfide, which is present in many steels), the hydrogen atoms combine to form hydrogen molecules. Hydrogen molecules are insoluble in the steel lattice and these molecules associate, producing high tensile stresses, which can initiate hydrogen cracking. When laminations are present in the steel, hydrogen molecules form at the interface between them, creating blisters. At high temperatures hydrogen atoms can react with carbides in the steel producing methane gas. Methane is insoluble and causes blistering or cracking. In clean steels containing none of these features the hydrogen atoms pass straight through the steel, causing no damage. Hydrogen can originate from several sources:

1. Electroplating (e.g. cadmium or hard chromium plating);
2. Acid pickling (e.g. prior to galvanizing);
3. Cathodic protection, especially if overprotected, through inadequate potential control;
4. Corrosion, where H^+ reduction is a cathodic reaction (Section 53.1.2);
5. The process inside the plant.

The amount of the hydrogen that is liberated on or near a metal surface, which then enters the metal, varies according to the environment and condition of the metal. The main factor that promotes the entry of hydrogen into a metal is the presence on the metal of a surface poison such as sulfide or other species, which inhibit the hydrogen recombination reaction.

Liquid metal embrittlement

The corrodent is a liquid metal in this form of stress corrosion cracking. Mercury at ambient temperature and metals including zinc (from galvanized steel-work) and copper (from electric cables) when melted during welding or in a fire cause rapid failure of certain metals.

Steels and austenitic stainless steels are susceptible to molten zinc, copper, lead and other metals. Molten mercury, zinc and lead attack aluminum and copper alloys. Mercury, zinc, silver and others attack nickel alloys. Other low-melting-point metals that can attack common constructional materials include tin, cadmium, lithium, indium, sodium and gallium.

53.1.4 Non-electrochemical corrosion

Although all the types of corrosion discussed in this section result in the oxidation of metal and some involve direct electron transfer, they can be understood without reference to electrochemistry.

Oxidation

In oxidizing atmospheres (i.e. in the presence of oxygen or a source of oxygen from which oxygen can be derived,

Table 53.4 Oxidation in air

Steels	480°C
Low-alloy steels	560°C
Ferritic stainless steel (type 410)	650°C
Austenitic stainless steel (type 316)	900°C
Nickel-based superlloys	1100°C
Superaustenitic stainless steel (type 310)	1150°C
Cobalt-based superalloys	>1150°C

1. Temperatures the those above which oxidation is generally too rapid to permit the metal's safe or economic use.
2. Atmospheres contaminated with acid or halogen gases reduce these values.
3. The mechanical properties including strength, ductility and creep resistance can be affected below these temperature.

such as water vapor), most metals are unstable with respect to their oxides. Oxidation to create the metal oxide depends on several factors, the most important being temperature (and hence the thermodynamics of the process), availability of oxygen and the protective action of any oxide that is created. Oxygen can migrate through the oxide, where it reacts with the metal. In other metal-oxide systems, metal ions migrate outwards through the oxide. The structure and the stability of the oxide, together with its semiconducting properties, control the mechanism of oxide. Oxidation rates can be linear, parabolic (tending to a steady state oxide thickness) or exponential (termed catastrophic, because the rate increases with time). Table 53.4 gives an indication of the temperature limits for some metals through oxidation in a cleaned atmosphere.

Sulfidation is analogous, but catastrophic sulfidation is common because of the generally lower melting points of sulfides than corresponding oxide. This is especially true in the case of nickel alloys, when a nickel/nickel sulfide eutectic is formed.

Fretting corrosion

Fretting corrosion occurs because of oscillating relative motion between touching surfaces. As little as 3×10^{-9} meter lateral movement is required. The amount of damage increases as the normal force between the surfaces is increased. In dry conditions, the corrosion product is usually the oxide. In the case of steel, this is wustite. This is otherwise the high-temperature form of the oxide, which infers that locally high temperatures are created on the fretting surfaces. The surfaces weld together in the high stress conditions at points of contact and are torn apart by the relative motion of fretting surfaces.

Materials with hard oxides, including stainless steels and aluminum and titanium alloys, are particularly susceptible to this form of attack. In steel, it is also known as *false Brinelling* because of the high surface hardness that can be created in work-hardening grades.

Molten salt corrosion

Any salt that is present on a hot metal surface can cause corrosion of that surface. This mechanism is often rapid and is due to straightforward dissolution of the metal and

any oxide which may be present on the surface. The mechanism is similar to that of aqueous corrosion. The high rate of attack is a consequence of the high activities of the ions present in the molten salt. The problem is greatest for salts of low melting point, since these are present over a wide temperature range and can be very fluid. Low melting-point salts include many chlorides, sulfides and sulfates.

53.2 The implications of corrosion

53.2.1 Economics

The principal economic implications of corrosion of a plant are the initial cost of construction, the cost of maintenance and replacement, and the loss of production through unplanned shutdowns. The initial cost of the plant is influenced by material selection, and a choice of material that is more corrosion resistant than is necessary for the safe operation of the plant over its design life is a very expensive error. This cost involves initial outlay of money, and plants have been built which could never be profitable because of the inappropriate materials selection.

Maintenance costs arising from corrosion can also prove to be unacceptably high. For plant that requires regular maintenance for other reasons (e.g. de-sludging or batch operation) it is often accepted that certain items are expendable, and annual or even monthly repair or replacement with low-cost materials is preferable to the use of expensive ones. Plant handling hot, strong hydrochloric acid is often regarded in this way, since the only possible construction materials that have acceptably low corrosion rates are too expensive to use. Plant which requires to be run continuously or maintenance-free for long periods necessitates high-integrity design and hence the use of more reliable corrosion-resistant materials.

Significant savings can be achieved by optimum material selection (guidance is given in Section 53.3), by considering corrosion at the design stage (Section 53.4), by employing corrosion-control techniques (Section 53.7) and corrosion monitoring (Section 53.8). In the industrialized countries of the Western world, corrosion is estimated to cost some 4 per cent of a country's GNP. This represents $6 billion per annum, of which a significant proportion could be saved by the use of existing knowledge. It cannot all be saved, since the gradual deterioration of a plant over its operational life is one of the costs incurred in the process.

53.2.2 Safety

The failure of plant by corrosion can be gradual or catastrophic. Gradual failure has few implications for safety providing it is monitored. Direct corrosion-monitoring techniques are described in Section 53.8. Indirectly, the correct interpretation of records relating to metal contamination of products or the loss of efficiency of heat exchangers, etc. can provide useful information.

Sudden or catastrophic failure of plant through corrosion can result in the loss of product at high velocity from a failed reactor vessel, high-energy steam from steam line, toxic or inflammable materials from storage vessels. Incidents of this nature from plants around the world have led to the death and injury of plant operatives and nearby householders. The majority of serious incidents in the chemical process industry, however, have been attributed to mechanical failure of plant.

53.2.3 Contamination of product

The metal lost from the inside of pumps, reaction vessels, pipework, etc. usually contaminates the product. The implications of this depend upon the product. Ppb levels of iron can discolor white plastics, though at this level the effect is purely cosmetic. Ppm levels of iron and other metals affect the taste of beer. Products sold to compositional requirements (such as reagent-grade acids) can be spoiled by metal pick-up. Pharmaceutical products for human use are often white tablets or powders and are easily discolored by slight contamination by corrosion products.

53.3 Materials selection

The materials selection procedure for new or replacement plant is crucial to the safe and economic operation of that plant. There is no one correct way to select the appropriate construction material, since for small plant handling toxic or inflammable products the integrity of the plant is high in priority, whereas large plant producing bulk material in a competitive market is more likely to be made of cheaper constructional materials, and the occasional leak or failure may have fewer safety implications.

Wherever possible, items should be made from materials which have proved themselves in similar service. Essential items of plant for which long delivery times are expected should be considered most carefully. The possible interactions between items fabricated from different materials should be considered. Section 53.1.2 discusses galvanic corrosion.

53.3.1 Sources of information

The first source of information for the behavior of a material in the proposed service environment is the potential supplier of the item of plant. Except for new (or significantly modified) processes, specialist suppliers or fabricators have relevant information and service experience. The supplier should be provided with all process or environmental details that are of possible relevant to corrosion. The most important are listed below:

1. *Environmental chemistry*: including complete typical and worst-case compositional analyses, pH, redox potential, dissolved gas content;
2. *Physical conditions*: temperature ranges in bulk and at surface, heat transfer, mass flow, fluid velocities, the presence of entrained particles and gases;
3. *Operating cycles*: commissioning procedures, heating, cooling and pressure cycling, stagnant periods, cleaning procedures;

4. *External environment*: corrosive gases, humidity temperature;
5. *Mechanical aspects*: the presence of tensile or cyclic stresses.

If the potential suppliers cannot provide documented service performance for similar plant on similar duties, materials suppliers should be consulted.

For all materials other than basic constructional steels and cast irons, reputable suppliers have information bases and applications laboratories from which information can be obtained. Trade organizations representing categories of materials suppliers are excellent sources of information; some are listed at the end of this chapter. The materials suppliers should be consulted in conjunction with equipment suppliers in order to ensure that the information generated is fully applicable to the end use to which the material is to be put. Fabrication techniques should be agreed between the two types of suppliers, since some materials cannot be cast or welded and forging cannot make some items.

For some new or modified processes, it will not be possible to obtain sufficient good information to permit material selection to be made. The following guidance is intended as a checklist for general use, since there is no universal, definitive guide to the selection of materials. It is based on corrosion and not on mechanical property requirements such as strength, toughness, hardness or fatigue resistance.

53.3.2 Aqueous systems

Water supply

For the purpose of corrosion, water is scaling or non-scaling. Scaling water tends to deposit generally protective hardness scales. Soft water does not scale and hence is potentially more corrosive; especially when it contains dissolved gases such as oxygen, carbon dioxide, ammonia and sulfur oxides.

For drinking water, copper has now replaced the use of lead. The local water authority provides information including analytical data. Stainless steels are being introduced especially for domestic systems.

For process water, steel pipes are used unless iron pick-up is to be minimized. Plastic pipes (polyethylene and polyvinylchloride) are used but they sometimes need external protection from solvents present in industrial atmospheres, ultraviolet radiation (including sunlight), freezing and mechanical damage.

For firewater, steel pipes are used but corrosion products can block sprinklers. Cement asbestos pipes are utilized but pressure limitations restrict their use. For critical applications, including offshore oil installations, cupronickel alloys and even duplex stainless steels are used. Fire-retardant grades of fiber-reinforced plastics are now available.

Borehole waters are generally very hard and cast iron pipes are still used because of the low internal corrosion rates permitted by the scaling which occurs naturally. Acidic waters cause graphitic attack on cast irons.

Boiler water, steam and condensate

The water supply for boilers is usually treated. Treatment depends on the quality of the water supply, the pressure of the boiler, the heat flux through the tube walls and the steam quality required. Most waters require de-alkalization. The water produced in this process is non-scaling and potentially corrosive (see above).

The quality of water thus produced is generally adequate for low-pressure boilers, but de-oxygenation is usually achieved through the addition of sodium sulfite in a feedwater tank. This should be designed to prevent further ingress of oxygen, employing a nitrogen blanket if necessary. Pipework from the feedwater tank (or 'hot well' if condensate is returned to it) to the boiler is steel, since the water is low in corrosivity.

For higher-pressure boilers demineralization is necessary to minimize total dissolved solids in the boiler. This water is normally carried in steel pipework, but if condensate is returned and the condensate has become contaminated (for example, with carbon dioxide or copper ions) more corrosion-resistant materials such as copper are required. Downstream of the boiler, steam pipework is usually steel with steel or stainless steel expansion bellows.

The presence of corrosive species in steam, however, creates the following corrosion problems:

1. Oxygen renders the steam corrosive towards steel. If the feed is not deoxygenated, a corrosion allowance should be added to the steam piping wall thickness, an oxygen scavenger such as hydrazine can be used or a volatile corrosion inhibitor could be employed. Candidate inhibitors include filming amines such as ethoxylated soya amines. Some amine-based inhibitors can break down and cause corrosion of copper in the steam or condensate system. Some products such as hydrazine are toxic or carcinogenic and cannot be used in food-processing plant, breweries and steam sterilizing equipment for hospitals.
2. Carbon dioxide, from the decomposition in the boiler of temporary hardness salts present in some waters, causes corrosion of steel steam pipework and cast iron valves and traps. Corrosion inhibitors may be used, but the choice of inhibitor must take into account the other materials in the system. Neutralizing amines such as morpholine or cyclohexylamine are commonly used.
3. Condensate returns lines are often copper. Copper has good corrosion resistance to oxygen and carbon dioxide individually. When both gases are present in the condensate, copper is susceptible to corrosion. Copper picked up in the condensate system and returned to the boiler causes serious corrosion problems in the boiler and any steel feedwater and steam pipework. Boiler tubes should last for 25 years but can fail within one year in a mismanaged or ill-designed boiler system suffering from these faults.
4. Boiler salts can contain chloride ions. When carried over into the steam (e.g. during priming) this can result in chloride stress corrosion cracking of austenitic stainless steel expansion bellows. In steam systems where freedom from chloride cannot be guaranteed, bellows

can be supplied in 9 per cent Cr. steel or austenitic alloys containing 40 per cent nickel.

Aqueous processes

The principal corrosive species as described in Section 53.1.2, combined with the reducing or oxidizing nature of the environment, can be used to select candidate materials for any aqueous system. The process variables discussed in Section 53.3.5 must also be considered.

Steel is the most common constructional material, and is used wherever corrosion rates are acceptable and product contamination by iron pick-up is not important. For processes at low or high pH, where iron pick-up must be avoided or where corrosive species such as dissolved gases are present, stainless steels are often employed. Stainless steels suffer various forms of corrosion, as described in Section 53.5.2. As the corrosivity of the environment increases, the more alloyed grades of stainless steel can be selected. At temperatures in excess of 60°C, in the presence of chloride ions, stress corrosion cracking presents the most serious threat to austenitic stainless steels. Duplex stainless steels, ferritic stainless steels and nickel alloys are very resistant to this form of attack. For more corrosive environments, titanium and ultimately nickel-molybdenum alloys are used.

The following brief summary of materials for acid duty is very simplified, and further guidance should be obtained from the sources of information outlined in Section 53.3.1:

1. *Sulfuric acid*: Plain carbon steel is used at strengths in the range 70–100% at temperatures up to 80°C. In flowing conditions the corrosion rate increases, thereby rendering steels unsuitable for pumps. Storage tanks for sulfuric acid require care in design to prevent the possible ingress of water vapor. This creates a layer of dilute acid on top of the bulk acid, creating rapid attack at the fill line. At strengths below 70°C chemical lead is the preferred material for tanks but at temperatures above 120°C corrosion becomes significant. For castings such as pumps, valves, fittings and some heat exchangers cast iron containing 15% silicon is used at strengths up to 100%, at temperatures up to the boiling point. Where iron pick-up must be avoided, conventional stainless steels can be used in certain ranges of strength and temperature. Anodically protected titanium is used at 70% (60°C) and 40% (up to 90°C). For higher temperatures and strengths above 100%, nickel-based alloys were molybdenum is used. Glass, gold and platinum are required for specific combinations of acid strength and temperature.
2. *Hydrochloric acid*: The selection of materials for handling hydrochloric is exceptionally difficult because of the sensitivity of corrosion to minor impurities. Oxidizing hydrochloric acid is highly corrosive; dissolved oxygen or trace levels of metal ions in high oxidation states (especially ferric or cupric) ions are the common oxidizing species. Dilute hydrochloric acid (2% at up to 70°C or 10% at 20°C) is handled by copper, nickel and some of their alloys. Moderate-strength acid (up to 10% at up to 70°C, up to 53% and 53°C and up to 40% at 15°C) requires the use of more resistant

materials. These materials include cast iron alloyed with silicon and molybdenum or, in the absence of oxidizing agents, silicon bronze. Strong hydrochloric acid at moderate temperatures (below 70°C) can be handled by some of the nickel-molybdenum alloys or non-metallic linings such as rubber. At higher temperatures silver, platinum, zirconium and tantalum are suitable across the whole range of acid strength.

3. *Nitric acid*: Stainless steels are the most commonly used materials for nitric acid duty. Because molybdenum is widely considered to be detrimental to corrosion behavior in nitric acid, type 304 is generally preferred to type 316L. It is used at temperatures up to 100°C and acid strengths up to 70%. At 100% strength the temperature limitation is 53°C. Higher alloys, based on 20% Cr, 25% Ni, extend the temperature range to nitric acid's boiling point at concentrations up to 50%. An austenitic stainless steel containing 18% chromium, 10% nickel and 2–3% silicon was developed specifically for this service. Cast iron with 15% silicon is used for valves, pumps, etc. in acid strengths of 0–40% (up to 70°C) and 40–80% (up to 110°C). Some aluminum alloys may be used for nitric acid above 80% strength at temperatures up to 35%. Titanium gold, platinum and tantalum are used to extend the temperatures range where required.

For alkaline duty, steel can sometimes be used up to approximately pH 11. Zinc, aluminum and similar metals and their alloys have limited use in alkaline conditions because they dissolve, giving complex anions. Iron and steel react in this way above about pH 12. Approximate limits of use for zinc are pH 6–12 and aluminum alloy pH 4–8. Stainless steels, including the lower grades, can be used even in the presence of chloride ions at pH levels of approximately In the absence of halides they can be used up to about pH 13.

Materials selection cannot be based on any simple combination of common corrosive species. There are many complicating factors, including the harmful or beneficial effects of contaminants at the ppm level, the relative proportion in which certain combinations of species are present (H^+ and Cl^- are often synergistic in their effect, whereas SO^{2-} and Cl^- often counter each other) and the presence of naturally occurring corrosion inhibitors.

53.3.3 Non-aqueous processes

Solvent systems

Most organic solvents, except for alcohol, have reasonably low ionic conductivity and hence do not support electrochemically corrosion to any significant extent. Steel is commonly used except in systems in which water can separate and where the conductivity is sufficient to permit the flow of ionic current.

Dry gases

Standard materials for handling dry gases include: for chlorine, UNS N10276 (type stainless steel is used for liquid chlorine); for bromine, UNS N10276 (below 60°C); for fluorine, copper; for hydrogen chloride, UNS N10276; for hydrogen sulfide (sulfur dioxide and trioxide), type 316 stainless steel. The presence of water in even trace quantities changes this significantly.

53.3.4 High-temperature environments

Chimneys, flues and ducts

The main corrosion processes that occur in these items arise from condensing liquids on the internal surface. Although often lagged, heat loss frequently causes internal skin temperatures to fall below the dewpoint of one or more components of the gas stream, albeit locally, such as at support points. Even at temperatures above its dewpoint a gas can dissolve in condensed water. Rapid corrosion can then occur in this thin film of corrosive liquid.

Condensing species of relevance to corrosion include water and all acid gases. The dewpoint of water is obtained from standard tables, requiring only the water content (i.e. relative humidity) of the gas stream. Above the water dewpoint, corrosion problems include condensing acids (Section 53.3.2), dry acid gases (Section 53.3.3) and erosion. Below the water dewpoint acid gases dissolve in the water film to create an acidic solution, including:

1. Carbon dioxide produces a solution of carbonic acid (as in boiler condensate, see Section 53.3.2). Carbon steel is often employed but corrosion rates of up to 1 mm/yr can be encountered. Coatings and non-metallic materials may be employed up to their temperature limits (Section 53.5.6). Basic austenitic stainless steels (type 534) are suitable up to their scaling temperatures.
2. Sulfur oxides produce sulphurous/sulfuric acid. Because oxygen and halides are often also present, high-nickel alloys are required.

   Current practice for some flue gas desulfurizer ducts operating in the temperature range 60–120°C involves lining the steel structure with 2-mm thick UNS N10276 or UNS N06022. Non-metallic linings, including glass flake epoxy materials, have limited application in the lower part of the temperature range.

Other acid gases such as hydrogen chloride and oxides of nitrogen produce similar corrosion problems. The corrosion effects produced by acid condensate are amplified by the motion of the gas stream (typically 20–53 m/s) and erosion effects due to entrained solids and impingement at bends, damper plates, reheaters, etc.

In the complete absence of acid gases, steel or galvanized ducting can be suitable but 80°C is often the operational temperature limit for galvanized steel).

Furnaces

Table 53.4 gives scaling data for classes of heat-resisting alloys in air. Mechanical properties, particularly creep strength, may be reduced significantly at temperatures well below scaling temperatures. In addition to scaling and loss of mechanical properties the third major factor in material performance in furnaces is the detailed chemical environment. Specific problems are encountered with the following:

1. Oxygen or steam in reformer tubes (typically up to 1000°C), steam superheater tubes (700–900°C), regenerators, etc. promote oxidation. Catastrophic oxidation occurs in some materials including molybdenum-bearing stainless steels because of the volatility of MoO_3. Alloys containing 25% Cr and 20% Ni are used up to 1100°C but it is sometimes necessary to employ 35% Cr, 25% Ni or even 45% Cr, 25% Ni types. Type 316 austenitic stainless steel can suffer catastrophic oxidation at temperatures as low as 760°C, although it is often used up to nearly 900°C provided that a free flow of gas is achieved over the entire surface.

2. Sulfur compounds, whether organic or inorganic in nature, cause sulfidation in susceptible materials. The sulfide film, which forms on the surface of much construction materials at low temperatures, becomes friable and melts at higher temperatures. The presence of molten sulfides (especially nickel sulfide) on a metal surface promotes the rapid conversion to metal sulfides at temperatures where these sulfides are thermodynamically stable. High-alloy materials such as 25% Cr, 20% Ni alloys are widely used, but these represent a compromise between sulfidation resistance and mechanical properties. Aluminum and similar diffusion coatings can be of use.

3. Carbon sources permit carbonization. In cracker furnaces, ethylene pyrolysis tubes, etc. carbonization results in rapid metal loss in the temperature range 850–1100°C. A lower-temperature form of this mechanism (sometimes termed 'metal dusting') occurs in the range 400–800°C in the carbon monoxide/carbon dioxide/hydrogen/hydrocarbon gas streams. Preferred materials include HK40 (25% Cr, 20% Ni, 0.4%C) or higher alloys (35% Ni, 25% Cr or 53% Cr, 53% Ni), depending on the severity of the environment.

4. Nitrogen containing compounds, including ammonia and amines, cause nitriding in susceptible materials. Ammonia synthesis (500–600°C, 250–400 bar pressure) requires the use of austenitic stainless steels or nickel alloys. The rate of penetration of nitriding, which causes severe embrittlement, can be 0.2 mm/year in basic austenitic grades (e.g. type 304), falling by more than an order of magnitude for 40% Ni base alloys.

5. Molten salts promote rapid corrosion of many constructional materials at relatively low temperatures. Low–melting-point salts include sodium salts from saline atmospheres, fireside ash, silicate insulation, contaminants in the feed, etc. Corrosion rates of several mm/year can be observed at temperatures as low as 520°C. High chromium- and nickel-containing alloys up to 50% Cr/50% Ni are employed.

Other detrimental factors which should to be taken into account in the materials selection process include temperature cycling and the presence of halide gases. Specialist alloys containing rare earth element additions such as cerium, lanthanum and yttrium have been developed for use in certain environments up to 130°C.

53.3.5 Influence of process variables on material selection

Flow

The main effects of flow on corrosion of most materials are deleterious. Increasing flow through the laminar region serves to create a thinner hydrodynamic boundary layer on the metal surface. This increases the rates of diffusion of cathodic species and anodic products, generally increasing corrosion rates. Protective deposits and scales are removed or prevented from forming. Damage is caused at locations such as those shown in Figure 53.6.

The most important exception to this is the behavior of stainless steels in sea water. Grades which are susceptible to pitting or crevice corrosion, including types 534 and 316, are liable to greatest pitting attack in stagnant conditions where debris settles onto upward-facing horizontal surfaces. In conditions of higher flow (greater than 1 m/s), the debris is less likely to settle and thus pitting (Section 53.5.2) is less likely.

If flow becomes turbulent, the corrosion rate increases even more rapidly. In practice, most engineering materials have a *critical velocity* above which the corrosion rate is unacceptably high. This does not correspond with the laminar-to-turbulence transition. Surface roughness is an important consideration.

Impingement attack (sometimes termed erosion corrosion) is a result of the combined effect of flow and corrosion on a metal surface and it occurs when metal is removed from the surface under conditions where passivation is insufficiently rapid. It is a function of flow, corrosion and passivation.

At higher flow rates cavitation is a serious degradation mechanism, where vapor bubbles created by pressure fluctuations brought about by the flow of liquid past the surface collapse on the metal surface with tremendous force. This damages any protective oxide which may be present, leading to pitting corrosion. It also causes mechanical damage to the metal.

The corrosive and mechanical effects of flow are observed in pipes, especially at bends and downstream of flow disturbances, tube and shell heat exchangers, valves and pumps. More corrosion and/or harder materials are used in such areas. Austenitic stainless steels work harden and hence are superior in flowing conditions to ferritic stainless steels of otherwise similar corrosion resistance. Hard

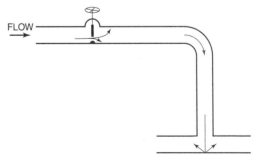

Figure 53.6 Areas affected by erosion corrosion

facings, such as hard chromium electroplate or cobalt based alloys, are used for local improvement of flow-assisted corrosion resistance where it is not required, practical or economic to fabricate the complete item of high resistant material. Hard facing by cobalt-based alloys can be achieved by weld overlay and plasma spray techniques. Correct choice of alloy is important to optimize corrosion resistance, though the resistance of most of the alloys in use is superior to the more common grades of stainless steel. Hard chromium plating contains micro cracks; this allows corrodents to gain access to the substrate. Layers of copper and/or nickel are often required to prevent the corrosion of underlying steel.

The alternative option for counteracting cavitation damage is the use of a resilient material such as rubber. The mechanical forces attendant on collapse of the bubbles are absorbed by elastic deformation of the resilient material.

Intermittent operation

This is generally harmful to plant, since deposits can form in periods of stagnation. The common grades of stainless steel, such as types 304 and 316 and other materials (including aluminum alloys and some copper alloys), are susceptible to under-deposit attack and are at risk in circumstances where the process liquid is left to stagnate in between operation cycles. Evaporation can also take place, creating more aggressive conditions than occur during operation.

Condensation can form on vessel walls, roofs and support points, as heat is lost. This condensation can absorb corrosive gases, creating localized corrosion effects of greater severity than the bulk environment that is normally present.

In rare circumstances, however, intermittent operation is beneficial. Plant that would pit if operated continuously can be operated successfully if the operational cycle is shorter than the initiation period for pitting attack and it is cleaned between batches. This is mainly the case for stainless steel plant operating with high enough chloride levels (or a low pH or high temperature) to initiate pitting if operated continuously. The cleaning regime between batches promotes re-passivation; hot water, steam, chloride-free caustic or sodium orthophosphate solution may be required to achieve the required rate of passivation. Plant is rarely designed on this principle, but a significant number of cooking vessels, especially in the food and drink industries, operate this way.

Commissioning

The first use of new plant, or start-up after a shutdown, poses corrosion hazards additional to those encountered in normal operation. New plant such as boilers requires special water treatment, involving boil-out, passivation and possible chemical cleaning. Actual requirements depend on the boiler type, the proposed service, the quality of water available during commissioning and the internal condition of the boiler. The condition of the boiler depends on for how long and in what conditions it has been stored. The presence of any salts, dirt or rust is harmful. An adherent, protective layer of magnetite in normal operation covers boiler internals. To create an intact layer of magnetite of the correct thickness, crystal type and morphology all debris must be removed from the boiler, manually or by means of a vacuum cleaner. If rusting is more than superficial, or if mill scale is present on the steel of the boiler tubes or shell, chemical cleaning is necessary. Such techniques are based on inhibited acids (e.g. citric acids). The most appropriate technique is determined by a number of factors, and the most important include the nature of deposits to be removed (i.e. rust, hardness scales, mill scale and metallic copper).

Degreasing may be necessary, especially on new plant. For optimum boiler life, passivation treatment is recommended.

Reliable contractors should carry out all boiler treatment and cleaning. Boilers are fired pressure vessels and are subject to mandatory insurance inspections. Significant benefits in safe and economic operation, particularly by reducing unnecessary chemical treatment, can be achieved by monitoring the condition of boilers.

Items of plant fabricated from stainless steels should be inspected before first use and after any maintenance work or unplanned shutdown. All materials that rely for their corrosion resistance on the presence of an oxide or similar passive layer are susceptible to localized attack where that layer is absent or damaged. Damage is most commonly caused by scratching, metallic contamination (nearby grinding or touching with ferrous tools), embedding of grit and weld spatter.

Serious damage of this nature should be rectified before the plant is used. Scratches, embedded grit and weld splatter can be removed by careful grinding. All grinding and abrading on stainless steel must be carried out using alumina or a similar abrasive, rather than silicon carbide consumables. Wire brushes must be stainless steel, rather than carbon steel. Local dressing should be fine-finish, since coarse scratches provide sites for the initiation of pitting attack.

In critical applications, if stainless steel is to be used near its limit (in terms of corrosion), and for cases such as welds, where a good finish cannot be otherwise achieved, additional passivation is required. Nitric acid (10–15 per cent by volume) is the best passivator. It also dissolves iron contamination. In circumstances where the use of nitric acid is not possible for safety or physical reasons (such as the underside of vessel roofs) passivation paste is appropriate. Both materials are used at ambient temperature and require a contact time of approximately 30 minutes. They must be removed by thorough rinsing with low chloride-content water.

New plant can be contaminated with corrosive species such as sea salt if it has been transported by sea as deck cargo. Corrosive species present on building sites can gain ingress into items of plant. Many metallic materials, particularly aluminum, zinc alloys and galvanized steel, are susceptible to attack by alkaline cement dust and acid floor cleaners. Significant corrosion can occur by any of these agents before or during commissioning. Titanium and, to a lesser extent, tantalum are at risk of hydrating attack when contaminated by iron smears (see Section 53.5.5).

The materials selection procedure should take into account the commissioning requirements of all parts of the plant in order not to include any materials that cannot safely be degreased or chemically cleaned.

Heat transfer

During the materials selection procedure isothermal corrosion testing may indicate the suitability of a material for handling a corrosive process fluid. In many cases where heat transfer is involved the metal wall temperature experienced in service is higher than the bulk process fluid temperature. This, and the actual heat transfer through the material, must be taken into account since both factors can increase corrosion rates significantly.

Liquid line

The corrosion conditions can be different at the fluid line from the bulk condition. Aqueous liquids have a concave meniscus, which creates a thin film of liquid on the vessel wall immediately above the liquid line. Some corrosion processes, particularly the diffusion of dissolved gases, are more rapid in these conditions. Additionally, the concentration of dissolved gases is highest near the liquid surface, especially when agitation is poor. Locally high corrosion rates can therefore occur at the liquid line, leading to thinning in a line around the vessel. This effect is reduced if the liquid level in the vessel varies with time. Any corrosion tests undertaken as part of the materials selection procedure should take this effect into account.

53.3.6 Influence of external environment

The external environment experienced by plant can be more corrosive than the internal process stream. Any construction material must be chosen to withstand or be able to be protected from external corrosive agents.

Corrosive atmospheres

Corrosive species in the atmospheres include water, salts and gases. Clean atmospheres contain little other than oxygen, nitrogen, water vapor and a small quantity of carbon dioxide. These species are virtually non-corrosive to any of the common constructional materials for plant at normal temperatures. Steel is susceptible to corrosion in even fairly clean air where water can exist as liquid. For plant operating at temperatures up to approximately 100°C coatings are employed to protect steel if required. In clean air corrosion rates are low, and corrosion is primarily a cosmetic problem, although it may be necessary to prevent rust staining of nearby materials.

Organic coatings (i.e. anti-corrosion paints) are used. Above 100°C the steel will be dry provided it is sheltered from rain. It is also likely to be lagged and therefore further protection from corrosion is not needed. Few organic coatings are suitable for use in this temperature range, but epoxy-based coatings are used up to 120°C and vinylesters up to 160°C. Silicones and inorganic and metallic coatings can be used at higher temperatures. At temperatures above 480°C, oxide scaling is a serious problem for steel, and

alternative materials such as stainless steels or nickel-base alloys are required.

Contaminated atmospheres create additional corrosion problems. In marine environments (e.g. coastal power stations, offshore oil platforms and above the splash zone on ships) the chief corrosive agents is sea salt. The chloride content of sea salt precludes the use of stainless steel and bare steel or aluminum in these areas. Steel requires to be coated as described in the previous section. Any coating scheme used for the protection of steel must be properly selected and applied. BS 5493 contains guidelines for coating systems for a variety of environments. The most common reasons for coating failure are incorrect application and inadequate surface preparation. The use of an independent coating inspector, as discussed in Section 53.7, is recommended. Aluminum alloys can be coated with plastics, such as U-PVC, epoxy finishes or conventional paints, with the correct etch primer where applicable to ensure adhesion. Some alloys are suitable for anodizing, but the corrosion protection afforded by anodizing is variable and requires tight specification and control to ensure correct anodized film thickness and sealing efficiency. The lower grades of stainless steels are not suitable for external exposure in marine atmospheric conditions unless they are cleaned regularly with fresh water. Higher grades of stainless steel, nickel alloys and most copper alloys do not normally benefit from additional protection. The development of the colored patina on copper alloys can, however, be avoided by coating with inhibited lacquer.

Industrial environments can be contaminated with sulfur oxides, other acid gases, ammonia, hydrogen sulfide, nitrogen oxides, chlorine, etc. Most of these are corrosive towards steel, which is normally protected as outlined above. Copper alloys are corroded by hydrogen sulfide at levels above 60 ppb although the corrosion product is generally stable and adherent and further protection is not usually required. Corroded in this way, copper is unsightly and the functioning of electronic equipment is affected. Ammonia corrodes and causes stress corrosion cracking of some copper alloys and if it is present in significant quantities, most alloys require protection by coating. Chlorine and most acid gases are sufficiently corrosive towards aluminum alloys and the lower grades of stainless steel that protection is required. Coatings are the only practical solution if these gases are present at high enough levels.

Corrosion under lagging

Thermal insulation of vessels and pipework usually employs glass fiber or foamed polyurethane products. In their pure forms these pose little corrosion risk. Generally, they are contaminated with leachable acids and/or chloride ions. Chloride-free lagging can be specified and this should be used for contact with metals which are susceptible to chloride pitting or chloride-induced stress corrosion cracking. The lower-grade austenitic stainless steels, including types 304, 321, 347, 316, 317 and their low-carbon variants, are at greatest risk. If the lagging remains dry, there is no corrosion problem. Lagging can become wet through rain ingress at damage sites, spillage around entry ports or leakage from steam coils or the lagged vessel wall. Chloride leached from the lagging or introduced with the water tends to concentrate on the pipe or the vessel, and even trace quantities create serious corrosion problems. If chloride-free lagging is not used, or the risk of water ingress cannot be removed, the vessel can be coated prior to lagging. Any barrier coating can be used but if it is damaged it will not prevent corrosion. Aluminum, in the form of foil, is used to prevent chloride-induced stress corrosion cracking of austenitic stainless steels. Applied to the vessel or pipe prior to lagging, it acts as a barrier to water and possibly by slight cathodic protection.

Corrosion in concrete

Intact, good-quality, void-free concrete creates an alkaline environment and protects embedded steelwork such as reinforcing steel. Penetration by chloride ions, carbon dioxide, oxygen and other corrosive species reduces this protection. Voids in the concrete created by inadequate compaction at the time of application accelerate this penetration. Carbon dioxide reacts chemically with the concrete, reducing its alkalinity. Chloride and oxygen reach the steel and initiate corrosion. The corrosion products occupy a greater volume than the steel from which they were formed. This expansion of some three to fourteen times creates a pressure greater than the tensile strength of the concrete. Spalling of the concrete allows further corrosion to take place. Footings for structural steelwork and vessel supports disintegrate and blocks of detached concrete fall off the roofs and walls of buildings. For reinforced concrete vessel bases, especially where acidic or chloride- or sulfate-containing spillage are anticipated, the following options should be considered at the materials selection process. Acid- and sulfate-resisting concrete are available or specialty coatings can be applied to concrete to prevent the ingress of species that would attack the concrete or corrode the reinforcing steel. In addition, these modifications will protect the process vessel that is encased in the concrete and reinforcing material can be galvanized steel, stainless steel or plastic coated; reinforced steel can be cathodically protected.

Corrosion in soil

Pipework is the item of plant that is most usually buried in soil. In addition to the internal environment, the soil can present a significant corrosion hazard.

Light, sandy, well-drained soil of high electrical resistivity is low in corrosivity and coated steel or bare stainless steels can be employed. It is unlikely that the whole pipe run would be in the same type of soil. In heavier or damp soils, or where the quality of back filling cannot be guaranteed, there are two major corrosion risks. Steel, copper alloys and most stainless steels are susceptible to sulfide attack brought about by the action of sulfate-reducing bacteria in the soil. SRB are ubiquitous but thrive particularly well in the anaerobic conditions which persist in compacted soil, especially clay. The mechanism of corrosion where SRB are involved is described in Section

53.1.2. Cathodic protection alone, described in Section 53.7.2, is not completely effective against this type of corrosion. Where SRB activity is likely or the resistivity of the soil is less than $4000\,\Omega\,cm$ the pipeline should be coated. Anti-corrosion paint systems at a minimum thickness of 0.2 mm and rubber, plastic or bituminous coatings are preferred protection systems. Close control of back-fill quality is necessary to ensure that sharp stones do not damage the coating, and full on-site inspection is required to ensure defect-free application of site-applied coatings and field joints. It is usually the case that a combination of coating and cathodic protection is the most economic alternative. For critical application or the upgrading of old pipes, cathodic protection is generally suitable. Whether or not cathodic protection is used, care is required to avoid stray current corrosion problems through interaction with nearby cathodic protection or other electrical systems such as electric railways.

Corrosion in timber

Metals in contact with or in the proximity of timber can suffer enhanced corrosion attack. Some species of timber, especially oak and Douglas fir, contain high levels of acetic acid. These are volatile and cause corrosion of nearby metals, especially iron, steel and lead alloys.

Metals in contact with timber can be corroded by the acetic acid of the timber and by treatment chemicals present in it. Treatment chemicals include ammonium sulfate and ammonium phosphate flame-retardants. These are particularly corrosive towards steel, aluminum and copper alloys. Preservative treatments include copper salts which, at high timber moisture contents, are corrosive towards steel, aluminum alloys and zinc-coated items.

53.4 Design and corrosion

The design of a plant has significant implications for its subsequent corrosion behavior. Good design minimizes corrosion risks whereas bad design promotes or exacerbates corrosion.

53.4.1 Shape

The shape of a vessel determines how well it drains (Figure 53.7). If the outlet is not at the very lowest point process liquid may be left inside. This will concentrate by evaporation unless cleaned out, and it will probably become more corrosive. This also applies to horizontal pipe runs and steam or cooling coils attached to vessels. Steam heating coils that do not drain adequately collect condensate. This is very often contaminated by chloride ions, which are soon concentrated to high enough levels (10–100 ppm) to pose serious pitting and stress corrosion cracking risks for 300-series austenitic stainless steel vessels and steam coils.

Flat-bottomed storage tanks tend to suffer pitting corrosion beneath deposits or sediments which settle out. Storage tanks may be emptied infrequently and may not experience sufficient agitation or flow to remove such deposits.

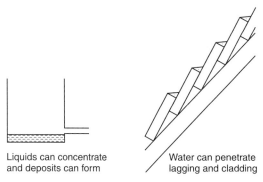

Liquids can concentrate and deposits can form

Water can penetrate lagging and cladding

Figure 53.7 Details of design creating corrosion problems

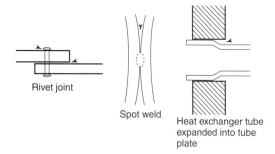

Rivet joint

Spot weld

Heat exchanger tube expanded into tube plate

Figure 53.8 Details of jointing processes creating additional corrosion risks (crevices and stress concentrations)

Flange face areas experience stagnant conditions. Additionally, some gasket materials, such as asbestos fiber, contain leachable chloride ions. This creates crevice and stress corrosion cracking problems on sealing surfaces. Where necessary, flange faces that are at risk can be overlaid with nickel-based alloys. Alternatively, compressed asbestos fiber gaskets shrouded in PTFE may be used.

Bends and tee-pieces in pipework often create locally turbulent flow. This enhances the corrosivity of the process liquid. These effects should be minimized by the use of flow straighteners, swept tees and gentle bends. Flow-induced corrosion downstream of control valves, orifice plates, etc. is sometimes so serious that pipework requires lining with resistant material for some twelve pipe diameters beyond the valve.

53.4.2 Stress

The presence of tensile stress in a metal surface renders that surface more susceptible to many kinds of corrosion than the same material in a non-stressed condition. Similarly, the presence of compressive stress in the surface layer can be beneficial for corrosion behavior.

Tensile stresses can be residual, from a forming or welding operation, or operational from heating–cooling, filling–emptying or pressurizing–depressurizing cycles. The presence of a tensile stress from whatever origin places some materials at risk from stress corrosion cracking, as described in Section 53.1.3. Some items of plant can be stress-relieved by suitable heat treatment, but this cannot prevent operational stress arising.

Cyclic stresses can also give rise to fatigue or corrosion fatigue problems. Information relating to the fatigue life of the material that in the service environment is required, together with the anticipated number of stress cycles to be experienced by the item over its operational life. The fatigue life (the number of cycles to failure) or the fatigue strength (the design strength below which it does not exhibit fatigue problems) is then used in the design.

The presence of stress raisers, including sharp corners and imperfect welds, produces locally high stress levels. These should be avoided where possible or taken into account when designing the materials for use in environments in which they are susceptible to stress corrosion cracking or corrosion fatigue.

53.4.3 Fabrication techniques

Most fabrication techniques have implications for corrosion performance. Riveted and folded seam construction creates crevices as shown in Figure 53.8. Those materials that are susceptible to crevice corrosion should be fabricated using alternative techniques (e.g. welding). Care should be taken to avoid lack of penetration or lack of fusion, since these are sites for crevice corrosion to initiate.

Welding should be continuous, employing fillets where possible, since tack welds create locally high stresses and leave crevice sites. Welding consumables should be chosen to create weld metals of similar corrosion resistance to the parent material. This often requires the use of a slightly over-alloyed consumable, to allow for loss of volatile alloying elements during the welding process and to compensate for the inherently poorer corrosion resistance of the weld metal structure. Strongly over-alloyed weld consumables can create galvanic corrosion problems if the weld metal is significantly more noble than the parent material. In all welds, the heat-affected zone is at risk. The new structure which forms because of the thermal cycle can be of lower corrosion resistance, in addition to the often poorer mechanical properties, than parent material. Austenitic steels such as type 304 and 316 are also susceptible to sensitization effects in the heat-affected zone. In these materials carbide precipitation during the welding thermal cycle denudes the parent material of chromium. This creates areas of significantly diminished corrosion resistance, resulting in knife-line attack in many corrosive environments. This is avoided by the use of the low-carbon equivalents (304L, 316L, etc.) or grades such as type 321 or 347 which are stabilized against sensitization. With correct welding techniques, however, this should be necessary only with thick sections (5 mm for 304 and 8 mm for 316). Some materials, particularly certain aluminum alloys, duplex stainless steels in certain reducing environments and most steel plate, are susceptible to end-grain attack. Penetration along the end grain can be very rapid, with corrosion exploiting the potential differences that exist between inclusions and ferrite crystals in steel and between austenitic and ferrite grains in duplex stainless steel. Where end-grain attack is significant, this should not be exposed to the corrosive environment. It can be covered by a fillet weld if necessary.

53.4.4 Design for inspection

Unseen corrosion can be the most damaging type of attack. Items should be designed to permit periodic inspection. This involves the provision of sufficiently large manways, the installation of inspection pits, the placing of flat-bottomed vessels on beams instead of directly onto concrete bases and the facility for removal of thermal insulation from vessel walls.

53.5 Uses and limitations of constructional materials

53.5.1 Steels and cast irons

Steel is essentially iron with a small amount of carbon. Additional elements are present in small quantities. Contaminants such as sulfur and phosphorus are tolerated at varying levels, depending on the use to which the steel is to be put. Since they are present in the raw material from which the steel is made it is not economic to remove them. Alloying elements such as manganese, silicon, nickel, chromium, molybdenum and vanadium are present at specified levels to improve physical properties such as toughness or corrosion resistance.

Steels are used primarily for their strength and other mechanical properties rather than for corrosion resistance. The corrosion rates of steels are very low only in a few environments, i.e. clean or hard water, certain non-electrolytes, clean atmospheres and properly constituted and compacted ordinary Portland cement. The economics of any high-volume chemical processes, including oil and gas production and transport systems, permits the widespread use of steels. By incorporating a corrosion allowance into the design of a plant, corrosion rates of up to 1 millimeter per year can be accommodated, particularly for plant designed for an operational life of only a few years. Steels generally corrode in a uniform manner. This is beneficial, since localized attack, as experienced by stainless steels, is very difficult to allow for in plant design. The use of corrosion inhibitors (Section 53.7.4) extends the economic range of use of steels. Steel is also available coated with zinc (galvanized) or aluminum-zinc.

Limitations on the use of steels include:

1. Acidic solutions below about pH 5 due to general corrosion;
2. Alkaline solutions above about pH 11 due to general dissolution and caustic cracking;
3. Marine and industrial atmospheres (without the use of protective coatings), because of high corrosion rates;
4. Atmospheric exposure at temperatures in excess of 480°C, because of oxide scaling;
5. Certain environments containing nitrate, cyanide, carbonate, amines, ammonia or strong caustic, due to the risk of stress corrosion cracking. Temperature is an important factor in assessment of each cracking environment;
6. Solutions of hydrogen sulfide, because of hydrogen-induced cracking. Grades of steel are available for certain ranges of pH and hydrogen sulfide partial pressure.

Cast irons are iron with high levels of carbon. Heat treatments and alloying element additions produce gray cast iron, malleable iron, ductile iron, spheroidal cast iron and other grades. The mechanical properties vary significantly. Nickel-containing cast irons have improved hardness and corrosion resistance. Copper or molybdenum additions improve strength.

Chromium, silicon and other alloying elements are used to create cast irons for corrosion resistance in specific environments. Silicon-containing cast irons are used for sulfuric acid duty.

Limitations on the use of cast irons are similar to those for steel, since in many environments most cast iron has poor corrosion resistance. Most grades are also susceptible to graphitization (the loss of iron, leaving a weak structure of graphite) in acidic environments below a pH of approximately 5.5. This attack occurs in soils.

53.5.2 Stainless steels

Stainless steels are iron-based alloys, which contain at least 11 per cent chromium. Hundreds of grades of stainless steel are available with alloying elements including nickel, molybdenum, manganese and copper. Stainless steels are not electrochemically inert, but are protected by a thin layer of oxide. This passive layer is unstable in oxidizing and reducing environments. If lost completely, the underlying steel corrodes rapidly in corrosive (particularly acidic) environments. Where the oxide is lost locally, pitting, crevice corrosion or stress corrosion cracking can proceed at rates of several millimeters per year. Chemical species including the halide ions assist in the breakdown of the passive film in even mild conditions such as neutral sodium chloride solution. For satisfactory service, stainless steels should be passivated continuously (or intermittently, see Section 53.3.5) with a suitable oxidizing species. The redox potential of an environment is of prime importance in the selection of a stainless steel. In reducing conditions, the protective oxide cannot form, and general corrosion takes place; in strongly oxidizing conditions, pitting corrosion is possible.

Stainless steels are used in a wide variety of applications and are most often selected because steel or cast iron would corrode at an unacceptably high rate or produce high levels of iron contamination in the proposed service environment. The main limitations on their uses are:

1. The lower grades of austenitic stainless steels suffer chloride-induced stress corrosion cracking at temperatures in excess of 60°C. Grades containing high levels of nickel are more resistant. Sensitization (Section 53.4.3) occurs near welds in certain non-stabilized grades. The maximum hardness limitations described in NACE MR-01-75 are necessary to prevent sulfide stress corrosion cracking in hydrogen sulfide-containing environments.
2. Ferritic stainless steels have inferior corrosion resistance compared with austenitic grades of equivalent chromium content, because of the absence of nickel. Stress corrosion cracking can occur in strong alkali.
3. Martensitic stainless steels are of limited use in chemical environments because of their inferior corrosion

resistance compared with ferritic and austenitic grades. Their mechanical properties allow them to be used where high hardness is required, but they are susceptible to hydrogen embrittlement.

4. Duplex stainless steels are mostly composed of alternate austenite and ferrite grains. Their structure improves resistance to chloride-induced stress corrosion cracking. In certain reducing acids, such as acetic and formic, preferential attack of the ferrite is a serious problem.

In all categories, the lower grades suffer pitting and crevice corrosion in even low-chloride environments. As the chloride content or temperature increases, or the pH falls, higher grades are required. Reducing acids, including acetic and formic, attack most grades, particularly at elevated temperatures. Free-cutting grades, containing deliberate additions of sulfur or selenium, have significantly poorer corrosion resistance than the corresponding standard grade.

53.5.3 Nickel alloys

Commercially pure nickel has good corrosion resistance to a variety of aggressive environments and is specially used for hot caustic service. It has moderate resistance to acid attack but cannot be used in oxidizing acidic conditions, i.e. in oxidizing acids or acidic media containing oxidizing agents. Nickel is widely used for electroplating steel. Electroless nickel plating is used, especially for items of complex shape that cannot be successfully electroplated. As with all coatings, the presence of defects such as pores allow the corrosive environment to reach the substrate. Since nickel is more noble than steel, rapid, galvanically assisted pitting of the substrate steel occurs at defect sites. Nickel is not resistant in the combined presence of ammonium compounds and oxidizing agents.

Nickel is usually alloyed with elements including copper, chromium, molybdenum and *then* for strengthening and to improve corrosion resistance for specific applications. Nickel-copper alloys (and copper-nickel alloys; see Section 53.5.4) are widely used for handling water. Pumps and valve bodies for fresh water, seawater and mildly acidic alkaline conditions are made from cast Ni-30% Cu type alloys. The wrought material is used for shafts and stems. In seawater contaminated with sulfide, these alloys are subject to pitting and corrosion fatigue. Ammonia contamination creates corrosion problems as for commercially pure nickel.

Nickel-chromium alloys can be used in place of austenitic stainless steels where additional corrosion resistance is required. These alloys are still austenitic but are highly resistant to chloride-induced stress corrosion cracking when their nickel content exceeds 40 per cent.

Molybdenum-containing nickel alloys (the Hastelloys[TM]) are a family of alloys which include grades for handling hot hydrochloric, sulfuric, oxidizing, reducing and organic acids. Some grades are susceptible to attack in ammonia- or sulfide-containing environments.

53.5.4 Copper alloys

Copper has excellent resistance to some corrosive environments, including fresh waters and fluoride-containing atmospheres. Alloying is necessary to achieve good strength, but copper limiting with steel for strength is an alternative (BS 5624). Copper and some of its alloys are susceptible to crevice corrosion, but the mechanism is different from that which affects stainless steels.

Alloying with zinc produces brasses. Brasses are generally suitable for fresh and potable waters but dezincification is a problem for grades containing more than 12 per cent zinc, especially in acidic/alkaline conditions. The addition of small amounts of tin, with arsenic or phosphorus, prevents dezincification and is particularly necessary for seawater applications. For handling flowing seawater or other corrosives, copper-nickel alloys are favored. Brasses are susceptible to stress corrosion cracking in the presence of ammonia or ammonium compounds. This includes lavatory fittings.

Copper-aluminum alloys (aluminum bronze) have good general corrosion resistance but de-aluminification is possible in some grades in certain environments. Copper-nickel alloys are used where flow-related corrosion is a problem, particularly in seawater heat exchanger tubes. The grades containing a small iron addition have the highest critical velocity. These alloys are all susceptible to pitting by sulfides (created in stagnant conditions in sea water systems), especially in the presence of chlorine and stress corrosion cracking in the presence of ammonia. To minimize their susceptibility to sulfide attack in estuarine or other polluted waters it can be advisable to pre-passivate and, if necessary, inhibit with ferrous sulfate.

53.5.5 Miscellaneous metallic materials

Aluminum alloys

Commercially pure aluminum has good resistance to atmospheric corrosion, except where chloride is present in significant quantities, i.e. within 1–2 km of the coast and in the vicinity of chemical plants. To achieve reasonable strength, alloying additions of silicon, manganese, zinc and copper are made. Aluminum-copper alloys have poor corrosion resistance and should not be used in corrosive environments. The other aluminum alloys have good corrosion resistance to most near-neutral media but are attacked in acidic and alkaline conditions. They are susceptible to pitting corrosion, especially where deposits form. Chloride ions also induce pitting of these alloys; chloride is often generated by the hydrolysis of chloride containing organic chemicals. Other organic materials can be handled safely when dry, except alcohols. Metal ions such as copper, which may be introduced by the corrosion of other plant, storage vessels or pipework, can plate out onto the aluminum and create a galvanic cell which produces intense pitting.

The use of anodic films on aluminum alloys is only applicable to some mildly corrosive environments, including architectural purposes and where abrasion resistance is required.

The presence of liquid mercury poses a liquid metal embrittlement problem for many alloys. This occurs by the spillage of mercury in aircraft and the condensation of mercury vapor from mercury pumps in cryogenic applications.

Titanium

Titanium is available unalloyed in several grades of purity, and alloyed. Each grade or alloy has specific uses. Generally, titanium has excellent resistance to seawater and oxidizing conditions, including acids and concentrated hypochlorite solution. It is susceptible to corrosion in reducing acids, rapid attack in dry chlorine gas, cracking in alcohols and embrittlement by hydriding. Hydriding is possible when titanium or a metal with which it is in contact corrodes, liberating hydrogen. For this reason, titanium plant should be inspected carefully for traces of metallic contamination (e.g. scratch marks from steel-nailed boots). Titanium can be used in the anodized form; a thick anodic coating greatly reduces hydrogen absorption. It is used with anodic protection for handling 40–70 per cent strength sulfuric acid and 36 per cent hydrochloric acid.

Other metals

Other metals are available which have very specific uses. Their cost is very high but their use is often justified in certain processes such as fine chemical manufacture. They are generally used as thin cladding or loose lining supported by steel or stainless steel.

1. *Gold* is resistant to many strong acids, but not cyanides or high levels of fluoride, hydrogen fluoride and chlorine.
2. *Platinum* is resistant to strong acids and most halogen gases.
3. *Silver* has reasonable resistance to inorganic and organic acids (not strongly oxidizing acids), hot alkaline conditions and some fluoride-containing environments. It tarnishes in the presence of sulfides.
4. *Tantalum* is inert to many environments including hydrochloric acid at strengths up to 25 per cent at 90°C. It is attacked by fluorine, hydrofluoric acid, fuming sulfuric acid containing free sulfur trioxide, and many alkalines. Above 300°C it reacts with many gases including air. It is subject to hydrogen embrittlement in a similar manner to titanium.

53.5.6 Linings and coatings

Many materials with good corrosion resistance have inadequate strength or other mechanical properties to enable their fabrication or economic use. Organic linings and coatings are used in a range of plant. Organic materials are generally susceptible to organic solvents and have use only within limited ranges of temperature. All coatings and linings must be applied in a manner to control defects within them, since the substrate on which they are applied will be corroded by any contact with the process fluid.

Inspection techniques are available for specific coating and lining systems. Proper bonding to substrate is necessary to avoid blistering and disbonding in case of a vacuum being created in the vessel.

Resins

Epoxy, polyester, phenolic and other resins are used as coatings and linings with or without reinforcement. Glass fiber, silica, carbon and many other materials can be used as filters or reinforcement to produce materials with specific properties of strength, flexibility, wear resistance and electrical conductivity.

Plastics

A large range of man-made polymeric materials is available, from polyethylene, which is attacked by most organic chemicals, to fluorinated products such as polytetrafluoroethylene and polyethyletherketones, which have exceptional resistance to virtually all chemicals. All polymers have their own adhesive, welding and fabrication limitations which must be taken into account in the design of the coated item. These materials can also be used in solid form.

Elastomers

Elastomers are used for their flexibility in seals, gaskets and hoses and to resist abrasion (through absorption of the kinetic energy of the impinging particles). The range of materials includes natural and synthetic rubbers and modern elastomers with chemical resistance.

Glass

Glass-lined reactor vessels are widely used in pharmaceutical and fine-drug manufacturing processes in which metallic contamination of the products has to be minimal. Most glasses are susceptible to attack in strong acid and alkali media. Manufacture and design of glass-lined vessels is a very specialized area. Such vessels require periodic inspection to ensure the integrity of the lining, which is susceptible to mechanical damage and repair is difficult.

53.6 Specifying materials

Having selected the appropriate material for an item of plant, it is imperative that this material is used in the required state of heat treatment, surface finishes, etc. for its construction.

53.6.1 Compositional aspects

1. *National standards*: British Standards, DIN (Deutsche Industries Normung, from Germany) and ASTM/AISI (American Society for Testing of Materials/American Iron and Steel Institute) are those in most common use for metallic materials. It is always preferable to select a material for which national or international specifications exist. Many materials, including steels and the

300-series austenitic stainless steels, have wide ranges of chemical composition. If an alloy or a grade of stainless steel is to be used in a marginal duty (i.e. where relatively small variations in alloying element content, heat treatment or surface condition could markedly affect its corrosion behavior) this should be taken into account when the material is specified. For stainless steels and some other materials of construction it is possible to purchase, at a premium rate, the material with specified minimums of alloying elements. Type 316 stainless steel, when supplied to AISI specification, can have 10–14 per cent nickel, 16–18 per cent chromium and 2–3 per cent molybdenum. This wide specification produces material across a range of corrosion resistance. It also creates latitude for mechanical and physical properties, including strength and magnetic permeability. For demanding applications, but where the more resistant type 317 material is not justified, type 316 with a minimum molybdenum content may be specified.

2. *Proprietary alloy designation*: Many alloys are the subject of patents and are the registered trademark of the producer. When produced by the holder of the patent, these alloys are generally of the highest standard, with tight control of composition, and exhibit consistent corrosion behavior. They are mostly originally developed for a specific corrosive service, but are subsequently marketed more widely. These materials can also be manufactured under license or, when the patent has expired, by other producers using a generic name.

   Produced in this way, these materials can be less reliable, with the introduction of unwanted tramp elements. Where proving tests have indicated the suitability of a proprietary alloy for specific application, the same source of material should be used for construction. Published compositional data are typical values; cast analyses should be requested for critical applications.

3. *Government standards*: These are developed by government departments, especially the military, and are primarily for aerospace and marine service materials.

4. *In-house specifications*: End users requiring larger amounts of material and fabricators of specialty plant benefit from preparing their own in-house specifications. These are often based on existing national standards modified to include additional requirements. They include tighter compositional ranges of alloying elements, mechanical properties such as strength and surface finish, inspection and certification. These specifications must be written by experts in order to satisfy the users' real requirements without compromising any aspect of the materials' performance.

53.6.2 Mechanical properties

Many available standards, particularly national standards, either do not cover the material, physical and mechanical properties or leave the user to specify from a range of options. The aspects of mechanical and physical properties which have specific implications for a material's corrosion behavior include:

1. Strength, particularly where the material will be subjected to fatigue loading or is susceptible to degrading mechanisms such as hydrogen-induced cracking;
2. Hardness, including surface hardness, especially for materials for sour service or environments in which stress corrosion cracking is expected. It is also important where erosion corrosion is likely;
3. Surface finish, since good surface finish retards the onset of certain types of corrosion attack, including pitting and stress corrosion cracking.

All relevant physical and mechanical property requirements should be included in the material specification.

53.6.3 Certification

All aspects of the material's chemical, mechanical and physical properties which are included in the specification should be capable of measurement and certification. For critical duties all material supplied should be fully tested and certified by competent approved, independent test laboratories. All items of plant should be purchased with material certification. Additional certification is required in cases where the fabricator, in manufacturing an item of plant, used techniques such as welding or heat treatment which may affect the corrosion behavior of the construction materials.

All bought-in items of plant, especially those supplied without adequate materials certification, should be subject to random inspection. Portable instruments are available for many types of non-destructive examination, chemical analysis and mechanical testing of fabricated items of plant.

53.7 Corrosion-control techniques

Economic reasons may dictate the use of an inexpensive constructional material (i.e. steel protected by one of the methods discussed below) in place of a more resistant but more expensive material. Although all these corrosion control techniques could be used on a variety of materials, with appropriate design and safeguards, they are only rarely employed to protect any material other than steel.

53.7.1 Painting

Paints are complex formulations of polymeric binders with additives including anti-corrosion pigments, colors, plasticizers, ultraviolet absorbers, flame-retardant chemicals, etc. Almost all binders are organic materials such as resins based on epoxy, polyurethanes, alkyds, esters, chlorinated rubber and acrylics. The common inorganic binder is the silicate used in inorganic zinc silicate primer for steel. Specific formulations are available for application to aluminum and for galvanized steel substrates.

Because of their generally poor resistance to solvents, acids, alkalis and other corrosive agents, paints are not normally used to protect plant internals handling anything other than waters. Even clean (i.e. potable or better) water is damaging to most epoxy-based coatings. Limitations on the use of paint systems include:

1. *Erosion resistance*: The mechanical properties of paints prevent their use in conditions where impingement or erosion by entrained solids is expected.
2. *Solvent resistance*: This varies from very poor (for chlorinated rubbers) to good (for polyurethanes). All paint binders have specific susceptibilities and the presence of small quantities of the appropriate solvent in the atmosphere in the region of an item of plant can cause rapid failure.
3. *Saponification*: Paints are most commonly used to protect steel from corrosion by seawater in marine applications and soil in the case of buried structures. Additional protection is often supplied by the application of cathodic protection to the steel. Any paint coating used in conjunction with cathodic protection must be resistant to the alkali which is produced on the steel at defect sites in the coating. The amount of alkali generated depends on the potential to which the steel is polarized. Some paint binders such as alkyds and vinyl ester are very susceptible to saponification, and should not be used on cathodically protected structures. Cathodic disbondment testing should be undertaken if the relevant information is not available.
4. *Temperature*: Thermal breakdown of the binder limits the service temperature for painted items to 70°C (e.g. chlorinated rubbers and polyvinylacetates), 120°C (many epoxies) and 160°C (for vinylesters). The commonest reason for coating failure is incorrect application. This is a skilled task, involving surface preparation (usually by grit blasting) and a tight control over environmental factors (temperature and relative humidity of the local atmosphere), the cleanliness of the steel. All coating should be supervised by an independent, qualified inspector certified to a recognized standard such as the National Association of Corrosion Engineers International Coating Inspector Training and Certification Scheme.

The selection of an anti-corrosion paint-coating scheme should be undertaken by a qualified expert.

53.7.2 Cathodic protection

Cathodic protection (CP) is an electrochemical technique of corrosion control in which the potential of a metal surface is moved in a cathodic direction to reduce the thermodynamic tendency for corrosion. CP requires that the item to be protected be in contact with an electrolyte. Only those parts of the item that are electrically coupled to the anode and to which the CP current can flow are protected. Thus, the inside of a buried pipe is not capable of cathodic protection unless a suitable anode is placed inside the pipe. The electrolyte through which the CP current flows is usually seawater or soil. Fresh waters generally have inadequate conductivity (but the interiors of galvanized hot water tanks are sometimes protected by a sacrificial magnesium anode) and the conductivity

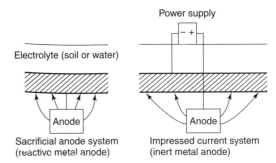

Figure 53.9 Cathodic protection

of air is far too low to permit the cathodic protection of buildings or vehicles.

The operating principle of CP is shown in Figure 53.9. For buried pipelines, the power source is most frequently D.C. voltage derived from a rectified mains power supply, or in suitable climates such as Australia and the Middle East, solar cells. Such systems operate by 'impressed current' and suitable anodes are platinized titanium or niobium wire, cast iron, and graphite. They are not consumed, but can be damaged by the chlorine generated in seawater CP systems. Reinforcing steel in concrete jetties, bridge and car park decks and some offshore structures are generally protected by means of impressed current systems.

Sacrificial anode systems operate without external power source. The anodes are reactive metals such as magnesium and zinc or aluminum alloys. The energy for the process is derived from the anode material. Careful design is required to match the output and lifetime of the anodes with the polarization and life-expectancy requirements of the plant. Sacrificial anode CP is used for offshore platforms, sub-sea pipelines and the inside of ballast tanks on tanker ships.

Correct design is necessary to achieve full protection without overprotection and to minimize wastage of power or anode materials. Overprotection is undesirable because alkali is generated at the cathode (this degrades many plant systems) and hydrogen can be evolved. Hydrogen is deleterious to some mechanical properties of steels and pipelines have failed through this mechanism. Steel is also susceptible to cracking in carbonate-containing environments in certain potential ranges. A reduction in the power requirements of a CP system can usually be achieved by coating the protected structure.

Stainless steel pipes (buried in the ground) and the interiors of stainless steel heat exchangers have been successfully cathodically protected, but CP is rarely used for materials other than steel. The protection potential usually adopted for steel is −850 mV to the saturated calomel reference electrode. This varies with temperature and the presence of other aggressive species in the environment.

53.7.3 Anodic protection

Anodic protection is possible only for material-environment combinations that exhibit fairly wide passive regions. Examples include type 304 stainless steel in phosphoric acid and titanium in sulfuric acid, steel and stainless steels in 98 per cent sulfuric acid. The effect of the anodic polarization is to shift the steel into a region of passivity. This promotes the formation of a protective film on the steel surface, preventing general corrosion. Inadequate or over-polarization creates corrosion and hydrogen evolution problems. As with CP, only wetted surfaces can be protected; this excludes condensed films or droplets on vessel roofs and walls above the liquid line. Such areas are subject to rapid corrosion attack.

53.7.4 Corrosion inhibitors

The use of cheap constructional materials such as steel can be tolerated in certain instances where corrosion inhibition is possible:

1. *Boilers and steam systems*: Steel steam lines can be inhibited by the use of a volatile amine-based inhibitor such as ammonia, morpholine or cyclohexylamine introduced with the feedwater. It passes through the boiler and into the steam system, where it neutralizes the acidic conditions in pipework. The inhibitor is chemically consumed and lost by physical means. Film-forming inhibitors such as heterocyclic amines and alkyl sulphonates must be present at levels sufficient to cover the entire steel surface, otherwise localized corrosion will occur on the bare steel. Inhibitor selection must take into account the presence of other materials in the system. Some amine products cause corrosion of copper. If copper is present and at risk of corrosion it can be inhibited by the addition of benzotriazole or tolutriazole at a level appropriate to the system (see also Section 53.3.2).
2. *Cooling waters*: Once-through cooling systems cannot usually be inhibited because of the expense of the chemicals required and the problems of the disposal of the treated water. Open recirculating systems lose water by evaporation from the tower and pick up atmospheric dust and other detritus. Inhibition of such water is part of the overall water treatment in which scaling tendency and other factors are also controlled. Phosphates and phosphonates together with zinc ions are commonly used. Biocide additions are also required and algal growth control in the case of systems into which sunlight penetrates. Closed recirculating systems, including those in motor vehicles, can be treated with the most efficient inhibitors such as chromates. Disposal problems have restricted the use of these chemicals, but borates, benzoates and nitrites are used. If there is copper in the system, this must be inhibited also. If aluminum is in the system, nitrite poses a pitting problem in certain pH ranges. Nitrite is also aggressive towards lead–tin soldered joints and is ineffective as an inhibitor for zinc and its alloys. Aluminum is only properly protected by benzoate in acidic solutions; in alkaline conditions, it is not fully effective (borates are preferred). The compatibility of all materials, including stainless steels and galvanized steel, must be ascertained before any corrosion-inhibition system is employed.
3. *Heating systems*: Where they are virtually sealed except for a make-up tank (as in the case of domestic central heating) systems can be treated as for closed recirculating cooling systems, and benzoates, nitrites and borates are used. In a correctly designed system such that oxygen ingress is minimal and little make-up is required the oxygen content diminishes as corrosion proceeds and the system can require no inhibitor treatment. The use of certain toxic corrosion inhibitors is not permitted in hospitals, food-manufacturing plant and breweries, where a leak could result in contamination of food or exposure of people to the chemical.
4. *Vapor phase inhibitors*: These are used for the temporary protection of new plant in transit or prior to commissioning. Volatile corrosion inhibitors such as cyclohexylamine derivatives are used. The plant must be sealed or contained to prevent rapid loss of the inhibitor. Sachets of these materials are placed in packing cases. Papers impregnated with them are available for wrapping steel items. These inhibitors are used primarily to protect steel.

53.8 Corrosion monitoring

Once a plant is in operation it is important to monitor the progress of any corrosion that might be taking place. The four approaches described below vary in sophistication and cost. The most appropriate for any plant is determined by a number of factors, including the mechanisms of corrosion that are anticipated and the implications of catastrophic or unexpected failure. Key areas of the plant require closer monitoring than ready replaceable items. The measures described below do not replace the mandatory inspections of pressure vessels, etc. for insurance purposes. The overall philosophy of corrosion monitoring is to improve the economics of the plant's operation by allowing the use of cheaper materials and generally reducing over-design that goes into plant to combat corrosion.

53.8.1 Physical examination

Full records of all constructional materials that are used in the plant should be maintained and updated when repairs are undertaken. The exteriors of all parts of the plant should be subjected to frequent visual examination and the results reported and stored for future reference. This maximizes the warning time before corrosion failures occur, since the majority of failure mechanisms cause leaks before bursting. Key items of plant, those in which some degree of corrosion is anticipated and those which might suffer catastrophic failure, should be examined in detail. Internal visual inspection during shutdowns is sufficient to identify most corrosion effects. Cracking can usually be seen with the naked eye but where cracking is considered a possible mechanism an appropriate non-destructive test method should be employed. In items of plant which are shut down only infrequently (relative to the time scale of possible corrosion or cracking failure) external non-destructive testing is often possible.

Candidate non-destructive test methods include:

1. *Ultrasonic techniques*: Wall thickness can be measured to monitor the progress of general corrosion, cracks can be detected and hydrogen blisters identified. Certain construction materials such as cast iron cannot be examined by ultrasound. Skilled operators and specialist equipment is required. Plant can be examined *in situ* except when it is above 80°C.
2. *Magnetic particle inspection*: Surface emergent and some sub-surface cracking can be detected in ferromagnetic materials. The technique must be used on the side of the material in contact with the corrodent.
3. *Dye penetration inspection*: This is a simple technique, requiring a minimum of operator training. In the hands of a skilled operator, it is capable of detecting fine cracks such as chloride stress corrosion cracks in austenitic stainless steels and fatigue cracks.

53.8.2 Exposure coupons and electrical resistance probes

If changes have been made to the process (e.g. if incoming water quality cannot be maintained or other uncertainties arise concerning the corrosion behavior of the construction materials) it is possible to incorporate coupons or probes of the material into the plant and monitor their corrosion behavior. This approach may be used to assist in the materials selection process for a replacement plant. Small coupons (typically, 25×50 mm) of any material may be suspended in the process stream and removed at intervals for weight loss determination and visual inspection for localized corrosion. Electrical resistance probes comprise short strands for the appropriate material electrically isolated from the item of plant. An electrical connection from each end of the probe is fed out of the plant to a control box. The box senses the electrical resistance of the probe. The probe's resistance rises as its cross-sectional area is lost through corrosion.

The materials should be in the appropriate form (i.e. cast/wrought/welded, heat treatment and surface condition). Metal coupons should be electrically isolated from any other metallic material in the system. They should be securely attached to prevent their being dislodged and causing damage downstream. Simple coupons and probes cannot replicate the corrosion effects due to heat transfer but otherwise provide very useful information. It should be noted that any corrosion they have suffered represents the integrated corrosion rate over the exposure time. Corrosion rates often diminish with time as scaling or filming takes place, thus short-term exposures can give values higher than the true corrosion rate.

53.8.3 Electrochemical corrosion monitoring

A number of corrosion-monitoring techniques based on electrochemical principles are available. These give an indication of the instantaneous corrosion rate, which is of use when changing process conditions create a variety of corrosion effects at different times in a plant. Some techniques monitor continuously, others take a finite time to make a measurement.

1. *Polarization resistance*: The current-potential behavior of a metal, externally polarized around its corrosion potential, provides a good indication of its corrosion rate. The technique has the advantage of being well established and hence reliable when used within certain limitations. This technique can only be used for certain metals, to give general corrosion rate date in electrolytes. It cannot be employed to monitor localized corrosion such as pitting, crevice corrosion or stress corrosion cracking, nor used in low-conductivity environments such as concrete, timber, soil and poor electrolytes (e.g. clean water and non-ionic solvents). Equipment is available commercially but professional advice should be sought for system design and location of probes.
2. *Impedance spectroscopy*: This technique is essentially the extension of polarization resistance measurements into low-conductivity environments, including those listed above. The technique can also be used to monitor atmospheric corrosion, corrosion under thin films of condensed liquid and the breakdown of protective paint coatings. Additionally, the method provides mechanistic data concerning the corrosion processes, which are taking place.
3. *Electrochemical noise*: A variety of related techniques are now available to monitor localized corrosion. No external polarization of the corroding metal is required, but the electrical noise on the corrosion potential of the metal is monitored and analyzed. Signatures characteristic of pit initiation, crevice corrosion and some forms of stress corrosion cracking is obtained.

53.8.4 Thin-layer activation

This technique is based upon the detection of corrosion products, in the form of dissolved metal ions, in the process stream. A thin layer of radioactive material is created on the process side of an item of plant. As corrosion occurs, radioactive isotopes of the elements in the construction material of the plant pass into the process stream and are detected. The rate of metal loss is quantified and local rates of corrosion are inferred. This monitoring technique is not yet in widespread use but it has been proven in several industries.

Further reading

Dillon, C. P., *Corrosion Control in the Chemical Process Industries*, McGraw-Hill, New York (1986).

Fontana, M. G. and Greene, N. D., *Corrosion Engineering*, 2nd edn, McGraw-Hill, New York (1978).

The Forms of Corrosion Recognition and Prevention, Corrosion Handbook No. 1, NACE, Houston, Texas (1982).

Lees, F. P., *Loss Prevention in the Process Industries*, 2nd edn, Butterworth-Heinemann, Oxford (1992).

Shreir, L. L. (ed.), *Corrosion*, 3rd edn, Butterworth-Heinemann, Oxford (1994).

Journals

British Corrosion Journal.

Industrial Corrosion is the main journal of the Institution of Corrosion Science and Technology, PO Box 253, Leighton Buzzard, LU7 7WB.

Corrosion is the scientific journal of the National Association of Corrosion Engineers, PO Box 21840, Houston, Texas, USA.

Materials Performance is the engineering journal of the National Association of Corrosion Engineers.

54

Shaft Alignment

R Keith Mobley
The Plant Performance Group

Contents

54.1 Introduction

Shaft alignment is the proper positioning of the shaft centerlines of the driver and driven components (i.e., pumps, gearboxes, etc.) that make up the machine drive train. Alignment is accomplished either through shimming and/or moving a machine component. Its objective is to obtain a common axis of rotation at operating equilibrium for two coupled shafts or a train of coupled shafts.

Shafts must be aligned as perfectly as possible to maximize equipment reliability and life, particularly for high-speed equipment. Alignment is important for directly coupled shafts, as well as coupled shafts of machines that are separated by distance – even those using flexible couplings. It is important because misalignment can introduce a high level of vibration, cause bearings to run hot, and result in the need for frequent repairs. Proper alignment reduces power consumption and noise level, and helps to achieve the design life of bearings, seals, and couplings.

Alignment procedures are based on the assumption that one machine-train component is stationary, level, and properly supported by its baseplate and foundation. Both angular and offset alignment must be performed in the vertical and horizontal planes, which is accomplished by raising or lowering the other machine components and/or moving them horizontally to align with the rotational centerline of the stationary shaft. The movable components are designated as 'machines to be moved,' or MTBM, or 'machines to be shimmed,' or MTBS. MTBM generally refers to corrections in the horizontal plane while MTBS generally refers to corrections in the vertical plane.

Too often, alignment operations are performed randomly and adjustments are made by trial and error, resulting in a time-consuming procedure.

54.2 Alignment fundamentals

This section discusses the fundamentals of machine alignment and presents an alternative to the commonly used method, trial-and-error. This section addresses exactly what alignment is and the tools needed to perform it, why it is needed, how often it should be performed, what is considered to be 'good enough,' and what steps should be taken prior to performing the alignment procedure. It also discusses types of alignment (or misalignment), alignment planes, and why alignment is performed on shafts as opposed to couplings.

What is alignment?

Shafts are considered to be in alignment when they are colinear at the coupling point. The term 'colinear' refers to the condition when the rotational centerlines of two mating shafts are parallel and intersect (i.e., join to form one line). When this is the case, the coupled shafts operate just like a solid shaft. Any deviation from the aligned or colinear condition, however, results in abnormal wear of machine-train components such as bearings and shaft seals.

Variations in machine-component configuration and thermal growth can cause mounting-feet elevations

and the horizontal orientations of individual drive-train components to be in different planes. Nevertheless, they are properly aligned as long as their shafts are colinear at the coupling point.

Note that it is important for final drive-train alignment to compensate for actual operating conditions because machines often move after start up. Such movement is generally the result of wear, thermal growth, dynamic loads, and support or structural shifts. These factors must be considered and compensated for during the alignment process.

Tools most commonly used for alignment procedures are dial indicators, adjustable parallels, taper gauges, feeler gauges, small-hole gauges, and outside micrometer calipers.

Why perform alignment and how often?

Periodic alignment checks on all coupled machinery are considered one of the best 'tools' in a preventive maintenance program. Such checks are important because the vibration effects of misalignment can seriously damage a piece of equipment. Misalignment of more than a few thousandths of an inch can cause vibration that significantly reduces equipment life.

Although the machinery may have been properly aligned during installation or during a previous check, misalignment may develop over a very short period of time. Potential causes include: foundation movement or settling, accidentally bumping the machine with another piece of equipment, thermal expansion, distortion caused by connected piping, loosened hold-down nuts, expanded grout, rusting of shims, etc.

Indications of misalignment in rotating machinery are shaft wobbling, excessive vibration (in both radial and axial directions), excessive bearing temperature (even if adequate lubrication is present), noise, bearing wear pattern, and coupling wear.

Trial-and-error versus calculation

Many alignments are done by the trial-and-error method. Although this method may eventually produce the correct answers, it is extremely time consuming and, as a result, it is usually considered 'good enough' before it really is. Rather than relying on 'feel' as with trial-and-error, some simple trigonometric principles allow alignment to be done properly with the exact amount of correction needed either measured or calculated, taking the guesswork out of the process. Such accurate measurements and calculations make it possible to align a piece of machinery on the first attempt.

What is good enough?

This is a difficult question to answer because there are vast differences in machinery strength, speed of rotation, type of coupling, etc. It also is important to understand that flexible couplings do not cure misalignment problems – a common myth in industry. Although they may somewhat dampen the effects, flexible couplings are not a total solution.

An easy (perhaps too easy) answer to the question of what is good enough is to align all machinery to comply exactly with the manufacturers' specifications. However, the question of which manufacturers' specifications to follow must be answered, as few manufacturers build entire assemblies. Therefore, an alignment is not considered good enough until it is well within all manufacturers' tolerances and a vibration analysis of the machinery in operation shows the vibration effects due to misalignment to be within the manufacturers' specifications or accepted industry standards. Note that manufacturers' alignment specifications may include intentional misalignment during 'cold' alignment to compensate for thermal growth, gear lash, etc. during operation.

54.3 Coupling alignment versus shaft alignment

If all couplings were perfectly bored through their exact center and perfectly machined about their rim and face, it might be possible to align a piece of machinery simply by aligning the two coupling halves. However, coupling eccentricity often results in coupling misalignment. This does not mean, however, that dial indicators should not be placed on the coupling halves to obtain alignment measurements. It does mean that the two shafts should be rotated simultaneously when obtaining readings, which makes the couplings an extension of the shaft centerlines whose irregularities will not affect the readings. Therefore, this chapter primarily addresses shaft alignment.

Although alignment operations are performed on coupling surfaces because they are convenient to use, it is extremely important that these surfaces and the shaft 'run true'. If there is any runout (i.e., axial or radial looseness) of the shaft and/or the coupling, a proportionate error in alignment will result. Therefore, prior to making alignment measurements, the shaft and coupling should be checked and corrected for runout.

54.3.1 Alignment conditions

There are four alignment conditions: perfect alignment, offset or parallel misalignment, angular or face misalignment, and skewed or combination misalignment (i.e., both offset and angular).

Perfect alignment

Two perfectly aligned shafts are colinear and operate as a solid shaft when coupled. This condition is illustrated in Figure 54.1. However, it is extremely rare for two shafts to be perfectly aligned without an alignment procedure

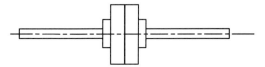

Figure 54.1 Perfect alignment

being performed on them. In addition, the state of alignment should be monitored on a regular basis in order to maintain the condition of perfect alignment.

Offset or parallel misalignment

Offset misalignment, also referred to as parallel misalignment, refers to the distance between two shaft centerlines, and is generally measured in thousandths of an inch. Offset can be present in either the vertical or horizontal plane. Figure 54.2 illustrates offset, showing two mating shafts that are parallel to each other, but not colinear. Theoretically, offset is measured at the coupling centerline.

Angular or face misalignment

A sound knowledge of angular, also called face, misalignment is needed for understanding alignment conditions and performing the tasks associated with machine-train alignment, such as drawing alignment graphics, calculating feet corrections, specifying thermal growth, obtaining target specifications, and determining spacer-shaft alignment.

Angular misalignment refers to the condition when the shafts are not parallel, but are in the same plane with no offset. This is illustrated in Figure 54.3. Note that with angular misalignment, it is possible for the mating shafts to be in the same plane at the coupling-face intersection, but to not have an angular relationship such that they are not colinear.

Angularity is the angle between the two shaft centerlines, which generally is expressed as a 'slope,' or 'rise over run,' of so many thousandths of an inch per inch (i.e., unit less), rather than as an angle in degrees. It must be determined in both the vertical and horizontal planes. Figure 54.4 illustrates the angles involved in angular misalignment.

From a practical standpoint, it is often difficult or undesirable to position the stems of the dial indicators at 90° angles to the rim and/or face surfaces of the coupling halves. For this reason, brackets are used to mount the devices on the shaft or a non-movable part of the coupling to facilitate taking readings and to insure greater accuracy. This is a valid method because any object that

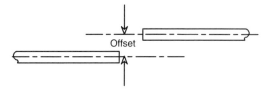

Figure 54.2 Offset misalignment

Figure 54.3 Angular misalignment (no offset)

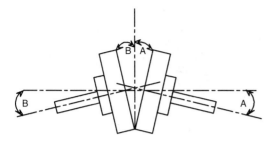

Figure 54.4 Angles are equal at the coupling or shaft centerline

is securely attached and rotated with the shaft or coupling hub becomes a radial extension of the shaft centerline and can be considered an integral part of the shaft. However, this somewhat complicates the process and requires right-triangle concepts to be understood and other adjustments (e.g., indicator sag) to be made to the readings.

Compare the two diagrams in Figure 54.5. Figure 54.5(a) is a common right triangle and Figure 54.5(b) is a simplified view of an alignment-measuring apparatus, or fixture, which incorporates a right triangle.

The length of side 'b' is measured with a tape measure and the length of side 'a' is measured with a device such as a dial indicator. Note that this diagram assumes the coupling is centered on the shaft and that its centerline is the same as the shaft's. Angle 'A' in degrees is calculated by:

$$A = \tan^{-1} \frac{a}{b}$$

This formula yields the angle 'A' expressed in degrees, which requires the use of a trigonometric table or a calculator that is capable of determining the inverse tangent.

While technically correct, however, alignment calculations do not require the use of an angle value in degrees. Note that it is common industry practice to refer to the following value as 'Angle A', even though it is not truly

an angle and is actually the tangent of 'Angle A':

$$\text{`Angle A'} = \frac{a}{b} = \frac{rise}{run}$$

Figures 54.6 and 54.7 illustrate the concept of rise and run. If one assumes that line O–A in Figure 54.6 represents a true, or target, shaft centerline, then side 'a' of the triangle represents the amount of offset present in the actual shaft, which is referred to as the rise.

Note that this 'offset' value is not the true theoretical offset. It is actually the theoretical offset plus one-half of the shaft diameter (see Figure 54.5) because the indicator dial is mounted on the outside edge of the shaft as opposed to the centerline. However, for the purposes of alignment calculations, it is not necessary to use the theoretical offset or the theoretical run that corresponds to it. Figure 54.7 illustrates why this is not necessary.

Figure 54.7 illustrates several rise/run measurements for a constant 'Angle A'. Unless 'Angle A' changes, an increase in rise results in a proportionate increase in run. This relationship allows the alignment calculations to be made without using the theoretical offset value and its corresponding run.

Therefore, the calculation of 'Angle A' can be made with any of the rise/run measurements:

$$\text{`Angle A'} = \frac{rise_1}{run_1} = \frac{rise_2}{run_2} = \frac{rise_3}{run_3} = \frac{rise_4}{run_4}$$

For example: If the rise at a machine foot is equal to 0.5 inches with a run of 12 inches, 'Angle A' is:

$$\text{`Angle A'} = \frac{0.5''}{12.0''} = 0.042$$

If the other machine foot is 12 inches away (i.e., run = 24 inches), the following relationship applies:

$$0.042 = \frac{X}{24.0''}, \text{ where X or rise} = 1 \text{ inch}$$

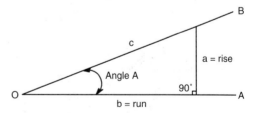

Figure 54.6 Concept of rise and run

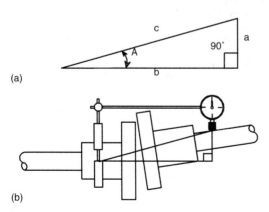

(a)

(b)

Figure 54.5 Common right triangle and simplified alignment-measuring apparatus

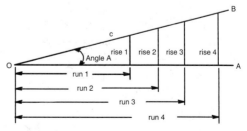

Figure 54.7 Rise/run measurements for constant angle

Combination or skewed misalignment

Combination or skewed misalignment occurs when the shafts are not parallel (i.e., angular), nor do they intersect at the coupling (i.e., offset). Figure 54.8 shows two shafts that are skewed, which is the most common type of misalignment problem encountered. This type of misalignment can occur in either the horizontal or vertical plane, or in both the horizontal and vertical planes.

For comparison, see Figure 54.3, which shows two shafts that have angular misalignment, but are not offset. Figure 54.9 shows how an offset measurement for non-parallel shafts can vary, depending upon where the distance between two shaft centerlines is measured. Again, note that theoretical offset is defined at the coupling face.

54.3.2 Alignment planes

There are two misalignment planes to correct: vertical and horizontal. Therefore, in the case where at least two machines make up a machine train, four types of misalignment can occur: vertical offset, vertical angularity, horizontal offset, and horizontal angularity. These can occur in any combination and, in many cases, all four are present.

Vertical

Both angular misalignment and offset can occur in the vertical plane. Vertical misalignment, which is corrected

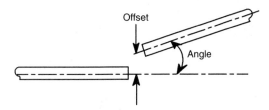

Figure 54.8 Offset and angular misalignment

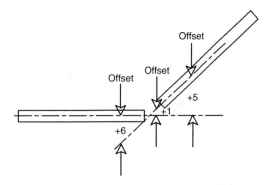

Figure 54.9 Offset measurement for angularly misaligned shafts

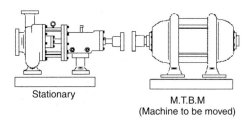

Stationary

M.T.B.M
(Machine to be moved)

Figure 54.10 Vertical misalignment

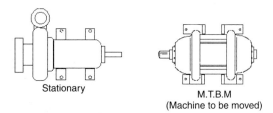

Stationary

M.T.B.M
(Machine to be moved)

Figure 54.11 Horizontal misalignment

by the use of shims, is usually illustrated in a side-view drawing as shown in Figure 54.10.

Horizontal

Both offset and angular misalignment can occur in the horizontal plane. Shims are not used to correct for horizontal misalignment, which is typically illustrated in a top-view drawing as shown in Figure 54.11. This type of misalignment is corrected by physically moving the machine to be moved (MTBM).

54.3.3 Actions to be taken before alignment

It is crucial that alignment procedures be performed correctly, regardless of what method is used. Actions to be taken before alignment are discussed in the following sections, which cover the preparatory steps as well as two major issues (i.e., soft-foot and indicator sag corrections) that must be resolved before alignment can be accomplished. This section provides procedures for making these corrections as well as the proper way to tighten hold-down nuts, an important procedure needed when correcting soft-foot.

Preparatory steps

The following preparatory steps should be taken before attempting to align a machine train:

1. Before placing a machine on its base, make sure that both the base and the bottom of the machine are clean, rust free, and do not have any burrs. Use a wire brush or file on these areas if necessary.
2. Common practice is to position, level, and secure the driven unit at the required elevation prior to adjusting the driver to align with it. Set the driven unit's shaft centerline slightly higher than the driver.

3. Make all connections, such as pipe connections to a pump or output shaft connections on a reducer, to the driven unit.
4. Use only clean shims that have not been 'kinked' or that have burrs.
5. Make sure the shaft does not have an indicated runout.
6. Before starting the alignment procedure, check for 'soft-foot' and correct the condition.
7. Always use the correct tightening sequence procedure on the hold-down nuts.
8. Determine the amount of indicator sag before starting the alignment procedure.
9. Always position the stem of the dial indicator so that it is perpendicular to the surface against which it will rest. Erroneous readings will result if the stem is not placed at a 90° angle to the surface.
10. Avoid lifting the machine more than is necessary to add or remove shims.
11. Jacking bolt assemblies should be welded onto the bases of all large machinery. If they are not provided, add them before starting the alignment procedure. Use jacking bolts to adjust for horizontal offset and angular misalignment and to hold the machine in place while shimming.

Correcting for soft-foot

Soft-foot is the condition when all four of a machine's feet do not support the weight of the machine. It is important to determine if this condition is present prior to performing shaft alignment on a piece of machinery. Not correcting soft-foot prior to alignment is a major cause of frustration and lost time during the aligning procedure.

The basis for understanding and correcting soft-foot and its causes is the knowledge that three points determine a plane. As an example, consider a chair with one short leg. The chair will never be stable unless the other three legs are shortened or the short leg is shimmed. In this example, the level floor is the 'plane' and the bottom tips of the legs are the 'points' of the plane. Three of the four chair tips will always rest on the floor. If a person is sitting with their weight positioned above the short leg, it will be on the floor and the normal leg diagonally opposite the short leg will be off the floor.

As in the chair example, when a machine with soft-foot is placed on its base, it will rest on three of its support feet unless the base and the bottoms of all of the feet are perfectly machined. Further, because the feet of the machine are actually square pads and not true points, it is possible that the machine can rest on only two support feet, ones that are diagonally opposite each other. In this case, the machine has two soft-feet.

Causes

Possible sources of soft-foot are shown in Figure 54.12.

Consequences

Placing a piece of machinery in service with uncorrected soft-foot may result in the following:

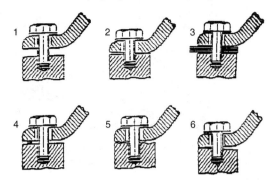

Figure 54.12 Diagrams of possible soft-foot causes (Ref. 3) 1. Loose foot, 2. Cocked foot, 3. Bad shims, 4. Debris under foot, 5. Irregular base surface, 6. Cocked foot

- Dial-indicator readings taken as part of the alignment procedure can be different each time the hold-down nuts are tightened, loosened, and retightened. This can be extremely frustrating because each attempted correction can cause a soft-foot condition in another location.
- The nuts securing the feet to the base may loosen, resulting in either machine looseness and/or misalignment. Either of these conditions can cause vibration, which can be dangerous to personnel as well as to the machine.
- If the nuts do not loosen, metal fatigue may occur at the source of soft-foot. Cracks can develop in the support base/frame and, in extreme cases, the soft-foot may actually break off.

Initial soft-foot correction

The following steps should be taken to check for and correct soft-foot:

- Before setting the machine in place, remove all dirt, rust, and burrs from the bottom of the machine's feet, the shims to be used for leveling, and the base at the areas where the machine's feet will rest.
- Set the machine in place, but do not tighten the hold-down nuts.
- Attempt to pass a thin feeler gauge underneath each of the four feet. Any foot that is not solidly resting on the base is a soft-foot. (A foot is considered 'soft' if the feeler gage passes beneath most of it and only contacts a small point or one edge.)
- If the feeler gage passes beneath a foot, install the necessary shims beneath that foot to make the 'initial' soft-foot correction.

Final soft-foot correction

The following procedure describes the final soft-foot correction:

- Tighten all hold-down nuts on both the stationary machine and the machine to be shimmed (MTBS).
- Secure a dial-indicator holder to the base of the stationary machine and the MTBS. The stem of the dial

indicator should be in a vertical position above the foot to be checked. A magnetic-base indicator holder is most suitable for this purpose.

- Set the dial indicator to zero. Completely loosen the hold-down nut on the foot to be checked. Watch the dial indicator closely for foot movement during the loosening process.
- If the foot rises from the base when the hold-down nut is loosened, place beneath the foot an amount of shim stock equal to the amount of deflection shown on the dial indicator.
- Retighten the hold-down nut and repeat the entire process once again to ensure that no movement occurs.
- Move the dial indicator and holder to the next foot to be checked and repeat the process. Note: The nuts on all of the other feet must remain securely tightened when a foot is being checked for a soft-foot condition.
- Repeat the above process on all of the feet.
- Make a three-point check on each foot by placing a feeler gage under each of the three exposed sides of the foot. This determines if the base of the foot is cocked.

Tightening hold-down nuts

Once soft-foot is removed, it is important to use the correct tightening procedure for the hold-down nuts. This helps ensure that any unequal stresses that cause the machine to shift during the tightening procedure remain the same throughout the entire alignment process. The following procedure should be followed:

- After eliminating soft-foot, loosen all hold-down nuts.
- Number each machine foot in the sequence in which the hold-down nuts will be tightened during the alignment procedure. The numbers (1, 2, 3 and 4) should be permanently marked on, or near, the feet.
- It is generally considered a good idea to tighten the nuts in an 'X' pattern as illustrated in Figure 54.13. Always tighten the nuts in the sequence in which the positions are numbered (1, 2, 3 and 4).
- Loosen nuts in the opposite sequences (4, 3, 2 and 1).
- Use a torque wrench to tighten all nuts with the same amount of torque.
- A similar procedure should be used for baseplates.
- Always tighten the nuts as though the final adjustment has been made, even if the first set of readings has not been taken.

54.3.4 Correcting for indicator sag

Indicator sag is the term used to describe the bending of the mounting hardware as the dial indicator is rotated from the top position to the bottom position during the alignment procedure. Bending can cause significant errors in the indicator readings that are used to determine vertical misalignment, especially in rim-and-face readings. The degree to which the mounting hardware bends depends on the length and material strength of the hardware.

To ensure that correct readings are obtained with the alignment apparatus, it is necessary to determine the amount of indicator sag present in the equipment and to correct the bottom or 6 o'clock readings before starting the alignment process.

Dial indicator mounting hardware consists of a bracket clamped to the shaft, which supports a rod extending beyond the coupling. When two shafts are perfectly aligned, the mounting rod should be parallel to the axis of rotation of the shafts. However, the rod bends or sags by an amount usually measured in mils (thousandths of an inch) due to the combined weight of the rod and the dial indicator attached to the end of the rod. Figure 54.14 illustrates this problem.

Indicator sag is best determined by mounting the dial indicator on a piece of straight pipe of the same length as in the actual application. Zero the dial indicator at the 12 o'clock, or upright, position and then rotate 180° to the 6 o'clock position. The reading obtained, which will be a negative number, is the measure of the mounting-bracket indicator sag for 180° of rotation and is called the sag factor.

All bottom or 6 o'clock readings should be corrected by subtracting the sag factor.

Example 1: Assume that the sag factor is −0.006 inch. If the indicator reading at 6 o'clock equals +0.010 inch, then the true reading is:

Indicator reading − sag factor

$$(+0.010'') - (-0.006'') = +0.016''$$

Example 2: If the indicator reading at 6 o'clock equals −0.010″, then the true reading is:

$$(-0.010'') - (-0.006'') = -0.004''$$

As shown by the above examples, the correct use of positive (+) and negative (−) signs is important in shaft alignment.

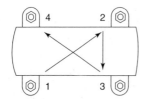

Figure 54.13 Correct bolting sequence for tightening nuts (Ref. 3)

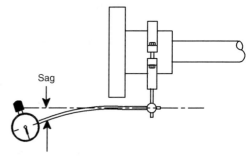

Figure 54.14 Dial indicator sag

Alignment equipment and methods

There are two primary methods of aligning machine trains: dial indicator alignment and optical, or laser, alignment. This section provides an overview of each, with an emphasis on dial-indicator methods.

Dial-indicator methods (i.e., reverse-dial indicator and the two variations of the rim-and-face method) use the same type of dial indicators and mounting equipment. However, the number of indicators and their orientations on the shaft are different. The optical technique does not use this device to make measurements, but uses laser transmitters and sensors.

While the dial-indicator and optical methods differ in the equipment and/or equipment setup used to align machine components, the theory on which they are based is essentially identical. Each method measures the offset and angularity of the shafts of movable components in reference to a pre-selected stationary component. Each assumes that the stationary unit is properly installed and that good mounting, shimming, and bolting techniques are used on all machine components.

54.4 Dial-indicator methods

There are three methods of aligning machinery with dial indicators. These methods are: (1) two-indicator method with readings taken at the stationary machine, (2) two-indicator method with readings taken at the machine to be shimmed, and (3) indicator reverse method. Methods 1 and 2 are often considered to be one method, which is referred to as rim-and-face.

Method selection

Although some manufacturers insist on the use of the reverse indicator method for alignment or, at least, as a final check of the alignment, two basic factors determine which method should be used. The determining factors in method selection are: (1) end play and (2) distance versus radius.

End play or float

Practically all machines with journal or sleeve bearings have some end play or float. It is considered manageable if sufficient pressure can be applied to the end of the shaft during rotation to keep it firmly seated against the thrust bearing or plate. However, for large machinery or machinery that must be energized and 'bumped' to obtain the desired rotation, application of pressure on the shaft is often difficult and/or dangerous. In these cases, float makes it impossible to obtain accurate face readings; therefore, the indicator reverse method must be used as float has a negligible affect.

Distance versus radius

If float is manageable, then there is a choice of which of the methods to use. When there is a choice, the best method is determined by the following rule:

If the distance between the points of contact of the two dial indicators set up to take rim readings for the indicator reverse method is larger than one-half the diameter of travel of the dial indicator set up to take face readings for the two-indicator method, the indicator reverse method should be used.

This rule is because misalignment is more apparent (i.e., dial indicator reading will be larger) under these circumstances and, therefore, corrections will be more accurate.

Equipment

Dial indicators and mounting hardware are the equipment needed to take alignment readings.

Dial indicators   Figure 54.15 shows a common dial indicator, which is also called a runout gage. A dial indicator is an instrument with either jeweled or plain bearings, precisely finished gears, pinions, and other precision parts designed to produce accurate measurements. It is possible to take measurements ranging from one-thousandth (0.001 inch or one mil) to 50 millionths of an inch.

The point that contacts the shaft is attached to a spindle and rack. When it encounters an irregularity, it moves. This movement is transmitted to a pinion, through a series of gears, and on to a hand or pointer that sweeps the dial of the indicator. It yields measurements in (+) or (−) mils.

Measurements taken with this device are based on a point of reference at the 'zero position,' which is defined as the alignment fixture at the top of the shaft – referred to as the 12 o'clock position. In order to perform the alignment procedure, readings also are required at the 3, 6, and 9 o'clock positions.

It is important to understand that the readings taken with this device are all relative, meaning they are dependent upon the location at which they are taken. Rim readings are obtained as the shafts are rotated and the dial indicator stem contacts the shaft at a 90° angle. Face readings, which are used to determine angular misalignment, are obtained as the shafts are rotated and the stem is parallel to the shaft centerline and touching the face of the coupling.

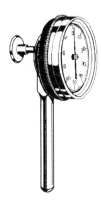

Figure 54.15 Common dial indicator

Mounting hardware Mounting hardware consists of the brackets, posts, connectors, and other hardware used to attach a dial indicator to a piece of machinery. Dial indicators can easily be attached to brackets and, because brackets are adjustable, they can easily be mounted on shafts or coupling hubs of varying size. Brackets eliminate the need to disassemble flexible couplings when checking alignments during predictive maintenance checks or when doing an actual alignment. This also allows more accurate 'hot alignment' checks to be made.

The brackets are designed so that dial indicators can easily be mounted for taking rim readings on the moveable machine and the fixed machine at the same time. This facilitates the use of the indicator reverse method of alignment. If there is not enough room on the shafts, it is permissible to attach brackets to the coupling hubs or any part of the coupling that is solidly attached to the shaft. Do not attach brackets to a movable part of the coupling, such as the shroud.

Note that misuse of equipment can result in costly mistakes. One example is the improper use of magnetic bases, which are generally designed for stationary service. They are not designed for direct attachment to a shaft or coupling that must be rotated to obtain the alignment readings. The shift in forces during rotation can cause movement of the magnetic base and erroneous readings.

54.5 Alignment methods

There are three primary methods of aligning machine trains with dial indicators: reverse-dial indicator method, also called indicator-reverse method, and two variations of the rim-and-face method.

With all three of these methods, it is usually possible to attach two dial indicators to the machinery in such a manner that both sets of readings can be taken simultaneously. However, if only one indicator can be attached, it is acceptable to take one set of readings, change the mounting arrangement, and then take the other set of readings.

There are advantages with the reverse-dial indicator method over the rim-and-face method – namely accuracy and the fact that the mechanic is forced to perform the procedure 'by the book,' as opposed to being able to use 'trial and error'. Accuracy is much better because only rim readings are used. This is because rim readings are not affected by shaft float or end play as are face readings. In addition, the accuracy is improved compared with rim-and-face methods because of the length of the span between indicators.

54.5.1 Reverse-dial indicator

Reverse-dial indicator method (also referred to as indicator-reverse method) is the most accurate form of mechanical alignment. This technique measures offset at two points and the amount of horizontal and vertical correction for offset and angularity is calculated. Rim readings are taken simultaneously at each of the four positions (12, 3, 6, and 9 o'clock) for the movable machine (MTBS/MTBM) and the stationary machine. The

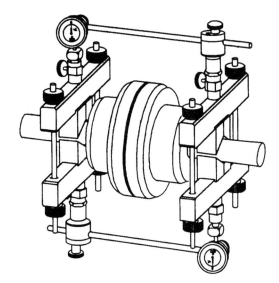

Figure 54.16 Typical reverse dial indicator fixture and mounting

measuring device for this type of alignment is a dual-dial indicator and the most common configuration is that shown in Figure 54.16.

Mounting configuration and readings

Dual runout gages are rigidly mounted on special fixtures attached to the two mating shafts. The runout gages are mounted so that readings can be obtained for both shafts with one 360° rotation.

When the reverse-dial fixture is mounted on mating shafts, the dials initially should be adjusted to their zero point. Once the dials are zeroed, slowly rotate the shafts in 90° increments. Record runout readings from both gages, being sure to record the positive or negative sign, when the fixture is at the 12, 3, 6, and 9 o'clock positions.

Limitations

There are potential errors or problems that limit the accuracy of this alignment technique. The common ones include data recording errors, failure to correct for indicator sag, mechanical looseness in the fixture installation, and failure to properly zero and/or calibrate the dial indicator.

Data recording One of the most common errors made with this technique is reversing the 3 and 9 o'clock readings. Technicians have a tendency to reverse their orientation to the machine train during the alignment process. As a result, they often reverse the orientation of the recorded data.

To eliminate this problem, always acquire and record runout readings facing away from the stationary machine component. In this orientation, the 3 o'clock data is taken with the fixture oriented at 90° (horizontal) to the right of the shafts. The 9 o'clock position is then horizontal to the left of the shafts.

Indicator sag The reverse dial indicator fixture is comprised of two mounting blocks, which are rigidly fixed to each of the mating shafts. The runout gages, or dial indicators, are mounted on long, relatively small diameter rods, which are held by the mounting blocks. Because of this configuration, there is always some degree of sag or deflection in the fixtures

Mechanical looseness As with all measurement instrumentation, proper mounting techniques must be followed. Any looseness in the fixture mounting or at any point within the fixture will result in errors in the alignment readings.

Zeroing and calibrating It is very important that the indicator dials be properly zeroed and calibrated before use. Zeroing is performed once the fixture is mounted on the equipment to be aligned at the 12 o'clock position. It is accomplished either by turning a knob located on the dial body or by rotating the dial face itself until the dial reads zero. Calibration is performed in the instrument lab by measuring known misalignments. It is important for indicator devices to be calibrated before each use.

54.5.2 Rim-and-face

There are two variations of the rim-and-face methods. One requires one rim reading and one face reading at the stationary machine, where the dial indicator mounting brackets and posts are attached to the machine to be shimmed. The other method is identical, except that the rim and face readings are taken at the machine to be shimmed, where the dial indicator mounting brackets and posts are attached to the stationary machine.

As with the reverse dial indicator method, the measuring device used for rim-and-face alignment is also a dial indicator. The fixture has two runout indicators mounted on a common arm as opposed to reverse-dial fixtures, which have two runout indicators mounted on two separate arms.

The rim-and-face gages measure both the offset and angularity for the movable machine train component only (as compared to the reverse-dial method, which measures offset and calculates angularity for both the stationary and movable components). With the rim-and-face method, one dial indicator is mounted perpendicular to the shaft, which defines the offset of the movable shaft. The second indicator is mounted parallel to the shaft, which registers the angularity of the movable shaft. Figure 54.17 illustrates the typical configuration of a rim-and-face fixture.

Mounting

As with the reverse dial alignment fixture, proper mounting of the rim-and-face fixture is essential. The fixture must be rigidly mounted on both the stationary and movable shafts. All mechanical linkages must be tight and looseness held to an absolute minimum. Any fixture movement will distort both the offset and angularity readings as the shafts are rotated through 360°.

Rim-and-face measurements are made in exactly the same manner as those of reverse-dial indicator methods.

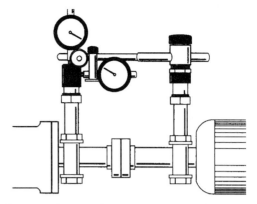

Figure 54.17 Typical configuration of a rim-and-face fixture (Ref. 2)

The shafts are slowly rotated in a clockwise direction in 90° increments. Measurements, including positive and negative signs, should be recorded at the 12, 3, 6 and 9 o'clock positions.

Limitations

Rim-and-face alignment is subject to the same errors as those of the reverse-dial indicator system. As with that system, care must be taken to insure proper orientation with the equipment and accurate recording of the data.

Note that rim-and-face alignment cannot be used when there is any end play, or axial movement, in the shafts of either the stationary or movable machine-train components. Since the dial indicator that is mounted parallel to the shaft is used to measure the angularity of the shafts, any axial movement or 'float' in either shaft will distort the measurement.

54.5.3 Optical or laser

Optical or laser alignment systems are based on the same principles as the reverse-dial method, but replace the mechanical components such as runout gages and cantilevered mounting arms with an optical device such as a laser. As with the reverse-dial method, offset is measured and angularity is calculated.

A typical system, which is shown in Figure 54.18, uses two transmitter/ sensors rigidly mounted on fixtures similar to the reverse-dial apparatus. When the shaft is rotated to one of the positions of interest (i.e., 12 o'clock, 3 o'clock, etc.), the transmitter projects a laser beam across the coupling. The receiver unit detects the beam and the offset and angularity are determined and recorded.

Advantages

Optical-alignment systems offer several advantages. Because laser fixtures eliminate the mechanical linkage and runout gages, there is no fixture sag. This greatly increases the accuracy and repeatability of the data obtained using this method.

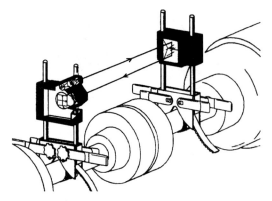

Figure 54.18 Typical optical or laser alignment system

Most of the optical-alignment systems incorporate a micro-processing unit, which eliminates recording errors commonly found with reverse-dial indicator and rim-and-face methods. Optical systems automatically maintain the proper orientation and provide accurate offset and angularity data, virtually eliminating operator error.

These microprocessor-based systems automatically calculate correction factors. If the fixtures are properly mounted and the shafts are rotated to the correct positions, the system automatically calculates and displays the appropriate correction for each foot of the movable machine-train component. This feature greatly increases the accuracy of the alignment process.

Disadvantages

Since optical-alignment systems are dependent on the transmission of a laser beam, which is a focused beam of light, they are susceptible to problems in some environments. Heat waves, steam, temperature variations, strong sunlight, and dust can distort the beam. When this happens, the system will not perform accurately.

One method that can be used to overcome most of the environment-induced problems is to use plastic tubing to shield the beam. This tubing can be placed between the transmitter and receiver of the optical-alignment fixture. It should be sized to permit transmission and reception of the light beam, but small enough to prevent distortion caused by atmospheric or environmental conditions. Typically, 2-inch, thin-wall tubing provides the protection required for most applications.

54.6 Alignment procedures

This section discusses the procedures for obtaining the measurements needed to align two classes of equipment: (1) horizontally installed units and (2) vertically installed units. The procedures for performing the initial alignment check for offset and angularity and for determining how much correction to make is presented.

Prior to taking alignment measurements, however, remember that it is necessary to remove any soft-foot that is present, making sure that the proper nut-tightening procedure is followed, and to correct for indicator sag (except when using the optical-alignment method).

Horizontal units

There are two parts to making alignment measurements on horizontally mounted units, which are typically taken using the reverse-dial indicator method. The first part of the procedure is to perform an initial alignment check by obtaining readings for the stationary and movable machines. The second part is to compare these values to the manufacturer's (i.e., desired) tolerances and to compute the difference between the actual readings and the desired readings.

The difference in the vertical readings is the amount of shim required to align the machine at the coupling for both vertical offset and angularity. The difference in the horizontal readings is the distance at the coupling to move the MTBM. These distances, however, must be converted to corrections to be made at the machine feet, computations that are made using rise and run concepts.

Initial alignment check

It is necessary to first obtain a complete set of indicator readings with the machines at ambient temperature, or non-operating condition. Figure 54.19 shows a hypothetical set of readings (i.e., top or 12 o'clock, right or 3 o'clock, bottom or 6 o'clock, and left or 9 o'clock) taken for the stationary machine shaft, 'A,' and the movable shaft, 'B'. The following is the procedure to be followed for obtaining these readings.

- The indicator bar either must be free of sags or compensated for in the readings.

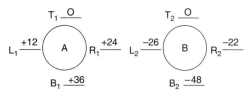

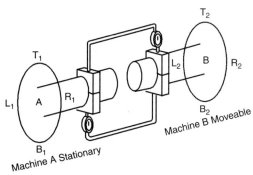

Figure 54.19 Hypothetical present state, or actual, dial-indicator readings

- Check the coupling for concentricity. If not concentric, replace the coupling.
- Zero the dial at top of the coupling.
- Record the readings at 90° increments taken clockwise as indicated in Figure 54.19.
- For any reading on a shaft, the algebraic sum of the left and right (9 o'clock and 3 o'clock) must equal the top and bottom (12 o'clock and 6 o'clock). The calculations below are for the example illustrated in Figure 54.19, in which shafts A and B are out of alignment as illustrated by the difference in the sums of the $(L + R)$ readings for shafts A and B and the difference in the sums of the $(T + B)$ readings for A and B.

Shaft A:
$$L1 + R1 = +12$$
$$+ (+24) = +36$$
$$T1 + B1 = 0$$
$$+ (+36) = +36$$

Shaft B:
$$L2 + R2 = -26$$
$$+ (-22) = -48$$
$$T2 + B2 = 0$$
$$+ (-48) = -48$$

Note, however, that this difference, which represents the amount of misalignment at the coupling, is not the amount of correction needed to be performed at the machine feet. This must be determined using rise and run concepts.

- The dial indicator should start at mid-range and not exceed the total range. In other words, do not peg the indicator. If misalignment exceeds the indicator span, it will be necessary to roughly align the machine before proceeding.

Determining corrections or amount of shim

With horizontally mounted units, it is possible to correct both angularity and offset with one adjustment. In order to compute the adjustments needed to achieve the desired alignment, it is necessary to establish three horizontal measurements. These measurements are critical to the success of any alignment and must be accurate to within $\frac{1}{16}$ inch (see Figure 54.20). Again, the procedure described here is for the reverse dial indicator method (see Figure 54.16).

1. Determine the distance, D_1, between the dial indicators.

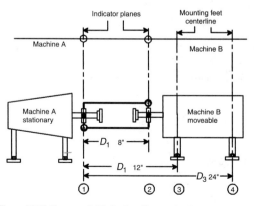

Figure 54.20 Reverse-dial indicator alignment setup

2. It is also necessary to know the distance from the indicator plane of the stationary machine, or Machine 'A', to the near adjustment plane of the MTBM, or Machine 'B'. This is the distance between the indicator plane of Machine 'A' to the near foot (Nf) of Machine 'B' and is referred to as D_2.
3. The distance between the indicator plane of Machine 'A' to the far adjustment plane is needed. This distance is referred to as D_3 and is the distance between the indicator plane of Machine 'A' to the far foot (Ff) of Machine 'B'.

The vertical and horizontal adjustments necessary to move Machine 'B' from the actual position (Figure 54.19 readings) to the desired state of alignment (Figure 54.21 readings) are determined using the equations below. Note that the desired state of alignment is obtained from manufacturer's tolerances. (When using manufacturer's tolerances, it is important to know if they compensate for thermal growth.)

For example: The shim adjustment at the near foot (Nf) and far foot (Ff) for the readings in Figures 54.19 and 54.21 can be determined by using the vertical movement formulas shown below. Since the top readings equal zero, only the bottom readings are needed in the calculation.

$$V_1 = \frac{B_3 - B_1}{2} = \frac{(-10) - (+36)}{2} = -23$$

$$V_2 = \frac{B_4 - B_2}{2} + V_1 = \frac{(+20) - (-48)}{2} + (-23) = +11$$

$$N_f = \frac{V_2 \times D_2}{D_1} - V_1 = \frac{(+11) \times (+12)}{8} - (-23) = +40$$

$$F_f = \frac{V_2 \times D_3}{D_1} - V_1 = \frac{(+11) \times (+24)}{8} - (-23) = +56$$

For Nf, at near foot of 'B', add 0.040-inch (40 mil) shims.

For Ff, at the far foot of 'B', add 0.056-inch (56 mil) shims.

For example: The side-to-side movement at Nf and Ff can be determined in the horizontal movement formula:

$$H_1 = \frac{(R_3 - L_3) - (R_1 - L_1)}{2}$$

$$= \frac{[(-15) - (+5)] - [(+24) - (+12)]}{2} = -16$$

$$H_2 = \frac{(R_4 - L_4) - (R_2 - L_2)}{2} + H_1$$

$$= \frac{[(+6) - (+14)] - [(-22) - (-26)]}{2} + (-16) = -22$$

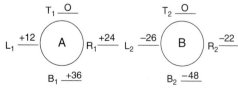

Figure 54.21 Desired dial indicator state readings at ambient conditions

$$N_f = \frac{H_2 \times D_2}{D_1} - H_1 = \frac{(-22) \times (+12)}{8} - (-16) = -17$$

$$F_f = \frac{H_2 \times D_3}{D_1} - H_1 = \frac{(-22) \times (+24)}{8} - (-16) = -50$$

For Nf, at near foot of 'B', move right 0.017 inch.

For Ff, at far foot of 'B', move right 0.050 inch.

Vertical units

The alignment process for most vertical units is quite different than that used for aligning horizontally mounted units. The major reason is that most vertical units are not designed to allow realignment to be performed under the assumption that they will always fit together perfectly. Field checks, however, have proven this assumption to be wrong in a vast majority of cases. Although it is quite difficult to correct misalignment on a vertical unit, it is essential that it be done to increase reliability and decrease maintenance costs.

Initial alignment check

The following procedure can be used on vertical units to obtain angularity and offset values needed to compare with recommended manufacturer's (i.e., desired) tolerances in order to determine if a unit is out of alignment. Perform an alignment check on the unit using the reverse-dial indicator method.

• Install brackets and dial indicators as illustrated in Figure 54.22.
• Check the alignment in two planes using the following directional designators: 'north/south' and 'east/west'.

Consider the point of reference nearest to you as being 'south,' which corresponds to the 'bottom' position of a horizontal unit. Note: Indicator sag does not occur when readings are taken as indicated below.

• Perform the 'north/south' alignment checks by setting the indicator dials to 'zero' on the 'north' side and take the readings on the 'south' side.
• Perform the 'east/west' alignment checks by setting the indicator dials to 'zero' on the 'west' side and take the readings on the 'east' side.
• Record the distance between the dial indicator center-lines, D_1.
• Record the distance from the centerline of the coupling to the top dial indicator.
• Record 'zero' for the distance, D_2, from the Indicator A to the 'top foot' of the movable unit.
• Record the distance, D_3, from Indicator A to the 'bottom foot' of the movable unit.
• Set the top dial indicator to 'zero' when it is in the 'north' position.

North/south alignment check

• Rotate shafts 180° until the top indicator is in the 'south' position and obtain a reading.
• Rotate shafts 180° again and check for repeatability of 'zero' on the 'north' side, then another 180° to check for repeatability of reading obtained on the 'south' side. Note: If results are not repeatable, check bracket and indicators for looseness and correct as necessary. If repeatable, record the 'south' reading.
• Rotate the shafts until the bottom dial indicator is in the 'north' position and set it to 'zero'.
• Rotate the shafts 180° and record 'south' side reading. Check for repeatability.

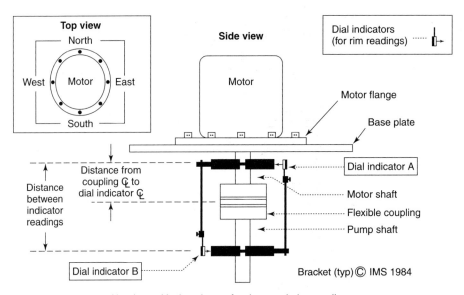

Figure 54.22 Proper dial indicator and bracket positioning when performing a vertical pump alignment

East/west alignment check

- Rotate the shafts until the top dial indicator is in the 'west' position and set it to 'zero'.
- Rotate the shafts 180° and obtain the reading on the 'east' side. Check for repeatability.
- Rotate the shafts until the bottom dial indicator is in the 'west' position and set it to 'zero'.
- Rotate the shafts 180° and again obtain the reading on the 'east' side. Check for repeatability.

Determining corrections

If the unit must be realigned, with vertical units it is necessary to use the rim-and-face method to obtain offset and angularity readings. Unlike horizontally mounted units, it is not possible to correct both angularity and offset with one adjustment. Instead, we must first correct the angular misalignment in the unit by shimming, and then correct the offset by properly positioning the motor base flange on the baseplate.

Because most units are designed in such a manner that realignment is not intended, it is necessary to change this design feature. Specifically, the 'rabbet fit' between the motor flange and the baseplate is the major hindrance to realignment.

Therefore, before proceeding with the alignment method, one should consider that the rabbet fit is designed to automatically 'center' the motor during installation. In theory, this should create a condition of perfect alignment between the motor and the driven-unit shafts. The rabbet fit is not designed to support the weight of the unit or resist the torque during start-up or operation; the motor flange and hold-down bolts are designed to do this. Since the rabbet fit is merely a positioning device, it is quite permissible to 'by-pass' it. This may be accomplished by either of the following:

- Machining off the entire male portion.
- Grinding off the male and/or female parts as necessary.

Angularity correction

There are three steps to follow when correcting for angularity. The first step is to obtain initial readings. The next step is to obtain corrected readings. The third step is to shim the machine.

Step 1: Initial readings The following procedure is for obtaining initial readings.

- Change the position of the bottom dial indicator so that it can obtain the 'face readings' of the lower bracket (see Figure 54.23).
- Looking from the 'south' side, identify the hold-down bolt at the 'north' position and label it #1. Proceeding clockwise, number each hold-down bolt until all are numbered (see Figure 54.24).
- Determine the largest negative reading, which occurs at the widest point, by setting the bottom dial indicator to 'zero' at point #1. This should be in line with centerline of hold-down bolt #1. Record the reading.
- Turn the shafts in a clockwise direction and record the data at each hold-down bolt centerline until readings have been taken at all positions.
- Use Figure 54.25 as an example of how the readings are taken. Remember that all readings are taken from the position of looking down on the lower bracket.

Note: We will always be looking for the largest negative (−) reading. If all readings are positive (+), the initial set point of zero will be considered the largest negative (−) reading. In Figure 54.25, the largest negative reading occurs at point #7.

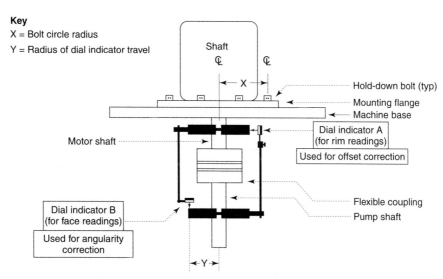

Key
X = Bolt circle radius
Y = Radius of dial indicator travel

Figure 54.23 Bottom dial indicator in position to obtain "face readings" (Ref. 1)

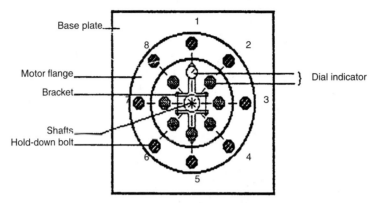

Note: Dial indicator B will be set up for taking face readings off of the lower bracket (as indicated by ⟳). Readings will then be taken at positions indicated by ⬢.

Figure 54.24 Diagram of a baseplate with hold-down bolts numbered

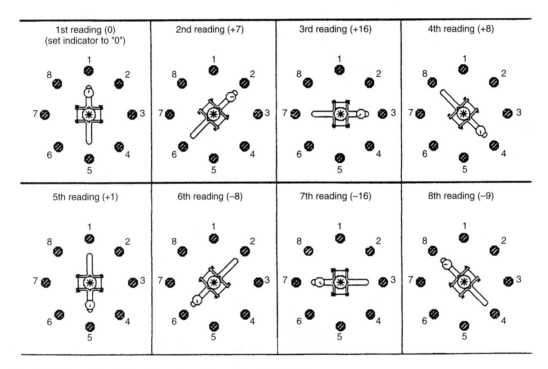

Figure 54.25 Determining the largest negative reading and the widest point

Step 2: Corrected readings Obtain corrected readings with the following procedure.

- Rotate the shafts until the indicator is again at the point where the largest negative reading occurs.
- Set the dial indicator to 'zero' at this point and take another complete set of readings. Using Figure 54.25 as an example, set the dial indicator to 'zero' at point

#7 (in line with centerline of bolt #7). The results of readings at the other hold-down bolt centerlines are as follows:

#1	+16
#2	+23
#3	+32
#4	+24

#5   +17
#6   +8
#7   0
#8   +7

Step 3: Shimming   Perform shimming with the following procedure:

• Measure the hold-down bolt circle radius and the radius of dial indicator travel as shown in Figure 54.26.
• Compute the shim multiplier, X/Y, where:

X = Bolt circle radius

Y = Radius of indicator travel

For example: If X = 9″ and Y = 4″, the shim multiplier is 9/4 = 2.25. The necessary shimming at each bolt equals the shim multiplier (2.25) times the bolt's corrected reading:

#1 – 2.25 × 16 = 36 mils = 0.036 inch
#2 – 2.25 × 23 = 52 mils = 0.052 inch
#3 – 2.25 × 32 = 72 mils = 0.072 inch
#4 – 2.25 × 24 = 54 mils = 0.054 inch
#5 – 2.25 × 17 = 38 mils = 0.038 inch
#6 – 2.25 × 8 = 18 mils = 0.018 inch
#7 – 2.25 × 0 = 0 mils = 0.000 inch
#8 – 2.25 × 7 = 16 mils = 0.016 inch

Offset correction

Once the angularity has been corrected by making the necessary shim adjustments at each of the hold-down bolts, it is necessary to correct the offset by sliding the moveable unit (i.e., motor in this example) on the baseplate. The top dial indicator is used to monitor the movements as

they are being made. 'North/south' and 'east/west' designations are used to describe the positioning of the unit.

North/south correction

The following is the procedure for making the 'north/south' corrections.

• Rotate shafts until the top dial indicator is in the 'north' position. Set it to 'zero'.
• Rotate the shafts 180° (until the top dial indicator is in the 'south' position) and record the reading.
• Determine movement necessary to correct the offset in this plane by dividing the reading by 2. This is the amount of movement (in mils) required. Direction of movement can be determined by the following rule: If the sign of the reading is positive (+), the motor must be moved toward the 'north'. If negative (−), it must be moved toward the 'south'.

East/west correction

The following is the procedure for making the 'east/west' corrections.

• Rotate the shafts until the top dial indicator is in the 'west' position. Set it to 'zero'.
• Rotate the shafts 180° (until the top dial indicator is in the 'east' position) and record the reading.
• Determine movement necessary to correct the offset in this plane by dividing the reading by 2. This value will be the amount of movement (in mils) required. Direction of movement can be determined by the following rule: If the sign of the reading is positive (+), the moveable unit (motor) must be moved toward the 'west'. If negative (−), it must be moved toward the 'east'.

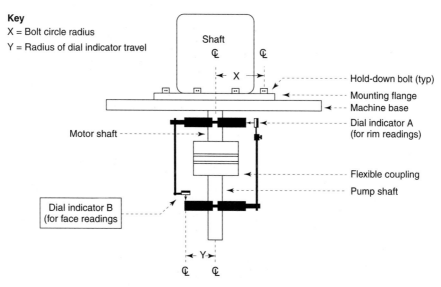

Figure 54.26 Determining bolt circle radius and radius of dial indicator travel

Making the offset corrections

After the amounts and directions of required offset adjustments have been obtained, the next step is to actually align the equipment. This is accomplished using two dial indicators with magnetic bases, which are installed on the south (or north) and west (or east) sides of the mounting flange of the moveable unit or motor. See Figure 54.27 for an illustration of this setup. It is important to zero both dial indicators before making adjustments and to watch both dial indicators while moving the unit.

Note: The motor position on the baseplate must be adjusted in order to align the equipment, which may require machining or grinding of the rabbet fit. Remember, however, the rabbet fit is only a positioning device and is not a structural support.

54.7 Computations, adjustments, and plots

Once initial alignment readings are obtained, they must be adjusted for changes in the machine train, which can be caused by process movement, vibration, or thermal growth. These adjustments must be made in order to achieve proper alignment at normal operating conditions. Once readings are obtained, the use of graphical plotting helps the technician visualize misalignment and the necessary corrections that must be made and to catch computation errors.

54.7.1 Adjustments for thermal growth

Thermal growth generally refers to the expansion of materials with increasing temperature. For alignment purposes, thermal growth is the shaft centerline movement associated with the change in temperature from the alignment process, which is generally performed at ambient conditions, to normal operating conditions. Such a temperature difference causes the elevation of one or both shafts to

change and misalignment to result. Temperature changes after alignment produce changes that may affect both offset and angularity of the shafts and can be in the vertical plane, horizontal plane, or any combination.

Proper alignment practices, therefore, must compensate for thermal growth. In effect, the shafts must be misaligned in the ambient condition so they will become aligned when machine temperatures reach their normal operating range. Generally, manufacturers supply dial indicator readings at ambient conditions, which compensate for thermal movement and result in colinear alignment at normal service conditions. When thermal rise information is supplied by the manufacturer or from machine history records, the necessary compensation may be made during the initial alignment procedure.

However, information concerning thermal rise is not available for all equipment. Generally, manufacturers of critical machinery, such as centrifugal air compressors and turbines, will include information relating to thermal rise in their installation manuals in the section dealing with alignment. When not available, the only method to determine the exact amount of compensation necessary to correct for thermal rise is referred to as a 'hot alignment check'.

Thermal compensation targets

A simple procedure for determining thermal compensation targets is to calculate the movement of the shaft due to temperature change at the bearings or feet, whichever is more convenient. Note that calculated thermal growth is highly dependent upon the accuracy of the temperature assumptions and is useful only for initial alignment estimates. Therefore, the targets developed from the following procedure should be revised when better data becomes available.

The formula for this calculation is very simple and very accurate. It requires three factors: (1) the difference in temperature of the machine housing between the feet and shaft bearings, (2) the distance between the shaft centerline and the feet, and (3) the coefficient of thermal expansion of the machine housing material. The thermal

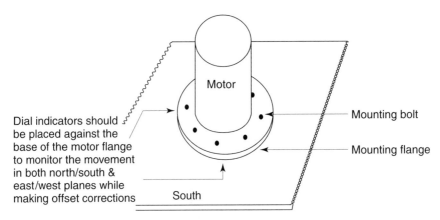

Dial indicators should
be placed against the
base of the motor flange
to monitor the movement
in both north/south &
east/west planes while
making offset corrections

Motor

Mounting bolt

Mounting flange

South

Figure 54.27 Placement of dial indicators to monitor offset corrections

growth between any two points of any metal can be predicted by the formula:

Growth = $\Delta T \times L \times C$

Where:

ΔT = Temperature difference between the feet and shaft bearings, °F
L = Length between points (often the vertical distance from the shim plane to the shaft centerline), inches
C = Growth factor (coefficient of thermal expansion)

Growth factors (mils/inch/°F) for common materials are:

Aluminum	0.0126
Bronze	0.0100
Cast iron, gray	0.0059
Stainless steel	0.0074
Mild steel, ductile iron	0.0063

Note: The thermal growth formula is usually applied only to the vertical components of the machine. While the formula can be applied to horizontal growth, this direction is often ignored.

For vertical growth, L is usually taken as the vertical height from the bottom of the foot where shims touch the machine to the shaft centerline. In the case where the machine is mounted on a base that has significant temperature variations along its length, L is the vertical distance from the concrete or other constant temperature baseline to the shaft centerline.

Hot alignment check

A hot alignment check is performed exactly like an ambient alignment check with the added safety precautions required for hot machinery. The accuracy of a hot alignment check depends on how soon after shutdown dial indicator readings can be taken. Readings may be taken within a few minutes with the use of shaft-mounted brackets that span a flexible coupling. To speed up the process, assemble the brackets to the fullest extent possible prior to shutdown so that they need only be bolted to the shafts once the machine stops rotating.

54.8 Graphical plotting

The graphical plotting technique for computing initial alignment can be performed with any of the three types of measurement fixtures (i.e. reverse dial indicator, rim-and-face, or optical). The following steps should be followed when plotting alignment problems:

1. Determine the following dimensions from the machine train, which are illustrated in Figure 54.28:

   FBS = Front-foot to back-foot of stationary train component
   CFS = Front-foot to coupling of stationary train component
   CF = Coupling to front-foot of movable train component
   FB = Front-foot to back-foot of movable train component
   CD = Coupling or working diameter

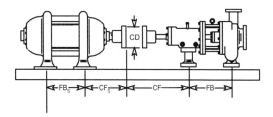

Figure 54.28 Graphical plotting measurements

2. On graph or grid paper, pick a horizontal line to be used as the baseline (also referred to as reference line or zero-line). This line usually crosses the center of the page from left to right and represents the rotational centerline of the stationary machine-train component.
3. Determine the number of inches or mils that each block on the graph paper represents by first finding the distance from the back-foot of the stationary component to the back-foot of the movable component. Then determine the inches or mils per square that will spread the entire machine train across the graph paper.
4. Plot inches or mils horizontally from left to right.
5. Plot mils from top to bottom vertically. As a rule, assign 0.5, 1, 2, 5, 10 mils to each vertical step. Note that this scale may need to be changed in cases where excessive misalignment is present.

Known foot correction values

The following steps should be followed to plot misalignment when foot correction values are known (see Figure 54.29):

1. On the baseline, start at the left end and mark the stationary back-foot. From the back-foot and moving right, count the number of squares along the baseline corresponding to FFS. Mark the stationary front-foot location.
2. Starting at the stationary front-foot and moving right, count the number of squares along the baseline corresponding to CFS and mark the coupling location.
3. Continue this process until the entire machine train is indicated on the graph.
4. To plot misalignment, locate the CFS or coupling on the horizontal baseline. From that point, count up or down on the vertical axis until the amount of offset is located on the mils scale. Mark this point on the graph. Use care to insure that the location is accurately located. Positive values should be above the horizontal baseline and negative values below the line.
5. Locate the FBM or back-foot of the movable component. Move either up or down vertically on the scale to the point of the offset measurement. Mark this point on the graph. Remember, positive values are above the horizontal baseline and negative values below the line.
6. Draw a line from the back-foot (FBM) of the movable component or MTBM through the front-foot of the movable component toward the vertical line where the stationary coupling is located. Draw a short vertical line at the coupling end of the line. Finish the

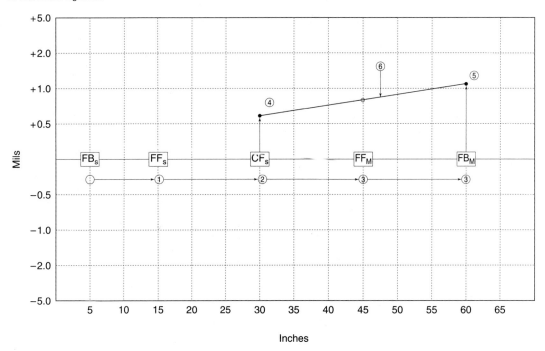

Figure 54.29 Graphical plotting of known foot correction

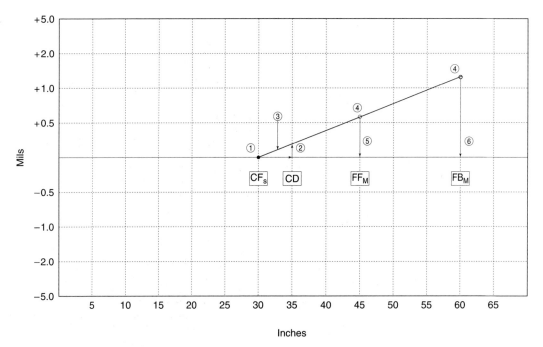

Figure 54.30 Graphical plotting of known coupling results

MTBM by drawing little squares to represent the feet, darkening the line from the back-foot to coupling, and darkening the coupling line.

Known coupling results

When plotting coupling misalignment, use the following steps instead of those from the preceding section (see Figure 54.30).

1. Start at the stationary coupling location and, moving up or down the vertical axis (mils), count the number of squares corresponding to the vertical or horizontal offset. Move up for positive offset and down for negative offset. Mark a point, which is the MTBM coupling location.
2. Start at the MTBM coupling center and, moving right on the horizontal line, count the number of squares corresponding to the CD dimension (see Figure 54.28) and lightly mark the point. From this point, move up or down vertically on the mils scale the number of squares corresponding to the total mils of angularity per diameter (CD) and mark lightly.

3. From the MTBM coupling center, draw a line through the point marked in the preceding step and extending past the MTBM back-feet location. This line is the MTBM centerline.
4. Now place the MTBM feet. Starting at the MTBM coupling and moving right along a horizontal line, count the number of squares corresponding to CF (see Figure 54.28). Then move straight vertically to the MTBM centerline and mark the location of the front-foot. Then starting at the MTBM front-foot and moving right, count the number of squares corresponding to FB. From this point, move vertically to the MTBM centerline and mark the location of the MTBM back-foot.
5. Draw a short line perpendicular to the shaft centerline to mark the MTBM coupling. Finish the MTBM by drawing little squares to represent the feet, and darkening the line from the back-foot to the coupling.
6. Correction of the MTBM machine-train component can now be measured directly from the graph. Locate the appropriate MTBM foot location and read the actual correction from the vertical or mils scale.

55

Rotor Balancing

R Keith Mobley

The Plant Performance Group

Contents

55.1 Introduction

Mechanical imbalance is one of the most common causes of machinery vibration and is present to some degree on nearly all machines that have rotating parts or rotors. Static, or standing, imbalance is the condition when there is more weight on one side of a centerline than the other. However, a rotor may be in perfect static balance and not be in a balanced state when rotating at high speed.

If the rotor is a thin disc, careful static balancing may be accurate enough for high speeds. However, if the rotating part is long in proportion to its diameter, and the unbalanced portions are at opposite ends or in different planes, the balancing must counteract the centrifugal force of these heavy parts when they are rotating rapidly.

This section provides information needed to understand and solve the majority of balancing problems using a vibration/balance analyzer, a portable device that detects the level of imbalance, misalignment, etc., in a rotating part based on the measurement of vibration signals.

55.2 Sources of vibration due to mechanical imbalance

Two major sources of vibration due to mechanical imbalance in equipment with rotating parts or rotors are: (1) assembly errors and (2) incorrect key length guesses during balancing.

55.2.1 Assembly errors

Even when parts are precision balanced to extremely close tolerances, vibration due to mechanical imbalance can be much greater than necessary due to assembly errors. Potential errors include relative placement of each part's center of rotation, location of the shaft relative to the bore, and cocked rotors.

Center of rotation

Assembly errors are not simply the additive effects of tolerances, but also include the relative placement of each part's center of rotation. For example, a 'perfectly' balanced blower rotor can be assembled to a 'perfectly' balanced shaft and yet the resultant imbalance can be high. This can happen if the rotor is balanced on a balancing shaft that fits the rotor bore within 0.5 mil (0.5 thousandths of an inch) and then is mounted on a standard cold-rolled steel shaft allowing a clearance of over 2 mils.

Shifting any rotor from the rotational center on which it was balanced to the piece of machinery on which it is intended to operate can cause an assembly imbalance four to five times greater than that resulting simply from tolerances. For this reason, all rotors should be balanced on a shaft having a diameter as nearly the same as the shaft on which it will be assembled.

For best results, balance the rotor *on its own shaft* rather than on a balancing shaft. This may require some rotors to be balanced in an overhung position, a procedure the balancing shop often wishes to avoid. However, it is better to use this technique rather than being forced to make too many balancing shafts. The extra precision balance attained by using this procedure is well worth the effort.

Method of locating position of shaft relative to bore

Imbalance often results with rotors that do not incorporate setscrews to locate the shaft relative to the bore (e.g., rotors that are end clamped). In this case, the balancing shaft is usually horizontal. When the operator slides the rotor on the shaft, gravity causes the rotor's bore to make contact at the 12 o'clock position on the top surface of the shaft. In this position, the rotor is end-clamped in place and then balanced.

If the operator removes the rotor from the balancing shaft without marking the point of bore and shaft contact, it may not be in the same position when reassembled. This often shifts the rotor by several mils as compared to the axis on which it was balanced, thus causing an imbalance to be introduced. The vibrations that result are usually enough to spoil what should have been a precision balance and produce a barely acceptable vibration level. In addition, if the resultant vibration is resonant with some part of the machine or structure, a more serious vibration could result.

To prevent this type of error, the balancer operators and those who do final assembly should follow the following procedure. The balancer operator should permanently mark the location of the contact point between the bore and the shaft during balancing. When the equipment is reassembled in the plant or the shop, the assembler should also use this mark. For end-clamped rotors, the assembler should slide the bore on the horizontal shaft, rotating both until the mark is at the 12 o'clock position and then clamp it in place.

Cocked rotor

If a rotor is cocked on a shaft in a position different from the one in which it was originally balanced, an imbalanced assembly will result. If, for example, a pulley has a wide face that requires more than one setscrew, it could be mounted on-center, but be cocked in a different position than during balancing. This can happen by reversing the order in which the setscrews are tightened against a straight key during final mounting as compared to the order in which the setscrews were tightened on the balancing arbor. This can introduce a pure couple imbalance, which adds to the small couple imbalance already existing in the rotor and causes unnecessary vibration.

For very narrow rotors (e.g., disc-shaped pump impellers or pulleys), the distance between the centrifugal forces of each half may be very small. Nevertheless, a very high centrifugal force, which is mostly counterbalanced statically by its counterpart in the other half of the rotor, can result. If the rotor is slightly cocked, the small axial distance between the two very large centrifugal forces causes an appreciable couple imbalance, which is often several times the allowable tolerance. This is due to the fact that the centrifugal force is proportional to half the rotor weight (at any one time, half of the rotor is pulling against the other half) times the radial distance from the axis of rotation to the center of gravity of that half.

To prevent this, the assembler should tighten each setscrew gradually – first one, then the other, and back again – so that the rotor is aligned evenly. On flange-mounted rotors such as flywheels, it is important to clean the mating surfaces and the bolt holes. Clean bolt holes are important because high couple imbalance can result from the assembly bolt pushing a small amount of dirt between the surfaces, cocking the rotor. Burrs on bolt holes also can produce the same problem.

Other

Other assembly errors can cause vibration. Variances in bolt weights when one bolt is replaced by one of a different length or material can cause vibration. For setscrews that are 90° apart, the tightening sequence may not be the same at final assembly as during balancing. To prevent this, the balancer operator should mark which was tightened first.

Key length

With a keyed-shaft rotor, the balancing process can introduce machine vibration if the assumed key length is different from the length of the one used during operation. Such an imbalance usually results in a mediocre or 'good' running machine as opposed to a very smooth running machine.

For example, a 'good' vibration level that can be obtained without following the precautions described in this section is amplitude of 0.12 in/sec. (3.0 mm/sec). By following the precautions, the orbit can be reduced to about 0.04 in/sec (1 mm/sec). This smaller orbit results in longer bearing or seal life, which is worth the effort required to make sure that the proper key length is used.

When balancing a keyed-shaft rotor, one half of the key's weight is assumed part of the shaft's male portion. The other half is considered part of the female portion that is coupled to it. However, when the two rotor parts are sent to a balancing shop for re-balancing, the actual key is rarely included. As a result, the balance operator usually *guesses* at the key's length, makes up a half key, and then balances the part. (Note: A 'half key' is of full-key length, but only half-key depth.)

In order to prevent an imbalance from occurring, *do not allow the balance operator to guess the key length*. It is strongly suggested that the actual key length be recorded on a tag that is attached to the rotor to be balanced. The tag should be attached in such a way that another device (such as a coupling half, pulley, fan, etc.) cannot be attached until the balance operator removes the tag.

55.3 Theory of imbalance

Imbalance is the condition when there is more weight on one side of a centerline than the other. This condition results in unnecessary vibration, which generally can be corrected by the addition of counterweights. There are four types of imbalance: (1) static, (2) dynamic, (3) couple, and (4) dynamic imbalance combinations of static and couple.

55.3.1 Static

Static imbalance is single-plane imbalance acting through the center of gravity of the rotor, perpendicular to the shaft axis. The imbalance also can be separated into two separate single-plane imbalances, each acting in-phase or at the same angular relationship to each other (i.e., 0° apart). However, the net effect is as if one force is acting through the center of gravity. For a uniform straight cylinder such as a simple paper machine roll or a multi-grooved sheave, the forces of static imbalance measured at each end of the rotor are equal in magnitude (i.e., the ounce-inches or gram-centimeters in one plane are equal to the ounce-inches or gram-centimeters in the other).

In static imbalance, the only force involved is weight. For example, assume that a rotor is perfectly balanced and, therefore, will not vibrate regardless of the speed of rotation. Also, assume that this rotor is placed on frictionless rollers or 'knife edges.' If a weight is applied on the rim at the center of gravity line between two ends, the weighted portion immediately rolls to the 6 o'clock position due to the gravitational force.

When rotation occurs, static imbalance translates into a centrifugal force. As a result, this type of imbalance is sometimes referred to as 'force imbalance' and some balancing machine manufacturers use the word 'force' instead of 'static' on their machines. However, when the term 'force imbalance' was just starting to be accepted as the proper term, an American standardization committee on balancing terminology standardized the term 'static' instead of 'force.' The rationale was that the role of the standardization committee was not to determine and/or correct right or wrong practices, but to standardize those currently in use by industry. As a result, the term 'static imbalance' is now widely accepted as the international standard and, therefore, is the term used in this chapter.

55.3.2 Dynamic

Dynamic imbalance is any imbalance resolved to at least two correction planes (i.e., planes in which a balancing correction is made by adding or removing weight). The imbalance in each of these two planes may be the result of many imbalances in many planes, but the final effects can be characterized to only two planes in almost all situations.

An example of a case where more than two planes are required is flexible rotors (i.e., long rotors running at high speeds). High speeds are considered to be revolutions per minute (rpm) higher than about 80 per cent of the rotor's first critical speed. However, in over 95 per cent of all run-of-the-mill rotors (e.g., pump impellers, armatures, generators, fans, couplings, pulleys, etc.), two-plane dynamic balance is sufficient. Therefore, flexible rotors are not covered in this document because of the low number in operation and the fact that balancing operations are almost always performed by specially trained people at the manufacturer's plant.

In dynamic imbalance, the two imbalances do not have to be equal in magnitude to each other, nor do they have to have any particular angular reference to each other. For example, they could be 0 (in-phase), 10, 80, or 180° from each other.

Although the definition of dynamic imbalance covers all two-plane situations, an understanding of the components of dynamic imbalance is needed so that its causes can be understood. Also, an understanding of the components makes it easier to understand why certain types of balancing do not always work with many older balancing machines for overhung rotors and very narrow rotors. The primary components of dynamic imbalance include: number of points of imbalance, amount of imbalance, phase relationships, and rotor speed.

55.4 Points of imbalance

The first consideration of dynamic balancing is the number of imbalance points on the rotor, as there can be more than one point of imbalance within a rotor assembly. This is especially true in rotor assemblies with more than one rotating element, such as a three-rotor fan or multi-stage pump.

Amount of imbalance

The amplitude of each point of imbalance must be known to resolve dynamic balance problems. Most dynamic balancing machines or in situ balancing instruments are able to isolate and define the specific amount of imbalance at each point on the rotor.

Phase relationship

The phase relationship of each point of imbalance is the third factor that must be known. Balancing instruments isolate each point of imbalance and determine their phase relationship. Plotting each point of imbalance on a polar plot does this. In simple terms, a polar plot is a circular display of the shaft end. Each point of imbalance is located on the polar plot as a specific radial, ranging from 0 to 360°.

Rotor speed

Rotor speed is the final factor that must be considered. Most rotating elements are balanced at their normal running speed or over their normal speed range. As a result, they may be out of balance at some speeds that are not included in the balancing solution. As an example, the wheel and tires on your car are dynamically balanced for speeds ranging from zero to the maximum expected speed (i.e., 80 miles per hour). At speeds above 80 miles per hour, they may be out of balance.

55.5 Coupled imbalance

Couple imbalance is caused by two equal non-colinear imbalance forces that oppose each other angularly (i.e., 180° apart). Assume that a rotor with pure couple imbalance is placed on frictionless rollers. Because the imbalance weights or forces are 180° apart and equal, the rotor is statically balanced. However, a pure couple imbalance

occurs if this same rotor is revolved at an appreciable speed.

Each weight causes a centrifugal force, which results in a rocking motion or rotor wobble. This condition can be simulated by placing a pencil on a table, then at one end pushing the side of the pencil with one finger. At the same time, push in the opposite direction at the other end. The pencil will tend to rotate end-over-end. This end-over-end action causes two imbalance 'orbits,' both 180° out of-phase, resulting in a 'wobble' motion.

Dynamic imbalance combinations of static and couple

Visualize a rotor that has only one imbalance in a single plane. Also, visualize that the plane is *not* at the rotor's center of gravity, but is off to one side. Although there is no other source of couple, this force to one side of the rotor not only causes translation (parallel motion due to pure static imbalance), but also causes the rotor to rotate or wobble end-over-end as from a couple. In other words, such a force would create a combination of both static and couple imbalance. This again is dynamic imbalance.

In addition, a rotor may have two imbalance forces exactly 180° opposite to each other. However, if the forces are not equal in magnitude, the rotor has a static imbalance in combination with its pure couple. This combination is also dynamic imbalance.

Another way of looking at it is to visualize the usual rendition of dynamic imbalance – imbalance in two separate planes at an angle and magnitude relative to each other not necessarily that of pure static or pure couple.

For example, assume that the angular relationship is 80° and the magnitudes are 8 units in one plane and 3 units in the other. Normally, you would simply balance this rotor on an ordinary two-plane dynamic balancer and that would be satisfactory. But for further understanding of balancing, imagine that this same rotor is placed on static balancing rollers, whereby gravity brings the static imbalance components of this dynamically out-of-balance rotor to the 6 o'clock position.

The static imbalance can be removed by adding counter-balancing weights at the 12 o'clock position. Although statically balanced, however, the two remaining forces result in a pure couple imbalance. With the entire static imbalance removed, these two forces are equal in magnitude and exactly 180° apart. The couple imbalance can be removed, as with any other couple imbalance, by using a two-plane dynamic balancer and adding counterweights.

Note that whenever you hear the word 'imbalance,' mentally add the word 'dynamic' to it. Then when you hear 'dynamic imbalance,' mentally visualize a 'combination of static and couple imbalance.' This will be of much help not only in balancing, but in understanding phase and coupling misalignment as well.

55.6 Balancing

Imbalance is one of the most common sources of major vibration in machinery. It is the main source in about

40 per cent of the excessive vibration situations. The vibration frequency of imbalance is equal to one times the rpm ($l \times$ rpm) of the imbalanced rotating part.

Before a part can be balanced using the vibration analyzer, certain conditions must be met:

- The vibration must be due to mechanical imbalance, and
- Weight corrections can be made on the rotating component.

In order to calculate imbalance units, simply multiply the amount of imbalance by the radius at which it is acting. In other words, one ounce of imbalance at a one-inch radius will result in one oz-in of imbalance. Five ounces at one-half inch radius results in $2\frac{1}{2}$ oz-ins of imbalance. (Dynamic imbalance units are measured in ounce-inches, oz-in, or gram-millimeters, g-mm.) Although this refers to a single plane, dynamic balancing is performed in at least two separate planes. Therefore, the tolerance is usually given in single-plane units for each plane of correction.

Important balancing techniques and concepts to be discussed in the sections to follow include: in-place balancing, single-plane versus two-plane balancing, precision balancing, techniques that make use of a phase shift, and balancing standards.

55.6.1 In-place balancing

In most cases, weight corrections can be made with the rotor mounted in its normal housing. The process of balancing a part without taking it out of the machine is called *in-place balancing*. This technique eliminates costly and time consuming disassembly. It also prevents the possibility of damage to the rotor, which can occur during removal, transportation to and from the balancing machine, and reinstallation in the machine.

Single-plane versus two-plane balancing

The most common rule of thumb is that a disc-shaped rotating part usually can be balanced in one correction plane only, whereas parts that have appreciable width require two-plane balancing. Precision tolerances, which become more meaningful for higher performance (even on relatively narrow face width), suggest two-plane balancing. However, the width should be the guide, not the diameter to width ratio.

For example, a 20-inch wide rotor could have a large enough couple imbalance component in its dynamic imbalance to require two-plane balancing. (Note: The couple component makes two-plane balancing important.) Yet, if the 20-inch width is on a rotor of large diameter to qualify as a 'disc-shaped rotor,' even some of the balance manufacturers erroneously would call for a single-plane balance.

It is true that the narrower the rotor, the less the chance for a large couple component and, therefore, the greater the possibility of getting by with a single-plane balance. For rotors over 4 to 5 inches in width, it is best to check for real dynamic imbalance (or for couple imbalance).

Unfortunately, you cannot always get by with a static- and couple-type balance, even for very narrow flywheels used in automobiles. Although most of the flywheels are only 1 inch to $1\frac{1}{2}$ inches wide, more than half have enough couple imbalance to cause excessive vibration. This obviously is not due to a large distance between the planes (width), but due to the fact that the flywheel's mounting surface can cause it to be slightly cocked or tilted. Instead of the flywheel being 90° to the shaft axis, it may be perhaps 85 to 95°, causing a large couple despite its narrow width.

This situation is very common with narrow and disc-shaped industrial rotors such as single-stage turbine wheels, narrow fans, and pump impellers. The original manufacturer often accepts the guidelines supplied by others and performs a single-plane balance only. By obtaining separate readings for static and couple, the manufacturer could and should easily remove the remaining couple.

An important point to remember is that static imbalance is always removed first. In static and couple balancing, remove the static imbalance first and then remove the couple.

55.6.2 Precision balancing

Most original-equipment manufacturers balance to commercial tolerances, a practice that has become acceptable to most buyers. However, due to frequent customer demands, some of the equipment manufacturers now provide precision balancing. Part of the driving force for providing this service is that many large mills and refineries have started doing their own precision balancing to tolerances considerably closer than those used by the original-equipment manufacturer. For example, the International Standards Organization (ISO) for process plant machinery calls for a G6.3 level of balancing in its balancing guide (refer to Table 55.1). This was calculated based on rotors running free in space with a restraint vibration of 6.3 mm/sec (0.25 in/sec) vibration velocity.

Precision balancing requires a G2.5 guide number, which is based on 2.5 mm/sec (0.1 in/sec) vibration velocity. As can be seen from this, 6.3 mm/sec (0.25 in/sec) balanced rotors will vibrate more than the 2.5 mm/sec (0.1 in/sec) precision balanced rotors. Many vibration guidelines now consider 2.5 mm/sec (0.1 in/sec) 'good,' creating the demand for precision balancing. Precision balancing tolerances can produce velocities of 0.01 in/sec (0.3 mm/sec) and lower.

It is true that the extra weight of non-rotating parts (i.e., frame and foundation) reduces the vibration somewhat from the free-in-space amplitude. However, it is possible to reach precision balancing levels in only two or three additional runs, providing the smoothest running rotor. The extra effort to the balance operator is minimal because he already has the 'feel' of the rotor and has the proper setup and tools in hand. In addition, there is a large financial pay off for this minimal extra effort due to decreased bearing and seal wear.

55.6.3 Techniques using phase shift

If we assume that there is no other source of vibration other than imbalance (i.e., we have perfect alignment, a perfectly straight shaft, etc.), it is readily seen that pure static imbalance gives in-phase vibrations and pure couple imbalance gives various phase relationships. Compare the *vertical* reading of a bearing at one end of the rotor with the *vertical* reading at the other end of the rotor to determine how that part is shaking vertically. Then compare the *horizontal* reading at one end with the *horizontal* reading at the other end to determine how the part is shaking horizontally.

If there is no resonant condition to modify the resultant vibration phase, then the phase for both vertical and horizontal readings are essentially the same even though the vertical and horizontal amplitudes do not necessarily correspond. In actual practice, this may be slightly off due to other vibration sources such as misalignment. In performing the analysis, what counts is that when the source of the vibration is *primarily from imbalance*, then the vertical reading phase differences between one end of the rotor and the other will be very similar to the phase differences when measured horizontally. For example, vibrations 60° out of phase vertically would show 60° out of phase horizontally within 20 per cent.

However, the horizontal reading on one bearing will not show the same phase relationship as the vertical reading on the same bearing. This is due to the pickup axis being oriented in a different angular position, as well as the phase adjustment due to possible resonance. For example, the horizontal vibration frequency may be below the horizontal resonance of various major portions of machinery, whereas the vertical vibration frequency may be above the natural frequency of the floor supporting the machine.

First, determine how the rotor is vibrating vertically by comparing *'vertical only'* readings with each other. Then, determine how the rotor is vibrating horizontally. If, the rotor is shaking horizontally and vertically and the phase differences are relatively similar, then the source of vibration is likely to be *imbalance*. However, before coming to a conclusion, be sure that other $l \times$ rpm sources (e.g., bent shaft, eccentric armature, misaligned coupling) are not at fault.

55.7 Balancing standards

The ISO has published standards for acceptable limits for residual imbalance in various classifications of rotor assemblies. Balancing standards are given in oz-ins or lb-ins per pound of rotor weight or the equivalent in metric units (g-mm/kg). The oz-ins are for each correction plane for which the imbalance is measured and corrected.

Caution must be exercised when using balancing standards. The recommended levels are for residual imbalance, which is defined as imbalance of any kind that remains *after* balancing.

Figure 55.1 and Table 55.1 are the norms established for most rotating equipment. Additional information can be obtained from ISO 5406 and 5343. Similar standards are available from the American National Standards Institute (ANSI) in their publication ANSI S2.43-1984.

So far, there has been no consideration of the angular positions of the usual two points of imbalance relative to each other or the distance between the two correction

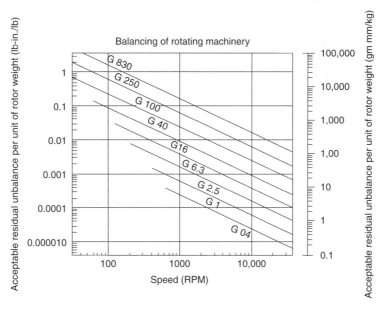

Figure 55.1 Balancing standards: residual imbalance per unit rotor weight (Ref. 5)

Table 55.1 Balance Quality Grades for Various Groups of Rigid Rotors

Balance quality grade	Type of rotor
G4,000	Crankshaft drives of rigidly mounted slow marine diesel engines with uneven number of cylinders.
G1,600	Crankshaft drives of rigidly mounted large two-cycle engines.
G630	Crankshaft drives of rigidly mounted large four-cycle engines; crankshaft drives of elastically mounted marine diesel engines.
G250	Crankshaft drives of rigidly mounted fast four-cylinder diesel engines.
G100	Crankshaft drives of fast diesel engines with six or more cylinders; complete engines (gasoline or diesel) for cars and trucks.
G40	Car wheels, wheel rims, wheel sets, drive shafts; crankshaft drives of elastically mounted fast four-cycle engines (gasoline and diesel) with six or more cylinders; crankshaft drives for engines of cars and trucks.
G16	Parts of agricultural machinery; individual components of engines (gasoline or diesel) for cars and trucks.
G6.3	Parts or process plant machines; marine main-turbine gears; centrifuge drums; fans; assembled aircraft gas-turbine rotors; fly wheels; pump impellers; machine-tool and general machinery parts; electrical armatures.
G2.5	Gas and steam turbines; rigid turbo-generator rotors; rotors; turbo-compressors; machine-tool drives; small electrical armatures; turbine-driven pumps.
G1	Tape recorder and phonograph drives; grinding-machine drives.
G0.4	Spindles, disks, and armatures of precision grinders; gyroscopes.

Source: 'Balancing Quality of Rotating Rigid Bodies,' *Shock and Vibration Handbook*, ISO 1940-1973; ANSI S2.19-1975.

planes. For example, if the residual imbalances in each of the two planes were in-phase, they would add to each other to create more static imbalance.

Most balancing standards are based on a *residual* imbalance and do not include multi-plane imbalance. If they are approximately 180° to each other, they form a couple. If the distance between the planes is small, the resulting couple is small; if the distance is large, the couple is large. A couple creates considerably more vibration than when the two residual imbalances are in-phase. Unfortunately, there is nothing in the balancing standards that considers this.

Another problem could also result in excessive imbalance-related vibration even though the ISO standards were met. The ISO standards call for a balancing grade of G6.3 for components such as pump impellers, normal electric armatures, and parts of process plant machines. This results in an operating speed vibration velocity of 6.3 mm/sec (0.25 in/sec) vibration velocity. However, practice has shown that an acceptable vibration velocity is 0.1 ins/sec and the ISO standard of G2.5 is required. Because of these discrepancies, changes in the recommended balancing grade are expected in the future.

References

1. Block, H. P. and Geitner, F. K., *Introduction to Machinery Reliability Assessment*, Van Nostrand Reinhold, New York (1990).
2. Block, H. P. and Geitner, F. K., *Machinery Failure Analysis and Troubleshooting*, Gulf Publishing Company Book Division, Houston, (1986).
3. Oberg, E., Jones, F. D., Horton, H. L. and Ryffel, H., *Machinery's Handbook 25*, 25th edn, Industrial Press Inc., New York (1996).
4. Nelson, C. A., *Millwrights and Mechanics Guide*, Macmillan Publishing Company, New York (1986).
5. Harris, C. M., *Shock and Vibration Handbook*, 3rd edn, McGraw-Hill Book Company (1988).

56

Packing and Seals

R Keith Mobley

The Plant Performance Group

Contents

56.1 Introduction

All machines such as pumps and compressors that handle liquids or gases must include a reliable means of sealing around their shafts so that the fluid being pumped or compressed does not leak. To accomplish this, the machine design must include a seal located at various points to prevent leakage between the shaft and housing. In order to provide a full understanding of seal and packing use and performance, this chapter discusses fundamentals, seal design, and installation practices.

56.2 Seal fundamentals

Shaft seal requirements and two common types of seals, packed stuffing boxes and simple mechanical seals, are described and discussed in this section. A packed box typically is used on slow- to moderate-speed machinery where a slight amount of leakage is permissible. A mechanical seal is used on centrifugal pumps or other type of fluid handling equipment where shaft sealing is critical.

56.2.1 Shaft seal requirements

Figure 56.1 shows the cross-section of a typical end-suction centrifugal pump where the fluid to be pumped enters the suction inlet at the eye of the impeller. Due to the relatively high speed of rotation, the fluid collected within the impeller vanes is held captive because of the close tolerance between the front face of the impeller and the pump housing.

With no other available escape route, the fluid is passed to the outside of the impeller by centrifugal force and into the volute where its kinetic energy is converted into pressure. At the point of discharge (i.e., discharge nozzle), the fluid is highly pressurized compared to its pressure at the inlet nozzle of the pump. This pressure drives the

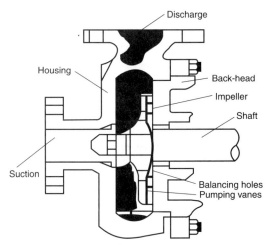

Figure 56.1 Cross-section of a typical end-suction centrifugal pump (Ref. 3)

fluid from the pump and allows a centrifugal pump to move fluids to considerable heights above the centerline of the pump.

This highly pressurized fluid also flows around the impeller to a lower pressure zone where, without an adequate seal, the fluid will leak along the drive shaft to the outside of the pump housing. The lower pressure results from a pump design intended to minimize the pressure behind the impeller. Note that this design element is specifically aimed at making drive shaft sealing easier.

Reducing the pressure acting on the fluid behind the impeller can be accomplished by two different methods, or a combination of both, on an open-impeller unit. One method is where small pumping vanes are cast on the backside of the impeller. The other method is for balance holes to be drilled through the impeller to the suction eye.

In addition to reducing the driving force behind shaft leakage, decreasing the pressure differential between the front and rear of the impeller using one or both of the methods described above greatly decreases the axial thrust on the drive shaft. This decreased pressure prolongs the thrust bearing life significantly.

56.2.2 Sealing devices

Two sealing devices are described and discussed in this section: packed stuffing boxes and simple mechanical seals.

Packed stuffing boxes

Before the development of mechanical seals, a soft pliable material or packing placed in a box and compressed into rings encircling the drive shaft was used to prevent leakage. Compressed packing rings between the pump housing and the drive shaft, accomplished by tightening the gland-stuffing follower, formed an effective seal.

Figure 56.2 shows a typical packed box that seals with rings of compressed packing. Note that if this packing is allowed to operate against the shaft without adequate lubrication and cooling, frictional heat eventually builds up to the point of total destruction of the packing and damage to the drive shaft. Therefore, all packed boxes must have a means of lubrication and cooling.

Lubrication and cooling can be accomplished by allowing a small amount of leakage of fluid from the machine or by providing an external source of fluid. When leakage from the machine is used, leaking fluid is captured in collection basins that are built into the machine housing or baseplate. Note that periodic maintenance to recompress the packing must be carried out when leakage becomes excessive.

Packed boxes must be protected against ingress of dirt and air, which can result in loss of resilience and lubricity. When this occurs, packing will act like a grinding stone, effectively destroying the shaft's sacrificial sleeve, and cause the gland to leak excessively. When the sacrificial sleeve on the drive shaft becomes ridged and worn, it should be replaced as soon as possible. In effect, this continuing maintenance program can readily be measured in terms of dollars and time.

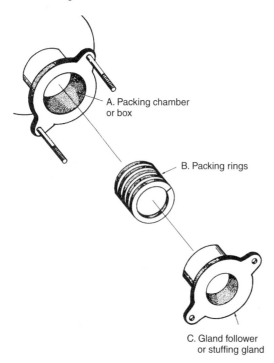

A. Packing chamber or box

B. Packing rings

C. Gland follower or stuffing gland

Figure 56.2 Typical packed stuffing box

Uneven pressures can be exerted on the drive shaft due to irregularities in the packing rings, resulting in irregular contact with the shaft. This causes uneven distribution of lubrication flow at certain locations, producing acute wear and packed-box leakages. The only effective solution to this problem is to replace the shaft sleeve or drive shaft at the earliest opportunity.

Simple mechanical seal

Mechanical seals, which are typically installed in applications where no leakage can be tolerated, are described and discussed in this section. Toxic chemicals and other hazardous materials are primary examples of applications where mechanical seals are used.

56.2.3 Components and assembly

Figure 56.3 shows the components of a simple mechanical seal, which is made up of the following:

- Coil spring;
- 'O' ring shaft packing;
- Seal ring.

The seal ring fits over the shaft and rotates with it. The spring must be made from a material that is compatible with the fluid being pumped so that it will withstand corrosion. Likewise, the same care must be taken in material selection of the 'O' ring and seal materials. The insert and insert 'O' ring mounting are installed in the bore cavity provided in the gland ring. This assembly is installed in a pump-stuffing box, which remains stationary when the pump shaft rotates.

A carbon graphite insertion ring provides a good bearing surface for the seal ring to rotate against. It is also resistive to attack by corrosive chemicals over a wide range of temperatures.

Figure 56.4 depicts a simple seal that has been installed in the pump's stuffing box. Note how the coil spring sits against the back of the pump's impeller, pushing the packing 'O' ring against the seal ring. By doing so, it remains in constant contact with the stationary insert ring.

As the pump shaft rotates, the shaft packing rotates with it due to friction. (In more complex mechanical seals, the shaft-packing element is secured to the rotating shaft by Allen screws.) There is also friction between the spring,

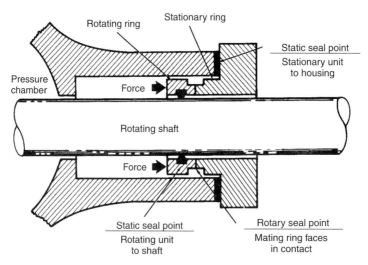

Rotating ring

Stationary ring

Static seal point

Stationary unit to housing

Pressure chamber

Force

Rotating shaft

Force

Static seal point

Rotating unit to shaft

Rotary seal point

Mating ring faces in contact

Figure 56.3 Simple mechanical seal

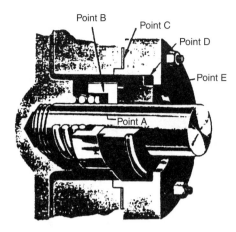

Figure 56.4 Pump stuffing box seal

the impeller, and the compressed 'O' ring. Thus, the whole assembly rotates together when the pump shaft rotates. The stationary insert ring is located within the gland bore. The gland itself is bolted to the face of the stuffing box. This part is held stationary due to the friction between the 'O' ring insert mounting and the inside diameter (ID) of the gland bore as the shaft rotates within the bore of the insert.

56.2.4 How to prevent leakage

Having discussed how a simple mechanical seal is assembled in the stuffing box, we must now consider how the pumped fluid is stopped from leaking out to the atmosphere.

In Figure 56.4, the path of the fluid along the drive shaft is blocked by the 'O' ring shaft packing at Point A. Any fluid attempting to pass through the seal ring is stopped by the 'O' ring shaft packing at Point B. Any further attempt by the fluid to pass through the seal ring to the atmospheric side of the pump is prevented by the gland gasket at Point C and the 'O' ring insert at Point D. The only other place where fluid can potentially escape is the joint surface around Point E, which is between the rotating carbon ring and the stationary insert. (Note: The surface areas of both rings must be machined-lapped perfectly flat, measured in light bands with tolerances of one-millionth of an inch.)

56.2.5 Sealing area and lubrication

The efficiency of all mechanical seals is dependent upon the condition of the sealing area surfaces. The surfaces remain in contact with each other for the effective working life of the seal and are friction-bearing surfaces.

As in the compressed packing gland, lubrication also must be provided in mechanical seals. The sealing area surfaces should be lubricated and cooled with pumped fluid (if it is clean enough) or an outside source of clean fluid. However, much less lubrication is required with this type of seal because the frictional surface area is smaller

than that of a compressed packing gland and the contact pressure is equally distributed throughout the interface. As a result, a smaller amount of lubrication passes between the seal faces to exit as leakage.

In most packing glands there is a measurable flow of lubrication fluid between the packing rings and the shaft. With mechanical seals, the faces ride on a microscopic film of fluid that migrates between them, resulting in leakage. However, leakage is so slight that if the temperature of the fluid is above its saturation point at atmospheric pressure, it flashes off to vapor before it can be visually detected.

56.2.6 Advantages and disadvantages

Mechanical seals offer a more reliable seal than compressed packing seals. Because the spring in a mechanical seal exerts a constant pressure on the seal ring, it automatically adjusts for wear at the faces. Thus, the need for manual adjustment is eliminated. Additionally, because the bearing surface is between the rotating and stationary components of the seal, the shaft or shaft sleeve does not become worn. Although the seal will eventually wear out and need replacing, the shaft will not experience wear.

However, much more precision and attention to detail must be given to the installation of mechanical seals compared to conventional packing. Nevertheless, it is not unusual for mechanical seals to remain in service for many thousands of operational hours if they have been properly installed and maintained.

56.3 Mechanical seal designs

Mechanical seal designs are referred to as friction drives, or single-coil spring seals, and positive drives.

56.3.1 Single-coil spring seal

The seal shown back in Figure 56.4 depicts a typical friction drive or single-coil spring seal unit. This design is limited in its use because the seal relies on friction to turn the rotary unit. Because of this, its use is limited to liquids such as water or other non-lubricating fluids. If this type of seal is to be used with liquids that have natural lubricating properties, it must be mechanically locked to the drive shaft.

Although this simple seal performs its function satisfactorily, two drawbacks must be considered. Both drawbacks are related to the use of a coil spring that is fitted over the drive shaft.

- One drawback of the spring is the need for relatively low shaft speeds because of a natural tendency of the components to distort at high surface speeds. This makes the spring push harder on one side of the seal than the other, resulting in an uneven liquid film between the faces. This causes excessive leakage and wears at the seal.
- The other drawback is simply one of economics. Because pumps come in a variety of shaft sizes and speeds, the use of this type of seal requires inventorying several sizes of spare springs, which ties up capital.

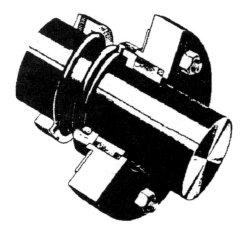

Figure 56.5 Conversion of a simple seal to positive drive

Nevertheless, the simple and reliable coil spring seal has proven itself in the pumping industry and is often selected for use despite its drawbacks. In regulated industries, this type of seal design far exceeds the capabilities of a compressed packing ring seal.

56.3.2 Positive drive seal

There are two methods of converting a simple seal to positive drive. Both methods, which use collars secured to the drive shaft by setscrews, are shown in Figure 56.5. In this figure, the end tabs of the spring are bent at 90° to the natural curve of the spring. These end tabs fit into notches in both the collar and the seal ring. This design transmits rotational drive from the collar to the seal ring by the spring. Figure 56.5 also shows two horizontally mounted pins that extend over the spring from the collar to the seal ring.

56.4 Installation procedures

This section describes the installation procedures for packed stuffing boxes and mechanical seals.

56.4.1 Packed stuffing box

This procedure provides detailed instructions on how to repack centrifugal pump packed stuffing boxes or glands. The methodology described here is applicable to other gland sealed units such as valves and reciprocating machinery.

Tool list

The following is a list of the tools needed to repack a centrifugal pump gland:

- Approved packing for specific equipment;
- Mandrel sized to shaft diameter;
- Packing ring extractor tool;
- Packing board;
- Sharp knife;
- Approved cleaning solvent;
- Lint-free cleaning rags.

Precautions

The following precautions should be taken in repacking a packed stuffing box:

- Ensure coordination with operations control.
- Observe site and area safety precautions at all time.
- Ensure equipment has been electrically isolated and suitably locked out and tagged.
- Ensure machine is isolated and depressurized with suction and discharge valves chained and locked shut.

Installation

The following are the steps to follow in installing a gland:

1. Loosen and remove nuts from the gland bolts.
2. Examine threads on bolts and nuts for stretching or damage. Replace if defective.
3. Remove the gland follower from the stuffing box and slide it along the shaft to provide access to the packing area.
4. Use packing extraction tool to carefully remove packing from the gland.
5. Keep the packing rings in the order they are extracted from the gland box. This is important in evaluating wear characteristics. Look for rub marks and any other unusual markings that would identify operational problems.
6. Carefully remove the lantern ring. This is a grooved, bobbin-like spool piece that is situated exactly on the centerline of the seal water inlet connection to the gland (Figure 56.6). *Note*: It is most important to place the lantern ring under the seal water inlet connection to ensure the water is properly distributed within the gland to perform its cooling and lubricating functions.

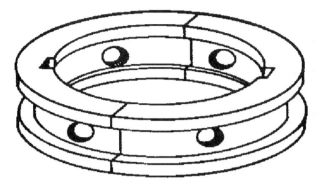

Figure 56.6 Lantern ring or seal cage

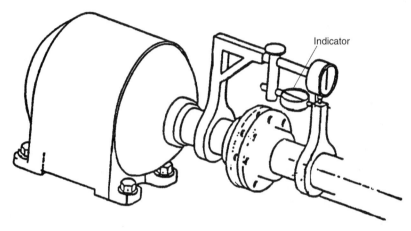

Indicator

Figure 56.7 Dial indicator check for run-out

7. Examine the lantern ring for scoring and possible signs of crushing. Make sure the lantern ring's outside diameter (OD) provides a sliding fit with the gland box's internal dimension. Check that the lantern ring's ID is a free fit along the pump's shaft sleeve. If the lantern ring does not meet this simple criterion, replace it with a new one.

8. Continue to remove the rest of the packing rings as previously described. Retain each ring in the sequence that it was removed for examination.

9. Do not discard packing rings until they have been thoroughly examined for potential problems.

10. Turn on the gland seal cooling water slightly to ensure there is no blockage in the line. Shut the valve when good flow conditions are established.

11. Repeat Steps 1 through 10 with the other gland box.

12. Carefully clean out the gland stuffing boxes with a solvent-soaked rag to ensure that no debris is left behind.

13. Examine the shaft sleeve in both gland areas for excessive wear caused by poorly lubricated or over-tightened packing.
   Note: If the shaft sleeve is ridged or badly scratched in any way, the split housing of the pump may have

to be split and the impeller removed for the sleeve to be replaced. Badly installed and maintained packing causes this.

14. Check total indicated runout (TIR) of the pump shaft by placing a magnetic base-mounted dial indicator on the pump housing and a dial stem on the shaft. Zero the dial and rotate the pump shaft one full turn. Record reading (Figure 56.7).
   Note: If the TIR is greater than ±0.002 inches, the pump shaft should be straightened.

15. Determine the correct packing size before installing using the following method (Figure 56.8):
   Measure the ID of the stuffing box, which is the OD at the packing (B), and the diameter of the shaft (A). With this data, the packing cross-section size is calculated by:

$$\text{Packing cross-section} = \frac{B - A}{2}$$

The packing length is determined by calculating the circumference of the packing within the stuffing box. The centerline diameter is calculated by adding the diameter of the shaft to the packing cross-section that was calculated in the preceding formula. For example,

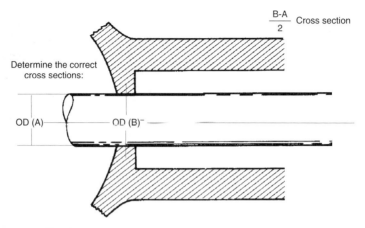

$$\frac{B\text{-}A}{2} \quad \text{Cross section}$$

Determine the correct
cross sections:

OD (A) ————— OD (B)⁻ ———

Figure 56.8 Selecting correct packing size

a stuffing box with a 4-inch ID and a shaft with a 2-inch diameter will require a packing cross-section of 1 inch. The centerline of the packing would then be 3 inches.

Therefore, the approximate length of each piece of packing would be:

Packing length = Centerline diameter × 3.1416

$$= 3.0 \times 3.1416$$

$$= 9.43 \text{ inches}$$

The packing should be cut approximately $\frac{1}{4}$ inch longer than the calculated length so that the end can be bevel cut.

16. Controlled leakage rates easily can be achieved with the correct size packing.

17. Cut the packing rings to size on a wooden mandrel that is the same diameter as the pump shaft. Rings can be cut either square (butt cut) or diagonally (approximately 30°). *Note*: Leave at least a $\frac{1}{16}$ inch gap between the butts regardless of the type of cut used. This permits the packing rings to move under compression or temperature without binding on the shaft surface.

18. Ensure that the gland area is perfectly clean and is not scratched in any way before installing the packing rings.

19. Lubricate each ring lightly before installing in the stuffing box.
    Note: When putting packing rings around the shaft, use an "S" twist. **Do not bend open**. See Figure 56.9.

20. Use a split bushing to install each ring, ensuring that the ring bottoms out inside the stuffing box. An offset tamping stick may be used if a split bushing is not available. **Do not use a screwdriver**.

21. Stagger the butt joints, placing the first ring butt at 12 o'clock; the second at 6 o'clock; the third at 3 o'clock; the fourth at 9 o'clock; etc., until the packing box is filled (Figure 56.10). *Note*: When the last ring has been installed, there should be enough room to insert the gland follower $\frac{1}{8}$ to $\frac{3}{16}$ inches into the stuffing box (Figure 56.11).

Figure 56.10 Stagger butt joints

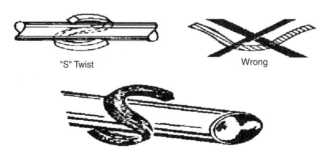

"S" Twist   Wrong

Figure 56.9 Proper and improper installation of packing

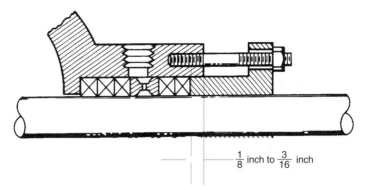

$\frac{1}{8}$ inch to $\frac{3}{16}$ inch

Figure 56.11 Proper gland follower clearance

22. Install the lantern ring in its correct location within the gland. Do not force the lantern ring into position (Figure 56.12).
23. Tighten up the gland bolts with a wrench to seat and form the packing to the stuffing box and shaft.
24. Loosen the gland nuts one complete turn and rotate the shaft by hand to get running clearance.
25. Retighten the nuts finger tight only. Again, rotate the shaft by hand to make sure the packing is not too tight.
26. Contact operations to start the pump and allow the stuffing box to leak freely. Tighten the gland bolts one flat at a time until the desired leakage is obtained and the pump runs cool.
27. Clean up the work area and account for all tools before returning them to the tool crib.
28. Inform operations of project status and complete all paperwork.
29. After the pump is in operation, periodically inspect the gland to determine its performance. If it tends to leak more than the allowable amount, tighten by turning the nuts one flat at a time. Give the packing enough time to adjust before tightening it more. If the gland is tightened too much at one time, the packing can be excessively compressed, causing unnecessary friction and subsequent burn-out of the packing.

56.4.2 Mechanical seals

A mechanical seal's performance depends on the operating condition of the equipment where it is installed. Therefore, inspection of the equipment before seal installation can potentially prevent seal failure and reduce overall maintenance expenses.

Equipment check points

The pre-installation equipment inspection should include the following: stuffing box space, lateral or axial shaft movement (end play), radial shaft movement (whip or deflection), shaft runout (bent shaft), stuffing box face squareness, stuffing box bore concentricity, driver alignment, and pipe strain.

Stuffing box space

To properly receive the seal, the radial space and depth of the stuffing box must be the same as the dimensions shown on the seal assembly drawing.

Lateral or axial shaft movement (end play)

Install a dial indicator with the stem against the shoulder of the shaft. Use a soft hammer or mallet to lightly tap the shaft on one end and then on the other. Total indicated end play should be between 0.001 and 0.004 inches. A mechanical seal cannot work properly with a large amount of end play or lateral movement. If the hydraulic condition changes (as frequently happens), the shaft could 'float,' resulting in sealing problems. Minimum end play is a desirable condition for the following reasons:

- Excessive end play can cause pitting, fretting, or wear at the point of contact between the shaft packing in the mechanical seal and the shaft or sleeve OD. As the mechanical seal driving element is locked to the shaft or sleeve, any excessive end play will result in either overloading or under-loading the springs causing excessive wear or leaks.
- Excessive end play as a result of defective thrust bearings can reduce seal performance by disturbing both the established wear pattern and the lubricating film.
- A floating shaft can cause chattering, which results in chipping of the seal faces, especially the carbon element. Ideal mechanical seal performance requires a uniform wear pattern and a liquid film between the mating contact faces.

Radial shaft movement (whip or deflection)

Install the dial indicator as close to the radial bearing as possible. Lift the shaft or exert light pressure at the impeller end. If more than 0.002 to 0.003 inches of radial movement occurs, investigate bearings and bearing fits (especially the bore) for the radial bearing fit. An oversized radial bearing bore caused by wear, improper machining, or corrosion will cause excessive radial shaft movement resulting in

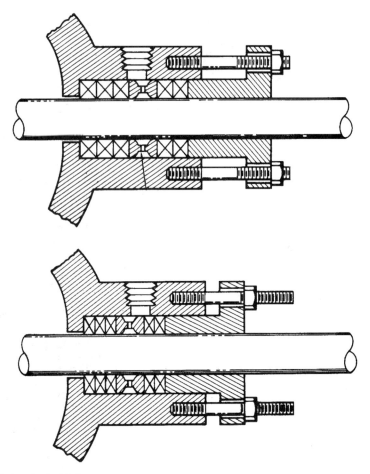

Figure 56.12 Proper lantern ring installation

shaft whip and deflection. Minimum radial shaft movement is important for the following reasons:

- Excessive radial movement can cause wear, fretting, or pitting of the shaft packing or secondary sealing element at the point of contact between the shaft packing and the shaft or sleeve OD.
- Extreme wear at the mating contact faces will occur when excessive shaft whip or deflection is present due to defective radial bearings or bearing fits. The contact area of the mating faces will be increased, resulting in increased wear and the elimination or reduction of the lubricating film between the faces, further shortening seal life.

Shaft runout (bent shaft)

A bent shaft can lead to poor sealing and cause vibration. Bearing life is greatly reduced and the operating conditions of both radial and thrust bearings can be affected.

Clamp the dial indicator to the pump housing and measure the shaft runout at two or more points on the OD of the impeller end of the shaft. Also measure the shaft runout at the coupling end of the shaft. If the runout exceeds 0.002 inches, remove the shaft and straighten or replace it.

Stuffing box face squareness

With the pump stuffing box cover bolted down, clamp the dial indicator to the shaft with the stem against the face of the stuffing box. The total indicator runout should not exceed 0.003 inches.

When the face of the stuffing box is 'out-of-square,' or not perpendicular to the shaft axis, the result can be serious malfunction of a mechanical seal for the following reasons:

- The stationary gland plate that holds the stationary insert or seat in position is bolted to the face of the stuffing box. Misalignment will cause the gland to cock, resulting in cocking of the stationary element. This results in seal wobble or operation in an elliptical pattern. This condition is a major factor in fretting, pitting, and wearing of the mechanical seal shaft packing at the point of contact with the shaft or sleeve.

- A seal that is wobbling on the shaft can also cause wear on the drive pins. Erratic wear on the face contact causes poor seal performance.

Stuffing box bore concentricity

With the dial indicator set up as described above, place the indicator stem well into the bore of the stuffing box. The stuffing box should be concentric to the shaft axis to within a 0.005 inch total indicator reading.

Eccentricity alters the hydraulic loading of the seal faces, reducing seal life and performance. If the shaft is eccentric to the box bore, check the slop, or looseness, in the pump bracket fits. Rust, atmospheric corrosion, or corrosion from leaking gaskets can cause damage to these fits, making it impossible to ensure a stuffing box that is concentric with the shaft. A possible remedy for this condition is welding the corroded area and re-machining to proper dimensions.

Driver alignment and pipe strain

Driver alignment is extremely important and periodic checks should be performed. Pipe strain can also damage pumps, bearings, and seals.

In most plants, it is customary to blind the suction and discharge flanges of inactive pumps. These blinds should be removed before the pump driver alignment is made or the alignment job is incomplete.

After the blinds have been removed and as the flanges on the suction and discharge are being connected to the piping, check the dial indicator reading on the OD of the coupling half and observe movement of the indicator dial as the flanges are being secured. Deviation indicates pipe strain. If severe strain exists, corrective measures should be taken or damage to the pump and unsatisfactory seal service can result.

Seal check points

The following are important seal check points:

- Ensure that all parts are kept clean, especially the running faces of the seal ring and insert.
- Check the seal rotary unit and make sure the drive pins and spring pins are free in the pin holes or slots.
- Check the setscrews in the rotary unit collar to see that they are free in the threads. Setscrews should be replaced after each use.
- Check the thickness of all gaskets against the dimensions shown on the assembly drawing. Improper gasket thickness will affect the seal setting and the spring load imposed on the seal.
- Check the fit of the gland ring to the equipment. Make sure there is no interference or binding on the studs or bolts or other obstructions. Be sure the gland ring pilot, if any, enters the bore with a reasonable guiding fit for proper seal alignment.
- Make sure all rotary unit parts of the seal fit over the shaft freely.

- Check both running faces of the seal (seal ring and insert) and be sure there are no nicks or scratches. Imperfections of any kind on either of these faces will cause leaks.

Installing the seal

The following steps should be taken when installing a seal:

- Instruction booklets and a copy of the assembly drawing are shipped with each seal. Be sure each is available and read the instructions before starting installation.
- Remove all burrs and sharp edges from the shaft or shaft sleeve, including sharp edges of keyways and threads. Worn shafts or sleeves should be replaced.
- Check the stuffing box bore and face to ensure they are clean and free of burrs.
- The shaft or sleeve should be lightly oiled before the seal is assembled to allow the seal parts to move freely over it. This is especially desirable when assembling the seal collar because the bore of the collar usually has only a few thousandths of an inch clearance. Care should be taken to avoid getting the collar cocked.
- Install the rotary unit parts on the shaft or sleeve in the proper order.
- Be careful when passing the seal gland ring and insert over the shaft. Do not bring the insert against the shaft because it might chip away small pieces from the edge of the running face.
- Wipe the seal faces clean and apply a clean oil film before completing the equipment assembly. A clean finger, which is not apt to leave lint, will do the best job when giving the seal faces the final wiping.
- Complete the equipment assembly, taking care when compressing the seal into the stuffing box.
- Seat the gland ring and gland ring gasket to the face of the stuffing box by tightening the nuts or bolts evenly and firmly. Be sure the gland ring is not cocked. Tighten the nuts or bolts only enough to affect a seal at the gland ring gasket, usually finger tight and $\frac{1}{2}$ to $\frac{3}{4}$ of a turn with a wrench. Excessively tightening the gland ring nut or bolt will cause distortion that will be transmitted to the running face, resulting in leaks.
- If the seal assembly drawing is not available, the proper seal setting dimension for inside seals can be determined as follows:
  - Establish a reference mark on the shaft or sleeve flush with the face of the stuffing box.
  - Determine how far the face of the insert will extend into the stuffing box bore. This dimension is taken from the face of the gasket.
  - Determine the compressed length of the rotary unit by compressing the rotary unit to the proper spring gap.
  - This dimension added to the distance the insert extends into the stuffing box will give the seal setting dimension from the reference mark on the shaft or sleeve to the back of the seal collar.
- Outside seals are set with the spring gap equal to the dimension stamped on the seal collar.

- Cartridge seals are set at the factory and installed as complete assemblies. These assemblies contain spacers that must be removed after the seal assembly is bolted into position and the sleeve collar is in place.

Installation of environmental controls

Mechanical seals are often chosen and designed to operate with environmental controls. If this is the case, check the seal assembly drawing or equipment drawing to ensure that all environmental control piping is properly installed. Before equipment startup, all cooling and heating lines should be operating and remain so for at least a short period after equipment shutdown.

Before startup, all systems should be properly vented. This is especially important on vertical installations where the stuffing box is the uppermost portion of the pressure-containing part of the equipment. The stuffing box area must be properly vented to avoid a vapor lock in the seal area that would cause the seal to run dry.

On double seal installations, be sure the sealing liquid lines are connected, the pressure control valves are properly adjusted, and the sealing liquid system is operating before starting the equipment.

Seal startup procedures

When starting equipment with mechanical seals, make sure the seal faces are immersed in liquid from the beginning so they will not be damaged from dry operation. The following recommendations for seal startup apply to most types of seal installations and will improve seal life if followed:

- Caution the electrician not to run the equipment dry while checking motor rotation. A slight turnover will not hurt the seal, but operating full speed for several minutes under dry conditions will destroy or severely damage the rubbing faces.
- The stuffing box of the equipment, especially centrifugal pumps, should always be vented before startup. Even though the pump has a flooded suction, it is still possible that air may be trapped in the top of the stuffing box after the initial liquid purge of the pump.
- Check installation for need of priming. Priming might be necessary in applications with a low or negative suction head.
- Where cooling or bypass re-circulation taps are incorporated in the seal gland, piping must be connected to or from these taps before startup. These specific environmental control features must be used to protect the organic materials in the seal and to ensure its proper performance. Cooling lines should be left open at all times or whenever possible. This is especially true when a hot product might be passing through standby

equipment while it is not on-line. Many systems provide for product to pass through the standby equipment so the need for additional product volume or an equipment change is only a matter of pushing a button.
- With hot operational equipment that is shut down at the end of each day, it is best to leave the cooling water on at least long enough for the seal area to cool below the temperature limits of the organic materials in the seal.
- Face lubricated-type seals must be connected from the source of lubrication to the tap openings in the seal gland before startup. This is another predetermined environmental control feature that is mandatory for proper seal function. Where double seals are to be operated, it is necessary that the lubrication feed lines be connected to the proper ports for both circulatory or dead-end systems before equipment startup. This is very important because all types of double seals depend on the controlled pressure and flow of the sealing fluid to function properly. Even before the shaft is rotated, the sealing liquid pressure must exceed the product pressure opposing the seal. Be sure a vapor trap does not prevent the lubricant from reaching the seal face promptly.
- Thorough warm-up procedures include a check of all steam piping arrangements to be sure that all are connected and functioning, as products that will solidify must be fully melted before startup. It is advisable to leave all heat sources on during shutdown to ensure a liquid condition of the product at all times. Leaving the heat on at all times further facilitates quick startups and equipment switchovers that may be necessary during a production cycle.
- Thorough chilling procedures are necessary on some installations, especially liquefied petroleum gases (LPG) applications. LPG must always be kept in a liquid state in the seal area and startup is usually the most critical time. Even during operation, the re-circulation line piped to the stuffing box might have to be run through a cooler in order to overcome frictional heat generated at the seal faces. LPG requires a stuffing box pressure that is greater than the vapor pressure of the product at pumping temperature (25 to 50 psi differential is desired).

References

1. *Basic Pumps, Power Plant Basics,* Overseas Bechtel Inc.
2. Neale, M. J., *Drives and Seals, A Tribology Handbook,* Society of Automotive Engineers, Inc. (1994).
3. Parker, S. (ed.), *Encyclopedia of Engineering,* 2nd edn, McGraw Hill, New York (1983).
4. Block, H. P. and Geitner, F. K., *Introduction to Machinery Reliability Assessment,* Van Nostrand Reinhold, New York (1990).
5. Block, H. P. and Geitner, F. K., *Machinery Failure Analysis and Troubleshooting,* Gulf Publishing Company Book Division, Houston (1986).
6. Nelson, C. A., *Millwrights and Mechanics Guide,* Macmillan Publishing Company, New York (1986).

57

Gears and Gear Drives

R Keith Mobley

The Plant Performance Group

Contents

57.1 Introduction

A gear is a disc or wheel with teeth around its periphery – either on the inside edge (i.e., internal gear) or on the outside edge (i.e., external gear). A gear is used for providing a positive means of power transmission. This transmission is effected by the teeth on one gear meshing with the teeth on another gear or rack (i.e., straight-line gear). Meshing teeth formed with special cutters provide a much more compact drive than either belts or chain drives and can operate at higher speeds and power.

57.2 Types of gears and their characteristics

There are several different types of gears used in industry, but the spur gear is the most commonly used. Many are complex in design and manufacture, and several have evolved directly from the spur gear, which is commonly referred to as the basic gear. Because gear design is extremely complex and the field of specialized gearing is beyond the scope of this text, only a general description and explanation of principles are given. The commonly used gears are discussed sufficiently to provide the millwright or mechanic with the basic information necessary to perform installation and maintenance work.

57.2.1 Gear types

Types of gears are: spur, helical, bevel (straight, spiral, zerol, and hypoid), and worm, each of which is discussed in the following sections. Table 57.1 summarizes the characteristics of each gear.

Table 57.1 Gear Characteristics Overview

Gear type	Characteristics Attributes/Positives	Negatives
Spur, external	Connects parallel shafts that rotate in opposite directions, inexpensive to manufacture to close tolerances, moderate peripheral speeds, no axial thrust, high mechanical efficiency	Noisy at high speeds
Spur, internal	Compact drive mechanism for parallel shafts rotating in same direction	
Helical, external	Connects parallel and non-parallel shafts; superior to spur gears in load-carrying capacity, quietness, and smoothness; high efficiency	Higher friction than spur gears
Helical, double (also referred to as herringbone)	Connects parallel shafts, overcomes high-end thrust present in single-helical gears, compact, quiet and smooth operation at higher speeds (1,000 to 12,000 fpm or higher), high efficiencies	
Helical, cross	Light loads with low power transmission demands	Narrow range of applications, requires extensive lubrication
Bevel	Connects angular or intersecting shafts	Gears overhang supporting shafts resulting in shaft deflection and gear misalignment
Bevel, straight	Peripheral speeds up to 1,000 fpm in applications where quietness and maximum smoothness not important, high efficiency	Thrust load causes gear pair to separate
Bevel, zerol	Same ratings as straight bevel gears and use same mountings, permits slight errors in assembly, permits some displacement due to deflection under load, highly accurate, hardened due to grinding	Limited to speeds less than 1,000 fpm due to noise
Bevel, spiral	Smoother and quieter than straight bevel gears at speeds greater than 1,000 fpm or 1,000 rpm, evenly distributed tooth loads, carry more load without surface fatigue, high efficiency, reduces size of installation for large reduction ratios, speed-reducing and speed-increasing drive	High tooth pressure, thrust loading depends on rotation and spiral angle
Bevel, miter	Same number of teeth in both gears, operate on shafts at 90°	
Bevel, hypoid	Connects non-intersecting shafts, high pinion strength, allows the use of compact straddle mounting on the gear and pinion, recommended when maximum smoothness required, compact system even with large reduction ratios, speed-reducing and speed-increasing drive	Lower efficiency, difficult to lubricate due to high tooth-contact pressures, materials of construction (steel) require use of extreme-pressure lubricants
Planetary or epicyclic	Compact transmission with driving and driven shafts in line, large speed reduction when required	
Worm, cylindrical	Provide high-ratio speed reduction over wide range of speed ratios (60:1 and higher from a single reduction, can go as high as 500:1), quiet transmission of power between shafts at 90°, reversible unit available, low wear, can be self-locking	Lower efficiency; heat removal difficult, which restricts use to low-speed applications
Worm, double-enveloping	Increased load capacity	Lower efficiencies

Source: Integrated Systems, Inc.

Spur

The spur gear is the least expensive of all gears to manufacture and is the most commonly used. It can be manufactured to close tolerances and is used to connect parallel shafts that rotate in opposite directions. It gives excellent results at moderate peripheral speeds and the tooth load produces no axial thrust. Because contact is simultaneous across the entire width of the meshing teeth, it tends to be noisy at high speeds. However, noise and wear can be minimized with proper lubrication.

There are three main classes of spur gears: external tooth, internal tooth, and rack-and-pinion. The external tooth variety shown in Figure 57.1 is the most common. Figure 57.2 illustrates an internal gear and Figure 57.3 shows a rack or straight-line spur gear. The smaller of the pair of spur gears illustrated in Figure 57.4 is a pinion, also called a sun gear in what is referred to as a planetary gearing system.

A planetary system consists of three sets of gears. A pinion meshes with two or more planet gears, which rotate within an internal gear. The planetary gears are connected to one shaft and the sun gear to another. This compact

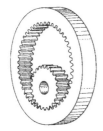

Figure 57.1 Example of a spur gear

Figure 57.2 Example of an internal spur gear

Figure 57.3 Rack or straight-line gear

Figure 57.4 Typical spur gears

system achieves a large speed reduction in a limited space. It also allows the input and output shaft to be on the same centerline.

The spur gear is cylindrical and has straight teeth cut parallel to its rotational axis. The diametric pitch establishes the tooth size of spur gears. Spur gear design accommodates mostly rolling, rather than sliding, contact of the tooth surfaces and tooth contact occurs along a line parallel to the axis. Such rolling contact produces less heat that results in high mechanical efficiency, often up to 99 per cent.

An internal spur gear, in combination with a standard spur gear pinion, provides a compact drive mechanism for transmitting motion between parallel shafts that rotate in the same direction. The internal gear is a wheel that has teeth cut on the inside of its rim and the pinion is housed inside the wheel. The driving and driven members rotate in the same direction at relative speeds inversely proportional the number of teeth. Figure 57.5 labels the parts of an internal spur gear.

Helical

Helical gears (shown in Figure 57.6) provide a means of connecting non-parallel shafts as well as provide an alternate means of connecting parallel shafts, serving the same purpose as spur gears. Cutters that produce an angle that allows several teeth to mesh simultaneously form helical gears. Helical gears are superior to spur gears in their load-carrying capacity, quietness, and smoothness of operation, which results from the sliding contact of the meshing teeth. A disadvantage, however, is higher friction and wear that accompanies this sliding action.

Single helical gears are manufactured with the same equipment as spur gears, but the teeth are cut at an angle to the axis of the gear and follow a spiral path. The angle at which the gear teeth are cut is called the helix angle, which is illustrated in Figure 57.7. This angle causes the position of tooth contact with the mating gear to vary at each section. Figure 57.8 shows the parts of a helical gear.

It is very important to note that the helix angle may be on either side of the gear's centerline. Or, if compared to the helix angle of a thread, it may be either a 'right-hand' or a 'left-hand' helix. The handedness of the helix is the same regardless of how viewed. Figure 57.9 illustrates a helical gear as viewed from opposite sides. Changing the

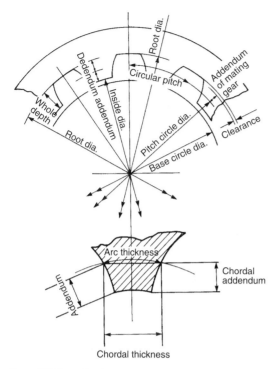

Figure 57.5 Parts of an internal spur gear

Figure 57.6 Typical set of helical gears

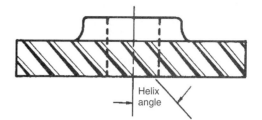

Figure 57.7 Illustrating the angle at which the teeth are cut

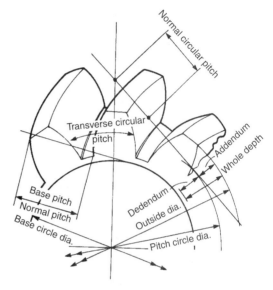

Figure 57.8 Helical gear and its parts

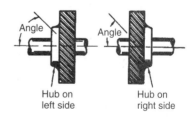

Figure 57.9 Helix angle of the teeth – the same no matter from which side the gear is viewed

position of the gear cannot change the handedness of the tooth's helix angle. A pair of helical gears such as the ones illustrated in Figure 57.9 must have the same pitch and helix angle, but must be of opposite handedness (one right hand and one left hand).

When helical gears are used to connect non-parallel shafts, they are often called 'spiral' or 'crossed-axis' helical gears. This style of helical gearing is shown in Figure 57.10.

The double-helical gear, also referred to as the Herringbone gear (Figure 57.11), is used for transmitting power between parallel shafts. It was developed to overcome

Figure 57.10 Typical set of spiral gears

Figure 57.11 Herringbone gear

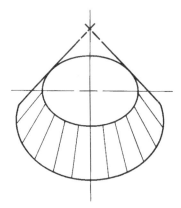

Figure 57.12 Basic cone shape of bevel gears

Figure 57.13 Typical set of bevel gears

the disadvantage of the high-end thrust that is present in single-helical gears.

The herringbone gear consists of two sets of gear teeth on the same gear, one right hand and one left hand. Having both hands of gear teeth causes the thrust of one set to cancel out the thrust of the other. Thus, an advantage of this gear type is quiet, smooth operation at higher speeds.

Bevel

Bevel gears are used most frequently for 90° drives, but other angles can be used. The most typical application is driving a vertical pump with a horizontal driver.

Two major differences between bevel gears and spur gears are their shape and the relation of the shafts on which they are mounted. A bevel gear is conical in shape while a spur gear is essentially cylindrical. Figure 57.12 illustrates the bevel gear's basic shape. Bevel gears transmit motion between angular or intersecting shafts, while spur gears transmit motion between parallel shafts.

Figure 57.13 shows a typical pair of bevel gears. As with other gears, the term 'pinion and gear' refers to the

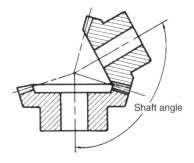

Figure 57.14 Shaft angle, which can be at any degree

members with the smaller and larger numbers of teeth in the pair, respectively. Special bevel gears can be manufactured to operate at any desired shaft angle, as shown in Figure 57.14.

As with spur gears, the tooth size of bevel gears is established by the diametric pitch. Because the tooth size varies along its length, measurements must be taken at a specific point. Note that, because each gear in a bevel gear set must have the same angle pressure, tooth lengths, and diametric pitch, they are manufactured and distributed only as mated pairs. Like spur gears, bevel gears are available in pressure angles of 14.5° and 20°.

Because there generally is no room to support bevel gears at both ends due to the intersecting shafts, one or both gears overhang their supporting shafts. This is referred to as an overhung load. It may result in shaft deflection and gear misalignment, causing poor tooth contact and accelerated wear.

Straight or plain

Straight bevel gears, also known as plain bevels, are the most commonly used and simplest type of bevel gear (Figure 57.15). They have teeth cut straight across the face of the gear. They are recommended for peripheral speeds up to 1,000 feet per minute in cases where quietness and maximum smoothness are not crucial. This gear type produces thrust loads in a direction that tends to cause the pair to separate.

Zerol

Zerol bevel gears are similar to straight bevel gears, carry the same ratings, and can be used in the same mountings. Zerol bevel gears, which should be considered as spiral bevel gears with zero spiral angles, have curved teeth that lie in the same general direction as straight bevel gears. This type of gear permits slight errors in assembly and some displacement due to deflection under load. Zerol gears should be used at speeds less than 1,000 feet per minute because of excessive noise at higher speeds.

Spiral

Spiral bevel gears (Figure 57.16) have curved oblique teeth that contact each other gradually and smoothly from one end of the tooth to the other, meshing with a rolling contact similar to helical gears. Spiral bevel gears are smoother and quieter in operation than straight bevel gears, primarily due to a design that incorporates two or more contacting teeth. Their design, however, results in high tooth pressure.

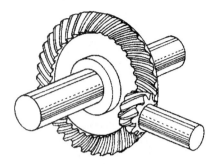

Figure 57.16 Spiral bevel gear

This type of gear is beginning to supersede straight bevel gears in many applications. They have the advantage of ensuring evenly distributed tooth loads and carry more load without surface fatigue. Thrust loading depends on the direction of rotation and whether the spiral angle of the teeth is positive or negative.

Miter

Miter gears are bevel gears with the same number of teeth in both gears, operating on shafts at right angles or at 90° as shown in Figure 57.17.

A typical pair of straight miter gears is shown in Figure 57.18. Another style of miter gears having spiral rather than straight teeth is shown in Figure 57.19.

Hypoid

Hypoid bevel gears are a cross between a spiral bevel gear and a worm gear (Figure 57.20). The axes of a pair of hypoid bevel gears are non-intersecting and the distance between the axes is referred to as the 'offset.' This configuration allows both shafts to be supported at both ends and provides high strength and rigidity.

Although stronger and more rigid than most other types of gears, they are less efficient and extremely difficult to lubricate because of high tooth-contact pressures. Further increasing the demands on the lubricant is the material of construction – both the driven and driving gears are made

Figure 57.15 Straight or plain bevel gear

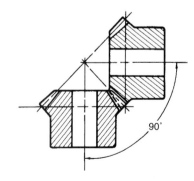

Figure 57.17 Miter gear shaft angle

Figure 57.18 Miter gears with straight teeth

Figure 57.19 Miter gears with spiral teeth

Figure 57.20 Hypoid bevel gear

of steel. This requires the use of special extreme-pressure lubricants that have both oiliness and anti-weld properties that can withstand the high contact pressures and rubbing speeds.

Despite its demand for special lubrication, this gear type is in widespread use in industrial and automotive applications. It is used extensively in rear axles of automobiles having rear-wheel drives and is increasingly being used in industrial machinery.

Worm

The worm and worm gear, which are illustrated in Figure 57.21, are used to transmit motion and power when a high-ratio speed reduction is required. They accommodate a wide range of speed ratios (60:1 and higher can be obtained from a single reduction and can go as high as 500:1). In most worm gear sets, the worm is most often the driver and the worm gear the driven member. They provide a steady, quiet transmission of power between shafts at right angles and can be self-locking. Thus, a torque on the gear will not cause the worm to rotate. A reversible worm-gear is also available.

The contact surface of the screw on the worm slides along the gear teeth. Because of the high level of rubbing between the worm and wheel teeth, however, slightly less efficiency is obtained than with precision spur gears. Note that large helix angles on the gear teeth produce higher efficiencies. Another problem with this gear type is heat removal, a limitation that restricts their use to low-speed applications.

One of the major advantages of the worm gear is low wear, which is due mostly to a full-fluid lubricant film. In addition, friction can be further reduced using metals having low coefficients of friction. For example, the wheel is typically made of bronze and the worm of highly finished hardened steel.

Most worms are cylindrical in shape with a uniform pitch diameter. However, a variable pitch diameter is used in the double-enveloping worm. This configuration is used when increased load capacity is required.

Figure 57.21 Typical set of worm gears

Table 57.2 Gear efficiencies

Gear type	Efficiency range, %
Bevel gear, hypoid	90–98
Bevel gear, miter	Not available
Bevel gear, spiral	97–99
Bevel gear, straight	97–99
Bevel gear, zerol	Not available
Helical gear, external	97–99
Helical gear–double, external (herringbone)	97–99
Spur gear, external	97–99
Worm, cylindrical	50–99
Worm, double-enveloping	50–98

Source: Adapted by Integrated Systems, Inc. from 'Gears and Gear Drives,' *1996 Power Transmission Design*, Penton Publishing Inc., Ohio, pp A199–A211.

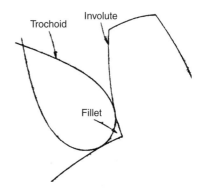

Figure 57.22 Involute tooth curve generated by straight-sided gear cutter

Like helical gears, worms and worm gears have handedness, which is determined by the direction of the angle of the teeth. The worm and worm gear must be the same hand in order to mesh correctly.

One revolution of the worm advances the gear teeth in direct proportion to the number of the worm threads. The most commonly used worms have either one, two, three, or four separate threads and are referred to as single-, double-, triple-, and quadruple- thread worms. The number of threads is determined by counting the number of starts or entrances at the end of the worm.

The worm thread is a major factor in worm ratios and is an important feature in worm design. The speed-reduction ratio is equal to the total number of worm-gear teeth divided by the number of worm threads.

57.2.2 Efficiencies

Table 57.2 provides a comparison of the approximate efficiency range of various gear types discussed in this chapter.

57.3 Gear teeth

Two important characteristics of gear teeth are the profile and the pressure angle. In particular, special tooth profiles must be used for higher gear speeds. As speeds increase, friction increases the heat released because of sliding contact between the teeth.

The *cycloidal* profile provides a rolling action, which minimizes friction. However, the gears and shafts must be very accurately aligned and spaced. The *involute* profile does not provide as good a rolling action as the cycloidal profile. As a result, some sliding between teeth occurs. However, the involute profile is less sensitive to shaft alignment and gear spacing, and is the shape most commonly used for gear teeth.

57.3.1 Involute profile

The involute tooth curve illustrated in Figure 57.22 is generated during the machining process using gear cutters

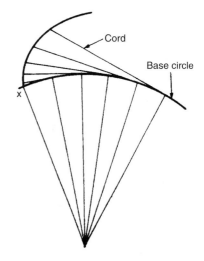

Figure 57.23 Theoretical generation of an involute tooth curve

with straight sides. Theoretically, the curve is generated by tracing a point on the end of a taut line that unwinds from a circle as shown in Figure 57.23.

The larger the circle, the straighter the curvature. For a rack, which is essentially a section of an infinitely large gear, the curve is straight or flat. The involute system of spur gearing is based on a straight- or flat-sided rack and all gears made for operation with such a rack will run with each other.

57.3.2 Pressure angle

The sides of each gear tooth incline at an angle called the pressure angle (Figure 57.24). For many years, the standard pressure angle was 14.5°. Today, the use of 14.5° gearing is generally limited to replacement work, while the 20° pressure angle is more commonly used. This switch has occurred because a 20° pressure angle results

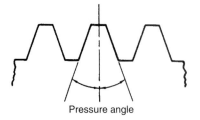

Figure 57.24 Pressure angle

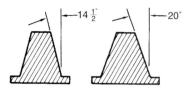

Figure 57.25 Effect of different pressure angles on gear teeth

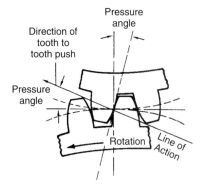

Figure 57.26 Relationship of the pressure angle to the line of action

in a higher strength gear tooth that offers better wear-resistance characteristics. The use of a 20° angle also permits pinions with fewer teeth to be used.

Figure 57.25 illustrates the effect of different pressure angles on a tooth. It is important to note that the *pressure angle must be the same for all mated gears* that run together.

The pressure angle of a gear is also the angle between the line of action and the line tangent to the pitch circles of mating gears. Figure 57.26 illustrates the relationship of the pressure angle to the line of action and the line tangent to the pitch circles.

57.4 Measurements and dimensions

All gear types are comprised of common gear tooth elements, which can be defined in terms of measurements and dimensions (linear, circular, and angular), and ratios.

57.4.1 Gear tooth elements

The elements of gear teeth common to all gears are tooth surface and profile, flank, top and bottom land, crown, root and pitch circle, gear center, line of centers, pitch point, line of action, line of contact, and point of contact. Figure 57.27 labels many of the common gear tooth elements. Figure 57.28 labels the common rack tooth elements.

57.4.2 Linear and circular measurements

Linear and circular measurements that define gears and are used in their specification and design are center distance, offset, pitch diameter, diametric pitch, axial pitch, base pitch, and axial base pitch, lead, and backlash.

Pitch diameter and center distance

A pitch circle is illustrated in Figure 57.29. The imaginary circle can be drawn to illustrate the motion of a gear in operation. The diameters of these imaginary circles are referred to as the pitch diameters of the gears. The center distance of two correctly meshed gears, illustrated in Figure 57.30, is equal to one half the sum of the two pitch diameters.

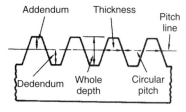

Figure 57.27 Names of gear parts

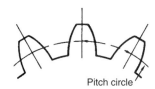

Figure 57.28 Names of rack parts

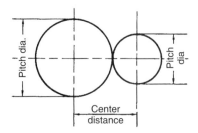

Figure 57.29 Pitch circle

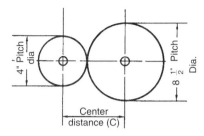

Figure 57.30 Pitch diameter and center distance

The center distance relationship also may be represented by the following equations:

$$C = \frac{D_1 + D_2}{2} \quad D_1 = 2C - D_s \quad D_2 = 2C - D_1$$

where:

C = Center distance, inches
D_1 = First pitch diameter, inches
D_2 = Second pitch diameter, inches

Example: Determine the center distance using the pitch diameters given in Figure 57.31.

$$C = \frac{D_1 + D_2}{2} = \frac{4 + 8.5}{2} = 6.25 \text{ inches}$$

Pitch measurements

The size and proportion of gear teeth are designated by a specific type of pitch. In gearing terms, there are two types of pitch: circular and diametric.

Circular   Circular pitch is the distance from a point on one tooth to the corresponding point on the next tooth measured along the pitch circle as shown in Figure 57.32. Its value is equal to the circumference of the pitch circle divided by the number of teeth in the gear. While most common-size gears are based on diametric pitch, large-diameter gears are frequently made to circular pitch dimensions.

Diametric   The most commonly used method of gear specification is based on diametric pitch. Practically all common-size gears are made to diametric pitch specifications, which also designate the size and proportions of gear teeth.

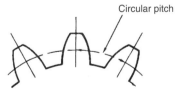

Figure 57.31 Determining center distance with known pitch diameter

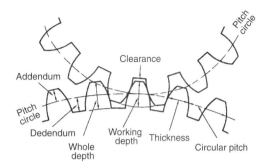

Figure 57.32 Circular pitch

Diametric pitch is a whole number used to specify the ratio of the number of teeth in a gear to its pitch diameter. Stated another way, it specifies the number of teeth in a gear per inch of pitch diameter. For each inch of pitch-circle diameter, there are pi ($\pi = 3.1416$) inches of pitch-circle circumference. Therefore, the diametric pitch provides the number of teeth for each 3.1416 inches of circumference along the pitch circle.

The pitch-circle diameter and the diametric pitch of a 4-inch pitch-circle diameter gear are illustrated in Figure 57.33. For this 4-inch gear, there are four 3.1416-inch circumference segments. Note that for a 3-inch gear, there are three 3.1416-inch segments.

These concepts may be better visualized and dimensions more easily obtained with the rack teeth presented in Figure 57.34. This clearly shows that there are 10 teeth in 3.1416 inches and, therefore, the rack illustrated is a 10 diametric-pitch rack.

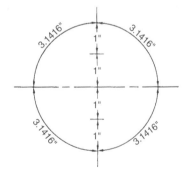

Figure 57.33 Pitch diameter and diametric pitch

Figure 57.34 Number of teeth in 3.1416 inches of a rack or straight-line gear

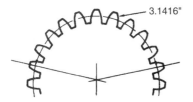

Figure 57.35 Number of teeth in 3.1416 inches on the pitch circle

Figure 57.35 illustrates a similar measurement along the pitch circle of a 10 diametric-pitch gear.

During the process of repairing a machine, a mechanic may need to quickly determine the diametric pitch of a gear. It is possible to do this easily without the use of precision measuring tools, templates, or gages; a ruler (preferably flexible) is all that is required to make the needed measurements. Because diametric pitch numbers are usually whole numbers, measurements need not be exact. An approximate calculation will usually result in a value close to a whole number, which is the diametric pitch of the gear. The following three methods may be used to determine the approximate diametric pitch of a gear.

Method 1

Count the number of teeth in the gear, add two to this number, and divide by the outside diameter (see Figure 57.36) of the gear. Rounding the gear diameter measurement to the closest fractional increment is sufficient. For example, assume Figure 57.36 illustrates a gear with 56 teeth (not all shown) and an outside diameter of $5\frac{13}{16}$ inches. Adding 2 to 56 gives 58; dividing 58 by $5\frac{13}{16}$ gives $9\frac{31}{32}$, which is approximately 10. Therefore, the diametric pitch of the gear is 10.

Method 2

Divide the number of gear teeth by the pitch diameter, which can be approximately measured from the root, or bottom, of a tooth to the top of a tooth on the opposite side of the gear. For example, assume Figure 57.37 illustrates a gear with 56 teeth (not all shown). The pitch diameter measured from the bottom of the tooth to the top of the opposite tooth is $5\frac{5}{8}$ inches. Dividing 56 by $5\frac{5}{8}$ gives $9\frac{15}{16}$ inches or approximately 10. As expected, this method also indicates that the gear is a 10 diametric-pitch gear.

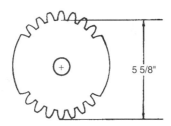

Figure 57.36 Using Method 1 to approximate the diametric pitch

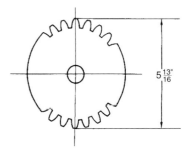

Figure 57.37 Using Method 2 to approximate the diametric pitch

Method 3

If a flexible ruler is available, measure approximately $3\frac{1}{8}$ inches along the gear's pitch line by bending the scale to match the curvature of the gear. Position the ruler about midway between the base and top of the teeth, the location of the imaginary pitch line. It might be helpful to mark this line on the gear, particularly if it is possible to rotate the gear. The diametric pitch is determined by counting the number of teeth in $3\frac{1}{8}$ inches. See Figure 57.38 for an illustration of this method.

Calculations

This section presents the equations used to make circular and diametric pitch calculations as well as tooth proportion calculations.

Circular pitch

The mathematical relationship of the circular pitch to the pitch-circle circumference, number of teeth, and the pitch diameter is shown in the following equations:

$$P = \frac{C}{N} \quad D = \frac{PN}{\pi} \quad N = \frac{C}{P}$$

where:

C = Pitch circle circumference (πD), inches
D = Pitch diameter, inches
N = Number of teeth
p = Circular pitch, inches
π = pi (3.1416)

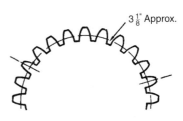

Figure 57.38 Using Method 3 to approximate the diametric pitch

Example 1   What is the circular pitch of a gear with 48 teeth and a pitch diameter of 6 inches?

$$P = \frac{\pi D}{N} \text{ or } \frac{3.1416 \times 6}{48} \text{ or } \frac{3.1416}{8} \text{ or } P = 0.3927 \text{ inches}$$

Example 2   What is the pitch diameter of a 0.500-inch circular-pitch gear with 128 teeth?

$$D = \frac{PN}{\pi} \text{ or } \frac{0.5 \times 128}{3.1416} \text{ or } D = 20.371 \text{ inches}$$

Diametric pitch

The mathematical relationship of pitch diameter, diametric pitch, and number of teeth is shown in the equations below. As with any equation, one of the variables can be calculated if any two are known.

$$P = \frac{N}{D} \quad D = \frac{N}{P} \quad N = D \times P$$

where:

D = Pitch diameter, inches
P = Diametric pitch, inches
N = Number of teeth

Example 1   What is the diametric pitch of a 40-tooth gear with a 5-inch pitch diameter?

$$P = \frac{N}{D} \text{ or } P = \frac{40}{5} \text{ or } P = 8 \text{ diametric pitch}$$

Example 2   What is the pitch diameter of a 12 diametric-pitch gear with 36 teeth?

$$D = \frac{N}{P} \text{ or } D = \frac{36}{12} \text{ or } D = 3\text{-inch pitch diameter}$$

Example 3   How many teeth are there in a 16 diametric-pitch gear with a pitch diameter of $3\frac{3}{4}$ inches?

$$N = D \times P \text{ or } N = 3\tfrac{3}{4} \times 16 \text{ or } N = 60 \text{ teeth}$$

Tooth proportion calculations

The most commonly used gear system is referred to as the full-depth involute. The tooth proportion formulas and symbols for full-depth, coarse-pitch involute gears (ANSI 136.1-1968, R1974) are shown below. All units are in inches.

Diametric pitch, P

$$\text{Addendum, } a = \frac{1}{P}$$

$$\text{Dedendum, } b = \frac{1.250}{P}$$

$$\text{Whole depth, } W_d = a + b = \frac{2.250}{P}$$

$$\text{Clearance, } c = b - a = \frac{0.250}{P}$$

$$\text{Tooth thickness, } t = \frac{1.5708}{P} \text{ on the pitch line}$$

Backlash

Gear backlash is the play between teeth, which is critical in preventing binding. Measured at the pitch circle, it is the distance between the involutes of the mating gear teeth, as illustrated in Figure 57.39.

Backlash is necessary to provide the running clearance needed to prevent binding of the mating gears, which can result in heat generation, noise, abnormal wear, overload, and/or failure of the drive. In addition to preventing binding, some backlash in gear systems is to be expected because of the dimensional tolerances needed for cost-effective manufacturing.

During the gear manufacturing process, backlash is achieved by cutting each gear tooth thinner by an amount equal to one-half of the backlash dimension required for the application. When two gears made in this manner are run together (i.e., mate), their allowances combine and provide the full amount of backlash required.

The increase in backlash that results from tooth wear does not adversely affect operation with non-reversing drives or drives with continuous load in one direction. However, for reversing drives and drives where timing is

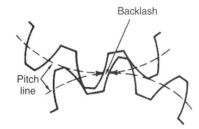

Figure 57.39 Backlash

Figure 57.40 Double-reduction single-helical drive

critical, excessive backlash that results from wear usually cannot be tolerated.

Angular dimensions

Angular dimensions that define gears and are used in their specification and design are helix angle, lead angle, shaft angle, and angular pitch.

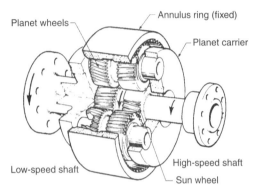

(a) Planetary gear drive

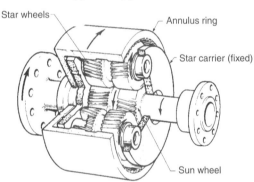

(b) Star gear drive

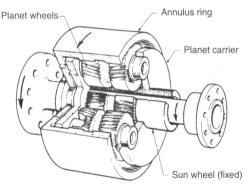

(c) Solar gear drive

Figure 57.41 Three double helical drives

Figure 57.42 High-speed gear drive

Ratios

Ratios that are used in defining and specifying gears are gear-tooth ratio, contact ratio, and hunting ratio. The *gear-tooth ratio* is the ratio of the larger to the smaller number of teeth in a pair of gears. The *contact ratio* is a measure of overlapping tooth action, which is necessary to assure smooth, continuous tooth action. For example, as one pair of teeth passes out of action, a succeeding pair of teeth must have already started action. The *hunting ratio* is the ratio of the number of gear and pinion teeth. It is a means of ensuring that each tooth in the pinion will contact every tooth in the gear before it contacts any gear tooth a second time.

57.5 Gear drives and speed reducers

Gear drives are packaged units used for a wide range of power-transmission applications. They are used to transmit power to a driven piece of machinery and to change or modify the power that is transmitted. Modifications include reducing speed and increasing output torque, increasing speed, changing the direction of shaft rotation, or changing the angle of shaft operation.

Motorized gear drives, which are commonly called gear motors or motor reducers, are used extensively throughout industry. Any of the basic gear drives can be manufactured as motorized units.

The following figures show typical gear-drive units. Figure 57.40 is a double-reduction single-helical drive. Figure 57.41 shows three types of double-helical drives. Figure 57.42(a) shows an enclosed high-speed gearbox, while Figure 57.42(b) shows the internal parts of this gear drive.

References

1. Stokes, A., *Gear Handbook – Design and Calculations*, Butterworth-Heinemann, Boston (1992).
2. 'Gears and Gear Drives,' *1996 Power Transmission Design*, Penton Publishing Inc., Ohio, pp. A199–A211.
3. Oberg, E., Jones, F. D., Horton, H. L. and Ryffel, H. H., *Machinery's Handbook 25*, 25th edn, Industrial Press Inc., New York (1996).
4. Higgins, and Mobley, K. R., *Maintenance Engineering Handbook*, McGraw-Hill, New York (1995).
5. Nelson, C. A., *Millwrights and Mechanics Guide*, Macmillan Publishing Company, New York (1986).

58

Flexible Intermediate Drives

R Keith Mobley
The Plant Performance Group

Contents

58.1 Introduction

There are two types of flexible intermediate drives used to transmit torsional power: belt drives and chain drives. Flexible belts are used in industrial power transmission applications primarily when the speeds of the driver and driven shafts must be different or when the shafts must be widely separated. The trend toward higher speed primary drivers and the need to achieve a slower, useful driven speed are additional factors favoring the use of belts. In addition to V-belts, there are round belts and flat belts. Chain drives are typically used in applications where space is limited or obstructions prevent direct coupling of machine-train components.

58.2 Belt drives

V-belts are the most commonly used belt in industrial power transmission applications. In addition to V-belts, there are round belts (e.g., O-rings) and flat belts, which are often reinforced with steel or Kevlar. Round belts are generally used in light-duty applications while reinforced flat belts are used in high-temperature applications such as automobiles.

Belts have numerous advantages, including overall economy, cleanliness, no need for lubrication, low maintenance costs, easy installation, dampening of shock loads, clutching, and variable speed power transmission between widely spaced shafts.

58.2.1 Power transmission

With belt drives, the force that produces work acts on the rim of a pulley or sheave, causing it to rotate. Since a belt on a drive must be tight enough to prevent slippage, there is a belt-pulling force on both sides of a driven wheel. When a drive is stationary or operating with no power transmission, the forces on both sides of the driven wheel are essentially equal. When the drive is transmitting power, however, the forces are not the same. There is a tight-side tension, T_T, and a slack-side tension, T_S. The difference between these two forces (T_T minus T_S) is called the working tension or effective or net pull. The effective pull is applied at the rim of the pulley and is the force that produces work.

$$\text{Net pull (pounds)} = \text{Horsepower (Hp)} \times \frac{33,000}{\text{Belt speed (fpm)}}$$

Belt speed is affected if the pulley or sheave diameter is changed. When the driven-shaft speed is constant, doubling the diameter of the pulley cuts the total belt speed in half.

Very often, in high horsepower applications, designers employ multiple belts in the drive components.

58.2.2 Belt tensions

A belt experiences three types of tension as it rotates around a pulley: working tension, bending tension, and centrifugal tension. The combination of tight side, bending, and centrifugal tensions is referred to as the peak tension of the belt drive system, which is calculated by:

$$T_{\text{Peak}} \text{ (pounds)} = T_T + T_B + T_C$$

Peak tension is the dominant factor that determines belt drive performance and service life. It should be noted that only the working tension components, T_T and T_S, have a direct impact on the pulley, shafts, and bearings of the drive train. Bending and centrifugal tension affects the belt, but should not transmit tension to the drive-train system.

The tension ratio, T_R, is equal to the tight-side tension divided by the slack-side tension. As the tension ratio increases, so does the potential for belt slippage.

$$T_R = \frac{\text{Tightside tension}}{\text{Slackside tension}}$$

Working tension

Working tension or effective pull, T_P, is the difference between the tight-side tension and the slack-side tension:

$$T_P \text{ (pounds)} = \text{Tightside tension} - \text{Slackside tension}$$

Bending tension

Bending tension, T_B, occurs when the belt bends around the pulley. The outside of the belt is in tension and the inside is in compression. The amount of tension depends on the belt's construction and the pulley diameter.

Centrifugal tension

Centrifugal tension, T_C, occurs as the belt rotates around the drive and is calculated by:

$$T_C = MV^2$$

Where:

T_C = Centrifugal tension, pounds
M = Constant dependent on the belt weight and the pulley diameter
V = Belt velocity, feet per minute

58.2.3 Selection of belts and pulleys

Belt and pulley selection is based on the design requirements of a specific machine train and the machine manufacturer will generally provide the belt system specifications. The following information is the minimum required for proper selection: horsepower requirements, center-to-center dimensions of the drive and driven shafts, and the process envelope.

The process envelope must define the full range of operating conditions that will be expected of the belt-drive system. For example, will the drive operate continuously in one direction or will it be subjected to start/stops or direction reversals? Will the drive system be subjected to radical speed changes or be used for braking the driven unit? Such factors are important in that they help define the full range of load dynamics that the belt system must accommodate. From this data, the appropriate service factor that must be applied as part of the selection process can be determined.

Belts

In the selection of V-belts, they can be completely specified by defining their cross-section and length. Nominal dimensions are shown in Figure 58.1 as an aid in identifying belt cross-sections.

Length

There are three ways to measure V-belt length: outside circumference (OC), datum length (DL) and effective length (EL).

Outside circumference   Outside circumference is measured by wrapping a tape measure around the outside surface of the belt. This method is useful for obtaining nominal dimensions, but does not give a truly accurate measurement.

Datum length   Datum length is a recent designation adopted by all belt manufacturers in order to retain standard belt and pulley designations. It more accurately reflects the changes that have occurred in belt-pitch length and pitch-line location within the belt. Standard V-belt datum lengths are obtained from ANSI/RMA IP-20 (1988) and are shown in Table 58.1.

Effective length   The effective length is measured on a length-inspection machine. The machine consists of two parallel shafts on movable centers with a scale to accurately measure the center distance between shafts. Inspection pulleys of equal diameter and grooved in accordance with industry standards are mounted on these shafts. A belt is mounted on these pulleys and tensioned to a specified force. The belt is rotated through at least three complete revolutions to ensure that the tension is equalized around the belt and that the belt is properly seated in the grooves.

The effective length is defined as the measured center distance plus the outside circumference of one of the inspection pulleys. This measurement method accounts for the modulus of elasticity, or stretch ability, and dimensional variations among belts with the same cross-section.

Number of belts and service factor

Few belt-drive systems use a single belt for power transmission. The number of belts is determined by the horsepower requirements of the machine train. Determination of these requirements must be based on the full operating range of the system. Rapid speed changes, direction reversal, and other operating practices have a direct impact on the requirements of the belt system. Therefore, the selection process should be based on the most demanding operating conditions.

Once the maximum horsepower required is determined, the number of belts required for the application can be determined by dividing the total horsepower by the corrected horsepower rating of a single belt. This corrected rating is obtained by applying a service factor to the horsepower rating of a single belt to adjust its acceptable limit to the variables of the process (e.g., speed and direction changes, start/stop operation, side tension, and a variety of other factors).

Cross-section

Always select belts to match the pulley grooves, using a pulley-groove gage to determine the correct replacement belt cross-section. On new applications, the design specifications will indicate the proper belt size, both length and cross-section (see Figure 58.1).

Use of matching belts in a set

In all applications, matching belts should be used in a belt set. This practice ensures the best service life of the belts and proper power transmission. Because of the need for uniformity in cross-section and length, belt manufacturers have adopted industry standards intended to assure such uniformity. However, no two belts are ever exactly alike due to manufacturing tolerances. Therefore, a matched set is considered the closest match of belts available from a manufacturer. As part of the quality control process, belt manufacturers group them based on variations in cross-section and length in order to provide the best possible belt set.

For maximum service, replace V-belts with a complete new set, **even if only one belt in the set breaks**. This is important because a new V-belt will ride much higher in the pulley groove than a worn belt. This is referred to as 'differential driving,' meaning that the belts riding higher in their respective grooves are actually traveling faster than their lower-riding counterparts. This creates the situation where one belt on a pulley is working against the others. Therefore, replacing only one belt in the set will cause the new belt to stretch much more rapidly than if the entire set had been replaced. **Note, however, that used belts may be used in a much lighter-loaded machine as a single belt or with belts from the same set.**

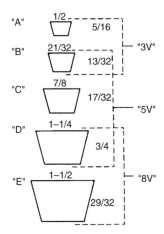

Figure 58.1 Nominal cross-section dimensions

Table 58.1 Classical V-belt standard datum length (ANSI/RMA PI-20, 1988)

Standard length designation[a]	Standard datum lengths cross-section				Permissible deviations from std. dantum length	Matching limits for one set
	A, AX	B, BX	C, CX	D		
26	27.3	–	–	–	±0.06	0.15
31	32.3	–	–	–	±0.06	0.15
35	36.3	36.8	–	–	±0.06	0.15
38	39.3	39.8	–	–	±0.07	0.15
42	43.3	43.8	–	–	±0.07	0.15
46	47.3	47.8	–	–	±0.07	0.15
51	52.3	52.8	53.9	–	±0.07	0.15
55	56.3	56.8	–	–	±0.07	0.15
60	61.3	61.8	62.9	–	±0.07	0.15
68	69.3	69.8	70.9	–	±0.07	0.30
75	75.3	76.8	77.9	–	±0.07	0.30
80	81.3	–	–	–	±0.07	0.30
81	–	82.8	83.9	–	±0.07	0.30
85	86.3	86.8	87.9	–	±0.07	0.30
90	91.3	91.8	92.9	–	±0.08	0.30
96	97.3	–	98.9	–	±0.08	0.30
97	–	98.8	–	–	±0.08	0.30
105	106.3	106.8	107.9	–	±0.08	0.30
112	113.3	113.8	114.9	–	±0.08	0.30
120	121.3	121.8	122.9	123.3	±0.08	0.30
128	129.3	129.8	130.9	131.3	±0.08	0.30
144	–	145.8	146.9	147.3	±0.08	0.30
158	–	159.8	160.9	161.3	±1.0	0.45
173	–	174.8	175.9	176.3	±1.0	0.45
180	–	181.8	182.9	183.3	±1.0	0.45
195	–	196.8	197.9	198.3	±1.1	0.45
210	–	211.8	212.9	213.3	±1.1	0.45
240	–	240.3	240.9	240.8	±1.3	0.45
270	–	270.3	270.9	270.8	±1.6	0.60
300	–	300.3	300.0	300.8	±1.6	0.60
330	–	–	330.9	330.8	±2.0	0.60
360	–	–	380	360.8	±2.0	0.60
540	–	–	–	540.8	±3.3	0.90
390	–	–	390.9	390.8	±2.0	0.75
420	–	–	420.9	420.8	±3.3	0.75
480	–	–	–	480.8	±3.3	0.75
600	–	–	–	600.8	±3.3	0.90
660	–	–	–	660.8	±3.3	0.90

Source: Machinery's Handbook 25, 25th edn, R. E. Green, editor, Industrial Press, Inc., New York, 1996.

Pulleys

Standard V-belt pulleys are specified with exact dimensions as opposed to nominal. The industry adopted the datum system in 1988 as the standard for specifying classic V-belt pulleys. Therefore, all pulley specifications should be in datum dimensions rather than pitch. Individual pulley dimensions are shown in Figure 58.2.

In applications where extreme vibration, belt twist, or extreme misalignment is encountered, deep-groove pulleys are used to increase belt stability.

The magnitudes of the forces on a belt have significant impact on belt life. When a V-belt bends around a pulley, compressive forces develop in the bottom of the belt and tension forces develop in the top of the belt. The magnitude of each force is a function of the diameter of the pulley and the cross-section of the belt.

Forces increase with smaller diameters and larger cross-sections. Therefore, minimum recommended diameters were developed for each belt cross-section. These minimum diameters can be found in V-belt selection guides provided by belt manufacturers. Using pulleys that are below the recommended minimum will always result in shorter belt life.

The National Electrical Manufacturers Association (NEMA) has developed a standard that specifies minimum pulley diameters that should be used on electric motors. Since motor bearings are specified using a relatively small maximum overhung load, minimum pulley diameters can have a dramatic effect on motor bearings. The increase in tension that is created by smaller pulleys can transmit a potentially destructive side-load on the bearings.

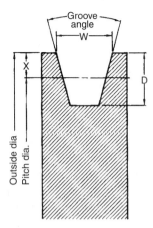

Figure 58.2 Typical pulley dimensions

58.3 Installation

Proper installation of V-belts is necessary to ensure trouble-free operation and maximum service life of the belts. The procedures and practices that should be followed in a proper installation can be broken down into three major categories: pre-installation inspections of the belt and pulley, installation practices, and post-installation practices (belt tension inspection and adjustment).

58.3.1 Pre-installation inspections

A thorough inspection of the belt-drive system components (i.e., belt and pulleys) should be performed prior to installing new or replacement belts and pulleys. However, before conducting any inspection or maintenance, always determine, understand, and follow all plant safety rules.

Safety

Before carrying out any inspections or maintenance procedures on belt drives, make sure they are not in operation. Equipment operators should electrically isolate the drive system and lock-out and tag the machine that is to undergo maintenance. Once this precaution has been taken, the belts and pulleys can be removed from the machine for inspection, cleaning, and repair.

Pulleys

Pulleys should be inspected prior to replacing belts to ensure they are in good operating condition. Damaged, worn, or dirty pulleys will reduce the effective lifetime of belts significantly. Wipe clean any oil or grease that has accumulated on the pulley and use a stiff brush with bristles that are softer than the pulley material to clean off rust and dirt.

Never reinstall damaged or worn pulleys on equipment; it is important to always repair or replace them. Always select the proper pulley-groove gage and template for the pulley diameter as shown in Figure 58.3. Inspect the

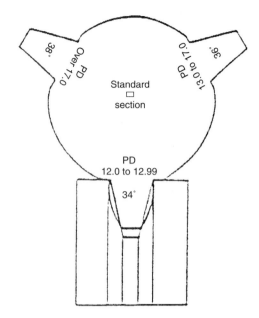

Figure 58.3 Illustration of a pulley groove gage

trueness of the sidewalls by inserting the gage into the groove. Look for voids, which indicate 'dishing' or other uneven and abnormal wear.

When the wear measurement reaches 0.015″, the pulley is unacceptable for steel-cable-reinforced belts and at 0.025″ the pulley is unacceptable for use with any belt.

Belts

Nicks, scratches, or gouge marks on a pulley can cut into a belt. Dirt abrades the belting material and wear grooves cause a belt to 'bottom-out' in the groove, resulting in slippage and ultimate damage to the belt.

Before installing onto a pulley, it is always a good idea to place each belt into its pulley groove. Figure 58.4 is an illustration of how a V-belt should sit in the pulley groove. The top of the belt should be within $\frac{1}{16}$-inch of the outer surface of the pulley. If the belt top is deeper in the groove, the groove or belt is worn, which can shorten the remaining belt life by as much as 50 per cent. Figure 58.5 shows a belt gage, an alternate means of determining belt wear, which measures belt total width (TW). For example, an 8V belt should measure $\frac{8}{8}″$ (i.e., 1 inch), a 5V belt should measure $\frac{5}{8}″$, and a 3V belt should measure $\frac{3}{8}″$.

58.3.2 Installation practices

Although most belt manufacturers have stringent quality assurance and control procedures in place to ensure that belts will operate as designed, this does not guarantee their performance if they are not properly installed and operated under the conditions for which they are designed. Therefore, it is important to consult and follow the installation

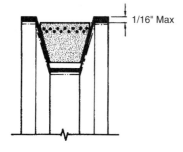

1/16" Max

1/32" Max
dished-out wear

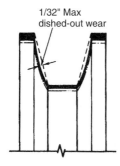

Figure 58.4 Illustration of how V-belts *should* sit in a pulley groove

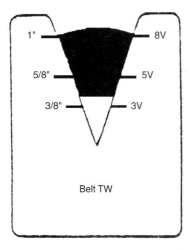

Figure 58.5 Belt gage

procedure prescribed by the belt manufacturer, as well as the one provided by the manufacturer of the equipment on which it will be installed.

Today's V-belts stretch as little as 1 per cent and, therefore, force should never be used during installation. If belts are forced onto a pulley with a pry bar or by cranking, damage will certainly occur. Therefore, if a belt fails within the first few hours of operation, the most probable cause is damage during installation.

On most drives, a pulley can be moved or opened (e.g., a split pulley) to obtain the slack necessary to install a belt. However, it is sometimes necessary to completely remove a pulley from the shaft in order to perform the installation without damaging the belt.

In addition, when installing multiple belts, it is imperative that the slack side of each belt be on the same side, either top or bottom. Tables 58.2 and 58.3 show the minimum movement below and above the standard center distance required between the pulleys. This movement allows the center distance to be shortened for new belt installation. It also allows the center distance to be increased to take up the slack and to maintain proper tension as the belt stretches (Figure 58.6).

58.3.3 Post-installation practices

Post-installation practices are belt tension inspection and adjustment.

Belt tension inspection

The following are the steps involved in checking the tension in conventional V-belt drives:

- Measure the belt span, 't'.
- Mark the center of the span, 't'. At the center mark, use a tension tester and apply a force perpendicular to the span that will be large enough to deflect the belt $\frac{1}{64}''$ for every inch of span length. For example, a $100''$ span will require a deflection of $\frac{100}{64}$ or $1\frac{9}{16}''$.
- This is the correct deflection force for operating speeds of 1000–3000 feet per minute. For belt speeds in excess of 3000 feet per minute, reduce the deflection force by 20 per cent. Check factory recommendations for operating speeds less than 1000 fpm.
- Compare the deflection force applied with the values in the detection Tables 58.4 and 58.5. A force between the value shown and 1.5 times (1.5×) the value shown should be satisfactory. A force below this value indicates an under-tensioned drive, which may result in slippage. A force above the tension value indicates an over-tensioned drive, which may result in belt and drive damage.
  Note: A drive with newly installed belts can be tightened initially to two times (2×) the normal tension to allow for the drop that occurs during the belt break-in period.

Tension adjustments

Prolonged slippage results in excessive belt surface wear, friction burn spots, and general overheating – all of which are very damaging to the belt. Therefore, tension should be applied until slippage no longer occurs, which can be verified using speed-ratio information. When the drive is correctly set, the belts will appear snug. However, there should be slight sag noticed on the loose side when in motion. After the initial tension adjustment, some additional tension should be applied to the belt in order to partially compensate for:

Table 58.2 Minimum pulley center distance allowances for belt installation

Standard length designation	Minimum allowance below standard center distance for installation of belts								Minimum allowances above standard center distance for maintaining tension (all cross-sections)
	A	B	B Torque team	C	C Torque team	D	D Torque team	E	
Up to and incl. 35	0.75	1.00	1.50						1.00
Over 35 to and incl. 55	0.75	1.00	1.50	1.50	2.00				1.50
Over 55 to and incl. 65	0.75	1.25	1.50	1.50	2.00				2.00
Over 65 to and incl. 112	1.00	1.25	1.60	1.50	2.00				2.50
Over 112 to and incl.144		1.25	1.80	1.50	2.10	2.00	2.90		3.00
Over 144 to and incl.180	1.00	1.25	1.90	2.00	2.20	2.00	3.00	2.50	3.50
Over 180 to and incl. 210		1.50	1.90	2.00	2.30	2.00	3.20	2.50	4.00
Over 210 to and incl. 240		1.50	2.00	2.00	2.50	2.50	3.20	2.50	4.50
Over 240 to and incl. 300		1.50	2.20	2.00	2.50	2.50	3.50	3.00	5.00
Over 300 to and incl. 390				2.00	2.70	2.50	3.60	3.00	6.00
Over 390				2.50	2.90	3.00	4.10	3.50	1.5% or belt length

Table 58.3 Minimum pulley center distance allowances for HY-T wedge V-belt installation

Standard length designation	Minimum allowance below standard center distance for installation of belts						Minimum allowances above standard center distance for maintaining tension (all cross-sections)
	A	B	B Torque team	C	C Torque team	D	
Up to and incl.475	0.5	1.2					1.0
Over 475 to and incl. 710	0.8	1.4	1.0	2.1			1.2
Over 710 to and incl. 1060	0.8	1.4	1.0	2.1	1.5	3.4	1.5
Over 1060 to and incl. 1250	0.8	1.4	1.0	2.1	1.5	3.4	1.6
Over 1250 to and incl. 1700	0.8	1.4	1.0	2.1	1.5	3.4	2.2
Over 1700 to and incl. 2000			1.0	2.1	1.8	3.6	2.5
Over 2000 to and incl. 2360			1.2	2.4	1.8	3.6	3.0
Over 2360 to and incl. 2650			1.2	2.4	1.8	3.6	3.2
Over 2650 to and incl. 3000			1.2	2.4	1.8	3.6	3.5
Over 3000 to and incl. 3550			1.2	2.4	2.0	4.0	4.0
Over 3550 to and Incl. 3750					2.0	4.0	4.5
Over 3750 to and incl. 5000					2.0	4.0	5.5

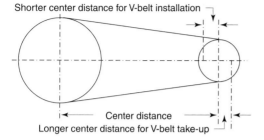

Shorter center distance for V-belt installation

Center distance
Longer center distance for V-belt take-up

Figure 58.6 Distance required between pulleys

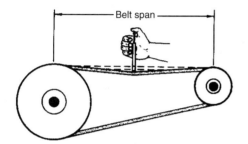

Belt span

Figure 58.7 Belt deflection force

- Initial belt stretch, and
- Slack caused by belt settling due to the soft rubber surface of the outer envelope being abraded away, which causes the belt to settle lower in the groove.

These changes occur during what is referred to as the run-in period. Once run-in, the slack on the new belts must be taken up to eliminate slippage, burning, or other irreparable damage. The following are common sense rules of V-belt tensioning:

- Check the belt tension frequently during the first 24 to 48 hours of run-in operation.

Table 58.4 Deflection chart for V-belts (tension in pounds force)

Cross-section	small diameter range	For normal tension	For $1\frac{1}{2}\times$ normal tension
3V	2.65 to 3.65	3.1	4.5
3V	4.12 to 6.90	4.5	6.5
5V	7.1 to 10.9	12	18
5V	11.8 to 16.0	14	21
8V	12.5 to 17.0	28	42
8V	18.0 to 22.4	33	50

- Maintain pulley alignment with a strong straightedge tool or, at a minimum, string while tensioning V-belts (Figure 58.8).
- Do not over tension belts as it shortens belt and bearing life. The ideal tension is the lowest at which the belt will not slip under peak load conditions.
- Inspect the V-belt drive periodically and re-tension the belts if slipping occurs.
- Use a strobe light to inspect belts in operation by timing the strobe to appear to slow-roll the drive. The belts that are slipping will appear to be running at a slower speed than the ones that are correctly set.
- Keep belts free from foreign material that may cause slippage.

58.4 Problem areas

Problem areas in belt drives include noises, such as squealing and squeaking (sometimes referred to as chirping); contamination from dirt, oil, and grease; belt damage, such as cracking, heat damage, stretching, and added load damage; and excessive belt whipping.

58.4.1 Belt noise

Belt noise can take the form of squeals and squeaks or chirps, both of which are generally more annoying than

Table 58.5 Deflection chart for CVS, HY-T, and torque-flex belts

Cross-section	Small diameter range	CVS Pounds force for normal horsepower	CVS Pounds force for $1\frac{1}{2}$ normal horsepower	HY-T Pounds force for normal horsepower	HY-T Pounds force for $1\frac{1}{2}$ normal horsepower	Torque-flex Pounds force for normal horsepower	Torque-flex Pounds force for $1\frac{1}{2}$ normal horsepower
A	3.0–3.5	3.2	4.5	3.6	5.2	5.0	7.0
A	3.8–4.8	3.8	5.2	4.3	6.2	5.9	8.4
A	5.0–7.0	4.4	6.0	5.0	7.2	6.7	9.6
B	3.4–4.2	4.6	6.1	4.9	6.9	7.1	10.1
B	4.4–5.6	5.8	8.0	6.5	9.3	7.7	11.0
B	5.8–8.6	7.1	10.0	8.2	11.8	9.6	13.8
C	7.0–9.0	11.5	15.4	15.5	22.1	16.9	23.6
C	9.5–16.0	14.4	20.0	16.9	24.3	18.8	27.4
D	12.0–16.0	24.4	33.9	28.1	40.9		
D	18.0–27.0	29.5	41.8	34.7	50.4		
E	20.0–32.0						

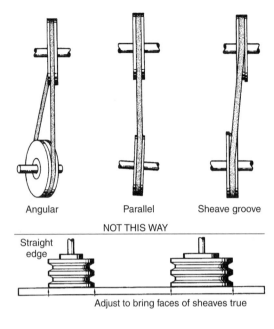

Angular   Parallel   Sheave groove

NOT THIS WAY

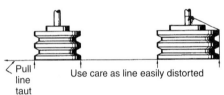

Straight edge

Adjust to bring faces of sheaves true

Pull line taut   Use care as line easily distorted

Figure 58.8 Proper way to tension belts

damaging. Noise commonly occurs with all types and makes of belts. Because of this, maintenance people often regard them as an operational noise and do not show much concern.

Important: Squeals and squeaks are not corrected by applying grease or oil to the surface of belts. This can cause serious belt damage!

Squeals

Although squeals are annoying and will not cause immediate harm to the belts, it does indicate that there may be a problem and, therefore, they should not be ignored. The following are some of the causes of belt squealing.

- Insufficient belt tension. This is a common cause and if squealing persists after belts are tensioned correctly, the drive itself should be examined for over-loading.
- Motors operating near or at full load
- Motor acceleration
- Misaligned idler pulley
- Dry and dusty belts

Squeaks

The following are some of the reasons for belt squeaking, which is often described as sounding like a bird chirping:

- Dry bearing
- Belts working in a dusty environment
- Misaligned idler pulley

Contamination

Common sources of belt contamination are dirt, oil, and grease, all of which should be removed from belts and pulleys. Dirt accelerates belt wear and belt traction is impaired when dirt is allowed to build up.

Do not expose belt drives to oil or grease under any circumstances! Potential sources of oil and grease include leaking bearings, which should be repaired immediately. It is important to avoid over-applying lubricants to bearings, but note that under-application can contribute to early failure with resulting damage to belts due to friction build-up and drag. If oily conditions cannot be avoided, special oil-resistant belts should be used.

Belt damage

Types of belt damage to watch for are cracking, which can be caused by excessive heat and dust; stretching; added load damage; and damage as a result of belt whipping.

Cracking   Cracking can be caused by exposure to high temperature and/or dust. Cracking on the bottom sections of belts does not cause a loss of strength or operating efficiency, but can lead to eventual failure. Nevertheless, there is no need to replace belts simply because bottom cracking has been detected.

Extreme heat   In the manufacturing process, belts are cured with scientifically controlled heat for given periods of time. If standard belts are operated below 140°F, their materials of construction are not affected. However, at temperatures above 140°F, over-curing will occur and belt life will be shortened. Therefore, the use of standard V-belts above this temperature should be avoided. Often, adequate shielding between the heat source and belts can be provided.

Dust   Dust can be very detrimental to belts and is often responsible for bottom cracks. The effects of dust can be slowed down greatly and possibly eliminated by installing larger pulleys and larger reverse idler pulleys.

Added load damage   Added loads shorten effective belt life. Always check to see if any other load has been added to the belt drive since original installation. Figure 58.9 illustrates the possible effects an added load may have on the motor belt life.

Stretching   All belts stretch, giving the appearance of flopping up and down when running. When this occurs, check bearings to ensure that they are free to turn and that there is no overload resulting from obstructions. Also

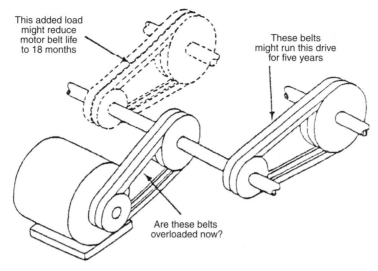

This added load
might reduce
motor belt life
to 18 months

These belts
might run this drive
for five years

Are these belts
overloaded now?

Figure 58.9 Added load

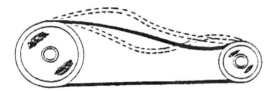

Figure 58.10 Belt whipping

check for pulley wear; worn pulleys create the impression that the belts are too long.

Excessive belt whip   Excessive belt whip is usually more common on long-center applications (applications where the distance between the two pulley shafts is great). Pulsating loads in the drive system can produce this phenomenon. Damage to the system is often in the form of belt breakage and/or bearing failure. Belt whip can be corrected by installing an idler pulley to dampen the vibration.

58.4.2 Troubleshooting V-belts

Several causes of V-belt failure and the action required to correct the problem are described in this section. Table 58.6 provides a troubleshooting overview.

58.5 Summary

The first rule for long belt life is to maintain correct belt tension. Loose belts slip, causing belt and pulley wear. The snapping action of loose belts adds sudden stress and often breaks the belt. To test for tension, press down firmly

Table 58.6   Troubleshooting V-belts

● *Cause of failure*: Ply separation caused by substandard pulley diameter (Figure 58.11).

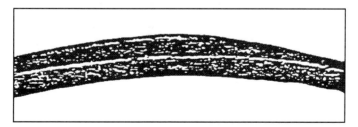

Figure 58.11 Failure due to substandard pulley diameter

(*continued overleaf*)

Table 58.6 (*continued*)

- *Action required:* Redesign drive to utilize proper size pulleys.
- *Cause of failure*: Rough pulley sidewalls cause the cover to wear off in an uneven pattern (Figure 58.12).

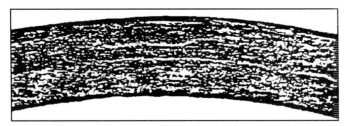

Figure 58.12 Failure due to rough pulley sidewalls

- *Action required:* File or machine out the rough spot on the pulley groove. If it is beyond repair, replace the pulley.
- *Cause of failure*: Belt has evenly spaced deep bottom cracks from use of substandard backside idler (Figure 58.13).

Figure 58.13 Failure due to use of substandard backside idler

- *Action required:* Replace backside idler with one that is in accordance with the minimum size recommendation.
- *Cause of failure*: Back of the belt rubbing on a belt guard or other component (Figure 58.14).

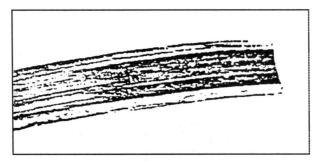

Figure 58.14 Failure due to belt rubbing guard

- *Action Required:* Provide adequate clearance between belt and guard of other component.
- *Cause of failure*: Tensile breaks caused by high shock loads, foreign objects lodged under the belt, or damage sustained during installation (Figure 58.15).

Table 58.6 (*continued*)

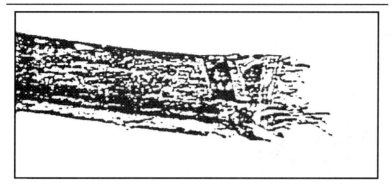

Figure 58.15 Failure due to tensile breaks

- *Action required:* Maintain proper drive tension and installation procedures. Provide guards to keep foreign material from coming in contact with the drive.
- *Cause of failure:* Excessive exposure to oil or grease causing the belt to become soft and swell, resulting in the bottom envelope seam splitting open (Figure 58.16).

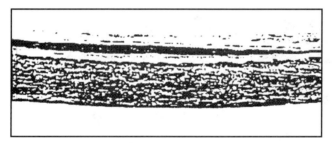

Figure 58.16 Failure due to excessive exposure to oil or grease

- *Action required:* Provide splashguards, do not over lubricate or clean belts with solvent.
- *Cause of failure:* Weathering or 'crazing' caused by the elements and aggravated by small pulleys (Figure 58.17).

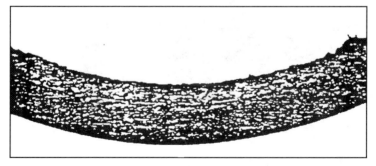

Figure 58.17 Failure due to a combination of weathering and small pulleys

(*continued overleaf*)

Table 58.6 (*continued*)

- *Action required:* Replace belt set and provide weather protection for the drive.
- *Cause of failure*: Cut bottom and sidewall indicates that the belt was pried over the pulley and damaged during installation (Figure 58.18).

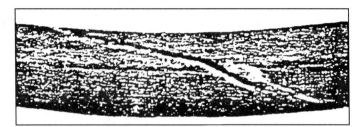

Figure 58.18 Failure due to installation damage

- *Action required:* Use the proper length of belt and move tensioning all the way 'in' when installing the belt.
- *Cause of failure*: Spin burn caused by a frozen or locked driven pulley (Figure 58.19).

Figure 58.19 Failure due to frozen or locked driven pulley

- *Action required:* Determine that the drive components turn freely and tighten the belt if necessary.
- *Cause of failure*: Worn pulley grooves allow the joined belt to ride too low thus cutting through to the top band (Figure 58.20).

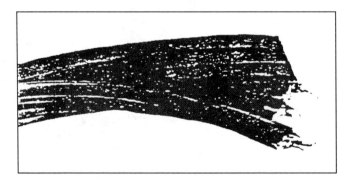

Figure 58.20 Failure due to worn pulley grooves

- *Action required:* Replace pulleys and maintain proper belt tension and pulley alignment.
- *Cause of failure*: Split on side at the belt pitch-line indicates use of a pulley with a substandard diameter (Figure 58.21).

Table 58.6 (*continued*)

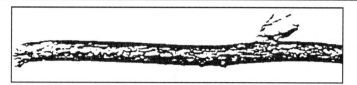

Figure 58.21 Failure due to use of pulley with substandard diameter

- *Action required:* Redesign drive to utilize proper size pulleys.
- *Cause of failure*: The load-carrying member has been broken by a shock load or damaged during operation (Figure 58.22).

Figure 58.22 Failure due to shock load or damage during operation

- *Action required:* Maintain proper tensioning and observe proper installation procedures.
- *Cause of failure*: Web fabrics wear caused by improper belt and pulley fit (Figure 58.23).

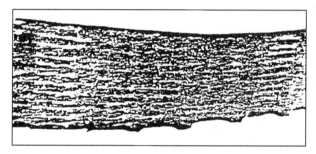

Figure 58.23 Failure due to improper belt and pulley fit

- *Action required:* Check out belt/pulley fit and replace worn or out-of-spec pulleys.
- *Cause of failure:* Flange wear on belt (Figure 58.24).

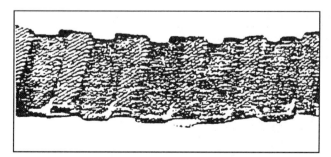

Figure 58.24 Failure due to flange wear

(*continued overleaf*)

Table 58.6 (*continued*)

- *Action required:* Adjust and maintain proper pulley alignment.
- *Cause of failure:* Tooth shear caused by belt overload condition from improper application or shock loads (Figure 58.25).

Figure 58.25 Failure due to belt overload

- *Action required:* Consult engineering manual for proper application and maintain proper belt tension.

on each belt. If there is proper tension, you can depress the belt an amount equal to the thickness of the belt for each 4 feet of center-to-center distance.

Keep pulleys aligned or excessive belt and pulley wear will occur. Shafts that are not parallel are very common and are the biggest contributor to early belt failure. In such cases, belts work harder on one side than on the other and wear faster. As a result, the entire belt set will need to be replaced. Misalignment in the pulley only is indicated by belt cover and pulley wear.

In addition to the need for pulleys to be aligned, pulley grooves must be smooth. Dust, oil and other foreign matter cause pitting and rust and 'dished-out' sidewalls ruin belts very quickly. If a shiny pulley bottom is showing, it indicates wear on the pulley groove, belt, or both. The biggest problem with badly worn grooves is that they cause the belt to ride lower than other belts on the same pulley. This results in 'differential driving,' meaning that the belts riding higher in their respective grooves are actually traveling faster than their lower-riding counterparts. This creates the situation where one belt on a pulley is working against the others.

58.6 Chain drives and sprockets

This section is an overview of flexible drive chains and sprockets. It provides the basic knowledge required to install, operate, and service this means of power transmission.

58.6.1 Chain drives

Like V-belts, chain drives are used to transmit power from a driver, such as an electric motor, to a driven unit.

Typically, chain drives are used in applications where space is limited or obstructions prevent direct coupling of machine-train components (e.g., timing chain in an engine).

Unlike V-belts, chain drives do not rely on friction to deliver power. As a result, transmission of power is positive (i.e., no slippage) and in many ways resembles the action of gears. A chain is constructed in such a way that it provides a connection between the teeth of the drive pinion and driven sprockets.

Operating characteristics

Since chain drives are designed to provide positive transmission of power from the pinion to the driven sprocket, there is little loss of efficiency. When properly installed, chain drives can approach 98 per cent efficiency. However, they are somewhat limited in speed and span.

Speed limitations

The dynamics of power transmission chain drives are not conducive to high-speed operation. Since proper installation requires some looseness in the chain, they tend to separate from the sprockets at higher speeds. In general, transmission chains are limited to a maximum sprocket speed of 3,600 revolutions per minute (rpm).

Span limitations

The span or center-to-center distance between sprockets is another limitation. Chain sag is difficult to control with long spans. Therefore, the total distance between sprockets must be limited to insure proper operation.

Horsepower

Properly designed chain drives can transmit almost unlimited power. As a result, this type of drive is suitable for low- to moderate-speed applications where the transmission of high force is needed.

Chain tension

Like V-belts, chains tend to stretch during prolonged operation. When this occurs, excessive looseness permits the chain to separate from the sprockets. Therefore, periodic inspection of chain tension is essential for proper operation of chain drives. Excessive looseness can be corrected by removing one or more links from the chain.

Types of chains

Types of chains that will be discussed in this section are: roller, leaf, and silent chains.

Roller chains   Probably the most common of all chains is the roller chain, which is found in applications such as:

• Timing drives in combustion engines
• Machinery drives
• Conveyor drives
• Fork lift trucks
• Most photocopiers
• Bicycles and motorcycles

Because of the versatility of roller chain drives, certain standards have been established. With these standards, interchangeability of chains is possible between one manufacturer and another. As long as chains are identifiable, they can be cross-referenced easily without any serious operational problems arising. Table 58.7 shows the standard roller chain American Standards Association (ASA) number by pitch length for single, double, and triple strand chains.

The roller chain uses a steel roller to engage the sprocket. The rolling action between the roller and the profile of the sprocket teeth causes the roller to rotate on a bushing, creating less friction than the traditional belt drives. There is some axial sliding action between the roller and bushing.

Table 58.7 Roller chain numbers by chain pitch

Chain pitch (inches)	ASA Number		
	Single strand	Double strand	Triple strand
1/4	25	25-2	25-3
3/8	35	35-2	35-3
1/2	40	40-2	40-3
1/2	41		
5/8	50	50-2	50-3
3/4	60	60-2	60-3
1	80	80-2	80-3
1 1/4	100	100-2	100-3
1 1/2	120	120-2	120-3
1 3/4	140	140-2	140-3
2	160	160-2	160-3
2 1/4	180	180-2	180-3
2 1/2	200	200-2	200-3
3	240	240-2	240-3

Source: American Standards Association

As with high-horsepower belt drive applications, chain drive designers often employ multiple chains. As illustrated in Figure 58.26, the chains in a multiple chain drive are all of the same size as a single roller chain (i.e., the pitch and roller sizes are all the same).

Roller chains are constructed by connecting two side plates together with pins that have bushings and rollers attached. The side plates are classified as either the pin link or roller link. The pin links are located outside the roller links and connect the roller links together. This provides an alternate pin and roller link combination.

As shown in Figure 58.27, the pins extend through the pin links. They are then usually riveted to one side plate, passed through the roller links, and then riveted to the outside of the other side plate. The roller link is formed by placing the rollers onto the bushings and, finally, pressing the roller plate links onto the bushings. This makes the roller link a solid unit.

Leaf chains   Leaf chains are usually not considered to be power transmission chains (see Figure 58.28). Instead, they are most often used in applications, such as hoisting

Figure 58.26 Single and quadruple roller chains

Figure 58.27 Types of pins

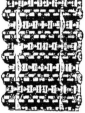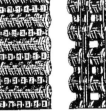

| Double guide | Non-flange silent chain | Duplex silent chain |

Figure 58.29 Three different silent chains

devices on fork lift trucks, where linear force must be transmitted.

They are designed to provide a flexible link between a power supply, such as a hydraulic cylinder, and another machine component. Leaf chains do not have any rollers or other sprocket-engaging device. They are designed with connector links located at each end of the chain that can be used to connect the drive and driven machine components.

Silent chains Silent chains (Figure 58.29) are similar to leaf chains in that neither incorporate rollers to facilitate mating to sprockets. However, a silent chain is designed for power transmission as opposed to linear force transmission. It has a tooth contour on its driving face that is designed to mate with sprockets and provide positive power transmission. This tooth contour is radically different than that of the roller-type chain in that it makes contact with both the leading *and* trailing edges

of the sprocket during operation. The close tolerance contact with the leading and trailing edges result in 'silent' operation as well as provide the ability for reverse driving of a sprocket.

Drives such as pumps, fans, blowers, and many other types of machinery commonly employ silent chains as their prime source of power transmission. Because of their unique design, silent chains provide the most positive means of traction. Because of this, they often are used in important applications such as timing drives on critical equipment.

Silent chains are similar to roller chains in that each has a master or connecting link to facilitate ease of installation and removal from a machine. They are usually made of high-carbon steel or heat-treated steel alloys. As a result, this type of chain provides trouble-free service for extended periods.

Chain length

Proper sizing of a chain is extremely important. The following procedure can be used to determine the proper chain length:

- Remove the chain from the sprockets and lay it on a smooth, horizontal surface or suspend it vertically.
- To remove slack from the chain measured in a horizontal position, use Table 58.8 and apply the

Figure 58.28 Typical leaf chain

Table 58.8 Chain measuring loads

Chain Number		Measuring load (pounds per strand)
Link-belt	ANSI	
RC 35	35	18
RC 40	40	31
RC 50	50	49
RC 60	60	70
RC 80	80	125
RC 100	100	195
RC 120	120	281
RC 140	140	383
RC 160	160	500
RC 180	180	633
RC 200	200	781
RC 240	240	1125

Source: Original source unknown

load recommended for the chain size being checked. To determine the measuring load for multiple-strand chains, multiply the load from Table 58.8 by the number of strands in the total chain width:

Measuring load = Pound/strand × Number of strands

For example, from Table 58.8, an RC120 chain should have a measuring load of 281 pounds per strand. Therefore, a four-strand chain should have the following measuring load:

Measuring load = 281 pound/strand × 4 strands
= 1124 pounds

- If the chain has to be measured on the sprockets, take up the slack on the chain and apply enough tension to keep it taut. See Figure 58.30 for guidance. Several measurements will have to be made until the entire chain is measured.
- Now you have the measured length. To calculate the proper length, L:

$$L = 2C + \frac{N}{2} + \frac{n}{2} + \left(\frac{N-n}{2\pi}\right) \times \frac{1}{C}$$

Where:

L = Proper chain length in number of pitches
C = Center distance between the sprockets in pitches

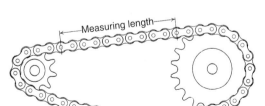

Figure 58.30 Illustration of recommended procedure for measuring chain length

N = Number of teeth in larger sprocket
n = Number of teeth in smaller sprocket

- The length obtained by this formula should be rounded upward to the nearest whole number. For a roller chain, the resultant number must be raised to the nearest *even* whole number or an offset link must be used to obtain the proper installed length. The proper chain length in inches is found by multiplying L by the pitch for the ASA type from Table 58.7.
- Compare the calculated proper chain length with the measurement taken. If the measurement exceeds the calculated proper chain length, replace the chain.

If a chain breaks or fails due to broken pins, sidebars, or rollers, temporary emergency repairs may be required to prevent a lengthy shutdown. Please note, however, the entire replacement of the chain is preferred for the following reasons:

1. If a section has broken due to fatigue, it is highly probable that other sections will have suffered similar fatigue and could be subject to early failure.
2. If the chain has been broken by a single high overload, other components besides those in the immediate region of the failure will usually be bent or severely weakened.

Chain installation and tensioning

Before installing a new chain, carefully check all sprocket teeth. If the teeth are worn to a hooked shape, the sprockets should be replaced to assure full-capacity performance and satisfactory life from the new chain. Because chain drives are not as flexible as belt drives, caution must be exercised during installation. The following steps should be observed during installation:

- Install the chain with a slight amount of sag on the slack side with the driving side in tension. Chain sag should be approximately $\frac{1}{4}$ inch for every 10 inches of sprocket centers.
- Install chain guards to ensure protection of personnel.
- Ensure that the chain is properly lubricated before start-up.
- Periodically inspect the chain drive for signs of wear and the condition of the oil.
- Follow plant safety guidelines and maintenance procedures when installing a chain drive.

Tensioning

Proper tension is essential when installing a new chain. A tight chain causes an additional load, which increases wear on chain joints, sprocket teeth, and shaft bearings. A slack chain produces vibration, which may result in excessive chain wear, noise, or shock loading.

Proper chain tension is obtained by adjusting the sag (catenary) in the unloaded span. For most horizontal and inclined drives, the chain should be installed with a depth of sag amounting to approximately 2 per cent of the sprocket centers (Table 58.9).

Table 58.9 Various sprocket centers and sag

Sprocket centers (inches)	Sag (inches)
20	0.50
30	0.63
40	0.88
60	1.25
80	1.63
100	2.00
125	2.50

Source: Original source unknown

To measure sag (Figure 58.31), pull the bottom span of the chain taut, allowing all of the excess chain to accumulate in the upper span. Place a straightedge on top of the sprockets and use a scale to measure sag.

For drives on vertical centers or those subject to conditions such as shock loads, rotation reversals, or dynamic braking, install the chain almost taut. It is essential to inspect such drives regularly for correct tension.

Alignment

Like all drive components, proper alignment is crucial to the smooth operation of a chain drive. If the sprockets are not parallel and in the same plane, the chain will lose contact with the sprocket teeth and drive-train failure will occur.

Sprocket alignment should be verified by placing a straightedge along the side face of both the driver and driven sprockets. The straightedge should fully contact the faces of both sprockets. If a gap is observed on either face, re-align the shafts.

Figure 58.31 Proper technique to measure sag (Ref. 2)

Lubrication

With the increased speed and horsepower capabilities of modern chain drives, the role of lubrication has increased in importance. The precision roller chain is actually a series of connected journal bearings and it is essential that lubrication minimizes the metal-to-metal contact of the pin/bushing joints of the chain. Many factors affect lubrication performance and chain life including heat, improper lubrication, windage, contamination, and oil viscosity.

Indications of improper lubrication

A lubricant-starved chain drive shows a brownish or rusty coloration around the joints and in the roller-bushing areas when the link is disassembled and the pin inspected. The normal highly polished surface of the pin will have deteriorated to a roughened, grooved, or galled surface that can eventually destroy the hardened surfaces of the chain parts and increase wear until the drive is completely destroyed. This also will be true if a lubricant that does not meet the chain's technical specifications is used.

Heat

Proper chain-drive lubrication increases drive life by dissipating frictional heat generated in the joint area. Heat varies according to the chain speed, horsepower transmitted, center distance, sprocket ratio, drive size, amount of lubricant, and viscosity. It generally ranges from surrounding temperature to 60 to 70°F above the ambient temperature. Normal chain drive temperatures should not exceed 180°F.

Windage

A chain drive can run through a sump of good lubricant and still be destroyed from lack of lubrication. This can occur at chain speeds approaching 3,000 feet per minute (fpm), which blows the lubricant out of the chain's path.

In high-horsepower, high-speed drives, it is necessary to use pressurized lubrication streams to ensure proper lubrication of the articulating components and to dissipate the heat generated. In this situation, the lubricant is sprayed onto the inside of the chain as it enters the sprockets, allowing centrifugal force to carry the lubricating material through the joints.

Contamination

Lubricants must be protected from dirt and moisture. A filtration system should be utilized to remove such abrasive particles and minimize wear on the drive chain.

Oil viscosity

A good grade of lubricant should be used to maximize the life of a chain drive due to wear. Lubricants containing anti-foam, anti-rust, and/or film-strength-enhancing additives may be useful. It is essential that the lubricant reach the sideplate-wearing surfaces and pin/bushing areas. Normally, heavy oils and greases are not recommended. The lubricant should be free-flowing

Table 58.10 Lubricant recommended by temperature

Temperature (°F)	Recommended lubricant
20–40	SAE 20
40–100	SAE 30
100–120	SAE 40
120–140	SAE 50

Source: Original source unknown.

at the prevailing temperature (see Table 58.10). The type of lubricant is dictated by the speed of the chain and the amount of power transmitted. The choice of lubrication method is determined by the drive itself.

58.6.2 Sprockets

There is a wide variety of commercially available sprockets. While they may vary in design, methods of manufacture, and materials of construction, they all have some common features. They will all have hardened teeth designed to mate with a specific type of chain, sufficient web strength to effectively transmit their rated horsepower or torsional forces, and a boss or hub that can be bored to the mating shaft's diameter.

Sprocket types

Sprockets can be either a solid-type body or cast with spokes. They also may be split into two halves to facilitate easy access during maintenance activities.

Fabricated Sprockets can be fabricated from machined parts such as hubs, webs, and rings. The actual material used in their construction is determined by the specific application. The teeth on most of the fabricated sprockets are hardened after the machining process to provide longer wear life.

Cast The teeth on cast iron sprockets are formed in a special pattern and chilled to create a hardened wearing surface at the tooth area.

Mounting configurations

A variety of mounting configurations are used to attach sprockets to their mating shafts. The most common has a tapered keyway that acts as a wedge when the key is driven in. No setscrews are provided with this type of key installation.

Hub diameters are larger when keys are used to ensure extra strength for power transmission purposes. In addition, some hubs will have tapered bushings instead of straight bores and key seats.

References

1. Oberg, E., Jones, F. D., Horton, H. L. and Ryffel, H. H., *Machinery's Handbook 25*, 25th edn, Industrial Press Inc., New York (1996).
2. Nelson, C. A., *Millwrights and Mechanics Guide*, Macmillan Publishing Company, New York (1986).
3. Mobley, R. K., *Maintenance Fundamentals*, Butterworth-Heinemann, Woburn, MA (1999).

59

Couplings and Clutches

R Keith Mobley

The Plant Performance Group

Contents

59.1 Introduction

This module discusses two important drive-train components: couplings and clutches. Each of these components is used to connect a driver (i.e., power source) shaft to the shaft of the driven unit. Such a connection allows torsional force to be converted into work in the driven unit. Keys and keyways, which are required to prevent slippage and to guarantee positive power with such connections, are also discussed.

Couplings

Couplings are mechanical devices used to connect the shaft of a driver (e.g., motor, turbine, etc.) to the shaft of a driven unit (e.g., fan, pump, etc.). The purpose of a coupling is to transmit rotary motion and/or torque on a continuous basis without slippage.

Clutches

Clutches are designed to transmit intermittent power to a driven unit. Unlike a normal coupling, which maintains constant power transmission between the driver and the driven unit, a clutch is designed to alternately engage and disengage. When engaged, the clutch transmits full torsional power to the driven unit. When disengaged, the clutch disconnects the driver from the driven unit.

59.2 Fundamentals of couplings

Couplings are designed to provide two functions: (1) to transmit torsional power between a power source and driven unit and (2) to absorb torsional variations in the drive train. They are not designed to correct misalignment between two shafts. While certain types of couplings provide some correction for slight misalignment, reliance on these devices to obtain alignment is not recommended.

59.2.1 Types

The sections to follow provide overviews of the more common coupling types: rigid and flexible. Also discussed are couplings used for special applications: floating-shaft (spacer) and fluid (hydraulic).

Rigid

A rigid coupling permits neither axial nor radial relative motion between the shafts of the driver and driven unit. When the two shafts are connected solidly and properly, they operate as a single shaft. A rigid coupling is primarily used for vertical applications, e.g., vertical pump. Types of rigid couplings discussed in this section are flanged, split, and compression.

Flanged couplings are used where there is free access to both shafts. Split couplings are used where access is limited on one side. Both flanged and split couplings require the use of keys and keyways. Compression couplings are used when it is not possible to use keys and keyways (e.g., brass shafts).

Flanged A flanged rigid coupling is comprised of two halves, one located on the end of the driver shaft and the other on the end of the driven shaft. These halves are bolted together to form a solid connection. To positively transmit torque, the coupling incorporates axially fitted keys and split circular key rings or dowels, which eliminate frictional dependency for transmission. The use of flanged couplings is restricted primarily to vertical pump shafts. A typical flanged rigid coupling is illustrated in Figure 59.1.

Split A split rigid coupling, also referred to as a clamp coupling, is basically a sleeve that is split horizontally along the shaft and held together with bolts. It is clamped over the adjoining ends of the driver and driven shafts, forming a solid connection. Clamp couplings are used primarily on vertical pump shafting. A typical split rigid coupling is illustrated in Figure 59.2. As with the flanged coupling, the split rigid coupling incorporates axially fitted keys and split circular key-rings to eliminate frictional dependency in the transmission of torque.

Compression A rigid compression coupling is comprised of three pieces: a compressible core and two encompassing coupling halves that apply force to the core. The core is comprised of a slotted bushing that has been machine bored to fit both ends of the shafts. It also has been taper machined on its external diameter from the center outward

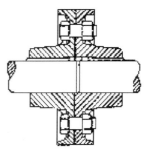

Figure 59.1 Typical flanged rigid coupling

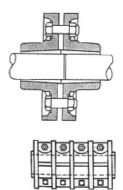

Figure 59.2 Typical split rigid coupling

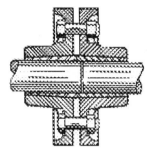

Figure 59.3 Typical compression rigid coupling

to both ends. The coupling halves are finish bored to fit this taper. When the coupling halves are bolted together, the core is compressed down on the shaft by the two halves and the resulting frictional grip transmits the torque without the use of keys. A typical compression coupling is illustrated in Figure 59.3.

Flexible

Flexible couplings, which are classified as mechanical flexing, material flexing, or combination, allow the coupled shafts to slide or move relative to each other. Although clearances are provided to permit movement within specified tolerance limits, flexible couplings are not designed to compensate for major misalignments. (Shafts must be aligned to less than 0.002 inches for proper operation.) Significant misalignment creates a whipping movement of the shaft, adds thrust to the shaft and bearings, causes axial vibrations, and leads to premature wear or failure of equipment.

Mechanical flexing

Mechanical-flexing couplings provide a flexible connection by permitting the coupling components to move or slide relative to each other. In order to permit such movement, clearance must be provided within specified limits. It is important to keep cross-loading on the connected shafts at a minimum. This is accomplished by providing adequate lubrication to reduce wear on the coupling components. The most popular of the mechanical-flexing type are the chain and gear couplings.

Chain

Chain couplings provide a good means of transmitting proportionately high torque at low speeds. Minor shaft misalignment is compensated for by means of clearances between the chain and sprocket teeth and the clearance that exists within the chain itself.

The design consists of two hubs with sprocket teeth connected by a chain of the single-roller, double-roller, or silent type.

Special purpose components may be specified when enhanced flexibility and reduced wear is required. Hardened sprocket teeth, special tooth design, and barrel shaped rollers are available for special needs. Light-duty drives

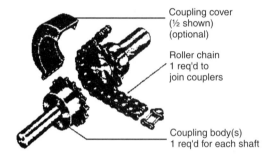

Coupling cover
(½ shown)
(optional)

Roller chain
1 req'd to
join couplers

Coupling body(s)
1 req'd for each shaft

Figure 59.4 Typical chain coupling

are sometimes supplied with non-metallic chains on which no lubrication should be used.

Gear

Gear couplings are capable of transmitting proportionately high torque at both high and low speeds. The most common type of gear coupling consists of two identical hubs with external gear teeth and a sleeve, or cover, with matching internal gear teeth. Torque is transmitted through the gear teeth, whereas the necessary sliding action and ability for slight adjustments in position comes from a certain freedom of action provided between the two sets of teeth.

Slight shaft misalignment is compensated for by the clearance between the matching gear teeth. However, any degree of misalignment decreases the useful life of the coupling and may cause damage to other machine-train components such as bearings. A typical example of a gear-tooth coupling is illustrated in Figure 59. 5.

Material-flexing

Material-flexing couplings incorporate elements that accommodate a certain amount of bending or flexing.

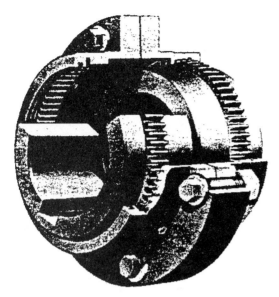

Figure 59.5 Typical gear-tooth coupling

Figure 59.6 Typical laminated disk-ring couplings

The material-flexing group includes laminated disk-ring, bellows, flexible shaft, diaphragm, and elastomeric couplings.

Various materials (e.g., metal, plastics, or rubber) are used to make the flexing elements in these couplings. The use of the couplings is governed by the operational fatigue limits of these materials. Practically all metals have fatigue limits that are predictable, therefore, they permit definite boundaries of operation to be established. Elastomers such as plastic or rubber, however, usually do not have a well-defined fatigue limit. Their service life is determined primarily by conditions of installation and operation.

Laminated disk-ring

The laminated disk-ring coupling consists of shaft hubs connected to a single flexible disk, or a series of disks, that allows axial movement. The laminated disk-ring coupling also reduces heat and axial vibration that can transmit between the driver and driven unit. Figure 59.6 illustrates some typical laminated disk-ring couplings.

Bellows

Bellows couplings consist of two shaft hubs connected to a flexible bellows. This design, which compensates for minor misalignment, is used at moderate rotational torque and shaft speed. This type of coupling provides flexibility to compensate for axial movement and misalignment caused by thermal expansion of the equipment components. Figure 59.7 illustrates a typical bellows coupling.

Flexible shaft or spring

Flexible shaft or spring couplings are generally used in small equipment applications that do not experience high

Figure 59.7 Typical bellows coupling

torque loads. Figure 59.8 illustrates a typical flexible shaft coupling.

Diaphragm

Diaphragm couplings provide torsional stiffness while allowing flexibility in axial movement. Typical construction consists of shaft hub flanges and a diaphragm spool, which provides the connection between the driver and driven unit. The diaphragm spool normally consists of a center shaft fastened to the inner diameter of a diaphragm on each end of the spool shaft. The shaft hub flanges are fastened to the outer diameter of the diaphragms to complete the mechanical connection. A typical diaphragm coupling is illustrated in Figure 59.9.

Elastomeric

Elastomeric couplings consist of two hubs connected by an elastomeric element. The couplings fall into two basic categories, one with the element placed in shear and the other with its element placed in compression. The coupling compensates for minor misalignments because of the flexing capability of the elastomer. These couplings

Figure 59.8 Typical flexible shaft coupling

Figure 59.9 Typical diaphragm coupling

Figure 59.11 Typical metallic-grid couplings

are usually applied in light- or medium-duty applications running at moderate speeds.

With the shear-type coupling, the elastomeric element may be clamped or bonded in place, or fitted securely to the hubs. The compression-type couplings may be fitted with projecting pins, bolts, or lugs to connect the components. Polyurethane, rubber, neoprene, or cloth and fiber materials are used in the manufacture of these elements.

Although elastomeric couplings are practically maintenance free, it is good practice to periodically inspect the condition of the elastomer and the alignment of the equipment. If the element shows signs of defects or wear, it should be replaced and the equipment realigned to the manufacturer's specifications. Typical elastomeric couplings are illustrated in Figure 59.10.

Combination (metallic-grid)

The metallic-grid coupling is an example of a combination of mechanical-flexing and material-flexing type couplings. Typical metallic-grid couplings are illustrated in Figure 59.11.

Figure 59.10 Typical elastomeric couplings

The metallic-grid coupling is a compact unit capable of transmitting high torque at moderate speeds. The construction of the coupling consists of two flanged hubs, each with specially grooved slots cut axially on the outer edges of the hub flanges. The flanges are connected by means of a serpentine-shaped spring grid that fits into the grooved slots. The flexibility of this grid provides torsional resilience.

Special application couplings

Two special application couplings are discussed in this section: (1) floating-shaft or spacer coupling and (2) hydraulic or fluid coupling.

Floating-shaft or spacer coupling   Regular flexible couplings connect the driver and driven shafts with relatively close ends and are suitable for limited misalignment. However, allowances sometimes have to be made to accommodate greater misalignment or when the ends of the driver and driven shafts have to be separated by a considerable distance.

Such is the case, for example, with end-suction pump designs in which the power unit of the pump assembly is removed for maintenance by being axially moved toward the driver. If neither the pump nor the driver can be readily removed, they should be separated sufficiently to permit withdrawal of the pump's power unit. An easily removable flexible coupling of sufficient length (i.e., floating-shaft or spacer coupling) is required for this type of maintenance. Examples of couplings for this type of application are shown in Figure 59.12.

In addition to the maintenance application described above, this coupling (also referred to as extension or spacer sleeve coupling) is commonly used where equipment is subject to thermal expansion and possible misalignment because of high process temperatures. The purpose of this

Figure 59.12 Typical floating-shaft or spacer couplings

Figure 59.13 Typical floating-shaft or spacer couplings for high-temperature applications

type of coupling is to prevent harmful misalignment with minimum separation of the driver and driven shaft ends. An example of a typical floating-shaft coupling for this application is shown in Figure 59.13.

The floating-shaft coupling consists of two support elements connected by a shaft. Manufacturers use various approaches in their designs for these couplings. For example, each of the two support elements may be of the single-engagement type, may consist of a flexible half-coupling on one end and a rigid half-coupling on the other end, or may be completely flexible with some piloting or guiding supports.

Floating-shaft gear couplings usually consist of a standard coupling with a two-piece sleeve. The sleeve halves are bolted to rigid flanges to form two single-flex couplings. These, in turn, are connected by an intermediate shaft that permits the transmission of power between widely separated drive components.

Hydraulic or fluid   Hydraulic couplings provide a soft start with gradual acceleration and limited maximum torque for fixed operating speeds. Hydraulic couplings are typically used in applications that undergo torsional shock from sudden changes in equipment loads (e.g., compressors). Figure 59.14 is an illustration of a typical hydraulic coupling.

59.2.2 Selection

Periodically, worn or broken couplings must be replaced. One of the most important steps in performing this maintenance procedure is to ensure that the correct replacement

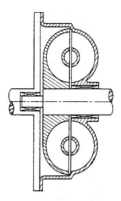

Figure 59.14 Typical hydraulic coupling

parts are used. After having determined the cause of failure, it is crucial to identify the correct type and size of coupling needed. Even if practically identical in appearance to the original, a part still may not be an adequate replacement.

The manufacturer's specification number usually provides the information needed for part selection. If the part is not in stock, a cross-reference guide will provide the information needed to verify ratings and to identify a coupling that meets the same requirements as the original.

A criterion that must be considered in part selection includes: equipment type, mode of operation, and cost. Each of these criteria are discussed in the sections to follow.

Equipment type

Coupling selection should be application specific and, therefore, it is important to consider the type of equipment, which it connects. For example, demanding applications such as variable, high-torque machine trains require couplings that are specifically designed to absorb radical changes in speed and torque (e.g., metallic-grid). Less demanding applications such as run-out table rolls can generally get by with elastomeric couplings. Table 59.1 lists the coupling type commonly used in a particular application.

Mode of operation

Coupling selection is highly dependent on the mode of operation, which includes torsional characteristics, speed, and the operating envelope.

Torsional characteristics

Torque requirements are a primary concern during the selection process. In all applications where variable or high torque is transmitted from the driver to the driven unit, a flexible coupling rated for the maximum torque requirement must be used. Rigid couplings are not designed to absorb variations in torque and should not be used.

Table 59.1 Coupling application overview

Application	Coupling* Selection Recommendation
Limited misalignment compensation	
Variable, high-torque machine trains operating at moderate speeds	Metallic-grid combination couplings
Run-out table rolls	Elastomeric flexible couplings
Vertical pump shafting	Flanged rigid couplings, split rigid or clamp couplings
Keys and keyways not appropriate (e.g., brass shafts)	Rigid compression couplings
Transmission of proportionately high torque at low speeds	Chain couplings (mechanical-flexing)
Transmission of proportionately high torque at both high and low speeds	Gear couplings (mechanical-flexing)
Allowance for axial movement and reduction of heat and axial vibration	Laminated disk-ring couplings (material-flexing)
Moderate rotational torque and shaft speed	Bellows couplings (material-flexing)
Small equipment that does not experience high torque loads	Flexible shaft or spring couplings (material-flexing)
Torsional stiffness while allowing flexibility in axial movement	Diaphragm material-flexing couplings
Light- or medium-duty applications running at moderate speeds	Elastomeric couplings (material-flexing)
Gradual acceleration and limited maximum torque for fixed operating speeds (e.g., compressors).	Hydraulic or fluid couplings
Variable or high torque and/or speed transmission	Flexible couplings rated for the maximum torque requirement
Greater misalignment compensation	
1) Maintenance requiring considerable distance between the driver and driven shaft ends. 2) Misalignment results from expansion due to high process temperatures.	Floating-shaft or spacer couplings

*See Table 59.6 for an application overview for clutches.
Note: Rigid couplings are not designed to absorb variations in torque and speed and should not be used in such applications. Maximum in-service coupling speed should be at least 15% below the maximum coupling speed rating.
Source: Integrated Systems, Inc.

Speed

Two speed-related factors should be considered as part of the selection process: maximum speed and speed variation.

Maximum speed When selecting coupling type and size, the maximum speed rating must be considered, which can be determined from the vendor's catalog. The maximum in-service speed of a coupling should be well below (at least 15 per cent) the maximum speed rating. The 15 per cent margin provides a service factor that should be sufficient to prevent coupling damage or catastrophic failure.

Speed variation Variation in speed equates to a corresponding variation in torque. Most variable-speed applications require some type of flexible coupling capable of absorbing these torsional variations.

Operating envelope

The operating envelope defines the physical requirements, dimensions, and type of coupling needed in a specific application. The envelope information should include: shaft sizes, orientation of shafts, required horsepower, full-range of operating torque, speed ramp rates, and any other data that would directly or indirectly affect the coupling.

Cost

Coupling cost should not be the deciding factor in the selection process, although it will certainly play a part in

it. Although higher performance couplings may be more expensive, they actually may be the cost effective solution in a particular application. Selecting the most appropriate coupling for an application not only extends coupling life, but also improves the overall performance of the machine train and its reliability.

59.2.3 Installation

Couplings must be installed properly if they are to operate satisfactorily. This section discusses shaft and coupling preparation, coupling installation, and alignment.

Shaft preparation

A careful inspection of both shaft ends must be made to ensure that no burrs, nicks, or scratches are present that will damage the hubs. Potentially damaging conditions must be corrected before coupling installation. Emery cloth should be used to remove any burrs, scratches, or oxidation that may be present. A light film of oil should be applied to the shafts prior to installation. Keys and keyways (discussed in Section 59.3) also should be checked for similar defects and to ensure that the keys fit properly. Properly sized key stock must be used with all keyways; do not use bar stock or other material.

Coupling preparation

The coupling must be disassembled and inspected prior to installation. The location and position of each component should be noted so that it can be reinstalled in the correct order. When old couplings are removed for inspection, bolts and bolt holes should be numbered so that they can be installed in the same location when the coupling is returned to service.

Any defects, such as burrs, should be corrected before the coupling is installed. Defects on the mating parts of the coupling can cause interference between the bore and shaft, preventing proper operation of the coupling.

Coupling installation

Once the inspection shows the coupling parts to be free of defects, the hubs can be mounted on their respective shafts. If it is necessary to heat the hubs to achieve the proper interference fit, an oil or water bath should be used. Spot-heating using a flame or torch should be avoided as it causes distortion and may adversely affect the hubs.

Care must be exercised during installation of a new coupling or the reassembly of an existing unit. Keys and keyways should be coated with a sealing compound that is resistant to the lubricant used in the coupling. Seals should be inspected to ensure that they are pliable and in good condition. They must be installed properly in the sleeve with the lip in good contact with the hub. Sleeve flange gaskets must be whole, in good condition, clean, and free of nicks or cracks. Lubrication plugs must be cleaned before being installed and must fit tightly.

The specific installation procedure is dependent on the type and mounting configuration of the coupling. However, common elements of all coupling installations include: spacing, bolting, lubrication, and the use of matching parts. The sections to follow discuss these installation elements.

Spacing

Spacing between the mating parts of the coupling must be within manufacturer's tolerances. For example, an elastomeric coupling must have a specific distance between the coupling faces. This distance determines the position of the rubber boot that provides transmission of power from the driver to the driven machine component. If this distance is not exact, the elastomer will attempt to return to its relaxed position, inducing excessive axial movement in both shafts.

Bolting

Couplings are designed to use a specific type of bolt. Coupling bolts have a hardened cylindrical body sized to match the assembled coupling width. Hardened bolts are required because standard bolts do not have the tensile strength to absorb the torsional and shearing loads in coupling applications and may fail resulting in coupling failure and machine-train damage.

Lubrication

Most couplings require lubrication and care must be taken to ensure that the proper type and quantity is used during the installation process. Inadequate or improper lubrication reduces coupling reliability and reduces its useful life. In addition, improper lubrication can cause serious damage to the machine train. For example, when a gear-type coupling is over-filled with grease, the coupling will lock. In most cases, its locked position will increase the vibration level and induce an abnormal loading on the bearings of both the driver and driven unit, resulting in bearing failure.

Matching parts

Couplings are designed for a specific range of applications and proper performance depends on the total design of the coupling system. As a result, it is generally not a good practice to mix coupling types. Note, however, that it is common practice in some steel industry applications to use coupling halves from two different types of couplings. For example, a rigid coupling half is sometimes mated to a flexible coupling half, creating a hybrid. While this approach may provide short-term power transmission, it can result in an increase in the number, frequency, and severity of machine-train problems.

Coupling alignment

The last step in the installation process is verifying coupling and shaft alignment. With the exception of special application couplings such as spindles and jackshafts, all couplings must be aligned within relatively close tolerances (i.e., 0.001–0.002 inch).

59.2.4 Lubrication and maintenance

Couplings require regular lubrication and maintenance to ensure optimum trouble-free service life. When proper maintenance is not conducted, premature coupling failure and/or damage to machine-train components such as bearings can be expected.

Determining cause of failure

When a coupling failure occurs, it is important to determine the cause of failure. Failure may result from a coupling defect, an external condition, or workmanship during installation.

Most faults are attributed to poorly machined surfaces causing out-of-specification tolerances, although defective material failures also occur. Inadequate material hardness and poor strength factors contribute to many premature failures. Other common causes are improper coupling selection, improper installation, and/or excessive misalignment.

Lubrication requirements

Lubrication requirements vary depending on application and coupling type. Because rigid couplings do not require lubrication, this section discusses lubrication requirements for mechanical-flexing, material-flexing, and combination flexible couplings only.

Mechanical-flexing couplings   It is important to follow the manufacturer's instructions for lubricating mechanical-flexing couplings, which must be lubricated internally.

Lubricant seals must be in good condition and properly fitted into place. Coupling covers contain the lubricant and prevent contaminants from entering the coupling interior. The covers are designed in two configurations, split either horizontally or vertically. Holes are provided in the covers to allow lubricant to be added without coupling disassembly.

Gear couplings are one type of mechanical-flexing coupling and there are several ways to lubricate them: grease pack, oil fill, oil collect, and continuous oil flow. Either grease or oil can be used at speeds of 3,600 rpm to 6,000 rpm. Oil is normally used as the lubricant in couplings operating over 6,000 rpm. Grease and oil lubricated units have end gaskets and seals, which are used to contain the lubricant and seal out the entry of contaminants. The sleeves have lubrication holes, which permit flushing and re-lubrication without disturbing the sleeve gasket or seals.

Material-flexing couplings Material-flexing couplings are designed to be lubrication free.

Combination couplings Combination (metallic-grid) couplings are lubricated in the same manner as mechanical-flexing couplings.

Periodic inspections

It is important to perform periodic inspections of all mechanical equipment and systems that incorporate rotating parts, including couplings and clutches.

Mechanical-flexing couplings To maintain coupling reliability, mechanical-flexing couplings require periodic inspections on a time- or condition-based frequency established by the history of the equipment's coupling life or a schedule established by the predictive maintenance engineer. Items to be included in an inspection are listed below. If any of these items or conditions are discovered, the coupling should be evaluated to determine its remaining operational life or should be repaired/replaced.

- Inspect lubricant for traces of metal (indicating component wear).
- Visually inspect coupling mechanical components (roller chains and gear teeth, and grid members) for wear and/or fatigue.
- Inspect seals to ensure they are pliable and in good condition. They must be installed properly in the sleeve with the lip in good contact with the hub.
- Sleeve flange gaskets must be whole, in good condition, clean, and free of nicks or cracks.
- Lubrication plugs must be clean (to prevent the introduction of contaminants to the lubricant and machine surfaces) before being installed and must be torqued to the manufacturer's specifications.
- Set-screws and retainers must be in place and tightened to manufacturer's specifications.
- Inspect shaft hubs, keyways, and keys for cracks, breaks, and physical damage.
- Under operating conditions, perform thermographic scans to determine temperature differences on the

coupling (indicates misalignment and/or uneven mechanical forces).

Material-flexing couplings Although designed to be lubrication-free, material-flexing couplings also require periodic inspection and maintenance. This is necessary to ensure that the coupling components are within acceptable specification limits. Periodic inspections for the following conditions are required to maintain coupling reliability. If any of these conditions are found, the coupling should be evaluated to determine its remaining operational life or repaired/replaced.

- Inspect flexing element for signs of wear or fatigue (cracks, element dust or particles).
- Set-screws and retainers must be in place and tightened to manufacturer's specifications.
- Inspect shaft hubs, keyways, and keys for cracks, breaks, and physical damage.
- Under operating conditions, perform thermographic scans for temperature differences on the coupling, which indicates misalignment and/or uneven mechanical forces.

Combination couplings Mechanical components (e.g., grid members) should be visually inspected for wear and/ or fatigue. In addition, the grid members on metallic-grid couplings should be replaced if any signs of wear are observed.

Rigid couplings The mechanical components of rigid couplings (e.g., hubs, bolts, compression sleeves and halves, keyways, and keys) should be visually inspected for cracks, breaks, physical damage, wear, and/or fatigue. Any component having any of these conditions should be replaced.

59.3 Keys, keyways and keyseats

A key is a piece of material, usually metal, placed in machined slots or grooves cut into two axially oriented parts in order to mechanically lock them together. For example, keys are used in making the coupling connection between the shaft of a driver and a hub or flange on that shaft. Any rotating element whose shaft incorporates such a keyed connection is referred to as a keyed-shaft rotor. Keys provide a positive means for transmitting torque between the shaft and coupling hub when a key is properly fitted in the axial groove.

The groove into which a key is fitted is referred to as a keyseat when referring to shafts and a keyway when referring to hubs. Keyseating is the actual machine operation of producing keyseats. Keyways are normally made on a keyseater or by a broach. Keyseats are normally made with a rotary or end mill cutter.

Figure 59.15 is an example of a keyed shaft that shows the key size versus the shaft diameter. Because of standardization and interchangeability, keys are generally proportioned with relation to shaft diameter instead of torsional load.

Top view

Top view

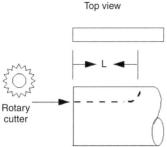

Rotary
cutter

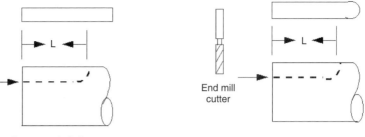

End mill
cutter

Figure 59.15 Keyed shaft: key size versus shaft diameter

The effective key length, 'L' is that portion of key having full bearing on hub and shaft. Note that the curved portion of the keyseat made with a rotary cutter does not provide full key bearing, so 'L' does not include this distance. The use of an end mill cutter results in a square-ended keyseat.

Figure 59.16 shows various key shapes: square ends, one square end and one round end, rounded ends, plain taper, and gib head taper. The majority of keys are square in cross-section, which are preferred through $4\frac{1}{2}$-inch diameter shafts. For bores over $4\frac{1}{2}$ inches and thin wall section of hubs, the rectangular (flat) key is used.

The ends are either square, rounded or gib-head. The gib-head is usually used with taper keys. If special considerations dictate the use of a keyway in the hub shallower than the preferred square key, it is recommended that the standard rectangular (flat) key be used.

Hub bores are usually straight, although for some special applications taper bores are sometimes specified. For smaller diameters, bores are designed for clearance fits and a set screw is used over the key. The major advantage of a clearance fit is that hubs can be easily assembled and disassembled. For larger diameters, the bores are designed for interference fits without setscrews. For rapid-reversing applications, interference fits are required.

The sections to follow discuss determining keyway depth and width, keyway manufacturing tolerances, key stress calculations, and shaft stress calculations.

59.3.1 Determining keyway depth and width

The formula given below and Figure 59.17, Table 59.1 (square keys), and Table 59.2 (flat keys) illustrate how the depth and width of standard square and flat keys and keyways for shafts and hubs are determined.

$$Y = \frac{D - \sqrt{D^2 - W^2}}{2}$$

where:

C = Allowance or clearance for key, inches
D = Nominal shaft or bore diameter, inches

Top Views   **Side Views**

Square Ends

Plain Taper

Square and Round

Rounded Ends

Gib Head Taper

Figure 59.16 Key shapes

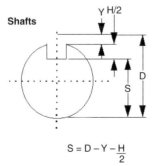

Shafts

$$S = D - Y - \frac{H}{2}$$

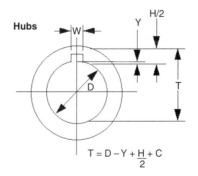

Hubs

$$T = D - Y + \frac{H}{2} + C$$

Figure 59.17 Shaft and hub dimensions

Table 59.2 Standard square keys and keyways (inches)*

Diameter of holes (inclusive)	Keyways		Key stock
	Width	Depth	
$\frac{5}{16}$ to $\frac{7}{16}$	$\frac{3}{32}$	$\frac{3}{64}$	$\frac{3}{32} \times \frac{3}{32}$
$\frac{1}{2}$ to $\frac{9}{16}$	$\frac{1}{8}$	$\frac{1}{16}$	$\frac{1}{8} \times \frac{1}{8}$
$\frac{5}{8}$ to $\frac{7}{8}$	$\frac{3}{16}$	$\frac{3}{32}$	$\frac{3}{16} \times \frac{3}{16}$
$1\frac{5}{16}$ to $1\frac{1}{4}$	$\frac{1}{4}$	$\frac{1}{8}$	$\frac{1}{4} \times \frac{1}{4}$
$1\frac{5}{16}$ to $1\frac{3}{8}$	$\frac{5}{16}$	$\frac{5}{32}$	$\frac{5}{16} \times \frac{5}{16}$
$1\frac{7}{16}$ to $1\frac{3}{4}$	$\frac{3}{8}$	$\frac{3}{16}$	$\frac{3}{8} \times \frac{3}{8}$
$1\frac{13}{16}$ to $2\frac{1}{4}$	$\frac{1}{2}$	$\frac{1}{4}$	$\frac{1}{2} \times \frac{1}{2}$
$2\frac{5}{16}$ to $2\frac{3}{4}$	$\frac{5}{8}$	$\frac{5}{16}$	$\frac{5}{8} \times \frac{5}{8}$
$2\frac{13}{16}$ to $3\frac{1}{4}$	$\frac{3}{4}$	$\frac{3}{8}$	$\frac{3}{4} \times \frac{3}{4}$
$3\frac{5}{16}$ to $3\frac{3}{4}$	$\frac{7}{8}$	$\frac{7}{16}$	$\frac{7}{8} \times \frac{7}{8}$
$3\frac{13}{16}$ to $4\frac{1}{2}$	1	$\frac{1}{2}$	1×1

*Square keys are normally used through shaft diameter $4\frac{1}{2}''$; larger shafts normally use flat keys.
Source: The Falk Corporation

Table 59.3 Standard flat keys and keyways (inches)

Diameter of holes (inclusive)	Keyways		Key stock
	Width	Depth	
$\frac{1}{2}$ to $\frac{9}{16}$	$\frac{1}{8}$	$\frac{3}{64}$	$\frac{1}{8} \times \frac{1}{32}$
$\frac{5}{8}$ to $\frac{7}{8}$	$\frac{3}{16}$	$\frac{1}{16}$	$\frac{3}{16} \times \frac{1}{8}$
$1\frac{5}{16}$ to $1\frac{1}{4}$	$\frac{1}{4}$	$\frac{3}{32}$	$\frac{1}{4} \times \frac{3}{16}$
$1\frac{5}{16}$ to $1\frac{3}{8}$	$\frac{5}{16}$	$\frac{1}{8}$	$\frac{5}{16} \times \frac{1}{4}$
$1\frac{7}{16}$ to $1\frac{3}{4}$	$\frac{3}{8}$	$\frac{1}{8}$	$\frac{3}{8} \times \frac{1}{4}$
$1\frac{13}{16}$ to $2\frac{1}{4}$	$\frac{1}{2}$	$\frac{3}{16}$	$\frac{1}{2} \times \frac{3}{8}$
$2\frac{5}{16}$ to $2\frac{3}{4}$	$\frac{5}{8}$	$\frac{7}{32}$	$\frac{5}{8} \times \frac{7}{16}$
$2\frac{13}{16}$ to $3\frac{1}{4}$	$\frac{3}{4}$	$\frac{1}{4}$	$\frac{3}{4} \times \frac{1}{2}$
$3\frac{5}{16}$ to $3\frac{3}{4}$	$\frac{7}{8}$	$\frac{5}{16}$	$\frac{7}{8} \times \frac{5}{8}$
$3\frac{13}{16}$ to $4\frac{1}{2}$	1	$\frac{3}{8}$	$1 \times \frac{3}{4}$
$4\frac{9}{16}$ to $5\frac{1}{2}$	$1\frac{1}{4}$	$\frac{7}{16}$	$1\frac{1}{4} \times \frac{7}{8}$
$5\frac{9}{16}$ to $6\frac{1}{2}$	$1\frac{1}{2}$	$\frac{1}{2}$	$1\frac{1}{2} \times 1$
$6\frac{9}{16}$ to $7\frac{1}{2}$	$1\frac{3}{4}$	$\frac{5}{8}$	$1\frac{3}{4} \times 1\frac{1}{4}$
$7\frac{9}{16}$ to 9	2	$\frac{3}{4}$	$2 \times 1\frac{1}{2}$
$9\frac{1}{16}$ to 11	$2\frac{1}{2}$	$\frac{7}{8}$	$2\frac{1}{2} \times 1\frac{3}{4}$
$11\frac{1}{16}$ to 13	3	1	3×2
$13\frac{1}{16}$ to 15	$3\frac{1}{2}$	$1\frac{1}{4}$	$3\frac{1}{2} \times 2\frac{1}{2}$
$15\frac{1}{16}$ to 18	4	$1\frac{1}{2}$	4×3
$18\frac{1}{16}$ to 22	5	$1\frac{3}{4}$	$5 \times 3\frac{1}{2}$
$22\frac{1}{16}$ to 26	6	4	
$26\frac{1}{16}$ to 30	7	5	

Source: The Falk Corporation

H = Nominal key height, inches
W = Nominal key width, inches
Y = Chordal height, inches

Note: Tables shown above are prepared for manufacturing use. Dimensions given are for standard shafts and keyways.

59.3.2 Keyway manufacturing tolerances

Keyway manufacturing tolerances (illustrated in Figure 59.18) are referred to as offset (centrality) and lead (cross axis). Offset or centrality is referred to as Dimension 'N'; lead or cross axis is referred to as Dimension 'J'. Both must be kept within permissible tolerances, usually 0.002 inches.

59.3.3 Key stress calculations

Calculations for shear and compressive key stresses are based on the following assumptions:

1. The force acts at the radius of the shaft.
2. The force is uniformly distributed along the key length.
3. None of the tangential load is carried by the frictional fit between shaft and bore.

The shear and compressive stresses in a key are calculated using the following equations (see Figure 59.19):

$$S_s = \frac{2T}{(d) \times (w) \times (L)} \qquad S_c = \frac{2T}{(d) \times (h_1) \times (L)}$$

where:

d = Shaft diameter, inches (use average diameter for taper shafts)
h_1 = Height of key in the shaft or hub that bears against the keyway, inches. Should equal h_2 for square keys. For designs where unequal portions of the key are in the hub or shaft, h_1 is the minimum portion.
Hp = Power, horsepower
L = Effective length of key, inches
RPM = Revolutions per minute
S_s = Shear stress, psi
S_c = Compressive stress, psi

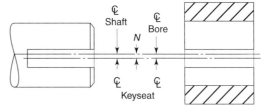

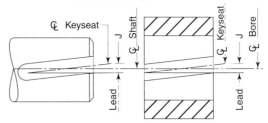

Figure 59.18 Manufacturing tolerances: offset and lead

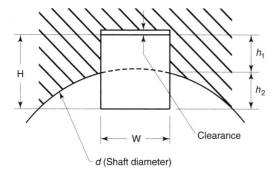

Figure 59.19 Measurements used in calculating shear and compressive key stress

$$T = \text{Shaft torque, lb-in or } \frac{\text{Hp} \times 63000}{\text{RPM}}$$

$w = $ Key width, inches

Key material is usually AISI 1018 or AISI 1045. Table 59.4 provides the allowable stresses for these materials.

Example: Select a key for the following conditions: 300 Hp at 600 RPM; 3″ diameter shaft, $\frac{3}{4}'' \times \frac{3}{4}''$ key, 4″ key engagement length.

$$T = \text{Torque} = \frac{Hp \times 63,000}{RPM} = \frac{300 \times 63,000}{600} = 31,500 \text{ in-lbs}$$

$$S_s = \frac{2T}{d \times w \times L} = \frac{2 \times 31,500}{3 \times \frac{3}{4} \times 4} = 7,000 \text{ psi}$$

$$S_c = \frac{2T}{d \times h1 \times L} = \frac{2 \times 31,500}{3 \times \frac{3}{8} \times 4} = 14,000 \text{ psi}$$

The AISI 1018 key can be used since it is within allowable stresses listed in Table 59.4 (Allowable $S_s = 7,500$, Allowable $S_c = 15,000$). Note: If shaft had been $2\frac{3}{4}''$ diameter (4″ hub), the key would be $\frac{5}{8}'' \times \frac{5}{8}''$, $S_s = 9,200$ psi, $S_c = 18,400$ psi, and a heat-treated key of AISI 1045 would have been required (Allowable $S_s = 15,000$, Allowable $S_c = 30,000$).

59.3.4 Shaft stress calculations

Torsional stresses are developed when power is transmitted through shafts. In addition, the tooth loads of gears mounted on shafts create bending stresses. Shaft design, therefore, is based on safe limits of torsion and bending.

Table 59.4 Allowable stresses for AISI 1018 and AISI 1045

Material	Heat treatment	Allowable stresses (psi)	
		Shear	Compressive
AISI 1018	None	7,500	15,000
AISI 1045	255–300 Bhn	15,000	30,000

Source: The Falk Corporation

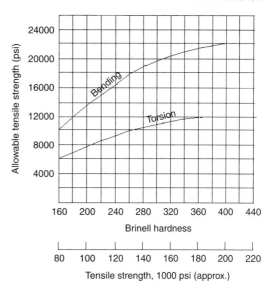

Figure 59.20 Allowable stress as a function of Brinell hardness

To determine minimum shaft diameter in inches:

$$\text{Minimum shaft diameter} = \sqrt[3]{\frac{Hp \times 321000}{RPM \times \text{Allowable stress}}}$$

Example:

Hp = 300
RPM = 30
Material = 225 Brinell

From Figure 59.20 at 225 Brinell, Allowable Torsion = 8000 psi

$$\text{Minimum shaft diameter} = \sqrt[3]{\frac{300 \times 321000}{30 \times 8000}}$$

$$= \sqrt[3]{402} = 7.38 \text{ inches}$$

From Table 59.5, note that the cube of $7\frac{1}{4}''$ is 381, which is too small (i.e., <402) for this example. The cube of $7\frac{1}{2}''$ is 422, which is large enough.

To determine shaft stress, psi:

$$\text{Shaft stress} = \frac{Hp \times 321,000}{RPM \times d^3}$$

Example: Given $7\frac{1}{2}''$ shaft for 300 Hp at 30 RPM

$$\text{Shaft stress} = \frac{300 \times 321,000}{30 \times (7\frac{1}{2})^3} = 7,600 \text{ psi}$$

Note: The $7\frac{1}{4}''$ diameter shaft would be stressed to 8420 psi.

59.4 Fundamentals of clutches

Many machine systems require clutches, devices that engage and disengage a driven unit from its rotating power

Table 59.5 Shaft diameters (inches) and their cubes (cubic inches)

D	D_3	D	D_3	D	D_3
1	1.00	5	125.0	9	729
$1\frac{1}{4}$	1.95	$5\frac{1}{4}$	145	$9\frac{1}{2}$	857
$1\frac{1}{2}$	3.38	$5\frac{1}{2}$	166.4	10	1000
$1\frac{3}{4}$	5.36	$5\frac{3}{4}$	190.1	$10\frac{1}{2}$	1157
2	8.00	6	216	11	1331
$2\frac{1}{4}$	11.39	$6\frac{1}{4}$	244	$11\frac{1}{2}$	1520
$2\frac{1}{2}$	15.63	$6\frac{1}{2}$	275	12	1728
$2\frac{3}{4}$	20.80	$6\frac{3}{4}$	308	$12\frac{1}{2}$	1953
3	27.00	7	343	13	2197
$3\frac{1}{4}$	34.33	$7\frac{1}{4}$	381	14	2744
$3\frac{1}{2}$	42.88	$7\frac{1}{2}$	422	15	3375
$3\frac{3}{4}$	52.73	$7\frac{3}{4}$	465	16	4096
4	64.00	8	512	17	4913
$4\frac{1}{4}$	76.77	$8\frac{1}{4}$	562	18	5832
$4\frac{1}{2}$	91.13	$8\frac{1}{2}$	614	19	6859
$4\frac{3}{4}$	107.2	$8\frac{3}{4}$	670	20	8000

Source: The Falk Corporation

source. While this is the primary role of clutches in machine trains, they also can be designed to accommodate changes in axial length of the attached shaft that result from thermal expansion or contraction.

A clutch acts like an on-demand spacer coupling. In the disengaged position there is space between the two shafts. When the clutch is engaged, the mechanism fills the void to reconnect the shaft.

59.4.1 Types of clutches

There is an almost infinite variety of clutches, which are designed to cover a wide range of applications. The more common clutch types are classified as positive and friction. Less common types are hydraulic and electromagnetic. Table 59.6 lists the clutches commonly used in a particular application.

Positive

Clutches in this classification transmit power by means of two mating hubs, one mounted on the driver and the

Table 59.6 Clutch Application Overview

Application	Clutch Selection Recommendation
Positive intermittent power transmission, transmitting torque in one direction, disengaging in the opposite direction	One-way positive clutches
Positive power transmission in both directions of rotation	Two-way positive clutches

Source: Integrated Computer Systems, Inc.

other mounted on the driven shaft. The interlocking hub teeth guarantee positive (i.e., no slippage) transmission of power due to the direct mechanical link they create between the drive and driven shaft.

The primary advantage of this type of clutch is its ability to transmit full torsional force without any possibility of slip. Its major disadvantage is that the two shafts are instantaneously coupled when the clutch engages. This results in abrupt starts, which may cause excessive torsional shock loads that damage drive-train components. Figure 59.21 shows a positive clutch.

There are numerous positive clutch and tooth configuration designs, but they generally can be broken down into two sub-classes: one- and two-way.

One-way   A one-way or flywheel clutch is a device that transmits torque in one direction, but disengages in the opposite direction. The typical design of flywheel clutches incorporates a tooth profile that engages in one direction only. If rotation of the drive shaft is reversed, the drive hub will freewheel and cannot engage the driven hub.

The most familiar examples of one-way clutches are the bicycle freewheel, which enables the cyclist to stop pedaling without stopping the bicycle, and the non-reversing ratchet used on vertical turbine pumps.

Two-way   The two-way clutch uses square or tapered teeth to provide positive mating between its drive and driven halves. This tooth profile provides positive power transmission in both directions of rotation.

Friction

The primary advantage of friction clutches is their ability to minimize engaging torque. Figure 59.22 shows an example of this type of clutch, which transmits torsional power from a driver to a driven unit as a result of friction between two or more mating parts. By relying on friction to transmit power, the transition from disengaged

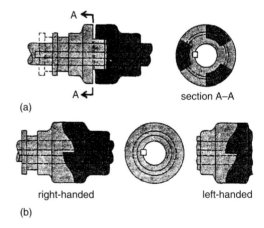

(a)

section A–A

right-handed   left-handed

(b)

Figure 59.21 Positive clutch

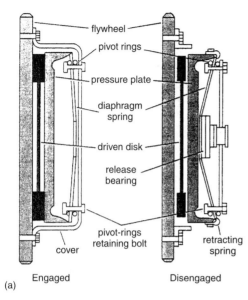

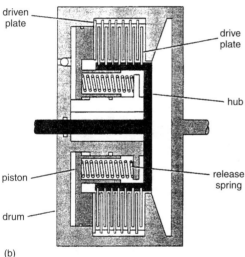

Figure 59.22 Friction clutch

to engaged is more gradual than with positive clutches, which incorporate mating hubs with interlocking teeth.

Although not a design that guarantees positive power transmission, a satisfactorily performing friction clutch will transmit the maximum torque output of the driver to the driven shaft without slip when fully engaged. The automobile clutch is the most common example, transmitting the maximum motor torque to the wheels.

The primary disadvantage of friction clutches is the regular maintenance and adjustment that are required: mating parts will wear due to friction ultimately resulting in slippage, springs will lose memory, and other parts will need replacement or adjustment. Although slippage caused by

wear due to friction is a primary mode of failure, proper maintenance can extend the lifetime of the clutch.

59.4.2 Maintenance

Like any other type of equipment, clutches require regular inspection and maintenance. The sections to follow describe the general maintenance practices that are required to keep positive and friction clutches functioning properly.

Positive clutches

Most reliability problems associated with positive clutches fall into two categories: mechanical damage and fatigue failures.

Mechanical damage   In applications where cycling occurs, damage of the tooth profile on the mating hubs is a common problem with the positive clutch. Since the hubs are designed to make near instantaneous engagement, the teeth in both the male and female halves are prone to mechanical damage. This damage is caused by the repeated insertion of the rotating set of teeth into the stationary set.

Regular inspection and lubrication of the teeth is an absolute requirement when this type of clutch is used. If tooth profile damage is observed, it should be repaired before being returned to service.

Fatigue failure   While tooth profile failure does occur in positive clutches, the more common failure is the actuator system. This system provides the means of engaging and disengaging the mating teeth that provide power transmission. Actuators range from compression springs to hydraulic or pneumatic devices. Repeated cycling results in fatigue failures in the actuator system components. Regular inspection and maintenance per the manufacturer's procedures are necessary to ensure proper operation and to achieve system-design lifetime.

Friction   Friction clutches tend to be more complex in design than positive clutches. Their complexity, coupled with reliance on friction for power transmission, demands constant care and maintenance. As a result, they require regular adjustment and good preventive maintenance to ensure trouble-free operation.

Clutch manufacturers generally provide complete inspection and maintenance instructions in their operations and maintenance manual. These instructions should be carefully followed.

References

1. Neale, M. J. *Drives and Seals – A Tribology Handbook*, Society of Automotive Engineers, Inc. (1994).
2. Parker, S. P. ed., *Encyclopedia of Engineering*, 2nd edn, McGraw Hill Publishing, New York (1993).
3. Sax, N. I. and Lewis, Sr, J. R., *Hawley's Condensed Chemical Dictionary*, 11th edn, Van Nostrand Reinhold Co, New York (1987).
4. Block, H. P. and Geitner, F. K., *Introduction to Machinery Reliability Assessment*, Van Nostrand Reinhold, New York (1990).

5. Block, H. P. and Geitner, F. K., *Machinery Failure Analysis and Troubleshooting*, Gulf Publishing Company Book Division, Houston (1986).

6. Oberg, E., Jones, F. D., Horton, H. L. and Ryffel, H. H., *Machinery's Handbook 25*, 25th edn, Industrial Press Inc., New York (1996).

7. Baumeister T. ed., *Marks' Standard Handbook for Mechanical Engineers*, 8th edn, McGraw-Hill, New York (1978).

8. McMaster-Carr Supply Co. *Catalog #99* (1993).

9. Nelson, C. A., *Millwrights and Mechanics Guide*, Macmillan Publishing Company, New York (1986).

60

Bearings

R Keith Mobley
The Plant Performance Group

Contents

60.1 Introduction

A bearing is a machine element that supports a part, such as a shaft, that rotates, slides, or oscillates in or on it. There are two broad classifications of bearings: plain and rolling element (also called anti-friction). Plain bearings are based on sliding motion made possible through the use of a lubricant. Anti-friction bearings are based on rolling motion, which is made possible by balls or other types of rollers. In modern rotor systems operating at relatively high speeds and loads, the proper selection and design of the bearings and bearing-support structure are key factors affecting system life.

Types of movement

The type of bearing used in a particular application is determined by the nature of the relative movement and other application constraints. Movement can be grouped into the following categories: rotation about a point, rotation about a line, translation along a line, rotation in a plane, and translation in a plane. These movements can be either continuous or oscillating.

Although many bearings perform more than one function, they can generally be classified based on types of movement and there are three major classifications of both plain and rolling element bearings: radial, thrust, and guide. Radial bearings support loads that act radially and at right angles to the shaft centerline. These loads may be visualized as radiating into or away from a center point like the spokes on a bicycle wheel. Thrust bearings support or resist loads that act axially. These may be described as endwise loads that act parallel to the centerline towards the ends of the shaft. This type of bearing prevents lengthwise or axial motion of a rotating shaft. Guide bearings support and align members having sliding or reciprocating motion. This type of bearing guides a machine element in its lengthwise motion, usually without rotation of the element.

Table 60.1 gives examples of bearings that are suitable for continuous movement; Table 60.2 shows bearings that are appropriate for oscillatory movement only. For the bearings that allow movements in addition to the one listed, the effect on machine design is described in the column, 'Effect of the other degrees of freedom.' Table 60.3 compares the characteristics, advantages, and disadvantages of plain and rolling element bearings.

Table 60.1 Bearing selection guide (continuous movement)

Bearing type	Accurate radial location	Axial load capacity as well	Low starting torque	Silent running	Standard parts available	Simple Lubrication
Plain, externally pressurized	1	No (Need separate thrust bearing)	1	1	No	4 (Need special system)
Plain, fluid film	3	No (Need separate thrust bearing)	2	1	Some	2 (Usually requires circulation system)
Plain, porous metal (oil impregnated)	2	Some	2	1	Yes	1
Plain, rubbing (non-metallic)	4	Some, in most instances	4	3	Some	1
Rolling	2	Yes, in most instances	1	Usually satisfactory	Yes	2 (When grease lubricated)

Rating: 1 –Excellent, 2 –Good, 3 –Fair, 4 –Poor

Table 60.2 Bearing selection guide (oscillatory movement)

Bearing type	High temperature	Low temperature	Low friction	Wet/ humid	Dirt/ dust	External vibration
Knife edge pivots	2	2	1	2 Watch corrosion	2	4
Plain, porous metal (oil impregnated)	4 Lubricant oxidizes	3 Friction can be high	2	2	Sealing essential	2
Plain, rubbing	2 Up to temperature limit of material	1	2 With PTFE	2 Shaft must not corrode	2 Sealing helps	1
Rolling element	Consult manufacturer above 150°C	2	1	2 With seals	Sealing essential	4
Rubber bushings	4	4	Elastically stiff	1	1	1
Strip flexures	2	1	1	2 Watch corrosion	1	1

Rating: 1 –Excellent, 2 –Good, 3 –Fair, 4 –Poor

Table 60.3 Comparison of plain and rolling element bearings

Rolling element	Plain
Assembly on crankshaft is virtually impossible, except with very short or built-up crankshafts	Assembly on crankshaft is no problem as split bearings can be used
Cost relatively high	Cost relatively low
Hardness of shaft unimportant	Hardness of shaft important with harder bearings
Heavier than plain bearings	Lighter than rolling element bearings
Housing requirement not critical	Rigidity and clamping most important housing requirement
Less rigid than plain bearings	More rigid than rolling element bearings
Life limited by material fatigue	Life not generally limited by material fatigue
Lower friction results in lower power consumption	Higher friction causes more power consumption
Lubrication easy to accomplish, the required flow is low except at high speed	Lubrication pressure feed critically important, required flow is large, susceptible to damage by contaminants and interrupted lubricant flow
Noisy operation	Quiet operation
Poor tolerance of shaft deflection	Moderate tolerance of shaft deflection
Poor tolerance of hard dirt particles	Moderate tolerance of dirt particles, depending on hardness of bearing
Requires more overall space: Length: smaller than plain Diameter: larger than plain	Requires less overall space: Length: larger than rolling element Diameter: smaller than rolling element
Running friction: Very low at low speeds May be high at high speeds	Running Friction: Higher at low speeds Moderate at usual crank speeds
Smaller radial clearance than plain	Larger radial clearance than rolling element

Source: Integrated Systems, Inc.

About a point (rotational) Continuous movement about a point is rotation, a motion that requires repeated use of accurate surfaces. If the motion is oscillatory rather than continuous, some additional arrangements must be made in which the geometric layout prevents continuous rotation.

About a line (rotational) Continuous movement about a line is also referred to as rotation, and the same comments apply as for movement about a point.

Along a line (translational) Movement along a line is referred to as translation. One surface is generally long

and continuous and the moving component is usually supported on a fluid film or rolling contact in order to achieve an acceptable wear rate. If the translational movement is reciprocation, the application makes repeated use of accurate surfaces and a variety of economical bearing mechanisms are available.

In a plane (rotational/translational) If the movement in a plane is rotational or both rotational and oscillatory, the same comments apply as for movement about a point.

If the movement in a plane is translational or both translational and oscillatory, the same comments apply as for movement along a line.

60.2 Commonly used bearing types

As mentioned before, major bearing classifications are plain and rolling element. These types of bearings are discussed in the sections to follow. Table 60.4 is a bearings characteristics summary.

Table 60.5 is a selection guide for bearings operating with continuous rotation and special environmental conditions. Table 60.6 is a selection guide for bearings operating with continuous rotation and special performance requirements. Table 60.7 is a selection guide for oscillating movement and special environment or performance requirements.

60.2.1 Plain bearings

All plain bearings also are referred to as fluid-film bearings. In addition, radial plain bearings also are commonly referred to as journal bearings. Plain bearings are available in a wide variety of types or styles and may be self-contained units or built into a machine assembly. Table 60.8 is a selection guide for radial and thrust plain bearings.

Plain bearings are dependent on maintaining an adequate lubricant film to prevent the bearing and shaft surfaces from coming into contact, which is necessary to prevent premature bearing failure. However, this is difficult to achieve and some contact usually occurs during operation. Material selection plays a critical role in the amount of friction and the resulting seizure and wear that occurs with surface contact. Refer to Section 60.3 for a discussion of common bearing materials. Note that fluid-film bearings do not have the ability to carry the full load of the rotor assembly at any speed and must have turning gear to support the rotor's weight at low speeds.

Thrust or fixed

Thrust plain bearings consist of fixed shaft shoulders or collars that rest against flat bearing rings. The lubrication state may be semi-fluid and friction is relatively high. In multi-collar thrust bearings, allowable service pressures are considerably lower because of the difficulty in distributing the load evenly between several collars. However, thrust ring performance can be improved by introducing tapered grooves. Figure 60.1 shows a mounting half-section for a vertical thrust bearing.

Table 60.4 Bearing characteristics summary

Bearing type	Description
Plain	
Lobed	See Radial, elliptical.
Radial or journal	
Cylindrical	Gas lubricated, low-speed applications.
Elliptical	Oil lubricated, gear and turbine applications, stiffer and somewhat more stable bearing.
Four-axial grooved	Oil lubricated, higher-speed applications than cylindrical.
Partial arc	Not a bearing type, but a theoretical component of grooved and lobed bearing configurations.
Tilting pad	High-speed applications where hydrodynamic instability and misalignment are common problems.
Thrust	Semi-fluid lubrication state, relatively high friction, lower service pressures with multi-collar version, used at low speeds.
Rolling element	Radial and axial loads, moderate- to high-speed applications.
Ball	Higher speed and lighter load applications than roller bearings.
Single-row Radial non-filling slot	Also referred to as Conrad or deep-groove bearing. Sustains combined radial and thrust loads, or thrust loads alone, in either direction, even at high speeds. Not self-aligning.
Radial filling slot	Handles heavier loads than non-filling slot.
Angular contact radial thrust	Radial loads combined with thrust loads alone. Axial deflection must be limited.
Ball-thrust	Very high thrust loads in one direction only, no radial loading, cannot be operated at high speeds.
Double-row	Heavy radial with minimal bearing deflection and light thrust loads.
Double-roll, self-aligning	Moderate radial and limited thrust loads.
Roller	Handles heavier loads and shock better than ball bearings, but are more limited in speed than ball bearings.
Cylindrical	Heavy radial loads, fairly high speeds, can allow free axial shaft movement.
Needle-type cylindrical or barrel	Does not normally support thrust loads, used in space-limited applications, angular mounting of rolls in double-row version tolerates combined axial and thrust loads.
Spherical	High radial and moderate-to-heavy thrust loads, usually comes in double-row mounting that is inherently self-aligning.
Tapered	Heavy radial and thrust loads. Can be preloaded for maximum system rigidity.

Radial or journal

Plain radial, or journal, bearings also are referred to as sleeve or Babbit bearings. The most common type is the full journal bearing, which has 360° contact with its mating journal. The partial journal bearing has less than 180° contact and is used when the load direction is constant. The sections to follow describe the major types of fluid-film journal bearings: plain cylindrical, four-axial groove, elliptical, partial arc, and tilting-pad.

Plain cylindrical   The plain cylindrical journal bearing (Figure 60.2) is the simplest of all journal bearing types. The performance characteristics of cylindrical bearings are well established and extensive design information is available. Practically, use of the unmodified cylindrical bearing is generally limited to gas-lubricated bearings and low-speed machinery.

Four-axial groove bearing   To make the plain cylindrical bearing practical for oil or other liquid lubricants, it is necessary to modify it by the addition of grooves or holes through which the lubricant can be introduced. Sometimes, a single circumferential groove in the middle of the bearing is used. In other cases, one or more axial grooves are provided.

The four-axial groove bearing is the most commonly used oil-lubricated sleeve bearing. The oil is supplied at a nominal gage pressure that ensures an adequate oil flow and some cooling capability. Figure 60.3 illustrates this type of bearing.

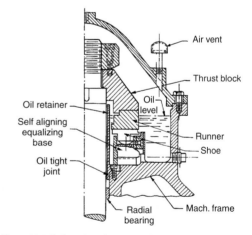

Figure 60.1 Half section of mounting for vertical thrust bearing

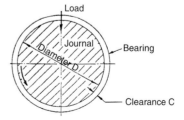

Figure 60.2 Plain cylindrical bearing

Table 60.5 Bearing selection guide for special environmental conditions (continuous rotation)

Bearing type	High temp.	Low temp.	Vacuum	Wet/humid	Dirt/dust	External vibration
Plain, externally pressurized	1 (With gas lubrication)	2	No (Affected by lubricant feed)	2	2 (1 when gas lubricated)	1
Plain, porous metal (oil impregnated)	4 (lubricant oxidizes)	3 (May have high starting torque)	Possible with special lubricant	2	Seals essential	2
Plain, rubbing (non-metallic)	2 (Up to temp. limit of material)	2	2	2 (Shaft must not corrode)	2 (Seals help)	2
Plain, fluid film	2 (Up to temp. limit of lubricant)	2 (May have high starting torque)	Possible with special lubricant	2	2 (With seals and filtration	2
Rolling	Consult manufacturer above 150 °C	2	3 (With special lubricant)	3 (With seals)	Sealing essential	3 (Consult manufacturers)
Things to watch with all	Effect of thermal expansion on fits	Effect of thermal expansion of fits		Corrosion		Fretting

Rating: 1–Excellent, 2–Good, 3–Fair, 4–Poor

Table 60.6 Bearing selection guide for particular performance requirements (continuous rotation)

Bearing type	Accurate radial location	Axial load capacity as well	Low starting torque	Silent running	Standard parts Available	Simple lubrication
Plain, externally pressurized	1	No (Need separate thrust bearing	1	1	No	4 (Need special system)
Plain, fluid film	3	No (Need separate thrust bearing	2	1	Some	2 (Usually requires circulation system)
Plain, porous metal (oil impregnated)	2	Some	2	1	Yes	1
Plain, rubbing (non-metallic)	4	Some in most instances	4	3	Some	1
Rolling	2	Yes in most instances	1	Usually satisfactory	Yes	2 (When grease lubricated)

Rating: 1–Excellent/very good, 2–Good, 3–Fair, 4–Poor

Table 60.7 Bearing selection guide for special environments or performance (oscillating movement)

Bearing type	High temp.	Low temp.	Low friction	Wet/humid	Dirt/dust	External vibration
Knife edge pivots	2	2	1	2 (Watch corrosion)	2	4
Plain, porous metal (oil impregnated)	4 (Lubricant oxidizes)	3 (Friction can be high)	2	2	Sealing essential	2
Plain, rubbing	2 (Up to temp. limit of material)	1	2 (With PTFE)	2 (Shaft must not corrode)	2 (Sealing helps)	1
Rolling	Consult manufacturer above 150 °C	2	1	2 (With seals)	Sealing essential	4
Rubber brushes	4	4	Elastically stiff	1	1	1
Strip flexures	2	2	1	2 (Watch corrosion)	1	1

Rating: 1–Excellent, 2–Good, 3–Fair, 4–Poor

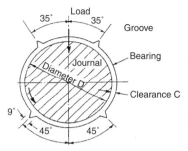

Figure 60.3 Four-axial groove bearing

Elliptical bearing The elliptical bearing is oil-lubricated and typically is used in gear and turbine applications. It is classified as a lobed bearing in contrast to a grooved bearing. Where the grooved bearing consists of a number of partial arcs with a common center, the lobed bearing is made up of partial arcs whose centers do not coincide. The elliptical bearing consists of two partial arcs where the bottom arc has its center a distance above the bearing center. This arrangement has the effect of preloading the bearing, where the journal center eccentricity with respect to the loaded arc is increased and never becomes zero. This results in the bearing being stiffened, somewhat improving its stability. An elliptical bearing is shown in Figure 60.4.

Partial-arc bearings A partial-arc bearing is not a separate type of bearing. Instead, it refers to a variation of previously discussed bearings (e.g., grooved and lobed bearings) that incorporates partial arcs. It is necessary to use partial-arc bearing data to incorporate partial arcs in a variety of grooved and lobed bearing configurations. In all cases, the lubricant is a liquid and the bearing film is laminar. Figure 60.5 illustrates a typical partial-arc bearing.

Tilting-pad bearings Tilting-pad bearings are widely used in high-speed applications where hydrodynamic instability and misalignment are common problems. This bearing consists of a number of shoes mounted on pivots, with each shoe being a partial-arc bearing. The shoes adjust and follow the motions of the journal, ensuring inherent stability if the inertia of the shoes does not interfere with the adjustment ability of the bearing. The load direction may either pass between the two bottom shoes

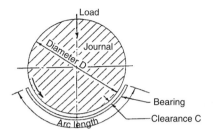

Figure 60.5 Partial arc bearing

or it may pass through the pivot of the bottom shoe. The lubricant is incompressible (i.e., liquid) and the lubricant film is laminar. Figure 60.6 illustrates a tilting-pad bearing.

60.3 Rolling element or anti-friction bearings

Rolling element anti-friction bearings are one of the most common types used in machinery. Anti-friction bearings are based on rolling motion as opposed to the sliding motion of plain bearings. The use of rolling elements between rotating and stationary surfaces reduces the friction to a fraction of that resulting with the use of plain bearings.

Use of rolling element bearings is determined by many factors, including load, speed, misalignment sensitivity, space limitations, and desire for precise shaft positioning. They support both radial and axial loads and are generally used in moderate- to high-speed applications.

Unlike fluid-film plain bearings (discussed in Section 60.2.1), rolling element bearings have the added ability to carry the full load of the rotor assembly at any speed. Where fluid-film bearings must have turning gear to support the rotor's weight at low speeds, rolling element bearings can maintain the proper shaft centerline through the entire speed range of the machine.

60.3.1 Grade classifications

Rolling element bearings are available in either commercial- or precision-grade classifications.

Most commercial-grade bearings are made to nonspecific standards and are not manufactured to the same precise standards as precision-grade bearings. This limits

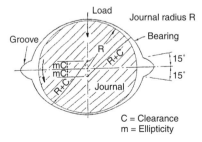

Figure 60.4 Elliptical bearing

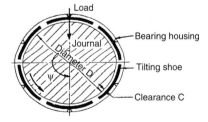

Figure 60.6 Tilting-pad bearing

Table 60.8 Plain bearings selection guide (a) Journal bearings

Characteristics	Direct lined	Insert liners
Accuracy	Dependent upon facilities and skill available	Precision components
Quality (consistency)	Doubtful	Consistent
Cost	Initial cost may be lower	Initial cost may be higher
Ease of repair	Difficult and costly	Easily done by replacement
Condition upon extensive use	Likely to be weak in fatigue	Ability to sustain higher peak loads
Materials used	Limited to white metals	Extensive range available

(b) Thrust Bearings

Characteristics	Flanged journal bearings	Separate thrust washer
Cost	Costly to manufacture	Much lower initial cost
Replacement	Involves whole journal/ thrust component	Easily replaced without moving journal bearing
Materials used	Thrust face materials limited in larger sizes	Extensive range available
Benefits	Aids assembly on a production line	Aligns itself with the housing

the speeds at which they can operate efficiently and given brand bearings may or may not be interchangeable.

Precision bearings are used extensively on many machines such as pumps, air compressors, gear drives, electric motors, and gas turbines. The shape of the rolling elements determines the use of the bearing in machinery. Because of standardization in bearing envelope dimensions, precision bearings were once considered to be interchangeable, even if manufactured by different companies. It has been discovered, however, that interchanging bearings is a major cause of machinery failure and should be done with extreme caution.

60.3.2 Rolling element types

There are two major classifications of rolling elements: ball and roller. Ball bearings function on point contact and are suited for higher speeds and lighter loads than roller bearings. Roller element bearings function on line contact and generally are more expensive than ball bearings, except for the larger sizes. Roller bearings carry heavy loads and handle shock more satisfactorily than ball bearings, but are more limited in speed. Figure 60.7 provides general guidelines to determine if a ball or roller bearing should be selected. This figure is based on a rated life of 30,000 hours.

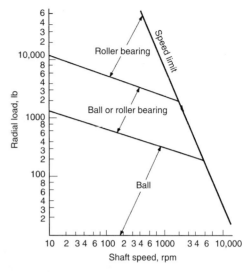

Figure 60.7 Guide to selecting ball or roller bearings

Although there are many types of rolling elements, each bearing design is based on a series of hardened rolling elements sandwiched between hardened inner and outer rings. The rings provide continuous tracks or races for the rollers or balls to roll in. Each ball or roller is separated from its neighbor by a separator cage or retainer, which properly spaces the rolling elements around the track and guides them through the load zone. Bearing size is usually given in terms of boundary dimensions: outside diameter, bore, and width.

Ball

Common functional groupings of ball bearings are radial, thrust, and angular-contact bearings. Radial bearings carry a load in a direction perpendicular to the axis of rotation. Thrust bearings carry only thrust loads, a force parallel to the axis of rotation tending to cause endwise motion of the shaft. Angular-contact bearings support combined radial and thrust loads. These loads are illustrated in Figure 60.8. Another common classification of ball bearings is single row (also referred to as Conrad or deep-groove bearing) and double row.

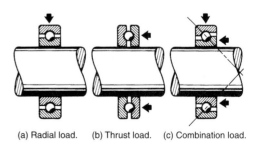

(a) Radial load. (b) Thrust load. (c) Combination load.

Figure 60.8 Three principal types of ball bearing loads

Single-row

Types of single-row ball bearings are: radial non-filling slot bearings, radial filling slot bearings, angular contact bearings, and ball thrust bearings.

Radial, non-filling slot bearings   This ball bearing is often referred to as the Conrad-type or deep-groove bearing and is the most widely used of all ball bearings (and probably of all anti-friction bearings). It is available in many variations, with single or double shields or seals. They sustain combined radial and thrust loads, or thrust loads alone, in either direction – even at extremely high speeds. This bearing is not designed to be self-aligning, therefore it is imperative that the shaft and the housing bore be accurately aligned.

Figure 60.10 labels the parts of the Conrad anti-friction ball bearing. This design is widely used and is versatile because the deep-grooved raceways permit the rotating balls to rapidly adjust to radial and thrust loadings, or a combination of these loadings.

Radial, filling slot bearings   The geometry of this ball bearing is similar to the Conrad bearing, except for the filling slot. This slot allows more balls in the complement and thus can carry heavier radial loads. The bearing is assembled with as many balls that fit in the gap created by eccentrically displacing the inner ring. The balls are evenly spaced by a slight spreading of the rings and heat expansion of the outer ring. However, because of the filling slot, the thrust capacity in both directions is reduced. In combination with radial loads, this bearing

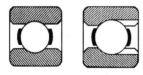

Figure 60.9 Single-row radial, non-filling slot bearing

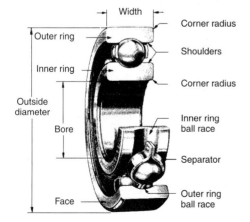

Figure 60.10 Conrad anti-friction ball bearing parts

design accomodates thrust of less than 60 per cent of the radial load.

Angular contact radial thrust bearings   This ball bearing is designed to support radial loads combined with thrust loads, or heavy thrust loads (depending on the contact-angle magnitude). The outer ring is designed with one shoulder higher than the other, which allows it to accommodate thrust loads. The shoulder on the other side of the ring is just high enough to prevent the bearing from separating. This type of bearing is used for pure thrust load in one direction and is applied either in opposed pairs (duplex), or one at each end of the shaft. They can be mounted either face-to-face or back-to-back and in tandem for constant thrust in one direction. This bearing is designed for combination loads where the thrust component is greater than the capacity of single-row deep-groove bearings. Axial deflection must be confined to very close tolerances.

Ball-thrust bearings   The ball-thrust bearing supports very high thrust loads in one direction only, but supports no radial loading. To operate successfully, this type of bearing must be at least moderately thrust-loaded at all times. It should not be operated at high speeds, since centrifugal force causes excessive loading of the outer edges of the races.

Double-row

Double-row ball bearings accommodate heavy radial and light thrust loads without increasing the outer diameter of the bearing. However, it is approximately 60 to 80 per cent wider than a comparable single-row bearing. The double-row bearing incorporates a filling slot, which requires the thrust load to be light. Figure 60.11 shows a double-row type ball bearing.

This unit is, in effect, two single-row angular contact bearings built as a unit with the internal fit between balls and raceway fixed during assembly. As a result, fit and internal stiffness are not dependent upon mounting methods. These bearings usually have a known amount of

Figure 60.11 Double row-type ball bearing

The labels in Figure 60.10 read:

Width

Outer ring

Inner ring

Outside diameter

Bore

Face

Corner radius

Shoulders

Corner radius

Inner ring ball race

Separator

Outer ring ball race

internal preload, or compression, built in for maximum resistance to deflection under combined loads with thrust from either direction. As a result of this compression prior to external loading, the bearings are very effective for radial loads where bearing deflection must be minimized.

Another double-row ball bearing is the internal self-aligning type, which is shown in Figure 60.12. It compensates for angular misalignment, which can be caused by errors in mounting, shaft deflection, misalignment, etc. This bearing supports moderate radial loads and limited thrust loads.

Roller

As with plain and ball bearings, roller bearings also may be classified by their ability to support radial, thrust, and combination loads. Note that combination load-supporting roller bearings are not called angular-contact bearings as they are with ball bearings. For example, the taper-roller bearing is a combination load-carrying bearing by virtue of the shape of its rollers.

Figure 60.13 shows the different types of roller elements used in these bearings. Roller elements are classified as cylindrical, barrel, spherical, and tapered. Note that barrel rollers are called needle rollers when less than $\frac{1}{4}$-inch in diameter and have a relatively high ratio of length to diameter.

Cylindrical

Cylindrical bearings have solid or helically wound hollow cylindrically shaped rollers, which have an approximate length:diameter ratio ranging from $1:1$ to $1:3$. They normally are used for heavy radial loads beyond the capacities of comparably sized radial ball bearings.

Cylindrical bearings are especially useful for free axial movement of the shaft. The free ring may have a restraining flange to provide some restraint to endwise movement in one direction. Another configuration comes without a flange, which allows the bearing rings to be displaced axially.

Either the rolls or the roller path on the races may be slightly crowned to prevent edge loading under slight shaft misalignment. Low friction makes this bearing type suitable for fairly high speeds. Figure 60.14 shows a typical cylindrical roller bearing.

Figure 60.15 shows separable inner-ring cylindrical roller bearings. Figure 60.16 shows separable inner-ring cylindrical roller bearings with a different inner ring.

The roller assembly in Figure 60.15 is located in the outer ring with retaining rings. The inner ring can be omitted and the roller operated on hardened ground shaft surfaces.

The style in Figure 60.16 is similar to the one in Figure 60.15, except the rib on the inner ring is different. This prohibits the outer ring from moving in a direction toward the rib.

Figure 60.17 shows separable inner ring-type cylindrical roller bearings with elimination of a retainer ring on one side.

The style shown in Figure 60.17 is similar to the two previous styles except for the elimination of a retainer

Figure 60.12 Double-row internal self-aligning bearing

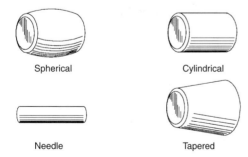

Spherical   Cylindrical

Needle   Tapered

Figure 60.13 Types of roller elements

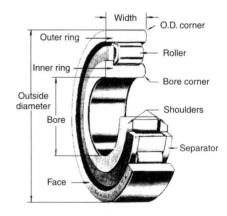

Figure 60.14 Cylindrical roller bearing

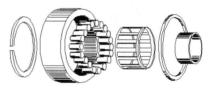

Figure 60.15 Separable inner ring-type cylindrical roller bearings

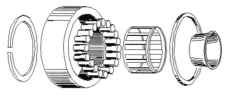

Figure 60.16 Separable inner ring-type cylindrical roller bearings with different inner ring

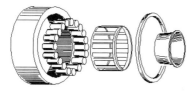

Figure 60.17 Separable inner ring-type cylindrical roller bearings with elimination of a retainer ring on one side

ring on one side. It can carry small thrust loads in only one direction.

Needle-type cylindrical or barrel

Needle-type cylindrical bearings (Figure 60.18) incorporate rollers that are symmetrical with a length at least

Figure 60.18 Needle bearings

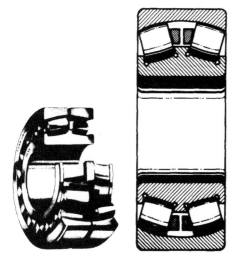

Figure 60.19 Spherical roller bearing assembly

four times their diameter. They are sometimes referred to as barrel rollers. These bearings are most useful where space is limited and thrust-load support is not required. They are available with or without an inner race. If a shaft takes the place of an inner race, it must be hardened and ground. The full-complement type is used for high loads and oscillating or slow speeds. The cage type should be used for rotational motion.

They come in both single-row and double-row mountings. As with all cylindrical roller bearings, the single-row mounting type has a low thrust capacity, but angular mounting of rolls in the double-row type permits its use for combined axial and thrust loads.

Spherical bearings are usually furnished in a double-row mounting that is inherently self-aligning. Both rows of rollers have a common spherical outer raceway. The rollers are barrel-shaped with one end smaller to provide a small thrust to keep the rollers in contact with the center guide flange.

This type of roller bearing has a high radial and moderate-to-heavy thrust load-carrying capacity. It maintains this capability with some degree of shaft and bearing housing misalignment. While their internal self-aligning feature is useful, care should be taken in specifying this type of bearing to compensate for misalignment. Figure 60.19 shows a typical spherical roller bearing assembly. Figure 60.20 shows a series of spherical roller bearings for a given shaft size.

Tapered bearings are used for heavy radial and thrust loads. They have straight tapered rollers, which are held in accurate alignment by means of a guide flange on the inner ring. Figure 60.21 shows a typical tapered-roller bearing. Figure 60.22 shows necessary information to identify a taper-roller bearing. Figure 60.23 shows various types of tapered roller bearings.

True rolling occurs because they are designed so all elements in the rolling surface and the raceways intersect at a common point on the axis. The basic characteristic of these bearings is that if the apexes of the tapered working surfaces of both rollers and races were extended, they would coincide on the bearing axis. Where maximum system rigidity is required, they can be adjusted for a preload. These bearings are separable.

60.3.3 Bearing materials

Because two contacting metal surfaces are in motion in bearing applications, material selection plays a crucial role in their life. Properties of the materials used in bearing construction determine the amount of sliding friction that occurs, a key factor affecting bearing life. When two similar metals are in contact without the presence of adequate lubrication, friction is generally high and the surfaces will seize (i.e., weld) at relatively low pressures or surface loads. However, certain combinations of materials support substantial loads without seizing or welding as a result of their low frictional qualities.

In most machinery, shafts are made of steel. Bearings are generally made of softer materials that have low frictional as well as sacrificial qualities when in contact with steel. A softer, sacrificial material is used for bearings

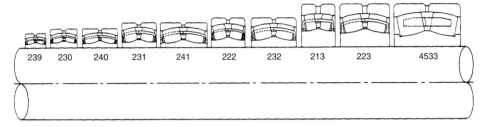

Figure 60.20 Series of spherical roller bearings for a given shaft size (available in several series)

because it is easier and cheaper to replace a worn bearing as opposed to a worn shaft. Common bearing materials are cast iron, bronze, and Babbitt. Other less commonly used materials include wood, plastics, and other synthetics.

There are several important characteristics to consider when specifying bearing materials, including: (1) strength

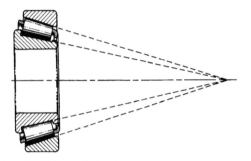

Figure 60.21 Taper-roller bearing

or the ability to withstand loads without plastic deformation, (2) ability to permit embedding of grit or dirt particles that are present in the lubricant, (3) ability to elastically deform in order to permit load distribution over the full bearing surface, (4) ability to dissipate heat and prevent hot spots that might seize, and (5) corrosion resistance.

Plain

As indicated above, dissimilar metals with low frictional characteristics are most suitable for plain bearing applications. With steel shafts, plain bearings made of bronze or Babbitt are commonly used. Bronze is one of the harder bearing materials and is generally used for low speeds and heavy loads.

A plain bearing may sometimes be made of a combination of materials. The outer portion may be constructed of bronze, steel, or iron to provide the strength needed to provide a load-carrying capability. The bearing may be lined with a softer material such as Babbitt to provide the sacrificial capability needed to protect the shaft.

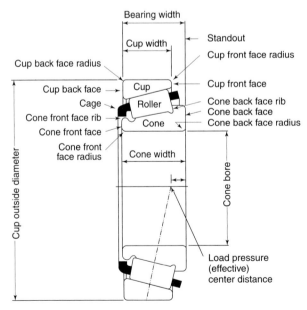

Figure 60.22 Information needed to identify a taper-roller bearing

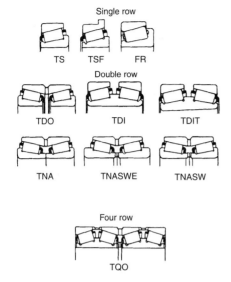

Single row

TS   TSF   FR

Double row

TDO   TDI   TDIT

TNA   TNASWE   TNASW

Four row

TQO

Figure 60.23 Various types of tapered roller bearings

Rolling element

A specially developed steel alloy is used for an estimated 98 per cent of all rolling element bearing uses. In certain special applications, however, materials such as glass, plastic, and other substances are sometimes used in rolling element construction.

Bearing steel is a high-carbon chrome alloy with high hardenability and good toughness characteristics in the hardened and drawn state. All load-carrying members of most rolling contact bearings are made with this steel.

Controlled procedures and practices are necessary to ensure specification of the proper alloy, maintain material cleanliness, and ensure freedom from defects – all of which affect bearing reliability. Alloying practices that conform to rigid specifications are required to reduce anomalies and inclusions that adversely affect a bearing's useful life. Magnaflux inspections ensure that rolling elements are free from material defects and cracks. Light etching is used between rough and finish grinding processes to stop burning during heavy machining operations.

60.4 Lubrication

It is critical to consider lubrication requirements when specifying bearings. Factors affecting lubricants include relatively high speeds, difficulty in performing relubrication, non-horizontal shafts, and applications where leakage cannot be tolerated. This section briefly discusses lubrication mechanisms and techniques for bearings.

60.4.1 Plain bearings

In plain bearings, the lubricating fluid must be replenished to compensate for end leakage in order to maintain their load-carrying capacity. Pressure lubrication from a pump- or gravity-fed tank, or automatic lubricating devices such as oil rings or oil disks, are provided in self-contained bearings. Another means of lubrication is to submerge the bearing (in particular, thrust bearings for vertical shafts) in an oil bath.

Lubricating fluids

Almost any process fluid may be used to lubricate plain bearings if parameters, such as viscosity, corrosive action, toxicity, change in state (where a liquid is close to its boiling point), and in the case of a gaseous fluid its compressibility, are appropriate for the application. Fluid-film journal and thrust bearings have run successfully, for example, on water, kerosene, gasoline, acid, liquid refrigerants, mercury, molten metals, and a wide variety of gases.

Gases, however, lack the cooling and boundary-lubrication capabilities of most liquid lubricants. Therefore, the operation of self-acting gas bearings is restricted by start/stop friction and wear. If start/stop is performed under load, then the design is limited to about seven pounds per square inch (lb/in^2) or 48 kilo-Newtons per square meter (kN/m^2) on the projected bearing area, depending upon the choice of materials. In general, the materials used for these bearings are those of dry rubbing bearings (e.g., either a hard/hard combination such as ceramics with or without a molecular layer of boundary lubricant, or a hard/soft combination with a plastic surface).

Externally pressurized gas journal bearings have the same principle of operation as hydrostatic liquid-lubricated bearings. Any clear gas can be used, but many of the design charts are based on air. There are three forms of external flow restrictors in use with these bearings: pocketed (simple) orifice, unpocketed (annular) orifice, and slot.

State of lubrication

Fluid or complete lubrication, the condition where the surfaces are completely separated by a fluid film, provides the lowest friction losses and prevents wear.

The semi-fluid lubrication state exists between the journal and bearing when a load-carrying fluid film does not form to separate the surfaces. This ocurrs at comparatively low speed with intermittent or oscillating motion, heavy load, and insufficient oil supply to the bearing. Semi-fluid lubrication also may exist in thrust bearings with fixed parallel-thrust collars; guide bearings of machine tools; bearings with plenty of lubrication, but have a bent or misaligned shaft; or where the bearing surface has improperly arranged oil grooves. The coefficient of friction in such bearings may range from 0.02 to 0.08.

In situations where the bearing is well lubricated, but the speed of rotation is very slow or the bearing is barely greasy, boundary lubrication takes place. In this situation, which occurs in bearings when the shaft is starting from rest, the coefficient of friction may vary from 0.08 to 0.14.

A bearing may run completely dry in exceptional cases of design or with a complete failure of lubrication. Depending on the contacting surface materials, the coefficient of friction will be between 0.25 and 0.40.

60.4.2 Rolling element bearings

Rolling element bearings also need a lubricant to meet or exceed their rated life. In the absence of high temperatures, however, excellent performance can be obtained with a very small quantity of lubricant. Excess lubricant causes excessive heating, which accelerates lubricant deterioration.

The most popular type of lubrication is the sealed grease ball-bearing cartridge. Grease is commonly used for lubrication because of its convenience and minimum maintenance requirements. A high-quality lithium-based NLGI 2 grease is commonly used for temperatures up to 180°F (82°C). Grease must be replenished and relubrication intervals in hours of operation are dependent on temperature, speed, and bearing size. Table 60.9 is a general guide to the time after which it is advisable to add a small amount of grease.

Some applications, however, cannot use the cartridge design, for example, when the operating environment is too hot for the seals. Another example is when minute leaks or the accumulation of traces of dirt at the lip seals cannot be tolerated (e.g., food processing machines). In these cases, bearings with specialized sealing and lubrication systems must be used.

In applications involving high speed, oil lubrication is typically required. Table 60.10 is a general guide in selecting oil of the proper viscosity for these bearings. For applications involving high-speed shafts, bearing selection must take into account the inherent speed limitations

Table 60.9 Ball-bearing grease relubrication intervals (hours of operation)

Bearing bore, mm	Bearing speed, rpm				
	5,000	3,600	1,750	1,000	200
10	8,700	12,000	25,000	44,000	220,000
20	5,500	8,000	17,000	30,000	150,000
30	4,000	6,000	13,000	24,000	127,000
40	2,800	4,500	11,000	20,000	111,000
50		3,500	9,300	18,000	97,000
60		2,600	8,000	16,000	88,000
70			6,700	14,000	81,000
80			5,700	12,000	75,000
90			4,800	11,000	70,000
100			4,000	10,000	66,000

Source: Maintenance Fundamentals, R. K. Mobley, Butterworth-Heinemann, 1999.

Table 60.10 Oil lubrication viscosity (ISO identification numbers)

Bearing Bore, mm	Bearing speed, rpm				
	10,000	3,600	1,800	600	50
4–7	68	150	220		
10–20	32	68	150	220	460
25–45	10	32	68	150	320
50–70	7	22	68	150	320
75–90	3	10	22	68	220
100	3	7	22	68	220

Source: Maintenance Fundamentals, R. K. Mobley, Butterworth-Heinemann, 1999.

of certain bearing designs, cooling needs, and lubrication issues such as churning and aeration suppression. A typical case is the effect of cage design and roller-end thrust-flange contact on the lubrication requirements in taper roller bearings. These design elements limit the speed and the thrust load that these bearings can endure. As a result, it is important to always refer to the bearing manufacturer's instructions on load-carrying design and lubrication specifications.

60.5 Installation and general handling precautions

Proper handling and installation practices are crucial to optimal bearing performance and life. In addition to standard handling and installation practices, the issue of emergency bearing substitutions is an area of critical importance. If substitute bearings are used as an emergency means of getting a machine back into production quickly, the substitution should be entered into the historical records for that machine. This documents the temporary change and avoids the possibility of the substitute bearing becoming a permanent replacement. This error can be extremely costly, particularly if the incorrectly specified bearing continually fails prematurely. It is important that an inferior substitute be removed as soon as possible and replaced with the originally specified bearing.

60.5.1 Plain bearing installation

It is important to keep plain bearings from shifting sideways during installation and to ensure an axial position that does not interfere with shaft fillets. Both of these can be accomplished with a locating lug at the parting line. Less frequently used is a dowel in the housing, which protrudes partially into a mating hole in the bearing.

The distance across the outside parting edges of a plain bearing are manufactured slightly greater than the housing bore diameter. During installation, a light force is necessary to snap it into place and, once installed, the bearing stays in place because of the pressure against the housing bore.

It is necessary to prevent a bearing from spinning during operation, which can cause a catastrophic failure. Spinning is prevented by what is referred to as 'crush.' Bearings are slightly longer circumferentially than their mating housings and upon installation this excess length is elastically deformed or 'crushed.' This sets up a high radial contact pressure between the bearing and housing, which ensures good back contact for heat conduction and, in combination with the bore-to-bearing friction, prevents spinning. It is important that under no circumstances should the bearing parting lines be filed or otherwise altered to remove the crush.

60.5.2 Rolling element bearing installation

A basic rule of rolling element bearing installation is that one ring must be mounted on its mating shaft or in its housing with an interference fit to prevent rotation. This is necessary because it is virtually impossible to prevent rotation by clamping the ring axially.

Mounting hardware

Bearings come as separate parts that require mounting hardware or as pre-mounted units that are supplied with their own housings, adapters, and seals.

Bearing mountings   Typical bearing mountings, which are shown in Figure 60.24, locate and hold the shaft axially and allow for thermal expansion and/or contraction of the shaft. Locating and holding the shaft axially is generally accomplished by clamping one of the bearings on the shaft so that all machine parts remain in proper relationship dimensionally. The inner ring is locked axially relative to the shaft by locating it between a shaft shoulder and some type of removable locking device once the inner ring has a tight fit. Typical removable locking devices are specially designed nuts, which are used for a through shaft, and clamp plates, which are commonly used when the bearing is mounted on the end of the shaft. For the locating or held bearing, the outer ring is clamped axially, usually between housing shoulders or end-cap pilots.

With general types of cylindrical roller bearings, shaft expansion is absorbed internally simply by allowing one ring to move relative to the other (Figure 60.24(a) and 60.24(c), non-locating positions). The advantage of this type of mounting is that both inner and outer rings may have a tight fit, which is desirable or even mandatory if significant vibration and/or imbalance exists in addition to the applied load.

Premounted bearing   Premounted bearings, referred to as pillow-block and flanged-housing mountings, are of considerable importance to millwrights. They are particularly adaptable to 'line-shafting' applications, which are a

Figure 60.26 Flanged bearing unit

series of ball and roller bearings supplied with their own housings, adapters, and seals. Premounted bearings come with a wide variety of flange mountings, which permit them to be located on faces parallel or perpendicular to the shaft axis. Figure 60.25 shows a typical pillow block. Figure 60.26 shows a flanged bearing unit.

Inner races can be mounted directly on ground shafts or can be adapter-mounted to 'drill-rod' or to commercial-shafting. For installations sensitive to imbalance and vibration, the use of accurately ground shaft seats is recommended.

Most pillow-block designs incorporate self-aligning bearing types and do not require the precision mountings utilized with other bearing installations.

Mounting techniques

When mounting or dismounting a roller bearing, the most important thing to remember is to apply the mounting or dismounting force to the side face of the ring with the interference fit. This force should not pass from one ring to the other through the ball or roller set, as internal damage can easily occur.

Mounting tapered-bore bearings can be accomplished simply by tightening the locknut or clamping plate. This locates it on the shaft until the bearing is forced the proper distance up the taper. This technique requires a significant amount of force, particularly for large bearings.

Cold mounting

Cold mounting, or force fitting a bearing onto a shaft or into a housing, is appropriate for all small bearings (i.e., 4-inch bore and smaller). The force, however, must be applied as uniformly as possible around the side face of the bearing and to the ring to be press-fitted. Mounting fixtures, such as a simple piece of tubing of appropriate size and a flat plate, should be used. It is not appropriate to use a drift and hammer to force the bearing on, which will cause the bearing to cock. It is possible to apply force by striking the plate with a hammer or by an arbor press. However, before forcing the the bearing on the shaft, a coat of light oil should be applied to the bearing seat on the shaft and the bearing bores. All sealed and shielded ball bearings should be cold mounted in this manner.

Temperature mounting

The simplest way to mount any open straight-bore bearing regardless of its size is temperature mounting, which

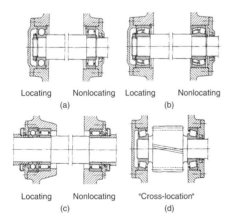

Locating	Nonlocating	Locating	Nonlocating
(a)		(b)	

Locating	Nonlocating	"Cross-location"
(c)		(d)

Figure 60.24 Typical bearing mounting

Figure 60.25 Typical pillow block

entails heating the entire bearing, pushing it on its seat, and holding it in place until it cools enough to grip the shaft. The housing may be heated if practical for tight outside-diameter fits, however, temperatures should not exceed 250°F. If heating of the housing is not practical, the bearing may be cooled with dry ice. The risk of cooling is that if the ambient conditions are humid, moisture is introduced and there is a potential for corrosion in the future. Acceptable ways of heating bearings are by hot plate, temperature-controlled oven, induction heaters, and hot-oil bath.

With the hot plate method, the bearing is simply laid on the plate until it reaches the approved temperature, using a pyrometer or Tempilstik to make certain it is not overheated. Difficulty in controlling the temperature is the major disadvantage of this method.

When using a temperature-controlled oven, the bearings should be left in the oven long enough to heat thoroughly, but they should never be left overnight.

The use of induction heaters is a quick method of heating bearings. However, some method of measuring the ring temperature (e.g., pyrometer or a Tempilstik) must be used or damage to the bearing may occur. Note that bearings must be demagnetized after the use of this method.

The use of a hot-oil bath is the most practical means of heating larger bearings. Disadvantages are that the temperature of the oil is hard to control and may ignite or overheat the bearing. The use of a soluble oil and water mixture (10 to 15 per cent oil) can eliminate these problems and still attain a boiling temperature of 210°F. The bearing should be kept off the bottom of the container by a grate or screen located several inches off the bottom. This is important to allow contaminants to sink to the bottom of the container and away from the bearing.

Dismounting

Commercially available bearing pullers allow rolling element bearings to be dismounted from their seats without damage. When removing a bearing, force should be applied to the ring with the tight fit, although sometimes it is necessary to use supplementary plates or fixtures. An arbor press is equally effective at removing smaller bearings as well as mounting them.

Ball installation

Figure 60.27 shows the ball installation procedure for roller bearings. The designed load carrying capacity of Conrad-type bearings is determined by the number of balls that can be installed between the rings. Ball installation is accomplished by the following procedure:

- Slip the inner ring slightly to one side.
- Insert balls into the gap, which centers the inner ring as the balls are positioned between the rings.
- Place stamped retainer rings on either side of the balls before riveting together. This positions the balls equidistant around the bearing.

60.5.3 General rolling element bearing handling precautions

In order for rolling element bearings to achieve their design life and perform with no abnormal noise, temperature rise, or shaft excursions, the following precautions should be taken:

- Always select the best bearing design for the application and not the cheapest. The cost of the original bearing is usually small by comparison to the costs of replacement components and the down-time in production when premature bearing failure occurs because an inappropriate bearing was used.
- If in doubt about bearings and their uses, consult the manufacturer's representative and the product literature.
- Bearings should always be handled with great care. Never ignore the handling and installation instructions from the manufacturer.
- Always work with clean hands, clean tools, and the cleanest environment available.
- Never wash or wipe bearings prior to installation unless the instructions specifically state that this should be done. Exceptions to this rule are when oil-mist lubrication is to be used and the slushing compound has hardened in storage or is blocking lubrication holes in the bearing rings. In this situation, it is best to clean the bearing with kerosene or other appropriate petroleum-based solvent. The other exception is if the slushing compound has been contaminated with dirt or foreign matter before mounting.
- Keep new bearings in their greased paper wrappings until they are ready to install.
- Place unwrapped bearings on clean paper or lint-free cloth if they cannot be kept in their original containers. Wrap bearings in clean, oil-proof paper when not in use.
- Never use wooden mallets, brittle or chipped tools, or dirty fixtures and tools when bearings are being installed.
- Do not spin bearings (particularly dirty ones) with compressed service air.
- Avoid scratching or nicking bearing surfaces. Care must be taken when polishing bearings with emery cloth to avoid scratching.
- Never strike or press on race flanges.
- Always use adapters for mounting that ensure uniform steady pressure rather than hammering on a drift or sleeve. **Never** use brass or bronze drifts to install bearings as these materials chip very easily into minute particles that will quickly damage a bearing.
- Avoid cocking bearings onto shafts during installation.
- Always inspect the mounting surface on the shaft and housing to insure that there are no burrs or defects.
- When bearings are being removed, clean housings and shafts before exposing the bearings. Dirt is abrasive and detrimental to the designed life-span of bearings.
- Always treat used bearings as if they are new, especially if they are to be reused.
- Protect dismantled bearings from moisture and dirt.
- Use clean filtered, water-free Stoddard's solvent or flushing oil to clean bearings.

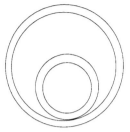

1. The inner ring is moved to one side.

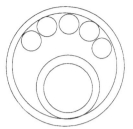

2. Balls are installed in the gap.

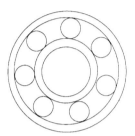

3. The inner ring is centered as the balls
are equally positioned in place.

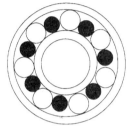

4. A retainer is installed.

Figure 60.27 Ball installation procedures

- When heating is used to mount bearings onto shafts, follow the manufacturer's instructions.
- When assembling and mounting bearings onto shafts, **never** strike the outer race or press on it to force the inner race. Apply the pressure on the inner race only. When dismantling, follow the same procedure.
- Never press, strike, or otherwise force the seal or shield on factory-sealed bearings.

60.6 Bearing failures, deficiencies, and their causes

The general classifications of failures and deficiencies requiring bearing removal are overheating, vibration, turning on the shaft, binding of the shaft, noise during operation, and lubricant leakage. Table 60.11 is a troubleshooting guide that lists the common causes for each of these failures and deficiencies. As indicated by the causes of failure listed, bearing failures are rarely caused by the bearing itself.

Many abnormal vibrations generated by actual bearing problems are the result of improper sizing of the bearing liner or improper lubrication. However, numerous machine and process-related problems generate abnormal vibration spectra in bearing data. The primary contributors to abnormal bearing signatures are: (1) imbalance, (2) misalignment, (3) rotor instability, (4) excessive or abnormal loads, and (5) mechanical looseness.

Defective bearings that leave the manufacturer are very rare and it is estimated that defective bearings contribute to only 2 per cent of total failures. The failure is invariably linked to symptoms of misalignment, imbalance, resonance and lubrication – or the lack of it. Most of the problems

that occur result from the following reasons: dirt, shipping damage, storage and handling, poor fit resulting in installation damage, wrong type of bearing design, overloading, improper lubrication practices, misalignment, bent shaft, imbalance, resonance, and soft foot. Any one of these conditions will eventually destroy a bearing – two or more of these problems can result in disaster!

Although most industrial machine designers provide adequate bearings for their equipment, there are some cases where bearings are improperly designed, manufactured, or installed at the factory. Usually, however, the trouble is caused by one or more of the following reasons: (1) improper on-site bearing selection and/or installation, (2) incorrect grooving, (3) unsuitable surface finish, (4) insufficient clearance, (5) faulty relining practices, (6) operating conditions, (7) excessive operating temperature, (8) contaminated oil supply, and (9) oil-film instability.

60.7 Improper bearing selection and/or installation

There are several things to consider when selecting and installing bearings, including the issue of interchangeability, materials of construction, and damage that might have occurred during shipping, storage, and handling.

Interchangeability

Because of the standardization in envelope dimensions, precision bearings were once regarded as interchangeable among manufacturers. This interchangeability has since been considered a major cause of failures in machinery and the practice should be used with extreme caution.

Table 60.11 Troubleshooting guide

Overheating	Vibration	Turning on the shaft	Binding of the shaft	Noisy bearing	Lubricant leakage
Inadequate or insufficient lubrication	Dirt or chips in bearing	Growth of race due to overheating	Lubricant breakdown	Lubricant breakdown	Overfilling of lubricant
Excessive lubrication	Fatigued race or rolling elements	Fretting wear	Contamination by abrasive or corrosive materials	Inadequate lubrication	Grease churning due to too soft consistency
Grease liquification or aeration	Rotor unbalance	Improper initial fit	Housing distortion or out-of-round pinching bearing	Pinched bearing	Grease deterioration due to excessive operating temperature
Oil foaming	Out-of-round shaft	Excessive shaft deflection	Uneven shimming of housing with loss of clearance	Contamination	Operating beyond grease life
Abrasion or corrosion due to contaminants	Race misalignment	Initial coarse finish on shaft	Tight rubbing seals	Seal rubbing	Seal wear
Housing distortion due to warping or out-of-round	Housing resonance	Seal rub on inner race	Preloaded bearings	Bearing slipping on shaft or in housing	Wrong shaft attitude (bearing seals designed for horizontal mounting only)
Seal rubbing or failure	Cage wear		Cocked races	Flatted roller or ball	Seal failure
Inadequate or blocked scavenge oil passages	Flats on races or rolling elements		Loss of clearance due to excessive adapter tightening	Brinelling due to assembly abuse, handing, or shock loads	Clogged breather
Inadequate bearing clearance or bearing preload	Race turning		Thermal shaft expansion	Variation in size of rolling elements	Oil foaming due to churning or air flow through housing
Race turning	Excessive clearance		Thermal shaft expansion	Variation in size of rolling elements	Oil foaming due to churning or air flow through housing
Cage wear	Corrosion			Housing bore waviness	Porous housing or closure
	False-brinelling or indentation of races			Chips or scores under bearing seat	Lubricator set at the wrong flow rate
	Electrical arcing				
	Mixed rolling element diameters				
	Out-of-square rolling paths in races				

Most of the problems with interchangeability stem from selecting and replacing bearings based only on bore size and outside diameters. Often, very little consideration is paid to the number of rolling elements contained in the bearings. This can seriously affect the operational frequency vibrations of the bearing and may generate destructive resonance in the host machine or adjacent machines.

More bearings are destroyed during their installation than fail in operation. Installation with a heavy hammer is the usual method in many plants. Heating the bearing with an oxy-acetylene burner is another classical method. However, the bearing does not stand a chance of reaching its life expectancy when either of these installation practices are used. The bearing manufacturer's installation instructions should always be followed.

Shipping damage

Bearings and the machinery containing them should be properly packaged to avoid damage during shipping. However, many installed bearings are exposed to vibrations, bending, and massive shock loadings through bad handling practices during shipping. It has been estimated that approximately 40 per cent of newly received machines have 'bad' bearings.

Because of this, all new machinery should be thoroughly inspected for defects before installation. Acceptance criteria should include guidelines that clearly define acceptable design/operational specifications. This practice pays big dividends by increasing productivity and decreasing unscheduled downtime.

Storage and handling

Stores and other appropriate personnel must be made aware of the potential havoc they can cause by their mishandling of bearings. Bearing failure often starts in the storeroom rather than the machinery. Premature opening of packages containing bearings should be avoided whenever possible. If packages must be opened for inspection, they should be protected from exposure to harmful dirt sources and then resealed in the original wrappings. The bearing should never be dropped or bumped as this can cause shock loading on the bearing surface.

Incorrect placement of oil grooves

Incorrectly placed oil grooves can cause bearing failure. Locating the grooves in high-pressure areas causes them to act as pressure-relief passages. This interferes with the formation of the hydrodynamic film, resulting in reduced load-carrying capability.

Unsuitable surface finish

Smooth surface finishes on both the shaft and the bearing are important to prevent surface variations from penetrating the oil film. Rough surfaces can cause scoring, overheating, and bearing failure. The smoother the finishes, the closer the shaft may approach the bearing without danger of surface contact. Although important in all bearing applications, surface finish is critical with the use of harder bearing materials such as bronze.

Insufficient clearance

There must be sufficient clearance between the journal and bearing in order to allow an oil film to form. An average diametral clearance of 0.001 inches per inch of shaft diameter is often used. This value may be adjusted depending on the type of bearing material, the load, speed, and the accuracy of the shaft position desired.

Faulty relining

Faulty relining occurs primarily with babbitted bearings rather than precision machine-made inserts. Babbitted bearings are fabricated by a pouring process that should be performed under carefully controlled conditions. Some reasons for faulty relining are: (1) improper preparation of the bonding surface, (2) poor pouring technique, (3) contamination of Babbitt, and (4) pouring bearing to size with journal in place.

60.8 Operating conditions

Abnormal operating conditions or neglecting necessary maintenance precautions cause most bearing failures. Bearings may experience premature and/or catastrophic failure on machines that are operated heavily loaded, speeded up, or being used for a purpose not appropriate for the system design. Improper use of lubricants can also result in bearing failure. Some typical causes of premature failure include: (1) excessive operating temperatures, (2) foreign material in the lubricant supply, (3) corrosion, (4) material fatigue, and (5) use of unsuitable lubricants.

Excessive temperatures

Excessive temperatures affect the strength, hardness, and life of bearing materials. Lower temperatures are required for thick Babbitt liners than for thin precision Babbitt inserts. Not only do high temperatures affect bearing materials, they also reduce the viscosity of the lubricant and affect the thickness of the film, which affects the bearing's load-carrying capacity. In addition, high temperatures result in more rapid oxidation of the lubricating oil, which can result in unsatisfactory performance.

Dirt and contaminations in oil supply

Dirt is one of the biggest culprits in the demise of bearings. Dirt makes its appearance in bearings in many subtle ways and it can be introduced by bad work habits. It also can be introduced through lubricants that have been exposed to dirt, which is responsible for approximately half of bearing failures throughout the industry.

To combat this problem, soft materials such as Babbitt are used when it is known that a bearing will be exposed to abrasive materials. Babbitt metal embeds hard particles, which protects the shaft against abrasion. When harder materials are used in the presence of abrasives, scoring and galling occurs as a result of abrasives caught between the journal and bearing.

In addition to the use of softer bearing materials for applications where abrasives may potentially be present, it is important to properly maintain filters and breathers, which should regularly be examined. In order to avoid oil supply contamination, foreign material that collects at the bottom of the bearing sump should be removed on a regular basis.

Oil-film instability

The primary vibration frequency components associated with fluid-film bearings problems are in fact displays of turbulent or non-uniform oil film. Such instability problems are classified as either oil whirl or oil whip depending on the severity of the instability.

Machine-trains that use sleeve bearings are designed based on the assumption that rotating elements and shafts operate in a balanced and, therefore, centered position. Under this assumption, the machine-train shaft will operate with an even, concentric oil film between the shaft and sleeve bearing.

For a normal machine, this assumption is valid after the rotating element has achieved equilibrium. When the forces associated with rotation are in balance, the rotating element will center the shaft within the bearing. However, several problems directly affect this self-centering operation. First, the machine-train must be at designed operating speed and load to achieve equilibrium. Second, any imbalance or abnormal operation limits the machine-train's ability to center itself within the bearing.

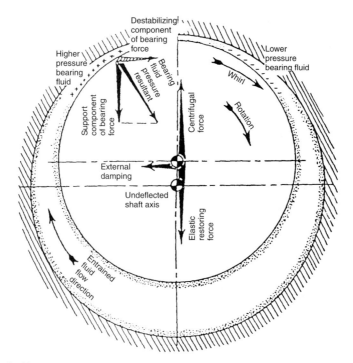

Figure 60.28 Oil whirl, oil whip

A typical example is a steam turbine. A turbine must be supported by auxiliary running gear during start-up or shut-down to prevent damage to the sleeve bearings. The lower speeds during the start-up and shut-down phase of operation prevent the self-centering ability of the rotating element. Once the turbine has achieved full speed and load, the rotating element and shaft should operate without assistance in the center of the sleeve bearings.

Oil whirl   In an abnormal mode of operation, the rotating shaft may not hold the centerline of the sleeve bearing. When this happens, an instability called oil whirl occurs. Oil whirl is an imbalance in the hydraulic forces within a sleeve bearing. Under normal operation, the hydraulic forces such as velocity and pressure are balanced. If the rotating shaft is offset from the true centerline of the bearing, instability occurs.

As Figure 60.28 illustrates, a restriction is created by the offset. This restriction creates a high pressure and another force vector in the direction of rotation. Oil whirl accelerates the wear and failure of the bearing and bearing support structure.

Oil whip   The most severe damage results if the oil whirl is allowed to degrade into oil whip. Oil whip occurs when the clearance between the rotating shaft and sleeve bearing is allowed to close to a point approaching actual metal-to-metal contact. When the clearance between the shaft and

bearing approaches contact, the oil film is no longer free to flow between the shaft and bearing. As a result, the oil film is forced to change directions. When this occurs, the high-pressure area created in the region behind the shaft is greatly increased. This vortex of oil increases the abnormal force vector created by the offset and rotational force to the point that metal-to-metal contact between the shaft and bearing occurs. In almost all instances where oil whip is allowed, severe damage to the sleeve bearing occurs.

References

1. Mobley, R. K., *Advanced Diagnostics and Analysis*, Plant Performance Group (1989).
2. Neale, M. J., *Bearings – A Tribology Handbook*, Society of Automotive Engineers, Inc., Butterworth-Heinemann, Oxford, Great Britain (1993).
3. *Bearings and Power Transmission Products Catalog*, Bearings, Inc., Greenville, South Carolina (1991).
4. Mobley, R. K., *Introduction to Predictive Maintenance*, Van Nonstrand-Reinhold, New York (1990).
5. Higgins, and Mobley, R. K., *Maintenance Engineering Handbook*, McGraw-Hill, New York (1995).
6. T. Baumeister, ed., *Marks' Standard Handbook for Mechanical Engineers*, 8th edn, McGraw-Hill, New York (1978).
7. Nelson, C. A., *Millwrights and Mechanics Guide*, Macmillan Publishing Company, New York (1986).
8. *SKF Product Service Guide*, SKF Bearing Services Co., USA (1992).

61

Finance for the Plant Engineer

Leon Turrell
Financial Consultant

Contents

61.1 Accounting

Ultimately, all activities of an organization will be expressed in money, the common denominator within the accounts of the proprietor, be it a company, a partnership or a statutory organization. The accounting for these activities will be under the control of the finance department, which, in turn, will be controlled, through the financial director, by the board of directors.

A typical accounting organization is shown in Figure 61.1. These divisions, depending on the size of the organization, may be broken down further to provide accounting services at each plant or operating facility. There will normally be an accountant at each plant, or, in the case of very large organizations, at each major profit or cost center at plant level. By this means, close liaison can be maintained with and management services information provided for the locally based operating management, in addition to providing the flow of information and appropriate documentation to enable the organization to prepare timely and informative accounts.

Many of the financial functions will tend to be centralized in order to enjoy economies of scale available using specially qualified staff and/or computers.

Prompt and accurate accounting is vital for the well-being of the organization and for the early detection of problem areas needing corrective treatment by management. The achievement of this promptitude and accuracy depends largely on the cooperation and attention to detail by staff in non-accounting departments responsible for the recording of information (e.g. fuel and water consumption, meter readings, the taking of stocks, etc.). Such staff

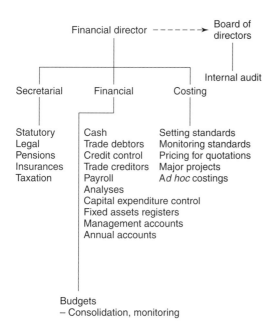

Figure 61.1 A typical accounting organization chart for a large company

may not recognize their involvement in these areas as one of their prime concerns. Each of these departmental chains will normally be staffed and headed by appropriately qualified people.

61.2 Types of organization

The most common of the organizations operating in today's industrial scene is the joint stock company with limited liability. In different countries there exist variants of this format, but there is a wide spread of organizations based upon broadly similar principles.

The partnership and sole proprietor types of organization are suitable for businesses while they remain small and can be managed and financed by the owners, but as expansion takes place and the business demands more capital, then the ability of relatively small numbers of proprietors to provide additional capital from their own resources becomes increasingly difficult. In the search for additional funds recourse may be had to a variety of institutions such as banks and the specialist companies which have been set up, often by the banks themselves in conjunction with insurance companies and merchant banks, to cater for this need.

These specialist lenders will often wish to take a share in the capital or 'equity' of the business as part of the arrangement by which they will advance capital. In order to achieve this the business will normally be required to incorporate.

The share capital of a company is expressed as a sum of money, the *authorized share capital*, made up of a stated number of shares, each of which has a 'face' or nominal value.

On incorporation, a small, family-owned firm may start with an initial capital of $10,000 divided into 10,000 shares of $1 each. A large public company, on the other hand, may have a share capital of many millions of dollars, sub-divided into many more millions of shares or stock units, of perhaps as little as 25 cents.

Not all the shares in a company may, or need to be, issued. By the issue of more shares, at an appropriate stage in the company's growth, capital may be raised from the institutions mentioned above or, through a flotation on the Stock Exchange, from the investing public at large. When this happens, the price paid per share by the purchasers may be many times greater than the nominal value.

Example

Company A has an authorized and issued share capital of 300,000 ordinary shares of $1 each held by five shareholders. After many years of successful trading, it is decided to seek a listing on the Stock Exchange. The capital is increased to $3 million and is then divided into 12 million shares of $0.25. Of these, 8 million are offered for sale to the public at $1 each. Subject to the costs of the issue, this will raise $8 million of fresh capital, some of which will be used in the business and some will go to the original owners of the business.

Before the issue:

	Shares of $1	
	No.	Value
Original shareholders	300,000	$300,000

After the issue:

	Shares of $0.25	
	No.	Value
Original shareholders	4 million	$1 million
Outside shareholders	8 million	$2 million

The original shareholders' ownership of the business has declined from 100 per cent to 33.3 per cent, and it may be necessary for the owners to accept the appointment of further directors from outside the business, possibly in a non-executive capacity.

A price for the shares will be set by the market, based upon its assessment of the prospective performance of the company and the perceived quality of its management, and trading in the shares will begin. This price does not affect the money value received by the company for the sale of the shares at flotation.

61.3 Definitions

61.3.1 Capital

Capital is variously described, but is generally taken to be the amount of finance provided by the proprietors or outside lenders to enable the business to operate and earn a profit, from which the lenders will expect to receive an income. Share capital is that part which is regarded as fixed in the sense that it changes infrequently, and is increased only when the business requires additional permanent finance. In a limited company, the amount of accumulated profits left in the business by the proprietors (i.e. not drawn in dividends) is grouped with the share capital and termed *shareholders' funds*. Under present-day legislation it is also possible for a company to reduce its share capital by buying back its own shares or, by applying to the court to have its capital reduced to enable a reorganization of the company's structure to take place. In this context, proprietors may be shareholders, stock-holders, partners or sole traders. Capital is often termed the *equity*.

61.3.2 Capital and revenue expenditure

The difference between capital and revenue expenditure can be compared broadly with that between expenditure on fixed and current assets. Capital expenditure is incurred in the acquisition of fixed assets and positioning the organization in order to start or continue its operations. Revenue expenditure relates to the purchase of goods and materials for manufacture or resale and in bringing those goods or materials to a condition suitable for sale, providing the facilities to bring about the sales and in the general administration and running of the business.

61.3.3 Cash flow

This is the amount of cash passing through the hands of an organization in an accounting period. The cash flow statement analyses the sources and the disposition of cash during a given period.

In addition to their historical use, cash flow statements are prepared as part of the budgeting process in order to identify the effects upon the cash facilities of the proposed activities for the period under review. A typical, simplified, statement would give the following information.

Cash flow statement

Operating profit	X
Depreciation	X
Cash flow from operations	X
Fixed assets bought	(X)
Fixed assets sold	X
Loans received	X
Loans repaid	(X)
Corporate taxes paid	(X)
Interest paid	(X)
Interest received	X
Increases in working capital*	(X)
Decreases in working capital*	X
Dividends paid	(X)
Net cash inflow (outflow)	X
Opening cash balances	X
Closing cash balances	X

*Working capital normally includes stocks, trade debtors, prepayments, trade creditors, accruals, current taxation, etc.

61.3.4 Liquidity

This is usually defined as the ratio that liquid assets (debtors + cash) bear to current liabilities. The ratio is a measure of the relation of short-term obligations to the funds likely to be available to meet them.

Example

Liquid assets	$300,000	
Current liabilities	$250,000	Ratio 1.2:1

Ratios well below 1 may indicate financial problems ahead while those substantially greater than 1 may point to poor credit control or under-utilization of cash. This ratio is sometimes known as the 'acid test'. The principal profitability ratio in use is the net profit before interest and tax (NPBIT) to net assets or return on capital employed.

61.3.5 Financial and operating ratios

Ratios are widely used to compare the performance of a business with predetermined objectives set by the business itself, with standards used by banks and other lenders, with other businesses in the same segment of industry (inter-firm comparison) and by the Stock Exchange and its attendant analysts. The principal criteria are profitability, return on capital, liquidity and growth. To undertake growth by acquisition a high standing in the market is required. This is often a reflection of the performance against these criteria.

The ratios which follow are illustrated by reference to the following balance sheet and profit and loss account.

Balance sheet	$000
Current assets	
Cash	1,175
Stocks	4,700
Debtors and prepayments	4,935
	10,810
Current liabilities	
Creditors	3,170
Short-term loans	600
Current taxation	1,056
Dividends	354
	5,180
Net current assets	5,630
Fixed assets	6,250
	11,880
Financed by:	
Share capital – Preference	825
– Ordinary	3,050* 3,875
Reserves	4,700
Shareholders' funds (net worth)	8,575
Long-term loans	2,500
Deferred taxation	805
	11,880

*12,200,000 shares of $0.25 each.

Profit and loss account		$000
Sales (or revenue)		22,000
Cost of sales (72%)		15,800
Net profit before interest and taxation (NPBIT)		2,555
Loan interest		200
Taxation		980
Net profit after taxation (NPAT)	1,375	
Dividends – Preference	47	
– Ordinary	330	377
Retained profits		998

1. Gearing – the ratio of fixed rate capital and borrowings to other capital, that is:

$$\frac{\text{Preference capital} + \text{long-term loans}}{\text{Ordinary capital} + \text{reserves}}$$

$825 + 2500 : 3050 + 4700 = 43\%.$

If the gearing is high, the capacity to borrow may be affected, since the risk to creditors is high and the company may already be burdened with heavy interest charges.

2. Outside liabilities to shareholders' funds

	$000
Long-term loans	2,500
Taxation – deferred	805
Current liabilities	5,180
	8,485
Shareholders' funds	8,575
Ratio	0.99:1

The total of capital and reserves is the amount by which the assets can fall below the balance sheet value without depleting the amount available for creditors. A high ratio will reduce borrowing capacity.

3. Current assets to current liabilities

$10,810 : 5,180 = 2.1 : 1$

Shows the margin of safety available to short-term creditors. Significantly higher ratios than 2:1 may indicate excess stocks, poor credit control or inadequate control of cash resources.

4. Stocks, debtors, creditors to sales

The ratio of the minimum net assets required supporting the sales volume of the business. From these ratios can be calculated the additional working capital needed if sales are to be increased:

$4,700 + 4,935 + 3,170 = 0.294 : 1$ or

Stocks plus debtors minus creditors = 29.4% of sales

The average for manufacturing industries is usually around 25% of sales. Changes in sales levels will give rise to movements in the working capital required to support them. The amounts can be calculated:

Increase in sales 15% of $22,000 = 3,300$

Increase in working capital

15% of $(4,700 + 4,935 + 3,170) = 970$

or 29.4% of $3,300 = 970$

5. Profitability

The most widely used gage of profitability is the ratio of profit to net assets (or return on capital employed):

$$\frac{\text{NPBIT} \times 100}{\text{Shareholders' funds} + \text{Long-term loans} + \text{Deferred taxation}}$$

$2,555 \times 100/11,880 = 21.5\%$

Two ratios amalgamate to provide the profitability ratio.

	$000
Sales	22,000
Net assets	11,880
NPBIT	2,555
Return on capital	$255,500 : 11,880 = 21.5\%$
Profit/Sales	$255,500 : 22,000 = 11.6\%$
Sales/Net assets	$22,000 : 11,880 = 1.85$
	$1.85 \times 11.6\% = 21.5\%$

These two ratios indicate funds usage efficiency and provide a basis for deciding the order in which improvements can be made.

6. Stock ratios

Expressed as a 'Stock turn', representing how many times stocks are turned round in a given period and/or as the number of days' stock of finished goods available to meet sales demand.

		$000
Stocks	– Raw materials	1,530
	– Work in progress	1,300
	– Finished goods	1,740
	– Other	130
		4,700

$$\text{Stock turn} = \frac{\text{Cost of sales}}{\text{Finished goods}}$$

$$= 15,800/1,740 = 9.1 \text{ times}$$

Days' stock $= 1,740 \times 365/15,800 = 40$ days

7. Debtors' ratio

Expressed as days of sales outstanding or 'debtors' days'. Used in conjunction with prior figures for measurement of credit control efficiency and trends:

	$000
Debtors	4,935
Sales	22,000

$$\text{Ratio } \frac{4,935 \times 365}{22,000} = 82 \text{ days}$$

This is sometimes refined by deducting the sales for the current period as being not yet payable and therefore leaving figure to represent *overdue* debt only. The ratio is also known as *collection period*.

8. Liquidity

Liquidity can be expressed as the ratio of liquid assets (cash plus debtors) to current liabilities. Such assets are also known as 'Quick assets', i.e. capable of swift realization.

$$\frac{(1,175 + 4,935)}{5,180} = 1.2 : 1$$

The ratio shows the ability to settle short-term liabilities and should not normally be lower than 1:1, though it very often is! The lower the ratio, the greater indication of possible financial strain.

A high ratio could mean poor credit control or underutilized cash.

9. Earnings per Share (net basis)

$$\frac{\text{NPAT} - \text{Preference dividend}}{\text{Number of issued shares}}$$

$$\frac{1,375 - 47}{12,200} = \$0.109$$

Growth in earnings per share is used widely by analysts as a measure of a company's success. Adjustments must be made where issues of shares have taken place to satisfy acquisition purchase considerations.

10. Yield

In this example, Advance Corporation Tax has been calculated at 25:75.

$$\frac{\text{Earnings per share} \times 100/75 \times 100}{\text{Market Price}}$$

$$\frac{0.109 \times 100/75 \times 100}{1.40} = 10.4\%$$

This demonstrates the current return on the stock market price.

11. Dividend yield

$$\frac{\text{Dividend} \times 100}{\text{Market price}} = \frac{0.036 \times 100}{1.40} = 2.58\%$$

The ratio gives the return, which the dividend provides on the market price.

12. Dividend cover

$$\frac{\text{Gross earnings}}{\text{Gross dividend}} = \frac{0.145}{0.036} = 4.0 \text{ times}$$

The ratio indicates by how much earnings can fall before the dividend must be reduced and shows the company policy towards the payment of dividends and profit retention.

13. Price/Earnings ratio (P/E ratio)

$$\frac{\text{Market price}}{\text{Earnings (net)}} = \frac{1.40}{0.109} = 12.8$$

This shows the number of years' earnings represented by the current market price. It evidences the Stock Market's assessment of the company's ability to maintain or increase earnings. A low P/E ratio will normally indicate a high-risk business, a high ratio a company with potential for growth.

61.4 Budgetary control

61.4.1 Preparation

Financial data on their own provide only an historical record of the transactions of a business. The accounts of a company published annually are mainly of historical value but must, by law, include comparisons with the figures for a prior corresponding period.

Until information can be compared with similarly classified data its use must be limited. In order to plan ahead, a business will prepare a strategy or budget for the next trading year and probably several years thereafter, with that for the next year broken down into the business's scheme of accounting periods.

The final form of the budget will include profit and loss account, balance sheet and cash flow statement for the planning periods together with such supporting statements as are deemed necessary (for example, proposed capital expenditure). These will provide management with advance warning of points of stress upon resources, and enable steps to be taken to ease their effects or to avoid them altogether. The financial controller or chief accountant will have the responsibility for bringing together the relevant information in the final budgets to be presented to senior management.

Budgeting practices vary from one business to another. Imposed budgets will be drawn up by senior management and, as the name implies, be imposed on those lower down the hierarchy. Participation budgets require the input of data and information at the formative stages by the people who will be responsible for bringing about the achievement of the results envisaged in the budgets.

The process of compiling the budgets will start with the preparation of estimates of the physical requirements of the plan, be they manpower, materials, tons of fuel, cubic meters or therms of gas, units of electricity or cubic meters of water. These estimates will be prepared in relation to the estimates of output by manufacturing or process departments, which, in turn, will have been based on the quantities, or other measures forecast in the sales budgets.

It is therefore vital that detailed and accurate records are kept by the operating and engineering departments of the usage and consumption of fuel, water and other services in order that performance against budget can be properly measured and so that data are available for use in compiling future estimates.

The budgets will also require estimates of expenditure upon equipment of a capital nature in the plan periods. This expenditure will normally fall into the following categories:

1. Replacement of existing machinery and equipment at the end of its useful life. (Straight replacement is rarely possible, since improvements in technology will usually have taken place.)
2. Items requiring replacement because of changes in the law (for example, those relating to fire prevention, safety measures for the protection of employees and public).
3. Items which will improve profitability by saving costs or by carrying out processes faster or by using less manpower (see Section 61.5).
4. Expenditure in connection with new projects or necessary to provide increased throughput.

Motor vehicles can also fall within the above categories but businesses will usually have specific policies for their replacement.

As said elsewhere, the importance of realistic estimates of costs cannot be over-stressed, if, in turn, realistic comparisons can be made as the budget periods progress.

The compilation of figures for the budget should almost always be done from 'the bottom up', that is, by calculating the costs for the individual accounting periods by taking estimated quantities and prices for each period and summing for the whole of the budget. Thus, for example, holiday periods and known peaks and troughs can be recognized and catered for within the figures. Estimates made in the first instance on an annual basis can rarely be analyzed satisfactorily to individual periods except where the costs themselves are expressed in annual terms (e.g. rents and rates).

The sources of information from which the data are obtained are infinitely variable, but will include manufacturers' and suppliers' price lists, direct contact with

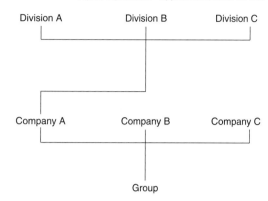

Figure 61.2 A typical group's profit centers

suppliers and long-term contracts with sources of fuel and power.

At all times the interrelationship between budgets must be maintained – sales with finished products – products with consumption of materials – consumption of materials with labor using them, machine time and power. Machine time will influence maintenance, which will have its own content of labor, materials and work by outside specialist contractors.

When all the budgets have been prepared they will be consolidated by the finance department for submission to the approving authority. In large businesses there will be several levels at which authorization will be made before the final total plan for the business and its component divisions is agreed.

In group organizations, company budgets will themselves be consolidated into group form for final approval by the group board. A typical budget program might be based on profit centers taking the form shown in Figure 61.2. (In multinational organizations this consolidation will continue worldwide.)

61.4.2 Control

During the progress of the financial year, budgets will be compared at intervals with the actual performance to enable adjustments to be made to ensure that the planned outcome is achieved as nearly as possible. Ideally, because of changing circumstances after the start of the budget year, the budgets should be revised, possibly at three-monthly intervals. In practice, particularly in large and complex organizations, this is rarely practicable, given limitations of staff and time. An alternative is to require to be given a forecast of the immediately following period(s) to be included with each report to identify areas diverging from planned performance and to alert management to the need for correction.

61.5 Capital expenditure — appraisal methods

There are a number of accepted methods available for the comparison and appraisal of the virtues of proposed capital expenditure projects. Those considered here are as follows:

1. *Pay-back period.* This consists of calculating how long it will take for the profits generated by the capital outlay to equal the outlay itself, such profits usually being calculated after taking into account tax and any grants receivable. The defects of this method are that it takes no account of the profitability of the schemes after the break-even point is reached and the same value is placed upon each pound of profit, whether it is earned in year 1 or year 10. This latter shortcoming is avoided by the use of discounted cash flow assessment.
2. *Rate of return.* A rate of return is calculated on the profits remaining after the initial outlay has been written off. This method suffers from the same defect mentioned in (1) above and from the use of arbitrary periods for the writing down of the initial expenditure and of arbitrary rates of interest for the calculation of the rate of return.

Table 61.1 Application of NPV. To illustrate the uses of the two DCF methods the following example assumes the following data relating to two competing projects

Year	Net cash flows	
	Project A $	Project B $
0	−100 000	−200 000
1	−15 000	−50 000
2	20 000	5 000
3	25 000	10 000
4	35 000	25 000
5	40 000	40 000
6	50 000	75 000
7	−5 000	75 000
8	65 000	90 000
9	85 000	110 000
10	100 000	130 000
Totals	300 000	310 000

Year	Net present values at 10%	
	Project A $	Project B $
0	−100 000	−200 000
1	−13 367	−45 455
2	16 528	4 132
3	18 783	7 513
4	23 905	17 075
5	24 836	24 836
6	28 225	42 338
7	−2 566	38 490
8	30 323	41 985
9	36 049	46 651
10	38 550	50 115
Totals	101 266	27 680

Although Project B shows a greater total of net cash inflows over the whole period, at net present values Project A indicates a more satisfactory return, all other factors being ignored.

3. *Discounted cash flow* (DCF). This method recognizes that $1,000 income in five years' time is worth less than $1,000 receivable this year. The use of DCF in appraising two or more competing projects offers two methods of assessment: the net present value (NPV) or DCF rate of return.

The principle of NPV is best understood by applying an agreed discounting rate, that is, the best investment rate obtainable by the company, to the sum to be invested. A discounting rate of 10 per cent assumes that $100 invested now will be worth $110 in a year's time. Conversely, it is assumed that $110 in a year's time is, at 10 per cent, worth $100 today. From these assumptions, it is possible to construct DCF tables for varying numbers of years and discounting rates (see Table 61.1).

To use Table 61.2, which is based on the formula $1/(1 + i)^n$, where i is the rate of interest and n is the number of years, the relevant factor is found for the rate and number of years and multiplied by the amount for which the NPV is required. Thus to find the NPV for $1500 receivable in 5 years at 10 per cent from the table is found the factor of 0.621 and this, multiplied by $1500, gives an NPV of $931.5. Most spreadsheets used on personal computers include a formula for calculating NPV, so avoiding the need to construct tables.

The use of NPV (or DCF) leaves unresolved an important problem, that of determining the rate of interest or return to be used. Different rates of return could alter the

Table 61.2

n	5%	10%	15%	20%	25%	30%
1	0.9524	0.9091	0.8696	0.8333	0.8000	0.7692
2	0.9070	0.8264	0.7561	0.6944	0.6400	0.5917
3	0.8638	0.7513	0.6575	0.5787	0.5120	0.4552
4	0.8227	0.6830	0.5718	0.4823	0.4096	0.3501
5	0.7835	0.6209	0.4972	0.4019	0.3277	0.2693
6	0.7462	0.5645	0.4323	0.3349	0.2621	0.2072
7	0.7107	0.5132	0.3759	0.2791	0.2097	0.1594
8	0.6768	0.4665	0.3269	0.2326	0.1678	0.1226
9	0.6446	0.4241	0.2843	0.1938	0.1342	0.0943
10	0.6139	0.3855	0.2472	0.1615	0.1074	0.0725
11	0.5847	0.3505	0.2149	0.1346	0.0859	0.0558
12	0.5568	0.3186	0.1869	0.1122	0.0687	0.0429
13	0.5303	0.2897	0.1625	0.0935	0.0550	0.0330
14	0.5051	0.2633	0.1413	0.0779	0.0440	0.0253
15	0.4810	0.2394	0.1229	0.0649	0.0352	0.0195
16	0.4581	0.2176	0.1069	0.0541	0.0281	0.0150
17	0.4363	0.1978	0.0929	0.0451	0.0225	0.0116
18	0.4155	0.1799	0.0808	0.0376	0.0180	0.0089
19	0.3957	0.1635	0.0703	0.0313	0.0144	0.0068
20	0.3769	0.1486	0.0611	0.0261	0.0115	0.0053
21	0.3589	0.1351	0.0531	0.0217	0.0092	0.0040

The cash flow expected for each period is discounted by the factor for the rate of interest chosen and the number of periods in which the cash flows will occur. The number of periods is calculated from the commencement of the capital expenditure. The factors are arrived at from the formula $1/(1 + i)^n$, where i is the rate of interest expressed as a decimal and n is the number of periods. In reality, the factors assume that the cash flow passes on the last day of each period but can be adopted where the flow is roughly even throughout the period.

ranking of the projects by changing the point at which the returns shown by the projects are in balance. If the company's own rate of return on capital is higher than that revealed by the NPV calculation then the apparently more viable scheme may not prove to be the more acceptable.

An alternative is therefore to use another method, using the same principles, by calculating a DCF rate of return. This has the advantage of not involving any assumptions as to interest rates, but calculates the effective rate of return on each project. The DCF rate of return is defined as the rate which reduces the NPV to zero. This method is more difficult to calculate in that it necessitates taking several trial values until two are found giving values on either side of zero. A weighted average can then be applied to 'fine tune' the result.

Where there are constraints upon the provision of funds then the DCF rate of return method will be the more appropriate. Where the organization has ready access to finance then the NPV method, using the known long-term borrowing rate, should be used.

61.6 Control of capital expenditure

Significant capital expenditure usually represents a substantial commitment of the resources of a business, both financially and in terms of man-hours. It is therefore incumbent upon management to ensure that proposals for such outlays receive proper and full consideration of all the relevant implications before implementation. Once policies as to levels of authorization and commitment are laid down, there should follow the formal appraisal of the financial effects of the proposal. These can be formulated only after detailed discussion with the appropriate departments as to all the physical, technical and environmental factors involved in making the final decision. There will also be brought into consideration, where pertinent, the marketing and sales effects.

Most organizations will have sets of forms for use in the authorization and control of capital expenditure and these will vary in design and content. A basic guide for such documents would include the following:

1. *Capital variation proposal* (Figure 61.3). This should embrace the following:
   (a) Description;
   (b) Amount for which authority is requested;
   (c) Reasons for application (e.g. to increase production, to maintain production, to reduce costs, for the introduction of a new product, etc.);
   (d) Expenditure or losses not included in the proposal, such as staff amenities, transport requirements, self-competition, etc.;
   (e) Summary of the cost;
   (f) Disposal or modification of existing assets;
   (g) Details and timing of the outlays;
   (h) Extra working capital demands which will arise from the implementation of the project;
   (i) Index of the documents supporting the application.
2. *DCF calculations* (Figure 61.4). These will normally be in the form of working sheets and/or graphs, the latter being used for interpolation where the calculations

are too numerous or too detailed for tabulation. Calculations will normally be carried out using computers but where this is not possible, DCF tables should be utilized.

3. *Calculation document* (Figure 61.5). This document provides a means of calculating the DCF rate of return or a net present value where these are required, usually where profitability of the project is of major concern.
4. *DCF rate of return graph* (Figure 61.6). This may not be necessary in all cases but would be used for the interpolation of rates of return where a precise answer is not obtained from the detailed DCF workings described above.
5. *Payback graph* (Figure 61.7). This provides a view of the profile of the cash flows emanating from the project and forms a useful adjunct to the DCF information.
6. *Summary of cash flows* (Figure 61.8). This document is an essential part of every capital expenditure project evaluation. The forecast thus provided will be needed in the preparation of the overall cash budget.

61.6.1 Current and post-event monitoring

Large and complex projects will be monitored as the expenditure proceeds while smaller outlays will be looked at after completion. The purpose of such monitoring is to provide a comparison of actual expenditure with that estimated when the project was sanctioned. The accounting or internal audit departments will normally carry this out. However, for a sensible comparison to be achieved it is essential that the planned expenditure demonstrated in the project be classified in the same way that the actual expenditure will be analyzed. While somewhat obvious, this is an area where very great difficulty is often met in practice, especially where large projects are involved.

Considerable thought should be given to this aspect at the planning stage, with consultation taking place between the engineering and accounting functions. Apart from its obvious use to prevent serious over-runs of expenditure, 'post-event monitoring' provides useful lessons for the future preparation of capital projects.

61.7 Standards and standard costing

Standard costing is usually thought of in connection with manufacturing production but can be used with advantage in the measurement of the efficiency of supporting plant and equipment. Most readers of this book will already be familiar with the measurement of efficiencies against, for example, manufacturers' standards for a specific item of equipment. The standards related to such plant will themselves play a part in setting the production standards mentioned above.

It is not proposed here to provide instruction in the techniques of standard costing but merely to illustrate the uses to which certain of the methods can be put. This can best be shown by the use of a worked example. The following is the statement for a service department supplying a factory, which uses both fuel and materials.

Description   Amount $

New
Other

Total

Budgeted

Reason
New product
Cost reduction
Legal requirement

RELATED EXPENDITURE not included here
Staff amenities

Transport and distribution costs

SUMMARY OF EXPENDITURE $ ASSETS TO BE DISPOSED OF Value $
Buildings
Other
Sub-total
Working capital

Grand total

Prepared by   Recommended by

Approved by Board   Date

Details of expenditure for approval
                      $ Comments   Timing

Details of additional working capital required
                      $ Comments   Timing
Stocks increase
Debtors increase

Minus creditors increase

Total

Figure 61.3 A capital variation proposal

	Budget	Standard	Actual
Output units	32,000	33,600	29,400
Operating days	20	21	21
Cost per unit	83.5c	83.5c	90.0c
Fuel – Consumption	200	210	213
– Unit price	$50	$50	$49
– Cost	$10,000	$10,500	$10,437
Materials – Consumption	800	840	700
– Unit price	$2	$2	$2.10
– Cost	$1,600	$1,680	$1,470
Labor – Hours	80	84	105
– Hourly rate	$5	$5	$7
– Cost	$400	$420	$735

Repairs			
– Rate per output unit	10c	10c	11c
– Cost	$3,200	$3,360	$3,234
Variable overheads			
– Rate per output unit	20c	20c	22c
– Cost	$6,400	$6,720	$6,468
Fixed overheads			
– Rater per output unit	16c	16c	14c
– Cost	$5,120	$4,704*	$4,116
Total cost	$26,720	$27,384	$26,460

*Actual output × standard cost.

Cash outflow

Year	Cash	0% Factor	NPV	5% Factor	NPV	10% Factor	NPV	15% Factor	NPV	20% Factor	NPV
0		1.0		1.0		1.0		1.0		1.0	
1		0.952		0.909		0.870		0.833			
2		0.907		0.826		0.756		0.694			
Total $		$		$		$		$		$	

Cash inflow

		Factor	NPV	Factor	NPV	Factor	NPV	Factor	NPV		NPV
1		0.952		0.909		0.870		0.833			
2		0.907		0.826		0.756		0.694			
3		0.986		0.751		0.658		0.579			
4		0.823		0.683		0.572		0.482			
5		0.784		0.621		0.497		0.402			
6		0.746		0.564		0.432		0.335			
7		0.711		0.513		0.376		0.279			
8		0.677		0.467		0.327		0.233			
9		0.645		0.424		0.284		0.194			
10		0.614		0.386		0.247		0.162			
Total $		$		$		$		$		$	

Figure 61.4 DCF calculations

Project —————————————————

Year ending	1	2	3	4	5	6	7	8	9	10
Revenue from project										
Other income arising										
Total income										
COSTS										
Operating profit (loss)										
Capital allowances										
Taxable profits										

	1	2	3	4	5	6	7	8	9	10	11
Tax											

Figure 61.5 Profit calculation

Variances from standards are then calculated as follows:

			$	
Fuel				
Usage variance	$SP(SQ-AQ)$	$= 50(210 - 213)$	$= 150A$	
Price variance	$AQ(SP-AP)$	$= 213(50 - 49)$	$= \underline{213F}$	63F
Materials				
Usage variance	$SP(SQ-AQ)$	$= 2(840 - 700)$	$= 280F$	
Price variance	$AQ(SP-AP)$	$= 700(2.00 - 2.10)$	$= \underline{70A}$	210F
Labor				
Rate variance	$AH(SR-AR)$	$= 105(5 - 7)$	$= 210A$	
Efficiency	$SR(SH-AH)$	$= 5(84 - 105)$	$= \underline{105A}$	315A
Repairs				
Volume variance	$SC(SV-AV)$	$= 0.10(33,600 - 29,400)$	$= 420F$	
Price variance	$AV(SC-AC)$	$= 29,400(0.10 - 0.11)$	$= \underline{294A}$	126F
Variable overheads				
Volume variance	$SC(SV-AV)$	$= 0.20(33,600 - 29,400)$	$= 840F$	
Price variance	$AV(SC-AC)$	$= 29,400(0.20 - 0.22)$	$= \underline{588A}$	252F
Fixed overheads				
Expenditure variance	$AFO-BFO$	$= 4116 - 5120$	$= 1004F$	
Volume efficiency	$SC(AQ-SQ)$	$= 0.16(32,000 - 33,600)$	$= 256A$	
Yield	$SC(SY-AY)$	$= 0.16(32,000 - 29,400)$	$= 416A$	
Capacity	$SC(RBQ-SQ)$	$= 0.16(33,600 - 33,600)$	$= NIL$	
Calendar variance	$SC(RBQ-BQ)$	$= 0.16(33,600 - 32,000)$	$= \underline{256F}$	$\underline{588F}$ 924F

SC	=	Standard cost	BQ	=	Budgeted quantity
SQ	=	Standard quantity	AQ	=	Actual quantity
SP	=	Standard unit price	AP	=	Actual unit price
AH	=	Actual hours			
SH	=	Standard hours	AH	=	Actual hours

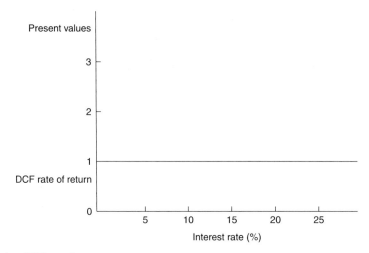

Figure 61.6 Interpolation of DCF rate of return

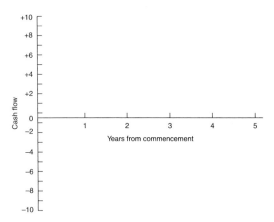

Figure 61.7 Cumulative cash flow

SV	= Standard volume	AV	= Actual volume
AFO	= Actual fixed overheads	BFO	= Budgeted fixed overheads
SR	= Standard rate per hour	AR	= Actual rate per hour
SY	= Standard yield	AY	= Actual yield
BQ	= Budgeted quantity	RBQ	= Revised budgeted quantity
A	= Adverse	F	= Favorable

Fixed overheads require some amplification. Budgeted output for the year is 400,000 units, budgeted fixed overheads for the year are $40,000. There are 50 working weeks in the year with 4 weeks in the period being reviewed, with one extra day being worked. The actual fixed overheads for the period amounted to $4,116.

Actual output for the period	29,400 units
Budgeted output for the period	32,000 units
Standard output for the period	33,600 units

(Budgeted output is 1600 (400,000/250) units per day, so 21 days' standard output is 33,600 units.)

It is necessary to present this information in an understandable way, and this may take the following form:

		$
Budgeted cost of output		26 720
Output variances		
Volume	2 171F	
Unit cost	1 911A	260F
Actual cost of output		26 460
Less: Standard cost of output		
Fuel	10 500	
Materials	1 680	
Labor	420	
Repairs	3 360	
Variable overheads	6 720	
Fixed overheads	4 704	27 384
Favourable variance		924

Analysis of variances:

Fuel	– Usage	150A	
	– Price	213F	63F
Materials	– Usage	280F	
	– Price	70A	210F
Labor	– Rate	210A	
	– Efficiency	105A	315A
Repairs	– Expenditure		126F
Variable overhead			252F
Fixed overhead			
	– Expenditure	1 004F	
	– Volume efficiency	256A	
	– Yield	416A	
	– Calendar	256F	588F
			924F

Comments

Fuel suffered from excessive usage, which was partly offset by a lower unit price.

Year	Fixed assets $	Working capital $	Other expenses $	Operating profit or loss $	Taxation $	Net profit or loss $	Cumulative totals $
0							
1							
2							
3							
4							
5							
6							
7							
8							
9							
10							
Totals							

Figure 61.8 Summary of cash flows

Materials showed lower usage but a higher than expected price was paid.

Labor both cost more per hour and were utilized longer.

Repairs and variable overheads both benefited from lower than expected expenditure.

Fixed overheads were less than forecast but this saving was partially offset by lower volume and yields. The extra day's working increased the net advantage gained from these factors.

61.8 Capital

Businesses require funds for day-to-day operations ('working capital') and for expansion by acquisition and for the provision of plant and machinery, buildings, etc. Most working capital needs are normally (and should be) met from the company's own cash generated from its own operations. Indeed, the need to meet this criterion serves as a discipline upon the company's standard of cash management in relation to credit control, payment of suppliers, etc.

Short-term needs will be met mainly from the company's own cash flow and from overdraft facilities provided by the company's bankers. Entering into arrangements for the hire purchase or leasing of specific items of equipment can provide medium-term facilities. The facilities available are many and varied. Long-term resources will normally be met by the raising of further capital by way of Stock Exchange placing or flotation or by the securing of loans through debentures secured upon the assets of the company, bearing fixed rates of interest and redeemable at a date in the future.

61.8.1 Short term

The main sources include:

1. *Overdrafts*. Usually the least expensive; interest is charged on the daily balance usually at a premium (often substantial) over the bank's base lending rate.
2. *Factoring*. Usually applied to sales invoices (effectively, the debtors). Specialist finance houses provide a service whereby the company is paid promptly a percentage of its sales value for a period and the finance house then collects the full amount from the debtor.

   In some cases the finance house will undertake the invoicing of the customers direct, providing a complete sales ledger service. Confidentiality as to the arrangement is lost in this case. There are many schemes with considerable variations available.
3. *Acceptance credits*. These are provided mainly for exports by the major banks and specialist accepting houses.

61.8.2 Medium term

This may be provided by:

1. *Hire-purchase*. Normally for periods of up to three or five years. The ownership of the goods remains with the finance company until the installment payments are complete and a final option to buy is exercised. Payments remain fixed throughout the term of the agreement with a predetermined unchanging rate of interest included.
2. *Leasing*. The asset is leased for a term of years at a fixed rental. The term can then be extended

for secondary or tertiary periods at nominal rentals. Recently-introduced accounting standards require that assets acquired under contracts of this sort must appear in the balance sheet at cost and be depreciated even though not legally owned by the company.

3. *Lease purchase*. There are on offer many variants of the two previously illustrated sources, giving choices of both methods in combination with each other. The underlying principles are the same.

4. *Contract hire*. Although not strictly a source of outside finance, this is a method of avoiding capital outlay, especially favored for vehicles. Again, there are many varying systems to be found.

5. *Sale and lease-back*. Once considered only for land and buildings, schemes are now available for other assets, notably sizeable fleets of vehicles.

61.8.3 Long term

The principal sources include:

1. *Flotation on stock markets*. This involves transferring in whole or in part the ultimate ownership of the business to persons outside the company and, dependent upon the proportion sold, may leave the business vulnerable to takeover by competitors or unwelcome bidders.

2. *Share placings*. Shares are issued for a consideration to known and specific parties at an agreed price, usually through the medium of a merchant bank. This reduces the element of vulnerability to takeover bids but does not eliminate the possibility, should any of the 'placees' be prepared to sell their holdings at a future date.

3. *Issue of debentures*. These will normally carry a fixed rate of interest and have a predetermined date of redemption, possibly at a premium. The holders of debentures will usually require security perhaps by means of a fixed charge over specific assets (or all the assets), and will have a right of prior payment in the event of a liquidation. Debenture holders can also sometimes exercise their rights on the occurrence of certain events. Widespread security given to one class of lender can militate against the provision of short-term finance from other lenders who require collateral.

4. *Loans from specialist companies*. There is a number of companies and institutions who specialize in providing long-term finance to industry. The forms of loans offered vary from one deal to another and, in some cases, will be accompanied by a request for some degree of participation in the equity of the borrower.

61.9 Value added tax

The calculation of VAT is straightforward when the rate at present in force is applied to a value to which it is to be added. The formula is:

Value + (Value × Rate/100)

Where the amount of value added tax included in a value is required the calculation is as follows:

Value × Rate/(100 + Rate)

Where the amount exclusive of value added tax included in a value is required the calculation is:

Value/[(100 + Rate)/100]

61.10 Break-even charts

It is often useful to prepare break-even charts in order to illustrate and give a clear picture of the position of a business. They can be adapted to help in showing the viability and profiles of individual projects.

The two most common designs of break-even charts are shown in Figures 61.9 and 61.10. In both cases the *y* (vertical) axis is used for sales (output) and costs while the *x* (horizontal) axis is used for volumes, capacity or time.

Illustration

Output (000s)	10	20	30	40	50	610
Fixed costs	5,000	5,000	5,000	5,000	5,000	5,000
Variable @ $0.30	3,000	6,000	9,000	12,000	15,000	18,000
Sales @ $0.50	5,000	10,000	15,000	20,000	25,000	30,000

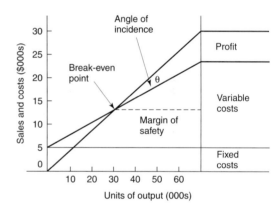

Figure 61.9 Break-even chart (1)

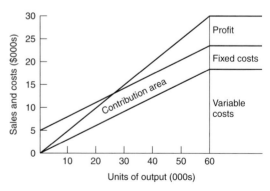

Figure 61.10 Break-even chart (2)

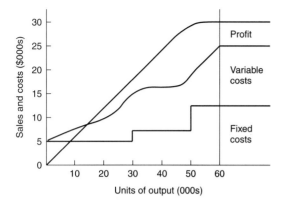

Figure 61.11 Break-even chart (3)

The angle of incidence, if wide, will indicate a high rate of profitability. If this is coupled with a high margin of safety an extremely advantageous investment is demonstrated.

The position of the break-even point serves as an indicator. If it is well to the right of the chart then the margin of safety will be poor. Conversely, a position to the right of the chart will hold out opportunities for expansion.

The guidance obtainable from break-even charts is limited and should be used only with other criteria in formulating decisions. The effects of policy decisions may alter the data used and vitiate or amend the results disclosed.

In Figure 61.10 fixed costs are shown above the variable costs and the non-recovery of the fixed costs below the break-even point is more clearly demonstrated. The *contribution* to fixed costs is of significance in the consideration of marginal costing.

In practice, costs do not increase smoothly or remain constant. Fixed costs frequently move in a series of steps and variable costs change unevenly. There may be several break-even points at different levels of sales and outputs (Figure 61.11).

61.11 Supply of steam, power, water, etc. to other departments

The plant engineer is frequently called upon to assist and advise on the bases for charging to other departments within the plant or to outside purchasers where there is a central generation plant for a variety of users, as may occur in an industrial estate or where the plant providing the service has a surplus. The elements to be considered and brought together in formulating a basis for charge can be summarized as follows, always bearing in mind individual needs and circumstances:

1. *Expenses associated with the capital cost of equipment.* Depreciation will be provided over the useful life of the equipment, varying according to the engineer's estimate. Present-day custom is to write assets down on a straight-line basis, but there are several accepted ways of providing the necessary depletion of the asset value.
2. *Costs relating to the space occupied by the equipment.* These will usually include rent and rates or similar

taxes, together with the upkeep of the buildings housing the machinery. Where the operator owns the buildings a notional rent should be included to reflect the full cost of occupancy. Such costs will normally need to be allocated to the department by reference to the area occupied, but may require more complex calculations where the buildings on a site differ in age, or in standards of construction or maintenance.

3. *Costs of gas, electricity, etc.* These are usually readily ascertainable from the suppliers' accounts, but for charging purposes, great reliance will be placed on results from regular meter readings. The importance of an efficient routine for reading meters has been stressed elsewhere. It will be essential for accurate and fair apportionment between users for there to have been a recognition of the need for and acceptance of the cost of metering equipment for all user departments, down to individual machines if necessary. Where the main source of energy is gas then it will be usual to have a stand-by oil-fired system, especially where the gas supply is interruptible.

The tanks for this reserve will form part of the capital equipment and the oil itself will need to be included when considering the capital employed. Where electricity is produced on site as a by-product in periods of low demand for steam, more cheaply than that obtained from the utility company, adjustments to the charges on a seasonal basis will be required.

4. *Water.* Costs are available from the supplier's invoices but where the operator has reservoirs and/or licenses for extraction from canals or rivers the annual fees and penalties, which are sometimes taxable, should not be forgotten. Projections of future demands will need to be carefully considered where this type of arrangement is made, since such contracts are often available only on long-term bases. Water-treatment plant will produce its own range of costs across the whole field of depreciation, materials, electricity, labor, etc. and these will need to be apportioned to the steam cost departments before final allocation to the user departments.
5. *Labor costs.* In arriving at the total labor costs, there must be included social security costs, pension costs and any provisions for holiday pay. It may be desirable, where users take steam overnight, to apportion costs in such a way as to reflect the higher cost of labor employed on night work.
6. *Maintenance expenses.* The nature of these is generally self-evident, but long-term maintenance contracts, insurance inspection fees and any subsequent requirements arising therefrom should not be overlooked. Inspections are commonly undertaken at weekends, at premium labor rates.
7. *Recovery of appropriate proportion of overheads.* The overhead charge to the supplying department will usually be provided by central finance in accordance with the company's policy for allocation of overheads.
8. *Recovery of proportion of exceptional financing and interest charges.* This will normally depend upon the policy of individual companies and will probably have been included, if at all, in the general overheads referred to in (7). It should not be left out of

consideration where provision has been made for a particular user's requirements involving, say, special extension or modification of existing plant.

61.11.1 Computation of charges to users

Assumptions

For a site operating four gas-fired boilers each of 10,000-kg/h output, operating at 70 per cent efficiency for 60 hours per week, employing three people directly in the department, annual costs may be summarized as follows:

Annual cost of producing steam	$
Depreciation of boilers and equipment	100,000
Gas	850,000
Electricity (for steam production)	30,000
Water and water treatment	12,500
Labor for steam production	36,000
Repairs, maintenance and insurance	35,000
Stand-by oil	1,500
Proportion of site overhead	25,000
Proportion of finance overhead	6,000
	1,096,000

These costs would then be allocated to the user departments in proportion to actual use, using metered records and adjusted for the incidence of peak-time usage, overnight working, weekends or early starts.

Given that metering equipment is in place, the allocation of the steam production costs to individual users is straightforward and needs no amplification here. If the allocation has to be made employing estimates of usage, perhaps by using the ratings of individual pieces of equipment, then everything will turn upon the general and continuing acceptance of those estimates by all the users concerned. Should these conditions not be met or maintained then the life of the plant engineer will be an uneasy one!

61.12 Charges for effluent and environmental services

These services are now very often a significant part of the costs of a factory and are worth a note here. Projecting accurate costs can be difficult, since water authorities commonly render accounts quarterly with charges fluctuating from month to month in relation to both volumes and chemical content ('toxicity') of the effluent, all of which can be highly variable. Monthly liaison with the authority's representative and inspection of sampling reports will often enable more accurate information to be available, both for the assessment of costs where charges are to be made to individual departments and for the accruing of costs by the management accountant.

62

Statistical Approaches in Machinery Problem Solving

Ricky Smith

Life Cycle Engineering, Inc

Contents

62.1 Introduction

Machinery failure and distress behavior can be influenced by appropriate design techniques. It would be wonderful to verify low failure during the specification, design, installation, and start-up phases of each machine within our PdM program. However, experience shows that good, basic design alone cannot prevent failures from occurring during the operational life span of each machine.

Failure analysis statistics have consistently shown that many machinery components failures can be directly attributed the equipment being operated outside of design parameters or unintended conditions. Most failure analysis and trouble-shooting activities are usually post-mortem and commence after installation and start-up of the equipment. The maintenance phase is now in motion, and failure analysis and trouble-shooting is now an integral part of that phase.

The failure fighting process includes all machinery maintenance activities, failure analysis, and trouble-shooting. Fighting failures involves using every strategy that is appropriate to the situation if the process is to be successful. However, before any strategies can be applied, as much information must be gathered as possible regarding failure modes and their frequencies. Appropriate statistical approaches will accomplish this task and therefore have to be part of the failure fighting system.

62.2 Machinery failure modes and maintenance strategies

Most plants have maintenance departments for the sole purpose of taking care of all types of machinery. Typical types of equipment deterioration are machinery component failure events, their malfunctions, and system troubles.

As in most cases, maintenance departments would like to perform machinery maintenance in a rational and planned manner. The fact that we plan our maintenance activities implies that we too behave in a rational and planned way, and – hopefully – respond favorably to planned maintenance actions.

However, in reality this sometimes does not always work that way. If we are to be realistic, we must examine how machinery component failure modes appear and behave as a function of time. To understand this fully, we must concern ourselves with the concept of machinery component life.

62.3 Useful life concepts

Failure is one of the ways in which engineered devices attain the end of their useful life. Table 62.1 shows some of the possible ways.

Usually most machinery failure analysis is performed at component level, therefore, a definition is required to state what constitutes life attainment in connection with component failure. 'Defect limit' is a term that is used in this context and needs to be expanded upon to understand it fully.

Defect limit describes a parameter which we are all familiar with. For example, when you buy original automotive equipment tires there are built-in tread wear indicators. These wear indicators are molded into the bottom

of the tread grooves and appear as approximately $\frac{1}{2}$-inch wide bands when the tire tread depth is down to $\frac{1}{16}$ inch. The wear indicators are 'defect limits'.

Other typical examples of defect limits are clearance tolerances of turbo-machinery assemblies, which are measured and recorded during machinery overhauls. Defect limits are not always as obvious as those stated above. Mechanical shaft couplings are very difficult to determine when serviceability is questionable, as they have no wear indicators to provide guidance.

There are defect limits that are associated with 'random failure' modes. For example, if there is a leak from a mechanical seal on a pump, where do we decide that the leakage is excessive and requires immediate maintenance? Vibration analysis severity levels are also typical examples of when do we have severe enough conditions to warrant equipment shutdown and overhaul. In such circumstances, the defect limit is dependent upon individual subjective judgment.

This is sometimes referred to as 'user tolerance' to the distress symptoms displayed by the machine. 'User tolerance' is really a catch-phrase to describe those people who, based upon their individual level of 'expertise', will determine that the defective machine can keep on running at a level of severe vibration as opposed to those who would take the machine out of service immediately.

Regardless of how service life is determined, the 'life' of a machinery component must be defined as the time interval between putting the component into operation and the point in time when the component reaches its defect limit.

'Standard life' is described as the average lifetime that is acceptable to any plant failure analyst or troubleshooter. Therefore, if we arrive at defect limits in machinery within the maintenance program, we have also reached the 'standard life' of all the failure modes in the plant. Do we now:

1. Live with the known defects?
2. Look for improvements to the program?

These questions must be answered by setting defect limits objectively.

If there are no baselines established, or any guidelines upon which to improve, and allow subjective judgment to prevail, we can guarantee some unpleasant failures because those judgment determinations will range from the conservative to the optimistic. The conservative approach reduces component life arbitrarily and the optimistic approach exposes the machine to unjustifiable high risks. Therefore, it is highly recommended that the team of

Table 62.1 Reaching end of useful machinery life

Way in which the end of useful life is reached	Example
Failure Slow	Mechanical seal leakage
Sudden	Motor winding failure
Obsolescence	Recip. steam pump/engine
Completion of mission	Oil mist
	Packaging
	Wear pads
Depletion	Electric battery

analysts or troubleshooters set defect limits, and the final set points fully documented to ensure compliance.

Experience shows that some machines have more frequent failures than do others. Obviously, different failure modes have different frequencies of occurrence. This is usually described as 'mean-time-between-failure' (MTBF) and expresses the probability of machinery failure and breakdown events as a function of time. This is of particular interest to the maintenance failure analyst and troubleshooter who have to grapple with the realization that some machinery failure modes appear slowly and predictably whilst others occur randomly and unpredictably. In most cases, both types of failures have been encountered.

Predictive maintenance

While preventive maintenance is concerned with regularly testing, and reconditioning equipment to prevent failures in service and premature deterioration, it follows that predictive maintenance procedures are concerned with the ability to predict when the equipment will fail and then developing schedules to implement timely repairs. Predictive maintenance does not imply that with the use of these techniques, failure modes in equipment can be prevented; rather, it suggests that the occurrence of failure can be predicted and thus planned for. An appropriate example would be the inspection and change of a major compressor face-type oil seal where random heat checking (FM) has been observed over the years.

The ideal predictive maintenance program will continuously search for defects in equipment by regularly monitoring the condition and performance of equipment, as well as maintaining a system for ongoing feedback. While the implementation of a program may depend on management philosophy, risk evaluations and cost/benefit ratios to name a few, the success of a program will depend largely on the monitoring devices provided and the willingness of operators and maintenance personnel to be involved in the program. Unfortunately, predictive maintenance activities are not performed as frequently as desired and usually with less sophisticated instruments such as the observer's sense of touch, sight, smell and sound.

Breakdown maintenance

Although there is widespread acceptance of the possibilities of machinery preventive maintenance, it should not be concluded that its concepts are without limitation.

The first limitation is set by the nature of random failure events. If we believe that failures occur periodically and not with equal probability in time, the premise of PM or periodic machinery maintenance is incorrect. Therefore, the only alternative is continuous monitoring.

The second limitation is the life dispersion of machinery components. It is difficult to predict time-dependent failure modes because even they do not occur at the exact same operating intervals. Consider the life dispersion of mechanical gear couplings on process compressors. Both components are clearly subject to wear. If we conclude that their MTBF (mean-time between failure), or mean-time-between-reaching-of-detect-limit is 7.5 years, it is possible to have an early failure after 3 years and another

after 15 years. The premise is that there are no undue extraneous influences such as excessive misalignment or lubricant loss on the machinery. Also, in this example, the longest life span will be five times as long as the shortest life span. Figure 62.1 describes this relationship.

Since most time-dependent failures have larger life dispersions, we must consider the maximum and minimum ratios of 4:1 and 40:1. Generally, relative life dispersion increases with the absolute value of MTBF. That is, wear items with a relatively short life expectancy such as rider rings on reciprocating compressors will have a comparatively smaller dispersion than components such as gear tooth flanks, which can be expected to remain serviceable for long periods of time.

Finally, the third limitation of predictive maintenance in process machinery is that downtime must be scheduled to allow for inspection of the equipment, which in itself poses risk. It is not an exaggeration to say that the source of the majority of equipment failures, which were studied, began shortly after predictive maintenance inspections made disassembly of equipment necessary. Figure 62.2 shows how trouble incident frequencies can increase contrary to expectations after inspection and overhaul.

The only thing we can say for sure is that even if the optimal combination of preventive/predictive maintenance were found, there would always be certain situations where breakdown maintenance would need to be performed. For instance, lube-oil changes will have failure modes that do not always respond to periodic servicing and may not even be noticed through inspection or monitoring. Other failure modes that are not easy to predict are motor winding failures, the insulation breakdown of a feeder cable for a large compressor motor driver, or pump shaft fracture caused by fatigue.

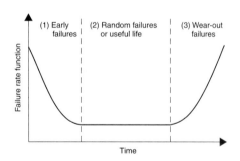

Figure 62.1 Bathtub curve

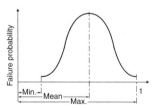

Figure 62.2 Failure probability and minimum–maximum life of machinery components

The existence of breakdown maintenance unquestionable. The maintenance activity is necessary to restore machinery equipment back to service after failure modes developed that were:

1. Preventable but not prevented.
2. Predictable but not predicted.
3. Predicted but not acted upon.
4. Not preventable or predictable.

62.4 Bad actor management

When certain components of the process machinery consistently malfunction, we call it bad actor management.

Another way to think of bad actor management is to liken it to Pareto's law, where 20 per cent of the components are causing 80 per cent of the problems. Many of the causes previously identified in connection with machinery component failure modes lend hand to bad actor management; however, some of the more significant are listed as follows:

1. Not suited for service due to:
   • wrong design assumptions
   • incorrect material
   • change in operation conditions.
2. Maintenance:
   • improper repair
   • design and/or selection errors.

Bad actor management is preoccupied with the following necessary problem solving steps:

1. Bad actor identification and tracking.
2. Failure analysis and documentation.
3. Follow-up.
4. Organizational aspects.

62.4.1 Bad actor identification and tracking

Similar to reliability work, bad actors are also identified through knowledge of the machinery MTBF by type, unit and service. Resources are likely to be limited, which means we will not be able to investigate all resources that deviate from an acceptable or expected time to failure (TTF). Therefore, the next step is to carefully choose an acceptable TTF for machine breakdown, keeping these limitations in mind. We don't want to start at the top, but it is imperative that we remember to use this process to cover all failures, even if we feel we know all of the bad actors. Later, the importance of this step will become obvious because it will not be as easy for us to tell the bad actors from acceptable actors. Figure 62.3 shows the methodology of this process.

The third step is to proceed to the component level because more than likely, the cause will be uncovered here. Table 62.2 shows machinery component failure modes commonly encountered in machinery failure analysis together with suggested standard life values, Weibull indices (β), and responsiveness to preventable or predictable maintenance strategies. Referring to this table will help decision-making. To assist the analyst in documenting bad actor failure modes on the component level,

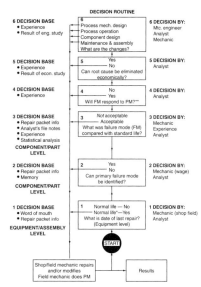

Figure 62.3 Machinery failure analysis methodology

Figure 62.4 is provided as an example of what a typical tracking sheet looks like.

Even though bad actor identification and tracking can be a complicated process, it is the first step in an effort to eliminate recurring and costly machinery failures. There are many other steps which are necessary in order to identify bad actors, and we will discuss them later.

The following two sections explore the use of statistical tools, which may help define failure experience trends in, given machinery.

62.4.2 Hazard plotting for incomplete failure data

In order to plot and analyze data, you must first take into account the form the data is in. There are two possibilities; failure data can either be complete or incomplete. Failure data is complete if the data consist of failure times for all the units in the sample. However, if only the failure times of the failed units and the running times of the unfailed units are collected, the data are incomplete. This type of data is also known as 'censored,' and the running times are known as 'censoring times.' If the unfailed units all have the same censoring time (which is greater than the failure times), the data are singly censored. If unfailed units have different censoring times, the data are multiply censored.

When all units have failed, the result is complete data. Singly-censored data result in life testing when testing is terminated before all units fail. And multiply-censored data result from:

1. Removal of units from use before failure.
2. Loss or failure of units due to extraneous causes.
3. Collection of data while units are still operating.

Table 62.2 Primary machinery component failure

Failure Mode	*Weibull Index*	*Standard Life*	*Maintenance Strategy* preventive	*Maintenance Strategy* predictive
Deformation				
Brinelling	1.0	Inf		•
Cold flow	1.0	Inf		•
Contracting	2.0	Inf	•	
Creeping	2.0	Inf		
Bending	1.0	Inf		•
Bowing	1.0	Inf		
Buckling	1.0	Inf		
Bulging	1.0	Inf		
Deformation	1.0	Inf		
Expanding	1.0	Inf		
Extruding	1.0	Inf		
Growth	1.0	Inf		
Necking	1.0	Inf		
Setting	2.0	Inf		
Shrinking	2.0	Inf		
Swelling	3.0	Inf		
Warping	1.0	Inf		
Yielding	1.0	Inf		
Examples:				
Deformation of springs	1.0	Inf		•
Extruding of elastomeric seals	1.0	4.0Y		•
Force-induced deformation	1.0	Inf		
Temperature-induced deformation	2.0	Inf		•
Yielding	1.0	Inf		
Fracture/Separation				
Blistering	1.0	Inf		
Brittle fracture	1.0	Inf		
Checking	1.0	Inf		
Chipping	1.0	Inf		
Cracking	1.0	Inf		
Caustic cracking	1.0	Inf		
Ductile rupture	1.0	Inf		
Fatigue fracture	1.0	Inf		
Flaking	1.0	Inf		
Fretting fatigue cracking	1.0	Inf		
Heat checking	1.0	Inf		
Pitting	1.0	Inf		
Spalling	1.0	Inf		
Splitting	1.0	Inf		

The following are examples of situations with these reasons for multiply-censored data:

1. In some life-testing work, units are put on test at various times but are all taken off together; this is done to terminate testing by some chosen time to meet a schedule.
2. In medical follow-up studies on the length of survival of patients after treatment, contact with some patients may be lost or some may die from causes unrelated to the disease treated. Similarly, in engineering studies, units may be lost or fail for some reason unrelated to the failure mode of interest; for example, a unit on test may be removed from a test or destroyed accidentally.
3. In many engineering studies, field data are used which consist of failure times for failed units and differing current running times for unfailed units. Such data arise because units are put into service at different times and receive different amounts of use.

Graphical analysis of failure data is most commonly plotted using probability. However, in order to understand the hazard plotting method presented here, is not necessary to understand probability plotting. While it is difficult to utilize probability plotting for multiply-censored data, it is relatively easy to use it for complete and singly-censored data. One simple expedient that is used to plot multiply-censored data is to ignore all failures that occurred after the earliest censoring time and plot only those that occurred before the first censoring time. This procedure may be acceptable if there are relatively few failures after the first censoring time and thus little information is lost by excluding them form the plot, although considerable information is lost if there are many failures after the earliest censoring time. However, if the earliest censoring time is in the lower tail of the distribution and only the failure times before that time are used, the distribution function must be estimated by considerable extrapolation beyond the earliest censoring time. The estimate of the distribution may be in error if you consider that a theoretical distribution, which agrees with the data below the earliest censoring time, may differ considerably from the true distribution function above the earliest censoring time. You take this risk if you extrapolate beyond the data that is there. To be as accurate as possible, it is advisable to plot all failure times occurring both before and after the earliest censoring time to make full use of the information in the data.

BAD ACTOR TRACKING		FAILURE ANALYSIS		NO: 023
DESCRIPTION: P–101 , Hightower Feed pump				
INFORMATION	FAILURE	FAILURE	FAILURE	FAILURE
DATE	80/11/05	81/01/15	81/05/10	
COMPONENT / PART	SEAL	SEAL	SEAL	
VENDOR DRWG. NO.				
ELEMENT/PART	Secondary sealing o–ring part # 13	Same Part # 13	(3) Springs Part # 5	
FAILURE MODE	Swelling	Swelling	Breakage	
TTF #	?	1.5M	5.0M	
CORRECTIVE ACTION	Replaced	Changed to viton	Corrected internal alignment	
BAD ACTOR PART (Check)	?	✓		
FAILURE CAUSE	Changed pump service	Changed pump service	Internal Mis-alignment due to flange loading	
FAILURE RESISTANCE	O–ring material	O–ring material	Pump casting strength	
ANALYSED BY:	J.Pickel		DATE: 81/05/30	

REVISION NO.	REMARKS	DATE
1		
2		
3		
4		

Time–To–Failure

Figure 62.4 Bad actor tracking sheet

Plotting data on hazard paper requires less effort, and at the same time, it succeeds in using all of the failures and gives the same information. Hazard papers are shown here for exponential, Weibull, normal, log normal, and extreme value distributions.

Generally, failure is described in terms of time to failure. Although in certain instances a more relevant measure for exposure before failure is necessary such as cycles to failure or miles to failure. It should also be noted that the word failure is used in a general sense to indicate any specific deterioration in performance of a unit. Additionally, if units are put into service at different times, the time in use for each unit is figured from its starting time.

The hazard plotting method is presented in detail in the next section followed by the step-by-step instructions on how to make a hazard plot of multiply-censored data with examples based on actual failure data to help illustrate the instructions. The next section describes how to obtain engineering information on the distribution of time to failure by using and interpreting a hazard plot. This section also contains plots and analyses of simulated data from different known theoretical distributions, which allows us to assess the performance of the hazard plotting method. Practical advice on how to use hazard paper finishes up the section.

62.4.3 How to make a hazard plot

The following are step-by-step instructions for making a hazard plot for multiply-censored failure data. Table 62.3 provides data to illustrate the hazard plotting method. These data for 70 diesel generators consist of hours of use on 12 generators at the time of fan failure, and hours on 58 generators without a failure. Figure 62.5 shows the ordered times to failure for the failed fans and the running times for the unfailed fans. The engineering problem was to determine if the failure rate of the fans was decreasing with time. Then based on the failure rate information, a decision by management and engineers would be made as to whether or not the unfailed fans should be replaced with better fans to prevent future failures. It is also possible that the weak fans would be removed because of failure, and that the remaining fans would have a failure rate low enough to be acceptable. Steps in the

hazard-plotting method follow with the fan failure data used as an example:

1. Suppose that the failure data on n units consist of the failure times for the failed units and the running (censoring) times for the unfailed units. Order the n times in the sample from smallest to largest without regard to whether they are censoring or failure times. In the list of ordered times the failures are each marked with an asterisk to distinguish them from the censoring times. This marking is done in Table 62.3 for the generator fan failure data. If some censoring and failure times have the same value, they should be put into the list in a well-mixed order.

2. Obtain the corresponding hazard value for each failure and record it as shown in figure 62.6. The hazard value for a failure time is 100 divided by the number K of units with a failure or censoring time greater than (or equal to) that failure time. The hazard value is the observed conditional probability of failure at a failure time, that is, the percentage $100 (1/K)$ of the K units that ran that length of time and failed then. This observed conditional failure probability for a failure time is readily apparent in Figure 62.6, where the number of units in use that length of time and surviving is easily seen. If the times in the ordered list are numbered backward with the first time labeled n, the second labeled $n - 1, \ldots$ and the nth labeled 1, then the K value is given by the label. For example, these K values are shown in parentheses in Table 62.3. Three-figure accuracy for K is sufficient. Note that the hazard value of a failure is determined by the number of failure and censoring times following it; in this way, hazard plotting takes into account the censoring times of unfailed units in determining the plotting position of a failure time.

3. For each failure time, calculate the corresponding cumulative hazard value, which is the sum of its hazard value and the hazard values of all preceding failure times. This calculation is done recursively by simple addition. For example, for the generator fan failure at 16,000 hours, the cumulative hazard value 5.93 is the hazard value 1.54 plus the cumulative hazard value 4.39 of the preceding failure. The cumulative hazard values are shown in Table 62.3 for the generator fan failures. Cumulative hazard values can be larger than 100 per cent.

4. Choose a theoretical distribution of time to failure and use the corresponding hazard paper to plot the data on. Various hazard papers are shown in the figure. The theoretical distribution should be chosen on the basis of engineering knowledge of the units and their failure modes. On the vertical axis of the hazard paper, mark a time scale that includes the range of the sample chosen, and mark the vertical scale off from 1000 to 100,000 hours, since the range of failure times for the data is from 4500 to 87,500 hours. This is shown in Figure 62.6.

5. Plot each failure time vertically against its corresponding cumulative hazard value on the horizontal axis of the hazard paper. This was done for the 12 generator fan failures and is shown in Figure 62.7.

Table 62.3 Generator fan failure data and hazard calculations

n(k)	Hours	Hazard	Cumulative Hazard	n(k)	Hours	Hazard	Cumulative Hazard
1 (70)	4,500*	1.43	1.43	36 (35)	43,000		
2 (69)	4,600			37 (34)	46,000*	2.94	18.78
3 (68)	11,500*	1.47	2.90	38 (33)	48,500		
4 (67)	11,500*	1.49	4.39	39 (32)	48,500		
5 (66)	15,600			40 (31)	48,500		
6 (65)	16,000*	1.54	5.03	41 (30)	48,500		
7 (64)	16,600			42 (29)	50,000		
8 (63)	18,500			43 (28)	50,000		
9 (62)	18,500			43 (27)	50,000		
10 (61)	18,500			45 (26)	61,000		
11 (60)	18,500			46 (25)	61,000*	4.00	22.78
12 (59)	18,500			47 (24)	61,000		
13 (58)	20,300			48 (23)	61,000		
14 (57)	20,300			49 (22)	63,000		
15 (56)	20,300			50 (21)	64,500		
16 (55)	20,700*	1.82	7.75	51 (20)	64,000		
17 (54)	20,700*	1.85	9.60	52 (19)	67,000		
18 (53)	20,800*	1.89	11.49	53 (18)	74,500		
19 (52)	22,000			54 (17)	78,000		
20 (51)	30,000			55 (16)	78,000		
21 (50)	30,000			56 (15)	81,000		
22 (49)	30,000			57 (14)	81,000		
23 (48)	30,000			58 (13)	82,000		
24 (47)	31,000*	2.13	13.62	59 (12)	85,000		
25 (46)	32,000			60 (11)	85,000		
26 (45)	34,500*	2.22	15.84	61 (10)	85,000		
27 (44)	37,500			62 (9)	87,500		
28 (43)	37,500			62 (8)	87,500*	12.50	35.28
29 (42)	41,500			64 (7)	87,500		
30 (41)	41,500			65 (6)	94,000		
31 (40)	41,500			66 (5)	99,000		
32 (39)	41,500			67 (4)	101,000		
33 (38)	43,000			68 (3)	101,000		
34 (37)	43,500			69 (2)	101,000		
35 (36)	43,000			70 (1)	115,000		

*Denotes failure

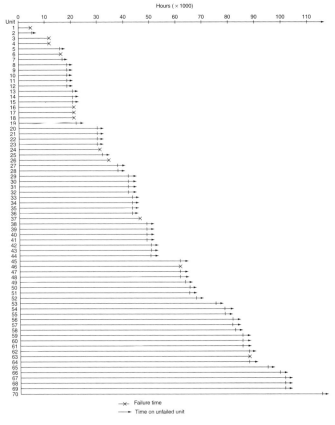

Figure 62.5 Graph of fan failure data

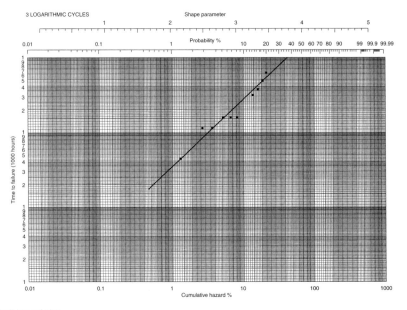

Figure 62.6 Weibull hazard plot

6. Check whether the failure data on the hazard plot follow a reasonably straight line. This can be done by holding the paper up to the eye and sighting along the data points. If the plot is reasonably straight, draw a best-fit straight line through the data points. This line is an estimate of the cumulative hazard function of the distribution that the data are from. This was done in Figure 62.5 for the generator fan failure data. The straight-line estimate of the cumulative hazard function is used to obtain information on the distribution of time to failure. It provides estimates of the cumulative distribution function and parameters of the distribution of time to failure. If the data do not follow a reasonably straight line, then plot the data on hazard paper for some other theoretical distribution to see if a better fit can be obtained. Curvature of a data plot such as shown in Figure 62.7 indicates poor fit.

When plotting data on two different hazard papers, it is possible that each will result in relatively straight plot. Generally, it doesn't matter which plot is used to interpolate within the range of data. There is an exception however, when extrapolation beyond the given range of data is necessary. In that case, the decision of which plot to use will be determined by the engineer.

Figures 62.8, 62.9, 62.10 show the data for generator fan failure plotted on exponential, normal and log normal hazard paper respectively. The exponential plot is a reasonably straight line which indicates that the failure rate is relatively constant over the range of the data. It should be noted that the reason the probability scale on the exponential hazard plot is crossed out is because that is not the proper way to plot data. (This will be discussed later.) The normal plot is curved concave upward which indicates that a normal distribution is not appropriate to the data. The log normal plot is a reasonably straight line. The log normal σ parameter of a line fitted to the data is approximately 0.7. For σ values near 0.7, a log normal distribution has a relatively constant failure rate over a large percentage of the distribution, particularly in the lower tail. The exponential, Weibull, and log normal plots are all comparable over the range of the data for prediction purposes. However, the best distribution to use when predicting times beyond the data, is a decision that should be based on engineering knowledge.

Like all other methods for analyzing censored failure data, the hazard plotting method is also based on a certain assumption that must be satisfied if we are going to rely on the results. The assumption is that if the unfailed units were run to failure, their failure times would be statistically independent of their censoring times. In other words, there is no relationship or correlation between the censoring time of a unit and the failure time. For example,

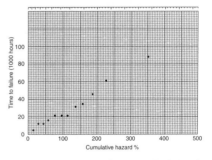

Figure 62.8 Normal hazard plot of generator fan failure data

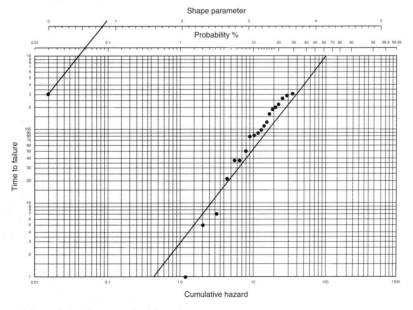

Figure 62.7 Exponential hazard plot of generator fan failure data

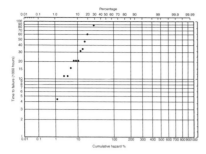

Figure 62.9 Generator fan failure data plotted on log normal hazard data

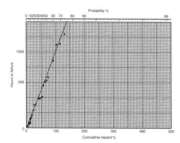

Figure 62.10 Exponential hazard plot of simulated exponential data

suppose that a certain type of unit has two competing modes of failure, which can be identified after failure, but there is interest in only one of them. Then failure of a unit by the other mode might be regarded as a censoring of the time to failure for the mode of interest. The two times to failure that a unit potentially has may not be independent, but correlated, for physical reasons. If this is so, the hazard-plotting method fails to give valid results on the distribution of time to failure for the mode of interest.

62.4.4 How to use hazard plots

The line the data supports on a hazard plot determines engineering information relating to the distribution of time to failure. Fan failure data and simulated data are illustrated here to explain how the information is obtained. The methods provide estimates of distribution parameters, percentiles, and probabilities of failure. The methods that give estimates of distribution parameters differ slightly from the hazard paper of one theoretical distribution to another and are given separately for each distribution. The methods that give estimates of distribution percentiles and failure probabilities are the same for all papers and are given first.

62.5 Estimates of distribution percentiles and probabilities of failure

For any distribution, the cumulative hazard function and the cumulative distribution junction are connected by a simple relationship. The probability scale for the cumulative distribution function appears on the horizontal axis at the top of hazard paper and is determined from that relationship. Thus, the line fitted to data on hazard paper

provided an estimate of the cumulative distribution function. The fitted line and the probability scale are used to obtain estimates of distribution percentiles and probabilities of failure. This procedure is explained next.

Suppose, for example, that an estimate based on a Weibull fit to the fan data is desired of the fifth percentile of the distribution of time to fan failure. Enter the Weibull plot, Figure 62.6, on the probability scale at the chosen percentage point, 5 per cent. Go vertically down to the fitted line and then horizontally to the time scale where the estimate of the percentile is read and is 14,000 hours.

An estimate of the probability of failure before some chosen specific time is obtained by the following. Suppose that an estimate is desired of the probability of fan failure before 100,000 hours, based on a Weibull fit to the fan data. Enter the Weibull plot on the vertical time scale at the chosen time, 100,000 hours. Go horizontally to the fitted line and then up to the probability scale where the estimate of the probability of failure is read and is 38 per cent. In other words, an estimated 38 per cent of the fans will fail before they run for 100,000 hours.

Given next is an estimate of the conditional probability that an unfailed unit of a given age will fail by a certain time. Such estimates for units operating in the field are useful for planning. An example of this is in the next paragraph. Suppose that an estimate based on the Weibull plot is desired for the probability that the unfailed fan with 32,000 hours will fail by 40,000. Enter the Weibull plot on the time scale at 32,000 hours. Go horizontally to the fitted line and then down to the cumulative hazard scale where the value is read as a percentage and is 13 per cent. Similarly, obtain the corresponding cumulative hazard value for 40,000 hours, which is 16.8 per cent. Take the difference between the two values, that is, $16.8 - 13 = 3.8$ per cent. Enter the plot on the cumulative hazard scale at the value of the difference, and go directly up to the probability scale to read the estimate of the conditional probability of failure, which is 3.7 per cent . For small percentages, cumulative probabilities and corresponding hazard values are approximately equal; this can be seen on the hazard papers. Thus, one could use the difference, if it is small, of the two cumulative hazard values as an approximation to the conditional probability of failure; this is 3.8 per cent here.

Conditional probabilities of failure can be used to predict the number of unfailed units that will fail within a specified period on each of the units. For each unit, the estimate of the conditional probability of failure within a specified period of time (8000 hours here) must be calculated. If there is a large number of units and the conditional probabilities are small, then the number of failures in that period will be approximately Poisson distributed (a special form of the normal distribution), with mean equal to the sum of the conditional probabilities, which must be expressed as decimals rather than percentages. The Poisson distribution allows us to make probability statements about the number of failures that will occur within a given period of time.

It should be emphasized that the estimation methods presented previously apply to any hazard paper and, in addition, to a nonparametric fit to the data obtained by drawing a smooth curve through data on any hazard paper.

Given next are the different methods for estimating distribution parameters on exponential, Weibull, normal, log normal, and extreme-value hazard papers. The methods are explained with the aid of simulated data from known distributions. Thus, we can judge from the hazard plots how well the hazard-plotting method does.

62.5.1 Exponential parameter estimate

Simulated data are shown in Table 62.4 and are a sample from an exponential distribution with a mean 1000 hours. The data are failure censored, which is a special case of multiply-censored data. That is, the data simulate a situation where 50 units were put on test at the same time and, according to a pre-specified plan, five unfailed units were removed from test when the fifth failure occurred, five more unfailed units were removed when the tenth failure occurred, and so on. Unfailed units would be removed for inspection for deterioration and to hasten the life test. A plot of these data on exponential hazard paper is shown in Figure 62.9. In Figure 62.9 the theoretical cumulative hazard function for the distribution is the straight line through the origin. On exponential hazard paper only, the time scale must start with time zero, and the straight line fitted to the data must pass through the origin.

For the purpose of showing how to obtain from an exponential hazard plot an estimate of the exponential mean time to failure, assume that the straight line on Figure 62.9 is the one fitted to the data. Enter the plot at the 100 per cent point on the horizontal cumulative hazard scale at the bottom of the paper. Go up to the fitted line and then across horizontally to the vertical time scale where the estimate of the mean time to failure is read and is 1000 hours. The corresponding estimate of the failure rate is the reciprocal of the mean time to failure and is $1/100 = 0.001$ failures per hour.

62.5.2 Weibull parameter estimates

Table 62.5 contains a randomly censored sample of 90 times from a Weibull distribution with shape parameter equal to 0.8 and scale parameter equal to 1000 hours. The data are randomly censored with independent censoring times from a Weibull distribution with shape parameter equal to 2.0 and scale parameter equal to 300 hours. The data were obtained by generating an observation from the failure distribution and an independent observation from the censoring distribution. If the censoring time was smaller than the failure time, the censoring time was used and vice versa. Such data arise if there are independent, competing modes of failure for a unit and the mode of failure can be determined on a failed unit. The 20 failure times are plotted in Figure 62.11 on Weibull hazard paper. The data follow the cumulative hazard function of the distribution of time to failure but do not fall on a particularly straight line.

62.5.3 Some practical aspects of data plotting

This section will cover the potential difficulties that may be encountered when using hazard and probability plotting paper. It will also look at how to use the plotting paper for the most effective results. Much of the discussion applies equally to both hazard or probability plotting, especially where good plotting techniques are concerned.

To determine which hazard paper is appropriate to use when plotting a set of multiply censored data, first rely on engineering experience. If that is not an option, try different papers until you find one that is suitable. To save time, it may be a good idea to try plotting the sample cumulative hazard function on exponential hazard paper, since it is just

Table 62.4 Simulated exponential data and hazard calculations

n(k)	Hours	Hazard	Cumulative hazard	n(k)	Hours	Hazard	Cumulative hazard
1 (50)	3*	2.00	2.00	26 (25)	385		
2 (49)	52*	2.04	4.04	27 (24)	385		
3 (48)	58*	2.08	6.12	28 (23)	385		
4 (47)	71*	2.13	8.25	29 (22)	385		
5 (46)	77*	2.17	10.42	30 (21)	385		
6 (45)	77			31 (20)*	391*	5.00	46.49
7 (44)	77			32 (19)*	410*	5.26	51.75
8 (43)	77			33 (18)*	562*	5.56	57.31
9 (42)	77			34 (17)*	611*	5.88	63.19
10 (41)	77			35 (16)*	621*	6.25	69.44
11 (40)	101*	2.50	12.92	36 (15)	621		
12 (39)	130*	2.56	15.48	37 (14)	621		
13 (38)	161*	2.63	18.11	38 (13)	621		
14 (37)	180*	2.70	20.81	39 (12)	621		
15 (36)	235*	2.78	23.59	40 (11)	621		
16 (35)	235			41 (10)	662*	10.00	79.44
17 (34)	235			42 (9)	884*	11.11	90.55
18 (33)	235			43 (8)	1101*	12.50	103.05
19 (32)	235			44 (7)	1110*	14.29	117.34
20 (31)	235			45 (6)	1232*	16.67	134.01
21 (30)	301*	3.33	26.92	46 (5)	1232		
22 (29)	306*	3.45	30.37	47 (4)	1232		
23 (28)	309*	3.57	33.94	48 (3)	1232		
24 (27)	334*	3.70	37.64	49 (2)	1232		
25 (26)	385*	3.85	41.49	50 (1)	1232		

*Denotes failure

Table 62.5 Randomly censored simulated Weibull data

Number	Time	Cumulative hazard	Number	Time	Cumulative hazard	Number	Time	Cumulative hazard
1	1*	1.11	31	199*	20.26	61	264	
2	5*	2.23	32	199		62	276	
3	7*	3.37	33	200		63	279	
4	21*	4.52	34	201		64	279	
5	37*	5.68	35	205		65	281	
6	38*	6.86	36	206		66	283	
7	49*	8.05	37	207		67	283	
8	78*	9.25	38	212		68	293*	29.59
9	85*	10.47	39	213		69	298	
10	91*	11.70	40	214		70	301*	34.35
11	96		41	217		71	301	
12	98*	12.97	42	217*	22.30	72	309	
13	111*	14.25	43	217		73	315	
14	125*	15.55	44	218		74	318	
15	141		45	220		75	322	
16	145		46	222		76	329	
17	156		47	223		77	330	
18	158		48	224		78	343	
19	160		49	226		79	344	
20	162		50	231		80	345	
21	163		51	232		81	349	
22	165*	17.00	52	239		82	354	
23	168		53	244		83	355	
24	176		54	244		84	363	
25	181		55	245		85	363	
26	182		56	251		86	370	
27	187		57	254*	25.24	87	380	
28	188*	18.59	58	256		88	385	
29	191		59	261		89	407	
30	198		60	262		90	412	

*Denotes failure time

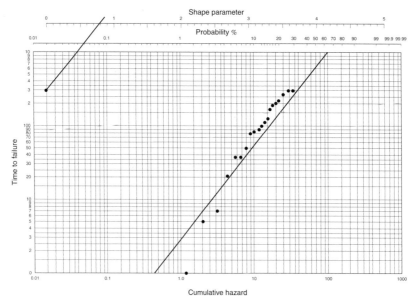

Figure 62.11 Weibull hazard plot of simulated Weibull data

a square grid. Then compare the plot with the theoretical cumulative hazard functions in Figures 62.12–62.15 and choose the distribution with the cumulative hazard function that agrees most closely with the shape of the sample cumulative hazard function over the range of cumulative hazard values in the sample. Plot the data on the corresponding hazard paper. The cumulative hazard scale on the horizontal axis of exponential hazard paper shown runs from 0 to 500 per cent, which may be too big a range, particularly if the data consist of some early failure times and many censoring times and the sample cumulative hazard values are small. Then the scale should be changed to run from 0 to 50 per cent, say, by adding decimal points. The probability scale then no longer corresponds to the hazard scale and must not be used. This was done in Figure 62.7 where the probability scale has been deleted.

If a data plot on a chosen hazard paper departs significantly from a straight line by being bowed up or down, it is an indication that the data should be re-plotted on a

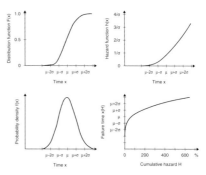

Figure 62.13 The normal distribution

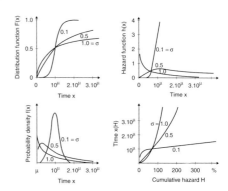

Figure 62.14 Long normal distributions

different hazard paper. Whether a plot of points is curved can best be judged by laying a transparent straight edge

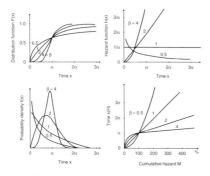

Figure 62.12 Weibull distributions

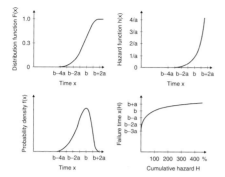

Figure 62.15 Smallest extreme-value distributions

along the points. If a plot has a definite double curvature on exponential paper, log normal hazard paper should be tried. If the plot is still not straight, then this may be an indication that the units have two or more failure modes with different distributions of time to failure. Then a mixture of such data from different distributions usually will not plot as a straight line on hazard or probability paper but will be a line with two or more curves to it.

When a set of data does not plot as a straight line on any of the available papers, then one may wish to draw a smooth curve through the data points on one of the plotting papers, and use the curve to obtain estimates of distribution percentiles and probabilities of failure for various given times. With such a nonparametric fit to the data, it is usually unsatisfactory to extrapolate beyond the data because it is difficult to determine how to extend how to extend the curve. Nonparametric fitting is best used only if the data contain a reasonably large number of failures.

The behavior of the failure rate as a function of time can be gaged from a hazard plot. If data are plotted on exponential hazard paper, the derivative of the cumulative hazard function at some time is the instantaneous failure rate at that time. Since time to failure is plotted as a function of the cumulative hazard, the instantaneous failure rate is actually the reciprocal of the slope of the plotted data, and the slope of the plotted data corresponds to the instantaneous mean time to failure. For the data that are plotted on one of the other hazard papers and that give a curved plot, one can determine from examining the changing slope of the plot whether the true failure rate is increasing or decreasing relative to the failure rate of the theoretical distribution for the paper. Such information on the behavior of the failure rate cannot be obtained from probability plots.

On exponential hazard paper, if a smooth curve through the data points clearly passes through the time axis above the origin instead of through the origin, it is an indication that the true distribution may have no failures before some fixed time. This may arise, for example, if time to failure is figured from date of manufacture, and there is a minimum possible time it takes to get a unit into service. The minimum time is estimated as the time on the axis that the smooth curve passes through. This minimum time is then treated as time zero, and all times to failure is calculated from it for purposes of plotting the data on another hazard paper.

In a data plot on exponential hazard paper, the early failure times are crowded together by the origin in the lower left-hand corner of the plot and the later failure times are space farther apart (see Figure 62.7, for example). Thus, on exponential hazard paper, the later failures influence the eye and placement of the best-fit line more than the early failures. If one is interested in the early failure part of the distribution, which is typical of reliability work, then the early failure should be emphasized more in the analysis. Plotting data on Weibull hazard paper can do this. The data points are then spaced out more equally throughout the range of the data, and the early and later failures are weighed about equally by eye (see Figure 62.5). If an exponential fit to the data is still desired, then a best-fit line with 45° slopes should be drawn through the data points. This gives an exponential fit to the data, since the exponential is a special case of the Weibull with the shape parameter equal to one. If a nonparametric fit to the data such as described previously is desired that emphasizes early times to failure, then the data should be plotted on Weibull hazard paper and a smooth curve drawn through the data to provide a nonparametric estimate of the cumulative hazard function of the distribution. Also, percentiles and probabilities of failure for various given times can then be estimated from the smooth curve by the methods given earlier.

If estimated of distribution parameters are desired from data plotted on a hazard paper, then the straight line drawn through the data should be based primarily on a fit to the data points near the center of the distribution the sample is from and not be influenced overly by data points in the tails of the distribution. This is suggested because the smallest and largest times to failure in a sample tend to vary considerably from the true cumulative hazard function, and the middle times tend to lie close to it. Similar comments apply to the probability plotting.

Failure times are typically recorded to the nearest hour, day, month, hundred miles, etc. This is so because the method of measurement has limited accuracy or because the data are rounded. For example, if units are inspected periodically for failure, the exact time of failure is not known but only the period in which the failure occurred. For data plotting, the amount that failure times are rounded off should be considerably smaller than the spread in the distribution of time to failure.

A rough rule of thumb for tolerable rounding for data plotting is that the amount of rounding should be no greater than one-fifth of the standard deviation of the distribution. Another way of saying this is that data are coarse if a large number of the failures are recorded for just a few different times. If data are too coarse, a plot on hazard or probability paper will have a staircase appearance with flat spots at those times with a number of failures. Plots of coarse data can be misleading unless the coarseness is taken into account; for example, estimated of failure probabilities and distribution parameters and percentiles can be somewhat in error if coarseness of the data in not taken into account.

Methods for analyzing coarse data can be found in the statistical literature where it is called grouped data. One method of taking into account the coarseness of data in plotting is to plot the failures with a common failure time at equal spaced times over the time interval associated with that time. For example, if failure times are recorded to the nearest 100 hours and five failures are at 1000 hours, then the corresponding interval is 950 to 1050, and the corresponding plotting times assigned to the five failures are 960, 980, 1000, 1020, and 1040 hours. This smoothes the steps out of the plot and makes the plot more accurate.

When a set of data consists of a large number of failure times, one may not wish to plot all failures but just selected ones. This is done to reduce the labor in making the data plot or to avoid a clutter of overlapping data points on the plot. Any selected set of points might be plotted. For example, one might plot every kth point, all points in a tail of the distribution, or only some of the points near the center of the distribution. Care must be taken in plotting just a subset of the data to ensure that the plot of the subset leads to the same conclusions as the plot of the entire set. In practice, the choice of the selected points to be plotted depends on how the plot is going to be used. For example, if one is interested in the lower tail of the distribution, that is, early failures, then all of the early failures should be plotted. In general, one plots the data points that are most relevant to the information wanted from the data.

Ideally, no fewer than 20 failure times, if available, should be plotted from a set of data. Often, in engineering practice there are so few failures that all should be kept in mind so that conclusions drawn from a plot are based on a limited amount of information. Note that if only selected failures from a sample are to be plotted on hazard paper, it is necessary to use all of the failures in the sample to calculate the appropriate cumulative hazard values for the plotting positions. Wrong plotting positions will result if some failures in the data are not included in the cumulative hazard calculations. A similar comment applies to the calculation of plotting positions for probability plotting.

For plots of some data, the ranges of the scales of the available hazard and probability papers shown throughout this chapter may not be large enough. To enlarge the time scale on any of the papers, join two or more pages together. To enlarge the hazard scale on Weibull or extreme-value hazard paper to get smaller hazard and probability values, add on extra log cycles for the lower values, and label them in the obvious manner; the cumulative hazard and cumulative probability scales are equivalent for the small probabilities in the added cycles. Weibull hazard papers with two different ranges are available to allow a user to choose the ranges that best suit his data (see Figures 62.5 and 62.6).

62.6 Concluding remarks

The cumulative hazard plotting method and papers presented here provide simple means for statistical analyses of multiply censored failure data to obtain engineering information. The hazard-plotting method is simpler to use for multiply censored data than other plotting methods given in the literature and directly gives failure-rate information not provided by others.

Method to identify bad repairs from bad designs

Earlier we dealt with Weibull analysis and noticed how data demanding that method can become. This is especially true when one wants to examine a sample of data from a time continuum such as in an established process plant. The sample of data may be a one-year period where the failure pattern can be considered similar to that in the previous year and in the succeeding year. However, this analysis method involves only a knowledge of the total sample population, total number of failures within the time period, the number of units on which these failures occurred, and time intervals between failures on units which failed more than once within the sample period. A good analogy would be a hospital operation with a yearly number of patients, of whom some are returning within that period for repeated treatment. Some of the returning patients only return once, others more than once during that year, and so forth.

In an earlier section, we acquainted ourselves with commonly used reliability terms: mean time between repairs (MTBR), mean time between failures (MTBF), etc. We saw that these terms have similar meanings but often include minor deviations. To avoid confusion, mean time to failure (MTTF) will be used in the following discussion. It can be expressed in terms of any time periods, i,e, days, months, years, etc.:

$$MTTF = P \times t$$

$$F_\tau$$

where:

$MTTF$ = mean time to failure

P = population

t = time periods in sample time

F_τ = failures in sample time

If t is unity, F_τ is failures

$MTTF = P$, years or other period

$$F_\tau$$

While such a number is very useful as an index of performance, it gives no indication of what needs to be done to correct the situation if the index is low.

Inherent defects in either the system or the equipment, or 'infant mortality' (see below) cause a low MTTF. Inherent defects will result in units failing more frequently than is desirable. These defects can be corrected by locating and correcting them or by taking action to reduce their effects, thus improving machine life.

Infant mortality results in repaired units failing shortly after their return to service. It can be corrected by simplifying repair techniques, quality control of repairs and repair parts, improved starting techniques, etc. Units which have persistent abnormally short lives may have an inherent defect which cannot be corrected by the previous methods. However, the usual characteristic of infant mortality is its variability. Very good pumps of last year become the very bad ones this year and vice versa. Any pump is a potential bad actor.

63

Health and Safety in the UK

George Pitblado

Support Services

Contents

63.1 Introduction

The initial introduction of the Health and Safety at Work Act 1974 enabled provisions in the Act to supersede legislation contained in 31 relevant Acts and 500 subsidiary regulations in use at that time. The 1974 Act is the framework which, over the years, will enable the gradual improvement of existing health and safety requirements. Revising and updating current provisions in the form of regulations and approved codes of practice prepared in consultation with industry will achieve this. These provisions will maintain or improve the health and safety standards established by existing legislation. The objective of the Act is not only that it is introduced to rationalize existing provisions but also to ensure that standards of health and safety are improved, thereby providing greater protection to persons at work and to the general public.

63.2 Legislation

Health and safety legislation is in four separate categories as follows.

63.2.1 Acts of parliament

The Factories Act 1961 and the Health and Safety at Work, etc. Act 1974 are the result of Bills, which, after being debated in Parliament, have received the Royal Assent, and now from part of criminal law.

63.2.2 Regulations

The 1974 Act is all encompassing. It lays down only general duties and grants the appropriate minister the power to make detailed Regulations to cover relevant aspects referred to in general legislation. That is, the Act, being an up-to-date form of legislative control, differs from that found in previous formats of the Factories Act 1961.

The Control of Substances Hazardous to Health Regulations 1988 (COSHH), the Noise at Work Regulations 1989, the Pressure Systems and Transportable Gas Containers Regulations 1989 and the Electricity at Work Regulations 1989 are examples of the Regulations being introduced. For example, the Electricity at Work Regulations which came into effect from 1 April 1990 revoke the Electricity Regulations 1908 and the Electricity (Factories Act) Special Regulations 1944. Within the structure of Regulations, provisions may be granted for exemptions from their requirements or application in some form, with respect to specific premises, processes or industries.

63.2.3 Codes of practice

Codes of Practice, where appropriate, will be issued to supplement Regulations. These Codes, although having a special legal status, are not statutory requirements but may be used in criminal proceedings as evidence that statutory requirements have been contravened. Prior to replacing or amending existing Codes of Practice discussions are normally held with interested parties before the Health and Safety Commission approves them for use under the Health and Safety at Work, etc. Act 1974. In some cases, these discussions could be supported by comments from interested bodies on a previously issued draft Code of Practice.

63.2.4 Guidance notes

Guidance notes relevant to sound health and safety practices are published from time to time by the Health and Safety Executive. They have no legal status or significance such as found in Codes of Practice. However, as their content is normally based on a wealth of practical experience, it would be expected by the enforcing authorities that employers would follow the advice contained in them.

63.3 Administration of the Health and Safety at Work Act

The bodies set up to ensure the satisfactory implementation and operation of the Health and Safety at Work, etc. Act 1974 (and the Employment Medical Advisory Service Act 1972) under Section 10 of the Act are the Health and Safety Commission and the Health and Safety Executive.

63.3.1 The Health and Safety Commission

This consists of a chairman appointed by the Secretary of State for Employment and a committee of not less than six and not more than nine members. The composition of the Commission is drawn from different groups: three members may be from employers' organizations and three from employees' organizations. The three remaining vacancies may be filled from such organizations as local authorities and professional bodies, as the Secretary of State considers appropriate.

63.3.2 The Health and Safety Executive

Whereas the Health and Safety Commission is responsible for the development of policies in the health and safety field, the Health and Safety Executive, being a separate statutory body appointed by the Commission, will work in accordance with directions and guidance given by the Commission. The Executive will provide advice on health and safety to both sides of industry and will enforce legal requirements.

63.3.3 Local authorities

Under the Commission's guidance, these will enforce legislation in certain areas of employment. In general, theirs will not be industrial activities.

63.3.4 Fire authorities

The Health and Safety at Work Act 1974 gives the relevant fire authority the power to designate a place of work as premises requiring a fire certificate. This certificate will indicate the measures to be taken with respect to means of escape in the event of a fire when the structural design of the premises is taken into consideration. The fire authority in most areas has the responsibility for enforcing the provisions of the Act, but the Health and Safety Executive are responsible for issuing fire certificates in certain classified premises (e.g. those used for storage, manufacture and use of highly flammable liquid).

63.4 General duties

Employers and others who now have duties under the 1974 Act will have had some form of responsibilities under previous legislation (e.g. the Factories Act 1961) to ensure the health and safety and welfare of people at work. However, a number of employers will be subject to such legislation for the first time.

Much of the earlier legislation has been retained, although as the respective regulations are amended they are presented within the remit of the 1974 Act, such as the Electricity at Work Regulations 1989, which revoked the Electricity Regulations 1908 and the Electricity (Factories Act) Special Regulations 1944. The earlier legislation, where it remains, has been brought under the same method of enforcement as those under the Act, and is applied universally.

63.4.1 Duties of employers

It is the duty of employers, as far as is reasonably practicable, to safeguard the health, safety and welfare of the people who work for them and others who may be affected. This applies in particular to the provision and maintenance of safe plant and equipment and methods of work, and covers all plant, equipment and substances used. Specific areas that require attention include:

1. Is the plant and equipment designed, installed and operated to the accepted standards?
2. Are there safe systems of work implemented during the operation and maintenance of the plant and equipment? (This may necessitate a permit to work to be issued by an authorized competent person.)
3. Is the work environment monitored, as required under the COSHH Regulations, to ensure that there is no hazard from, for example, toxic contaminants?
4. Has the monitoring of any control measure been implemented?
5. Is there a program in place whereby all equipment and appliances found necessary to ensure the safety and health of those likely to be affected are regularly inspected?
6. Have the risks to health from the use, storage, handling or transport of articles or substances been kept to the minimum? (Expert advice should be sought in this area if there is no sufficient in-house knowledge.)

63.4.2 Safety policies: organization and arrangements

It is a statutory requirement that an employer with five or more employees prepares a written statement of the company's general policy, organization and arrangements for health and safety at work. This policy statement should be revised at regular intervals and, where appropriate, amended.

63.4.3 Safety training and information

Employers have a duty under the 1974 Act to provide, as necessary, training and information to ensure that there is no risk to the health and safety of their employees. It may be found necessary to provide operators and maintenance employees with specific training to carry out certain processes or work tasks.

63.4.4 Duties to others

Employers must, as far as is reasonably practicable, have regard for the health and safety of contractors' employees or the self-employed who may be affected by the company's operations and for the health and safety of the general public. This covers, for example, the emission of noxious or offensive gases and dust into the atmosphere, or danger from plant and equipment to which the public or those not directly employed by the company have access.

63.4.5 Responsibilities of the self-employed

The self-employed have a similar duty to that of employers to ensure that there is no risk or danger to the health and safety of themselves or any other persons.

63.4.6 Duties of manufacturers and suppliers

Section 6 of the Act has been amended by the Consumer Protection Act 1987 and now imposes specific duties on manufacturers, importers, designers and suppliers to ensure that articles and substance supplied for use at work are safe and without risk to health.

To assist manufactures and suppliers attain the required level of health and safety, the Health and Safety Commission have commenced approving standards laid down by such bodies as the British Standards Institution.

63.4.7 Duties of employees

The Act also places a responsibility on the employees in that they must take reasonable care to avoid injury to themselves or to others by their work activities, and cooperate with their employer and others in meeting all statutory requirements. The Act also requires that employees do not interfere with or misuse anything designed to protect their health, safety or welfare.

63.4.8 Enforcement

On finding that there has been a contravention of either an existing Act or Regulation, or of a provision of the Health and Safety at Work, etc. Act 1974, an inspector can:

1. Issue a prohibition notice, if there is a risk of serious personal injury, to stop the work or process giving rise to this risk until remedial action specified by the inspector in the notice has been undertaken. This notice

can be served on the person carrying out the activity or in control of it at the time the notice is served.

2. Issue an improvement notice if there is a legal contravention of any statutory legislation relevant to the activity in that the activity cannot be continued until the remedial action specified in the notice is implemented. This notice can be served on the person who is deemed to be contravening the legal provision or who is responsible for the activity, whether they are an employer, employee or a supplier of equipment or materials.
3. Prosecute any person who may be contravening a relevant statutory provision instead of (or in addition to) serving a notice.

63.5 Safety policy

The safety policy for the company, although common in most parts with the safety policy for every company, may require being prepared in detail. This would naturally cover any process, hazardous product or procedure specific to the organization. In preparing the policy, it is essential to ensure that parts referred to in Section 2(3) of the 1974 Act are covered, i.e.:

1. The general policy statement;
2. The organization and arrangements for carrying out the policy.

In large companies, these two aspects may be dealt with separately. In larger organizations, it can be found that it is of greater benefit if the policy document is produced in two separate sections. These would include:

1. A concise statement of the company's general policy, organization and arrangements in a single document. It follows that this document can then be issued individually to all members of staff and contractors who work on the premises.
2. A more detailed document in the form of a health and safety manual. This manual would include the company's policy statement, company rules, safe working procedures, etc., and would normally be located in a nominated office. It is necessary in using this method that all staff are made aware of its contents and its location.

63.5.1 Policy statement

The company's general policy statement should be a declaration of the employer's intent to provide a safe and healthy workplace for all employees and should include the request that the employees provide the necessary support towards achieving the company's aims.

63.5.2 Levels of responsibility

The policy statement should give the name, designation and office location of the nominated senior member of the company designated as the responsible person within the organization for ensuring that the company's policy

statement is complied with. This nominated person should have high position within the company (e.g. director, senior manager, company secretary).

While the overall policy responsibility for health and safety rests with senior managers, all employees within the organization, irrespective of their duties, have some degree of responsibility for carrying out the policy. Where appropriate, nominated persons with specific responsibilities for health and safety should be named with a summary of their responsibilities defined. It follows that there should be a procedure established whereby deputies are available during the nominated person's absences.

Where specialist knowledge is required, the relevant aspects should be clearly established and the respective persons made aware of such. This expertise can be obtained, for example, from the company's safety officer, chemist, etc. Finally, the policy statement should make clear the level of responsibility of every employee.

63.5.3 Safety representatives and joint safety committees

Where health and safety consultation is in place (e.g. joint safety committees) the structure with regards to members' representation should be described.

63.5.4 Training and supervision

The policy statement should indicate how the company proposes to carry out training with respect to health and safety. Equally important is the responsibility placed on managers, supervisors, etc. to ensure that the individual who has been given the work task has the knowledge to carry it out without risk to themselves or others. This task may be operating a piece of equipment on which all safety measures must be in order or carrying out a maintenance task where there is an acknowledged hazard. It is therefore equally important that managers and supervisors are suitably trained in both the technical and safety aspects of the work.

63.5.5 Hazard details

Many accidents occur because the operator or maintenance person does not understand the hazards involved or has not been instructed on the precautions to be taken. The policy document should identify the main hazards within the company with advice on which rules must be obeyed while carrying out a hazardous task. General rules should also be included to cover items such as untidy work areas, replacement of guards, the use of protective clothing/equipment where appropriate, safe working practices in handling goods and materials, etc. It is essential that on the introduction of new products, processes, operations or plant and equipment that any hazards associated with these are brought to the attention of all concerned.

63.5.6 Reporting accidents

Procedures must be prepared to ensure that the reporting of accidents meets the requirements as set out in the Reporting of Injuries as prescribed in the Diseases and Dangerous Occurrences Regulations 1985. Accidents should also be recorded in the Accident Book (Form BI 510). The procedure adopted should include a measure of recording and reporting of accidents, which should then form the basis of discussion by management and/or joint safety committees, with a view to identifying any hazard that gives rise for concern and introducing the appropriate corrective action where necessary. The number of accidents could be taken as a measure of how the organization is performing overall with respect to the health and safety of its workforce and others.

63.5.7 Policy review

The policy document should be reviewed at regular intervals to ensure that any changes from the introduction of new legislation (i.e. regulations, codes of practice, information concerning safe working procedures) are implemented.

63.6 Information

Health and safety information can be obtained from the following Health and Safety Executive offices:

Public enquiry points: HSE library and information services

St Hugh's House
Stanley Precinct
Trinity Road
Bootle
Merseyside L20 3QY
Tel. 0151 951 4381

Broad Lane
Sheffield S3 7HQ
Tel. 0870 1545500

Baynards House
1 Chepstow Place
Westbourne Grove
London W2 4TF
Tel. 020 7221 0870

Inquiries with respect to health and safety at the workplace should be directed to the local area office of the HSE or the local authority's Environmental Health Department. The name of the authority responsible for the health and safety in the organization should be displayed prominently within the premises on the poster *Health and Safety Law, What You Should Know*. This poster gives a brief guide to health and safety law and specific details on key points must be obtained from the relevant legislative documents.

63.7 HSE inspectorates

HSE inspectors systematically visit and review a wide range of work activities, giving expert advice and guidance where necessary. The Inspectorates, within the HSE, who monitor or enforce standards and provide advice in specific sectors of industry or employment, are as follows:

1. *HM Factory Inspectorate*: Manufacturing and heavy industrial premises and processes, as well as construction activities, local authority undertakings, hospitals, schools, universities and fairgrounds;
2. *HM Agriculture Inspectorate*: Farming, horticulture and forestry;
3. *HM Explosives Inspectorate*: The manufacture, transport, handling and security of explosives;
4. *HM Mines and Quarries Inspectorate*: All mines, quarries and landfill sites;
5. *HM Nuclear Installations Inspectorate*: On HSE's behalf licenses nuclear installations ranging from nuclear power stations and chemical works to research reactors.

63.8 The Employment Medical Advisory Service (EMAS)

This is an organization of doctors and nurses based in area offices and form the medical arm of the HSE. The EMAS provides advice, at the request of the employer, employee, self-employed, trade union representative or medical practitioner, on the effects of work on health.

63.9 HSE area offices

The Health and Safety Executive operate from area offices as follows.
South-West
Inter City House
Mitchell Lane
Victoria Street
Bristol BS1 6AN
Tel. 0272 290681
(Avon, Cornwall, Devon, Gloucestershire, Somerset, Isles of Scilly)
South
Priestly House
Priestly Road
Basingstoke RG24 9NW
Tel. 0256 473181
(Berkshire, Dorset, Hampshire, Isle of Wight, Wiltshire)
South-East
3 East Grinstead House
London Road
East Grindstead
West Sussex RH19 1RR
Tel. 0342 26922
(Kent, Surrey, East Sussex, West Sussex)
London North
Maritime House
1 Linton Road
Barking
Essex IG11 8HF
Tel. 081-594 5522
(Barking and Dagenham, Barnet, Brent Camden, Ealing, Hackney, Haringey, Harrow, Havering, Islington, Newham, Redbridge, Tower Hamlets, Waltham Forest)

London South
1 Long Lane
London SE1 4PG
Tel. 071-407 8911
(Bexley, Bromley, City on London, Croydon, Greenwich, Hammersmith and Fulham, Hillingdon, Hounslow, Kensington and Chelsea, Kingston, Lambeth, Lewisham, Merton, Richmond, Southwark, Sutton, Wandsworth, Westminster)
East Anglia
39 Baddow Road
Chelmsford
Essex CM2 0H1
Tel. 0245 284661
(Essex)
Northern Home Counties
14 Cardiff Road
Luton
Beds LU1 1PP
Tel. 0582 34121
(Bedfordshire, Buckinghamshire, Cambridgeshire, Hertfordshire)
East Midlands
Belgrave House
1 Greyfriars
Northampton NN1 2BS
Tel. 0604 21233
(Leicestershire, Northamptonshire, Oxfordshire, Warwickshire)
West Midlands
McLaren Building
2 Masshouse Circus
Queensway
Birmingham B4 7NP
Tel. 021 200 2299
(West Midlands)
Wales
Brunel House
2 Fitzalan Road
Cardiff CF2 1SH
Tel. 0222 497777
(Clwyd, Dyfed, Gwent, Gwynedd, Mid-Glamorgan, Powys, South Glamorgan, West Glamorgan)
Marches
The Marches House
Midway
Newcastle-under-Lyme
Staffs ST5 1DT
Tel. 0782 717181
(Hereford and Worcester, Shropshire, Newcastle-under-Lyme, Staffordshire)
North Midlands
Birbeck House
Trinity Square
Nottingham HB1 4AU
Tel. 0602 470712
(Derbyshire, Lincolnshire, Nottinghamshire)
South Yorkshire
Sovereign House
40 Silver Street
Sheffield S1 2ES

Tel. 0742 739081
(Humberside, South Yorkshire)
West and North Yorkshire
8 St Pauls Street
Leeds LS1 2LE
Tel. 0532 446191
(North Yorkshire, West Yorkshire)
Greater Manchester
Quay House
Quay Street
Manchester M3 3JB
Tel. 061 831 7111
(Greater Manchester)
Merseyside
The Triad
Stanley Road
Bootle L20 3PG
Tel. 051 922 7211
(Cheshire, Merseyside)
North-West
Victoria House
Ormskirk Road
Preston PR1 1HH
Tel. 0772 59321
(Cumbria, Lancashire)
North-East
Arden House
Regent Centre
Regent Farm Road
Gosforth
Newcastle-upon-Tyne NE3 3JN
Tel. 091 284 8448
(Cleveland, Durham, Northumberland, Tyne and Wear)
Scotland East
Belford House
59 Belford Road
Edinburgh EH4 3UE
Tel. 031 225 1313
(Borders, Central, Fife, Grampian, Highland, Lothian, Tayside, and the island area of Orkney and Shetland)
Scotland West
314 St Vincent Street
Glasgow G3 8XG
Tel. 041 204 2646
(Dumfries and Galloway, Strathclyde, and the Western Isles)

63.10 Health and safety procedures

Note that the following procedures only provide a framework in which specific health and safety procedures for the individual works or premises may be prepared. A senior member of the company or the person nominated as responsible for health and safety in the organization should write the foreword to a company's procedure. To emphasize its importance it should also include a part of the company's health and safety statement. It should be followed by an instruction such as 'You are required to read the following procedures carefully and comply with the sections relevant to your workplace'.

63.11 Fire and first-aid instructions

63.11.1 Fire

Fire bells are installed in works or premises:

1. To alert all staff in the event of a fire alarm;
2. To order complete evacuation of the works or premises (in a number of workplaces such evacuation may not be necessary).

On hearing the alarm all personnel should leave the building by the nearest safe fire exit. They should not use the lifts or re-enter the building until given permission by the person controlling the incident. Staff should be aware of all fire-precaution notices and procedures (including the fire brigade's telephone number), fire call points, extinguishers, fire exits, escape routes and assembly points.

63.11.2 First-aid/medical services

Staff should make themselves familiar with the first-aid procedures in the event of an accident and know the location of the nearest first aid officers and the first-aid room (if there is one on the premises). They should also be aware of all first-aid/medical notices and procedures and the telephone number to call for assistance during and outside normal hours. (These are likely to differ unless the works or premises operate 24 hours a day.)

63.12 Good housekeeping

Good housekeeping can play a major part in maintaining a safe and environmentally sound place of work. Tripping over material not tidied away causes many accidents. Another source of potential injury is in the lack of secure storage of cleaning equipment, tools, etc.

Liquids allowed to overflow and those not adequately cleaned off the floor present potential hazards by persons slipping.

General hints on keeping the workplace clean are:

1. Use waste bins. Special containers may be required for specific hazardous waste material.
2. Return everything to its correct place.
3. Always tidy up after completing the task.
4. Keep passageways and footpaths clear at all times.
5. Store materials and equipment in a safe manner.
6. Store hazardous material in its correct location.
7. Encourage everyone to develop *clean* habits.

63.13 Protective clothing and equipment

1. Eye protection should comply with the Protection of Eyes Regulations 1974 and relevant British Standard (e.g. BS 2092, BS 1542, BS 679).
2. Safety helmets should comply with the relevant British Standards (e.g. BS 5240, BS 4033). If engaged in construction, the Construction (Head Protection) Regulations 1989 must be applied.

3. The Noise at Work Regulations 1989 has set action levels which must be complied with:
   (a) Where employees are exposed to between 85 dB(A) and 90 dB(A) they are entitled to ask for ear protectors and the employer must provide them. There is no obligation on the employer to ensure that protectors are used or on the employee to use them.
   (b) At or above 90 dB(A) and the peak action level of 200 Pa or 140 dB re 20 µPa the employer must provide suitable ear protectors and enforce their use, and employees are under an obligation to use them. Hearing protectors should comply with BS 6344, with sound-attenuation measurement to BS 5108.
   (c) Where exposure is at or above 90 dB(A) the employer is required to reduce exposure to noise as far as is reasonably practicable by means other than ear protectors. As with other regulations, it is now a duty of the employer to assess the noise exposure and advise the employee as appropriate. In a noisy environment, the employer may have to seek expertise from a competent person to carry out the necessary assessments and monitoring of the premises.
4. Safety footwear *must* be worn where applicable (moving heavy objects, working with hazardous materials (e.g. chemicals), working in wet conditions). Footwear should comply with the following standards as appropriate: BS 1870, BS 4972, BS 5131, BS 5145, BS 5451, BS 5462, BS 6159.
5. Safety belts should comply with BS 1397 and BS 6868.
6. Protective clothing (including gloves) *must* be used where applicable (e.g. normal duties; welding/cutting hot work; water treatment; application/use of chemicals; during servicing of batteries; descaling/chemical cleaning). Clothing should comply with BS 1542, BS 5426, BS 2653, BS 4171, BS 6408 and BS 4724 and gloves with BS 1651, BS 697 and BS 6526.
7. Respirators should comply with BS 2091, BS 4555, BS 4558, BS 4771 and BS 4275. (Refer to Form F2486, published by HSE for suppliers of respirators suitable for use against aerosols containing *Legionella*.)

When in doubt as to which standard of protection is necessary, consult the manufactures, suppliers, HSE, EMAS, the local authority or the Environmental Health Officer.

63.14 Safe working areas

It is essential when carrying out any operation or task that cannot be undertaken standing on a floor that an appropriate working platform is used. This can take many forms, but the methods that should be used fall within the following types: stepladders; trestles with suitable boards; towers (mobile); proprietary scaffolding or general-access scaffolding. Such access equipment should be to the relevant British Standard, e.g.

Stepladders: BS 1129, BS 2037;
Trestles: BS 1129, BS 2037, BS 1139;

Scaffolds; BS 1139, DD72, CP 5973, BS 2482, BS 6037, BS 5974, BS 6289.

The use of access equipment should meet the acceptable standards as laid down in the Construction (Working Places) Regulations 1966.

Items that should be checked on access equipment before use include:

1. General condition – is there any damaged section or parts?
2. Is it the correct access equipment to carry out the task safely? For example, is a ladder adequate or should a tower be used?
3. Is it positioned correctly (e.g. ladders at the correct ratio of 4 : 1; scaffold on a sound footing)?
4. Is the ladder or scaffold secured safely?
5. What is the condition of access equipment after inclement weather?
6. Is there any risk to its use in a specific area (e.g. close to overhead power lines)?
7. Does it have handrails and toeboards fitted? (Regulations insist on these when working height is at 2 meters or more. This should not preclude taking the same measures at lower heights.)
8. Does it provide a safe working area? If in doubt, seek advice, as there are many injuries each year caused by falls or by material dropping from heights. The wearing of safety helmets is a secondary protection and should not be allowed to encourage carelessness.

63.15 Materials handling

Whenever possible, lifting of excessive weights should be by mechanical means and should only be done manually if there is no practicable method of obtaining access for mechanical equipment. It follows that material must only be moved manually if it is within the capability of the persons involved and that there is no risk of injury.

63.15.1 Manual lifting

Correct manual lifting and handling of material prevents strains and injury and also reduces the effort required. Persistently incorrect lifting and handling may lead to the person suffering from permanent back strain or other health problems. Points to consider when preparing to lift material are:

1. Can you lift it yourself?
2. Does it have sharp corners or edges?
3. Is there any obstruction where you have to walk or place the material?

When lifting the material:

1. Stand with your feet apart, with one foot in the direction you intend to move.
2. Grip the object with the palms of your hands.
3. Keep a straight back and, bending slightly at the knees, use your thigh muscles to lift.
4. With elbows in and arms close to the body, slowly rise.

5. Keeping your chin in and raising your head, lift the object to the height required, adjusting the position of your feet to ensure that you maintain your balance.
6. Take care when a hazardous substance (e.g. acid) is lifted.
7. Always wear the correct protective clothing.
8. Get help with heavy loads.

63.15.2 Mechanical lifting

In addition to the general rules of the Health and Safety at Work, etc. Act 1974 there are specific requirements under the following legislation with respect to chains, ropes and lifting tackle:

Factories Act 1961
The Construction (Lifting Operations) Regulations 1961
The Docks Regulations 1934
The Shipbuilding and Ship-Repairing Regulations 1960

Depending on the location of the workplace, the regulations will indicate the frequency of tests and examination of the equipment.

63.16 Portable tools and equipment

1. Portable electrical hand tools and equipment shall be properly grounded and wound to operate on 110V a.c. center tapped to earth supply, and shall only be connected to the system by permanent joints or proper connections.
2. Portable lighting, when used in wet conditions or confined spaces, should operate at no more than 25V a.c. single phase, and must be fitted with the correct guard.
3. Persons operating abrasive wheels (e.g. bench grinders, machine grinders, disc grinders) should be trained in their use. To dress or mount (change) a wheel or a disc, the operative must have attended a training course in compliance with the Abrasive Wheels Regulations 1970. Attendees at such course must be recorded in the company register (F2346) for the purpose of the Abrasive Wheels Regulations 1970.
4. It follows that with any tool or equipment that is potentially dangerous or operators who may suffer injury from its use must be protected. To meet this requirement, training must be provided.

63.17 Confined spaces

Some plant and equipment are immediately classified as confined spaces, but extreme caution is necessary in the assessment of other areas. Section 2 of the Health and Safety at Work, etc. Act 1974 requires employers to ensure the health and safety at work of their employees; this duty is *so far as is reasonably practicable*. Therefore, as work in confined spaces is potentially dangerous this Section of the Act clearly requires employers to ensure that there is no risk to their employees when working in such an area.

Where work is to be carried out in a confined space within a factory, Section 30 of the Factories Act 1961 lays down specific requirements when work is performed on or in a number of items of plant and equipment. No person must enter or be in a confined space unless he or she has been authorized to do so, and all the appropriate safety precautions have been strictly adhered to prior to entry and while within it.

Points that must be checked prior to the issue of the appropriate permit to work should include the following:

1. Are there fumes present?
2. Is there an adequate supply of maintainable oxygen?
3. Will breathing apparatus be required? Has the operative had training in its use and is there adequate support?
4. Is there any hazardous residue in the 'space'?
5. Has the space been tested for being clean and gas free?
6. Is there a supporting person standing by in the event of an emergency?
7. Is there a safety harness and rope available for use when the work is to be carried out? The operative entering the space will wear a harness while the operative standing by the entrance will retain the free end of the rope attached to the safety harness.

63.18 Electricity

The Electricity at Work Regulations 1989 revoke, among others, the Electricity Regulations 1908 and the Electricity (Factories Act) Special Regulations 1944, which have been the most important electricity regulations for many years. These regulations, effective from 1 April 1990, place a responsibility on both employer and employee to comply as far as it is reasonable and practicable within their control. Major changes from the previous regulations are as follows.

Regulation 13: Precautions for work on equipment made dead

In order to prevent danger while work is carried out on or near electrical equipment which has been made dead, adequate precautions shall be taken to prevent it from becoming electrically charged during that work if danger may arise thereby.

Regulation 14: Work on or near live conductors

No person shall be engaged in any work activity on or near any live conductor (other than one suitably covered with insulating material) so that danger may arise unless:

1. It is unreasonable in all circumstances for it to be live;
2. It is reasonable in all circumstances for the person to be at work on or near it while it is live; and
3. Suitable precautions (including, where necessary, the provision of suitable protective equipment) are taken to prevent injury.

Regulation 16: Persons to be competent to prevent danger and injury

No person shall be engaged in any work activity where technical knowledge or experience is necessary to prevent danger. Where appropriate, injury, unless he or she possesses such knowledge or experience, or is under such degree of supervision as may be appropriate having regard to the nature of the work.

Additional safety precautions

1. You must not work on or interfere with any electrical equipment unless you have been authorized to do so.
2. Do not use electrical equipment or switch rooms as a storage area.
3. Assume that all electrical distribution circuits are live.
4. Obtain the necessary authority/permit to work before commencing any work on electrical systems.
5. Do not improvize: electricity can kill.

63.19 Plant and equipment

63.19.1 Compressed air

Compressed air can be dangerous if used incorrectly:

1. Never use compressed air for cleaning. The pressure from nozzles may blow particles of dirt and dust into the eyes, ears or skin of the person using the nozzle for cleaning or of others in the vicinity.
2. Do not dust yourself down with the compressed air nozzle.
3. Indulging in horseplay with compressed air can have disastrous results.

Persons employed to use compressed air in carrying out a specific work task must be correctly equipped to prevent injury. The minimum requirement is to wear eye protectors, which conform to BS 2092.

63.19.2 Steam boilers

Section 38 of the Factories Act 1961 defines a steam boiler as a 'any closed vessel in which for any purpose steam is generated under pressure greater than atmospheric pressure'. Economizers used to heat water being fed to such a vessel and superheater for heating steam are also included. Every boiler must be fitted with the recommended safety measures (e.g. safety valve, stop valve, water gauge, low-water alarms, pressure gages, etc.).

Prior to being brought into use in a factory, boilers must have been thoroughly examined (when cold) by a competent person in accordance with the Examination of Steam Boilers Regulations 1964. Following subsequent examination at the recommended frequency, the results on the prescribed forms must be attached to the General Register (F31) within 28 days.

Entry into boilers are controlled by Section 34 of the Factories Act 1961, which states that no person shall enter or be in any steam boiler which is one of a range of two or more boilers unless:

1. All inlets through which steam or hot water might otherwise enter the boiler from any part of the range are disconnected from that part;

2. All valves or taps controlling the entry of steam or hot water are closed and securely locked. Where the boiler has a blow-off pipe in common with one or more other boilers or delivering into a common blow-off vessel or sump, the blow-off valve or tap on each boiler is so constructed that it can only be opened by a key which cannot be removed until the valve or tap is closed. There can be only one key in use for that set of blow-off valves or taps.

63.19.3 Pressure systems

The Pressure Systems and Transportable Gas Regulations 1989, effective from 1 July 1990 in parts, will be fully effective from 1 July 1994. Aspects that became effective from 1 July 1990 are:

1. The supply of information for new plant;
2. The proper installation of plant; and
3. The establishment of safe operating limits on existing plants.

The Regulations relating to gas cylinders apply to suppliers, importers, fillers and owners of cylinders from 1 January 1991 if:

1. The plant contains a compressed gas (such as compressed air or liquefied gas) at a pressure greater than 0.5 bar (about 7 psi) above atmospheric;
2. If steam systems with a pressure vessel are used at work, irrespective of the pressure;
3. Employees and/or the self-employed use the plant at work. For full details of the requirements under these Regulations reference should be made to the following: The Pressure Systems and Transportable Gas Containers Regulations 1981
Approved Code of Practice, Safety of pressure systems
Approved Code of Practice, Safety of transportable gas containers
HS(R) 30, A Guide to the Pressure Systems and Transportable Gas
Containers Regulations 1989
The Approved Codes of Practice list relevant HSE Guidance Notes.

63.19.4 Machinery

The requirement for safety precautions when operating or working on all types of machinery cannot be overstressed, in that the likelihood of an accident occurring on moving machinery is invariably high. It is essential that all those who work on such machines are adequately trained as to the operation and the safeguards fitted. Regulations that must be complied with include:

Health and Safety at Work, etc. Act 1974, Sections 2 and 33
Factories Act 1961, Section 14
Offices, Shops and Railway Premises Act 1963, Section 17

Other information can be found in BS 5304, Code of Practice for safety of machinery.

In premises in which the Factories Act 1961 apply, Section 15 of the Act acknowledges that certain dangerous parts of machinery can only be adjusted or lubricated while in motion. Employees carrying out this work must be properly trained and their employers must specify their tasks in writing.

Details of the appointed persons must be entered in Register F31. Also, such persons must be instructed in the requirements of the Operations at Unfenced Machinery Regulations 1938 and issued with the precautionary leaflet F2487. General precautions to observe include:

1. Machines must not be used unless properly guarded (or operatives are trained). All damaged guards/security measures must be reported.
2. Never reach over operating machinery or wear loose clothing when working at or on machinery.
3. Never start up a machine until you are certain that doing so can injure no one.
4. Keep the floor area around the machine clean.
5. Do not clean the machine when it is in operation.
6. Wear appropriate protective clothing.

63.20 Safety signs and pipeline identification

63.20.1 Safety signs

The Safety Signs Regulations 1980 apply to signs that give a health and safety message to people at work by using certain shapes, colours and pictorial symbols. There are four basic types of safety signs which the Regulations require to conform to BS 5378: 1980, Safety signs and colours:

1. Solid blue circle: mandatory ('You must do');
2. Red circle with a red band across the diameter: prohibition ('Do not do');
3. Yellow triangle with black border: caution ('Warning of danger');
4. Solid green square: information ('Safe conditions').

Graphic symbols may be placed in the above shapes to give them more meaning and there may also be text affixed to them.

63.20.2 Pipeline identification

Pipelines should be marked clearly as to the contents. This can be achieved by marking them in accordance with BS 1710, System (contents) and BS 4800, Colours (indicating contents).

63.21 Asbestos

In the 1987 Regulations, asbestos is defined as any of the following minerals; crocidolite, amosite, chrysotile, fibrous anthophyllite, fibrous actinolite, fibrous tremolite and any mixture containing any of these. Before carrying out work on any substance suspected of being asbestos, a competent person must be called to advise on its possible composition. Such a person will advise on the measures that must be taken to avoid any hazard to the occupants or others likely to be affected. Companies with asbestos on their premises should hold the appropriate documents. Current Regulations include:

Asbestos (Licensing) Regulations 1983
Asbestos (Prohibitions) Regulations (amended) 1985
Control of Asbestos at Work Regulations 1987

63.22 COSHH

The Control of Substances Hazardous to Health (COSHH) Regulations 1989 covers virtually all substances hazardous to health. Only asbestos, lead, materials producing ionizing radiation and substances below ground in mines (which all have their own legislation) are excluded. The Regulations set out measures that employers must implement. Failure to comply with COSHH, in addition to exposing employees and others to risk, constitutes an offence and is subject to penalties under the Health and Safety at Work Act, etc. 1974.

Substances that are hazardous to health include:

1. Those labeled as dangerous (i.e. very toxic, toxic, harmful, irritant or corrosive);
2. Agricultural pesticides and other chemicals used in farming;
3. Those with occupational exposure limits;
4. Harmful microorganisms;
5. Substantial quantities of dust;
6. Any material, mixture or compound used at work, or arising from work activities, which can harm people's health.

In works, premises, factories, etc. there will be substances in use that come within the control of COSHH. Seek advice!

The following publications give detailed information on COSHH and its requirements:

1. Control of Substances Hazardous to Health Regulations 1988, Approved Code of Practice Control of substances hazardous to health and Approved Code of Practice Control of carcinogenic substances;
2. Occupational Exposure Limits: Guidance Note EH40 (revised and issued yearly);
3. Control of Substances Hazardous to Health Regulations 1988 and Approved Code of Practice Control of substances hazardous to health in fumigation operations;
4. Approved Code of Practice Control of vinyl chloride at work;
5. COSHH Assessments (a step-by-step guide to assessment and the skills needed for it);
6. Also available are free leaflets from the HSE.

63.23 Lead

The Control of Lead at Work Regulations 1980 apply to work which exposes persons (both employees and others) to lead as defined in Regulation 2(1) (i.e. to the metal

and its alloys, to compounds of lead, which will include both organic and inorganic compounds, and to lead as a constituent of any substance or material) when lead is in a form in which it is liable to be inhaled, ingested or otherwise absorbed by persons. Relevant publications are:

1. Approved Code of Practice Control of lead at work;
2. Guidance Notes;
    EH 28 Control of lead: air sampling techniques and strategies
    EH 29 Control of lead: outside workers
    EH 30 Control of lead: pottery and related industries;
3. Also available are free leaflets from the HSE.

63.24 Other information

Numerous leaflets can be obtained from the local offices of the Health and Safety Executive or HSE Information Points. Remember, if you have any doubts about the health and safety risks involved with respect to any substance you may consider using, or any work you are planning to undertake, seek advice. This information may be obtained from your company safety officer, consultants, manufacturer/supplier or equipment/material/substance, HSE or local Environmental Health Officer. Use their expertise and knowledge. Do not be a statistic.

63.25 Assessment of potential hazards

63.25.1 Purpose

An assessment of potential hazards (APH) shall be carried out prior to work tasks being issued to determine:

1. Physical hazards;
2. Access/exit;
3. Potentially unsafe practices likely to be implemented in carrying out the work;
4. Safe working procedures;
5. Whether a permit to work is required.

63.25.2 Working practices

An assessment of each element of the program of work is carried out using the Assessment of Potential Hazards Form APH1 (see below), which is to be completed by the person responsible for the task to be carried out. This may include 'hot' work; cutting/grinding; work on electrical systems; working alone; use of hazardous materials (e.g. chemicals, solvents, etc.); confined spaces or any other work which may be of a hazardous nature.

In establishing potential hazards in carrying out specific tasks, other sources may have to be consulted (e.g. manufacturer, supplier, HSE, environmental health officer, consultants, etc.) and protective clothing and equipment must be considered.

Prior to implementing any APH within the Method Statement for the task the APH and Method Statement must first be approved by the responsible person's senior and/or the manager of the department responsible for the plant and equipment. The approved APH and Method

Statements will be brought to the attention of all persons called upon to carry out the specific task. Copies of the APHs will be issued to:

The department responsible for the person carrying out work;
The department in which work is being done;
Health and safety/security departments.

APHs will be reviewed annually to ensure compliance with current plant and equipment, procedures and changes in health and safety recommendations.

In the event of accidents, an investigation should include examination of all aspects of the APH.

Assessment of Potential Hazards - Analysis Form APH1
Job Description/Classification   Department Date

Task Elements	Hazards	Safe Working Practices

Completed by: . . . .   Department . . . .
Approved by: . . . .   Department . . . .

63.26 Alternative method of assessing hazards

An alternative method of assessing the hazards of a specific work task may be as follows. The work task envisaged, to be carried out by a contractor, is that of connecting an additional outlet to fuel oil tanks situated within an oil tank room, which is located below ground level.

63.26.1 Work on fuel oil tanks

Relevant information:
Guidance Note CS1: Industrial use of flammable gas detectors (HMSO)
Guidance Note CS2: Storage of highly flammable liquids (HMSO)
Guidance Note CS15: The gas-freeing and cleaning of tanks containing flammable residues (HMSO)
Guidance Note GS5: Entry into confined spaces (HMSO)
Hot Work on Tanks and Drums: IND(G)35(L) M20 (HSE)

63.26.2 Procedures: 1

1. Identify the work task.
2. Examine alternative methods for carrying out the task.

3. Select the appropriate method.
4. Plan how the work will be carried out.
5. Who is to carry out the work task and when?
6. Carry out the work task.
7. Continually monitor the work task and the safety measures taken.
8. Review the procedures.

63.26.3 Procedures: 2

1. Identify the work task. Can an alternative method be suggested to the client, hence eliminating potential hazards from the work task?
   (a) Task 1: Ensure that there is no hazard from tanks, etc. that have to be worked on or from the locality of work (i.e. tanks have been emptied, degreased and gas-freed and checked and tested);
   (b) Task 2: As per client's instructions (client provides Method of Work Statement).
2. Examine alternative methods for carrying out the task.
   (a) Can the work be carried out without the use of heat (i.e. welding, burning, electrical equipment (arcing), cutting causing heat or sparks), assembled externally and flanged rather than welded?
   (b) Will the work be carried out in stages, i.e. will some of the adjacent tanks remain charged?
3. Select the appropriate method.
4. Plan how the work will be carried out.
5. Discuss with the client:
   (a) The period during which work can be carried out;
   (b) Their involvement and responsibility with respect to fire precautions, extinguishers, permit to work, liaison with fire brigade, fire officer attendance;
   (c) Who is responsible for emptying the fuel tanks;
   (d) Cleaning, degreasing and gas freeing the tanks;
   (e) Testing for explosions or oxygen deficiency.
6. Nominate a person to be in charge.
7. Prepare a program of work.
8. Is training required on use of respirators, entry into confined spaces, fire prevention and use of extinguishers?
9. Who is to carry out the work task and when?
   Task 1:
   (a) Contractor;
   (b) Nominated specialist contractor;
   (c) Client;
   Task 2:
   (a) Contractor;
   (b) Nominated specialist contractor.
10. Carry out the work task *only* when a permit to work has been issued which states that the work area is safe for working and details the safety checks that have been made and precautions to be taken.
11. Continually monitor the work task and the safety measures taken.
    (a) Ensure that Task 1 has been carried out satisfactorily and check report/record;
    (b) Continually monitor working environment for explosion or fire risk and/or oxygen deficiency.

12. Review the procedures.
  (a) Ensure that the company's safety officer checks the procedures;
  (b) Have the procedures been effective or should they be amended? If so, why?

If specialist contractors are to be engaged, a Method Statement should first be requested from them on how they plan to carry out the work.

63.27 Permits to work

Where there is a risk to the employee carrying out work, in addition to a Method Statement a permit to work should be used. A permit to work is a formal written means of making sure that potentially dangerous jobs are examined first, before authorizing the work to commence. Its task is twofold: it ensures that the person making the system safe and the person who is to carry out the work have both checked that it is safe to carry out the work task. Permits to work cannot be transferred to other parties. If any circumstances change from the issue of the original permit to work, it must be canceled and a new one prepared.

Permits to work can be of differing formats. Therefore, it is essential that when a permit to work has been issued it should be read carefully and understood and all possible measures taken to reduce or eliminate the known danger. Permits can be issued for a variety of work tasks (e.g. work on high-voltage electrical systems, steam boilers, hot work, confined spaces, etc.). The permit below is an example of differing types of work tasks.

63.27.1 Permission to Work on Electrical Systems Certificate (not exceeding 1000 V

The authorized person and sections 1 to 3 inclusive must always raise this certificate completed before any work is

Company   **Serial No** . . . .

PERMISSION TO WORK ON ELECTRICAL SYSTEMS CERTIFICATE
(not exceeding 1000 V)

1. *This Certificate is issued for the following Authorized Work. No other work may be carried out. Entrance to work area is limited to those authorized. Tools/access equipment etc. must be approved by authorized person prior to them being taken into the work area.*
2. **Work Details/Job No./Method Statement Ref. No**
. . . .
. . . .
. . . .
3. **Work Location:**
. . . .
. . . .
4. **Method of Isolation:**
The equipment on which the work is to be carried out has been proven dead after isolation at: . . . .
. . . .
Care must be exercised as the equipment is still live at: . . . .
. . . .
Safety screening has been applied at: . . . .
. . . . . Lock-Off Key Nos . . . . Danger Notices Posted: Y/N
Warning Signs Displayed: Y/N Safety Barriers Erected: Y/N
5. **Authorized Person issuing Certificate:**
. . . . Designation . . . .
Date . / . / . Time . . . . Signature . . . .
6. **Person Responsible for Work being carried out:**
Company/Dept . . . . Name . . . .
Designation . . . . Signature . . . .
7. **Work Completed/Suspended:** Date . / . / . Time . . . .
Person Responsible for Work being carried out:
Company/Dept . . . . Name . . . .
Signature . . . .
8. **Cancellation of Certificate by Authorized Person:**
Receipt of Certificate indicating work is
Complete/Suspended:
Name . . . . Signature . . . .
Designation . . . . Date . / . / . Time . . . .

On cancellation of this certificate, no other work must be carried out on the plant/ equipment without a further certificate being issued.

carried out. A consecutive serial number must be entered on the certificate.

Where the authorized person is to carry out the work, he must complete section as the person responsible for work being carried out and retain the top copy on him throughout the period the work is being carried out. The bottom control copy must remain in the register.

Where a competent person is to carry out the work, he must complete section 3 as the certificate holder. The authorized person then issues him the top copy, which he must retain on him throughout the period the work is being carried out. The bottom copy is retained in the register.

On completion or suspension of the work, the responsible person must return the top copy to the register, so that the authorized person ensures that sections 7 and 8 are completed on both top and bottom copies of the certificate. The bottom copy must always be retained in the register for record purposes. Before any suspended work is restarted, a new certificate must be raised and processed by the authorized person as above.

Examples of tasks that will require the issuing of the Permission to Work on Electrical Systems Certificate are indicated in Method Statements for Work on Electrical Systems.

Note

1. The company nominates authorized and competent persons after they have been given the appropriate training in both the tasks to be carried out and their responsibility as nominated persons.
2. The electrical engineer would normally stipulate the tasks that require the use of Method Statements for Work on Electrical Systems and would also prepare the Method Statements.
3. A similar format of certificate would suffice as a permit to work for high-voltage systems (i.e. exceeding 1000 V).

An example of this certificate as well as a Hot Work Permit is given below.

63.28 Working alone

63.28.1 Company and contractor's responsibilities for the lone worker

Managers responsible for services within works, offices or premises have an additional task in that when engaging service contractors they then have joint responsibility under the Health and Safety at Work, etc. Act 1974 for the health and safety of the contractor's employees while on their premises. When engaging contractors to carry out work within the premises, systems must be implemented by which the contractor's employee works in a safe manner and does not create a hazard to the premises' occupants or staff while carrying out this work. This responsibility is greater when there is an employee or service contractor working alone, as in most instances the premises' communications do not allow for such circumstances (e.g. the lone employee may be working in remote areas such as plant rooms).

To enable both company and contractor to comply with their specific responsibilities under the Health and Safety at Work, etc. Act 1974, a Lone Workers' policy, supported by a method of assessment of potential hazards relating to the work to be carried out, should be prepared. Such a system would prove beneficial when, for any reason, the regular lone worker cannot be found.

63.29 Safety policy for lone workers

In the Health and Safety at Work, etc. Act 1974 the following Sections apply:

> Section 2: General duties of employers to their employees
> Section 4: General duties of persons concerned with premises to persons other than their employees.

Working alone presents a specific health and safety problem, and companies, department managers and contractors must ensure that there is a safe system whereby:

1. The person working alone is adequately trained for the work that is to be carried out;
2. Having carried out an assessment of potential hazards (APH), a Method Statement (including advice on whether a permit to work is required) is prepared;
3. The correct tools, equipment and manufacturers instructions are available;
4. Contact is maintained by:
   (a) Means of telephone at regular intervals;
   (b) Visits by others at regular intervals;
   (c) Persons passing through a 'Lone Worker' area are made aware of the lone worker's presence;
   (d) The intervals between contacts being made should not exceed one hour.

A lone worker must:

1. Comply with any Method Statement or safe system introduced for his or her health and safety;
2. Work safely at all times, and not take risks;
3. Understand the hazards/risks associated with the work to be carried out;
4. Understand the hazards/risks when *working alone*;
5. Ensure, where necessary, that a permit to work has been requested and issued.

63.30 Contractor's conditions and safe working practices

The use of contractors to carry out work that may in the past have been done by direct labor or during a normal sub-contract does not relieve the company from the responsibility of ensuring that all those working on their premises do carry out their work in a safe manner. The following contractor's conditions and safe working practices should form a part of the company's health and safety policy manual.

63.30.1 Introduction

These conditions and safe working practices form part of the contract and are to be read in conjunction with

Company   **Serial No . . . . .**

AUTHORIZATION TO WORK ON ELECTRICAL/MECHANICAL APPARATUS CERTIFICATE

1. This Certificate is issued for the following Authorized Work. *No other work other than that detailed must be carried out.*

Work Details:

....

....

Work Location:

....

....

Method of Isolation/Making Safe:

....

....

Lock-Off Key Nos ....

Danger Noticed Posted: Yes/No ....

Warning Signs Displayed: Yes/No ....

Authorized Person issuing Certificate ....

Designation .... Date . / . / . Time ....

Person Responsible for Work being carried out:

Company .... Name ....

Designation .... Signature ....

2. Work Completed: Date . / . / . Time ....

Person Responsible for Work carried out:

Company .... Name ....

Designation .... Signature ....

3. Authorized Person:

Receipt of Certificate indicating work is complete:

Designation .... Date . / . / . Time ....

Name .... Signature ....

N.B. Copy of this Certificate to be issued to Person responsible for carrying out Work.

Signature of Receipt ....

---

the general conditions of the contract form. The company reserves the right to make alterations or additions to those conditions and practices as and when necessary. Contractors, sub-contractors or their employees must not communicate to persons other than the company staff any information relating to the work being carried out by the contractor or sub-contractors. Information, drawings, etc. must not be removed from company premises without prior permission from the company contact, who is the responsible person for arranging the appropriate clearance relating to that information.

63.30.2 Contents

Local conditions

1. On arrival at company premises
2. Prior to commencement of work
3. Attendance at company premises for more than one day.

Safe working practices and procedures

1. Health and safety
2. Emergency procedures

63.31 Safe working practices and procedures

63.31.1 Health and safety

1. Health and safety is each employer's responsibility with respect to their employees under the Health and Safety at Work, etc. Act 1974.
2. Contractors must comply with all statutory and site regulations.
3. It is the contractor's responsibility to ensure that all work is carried out safely and by competent persons.
4. Method statements as to how the work is to be carried out, clearly indicating matters relating to health and safety, must be passed for approval to the

HOT WORK PERMIT

Required when any of the following tasks are to be carried out:
Cutting; Welding; Soldering, Brazing and also during the use of any equipment which can produce heat, naked flames or dust (i.e. dust likely to affect Detector Heads)

Period of Work:
Commence: .... hrs .... day .... month .... year
Completion / .... hrs .... day .... month .... year
Expiry

Location of Work:
....
....
....
....

Description of Work:
....
....
....
....

Precautions:
Department in which work is to be carried out has been advised
Cutting/Welding/Brazing/Hot Work Equipment is in good order

Fire Alarm: ON/OFF   **Halon System:** ON/OFF   **Sprinkler System:** ON/OFF
Fire Notices Posted: YES/NO   Insurance Informed YES/NO
Fire Detectors are protected

Precautions within 35 ft (10 m) of work area:
Floors and areas clear of combustibles
Walls and floor openings covered with e.g. Fire Blanket
Combustibles if not moved protected by Fire Blanket
Fire Watch will be provided for at least 30 minutes after work (incl. breaks)
Fire Watch is supplied with suitable extinguishers
Fire Watch is trained in use of extinguishers and method of raising alarm

I have examined the location of the work, checked that all precuations and preventive measures have been implemented.
Authorizing Signature .... **Name** ....
Designation .... hrs 90

I hereby declare that no work other than that stated above will be carried out.
Signed .... **Name** .... **Design'n/Company** ....

I hereby declare that the work has been completed and that the work and adjacent areas have been checked for fire 30 minutes after the completion of the work task.
Signed .... **Name** .... **Design'n/Company** ....
Authorizing Signature .... **Name** ....
Designation .... hrs 90

nominated company officer prior to the commencement of work.
5. The successful contractor is responsible for ensuring that all items above are brought to the attention of any contractors to whom work is sub-contracted.

63.31.2 Protective clothing and equipment

1. Eye protection must comply with the Protection of Eyes Regulations 1974 and BS 2092, BS 1542 and BS 769.

2. Helmets must comply with BS 5240 and BS 4033.
3. Noise must be comply with the Noise at Work Regulations and ear protectors with BS 6344, Industrial hearing protectors.
4. Safety footwear *must* be worn where applicable (e.g. moving heavy objects; working with hazardous materials and in wet conditions). Footwear should comply with the following as appropriate: BS 1870, BS 4792, BS 5131, BS 5145, BS 5451, BS 5462 and BS 6159.
5. Safety belts must comply with BS 1397 and BS 6858.
6. Protective clothing (including gloves) *must* be used where applicable (e.g. normal duties; welding/cutting

hot work; water treatment; application/use of chemicals; during servicing of batteries; descaling/chemical cleaning). Clothing: BS 1542, BS 5426, BS 2653, BS 4171, BS 6408, BS 4724; Gloves: BS 1651, BS 697, BS 6526.
7. Respirators must comply with BS 2091, BS 4555, BS 4558, BS 4771 and BS 4275. Also refer to Form F2486, published by HSE for suppliers of respirators suitable for use against aerosols containing *Legionella*.

When in doubt, consult with manufactures, suppliers, HSE, EMAS, the local authority or the environmental health officer.

63.31.3 Access equipment

1. Stepladders: BS 1129, BS 2037
2. Trestles: BS 1129, BS 2037, BS 1139
3. Scaffolds: BS 1139, DD72, CP 5973, BS 2482, BS 6037, BS 5974, BS 6289

Scaffold and other access equipment should comply with a standard as for the Construction (Working Places) Regulations 1966.

63.31.4 Warning signs and good housekeeping

All appropriate warning signs must comply with the Safety Signs Regulations 1980 and must be displayed prior to commencement and for the duration of the work to be carried out (see BS 5378). Workplaces must be kept safe and tidy and all access/exit areas clear of obstruction. Additional information is available from HSE publications.

63.31.5 Tools and equipment

The contractor will be responsible for the supply of all tools and equipment, unless otherwise agreed by company representatives:

1. Portable electrical hand tools and equipment shall be properly grounded and wound to operate on 110V a.c. center tapped to earth supply, and shall only be connected to the system by permanent joints or proper connections.
2. Portable lighting must operate at no more than 25V a.c. single phase, and must be fitted with the correct guard.
3. Equipment issued by the company is provided on the understanding that competent workers under adequate supervision will properly use it. Also that the contractor shall be solely responsible for and indemnify the company against loss or damage to such equipment and all or any loss, injury, death of any person or damage to any property whatsoever arising out of the use of such equipment.

63.31.6 Materials/parts/spares

1. Materials used must be of the highest standard, complying as necessary with the appropriate British Standard.

2. Materials supplied by the company must only be used for the work task for which they have been issued.

63.32 Emergency procedures

63.32.1 Fire

When arriving at the Company premises and being shown their work area the contractor's employees must:

1. Check the location of the nearest fire exit.
2. Make themselves familiar with fire alarm procedures.
3. Ensure that means of escape are kept clear at all times.

63.32.2 First-aid

The contractor's employees must:

1. Make themselves familiar with the first-aid procedures in the event of an accident.
2. Check the location of the nearest first-aid persons and the first-aid room if there is one on the premises.

63.33 Contractor's guide

This is to be issued to an individual on arrival at a company's premises. All contractors' employees must adhere to the following procedures.

63.33.1 On arrival at company premises

1. Report to reception
2. Advise reception/office of:
   (a) Company contact who requested attendance;
   (b) Work to be carried out, stating department location;
3. On being signed in/issued identity badge be advised of:
   (a) Whereabouts of work location; or
   (b) Be escorted to work location by member of staff; or
   (c) Be escorted by company contact;
4. Be issued with a copy of the *Contractor's Guide*.

63.33.2 Prior to commencement of work

The company contact/coordinator will:

1. Discuss items in the *Contractor's Guide* relevant to the work to be carried out;
2. Ensure that adequate notice has been issued to all other parties, e.g.:
   (a) Members of department, including deputy;
   (b) Others who may be carrying out work in that area;
   (c) Contact security/loss prevention coordinator for issue of permits to work, advice/instructions on health and safety, specific hazards relevant to work and work area and issue of fire extinguishers/fire blankets;
3. Give permission to commence work.

63.33.3 Security

If attendance at company premises is required on more than one day to complete the work the contractor's employees must:

1. Sign in/out at reception/security;
2. Collect/return;
   (a) Identity badge;
   (b) Permits to work (not to be removed from premises);
   (c) Keys (not to be removed from premises).
3. Report proposed absence from/return to premises to:
   (a) Contact;
   (b) Security/loss prevention coordinator if work requires a permit to work;
   (c) Facilities coordinator/department head if work involves general premises/site facilities/services.

63.33.4 Welfare/canteen

Contractors subject to discussion with the company contact will use these facilities.

63.33.5 Useful contacts

These include the telephone numbers of contact, security, first-aid, etc.

Regulatory Compliance Issues in the US

R Keith Mobley
The Plant Performance Group

Contents

64.1 Introduction

Regulatory compliance problems associated with work-related incidents generally fall into the domain of safety, which is regulated by OSHA, and environment, which is regulated by the EPA and state (although a state may defer to the EPA through a formal agreement) and local governments. This section provides guidelines for investigating incidents that fall under the jurisdiction of these regulations.

Note: The material presented in this chapter does not constitute legal advice and is provided for information purposes only. Please consult the appropriate agencies directly and seek qualified legal council for this type of assistance.

Spill hazards

Be certain that spill hazards are reviewed as part of the plant's internal audit program. Audit-team members should be trained to look for deficient containers, fall hazards that also can contribute to a spill, improper container labeling, and other potential hazards. Major causes of chemical spills are:

- Mishandling of equipment and supplies in the work area (e.g., use of improper containers),
- Mishandling during transportation (even within a facility),
- Leaks from process equipment such as flanges or valves, and
- Improperly stored containers (e.g., storage in a damp location leads to rusty, deteriorating containers).

Worker education and communication

Employers must ensure that their workers know the potential hazards of the chemicals they work with, how to protect themselves against those hazards (e.g., safe practices, personal protection equipment, etc.), and what to do in case of an emergency. Accordingly, OSHA has established basic communication requirements under the *Hazard Communication Standard* to inform workers about chemicals in use in the workplace. Under this standard, chemical makers must meet the following requirements:

- Manufacturers must determine the physical and health hazards of each chemical they produce.
- Users must be informed of the hazards through container labels and Material Safety Data Sheets (MSDSs).
- A written hazard-communication program must describe the employer's efforts to tell employees about the standard and how it is being implemented at the worksite.

Prevention

Preventive measures include everything from safe work practices to training programs, good housekeeping, regular audits, administrative and engineering controls, and chemical-protection equipment.

Workers must be properly trained to handle the responsibilities assigned to them, with members of a designated spill team undergoing more specialized training than the average employee. All workers training should include the basic instruction that *employees not involved in cleanup must stay away from a spill.* This is important because of risks such as inhaling chemical vapors or fire or explosion hazards. Another important lesson to teach workers is how to prevent a spill or leak from becoming worse. Proper containment can mean the difference between a small problem and a huge, dangerous mess.

64.2 Recommended spill response

The following steps are recommended responses when a chemical spill occurs:

1. If necessary, evacuate the area of untrained spill-response personnel.
2. Check labels, MSDSs, and other key documentation to identify what has spilled.
3. Assign at least two qualified individuals to the cleanup. They may be internal staff or from an outside firm that you have prearranged to handle this task. Two workers are needed in case one is overcome or injured.
4. Promptly clear the spill area by ventilation.
5. If the spilled material is on fire, douse the flames in a safe manner.
6. Limit the spread of the spilled substance by containing it with a dike. Use an absorbent material appropriate for the type of spill (e.g., polymer-type absorbents such as spill booms, granular materials such as kitty litter, ground corncobs, etc.).

However, when absorbent materials are used, they become as hazardous as the material absorbed. Therefore, federal, state, and local regulations must be consulted before disposing of these contaminated materials. In some cases, a neutralizer can be used to convert one material into another (e.g., an acid neutralized with a base produces salt and water).

Caution: Specialized worker training is legally required to use absorbent materials on spills because of the chemical reactions that can occur. For example, organic materials should never be used on nitric acid spills because a fire will result!

Workers' responsibilities

Employees must take most of the responsibility for protecting themselves from chemical hazards. However, adequate training and frequent reminders from the employer can help ensure that they take that responsibility. The following are some basic chemical-safety tips to incorporate in a worker-safety-training program or to post on the bulletin board:

- Pay attention to training that is provided.
- Know what hazardous chemicals are used in your work area.
- Read labels and MSDSs before starting a job.
- Do not use chemicals from unlabeled containers.

- Follow manufacturers' instructions for chemicals and equipment.
- Follow company rules and procedures and avoid short-cuts.
- Wear all personal protection equipment (PPE) required by your organization and ask your supervisor if you are not sure.
- Keep containers closed when not in use.
- Check containers regularly for leaks.
- Make sure that equipment is in good working condition before use.
- Keep incompatible materials apart.
- Keep flammable and explosive materials away from heat sources.
- Make sure the work area is adequately ventilated.
- Do not bring food or drinks into a work area where chemicals are used.
- Wash before eating or drinking.
- Cleanse tools, equipment, and clothing that have been exposed to hazardous materials before storing or reusing them.
- Dispose of contaminated materials properly.
- Ask a supervisor what to do with old or unused chemicals in your work area.

Legislation and reporting requirements

In any industrial facility, from offices to factories and laboratories, spills happen and create a variety of risks to workers. Inside a plant, spills result in chemicals on the floor, in the air, or on the workers themselves. When releases occur outside the plant (e.g., chemical releases from tank cars or trucks, the spread of noxious fumes from an internal spill), the potential for harm extends far beyond the facility, particularly with major catastrophes such as the Bhopal chemical release, the *Exxon Valdez* oil spill, New York's Love Canal, and dioxin-contaminated Times Beach in Missouri, have led several federal departments and agencies to enact protective regulations. These protections are aimed at protecting a much broader range of people, property, and the environment than most regulations administered by OSHA.

Spills are covered by a variety of federal, state, and local reporting requirements and substantial penalties can result to a company and its employees for failing to report certain spills. Initial release notification is usually required immediately or within 24 hours of the release and, in some cases, written follow-up reports are required. Some of the applicable legislation is listed below and Table 64.1 lists some of the major reporting requirements for chemical spills that are specified by these Acts. However, refer to Hoechst Celanese's procedures for plant-specific requirements.

- OSHA's Process Safety Management (PSM) Standard
- OSHA's Hazardous Waste Operations and Emergency Response (HAZWOPER)
- Superfund: Comprehensive Environmental Response, Compensation, and Liability Act (CERCLA)
- Superfund Amendments and Reauthorization Act (SARA): SARA Title II contains the Emergency Planning and Community Right-to-know Act (EPCRA)

- Resource Conservation and Recovery Act (RCRA)
- Toxic Substances Control Act (TSCA)
- Clean Water Act (CWA)
- Department of Transportation (DOT) Rules for packing and shipping
- Hazardous Materials Transportation Act (HMTA)

64.3 OSHA Process Safety Management Standard

OSHA legislation focuses primarily on individual workplaces and is intended to prevent explosions, spills, and other disasters. The Process Safety Management (PSM) Standard covers large-scale makers and users of highly hazardous chemicals and other chemical manufacturers. However, small companies whose core business has nothing to do with chemicals also are vulnerable to spills (e.g., cleaning products, toner for the copying machine, etc.). Although the PSM standard does not specifically apply to smaller chemical spills, its principles are still valid. Note that RCRA regulations apply to these small spills. This section summarizes the requirements for OSHA's *PSM of Highly Hazardous Chemicals, Explosives, and Blasting Agents* procedures for incident investigation (CFR 1910, Part 11.9, Section m).

64.4 Incident investigation requirements

The regulation states: 'The employer shall investigate each incident which results in, or could reasonably have resulted in, a catastrophic release of highly hazardous chemicals in to the workplace.'

To meet this requirement, a company must define an incident in specific terms for their facility. This includes an operational definition that indicates the number of pounds of the substance used in a particular process that would qualify as a 'catastrophic event'. Defining an incident in site-specific terms also includes defining the terms '*could reasonably have resulted in*'. Appendix C of the regulation provides guidelines for clarifying this point. It includes definitions of '*near misses*' in which a catastrophic failure occurred, but a chemical release did not occur. Clear guidelines should be established that provide the employee with a quantifiable means of defining those incidents that require a violation report.

Table 64.2 provides examples of hazardous chemicals that require investigation when a catastrophic release occurs or when one could have happened. These OSHA guidelines should be used in conjunction with site-specific procedures. For a complete listing of the reportable chemical used in your plant, refer to the site Hazardous Materials Policy and Procedure Manual.

Required scope

OSHA 1910.119 does not mandate the specific type of investigation a plant must conduct when a reportable incident occurs. However, it provides stipulations that must be met for the following: investigator qualifications, time requirements, report content, review process, and corrective actions.

Table 64.1 Major Regulatory Reporting Requirements for Chemical Spills

Regulation	Reference	Reporting requirements
Super/fund	40 CFR 302.4	Immediately report to the Coast Guard's National Response Center (NRC) the release of CERCLA hazardous substances in quantities equal to or greater than its reportable quantity.
EPCRA	40 CFR 355	Report releases of a reportable quantity of a hazardous substance to the state Emergency Response Commission (SERC) for each state likely to be affected. Also provide notice to the local Emergency Planning Committee (LEPC) for any area affected by the release.
RCRA	40 CFR 240–281	Notification to the NRC is required for releases equal to or greater than the reportable quantity of a RCRA hazardous waste. If the waste also is on the CERCLA list, that reportable quantity applies. If not, the reportable quantity is 100 pounds if the waste is ignitable, corrosive, reactive, or toxic.
TSCA	40 CFR 761.120 et seq. Section 8(e)	Immediately report by telephone to the EPA regional office any spill of a hazardous chemical that "seriously threatens humans with cancer, birth defects, mutation, death, or serious prolonged incapacitation, or seriously threatens the environment with large-scale or ecologically significant population destruction." A written follow-up report is required within 15 days.
CWA	Oil: 40 CFR 110–114 Haz: 40 CRF 116–117	Report any *oil spill* that occurs into navigable waters or adjoining shorelines to your regional EPA office and state water pollution-control agency if it violates water-quality standards, causes a sheen or discoloration of the water or shoreline, or causes a sludge or emulsion to be deposited beneath the surface of the water or on the shoreline. Immediate notification also is required to the NRC for the release of a *designated hazardous substance* in a reportable quantity during a 24-hour period if the spill is in or alongside navigable waters.
HMTA		Generally, the transporter of hazardous materials (including waters) must immediately report to the NRC and the state response center a release during transport if the release meets any of the following criteria: causes death or serious injury, involves more than $50,000 in property damage, involves the release of radioactive materials or etiological agents, requires public evacuation lasting at lease one hour, closes one or more "major transportation artery or facility" for at least one hour, alters the flight pattern or routine of an aircraft. Even if none of these criteria are met, it should be reported if the carrier believes a spill or incident poses "such a danger" that it should be reported. Follow-up written reports are due within 30 days. The carrier also must file DOT Form 5800.1.

For all releases:
  Always call 911 first to assure that first responders are dispatched to the scene to stabilize the release, render first aid, establish a perimeter, and extinguish/minimize the threat of fire or explosion.
Useful telephone numbers:
  Coast Guard's National Response Center (NRC): 800-424-8802
  EPA's national database of all toxic chemical release information: 800-638-8480
  EPA's Emergency Planning and Community Right-to-know Information Hotline: 800-535-0202

CFR = Code of Federal Regulations, CWA = Clean Water Act, EPCRA = Emergency Planning and Community Right-to-know Act; TSCA = Toxic Substance Control Act, RCRA = Resource Conservation and Recovery Act, HMTA = Hazardous Materials Transportation Act

Source: Adapted by Integrated Systems, Inc. from *Environmental Compliance National Edition* (Issue No. 181, Business & Legal Reports, Inc., Madison, CR, Dex. 1996) and other sources.

Qualified investigator

The regulation clearly defines the investigating team: '…at least one person knowledgeable in the process involved, including a contract employee if the incident involved work of the contractor, and other persons with appropriate knowledge and experience to thoroughly investigate and analyze the incident.'

Time requirements

OSHA regulations define specific time requirements for investigating any release or potential release of any chemical that is within the scope of 29 CFR 1910.

The regulation states: 'An incident investigation shall be initiated as promptly as possible, but not later than 48 hours following the incident.' In part, the reason for this quick response to an incident is to assure that all pertinent evidence and facts can be preserved to facilitate the investigation.

Report content

The regulation also defines specific topics that must be addressed in the report. These include:

- Date of incident,
- Date investigation began,

Table 64.2 Examples of OSHA-listed chemicals

Chemical name	Chemical abstract service	Threshold quantity (pounds)
Ammonia, anhydrous	7664-41-7	10,000
Bromine	7726-95-6	1,500
Chlorine	7782-50-5	1,500
Ethylene oxide	75-21-8	5,000
Hydrogen chloride	7647-01-0	5,000
Hydrogen sulfide	7783-06-4	1,500
Isopropylamine	75-31-0	5,000
Ketene	463-51-4	100
Methylamine, anhydrous	74-89-5	1,000
Methyl chloride	74-87-3	15,000
Methyl isocyanate	624-83-9	250
Nitric acid ($\geq$94.5% by weight)	7697-37-2	500
Perchloromethyl mercaptan	594-42-3	150
Perchloryl floride	7616-94-6	5,000
Trifluorochloroethylene	79-38-9	10,000

Source: OSHA 1910.119, Appendix A

- A complete description of the incident,
- Contributing factors, and
- Any recommendations resulting from the investigation.

Review process

The regulation states: 'The report must be reviewed with all affected personnel whose job tasks are relevant to the incident findings, including contract employees where applicable.' The aim of this clause is to ensure that all affected employees understand why the incident occurred and what actions could prevent a recurrence. While the regulation does not specifically define the methods to comply with this requirement, it is imperative that the review is prompt and complete. In almost all cases, the review requires personal meetings, either individually or in small groups, to thoroughly review the incident and recommend corrective actions.

Corrective actions

Regarding corrective actions, the regulation states: 'The employer shall establish a system to promptly address and resolve the incident-report findings and recommendations. Resolution and corrective actions shall be documented.' The regulation does not define *promptly* in definitive terms, but the intent is that all corrective actions must be implemented immediately.

The major difference between an OSHA-mandated investigation and other RCFA is that an appropriate corrective action or actions *must be implemented* as quickly as possible. In the non-OSHA-mandated RCFA process, a corrective action may or may not be implemented, depending on the results of the cost-benefit analysis.

The cost of corrective actions is not a consideration in the OSHA regulations, but it must be considered as part of the analysis. Because of the critical timeline that governs an OSHA-mandated investigation, a full cost-benefit analysis may not be possible. *However, the investigating*

team should consider the cost-benefit impact of potential corrective actions. The guidelines provided in Section 3.9 should be followed as much as possible within the time constraints of the investigation.

64.5 OSHA's investigation process

Figure 64.1 illustrates the logic tree to follow for an OSHA-mandated investigation. While it is similar to other non-mandated investigations, there are distinct differences.

OSHA's Hazardous Waste Operations and Emergency Response (HAZWOPER) legislation protects workers who respond to emergencies, such as serious spills, involving hazardous materials. It also covers those employed in cleanup operations at uncontrolled hazardous waste sites and at EPA-licensed waste treatment, storage, and disposal facilities.

Emergency planning and community right-to-know act

The Emergency Planning and Community Right-to-know Act (EPCRA) is administered by the EPA and state and local agencies. It affects virtually all facilities that manufacture, use, or store hazardous chemicals. The following are the reporting requirements of the Act:

- An inventory that includes the amount, nature, and location of any hazardous or extremely hazardous chemical present at a facility in an amount equal to or greater than its assigned 'threshold level'.
- Reports on releases of a 'reportable quantity' of a listed hazardous substance, including the total annual releases during normal operations.

Resource conservation and recovery act

The Resource Conservation and Recovery Act (RCRA) has a considerable number of regulations affecting spills in the workplace, including training of workers who might be expected to respond to them. The EPA administers RCRA.

US department of transportation

Title 49 of the Code of Federal Regulations (CFR) addresses the US DOT rules for packing and shipping of hazardous materials by air, road, rail, or water. CFR 49 covers issues such as appropriate containers, labeling, truck and rail car placarding, and providing essential information that can aid in emergency response in case of an incident involving hazardous materials.

Hazardous materials transportation act

The Hazardous Materials Transportation Act (HMTA) defines transportation releases to be those that occur during loading, unloading, transportation or temporary storage of hazardous materials or waste. Releases that meet certain criteria (see Table 64.1) should be reported to the National Response Center (NRC) and the state response center. Most states also require calls to the local police or response agencies (often by calling 911). Follow-up writ-

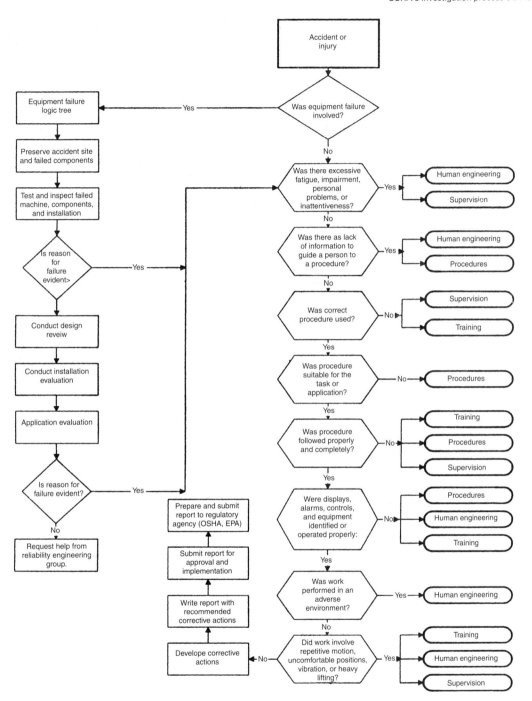

Figure 64.1 Logic tree for an OSHA-mandated investigation

ten reports are due within 30 days. If the carrier is required to report releases to the DOT, Form 5800.1 should be completed and sent to the following address: Systems Manager, DHM-63

Information
Research and Special Programs Administration
US Department of Transportation
Washington, DC 20590-0001

CERCLA also requires certain spills to be reported by the owner/shipper to the NRC. To alert drivers and emergency responders to this requirement, the letters 'RQ' must appear on shipping papers if the transporter is carrying, in one package, a substance on DOT's Hazardous Materials Table in an amount equal to or greater than the RQ shown in the table. Two additional requirements apply if hazardous waste is involved:

- Attach to the report a copy of the hazardous waste manifest.
- Include in the report an estimate of the quantity of the waste removed from the scene, the name and address of the facility to which it was taken, and the manner of disposition (Section IX of DOT Form F5800.1).

Appendix 64.1 List of abbreviations

CERCLA	Comprehensive Environmental Response, Compensation, and Liability Act
CFR	Code of Federal Regulations
CWA	Clean Water Act
DOT	Department of Transportation
EPA	Environmental Protection Act
EPCRA	Emergency Planning and Community Right-to-know Act
FMEA	Failure Mode and Effects Analysis
HAZWOPER	Hazardous Waste Operations and Emergency Response
HMTA	Hazardous Materials Transportation Act
MSDS	Material Safety Data Sheets
NRC	National Response Center
OSHA	Occupational Safety and Health Act
PPE	Personal protection equipment
PSM	Process safety management
RCRA	Resource Conservation and Recovery Act
SARA	Superfund Amendments and Reauthorization
TSCA	Toxic Substances Control Act

Index